ASTRONOMY AND ASTROPHYSICS ABSTRACTS

A Publication of the Astronomisches Rechen-Institut Heidelberg
Member of the Abstracting Board
of the International Council of Scientific Unions
Astronomy and Astrophysics Abstracts is Prepared
Under the Auspices of the International Astronomical Union

Volume 30
Literature 1981, Part 2

Edited by
S. Böhme W. Fricke I. Heinrich W. Hofmann D. Krahn
V. R. Matas D. Rosa L. D. Schmadel G. Zech

Springer-Verlag Berlin Heidelberg GmbH 1982

Astronomisches Rechen-Institut Heidelberg
Director: Professor Dr. Walter Fricke

Astronomy and Astrophysics Abstracts
Editors-in-Chief: Inge Heinrich, Dr. Lutz D. Schmadel

ISBN 978-3-662-12333-1   ISBN 978-3-662-12331-7 (eBook)
DOI 10.1007/978-3-662-12331-7

2153/3130-543210

Preface

Astronomy and Astrophysics Abstracts, which has appeared in semi-annual volumes since 1969, is devoted to the recording, summarizing and indexing of astronomical publications throughout the world. It is prepared under the auspices of the International Astronomical Union (according to a resolution adopted at the 14th General Assembly in 1970).

Astronomy and Astrophysics Abstracts aims to present a comprehensive documentation of literature in all fields of astronomy and astrophysics. Every effort will be made to ensure that the average time interval between the date of receipt of the original literature and publication of the abstracts will not exceed eight months. This time interval is near to that achieved by monthly abstracting journals, compared to which our system of accumulating abstracts for about six months offers the advantage of greater convenience for the user.

Volume 30 contains literature published in 1981 and received before February 12, 1982; some older literature which was received late and which is not recorded in earlier volumes is also included.

Dr. Herbert Hefele joined our staff in October 1981. We want to acknowledge his valuable contributions to this volume. We acknowledge with thanks contributions to this volume by Dr. J. Bouška, Prague, who surveyed journals and publications in Czech and supplied us with abstracts in English.

We express our warmest thanks again to Ms. Helga Ballmann, Ms. Mona El-Choura, Ms. Nadja Hodges, Ms. Monika Kohl, Ms. Sylvia Matyssek and Ms. Angelika Meßmer for typing the text of this volume on IBM 72 Composers, for compiling the pages from abstract slips in a perfect form for offset reproduction, and for punching material for the author index and for the subjet index, which finally were printed with a TN chain on a 1403 IBM high-speed printer. Finally, we have to thank Mr. Claus Leitherer, Mr. Uwe Reichert and Mr. Werner Sanns who supported our task by careful proofreading.

Heidelberg, April 1982

Siegfried Böhme   Vladimir R. Matas
Walter Fricke   Dorothea Rosa
Inge Heinrich   Lutz D. Schmadel
Wilfried Hofmann   Gert Zech
Dietlinde Krahn

Contents

Introduction

Astronomical bibliographies

Astronomy and Astrophysics Abstracts begins documentation and abstracting from the year 1969. For information on astronomical literature before this date consultation of one of the following bibliographies is suggested:

(1) J. J. de Lalande, Bibliographie Astronomique, Paris 1803 (this work covers the time from 480 B. C. to the year 1803, VIII + 966 pages).

(2) J. C. Houzeau, A. Lancaster, Bibliographie générale de l'astronomie, Volume I (in two parts), Bruxelles 1882, 1887, Volume II, Bruxelles 1889. The complete title of Volume II is "Bibliographie générale de l'astronomie ou catalogue méthodique des ouvrages, des mémoires et des observations astronomiques, publiés depuis l'origine de l'imprimerie jusqu'en 1880". A new edition of these volumes was prepared by D. W. Dewhirst in 1964.

(3) Bibliography of Astronomy, 1881 - 1898. The literature of this period was recorded on standard slips by the Observatoire Royal de Belgique. From the material (some 52,000 items) a microfilm version was produced by University Microfilms Limited, Tylers Green, High Wycombe, Buckinghamshire, England, in 1970.

(4) Astronomischer Jahresbericht, 1899 gegründet von Walter Wislicenus, herausgegeben vom Astronomischen Rechen-Institut in Heidelberg (formerly in Berlin), Verlag W. de Gruyter, Berlin. For the period from 1899 to 1968 sixty-eight volumes were published, each of which, in general, covers the literature of one year.

(5) Bulletin Signalétique – Section 120: Astronomie, Physique Spatiale, Géophysique. Published by Centre de Documentation du Centre National de la Recherche Scientifique, Paris. This publication is a continuation of "Bibliographie Mensuelle de l'Astronomie" founded in 1933 by the Société Astronomique de France. The publication is continued.

(6) Referativnyj Zhurnal. Founded in 1953 and published by Vsesoyuznyj Institut Nauchnoj i Tekhnicheskoj Informatsii, Akademiya Nauk, Moskva. The publication is continued.

Concept of Astronomy and Astrophysics Abstracts

This abstracting service aims to present a comprehensive documentation of the literature in all fields of astronomy and astrophysics and their border fields. It appears in semi-annual volumes. Two of these volumes cover the literature of one calendar year. The half-yearly period of issue is regarded as an optimal period for summarizing papers into subject categories and for the presentation of abstracts as quickly as possible after the publication of the original literature. The recording summarizing and indexing of astronomical publications of the year 1981 received from July 1981 to February 1982 are subjects of **Volume 30**. It also records a number of papers issued before 1981 but received within this period.

The main characteristics of the concept of Astronomy and Astrophysics Abstracts may be summarized as follows:

(1) The subdivision of astronomy and its border fields into subject categories is facilitated by the fact that the astronomical objects appear to be particularly well suited for the formation of categories. It may be assumed that such subdivisions can be maintained for a long period. Experience shows, however, that progress in research might imply minor changes in the classification scheme.

(2) Each paper has been classified into one of 108 numbered subject categories and given a serial number within the category. In this way each item is numbered by six figures: the first three indicate the number of the category, the following three the serial number within the category. Reference to an abstract in Volume 1 is indicated by "01" before the number of the category; for example: 01.074.028, denotes Volume 1, category 074, abstract 028.

A paper might be classified into more than one category. In this case, its abstract is placed only in one category, whereas in the other categories only cross references are given. These are listed at the end of each category.

(3) Authors' abstracts are used whenever possible. Popular articles are not abstracted.

(4) If possible, titles of papers and abstracts are given in English. A special reference is made to titles which we have not taken in the original language.

Transliteration scheme for the Russian alphabet

The transliteration of the Russian alphabet in use in Astronomy and Astrophysics Abstracts is presented here.

А	а	a	П	п	p
Б	б	b	Р	р	r
В	в	v	С	с	s
Г	г	g	Т	т	t
Д	д	d	У	у	u
Е	е	e	Ф	ф	f
Ё	ё	e	Х	х	kh
Ж	ж	zh	Ц	ц	ts
З	з	z	Ч	ч	ch
И	и	i	Ш	ш	sh
Й	й	j	Щ	щ	shch
К	к	k	Ы	ы	y
Л	л	l	Ь	ь	'
М	м	m	Э	э	eh
Н	н	n	Ю	ю	yu
О	о	o	Я	я	ya

This transliteration was recommended by the Abstracting Board of the International Council of Scientific Unions in 1969. It corresponds essentially to the transliteration proposed by the Academy of Sciences, Moscow, which is used by the Referativnyj Zhurnal. In this case the letters can be read and printed by usual data processing machines.

If the names of Russian authors in the literature are transliterated in a different scheme, we present the names as they are given in the references cited and in addition in brackets according to our transliteration table.

Sources of information

The majority of sources of information for this volume is given in section 001 Periodicals and in section 008 Observatories, Institutes. Section 001 records 614 periodicals indicating full titles and publishers. It may be noted that the titles of the periodicals are given in the original languages, and that Russian titles have been transliterated applying the transliteration scheme given above. Section 008 records 108 periodicals; these are publication series of observatories and astronomical institutes. Titles of the periodicals have been given following the recommendations of the "International List of Periodical Title Word Abbreviations" and its additions (see also **Abbreviations**, p. 10). In most cases they permit recognition of the full title without recourse to the key in section 001.

If other abstracting journals have been consulted in order to examine the degree of completeness of our service, we cite these papers and give reference to the abstracting service.

Author index and subject index

The subject category and the serial number have been used as a reference both in the author index and the subject index. These references are more precise than page references. They offer considerable advantages in indexing by means of data processing machines, and are more convenient for the user.

The author index of this volume contains 10140 names. A complete reference comprises six figures, three for the subject category and three for the serial number within the category. In the case of more than one reference to abstracts in one category, the number of the category is given only once and not repeated in the immediately following references. The total number of papers (some do not give names of authors) recorded in this volume amounts to 8550.

We consider the subject index as an approximation to an optimal index covering all fields of astronomy and astrophysics and their border fields. The assigning of one or more key words to a paper is, undoubtedly, a difficult task. Some journals have started giving key words together with the titles of papers. These key words are chosen by the authors themselves. Starting with Volume 18, the subject index was enlarged to a certain extent in order to provide a thesaurus of astronomical and astrophysical terms. This is done not only for the users' convenience, but also with the intention to propose the use of special key words to authors and publishers.

While each volume is scheduled to contain an author index and a subject index, the magnetic tapes containing the index information will be used to produce separate index volumes (authors and subjects) at intervals of five years.

The sorting program for the author and subject indexes is based on the IBM SORT/MERGE Program. This program sorts blank before hyphen (−) and before letters. Apostrophes are ignored by a special routine. The computations and printing were carried out on an IBM 360/44.

The two most common and widely used classification systems in astronomy and astrophysics are given by Class 9 of the International Classification System for Physics, published by the International Council of Scientific Unions Abstracting Board (Second edition 1978. ICSU-AB, 17 Rue Mirabeau, 75017 Paris, France, ISSN 0305-9618), and the Astronomy and Astrophysics Abstracts classification. In order to facilitate literature searches, we introduce a concordance relation between these two very different systems. This solution is only a unilateral one. Starting from the fourth hierarchical level of the ICSU-AB system, the appropriate Astronomy and Astrophysics Abstracts chapter numbers are listed. This cannot imply an identical content of the respective chapters in both systems. In many cases there is only a rather partial concordance, and therefore the Astronomy and Astrophysics Abstracts numbers are enclosed in parentheses. Due to the fact that our service only aims to present a comprehensive documentation of the literature in all fields of astronomy and astrophysics, only a certain part of Class 9 of the ISCU-AB scheme is covered.

The users are requested to inform us on spelling errors within the author and subject indexes in order to assist us in eliminating mistakes in future cumulative indexes.

Concordance Relation

between the ICSU-AB International Classification System for Physics
and the *Astronomy and Astrophysics Abstracts* Classification Scheme

ICSU-AB International Classification System for Physics	Astronomy and Astrophysics Abstracts Classification Scheme
0 General	
01.10 Announcements, news, and organizational activities	
01.10.C	006 (010)
01.10.F	011
01.10.H	013 (010)
01.30 Physics literature and publications	
01.30.B	012 (014)
01.30.C	012
01.30.E	003
01.30.K	002 (003)
01.30.M	003
01.30.P	003
01.30.R	014
01.30.T	002
01.40 Education	014
01.50 Educational aids	014
01.60 Biographical, historical, and personal notes	004 (005, 006, 007)
01.65 History of science	004
01.90 Other topics of general interest	015

ICSU-AB International Classification System for Physics	Astronomy and Astrophysics Abstracts Classification Scheme

9 Geophysics, Astronomy, and Astrophysics

91.10   Geodesy and gravity

91.10.B	046
91.10.N	044 (045)
91.10.Q	081

91.25 Geomagnetism and paleomagnetism; **geoelectricity**	084

91.35 Earth's interior structure and properties	081

91.90 Other topics in solid Earth physics	081

92.60 Meteorology	082

92.65 Atmospheric optics	082

94.10   Physics of the neutral atmosphere

94.10.B	082
94.10.D	082
94.10.F	082
94.10.G	082 (063)
94.10.H	082
94.10.L	082
94.10.N	106
94.10.Q	082
94.10.S	084

94.20   Physics of the ionosphere

94.20.B	083
94.20.D	083
94.20.M	083 (084)
94.20.P	083 (084)
94.20.W	083 (062)
94.20.Y	083 (084)

94.30   Physics of the magnetosphere

94.30.C	084
94.30.D	084

ICSU-AB International Classification System for Physics	Astronomy and Astrophysics Abstracts Classification Scheme
94.30.E	084
94.30.F	084 (062)
94.30.G	084 (062)
94.30.H	084
94.30.L	084
94.30.M	084
94.30.S	084
94.30.V	084 (074)
94.30.W	084 (078, 143)

94.40   Cosmic rays

94.40.C	143
94.40.E	143 (078, 106)
94.40.H	078
94.40.K	143 (085)
94.40.L	143 (078)
94.40.V	105 (143)

94.60   Interplanetary space

94.60.D	074
94.60.F	074
94.60.G	074 (062)
94.60.K	106
94.60.M	106
94.60.Q	074 (091, 094.0)
94.60.R	106 (062)

**94.80   Aerospace facilities and techniques,
space research**

94.80.P	053 (051)
94.80.R	054 (051)
94.80.W	032.5

95.10   Fundamental astronomy

95.10.C	042 (043, 052)
95.10.E	042 (052)
95.10.G	041 (079, 095, 096)
95.10.J	041

ICSU-AB International Classification System for Physics	Astronomy and Astrophysics Abstracts Classification Scheme
95.30 **Fundamental aspects of astrophysics**	
95.30.C	061 (022)
95.30.E	022
95.30.G	022 (061)
95.30.J	063
95.30.L	062
95.30.Q	062
95.30.S	066.0 (162)
95.45 **Observatories**	008 (009)
95.55 **Astronomical instruments**	
95.55.B	032.0
95.55.C	032.0 (031.0)
95.55.E	032.0
95.55.J	033
95.55.L	032.5
95.65 **Auxiliary and recording instruments**	034
95.70 **Other instrumentation and techniques (including clocks, frequency standards, etc.)**	035 (031.5, 034, 036)
95.75 **Techniques of observation and reduction**	
95.75.D	031.5
95.75.F	031.5
95.75.H	031.5
95.75.K	031.5
95.75.M	031.5 (021)
95.75.P	021
95.80 **Catalogues, atlases, etc.**	002 (047)
96.10 **General, solar nebula, and cosmogony**	107 (091)
96.20 **Moon**	
96.20.B	094.0
96.20.D	094.0 (094.5)
96.20.J	094.0

ICSU-AB International Classification System for Physics	Astronomy and Astrophysics Abstracts Classification Scheme

96.30 Planets and satellites (excluding the moon)

96.30.D	092
96.30.E	093
96.30.G	097
96.30.H	098
96.30.K	099
96.30.M	100
96.30.T	101

96.50 Other objects in the planetary system

96.50.D	106
96.50.G	102 (103)
96.50.K	104
96.50.M	105

96.60 Solar physics

96.60.C	080 (075)
96.60.F	071 (080)
96.60.K	080
96.60.M	071
96.60.N	073 (074)
96.60.P	074
96.60.Q	072
96.60.R	073 (076, 077)
96.60.S	073
96.60.V	078 (074)

97.10 Stellar characteristics and properties

97.10.B	131
97.10.C	065 (061)
97.10.E	064 (063)
97.10.F	112 (064)
97.10.H	064 (112)
97.10.K	116 (065)
97.10.L	116 (065)
97.10.N	115
97.10.Q	115
97.10.R	115 (113, 114).
97.10.T	114
97.10.V	111
97.10.W	111

ICSU-AB International Classification System for Physics	Astronomy and Astrophysics Abstracts Classification Scheme

97.20   Normal stars (by class): general or individual

97.20.D	121
97.20.R	126

97.30   Variable and peculiar stars (including novae)

97.30.E	114
97.30.F	116
97.30.G	122 (123)
97.30.J	122 (123)
97.30.K	122 (123)
97.30.N	122 (123)
97.30.Q	124 (122, 123)
97.30.S	122 (123)

97.60   Late stages of stellar evolution (including black holes)

97.60.B	125
97.60.G	141.5
97.60.J	066.5
97.60.L	066.0
97.60.S	066.0 (065)

97.80   Binary and multiple stars (including extrasolar planetary systems)

97.80.D	118 (002)
97.80.F	120
97.80.H	119
97.80.J	142.0 (117)
97.80.K	118
97.80.M	118

98.10   Stellar dynamics — 151

98.20   Stellar clusters and associations

98.20.C	152
98.20.E	153
98.20.H	154

ICSU-AB International Classification System for Physics	Astronomy and Astrophysics Abstracts Classification Scheme

98.40 Interstellar matter and nebulae

98.40.B	131
98.40.C	131
98.40.F	132
98.40.H	132 (134)
98.40.J	133 (112)
98.40.K	134 (131)
98.40.M	134
98.40.N	125

98.50 The Galaxy; extragalactic objects and systems

98.50.C	158
98.50.E	151 (158)
98.50.H	158 (162)
98.50.K	160
98.50.L	155 (156, 157)
98.50.M	160 (159)
98.50.R	158
98.50.T	161

98.70 Other objects and background radiations of
unknown origin or distances

98.70.D	141.0
98.70.J	141.0
98.70.L	133
98.70.Q	142.0 (142.5)
98.70.S	143
98.70.V	066.0 (142.0, 142.5, 162)

98.80 Cosmology

98.80.B	162
98.80.D	162 (066.0)
98.80.F	061 (162)

Abbreviations

Abbreviations used in *Astronomy and Astrophysics Abstracts* are primarily based on the 'International List of Periodical Title Word Abbreviations', prepared for the UNISIST/ICSU-AB Working Group on Bibliographic Descriptions (1970).

A.A.B.	Associazione Astrofili Bolognesi	Atmos.	Atmosf−, Atmosph−
Aarg.	Aargang	BAA	British Astronomical Association
AAS	American Astronomical Society	Bayer.	Bayerisch−
AAVSO	American Association of Variable Star Observers	Beitr.	Beitrag, Beiträge
		Beob.	Beobacht−
Abh.	Abhandlung−	Ber.	Bericht−
Abstr.	Abstract−	Bibl.	Bibliot−
Abt.	Abteilung	Bibliogr.	Bibliograf−, Bibliograph−
Acad.	Academi−, Academy	BIH	Bureau International de l'Heure (Paris)
Accad.	Accademi−	Bimest.	Bimestr−
Act.	Active, Activit−	Bl.	Blatt, Blätter
Adm.	Administr−	Bol.	Boletin
Adv.	Advanc−	Boll.	Bolletino
Aehron.	Aehronomi−	Bul.	Buleten−, Buletin−, Bulten
Aeron.	Aeronom−	Bull.	Bulletin−, Bullettino
Aeronaut.	Aeronauti−	Bur.	Bureau−
Aerosp.	Aerospace	Byul.	Byuleten−, Byuletin−
AG	Astronomische Gesellschaft	Byull.	Byulleten−
AIAA	American Institute of Aeronautics and Astronautics	C.R.	Comptes Rendus
		Cah.	Cahier−
AJB	Astronomischer Jahresbericht	Calif.	California
Akad.	Akadem−	Cas.	Casopis
Ala.	Alabama	Cent.	Center−, Central, Centrale, Centrally, Centre
Alm.	Almanac−, Almanak−		
Amat.	Amateur−	Cercet.	Cercetary
An.	Anais, Anale−, Anali−, Anals	Chem.	Chemi−
Anal.	Analis−, Analit−, Analys−, Analyt−	Chim.	Chimi−
Angew.	Angewandt−	Chron.	Chronic−, Chronik, Chronique
Ann.	Annaes, Annal−	Chronom.	Chronometr−
Annu.	Annu−	Cie.	Compagnie
Anst.	Anstalt	Cienc.	Ciencia−
Anu.	Anual−, Anuar−	Cient.	Cientific−
Anz.	Anzeiger	Circ.	Circolar−, Circolo, Circolaire−, Circular−, Circulo
Appl.	Applied		
Arb.	Arbeit	Cirk.	Cirkulaer−
Arch.	Archiv−	Cl.	Clasa, Classe−
Årg.	Årgang	Co.	Companies, Company
Ariz.	Arizona	Coll.	College
Ark.	Arkiv−	Collect.	Collect−
Arkh.	Arkhiv−	Colloq.	Colloqui−
Artif.	Artifici−	Colo.	Colorado
ASA	Astronomical Society of Australia	Comet.	Cometary
Asoc.	Asocia−	Commentat.	Commentat−
ASP	Astronomical Society of the Pacific	Commun.	Communica−
ASSA	Astronomical Society of Southern Africa	Comput.	Computation, Computer−, Computing
Assem.	Assembl−	Comun.	Comunica−
Assoc.	Associ−	Conf.	Conferen−
Assoz.	Assozi−	Congr.	Congres−
Astrofis.	Astrofisic−	Conn.	Connecticut
Astrofiz.	Astrofizi−	Contract.	Contract−
Astrometr.	Astrometr−	Contrib.	Contribu−
Astron.	Astronom−	Cosm.	Cosmic−
Astronaut.	Astronauti−, Astronauty−	Cosmochim.	Cosmochimi−
Astrophys.	Astrophys−	COSPAR	Committee on Space Research
ASV	Astronomical Society of Victoria	Crystallogr.	Crystallograph−
ASWA	Astronomical Society of Western Australia	CSIRO	Commonwealth Scientific and Industrial Research Organization
At.	Atom−	Cult.	Cultur−, Cultuur

Curr.	Current
D.C.	District of Columbia
DDR	Deutsche Demokratische Republik
Del.	Delaware
Dep.	Departament, Département, Department
Dev.	Development–, Développement–
Diss.	Disserta–
Div.	Divis–
Doc.	Document–
Dok.	Dokument–
Dokl.	Doklad–
Ehksp.	Ehksperiment–
Eidg.	Eidgenössisch–
Eksp.	Eksperiment–
Electron.	Electroni–
Eng.	Engineer–
Environ.	Environment–
Equip.	Equipement, Equipment
Ergeb.	Ergebnis–
ESA	European Space Agency
ESO	European Southern Observatory
ESRO	European Space Research Organization
Eval.	Evaluation–
Exp.	Experiment–
Extraterr.	Extraterrestr–
F. R. Germany	Federal Republic of Germany
Fac.	Facolt–, Faculd–, Facult–
Fak.	Fakult
Fasc.	Fascicul–
Fenn.	Fenni–
Fis.	Fisic–, Fisik–
Fiz.	Fizic–, Fizik–, Fizyk–
Fla.	Florida
Fluid.	Fluidi–
Fond.	Fondation–, Fondazione
Fortschr.	Fortschritt–
Fotogr.	Fotograf–
Found.	Foundation–
Freq.	Frequen–
Fundam.	Fundamenta–
Fys.	Fysik–, Fysisch, Fysisk–
Fyz.	Fyzik–
G.	Giornale
Ga.	Georgia
Gaz.	Gazeta, Gazette
Gazz.	Gazzetta
Gen.	General
Geochem.	Geochem–
Geochim.	Geochim–
Geod.	Geodaes–, Geodaet–, Geodes–, Geodet–, Geodez–
Geofis.	Geofis–
Geofiz.	Geofiz–
Geofys.	Geofys–
Geogr.	Geograf–, Geograph–
Geokhim.	Geokhim–
Geol.	Geolog–, Geolosk–
Geomagn.	Geomagneti–
Geophys.	Geophys–
Ges.	Gesellschaft
Gesch.	Geschichte
Gl.	Glavno–
Glas.	Glasnik
Gos.	Gosudarst–
Gov.	Government–
Grenzgeb.	Grenzgebiet–
GSFC	Goddard Space Flight Center
H. M.	Her Majesty's, His Majesty's
Handb.	Handbook, Handbuch
Her.	Herald–
Hist.	History
Hochsch.	Hochschule
Hoegsk.	Hoegskol–
HR-diagram	Hertzsprung-Russell diagram
Hydrogr.	Hydrograf–, Hydrograph–
IAF	International Astronautical Federation
IAU	International Astronomical Union
IBM	International Business Machines Corporation
ICSU	International Council of Scientific Unions
ICSU-AB	International Council of Scientific Unions– Abstracting Board
IEEE	Institute of Electrical and Electronics Engineers
Ill.	Illinois
Inc.	Incorporated
Ind.	Industr–
Inf.	Informat–, Informaz–, Informe–
Ing.	Ingenieur
INIS	International Nuclear Information System
INSPEC	International Information Services for the Physics and Engineering Communities
Inst.	Institut–, Instytut–
Instn.	Institution
Instrum.	Instrument–
Int.	Internationa–, Internazional–
Inter.	Intérieur–, Interior
Interplanet.	Interplanetary
Intez.	Intezet–
Invest.	Investiga–
Ionos.	Ionosfer–, Ionospher–
Iskusstv.	Iskusstvenn–
Issled.	Issledovan–
Ist.	Istitut–
Izd.	Izdatel–
Izv.	Izvesti–
J.	Joernaal–, Jornal–, Journal–
Jaarb.	Jaarboek–
Jahrb.	Jahrbuch, Jahrbücher
Jahresber.	Jahresbericht–
Jahresschr.	Jahresschrift
Jahrg.	Jahrgang
JPL	Jet Propulsion Laboratory
K.	Königlich–, Koninklijk–, Kunglig–
Kans.	Kansas
Kartogr.	Kartograf–
Kernforsch.	Kernforschung
Kernphys.	Kernphysik–
Khem.	Khemyi–
Khim.	Khimi–
Kim.	Kimija–, Kimya
Kl.	Klass–
Kolloq.	Kolloquium–
Komet.	Kometnyj
Komm.	Kommission–
Konf.	Konfer–
Kongr.	Kongress
Kosm.	Kosmich–
Kosmog.	Kosmogon–
Kozp.	Kozponti
KPNO	Kitt Peak National Observatory
Kut.	Kutato
Ky.	Kentucky
La.	Louisiana
Lab.	Laborato–
Lett.	Letter–, Lettra, Lettre
Libr.	Librair–, Librar–
Mag.	Magasin, Magazin–
Magn.	Magneti–, Magnitn–
Mass.	Massachusetts
Mat.	Matemaat–, Matemat–
Mater.	Material–
Math.	Mathemat–
Md.	Maryland

Meas.	Measur–	Period.	Periodi–
Mec.	Mecani–	Petrol.	Petrolog–
Mech.	Mechani–	Philos.	Philosoph–
Medd.	Meddelande–, Meddelelse	Photogr.	Photograf–, Photograph–
Meded.	Mededeeling–, Mededeling–	Photogramm.	Photogrammetr–
Mekh.	Mekhani–	Photom.	Photometr–
Mem.	Memento–, Memoir–, Memori–,	Phys.	Physic–, Physik–, Physique–, Physisch–
	Memory–, Memuary	Pict.	Picture–
Memo.	Memorand–	Planet.	Planetary
Mens.	Mensile, Mensual–, Mensuel–	Pr.	Prac–
Messtech.	Messtechni–	Prelim.	Prelimin–
Meteorol.	Meteorolog–	Prepr.	Preprint
Mich.	Michigan	Prib.	Pribor–
Micromec.	Micromecaniq–	Prikl.	Prikladnoj
Miner.	Mineral, Minerale–, Minerali–	Prir.	Prirodn–
Mineral.	Mineralog–	Prirodoved.	Prirodoved–
Minn.	Minnesota	Probl.	Problem–
Miss.	Mississippi	Proc.	Proceedings
MIT	Massachusetts Institute of Technology	Prod.	Prodott–, Produc–, Produkt,
Mitt.	Mitteilung–	Prog.	Progres–
Mo.	Missouri	Propag.	Propagation
Mod.	Modern–	Prov.	Provinc–, Provints–, Provinz–
Mol.	Molecul–, Molekul–	Pubbl.	Pubblicazion–
Mon.	Monat, Monatlich–, Month–	Publ.	Publicac–, Publicas–, Publicat–,
Monogr.	Monograph–		Publikas–, Publikat–
Mont.	Montana	Q.	Quarterly
MPI	Max-Planck-Institut	Quant.	Quantit–
Mus.	Museum	R.	Royal
N. C.	North Carolina	R. I.	Rhode Island
N. D.	North Dakota	Radiat.	Radiati–
N. H.	New Hampshire	Radioact.	Radioactiv–, Radioaktiv–
N. J.	New Jersey	Radioisot.	Radioisotop–
N. M.	New Mexico	Rap.	Raport–
N. Y.	New York	Rapp.	Rapport–
Nablyud.	Nablyudeni–	RAS	Royal Astronomical Society
Nac.	Nacion–	Rec.	Record–
Nachr.	Nachricht–	Rech.	Recherche–
NASA	National Aeronautics and Space	Ref.	Referat–, Reference–, Referieren
	Administration	Relat.	Related, Relation–
Nat.	Natur–	Relativ.	Relativit–
Natl.	National–	Rend.	Rendicont–
Naturforsch.	Naturforsch–	Rep.	Report–
Naturwiss.	Naturwissenschaft–	Repr.	Reprint–
Natuurkd.	Natuurkunde	Repub.	Republi–
Nauchn.	Nauchny–	Res.	Research–
Nauk.	Nauka, Naukite, Naukov–, Naukow–	Result.	Resultad–, Resultat–
Naut.	Nautic–	Rev.	Review–, Revisio, Revista, Revue–
Nav.	Naval–	Rezul't.	Rezul'tat–
Navig.	Navigat–	Ric.	Ricerca, Ricerche
Naz.	Nazion–	Riv.	Rivist–
Nebr.	Nebraska	Rundsch.	Rundschau
Nev.	Nevada	S. C.	South Carolina
Newsl.	Newsletter–	S. D.	South Dakota
Not.	Notationes, Notic–, Notise–, Notizi–	SAF	Société Astronomique de France
Nouv.	Nouveau–, Nouvell–	SAI	Società Astronomica Italiana
Nov.	Novoe	Samml.	Sammlung–
Nucl.	Nucléaire–, Nuclear–, Nucl–	SAO	Smithsonian Astrophysical Observatory
Nukl.	Nukle–	SAS	Société Astronomique de Suisse
Numer.	Numeri–	Satell.	Satellite
O-va	Obshchestva	Sb.	Sbornik–
O-vo	Obshchestvo	Schr.	Schrift–
Obs.	Observ–	Schriftenr.	Schriftenreihe
Obz.	Obzor–	Sci.	Scienc–, Scient–, Scienz–
Okla.	Oklahoma	Scr.	Scripta, Scritt–
Opt.	Optic–, Optik–, Optique	Secc.	Seccion–
Oreg.	Oregon	Sect.	Secti–
Oss.	Osserva–	Sekc.	Sekci–, Sekcj–
Pa.	Pennsylvania	Sekt.	Sektion–, Sektor–
Paleontol.	Paleontolog–	Sekts.	Sektsi–
Pap.	Paper–, Papier	Sel.	Seleccion–, Select–, Selek–, Selezione
Part.	Particle	Selsk.	Selskab–, Selskap–
Perem.	Peremenn–	Semin.	Séminair–, Seminar–

Sep.	Separat–		Tidskr.	Tidskrift–
Ser.	Seria–, Serie–, Seriya		Tidsskr.	Tidsskrift–
Serv.	Servic–, Serviz–		Top.	Topic–
Sess.	Sessi–		Tr.	Trudy
Signal.	Signalétique–		Trans.	Transactions, Transazione
Simp.	Simpoz–		Tsentr.	Tsentral–
Sitzungsber.	Sitzungsbericht–		Tsirk.	Tsirkulyar–
Skr.	Skrift–		TU	Technical University
Soc.	Sociedad–, Societ–		Uch.	Uchen–
Sol.	Solar		Uchebn.	Uchebn–
Soln.	Solnechn–		UK	United Kingdom
Sonderdr.	Sonderdruck–		Umsch.	Umschau
Soobshch.	Soobshchen–		UN	United Nations
South.	Southern		Univ.	Universidad–, Universit–, Univerzitet–
Spacecr.	Spacecraft		US	United States
Spat.	Spatial–		USA	United States of America
Spec.	Special–		USSR	Union of Soviet Socialist Republics
Spectrosc.	Spectroscop–		Va.	Virginia
Spectrosk.	Spectroskop–		Var.	Various
Spets.	Spetsial–		Ver.	Verein–, Verenig–
Spez.	Spezial–, Speziell–		Veränderl.	Veränderlich–
SSR	Sovetskaya Sotsialisticheskaya Respublika		Verh.	Verhandl–
			Vermess.	Vermessung–
SSSR	Soyuz Sovetskikh Sotsialisticheskikh Respublik		Vermessungswes.	Vermessungswesen
			Veröff.	Veröffentlich–
St.	Saint–, Sankt–, Sant–		Vesn.	Vesnik
–St.	–Straße, Street		Vestn.	Vestnik
Stand.	Standard–, Standart–		Vetensk.	Vetenskap–
Sternw.	Sternwarte–		Vidensk.	Videnskab–, Videnskap
Stiint.	Stiintific–		Vierteljahresschr.	Vierteljahresschrift–
Stn.	Station, Stazione		Vierteljahrsschr.	Vierteljahrsschrift–
Stud.	Studia, Studie–, Studii		VLB	Very Long Baseline
Supl.	Suplement–, Supliment–		Volcanol.	Volcanolog–
Suppl.	Supplement–		Vopr.	Vopros–
Surv.	Survey–		Vortr.	Vorträge
Symp.	Sympos–, Sympoz–		Vses.	Vsesoyuzn–
Syst.	System–		Vt.	Vermont
Sz.	Szemle		Vyp.	Vypusk–
Teach.	Teacher–, Teaching		Vyssh.	Vyssh–
Tec.	Tecni–		Vyzk.	Vyzkum–
Tech.	Techni–		W. Va.	West Virginia
Technol.	Technolog–		Wash.	Washington
Tecnol.	Tecnolog–		West.	Western
Teh.	Tehnic–, Tehnika, Tehnisk–		Wet.	Wetenschap–, Wetenskap–
Tehnol.	Tehnolog–, Tehnolosk–		Wis.	Wisconsin
Tek.	Tekni–		Wiss.	Wissenschaft–
Tekh.	Tekhni–		Wyo.	Wyoming
Tekhnol.	Tekhnolog–		Yad.	Yadern–
Teknol.	Teknolog–		Z.	Zeitschrift–
Telesc.	Telescop–		ZA	Zero Age
Telev.	Television–		ZAED	Zentralstelle für Atomkernenergie-Dokumentation
Tenn.	Tennessee			
Teor.	Teoret–, Teori–		Zap.	Zapisk–, Zapyisk–
Terr.	Terrestr–		Zaved.	Zaveden–
Test.	Testing		Zent.	Zentral
Tex.	Texas		Zentralbl.	Zentralblatt
TH	Technische Hochschule		Zesz.	Zeszyt
Theor.	Theoret–, Theori–		Zh.	Zhurnal–
Tidschr.	Tidschrift–		Zirk.	Zirkular

Periodicals, Proceedings, Books, Activities

001 Periodicals

A. A. O. Newsl.
Anglo-Australian Observatory Newsletter. Published by the Anglo-Australian Observatory, PO Box 296, Epping, NSW 2121, Australia.

AAS Photo-Bull.
AAS (American Astronomical Society) Photo-Bulletin. Published by the Working Group on Photographic Materials. Produced by Eastman Kodak Co., Rochester, N.Y.

AAVSO Bull.
American Association of Variable Star Observers Bulletin. 187 Concord Avenue, Cambridge, Mass., 02138, U.S.A.

Acad. R. Belgique, Bull. Cl. Sci.
Académie Royale de Belgique, Bulletin de la Classe des Sciences (Koninklijke Academie van België, Mededelingen van de Klasse der Wetenschappen). 5ᵉ Série, Palais des Académies, Bruxelles.

Acta Astron.
Acta Astronomica. An international quarterly journal. Publisher: Polska Akademia Nauk, Komitet Astronomii (Polish Academy of Sciences, Committee of Astronomy), Warszawa — Wrocław.

Acta Astron. Sinica
Acta Astronomica Sinica. Published by Purple Mountain Observatory, Academia Sinica, Nanking, China.

Acta Astronaut.
Acta Astronautica. Journal of the International Academy of Astronautics. Publisher: Pergamon Press Inc., Elmsford, New York, U.S.A.; Pergamon Press Ltd., Oxford, England.

Acta Astrophys. Sinica
Acta Astrophysica Sinica. Editorial Board "Acta Astrophysica Sinica", P. O. Box 399, Beijing, China.

Acta Cosmologica
Acta Cosmologica. Published by Obserwatorium Astronomiczne Uniwersytetu Jagiellońskiego, Kraków, Poland.

Acta Fac. Rerum Nat. Univ. Comenianae Phys.
Acta Facultatis Rerum Naturalium Universitatis Comenianae, Physica. Ústredná knižnica PFUK, 886 11 Bratislava, U1. 29 Augusta č.5, Czechoslovakia.

Acta Geod. Geophys. Montan.
Acta Geodaetica, Geophysica et Montanistica. Akademiai Kiado, 1054 Budapest, Alkotmany utca 21, Hungary.

Acta Geophys. Polonica
Acta Geophysica Polonica. ARS Polona-Ruch, 00-068 Warszawa, Krakowskie Przedmiescie 7, P.O. Box 1001, Poland.

Acta Geophys. Sinica
Acta Geophysica Sinica. Chinese Academy of Sciences, Department of Geophysical Research. Published by Science Press, Peking, People's Republic of China.

Acta Mech. Sinica
Acta Mechanica Sinica, Science Press, Peking, People's Republic of China. Subscription address: Guozi Shudian, P.O. Box 399, Peking.

Acta Phys. Acad. Sci. Hungaricae
Acta Physica Academiae Scientiarum Hungaricae. Postafiok 24, Budapest 502, Hungary.

Acta Phys. Austriaca
Acta Physica Austriaca. Springer-Verlag, A-1011 Wien, Molkerbastei 5, Postfach 367, Austria.

Acta Phys. Polonica B
Acta Physica Polonica B. ARS Polona-Ruch, Warszawa 1, P.O. Box 154, Poland.

Acta Phys. Sinica
Acta Physica Sinica. Chinese Academy of Sciences, Institute of Physics, Peking, People's Republic of China. [English translation in: Chinese J. Phys. (*USA*)].

Acta Phys. Slovaca
Acta Physica Slovaca. VEDA Publishing House of the Slovak Academy of Sciences, 895 30 Bratislava, Klemensova 27, Czechoslovakia.

Acta Polytech. III
Acta Polytechnica. Series III. Elektrotechnická fakulta ČVUT v Praze, Technická ul. 2, Praha 6-Dejvice, Czechoslovakia.

Acta Sci. Nat. Univ. Pekinensis
Acta Scientiarum Naturalium Universitatis Pekinensis. Peking, People's Republic of China.

Acta Sci. Nat. Univ. Sunyatseni
Acta Scientiarum Naturalium Universitatis Sunyatseni (Zhongshandaxue Xuebao). Canton Post Office, Canton, People's Republic of China.

Acta Tech. Acad. Sci. Hungaricae
Acta Technica Academiae Scientiarum Hungaricae. Akademiai Kiado, Budapest 1363, P. O. Box 24, Hungary.

Acta Tech. CSAV
Acta Technica Československá akademie věd. Academia, Publishing House of the Czechoslovak Academy of Sciences, Vodickova 40, 112 29 Praha 1, Czechoslovakia. John Benjamins N.V., Periodical Trade, Warmoesstraat 54, Amsterdam, Netherlands.

Acta Univ. Carolinae Math. Phys.
Acta Universitatis Carolinae, Mathematica et Physica. Administrace: Matematicko-fyzikální fakulta University Karlovy, Praha.

Adv. Phys.
Advances in Physics. Taylor & Francis Ltd., 10–14 Macklin Street London, WC2B 5NF, England.

Adv. Space Res.
Advances in Space Research. The official journal of the Committee on Space Research (COSPAR). Pergamon Press, Oxford – New York – Toronto – Sydney – Paris – Frankfurt. ISSN 0273-1177.

Aeronaut. Astronaut.
L'Aeronautique et l'Astronautique. Editions Air et Cosmos, 6 Rue Anatole de la Forge, 75017 Paris, France.

AIAA J.
AIAA Journal. A Publication of the American Institute of Aeronautics and Astronautics devoted to Aerospace Research and Development. Published by the American Institute of Aeronautics and Astronautics, New York, N.Y.

AIP Conf. Proc.
AIP Conference Proceedings. American Institute of Physics, 335 East 45th Street, New York, N.Y. 10017, USA.

Alta Freq.
Alta Frequenza.Ufficio Centrale AEI-CEI, Viale Monza 259, 20126 Milano, Italy.

American J. Phys.
American Journal of Physics. Published for the American Association of Physics Teachers by the American Institute of Physics, 335 East 45th Street, New York, N. Y. 10017, USA.

American Sci.
American Scientist. Society of Sigma XI, 345 Whitney Avenue, New Haven, CT 06510, USA.

An. Acad. Brasil. Cienc.
Anais da Academia Brasileira de Ciencias. Caixa Postal 229, ZC-00 Rio de Janeiro gb, Brazil.

An. Fis.
Anales de Física. Real Sociedad Española de Física y Química (Facultad de Ciencias), Ciudad Universitaria, Madrid-3, Spain.

Ann. Acad. Sci. Fennicae, Ser. A. VI
Annales Academiae Scientiarum Fennicae, Series A VI (Physica). Snellmaninkato 9-11, 00170 Helsinki-17, Finnland.

Ann. Geofis.
Annali di Geofisica. Istituto Nazionale di Geofisica, Citta Universitaria, Via Ruggero Bonghi 11/B, 00184 Roma, Italy.

Ann. Géophys.
Annales de Géophysique. Service des Publications du CNRS, 15 Quai Anatole-France, 75700 Paris, France.

Ann. Inst. Henri Poincaré A
Annales de l'Institut Henri Poincaré, Section A (Physique Théorique). 11 Rue Pierre-Curie, Paris 5, France.

Ann. Nucl. Energy
Annals of Nuclear Energy. Pergamon Press Ltd.,
Headington Hill Hall Oxford, OX3 0BW, England.

Ann. Physics
Annals of Physics. Academic Press Inc., 111 Fifth Avenue, New York, NY 10003, USA.

Ann. Physik
Annalen der Physik. 7. Folge. Publisher: Johann Ambrosius Barth, Salomonstr. 18B, Leipzig 701, German Democratic Republic.

Ann. Physique
Annales de Physique. Publisher: Masson et Cie., 120 Boulevard Saint-Germain, Paris 6, France.

Ann. Sci.
Annals of Science. Taylor & Francis Ltd., 10-14 Macklin Street, London, WC2B 5NF, England.

Ann. Shanghai Obs., Acad. Sinica
Annuals of Shanghai Observatory, Academia Sinica. Published by Shanghai Scientific and Technical Publishers, Shanghai, Rei Jing Er Street, No. 450.

Ann. Soc. Sci. Bruxelles I
Annales de la Société Scientifique de Bruxelles. Série I: Sciences Mathématiques, Astronomiques et Physiques. Rue de Bruxelles 61, B 5000 Namur, Belgium.

Ann. Télécommun.
Annales des Télécommunications. Centre National d'Études des Télécommunications, 38 rue du Général Leclerc, 92 Issy-les-Moulineaux, France.

Ann. Tokyo Astron. Obs.
Annals of the Tokyo Astronomical Observatory. University of Tokyo, Mitaka, Tokyo, Japan.

Annu. Rep. Astron. Inst. Greece
Annual Reports of the Astronomical Institutes of Greece. Published by the Greek National Committee for Astronomy. Academy of Athens, Research Center for Astronomy and Applied Mathematics.

Annu. Rev. Astron. Astrophys.
Annual Review of Astronomy and Astrophysics. Annual Reviews Inc., 4139 El Camino Way, Palo Alto, Calif. 94306, USA. ISSN 0066-4146.

Annu. Rev. Earth Planet. Sci.
Annual Review of Earth and Planetary Sciences. Annual Reviews Inc., 4139 El Camino Way, Palo Alto, Calif. 94306, USA. ISSN 0084-6597.

Annu. Univ. Sofia Fac. Phys.
Annuaire de l'Université de Sofia Faculté de Physique, Sofiya, Bulgaria.

Antenna
L'Antenna. Via Monte Generoso 6/a, 20155 Milano, Italy.

Anz. Österreich. Akad. Wiss. Math.-Naturwiss. Kl.
Anzeiger. Österreichische Akademie der Wissenschaften. Mathematisch-Naturwissenschaftliche Klasse. Publisher: Springer-Verlag, Wien.

APL Tech. Dig.
APL Technical Digest. Applied Physics Laboratory, The John Hopkins University, 8621 Georgia Avenue, Silver Spring, MD 20910, USA.

Appl. Opt.
Applied Optics. A monthly publication of the Optical Society of America. Published for the Optical Society of America by the American Institute of Physics, 335 East 45th Street, New York, NY 10017, USA.

Appl. Phys.
Applied Physics. Springer-Verlag, Heidelberger Platz 3, D-1000 Berlin 33, F. R. Germany.

Appl. Phys. Lett.
Applied Physics Letters. American Institute of Physics, 335 East 45th Street, New York, N.Y. 10017, USA.

Appl. Spectrosc.
Applied Spectroscopy. 428 East Preston Street, Baltimore, MD 21202, USA.

Appl. Spectrosc. Rev.
Applied Spectroscopy Reviews. Marcel Dekker Inc., 95 Madison Avenue, New York, NY 10016, USA.

Arch. Hist. Exact Sci.
Archive for History of Exact Sciences. Springer-Verlag, Berlin - Heidelberg - New York. ISSN 0003 - 9519.

Arch. Mech.
Archives of Mechanics (Archiwum Mechaniki Stosowanej). Polish Scientific Publishers, Swietokrzyska 21, Warszawa, Poland.

Arch. Sci.
Archives des Sciences, éditées par la Société de Physique et d'Histoire Naturelle de Genève. Publisher: Imprimerie Kundig, Genève. Subscription address: Librairie Payot, Genève.

Archaeoastronomy (*England*)
Archaeoastronomy. Supplement to Journal for the History of Astronomy. Published by Science History Publications Ltd, Halfpenny Furze, Mill Lane, Chalfont St Giles, Bucks, England, HP8 4NR. ISSN 0142–7253.

Archaeoastronomy (*U.S.A.*)
Archaeoastronomy. The Bulletin of the Center for Archaeoastronomy, Space Sciences Building, University of Maryland, College Park, MD 20742 (301) 454-4460. ISSN 0190–9940.

Archaeometry
Archaeometry. Cambridge University Press, P.O. Box 92, London, NW1 2DB, England.

Ark. Fys. Semin. Trondheim
Arkiv for det Fysiske Seminar i Trondheim. c/o Institutt for Teoretisk Fysikk, Universitetet i Trondheim, NTH, N-7034 Trondheim, Norway.

Ark. Mat.
Arkiv för Matematik. Published by Institut Mittag-Leffler, Auravägen 17, S-182 62 Djursholm, Sweden.

Artif. Satell.
Artificial Satellites. Publication of Polish Scientific Institutions. Polish Academy of Sciences, National Committee of Geophysics and Geodesy, National Committee for Space Research, Warsaw. Space Research Centre, Pałac Kultury i Nauki 2301, 00-901 Warszawa, Poland.

Astrofiz. Issled. Izv. Spets. Astrofiz. Obs.
Astrofizicheskie Issledovaniya. Izvestiya Spetsial'noj

Astrofizicheskoj Observatorii. Akademiya Nauk SSSR. Publishers: Izdatel'stvo "Nauka", Leningradskoe Otdelenie, Leningrad.

Astrofizika
Astrofizika. Izdatel'stvo Akademii Nauk Armyanskoj SSR, Erevan. [An English translation is published in "Astrophysics"].

Astrometr. Astrofiz.
Astrometriya i Astrofizika. Respublikanskij Mezhvedomstvennyj Sbornik. Akademiya Nauk Ukrainskoj SSR, Glavnaya Astronomicheskaya Observatoriya. Naukova Dumka, Kiev.

Astron. Astrophys.
Astronomy and Astrophysics. A European Journal. Published by Springer-Verlag, Berlin–Heidelberg–New York.

Astron. Astrophys., Suppl. Ser.
Astronomy and Astrophysics. Supplement Series. A European Journal. Published by les Editions de Physique, Orsay, France, on behalf of the Board of Directors of the European Southern Observatory. ISSN 0365-0138.

Astron. Circ.
Astronomical Circular. Compiled by the editor section of Acta Astronomica Sinica, Purple Mountain Observatory, Nanking, China. Edited by the Chinese Astronomical Society.

Astron. Data Cent. Bull.
Astronomical Data Center Bulletin. National Space Science Data Center/World Data Center A for Rockets and Satellites. National Aeronautics and Space Administration, Goddard Space Flight Center, Greenbelt, Maryland 20771, USA.

Astron. J.
The Astronomical Journal. Published for the American Astronomical Society by the American Institute of Physics, New York, N.Y. Editorial Office: Department of Astronomy, Columbia University, New York, N.Y.

Astron. Mitt. Eidg. Sternw. Zürich
Astronomische Mitteilungen der Eidgenössischen Sternwarte Zürich, Switzerland.

Astron. Nachr.
Astronomische Nachrichten. Publisher: Akademie-Verlag, Berlin.

Astron. Pap.
Astronomical Papers prepared for the use of the American Ephemeris and Nautical Almanac. Published by the Nautical Almanac Office, U.S. Naval Observatory by direction of the Secretary of the Navy and under the authority of Congress. U.S. Government Printing Office, Washington, D.C.

Astron. Q.
The Astronomy Quarterly. Pachart Publishing House, P.O. Box 35549, Tucson, Ariz. 85740. ISSN 0364-9229.

Astron. Rep.
The Astronomical Reports. Polish Amateur Astronomical Society. Polskie Towarzystwo Miłośników Astronomii, Kraków, Poland.

Astron. Schule
Astronomie in der Schule. Zeitschrift für die Hand des Astronomielehrers. Herausgegeben vom Verlag Volk und Wissen, Berlin. Redaktion: Sternwarte Bautzen.

Astron. Soc. Western Australia, Circ.
The Astronomical Society of Western Australia, Circular.

Astron. Tidsskr.
Astronomisk Tidsskrift. Edited by Astronomisk Selskab, København; Norsk Astronomisk Selskap, Oslo; Svenska Astronomiska Sällskapet, Stockholm. Printed by John Griegs Boktrykkeri, Bergen.

Astron. Tsirk.
Astronomicheskij Tsirkulyar, izdavaemyj Byuro Astronomicheskikh Soobshchenij Akademii Nauk SSSR. Tipografiya Astrosoveta AN SSSR, Moskva.

Astron. Vestn.
Astronomicheskij Vestnik. Publishers: Izdatel'stvo "Nauka", Moskva.

Astron. Zh.
Astronomicheskij Zhurnal. Akademiya Nauk SSSR. Publishers: Izdatel'stvo "Nauka", Moskva. [An English translation is published in "Soviet Astronomy"].

Astronomia
Astronomia. Periodico trimestrale dell'Unione Astrofili Italiani.

Astronomie
L'Astronomie et Bulletin de la Société Astronomique de France. Société Astronomique de France, Paris.

Astronomy
Astronomy. AstroMedia Corp., 757 North Broadway, Suite 204, Milwaukee, WI 53202, USA.

Astrophys. J.
The Astrophysical Journal. Published for the American Astronomical Society by the University of Chicago Press, Chicago, Illinois.

Astrophys. J., Lett.
The Astrophysical Journal. Letters to the Editors. Published for the American Astronomical Society by the University of Chicago Press, Chicago, Illinois.

Astrophys. J., Suppl. Ser.
The Astrophysical Journal. Supplement Series. Published for the American Astronomical Society by the University of Chicago Press, Chicago, Illinois.

Astrophys. Lett.
Astrophysical Letters. Published by NASA–Goddard Space Flight Center. Gordon and Breach Science Publishers Ltd., New York–London–Paris.

Astrophys. Space Sci.
Astrophysics and Space Science. An International Journal of Cosmic Physics. Published by D. Reidel Publishing Company, Dordrecht, Holland.

Astrophysics
Astrophysics. A cover-to-cover translation of Astrofizika (USSR). Consultants Bureau, New York, N. Y.

Atmos. Environ.
Atmospheric Environment. Pergamon Press Ltd., Headington Hill Hall, Oxford, OX3 OBW, England.

Atomkernenergie
Atomkernenergie. Verlag Karl Thiemig, Pilgersheimerstrasse 38, 8 München 90, Postfach 900740, F.R. Germany.

Atti Accad. Naz. Lincei, Mem. Ser. Ottava
Atti della Accademia Nazionale dei Lincei. Serie Ottava. Memorie. Classe di Scienze fisiche, matematiche e naturali. Sezione I: Matematica, Meccanica, Astronomia, Geodesia e Geofisica. Published by Accademia Nazionale dei Lincei, Roma.

Atti Accad. Naz. Lincei, Rend. Ser. Ottava
Atti della Accademia Nazionale dei Lincei. Serie Ottava. Rendiconti. Classe di Scienze fisiche, matematiche e naturali. Published by Accademia Nazionale dei Lincei, Roma.

Atti Accad. Sci. Torino I
Atti della Accademia delle Scienze di Torino. I. Classe di Scienze Fisiche, Mathematiche e Naturali. Via Accademia delle Scienze 6, Via Maria Vittoria 3, Torino (208), Italy.

Atti Fond. Giorgio Ronchi
Atti della Fondazione Giorgio Ronchi. Largo Enrico Fermi 1, 50125 Arcetri-Firenze, Italy.

Australian J. Phys.
Australian Journal of Physics. Published by the Commonwealth Scientific and Industrial Research Organization, 372 Albert Street, East Melbourne, Victoria 3002, Australia.

Australian J. Phys., Astrophys. Suppl.
Australian Journal of Physics, Astrophysical Supplement. Published by Commonwealth Scientific and Industrial Research Organization, 372 Albert Street, East Melbourne, Victoria 3002, Australia.

Autom. Strum.
Automazione e Strumentazione. Associazione Nazionale Italiana per l'Automazione, Via Le Premuda 2, 21029 Milano, Italy.

B. I. H., Paris, Circ.
Bureau International de l'Heure, B. I. H., Paris, Circulars. 61, Avenue de l'Observatoire, 75014-Paris.

BAV Rundbrief
BAV Rundbrief. Mitteilungsblatt der Berliner Arbeitsgemeinschaft für Veränderliche Sterne. Editor: BAV Berliner Arbeitsgemeinschaft für Veränderliche Sterne eV., Berlin.

BBSAG Bull.
Bedeckungsveränderlichen Beobachter der Schweizerischen Astronomischen Gesellschaft, [Swiss Astronomical Society's Eclipsing Variable Observers], Bulletin. To be obtained from R. Diethelm, Winterthur, Switzerland.

Biul. Obs. Astron. Uniw. M. Kopernika Toruniu
Biuletyn Obserwatorium Astronomicznego Uniwersytetu M. Kopernika w Toruniu.

Blick Weltall
Blick in das Weltall. Monatsprogramm und Mitteilungen für Sternfreunde. Archenhold-Sternwarte, Berlin-Treptow.

Bol. Acad. Cienc. Fis. Mat. Nat.
Boletin de la Academia de Ciencias Fisicas Matematicas y Naturales. Printed by Italgrafica, S. R. L. Republica de Venezuela.

Bol. Asoc. Argentina Astron.
Boletin de la Asociación Argentina de Astronomía,
San Juan, Argentina. ISSN 0571-3285.

Bol. Astron.
Boletin Astronômico. Observatório do Capricórnio,
Prefeitura Municipal de Campinas–SP, Brazil.

Bol. Astron. Obs. Madrid
Boletín Astronómico del Observatorio de Madrid.
Instituto Geografico Nacional, General Ibáñez de Ibero,
3. Madrid 3. Spain.

Bol. Inst. Tonantzintla
Boletin del Instituto de Tonantzintla. Instituto Nacional
de Astrofisica, Optica y Electronica, Apartados Postales
Nos. 216 y 51, Puebla, Pue, Mexico.

Bol. Obs. Ebro
Boletín del Observatorio del Ebro, Tortosa. Printed by
Cooperativa Gráfica Dertosense, Tortosa.

Boll. Geod. Sci. Affini
Bolletino di Geodesia e Scienze Affini. Pubblicazione
dell'Istituto Geografico Militare, Firenze.

Boundary-Layer Meteorol.
Boundary-Layer Meteorology. D. Reidel Publishing Co.,
P.O. Box 17, Dordrecht, Netherlands.

British Astron. Assoc. Circ.
British Astronomical Association, Circular. Editorial
Office: S. W. Milbourn, Brookhill Road, Copthorne Bank,
Crawley, West Sussex, RH10 3QJ.

British J. Philos. Sci.
British Journal for the Philosophy of Science. Cambridge
University Press, Bentley House, 200 Euston Road,
London, NW1 2DB, England.

British J. Photogr.
British Journal of Photography. Henry Greenwood &
Co., 24 Wellington Street, London, WC2E 7DH, England.

Bul. Inst. Politeh. 'Gheorghe Gheorghiu-Dej' Bucuresti.
Buletinul Institutului Politehnic 'Gheorghe Gheorghiu-
Dej' Bucuresti.Calea Grivitei 132, Bucuresti, Rumania.
Journal split into three series, Bul. Inst. Politeh.
'Gheorghe Gheorghiu-Dej' Bucuresti Ser. Chim. – Metal.,
Ser. Electroteh. and Ser. Mec. (Rumania).

Bulgarian J. Phys.
Bulgarian Journal of Physics. Bulgarian Academy of
Sciences, Faculty of Physics, 5 Anton Ivanov Blvd.,
1126 Sofia, Bulgaria.

Bull. Acad. Polonaise Sci., Ser. Sci. Tech.
Bulletin de l'Académie Polonaise des Sciences. Série des
Sciences Techniques. 00-901 Warszawa, Palac Kultury i
Nauki, P. O. Box 20, Poland.

Bull. AFOEV
Bulletin de l'Association Française des Observateurs
d'Etoiles Variables. Rédaction et publication:
E. Schweitzer, "La Moineaudière", 16, rue de Plobsheim,
67100 Strasbourg–Neudorf, France.

Bull. American Astron. Soc.
Bulletin of the American Astronomical Society. Published
for the American Astronomical Society by the American
Institute of Physics, 335 East 45th Street, New York,
N.Y. 10017, USA.

Bull. Astron. Inst. Czechoslovakia
Bulletin of the Astronomical Institutes of Czechoslovakia.
Published under the auspices of the Czechoslovak Acade-
my of Sciences by Academia, Praha. Editor: Astronomi-
cal Institute of the Czechoslovak Academy of Sciences,
Praha.

Bull. Astron., Obs. R. Belgique
Bulletin Astronomique, Observatoire Royal de Belgique.
(Astronomisch Bulletin, Koninklijke Sterrenwacht van
België).

Bull. Astron. Soc. India
Bulletin of the Astronomical Society of India. Edited and
published by M. S. Vardya, Tata Institute of Fundamental
Research, Bombay on behalf of the Astronomical Society
of India, Osmania University, Hyderabad.

Bull. Geod.
Bulletin Géodésique. The Journal of the International
Association of Geodesy. Publié par le Bureau Central
de l'Association Internationale de Géodésie, 39 Rue Gay-
Lussac, 75005 Paris, France.

Bull. Inf. Cent. Données Stellaires
Bulletin d'Information du Centre de Données Stellaires.
Compiled at Observatoire de Strasbourg, 11, rue de
l'Université, 67000-Strasbourg, France.

Bull. Inst. Space Aeronaut. Sci. Univ. Tokyo A
Bulletin of the Institute of Space and Aeronautical
Science, University of Tokyo A. Tokyo, Japan.

Bull. Inst. Space Aeronaut. Sci., Univ. Tokyo B
Bulletin of the Institute of Space and Aeronautical
Science, University of Tokyo B. Tokyo, Japan.

Bull. Obs. Astron. Belgrade
Bulletin de l'Observatoire Astronomique de Belgrade.
Editor: Observatoire Astronomique de Belgrade. Printed
by Naucna delo, Belgrade.

Bull. Res. Inst. Sci. Meas. Tôhoku Univ.
Bulletin of the Research Institute for Scientific Measurements, Tôhoku University, Sendai, Japan.

Bull. Sci. Yougoslavie
Bulletin Scientifique. Conseil des Academies des Sciences
et des Arts de la RSF de Yougoslavie. Section A: Sciences
Naturelles, Techniques et Médicales. Rédaction et Administration: Opaticka ul. 18/II, Zagreb, Yougoslavie.

Bull. Signal.
Bulletin Signalétique. Section 120: Astronomie, physique
spatiale, geophysique. Centre Nationale de la Recherche
Scientifique, Informascience, Centre de Documentation
Scientifique et Technique, 26, rue Boyer, 75971 Paris.
ISSN 0007-5337.

Bull. Soc. R. Sci. Liège
Bulletin de la Société Royale des Sciences de Liège.
L'Université, 15 Avenue des Tilleurs, Liège, Belgium.

Bull. Tokyo Gakugei Univ., Ser. IV
Bulletin of Tokyo Gakugei University. Series IV (Mathe-
matics and Natural Sciences) 4-1-1 Nukui-kita-machi,
Koganei, Tokyo, Japan.

Bull. Yamagata Univ. (Nat. Sci.)
Bulletin of the Yamagata University (Natural Science).
Yamagata, Japan.

Byull. Abastumanskaya Astrofiz. Obs.
Abastumanskaya Astrofizicheskaya Observatoriya, Gora
Kanobili. Byulleten'. Akademiya Nauk Gruzinskoj SSR.
Publishers: Izdatel'stvo "Metsniereba", Tbilisi.

Byull. Inst. Astrofiz.
Byulleten' Instituta Astrofiziki, Akademiya Nauk
Tadzhikskoj SSR. Izdatel'stvo Donish, Dushanbe.

Byull. Inst. Teor. Astron.
Byulleten' Instituta Teoreticheskoj Astronomii. Izdatel'-
stvo Nauka, Leningradskoe Otdelenie, Leningrad.

C. R. Acad. Sci. Paris
Comptes Rendus des Séances de l'Académie des Sciences.
Série II: mécanique, physique, chimie, sciences de
l'univers, sciences de la terre. Publiés par MM. les
Secrétaire perpétuels. Gauthier-Villars, C.D.R., Centrale
des Revues, B.P. No. 119, 93104 Montreuil Cedex,
France. ISSN 0567-6541.

Canadian J. Earth Sci.
Canadian Journal of Earth Sciences. National Research
Council of Canada, Ottawa KIA OR6, Canada.

Canadian J. Phys.
Canadian Journal of Physics. Published by the National
Research Council of Canada, Ottawa. Printed in Canada
by the University of Toronto Press, Toronto, Ont.

Carter Obs., Astron. Bull.
Carter Observatory, Astronomical Bulletin. Carter Observatory, P.O. Box 2909, Wellington 1, New Zealand.

Celestial Mech.
Celestial Mechanics. An International Journal of Space
Dynamics. Publishers: D. Reidel Publishing Company,
Dordrecht, Holland.

Cent. Astrophys. Prepr. Ser.
Center for Astrophysics, Preprint Series. Harvard College
Observatory, Smithsonian Astrophysical Observatory.
Center for Astrophysics, 60 Garden St., Cambridge,
Mass. 02138.

Centaurus
Centaurus. International magazine of the history of
mathematics, science, and technology. Munksgaard,
Copenhagen.

Ceskoslovensky Cas. Fyz., A.
Československý časopis pro fyziku. Sekce A. Academia
Publishing House of the Czechoslovak Academy of
Sciences, Vodičkova 40, 112 29 Praha 1, Czechoslovakia.

Chinese Astron.
Chinese Astronomy. A cover-to-cover translation of
Acta Astron. Sinica and Stud. Astron. Sinica. Published
by Pergamon Press, Headington Hill Hall, Oxford,
OX3 0BW, England — Maxwell House, Fairview Park,
Elmsford, N.Y. 10523, USA.

Chinese J. Phys.
Chinese Journal of Physics. Physical Society of the
Republic of China, Physics Department, National Taiwan
University, Taipei, Taiwan, China.

Ciel
Le Ciel. Bulletin de la Société Astronomique de Liège.
Éditeur responsable: Françoise Rameau, 19 - 21, rue
des genêts, 4310 St. Nicolas.

Ciel Terre
Ciel et Terre. Bulletin de la Société Belge d'Astronomie,
de Météorologie et de Physique du Globe. Administration: Avenue Circulaire, 3, Bruxelles. Printed by
Imprimerie R. Louis, Bruxelles.

Circ. Czechoslovak Obs., Time and Latitude
Circular of the Czechoslovak Observatories, Time and
Latitude. Czechoslovak Academy of Sciences, Astro-
nomical Institute, Prague, Czechoslovakia.

Circ. Inf.
Circulaire d'Information. Union Astronomique Interna-
tionale. Commission des Etoiles Doubles. Address:
Observatoire de Meudon, Meudon, France.

Circ. Stn. Astron. Int. Latitudine, Carloforte-Cagliari
Circolari della Stazione Astronomica Internazionale di
Latitudine, Carloforte-Cagliari. Serie A printed by Tipo-
Offset "3T", Cagliari. Serie B printed by Multi Copy,
Milano.

Circ. Time and Latitude Serv.
Circular Time and Latitude Service. Polish Academy of
Sciences, Astronomical Latitude Observatory, Borowiec,
Poland.

Coelum
Coelum. Periodico bimestrale per la Divulgazione dell'
Astronomia. Editor: Osservatorio Astronomico Univer-
sitario di Bologna.

Comments Astrophys.
Comments on Astrophysics. A Journal of Critical Dis-
cussion of the Current Literature. Comments on Modern
Physics: Part C. Publishers: Gordon and Breach, Science
Publishers Ltd., 42 William IV Street, London WC2,
England.

Comments At. Mol. Phys.
Comments on Atomic and Molecular Physics. Gordon
& Breach Science Publishers Ltd., 41 and 42 William IV
Street, London, WC2, England.

Comments Nucl. Part. Phys.
Comments on Nuclear and Particle Physics. Gordon &
Breach Science Publishers Ltd., 41 and 42 William IV
Street, London, WC2, England.

Comments Plasma Phys. Controlled Fusion
Comments on Plasma Physics and Controlled Fusion.
Gordon & Breach Science Publishers Ltd., 41 and 42
William IV Street, London, WC2, England.

Commun. Fac. Sci. Univ. Ankara
Communications de la Faculté des Sciences de l'Université
d'Ankara, Série A_2: Physique, Série A_3: Astronomie.

Commun. Math. Phys.
Communications in Mathematical Physics, Springer-
Verlag, Postfach 105280, 6900 Heidelberg 1, F.R.
Germany.

Comput. Phys. Commun.
Computer Physics Communications. North-Holland
Publishing Co., P. O. Box 211 Amsterdam, Netherlands.

Comun. Obs. Astron. Univ. Coimbra
Comunicações do Observatório Astronomico da Universidade de Coimbra, Portugal.

Contemp. Phys.
Contemporary Physics.Taylor and Francis Ltd., 10 - 14
Macklin Street, London, WC2B 5NF, England.

Contrib. Astron. Obs. Skalnaté Pleso.
Contributions of the Astronomical Observatory Skalnaté
Pleso. VEDA, vydavateľstvo Slovenskej akadémie vied,
Bratislava, Czechoslovakia.

Contrib. Atmos. Phys.
Contributions to Atmospheric Physics – Beiträge zur
Physik der Atmosphäre. Publisher: Friedrich Vieweg &
Sohn, Braunschweig.

Czechoslovak J. Phys. B
Czechoslovak Journal of Physics, Section B. Czechoslovak
Academy of Science, Akademia, Vodičkova 40, 112 29
Praha 1, Czechoslovakia.

Data Rep. Hydrogr. Obs., Ser. Astron. Geod., Tokyo
Data Report of Hydrographic Observations. Series of
Astronomy and Geodesy. Maritime Safety Agency,
Hydrographic Department Tsukiji-5, Chuo-ku, Tokyo,
104 Japan.

Debrecen Heliophys. Obs. Hungarian Acad. Sci., Repr.
Debrecen Heliophysical Observatory of the Hungarian
Academy of Sciences, Reprints.

Deutsche Geod. Komm. Bayerisch. Akad. Wiss.
Deutsche Geodätische Kommission bei der Bayerischen
Akademie der Wissenschaften. Reihe A: Höhere Geo-
däsie; Reihe B: Angewandte Geodäsie; Reihe C: Disser-
tationen; Reihe D: Tafelwerke; Reihe E: Geschichte und
Entwicklung der Geodäsie. Published by Verlag der
Bayerischen Akademie der Wissenschaften, München.

Dokl. Akad. Nauk SSSR
Doklady Akademii Nauk SSSR. Seriya Matematika,
Fizika. Publishers: Izdatel'stvo "Nauka", Moskva.

Dokl. Bolg. Akad. Nauk
Doklady Bolgarskoj Akademii Nauk. Sofiya, Bulgaria.

Dudley Obs. Rep.
Dudley Observatory Reports. Dudley Observatory,
Albany, N.Y., USA.

Dunsink Obs. Publ.
Dunsink Observatory Publications. The Observatory of
the School of Cosmic Physics, Dublin Institute for Advanced Studies, Dublin.

Earth Planet. Sci. Lett.
Earth and Planetary Science Letters. A Letter Journal
devoted to the Development in Time of the Earth and
Planetary System. Publisher: North-Holland Publishing
Company, Amsterdam, Netherlands.

Electro-Opt. Syst. Des.
Electro-Optical Systems Design. Milton S. Kiver Publications Inc., 222 West Adams, Chicago, IL 60606, USA.

Electron. Lett.
Electronics Letters. Institution of Electrical Engineers,
Savoy Place, London, WC2R OBL, England.

Electronics
Electronics. McGraw-Hill Publishing Co., 1221 Avenue of
the Americas, New York, N.Y. 10020, USA.

Elektroteh. Vestn.
Elektrotehniski Vestnik. Editorial address: YU-61001
Ljubljana, P. O. Box 92 - II, Trzaska 25, Yugoslavia.

Endeavour New Ser.
Endeavour New Series. Pergamon Press Ltd., Headington
Hill Hall, Oxford, OX3 0BW, England.

EOS Trans. American Geophys. Union
EOS Transactions of the American Geophysical Union.
1707 L Street, N.W., Washington, DC 20036, USA.

ESA Bull.
ESA Bulletin. Editorial Office: ESA Scientific and
Technical Publications Branch, ESTEC, Noordwijk, The
Netherlands. ISSN 0376-4265.

ESA IUE Newsl.
ESA IUE Newsletter. Published by The ESA IUE
Observatory, Villafranca Satellite Tracking Station,
Apartado 54065, Madrid, Spain.

ESA J.
ESA Journal. Editorial Office: ESA Scientific and
Technical Publications Branch, ESTEC, Noordwijk, The
Netherlands. ISSN 0379-2285.

ESO Sci. Prepr.
European Southern Observatory, Scientific Preprints.
Published by European Southern Observatory, Karl-
Schwarzschild-Straße 2, D-8046 Garching bei München.

ESO Tech. Rep.
European Southern Observatory, Technical Report.
Published by European Southern Observatory, Karl-
Schwarzschild-Straße 2, D-8046 Garching bei München.

European J. Phys.
European Journal of Physics. Institute of Physics,
47 Belgrave Square, London SW1X 8QX.

Exp. Tech. Phys.
Experimentelle Technik der Physik, VEB Deutscher
Verlag der Wissenschaften, Traubenstrasse 10,
108 Berlin 8, German Democratic Republic.

Feingerätetechnik
Feingerätetechnik. VEB Verlag Technik, Oranienburger
Strasse 13/14, 1020 Berlin, DDR.

Feinwerktech. Messtech.
F & M. Feinwerktechnik und Messtechnik. Fusion of
"Feinwerktechnik" and "Messtechnik" (formerly Zeit-
schrift für Instrumentenkunde) beginning with Jahrgang
82, No. 5 (1974). Publishers: Karl Hanser Verlag,
Kolbergerstr. 22, D-8000 München 80. F. R. Germany.

Fis. Tecnol.
Fisica e Tecnologia. Societa Italiana di Fisica, Via
Loderingo Degli Andalo 2, 40124 Bologna, Italy.

Fiz. Sz.
Fizikai Szemle. Kiadja a Lapkiado Vallalat, Budapest VII,
Lenin korut 9–11, Hungary.

Fizika
Fizika. 'Mladost' Export-Import, Zagreb, Ilica 30,
Yugoslavia.

Folia Fac. Sci. Nat. Univ. Purkynianae Brunensis Phys.
Folia Facultatis Scientiarum Naturalium Universitatis

Purkynianae Brunensis, Physica. University J. E. Purkyne, 61137 Brno-Kotlarska 2, Czechoslovakia.

Fortschr. Phys.
Fortschritte der Physik. Akademie-Verlag. DDR-108 Berlin, Leipzigerstrasse 3 - 4, Germany.

Found. Phys.
Foundations of Physics. Plenum Publishing Co., 8 Scrubs Lane, Harlesden, London, NW10 6SE, England.

Fra Fys. Verden
Fra Fysikkens Verden. Fysisk Institutt, Universitetet i Trondheim, Norges Laererhogskole, 7000 Trondheim, Norway.

Fundam. Cosmic Phys.
Fundamentals of Cosmic Physics. Gordon and Breach Science Publishers Ltd., New York–London–Paris.

Funkschau
Funkschau. Francis-Verlag, 8 München 37, Postfach 37 01 20, Karlstrasse 37, Germany.

Fys. Tidsskr.
Fysisk Tidsskrift. Subscription address: Jul. Gjellerups Boghandel, Solvgade 87, 1307 Kobenhavn, Denmark.

G. A.A.B.
Giornale dell'A.A.B. Notiziario trimestrale delle attività culturali e scientifiche della Associazione Astrofili Bolognesi, Bologna, Italy.

G. Astron.
Giornale di Astronomia, Pubblicazione della Società Astronomica Italiana. Printed by Tipolitografia Lodigraf S.p.A. Lodi (MI).

Gen. Relativ. Gravitation
General Relativity and Gravitation. Published under the auspices of the International Committee on General Relativity and Gravitation GRG. Publishing Office: Plenum Publishing Corporation, 233 Spring Street, New York, N. Y. 10013, USA.

Geochim. Cosmochim. Acta
Geochimica et Cosmochimica Acta. Journal of the Geochemical Society. Publishing House: Pergamon Press, Ltd., Oxford.

Geod. Geophys. Veröff., Reihe III
Geodätische und Geophysikalische Veröffentlichungen. Reihe III: Physik der festen Erde. Herausgegeben vom Nationalkomitee für Geodäsie und Geophysik bei der Akademie der Wissenschaften der Deutschen Demokratischen Republik.

Geod. Kartogr.
Geodezja i Kartografia. Komitet Geodezji Polskiej Akademii Nauk. Publisher: Państwowe Wydawnictwo Naukowe, Warszawa.

Geomagn. Aehron.
Geomagnetizm i Aehronomiya. Akademiya Nauk SSSR. Izdatel'stvo "Nauka", Moskva [An English translation is published in "Geomagnetism and Aeronomy", American Geophysical Union, Washington, D.C.].

Geomagn. Ser. Earth Phys. Branch
Geomagnetic Series, Earth Physics Branch. Energy, Mines and Resources Canada, 1 Observatory Crescent, Ottawa K1A OE4, Canada.

Geophys. Astrophys. Fluid Dyn.
Geophysical and Astrophysical Fluid Dynamics. Gordon and Breach Science Publishers Ltd., 41/42 William IV Street, London, WC2, England.

Geophys. J. R. Astron. Soc.
The Geophysical Journal of the Royal Astronomical Society. Published for the Royal Astronomical Society by Blackwell Scientific Publications, Oxford–Edinburgh. American Office of the Geophys. J., US Geological Survey, Stop 967, Box 25046, Federal Center, Denver, Colorado 80225, USA.

Geophys. Res. Lett.
Geophysical Research Letters. Published monthly by the American Geophysical Union, Washington, D.C., U.S.A.

Geophys. Surv.
Geophysical Surveys. D. Reidel Publishing Co., P.O. Box 17, Dordrecht, Netherlands.

Geophysics
Geophysics. Society of Exploration Geophysicists, P.O. Box 3098, Tulsa, OK 74101, USA.

GEOS
GEOS. Department of Energy, Mines and Resources, Ottawa, Canada.

GEOS Circ.
GEOS (Groupe: Etude et Observation Stellaire and Gruppo Europeo di Osservazione Stellare) Circulars, Series: RR (RR Lyrae type variables), SR (red variables), EB (eclipsing binaries), SA (small-amplitude variables). Published by A. Figer, GEOS, 12 rue Bezout, 75014 Paris, France.

Gerlands Beitr. Geophys.
Gerlands Beiträge zur Geophysik. Publisher: Akademische Verlagsgesellschaft Geest & Portig K.-G., Leipzig.

Glasnik Mat.
Glasnik Matematicki. Published by the Society of Mathematicians and Physicists of the S. R. of Croatia. Publisher: Drustvo Matematicara i Fizicara S. R. Hrvatske, Zagreb.

Heavens
The Heavens. The Oriental Astronomical Association, Ōtsu-shi, Shiga-ken, Japan. In Japanese.

HHI Sol. Data
HHI Solar Data. Heinrich-Hertz-Institut, Solare Beobachtungsergebnisse. Akademie der Wissenschaften der DDR, Zentralinstitut für Solar-Terrestrische Physik (Heinrich-Hertz-Institut), DDR-1199 Berlin-Adlershof.

HHI-STP-Rep.
HHI Solar-Terrestrische Physik Reports. Heinrich-Hertz-Institut. Akademie der Wissenschaften der DDR, Zentralinstitut für Solar-Terrestrische Physik (Heinrich-Hertz-Institut), DDR-1199 Berlin-Adlershof.

Hvar Obs. Bull.
Hvar Observatory Bulletin. Faculty of Geodesy. 41000 Zagreb, Kačićeva 26, Yugoslavia.

I. A. P. P. P. Commun.
International Amateur-Professional Photoelectric Photometry, Communication. R. M. Genet, Fairborn Observatory, 1247 Folk Road, Fairborn, Ohio 45324, USA.

I.U.A.A. Bull.
I.U.A.A. Bulletin. International Union of Amateur Astronomers, Contributions. I.U.A.A. c/o Achille Leani, via Bertesi 15, 26100 Cremona, Italy.

IAU Circ.
International Astronomical Union, Circular. Central Bureau for Astronomical Telegrams, Smithsonian Astrophysical Observatory, Cambridge, Mass.

Icarus
Icarus. International Journal of Solar System Studies. Publisher: Academic Press, New York—London.

IEE J. Microwave Opt. Acoust.
IEE Journal on Microwave, Optics and Acoustics. Institution of Electrical Engineers, Publishing Department, P.O. Box 8, Southgate House, Stevenage, Herts. SG1 1HQ, England.

IEEE Spectrum
IEEE Spectrum. Published monthly by the Institute of Electrical and Electronics Engineers, 345 East 47th Street. New York, N.Y. 10017, USA.

IEEE Trans. Aerosp. Electron. Syst.
IEEE Transactions on Aerospace and Electronic Systems. Published by the Institute of Electrical and Electronics Engineers, 345 East 47th Street, New York, N.Y. 10017, USA.

IEEE Trans. Antennas Propag.
IEEE Transactions on Antennas and Propagation. Published by the Institute of Electrical and Electronics Engineers, 345 East 47th Street, New York, N.Y. 10017, USA.

IEEE Trans. Commun.
IEEE Transactions on Communications. Institute of Electrical and Electronics Engineers, 345 East 47th Street, New York, NY 10017, USA.

IEEE Trans. Consum. Electron.
IEEE Transactions on Consumer Electronics. Institute of Electrical and Electronics Engineers, 345 East 47th Street, New York, N. Y. 10017, USA.

IEEE Trans. Electromagn. Compat.
IEEE Transactions on Electromagnetic Compatibility. Institute of Electrical and Electronics Engineers, 345 East 47th Street, New York, NY 10017, USA.

IEEE Trans. Electron Devices
IEEE Transactions on Electron Devices. Published by the Institute of Electrical and Electronics Engineers, 345 East 47th Street, New York, N. Y. 10017, USA.

IEEE Trans. Geosci. Electron.
IEEE Transactions on Geoscience Electronics. Published by the Institute of Electrical and Electronics Engineers, 345 East 47th Street, New York, N.Y. 10017, USA.

IEEE Trans. Instrum. Meas.
IEEE Transactions on Instrumentation and Measurement. Published by the Institute of Electrical and Electronics Engineers, 345 East 47th Street, New York, N.Y. 10017, USA.

IEEE Trans. Magn.
IEEE Transactions on Magnetics. Institute of Electrical and Electronics Engineers, 345 East 47th Street, New York, N. Y. 10017, USA.

IEEE Trans. Microwave Theory Tech.
IEEE Transactions on Microwave Theory and Techniques. Published by the Institute of Electrical and Electronics Engineers, 345 East 47th Street, New York, N.Y. 10017, USA.

IEEE Trans. Nucl. Sci.
IEEE Transactions on Nuclear Science. Institute of Electrical and Electronics Engineers, 345 East 47th Street, New York, N.Y. 10017, USA.

IEEE Trans. Plasma Sci.
IEEE Transactions on Plasma Science. Institute of Electrical and Electronics Engineers, 345 East 47th Street, New York, NY 10017, USA.

Indian J. Hist. Sci.
Indian Journal of History of Science. Published and printed by Indian National Science Academy, Bahadur Shah Zafar Marg, New Delhi 110002, at Mudranika, 13-A Bepin Pal Road, Calcutta 700026.

Indian J. Phys. Part B
Indian Journal of Physics Part B. Indian Association for the Cultivation of Science, 2 & 3 Raja Subodh Chandra Mallik Road, Calcutta 700032, India.

Indian J. Pure Appl. Math.
Indian Journal of Pure and Applied Mathematics. National Institute of Sciences India, Bahadur Shah Zafar Marg, New Delhi 1, India.

Indian J. Pure Appl. Phys.
Indian Journal of Pure and Applied Physics. Council of Scientific and Industrial Research, Hillside Road, New Delhi 110012, India.

Indian J. Radio Space Phys.
Indian Journal of Radio & Space Physics. Council of Scientific & Industrial Research. Editorial address: Publications & Information Directorate, Hillside Road, New Delhi 110012, India.

Indian J. Theor. Phys.
Indian Journal of Theoretical Physics. Institute of Theoretical Physics, Bognan Kutir, 4 - 1 Mohan Bagan Lane, Calcutta 700004, India

Inf. Bull. Variable Stars
Commission 27 of the I.A.U. Information Bulletin on Variable Stars. Konkoly Observatory, Budapest.

Informeto Astron. Obs. Univ. Turku
Informeto Astronomia Observatorio Universitato de Turku, Finnlando.

Infrared Phys.
Infrared Physics. An International Research Journal. Publisher: Pergamon Press Ltd., Oxford, England.

Inst. Theor. Astrophys., Blindern—Oslo, Rep.
Institute of Theoretical Astrophysics, Blindern—Oslo, Report. Universitetsforlagets trykningssentral, Oslo.

Int. Comet Q.
The International Comet Quarterly, Physics Department, Appalachian State University, Boone, NC 28608.

Int. J. Electron.
International Journal of Electronics. Taylor and Francis Ltd., 10—14 Macklin Street, London, WC2B 5BF, England.

Int. J. Heat Mass Transfer
International Journal of Heat and Mass Transfer.
Pergamon Press Ltd., Headington Hill Hall, Oxford,
OX3 OBW, England.

Int. J. Mass Spectrom. Ion Phys.
International Journal of Mass Spectrometry and Ion
Physics. Elsevier Scientific Publishing Co., P.O. Box 211
Amsterdam, Netherlands.

Int. J. Theor. Phys.
International Journal of Theoretical Physics. Plenum
Publishing Co. Ltd., Davis House, 8 Scrubs Lane, London,
NW10 6SE, England.

Interdisciplinary Sci. Rev.
Interdisciplinary Science Reviews. Heyden & Son Ltd.,
Spectrum House, Alderton Crescent, London NW4
3XX, England.

Irish Astron. J.
The Irish Astronomical Journal. A Quarterly Publication
under the auspices of the Observatories of Armagh and
Dunsink. Armagh Observatory, Northern Ireland.

ISIS
ISIS. An international review devoted to the history of
science and its cultural influences. Publication and
Editorial Office, Department of History and Sociology
of Science, University of Pennsylvania, Philadelphia
19104.

Istanbul Üniv. Fen Fak. Mec., Ser. C
Istanbul Üniversitesi Fen Fakultesi Mecmuasi, Serie C
(Astronomie, Physique, Chimie). Istanbul, Turkey.

Izv. Akad. Nauk Armyansk. SSR
Izvestiya Akademii Nauk Armyanskoj SSR. Fizika. Publisher: Izdatel'stvo AN Armyanskoj SSR, Erevan.

Izv. Astron. Ehngel'gardt. Obs.
Izvestiya Astronomicheskoj Ehngel'gardtovskoj Observatorii. Izdatel'stvo Kazanskogo Universiteta, Kazan.

Izv. Glav. Astron. Obs. Pulkovo
Izvestiya Glavnoj Astronomicheskoj Observatorii v Pulkove. Akademiya Nauk SSSR. Izdanie Glavnoj astronomicheskoj observatorii v Pulkove, Leningrad.

Izv. Krymskoj Astrofiz. Obs.
Izvestiya Krymskoj Astrofizicheskoj Observatorii.
Akademiya Nauk SSR. Publishers: Izdatel'stvo "Nauka",
Moskva.

J. Acoust. Soc. America
Journal of the Acoustical Society of America. Published
for the Acoustical Society of America by the American
Institute of Physics, 335 East 45th Street, New York,
NY 10017, USA

J. American Assoc. Variable Star Obs.
The Journal of the American Association of Variable Star
Observers. Published by The American Association of
Variable Star Observers, 187 Concord Avenue, Cambridge,
Mass. 02138, USA.

J. Appl. Meteorol.
Journal of Applied Meteorology. American Meteorological
Society, 45 Beacon Street, Boston, MA 02108, USA.

J. Appl. Photogr. Eng.
Journal of Applied Photographic Engineering. Society
of Photographic Scientists and Engineers, Suite 204,
1330 Massachusetts Avenue, N.W. Washington, D.C.
20005, USA.

J. Appl. Phys.
Journal of Applied Physics. American Institute of
Physics, 335 East 45th Street, New York, NY 10017,
USA.

J. Astron. Soc. Egypt
Journal of the Astronomical Society of Egypt.
Published by Helwan Observatory, Helwan, Egypt.

J. Astron. Soc. Western Australia
The Journal of the Astronomical Society of Western
Australia. Edited by the Astronomical Society of Western
Australia, Perth, W. A.

J. Astronaut. Sci.
Journal of the Astronautical Sciences. American Astronautical Society, 6060 Duke Street, Alexandria, VA
22304, USA.

J. Astrophys. Astron.
Journal of Astrophysics and Astronomy. Published by
Indian Academy of Sciences, Post Box No. 8005,
Bangalore 560080, India.

J. Atmos. Sci.
Journal of the Atmospheric Sciences. American Meteoro-
logical Society, 45 Beacon Street, Boston, MA 02108,
USA.

J. Atmos. Terr. Phys.
Journal of Atmospheric and Terrestrial Physics. Pergamon
Press Ltd., Oxford, England.

J. British Astron. Assoc.
Journal of the British Astronomical Association.
Burlington House, Piccadilly, London, W1V ONL, Eng-
land.

J. British Interplanet. Soc.
Journal of the British Interplanetary Society. British
Interplanetary Society, 12 Bessborough Gardens, London.
SW1V 2JJ, England.

J. Colloid Interface Sci.
Journal of Colloid and Interface Science. Academic Press
Inc., 111 Fifth Avenue, New York, N.Y. 10003, USA.

J. Comput. Phys.
Journal of Computational Physics. Academic Press Inc.,
111 Fifth Avenue, New York, NY 10003, USA.

J. Fluid Mech.
Journal of Fluid Mechanics. Cambridge University Press,
Bentley House, 200 Euston Road, London, NW1 2DB,
England.

J. Geomagn. Geoelectr.
Journal of Geomagnetism and Geoelectricity. Society of
Terrestrial Magnetism and Electricity of Japan, Geophysi-
cal Institute, Tokyo University, Tokyo 113, Japan.

J. Geophys.
Journal of Geophysics / Zeitschrift für Geophysik.
Springer Verlag, D-6900 Heidelberg 1, Postfach 105280,
F. R. Germany.

J. Geophys. Res.
Journal of Geophysical Research. Published by American

Geophysical Union, 1909 K Street, N.W. Washington D.C.
First section: Space physics; Second section: Physics and
chemistry of the solid earth, planetology, geodesy; Third
section: Oceans and atmospheres.

J. Guid. Control
Journal of Guidance and Control. American Institute
of Aeronautics and Astronautics, 1290 Avenue of the
Americas, New York, NY 10019, USA.

J. Hist. Astron.
Journal for the History of Astronomy. Published by
Science History Publications Ltd., Halfpenny Furze,
Mill Lane, Chalfont St Giles, Buckinghamshire, England.

J. Inst. Math. Appl.
Journal of the Institute of Mathematics and its Applica-
tions. Academic Press Inc. (London) Ltd., 24 - 28 Oval
Road, London NW1 7DX, England.

J. Magn. Magn. Mater.
Journal of Magnetism and Magnetic Materials. North-
Holland Publishing Co., P.O. Box 211, Amsterdam,
Netherlands.

J. Math. Phys.
Journal of Mathematical Physics. American Institute of
Physics, 335 East 45th Street, New York, N.Y. 10017,
USA.

J. Mol. Spectrosc.
Journal of Molecular Spectroscopy. Academic Press Inc.,
111 Fifth Avenue, New York, NY 10003, USA.

J. Nanjing Univ.
Journal of Nanjing University. Nanjing Daxue Xuebao.
(Natural Science Edition). Nanking University, Nanking,
China.

J. Navig.
The Journal of Navigation. The Royal Institute of Naviga-
tion at the Royal Geographical Society, Kensington Gore,
London, SW7 2AT. Scottish Academic Press Ltd., 33
Montgomery Street, Edinburgh EH7 5JX.

J. Opt. (France)
Journal of Optics. Masson Editeur, 120 Boulevard
Saint-Germain, 75280 Paris Cedex 06, France.

J. Opt. (India)
Journal of Optics. Optical Society of India, Department
of Applied Physics, University of Calcutta, 92 Acharya
Prafulla Chandra Road, Calcutta-9, India.

J. Opt. Soc. America
Journal of the Optical Society of America. American
Institute of Physics, 335 East 45th Street, New York,
N.Y. 10017, USA.

J. Phys. A
Journal of Physics A, (Mathematical, Nuclear and
General). Institute of Physics, 47 Belgrave Square,
London, SW1X 8QX, England.

J. Phys. B
Journal of Physics B, (Atomic and Molecular Physics).
Institute of Physics, 47 Belgrave Square, London,
SW1X 8QX, England.

J. Phys. Chem. Ref. Data
Journal of Physical and Chemical Reference Data.

American Chemical Society, 1155 Sixteenth Street,
N.W., Washington, DC 20036, USA.

J. Phys. Colloq.
Journal de Physique Colloque. Société Française de
Physique, 87 bis Avenue du Général Leclerc, 75014 Paris,
France.

J. Phys. E
Journal of Physics E, (Scientific Instruments). Formerly:
J. Sci. Instrum. (GB). Institute of Physics, 47 Belgrave
Square, London, SW1X 8QX, England.

J. Phys. F
Journal of Physics F, (Metal Physics). Institute of Physics,
47 Belgrave Square, London, SW1X 8QX, England.

J. Phys. G
Journal of Physics G, (Nuclear Physics). Institute of
Physics, 47 Belgrave Square, London, SW1X 8QX,
England.

J. Phys. Soc. Japan
Journal of the Physical Society of Japan. Room 211,
Kikai Shinko Building, Shiba Koen, Minato-ku, Tokyo
105, Japan.

J. Physique
Journal de Physique. Z. I. de Courtaboeuf, B. P. 112,
91402 Orsay, France.

J. Plasma Phys.
Journal of Plasma Physics. Cambridge University Press,
Bentley House, 200 Euston Road, London, NW1 2DB,
England.

J. Proc. R. Soc. New South Wales
Journal and Proceedings of the Royal Society of New
South Wales. Science Centre, 35 Clarence Street, Sydney,
N.S.W. 2000, Australia

J. Quant. Spectrosc. Radiat. Transfer
Journal of Quantitative Spectroscopy & Radiative Trans-
fer. Pergamon Press Ltd., Headington Hill Hall, Oxford,
OX3 OBW, England.

J. R. Astron. Soc. Canada
The Journal of the Royal Astronomical Society of Canada, devoted to the advancement of astronomy and allied
sciences. The Royal Astronomical Society of Canada,
124 Merten Street, Toronto, Ontario, Canada.

J. Radio Res. Lab.
Journal of the Radio Research Laboratories. Chief Planning Section, Radio Research Laboratories, Ministry of
Posts & Telecommunications, Nukui-Kitamachi, Konganei-shi, Tokyo 184, Japan.

J. Res. Natl. Bur. Stand. B
Journal of Research of the National Bureau of Standards.
Section B (Mathematics and Mathematical Physics). US
Government Printing Office, Division of Public
Documents, Washington, DC 20402, USA.

J. Sci. Ind. Res.
Journal of Scientific and Industrial Research. Sales &
Distribution Office, Publications & Information Director-
ate, Hillside Road, New Delhi 110012, India.

J. Spacecr. Rockets
Journal of Spacecraft and Rockets. American Institute

of Aeronautics and Astronautics, 1290 Avenue of the Americas, New York, N. Y. 10019, USA.

J. Spectrosc. Soc. Japan
Journal of the Spectroscopical Society of Japan. (Bunkyo Kenkyu). 2-15-1, Nakai, Shinjuku-ku, Tokyo 161, Japan.

J. Toyo Univ., Gen. Educ., Nat. Sci.
Journal of the Toyo University, General Education, Natural Science. Published by The Toyo University, 28, Hakusan 5-chôme, Bunkyo-ku, Tokyo, Japan.

Japanese J. Appl. Phys.
Japanese Journal of Applied Physics. Publication Office, 2nd Toya Kaiji Building, 24-8 Shinbashi, Minato-ku, Tokyo 105, Japan.

Jenaer Rundsch. (Jena Rev.)
Jenaer Rundschau (Jena Review). Publisher: VEB Verlag Technik, Berlin, German Democratic Republic.

JETP Lett.
JETP Letters. A translation of JETP Pis'ma v Redaktsiyu of the Academy of Sciences in the USSR. American Institute of Physics, 335 East 45th Street, New York, NY 10017, USA.

Kexue Tongbao
Kexue Tongbao. Academia Sinica, Peking, People's Republic of China [English translation in: Kexue Tongbao (Scientia)(USA)].

Kodaikanal Obs. Bull.
Kodaikanal Observatory Bulletins, Series A, Indian Institute of Astrophysics, Bangalore, India.

Komet. Tsirk.
Kometnyj Tsirkulyar. Gruppa po Issledovaniyu Komet Astrosoveta i Mezhduvedomstvennyj Geofizicheskij Komitet Akademii Nauk SSSR. Kievskij Universitet im. T. G. Shevchenko.

Komety i Meteory
Komety i Meteory. Akademiya Nauk Tadzhikskoj SSR. Astronomicheskij Sovet Akademii Nauk SSSR. Publishers: Izdatel'stvo "Donish", Dushanbe.

Kosm. Issled.
Kosmicheskie Issledovaniya. Akademiya Nauk SSSR. Publishers: Izdatel'stvo "Nauka", Moskva [An English translation is published as "Cosmic Research", Consultants Bureau, New York, N. Y.].

Kozmos
Kozmos. Popular Astronomical Journal of the Slovak Central Observatory in Hurbanovo. Publisher: Slovenská ústredná hvezdáreň v Hurbanove.

Lett. Math. Phys.
Letters in Mathematical Physics. D. Reidel Publishing Co., P.O. Box 17, Dordrecht, Netherlands.

L'Universo
L'Universo. Rivista dell'Istituto Geografico Militare. Direzione, Redazione e Amministrazione: Istituto Geografico Militare, Firenze.

Mada
Mada (Science). Published by The Weizmann Science Press of Israel, Jerusalem.

Magn. Polya Soln. Pyaten
Magnitnye Polya Solnechnykh Pyaten. (Supplements to Solnechnye Dannye. Byulleten' (*Solar Data*)). Publishers: Izdatel'stvo "Nauka", Leningrad.

Magy. Geofiz.
Magyar Geofizika. Lapkiado Vallalat, 1073 Budapest, Lenin korut 9 - 11, Hungary

Math. Intelligencer
The Mathematical Intelligencer. Springer-Verlag, Berlin–Heidelberg–New York, 175 Fifth Avenue, New York, NY 10010, USA.

Math. Proc. Cambridge Philos. Soc.
Mathematical Proceedings of the Cambridge Philosophical Society. Formerly: Proceedings of the Cambridge, Philosophical Society (Mathematical and Physical Sciences). Cambridge University Press, Bentley House, 200 Euston Road, London, NW1 2DB, England.

Mem. Astron. Soc. India
Memoirs of the Astronomical Society of India. Edited and published by M.S. Vardya, Tata Institute of Fundamental Research, Bombay 400005 on behalf of the Astronomical Society of India, Osmania University, Hyderabad 500007.

Mem. Fac. Eng. Kyoto Univ.
Memoirs of the Faculty of Engineering, Kyoto University, Kyoto, Japan.

Mem. Fac. Eng. Osaka City Univ.
Memoirs of the Faculty of Engineering, Osaka City University. 459 Sugimoto-cho, Sumi Yoshi-kum, Osaka, Japan.

Mem. Fac. Sci. Kyoto Univ.
Memoirs of the Faculty of Science, Kyoto University. Series of Physics, Astrophysics, Geophysics, and Chemistry. Printed by Yamashiro Printing Publishing Co. Ltd., Kamigyo, Kyoto.

Mem. Japan Astron. Study Assoc.
Memoirs of the Japan Astronomical Study Association. c/o National Science Museum, Ueno Park, Taito-ku, Tokyo, Japan.

Mem. Soc. Astron. Italiana
Memorie della Società Astronomica Italiana. Presso Laboratorio di Astrofisica Spaziale, Castella Postale 67, 00044 Frascati, Italy.

Mercury
Mercury. The Journal of the Astronomical Society of the Pacific. Published by the Astronomical Society of the Pacific, 1290 24th Avenue, San Francisco, California 94122, USA. (415) 661 - 8660.

Messenger
The Messenger – El Mensajero. Published by European Southern Observatory, Karl-Schwarzschild-Straße 2, D-8046 Garching bei München.

Meteoritics
Meteoritics. The Journal of the Meteoritical Society. Published quarterly by The Meteoritical Society and Arizona State University Bureau of Publications. Editorial address: Center for Meteorite Studies, The Arizona State University, Tempe, Ariz. 85281, USA.

Meteoritika
Akademiya Nauk SSSR. Komitet po Meteoritam. Publishers: Izdatel'stvo "Nauka", Moskva.

Meteorol. Rundsch.
Meteorologische Rundschau. Springer-Verlag, D-1000 Berlin 33, Heidelberger Platz 3, Germany.

Metrologia
Metrologia. Springer-Verlag, Heidelberger Platz 3, D-1000 Berlin 33, F. R. Germany.

Microwave J.
Microwave Journal. To be obtained from 610 Washington Street, Dedham Plaza, Dedham, Massachusetts, U.S.A.

Minor Planet Bull.
The Minor Planet Bulletin. Bulletin of the Minor Planets Section of the Association of Lunar and Planetary Observers. Editorial Office: R. G. Hodgson, Dordt College, Sioux Center, Iowa, U.S.A.

Minor Planet Circ., (M. P. C.)
The Minor Planet Circulars/Minor Planets and Comets. Edited under the supervision of B. G. Marsden. Published by Minor Planet Center, Smithsonian Astrophysical Observatory, Cambridge, Mass. 02138, USA.

Mitt. Astron. Ges.
Mitteilungen der Astronomischen Gesellschaft, Hamburg. Available from Astron. Instit. Univ. Bochum, Postfach 10 21 48, D-4630 Bochum. ISSN 0172–5483.

Mitt. Inst. Theor. Geod. Univ. Bonn
Mitteilungen aus dem Institut für Theoretische Geodäsie der Universität Bonn, Nußallee 17, 5300 Bonn 1, F. R. Germany.

Mitt. Karl-Schwarzschild-Obs. Tautenburg
Mitteilungen des Karl-Schwarzschild-Observatoriums Tautenburg der Deutschen Akademie der Wissenschaften der DDR. Zentralinstitut für Astrophysik.

Mitt. Satell.-Beobachtungsstn. Zimmerwald
Mitteilungen der Satelliten-Beobachtungsstation Zimmerwald. Hausdruckerei Institut für Exakte Wissenschaften, Bern, Switzerland.

Mitt. Sternw. Sonneberg
Akademie der Wissenschaften der DDR, Zentralinstitut für Astrophysik, Sternwarte Sonneberg. Mitteilungen der Sternwarte zu Sonneberg.

Mitt. Sternw. Ungar. Akad. Wiss.
Mitteilungen der Sternwarte der Ungarischen Akademie der Wissenschaften. A Magyar Tudományos Akadémia Csillagvizsgáló Intézetének Közleményei, Budapest–Szabadsághegy, HU ISSN 0324-2234.

Mitt. Veränderl. Sterne (MVS)
Mitteilungen über Veränderliche Sterne. Herausgegeben von der Sternwarte Sonneberg der Akademie der Wissenschaften der DDR, Sonneberg, German Democratic Republic.

Mod. Geol.
Modern Geology. Gordon & Breach Science Publishers Ltd., 41 and 42 William IV Street, London WC2, England.

Mon. Not. R. Astron. Soc.
Monthly Notices of the Royal Astronomical Society. Published for the Royal Astronomical Society by Blackwell Scientific Publications, Oxford – London – Edinburgh – Melbourne.

Mon. Notes Astron. Soc. South. Africa
Monthly Notes of the Astronomical Society of Southern Africa. Published by the Astronomical Society of Southern Africa, S. A. Astronomical Observatory, Cape Province, South Africa.

Mon. Notes Int. Polar Motion Serv.
Monthly Notes of the International Polar Motion Service. Published by the Central Bureau, International Latitude Observatory of Mizusawa, Mizusawa-shi, Iwate-ken, Japan.

Moon Planets
The Moon and the Planets. An International Journal of Comparative Planetology. Publisher: D. Reidel Publishing Company, Dordrecht, Holland – Boston, USA. Formerly: Moon.

Nablyud. Iskusstv. Nebesn. Tel
Nablyudeniya Isskusstvennykh Nebesnykh Tel. Published by Astronomicheskij Sovet Akademii Nauk SSSR, Moskva.

Nachr. Akad. Wiss. Göttingen II
Nachrichten der Akademie der Wissenschaften in Göttingen. II. Mathematisch-Physikalische Klasse. Vandenhoeck & Ruprecht, Göttingen.

Nachr. Karten-, Vermessungswesen
Nachrichten aus dem Karten- und Vermessungswesen. Editor: Institut für Angewandte Geodäsie (Abt. II des Deutschen Geodätischen Forschungsinstituts). Published by Verlag des Instituts für Angewandte Geodäsie, Frankfurt a. M.

Nachr. Olbers-Ges. Bremen
Nachrichten der Olbers-Gesellschaft Bremen. Werderstraße 73, Bremen.

NASA Conf. Publ.
NASA Conference Publication. National Aeronautics and Space Administration. Scientific and Technical Information Branch, Washington, D.C. For sale by the National Technical Information Service, Springfield, Virginia 22161.

NASA Contract. Rep.
NASA Contractor Report. National Aeronautics and Space Administration, Washington, D.C. For sale by the National Technical Information Service, Springfield, Virginia 22161.

NASA Ref. Publ.
NASA Reference Publication. National Aeronautics and Space Administration. Scientific and Technical Information Office. Washington, D.C. 20546. For sale by the National Technical Information Service, Springfield, Virginia 22161.

NASA Tech Briefs
NASA Tech Briefs. NASA Technology Utilization Program, Technology Transfer Division, P. O. Box 8757, Baltimore/Washington International Airport, MD 21240, USA.

NASA Tech. Memo.
NASA Technical Memorandum. National Aeronautics and Space Administration, Washington, D.C. For sale by the National Technical Information Service, Springfield, Virginia 22161.

NASA Tech. Note
NASA Technical Note. National Aeronautics and Space Administration, Washington, D.C. For sale by the National Technical Information Service, Springfield, Virginia 22161.

NASA Tech. Pap.
NASA Technical Paper. National Aeronautics and Space Administration, Washington, D.C. For sale by the National Technical Information Service, Springfield, Virginia 22161.

Natl. Geogr.
National Geographic. Official Journal of the National Geographic Society, Washington, D.C. 17th and M Sts. N.W., Washington, D.C. 20036.

Nature
Nature. Editorial and Publishing Offices: Macmillan Journals Limited, 4 Little Essex Street, London WC2R 3LF, 711 National Press Building, Washington, D.C. 20045.

Naturwissenschaften
Die Naturwissenschaften. Publisher:Springer-Verlag, Berlin – Heidelberg – New York.

Nauchn. Inf.
Nauchnye Informatsii. Astronomicheskij Sovet Akademii Nauk SSSR, Moskva.

Naučna Misao
Naučna Misao. Društvo za Unapredivanje i Širenje Nauke, Zagreb, Babonićeva 54, Yugoslavia. Scientific Idea. Society for Promotion and Propagation of Science.

Navigation *(France)*
Navigation. Institut Francaise de Navigation, 3 avenue Octave-Greard, Paris 7, France.

Navigation *(USA)*
Navigation. Journal of the Institute of Navigation, Institute of Navigation, Suite 832, 815 15th Street, N. W., Washington, DC 20005, USA.

NBS Monogr.
National Bureau of Standards Monograph. U.S. Government Printing Office, Washington, D.C. 20402.

Nederlands Tijdschr. Natuurkd. A
Nederlands Tijdschrift voor Natuurkunde, Publisher: Martinus Nijhoff, Lange Voorhout 9, Den Haag, Netherlands.

New Scientist
New Scientist. New Science Publications, 128 Long Acre, London, WC2E 9QH, England.

New Zealand J. Sci.
New Zealand Journal of Science. Department of Scientific and Industrial Research, Private Bag, Wellington, New Zealand.

News Lett. Astron. Soc. N.Y.
News Letter of the Astronomical Society of New York. A. G. D. Philip (Editor). Astronomical Society of New York, Dudley Observatory, 69 Union Avenue, Schenectady, New York 12308.

Nucl. Instrum. Methods
Nuclear Instruments and Methods. North-Holland Publishing Co., P. O. Box 211, Amsterdam, Netherlands.

Nucl. Phys. A
Nuclear Physics, Volume A. North-Holland Publishing Co., P. O. Box 211, Amsterdam, Netherlands.

Nucl. Tracks Methods Instrum. Appl.
Nuclear Tracks, Methods, Instruments and Application. Formerly: Nucl. Track Detect. (GB) Pergamon Press Ltd., Headington Hill Hall, Oxford OX3 0BW, England.

Numer. Math.
Numerische Mathematik. Springer-Verlag, Berlin–Heidelberg–New York.

Nuovo Cimento A
Il Nuovo Cimento. A. Societa Italiana di Fisica, vialle XII Guigno 1, 40124 Bologna, Italy.

Nuovo Cimento B
Il Nuovo Cimento. B. Societa Italiana di Fisica, vialle XII Guigno 1, 40124 Bologna, Italy.

Nuovo Cimento C
Il Nuovo Cimento. C. Societa Italiana di Fisica, vialle XII Guigno 1, 40124 Bologna, Italy.

Nuovo Cimento, Lett.
Lettere al Nuovo Cimento, a Cura della Società Italiana di Fiscia. Via Degli Andalo 2, 40124 Bologna, Italy.

Nuovo Cimento, Riv.
Rivista del Nuovo Cimento. Società Italiana di Fisica, Via Degli Andalo 2, 40124 Bologna, Italy.

Nuovo Cimento, Suppl.
Supplemento al Nuovo Cimento. Società Italiana di Fisica. Via Degli Andalo 2, 40124 Bologna, Italy.

Obs. Artif. Earth Satell.
Observations of Artificial Satellites of the Earth (Nablyudeniya Iskusstvennykh Sputnikov Zemli). Magyar Tudományos Akadémia Csillagvizsgáló Intézete. Budapest.

Obs. Astron. Antares, Contrib. Cient.
Universidade Estadual de Feira de Santana, Observatório Astronômico Antares, Contribuição Cientifica. Feira de Santana, Brazil.

Obs. Astrophys. Lab., Univ. Helsinki.Rep.
Observatory and Astrophysics Laboratory, University of Helsinki. Report. Helsinki, Finland.

Observatory
The Observatory. A Review of Astronomy. Publishers: The Editors of 'The Observatory', Royal Greenwich Observatory, Herstmonceux Castle, Hailsham, Sussex, England, BN27 1RP.

Occas. Rep. R. Obs. Edinburgh
Occasional Reports of the Royal Observatory, Edinburgh, Blackford Hill, Edinburgh EH9 3HJ, Scotland.

Occultation Newsl.
Occultation Newsletter. Published by the International

Occultation Timing Association (I.O.T.A.). 6 N 106 White Oak Lane, St. Charles, Ill. 60174, USA.

Österreich. Z. Vermessungswes. Photogramm.
Österreichische Zeitschrift für Vermessungswesen und Photogrammetrie. Editor and Publisher: Österreichischer Verein für Vermessungswesen und Photogrammetrie, Wien, Austria.

Opt. Acta
Optica Acta. Taylor and Francis Ltd., 10 - 14 Macklin Street, London, WC2B 5NF, England.

Opt. Commun.
Optics Communications. North-Holland Publishing Co., P.O. Box 211, Amsterdam, Netherlands.

Opt. Eng.
Optical Engineering. Society of Photo-Optical Instrumentation Engineers, 337 Tejon Place, Palos Verdes Estates, CA 90274, USA.

Opt. Lett.
Optics Letters. A publication of the Optical Society of America. American Institute of Physics, 335 East 45th Street, New York, N. Y. 10017, USA.

Opt. Spectra
Optical Spectra. The Magazine of Optical/Electro-Optical/Laser Technology. Published by The Optical Publishing Co., Inc., 59 Bartlett Ave., P.O. Box 1146, Pittsfield, Mass. 01201, USA.

Optik
Optik. Zeitschrift für das gesamte Gebiet der Licht- und Elektronenoptik. Publishers: Wissenschaftliche Verlagsgesellschaft mbH, Postfach 40, D-7000 Stuttgart, F. R. Germany.

Origins of Life
Origins of Life (Formerly Space Life Sciences). An International Journal. Publisher: D. Reidel Publishing Company, Dordrecht, Holland.

Orion
Orion. Zeitschrift der Schweizerischen Astronomischen Gesellschaft (SAG). Revue de la Société Astronomique de Suisse (SAS). Printed by A. Schudel & Co. AG, 4125 Riehen, Switzerland.

Orione
Orione. Rivista Trimestrale di Divulgazione Astronomica. Via Roma 6, 10025 Pino Torinese (Torino).

Oss. Astrofis. Catania, Pubbl.
Osservatorio Astrofisico di Catania, Pubblicazione. Printed by Scuola Salesiana del Libro, Catania.

Oss. Mem. Oss. Astrofis. Arcetri
Osservazioni e Memorie dell'Osservatorio Astrofisico di Arcetri. Università Degli Studi di Firenze, Firenze, Italy.

Oyo Buturi
Oyo Buturi. Japan Society of Applied Physics, Room No. 209-2, Kikai-Shinko Building, 21 Shiba-Koen Minato-ku, Tokyo, Japan.

Perem. Zvezdy
Peremennye Zvezdy, Byulleten', izdavaemyj Astronomicheskim Sovetom Akademii Nauk SSSR. Published by Astronomicheskij Sovet Akademii Nauk SSSR, Moskva.

Perem. Zvezdy, Prilozhenie
Peremennye Zvezdy, Prilozhenie (The Variable Stars, Supplement). Astronomicheskij Sovet Akademii Nauk SSSR, Moskva.

Philos. Trans. R. Soc. London, Ser. A
Philosophical Transactions of the Royal Society of London. Series A, Mathematical and Physical Sciences. Carlton House Terrace, London, SW1Y 5 AG England.

Photogr. J. Sun
Photographic Journal of the Sun. Supplement to Monthly Bulletin, Solar Phenomena. Osservatorio Astronomico di Roma.

Photogr. Sci. Eng.
Photographic Science and Engineering. Society of Photographic Scientists and Engineers, Suite 204, 1330 Massachusetts Avenue N.W., Washington, DC 20005, USA.

Photogramm. Eng. Remote Sensing
Photogrammetric Engineering and Remote Sensing. Formerly: Photogramm. Eng. American Society of Photogrammetry, 105 North Virginia Avenue, Falls Church, VA 22046, USA.

Phys. Abstr.
Physics Abstracts. Science Abstracts, Series A. An INSPEC Publication, published by The Institution of Electrical Engineers in Association with the Institute of Electrical and Electronics Engineers Inc. Printed by Pindar & Son Ltd., Scarborough, N. Yorkshire, England.

Phys. Bl.
Physikalische Blätter. Physik-Verlag GmbH, Pappelallee 3, Postfach 1260/1280, D-6940 Weinheim, F. R. Germany.

Phys. Briefs
Physics Briefs. Physikalische Berichte. Edited by Deutsche Physikalische Gesellschaft and Fachinformationszentrum Energie, Physik, Mathematik in cooperation with American Institute of Physics. Published by Physik Verlag GmbH, Postfach 1260/1280, D-6940 Weinheim, F. R. Germany. ISSN 0170-7434.

Phys. Chem. Miner.
Physics and Chemistry of Minerals. Springer-Verlag, D-1000 Berlin 33, Heidelberger Platz 3, Germany.

Phys. Earth Planet. Inter.
Physics of the Earth and Planetary Interiors. A journal devoted to observational and experimental studies of the Earth and Planetary interiors and their theoretical interpretation by the physical sciences. Publisher: North-Holland Publishing Company, Amsterdam, Netherlands.

Phys. Fluids
The Physics of Fluids. Published by the American Institute of Physics, 335 East 45th Street, New York, NY 10017, USA.

Phys. Lett.
Physics Letters. Volumes A and B. Publisher: North-Holland Publishing Company, Amsterdam.

Phys. Rep.
Physics Reports. North-Holland Publishing Co., P. O. Box 211, Amsterdam, Netherlands.

Phys. Rev. A
Physical Review A, General Physics. Published for the

American Physical Society by the American Institute
of Physics, 335 East 45th Street, New York, N. Y. 10017,
USA.

Phys. Rev. B

Physical Review B, Solid State. Published for the American Physical Society by the American Institute of Physics,
335 East 45th Street, New York, NY 10017, USA.

Phys. Rev. C

Physical Review C, Nuclear Physics. Published for the
American Physical Society by the American Institute of
Physics, 335 East 45th Street, New York, NY 10017, USA.

Phys. Rev. D

Physical Review D, Particles and Fields. Published for the
American Physical Society by the American Institute of
Physics, 335 East 45th Street, New York, NY 10017, USA.

Phys. Rev. Lett.

Physical Review Letters. Published for the American
Physical Society by the American Institute of Physics,
335 East 45th Street, New York, NY 10017, USA.

Phys. Scr.

Physica Scripta. (Formerly Arkiv för Fysik). Published by
the Royal Swedish Academy of Sciences, S-104 05
Stockholm 50, Sweden.

Phys. Teach.

Physics Teacher. American Institute of Physics, 335 East
45th Street, New York, NY 10017, USA.

Phys. Today

Physics Today. Published by the American Institute of
Physics, 335 East 45th Street, New York, NY 10017,
USA.

Phys. unserer Zeit

Physik in unserer Zeit. Verlag Chemie GmbH, Pappelallee 3, Postfach 1260/1280, D-6940 Weinheim, F. R.
Germany.

Physica B, C

Physica B & C. Subscription address: North-Holland
Publishing Co., P.O. Box 211, Amsterdam, Netherlands.

Pis'ma Astron. Zh.

Pis'ma v Astronomicheskij Zhurnal. Akademiya Nauk
SSSR. Publishers: Izdatel'stvo 'Nauka', Moskva. Translation in Soviet Astron. Lett.

Planet. Space Sci.

Planetary and Space Science. Pergamon Press Ltd.,
Headington Hill Hall, Oxford, OX3 OBW, England.

Postępy Astron.

Postępy Astronomii. Czasopismo Poświecone Upowszech-
nianiu Wiedzy Astronomicznej. Polskie Towarzystwo
Astronomiczne, Warszawa. Printed in Poland by Pánstwo-
we Wydawnictwo Naukowe, Lódź.

Postępy Fiz.

Postępy Fizyki. Polskie Towarzystwo Fizyczne,
00-681 Warszawa, ul. Hoza 69, Poland.

Pramăna

Pramăna. Indian Academy of Sciences, Bangalore
560006, India.

Priroda

Priroda. Publishers: Izdatel'stvo "Nauka", Moskva.

Probl. Kosm. Fiz.

Problemy Kosmichskoj Fiziki. Mezhvedomstvennyj
Nauchnoj Sbornik. Izdatel'skoe Obedinenie Vishcha
Shkola. Izdatel'stvo pri Kievskom Universitete, Kiev.

Proc. Astron. Soc. Australia

Proceedings of the Astronomical Society of Australia.
Published for the Society by Sydney University Press,
Sydney.

Proc. IEEE

Proceedings of the IEEE. Published by the Institute of
Electrical and Electronics Engineers, 345 East 47th Street,
New York, NY 10017, USA.

Proc. Indian Acad. Sci., Sect. A

Proceedings of the Indian Academy of Sciences,
Section A. Bangalore 560006, India.

Proc. Indian Acad. Sci., Sect. C

Proceedings of the Indian Academy of Sciences,
Section C: Engineering Sciences, Bangalore 560006,
India.

Proc. Indian Natl. Sci. Acad., Part A.

Proceedings of the Indian National Science Academy,
Part A, Bahadur Shah Zafar Marg, New Delhi 1, India.

Proc. Int. Latitude Obs. Mizusawa

Proceedings of the International Latitude Observatory of
Mizusawa. Published by the International Latitude Observatory of Mizusawa, Japan.

Proc. Japan Acad., Ser. B

Proceedings of the Japan Academy. Series B, Physical
and Biological Sciences, Ueno Park, Tokyo 110, Japan.

Proc. K. Nederlandse Akad. Wet. B

Koninklijke Nederlandse Akademie van Wetenschappen.
Proceedings. Series B, Physical Sciences. Publisher:
North-Holland Publishing Company, Amsterdam,
Netherlands.

Proc. Natl. Acad. Sci. India, Sect. A

Proceedings of the National Academy of Sciences of
India. Section A (Physical Sciences). Lajpatrai Road,
Allahabad-2, India

Proc. Natl. Acad. Sci. U.S.A.

Proceedings of the National Academy of Sciences of the
United States of America. National Academy of Sciences,
2101 Constitution Avenue, Washington, DC 20418, USA.

Proc. R. Soc. London, Ser. A

Proceedings of the Royal Society of London. Series A:
Mathematical and Physical Sciences. Published by the
Royal Society, 6 Carlton House Terrace, London,
SW1Y 5AG, England.

Proc. Res. Inst. Atmos. Nagoya Univ.

Proceedings of the Research Institute of Atmospherics
Nagoya University. Nagoya University, 13 Honohara,
3 Chrome, Toyokawa 442, Japan.

Proc. Soc. Photo-Opt. Instrum. Eng.

Proceedings of the Society of Photo-Optical Instrumentation Engineers, Bellingham, WA 98225, USA.

Prog. Part.Nucl. Phys.

Progress in Particle and Nuclear Physics. Pergamon
Press Ltd., Headington Hill Hall, Oxford OX3 OBW,
England.

Prog. Theor. Phys.
Progress of Theoretical Physics. Published for the Research Institute for Fundamental Physics and the Physical Society of Japan. Publication Office: Progress of Theoretical Physics, Yukawa Hall, Kyoto University, 606 Kyoto, Japan.

Prog. Theor. Phys. Suppl.
Supplement of the Progress of Theoretical Physics. Published for the Research Institute for Fundamental Physics and The Physical Society of Japan. Publication Office: Progress of Theoretical Physics. Yukawa Hall, Kyoto University, 606 Kyoto, Japan.

PTB Mitt.
PTB Mitteilungen. Forschen + Prüfen. Fachorgan für Wirtschaft und Wissenschaft. Amts- und Mitteilungsblatt der Physikalisch-Technischen Bundesanstalt. Braunschweig–Berlin. Deutscher Eichverlag, Postfach 3367, D-3300 Braunschweig, F.R. Germany.

Publ. Astron. Soc. Japan
Publications of the Astronomical Society of Japan. Published by the Astronomical Society of Japan. Office of the Society: Tokyo Astronomical Observatory, Mitaka, Tokyo. Agent: Maruzen Co. Ltd. (Export Department), Nihonbashi, Tokyo, Japan.

Publ. Astron. Soc. Pacific
Publications of the Astronomical Society of the Pacific. Published by the Astronomical Society of the Pacific, 1290 24th Avenue, San Francisco, California 94122, USA. (415) 661 - 8660.

Publ. Beijing Astron. Obs.
Publications of the Beijing Astronomical Observatory. Beijing Astronomical Observatory, Academia Sinica, China.

Publ. Bosscha Obs.
Publications of the Bosscha Observatory. Bandung Institute of Technology, Department of Science. Lembang, Indonesia.

Publ. Debrecen Heliophys. Obs. Hungarian Acad. Sci.
Publications of Debrecen Heliophysical Observatory of the Hungarian Academy of Sciences.

Publ. Dep. Astron., Univ. Beograd
Publications of the Department of Astronomy, University of Beograd, Faculty of Sciences (Publications de la Chaire d'Astronomie, Université de Beograd, Faculté des Sciences) Beograd. YU ISSN 0350-3283.

Publ. Dep. Geod. Astron., Univ. Thessaloniki
Publications of the Department of Geodetic Astronomy, University of Thessaloniki.

Publ. Dominion Astrophys. Obs.
Publications of the Dominion Astrophysical Observatory, Victoria, B. C. National Research Council of Canada.

Publ. Eidg. Sternw. Zürich
Publikationen der Eidgenössischen Sternwarte Zürich. Schulthess Polygraphischer Verlag, Zürich.

Publ. Inst. Geophys., Polish Acad. Sci.
Publications of the Institute of Geophysics, Polish Academy of Sciences. Państwowe Wydawnictwo Naukowe, Warszawa-Łódź. ISBN 83-01-02210-8. ISSN 0138-0214.

Publ. Inst. R. Meteorol. Belgique A
Publications, Institut Royal Meteorologique de Belgique. Serie A. 3 Avenue Circulaire, Uccle-Bruxelles 1180, Belgium.

Publ. Int. Latitude Obs. Mizusawa
Publications of the International Latitude Observatory of Mizusawa. Published by the International Latitude Observatory of Mizusawa, Japan.

Publ. Korean Natl. Astron. Obs.
Publications of the Korean National Astronomical Observatory. Published by the Korean National Astronomical Observatory, Seoul, Korea.

Publ. Obs. Astron. Beograd
Publications de l'Observatoire Astronomique de Beograd Editeur: Observatoire Astronomique de Belgrade, 11050 Beograd, Volgina 7, Yougoslavie.

Publ. R. Obs. Edinburgh
Publications of the Royal Observatory, Edinburgh. Published by The Royal Observatory, Edinburgh, Scotland.

Publ. Shaanxi Astron. Obs.
Publications of the Shaanxi Astronomical Observatory. Academia Sinica, P. O. Box 18, Lintong, near Xian, China.

Publ. United States Naval Obs.
Publications of the United States Naval Observatory. Department of the Navy, U.S. Naval Observatory, Washington. U.S. Government Printing Office, Washington, D.C.

Publ. Variable Star Sect.,R. Astron. Soc. New Zealand
Publications of the Variable Star Section, Royal Astronomical Society of New Zealand. Director: F. M. Bateson, Greerton, Tauranga, New Zealand.

Publ. Warner Swasey Obs.
Publications of the Warner and Swasey Observatory, Case Western Reserve University, Cleveland, Ohio 44106.

Q. Appl. Math.
Quarterly of Applied Mathematics. American Mathematical Society, P. O. Box 1571, Providence, RI 02901, USA.

Q. Bull. Sol. Act.
International Astronomical Union, Quarterly Bulletin on Solar Activity. Published by the Tokyo Astronomical Observatory. Beginning with No. 197; formerly Zürich.

Q. J. R. Astron. Soc.
Quarterly Journal of the Royal Astronomical Society. Burlington House, London, W1V ONL, England.

R. Greenwich Obs., Time Latitude Serv.
Royal Greenwich Observatory, Time and Latitude Service Royal Greenwich Observatory, Herstmonceux Castle, Hailsham, East Sussex BN27 1RP, England.

R Muscae
R Muscae. Revista de Estrellas Variables. Published by Departamento de Astronomía, Instituto Copérnico, Buenos Aires, Argentina. Editorial address: Montevideo 724, 1° "4", (1019), Buenos Aires, Argentina.

R. Obs. Ann.
Royal Observatory Annals, Royal Greenwich Observato-

ry, Herstmonceux Castle, Hailsham, East Sussex,
BN 27 1RP, England.

Radiotekh. Ehlektron.
Radiotekhnika i Ehlektronika. Moskva TSP-3, Pr. Karl
Marx 18, USSR.

Rech. Aerosp.
Recherche Aerospatiale. Office National d'Études et de
Recherches Aerospatiales, 29 Avenue de la Division
Leclerc, 92320-Chatillon, France.

Recherche
Recherche, 4 Place de l'Odéon, Paris 6, France.

Ref. Zh., 51. Astron.
Referativnyi Zhurnal, 51. Astronomiya. Vsesoyuznyj
Institut Nauchnoj i Tekhnicheskoj Informatsii. Moskva.

Ref. Zh., 52. Geod. Aehrosemka
Referativnyj Zhurnal, 52. Geodeziya i Aehrosemka.
Vsesoyuznyj Institut Nauchnoj i Tekhnicheskoj Infor-
matsii. Moskva.

Ref. Zh., 62. Issled. kosm. prostranstva
Referativnyj Zhurnal, 62. Issledovanie Kosmicheskogo
Prostranstva. Vsesoyuznyj Institut Nauchnoj i Tekhni-
cheskoj Informatsii. Moskva.

Rep. Finnish Geod. Inst.
Reports of the Finnish Geodetic Institute. Suomen
Geodeettisen Laitoksen Tiedonantoja. Helsinki, Finland.

Rep. Math. Phys.
Reports on Mathematical Physics, Pergamon Press Ltd.,
Headington Hill Hall, Oxford OX3 OBW, England.
Subscription address: ARS Polona-Ruch Foreign Trade
Enterprise, Krakowskie Przedmiescie 7, 00-068
Warszawa.

Rep. Obs. Lund
Reports from the Observatory of Lund. Lunds Universitet,
Institutionen för Astronomie, S-222 24 Lund, Sweden.

Rep. Prog. Phys.
Reports on Progress in Physics. Published by the Institute
of Physics, 47 Belgrave Square, London, SW1X 8QX,
England.

Rep. Ser. Dep. Phys. Sci., Univ. Turku
Report Series, Department of Physical Sciences,
Institute of Astronomy, University of Turku, SF-20500
Turku 50, Finland.

Res. Lab. Electron. Onsala Space Obs.
Research Laboratory of Electronics and Onsala Space
Observatory, Chalmers University of Technology,
Gothenburg, Sweden. Research Report.

Rev. Astron.
Revista Astronomica. Organo de la Asociación Argentina
Amigos de la Astronomia, Avenida Patricias Argentinas
550, Buenos Aires 5, Argentina.

Rev. Brasil. Fis.
Revista Brasileira de Fisica. Sociedade Brasileira de Fisica,
Cx. Postal 20553, Sao Paolo SP, Brazil.

Rev. Geophys. Space Phys.
Reviews of Geophysics and Space Physics (formerly Reviews of Geophysics). Published by the American Geo-

physical Union, 1909 K Street, N.W., Washington,
DC 20006, USA.

Rev. Hist. Sci.
Revue d'Histoire des Sciences. Revue trimestrielle publiée
avec le concours du C.N.R.S. Centre International de
Synthèse, Section d'Histoire des Sciences.

Rev. Mexicana Astron. Astrofis.
Revista Mexicana de Astronomia y Astrofisica. Dirección:
Instituto de Astronomia, Universidad Nacional Autóno-
ma de México, Apartado Postal 70-264, Mexico 20,
D. F., Mexico.

Rev. Mexicana Fis.
Revista Mexicana de Fisica. Sociedad Mexicana de Fisica,
Apartado Postal No. 20-364, Mexico 20, D. F. Mexico.

Rev. Mod. Phys.
Reviews of Modern Physics. Published for the American
Physical Society by the American Institute of Physics,
335 East 35th Street, New York, NY 10017, USA.

Rev. Radio Res. Lab.
Review of the Radio Research Laboratories. Ministry of
Posts & Telecommunications, Nukui-Kitamachi,
Konganei-shi, Tokyo 184, Japan.

Rev. Roumaine Phys.
Revue Roumaine de Physique. Academie Republicii
Populare Romine, Boite Postale 134 - 135, Bucuresti,
Rumania.

Rev. Sci. Instrum.
Review of Scientific Instruments. American Institute of
Physics, 335 East 45th Street, New York, NY 10017,
USA.

Rezul't. Nablyud. Iskusstv. Sputnikov Zemli
Rezul'taty Nablyudenij Iskusstvennykh Sputnikov
Zemli. Published by Astronomicheskij Sovet Akademii
Nauk SSSR, Ryazanskij Gosudarstvennyj Pedagogicheskij
Institut, Ryazan'.

Ric. Astron.
Ricerche Astronomiche. Specola Vaticana, Città del
Vaticano.

Ric. Spettrosc.
Ricerche Spettroscopiche. Specola Vaticana, Città del
Vaticano.

Říše hvězd
Říše hvězd. Czech popular astronomical journal.
Publisher: Panorama, Praha.

Sci. American
Scientific American. 415 Madison Avenue, New York,
NY 10017, USA.

Sci. Atmos. Sinica
Scientia Atmospherica Sinica. Science Press, Peking.
Subscription address: Guozi Shudian, P.O. Box 399,
Peking, Peoples's Republic of China.

Sci. Dimension
Science Dimension. National Research Council of Canada,
Ottawa K1A 0R6, Canada.

Sci. Pap. Inst. Phys. Chem. Res.
Scientific Papers of the Institute of Physical and Chemical

Research. Rikagaku Kenkyusho, Wako-shi, Saitama 351, Japan.

Sci. Prog.
Science Progress. Blackwell Scientific Publications, Oxford, England.

Sci. Rep. Tôhoku Univ., Ser. 1.
The Science Reports of the Tôhoku University. First Series (Physics, Chemistry, Astronomy). Published by the Faculty of Science, Tôhoku University, Sendai, Japan. ISSN 0040-8778.

Sci. Rep. Tôhoku Univ., Ser. 8.
The Science Reports of the Tôhoku University. Eighth Series (Physics and Astronomy). Published by the Faculty of Science, Tôhoku University, Sendai, Japan. ISSN 0388-5607.

Sci. Rev.
Scienca Revuo. Prof. B. Popovic, Ognjena Price 80, Beograd, Yugoslavia.

Sci. Sinica
Scientia Sinica. Science Press, Peking. Subscription address: Guozi Shudian, P.O. Box 399, Peking, China.

Science
Science. American Association for the Advancement of Science, 1515 Massachusetts Avenue, N. W., Washington, D. C. 20005, USA.

Scr. Fac. Sci. Nat. Univ. Purkynianae Brunensis Phys.
Scripta Facultatis Scientiarum Naturalium Universitatis Purkynianae Brunensis, Physica. University J. E. Purkyne, 61137 Brno-Kotlarská 2, Czechoslovakia.

Shaanxi Astron. Obs. Repr.
Shaanxi Astronomical Observatory Reprints. Academia Sinica, P. O. Box 18, Lintong, near Xian, China.

SIAM J. Appl. Math.
SIAM Journal on Applied Mathematics. Society for Industrial and Applied Mathematics, 33 South 17th Street, Philadelphia, PA 19103, USA.

Sitzungsber. Akad. Wiss. DDR
Sitzungsberichte der Akademie der Wissenschaften der DDR. Mathematik-Naturwissenschaften-Technik. Akademie-Verlag, Berlin.

Sitzungsber. Bayerische Akad. Wiss.
Bayerische Akademie der Wissenschaften. Mathematisch-Naturwissenschaftliche Klasse. Sitzungsberichte. Publisher: Verlag der Bayerischen Akademie der Wissenschaften, München.

Sitzungsber. Heidelberger Akad. Wiss.
Sitzungsberichte der Heidelberger Akademie der Wissenschaften. Mathematisch-Naturwissenschaftliche Klasse. Publisher: Springer-Verlag, Heidelberg.

Sitzungsber. Österreich. Akad. Wiss.
Sitzungsberichte. Österreichische Akademie der Wissenschaften. Mathematisch-Naturwissenschaftliche Klasse. Abteilung II: Mathematik, Astronomie, Meteorologie und Technik. Publisher: Springer-Verlag, Wien.

Sky Telesc.
Sky and Telescope. Published by Sky Publishing Corporation, 49-50-51 Bay State Road, Cambridge, Mass. 02138, USA.

Smithsonian Astrophys. Obs., Spec. Rep.
Smithsonian Astrophysical Observatory, Special Report. Available from the Publications Division, Distribution Section, Smithsonian Astrophysical Observatory, Cambridge, Mass. 02138.

Smithsonian Contrib. Astrophys.
Smithsonian Contributions to Astrophysics. Smithsonian Institution Astrophysical Observatory, Cambridge, Mass. Printed by Smithsonian Institution Press, City of Washington. For sale by the Superintendent of Documents, U.S. Government Printing Office, Washington, D.C.

Sol. Energy
Solar Energy, Pergamon Press, Maxwell House, Fairview Park, Elmsford, NY 10523, USA.

Sol. Phenom.
Solar Phenomena. Osservatorio Astronomico di Roma.

Sol. Phys.
Solar Physics. A Journal for Solar Research and the Study of Solar Terrestrial Physics. Publisher: D. Reidel Publishing Company, Dordrecht, Holland.

Soln. Dannye, Byull.
Solnechnye Dannye. Byulleten'. (*Solar Data*). Publishers: Izdatel'stvo "Nauka", Leningradskoe Otdelenie, Leningrad.

Sonne
Sonne. Mitteilungsblatt der Amateursonnenbeobachter. Peter Völker, c/o Wilhelm-Foerster-Sternwarte, Munsterdamm 90, 1000 Berlin 41.

Soobshch. Byurakan. Obs.
Soobshcheniya Byurakanskoj Observatorii. Akademiya Nauk Armyanskoj SSR, Erevan.

Soobshch. Gos. Astron. Inst. Shternberg
Soobshcheniya Gosudarstvennogo Astronomicheskogo Instituta im. P. K. Shternberga. Publishers: Izdatel'stvo Moskovskogo Universiteta, Moskva.

Soobshch. Spets. Astrofiz. Obs.
Soobshcheniya Spetsial'noj Astrofizicheskoj Observatorii. Izdanie Spetsial'noj Astrofizicheskoj Observatorii AN SSSR.

South African Astron. Obs. Circ.
South African Astronomical Observatory, Circulars. S.A. Astronomical Observatory, Observatory, Cape.

South African J. Phys.
South African Journal of Physics (Suid-Afrikaanse Tydskrif vir Fisika). Bureau for Scientific Publications, P.O. Box 1758, Pretoria 0001, South Africa.

South. Stars
Southern Stars. The Journal of the Royal Astronomical Society of New Zealand (Inc.). Address of the Society: P.O. Box 3181, Wellington C1, New Zealand.

Soviet Astron.
Soviet Astronomy. A translation of Astronomicheskij Zhurnal (Astronomical Journal). Published by the American Institute of Physics, New York, N.Y.

Soviet Astron. Lett.
Soviet Astronomy Letters. A translation of "Pis'ma v

Astronomicheskij Zhurnal". Published by the American Institute of Physics.

Space Educ.
Space Education. Supplement to Spaceflight. Published by the British Interplanetary Society, 27/29 South Lambeth Road, London SW8 1SZ, England.

Space Sci. Instrum.
Space Science Instrumentation. An International Journal of Scientific Instruments for Aircraft, Balloons, Sounding Rockets, and Spacecraft. Published by D. Reidel Publishing Company, Dordrecht, Holland.

Space Sci. Rev.
Space Science Reviews. Publishers: D. Reidel Publishing Company, Dordrecht, Holland.

Spaceflight
Spaceflight. Published by the British Interplanetary Society. Printed by Unwin Brothers Ltd., at the Gresham Press, Old Woking, England.

Spaceworld
Spaceworld. Palmer Publications Inc., Amherst, WI 54406, USA.

Sterne
Die Sterne. Zeitschrift für alle Gebiete der Himmelskunde. Johann Ambrosius Barth, Leipzig, German Democratic Republic.

Sterne Weltraum
Sterne und Weltraum. Astronomische Monatsschrift. Publisher: Verlag Sterne und Weltraum Dr. Vehrenberg, Düsseldorf, F. R. Germany.

Sternenbote
Sternenbote. Monatsschrift für Österreichs Amateurastronomen. Publisher: Astronomisches Büro. Hermann Mucke, Wien, Austria.

Stockholms Obs. Ann.
Stockholms Observatorium Annaler. Printed by Almquist & Wiksell, Stockholm, Sweden.

Stockholms Obs. Rep.
Stockholms Observatorium, Saltsjöbaden, Sweden, Report.

Strolling Astron.
The Strolling Astronomer. The Journal of The Association of Lunar and Planetary Observers, Publication Office: The Strolling Astronomer, Box 3 AZ, University Park, New Mexico, USA.

Stud. Astron. Sinica
Studia Astronomica Sinica. Published by the Purple Mountain Observatory, Academia Sinica, Nanking, People's Republic of China.

Stud. Geophys. Geod.
Studia geophysica et geodaetica. Published for the Geophysical Institute of the Czechoslovak Academy of Sciences by Academia, Praha.

Stud. Hist. Philos. Sci.
Studies in History and Philosophy of Science. Pergamon Press Ltd., Headington Hill Hall, Oxford OX3 0BW, England.

Stud. Soc. Sci. Torunensis
Studia Societatis Scientiarum Torunensis, Toruń – Polonia. Sectio F (Astronomia).

Stud. Univ. Babeş-Bolyai
Studia Universitatis Babeş-Bolyai. Series Mathematica-Physica. Publishers: Intreprinderea Poligrafica, Cluj.

Surv. High Energy Phys.
Surveys in High Energy Physics. Harwood Academic Publishers GmbH, P. O. Box 786, Cooper Station, New York, NY 10003, USA.

Tartu Astrofüüs. Obs. Publ.
W. Struve nimelise Tartu Astrofüüsika Observatooriumi, Publikatsioonid. Eesti NSV Teaduste Akadeemia, Tartu.

Tartu Astrofüüs. Obs. Teated
Tartu Astrofüüsika Observatoorium Teated. Eesti NSV Teaduste Akadeemia W. Struve nim. Tartu Astrofüüsika Observatoorium, Tartu.

Telecommun. J.
Telecommunication Journal.(English Edition). International Telecommunications Union, Place des Nations, 1211 Genève 20, Switzerland.

Tellus
Tellus, a bi-monthly Journal of Geophysics. Svenska Geofysiska Foreningen, Arrhenius laboratoriet, Fack, S - 104 05 Stockholm, Sweden.

Time Freq. Serv. Bull.
Time and Frequency Services Bulletin. Shaanxi Astronomical Observatory, Chinese Academy of Sciences, Lintong, Xian, China.

Tokyo Astron. Bull., Second Ser.
Tokyo Astronomical Observatory, Japan. Tokyo Astronomical Bulletin, Second Series.

Tokyo Astron. Obs. Rep.
University of Tokyo, Tokyo Astronomical Observatory, Japan. Report. ISSN 0374-4639.

Tokyo Astron. Obs., Time and Latitude Bull.
Tokyo Astronomical Observatory, Time and Latitude Bulletins. Mitaka, Tokyo, Japan.

Tr. Astrofiz. Inst. Alma-Ata
Trudy Astrofizicheskogo Instituta, Alma-Ata. Akademiya Nauk Kazakhskoj SSR. Publishers: Izdatel'stvo "Nauka" Kazakhskoj SSR, Alma-Ata.

Tr. Astron. Obs., *Leningrad*
Uchenye Zapiski Gosudarstvennogo Universiteta im. A. A. Zhdanova, Seriya matematicheskikh nauk = Trudy Astronomicheskoj Observatorii. Izdatel'stvo Leningradskogo Universiteta, Leningrad.

Tr. Glav. Astron. Obs. Pulkovo
Trudy Glavnoj Astronomicheskoj Observatorii v Pulkove. Akademiya Nauk SSR. Izdanie Glavnoj astronomicheskoj observatorii v Pulkove, Leningrad.

Tr. Inst. Teor. Astron., *Leningrad*
Trudy Instituta Teoreticheskoj Astronomii. Akademiya Nauk SSSR. Publishers: Izdatel'stvo "Nauka", Leningrad.

Tr. Kazan. Gorod. Astron. Obs.
Trudy Kazanskoj Gorodskoj Astronomicheskoj Observatorii. Izdatel'stvo Kazanskogo Universiteta, Kazan.

Tr. Tashkent. Astron. Obs.
Trudy Tashkentskoj Astronomicheskoj Observatorii.
Akademiya Nauk Uzbekskoj SSR. Publishers: Izdatel'-
stvo "FAN" Uzbekskoj SSR, Tashkent.

Trans. Astron. Obs. Yale Univ.
Transactions of the Astronomical Observatory of Yale
University. Published by the Observatory, New Haven.

Trans. IAU
Transactions of the International Astronomical Union.
Published and distributed for the IAU (UAI) by D. Rei-
del Publishing Company, Dordrecht, Holland – Boston,
U.S.A.

Tsirk. Astron. Inst. Tashkent
Tsirkulyar Astronomicheskogo Instituta. Akademiya
Nauk Uzbekskoj SSR. Izdatel'stvo "FAN" Uzbekskoj
SSR, Tashkent.

Tsirk. Astron. Obs. L'vov
Tsirkulyar. Astronomicheskaya Observatoriya. L'vovskij
Ordena Lenina Gosudarstvennyj Universitet imeni Ivana
Franko. Publisher: Izdatel'stvo L'vovskogo Universiteta,
L'vov.

UKIRT Rep.
United Kingdom Infrared Telescope Report. ISSN 0260-
9983.

Umschau
Umschau in Wissenschaft und Technik. Umschau Verlag,
Stuttgarter Str. 18 - 24, D-6000 Frankfurt/M., F. R.
Germany.

United States Naval Obs., Circ.
United States Naval Observatory, Circular. U.S. Naval
Observatory, Washington, D.C. 20390.

Univ. Chile, Dep. Astron., Publ.
Universidad de Chile, Facultad de Ciencias Fisicas y
Matematicas, Departamento de Astronomía, Publica-
ciones. Observatorio Astronómico Nacional, Cerro
Calán, Santiago de Chile.

Urania Barcelona
Urania. Revista de Astronomía y Ciencias Afines. Órgano
de la Sociedad Astronómica de España y América, Barcelona; Unión Nacional de Astronomía y Ciencias Afines,
Madrid, Spain.

Urania Kraków
Urania. Miesięcznik Polskiego Towarzystwa Miłośników
Astronomii, Kraków. Publisher: Z. N. im Ossolińskich,
Kraków, Poland.

Vasiona
Vasiona. Revue d'Astronomie et d'Astronautique. Bulletin de la Société Astronomique "R. Bosković", Beograd.

Vatican Obs. Publ.
Vatican Observatory Publications, Specola Vaticana,
Città del Vaticano.

Veröff. Astron. Rechen-Inst. Heidelberg
Veröffentlichungen des Astronomischen Rechen-Instituts
Heidelberg. Verlag G. Braun, Karlsruhe, F.R. Germany.

**Veröff. Bayer. Komm. Int. Erdmessung, Bayer. Akad. Wiss.,
Astron.-Geod. Arb.**
Veröffentlichungen der Bayerischen Kommission für die
Internationale Erdmessung der Bayerischen Akademie der

Wissenschaften. Astronomisch-Geodätische Arbeiten.
Published by Verlag der Bayerischen Akademie der
Wissenschaften, München, F. R. Germany. ISSN 0340-
7691, ISBN 3-7696-9782-0.

Veröff. Remeis-Sternw. Bamberg
Veröffentlichungen der Remeis-Sternwarte Bamberg,
Astronomisches Institut der Universität Erlangen-Nürnberg.

Veröff. Sternw. Sonneberg
Akademie der Wissenschaften der DDR, Zentralinstitut
für Astrophysik, Veröffentlichungen der Sternwarte in
Sonneberg. Publisher: Akademie-Verlag, Berlin, German
Democratic Republic.

Veröff. Zentralinst. Phys. Erde
Akademie der Wissenschaften der DDR, Forschungsbe-
reich Geo- und Kosmowissenschaften. Veröffentlichungen des Zentralinstituts für Physik der Erde, Potsdam,
German Democratic Republic.

Vesmír
Vesmír. Přírodovědecký časopis Čs. akademie věd. Publisher: Academia, Praha.

Vestn. Khar'kov. Univ.
Vestnik Khar'kovskogo Universiteta. Seriya Astronomicheskaya. Publishers: Izdatel'stvo Khar'kovskogo Universiteta, Khar'kov.

Vestn. Kiev. Univ.
Vestnik Kievskogo Universiteta. Seriya Astronomii.
Publishers: Izdatel'stvo Kievskogo Universiteta, Kiev.

Vistas Astron.
Vistas in Astronomy. An international review journal.
Pergamon Press, Oxford – New York – Braunschweig.

Weather
Weather. James Glaisher House, Grenville Place, Brack-
nell, Berks RG12 1BX, England.

Wiss. Z. Friedrich-Schiller-Univ. Jena
Wissenschaftliche Zeitschrift der Friedrich-Schiller-Uni-
versität. Jena. Mathematisch-Naturwissenschaftliche
Reihe; Edited by the Rektor der Friedrich-Schiller-Uni-
versität Jena, Am Anger 24, Jena, German Democratic
Republic.

Wiss. Z. Humboldt-Univ. Berlin
Wissenschaftliche Zeitschrift der Humboldt-Universität
zu Berlin. Mathematisch-Naturwissenschaftliche Reihe.
Edited by the Rektor der Humboldt-Universität Berlin,
Unter den Linden 6, 108 Berlin, German Democratic
Republic.

Wiss. Z. Tech. Univ. Dresden
Wissenschaftliche Zeitschrift der Technischen Universität
Dresden. Mommsenstraße 13, Dresden A227, Germany.

Wuli
Wuli. Science Press, Peking, People's Republic of China.
Subscription address: Guozi Shudian, P.O. Box 399,
Peking.

Yamamoto Circ.
Yamamoto Circular. Published by the Yamamoto
Observatory, Kamitanakami-Kiryutyo, Otu, Siga-Ken,
[520-21] Japan.

Z. angew. Math. Mech.
Zeitschrift für angewandte Mathematik und Mechanik.

Akademie-Verlag GmbH, 108 Berlin, Leipziger Strasse 3–4, German Democratic Republic.

Z. angew. Math. Phys.
Zeitschrift für angewandte Mathematik und Physik.
Verlag Birkhauser, Postfach 4000, Basel 24, Switzerland.

Z. Naturforsch.
Zeitschrift für Naturforschung. Teil A. A Europhysics Journal. Physik–Physikalische Chemie–Kosmophysik.
Verlag der Zeitschrift für Naturforschung, Tübingen.
P. O. Box 2645, D–7400 Tübingen, F. R. Germany.

Z. Phys. A
Zeitschrift für Physik A. Atoms and Nuclei. Springer-Verlag, Berlin–Heidelberg–New York.

Z. Phys. B
Zeitschrift für Physik B. Condensed Matter and Quanta.
Springer-Verlag, Berlin–Heidelberg–New York.

Z. Phys. C
Zeitschrift für Physik C. Particles and Fields. Springer-Verlag, P.O. Box 105280, D-6900 Heidelberg 1, Germany.

Z. Vermessungswes.
Zeitschrift für Vermessungswesen. Verlag Konrad Wittwer,

7000 Stuttgart 1, Nordbahnhofstrasse 16, Postfach 147, F. R. Germany.

Zeiss Inf.
Zeiss Information. Carl Zeiss, Oberkochen. F. R. Germany.

Zemlya i Vselennaya
Zemlya i Vselennaya. Astronomiya, Geofizika. Issledovaniya Kosmicheskogo Prostranstva, Nauchno-Populyarnyj Zhurnal Akademii Nauk SSSR. Publishers: Izdatel'-stvo "Nauka", Moskva.

Zenit
Populair wetenschappelijk maandblad over sterrenkunde/weerkunde/ruimtevaart/ruimte-onderzoek/aanverwante wetenschappen en technieken. Bureau: Stichting De Koepel, Utrecht.

Zentralbl. Math. Grenzgeb. – Math. Abstr.
Zentralblatt für Mathematik und ihre Grenzgebiete – Mathematics Abstracts. Publisher: Springer-Verlag, Berlin–Heidelberg–New York.

Zvaigžņota Debess
Latvijas PSR Zinātņu Akadēmijas Radioastrofizikas Observatorijas Populārzinatnisks Gadalaiku Izdevums. Izdevnieciba "Zinātne", Riga.

Journals abstracted completely.

A selected number of periodicals listed in category 001 are central to the subject scope of *Astronomy and Astrophysics Abstracts*. Depending on their relevance, almost all papers of the journals listed below are abstracted in our service.

AAVSO Bull.
Acta Astron.
Acta Astron. Sinica
Acta Astrophys. Sinica
Acta Cosmologica
Ann. Shanghai Obs., Acad. Sinica
Ann. Tokyo Astron. Obs.
Annu. Rep. Astron. Inst. Greece
Annu. Rev. Astron. Astrophys.
Astrofiz. Issled. Izv. Spets. Astrofiz. Obs.
Astrofizika
Astrometr. Astrofiz.
Astron. Astrophys.
Astron. Astrophys., Suppl. Ser.
Astron. J.
Astron. Nachr.
Astron. Pap.
Astron. Q.
Astron. Tidsskr.
Astron. Tsirk.
Astron. Vestn.
Astron. Zh.
Astronomia
Astronomie
Astrophys. J.
Astrophys. J., Lett.
Astrophys. J., Suppl. Ser.
Astrophys. Lett.
Astrophys. Space Sci.
Australian J. Phys., Astrophys. Suppl.
BAV Rundbrief
BBSAG Bull.
Bol. Astron. Obs. Madrid
Bol. Inst. Tonantzintla
British Astron. Assoc. Circ.
Bull. AFOEV
Bull. American Astron. Soc.
Bull. Astron. Inst. Czechoslovakia
Bull. Astron., Obs. R. Belgique
Bull. Astron. Soc. India
Bull. Inf. Cent. Données Stellaires
Bull. Obs. Astron. Belgrade
Byull. Abastumanskaya Astrofiz. Obs.
Byull. Inst. Astrofiz.
Byull. Inst. Teor. Astron.
Carter Obs., Astron. Bull.
Celestial Mech.
Circ. Inf.
Circ. Stn. Astron. Int. Latitudine, Carloforte-Cagliari
Coelum
Comments Astrophys.
Comun. Obs. Astron. Univ. Coimbra
Dudley Obs. Rep.
Dunsink Obs. Publ.
ESO Sci. Prepr.
ESO Tech. Rep.
G. Astron.
IAU Circ.
Icarus
Inf. Bull. Variable Stars
Inst. Theor. Astrophys., Blindern–Oslo, Rep.
Irish Astron. J.
Izv. Astron. Ehngel'gardt. Obs.
Izv. Glav. Astron. Obs. Pulkovo
Izv. Krymskoj Astrofiz. Obs.

J. American Assoc. Variable Star Obs.
J. British Astron. Assoc.
J. Hist. Astron.
J. R. Astron. Soc. Canada
Komet. Tsirk.
Komety i Meteory
Mem. Soc. Astron. Italiana
Mercury
Messenger
Meteoritics
Meteoritika
Minor Planet Bull.
Minor Planet Circ.
Mitt. Astron. Ges.
Mitt. Sternw. Sonneberg
Mitt. Veränderl. Sterne (MVS)
Mon. Not. R. Astron. Soc.
Mon. Notes Astron. Soc. South Africa
Mon. Notes Int. Polar Motion Serv.
Moon Planets
Nablyud. Iskusstv. Nebesn. Tel
Nauchn. Inf.
News Lett. Astron. Soc. N.Y.
Obs. Astrophys. Lab., Univ. Helsinki. Rep.
Observatory
Occas. Rep. R. Obs. Edinburgh
Occultation Newsl.
Orion
Oss. Astrofis. Catania, Pubbl.
Oss. Mem. Oss. Astrofis. Arcetri
Perem. Zvezdy
Perem. Zvezdy, Prilozhenie
Pis'ma Astron. Zh.
Postępy Astron.
Probl. Kosm. Fiz.
Proc. Astron. Soc. Australia
Proc. Int. Latitude Obs. Mizusawa
Publ. Astron. Soc. Japan
Publ. Astron. Soc. Pacific
Publ. Dominion Astrophys. Obs.
Publ. Eidg. Sternw. Zürich
Publ. Int. Latitude Obs. Mizusawa
Publ. R. Obs. Edinburgh
Publ. United States Naval Obs.
Publ. Variable Star Sect. R. Astron. Soc. New Zealand
Q. Bull. Sol. Act.
Q. J. R. Astron. Soc.
R. Obs. Ann.
Rep. Obs. Lund
Rev. Mexicana Astron. Astrofis.
Rezul't. Nablyud. Iskusstv. Sputnikov Zemli
Sky Telesc.
Smithsonian Astrophys. Obs., Spec. Rep.
Smithsonian Contrib. Astrophys.
Sol. Phys.
Soln. Dannye, Byull.
Soobshch. Byurakan. Obs.
Soobshch. Gos. Astron. Inst. Shternberg
Soobshch. Spets. Astrofiz. Obs.
South African Astron. Obs. Circ.
South. Stars
Space Sci. Instrum.
Space Sci. Rev.
Sterne
Sterne Weltraum

Stockholms Obs. Ann.
Stockholms Obs. Rep.
Strolling Astron.
Tartu Astrofüüs. Obs. Publ.
Tartu Astrofüüs. Obs. Teated
Tokyo Astron. Bull.
Tokyo Astron. Obs. Rep.
Tr. Astrofiz. Inst. Alma-Ata
Tr. Astron. Obs., *Leningrad*
Tr. Glav. Astron. Obs. Pulkovo
Tr. Inst. Teor. Astron., *Leningrad*
Tr. Kazan. Gorod. Astron. Obs.
Tr. Tashkent. Astron. Obs.

Trans. IAU
Tsirk. Astron. Inst. Tashkent
Tsirk. Astron. Obs. L'vov
United States Naval Obs., Circ.
Urania Barcelona
Urania Kraków
Vatican Obs. Publ.
Veröff. Astron. Rechen-Inst. Heidelberg
Veröff. Sternw. Sonneberg
Vestn. Khar'kov. Univ.
Vestn. Kiev. Univ.
Vistas Astron.
Yamamoto Circ.

002 Bibliographical Publications, Catalogues, Atlases

002.001 A catalogue of [Fe/H] determinations. (Magnetic tape version, updated to 31.12.1980).
G. Cayrel de Strobel, C. Bentolila, B. Hauck, D. Lovy.
Astron. Astrophys., Suppl. Ser., Vol. 45, 97 (1981).
   This catalogue is an updated version of that published recently by G. Cayrel de Strobel et al. (1980) since many new determinations were available. This version contains 1298 [Fe/H] determinations for 703 stars, being an increase of 17% and 12% respectively. It is also possible to obtain a copy of this catalogue on magnetic tape from the Stellar Data Centre of Strasbourg.

002.002 A catalogue of observations in Hα.
   J. R. Ducati.
Astron. Astrophys., Suppl. Ser., Vol. 45, 119 (1981).
   A catalogue of all quantitative observations of Hα published between 1949 and 1980 has been prepared. It contains 4095 measurements concerning 2700 stars.

002.003 Untrivial redshifts : a bibliographical catalogue.
   H. J. Reboul.
Astron. Astrophys., Suppl. Ser., Vol. 45, 129 - 144 (1981).
   The author presents 780 coded references covering 70 years in the problem of anomalous – hereafter untrivial (NT) – redshifts (Z). An arbitrary definition is primarily settled for trivial Z and classical theories. 17 classes of untrivial redshifts (NTZ) (part 1) and 19 classes of unclassical theories (part 2) are investigated in that frame.

002.004 Catalogue of apparent diameters and absolute radii of stars (CADARS).
M. Fracassini, L. E. Pasinetti, F. Manzolini.
Astron. Astrophys., Suppl. Ser., Vol. 45, 145 - 174 (1981).
   A catalogue of stellar apparent diameters and/or absolute radii, listing 6313 data files for 4266 stars, is given.

002.005 Sky Atlas 2000.0. W. Tirion.
Zenit, 8. Jaarg., 370 - 373 (1981).

002.006 Some trends in American astronomical publications.
H. A. Abt.
Publ. Astron. Soc. Pacific, Vol. 93, 269 - 272 (1981).
   The author has studied the papers published in 1910, 1920, . . . 1980 in the three principal American astronomical journals to derive the following results: the total annual publication rate was relatively constant until World War II, followed by a doubling time of 7.8 years and with a recent slower rate. The average paper lengths, normalized to the same page content, has tripled since 1940, with a much larger increase in information. But the most surprising result is the virtual disappearance of the very short papers (less than one page) that accounted for one-third of the papers published earlier in the century. While papers are becoming longer, authors are writing fewer papers per person (by a factor of 0.6). The fraction of theoretical papers has increased from 5% early in the century to a current 33%. The strong tendency for single author papers early in the century has been replaced by a current average of double author theoretical papers and triple author observational papers.

002.007 Graphic catalogue of spot groups, Carrington longitude – time – latitude, for cycle No. 20. I. Interaction of groups and flare occurrence with type II radio bursts.   J. Klimeš, L. Křivský.
Bull. Astron. Inst. Czechoslovakia, Vol. 32, 222 - 227 (1981).
   The author designed a graphic catalogue with all spot groups plotted in daily values of their areas in the system of Carrington longitude – time – latitude position, for the entire cycle No. 20 (1965 - 1976). Part I contains a classification of the interactions (close approaching or spot group merging) from the point of view of the evolution of groups during solar rotations. A first attempt is then made to use the catalogue in investigating the connection between type II radio flares and the interacting spot groups.

002.008 Third catalogue of stars measured in the Geneva Observatory Photometric System.
F. Rufener.
Astron. Astrophys., Suppl. Ser., Vol. 45, 207 - 366 (1981).
   A compilation of the photoelectric measurements in the Geneva Observatory seven colour system ($U, B, V, B1, B2, V1, G$) is presented for 14633 stars. The catalogue, besides the coordinates and principal bibliographic references of each star, gives the colours (normalized magnitudes), the colour indices $B1$-$B2$, $B2$-$V1$ and $V1$-G, the parameters d, Δ, g and $m2$, the V magnitude. The number of accurate measurements and the standard deviation of their mean value is given separately for the colours and the V magnitude. An estimation procedure of the uncertainties associated with the photometric parameters is described in the preface.

002.009 A comparison of the AGK 3 with the SAOC – Part I.
   L. G. Taff, S. A. Stansfield.
Bull. American Astron. Soc., Vol. 13, 570 (1981). – Abstract.

002.010 A comparison of the AGK3 with the SAOC – Part II.
   L. G. Taff, S. A. Stansfield.
Bull. American Astron. Soc., Vol. 13, 510 (1981). – Abstract.

002.011 A catalogue of extragalactic radio sources having flux densities greater than 1 Jy at 5 GHz.
H. Kühr, A. Witzel, I. I. K. Pauliny-Toth, U. Nauber.
Astron. Astrophys., Suppl. Ser., Vol. 45, 367 - 430 (1981).
   A catalogue of 518 extragalactic radio sources is presented. It contains sources from the combined NRAO-MPI 5 GHz Strong Source Surveys and the Parkes 2.7 GHz Surveys which altogether cover 9.811 sr of the sky. Within this area the catalogue is essentially complete for sources with flux densities $S \geqslant 1$ Jy at 5 GHz. For most sources the radio spectra are given, as well as the radio positions and basic optical information.

002.012 The Sixth Catalogue of galactic Wolf-Rayet stars, their past and present.
K. A. van der Hucht, P. S. Conti, I. Lundström, B. Stenholm.
Space Sci. Rev., Vol. 28, 227 - 306 (1981).
   This paper presents the Sixth Catalogue of galactic Wolf-Rayet stars (Pop. I), a short history on the five earlier WR catalogues, improved spectral classification, finding charts, a discussion on related objects, and a review of the current status of Wolf-Rayet star research.

002.013 A bibliography on galactic Wolf-Rayet literature 1867–1980. Appendix to the Sixth Catalogue of galactic Wolf-Rayet stars, their past and present.
K. A. van der Hucht, P. S. Conti, I. Lundström, B. Stenholm.
Space Sci. Rev.,Vol. 28, 307 - 382 (1981).

002.014 Mapping the planets and moons.   G. Day.
Spaceflight, Vol. 23, 320 (1981).

002.015 I.A.U. archives of unpublished observations of variable stars: 1979–81 data.   M. Breger.
Publ. Astron. Soc. Pacific, Vol. 93, 528 - 531 (1981).

002.016   **The AAVSO Variable Star Atlas.**   C. E. Scovil.
J. R. Astron. Soc. Canada, Vol. 75, 164 - 168
(1981).

002.017   **Accurate positions of Markarian galaxies.**
M. N. Joshi, R. A. Kandalian.
Bull. Astron. Soc. India, Vol. 9, 24 - 30 (1981).
Optical positions of 304 Markarian galaxies, MRK 1096
to 1399, are given with accuracies better than 1 arcsec.

002.018   **Comets and the origin of life: bibliography.**
L. G. Pleasant.
Comets and the origin of life, (see 012.022), p. 255 - 268
(1981).

002.019   **Viking Lander imaging investigation during**
extended and continuation automatic missions.
Volume I – Lander 1 Picture Catalog of Experiment Data
Record, Volume II – Lander 2 Picture Catalog of Experiment
Data Record.   K. L. Jones, M. Henshaw, C. McMenomy,
A. Robles, P. C. Scribner, S. D. Wall, J. W. Wilson.
NASA Ref. Publ., NASA RP 1068, Vol. I: 6 + 648 pp.,
Vol. II: 6 + 559 pp. (1981).
All the images returned by the two Viking landers
during the extended and continuation automatic phases of
the Viking Mission are presented in the two volumes. Listings
of supplemental information which describe the conditions
under which the images were acquired are included together
with skyline drawings which show where the images are
positioned in the field of view of the cameras. Subsets of the
images are listed in a variety of sequences to aid in locating
images of interest. The format and organization of the digital
magnetic tape storage of the images are described. A brief
description of the mission and the camera system is also
included.

002.020   **Bibliography of non-commercial publications of**
Observatories and Astronomical Societies.
P. A. H. Smulders (Editor).
4th completely revised edition. "Sonnenborgh" Observatory,
Zonnenburg 2, 3512 NL Utrecht, The Netherlands. (1980/81).

002.021   **Documentation for the machine-readable version of**
the Thirteen-Color Photometry of 1380 Bright
Stars.   W. H. Warren, Jr., N. G. Roman.
National Space Science Data Center/World Data Center A for
Rockets and Satellites, National Aeronautics and Space
Administration, Goddard Space Flight Center, Greenbelt,
Maryland 20771, NSSDC/WDC-A-R & S 81-02. 11 pp. (1981).

002.022   **Documentation for the machine-readable character**
coded version of the SKYMAP catalogue.
W. H. Warren, Jr.
National Space Science Data Center/World Data Center A for
Rockets and Satellites, National Aeronautics and Space
Administration, Goddard Space Flight Center, Greenbelt,
Maryland 20771, NSSDC/WDC-A-R & S 81-05. 3 + 106 pp.
(1981).

002.023   **Documentation for the machine-readable version of**
the catalog of 5,268 standard stars, 1950.0 based on
the normal system N30.   W. H. Warren, Jr.
National Space Science Data Center/World Data Center A for
Rockets and Satellites, National Aeronautics and Space
Administration, Goddard Space Flight Center, Greenbelt,
Maryland 20771, NSSDC/WDC-A-R & S 81-08. 3 + 11 pp.
(1981).

002.024   **Bibliography of articles concerning the solar system**
from journals received in 1980.
Z. Kopal, M. Moutsoulas, F. B. Waranius (Editors).
Moon Planets, Vol. 25, 139 - 356 (1981).

002.025   **Refined data for parallax stars.**   W. Buscombe.
Bull. Inf. Cent. Données Stellaires, No. 21, p. 26 -
34 (1981).

002.026   **Photometric data for the nearby stars.**
B. Hauck, M. Mermilliod.
Bull. Inf. Cent. Données Stellaires, No. 21, p. 35 - 39 (1981).

002.027   **A proposal for a glossary of terms relating to the**
storage, retrieval and analysis of astronomical data.
G. A. Wilkins.
Bull. Inf. Cent. Données Stellaires, No. 21, p. 47 - 49 (1981).

002.028   **A master list of nonstellar optical astronomical**
objects.   R. S. Dixon, G. Sonneborn.
Bull. Inf. Cent. Données Stellaires, No. 21, p. 50 (1981).
This compendium of approximately 185,000 listings
from all known catalogues of nonstellar objects has been compiled by entering all of the items in each individual component
onto punched cards, converting them to a standard form of
notation, and combining them in right ascension order. Every
object in every catalogue is included, so that multiple listings
of the same item may be compared. The data given for each
object include 1950.0 position, angular diameter, magnitude,
and description.

002.029   **Cross-identification tables CoD-CPD.**
R. Bonnet, F. Ochsenbein.
Bull. Inf. Cent Données Stellaires, No. 21, p. 51 (1981).
Cross-identification tables between Cordoba (CoD) and
Cape Photographic (CPD) Durchmusterungen have been built
up in the zones −22° to −39° (144 600 stars). The catalog
includes all CoD stars in the mentioned zones and is sorted by
CoD numbers.

002.030   **Star catalogs and files available at the Stellar Data**
Center. June 1981.
Bull. Inf. Cent. Données Stellaires, No. 21, p. 52 - 107 (1981).

002.031   **Availability of an interactive computer reference**
search of the astronomical literature 1950 - 1976.
T. A. Nagy, R. S. Hill, J. M. Mead.
Astron. Data Cent. Bull., Vol. 1, 62 - 68 (1981).

002.032   **Suggestions for formatting, checking and document-**
ing machine-readable astronomical catalogues.
R. S. Hill.
Astron. Data Cent. Bull., Vol. 1, 69 - 77 (1981).
The purpose of this paper is to describe ADC practice in
modifying and documenting machine-readable catalogues in
order to provide guidelines for others who make such catalogues. The following sections describe practices followed by
the ADC to make machine-readable catalogues transportable
to many computers, to insure the integrity of the data in the
catalogues, and to make the catalogues easy to use for input
to computer programs.

002.033   **A catalogue of ultraviolet, optical, and H I data for**
201 Virgo cluster galaxies.
R. H. Cornett, A. M. Smith.
Astron. Data Cent. Bull., Vol. 1, 78 (1981).

002.034   **The machine-readable version of the General Cata-**
logue of Variable Stars, third edition.
R. S. Hill, T. A. Nagy.
Astron. Data Cent. Bull., Vol. 1, 79 - 80 (1981).

002.035   **Galactic latitude and magnitude distribution of two**
astronomical catalogues.
T. A. Nagy, R. S. Hill, J. M. Mead.
Astron. Data Cent. Bull., Vol. 1, 81 - 89 (1981).

002.036   **The Lund-Strasbourg catalogue of open cluster data.**
G. Lyngå.
Astron. Data Cent.Bull., Vol. 1, 90 - 93 (1981).
The intention is to provide a selection of fundamental data values for open clusters. This has entailed and will continue to entail a critical examination of the large volume of available information. The best available value for each parameter should be presented together with a reference.

002.037   **Infrared Astronomical Data Base and Catalog of Infrared Observations.**
M. Schmitz, D. Y. Gezari, J. M. Mead.
Astron. Data Cent. Bull., Vol. 1, 94 - 104 (1981).
A computer data base of infrared astronomical observations has been established at NASA/Goddard Space Flight Center. It contains a summary of all infrared (1 μm - 1 000 μm) observations of celestial sources outside the solar system, published in the major scientific journals since 1960, as well as the contents of infrared surveys and catalogs.

002.038   **Three short notes on the Air Force Geophysics Laboratory (AFGL) catalogue.**
T. A. Nagy, R. S. Hill, J. M. Mead.
Astron. Data Cent. Bull., Vol. 1, 105 - 107 (1981).

002.039   **Revised magnetic tape of the N30 catalog of 5,268 standard stars.**   W. H. Warren, Jr.
Astron. Data Cent. Bull., Vol. 1, 108 - 112 (1981).

002.040   **Positional correlation of the Two-Micron Sky Survey and Smithsonian Astrophysical Observatory catalogue sources.**   T. A. Nagy, R. S. Hill.
Astron. Data Cent. Bull., Vol. 1, 113 (1981).

002.041   **Search capabilities for the Catalog of Stellar Identifications, edition 1979.**
W. H. Warren, Jr., W. T. Sheridan.
Astron. Data Cent. Bull., Vol. 1, 114 - 120 (1981).

002.042   **CSI statistics: blue magnitude versus spectral type.**
T. A. Nagy, R. S. Hill, J. M. Mead.
Astron. Data Cent.Bull., Vol. 1, 121 - 138 (1981).

002.043   **New and revised catalogues available from the Astronomical Data Center (ADC).**
W. H. Warren, Jr., T. A. Nagy, R. S. Hill.
Astron. Data Cent. Bull., Vol. 1, 139 - 145 (1981).

002.044   **Status report on machine-readable astronomical catalogues (1 July 1981): Astronomical Data Center, NASA/Goddard Space Flight Center.**
T. A. Nagy, J. M. Mead, W. H. Warren, Jr.
Astron. Data Cent. Bull., Vol. 1, 146 - 169 (1981).

002.045   **A photometric catalogue of stars in the direction of the bright cloud B in Sagittarius.**
A. Terzan, A. Bernard.
Astron. Astrophys.,Suppl. Ser., Vol. 46, 49 - 56 (1981).
The present catalogue includes photometric data on stars in a 5.5×5.5 area in the direction of the bright cloud B in Sagittarius, centred at $l = 357°.7$, $b = +5°.6$. Magnitudes V and colours B-V and U-B are reported for nearly all HD and HDE stars in the field and also for a number of interesting stars (402 stars), 21 proper motion stars, and 45 stars in a 15' radius circular region centred on the galactic cluster Trumpler 26 (Tr 26). Furthermore, photographic B, V, R, and IR magnitudes are listed for 274 stars in the field of Tr 26.

002.046   **Spectroscopic and photometric observations of galaxies from the ESO/Uppsala list. Third catalogue.**
R. M. West, J. Surdej, H.-E. Schuster, A. B. Muller, S. Laustsen, T. M. Borchkhadze.

Astron. Astrophys.,Suppl. Ser., Vol. 46, 57 - 77 (1981).
Spectroscopic and photometric observations are presented for a total of about 275 southern galaxies, selected from ESO/Uppsala lists Nos. 1 - 8. Many of the galaxies have emission lines and several are members of multiple systems.

002.047   **A catalogue of Jovian decametric radio observations from January 1978 to December 1979.**
Y. Leblanc, J. de la Noë, F. Genova, A. Gerbault, A. Lecacheux.
Astron. Astrophys.,Suppl. Ser., Vol. 46, 135 - 149 (1981).
This catalogue contains a complete list of all Jupiter decameter emissions recorded on the Nançay Spectrograph, over a period January 1978 to December 1979. A brief description of the equipment and its characteristics are presented. The observing program and the limitations arising from daytime observations are developed. A list of observing times as well as histograms showing the occurrence probability of emission are also given.

002.048   **The second COS-B catalogue of high-energy γ-ray sources.**   W. Hermsen.
Philos. Trans. R. Soc. London, Ser. A, Vol. 301, 519 - 521 (1981). – Abstr. in Phys. Abstr., Vol. 84, Abstr. 94202 (1981).

002.049   **UBV photometry of double galaxies. Catalogue.**
V. V. Demin, Eh. A. Dibaj, A. N. Tomov.
Astron. Zh., Tom 58, 925 - 932 (1981). In Russian. English translation in Soviet Astron., Vol. 25, No. 5.
A catalogue of photometric UBV magnitudes and colours of 105 double galaxies is presented.

002.050   **Absolute proper motions of 117 O-type stars.**
D. K. Karimova, E. D. Pavlovskaya.
Pis'ma Astron. Zh., Tom 7, 627 - 631 (1981). In Russian. English translation in Soviet Astron. Lett., Vol. 7.

002.051   **A catalogue of high-speed plasma streams in the solar wind.**   B. A. Lindblad, H. Lundstedt.
Sol. Phys., Vol. 74, (see 012.029), 197 - 206 (1981).
A catalogue of 346 well defined high-speed plasma streams in solar wind observations 1964 - 1975 is presented. The data base for the study is the compilation of interplanetary plasma/magnetic field data prepared by J. King. It is believed that the catalogue may be found useful for studies of various solar-interplanetary and solar-terrestrial phenomena.

002.052   **The early-type chemically peculiar stars in the catalogue of stellar groups.**
D. Egret, M. Jaschek.
Upper main sequence chemically peculiar stars, (see 012.033), p. 495 - 502, microfiche (1981).

002.053   **Le catalogue d'étoiles Ap et Am.**   P. Renson.
Upper main sequence chemically peculiar stars, (see 012.033), p. 503 - 504 (1981).

002.054   **A search for S stars in the IRC and AFGL Catalogues.**
A. Borghesi, E. Bussoletti, L. Rossi, S. Uras.
Astrophys. Space Sci., Vol. 79, 435 - 441 (1981).
Seventeen unidentified sources contained in the 2μ IR Catalogue, and AFGL-AFGLS Catalogues have been tentatively indicated as S stars from a comparison with the General Catalogue of S stars. An analysis based on the I–K and K–[4.2 μ] behavior has been performed for all the sources which have been found to be in common with all the above mentioned catalogues. The results lead to the conclusion that the presence of circumstellar shells seems to be very unlike for these objects in spite of their evolutionary status which is intermediate between M and C type stars.

002.055 **The ESO/Uppsala survey of the ESO (B) Atlas of the southern sky. IX.**
A. Lauberts, E. B. Holmberg, H.-E. Schuster, R. M. West.
Astron. Astrophys., Suppl. Ser., Vol. 46, 311 - 346 (1981).

A systematic search for certain objects (NGC + IC galaxies, all galaxies with a diameter larger than about 1ʹ.0, all disturbed galaxies, all star clusters in the Budapest Catalogue, and all listed planetary nebulae) has been carried out by means of the ESO (B) Atlas, covering the southern sky from $-90°$ to $-20°$. The present paper contains the ninth list of objects in 64 fields. A total of 1881 objects is listed; of these, about 50% for the first time. This list concludes the search; the nine lists of the ESO/Uppsala Survey of the ESO (B) Atlas comprise about 18500 entries.

002.056 **Catalog of cosmic gamma-ray bursts from the KONUS experiment data.** E. P. Mazets, S. V. Golenetskii *(Golenetskij)*, V. N. Il'inskii *(Il'inskij)*, V. N. Panov, R. L. Aptekar *(Aptekar')*, Yu. A. Gur'yan, M. P. Proskura, I. A. Sokolov, Z. Ya. Sokolova, T. V. Kharitonova, A. V. Dyatchkov *(Dyachkov)*, N. G. Khavenson, with a preface by K. Hurley.
Astrophys. Space Sci., Vol. 80, 1 - 143 (1981).

Data are presented on the temporal structure, fluxes, energy spectra and coordinates of the sources of gamma-ray bursts detected in the KONUS experiment on Venera 11 and Venera 12 space probes in the period September 1978 to February 1980. The statistical distributions of gamma bursts in duration, intensity, and peak power, as well as the distribution of the burst sources over the celestial sphere are given.

002.057 **Annotations on papers on geomagnetism and aeronomy published in "News of Universities. Radiophysics", Tom 23, No. 12 (1980); Tom 24, No. 4 (1981). In Russian.**
Geomagn. Aehron., Tom 21, 956 - 957 (1981). In Russian.

002.058 **BAA star charts.**
W. Tirion.
British Astronomical Association, London, England. 5 pp. (1981). — Review in Phys. Abstr., Vol. 84, Abstr. 108082 (1981).

002.059 **SS 433 – essai de bibliographie.** E. Schweitzer.
Bull. AFOEV, No. 18, p. 5 - 11 (1981).

002.060 **A brief subject index for N. A. S. A.'s special publications relating to astronomy.** A. Fraknoi.
Mercury, Vol. 10, 152 - 158 (1981).

002.061 **The Ariel V (3 A) catalogue of X-ray sources – I. Sources at low galactic latitude ($|b| < 10°$).**
R. S. Warwick, N. Marshall, G. W. Fraser, M. G. Watson, A. Lawrence, C. G. Page, K. A. Pounds, M. J. Ricketts, M. R. Sims, A. Smith.
Mon. Not. R. Astron. Soc., Vol. 197, 865 - 891 (1981).

A catalogue is presented of 109 X-ray sources at low galactic latitude ($|b| < 10°$) observed by the Leicester Sky Survey Instrument on Ariel V. The catalogue is based on observations extending over a $5^1/_2$-yr period and gives details of the position, 2–10 keV flux, X-ray variability and the proposed optical identification of each source. The distribution, $\log N - \log S$ relation, X-ray variability and optical identifications of the sources are then briefly discussed.

002.062 **The Ariel V (3 A) catalogue of X-ray sources – II. Sources at high galactic latitude ($|b| > 10°$).**
I. M. McHardy, A. Lawrence, J. P. Pye, K. A. Pounds.
Mon. Not. R. Astron. Soc., Vol. 197, 893 - 919 (1981).

This paper forms the second part of the Ariel V 3 A catalogue and covers all the sky outside the galactic plane ($|b| > 10°$). It is based on all the data collected by the Sky Survey Instrument on the Ariel V satellite during its $5^1/_2$ yrs in orbit. It covers 90 per cent of the high latitude sky down to ~ 0.8 SSI count s^{-1} and contains 142 sources. The main difference from the 2A catalogue is the far greater number of identifications with stars, and, to a lesser extent, with active galaxies. The authors suspect that many of the unidentified sources will eventually be identified with cataclysmic variables.

002.063 **Planetary mission data available at NSSDC *(National Space Science Data Center)*.** W. S. Cameron.
Bull. American Astron. Soc., Vol. 13, 700 (1981). – Abstract.

002.064 **Neutral hydrogen observations of a large sample of galaxies.** J. R. Fisher, R. B. Tully.
Astrophys. J., Suppl. Ser., Vol. 47, 139 - 200 (1981).

A sample of 1787 nearby galaxies has been observed in the 21 cm line with the NRAO 91 m and 43 m telescopes and the Bonn 100 m telescope. A total of 1171 galaxies were detected. The radio observations provide an accurate heliocentric velocity, an H I flux, and a line profile width for each detection. Literature optical magnitudes, dimensions, and morphological types are reduced to common systems and tabulated. Intrinsic luminosities, dimensions, H I masses, and total masses are determined, assuming distances derived from redshifts.

002.065 **Catalogue of star formation regions in the Galaxy: Part 1. Observational data. Supplement to the collection "Nauchnye Informatsii" iss. 47 [Vyp. 47].**
V. S. Avedisova.
Astronomicheskij Sovet Akademii Nauk SSSR, Radioastrofizicheskaya Observatoriya Akademii Nauk Latvijskoj SSR.
Zinatne, Riga. 139 pp. Price 65 Kop. (1981). In Russian.

002.066 **Astronomical data bases and retrieval systems.**
J. M. Mead, T. A. Nagy, W. H. Warren, Jr.
Modern observational techniques for comets, (see 012.047), p. 232 - 236 (1981).

The status of the development of machine-readable stellar and extragalactic data bases is summarized, including several examples of astronomical applications using these data sets. The creation of a computerized bibliographical data base for cometary research is described.

002.067 **Catalogue of clusters of galaxies with measured redshifts.** T. S. Fetisova.
Astron. Zh., Tom 58, 1137 - 1157 (1981). In Russian.
English translation in Soviet Astron., Vol. 25, No. 6.

The paper lists 452 clusters of galaxies with measured radial velocities. The 1950 right ascension and declination of the clusters, their richness, compactness and Bautz-Morgan classification are also given.

002.068 **Bibliographie générale des marées terrestres. Supplément V, 1980 - 1981.**
P. Melchior (Editor).
Obs. R. Belgique, Bruxelles. 12 pp. (1981).

002.069 **Charts for southern variables. Series No. 13.**
F. M. Bateson, M. Morel, B. Sumner, R. Winnett.
Published by Astronomical Research Ltd., P. O. Box 3093, Greerton, Tauranga, New Zealand. 28 pp. + charts 552-600 (1981).

002.070 **Testing of optical surfaces: a bibliography.**
A. Cornejo-Rodriguez, H. J. Caulfield, W. Friday.
Appl. Qpt., Vol. 20, 4148 (1981).

002.071 **Solar spectrum synthesis. I. A sample atlas from 224 to 300 nm.**
R. L. Kurucz, E. H. Avrett.
Smithsonian Astrophys. Obs., Spec. Rep. No. 391, 4 + 139 pp. (1981).

The authors have developed sophisticated computer programs for determining solar and stellar atmospheric structure through the analysis of spectra. These programs allow to treat the spectrum as a whole and to draw much stronger conclusions than would be apparent from individual spectral features.

002.072 Astronomy and Astrophysics Abstracts. Vol. 29.
            Literature 1981, Part 1.
S. Böhme, W. Fricke, I. Heinrich, W. Hofmann, D. Krahn, D. Rosa, L. D. Schmadel, G. Zech (Editors).
Published for Astronomisches Rechen-Institut by Springer-Verlag, Berlin—Heidelberg—New York. 10 + 853 pp. Price DM 118.00; ca. US $ 54.90 [Subscription price DM 94.40; ca. US $ 43.90] (1981). ISBN 3-540-11264-2. ISBN 0-387-11264-2.

002.073 A near-infrared atlas of spiral galaxies.
            D. M. Elmegreen.
Astrophys. J., Suppl. Ser., Vol. 47, 229 - 233, plates 1 - 19 (1981).
        A near-infrared photographic atlas of 54 spiral galaxies is presented. Blue photographs are included so that features may be compared in the two passbands. Hubble types range from Sa through Sd for SA, SAB, and SB galaxies. Most are nearly face-on and larger than 2'. These galaxies have been subdivided into the subjective categories of grand design and flocculent spirals, depending upon the appearance of their spiral arms. Photographic procedures and their effects are discussed.

002.074 Preliminary orbital parallax catalog.
            M. Halliwell.
Astrophys. J., Suppl. Ser., Vol. 47, 243 - 278 (1981).
        Thirty five orbital parallaxes calculated from comparison of spectroscopic and visual or astrometric elements are compiled. Thirty-six additional systems with incomplete or discordant spectroscopic and visual data are also discussed. Also listed, but not discussed, are 57 more systems which are good eventual prospects for orbital parallaxes. An analysis of seven methods of computing orbital parallaxes for varying amounts of available data is provided. Summary tables for orbital elements and derived system properties are included for the 35 systems with the best orbital parallaxes.

002.075 An atlas of southern and equatorial dwarf novae.
            N. Vogt, F. M. Bateson.
ESO Sci. Prepr., No. 161, 36 pp. (1981). – Submitted to Astron. Astrophys., Suppl. Ser.

002.076 Atlas and catalogue of stellar magnitudes and
            photoelectric standards.
M. S. Kazanasmas, L. A. Zavershneva, L. F. Tomak.
Naukova dumka, Kiev. 219 pp. (1981). In Russian. – Abstr. in Ref. zh., 51. Astron., 12.51.82 (1981).

002.077 A catalogue of homogeneous photometry of bright
            stars on the DDO system.
R. D. McClure, W. T. Forrester.
Publ. Dominion Astrophys. Obs., Vol. 15, 439 - 457 (1981) = NRC No. 18853.
        DDO photometry for 2196 bright G and K stars is presented. This photometry was compiled from observations made at Kitt Peak and Cerro Tololo with filter sets well matched to the standard system. Mean values of DDO indices for MK spectral classes are listed.

002.078 Double galaxy investigations. I. Observations.
            W. G. Tifft.
Prepr. Steward Obs., No. 344, 102 pp. (1981).
        Redshift information from 240 Å mm^{-1} spectrograms is presented for 370 double galaxy systems from Karachentsev

including all pairs in that catalog with separation less than 80 arc-seconds. An extensive error discussion utilizing internal and external (21-cm) comparisons provide calibration of systematic error and determines the uncertainty for a typical high weigth optical redshift to be ±65 km s^{-1}. Internal differential redshifts within single spectra using common lines achieve accuracies of 18 - 30 km s^{-1}, depending upon separation, and are available for about 200 pairs. Extensive information on emission and other properties is also provided.

002.079 An improvement of the Tokyo PZT star catalogue.
            S. Fujii.
Ann. Tokyo Astron. Obs., Second Ser., Vol. 18, 175 - 190 (1981).
        The author calculates a revised star catalogue for the PZT observations at the Tokyo Astronomical Observatory, by using 23 years of observational data. By using this catalogue, one can reduce the large annual term in the local z, which has appeared since the adoption of the previous α_{75}/δ_{75} star system.

002.080 Right ascension and proper motion in R. A. of
            904 bright stars.   C.-q. Fu, S.-j. Gong.
Ann. Shanghai Obs. Acad. Sinica, No. 1, p. 135 - 167 (1979).
        A catalogue of right ascension and proper motion in R. A. of 904 bright stars with declination from −30° to +65°, which is called the ZPC catalogue, has been compiled by using about 70000 individual star observations obtained with the transit instruments at Shanghai Observatory and Hainan Island Station during eighteen years from 1957 to 1974. The comparisons of the catalogue ZPC with the FK4, N30 and KCB show that the systematic accuracy in R. A. and proper motion of the ZPC is comparable to KCB.

002.081 A catalogue of proper motions of RR Lyrae
            variables.   L. Wan, Y.-q. Mao, D.-s. Ji.
Ann. Shanghai Obs. Acad. Sinica, No. 2, p. 1 - 9 (1980).

002.082 The photoelectric astrolabe catalogue of Shanghai
            Observatory (1975.9—1978) — corrections to
individual FK4 positions.   T.-q. Xu, P.-z. Lu.
Ann. Shanghai Obs. Acad. Sinica, No. 2, p. 10 - 16 (1980).

002.083 The third astrolabe catalogue of Shanghai
            Observatory (1973.0—1978.0).
H.-j. Yan, Y.-f. Chen, X.-m. Tong.
Ann. Shanghai Obs. Acad. Sinica, No. 2, p. 17 - 29 (1980).
        Using the observational data with the Danjon astrolabe at the Shanghai Observatory (O.P.L. No. 14) during 1973.0—1978.0, the third astrolabe catalogue of this instrument is given. It includes the corrections of the individual positions $\Delta\alpha$, $\Delta\delta$ of 92 stars observed at both transits, $\Delta\delta$ of 28 stars and $\Delta\alpha$ of 37 stars at one transit only. The systematic corrections of catalogue $(\Delta\alpha)_\delta$, $(\Delta\delta)_\delta$ are also given.

002.084 NSSDC Data Listing.
        National Space Science Data Center/World Data Center A for Rockets and Satellites, National Aeronautics and Space Administration, Goddard Space Flight Center, Greenbelt, Maryland, NSSDC/WDC-A-R & S 81-11, 64 pp. (1981).
        The NSSDC Data Listing provides a convenient reference to space science and supportive data available from the National Space Science Data Center (NSSDC). The first part of this listing, Satellite Data, is in an abbreviated form compared to the data catalogs published by NSSDC. The second part, Supplementary Data, contains a listing of ground-based data, models, computer routines, and composite spacecraft data that are available from NSSDC.

002.085 Verzeichnis der Publikationen zur solaren Radio-

astronomie 1954 - 1979.
HHI-STP-Rep. No. 12, 32 pp. (1979).

002.086 The star of Bethlehem: a list of references.
R. S. Freitag.
Library of Congress, Washington, D.C., U.S. Superintendent
of Documents No. 030-000-00112-5. 12 + 44 pp. Price
$ 2.50 (1979). – Review in J. Hist. Astron., Vol. 12, 212 -
213; 1981 (*O. Gingerich*).

002.087 Index of galaxy spectra.
G. R. Gisler, E. D. Friel.
Astronomy and Astrophysics Series, Vol. 10. Pachart Publish-
ing House, Tucson. 5 + 190 pp. Price $ 38.00, DM 133.90
(1979). ISBN 0-912918-19-5. – Review in Astrophys. Lett.,
Vol. 22, 210; 1981 (*S. Wyckoff*).

002.088 Jugendlexikon "Astronomie und Raumfahrt".
K. Lindner, K.-H. Neumann.
VEB Bibliographisches Institut, Leipzig. 256 pp. Price M 7.50
(1980). – Reviews in Astron. Schule, 18. Jahrg., 94; 1981
(*H. Kühnhold*); Sterne, 57. Band, 327 - 328; 1981
(*H. Meusinger*).

002.089 Atlas des étoiles (plus 4.000 étoiles localisées et
identifiées).   J. Mitton, S. Mitton.
Editions du Fanal, Paris (1980). – Review in Ciel Terre,
Vol. 97, 338; 1981 (*A. Koeckelenbergh*).

002.090 True visual magnitude photographic star atlas.
Three volumes.   C. Papadopolous, C. Scovil.
Pergamon Press, New York. Price $ 675.00 (1981). From
Phys. Today, Vol. 34, No. 12, p. 59 (1981).

002.091 A revised Shapley-Ames catalog of bright galaxies.
A. Sandage, G. A. Tammann.
Carnegie Institution of Washington Publication No. 635.
Carnegie Institution, Washington, D.C. 8 + 157 pp. Price
$ 29.00 (1981). ISBN 0-87279-652-3. – Reviews in J. R.
Astron. Soc. Canada, Vol. 75, 267; 1981 (*S. van den Bergh*).
Nature, Vol. 294, 676; 1981 (*B. F. Madore*); Sky Telesc.,
Vol. 62, 476 (1981).

**002.092 La galaxie, l'univers extragalactique. Tome no. 3
de l'encyclopédie scientifique de l'univers.**
Compiled by J. Terrien.
Gauthier – Villars, Paris. 10 + 278 pp. Price Can. $ 28.00
(1980). – Review in J. R. Astron. Soc. Canada, Vol. 75,
270 - 271; 1981 (*J. P. Vallee*).

002.093 Vocabulaire d'astronomie. Par le Conseil Inter-
national de la Langue Francaise.
Editions Hachette, Paris. 330 pp. (1980). – Review in
Astronomie, Vol. 95, 462; 1981 (*G. Oudenot*).

002.094 Chung-Kuo Ta Pai K'o Ch'üan Shu (The Greater
Encyclopaedia of China). Vol. 1. T'ien Wên Hsüeh
(Astronomy).
Greater Encyclopaedia of China Publishing House, 1–A Wai
Guan Dong Jie, Beijing, 27 (Peking), China. 650 pp. Price
$ 30.00 (1981). – Review in Nature, Vol. 292, 276 - 277;
1981 (*M. Salt, J. Needham*).

002.095 Catalogs currently available on microfiche.
Bull. Inf. Cent. Données Stellaires, No. 21, p. 108 -
114 (1981).

Astronomical data center operations.
See Abstr. 013.012.

Automated astronomical data retrieval at GSFC.
See Abstr. 013.013.

Réduction des catalogues photographiques: zones
+31° à −2° (R. C. P. 164).   See Abstr. 031.569.

Investigation of systematic differences between the
new Washington catalogs W5-50 and WL50, the Perth 70 and
the AGK 3 R in their common zone of overlap, declinations
−5° to +5°.   See Abstr. 041.016.

The magnitude equations between the fundamental
coordinate systems N30, FK3, FK4.
See Abstr. 041.017.

Right ascensions of 586 FKSZ stars observed at the
Nikolaev Observatory in 1974 - 1976.
See Abstr. 041.021.

A catalogue of Jupiter's decametric emission observed by Voyager-1 and by Voyager-2 in the range
15-40 MHz.   See Abstr. 099.064.

Interplanetary gas. XXVII. A catalog of disconnection events in cometary plasma tails.
See Abstr. 102.007.

Proper motions of stars in the Orion association.
See Abstr. 111.019.

On the estimation of photometric spectral types.
See Abstr. 114.081.

Near-infrared observations of trapezium-type
multiple systems. Catalogue of observations and a new
determination of the reddening law.   See Abstr. 118.023.

Preliminary results of the Air Force Infrared Sky
Survey.   See Abstr. 133.008.

Identification of objects discovered in the AFGL
four-color infrared sky survey.   See Abstr. 133.014.

An atlas of emission line fluxes of planetary nebulae
in the 1150–3200 Å region.   See Abstr. 135.020.

Positions and flux densities at 5000 MHz for
1133 weak radio sources.   See Abstr. 141.129.

New optical identifications for weak sources from
the Parkes 2700 MHz catalogue.   See Abstr. 141.130.

Flux densities at 8.87 GHz of 347 small diameter
radio sources.   See Abstr. 141.131.

Observed and derived parameters for 330 pulsars.
See Abstr. 141.556.

Integrated magnitudes and mean colors of the
DDO dwarf galaxies in the *UBV* system. I. Observations and
catalog.   See Abstr. 158.170.

Multidimensional statistical analysis of normal galaxies.   See Abstr. 158.215.

Erratum

**002.901 Erratum: "A revised optical catalog of quasi-stellar
objects"** [Astrophys. J., Suppl. Ser., Vol. 43, 57 -
158 (1980)].   A. Hewitt, G. Burbidge.
Astrophys. J., Suppl. Ser., Vol. 46, 113 - 115 (1981). – See
Abstr. 28.002.013.

003 Books

003.001   **Some strangeness in the proportion. Centennial symposium to celebrate the achievements of Albert Einstein.**   H. Woolf (Editor).
Addison-Wesley, Reading, Mass., USA. 31 + 539 pp. Price $ 43.50 (1980). ISBN 0-201-09924-1. – Review in Phys. Abstr., Vol. 84, Abstr. 71366 (1981). – See abstracts 022.046, 066.035, 066.036, 066.042, 066.043, 161.002, 162.022, 162.028, 162.029, 162.036.

003.002   **Solar flare magnetohydrodynamics.**   E. R. Priest (Editor).
The fluid mechanics of astrophysics and geophysics, Vol. 1. Gordon and Breach Science Publishers, New York–London–Paris. 12 + 563 pp. Price DM 238.20 (1981). ISBN 0-677-05530-7. – The individual contributions are included in their corresponding subject categories – see abstracts 062.047, 073.032 - 073.036, 074.036, 074.037.

003.003   **Annual Review of Earth and Planetary Sciences. Volume 9.**
G. W. Wetherill, A. L. Albee, F. G. Stehli (Editors). Annual Reviews Inc., 4139 El Camino Way, Palo Alto, Calif. 94306, USA. 9 + 497 pp. Price $ 21.00 (1981). ISBN 0-8243-2009-3. – The individual contributions within the subject scope of Astronomy and Astrophysics Abstracts are included in their corresponding categories – see abstracts 081.015, 081.016, 082.025, 082.026, 102.010.

003.004   **The sun as a star.**   S. Jordan (Editor), with an introduction by R. J. Rutten, L. E. Cram.
Monograph Series on Nonthermal Phenomena in Stellar Atmospheres. NASA SP-450. 54 + 518 pp. Price $ 39.50 (1981). For sale by the National Technical Information Service, Springfield, Virginia 22161. – The individual contributions are included in their corresponding subject categories – see abstracts 062.048 - 062.050, 064.031, 071.019, 073.038 - 073.042, 074.038 - 074.041, 075.008, 080.021 - 080.023.

003.005   **Problems of cosmic physics. Vypusk 16.**   S. K. Vsekhsvyatskij (Editor).
Respublikanskij Mezhvedomstvennyj Nauchnyj Sbornik. Izdatel'stvo pri Kievskom Gosudarstvennom Universitete Izdatel'skogo Obedineniya "Vishcha Shkola", Kiev. 140 pp. Price 1 Rbl. 50 Kop. (1981). In Russian. ISSN 0555-2796. The individual contributions are included in their corresponding subject categories – see abstracts 022.060 - 022.062, 034.022, 036.006, 042.030, 062.051, 062.052, 071.021, 071.022, 077.027, 082.031, 082.032, 084.033 - 084.035, 102.012, 104.005, 106.013, 107.002.

003.006   **Stellar aggregates.**   A. E. Vasilevskij (Editor).
Ural'sk. univ., Sverdlovsk. 160 pp. (1980). In Russian. – Review in Ref. zh., 51. Astron., 8.51.49 (1981). See abstracts 013.014, 034.024, 119.038, 153.013, 153.014, 153.016, 153.019 - 153.022, 154.026, 155.034, 155.035.

003.007   **Histoire de l'univers.**   A. Hayli (Editor).
Hachette, Littérature, Paris. 448 pp. (1980). ISBN 2-01-007005-4. – The individual contributions are included in their corresponding subject categories – see abstracts 015.018, 031.560, 065.036, 080.029, 081.020, 091.017, 131.105, 151.027, 155.028, 158.128, 158.129, 162.051 - 162.054.

003.008   **Determination of the coordinates of celestial bodies.**   L. Lauzenieks (Editor).
Mezhvedomstvennyj Sbornik Nauchnykh Trudov. Latv. univ., Riga. 195 pp. (1981). In Russian. – Review in Ref. zh., 51.

Astron., 9.51.24 (1981). – See abstracts 021.025, 021.028, 034.031, 034.032, 041.007, 042.046, 043.001, 052.029, 052.030, 098.030, 098.031, 102.026 - 102.031, 103.721.

003.009   **Observations of the total solar eclipse of 16 February 1980.** (Preliminary results). With a foreword by
S. K. Trehan and a preface by J. C. Bhattacharyya. Published by the Executive Secretary, Indian National Science Academy, 1, Bahadur Shah Zafar Marg, New Delhi-110 002, India. 8 + 133 pp. Price $ 15.00 (1981). – The individual contributions within the subject scope of Astronomy and Astrophysics Abstracts are included in their corresponding categories – see abstracts 071.025, 073.065, 073.066, 074.053 - 074.070, 077.053, 079.302 - 079.309, 082.039 - 082.041, 083.028 - 083.033, 084.043, 084.044, 085.018.

003.010   **Advances in lunar research. Luna-24 samples.**
S. K. Trehan, H. Y. Mohan Ram (Editors). Published by Indian Natl. Sci. Acad., New Delhi, India. 69 pp. Price Rs. 30.00 (1979). – Review in Phys. Abstr., Vol. 84, Abstr. 89173 (1981). – See abstracts 094.595 - 094.604.

003.011   **Annual Review of Fluid Mechanics. Vol. 13.**
M. Van Dyke, J. V. Wehausen, J. L. Lumley. Annual Reviews Inc., Palo Alto, Calif., USA. 530 pp. (1981). ISBN 0-8243-0713-5. – Review in Phys. Abstr., Vol. 84, Abstr. 91210 (1981). – See Abstract 151.045.

003.012   **Annual Review of Astronomy and Astrophysics. Volume 19.**   G. Burbidge, D. Layzer,
J. G. Phillips (Editors). Annual Reviews Inc., 4139 El Camino Way, Palo Alto, Calif. 94306, USA. 8 + 476 pp. Price $ 21.00 (1981). ISBN 0-8243-0919-7. – The individual contributions are included in their corresponding subject categories – see abstracts 005.007, 062.077, 062.078, 064.057, 074.077, 114.106, 117.070, 131.132, 133.008, 141.110, 141.111, 154.028, 155.039, 162.089.

003.013   **Cosmic ray variations and solar wind.** Collection of scientific papers. Institute of cosmophysical investigations and aeronomy. Yakutsk Department of the SO of the USSR Academy of Sciences.
N. P. Chirkov (Editor). Yakutsk. 109 pp. (1980). In Russian. – Review in Ref. zh., 51. Astron., 10.51.115 (1981). – See abstracts 021.040, 078.020, 143.094.

003.014   **Investigating the universe.** Papers presented to Zdeněk Kopal on the occasion of his retirement, September 1981.
F. D. Kahn (Editor), with a foreword by M. K. V. Bappu. Astrophysics and space science library, Vol. 91. D. Reidel Publishing Company, Dordrecht, Holland – Boston, U.S.A. – London, England. 10 + 458 pp. Price Dfl. 125.00, $ 54.50 (1981). ISBN 90-277-1325-1. – The individual contributions are included in their corresponding subject categories – see abstracts 042.075, 091.066, 117.098 - 117.101, 121.031, 131.175, 131.176, 141.144, 151.060, 151.061, 155.052, 159.018.

003.015   **Radio investigations of the moon and terrestrial planets.**   N. N. Krupenio.
Edited by I. S. Shcherbina-Samojlova. Itogi Nauki i Tekhniki. Seriya Astronomiya, Tom 17. Part 2. Vsesoyuznyj Institut Nauchnoj i Tekhnicheskoj Informatsii. Moskva. 167 pp. Price 1 Rbl. 5 Kop. (1981). ISSN 0202-0742. In Russian. See 27.003.010, Part 1.

003.016 **Evolution of stars and stellar aggregates.**
A. V. Tutukov, Eh. K. Grasberg, A. G. Masevich,
E. I. Popova, Yu. L. Frantsman (Editors).
Nauchn. Inf., Vyp. (No.) 47. Zinatne, Riga. 116 pp. Price
60 Kop. (1981). In Russian. – The individual contributions
are included in their corresponding subject categories – see
abstracts 065.089, 065.090, 114.170, 122.132 - 122.134,
131.191, 131.192, 151.070, 151.071.

003.017 **Evolution of stars and stellar aggregates.**
A. V. Tutukov, Eh. K. Grasberg, A. G. Masevich,
E. I. Popova, Yu. L. Frantsman (Editors).
Nauchn. Inf., Vyp. (No.) 49. Zinatne, Riga. 176 pp. Price 1 Rbl.
(1981). In Russian. – The individual contributions are included
in their corresponding subject categories – see abstracts
064.091, 065.092 - 065.094, 066.521, 117.108, 131.193,
131.194, 151.072, 155.056, 155.057.

003.018 **Problems of extragalactic astronomy.**
I. S. Shcherbina-Samojlova (Editor).
Itogi nauki i tekhniki. Seriya Astronomiya, Tom 18.
Vsesoyuznyj Institut Nauchnoj i Tekhnicheskoj Informatsii
(VINITI). Moskva. 124 pp. Price 1 Rbl. 5 Kop. (1981).
ISSN 0202-0742. In Russian. – The individual contributions
are included in their corresponding subject categories – see
abstracts 151.077, 151.078, 160.052.

003.019 **Dynamics Explorer.**
R. A. Hoffman (Editor).
Space Sci. Instrum., Vol. 5, No. 4, p. 339 - 584 (1981). – The
individual contributions within the subject scope of Astronomy and Astrophysics Abstracts are included in their corresponding categories – see abstracts 021.046, 032.579 -
032.589, 051.034, 051.035.

003.020 **Automation of astronomical observations.**
V. G. Tejfel' (Editor).
Nauka, Alma-Ata. 114 pp. (1981). In Russian. – Review in
Ref. zh., 51. Astron., 11.51.79 (1981). – See abstracts
021.047 - 021.049, 031.620 - 031.622, 034.082.

003.021 **Investigations of the sun and red stars. 11.**
A. Balklavs (Editor).
Latvijas PSR Zinātņu akademija, Radioastrofizikas observatorija. Akademiya nauk Latvijskoj SSR, Radioastrofizicheskaya
observatoriya. Zinatne, Riga. 94 pp. Price 35 Kop. (1980).
ISSN 0135-1303. In Russian. – The individual contributions
are included in their corresponding subject categories – see
abstracts 077.064, 080.069, 113.070, 114.188, 122.190.

003.022 **Investigations of the sun and red stars. 12.**
A. Balklavs (Editor).
Latvijas PSR Zinātņu akademija, Radioastrofizikas observatorija. Akademiya nauk Latvijskoj SSR, Radioastrofizicheskaya
observatoriya. Zinatne, Riga. 76 pp. Price 30 Kop. (1981).
ISSN 0135-1303. In Russian. – The individual contributions
are included in their corresponding subject categories – see
abstracts 021.054, 064.093, 074.110, 114.189, 114.190,
122.191.

003.023 **Investigations of the sun and red stars. 13.**
A. Balklavs (Editor).
Latvijas PSR Zinātņu akademija, Radioastrofizikas observatorija. Akademiya nauk Latvijskoj SSR, Radioastrofizicheskaya
observatoriya. Zinatne, Riga. 132 pp. Price 45 Kop. (1981).
ISSN 0135-1303. In Russian. – The individual contributions
are included in their corresponding subject categories – see
abstracts 031.630, 033.016, 033.017, 064.094, 074.111,
113.071 - 113.073, 114.191.

003.024 **Investigations of the sun and red stars. 14.**
A. Balklavs (Editor).

Latvijas PSR Zinātņu akademija, Radioastrofizikas
observatorija. Akademiya nauk Latvijskoj SSR,
Radioastrofizicheskaya observatorija. Zinatne, Riga. 70 pp.
Price 25 Kop. (1981). ISSN 0135-1303. In Russian. – The
individual contributions are included in their corresponding
subject categories – see abstracts 113.074, 113.075, 122.192 -
122.194.

003.025 **Researches in the history of astronomy.** 15th issue.
L. E. Majstrov (Editor).
Glavnaya Redaktsiya Fiziko-Matematicheskoj Literatury,
Nauka, Moskva. 368 pp. Price 3 Rbl. 10 Kop. (1980).
In Russian. – The individual contributions are included in
their corresponding subject categories – see abstracts
004.070 - 004.079, 005.019 - 005.021, 015.050 - 015.054,
035.019, 047.047, 106.057.

003.026 **The new solar system.**
J. K. Beatty, B. O'Leary, A. Chaikin (Editors),
with an introduction by C. Sagan.
Cambridge University Press, Cambridge - London - New York -
New Rochelle - Melbourne - Sydney and Sky Publishing Corporation, Cambridge, Mass. 8 + 224 pp. Price $ 19.95, £ 9.95,
DM 58.80 (1981). ISBN 0-521-23881-1. – Reviews in J. British Astron. Assoc., Vol. 91, 604; 1981 (*C. A. Ronan*); Nature,
Vol. 294, 21 - 22; 1981 (*D. W. Hughes*); Science, Vol. 214,
174 - 175; 1981 (*T. Owen*). – The individual contributions
are included in their corresponding subject categories – see
abstracts 051.061, 051.062, 080.082, 091.096 - 091.101,
094.050, 097.086, 097.087, 098.109, 099.232, 099.233,
100.148, 101.040, 102.089, 105.123, 107.032.

003.027 **Carbon stars.** Z. Alksne, J. Ikaunieks.
Pachart, Tucson. 182 pp. Price $ 24.00 (1981).
From Phys. Today, Vol. 34, No. 12, p. 59 (1981).

003.028 **Probleme der modernen Kosmogonie.**
V. A. Ambarzumjan (*Ambartsumyan*).
Akademie-Verlag, Berlin, GDR. 341 pp. Price M 28.00 (1980).
Review in Sterne, 57. Band, 258 - 259; 1981 (*R. Kippenhahn*).

003.029 **The creation.** P. W. Atkins.
W. H. Freeman & Company, Oxford–San Francisco.
8 + 132 pp. Price £ 5.95 (1981). ISBN 0-7167-1350-0.
Review in Astron. Tidsskr., Årg. 14, 179 (1981).
The book is an account of the nature and the origin of
the universe, but it is not just another book about astronomy
or elementary particles. There are central aspects of the
universe – among them its beginning, the nature of time, and
consciousness – that are now open to scientific elucidation.
The author sets them centrally on the stage. He avoids details
and selects from modern science the broad features of its
explanations and implications.

003.030 **Atmospheres.** A view of the gaseous envelopes
surrounding members of our solar system.
J. P. Barbato, E. A. Ayer.
Pergamon Press, New York. 20 + 266 pp. Price $ 35.00 cloth;
$ 12.95 paper (1981). – From Science, Vol. 214, 1235 (1981).

003.031 **Image et information.** Introduction au traitement
numérique des images.
A. Bijaoui, with a preface by C. Fehrenbach.
Masson, Paris–New York–Barcelone–Milan–Mexico–Rio de
Janeiro. 242 pp. Price FF 140.00 (1981). ISBN 2-225-
74943-4.
Contents: Propriétés générales d'un récepteur d'images.
La photographie. Récepteurs photoélectriques. Le matériel du
traitement numérique des images. Introduction à la théorie de
l'information. Information continue dans une image. Aspects
statistiques. Introduction à la représentation de l'image.
Filtrage et transformée de Fourier. Utilisation des corrélations.

Traitement par représentation locale. La déconvolution. Autres types de représentations. Structure générale des images. La compression numérique des images. Système de traitement des images.

003.032 **Astronomia Nautica.** M. Bini.
Accademia Navale, Livorno. 396 pp. – Review in J. Navig., Vol. 34, 481; 1981 (*D. H. Sadler*).

003.033 **Update on space. Vol. 1.**
B. J. Bluth, S. R. McNeal (Editors).
National Behavior Systems, 11601 Balboa Blvd., Granada Hills, Calif. 91344. 196 pp. Price $ 7.95 (1981). ISBN 0-937654-00-0. – Review in Spaceflight, Vol. 23, 327 (1981).

003.034 **Climate of the planets.**
E. P. Borisenkov, O. G. Vakhmistrova, K. Ya. Kondrat'ev, N. I. Moskalenko, S. L. Khess.
Gidrometeoizdat, Leningrad. 96 pp. (1981). In Russian. Review in Ref. zh., 51. Astron., 1.51.49 (1982).

003.035 **Calcul astronomique pour amateurs adapté à l'emploi d'un calculateur ou d'un micro-ordinateur.**
S. Bouiges.
3., revised and extended edition. Masson. Paris–New York– Barcelone–Milan–Mexico–Rio de Janeiro. 154 pp. (1981). ISBN 2-225-74514-5.
Contents: Les mouvements dans le système solaire. Systèmes de coordonnées. L'heure. Calcul des coordonnées écliptiques héliocentriques d'une planète. Coordonnées géocentriques. Constantes planétaires. Mouvement de la Lune. Amélioration de la précision. Coordonnées apparentes. Satellites de Jupiter et de Saturne. Cycles. Détermination de l'orbite d'une comète. Temps sidéral. Calcul des coordonnées locales. Lever et coucher des astres. Le triangle parallactique. Visibilité d'un astre. Calcul d'un cadran solaire plan. Coordonnées héliographiques. Qu'est-ce que la programmation? Les registres mémoire. Fonctions logiques. Boucles et sous-programmes. L'adressage indirect. Apprendre à programmer sa machine.

003.036 **Comets.** Readings from Scientific American.
J. C. Brandt (Editor).
W. H. Freeman, San Francisco. 92 pp. Price $ 11.95, £ 7.50; $ 5.95, £ 3.50 pbk (1981). ISBN 0-7167-1319-5; ISBN 0-7167-1320-9 pbk. – Review in Strolling Astron., Vol. 29, 80; 1981 (*J. D. Sabia*).

003.037 **Physics of meteor phenomena.** V. A. Bronshtehn.
Nauka, Moskva. 416 pp. (1981). In Russian.
Review in Ref. zh., 51. Astron., 7.51.31 (1981).

003.038 **The Milky Way.** B. J. Bok, P. F. Bok.
Fifth Edition. Harvard University Press, Cambridge, Mass. 12 + 356 pp. Price $ 20.00 (1981). ISBN 0-674-5703-2. Reviews in Sky Telesc., Vol. 62, 476 (1981); Strolling Astron., Vol. 29, 83; 1981 (*J. R. Smith, C. S. Morris*).

003.039 **Introduction to comets.**
J. C. Brandt, R. D. Chapman.
Cambridge University Press, Cambridge–London–New York– New Rochelle–Melbourne–Sydney. 8 + 246 pp. Price DM 106.00 (1981). ISBN 0-521-23906-0.
Contents: Comets in history. Development of modern ideas on the physics of comets. Dynamics of comets. Structure of comets. Model and origins of comets. Comets and the solar system. Recent comets and some current developments. Missions to comets. Are comets dangerous? Comet lore.

003.040 **Nomenclature of features of the Martian relief.**
G. A. Burba.
Nauka, Moskva. 85 pp. (1981). In Russian. – Review in Ref.

zh., 51. Astron., 9.51.28; 62. Issled. kosm. prostranstva, 9.62.402 (1981).

003.041 **Telescopes for the 1980s.**
G. Burbidge, A. Hewitt (Editors).
Annual Reviews, Inc., Palo Alto, Calif. 10 + 278 pp. Price $ 27.00 (1981). – Review in Sky Telesc., Vol. 62, 590 - 593; 1981 (*L. J. Robinson*).

003.042 **The moon – our sister planet.** P. Cadogan.
Cambridge University Press, 391 pp. Price £ 27.50, £ 12.50 pbk. (1981). – Review in Space Educ., Vol. 1, 90 (1981).

003.043 **The summer stargazer.** Astronomy for beginners.
R. Claiborne.
Revised Edition. Penguin, New York. 222 pp. Price $ 4.95 (1981). – From Science, Vol. 214, 218 (1981).

003.044 **Elementary general relativity.** C. Clarke.
John Wiley and Sons, New York. 131 pp. Price $ 18.95 (1980). ISBN 0-470-26930-8. – Review in Astrophys. Lett., Vol. 21, 115 - 116; 1981 (*H. L. Shipman*).

003.045 **The Newtonian revolution.** I. B. Cohen.
Cambridge University Press, New York. Price $ 37.50 (1981). – Review in Phys. Today, Vol. 34, No. 8, p. 57 - 58; 1981 (*C. A. Wilson*).

003.046 **The first stargazers.** An introduction to the origins of astronomy. J. Cornell.
Charles Scribner's Sons/Athlone. 262 pp. Price £ 7.95, $ 15.95 (1981). ISBN US 0-684-16799-9; ISBN UK 0-485-30004-4. – Review in Nature, Vol. 294, 485 - 486; 1981 (*C. Ruggles*).

003.047 **L'observation des étoiles doubles visuelles.**
P. Couteau.
Translated from the French edition. Mir, Moskva. 238 pp. (1981). In Russian. – From Ref. zh., 51. Astron., 1.51.53 (1982).

003.048 **Observing visual double stars.** P. Couteau.
Translated from the French edition by A. H. Batten. MIT Press, Cambridge, Mass. 18 + 258 pp. Price $ 19.95 (1981). ISBN 0-262-03077-2. Available from Sky Publishing Corp. – Review in Sky Telesc., Vol. 62, 593 (1981).

003.049 **X-ray astronomy.** J. L. Culhane, P. W. Sanford.
Faber & Faber, London/Scribner's, New York. 192 pp. Price $ 25.00 (1981). ISBN 0-571-11550-0. – Review in Phys. Abstr., Vol. 85, Abstr. 3344 (1982).

003.050 **Am Ende ein neuer Anfang.** Die Biographie des Universums. P. Davies.
Eugen Diederichs Verlag, Düsseldorf-Köln. 208 pp. Price DM 24.80 (1979). – Review in Sterne, 57. Band, 257; 1981 (*B. Stecklum*).

003.051 **The edge of infinity: naked singularities and the destruction of spacetime.** P. Davies.
J. M. Dent/Simon & Schuster. 10 + 194 pp. Price £ 7.95, $ 14.50 (1981). ISBN UK 0-460-04490-7; ISBN US 0-671-44063-2. Reviews in Astron. Tidsskr., Årg. 14, 179 (1981); Nature, Vol. 294, 488 - 489; 1981 (*W. H. Press*).

003.052 **Master optical techniques.** A. S. De Vany.
Wiley Series in Pure and Applied Optics. Wiley, New York. 8 + 600 pp. Price $ 55.00 (1981). ISBN 0-471-07720-8. – Review in Sky Telesc., Vol. 62, 595 (1981).

003.053 **Briefwechsel zwischen Alexander von Humboldt und**

Heinrich Christian Schumacher. S. Débarbat.
Akademie-Verlag, Berlin. 192 pp. (1979). – Review in
Astronomie, Vol. 95, 517; 1981 (*P. Wallach*).

003.054 Exploration of the polar upper atmosphere.
C. S. Deehr, J. A. Holtet (Editors).
NATO Advanced Study Institutes Series. D. Reidel Publishing
Company, Dordrecht, Holland. 16 + 498 pp. Price US $ 58.00
(1980). – Review in Planet. Space Sci., Vol. 29, 1019; 1981
(*H. Rishbeth*).

003.055 Space activities in the eighties. Vol. 1: Survey of
space programmes. Vol. 2: Detailed presentation of
the European Space Industry (1981). **Vol. 3: Outlook of space**
industry outside Europe (1981). G. Dondi.
European Space Agency, Paris. ESA SP 1012. Vol. 1: 289 pp.
Vol. 2: 533 pp. Vol. 3: 258 pp. Price Vol. 1: 80.00 FF, Vol. 2:
160.00 FF, Vol. 3: 80.00 FF. (1981). Available from ESA
Information Retrieval Service, Rue Mario Nikis 8 - 10,
75738 Paris Cedex 15, France.

003.056 Dynamics of flights between the earth and the
moon. V. A. Egorov, L. I. Gusev.
Nauka, Moskva. 543 pp. (1980). In Russian. – Review in
Ref. zh., 62. Issled. kosm. prostranstva, 7.62.61 (1981).

003.057 A reader/study guide for cosmos and a viewer's
guide to cosmos. A. Elwood, L. Wood.
Random House, New York. 278 + 71 pp. (1980). – From
Mercury, Vol. 10, 90 (1981).

003.058 Le viscere della terra – dinamica e struttura dell'-
interno terrestre. J. Elder.
Zanichelli, Bologna. 249 pp. Price L. 10,000 (1980).
Review in G. Astron., Vol. 7, 243; 1981 (*P. Picciafuoco*).

003.059 Twenty-five years of the American Astronautical
Society 1954–1979. E. M. Emme (Editor).
American Astronautical Society, 235 pp. (1981). – Review in
Spaceflight, Vol. 23, 327 (1981).

003.060 The map of the universe. T. J. Filsinger.
Celestial Arts, 231 Adrian Road, Millbrae,
CA 94030. Sky map and 15 pp. booklet. Price $ 9.95 (1981).
From Astron. Q., Vol. 4, No. 13, p. 47 (1981).

003.061 The cosmology of infinity. A. Firth.
Sono Niss Press, Victoria B.C., Canada. 109 pp.
Price Can. $ 10.95. – From J. British Astron. Assoc., Vol. 92,
50 (1981).

003.062 Tobias Mayer (1723 - 1762), pioneer of enlightened
science in Germany. E. G. Forbes.
Arbeiten a. d. niedersächsischen Staats- und Universitäts-
bibliothek, Band 17. Vandenhoeck & Ruprecht, Göttingen,
248 pp. Price DM 38.00 (1980). – Review in Phys. Bl.,
Vol. 37, 354; 1981 (*H. Rechenberg*).

003.063 The planets, a decade of discovery. P. Francis.
Penguin Books, Harmondsworth–New York–
Ringwood–Markham–Auckland. 411 pp. Price £ 3.95 (1981).
Reviews in Astron. Tidsskr., Årg. 14, 138 - 139 (1981);
J. British Astron. Assoc., Vol. 91, 517 - 518; 1981
(*D. W. Hughes*); Observatory, Vol. 101, 185; 1981 (*A. Firsoff*);
Planet. Space Sci., Vol. 29, 1019; 1981 (*P. S. Butterworth*);
Spaceflight, Vol. 23, 328 (1981).

003.064 The computer in optical research, methods and
applications. B. R. Frieden (Editor).
Springer-Verlag, Berlin–Heidelberg–New York. 13 + 371 pp.
Price $ 58.00 (1980). – Review in J. Opt. Soc. America,
Vol. 71, 1413; 1981 (*D. C. Sinclair*).

003.065 Theoretical physics and astrophysics.
V. L. Ginzburg.
Nauka, Glavnaya redaktsiya fiziko-matematicheskoj literatury.
2nd edition. Moskva. 504 pp. Price 3 Rbl. 40 Kop. (1981).
In Russian. – Review in Priroda, 1981, No. 11, p. 125.

003.066 Theoretical physics and astrophysics. Supplemen-
tary chapters. V. L. Ginzburg.
Second revised edition. Nauka, Moskva. 503 pp. (1981).
In Russian. – Review in Ref. zh., 51. Astron., 11.51.65
(1981).

003.067 Genesis: the origins of man and the universe.
J. Gribbin.
Dent/Delacorte. 360 pp. Price £ 7.95, $13.95 (1981). ISBN
UK 0-460-04505-9; ISBN US 0-440-2832-9. – Review in
Nature, Vol. 294, 491; 1981 (*J. N. Wilford*).

003.068 Brazilian stone meteorites.
C. B. Gomes, K. Keil.
University of New Mexico Press. 161 pp. Price $ 20.00 (1980).
Review in Geochim. Cosmochim. Acta, Vol. 45, 2296; 1981
(*R. S. Clarke, Jr.*).

003.069 The camera obscura: a chronicle. J. H. Hammond.
Adam Hilger, Bristol/Heyden, Philadelphia. 182 pp.
Price £ 13.50, $ 33.50 (1981). ISBN 0-85274-451-X. – Review
in Nature, Vol. 294, 492; 1981 (*S. A. Bedini*).

003.070 Cosmology. The science of the universe.
E. R. Harrison.
Cambridge University Press, Cambridge–London–New York–
New Rochelle–Melbourne–Sydney. 11 + 430 pp. Price
$ 24.95, DM 69.80 (1981). ISBN 0-521-22981-2.
Contents: What is cosmology? Stars. Galaxies. Location
and the cosmic center. Containment and the cosmic edge.
Space and time. Curved space. General relativity. Black holes.
Expansion. Redshifts. Darkness at night. The universe in a
nutshell, Newtonian cosmology. The many universes. Theories
of the universe. The cosmic numbers. The early universe.
Horizons in the universe. Life in the universe.

003.071 Cosmic discovery: the search, scope, and heritage
of astronomy. M. Harwit.
Harvester Press, Brighton/Basic Books, New York. 12 +
334 pp. Price £ 12.95, $ 25.00 (1981).ISBN UK 0-7108-
0089-4; ISBN US 0-465-01428-3. – Reviews in Nature, Vol.
294, 22 - 23; 1981 (*D. S. Evans*); Sci. American, Vol. 245,
No. 5, p. 33 - 34 (1981).

003.072 Early scientific instruments. N. Hawkes.
Abbeville Press, New York. 168 pp. Price $ 29.95
(1981). ISBN 0-89659-192-1. – Review in Nature, Vol. 294,
491 - 492; 1981 (*W. D. Hackmann*).

003.073 International Directory of Amateur Astronomical
Societies. – IDAAS.
A. Heck, J. Manfroid (Editors).
Heck, ESA Satellite Tracking Station, Apartado 54065,
Madrid, Spain. Manfroid, Institut d'Astrophysique, avenue de
Cointe 5, B-4200 Cointe-Ougrée, Belgium. 308 pp. (1981).

003.074 Das Sternguckerbuch. D. B. Herrmann.
Verlag Neues Leben, Berlin. 287 pp. Price M 9.80
(1981). – Review in Sterne Weltraum, Jahrg. 20, 434; 1981
(*K. Meisenheimer*).

003.075 Evolution from space. F. Hoyle,
C. Wickramasinghe.
J. M. Dent, London. 176 pp. Price £ 7.95 (1981). ISBN 0-
460-04535-0. – Reviews in Nature, Vol. 294, 489 - 490; 1981
(*D. A. J. Tyrrell*); Space Educ., Vol. 1, 93 (1981).

003.076 **Space travellers, the bringers of life.**
F. Hoyle, C. Wickramasinghe.
University College Cardiff Press, Cardiff. 197 pp. Price
£ 8.95 (1981). − Reviews in Astron. Tidsskr.,Årg. 14, 180
(1981); Space Educ., Vol. 1, 92 (1981).

003.077 **Space.**   M. Ince.
Sphere Books Ltd., London. 215 pp. Price
£ 1.50 (1981). − Review in J. British Astron. Assoc., Vol. 91,
609 - 610; 1981 (*H. Miles*).

003.078 **Mechanics of relative motion and inertial forces.**
A. Yu. Ishlinskij.
Nauka, Moskva. 191 pp. (1981). In Russian. − Review in
Ref. zh., 51. Astron., 10.51.110 (1981).

003.079 **Pictorial guide to the planets.**
J. H. Jackson, J. H. Baumert.
Harper and Row, New York. 10 + 246 pp. Price $ 22.50
(1981). From Science, Vol. 214, 217 (1981).

003.080 **Discovering astronomy.**
W. H. Jefferys, R. R. Robbins.
John Wiley & Sons, New York − Chichester − Brisbane −
Toronto. 466 pp. Price £ 15.85 (1981). − Review in Astron.
Tidsskr., Årg. 14, 180 (1981).

003.081 **Webb Society deep-sky observer's handbook.**
    Vol. 4: Galaxies.   K. G. Jones (Editor).
Enslow Publishers, Hillside,N. J. 238 pp. Price $ 15.95 (1981).
ISBN 0-89490-050-1. − Review in Strolling Astron., Vol. 29,
82; 1981 (*J. L. Benton, Jr.*).

003.082 **Dynamics of the upper atmosphere.**   S. Kato.
    D. Reidel, Dordrecht, Holland. 233 pp. Price
Dfl. 57.00, $ 29.95 (1980). ISBN 90-277-1132-1. − Reviews
in Nature, Vol. 294, 782; 1981 (*R. Hide*); Planet. Space Sci.,
Vol. 29, 1135; 1981 (*H. Volland*).

003.083 **Geophysik in Heidelberg.**   T. Kirsten (Editor).
    Springer-Verlag, Berlin-Heidelberg−New York.
112 pp. Price DM 52.00 (1980). (Sitzungsber. der Heidelberger Akademie der Wissenschaften, math.-naturwiss. Klasse,
Jg. 1979/80, 4. Abh.). − Review in Sterne, 57. Band, 383;
1981 (*U. Walzer*).

003.084 **Spectral investigations of cosmic and atmospheric**
    radiation.   A. G. Kislyakov (Editor).
Sb. nauch. tr. Inst. prikl. fiz. AN SSSR. Gor'kij. 171 pp.
(1979). In Russian. − Review in Ref. zh., 51. Astron.,
7.51.35 (1981).

003.085 **Calendar and chronology.**   I. A. Klimishin.
    Nauka, Moskva. 191 pp. (1981). In Russian.
Review in Ref. zh., 51. Astron., 11.51.61 (1981).

003.086 **Nuclear-physical investigations of the moon and**
    planets.
G. E. Kocharov, S. V. Viktorov, V. I. Chesnokov.
Ehnergoizdat, Moskva. 184 pp. (1981). In Russian. − Review
in Ref. zh., 51. Astron., 12.51.84 (1981).

003.087 **Automation of control processes and reduction of**
    information.   S. V. Korotkov (Editor).
Leningrad. 157 pp. (1980). In Russian. − Review in Ref. zh.,
51. Astron., 10.51.118 (1981).

003.088 **Mean-field magnetohydrodynamics and dynamo**
    theory.   F. Krause, K.-H. Rädler.
Pergamon Press, Oxford/Akademie-Verlag, Berlin, GDR.
271 pp. Price £ 15.00, M 48.00 (1980). − Reviews in Astron.

Tidsskr., Årg. 14, 139 (1981); Astron. Nachr., Band 302, 318 -
319; 1981 (*P. H. Roberts*).

003.089 **In search of ancient astronomies.**
    E. C. Krupp (Editor).
Chatto and Windus, London. 276 pp. Price £ 8.95 (1979).
Review in Irish Astron. J., Vol. 14, 191 - 192; 1981 (*I. Elliott*).

003.090 **Astronomy in al-Farabi's papers.**   A. Kubesov.
    Nauka, Alma-Ata. 124 pp. (1981). In Russian.
Review in Ref. zh., 51. Astron., 11.51.59 (1981).

003.091 **Free oscillations of the earth.**
    E. R. Lapwood, T. Usami.
Cambridge University Press, Cambridge. 12 + 243 pp. Price
£ 25.00 (1981). ISBN 0-521-23536-7. − Review in Phys.
Earth Planet. Inter., Vol. 27, 230 - 231; 1981 (*S. Crampin*).

003.092 **Stones from the stars: the unresolved mysteries of**
    meteorites.   T. R. LeMaire.
Prentice Hall Inc., Englewood Cliffş, N.J. 07632. 185 pp.
Price $ 9.95 (1980). − Review in Strolling Astron., Vol. 29,
30; 1981 (*J. R. Smith*).

003.093 **High energy astrophysics.** An informal introduction
    for students of physics and astronomy.
M. S. Longair.
Cambridge University Press, Cambridge−London−New York−
New Rochelle−Melbourne−Sydney. 10 + 412 pp. Price
DM 49.20 (1981). ISBN 0-521-23513-8, ISBN 0-521-28013-3
(pbk.).
    Contents: The origin of cosmic rays as an introduction
to high energy astrophysics. Interaction of high energy
particles with matter I: ionisation losses. Interaction of high
energy particles with matter II: electrons − ionisation losses
and bremsstrahlung. Interactions of high energy photons and
electron-photon cascades. Nuclear interactions. Detectors for
cosmic rays, X-rays and γ-rays. Cosmic ray, X-ray and γ-ray
telescopes. Plastics and meteorites. Cosmic ray at the top of
the atmosphere. Extensive air showers − the highest energy
cosmic rays. The influence of the Earth's magnetic field. The
solar wind and its influence upon the local flux of cosmic
rays. The dynamics of cosmic rays in the solar wind. Introduction to observational optical and radio astronomy. A guided
tour of the Galaxy and beyond. The evolution of stars and
our Galaxy. The interstellar gas and magnetic field. Synchrotron radiation and the radio emission from the Galaxy. The
energy losses and diffusion of cosmic ray electrons under
cosmic conditions − sources of cosmic ray electrons. Cosmic
ray protons and nuclei − their distribution in the Galaxy and
the origin of the light elements. The diffusion and confinement of cosmic rays in the Galaxy. Extragalactic sources of
cosmic rays − active galactic nuclei.The origin of cosmic rays −
Galactic or extragalactic. The acceleration of cosmic rays.
Concluding remarks.

003.094 **Emerging cosmology.**   B. Lovell.
    Columbia University Press, New York. 8 + 208 pp.
Price $ 14.95, £ 10.80 (1981). ISBN 0-231-05304-5. − Review
in Nature, Vol. 294, 21; 1981 (*J. Silk*).

003.095 **Impactites.**   A. A. Marakushev (Editor).
    Moskovskij Gosudarstvennyj Universitet, Moskva.
240 pp. (1981). In Russian. − Review in Ref. zh., 51. Astron.,
12.51.86 (1981).

003.096 **Planets of the solar system.**   M. Ya. Marov.
    Nauka, Moskva. 256 pp. (1981). In Russian.
Review in Ref. zh., 51. Astron., 12.51.83 (1981).

003.097 **Astronomia spaziale − Nuove immagini dell'universo.**
    M. Martin, J. Chesterman.

Mondadori, Milano. 128 pp. Price L. 25,000 (1980). – Review in G. Astron., Vol. 7, 122 - 123; 1981 (*M. Rigutti*).

003.098 Stars and stellar systems.
D. Ya. Martynov (Editor).
Nauka, Moskva. 416 pp. (1981). In Russian. – Review in Ref. zh., 51. Astron., 11.51.72 (1981).

003.099 Geology of astroblemes.
V. L. Masajtis, A. N. Danilin, M. S. Mashchak, A. I. Rajkhlin, T. V. Selivanovskaya, E. M. Shadenkov.
Nedra, Leningrad. (1980). In Russian. – Review in Zemlya Vselennaya, 1981, No. 6, p. 39.

003.100 Physics and evolution of stars.
A. G. Masevich, A. V. Tutukov.
Itogi nauk. i tekh. VINITI. Issled. kosm. prostranstva, Tom 17, 90 pp. (1981). In Russian. – Review in Ref. zh., 51. Astron. 1.51.52 (1982).

003.101 Atomic and molecular collisions. H. Massey.
Taylor & Francis Ltd., London. 309 pp. Price $ 34.95 (1979). ISBN 0-470-26742-9. – Review in Astrophys. Lett., Vol. 21, 127 - 128; 1981 (*D. A. Landman*).

003.102 Mathematical and astronomical manuscripts of Central Asia scientists from the Xth to the XVIIIth century. G. P. Matvievskaya, Kh. Tllashev.
Fan, Tashkent. 147 pp. (1981). In Russian. – Review in Ref. zh., 51. Astron., 9.51.21 (1981).

003.103 Galactic astronomy. Structure and kinematics.
D. Mihalas, J. Binney.
Second Edition. W. H. Freeman and Company, San Francisco. 13 + 597 pp. Price $ 29.95, £ 19.50 (1981). ISBN 0-7167-1280-6. – Review in Sky Telesc., Vol. 62, 593 (1981).
Contents: The Galaxy: an overview. Astronomical background. Physical properties of stars and the interstellar medium. The space distribution of stars and the chemical elements in our galaxy. Large-scale structure and stellar content of galaxies. Stellar kinematics: the solar motion. Stellar kinematics: the stellar residual-velocity distribution. The rotation of galaxies. The large-scale distribution of gas in galaxies.

003.104 Instationarity and evolution of stars.
L. V. Mirzoyan.
Edited by V. A. Ambartsumyan.
Akademiya Nauk Armyanskoj SSR. Byurakanskaya Astrofizicheskaya Observatoriya. Izdatel'stvo Akademii Nauk Armyanskoj SSR. Erevan. 379 pp. Price 4 Rbl. 30 Kop. (1981). In Russian.

003.105 Daytime star: the story of our sun. S. Mitton.
Charles Scribner's Sons/Faber & Faber. 191 pp. Price $ 14.95, £ 10.00 (1981). ISBN US 0-684-16840-5; ISBN UK 0-571-11659-0. – Reviews in Nature, Vol. 294, 21 - 22; 1981 (*D. W. Hughes*); Sky Telesc., Vol. 62, 476 (1981); Strolling Astron., Vol. 29, 81 - 82; 1981 (*F. D. Miller*).

003.106 Astronomy: a self-teaching guide. D. Moché.
Second Edition. Wiley, New York. 284 pp. Price $ 7.95 (1981). – From Phys. Today, Vol. 34, No. 12, p. 59 (1981).

003.107 Physics of solar activity.
Eh. I. Mogilevskij (Editor).
Inst. zemn. magn., ionos. i rasprostr. radiovoln AN SSSR. 189 pp. (1980). In Russian. – Review in Ref. zh., 51. Astron., 8.51.44 (1981).

003.108 1982 yearbook of astronomy.
P. Moore (Editor).
Sidgwick & Jackson Ltd., London. 248 pp. Price £ 6.50; £ 4.50 pbk (1981). – Review in J. British Astron. Assoc., Vol. 92, 48; 1981 (*I. Nicolson*).

003.109 Astronomi. P. Moore.
Politikens Forlag, Köpenhamn. 144 pp. Price D. kr. 68.00 (1981). – Review in Astron. Tidsskr., Arg. 14, 180 (1981).

003.110 Astronomy for O-level. P. Moore.
G. Duckworth, London. 216 pp. Price £ 2.95 pbk. (1979). – Review in Space Educ., Vol. 1, 90 (1981).

003.111 The astronomy of Birr Castle. P. Moore.
Second Edition. The Tribune Printing and Publishing Group, Birr. 14 + 81 pp. Price £ 1.60 (Irish £ 2.00) (1981). – Review in J. British Astron. Assoc., Vol. 92, 49 - 50; 1981 (*D. Howse*).

003.112 The development of astronomical thought. P. Moore.
Ian Henry Publications. 108 pp. Price £ 95.00 (1981). Review in Space Educ., Vol. 1, 90 (1981).

003.113 Das Geheimnis der Planeten. Ein Roman um Johannes Kepler. L. Moritzberger.
Georg Bitter Verlag, Recklinghausen, FRG. 154 pp. Price DM 16.90 (1980). – Review in Sterne Weltraum, Jahrg. 20, 348; 1981 (*I. Richter*).

003.114 Notre univers. J. Muirden, traduit par M.-A. Sevin.
Hatier éditeur, Paris. (1980). – Review in Astronomie, Vol. 95, 462; 1981 (*R. Servajean*).

003.115 Earthlike planets. Surfaces of Mercury, Venus, Earth, Moon, Mars.
B. Murray, M. C. Malin, R. Greeley.
W. H. Freeman and Company, San Francisco. 387 pp. Price $ 24.95 cloth, $ 14.95 paper (1981). ISBN 0-7167-1148-6; ISBN 0-7167-1149-4 pbk. – Reviews in Sky Telesc., Vol. 62, 588 - 589; 1981 (*J. A. Wood*); Strolling Astron., Vol. 29, 79; 1981 (*W. S. Cameron*).

003.116 Evolution and stability of satellite systems.
A. I. Nazarenko, B. S. Skrebushevskij.
Mashinostroenie, Moskva. 284 pp. (1981). In Russian. Review in Ref. zh., 62. Issled. kosm. prostranstva, 12.62.41 (1981).

003.117 Beyond the atmosphere. Early years of space science. H. E. Newell.
National Aeronautics and Space Administration, Washington, D.C. The NASA History Series. NASA SP-4211. 18 + 502 pp. Price $ 11.00 (1980). Available from the Superintendent of Documents, U.S. Government Printing Office, Washington, D.C. 20402. – Review in Sky Telesc., Vol. 62, 474 - 476; 1981 (*L. Goldberg*).

003.118 Red star in orbit. J. E. Oberg.
Random House, New York. 272 pp. Price $ 12.95 (1981). ISBN 0-394-51429-7. Available from Sky Publishing Corp. – Review in Sky Telesc., Vol. 62, 593 (1981).

003.119 Determination of the positions of major planets with the photographic method. A. B. Onegina.
Naukova dumka, Kiev. 123 pp. (1981). In Russian. – Review in Ref. zh., 51. Astron., 9.51.25 (1981).

003.120 **The earth: its birth and growth.**   M. Ozima.
Cambridge University Press. 117 pp. Price £ 10.50,
$ 22.50; £ 3.95, $ 8.95 pbk (1981). ISBN 0-521-23500-6;
ISBN 0-521-28005-2 pbk. – Review in Nature, Vol. 294,
490 - 491, 1981 (*P. J. Smith*).

003.121 **Radio galaxies. Radiative transfer, dynamics,
stability and evolution of a synchrotron plasmon.**
A. G. Pacholczyk.
Translated from the English edition. Mir, Moskva. 239 pp.
(1980). In Russian. – Review in Ref. zh., 51. Astron.,
8.51.50 (1981).

003.122 **Hilfsmaterialien für Übungen in Astronomie und
Astrophysik, 2. Teil.**   M. Pańków.
Uniwersytet Slaski, Katowice. 124 pp. Price zł. 45.00 (1980).
In Polish. – Review in Astron. Schule, 18. Jahrg., 93 - 94;
1981 (*A. Müssiggang*).

003.123 **Systematics of global cycles of natural processes.
Geological aspects.**   L. I. Pankul'.
Nauka, Alma-Ata, 146 pp. (1981). In Russian. – Review in
Ref. zh., 51. Astron., 7.51.28 (1981).

003.124 **The theory of cosmic aberration – a new interpretation of the Hubble redshift.**   L. Parish.
Cortney Publ., Luton, England. 44 pp. Price £ 3.50 (1981).
ISBN 0-904378-11-X. – Review in Observatory, Vol. 101,
223; 1981 (*R. C. Smith*).

003.125 **Invitation to physics.**   J. M. Pasachoff,
M. L. Kutner.
Norton. 536 pp. Price $ 17.95 (1981). ISBN 0-393-95152-9.
Review in Sky Telesc., Vol. 62, 156 (1981).

003.126 **America's Stonhenge.**   E. W. Piini.
Sarsen Press, P.O. Box 3361, Redwood City, Calif.
94063. 29 pp. Price $ 3.50(1980). – Review in Sky Telesc.,
Vol. 62, 595 (1981).

003.127 **Equatoires et Horlogerie Planétaire du XIIIᵉ au
XVIᵉ Siècle.**   E. Poulle.
Librarie Droz, Geneva. 1162 pp. (two volumes), Price
SF 150 (1980). – From J. Hist. Astron., Vol. 12, 215 (1981).

003.128 **Voyager – the story of a space mission.**
M. Poynter, A. L. Lane.
Atheneum, New York. 152 pp. Price $ 9.95 (1981). – Review
in Sky Telesc., Vol. 62, 476 (1981).

003.129 **The isotropic universe. An introduction to cosmolo-
gy.**   D. J. Raine.
Monographs on Astronomical Subjects: 7. Adam Hilger Ltd.,
Bristol. 14 + 253 pp. Price £ 19.50, $ 49.00, DM 97.70 (1981).
ISBN 0-85274-370-X. – Review in Astron. Tidsskr., Årg. 14,
139 (1981).
 Contents: The quality of matter. The expanding Universe.
The quality of radiation. The quantity of matter. The general
theory of relativity. Cosmological models. Cosmological tests.
Matter and radiation. The limits of isotropy. Why is the
Universe isotropic? Singularities. The evolution of structure.

003.130 **The science of space-time.**
D. J. Raine, M. Heller.
Astronomy and Astrophysics Series, Vol. 9. Pachart Publish-
ing House, Tucson. 11 + 244 pp. Price $ 24.00 (1981).
ISBN 0-912918-12-8.

003.131 **Scattering theory. Methods of modern mathematical
physics, Vol. 3.**   M. Reed, B. Simon.
American Press, Inc., New York. 15 + 463 pp. Price $ 42.00

(1979). – Review in J. Opt. Soc. America, Vol. 71, 1289;
1981 (*R. Barakat*).

003.132 **The young astronomer's handbook.**   I. Ridpath.
Hamlyn, London. 224 pp. Price £ 3.50 (1981).
Review in Space Educ., Vol. 1, 92 (1981).

003.133 **Atmospherically-optical phenomena from observations with the Salyut scientific orbital stations.**
L. Riives (Editor).
Inst. astrofiz. i fiz. atmos. AN EhSSR, Gos. opt. inst. Tartu.
188 pp. (1981). In Russian. – Review in Ref. zh., 62. Issled.
kosm. prostranstva, 11.62.56 (1981).

003.134 **Amatörastronomen.**   C. A. Ronan.
P. A. Norstedt & Söners Förlag, Stockholm. 208 pp.
Price S. kr. 150.00 (1981). – Review in Astron. Tidsskr.,
Årg. 14, 180 (1981).

003.135 **Are we alone? The possibility of extraterrestrial
civilizations.**   R. T. Rood, J. S. Trefil.
Scribner's, New York. 262 pp. Price $ 14.95 (1981). ISBN
0-684-16826-X. – Review in Sky Telesc., Vol. 62, 476 (1981).

003.136 **Le stelle variabili.**   L. Rosino.
Collana Rivista Coelum n. 1. Compositori, Bologna.
211 pp. Price L. 8.000 (1981). – Review in G. Astron., Vol. 7,
119 - 120; 1981 (*G. Romano*).

003.137 **Orbital motion.**   A. E. Roy.
Translated from the English edition. Mir, Moskva.
544 pp. (1981). In Russian. – Reviews in Ref. zh., 51. Astron.,
11.51.62 (1981); Ref. zh., 62. Issled. kosm. prostranstva,
12.62.40 (1981).

003.138 **Cosmology.**   M. Rowan-Robinson.
Second Edition. Oxford Physics Series, 15.
Clarendon Press/Oxford University Press, Oxford. 14 + 152 pp.
Price £ 12.50, £ 5.95 pbk; $ 34.50, $ 17.95 pbk. (1981).
Reviews in Astron. Tidsskr., Årg. 14, 180 (1981); Space Educ.,
Vol. 1, 91 (1981).

003.139 **Die Magnetosphäre der Erde und ihre Dynamik.**
K. Schindler.
Westdeutscher Verlag, Opladen, FRG. 32 pp. Price DM 15.00
(1980). (Rheinisch-Westfälische Akademie der Wissenschaften,
Vorträge Nr. 295). – Review in Sterne, 57. Band, 384; 1981
(*H.-R. Lehmann*).

003.140 **Horizons: exploring the universe.**   M. A. Seeds.
Wadsworth Publishing Co., Belmont, Calif. 94002.
12 + 414 pp. Price $ 19.95 (1981). ISBN 0-534-00888-7.
Review in Sky Telesc., Vol. 62, 476 (1981).

003.141 **The satellites of Mars.**   P. Seidelmann (Editor).
Translated from the English edition. Mir, Moskva.
100 pp. (1981). In Russian. – Review in Ref. zh., 51. Astron.,
12.51.85 (1981).

003.142 **Spaziotempo.**   R. Sexl, K. Schmidt.
Universale Scientifica Boringhieri, Torino. 314 pp.
(1980). – Review in G. Astron., Vol. 7, 121; 1981 (*G. Longo*).

003.143 **The UFO verdict – examining the evidence.**
R. Sheaffer.
Prometheus Books, Buffalo, N.Y. 242 pp. Price $ 15.95(1981).
Review in Sky Telesc., Vol. 62, 472; 1981 (*R. A. Schorn*).

003.144 **A complete manual of amateur astronomy. Tools
and techniques for astronomical observations.**
P. C. Sherrod.
Prentice-Hall. Englewood Cliffs, N.J. 16 + 320 pp. Price

$ 24.95, $ 10.95 pbk; £ 18.70, £ 8.20 pbk (1981). ISBN
0-13-162115-7; ISBN 0-13-162107-6 pbk. – Reviews in
Nature, Vol. 294, 488; 1981 (*P. Moore*); Sky Telesc., Vol. 62,
595 (1981).

003.145 Observational astronomy for amateurs.
J. B. Sidgwick.
Third edition prepared by R. C. Gamble. Dover Publications,
New York. 358 pp. Price $ 4.50 (1981). – From Science,
Vol. 214, 1022 (1981).

003.146 La vita intelligente nell'Universo.
I. S. Shklovskij, C. Sagan.
Ed. Feltrinelli, Milano. 356 pp. Price L. 28,000 (1980).
Review in Orione, Vol. 2, 246; 1981 (*S. Baroni*).

003.147 Remote sensing. Optics and optical systems.
P. N. Slater.
Addison-Wesley Publ. Co., Reading, Mass. 575 pp. Price
$ 34.50 (1980). – Review in Space Sci. Rev., Vol. 28, 451;
1981 (*H. Bokhove*).

003.148 Radiation transfer and stellar atmospheres.
T. L. Swihart.
Astronomy and Astrophysics Series, Vol. 12. Pachart Pub-
lishing House, Tucson. 11 + 130 pp. Price $ 24.00, DM 89.90
(1981). ISBN 0-912918-18-7.
Contents: The equation of transfer. The gray atmosphere.
The non-gray atmosphere. Line formation. Polarization.

003.149 Astrophysics and space physics reviews. Soviet
scientific reviews, Section E, Vol. 1.
R. A. Syunyaev (Editor).
Harwood, New York. 314 pp. Price $ 82.00 (1981). – From
Phys. Today, Vol. 34, No. 12, p. 59 (1981).

003.150 Galaxies: structure and evolution.
R. J. Tayler.
Translated from the English edition. Mir, Moskva. 223 pp.
(1981). In Russian. – Review in Ref. zh., 51. Astron.,
1.51.54 (1982).

003.151 Problem of search for extraterrestrial intelligence.
V. S. Troitskij, N. S. Kardashev (Editors).
Nauka, Moskva. 263 pp. (1981). In Russian. – Review in Ref.
zh., 51. Astron., 12.51.79 (1981).

003.152 Evolution kosmischer, biologischer und geistiger
Strukturen. A. Unsöld.
Wissenschaftliche Verlagsgesellschaft, Stuttgart. 150 pp. Price
DM 32.00 (1981). – Review in Phys. Bl., Vol. 37, 329; 1981
(*W. A. P. Luck*).

003.153 Devenez astronome amateur. J. Vallières.
La Fédération Québecoise du Loisir Scientifique.
239 pp. (1980). – Review in J. R. Astron. Soc. Canada,
Vol. 75, 218; 1981 (*R. Auclair*).

003.154 Stellar paths. Photographic astrometry with long-
focus instruments.
P. van de Kamp, with an introduction by J.-C. Pecker.
Astrophysics and space science library, Vol. 85. D. Reidel
Publishing Company, Dordrecht, Holland–Boston, U.S.A.–
London, England. 22 + 155 pp. Price Dfl. 70.00, $ 34.95
(1981). ISBN 90-277-1256-5.
Contents: Astrometry: historical highlights. Long-focus
photographic astrometry. Telescope; measuring machine.
Observational errors. Instrumental equation. Stellar paths.
Reduction of measurements. Path of single star. Relative
parallax, proper motion, quadratic time effect. Reduction to
absolute. Accuracy: cosmic errors. Parallax results for nearest
stars. H-R diagrams. Perspective secular changes in proper

motion, radial velocity, and parallax. Reduction from heliocentric to barycentric. Visual binaries. Orbital elements. Path
of star with orbital motion. Photocenter. Mass-ratio and
masses. Harmonic relation. Perturbations in stellar paths.
History. Analysis. Unseen astrometric companions. Illustrations. Unseen astrometric companions. General. Planetary
companions. Barnard's star. Long-period eclipsing binaries:
VV Cephei and Epsilon Aurigae. Epilogue. Attainable accuracy. Substellar and planetary detectability.

003.155 Relativistic theory of gravity. Sources and forma-
tion. 1900 - 1915. V. P. Vizgin.
Nauka, Moskva. 352 pp. (1981). In Russian. – Review in
Ref. zh., 51. Astron., 9.51.34 (1981).

003.156 Distribution of meteor bodies near the earth's orbit.
Yu. I. Voloshchuk, B. L. Kashcheev.
Nauka, Moskva. 187 pp. (1981). In Russian. – Review in Ref.
zh., 51. Astron., 7.51.30 (1981).

003.157 Space, time, and gravity. The theory of the big-bang
and black holes. R. M. Wald.
Reprint of the 1977 edition. University of Chicago Press,
Chicago. 8 + 132 pp. Price $ 3.95 (1981). – From Science,
Vol. 213, 369 (1981).

003.158 Prelude to Galileo. Essays on Medieval and
Sixteenth-Century Sources of Galileo's thought.
W. A. Wallace.
Reidel, Boston. 16 + 372 pp. Price $ 49.95, $ 23.50 pbk
(1981). – Review in Science, Vol. 214, 55 - 56; 1981
(*E. Grant*).

003.159 Astronomin i vår tid. Å. Wallenquist.
Revised edition. Bokförlaget Prisma, Stockholm.
215 pp. Price S. kr. 115.00 (1980). – Review in Astron.
Tidsskr., Årg. 14, 137; 1981 (*G. Lyngå*).

003.160 In the light of the Sun. M. Washburn.
Harcourt Brace Jovanovich, New York. 207 pp.
Price $ 14.95 (1981). ISBN 0-15-186737-2. – Review in Sky
Telesc., Vol. 62, 257 (1981).

003.161 Orbiting the sun: planets and satellites of the solar
system. F. L. Whipple.
Harvard University Press, Cambridge, Mass. 12 + 338 pp.
Price $ 20.00, £ 14.00 (1981). ISBN 0-674-64125-6. –
Reviews in Nature, Vol. 294, 21 - 22; 1981 (*D. W. Hughes*);
Sky Telesc., Vol. 62, 156 (1981); Strolling Astron., Vol. 29,
80 - 81; 1981 (*J. Olivarez*).

003.162 Sun, moon and standing stones. J. E. Wood.
Translated from the English edition. Mir, Moskva.
268 pp. (1981). In Russian. – Review in Ref. zh., 51. Astron.,
12.51.80 (1981).

003.163 Principles of astronomy. A short version.
S. P. Wyatt, J. B. Kaler.
Allyn and Bacon, Boston. 14 + 516 pp. Price $ 20.95, £ 10.95
(1981). ISBN 0-205-07315-8. – Reviews in Space Educ.,
Vol. 1, 93 (1981); Strolling Astron., Vol. 29, 79 - 80; 1981
(*R. C. Maag*).

003.164 Digital spectral analysis. C. K. Yuen, D. Fraser.
Pitman Publishing Ltd., London; SCIRO, East
Melbourne. 156 pp. Price £ 8.50 (1979). ISBN 0-273-08439-9.
Review in Astrophys. Lett., Vol. 22, 210; 1981 (*E. J. Groth*).

003.165 Problems of cosmic ray physics. Volume 1.
G. T. Zatsepin.
Inst. yader. issled., Moskva. 336 pp. (1978). In Russian.
Review in Ref. zh., 51. Astron., 12.51.90 (1981).

003.166 **Problems of neutrino astrophysics. Volume 2.**
G. T. Zatsepin.
Inst. yader. issled., Moskva. 218 pp. (1980). In Russian.
Review in Ref. zh., 51. Astron., 12.51.91 (1981).

003.167 **Breve storia dell'infinito.**   P. Zellini.
Ed. Adelphi, Milano. 256 pp. Price L.8,000 (1980).
Review in G. Astron., Vol. 7, 244; 1981 (*M. A. Santaniello*).

003.168 **Cosmic investigations made in the USSR in 1980.**
Nauka, Moskva. 47 pp. (1981). In Russian.
Review in Ref. zh., 62. Issled. kosm. prostranstva, 12.62.39
(1981).

003.169 **Le grand livre du ciel (Tout en couleurs).**
Les Editions des deux Coqs d'Or, Paris. 384 pp.
Price F 649.00 (1981). – Review in Ciel Terre, Vol. 97, 413;
1981 (*J. C. Jodogne*).

003.170 **Physics of the atomic nucleus and cosmic rays.**
Kazakhsk. univ., Alma-Ata. 137 pp. (1980).
In Russian. Review in Ref. zh., 51. Astron., 9.51.32 (1981).

003.171 **L'astronomie en 4°.** Publication du Centre National
de Documentation Pédagogique (CNDP) de Stras-
bourg (Université Louis-Pasteur). 53 pp. – Review in Astronomie, Vol. 95, 463 (1981).

003.172 **Landolt-Börnstein.** Numerical data and functional
relationships in science and technology. New series.

Group VI: Astronomy, astrophysics and space research.
Volume 2. Astronomy and astrophysics. Extension and
supplement to Volume 1. Subvolume a: methods—constants—
solar system.
K.-H. Hellwege (Editor-in-chief), K. Schaifers, H. H. Voigt
(Editors).
Springer-Verlag, Berlin—Heidelberg—New York. 17 + 305 pp.
Price DM 510.00, $ 217.10 (1981). ISBN 3-540-10054-7.
   Contents: Optical telescopes (*R. Wolf*). Solar telescopes
(*C. J. Durrant*). Photoelectric photometry (*A. Behr*). Photo-
graphic emulsions (*R. M. West*). Spectrometers and spectro-
graphs (*J. Solf*). Optical high resolution methods (*G. Weigelt,
G. Schmahl*). X-ray and γ-ray instruments (*R. Staubert,
J. Trümper*). Infrared techniques (*D. Lemke*). Radio astronom-
ical receiver systems (*O. Hachenberg*). Determination of
astronomical latitude and longitude (*H. Enslin*). Time deter-
mination (*H. Enslin*). The system of astronomical constants
(*W. Fricke*). The sun (*C. J. Durrant, O. Hachenberg, A. Bruzek*).
The planets and their satellites (*F. Gondolatsch, W.-H. Ip,
W. I. Axford*). Small bodies in the solar system (*J. Schubart,
E. K. Jessberger, J. Rahe, H. Fechtig, C. Leinert, E. Grün,
W. Pilipp, M. Scholer*). Abundances of the elements in the
solar system (*H. Palme, H. E. Suess, H. D. Zeh*). Chronology
of the solar system (*T. Kirsten*).

003.173 **Some results and prospects of an investigation of
physical properties of the moon.**
Yu. V. Solov'ev, E. G. Piskunov, V. V. Tikshaev.
Saratov. univ., Saratov. 104 pp. (1980). In Russian.
Reviews in Ref. zh., 51. Astron., 9.51.29 (1981); Ref. zh.,
52. Geod. Aehrosemka, 9.52.175 (1981).

004 History of Astronomy

004.001 **Christoph Scheiner – Entdecker der Sonnenflecken?**
E. Goercke.
Sonne, Jahrg. 5, 109 - 116 (1981).

004.002 **Eisinga's planetarium.** H. Terpstra.
Zenit, 8. Jaarg., 358 - 365 (1981).

004.003 **Die Entwicklung der Astronomie in den letzten
50 Jahren.** M. Schürer.
Orion, 39. Jahrg., 110 - 114 (1981).

004.004 **Tropical archeoastronomy.** A. F. Aveni.
Science, Vol. 213, 161 - 171 (1981).

004.005 **The great moon hoax – I, II.** D. S. Evans.
Sky Telesc., Vol. 62, 196 - 198, 308 - 311 (1981).

004.006 **Herschel and the construction of the heavens.**
M. A. Hoskin.
J. British Astron. Assoc., Vol. 91, 440 - 448 (1981).

004.007 **The great comet of 1843.** B. Warner.
Mon. Notes Astron. Soc. South. Africa, Vol. 39,
69 - 75 (1980).

004.008 **Early astronomical observations in South Africa.**
A. H. Jarrett.
Mon. Notes Astron. Soc. South. Africa, Vol. 39, 76 - 80 (1980).

004.009 **Premières photographies d'une comète et de son
spectre.** J. Sauval.
Ciel Terre, Vol. 97, 334 - 335 (1981).

004.010 **The star list of Huang You – a Chinese star list in
the early Middle Ages.** N. Pan, D.-c. Wang.
Acta Astron. Sinica, Vol. 22, 107 - 119 (1981). In Chinese.

004.011 **Towards an impressionist picture of the history of
astronomy.** L. M. Celnikier.
American J. Phys., Vol. 49, 473 - 476 (1981). – Abstr. in Phys.
Abstr., Vol. 84, Abstr. 63127 (1981).

004.012 **A note on Galileo's observation of Neptune.**
E. M. Standish.
Bull. American Astron. Soc., Vol. 13, 568 (1981). – Abstract.

004.013 **Sir William Huggins, la raie F de l'hydrogène et le
"parcours".** D. Bardin.
Astronomie, Vol. 95, 400 - 402 (1981).

004.014 **Ein Tempel als Kalender. Astronomie im alten Peru.**
R. Breitinger.
Umschau, 81. Jahrg., 557 - 559 (1981).

004.015 **On the history of mathematics and astronomy of
Medieval East.** G. P. Matvievskaya.
Izv. AN UzSSR. Ser. fiz.-mat. nauk, 1981, No. 1, p. 23 - 26. In
Russian. – Abstr. in Ref. zh., 51. Astron., 7.51.2. (1981).

004.016 **Evolution of the astronomical picture of the World
in the XIXth and at the beginning of the XXth
centuries and development of ideas on the nature of the sun.**
N. E. Khovanova.
Tr. 21−22-j nauchn. konf. aspirantov i molod. spets. po istor.
estestvozn. i tekh. Sekts. istor. fiz. Moskva, 1978 - 1979,
p. 104 - 117. In Russian. – Abstr. in Ref. zh., 51. Astron.,
8.51.7 (1981).

004.017 **Development of ideas on a model of the sun in the
XIXth and at the beginning of the XXth centuries.**
N. E. Khovanova.
Tr. 21−22-j nauchn. konf. aspirantov i molod. spets. po istor.
estestvozn. i tekh. Sekts. istor. fiz. Moskva, 1978 - 1979, p.
35 - 46. In Russian. – Abstr. in Ref. zh., 51. Astron., 8.51.8
(1981).

004.018 **Periodicity of comets as known to Indian
astronomers before eleventh century A. D.**
B. P. Bhador, S. D. Sharma.
Bull. Astron. Soc. India, Vol. 9, 87 (1981). – Abstract.

004.019 **V. B. Ketakara's independent approach to some
important problems in astronomy.** S. D. Sharma.
Bull. Astron. Soc. India, Vol. 9, 87 (1981). – Abstract.

004.020 **Astronomical content of the Rig Veda.** U. Sen.
Bull. Astron. Soc. India, Vol. 9, 87 - 88 (1981).
Abstract.

004.021 **Astronomy in seventeenth-century Canada.**
P. Broughton.
J. R. Astron. Soc. Canada, Vol. 75, 175 - 208 (1981).

004.022 **The view from the tropics.** A. F. Aveni.
News Lett. Astron. Soc. N. Y., Vol. 1, No. 10,
p. 5 - 22 (1981).

004.023 **The chronology of archbishop James Ussher.**
R. L. Reese, S. M. Everett, E. D. Craun.
Sky Telesc., Vol. 62, 404 - 405 (1981).

004.024 **New light on the moon hoax.** M. J. Crowe.
Sky Telesc., Vol. 62, 428 - 429 (1981).

004.025 **Comet observations and the quest for precision in
the seventeenth century.** N. S. Hetherington.
J. British Astron. Assoc., Vol. 91, 588 - 592 (1981).

004.026 **A propos de la constellation du Cocher.**
J.- C. Mathelin.
Astronomie, Vol. 95, 443 - 444 (1981).

004.027 **Por que el cielo es negro por la noche?**
J. R. García.
R Muscae, Vol. 6, No. 1/2, p. 6 - 19 (1981).

004.028 **Astronomie im Zeitalter der Kuenringer.**
M. G. Firneis.
Die Kuenringer, Niederösterreichische Landesausstellung,
Stift Zwettl, 16. Mai - 26. Oktober 1981. Katalog des Nieder-
österreichischen Landesmuseums, Neue Folge, Nr. 110, Wien,
Austria. 31 + 748 pp. (1981). p. 656 - 668, Abb. 54 - 58.

004.029 **Die Astronomie im mittelalterlichen Österreich.**
G. Hamann.
Die Kuenringer, Niederösterreichische Landesausstellung,
Stift Zwettl, 16. Mai - 26. Oktober 1981. Katalog des Nieder-
österreichischen Landesmuseums, Neue Folge, Nr. 110, Wien,
Austria. 31 + 748 pp. (1981). p. 668 - 685.

004.030 **Kepler, Tycho, and the "optical part of astronomy":
the genesis of Kepler's theory of pinhole images.**
S. Straker.
Arch. Hist. Exact Sci., Vol. 24, 267 - 293 (1981).

004.031   Atmospheric refraction and extinction near the
horizon.   F. Bruin.
Arch. Hist. Exact Sci., Vol. 25, 1 - 17 (1981).

004.032   Nicolaus Copernicus in the history of science.
V. M. Verkhunov.
Vopr. metodiki prepodavaniya fiz. v sredn. shk., Cheboksary,
1980, p. 118 - 127. In Russian. — Abstr. in Ref. zh., 51.
Astron., 10.51.17 (1981).

004.033   Early days in cosmic rays.   B. Rossi.
Phys. Today, Vol. 34, No. 10, p. 34 - 41 (1981).

004.034   A possible identification of Cygnus X-1 with a guest
star observed in China and Japan in 1408.
T. Kiang, K. Imaeda.
Irish Astron. J., Vol. 14, (see 012.035), 138 - 141 (1980).

004.035   The first time balls.   I. R. Bartky, S. J. Dick.
J. Hist. Astron., Vol. 12, 155 - 164 (1981).

004.036   Cosmology and the magnetical philosophy, 1640 -
1680.   J. A. Bennett.
J. Hist. Astron., Vol. 12, 165 - 177 (1981).

004.037   Durées de lever et de coucher des signes du zodiaque.
J.-P. Brunet, R. Nadal.
J. Hist. Astron., Vol. 12, 178 - 194 (1981).

004.038   A surviving flat from William Lassell's four-foot
equatorial Newtonian telescope.   J. A. Bennett.
J. Hist. Astron., Vol. 12, 195 - 197 (1981).

004.039   "In the beginning..." the Ussher chronology and
other Renaissance ideas dating the creation.
R. L. Reese, S. M. Everett, E. D. Craun.
Publ. Astron. Soc. Pacific, Vol. 93, 549 (1981). — Abstract.

004.040   Maya planetary astronomy.   A. F. Aveni.
Bull. American Astron. Soc., Vol. 13, 720 (1981).
Abstract.

004.041   The recumbent stone circles of Scotland.
A. Burl.
Sci. American, Vol. 245, No. 6, p. 50 - 56 (1981).
    Megalithic monuments, like many others, have been seen
as ancient astronomical observatories. It seems clear that
although they are astronomically aligned, their purpose was
purely ritual.

004.042   On events in the sky in two new Arabic sources.
Z. M. Buniyatov.
Tsirk. Shemakhinsk. Astrofiz. Obs., No. 68, p. 33 - 34 (1981).
In Russian.

004.043   Prehistoric astronomy: how far did it go? .
C. Ruggles.
New Scientist, Vol. 90, 750 - 753 (1981). — Abstr. in Phys.
Abstr., Vol. 85, Abstr. 6547 (1982).

004.044   Mathematical and physical problems in the
manuscript of the Turkmenian astronomer of the
XIVth century Kamal-ad-Din at-Turkmani written in the
capital of the Golden Horde.
M. Atagarryev, N. Khalimov.
Tr. 20-22 Nauchn. konf. aspirantov i molod. spets. po istor.
estestvozn. i tekh. Sekts. istor. mat. i mekh. Moskva, 1977 -
1979, p. 23 - 27. In Russian. — Abstr. in Ref. zh., 51. Astron.,
11.51.9 (1981).

004.045   On the history of dissemination of heliocentrism in
Moldavia.   A. I. Babij.

Probl. sovrem. nauchn. deyat-sti i istor. nauk. Kishi-nev, 1981,
p. 27 - 31. In Russian. — Abstr. in Ref. zh., 51. Astron.,
11.51.10 (1981).

004.046   Analyzing Galileo's scientific work.   V. M. Rozin.
Vopr. filos. 1981, No. 5, p. 73 - 85, 187 - 188. In
Russian. — Abstr. in Ref. zh., 51. Astron., 11.51.20 (1981).

004.047   L'histoire du télescope de 1600 à 1900.
M. Gabriel.
Ciel, Vol. 43, 186 - 191 (1981).

004.048   Bemerkenswerte Daten aus der Geschichte himmelskundlicher Forschung: Der Mond.   H. Mucke.
Sternenbote, 24. Jahrg., 114 - 119, 122 (1981).

004.049   Bicentennial of the discovery of Uranus.
B. Gezeman.
Vasiona, Année 29, 64 - 65 (1981). In Serbo-Croatian.

004.050   The cosmological model of R. Bošković.
B. Jovanović.
Vasiona, Année 29, 67 - 70 (1981). In Croatian.

004.051   Astronomical information in Jelgava Latvian
calendars of the 18th century.
J. Kožankova, L. Roze.
Zvaigžŋota debess, 1980/81. gada ziema, p. 62 - 66. In
Latvian.

004.052   N. I. Lobashevskij and I. M. Simonov in the
rectorate elections of the years 1845 - 1846.
S. N. Korytnikov.
Tr. Kazan. Gorod. Astron. Obs., Vyp. 45, p. 98 - 125 (1980).
In Russian.

004.053   Early days of astronomy at Toronto — Part I: The
Toronto Astronomical Club.   H. Sawyer Hogg.
J. R. Astron. Soc. Canada, Vol. 75, 281 - 288 (1981).

004.054   Solar-system astronomy in volumes 1 - 25 of the
Journal of the R. A. S. C. (1907 - 1931).
I. Halliday.
J. R. Astron. Soc. Canada, Vol. 75, 299 - 304 (1981).

004.055   La recherche d'une planète trans-neptunienne ou la
découverte de Pluton.   C. W. Tombaugh.
Astronomie, Vol. 95, 527 - 545 (1981).

004.056   L'observation de la planète Pluton à l'Observatoire
de Meudon en 1930.   A. Dollfus.
Astronomie, Vol. 95, 546 - 547 (1981).

004.057   Airy, Christie, and 1881.   J. Burnett.
J. British Astron. Assoc., Vol. 92, 11 - 12 (1981).
    Some reflections on Airy and Christie while Astronomers
Royal at Greenwich, and their attitudes to amateur astronomy.

004.058   Prähistorische Kalenderastronomie. I.
H. Hindrichs.
Orion, 39. Jahrg., 174 - 179 (1981).

004.059   La astronomia en Grecia antigua.   C. A. Altavista.
Rev. Astron., Tomo 53, No. 218, p. 6 - 7 (1981).

004.060   On the change of rate of the water clock detected
by Shen Kuo and the non-uniformity of the
apparent annual motion of the sun.   S.-c. Guo.
Ann. Shanghai Obs. Acad. Sinica, No. 2, p. 202 - 210 (1980).

004.061   Further arguments on no discovery of the proper
motion of stars by Yi Xing of the Tang Dynasty.

D.-s. Pan.
Ann. Shanghai Obs. Acad. Sinica, No. 2, p. 211 - 219 (1980).

004.062 **Jesuiten-Mathematiker in der deutschen Assistenz bis 1773.** K. Fischer.
Arch. Hist. Soc. Iesu, Vol. XLVII, 159 - 224 (1978).

004.063 **Ein Beitrag zur Geschichte der Sternatlanten.** K. Fischer.
Bohemia, Jahrbuch des Collegium Carolinum 11, 338 - 347 (1970).

004.064 **Astronomy in Transylvanian in feudalism.** K. Fischer.
Extrait des actes du XIe congrès international d'histoire des sciences, p. 117 - 126.

004.065 **The history of astronomy in Slovakia through the 19th century.** K. Fischer.
Proceedings of the Xth international congress of History of Science, Ithaca, USA, 1962, p. 769 - 771.

004.066 **Beiträge zur Geschichte der Mondgloben.** K. Fischer.
Veröff. Staatl. Math.-Phys. Salons, Band 5, 103 - 122.

004.067 **The great books of astronomy (and some others).** B. Warner.
Bibliophilia Africana IV, South African Library, Cape Town, p. 125 - 136 (1981).

004.068 **Astronomy and kinematics in Plato's project of rationalist explanation.** A. P. D. Mourelatos.
Stud. Hist. Philos. Sci., Vol. 12, No. 1, p. 1 - 32 (1981).

004.069 **Solar and lunar velocities in the Alfonsine Tables.** B. R. Goldstein.
Hist. Math., Vol. 7, 134 - 140 (1980).

004.070 **The history of the creation of a telescope.** V. A. Gurikov.
Researches in the history of astronomy. 15th issue, (see 003.025), p. 109 - 141 (1980). In Russian.

004.071 **The history of the creation of a telescope by M. V. Lomonosov.** D. N. Ponomarev.
Researches in the history of astronomy. 15th issue, (see 003.025), p. 143 - 155 (1980). In Russian.

004.072 **The first Copernican planetarium,** with a supplement from the editorial board.
W. Scheffer.
Researches in the history of astronomy. 15th issue, (see 003.025), p. 157 - 160 (1980). In Russian.

004.073 **The astronomical tools in physical studies of the Museum-Lyceum in Pushkin.**
K. G. Bol'shakova, L. E. Majstrov.
Researches in the history of astronomy. 15th issue, (see 003.025), p. 163 - 176 (1980). In Russian.

004.074 **On the source of image of astronomical instruments in the Russian mathematical manuscript of the beginning of the XVIIth century.** Yu. A. Belyj.
Researches in the history of astronomy. 15th issue, (see 003.025), p. 181 - 185 (1980). In Russian.

004.075 **On the origin of the results of determining the size of the earth mentioned by Aristotle and Archimedes.**
A. V. Klimenko.
Researches in the history of astronomy. 15th issue, (see 003.025), p. 189 - 197 (1980). In Russian.

004.076 **From the history of discovery of the Martian satellites.** G. G. Georgobiani.
Researches in the history of astronomy. 15th issue, (see 003.025), p. 199 - 206 (1980). In Russian.

004.077 **On a statement by Laplace.** A. A. Mikhajlov.
Researches in the history of astronomy. 15th issue, (see 003.025), p. 207 - 211 (1980). In Russian.

004.078 **On the first extra-eclipse observations of solar prominences.** S. V. Al'tshuler.
Researches in the history of astronomy. 15th issue, (see 003.025), p. 213 - 218 (1980). In Russian.

004.079 **The treatises about burning mirrors,** by Ibn al-Haysam with an introduction by B. A. Rozenfel'd, translated by I. O. Mohammed and N. V. Orlova.
Researches in the history of astronomy. 15th issue, (see 003.025), p. 305 - 338 (1980). In Russian.

Researches in the history of astronomy.
See Abstr. 003.025.

The Newtonian revolution.
See Abstr. 003.045.

The first stargazers. An introduction to the origins of astronomy. See Abstr. 003.046.

Briefwechsel zwischen Alexander von Humboldt und Heinrich Christian Schumacher. See Abstr. 003.053.

Astronomy in al-Farabi's papers.
See Abstr. 003.090.

Atomic and molecular collisions.
See Abstr. 003.101.

Mathematical and astronomical manuscripts of Central Asia scientists from the Xth to the XVIIIth century.
See Abstr. 003.102.

The development of astronomical thought.
See Abstr. 003.112.

America's Stonhenge. See Abstr. 003.126.

Prelude to Galileo. Essays on Medieval and Sixteenth-Century Sources of Galileo's thought.
See Abstr. 003.158.

Sun, moon and standing stones. See Abstr. 003.162.

Christoph Scheiner, with supplements from the editorial board.   See Abstr. 005.021.

Ancient astronomers of the new and old worlds.
See Abstr. 011.010.

Demonstrating the motion of comets.
See Abstr. 014.007.

Astrophysical investigations by G. A. Tikhov and the problems of exobiology. See Abstr. 015.050.

Copernicus' teaching and practical needs of the society. See Abstr. 015.051.

On Soldner's value of Newtonian deflection of light.
See Abstr. 066.195.

The zodiacal light. See Abstr. 106.057.

005 Biography

005.001   The Brydone Jack lectures on astronomy and
related topics.   J. E. Kennedy.
J. R. Astron. Soc. Canada, Vol. 75, 132 - 138 (1981).

005.002   Poet and observer: Gerard Manley Hopkins and
some mid-19th century comets.   D. H. Levy.
J. R. Astron. Soc. Canada, Vol. 75, 139 - 150 (1981).

005.003   Confessioni di un astronomo (1930-1960).
L. Gratton.
Mem. Soc. Astron. Italiana, Vol. 52, (see 012.019), 165 - 179
(1981).

005.004   Sir Isaac Newton: mad as a hatter.   W. J. Broad.
Science, Vol. 213, 1341 - 1342, 1344 (1981).

005.005   Alfred Wegener and the physics of high atmosphere.
W. Schröder.
Astron. Nachr., Band 302, 197 - 201 (1981). In German.
     Extracts from the papers of Alfred Wegener relating to
noctilucent clouds, aurora (including the so-calling geocorona)
and other upper atmospheric phenomena are presented so as to
give an idea of his contribution to this subject made during the
years of 1900–1930.

005.006   In defense of Tycho Brahe.   E. Rosen.
Arch. Hist. Exact Sci., Vol. 24, 257 - 265 (1981).

005.007   Some notes on my life as an astronomer.
J. H. Oort.
Annu. Rev. Astron. Astrophys., Vol. 19, (see 003.012), 1 - 5
(1981).

005.008   The Rosse papers and instruments.
J. A. Bennett, M. Hoskin.
J. Hist. Astron., Vol. 12, 216 - 229 (1981).

005.009   Dorothea Klumpke Roberts: a forgotten astronomer.
K. Bracher.
Mercury, Vol. 10, 139 - 140 (1981).

005.010   Edmond Halley (1656 - 1742).
Vesmír, Vol. 60, 286 (1981). In Czech.

005.011   F. Nušl (1867 - 1951).   L. Křivský.
Říše hvězd, Vol. 62, 208, 213 (1981). In Czech.

005.012   V. G. Fesenkov (1889 - 1972).
Vesmír, Vol. 60, 350 (1981). In Czech.

005.013   Joel Stebbins: pioneer of astronomical photoelectric
photometry.   R. M. Genet.
I. A. P. P. P. Commun., No. 6, p. 1 - 3 (1981).

005.014   On the occasion of the 80th birthday of
K. Ogorodnikov in 1980.   J. I. Straume.
Zvaigžņota debess, 1981. gada vasara, p. 38 - 40. In Latvian.

005.015   Sir William Herschel as a solar observer.
R. B. Ammons, D. Ammons, S. Ammons.
J. American Assoc. Variable Star Obs., Vol. 10, 33 (1981).
Abstract.

005.016   Henrietta Hill Swope, 1902 - 1980. Astronomer,
educator, benefactor.   B. L. Welther.
J. American Assoc. Variable Star Obs., Vol. 10, 37 (1981).
Abstract.

005.017   The scientific personality of Kepler.   S. Sambursky.
Mada, Vol. 25, 12 - 15 (1981). In Hebrew.

005.018   Der Astronom Pater Max Hell SJ.   A. Pinsker.
Freinberger Stimmen, Vol. 41, No. 2, p. 98 - 111
(1971).

005.019   Nikolai Ivanovich Dneprovsky (1887 - 1944).
M. S. Zverev.
Researches in the history of astronomy. 15th issue, (see
003.025), p. 15 - 60 (1980). In Russian.

005.020   Anatoly Alexeyevich Nefedyev (1910 - 1976).
S. N. Korytnikov, A. I. Nefed'eva.
Researches in the history of astronomy. 15th issue, (see
003.025), p. 61 - 72 (1980). In Russian.

005.021   Christoph Scheiner, with supplements from the
editorial board.   I. M. Rabinovich.
Researches in the history of astronomy. 15th issue, (see
003.025), p. 85 - 92 (1980). In Russian.

     Briefwechsel zwischen Alexander von Humboldt und
Heinrich Christian Schumacher.   See Abstr. 003.053.

     Tobias Mayer (1723 - 1762), pioneer of enlightened
science in Germany.   See Abstr. 003.062.

     Das Geheimnis der Planeten. Ein Roman um
Johannes Kepler.   See Abstr. 003.113.

     Prelude to Galileo. Essays on Medieval and Sixteenth
Century Sources of Galileo's thought.
See Abstr. 003.158.

006 Personal Notes

G. de Vaucouleurs received the 1980 Herschel Medal.   A. W. Wolfendale.
Q. J. R. Astron. Soc., Vol. 22, 352 (1981).

J. F. Gilbert received the 1981 Gold Medal of the Royal Astronomical Society.   A. W. Wolfendale.
Q. J. R. Astron. Soc., Vol. 22, 349 (1981).

H. Jeffreys, 90th birthday.
Q. J. R. Astron. Soc., Vol. 22, 467 (1981).

Sir Bernard Lovell received the Gold Medal of the Royal Astronomical Society 1981.   M. J. Seaton.

Q. J. R. Astron. Soc., Vol. 22, 225 - 226 (1981).

R. E. Nather received the Boyden Premium of the Franklin Institute of Philadelphia.
Observatory, Vol. 101, 132 (1981).

P. J. E. Peebles received the 1981 Eddington Medal. A. W. Wolfendale.
Q. J. R. Astron. Soc., Vol. 22, 350 - 351 (1981).

B. Warner received the Boyden Premium of the Franklin Institute of Philadelphia.
Observatory, Vol. 101, 132 (1981).

007 Obituaries

Z. B. Brkić, 1910 - 1979.
Bull. Obs. Astron. Belgrade, No. 131, p. 1 (1981).

Balfour Watson Currie, 1902 - 1981.
J. E. Kennedy.
J. R. Astron. Soc. Canada, Vol. 75, 219 - 221 (1981).

Alexander E. Douglas died 1981, July 26.
G. Herzberg.
Phys. Today, Vol. 34, No. 10, p. 116 - 117 (1981).

In memoriam: Pero M. Đurković (1908 - 1981).
G. M. Popović, B. Popović, A. Tomić.
Vasiona, Année 29, 21 - 25 (1981). In Serbo-Croatian.

Paul Herget, 1908 - 1981.
B. G. Marsden.
Minor Planet. Circ., (M.P.C.), No. 6221 (1981).

Paul Herget: tracker of the skies.
P. K. Seidelmann.
Sky Telesc., Vol. 62, 531 (1981).

C. M. Huffer (1894 - 1981): pioneer of photoelectric photometry.   R. M. Genet.
I. A. P. P. P. Commun., No. 5, p. 1 - 3 (1981).

Morris James Sage Innes, 1907 - 1980, October 28.
M. R. Dence, J. G. Tanner.
J. R. Astron. Soc. Canada, Vol. 75, 117 - 123 (1981).

Kazimierz Kordylewski, 1903 October 11 - 1981 March 11.   P. Flin.
Rocznik Astronomiczny Obserwatorium Krakowskiego 1982, (see 047.025), 2 pp. (1981).

In memoriam: Henri Michel, 1885 - 1981.
E. Hoge.
Ciel Terre, Vol. 97, 349 - 351 (1981).

Nikolaus Benjamin Richter, 5. 2. 1910 - 26. 11. 1980.   S. Marx.
Astron. Nachr., Band 302, 203 (1981).

Nikolaus B. Richter, 1910 February 5 - 1980 November 26.   J. Rahe.
Mitt. Astron. Ges., Nr. 53, p. III - V (1981).

N. B. Richter, 5.2. 1910 - 26.11. 1980.
K.-H. Schmidt.
Sterne, 57. Band, 182 - 183 (1981).

Guglielmo Righini (1908 - 1978).
L. Rosino.
Atti Accad. Naz. Lincei, Rend. Ser. Ottava, Vol. 68, 463 - 480 (1980).

Martha Betz Shapley, 1890 August 3 - 1981 January 24.   Z. Kopal.
Astrophys. Space Sci., Vol. 79, 261 - 264 (1981).

Beatrice M. Tinsley died on 23 March 1981.
S. Faber.
Phys. Today, Vol. 34, No. 9, p. 110, 112 (1981).

Harold C. Urey, 1893 April 29 - 1981 January 6.
J. R. Arnold.
Geochim. Cosmochim. Acta, Vol. 45, 1974 (1981).

Johann Wempe, 1906 December 31 - 1980 May 29.
H. Scheffler.
Mitt. Astron. Ges., Nr. 53, p. I - II (1981).

008 Observatories, Institutes

Reports, communications and publications of observatories and astronomical institutes are recorded in this section; included are numbered series of reprints. Whenever possible, the numbers of the abstracts referring to the publications are given. Observatories and institutes are listed in alphabetical order of their towns. In some cases observatory publications do not give the name of the town; the following list which gives names and towns of some institutions may serve as an aid in such cases.

Aarne Karjalainen Observatory	Oulu, Finland	Dudley Observatory	Schenectady, New York, USA
Algonquin Radio Observatory	Lake Traverse, Ontario, Canada	Dunsink Observatory	Dublin, Ireland
Allegheny Observatory	Pittsburgh, Pennsylvania, USA	Dyer Observatory, Vanderbilt University	Nashville, Tennessee, USA
Anglo-Australian Observatory	Epping, N. S. W., Australia	Ege University Observatory	Izmir, Turkey
Archenhold-Sternwarte	Berlin-Treptow, German Democratic Republic	Engelhardt Observatory	Kazan, USSR
Argentine Radioastronomy Institute	Pereyra, Iraola, Argentina	Erwin W. Fick Observatory, Iowa State University	Ames, Iowa, USA
Arizona State University	Tempe, Arizona, USA	European Southern Observatory	La Silla, Chile
Arthur J. Dyer Observatory	Nashville, Tennessee, USA	Fabra Observatory	Barcelona, Spain
Astronomical Latitude Station, Polish Academy of Sciences	Borowiec, Poland	Felix Aguilar Observatory	San Juan, Argentina
		Fernbank Observatory	Atlanta, Georgia, USA
Astronomisches Rechen-Institut	Heidelberg, F. R. Germany	Five College Observatories	Amherst, Massachusetts, USA
Bell Laboratories	Murray Hill, New Jersey, USA	Florida State University Radio Observatory	Tallahassee, Florida, USA
Bell Telephone Laboratories	Holmdel, New Jersey, USA	Flower and Cook Observatories, University of Pennsylvania	Philadelphia, Pennsylvania, USA
Bosscha Observatory	Lembang, Indonesia		
Boyden Observatory	Bloemfontein, South Africa		
Bureau International de l'Heure	Paris, France	George R. Wallace Jr. Astrophysical Observatory	Cambridge, Massachusetts, USA
Cajigal Observatory	Caracas, Venezuela		
California Institute of Technology	Pasadena, California, USA	Georgetown Observatory	Washington, D.C., USA
Carter Observatory	Wellington, New Zealand	Glavnaya Astronomicheskaya Observatoriya AN SSSR	Pulkovo, USSR
Cavendish Laboratory	Cambridge, England	Goddard Space Flight Center	Greenbelt, Maryland, USA
Centre de Données Stellaires	Strasbourg, France	Goethe Link Observatory, Indiana University	Bloomington, Indiana, USA
Centro Astronomico Hispano-Aleman (Max-Planck Institut für Astronomie Station)	Calar Alto, Almeria, Spain	H. M. Nautical Almanac Office, Royal Greenwich Observatory	Greenwich, England
Cerro Tololo Interamerican Observatory	La Serena, Chile	Hale Observatories	Pasadena, California, USA
Československá akademie věd, Astronomický ústav	Prague, Czechoslovakia	Harvard Radio Astronomy Station	Cambridge, Massachusetts, USA
Chamberlin Observatory, University of Denver	Denver, Colorado, USA	Haute Provence Observatory	Saint Michel, France
Columbia University, Department of Astronomy	New York, New York, USA	Haystack Observatory	Westford, Massachusetts, USA
Commonwealth Observatory	Canberra, Australia	Heinrich-Hertz Institut	Berlin-Adlershof, German Democratic Republic
Cornell University, Center for Radiophysics and Space Research	Ithaca, New York, USA	Herzberg Institute of Astrophysics	Victoria, B. C., Ottawa, Canada
Corralitos Observatory	Las Cruces, New Mexico, USA		
Crawford Hill Laboratory	Holmdel, New Jersey, USA	High Altitude Observatory, University of Colorado	Boulder, Colorado, USA
CSIRO, Radiophysics Division	Sydney, Australia	Hopkins Observatory	Williamstown, Massachusetts, USA
David Dunlap Observatory, University of Toronto	Richmond Hill, Ontario, Canada	Horn d'Arturo Observatory	Bologna, Italy
Dearborn Observatory	Evanston, Illinois, USA	Hvar Observatory	Zagreb, Yugoslavia
Department of Astronomy and Observatory, University of California	Los Angeles, California, USA	IBM Thomas J. Watson Research Center	Yorktown Heights, New York, USA
Department of Astronomy Swarthmore College	Swarthmore, Pennsylvania, USA	Indian Institute of Astrophysics	Bangalore, India
		Infrared Telescope Facility	Honolulu, Hawaii, USA
Department of Astronomy, University of Texas	Austin, Texas, USA	Institute for Astronomy, University of Hawaii	Honolulu, Hawaii, USA
Deutsches Hydrographisches Institut (DHI)	Hamburg, F. R. Germany	Institute for Theoretical Astronomy	Leningrad, USSR
Dominion Astrophysical Observatory	Victoria, B. C., Canada	Institute of Astronomy and Space Science, University of British Columbia	Vancouver, B. C., Canada
Dominion Observatory	Ottawa, Ontario, Canada	Institute of Theoretical Astrophysics, Blindern	Oslo, Norway
Dominion Radio Astrophysical Observatory	Penticton, B. C., Canada		

Instituto Argentino de Radioastronomía	Villa Elisa, Provincia de Buenos Aires, Argentina
Instituto de Astronomía y Física del Espacio (IAFE)	Buenos Aires, Argentina
Instituto Venezolano de Astronomia	Merida, Venezuela
Instituto y Observatorio de Marina	San Fernando (Cádiz), Spain
International Latitude Observatory	Mizusawa, Japan
IUE Observatory, European Space Agency	Villafranca, Madrid, Spain
Jet Propulsion Laboratory, California Institute of Technology	Pasadena, California
Joint Institute for Laboratory Astrophysics (JILA)	Boulder, Colorado, USA
Judson B. Coit Observatory	Boston, Massachusetts, USA
Kandilli Observatory	Istanbul, Turkey
Kansas University Observatory	Lawrence, Kansas, USA
Kapteyn Astronomical Laboratory	Groningen, Netherlands
Karl-Schwarzschild-Observatorium	Tautenburg, German Democratic Republic
Kenneth Mees Observatory	Rochester, New York, USA
Kiepenheuer-Institut für Sonnenphysik, formerly Fraunhofer-Institut	Freiburg, F. R. Germany
Kitt Peak National Observatory	Tucson, Arizona, USA
Kodaikanal Observatory	Bangalore, India
Korean National Astronomical Observatory	Seoul, Korea
Kwasan and Hida Observatories	Kyoto, Japan
Lamont-Hussey Observatory	Bloemfontein, South Africa
Landessternwarte Heidelberg-Königstuhl	Heidelberg, F. R. Germany
Las Campanas Observatory	Pasadena, California, USA
Lawrence Livermore Laboratory, University of California	Livermore, California, USA
Leander McCormick Observatory, University of Virginia	Charlottesville, Virginia, USA
Lee Observatory	Beirut, Lebanon
Leopold-Figl-Observatorium	Vienna, Austria
Leuschner Observatory	Berkeley, California, USA
Lick Observatory	Santa Cruz, (Mount Hamilton), California, USA
Lindheimer Astronomical Research Center	Evanston, Illinois, USA
Lockheed Palo Alto Research Laboratory	Palo Alto, California, USA
Lockheed Solar Observatory	Saugus, California, USA
Lohrmann-Observatorium der Technischen Universität Dresden	Dresden, German Democratic Republic
Louisiana State University Observatory	Baton Rouge, Louisiana, USA
Lowell Observatory	Flagstaff, Arizona, USA
Lunar and Planetary Laboratory, Catalina Station	Tucson, Arizona, USA
Max-Planck-Institut für Astronomie	Heidelberg, F. R. Germany
Max-Planck-Institut für Physik und Astrophysik	Munich, F. R. Germany
Max-Planck-Institut für Radioastronomie	Bonn, F. R. Germany
McDonald Observatory	Fort Davis, Texas, USA
McMath Hulbert Observatory	Pontiac, Michigan, USA
Michigan State University Observatory	East Lansing, Michigan, USA

Molonglo Radio Observatory, University of Sydney	Sydney, Australia
Monterey Institut for Research in Astronomy	Carmel Valley, California, USA
Mount Cuba Observatory	Wilmington, Delaware, USA
Mount John Observatory	Lake Tekapo, New Zealand
Mount Stromlo Observatory	Canberra, Australia
Mount Wilson Observatory	Pasadena, California, USA
Mt. Laguna Observatory	San Diego, California, USA
Mullard Radio Astronomy Observatory	Cambridge, England
Mullard Space Science Laboratory	London, England
Multiple Mirror Telescope Observatory	Tucson, Arizona, USA
Narrabri Observatory, University of Sydney	Sydney, Australia
National Bureau of Standards	Washington, D.C., USA
National Radio Astronomy Observatory	Charlottesville, Virginia, USA Green Bank, West Virginia, USA Socorro, New Mexico, USA Tucson, Arizona, USA
National Research Council of Canada	Ottawa, Ontario, Canada
New Mexico State University Observatory	Las Cruces, New Mexico, USA
Nicholas Copernicus Observatory	Brno, Czechoslovakia
Nizamiah & Rangapur Observatories	Hyderabad, India
Nuffield Radio Astronomy Laboratories, Jodrell Bank University of Manchester	Manchester, England
Oak Ridge Observatory (formerly Harvard College Observatory)	Cambridge, Massachusetts, USA
Observatoire Royal de Belgique	Uccle, Belgium
Observatories of the University of Western Ontario	London, Canada
Observatório Astronômico Antares	Feira de Santana, Brazil
Observatório Astronômico do Instituto de Física da Universidade Federal do Rio Grande do Sul	Porto Alegre, Rio Grande do Sul, Brazil
Observatorio de Cartuja	Granada, Spain
Observatorio del Ebro	Tortosa, Spain
Observatorio Nacional	Rio de Janeiro, Brazil
Observatorio Nacional de Física Cósmica	San Miguel, Argentina
Observatory, University of Michigan	Ann Arbor, Michigan, USA
Ohio State University Radio Observatory	Columbus, Ohio, USA
Ole Roemer-Observatoriet	Aarhus, Denmark
Onsala Space Observatory	Göteborg, Sweden
Owens Valley Radio Observatory	Big Pine, California, USA
Palomar Observatory	Pasadena, California, USA
Perkins Observatory, Ohio State and Wesleyan Universities	Delaware, Ohio, USA
Physical Research Laboratory	Ahmedabad, India
Purple Mountain Observatory	Nanking, China
Radcliffe Observatory	Pretoria, South Africa
Raman Research Institute	Bangalore, India
Rattlesnake Mountain Observatory	Richland, Washington, USA

Remeis-Sternwarte	Bamberg, F. R. Germany
Ritter Astrophysical Research Center of the University of Toledo	Toledo, Ohio, USA
Rosemary Hill Observatory	Gainesville, Florida, USA
Rothney Astrophysical Observatory	Calgary, Canada
Royal Aircraft Establishment, Geophysical Studies in Space Department	Farnborough, England
Royal Radar Establishment, Radio Astronomy Division	Malvern, England
Sacramento Peak Observatory	Sunspot, New Mexico, USA
Sagamore Hill Radio Observatory	Hamilton, Massachusetts, USA
San Fernando Observatory	Northridge, California, USA
Shaanxi Astronomical Observatory	Lintong, Sian, China
Siding Spring Observatory	Siding Spring, N. S. W., Australia
Smithsonian Astrophysical Observatory	Cambridge, Massachusetts, USA
Sonnenobservatorium Kanzelhöhe	Graz, Austria
South African Astronomical Observatory	Cape Town, South Africa
Specola Vaticana	Castel Gandolfo, Vatican
Specola di Padova	Asiago, Italy
Sproul Observatory	Swarthmore, Pennsylvania, USA
Sternberg Astronomical Institute	Moscow, USSR
Steward Observatory, University of Arizona	Tucson, Arizona, USA
W. Struve Tartu Astrophysical Observatory	Tartu, USSR
Tata Institute of Fundamental Research	Bombay, India
United States Naval Observatory	Washington, D. C., USA
University of Alabama	University, Alabama, USA
University of California	Berkeley, California, USA
University of Florida Observatories	Bronson, Florida, USA

University of Florida, Radio Observatory	Old Town, Florida, USA
University of Hawaii	Honolulu, Hawaii, USA
University of Illinois Observatory	Urbana, Illinois, USA
University of Kansas Observatory	Lawrence, Kansas, USA
University of Maryland	College Park, Maryland, USA
University of Michigan Observatories	Ann Arbor, Michigan, USA
University of Minnesota	Minneapolis, Minnesota, USA
University of South Florida Observatory	Tampa, Florida, USA
University of Texas, Department of Astronomy	Austin, Texas, USA
University of Washington, Astronomy Department	Seattle, Washington, USA
Uttar Pradesh State Observatory	Nainital, India
Van Vleck Observatory	Middletown, Connecticut, USA
Vatican Observatory	Castel Gandolfo, Vatican
Venezuelan Astronomical Institute	Merida, Venezuela
Wallace Astrophysical Observatory	Cambridge, Massachusetts, USA
Warner and Swasey Observatory	Cleveland, Ohio, USA
Washburn Observatory University of Wisconsin	Madison, Wisconsin, USA
West Melton Observatory	Christchurch, New Zealand
Wilhelm-Foerster Sternwarte	Berlin, F. R. Germany
Yale University Observatory	New Haven, Connecticut, USA
Yerkes Observatory	Williams Bay, Wisconsin, USA
Zentralinstitut für Astrophysik, Sternwarte Babelsberg	Potsdam-Babelsberg, German Democratic Republic
Zentralinstitut für Astrophysik, Sternwarte in Sonneberg	Sonneberg, German Democratic Republic
Zentralinstitut für solar-terrestrische Physik	Berlin-Adlershof, German Democratic Republic

008.001   Athens

National Observatory of Athens, Astronomical Institute. Annual report 1979.   G. Contopoulos. Annu. Rep. Astron. Inst. Greece, 1979, p. 3 - 5 (1980).

National Observatory of Athens, Astronomical Institute. – Annual report 1980.   G. Contopoulos. Annu. Rep. Astron. Inst. Greece 1980, p. 3 - 6 (1981).

Department of Astronomy, University of Athens. Annual report 1979. G. Contopoulos. Annu. Rep. Astron. Inst. Greece 1979, p. 6 - 10 (1980).

Department of Astronomy, University of Athens. Annual report 1980.   G. Contopoulos. Annu. Rep. Astron. Inst. Greece 1980, p. 7 - 11 (1981).

Department of Astronomy, Technical University of Athens. – Annual report 1979.   J. Argyrakos. Annu. Rep. Astron. Inst. Greece, 1979, p. 16 (1980).

Department of Astronomy, Technical University of Athens. – Annual report 1980.   J. Argyrakos. Annu. Rep. Astron. Inst. Greece 1980, p. 17 (1981).

Department of Astrophysics, University of Athens. Annual report 1979.   S. N. Svolopoulos. Annu. Rep. Astron. Inst. Greece 1979, p. 29 - 32 (1980).

Department of Astrophysics, University of Athens. Annual report 1980.   S. N. Svolopoulos. Annu. Rep. Astron. Inst. Greece 1980, p. 32 - 36 (1981).

Research Center for Astronomy and Applied Mathematics, Academy of Athens. – Annual report 1979. J. N. Xanthakis. Annu. Rep. Astron. Inst. Greece, 1979, p. 17 - 19 (1980).

Research Center for Astronomy and Applied Mathematics, Academy of Athens. – Annual Report 1980. J. N. Xanthakis. Annu. Rep. Astron. Inst. Greece 1980, p. 18 - 21 (1981).

Research Center for Astronomy and Applied
Mathematics, Academy of Athens, Contributions Series I
(Astronomy), Nos. 70 (30.077.066), 71, 72, 75 (30.093.
067), 76 (27.093.012), 77, 78 (28.073.103), 79 (30.085.
058), 80 (29.143.004), 81 (30.093.068), 82 (30.075.023),
84 (30.106.041), 86 (29.085.005), 87 (30.085.059).

008.002 Austin, Tex.

Department of Astronomy and McDonald
Observatory of the University of Texas, Austin, Texas,
Reprints Nos. 901 (28.120.022), 902 (28.122.146), 903
(28.141.165), 904 (28.131.158),905 (28.158.216), 906
(28.034.047), 907 (28.131.156), 908 (28.100.120), 909
(29.021.006), 910 (29.131.034), 911 (29.091.007), 912
(28.158.325), 913 (29.158.080), 914, 915 (29.118.008),
916 (29.114.046), 917 (29.103.221), 918 (29.131.003),
919 (30.101.011), 920 (30.091.079), 921 (29.101.006),
922 (29.101.007), 923, 924 (29.114.034), 925 (28.065.094),
926 (29.114.035), 927 (29.022.128), 928 (29.091.040),
929 (29.014.027), 930 (29.099.075), 931 (29.125.203),
932 (29.122.173), 933 (29.125.221), 934 (29.022.045),
935 (29.122.008), 936 (29.158.286), 937 (29.102.009),
938 (30.100.043), 939 (30.091.091), 940 (29.131.058),
941 (29.117.023), 942 (29.107.029), 943 (28.135.032),
944 (30.065.011), 945 (29.122.045), 946 (29.034.067),
947 (29.119.039), 948 (29.114.161), 949 (29.131.251),
950 (29.125.201).

The University of Texas, Publications in Astronomy,
No. 18 (30.154.043).

008.003 Bamberg

Dr.-Remeis-Sternwarte Bamberg, Astronomisches
Institut der Universität Erlangen-Nürnberg. – Report for 1980.
J. Rahe.
Mitt. Astron. Ges., Nr. 53, p. 7 - 14 (1981).

008.004 Basel

Astronomisches Institut der Universität Basel.
Report for 1980.   G. A. Tammann.
Mitt. Astron. Ges., Nr. 53, p. 15 - 21 (1981).

008.005 Baton Rouge, La.

Contribution of the James-Mims Observatory,
Baton Rouge, La., No. 5 (30.104.036).

Contribution of the Institute for Astronomical
Research, Baton Rouge, La., No. 6 (30.124.402).

008.006 Belgrade

Bulletin de l'Observatoire Astronomique de
Belgrade, No. 131 (30.007, 30.082.103, 30.032.045,
30.116.044, 30.118.030, 30.118.031, 30.118.032, 30.118.
033, 30.098.099, 30.096.018, 30.096.019, 30.045.014,
30.044.019).

Publications de l'Observatoire Astronomique de
Beograd, No. 28 (30.116.046).

008.007 Berlin

Institut für Astronomie und Astrophysik der
Technischen Universität Berlin. – Report for 1980.
R. Wielen.
Mitt. Astron. Ges., Nr. 53, p. 23 - 30 (1981).

Institut für Astronomie und Astrophysik der
Technischen Universität Berlin, Mitteilungen, No. 2(30.008.
007).

Veröffentlichungen der Wilhelm-Foerster-Sternwarte
Berlin, Nr. 53 (30.011.051).

Heinrich-Hertz-Institut, Solare Beobachtungsergebnisse. Akademie der Wissenschaften der DDR, Zentralinstitut
für Solar-Terrestrische Physik (Heinrich-Hertz-Institut),
Berlin-Adlershof, HHI Solar Data, Vol. 32, 1981 January -
August (30.072.059).

Heinrich-Hertz-Institut, Solar-Terrestrische Physik,
Reports. Akademie der Wissenschaften der DDR, Zentralinstitut für Solar-Terrestrische Physik (Heinrich-Hertz-Institut),
Berlin-Adlershof, HHI-STP-Report, Nos. 6 (30.077.067),
12 (30.002.085).

008.008 Bloemfontein

University of the Orange Free State: Boyden
Observatory, Department of Astronomy.
A. H. Jarrett.
Mon. Notes Astron. Soc. South. Africa, Vol. 40, 5 - 6 (1981).

Boyden Observatory Reprint Nos. 45 (30.073.102),
46 (30.004.008), 47 (30.073.103), 48 (30.073.104), 49
(30.158.036).

008.009 Bochum

Astronomisches Institut der Ruhr-Universität
Bochum. – Report for 1980.   K. Rohlfs.
Mitt. Astron. Ges., Nr. 53, p. 31 - 41 (1981).

Bereich Extraterrestrische Physik in der Abteilung
XII der Ruhr-Universität Bochum. – Report for 1980.
R. H. Giese.
Mitt. Astron. Ges., Nr. 53, p. 41 - 43 (1981).

Institut für Theoretische Physik, Lehrstuhl IV der
Ruhr-Universität Bochum. – Report for 1980.
K. Schindler.
Mitt. Astron. Ges., Nr. 53, p. 44 - 45 (1981).

008.010 Bonn

Astronomische Institute der Universität Bonn.
Sternwarte mit Observatorium Hoher List; Radioastronomi-
sches Institut mit Radioobservatorium Stockert; Institut für
Astrophysik und Extraterrestrische Forschung; Sonderfor-
schungsbereich Radioastronomie. – Reports for 1980.

U. Mebold, W. Priester, H. Schmidt.
Mitt. Astron. Ges., Nr. 53, p. 47 - 81 (1981).

Veröffentlichungen der Astronomischen Institute
Bonn, Nr. 94 (30.082.104).

Max-Planck-Institut für Radioastronomie. – Report
for 1980.   I. Pauliny-Toth, R. Schwartz.
Mitt. Astron. Ges., Nr. 53, p. 83 - 114 (1981).

Max-Planck-Gesellschaft zur Förderung der Wissen-
schaften, M.P.I. für Radioastronomie, Bonn, Sonderdruck,
Ser. A, Nos. 408 (27.021.036), 409 (28.131.065), 410
(28.141.502), 411 (30.022.170), 412 (30.022.171), 413
(28.141.070), 414 (28.132.006), 415 (30.033.007), 416
(28.134.005), 417 (28.125.001), 418 (28.131.015), 419
(27.141.133), 420 (27.141.537), 421 (28.141.524), 422
(30.022.172), 423 (27.141.168), 424 (28.134.021), 425
(28.132.019), 426 (28.141.065), 427 (28.158.006), 428
(28.141.064), 429 (28.158.142), 430 (29.131.132), 431
(28.141.139), 432 (28.141.071), 433 (28.125.041), 434
(28.160.003), 435 (28.071.031), 436 (28.071.032), 437
(28.114.081), 438 (28.031.029), 439 (28.141.115), 440
(28.131.146), 441 (28.141.022), 442 (28.131.254), 443,
444 (28.141.024), 445 (28.141.128), 446 (28.131.282),
447 (29.033.001), 448 (28.158.175), 449 (28.131.235),
450 (28.125.061), 451 (28.125.060), 452 (29.141.505),
453 (30.033.018), 454 (28.131.150), 455 (28.158.134),
456 (29.158.079), 457 (28.141.102), 458 (28.116.024),
459 (29.064.002), 460 (29.158.057), 461 (28.131.256),
462 (28.141.092), 463 (29.131.022), 464 (29.131.023),
465 (29.064.107), 466 (29.112.002), 467 (28.141.124),
468 (29.133.023), 469 (29.121.001), 470 (28.125.062),
471 (29.125.002), 472 (29.031.507), 473 (29.141.504),
474 (29.131.028), 475 (29.141.011), 476 (29.131.021),
477 (29.158.009), 478 (29.064.006), 479 (29.160.007),
480 (29.142.508), 481 (29.131.077), 482 (29.141.064),
483 (29.158.103), 484 (29.131.116), 485 (29.141.071),
486 (29.141.048), 487 (29.141.520), 488 (29.131.111),
489 (29.141.027), 491 (29.034.020), 492 (29.114.013),
493 (29.112.005), 494 (29.155.032), 495 (29.158.087),
496 (29.135.022), 497 (29.141.092), 498 (29.158.247),
499 (29.141.138), 500 (29.131.220), 501 (29.158.163),
502 (29.158.150), 503 (29.141.090), 504 (29.131.154),
505 (29.156.008), 506 (30.158.055), 507 (30.141.085),
508 (29.131.271), 509 (30.141.014), 510 (29.125.092),
511 (29.131.267), 512 (29.131.152), 513 (30.132.007),
515 (29.141.123), 516 (30.158.022), 517 (30.141.502),
518 (30.141.111).

008.011 Borowiec

Circular, Time and Latitude Service, No. 157
(30.044.018).

008.012 Bucharest

Observations solaires. Rotations 1690 - 1702
(30.072.070).

008.013 Cambridge, U.K.

Institute of Astronomy. Annual Report 1980 - 1981.
M. J. Rees.
Published by Univ. Cambridge, The Observatories, Madingley
Road, Cambridge CB3 OHA. 43 pp. (1981).

Mullard Radio Astronomy Observatory, Cavendish
Laboratory, University of Cambridge. Report for the period
1979 October 1 to 1980 September 30.   M. Ryle.
Q. J. R. Astron. Soc., Vol. 22, 452 - 466 (1981).

008.014 Cambridge, Mass.

Smithsonian Astrophysical Observatory, Special
Report, Nos. 390 (30.022.173), 391 (30.002.071).

IAU Circulars, Nos, 3615 - 3655 (1981).

Minor Planet Circulars, (M.P.C.), Nos. 6065 - 6572
(1981).

008.015 Cape Town

South African Astronomical Observatory. Report
for the year ending 31 December 1980.   M. W. Feast.
Council for Scientific and Industrial Research (South Africa);
Science Research Council (United Kingdom). 48 pp. (1981).

South African Astronomical Observatory Circulars,
No. 6 (30.119.109, 30.113.079, 30.113.080, 30.154.044,
30.122.199, 30.122.200, 30.082.107).

University of Cape Town: Department of Astrono-
my.   B. Warner.
Mon. Notes Astron. Soc. South. Africa, Vol. 40, 2 - 4 (1981).

University of Cape Town: Department of Applied
Mathematics.   G. F. R. Ellis.
Mon. Notes Astron. Soc. South. Africa, Vol. 40, 4 - 5 (1981).

008.016 College Park, Md.

Astronomy Program, University of Maryland,
College Park, Maryland 20742. – Report for the period
1 September 1979 - 31 August 1980.
Bull. American Astron. Soc., Vol. 13, 669 - 682 (1981).

008.017 Córdoba

Observatorio Astronómico (Universidad Nacional de
Córdoba, Córdoba, Argentina), Tirada Aparte, Nos. 231,
232 (28.153.013), 233 (28.158.180), 234.

008.018 Cracow

Cracow Observatory, Reprints No. 129 (30.106.042).

008.019 Crimea

Izvestiya Ordena Trudovogo Krasnogo Znameni
Krymskoj Astrofizicheskoj Observatorii, Akademiya Nauk
SSSR, Tom 63 (30.075.012, 30.080.054, 30.072.038,
30.072.039, 30.114.131, 30.120.021, 30.114.132, 30.117.072,
30.126.026, 30.141.123, 30.142.533, 30.142.534, 30.142.062,
30.032.023, 30.031.006, 30.031.007, 30.031.008, 30.033.006,
30.032.548, 30.034.044).

008.020 Dresden

Mitteilungen des Lohrmann-Observatoriums der Technischen Universität Dresden, No. 38 (30.096.021).

Technische Universität Dresden, Lohrmann-Observatorium, Zirkular, Nos. 94 - 95 (30.045.015).

008.021 Epping

Anglo-Australian Telescope 1980/81. Report of the Anglo-Australian Telescope Board, 1 July 1980 to 30 June 1981.   H. Massey.
Australian Government Publishing Service, Canberra. 6 + 39 pp. (1981).

Anglo-Australian Observatory, Preprint Nos. 150 (30.034.100), 152 (30.141.198; 30.158.271; 30.158.272), 153 (30.158.273), 154 (30.135.065), 155 (30.036.019), 156 (30.142.152), 157 (30.158.274), 158 (30.158.275), 159 (30.034.101).

A. A. O. Newsletter, Nos. 18, 19 (1981).

008.022 Erlangen

Physikalisches Institut, Angewandte Optik.
Report for 1980.   G. P. Weigelt.
Mitt. Astron. Ges., Nr. 53, p. 115 - 116 (1981).

008.023 Farnborough

Royal Aircraft Establishment, Farnborough, Geophysical studies in space department. Report for the year ending 1981 March 31.
Q. J. R. Astron. Soc., Vol. 22, 432 - 435 (1981).

008.024 Frankfurt

Institut für Theoretische Physik/Astrophysik.
Report for 1980.   W. H. Kegel.
Mitt. Astron. Ges., Nr. 53, p. 117 (1981).

008.025 Freiburg

Kiepenheuer-Institut für Sonnenphysik. – Report for 1980.   E. H. Schröter.
Mitt. Astron. Ges., Nr. 53, p. 119 - 129 (1981).

008.026 Gainesville, Fla.

University of Florida, Department of Astronomy, Gainesville, Florida 32611.-Report.   H. Eichhorn.
Bull. American Astron. Soc., Vol. 13, 588 - 592 (1981).

008.027 Geneva

Publications de l'Observatoire de Genève, Série A, Fasc. 84 (30.151.043; 30.114.099; 30.031.640).

008.028 Glasgow

Department of Astronomy, University of Glasgow. Report for the period 1979 October to 1980 September. P. A. Sweet.
Q. J. R. Astron. Soc., Vol. 22, 299 - 302 (1981).

008.029 Göteborg

Research Laboratory of Electronics and Onsala Space Observatory, Chalmers University of Technology, Gothenburg, Sweden. Research Report, Nos. 142 (30.112. 052), 143 (30.131.209).

008.030 Göttingen

Universitäts-Sternwarte Göttingen und Institut für Sonnenforschung Locarno-Orselina (Tessin). – Report for 1980.   K. J. Fricke, H. H. Voigt.
Mitt. Astron. Ges., Nr. 53, p. 131 - 141 (1981).

008.031 Graz

Institut für Astronomie der Universität Graz; Institut für Astronomie (Universitäts-Sternwarte); Observatorium Lustbühel; Sonnenobservatorium Kanzelhöhe. – Report for 1980.   H. Haupt.
Mitt. Astron. Ges., Nr. 53, p. 143 - 148 (1981).

Mitteilungen der Universitätssternwarte Graz, Nos. 52 (26.042.012), 55 (26.098.028), 56 (26.098.029), 58 (27.098.002), 59 (27.098.075), 60 (27.098.076), 62 (28.098.001), 63 (28.021.004), 64 (28.107.007), 65 (28.042.023), 66 (28.098.028), 67 (28.098.087), 71 (29.098.074), 72 (29.042.059), 74 (29.098.023), 75 (29.098.092), 76 (30.098.005), 77 (30.098.006).

Mitteilungen des Sonnenobservatoriums Kanzelhöhe, Nos. 28 (26.075.018), 30 (29.080.034).

Austrian Papers on Asteroids (APAs) Nos. 42 (26.098.028), 43 (26.098.029), 44 (27.098.002), 45 (27.098.075), 47 (28.098.001), 48 (28.098.028), 49 (28.098.087), 53 (29.098.074), 54 (29.098.023), 55 (29.098.092), 56 (30.098.005), 57 (30.098.006).

008.032 Green Bank, W. Va.

National Radio Astronomy Observatory, Green Bank, Reprints, Series A, Nos. 1194 (29.112.004), 1195 (28.141.159), 1196 (28.158.227), 1197 (28.158.228), 1198 (28.131.295), 1199 (28.131.298), 1200 (28.131.300), 1201 (28.131.302), 1202 (28.158.324), 1203 (28.141.183), 1204 (29.063.029), 1205 (28.132.001), 1206 (28.132.002), 1207 (29.158.048), 1208 (30.033.019), 1209 (30.033.020), 1210 (28.141.178), 1211 (28.131.332),* 1213 (29.141.059), 1214 (29.141.048), 1215 (29.141.512), 1216 (28.141.179), 1217 (28.158.320), 1218 (28.112.054), 1219 (29.158.057), 1220 (29.141.052), 1221 (29.131.038), 1222 (29.141.028), 1223 (29.132.040), 1224 (29.116.042), 1225 (29.131.046), 1226 (29.116.007), 1227 (29.077.011), 1228 (29.141.064), 1229 (29.125.015), 1230 (29.141.027), 1231 (29.141.070), 1232 (29.155.021), 1233 (29.158.070), 1234 (29.158.073), 1235 (29.131.051), 1236 (29.131.052), 1237 (29.143.010), 1238 (30.033.021),

1239 (29.141.203), 1240 (29.131.251), 1241 (29.141.066),
1242 (29.141.067), 1243 (29.131.055), 1244 (29.131.057),
1245 (29.103.162), 1246 (29.158.086), 1247 (29.158.088),
1248 (29.112.008), 1249 (29.158.094), 1250 (29.141.088),
1251 (29.141.090), 1252 (29.141.181), 1253 (29.141.183),
1254 (29.131.072), 1255 (29.131.074), 1256 (29.160.028),
1257 (29.131.089), 1258 (29.133.006), 1259 (29.131.086),
1260 (29.155.031), 1261 (29.131.107), 1262 (29.133.009),
1263 (29.132.025), 1264 (29.131.115), 1265 (29.116.020),
1266 (29.131.221), 1267 (29.141.189), 1268 (30.133.004),
1269 (29.008.029), 1270 (29.158.230), 1271 (29.131.217),
1272 (29.131.218), 1273 (29.131.219), 1274 (29.131.228),
1275 (29.141.194), 1276 (29.151.086), 1277 (29.158.247),
1278 (29.131.237), 1279 (29.133.022), 1280 (29.142.552),
1281 (29.141.195), 1282 (29.131.240), 1283 (29.131.153),
1284 (29.141.193), 1285 (29.158.278), 1286 (29.131.252),
1287 (29.131.253), 1288 (29.131.254), 1289 (29.077.062),
1290 (29.077.063), 1291 (30.158.003), 1292 (30.158.004),
1293 (30.064.001), 1294 (30.141.002), 1295 (30.141.004),
1296 (30.133.002), 1297 (30.132.001), 1298 (30.141.104),
1299 (29.021.035), 1300 (29.021.036), 1301 (30.160.008),
1302 (30.158.028), 1303 (30.141.013), 1304 (30.141.015),
1305 (30.112.004), 1306 (30.096.008), 1307 (30.033.022),
1308 (30.158.034), 1309 (30.141.033), 1310 (30.141.035),
1311 (30.158.042), 1312 (30.132.010), 1313 (30.141.036),
1314 (30.131.016), 1315 (30.141.043), 1316 (30.141.044),
1317 (30.102.005), 1318 (30.125.009), 1319 (30.131.032),
1320 (30.141.047), 1321 (30.141.048), *1212 (29.161.004).

National Radio Astronomy Observatory, Green
Bank, Reprints, Series B, Nos. 518 (29.131.005), 519
(29.141.122), 520 (29.141.112), 521 (29.154.015), 522
(29.141.137), 523 (29.093.038), 524 (30.141.111).

008.033 Greenwich

Royal Greenwich Observatory Bulletins, No. 185
(30.047.045).

Greenwich Time Report. Royal Greenwich Observatory, Time and Latitude Service, 1980 October - 1981 March
(30.044.020).

008.034 Groningen

Nederlandse Vereniging voor Weer- en Sterrenkunde.
Kapteyn Astronomical Laboratory, Groningen Netherlands.
Observations of Variable Stars. Report. No. 35 (30.123.021).

008.035 Hamburg

Hamburger Sternwarte. – Report for 1980.
A. Weigert.
Mitt. Astron. Ges., Nr. 53, p. 149 - 155 (1981).

Deutsches Hydrographisches Institut. 35. Jahresbericht 1980.   G. Zickwolff.
Deutsches Hydrographisches Institut, Hamburg. 113 pp. (1981)

Deutsches Hydrographisches Institut, Hamburg.
Zeit- und Breitendienst, 1980 April - December, 1981 January -
June (30.044.021).

008.036 Heidelberg

Astronomisches Recheninstitut. – Report for 1980.
W. Fricke.
Mitt. Astron. Ges., Nr. 53, p. 157 - 166 (1981).

Astronomisches Rechen-Institut, Heidelberg,
Mitteilungen Serie B, Nos. 97 (29.111.009), 98 (29.151.033),
99 (29.151.034), 100 (29.111.023), 101 (29.002.013),
102 (29.031.549).

Apparent Places of Fundamental Stars 1983.
(30.047.039).

Astronomische Grundlagen für den Kalender 1983
(30.047.011).

Astronomy and Astrophysics Abstracts, Vol. 29
(30.002.072).

Institut für Theoretische Astrophysik. – Report for
1980.   B. Baschek, G. Traving.
Mitt. Astron. Ges., Nr. 53, p. 167 - 170 (1981).

Landessternwarte. – Report for 1980.
I. Appenzeller.
Mitt. Astron. Ges., Nr. 53, p. 171 - 180 (1981).

Max-Planck-Institut für Astronomie. Deutsch-
Spanisches Astronomisches Zentrum. – Report for 1980.
H. Elsässer.
Mitt. Astron. Ges., Nr. 53, p. 181 - 206 (1981).

008.037 Helsinki

Observatory and Astrophysics Laboratory, University of Helsinki. Report 1/1981 (30.031.641), 2/1981
(30.012.057).

008.038 Honolulu, Hawaii

UKIRT Report, Nos. 1 (30.031.035), 2 (30.032.046),
3 (30.032.047), 4 (30.032.048), 5 (30.032.049), 6 (30.032.
050), 7 (30.032.051), 8 (30.032.052), 9 (30.032.053), 10
(30.031.036), 11 (30.031.037), 12 (30.021.062), 13 (30.031.
038), 14 (30.031.039), 15 (30.032.054), 16 (30.082.108).

008.039 Innsbruck

Institut für Astronomie. – Report for 1980.
J. Pfleiderer.
Mitt. Astron. Ges., Nr. 53, p. 207 - 210 (1981).

008.040 Ioannina

Department of Astronomy, University of Ioannina.
Annual report 1979.   G. Banos.
Annu. Rep. Astron. Inst. Greece 1979, p. 25 - 26 (1980).

Department of Astronomy, University of Ioannina.
Annual report 1980.   G. Banos.
Annu. Rep. Astron. Inst. Greece 1980, p. 28 - 29 (1981).

008.041 Ithaca, N.Y.

National Astronomy and Ionosphere Center, Cornell University, Ithaca, Astronomy Publications, Nos. A81-1 (29.158.074), A81-2 (29.141.568), A81-3 (29.141. 570), A81-4 (29.132.017), A81-5 (29.141.558), A81-6 (29.141.530), A81-7 (29.131.219), A81-8 (29.131.221), A81-9, A81-10, A81-11, A81-12, A81-13, A81-14 (29.141. 506), A81-15 (29.158.051), A81-16 (29.158.048), A81-17 (29.158.087), A81-18 (29.158.248), A81-19 (29.158.230), A81-20 (29.131.217), A81-21 (29.099.083), A81-22 (30.097.076), A81-23, A81-24 (29.135.022).

008.042 Izmir

Publications of the Ege University Observatory, Izmir, Nos. 20 (25.118.001, 26.119.035, 25.119.097, 25.119.107, 25.119.115), 21 (27.118.003, 27.119.025, 28.119.033, 27.119.088, 27.119.090, 27.119.093, 27.117. 057, 28.122.126, 28.119.069, 28.119.071, 28.119.083, 28.119.093).

008.043 Kazan

Trudy Kazanskoj Gorodskoj Astronomicheskoj Observatorii, Izdatel'stvo Kazanskogo Universiteta, Vyp. 45 (30.116.043; 30.065.096; 30.064.095; 30.158.267; 30.066. 190; 30.094.039; 30.046.006; 30.004.052), 46 (30.065.097; 30.064.096; 30.062.120; 30.012.054; 30.022.161; 30.094. 040; 30.094.041).

008.044 Kiel

Institut für Theoretische Physik und Sternwarte der Universität. – Report for 1980.   V. Weidemann. Mitt. Astron. Ges., Nr. 53, p. 211 - 216 (1981).

Sonderdrucke der Sternwarte Kiel, Nos. 271 (30.117.036), 272 (30.126.013), 273 (30.064.029), 274 (30.112.019), 275 (30.126.011), 276 (30.126.012), 277 (30.126.014), 278 (30.126.015), 287 (30.114.076), 290 (30.126.017), 297 (30.122.053).

Institut für Reine und Angewandte Kernphysik, Arbeitsgruppe Mathematische Physik. – Report for 1980. K. O. Thielheim. Mitt. Astron. Ges., Nr. 53, p. 217 - 219 (1981).

008.045 Kodaikanal

Indian Institute of Astrophysics. Report for the period 1979 April 1 - 1980 March 31. M. K. V. Bappu. Bull. Astron. Soc. India, Vol. 9, 151 - 163 (1981).

Raman Research Institute. Report for the period 1979 April 1 - 1980 March 31.   V. Radhakrishnan. Bull. Astron. Soc. India, Vol. 9, 164 - 167 (1981).

008.046 La Serena

Cerro Tololo Inter-American Observatory, La Serena, Chile. – Report for the period 1 October 1979 - 30 September 1980.   V. M. Blanco. Bull. American Astron. Soc., Vol. 13, 575 - 587 (1981).

008.047 La Silla

Sonia (1967 - 1981). Messenger, No. 26, p. 31 (1981). During a thundery evening, on May 13, 1967, a little female with fiery eyes and an independent character was born on La Silla. She was called Sonia. It was said that she was very pretty. Coquettish until the end, she enjoyed her young life as a perfect courtesan. Later, perhaps as a result of the influence of her protector, H. E. Schuster, she became a paragon of virtue. Intelligent and discrete, she devoted her life to the Observatory where, on September 28, 1981, she passed away quietly.

ESO Scientific Preprint, Nos. 159 (30.151.086), 160 (30.158.276), 161 (30.002.075), 162 (30.159.022), 163 (30.151.087), 164 (30.141.199), 165 (30.141.200), 166 (30.132.057), 167 (30.125.064), 168 (30.135.040), 169 (30.158.277), 170 (30.154.045), 171 (30.121.035), 172 (30.141.201), 173 (30.125.065), 174 (30.162.161), 175 (30.132.058), 176 (30.132.059), 177 (30.119.111), 178 (30.151.088), 179 (30.158.278), 180 (30.132.060).

ESO user's manual.   A. C. Danks (Editor). Published by European Southern Observatory, Karl-Schwarzschild-Str. 2, D-8046 Garching bei München (1981). Contents: I. General information. II. Telescopes. III. Instrumentation. IV. Photomaterials/detectors/calibration techniques. V. Data analysis. VI. Auxiliaries.

The Messenger – El Mensajero, Nos. 25, 26 (1981).

008.048 Lake Tekapo, New Zealand

Mt. John University Observatory. Report for 1980. J. B. Hearnshaw. South. Stars, Vol. 29, 21 - 23 (1981).

008.049 Lawrence, Kans.

University of Kansas, Lawrence, Kansas 66045. Report. Bull. American Astron. Soc., Vol. 13, 593 - 594 (1981).

008.050 Leningrad

Trudy Astronomicheskoj Observatorii, (Transactions of the Astronomical Observatory), Leningrad, Tom 36 (30.063.002; 30.061.003; 30.021.002; 30.122.012; 30.153.005; 30.160.014; 30.155.003; 30.042.003; 30.021.003; 30.041.001; 30.031.512; 30.071.004; 30.052.001) Tom 37 (30.009.006 - 30.009.016; 30.116.005).

008.051 Liège

Institut d'Astrophysique, Université de Liège, Cointe-Ougrée (Belgique), Collection in 4°, Nos. 319 (20.071.024), 320 (21.162.178), 321 (22.065.021), 322 (22.122.099), 323 (25.162.007), 324 (25.151.070), 325 (27.065.017), 326 (27.162.017), 327 (26.065.029), 328 (26.113.014), 329 (27.065.053), 330 (27.114.141), 331 (28.066.070), 332 (30.162.160), 333 (28.065.071), 334 (28.123.021).

Institut d'Astrophysique, Université de Liège, Cointe-Ougrée (Belgique), Collection in 8°, Nos. 675 (21.042.035), 676 (22.032.579), 677 (25.065.060), 678 (30.102.078), 679 (28.162.009), 680 (30.065.099), 681 (30.162.159), 682 (29.080.015).

008.052 Lintong

Publications of the Shaanxi Astronomical Observatory, 1981, No. 1 (30.035.008; 30.033.023; 30.035.009; 30.045.016; 30.052.055; 30.052.056; 30.035.010).

Time and Frequency Services Bulletin, April - December 1981 (30.044.023).

008.053 L'vov

L'vovskij Ordena Lenina Gosudarstvennyj Universitet imeni Ivana Franko, Astronomicheskaya Observatoriya, Tsirkulyar, No. 54 (30.135.063; 30.135.064; 30.158.268; 30.141.561; 30.114.194; 30.071.036; 30.022.168; 30.022.169; 30.035.006; 30.113.076).

008.054 Madrid

Boletín Astronómico del Observatorio de Madrid, Vol. 10, No. 4 (30.072.069).

Universidad Complutense – Facultad de Ciencias, Madrid. Seminario de Astronomia y Geodesia, Publicación, Nos. 101, 102 (30.046.007), 103 (30.045.017), 104 (30.045.018), 105, 106 (30.021.064), 107 (30.119.112).

008.055 Manchester

University of Manchester, Nuffield Radio Astronomy Laboratories, Jodrell Bank. Report on the researches in radio astronomy at Jodrell Bank for the year ending 1980 September 30.   B. Lovell.
Q. J. R. Astron. Soc., Vol. 22, 303 - 318 (1981).

Astronomical Contributions from the University of Manchester, Series II, Jodrell Bank Reprints, Nos. 626 (28.141.004), 627 (28.008.058), 628 (28.156.010), 629 (29.131.010), 630 (28.132.028), 632 (29.141.007), 633 (29.132.008), 634 (28.158.022), 635 (28.033.010), 636 (28.141.096), 637 (29.158.011), 638 (29.141.035), 639 (29.160.004), 640 (28.141.097), 641 (28.131.141), 642 (29.141.528), 644 (29.141.027), 647 (29.112.003), 649 (30.008.055), 650 (29.141.029).

008.056 Manila

Manila Observatory, Solar Division. Solar maps and activity, 1981 March - 1981 August (30.072.062).

008.057 Mizusawa

Annual Report of the International Polar Motion Service 1979 (30.045.011).

Monthly Notes of the International Polar Motion Service, 1981, Nos. 5 - 11 (30.045.021).

Proceedings of the International Latitude Observatory of Mizusawa, No. 20 (30.082.112; 30.046.008; 30.033.027; 30.031.040; 30.035.011; 30.045.019; 30.045.020; 30.082.113).

Publications of the International Latitude Observatory of Mizusawa, Vol. 14, No. 1 (30.045.022; 30.045.023), 2 (30.044.032).

008.058 Mons

Communications du Département d'Astrophysique de la Faculté des Sciences de Mons. Mons Astrophysical Papers. Nos. 83 (30.034.011), 84 (30.031.524), 85 (30.031.525), 86 (30.031.526), 87, 88 (29.022.100), 89 (29.076.048).

008.059 Münster

Astronomisches Institut der Universität. Außenstation Schalkenmehren. – Report for 1980.
D. Stöffler, W. Seitter.
Mitt. Astron. Ges., Nr. 53, p. 261 - 268 (1981).

008.060 Munich

Institut für Astronomie und Astrophysik der Universität München. Universitäts-Sternwarte. – Report for 1980.   P. Wellmann.
Mitt. Astron. Ges., Nr. 53, p. 253 - 260 (1981).

Institut für Astronomische und Physikalische Geodäsie der Technischen Universität München, aus den Mitteilungen, Nos. 153, 154 (30.046.010), 155 (30.094.046).

Max-Planck-Institut für Physik und Astrophysik. Institut für Astrophysik und Institut für extraterrestrische Physik. – Reports for 1980.   R. Kippenhahn, H. Fink.
Mitt. Astron. Ges., Nr. 53, p. 221 - 252 (1981).

Max-Planck-Institut für Physik und Astrophysik, Institut für Extraterrestrische Physik, Garching bei München, MPI-PAE/Extraterr. 167 (30.131.210).

008.061 Northridge, Calif.

San Fernando Observatory, Department of Physics and Astronomy, California State University, Northridge, California 91130.-Report.
Bull. American Astron. Soc., Vol. 13, 683 - 684 (1981).

008.062   Oxford

Department of Astrophysics, University Observatory, Oxford, Publications Nos. 295 (30.155.019), 296 (30.114.077).

008.063   Paris

Bureau International de l'Heure, (B.I.H.), Circular A (30.044.025), D177 - D181 (30.044.026).

008.064   Patras

Department of Astronomy, University of Patras. Annual report 1979.   G. Antonacopoulos. Annu. Rep. Astron. Inst. Greece 1979, p. 27 - 28 (1980).

Department of Astronomy, University of Patras. Annual report 1980.   G. Antonacopoulos. Annu. Rep. Astron. Inst. Greece 1980, p. 30 - 31 (1981).

008.065   Potsdam

Zentralinstitut für Astrophysik, Sternwarte Babelsberg, Mitteilungen. Neue Folge, Nos. 234 (25.160.025), 235 (25.022.042), 238 (25.141.150), 239 (25.160.050), 240 (25.162.078), 245 (26.162.031), 246 (26.141.061), 247 (26.066.116), 248 (26.160.057), 249 (26.113.042), 251 (27.066.049), 252 (27.066.050), 253 (27.160.031), 254 (27.158.084), 255 (28.160.004), 256 (28.031.510), 257 (28.160.005), 258 (28.158.014), 259 (28.158.097), 260, 261 (28.066.056), 262 (28.066.150).

Mitteilungen des Astrophysikalischen Observatoriums Potsdam, Nos. 229 (21.114.553), 230 (21.063.019), 231 (21.063.020), 232 (21.022.076), 233 (28.062.100), 234 (25.004.021), 235 (22.114.570), 236 (25.116.006), 237 (26.116.017), 238 (26.061.030), 239 (26.064.049), 240 (27.113.027), 241 (27.062.129), 242 (28.062.006), 243 (28.122.060), 244 (28.116.006).

Mitteilungen des Zentralinstituts für Physik der Erde, Potsdam, No. 825 (30.046.009).

Zeit- und Breitenbestimmungen, Zeitsysteme, Präzisionszeitvergleiche, Jahrg. 1980, Nos. 5 - 6; Jahrg. 1981, Nos. 1 - 2 (30.044.027).

008.066   Prague

Czechoslovak Academy of Sciences, Astronomical Institute, Circular of the Czechoslovak Observatories, Time and Latitude, 1980 October - December, 1981 January - March (30.044.024).

008.067   Pulkovo

Solar research at the Pulkovo Astronomical Observatory.   V. A. Krat. Sol. Phys., Vol. 73, 405 - 409 (1981).
A brief survey of the progress in the solar physics research at the Pulkovo Observatory during the recent years is given.

The Main Astronomical Observatory of the Academy of Sciences of the USSR at Pulkovo.   See Abstr. 009.045.

008.068   San Fernando

Memoria de las actividades en 1980. Inst. Obs. Marina, San Fernando (Cádiz). 21 pp. (1981).

Instituto y Observatorio de Marina, San Fernando (Cádiz), España, Boletin Astronomico, No. 5 (30.044.036).

Instituto y Observatorio de Marina, San Fernando (Cádiz), España, Serie B, No. 5 (30.098.101).

Instituto y Observatorio de Marina, San Fernando (Cádiz), España, Serie C, Nos. 82 (30.044.034), 83 (30.044. 035).

008.069   Santa Cruz

Lick Observatory Bulletin, Nos. 824 (29.141.023), 825 (29.143.006), 844 (29.153.007), 848 (29.066.022), 856 (28.131.290), 857 (27.158.233), 862 (28.158.099), 863 (28.158.079), 865 (28.158.078), 867 (28.122.001), 868 (28.125.002), 869 (28.134.020), 870 (28.034.069), 872 (29.158.056), 873 (28.121.014), 874 (29.158.068), 875 (28.121.012), 878 (29.096.005), 879 (29.096.004), 880 (29.111.017).

Lick Observatory Contributions No. 423 (28.141. 089).

008.070   Sendai

Annual report of research activities April 1980 - March 1981. Astronomy group. Sci. Rep. Tôhoku Univ., Ser. 8, Vol. 2, 100 - 105 (1981).

Sendai Astronomiaj Raportoj, Nos. 219 (29.065.015) 220 (29.062.019), 222 (30.122.204), 223 (30.062.125), 224 (30.021.065), 226 (28.131.275), 227 (28.064.089).

008.071   Shanghai

Annals of Shanghai Observatory Academia Sinica Nos. 1 (30.045.026, 30.081.055, 30.045.027, 30.044.037, 30.031.644, 30.045.028, 30.082.116, 30.141.206 - 30.141. 208, 30.111.018, 30.118.035, 30.032.058, 30.032.059, 30.031.645, 30.021.069, 30.082.117, 30.033.029, 30.031. 042, 30.031.043, 30.032.057, 30.002.080), 2 (30.002.081 - 30.002.083, 30.045.029, 30.045.030, 30.044.039, 30.045.031, 30.099.212, 30.046.012, 30.052.061, 30.021.070, 30.098.102, 30.021.071, 30.031.646, 30.131.216, 30.151.093, 30.111.019, 30.033.030, 30.032.060, 30.021.072, 30.032.061 - 30.032.063, 30.035.012 - 30.035.016, 30.004.060, 30.004.061).

008.072   Shemakha

Shemakhinskaya Astrofizicheskaya Observatoriya, Tsirkulyar, No. 68 (30.073.097; 30.077.059; 30.033.009; 30.077.060; 30.031.600; 30.072.047; 30.004.042).

008.073 Skalnaté Pleso

Contributions of the Astronomical Observatory Skalnaté Pleso, Vol. 10 (30.074.071, 30.104.013, 30.119.036, 30.119.037, 30.104.014).

008.074 Sonneberg

Zentralinstitut für Astrophysik, Sonneberg, Mitteilungen über Veränderliche Sterne, Band 9, Heft 1 (30.113. 082, 30.122.206 - 30.122.209, 30.123.022, 30.123.023, 30. 122.210, 30.124.222, 30.123.024, 30.119.114, 30.119.115, 30.123.025, 30.123.026, 30.123.027), 2 (30.122.211, 30.121.039, 30.119.116, 30.122.212, 30.122.213, 30.123. 028 - 30.123.032).

Akademie der Wissenschaften der DDR, Zentralinstitut für Astrophysik. Veröffentlichungen der Sternwarte in Sonneberg, Band 9, Heft 4 (30.121.038).

008.075 Strasbourg

Bulletin d'Information du Centre de Données Stellaires. No. 21 (1981).

008.076 Sydney

Sydney Observatory. Report for the year ending 1980 December 31.   W. H. Robertson. Q. J. R. Astron. Soc., Vol. 22, 319 (1981).

008.077 Tartu

W. Struve nimeline Tartu Astrofüüsika Observatoorium, Teated, Nos. 63 (30.162.090), 64 (30.114.107; 30.064.060; 30.021.039), 65 (30.063.026; 30.131.135; 30.031.573).

008.078 Thessaloniki

Astronomy Department, University of Thessaloniki. Annual report 1979.   B. Barbanis. Annu. Rep. Astron. Inst. Greece, 1979, p. 11 - 15 (1980).

Astronomy Department, University of Thessaloniki. Annual report 1980.   B. Barbanis. Annu. Rep. Astron. Inst. Greece 1980, p. 12 - 16 (1981).

Department of Geodetic Astronomy, University of Thessaloniki. – Annual report 1980.   L. N. Mavridis. Annu. Rep. Astron. Inst. Greece, 1980, 22 - 27 (1981).

Contributions from the Department of Geodetic Astronomy, University of Thessaloniki, Nos. 25 (25.113.086), 26 (27.122.200), 27 (27.122.201), 28 (27.122.202), 29 (27.122.203), 30 (27.122.204), 31 (27.122.205), 32 (27.122. 206), 33 (28.124.801), 34 (28.122.166), 35 (29.122.118), 36 (29.122.119).

008.079 Tokyo

Annals of the Tokyo Astronomical Observatory, University of Tokyo, Second Series, Vol. 18, No. 3 (30.114. 201; 30.122.205; 30.151.092; 30.002.079).

Tokyo Astronomical Bulletin, Second Series, No. 265 (30.033.028).

University of Tokyo, Tokyo Astronomical Observatory, Report (No. 74), Vol. 19, No. 3 (30.082.114, 30.034. 104, 30.034.105, 30.034.106, 30.021.066, 30.021.067, 30.034.107, 30.042.085, 30.021.068, 30.036.020, 30.034. 108, 30.082.115).

Tokyo Astronomical Observatory, Reprints Nos. 595 (29.114.099), 596 (26.077.055), 597 (29.122.172), 598 (29.158.131), 599 (29.114.148), 600 (30.151.023), 601 (30.134.016), 602 (30.131.078), 603 (30.034.057), 604 (29.114.168), 605 (30.117.100), 606 (30.071.038).

Tokyo Astronomical Observatory, Time and Latitude Bulletins, Vol. 55, Nos. 1 - 3 (30.044.033).

Tokyo Astronomical Observatory, Kiso Information Bulletin, Vol. 1, No. 5 (1981).

Quarterly Bulletin on Solar Activity, Vol. 21, Part I (30.072.071); Part II (30.075.025); Part III (30.073. 107); Part IV (30.074.116).

Contributions from the Department of Astronomy, University of Tokyo, Nos. 280 (27.151.079), 281 (27.064. 065), 282 (27.119.069), 283 (28.158.030), 284 (28.065.012), 285 (28.065.013), 286 (27.077.058), 287 (28.065.048), 288 (28.119.043), 289 (28.120.010), 290 (29.155.015).

008.080 Tortosa

Publicaciones del Observatorio del Ebro, Miscelánea, No. 35 (26.083.051).

008.081 Torun

Biuletyn Obserwatorium Astronomicznego Uniwersytetu M. Kopernika w Toruniu, Nr. 62 (28.112.003; 28.122.015; 28.114.125; 28.155.040; 28.116.022; 28.131. 279; 28.103.302), 63 (30.131.214; 30.131.215; 30.098.100).

008.082 Trieste

Publications of the Astronomical Observatory of Trieste, Nos. 676 (30.114.066), 677 (30.031.555), 678 (30.112.017), 679 (30.114.071), 680 (30.114.073), 681 (30.118.012), 682 (30.119.023), 714 (28.077.035), 715 (29.119.025), 716 (30.119.012), 717 (29.119.018), 718 (29.160.024), 719 (29.119.034), 720 (29.119.021), 721 (29.119.005), 722 (29.119.006), 723 (29.119.044), 724 (30.072.063), 725 (29.119.043), 727 (29.119.045), 728 (30.119.046), 729 (29.119.118), 730 (29.119.117), 731 (30.117.062), 734 (29.031.623), 735 (30.119.001), 736 (29.119.004), 737 (29.112.013), 738 (30.064.046), 739 (30.119.005), 740 (29.033.044), 741 (30.119.025), 742 (30.065.022), 743 (30.117.044), 744 (30.124.202), 748 (30.115.013), 749 (30.072.063), 750 (30.114.109), 752 (30.122.083), 753 (30.114.124), 754 (30.114.113),

755 (30.114.018), 756 (30.113.034), 757 (30.116.026), 758 (30.113.039), 759 (30.114.152), 760 (29.021.046), 761 (29.114.152), 762 (29.031.021), 763 (29.031.643), 764 (30.072.063), 765 (30.114.005), 766 (30.114.129), 768 (29.031.021), 771 (30.031.595), 772 (30.064.017), 773 (30.114.033), 774 (30.114.034), 775 (30.141.516), 776 (30.124.301).

008.083   Tucson, Ariz.

Kitt Peak National Observatory, Tucson, Arizona 85726. – Report for the period 1 October 1979 - 30 September 1980.   G. Burbidge.
Bull. American Astron. Soc., Vol. 13, 595 - 668 (1981).

Preprints of the Steward Observatory, University of Arizona, Tucson, Nos. 327 (30.034.103), 328 (30.117.137), 329 (30.158.281), 330 (30.160.056), 331 (30.131.212), 332 (30.154.046), 333 (30.101.035), 334 (30.141.203), 335 (30.158.282), 336 (30.119.113), 337 (30.141.204), 338 (30.158.283), 339 (30.121.036), 340 (30.131.213), 341 (30.158.284), 342 (30.031.643), 343 (30.158.285), 344 (30.002.078), 345 (30.158.286), 346 (30.031.041), 347 (30.032.055), 348 (30.158.287), 349 (30.113.081), 350 (30.117.138), 351 (30.160.057), 352 (30.151.091), 353 (30.158.288), 354 (30.121.037), 355 (30.101.036), 356 (30.032.056).

The Steward Observatory speckle interferometry program.   See Abstr. 031.529.

008.084   Tübingen

Astronomisches Institut der Universität und Lehrstuhl für Theoretische Astrophysik. – Report for 1980.
G. Elwert, M. Grewing.
Mitt. Astron. Ges., Nr. 53, p. 269 - 278 (1981).

008.085   Turku

Turku University Observatory Informo, Nos. 48 (30.158.289, 49 (30.102.079), 50 (30.159.023), 51 (30.042.084), 52 (30.141.205), 53 (30.142.153), 54 (29.082.007), 55 (29.141.131), 56 (29.155.039), 57 (30.045.024), 58 (30.142.154).

Report Series, Department of Physical Sciences, Institute of Astronomy, University of Turku, Nos. R19 (30.158.289), R20 (30.102.079), R21 (30.159.023).

008.086   Uccle

Observatoire Royal de Belgique, Communications. Koninklijke Sterrenwacht van België, Mededelingen, Série A, Nos. 61, 62 (30.032.577), 63 (30.081.054), 64 (30.099.037), 65 (30.098.002).

Observatoire Royal de Belgique, Communications, Koninklijke Sterrenwacht van België, Mededelingen Série B, Nos. 119 (30.118.014), 120 (29.072.090), 121 (29.077.053), 122 (29.033.045).

008.087   Utrecht

Utrechtse Sterrekundige Overdrukken, No. 541 (30.122.051).

008.088   Victoria

Dominion Astrophysical Observatory, Victoria, British Columbia. Herzberg Institute of Astrophysics National Research Council, Canada. Report for the year 1980 April 1 to 1981 March 31.   S. van den Bergh.
Q. J. R. Astron. Soc., Vol. 22, 436 - 451 (1981).

Publications of the Dominion Astrophysical Observatory, Victoria, B.C. Vol. 15, No. 14 (30.002.077).

008.089   Vienna

Institut für Astronomie der Universität Wien.
Report for 1980.   K. D. Rakos.
Mitt. Astron. Ges., Nr. 53, p. 279 - 293 (1981).

008.090   Villa Elisa, Provincia de Buenos Aires

Contribuciones del Instituto Argentino de Radioastronomia, Nos. 95 (27.131.167), 96 (29.131.168), 99 (28.131.293), 100 (29.131.062), 101 (29.131.064), 102 (30.116.002).

008.091   Vilnius

Vilniaus Astronomijos Observatorijos Biuletenis (Bulletin of the Vilnius Astronomical Observatory), Nr. 56 (30.113.054; 30.113.055; 30.114.175; 30.103.881), Nr. 57 (30.153.029; 30.113.056; 30.114.176; 30.152.007).

008.092   Warsaw

Warsaw University Observatory and Polish Academy of Sciences, N. Copernicus Astronomical Center, Reprint Nos. 437 (29.117.126), 438 (29.158.225), 439 (29.062.092), 440 (29.064.082), 441 (29.119.114), 442 (29.066.154), 443 (29.117.127), 444 (29.119.113).

Politechnika Warszawska, Obserwatorium Astronomiczno-Geodezyjne w Józefosławiu, (Warsaw Technical University, Astronomic-Geodetical Observatory at Józefosław), **Latitude Circular,** Nos. 78, 79 (30.045.025).

008.093   Washington

United States Naval Observatory, Washington, D.C., Circular, Nos. 161 (30.099.211), 162 (30.101.037).

,   **U. S. Naval Observatory, Washington, D.C. Time Service Publications.** Series 4, Nos. 753 - 778; Series 6, Nos. 65 - 70; Series 7, Nos. 706 - 731; Series 14, No. 30 (30.044.028 - 30.044.031).

008.094 **Wellington**

Annual report of the Carter Observatory Board for the year ended 1981 March 31.   J. B. Mackie. Carter Obs., Astron. Bull., No. 96, 8 pp. (1981).

Carter Observatory, Astronomical Bulletin, Nos. 94, (30.047.046), 96 (30.008.094).

008.095 **Wroclaw**

Wrocław Astronomical Observatory, Reprint, No. 111 (29.122.152).

008.096 **Würzburg**

Institut für Astronomie und Astrophysik, Lehrstuhl Astronomie. – Report for 1980.   F. - L. Deubner. Mitt. Astron. Ges., Nr. 53, p. 295 - 299 (1981).

008.097 **Zürich**

Institut für Astronomie. – Report for 1980. J. O. Stenflo. Mitt. Astron. Ges., Nr. 53, p. 301 - 307 (1981).

009 Notes on Observatories, Planetaria, Exhibitions

009.001 L'attività dell'Osservatorio Astronomico del Sud Africa.   M. Feast. Orione, Vol. 2, 214 - 223 (1981).

009.002 L'observatoire du Pic-du-Midi et les astronomes amateurs.   C. Boyer. Astronomie, Vol. 95, 333 - 341 (1981).

009.003 L'Osservatorio di Arcetri.   G. Tagliaferri. Coelum, Vol. 50, 145 - 155 (1981).

009.004 Einweihung der ESO-Zentrale in Garching. J. Krautter. Sterne Weltraum, Jahrg. 20, 264 (1981).

009.005 Das ZKP2 aus Jena. Eine Nova am Himmel der Planetariumsprojektoren. Orion, 39. Jahrg., 114 - 115 (1981).

009.006 Astronomy at the St. Petersburg – Leningrad University in 1881 - 1930. V. V. Ivanov, T. M. Maksimova. Tr. Astron. Obs., Leningrad, Tom 37 = Uch. Zap. Leningr. Univ., No. 406 = Ser. mat. nauk, Vyp. 59, 7 - 26 (1981). In Russian.

009.007 Astronomy at the Leningrad University since 1931 up to the present time.   V. V. Sobolev. Tr. Astron. Obs., Leningrad, Tom 37 = Uch. Zap. Leningr. Univ., No. 406 = Ser. mat. nauk, Vyp. 59, 26 - 42 (1981). In Russian.

009.008 My work at the Astronomical Observatory at the Leningrad University.   V. A. Ambartsumyan. Tr. Astron. Obs., Leningrad, Tom 37 = Uch. Zap. Leningr. Univ., No. 406 = Ser. mat. nauk, Vyp. 59, 42 - 49 (1981). In Russian.

009.009 My student years (memoirs).   A. N. Deutsch. Tr. Astron. Obs. Leningrad, Tom 37 = Uch. Zap. Leningr. Univ., No. 406 = Ser. mat. nauk, Vyp. 59, 50 - 53 (1981). In Russian.

009.010 The development of science at the Astronomical Observatory of the Leningrad University. Theoretical astrophysics.   V. G. Gorbatskij. Tr. Astron. Obs., Leningrad, Tom 37 = Uch. Zap. Leningr.

Univ., No. 406 = Ser. mat. nauk, Vyp. 59, 54 - 70 (1981). In Russian.

009.011 The development of science at the Astronomical Observatory of the Leningrad University. Theory of radiative transfer.   V. V. Ivanov. Tr. Astron. Obs., Leningrad, Tom 37 = Uch. Zap. Leningr. Univ., No. 406 = Ser. mat. nauk, Vyp. 59, 70 - 90 (1981). In Russian.

009.012 The development of science at the Astronomical Observatory of the Leningrad University. Observational astrophysics.   V. A. Hagen-Thorn. Tr. Astron. Obs., Leningrad, Tom 37 = Uch. Zap. Leningr. Univ., No. 406 = Ser. mat. nauk, Vyp. 59, 90 - 106 (1981). In Russian.

009.013 The development of science at the Astronomical Observatory of the Leningrad University. Radioastronomy.   V. G. Nagnibeda. Tr. Astron. Obs., Leningrad, Tom 37 = Uch. Zap. Leningr. Univ., No. 406 = Ser. mat. nauk, Vyp. 59, 106 - 112 (1981). In Russian.

009.014 The development of science at the Astronomical Observatory of the Leningrad University. Stellar astronomy.   T. A. Agekyan. Tr. Astron. Obs., Leningrad, Tom 37 = Uch. Zap. Leningr. Univ., No. 406 = Ser. mat. nauk, Vyp. 59, 113 - 144 (1981). In Russian.

009.015 The development of science at the Astronomical Observatory of the Leningrad University. Celestial mechanics, geodesy and gravimetry. R. A. Lyakh, K. V. Kholshevnikov. Tr. Astron. Obs., Leningrad, Tom 37 = Uch. Zap. Leningr. Univ., No. 406 = Ser. mat. nauk, Vyp. 59, 145 - 160 (1981). In Russian.

009.016 The development of science at the Astronomical Observatory of the Leningrad University. Astrometry.   M. S. Zverev, M. P. Mishchenko. Tr. Astron. Obs., Leningrad, Tom 37 = Uch. Zap. Leningr. Univ., No. 406 = Ser. mat. nauk, Vyp. 59, 160 - 180 (1981). In Russian.

009.017 **The Canary Islands – an astronomer's experiment.**
A. W. Jones.
Sky Telesc., Vol. 62, 199 - 201 (1981).

009.018 **The new observatory on La Palma.**
F. G. Smith.
J. British Astron. Assoc., Vol. 91, 449 - 452 (1981).

009.019 **The new observatory on La Palma.** F. G. Smith.
Q. J. R. Astron. Soc., Vol. 22, 254 - 265 (1981).

009.020 **Stabilité ou dégradation du ciel nocturne à l'Observatoire de Nice?**
M. Fulconis, M. Fulconis.
Astronomie, Vol. 95, 371 - 379 (1981).

009.021 **Crisis at Kitt Peak.** L. J. Robinson.
Sky Telesc., Vol. 62, 413 (1981).

009.022 **Observing at Kitt Peak National Observatory.**
M. Cohen.
Mercury, Vol. 10, 98 - 106 (1981).

009.023 **Le Centre de Données Stellaires de Strasbourg.**
F. Ochsenbein.
Astronomie, Vol. 95, 435 - 442 (1981).

009.024 **W. Struve Astrophysical Observatory.**
J. I. Straume.
Zvaigźnotā debess, 1980. gada vasara, p. 24 - 29. In Latvian.

009.025 **Black Birch Observatory, Awatere Valley, Blenheim.**
F. P. Andrews.
South. Stars, Vol. 29, 38 - 40 (1981).

009.026 **Hundred years of the Astronomical Observatory of the Leningrad State University.** V. V. Sobolev.
Vestn. LGU, 1981, No. 7, p. 5 - 9. In Russian. – Abstr. in Ref. zh., 51. Astron., 10.51.15 (1981).

009.027 **Anniversary of the Moscow astronomical school.**
E. P. Aksenov.
Zemlya Vselennaya, 1981, No. 6, p. 43 - 49. In Russian.

009.028 **L'observatoire de la Société Astronomique de France.** B. Clouet, M. Dumont.
Astronomie, Vol. 95, 475 - 484 (1981).

009.029 **La Société Astronomique de France visite l'Observatoire de Paris.** A. Dollfus.
Astronomie, Vol. 95, 485 - 487 (1981).

009.030 **L'observatoire du Dr. Vehrenberg.**
E. Schweitzer.
Astronomie, Vol. 95, 509 - 512 (1981).

009.031 **Rhodes University: Department of Physics and Electronics.** E. E. Baart.
Mon. Notes Astron. Soc. South. Africa, Vol. 40, 6 - 7 (1981).

009.032 **University of South Africa: Department of Mathematics, Applied Mathematics and Astronomy.**
P. D. Bennewith, J. Wolterbeek.
Mon. Notes Astron. Soc. South. Africa, Vol. 40, 7 - 8 (1981).

009.033 **L'observatoire populaire MIRA.** T. Pieraerts.
Ciel Terre, Vol. 97, 375, 378 - 382 (1981).

009.034 **Showcase for Scotland's Royal Observatory.**
F. Watson.
Sky Telesc., Vol. 62, 535 - 537 (1981).

009.035 **The largest Schmidt's first 20 years.**
Sky Telesc., Vol. 62, 554 - 557 (1981).

009.036 **110 years of the Hurbanovo Observatory (Ó Gyalla).** L. Druga.
Kozmos, Vol. 12, 138 - 141 (1981). In Slovak.

009.037 **The Zelenchuk Observatory.** J. Zverko.
Kozmos, Vol. 12, 112 (1981). In Slovak.

009.038 **Astronomische Observatorien.**
H. G. Beck, H. Groth.
Jenaer Rundsch. (Jena Rev.), 26. Jahrg., 234 - 235, 238 (1981).

009.039 **Planetarien – schlüsselfertige Anlagen aus Jena.**
G. Schille.
Jenaer Rundsch. (Jena Rev.), 26. Jahrg., 236 - 238 (1981).

009.040 **L'Osservatorio Astronomico di Palermo.**
G. Foderà Serio.
Coelum, Vol. 50, 250 - 257 (1981).

009.041 **Mauna Kea (I): Halfway to space. (II): Coming of age.** M. M. Waldrop.
Science, Vol. 214, 1010 - 1013, 1110 - 1114 (1981).

009.042 **Observing facilities at the European Southern Observatory (ESO) in Chile for cometary observations.**
G. F. O. Schnur, L. Kohoutek, J. Rahe.
Modern observational techniques for comets, (see 012.047), p. 115 - 128 (1981).

009.043 **Calar Alto – bisherige Erfahrungen, künftige Entwicklung.** H. Elsässer.
Mitt. Astron. Ges., Nr. 54, (see 012.050), p. 15 - 19 (1981).

009.044 **Centenary of the Astronomical Observatory of the Leningrad University.** J. I. Straume.
Zvaigźnotā debess, 1981. gada pavasaris, p. 24 - 33. In Latvian.

009.045 **The Main Astronomical Observatory of the Academy of Sciences of the USSR at Pulkovo.**
Yu. I. Vitinskij; edited by A. A. Mikhajlov, translated by I. N. Voronina.
Academy of Sciences of the USSR, Main Astronomical Observatory. Leningrad Section, Nauka, Leningrad. 40 pp. Price 15 Kop. (1981).
    The booklet deals with the history, main problems and results of investigations of the Pulkovo Observatory (the Main Astronomical Observatory of the USSR Academy of Sciences). It also gives a description of the observatory and can be used as a guide-book.

009.046 **Astrophysikalisches Observatorium Arosa "Tschuggen".** P. Altermatt.
Orion, 39. Jahrg., 180 - 182 (1981).

Eisinga's planetarium. See Abstr. 004.002.

The Astronomical Institute of the Czechoslovak Academy of Sciences. See Abstr. 013.024.

Det danske 1,5 m teleskop på La Silla. See Abstr. 032.016.

Drei Jahre IUE-Observatorium. See Abstr. 032.549.

010 Societies, Associations, Organizations

010.001   American Association of Variable Star Observers
(AAVSO)

**Meetings and activities of the Society, committee
reports.**
J. American Assoc. Variable Star Obs., Vol. 9, 94 - 102, 103 -
112 (1980); Vol. 10, 39 - 46 (1981).

**Annual report of the director for fiscal year 1979 -
1980.**   J. A. Mattei.
J. American Assoc. Variable Star Obs., Vol. 9, 103 - 112 (1980).

**Abstracts of papers presented at the spring meeting
of the AAVSO in Tucson, Arizona, April 25, 1981.**
J. American Assoc. Variable Star Obs., Vol. 10, 33 - 37
(1981).

**The Journal of the American Association of Variable
Star Observers**, Vol. 9, No. 2 (1980); Vol. 10, No. 1 (1981).

010.002   American Astronautical Society

**Twenty-five years of the American Astronautical
Society 1954 - 1979.**   See Abstr. 003.059.

010.003   American Astronomical Society (AAS)

**The 158th meeting of the American Astronomical
Society held 28 June - 1 July 1981 at Calgary, Alberta.**
Abstracts of papers presented.
Bull. American Astron. Soc., Vol. 13, 493 - 565 (1981).

**The 12th regular meeting of the Dynamical Astron-
omy Division held 23 - 25 March 1981 at Tuscaloosa, Alabama.**
Abstracts of papers presented.
Bull. American Astron. Soc., Vol. 13, 566 - 574 (1981).

**The 13th annual meeting of the Division for
Planetary Sciences of the American Astronomical Society,
held 13 - 16 October 1981 at Pittsburgh, Pennsylvania.**
Abstracts of papers presented.
Bull. American Astron. Soc., Vol. 13, 687 - 750 (1981).

AAS Photo-Bulletin, Issue 25, No. 3 (1980).

Bulletin of the American Astronomical Society,
Vol. 13, Nos. 2, 3 (1981).

010.004   Association Française des Observateurs d'Etoiles
Variables (A.F.O.E.V.)

La vie de l'Association.   E. Schweitzer.
Bull. AFOEV, No. 17, p. 116; No. 18, p. 12 (1981).

Activité de l'A.F.O.E.V.   E. Schweitzer.
Astronomie, Vol. 95, 351 - 353, 554 - 556 (1981).

**Bulletin de l'Association Française des Observateurs
d'Etoiles Variables**, Nos. 17, 18 (1981).

010.005   Association of Lunar and Planetary Observers
(A.L.P.O)

What's ahead for the Minor Planets Section?
R. G. Hodgson.
Minor Planet Bull., Vol. 8, 34 (1981).

Meetings of the Association.
Strolling Astron., Vol. 29, 60 - 61 (1981).

The Minor Planet Bulletin. Bulletin of the Minor
Planets Section of the Association of Lunar and Planetary
Observers, Vol. 8, No. 3 (1981).

The Strolling Astronomer. The Journal of the
Association of Lunar and Planetary Observers, Vol. 29,
Nos. 1 - 4 (1981).

010.006   Astronomical Society of Egypt

Journal of the Astronomical Society of Egypt,
Vol. 2 (1980).

010.007   Astronomical Society of India

**Abstracts of papers presented at the sixth annual
meeting 1980 November 25 - 29 at Ahmedabad.**
Bull. Astron. Soc. India, Vol. 9, 64 - 88 (1981).

Bulletin of the Astronomical Society of India,
Vol. 9, Nos. 1 - 3 (1981).

010.008   Astronomical Society of Japan

Publications of the Astronomical Society of Japan,
Vol. 33, Nos. 2, 3 (1981).

010.009   Astronomical Society of New York

**Abstracts of papers presented at the Spring Meeting
of the Astronomical Society of New York held at Colgate
University, Hamilton, N. Y., 1981 April 25.**
News Lett. Astron. Soc. N. Y., Vol. 1, No. 10, p. 5 - 42 (1981).

**News Letter of the Astronomical Society of New
York**, Vol. 1, No. 10 (1981).

010.010   Astronomical Society of the Pacific (ASP)

**Abstracts of papers presented at the University of
Washington, Seattle, Meeting of the Astronomical Society of
the Pacific, 18 - 23 July 1981.**
Publ. Astron. Soc. Pacific, Vol. 93, 544 - 551 (1981).

**Publications of the Astronomical Society of the
Pacific**, Vol. 93, Nos. 553 - 555 (1981).

Mercury. The Journal of the Astronomical Society of the Pacific, Vol. 10, Nos. 3 - 5 (1981).

010.011 Astronomical Society of Southern Africa (ASSA)

Centre reports, 1979 - 1980. Cape Centre *(J. F. Dean)*; Natal Centre *(J. Barker)*; Natal Midlands Centre *(G. L. Prosser)*; O. F. S. Centre *(G. N. Walker)*; Pretoria Centre *(J. R. Starkey)*; Salisbury Centre *(C. J. Armstrong)*; Transvaal Centre *(G. R. Marshall)*.
Mon. Notes Astron. Soc. South. Africa, Vol. 39, 60 - 69 (1980).

The Astronomical Society of Southern Africa. Proceedings of the Annual General Meeting, 1981.
Mon. Notes Astron. Soc. South. Africa, Vol. 40, 24 - 30 (1981).

The Society — where does it stand today? Presidential address.   M. A. Gray.
Mon. Notes Astron. Soc. South. Africa, Vol. 40, 33 - 37 (1981).

Notices.
Mon. Notes Astron. Soc. South. Africa, Vol. 39, 59 (1980); Vol. 40, 1, 23 (1981).

Section reports.
Mon. Notes Astron. Soc. South. Africa, Vol. 40, 30 - 33 (1981).

Monthly Notes of the Astronomical Society of Southern Africa, Vol. 39, Nos. 11, 12 (1980); Vol. 40, Nos. 1- 9 (1981).

010.012 Astronomical Society of Western Australia (ASWA)

The Journal of the Astronomical Society of Western Australia, Vol. 28, Nos. 11 - 12 (1981).

010.013 Astronomische Gesellschaft (AG)

Mitteilungen der Astronomischen Gesellschaft, Nr. 53, 54 (1981).

010.014 British Astronomical Association (BAA)

Meteor section meeting, held at Hawkstone Hall, London, 1981 March 7.   G. H. Spalding.
J. British Astron. Assoc., Vol. 91, 435 - 438 (1981).

Report on the exhibits at the meeting of 1981 May 30.   S. R. Dunlop.
J. British Astron. Assoc., Vol. 91, 432 - 434 (1981).

Spaceprobe images and the terrestrial planets section.   G. J. Day.
J. British Astron. Assoc., Vol. 92, 35 - 38 (1981).

Meetings and activities of the Association.
J. British Astron. Assoc., Vol. 91, 426 - 431, 438 - 439, 529 - 554, 558 - 560 (1981).

Notices.
J. British Astron. Assoc., Vol. 91, 421 - 425, 555 - 558 (1981).

Section reports.
J. British Astron. Assoc., Vol. 91, 583 - 587; Vol. 92, 22 - 32, 35 - 40 (1981).

Journal of the British Astronomical Association, Vol. 91, Nos. 5, 6; Vol. 92, No. 1 (1981).

010.015 British Interplanetary Society (BIS)

JBIS. Journal of the British Interplanetary Society, Vol. 34, Nos. 8 - 12 (1981).

Spaceflight. A publication of the British Interplanetary Society, Vol. 23, Nos. 8, 9 (1981).

Space Education. A publication of the British Interplanetary Society, Vol. 1, No. 2 (1981).

010.016 Canadian Astronomical Society

Twelfth meeting of the Canadian Astronomical Society held at Université Laval, Québec, Québec, May 27 - 29, 1981. Abstracts of papers presented.
J. R. Astron. Soc. Canada, Vol. 75, 243 - 254 (1981).

010.017 European Space Agency (ESA)

ESA IUE Newsletter, Nos. 11 - 12 (1981).

010.018 International Amateur-Professional Photoelectric Photometry (I.A.P.P.P.)

I. A. P. P. P. gegründet.   M. Fernandes.
BAV Rundbrief, 30. Jahrg., 57 (1981).

Photoelectric photometry of asteroids. R. M. Genet.
Minor Planet Bull., Vol. 8, 27 (1981).

I. A. P. P. P. Communication, Nos. 5 - 6 (1981).

010.019 International Astronomical Union (IAU)

Circulaire d'Information, No. 85 (1981).

IAU Circulars, Nos. 3615 - 3655 (1981).

Commission 27 of the I. A. U. Information Bulletin on Variable Stars, Nos. 1985 - 2061 (1981).

Minor Planet Circulars, (M.P.C.), Nos. 6065 - 6572 (1981).

010.020 Meteoritical Society

Meteoritics. The Journal of the Meteoritical Society, Vol. 16, Nos. 2, 3 (1981).

010.021 **Nantucket Maria Mitchell Association**

The Nantucket Maria Mitchell Association.
Seventy-ninth annual report for the year ending December 31, 1980.
Edited by the Nantucket Maria Mitchell Assoc., Nantucket, Mass., 47 pp. (1981).

010.022 **Oriental Astronomical Association**

The Heavens, Nos. 673 - 679, Vol. 62, Nos. 6 - 12 (1981).

010.023 **Royal Astronomical Society (RAS)**

New techniques and telescopes in optical astronomy.
Summaries of papers presented at the RAS specialist discussion, held 1980 December 12, with an opening address by M. J. Disney and a review of the discussion by P. B. Fellgett.
Observatory, Vol. 101, 133 - 142 (1981).

Quasars. RAS specialist discussion, held 1981 February 13.
Observatory, Vol. 101, 143 - 149 (1981).

Meetings of the Society.
Observatory, Vol. 101, 93 - 105, 189 - 197 (1981).

Meetings and activities of the Society.
Q. J. R. Astron. Soc., Vol. 22, 324 - 332, 470 (1981).

Geophysical Journal of the Royal Astronomical Society, Vol. 66, Nos. 2, 3; Vol. 67, Nos. 1 - 3 (1981).

Monthly Notices of the Royal Astronomical Society, Vol. 196, Nos. 2, 3; Vol. 197, Nos. 1 - 3 (1981).

The Quarterly Journal of the Royal Astronomical Society, Vol. 22, Nos. 3, 4 (1981).

010.024 **Royal Astronomical Society of Canada**
         (RAS Canada)

Meetings of the Society.
J. R. Astron. Soc. Canada, Vol. 75, 261 - 266 (1981).

The Journal of the Royal Astronomical Society of Canada, Vol. 75, Nos. 3 - 6 (1981).

National Newsletter. Supplement to the Journal of the Royal Astronomical Society of Canada, Vol. 75, Nos. 3 -6 (1981).

010.025 **Royal Astronomical Society of New Zealand**
         (RAS New Zealand)

The Royal Astronomical Society of New Zealand (Inc.). 58th annual report of Council, for the year ended 1980 September 30.
South. Stars, Vol. 29, 24 - 37 (1981).

Southern Stars. Journal of the Royal Astronomical Society of New Zealand, Vol. 29, Nos. 1, 2 (1981).

010.026 **Schweizerische Astronomische Gesellschaft (SAG)**

Mitteilungen.
Orion, 39. Jahrg., 121 - 124, 153 - 156, 185 - 188 (1981).

BBSAG Bulletin, Nos. 56 - 57 (1981).

Orion. Zeitschrift der Schweizerischen Astronomischen Gesellschaft. Revue de la Société Astronomique de Suisse, 39. Jahrg., Nr. 185 - 187 (1981).

010.027 **Società Astronomica Italiana (S.A.It.)**

Giornale di Astronomia, Vol. 7, N. 2, 3 (1981).

Memorie della Società Astronomica Italiana, Vol. 52, Nos. 1 - 3 (1981).

010.028 **Société Astronomique de France (SAF)**

Séances, commissions, activités de la Société.
Astronomie, Vol. 95, 343 - 346, 397 - 399, 423 - 434, 456 - 460, 561 - 569 (1981).

Société Astronomique de France, Groupe Alsace.
E. Schweitzer.
Astronomie, Vol. 95, 453 - 455 (1981).

L'Astronomie et Bulletin de la Société Astronomique de France, Vol. 95, juillet - décembre (1981).

010.029 **Société Astronomique de Liège**

Le Ciel, Vol. 43, 171 - 178, septembre - décembre (1981).

010.030 **Société Belge d'Astronomie, de Météorologie et de**
         Physique du Globe

Meetings and activities of the Society.
Ciel Terre, Vol. 97, 321 - 333 (1981).

Ciel et Terre. Bulletin de la Société Belge d'Astronomie, de Météorologie et de Physique du Globe, Vol. 97, Nos. 4 - 6 (1981).

010.031 **Vereinigung der Sternfreunde e.V.**

Nachrichten der Vereinigung der Sternfreunde e.V.
Sterne Weltraum, Jahrg. 20, 292 - 297, 339 - 346, 380, 432 - 433, 471 - 473 (1981).

Sonne. Mitteilungsblatt der Amateursonnenbeobachter, Jahrg. 5, Nr. 19, 20 (1981).

011 Reports on Colloquia, Congresses, Meetings, Symposia, Expeditions

011.001 **Solar physics at Oxford.**
I. W. Roxburgh, C. Jordan.
Nature, Vol. 292, 194 - 195 (1981).
Report on the Solar Physics Section of the European Physical Society meeting on 13 - 15 April, 1981. The meeting was preceded by a two-day workshop on 'Near future and plans for solar research'.

011.002 **Cosmochemistry and the origin of life.**
G. Eglinton, A. Henderson-Sellers, S. Moorbath.
Nature, Vol. 292, 669 (1981).
Report on the NATO Advanced Study Institute held in Maratea, Italy on 1 - 12 June 1980.

011.003 **Symbiotic stars.** D. Allen.
Nature, Vol. 293, 99 - 100 (1981).

011.004 **The physics of sunspots.**
L. E. Cram, J. H. Thomas.
Nature, Vol. 293, 101 - 102 (1981).
Report of the workshop 'The physics of sunspots', held at Sacramento Peak Observatory, Sunspot, New Mexico, 14 - 17 July 1981.

011.005 **Report on the Tenth Texas Symposium on Relativistic Astrophysics.** J. F. Dolan.
Comments Astrophys., Vol. 9, 171 - 177 (1981).

011.006 **IAU/RAS Colloquium: Uranus and the outer planets.** P. Moore.
J. British Astron. Assoc., Vol. 91, 499 - 500 (1981).

011.007 **Autumn 1980 MIST meeting.**
P. A. Hadjiry, M. J. Laird.
Q. J. R. Astron. Soc., Vol. 22, 293 - 298 (1981).

011.008 **Supernovae.** V. Trimble.
Nature, Vol. 293, 186 - 187 (1981).
Report on the NATO Advanced Study Institute on supernovae, held 29 June - 10 July 1981 at the Institute of Astronomy, Cambridge.

011.009 **School on cosmophysics.** – Riga, 1981, Jan. 26 - Feb. 4.
V. S. Imshennik, P. V. Sasorov, A. G. Frank.
Priroda, 1981, No. 7, p. 99 - 101. In Russian.

011.010 **Ancient astronomers of the new and old worlds.**
A. Burl.
Nature, Vol. 293, 335 (1981).
Report on an archaeo-astronomical symposium, held at Queen's College, Oxford, from September 4 to 9, 1981.

011.011 **Extragalactic jets: facts and fancies.** P. Scheuer.
Nature, Vol. 293, 336 - 337 (1981).
Report on the IAU symposium No. 97 held at Albuquerque, New Mexico on 3 - 7 August 1981.

011.012 **Hot and noisy stars.** M. G. Edmunds.
Nature, Vol. 293, 432 - 433 (1981).

011.013 **Star Gazing Workshop.** E. J. Grayzeck.
Bull. American Astron. Soc., Vol. 13, 536 - 537 (1981). – Abstract.

011.014 **Future directions in ground-based optical observations.** R. B. Dunn.
Bull. American Astron. Soc., Vol. 13, 548 (1981). – Abstract.

011.015 **The MHD of sunspots.** Brief report on the Joint Meeting of Commisions 10, 12, and 44 during the IAU Assembly in Montreal. C. Zwaan.
Space Sci. Rev., Vol. 28, 385 - 386 (1981).

011.016 **Relativistische Astrophysik.** Ein Bericht über das X. Texas Symposium in Baltimore. G. Börner.
Naturwissenschaften, 68. Jahrg., 458 - 463 (1981).

011.017 **Scientific session of the Department of General Physics and Astronomy and the Department of Nuclear Physics of the USSR Academy of Sciences, September 24 - 25, 1980.**
Usp. fiz. nauk, Tom 133, 543 - 547 (1981). In Russian. – From Ref. zh., 51. Astron., 8.51.16 (1981).

011.018 **Heads of the Departments of Physics and Astronomy are consulting.**
R. N. Kuz'min, A. P. Sukhorukov.
Vestn. vyssh. shkoly, 1981, No. 4, p. 48 - 50. In Russian. Abstr. in Ref. zh., 51. Astron., 8.51.17 (1981).

011.019 **IAU symposium on pulsars.** C. S. Shukre.
Bull. Astron. Soc. India, Vol. 9, 90 - 92 (1981).

011.020 **Symposium on the total solar eclipse of 1980 February 16.**
Bull. Astron. Soc. India, Vol. 9, 92 - 94 (1981).

011.021 **I. A. U. Colloquium No. 64. "Automated data retrieval in astronomy".** C. Jaschek.
Bull. Inf. Cent. Données Stellaires, No. 21, p. 46 (1981).

011.022 **Kernexplosionen an Sternoberflächen.**
W. Kundt.
Phys. Bl., Vol. 37, 324 - 325 (1981).

011.023 **Expedition zur streifenden Bedeckung der Venus durch den Mond.** W. Rothe.
Sterne, 57. Band, 231 - 237 (1981).

011.024 **Relativistic astrophysics: the view from Texas in Baltimore.** V. L. Trimble, S. P. Maran.
Astrophys. Lett., Vol. 22, 151 - 160 (1981).
The authors outline some of the recent observations and theoretical work incorporated into the Tenth Texas Symposium on Relativistic Astrophysics, held at Baltimore, U.S.A., December 15 - 19, 1980.

011.025 **Meteoritics – more facts, more complexity.**
R. Hutchison, C. T. Pillinger.
Nature, Vol. 294, 15 - 16 (1981).
Report on the 44th meeting of the Meteoritical Society, held at Berne, Switzerland, August 17 - 22, 1981.

011.026 **All-Union school on cosmic physics.** Riga, 1981, January 26 - February 5.
Eh. E. Dubov.
Astron. Zh., Tom 58, 1133 (1981). In Russian. English translation in Soviet Astron., Vol. 25, No. 5.

011.027 **Scientific session of the Department of General Physics and Astronomy and the Department of Nuclear Physics of the USSR Academy of Sciences, October 22 - 23, 1980.**
Usp. fiz. nauk, Tom 134, 153 - 164 (1981). In Russian. Abstr. in Ref. zh., 51. Astron., 10.51.22 (1981).

011.028   Scientific session of the Department of General
          Physics and Astronomy and the Department of
Nuclear Physics of the USSR Academy of Sciences,
November 26 - 27, 1980.
Usp. fiz. nauk, Tom 134, 164 - 172 (1981). In Russian.
Abstr. in Ref. zh., 51. Astron., 10.51.23 (1981).

011.029   Symposium dedicated to Johannes Kepler.
          Leningrad, 1980, December 23.
N. I. Nevskaya.
Vopr. istor. estestvozn. i tekh., 1981, No. 2, p. 187. In Russian.
From Ref. zh., 51. Astron., 10.51.30 (1981).

011.030   The moon from different points of view.
          Conference in Kazan, 1981, May 18 - 22.
V. V. Shevchenko.
Zemlya Vselennaya, 1981, No. 6, p. 40 - 42. In Russian.

011.031   Twenty-first meeting of the NRC Associate
          Committee on Astronomy.   A. H. Batten.
J. R. Astron. Soc. Canada, Vol. 75, 259 (1981).

011.032   Impact looks real, the catastrophe smaller.
          R. A. Kerr.
Science, Vol. 214, 896 - 898 (1981).
    Report on the conference on large body impacts and ter-
restrial evolution: geological, climatological, and biological
implications, 19 to 22 October 1981 at Snowbird, Utah.
Diverse specialists agree that the evidence for a huge asteroid
(or comet) impact is impressive, but they have scaled down its
effects.

011.033   Conference on particle interactions and astrophysics,
          held in Mysore 1981 February 4 - 8.
T. S. Santhanam, R. Parthasarathy.
Bull. Astron. Soc. India, Vol.9, 168 - 169 (1981).

011.034   Die Tagung der Astronomischen Gesellschaft in
          Innsbruck 1981.   R. Lukas.
Sterne Weltraum, Jahrg. 20, 444 - 445 (1981).

011.035   XXXIInd congress of the International Astronautical
          Federation.   P. Lála.
Vesmír, Vol. 60, 347 (1981). In Czech.

011.036   Conference on stellar atmospheres, June 1981, Brno.
          P. Heinzel.
Říše hvězd, Vol. 62, 203 - 204 (1981). In Czech.

011.037   Requirements for future SN observations: γ-ray,
          X-ray, UV, visible, IR.   E. M. Burbidge.
AIP Conf. Proc., No. 63, (see 012.001), p. 167 - 170 (1980).
Report of a La Jolla Institute workshop working group
meeting.

011.038   Spectroscopic data needs for the first five spectra of
          Fe, Co, and Ni.   W. L. Wiese.
AIP Conf. Proc., No. 63, (see 012.001), p. 167, 171 (1980).
Report of a La Jolla Institute workshop working group
meeting.

011.039   Recombination rates and recombination spectra.
          V. L. Jacobs.
AIP Conf. Proc., No. 63, (see 012.001), p. 167, 172 (1980).
Report of a La Jolla Institute workshop working group meeting.

011.040   Atomic physics and spectroscopic data needs for
          improved hydrodynamic predictions of composition,
temperatures, and densities of SN envelopes.
R. A. Chevalier.
AIP Conf. Proc., No. 63, (see 012.001), p. 167, 173 (1980).
Report of a La Jolla Institute workshop working group meeting.

011.041   Workshop on acceleration.
          R. Kulsrud, C. Cesarsky.
Plasma astrophysics, (see 012.042), p. 307 - 308 (1981).

011.042   Workshop on large scale collective instabilities in the
          universe.   A. Ferrari.
Plasma astrophysics, (see 012.042), p. 345 - 347 (1981).

011.043   The early Universe.   E. W. Kolb, M. S. Turner.
          Nature, Vol. 294, 521 - 526 (1981).
    In the past few years one of the most exciting areas of
research in physics has been the interdisciplinary field of
cosmology and particle physics. The NSF's Institute for
Theoretical Physics in Santa Barbara devoted a 6-month
program and an intensive 1-week workshop to the subject. A
brief review is given of both the workshop and this field which
is attracting attention, in part, because the early Universe
seems to be the only laboratory in which to study grand
unification.

011.044   Il convegno del coassi su «Educazione scientifica di
          base».   M. A. Santaniello.
G. Astron., Vol. 7, 233 - 234 (1981).

011.045   Historien om en solförmörkelse, som började dagen
          efter den slutade. Tycho Brahe-sällskapets resa för
att observera densamma.   P.-Å. Björklund.
Astron. Tidsskr., Årg. 14, 141 - 153 (1981).

011.046   Significant achievements in the Planetary Geology
          Program, 1981.   H. E. Holt (Editor), with
contributions by R. E. Arvidson, E. A. King, J. S. King,
M. C. Malin, D. Nummedal, D. J. Roddy, J. Veverka.
NASA Tech. Memo., NASA TM 83809, 37 pp. (1981).
    Recent developments in planetology research as
reported at the 1981 NASA Planetary Geology Principal
Investigators meeting are summarized. Important developments are summarized in topics ranging from solar system
evolution and comparative planetology to geologic processes
active on other planetary bodies.

011.047   Seminar-conference "Program of complex investiga-
          tion of the main meridional section of the Galaxy",
Tartu, 17 - 18 March 1981.
J. Einasto, V. Malyuto.
Astron. Zh., Tom 58, 1331 - 1332 (1981). In Russian.
English translation in Soviet Astron., Vol. 25, No. 6.

011.048   All-Union seminar on physics of the interstellar
          medium. Moscow, 1981, April 20.
N. G. Bochkarev.
Astron. Zh., Tom 58, 1332 - 1333 (1981). In Russian.
English translation in Soviet Astron., Vol. 25, No. 6.

011.049   The Tucson workshop on active galaxies.
          P. Véron.
Optical jets in galaxies, (see 012.052), p. 7 - 8 (1981).

011.050   All-Union seminar on problems of the filtration
          theory. Tashkent, 1980, 23 - 25 September.
A. Buiķis.
Zvaigžņotā debess, 1981. gada pavasaris, p. 34 - 35. In Latvian.

011.051   Tagebuch einer Sonnenfinsternisexpedition nach
          Kenia.   B. Wedel.
Veröff. Wilhelm-Foerster-Sternw., Berlin, Nr. 53, 32 pp.

011.052   Scientific session of the Department of General
          Physics and Astronomy and the Department of
Nuclear physics of the USSR Academy of Sciences, 24 - 25
December 1980.
Usp. fiz. nauk, Tom 134, 553 - 556 (1981). In Russian.

Abstr. in Ref. zh., 51. Astron., 12.51.18 (1981).

011.053 **Scientific session of the Department of General Physics and Astronomy and the Department of Nuclear Physics of the USSR Academy of Sciences, 21 - 22 January 1981.**
Usp. fiz. nauk, Tom 134, 739 - 743 (1981). In Russian. From Ref. zh., 51. Astron., 12.51.19 (1981).

011.054 **Scientific session of the Department of General Physics and Astronomy and the Department of Nuclear Physics of the USSR Academy of Sciences, 25 - 26 February 1981.**
Usp. fiz. nauk, Tom 134, 743 - 750 (1981). In Russian. Abstr. in Ref. zh., 51. Astron., 12.51.20 (1981).

011.055 **Spring 1981 MIST meeting at Fitzwilliam College, Cambridge.** P. A. Hadjiry, M. J. Laird.

Q. J. R. Astron. Soc., Vol. 22, 423 - 431 (1981).

011.056 **The new constitution of the Scientific Committee on Solar-Terrestrial Physics (SCOSTEP).**
Izv. Soveta "Solntse-Zemlya" AN SSSR, 1981, No. 24, p. 6 - 14. In Russian. – Abstr. in Ref. zh., 51. Astron., 1.51.22 (1982).

011.057 **The report on the organization activity of the Solar-Terrestrial Council in 1980.**
Izv. Soveta "Solntse-Zemlya" AN SSSR, 1981, No. 24, p. 14 - 22. In Russian. – Abstr. in Ref. zh., 51. Astron., 1.51.23 (1982).

011.058 **Reports on the activity of the Solar-Terrestrial Council sections in 1980.**
Izv. Soveta "Solntse-Zemlya" AN SSSR, 1981, No. 24, p. 22 - 32. In Russian. – From Ref. zh., 51. Astron., 1.51.24 (1982).

012 Proceedings of Colloquia, Congresses, Meetings, Symposia

012.001 **Supernovae spectra.** The La Jolla Institute workshop on atomic physics and spectroscopy for supernovae spectra, held at La Jolly, Calif., USA, 10 - 12 January 1980. R. Meyerott, G. H. Gillespie (Editors).
AIP Conf. Proc., No. 63, 7 + 173 pp. Price $ 18.25 (1980). ISBN 0-88318-162-2. – The individual contributions are included in their corresponding subject categories – see abstracts 011.037 - 011.040, 022.027 - 022.033, 062.025, 125.010 - 125.015.

012.002 **Applications of speckle phenomena.** Conference held at San Diego, Calif., USA, 29 - 30 July 1980.
Proc. Soc. Photo-Opt. Instrum. Eng., Vol. 243 (1980). – Review in Phys. Abstr., Vol. 84, Abstr. 63090 (1981). – See abstracts 031.522, 031.523, 031.527 - 031.533.

012.003 **Nuclear power systems and nuclear science symposium,** held at Orlando, Fla., USA, 5 - 7 November 1980.
IEEE Trans. Nucl. Sci., Vol. NS-28, No. 1 (1981). – See abstracts 032.502 - 032.507, 034.007.

012.004 **Optomechanical systems design.** Conference held at San Diego, Calif., USA, 31 July 1980.
Proc. Soc. Photo-Opt. Instrum. Eng., Vol. 250 (1980). – Review in Phys. Abstr., Vol. 84, Abstr. 67151 (1981). – See abstracts 032.004, 032.005, 032.510.

012.005 **Advances in atomic and molecular physics.** Proceedings of a symposium held at London, England, 20 - 22 September 1978. D. R. Bates, B. Bederson (Editors).
Academic Press, London, England. 18 + 544 pp. Price $ 55.00 (1979). ISBN 0-12-003815-3. – Review in Phys. Abstr., Vol. 84, Abstr. 67154 (1981). – See abstracts 022.035 - 022.036.

012.006 **Magnetohydrodynamic aspects of the solar corona.** Conference held at Paris, France, 10 - 14 December 1979.
Ann.Physique, Vol. 5, Nos. 5-6 (1980). – Review in Phys. Abstr., Vol. 84, Abstr. 71349 (1981). – See abstracts 062.028, 062.029, 062.032, 074.018 - 074.021.

012.007 **V high energy physics symposium.**
Conference held at Tripunithura, India, 26 - 30 December 1980. Papers in summary form.
Published by Dep. Atomic Gov. India, Tripunithura, India. 146 pp. (1980). – Review in Phys. Abstr., Vol. 84, Abstr. 79885 (1981). – See Abstr. 162.038.

012.008 **Equatorial aeronomy.** Sixth international symposium on equatorial aeronomy, held at Aguadilla, Puerto Rico, 17 - 23 July 1980. Parts I, II. S.Matsushita, B. B. Balsley, H. Rishbeth (Editors).
J. Atmos. Terr. Phys., Vol. 43, No. 5/6, p. 375 - 632 (Part I); No. 8, p. 737 - 872 (Part II) (1981). – The individual contributions within the subject scope of Astronomy and Astrophysics Abstracts are included in their corresponding categories – see abstracts 062.040, 082.022 - 082.024, 083.016 - 083.025, 084.029.

012.009 **The Universe at ultraviolet wavelengths.** The first two years of International Ultraviolet Explorer. Proceedings of a symposium held at NASA Goddard Space Flight Center, Greenbelt, Maryland, May 7 - 9, 1980. R. D. Chapman (Editor).
NASA Conf. Publ., NASA CP 2171, 13 + 824 pp. (1981). For sale by the National Technical Information Service, Springfield, Virginia 22161. – The individual contributions are included in their corresponding subject categories – see abstracts 021.008, 021.009, 031.543 - 031.546, 064.016 - 064.020, 065.015, 091.008, 093.017, 097.013, 099.040 - 099.043, 101.005, 103.102, 103.103, 103.501, 112.010, 112.011, 114.028 - 114.044, 117.019 - 117.026, 119.015 - 119.017, 120.007, 121.007, 121.008, 122.032 - 122.038, 124.002, 124.301, 125.021, 125.022, 125.101, 126.008, 126.009, 131.040 - 131.045, 132.015 - 132.017, 134.009, 134.010, 135.014 - 135.020, 141.516, 158.064 - 158.071.

012.010 **The structure and evolution of normal galaxies.** North Atlantic Treaty Organisation, Advanced Study Institute held at the Institute of Astronomy and Clare College, Cambridge, 3 - 15 August, 1980. S. M. Fall, D. Lynden-Bell (Editors).
Cambridge University Press, Cambridge–London–New York– New Rochelle–Melbourne–Sydney. 13 + 272 pp. Price

DM 81.00 (1981). ISBN 0-521-23907-9. – The individual contributions are included in their corresponding subject categories – see abstracts 131.072, 151.019 - 151.022, 154.018, 158.100 - 158.108, 160.028.

012.011   **Second European IUE Conference.** Proceedings of an International Conference held at Tübingen, Germany, 26 - 28 March 1980.
Compiled by B. Battrick, J. Mort, with a foreword by B. Fitton and M. Grewing.
ESA SP – 157, ESA Scientific and Technical Publications Branch, c/o ESTEC, Noordwijk, Netherlands. 80 + 368 pp. Price FF 160 (1980). ISSN 0379-6566. – The individual contributions are included in their corresponding subject categories – see abstracts 031.555, 031.556, 064.029, 099.051, 102.009, 103.104, 106.012, 112.016 - 112.019, 114.057 - 114.078, 116.010, 117.034 - 117.038, 118.012, 119.022, 119.023, 120.008, 121.016, 122.050 - 122.054, 124.003, 124.103, 125.027, 125.028, 126.011 - 126.017, 131.073 - 131.076, 132.026 - 132.028, 134.015, 135.028 - 135.032, 141.070 - 141.072, 141.520, 142.040 - 142.042, 154.019, 154.020, 155.019, 158.109 - 158.120.

012.012   **Activity and outer atmospheres of the sun and stars.** Eleventh advanced course of the Swiss Society of Astronomy and Astrophysics, held in Saas-Fee, Switzerland, March 30 - April 4, 1981.
A. O. Benz, Y. Chmielewski, M. C. E. Huber, H. Nussbaumer (Editors).
Published and sold by Observatoire de Genève, Chemin des Maillettes, CH-1290 Sauverny, Switzerland. 9 + 261 pp. (1981). The individual contributions are included in their corresponding subject categories – see abstracts 064.030, 073.037, 080.020.

012.013   **Proceedings of the conference on the lunar highlands crust,** Houston, Texas, November 14 - 16, 1979.
Compiled by the Lunar and Planetary Institute,Houston, Texas.
R. B. Merrill (Managing Editor), J. J. Papike (Science Editor).
Geochim. Cosmochim. Acta, Suppl. 12. Pergamon Press, New York–Oxford–Toronto–Sydney–Frankfurt–Paris.
10 + 505 pp. Price $52.00 (1980). ISBN 0-08-026304-6. – The individual contributions are included in their corresponding subject categories – see abstracts 022.058, 091.009, 094.006 - 094.008, 094.506 - 094.522, 105.020.

012.014   **The MHD of sunspots.** Invited reviews presented at the Joint Meeting of IAU Commissions 10, 12, and 44 during the IAU Assembly in Montreal, August 20, 1979.
C.Zwaan (Editor).
Space Sci. Rev., Vol. 28, No. 4, p. 387 - 448 (1981). – The individual contributions are included in their corresponding subject categories – see abstracts 072.016, 072.017, 073.047.

012.015   **Effects of mass loss on stellar evolution.** IAU Colloquium No. 59, held in Miramare, Trieste, Italy, September 15 - 19, 1980.
C. Chiosi, R. Stalio (Editors), with an introductory address by M. Hack and concluding remarks by N. Dallaporta.
Astrophysics and space science library, Vol. 89. D. Reidel Publishing Company, Dordrecht, Holland – Boston, U.S.A. – London, England. 22 + 566 pp. Price Dfl. 140.00, US $ 73.50 (1981). ISBN 90-277-1292-1. – The individual contributions are included in their corresponding subject categories – see abstracts 064.032 - 064.043, 065.022 - 065.034, 112.020 - 112.038, 114.083 - 114.087, 115. 007 - 115.009, 116.012, 117.040 - 117.049, 118.014, 119.027, 122.059 - 122.061, 124.202, 132.030, 135.033, 135.034, 142.043, 143.032, 155.020 - 155.023.

012.016   **Problems of solar-terrestrial relations. International symposium on solar-terrestrial physics, Ashkhabad,**

23 - 27 Okt. 1979.
Fiy.-tekh. inst. AN TurkmSSR, Ashkhabad. 173 pp. (1981). In Russian. – From Ref. zh., 51. Astron., 8.51.46 (1981).

012.017   **Proceedings of the VIIth European symposium on cosmic rays, Leningrad, 15 - 19 September, 1980.** Izv. AN SSSR. Ser. fiz., Tom 45, 449 - 666 (1981). In Russian. – From Ref. zh., 51. Astron., 8.51.48 (1981).

012.018   **The phases of the interstellar medium.** Proceedings of a workshop held at the National Radio Astronomy Observatory, Green Bank, West Virginia, May 10 - 13, 1981.   J. M. Dickey (Editor).
Publications Division, NRAO, P. O. Box 2, Green Bank, WV 24954, 6 + 200 pp. (1981). – The individual contributions are included in their corresponding subject categories – see abstracts 062.053, 131.084 - 131.104, 132.031, 132.032, 135.037, 151.026, 155.024 - 155.027, 156.004, 157.007, 157.008, 158.127.

012.019   **Proceedings of the meeting in honour of Prof. Livio Gratton,** held in Rome, May 30 and 31, 1980.
V. Caloi (Editor), with an introduction by M. Rigutti.
Mem Soc. Astron. Italiana, Vol. 52, 1 - 179 (1981). – The individual contributions are included in their corresponding subject categories – see abstracts 005.003, 013.009, 013.010, 051.009, 064.046, 065.040, 065.041, 091.018, 115.010, 125.102, 142.046, 151.034, 151.035, 154.024, 158.132.

012.020   **Proceedings of the Eleventh Lunar and Planetary Science Conference,** Houston, Texas, March 17 - 21, 1980.
Compiled by the Lunar and Planetary Institute, Houston, Texas.
R. B. Merrill (Managing Editor), F. Hörz, J. Minear, D. Phinney, W. C. Phinney (Editors).
Vol. 1: Igneous processes and remote sensing. Vol. 2: Meteorite and regolith studies. Vol. 3: Physical processes.
Geochim. Cosmochim. Acta, Suppl. 14. Pergamon Press, New York–Oxford–Toronto–Sydney–Frankfurt–Paris.
33 + 10 + 9 + 2502 + 21 + 21 + 23 pp. Price $ 200.00 (1980). ISBN 0-08-026314-3 (set). – The individual contributions are included in their corresponding subject categories – see abstracts 015.019, 022.064 - 022.069, 078.008, 081.022, 091.019 - 091.025, 092.003 - 092.005, 093.023, 093.024, 094.012 - 094.018, 094.523 - 094.592, 097.019 - 097.024, 099.059, 099.060, 101.011, 105.031 - 105.059, 106.016, 106.017, 107.004, 125.033.

012.021   **Spacecraft flight dynamics.** Proceedings of an International Symposium, held at Darmstadt, Germany, 18 - 22 May 1981.
T. D. Guyenne, G. Lévy (Editors), with an introduction by R. E. Münch.
Published by ESA Scientific and Technical Publication Branch, Noordwijk, The Netherlands. ESA SP-160. 9 + 494 pp. Price FF 125.00 (1981). ISSN 039-6566. – The individual contributions within the subject scope of Astronomy and Astrophysics Abstracts are included in their corresponding categories – see abstracts 046.001, 051.010, 051.011, 052.015 - 052.022, 082.035, 094.019.

012.022   **Comets and the origin of life.** Proceedings of the Fifth College Park Colloquium on Chemical Evolution, University of Maryland, College Park, Maryland, U.S.A., October 29 - 31, 1980.   C. Ponnamperuma (Editor).
D. Reidel Publishing Company, Dordrecht, Holland – Boston, U.S.A. – London, England. 7 + 282 pp. Price Dfl. 85.00, US $ 39.50 (1981). ISBN 90-277-1318-9. – The individual contributions are included in their corresponding subject categories – see abstracts 002.018, 015.020, 015.021, 051.012, 081.036, 102.014 - 102.024, 106.053, 131.108.

012.023 **Comparative study of the planets.** Proceedings of the NATO Advanced Institute held at Vulcano, Eolie, 14 - 25 September 1981.
With a foreword by A. Coradini, M. Fulchignoni.
Mem. Soc. Astron. Italiana, Vol. 52, 323 - 597 (1981). — The individual contributions within the subject scope of Astronomy and Astrophysics Abstracts are included in their corresponding categories — see abstracts 022.072 - 022.074, 042.041, 042.042, 045.006, 051.014, 080.030, 081.024 - 081.027, 091.026 - 091.042, 093.025, 094.021, 094.593, 097.027 - 097.029, 099.062, 100.064, 102.025, 105.062, 107.005, 107.006.

012.024 **XIIIth All-Union conference on radio astronomical investigations of the solar system.** Kiev, 20 - 23 April 1981.
Naukova dumka, Kiev. 89 pp. (1981). In Russian. — From Ref. zh., 51. Astron., 9.51.27 (1981).

012.025 **Cosmic rays.** Proceedings of the VIIth European cosmic ray symposium. 15 - 19 September 1980.
V. A. Dergachev, G. E. Kocharov (Editors).
Fiz.-tekh. inst. AN SSSR. Leningrad. 401 pp. (1980). In Russian. — From Ref. zh., 51. Astron., 9.51.33 (1981).
See abstracts 072.037, 073.062, 078.013, 078.015, 082.059, 085.042, 105.067, 106.022, 142.517, 143.044, 143.045, 143.047 - 143.049, 143.097 - 143.101.

012.026 **Colors and populations of galaxies.** Conference held at Paris, France, 23 - 25 September 1980.
Ann. Physique, Vol. 6, No. 1 - 2 (1981). — Review in Phys. Abstr., Vol. 84, Abstr. 89134 (1981). — See abstracts 064.054, 065.046, 065.047, 080.033, 113.022, 113.023, 114.099, 115.013, 131.118, 151.043, 151.044, 156.006, 158.147 - 158.150.

012.027 **Optical alignment.** Conference held at San Diego, Calif., USA, 29 - 31 July 1980.
Proc. Soc. Photo-Opt. Instrum. Eng., Vol. 251 (1980).
Review in Phys. Abstr., Vol. 84, Abstr. 89145 (1981). — See abstracts 032.020 - 032.022, 032.533 - 032.535.

012.028 **Cryogenically cooled sensor technology.** Conference held at San Diego, Calif., USA, 29 - 30 July 1980.
Proc. Soc. Photo-Opt. Instrum. Eng., Vol. 245 (1980).
Review in Phys. Abstr., Vol. 84, Abstr. 89147 (1981). — See abstracts 032.536, 032.537, 034.035.

012.029 **Physics of solar variations.** Proceedings of the 14th ESLAB Symposium held in Scheveningen, The Netherlands, 16 - 19 September 1980.
V. Domingo (Editor), with an opening address by D. E. Page.
Sol. Phys., Vol. 74, Nos. 1 - 2, p. 1 - 555 (1981). — The individual contributions are included in their corresponding subject categories — see abstracts 002.051, 031.572, 032.541 - 032.544, 034.037, 034.038, 072.033 - 072.036, 073.071, 074.078 - 074.080, 075.009, 075.010, 076.013, 076.014, 080.037 - 080.051, 085.029 - 085.040.

012.030 **Group theoretical methods in physics.** Proceedings of the IXth international colloquium on group theoretical methods in physics, held at Cocoyoc, Mexico, 23 - 27 June 1980.   K. B. Wolf (Editor).
Springer-Verlag, Berlin, Germany. 26 + 629 pp. Price $ 42.50 (1980). ISBN 3-540-10271-X. — Review in Phys. Abstr., Vol. 84, Abstr. 89156 (1981). — See abstracts 066.107, 066.110.

012.031 **Infrared systems.** Conference held at Huntsville, Ala., USA, 30 September - 1 October 1980.
Proc. Soc. Photo-Opt. Instrum. Eng., Vol. 256 (1980).
Review in Phys. Abstr., Vol. 84, Abstr. 94593 (1981). — See Abstract 133.009.

012.032 **6e session d'études biennale de physique nucléaire. (LYCEN 8101).** Conference held at Lyon, France, 2 - 6 February 1981.
Inst. Phys. Nucl., Lyon, France. 452 pp. (1981). Review in Phys. Abstr., Vol. 84, Abstr. 94598 (1981). — See Abstract 061.032.

012.033 **Les étoiles de composition chimique anormale du début de la séquence principale.** Upper main sequence chemically peculiar stars.
Avec un avant-propos de P. Renson.
Published by Institut d'Astrophysique, Université de Liège, Belgique. 6 + 512 pp. Price 800 Belgian Francs, 23 U. S. $ (1981). — The individual contributions are included in their corresponding subject categories — see abstracts 002.052, 002.053, 022.100, 022.101, 031.575 - 031.577, 064.061 - 064.067, 065.055 - 065.061, 113.028 - 113.035, 114.108 - 114.126, 115.016, 115.017, 116.019 - 116.026, 120.018, 122.083 - 122.092, 126.022.

012.034 **Proceedings of the meeting held by the Astronomical Science Group of Ireland,** held at the New University of Ulster, Coleraine, May 23rd, 1979.
Irish Astron. J., Vol. 14, 114 - 137 (1980). — The individual contributions are included in their corresponding subject categories — see abstracts 080.056, 114.139, 121.025.

012.035 **Proceedings of the meeting held by the Astronomical Science Group of Ireland,** held at the University College, Dublin, December 20th, 1979.
Irish Astron. J., Vol. 14, 138 - 155 (1980). — The individual contributions are included in their corresponding subject categories — see abstracts 004.034, 013.023, 032.024, 066.135.

012.036 **Nuclear astrophysics.** Proceedings of the International School of Nuclear Physics, Erice, 25 March - 6 April, 1980.   D. H. Wilkinson (Editor).
Progress in particle and nuclear physics. Vol. 6. Pergamon Press, Oxford—New York—Toronto—Sydney—Paris—Frankfurt 6 + 345 pp. Price DM 210.00 (1981). ISBN 0-08-027117-0.
The individual contributions are included in their corresponding subject categories — see abstracts 022.113, 061.037 - 061.045, 065.071, 066.514, 066.515, 080.057, 124.004, 132.049, 162.118 - 162.121.

012.037 **La géodésie spatiale et ses applications.** Conference held at Cannes, France, 18 - 21 November 1980.
Ann. Géophys., Vol. 37, No. 1 (1981). — Review in Phys. Abstr., Vol. 84, Abstr. 102289 (1981). — See abstracts 021.043, 032.577, 035.003, 046.003, 046.004, 081.038, 081.039, 081.047, 091.056, 091.057, 091.067.

012.038 **Unification of the fundamental particle interactions.** Proceedings of the Europhysics Conference, held at Erice, Sicily, Italy, 17 - 24 March 1980.
S. Ferrara, J. Ellis, P. van Nieuwenhuizen (Editors).
Plenum Press, New York, USA. 12 + 727 pp. (1980). ISBN 0-306-40575-X. — Review in Phys. Abstr., Vol. 84, Abstr. 102298 (1981). — See abstracts 022.114, 162.122, 162.123, 162.129.

012.039 **Nuclear cross sections for technology.** Proceedings of the international conference on nuclear cross sections for technology, held at Knoxville, Tenn., USA, 22 - 26 October 1979.
J. L. Fowler, C. H. Johnson, C. D. Bowman (Editors).
NBS, Washington, D.C. 16 + 1039 pp. (1980). — Review in Phys. Abstr., Vol. 84, Abstr. 102299 (1981). — See abstracts 022.148, 061.047, 061.049.

012.040 **SS 433: a new extraordinary object in astrophysics.** Proceedings of an International Conference, held at

Rome, Italy on 8 - 10 October, 1980.
P. Beer (Editor), with opening remarks by G. Salvini, M. Hack.
Vistas Astron., Vol. 25, 1 - 233 (1981). – The individual
contributions are included in their corresponding subject
categories – see abstracts 064.082, 064.083, 065.078,
066.153, 066.154, 113.048, 116.037 - 116.039, 117.082 -
117.092, 125.051, 125.052, 141.139, 141.140, 142.069 -
142.071.

012.041   Proceedings of the 5th Göttingen-Jerusalem-
           Symposium held at Göttingen, October 6 - 9, 1980
on Astrophysics.   K. J. Fricke, J. Shaham (Editors).
Abhandlungen der Akademie der Wissenschaften in Göttingen,
Mathematische-Physikalische Klasse, Dritte Folge, Nr. 33.
Vandenhoeck & Ruprecht, Göttingen. 11 + 298 pp. Price
DM 44.00 (1981). ISBN 3-525-82107-7. – The individual
contributions are included in their corresponding subject
categories – see abstracts 061.056, 061.057, 062.097,
064.088, 065.083 - 065.087, 066.166, 066.167, 066.520,
117.106, 131.185 - 131.188, 135.058, 141.157, 142.081,
151.067, 154.039, 155.054, 158.221 - 158.224, 160.050.

012.042   Plasma astrophysics. Course and workshop held at
           Varenna (Como), Italy, 27 August - 7 September
1981.   T. D. Guyenne, G. Lévy (Editors), with an opening
address by P. Caldirola and an introduction and concluding
remarks by B. Coppi.
European Space Agency, ESA SP-161. Available from ESA
Information Retrieval Service, Rue Mario Nikis 8 - 10,
75738 Paris Cedex 15, France. 12 + 458 pp. Price FF 125.00
(1981). ISSN 0379-6566. – The individual contributions are
included in their corresponding subject categories – see
abstracts 011.041, 011.042, 022.144, 051.030, 062.099 -
062.115, 063.052, 064.089, 064.090, 066.168, 072.046,
073.092, 074.098 - 074.101, 076.020, 084.077, 084.078,
099.194, 107.017, 125.056, 131.189, 131.190, 141.161 -
141.167, 141.551, 141.552, 142.082, 142.083, 143.079 -
143.083, 151.068, 151.069, 158.227, 158.228, 162.144.

012.043   Solar physics from space. Proceedings of a con-
           ference held at ETH Zurich, November 11 - 14,
1980.   M. C. E. Huber (Editor).
Space Sci. Rev., Vol. 29, No. 4, p. 295 - 496 (1981). – The
individual contributions are included in their corresponding
subject categories – see abstracts 022.146, 022.147,
032.553 - 032.566, 064.092, 071.034, 072.048, 072.049,
073.098, 074.106, 074.107, 075.017, 076.021, 080.065,
114.172.

012.044   X-ray astronomy. Proceedings of the XV ESLAB
           Symposium held in Amsterdam, The Netherlands, 22 -
26 June 1981.
R. D. Andresen (Editor), with an opening address by
D. E. Page.
Space Sci. Rev., Vol. 30, Nos. 1 - 4, p. 1 - 642 (1981). – The
individual contributions are included in their corresponding
subject categories – see abstracts 013.035, 032.567 - 032.576,
051.031, 051.032, 062.116, 066.522 - 066.524, 113.052,
117.111 - 117.113, 122.145, 122.146, 125.059, 126.030,
134.025, 141.553, 142.085 - 142.132, 142.539, 157.016,
157.017, 158.229 - 158.236, 159.020.

012.045   Comets: gases, ices, grains and plasma.
           International Astronomical Union. Colloquium
No. 61, held in Tucson, Arizona, March 11 - 14, 1981. Some
of the contributed papers.   J. A. Burns (Editor).
Icarus, Vol. 47, No. 3, p. 301 - 522 (1981). – The individual
contributions are included in their corresponding subject
categories – see abstracts 022.152, 031.601, 098.074,
098.075, 102.051 - 102.059, 103.015, 103.016, 103.109,
103.224, 103.242, 103.602, 103.603, 103.806, 104.021,
104.022, 107.018.

012.046   Goddard contributions to the La Jolla workshop on
           gamma-ray transients, 1981 August 5 - 8.
NASA Tech. Memo., NASA TM 83816, 83 pp. (1981). – The
individual contributions are included in their corresponding
subject categories – see abstracts 063.057, 076.022, 142.541 -
142.544.

012.047   Modern observational techniques for comets.
           Proceedings of a workshop held at Goddard Space
Flight Center, Greenbelt, Maryland, on October 22 - 24, 1980.
J. C. Brandt, B. Donn, J. M. Greenberg, J. Rahe (Editors),
with a summary by F. L. Whipple.
National Aeronautics and Space Administration, Jet Propulsion Laboratory. California Institute of Technology, Pasadena,
California, JPL Publication 81 - 68, 5 + 319 pp. (1981). – The
individual contributions are included in their corresponding
subject categories – see abstracts 002.066, 009.042, 013.037,
022.153, 031.602 - 031.608, 032.590, 051.036 - 051.041,
102.060 - 102.072, 103.017 - 103.021, 103.110, 103.604,
103.807 - 103.809, 103.901.

012.048   Astrometric investigations. Proceedings of the 21st
           Astrometric conference of the USSR. Tashkent,
1978, November 14 - 17.   A. S. Kharin (Editor).
Naukova dumka, Kiev. 144 pp. (1981). In Russian. – Abstr.
in Ref. zh., 51. Astron., 11.51.64 (1981). – The individual
contributions are included in their corresponding subject
categories – see abstracts 032.028, 034.081, 034.083,
041.018 - 041.032, 082.072 - 082.075, 111.014, 111.015.

012.049   Modern theoretical and experimental problems of
           relativity theory.   A. A. Sokolov (Editor).
Fifth All-Union gravitational conference, Moscow, June, 1981.
306 pp. In Russian. – From Ref. zh., 51. Astron., 11.51.78
(1981).

012.050   Astronomische Optik, Instrumentierung und Technik.
           Wissenschaftliche Tagung der Astronomischen
Gesellschaft in Oberkochen, 1. - 3. April 1981. Übersichts-
berichte, Review Articles III, Kurzberichte.
W. Seggewiß (Editor).
Mitt. Astron. Ges., Nr. 54, p. 1 - 280 (1981). – The individual
contributions are included in their corresponding subject
categories – see abstracts 009.043, 015.036, 022.154,
022.155, 031.012 - 031.015, 031.610 - 031.618, 032.032 -
032.038, 032.591 - 032.606, 033.012, 034.065 - 034.080,
036.012, 051.042, 062.117, 131.202.

012.051   Cosmology and gravitation. Spin, torsion, rotation,
           and supergravity. Proceedings of the NATO Advanc-
ed Study Institute, held at the Ettore Majorana International
Center for Scientific Culture, Erice, Italy, May 6 - 18, 1979.
P. G. Bergmann, V. de Sabbata (Editors).
NATO Advanced Study Institutes, Series B, Vol. 58.
Plenum Press, New York - London. Published in cooperation
with NATO Scientific Affairs Division. 9 + 510 pp. Price
DM 205.60, $ 65.00 (1980). ISBN 0-306-40478-8. – The indi-
vidual contributions within the subject scope of Astronomy and
Astrophysics Abstracts are included in their corresponding
categories – see abstracts 066.180 - 066.184, 066.193,
151.079, 162.154.

012.052   Optical jets in galaxies. Proceedings of the second
           ESO/ESA workshop on the use of the Space Tele-
scope and co-ordinated ground based research, held in Munich,
18 - 19 February 1981.   Compiled by B. Battick, J. Mort,
with a welcoming address by P. O. Lindblad and concluding
remarks by M. S. Longair.
ESA SP-162. ESA Scientific & Technical Publications Branch,
c/o ESTEC Nordwijk, Netherlands. 4 + 151 pp. Price FF 80.00
(1981). ISSN 0379-6566. Available from ESA Information
Retrieval Service, Rue Mario Nikis 8 - 10, 75738 Paris CEDEX

15, France. — The individual contributions are included in their corresponding subject categories — see abstracts 011.049, 032.609, 141.183 - 141.197, 158.258 - 158.266.

012.053 Proceedings of the symposium on astrophysics, held at Huang Mountain, China, August 1977. With an introduction by S. Q. Qian.
Science Press, Beijing, Zao Yang Man Nei Street No. 137, People's Republic of China. 4 + 302 pp. Price ¥ 2.95 (1981). In Chinese.
Contents: I. Stellar systems and cosmology. II. Stars. III. Sun and solar system.

012.054 Materials of the All-Union conference "Use of model atmospheres for the interpretation of stellar spectra". Kazan, 1978, June 12 - 13.
Program of the conference and opening address by N. A. Sakhibullin.
Tr. Kazan. Gorod. Astron. Obs., Vyp. 46, p. 72 - 100 (1980). In Russian. — The individual contributions are included in their corresponding subject categories — see abstracts 021.055 - 021.057, 064.097, 064.098, 114.192, 114.193, 135.062.

012.055 Proceedings of the 14th ISAS Lunar and Planetary Symposium, held at Tokyo, July 20 - 22, 1981.
H. Hasegawa, M. Shimizu (Editors).
Published by the Institute of Space and Astronautical Science, Tokyo. 7 + 387 pp. (1981). — The individual contributions within the subject scope of Astronomy and Astrophysics Abstracts are included in their corresponding categories — see abstracts 015.038, 022.162 - 022.167, 051.047 - 051.049, 063.059, 075.020, 075.021, 080.070, 085.056, 091.072, 091.073, 092.008, 093.064, 093.065, 094.042, 094.043, 094.615, 097.071 - 097.073, 098.095 - 098.097, 099.204 - 099.205, 102.076, 105.091 - 105.102, 106.040, 107.019 - 107.024, 131.205, 131.206.

012.056 Colloque entre la Classe des Sciences de l'Académie royale de Belgique et l'Académie des Sciences de l'Institut de France, Astronomie, Paris, 28 avril 1981.
C. R. Acad. Sci. Paris, Vie Acad., Tome 293, I - XIX (1981). The individual contributions are included in their corresponding subject categories — see abstracts 071.037, 076.025, 080.071, 080.072, 093.066.

012.057 Proceedings of the Third Finnish-Soviet Astronomical Symposium, held at Helsinki and Kevo, June 1 - 8, 1979.
Published by the Astronomical Society of Finland, Helsinki. Obs. Astrophys. Lab., Univ. Helsinki, Rep. 2/1981. 141 pp. (1981). ISBN 951-45-2417-9. — The individual contributions are included in their corresponding subject categories — see abstracts 034.102, 072.061, 091.075, 099.209, 114.197 - 114.200, 116.045, 117.133 - 117.136, 119.110, 122.201 - 122.203, 151.085, 155.061, 156.011.

012.058 Proceedings of joint scientific readings on cosmonautics dedicated to the memory of eminent Soviet scientists — pioneers of cosmic space exploration. Moscow, 28 January - 1 February, 1980. Applied celestial mechanics and motion control.
D. E. Okhotsimskij, V. A. Sarychev (Editors).
Moskva, 226 pp. (1980). In Russian. — From Ref. zh., 62. Issled. kosm. prostranstva, 12.62.51 (1981). — See abstracts 042.081, 052.058, 099.210.

012.059 Cosmic rays in the heliosphere. Proceedings of Symposium 2 of the COSPAR Twenty-third Plenary Meeting held in Budapest, Hungary, 2 - 14 June 1980.
A. J. Somogyi, J. Kota, K. Kecskeméty (Editors).
Adv. Space Res., Vol. 1, No. 3, 8 + 178 pp. (1981).

ISBN 0-08-027159-6. — The individual contributions are included in their corresponding subject categories — see abstracts 031.650, 032.618, 032.619, 062.126, 072.072, 073.108 - 073.111; 074.117 - 074.120, 078.021 - 078.023, 106.046 - 106.052, 143.109 - 143.115.

012.060 Satellite perturbations and orbital determination. Proceedings of the Topical Meeting of the COSPAR Technical Panel on Dynamics of Artificial Satellites and Space Probes of the COSPAR Twenty-third Plenary Meeting, held in Budapest, Hungary, 2 - 14 June 1980.
P. Lála (Editor).
Adv. Space Res., Vol. 1, No. 6, 6 + 95 pp. Price $ 12.00 (1981). ISBN 0-08-028380-2. — The individual contributions are included in their corresponding subject categories — see abstracts 042.086, 042.087, 052.062 - 052.067, 081.056, 081.057.

012.061 Planetary interiors. Proceedings of the Workshop on Comparative Studies of Planetary Interiors, from the COSPAR Twenty-third Plenary Meeting held in Budapest, Hungary, 2 - 14 June 1980.
H. Stiller, R. Z. Sagdeev (Editors).
Adv. Space Res., Vol. 1, No. 7, 6 + 265 pp. Price $ 34.00 (1981). ISBN 0-08-028382-9. The individual contributions are included in their corresponding subject categories — see abstracts 022.176, 022.177, 065.101, 081.058, 084.095, 091.076 - 091.085, 093.077, 094.619, 099.213, 107.026 - 107.029, 118.037.

012.062 Progress in planetary exploration. Proceedings of Symposium 1 and the Topical Meeting of the COSPAR Interdisciplinary Scientific Commission B of the COSPAR Twenty-third Plenary Meeting held in Budapest, Hungary, 2 - 14 June 1980.
R. W. Shorthill, M. Ya. Marov, J. A. M. McDonnell (Editors).
Adv. Space Res., Vol. 1, No. 8, 7 + 224 pp. Price $ 29.00 (1981). ISBN 0-08-028384-5. — The individual contributions are included in their corresponding subject categories — see abstracts 031.651, 051.050, 051.051, 091.086 - 091.090, 093.078 - 093.081, 094.620, 094.621, 097.079, 097.080, 098.105, 098.106, 099.214 - 099.216, 100.136 - 100.138, 102.082, 106.054 - 106.056, 107.030, 131.217.

012.063 Planetary aeronomy and astronomy. Proceedings of the Topical Meeting of the COSPAR Interdisciplinary Scientific Commission C (Sessions C2 and C4) of the COSPAR Twenty-third Plenary Meeting held in Budapest, Hungary, 2 - 14 June 1980.
S. K. Atreya, J. J. Caldwell (Editors).
Adv. Space Res., Vol. 1, No. 9, 6 + 215 pp. Price $ 28.00 (1981). ISBN 0-08-028385-3. — The individual contributions are included in their corresponding subject categories — see abstracts 032.620, 076.026, 082.118, 082.119, 091.091, 093.082 - 093.089, 099.217 - 099.225, 100.139, 101.038, 103.025.

012.064 International symposium on physics of the earth's ionosphere and magnetosphere, and solar wind, Tsakhkadzor, 27 - 30 May 1981. Program and abstracts.
Moskva, Inst. kosm. issled. AN SSSR, 1981. 64 pp. In Russian. Abstr. in Ref. zh., 51. Astron., 1.51.51 (1982).

012.065 Scientific ballooning—II. Proceedings of Symposium 6 on Perspectives for Scientific Ballooning During the 1980s, of the COSPAR Twenty-third Plenary Meeting held in Budapest, Hungary, 2 - 14 June 1980.
W. Riedler, M. Friedrich (Editors).
Adv. Space Res., Vol. 1, No. 11, 7 + 274 pp. (1981). ISBN 0-08-028390-X, ISSN 0273-1177. — The individual contributions within the subject scope of Astronomy and Astrophysics Abstracts are included in their corresponding categories — see

abstracts 013.050 - 013.053, 032.621 - 032.630, 046.011, 051.054 - 051.057, 082.121, 084.097, 084.098, 158.291.

012.066 **The mesosphere and thermosphere.** Proceedings of the Topical Meeting of the COSPAR Interdisciplinary Scientific Commission C (Sessions C1 and C3) of the COSPAR Twenty-third Plenary Meeting held in Budapest, Hungary, 2 - 14 June 1980.
G. Schmidtke, K. S. W. Champion (Editors).
Adv. Space Res., Vol. 1, No. 12, 7 + 238 pp. Price $ 30.00 (1981). ISBN 0-08-028393-4.— The individual contributions within the subject scope of Astronomy and Astrophysics Abstracts are included in their corresponding categories — see abstracts 074.121, 076.027, 082.122, 082.123, 083.067, 084.099.

012.067 **High-energy astrophysics.** Proceedings of Symposium 5 and the Topical Meeting of the COSPAR Interdisciplinary Scientific Commission E of the COSPAR Twenty-third Plenary Meeting held in Budapest, Hungary, 2 - 14 June 1980.
H. S. Hudson (Editor).
Adv. Space Res., Vol. 1, No. 13, 7 + 290 pp. Price $ 37.50 (1981). ISBN 0-08-028395-0. — The individual contributions are included in their corresponding subject categories — see abstracts 032.631, 062.127, 066.197, 066.527, 066.528, 073.112 - 073.114, 076.028, 076.029, 078.024, 080.078, 117.093, 125.066 - 125.068, 126.031, 131.219, 131.220, 141.211 - 141.213, 141.562, 141.563, 142.155 - 142.164, 142.547 - 142.549, 143.116, 157.019, 157.020.

012.068 **Life sciences and space research XIX.** Proceedings of the Topical Meeting of the COSPAR Interdisciplinary Scientific Commission F of the COSPAR Twenty-third Plenary Meeting held in Budapest, Hungary, 2 - 14 June 1980.
W. R. Holmquist (Editor).

Adv. Space Res., Vol. 1, No. 14, 8 + 232 pp. Price $ 29.00 (1981). ISBN 0-08-028396-9. — The individual contributions within the subject scope of Astronomy and Astrophysics Abstracts are included in their corresponding categories — see abstracts 015.048, 015.049, 032.632, 082.124, 143.117.

012.069 **Rôle des nuages moléculaires dans la formation des étoiles.** Compte-rendu de l'école de Goutelas, France, 5 - 9 mai 1980.
Y. Viala (Editor), with an introduction by E. Schatzman.
Published by Société Française des Specialistes d'Astronomie , Observatoire de Meudon, France, Price FF 50.00. — The individual contributions are included in their corresponding subject categories —see abstracts 131.221 - 131.226, 133.016, 155.062, 155.063.

012.070 **Superspace and supergravity.** Proceedings of a workshop, Cambridge, England, July 1980.
S. W. Hawking M. Roček (Editors).
Cambridge University Press, New York. 12 + 528 pp. Price $ 49.95, £ 24.00 (1981). — Review in Observatory, Vol. 101, 218; 1981 (*D. Bailin*).

012.071 **The large-scale structure of the universe.**
M. Longair, J. Einasto (Editors).
Translated from the English edition.
Mir, Moskva. 515 pp. (1981). In Russian. — Proceedings of the 1977 Tallin symposium of the International Astronomical Union. — From Ref. zh., 51. Astron., 11.51.76 (1981).

012.072 **Image analysis techniques and applications. Technical digest.** Conference held at Tucson, Ariz., USA, 6 - 9 January 1981.   P. N. Slater, R. F. Wagner (Editors).
Photographic Sci. Eng., Washington, D.C., USA. 167 pp. (1981). ISBN 0-89208-102-3. — Review in Phys. Abstr., Vol. 84, Abstr. 94595 (1981).

013 Reports on Astronomy in Various Countries and Particular Fields, International Cooperation

013.001 **Entwicklung und Perspektiven der Kosmosforschungen.**   B. N. Petrow.
Sitzungsber. Akad. Wiss. DDR, No. 20N, p. 5 - 21 (1980).

013.002 **Das Zusammenwirken von Naturwissenschaften und Technik bei Interkosmosexperimenten.**
H. Stiller, K.-H. Schmelovsky.
Sitzungsber. Akad. Wiss. DDR, No. 20N, p. 22 - 38 (1980).

013.003 **The definition and implementation of the Science Program for the Space Telescope.**   B. M. Lasker.
Bull. American Astron. Soc., Vol. 13, 535 (1981). — Abstract.

013.004 **The ST data analysis software development.**
R. Albrecht.
Bull. American Astron. Soc., Vol. 13, 535 - 536 (1981). Abstract.

013.005 **The Air Force RSTN *(Radio Solar Telescope Network)* system.**
D. A. Guidice, E. W. Cliver, W. R. Barron, S. Kahler.
Bull. American Astron. Soc., Vol. 13, 553 (1981). — Abstract.

013.006 **Known infrared data base for IRAS.**
T. A. Nagy, R. S. Hill, J. M. Mead.
Bull. American Astron. Soc., Vol. 13, 556 (1981). — Abstract.

013.007 **Realization of B. P. Konstantinov's ideas on cosmic investigations.**   G. E. Kocharov.
Fiz.-tekh. inst. AN SSSR. Prepr., 1980, No. 643, 28 pp. In Russian. — Abstr. in Ref. zh., 62. Issled. kosm. prostranstva, 7.62.55 (1981).

013.008 **Development of radar investigations of planets in the Soviet Union.**   V. A. Kotel'nikov, O. N. Rzhiga, Yu. N. Aleksandrov, V. M. Dubrovin, V. A. Morozov, G. M. Petrov, A. M. Shakhovskoj, A. V. Frantsesson.
Probl. sovrem. radiotekh. i ehlektron. Moskva, 1980, p. 32 - 57. In Russian. — Abstr. in Ref. zh., 52. Geod. Aehrosemka, 8.52.185 (1981).

013.009 **L'astronomia italiana negli ultimi 50 anni (cenni).**
M. G. Fracastoťo.
Mem. Soc. Astron. Italiana, Vol. 52, (see 012.019), 17 - 22 (1981).

013.010 **Hard X-ray astronomy at IAS (*Istituto di Astrofisica Spaziale*).**   P. Ubertini.
Mem. Soc. Astron. Italiana, Vol. 52, (see 012.019), 117 - 135 (1981).

013.011 **Die Astronomie in Bulgarien.**   A. Nicolov.
Sterne Weltraum, Jahrg. 20, 367 - 368 (1981).

013.012 **Astronomical data center operations.**
W. H. Warren, Jr., J. M. Mead, T. A. Nagy.
Astron. Data Cent. Bull., Vol. 1, 52 - 56 (1981).

013.013 **Automated astronomical data retrieval at GSFC.**
J. M. Mead, T. A. Nagy, R. S. Hill, W. H. Warren, Jr.
Astron. Data Cent. Bull., Vol. 1, 57 - 61 (1981).
More than 250 machine-readable catalogs of stars and extended celestial objects are now available at the NASA/ Goddard Space Flight Center (GSFC) as the result of over a decade of catalog acquisition, verification and documentation. Retrieval programs are described which permit the user to obtain from a remote terminal bibliographical listings for stars.

013.014 **Astronomy in the Ural.**   V. V. Syrovoj.
Stellar aggregates, Sverdlovsk, 1980, (see 003.006), p. 3 - 7. In Russian. − Abstr. in Ref. zh.; 51. Astron., 9.51.12 (1981).

013.015 **Investigations on solar-terrestrial physics in Turkmenistan.**
O. G. Ovezgel'dyev, A. Kh. Khanberdyev.
Probl. soln.-zemn. svyazej. Dokl. simpoz. KAPG, Ashkhabad, 1979. Ashkhabad, 1981, p. 8 - 20. In Russian. − Abstr. in Ref. zh., 51. Astron., 9.51.14 (1981).

013.016 **Prospects of observational gamma-astronomy.**
V. G. Kirillov-Ugryumov, R. Z. Sagdeev, Yu. P. Semenov.
Zemlya Vselennaya, 1981, No. 4, p. 4 - 8. In Russian.

013.017 **Cosmic investigations in Sweden.**
N. S. Novikov.
Zemlya Vselennaya, 1981, No. 4, p. 61 - 65. In Russian.

013.018 **Decametric radio astronomy.**   S. Ya. Braude.
Zemlya Vselennaya, 1981, No. 5, p. 24 - 31. In Russian.

013.019 **International scientific research program "The year of solar maximum".**   A. Balklavs.
Zvaigžnotā debess, 1980. gada vasara, p. 14 - 16. In Latvian.

013.020 **Practical astronomy, geodesy and cartography in Lithuania.**   A. Ražinsks.
Zvaigžnotā debess, 1980. gada vasara, p. 55 - 62. In Latvian.

013.021 **The future of radio astronomy.**
H. Tamaka.
J. Inst. Electron. Commun. Eng. Japan, Vol. 64, No. 1, p. 74 - 76 (1981). In Japanese. − Abstr. in Phys. Abstr., Vol. 84, Abstr. 98686 (1981).

013.022 **The Astronomical Science Group of Ireland 1974 - 1980.**   P. A. Wayman, M. de Groot.
Irish Astron. J., Vol. 14, 111 - 113 (1980).

013.023 **Status of research on ultra-heavy nuclei in cosmic rays.**   D. O'Sullivan.
Irish Astron. J., Vol. 14, (see 012.035), 147 - 152 (1980).

013.024 **The Astronomical Institute of the Czechoslovak Academy of Sciences.** Preface by V. Bumba.
Academia, Publishing House of the Czechoslovak Academy of Sciences, Praha. 112 pp. Price 20 Kčs paperback (1979). In Czech, Russian, English.

013.025 **The International Halley Watch.**
R. L. Newburn, Jr., D. K. Yeomans.
Bull. American Astron. Soc., Vol. 13, 699 (1981). − Abstract.

013.026 **The role of photoelectric photometry in small observatories.**   F. M. Bateson.
I. A. P. P. P. Commun., No. 5, p. 7 - 9 (1981).

013.027 **The Association for Pulsating Variables.**
A. Gautschy.
I. A. P. P. P. Commun., No. 5, p. 15 - 16 (1981).

013.028 **Physicists and astronomy − will you join the dance?**
M. Harwit.
Phys. Today, Vol. 34, No. 11, p. 172 - 175, 177 - 183, 185, 187 (1981).

013.029 **On the problem of planning Soviet astronomical researches.**   G. S. Khromov.
Nauchn. Inf., Vyp. (No.) 36, p. 3 - 7 (1975). In Russian.

013.030 **Characteristic features of present-day astronomical researches.**   V. S. Bojchenko, V. A. Minin.
Nauchn. Inf., Vyp. (No.) 36, p. 8 - 12 (1975). In Russian.

013.031 **Perspective planning of basic researches.**
V. S. Bojchenko.
Nauchn. Inf., Vyp. (No.) 36, p. 13 - 26 (1975). In Russian.

013.032 **On modelling of logical connections between the elements of the process of astronomical research.**
V. A. Minin.
Nauchn. Inf., Vyp. (No.) 36, p. 27 - 41 (1975). In Russian.

013.033 **Classification system of astronomy as a basis for "a model of internal connections of astronomical research".**   V. A. Minin.
Nauchn. Inf., Vyp. (No.) 36, p. 42 - 60 (1975). In Russian.

013.034 **Principles of multi-criterion expert estimates (applied to the model of internal connections of astronomical research).**   V. S. Bojchenko, V. A. Minin.
Nauchn. Inf., Vyp. (No.) 36, p. 61 - 71 (1975). In Russian.

013.035 **Future programme of X-ray astronomy in Japan.**
Y. Tanaka.
Space Sci. Rev., Vol. 30, (see 012.044), 565 - 568 (1981).
The author briefly describes the X-ray astronomy programme in Japan in the next decade. The main items are the ASTRO-B mission, scheduled for launch in February 1983, the ASTRO-C mission in the late 80s and considerations for a large X-ray satellite CXGT.

013.036 **Enträtselung der Sonnenaktivität und ihrer Auswirkungen − ein internationales Forschungsprogramm.**   M. Kopecký.
Sterne, 57. Band, 347 - 354 (1981).

013.037 **The international Halley watch: a program of coordination, cooperation and advocacy.**
L. Friedman, R. L. Newburn.
Modern observational techniques for comets, (see 012.047), p. 313 - 314 (1981).

013.038 **Promising projects of scientific instruments.**
A. Balklavs.
Zvaigžnotā debess, 1981. gada pavasaris, p. 2 - 7. In Latvian.

013.039 **Quelques travaux en astrophysique théorique au Canada.**   G. Michaud.
J. R. Astron. Soc. Canada, Vol. 75, 289 - 296 (1981).

013.040 **Canadian radio astronomy – past, present, future?**   J. A. Galt.
J. R. Astron. Soc. Canada, Vol. 75, 297 - 298 (1981).

013.041 **Le télescope Canada - France - Hawaii; gestation, enfance et avenir.**   R. Racine.
J. R. Astron. Soc. Canada, Vol. 75, 305 - 315 (1981).

013.042 **Activities in the centres: Calgary, Halifax, Kingston, Ottawa, Québec, Toronto, Vancouver, Victoria.**
J. R. Astron. Soc. Canada, Vol. 75, 323 - 329 (1981).

013.043 **Real-time monitoring of solar activity: the NOAA Space Environment Laboratory.**   W. R. Winkler.
J. American Assoc. Variable Star Obs., Vol. 10, 37 (1981). Abstract.

013.044 **Urania e informática.**   H. Marraco.
Rev. Astron., Tomo 53, No. 218, p. 8 - 11 (1981).

013.045 **N. D. Papaleksi and Soviet radio astronomy.**   A. E. Salomonovich.
Usp. fiz. nauk, Tom 134, 541 - 550 (1981). In Russian. Abstr. in Ref. zh., 51. Astron., 12.51.14 (1981).

013.046 **The roles of the ICQ and individual observers in the improvement of visual cometary photometry.**
D. W. E. Green, C. S. Morris.
Int. Comet Q., Vol. 3, 67 - 69 (1981).
The roles of the ICQ and individual observers are discussed, and steps are suggested to improve the quality of total visual magnitude estimates of comets.

013.047 **The most important results in solar-terrestrial physics obtained in the USSR in 1980.**
Izv. Soveta "Solntse-Zemlya" AN SSSR, 1981, No. 24, p. 32 - 35. In Russian. – Abstr. in Ref. zh., 51. Astron., 1.51.25 (1982).

013.048 **Accounts on the meetings during 1980.**
Izv. Soveta "Solntse-Zemlya"AN SSSR, 1981, No. 24, p. 35 - 41. In Russian. – Abstr. in Ref. zh., 51. Astron., 1.51.26 (1982).

013.049 **Results of realization of "The Solar Maximum Year" program in the USSR and abroad in 1979 - 1980.**
Izv. Soveta "Solntse-Zemlya" AN SSSR, 1981, No. 24, p. 42 - 65. In Russian. – Abstr. in Ref. zh., 51. Astron., 1.51.27 (1982).

013.050 **Soviet-Indian research of cosmic radiation by high altitude balloons.**   A. M. Galper (Gal'per),
V. G. Kirillov-Ugryumov, Yu. D. Kotov, L. V. Kurnosova, A. V. Kurochkin, N. G. Leikov (Lejkov), V. I. Logachev, L. A. Razorenov, Yu. V. Smirnov, M. I. Fradkin, S. V. Damle, P. K. Kunte, B. V. Sreekantan.

Scientific ballooning–II, (see 012.065), p. 101 - 106 (1981).
The report presents some results on the cosmic radiation intensity research carried out with high-altitude balloons for the period of 1977-1979. The intensity of gamma-radiation with energies above 40 MeV was measured in two balloon flights at an altitude of 4-7 g/cm^2 of residual atmosphere in the vicinity of the geomagnetic equator. A temporal analysis of the intensity to discover fluctuations with periods in the range of 4-60 min was made. Quasi-periodic fluctuations of gamma-radiation intensity with 5 min periods, amplitude ~20% and duration of several hours were discovered. Possible mechanisms of such fluctuation appearance are discussed.

013.051 **Balloon research from Scandinavia.**   W. Riedler, M. Friedrich.
Scientific ballooning–II, (see 012.065), p. 251 - 260 (1981).
The scientific and technical facilities for balloon launches from Scandinavia are described and the advantage of this region for various types of research is given.

013.052 **Transmediterranean flights.**   A. Soubrier, M. Malavasi, J. Puyeo.
Scientific ballooning–II, (see 012.065), p. 261 - 264 (1981).

013.053 **Program of balloon launchings in Indonesia in the 1980s.**   J. Soegijo.
Scientific ballooning–II, (see 012.065), p. 265 - 271 (1981).

013.054 **Astronomy in Egypt past and present.**   A. S. Assad.
Astronomy in Egypt.XIIth international astronomical summer school, Cairo 22. August - 9. September 1981, p. 3 - 13 (1981).

Space activities in the eighties. Vol. 1: Survey of space programmes. Vol. 2: Detailed presentation of the European Space Industry (1981). Vol. 3: Outlook of space industry outside Europe (1981).   See Abstr. 003.055.

Cosmic investigations made in the USSR in 1980. See Abstr. 003.168.

Early days of astronomy at Toronto – Part I: The Toronto Astronomical Club.   See Abstr. 004.053.

Data link between CDS and ESA IUE Observatory. See Abstr. 021.037.

A view of satellite development in Japan. See Abstr. 051.026.

Ultraviolet astronomy with the International Ultraviolet Explorer.   See Abstr. 051.028.

The International Solar Polar Mission (ISPM). See Abstr. 051.030.

X-ray astronomy in the future. See Abstr. 051.031.

014 Teaching in Astronomy

014.001 Aspetti problematici delle attività di divulgazione e sostegno alla didattica degli astrofili: una esperienza. L. Orazzo.
Astronomia, N. 2, p. 39 - 41 (1981).

014.002 Second-year introductory courses in astrophysics. E. Schatzman.
European J. Phys., Vol. 1, 219 - 221 (1980). − Abstr. in Phys. Abstr., Vol. 84, Abstr. 71378 (1981).

014.003 Computer projects for intermediate astronomy courses. H. J. Augensen.
Bull. American Astron. Soc., Vol. 13, 536 (1981). − Abstract.

014.004 Women in astronomy. J. R. Percy.
J. R. Astron. Soc. Canada, Vol. 75, 210 - 214 (1981).

014.005 Flexible, mastery-oriented astrophysics sequence. M. Zeilik.
American J. Phys., Vol. 49, 827 - 829 (1981). − Abstr. in Phys. Abstr., Vol. 84, Abstr. 89175 (1981).

014.006 Polytropic, differentially rotating cylinders. S. A. Simon, M. F. Czysz III, K. Everett, C. Field.
American J. Phys., Vol. 49, 662 - 665 (1981). − Abstr. in Phys. Abstr., Vol. 84, Abstr. 89185 (1981).

014.007 Demonstrating the motion of comets. J. R. Milburn.
Space Educ., Vol. 1, 55 - 58 (1981).

014.008 The stellarium. A three-dimensional model of the stars. M. J. H. Mogridge.
Space Educ., Vol. 1, 76 - 77 (1981).

014.009 Teaching "Life on other worlds". R. F. Garrison.
J. R. Astron. Soc. Canada, Vol. 75, 255 - 258 (1981).

014.010 Eine unzeitgemäße Mondbetrachtung. Zum Widerspruch zwischen Wissen und Erleben. T. Schmidt.
Sterne Weltraum, Jahrg. 20, 419 - 422 (1981).

014.011 Commento a libri di testo di osservazioni scientifiche relativi ai nuovi programmi per la scuola media inferiore. M. L. Bruni Ciampini.
G. Astron., Vol. 7, 149 - 154 (1981).

014.012 Il sistema solare. U. Moncharmont.
G. Astron., Vol. 7, 157 - 166 (1981).

014.013 Tavola del sistema planetario. U. Moncharmont.
G. Astron., Vol. 7, 167 - 169 (1981).

014.014 Lavori pratici per la didattica del sistema solare in generale. U. Moncharmont.
G. Astron., Vol. 7, 171 - 178 (1981).

014.015 Moti apparenti degli oggetti del sistema solare. A. Vittone.
G. Astron., Vol. 7, 183 - 194 (1981).

014.016 Leggi di Keplero e di Newton. M. A. Santaniello.
G. Astron., Vol. 7, 195 - 212 (1981).

014.017 La determinazione delle distanze nell'ambito del sistema solare. M. Rigutti.
G. Astron., Vol. 7, 213 - 223 (1981).

014.018 Esercitazione − Misura dell'altezza del sole. M. Rigutti.
G. Astron., Vol. 7, 225 - 227 (1981).

014.019 La formazione del sistema solare. M. Rigutti.
G. Astron., Vol. 7, 229 - 232 (1981).

014.020 Bagaglio culturale e professionale di scienze della terra e dell'universo nella scuola secondaria superiore. E. Proverbio.
G. Astron., Vol. 7, 235 - 242 (1981).

014.021 Astronomische Bildung und Bildungseinrichtungen. H. Groth, J. Krahner.
Jenaer Rundsch. (Jena Rev.), 26. Jahrg., 239 - 241 (1981).

014.022 Teaching constellations in an introductory astronomy course. J. N. Scrimger.
J. R. Astron. Soc. Canada, Vol. 75, 330 - 336 (1981).

014.023 Education of astronomy in Egypt since 1937. A. Aiad.
Astronomy in Egypt. XIIth international astronomical summer school, Cairo 22. August - 9. September 1981, p. 14 - 26 (1981).

Jugendlexikon "Astronomie und Raumfahrt". See Abstr. 002.088.

Astronomy for O-level. See Abstr. 003.110.

Hilfsmaterialien für Übungen in Astronomie und Astrophysik, 2. Teil. See Abstr. 003.122.

L'astronomie en 4°. See Abstr. 003.171.

Star Gazing Workshop. See Abstr. 011.013.

015 Miscellaneous Papers (Philosophical Aspects, Extraterrestrial Civilizations, etc.)

015.001   The SETI ellipsoid.   W. F. Hilton.
J. British Interplanet. Soc., Vol. 34, 400 - 401
(1981).

015.002   The anthropic principle and the early universe.
P. Davies.
Mercury, Vol. 10, 66 - 77 (1981).

015.003   Modelli epistemologici, costrutti scientifici e mondo
del senso comune.   M. Bianca.
G. Astron., Vol. 7, 69 - 86 (1981).

015.004   Galactic civilizations: population dynamics and
interstellar diffusion.   W. I. Newman, C. Sagan.
Icarus, Vol. 46, 293 - 327 (1981).
The interstellar diffusion of galactic civilizations is
reexamined by potential theory. The authors conclude that the
Earth is uncolonized not because interstellar spacefaring
societies are rare, but because there are too many worlds to be
colonized in the plausible lifetime of the colonization phase of
nearby galactic civilizations. They also conclude that there are
no very old galactic civilizations with a consistent policy of
conquest of inhabited worlds; there is no Galactic Empire.
There may, however, be abundant groups of $\sim 10^5$ to 10^6 worlds
linked by a common colonial heritage.

015.005   Discrete calculations of interstellar migration and
settlement.   E. M. Jones.
Icarus, Vol. 46, 328 - 336 (1981).
Monte Carlo calculations of the expansion of space-faring
civilizations are presented for a wide range of values of the
population growth coefficient (α) and emigration coefficient (γ).
Even for the very low values proposed by Newman and Sagan
($\alpha = 10^{-4}$ per year; $\gamma = 10^{-8}$ per year) the migration wavefront
expands at 1.4×10^{-5} pc per year. Even with this low expansion
velocity, such a civilization would fill the Galaxy in about 10^9
years. Filling times of the order of 60 million years seem
probable.

015.006   Extraterrestrial beings don't exist.   R. A. Schorn.
Sky Telesc., Vol. 62, 207 (1981).

015.007   Additional remarks on extraterrestrial intelligence.
F. J. Tipler.
Q. J. R. Astron. Soc., Vol. 22, 279 - 292 (1981).

015.008   Astronomers working in industry.
R. C. Bless, I. R. King.
Phys. Today, Vol. 34, No. 7, p. 24 - 30 (1981).

015.009   Motion of the Earth as viewed from the Moon and
the Y-suspended pendulum.   A. D. Crowell.
American J. Phys., Vol. 49, 452 - 454 (1981). - Abstr. in Phys.
Abstr., Vol. 84, Abstr. 63123 (1981).

015.010   Space for everyone.   S. Preston.
Bull. American Astron. Soc., Vol. 13, 536 (1981).
Abstract.

015.011   Les comètes et l'origine de la vie (II).
A. H. Delsemme.
Astronomie, Vol. 95, 381 - 395 (1981).

015.012   Geomagnetic field — its role in the evolution of life
and intelligence on earth.   U. R. Rao.
J. British Interplanet. Soc., Vol. 34, 459 - 465 (1981).

015.013   Correlation between microcosmos and macrocosmos
as a philosophical problem: history and present.
M. S. Asimov, A. Tursunov.
Filos., estestvozn., sovremennost'. Itogi i perspektivy issled.
1970 - 1980. Moskva, 1981, p. 44 - 59. In Russian. — Abstr. in
Ref. zh., 51. Astron., 8.51.1 (1981).

015.014   Philosophical problems of investigation of the
universe.   V. V. Kazyutinskij.
Filos., estestvozn., sovremennost'. Itogi i perspektivy issled.
1970 - 1980. Moskva, 1981, p. 258 - 281. In Russian. — Abstr.
in Ref. zh., 51. Astron., 8.51.2 (1981).

015.015   Scientific revolutions and investigations of the
universe.   V. A. Ambartsumyan, V. V. Kazyutinskij.
Filos., estestvozn., sovremennost'. Itogi i perspektivy issled.
1970 - 1980. Moskva, 1981, p. 26 - 43. In Russian. — Abstr. in
Ref. zh., 51. Astron., 8.51.3 (1981).

015.016   Dialectics of knowledge of evolutionary processes in
the universe.
V. A. Ambartsumyan, V. V. Kazyutinskij.
Vopr. filos., 1981, No. 4, p. 52 - 70. In Russian. — Abstr. in
Ref. zh., 51. Astron., 8.51.4 (1981).

015.017   Socially-philosophical problems of space exploration.
A. D. Ursul, Yu. A. Shkolenko.
Filos., estestvozn., sovremennost'. Itogi i perspektivy issled.
1970 - 1980. Moskva, 1981, p. 282 - 302. In Russian. — Abstr.
in Ref. zh., 51. Astron., 8.51.5 (1981).

015.018   Vie, intelligence et communication dans l'univers.
A. Hayli.
Histoire de l'univers, (see 003.007), p. 409 - 418 (1980).
Contents: L'apparition de la vie à la surface de la Terre.
Vie et intelligence extra-terrestres. Peut-on espérer
communiquer avec d'éventuelles civilisations extra-terrestres?

015.019   Introduction: the Vatican's astronomical paintings
and the Institute of the Sciences of Bologna.
S. A. Bedini.
Proc. Eleventh Lunar Planet. Sci. Conf., (see 012.020),
p. XIII - XXXIII (1980).

015.020   Comets — a vehicle for panspermia.
F. Hoyle, C. Wickramasinghe.
Comets and the origin of life, (see 012.022), p. 227 - 239
(1981).
Arguments are given for life being a cosmic phenomenon.
The physical and chemical conditions associated with comets
favour the hypothesis that comets carry, amplify and disperse
life throughout the universe.

015.021   Extreme environments: are there any limits to life?
D. Kushner.
Comets and the origin of life, (see 012.022), p. 241 - 248
(1981).
Many microorganisms can grow in conditions that seem
extreme to us. There do seem to be limits, however, to the
environments in which microorganisms can grow. All need
relatively abundant amounts of liquid water. Unless comets
contain substantial amounts of liquid water, life, similar to
any that we know, probably cannot exist or arise in them.

015.022   Mit Flußkrebs-Augen ins Weltall blicken. Augen mit
Spiegeloptik — Ein biologisches Vorbild für Röntgen-

teleskope.   K. Kirschfeld.
Sterne Weltraum, Jahrg. 20, 357 - 358 (1981).

015.023 Extraterrestrials – where are they?
B. Zuckerman.
Nature, Vol. 294, 10 - 11 (1981).

015.024 On the problem of population of the Galaxy.
V. S. Troitskij.
Astron. Zh., Tom 58, 1121 - 1130 (1981). In Russian. English
translation in Soviet Astron., Vol. 25, No. 5.
A new theory of the population of the Galaxy is considered, based on the hypothesis of the "blast" origin, simultaneous
and one-fold, of life in the Universe in a certain moment of its
development. According to the suggested theory, civilizations
began to originate around the present moment of the history
of the Universe. Their possible number is restricted even when
their life-time is unlimited.

**015.025 Cooperation – the aim of contacts between galactic
civilizations.**   Z. Alksne.
Zvaigžnotā debess, 1980. gada vasara, p. 2 - 11. In Latvian.

015.026 The anthropic principle.   G. Gale.
Sci. American, Vol. 245, No. 6, p. 114 - 122 (1981).
Certain conditions, such as temperature, were favorable
to the emergence of life on the earth. The anthropic principle
argues the reverse: the presence of life may "explain" the
conditions.

015.027 Planetenspezifische astronomische Kalendertypen.
M. Gossler.
Sterne, 57. Band, 298 - 300 (1981).

**015.028 On the place of organic life in the universe and the
possibility of extraterrestrial life.**
R. N. Butler.
Irish Astron. J., Vol. 14, 177 - 188 (1980).

015.029 Starscapes near the speed of light.
Sky Telesc., Vol. 62, 530 - 531 (1981).

015.030 Sun, moon, eclipse: all at once.   W. M. Sinton.
Sky Telesc., Vol. 62, 551 (1981).

015.031 Polarität und Steigerung.   T. Schmidt.
Sterne Weltraum, Jahrg. 20, 465 - 467 (1981).

015.032 About the Olbers' paradox.   P. Andrle.
Vesmír, Vol. 60, 331 - 332 (1981). In Czech.

**015.033 Das Phänomen Halley. Kometenfieber – eine
Erkrankung besonderer Art.**   N. Calder.
Umschau, 81. Jahrg., 742 - 744 (1981).

015.034 Julestjernen eller Betlehemsstjernen.
R. Brahde.
Astron. Tidsskr., Årg. 14, 161 - 166 (1981).

015.035 Information on Vinchucas and Chagas disease.
H. Schenone.
Messenger, No. 26, p. 1 - 2 (1981).

015.036 Technik, astronomische Forschung und Philosophie.
T. Schmidt-Kaler.
Mitt. Astron. Ges., Nr. 54, (see 012.050), p. 11 - 14 (1981).

**015.037 What astronomers have discovered in other
sciences. II.**   B. Ševarlić.
Vasiona, Année 29, 42 - 44 (1981). In Serbo-Croatian.

015.038 To the scientific exobiology.
M. Shimizu.
Proceedings of the 14th ISAS Lunar and Planetary Symposium,
(see 012.055), p. 118 - 130 (1981).
Comparative molecular biology should provide insight
into the prerequisites for the occurrence of extraterrestrial life.
The author discusses in some detail the molecular basis of the
genetic code.

015.039 The joys of a journal editor.   L. A. Higgs.
J. R. Astron. Soc. Canada, Vol. 75, 316 - 318
(1981).

015.040 Sur la réalité astronomique de l'étoile de Bethléem.
J.-P. Parisot.
Astronomie, Vol. 95, 549 - 553 (1981).

**015.041 Interaction between the general scientific and the
astronomical picture of the world.**
I. V. Chernikova.
Zakonomernosti razvitiya sovrem. nauk. Tomsk, 1981, p. 125 -
132. In Russian. – Abstr. in Ref. zh., 51. Astron., 12.51.1
(1981).

**015.042 Heuristic role of philosophical principles and the
picture of the world in history of astronomy.**
I. V. Chernikova.
Dialektika svyazi filos. i konkretno-nauch. znaniya. Irkutsk,
1980, p. 107 - 114. In Russian. – Abstr. in Ref. zh., 51.
Astron., 12.51.2 (1981).

**015.043 General picture of the world and the problem of
life in the universe.**   L. V. Fesenkova.
Nauch. dokl. vyssh. shkoly. Filos. nauk., 1981, No. 3, p. 57 -
63. In Russian. – Abstr. in Ref. zh., 51. Astron., 12.51.3
(1981).

015.044 The Drake equation re-examined.
S. G. Wallenhorst.
Q. J. R. Astron. Soc., Vol. 22, 380 - 387 (1981).
The Drake equation, which predicts the number of
extant technological civilizations in the Galaxy, is considered.
A formulation originally due to Kreifeldt, and assumptions
about civilization development time and lifetime, are used to
set an upper limit to the number of extant civilizations. This
number $N_c(T)$, is found to be of order $10^0 - 10^2$, rather than
the 10^6 suggested previously by other authors. The implications of this much-reduced galactic population on the SETI
program and the arguments of Tipler against the existence of
extraterrestrial life are discussed.

**015.045 The problem of idealization of time in philosophy
and physics.**   I. B. Shamshev.
Metod. probl. sovrem. nauk. Yaroslavl', 1980, No. 2, p. 65 -
80. In Russian. – Abstr. in Ref. zh., 51. Astron., 1.51.1 (1982)

015.046 The idea of the universe.
V. V. Kazyutinskij.
Filos. i mirovozzrenchesk. probl. sovrem. nauk. 16-j Vsemirn.
filos. kongr., Moskva, 1981, p. 49 - 96. In Russian. – Abstr.
in Ref. zh., 51. Astron., 1.51.2 (1982).

**015.047 Cosmology and the development of the scientific
picture of the world.**   A. S. Karmin.
Filos. i razvitie estestv.-nauchn. kartiny mira. Leningrad, 1981,
p. 97 - 116. In Russian. – Abstr. in Ref. zh., 51. Astron.,
1.51.3 (1982).

015.048 Biological studies of Martian soil analogues.
A. A. Imshenetsky (*Imshenetskij*), B. G. Murzakov,
M. D. Evdokimova, I. K. Dorofeyeva (*Dorofeeva*).
Life sciences and space research XIX, (see 012.068), p. 21 - 26

(1981).

Results of the study of the influence of Martian soil analogues and of hydrogen peroxide on the viability of microorganisms are presented. In view of the fact that Martian soil cannot contain microorganisms in great quantities the authors suggest using electroadsorption to make detection reliable. A device was designed for this purpose, using the principle of electroadsorption on a polarisable carrier.

015.049 **Survival of microorganisms in space: a review.**
G. Horneck.
Life sciences and space research XIX, (see 012.068), p. 39 - 48 (1981).

Spores of Bacillus subtilis were exposed to selected factors of space (vacuum, solar UV radiation, heavy ions of cosmic radiation), and their response was studied after recovery. These investigations were supplemented by ground-based studies under simulated space conditions. A spore travelling through space and protected from ultraviolet radiation could possibly survive an interplanetary journey.

015.050 **Astrophysical investigations by G. A. Tikhov and the problems of exobiology.**   A. K. Suslov.
Researches in the history of astronomy. 15th issue, (see 003.025), p. 73 - 83 (1980). In Russian.

015.051 **Copernicus' teaching and practical needs of the society.**   N. I. Nevskaya.
Researches in the history of astronomy. 15th issue, (see 003.025), p. 93 - 106 (1980). In Russian.

015.052 **Memorial astronomical medals.**   Yu. A. Barshtejn.
Researches in the history of astronomy. 15th issue, (see 003.025), p. 219 - 230 (1980). In Russian.

015.053 **From K. E. Tsiolkovsky's scientific heritage: "The property of space". "The necessity of a cosmic point of view".**   Published with notes by N. K. Gavryushin.
Researches in the history of astronomy. 15th issue, (see 003.025), p. 293 - 304 (1980). In Russian.

015.054 **Fragments from the dialogue "About the nature of gods"** by M. T. Cicero.

Translation with introduction and comments by T. A. Lapina and A. Ch. Kozarzhevskij.
Researches in the history of astronomy. 15th issue, (see 003.025), p. 339 - 360 (1980). In Russian.

Comets and the origin of life: bibliography.
See Abstr. 002.018.

The star of Bethlehem: a list of references.
See Abstr. 002.086.

Evolution from space.   See Abstr. 003.075.

Space travellers, the bringers of life.
See Abstr. 003.076.

Are we alone? The possibility of extraterrestrial civilizations.   See Abstr. 003.135.

La vita intelligente nell'Universo.
See Abstr. 003.146.

Problem of search for extraterrestrial intelligence.
See Abstr. 003.151.

Evolution kosmischer, biologischer und geistiger Strukturen.   See Abstr. 003.152.

SI and Gaussian CGS units, conversions and equations for use in geomagnetism.   See Abstr. 084.100.

Comets and the origin of life — the stable isotope approach.   See Abstr. 102.022.

Are comets connected to the origin of life?
See Abstr. 102.023.

Cometary material and the origins of life on earth.
See Abstr. 102.024.

Chemical evolution of interstellar dust — a source of prebiotic material?   See Abstr. 131.108.

Applied Mathematics, Physics

021 Mathematical Papers Related to Astronomy and Astrophysics, Computing, Data Processing

021.001 The analysis of periodicities in irregularly sampled data. T. Ponman.
Mon. Not. R. Astron. Soc., Vol. 196, 583 - 596 (1981).

The use of standard Fourier techniques to produce power spectra is limited to data which are sampled at regular intervals. This is not always easily achieved in astronomy, nor is it entirely desirable since it introduces ambiguities into the interpretation of the resulting power spectra. A method is presented for producing spectral estimates from data which may be weighted and sampled at any intervals. A Green's function approach is introduced to help identify aliases, and a formula given which enables the significance of peaks to be ascertained. Finally, an analysis of the X-ray intensity of Cygnus X-3 is used to illustrate the method.

021.002 On the comparison of Yurkevich's method of searching for a periodicity with the Fourier analysis in the case of unequally spaced data.
V. A. Hagen-Thorn, V. A. Yakovleva.
Tr. Astron. Obs., Leningrad, Tom 36 = Uch. Zap. Leningr. Univ., No. 402 = Ser. mat. nauk, Vyp. 58, 26 - 40 (1981). In Russian.

It is shown that the Fourier analysis is preferable.

021.003 Short-periodic perturbations of the state vector in the method of Lie transforms.
K. V. Kholshevnikov.
Tr. Astron. Obs., Leningrad, Tom 36 = Uch. Zap. Leningr. Univ., No. 402 = Ser. mat. nauk, Vyp. 58, 124 - 134 (1981). In Russian.

021.004 Software for VLBI experiments.
L. R. Kogan, L. S. Chesalin.
Astron. Zh., Tom 58, 898 - 903 (1981). In Russian. English translation in Soviet Astron., Vol. 25, No. 4.

021.005 Predictions of occultations by the Earth as seen from a geostationary satellite. G. E. Taylor.
J. British Astron. Assoc., Vol. 91, 473 - 482 (1981).

A simple method is described for calculating the times when the Moon and some stars can be photographed close to the limb of the Earth, as seen from a geostationary satellite.

021.006 Satellite orbit parameters and visibility characteristics calculation by pocket calculator.
J. Molnar.
Hiradastechnika, Vol. 32, No. 2, p. 47 - 58 (1981). In Hungarian. – Abstr. in Phys. Abstr., Vol. 84, Abstr. 71057 (1981).

021.007 Tracking the planets. E. S. Maxey.
Kilobaud Microcomput., Vol. 5, 130 - 134 (1981). Abstr. in Phys. Abstr., Vol. 84, Abstr. 75003 (1981).

021.008 Analysis of IUE spectra using the Interactive Data Language. C. L. Joseph.
The Universe at ultraviolet wavelengths, (see 012.009), p. 801 - 803 (1981).

In recent years, the use of interactive computer languages has made a major impact on data handling. One language in particular, the Interactive Data Language (IDL), has been extensively used to analyze high-resolution spectra from the IUE. The author confines the attention to a package of programs created to analyze interstellar absorption lines as an example of the graphical power of IDL.

021.009 Problems and programming for analysis of IUE high resolution data for variability.
C. A. Grady.
The Universe at ultraviolet wavelengths, (see 012.009), p. 805 - 810 (1981).

In the course of analysis of IUE high resolution data of α Cam and other O, B and Wolf-Rayet stars several effects were found which cause spurious variability or spurious spectral features in the data. Programming has been developed to partially compensate for these effects using the Interactive Data Language (IDL) on the LASP PDP 11/34 at the University of Colorado.

021.010 Instability of an area preserving mapping.
J. H. Bartlett.
Bull. American Astron. Soc., Vol. 13, 572 (1981). – Abstract.

021.011 On a mathematical model of variation of the atmospherical parameters under conditions of their indeterminacy. G. I. Sakharov, V. E. Usachev.
Nauchn. chteniya po aviats. i kosmonavt. 10-e Gagarinsk. chteniya, 1980. Moskva, 1981, p. 197. In Russian. – Abstr. in Ref. zh., 62. Issled. kosm. prostranstva, 8.62.477 (1981).

021.012 Microcomputer system for a digital correlation receiver. D. K. Ravindra, S. Krishnan.
Bull. Astron. Soc. India, Vol. 9, 77 (1981). – Abstract.

021.013 Microcomputer on-line data reduction for satellite experiments. S. Beros.
Space Sci. Instrum., Vol. 5, 247 - 256 (1981).

A microcomputer data reduction system based on Motorola's MC6802 microprocessor is presented. This system was developed to be used on board satellites and balloons. After processing, only statistically relevant data have to be stored on board or transmitted to Earth. The performances were tested with real data from X-ray burst sources. The rate reduction depends on the structure of the processed data and may be a factor of ten or even more.

021.014 The contact transformation groups of the extended Hamiltonian system. J. Bryant.
Celestial Mech., Vol. 25, 41 - 49 (1981).

021.015 On the number of isolating integrals in resonant systems with 3 degrees of freedom.
L. Martinet, P. Magnenat, F. Verhulst.
Celestial Mech., Vol. 25, 93 - 99 (1981).

The 2:2:1-resonance case for a potential problem with three degrees of freedom is characterized by the existence of two isolating approximate integrals apart from the energy. This result completes a statement by Gustavson concerning the number of formal integrals in resonant Hamiltonian systems.

021.016 **The computation of tables of Hansen coefficients.**
S. Hughes.
Celestial Mech., Vol. 25, 101 - 107 (1981).

021.017 **Sur une étude de stabilité par analogie avec la collision triple.**   L. Losco.
Celestial Mech., Vol. 25, 159 - 167 (1981).

021.018 **The Lagrangian theory of Stäckel systems.**
R. Broucke.
Celestial Mech., Vol. 25, 185 - 193 (1981).
A purely Lagrangian formulation and a direct proof of the separation of variables theorem is given for what is called Stäckel systems in dynamics and celestial mechanics. The proof is essentially based on some properties of determinants and minors. In contrast with the standard literature on the subject, the author avoids the use of the Hamiltonian, canonical transformations or the Hamilton-Jacobi equation, by using instead a more elementary approach based on the Lagrangian.

021.019 **On the possibility of using the program system "Auto-Analyzer" for the solution of problems of celestial mechanics.**   L. A. Moskovkina, Yu. B. Shmidt.
Astron. i geod., Tomsk, 1980, No. 8, p. 92 - 96. In Russian.
Abstr. in Ref. zh., 51. Astron., 9.51.51 (1981).

021.020 **Some algorithms for constructing analytical solutions in problems of celestial mechanics.**
K. V. Kholshevnikov.
Astron. i geod., Tomsk, 1980, No. 8, p. 16 - 26. In Russian.
Abstr. in Ref. zh., 51. Astron., 9.51.52 (1981).

021.021 **Algorithms of evaluation in the restricted three-body problem.**   L. L. Sokolov.
Astron. i geod., Tomsk, 1980, No. 8, p. 32 - 34. In Russian.
Abstr. in Ref. zh., 51. Astron., 9.51.56 (1981).

021.022 **Review of modern methods for increasing the accuracy of numerical integration of differential equations of the motion of celestial bodies.**
T. V. Bordovitsyna.
Astron. i geod., Tomsk, 1980, No. 8, p. 54 - 75. In Russian.
Abstr. in Ref. zh., 51. Astron., 9.51.70 (1981).

021.023 **A semi-analytical algorithm for calculating the first-order perturbations in regularizing variables.**
T. V. Ivanova.
Astron. i geod., Tomsk, 1980, No. 8, p. 40 - 44. In Russian.
Abstr. in Ref. zh., 51. Astron., 9.51.71 (1981).

021.024 **Investigation of the efficiency of numerical algorithms using the equations of motion of AES in stabilized and regularized form.**
T. V. Bordovitsyna, L. E. Sukhoplyueva.
Astron. i geod., Tomsk, 1980, No. 8, p. 76 - 80. In Russian.
Abstr. in Ref. zh., 51. Astron., 9.51.92 (1981).

021.025 **Computation of ephemerides in the case of ill-conditioned normal equations.**
V. A. Izvekov.
Determination of the coordinates of celestial bodies, Riga, 1981, (see 003.008), p. 11 - 18. In Russian. – Abstr. in Ref. zh., 51. Astron., 9.51.99 (1981).

021.026 **Programmes for measurement processing of linear polarized galactic radio emission by the NAIRI-K computer.**   V. G. Arkhangel'skij, I. P. Kuznetsova.
Gor'kov. nauchn.-issled. radiofiz. inst. Prepr., 1981, No. 145, 43 pp. In Russian. – Abstr. in Ref. zh., 51. Astron., 9.51.1028 (1981).

021.027 **UPP – universal system for analytical operations with Poisson series.**   I. O. Babaev, V. A. Brumberg, N. N. Vasil'ev, T. V. Ivanova, V. I. Skripnichenko, S. V. Tarasevich.
Astron. i geod., Tomsk, 1980, No. 8, p. 49 - 53. In Russian.
Abstr. in Ref. zh., 51. Astron., 9.51.1030 (1981).

021.028 **A new criterion for rough error selection in astronomical observations.**   S. D. Shaporev.
Determination of the coordinates of celestial bodies, Riga, 1981, (see 003.008), p. 104 - 112. In Russian. – Abstr. in Ref. zh., 51. Astron., 9.51.1031 (1981).

021.029 **On the possibility of a new approach to processing of cosmic pulsed radio signals.**   A. A. Bocharov.
Inst. kosm. issled. AN SSSR. Prepr., 1980, No. 605, 18 pp. In Russian. – Abstr. in Ref. zh., 51. Astron., 9.51.1032 (1981).

021.030 **SPORA – a special program for solving astronomical problems.**   S. S. Lavrov, V. I. Skripnichenko.
Astron. i geod., Tomsk, 1980, No. 8, p. 3 - 15. In Russian.
Abstr. in Ref. zh., 51. Astron., 9.51.1036 (1981).

021.031 **Storing means of observational results and provision for their availability.**   N. V. Emel'yanov.
Astron. i geod., Tomsk, 1980, No. 8, p. 45 - 48. In Russian.
Abstr. in Ref. zh., 51. Astron., 9.51.1037 (1981).

021.032 **On the determination of the number of gradations distinguishable by an information-measurement system.**   A. B. Bukach, L. V. Granitskij, N. I. Bukach.
Krym. astrofiz. obs. AN SSSR. pos. Nauchnyj, 1981. 20 pp. In Russian. – Abstr. in Ref. zh., 51. Astron., 9.51.1040 (1981).

021.033 **A program for the search of periods in large series of observational data.**   W. Schöneich, D. Lange.
Astron. Nachr., Band 302, 153 - 154 (1981). In German.

021.034 **What time does the sun rise and set?**
B. Barkstrom.
BYTE, Vol. 6, 94 - 114 (1981). – Abstr. in Phys. Abstr., Vol. 84, Abstr. 88715 (1981).

021.035 **Data processing in synthetic-aperture radio astronomy.**   G. W. Swenson, Jr.
1981 IEEE international symposium on circuits and systems. Vol. 1. Conference held at Chicago, Ill., USA, 27 - 29 April 1981. IEEE, New York, USA (1981). p. 62 - 65. – Abstr. in Phys. Abstr., Vol. 84, Abstr. 94194 (1981).

021.036 **The origin of the Julian Period: an application of congruences and the Chinese Remainder Theorem.**
R. L. Reese, S. M. Everett, E. D. Craun.
American J. Phys., Vol. 49, 658 - 661 (1981). – Abstr. in Phys. Abstr., Vol. 84, Abstr. 94204 (1981).

021.037 **Data link between CDS and ESA IUE Observatory.**
A. Heck.
Bull. Inf. Cent. Données Stellaires, No. 21, p. 45 (1981).

021.038 **On stationary solutions of some canonical systems of differential equations.**   S. G. Zhuravlev.
Celestial Mech., Vol. 25, 297 - 315 (1981).

021.039 **On a rapid algorithm for evaluation of the Voigt function.** A. Heinlo, T. Viik.
Tartu Astrofüüs. Obs., Teated, Nr. 64, p. 24 - 31 (1981).
In Russian.

021.040 **On a method for solving the Fokker-Planck equation of general mode.** N. G. Skryabin.
Cosmic ray variations and solar wind, (see 003.013), p. 9 - 12 (1980). In Russian. – Abstr. in Ref. zh., 51. Astron., 10.51.178 (1981).

021.041 **QIKAIM, a fast semi-numerical algorithm for the generation of minute-of-arc accuracy satellite predictions.** M. Vermeer.
Rep. Finnish Geod. Inst., 81:1, 17 + 11 pp. (1981). ISBN 951-711-068-5, ISSN 0355-1962.

021.042 **A versatile interactive computer program for computation and automatic optimization of gravity models.** T. Enmark.
Geoexploration, Vol. 19, No. 1, p. 47 - 66 (1981). – Abstr. in Phys. Abstr., Vol. 84, Abstr. 107575 (1981).

021.043 **Orbit computation comparisons using the method of fictitious observations.**
A. Drozyner, C. Reigber.
Ann. Géophys., Vol. 37, (see 012.037), 11 - 14 (1981). Abstr. in Phys. Abstr., Vol. 84, Abstr. 108070 (1981).

021.044 **Second order solution of Lindstedt's equation with constant coefficients.** F. Tong, Z. Chen.
Acta Astron. Sinica, Vol. 22, 213 - 222 (1981). In Chinese.

021.045 **A Monte Carlo approach to surface and interior scattering of tri-axial ellipsoids.**
W. T. Thompson, D. Van Blerkom.
Bull. American Astron. Soc., Vol. 13, 727 (1981). – Abstract.

021.046 **Dynamics Explorer science data processing system.** P. H. Smith, C. H. Freeman, R. A. Hoffman.
Space Sci. Instrum., Vol. 5, (see 003.019), 561 - 573 (1981).
The Dynamics Explorer project has acquired the ground data processing system from the Atmosphere Explorer project to provide a central computer facility for the data processing, data management and data analysis activities of the investigators.

021.047 **Complex of algorithms for detection of astronomical objects and determination of their coordinates.**
S. M. Kaganov, V. D. Levchenko, V. S. Matyagin, L. A. Usol'tseva, L. I. Shchavlev.
Automation of astronomical observations, (see 003.020), p. 41 - 56 (1981). In Russian. – Abstr. in Ref. zh., 51. Astron., 11.51.1065 (1981).

021.048 **Device for coding and input of a television image into a M-220 computer.**
E. S. Agapov, V. A. Kachmin, G. V. Motarygina, V. L. Rymarenko.
Automation of astronomical observations, (see 003.020), p. 64 - 69 (1981). In Russian. – Abstr. in Ref. zh., 51. Astron., 11.51.1066 (1981).

021.049 **Automation of search for distant cosmic objects and their coordinate measuring.**
B. I. Kontorovich, V. N. Medvedev.
Automation of astronomical observations, (see 003.020), p. 70 - 82 (1981). In Russian. – Abstr. in Ref. zh., 51. Astron., 11.51.1067 (1981).

021.050 **Comparison of two methods for calculation of differences between star positions.** V. S. Borovskikh.

Kazan. inzh.-stroit. inst. Kazan', 1981. 6 pp. In Russian. Abstr. in Ref. zh., 51. Astron., 11.51.1071 (1981).

021.051 **Mathematical modelling of electrophotometry of celestial bodies moving on the background of stars.**
A. M. Androsov, V. G. Vygon, S. N. Malikov.
Tr. 6-j konf. molod. uchenykh. Mosk. fiz.-tekh. inst. Moskva, 1981, p. 231 - 233. In Russian. – Abstr. in Ref. zh., 51. Astron., 11.51.1079 (1981).

021.052 **High-order Runge-Kutta method and stabilizing transformations in the problems of forecasting the motion of AES.** T. V. Bordovitsyna.
Kosm. Issled., Tom 19, 941 - 943 (1981). In Russian.

021.053 **Application of the Buys-Ballot scheme to observations of pulsating variable stars.**
B. N. Irkaev.
Astron. Tsirk., No. 1139, p. 3 - 4 (1980). In Russian.

021.054 **On some possibilities to use bispectral analysis in the treatment of observations of solar radio emission fluctuations.** E. Aver'yanikhina.
Investigations of the sun and red stars. 12, (see 003.022), p. 57 - 71 (1981). In Russian.
The bispectra of artificial time series are calculated and analysed for the illustration how to use such an analysis in the investigation of solar radio emission fluctuations. Some practical recommendations are given.

021.055 **Digital method for solving equations of the state of matter of a stellar atmosphere.**
V. V. Tsymbal, N. S. Komarov.
Use of model atmospheres for the interpretation of stellar spectra, (see 012.054), p. 80 - 83 (1980). In Russian.

021.056 **Functional structure and information provision of an automatic dialogue system for high-dispersion astronomical spectra processing.** Yu. V. Borisov.
Use of model atmospheres for the interpretation of stellar spectra, (see 012.054), p. 89 - 96 (1980). In Russian.

021.057 **ATLAS-5 program on the BESM-6 computer.**
A. K. Gal'dikas.
Use of model atmospheres for the interpretation of stellar spectra, (see 012.054), p. 96 - 97 (1980). In Russian.

021.058 **Numerical simulation of the signal and data processing of the HIPPARCOS satellite.**
M. Froeschle. F. Mignard.
Appl. Opt., Vol. 20, 3251 - 3258 (1981).
This paper presents a theoretical study of the diffraction pattern which will be obtained with the telescope launched aboard the HIPPARCOS satellite. The output signal is studied by simulating observations of various stars of different spectral types and classes. In the case of very faint stars, the signal will consist of only a few photons during a sampling interval, so that a statistical recovery of the signal will be needed. A simulation of such a signal is described. The last part of this paper deals with the data processing.

021.059 **Algorithms for the calculation of scattering by stratified spheres.** O. B. Toon, T. P. Ackerman.
Appl. Opt., Vol. 20, 3657 - 3660 (1981).
Efficient, numerically stable, methods for the calculation of light-scattering intensity functions for concentrically coated spheres are discussed. Earlier forms of these equations are subject to various numerical difficulties which give rise to significant errors, especially for thin absorbing shells. The present equations are accurate for all refractive indices, for large and small particles, and for cores with any relative size.

021.060 A computer program for converting local time
observations of variable stars to Julian Date and
decimal.   K. Wilson.
J. American Assoc. Variable Star Obs., Vol. 10, 33 (1981).
Abstract.

021.061 Sternzeitberechnung mit TI-58 oder TI-59.
P. Weber.
Orion, 39. Jahrg., 182 - 183 (1981).

021.062 UKIRT instrumentation computer.   J. M. Stewart.
UKIRT Rep., No. 12, 41 + 26 + 11 pp. (1979).

021.063 On the construction of a compiled selenodetic
catalogue.   S. G. Valeev.
Izv. vuzov. Geod. i aehrofotosemka, No. 3, p. 38 - 46. In
Russian. − Abstr. in Ref. zh., 52. Geod. Aehrosemka, 12.52.
308 (1981).

021.064 Inversion de matrices simétricas en el metodo de
minimos cuadrados.   M. J. Sevilla.
Bol. Inf., Núm. 44, Serv. Geogr. Ejército, 16 + 3 pp. (1979) =
Univ. Complutense - Fac. Cienc., Madrid, Semin. Astron.
Geod., Publ. núm. 106.

021.065 Researches in commutative hyper complex systems.
S. Kikuchi.
Sci. Rep. Tôhoku Univ., Ser. 8, Vol. 1, 187 - 212 (1980).

021.066 Automated data acquisition system for the time
keeping.   T. Yamazaki.
Tokyo Astron. Obs. Rep. (No. 74), Vol. 19, 379 - 385 (1981).
In Japanese.

021.067 HP-IB universal board and its application to the
clock data acquisition system.   T. Yamazaki.
Tokyo Astron. Obs. Rep. (No. 74), Vol. 19, 386 - 398 (1981).
In Japanese.

021.068 Data transfer from floppy disc to UNIVAC
1100/80B.   Y. Nishino.
Tokyo Astron. Obs. Rep. (No. 74), Vol. 19, 425 - 440 (1981).
In Japanese.

021.069 Using computers for literal developments: the lunar
disturbing function arising from the sun.
Q.-c. Lin.
Ann. Shanghai Obs. Acad. Sinica, No. 1, p. 97 - 100 (1979).

021.070 Predicting ephemerides and visible periods of
artificial satellites using an electronic computer.
Q.-c. Lin.
Ann. Shanghai Obs. Acad. Sinica, No. 2, p. 73 - 79 (1980).

021.071 Estimation of the standard error from correlated
observations.   S.-y. Zhu.
Ann. Shanghai Obs. Acad. Sinica, No. 2, p. 88 - 89 (1980).

021.072 Design of the real time display of encoder output.
C.-k. Xiao, S.-n. Huang.
Ann. Shanghai Obs. Acad. Sinica, No. 2, p. 139 - 144 (1980).

021.073 On the addition theorem of spherical functions.
V. G. Shkodrov.
Dokl. Bolg. AN, Vol. 34, 605 - 608 (1981). − Abstr. in Ref.
zh., 51. Astron., 1.51.95 (1982).

021.074 Latitude and longitude from van der Grinten grid
coordinates.   D. P. Rubincam.
American Cartogr., Vol. 8, 177 - 180 (1981).

021.075 Interplanetary two point boundary value problems.
(Extension of the Unified System for Orbit Computation, USOC, Vol. IV).   With a preface by G. Scheifele.
Analytical and Computational Mathematics, ACM Schänis AG,
Switzerland. Prepared under ESOC Contract No. 3488/78/D/
IM/SC of ACM with European Space Agency. ESA CR(P)-
1347. 241 pp. (1979).
The Unified System for Orbit Computation (USOC) has
been extended to allow the resolution of two point boundary
value problems for interplanetary flight computation. The
implementation is done in three different stages corresponding to three levels of accuracy: unperturbed patched conics
method, method of matched asymptotic expansions and
numerical quasilinearization method.

021.076 Calculs astronomiques pour amateurs.
J. Meeus.
Astronomie, Vol. 95, 356 - 357, 396, 449 - 452, 513 - 514,
557 - 558 (1981).

A proposal for a glossary of terms relating to the
storage, retrieval and analysis of astronomical data.
See Abstr. 002.027.

Availability of an interactive computer reference
search of the astronomical literature 1950 - 1976.
See Abstr. 002.031.

Suggestions for formatting, checking and documenting machine-readable astronomical catalogues.
See Abstr. 002.032.

Astronomical data bases and retrieval systems.
See Abstr. 002.066.

Calcul astronomique pour amateurs adapté à
l'emploi d'un calculateur ou d'un micro-ordinateur.
See Abstr. 003.035.

The computer in optical research, methods and
applications.   See Abstr. 003.064.

Automation of control processes and reduction of
information.   See Abstr. 003.087.

The ST data analysis software development.
See Abstr. 013.004.

Optical aberration coefficients: FORTRAN subroutines for symmetrical systems.
See Abstr. 031.031.

On the computation of acceleration parallaxes.
See Abstr. 031.506.

An improved IUE image processing system − the
case of absorption lines in high dispersion spectra.
See Abstr. 031.556.

Photometric observations of faint objects and their
processing on the M-220M computer.
See Abstr. 031.622.

Datensysteme am 2.2 m-Teleskop.
See Abstr. 032.038.

The Camac control system used on the UKIRT
3.8 m telescope.   See Abstr. 032.050.

Examples of interactive instrument control and
data acquisition.   See Abstr. 034.078.

Derivation of positions and parallaxes from simulated observations with a scanning astrometry satellite. See Abstr. 041.002.

Expansion theory for the elliptic motion of arbitrary eccentricity and semi-major axis. II. Analyses of the function

$$G^{(N, n)}(\theta, x_n, \xi_n) = \prod_{i=1}^{n} (1 - \xi_i \sin^{2N-1}\theta)^{x_i}.$$

See Abstr. 042.028.

On the construction of Roche harmonics. See Abstr. 042.067.

Determinación de la latitud por el metodo de Sterneck. Programas de cálculo automático. See Abstr. 045.017.

Determinación de la latitud y la longitud por el metodo de alturas iguales. Prográmas de cálculo automático. See Abstr. 045.018.

Determinación de la longitud por el metodo de Mayer. Programas de cálculo automático. See Abstr. 046.007.

Almanac for computers 1982. See Abstr. 047.015.

Collisions of artificial earth orbiting bodies. See Abstr. 052.014.

Algorithms and programs for the construction of an analytical theory of motion of AES in coordinates. See Abstr. 052.027.

Radius of convergence of Lie series for some elliptic elements.   See Abstr. 052.038.

Numerical computations of refraction. See Abstr. 082.115.

Coordinated ionospheric and magnetospheric observations from the ISIS 2 satellite by the ISIS 2 experimenters. Volume 4. A. Large storms. B. Airglow and related measurements. C. VLF observations.   See Abstr. 084.096.

Erratum

021.901 Erratum: "Prediction of minima – a new BASIC-program for small computers" [BBSAG Bull., No. 55, p. 3 - 7 (1981)].   N. Hasler. BBSAG Bull., No. 56, p. 7 (1981). – See Abstr. 29.021.038.

022 Physical Papers Related to Astronomy and Astrophysics

022.001 Calculated X-radiation from optically thin plasmas IV. Atomic data and rate coefficients for spectra in the range 1 - 270 Å.
R. Mewe, E. H. B. M. Gronenschild.
Astron. Astrophys., Suppl. Ser., Vol. 45, 11 - 52 (1981).

Rate coefficients of ionization, recombination, and excitation, used in recent computations of the soft X-ray (1 - 270 Å) spectra of transient optically thin plasmas are described in detail. The authors considered 870 spectral lines containing about 100 satellites to resonance lines in hydrogen- and helium-like ions. These satellite lines are produced by innershell excitation, dielectronic recombination, and innershell ionization processes that can be very sensitive to non-equilibrium effects in the ionization balance. The elements included in the calculations are H, He, C, N, O, Ne, Na, Mg, Al, Si, S, Ar, Ca, Fe and Ni. Some resulting spectra are shown in graphical form and the intensities of the emission lines of a stationary plasma are tabulated over the temperature range 10^5 - 10^8 K. The effects on the emergent X-ray spectrum and on the satellite line formation processes due to deviations from the ionization equilibrium are illustrated for the model of a transient plasma in which the electron temperature strongly increases with a jump.

022.002 Anomalous $^{12}CH_4$: $^{13}CH_4$ strengths in $3\nu_3$.
J. W. Brault, K. Fox, D. E. Jennings, J. S. Margolis.
Astrophys. J., Lett., Vol. 247, L101 - L104 (1981).

The vibration-rotation band near 9050 cm^{-1} has been useful in determinations of methane abundances and ^{12}C: ^{13}C ratios in the atmospheres of the outer planets. Absolute strengths of corresponding R-branch lines in isotopically pure samples have now been measured in the laboratory under identical conditions. Line strengths for $^{12}CH_4$ were determined here to be larger than those for $^{13}CH_4$ by 3 - 87%. The ^{12}C: ^{13}C abundance ratios deduced from $3\nu_3$ spectra of the atmospheres of Jupiter and Saturn have been reevaluated on the basis of the new laboratory data.

022.003 Density sensitive C II lines in cool stars of low gravity. R. E. Stencel, J. L. Linsky, A. Brown, C. Jordan, K. G. Carpenter, R. F. Wing, S. Czyzak.
Mon. Not. R. Astron. Soc., Vol. 196, 47P - 53P (1981).

It is shown that the relative intensities of emission lines within the multiplet UV 0.01 of C II, around 2325 Å, are sensitive to electron density in the range $10^9 > N_e > 10^7$ cm^{-3}. The lines therefore offer a valuable method for measuring N_e in the chromospheres of late-type giants and supergiants. Calculated line ratios are compared with those observed in a range of objects.

022.004 Band analysis of the $A-X$ transition in $^{16}O_2^+$ and $^{18}O_2^+$ molecules.
D. S. Pesic, D. Rakotoarijimy, S. Weniger.
Astrophys. J., Suppl. Ser., Vol. 45, 507 - 515, plate 26, 27 (1981).

Using a low-pressure hollow cathode discharge cooled by liquid nitrogen, the authors have obtained the bands of the second negative system $A-X$ of $^{16}O_2^+$ and $^{18}O_2^+$ in emission between 4800 and 2600 Å. Photographed with high-resolution spectrograph, the complex band structure was available for analysis.

022.005 Scaled collision strengths for hydrogenic ions.
L. B. Golden, R. E. H. Clark, S. J. Goett, D. H. Sampson.
Astrophys. J., Suppl. Ser., Vol. 45, 603 - 612 (1981).

Scaled hydrogenic ion collision strengths $Z^2 \Omega_H$ and $Z^2 \Omega_H^e$ have been calculated using the Coulomb-Born-

Oppenheimer approximation for the limit $Z \to \infty$ or, equivalently, time-dependent perturbation theory with the perturbation consisting of the entire electrostatic interaction between the electrons.

022.006 Polynomial partition function approximations of 344 atomic and molecular species. A. W. Irwin.
Astrophys. J., Suppl. Ser., Vol. 45, 621 - 633 (1981).

Approximations of 344 atomic and molecular partition functions are given for a temperature range from 1000 to 16000 K. These approximations are useful for the calculation of the equation of state and the line and continuum opacity of intermediate and late-type stellar atmospheres.

022.007 Accurate ab initio calculation of the HCS^+ interstellar ion.
S. Chekir, F. Pauzat, G. Berthier.
Astron. Astrophys., Vol. 100, L14 - L15 (1981).

The equilibrium structure of the thioformyl ion HCS^+ is determined by SCF + CI calculations, giving a value of 21.37 GHz for the rotational constant B_e. Taking into account the rotation-vibration corrections deduced from experimental data for HCP leads to B_0 = 21.27 GHz and $\nu(1 \leftarrow 2)$ = 85.08 GHz. The error bar of this calculation is estimated at 0.60 GHz, which supports the assignment of the B-line observed around 85.34 GHz in interstellar space. A value of μ = 2.2D is found for the corresponding dipole moment.

022.008 Collisional l-mixing of Rydberg states of carbon due to thermal energy charged particles.
A. S. Dickinson.
Astron. Astrophys., Vol. 100, 302 - 306 (1981).

Cross sections for thermal energy charged-particle collisions with Rydberg levels of carbon are of importance in the new dielectronic-like recombination process proposed recently by Watson et al. (1980). The author reexamines such cross sections and derives a new approximation applicable to the adiabatic proton collisions of interest in this process. His cross sections are significantly smaller than those used by Watson et al. so that recombination in carbon is likely to be more hydrogen-like than their estimate.

022.009 $^{45}Sc(p, p')^{45}Sc^*$ and the attainment of a thermal distribution of states in an astrophysical environment.
M. R. Anderson, S. R. Kennett, M. E. Sevior, D. G. Sargood.
Australian J. Phys., Vol. 34, 105 - 112 (1981).

Cross sections for inelastic scattering of protons by ^{45}Sc have been measured over the energy range E_p = 1.2 - 2.5 MeV for protons leading to the second, third, fourth, sixth and eighth excited states of ^{45}Sc. Statistical model calculations are in good agreement with the data.

022.010 Existiert eine kosmische Quantelung?
K. Volkamer.
Sterne Weltraum, Jahrg. 20, 273 - 275 (1981).

022.011 Vibrational disequilibrium in low pressure clouds.
J. A. Nuth, B. Donn.
Astrophys. J., Vol. 247, 925 - 935 (1981).

In this paper the authors calculate the vibrational distribution and corresponding vibrational temperatures for the diatomic molecules carbon monoxide and silicon monoxide over a wide range of gas kinetic temperatures and pressures for several values of a diluted blackbody radiation field. They then estimate the vibrational temperature of an idealized polyatomic molecule under similar conditions. Carbon monoxide was chosen because of its abundance in most regions of interest and silicon monoxide because it plays a major role in the condensation of numerous predicted grain materials. Most of the necessary molecular data are available for both.

022.012 Rotational excitation of molecular ions by electron impact under interstellar conditions.
S. S. Bhattacharyya, B. Bhattacharyya, M. V. Narayan.
Astrophys. J., Vol. 247, 936 - 940 (1981).

A generalized expression for the rate coefficient for rotational excitation of molecular ions by electron impact under interstellar conditions has been obtained from the first order perturbation theory. The expression has been obtained by considering only the electron-dipole term of the interaction potential which is the most dominant term. The effect of short range and electron-quadrupole interactions has been assessed. The importance of electron-ion collisions in relation to H_2-ion collisions in the interpretation of the spectral data obtained for ions from interstellar sources has also been assessed.

022.013 $^{14}N/^{15}N$ isotope fractionation in the reaction $N_2H^+ + N_2$: interstellar significance.
N. G. Adams, D. Smith.
Astrophys. J., Lett., Vol. 247, L123 - L125 (1981).

The authors' determination of the forward and reverse rate coefficients for the proton transfer reactions between N_2H^+ and N_2 in their various isotopic forms suggests that, under typical interstellar cloud conditions, isotopic fractionation can occur in the reactions and that equilibrium will be established in reaction $^{14}N_2H^+ + {}^{15}N_2 \rightleftharpoons {}^{15}N_2H^+ + {}^{14}N_2$ in a short time compared with cloud lifetimes. Thus, within the limitations of the ion-chemical model, the $[^{14}N^{15}NH^+]/[^{14}N_2H^+]$ ratio is seen to be directly proportional to the $[^{14}N^{15}N]/[^{14}N_2]$ ratio, but with an enhancement of ^{15}N in the N_2H^+ of about a factor of 2 under cloud conditions. This enhancement improves the possibility of $^{14}N^{15}NH^+$ detection.

022.014 Aspects of isochronism in Pb isotope systematics – application to planetary evolution.   F. Tera.
Geochim. Cosmochim. Acta, Vol. 45, 1439 - 1448 (1981).

Some interrelated aspects of lead isotope systematics are exposed and their bearing on the problem of planetary evolution is discussed.

022.015 A determination of the nitric oxide Einstein coefficient ratios.
B. D. Green, G. E. Caledonia, R. E. Murphy.
J. Quant. Spectrosc. Radiat. Transfer, Vol. 26, 215 - 221 (1981).

The ratios of the fundamental and first overtone A-values for NO vibrational levels 2–9 have been measured experimentally by observing the i.r. fluorescence from electron irradiated mixtures of N_2 and O_2.

022.016 An analysis of Markovian model microfield methods for Stark broadening.
E. W. Smith, B. Talin, J. Cooper.
J. Quant. Spectrosc. Radiat. Transfer, Vol. 26, 229 - 242 (1981).

022.017 Electron impact ionization cross sections and rates for highly ionized atoms.   S. M. Younger.
J. Quant. Spectrosc. Radiat. Transfer, Vol. 26, 329 - 337 (1981).

Electron impact ionization cross sections and rates for hydrogen-, helium- and lithium-like ions are presented in simple analytical form, based on a set of parameters derived from distorted wave exchange cross section calculations.

022.018 Tables for direct determination of spectral line parameters, from spectrometric data – I. Pure Lorentz broadening.   M. L. Claude, A. Valentin.
J. Quant. Spectrosc. Radiat. Transfer, Vol. 26, 349 - 363 (1981).

Numerical tables and curves have been prepared for the precise calculation of spectral line parameters from observed spectra with a spectrometer of known instrument response function.

022.019 Exact random walk definition of the collisional-radiative ionization and recombination coefficients.
S. O. Kastner.
J. Quant. Spectrosc. Radiat. Transfer, Vol. 26, 371 - 375 (1981).

The collisional-radiative ionization and recombination coefficients are shown to be expressible in terms of the "total probability" t_{ij} from initial level i to final level j, which excludes intermediate returns to the initial level. Applications are made to three-level and four-level systems to illustrate the method of calculation and compare with an approximation currently used.

022.020 A new method for the rapid calculation of infrared transmittances of atmospheric gases.
V. Oinas.
J. Quant. Spectrosc. Radiat. Transfer, Vol. 26, 381 - 383 (1981).

A new procedure is outlined for the rapid calculation of transmittances of atmospheric gases in the infrared. A feature of the method is the relatively high accuracy (better than 1%) and the fact that any line shape can be handled. The highest gain in speed over direct line-by-line integration is in the low pressure region (Voigt or Doppler profiles).

022.021 The auroral 2145 Å feature.
A. Dalgarno, G. A. Victor, T. W. Hartquist.
Geophys. Res. Lett., Vol. 8, 603 - 605 (1981).

022.022 Broad-band polarization in molecular spectra.
R. M. E. Illing.
Astrophys. J., Vol. 248, 358 - 372 (1981).

The rotational lines of the CN(0,0) red system have been observed to show a strongly asymmetric Zeeman profile. Certain molecules are very susceptible to magnetic perturbation because of the weakness of their spin-rotation coupling; a fairly weak magnetic field can cause a complete Paschen-Back effect. The calculation of transition probabilities incorporating this effect into the Hamiltonian is discussed; the detailed calculation is then given. The resulting transition probabilities are transformed into synthetic line profiles by using the Unno model of polarized radiation transfer. The dependence of the net polarized flux on magnetic field and equivalent width is investigated. Broad-band circular polarization of sunspots may be due, in part, to molecular bands. Analysis of the CH G band indicates a magnetic field of $0.25-0.50 \times 10^6$ gauss in the white dwarf G99–37.

022.023 A laboratory study of the reaction $H_3^+ + HD \rightleftharpoons H_2D^+ + H_2$: the electron densities and the temperatures in interstellar clouds.   N. G. Adams, D. Smith.
Astrophys. J., Vol. 248, 373 - 379 (1981).

The rate coefficients k_1 and k_{-1} have been determined at 80, 200, and 295 K for the astrophysically important reaction: $H_3^+ + HD \underset{k_{-1}}{\overset{k_1}{\rightleftharpoons}} H_2D^+ + H_2$. The data have been used in an ion-chemical model, first proposed by Watson, to reestimate the upper limits to the relative electron density, $n_e/[H_2]$, and the temperature, T, in interstellar clouds. Thus, $n_e/[H_2]$ is increased by a factor of ~ 6, and T is reduced by a factor of ~ 2 compared to the previous estimates.

022.024 Pressure broadening of oxygen and its implications for cosmic background measurements.
H. M. Pickett, E. A. Cohen, D. E. Brinza.
Astrophys. J., Lett., Vol. 248, L49 - L51 (1981).

The pressure broadening of oxygen transitions at 119 and 425 GHz have been measured in the laboratory for nitrogen and oxygen broadening as a function of temperature. The measured values are significantly different from estimates used by Woody and Richards to obtain a secondary calibration in their cosmic background measurements. The effect of the new determination is to move the calibrated cosmic background further from a Planck function.

022.025   **Laboratory measurements of millimeter and sub-millimeter transitions of sodium hydride and sodium deuteride.**   K. V. L. N. Sastry, E. Herbst, F. C. De Lucia.
Astrophys. J., Lett., Vol. 248, L53 - L55 (1981).
The frequencies of several rotational transitions of sodium hydride and sodium deuteride have been measured by means of millimeter and submillimeter microwave spectroscopy. Sodium hydride is potentially a repository of sodium in dense interstellar clouds.

022.026   **Quark matter in the Hartree-Fock approximation.**   E. Alvarez.
Phys. Rev. D, Vol. 23, 1715 - 1722 (1981). — Abstr. in Phys. Abstr., Vol. 84, Abstr. 63500 (1981).

022.027   **Energy levels, wavelengths and transition probabilities for the first five spectra of Fe, Co and Ni.**
W. L. Wiese.
AIP Conf. Proc., No. 63, (see 012.001), p. 103 - 117 (1980).
Atomic energy level, wavelength and transition probability data for the first five spectra of Fe, Co and Ni are reviewed, and lists of recent comprehensive data tables are presented. The source material for transition probabilities, both for allowed and forbidden lines, is critically discussed, since these data contain large uncertainties.

022.028   **Excitation and ionization of moderately heavy ions.**   R. H. Garstang.
AIP Conf. Proc., No. 63, (see 012.001), p. 119 - 123 (1980).
A brief review is given of some recent work on excitation, deexcitation and ionization of moderately heavy ions (iron, zinc and gallium).

022.029   **Dielectronic recombination, ionization equilibrium, and radiative emission for astrophysically abundant elements.**   V. L. Jacobs, J. Davis.
AIP Conf. Proc., No. 63, (see 012.001), p. 139 - 144 (1980).
The authors have carried out systematic calculations of the total dielectronic recombination rates, the corona ionization equilibrium abundances, and the radiative emission rates for all of the astrophysically important elements, including Fe and Ni. These calculations demonstrate that the inclusion of the dielectronic recombination rates produces a substantial shift in the corona equilibrium abundances of certain charge-states.

022.030   **Photoionization cross sections calculated by many body theory.**   H. P. Kelly.
AIP Conf. Proc., No. 63, (see 012.001), p. 145 - 161 (1980).
The use of many body perturbation theory to calculate photoionization cross sections of atoms including electron correlations is discussed. Results are presented for neutral argon, zinc, iron, and chlorine.

022.031   **Semiempirical calculation of gf values.**   R. L. Kurucz.
AIP Conf. Proc., No. 63, (see 012.001), p. 163 - 166 (1980).

022.032   **Recent advances in charged particle energy deposition and applications to supernova spectra.**   A. E. S. Green.
AIP Conf. Proc., No. 63, (see 012.001), p. 75 - 94 (1980).
The author explores the possibility of applying the continuous slowing down approximation and the yield spectrum method to the explanation of late time supernova spectra. He is particularly concerned with excitations induced by incident electrons. Calculation of the excitations induced by gamma rays follows similar methods after allowing for pair production, the Compton effect, and the photoelectric effect.

022.033   **Charge and energy transfer in heavy particle collisions.**   R. E. Olson.
AIP Conf. Proc., No. 63, (see 012.001), p. 95 - 101 (1980).
A short discussion of the general classes of excitation and charge transfer reactions is presented. Specific emphasis is placed on collision mechanisms applicable to low energy ($E \lesssim 10$ eV) scattering between ions and atoms and the resulting scaling of the cross sections on parameters such as collision energy and ion charge state.

022.034   **High-resolution infrared spectrum and analysis of the ν_9 band of ethane at 12.17 μm.**
S. J. Daunt, W. E. Blass, G. W. Halsey, K. Fox, R. J. Lovell.
J. Mol. Spectrosc., Vol. 86, 327 - 343 (1981). — Abstr. in Phys. Abstr., Vol. 84, Abstr. 68177 (1981).

022.035   **Atomic physics from atmospheric and astrophysical studies.**   A. Dalgarno.
Advances in atomic and molecular physics, (see 012.005), p. 37 - 76 (1979). — Abstr. in Phys. Abstr., Vol. 84, Abstr. 68271 (1981).

022.036   **Aspects of recombination.**
D. R. Bates.
Advances in atomic and molecular physics, (see 012.005), p. 235 - 262 (1979). — Abstr. in Phys. Abstr., Vol. 84, Abstr. 68272 (1981).

022.037   **Oscillator strengths for Zr I and Zr II and a new determination of the solar abundance of zirconium.**
E. Biémont, N. Grevesse, P. Hannaford, R. M. Lowe.
Astrophys. J., Vol. 248, 867 - 873 (1981).
A new determination of the solar abundance of zirconium has been made using equivalent-width data measured on the Jungfraujoch solar atlas together with new oscillator strengths derived from measurements of atomic lifetimes and branching ratios for 34 lines of Zr I and 24 lines of Zr II. Excellent agreement is found between the results derived from Zr I and Zr II lines and also with recent meteoritic results. The mean abundance of zirconium in the sun is found to be $A_{Zr} = 2.56 \pm 0.05$.

022.038   **Sum rules for hydrogenic atoms in an arbitrary magnetic field.**   J. M. Wadehra.
Astrophys. J., Vol. 248, 874 - 879 (1981).
A general formulation of energy-weighted sum rules for one electron atoms in an arbitrary magnetic field is presented. These sum rules are evaluated in the $B \to 0$ and $B \to \infty$ limits using a trial wave function that is product of a Landau function and a hydrogenic function. As an application, the present sum rules are used to estimate other useful sum rules which cannot be evaluated easily.

022.039   **Cross sections relevant to gamma-ray astronomy: proton induced reactions.**   P. Dyer, D. Bodansky, A. G. Seamster, E. B. Norman, D. R. Maxson.
Phys. Rev. C, Vol. 23, 1865 - 1882 (1981). — Abstr. in Phys. Abstr., Vol. 84, Abstr. 72035 (1981).

022.040   **Spin-orbit coupling in the butadiyne ion.**   D. L. Cooper.
J. Phys. B, Vol. 14, L397 - L399 (1981). — Abstr. in Phys. Abstr., Vol. 84, Abstr. 76707 (1981).

022.041 High-resolution rotation-vibration spectra of D_2O in the region of the ν_1 and ν_3 bands.   V. D. Gupta.
J. Phys. B, Vol. 14, 1761 - 1770 (1981). − Abstr. in Phys. Abstr., Vol. 84, Abstr. 76798 (1981).

022.042 Some studies on the astrophysically significant BN molecule.   V. M. Mummigatti, B. G. Jyoti.
Curr. Sci., Vol. 49, 897 - 898 (1980). − Abstr. in Phys. Abstr., Vol. 84, Abstr. 76874 (1981).

022.043 Rotational cross sections and rate coefficients for e-CO and e-HCN collisions under interstellar conditions.   S. Saha, S. Ray, B. Bhattacharyya, A. K. Barua.
Phys. Rev. A, Vol. 23, 2926 - 2932 (1981). − Abstr. in Phys. Abstr., Vol. 84, Abstr.76963 (1981).

022.044 14 MeV neutron activation analysis of geological and lunar samples.   J. C. Laul, N. A. Wogman.
IEEE Trans. Nucl. Sci., Vol. NS-28, 1703 - 1705 (1981). Abstr. in Phys. Abstr., Vol. 84, Abstr. 79632 (1981).

022.045 Uranium determinations by ^{133}Xe in terrestrial zircon, apatite and chromite; comparative study of thermal releases of fissiogenic xenon from terrestrial and meteoritic minerals.   K. Sakamoto, Y. Hamajima, K. Itoh, K. Yamazaki.
J. Radioanal. Chem., Vol. 60, 343 - 352 (1980). − Abstr. in Phys. Abstr., Vol. 84, Abstr. 79640 (1981).

022.046 Beyond the black hole.
J. A. Wheeler.
Some strangeness in the proportion. Centennial symposium to celebrate the achievements of Albert Einstein, (see 003.001), p. 341 - 375 (1980). − Abstr. in Phys. Abstr., Vol. 84, Abstr. 79719 (1981).

022.047 Annihilations of stationary particles on a lattice.   V. M. Kenkre, H. M. Van Horn.
Phys. Rev. A, Vol. 23, 3200 - 3206 (1981). − Abstr. in Phys. Abstr., Vol. 84, Abstr. 80060 (1981).

022.048 On relativistic kinetic theory: neutrino-antineutron systems. II. The transport coefficients. Separation of matter and anti-matter.   S. R. de Groot.
Proc. K. Nederlandse Akad. Wet., Ser. B, Vol. 84, 41 - 45 (1981). − Abstr. in Phys. Abstr., Vol. 84, Abstr. 80201 (1981).

022.049 A measurement of the hyperfine structure of $C^{17}O$.   M. A. Frerking, W. D. Langer.
J. Chem. Phys., Vol. 74, 6990 - 6991 (1981). − Abstr. in Phys. Abstr., Vol. 84, Abstr. 81110 (1981).

022.050 Measurement of the oscillator strengths and auto-ionization widths of the neutral-aluminum multiplet $3s^2 3p^2 P^o − 3s3p^2 \, ^2P$.
G. G. Lombardi, B. L. Cardon, R. L. Kurucz.
Astrophys. J., Vol. 248, 1202 - 1208 (1981).
The oscillator strengths and line widths of the Al I multiplet $3s^2 3p \, ^2P^o − 3s3p^2 \, ^2P$ at 176 nm were determined by the hook method in conjunction with absorption equivalent width measurements. The measured oscillator strengths agree well with the calculations of Kurucz and Peytremann. Auto-ionization is found to be the dominant decay channel for both the levels of the $3p^2 \, ^2P$ term. The van der Waals scattering cross section for the Al I transition at 176.91 nm by He was found to be $(1.3 \pm 0.3) \times 10^{-14}$ cm^2 at 2400 K. A solar spectrum calculation, centered at 176 nm, showed that the Al I features are possible diagnostic probes of solar non-LTE processes and the temperature structure of the lower chromosphere and temperature minimum.

022.051 Radiative lifetimes of excited electronic states in molecular ions.   B. H. Mahan, A. O'Keefe.

Astrophys. J., Vol. 248, 1209 - 1216 (1981).
Radiative lifetimes of excited electronic states of several molecular ions have been measured using a technique which permits the mass selective storage of ions for periods of many ms. This technique is used to record radiative lifetimes ranging from 60 ns to $\sim 10 \, \mu$s. Careful measurements of the $B \, ^2\Sigma_u^+$ state of N_2^+ and the $A \, ^2\Pi$ states of N_2^+ and CO^+ have been made. The results obtained are in good agreement with previous studies. The analysis of the radiative lifetime for the CH^+ and $CD^+ A \, ^1\Pi(v'=0)$ state results in a value significantly larger than earlier determinations. The new results are used to make revised estimates of interstellar abundances of CH^+.

022.052 The X-ray spectrum of a hot interstellar plasma.   J. M. Shull.
Astrophys. J., Suppl. Ser., Vol. 46, 27 - 40 (1981).
The author calculates the soft X-ray spectrum of a hot, low density, optically thin plasma in collisional ionization equilibrium, using current atomic data. Many Fe collision strengths differ from previous calculations. The spectra from individual elements are presented for use in abundance determinations from X-ray observations. The author tabulates collision strengths and line energies, sorted both according to energy and according to ion.

022.053 Oscillator strengths in the first spectrum of technetium.   R. H. Garstang.
Bull. American Astron. Soc., Vol. 13, 516 (1981). − Abstract.

022.054 Partition functions for uranium at elevated temperature.   J. M. Mack, L. J. Radziemski.
Bull. American Astron. Soc., Vol. 13, 527 (1981). − Abstract.

022.055 The laser-induced breakdown spark − a simple source of laboratory plasmas.   L. J. Radziemski.
Bull. American Astron. Soc., Vol. 13, 556 (1981). − Abstract.

022.056 Motions of a chain of oscillators.   R. Broucke.
Bull. American Astron. Soc., Vol. 13, 571 (1981). − Abstract.

022.057 Some positive ion reactions with H_2 : interstellar implications.   D. Smith, N. G. Adams.
Mon. Not. R. Astron. Soc., Vol. 197, 377 - 384 (1981).
The rate coefficients, k, for the reactions of NH_3^+, Cl^+, S^+, SH^+ and C^{2+} with H_2 have been measured at various temperatures down to 80 K in a SIFT experiment. The problem of explaining NH_3 production in interstellar clouds (ISC) is alleviated, the absence of HCl from ISC spectra remains uncertain. It is argued that radiative association does not produce H_2S^+ in ISC. Also the C^{2+} reaction does not contribute to the production of CH^+ in diffuse ISC. Possible alternative sources of H_2S and CH^+ in ISC are tentatively proposed.

022.058 Laboratory verification of the lunar orbital X-ray fluorescence experiment: initial results.
N. Hubbard, B.-S. King.
The lunar highlands crust, (see 012.013), p. 457 - 466 (1980).

022.059 Rossby soliton.   S. V. Antipov, M. V. Nezlin, E. N. Snezhkin, A. S. Trubnikov.
Pis'ma v ZhEhTF, Tom 33, 368 - 372 (1981). In Russian. Abstr. in Ref. zh., 51. Astron., 7.51.161 (1981).

022.060 Investigation of the light curves of cylindrical artificial satellites.
M. V. Bratijchuk, V. M. Mikhajlets, I. F. Najbauer.
Problems of cosmic physics. Vyp. 16, (see 003.005), p. 105 - 108 (1981). In Russian.
Some characteristics of brightness curves of cylindrical objects are given for comparison with experimental curves of artificial satellites.

022.061 Possibility of the existence of comet-like bodies with small average density in the solar system.
V. A. Dranevich, E. A. Kajmakov, I. S. Lizunkova,
I. N. Matveev, Yu. I. Svetov.
Problems of cosmic physics. Vyp. 16, (see 003.005), p. 127 - 131 (1981). In Russian.
    The suggestion on the existence of small bodies of average density in the solar system is put forward. It is based on the results of numerous laboratory experiments on the sublimation of frozen solutions into a vacuum. Genetically these objects are related to comets.

022.062 Modelling the interaction between solar radiation and cometary matter.
V. A. Dranevich, E. A. Kajmakov, I. S. Lizunkova,
I. N. Matveev, Yu. I. Svetov.
Problems of cosmic physics. Vyp. 16, (see 003.005), p. 131 - 135 (1981). In Russian.
    The question of possible parental molecules of comets is discussed.

022.063 Near-infrared spectral reflectance of mineral mixtures: systematic combinations of pyroxenes, olivine, and iron oxides.   R. B. Singer.
J. Geophys. Res., Vol. 86, 7967 - 7982 (1981).
    Near-infrared spectral reflectance data are presented for systematic variations in weight percent of two component mixtures of ferromagnesian and iron oxide minerals. Mixtures were chosen for application to the study of the dark materials on Mars, but the results are equally applicable to mafic and ultramafic assemblages elsewhere in the solar system.

022.064 High temperature crystal field spectra of transition metal-bearing minerals: relevance to remote-sensed spectra of planetary surfaces.   K. M. Parkin, R. G. Burns.
Proc. Eleventh Lunar Planet. Sci. Conf., (see 012.020), p. 731 - 755 (1980).

022.065 Amorphous gels as possible analogs to martian weathering products.
D. L. Evans, J. B. Adams.
Proc. Eleventh Lunar Planet. Sci. Conf., (see 012.020), p. 757 - 763 (1980).

022.066 The effects of scattering geometry on the spectrophotometric properties of powdered material.
J. Gradie, J. Veverka, B. Buratti.
Proc. Eleventh Lunar Planet. Sci. Conf., (see 012.020), p. 799 - 815 (1980).

022.067 Laboratory studies of actinide metal-silicate fractionation.   J. H. Jones, D. S. Burnett.
Proc. Eleventh Lunar Planet. Sci. Conf., (see 012.020), p. 995 - 1001 (1980).

022.068 Impact cratering in viscous targets: laboratory experiments.   R. Greeley, J. Fink, D. E. Gault,
D. B. Snyder, J. E. Guest, P. H. Schultz.
Proc. Eleventh Lunar Planet. Sci. Conf., (see 012.020), p. 2075 - 2097 (1980).

022.069 The equivalent depth of burst for impact cratering.
K. A. Holsapple.
Proc. Eleventh Lunar Planet. Sci. Conf., (see 012.020), p. 2379 - 2401 (1980).

022.070 Semiempirical Stark linewidths of alkali like ions.
M. S. Dimitrijević, N. Konjević.
Astron. Astrophys., Vol. 102, 93 - 96 (1981).
    In this paper the authors have tested the applicability of modified semi-empirical formula with newly proposed Gaunt factors for evaluation of Stark linewidths of alkali like ions.

The comparison with experiments and other, already tested theoretical approaches indicates that these new Gaunt factors can be used successfully for Stark linewidth calculations.

022.071 About the ergodicity hypothesis in random propagation studies.   G. Bourgois.
Astron. Astrophys., Vol. 102, 212 - 222 (1981).
    The theoretical study of the propagation of a wave through a random medium has as a motive to connect the statistical properties of the medium to the statistical properties of the received wave. The theory assumes the stationarity of the medium and uses ensemble averaged mean values. Under the ergodicity hypothesis, the theory requires a duration of observation long enough to establish an equivalence between a temporal average and the ensemble average. This leads to an equivalence condition which involves the correlation length of the random refractive index of the medium, its velocity with respect to the observer and the duration of the observation.

022.072 Hypervelocity impact craters in leucitite targets.
R. Bianchi, M. Coradini, G. Martelli.
Mem. Soc. Astron. Italiana, Vol. 52, (see 012.023), 349 (1981). Abstract.

022.073 A possible mechanism for the magnification of magnetic fields during crater formation.
P. Cerroni, G. Martelli.
Mem. Soc. Astron. Italiana, Vol. 52, (see 012.023), 375 - 377 (1981).

022.074 Hypervelocity impact cratering experiments.
G. Martelli, P. Cerroni, P. Hurren, P. N. Smith,
M. Coradini, R. Bianchi, M. Fulchignoni, F. Waldner, R. Flavill.
Mem. Soc. Astron. Italiana, Vol. 52, (see 012.023), 447 (1981). Abstract.

022.075 On the foundation of the method of virtual contacts.   N. N. Kozlov.
Inst. prikl. mat. AN SSSR. Prepr., 1980 (1981), No. 4, p. 28.
In Russian. — Abstr. in Ref. zh., 51. Astron., 9.51.182 (1981).

022.076 Detection of the infrared spectrum of cyanobutadiyne, HC_5N.   K. Yamada, G. Winnewisser.
Zeitschr. Naturforsch., Band 36a, 1052 - 1056 (1981).

022.077 f values for isoelectronic ions of carbon.
P. S. Ganas.
Astron. Astrophys., Suppl. Ser., Vol. 46, 101 - 103 (1981).

022.078 Phase picture of the SU(5) grand unified model.
V. A. Kuzmin (Kuz'min), M. E. Shaposhnikov,
I. I. Tkachev.
Phys. Lett. B, Vol. 102B, 397 - 400 (1981). — Abstr. in Phys. Abstr., Vol. 84, Abstr. 84219 (1981).

022.079 Grand unified theories and proton decay.
P. Langacker.
Phys. Rep., Vol. 72, 185 - 385 (1981). — Abstr. in Phys. Abstr., Vol. 84, Abstr. 84227 (1981).

022.080 Measurement of the absolute photoionization cross section of the $3p^3 \, P^0$ term of neutral magnesium.
G. G. Lombardi, P. L. Smith, W. H. Parkinson.
Phys. Rev. A, Vol. 24, 326 - 330 (1981). — Abstr. in Phys. Abstr., Vol. 84, Abstr. 85120 (1981).

022.081 Collisional excitation of OH by H_2: transitions within the groundstate Λ doublet.
D. P. Dewangan, D. R. Flower.

J. Phys. B, Vol. 14, L425 - L429 (1981). – Abstr. in Phys. Abstr., Vol. 84, Abstr. 85248 (1981).

022.082 Rotational excitation of OH by H_2 at thermal energies.   D. P. Dewangan, D. R. Flower.
J. Phys. B, Vol. 14, 2179 - 2190 (1981). – Abstr. in Phys. Abstr., Vol. 84, Abstr. 85250 (1981).

022.083 Theory of the pressure broadening and shift of spectral lines.   G. Peach.
Adv. Phys., Vol. 30, 367 - 474 (1981). – Abstr. in Phys. Abstr., Vol. 84, Abstr. 85873 (1981).

022.084 Interference between ordinary and higher-order amplitudes in O II magnetic dipole transitions.
W. Eissner, C. J. Zeippen.
J. Phys. B, Vol. 14, 2125 - 2137 (1981). – Abstr. in Phys. Abstr., Vol. 84, Abstr. 88733 (1981).

022.085 Some comments on the relativistic Thomas-Fermi model and the Vallarta-Rosen equation.
R. Ruffini, L. Stella.
Phys. Lett. B, Vol. 102B, 442 - 444 (1981). – Abstr. in Phys. Abstr., Vol. 84, Abstr. 88738 (1981).

022.086 The extinction coefficients in mid- and far-infrared of silicate and iron-oxide minerals of interest for astronomical observations.
C. Koike, H. Hasegawa, N. Asada, T. Hattori.
Astrophys. Space Sci., Vol. 79, 77 - 85 (1981).
Extinction measurements were made for some silicate and iron-oxide mineral grains in mid- and far-infrared region. For far-infrared region, high temperature magnesium silicates such as olivine and pyroxenes show the absorption spectra of steep dependence as λ^{-3} (λ being the wavelength) with some peak structure, but the spectrum of magnetite shows λ^{-1} dependence.

022.087 Comparison of band model and integrated line-by-line synthetic spectra for methane in the 2.3 μm region.   J. Apt, J. V. Martonchik, L. R. Brown.
J. Quant. Spectrosc. Radiat. Transfer, Vol. 26, 431 - 442 (1981).
The 2.3 μm spectral region of methane can be used to retrieve cloud properties of planetary spectra, provided parameters for the methane spectrum are known. Two standard techniques for calculating absorption spectra in this region are compared. A Voigt profile Mayer-Goody random band model is applied, using coefficients empirically fitted by Fink et al., to CH_4 spectra recorded with high absorbing amounts at 10 cm^{-1} resolution. Calculation of the absorption is also done with a line-by-line direct integration method for the same gas conditions using molecular parameters obtained by combining an older unpublished list of observed positions and estimated line strengths with quantum assignments from the literature. The molecular parameters have been evaluated for the 4180 - 4590 cm^{-1} region by comparing new laboratory spectra with 0.01 cm^{-1} resolution recorded at 296 and 153 K with synthetic spectra calculated at the same conditions. Curves of growth of the total equivalent width are calculated at 296 and 55 K for a pathlength of 50 cm and pressures up to 10 atm.

022.088 Rotational structure of CN emission from active nitrogen flames.   M. R. Gorbal, M. I. Savadatti.
J. Quant. Spectrosc. Radiat. Transfer, Vol. 26, 457 - 462 (1981).
The CN violet bands from active nitrogen flames have been studied. Rotational line intensities and rotational temperatures have been measured for a number of bands (0-1, 4-6, 5-7, 11-11, 12-12, and13-13) at pressures ranging from 3 to 30 torr.

022.089 Magnetic susceptibility of relativistic Fermi gas.
E. M. Chudnovsky.

J. Phys. A, Vol. 14, 2091 - 2094 (1981). – Abstr. in Phys. Abstr., Vol. 84, Abstr. 89403 (1981).

022.090 Topological symmetry restoration.
G. Kennedy.
Phys. Rev. D, Vol. 23, 2884 - 2900 (1981). – Abstr. in Phys. Abstr., Vol. 84, Abstr. 89720 (1981).

022.091 Observation of the infrared spectrum of the triatomic molecular ion $H_2 D^+$.
J.-T. Shy, J. W. Farley, W. H. Wing.
Phys. Rev. A, Vol. 24, 1146 - 1149 (1981). – Abstr. in Phys. Abstr., Vol. 84, Abstr. 90575 (1981).

022.092 The emission spectrum of H_2 from associative detachment and ultraviolet pumping.
J. H. Black, A. Porter, A. Dalgarno.
Astrophys. J., Vol. 249, 138 - 144 (1981).
The infrared emission spectrum of H_2 resulting from the associative detachment of H and H^- is calculated. It is characterized by pure rotational transitions in excited vibrational states and is readily distinguished from that expected from ultraviolet pumping or from shock excitation. The combined effects of associative detachment and of ultraviolet pumping are discussed with reference to H_2 line emission from planetary nebulae.

022.093 Recombination coefficients for iron ions.
D. T. Woods, J. M. Shull, C. L. Sarazin.
Astrophys. J., Vol. 249, 399 - 401 (1981).
The authors present new calculations of radiative recombination rate coefficients for Fe ions, based on photoionization cross sections computed by Reilman and Manson. They provide analytic fits to these radiative rates and to dielectronic recombination rates recently computed by Jacobs et al. For Fe I through Fe VII, the authors compute dielectronic rates with the Burgess general formula.

022.094 Quasi-static Stark profiles of hydrogen H$n\alpha$ lines in a plasma.   S. A. Gulyaev, G. V. Sholin.
Astron. Zh., Tom 58, 1057 - 1062 (1981). In Russian. English translation in Soviet Astron., Vol. 25, No. 5.
On the basis of numerical calculations of quasistatic contours of the hydrogen H$n\alpha$-type lines, new formulae are derived for Stark halfwidths of these lines. The new formulae predict a difference by a factor of 7.5 from the commonly used asymptotic halfwidths. The universal function $T_\alpha(x)$ describing the quasistatic Stark contour of an H$n\alpha$ line has been redetermined. The behaviour of the line intensity in the wing and in the central part of the contour is investigated. A comparison is performed with the results of observations of Stark contours of the lines of hydrogen-like ions emitted by a laser plasma.

022.095 A comparison of methods for the calculation of Voigt profiles.   A. Klim.
J. Quant. Spectrosc. Radiat. Transfer, Vol. 26, 537 - 545 (1981).
Six algorithms and approximations for the calculation of Voigt profiles are compared for accuracy and time required for calculation. For high-precision claculations, a GAUTSCHI program is recommended; for calculations with a few per cent of error, the Matvejev and the Kielkopf approximations are recommended.

022.096 Hydroxycarbene (HCOH) and protonated formaldehyde: two potentially observable interstellar molecules.   M. R. Hoffmann, H. F. Schaefer III.
Astrophys. J., Vol. 249, 563 - 565 (1981).
It is known that the lowest energy isomer of protonated formaldehyde has the proton attached to the oxygen atom. Therefore, a purely statistical argument would suggest that the hydroxycarbene molecule HCOH is twice as likely as formal-

dehyde to be formed from H_3CO^+ via dissociative recombination, a process generally considered to be very important in interstellar clouds. Theoretical studies show that, while HCOH has never been observed in the laboratory, it should nevertheless be a relatively stable molecule. Theoretical equilibrium structures for H_3CO^+ and HCOH suggest that they may be observed at 63.5 and 68.4 GHz respectively.

022.097　**A mechanism for producing ground vibrational state SiO masers.**
S. E. Robinson, D. J. Van Blerkom.
Astrophys. J., Vol. 249, 566 - 571 (1981).

The authors propose a process by which $v = 0$ SiO masers may be produced in slowly accelerating circumstellar envelopes associated with Mira variables. A numerical model is used to demonstrate the feasibility of this mechanism and to investigate its properties. The authors conclude that ground vibrational state SiO masers will be strong only when $v = 1$ masers are absent or weak. The ground state $J = 1 \rightarrow 0$ transition is found to be the strongest $v = 0$ maser transition.

022.098　**Density and temperature diagnostics of X-ray sources: line ratios for helium-like ions.**
A. K. Pradhan, J. M. Shull.
Astrophys. J., Vol. 249, 821 - 830 (1981).

Improved calculations are made for the line ratios $R = f/i$ and $G = (i + f)/r$ involving the forbidden (f), the intercombination (i), and the resonance (r) lines of helium-like ions: C V, O VII, Ne IX, Si XIII, Ca XIX, and Fe XXV. The ratios in ionization equilibrium are obtained over a range of electron temperatures. Departures from ionization equilibrium are also studied. For the line ratios observed in the Sun and in supernova remnants, the density and temperature diagnostics based on previous works may not be entirely correct.

022.099　**The infrared spectrum of SiO near 1240 cm^{-1} and its relation to the circumstellar SiO maser.**
F. J. Lovas, A. G. Maki, W. B. Olson.
J. Mol. Spectrosc., Vol. 87, 449 - 458 (1981). – Abstr. in Phys. Abstr., Vol. 84, Abstr. 95713 (1981).

022.100　**Absolute oscillator strengths of Eu$^+$ spectral lines.**
C. Karner, G. Meyer, W. Ruland, A. Sahm, F. Träger, G. zu Putlitz.
Upper main sequence chemically peculiar stars, (see 012.033), p. 223 - 227 (1981).

022.101　**New spectroscopic data for the iron group elements.**
S. Johansson.
Upper main sequence chemically peculiar stars, (see 012.033), p. 229 - 231 (1981).

022.102　**The 11.5 micrometer emission from carbon stars: comparison with IR spectra of submicrometer-sized silicon carbide grains.**
C. Friedemann, J. Gürtler, R. Schmidt, J. Dorschner.
Astrophys. Space Sci., Vol. 79, 405 - 417 (1981).

In this paper the results of an experimental investigation of the spectra of submicrometer-sized silicon carbide grains are presented. The theoretical profile of the 11.5 μm band shows a striking similarity with the observed profile in the spectrum of the carbon star Y CVn. The total amount of SiC dust in the envelope of this star has been estimated at about 10^{24} g.

022.103　**Sextet transitions in Fe II.**
H. Nussbaumer, M. Pettini, P. J. Storey.
Astron. Astrophys., Vol. 102, 351 - 358 (1981).

022.104　**Oscillator strengths for lines of the $\widetilde{F}$ (0, 0, 0) – $\widetilde{X}$ (0, 0, 0) band of H_2O at 111.5 nanometers and the abundance of H_2O in diffuse interstellar clouds.**
P. L. Smith, K. Yoshino, H. E. Griesinger, J. H. Black.
Astrophys. J., Vol. 250, 166 - 174 (1981).

Absolute oscillator strengths (f-values) for rotational lines of the $\widetilde{F}$ (0, 0, 0) – $\widetilde{X}$ (0, 0, 0) band of H_2O at 111.5 nm have been measured with an uncertainty of ±40% using quantitative photographic techniques. The f-value for the $1_{11} - 0_{00}$ line, the one most likely to be seen in absorption in interstellar clouds, is $(3.0 \pm 1.2) \times 10^{-2}$. This value is combined with a measured upper limit for absorption by this line in the ζ Oph cloud to show that the column density of H_2O in the 0_{00} level is less than 5.3×10^{12} cm^{-2}. This upper limit is larger than the predictions of gas-phase, chemical models of this cloud. The authors show that definite detection of H_2O in diffuse clouds at the level of this upper limit would be in conflict with models of molecular formation in shock-heated gas.

022.105　**On the establishment of internally consistent solar scales of oscillator strengths and abundances of chemical elements. I. Oscillator strengths for 865 Fe I lines iron abundance.**
E. A. (*Eh. A.*) Gurtovenko, R. I. Kostik (*Kostyk*).
Astron. Astrophys., Suppl. Ser., Vol. 46, 239 - 248 (1981).

The authors use the observed central intensities of 865 solar Fe I lines between 4000 Å and 8000 Å to derive the photospheric iron abundance and the values of the oscillator strengths of these lines on a self-consistent scale, which is calibrated using the precise laboratory data of Blackwell et al. (1976, 1979) for 20 lines. The authors' value of the solar iron abundance is log A_{Fe} (H = 12) = 7.57 ± 0.01.

022.106　**Present state and prospects of high-energy physics.**
L. B. Okun'.
Usp. fiz. nauk, Tom 134, 3 - 44 (1981). In Russian. – Abstr. in Ref. zh., 51. Astron., 10.51.102 (1981).

022.107　**Investigation of the phenomena originating in a collision with different surfaces of solid particles with high velocities.**　Yu. G. Mallama.
Inst. kosm. issled. AN SSSR. Prepr., 1981, No. 642, 67 pp. In Russian. – Abstr. in Ref. zh., 62. Issled. kosm. prostranstva, 10.62.242 (1981).

022.108　**Can silicon monoxide grains be responsible for the interstellar 9.7 micrometer absorption band?**
J. Gürtler, R. Schmidt, J. Dorschner, C. Friedemann.
Astrophys. Space Sci., Vol. 80, 249 - 255 (1981).

In this paper results of an experimental investigation of the spectra of submicrometer-sized silicon monoxide grains are presented. The sensitivity against heating by some hundred K rules out that SiO grains can be responsible for the 9.7 μm circumstellar as well as interstellar spectral feature.

022.109　**Lifetime measurements in Cr I by laser excitation from metastable states.**　M. Kwiatkowski, G. Micali, K. Werner, P. Zimmermann.
Astron. Astrophys., Vol. 103, 108 - 110 (1981).

A combination of collisional and laser excitation was used to measure radiative lifetimes in Cr I. By a discharge an atomic beam of metastable atoms in the $3d^5 4s a^5 S$, $a^5 G$, $b^5 D$, $a^3 I$, $b^1 I$ and $3d^4 4s^2 a^5 D$ terms was produced. Spatially separated from the place of collisional excitation laser radiation selectively populated levels belonging to the $3d^5 4pz^5 P$, $y^5 P$, $u^5 F$, $u^5 D$, $x^3 I$, $y^1 I$, $3d^5 5pz^5 G$ and $3d^4 4s4px^5 G$ terms. Time-resolved observation of the reemitted resonance fluorescence yielded the lifetimes of 28 levels. The values are compared with other experimental and theoretical results.

022.110　**Atomic calculation for Fe XXIII, UV, and X-ray lines.**　A. K. Bhatia, H. E. Mason.
Astron. Astrophys., Vol. 103, 324 - 330 (1981).

Electron-ion scattering data have been obtained for the configurations $2s^2$, $2s\,2p$, $2p^2$, $2s\,\overline{3}s$, $2s\,3p$, and $2s\,3d$ of

Fe XXIII in the "distorted wave" approximation. Oscillator and collision strengths are compared with other calculations. Wavelengths for various UV and X-ray lines are compared with laboratory measurements. New identifications are given for recent solar spectra based on intensity estimates. A line at 11.737 Å is identified as the strongest Fe XXIII line in the solar X-ray spectrum.

022.111 **HOC$^+$: an observable interstellar species? A comparison with the isomeric and isoelectronic HCO$^+$, HCN and HNC.** R. H. Nobes, L. Radom.
Chem. Phys., Vol. 60, 1 - 10 (1981). − Abstr. in Phys. Abstr., Vol. 84, Abstr. 99703 (1981).

022.112 **Reactions of Si$^+$ with H$_2$O and O$_2$ and SiO$^+$ with H$_2$ and D$_2$.**
D. W. Fahey, F. C. Fehsenfeld, E. E. Ferguson, L. A. Viehland.
J. Chem. Phys., Vol. 75, 669 - 674 (1981). − Abstr. in Phys. Abstr., Vol. 84, Abstr. 101487 (1981).

022.113 **The neutron lifetime.** D. H. Wilkinson.
Nuclear astrophysics, (see 012.036), p. 325 - 332 (1981).

022.114 **Magnetic monopoles and grand unified theories.** D. I. Olive.
Unification of the fundamental particle interactions, (see 012.038), p. 451 - 459 (1980). − Abstr. in Phys. Abstr., Vol. 84, Abstr. 102878 (1981).

022.115 **Vibration-rotation intensities of SiO.**
R. H. Tipping, C. Chackerian, Jr.
J. Mol. Spectrosc., Vol. 88, 352 - 363 (1981). − Abstr. in Phys. Abstr., Vol. 84, Abstr. 103658 (1981).

022.116 **Simulating the transparent gravitational lens.**
C. C. Dyer, R. C. Roeder.
J. R. Astron. Soc. Canada, Vol. 75, 227 - 236 (1981).

022.117 **Oscillator strengths for neutral technetium.**
R. H. Garstang.
Publ. Astron. Soc. Pacific, Vol. 93, 641 - 646 (1981).
Oscillator strengths have been calculated for most of the spectral lines of Tc I which are of interest in the study of stars of spectral type S. Oscillator strengths have been computed for the corresponding transitions in Mn I as a partial check of the technetium calculations.

022.118 **Etch a meteorite.** E. Fortier.
Sky Telesc., Vol. 62, 527 - 529 (1981).

022.119 **Dust-sensitive forbidden line ratios.**
B. T. Draine, J. N. Bahcall.
Astrophys. J., Vol. 250, 579 - 589 (1981).
The differential extinction toward distant objects such as quasars, or galactic objects such as planetary nebulae or H II regions, may be measured by using the known intrinsic ratios of forbidden emission lines. A compilation is made of all line pairs of cosmically abundant elements that are suitable for this purpose. [O III] 2322, 4364 is the best line pair for the (ground-based) detection of dust toward moderate-redshift quasars ($0.51 < z < 0.95$).

022.120 **Solar observations and atomic data for the $3s^2\,^1S_0-3s3p\,^3P_1$ transition in S V.**
U. Feldman, G. A. Doschek, A. K. Bhatia.
Astrophys. J., Vol. 250, 799 - 804 (1981).
High resolution solar observations of the S V intersystem line at 1199.18 Å are available from Skylab. This line is potentially useful as a density diagnostic for high density plasmas expected in solar flares. S V lines are also prominent in solar spectra at wavelengths below 1000 Å. Collision strengths and radiative decay rates are calculated for levels of the configurations $3s^2$, $3s3p$, $3p^2$, and $3s3d$. Level populations for the five lowest energy levels have been calculated as a function of electron density. These calculations are carried out assuming the temperature at which S V is most abundant in solar plasmas. The calculated population of the $3s3p\,^3P_1$ level reaches a pseudo-Boltzmann equilibrium at a density which is ~5 times higher than is derived from solar spectra and previous density determinations.

022.121 **Laboratory identification of Keenan and Wing bands.**
S. P. Davis, P. D. Hammer.
Astrophys. J., Vol. 250, 805 - 810 (1981).
The spectrum of ZrO produced in a furnace at 2700 K is compared with astronomical observations of the Keenan and Wing bands. Bands in the laboratory spectrum at 929.9, 931.6, 973.3, 1051.5, and 1052.8 nm are identified as belonging to the ZrO molecule, through analysis of the rotational structure. The bands at 827.4, 828.3, 847.6, 848.1, 861.0, 882.1, and 924.6 nm appear to belong to ZrO as well, although an analysis has not been made.

022.122 **Laboratory millimeter and submillimeter spectra of CO$^+$.**
K. V. L. N. Sastry, P. Helminger, E. Herbst, F. C. De Lucia.
Astrophys. J., Lett., Vol. 250, L91 - L92 (1981).
The strong electric dipole-allowed transitions of the molecular ion CO$^+$ in the region 235 - 470 GHz have been measured in the laboratory. The laboratory spectra at 235 GHz appear to confirm the claim by Erickson et al. to have observed CO$^+$ in OMC-1.

022.123 **Dynamical motions of charged particles. Equilibrium conditions.** D. D. Dionysiou.
Astrophys. Space Sci., Vol. 80, 477 - 482 (1981).
A new approach to the problem of the equilibrium conditions of self-interacting massive charged particles is presented.

022.124 **Suppression of the firehouse instability by hard energy spectra.** P. D. Noerdlinger.
Astrophys. Space Sci., Vol. 80, 513 - 516 (1981).
The growth rate of the relativistic firehouse instability is an order of magnitude smaller, near marginal instability, for plasmas with particle energy spectra $N(E)dE \propto E^{-2.5}\,dE$ than for those with $N(E)dE \propto E^{-3}\,dE$. If cosmic rays are accelerated impulsively in limited regions, with initially flat spectra, the sudden increase in the firehouse growth rate could lead to release of the particles when the spectrum has steepened beyond exponent 2.5.

022.125 **Spectral properties of ice mineral mixtures: implications on the composition of the Galilean satellites and other icy bodies.** R. N. Clark, P. G. Lucey.
Bull. American Astron. Soc., Vol. 13, 700 (1981). − Abstract.

022.126 **Reflection spectra of simulated Titan organic clouds.**
B. N. Khare, C. Sagan, J. Gradie.
Bull. American Astron. Soc., Vol. 13, 701 (1981). − Abstract.

022.127 **Particle size, absorption strength and mineral mixtures.** T. V. V. King, C. M. Pieters.
Bull. American Astron. Soc., Vol. 13, 709 (1981). − Abstract.

022.128 **Microtektite spherules vs. volcanic shards: a question of foamability.**
J. A. O'Keefe, A. N. Thorpe.
Bull. American Astron. Soc., Vol. 13, 711 (1981). − Abstract.

022.129 **A new interpretation of Venera 11 spectra. I. Comparison with laboratory data.** L. D. G. Young.
Bull. American Astron. Soc., Vol. 13, 715 (1981). − Abstract.

022.130 **Laboratory measurements of the microwave opacity of cloud related gases under simulated conditions for the Venus atmosphere.**   P. G. Steffes, V. R. Eshleman.
Bull. American Astron. Soc., Vol. 13, 716 (1981). – Abstract.

022.131 **Monte Carlo simulations of the charged-particle irradiation of asteroidal regoliths.**
K. R. Housen, L. L. Wilkening.
Bull. American Astron. Soc., Vol. 13, 718 - 719 (1981). Abstract.

022.132 **Experimental impacts: recent progress on regolith velocity distributions.**   W. K. Hartmann.
Bull. American Astron. Soc., Vol. 13, 719 (1981). – Abstract.

022.133 **Laser induced photodissociation of cometary and planetary molecules: the production and quenching of NH ($A^3\pi_i$).**   R. J. Cody, J. E. Allen, Jr., W. F. Rowe.
Bull. American Astron. Soc., Vol. 13, 724 (1981). – Abstract.

022.134 **Strength and temperature dependence of infrared spectrum of methane.**   K. Fox.
Bull. American Astron. Soc., Vol. 13, 733 (1981). – Abstract.

022.135 **Laboratory spectra of cold methane from 1.0 to 2.7 μ with abundances up to 1 km-amagat.**
D. C. Benner, L. P. Giver, R. W. Boese, D. Goorvitch, U. Fink.
Bull. American Astron. Soc., Vol. 13, 733 - 734 (1981). Abstract.

022.136 **Line strengths versus band strengths for the v_2 and v_4 bands of methane.**
B. L. Lutz, C. Pierre, G. Pierre, J. P. Champion.
Bull. American Astron. Soc., Vol. 13, 734 (1981). – Abstract.

022.137 **Sulfur-containing glass: a possible colorant on Io and Amalthea.**
J. Gradie, S. J. Ostro, J. Veverka.
Bull. American Astron. Soc., Vol. 13, 740 (1981). – Abstract.

022.138 **Reflection spectra of molten sulfur: could sulfur liquid be on Io's surface?**
R. M. Nelson, D. C. Pieri, S. M. Baloga, C. Sagan, D. B. Nash.
Bull. American Astron. Soc., Vol. 13, 740 - 741 (1981). Abstract.

022.139 **A review of laboratory data on charged particle erosion of condensed gas frosts.**
W. L. Brown, L. J. Lanzerotti, R. E. Johnson, J. W. Boring, C. Reimann, V. Pirronello.
Bull. American Astron. Soc., Vol. 13, 743 - 744 (1981). Abstract.

022.140 **Statistical equilibrium in cometary C_2. III. Triplet-singlet, Phillips, Ballik-Ramsay, and Mulliken bands.**
K. S. Krishna Swamy, C. R. O'Dell.
Astrophys. J., Vol. 251, 805 - 808 (1981).
A new series of vibrational bands is predicted. These bands arise from transitions between the lowest electronic states of the triplet and singlet states of homonuclear C_2. Both intensities and wavelengths are predicted. Detailed predictions for Mulliken, Phillips, and Ballik-Ramsay band sequences are also given.

022.141 **Cross section for the $^{26}Mg + p \rightleftharpoons ^{26}Al + n$ reactions.**
E. B. Norman, T. E. Chupp, K. T. Lesko, P. J. Grant.
Astrophys. J., Vol. 251, 834 - 837 (1981).
A thick-target yield technique has been used to determine cross sections for the $^{26}Mg(p, n)^{26}Al$ reaction from threshold to $E_p = 7.0$ MeV. Using these cross sections for proton energies below the $^{26}Al^m$ threshold and the principle of detailed balance, cross sections for the $^{26}Al_{g.s.}(n, p_0)^{26}Mg_{g.s.}$ reac-

tion have been calculated. The significance of these cross sections for the nucleosynthesis of ^{26}Al is discussed.

022.142 **Laboratory millimeter and submillimeter spectrum of CCH.**   K. V. L. N. Sastry, P. Helminger, A. Charo, E. Herbst, F. C. De Lucia.
Astrophys. J., Lett., Vol. 251, L119 - L120 (1981).
The laboratory millimeter and submillimeter spectrum of the interstellar radical CCH has been observed and measured in the 174 - 350 GHz region. The fitted spectral constants are in excellent agreement with those determined from radioastronomical measurement.

022.143 **The infrared $B^1\Pi - A^1\Delta$ system of ZrO.**
P. D. Hammer, S. P. Davis.
Astrophys. J., Suppl. Ser., Vol. 47, 201 - 228 (1981).
The emission spectrum of gaseous ZrO was obtained using a King-type furnace and a Fourier transform spectrometer. A rotational and vibrational analysis has been made of a system extending from 0.96 to 1.21 μm, verifying the identity of the new $B^1\Pi - A^1\Delta$ system and establishing $T_e(A^1\Delta) = 5904.19$ cm^{-1} above the ground state for ^{90}ZrO.

022.144 **Particle acceleration.**   R. M. Kulsrud.
Plasma astrophysics, (see 012.042), p. 105 - 109 (1981).
The general theory of particle acceleration is illustrated by a discussion of three particular types of acceleration. These types are Fermi acceleration, acceleration by strong electromagnetic waves such as those assumed to be produced by pulsars, and acceleration by collisionless shocks. The physical processes involved as well as the amount of acceleration produced in each type is discussed.

022.145 **Laboratory experiments and space phenomena.**
I. Podgorny (Podgornyj).
Active experiments in space plasmas. Proceedings of Symposium 9 of the COSPAR Twenty-third Plenary Meeting held in Budapest, Hungary, 2 - 14 June 1980. C. T. Russell, M. J. Rycroft (Editors). Adv. Space Res., Vol. 1, No. 2, p. 385 - 401 (1981).
Two types of convection were observed in the laboratory model of the magnetosphere: viscous convection and convection due to field lines common to both the magnetosphere and artificial solar wind. The formation of an induced magnetosphere with a magnetic tail is shown in the experiments of the simulated conditions near non-magnetic bodies with a plasma shell (Venus, comets). A combined induced-intrinsic magnetosphere also was investigated.

022.146 **The relevant atomic data.**
D. L. Moores, H. Nussbaumer.
Space Sci. Rev., Vol. 29, (see 012.043), 379 - 386 (1981).
The current status of the theoretical methods for producing the relevant atomic data is surveyed.

022.147 **Atomic spectroscopy for astrophysical applications.**
S. Johansson.
Space Sci. Rev., Vol. 29, (see 012.043), 387 - 390 (1981).
The recent work in atomic spectroscopy at the Department of Physics, University of Lund, is briefly reviewed and some examples of application to astrophysics are presented.

022.148 **keV neutron capture cross sections for the s-process isotopes of Se, Br and Kr and the abundance of krypton in the solar system.**
B. Leugers, F. Kappeler, F. Fabbri, G. Reffo.
Nuclear cross sections for technology, (see 012.039), p. 857 - 861 (1980). – Abstr. in Phys. Abstr., Vol. 85, Abstr. 3935 (1982).

022.149 **Millimeter and submillimeter spectra of HCO⁺ and DCO⁺.** K. V. L. N. Sastry, E. Herbst, F. C. De Lucia.
J. Chem. Phys., Vol. 75, 4169 - 4170 (1981). − Abstr. in Phys. Abstr., Vol. 85, Abstr. 4264 (1982).

022.150 **Negative masses and the energy-sources of the Universe.** Ya. P. Terletskij.
Exp. Tech. Phys., Vol. 29, 331 - 332 (1981). − Abstr. in Phys. Abstr., Vol. 85, Abstr. 6689 (1982).

022.151 **On the MgH oscillator strengths.** K. Sinha.
J. Astrophys. Astron., Vol. 2, 285 - 288 (1981).

022.152 **Infrared spectroscopy of interplanetary dust in the laboratory.**
P. Fraundorf, R. I. Patel, J. J. Freeman.
Icarus, Vol. 47, (see 012.045), 368 - 380 (1981).
Diagnostic infrared spectra of individual nanogram-sized interplanetary dust particles (IDPs) collected in the Earth's stratosphere have been obtained. The structural diversity of " "chondritic" IDPs and possible effects of atmospheric heating must be considered when comparing this spectrum with astrophysical spectra of interplanetary and cometary dust. Transmission electron microscope (TEM) and infrared observations are also reported on one member of the rare subset of IDPs which resemble hydrated carbonaceous chondrite matrix material.

022.153 **Laboratory measurements of cometary photochemical phenomena.** W. M. Jackson.
Modern observational techniques for comets, (see 012.047), p. 257 - 274 (1981).
Laboratory experiments are described that provide fundamental information about photochemical processes in comets. The yield of cometary radicals such as CN, OH, etc. can be determined as a function of photolyzing wavelength. Quantum state distributions of the internal energy of the cometary radicals can also be measured as a function of wavelength permitting one to define the recoil velocity of the fragments. This type of information supplies the data needed for more elaborate models to interpret the data being obtained on comets.

022.154 **Eisenlinien in superstarken Magnetfeldern.** H. Ruder, G. Wunner, H. Herold, M. Reinecke, J. Trümper.
Mitt. Astron. Ges., Nr. 54, (see 012.050), p. 271 - 273 (1981).

022.155 **Photoionisation von wasserstoffähnlichen Atomen in starken Magnetfeldern.** G. Wunner, H. Ruder, H. Herold, M. Reinecke, W. Schmitt.
Mitt. Astron. Ges., Nr. 54, (see 012.050), p. 273 - 275 (1981).

022.156 **Electron-impact excitation cross-sections for the silicon atom.**
P. A. Kolosov, A. Yu. Krasavin, Yu. M. Smirnov.
Astron. Zh., Tom 58, 1213 - 1216 (1981). In Russian. English translation in Soviet Astron., Vol. 25, No. 6.
The excitation of spectral lines of the silicon atom has been investigated by the method of crossing beams. The absolute values of excitation cross-sections of 14 ultraviolet lines of the silicon atom have been measured.

022.157 **On the problem of estimate of the neutrino mass.** K. P. Stanyukovich, V. N. Mel'nikov, Yu. M. Nikolaev.
Astron. Tsirk., No. 1139, p. 5 - 7 (1980). In Russian.

022.158 **On the kinetics of nuclear reactions in quasi-equilibrium approximation.** A. M. Khokhlov.

Pis'ma Astron. Zh., Tom 7, 741 - 747 (1981). In Russian. English translation in Soviet Astron. Lett., Vol. 7.
Nuclear statistical quasi-equilibrium is considered under variable temperature, density and neutron excess.

022.159 **Radiative-lifetime measurements of the $4p\,^5P$, $4p\,^3P$, and $4d\,^5D°$ multiplets of O I.**
R. L. Day, R. J. Anderson, G. J. Salamo.
J. Opt. Soc. America, Vol. 71, 851 - 855 (1981).

022.160 **Optical oscillator strengths for the phosphorus isoelectronic sequence.** P. S. Ganas.
J. Opt. Soc. America, Vol. 71, 908 - 909 (1981).

022.161 **Oscillator strengths of electric dipole transitions $2ln_1l_1 - 2ln_2l_2$ in C III.**
A. V. Kupliauskiene, Z. J. Kupliauskis.
Tr. Kazan. Gorod. Astron. Obs., Vyp. 46, p. 100 - 105 (1980). In Russian.

022.162 **The mid-infrared spectrum of high temperature silicate minerals cooling on low temperature.**
C. Koike, T. Maihara, H. Hasegawa.
Proceedings of the 14th ISAS Lunar and Planetary Symposium, (see 012.055), p. 149 - 155 (1981).
The authors measured absorption spectra of the minerals olivine, clinopyroxene and orthopyroxene in the mid infrared region at low temperatures with a double beam spectrophotometer.

022.163 **Thermal metamorphism of synthetic silicates.** N. Asada.
Proceedings of the 14th ISAS Lunar and Planetary Symposium, (see 012.055), p. 156 - 161 (1981).
IR absorption spectra of thermally metamorphosed synthetic silicates were measured. The results are compared with the 10 μm feature in astronomical objects.

022.164 **Formation of Ca-Al mineral grains through gas-solid reactions.** H. Hasegawa, J. Seki.
Proceedings of the 14th ISAS Lunar and Planetary Symposium, (see 012.055), p. 249 - 256 (1981).
Direct condensates from the cooling gas of the cosmic abundance are Al_2O_3, $CaTiO_3$, $MgSiO_3$ and Fe(Ni). The other high temperature Ca-Al minerals are the products of the gas-solid reactions between the pre-existing Al_2O_3 grains and the ambient gas. A diffusion mechanism governs the reactions and the reaction rates depend on the diffusion coefficients.

022.165 **Low-velocity impact experiment on rocks − comparison with high-velocity impact experiment.**
T. Matsui, T. Waza, K. Kani, S. Suzuki.
Proceedings of the 14th ISAS Lunar and Planetary Symposium, (see 012.055), p. 258 - 266 (1981).
Low-velocity impact experiments on rocks were performed to reveal the nature of collision processes of planetesimals. Projectiles of steel (15CK, carbon content 0.15%) with velocities of 12 to 270 m/sec were impacted against four kinds of rocks with various shapes and sizes. Imparted energies to unit target mass range from 10^6 to 10^8 erg/g. Collisional phenomena associated with low-velocity impact experiment are shown to be different from those observed in the high-velocity impact experiment (a few km/sec).

022.166 **Performance test of the high velocity shock gun with a novel sabot stopper.** H. Mizutani, M. Kumazawa, M. Kato, T. Masuda, S. Kawakami, Y. Takagi, K. Kani.
Proceedings of the 14th ISAS Lunar and Planetary Symposium, (see 012.055), p. 267 - 277 (1981).
The design and operation of a gas-driven gun capable of operation to velocity of 1000 m/sec is described. The gun

utilizes a unique sabot stopper device that captures the sabot and prevents gas discharge from the propellant explosive. This feature makes it possible to study the collision process for various materials in controlled environment.

022.167 Sr/Ca–Ba/Ca systematics on eucrite parent body, the moon and the earth.   N. Onuma.
Proceedings of the 14th ISAS Lunar and Planetary Symposium, (see 012.055), p. 334 - 340 (1981).
    A new diagram, "Sr/Ca–Ba/Ca diagram", has been proposed for elucidation of magma genesis. In this paper, three "Sr/Ca–Ba/Ca diagrams" for eucrite parent body, the moon and the earth are presented. A comparative study of these diagrams reveals that primitive source materials for these three planetary bodies are very similar to one another in terms of non-volatile element ratios Sr/Ca and Ba/Ca, and the ratios correspond to chondritic composition or solar composition. The result suggests that accretion processes in the proto solar system did not change the refractory element ratios.

022.168 On the mechanism of formation of weak lines of rare-earth elements.
B. T. Babij, M. M. Koval'chuk.
Tsirk. Astron. Obs., L'vov, No. 54, p. 34 - 37 (1979). In Russian.

022.169 Source functions for weak lines of rare-earth elements.   M. M. Koval'chuk.
Tsirk. Astron. Obs., L'vov, No. 54, p. 38 - 40 (1979). In Russian.

022.170 A precise study of the rotational spectrum of formaldehyde $H_2{}^{12}C^{17}O$ and $H_2{}^{13}C^{17}O$.
R. Cornet, B. M. Landsberg, G. Winnewisser.
J. Mol. Spectrosc., Vol. 82, 253 - 263 (1980) = Max-Planck-Inst. Radioastron., Bonn, Sonderdr. Ser. A, Nr. 411.

022.171 A precise study of the rotational spectrum of formaldehyde $H_2{}^{12}C^{16}O$, $H_2{}^{13}C^{16}O$, $H_2{}^{12}C^{18}O$, $H_2{}^{13}C^{18}O$.
R. Cornet, G. Winnewisser.
J. Mol. Spectrosc., Vol. 80, 438 - 452 (1980) = Max-Planck-Inst. Radioastron., Bonn, Sonderdr. Ser. A, Nr. 412.

022.172 Millimeter wave spectrum of barium sulfide in a low-pressure flame. Current millimeter wave measurements of high-temperature species.
D. A. Helms, M. Winnewisser, G. Winnewisser.
J. Phys. Chem., Vol. 84, 1758 - 1765 (1980) = Max-Planck-Inst. Radioastron., Bonn, Sonderdr. Ser. A, Nr. 422.

022.173 Semiempirical calculation of gf values, IV: Fe II.   R. L. Kurucz.
Smithsonian Astrophys. Obs., Spec. Rep. No. 390, 5 + 314 + 5 pp. (1981).
    Slater parameters, eigenvalues, eigenvectors, and scaled Thomas-Fermi-Dirac wavefunctions have been computed for all known configurations of Fe II. The complete transition array has been computed, together with partial sums required for radiative, Stark, and van der Waals damping constants. A listing of the 431,933 strongest lines is available on magnetic tape. A subset of that listing containing 22,547 lines between observed energy levels is presented here. Comparisons are made with other published lifetime and gf value data.

022.174 Franck-Condon factor formulae for astrophysical and other molecules.   R. W. Nicholls.
Astrophys. J., Suppl. Ser., Vol. 47, 279 - 290 (1981).
    Simple closed-form, approximate, analytic expressions for Franck-Condon factors are given. They provide reliable estimates for Franck-Condon factor arrays for molecular band systems for which only ω_e, r_e, μ_A values are known, as is often the case for astrophysically interesting molecules such as

CeO, CoH, CrH, CrO, CuH, GeH, LaO, NiH, SnH, and ZnH for band systems of which Franck-Condon arrays have been calculated.

022.175 On the distribution function of electrons in an electric field in the presence of a strong ion-sound turbulence.
M. A. Livshits, V. A. Liperovskij, V. M. Tomozov.
Uzv. vuzov. Radiofiz., Tom 23, 1399 - 1405 (1980). In Russian. Abstr. in Ref. zh., 51. Astron., 12.51.244 (1981).

022.176 Critical temperatures for dielectric-metal transition in hydrogen and implications for Jupiter and Saturn.
S. Franck.
Planetary interiors, (see 012.061), p. 203 - 209 (1981).
    The dielectric-metal transition in hydrogen is investigated within a microscopic model.

022.177 Adiabats and phase boundaries in planets.   U. Schmit.
Planetary interiors, (see 012.061), p. 211 - 215 (1981).
    The following result is derived: If the temperature distribution in a planet follows an adiabat and crosses a phase boundary, then the phase with the lower specific volume lies in the deeper part. The author's conclusion is the rejection of the so-called earth core paradox.

022.178 The lore of large numbers: some historical background to the anthropic principle.   J. D. Barrow.
Q. J. R. Astron. Soc., Vol. 22, 388 - 420 (1981).
    The author describes how the study of numerological coincidences in physics and cosmology led first to the large numbers hypothesis of Dirac and then to the suggestion of the anthropic principle in a variety of forms. The early history of 'coincidences' is discussed together with the work of Weyl, Eddington and Dirac.

022.179 On Sommerfeld's "tachyonic" solutions.   J. L. Agudin, A. M. Platzeck.
Nuovo Cimento, Lett., Vol. 31, 421 - 428 (1981).

022.180 On a method of estimating the spectral characteristics of the electromagnetic field scattered on ionospheric inhomogeneities.   I. V. Krasheninnikov.
Rasprostr. dekametr. radiovoln, Moskva, 1980, p. 162 - 166. In Russian. – Abstr. in Ref. zh., 62. Issled. kosm. prostranstva, 1.62.540 (1982).

022.181 Spectral characteristics of the electromagnetic field scattered on ionospheric inhomogeneities.
Yu. V. Berezin, I. V. Krasheninnikov.
Rasprostr. dekametr. radiovoln, Moskva, 1980, p. 167 - 173. In Russian. – Abstr. in Ref. zh., 62. Issled. kosm. prostranstva, 1.62.541 (1982).

022.182 A formula for calculating negative polarization of light reflected by atmosphereless planets.
Yu. G. Shkuratov, L. Ya. Melkumova.
Astron. Tsirk., No. 1159, p. 1 - 3 (1981). In Russian.

022.183 Laboratory measurements of the microwave opacity of sulfur dioxide and other cloud-related gases under simulated conditions for the middle atmosphere of Venus.   P. G. Steffes, V. R. Eshleman.
Icarus, Vol. 48, 180 - 187 (1981).
    Recent papers attributing the observed microwave opacity of the middle atmosphere of Venus to gaseous sulfur dioxide (SO_2) and other cloud-related gases have motivated laboratory measurements of their microwave absorbing properties under simulated conditions for this region. The results reduce the amount of SO_2 in the Venus middle atmosphere required to explain the opacity measured by radio occultation,

but this amount still exceeds the abundance measured in situ by atmospheric probes, suggesting that there must be another important source of opacity. Preliminary measurements of the 13-cm absorptivity of gaseous sulfuric acid (H_2SO_4) show it likely to be responsible for a significant part of the observed opacity. Sulfur trioxide (SO_3), was also tested in a CO_2 atmosphere and found to be relatively transparent.

022.184 Organic solids produced by electrical discharge in reducing atmospheres: tholin molecular analysis.
B. N. Khare, C. Sagan, J. E. Zumberge, D. S. Sklarew, B. Nagy.
Icarus, Vol. 48, 290 - 297 (1981).

The complex dark brown solid of a class called tholins, produced on passage of an electrical discharge through a roughly equimolar mixture of methane and ammonia with 2.6% water vapor, is analyzed by vacuum pyrolysis followed by gas chromatography and mass spectrometry. Pyrolyzates include a wide range of aliphatic and aromatic nitriles, alkanes, alkenes, aromatic hydrocarbons, pyrrole, and pyridine. This tholin is remarkably stable to 950°C. It and its degradation products are candidate constituents of planetary aerosols in the outer solar system and of the grains in the interstellar medium.

022.185 Rotation of fragments in catastrophic impact.
A. Fujiwara, A. Tsukamoto.
Icarus, Vol. 48, 329 - 334 (1981).

The rotation of fragments produced by catastrophic impacts into basalt targets was investigated using framing camera records taken by Fujiwara and Tsukamoto (1980). Most of the cores have low rotation rates, of the order 1 rev . sec⁻¹ or less. Many spall fragments have high rotation rates in the strong shear field produced in the target material by impact.

022.186 Lifetime constraints on massive neutrinos from ultraviolet observations of clusters of galaxies.
R. C. Henry, P. D. Feldman.
Phys. Rev. Lett., Vol. 47, 618 - 619 (1981).

Ultraviolet observations are presented that for the first time provide a better lower limit on the lifetimes of 16-20-eV/c^2 neutrinos than do laboratory muon-decay experiments.

022.187 Unresolved dielectronic satellites of the resonance line of heliumlike iron (Fe XXV).
M. Bitter, S. von Goeler, K. W. Hill, R. Horton, D. Johnson, W. Roney, N. Sauthoff, E. Silver, W. Stodiek.
Phys. Rev. Lett., Vol. 47, 921 - 924 (1981).

022.188 Astrophysical production of fractional charge in broken quantum chromodynamics.
E. W. Kolb, G. Steigman, M. S. Turner.
Phys. Rev. Lett., Vol. 47, 1357 - 1360 (1981).

The production of states of fractional charge in the early universe and by cosmic rays is calculated when quantum chromodynamics is broken at a scale less than that of the quantum chromodynamics scale parameter Λ. It is shown that it is possible to produce a detectable abundance of fractional charge states by both mechanisms.

Mechanics of relative motion and inertial forces.
See Abstr. 003.078.

Laboratory research. See Abstr. 031.608.

On the establishment of internally consistent solar scales of oscillator strengths and abundances of chemical elements. II. On the errors of the oscillator strengths of Fe I lines in the Kurucz-Peytremann gf-scale.
See Abstr. 071.003.

Isotopes of nickel in the sun. See Abstr. 071.026.

Non-resonance lines of neutral calcium in the spectra of the Sun and Procyon. See Abstr. 071.027.

Nonthermal line profiles from O(^{1}S) in the thermospheric nightglow. See Abstr. 082.097.

Errata

022.901 Erratum: "Photoabsorption cross sections for positive ions with $Z \leqslant 30$" [Astrophys. J., Suppl. Ser., Vol. 40, 815 - 880 (1979)]. R. F. Reilman, S. T. Manson. Astrophys. J., Suppl. Ser., Vol. 46, 115 (1981). − See Abstr. 26.022.054.

022.902 Errata: "Spectral line parameters for the $A^2\Sigma - X^2\Pi$ (0, 0) band of OH for atmospheric and high temperatures [J. Quant. Spectrosc. Radiat. Transfer, Vol. 25, 111 - 135 (1981)]. A. Goldman, J. R. Gillis. J. Quant. Spectrosc. Radiat. Transfer, Vol. 26, 547 (1981). See Abstr. 29.022.010.

022.903 Errata: "Spectral line parameters for the $X^2\Pi - X^2\Pi$ (1, 0) bands of OH and ClO for atmospheric applications [J. Quant. Spectrosc. Radiat. Transfer, Vol. 26, 23 - 31 (1981)]. J. R. Gillis, A. Goldman. J. Quant. Spectrosc. Radiat. Transfer, Vol. 26, 547 (1981). See Abstr. 29.022.084.

Astronomical Instruments and Techniques

031 Astronomical Optics, Methods of Observation and Reduction

Astronomical Optics

031.001 **The optical system of the Space Telescope.**
S. Dunlop.
J. British Astron. Assoc., Vol. 91, 504 - 505 (1981).

031.002 **On the optical design of photoelectric stellar photometers.** J. F. James.
Q. J. R. Astron. Soc., Vol. 22, 244 - 253 (1981).
The design criteria for an astronomical stellar photometer are discussed and proposals are made for implementing them. Some of the pitfalls are described together with the means of avoiding them. A description is given of an instrument that has been constructed with these criteria in mind.

031.003 **Specifying the figure of telescope mirrors.**
R. V. Willstrop.
Observatory, Vol. 101, 134 - 135 (1981).

031.004 **The flexure of thin monolithic mirrors.**
B. Mack.
Observatory, Vol. 101, 135 (1981).

031.005 **The optical performance of mosaic-mirror telescopes.**
C. M. Humphries, T. E. Purkins.
Observatory, Vol. 101, 135 - 136 (1981).

031.006 **A possible optical scheme for a telescope with a main spherical mirror of 20 - 25 meters in diameter.**
B. A. Burnasheva, R. E. Gershberg, V. I. Pronik, N. V. Steshenko.
Izv. Krymskoj Astrofiz. Obs., Tom 63, 166 - 170 (1981). In Russian. English translation in Bull. Crimean Astrophys. Obs., Vol. 63.

031.007 **Control of concave and convex aspherical mirrors with a T-shaped spherometer.**
R. E. Gershberg, V. I. Pronik.
Izv. Krymskoj Astrofiz. Obs., Tom 63, 171 - 179 (1981). In Russian. English translation in Bull. Crimean Astrophys. Obs., Vol. 63.
The main relations used to construct the program SFERA to compute the data necessary for making and controlling astronomical mirrors are given.

031.008 **Method for computation of the profile of a surface which corrects exactly the spherical aberration on the axis in a centered optical system.** G. M. Popov.
Izv. Krymskoj Astrofiz. Obs., Tom 63, 180 - 188 (1981). In Russian. English translation in Bull. Crimean Astrophys. Obs., Vol. 63.

031.009 **Use of modelling for estimating the image quality of a composite telescope mirror.**
I. A. Lapshina, Yu. A. Sabinin.
Leningr. inst. tochn. mekh. i opt. Leningrad, 1981. 24 pp. In Russian. − Abstr. in Ref. zh., 51. Astron., 1.51.850 (1982).

031.010 **Zes oculairen vergeleken door middel van optische doorrekening.**
H. G. J. Rutten, M. A. M. van Venrooij.
Zenit, 8. Jaarg., 513 - 519 (1981).

031.011 **Test of light energy concentration for the 1 m reflector at Yunnan Observatory.**
Y.-m. Wang, S.-n. Qin.
Acta Astron. Sinica, Vol. 22, 310 - 314 (1981). In Chinese.

031.012 **Hartmann-Tests am MPIA-2.2 m-Teleskop mit Blende im Pupillenbild.** B. Loibl.
Mitt. Astron. Ges., Nr. 54, (see 012.050), p. 139 - 142 (1981).

031.013 **Aktiver Spiegel zum Einsatz am astronomischen Teleskop.**
K. Freischlad, H.-L. Reischmann, F. Merkle, J. Bille.
Mitt. Astron. Ges., Nr. 54, (see 012.050), p. 149 - 156 (1981).

031.014 **Bemerkungen zu astronomischen Anwendungen von Fokalreduktoren.** E. H. Geyer.
Mitt. Astron. Ges., Nr. 54, (see 012.050), p. 156 - 158 (1981).

031.015 **Vorschlag einer Weitwinkel-Röntgen-Optik mittlerer Auflösung für astronomische Anwendungen.**
W. K. H. Schmidt.
Mitt. Astron. Ges., Nr. 54, (see 012.050), p. 209 - 221 (1981).

031.016 **Determination of the effective focal length in Ritchey-Chrétien systems by the Hartmann method.**
L. I. Snezhko.
Astron. Tsirk., No. 1117, p. 2 - 4 (1980). In Russian.

031.017 **Strehl ratio for primary aberrations: some analytical results.** V. N. Mahajan.
J. Opt. Soc. America, Vol. 71, 1561 (1981). − Abstract.

031.018 **Fabrication and testing of a mirror for NASA.**
J. A. Magner, C. A. Martin, D. R. Eastman.
J. Opt. Soc. America, Vol. 71, 1580 (1981). − Abstract.

031.019 **Polishing and testing of aspheric diamond-turned surfaces.** R. E. Parks.
J. Opt. Soc. America, Vol. 71, 1580 - 1581 (1981). − Abstract

031.020 **Subaperture test of a large flat or a fast aspheric surface.** C.-J. Kin, J. C. Wyant.
J. Opt. Soc. America, Vol. 71, 1587 (1981). − Abstract.

031.021 **Exact ray trace analysis of the Ritchey-Common test.** K.-L. Shu, R. E. Parks, R. R. Shannon.
J. Opt. Soc. America, Vol. 71, 1587 (1981). − Abstract.

031.022 **Active optical mirror with model control system for astronomical application.**
F. Merkle, K. Freischlad, J. Bille, G. Jahn.
J. Opt. Soc. America, Vol. 71, 1594 (1981). − Abstract.

031.023 **Reflectance and preparation of front-surface mirrors for use at various angles of incidence from the ultraviolet to the far infrared.** G. Hass.
J. Opt. Soc. America, Vol. 71, 1621 (1981). – Abstract.

031.024 **Aspheric production and testing.** E. Heynacher.
J. Opt. Soc. America, Vol. 71, 1624 (1981).
Abstract.

031.025 **Computer-controlled grinding of optical surfaces.** R. A. Jones.
J. Opt. Soc. America, Vol. 71, 1634 (1981). – Abstract.

031.026 **All-spherical catadioptic telescope with subaperture correctors.** R. D. Sigler.
J. Opt. Soc. America, Vol. 71, 1647 (1981). – Abstract.

031.027 **Thick meniscus field correctors.** T. H. Jamieson.
J. Opt. Soc. America, Vol. 71, 1647 (1981).
Abstract.

031.028 **Accelerating convergence in automatic lens design.** B. Brixner.
Appl. Opt., Vol. 20, 2452 - 2456 (1981).
Among the various factors that slow lens optimization—insufficient performance targets, the absence of a unique solution, false local minima, a poorly scaled change vector, failure to find the optimum damping number, and failure to equalize the parameter gradients – the importance of parameter gradient equalization has been insufficiently recognized. Gradients can be approximately equalized by scaling the lens to a suitable size while it is being optimized. For best results, the size of the damping number should also be optimized during each iteration.

031.029 **Single sideband Ronchi test.** J. Schwider.
Appl. Opt., Vol. 20, 2635 - 2642 (1981).
A new lateral shear interferometer based on the Ronchi grating is proposed. The normal Ronchi test is impaired by multiple beam interference (Talbot effect). Furthermore, the fringe pattern is somewhat restricted by fringe number and orientation. The new shear interferometer suppresses the Talbot effect by spatial filtering and by using a second grating enables arbitrary fringe orientation and number.

031.030 **Automatic computation of optical focal surfaces.** T. B. Andersen.
Appl. Opt., Vol. 20, 2754 - 2760 (1981).
The use of automatically computed aberration coefficients of arbitrary order for an exact determination of the surface of best focus of a symmetrical system, by minimizing the gyration radius of the spot diagram is demonstrated. For the two optical systems studied, a Schmidt camera and a Ritchey-Chrétien telescope, the results from the exact model are found to be significantly different from the paraxial approximation predictions.

031.031 **Optical aberration coefficients: FORTRAN subroutines for symmetrical systems.**
T. B. Andersen.
Appl. Opt., Vol. 20, 3263 - 3268 (1981).
FORTRAN computer subroutines for the automatic computation of the optical aberration functions S, T, V, W, and K to the 15th order for rotationally symmetric systems are presented. The routines may be conveniently extended toward higher orders.

031.032 **Optimum solution for spherical primary mirror with two and three aspheric corrector plates located near focus.** A. B. Meinel, M. P. Meinel.
Appl. Opt., Vol. 20, 3627 - 3629 (1981).
Aspheric corrector plates can provide good aberration control for large spherical primary mirror telescopes, but the regions where good solutions can be obtained are very limited. Two isolated solution regions exist for the two-corrector configuration.

031.033 **Optical aberration functions: computation of caustic surfaces and illuminance in symmetrical systems.**
T. B. Andersen.
Appl. Opt., Vol. 20, 3723 - 3728 (1981).
The use of automatically computed aberration functions for an exact determination of the geometric illuminance and caustic surfaces is demonstrated. Numerical results for two optical systems studied using aberration coefficients to the thirteenth order are presented.

031.034 **Annular-aperture wide-field camera.** G. R. Carruthers.
Appl. Opt., Vol. 20, 4094 - 4096 (1981).
An annular-aperture wide-field camera, consisting of a spherical mirror and an annular aperture stop at its center of curvature (essentially a correctorless Schmidt camera), has been shown through analysis and laboratory tests to be a useful imaging device in the far and extreme UV, where the efficiencies of conventional refractive correctors and of reflective mirror coatings are relatively low.

031.035 **The initial aluminising of the UKIRT (*UK Infrared Telescope*) 3.8 m primary mirror.** T. E. Purkins.
UKIRT Rep., No. 1, 2 + 6 pp. (1978).

031.036 **Alignment of the Cassegrain and coudé optics of the 3.8 m UK Infrared Telescope; assessment of the optical performance and results of aberration calculations.**
D. S. Brown, C. M. Humphries, T. E. Purkins, R. V. Willstrop.
UKIRT Rep., No. 10, 14 + 3 pp. (1981).

031.037 **The removal, mechanical handling and installation procedures for re-aluminising the UKIRT 3.8 m primary mirror.** A. A. Neild.
UKIRT Rep., No. 11, 34 pp. (1980).

031.038 **Metallic film thickness requirements for mirrors in the infrared.** C. M. Humphries, H. Seddon.
UKIRT Rep., No. 13, 6 + 4 + 2 pp. (1980).

031.039 **Thin mirror telescopes: experience gained with the U. K. Infrared Telescope.** C. M. Humphries.
UKIRT Rep., No. 14, 4 pp. (1980).

031.040 **Chromatic aberration of the Mizusawa PZT No. 1.** G. Murakami, H. Kitago, K. Iwadate.
Proc. Int. Latitude Obs. Mizusawa, No. 20, p. 53 - 60 (1981). In Japanese.
Chromatic aberration of the Mizusawa PZT No. 1 is tested by taking account of variations of the scale value, determined by the measured E/W distance. The focal length is found to be longer for K and M types than those for earlier types. This may correspond to the fact that the objective of the Mizusawa PZT No. 1 is achromatized for visual wavelength.

031.041 **Honeycomb mirrors of borosilicate glass: current results and plans for 7 - 8 m diameter.**
J. R. P. Angel, D. Arganbright, L. Harmonson, J. M. Hill, N. Woolf.
Prepr. Steward Obs., No. 346, 5 pp. (1981).

031.042 **A design of the corrector system for the 1.5 meter telescope.** L.-j. Wang, Z.-f. Ye.
Ann. Shanghai Obs. Acad. Sinica, No. 1, p. 119 - 125 (1979).

031.043 **Calculating the elastic deformation of mirrors by the finite element method.**
S.-y. Zhu, Y.-l. Dai.
Ann. Shanghai Obs. Acad. Sinica, No. 1, p. 126 - 129 (1979).

031.044 **Optical pantograph.**   E. M. Makhov.
Issled., raschet i konstruir. izmerit. priborov.
Leningrad, 1981, p. 20 - 23. In Russian. − Abstr. in Ref, zh.,
51. Astron., 1.51.853 (1982).

031.045 **Approximate estimate of frequency-contrast characteristics of objectives.**   R. A. Kraskovskij.
Tr. Gos. opt. inst., Tom 46, No. 180, p. 61 - 65 (1980). In
Russian. − Abstr. in Ref. zh., 51. Astron., 1.51.854 (1982).

031.046 **Bemerkungen zur Herstellung und Justierung von Schmidt-Spiegeln.**   M. Schürer.
Orion, 39. Jahrg., 196 - 198 (1981).

Testing of optical surfaces: a bibliography.
See Abstr. 002.070.

Master optical techniques.   See Abstr. 003.052.

The computer in optical research, methods and applications.   See Abstr. 003.064.

Das System des 2.2 m-Teleskops.
See Abstr. 032.034.

Report on optical work carried out at the U. K. Infrared Telescope installation at Mauna Kea, Hawaii from mid-July until early August, 1978.   See Abstr. 032.047.

Automatic adjustment of a multi-element telescope mirror from the criterion of maximum intensity.
See Abstr. 032.066.

Demonstration of a phased segmented mirror for space applications.   See Abstr. 032.613.

Methods of Observation and Reduction

031.501 **Local gas without reddening: the contribution of stray radiation to 21 centimeter line measurements.**
C. Heiles, A. A. Stark, S. Kulkarni.
Astrophys. J., Lett., Vol. 247, L73 - L76 (1981).
Previous studies of the relationship between H I column density derived from 21 cm line measurements and reddening $E(B-V)$ have disagreed concerning the presence of a zero-point offset, i. e., a component of gas that produces no reddening. It was suspected that the putative effect may have been a result of errors produced by the stray radiation of paraboloidal radio telescopes. The authors have repeated a number of H I measurements with the Bell Telephone Laboratories horn reflector antenna and confirm the presence of the zero-point offset in the gas-to-reddening relationship.

031.502 **Analysis of lunar occultations − II. Personal equation.**   L. V. Morrison, G. M. Appleby.
Mon. Not. R. Astron. Soc., Vol. 196, 1005 - 1011 (1981).
From an analysis of 42 000 timings of occultations of stars by the Moon in the years 1943 to 1979 it is found that the observers' personal equation is dependent on the phase of the Moon and the magnitude of the star. Expressions for this dependence are derived for two different methods of timing.

031.503 **Analysis of lunar occultations − III. Systematic corrections to Watts' limb-profiles for the Moon.**
L. V. Morrison, G. M. Appleby.
Mon. Not. R. Astron. Soc., Vol. 196, 1013 - 1020 (1981).
Timings of 66 000 occultations of stars by the Moon are analysed for systematic corrections to the limb-profile heights taken from Watts' charts of the marginal zone of the Moon. The radius, shape and location of the centre of the datum implicit in the charts are found to vary with libration and these variations produce systematic errors in the limb-profile heights which attain 0.4 arcsec in some position angles.

031.504 **A new method for making maps with unstable radio interferometers.**   T. J. Cornwell, P. N. Wilkinson.
Mon. Not. R. Astron. Soc., Vol. 196, 1067 - 1086 (1981).
The authors present a new method for making "hybrid" maps using data from unstable interferometers. Their approach is based on correcting errors occurring at individual telescopes, and is more general than previous ones in that it allows for different degrees of instability at each telescope as well as for varying signal-to-noise ratios in the visibility data on each baseline. The method was developed, and will be used, for the analysis of the data from the Jodrell Bank Multi-Telescope Radio-Linked Interferometer (MTRLI). The authors show the results of tests on simulated data to indicate the quality of maps which can be made with the MTRLI using the new method, and present a hybrid map of the quasar 3C 309.1 made from actual MTRLI data at 1666 MHz.

031.505 **Observing technique for photoelectric photometry: analytical expressions for the optimum choice of integration.**   M. Claudius, R. Florentin-Nielsen.
Astron. Astrophys., Vol. 100, 186 - 188 (1981).
Analytical expressions are derived for the optimum choice of integration times for photon counting photoelectric photometers. The expressions are applied to simultaneous photometry of star and sky background as well as to conventional photometry in which star plus background and background alone are observed sequentially. The gain in efficiency over commonly adopted ways of choosing the integration times is discussed.

031.506 **On the computation of acceleration parallaxes.**   H. Eichhorn.
Astron. J., Vol. 86, 915 - 917 (1981) = Contrib. Dep. Astron., Univ. Florida, No. 27.
Rigorous formulas are derived with which one can compute from three observed (coplanar) directions the proper-motion components and the product of parallax and radial velocity of an object (star) which moves uniformly with respect to the observer (Earth). The errors of the determination of distance by parallactic acceleration are analyzed and the possibility of determining distances by this method is discussed.

031.507 **Automated detection of variable objects on Schmidt plates.**   M. R. S. Hawkins.
Nature, Vol. 293, 116 - 120 (1981).
The author reports a procedure for automatically detecting variable objects and gives a preliminary description of the type of object to be found. This technique should make it

possible to extend the work on RR Lyraes and quasars down to a limit of at least $B = 21$, and to investigate the distribution of other classes of faint variable objects.

031.508 Observations astrométriques à Paris: gageure ou réalité? F. Chollet, S. K. Lam.
Astronomie, Vol. 95, 347 - 350 (1981).

031.509 Astronomia infrarossa – parte seconda: le osservazioni. A. D'Ercole.
Coelum, Vol. 50, 163 - 172 (1981).

031.510 Die Beobachtung sporadischer Meteore und Meteorschauer mittels radiotechnischer Meßeinrichtungen.
W. Schippke.
Sterne Weltraum, Jahrg. 20, 287 - 291 (1981).

031.511 Studio su un possibile effetto Schröter lunare.
G. Adamoli.
Astronomia, N. 2, p. 3 - 7 (1981).

031.512 Measurement of small temperature differences in astronomical observations with piezoquartz frequency thermometers. S. B. Volodin.
Tr. Astron. Obs., Leningrad, Tom 36 = Uch. Zap. Leningr. Univ., No. 402 = Ser. mat. nauk, Vyp. 58, 143 - 152 (1981). In Russian.

It is shown that temperature sensors for measurement of small temperature differences during astronomical observations may be developed on the basis of quartz resonators.

031.513 On the application of numerical models to the inverse mapping of solar wind flow structures.
V. J. Pizzo.
J. Geophys. Res., Vol. 86, 6685 - 6690 (1981).

The widely used constant velocity technique is capable of mapping solar wind structures seen near 1 AU back to the vicinity of the sun with only limited accuracy. This report points out that relatively sophisticated MHD numerical models can be applied to this 'inverse' mapping problem, promising better accuracy, especially in the region of the stream front.

031.514 The perils of high-speed stellar photometry.
Y. P. Elsworth, J. F. James.
Observatory, Vol. 101, 120 - 122 (1981).

There are several factors that limit the data which can be collected in high-speed photometry of variable stars; not the least of these factors is mechanical vibration of the telescope itself.

031.515 Une méthode pour estimer l'équation personnelle lors de chaque observation d'occultation d'étoile par la lune. J. Bourgeois.
Ciel Terre, Vol. 97, 291 - 296 (1981).

The author presents a method for measuring the observer's personal equation after each lunar occultation. He describes an apparatus producing simulated occultations, which has been especially built, and he shows the preliminary results as yet obtained.

031.516 On the essentiality of the Whittaker-Vondrak method as a filter and estimations of the standard deviation and the correlation for a digital filter.
K.-y. Huang, X. Zhou.
Acta Astron. Sinica, Vol. 22, 120 - 130 (1981). In Chinese.

031.517 Whether the equal altitude method will make new contributions in modern astrometry? D.-m. Li.
Acta Astron. Sinica, Vol. 22, 131 - 137 (1981). In Chinese.

031.518 Discussion on the maximum-likelihood method for the determination of the membership in open
clusters. J.-l. Zhao, K.-p. Tian, Z.-h. Xu, M.-g. Yin.
Acta Astron. Sinica, Vol. 22, 180 - 187 (1981). In Chinese.

031.519 Fourier analysis of differential rotation in late-type stars. D. H. Bruning.
Astrophys. J., Vol. 248, 274 - 278 (1981).

A numerical model is used to describe the modification of the rotational broadening function by differential rotation in late-type stars. It is shown that the differences between the Fourier transforms for solid body and differential rotation for stars with small polar inclinations ($i < 60°$) are more conspicuous than the differences described by Gray for stars seen equator-on ($i = 90°$). Furthermore, the signature of equatorial acceleration may be distinguished from that of polar acceleration, since the former acts to decrease the amplitude of the first sidelobe, while the latter increases the first sidelobe amplitude. An observational scheme for the detection of differential rotation is outlined.

031.520 Automatic resonance recognition method in solar wind. J. P. Thouvenin, J. G. Trotignon.
Nuovo Cimento C, Ser. 1, Vol. 3C, 696 - 710 (1980). – Abstr. in Phys. Abstr., Vol. 84, Abstr. 66839 (1981).

031.521 Optical polarimeters for solar research.
T. G. Baur.
Opt. Eng., Vol. 20, 2 - 13 (1981). – Abstr. in Phys. Abstr., Vol. 84, Abstr. 66840 (1981).

031.522 New trends in stellar speckle interferometry.
F. Roddier.
Proc. Soc. Photo-Opt. Instrum. Eng., Vol. 243, (see 012.002), p. 83 - 87 (1980). – Abstr. in Phys. Abstr., Vol. 84, Abstr. 66841 (1981).

031.523 Speckle imaging under non-isoplanatic conditions.
J. W. Sherman.
Proc. Soc. Photo-Opt. Instrum. Eng., Vol. 243, (see 012.002), p. 51 - 57 (1980). – Abstr. in Phys. Abstr., Vol. 84, Abstr. 68401 (1981).

031.524 Réduction des spectres stellaires obtenus sur plaques photographiques. I. Réductions en intensité et en longueur d'onde. A. Delcroix.
Bull. Soc. R. Sci. Liège, 49e année, p. 244 - 257 (1980) = Commun. Dép. Astrophys. Mons, No. 84.

The author discusses the procedures created to reduce data produced by the photometer described in a preceding paper. The reduction is realized in intensity as well as in wavelength. The accuracy reached for the wavelength scale is about 0.03 Å, comparable to that obtained with much more sophisticated machines.

031.525 Réduction des spectres stellaires obtenus sur plaques photographiques. II. Traitement du bruit de fond. A. Delcroix.
Bull. Soc. R. Sci. Liège, 49e année, p. 258 - 272 (1980) = Commun. Dép. Astrophys. Mons, No. 85.

The noise due to photographic plates granulation is a factor of degradation of spectrum quality. The author presents here a statistical study of that noise. He shows that its probability distribution function may be considered as gaussian, at the 5% probability level.

031.526 Réduction des spectres stellaires obtenus sur plaques photographiques. III. Moyennes de spectres.
A. Delcroix.
Bull. Soc. R. Sci. Liège, 49e année, p. 273 - 277 (1980) = Commun. Dép. Astrophys. Mons, No. 86.

The author presents here a method based on spectra addition, which leads to a dramatic increase of spectrum quality in the definition of the continuum, in the shape of lines and the separation of blends components.

031.527   Speckle interferometric techniques applied to the observation of the solar photosphere.
C. Aime, G. Ricort.
Proc. Soc. Photo-Opt. Instrum. Eng., Vol. 243, (see 012.002), p. 58 - 64 (1980). − Abstr. in Phys. Abstr., Vol. 84, Abstr. 71091 (1981).

031.528   High angular resolution astronomical techniques. Speckle interferometry and related methods.
S. P. Worden.
Proc. Soc. Photo-Opt. Instrum. Eng., Vol. 243, (see 012.002), p. 66 - 73 (1980). − Abstr. in Phys. Abstr., Vol. 84, Abstr. 71092 (1981).

031.529   The Steward Observatory speckle interferometry program.   P. A. Strittmatter.
Proc. Soc. Photo-Opt. Instrum. Eng., Vol. 243, (see 012.002), p. 75 - 79 (1980). − Abstr. in Phys. Abstr., Vol. 84, Abstr. 71093 (1981).

031.530   Observational speckle interferometry.
D. Bonneau, M. Faucherre, L. Koechlin, F. Vakili.
Proc. Soc. Photo-Opt. Instrum. Eng., Vol. 243, (see 012.002), p. 80 - 81 (1980). − Abstr. in Phys. Abstr., Vol. 84, Abstr. 71094 (1981).

031.531   Data recording and processing for speckle image reconstruction.
P. Nisenson, R. Stachnik, C. Papaliolios, P. Horowitz.
Proc. Soc. Photo-Opt. Instrum. Eng., Vol. 243, (see 012.002), p. 88 - 94 (1980). − Abstr. in Phys. Abstr., Vol. 84, Abstr. 71095 (1981).

031.532   Astronomical imaging by processing stellar speckle interferometry data.   J. R. Fienup,
G. B. Feldkamp.
Proc. Soc. Photo-Opt. Instrum. Eng., Vol. 243, (see 012.002), p. 95 - 102 (1980). − Abstr. in Phys. Abstr., Vol. 84, Abstr. 71096 (1981).

031.533   Stellar speckle interferometry and speckle holography at low light levels.   G. P. Weigelt.
Proc. Soc. Photo-Opt. Instrum. Eng., Vol. 243, (see 012.002), p. 103 - 111 (1980). − Abstr. in Phys. Abstr., Vol. 84, Abstr. 71097 (1981).

031.534   New method for measuring Earth albedo.
C. A. Coombes, A. W. Harrison.
Canadian J. Phys., Vol. 59, 635 - 638 (1981). − Abstr. in Phys. Abstr., Vol. 84, Abstr. 74924 (1981).

031.535   Remote investigation of tropospheric turbulence by two-dimensional analysis of stellar scintillation.
M. Azouit, J. Vernin.
J. Atmos. Sci., Vol. 37, 1550 - 1557 (1980). − Abstr. in Phys. Abstr., Vol. 84, Abstr. 74949 (1981).

031.536   Geometrical rectification of spin-scan images from Pioneer 11.   R. N. Strickland, J. J. Burke.
Proc. Soc. Photo-Opt. Instrum. Eng., Vol. 249, 47 - 52 (1980). Abstr. in Phys. Abstr., Vol. 84, Abstr. 75026 (1981).

031.537   Imaging of star clusters from speckle interferometry.
J. Christou.
Opt. Commun., Vol. 37, 331 - 334 (1981). − Abstr. in Phys. Abstr., Vol. 84, Abstr. 79728 (1981).

031.538   On the influence of scanning speed in turbulence degraded one-dimensional stellar images.

G. Lund, F. Martin, C. Aime.
J. Opt. (*France*), Vol. 12, 207 - 215 (1981). − Abstr. in Phys. Abstr., Vol. 84, Abstr. 83491 (1981).

031.539   A new method for studying celestial emission spectrum.   C. Zhu, Z. Shi.
Kexue Tongbao, Vol. 26, 48 - 52 (1981). − Abstr. in Phys. Abstr., Vol. 84, Abstr. 83496 (1981).

031.540   A study of sky noise, 1.5 μm - 5 μm.
D. A. Allen, J. R. Barton.
Publ. Astron. Soc. Pacific, Vol. 93, 381 - 384 (1981).
Power spectra of the sky noise at Siding Spring Observatory are given for wavelengths from 1.6 μm to 5 μm. These data indicate that sky-chopping infrared photometers, which conventionally operate near 10 Hz could be used at frequencies below 1 Hz in many applications.

031.541   Magnitude-image diameter calibration of the UKSTU J-survey films.
D. J. King, C. J. Birch, C. Johnson, K. N. R. Taylor.
Publ. Astron. Soc. Pacific, Vol. 93, 385 - 387 (1981).
A calibration curve and table are presented for deriving approximate magnitudes from the UKSTU J-survey.

031.542   Frequency shift in Fourier analysis.   G. Kovács.
Astrophys. Space Sci., Vol. 78, 175 - 188 (1981).
Error analysis of the frequency determination by conventional Fourier method was performed. Taking into consideration only equidistant data, estimates were obtained on the accuracy of the frequencies deduced by analysing a noise-free signal containing not more than two sinusoidal components with closely spaced frequencies. An important consequence of these results was the empirical distribution function of the frequencies for signals contaminated by noise and containing sufficiently distant frequencies.

031.543   Washburn extraction and width of IUE point spread function.
K. S. de Boer, J. Koornneef, M. R. Meade.
The Universe at ultraviolet wavelengths, (see 012.009), p. 771 - 775 (1981).
The authors review the Washburn extraction routine for low dispersion IUE spectra. The shape of the point spread function (PSF) in low dispersion spectra is sufficiently well described by a gaussian function. The PSF is in large and small aperture essentially identical and the authors present values of σ. Several advantages of the extraction routine are mentioned.

031.544   Results of basic improvements to the extraction of spectra from IUE images.
D. J. Lindler, R. C. Bohlin.
The Universe at ultraviolet wavelengths, (see 012.009), p. 777 - 785 (1981).
Results of two methods of extracting spectra from IUE images are compared. The first method, which is presently implemented, performs a geometric correction of the image followed by a photometric correction. The spectral data are then extracted using a slit with an effective width and sampling interval of 2.4 Å for the SWP camera and 3.7 Å for the LWR camera in low dispersion. The second method performs the photometric correction without doing a geometric correction. The spectral data are then extracted from the photometrically corrected image by an extraction slit, which follows the spectral orders in the non-geometrically corrected space, with an effective width and sampling interval 1/2 that of the present method.

031.545   Extracting spatial information from large aperture exposures of diffuse sources.
J. T. Clarke, H. W. Moos.

The Universe at ultraviolet wavelengths, (see 012.009), p. 787 - 794 (1981).

The spatial properties of large aperture exposures of diffuse emission can be used both to investigate spatial variations in the emission and to filter out camera noise in exposures of weak emission sources. Spatial imaging can be accomplished both parallel and perpendicular to dispersion with a resolution of 5-6 arc sec, and a narrow median filter running perpendicular to dispersion across a diffuse image selectively filters out point source features, such a reseaux marks and fast particle hits. Spatial information derived from observations of solar system objects are presented.

031.546   Effects of temperature fluctuations on IUE data quality.
R. W. Thompson, B. E. Turnrose, R. C. Bohlin.
The Universe at ultraviolet wavelengths, (see 012.009), p. 811 - 819 (1981).

Analysis of IUE calibration lamp images has shown that variation in the temperature of the scientific instrument causes shifts in the location of the spectral format with respect to the reseau grid on the detector and in the location of the reseaux themselves. Procedures are under development for utilizing these temperature correlations to correct the dispersion-relation and reseau-position files to the temperature of each target image and thereby achieve improved wavelength and photometric accuracy in reduced IUE spectra.

031.547   Real-time excising of narrowband interference in radio interferometry.
D. Routledge, B. L. Kasper, F. S. Chute.
Bull. American Astron. Soc., Vol. 13, 510 (1981). – Abstract.

031.548   High resolution image reconstruction of binary stars and other objects.   P. Nisenson, R. S. Stachnik.
Bull. American Astron. Soc., Vol. 13, 514 (1981). – Abstract.

031.549   The inapplicability of the Fourier convolution method to the analysis of late-type stellar spectra.
D. H. Bruning.
Bull. American Astron. Soc., Vol. 13, 516 (1981). – Abstract.

031.550   Analysis of cryogenic FTS as a survey instrument to discover new quasars.   P. R. Silverglate.
Bull. American Astron. Soc., Vol. 13, 517 (1981). – Abstract.

031.551   New survey technique for halo G-dwarfs.
C. J. Corbally, R. F. Garrison.
Bull. American Astron. Soc., Vol. 13, 534 (1981). – Abstract.

031.552   Simultaneous echelle spectroscopy of 15 or more giants in globular clusters using an aperture plate.
R. C. Peterson, C. A. Pilachowski, D. W. Willmarth.
Bull. American Astron. Soc., Vol. 13, 545 (1981). – Abstract.

031.553   A method to use ground-based observations to determine the Lα flux in prominences.
A. P. Bernat, D. A. Landman.
Bull. American Astron. Soc., Vol. 13, 552 (1981). – Abstract.

031.554   Overlapping plate reduction using orthogonal transformations.   W. H. Jefferys.
Bull. American Astron. Soc., Vol. 13, 569 (1981). – Abstract.

031.555   Parasite light effects in IUE low resolution spectra.
L. Crivellari, C. Morossi.
Second European IUE Conference, (see 012.011), p. 99 - 102 (1980).

The problem of parasite light is a serious one for low resolution IUE spectra. An approach has been made to minimize spurious contributions due to halation effect and scatter-

ing of light along the direction of dispersion. Some results are presented.

031.556   An improved IUE image processing system – the case of absorption lines in high dispersion spectra.
C. Laurent, A. Vidal-Madjar, J. Paul.
Second European IUE Conference, (see 012.011), p. 103 - 107 (1980).

Among the problems which remain in the IUE data reduction system, some are of crucial importance in the case of study of absorption lines. Firstly, the background correction, which is taken as the interorder value, is obviously overestimated where the echelle orders crowd together. Also the long exposure spectra, often required for interstellar studies, are affected by cosmic-ray induced spots which contribute either to the observed intensity or to the background. Most of these difficulties could be overtaken if one substitutes to the pseudo-slit passing along the orders a step-by-step determination of both background profile and intensity profile across the orders. In this contribution, the authors describe such a procedure and present preliminary results derived from a 7 1/2 hours exposure spectrum of the SMC star S-159.

031.557   L'exploitation des observations d'étoiles variables.
E. Schweitzer.
Bull. AFOEV, No. 17, p. 91 - 105 (1981).

031.558   Photoelectric photometry: an approach to data reduction.   W. E. Harris, M. P. FitzGerald, B. C. Reed.
Publ. Astron. Soc. Pacific, Vol. 93, 507 - 517 (1981) = Contrib. Univ. Waterloo Obs., No. 82.

The overall problem of reducing photoelectric photometry data is reviewed, and a general reduction method that differs significantly from the traditional approach of Hardie and others is summarized and analyzed. The technique, essentially a multilinear solution for all the transformation model parameters simultaneously, is especially suitable for modern computing facilities, and is both faster and more reliable than older approaches. This particular technique is also shown to be sufficiently powerful and versatile that it gives the observer several important advantages and flexibilities at the telescope, and can be set up to apply to many general situations. Examples and comparisons of its use for single- and multiple-night reductions are presented, and the role of the reduction problem in the whole area of photoelectric photometry is discussed.

031.559   Estimate of the possibility of continuous recording of the solar electromagnetic radiation with an instrument aboard a satellite flying at lower heights in a polar circular orbit.   Yu. M. Kulagin, V. F. Samolin.
Tr. Inst. prikl. geofiz. Gos. kom. SSSR po gidrometeorol. i kontrolyu prirod. sredy, 1981, No. 50, p. 84 - 91. In Russian. Abstr. in Ref. zh., 62. Issled. kosm. prostranstva, 8.62.67 (1981).

031.560   L'observation et les propriétés des étoiles.
B. Hauck, A. Hayli.
Histoire de l'univers (see 003.007), p. 17 - 51 (1980).

Contents: L'observation des étoiles. La distance et les mouvements des étoiles. La lumière des étoiles. Les masses et les dimensions des étoiles. Les populations stellaires.

031.561   Contact timings of solar eclipse from photographs.
J. E. S. Singh.
Bull. Astron. Soc. India, Vol. 9, 72 (1981). – Abstract.

031.562   Parallax and proper motion analyses of 21 plate series taken with the Sproul 61-cm refractor.
S. L. Lippincott.
Astron. J., Vol. 86, 1401 - 1403 (1981).

Results are presented from 21 plate series with observational coverage predominantly over the last decade; 17 series extend back to ~1940. No significant variable proper motion was found. All but two appear to be dwarf or subdwarf K or M stars.

031.563   A comparison of optical and digital Fourier transformation of solar granulation.
O. von der Lühe.
Astron. Astrophys., Vol. 101, 277 - 283 (1981) = Mitt. Kiepenheuer-Inst. Nr. 200.

The application of an optical Fourier transformation setup on solar granulation transparencies has been studied. High quality pictures have been analyzed without using liquid gate techniques. An analysis of the photographic process allows to derive calibrated radial power spectra of the granulation intensity fluctuations. The author compared his results with power spectra of the same pictures obtained via micro densitometry and fast Fourier algorithms. The two independent methods lead to essentially the same quantitative results.

031.564   A measurement of the resemblance between two contours.   Z. Mouradian.
Astron. Astrophys., Vol. 101, 292 - 294 (1981).

A criterion is given which allows to select from a group of contours on an image that most closely resembles a defined contour on another image. This method is suitable to the study of images of large objects (the sun, planets, galaxies, nebulae etc.).

031.565   Reduction of double-beam observations of extended radio sources.   J. Pfleiderer.
Astron. Astrophys., Vol. 101, 320 - 322 (1981).

A direct method for reducing double-beam observations of extended radio sources to the equivalent single-beam data is proposed. It involves the solution of a set of linear equations which becomes unique by applying suitable boundary conditions. The method is also compared to the Fourier transform method.

031.566   Recursive estimation of the reduction parameters of an astrometric plate.   A. Fresneau.
Astron. Astrophys., Vol. 102, 143 - 146 (1981).

Automatic systems of data acquisition provide the possibility to use sequential techniques of parameters estimation during the process of measurements. The estimates for the astrometric plate constants can be updated when the measurement of each of the reference stars becomes available. A material of 255 plates is intended to demonstrate the reliability and efficiency of a recursive estimation procedure. The method seems to be appropriate in checking the quality of the on-line processing.

031.567   Surveying velocity fields in galaxies.
G. de Vaucouleurs.
Sky Telesc., Vol. 62, 406 - 410 (1981).

031.568   Über die Führung eines Beobachtungsbuches.
W. Braune.
BAV Rundbrief, 30. Jahrg., 70 - 78 (1981).

031.569   Réduction des catalogues photographiques: zones +31° à −2° (R. C. P. 164).   P. Lacroute.
Bull. Inf. Cent. Données Stellaires, No. 21, p. 2 - 25 (1981).

The reduction of the Photographic Catalog to the system AGK2 - AGK3 has been performed for all stars of the zones +31° to −2°. The work done is described in detail.

031.570   Application of technical diagnostics methods at remote sounding of scattered cosmic objects.
B. N. Petrov, V. V. Bugrovskij, E. B. Dudin.

Dokl. AN SSSR, Tom 257, 553 - 558 (1981). In Russian.
Abstr. in Ref. zh., 51. Astron., 9.51.1039 (1981).

031.571   Statistical accuracy in stellar speckle interferometry at low light levels.   J. G. Walker.
Opt. Acta, Vol. 28, 885 - 905 (1981). − Abstr. in Phys. Abstr., Vol. 84, Abstr. 88764 (1981).

031.572   The observation of the solar irradiance and its variations, challenging space metrology.
D. Crommelynck.
Sol. Phys., Vol. 74, (see 012.029), 509 - 519 (1981).

The problems associated with the accurate determination of the total and spectral irradiances of the Sun are discussed. It is estimated that an ultimate accuracy of the order of 2 to 5×10^{-4} should be aimed at and be feasible for total solar irradiance measurements made with second generation objectively characterised absolute radiometers.

031.573   On recording of black body radiation.
J. Ojaste.
Tartu Astrofüüs. Obs., Teated, Nr. 65, p. 16 - 30 (1981). In Russian.

The response of photon counts to thermal light is studied following the ideas of Bothe, Bose and Einstein about multiple quanta. Formulae for calculating the angular diameter and temperature of a star are derived from the statistical interferometer formula.

031.574   Electromagnetic isotope separation as a method of improving the detection limits of accelerator spectrometry.   G. M. Raisbeck, F. Yiou.
Nucl. Instrum. Methods Phys. Res., Vol. 186, 483 - 486 (1981). − Abstr. in Phys. Abstr., Vol. 84, Abstr. 94939 (1981).

031.575   Digital image processing of Ap-star coudé Zeeman plates.   H. J. Wood, R. Albrecht.
Upper main sequence chemically peculiar stars, (see 012.033), p. 199 - 210 (1981).

Plots as well as two-dimensional video displays of short spectral regions around relatively blend-free lines of high magnetic sensitivity are presented. The clear case of the resolved Zeeman patterns of HR 5747 (β CrB) is contrasted to the similar but mixed Doppler/Zeeman spectrum of HR 183 (HD 3980). The discovery of a variable magnetic field in the double-line Ap-binary HR 5303 (η Aps) is announced.

031.576   Preliminary results of a statistical study of some period determination methods.
G. Mersch, A. Heck.
Upper main sequence chemically peculiar stars, (see 012.033), p. 299 - 305 (1981).

The purpose of the present paper is to compare the performance and the accuracy of some methods of period determination. As it was not possible to make an exhaustive discussion of all existing methods the authors have restricted their investigations to the field of the Fourier analysis on one hand and, on the other hand, to the non-parametric techniques using the θ-criterion inspired by Lafler and Kinman (1965).

031.577   Measurements of stellar magnetic field.
K. Nariai.
Upper main sequence chemically peculiar stars, (see 012.033), p. 321 - 322 (1981).

031.578   Selection effects in spectral searches for quasars.
R. G. Clowes.
Mon. Not. R. Astron. Soc., Vol. 197, 731 - 738 (1981).

The paper shows how the selection effects in detection that operate in spectral searches for quasars may be quantified and the quantification is tested with data from the CTIO-4m grism and the Curtis Schmidt objective-prism surveys. The quantification leads to an explanation of the apparent discrepancy in the surface densities of quasars from the two surveys.

031.579 **A method for unambiguous determination of starspot temperatures and areas: application to II Pegasi, BY Draconis, and HD 209813.** S. S. Vogt.
Astrophys. J., Vol. 250, 327 - 340 (1981) = Lick Obs. Bull. No. 893.

A method is presented for unambiguously determining both starspot effective temperatures and areas from standardized V and R light curves and the Barnes-Evans surface brightness relation. The method is applied to the 1977 spot of II Peg and describes a spot covering 37–45% of the visible hemisphere at an effective temperature of $T_e = 3400 \pm 100$ K. This spot contributed 7% and 12% of the total observed flux from the system at V and R, respectively. New high-precision V and R photometry is presented for BY Dra and HD 209813, and spot temperatures are derived to be 3500 ± 450 K and $\leqslant 3840 \pm 200$ K, respectively. Comparison of all these temperatures with the Sun reveals that T_{spot}/T_{star} is remarkably similar between starspots and the umbrae of sunspots; however, it is also shown that if the sunspot analogy is valid, the derived temperatures must refer predominantly to the starspots' penumbrae. This contradiction implies that either the temperature agreement is merely coincidental, or starspots are not morphologically similar to sunspots. Assuming the former, these penumbral temperatures imply umbral temperatures of 2600–2900 K, in good agreement with the value of 2760 K predicted by the 3 kilogauss convective starspot model of Mullan.

031.580 **Recommendations for calibration of millimeter-wavelength spectral line data.**
M. L. Kutner, B. L. Ulich.
Astrophys. J., Vol. 250, 341 - 348 (1981).

The authors examine the methods currently used to calibrate millimeter-wavelength photometric data and make recommendations on procedures to be followed. Since confusing references to the quantity $T_A{}^*$ have appeared in the literature, they introduce a series of formal definitions of quantities related to intensity measurements. A new quantity called $T_R{}^*$ is introduced which is the source antenna temperature corrected for atmosheric, ohmic, and all spillover losses. Physically, $T_R{}^*$ corresponds to the source brightness distribution convolved with the diffraction and error beam patterns of the telescope. The authors also reexamine the relative merits of the so-called "direct" and the "chopper-wheel" methods of correcting for atmospheric absorption and conclude that the chopper-wheel technique is generally preferable.

031.581 **On the problem of detection of the circumstellar clouds from 2800 Mg II observations.**
G. A. Gurzadyan.
Astrophys. Space Sci., Vol. 80, 197 - 209 (1981).

A new method for the detection of the circumstellar clouds around the hot stars of O–B classes is developed. The method is based on a fact connected with the large dispersion in the observed equivalent widths. $W_*(2800$ Mg II) of non-stellar origin, for a selected group of stars with $E(B-V) = 0$. It is shown that the circumstellar clouds really existed around ~90% of hot stars analysed in the present paper.

031.582 **A method for searching for optical pulsars.**
Y. P. Elsworth, J. F. James.
Astron. Astrophys., Vol. 103, 131 - 134 (1981).

A method is described for searching for optical pulsars and other short-period phenomena in astronomical objects. No knowledge of the period is required and the search can be carried out with portable equipment at any reasonably large

telescope. A photon counter collects samples from the object and a real-time Fast Fourier Transform is carried out on successive sets of 1024 samples. The result is stored and displayed graphically and the displays updated continuously as new data arrive, are transformed and added into the store. The sensitivity of the method is derived theoretically, and the results of laboratory and field tests are given, showing satisfactory agreement with this theory.

031.583 **Extended radio sources: a method for baseline improvement.** J. Pfleiderer.
Astron. Astrophys., Vol. 103, 220 - 222 (1981).

The author describes a method of determining and correcting baselines of individual scans in an observed map of an extended radio source. Initial baselines are determined by selecting only those data of a single scan which lie closest to a straight line. A correction of these initial baselines is achieved by comparing different scans. This requires having scanned the source in two perpendicular directions. The baselines are adjusted such that the differences of the data at the cross-over points between perpendicular scans are minimized over the whole map using the method of least squares.

031.584 **Photographic polarization survey with a Savart plate.** H.-J. Röser.
Astron. Astrophys., Vol. 103, 374 - 381 (1981).

A reduction procedure for Savart plate measurements to survey for high linear polarization in point-like galactic and extragalactic objects is described. The detection limit on a single plate is shown to be ~8% in linear polarization for the range 13–18 mag (1 h exposure at a 1.2-m-telescope on IIaO). Possible applications are discussed and first results of a survey near M 17 are given.

031.585 **The use of VLBI in observing astronomical and terrestrial survey parameters.** S. Kwase.
Rev. Radio Res. Lab., Vol. 26, 479 - 484 (1980). In Japanese. Abstr. in Phys. Abstr., Vol. 84, Abstr. 101939 (1981).

031.586 **Images in radioastronomy.** T. W. Cole.
J. Electr. Electron. Eng. Australia, Vol. 1, 123 - 129 (1981). – Abstr. in Phys. Abstr., Vol. 84, Abstr. 102044 (1981).

031.587 **Analysis of a satellite interferometer system (SIS).** P. R. Silverglate.
Opt. Eng., Vol. 20, 621 - 624 (1981). – Abstr. in Phys. Abstr., Vol. 84, Abstr. 108073 (1981).

031.588 **Recherche d'objets faibles fortement polarisés.** G. Corriveau, E. Borra, C. Rheault.
J. R. Astron. Soc. Canada, Vol. 75, 246 (1981). – Abstract.

031.589 **A new asteroid observation and search technique.** L. G. Taff.
Publ. Astron. Soc. Pacific, Vol. 93, 658 - 660 (1981).

An observatory designed for artificial satellite work can be reconfigured to search for and observe asteroids. It will be used to conduct searches for Earth-approaching asteroids.

031.590 **An investigation of autoradiographically intensified electrographic imagery.**
H. M. Heckathorn, G. R. Carruthers.
Publ. Astron. Soc. Pacific, Vol. 93, 672 - 680 (1981).

The potential benefit of post-development intensification of electrographic film negatives using the technique of Thiourea-S^{35} autoradiography is investigated using microphotometric scans of uv laboratory spectra. It is shown that the technique improves the effective signal-to-noise ratio on weakly exposed regions of emulsion by increasing image density to the point where microphotometric noise becomes a minor contributor to total system noise.

031.591 **Image processing with COSMOS.**
R. S. Stobie, R. J. Dodd, H. T. MacGillivray.
Sky Telesc., Vol. 62, 538 - 539 (1981).

031.592 **The microwave solar radar experiment. I. Observations.**   H. R. Fitze, A. O. Benz.
Astrophys. J., Vol. 250, 782 - 790 (1981).
This is a report on the first solar radar experiment in microwaves. It was carried out with the 300 m dish in Arecibo using a 250 kW transmitter. Receiving at a displaced frequency from the transmitted radar frequency allowed to probe the Langmuir (plasma) wave energy density of the corona in the 170 - 270 MHz range. The authors have not found any echo in various regions on the Sun: quiet regions, active regions, type I radio sources and even a possible type IV radio source. This contradicts some models of type I radio bursts and a proposed scattering mechanism of metric solar radar echos. An alternative experiment, in which they received at the transmitted frequency, did not produce any echo either.

031.593 **Measurement and reduction of photographic plates.**
R. Albrecht.
AAS Photo-Bull., No. 25, p. 9 - 12 (1980).

031.594 **One-dimensional infrared speckle interferometry.**
R. R. Howell, D. W. McCarthy, F. J. Low.
Astrophys. J., Lett., Vol. 251, L21'- L25 (1981).
An improved technique of one-dimensional speckle interferometry yields high sensitivity at wavelengths from 2.2 to 11.6 μm. The "frozen" images at 11.6 μm are essentially diffraction limited under good "seeing" at the 2.3 m telescope. The binary character of the protostellar source W3 IRS 5 is confirmed at 5 μm; the separation, position angle, and brightness ratio are $1''.26$, $37°$, and 1.9, respectively. The individual components are $\leqslant 0''.25$ in diameter. S140 IRS 1 and Mon R2 IRS 3 are extended, $\geqslant 1''$, at 2.2 μm, while GL 2591, NGC 2264 IRS, and the Becklin-Neugebauer source are unresolved, $\leqslant 0''.2$. At 11.6 μm the size of the dust shell around α Ori is measured for the first time; assuming a Gaussian intensity distribution, the $1/e$ intensity points are separated by $3''.4$, more than 60 stellar diameters.

031.595 **Automatic classification of galaxy images by Fourier structural analysis.**
G. Sedmak, M. L. Trujillo Lamas.
Astron. Astrophys., Vol. 104, 93 - 98 (1981).
The authors describe a procedure for the automatic classification of galaxy images based on a two-bits Fourier structural detector which monitors the radial and angular excess noises in Fourier domain. The relevant section of this procedure was successfully tested on numerical simulations of basic types of galaxy-like images and on true images of galaxies. Sample results of the Fourier morphology of basic galaxy-like images and of the operations of the Fourier structural detector are reported.

031.596 **Information treatment of astronomical images.**
J.-F. Lallier, E. Hardy.
Proceedings of the 7th Canadian Man-Computer Communications Conference, Waterloo, Ont., Canada, 10 - 12 June 1981. Canadian Man-Computer Comm. Soc., Toronto, Ont., Canada (1981). p. 335 - 341. In French. – Abstr. in Phys. Abstr., Vol. 85, Abstr. 3205 (1982).

031.597 **A study of the convergent solutions of Kopal's iterative method for solving the light curves of eclipsing binaries.**   J.-s. Chen, Z.-l. Zou.
Acta Astrophys. Sinica, Vol. 1, 213 - 221 (1981). In Chinese.

031.598 **The analysis of solar limb observations. I. Restoration of data in a tilted reference frame.**
A. Wiesmeier, C. J. Durrant.

Astron. Astrophys., Vol. 104, 207 - 210 (1981).
The authors derive expressions for the correction of the statistical properties of two-dimensional fields degraded by arbitrary (without axial symmetry) point-spread functions. They treat both one- and two-dimensional sampling. Several earlier estimates of corrections to observations of brightness and velocity fluctuations away from the centre of the solar disk are shown to be in error. Corrected values of the rms granular brightness fluctuations as a function of heliocentric angle are reported.

031.599 **Millimeter-wavelength continuum calibration sources.**   B. L. Ulich.
Astron. J., Vol. 86, 1619 - 1626 (1981).
High-precision observations of 18 radio sources have been made at wavelengths of 9.55, 3.33, and 2.00 mm between 1971 and 1980. The sources include the Sun, Moon, planets, planetary satellites, asteroids, H II regions, planetary nebulae, and a quasar. Analyses of these data provide accurate flux densities for the interpretation of source spectra and for the calibration of other radio astronomical observations. A brief description of the observational techniques is given and tables of absolute flux densities are presented.

031.600 **On the elimination of the errors due to non-linearity of converters of graphs.**   G. I. Abbasov.
Tsirk. Shemakhinsk. Astrofiz. Obs., No. 68, p. 24 - 27 (1981). In Russian.

031.601 **Faint comet searching.**   T. Gehrels.
Icarus, Vol. 47, (see 012.045), 518 - 522 (1981).
Comets are usually discovered to a magnitude limit of $B_{lim} \sim 19$ at best. This can be improved by one magnitude with a blink-search technique that is described here. Three faint comets were found this way. Another, future, method to find such inconspicuous comets is with electronically scanning cameras and computerized reduction.

031.602 **Introductory remarks. Photometry section.**
M. F. A'Hearn.
Modern observational techniques for comets, (see 012.047), p. 53 - 56 (1981).

031.603 **Correlated ground-based and IUE observations.**
M. F. A'Hearn.
Modern observational techniques for comets, (see 012.047), p. 138 - 140 (1981).

031.604 **Use of an image dissector scanner for spectrophotometry of faint comets.**
H. Spinrad, R. L. Newburn, Jr.
Modern observational techniques for comets, (see 012.047), p. 148 - 149 (1981).

031.605 **Observations of faint comets at McDonald Observatory: 1978 - 1980.**
E. S. Barker, A. L. Cochran, P. M. Rybski.
Modern observational techniques for comets, (see 012.047), p. 150 - 155 (1981).
The assumption that periodic comets are quiescent at large (> 3 a. u.) heliocentric distances is proving to be erroneous, based on the photometry of P/Tempel 2 at 3.1 a.u. and P/Encke at 3.9 a. u. Both showed outbursts of greater than a magnitude. The colors of both comets appear to be bluer during outburst than before or after outburst. Rotation periods of 19 - 20th magnitude comets can be measured using the DAP system, if care is taken to avoid many periods of time when near occultations occur with field stars which are usually brighter than the comet. The ability of an area photometer to detect other objects which are in the field and which could add significant numbers of photons to the light from the comet makes the DAP system so ideal for comet photometry.

031.606   **Spectral imagery: recent results with the SPIFI and their implications for cometary atmospheric studies.**
W. Hayden Smith.
Modern observational techniques for comets, (see 012.047), p. 156 - 160 (1981).

Spectral polarimetric imaging of comets is proposed with state-of-the-art methods. Related observational data are shown and sensitivity levels are given which indicate the range of objects which may be studied. Some proposed observational goals are also indicated.

031.607   **A possible technique for cometary studies with high angular and spectral resolution.**   T. R. Gull.
Modern observational techniques for comets, (see 012.047), p. 161 - 166 (1981).

The echelle spectrographs, designed for and used at the Cassegrain stations of the KPNO and CTIO 4-meter telescopes, are capable of cometary spectroscopy with seeing-limited angular resolution along the slit and with spectral resolving power $(\lambda/\Delta\lambda)$ ranging from 10^4 to 10^5. Various gratings, cameras and detectors can be used in combination for specific studies in the 3000 Å to 10,000 Å range.

031.608   **Laboratory research.**
B. Donn.
Modern observational techniques for comets, (see 012.047), p. 251 - 256 (1981).

031.609   **On the possibility of using cosmic radio emission sources for investigation of the moon and planets by a bistatic radar method.**
A. G. Pavel'ev, A. I. Kucheryavenkov.
13-ya Vses. konf. po radioastron. issled. soln. sistemy, Kiev, 1981. Kiev, 1981, p. 75 - 76. In Russian. − From Ref. zh., 51. Astron., 11.51.259 (1981).

031.610   **Digitale Bildverarbeitung in der Astronomie.**
R. Albrecht.
Mitt. Astron. Ges., Nr. 54, (see 012.050), p. 53 - 59 (1981).

The most important aspects of developing and maintaining a computer-controlled astronomical image processing system are addressed in the paper. These include considerations pertaining to system hardware and software structure, the nature of image data, and the operations to be carried out on an astronomy-oriented interactive image processing system. Special attention is devoted to the problem of system software maintainance and transportability of application programs.

031.611   **Photographische Photometrie mit Racine-Prisma in Fokalreduktoranordnung.**   F.-T. Lentes.
Mitt. Astron. Ges., Nr. 54, (see 012.050), p. 181 - 182 (1981).

031.612   **Radialgeschwindigkeitsbestimmungen mit Doppelgitterprismen mit spaltlosen Spektrographen.**
B. Nelles.
Mitt. Astron. Ges., Nr. 54, (see 012.050), p. 186 - 188 (1981).

031.613   **Beobachtungen von H II-Gebieten mit verschiedenen Fabry-Perot-Interferometern.**   A. Hänel.
Mitt. Astron. Ges., Nr. 54, (see 012.050), p. 189 - 193 (1981).

031.614   **Eine Fabry-Perot-Kamera zum Studium von Ionen in der Koma von Kometen.**   K. Jockers.
Mitt. Astron. Ges., Nr. 54, (see 012.050), p. 196 - 197 (1981).

031.615   **35 mm-Kameras zur Aufnahme und Nachverarbeitung astronomischer Speckle-Bilder.**
C. Ziegler, J. Beringer, F. Merkle, J. Bille.
Mitt. Astron. Ges., Nr. 54, (see 012.050), p. 256 - 260 (1981).

031.616   **Photographische Polarisationsdurchmusterung.**
H.-J. Röser.
Mitt. Astron. Ges., Nr. 54, (see 012.050), p. 260 (1981). Abstract.

031.617   **Ein Vergleich zwischen optischer und digitaler Fouriertransformation von Sonnengranulationsaufnahmen.**   O. von der Lühe.
Mitt. Astron. Ges., Nr. 54, (see 012.050), p. 261 (1981). Abstract.

031.618   **Neue Entwicklungen bei der Reduktion und Analyse von Radiokontinuumskartierungen.**   K. Reif.
Mitt. Astron. Ges., Nr. 54, (see 012.050), p. 269 (1981).

031.619   **Analysis of multi-ray methods for radio observations of meteor trains.**   V. A. Nechitajlenko.
Meteorn. issled., Moskva, 1981, No. 7, p. 90 - 99. In Russian. Abstr. in Ref. zh., 51. Astron., 11.51.1056 (1981).

031.620   **On the exclusion of distorted images in an electron-optical TV system.**
V. S. Matyagin, N. V. Sinyaeva, L. A. Usol'tseva.
Automation of astronomical observations, (see 003.020), p. 94 - 109 (1981). In Russian. − Abstr. in Ref. zh., 51. Astron., 11.51.1068 (1981).

031.621   **Analysis of some methods of TV astronomy in the problem of determination of the equatorial coordinates of astronomical objects.**   V. S. Melkumov.
Automation of astronomical observations, (see 003.020), p. 57 - 63 (1981). In Russian. − Abstr. in Ref. zh., 51. Astron. 11.51.1069 (1981).

031.622   **Photometric observations of faint objects and their processing on the M-220M computer.**
I. G. Babkin, A. V. Didenko, T. P. Nosova, G. V. Khonin.
Automation of astronomical observations, (see 003.020), p. 83 - 93 (1981). In Russian. − Abstr. in Ref. zh., 51. Astron., 11.51.1080 (1981).

031.623   **The main sources of errors in amplitude-phase field distribution of multi-element interferometers.**
B. B. Krissinel', T. A. Treskov.
Issled. po geomagn., aehron. i fiz. Solntsa, Moskva, 1981, No. 56, p. 176 - 184. In Russian. − Abstr. in Ref. zh., 51. Astron., 11.51.1083 (1981).

031.624   **Low-energy particle registration with a monocrystal scintillation radiometer with anticoincidence scheme.**
S. N. Fedorchenko.
Vopr. atom. nauk. i tekh. Yader. priborostr. Moskva, 1981, No. 2147, p. 35 - 39. In Russian. − Abstr. in Ref. zh., 51. Astron., 11.51.1107 (1981).

031.625   **A quick method of determining monochromatic absorption corrected stellar magnitudes.**
V. A. Chernobaj, V. F. Chernobaj.
Astron. Tsirk., No. 1117, p. 4 - 6 (1980). In Russian.

031.626   **Features of speckle interferometry at a moderate-diameter telescope.**   V. G. Chernyj.
Astron. Tsirk., No. 1133, p. 4 - 5 (1980). In Russian.

031.627   **On the possibility of using a sodar in site testing.**
M. A. Kallistratova.
Astron. Tsirk., No. 1124, p. 4 - 6 (1980). In Russian.

031.628   **Influence of long-period scintillations on the accuracy of spectrophotometric measurements.**
G. A. Alekseeva, L. A. Kamionko.
Astron. Tsirk., No. 1144, p. 4 - 6 (1980). In Russian.

031.629 **Determination of brightness and integral apparent magnitude of weak diffuse sources.**   A. Tomić.
Vasiona, Année 29, 75 - 77 (1981). In Croatian.

031.630 **Automatic method for determination of the variability index of the solar radio flux.**
A. Balklavs, M. Eliáss, I. Šmelds.
Investigations of the sun and red stars. 13, (see 003.023), p. 90 - 97 (1981). In Russian.
Amplitude and frequency of radiobursts can serve as indicators of the state before solar flares. The amplitude and number of such bursts could be characterized with the so-called variability index. An equipment for automatic determination of this index has been developed.

031.631 **Method for determining object intensity distributions in stellar speckle interferometry.**
B. J. Brames, J. C. Dainty.
J. Opt. Soc. America, Vol. 71, 1542 - 1545 (1981).
The authors describe a method for finding the intensity distribution of an object from a measurement, or measurements, of its cross-power spectrum. This technique, like the Knox-Thompson algorithm, makes use of the phase information contained in the cross spectrum; it is based on a systematic procedure for locating the complex zeros representing the spectrum of an object.

031.632 **Object reconstruction from stellar speckle interferometry data.**   J. G. Walker.
J. Opt. Soc. America, Vol. 71, 1562 - 1563 (1981). – Abstract.

031.633 **Maximum entropy estimation of spread function in astronomical imagery.**   F. Sabet-Peyman.
J. Opt. Soc. America, Vol. 71, 1563 (1981). – Abstract.

031.634 **On an improved method for measuring the variability of solar radio emission.**
A. Balklavs, M. Eliáss, I. Šmelds.
Zvaigžnota debess, 1981. gada vasara, p. 18 - 19. In Latvian.

031.635 **Alternate approach to the analysis of solar photometer data.**   B. M. Herman, M. A. Box,
J. A. Reagan, C. M. Evans, with comments by A. T. Young.
Appl. Opt., Vol. 20, 2925 - 2928, 3993 (1981).
The standard technique of analyzing solar photometer data to determine atmospheric optical depth and the spectral solar constant is shown to inadvertently weight the data unequally. A new approach is proposed which equally weights all the data. Assuming that the deviations of the data points result from real random variations of optical depth during the period of the measurements, this latter approach is shown to yield more reliable results.

031.636 **Rectification and enhancement of three severely distorted images of Jupiter's north polar region.**
R. N. Strickland, J. J. Burke.
Appl. Opt., Vol. 20, 3612 - 3618 (1981).
A number of high-resolution images of Jupiter's northern hemisphere were received from the imaging photopolarimeter (IPP) aboard Pioneer 11 in 1974. Erratic scanning of the IPP caused severe distortions in three scientifically important images, which until now have never been satisfactorily restored. The authors report new rectification and enhancement techniques, implemented on up-to-date image processing hardware, yielding images of sufficient quality to enable full scientific exploitation of the photometric data.

031.637 **A spirit of search.**   K. C. Beckmann.
J. American Assoc. Variable Star Obs., Vol. 10, 25 - 27 (1981).
A description is provided of a nova-search program that

is guided by studies of historical novae and their distribution on the Milky Way.

031.638 **Perceptual experiences in solar observing.**
D. Ammons, R. B. Ammons.
J. American Assoc. Variable Star. Obs., Vol. 10, 36 (1981). Abstract.

031.639 **Detecting brightness variations due to sunspot cycles in normal stars.**   R. D. Ferdie.
J. American Assoc. Variable Star Obs., Vol. 10, 37 (1981). Abstract.

031.640 **Spectrographie stellaire I. Optimisation du temps de pose.**   G. Goy.
Arch. Sci., Vol. 34, 251 - 258 (1981) = Publ. Obs. Genève, Sér. A, Fasc. 84.

031.641 **Calibration of spectral line observations at the Metsähovi Radio Observatory.**
L. J. Malkamäki.
Obs. Astrophys. Lab. Univ. Helsinki, Rep. 1/1981, 23 pp. (1981).

031.642 **Restoration errors of the directivity pattern due to measurement errors in scanning of the near antenna field.**   V. V. Arabadzhi, Yu. I. Belov, A. I. Knafel'.
Izv. vuzov. Radiofiz., Tom 24, 349 - 357 (1981). In Russian.
Abstr. in Ref. zh., 51. Astron., 12.51.1055 (1981).

031.643 **Automated star counts in the dark cloud L1454.**
R. Duerr, E. R. Craine.
Prepr. Steward Obs., No. 342, 36 pp. (1981).
The Near Infrared Photographic Sky Survey (NIPSS) is a data base with broad potential not only for optical identification of point infrared sources, but also for a variety of studies related to the distribution of red stars. It has been suggested that these applications could be greatly aided by digitization of the data base, for which a feasibility study is underway. As a part of this study a NIPSS visual and near infrared photographic pair, encompassing the dark cloud L1454, was digitized. Star counts in this region suggest the existence of two clouds along the line of sight and allowed the distances, extinctions, and masses of the clouds to be estimated. In addition, maps of the region as a function of the redness of the constituent stars were generated. These, when compared to the extinction maps, allow a discussion of the star formation properties of the region.

031.644 **Analyse de l'équation de magnitude sur l'astrolabe de Danjon à l'Observatoire de Shanghai.**
T.-q. Xu, Z.-g. Yang.
Ann. Shanghai Obs. Acad. Sinica, No. 1, p. 34 - 39 (1979).

031.645 **A method for acquiring the tracks of artificial satellites.**   Q.-c. Lin, X.-l. Niu.
Ann. Shanghai Obs. Acad. Sinica, No. 1, p. 92 - 96 (1979).

031.646 **Investigation on measurement errors of an astrophotogram.**   K.-p. Tian, M.-g. Yin, Z.-h. Xu, J.-l. Zhao.
Ann. Shanghai Obs. Acad. Sinica, No. 2, p. 90 - 94 (1980).

031.647 **Some procedures for comet discovery.**
W. A. Bradfield.
Int. Comet Q., Vol. 3, 71 - 75 (1981).

031.648 **A review of visual comet observing techniques – III.**
C. S. Morris.
Int. Comet Q., Vol. 3, 89 - 90 (1981).

031.649 Beitrag zur geodätischen und astronomischen
         Nutzung von Sternbedeckungen durch den Mond.
D. Böhme.
Geod. Geophys. Veröff., Reihe III, Heft 42, 84 pp. (1980).

031.650 Characteristics of cosmic ray cutoffs for a satellite
         orbiting at 400 km; the effect of the solid earth.
D. F. Smart, M. A. Shea, J. E. Humble.
Cosmic rays in the heliosphere, (see 012.059), p. 165 - 168
(1981).

Studies of galactic cosmic ray intensity and composition
by means of earth orbiting space vehicles often rely on the
earth's magnetic field as a momentum analyzer or threshold
energy indicator. The authors have determined, by the
trajectory-tracing process, cosmic ray cutoff characteristics
for a satellite at 400 km altitude. These calculations indicate
that cosmic rays have direct access to the satellite from the
magnetic west down to angles of 135° from the zenith.

031.651 Cometary dust observations by optical *in-situ*
         methods.   A. C. Levasseur-Regourd,
D. W. Schuerman, R. H. Zerull, R. H. Giese.
Progress in planetary exploration, (see 012.062), p. 113 - 120
(1981).

Remote optical observations of comets provide information only along the whole line of sight and require some
assumptions to be interpreted. Due to the advent of cometary
space missions, a two-step strategy has been defined to derive
without any assumption spatial distribution and physical
properties of dust by *in-situ* optical observations. First, an
Optical Probe Experiment, suitable for a fast flyby, should
provide passive *in-situ* measurements in the direction of the
approaching (or receding) comet near encounter. Secondly, a
Light Scattering Dust Analyzer, suitable for a rendez-vous mission, should permit the determination of the scattering properties of individual particles.

031.652 Restoration of the spectra of solar emission scatter-
         ed in the Venus atmosphere (according to Venera 11
and Venera 12 data).
A. V. D'yachkov, B. E. Moshkin, E. B. Shesterikova.
Inst. kosm. issled. AN SSSR. Prepr., 1981, No. 633, 31 pp. In
Russian. – Abstr. in Ref. zh., 51. Astron., 1.51.877 (1982).

031.653 Automatic quantitative spectral classification of
         stars. I. Observational data reduction.
V. Malyuto, J. Pelt.
Academy of Sciences of the Estonian SSR, Prepr. A-1, p. 1 - 15
(1981).

As a first step in the quantitative spectral classification of
the F-G-K stars with the help of the automated microphotometer at the Tartu Observatory a technique of observational
data reduction has been worked out.

Image et information.   See Abstr. 003.031.

Observing visual double stars.
See Abstr. 003.048.

The computer in optical research, methods and
applications.   See Abstr. 003.064.

Digital spectral analysis.   See Abstr. 003.164.

Image analysis techniques and applications. Technical digest.   See Abstr. 012.072.

The analysis of periodicities in irregularly sampled
data.   See Abstr. 021.001.

Two-star high-speed photometry.
See Abstr. 034.015.

Experiments with an SIT camera.
See Abstr. 034.057.

Reseau astrometry with Palomar Schmidt plates:
position-coincidence optical identification of radio sources.
See Abstr. 041.014.

Reseau astrometry with Palomar Schmidt plates: a
two-way analysis of variance.   See Abstr. 041.015.

Programme de latitude du tube zénithal visuel
de l'Observatoire Turku-Tuorla, système amélioré de 1976.
See Abstr. 045.024.

New problems of cometary observations from space.
See Abstr. 051.037.

Minimizing the calculation scope of spectrum
interpretation with fine structure of the solar plasma taken
into account.   See Abstr. 071.023

The analysis of solar limb observations. II. Geometrical smearing.   See Abstr. 071.030.

Physical properties of the solar chromosphere
deduced from optically thick lines: I. Observations, data
reduction, and modeling of an average plage.
See Abstr. 073.077.

Study of Hα emission from solar limb prominences
using Fabry-Perot interferometry.   See Abstr. 073.102.

Interferometry of solar limb prominences at
656.3 nm. Part 1.   See Abstr. 073.103.

Finite sun effect on the interpretation of solar
aureole.   See Abstr. 082.094.

Measurement of minor species (H_2, Cl, O_3, NO) in
the earth's atmosphere by the occultation technique.
See Abstr. 082.119.

Parallax and atmospheric scattering effects on the
inversion of satellite auroral observations.
See Abstr. 084.091.

Report on the A. L. P. O. - LTP *(Lunar Transient
Phenomena)* observing program.   See Abstr. 094.502.

On observing comets for nuclear rotation.
See Abstr. 102.071.

The interactive astronomical data analysis facility –
image enhancement techniques applied to comet Halley.
See Abstr. 103.809.

An RV digital measuring method applied to Lac OB1:
preliminary results.   See Abstr. 111.005.

Observations of stellar winds in early type stars.
See Abstr. 112.020.

La misura dei raggi stellari.   See Abstr. 114.138.

Statistical method for calculating parallaxes and
masses of binaries with unknown orbits.
See Abstr. 118.019.

On the possibility of detecting companions to
cepheids and their effect on the CORS method.
See Abstr. 122.074.

The multiple-point fringe-rate method of mapping spectral-line VLBI sources with application to H_2O masers in W3-IRS5 and W3(OH).   See Abstr. 131.109.

Distances to 14 molecular clouds (including two associated with supernova remnants) by a new technique. See Abstr. 131.200.

A 408 MHz all-sky continuum survey. I. Observations at southern declinations and for the north polar region. See Abstr. 141.017.

Imaging observations of optical emission from jets. See Abstr. 141.183.

On the optical appearance of distant galaxies. See Abstr. 158.242.

The optical jet in M87 — deconvolution by maximum entropy.   See Abstr. 158.259.

Neutral hydrogen in M31. I. The distribution of H I gas and spiral arms.   See Abstr. 158.269.

Errata

031.901 Erratum: 'Zernike annular polynomials for imaging systems with annular pupils: errata' [J. Opt. Soc. America, Vol. 71, 75 - 85 (1981)].   V. N. Mahajan. J. Opt. Soc. America, Vol. 71, 1408 (1981). – See Abstr. 29.031.036.

031.902 Erratum: 'Distortion of grating prisms, and their use in radial velocity determinations of astronomical objects with slitless field spectrographs.' [Appl. Opt., Vol. 20, 660 - 664 (1981)]. B. Nelles, E. H. Geyer. Appl. Opt., Vol. 20, 3999 (1981). – See Abstr. 029.031.025.

032 Astronomical Instruments, Space Instrumentation

Astronomical Instruments

032.001 Sul montaggio e sul calcolo della parte ottica in un telescopio di Newton visuale e fotografico (Parte seconda).   C. Giordanengo, U. Sisto. Orione, Vol. 2, 232 - 239, with a correction p. 289 (1981).

032.002 On the elimination of the diffraction halo in extra-eclipsing instruments with apodization of the objective.   V. P. Vasil'ev. Soln. Dannye 1981 Byull., No. 2, p. 86 - 90 (1981). In Russian.

The image formation in a coronograph with a uniform objective and with an objective whose amplitude transparency decreases limbwards as a Gaussian is considered. It is shown that such apodization of the objective leads to full extinction of the diffractional halo even near the inner corona with small increase of the diameter of the solar disc image.

032.003 Meine Erfahrungen mit Celestron-Teleskopen. U. Köhler. Sterne Weltraum, Jahrg. 20, 332 - 334 (1981).

032.004 Secondary mirror subassembly in the 2.4 m NASA optical telescope.   N. Ahmed, K. Doyle. Proc. Soc. Photo-Opt. Instrum. Eng., Vol. 250, (see 012.004), p. 24 - 26 (1980). – Abstr. in Phys. Abstr., Vol. 84, Abstr. 71079 (1981).

032.005 Wide-angle infrared telescope with stable performance.   D. Shafer. Proc. Soc. Photo-Opt. Instrum. Eng., Vol. 250, (see 012.004), 14 - 17 (1980). – Abstr. in Phys. Abstr., Vol. 84, Abstr. 75016 (1981).

032.006 Measuring electron density in coronal active regions. I: A K-coronameter with a reflex monitor at $\lambda 5303$ Å.   J.-C. Noens, J.-L. Leroy. Sol. Phys., Vol. 73, 81 - 87 (1981).

The coronameter described in this paper has been designed for the study of coronal condensations with a 30″ spatial resolution. The instrument associates measurements of the K-corona polarized light with simultaneous pictures of the coronal structures as seen in the light of the green emission line of Fe XIV ($\lambda 5303$ Å). It has allowed to engage in an extensive program of observation devoted to the study of electron density in active coronal regions. The authors present results concerning a coronal condensation observed on 1980 February 15.

032.007 Solar telescopes filled with helium? O. Engvold, R. Brahde, B. Fossum. Publ. Astron. Soc. Pacific, Vol. 93, 526 - 527 (1981).

The possibility of improving telescope seeing by replacing the air with helium in telescope tubes has earlier been proposed (Rösch 1955). Experimental evidence is given in support of this hypothesis.

032.008 Large-telescope projects in the United States. G. R. Burbidge, D. Hall. Observatory, Vol. 101, 133 - 134 (1981).

032.009 Versatility of the telescope array concept. M. J. Disney. Observatory, Vol. 101, 136 - 137 (1981).

032.010 Problems of superposition and guidance for a hundred-telescope array.   R. C. M. Learner. Observatory, Vol. 101, 137 - 138 (1981).

032.011 On maximum-area telescopes at given cost. D. Lynden-Bell. Observatory, Vol. 101, 138 - 139 (1981).

032.012 Multiple mirrors, multiple objects. J. R. P. Angel. Observatory, Vol. 101, 139 - 140 (1981).

032.013 The UMIST 7 × 15-inch multi-aperture telescope programme.   J. F. Grainger. Observatory, Vol. 101, 140 (1981).

032.014 **The advantages of a single large telescope.**
R. G. Bingham.
Observatory, Vol. 101, 167 - 169 (1981).

032.015 **Les coronographes.** F. Crifo.
Astronomie, Vol. 95, 319 - 328 (1981).

032.016 **Det danske 1,5 m teleskop på La Silla.**
J. Andersen.
Astron. Tidsskr., Årg. 14, 93 - 109 (1981).

032.017 **Cutter design in meniscus telescopes.**
N. A. Agapov.
Izmerenie opt.-meteorol. parametrov atmos. s ispol'z. lazer.
izlucheniya. Tomsk, 1980, p. 160 - 164. In Russian. – Abstr.
in Ref. zh., 51. Astron., 9.51.950 (1981).

032.018 **Automation of a photoelectric reflecting zenith**
tube. V. A. Gedrovits.
Determination of the coordinates of celestial bodies, Riga,
1981, (see 003.008), p. 173 - 180. In Russian. – Abstr. in
Ref. zh., 51. Astron., 9.51.962 (1981).

032.019 **Telescopes: a look to the future.**
N. Henbest.
New Scientist, Vol. 89, 796 - 799 (1981). – Abstr. in Phys.
Abstr., Vol. 84, Abstr. 88743 (1981).

032.020 **Alignment of a four meter Ritchey-Chrétien tele-**
scope.
J. E. Simmons, W. Schoening, D. Graham, L. Ott.
Proc. Soc. Photo-Opt. Instrum. Eng., Vol. 251, (see 012.027),
p. 138 - 145 (1980). – Abstr. in Phys. Abstr., Vol. 84, Abstr.
94171 (1981).

032.021 **Influence of alignment errors of a telescope system**
on its aberration field.
R. V. Shack, K. Thompson.
Proc. Soc. Photo-Opt. Instrum. Eng., Vol. 251, (see 012.027),
p. 146 - 153 (1980). – Abstr. in Phys. Abstr., Vol. 84, Abstr.
94172 (1981).

032.022 **Telescope coalignment system for the Multiple**
Mirror Telescope (MMT). D. McDonough.
Proc. Soc. Photo-Opt. Instrum. Eng., Vol. 251, (see 012.027),
p. 154 - 162 (1980). – Abstr. in Phys. Abstr., Vol. 84, Abstr.
94173 (1981).

032.023 **The optical telescope with a main mosaic mirror of**
25 m in diameter. N. V. Steshenko.
Izv. Krymskoj Astrofiz. Obs., Tom 63, 161 - 165 (1981). In
Russian. English translation in Bull. Crimean Astrophys. Obs.,
Vol. 63.

032.024 **The Chinese photoelectric astrolabe.**
P. A. Wayman.
Irish Astron. J., Vol. 14, (see 012.035), 142 - 146 (1980).

032.025 **New prospects for astronomy.**
F. G. Smith.
Contemp. Phys., Vol. 22, 533 - 544 (1981). – Abstr. in Phys.
Abstr., Vol. 84, Abstr. 108059 (1981).

032.026 **Proposals of large telescopes.** P. Mayer.
Říše hvězd, Vol. 62, 252 - 256 (1981). In Czech.

032.027 **On the setting errors of the angle standard of the**
circumzenithal instrument and of the astrolabe.
A. Niemi.
Astron. Astrophys., Vol. 104, 276 - 279 (1981).
General formulae are derived allowing the computation
of the instrumental zenith distance of the circumzenithal instrument or of the prismatic astrolabe. A comparison is made
between the two instruments concerning the influence of the
setting errors of the angle standard. It is pointed out that,
in some cases, the collimation error may be a significant error
source. The method used at the Turku Observatory to control
the position of the angle standard is explained.

032.028 **Choice of a new place for establishing the Repsold**
meridian circle of the Moscow Observatory.
V. G. Shamaev.
Astrometric investigations, (see 012.048), p. 117 - 119 (1981).
In Russian. – Abstr. in Ref. zh., 51. Astron., 11.51.181 (1981).

032.029 **Large azimuthal telescope: status, results, perspec-**
tives. I. M. Kopylov.
Vestn. AN SSSR, 1981, No. 6, p. 26 - 36. In Russian.
From Ref. zh., 51. Astron., 11.51.1039 (1981).

032.030 **Coordinate guiding and scanning system as a means**
of automatic control of a telescope. B. F. Osak.
Issled. po geomagn., aehron. i fiz. Solntsa, Moskva, 1981,
No. 56, p. 140 - 142. In Russian. – Abstr. in Ref. zh., 51.
Astron., 11.51.1046 (1981).

032.031 **Vibrations of the tower of the large solar vacuum**
telescope. A. K. Kitov.
Issled. po geomagn., aehron. i fiz. Solntsa, Moskva, 1981,
No. 56, p. 143 - 154. In Russian. – Abstr. in Ref. zh., 51.
Astron., 11.51.1053 (1981).

032.032 **Das 3,5 m-Teleskop.** K. Bahner.
Mitt. Astron. Ges., Nr. 54, (see 012.050), p. 21 - 25
(1981).

032.033 **Konzeptionelle technische Merkmale des 3,5 m-**
Teleskops. C. Kühne.
Mitt. Astron. Ges., Nr. 54, (see 012.050), p. 27 - 38 (1981).

032.034 **Das System des 2.2 m-Teleskops.** K. Bahner.
Mitt. Astron. Ges., Nr. 54, (see 012.050), p. 135 -
137 (1981).

032.035 **Fernsehleiteinrichtung für das 2.2 m-Teleskop.**
H. Bellemann, D. Fath, H. Unser, R. Wolf,
K. Zimmermann.
Mitt. Astron. Ges., Nr. 54, (see 012.050), p. 137 - 139 (1981).

032.036 **Eine neues Steuerungskonzept für das 1.5 m-**
Teleskop. I: Hardware-Aspekte.
M. Stoll, H. Jenkner.
Mitt. Astron. Ges., Nr. 54, (see 012.050), p. 142 - 144 (1981).

032.037 **Ein neues Steuerungskonzept für das 1.5 m-**
Teleskop. II: Software-Aspekte.
H. Jenkner, M. Stoll.
Mitt. Astron. Ges., Nr. 54, (see 012.050), p. 144 - 149 (1981).

032.038 **Datensysteme am 2.2 m-Teleskop.** D. Fath.
Mitt. Astron. Ges., Nr. 54, (see 012.050), p. 247 -
251 (1981).

032.039 **Solar instruments.**
V. M. Grigor'ev, V. N. Karpinskij.
Issled. po geomagn., aehron. i fiz. Solntsa, Moskva, 1981,
No. 56, p. 57 - 75. In Russian. – Abstr. in Ref. zh., 51.
Astron., 11.51.1072 (1981).

032.040 **Solar telescope for operative predictions (STOP).**
V. M. Grigor'ev, B. F. Osak, N. I. Kobanov,
N. V. Klochek, I. L. Maslov, M. F. Shtol'.
Issled. po geomagn., aehron. i fiz. Solntsa, Moskva, 1981,

No. 56, p. 129 - 139. In Russian. – Abstr. in Ref. zh., 51.
Astron., 11.51.1073 (1981).

032.041   On temperature compensation in the body of a
          transit instrument by means of a ventilator.
G. M. Blank.
Astron. Tsirk., No. 1129, p. 7 - 8 (1980). In Russian.

032.042   Discrete drives for telescopes.
          V. N. Ivanov, L. S. Paulin.
Astron. Tsirk., No. 1133, p. 7 - 8 (1980). In Russian.

032.043   Current status of the national 15-meter telescope.
          L. D. Barr.
J. Opt. Soc. America, Vol. 71, 1611 (1981). – Abstract.

032.044   L'osservazione astronomica ad alta risoluzione.
          P. Couteau.
Orione, Vol. 2, 254 - 262 (1981).

032.045   Investigation of the level division inequalities with
          an extrafocally adjusted collimator.
L. A. Mitić.
Bull. Obs. Astron. Belgrade, No. 131, p. 9 - 12 (1981).

032.046   Report on the UKIRT control system.
          R. L. Davey.
UKIRT Rep., No. 2, 14 pp. (1978).

032.047   Report on optical work carried out at the U. K.
          Infrared Telescope installation at Mauna Kea,
Hawaii from mid-July until early August, 1978.
R. V. Willstrop.
UKIRT Rep., No. 3, 13 pp. (1978).

032.048   Report on mechanical tests performed with the
          UKIRT 3.8 m telescope from August to mid-
September 1978.   R. J. Beetles.
UKIRT Rep., No. 4, 18 pp. (1978).

032.049   The United Kingdom 3.8 metre telescope in Hawaii.
          C. M. Humphries.
UKIRT Rep., No. 5, 13 pp. (1978).

032.050   The Camac control system used on the UKIRT
          3.8 m telescope.
B. D. S. Bell, G. R. Adam, C. L. Stephens.
UKIRT Rep., No. 6, 9 pp. (1978).

032.051   UKIRT polar axis: determination of alignment
          errors.   A. McLachlan.
UKIRT Rep., No. 7, 12 pp. (1978).

032.052   Report on observations at UKIRT in December 1978.
          P. M. Williams.
UKIRT Rep., No. 8, 8 pp. (1978).

032.053   Balancing of the UKIRT 3.8 metre telescope.
          A. A. Neild.
UKIRT Rep., No. 9, 7 pp. (1978).

032.054   Computer aided balancing of UKIRT: a preliminary
          report.   A. McLachlan.
UKIRT Rep., No. 15, 7 pp. (1980).

032.055   Transit telescope designs optimized for multiple
          object spectroscopy with fibers.   J. R. P. Angel.
Prepr. Steward Obs., No. 347, 7 pp. (1981).

032.056   Very large ground-based telescopes for optical and
          infrared astronomy.

J. R. P. Angel.
Prepr. Steward Obs., No. 356, 26 pp. (1981).

032.057   The features of the optical system of the photo-
          electric astrolabe type II.   L.-j. Wang.
Ann. Shanghai Obs. Acad. Sinica, No. 1, p. 130 - 134 (1979).

032.058   The satellite laser ranging system at the Shanghai
          Observatory.   F.-m. Yang, C.-k. Xiao, D.-t. Tan,
C.-p. Jiang, H.-m. Le, Z.-y. Li, J.-l. Sun, F.-g. Zhan, B.-s. Wei,
S.-n. Huang, S.-f. Cai.
Ann. Shanghai Obs. Acad. Sinica, No. 1, p. 83 - 87 (1979).

032.059   A preliminary analysis of the accuracy of the satel-
          lite laser ranging system at Shanghai Observatory.
S.-n. Huang, D.-t. Tan, F.-m. Yang, C.-k. Xiao.
Ann. Shanghai Obs. Acad. Sinica, No. 1, p. 88 - 91 (1979).

032.060   A semi-automatic photoelectric transit instrument.
          G. Zhao, Y.-d. Yang, C.-q. Fu, Y.-s. Ye, X.-l. Shi,
H.-s. Wang, H.-q. Liang.
Ann. Shanghai Obs. Acad. Sinica, No. 2, p. 130 - 138 (1980).

032.061   On the adjustment of the polar axis of the astro-
          graph of 400 mm aperture at the Zo-Se section.
L.-s. Yan.
Ann. Shanghai Obs. Acad. Sinica, No. 2, p. 145 - 149 (1980).

032.062   Effects of atmospheric refraction on locating and
          tracking of a telescope.   J.-l. Zhao.
Ann. Shanghai Obs. Acad. Sinica, No. 2, p. 150 - 155 (1980).

032.063   An automatic positioning method of equatorial
          telescopes utilizing resolvers.   Y.-j. Zheng.
Ann. Shanghai Obs. Acad. Sinica, No. 2, p. 156 - 162 (1980).

032.064   Revolutionary changes in the astronomical telescope.
          K. Pines.
Mada, Vol. 24, 41 - 43 (1980). In Hebrew.

032.065   Calculation of the permeability of a telescope and
          optimization of its parameters.   Yu. D. Pimenov.
Opt.-mekh. prom-st', 1981, No. 8, p. 9 - 13. – Abstr. in Ref.
zh., 51. Astron., 1.51.842 (1982).

032.066   Automatic adjustment of a multi-element telescope
          mirror from the criterion of maximum intensity.
V. V. Lavrent'ev, I. A. Lapshina, Yu. A. Sabinin.
Leningr. inst. tochn. mekh. i opt. Leningrad, 1981, 9 pp. In
Russian. – Abstr. in Ref. zh., 51. Astron., 1.51.851 (1982).

032.067   Frequency-stabilized modulator.
          A. P. Isaev, P. I. Prokof'ev, K. A. Zanina.
Issled., raschet i konstruir. izmerit. priborov. Leningrad, 1981,
p. 64 - 68. In Russian. – Abstr. in Ref. zh., 51. Astron.,
1.51.855 (1982).

032.068   Investigation of the horizontal flexure of the
          Engelhardt Astronomical Observatory's meridian
circle.   S. S. Peruanskij, Z. M. Abdullina, G. V. Auzin'.
Kazan. univ. Kazan', 1981. 7 pp. In Russian. – Abstr. in Ref.
zh., 51. Astron., 1.51.860 (1982).

032.069   On determination of the limb setting error for the
          Poltava Observatory's astrolabe OPL No. 21.
A. A. Slavinskaya, N. M. Zalivadnyj.
Vrashchenie i priliv. deformatsii Zemli, Kiev, 1981, No. 13,
p. 97 - 98. In Russian. – Abstr. in Ref. zh., 51. Astron.,
1.51.862 (1982).

          Telescopes for the 1980s.   See Abstr. 003.041.

Landolt-Börnstein.   See Abstr. 003.172.

The history of the creation of a telescope.
See Abstr. 004.070.

The history of the creation of a telescope by
M. V. Lomonosov.   See Abstr. 004.071.

New techniques and telescopes in optical astronomy.
See Abstr. 010.023.

Le telescope Canada - France - Hawaii; gestation,
enfance et avenir.   See Abstr. 013.041.

UKIRT instrumentation computer.
See Abstr. 021.062.

Use of modelling for estimating the image quality of
a composite telescope mirror.   See Abstr. 031.009

Test of light energy concentration for the 1 m
reflector at Yunnan Observatory.   See Abstr. 031.011.

Hartmann-Tests am MPIA-2.2 m-Teleskop mit
Blende im Pupillenbild.   See Abstr. 031.012.

Chromatic aberration of the Mizusawa PZT No. 1.
See Abstr. 031.040.

A design of the corrector system for the 1.5 meter
telescope.   See Abstr. 031.042.

Space Instrumentation

032.501   A glimpse of new results: the preliminary Solar
Maximum Mission data.   R. Rosner.
Nature, Vol. 292, 582 - 584 (1981).

032.502   A parallel plate imaging proportional counter with
high background rejection capability.
O. Siegmund, P. Sanford, I. Mason, L. Culhane, S. Kellock,
R. Cockshott.
IEEE Trans. Nucl. Sci., Vol. NS-28, (see 012.003), p. 478 - 482
(1981). − Abstr. in Phys. Abstr., Vol. 84, Abstr. 63829 (1981).

032.503   A high resolution sealed xenon filled position sensitive proportional counter.
M. R. Sims, H. D. Thomas, M. J. L. Turner.
IEEE Trans. Nucl. Sci., Vol. NS-28, (see 012.003), p. 825 - 829
(1981). − Abstr. in Phys. Abstr., Vol. 84, Abstr. 63860 (1981).

032.504   Spectroscopy in X-ray astronomy using gas scintillation cameras.
B. G. Taylor, J. Davelaar, G. Manzo, A. Peacock.
IEEE Trans. Nucl. Sci., Vol. NS-28, (see 012.003), p. 857 - 860
(1981). − Abstr. in Phys. Abstr., Vol. 84, Abstr. 63865 (1981).

032.505   A fluorescent gated imaging gas scintillation spec-
trometer for high energy X-ray astronomy.
B. G. Taylor, R. D. Andresen, J. Davelaar, G. Manzo,
A. Peacock, S. Re.
IEEE Trans. Nucl. Sci., Vol. NS-28, (see 012.003), p. 861 - 868
(1981). − Abstr. in Phys. Abstr., Vol. 84, Abstr. 63866 (1981).

032.506   In orbit performance of the Einstein Observatory
HEAO-2 imaging proportional counter.
P. Gorenstein, F. R. Harnden, Jr., D. G. Fabricant.
IEEE Trans. Nucl. Sci., Vol. NS-28, (see 012.003), p. 869 - 874
(1981). − Abstr. in Phys. Abstr., Vol. 84, Abstr. 63867 (1981).

032.507   EOSCOR III, a high energy solar neutron detector.
R. Koga, G. M. Frye, Jr., B. V. Denehy, O. Mace.
IEEE Trans. Nucl. Sci., Vol. NS-28, (see 012.003), p. 875 - 879
(1981). − Abstr. in Phys. Abstr., Vol. 84, Abstr. 63868 (1981).

032.508   A large area gas-flow proportional counter system
for soft X-ray astronomy.
K. P. Singh, R. K. Manchanda, J. Rodrigues, P. C. Agrawal,
S. Naranan.

Indian J. Radio Space Phys., Vol. 9, 201 - 208 (1980). − Abstr.
in Phys. Abstr., Vol. 84, Abstr. 66830 (1981).

032.509   Stellar and solar X-ray polarimetry.
R. Novick.
Opt. Eng., Vol. 20, No. 1, p.   31 - 38   (1981). − Abstr. in
Phys. Abstr., Vol. 84, Abstr. 66832 (1981).

032.510   Design of infrared astronomical satellite (IRAS)
primary mirror mounts.
M. Schreibman, P. Young.
Proc. Soc. Photo-Opt. Instrum. Eng., Vol. 250, (see 012.004),
p. 50 - 58 (1980). − Abstr. in Phys. Abstr., Vol. 84, Abstr.
71055 (1981).

032.511   Some characteristics of the Hadamard transform
X-ray telescope.   S. Miyamoto, H. Tsunemi,
K. Tsuno.
Nucl. Instrum. Methods, Vol. 180, No. 2 - 3, p. 55 - 72 (1981).
Abstr. in Phys. Abstr., Vol. 84, Abstr. 71077 (1981).

032.512   Small helium-cooled infrared telescope for
Spacelab 2.   T. N. Gautier, III, W. Poteet,
G. H. Rieke, F. J. Low, W. F. Hoffmann.
Opt. Eng., Vol. 20, 285 - 290 (1981). − Abstr. in Phys. Abstr.,
Vol. 84, Abstr. 75014 (1981).

032.513   Advanced X-ray Astrophysics Facility (AXAF).
M. V. Zombeck.
Opt. Eng., Vol. 20, 297 - 309 (1981). − Abstr. in Phys. Abstr.,
Vol. 84, Abstr. 75015 (1981).

032.514   Alignment and evaluation of the cryogenic correct-
ed Infrared Astronomical Satellite (IRAS) telescope.
N. Harned, R. Harned, R. Melugin.
Opt. Eng., Vol. 20, 195 - 200 (1981). − Abstr. in Phys. Abstr.,
Vol. 84, Abstr. 79720 (1981).

032.515   Design of Infrared Astronomical Satellite (IRAS)
primary mirror mounts.
M. Schreibman, P. Young.
Opt. Eng., Vol. 20, 190 - 194 (1981). − Abstr. in Phys. Abstr.,
Vol. 84, Abstr. 79724 (1981).

032.516   Measured response of a Wolter I X-ray telescope
figured by diamond turning.
R. C. Gatura, D. Bardas, W. A. Brown, D. T. Roethig,
J. L. Culhane, R. Berthelsdorf, A. Franks.
Bull. American Astron. Soc., Vol. 13, 510 (1981). − Abstract.

032.517 **Astrometric data reduction software for Space Telescope.**   W. H. Jefferys, J. T. Feo.
Bull. American Astron. Soc., Vol. 13, 536 (1981). – Abstract.

032.518 **A multiple-spacecraft optical very long baseline interferometer.**
R. V. Stachnik, D. Y. Gezari, E. McCormack, P. Melroy, D. Arnold.
Bull. American Astron. Soc., Vol. 13, 556 (1981). – Abstract.

032.519 **The contribution of OSO-8 to our knowledge of the chromosphere and transition region.**
R. M. Bonnet.
Space Sci. Rev., Vol. 29, 131 - 200 (1981).
    This review deals with the pointed instrumentation placed onboard the NASA Orbiting Solar Observatory No. 8 which was successfully launched in June 1975 and operated continuously for more than 3 years. A brief description of the instrumentation and of its capabilities is given together with an appreciation of how it performed into orbit. The most important scientific achievements are reviewed in detail. The high spectral resolution of the OSO-8 pointed instruments allowed to make a real break through in the investigation of the heating mechanisms of the chromosphere and corona, by showing that the energy contained in the 150–300 s oscillations is 2 to 3 orders of magnitude too low to accomodate the radiative losses of these layers. Line profile and power spectrum analysis, together with the variation over the disk and in height of the non thermal velocities lend more support to the concept of a chromosphere and transition region being magnetically heated. The dynamics of the upper solar atmosphere has been studied at altitudes never achieved before. The physical structure of sunspots, prominences and of a few active regions has been investigated through the analysis of the profiles observed for the first time in these features.

032.520 **X-ray sky monitor experiment onboard Bhaskara.**
    R. K. Manchanda, D. P. Sharma, P. C. Agrawal, S. V. Damle, K. Kasturirangan, U. R. Rao, B. V. Sreekantan.
Bull. Astron. Soc. India, Vol. 9, 65 - 66 (1981). – Abstract.

032.521 **A CCD camera for a possible Indian space mission for comet Halley observations.**
U. R. Rao, P. Durgaprasad, A. K. Jain, K. Kasturirangan, T. M. K. Marar.
Bull. Astron. Soc. India, Vol. 9, 73 - 74 (1981). – Abstract.

032.522 **Cosmic X-ray satellite "HAKUCHO".**
    I. Kondo, H. Inoue, K. Koyama, K. Makishima, M. Matsuoka, T. Murakami, M. Oda, Y. Ogawara, T. Ohashi, N. Shibazaki, Y. Tanaka, S. Hayakawa, H. Kunieda, F. Makino, K. Masai, F. Nagase, Y. Tawara, S. Miyamoto, H. Tsunemi, K. Yamashita, M. Yoshimori.
Space Sci. Instrum., Vol. 5, 211 - 228 (1981).
    Instrumentation and initial performance of the X-ray astronomy satellite "HAKUCHO" are described. Seven X-ray counters have fields of view almost parallel to the spin axis and cover the energy range 0.1 - 100 keV, whereas four counters have their optical axis perpendicular to the spin axis and cover the energy range 0.1 - 30 keV. These counters watch for appearance of X-ray bursts in a sky region around the direction of the spin-axis, and survey a large part of the sky for X-ray sources.

032.523 **The Signe 2 Franco-Soviet interplanetary gamma ray burst experiment network.**
C. Barat, G. Chambon, K. Hurley, M. Niel, G. Vedrenne, I. V. Estulin *(Ehstulin)*, A. V. Kuznetsov, V. M. Zenchenko.
Space Sci. Instrum., Vol. 5, 229 - 235 (1981).
    The Signe 2 project, consisting of French-built gamma ray detectors on two Soviet Venera spacecraft and one Prognoz satellite, is described. The Prognoz and Venera mis-sions are discussed, and the experiments, their operating modes, and their scientific objectives are explained.

032.524 **Very Large Area Multiwire Spectroscopic Proportional Counters.**   P. Ubertini, A. Bazzano, L. Boccaccini, M. Mastropietro, C. D. La Padula, R. Patriarca, V. F. Polcaro.
Space Sci. Instrum., Vol. 5, 237 - 246 (1981).
    The instrument is the last version of a new generation of Multiwire Spectroscopic Proportional Counters (MWSPC) succesfully employed in many balloon borne flights, devoted to hard X-ray astronomy. The sensitive area of this standard unit is 2700 cm^2 with an efficiency higher than 10% in the range 15 - 180 keV (80% at 60 keV).

032.525 **Calibration and in-flight performance of the zodiacal light experiment on Helios.**
C. Leinert, E. Pitz, H. Link, N. Salm.
Space Sci. Instrum., Vol. 5, 257 - 270 (1981).
    The zodiacal light experiment on Helios is a broad-band, low light level photometer measuring in U, B, and V. The calibration procedures and results for both laboratory calibration and calibration by star crossings are described. The authors found good agreement between these two independent methods. If one assumes that both calibration procedures are correct, the remaining differences between the two sets can be explained by larger solar (B–V) of 0.69 ± 0.02. Examples for the stability of the experiment in flight, the very satisfactory stray light suppression and the non-Poissonian dark current statistics are given.

032.526 **The supra-thermal plasma analysers on the ESA GEOS satellites.**
G. L. Wrenn, J. F. E. Johnson, J. J. Sojka.
Space Sci. Instrum., Vol. 5, 271 - 293 (1981).
    Two GEOS scientific spacecraft have been launched by ESA for the study of magnetospheric processes. This paper describes these Suprathermal Plasma Analysers (SPA) which measure differential fluxes of electrons and protons with energies less than 500 eV.

032.527 **Sheath effects on current collection by particle detectors with narrow acceptance angles.**
N. Singh, C. R. Baugher.
Space Sci. Instrum., Vol. 5, 295 - 305 (1981).

032.528 **The University College London balloon-borne ultraviolet telescope and high resolution echelle spectrograph.**
B. Y. Welsh, B. Anderson, A. Boksenberg, M. Pettini, W. Towlson.
Space Sci. Instrum., Vol. 5, 307 - 322 (1981).
    A description of the new University College London ultraviolet balloon platform is presented. The payload comprises a three-axes star-pointing platform, carrying a 310 mm ultraviolet telescope coupled to an echelle spectrograph and an image tube detector system. The instrument covers the whole balloon ultraviolet range (2000 - 3200 Å) at high resolution ($\sim$ 7 km s^{-1}). In the present configuration it can be used to observe stars brighter than $V = 7.5$, and in the near future it will be upgraded to acces fainter objects ($V < 9$). This instrumentation will be used in the coming year to obtain high quality ultraviolet spectra of stars and interstellar matter in and out of the galactic plane.

032.529 **Rocket-borne instrumentation for the measurement of atomic oxygen based on chemical release in the lower thermosphere.**
E. Van Hemelrijck, E. Van Ransbeeck.
Space Sci. Instrum., Vol. 5, 323 - 338 (1981).
    A rocket-borne instrument to measure mainly atomic oxygen density based on nitric oxide point releases in the

lower thermosphere has been developed. Many of the difficulties associated with these kinds of experiments have been avoided by a new technique ejecting the NO gas in the backward direction of the flight. As a result, shock wave problems are eliminated substantially facilitating measurements and analysis. Preliminary tests are discussed and construction and mode of operation of the instrument are described.

032.530 On perspectives of using compressed xenon for registration of gamma quanta with 0.1 - 10 MeV.
A. M. Gal'per, V. V. Dmitrenko, A. S. Romanyuk, Z. M. Uteshev.
Izv. AN SSSR. Ser. fiz., Tom 45, 649 - 652 (1981). In Russian. Abstr. in Ref. zh., 51. Astron., 9.51.1024 (1981).

032.531 Gamma-telescope characteristics on the AES Cosmos 561 and Cosmos 731.
V. L. Bokov, E. M. Kruglov.
Izv. AN SSSR. Ser. fiz., Tom 45, 653 - 656 (1981). In Russian. Abstr. in Ref. zh., 51. Astron., 9.51.1025; 62. Issled. kosm. prostranstva, 9.62.90 (1981).

032.532 TV pictures of another world.
A. Cantoni.
Antenna, Vol. 53, 71 - 74 (1981). In Italian. − Abstr. in Phys. Abstr., Vol. 84, Abstr. 94132 (1981).

032.533 Alignment and evaluation of the cryogenic corrected infrared astronomical satellite (IRAS) telescope.
N. Harned, R. Harned, R. Melugin.
Proc. Soc. Photo-Opt. Instrum. Eng., Vol. 251, (see 012.027), p. 163 - 170 (1980). − Abstr. in Phys. Abstr., Vol. 84, Abstr. 94174 (1981).

032.534 Alignment design for a cryogenic telescope.
P. Young, M. Schreibman.
Proc. Soc. Photo-Opt. Instrum. Eng., Vol. 251, (see 012.027), p. 171 - 178 (1980). − Abstr. in Phys. Abstr., Vol. 84, Abstr. 94175 (1981).

032.535 Modular alignment system for spacecraft.
M. Henrist, D. Malaise, A. Monfils.
Proc. Soc. Photo-Opt. Instrum. Eng., Vol. 251, (see 012.027), p. 197 - 202 (1980). − Abstr. in Phys. Abstr., Vol. 84, Abstr. 94176 (1981).

032.536 An infrared sensor designed to measure the diffuse zodiacal light.
T. Murdock, P. Tandy, P. Walters, D. Wang.
Proc. Soc. Photo-Opt. Instrum. Eng., Vol. 245, (see 012.028), p. 9 - 13 (1980). − Abstr. in Phys. Abstr., Vol. 84, Abstr. 94177 (1981).

032.537 Preliminary cryogenic performance of the Shuttle infrared telescope facility.   H. L. Gier, R. Stoll.
Proc. Soc. Photo-Opt. Instrum. Eng., Vol. 245, (see 012.028), p. 69 - 76 (1980). − Abstr. in Phys. Abstr., Vol. 84, Abstr. 94179 (1981).

032.538 The UH-nuclei cosmic ray detector on the third High Energy Astronomy Observatory.
W. R. Binns, M. H. Israel, J. Klarmann, W. R. Scarlett, E. C. Stone, C. J. Waddington.
Nucl. Instrum. Methods Phys. Res., Vol. 185, 415 - 426 (1981). − Abstr. in Phys. Abstr., Vol. 84, Abstr. 94183 (1981).

032.539 A new experiment for measuring the flux of high energy iron nuclei in cosmic rays.
R. K. Sood, J. Panettieri.
Nucl. Instrum. Methods Phys. Res., Vol. 185, 427 - 431

(1981). − Abstr. in Phys. Abstr., Vol. 84, Abstr. 94184 (1981).

032.540 Calibration of a gamma-ray telescope using tagged positron annihilation photons.
D. L. Bertsch, W. R. Dodge.
Nucl. Instrum. Methods Phys. Res., Vol. 185, 439 - 448 (1981). − Abstr. in Phys. Abstr., Vol. 84, Abstr. 94186 (1981).

032.541 The role of space techniques in the understanding of solar variability.   R. M. Bonnet.
Sol. Phys., Vol. 74, (see 012.029), 485 - 501 (1981).
The unique advantages of space observations are recalled, and the difficulties in performing reliable measurements of solar variability from space are stressed. The author shows with more details how space observations are crucial in determining the causes, the amplitude and the previsions of solar variability. He shows how they would permit to gain a better view of the solar interior, of the origin and of the effects of the solar magnetic field.

032.542 A high precision solar ultraviolet spectral irradiance monitor for the wavelength region 120 - 400 nm.
M. E. VanHoosier, J.-D. F. Bartoe, G. E. Brueckner, D. K. Prinz, J. W. Cook.
Sol. Phys., Vol. 74, (see 012.029), 521 - 530 (1981).
The authors describe the Solar Ultraviolet Spectral Irradiance Monitor (SUSIM) experiment under development at the Naval Research Laboratory for flight aboard the Space Shuttle and the Upper Atmospheric Research Satellite. SUSIM will monitor the solar flux in the 120 - 400 nm region with high precision, using an in-flight calibration system to reduce absolute error to $< 10\%$, and error relative to the 400 nm continuum to $< 1\%$.

032.543 An instrument to measure the solar spectrum from 170 to 3200 nm on board Spacelab.
G. Thuillier, P. C. Simon, D. Labs, R. Pastiels, H. Neckel.
Sol. Phys., Vol. 74, (see 012.029), 531 - 537 (1981).
The instrument described, at the present time in development, will fly on board Spacelab I in May 1983. Other flights are foreseen during the following missions. This instrument is composed by three double monochromators covering the range 170 to 3200 nm. The spectrometers have band-passes of 1 nm up to 900 nm and 20 nm from 850 to 3200 nm with an accuracy 10^{-2} nm. Calibration lamps are included in the instrument to monitor any change of its sensitivity and wavelength scale.

032.544 The ESA project for a Grazing Incidence Solar Telescope (GRIST).   M. C. E. Huber.
Sol. Phys., Vol. 74, (see 012.029), 539 - 542 (1981).
The Grazing Incidence Solar Telescope (GRIST) as it is being studied by the European Space Agency is described: a soft X-ray and extreme-ultraviolet facility for solar observations with 1 arc sec spatial resolution in a wavelength range extending from 9 to beyond 100 nm. The telescope, a 35° sector of a Wolter, type-II, configuration, will have a focal length of 4 m and a collection area of 280 cm^2. It is planned that GRIST be flown on joint Spacelab flights with the Solar Optical Telescope, a NASA facility for the wavelength range 110 nm to 1 μm.

032.545 Gravity sensors and the principle of equivalence.
R. L. Forward.
IEEE Trans. Aerosp. Electron. Syst., Vol. AES-17, 511 - 519 (1981). − Abstr. in Phys. Abstr., Vol. 84, Abstr. 94815 (1981).

032.546 Automatic control system of the spectrometer aboard the Proton 1 artificial earth satellite.
V. K. Georgiev, V. N. Lutsenko, K. M. Kazakov.

Dokl. Bolg. AN, Vol. 34, 33 - 41 (1981). – Abstr. in Ref. zh., 62. Issled. kosm. prostranstva, 10.62.54 (1981).

032.547 Telescope on compressed xenon for recording cosmic gamma radiation in the 0.1–10 MeV energy range. A. M. Gal'per, V. V. Dmitrenko, A. S. Lenin, A. S. Romanyuk, Z. M. Uteshev, E. I. Chujkin.
Fiz.-tekh. inst. AN SSSR, Prepr., 1981, No. 700. 25 pp. In Russian. – Abstr. in Ref. zh., 62. Issled. kosm. prostranstva, 10.62.55 (1981).

032.548 On the problem of coating mirrors of devices during extraatmospheric experiments.   A. V. Bruns.
Izv. Krymskoj Astrofiz. Obs., Tom 63, 197 - 203 (1981). In Russian. English translation in Bull.Crimean Astrophys. Obs., Vol. 63.

032.549 Drei Jahre IUE-Observatorium.   V. Weidemann.
Phys. Bl., Vol. 37, 344 - 347 (1981).

032.550 Radiometric performance of the Voyager cameras.
G. E. Danielson, P. N. Kupferman, T. V. Johnson, L. A. Soderblom.
J. Geophys. Res., Vol. 86, 8683 - 8689 (1981).
The Voyager Imaging Experiment provided high-quality data of Jupiter and the Galilean satellites with the two flyby trajectories in March and July of 1979. Moderately accurate radio-metric measurements have been made using these data. This paper evaluates the radiometric results and describes the inflight and ground geometric and radiometric correction factors. The radiometric quantities of intensity I and geometric albedo I/F are derived, and scaling factors for each of the filters are tabulated for correcting the 'calibrated' data from the Image Processing Laboratory at JPL. In addition, the key characteristics of both Voyager 1 and Voyager 2 cameras are tabulated.

032.551 New design aspects for an energetic particle telescope for space missions.
E. Keppler, A. Glasmachers, W. Winkelnkemper.
IEEE Trans. Instrum. Meas., Vol. IM-30, 177 - 181 (1981). Abstr. in Phys. Abstr., Vol. 84, Abstr. 108030 (1981).

032.552 The infrared experiment of the Venera-Halley mission.
J. F. Crifo, N. Coron, J. P. Bibring, M. Combes, J. Crovisier, T. Encrenaz, E. Gérard.
Bull. American Astron. Soc., Vol. 13, 699 (1981). – Abstract.

032.553 Solar physics assessment of future high-resolution observations in the grazing-incidence domain.
M. Malinovsky-Arduini.
Space Sci. Rev., Vol. 29, (see 012.043), 301 - 318 (1981).
The project for a Grazing Incidence Solar Telescope (GRIST) offers, for the first time, the combinations of high spatial ($1''$) and spectral resolution in the extreme-ultraviolet wavelength range. The 3-dimensional electron density and temperature structure of the transition region and corona will be determined. The dynamics of the structures which make up the corona will be studied. GRIST can be expected to provide definitive improvement in the understanding of the coronal heating problem.

032.554 GRIST instrumentation.   G. Tondello.
Space Sci. Rev., Vol. 29, (see 012.043), 319 - 326 (1981).
The Grazing Incidence Solar Telescope (GRIST) as it is being studied by the European Space Agency (ESA) is briefly described: it is an extreme ultraviolet facility for solar observation with 1 arc sec spatial resolution in the wavelength range from 9 to 120 nm.

032.555 The solar grazing incidence (GRIST) and optical (SOT) telescopes joint accommodation.
V. Domingo.
Space Sci. Rev., Vol. 29, (see 012.043), 327 - 331 (1981).
Some of the problems foreseen for the joint accommodation and operation of the Grazing Incidence Solar Telescope (GRIST) under study by ESA to operate in the extreme ultraviolet region ($90 < \lambda < 1700$ Å), and the Solar Optical Telescope (SOT), developed by NASA to operate in the ultraviolet, optical and infrared region ($\lambda > 1100$ Å) on a Spacelab mission are described.

032.556 The Solar Optical Telescope (SOT).
S. D. Jordan.
Space Sci. Rev., Vol. 29, (see 012.043), 333 - 340 (1981).
The Solar Optical Telescope (SOT), which NASA plans to operate on Spacelab, should provide resolution down to 0.1 arc sec, thus offering the capability for solving a number of fundamental problems in solar magnetism and in atmospheric heating and dynamics.

032.557 SOT instrumentation.   R. B. Dunn.
Space Sci. Rev., Vol. 29, (see 012.043), 341 - 345 (1981).
NASA has conducted numerous studies to define the concepts of a workable Solar Optical Telescope (SOT). The author describes SOT as it is envisioned by the U.S. solar community. The final configuration is not entirely defined, as it is being proposed on the basis of performance specifications and the instrument configurations for the first flight are just now being evaluated.

032.558 Magnetograph based on magneto-optical effects.
A. Cacciani.
Space Sci. Rev., Vol. 29, (see 012.043), 403 - 404 (1981).
Simple and low-cost instruments, reduced in dimensions and, possibly, weight, are always welcome and very convenient, when they are intended for space applications.

032.559 The Spacelab Lyman alpha and white light coronagraphs program.   J. L. Kohl, G. L. Withbroe, H. Weiser, R. M. MacQueen, R. H. Munro.
Space Sci. Rev., Vol. 29, (see 012.043), 419 - 424 (1981).
The Harvard-Smithsonian Center for Astrophysics and the High Altitude Observatory have defined a joint coronagraphs experiment for a future Spacelab mission. The instrumentation package would include an ultraviolet light coronagraph to measure the intensity and profiles of spectral lines formed between 1.2 and 8 solar radii from Sun center and a white light coronagraph to measure the intensity and polarization of visible light.

032.560 A solar extreme ultraviolet telescope and spectrograph for Shuttle/Spacelab.
W. M. Neupert, G. L. Epstein, R. J. Thomas, U. Feldman.
Space Sci. Rev., Vol. 29, (see 012.043), 425 - 429 (1981).
An instrument for advanced studies of the solar corona is described. Its optical system provides nearly stigmatic imaging of selected portions of the Sun over the spectral range from 22.5 to 44.0 nm. Both spectroheliograms and emission line profiles of coronal features will be obtained over a wide range of coronal temperatures.

032.561 The Coronal Helium Abundance Experiment on Spacelab 2.
B. E. Patchett, K. Norman, A. H. Gabriel, J. L. Culhane.
Space Sci. Rev., Vol. 29, (see 012.043), 431 - 437 (1981).
The Coronal Helium Abundance Spacelab Experiment, (CHASE), basically consists of a grazing incidence telescope and spectrometer sensitive over the range 150-1335 Å. Whilst aimed primarily at deriving the solar helium abundance from measurements of coronal resonance scattering, its specifica-

tion has been extended in order to provide a more general purpose solar XUV facility. The instrument will be flown on the Spacelab 2 Mission, currently scheduled for launch in November 1984.

032.562 High resolution grazing incidence telescopes for the EUV-regime. B. Aschenbach.
Space Sci. Rev., Vol. 29, (see 012.043), 443 - 449 (1981).

Much experience has been gathered over the past two decades in the building of grazing incidence X-ray telescopes. Based on the performance data of the best X-ray optics, the requirements for building an EUV-telescope with sub-arcsecond angular resolution are estimated.

032.563 The manufacture of high resolution X-ray grazing incidence telescopes in the U.K.
A. P. Vickery, A. Franks.
Space Sci. Rev., Vol. 29, (see 012.043), 451 - 453 (1981).

Research on materials and manufacturing methods are now aimed at producing 1 arc sec resolution X-ray telescopes.

032.564 X-ray and extended UV spectrometer designs based on off-plane grating mountings. W. Werner.
Space Sci. Rev., Vol. 29, (see 012.043), 455 - 459 (1981).

032.565 Photoelectric array detectors for use at XUV wavelengths. J. G. Timothy.
Space Sci. Rev., Vol. 29, (see 012.043), 461 - 476 (1981).

The characteristics of photoelectric detector systems for use at visible-light, ultraviolet, and X-ray wavelengths are briefly reviewed in the context of the needs of the Spacelab solar-physics facilities. Photoelectric array detectors for use at XUV wavelengths between 90 and 1500 Å are described, and their use in the ESA Grazing-Incidence Solar Telescope (GRIST) facility is discussed.

032.566 GRIST observations of non-solar objects: observational constraints.
J. F. Crifo, M. Malinovsky-Arduini.
Space Sci. Rev., Vol. 29, (see 012.043), 489 - 496 (1981).

The use of the Grazing Incidence Solar Telescope (GRIST) for the observation of celestial sources other than the Sun in the Extreme Ultraviolet (EUV) is discussed. By use of galactic point sources as sample objects, the capabilities offered by the presently proposed telescope and its focal plane instruments set are shown to be substantial, in spite of the short duration of a single Shuttle mission. This seems to legitimate complementary technical studies needed to demonstrate the feasibility of stellar pointing.

032.567 The EXOSAT mission.
B. G. Taylor, R. D. Andresen, A. Peacock, R. Zobl.
Space Sci. Rev., Vol. 30, (see 012.044), 479 - 494 (1981).

A review of the EXOSAT (European X-ray Observatory Satellite) project is given. Its status in mid-1981 is reported, the lunar occultation technique is outlined and distinctive features of the instrumentation are described. The scientific objectives of the mission and expected performance parameters are summarized.

032.568 The X-ray imaging telescopes on EXOSAT.
P. A. J. de Korte, J. A. M. Bleeker,
A. J. F. den Boggende, G. Branduardi-Raymont,
A. C. Brinkman, J. L. Culhane, E. H. B. M. Gronenschild,
I. Mason, S. P. McKechnie.
Space Sci. Rev., Vol. 30, (see 012.044), 495 - 511 (1981).

The instrument configuration and performance characteristics of the X-ray imaging telescopes on EXOSAT are described. The instrument comprises two fully independent Wolter I imaging telescopes. Each telescope can be used in either of two principal modes: (I) an imaging mode with either a position sensitive proportional counter or a channel multiplier array

plate in the focal plane, (II) a spectrometer mode which features a 500 lines/mm and/or a 1000 lines/mm transmission grating as dispersive element. Preliminary results indicate an ultimate angular resolution of 8.5 arc sec full width at half maximum or 17.5 arc sec half-power beam width. The ultimate wavelength resolution in the spectrometer mode ranges from 1 Å for wavelengths below 50 Å, to 5 Å at wavelengths near 300 Å.

032.569 The medium energy instrument on EXOSAT.
M. J. L. Turner, A. Smith, H. U. Zimmermann.
Space Sci. Rev., Vol. 30, (see 012.044), 513 - 524 (1981).

The medium energy instrument on EXOSAT, although conceived as the main instrument for occultations, has been made sufficiently versatile to provide a significant advance when used for individual source studies of timing and spectra. The energy range is 1.2 to 50 keV, with $\Delta E/E$ of 0.2 at 6 keV, sufficient to detect iron lines. The effective area of ~1800 cm^2 and narrow field of view ($3/4° \times 3/4°$) make it suitable for the detailed study of sources down to the 0.3 mCrab confusion limit.

032.570 The gas scintillation proportional counter on EXOSAT. A. Peacock, R. D. Andresen,
G. Manzo, B. G. Taylor, G. Villa, S. Re, J. C. Ives, S. Kellock.
Space Sci. Rev., Vol. 30, (see 012.044), 525 - 534 (1981).

The inclusion of a gas scintillation proportional counter (GSPC) within the EXOSAT payload significantly improves the spectroscopic capability of the mission. This broad-band medium energy spectrometer used in conjunction with the large area proportional counter array (ME) should provide additional spectroscopic details on strong X-ray sources at photon energies above ~2 keV.

032.571 Development of a 1 m-normal-incidence-EUV telescope. M. Grewing, G. Krämer, E. Schulz-Lüpertz,
C. Wulf-Mathies, S. Bowyer, R. Kimble.
Space Sci. Rev., Vol. 30, (see 012.044), 575 - 580 (1981).

Astrophysical plasmas at temperatures in the range $(0.5-5) \times 10^5$ K can best be studied at extreme ultraviolet wavelengths where they release the bulk of their energy. The authors report here the current development status of a 1m-normal-incidence-EUV-telescope that will be flown on an ARIES rocket to observe the spectra of nearby stars in the 350 - 700 Å range.

032.572 The use of diamond turned & replicated Wolter 1 telescopes for high sensitivity X-ray astronomy.
J. L. Culhane, R. C. Catura, K. A. Pounds, P. de Korte,
A. Franks, G. P. Garmire, A. Fabian, B. Margon.
Space Sci. Rev., Vol. 30, (see 012.044), 581 - 589 (1981).

Following the success of Einstein, it is clear that telescopes of very large area (~10^4 cm^2) with angular resolution ($\gtrsim 20''$) are needed for deep X-ray surveys and other observations. After a discussion of these objectives, which form the basis of the NASA LAMAR mission, the design & performance of a five mirror telescope is described. The system was studied for possible flight on Spacelab to act as a prototype module for LAMAR. Both diamond turning & replication methods of mirror production are discussed. The performance of a single Wolter 1 telescope with diamond turned mirrors will be described.

032.573 The capabilities of a gas scintillation camera for cosmic X-ray astronomy.
J. Davelaar, A. Peacock, B. G. Taylor, J. A. M. Bleeker.
Space Sci. Rev., Vol. 30, (see 012.044), 591 - 600 (1981).

The capabilities of a gas scintillator camera for use in X-ray astronomy are investigated. Detailed experimental results are presented on both the position and energy resolution over the energy range 0.28 to 6 keV. The energy resolution varies from 38% to 9.5% for 0.28 keV C−K and 6 keV

X-rays respectively. Position resolutions of ~1.8 mm and 3.5 mm for 6 keV and 1.5 keV Al-K X-rays were obtained. Applications of these cameras in conjunction with grazing incidence and coded mask X-ray optics are discussed.

032.574 A pseudo random mask telescope for Spacelab.
A. P. Willmore, G. K. Skinner, C. J. Eyles,
B. Ramsey.
Space Sci. Rev., Vol. 30, (see 012.044), 601 - 605 (1981).

A pair of pseudo-random mask telescopes is being constructed for Spacelab 2, mainly to observe the emission from galaxy clusters at energies from 2.5 to 25 keV. The main features and expected performance of the telescopes is described.

032.575 Spherical crystal cosmic X-ray spectrometer.
H. W. Schnopper, L. Koch, M. Cantin, B. Mougin,
R. Rocchia, R. F. Berthelsdorf, J. L. Culhane.
Space Sci. Rev., Vol. 30, (see 012.044), 607 - 614 (1981).

For spectral studies at energies ~3 keV, higher than those usually neglected by grazing incidence telescopes with high efficiency, freestanding, self-focussing, crystal arrays offer the most practical way to achieve adequate sensitivity through concentration. Such spectrometers can be designed for the entire range of energies that can be diffracted by crystals, ~500 eV to 10^4 eV, and, for energies below 3keV, can have sensitivities greater than or comparable with that of instruments at the focal plane of a large telescope.

032.576 Wide angle X-ray optics for use in astronomy.
W. K. H. Schmidt.
Space Sci. Rev., Vol. 30, (see 012.044), 615 - 621 (1981).

The author suggests a wide angle X-ray telescope. It consists of one or two sets of plane mirrors used in a grazing incidence configuration. The advantages of this type of X-ray optics over other systems for particular astronomical observations are discussed.

032.577 The use of micro-accelerometers for space geodetic experiments.   P. Paquet, F. Barlier, L. Bossy,
L. Mezzani, F. Nouel, C. Reigber.
Ann. Géophys., Tome 37, 1 - 5 (1981) = Obs. R. Belgique Commun., Sér. A, No. 62.

032.578 X-ray telescope on board K-10-14 for the observation of the Cygnus Loop.   S. Miyamoto,
K. Yamashita, H. Tsunemi, K. Tsuno, A. Kawabata, Y. Sato,
S. Kitamoto, M. Higuchi.
Bull. Inst. Space Aeronaut. Sci. Univ. Tokyo B, Vol. 16, 1469 - 1505 (1980). In Japanese. − Abstr. in Phys. Abstr., Vol. 85, Abstr. 6536 (1982).

032.579 Global auroral imaging instrumentation for the Dynamics Explorer mission.   L. A. Frank,
J. D. Craven, K. L. Ackerson, M. R. English, R. H. Eather,
R. L. Carovillano.
Space Sci. Instrum., Vol. 5, (see 003.019), 369 - 393 (1981).

The instrumentation for gaining global images of the auroral oval from the high-altitude spacecraft of the Dynamics Explorer Mission is described.

032.580 The Fabry-Perot Interferometer on Dynamics Explorer.
P. B. Hays, T. L. Killeen, B. C. Kennedy.
Space Sci. Instrum., Vol. 5, (see 003.019), 395 - 416 (1981).

The Fabry-Perot Interferometer (FPI) on Dynamics Explorer is a remote sensing instrument designed to measure the temperature, meridional wind and density of the metastable atoms $O(^1 S)$ and $O(^1 D)$ and the ion $O^+(^2 P)$ in the thermosphere.

032.581 The Dynamics Explorer Wind and Temperature Spectrometer.   N. W. Spencer, L. E. Wharton,

H. B. Niemann, A. E. Hedin, G. R. Carignan, J. C. Maurer.
Space Sci. Instrum., Vol. 5, (see 003.019), 417 - 428 (1981).

The Wind and Temperature Spectrometer (WATS) is designed to measure the concentration, kinetic temperature and motions (3 mutually perpendicular components of the wind) of the neutral particles. In addition, measurements of the concentration and velocity of the ambient thermal ions are possible.

032.582 The neutral mass spectrometer on Dynamics Explorer B.   G. R. Carignan, B. P. Block,
J. C. Maurer, A. E. Hedin, C. A. Reber, N. W. Spencer.
Space Sci. Instrum., Vol. 5, (see 003.019), 429 - 441 (1981).

A neutral gas mass spectrometer has been developed to satisfy the measurement requirements of the Dynamics Explorer mission. The mass spectrometer, a quadrupole, will measure the abundances of neutral species in the region 300 - 500 km in the earth's atmosphere.

032.583 The energetic ion composition spectrometer (EICS) for the Dynamics Explorer-A.   E. G. Shelley,
D. A. Simpson, T. C. Sanders, E. Hertzberg, H. Balsiger,
A. Ghielmetti.
Space Sci. Instrum., Vol. 5, (see 003.019), 443 - 454 (1981).

The energetic ion composition spectrometer (EICS) on the DE-A spacecraft is a high sensitivity (≈ 1 cm^2-sr-eV), high resolution ($M/\Delta M \gtrsim 10$ FWHM at focus) instrument. It covers the entire mass range from less than 1 amu/e to greater than 150 amu/e in 64 mass channels at each of 32 energy per charge steps covering the range from 0 to approximately 17 keV/e.

032.584 High-Altitude Plasma Instrument for Dynamics Explorer-A.   J. L. Burch, J. D. Winningham,
V. A. Blevins, N. Eaker, W. C. Gibson, R. A. Hoffman.
Space Sci. Instrum., Vol. 5, (see 003.019), 455 - 463 (1981).

The Dynamics Explorer A High-Altitude Plasma Instrument (HAPI) will make differential measurements of the velocity-space distributions of electrons and positive ions over the energy/charge range of 5 eV to 32 keV.

032.585 The Low Altitude Plasma Instrument (LAPI).
J. D. Winningham, J. L. Burch, N. Eaker,
V. A. Blevins, R. A. Hoffman.
Space Sci. Instrum., Vol. 5, (see 003.019), 465 - 475 (1981).

The Low Altitude Plasma Instrument on the Dynamics Explorer-B spacecraft provides high resolution velocity space measurements of positive ions and electrons from 5 eV to 32 keV and a monitor of electrons with energies above 35 keV.

032.586 The Retarding Ion Mass Spectrometer on Dynamics Explorer-A.   C. R. Chappell, S. A. Fields,
C. R. Baugher, J. H. Hoffman, W. B. Hanson, W. W. Wright,
H. D. Hammack, G. R. Carignan, A. F. Nagy.
Space Sci. Instrum., Vol. 5, (see 003.019), 477 - 491 (1981).

The thermal component of the magnetospheric plasma plays a key role in magnetosphere-ionosphere coupling processes acting as a strong influence on ionospheric structure at low altitudes and as a source and modifier of the hotter plasma population at high altitudes. The Retarding Ion Mass Spectrometer (RIMS) instrument on Dynamics Explorer-A is designed to measure this important thermal plasma component.

032.587 The Dynamics Explorer Langmuir probe instrument.   J. P. Krehbiel, L. H. Brace, R. F. Theis,
W. H. Pinkus, R. B. Kaplan.
Space Sci. Instrum., Vol. 5, (see 003.019), 493 - 502 (1981).

The Dynamics Explorer Langmuir probe instrument (DE-LANG) has been designed to perform in-situ measurements of electron temperature and electron and ion density in the Earth's ionosphere. It is a former spare unit from the

Pioneer Venus mission that has been modified to interface with the DE spacecraft.

032.588 The Ion Drift Meter for Dynamics Explorer-B.
R. A. Heelis, W. B. Hanson, C. R. Lippincott, D. R. Zuccaro, L. H. Harmon, B. J. Holt, J. E. Doherty, R. A. Power.
Space Sci. Instrum., Vol. 5, (see 003.019), 511 - 521 (1981).
The Ion Drift Meter on Dynamics Explorer-B measures two mutually perpendicular angles of arrival of thermal ions with respect to the sensor look direction. These measurements are used to derive two components of the ambient thermal ion drift velocity, which together with the third component from the Retarding Potential Analyzer instrument provide the total velocity.

032.589 Instrumentation for vector electric field measurements from DE-B.
N. C. Maynard, E. A. Bielecki, H. F. Burdick.
Space Sci. Instrum., Vol. 5, (see 003.019), 523 - 534 (1981).
Instrumentation has been developed using the symmetric double floating probe technique to make vector electric field measurements from the low altitude Dynamics Explorer (DE-B) spacecraft.

032.590 An opportunity for the observations of comets with wide-field cameras aboard the Saliout space station. P. L. Lamy, S. Koutchmy.
Modern observational techniques for comets, (see 012.047), p. 190 (1981).

032.591 Das deutsche Infrarot-Teleskop GIRL.
D. Lemke.
Mitt. Astron. Ges., Nr. 54, (see 012.050), p. 71 - 89 (1981).

032.592 The 1-m balloon-borne telescope "Golden Dragon" for far-infrared astronomy.
S. Drapatz, L. Haser, R. Hofmann, J. Niekerke, H. Rothermel, G. Sämann, J. Stöcker, H. Wutz.
Mitt. Astron. Ges., Nr. 54, (see 012.050), p. 91 - 99 (1981).
A three-axis stabilized balloon platform has been developed for far-infrared observations. It was flown successfully with a 1-m telescope and a Michelson interferometer in 1980. Design and overall properties of the gondola are presented. Specific features of the payload are discussed, including the two-dimensional chopping secondary mirror, the in-flight alignment procedure, and the step-and-integrate Fourier-interferometer.

032.593 Röntgenastronomie mit Ballonen und Satelliten.
R. Staubert.
Mitt. Astron. Ges., Nr. 54, (see 012.050), p. 101 - 116 (1981).
Two X-ray astronomy programs in Germany are described in some detail: the AIT/MPE balloon program with large area phoswich and solid state detectors. Observations of a number of galactic and extragalactic X-ray sources have been made in the photon energy range 20-200 keV. The second program, pursued by the MPE, Garching, is the development of a German X-ray satellite to be launched in 1986. It will carry a Wolter-Type I grazing incidence telescope of 80 cm diameter. Its scientific objective is to perform an unbiased sky survey in the photon energy range of 0.1-2 keV for the first six months. The second part of the mission will be devoted to detailed studies of individual sources.

032.594 Das Compton-Teleskop. V. Schönfelder.
Mitt. Astron. Ges., Nr. 54, (see 012.050), p. 117 - 133 (1981).
The paper describes the principle of measurement of the Compton telescope and two different telescopes of this type: the MPI balloon telescope and the GRO telescope COMPTEL. The properties of the telescopes are explained, especially their

angular resolution. Balloon flight data are used to illustrate how the data analysis is performed.

032.595 Das Steuer- und Datensystem von GIRL.
K. Proetel, H. Drexler, H. Eichel, F. Gliem.
Mitt. Astron. Ges., Nr. 54, (see 012.050), p. 158 - 160 (1981).

032.596 Zum Teleskop-Subsystem von GIRL.
R. Schlegelmilch, K. Proetel, E. D. Knohl, H.-J. Meier.
Mitt. Astron. Ges., Nr. 54, (see 012.050), p. 161 - 163 (1981).

032.597 Der Sekundärspiegelchopper für GIRL. A. Krabbe, K. Haussecker, D. Lemke, K. Proetel.
Mitt. Astron. Ges., Nr. 54, (see 012.050), p. 163 - 166 (1981).

032.598 Stand des 1 m-Teleskop-Projekts für den extremen Ultraviolettbereich. E. Schulz-Lüpertz, C. Wulf-Mathies, M. Grewing, G. Krämer.
Mitt. Astron. Ges., Nr. 54, (see 012.050), p. 166 - 169 (1981).

032.599 Eigenschaften der 32 cm-Röntgenteleskope.
B. Aschenbach, H. Bräuninger, G. Hasinger, A. Ondrusch.
Mitt. Astron. Ges., Nr. 54, (see 012.050), p. 169 - 172 (1981).

032.600 Entwicklung eines 80 cm-Röntgenteleskops für den Röntgensatelliten ROSAT.
B. Aschenbach, H. Bräuninger, G. Kettenring.
Mitt. Astron. Ges., Nr. 54, (see 012.050), p. 172 - 173 (1981).

032.601 Steuerung eines Ballon-Röntgenteleskops mit Prozeßrechnern.
E. Kendziorra, F. Laux, R. Staubert.
Mitt. Astron. Ges., Nr. 54, (see 012.050), p. 173 - 177 (1981).

032.602 Das IR-Photometer/Polarimeter für GIRL.
W. Martin, D. Lemke, J. Riedinger, H. Bauer, H. Bellemann.
Mitt. Astron. Ges., Nr. 54, (see 012.050), p. 198 - 200 (1981).

032.603 GIRL-Experiment E4: Ferninfrarot-Michelson-interferometer.
S. Drapatz, R. Hofmann, R. Katterloher.
Mitt. Astron. Ges., Nr. 54, (see 012.050), p. 200 - 203 (1981).

032.604 Entwicklung und Bau einer Infrarot-Kamera für den Einsatz auf dem Space-Shuttle-Experiment GIRL.
P. R. Preußner, R. Mühlheim, M. Grewing.
Mitt. Astron. Ges., Nr. 54, (see 012.050), p. 204 - 208 (1981).

032.605 Untersuchungen an höchstempfindlichen Infrarot-detektoren für GIRL.
J. Wolf, D. Lemke, K. Haussecker.
Mitt. Astron. Ges., Nr. 54, (see 012.050), p. 235 - 237 (1981).

032.606 Entwicklung eines ortsempfindlichen Proportional-zählers für röntgenastronomische Beobachtungen.
E. Pfeffermann, U. Briel.
Mitt. Astron. Ges., Nr. 54, (see 012.050), p. 242 - 246 (1981).

032.607 Device for estimating the energetic spectrum of mean energy protons.
L. S. Gorn, M. V. Iovlev, E. I. Morozova, N. F. Pisarenko, V. V. Pomogaev, A. Yu. Safronov, E. B. Usha, A. V. Shifrin.
Vopr. atom. nauk. i tekh. Yader. priborostr. Moskva, 1981, No. 2147, p. 58 - 63. In Russian. – Abstr. in Ref. zh., 51. Astron., 11.51.1106 (1981).

032.608 Calculation of the characteristics of a gamma-telescope with coded aperture. A. M. Gal'per,

S. A. Lazarev, B. I. Luchkov, Yu. V. Ozerov, O. F. Prilutskij.
Kosm. Issled., Tom 19, 947 - 951 (1981). In Russian.

032.609 The Space Telescope.
F. Macchetto.
Optical jets in galaxies, (see 012.052), p. 15 - 21 (1981).

The Space Telescope will be an extremely sophisticated
observatory that will make major contributions to astronomical research during the rest of this century. Its main characteristics and those of the scientific instruments that it carries are
briefly described.

032.610 Optical systems of the experiments in Spacelab 2.
E. W. Urban.
J. Opt. Soc. America, Vol. 71, 1566 (1981). — Abstract.

**032.611 Atmospheric research from the shuttle–cryogenic
limb-scanning interferometer and radiometer.**
V. Kunde, M. Coffey, M. Dubin, R. Drummond, J. Gille,
W. Mankin.
J. Opt. Soc. America, Vol. 71, 1566 (1981). — Abstract.

**032.612 Optical systems of the high-resolution telescope and
spectrograph on Spacelab 2.   J.-D. F. Bartoe.**
J. Opt. Soc. America, Vol. 71, 1567 (1981). — Abstract.

**032.613 Demonstration of a phased segmented mirror for
space applications.   P. A. Jones.**
J. Opt. Soc. America, Vol. 71, 1567 (1981). — Abstract.

**032.614 Grazing-incidence high-resolution stigmatic spec-
trograph with two optical elements.**
A. M. Malvezzi, L. Garifo, G. Tondello.
Appl. Opt., Vol. 20, 2560 - 2565 (1981).

Using two optical toroidal elements, a mirror and a grating, both working at grazing incidence, a spectrometer can be
built that is stigmatic in the XUV region at one wavelength.
Good compensation of the aberrations is achieved when the
intermediate sagittal image is nearly at infinity. By varying the
angle of incidence on the grating with simple movements, a
given couple of optical elements could cover stigmatically a
rather extended spectral range. If coupled with bidimensional
array detectors, such a spectrograph could find applications in
planned solar XUV telescopes.

**032.615 Extreme ultraviolet transmission grating mono-
chromator.**
P. J. Caldwell, E. T. Arakawa, T. A. Callcott.
Appl. Opt., Vol. 20, 3047 - 3049 (1981).

A compact monochromator has been designed to utilize
recently developed free-standing XUV transmission gratings.
Tests using a condensed spark source of far UV radiation show
that this monochromator, with 20 μm slits, is capable of
0.08 nm resolution at 30 nm in the first order. A physical description of the system and test results are presented.

032.616 Optical alignment of an X-ray collimator.
R. Hoekstra, T. E. d'Arnaud, H. F. van Beek.
Appl. Opt., Vol. 20, 3630 - 3634 (1981).

The hard X-ray imaging spectrometer (HXIS) is one of
the instruments aboard the NASA Solar Maximum Mission
satellite. For the imaging of solar flares in hard X rays HXIS
employs a grid collimator with an 8-sec of arc FWHM angular
resolution. The collimator consists of ten grid plates in a sequence. The position accuracy of the grids in their planes must
be better than a few microns. This paper describes a simple
optical alignment method developed to perform measurements
of the grid positions with a precision of 1 μm.

**032.617 Ultraviolet spectrometer and polarimeter for the
Solar Maximum Mission.**
M. S. Miller, A. J. Caruso, B. E. Woodgate, A. A. Sterk.

Appl. Opt., Vol. 20, 3805 - 3814 (1981).

The detailed optical design of the Solar Maximum
Mission-Ultraviolet Spectrometer and Polarimeter is discussed
in conjunction with the scientific objectives that led to the
design. The instrument consists of a 1.8-m effective focal
length aplanatic Gregorian telescope followed by a 1-m Ebert
spectrometer. The design of the Stokes polarimeter is also discussed.

032.618 Fast heavy ions in the heliosphere.
J. H. Adams, Jr., M. M. Shapiro, R. Silberberg,
C. H. Tsao.
Cosmic rays in the heliosphere, (see 012.059), p. 169 - 172
(1981).

The authors describe here a satellite experiment that
uses plastic track detectors to study the age of cosmic-ray
source material and the distribution of pathlengths over which
the heaviest cosmic rays travel to earth. The experiment will
also search for the singly charged particles, thought to make
up the anomalous component, and for low-energy heavy ions
deep in the magnetosphere.

**032.619 Two-dimensional analyzer with dE-dE silicon
detectors for cosmic ray experiments in space.**
J. Erő, Jr., K. Kecskeméty, T. Kovács, Jr., G. Kozma, I. Náday,
A. Somogyi, L. Szabó, I.T. Szücs, A. Zarándy.
Cosmic rays in the heliosphere, (see 012.059), p. 173 - 176
(1981).

A charged particle semiconductor-telescope is under
construction in the CRIP, Budapest, to measure fluxes of
4–30 MeV/AMU protons, alphas, and M nuclei in interplanetary space. Identification and counting of particles are made
on-board using microprocessor techniques. Details including
some new ideas on economical coding of information and
design of electronics are given.

032.620 Planetary science with Space Telescope.
J. Caldwell.
Planetary aeronomy and astronomy, (see 012.063), p. 199 -
213 (1981).

A discussion is given of the various capabilities and advantages of the Space Telescope observatory, scheduled for launch
into Earth-orbit in 1984. The first generation instruments are
described, and a detailed example for one specific observing
program, an intercomparison of Uranus and Neptune, is made.
The importance of an extra-solar planet search is emphasized.

032.621 Synopsis of troposphere and lower stratosphere.
D. Cadet.
Scientific ballooning–II, (see 012.065), p. 5 - 11 (1981).

The use of different types of balloons for the investigation of the troposphere and lower stratosphere is reviewed
with a special emphasis on the application for the next
10 years. The instrumentation currently flown aboard
balloons or under development is described. Some possible
scientific objectives of such balloon experiments are presented.
The specific applications of the different types of balloons
available within the next few years for scientific flights are
discussed.

032.622 Atmospheric sampling.   P. Fabian.
Scientific ballooning–II, (see 012.065), p. 17 - 27
(1981).

Four important sampling techniques are briefly reviewed:
Selective sampling on impregnated filters for measuring acidic
gases, the matrix isolation technique for measuring radicals,
whole air grabsampling and whole air cryogenic sampling for
measuring stable source gases. Vertical profiles of H_2, CH_4,
CO, N_2O, $CFCl_3$ and CF_2Cl_2 resulting from gas chromato-
graphic analysis of whole air samples collected with a cryogenic sampler are presented. The CO_2 mixing ratio and the
vertical distribution of methyl chloride are given.

032.623 Vertical sounding balloons for stratospheric photochemistry.   J. P. Pommereau.
Scientific ballooning–II, (see 012.065), p. 35 - 38 (1981).

Vertical sounding balloons are unique tools for stratospheric photochemistry investigations. An example of the use of a vertically piloted gas balloon for nitrogen dioxide diurnal variation search is presented. Improvements of such a method with the use of montgolfieres (hot air balloons) are proposed.

032.624 A balloon-borne sub-millimeter spectro-radiometer.   A. E. Salomonovich, V. I. Lapshin, A. V. Kiktev.
Scientific ballooning–II, (see 012.065), p. 89 - 92 (1981).

032.625 Design concept for a small, helium cooled, high-resolution interferometer for limb emission measurement.   F. Fergg, P. Burkert.
Scientific ballooning–II, (see 012.065), p. 93 - 96 (1981).

A helium cooled two-beam interferometer for measuring atmospheric emission spectra between 5 μm and 15 μm with a spectral resolution of 0.1 cm^{-1} has been designed. The unconventional arrangement of the interferometer optics results in very low heat generation by the mirror drive, high reliability under cryogenic conditions and suitable size and weight for space applications.

032.626 Experiment for the observation of short-term time variations in the hard X-ray region of Circinus X-1.
M. Nakagawa.
Scientific ballooning–II, (see 012.065), p. 97 - 100 (1981).

032.627 An imaging telescope for soft gamma ray astronomy.   J. N. Carter, P. Charalambous, A. J. Dean, D. Ramsden, M. Badiali, P. Ubertini, G. Boella, F. Perotti, G. Villa, G. di Cocco, G. Spada, A. Spizzichino.
Scientific ballooning–II, (see 012.065), p. 115 - 118 (1981).

A telescope capable of producing images of the gamma ray sky in the energy range 0.2-20 MeV with an angular resolution of a few tenths of a degree is presented. This capability is achieved by means of a large array of Sodium Iodide position sensitive elements together with a coded imaging mask. The expected performance, derived from calculations and preliminary laboratory tests, is described.

032.628 Medium energy gamma ray astronomy with transpacific balloon flights.
A. D. Zych, B. Dayton, M. C. Jennings, R. S. White.
Scientific ballooning–II, (see 012.065), p. 119 - 122 (1981).

Transpacific balloon flights with the University of California, Riverside double scatter telescope are discussed. With flight durations from 5 days up to perhaps 15 days the long observation times necessary for medium energy (1–30 MeV) gamma ray astronomy can be obtained.

032.629 A star tracker for use during daytime at balloon altitude.   S. Bowring, D. Ramsden.
Scientific ballooning–II, (see 012.065), p. 123 - 126 (1981).

Recent developments in X- and γ-ray astronomy indicate the need to know the absolute attitude of balloon-borne telescopes with a precision greater than can be achieved using magnetic sensors. The prospects for being able to construct a star tracker for use during both day and night time are reviewed.

032.630 A control system for a balloon-borne telescope.   J. Nishimura, N. Yajima, S. Kokaji, S. Hashino.
Scientific ballooning–II, (see 012.065), p. 127 - 133 (1981).

The paper deals with a star tracking system for a balloon-borne infrared telescope. A control moment gyro has been adopted for controlling the azimuth angle of the gondola. A CCD image sensor is used in a star field camera. Newly developed on-board microprocessor system is used for supervisory control of on-board equipment.

032.631 Solar Maximum Mission experiment: ultraviolet spectroscopy and polarimetry on the Solar Maximum Mission.   E. Tandberg-Hanssen, B. E. Woodgate, R. G. Athay, J. M. Beckers, J. C. Brandt, E. C. Bruner, R. D. Chapman, C. C. Cheng, J. B. Gurman, C. L. Hyder, P. J. Kenney, A. G. Michalitsianos, R. A. Rehse, S. A. Schoolman, R. A. Shine, W. Henze.
High-energy astrophysics, (see 012.067), p. 275 - 283 (1981).

The authors describe the Ultraviolet Spectrometer and Polarimeter (UVSP) on the Solar Maximum Mission spacecraft. The instrument, which operates in the wavelength range 1150 - 3600 Å, has a spatial resolution of 2 - 3 arc sec and a spectral resolution of 0.02 Å FWHM in second order. A Gregorian telescope, focal length 1.8 m, feeds a 1 m Ebert-Fastie spectrometer. A polarimeter comprising rotating MgF_2 waveplates can be inserted behind the spectrometer entrance slit and allows all four Stokes parameters to be determined.

032.632 A new thermoluminescent dosimeter system for space research.   I. Fehér, S. Deme, B. Szabó, J. Vágvölgyi, P. P. Szabó, A. Csőke, M. Ránky, Yu. A. Akatov.
Life sciences and space research XIX, (see 012.068), p. 61 - 66 (1981).

A small, portable, vibration and shock resistant thermoluminescent dosimeter system was developed to measure cosmic radiation dose on board a spacecraft. The system consists of a small battery-operated reader and a special bulb dosimeter. Doses from 10 μGy up to 100 mGy can be measured. The electrical power consumption of the reader is about 5 W, its volume is about 1 dm^3 and its mass is about 1 kg. Details are given for the construction and technical parameters of the dosimeter and reader.

Telescopes for the 1980s.   See Abstr. 003.041.

The definition and implementation of the Science Program for the Space Telescope.   See Abstr. 013.003.

Microcomputer on-line data reduction for satellite experiments.   See Abstr. 021.013.

The optical system of the Space Telescope.   See Abstr. 031.001.

Annular-aperture wide-field camera.   See Abstr. 031.034.

Cometary dust observations by optical _in-situ_ methods.   See Abstr. 031.651.

Design of cosmic radio telescopes: technological solutions.   See Abstr. 033.025.

Multi-anode microchannel arrays.   See Abstr. 034.007.

The ROSAT mission.   See Abstr. 051.032.

The ESA mission to comet Halley.   See Abstr. 051.041.

Recent developments in space-borne zodiacal light photometry.   See Abstr. 051.050.

Long duration hard X-ray transatlantic payload.   See Abstr. 051.057.

Near-millimeter spectrum of the microwave background.   See Abstr. 066.005.

Direct observation of charge state abundances of energetic He, C, O, and Fe emitted in solar flares. See Abstr. 073.108.

Observation of the solar H Lyman-α radiation by a ion-chamber with an O_2 gas filter.   See Abstr. 076.018.

Solar Maximum Mission experiment: early results of the hard X-ray imaging experiment.   See Abstr. 076.028.

Solar Maximum Mission experiment: early results from the soft X-ray polychromator experiment. See Abstr. 076.029.

Nitric oxide γ band airglow radiometer with a self-absorbing gas cell.   See Abstr. 082.092.

Quartz-glass chopping method of measuring solar irradiance.   See Abstr. 082.093.

Auroral X-ray images.   See Abstr. 084.009.

Coordinated ionospheric and magnetospheric observations from the ISIS 2 satellite by the ISIS 2 experimenters. Volume 4. A. Large storms. B. Airglow and related measurements. C. VLF observations.   See Abstr. 084.096.

Perspectives of auroral X-ray measurements in the 1980s.   See Abstr. 084.097.

Vikinig bistatic radar experiment: summary of first-order results emphasizing north polar data. See Abstr. 097.006.

Ultraviolet absorption studies of H_2O and other species in comet comae with satellite telescope-spectrometers. See Abstr. 102.058.

The astrophysical interest of EUV observations of extra-solar objects.   See Abstr. 114.172.

Galactic X rays observed with X-ray astronomy satellite 'Hakucho'.   See Abstr. 142.053.

Cosmic X-ray observations performed with a gas scintillation spectrometer.   See Abstr. 142.099.

Observations of gamma radiation between 0.4 MeV and 7 MeV at balloon altitudes using a Compton telescope. See Abstr. 142.511.

Search for time variations in 511 keV flux by ISEE−3 gamma-ray spectrometer.   See Abstr. 142.543.

High-energy gamma astronomy. See Abstr. 142.546.

The HEAO-3 French-Danish cosmic ray spectrometer: preliminary results on the elemental abundances of cosmic ray nuclei in the iron peak.   See Abstr. 143.116.

Jets and the Space Telescope − an introduction. See Abstr. 158.258.

033 Radio Telescopes and Equipment

033.001   A 200-350-GHz heterodyne receiver.
N. R. Erickson
IEEE Trans. Microwave Theory Tech., Vol. MTT-29, 557 - 561 (1981). − Abstr. in Phys. Abstr., Vol. 84, Abstr. 83480 (1981).

033.002   On the problem of constructing mathematical models of mirror systems of radio telescopes.
P. V. Belyanskij, Eh. Eh. Gasanov, A. B. Danilevich.
Izv. AN AzSSR. Ser. fiz.-tekh. i mat. nauk, 1980, No. 5, p. 86 - 93. In Russian. − Abstr. in Ref. zh., 51. Astron., 7.51.836 (1981).

033.003   The interplanetary scintillation observatory at Thaltej, Ahmedabad.
S. K. Alurkar, R. V. Bhonsle, R. Sharma, A. D. Bobra, Sohanlal, N. S. Nirman, P. Venat, G. Sethia.
Bull. Astron. Soc. India, Vol. 9, 76 (1981). − Abstract.

033.004   RT-7.5 radio telescope for millimeter wavelengths.
B. A. Rozanov.
Izv. vuzov. Radioehlektron., Tom 24, No. 3, p. 3 - 8 (1981). In Russian. − Abstr. in Ref. zh., 51. Astron., 9.51.1008 (1981).

033.005   Project of a ground-cosmic radiointerferometer with long base up to 1 million km and coherent radio communication between the telescopes.

V. V. Andreyanov, N. S. Kardashev.
Kosm. Issled., Tom 19, 763 - 772 (1981). In Russian.

033.006   Computation of polarized characteristics of systems by the symbolic method.
D. N. Rachkovskij, V. I. Abramenko, L. I. Tsvetkov.
Izv. Krymskoj Astrofiz. Obs., Tom 63, 189 - 196 (1981). In Russian. English translation in Bull. Crimean Astrophys. Obs., Vol. 63.

033.007   Metrology of the Effelsberg 100 meter radio reflector.   A. Greve.
Z. Vermessungswes., Vol. 106, 308 - 315 (1981). − Abstr. in Phys. Abstr., Vol. 84, Abstr. 108068 (1981).

033.008   Real-time excising of narrowband interference in radio interferometry.
B. L. Kasper, F. S. Chute, D. Routledge.
J. R. Astron. Soc. Canada, Vol. 75, 249 (1981). − Abstract.

033.009   The RT-6 radiotelescope of the Shemakha Astrophysical Observatory.
A. M. Gusejnov, N. M. Svikhnushin.
Tsirk. Shemakhinsk. Astrofiz. Obs., No. 68, p. 13 - 16 (1981). In Russian.

033.010   The Nançay radioheliograph.
J. Bonmartin, A. Kerdraon.

Onde Electr., Vol. 61, No. 8 - 9, p. 11 - 16 (1981). In French.
Abstr. in Phys. Abstr., Vol. 85, Abstr. 6537 (1982).

033.011 Numerical filter for an impulse meteor radar.
N. S. Andrianov, R. G. Khuzyashev.
Meteorn. rasprostr. radiovoln, Kazan', 1980, No. 16, p. 69 - 73.
In Russian. – From Ref. zh., 51. Astron., 11.51.1054 (1981).

033.012 The 30 meter millimeterwavelength radiotelescope.
J. W. M. Baars.
Mitt. Astron. Ges., Nr. 54, (see 012.050), p. 61 - 69 (1981).

**033.013 Automated equipment for radio meteor studies of
the upper atmosphere.**
V. V. Sidorov, A. N. Fakhrutdinova, V. A. Makarov,
A. M. Stepanov, V. Yu. Nesterov, V. A. Shuvarikov.
Meteorn. issled., Moskva, 1981, No. 7, p. 83 - 89. In Russian.
Abstr. in Ref. zh., 51. Astron., 11.51.1055 (1981).

**033.014 A multi-ray method for meteor radar observations
with electronic commutation of receiving antennas.**
Yu. I. Suvorov, O. A. Solyanik, V. V. Lizogub.
Meteorn. issled., Moskva, 1981, No. 7, p. 100 - 102. In Russian.
Abstr. in Ref. zh., 51. Astron., 11.51.1057 (1981).

**033.015 RATAN 600: status, results, perspectives. Scientific
report.**   Yu. N. Parijskij.
Vestn. AN SSSR. 1981, No. 6, p. 37 - 41. In Russian. – From
Ref. zh., 51. Astron., 11.51.1087 (1981).

033.016 A radiometric device for decimeter waves.
G. Ozoliņš, M. Eliāss.
Investigations of the sun and red stars. 13, (see 003.023),
p. 77 - 81 (1981). In Russian.

**033.017 Low-frequency circuit of a radiometer with high
dynamical range.**   A. Grišāns.
Investigations of the sun and red stars. 13, (see 003.023),
p. 82 - 89 (1981). In Russian.

**033.018 Modulated laser surface surveying system for
mm-wavelength radio telescopes.**
A. Greve, W. Harth.
Proc. Soc. Photo-Opt. Instrum. Eng., Vol. 36, 110 - 112
(1981) = Max-Planck-Inst. Radioastron., Bonn, Sonderdr.
Ser. A, Nr. 453.

033.019 The sector coupler – theory and performance.
J. W. Archer, M. Ogai, E. M. Caloccia.
IEEE Trans. Microwave Theory Tech., Vol. MTT-29, 202 - 208
(1981) = Natl. Radio Astron. Obs., Green Bank, Repr. Ser. A,
No. 1208.

033.020 Low noise single-ended mixer for 230 GHz.
J. W. Archer, R. J. Mattauch.
Electron. Lett., Vol. 17, No. 5, p. 180 - 181 (1981) = Natl.
Radio Astron. Obs., Green Bank, Repr. Ser. A, No. 1209.

033.021 Millimeter wavelength frequency multipliers.
J. W. Archer.
IEEE Trans. Microwave Theory Tech., Vol. MTT-29, 552 - 557
(1981) = Natl. Radio Astron. Obs., Green Bank, Repr. Ser. A,
No. 1238.

**033.022 A broad-band UHF mixer exhibiting high image
rejection over a multidecade baseband frequency
range.**   J. W. Archer, J. Granlund, R. E. Mauzy.
IEEE J. Solid-State Circuits, Vol. SC-16, 385 - 392 (1981) =
Natl. Radio Astron. Obs., Green Bank, Repr. Ser. A, No. 1307.

**033.023 Hydrogen masers in Shaanxi Astronomical Observa-
tory.**   R.-q. Tian, B.-t. Zhu.

Publ. Shaanxi Astron. Obs., No. 1, p. 17 - 23 (1981). In
Chinese.

**033.024 Use of cosmic radio sources for investigation of the
characteristics of the RATAN-600 radio telescope.**
B. V. Braude, N. A. Esepkina, Yu. K. Zverev, S. Ya. Golosova,
Yu. N. Kalikhevich, G. N. Pinchuk, D. V. Korol'kov,
O. I. Krat, M. N. Naugol'naya, Yu. N. Parijskij, N. S. Soboleva.
Radiotekh. i ehlektron., Tom 26, 1530 - 1544 (1981). In
Russian. – Abstr. in Ref. zh., 51. Astron., 11.51.1052 (1981).

**033.025 Design of cosmic radio telescopes: technological
solutions.**   A. G. Sokolov, A. S. Gvamichava.
Antenny, Moskva, 1981, No. 29, p. 3 - 10. In Russian.
From Ref. zh., 51. Astron., 12.51.1053 (1981).

033.026 "Circular periscope" antenna for radio telescopes.
N. L. Kajdanovskij.
Antenny, Moskva, 1981, No. 29, p. 32 - 44. In Russian.
From Ref. zh., 51. Astron., 12.51.1054 (1981).

**033.027 35 GHz 16-element solar radio interferometer at
Nagoya University.**   M. Fujishita.
Proc. Int. Latitude Obs. Mizusawa, No. 20, p. 38 - 52 (1981).
In Japanese.

033.028 Measurements on phase delay of a Loran-C antenna.
M.-K. Fujimoto, K. Fujiwara.
Tokyo Astron. Bull., Second Ser., No. 265, p. 3015 - 3020
(1981).

033.029 Shanghai Observatory's hydrogen maser.
Ann. Shanghai Obs. Acad. Sinica, No. 1, p.
106 - 118 (1979).

**033.030 Experimental very long baseline interferometry
system.**
Ann. Shanghai Obs. Acad. Sinica, No. 2, p. 117 - 129 (1980).

033.031 Selectivity of the meteor radar station Kama-2.
G. S. Kardonik.
Meteor. rasprostr. radiovoln, Kazan', 1981, No. 17, p. 89 - 95.
In Russian. – Abstr. in Ref. zh., 51. Astron., 1.51.858 (1982).

**033.032 Automatic meteor radar KGU-M5 for measurement
of coordinates of reflecting regions on meteor
trails.**   Yu. A. Pupyshev, V. A. Makarov,
V. Yu. Nesterov, V. V. Sidorov, A. M. Stepanov,
A. N. Fakhrutdinova, V. A. Shuvarikov.
Meteor. rasprostr. radiovoln, Kazan', 1981, No. 17, p. 96 - 100.
In Russian. – Abstr. in Ref. zh., 51. Astron., 1.51.859 (1982).

Telescopes for the 1980s.   See Abstr. 003.041.

**Data processing in synthetic-aperture radio astrono-
my.**   See Abstr. 021.035.

**Local gas without reddening: the contribution of
stray radiation to 21 centimeter line measurements.**
See Abstr. 031.501.

**A new method for making maps with unstable radio
interferometers.**   See Abstr. 031.504.

Images in radioastronomy.   See Abstr. 031.586.

**Calibration of spectral line observations at the
Metsähovi Radio Observatory.**   See Abstr. 031.641.

**A 408 MHz all-sky continuum survey. I. Observa-
tions at southern declinations and for the north polar region.**
See Abstr. 141.017.

034 Auxiliary Instrumentation

034.001 **A celestial pole locator for portable telescopes.**
M. M. Taylor.
J. British Astron. Assoc., Vol. 91, 576 - 582 (1981).

034.002 **Investigations of the birefringent filter of the Astronomical Observatory of the State Ural University.**
V. P. Kozhevnikov, T. P. Nikiforova.
Soln. Dannye 1981 Byull., No. 1, p. 82 - 84 (1981). In Russian.

034.003 **Use of an apparatus for visual observations of small-scale solar features in the Solar Service.**
A. M. Kumantsev, L. N. Kuznetsov.
Soln. Dannye 1981 Byull., No. 1, p. 92 - 93 (1981). In Russian.
    The use of a simple device with a rotating white screen in the Cassegrain focus for the Solar Service is described.

034.004 **Safe solar filters.**   B. R. Chou.
Sky Telesc., Vol. 62, 119 - 121 (1981).

034.005 **Nature's own particle accelerator.**
M. M. Waldrop.
Sky Telesc., Vol. 62, 208 - 215 (1981).

034.006 **Saturation and colour corrections for the SAAO Mk II JHKL photometer for the period 1975 to 1978.**   B. S. C. Robertson, R. M. Catchpole.
Mon. Notes Astron. Soc. South. Africa, Vol. 39, 82 - 88 (1980).
    The authors present the saturation and colour corrections required to relate the natural system of the Mk II JHKL (1.2 to 3.4 μ) photometer, using a PbS detector, to the Glass (1974) standards.

034.007 **Multi-anode microchannel arrays.**
J. G. Timothy, G. H. Mount, R. L. Bybee.
IEEE Trans. Nucl. Sci., Vol. NS-28, (see 012.003), p. 689 - 697 (1981). − Abstr. in Phys. Abstr , Vol. 84, Abstr. 63847 (1981).

034.008 **A new 10 micron astronomical polarimeter.**
P. A. Ekstrom, R. A. Stokes, G. M. Stokes, J. A. Hackwell.
Opt. Eng., Vol. 20, No. 1, p. 19 - 24 (1981). − Abstr. in Phys. Abstr., Vol. 84, Abstr. 66831 (1981).

034.009 **A cold camera for astrophotography.**
J. Newton.
Astronomy, Vol. 9, No. 2, p. 39 - 42 (1981). − Abstr. in Phys. Abstr., Vol. 84, Abstr. 67484 (1981).

034.010 **Fiber optics link to a solar spectrophotometer.**
C. V. Marques, J. S. Lemos, M. R. Teixeira, M. M. Leirao, F. C. Rodrigues, J. P. Peixoto.
Rev. Sci. Instrum., Vol. 52, 475 - 476 (1981). − Abstr. in Phys. Abstr., Vol. 84, Abstr. 71080 (1981).

034.011 **Conception et réalisation d'un photomètre pour la lecture des spectres stellaires sur plaque photographique.**   A. Delcroix.
Bull. Soc. R. Sci. Liège, 49e année, p. 238 - 243 (1980) = Commun. Dép. Astrophys. Mons, No. 83.
    The author presents here the modifications made to a Zeiss photometer in order to increase its precision, its speed, its security as well as a digital interface for the use of the data on a computer, all the changes having only very small budget impact.

034.012 **Sky photometer for ground-based coronal photography.**   J. E. Graves, A. L. Widener.

Proc. Soc. Photo-Opt. Instrum. Eng., Vol. 252, 82 - 84 (1980). Abstr. in Phys. Abstr., Vol. 84, Abstr. 74956 (1981).

034.013 **A study of the external-noise input in Weber-type gravitational-wave antennas.**   F. Fuligni, F. Ricci.
Nuovo Cimento C, Ser. 1, Vol. 4C, 93 - 102 (1981). − Abstr. in Phys. Abstr., Vol. 84, Abstr. 75021 (1981).

034.014 **An all-sky imaging photometer.**
S. K. Babey, L. L. Cogger, C. D. Anger.
Opt. Eng., Vol. 20, 216 - 219 (1981). − Abstr. in Phys. Abstr., Vol. 84, Abstr. 79650 (1981).

034.015 **Two-star high-speed photometry.**
A. D. Grauer, H. E. Bond.
Publ. Astron. Soc. Pacific, Vol. 93, 388 - 396 (1981) = Contrib. Louisiana State Univ. Obs., No. 164.
    The authors have constructed two dual-channel photometers that are capable of observing a rapid variable star and a nearby comparison star simultaneously. They are slightly modified versions of the two-star photometers used at McDonald Observatory. One of them is in use on the 0.9 m reflector of the Louisiana State University Observatory, and the other is used regularly at Kitt Peak National Observatory. The authors routinely employ the simultaneous comparison-star observations to remove sky transparency variations. Examples are given of data obtained with these photometers, demonstrating that precision high-speed photometry can be obtained under nonphotometric conditions. Further, even under photometric conditions the second channel gives the observer substantial confidence in the reality of observed variations in the brightness of the program star. Such an instrument is capable of increasing significantly the scientific productivity of telescopes located at nonphotometric sites.

034.016 **A new Cassegrain spectrograph for the Hale 5-meter telescope.**   J. B. Oke.
Bull. American Astron. Soc., Vol. 13, 509 (1981). − Abstract.

034.017 **Infrared observations with a 10 μm monolithic array camera.**
J. F. Arens, G. M. Lamb, M. C. Peck, G. G. Fazio, W. Hoffmann.
Bull. American Astron. Soc., Vol. 13, 509 - 510 (1981). Abstract.

034.018 **Automation of an iris astrophotometer.**
L. H. W. Bradfield, M. P. FitzGerald.
Bull. American Astron. Soc., Vol. 13, 517 (1981). − Abstract.

034.019 **A tube sensitometer for testing photographic hypersensitization techniques.**
R. L. Scott, R. L. Tonkel.
Bull. American Astron. Soc., Vol. 13, 556 (1981). − Abstract.

034.020 **First observations with CORAVEL at La Silla.**
M. Imbert, L. Prévot.
Messenger, No. 25, p. 6 - 7 (1981).

034.021 **An infrared speckle interferometer.**   C. Perrier.
Messenger, No. 25, p. 26 - 29 (1981).

034.022 **Calculation of illumination in the shadow cast by an external occulting shield of a coronograph. IV. Three-disk system. V. Comparison between efficiencies of various external occulting systems.**   A. V. Lenskij.
Problems of cosmic physics. Vyp. 16, (see 003.005), p. 116 - 122 (1981). In Russian.

034.023   A photoelastic-modulator polarimeter at Pine
Mountain Observatory.
J. C. Kemp, M. S. Barbour.
Publ. Astron. Soc. Pacific, Vol. 93, 521 - 525 (1981).
The history of astronomical photoelastic polarimeters
is briefly outlined. A description is given of a system of this
type which has been in continuous use at Pine Mountain
Observatory for six years. The instrument has produced data
at the sensitivity level of 0.01%, with long-term absolute re-
producibility.

034.024   Investigation of the Askorekord E-2 coordina-
tometer.
G. T. Kajzer, G. P. Khremli, T. I. Levitskaya, S. N. Timofeev.
Stellar aggregates, Sverdlovsk, 1980, (see 003.006), p. 146 -
154. In Russian. − Abstr. in Ref. zh., 51. Astron., 8.51.959
(1981).

034.025   The capabilities of charge-coupled device (CCD)
detectors.   D. J. Purll.
Observatory, Vol. 101, 138 (1981).

034.026   Charge and mass spectrometer for energetic charged
particles.   V. N. Lutsenko, J. Plch, S. Slabý,
J. Tomek, V. Vitouš, S. Fischer.
Bull. Astron. Inst. Czechoslovakia, Vol. 32, 257 - 270 (1981).
The TP-2 instrument which was used in the Soviet-
Czechoslovak experiments devoted to studies of the isotopic
and charge compositions of solar cosmic rays and of charged
particles of the terrestrial radiation belts in the Prognoz 6 and
Intercosmos 17 satellites is described. The instrument designed for the energy range of 1 - 20 MeV/nucleon employed dual
ΔE-E analysis, a system of priorities with a particle identifier,
automatic range selection of the linear sections, and an effective multichannel digital intensimeter.

034.027   A new photoelectric photometer for astronomical
use.   R. Burchi, S. Mancuso.
Mem. Soc. Astron. Italiana, Vol. 52, 187 - 193 (1981).
A new photoelectric photometer photon counting,
double beam, monochannel is described.

034.028   New photoelectric photometer.
A. I. Gamal Eldin, A. M. Osman, S. M. Hassan.
J. Astron. Soc. Egypt, Vol. 2, 6 - 10 (1980).

034.029   Un moderno fotómetro astronómico.
L. M. Weinstein.
R Muscae, Vol. 6, No. 1/2, p. 28 - 32 (1981).

034.030   Elektronografiske kameraer i drift på det danske
1,5 m teleskop.   R. F. Nielsen.
Astron. Tidsskr., Årg. 14, 110 - 115 (1981).

034.031   On the effect of nonlinear characteristics of a
photoelectric device on the determination of mean
moments.   M. Ogriňš.
Determination of the coordinates of celestial bodies, Riga,
1981, (see 003.008), p. 186 - 191. In Russian. − Abstr. in Ref.
zh., 51. Astron., 9.51.968 (1981).

034.032   On the synchronization of reference marks for
determination of mean moments.
K. Šteins, M. Ogriňš.
Determination of the coordinates of celestial bodies, Riga,
1981, (see 003.008), p. 181 - 185. In Russian. − Abstr. in Ref.
zh., 51. Astron., 9.51.971 (1981).

034.033   Stellar-solar electrophotometer.
A. I. Abramochkin, P. P. Vaulin, V. P. Galilejskij,
A. V. Isakov, G. V. Potemkin.
Izmerenie opt.-meteorol. parametrov atmos. s ispol'z. lazer.

izlucheniya. Tomsk, 1980, p. 14 - 19. In Russian. − Abstr. in
Ref. zh., 51. Astron., 9.51.982 (1981).

034.034   The cryogenic detector of gravitational waves in
Frascati.   U. Giovanardi, V. Iafolla, P. Napoleoni,
B. Pavan, S. Ugazio, F. Ricci.
J. Phys. E, Vol. 14, 1067 - 1072 (1981). − Abstr. in Phys.
Abstr., Vol. 84, Abstr. 89396 (1981).

034.035   Cryogenically cooled detector system for spectroscopic applications of solid-state arrays.
W. G. Robinson, J. E. Simmons, A. S. Brar, A. M. Fowler.
Proc. Soc. Photo-Opt. Instrum. Eng., Vol. 245, (see 012.028),
p. 14 - 33 (1980). − Abstr. in Phys. Abstr., Vol. 84, Abstr.
94178 (1981).

034.036   Cooled filters 16 - 42 µm wavelength for FIR
astronomy.
J. S. Seeley, R. Hunneman, A. Whatley.
Proc. Soc. Photo-Opt. Instrum. Eng., Vol. 253, p. 266 - 273
(1980). − Abstr. in Phys. Abstr., Vol. 84, Abstr. 94182
(1981).

034.037   A two-dimensional solar spectrometer.
J. R. Brookes, G. R. Isaak, H. B. van der Raay.
Sol. Phys., Vol. 74, (see 012.029), 503 - 508 (1981).
A precise two-dimensional positioning device has been
developed for use in conjunction with a resonant scattering
spectrometer to study the spatial distribution of solar velocity
fields. The principle of operation and constructional details are
discussed and the experimental performance is evaluated. As
an illustration of its use preliminary data obtained from a
meridional scan are presented.

034.038   Searching for l=1 modes of solar oscillations.
A. Cacciani, V. Croce, T. Fortini, M. Torelli.
Sol. Phys., Vol. 74, (see 012.029), 543 - 546 (1981).
An instrument to measure the non-full-disk low-order
solar oscillations that uses a magneto-optical filter in Na-D
lines is described. It has the advantage, over the resonant cells
used by other observers, that it gives an image of the Sun and
a higher photonic flux.

034.039   Design aspects at Fourier-infrared-spectrometers.
J. Puder.
Feingerätetechnik, Vol. 30, 99 - 101 (1981). In German.
Abstr. in Phys. Abstr., Vol. 84, Abstr. 94908 (1981).

034.040   Measurements of characteristics of the gas propor-
tional chamber in cosmic ray showers.
K. Mitsui.
Nucl. Instrum. Methods Phys. Res., Vol. 186, 579 - 584 (1981).
Abstr. in Phys. Abstr., Vol. 84, Abstr. 95570 (1981).

034.041   A high efficiency 4π neutron detector.
B. Holmqvist, E. Ramstrom.
Nucl. Instrum. Methods Phys. Res., Vol. 188, 153 - 157 (1981).
Abstr. in Phys. Abstr., Vol. 84, Abstr. 95575 (1981).

034.042   Superconducting tensor gravity gradiometer for
satellite geodesy and inertial navigation.
J. P. Ho.
J. Astronaut. Sci., Vol. 29, 1 - 18 (1981). − Abstr. in Phys.
Abstr., Vol. 84, Abstr. 98597 (1981).

034.043   A narrow gap, servo-controlled tunable Fabry-Pérot
filter for astronomy.
P. D. Atherton, N. K. Reay.
Mon. Not. R. Astron. Soc., Vol. 197, 507 - 511 (1981).
The authors have constructed an 85 mm clear aperture
servo-controlled Fabry-Pérot etalon with a nominal 4 µm gap.
The etalon can be operated between orders 7 and 34. It be-

haves as a single period interference filter tunable over the wavelength range 4300 - 6400 Å with a bandwidth at 5000 Å of between 30 and 7.3 Å.

034.044 Testing of the five-channel spectrometer on the AZT-8 telescope.   A. B. Bukach.
Izv. Krymskoj Astrofiz. Obs., Tom 63, 204 - 215 (1981). In Russian. English translation in Bull. Crimean Astrophys. Obs., Vol. 63.
The results of a full-scale test of the spectrophotometer are presented. The error of measurements and stability of the main characteristics of the device were evaluated and testified during the operational phase on the telescope.

034.045 Combinaisons optiques à réseaux aspheriques: le spectrographe UV-PRIM.   G. Lemaître.
Astron. Astrophys., Vol. 103, L14 - L16 (1981).
Aspherical diffraction gratings have been produced by the elastic relaxation method. These gratings lead to more nearly ideal mountings from the standpoints of a small number of surfaces, of a wide field and of a fast aperture ratio. The UV-PRIM spectrograph is the first instrument of this type to have been constructed.

034.046 An air shower array designed for cosmic ray variation measurements and high energy gamma ray astronomy.   C. Morello, G. Navarra.
Nucl. Instrum. Methods Phys. Res., Vol. 187, 533 - 538 (1981). — Abstr. in Phys. Abstr., Vol. 84, Abstr. 102040 (1981).

034.047 Sensitivity analysis of a resonant mass gravitational wave antenna with resonant transducer.
P. F. Michelson, R. C. Taber.
J. Appl. Phys., Vol. 52, 4313 - 4319 (1981). — Abstr. in Phys. Abstr., Vol. 84, Abstr. 102485 (1981).

034.048 Enhanced sensitivity of a gravitational wave detector.
D. F. Walls, P. Zoller.
Phys. Lett. A, Vol. 85A, 118 - 120 (1981). — Abstr. in Phys. Abstr., Vol. 84, Abstr. 102487 (1981).

034.049 Adjustable mosaic grating mounts.
G. A. Brealey, J. M. Fletcher, W. A. Grundmann, E. H. Richardson.
Proc. Soc. Photo-Opt. Instrum Eng., Vol. 240, 225 - 228 (1980). — Abstr. in Phys. Abstr., Vol. 84, Abstr. 104219 (1981).

034.050 Consideration on the counting rate and the path length distribution of a telescope composed of two rectangular parallelepiped scintillators.   H. Takeuchi.
Sci. Pap. Inst. Phys. Chem. Res., Vol. 75, No. 1, p. 1 - 32 (1981). — Abstr. in Phys. Abstr., Vol. 84, Abstr. 108064 (1981).

034.051 WATFAA — the automation of a Cuffey iris astrophotometer.   L. Bradfield, M. P. Fitzgerald.
J. R. Astron. Soc. Canada, Vol. 75, 245 - 246 (1981). Abstract.

034.052 Un système d'interférométrie à balayage pour l'imagerie et la spectrométrie.
J.-R. Roy, R. Arsenault, G. Joncas.
J. R. Astron. Soc. Canada, Vol. 75, 252 (1981). — Abstract.

034.053 Photometry with the Kitt Peak Reticon scanner.
J. V. Barnes, D. L. Crawford.
Publ. Astron. Soc. Pacific, Vol. 93, 544 (1981). — Abstract.

034.054 An automated system for photoelectric photometry, II.   A. P. Linnell.
Publ. Astron. Soc. Pacific, Vol. 93, 661 - 671 (1981).
A system of automated photometry described in an earlier publication has been revised extensively. The modified system is flexible and particularly well adapted to high-speed photometry of variable stars.

034.055 Der Einsatz eines zwei-dimensionalen Photodioden-arrays zur spektroskopischen Erfassung solarer Geschwindigkeitsfelder.   G. Küveler, H. Wöhl.
Sterne Weltraum, Jahrg. 20, 452 - 456, 458 (1981).

034.056 Test of a transit photoelectric micrometer and automatic observation.   H. Li, Z.-l. Zou, X.-f. Hua.
Acta Astron. Sinica, Vol. 22, 248 - 253 (1981). In Chinese.

034.057 Experiments with an SIT camera.
S. Isobe, T. Hirayama, G. Sasaki, K. Tomita, K. Saito, N. Oshima.
AAS Photo-Bull., No. 25, p. 5 - 8 (1980).

034.058 The 1P21.   G. E. Kron.
I. A. P. P. P. Commun., No. 5, p. 4 - 6 (1981).

034.059 Solid state photometric photometer.
Z. A. Saroch.
I. A. P. P. P. Commun., No. 5, p. 10 - 11 (1981).

034.060 An inexpensive photoelectric system.
K. Krisciunas.
I. A. P. P. P. Commun., No. 6, p. 24 - 27 (1981).

034.061 The KD*P modulator in the Solar Magnetic Field Telescope.   G.-x. Ai, Y.-f. Hu.
Acta Astrophys. Sinica, Vol. 1, 273 - 284 (1981). In Chinese.

034.062 The Roman experiment for the detection of gravitational waves.   G. V. Pallottino, G. Pizzella.
Fis. Tecnol., Vol. 4, No. 1, p. 15 - 29 (1981). In Italian. Abstr. in Phys. Abstr., Vol. 85, Abstr. 3467 (1982).

034.063 Installation and first results of the Coudé Echelle Spectrometer.   D. Enard.
Messenger, No. 26, p. 22 - 24 (1981).

034.064 New large interference filters for the 3.6-m triplet.
D. Enard, M. Tarenghi.
Messenger, No. 26, p. 24 - 27 (1981).

034.065 Spektroskopische Instrumentierung.   J. Solf.
Mitt. Astron. Ges., Nr. 54, (see 012.050), p. 39 - 41 (1981).

034.066 Bildröhren.   M. Beetz.
Mitt. Astron. Ges., Nr. 54, (see 012.050), p. 43 - 46 (1981).

034.067 Digitale Bildsysteme.   K.-H. Marien.
Mitt. Astron. Ges., Nr. 54, (see 012.050), p. 47 - 52 (1981).

034.068 Ein neues lichtelektrisches Photometer für das Deutsch-Spanische Astronomische Zentrum.
R. Lenzen, H. Bellemann, D. Fath, H. Schütz, K. Zimmermann.
Mitt. Astron. Ges., Nr. 54, (see 012.050), p. 179 - 180 (1981).

034.069 Ein Lichtleiter-gekoppelter Spektrograph.
J. G. Schiffer.
Mitt. Astron. Ges., Nr. 54, (see 012.050), p. 182 - 186 (1981).

034.070 Das Fabry-Perot-Interferometer des MPI für Astronomie.   H. Hippelein, G. Münch.
Mitt. Astron. Ges., Nr. 54, (see 012.050), p. 193 - 196 (1981).

034.071 Protuberanzenbeobachtung mit dem 20 cm-Korono-
graphen des Sonnenobservatoriums Wendelstein.
O. Bärnbantner, C. Spannagl.
Mitt. Astron. Ges., Nr. 54, (see 012.050), p. 197 - 198 (1981).

034.072 A multichannel photon counting detector system
for spectroscopy.   R. Östreicher.
Mitt. Astron. Ges., Nr. 54, (see 012.050), p. 223 - 226 (1981).

034.073 Bildverstärkerkamera für 35 mm-Film.   B. Wedel.
Mitt. Astron. Ges., Nr. 54, (see 012.050), p. 227
(1981).

034.074 Erste Erfahrungen zur astronomischen Verwendung
des panoramischen SIT-Vidicon-Detektorsystems
OMA 2.   E. H. Geyer.
Mitt. Astron. Ges., Nr. 54, (see 012.050), p. 228 - 231 (1981).

034.075 Solid state area scanner photometer.   K. Rakos.
Mitt. Astron. Ges., Nr. 54, (see 012.050), p. 231 -
233 (1981).

034.076 Kameras und Kassetten für hypersensibilisierte
Photoplatten.   J. D. Schumann.
Mitt. Astron. Ges., Nr. 54, (see 012.050), p. 233 (1981).

034.077 Development of a microchannelplate-detector for
laboratory applications in the extreme ultraviolet.
M. Grewing, J. Barnstedt, R. Streitenberger, C. Wulf-Mathies.
Mitt. Astron. Ges., Nr. 54, (see 012.050), p. 237 - 241 (1981).

034.078 Examples of interactive instrument control and
data acquisition.
H. Jenkner, P. L. Byard, C. B. Foltz, B. M. Peterson.
Mitt. Astron. Ges., Nr. 54, (see 012.050), p. 251 - 255 (1981).
In German.

034.079 Untersuchungen der photometrischen Eigenschaften
des PDS-Mikrodensitometers.   A. M. Quetsch.
Mitt. Astron. Ges., Nr. 54, (see 012.050), p. 261 - 263 (1981).

034.080 Genauigkeitsuntersuchung am Wiener Ascorecord.
M. Barylak, F. Prochazka.
Mitt. Astron. Ges., Nr. 54, (see 012.050), p. 263 - 267 (1981).

034.081 Two-coordinate stellar photoelectric micrometer.
V. V. Konin, A. D. Pogonij.
Astrometric investigations, (see 012.048), p. 120 - 123 (1981).
In Russian. – Abstr. in Ref. zh., 51. Astron., 11.51.1062
(1981).

034.082 Measuring-calculating system for detection of
astronomical objects and determination of their
coordinates.
V. F. Anisimov, S. M. Kaganov, V. A. Kachmin,
V. D. Levchenko, T. V. Omarov, S. M. Sinenok, L. I. Shchavlev.
Automation of astronomical observations, (see 003.020), p. 5 -
40 (1981). In Russian. – Abstr. in Ref. zh., 51. Astron.,
11.51.1064 (1981).

034.083 Perforating chronograph.
A. F. Belosvet, L. G. Karyakina, Eh. A. Pryanchikova.
Astrometric investigations, (see 012.048), p. 126 - 131 (1981).
In Russian. – Abstr. in Ref. zh., 51. Astron., 11.51.1070
(1981).

034.084 Increasing of reliability of work of multi-counter
spectrometers.   S. I. Babichenko,
A. F. Gritsenko, V. M. Dvojnishnikov, A. P. Utekhin.
Vopr. atom. nauk. i tekh. Yader. priborostr. Moskva, 1981,
No. 2147, p. 43 - 46. In Russian. – Abstr. in Ref. zh., 51.
Astron., 11.51.1102 (1981).

034.085 SKS-04 spectrometer for low-energy charged
particles.   O. L. Vajsberg, L. S. Gorn,
L. S. Zhurina, D. S. Zakharov, A. A. Klimashov,
A. N. Omel'chenko, L. G. Ol'dekop, V. V. Pomogaev,
B. I. Khazanov, A. V. Shifrin.
Vopr. atom. nauk. i tekh. Yader. priborostr. Moskva, 1981,
No. 2147, p. 51 - 57. In Russian. – Abstr. in Ref. zh., 51.
Astron., 11.51.1103 (1981).

034.086 Intensification of signals of secondary electron
multipliers.   L. S. Gorn, D. S. Zakharov,
A. A. Klimashov, V. V. Pomogaev, B. I. Khazanov.
Vopr. atom. nauk. i tekh. Yader. priborostr. Moskva, 1981,
No. 2147, p. 47 - 50. In Russian. – Abstr. in Ref. zh., 51.
Astron., 11.51.1104 (1981).

034.087 Integrating detector for low-energy charged particles.
G. N. Zastenker, A. B. Denin, Yu. I. Ermolaev,
L. S. Zhurina, A. A. Klimashov, Eh. L. Lein, A. E. Stefanovich,
B. I. Khazanov, A. V. Shifrin.
Vopr. atom. nauk. i tekh. Yader. priborostr. Moskva, 1981,
No. 2147, p. 64 - 69. In Russian. – Abstr. in Ref. zh., 51.
Astron., 11.51.1105 (1981).

034.088 On FEU-79 PM tube efficiency at low light-level
photometry.
B. M. Glukhovskoj, A. A. Tokovinin, P. V. Shcheglov.
Astron. Tsirk., No. 1120, p. 1 - 3 (1980). In Russian.

034.089 Investigation of FEU-115 PM tubes.
B. M. Glukhovskoj, Yu. V. Khan.
Astron. Tsirk., No. 1120, p. 3 - 5 (1980). In Russian.

034.090 Absorption of composite bolometers.
B. Carli, D. Iorio-Fili.
J. Opt. Soc. America, Vol. 71, 1020 - 1025 (1981).
    A general formula that gives the absorption of a dielectric
substrate with two metallic films is derived. The variation of
the absorption as a function of the relevant parameters
(refractive index of substrate, thickness of metallic films, and
angle of incidence) is analyzed in order to identify the criteria
for the optimization of this substrate as an absorber for
composite bolometers.

034.091 Argon branching ratios for spectral-intensity
calibration.   D. L. Adams, W. Whaling.
J. Opt. Soc. America, Vol. 71, 1036 - 1038 (1981).
    Branching ratios have been measured for 104 emission
lines from 24 levels in Ar I and Ar II. These branching ratios
may be used to calibrate the relative detection efficiency of a
spectrometer over the 2900–23000-Å wavelength range.

034.092 Low temperature performance of a metallic-mesh
low-pass interference filter in the far infrared.
B. W. Davis.
Appl. Opt., Vol. 20, 2345 - 2346 (1981).

034.093 Compliance of Si photodiodes with Talbot's law.
J. Priu, W. Budde.
Appl. Opt., Vol. 20, 2676 - 2682 (1981)
    For accurate measurements of chopped radiation, it is
important that the photoelectric instrumentation complies
with Talbot's law, that is, measures the time average. Two different photoelectric measuring systems were investigated for
their compliance with Talbot's law: (1) Si photodiode with an
integrating digital voltmeter (IDVM) system, and (2) Si photodiode with a lock-in amplifier. The test consisted in comparing
values predicted from the amplitude and the duty factor of
the chopper with values measured by both systems. It was
found that the IDVM system showed satisfactory compliance
with Talbot's law, whereas the lock-in system did not.

034.094 Infrared charge-injection-device array performance at low background.
C. R. McCreight, J. H. Goebel.
Appl. Opt., Vol. 20, 3189 - 3196 (1981).

Low-background tests of a 1 × 32 Si:Bi charge-injection-device (CID) IR detector array were carried out to evaluate its feasibility for space-based astronomical observations. Optimum performance was obtained at a temperature of 11 K. The device showed a peak responsivity of 4.4 A/W, an average noise level of ~ 670 electrons, and a minimum noise equivalent power of 3×10^{-17} W/$\sqrt{\text{Hz}}$ for 1 sec integration time. This sensitivity compares well with that of discrete extrinsic silicon photoconductors.

034.095 Echelle and holographic gratings compared for scattering and spectral resolution.
J. Kielkopf.
Appl. Opt., Vol. 20, 3327 - 3331 (1981).

An echelle grating with 316 grooves/mm and a 63° blaze angle is compared with a 3600 groove/mm holographic grating. The scattered light and resolution of the 10 × 20 cm gratings are photoelectrically evaluated in a 3 m spectrometer optimized to eliminate coma and baffled to minimize instrumental scattered light. Both gratings are ghost-free, but the holographic grating has substantially lower scattered light for test lines near 4000 Å and yields improved stray light rejection within absorption lines. It also exhibits slightly higher resolution. The throughput of the spectrometer with the holographic grating is about 5 times higher than the equivalent combination of spectrometer, echelle grating, and predisperser.

034.096 TESS: a high-luminosity high-resolution twin-étalon scanning spectrometer.
G. Hernandez, O. A. Mills, J. L. Smith.
Appl. Opt., Vol. 20, 3687 - 3688 (1981).

034.097 Imaging detectors for the ultraviolet.
C. I. Coleman.
Appl. Opt., Vol. 20, 3693 - 3703 (1981).

A review is presented of the available types of efficient middle- and far-UV imaging detectors. These include luminescent phosphors, semitransparent and opaque photocathodes, mesh-based photocathodes, microchannel plates, and solid-state devices. Certain important aspects of their performance and calibration are discussed, and mention is also made of some applications of these systems, particularly in UV astronomy.

034.098 Monochromator-interferometer combination for submillimeter astronomical spectrometry from aircraft.   M. Harwit, N. T. Kurtz, R. W. Russell, S. Smyers.
Appl. Opt., Vol. 20, 3792 - 3796 (1981).

A simple interferometer placed in front of a liquid helium cooled grating instrument was constructed; one has attained a spectral resolution of $\gtrsim 0.1$ cm^{-1} in the wavelength range of ~64 cm^{-1} (157 μm) during observations from the NASA Kuiper Airborne Observatory. Spectra obtained on the planetary nebula NGC 7027 are shown.

034.099 Star brightness measurement using photon counting.
R. H. Stanton.
J. American Assoc. Variable Star Obs., Vol. 10, 21 - 24 (1981).

A brief overview of the use of photon counting for star brightness measurement is presented.

034.100 Taurus: a wide-field imaging Fabry-Perot spectrometer for astronomy.   P. D. Atherton, K. Taylor,
C. D. Pike, C. F. W. Harmer, N. Parker, R. N. Hook.
Anglo-Australian Obs. Prepr. No. 150, 75 pp. (1981). – Submitted to Mon. Not. R. Astron. Soc.

034.101 Photographic calibration of astronomical photographic plates. II. – The Anglo-Australian Telescope prime focus spot sensitometer.
D. F. Malin, B. A. Peterson.
Anglo-Australian Obs. Prepr. No. 159, 13 pp. (1981). – Submitted to Publ. Astron. Soc. Pacific.

034.102 Using the PDS microdensitometer at the Observatory of Lund, Sweden.   N. Holsti.
Proceedings of the Third Finnish-Soviet Astronomical Symposium, (see 012.057), p. 43 - 44 (1981).

034.103 One-dimensional infrared speckle interferometry.
R. R. Howell, D. W. McCarthy, F. J. Low.
Prepr. Steward Obs., No. 327, 19 pp. (1981).

An improved technique of one-dimensional speckle interferometry yields high sensitivity at wavelengths from 2.2 to 11.6 μm. The binary character of the protostellar source W3 IRS 5 is confirmed at 5 μm; the separation, position angle, and brightness ratio are 1.26 arcsec, 37° and 1.7, respectively, The individual components are $\leqslant 0.25$ arcsec in diameter. S140 IRS 1 and Mon R2 IRS 3 are extended, $\geqslant 1$ arcsec, at 2.2 μm, while GL 2591, NGC 2264 IRS, and the Becklin-Neugebauer source are unresolved, $\leqslant 0.2$ arcsec. At 11.6 μm the size of the dust shell around α Ori is measured for the first time; assuming a Gaussian intensity distribution the 1/e intensity points are separated by 3.4 arcsec, more than 60 stellar diameters.

034.104 Photoelectric photometry system for the new 30 cm reflector at Mitaka, Tokyo Astronomical Observatory (I).   H. Sato, Y. Nishino.
Tokyo Astron. Obs. Rep. (No. 74), Vol. 19, 336 - 360 (1981). In Japanese.

034.105 Improvement of sidereal divider in the timing system of 91 cm telescope at the Dodaira station.
T. Yamaguchi.
Tokyo Astron. Obs. Rep. (No. 74), Vol. 19, 361 - 363 (1981). In Japanese.

034.106 The software of the Okayama 91 cm telescope and the photoelectric photometer.   K. Okida.
Tokyo Astron. Obs. Rep. (No. 74), Vol. 19, 364 - 378 (1981). In Japanese.

034.107 A multi-channel polarimeter IV. A TV guiding system.   S. Kikuchi, M. Konno, T. Yamaguchi, H. Shibasaki, Y. Mikami, M. Noguchi.
Tokyo Astron. Obs. Rep. (No. 74), Vol. 19, 399 - 404 (1981). In Japanese.

034.108 Performance test for the PDS micro-densitometer.
T. Noguchi, N. Miyauchi, T. Tsuji.
Tokyo Astron. Obs. Rep. (No. 74), Vol. 19, 452 - 465 (1981). In Japanese.

034.109 Practical experience with dichroic beamsplitters for visible wavelengths.   C. F. Velt, J. Tinbergen.
Astron. Astrophys., Vol. 103, 422 - 423 (1981).

Quality tests of off-the-shelf dichroic beamsplitters are presented. These components have many possible uses in astronomical instrumentation and some (but by no means all) are of sufficient quality. Prospective users must in general verify that the beamsplitters they obtain meet the manufacturer's specifications.

034.110 Narrowband filters for cometary photometry.
M. F. A'Hearn.
Int. Comet Q., Vol. 3, 91 (1981).

034.111 Device for uniform illumination of the photo-sensitive surface of astronomical detectors.
A. N. Abramenko.
Astron. Tsirk., No. 1153, p. 4 - 6 (1981). In Russian.

034.112 Photon counting Czerny-Turner spectrophotometer.
A. P. Ipatov.
Astron. Tsirk., No. 1160, p. 1 - 3 (1981). In Russian.

Remote sensing. Optics and optical systems.
See Abstr. 003.147.

Landolt-Börnstein. See Abstr. 003.172.

On the optical design of photoelectric stellar photometers. See Abstr. 031.002.

Optical polarimeters for solar research.
See Abstr. 031.521.

Surveying velocity fields in galaxies.
See Abstr. 031.567.

The observation of the solar irradiance and its variations, challenging space metrology. See Abstr. 031.572.

Recommendations for calibration of millimeter-wavelength spectral line data. See Abstr. 031.580.

A new asteroid observation and search technique.
See Abstr. 031.589.

An investigation of autoradiographically intensified electrographic imagery. See Abstr. 031.590.

A possible technique for cometary studies with high angular and spectral resolution. See Abstr. 031.607.

35 mm-Kameras zur Aufnahme und Nachverarbeitung astronomischer Speckle-Bilder.
See Abstr. 031.615.

Fernsehleiteinrichtung für das 2.2 m-Teleskop.
See Abstr. 032.035.

Untersuchungen an höchstempfindlichen Infrarotdetektoren für GIRL. See Abstr. 032.605.

Entwicklung eines ortsempfindlichen Proportionalzählers für röntgenastronomische Beobachtungen.
See Abstr. 032.606.

The atmospheric extinction at SAAO Sutherland, and the red leak of the Schott BG18 filter in the St. Andrews Scanner. See Abstr. 082.016.

Photometric and polarimetric observations of the Moon's surface using the multichannel polarimeter installed at the Dodaira Station of the Tokyo Astronomical Observatory. See Abstr. 094.042.

Multislit photoelectric magnetometer observations of Cepheids and supergiants: probable detections of weak magnetic fields. See Abstr. 116.001.

Small-scale structure of the core of M33 (NGC 598).
See Abstr. 158.045.

035 Clocks and Frequency Standards

035.001 Un tipo di orologio solare. R. Valentini.
G. Astron., Vol. 7, 87 - 96 (1981).

035.002 A non-skiaterical heliochronometer. J. Wall.
J. British Astron. Assoc., Vol. 91, 491 - 498 (1981).

035.003 A possibility of clock synchronization between two stations by the way of an artificial satellite.
N. Capitaine, D. Gambis.
Ann. Géophys., Vol. 37, (see 012.037), 139 - 142 (1981).
Abstr. in Phys. Abstr., Vol. 84, Abstr. 107882 (1981).

035.004 Experiments on extremely precise time synchronisation, using satellites with laser mirrors.
I. Kardos.
Fiz. Sz., Vol. 31, 176 - 180 (1981). In Hungarian. – Abstr. in Phys. Abstr., Vol. 85, Abstr. 6539 (1982).

035.005 Die astronomische Uhr in der Marienkirche zu Rostock. M. Schukowski.
Sterne, 57. Band, 331 - 341 (1981).

035.006 Chronograph for magnetic registration of time of astronomical field observations with quartz clocks.
A. A. Logvinenko, G. G. Krajnyuk, A. T. Dul'tsev.
Tsirk. Astron. Obs., L'vov, No. 54, p. 41 - 43 (1979). In Russian.

035.007 Another standard-time sundial. H. Brix.
J. British Astron. Assoc., Vol. 92, 16 - 21 (1981).

035.008 International comparison test of Shaanxi Observatory's clock.
Y.-r. Miao, J.-a. Song, H.-q. Zheng, H.-r. Xiang.
Publ. Shaanxi Astron. Obs., No. 1, p. 1 - 16 (1981). In Chinese.

035.009 Effects of solar X-ray bursts on the time and frequency calibration by using LF signals.
Y.-r. Miao, K.-j. Yang, L.-d. Pan.
Publ. Shaanxi Astron. Obs., No. 1, p. 24 - 30 (1981). In Chinese.

035.010 An automatic measurement system for time comparison. H.-q. Zheng, S.-f. Chen.
Publ. Shaanxi Astron. Obs., No. 1, p. 58 - 63 (1981). In Chinese.

035.011 The clock comparison using TV signals emitted by Broadcasting Satellite. K. Horiai, T. Hara, M. Aihara, K. Sato, M. Fujishita.
Proc. Int. Latitude Obs. Mizusawa, No. 20, p. 61 - 68 (1981). In Japanese.

035.012 Establishment of the Shanghai Observatory independent atomic time scale.
Ann. Shanghai Obs. Acad. Sinica, No. 2, p. 163 - 168 (1980).

035.013   The second order frequency drift of rubidium
clocks and its effects on time scale.   J.-l. Hu.
Ann. Shanghai Obs. Acad. Sinica, No. 2, p. 169 - 174 (1980).

035.014   Reliability research of rubidium clocks.
Y.-x. Zhuang.
Ann. Shanghai Obs. Acad. Sinica, No. 2, p. 175 - 180 (1980).

035.015   Frequency comparison by colour sub-carrier pulse.
J.-x. Cai.
Ann. Shanghai Obs. Acad. Sinica, No. 2, p. 181 - 185 (1980).

035.016   Design and development of a two-channel automatic frequency tuner for a hydrogen maser
cavity.   S.-j. Shi, J.-y. Su, Y.-d. Song.
Ann. Shanghai Obs. Acad. Sinica, No. 2, p. 186 - 201 (1980).

035.017   Investigation of the rates of the crystal-controlled
synchronometers "Ch7-15".   M. I. Malyshev.
Astron. Tsirk., No. 1156, p. 3 - 5 (1981). In Russian.

035.018   High-precision clocks of sidereal time.
M. I. Malyshev, E. N. Fedoseev.
Astron. Tsirk., No. 1156, p. 5 - 6 (1981). In Russian.

035.019   The sundial of 1556 from the Hermitage.
V. Yu. Matveev.
Researches in the history of astronomy. 15th issue, (see
003.025), p. 177 - 180 (1980). In Russian.

Equatoires et Horlogerie Planétaire du XIIIᵉ au
XVIᵉ Siècle.   See Abstr. 003.127.

Automated data acquisition system for the time
keeping.   See Abstr. 021.066.

HP-IB universal board and its application to the
clock data acquisition system.   See Abstr. 021.067.

Gravitational effects of sun and moon on atomic
time scales.   See Abstr. 044.007.

Time and Frequency Services Bulletin, April -
December 1981.   See Abstr. 044.023.

036 Photographic Materials and Techniques

036.001   Il concetto di efficienza quantica dei materiali
fotografici e i criteri di esposizione per la rivelazione
dei segnali deboli in astronomia.
F. Bònoli, L. Scaltriti.
Coelum, Vol. 50, 112 - 126 (1981).

036.002   Il trattamento di sviluppo delle emulsioni
fotografiche in bianco e nero.   A. D. Donne.
Coelum, Vol. 50, 156 - 162 (1981).

036.003   Die Bestimmung der Belichtungszeiten für Mond-
und Planetenaufnahmen durch die Berechnung von
Verlängerungsfaktoren.   D. Haesloop.
Sterne Weltraum, Jahrg. 20, 298 - 299 (1981).

036.004   The deep sky in color.   D. F. Malin.
Sky Telesc., Vol. 62, 216 - 219 (1981).

036.005   Non-atlas photographic work in the Sky Atlas
Laboratory.   C. Madsen.
Messenger, No. 25, p. 16 - 18 (1981).

036.006   Method of analytical approximation of the
characteristic curve of photomaterial.
Yu. S. Muzalevskij.
Problems of cosmic physics. Vyp. 16, (see 003.005), p. 122 -
127 (1981). In Russian.

036.007   The Orion nebulae in color.   D. F. Malin.
Sky Telesc., Vol. 62, 414 - 417 (1981).

036.008   Ermittlung der Belichtungszeit für Aufnahmen von
Sonne, Mond und Planeten. Teil I, II, III.
K. Schiefer, U. Schiefer.
Sterne Weltraum, Jahrg. 20, 378 - 380, 427 - 432, 474 - 479
(1981).

036.009   Hypersensitization of Kodak infrared plates: results
of some chemical methods and pre-exposure.
F. Bònoli, F. Fusi Pecci (Fusi-Pecci), E. Milandri.
Astrophys. Space Sci., Vol. 79, 191 - 201 (1981).
Tests have been carried out on Kodak Spectroscopic
Plates Type I-N, in order to find a treatment which could be
easily applied to the infrared plates to be used with the
Loiano 152 cm, f/8, Ritchey-Chrétien telescope. Different
techniques have been tested.

036.010   Die fotografische Sonnenbeobachtung. 4. Teil:
Beobachtungsprogramme und Auswertungsmöglich-
keiten.   E. Remmert.
Sonne, Jahrg. 5, 152 - 157 (1981).

036.011   Airglow photography with hypersensitized infrared
film.   E. B. Armstrong.
AAS Photo-Bull., No. 25, p. 3 - 4 (1980).

036.012   Ein einfacher Preflasher für Photoplatten.
J. D. Schumann.
Mitt. Astron. Ges., Nr. 54, (see 012.050), p. 234 - 235 (1981).

036.013   Results of an investigation of B-type plates.
N. M. Bronnikova.
Astron. Tsirk., No. 1127, p. 7 - 8 (1980). In Russian.

036.014   New experimental astronomical plates.
I. I. Brejdo, O. M. Mikhajlova, M. R. Shpol'skij,
B. A. Blekhman, G. I. Chistova.
Astron. Tsirk., No. 1136, p. 4 - 5 (1980). In Russian.

036.015   Photography of weak extended sources.
A. Tomić.
Vasiona, Année 29, 50 - 51 (1981). In Croatian.

036.016   Un grafico per la fotografia degli astri.   W. Ferreri.
Orione, Vol. 2, 272 - 274 (1981).

036.017 **Making films more efficient for astronomical photography.** C. R. Martys.
J. British Astron. Assoc., Vol. 92, 7 - 10 (1981).
This paper examines the various methods by which photographic emulsions may be made to record faint deep-sky objects more efficiently. The effects of lowering emulsion temperature during exposure, and gas and liquid treatment prior to exposure, are discussed and examples are given from the author's own work.

036.018 **Astro-films hypersensibilisés.** W. Maeder.
Orion, 39. Jahrg., 191 - 192 (1981).

036.019 **Photographic image intensification and reduction: a unified optical approach.** D . F. Malin.
Anglo-Australian Obs. Prepr. No. 155, 16 pp. (1981).
A simple, non destructive method of image intensification and reduction is described which can be applied to continuous-tone negative materials. The method employs a commercially available direct copy (reversal) film whose contrast can be continuously varied by pre-exposure from 0.8 to ~10.0. This film is then used with a diffuse-light contact printer to extract images from negatives which have been severely under- or over-exposed. Examples of images extracted from Plus-X negatives which have received from 8 stops over- to 7 stops under-exposure are shown.

036.020 **The standard wedge sensitometer for large photographic plates.**
T. Soyano, Y. Simizu, T. Noguchi.
Tokyo Astron. Obs. Rep. (No. 74), Vol. 19, 441 - 451 (1981). In Japanese.

036.021 **Lowering of the characteristic curve's gradient of astro-plates by FAH-treatment.**
W. Högner, R. Ziener.
Astron. Nachr., Band 302, 287 - 290 (1981). In German.
Photographic astro-plates were FAH (*Feinkorn-Ausgleich-Hypersensibilisierung*)-attended, in order to prevent development effects by a large gradient of the characteristic curve. The influence of the FAH attendance on the photometric error was researched too.

036.022 **Comparison of speed and noise of astronomical photographic emulsions Kodak 103aO, IIaO and ORWO ZU-21.** M. F. Shabanov.
Astron Tsirk., No. 1148, p. 6 - 8 (1981). In Russian.

Landolt-Börnstein. See Abstr. 003.172.

An investigation of autoradiographically intensified electrographic imagery. See Abstr. 031.590.

Measurement and reduction of photographic plates. See Abstr. 031.593.

Spectrographie stellaire I. Optimisation du temps de pose. See Abstr. 031.640.

A tube sensitometer for testing photographic hypersensitization techniques. See Abstr. 034.019.

Elektronografiske kameraer i drift på det danske 1,5 m teleskop. See Abstr. 034.030.

Positional Astronomy, Celestial Mechanics

041 Astrometry

041.001 **Determination of absolute right ascensions of stars from observations made at the equator.**
V. I. Kiyaev.
Tr. Astron. Obs., Leningrad, Tom 36 = Uch. Zap. Leningr. Univ., No. 402 = Ser. mat. nauk, Vyp. 58, 135 - 143 (1981). In Russian.

It is shown that observations of pairs of stars symmetric to the zenith made at the equator permit to determine absolute right ascensions of stars of the northern and southern hemispheres. The principle of the compilation of an observational program is discussed.

041.002 **Derivation of positions and parallaxes from simulated observations with a scanning astrometry satellite.** P. Høyer, K. Poder, L. Lindegren, E. Høg.
Astron. Astrophys., Vol. 101, 228 - 237 (1981).

The construction of a consistent set of positions and parallaxes from angular measurements obtained with a scanning astrometry satellite like Hipparcos has been studied by means of numerical experiments simulating the observation of up to 450 stars distributed on the whole sky. Extrapolation to full scale (10^5 stars) confirms the expected overall accuracy of the Hipparcos mission.

041.003 **A practical introduction to astrometry.**
B. M. Haisch.
J. Astronaut. Sci., Vol. 28, 205 - 230 (1980). — Abstr. in Phys. Abstr., Vol. 84, Abstr. 71058 (1981).

041.004 **Astrometric observations of planets, minor planets, and satellites: 1976—1980.** A. R. Klemola.
Astron. J., Vol. 86, 1108 - 1109 (1981) = Lick Obs. Bull., No. 885.

Equatorial coordinates for Uranus, Neptune, selected minor planets, and the Jovian satellites have been measured on direct photographs taken with the 0.51-m Carnegie double astrograph of the Lick Observatory during the period 1976—1980.

041.005 **Two differing definitions of the dynamical equinox and the mean obliquity.** E. M. Standish, Jr.
Astron. Astrophys., Vol. 101, L17 - L18 (1981).

When one computes (either implicitly or explicitly) the location of the dynamical equinox or the value of the mean obliquity in a rotating reference frame, one gets different results from those computed in an inertial frame. The differences must be accounted for when intercomparing results.

041.006 **Remark on the computation of position angle and distance from standard coordinates.**
H. Eichhorn.
Astron. Astrophys., Vol. 102, 35 (1981) = Contrib. Dep. Astron. Univ. Florida No. 32.

Rigorous formulas are given for computing position angles and distances from the standard coordinates of the objects concerned.

041.007 **Questions of organization of position observations of major planets with means of astrometry.**
A. S. Kharin.
Determination of the coordinates of celestial bodies, Riga, 1981, (see 003.008), p. 113 - 121. In Russian. — Abstr. in Ref. zh., 51. Astron., 9.51.103 (1981).

041.008 **Influence of changes of element orientation of a transit instrument in determination of right ascensions of stars with the pair method.** V. I. Kiyaev.
Vestn. LGU, 1981, No. 7, p. 104 - 106. In Russian. — Abstr. in Ref. zh., 51. Astron., 9.51.106 (1981).

041.009 **The N30 catalogue $\Delta\mu_\alpha$ proper motion corrections for 2305 stars.** E. V. Vityazeva.
Redkol. zh. Vestn. LGU. Mat., mekh., astron. Leningrad, 1980. 43 pp. In Russian. — Abstr. in Ref. zh., 51. Astron., 9.51.107 (1981).

041.010 **On the N30 catalogue $\Delta\mu_\alpha$ proper motion corrections. Comparison of the improved system of proper motions of the N30 catalogue with the FK4 and GC systems.** E. V. Vityazeva.
Redkol. zh. Vestn. LGU. Mat., mekh. astron. Leningrad, 1981. 18 pp. In Russian. — Abstr. in Ref. zh., 51. Astron., 9.51.108 (1981).

041.011 **Observations du soleil en 1979 à l'astrolabe du CERGA.** F. Laclare, M. Glentzlin.
Astron. Astrophys.,Suppl. Ser., Vol. 46, 1 - 2 (1981).

Results of observational campaign of solar positions are reported. Important improvements in instrumentation have been made in order to study the zenithal distance variation linked to variation of equilateral prism angle and also in order to extend observations at larger zenithal distances ($45°$ and $60°$) by using reflector prisms at steady angle. Therefore $\Delta\alpha$ of the theory used can be determined from an orbital arc of 9 months. Other works are in progress to determine orbital elements of the Earth.

041.012 **Construction of reference networks on planets and their satellites.** Yu. S. Tyuflin.
Geod. i kartogr., 1981, No. 4, p. 26 - 30. In Russian. — Abstr. in Ref. zh., 62. Issled. kosm. prostranstva, 9.62.399 (1981).

041.013 **On a relation between radio interferometric and equatorial reference systems.** E. P. Fedorov.
Pis'ma Astron. Zh., Tom 7, 632 - 635 (1981). In Russian. English translation in Soviet Astron. Lett., Vol. 7.

Observations with the meridian instruments being capable to give positions of celestial bodies only in the instantaneous equatorial system, reduction of these positions to a basic non-rotating reference system defined by a catalogue of adopted coordinates of extragalactic radio sources is considered.

041.014 **Reseau astrometry with Palomar Schmidt plates: position-coincidence optical identification of radio sources.** K. J. Mitchell, J. J. Condon, A. Warnock III, P. D. Usher.
Publ. Astron. Soc. Pacific, Vol. 93, 647 - 654 (1981).

The simple, inexpensive method for determining celestial

coordinates on original Palomar Schmidt plates developed previously is here inverted and shown to be a useful way of finding and identifying objects with known celestial coordinates. The method is especially suited for rapidly identifying large numbers of radio sources having accurate radio positions which are found in surveys made with large aperture-synthesis instruments. The average position error ($\simeq$ 0.9 arcsec) for stellar objects, galaxies, and faint objects near the plate limit is sufficient for making reliable optical identifications of objects as faint as m_J = + 24.

041.015 **Reseau astrometry with Palomar Schmidt plates: a two-way analysis of variance.**
A. Warnock III, D. Mattson, P. D. Usher.
Publ. Astron. Soc. Pacific, Vol. 93, 655 - 657 (1981).
Astrometric accuracy of about one arc second or better can be achieved over the entire area of 14-inch square Palomar Schmidt telescope plates with the use of a precision grid. A formal two-way analysis of variance has been performed to test for differences between positions generated by the use of (a) the coordinate transformation formulae of Dixon and of Smart, respectively, and (b) positional standards selected from the AGK3 and SAO star catalogs.

041.016 **Investigation of systematic differences between the new Washington catalogs W5-50 and WL50, the Perth 70 and the AGK 3 R in their common zone of overlap, declinations −5° to +5°.** T. Corbin, C. de Vegt.
Astron. Astrophys., Vol. 104, 88 - 92 (1981).
A detailed comparison of three of the main contributors to the SRS program, the Perth 70, W5-50 and WL50 is made with the AGK 3 R in the zone of common overlap, −5° to +5° declination.

041.017 **The magnitude equations between the fundamental coordinate systems N30, FK3, FK4.**
H. Schwan.
Astron. Astrophys., Vol. 104, 155 - 158 (1981).
The magnitude equations between the fundamental systems N30, FK3, FK4 at epoch and equinox 1950.0 have been determined by means of an analytical method. The results are presented by a set of significant functions from which tables have been derived giving a mean magnitude equation for regions of the sky where the magnitude equation changes numerically only a little. The systematic differences depending exclusively on α and δ have also been derived.

041.018 **On the catalogue of absolute right ascensions of 1960 bright and faint fundamental stars of the southern sky (SPu 71).**
M. P. Varin, V. A. Varina, V. S. Gubanov, L. I. Medvedeva, A. A. Nemiro, D. D. Polozhentsev, T. A. Polozhentseva, G. M. Timashkova.
Astrometric investigations, (see 012.048), p. 7 - 11 (1981). In Russian. − Abstr. in Ref. zh., 51. Astron., 11.51.138 (1981).

041.019 **On the catalogue of right ascensions of circumpolar stars.** A. P. Gulyaev.
Astrometric investigations, (see 012.048), p. 11 - 14 (1981). In Russian. − Abstr. in Ref. zh., 51. Astron., 11.51.139 (1981).

041.020 **Three-year series of observations of absolute right ascensions of stars on Spitsbergen island.**
G. M. Petrov.
Astrometric investigations, (see 012.048), p. 14 - 20 (1981). In Russian. − Abstr. in Ref. zh., 51. Astron., 11.51.142 (1981).

041.021 **Right ascensions of 586 FKSZ stars observed at the Nikolaev Observatory in 1974 - 1976.**
E. V. Khrutskaya.
Astrometric investigations, (see 012.048), p. 20 - 31 (1981). In Russian. − Abstr. in Ref. zh., 51. Astron., 11.51.143 (1981).

041.022 **Improvement of declinations and proper motions of the latitude programme stars compiled at the Engelhardt Astronomical and the Poltava Gravimetrical Observatories.** I. A. Urasina.
Astrometric investigations, (see 012.048), p. 31 - 41 (1981). In Russian. − Abstr. in Ref. zh., 51. Astron., 11.51.144 (1981).

041.023 **Results of using minor planet observations for fundamental astrometry.**
D. P. Duma, L. N. Kizyun.
Astrometric investigations, (see 012.048), p. 41 - 47 (1981). In Russian. − Abstr. in Ref. zh., 51. Astron., 11.51.145 (1981).

041.024 **Improvement of the FK4 zero points from photographic observations of asteroids.**
V. I. Orel'skaya.
Astrometric investigations, (see 012.048), p. 48 - 53 (1981). In Russian. − Abstr. in Ref. zh., 51. Astron., 11.51.146 (1981).

041.025 **On a possibility of improving the fundamental system of stellar coordinates from radio interferometric and photographic quasar observations.**
V. S. Gubanov, I. I. Kumkova.
Astrometric investigations, (see 012.048), p. 59 - 66 (1981). In Russian. − Abstr. in Ref. zh., 51. Astron., 11.51.148 (1981).

041.026 **On the choice of cosmic radio sources for constructing an inertial coordinate system by means of long base-line radiointerferometry.** V. A. Alekseev.
Astrometric investigations, (see 012.048), p. 71 - 75 (1981). In Russian. − Abstr. in Ref. zh., 51. Astron., 11.51.149 (1981).

041.027 **On an optimum programme of very long base-line interferometry observations for constructing an inertial coordinate system.** V. A. Alekseev.
Astrometric investigations, (see 012.048), p. 75 - 78 (1981). In Russian. − Abstr. in Ref. zh., 51. Astron., 11.51.150 (1981).

041.028 **On meridian observations according to the international programme (double stars).**
M. S. Zverev.
Astrometric investigations, (see 012.048), p. 79 - 82 (1981). In Russian. − Abstr. in Ref. zh., 51. Astron., 11.51.151 (1981).

041.029 **Determination of coordinates of stars using precise data on the earth's rotation (VLBI).**
S. A. Tolchel'nikova-Murri, E. I. Krejnin.
Astrometric investigations, (see 012.048), p. 82 - 87 (1981). In Russian. − Abstr. in Ref. zh., 51. Astron., 11.51.152 (1981).

041.030 **On the accuracy of differential coordinate measurements with RATAN-600.**
V. A. Fomin, P. M. Afanas'eva.
Astrometric investigations, (see 012.048), p. 66 - 71 (1981). In Russian. − Abstr. in Ref. zh., 51. Astron., 11.51.153 (1981).

041.031 **On the observation of the zodiacal star catalogue with the Nikolaev Observatory's zonal astrograph.**
F. F. Kalikhevich.
Astrometric investigations, (see 012.048), p. 87 - 90 (1981). In Russian. − Abstr. in Ref. zh., 51. Astron., 11.51.154 (1981).

041.032 **Studies of the declination system of the Pulkovo FVK according to results of observations of stars of the experimental programme.**
B. K. Bagil'dinskij, E. G. Zhilinskij, V. D. Shkutov.
Astrometric investigations, (see 012.048), p. 93 - 99 (1981). In Russian. − Abstr. in Ref. zh., 51. Astron., 11.51.156 (1981).

041.033 **On an investigation of magnitude equations by a statistical method.** G. G. Borzov,

N. R. Kasimov, O. I. Stal'bovskij, Yu. A. Shokin.
Astron. Tsirk., No. 1132, p. 1 - 4 (1980). In Russian.

A comparison of the AGK 3 with the SAOC − Part I.
See Abstr. 002.009.

A comparison of the AGK3 with the SAOC − Part II.
See Abstr. 002.010.

An improvement of the Tokyo PZT star catalogue.
See Abstr. 002.079.

Right ascension and proper motion in R. A. of
904 bright stars.   See Abstr. 002.080.

The photoelectric astrolabe catalogue of Shanghai
Observatory (1975.9−1978) − corrections to individual
FK4 positions.   See Abstr. 002.082.

The third astrolabe catalogue of Shanghai
Observatory (1973.0−1978.0).   See Abstr. 002.083.

Determination of the positions of major planets
with the photographic method.   See Abstr. 003.119.

Stellar paths. Photographic astrometry with long-
focus instruments.   See Abstr. 003.154.

The development of science at the Astronomical
Observatory of the Leningrad University. Astrometry.
See Abstr. 009.016.

Comparison of two methods for calculation of differences between star positions.   See Abstr. 021.050.

Numerical simulation of the signal and data processing of the HIPPARCOS satellite.
See Abstr. 021.058.

Observations astrométriques à Paris: gageure ou
réalité?   See Abstr. 031.508.

Whether the equal altitude method will make new
contributions in modern astrometry?   See Abstr. 031.517.

Réduction des catalogues photographiques: zones
+31° à −2° (R. C. P. 164).   See Abstr. 031.569.

Analysis of some methods of TV astronomy in the
problem of determination of the equatorial coordinates of
astronomical objects.   See Abstr. 031.621.

Analyse de l'equation de magnitude sur l'astrolabe
de Danjon à l'Observatoire de Shanghai.
See Abstr. 031.644.

On a mode of direct determination of the geodetic
azimuth from observations of faint stars in the first vertical.
See Abstr. 046.006.

The astrometric satellite HIPPARCOS.
See Abstr. 051.006.

Hipparcos, the European Space Agency's astrometric
satellite.   See Abstr. 051.023.

On the measurement of proper motions of star
clusters with the European astrometry satellite HIPPARCOS.
See Abstr. 051.042.

Hipparcos. Space astrometry mission. Answer to the
announcement of opportunity for the compilation of the
input catalogue.   See Abstr. 051.059.

Hipparcos. Proposal for scientific data processing.
See Abstr. 051.060.

The temperature characteristic for the zenith
distance correction of the Danjon astrolabe − discussion of a
method for studying the characteristic curve.
See Abstr. 082.116.

Astrolabe observations of Mars.
See Abstr. 097.036.

Photographic position observations of Mars and
Deimos at the Main Astronomical Observatory of the
Ukrainian Academy of Sciences in 1975 - 1976.
See Abstr. 097.077.

Positions of selected minor planets (1979 - 80).
See Abstr. 098.003.

Positions of Jupiter and Galilean satellites in 1978.
See Abstr. 099.195.

Observations of planets with the Danjon astrolabe
of the Shanghai Observatory during 1978.12−1979.4.
See Abstr. 099.212.

Cometary ephemerides − needs and concerns.
See Abstr. 102.064.

Corrections of the components of solar motion, of
the galactic rotation and variations of corrections for the
precessional constant in improved systems of proper motions
of the GC and N30 star catalogues.
See Abstr. 111.020.

An improved optical position of 3C 273B in the
FK4-system.   See Abstr. 141.038.

Optical positions of benchmark radio sources south
of +5° declination.   See Abstr. 141.121.

Optical position and "proper motion" of the radio
source OQ 208.   See Abstr. 141.126.

Erratum

041.901   Erratum: "On the simultaneous determination of
          differential right ascensions and declinations of
radio sources with RATAN-600" [Astron. Zh., Tom 57, 649 -
652 (1980)].   V. N. L'vov.
Astron. Zh., Tom 58, 1133 (1981). In Russian. − See Abstr.
27.041.027.

042 Celestial Mechanics, Figures of Celestial Bodies

042.001 Relativistic perturbations of planetary orbits in the generalized three-parametric Schwarzschild metric. The case of Mercury. J.-F. Lestrade.
Astron. Astrophys., Vol. 100, 143 - 155 (1981). In French.
In this paper, analytical formulas for relativistic perturbations of planetary orbits in the generalized three-parametric Schwarzschild metric are given and the orbital motion of Mercury is compared with the numerical integration of Oesterwinter and Cohen as well as with DE 102.

042.002 Apsidal precession of orbits about an oblate planet. R. Greenberg.
Astron. J., Vol. 86, 912 - 914 (1981).
Expressions for the apsidal precession rates to second order in J_2 appear in the literature in at least three apparently mutually contradictory forms. These expressions are reconciled by accounting for subtle differences in the definitions of orbital elements.

042.003 On close triple approaches in the three-body problem. Zh. P. Anosova, N. N. Zavalov.
Tr. Astron. Obs., Leningrad, Tom 36 = Uch. Zap. Leningr. Univ., No. 402 = Ser. mat. nauk, Vyp. 58, 109 - 123 (1981). In Russian.
Disruption of triple systems with negative total energy always takes place after close triple approaches. Preliminary classification of close triple approaches leading and not leading to disruption of triple systems has been made. An atlas of 150 close triple approaches is compiled.

042.004 A new class of stationary and conditionally periodic solutions in the three-dimensional restricted elliptic three-body problem. S. G. Zhuravlev, I. A. Gerasimov.
Astron. Zh., Tom 58, 879 - 885 (1981). In Russian. English translation in Soviet Astron., Vol. 25, No. 4.
The motion of a mass point is considered on the basis of the resonant three-dimensional restricted elliptic three-body problem. For commensurabilities 2/1, 5/2 and 4/1 between the mean motions of the mass point and the small primary the families of stationary and conditionally periodic solutions are found.

042.005 On the equations of motion of binary systems with variable mass. V. I. Kuryshev, N. I. Perov.
Astron. Zh., Tom 58, 886 - 887 (1981). In Russian. English translation in Soviet Astron., Vol. 25, No. 4.
An equation of the relative motion of a binary system with variable mass is obtained. It is shown that the equations of Gylden – Mestchersky, Mestchersky – Levi-Cività, the combined equation of Gylden – Mestchersky and Mestchersky – Levi-Cività can be derived from it. These equations can be used only for small intervals of time. It is also shown that these equations cannot be used in the problems of stellar evolution with an arbitrary law of mass loss.

042.006 Expansion of the perturbing function of a satellite restricted four-body problem.
S. G. Zhuravlev, V. V. Anikovsky.
Celestial Mech., Vol. 24, 237 - 254 (1981).
A satellite four-body problem is the problem of motion of an artificial satellite of a planet in a region of the space where perturbations due to the gravitational field of the planet are of the same order as perturbations due to influences of two perturbing bodies. In this paper an expansion of the perturbing function into a Fourier series in terms of angular Keplerian elements $(\Omega_j, \omega_j, M_j; j = 0, 1, 2)$ is obtained taking into account a sharp commensurability.

042.007 On periodic flybys of the Moon.
A. D. Bruno (*Bryuno*).
Celestial Mech., Vol. 24, 255 - 268 (1981).
This paper considers the plane circular restricted three-body problem for small μ. Symmetric periodic solutions of the second species (passing near the body of mass μ) and their distance from the center of the body of mass μ are studied by constructing perturbations of arc-solutions (solutions with consecutive collisions) existing for $\mu = 0$. Orbits which also pass near the body of mass $1-\mu$ are studied in detail. The results are applied to finding periodic orbits in the Earth-Moon system and in the Sun-Jupiter system.

042.008 Time elements in rectangular coordinates.
J. H. Kwok, P. Nacozy.
Celestial Mech., Vol. 24, 269 - 287 (1981).
In a paper by the second author (Nacozy, 1981), various time elements are presented for use with the Sundman time transformation. In that paper, the time elements are given in terms of Keplerian orbital elements. The authors give here the corresponding time elements in terms of rectangular coordinates. They present additional numerical experiments comparing the use of time elements and time transformations together with the use of time transformations alone. The results indicate a reduction in computational error when time elements are used.

042.009 An asymptotic solution for the stellar case of the non-planar three-body problem.
J. Degraeve, M. Pascal.
Celestial Mech., Vol. 24, 289 - 327 (1981).
A simplified model of the non-planar three-body problem is considered in which two particles, forming a close binary, orbit a distant point. A small parameter ϵ, related to the distance separating the binary and the remaining mass, is defined. The time is eliminated from the equations of motion and an angular variable is used instead. A three-variable expansion procedure is used to find an asymptotic solution of the problem. It is possible to obtain a solution up to the order six in ϵ without secular terms only if the mutual inclination i_0 of the unperturbed orbits is less than a critical inclination i_1 ($i_1 \simeq 39°$).

042.010 A note on a conjecture of Poincaré.
G. Gómez, J. Llibre.
Celestial Mech., Vol. 24, 335 - 343 (1981).
The authors prove the following weakened version of Poincaré's conjecture on the density of periodic orbits of the restricted three-body problem: The measure of Lebesgue of the set bounded orbits which are not contained in the closure of the set of periodic orbits approaches zero as the mass parameter approaches zero.

042.011 Properties of the moment of inertia in the problem of three bodies. K. Zare.
Celestial Mech., Vol. 24, 345 - 354 (1981).
Sundman's and Birkhoff's results are combined with a recently developed inequality and new qualitative results are given for the problem of three bodies.

042.012 The 4:1 resonance. G. Contopoulos.
Celestial Mech., Vol. 24, 355 - 366 (1981).
The author studies the resonance $\omega_1 : \omega_2 = 4 : 1$ and some near-resonance cases. The main peculiarity of this resonance is that for $\omega_1 : \omega_2 < 4$ the characteristic of the central periodic orbits is broken into two and each part is joined with a resonance characteristic. This behaviour is described theoretically by means of the 'third' integral. It seems that there are

infinite families of simple periodic orbits near the escape region. Finally, a comparison is made with the cases near the $\omega_1 : \omega_2 = 2 : 1$ resonance.

042.013 Reformulation of the Brouwer geopotential theory for improved computational efficiency.
F. R. Hoots.
Celestial Mech., Vol. 24, 367 - 375 (1981).
The theory, as derived by Brouwer and later modified by Lyddane, of the motion of an artificial Earth satellite, perturbed by the first five zonal harmonics is reformulated in terms of an alternate set of variables. This alternate set of variables produces an equivalent solution, has no small eccentricity or small inclination restrictions, and allows calculation of position and velocity with considerably fewer algebraic and trigonometric operations. In addition, the alternate set of variables avoids one solution of Kepler's equation.

042.014 Orbit calculations nearby the equilibrium points by a discrete mechanics method.
J. Albrycht, A. Marciniak.
Celestial Mech., Vol. 24, 391 - 405 (1981).
Basic formulas of discrete mechanics in a rotating frame of references are obtained. Their application to several computer examples of orbits nearby the equilibrium points of the Earth-Moon system is presented, and — for computational purposes — an algorithm is given.

042.015 Gylden—Meščerskii problem.
L. M. Berkovič (*Berkovich*).
Celestial Mech., Vol. 24, 407 - 429 (1981).
Classical non-stationary two-body problem, described by the equation of the form $\ddot{r} = -\mu(t)\frac{\mathbf{r}}{r^3}$ is investigated using differential equation transformation methods developed by the author. All laws of mass variation for which Gylden—Meščerskii problem is reduced to autonomous form are stated. The problem symmetry properties are investigated and reviews of integrable cases from point of view of the theory of groups are made.

042.016 On periodic solutions adjacent to the Lagrangian ones in the problem of translational-rotational motion of three solid bodies.   N. I. Gamarnik.
Pis'ma Astron. Zh., Tom 7, 442 - 448 (1981). In Russian. English translation in Soviet Astron. Lett., Vol. 7.
The existence of periodic solutions in the problem of the translational-rotational motion of three solid bodies is proved.

042.017 On a "correction" to the tidal force.
G. F. Sitnik, A. I. Khlystov.
Pis'ma Astron. Zh., Tom 7, 510 - 512 (1981). In Russian. English translation in Soviet Astron. Lett., Vol. 7.
It is shown that the resultant tidal force acting on some body from other two (or more) bodies is equal to the vectorial sum of tidal forces from each of that bodies, in contradiction with Avsyuk (1976, 1977, 1980) who proposed a correction to this sum.

042.018 The dependence of the Kolmogorov entropy of mappings on the system of coordinates.
Y.-s. Sun, C. Froeschlé.
Acta Astron. Sinica, Vol. 22, 159 - 168 (1981). In Chinese.

042.019 On Kepler's equation and strange attractors.
R. Broucke.
J. Astronaut. Sci., Vol. 28, 255 - 265 (1980). — Abstr. in Phys. Abstr., Vol. 84, Abstr. 71061 (1981).

042.020 Combination resonance in the three-body problem.
E. Mettler.
Z. Angew. Math. Mech., Vol. 61, 1 - 6 (1981). In German. Abstr. in Phys. Abstr., Vol. 84, Abstr. 71064 (1981).

042.021 Celestial mechanics — the potential torsion.
A. I. Klat, C. Marchal.
Acta Astronaut., Vol. 8, 101 - 104 (1981). In French. — Abstr. in Phys. Abstr., Vol. 84, Abstr. 75004 (1981).

042.022 Expansion of the inverse of mutual distance between two bodies raised to any power.
O. M. Kamel, A. A. Bakry.
Astrophys. Space Sci., Vol. 78, 3 - 26 (1981).
The authors calculate the expression for Δ^{-s} in terms of true anomalies and classical orbital elements, referring to a common fixed plane and working up to power four of eccentricities and tangents of inclinations.

042.023 Numerical investigation of symmetric and asymmetric periodic oscillations.   P. G. Kazantzis.
Astrophys. Space Sci., Vol. 78, 27 - 43 (1981).
Global information for the periodic solutions — symmetric and asymmetric — of the "gravitational" spring-pendulum problem is given for the first time. For two different sets of the parameters of this problem, the families of symmetric periodic solutions which emanate from the equilibrium point have been determined. Further families of asymmetric and symmetric solutions which bifurcate from them have also been examined and interesting results for their behaviour have been pointed out.

042.024 Evolution of rigid body oscillations relative to the mass centre near a libration point.
A. P. Markeev.
Pis'ma Astron. Zh., Tom 7, 572 - 576 (1981). In Russian. English translation in Soviet Astron. Lett., Vol. 7.
Using the theory of adiabatic invariants the evolution of rigid body oscillations is investigated. The orbit of the mass centre of a rigid body near a libration point is supposed to be a long-period one.

042.025 Semi-simple 1- 1 resonances.   A. Deprit.
Bull. American Astron. Soc., Vol. 13, 570 - 571 (1981). — Abstract.

042.026 An intermediate orbit for planetary theory.
D. L. Richardson.
Bull. American Astron. Soc., Vol. 13, 571 (1981). — Abstract.

042.027 Diffusion of ring particle orbits by a monolayer of gravitational scatterers.   G. R. Stewart.
Bull. American Astron. Soc., Vol. 13, 571 - 572 (1981). Abstract.

042.028 Expansion theory for the elliptic motion of arbitrary eccentricity and semi-major axis. II. Analyses of the
function $G^{(N, n)}(\theta, x_n, \xi_n) = \prod_{i=1}^{n}(1 - \xi_i \sin^{2N-1}\theta)^{x_i}$.
M. A. Sharaf.
Astrophys. Space Sci., Vol. 78, 359 - 400 (1981).
In this paper literal analytical expressions for the coefficients of the Fourier series representation of G are established. Moreover, the recurrence formulae satisfied by these coefficients are also established. Illustrative analytical examples and a full recursive computational algorithm, with its numerical results, are included. The applications of the recurrence formulae are also illustrated by their stencils. As by-products of the analyses are two important periodic integrals developed analytically and computationally.

042.029 Force function between two celestial bodies.
V. G. Shkodrov, T. G. Gechev.
Dokl. Bolg. AN, Vol. 33, 1025 - 1028 (1980). — Abstr. in Ref., zh., 51. Astron., 7.51.75 (1981).

042.030 **The region of resonant orbits in the 1 : 3 Kirkwood gap in the coordinate space a_0, e_0, M_0.**
V. G. Kruchinenko, L. M. Sherbaum, A. N. Simonenko.
Problems of cosmic physics. Vyp. 16, (see 003.005), p. 25 - 30 (1981). In Russian.

Model calculations are carried out for perturbed motion of material points in the vicinity of the commensurability 1/3 with Jupiter with reference values of the major semi-axes, eccentricities and the positions of points of the orbits varying considerably. From the results obtained the region of resonance is found and the dependence of the dimensions of orbit resonance section upon the semi-axes, eccentricities and initial position of the points is determined.

042.031 **Necessary and sufficient conditions of the existence of Lagrangian and Eulerian solutions of the general problem of three solid bodies.**
V. T. Kondurar', L. S. Troitskaya.
Mekh. tverd. tela, Kiev, 1981, No. 13, p. 3, 9. In Russian.
Abstr. in Ref. zh., 51. Astron., 8.51.107 (1981).

042.032 **Motion of a relativistic point of variable mass in a central force field.** B. A. Gordienko.
Nauchn. chteniya po aviats. i kosmonavt. 10-e Gagarinsk. chteniya, 1980. Moskva, 1981, p. 203. In Russian. – Abstr. in Ref. zh., 62. Issled. kosm. prostranstva, 8.62.135 (1981).

042.033 **Construction of a theory of the outer planets through an iterative method.** P. Bretagnon.
Astron. Astrophys., Vol. 101, 342 - 349 (1981). In French.

The author has undertaken the construction of a theory for the outer planets Jupiter, Saturn, Uranus, and Neptune through an iterative method. This method makes it possible to reach a high order with respect to the masses. The solutions are analytical functions of the mean longitudes of the eight planets and are numerical with respect to other integration constants. The precision is shown by comparisons to numerical integrations over 1000 yr.

042.034 **A simple method of orbit determination.**
W. Neutsch.
Astron. Astrophys., Vol. 102, 59 - 64 (1981).

In the following a simple and straightforward method of orbit determination is discussed. It may be used for all types of Kepler orbits and for any given number $N \geqslant 3$ of observations all of which are treated symmetrically. Furthermore the use of position and velocity at a fixed epoch as orbital elements makes it unnecessary to discuss some special cases, as e.g. small inclination or small eccentricity, separately. Hence the formulae of the text are easily transformed into a computer code.

042.035 **Bifurcations of planar to three-dimensional periodic orbits in the general three-body problem.**
V. V. Markellos.
Celestial Mech., Vol. 25, 3 - 31 (1981).

The author studies the generation of three-dimensional periodic orbits of the general three-body problem from special generating plane orbits, the vertical-critical orbits. The bifurcation process is examined analytically and geometrically. A method of obtaining numerically continuous sets of vertical-critical orbits is outlined, and applied for the determination of 16 monoparametric sets including all possible types of such orbits corresponding to all possible types of symmetry of the bifurcating three-dimensional orbits. The stability of all bifurcation orbits is assessed. Examples of three-dimensional periodic orbits generated from the bifurcation orbits are given.

042.036 **Theory of satellite orbit – orbit resonance.**
L. Blitzer, J. D. Anderson.
Celestial Mech., Vol. 25, 65 - 78 (1981).

On the basis of the strong mathematical and physical parallels between orbit – orbit and spin – orbit resonances, the dynamics of mutual orbit perturbations between two satellites about a massive planet are examined. The theory is applicable to Saturn's resonant pairs Titan–Hyperion and Mimas-Tethys, and in these cases the calculated libration periods are in reasonably good agreement with the observed periods.

042.037 **Saari's conjecture revisited.** J. I. Palmore.
Celestial Mech., Vol. 25, 79 - 80 (1981).

The author elaborates a variational method used recently in the proof of Saari's conjecture.

042.038 **Applications of Szebehely's equation.** S. Molnár.
Celestial Mech., Vol. 25, 81 - 88 (1981).

It is shown that the potential obtained from Joukovsky's formula, corresponding to a given family of orbits, is a general solution of Szebehely's equation. Then it is shown how a general solution of Szebehely's equation can be obtained from its particular solution. This method is applied to several examples. Potentials generating families of concentric elliptic orbits and families of orbits of conic sections are determined. Finally, the inverse Keplerian problem is solved using Szebehely's equation in polar coordinates.

042.039 **Comment on 'General relativity and satellite orbits: the motion of a test particle in the Schwarzschild Metric' by D. P. Rubincam.** E. M. Gaposchkin.
Celestial Mech., Vol. 25, 181 - 183 (1981).

The short periodic perturbations due to the Schwarzschild Metric can be ignored when analyzing earth satellite data. See also Abstr. 20.042.035.

042.040 **A new family of periodic orbits for the restricted problem.** E. A. Belbruno.
Celestial Mech., Vol. 25, 195 - 217 (1981).

A new family of periodic orbits of the three-dimensional restricted three-body problem which continue off from a consecutive collision orbit are numerically studied. Their behavior for varying energy is unexpected.

042.041 **Some new concepts in the plane n-body and 3-body problems.** A. Kyrala.
Mem. Soc. Astron. Italiana, Vol. 52, (see 012.023), 423 - 429 (1981).

042.042 **Dynamics of the asteroids.** H. Scholl.
Mem. Soc. Astron. Italiana, Vol. 52, (see 012.023), 515 - 522 (1981).

The large majority of the known asteroids is situated in a belt between the orbits of Mars and Jupiter. Obviously, the boundaries of the belt are due to these two planets. The dynamics of the belt asteroids is at present mainly determined by Jupiter. In the past, also collisions among asteroids played an important role which is indicated by the Hirayama families. Those asteroids which cross the orbits of a planet might have suffered or will suffer drastic changes in their orbits or will even collide with that planet unless particular protection mechanisms prevent such close approaches.

042.043 **Potential energy and moment of inertia of rapidly rotating bodies of arbitrary structure.**
M. Marie.
J. Astron. Soc. Egypt, Vol. 2, 11 - 21 (1980).

042.044 **Libration points in the restricted triangular circular problem of four bodies.**
G. V. Ufimtsev, S. M. Poleshchikov.
Astron. i geod., Tomsk, 1980, No. 8, p. 124 - 135. In Russian.
Abstr. in Ref. zh., 51. Astron., 9.51.61 (1981).

042.045 **An algorithm for numerical study of the motion of peculiar minor planets based on double regularization of the equations of motion.**   V. A. Shefer.
Astron. i geod., Tomsk, 1980, No. 8, p. 81 - 91. In Russian.
Abstr. in Ref. zh., 51. Astron., 9.51.62 (1981).

042.046 **Intermediate orbits for the initial part of motion.**   Yu. V. Batrakov.
Determination of the coordinates of celestial bodies, Riga, 1981, (see 003.008), p. 3 - 10. In Russian. − Abstr. in Ref. zh., 51. Astron., 9.51.65 (1981).

042.047 **Analytical theory of the sixth order motion of an outer satellite of a planet.**   T. S. Boronenko.
Astron. i geod., Tomsk, 1980, No. 8, p. 97 - 101. In Russian.
Abstr. in Ref. zh., 51. Astron., 9.51.77 (1981).

042.048 **The effect of Lorentz forces on the motion of a satellite relative to the mass center.**
G. V. Lyakhovka.
Vestn. LGU, 1981, No. 7, p. 77 - 82. In Russian. − Abstr. in Ref. zh., 51. Astron., 9.51.78 (1981).

042.049 **Kepler's third law.**
A. Tan, W. L. Chameides.
American J. Phys., Vol. 49, 691 - 692 (1981). − Abstr. in Phys. Abstr., Vol. 84, Abstr. 83826 (1981).

042.050 **Secular effects in the translatory-rotational motion of planets caused by their non-sphericity.**
Yu. V. Barkin.
Astron. Vestn., Tom 15, 105 - 112 (1981). In Russian.

042.051 **On the calculation of the first-order secular perturbations of orbital elements according to the Halphen-Goryachev method.**   Yu. V. Obrubov.
Astron. Vestn., Tom 15, 113 - 117 (1981). In Russian.
It is shown that the Halphen-Goryachev method is applicable in the cases of intersecting orbits, small inclinations and large aphelion distances. Some examples of the numerical calculations are given.

042.052 **On a four-body problem.**   A. Majorana.
Celestial Mech., Vol. 25, 267 - 270 (1981).
The author studies a particular four-body problem: three bodies revolve around their center of mass in circular orbits under the influence of their mutual gravitational attraction, while a fourth body moves in the plane defined by the three bodies but not influencing their motion. The linear stability of the eight equilibrium points is studied, and it is found that it depends on the values of the masses.

042.053 **The Lyapunov characteristic exponents as indicators of stochasticity in the restricted three-body problem.**
R. Gonczi, C. Froeschlé.
Celestial Mech., Vol. 25, 271 - 280 (1981).
The Lyapunov characteristic numbers are used to continue the study of a larger sample of orbits in the three-body three-dimensional restricted problem than that previously studied by Froeschlé. These numbers are found to be sensitive indicators of stochasticity and the results confirm those found by Froeschlé.

042.054 **On the problem of lateral modes of oscillations of equilibrium figures.**   A. S. Baranov.
Astron. Zh., Tom 58, 949 - 958 (1981). In Russian. English translation in Soviet Astron., Vol. 25, No. 5.
Canonical equations of motion of a liquid are derived for the case of conservation of the symmetry of the figure with respect to the equatorial plane. A preliminary analysis of the equations obtained has been made. Dynamical stability of the Maclaurin ellipsoid in nonlinear treatment has been investigated

by the variational method. The problem of orientation of the figure has been considered as well.

042.055 **Construction of conditionally periodic solutions of the restricted circular three-body problem in the three-dimensional case.**   S. G. Zhuravlev.
Astron. Zh., Tom 58, 1095 - 1100 (1981). In Russian. English translation in Soviet Astron., Vol. 25, No. 5.
The second stage of construction of conditionally periodic solutions of the problem is considered. A calculation of stationary solutions and construction of the first approximation are included in this stage.

042.056 **Semi-regular motions in the problem of the translatory-rotational motion of two axisymmetric bodies.**   V. V. Vidyakin.
Astron. Zh., Tom 58, 1115 - 1120 (1981). In Russian. English translation in Soviet Astron., Vol. 25, No. 5.
It is proved that, if in the problem considered one of Euler's angles (precession or nutation) describing the orientation of the body is constant, then the determination of the other angle can be reduced to the solution of a system of two differential equations of the first order.

042.057 **On the relationship between the total mass of a celestial body and the averaged mass of its constituent particles.**   V. I. Ferronsky, S. A. Denisik, S. V. Ferronsky.
Phys. Lett. A, Vol. 84A, 223 - 225 (1981). − Abstr. in Phys. Abstr., Vol. 84, Abstr. 98668 (1981).

042.058 **On oscillations of a satellite in the plane of an elliptical orbit.**   A. A. Zevin.
Kosm. Issled., Tom 19, 674 - 679 (1981). In Russian.

042.059 **On stationary solutions in the problem of translatory-rotational motion of an axisymmetric satellite of a triaxial planet.**   A. A. Zlenko.
Kosm. Issled., Tom 19, 688 - 694 (1981). In Russian.

042.060 **On the existence and structure of inhomogeneous analogs of the Dedekind and Jacobi ellipsoids.**
J. R. Ipser, R. A. Managan.
Astrophys. J., Vol. 250, 362 - 372 (1981).
A study of the existence and structure of nonaxisymmetric rotating configurations of inhomogeneous fluid in gravitational equilibrium is initiated. A vector-potential formalism is set up for describing configurations with constant density along velocity streamlines. The formalism yields necessary conditions for the existence of a Dedekind configuration, a nonaxisymmetric configuration stationary in the inertial frame, that is inhomogeneous and whose velocity lacks a meridional component and depends linearly on position.

042.061 **New periodic motions in the problem of two gravitating solid bodies one of which is a sphere.**
V. A. Proshkin.
MGU, Moskva. 1981. 17 pp. In Russian. − Abstr. in Ref. zh., 51. Astron., 10.51.131 (1981).

042.062 **Periodic motions of the axis of symmetry of a dynamically symmetric satellite under the action of a gravitational moment.**   V. V. Sazonov, M. E. Sidoryuk.
Inst. prikl. mat. AN SSSR. Prepr., 1981, No. 31, 29 pp. In Russian. − Abstr. in Ref. zh., 62. Issled. kosm. prostranstva, 10.62.146 (1981).

042.063 **Three-dimensional periodic motion of three finite masses around collinear equilibrium configurations.**
V. V. Markellos.
Celestial Mech., Vol. 25, 319 - 344 (1981).

Three-dimensional periodic motions of three bodies are shown to exist in the infinitesimal neighbourhood of their collinear equilibrium configurations. These configurations and some characteristic quantities of the emanating three-dimensional periodic orbits are given for many values of the two mass parameters, $\mu = m_2/(m_1 + m_2)$ and m_3, of the general three-body problem.

042.064 **On some applications of the problem of many fixed centres to geophysics.**   G. T. Arazov.
Celestial Mech., Vol. 25, 345 - 352 (1981).

In the present paper, using as examples the problems of four, five and six fixed centres, some applications of the problem of many fixed centres to geophysics are given.

042.065 **Sur les mouvements réguliers des satellites.**   G. N. Doubochine (*Duboshin*).
Celestial Mech., Vol. 25, 375 - 396 (1981).

Il est envisagé dans ce travail le problème du mouvement translatoire-rotatoire d'un corps solide invariable dans le champ centrale de la gravitation Newtonienne.

042.066 **A new regularization of the restricted three-body problem and an application.**   E. A. Belbruno.
Celestial Mech., Vol. 25, 397 - 415 (1981).

A new regularizing transformation for the three-dimensional restricted three-body problem is constructed. It is explicitly derived and is equivalent to a simple rational map. Geometrically it is equivalent to a rotation of the 3-sphere. Unlike the KS map it is dimension preserving and is valid in n dimensions. This regularizing map is applied to the restricted problem in order to prove the existence of a family of periodic orbits which continue from a family of collision orbits.

042.067 **On the construction of Roche harmonics.**   G. F. Roach.
Astrophys. Space Sci., Vol. 80, 237 - 247 (1981).

A method, involving boundary integral equations of the first kind is offered for obtaining exact representations for solutions to interior and exterior harmonic problems. The method is constructive and a representation can be obtained to any degree of accuracy. The determination of Roche harmonics appears as a special case.

042.068 **Horseshoe periodic orbits in the restricted problem of three bodies for a Sun-Jupiter mass ratio.**
D. B. Taylor.
Astron. Astrophys., Vol. 103, 288 - 294 (1981).

Segments of seven families of symmetric horseshoe periodic orbits of the restricted three body problem for a Sun-Jupiter mass ratio have been numerically determined. Each family is found to have a region consisting of smooth horseshoe shaped orbits with the family evolving to orbits acquiring loops on both sides of the smooth horseshoes. The general evolution of these families is discussed and one in particular (Rabe's horseshoe is a member of this family) is described in more detail with the aid of computer plots of orbits in this family.

042.069 **Another equilibrium sequence of a self-gravitating and rotating incompressible fluid.**
Y. Eriguchi, D. Sugimoto.
Prog. Theor. Phys., Vol. 65, 1870 - 1875 (1981). – Abstr. in Phys. Abstr., Vol. 84, Abstr. 98998 (1981).

042.070 **Quasi-equilibrium in collisional systems.**
K. A. Hämeen-Anttila.
Moon Planets, Vol. 25, 477 - 506 (1981).

Quasi-equilibrium states of collisional systems consist of two major types. In the first type the coefficient of restitution must depend on the velocity of impact to maintain the equilibrium. In the second type this is not necessary, but an appropriate perturbation is needed. Drag in a co-moving medium leads to this type. The accretion of matter in a protoplanet may trigger a jump from one stable region to another in the surrounding medium. If this happens, a dense zone is produced at the distance of 16/9 planetary radii. This phenomenon seems to explain the origin of planetary rings.

042.071 **On the evolution of the homogeneous ellipsoidal figures. II. Gravitational collapse and gravitational radiation.**   S. Detweiler, L. Lindblom.
Astrophys. J., Vol. 250, 739 - 749 (1981).

The authors use homogeneous ellipsoids to model the gravitational collapse of a stellar core and the subsequent emission of gravitational radiation. The growth of asymmetry caused by the collapse is computed along with the flux and energy spectrum of the gravitational radiation produced as the ellipsoid evolves toward an axisymmetric state following the collapse. In the very low angular momentum limit they find that a 1 $M_\odot$ core of radius $R=10^6$ cm and angular velocity $\Omega_p=10^3 \mathrm{s}^{-1}$ should emit gravitational radiation by this mechanism with a total energy of approximately $E_{\mathrm{GR}}=10^{-6} Mc^2$, at a frequency of 1600 Hz and a bandwidth of only about 1.6 Hz.

042.072 **Another numerical treatment of the evolution of circumplanetary disks.**   T. W. Lougheed.
Bull. American Astron. Soc., Vol. 13, 721 - 722 (1981). Abstract.

042.073 **Orbital stability constraints on the nature of planetary systems.**   F. Graziani, D. C. Black.
Astrophys. J., Vol. 251, 337 - 341 (1981).

A fully self-consistent, N-body computer code has been used to study conditions under which model planetary systems, each consisting of a star and two "planetary" companions, become orbitally unstable due to gravitational interactions between the companions.

042.074 **Numerical integration of the satellites of the outer planets.**   C. F. Peters.
Astron. Astrophys., Vol. 104, 37 - 41 (1981).

Equations of motion and variational equations, suitable for the construction of numerically integrated outer planet satellite ephemerides, are derived. The force model includes n oblate integrated satellites and m external perturbing planets. Variational equations are included for initial state, mass, oblateness, and pole parameters. This formulation was used to construct Saturn satellite ephemerides for the Voyager mission.

042.075 **Bifurcations of planar to three-dimensional periodic orbits in the restricted three-body problem.**
V. V. Markellos, C. L. Goudas, G. A. Katsiaris.
Investigating the universe, (see 003.014), p. 321 - 351 (1981).

The authors study the mechanism of bifurcation of three-dimensional periodic orbits from the plane of motion of the primaries. The branchings of simple and double periodic three-dimensional orbits occur at the plane periodic orbits of indifferent vertical stability. The authors describe suitable numerical techniques for their accurate determination and for their continuation through the range of the mass parameter μ of the problem. The authors use these techniques to determine the first few vertical-critical orbits of the basic families of periodic orbits of the problem, namely, families l, m, i, h, a, b and c, for the entire range of μ. The initial conditions of over three hundred vertical-critical orbits are given, classified in twenty-one series. Their horizontal stability is also examined and this leads to predictions of where in the range of μ stable three-dimensional periodic orbits may not branch from these basic families.

042.076 On the triple resonance case of disturbed motion of
the satellite of a spheroidal planet.
N. S. Mamedov, G. F. Sultanov.
Izv. 'AzSSR. Ser. fiz.-tekh. i mat. nauk, 1980, No. 4, p. 50 - 54.
In Russian. – Abstr. in Ref. zh., 51. Astron., 11.51.107
(1981).

042.077 Periodic solutions of the restricted problem of the
translatory-rotational motion of three rigid bodies.
N. I. Gamarnik.
Astron. Zh., Tom 58, 1296 - 1305 (1981). In Russian.
English translation in Soviet Astron., Vol. 25, No. 6.
   Periodic solutions of the Hamiltonian system of equations
close to Lagrange's triangular ones of the restricted circular
problem of three rigid bodies are found in first approximation.
By Lyapunov's method the existence is proved of periodic
solutions of nonlinear equations of the problem which correspond to zero or purely imaginary roots.

042.078 Estimate of the error in numerical integration of
the N-body problem.   L. K. Babadzhanyants.
Pis'ma Astron. Zh., Tom 7, 752 - 755 (1981). In Russian.
English translation in Soviet Astron. Lett., Vol. 7.
   Estimates of the radius of convergence and the remainder
of the Taylor solution in rectangular coordinates of the
N-body problem are given.

042.079 Final motions in the three-body problem and
symbolic dynamics.   V. M. Alekseev.
Usp. mat. nauk, Tom 36, No. 4, p. 161 - 176 (1981). In
Russian. – Abstr. in Ref. zh., 51. Astron., 12.51.107 (1981).

042.080 Investigation of the planar restricted elliptical twice
averaged three-body problem.
M. A. Vashkov'yak.
Inst. prikl. mat. AN SSSR. Prepr., 1981, No. 64, 26 pp. In
Russian. – Abstr. in Ref. zh., 51. Astron., 12.51.109 (1981).

042.081 On the evolution of orbits in the restricted twice
averaged three-body problem.
M. A. Vashkov'yak.
Proceedings of scientific lectures on cosmonautics, (see
012.058), 1980, p. 171 - 184. In Russian. – Abstr. in Ref. zh.,
51. Astron., 12.51.110 (1981).

042.082 Canonic equations of the perturbed translatory-
rotational motion of a planet and its satellite expressed in Delaunay-Andoyer elements.   D. Z. Koenov.
Izv. AN TadzhSSR. Otd. fiz.-mat., khim. i geol. nauk, 1981,
No. 2, p. 34 - 41. In Russian. – Abstr. in Ref. zh., 51. Astron.,
12.51.124 (1981).

042.083 The effect of gravitational and Lorentz forces on a
satellite's motion relative to the mass centre.
G. V. Lyakhovka.
Vestn. LGU, 1981, No. 13, p. 85 - 90. In Russian. – Abstr. in
Ref. zh., 51. Astron., 12.51.128 (1981).

042.084 Orbital eccentricity in a logarithmic potential.
K. A. Innanen, L. Tähtinen, M. J. Valtonen.
Turku Univ. Obs. Informo No. 51, 5 pp. (1981). – Submitted
to Astron. J.

042.085 Meteor position and velocity in the original orbit.
K. Nagasawa.
Tokyo Astron. Obs. Rep. (No. 74), Vol. 19, 405 - 424 (1981).
In Japanese.

042.086 Some possibilities for determining the influence of
Earth body tides on the motion of artificial satellites.
J. Kostelecký.
Satellite perturbations and orbital determination, (see

012.060), p. 29 - 35 (1981).
   The paper treats some aspects of the widespread concept
of a deforming potential and a more generalized approach to
the Dirichlet problem in defining the influence of Earth tides
on a satellite's movement. In considering theoretical, rather
than practical aspects, the solution of the Dirichlet problem is
discussed for various reference surfaces in place of the Earth.

042.087 A general time element for orbit integration in
Cartesian coordinates.   G. Janin, V. R. Bond.
Satellite perturbations and orbital determination, (see
012.060), p. 69 - 78 (1981).
   For increasing the accuracy of the numerical integration
of eccentric orbits in Cartesian coordinates, two techniques
are discussed: (1) the use of an independent variable different
from the time, which increases the efficiency of the numerical
integration, (2) the use of a time element, which reduces the
in-track error. A general expression of a time element valid
for an arbitrary independent variable is presented. Taking the
example of a transfer orbit for a geosynchronous mission, a
comparison between the use of the eccentric, true and elliptic
anomaly as independent variable is shown.

042.088 The effect of precession and nutation on the
harmonic coefficients of the gravitational potential
of a planet.   V. I. Umlenski, V. G. Shkodrov.
Dokl. Bolg. AN, Vol. 34, 613 - 616 (1981). – Abstr. in Ref.
zh., 51. Astron., 1.51.74 (1982).

042.089 Some qualitative characteristics of the force function between two celestial bodies.
V. G. Shkodrov, T. G. Gechev.
Dokl. Bolg. AN, Tom 34, 609 - 611 (1981). – Abstr. in Ref.
zh., 51. Astron., 1.51.96 (1982).

042.090 On secular effects in the movement of a planetary
satellite.   V. G. Ivanova.
Dokl. Bolg. AN, Vol. 34, 617 - 620 (1981). – Abstr. in Ref.
zh., 62. Issled. kosm. prostranstva, 1.62.143 (1982).

042.091 The dynamics of tadpole and horseshoe orbits.
I. Theory.   S. F. Dermott, C. D. Murray.
Icarus, Vol. 48, 1 - 11 (1981).
   The properties of the tadpole and the horseshoe orbit
solutions of both the circular and elliptic restricted three-body
problem with small mass ratio are examined using analytical
and numerical methods. The authors show how the trajectory
of a particle in a near-circular orbit is critically dependent on
the initial radial separation between the particle and satellite
orbits. By numerical integration of a number of horseshoe
orbits they have investigated the effects of changes in
eccentricity and longitude of pericenter over repeated satellite encounters. Such changes will determine the long-term
stability of particles in horseshoe orbits.

Orbital motion.   See Abstr. 003.137.

The development of science at the Astronomical
Observatory of the Leningrad University. Celestial mechanics,
geodesy and gravimetry.   See Abstr. 009.015.

Proceedings of joint scientific readings on cosmonautics dedicated to the memory of eminent Soviet scientists –
pioneers of cosmic space exploration. Moscow, 28 January -
1 February, 1980. Applied celestial mechanics and motion
control.   See Abstr. 012.058.

The contact transformation groups of the extended
Hamiltonian system.   See Abstr. 021.014.

The computation of tables of Hansen coefficients.
See Abstr. 021.016.

The Lagrangian theory of Stäckel systems.
See Abstr. 021.018.

On the possibility of using the program system "Auto-Analyzer" for the solution of problems of celestial mechanics.   See Abstr. 021.019.

Some algorithms for constructing analytical solutions in problems of celestial mechanics. See Abstr. 021.020.

Algorithms of evaluation in the restricted three-body problem.   See Abstr. 021.021.

Review of modern methods for increasing the accuracy of numerical integration of differential equations of the motion of celestial bodies.   See Abstr. 021.022.

A semi-analytical algorithm for calculating the first-order perturbations in regularizing variables. See Abstr. 021.023.

On stationary solutions of some canonical systems of differential equations.   See Abstr. 021.038.

Motions of a chain of oscillators. See Abstr. 022.056.

Note on the numerical expressions for precession quantities.   See Abstr. 043.004.

Radius of convergence of Lie series for some elliptic elements.   See Abstr. 052.038.

Elimination of secular terms generated by the coupling of perturbations.   See Abstr. 052.041.

Analysis of Lageos' altitude decrease. See Abstr. 052.062.

Explosive mass loss in binary stars: the two time-scale method.   See Abstr. 064.009.

On observable consequences of the three body problem in Einstein's gravitation theory. See Abstr. 066.059.

Planetary tides as a part of deformations of the Earth's body.   See Abstr. 081.041.

The positions of secular resonance surfaces. See Abstr. 091.004.

The stability of the solar system. See Abstr. 091.039.

Third order theory of the four large planets. See Abstr. 091.052.

Effects of the tides on the evolution of natural satellites.   See Abstr. 091.067.

The Earth-figure perturbations in the lunar theory. See Abstr. 094.028.

Comparison of ELP-2000 to a JPL numerical integration.   See Abstr. 094.029.

The perturbations of the orbital elements of Trojan asteroids.   See Abstr. 098.008.

On stability of asteroidal orbits near the 2/3 commensurability with Jupiter.   See Abstr. 098.092.

Dispersion of the minor planet inclinations as a function of time.   See Abstr. 098.100.

Tidal evolution of the Galilean satellites: a linearized theory.   See Abstr. 099.021.

Numerical and analytical theories of motion of the outer satellites of Jupiter.   See Abstr. 099.210.

Derivation of the collision probability between orbiting objects: the lifetimes of Jupiter's outer moons. See Abstr. 099.227.

The dynamics of tadpole and horseshoe orbits. II. The coorbital satellites of Saturn.   See Abstr. 100.143.

A Monte Carlo investigation of Jovian perturbations on short-period comet orbits.   See Abstr. 102.002.

When we may and need to use barycentric orbit of a comet?   See Abstr. 102.038.

Solar nebula dispersal and the stability of the planetary system. I. Scanning secular resonance theory. See Abstr. 107.011.

The mass-angular momentum-diagram of astronomical objects.   See Abstr. 151.079.

Invariant surfaces and orbital behaviour in dynamical systems of 3 degrees of freedom. II.   See Abstr. 151.087.

043 Astronomical Constants, Reference Systems

043.001   **The new IAU system of astronomical constants (1976, 1979).**   V. K. Abalakin.
Determination of the coordinates of celestial bodies, Riga, 1981, (see 003.008), p. 96 - 103. In Russian. — Abstr. in Ref. zh., 51. Astron., 9.51.120 (1981).

043.002   **Astronomical precession: a good and a bad first-order approximation.**   B. M. Haisch.
American J. Phys., Vol. 49, 636 - 640 (1981). — Abstr. in Phys. Abstr., Vol. 84, Abstr. 89180 (1981).

043.003   **On the new method of determining the gravitational constant.**   L. Fialovszky.
Gerlands Beitr. Geophys., Band 90, 448 (1981).
    Critical comments to a paper by Bleyer et al. — see Abstr. 19.043.004.

043.004   **Note on the numerical expressions for precession quantities.**   P. Bretagnon, J. Chapront.
Astron. Astrophys., Vol. 103, 103 - 107 (1981). In French.
    The precession quantities are completely determined by the two motions of the equatorial pole and the ecliptic pole. The precessional formulae derived by (Lieske et al., 1977) are based upon the use of the secular variations of the ecliptic pole from Newcomb's theory of the Sun. Taking advantage of the analytic formulae given by Lieske et al., the authors propose new expressions for the precession quantities, setting up the secular motion of the ecliptic pole from Bretagnon's theory of the Sun, which has been developed up to the third order of planetary mass. First, the authors discuss the proposed expressions for precession quantities, computed with the new value of general precession p_1, and the obliquity ϵ_0, at the epoch J2000, recommended by IAU. Secondly they present the results of a comparison of Bretagnon's theory of the Sun with the JPL numerical integration DE102.

043.005   **The system of astronomical constants.**   K. A. Kulikov.
Zemlya Vselennaya, 1981, No. 6, p. 25 - 27. In Russian.

043.006   **Amélioration des constantes fondamentales de l'astronomie.**   N. Capitaine.
Astronomie, Vol. 95, 489 - 502 (1981).

043.007   **The determination of the corrections of the equinox and equator by means of the observations of planets with the Danjon astrolabe.**   Z.-f. Li, H.-j. Yan.
Acta Astron. Sinica, Vol. 22, 230 - 240 (1981). In Chinese.
    Using the results of the observations of the Mars, Jupiter, and Saturn made with the Danjon astrolabe at Shanghai and the other six observatories and adopting the differential method to modify orbital elements, the corrections to the origin of the right ascension and the equator for the Fourth Fundamental Catalogue FK4 are calculated.

Landolt-Börnstein.   See Abstr. 003.172.

On a statement by Laplace.
See Abstr. 004.077.

A practical introduction to astrometry.
See Abstr. 041.003.

Independence of the gravitational constant from gross Earth data.   See Abstr. 081.061.

044 Time, Rotation of the Earth

044.001 Solar activity and Earth's rotation.
D. Djurović.
Astron. Astrophys., Vol. 100, 156 - 158 (1981).
In the spectrum of Wolf number W and geomagnetic index Ap the cyclic variations with periods of 0.5, 3.3, 6.6, and 11 yr were identified. The first three cyclic variations were earlier identified in the Earth's rotation angular velocity ω (Djurović, 1979). Since the corresponding phases of the mentioned variations are approximately equal, the author considers that the solar activity contributes to the variations of ω.

044.002 Comments on the paper by Y. Kubo: 'A core-mantle interaction in the rotation of the earth'.
M. G. Rochester, with a reply by Y. Kubo.
Celestial Mech., Vol. 24, 231 - 235 (1981). — See Abstr. 25.044.012.
A recently-published paper, concerning the effects of core-mantle coupling on the earth's precession and nutation, is critically appraised in the light of previously-published work in astronomy and geophysics.

044.003 Pleistocene deglaciation and the earth's rotation: implications for mantle viscosity.
R. Sabadini, W. R. Peltier.
Geophys. J. R. Astron. Soc., Vol. 66, 553 - 578 (1981).
The authors analyse the effect of deglaciation upon the earth's rotation and thereby show that the observed secular trend (polar wander) evident in the ILS—IPMS pole path, and measurements of the non-tidal acceleration of the length of day, are both consistent with the viscosity profile deduced from postglacial rebound.

044.004 The Earth's rotation.
F. Mignard.
Recherche, No. 119, p. 136 - 147 (1981). In French. — Abstr. in Phys. Abstr., Vol. 84, Abstr. 70821 (1981).

044.005 The effective time of Newcomb's theory of the sun.
S. J. Goldstein, Jr.
Bull. American Astron. Soc., Vol. 13, 571 (1981). — Abstract.

044.006 Earth rotation from a simultaneous reduction of LLR/LAGEOS data.
P. J. Shelus, R. J. Eanes, N. R. Zarate.
Bull. American Astron. Soc., Vol. 13, 573 (1981). — Abstract.

044.007 Gravitational effects of sun and moon on atomic time scales. L. Mereddu, E. Proverbio.
Mem. Soc. Astron. Italiana, Vol. 52, 233 - 248 (1981).
The analysis of frequency standard comparisons between the local atomic time scale and the time signals emitted in the Loran-C transmission system emphasizes the existence of diurnal variation in the relative phase of the frequency standards. The computation of the gravitational effects of the moon and sun upon atomic time scale shows a good agreement in the phase of the calculated and observed diurnal fluctuations. However, the amplitude of this last is about five times greater than the calculated gravitational effects. More careful and detailed investigations are required on this matter concerning the influence of gravity waves and other atmospheric effects on the diurnal propagation of radio waves.

044.008 Continental drift and the rotation of the Earth — a new and critical evaluation. V. J. Slabinski.
Celestial Mech., Vol. 25, 89 - 91 (1981).
Sconzo has proposed that the continental drift of a land mass like Greenland can make a detectable change in the Earth's rotation rate through a change in the Earth's moment of inertia. The correction of theoretical and numerical errors in his paper results in a change in the length of the day of only 3.5×10^{-8} s/100 yr, a rate which is 6 orders of magnitude less than the value he gives. Continental drift does not appear to cause an important change in the length of the day.

044.009 A new analytic expression for the equation of time.
L. O. Lamm.
Sol. Energy, Vol. 26, 465 (1981). — Abstr. in Phys. Abstr., Vol. 84, Abstr. 88255 (1981).

044.010 Time and latitude results of observations made at Merate Observatory with the astrolabe for the year 1980. L. Buffoni, F. Carta, F. Chlistovsky, A. Manara, F. Mazzoleni.
Astron. Astrophys., Suppl. Ser., Vol. 46, 179 - 180 (1981).
Results of the observations made with the astrolabe Danjon OPL No. 32 are given. The results are in the FK4 system.

044.011 Résultats des observations faites à Paris avec l'astrolabe. Temps et latitude 1979-1980.
F. Chollet, S. Débarbat.
Astron. Astrophys., Suppl. Ser., Vol. 46, 249 - 256 (1981).
Results are given for the observations made with the astrolabe APP during the years 1979 and 1980. No change has been made in the star catalogue since 1956. The observing technique and computation method have also been unchanged since 1972. The results are in the FK4 system.

044.012 Influence of the sector structure of the interplanetary magnetic field on the regime of the daily rotation of the earth. V. M. Kiselev.
Geomagn. Aehron., Tom 21, 798 - 802 (1981). In Russian.

044.013 The tidal effect on the observation with the astrolabe at Shanghai. Z.-g. Yang.
Acta Astron. Sinica, Vol. 22, 241 - 247 (1981). In Chinese.
Based on the observational results with the astrolabe of the Shanghai Observatory from 1960 - 1979, the lunar tidal effect on the vertical is analysed.

044.014 Variations in the angular velocity of the earth's rotation and the rate of change of the pulsar period.
Y.-f. Xia, J.-x. Rong.
Acta Astron. Sinica, Vol. 22, 299 - 304 (1981). In Chinese.
In this paper the effect of the variations in the angular velocity of the earth's rotation on the rate of change of the pulsar period is analysed.

044.015 Atmospheric angular momentum and the length of day: a common fluctuation with a period near 50 days. R. B. Langley, R. W. King, I. I. Shapiro, R. D. Rosen, D. A. Salstein.
Nature, Vol. 294, 730 - 733 (1981).
Last year, four astronomical measures of changes in the length of day obtained in 1979 were shown to exhibit a ~50-day fluctuation. To find out whether this fluctuation was persistent and of meteorological origin the authors analysed lunar laser ranging observations from the McDonald Observatory in Texas and zonal (east-west) wind data deduced from sources distributed over the globe. The relevant data were available for the common period 1976—79. The authors describe these analyses in turn and then compare the results.

044.016   **On the new system of Universal Time.**
G. P. Pil'nik, K. S. Sitnik.
Astron. Tsirk., No. 1123, p. 7 - 8 (1980). In Russian.

044.017   **On determination of the parameters of the earth's rotation from observations.**   E. P. Fedorov.
Astron. Zh., Tom 58, 1314 - 1322 (1981). In Russian.
English translation in Soviet Astron., Vol. 25, No. 6.

044.018   **Time and Latitude Service.** 1981 January - March.
Circ. Time Latitude Serv., No. 157 (1981).

044.019   **Détermination astronomique de l'heure.**
M. Jovanović, L. Đurović.
Bull. Obs. Astron. Belgrade, No. 131, p. 45 - 48 (1981).

044.020   **Greenwich Time Report.** 1980 October - December, 1981 January - March.   F. G. Smith.
R. Greenwich Obs., Time Latitude Serv., p. 225 - 254 (1981).

044.021   **Zeit- und Breitendienst.** April - December 1980, January - June 1981.
Deutsches Hydrogr. Inst., Hamburg. 10 + 9 + 8 + 5 + 6 pp. (1980/81).

044.022   **Earth rotation from lunar laser ranging.**
R. B. Langley, R. W. King, I. I. Shapiro.
J. Geophys. Res., Vol. 86, 11913 - 11918 (1981).
The authors have estimated the variation of latitude and UT0 from McDonald Observatory lunar laser ranging observations for the period October 1970 to October 1979. The typical formal uncertainties of the values are about 6 milliarcseconds and 0.5 milliseconds of time, respectively. The authors compare their values of variation of latitude with those derived from pole positions determined using classical astrometric, satellite Doppler, and satellite laser ranging techniques for the period July 1976 to November 1978 when data from all of these techniques were available.

044.023   **Time and Frequency Services Bulletin,** April - December 1981.
Published by Shaanxi Astron. Obs., Chinese Acad. Sci., Lintong, Xian, China (1981).

044.024   **Time and Latitude.** October 1980 - March 1981.
V. Ptáček, J. Vondrák, R. Weber.
Circ. Czechoslovak Obs., Time and Latitude (1980/81).

044.025   **UT2−UT1 for 1982.**
B.I.H., Paris, Circ. A (1981).

044.026   **Bureau International de l'Heure (B.I.H.) Circular D.** August - December 1981.
B.I.H., Paris, Circ. D177 - D181 (1981).
Coordinated Universal Time UTC. International Atomic Time. Universal Time and coordinates of the pole. Short term variation of UT1 and of the duration of the day. Data of individual networks. Informations on time signals.

044.027   **Zeit- und Breitenbestimmungen, Zeitsysteme, Präzisionszeitvergleiche.** 1980 September - 1981 April.
Published by Akad. Wiss. DDR, Zentralinst. Phys. Erde, Potsdam, Geod.-astron. Obs., Jahrg. 1980, Nos. 5 - 6 (1981); Jahrg. 1981, Nos. 1 - 2 (1981).

044.028   **Daily time differences and relative phase values.** 1981 July - December.
U. S. Naval Obs., Washington, D.C. Time Serv. Publ., Ser. 4, Nos. 753 - 778 (1981).

044.029   **A.1 − UT1 data.** 1981 July - December.
U. S. Naval Obs., Washington, D.C. Time Serv. Publ., Ser. 6, Nos. 65 - 70 (1981).

044.030   **Preliminary times and coordinates of the pole.** 1981 July - December.
U. S. Naval Obs., Washington, D.C. Time Serv. Publ., Ser. 7, Nos. 706 - 731 (1981).

044.031   **Time Service announcement. UTC time scale.**
G. M. R. Winkler.
U. S. Naval Obs., Washington, D.C. Time Serv. Publ., Ser. 14, No. 30 (1981).

044.032   **Variations of UT1−TAI and atmospheric excitations.**   T. Hara.
Publ. Int. Latitude Obs. Mizusawa, Vol. 14, 45 - 75 (1980).
Astronomical evidence for the fluctuation of the spinning rate of the Earth is presented with the use of data published by the BIH. The process of the angular momentum exchange between the atmosphere and the solid Earth is studied by looking into the latitude- and longitude-height distributions of the atmospheric angular momentum and the effects of mountains such as surface stress and mountain torque rather than by comparing the total angular momenta of the atmosphere and the solid Earth.

044.033   **Time and Latitude Bulletins.** January - March 1981.
Tokyo Astron. Obs., Time Latitude Bull., Vol. 55, Nos. 1 - 3 (1981).

044.034   **Rotacion de la tierra año 1979.** Resultados obtenidos en San Fernando con el Astrolabio Impersonal Danjon OPL no. 37. I.−Tiempo y Latitud. II.− Observaciones de Júpiter.
Published by Inst. Obs. de Marina, San Fernando (Cádiz), Ser. C, No. 82, 20 pp. (1981).
This bulletin contains definitive results of time and latitude observations made at San Fernando with the Danjon astrolabe during 1979. Corrections for internal regularization of stars inside each fundamental group (CL I) have been applied, but not "group corrections". In tables the BIH corrections and astronomical coordinates of the astrolabe are shown. Results of Jupiter observations during the first months of 1979 are given.

044.035   **Rotacion de la tierra año 1980.** Resultados obtenidos en San Fernando con el Astrolabio Impersonal Danjon OPL no. 37. I.− Tiempo y Latitud. II.− Observaciones de Marte y Júpiter.
Published by Inst. Obs. de Marina, San Fernando (Cádiz), Ser. C, No. 83, 23 pp. (1981).
This bulletin contains definitive results of time and latitude observations made at San Fernando with the Danjon astrolabe during 1980. Corrections for internal regularization of stars inside each fundamental group (CL I) have been applied, but not "group corrections". In tables the BIH corrections and astronomical coordinates of the astrolabe are shown. Results of Mars and Jupiter observations during the winter 1979−80 are given.

044.036   **Utilización de las distintas escalas de tiempo en los problemas de astronomía.**   A. Orte.
Inst. Obs. Marina, San Fernando (Cádiz), Bol. Astron. No. 5, 18 pp. (1981).

044.037   **Relation between the secular variation of longitude and the plate motion.**
Z.-n. Gu,,C.-x. Qian, X.-q. Liang.
Ann. Shanghai Obs. Acad. Sinica, No. 1, p. 26 - 33 (1979).

044.038 **Approximate function for the equation of time.**
H. Claussnitzer.
Astron. Nachr., Band 302, 311 - 316 (1981).

Starting from the equation stated by Schaub (1950) it is possible by adding suitable correction terms to find a simple mathematical expression, which allows to obtain $g = \alpha_\odot - \bar{\alpha}_\odot$ for any time with an accuracy sufficent for many purposes. The maximum deviations from the true values amount for the initial year to about $\pm 0^s3$, and for the two following years to about $\pm 0^s7$ and 1^s5. When later the true values of g are known, the deviations may be considerably reduced by adding generally only one trigonometric correction term. So it is possible to keep the approximate function at an accuracy of $\pm 1^s$ for several years.

044.039 **An analysis of non-polar longitude variations in the BIH system.** Z.-n. Gu.
Ann. Shanghai Obs. Acad. Sinica, No. 2, p. 48 - 51 (1980).

Landolt-Börnstein. See Abstr. 003.172.

Electromagnetic-gravitational energy systems. See Abstr. 066.161.

Beitrag zur geodätischen und astronomischen Nutzung von Sternbedeckungen durch den Mond. See Abstr. 096.021.

045 Latitude Determination, Polar Motion

045.001 **Investigation of controversial polar motion features using homogeneous International Latitude Service data.**   S. R. Dickman.
J. Geophys. Res., Vol. 86, 4904 - 4912 (1981).
Standard time series analysis techniques have been applied to the homogeneous polar motion data recently published by the ILS-IPMS (Yumi and Yokoyama, 1980) in order to study some of the more controversial features apparently possessed by the older ILS data. The magnitude and direction of the secular trend unbiased by the presence of harmonics in the data were determined, yielding a rate of polar wander ~3.52 × 10⁻³ arc sec/yr in direction 80.1°W longitude. The long-perid Markowitz wobble, which dominates the retrograde power spectrum of the data, has a signal to noise ratio in that spectrum of 21:1; its period is well-determined as 31 years. Variations with time of the annual wobble and Chandler wobble were investigated using complex demodulation.

045.002 **Excitation of the earth's polar motion: a reassessment with new data.**
C. R. Wilson, S. Gabay.
Geophys. Res. Lett., Vol. 8, 745 - 748 (1981).
A comparison of a revised ILS polar motion series for the years 1901 to 1970 with an atmospheric time series shows evidence for coherence near the period of the 14 month Chandler motion. The same ILS series, and two other BIH series, have also been used to compute shifts in the apparent pole position near the times of the 1960 Chilean and 1964 Alaskan earthquakes.

045.003 **True polar wander.**
D. M. Jurdy.
Tectonophysics, Vol. 74, No. 1 - 2, p. 1 - 16 (1981). — Abstr. in Phys. Abstr., Vol. 84, Abstr. 79421 (1981).

045.004 **On the study of the secular motion of the earth poles.**   I. I. Krasnorylov.
Redkol. zh. Izv. vuzov. Geod. i aehrofotosemka. Moskva, 1981. 25 pp. In Russian. — Abstr. in Ref. zh., 51. Astron., 7.51.85 (1981).

045.005 **Some "corrections" to the static theory of the polar tide.**   V. P. Karklin.
Probl. Arktiki i Antarktiki, Leningrad, 1981, No. 56, p. 83 - 86. In Russian. — Abstr. in Ref. zh., 51. Astron., 8.51.125 (1981).

045.006 **Polar wandering and the forced responses of a rotating, multilayered, viscoelastic planet.**
R. Sabadini, E. Boschi, D. A. Yuen.
Mem. Soc. Astron. Italiana, Vol. 52, (see 012.023), 513 (1981). Abstract.

045.007 **Study of the secular motion of the earth poles.**
I. I. Krasnorylov.
Izv. vuzov. Geod. i aehrofotosemka, Moskva, 1981. 25 pp. In Russian. — Abstr. in Ref. zh., 52. Geod. Aehrosemka, 9.52.85 (1981).

045.008 **On the cause of a wobble of polar motion oscillating with a half Chandler period.**   H. Jochmann.
Astron. Nachr., Band 302, 193 - 196 (1981) = Mitt. Zentralinst. Phys. Erde, Nr. 966.
It is proved that a wobble of polar motion with a half Chandler period, discussed recently in several publications, may be caused by lateral heterogeneities of rheological properties of the earth.

045.009 **Estimates of Chandler's component of polar motion as derived from various data sets.**
C. R. Wilson, R. O. Vicente.
Astron. Nachr., Band 302, 227 - 232 (1981).
Data sets consisting of several combinations of classical and modern techniques (Doppler and laser ranging to artificial satellites) have been analysed by autoregressive (AR), sinusoidal and autoregressive moving average (ARMA) methods in order to obtain estimates of the frequency and quality factor of the Chandler component of polar motion. The ARMA method shows advantages over the other methods and the results obtained do not furnish evidence of temporal variation in these polar motion parameters.

045.010 **Die Eingangs-Ausgangs-Analyse der Polbewegung mit Hilfe der meteorologischen Erregerfunktion.**
H. Jochmann.
Gerlands Beitr. Geophys., Band 90, 477 - 488 (1981).
The parameters of the polar motion system are determined by input-output analysis, using the meteorological excitation function. The meteorological excitation has been proved for different terms of polar motion. The transfer function for these terms has to be larger or equal to ten.

045.011 **Annual report of the International Polar Motion Service for the year 1979.**   K. Yokoyama.
Published for the International Council of Scientific Unions by Central Bureau of the International Polar Motion Service, Mizusawa, Japan. 4 + 117 pp. (1981). ISSN 0074-7432.

045.012 **Programme de latitude du tube zénithal visuel de l'Observatoire Turku-Tuorla système amélioré de 1976.**   L. Oterma, with a preface by J. Kakkuri.
Publ. Finnish Geod. Inst., No. 92, 22 pp. (1981).

045.013 **On Chandler wobble.**   H.-z. Zhang.
Acta Astrophys. Sinica, Vol. 1, 253 - 263 (1981). In Chinese.

045.014 **Observations à la lunettte zénithale (de 110 mm) du service de la latitude de l'Observatoire de Beograd en 1979.**   R. Grujić, M. Đokić.
Bull. Obs. Astron. Belgrade, No. 131, p. 43 - 44 (1981).

045.015 **Breitenbestimmungen. 1980 November - 1981 Juni.**
Tech. Univ. Dresden, Lohrmann - Obs., Zirk. Nr. 94 - 95 (1981).

045.016 **The solution of polar motion and its reference system.**   H. Zhang.
Publ. Shaanxi Astron. Obs., No. 1, p. 31 - 43 (1981). In Chinese.

045.017 **Determinación de la latitud por el metodo de Sterneck.** Programas de cálculo automático.
M. J. Sevilla, A. Nuñez.
Téc. Topogr., Vol. 7, Núms.27 - 28, 23 pp. (1979) = Univ. Complutense - Fac. Cienc., Madrid, Semin. Astron. Geod., Publ. núm. 103.

045.018 **Determinación de la latitud y la longitud por el metodo de alturas iguales.** Programas de cálculo automático.   M. J. Sevilla.
Bol. Inf., Núm. 43, Serv. Geogr. Ejército, 13 + 27 pp. (1979) = Univ. Complutense - Fac. Cienc., Madrid, Semin. Astron. Geod., Publ. núm. 104.

045.019 **Renewal of the ILS star pairs.**
S. Abe, S. Yumi, K. Yokoyama, K. Sato.
Proc. Int. Latitude Obs. Mizusawa, No. 20, p. 69 - 78 (1981).
In Japanese.

045.020 **On characteristics and changes of weights of the IPMS stations.**   H. Ishii.
Proc. Int. Latitude Obs. Mizusawa, No. 20, p. 79 - 103 (1981).
In Japanese.

045.021 **Monthly Notes of the International Polar Motion Service.**
Mon. Notes Int. Polar Motion Serv., Nos. 5 - 11, p. 45 - 119 (1981).
Announces the values of latitudes observed at the collaborating stations during May 1981 - October 1981.

045.022 **An observing program for the photographic zenith tubes at 39°8′N latitude.**   D. D. McCarthy.
Publ. Int. Latitude Obs., Vol. 14, 1 - 34 (1980).
Past observational results of the Mizusawa, Richmond, and Washington photographic zenith tubes were investigated to determine criteria to be used in the selection of an observing program for the chain of photographic zenith tubes located at 39°8′N latitude. Following these criteria, an observing program is suggested and the results to be expected from a chain of photographic zenith tubes are estimated.

045.023 **An observing list of the ILS PZT stars – a proposal.**
H. Kitago.
Publ. Int. Latitude Obs., Vol. 14, 35 - 43 (1980).
A list of stars for the cooperative ILS PZT observation is proposed. The list contains 127 stars which are divided into 12 groups. All the stars are selected from the Smithsonian Astrophysical Observatory Catalogue and observable for all ILS PZT's including the small field ones of Ukiah and Kitab.

045.024 **Programme de latitude du tube zénithal visuel de l'Observatoire Turku-Tuorla, système amélioré de 1976.**   L. Oterma.
Veröff. Finnischen Geod. Inst., No. 92, 22 pp. (1981) = Turku Univ. Obs. Informo, No. 57.

045.025 **Results of determination of latitude in Józefosław by observations of the Horrebow-Talcott pairs.**
1981 January - June.   L. Pieczyński.
Warsaw Tech. Univ., Astron.-Geod. Obs. Józefosław.
Latitude Circ. Nos. 78 - 79 (1981).

045.026 **On the free wobble of the earth's pole.**
M. Zhao.
Ann. Shanghai Obs. Acad. Sinica, No. 1, p. 1 - 14 (1979).
As many authors have pointed out, the free wobble of the earth's pole has undergone considerable changes during this century. Some various hypotheses have been suggested in order to explain the changes. The present author discusses the hypotheses, points out the contradictions existing between the hypotheses and the observed facts, and suggests a new hypothesis to explain the changes of the wobble.

045.027 **The stability of the polar coordinate systems.**
S.-y. Zhu, J.-l. Zhao.
Ann. Shanghai Obs. Acad. Sinica, No. 1, p. 20 - 25 (1979).

Taking the ILS and BIH systems as examples, the authors discuss the effects on polar coordinate systems influenced by various errors in latitude observations. The disadvantage of the ILS system and some disadvantages of the BIH system have been pointed out. Some proposals to improve the BIH system have been suggested.

045.028 **Periodic analysis of the Z-term.**   S.-Z. Ji.
Ann. Shanghai Obs. Acad. Sinica, No. 1, p. 40 - 43 (1979).

045.029 **Analysis of the nearly diurnal free wobble using data observed with the Danjon astrolabe at the Shanghai Observatory.**   Z.-g. Yang.
Ann. Shanghai Obs. Acad. Sinica, No. 2, p. 30 - 37 (1980).

045.030 **The non-polar long period terms in latitude and their influence on the study of secular polar motion.**
Z.-x. Li, P.-z. Lu.
Ann. Shanghai Obs. Acad. Sinica, No. 2, p. 38 - 47 (1980).

045.031 **Reduction of latitude data obtained with the Danjon astrolabe at the Shanghai Observatory during 1960–1979.**   S.-z. Ji, D.-w. Zheng, J.-z. Zhang.
Ann. Shanghai Obs. Acad. Sinica, No. 2, p. 52 - 56 (1980).

045.032 **On the law of distribution of random errors in long series of latitude observations.**   S. N. Ped'ko.
Astron. Tsirk., No. 1163, p. 3 - 5 (1981). In Russian.

Landolt-Börnstein.   See Abstr. 003.172.

Time and latitude results of observations made at Merate Observatory with the astrolabe for the year 1980.
See Abstr. 044.010.

Résultats des observations faites à Paris avec l'astrolabe. Temps et latitude 1979–1980.   See Abstr. 044.011.

Time and Latitude Service.   See Abstr. 044.018.

Greenwich Time Report.   See Abstr. 044.020.

Zeit- und Breitendienst.   See Abstr. 044.021.

Time and Latitude. October 1980 - March 1981.
See Abstr. 044.024.

Zeit- und Breitenbestimmungen, Zeitsysteme, Präzisionszeitvergleiche. 1980 September - 1981 April.
See Abstr. 044.027.

Time and Latitude Bulletins. January - March 1981.
See Abstr. 044.033.

Rotacion de la tierra año 1979.
See Abstr. 044.034.

Rotacion de la tierra año 1980.
See Abstr. 044.035.

A SIRIO-2 laser ranging experiment.
See Abstr. 052.015.

046 Astronomical Geodesy, Satellite Geodesy, Navigation

046.001 Navstar Global Positioning System (GPS): future
enhanced capabilities.   M. Ananda.
Spacecraft flight dynamics, (see 012.021), p. 61 - 71 (1981).
This paper reviews and addresses three specific areas of
research which are currently being done in support of en-
hancing future capabilities of the Global Positioning System.
These areas are 1) navigation using radio interferometric
techniques, 2) analyses leading to more accurate orbit prediction of GPS satellites and 3) high precision application of GPS,
beyond the current 16-meter spherical error probability
capability.

046.002 Relations between disturbances of the local temperature field and the results of geodetic-astronomical
observations with the Danjon astrolabe.   J. Dittrich.
Astron. Nachr., Band 302, 145 - 151 (1981) = Mitt. Zentralinst. Phys. Erde Nr. 955. In German.

046.003 Performance and results of the satellite ranging laser
station at Kavalur, India, in 1980.   P. S. Dixit,
P. K. Rao, K. Elango, K. Gopalakrishnan, K. N. Rao,
S. Schillak, W. Kielek, M. Abele.
Ann. Géophys., Vol. 37, (see 012.037), 59 - 62 (1981).
Abstr. in Phys. Abstr., Vol. 84, Abstr. 107568 (1981).

046.004 Space geodesy, paleogeodesy and paleogeophysics.
N.-A. Morner.
Ann. Géophys., Vol. 37, (see 012.037), 69 - 76 (1981).
Abstr. in Phys. Abstr., Vol. 84, Abstr. 107569 (1981).

046.005 Space techniques to monitor movements in the
Earth's crust.   S. Hieber.
ESA Bull., No. 28, p. 62 - 67 (1981).

046.006 On a mode of direct determination of the geodetic
azimuth from observations of faint stars in the first
vertical.   M. A. Vajsov.
Tr. Kazan. Gorod. Astron. Obs., Vyp. 45, p. 81 - 97 (1980).
In Russian.

046.007 Determinación de la longitud por el metodo de
Mayer. Programas de cálculo automático.
M. J. Sevilla, A. Nuñez.
Bol. Inf., Núm. 42, Serv. Geogr. Ejército, 23 + 32 pp.
(1979) = Univ. Complutense - Fac. Cienc., Madrid, Semin.
Astron. Geod., Publ. núm. 102.

046.008 Mutual comparison between two NNS in satellite
Doppler positioning.   K. Sato.
Proc. Int. Latitude Obs. Mizusawa, No. 20, p. 32 - 37 (1981).
In Japanese.

046.009 Satelliten-Dopplermessungen in Potsdam und
Dresden.
I. Fejes, R. Dietrich.
Vermessungstechnik, Heft 3, 4 pp. (1980) = Mitt. Zentralinst.
Phys. Erde Nr. 825.

046.010 On the establishment of a unified astronomical
longitude reference system in Europe.   R. Sigl.
Mitt. Inst. Astron. Phys. Geod. Tech. Univ. München, Nr.
154 (1980).

046.011 Balloon-borne beacons for stellar triangulation.
Á. Czobor, Z. Németh.
Scientific ballooning—II, (see 012.065), p. 29 - 34 (1981).
The method of stellar triangulation proved to be one of
the best to check the orientation of national geodetic networks. The authors suggest that if using laser switched
balloon-borne flash sondes at the 35-36 km ceiling altitudes
these sondes might be used for international purposes as well as
combining national networks in Europe.

046.012 Some discussion on two-dimensional satellite
Doppler positioning.   W.-y. Zhu.
Ann. Shanghai Obs. Acad. Sinica, No. 2, p. 60 - 64 (1980).

Astronomia Nautica.   See Abstr. 003.032.

The development of science at the Astronomical
Observatory of the Leningrad University. Celestial mechanics,
geodesy and gravimetry.   See Abstr. 009.015.

Beitrag zur geodätischen und astronomischen
Nutzung von Sternbedeckungen durch den Mond.
See Abstr. 031.649.

The use of micro-accelerometers for space geodetic
experiments.   See Abstr. 032.577.

Superconducting tensor gravity gradiometer for
satellite geodesy and inertial navigation.
See Abstr. 034.042.

Determinación de la latitud y la longitud por el
metodo de alturas iguales. Programas de cálculo automático.
See Abstr. 045.018.

Resonance orbits of AES and their use in dynamical
problems of space geodesy.   See Abstr. 052.047.

Least squares combination of satellite and terrestrial
data in physical geodesy.   See Abstr. 081.039.

On the use of point mass models of the geopotential
for orbit predictions.   See Abstr. 081.056.

Beitrag zur geodätischen und astronomischen
Nutzung von Sternbedeckungen durch den Mond.
See Abstr. 096.021.

047 Ephemerides, Almanacs, Calendars, Chronology

047.001 1982 Abridged Nautical Almanac. Pub. No. 683.
Published by Hydrographic Office of Japan,
Maritime Safety Agency, Tokyo, Japan. 3 + 243 + 8 pp. (1981).

047.002 Nautisches Jahrbuch oder Ephemeriden und Tafeln
für das Jahr 1982, zur Bestimmung der Zeit, Länge
und Breite auf See nach astronomischen Beobachtungen.
Edited by Deutsches Hydrographisches Institut, Hamburg.
131. Jahrg. 45 + 365 + 30 pp. (1981). ISSN 0077-6211.

047.003 The Handbook of the British Astronomical Associa-
tion 1982.
Prepared by the Computing Section of the Association under
the supervision of G. E. Taylor.
Office of the Association: Burlington House, Piccadilly,
London, W1V 0NL. 97 pp. Price £ 3.00 (1981).

047.004 Almanaque Náutico, 1982. Con suplemento para
la navegación aerea.
Published by Instituto y Observatorio de Marina, San Fernando
(Cádiz). 420 + 30 + 5 pp. (1981). ISBN 84-7469-008-0;
ISSN 0210-735X.

047.005 1982 Nautical Almanac. Pub. No. 681.
Published by Hydrographic Office of Japan,
Maritime Safety Agency, Tokyo, Japan. 6 + 467 + 12 pp.
(1981).

047.006 Astronomical Yearbook for 1981.
Metsniereba, Tbilisi. 175 pp. (1980). In Russian and
Georgian. – Abstr. in Ref. zh., 51. Astron., 7.51.27 (1981).

047.007 Himmelskalender 1982. Ein kleines astronomisches
Jahrbuch für Österreich. H. Mucke.
Astronomisches Büro, Hermann Mucke, Hasenwartgasse 32,
A-1238 Wien, Österreich. 111 pp. (1981).

047.008 Éphémérides Nautiques pour l'an 1982. Ouvrage
publié par le Bureau des Longitudes spécialement
à l'usage des marins.
Gauthier-Villars, Paris, France. 485 pp. (1981). ISBN 2-04-
010678-2.

047.009 Triple conjunctions of the planets.
J. G. Porter.
J. British Astron. Assoc., Vol. 91, 567 - 575 (1981).
This paper gives a general discussion of the conditions
necessary for the occurrence of triple conjunctions, and of
the ten possible cases which may arise with the planets Mars
to Neptune.

047.010 1982 Polaris Almanac for Azimuth Determination.
Pub. No. 685.
Published by Hydrographic Office of Japan. Maritime Safety
Agency, Tokyo, Japan. 15 pp. (1981).

047.011 Astronomische Grundlagen für den Kalender 1983.
Compiled by T. Lederle, edited by Astronomisches
Rechen-Institut, Heidelberg.
Verlag G. Braun, Karlsruhe. 88 pp. Price DM 52.00 (1981).

047.012 Basic formulas for computer programmed computation of cyclic fixed lunisolar types of calendars.
M. Gossler.
Astron. Nachr., Band 302, 161 - 164 (1981). In German.
Direct formulas for mutual transformations of Jewish
calendar dates and Julian Days are given as well as for calculations of Julian and Gregorian Easter Days.

047.013 Annuaire de l'Observatoire Royal de Belgique,1982.
[Jaarboek van de Koninklijke Sterrenwacht van
België].
Imprimerie Hayez, Rue Fin 4, 1080 Bruxelles. 149e année,
227 pp. (1981).

047.014 Kalender für Sternfreunde 1982. Kleines astronomi-
sches Jahrbuch. P. Ahnert.
Johann Ambrosius Barth, Leipzig, German Democratic
Republic. 184 pp. Price M 4.80, DM 8.50 (1981).

047.015 Almanac for computers 1982.
Nautical Almanac Office, United States Naval
Observatory, Washington, D.C. 20390. A14 + B17 + C23 +
D33 + E10 pp. (1981).

047.016 Philippine Astronomical Handbook 1981.
Prepared by the Astronomical Observation Division
of the National Geophysical and Astronomical Office under
the supervision of J. P. Enrile, R. L. Kintanar, E. V. Calpo.
Published by the Philippine Atmospheric, Geophysical and
Astronomical Services Administration. 12 + 62 pp. (1980).
ISSN 0115-1207.

047.017 Tables of sunrise, sunset, twilight, moonrise and
moonset 1981.
Prepared under the supervision of J. P. Enrile, R. L. Kintanar,
E. V. Calpo.
Published by the Philippine Atmospheric, Geophysical and
Astronomical Services Administration.14 + 57 pp. ISSN
0115-3307.

047.018 The frequency distribution of the dates of Easter.
M. E. Davison.
Irish Astron. J., Vol. 14, 156 - 164 (1980).

047.019 Hvězdářská ročenka 1981.
Compiled by V. Guth, B. Onderlička, P. Příhoda,
J. Ruprecht.
Ročník 57, svazek 2; Přehled pokroků v astronomii. Academia,
nakladatelství Československé akademie věd, Praha. 175 pp.
24.00 Kčs (1981).

047.020 Hvězdářská ročenka 1982.
Compiled by P. Příhoda, J. Vondrák, B. Onderlička,
Z. Pokorný.
Ročník 58. Academia, nakladatelstvi Československé akademie
věd, Praha. 149 pp. Price 21.00 Kčs. (1981).

047.021 Observer's Handbook 1982.
R. L. Bishop (Editor).
Published by Royal Astronomical Society of Canada,
124 Merton Street, Toronto, Ontario, M4S 2Z2. 168 pp.
(1981). ISSN 0080-4193.

047.022 Efemérides Astronómicas 1982.
Published by Instituto y Observatorio de Marina,
San Fernando (Cádiz), Spain. Vol. 191, 16 + 404 pp. (1981).
ISBN 84-7469-010-2, ISSN 0080-5971.

047.023 1983 Japanese Ephemeris. Pub. No. 684.
Compiled by A. Yamazaki, T. Mori, Y. Tano,
A. Senda, Y. Kubo, Y. Ganeko, T. Takemura, F. Ono,
Y. Harada, K. Inoue, T. Uchiyama, M. Sasaki.
Published by Hydrographic Department, Maritime Safety
Agency, Tokyo, Japan. 6 + 466 + 20 pp. (1981).

047.024   Ephémérides astronomiques et calendrier des
          événements célestes pour 1982.
P. Cugnon, A. Koeckelenbergh.
Ciel Terre, Vol. 97, No. 6, 90 pp. (1981).

047.025   Rocznik Astronomiczny Obserwatorium Krakows-
          kiego 1982. International Supplement Nr. 53.
K. Rudnicki, P. Flin (Editors).
Państwowe Wydawnictwo Naukowe, Kraków. 7 + 137 pp.
Price zł 80.00 (1981). ISBN 83-01-03969-8, ISSN 0075-7047.
See abstracts 047.026, 119.049 - 119.051, 122.112.

047.026   Geocentric ephemerides for the year 1982 of the
          libration points L_4 and L_5 in the Earth-Moon
system and in the Sun-Venus system.
K. Kordylewski, R. Szafraniec.
Rocznik Astronomiczny Obserwatorium Krakowskiego 1982,
(see 047.025), p. 131 - 136 (1981).

047.027   Anuarul Astronomic 1982.
          G. Stănilă, M. Stavinschi, S. Dinulescu (Editors).
Centrul de Astronomie şi ştiinţe spaţiale, Bucureşti. Editura
Academiei Republicii Socialiste România, R 79717 Calea
Victoriei 125, Bucureşti. 319 pp. Price Lei 15.50 (1981).

047.028   The Astronomical Almanac for the year 1982.
          Data for astronomy, space sciences, geodesy,
surveying, navigation and other applications.
Issued by the Nautical Almanac Office, United States Naval
Observatory, Washington; and Her Majesty's Nautical Almanac
Office, London. For sale by the Superintendent of Documents,
U.S. Government Printing Office, Washington, D.C. 20402 and
Her Majesty's Stationery Office, London. 9 + 554 pp. (1981).
ISBN 0-11-886904-3.

047.029   The Air Almanac 1982, January - June.
          Air Publication 1602, issued by Her Majesty's
Nautical Almanac Office, London; and Nautical Almanac
Office, United States Naval Observatory, Washington. Her
Majesty's Stationery Office, London. For sale by the Super-
intendent of Documents, US Government Printing Office,
Washington, D. C. 20402. p. 1 - 364, A 104 + F 4 pp.
Price £ 17.50 (1981). ISBN 0-11-772266-9.

047.030   Astronomický kalendár na rok 1982.
          E. Pittich (Editor).
Krajská hvezdáreň, Hlohovec. 200 pp. (1981).

047.031   About the calculation of phenomena predictions of
          Jupiter's satellites.   J. Vondrák.
Říše hvězd, Vol. 62, 225 - 230 (1981). In Czech.

047.032   Vår tidsregning og Dionysius Exiguus.
          G. Teres.
Astron. Tidsskr., Årg. 14, 154 - 160 (1981).

047.033   Almanacco Astronomico della Rivista Coelum per
          l'anno 1982.   P. Battistini.
Coelum, Suppl. al fasc. 11 - 12, 43 + 38 pp. Price L 8,000
(1981).

047.034   Muster-Kalender für das Jahr 1982.
          Compiled by H. Haupt.
Published by Verlag Georg Fromme & Co., Wien, Austria.
36 pp. (1980).

047.035   Nautisches Jahrbuch 1982.
          Edited by Seehydrographischer Dienst der
Deutschen Demokratischen Republik, Rostock. 32. Jahrg.
45 + 366 pp. (1981).

047.036   Annuaire du Bureau des Longitudes. Éphémérides
          1982. Calendriers–soleil–lune–planètes–satellites–
étoiles–marées–déclinaison magnétique.
Gauthier-Villars, Paris, France. 15 + 256 pp. (1981). ISBN
2-04-011368-1 = Supplément à l'Astronomie de Janvier 1982.

047.037   Satellites galiléens de Jupiter. Phénomènes et
          configurations pour 1982.
B. Morando (Directeur du Service des Calculs), J.-E. Arlot,
Y. Jannot, W. Thuillot, D. T. Vu (Rédaction et calculs).
Supplément à la Connaissance des Temps, Bureau des
Longitudes, Paris, France. 61 pp. (1981).

047.038   Éphémérides des huit premiers satellites de Saturne
          pour 1982.
Supplément à la Connaissance des Temps, Bureau des
Longitudes, Paris, France. 46 pp. (1981).

047.039   Apparent Places of Fundamental Stars 1983.
          Containing the 1535 stars in the Fourth Funda-
mental Catalogue (FK4).
Edited by Astronomisches Rechen-Institut, Heidelberg,
under the supervision of W. Fricke, T. Lederle, H. Schwan.
Published and produced by G. Braun, 7500 Karlsruhe,
F. R. Germany. 44 + 510 pp. Price DM 52.00 (1981).
ISBN 3-7650-0083-3; ISSN 0174-254-X.

047.040   Chinese Astronomical Ephemeris 1982.
          Published by Purple Mountain Observatory,
Academia Sinica, Nanking, China. 4 + 541 pp. (1981).
In Chinese.

047.041   Astronomical ephemerides for the year 1982.
          Z. Ivanović, N. Čabrić, A. Tomić.
Supplement to Vasiona, Année 29, No. 4, 85 - 106 (1981).
In Croatian.

047.042   Anuario del Observatorio Astronomico de Madrid
          para 1982.
Published by Instituto Geografico Nacional, Madrid. 449 pp.
Price 300 pesetas (1981). ISBN 84-500-3470-1, ISSN 0373-
5125.

047.043   Dados Astronómicos para os Almanaques de 1982
          para Portugal.   E. M. L. Cabrita.
Published by Observatório Astronómico de Lisboa. 56 pp.
(1981).

047.044   Almanacco 1982.
          Astronomia, N. 4, p. 1 - 124 (1981).

047.045   Compact data for navigation and astronomy for
          1981 to 1985.   B. D. Yallop.
R. Greenwich Obs. Bull., No. 185, 93 pp. (1981).

047.046   Astronomical Handbook for 1981.   B. M. Lewis.
          Carter Obs., Astron. Bull., No. 94, 32 pp. (1980).

047.047   On the simplest calendar.   L. E. Majstrov.
          Researches in the history of astronomy. 15th issue,
(see 003.025), p. 361 - 364 (1980). In Russian.

          Calendar and chronology.   See Abstr. 003.085.

          Ein Tempel als Kalender. Astronomie im alten Peru.
See Abstr. 004.014.

          What time does the sun rise and set?
See Abstr. 021.034.

The origin of the Julian Period: an application of congruences and the Chinese Remainder Theorem.
See Abstr. 021.036.

Contact timings of solar eclipse from photographs.
See Abstr. 031.561.

Geocentric ephemeris of Pluto 1975 - 1985.
See Abstr. 101.037.

Space Research

051 Extraterrestrial Research, Spaceflight Related to Astronomy and Astrophysics

051.001 **Die 54 Stunden der Columbia.** J. von Puttkamer.
Umschau, 81. Jahrg., 434 - 436 (1981).

051.002 **Zwanzig Jahre bemannte Raumfahrt.**
S. Jähn.
Sterne, 57. Band, 139 - 141 (1981).

051.003 **Some modern astronomy from space.**
R. L. F. Boyd.
J. British Astron. Assoc., Vol. 91, 453 - 461 (1981).

051.004 **Spaceflight of a Soviet-Mongolian crew.**
S. A. Nikitin.
Priroda, 1981, No. 7, p. 103 - 105. In Russian.

051.005 **VOIR mission: the global mapping of Venus.**
S. J. Kerridge.
Bull. American Astron. Soc., Vol. 13, 511 (1981). − Abstract.

051.006 **The astrometric satellite HIPPARCOS.**
H. Scholl.
Bull. American Astron. Soc., Vol. 13, 569 (1981). − Abstract.

051.007 **The fifth main experiment aboard Salyut 6.**
S. A. Nikitin.
Priroda, 1981, No. 8, p. 103 - 104. In Russian.

051.008 **Halley and contemporary space science.**
R. L. F. Boyd.
Observatory, Vol. 101, 149 - 164 (1981).

051.009 **Space astronomy in the European Space Agency.**
F. Macchetto.
Mem. Soc. Astron. Italiana, Vol. 52, (see 012.019), 95 - 116 (1981).

051.010 **Midcourse navigation for the European Comet Halley Mission.** M. Hechler, F. W. Hechler.
Spacecraft flight dynamics, (see 012.021), p. 189 - 195 (1981).
The European Comet Halley Mission involves a sequence of mid-course orbit corrections to remove deviations from the nominal flight path, as indicated by orbit determination. As a function of the spacecraft design, several manoeuvre execution modes are studied, and the corresponding optimization problems are discussed.

051.011 **The observation planning facility for the Infra-Red Astronomical Satellite (IRAS).**
E. Oord, P. J. de Pagter, R. C. van Holtz, J. R. Mcdougall.
Spacecraft flight dynamics, (see 012.021), p. 269 - 280 (1981).
The mission objective of the Infrared Astronomical Satellite (IRAS) is to perform an all-sky survey in the infrared and to carry out additional observations on specific sources. This paper describes the software system that will generate and schedule observations to perform these two tasks.

051.012 **Missions to comets: the perspective in 1980.**
L. L. Wilkening.
Comets and the origin of life, (see 012.022), p. 249 - 254 (1981).
The impending return of Comet Halley to the inner solar system in 1985 - 1986 has prompted several countries to design missions to take advantage of this rare opportunity. The European Space Agency, Soviet Union and Japan have missions to study Comet Halley. The United States has not committed itself to a Halley mission, but a rendezvous mission with a short period comet remains among the National Aeronautics and Space Administration's high priorities for the 1980's.

051.013 **Letzte Runde vor der langen Pause. Voyager 2 nach dem Saturn auf dem Weg zum Planeten Uranus.**
W. Engelhardt.
Umschau, 81. Jahrg., 634 (1981).

051.014 **A pictoral history of the NASA Lunar and Planetary Exploration Program.** S. E. Dwornik.
Mem. Soc. Astron. Italiana, Vol. 52, (see 012.023), 399 - 400 (1981).

051.015 **HIPPARCOS: space astrometry mission. A report on the project − I.** M. A. C. Perryman.
Bull. Inf. Cent. Données Stellaires, No. 21, p. 40 - 44 (1981).

051.016 **A new trajectory concept for exploring the Earth's geomagnetic tail.** R. W. Farquhar, D. W. Dunham.
J. Guid. Control, Vol. 4, 192 - 196 (1981). − Abstr. in Phys. Abstr., Vol. 84, Abstr. 88706 (1981).

051.017 **Probing the solar poles.**
S. A. Haines.
Sci. Dimension, Vol. 13, No. 1, p. 12, 14, 16 (1981). − Abstr. in Phys. Abstr., Vol. 84, Abstr. 94143 (1981).

051.018 **Future prospects for γ-ray astronomy.**
C. Fichtel.
Philos. Trans. R. Soc. London, Ser. A, Vol. 301, 693 - 701 (1981). − Abstr. in Phys. Abstr., Vol. 84, Abstr. 94205 (1981).

051.019 **Cosmic velocities of the present and future.**
A. A. Sternfeld.
Zemlya Vselennaya, 1981, No. 4, p. 40 - 43. In Russian.

051.020 **Entry and landing probe for Titan.**
J. P. Murphy, J. N. Cuzzi, A. J. Butts, P. C. Carroll.
J. Spacecr. Rockets, Vol. 18, 157 - 163 (1981). − Abstr. in Phys. Abstr., Vol. 84, Abstr. 98660 (1981).

051.021 **Project "Bolgariya 1300".** S. A. Nikitin.
Priroda, 1981, No. 11, p. 104 - 106. In Russian.

051.022 **The Voyager mission through the Jupiter encounters.**
E. C. Stone.
J. Geophys. Res., Vol. 86, 8123 - 8124 (1981).

051.023 **Hipparcos, the European Space Agency's astrometric satellite.**   J. Kovalevsky, R. Pacault.
Aeronaut. Astronaut., No. 89, p. 21 - 26 (1981). In French.
Abstr. in Phys. Abstr., Vol. 84, Abstr. 102008 (1981).

051.024 **Assessment of the Cosmos Programme 1980.**
D. Laurent.
Aeronaut. Astronaut., No. 89, p. 44 - 50 (1981). In French.
Abstr. in Phys. Abstr., Vol. 84, Abstr. 102009 (1981).

051.025 **Satellites for the detection of nonsolar planets.**
L. W. Bandermann, R. H. Pohle, J. P. Murphy,
J. W. Vorreiter, D. C. Black.
J. Spacecr. Rockets, Vol. 18, 164 - 171 (1981). – Abstr. in
Phys. Abstr., Vol. 84, Abstr. 102012 (1981).

051.026 **A view of satellite development in Japan.**
M. Hirai.
J. Inst. Electron. Commun. Eng. Japan, Vol. 64, 77 - 81
(1981). In Japanese. – Abstr. in Phys. Abstr., Vol. 84,
Abstr. 102016 (1981).

051.027 **Japan aims for Moon – literally.**
MSN Microwave Syst. News, Vol. 11, No. 3, p. 19 -
20 (1981). – Abstr. in Phys. Abstr., Vol. 84, Abstr. 102017
(1981).

051.028 **Ultraviolet astronomy with the International Ultraviolet Explorer.**   P. M. Gondhalekar.
Bull. Astron. Soc. India, Vol. 9, 112 - 138 (1981).

051.029 **Astronautics in the year 1980.**
M. Grün, P. Koubský.
Říše hvězd, Vol. 62, 157 - 160 (1981). In Czech.

051.030 **The International Solar Polar Mission (ISPM).**
R. G. Marsden, K.-P. Wenzel.
Plasma astrophysics, (see 012.042), p. 167 - 175 (1981).
The paper presents an overview of the scientific objectives of the instrumentation chosen jointly by ESA and NASA to carry out the exploratory investigations unique to ISPM. A brief description of the mission time-lines and trajectory will also be given.

051.031 **X-ray astronomy in the future.**
J. L. Culhane.
Space Sci. Rev., Vol. 30, (see 012.044), 537 - 564 (1981).
After a brief outline of major scientific achievements of X-ray astronomy in the past decade the author reviews the state of various future projects in this field. Scientific objectives and expected performance parameters of the European X-80 and NASA's Advanced X-ray Astronomy Facility (AXAF) are discussed in detail. Preliminary considerations for the ultimate sensitivity X-ray telescope (LAMAR) studied by NASA are described.

051.032 **The ROSAT mission.**   B. Aschenbach,
H. Bräuninger, U. Briel, W. Brinkmann, H. Fink,
N. Heinecke, H. Hippmann, G. Kettenring, G. Metzner,
A. Ondrusch, E. Pfeffermann, P. Predehl, G. Reger,
K.-H. Stephan, J. Trümper, H. U. Zimmermann.
Space Sci. Rev., Vol. 30, (see 012.044), 569 - 573 (1981).
The primary scientific objective of the ROSAT mission is to perform the first all sky survey with an imaging X-ray telescope leading to an improvement in sensitivity by several orders of magnitude compared with previous surveys. Consequently a large number of new sources ($> 10^5$) will be discovered and located with an accuracy of 1 arcmin. The X-ray telescope consists of a fourfold nested Wolter type I mirror system with 80 cm aperture and 240 cm focal length, and three focal plane detectors. In the baseline version these will be imaging

proportional counters (0.1 - 2 keV) providing a field of view of $2° \times 2°$.

051.033 **Six years of gamma-ray astronomy with Cos-B.**
B. G. Taylor, R. D. Wills.
ESA Bull., No. 28, p. 48 - 61 (1981).

051.034 **Dynamics Explorer program: an overview.**
R. A. Hoffman, E. R. Schmerling.
Space Sci. Instrum., Vol. 5, (see 003.019), 345 - 348 (1981).

051.035 **Dynamics Explorer spacecraft and ground operations systems.**   R. A. Hoffman, G. D. Hogan,
R. C. Maehl.
Space Sci. Instrum., Vol. 5, (see 003.019), 349 - 367 (1981).
The Dynamics Explorer program of NASA is designed to study the coupling of energy, electric currents, electric fields, and plasmas between the atmosphere, ionosphere, and the magnetosphere. The program is composed of two well instrumented polar orbiting satellites and ground operations and data handling systems which have been designed to acquire, process, and make readily available to the scientists of the program the data sets necessary for analysis to accomplish the science objectives. These systems are described with emphasis placed on the relationship of the two instrumented spacecraft and the ground systems to their roles in meeting the program requirements. In addition, the characteristics of the orbits selected are specified.

051.036 **Anticipated results from dust experiments on cometary missions.**
J. Kissel, H. Fechtig, E. Grün.
Modern observational techniques for comets, (see 012.047),
p. 33 - 38 (1981).

051.037 **New problems of cometary observations from space.**
O. V. Dobrovolsky (*Dobrovol'skij*), S. I. Ibadov.
Modern observational techniques for comets, (see 012.047),
p. 39 - 41 (1981).

051.038 **Results to be expected from light scattering dust analyzer during a rendezvous mission.**
R. H. Zerull, R. H. Giese, B. Kneissel.
Modern observational techniques for comets, (see 012.047),
p. 72 - 76 (1981).

051.039 **Near-perihelion observations of comet Halley from Shuttle Orbiter.**   J. T. Bergstralh.
Modern observational techniques for comets, (see 012.047),
p. 241 - 248 (1981).
The goals of cometary research, articulated by several comet science working groups, imply that it would be desirable to (1) observe comets from space, and (2) to make synoptic sequences of comet observations. Intercept missions and the Space Telescope will return unique data on comet Halley, but will leave important gaps in the observational coverage of the comet's activity, especially around the time of perihelion passage. A cometary instrument package of modest size could be assembled to share space in the Shuttle cargo bay with other payloads; this approach should be economical enough to permit scheduling such a package for several flights during Halley's apparition, and thus partially fill the observational gaps left by ST and the intercept missions.

051.040 **Plans for comet Halley.**
J. Rahe.
Modern observational techniques for comets, (see 012.047),
p. 277 - 283 (1981).

051.041 **The ESA mission to comet Halley.**
R. Reinhard.
Modern observational techniques for comets, (see 012.047),

p. 284 - 312 (1981).

The mission scientific objectives are given, followed by a somewhat more detailed description of the 10 Giotto scientific instruments. The principles are explained on which the experiments are based, and the experiment key performance data are summarized. "Mission analysis" focusses on the launch constraints, describes the heliocentric transfer trajectory and, in more detail, the encounter scenario. The Giotto spacecraft major design criteria are explained leading to a brief discussion of the spacecraft subsystem. The ground system design as an integral part of the mission is described. Particular attention, finally, is drawn to the problem of hypervelocity dust particle impacts in the innermost part of the coma and the problem of spacecraft survival, and the adverse effects of impact-generated plasma around the spacecraft.

051.042   **On the measurement of proper motions of star clusters with the European astrometry satellite HIPPARCOS.**   R. Wielen.
Mitt. Astron. Ges., Nr. 54, (see 012.050), p. 268 (1981).

051.043   **Astronomische Beobachtungen mit dem IUE-Satelliten.**   R. Albrecht, H. M. Maitzen, K. Rakos.
Sternenbote, 24. Jahrg., 146 - 152 (1981).

051.044   **Visit to space.**   M. M. Novaković.
Vasiona, Année 29, 35 - 42 (1981). In Serbo-Croatian.

051.045   **Rendezvous with comet Halley.**   U. Dzērvītis.
Zvaigžnota debess, 1980/81. gada ziema, p. 12 - 19. In Latvian.

051.046   **Viking – end and continuation.**   E. Mūkins.
Zvaigžnotā debess, 1981. gada pavasaris, p. 19 - 23. In Latvian.

051.047   **A proposal to PLANET–B drag-free satellite and Mercury observation.**
N. Kawashima, S. Sasaki, M. Yanagisawa.
Proceedings of the 14th ISAS Lunar and Planetary Symposium, (see 012.055), p. 350 - 363 (1981).

A verification experiment of general relativity and Mercury observation is proposed for ISAS PLANET–B mission. A drag-free system will be employed to perform a very accurate determination of the group delay of tracking waves of the artificial planet by the solar gravity. CCD imaging system will be employed for Mercury observation. The present status of hardware development for both drag-free detection subsystem and CCD subsystem is described.

051.048   **Laser ranging from satellites orbiting around the moon and Mercury – a proposal to the solid planets exploration campaign in Japan.**   T. Nakamura.
Proceedings of the 14th ISAS Lunar and Planetary Symposium, (see 012.055), p. 372 - 379 (1981).

The possibility of laser ranging from an orbiter to the retroreflectors thrown down onto the surfaces of the moon and Mercury is explored. The scientific significance of this mission is discussed specifically from the viewpoint of solid planets dynamics. Brief specifications of a laser ranging system are proposed and several necessary items to materialize this mission are discussed in connection with the space projects now proposed in Japan.

051.049   **System study of lunar exploration.**
H. Koshiishi, T. Matsuda.
Proceedings of the 14th ISAS Lunar and Planetary Symposium, (see 012.055), p. 380 - 387 (1981).

System study of lunar exploration by a polar orbiter was done. The study consists of four parts: (1) Set up of mission (2) Orbit analysis (3) Launch vehicle, it's guidance and control

(4) Orbiter and remote sensors. As a result of the study it was concluded that an orbiter weighing about 750 kg can be put into polar orbit with 100 km altitude by a candidate launch vehicle H–IA rocket.

051.050   **Recent developments in space-borne zodiacal light photometry.**   G. H. Schwehm, R. H. Giese, F. Giovane, D. W. Schuerman, J. L. Weinberg.
Progress in planetary exploration, (see 012.062), p. 121 - 125 (1981).

The 1985 International Solar Polar Mission will provide the first opportunity to perform measurements from out of the ecliptic and from above the solar poles. Included in this mission is the Zodiacal Light/Background Starlight Experiment of the Ruhr-University Bochum, FRG (in collaboration with the State University of New York at Albany). The experiment is based on the use of a multicolor, sky-scanning photopolarimeter, which will be carried on the NASA S/C. The scientific objectives of the experiment are to investigate the intensity, polarization, and color of the diffuse sky brightness and determine the spatial distribution and physical properties of the interplanetary dust, including a possible interstellar component, as a function of S/C position in and out of the ecliptic.

051.051   **A European probe to comet Halley.**
W. Trogus, R. Ockert, R. D. Auer.
Progress in planetary exploration, (see 012.062), p. 131 - 136 (1981).

A European probe to comet Halley is proposed. The probe's model payload consists of 8 scientific instruments, viz. neutral, ion and dust impact mass spectrometers, magnetometer, medium energy ion and electron analyzer, camera, dust impact detectors and plasma wave experiment. Fly-by of the comet Halley nucleus will take place on November 28th, 1985, at about 500 km miss distance. The main spacecraft serves as relay link to transmit the observed data to Earth. As probe, a modified ISEE 2 design is proposed. Because of the cometary dust hazard expected in the coma a heavy dust shield (27 kg) is required, consisting of a thin front sheet and a 3 layer rear sheet.

051.052   **DISCO. Re-assessment study.**
European Space Agency, SCI (81)6, Paris. 5 + 55 pp. (1981).

Revised scientific objectives for DISCO and a corresponding model payload are presented and discussed. It is planned to investigate the physics of the Sun's interior by measuring global oscillations in the visible and by measuring variations of the solar constant and the spectral irradiance. It is proposed to study large-scale temperature and density distributions in the source regions of the solar wind by means of a far-UV spectroheliograph and to monitor the interplanetary plasma, energetic charged particles and solar radio emissions in support of the ISPM mission.

051.053   **Spacecraft discover the nature of planets.**
V. S. Avduevskij.
Nauk. i chelovechestvo, 1981. Moskva, 1981, p. 191 - 203. In Russian. – From Ref. zh., 51. Astron., 1.51.214 (1982).

051.054   **Synoptic studies of the middle atmosphere using balloons.**   S. A. Bowhill.
Scientific ballooning–II, (see 012.065), p. 13 - 16 (1981).

During the Middle Atmosphere Program, 1982–1985, balloons will play a central role in gathering stratospheric, chemical, and dynamical data. The paper discusses some of the unique aspects of balloons as tools for probing the middle atmosphere, and suggests some directions for future projects and enhanced capabilities that would improve their utility in future investigations.

051.055 **Balloons on other planets.**   J. Blamont.
Scientific ballooning–II, (see 012.065), p. 63 - 69 (1981).

A description of the balloon systems designed for the exploration of Venus (constant levels of 55 km and 18 km, and aerostats able to take off from the ground and reach 60 km of altitude) and of balls to be used as small rovers on the soil of Mars is given.

051.056 **UV astronomy with balloons.**   M. Golay.
Scientific ballooning–II, (see 012.065), p. 71 - 79 (1981).

A stabilized astronomical gondola was carrying a Schmidt photographic camera with UV microchannel plate image converter-intensifier. Systematic surveys of the galactic plane and the galactic poles are in progress.

051.057 **Long duration hard X-ray transatlantic payload.**
C. D. LaPadula, A. Bazzano, L. Boccaccini, M. Mastropietro, R. Patriarca, V. F. Polcaro, P. Ubertini.
Scientific ballooning–II, (see 012.065), p. 111 - 114 (1981).

The note describes the HXR 80M large area hard X-ray Astronomy experiment. The payload is scheduled for a flight on board of a transatlantic balloon in the framework of the CNR experimental transatlantic campaign.

051.058 **The European X-ray astronomy satellite EXOSAT observer's guide. Part I: Executive summary.**
Part II: Observer's handbook.
Published by Space Science Department of the European Space Agency, 12 + 153 pp. (1981).

Part I provides a brief summary of the EXOSAT mission, the mission objectives, and the various sensitivities of the payload. Part II represents the detailed observer's handbook to the observatory, including proposal submission.

051.059 **Hipparcos.** Space astrometry mission. Answer to the announcement of opportunity for the compilation of the input catalogue.
Published by Observatoire de Paris–Meudon. 288 pp. (1981).

051.060 **Hipparcos.** Proposal for scientific data processing.
Fundamental Astronomy by Space Techniques Consortium. European Space Agency, ESA HIP 81/02 (1981).

051.061 **The golden age of solar system exploration.**
N. W. Hinners.
The new solar system, (see 003.026), p. 3 - 10 (1981).

051.062 **The Voyager encounters.**   B. A. Smith.
The new solar system, (see 003.026), p. 105 - 116 (1981).

051.063 **Space report.**
Spaceflight, Vol. 23, 252 - 256, 297 - 301 (1981).

NSSDC Data Listing.   See Abstr. 002.084.

Space activities in the eighties. Vol. 1: Survey of space programmes. Vol. 2: Detailed presentation of the European Space Industry (1981). Vol. 3: Outlook of space industry outside Europe (1981).   See Abstr. 003.055.

Beyond the atmosphere. Early years of space science.   See Abstr. 003.117.

Future programme of X-ray astronomy in Japan. See Abstr. 013.035.

Soviet-Indian research of cosmic radiation by high altitude balloons.   See Abstr. 013.050.

Survival of microorganisms in space: a review. See Abstr. 015.049.

Numerical simulation of the signal and data processing of the HIPPARCOS satellite. See Abstr. 021.058.

The EXOSAT mission.   See Abstr. 032.567.

Experimental gravitation with measurements made from within a planetary system.   See Abstr. 066.182.

Seismic exploration of Mercury: an assessment. See Abstr. 092.008.

Millimeter-wave experiments for cometary space missions.   See Abstr. 102.057.

Comet Halley 1986 as a target of space probes. See Abstr. 103.805.

Zodiacal light and space observation of faint objects. See Abstr. 106.056.

Results of cosmic radiation dose field measurements aboard the "Salyut-6" orbital station.   See Abstr. 143.117.

052 Astrodynamics, Navigation of Space Vehicles

052.001 The influence of earth-reflected solar radiation on the motion of artificial satellites: the problem of perturbations.   E. N. Polyakhova.
Tr. Astron. Obs., Leningrad, Tom 36 = Uch. Zap. Leningr. Univ., No. 402 = Ser. mat. nauk, Vyp. 58, 220 - 238 (1981). In Russian.
The effects of terrestrial albedo radiation pressure on the motion of artificial satellites are described. The general aspects of the last 15 years are discussed.

052.002 On the third order artificial satellite theory.
F. Tong, L.-d. Wu.
Acta Astron. Sinica, Vol. 22, 138 - 144 (1981). In Chinese.
The authors investigate in this paper a principal approach to solve the perturbation equations of the Vinti's intermediary and propose a practical semi-analytical and seminumerical method based on the average elements method.

052.003 Combined perturbations on the orbit of close earth satellites.   L. Liu, D.-z. Zhao.
Acta Astron. Sinica, Vol. 22, 145 - 158 (1981). In Chinese.
Using an effective semi-analytic method, the authors obtained the first order solution of orbital perturbations for a near earth satellite, including the disturbing effects of the earth's figure and air drag.

052.004 A second-order satellite orbit theory, with compact results in cylindrical coordinates.   R. H. Gooding.
Philos. Trans. R. Soc. London, Ser. A, Vol. 299, 426 - 474 (1981). − Abstr. in Phys. Abstr., Vol. 84, Abstr. 71062 (1981).

052.005 Definition and analysis of properties of optimal three-impulse point-to-orbit transfers under time constraint.   V. V. Ivashkin, A. P. Skorokhodov.
Acta Astronaut., Vol. 8, 11 - 23 (1981). − Abstr. in Phys. Abstr., Vol. 84, Abstr. 74995 (1981).

052.006 Unmodeled perturbation effect on the orbit determination accuracy.
B. T. Bakshian (B. Ts. Bakhshiyan), R. R. Nazirov, P. E. Eliasberg (Ehl'yasberg).
Acta Astronaut., Vol. 8, 25 - 29 (1981). − Abstr. in Phys. Abstr., Vol. 84, Abstr. 74996 (1981).

052.007 A new treatment of Lambertian mechanics.
F. T. Sun.
Acta Astronaut., Vol. 8, 105 - 122 (1981). − Abstr. in Phys. Abstr., Vol. 84, Abstr. 83459 (1981).

052.008 Analyses of the US Navy orbits of 1963 − 24B and 1974 − 34A at 15th-order resonance.
D. M. C. Walker.
Geophys. J. R. Astron. Soc., Vol. 67, 1 - 18 (1981).

052.009 Third order solution of orbital perturbations of artificial earth satellites.   L. Liu, D. Zhao.
Sci. Sinica, Vol. 24, 813 - 828 (1981).

052.010 Relativistic astrodynamics: non-rectilinear trajectories for star exploration flights.   G. Vulpetti.
J. British Interplanet. Soc., Vol. 34, 477 - 485 (1981).

052.011 Determination of the parameters of precession from a light curve.   A. S. Kuznetsov, A. V. Ehkalo.
Leningr. ehlektrotekh. inst. im. V. I. Ul'yanova (Lenina).
Leningrad, 1980. 8 pp. In Russian. − Abstr. in Ref. zh., 51. Astron., 7.51.74; 62. Issled kosm. prostranstva, 7.62.125 (1981).

052.012 Some questions of the dynamics of strongly charged bodies in the cosmic space.
S. A. Vavilov, E. K. Kolesnikov.
Fiz. mekh., Leningrad, 1980, No. 4, p. 168 - 180. In Russian. Abstr. in Ref. zh., 62. Issled. kosm. prostranstva, 7.62.120 (1981).

052.013 Flights to short-period comets in one-pulse flight trajectories.
A. V. Leshchenko, V. A. Kotin, O. V. Papkov.
Nauchn. chteniya po aviats. i kosmonavt. 10-e Gagarinsk. chteniya, 1980. Moskva, 1981, p. 604. In Russian. − Abstr. in Ref. zh., 62. Issled. kosm. prostranstva, 8.62.141 (1981).

052.014 Collisions of artificial earth orbiting bodies.
L. Sehnal, L. Pospíšilová.
Bull. Astron. Inst. Czechoslovakia, Vol. 32, 310 - 315 (1981).
The study concerns a statistical treatment of a probable collision frequency between man-made objects in the earth's environment.

052.015 A SIRIO-2 laser ranging experiment.
J. M. Dow, W. Flury.
Spacecraft flight dynamics, (see 012.021), p. 13 - 23 (1981).
On the basis of existing orbit determination software, ESOC is preparing to make use of laser ranging data from the SIRIO-2 LASSO experiment (Laser Synchronisation from Sationary Orbit). Four areas have been proposed for investigation: precise orbit determination, improvement of resonant gravity coefficients, polar motion, and calibration of the ESA VHF ranging system. The paper outlines the aims of this offline software experiment and the preparations under way in support of them.

052.016 Accurate orbit determinations from laser range observations of LAGEOS, STARLETTE and GEOS-3.
K. F. Wakker, B. A. C. Ambrosius.
Spacecraft flight dynamics, (see 012.021), p. 25 - 35 (1981).

052.017 Modelling of orbital perturbations due to radiation pressure for high earth satellites.
L. Anselmo, P. Farinella, A. Milani, A. M. Nobili.
Spacecraft flight dynamics, (see 012.021), p. 47 - 52 (1981).
The force model of radiation pressure is discussed for spacecraft of complex shape and structure. By general perturbations techniques, orbital effects in semimajor axis and in longitude are computed and divided in long-and short-periodic ones. As an example the perturbations on the orbit of ESA SIRIO 2 satellite are computed.

052.018 Present status and future trends in near-Earth satellite orbit determination.   A. J. Fuchs.
Spacecraft flight dynamics, (see 012.021), p. 287 - 292 (1981).
The scope of the paper is limited to unmanned artificial satellites and near-Earth is defined to include satellites within the geosynchronous distance of the Earth. The major components of an orbit determination system and the evolution of the elements making up each component is reviewed. Typical accuracies presently achievable in the orbit determination process are summarized as well as the factors limiting the accuracies and some interesting examples of recent improvements made in the dynamic models used in the process.

052.019 An analytical singularity-free orbit predictor for near-earth satellites.   G. Scheifele.
Spacecraft flight dynamics, (see 012.021), p. 299 - 305 (1981).
A completely analytical, first order satellite theory intended for low earth orbits is presented. Perturbations are

unified under the non-singular Poincaré-Similar element formation. The theory includes short period, long period and secular effects of J_2 and all higher zonal harmonics; secular and quadratic effects of atmospheric drag; and the average mean motion considering all harmonics of the geopotential.

052.020 **Some aspects in measuring and correcting of satellite orbits.** Z.-y. Liu, H.-b. Gao.
Spacecraft flight dynamics, (see 012.021), p. 307 - 309 (1981).
This article deals with methods of orbit determination and orbit correction.

052.021 **SIRIO-2 L.A.S.S.O. mission: geosynchronous satellite laser station ranging-data orbit determination simulations.** G. Vulpetti.
Spacecraft flight dynamics, (see 012.021), p. 313 - 324 (1981).

052.022 **Orbit and attitude determination of Bhaskara.** R. S. Bhat, P. Rajendra Prasad, N. Ramani, P. Padmanabhan, K. Harendranath.
Spacecraft flight dynamics, (see 012.021), p. 325 - 332 (1981).
An orbit determination program based on R. H. Merson's analytical theory of satellite motion was developed. The post facto nominal position accuracy achieved during the initial period of the mission was of the order of 500 meters.

052.023 **Orbits near critical inclinations, including lunisolar perturbations.** M. E. Hough.
Celestial Mech., Vol. 25, 111 - 136 (1981).
An improved theory is presented of long period perigee motion for orbits near the critical inclinations 63.4° and 116.6°. Inclusion of lunisolar perturbations and all measured zonal harmonic coefficients from a recent Earth model are significant improvements over existing theories. Numerical studies show that lunisolar perturbations control stability of equilibria for orbits with semimajor axes exceeding 1.4 Earth radii. Moreover, a theory which includes lunisolar perturbations predicts larger maximum fluctuations in eccentricity and faster oscillations near stable equilibria compared to a theory which models only the zonal harmonics.

052.024 **Sun-synchronous orbits near critical inclination.** M. E. Hough.
Celestial Mech., Vol. 25, 137 - 157 (1981).
The long period dynamics of Sun-synchronous orbits near the critical inclination 116.6° are investigated. It is known that, at the critical inclination, the average perigee location is unchanged by Earth oblateness. For certain values of semimajor axis and eccentricity, orbit plane precession caused by Earth oblateness is synchronous with the mean orbital motion of the apparent Sun (a Sun-synchronism). It is shown that the total number and infinitesimal stability of equilibrium solutions can change abruptly over the Sun-synchronous range of semimajor axis values (1.54 to 1.70 Earth radii). The effect of direct solar radiation pressure upon certain stable equilibria is investigated.

052.025 **Effects of the terrestrial infrared radiation pressure on the motion of an artificial satellite.** L. Sehnal.
Celestial Mech., Vol. 25, 169 - 179 (1981).
The changes of the orbital motion of an artificial satellite caused by the terrestrial infrared radiation pressure are studied. The infrared radiation is described as a series of spherical harmonics and only the zero-th and second-order harmonics are taken into account. The expressions for the disturbing components and for the changes of the orbital elements are given together with numerical examples. The comparison of the computed magnitudes of the disturbing force with observed data shows good agreement.

052.026 **On program realization of a semi-analytical method for computation of the motion of a stationary artificial earth satellite.** M. A. Vashkov'yak, V. A. Lyakhova, O. S. Ryzhina.
Astron. i geod., Tomsk, 1980, No. 8, p. 29 - 31. In Russian. Abstr. in Ref. zh., 51. Astron., 8.51.90 (1981).

052.027 **Algorithms and programs for the construction of an analytical theory of motion of AES in coordinates.** V. B. Titov.
Astron. i geod., Tomsk, 1980, No. 8, p. 35 - 39. In Russian. Abstr. in Ref. zh., 51. Astron., 9.51.91 (1981).

052.028 **Stable orbits of circular resonant AES.** L. L. Sokolov, V. B. Titov.
Astron. i geod., Tomsk, 1980, No. 8, p. 111 - 123. In Russian. Abstr. in Ref. zh., 51. Astron., 9.51.93 (1981).

052.029 **On the efficiency of applying minimizing methods to determination of orbital elements of AES.** M. Abele, Ya. V. Vyater, J. Balodis, L. Lauzenieks.
Determination of the coordinates of celestial bodies, Riga, 1981, (see 003.008), p. 129 - 136. In Russian. – Abstr. in Ref. zh., 62. Issled. kosm. prostranstva, 9.62.164 (1981).

052.030 **On the variation of the parameters of the apparent motion of AES.** M. Abele, Ya. V. Vyater, J. Balodis, L. Lauzenieks.
Determination of the coordinates of celestial bodies, Riga, 1981, (see 003.008), p. 137 - 146. In Russian. – Abstr. in Ref. zh., 62. Issled. kosm. prostranstva, 9.62.165 (1981).

052.031 **Drag perturbations of artificial satellite orbits in a spherically symmetrical atmosphere.** L. Burs, A. Pál.
Astron. Nachr., Band 302, 189 - 192 (1981).
The equations characterizing the motion of an artificial satellite in a non-rotating spherically symmetrical atmosphere are integrated in the assumption of a linear variation of the density scale height with height, and using a new variable instead of the true anomaly. The secular perturbations in the semi-major axis and eccentricity are deduced.

052.032 **On-orbit control system performance of the HEAO-2 Observatory.** R. E. Rose, E. A. Berkery.
J. Guid. Control, Vol. 4, 148 - 156 (1981). – Abstr. in Phys. Abstr., Vol. 84, Abstr. 88705 (1981).

052.033 **The Seasat Precision Orbit Determination Experiment.** B. D. Tapley, G. H. Born.
J. Astronaut. Sci., Vol. 28, 315 - 326 (1980). – Abstr. in Phys. Abstr., Vol. 84, Abstr. 88711 (1981).

052.034 **Precision orbit analyses in support of the Seasat altimeter experiment.** J. G. Marsh, R. G. Williamson.
J. Astronaut. Sci., Vol. 28, 345 - 369 (1980). – Abstr. in Phys. Abstr., Vol. 84, Abstr. 88712 (1981).

052.035 **Precision orbit determination software validation experiment.** B. E. Schutz, B. D. Tapley, R. J. Eanes, J. G. Marsh, R. G. Williamson, T. V. Martin.
J. Astronaut. Sci., Vol. 28, 327 - 343 (1980). – Abstr. in Phys. Abstr., Vol. 84, Abstr. 88721 (1981).

052.036 **Doppler computed Seasat orbits.** E. S. Colquitt, C. W. Malyevac, R. J. Anderle.
J. Astronaut. Sci., Vol. 28, 391 - 403 (1980). – Abstr. in Phys. Abstr., Vol. 84. Abstr. 88722 (1981).

052.037 **On the optimum controlled movement around the Lagrangian points in the Earth-Moon system.**
M. Popescu.
Rev. Roumaine Sci. Tech. Ser. Mec. Appl., Vol. 26, 301 - 323 (1981). − Abstr. in Phys. Abstr., Vol. 84, Abstr. 88724 (1981).

052.038 **Radius of convergence of Lie series for some elliptic elements.**   C. J. Cohen, R. H. Lyddane.
Celestial Mech., Vol. 25, 221 - 234 (1981).
    For equatorial orbits about an oblate body, the authors show that the Lie series for the elliptic elements e, f, l, and ϖ diverge when the oblateness exceeds a critical multiple of the transformed eccentricity constant. The use of similar truncated series expansions for such elliptic elements by Brouwer accounts for the first-order errors at low eccentricity in his derived coordinates for an artificial satellite.

052.039 **The 'critical inclination': another look.**   S. Hughes.
Celestial Mech., Vol. 25, 235 - 266 (1981).
    The occurrence and uniqueness of the 'critical inclination' in satellite theory is discussed. An infinite set of canonical transformations in Hill variables are shown to exist whereby the first order secular part of the disturbing function can be changed into an alternative form. As a result of such a transformation the 'critical inclination' can become (1) any other real or complex inclination or a function independent of the satellite's orbital inclination and (2) a function dependent on the semi-major axis, eccentricity and inclination of the satellite's orbit. It is also shown that all transformations of types (1) and (2) are only valid for short intervals of time of the order of a few satellite revolutions. Furthermore if such transformations are modified so that they become valid for greater intervals of time, then the resulting solutions in all cases contain no singular divisor other than the 'critical inclination'.

052.040 **Combined solar radiation pressure and drag effects on the orbits of artificial satellites.**
R. Vilhena de Moraes.
Celestial Mech., Vol. 25, 281 - 292 (1981).
    A semi-analytical theory is proposed to study the joint effects of direct solar radiation pressure and atmospheric drag on the orbit of an artificial Earth satellite. Making the solar radiation pressure equal to zero the problem is reduced to one already solved by Brouwer and Hori. The solutions are not equivalent, however, since in the Brouwer and Hori theory one has spurious Poisson terms.

052.041 **Elimination of secular terms generated by the coupling of perturbations.**   S. Ferraz-Mello.
Celestial Mech., Vol. 25, 293 - 296 (1981).
    The idea of the transformation of Tisserand is used to avoid the appearance of spurious secular terms in the study of the simultaneous action of two disturbing forces of different type.

052.042 **Gravitational orientation of a rotating satellite.**   V. A. Sarychev, V. V. Sazonov.
Kosm. Issled., Tom 19, 499 - 512 (1981). In Russian.

052.043 **On the problem of determination of the orbit of a satellite in the case of restricted measurement errors.**   E. O. Kotov.
Kosm. Issled., Tom 19, 513 - 517 (1981). In Russian.

052.044 **Collision between satellites in stationary orbits.**   K. Takahashi.
IEEE Trans. Aerosp. Electron. Syst., Vol. AES-17, 591 - 596 (1981). − Abstr. in Phys. Abstr., Vol. 84, Abstr. 98658 (1981).

052.045 **Uniaxial gravitational orientation of artificial satellites.**   V. A. Sarychev, V. V. Sazonov.
Kosm. Issled., Tom 19, 659 - 673 (1981). In Russian.

052.046 **On the determination of the methodical error of numerical integration of equations of the perturbed motion of AES in the orbital method of space geodesy.**
Yu. D. Stepin.
Izv. vuzov. Geod. i aehrofotosemka, 1981, No. 2, p. 42 - 45. In Russian. − Abstr. in Ref. zh., 52. Geod. Aehrosemka, 10.52.73; 62. Issled. kosm. prostranstva, 10.62.418 (1981).

052.047 **Resonance orbits of AES and their use in dynamical problems of space geodesy.**   S. N. Yashkin.
Izv. vuzov. Geod. i aehrofotosemka, 1981, No. 2, p. 36 - 42. In Russian. − Abstr. in Ref. zh., 52. Geod. Aehrosemka, 10.52.74; 62. Issled. kosm. prostranstva, 10.62.74 (1981).

052.048 **Perturbation of the rotational motion of a body in a central gravitational field by Lorentz forces.**
N. V. Chikova.
Prikl. mekh., Leningrad, 1981, No. 5, p. 38 - 47. In Russian. Abstr. in Ref. zh., 62. Issled. kosm. prostranstva, 10.62.149 (1981).

052.049 **Stabilization due to gyroscopic coupling in dual-spin satellites subject to gravitational torques.**
R. Longman, P. Hagedorn, A. Beck.
Celestial Mech., Vol. 25, 353 - 373 (1981).
    The stability of attitude equilibria relative to gravitational torques for a rigid satellite in a circular orbit has been divided into three inertia regions, the Lagrange region of assured Liapunov stability, the Beletskii-Delp region which is often described as stabilized due to gyroscopic coupling, and an assured instability region. The stability boundaries are obtained for all possible equilibrium orientations for such vehicles, and the variations of these boundaries corresponding to changes in the internal momentum magnitude, or to aligning the momentum with a different principal axis, are determined.

052.050 **Vie et mort des satellites artificiels.**   J. Vercheval.
Ciel Terre, Vol. 97, 353 - 370 (1981).
    The characteristics of the orbital motion of artificial earth satellites are briefly presented. In particular, a description is made of the most marked effects of the main perturbing forces acting upon an artificial earth satellite, as the atmospheric drag, the irregularities of the earth's gravitational field, the luni-solar gravitational perturbations and the solar radiation pressure. The lifetime of satellites is related to the behaviour of the perigee height in presence of these perturbations.

052.051 **The variations of the orbit of 1976 - 87A and determination of air density.**
Y.-y. Liu, Y.-b. Wang, Z.-l. Feng.
Acta Astron. Sinica, Vol. 22, 223 - 229 (1981). In Chinese.
    The variations of the orbit of 1976 - 87A (China 6) have been studied in this paper. Using the orbital decay rate the air density at hight of 205 - 220 km is determined. The average values of density are over 20 per cent higher than that computed by CIRA 1972 model.

052.052 **Influence of atmospheric density fluctuations on the accuracy of determination and prediction of orbits of artificial earth satellites.**   P. E. Ehl'yasberg.
Kosm. Issled., Tom 19, 803 - 812 (1981). In Russian.

052.053 **On optimum stabilization of stationary motions of a space vehicle relative to the mass center in a triangular point of libration.**   A. P. Blinov, Yu. G. Markov.
Kosm. Issled., Tom 19, 829 - 834 (1981). In Russian.

052.054 **Disturbed satellite rotation around the mass centre in the region of a triangular point of libration.**
Yu. G. Markov.
Astron. Zh., Tom 58, 1306 - 1313 (1981). In Russian. English translation in Soviet Astron., Vol. 25, No. 6.
The paper describes the disturbed rotation of an axisymmetric satellite with its mass centre located at the triangular point of libration L_4 of the restricted elliptic three-body problem. The first-order disturbances based on the intermediate satellite rotation are calculated.

052.055 **A simple method of prediction of the "TRANSIT" satellite position.** A.-x. Ma.
Publ. Shaanxi Astron. Obs., No. 1, p. 44 - 53 (1981). In Chinese.

052.056 **A low-cost decoder for TRANSIT satellite ephemeris.**
X.-z. Zhang.
Publ. Shaanxi Astron. Obs., No. 1, p. 54 - 57 (1981). In Chinese.

052.057 **On the stability of a gyro-horizon compass in the Newtonian field of gravitation.**
Zh. G. Zavozin.
Izv. AN SSSR. Mekh. tverd. tela, 1981, No. 3, p. 114 - 117. In Russian. − Abstr. in Ref. zh., 51. Astron., 12.51.123; 62. Issled. kosm. prostranstva, 12.62.132 (1981).

052.058 **On the motion of a dynamically symmetric satellite relative to the mass centre near a collinear libration point.**
A. P. Markeev, N. K. Moshchuk, A. A. Saitbattalov.
Proceedings of scientific lectures on cosmonautics, (see 012.058), 1980, p. 105 - 114. In Russian. − Abstr. in Ref. zh., 51. Astron., 12.51.139 (1981).

052.059 **On the influence of an electric charge on the rotational motion of an earth satellite.**
L. I. Kuznetsov.
Prikl. mekh. Leningrad, 1981, No. 5, p. 78 - 83. In Russian. Abstr. in Ref. zh., 51. Astron., 12.51.144; 62. Issled. kosm. prostranstva, 12.62.134 (1981).

052.060 **Periodic oscillations of a satellite-gyrostat relative to the mass center.** V. V. Sazonov.
Inst. prikl. mat. AN SSSR, Prepr., 1981, No. 62, 28 pp. In Russian. − Abstr. in Ref. zh., 62. Issled. kosm. prostranstva, 12.62.131 (1981).

052.061 **Computing the coordinates and velocities of satellites by means of the broadcast ephemeris of the "Transit" satellites.** H.-g. Xu, W.-y. Zhu, Y.-l. Zhu.
Ann. Shanghai Obs. Acad. Sinica, No. 2, p. 65 - 72 (1980).

052.062 **Analysis of Lageos' altitude decrease.**
V. Szebehely.
Satellite perturbations and orbital determination, (see 012.060), p. 7 - 11 (1981).
The paper treats the inverse problem of celestial mechanics which consists of determining the force field or potential from given or observed orbit(s). From the observational information, according to which Lageos loses approximately 1 mm altitude per day, a linear partial differential equation is formulated. The solution of this equation gives the field responsible for the above-mentioned, as yet unexplained, small but well established secular decrease in the semi-major axis. The altitude-loss is not due to air-drag because of the very high altitude of this satellite.

052.063 **The motion of an artificial satellite under the terrestrial radiation pressure.** L. Sehnal.
Satellite perturbations and orbital determination, (see 012.060), p. 37 - 41 (1981).
The terrestrial infrared radiation pressure effects on the motion of an artificial satellite are investigated. The radiative field is described by a series of spherical harmonics. The equations for the changes of the elements are found and numerical examples for the case of the satellite D−5−B (1975 39 B) are given.

052.064 **The program system "Potsdam-4" for differential improvement of orbital elements and other parameters.** G. Gendt, H. Montag.
Satellite perturbations and orbital determination, (see 012.060), p. 43 - 56 (1981).
The "Potsdam-4" program system can model satellite orbits with a high accuracy by means of numerical integration. It takes into account gravitational forces of the earth, the moon, the sun, and non-gravitational forces, and realizes an inertial system with a high accuracy. On the basis of this orbit calculation, geodetic and geodynamic parameters can be determined by means of photographic, laser and Doppler observations.

052.065 **Computer program PRIOR used for orbit determination at the Ondřejov Observatory.** P. Lála.
Satellite perturbations and orbital determination, (see 012.060), p. 57 - 67 (1981).
The computer program used mainly for determination of orbits of the Interkosmos satellites is described. Examples of results are given for Geos B and Interkosmos 17 satellites.

052.066 **Kalman filter orbit improvement from Kootwijk laser range observations.**
K. F. Wakker, B. A. C. Ambrosius, J. J. P. van Hulzen.
Satellite perturbations and orbital determination, (see 012.060), p. 79 - 82 (1981).
Modern satellite ranging lasers emit short pulses at a low beam divergence and therefore require accurate satellite position predictions. To reach these accuracies the application of a Kalman filter orbit improvement technique has been investigated. Using laser observations acquired at only one ground-station the filter scheme provides real-time satellite position prediction updates, and also yields better predictions for subsequent passes over that station.

052.067 **On the generation of satellite position (and velocity) by a mixed analytical-numerical procedure.**
R. H. Gooding.
Satellite perturbations and orbital determination, (see 012.060), p. 83 - 93 (1981).
The generation of accurate Earth-satellite ephemerides by numerical integration, over a period of perhaps weeks, can consume an inordinate amount of computer time. It is best if short-period perturbations are not applied to the orbital elements themselves but to the satellite's position (and velocity if required), expressed in a system of cylindrical polar coordinates, and the paper shows how mean elements can be recovered from position and velocity. A computer program has been written to test the proposed procedure for generating ephemerides, using a truncated potential field. Some results from this program are presented.

Dynamics of flights between the earth and the moon. See Abstr. 003.056.

Evolution and stability of satellite systems. See Abstr. 003.116.

Investigation of the efficiency of numerical algorithms using the equations of motion of AES in stabilized and regularized form. See Abstr. 021.024.

QIKAIM, a fast semi-numerical algorithm for the generation of minute-of-arc accuracy satellite predictions. See Abstr. 021.041.

High-order Runge-Kutta method and stabilizing transformations in the problems of forecasting the motion of AES.   See Abstr. 021.052.

Predicting ephemerides and visible periods of artificial satellites using an electronic computer. See Abstr. 021.070.

Interplanetary two point boundary value problems. (Extension of the Unified System for Orbit Computation, USOC, Vol. IV).   See Abstr. 021.075.

Reformulation of the Brouwer geopotential theory for improved computational efficiency. See Abstr. 042.013.

Sur les mouvements réguliers des satellites. See Abstr. 042.065.

Some possibilities for determining the influence of Earth body tides on the motion of artificial satellites. See Abstr. 042.086.

Earth and ocean tides from long-term analysis of satellite orbits.   See Abstr. 081.057.

Research results from analysis of satellite orbits. See Abstr. 082.035.

053 Lunar and Planetary Probes and Satellites

053.001 How Voyager 2 has been reprogrammed.
E. C. Stone.
Nature, Vol. 292, 675 - 676 (1981).

053.002 Voyager Saturn: preview of the August encounter.
D. Morrison.
Bull. American Astron. Soc., Vol. 13, 526 (1981). — Abstract.

053.003 Robots for the longest voyage.   A. Chaikin.
Sky Telesc., Vol. 62, 328 (1981).

053.004 Pioneer 11 to Saturn.   E. Mūkins.
Zvaigžnotā debess, 1980. gada pavasaris, p. 28 - 33.
In Latvian.

Voyager — the story of a space mission. See Abstr. 003.128.

Mission objectives for geological exploration of the Apollo 16 landing site.   See Abstr. 094.506.

054 Artificial Earth Satellites

054.001 **The eighth from the Prognoz family.**
M. A. Rimsha.
Zemlya Vselennaya, 1981, No. 4, p. 8. In Russian.

054.002 **Development of the Salyuts.**
K. P. Feoktistov, M. M. Markov.
Zemlya Vselennaya, 1981, No. 5, p. 10 - 16. In Russian.

054.003 **Intercosmos 21.** M. A. Rimsha.
Zemlya Vselennaya, 1981, No. 5, p. 39 - 41. In Russian.

054.004 **Poljot 1 – 1963-43-1. July - November 1978.**
Poljot 1 – 1963-43-1. May - August 1979.
Explorer 19 – 1963-63-1. May - August 1979. Intercosmos
17 – 1977-96-1. April - August 1979. Intercosmos 18 –
1978-99-1. April - July 1979. Intercosmos 19 – 1979-20-1.
May - August 1979. Visual observations. Horizontal coordinates.
Rezul'taty Nablyud. Iskusstv. Sputnikov Zemli, Vyp. (No.)
85 (225), 66 pp. (1979). In Russian.

054.005 **Poljot 1 – 1963-43-1. June - November 1979.**
Explorer 19 – 1963-53-1. June - November 1979.
Intercosmos 18 – 1978-99-1. June - September 1979.
Intercosmos 19 – 1979-20-1. June - November 1979. Visual
observations. Equatorial coordinates (1950.0).
Rezul'taty Nablyud. Iskusstv. Sputnikov Zemli, Vyp. (No.)
86 (226), 68 pp. (1979). In Russian.

054.006 **Poljot 1 – 1963-43-1. August - December 1979.**
Explorer 19 – 1963-53-1. August - December 1979.
Intercosmos 18 – 1978-99-1. September - December 1979.
Intercosmos 19 – 1979-20-1. September - December 1979.
Visual observations. Horizontal coordinates. Poljot
1 – 1963-43-1. September - December 1979. Explorer
19 – 1963-53-1. September - December 1979. Intercosmos
18 – 1978-99-1. September - December 1979. Intercosmos
19 – 1979-20-1. September - December 1979. Visual observations. Equatorial coordinates (1950.0).
Rezul'taty Nablyud. Iskusstv. Sputnikov Zemli, Vyp. (No.)
87 (227), 60 pp. (1980). In Russian.

054.007 **Poljot 1 – 1963-43-1. January - March 1980.**
Explorer 19 – 1963-53-1. January - March 1980.
Intercosmos 18 – 1978-99-1. February - March 1980. Intercosmos 17 – 1979-96-1. January - March 1980. Visual
observations. Equatorial coordinates (1950.0).
Rezul'taty Nablyud. Iskusstv. Sputnikov Zemli, Vyp. (No.)
88 (228), 61 pp. (1980). In Russian.

054.008 **Poljot 1 – 1963-43-1. January - March 1980.**
Explorer 19 – 1963-53-1. January - April 1980.
Intercosmos 18 – 1978-99-1. February - March 1980.
Intercosmos 19 – 1979-20-1. January - April 1980.
Intercosmos 17 – 1979-96-1. January - April 1980. Visual
observations. Horizontal coordinates.
Rezul'taty Nablyud. Iskusstv. Sputnikov Zemli, Vyp. (No.)
89 (229), 58 pp. (1980). In Russian.

054.009 **Poljot 1 – 1963-43-1. March 1980. Explorer**
19 – 1963-53-1. March - April 1980. Intercosmos
18 – 1978-99-1. March - April 1980. Intercosmos
19 – 1979-20-1. January - April 1980. Intercosmos
17 – 1979-96-1. March - April. Visual observations. Equatorial coordinates (1950.0).
Rezul'taty Nablyud. Iskusstv. Sputnikov Zemli, Vyp. (No.)
90 (230), 64 pp. (1980). In Russian.

054.010 **Poljot 1 – 1963-43-1. May - June 1980. Explorer**
19 – 1963-53-1. March - July 1980. Intercosmos
18 – 1978-99-1. March - July 1980. Intercosmos
19 – 1979-20-1. March - July 1980. Intercosmos
17 – 1979-96-1. April - July 1980. Visual observations.
Horizontal coordinates.
Rezul'taty Nablyud. Iskusstv. Sputnikov Zemli, Vyp. (No.)
91 (231), 60 pp. (1980). In Russian.

054.011 **Poljot 1 – 1963-43-1. May - August 1980.**
Explorer 19 – 1963-53-1. April - July 1980.
Intercosmos 18 – 1978-99-1. May - August 1980. Intercos-
mos 19 – 1979-20-1. April - August 1980. Intercosmos
17 – 1979-96-1. May - September 1980. Visual observations.
Equatorial coordinates (1950.0).
Rezul'taty Nablyud. Iskusstv. Sputnikov Zemli, Vyp. (No.)
92 (232), 70 pp. (1980). In Russian.

054.012 **Poljot 1 – 1963-43-1. June - October 1980.**
Explorer 19 – 1963-53-1. June - September 1980.
Intercosmos 18 – 1978-99-1. July - August 1980. Inter-
cosmos 19 – 1979-20-1. June - October 1980. Intercosmos
17 – 1979-96-1. July - October 1980. Visual observations.
Horizontal coordinates. Explorer 19 – 1963-53-1. August -
September 1980. Visual observations. Equatorial coordinates
(1950.0).
Rezul'taty Nablyud. Iskusstv. Sputnikov Zemli, Vyp. (No.)
93 (233), 59 pp. (1980). In Russian.

054.013 **Poljot 1 – 1963-43-1. August - November 1980.**
Explorer 19 – 1963-53-1. October - November
1980. Intercosmos 18 – 1978-99-1. September 1980. Intercosmos 17 – 1979-96-1. September - November 1980. Intercosmos 19 – 1979-20-1. September - November 1980. Visual
observations. Equatorial coordinates (1950.0). Poljot
1 – 1963-43-1. June - November 1980. Explorer
19 – 1963-53-1. June - November 1980. Intercosmos
19 – 1979-20-1. June - November 1980. Intercosmos
17 – 1979-96-1. June - October 1980. Visual observations.
Horizontal coordinates.
Rezul'taty Nablyud. Iskusstv. Sputnikov Zemli, Vyp. (No.)
94 (234), 66 pp. (1981). In Russian.

054.014 **Satellite digest – 148, 149.**
Compiled by R. D. Christy.
Spaceflight, Vol. 23, 273 - 274, 319 (1981).

Dynamics Explorer science data processing system.
See Abstr. 021.046.

A method for acquiring the tracks of artificial
satellites. See Abstr. 031.645.

Theoretical Astrophysics

061 General Aspects (Nucleosynthesis, Neutrino Astronomy, etc.)

061.001 The elements just beyond iron: formation during explosive carbon burning.
J. P. Wefel, D. N. Schramm, J. B. Blake, D. Pridmore-Brown.
Astrophys. J., Suppl. Ser., Vol. 45, 565 - 584 (1981).
   The synthesis of the elements just beyond iron during explosive carbon burning has been studied using an exact n (neutron)-process calculation incorporating neutron captures, charged particle reactions, and beta-decay in a network covering the elements Cr-Zr. The temperature, density, and free particle abundances were taken directly from explosive carbon-burning calculations. Heavy element seed distributions incorporating the s-process enhancements from core helium burning were employed, along with realistic postshock conditions, to obtain an evolution characteristic of the average massive star contributing material to the solar system. The results for the heavy isotopes are significantly different from earlier calculations.

061.002 Existing and potential limits to lifetimes of massive neutrinos. H. L. Shipman, R. Cowsik.
Astrophys. J., Lett., Vol. 247, L111 - L114 (1981).
   The most stringent lower limits to the lifetime of neutrinos against radiative decay processes can be set by observations of discrete sources rather than from background measurements. Analysis of existing data places lower limits of a few times 10^{23} to 10^{25} s on the neutrinos' radiative lifetimes, depending on the wavelength of the emitted photon. These limits are significantly more stringent than the 10^{22} to 10^{23} s limits obtained from analyses of background measurements by Cowsik and most recently by Stecker and by Kimble, Bowyer, and Jakobsen. The authors demonstrate that existing or proposed instruments could be used to improve this limit significantly, with values of 10^{26} to 10^{27} s as realistic possibilities.

061.003 On a method of investigation of polarization variability of cosmic objects. V. A. Hagen-Thorn.
Tr. Astron. Obs., Leningrad, Tom 36 = Uch. Zap. Leningr. Univ., No. 402 = Ser. mat. nauk, Vyp. 58, 20 - 26 (1981). In Russian.
   Simultaneous polarization and brightness changes may sometimes be caused by a source of variable intensity with constant polarization parameters. The paper gives a detailed study of this possibility. The case of two variable sources is also briefly discussed.

061.004 The formation and the structure of neutrino astronomical objects. T. Lu, L.-f. Luo, G.-c. Yang.
Acta Astron. Sinica, Vol. 22, 207 - 212 (1981). In Chinese.

061.005 Nucleosynthesis of neutron-rich heavy nuclei during explosive helium burning in massive stars.
J. B. Blake, S. E. Woosley, T. A. Weaver, D. N. Schramm.
Astrophys. J., Vol. 248, 315 - 320 (1981).
   The production of heavy nuclei during explosive helium burning has been calculated using a hydrodynamical model of a 15 $M_\odot$ (Type II) supernova and an "n-process" nuclear reaction network. The authors find that the resulting neutron-rich heavy nuclei are not produced in the relative abundances of solar-system r-process material, especially in the vicinity of Pt,

nor are any actinides produced. These deficiencies reflect an inadequate supply of neutrons. However, some neutron-rich isotopes, normally associated with the r-process, are produced which may be significant for the production of isotopic anomalies in meteorites.

061.006 Neutron-capture cross sections for osmium isotopes and the age of the Universe.
J. C. Browne, B. L. Berman.
Phys. Rev. C, Vol. 23, 1434 - 1445 (1981). – Abstr. in Phys. Abstr., Vol. 84, Abstr. 63650 (1981).

061.007 Effects of neutrino degeneracy on the transport properties of presupernovae.
L. J. van den Horn, C. G. van den Weert.
Phys. Lett. A, Vol. 83A, 88 - 90 (1981). – Abstr. in Phys. Abstr., Vol. 84, Abstr. 67014 (1981).

061.008 Limits on new superweakly interacting particles from primordial nucleosynthesis.
K. A. Olive, D. N. Schramm, G. Steigman.
Nucl. Phys. B, Vol. B180, 497 - 515 (1981). – Abstr. in Phys. Abstr., Vol. 84, Abstr. 71073 (1981).

061.009 The beta strength function and the astrophysical site of the r-process.
H. V. Klapdor, T. Oda, J. Metzinger, W. Hillebrandt, F. K. Thielemann.
Z. Phys. A, Vol. 299, 213 - 229 (1981). – Abstr. in Phys. Abstr., Vol. 84, Abstr. 71075 (1981).

061.010 Magnetic moment of massive neutrinos and the cosmic helium abundances. B. W. Lynn.
Phys. Rev. D, Vol. 23, 2151 - 2156 (1981). – Abstr. in Phys. Abstr., Vol. 84, Abstr. 71875 (1981).

061.011 Elastic and inelastic scattering of the relic neutrinos by high energy cosmic rays. T. Hara, H. Sato.
Prog. Theor. Phys., Vol. 65, 477 - 488 (1981). – Abstr. in Phys. Abstr., Vol. 84, Abstr. 74993 (1981).

061.012 Possible existence of massive neutrino halos.
C.-G. Kallman.
Phys. Lett. A, Vol. 83A, 179 - 180 (1981). – Abstr. in Phys. Abstr., Vol. 84, Abstr. 75202 (1981).

061.013 Transport properties of degenerate neutrinos and photons. L. J. van den Horn, C. G. van Weert.
Phys. Lett. B, Vol. 101B, 447 - 449 (1981). – Abstr. in Phys. Abstr., Vol. 84, Abstr. 79717 (1981).

061.014 Selfconsistent equation of state.
R. K. Tripathi.
Phys. Lett. B, Vol. 101B, 369 - 374 (1981). – Abstr. in Phys. Abstr., Vol. 84, Abstr. 79787 (1981).

061.015 Partition functions of nuclei in Hartree-Fock approximation. B. Banerjee, S. M. Chitre, B. Datta.

Phys. Lett. B, Vol. 101B, 450 - 452 (1981). – Abstr. in Phys. Abstr., Vol. 84, Abstr. 79788 (1981).

061.016 Nucleosynthesis with nonzero lepton numbers: is there a limit on the neutrino flavors?
A. S. Szalay.
Phys. Lett. B, Vol. 101B, 453 - 456 (1981). – Abstr. in Phys. Abstr., Vol. 84, Abstr. 79875 (1981).

061.017 50 years of neutrino physics – some episodes.
B. Pontecorvo.
Fiz. Sz., Vol. 30, 443 - 454 (1980). In Hungarian. – Abstr. in Phys. Abstr., Vol. 84, Abstr. 80345 (1981).

061.018 Direct capture cross sections at low energy.
R. D. Williams, S. E. Koonin.
Phys. Rev. C, Vol. 23, 2773 - 2774 (1981). – Abstr. in Phys. Abstr., Vol. 84, Abstr. 80531 (1981).

061.019 Nucleocosmochronology.   E. M. D. Symbalisty, D. N. Schramm.
Rep. Prog. Phys., Vol. 44, 293 - 328 (1981). – Abstr. in Phys. Abstr., Vol. 84, Abstr. 83479 (1981).

061.020 Beta decay rates for s-process studies.
K. Cosner, J. W. Truran.
Astrophys. Space Sci., Vol. 78, 85 - 94 (1981).
The rates for a variety of beta decay processes have been determined as a function of temperature for nuclei which can participate in the s-process production of heavy elements, occurring in the presence of the ^{22}Ne $(\alpha, n)^{25}$Mg neutron source operating in the convective helium shells of thermally pulsing stars. Specifically: calculated half-lives are presented for electron emission, positron emission, and electron capture over the temperature range $10^8 - 10^9$ K.

061.021 Beta decay rates for astrophysical environments.
M. J. Newman, G. M. Fuller, W. A. Fowler.
Bull. American Astron. Soc., Vol. 13, 516 (1981). – Abstract.

061.022 An unconvential site for the r-process.
J. J. Cowan, A. G. W. Cameron, J. W. Truran.
Bull. American Astron. Soc., Vol. 13, 528 (1981). – Abstract.

061.023 Neutrino transport with isoenergetic and neutrino-electron scattering.
S. W. Bruenn, J. L. Ballester.
Astrophys. Space Sci., Vol. 78, 247 - 272 (1981).
Solutions to steady-state, bigroup neutrino transport with isoenergetic and neutrino-electron scattering have been obtained for idealized conditions representative of the outer core or envelope of a star whose inner core is undergoing gravitational collapse. The incident neutrino distribution was chosen to be non-equilibrium, and the solutions clearly show the tendency of the neutrinos to equilibrate via neutrino-electron scattering as they propagate through the material.

061.024 A new mechanism for energy release in astrophysical objects.   V. I. Denisov, A. A. Logunov.
Inst. yader. issled. AN SSSR. Prepr., 1981, No. 0206, 10 pp. In Russian. – Abstr. in Ref. zh., 51. Astron., 7.51.134 (1981).

061.025 High-energy neutrino astronomy: sources and fluxes.   V. S. Berezinskij.
Usp. fiz. nauk, Tom 133, 545 - 546 (1981). In Russian. Abstr. in Ref. zh., 51. Astron., 7.51.544 (1981).

061.026 Current rate of nucleosynthesis and its implications.   D. C. V. Mallik.
J. Astrophys. Astron., Vol. 2, 171 - 185 (1981).
The current rate of nucleosynthesis in the solar neighbourhood is re-evaluated. If massive stars are held responsible for most of the metals we observe, a higher birthrate of these stars in the past is indicated in view of the low current rate of nucleosynthesis. The intermediate mass stars may not supply the bulk of the metals unless total disruption of their carbon core takes place. While a declining birthrate is in conflict with the result obtained from the age-metallicity relation of stars, it is supported by some galactic evolution models. If the constraint of a nearly time-invariant birthrate were strictly accepted, then models of the prompt initial enrichment type are required to explain the observed abundances in terms of nucleosynthesis in massive stars.

061.027 Gravitational energy release induced by the nuclear energy generation processes: the resolution of the solar neutrino dilemma.   C. A. Rouse.
Astron. Astrophys., Vol. 102, 8 - 11 (1981).
It is proposed that a source of solar (stellar) energy emission on the nuclear time scale is gravitational. This gravitational energy release is caused by a slow steady-state and/or a more rapid recurring transient contraction of the sun (star) induced by thermodynamic perturbations produced by the nuclear energy generation processes. The basic reference steady-state results are obtained. The resolution of the solar neutrino dilemma is described as being obtained with a model of the sun with a central temperature less than 14×10^6 K and with the sum of the nuclear luminosity and gravitational luminosity equal to the observed luminosity without significant production of B^8 neutrinos.

061.028 Peculiarities of solutions of hyperbolic equations and the problem of twins in astronomy.
V. P. Palamodov.
Dokl. AN SSSR, Tom 257, 849 - 852 (1981). In Russian. Abstr. in Ref. zh., 51. Astron., 9.51.171 (1981).

061.029 Cosmological upper limit to neutrino magnetic moments.   J. A. Morgan.
Phys. Lett. B, Vol. 102B, 247 - 250 (1981). – Abstr. in Phys. Abstr., Vol. 84, Abstr. 84300 (1981).

061.030 Neutrinos and cosmology.
M. S. Turner.
AIP Conf. Proc., No. 72, p. 335 - 354 (1980). – Abstr. in Phys. Abstr., Vol. 84, Abstr. 89129 (1981).

061.031 Silicon in the outer space and on the earth; its biological role.   G. Zelčāns.
Zvaigžnotā debess, 1980. gada rudens, p. 2 - 9. In Latvian.

061.032 Equations of state for hot, dense nuclear matter.
P. Bonche, D. Vautherin.
6e session d'études biennale de physique nucléaire, (see 012.032), p. C. 1/1-2 (1981). In French. – Abstr. in Phys. Abstr., Vol. 84, Abstr. 95225 (1981).

061.033 On the decay law of external fields of a collapsing body.   L. A. Kofman.
Phys. Lett. A, Vol. 84A, 232 - 234 (1981). – Abstr. in Phys. Abstr., Vol. 84, Abstr. 98679 (1981).

061.034 Bulk viscosity of a reactive $(\nu_e npe^-)$ system.
L. J. van den Horn, C. G. van Weert.
Phys. Lett. A, Vol. 84A, 226 - 228 (1981). – Abstr. in Phys. Abstr., Vol. 84, Abstr. 98782 (1981).

061.035 Analysis of background events of $\bar{\nu}$-flares of collapsing stars of our Galaxy recorded on the Baksan underground scintillation telescope.
E. N. Alekseev, L. N. Alekseeva, B. Sh. Borchaeva, V. N. Kalinchenko, G. P. Kejdan, A. I. Pavlov, A. E. Chudakov.
Inst. yader. issled. AN SSSR. Prepr., 1981, No. 0196, 20 pp. In Russian. – Abstr. in Ref. zh., 51. Astron., 10.51.210 (1981).

061.036 On the origin of the solar-system abundances of
^{113}In, ^{114}Sn, and ^{115}Sn.   R. A. Ward, H. Beer.
Astron. Astrophys., Vol. 103, 189 - 196 (1981).

The neutron-capture cross section of ^{114}Cd to the 53.38 h
ground state in ^{115}Cd has been measured via neutron activation. Using this result in conjunction with the total neutron-
capture rate of ^{114}Cd, the relative population of
^{115}Cd^0 (53.38 h) was found to be 0.79 ± 0.13 at an energy
appropriate to 30 keV stellar neutrons. In addition, the
authors quantitatively examined the isomeric structure
of the key nuclei: ^{113}Cd, ^{114}In, ^{115}Cd, and ^{115}In which all
crucially influence the neutron-capture flows of the s-process
as well as the final beta-decays of the r-process in the Cd-In-Sn
region of atomic weights.

061.037 Nuclear astrophysics. Introductory note.
D. H. Wilkinson.
Nuclear astrophysics, (see 012.036), p. 1 - 4 (1981).

061.038 Nuclear and particle physics in the early universe.
D. N. Schramm.
Nuclear astrophysics, (see 012.036), p. 69 - 86 (1981).

The author reviews the arguments favoring the big bang
model of the universe. He describes the big bang in chronologi-
cal order beginning with the Planck time and quantum gravity
and continues on through the era of grand unification where
one expects the generation of baryonic matter to have
occurred. The chronologic history continues on into the quark-
hadron phase transition where the free quarks of the early
universe combine to make hadrons. The decoupling of
neutrinos at $\sim 10^{10}$ K will be described and the possible consequence of these relic neutrinos on the future dynamics of the
universe. The author discusses the epoch of big bang nucleosynthesis where the observed abundance of ^{4}He and ^{2}D were
produced. It is shown that the He abundance can be used to
set a limit on the number of neutrino types and thus also on
the number of quark flavors.

061.039 The quark matter.   M. Rho.
Nuclear astrophysics, (see 012.036), p. 87 - 110
(1981).

The author discusses the present status of our understanding of the physics of hadronic (nuclear or neutron)
matter under extreme conditions, in particular at high densities. This is a problem which challenges three disciplines of
physics: nuclear physics, astrophysics and particle physics. In
astrophysics, the properties of neutron stars will be properly
understood only when the equation of state of "neutron"
matter at densities exceeding that of nuclear matter can be
reliably calculated. Most fascinating is the possibility of quark
stars existing in Nature, not entirely an absurd idea. Finally
the quark matter – nuclear matter phase transition must have
occurred in the early stage of universe when matter expanded
from high temperature and density; this could be an essential
ingredient in the big-bang cosmology.

061.040 Some aspects of the nucleosynthesis of the light
elements.   J. Audouze.
Nuclear astrophysics, (see 012.036), p. 125 - 157 (1981).

The author reviews the current situation regarding the
nucleosynthesis of the light elements. He concentrates mainly
on the formation of the light elements lithium, beryllium and
boron, on the explosive hydrogen and helium-burning nucleosynthesis occurring in novae or in (still hypothetical) supermassive stars and on some consequences of the discoveries of
isotope "anomalies" in carbonaceous chondrites and galactic
cosmic rays (especially those concerning the formation of the
solar system and the origin of the cosmic rays).

061.041 Synthesis of the heavy elements.
J. W. Truran.
Nuclear astrophysics, (see 012.036), p. 161 - 176 (1981).

The author's aim is to identify the basic mechanisms of
heavy element nucleosynthesis and to review the important
underlying physical conditions. He concentrates on those
results of nucleosynthesis studies which he believes to be well
established, independent of the details of the appropriate
stellar or supernova hydrodynamic models.

061.042 Laboratory approaches to nuclear astrophysics.
C. A. Barnes.
Nuclear astrophysics, (see 012.036), p. 235 - 277 (1981).

The author discusses the hypothesis that most of the
elements one observes are produced in stars by nuclear reactions for which the reaction rates can be measured, or at least
estimated by semi-empirical methods. One must also ask
whether a significant fraction of the matter one studies by
direct measurement, or by astronomical observations, could
have been made in other ways than in stars. The consensus
opinion arising from the study of "big-bang" nucleosynthesis
is that this universal process could have made most of the
observed helium (roughly 25% by mass), and possibly also the
tiny traces of deuterium, helium-3, and lithium-7. The author
adopts this hypothesis and concludes that the remainder of the
chemical elements and their various isotopes were made in
stars – either the kinds of stars that one observed now:
"population 1" stars with about two percent by mass of
elements heavier than helium, and "population 2" stars, with
very small abundances of the "trans-helium" elements, which
are presumably older than the population 1 stars.

061.043 Nucleosynthesis of odd-odd nuclei.
E. B. Norman.
Nuclear astrophysics, (see 012.036), p. 285 - 294 (1981).

The author discusses the results of recent work involving
the nuclei ^{26}Al, ^{176}Lu, and ^{180}Ta.

061.044 Neutrinos from electron-positron pair annihilation
in supernovae.   T. J.-L. Courvoisier.
Nuclear astrophysics, (see 012.036), p. 305 - 310 (1981).

Neutrino reactions play an important role in the physics
of supernova explosions: Neutrinos are the most effective
carrier of the gravitational energy released during the collapse
of the core. Many reactions involving neutrinos occur during
the collapse. In this paper the author takes only the electron-
positron pair annihilation into account. It is shown that, at the
end of the collapse, the electron-positron pair annihilation
gives rise to a degenerate sea of mu and tau neutrinos. The
Fermi energy of the neutrino seas is about 100 MeV. This
happens despite the degeneracy of the electron gas, because
both the Fermi energy of the electron gas and the temperature
of the core are high when the collapse is stopped.

061.045 Limits on neutrino degeneracy from nucleosynthesis.
A. S. Szalay.
Nuclear astrophysics, (see 012.036), p. 319 - 324 (1981).

The abundances of ^{4}He and ^{2}H are calculated in terms of
the baryon density and neutrino degeneration, under the
assumption, that the present dynamical properties of the
Universe are determined by neutrinos, as it is suggested by
recent measurements. It is shown, that only a limited range of
these parameters gives the abundances in agreement with
observations. This provides a limit on both the baryon
density of the Universe and the neutrino degeneration parameter. These limits are in agreement with the observed amount
of visible matter.

061.046 s-process branch at ^{176}Lu.
B. J. Allen, G. L. Lowenthal, J. R. de Laeter.
J. Phys. G, Vol. 7, 1271 - 1284 (1981). – Abstr. in Phys.
Abstr., Vol. 84, Abstr. 103187 (1981).

061.047 Stellar nucleosynthesis and the 24-keV neutron
capture cross sections of some heavy nuclei.

T. Bradley, Z. Parsa, M. L. Stelts, R. E. Chrien.
Nuclear cross sections for technology, (see 012.039), p. 344 -
347 (1980). – Abstr. in Phys. Abstr. Vol. 84, Abstr. 103215
(1981).

**061.048 Time delay between gravitational waves and neutrino
bursts from a supernova explosion: a test for the
neutrino mass.** D. Fargion.
Nuovo Cimento Lett., Ser. 2, Vol. 31, 499 - 500 (1981).
Abstr. in Phys. Abstr., Vol. 84, Abstr. 108049 (1981).

**061.049 Neutron cross sections of importance to astro-
physics.** J. C. Browne.
Nuclear cross sections for technology, (see 012.039), p. 627 -
633 (1980). – Abstr. in Phys. Abstr., Vol. 84, Abstr. 108058
(1981).

061.050 Barium nucleosynthesis in the disk.
B. A. Twarog.
Astrophys. J., Vol. 250, 753 - 757 (1981).
The history of Ba production in the disk is discussed,
particularly with regard to the apparent constancy of the production rate of Ba relative to Fe over the lifetime of the disk.
An infall model of the chemical evolution of Ba/Fe within the
disk is constructed under the assumption that the mass function and star formation rate are independent of time and Ba is
produced as purely a secondary element. The model not only
satisfies the present observational constraints for the disk, but
produces a [Ba/H]-[Fe/H] relation which is consistent with the
available observational data.

**061.051 Neutrino and photon emission from a dense, high
temperature atmosphere.**
E. E. Salpeter, S. L. Shapiro.
Astrophys. J., Vol. 251, 311 - 324 (1981).
The thermal evolution of a hot, dense medium, cooling
via neutrino and photon emission, is analyzed. The medium
is assumed to be a plane-parallel, semi-infinite atmosphere in
hydrostatic equilibrium and initially isothermal. The thermal
history of the neutrino photosphere is governed by a nonlinear
diffusion equation. At late times the solution of this equation
acquires a self-similar form. The temperature profile, together
with the neutrino and photon fluxes, are obtained as functions
of depth and time by a combination of numerical and analytic
calculations. Key results are summarized by a convenient set
of compact, closed-form similarity formulae. These are employed to examine the cooling of a young neutron star during
the first ~15 s of its lifetime, when the star is hot, optically
thick to photons and neutrinos, and quasi-static out to the
neutrino photosphere.

061.052 Neutrinos, Sun and universe. A. D. Černin.
Pokroky, Vol. 26, 155 - 165 (1981). In Czech.

**061.053 s-process nucleosynthesis, stellar abundances, and
galactic evolution.** J. W. Truran.
Nukleonika, Vol. 25, 1463 - 1476 (1980). – Abstr. in Phys.
Abstr., Vol. 85, Abstr. 3191 (1982).

061.054 ^{176}Lu: cosmic clock or stellar thermometer?
H. Beer, F. Käppeler, K. Wisshak, R. A. Ward.
Astrophys. J., Suppl. Ser., Vol. 46, 295 - 317 (1981).
The authors quantitatively examine the various experimental and theoretical aspects of the stellar synthesis of the
long-lived ground state of ^{176}Lu (3.6×10^{10} yr). They discuss
the various regimes of stellar temperature and free-neutron
density in which either the internal electromagnetic couplings
between ^{176}Lu^o and ^{176}Lu^m (3.68 hr) are sufficiently slow that
they may be treated as separate nuclei, or the internal couplings are rapidly able to establish thermal equilibrium between
^{176}Lu^o and ^{176}Lu^m.

**061.055 On neutrino thermal conduction and viscosity in
stellar collapse.**
L. J. van den Horn, C. G. van Weert.
Astrophys. J., Lett., Vol. 251, L97 - L100 (1981).
The authors point out that Weinberg's formulae for heat
conduction and viscosity, as adapted by several authors to
contain the degenerate neutrino energy density, do not properly account for dissipative phenomena in stellar collapse. It appears that the transport coefficients are not simply proportional to the neutrino energy density, owing to the energy dependence of the weak-interaction cross section. By using the correct expressions, it follows that the relative importance of
viscous dissipation is much greater than has been previously
claimed.

061.056 Electron capture as supernova killer. N. Sack.
Proceedings of the 5th Göttingen-Jerusalem-
Symposium on Astrophysics, (see 012.041), p. 257 - 260 (1981).
Preliminary model calculations indicate that the process
of electron capture, regardless of the details of neutrino transport, may be responsible for the inability of stellar collapse to
end in motion reversal and mass ejection.

061.057 Linear series and "catastrophies" in astronomy.
J. Katz.
Proceedings of the 5th Göttingen-Jerusalem-Symposium on
Astrophysics, (see 012.041), p. 281 - 289 (1981).
It is indicated at which point Thom's classification of
elementary catastrophies is related to the well known method
of linear series for finding stability limits. So far, there has been
no pertinent application in astronomy.

061.058 On some trends in the development of astrophysics.
V. A. Ambartsumyan.
Izv. Akad. Nauk Arm. SSR, Fizika, Tom 16, 239 - 251 (1981).
In Russian.
Main results on the following topics of astrophysics are
discussed: 1. The invariance principles in the theory of radiative
transfer. 2. The inverse problems of astrophysics. 3. The
empirical approach to the problem of the origin and evolution
of stars and galaxies.

061.059 Thermal cyclotron emission in astrophysics.
V. V. Zheleznyakov.
Inst. prikl. fiz. AN SSSR. Prepr., 1981, No. 15, 45 pp. In
Russian. – Abstr. in Ref. zh., 51. Astron., 11.51.227 (1981).

061.060 Theory of the nucleosynthesis in stars: the s-process.
V. P. Chechev, Ya. M. Kramarovskij.
Usp. fiz. nauk, Tom 134, 431 - 467 (1981). In Russian.
Abstr. in Ref. zh., 51. Astron., 11.51.531 (1981).

**061.061 Which problems of physics and astrophysics seem to
be now most important and interesting? (Ten years
afterwards).** V. L. Ginzburg.
Usp. fiz. nauk, Tom 134, 469 - 517 (1981). In Russian.
From Ref. zh., 51. Astron., 12.51.72 (1981).

Theoretical physics and astrophysics.
See Abstr. 003.065.

Problems of neutrino astrophysics. Volume 2.
See Abstr. 003.166.

**The development of science at the Astronomical
Observatory of the Leningrad University. Theoretical astrophysics.** See Abstr. 009.010.

**The development of science at the Astronomical
Observatory of the Leningrad University. Observational astrophysics.** See Abstr. 009.012.

Existiert eine kosmische Quantelung?
See Abstr. 022.010.

The neutron lifetime.   See Abstr. 022.113.

La structure, l'énergie et l'évolution des étoiles.
See Abstr. 065.036.

Evolution and nucleosynthesis in massive stars with mass loss: the yields in helium and heavy elements and constraints on the past star formation rate.
See Abstr. 065.042.

The evolution and nucleosynthesis of massive pop III stars.   See Abstr. 065.087.

Deflection of massive neutrinos by gravitational fields.   See Abstr. 066.028.

Nucleosynthesis in solar flares as a source of solar gamma radiation.   See Abstr. 076.002.

Systematics of r-process enrichment factors for barium, neodymium, and samarium isotopic anomalies in the Allende meteorite.   See Abstr. 105.079.

Neutrino mass and detection of neutrino supernova bursts.   See Abstr. 125.034.

Heavy neutrinos and supernova explosions.
See Abstr. 125.050.

Gravitational-collapse supernovae.
See Abstr. 125.067.

Radio astronomy confronts elementary particle cosmology: radio measurements of helium abundances.
See Abstr. 132.049.

High-energy neutrinos from powerful radio galaxies.
See Abstr. 141.213.

Expanding shells of young pulsars as sources of high-energy neutrinos.   See Abstr. 141.563.

Cosmic-ray abundances of elements with atomic number $26 \lesssim Z \lesssim 40$ measured on *HEAO* 3.
See Abstr. 143.008.

Indicators of nucleosynthesis and acceleration processes in the ultraheavy cosmic rays: $24 < Z < 59$.
See Abstr. 143.076.

Clustering of the relic neutrinos and the velocity dispersion in a cluster of galaxies.   See Abstr. 160.018.

Age of a neutrino dominated universe.
See Abstr. 162.032.

Neutrinos of non-zero mass in Friedmann universes.
See Abstr. 162.062.

Production of massless particles in the de Sitter-Schwarzschild Universe.   See Abstr. 162.113.

Nuclear astrophysics and particles.
See Abstr. 162.118.

Cosmological implications of Grand Unified Theories.
See Abstr. 162.119.

The formation of galaxies from massive neutrinos.
See Abstr. 162.133.

The rest mass of neutrinos and clustering in the early universe.   See Abstr. 162.140.

Upper bounds on neutrino masses from the large-scale structure of space-time.   See Abstr. 162.148.

062 Hydrodynamics, Magnetohydrodynamics, Plasma

062.001 Disk models with convective viscosity.
S. C. Vila.
Astrophys. J., Vol. 247, 499 - 502 (1981).

This paper presents disk models with convection in the central layers. The calculation of the vertical structure includes the viscosity produced by convection. The models have a central mass of $10^6 M_\odot$ and matter fluxes from 10^{19} to 10^{23} g s^{-1}. The disk luminosities are in the range $10^{39} - 10^{43}$ ergs s^{-1}. They are too low to account for the energy output of quasars if these are at cosmological distances but are adequate if they are at noncosmological ones.

062.002 On charge neutrality and dissipative effects in pulsar systems.   E. A. Jackson.
Astrophys. J., Vol. 247, 650 - 663 (1981).

In the search for the state of minimum energy and minimum energy dissipation in aligned pulsars, force-free states are considered which have no net charge to the star plus its atmosphere (neutral system). All attempts to obtain finite, force-free atmospheres for neutral systems have failed to date, and some necessary complexities of such systems (if they exist) are discussed. Specific models are presented. Several dissipative mechanisms in the atmosphere are discussed and contrasted with the dissipation in the crust. Some of the results of this analysis support the possibility that the nonaligned pulsar atmosphere may be composed of two disjoint plasma regions.

062.003 Evolution of magnetic field and atmospheric response. I. Three-dimensional formulation by the method of projected characteristics.   Y. Nakagawa.
Astrophys. J., Vol. 247, 707 - 718 (1981).

In a previous paper, a method of examining the evolution of a magnetic field together with atmospheric responses in a physically self-consistent manner was described. With the full use of properties of the projected characteristics, a new and simpler formulation is developed. Therefore, the formulation for the examination of the general three-dimensional problems is presented. In practice, numerical solutions must be obtained; hence, the final formulation is given in the form of difference equations. The possibility of including effects of viscous and ohmic dissipations in the formulation is considered, together with discussion of physical interpretation.

062.004 Evolution of magnetic field and atmospheric responses. II. Formulation of proper boundary equations.   Y. Nakagawa.
Astrophys. J., Vol. 247, 719 - 733 (1981).

A systematic manner of deriving physically self-consistent, time-dependent boundary equations for MHD initial boundary problems is presented. The formulation is based on the use of the method of projected characteristics. It is shown that the full use of the compatibility equations is required in determining the time-dependent boundary conditions. To provide a clear physical picture as an example, the evolution of axisymmetric global magnetic field by photospheric differential rotation is considered. The details of the formulation are described together with other considerations − in particular, the possible cause of the observed time difference for the reversal of the polar magnetic field on the Sun.

062.005 Internal motions in magnetic stars.
J. Nittmann, W. P. Wood.
Mon. Not. R. Astron. Soc., Vol. 196, 491 - 506 (1981).

The dynamically driven, adiabatic, internal motions for an obliquely rotating star are estimated for the case where the toroidal component of the magnetic field dominates. The consequent mixing due to these motions is discussed in relation to: (1) the diffusion theory for the Ap phenomena, and (2) the

blue straggler stars. The authors report the behaviour of the time-scales for the obliquity angle to reach its asymptotic state for a range of stellar masses for stars with either dominant poloidal or toroidal magnetic fields.

062.006 A discussion of theoretical ionization equilibrium calculations based on solar flare X-ray spectra.
U. Feldman, G. A. Doschek, R. D. Cowan.
Mon. Not. R. Astron. Soc., Vol. 196, 517 - 526 (1981).

Several sets of ionization equilibrium calculations exist for use in interpreting X-ray and EUV spectra of astrophysical plasmas. In particular, the calculations of Jordan (1969, 1970), Jacobs et al. (1977, 1978) and Summers (1974) are well known. The temperatures of maximum fractional abundance calculated by Summers for the more highly ionized and heavier elements such as iron are about a factor of 2 higher than the temperatures calculated by Jordan and Jacobs et al. Using recently obtained X-ray spectra of solar flares, the authors show that the temperatures calculated by Summers (1974) for iron are incorrect. In their opinion the temperatures calculated by Jordan or Jacobs et al. should be used until further improvements become available.

062.007 Some exact axisymmetric force-free magnetospheres − I.   R. Buckley.
Mon. Not. R. Astron. Soc., Vol. 196, 1021 - 1049 (1981).

In this paper, and another to follow, classes of globally exact magnetospheres in the force-free approximation are derived and discussed. These classes include all previously known exact solutions. The method of solution allows for differential rotation of the field lines, and so the models are not applicable to magnetospheres attached to rigidly rotating stars. Several of the models do, however, seem promising candidates for magnetospheres associated with non-uniformly rotating discs, and it is especially interesting that the focusing of electromagnetic energy flux along the axis of rotation, an essential ingredient of Blandford's model, is a property of several of the models considered here.

062.008 Magnetohydrodynamic Kelvin-Helmholtz instabilities in astrophysics − II. Cylindrical boundary layer in vortex sheet approximation.
A. Ferrari, E. Trussoni, L. Zaninetti.
Mon. Not. R. Astron. Soc., Vol. 196, 1051 - 1066 (1981).

This second paper of the series is devoted to Kelvin-Helmholtz instabilities in cylindrical boundary layer flows (jets). The vortex-sheet approximation is still used, and compressible flows are studied in subsonic, transonic, supersonic and relativistic regimes. Magnetic field effects are analysed, together with density contrast inside and outside the jet. The general result is that, due to the onset of a so-called reflection branch of resonant modes, jets are always unstable, both to pinching and helical perturbations with wavelengths of the order of the jet circumference. In particular the time-scales for instability are such that this certainly plays a significant part in the morphology and energetics of extended radio sources.

062.009 The effects of cold plasma on the Kelvin-Helmholtz instability.   B. G. Melander, G. K. Parks.
J. Geophys. Res., Vol. 86, 4697 - 4707 (1981).

The article studies the effect of a two-component plasma (hot and cold) on the shear driven Kelvin-Helmholtz instability. The theory incorporates an ion distribution function with a shear flow parallel to the ambient magnetic field and a density gradient perpendicular to both. The electrostatic and electromagnetic modes of the instability are studied in the limit of hydromagnetic frequencies. The dispersion relation for the electrostatic case is obtained by solving the Vlasov equation for the perturbed ion and electron densities that are then used in the quasi-neutrality condition. The electromagnetic dispersion relation is obtained by solving the coupled Vlasov and Maxwell equations. The results applied to the magnetosphere show that the ionosphere plays an important role in determining the type of waves that are generated by a sheared plasma flow and indicate whether the flow boundary is unstable to the Kelvin-Helmholtz instability.

062.010   Large-amplitude waves in an anisotropic plasma.
A. T. Granik.
J. Geophys. Res., Vol. 86, 5431 - 5437 (1981).

The propagation of large-amplitude hydromagnetic waves of a collisionless anisotropic plasma is treated by using the Chew-Goldberg-Low equations. The generalization of the usual theory of characteristics as applied to this treatment has been developed. It is shown that in such a plasma there are six different types of waves. They incorporate all modes considered in a small-amplitude theory.

062.011   Self-consistent solutions for electromagnetic plane waves in a relativistic plasma.
A. Che, H. Herold, M. Reinecke, H. Ruder, G. Wunner.
Astron. Astrophys., Vol. 100, 164 - 168 (1981).

Using numerical methods, self-consistent solutions for electromagnetic waves propagating in a cold electron – ion plasma are obtained for amplitudes covering the whole range from the classical limit to the extreme relativistic region. For the three principal modes of strictly periodic solutions – longitudinal, linearly polarized and circularly polarized waves – the dependence of the cutoff frequencies on the amplitudes is discussed for different ion masses and several values of the parameter u_0, which is related to the drift velocity of the plasma. The domain of intermediate wave strengths is here investigated quantitatively for the first time. A possible relevance to the propagation of pulsar waves is pointed out.

062.012   Physical conditions in an optically thin relativistic gas irradiated by γ-rays.
I. Kovner, M. Milgrom.
Astron. Astrophys., Vol. 100, 271 - 276 (1981).

The temperature T_e and the positron density n_+ in an optically thin relativistic gas cloud irradiated by γ-rays are calculated. A power-law spectrum with an energy power index β, and a low energy cutoff E_0 is assumed. It is assumed that both E_0 and kT_e are much larger than mc^2 (m is the electron mass). The behaviour of T_e and n_+, as functions of β, E_0, and the γ-ray flux I, is rather regular. The approximate expressions in the extreme cases of large or small γ-ray fluxes or equivalently low and high nucleon densities respectively are given. The cooling effect of X-rays and softer radiation is discussed.

062.013   Collisionless perpendicular shocks: applications to solar type II radio bursts and the Antares (α Sco) B radio emission.
F. R. Klinkhamer, J. Kuijpers.
Astron. Astrophys., Vol. 100, 291 - 301 (1981).

The authors propose a model of a collisionless perpendicular ($v \perp B$) shock. They calculate a mean turbulence energy density in the shock. Type II radio bursts of the Sun and their characteristic fine structures can be explained. Using the turbulent bremsstrahlung mechanism for Langmuir waves the radio luminosity can be accounted for. The authors consider the observed radio emission from the Antares B companion star which is embedded in a stellar wind from the M supergiant.

062.014   Radiation pressure in a superdense electron plasma.
C. K. Chou.
Bull. Geophys., No. 20, p. 114 - 123 (1980). – Abstr. in Phys. Abstr., Vol. 84, Abstr. 88732 (1981).

062.015   The potentials for the g-, p-, and the toroidal-modes of self-gravitating fluids.   Y. Sobouti.
Astron. Astrophys., Vol. 100, 319 - 322 (1981) = Biruni Obs., Contrib. No. 9.

In a convectively neutral fluid the g-modes are derived from a vector potential and the p-modes from a scalar potential. In a convectively non-neutral fluid the two potentials are coupled. For small and moderate deviations from convective neutrality, however, the solenoidal character of the g-modes and the irrotational nature of the p-modes persist.

062.016   Stability of galactic radio jets.
G. Benford.
Astrophys. J., Vol. 247, 792 - 802 (1981).

The author considers a simple cold flowing beam undergoing sidewise perturbations. He ignores internal dynamics and uses only Newton's laws. He finds that large scale instability of the "firehose" type can be avoided, or its effect diminished, by (a) sizable internal magnetic fields, (b) a dense cocoon of plasma which increases in pressure with distance from the beam, (c) linking a large, dense cocoon to the beam, which adds inertia to the system. Exponential growth can be stopped by two nonlinear effects: (a) when $k\xi > 1$, growth shifts to longer wavelengths; (b) when the beam boundary sidewise velocity exceeds the sound speed, a bow shock forms, allowing only algebraic growth.

062.017   The effect of turbulence on the $K\|B$ relativistic beam instability.   S. R. Spangler, C. K. Goertz.
Astrophys. J., Vol. 247, 1078 - 1088 (1981).

The authors have considered the effect of magnetic turbulence, causing electron gyrophase diffusion, on the instability of an anisotropic, field-aligned relativistic electron beam. Formally, inclusion of gyrophase diffusion results in a change in the formula for the wave growth rate whereby the delta function describing the wave-particle resonance is replaced by a Lorentzian function.

062.018   On the acceleration of fast particles at a spherical shock wave front.
V. L. Prishchep, V. S. Ptuskin.
Astron. Zh., Tom 58, 779 - 789 (1981). In Russian. English translation in Soviet Astron., Vol. 25, No. 4.

The mechanism of particle acceleration at a shock wave front is discussed, the scattering of particles in front of the shock being taken into account. The possibility of cosmic rays acceleration in supernova shells is discussed.

062.019   On the parametric generation of acousto-gravity waves in an isothermal atmosphere.
L. A. Ostrovskij, N. S. Petrukhin.
Astron. Zh., Tom 58, 848 - 858 (1981). In Russian. English translation in Soviet Astron., Vol. 25, No. 4.

Nonlinear interaction of acousto-gravity waves (AGW) in an isothermal layer of finite thickness is considered. Averaged equations for a connected triple of waves are obtained by means of the Lagrangian description. It is shown that due to the feedback caused by the direction difference of phase and group velocities of the internal gravity mode such a layer excited by an acoustic "pump" can serve as a parametric generator of AGW which do not get into it from outside. The threshold values of the layer length as well as of the amplitudes of the generated waves are found. Quantitative evaluations for the sun's chromosphere are made.

062.020   The hydrogen two-photon emission spectrum of moderately dense plasmas.
S. A. Drake, R. K. Ulrich.
Astrophys. J., Vol. 248, 380 - 383 (1981).

The physical conditions in which two-photon emission is a nonnegligible contributor to the continuous spectrum of hydrogen have been examined. This process can make a significant contribution to the continuous spectral energy distribution for electron densities as high as 10^{11} cm^{-3} and for temperatures and geometrical thicknesses that are of astrophysical interest. The authors present quantitative results which define the region of parameter space over which the emission from this process might be detectable.

062.021 Chaos in the segmented disc dynamo.
E. Knobloch.
Phys. Lett. A, Vol. 82A, 439 - 440 (1981). — Abstr. in Phys. Abstr., Vol. 84, Abstr. 66827 (1981).

062.022 Taylor vortices and the Goldreich-Schubert instability.   D. G. Economides, G. Moir.
Geophys. Astrophys. Fluid Dyn., Vol. 16, 299 - 317 (1981). Abstr. in Phys. Abstr., Vol. 84, Abstr. 66932 (1981).

062.023 On the maintenance of magnetic fields by compressible flows and the Nernst-Ettingshausen effect.
D. J. Ivers, R. W. James.
Geophys. Astrophys. Fluid Dyn., Vol. 16, 319 - 323 (1981). Abstr. in Phys. Abstr., Vol. 84, Abstr. 66933 (1981).

062.024 Interpretation of the spectral proof of Cowling's theorem.   D. J. Ivers, R. W. James.
Geophys. Astrophys. Fluid Dyn., Vol. 16, 325 - 328 (1981). Abstr. in Phys. Abstr., Vol. 84, Abstr. 66934 (1981).

062.025 The opacity of an expanding medium.
A. H. Karp.
AIP Conf. Proc., No. 63, (see 012.001), p. 125 - 138 (1980).
After reviewing the expansion opacity for a homogeneous, isotropically expanding medium, the author relaxes the assumptions. The expansion opacity in a homogeneous, non-isotropically expanding gas is considered. The assumption of homogeneity is dropped, attention is restricted to the case of an optically thick gas. The author considers the general problem in which the photons are allowed to have an arbitrarily large mean free path.

062.026 On the construction of models of rotating stars and stellar systems.
P. O. Vandervoort, D. E. Welty.
Astrophys. J., Vol. 248, 504 - 515 (1981).
A new method is presented for the construction of the equilibrium configurations of rotating stars and stellar systems. A configuration is represented in a first approximation as stratified on similar and similarly situated ellipsoids. The stratification is determined by solving a spherically symmetric model of the governing equations, whereas the ellipsoidal geometry is determined by solving the tensor virial equations of the second order. The structure of the configuration is determined in a second approximation by solving exactly the equations of mechanical equilibrium in which the prevailing gravitational field is given by the solution for the field obtained in the first approximation. The method is formulated in detail for barotropic configurations and applied to the construction of both axisymmetric and nonaxisymmetric configurations of uniformly rotating polytropes. Generalizations and wider applications of the method are briefly described.

062.027 Enhancement of passive diffusion and suppression of heat flux in a fluid with time-varying shear.
W. H. Press, G. B. Rybicki.
Astrophys. J., Vol. 248, 751 - 766 (1981).
Diffusion and heat conductivity within stars are shown to be substantially affected by fluid shear motions, even when these motions do not mix fluid elements. Internal gravity waves are one example of such a nonmixing flow. A formalism, based on Lagrangian flow coordinates, is developed to describe the effect in Boussinesq approximation. The diffusion

of passive contaminants and of specific entropy is found to be generally enhanced, while the conduction of heat in stable stratified layers is suppressed. Explicit formulae are derived for flows which can be approximated as having spatially constant velocity gradients, and also for interval gravity waves. In the latter case the mechanical dissipation and thermodynamic efficiencies are calculated.

062.028 Resistive stability.   J. Heyvaerts.
Ann. Physique, Vol. 5, (see 012.006), 379 - 407 (1980). In French. — Abstr. in Phys. Abstr., Vol. 84, Abstr. 73333 (1981).

062.029 Recent development in the theory of perfect MHD waves in magnetically structured media.
B. Roberts.
Ann. Physique, Vol. 5, (see 012.006), 453 - 481 (1980). In French. — Abstr. in Phys. Abstr., Vol. 84, Abstr. 73335 (1981).

062.030 Inductive interaction of electroconductive bodies with ionospheric and interplanetary plasma.
A. V. Gurevich, N. T. Paschenko.
Acta Astronaut., Vol. 8, 47 - 56 (1981). — Abstr. in Phys. Abstr., Vol. 84, Abstr. 74975 (1981).

062.031 Monte Carlo simulation of the unsaturated Comptonization in a semi-relativistic plasma.   F. Takahara.
Prog. Theor. Phys., Vol. 65, 883 - 893 (1981). — Abstr. in Phys. Abstr., Vol. 84, Abstr. 75009 (1981).

062.032 Kinetic effects on the propagation of surface waves and their relevance to the heating of the solar corona.
M. Kuperus, J. Heyvaerts.
Ann. Physique, Vol. 5, (see 012.006), 483 - 502 (1980). In French. — Abstr. in Phys. Abstr., Vol. 84, Abstr. 75089 (1981).

062.033 Comment on 'First phase acceleration mechanisms and implications for hard X-ray burst models in solar flares'.   G. Van Hoven.
Sol. Phys., Vol. 73, 205 (1981).

062.034 Introduction to high temperature plasma physics.
R. J. Bickerton.
Philos. Trans. R. Soc. London, Ser. A, Vol. 300, 475 - 488 (1981). — Abstr. in Phys. Abstr., Vol. 84, Abstr. 81692 (1981).

062.035 Experiment and m.h.d. theory of stability and relaxation in toroidal discharges.   D. C. Robinson.
Philos. Trans. R. Soc. London Ser. A, Vol. 300, 525 - 533 (1981). — Abstr. in Phys. Abstr., Vol. 84, Abstr. 81713 (1981).

062.036 Dynamically consistent nonlinear dynamos driven by convection in a rotating spherical shell.
P. A. Gilman, J. Miller.
Astrophys. J., Suppl. Ser., Vol. 46, 211 - 238 (1981).
The authors present extensive calculations from a convectively driven hydromagnetic dynamo for rotating spherical shell of fluid and compare the results with observations of the solar dynamo. What distinguishes this dynamo model from previous models applied to the sun is that the motion fields used are themselves solutions to the nonlinear equations of momentum, thermodynamics, and mass continuity, albeit for a Boussinesq fluid. The full feedbacks of the induced magnetic fields on these motions are included. The motions take the form of a time-evolving spectrum of convection patterns driven by uniform heating from below, which in turn drive a differential rotation whose outer boundary amplitude and profile are similar to that of the equatorial acceleration of the sun. Despite the similarity of the calculated differential rotation to the observed one, the authors find the calculated dynamo behaves much differently than the sun.

062.037 A model of flare-produced magneto-radiative shock with increasing energy.
B. G. Verma, R. C. Srivastava.
Astrophys. Space Sci., Vol. 78, 95 - 103 (1981).

An exact solution for a spherically-symmetric model of a magneto-radiative shock wave in the solar wind caused by the explosive energy release of a solar flare has been obtained in the case when energy released is an increasing function of the time. It has been shown that due to increasing energy, density, pressure, radiation flux, magnetic field and shock velocity change considerably.

062.038 Free convection effects on the hydromagnetic oscillatory flow in the Stokes problem past an infinite porous vertical limiting surface with constant suction. II.
N. G. Kafousias, A. A. Raptis, G. J. Tzivanidis, C. V. Massalas.
Astrophys. Space Sci., Vol. 78, 157 - 168 (1981).

Unsteady hydromagnetic boundary layer flow of a viscous incompressible and electrically conducting fluid past an infinite vertical non-conducting porous limiting surface in presence of a transverse magnetic field is considered when the limiting surface is moving impulsively in its own plane and is subjected to a constant suction. The solutions for the transient velocity profiles, transient temperature profiles, the amplitude and the phase of the skin friction and the rate of the heat transfer are presented.

062.039 Plasma diagnostics in an astrophysical setting.
D. J. Helfand.
Nature, Vol. 293, 338 - 339 (1981).

062.040 Physical mechanism of the lower-hybrid-drift instability in a collisional plasma.
J. D. Huba, S. L. Ossakow.
J. Atmos. Terr. Phys., Vol. 43, (see 012.008), 775 - 778 (1981).

The authors present a physical discussion of the lower-hybrid-drift instability in both collisionless and collisional plasmas. The instability is important since it is the most promising explanation of small-scale irregularities (i.e. $\lesssim 1$ m) observed during equatorial spread F.

062.041 Differential rotation induced by compressible convection.
G. A. Glatzmaier, P. A. Gilman.
Bull. American Astron. Soc., Vol. 13, 527 (1981). – Abstract.

062.042 Thermodynamic transport properties for relativistic magnetic imperfect fluids.
T. W. Edwards.
Bull. American Astron. Soc., Vol. 13, 534 (1981). – Abstract.

062.043 Synchrotron radiation at mildly relativistic energies.
V. Petrosian.
Bull. American Astron. Soc., Vol. 13, 551 (1981). – Abstract.

062.044 X-ray and gamma-burst radiation stress limits and diamagnetic accretion.
A. G. Petschek, S. A. Colgate.
Bull. American Astron. Soc., Vol. 13, 551 (1981). – Abstract.

062.045 Solutions for force-free magnetic fields.
H. M. Chang, R. L. Carovillano.
Bull. American Astron. Soc., Vol. 13, 552 (1981). – Abstract.

062.046 Effects of mass transfer, free-convection currents and heat sources on the Stokes' problem for an infinite vertical plate.
A. A. Raptis, G. J. Tzivanidis.
Astrophys. Space Sci., Vol. 78, 351 - 357 (1981).

An exact analysis of the effects of heat sources, mass transfer and free-convection currents on the flow past an impulsively started plate is investigated. Closed-form solutions to the velocity and temperature field have been derived by using Laplace transform and expressions are given for the skin-friction and the rate of heat transfer. Variations of the above quantities are presented graphically, and the paper is concluded with a quantitative discussion.

062.047 Current sheets.
E. R. Priest.
Solar flare magnetohydrodynamics, (see 003.002), p. 139 - 215 (1981).

Contents: Introduction. Means of formation. Behaviour of current sheets. Role of current sheets in flare models. The emerging (or evolving) flux model. Conclusion.

062.048 Penetrative convection.
D. R. Moore.
The sun as a star, (see 003.004), p. 253 - 262 (1981).

This paper attempts to summarize the current state of understanding of the most directly observable solar convection, the granulation and supergranulation. It approaches the subject from a fluid mechanics point of view.

062.049 Wave generation.
R. F. Stein, J. W. Leibacher.
The sun as a star, (see 003.004), p. 289 - 300 (1981).

This paper treats mainly the generation of relatively short period waves by turbulence in the lower solar photosphere at the top of the convection zone. Periods are typically of the order of 100 seconds and less, although the generation of internal gravity waves by penetrative convection in the upper photosphere is also considered, and these waves generally have much longer periods. Other than gravity waves, the chief modes considered are the sound wave and also the Alfvén wave, which will be generated in the highly conductive solar plasma wherever there is a magntic field and a dynamical perturbation of the medium.

062.050 Magnetic flux tubes.
H. C. Spruit.
The sun as a star, (see 003.004), p. 385 - 412 (1981).

In the sections on photospheric and coronal flux tubes, the general environment of flux tubes in the convection zone and in the corona is discussed, as well as the reasons for their existence as individual structures. In the sections equilibrium of flux tubes and stability and waves, the problems relating to the equilibrium and stability of flux tubes in the photosphere and convection zone are treated. The corresponding problems in coronal tubes are discussed briefly in the section on coronal flux tubes, while in the section evolution of active regions and sunspots, current ideas on the evolution of active regions are discussed from a flux tube point of view.

062.051 Expansion of plasma clouds with different geometry into a vacuum.
Yu. I. Chutov, A. Yu. Kravchenko.
Problems of cosmic physics. Vyp. 16, (see 003.005), p. 89 - 104 (1981). In Russian.

062.052 On the evolution of thermal disturbances in stars.
I. A. Klimishin, B. I. Gnatyk.
Problems of cosmic physics. Vyp. 16, (see 003.005), p. 109 - 115 (1981). In Russian.

The relation between the velocities of hydrodynamic processes and those of thermal conduction in a medium of constant density is studied. At the energy concentration during the outburst of nova- or supernova-type the evolution of temperature disturbance occurs mainly through hydrodynamic energy transfer.

062.053 Diffuse clouds and Alfvén waves.
J. C. Higdon.
The phases of the interstellar medium, (see 012.018), p. 169 - 174 (1981),

21 cm emission-absorption measurements show that most diffuse clouds possess supersonic motions. It is difficult to understand this phenomenon in the context of acoustic turbulence models, since shock waves would form that would radiate the initial kinetic energy relatively quickly. It is suggested here that these supersonic motions in diffuse clouds

are produced by large-scale Alfvén waves. In a partially ionized medium the neutral gas component can be forced to move at speeds approaching $H_o/(4\pi\rho c)1/2$ without generating shocks, where H_o and ρc are magnetic field strength, and neutral gas density, respectively. Field strengths between $\approx (3.5$ to $5) \times 10^{-6}$ gauss can reproduce the observed broadening of the 21 cm profiles in diffuse clouds.

062.054 **Non-linear electromagnetic waves in relativistic plasma.** G. S. Lakhina, B. Buti.
Bull. Astron. Soc. India, Vol. 9, 64 (1981). − Abstract.

062.055 **Wave interactions in astrophysical plasma.** G. L. Kalra.
Bull. Astron. Soc. India, Vol. 9, 64 - 65 (1981). − Abstract.

062.056 **The equilibrium statistical mechanics of self-gravitating systems.** P. S. Cally.
Australian J. Phys., Vol. 34, 267 - 278 (1981).

062.057 **Mathematical theory of cylindrical isothermal blast waves in a magnetic field.** I. Lerche.
Australian J. Phys., Vol. 34, 279 - 301 (1981).

062.058 **Influence of a dense plasma on ions with K-electrons.** H. Pilkuhn, T. Reddmann.
Astron. Astrophys., Vol. 101, 350 - 351 (1981).
Wave functions and ionization energies of ions with bound K-electrons in a shielding plasma are calculated, using a variational method.

062.059 **Equations for thin flux tubes in ideal MHD.** H. C. Spruit.
Astron. Astrophys., Vol. 102, 129 - 133 (1981).
A Hamilton-Lagrange formalism is used to derive the equation of motion for an isolated magnetic flux tube which is embedded in a static gravitating fluid. It is shown that the force operator corresponding to this equation is self-adjoint. An equation of continuity for thin flux tubes is also given. The equations can be used for the study of wave and stability problems involving flux tubes, and possibly for the study of turbulent dynamo models in which the Lorentz force is taken into account.

062.060 **Charged particle acceleration in shearing flows of cosmic plasma.** E. G. Berezhko.
Pis'ma v ZhEhTF, Tom 33, 416 - 419 (1981). In Russian. Abstr. in Ref. zh., 51. Astron., 9.51.162 (1981).

062.061 **Shock wave detection in cosmic plasma on the basis of slipping statistics.**
V. P. Evdokimov, M. Ya. Natenzon.
Inst. kosm. issled. AN SSSR. Prepr., 1981, No. 613, 41 pp. In Russian. − Abstr. in Ref. zh., 51. Astron., 9.51.163 (1981).

062.062 **Resonant absorption of Alfvén waves in a homogeneous magnetic tube.** Yu. A. Sukovatov.
Fiz. soln. aktivnosti. Moskva, 1980, p. 131 - 144. In Russian. Abstr. in Ref. zh., 51. Astron., 9.51.358 (1981).

062.063 **On parametric generation of acoustic-gravitational waves in the solar atmosphere.**
L. A. Ostrovskij, N. S. Petrukhin.
13-ya Vses. konf. po radioastron. issled. soln. sistemy, Kiev, 1981. Tez. dokl. Kiev, 1981, p. 8 - 9. In Russian. − From Ref. zh., 51. Astron., 9.51.359 (1981).

062.064 **Propagation of fast MHD waves in stratified medium.** W.-r. Hu.
Acta Mech. Sinica, No. 2, p. 190 - 193 (1981). In Chinese. Abstr. in Phys. Abstr., Vol. 84, Abstr. 85830 (1981).

062.065 **Modulation instabilities in astrophysics.** D. ter Haar, V. N. Tsytovich.
Phys. Rep., Vol. 73, 175 - 236 (1981). − Abstr. in Phys. Abstr., Vol. 84, Abstr. 88740 (1981).

062.066 **Rayleigh instability of compressible plasma in a vertical magnetic field.** B. K. Shivamoggi.
Astrophys. Space Sci., Vol. 79, 3 - 9 (1981).
A study of the Rayleigh instability of a compressible plasma of density stratified in horizontal planes and subjected to a vertical magnetic field is made. The special case of a plane interface separating two superposed uniform plasmas of different densities and speeds of sound is treated as an example to illustrate the compressibility effects on the hydromagnetic Rayleigh instability. It is found that the hydromagnetic-compressibility effects act toward reducing the growth rate in a hydrodynamically unstable situation.

062.067 **Transport phenomena and abundance anomalies in cosmic plasma.** V. A. Urpin.
Astrophys. Space Sci., Vol. 79, 11 - 24 (1981).
Hydrodynamical equations for a fully ionized hydrogen-helium plasma are derived by the Chapman-Enskog method. The electron and ion transport coefficients are found as the functions of electron and ion temperatures and number densities as well as of the magnetic field strength. The presented equations are needed for describing transport phenomena in laboratory and cosmic plasmas. It is shown that transport phenomena can produce abundance anomalies. Various astrophysical consequences of the theory are discussed.

062.068 **Generation of a d.c. field by nonlinear electromagnetic waves in relativistic plasmas.**
G. S. Lakhina, B. Buti.
Astrophys. Space Sci., Vol. 79, 25 - 36 (1981).
A finite amplitude linearly polarized electromagnetic wave propagating in a relativistic plasma, is found to generate the longitudinal d.c. as well as the oscillating electric field at the second harmonic. In a plasma consisting of only electrons and positrons, these fields cannot be generated. The evolution of the electromagnetic waves is governed by the non-linear Schrödinger equation. The model has direct relevance to some plasma processes occurring in pulsars.

062.069 **Hydromagnetic wave propagation and the Kelvin-Helmholtz instability in an arbitrary-β collisional inhomogeneous plasma.** K. M. Srivastava, R. Singh.
Astrophys. Space Sci., Vol. 79, 37 - 54 (1981).
The instability of an inhomogeneous arbitrary-β plasma occurring due to the transverse velocity shear, has been studied to analyse the effects of collisional thermal transfer. The dissipation of hydromagnetic waves in such a plasma has also been discussed. It has been found that the thermal forces modify the instability criteria in several limiting cases. Numerical solutions have also been obtained to investigate the effects of various physical parameters for a non-isothermal plasma with different adiabaticity of two species, viz., electrons and ions.

062.070 **The Schwarzschild criterion for convection in the presence of a magnetic field.**
K. H. Schatten, S. Sofia.
Astrophys. Lett., Vol. 21, 93 - 96 (1981).
The Schwarzschild criterion governing the onset of convective instability has been modified to include magnetic field. This may be of importance for solar variability. The revised condition suggests that the underside of field layers are stabilizing and the upper side destabilizing. Absolute instability can be reached to achieve conventional magnetic buoyancy. This may explain the inverse correlation between the time intervals between sunspot minima and sunspot maxima with the maximum values of sunspot number.

062.071 **The effect of losses on acceleration of energetic particles by diffusive scattering through shock waves.**   H. J. Völk, G. E. Morfill, M. A. Forman.
Astrophys. J., Vol. 249, 161 - 175 (1981).

The authors discuss the problem of shock acceleration in a lossy medium in a number of illustrative examples . They do this always in the time asymptotic limit. The effect of losses on the build-up time of the scattering wave field, for example, would be even more pronounced in a situation where a freshly created shock starts to accelerate an existing' population of cosmic rays over its finite lifetime. The loss process itself is described in terms of a (energy-dependent) loss time τ for analytical simplicity. Since the authors are concerned with acceleration rather than, for example, modulation, this approximation seems acceptable. They first consider steady state acceleration by a shock in a medium with spatially uniform losses and a background source of the cosmic rays. The conditions on the local diffusion properties of the medium are discussed, and the requirements on the wave field in the face of damping and energy losses are determined for several characteristic environments. Up to this point, the ionization structure of the medium was assumed to be unaffected by the presence of the shock. The authors discuss the possibility of local acceleration within the ionization layer accompanying strong shocks in essentially neutral interstellar clouds.

062.072 **A criterion for the onset of instability at a turning point.**   R. Sorkin.
Astrophys. J., Vol. 249, 254 - 257 (1981).

A very general sufficient condition for the onset of "thermodynamic" instability at a turning point is proved, assuming essentially nothing beyond the smoothness of the equilibrium-sequence in question.

062.073 **The effects of *in situ* particle acceleration on the appearance of relativistically moving plasmons.**   R. Duerr.
Astrophys. J., Vol. 249, 263 - 269 (1981).

The effects of *in situ* particle acceleration in a relativistically moving ram-pressure confined radio emitting plasmon (plasmoid) are discussed. Although the inclusion of these effects is found to be insufficient to explain the observation of superluminal events with reasonable values for the acceleration parameters, the situation of intermediate values of beta is improved. Indeed, with values of the acceleration parameters within the ranges proposed by other authors, plasmons with β as high as 0.6 can be simultaneously observed.

062.074 **On the function of size distribution of solid particles in a turbulent gas.**   V. S. Kessel'man.
Astron. Zh., Tom 58, 1011 - 1016 (1981). In Russian. English translation in Soviet Astron., Vol. 25, No. 5.

A stationary solution of the coagulation equation for solid particles in a turbulent gas is obtained.

062.075 **Kinetic consideration of charged particle acceleration process in shearing flows of collisionless plasma.**
E. G. Berezhko, G. F. Krymskij.
Pis'ma Astron. Zh., Tom 7, 636 - 640 (1981). In Russian. English translation in Soviet Astron. Lett., Vol. 7.

The equation for the density of accelerated charged particles is derived from Boltzmann's kinetic equation for the case of collisionless plasma with shearing flows. The stationary impulse spectrum of accelerated particles is shown to be of power-law type provided there is a power-law growth of the mean scattering time as the impulse increases. The role of the considered process of acceleration in formation of non-thermal spectra of charged particles in the interstellar and interplanetary space is discussed.

062.076 **MHD instabilities of atmospheres with magnetic fields.**   E. G. Zweibel.
Astrophys. J., Vol. 249, 731 - 745 (1981).

The problem of determining the stability of an atmosphere with a magnetic field is important for a variety of astrophysical problems. The author uses the MHD energy principle to study the stability of two-dimensional configurations, and develops a criterion for demonstrating local instability. The author then applies this criterion to the family of equilibria recently described by Zweibel and Hundhausen, and finds that all members of this family except the potential field are locally unstable. The instabilities are explained in terms of pinching, interchange, and Rayleigh-Taylor effects.

062.077 **The present status of dynamo theory.**   T. G. Cowling.
Annu. Rev. Astron. Astrophys., Vol. 19, (see 003.012), 115 - 135 (1981).

This review does not consider fields explicable in terms of the compression of preexistent fields (fields of early-type stars, white dwarfs, and pulsars). Of the three main aspects of dynamo theory (planetary, stellar, and galactic) the author concentrates on the fields of the Sun and similar stars, with only passing references to other bodies. No attempt will be made to review "pure" dynamo theory.

062.078 **Pinch sheets and reconnection in astrophysics.**   S. I. Syrovatskii (*Syrovatskij*).
Annu. Rev. Astron. Astrophys., Vol. 19, (see 003.012), 163 - 229 (1981).

Contents: Introduction. Singular lines in magnetized plasma. Current sheet formation. Quasi-stationary current sheet. Current sheet stability and flare processes. Numerical simulation and laboratory experiments. Current sheets in astrophysics. Conclusion.

062.079 **Temperature anisotropy instabilities driven by Whang's ion model.**   M. P. Leubner.
J. Geophys. Res., Vol. 86, 7787 - 7791 (1981).

Four plasma instabilities driven by proton temperature anisotropies are investigated theoretically by using Whang's ion model, i.e., heat flux modified non-Maxwellian velocity distribution functions. The linear dispersion properties are studied in an infinite Vlasov plasma where the proton distribution is assumed to consist of one warm slow-speed component and the electron background is isotropic. The parameters are chosen according to Whang's solar wind ion model.

062.080 **Similarity solutions for an axially symmetric explosion model in magnetogasdynamics. I.**
J. B. Singh, S. K. Srivastava.
Astrophys. Space Sci., Vol. 79, 345 - 353 (1981).

Similarity solutions, for a point explosion in a spheroid with axially symmetric density distribution obeying power laws in the presence of magnetic field, are obtained. A new technique suggested by Bhowmick (1978) has been utilised to study the character of flow variables behind the shock front in an axisymmetric model. The total energy of the wave is constant.

062.081 **An axially symmetric explosion model in magnetogasdynamics. II.**   J. B. Singh, S. K. Srivastava.
Astrophys. Space Sci., Vol. 79, 355 - 357 (1981).

A method using Eulerian coordinate system is developed under a local radiality assumption to study a point explosion in a spheroid with axially symmetric exponential density distribution, including the effect of azimuthal magnetic field.

062.082 **Effect of a helical magnetic field on the stability of a gravitating cylinder.**   A. M. Karnik, S. P. Talwar.
Astrophys. Space Sci., Vol. 79, 379 - 386 (1981).

The effect of a helical magnetic field on the stability of

an infinitely conducting, inviscid, incompressible and infinitely long self-gravitating cylinder is studied for axisymmetric perturbations. The effect of helicity is also examined.

062.083 Annihilation radiation from a hot e^+-e^- plasma.
R. Ramaty, P. Mészáros.
Astrophys. J., Vol. 250, 384 - 388 (1981).
The authors have studied the details of pair annihilation in hot e^+-e^- plasmas. They have calculated the annihilation rate, luminosity, and spectrum of optically thin plasmas of temperatures above 10^8 K by means of a Monte Carlo simulation. For a given temperature, the spectrum is peaked at an energy equal to 0.511 MeV plus a positive definite quantity of order kT. In high temperature sources, such as γ-ray bursts, this blueshift can amount to a significant fraction of 0.511 MeV.

062.084 The magnetic barrier effect in non-dissipative magneto-gasdynamics.   N. V. Erkaev.
VTs Krasnoyar. fil. SO AN SSSR. Krasnoyarsk, 1981, 55 pp. In Russian. – Abstr. in Ref. zh., 51. Astron., 10.51.205 (1981).

062.085 Magnetic field in a reflectionally invariant turbulence.
A. A. Ruzmajkin, D. D. Sokolov.
Pis'ma Astron. Zh., Tom 7, 701 - 704 (1981). In Russian. English translation in Soviet Astron. Lett., Vol. 7.
Exponential growth of a magnetic field in a reflectionally invariant (on an average) turbulence is shown to be possible. The excitation criterium and the spectrum at low wave numbers are obtained.

062.086 Asymptotic properties of a disk dynamo.
R. V. Isakov, A. A. Ruzmaikin *(Ruzmajkin)*,
D. D. Sokoloff *(Sokolov)*, M. V. Faminskaya.
Astrophys. Space Sci., Vol. 80, 145 - 155 (1981).
The asymptotic properties of a turbulent disk dynamo at large dimensionless numbers R_α, and R_ω, characterizing the helicity and the differential rotation are analysed. Three types of generations in the dependence of the relations between R_α and R_ω are found: α^2-dynamo and two types of $\alpha\omega$-dynamo. For each of these types the rates of growth are obtained and the forms of solution are pointed out. Boundaries of the disk dynamo approximation are given.

062.087 Diffusion models for magnetic Ap–Bp stars.
G. Michaud, C. Mégessier, Y. Charland.
Astron. Astrophys., Vol. 103, 244 - 262 (1981).
Using the simplest magnetic field configurations that can explain the observed magnetic fields on Ap stars (H_e and H_s), diffusion is shown to lead to the kind of surface inhomogeneities observed on those stars. It is shown, for instance, where the rare earths and iron should be expected to migrate. Turbulence is assumed to be negligible. It is also shown how the observed surface inhomogeneities imply vertical inhomogeneities. The results are compared to observations mainly for 53 Cam. It is shown that elements concentrated where magnetic field lines are horizontal should lead to an H_e of opposite sign to the H_e measured using lines of hydrogen or of elements concentrated where the magnetic field is vertical.

062.088 Stabilization of the electrostatic Kelvin-Helmholtz instability in high β plasmas.   J. D. Huba.
J. Geophys. Res., Vol. 86, 8991 - 9000 (1981).
A fully self-consistent theory of the electrostatic Kelvin-Helmholtz instability in finite β plasmas is presented. The results are applied to several space plasmas (i. e., polar cusp, cometary tails, magnetopause) and estimates of the anomalous diffusion coefficient associated with this instability are presented.

062.089 Laboratory study of coherent curvature radiation as a pulsar emission mechanism.
G. Benford, D. Tzach.
Astrophys. Space Sci., Vol. 80, 307 - 321 (1981).
To simulate some of the major physical processes occurring in pulsars, the authors performed experiments using a relativistic electron beam propagating helically through a magnetized plasma. Microwave radiation with $\lambda \approx 1$ cm emerged when the predicted resonance conditions were satisfied. Power exceeded 1 MW and radiation lasted as long as the electron beam pulse. The spectrum, harmonics, power and scalings were consistent with a model of coherent curvature radiation from electrons which are bunched by a beam-plasma streaming instability. Brightness temperature was $\sim 10^{20}$ degrees.

062.090 Convective instability in bounded uniform self-gravitating spherical clouds.
I. B. Bernstein, D. L. Book.
Astrophys. J., Vol. 251, 271 - 277 (1981).
A new type of instability, similar to thermal convective instabilities, has been found to occur in self-gravitating spherical clouds with finite pressure homologously expanding or collapsing in a vacuum. Unlike the Jeans instability, it has a growth rate which increases with zonal mode number l, so that objects of arbitrarily small size are unstable. The modes are compressional and have nonvanishing vorticity. The perturbation amplitudes grow algebraically at first, then at late times become proportional to the unperturbed cloud radius $R(t)$. For a collapsing object such as a protogalaxy the theory predicts fragmentation and enhanced condensation in the peripheral regions.

062.091 Higher order fluid equations for multicomponent nonequilibrium stellar (plasma) atmospheres and star clusters. II. Effects of nonzero relative flow velocities and skewing of velocity distribution functions.
S. Cuperman, I. Weiss, M. Dryer.
Astrophys. J., Vol. 251, 297 - 310 (1981).
Further development of a six-moments fluid theory required for the description of spherical systems of particles obeying an inverse-square law of interaction (e.g., stellar atmospheres and winds, stellar systems, inertial confinement fusion systems, etc.) is presented.

062.092 On the magnetospheric structure of pulsars.
W.-r. Hu.
Acta Astrophys. Sinica, Vol. 1, 222 - 233 (1981). In Chinese.

062.093 Comments on the paper "Radiation function $P_\lambda(T, \Delta E_f)_G$ of an open system at definite states."
S.-h. Ye, with a reply by C.-s Zhu.
Acta Astrophys. Sinica, Vol. 1, 322 - 332 (1981). In Chinese.

062.094 Some properties of adiabatic blast waves in pre-existing cavities.
D. P. Cox, J. Franco.
Astrophys. J., Vol. 251, 687 - 694 (1981).
The authors review Sedov's self-similar blast wave structures for power law density distributions, simplifying to the $\gamma = 5/3$ case. They performed numerical energy integrals over the structures and provide a good approximation to the total energy contents, from which they derive the shock radius $R_s(E_0, t)$ for the cases of outwardly increasing density. They furthermore utilize an approximation due to Itoh to follow the evolution of the electron temperature distribution (given only Coulomb heating of the electrons) and to define the electron temperature equilibration time scale. They also present compaction-parameter integrals over the structure to simplify the estimation of luminosities of such blast waves.

062.095 Analytical description of charged particle transport along arbitrary guiding field configurations.
J. A. Earl.
Astrophys. J., Vol. 251, 739 - 755 (1981).

A new description of focused transport has been developed which is valid for arbitrary spatial dependence of both the scattering mean free path and the focusing length. This description is formulated in terms of partial differential equations, which were derived with the aid of perturbation methods, and whose solutions define first-order approximate solutions of the Boltzmann equation in terms of relatively simple functions of pitch angle. The new formulation describes the "supercoherent transition" from coherent modes that occur when focusing is strong compared to scattering to the diffusive mode that occurs when focusing is weak. In the steady state, it predicts that the angular distribution consists of a collimated "strahl" which is superposed on an isotropic "halo".

062.096 Free and forced oscillations of a flux tube.
P. R. Wilson.
Astrophys. J., Vol. 251, 756 - 767 (1981).

The free oscillations of a magnetic flux tube of finite dimension, embedded in a nonmagnetic plasma, are studied for models in which energy may propagate laterally away from the tube. All solutions for effectively thick tubes exhibit real frequencies and imaginary external wavenumbers. Thus the waves propagate parallel to the flux tube with constant amplitudes and are evanescent in the transverse direction. The same general formalism is applied to oscillations set up by an oscillatory disturbance applied at a point on the axis of the tube. Here it is found that for weak tubes the wave amplitudes decrease along the tube away from the disturbance, whereas for strong tubes the amplitudes increase away from the disturbance.

062.097 Stability analysis of non-steady MHD-equilibria.
D. Schmitt.
Proceedings of the 5th Göttingen-Jerusalem-Symposium on Astrophysics, (see 012.041), p. 291 - 298 (1981).

Following the work of Bernstein et al. (1958), Frieman and Rotenberg (1960) and Unno (1968) a formalism is developed which allows to examine the stability of a perfectly conducting, isothermal, non-viscous rotating and self-gravitating plasma in non-steady equilibrium. Using this the stability of a plasma in a dynamical phase of its evolution can be predicted. Global stability investigations are carried out which are based on a variation of the total energy of the system and, in general, lead to sufficient stability conditions. The formalism is applied to the stability of a horizontal magnetic field in a medium stratified by a gravitational field.

062.098 Compressible convection in a rotating spherical shell. IV. Effects of viscosity, conductivity, boundary conditions, and zone depth.
G. A. Glatzmaier, P. A. Gilman.
Astrophys. J., Suppl. Ser., Vol. 47, 103 - 116 (1981).

The authors investigate the effects of certain variations on the stability and structure of the linear solutions. In this paper they examine the effects of several different eddy diffusivities. Here they study the influence of several different sets of temperature boundary conditions. Solutions for a 20% convection zone depth are compared to those for a 40% depth which was assumed in a previous paper.

062.099 The new paradigm in cosmic plasma physics.
H. Alfvén.
Plasma astrophysics, (see 012.042), p. 3 - 10 (1981).

062.100 The hydromagnetic dynamo. E. N. Parker.
Plasma astrophysics, (see 012.042), p. 23 - 29 (1981).

It appears that the cyclonic convection and nonuniform rotation within the liquid metal cores of planets, the convective zones of stars, and the gaseous disks of galaxies provide an efficient dynamo mechanism for the production of their fields. The dominant field within the body is azimuthal, produced by the nonuniform rotation from the poloidal field. The interaction of the cyclonic convection within the azimuthal field produces loops of flux in meridional planes whose net effect is a meridional circulation of field in the same sense as the original poloidal field. Both stationary and oscillatory fields can be generated in this way.

062.101 The non-equilibrium of magnetic fields.
E. N. Parker.
Plasma astrophysics, (see 012.042), p. 31 - 35 (1981).

A magnetic field embedded in a conducting fluid possesses a static equilibrium only if the topology of the lines of force is suitably invariant along the direction of the field. The basic principles are enumerated. It is shown that when the field does not possess the necessary invariance, it is subject to internal neutral point reconnection. The magnetic fields of most stars suffer such nonequilibrium, and consequent enhanced dissipation, providing an important source of heat in their outer atmospheres.

062.102 Modulational instabilities in astrophysics.
D. ter Haar.
Plasma astrophysics, (see 012.042), p. 45 - 49 (1981).

A brief discussion is given of (1) the nature and importance of modulational instabilities in plasmas, (2) the equations describing such instabilities, (3) applications to type I solar bursts, type III solar bursts, auroral kilometric radiation, and pulsar radiation, and (4) applications to galactic density waves.

062.103 Plasma mechanisms of emission and acceleration of relativistic particles in cosmic plasmas.
V. N. Tsytovich.
Plasma astrophysics, (see 012.042), p. 61 - 75 (1981).

A review of the new approaches and new results on the plasma mechanisms of emission and acceleration of relativistic particles in cosmic plasma is given.

062.104 Weak turbulence theory of an enhanced gas ionisation by the plasma flow. A. A. Galeev.
Plasma astrophysics, (see 012.042), p. 77 - 82 (1981).

062.105 Ideal MHD fluids. R. M. Kulsrud.
Plasma astrophysics, (see 012.042), p. 83 - 87 (1981).

The general properties of ideal hydromagnetic fluids are discussed. In particular the definition and significance of flux freezing is emphasized. Also energy conservation is demonstrated and its importance for plasma motion and equilibrium is also emphasized. These concepts are illustrated by a discussion of magnetic fields above the solar photosphere.

062.106 Plasma lasers (a strong source of coherent radiation in astrophysics). K. Papadopoulos.
Plasma astrophysics, (see 012.042), p. 89 - 95 (1981).

There are many collective plasma radiation processes of interest in astrophysics. The author reviews the fundamental principles involved in a particular class of coherent plasma radiation sources: the plasma lasers.

062.107 Plasma processes at collisionless shock waves.
V. Formisano.
Plasma astrophysics, (see 012.042), p. 145 - 165 (1981).

062.108 On the possibility of existence of large amplitude waves around pulsars. E. Asseo.
Plasma astrophysics, (see 012.042), p. 309 - 312 (1981).

Radiative damping effects compete with Weibel instability to prevent the propagation of large amplitude, low frequency, linearly polarized plasma waves. These inhibition phenomena are important for understanding the eventual acceleration of particles in the strong waves emitted by pulsars beyond the light cylinder distance.

062.109 Electron acceleration in magnetosonic shock fronts.
     K. Papadopoulos.
Plasma astrophysics, (see 012.042), p. 313 - 315 (1981).

It is shown that the low frequency turbulence driven by ions reflected in the foot of magnetosonic shocks can accelerate electrons to large energies. The mechanism seems to account for the energetic electron fluxes and the associated wave activity observed in the earth's bow shock.

062.110 Ion pressure-supported accretion tori and the origin of radio jets.   E. S. Phinney.
Plasma astrophysics, (see 012.042), p. 337 - 341 (1981).

Despite their impressive outputs of power, quasars probably do not provide us with diagnostics of any exotic relativistic physics. The author argues that radio galaxies may be cleaner laboratories which give us a direct view of purely nonthermal electrodynamic processes occurring near black holes. A review of the observational facts which motivate the discussion is given. It is described how inadequate cooling and particle coupling can give rise to a pressure-supported accretion torus, its structure in the absence of large-scale magnetic fields is discussed, and it is described how it can be modified to extract sufficient energy from a rotating black hole to power a double radio source. Plasma parameters are given in an appendix.

062.111 Second-order nonlinear conductivity tensor for an unmagnetized relativistic turbulent plasma.
H. E. Brandt.
Plasma astrophysics, (see 012.042), p. 361 - 364 (1981).

062.112 On the existence of the ion tearing mode.
H. Goldstein.
Plasma astrophysics, (see 012.042), p. 383 - 384 (1981).

062.113 Comments on high Mach number magnetosonic shocks.   K. Papadopoulos.
Plasma astrophysics, (see 012.042), p. 409 - 411 (1981).

062.114 A plasma current sheet studied by means of the particle motion.   R.-l. Xu.
Plasma astrophysics, (see 012.042), p. 417 - 419 (1981).

062.115 The analytical trajectory of the charged particle moving in a neutral sheet magnetic field.   R.-l. Xu.
Plasma astrophysics, (see 012.042), p. 421 - 423 (1981).

062.116 High X-ray luminosity from dynamo stars.
     G. Belvedere, D. Molteni.
Space Sci. Rev., Vol. 30, (see 012.044), 181 - 184 (1981).

The authors intend to show that a stellar dynamo mechanism can produce high X-ray luminosities and also give account for modulation periods of the order thousand seconds or larger. They intend to show that faint late main sequence stars sufficiently fast rotating can give rise by dynamo action to sufficiently high magnetic fields to give account for the strong X-ray emission of some galactic X-ray sources.

062.117 Relativistische Effekte bei der Photonenpropagation in stark magnetisierten Plasmen.
H. Herold, M. Reinecke, H. Ruder, G. Wunner.
Mitt. Astron. Ges., Nr. 54, (see 012.050), p. 275 - 278 (1981).

062.118 Radiation reflection from a semi-infinite plasma layer with homogeneous magnetic field.

N. A. Silant'ev.
Astron. Zh., Tom 58, 1217 - 1225 (1981). In Russian.
English translation in Soviet Astron., Vol. 25, No. 6.

From the radiative transfer equation and the invariance principle an expression is derived for the density matrix of the reflected radiation. Numerical solutions are given for the Stokes parameters of the radiation for the case when the magnetic field is perpendicular to the surface. It is shown that the presence of a magnetic field may significantly decrease the albedo.

062.119 On refraction of acousto-gravity waves in the solar atmosphere.   N. S. Petrukhin.
Astron. Zh., Tom 58, 1257 - 1261 (1981). In Russian.
English translation in Soviet Astron., Vol. 25, No. 6.

The refraction of acousto-gravity waves (AGW) on the sharp boundary separating two isothermal media with different temperatures is investigated. It is shown that long and short AGW are refracted in opposite directions. The energetic factor of wave transmission is found. The results of the investigation are used for the evaluation of the wave flow crossing the transition layer of the sun's atmosphere.

062.120 Nonlinear electric field in a plasma due to a plane gravitational wave.   A. B. Balakin, V. I. Bashkov.
Tr. Kazan. Gorod. Astron. Obs., Vyp. 46, p. 66 - 71 (1980). In Russian.

062.121 Wave-particle transport by weak electrostatic flow shear fluctuations.   S. P. Gary, S. J. Schwartz.
J. Geophys. Res., Vol. 86, 11139 - 11147 (1981).

062.122 The interplay between magnetic fields and convection.   N. O. Weiss.
J. Geophys. Res., Vol. 86, 11689 - 11694 (1981).

The interaction between magnetic fields and convection was first modeled by considering the effect of a solid rotor on an otherwise uniform magnetic field. This approach has subsequently been extended by considering convection in an electrically conducting Boussinesq fluid in the presence of an imposed magnetic field. The results of a series of numerical experiments at a fixed Rayleigh number of 10^5 are presented and contrasted with predictions based on linear theory.

062.123 Propagation of large-scale disturbances in an atmosphere with anisotropic conductivity.
Eh. G. Doil'nitsyna.
Izv. vuzov. Radiofiz., Tom 24, 267 - 275 (1981). In Russian.
Abstr. in Ref. zh., 51. Astron., 12.51.245 (1981).

062.124 Magnetic field dynamo in space.
     S. I. Vajnshtejn.
Usp. fiz. nauk, Tom 134, 744 - 745 (1981). In Russian.
Abstr. in Ref. zh., 51. Astron., 12.51.259 (1981).

062.125 On the global structure of stellar magnetospheres with stellar winds.
S. Shibata, O. Kaburaki, M. Okazaki.
Sci. Rep. Tôhoku Univ., Ser. 8, Vol. 1, 161 - 186 (1980).

The solutions for the magnetic field and stellar wind in axially symmetric stellar magnetospheres are obtained through perturbational method in three artificial extreme situations, i.e., the limits of weak magnetic field, strong magnetic field and weak electromagnetic coupling.

062.126 A transient MHD model applicable for the source of solar cosmic ray acceleration.
M. Dryer, S. T. Wu.
Cosmic rays in the heliosphere, (see 012.059), p. 85 - 88 (1981).

A two-dimensional, time-dependent magnetohydrodynamic (MHD) model is used to describe the possible mechanisms for the source of solar cosmic ray acceleration following a solar flare. The hypothesis is based on the propagation of fast mode MHD shocks following a sudden release of energy.

062.127   New magnetodynamic principles in cosmic physics.
      E. Woyk (Chvojková).
High-energy astrophysics, (see 012.067), p. 167 - 169 (1981).
      The effect of gravity on super-escape particles spiralling along magnetic field lines need not be negligible when the field lines are long enough and the field-strength variation small. If the magnetic field strength decreases with altitude but only very slowly, some unexpected phenomena may occur owing to gravity: some super-high-velocity particles can possess an upper level of reflection which impedes their escape into higher regions.

062.128   On the law of motion of a strong shock wave in an inhomogeneous medium.
B. I. Gnatyk, I. A. Klimishin.
Astron. Tsirk., No. 1149, p. 1 - 2 (1981). In Russian.

Mean-field magnetohydrodynamics and dynamo theory.   See Abstr. 003.088.

Calculated X-radiation from optically thin plasmas IV. Atomic data and rate coefficients for spectra in the range 1 - 270 Å.   See Abstr. 022.001.

The X-ray spectrum of a hot interstellar plasma.
See Abstr. 022.052.

About the ergodicity hypothesis in random propagation studies.   See Abstr. 022.071.

On the existence and structure of inhomogeneous analogs of the Dedekind and Jacobi ellipsoids.
See Abstr. 042.060.

Radiative transfer in a strongly magnetized plasma. I. Effects of anisotropy.   See Abstr. 063.041.

Radiative transfer in a strongly magnetized plasma. II. Effects of Comptonization.   See Abstr. 063.042.

Compton losses, Compton rockets.
See Abstr. 063.043.

Effects of turbulent plasma scattering on the X-ray spectra of celestial bodies.   See Abstr. 063.045.

On the transport and propagation of relativistic electrons in galaxies.   See Abstr. 063.046.

Relativistic thermal plasmas: radiation mechanisms.
See Abstr. 063.049.

Element stratification in the atmospheres of main sequence stars: the silicon accumulation.
See Abstr. 064.008.

Adiabatic oscillations of accretion disks.
See Abstr. 064.011.

Photospheric flow and stellar winds.
See Abstr. 064.086.

Accretion in two phases.   See Abstr. 064.088.

Stellar core collapse. I. Infall epoch.
See Abstr. 065.050.

Mass loss during the core helium flash.
See Abstr. 065.051.

The origin and structure of the fields of the upper main sequence magnetic stars.   See Abstr. 065.055.

Does γ Equ have a magnetic cycle like the sun?
See Abstr. 065.056.

Hydromagnetic rotational deceleration of magnetic Ap stars.   See Abstr. 065.057.

Comments on theories of the chemically-peculiar stars of the upper main sequence.   See Abstr. 065.058.

A parameter free model for HgMn stars.
See Abstr. 065.059.

Transfer equation of arbitrarily polarized radiation in the presence of a magnetic field.   See Abstr. 065.060.

Diffusion in the presence of a magnetic field.
See Abstr. 065.061.

Comments on the equation of state of supernova matter.   See Abstr. 065.084.

Hydrodynamic collapse calculations for nonrotating and rotating supernova models.   See Abstr. 065.085.

On the collapse instability of rapidly rotating pre-supernova cores having finite entropy.   See Abstr. 065.086.

Homothetic solutions of Einstein's equations and shock waves.   See Abstr. 066.046.

The equation of state of a cosmological self-gravitating fluid influenced by thermal or turbulent motion and the primeval background radiation.
See Abstr. 066.111.

Restrictions on relativistically rotating fluids.
See Abstr. 066.143.

SS 433: background for a relativistic model.
See Abstr. 066.153.

Coordinate-dependent 3+1 formulation of the general relativity equations for a perfect fluid.
See Abstr. 066.162.

Models for accretion flow on to neutron stars and X-ray photon-electron interactions in ultra-high magnetic fields.   See Abstr. 066.523.

Slow magnetohydrodynamic shocks as transformer of the magnetic energy in solar active regions.
See Abstr. 072.047.

A class of analytic solutions for the thermally balanced magnetostatic prominence sheet.
See Abstr. 073.009.

Two-ribbon flares: magnetostatic equilibria.
See Abstr. 073.034.

Two-ribbon flares: (post)-flare loops.
See Abstr. 073.035.

Particle acceleration in solar flares.
See Abstr. 073.036.

On the importance of reverse current ohmic losses in electron-heated solar flare atmospheres. See Abstr. 073.070.

Preacceleration in collapsing magnetic neutral sheets and anomalous abundances of solar flare particles. See Abstr. 073.091.

Generation of coronal electric currents due to convective motions on the photosphere.   See Abstr. 074.017.

Simple-loop flares: magnetic instabilities. See Abstr. 074.036.

Simple-loop flares: thermal evolution. See Abstr. 074.037.

The polarization of second harmonic radio emission in type III bursts.   See Abstr. 077.013.

The Voyager 1/Saturn encounter and the cosmogonic shadow effect.   See Abstr. 107.017.

Broad-band linear polarization and magnetic intensification in rotating magnetic stars.   See Abstr. 116.017.

A theoretical review of SS 433. See Abstr. 117.093.

Analytical methods for the hydrodynamical evolution of supernova remnants. II. Arbitrary form of boundary conditions.   See Abstr. 125.040.

Axisymmetric gravitational collapse of an isothermal rotating interstellar gas cloud. A semi-analytical approach. See Abstr. 131.034.

Collapse and fragmentation of rotating, adiabatic clouds.   See Abstr. 131.164.

Star formation in molecular clouds and the formation of compact H II regions.   See Abstr. 131.185.

Quelques aspects du rôle joué par le champ magnétique dans l'évolution des nuages interstellaires. See Abstr. 131.225.

Extra-galactic double radio sources. See Abstr. 141.162.

Production and propagation of jets. See Abstr. 141.163.

Plasma beams from doughnuts. See Abstr. 141.165.

Large scale Kelvin-Helmholtz instabilities in extragalactic radio jets.   See Abstr. 141.166.

Inertial development of vorticity in pulsar magnetospheres.   See Abstr. 141.506.

Sub-rotation and super-rotation in pulsar magnetospheres.   See Abstr. 141.507.

Pulsar disk systems.   See Abstr. 141.548.

Pulsars: polar pumps, interpolar currents, and induced Landau radiation.   See Abstr. 141.549.

Pulsar electrodynamics.   See Abstr. 141.550.

Magnetic alignment and counter-alignment in binary X-ray pulsars.   See Abstr. 142.107.

The power-law spectrum of shock-accelerated relativistic particles.   See Abstr. 143.001.

Some problems with the evolution of gas in clusters of galaxies.   See Abstr. 151.037.

The role of magnetic fields in extragalactic astronomy.   See Abstr. 151.057.

A hydrodynamic interaction between the galactic disk and halo, and its application to the origin of warping gaseous disks.   See Abstr. 151.059.

Thermodynamic equilibrium hard binary models for globular star clusters.   See Abstr. 154.039.

Radiation processes in active galactic nuclei. See Abstr. 158.227.

Hot plasmas in active galactic nuclei. See Abstr. 158.228.

X-ray emission from galactic jets. See Abstr. 158.235.

A direct hydrodynamical approach to the evolution of the density correlations in an expanding flat Friedmann universe.   See Abstr. 162.042.

Erratum

062.901   Erratum: "Radiation force on a relativistic plasma and the Eddington limit" [Astrophys. J., Lett., Vol. 243, L147 - L149 (1981)].   S. L. O'Dell. Astrophys. J., Lett., Vol. 251, L55 (1981). – See Abstr. 29.062.034.

063 Radiative Transfer, Scattering

063.001 Compton scattering in a converging fluid flow – III. Spherical supercritical accretion.
D. G. Payne, R. D. Blandford.
Mon. Not. R. Astron. Soc., Vol. 196, 781 - 795 (1981).
Radiative transfer in spherical, supercritical accretion on to a massive black hole is considered. Particular emphasis is placed on the case of non-adiabatic flow in which electron scattering in the converging flow is the dominant source of opacity and photon heating.

063.002 The light regime in a medium consisting of two semi-infinite layers with different refraction indices.
A. K. Kolesov.
Tr. Astron. Obs., Leningrad, Tom 36 = Uch. Zap. Leningr. Univ., No. 402 = Ser. mat. nauk, Vyp. 58, 3 - 19 (1981). In Russian.
Anisotropic scattering of radiation in a medium consisting of two adjacent half-spaces with different optical properties is investigated. The intensities of radiation in the medium are expressed in terms of the auxiliary functions. A system of linear integral equations for these functions is derived.

063.003 Expressions for the computer-evaluation of the four kernel functions for line formation with Doppler and Lorentz profiles.   D. G. Hummer.
J. Quant. Spectrosc. Radiat. Transfer, Vol. 26, 187 - 195 (1981).
Rational approximations for the kernel functions $K_1(\tau)$, $K_2(\tau), M_1(\tau)$ and $M_2(\tau)$ that describe the transfer of radiation scattered with complete redistribution over Doppler and Lorentz profiles have been obtained from their series and asymptotic expressions by the techniques of Padé approximants with a maximum relative error of less than 10^{-4}.

063.004 On the use of mean absorption coefficients in the presence of strong temperature gradients.
G. C. Pomraning, A. K. Prinja, I. R. Shokair.
J. Quant. Spectrosc. Radiat. Transfer, Vol. 26, 199 - 213 (1981).
The authors compare results obtained by using the Planck and Rosseland mean absorption coefficients with exact multifrequency transport results in a purely absorbing halfspace problem in the presence of strong temperature gradients. They give numerical results for both a prescribed temperature distribution in the presence of an Elsässer band absorption coefficient and a self-consistent radiative equilibrium distribution corresponding to a source deep within the halfspace transporting through a picket fence absorption coefficient.

063.005 Particular solutions of the equation of transfer.
C. E. Siewert, F. O. Oruma.
J. Quant. Spectrosc. Radiat. Transfer, Vol. 26, 223 - 227 (1981).
Particular solutions corresponding to various forms for the inhomogeneous source term are established for the equation of transfer with Lth order anisotropic scattering .

063.006 Evaluation and universal curves of the photon loss probability ϵ.   S. O. Kastner.
J. Quant. Spectrosc. Radiat. Transfer, Vol. 26, 377 - 380 (1981).
Convenient expressions are developed for the photon loss probability per scattering event, for lines of neutral atoms and ions, which depend only on electron density, temperature and transition energy.

063.007 Estimate of self-absorption in a spectral line for an inhomogeneous medium. I.   V. A. Ostapenko.

Soln. Dannye 1981 Byull., No. 2, p. 90 - 96 (1981). In Russian.
The probability expression for the slipping away of photons from an optically thick inhomogeneous object is obtained. Calculation results of the probabilities for a wide range of medium parameters are given in a table.

063.008 A flux-limited diffusion theory.
C. D. Levermore, G. C. Pomraning.
Astrophys. J., Vol. 248, 321 - 334 (1981).
A diffusion theory for radiative transfer is derived which is naturally flux limited, i.e., the magnitude of the flux can be no greater than the density times the maximum transport speed. Numerical comparisons with exact solutions of the equation of transfer indicate that this approximate theory is significantly more accurate than classical isotropic diffusion theory (the Eddington approximation) and asymptotic diffusion theory.

063.009 Redistribution of radiation for the wings of Lyman-alpha.   J.-B. Yelnik, K. Burnett, J. Cooper,
R. J. Ballagh, D. Voslamber.
Astrophys. J., Vol. 248, 705 - 716 (1981).
Earlier work on redistribution of radiation by collisions for isolated lines is extended to overlapping lines, and an explicit expression for the frequency redistribution is given for Lyman-α. This expression is valid, even when the emitted photon is in the (non-impact) line wings. A simple physical explanation of the result is possible.

063.010 Comptonization by cold electrons.
A. P. Lightman, D. Q. Lamb, G. B. Rybicki.
Astrophys. J., Vol. 248, 738 - 750 (1981).
The authors derive analytic approximations to the Green's functions for scattering of photons by cold electrons in finite media of various geometries. They first compute the probabilities of a photon to escape spherical and plane-parallel, semi-infinite media after n scatterings, and compare these results to those previously obtained by the "scattering time" method. Green's functions are then obtained by combining these probabilities with the wavelength distribution after n scatterings. To illustrate their results, the authors compute the emergent spectra from input blackbody and bremsstrahlung spectra in the geometries mentioned above.

063.011 Quantum theory of radiative transfer.
E. C. G. Sudarshan.
Phys. Rev. A, Vol. 23, 2802 - 2809 (1981). – Abstr. in Phys. Abstr., Vol. 84, Abstr. 79718 (1981).

063.012 Radiation from an asteroid-neutron star collision.
W. M. Howard, J. R. Wilson, R. T. Barton.
Bull. American Astron. Soc., Vol. 13, 557 (1981). – Abstract.

063.013 On the importance of convective transport of excited atoms in stellar atmospheres.
I. Hubený.
Astron. Astrophys., Vol. 100, 314 - 318 (1981).
The question of the importance of the convective or diffusion transport of excited atoms due to their density gradients generated by non-LTE line transfer, is reconsidered. It is shown that such kinetic effects are practically never important under stellar atmospheric conditions, contrary to the previous study of Oxenius (1979). This different conclusion follows from the discussion of an explicit form of the coupled set of kinetic equations for excited atoms and the radiative transfer equation, as well as from the proper parametrization of the physical state of stellar atmospheres.

063.014 **Study of interstellar scattering by interplanetary scintillations.**
A. Pramesh Rao, S. Ananthakrishnan.
Bull. Astron. Soc. India, Vol. 9, 76 - 77 (1981). − Abstract.

063.015 **Non-LTE line transfer with partial redistribution.**
I. General emission profile. I. Hubený.
Bull. Astron. Inst. Czechoslovakia, Vol. 32, 271 - 286 (1981).
An attempt to formulate, in a general way, the problem of the transfer of line radiation by multilevel atoms, allowing for the effects of partial redistribution, is presented.

063.016 **On the transfer equation for the cyclotron line in Her X-1.** D. B. Melrose.
Astron. Astrophys., Vol. 101, 284 - 288 (1981).
A scattering-type equation is derived for cyclotron photons absorbed and re-emitted by electrons in their lowest Landau orbital. An adjunct equation describing the effect on the parallel momentum of the electrons is also derived. The effects of induced emission are taken into account fully. It is pointed out that the parallel velocity spread of the electrons and the bandwidth of the cyclotron line are strongly coupled in Her X-1.

063.017 **Formation and transfer of permitted Si II emission lines in Seyfert I galaxies and quasars.**
A. M. Dumont, G. Mathez.
Astron. Astrophys., Vol. 102, 1 - 7 (1981).
Results on the transfer of Si II emission lines in finite, optically thick, atmospheres of quasars and Seyfert I galaxies are given. Intensities have been computed solving exactly the equations of transfer, ionization equilibrium, and statistical equilibrium in a 11-level model ion. By comparison to observational data, the absence of observed λλ 3858 and 6355 lines sets upper limits to the total hydrogen column density. In homogeneous models, permitted ranges of the electron temperatures, density, and the hydrogen column density are deduced. Photoionized models are also examined.

063.018 **Maximum entropy Eddington factors and flux limited diffusion theory.** G. C. Pomraning.
J. Quant. Spectrosc. Radiat. Transfer, Vol. 26, 385 - 388 (1981).
The relationship between the maximum entropy Eddington factor and flux limited diffusion theory is established. Although these two approaches to radiative transfer are philosophically very different (one treats an ensemble, while the other treats a specific system), the resulting variable Eddington factors are in semiquantitative agreement. It is shown that a simple Padé approximant to the maximum entropy angular distribution yields precisely the same angular distribution extant in a recently reported flux limited diffusion theory.

063.019 **A simple computational method for internal polarized radiation fields of finite slab atmospheres.**
H. Domke, E.(Eh.) G. Yanovitskij.
J. Quant. Spectrosc. Radiat. Transfer, Vol. 26, 389 - 396 (1981).
An extended doubling method is formulated, which provides together with the emergent radiation also the internal polarized radiation field without additional iterations. Two sets of linear regular integral relations are derived, which have to be fulfilled by the surface Green's function matrix or, equivalently, by the Stokes vector of the slab albedo problem radiation field. The integral relations refer to the half range angular variable of the direction of incidence and to the full range angular variable of the direction of light propagation, respectively.

063.020 **Modification to Scott's method for direct integration of gaseous transmission to improve speed and accuracy.** G. S. Orton.
J. Quant. Spectrosc. Radiat. Transfer, Vol. 26, 463 - 466 (1981).

Two limitations to the method of Scott for rapid direct integration of gaseous transmission in a finite spectral interval over an inhomogeneous optical path are discussed.

063.021 **Electromagnetic wave propagation in moving media with special regard to frequency-shifts ('anomalous' frequency shifts in astronomy). I. Basic problems in the determination of a propagating wave pattern in moving media.**
C. Ferencz.
Acta Tech. Acad. Sci. Hungaricae, Vol. 89, 445 - 471 (1979). Abstr. in Phys. Abstr., Vol. 84, Abstr. 90774 (1981).

063.022 **Electromagnetic wave propagation in moving media with special regard to frequency-shifts in astronomy ('anomalous' frequency-shifts in astronomy). II. The relativistic ray tracing method.** C. Ferencz.
Acta Tech. Acad. Sci. Hungaricae, Vol. 90, No. 1 - 2, p. 29 - 58 (1980). − Abstr. in Phys. Abstr., Vol. 84, Abstr. 94156 (1981).

063.023 **Directionality effects in the transfer of X-rays from an accreting magnetized neutron star: beam and pulse shapes.** P. Mészáros, S. Bonazzola.
NASA Tech. Memo., NASA TM 82094, 45 pp. (1981).
The authors discuss the direction dependent transfer of X-rays in a plane parallel atmosphere with a strong magnetic field perpendicular to the surface. They treat the problem of a slab illuminated from below, and of a semi-infinite medium at constant temperature and density, and present numerical results for parameters typical of the hot polar caps of accreting magnetized neutron stars. Theoretical beam and X-ray pulse shapes are obtained for various models of X-ray pulsars, and the frequency and phase dependence of the pulse structure is briefly compared with observations.

063.024 **Chandrasekhar's X- and Y-functions.**
J. Caldwell, A. J. Perks.
Astrophys. J., Vol. 249, 258 - 262 (1981).
The X- and Y-functions of Chandrasekhar are defined in terms of two coupled integral equations whose solution for certain values of optical depth and albedo have already been found by means of iterative methods. This paper describes a method which attempts to overcome the difficulty which has arisen in previous iterative methods in obtaining accurate solutions $X(\mu)$, $Y(\mu)$ for values of μ close to unity. Reasonable accuracy is achieved for certain cases at the expense of little computing time.

063.025 **Solutions to radiative transfer problems using approximate lambda operators.**
G. B. Scharmer.
Astrophys. J., Vol. 249, 720 - 730 (1981).
Two new techniques for obtaining solutions to non-LTE problems are developed. The essence of the first technique is that the core-saturation approximation, the Eddington-Barbier relation, and the upper boundary condition are used to derive a simple first-order differential equation for S_l. This approximate equation is solved analytically for a two-level atom in an atmosphere with constant properties and zero continuum opacity. Second, the author develops a new efficient perturbation technique for numerical solutions to non-LTE problems. Using a linear test function, a one-point quadrature relation between I_ν and S_ν is derived. From this relation it is possible to obtain an approximate solution of the transfer equation with a small amount of computing time.

063.026 **The radiation field in an inhomogeneous atmosphere.** T. Viik.
Tartu Astrofüüs. Obs., Teated, Nr. 65, p. 3 - 10 (1981). In Russian.
The problem of determining the radiation field in an inhomogeneous atmosphere by solving the equation of radiative

transfer is a boundary problem. Changing the upward intensity into the downward intensity by means of Chandrasekhar's principle of invariance, one can convert the boundary value problem into a respective Cauchy problem. This problem may be solved in a straight-forward manner provided we know the scattering function for every level of the atmosphere. These functions are easy to find using an approach of Bellmann and Vasudevan (1974).

063.027 Electromagnetic wave propagation in moving media with special regard to frequency-shifts.'Anomalous' frequency-shifts in astronomy. III. Examples of application.
C. Ferencz.
Acta Tech. Acad. Sci. Hungaricae, Vol. 90, No. 3 - 4, p. 303 - 319 (1980). − Abstr. in Phys. Abstr., Vol. 84, Abstr. 98670 (1981).

063.028 A new look at the discrete ordinate method for radiative transfer calculations in anisotropically scattering atmospheres.   K. Stamnes, R. A. Swanson.
J. Atmos. Sci., Vol. 38, 387 - 399 (1981). − Abstr. in Phys. Abstr., Vol. 84, Abstr. 98678 (1981).

063.029 Compton scattering of relativistic electrons in compact X-ray sources.
F. A. Aharonian (*Agaronyan*), A. M. Atoyan.
Astrophys. Space Sci., Vol. 79, 321 - 336 (1981).
The problem of single Compton scattering is considered and the resulting spectrum, angular distribution and polarization of scattered photons in a general case are obtained. The inverse Compton scattering (ICS) for arbitrary energies of electrons E and photons ω_0 is investigated in detail. The radiation spectrum of ICS, as well as the energy-losses of relativistic electrons distributed by power-law $E^{-\alpha}$, are calculated.

063.030 Inverse Compton scattering in anisotropic synchrotron sources.   G. Gilmore.
Astrophys. Space Sci., Vol. 79, 483 - 490 (1981).
Models of compact, high brightness temperature sources such as quasars and active galactic nuclei often predict substantial inverse Compton scattered flux to be produced at high frequencies. The often used equations to include these effects are derived with several simplifying assumptions, which may not always be consistent with the derived source parameters. The author therefore presents here an explicit derivation of the dependence of the rate of inverse Compton scattering on anisotropies in the source, to emphasise the importance and number of assumptions required in the derivation.

063.031 Numerical solution of time dependent Compton scattering problems by means of an integral equation.   P. W. Guilbert.
Mon. Not. R. Astron. Soc., Vol. 197, 451 - 460 (1981).
Compton scattering is important in determining the spectrum of certain compact high energy X-ray sources and so to understand these sources one must be able to calculate the effect of Compton scattering on a radiation field. A method is presented in this paper which considerably extends the range of problems which can be solved numerically. In particular, time dependent problems where heating and cooling are important can be solved for gas temperatures and photon energies up to and somewhat beyond $m_e c^2$.

063.032 A comment on "Radiative transfer in spherically symmetric flows" by J. I. Castor.
D. Mihalas.
Astrophys. J., Vol. 250, 373 - 375 (1981).
Castor's comoving-frame equation of transfer is rederived retaining a term omitted from one of the coefficients in his metric. This term is important when the flow changes significantly on time scales $\Delta t \sim \Delta r/c$, so that the fluid ac-

celeration is comparable with cv/r or $c(dv/dr)$. When the term is retained, terms that depend on the acceleration now appear in the coefficients of Castor's transfer equation, in agreement with independent derivations by Buchler and Mihalas.

063.033 Angle-dependent frequency redistribution: internal source case.   R. R. Meier, J.-S. Lee.
Astrophys. J., Vol. 250, 376 - 383 (1981).
The effects of various approximations of the theory of frequency redistribution in resonant scattering are investigated for a plane-parallel layer of two-level Maxwell-Boltzmann atoms with natural broadening of the upper level. The excitation source is inside the medium. Two limiting cases are examined: a uniform source and a discrete source located at the center of the medium. The authors find that for large optical depths (10^4), the "exact" angle-dependent partial redistribution function is required if high accuracy is desired for the emission rate inside the medium or for line profiles. Complete frequency redistribution, both with and without inclusion of the natural wings, is often a poor approximation. However, the integrated emission rate as viewed from outside the medium is virtually independent of the details of the frequency redistribution for the cases considered.

063.034 Intensity fluctuations of an optical wave in a turbulent medium. Effect of source coherence.
R. L. Fante.
Opt. Acta, Vol. 28, 1203 - 1207 (1981). − Abstr. in Phys. Abstr., Vol. 84, Abstr. 100007 (1981).

063.035 Inverse scattering − exact solution of the Gel'fand− Levitan equation (*ionosphere structure measurement*).   K. R. Pechenick, J. M. Cohen.
J. Math. Phys., Vol. 22, 1513 - 1516 (1981). − Abstr. in Phys. Abstr., Vol. 84, Abstr. 101972 (1981).

063.036 Effects of scattering on continuum radiation from supernovae and determination of their distances.
R. V. Wagoner.
Astrophys. J., Lett., Vol. 250, L65 - L69 (1981).
The continuum radiation from supernovae is calculated from simple models in which scattering can dominate the opacity. It is shown that previous analyses of supernovae expansion rates may have led to underestimates of the Hubble constant.

063.037 Effects of partial frequency redistribution on the level population densities in a resonance line.
A. Peraiah, D. M. Rao.
Astrophys. Space Sci., Vol. 80, 437 - 442 (1981).
The authors have obtained a simultaneous solution of the statistical equilibrium equation for a non-LTE two-level atom and the radiative transfer equation in the comoving frames by employing the angle-averaged partial frequency redistribution R_I with isotropic scattering. In the first iteration they have set the population density of the upper level equal to zero and allow it to be populated in the subsequent iterations. The solution converges within two to four iterations. The process of iteration is terminated when the ratios of population densities in two successive iterations at each radial point, attain an accuracy of 1%. The effects of partial frequency redistribution is to increase the population density of the upper level. Radial gas motions do not seem to have significant effects, although in highly extended geometries, velocity gradients change the population densities considerably.

063.038 Orders of scattering contributing to each Fourier term representing radiative intensity.
K. Kawabata.
Bull. American Astron. Soc., Vol. 13, 734 (1981). − Abstract.

063.039 Two photometric functions that reproduce photometry of particulate surfaces of any albedo.

J. Goguen, J. Veverka.
Bull. American Astron. Soc., Vol. 13, 742 (1981). – Abstract.

063.040 **The role of scattering in planetary surface reflec-**
tance spectra. R. V. Morris, W. W. Mendell.
Bull. American Astron. Soc., Vol. 13, 743 (1981). – Abstract.

063.041 **Radiative transfer in a strongly magnetized plasma.**
I. Effects of anisotropy. W. Nagel.
Astrophys. J., Vol. 251, 278 - 287 (1981).

The author presents results of radiative transfer calculations for radiating slabs and columns of strongly magnetized
plasma. The angular dependence of the escaping radiation was
found numerically by Feautrier's method, using the differential scattering cross sections derived by Ventura. The author
also gives an approximate analytical expression for the anisotropy of the outgoing radiation, based on a system of two
coupled diffusion equations for ordinary and extraordinary
photons. Giving the polarization dependence of the beaming
pattern of radiating slabs as well as columns, the author
generalizes previous results of Basko and Kanno. Some implications for models of the pulsating X-ray source Her X-1 are
discussed.

063.042 **Radiative transfer in a strongly magnetized plasma.**
II. Effects of Comptonization. W. Nagel.
Astrophys. J., Vol. 251, 288 - 296 (1981).

The author presents results of numerical calculations of
radiative transfer in hot, strongly magnetized plasma, including Comptonization and redistribution of photons in the
cyclotron line. Polarization, angle, and frequency dependent
differential scattering cross sections were obtained in the non-
relativistic approximation. An approximate, two-stream
transfer equation was solved for homogeneous radiating slabs
of various optical depths using Feautrier's method. For intermediate optical depths the author obtained the asymmetric,
self-reversed emission line recently predicted by Wasserman
and Salpeter.

063.043 **Compton losses, Compton rockets.**
A. Y. S. Cheng, S. L. O'Dell.
Astrophys. J., Lett., Vol. 251, L49 - L54 (1981).

Near a luminous object, a relativistic plasma suffers
severe Compton losses. Bulk relativistic motion alleviates these
losses, but hydrodynamic acceleration processes are too slow.
However, the Compton-rocket radiation force due to anisotropic Compton losses can accelerate the plasma on a time
scale comparable to the loss time scale. In an anisotropic radiation field, the Compton-rocket thrust exceeds the hydrodynamic force when the light travel time across a proper pressure scale length exceeds the cooling time for the total internal
kinetic energy of the plasma. The Compton-rocket thrust can
effect relativistic velocities if the plasma radiates away most of
its total relativistic enthalpy. Consequently, bulk relativistic
motion not only alleviates Compton losses, but can actually
result from such losses.

063.044 **Radiation transfer in circumstellar envelopes with**
spherical symmetry – Henyey's method of solution.
M. Šolc.
Acta Univ. Carolinae Math. Phys., Vol. 22, No. 1, p. 63 - 75
(1981).

A numerical method solving the radiative transfer is developed on the basis of Henyey's approach to the multiple
scattering problem, giving for model dust nebulae primarily the
surface brightness distribution. Generally polarized radiation
and anisotropic scattering are taken into account. The method
proves efficient only if the model of circumstellar dust envelope has spherical symmetry.

063.045 **Effects of turbulent plasma scattering on the X-ray**
spectra of celestial bodies. X.-q. Li, Z.-d. Zhang.
Acta Astrophys. Sinica, Vol, 1, 301 - 310 (1981). In Chinese.

063.046 **On the transport and propagation of relativistic**
electrons in galaxies. I. Lerche, R. Schlickeiser.
Astrophys. J., Suppl. Ser., Vol. 47, 33 - 85 (1981).

The authors present analytic solutions to the time-
dependent equation describing the transport of relativistic
electrons, which are being convected and diffused and may be
either gaining or losing energy. A body of solutions is available
from which the interested reader can choose that particular
solution which would seem to model most closely observations
of (a) synchrotron radiation in other galaxies both in respect
of its spatial, frequency, and temporal dependence, and (b)
synchrotron radiation in our own Galaxy, both in respect of
the global background of such radiation and also in respect of
discrete sources of such radiation (e.g., supernova) observed
in our Galaxy.

063.047 **Theoretical quasar emission line ratios. IV. General**
asymptotic escape probabilities and the effects of
linear Stark broadening. R. C. Puetter.
Astrophys. J., Vol. 251, 446 - 450 (1981).

The extensive use of photon escape probabilities in studies of radiative transfer effects in QSOs makes it very important that these quantities be on a firm physical and mathematical basis. In this paper the author established techniques
that allow evaluation of exact asymptotic forms of photon
escape probabilities for a wide range of types of absorption
coefficient profiles. He derives a simple computationally ex-
pedient approximate expression for the asymptotic form of
the photon escape probability when linear Stark broadening
contributes to the line absorption coefficient profile in hydrogenic lines. In particular, special attention is paid to Lyα for
the case in which partial redistribution over the linear Stark
component of the line absorption coefficient profile dominates single flight photon escape.

063.048 **Directionality effects in the transfer of X-rays from**
an accreting magnetized neutron star: beam and
pulse shapes. P. Mészáros, S. Bonazzola.
Astrophys. J., Vol. 251, 695 - 712 (1981).

The authors discuss the direction-dependent transfer of
X-rays in a plane-parallel atmosphere with a strong magnetic
field perpendicular to the surface. They present a transfer
formalism incorporating the full angular and polarization dependence of the cross sections, including vacuum polarization,
for frequencies not too close to the cyclotron resonance. They
treat the problem of a slab illuminated from below and of a
semi-infinite medium at constant temperature and density and
present numerical results for parameters typical of the hot
polar caps of accreting magnetized neutron stars. Theoretical
beam and X-ray pulse shapes are obtained for various models
of X-ray pulsars, and the frequency and phase dependence of
the pulse structure is briefly compared with observations.

063.049 **Relativistic thermal plasmas: radiation mechanisms.**
A. P. Lightman, D. L. Band.
Astrophys. J., Vol. 251, 713 - 726 (1981).

The authors investigate several important radiation mechanisms in relativistic thermal plasmas, obtaining approximate
analytic expressions rather than very detailed, numerical results. They obtain approximate spectra for pair annihilation
radiation and double Compton radiation. They also calculate
spectra for Comptonized bremsstrahlung radiation as a function of temperature and scattering depth of the medium, including the factor by which Comptonization amplifies the
total photon energy. They discuss the conditions under which
various of these processes dominate and make some comparisons to the nonrelativistic regime. These radiation processes will be important for determining the electron-positron

pair densities in relativistic plasmas, as well as for the emergent spectra from such plasmas.

063.050 Synchrotron emissivity from mildly relativistic particles.   V. Petrosian.
Astrophys. J., Vol. 251, 727 - 738 (1981).

The author presents approximate analytic expressions for the evaluation of the frequency and angular dependence of the synchrotron emissivity from mildly relativistic particles with an arbitrary energy spectrum and pitch angle distribution in a given magnetic field. He presents some general results for particle distributions which are not extremely anisotropic and simplify the equations empirically. He also considers an example of an extremely anisotropic distribution, namely that of particles with small pitch angles streaming along magnetic field lines. He compares the result from the analytic expression with results from detailed numerical evaluations and finds excellent agreement not only at frequencies large compared to the gyrofrequency but also at lower frequencies.

063.051 Radiative transfer in the surfaces of atmosphereless bodies. I. Theory.   K. Lumme, E. Bowell.
Astron. J., Vol. 86, 1694 - 1704 (1981) = Contrib. Five Coll. Obs., No. 437.

The authors present a new, generalized theory of radiative transfer, applicable both to the surfaces of atmosphereless bodies in the solar system and to laboratory samples. The theory includes single and multiple scattering of light in particulate surfaces. Single scattering, formulated by means of a probabilistic method, is described in terms of the effects of porosity and roughness. They demonstrate that, for low-albedo surfaces, these effects are distinctly separable: the opposition effect is caused by porosity, and the slope of the linear part of the phase curve is mainly controlled by roughness.

063.052 Electron kinetic equations in comptonisation by isotropic photons.   N. Iwamoto.
Plasma astrophysics, (see 012.042), p. 331 - 332 (1981).

The author reports the derivation of the kinetic equations that describe the time evolution of the distribution of non-degenerate nonrelativistic electrons interacting with isotropic thermal photons via Compton scattering. Two cases, when the electrons are isotropic and when the electron motion is restricted to one spatial dimension due to strong magnetic field, are considered.

063.053 Role of scattering in emergent planetary radiation.   R. K. Bhatia, K. D. Abhyankar.
Bull. Astron. Soc. India, Vol. 9, 181 - 213 (1981).

Starting from the basic concepts, the role played by multiple scattering in modifying the intensity and polarization of radiation incident on a planetary atmosphere is brought out. Formation of spectral lines in such an atmosphere is discussed in detail. Various ground-based methods of observation are given and the spectroscopic observational data for Jupiter and Venus reviewed till 1981.

063.054 Effects of redistribution with dipole scattering on line source functions.
A. Peraiah, K. E. Rangarajan.
J. Astrophys. Astron., Vol. 2, 245 - 252 (1981).

The partial frequency redistribution function for zero natural line width with dipole scattering (R_I) has bee considered in obtaining the simultaneous solution of the statistical equilibrium and line transfer equations in the comoving frame of the expanding gas. The authors have considered a non-LTE two level atom in an expanding spherical medium whose outer radii are 3, 10 and 20 times the stellar radius with a total optical depth $T \simeq 2 \times 10^3$. In all the cases, the authors have calculated the population ratio of the two levels N_2/N_1 and

compared these results with those obtained by using different expansion velocities and geometrical extensions.

063.055 Optical depth effects on the formation of spectral lines in rotating and expanding spherical atmospheres.
A. Peraiah, G. Raghunath, K. N. Nagendra.
J. Astrophys. Astron., Vol. 2, 277 - 284 (1981).

The authors have investigated the effects of increasing optical depths on spectral lines formed in a rotating and expanding spherical shell.

063.056 On the theory of gamma ray amplification through stimulated annihilation radiation (grasar).
R. Ramaty, J. M. McKinley, F. C. Jones.
NASA Tech. Memo., NASA TM 83857, 34 pp. (1981).

The theory of photon emission, absorption and scattering in a relativistic plasma of positrons, electrons and photons is studied. In a system of photons and e^+-e^- pairs, an emission line at ~0.43 MeV can be produced by grasar action provided that the pair chemical potential exceeds ~1 MeV. This emission line could account without a gravitational redshift for the observed lines at this energy from gamma ray bursts.

063.057 On the theory of gamma ray amplification through stimulated annihilation radiation (grasar).
R. Ramaty, J. M. McKinley, F. C. Jones.
Gamma-ray transients, (see 012.046), 10 pp. (1981).

063.058 Depolarization effects in Rayleigh scattering.
A. T. Young.
J. Opt. Soc. America, Vol. 71, 1142 (1981).

063.059 Light scattering by dust particles.
S. Mukai, T. Mukai.
Proceedings of the 14th ISAS Lunar and Planetary Symposium, (see 012.055), p. 162 - 167 (1981).

Light scattering by a large particle bounded by a rough surface is investigated. The multiple reflection of light by such a rough surface is examined based on multiple scattering theory, because each surface element, where a reflection occurs, behaves itself like a grain. It is found that, for a compact particle with large radius, the method gives a powerful tool to predict the phase angle variation of intensity and polarization of scattered light.

063.060 Computing diffuse reflection from particulate planetary surface with a new function.   M. Wolff.
Appl. Opt., Vol. 20, 2493 - 2498 (1981).

An equation is derived to compute the amount of diffuse light reflected by a particulate surface such as on Mars or an asteroid. The method traces the paths of rays within an ensemble of randomly shaped grains and finds the eventual probability of emission. The amount of diffuse, unpolarized emitted light is obtained in terms of the real index of refraction, the imaginary index, and the average diameter of particles making up the surface. The equation is used to compute the empirical rule for obtaining the planetary albedo from the slope of its polarization curve. Results for the indices of the moon, Mars, Io, and Europa are obtained and compared with other data.

063.061 Systematic studies of light scattering. 1: Particle shape.
D. W. Schuerman, R. T. Wang, B. Å. S. Gustafson, R. W. Schaefer.
Appl. Opt., Vol. 20, 4039 - 4050 (1981).

The light scattering properties of twenty-eight particles, spanning four sizes (near the resonance region) and seven related shapes (a 4:1 cylinder, 4:1 and 2:1 prolate spheroids, a sphere, 2:1 and 4:1 oblate spheroids, and a 4:1 disk), are presented for a common index of refraction, $m = 1.61 - i0.004$, representing silicates. All results refer to an ensemble or a cloud of identical particles because averages have been taken over random particle orientations. The degree of polarization, backscatter, and the radiation-pressure cross section are most sensitive to particle shape, implying that the use of Mie theory may be inappropriate for many applications.

063.062   Experimental results of dependent light scattering by two spheres.
R. T. Wang, J. M. Greenberg, D. W. Schuerman.
Opt. Lett., Vol. 6, 543 - 545 (1981).
Microwave analog measurements of the forward scattering produced by two spheres yield the extinction as a function of sphere separation for the case of one sphere's being shadowed by the other. The effects of dependent scattering are obvious up to a separation distance of about ten sphere diameters. Side-scattering measurements show a resonance when the axis of the two-sphere system is in the scattering plane and bisects the scattering angle. The magnitude of at least one measured resonance is a factor of 44 larger than the scattering that is due to a single sphere.

063.063   On radiative transfer in a homogeneous spherical medium with anisotropic scattering.
A. K. Kolesov.
Vestn. LGU, 1981, No. 13, p. 97 - 103. In Russian. – Abstr. in Ref. zh., 51. Astron., 12.51.236 (1981).

063.064   Correlation function of intensity fluctuations of a Gaussian optical beam in a random medium with large-scale discrete scatterers.   V. A. Krutikov.
Izv. vuzov. Radiofiz., Tom 24, 314 - 321 (1981). In Russian. Abstr. in Ref. zh., 51. Astron., 12.51.237 (1981).

063.065   Saturation effect of a coherence function of a wave reflected from discrete scatterers in a turbulent atmosphere.   A. B. Krupnik, A. I. Saichev.
Izv. vuzov. Radiofiz., Tom 24, 322 - 325 (1981). In Russian. Abstr. in Ref. zh., 51. Astron., 12.51.238 (1981).

063.066   Condition of adiabatic invariant conservation in the theory of radiative transfer.   L. Apresyan.
Izv. vuzov. Radiofiz., Tom 24, 308 - 313 (1981). In Russian. Abstr. in Ref. zh., 51. Astron., 12.51.240 (1981).

063.067   Penumbra diffraction fields in inhomogeneous media.   Yu. I. Orlov, S. K. Tropkin.
Izv. vuzov. Radiofiz., Tom 24, 334 - 342 (1981). In Russian. Abstr. in Ref. zh., 51. Astron., 12.51.279 (1981).

Scattering theory. Methods of modern mathematical physics, Vol. 3.   See Abstr. 003.131.

Radiation transfer and stellar atmospheres. See Abstr. 003.148.

The development of science at the Astronomical Observatory of the Leningrad University. Theory of radiative transfer.   See Abstr. 009.011.

Algorithms for the calculation of scattering by stratified spheres.   See Abstr. 021.059.

A comparison of methods for the calculation of Voigt profiles.   See Abstr. 022.095.

On some trends in the development of astrophysics. See Abstr. 061.058.

Monte Carlo simulation of the unsaturated Comptonization in a semi-relativistic plasma.   See Abstr. 062.031.

Annihilation radiation from a hot e^+-e^- plasma. See Abstr. 062.083.

Plasma mechanisms of emission and acceleration of relativistic particles in cosmic plasmas. See Abstr. 062.103.

Plasma lasers (a strong source of coherent radiation in astrophysics).   See Abstr. 062.106.

Radiation reflection from a semi-infinite plasma layer with homogeneous magnetic field. See Abstr. 062.118.

Mass-loss rates in early-type stars determined by fitting Balmer alpha profiles.   See Abstr. 064.014.

Collisional and radiative excitation of SiO masers. See Abstr. 064.048.

Radiation transfer in extended atmospheres. See Abstr. 064.052.

Infrared atomic hydrogen line formation in luminous stars.   See Abstr. 064.056.

Self-focussed electromagnetic waves and SS 433. See Abstr. 066.154.

Cyclotron radiation of magnetized degenerate dwarfs and neutron stars: I – theories. See Abstr. 066.527.

A theoretical interpretation of the phase curves of atmosphereless solar system bodies.   See Abstr. 091.075.

On the definition of albedo and application to irregular particles.   See Abstr. 106.034.

Study of stellar polarization with the CERGA interferometer.   See Abstr. 116.016.

Cyclotron emission from AM Herculis. See Abstr. 117.026.

A theoretical review of SS 433. See Abstr. 117.093.

Inverse Compton effect and the colorimetric characteristics of flare stars.   See Abstr. 122.102.

Interstellar masers: the influence of the geometrical shape on the radiation properties.   See Abstr. 131.079.

Masers.   See Abstr. 131.132.

Far UV radiation transfer and H_2CO lifetime in dense interstellar clouds.   See Abstr. 131.152.

Star formation in molecular clouds and the formation of compact H II regions.   See Abstr. 131.185.

Theoretical quasar emission-line ratios. III. Flux divergence and photon escape.   See Abstr. 141.034.

Broad emission line profiles from discs and quasars. See Abstr. 141.069.

The formation of emission lines in quasars and Seyfert nuclei.   See Abstr. 141.136.

Hydrogen line spectrum in quasars. I. Approximation procedures for line transfer versus an exact treatment.
See Abstr. 141.156.

The gamma-ray spectra of radio pulsars.
See Abstr. 141.534.

A revised calculation of the bremsstrahlung cross-section in the high magnetic field of pulsars.
See Abstr. 141.536.

Radiation from an asteroid–neutron star collision.
See Abstr. 142.524.

Radiation processes in active galactic nuclei.
See Abstr. 158.227.

Erratum

063.901   **Erratum: "Radiative transfer through an arbitrarily thick, scattering atmosphere [J. Quant. Spectrosc. Radiat. Transfer, Vol. 24, 391 - 406 (1980)].**
A. H. Karp, J. Greenstadt, J. A. Fillmore.
J. Quant. Spectrosc. Radiat. Transfer, Vol. 26, 547 (1981).
See Abstr. 28.063.031.

064 Stellar Atmospheres, Stellar Envelopes, Mass Loss, Accretion

064.001 A new look at the dynamics of twisted accretion disks. S. P. Hatchett, M. C. Begelman, C. L. Sarazin.
Astrophys. J., Vol. 247, 677 - 685 (1981).

The authors reexamine the dynamic response of a thin, viscous accretion disk to twisting torques, guided by the earlier analyses by Bardeen and Petterson. They make several corrections to this earlier work, and present a new version of the twist equations consistent with their physical assumptions. They find exact, analytic solutions for the steady state structure of a disk subject to Lense-Thirring torques by a non-aligned central Kerr black hole and also for the time-dependent problem of the structure of a slaved disk with its oscillating boundary conditions and show that undriven disks and disks subject to time-independent driving torques are stable.

064.002 The sizes and edge temperatures of optically thick accretion discs. J. Frank, A. R. King.
Mon. Not. R. Astron. Soc., Vol. 196, 507 - 515 (1981).

The vertical optical depth through a steady-state accretion disc is calculated using a simple Rosseland mean technique. The authors show that for accretion rates $\dot{M} \sim 10^{16}$ g s^{-1} the edge temperature T_{out} of the optically thick region of the disc will be $\lesssim 5000$ K for values of the viscosity parameter $\alpha \leqslant 1$, and can be considerably cooler, in agreement with recent observations of cataclysmic variables. For a wide range of conditions they find $T_{out} \simeq \dot{M}^{-0.056}$ with corresponding disc radius $R_{out} \simeq \dot{M}^{0.048}$, although in some regimes the edge temperature is shown to be very sensitive to the accretion rate.

064.003 On the stability of stellar winds from early-type stars. F. D. Kahn.
Mon. Not. R. Astron. Soc., Vol. 196, 641 - 657 (1981).

Early-type stars are known to emit powerful stellar winds. Pressure effects are significant only within a few scale heights of the surface unless the gas becomes very hot. Once the wind has become highly supersonic it is driven by radiation pressure acting on suitable resonance lines. The body force on the gas depends on the local velocity gradient. If the flow is everywhere radial the motion remains laminar. Especially high wind speeds are predicted when the mass-loss rate $\dot{M}$ is close to the maximum allowable value, λ.

064.004 Emission, absorption and self-absorption in circumstellar dust shells. R. M. Mitchell, G. Robinson.
Mon. Not. R. Astron. Soc., Vol. 196, 801 - 822 (1981).

An extended spherical circumstellar dust shell model, in which the radiative transfer problem is treated accurately, is used to construct an extensive grid of model results in order to investigate quantitatively the dependence of the 10 and 20 μm silicate features on model parameters. A wide range of results is presented graphically, and these results demonstrate that compact shells characterized by inward peaking density distributions exhibit stronger emission features and weaker absorption features than those of the reverse characteristics.

064.005 The temperature scale of solar-type stars. T. Gehren.
Astron. Astrophys., Vol. 100, 97 - 106 (1981).

Effective temperatures for the atmospheres of solar-type stars have been determined by comparison of observed hydrogen lines with profiles computed from scaled solar model atmospheres. The author's spectroscopic calibration yields effective temperatures for G-type stars that are about 200 - 300 K higher than those predicted from comparison of observed and synthesized colour indices.

**064.006 Line blanketed model atmospheres of Ap-stars.
V. The Hg—Mn stars 53 Tau and κ Cnc.**
K. Stępień, H. Muthsam.
Astron. Astrophys., Vol. 100, 159 - 163 (1981).

All available observations of the energy distributions of 53 Tau and κ Cnc from the wavelength interval between $L\alpha$ and the Paschen jump are compared with a series of model atmospheres in which temperature, microturbulent velocity and chemical composition were varied. Good agreement is obtained if effective temperatures of 53 Tau and κ Cnc are 11,250 K and 12,800 K, respectively, and the chemical compositions are taken from Strom (1969) and Kodaira (1965). For 53 Tau the set with $v_{mic} = 0$ gives equally good fit as the old abundances and the resulting temperature is 11,400 K. Observations of both stars show a slight excess of radiation in the red. It is suggested that this excess may be caused by the circumstellar matter.

064.007 The expanding envelope of Tau Scorpii: a detailed UV-line fit. W.-R. Hamann.
Astron. Astrophys., Vol. 100, 169 - 174 (1981).

The comoving-frame technique is applied to the UV resonance lines of τ Sco. A detailed fit is achieved by construction of a simple empirical wind model. The profiles are reproduced over the whole range including the line center regions and the overlapping doublets. The results confirm that the ionization decreases outwards in the envelope of τ Sco. On the basis of the "warm wind model" an electron temperature from 250,000 K to 140,000 K is required for the observed ionization. The derived mass loss rate is log $[\dot{M}/(M_{\odot}/\text{yr})] = -8.9 \pm 0.5$. A high turbulence velocity (100 km s^{-1}) is observed even in the low-velocity layers.

064.008 Element stratification in the atmospheres of main sequence stars: the silicon accumulation.
G. Alecian, S. Vauclair.
Astron. Astrophys., Vol. 101, 16 - 25 (1981).

Radiative accelerations on silicon atoms in magnetic stars have been precisely computed, including Si II autoionization lines, for two stellar models ($T_{eff} = 12000$ K and $T_{eff} = 14000$ K). The main conclusion is that silicon overabundances by factors up to 100 as observed in magnetic stars cannot be obtained without magnetic effects: horizontal, or nearly horizontal magnetic fields are needed to help supporting silicon up to the observed values. The abundance anomalies which can be expected in case of horizontal magnetic fields are evaluated, and precisely computed in one case ($T_{eff} = 12000$ K, H = 10000 G). The effect of oblique lines is also discussed. It is shown that accumulation proceeds as long as the angle between the magnetic lines and the stellar surface is smaller than an angle θ_0 which is computed for the same case. Silicon spots are definitely expected in magnetic stars at places where the magnetic lines in the atmosphere are nearly parallel to the stellar surface.

064.009 Explosive mass loss in binary stars: the two time-scale method. P. Hut, F. Verhulst.
Astron. Astrophys., Vol. 101, 134 - 137 (1981).

Quick loss of mass, as happens in a supernova explosion, disrupts a binary star with an initially circular orbit if more than roughly half the total mass is ejected. This critical mass fraction is determined by the form and duration of the process of mass loss. Asymptotic expansions by the two time-scale method are introduced and illustrated by a simple example. The method is then used to obtain approximate solutions to this boundary layer type problem of jumps between Kepler orbits.

064.010 Line formation in the wind of Alpha Cygni.
P. B. Kunasz, F. Praderie.
Astrophys. J., Vol. 247, 949 - 959 (1981).

The authors summarize the results of recent *IUE* observations of the spectra of α Cyg and discuss possible explanations for the Doppler-shifted deep absorption profiles in the resonance lines of the singly ionized metals. The results of detailed modeling for Mg II λ 2802 are presented and discussed in the context of observations of α Cyg in the radio, infrared, visible, ultraviolet and X-ray parts of the spectrum. Extensive semi-empirical modeling was carried out in which the Mg$^+$ resonance transition was treated as a NLTE two-level atom in an expanding, spherically symmetric atmosphere. Supersonic microturbulence in a cool wind is required for reasonable agreement with observations.

064.011 Adiabatic oscillations of accretion disks.
J. P. Cox.
Astrophys. J., Vol. 247, 1070 - 1077 (1981).

The author has used a variational principle to derive an integral expression for the frequencies of adiabatic oscillations of an accretion disk and has applied this expression to vertical oscillations. A number of examples show that this expression generally gives fairly good results with a simple trial function. In particular, the oscillatory periods of vertical oscillations in some of the low modes are found to be of the order of the corresponding Keplerian periods (or less, if self-gravity is taken into account), regardless of the detailed structure of the disk model.

064.012 Massenverlust normaler Sterne.
H. - A. Ott.
Sterne Weltraum, Jahrg. 20, 312 - 318 (1981).

064.013 On thermohaline-convective instability with finite Larmor radius and Hall effects (*stellar atmosphere model*).   R. C. Sharma, K. N. Sharma.
Acta Phys. Acad. Sci. Hungaricae, Vol. 48, 349 - 357 (1980). Abstr. in Phys. Abstr., Vol. 84, Abstr. 66925 (1981).

064.014 Mass-loss rates in early-type stars determined by fitting Balmer alpha profiles.
G. L. Olson, D. Ebbets.
Astrophys. J., Vol. 248, 1021 - 1030 (1981).

Theoretical Hα profiles are computed to match the observed profiles in 10 stars ranging in spectral type from O4f to B3 Ia. The theoretical profiles assume spherically symmetric mass loss and use the Sobolev approximation in the radiative transfer. Therefore, it is not possible to fit exactly the region near line center, where stellar rotation causes deviations from spherical symmetry and low velocities cause a larger error in the radiative transfer. However, by obtaining a good fit to the high-velocity wings of the observed profiles, it is possible to get a good estimate of the mass-loss rate and to find out how the flow velocity increases with distance from the star. Variability in the observed profiles is discussed and a comparison of the derived mass-loss rates with other rate determinations presented.

064.015 Stellar convection theory. III. Dynamical coupling of the two convection zones in A-type stars by penetrative motions.   J. Latour, J. Toomre, J.-P. Zahn.
Astrophys. J., Vol. 248, 1081 - 1098 (1981).

Anelastic modal equations are used to examine thermal convection occurring over many density scale heights in the entire outer envelope of an A-type star, encompassing both the hydrogen and helium convectively unstable zones. Single-mode anelastic solutions for such compressible convection display strong overshooting of the motions into adjacent radiative zones. Such mixing would preclude diffuse separation of elements in the supposedly quiescent region between the two unstable zones. Indeed, the anelastic solutions reveal that the two zones of convective instability are dynamically coupled by the overshooting motions. The nonlinear single-mode equations admit two solutions for the same horizontal wavelength, and these are distinguished by the sense of the vertical velocity at the center of the three-dimensional cell.

064.016 Outlook for ultraviolet astronomy.
E. Böhm-Vitense.
The Universe at ultraviolet wavelengths, (see 012.009), p. 3 - 17 (1981).

064.017 Proposed three-phase modeling of Be stars from combined UV and visual observations.
V. Doazan, R. Stalio, R. N. Thomas.
The Universe at ultraviolet wavelengths, (see 012.009), p. 149 - 155 (1981).

Far UV observations of the behavior of (wind-velocity, superionization) values as a function of the phase of the (Be, B-shell, B-normal) pattern established by visual observations for γ Cas and 59 Cyg are translated into a crude atmospheric model for the Be phase and several kinds of mass-flux variability across the three phases.

064.018 Comments on the origin of heating in the mantles of early-type stars.   A. B. Underhill.
The Universe at ultraviolet wavelengths, (see 012.009), p. 225 - 226 (1981).

It is shown that quiescent heating by magneto-dynamic effects can produce a range in injected non-radiative energy that is probably adequate to account for the range in physical properties that is observed for the mantles of early-type supergiants.

064.019 An IUE's eye view of cool-star outer atmospheres.
T. Ayres.
The Universe at ultraviolet wavelengths, (see 012.009), p. 237-278 (1981).

The author describes how the IUE is used in practice to deduce the physical properties of stellar outer atmospheres. He discusses the sorts of information that one can extract from low-dispersion and echelle-mode spectra with the SWP and LWR cameras. The author illustrates the discussion with spectra of two bright binary systems, α Centauri (G2 V+K1 V) and Capella (G6 III+F9 III). In fact, the comparison of these two systems with the sun has revealed unexpected clues to the nature of the chromosphere-corona phenomena.

064.020 The structure of chromospheres around late-type giants and supergiants.
A. Brown, M. Ferraz, C. Jordan.
The Universe at ultraviolet wavelengths, (see 012.009), p. 297 - 302 (1981).

Observations α Tau (K5 III) and β Gru (M2 II) made at high resolution have been used to confirm line identifications of features blended at low resolution. The high resolution spectra allow selected pairs of lines to be used to find the electron density, N_e, and the opacity, τ. These can be used together with the emission measure to place constraints on the structure of the atmosphere. The line formation processes are briefly discussed. Photo-excitation by strong lines appears to be important in these late-type atmospheres.

064.021 Large macroturbulence in the chromosphere of an RS CVn star.   L. E. Cram, D. J. Mullan.
Bull. American Astron. Soc., Vol. 13, 514 - 515 (1981). Abstract.

064.022 Autoionization and dielectronic recombination of Si II in B-star atmospheres.   A. B. Underhill.
Bull. American Astron. Soc., Vol. 13, 527 (1981). – Abstract.

064.023 A model of the outer atmosphere of β Ceti.
K. Eriksson, T. Simon, J. L. Linsky.
Bull. American Astron. Soc., Vol. 13, 547 (1981). – Abstract.

064.024 Variable mass loss and magnetic topology in cool
giant stars.   D. J. Mullan, R. E. Stencel.
Bull. American Astron. Soc., Vol. 13, 547 (1981). – Abstract.

064.025 Stellar chromospheres and coronae.   R. G. Athay.
Bull. American Astron. Soc., Vol. 13, 547 - 548
(1981). – Abstract.

064.026 Time-dependent models of grain-forming stellar
atmospheres.   J. E. J. Woodrow.
Bull. American Astron. Soc., Vol. 13, 548 (1981). – Abstract.

064.027 Structure and stability of heated accretion flow onto
compact objects.
R. F. Stellingwerf, J. Buff, A. Gillman.
Bull. American Astron. Soc., Vol. 13, 557 (1981). – Abstract.

064.028 Mechanical wave-energy flux in magnetoatmospheres:
discrete and continuous spectra.   J. A. Adam.
Astrophys. Space Sci., Vol. 78, 293 - 350 (1981).
This paper deals with a rather general class of magneto-atmospheres. After discussion of the equilibrium and stability of such systems, and certain mathematical properties of a particular system, theory is developed which enables expressions to be written down for the mechanical wave energy flux associated with wave motion due to a transient source. These analytic expressions are very general and contain contributions from the continuous and discrete frequency spectra. The relative distribution of wave energy flux in the various modes is discussed in the context of solar physics parameters.

064.029 Mass loss and coronae of cool stars.
D. Reimers.
Second European IUE Conference, (see 012.011), p. XXXIII - XXXVIII (1980).
The current knowledge about the occurrence of coronae and mass loss in cool stars is reviewed. Observational arguments are collected that – contrary to recent claims of Mullan and Linsky & Haisch – both mass-loss phenomena and coronae around cool stars vary in a continuous way from solar-type stars via G and K giants to luminous M supergiants. It is pointed out that IUE observations of cool giants with hot companions are important.

064.030 Activity and outer atmospheres of the stars.
F. Praderie.
Activity and outer atmospheres of the sun and stars, (see 012.012), p. 157 - 261 (1981).
Contents: The chromosphere/corona/wind complex. How to recognize the presence of a stellar chromosphere or corona: problems of diagnostic procedures (static and non static case). Stellar chromospheres, transition regions and coronae: detection from visible, UV, X and radio observations. Semi empirical models of stellar chromospheres and transition regions. Radiative losses in stellar outer layers. Towards theoretical models. Ensemble properties of the outer layers.

064.031 Oscillations and pulsations.
J. W. Leibacher, R. F. Stein.
The sun as a star, (see 003.004), p. 263 - 287 (1981).
The paper begins by discussing the properties of waves in stars: their restoring forces, periods and wavelengths, their propagation and motions. It continues by summarizing the characteristics of the eigenmodes of oscillation of a cavity. This is applied to the solar oscillations, to describe their nature and to point out the diagnostic capabilities afforded by them.

064.032 On the stellar gravity and effective temperature
dependence of the ratio of terminal to escape veloc-
ities in stellar winds.   L. Carrasco.
Effects of mass loss on stellar evolution, (see 012.015), p. 75 - 78 (1981).
Contrary to the results of some investigators, the ratio of terminal to escape velocities ($V\infty/V_{esc}$) observed for the winds in early-type stars is found to be linearly correlated with log Γ, Γ being the ratio of stellar to Eddington's luminosities.

064.033 The theory of winds in early type stars.
A. G. Hearn.
Effects of mass loss on stellar evolution, (see 012.015), p. 125 - 130 (1981).

064.034 Empirical wind models from detailed UV-line fits:
Tau Scorpii.   W.-R. Hamann.
Effects of mass loss on stellar evolution, (see 012.015), p. 161 - 165 (1981).
The main-sequence B0 star Tau Sco has been studied previously by Lamers and Rogerson (1978). However, their line fit was restricted to the blue wings of the UV resonance lines because the line formation was calculated in Sobolev approximation. The author now repeats this investigation by means of the comoving-frame (CMF) method, which he has extended to the treatment of overlapping doublets. From a systematic comparison the author knows that the results may deviate considerably from those of the Sobolev method.

064.035 Can hot star winds be driven by radiation pressure?
M. Leroy, J.-P. J. Lafon.
Effects of mass loss on stellar evolution, (see 012.015), p. 167 - 171 (1981).
The mechanism by which early type (O and B) stars lose mass is investigated. The phenomena occurring close to the photosphere are crucial for the structure of the wind. It is shown that the effects of radiation are not sufficient to explain the observed winds. A general scheme for energy balance including a corona is proposed.

064.036 Radiative wind acceleration in early type stars.
N. Panagia, F. Macchetto.
Effects of mass loss on stellar evolution, (see 012.015), p. 173 - 178 (1981).
The processes of radiative acceleration of stellar winds in OB stars by single and multiple photon scattering are considered. Single scattering can be the dominant accelerating process for stars later than B2 and can account for terminal velocities up to 500 - 1000 km s^{-1}. Multiple scattering of photons in the approximate range 200 - 500 Å provides additional wind acceleration for stars earlier than B2 to reach terminal velocities of up to 2000 - 4000 km s^{-1}. A systematic increase of the terminal velocity as a function of the effective temperature is predicted.

064.037 Winds in late-type stars: mechanisms of mass
outflow.   J. L. Linsky.
Effects of mass loss on stellar evolution, (see 012.015), p. 187 - 212 (1981).
Four basic mechanisms have been proposed to explain the acceleration of winds in late-type stars – thermal pressure gradients, radiation pressure on circumstellar dust grains, momentum addition by Alfvén waves, and momentum addition by periodic shock waves. In the review the author describes recent work in applying these mechanisms to stars, and considers whether these mechanisms can work even in principle and whether they are consistent with recent ultraviolet and X-ray data from the IUE and Einstein spacecraft.

064.038 The fluctuation theory of the stellar mass loss.
C. D. Andriesse.
Effects of mass loss on stellar evolution, (see 012.015),

p. 213 - 227 (1981).

The idea that fluctuations in the mass flow are as significant as the very existence of the flow has led to the development of a fluctuation theory of the stellar mass loss. A general theory for fluctuations in non-equilibrium systems – and such are stellar atmospheres – has been developed long ago. In developing the general theory to a specific stellar theory, however, the arguments have not come up in their logical order. The present sketch of this theory improves on that order and is offered as a framework for further study.

064.039 Theoretical evidence of mass loss from globular cluster stars.   V. Castellani, A. Tornambè.
Effects of mass loss on stellar evolution, (see 012.015), p. 357 - 359 (1981).

The total amount of mass lost during the first giant phase of Population II red giants is derived from Horizontal Branch properties: it is found to be proportional to the metal content.

064.040 Effect of mass gain on stellar evolution.   R. Ebert, H. Zinnecker.
Effects of mass loss on stellar evolution, (see 012.015), p. 361 - 371 (1981).

The authors present a fully hydrodynamical treatment of the stationary isothermal accretion problem onto a moving gravitating point mass. The derivation is purely analytical. The authors find that the accretion rate is more than a factor of 50 higher than the accretion rate derived from the partially non-hydrodynamical treatment by Hoyle and Lyttleton (1939) or Bondi and Hoyle (1944). This result may have some bearing on the evolutionary tracks of young pre-Main Sequence stars still embedded in their parent protocluster cloud.

064.041 Response of low-mass main sequence stars to accretion.   M. Y. Fujimoto, I. Iben, Jr., S. A. Becker.
Effects of mass loss on stellar evolution, (see 012.015), p. 401 - 403 (1981).

064.042 Properties of optically thick winds driven by radiation pressure.   M. Friedjung.
Effects of mass loss on stellar evolution, (see 012.015), p. 495 - 497 (1981).

Conditions are derived for supercritical winds, accelerated by the radiation pressure of an object above the Eddington limit. A relation between the different energy fluxes is found. The conditions seem to be satisfied by novae, and perhaps by η Car. They are not satisfied however by the symbiotic star Z And.

064.043 Subionization and decelerated-flow in the vicinity of a B-shell star.   J. Zorec.
Effects of mass loss on stellar evolution, (see 012.015), p. 539 - 541 (1981).

064.044 On the applicability of the Kurucz – Peytremann oscillator strengths for the analysis of stellar atmospheres.   V. Dobrichev, D. Rajkova.
Astrofiz. issled., NRB, Vol. 3, 18 - 35 (1981). In Russian. Abstr. in Ref. zh., 51. Astron., 8.51.150 (1981).

064.045 On the initial phase of interaction between an expanding stellar envelope and the surrounding medium.   D. K. Nadezhin.
Inst. teor. i ehksp. fiz. Moskva. Prepr., 1981, No. 1, 44 pp. In Russian. – Abstr. in Ref. zh., 51. Astron., 8.51.537 (1981).

064.046 Mass-loss from stars: a review of the observational evidence.   M. Hack.
Mem. Soc. Astron. Italiana, Vol. 52, (see 012.019), 37 - 48 (1981).

064.047 Mass loss and overshooting in massive stars.   A. G. Bressan, G. Bertelli, C. Chiosi.
Astron. Astrophys., Vol. 102, 25 - 30 (1981).

In this paper the authors examine the problem of the boundary of convective cores in massive stars (M > $10M_\odot$). A new method is developed, which enables to calculate the overshooting from convective cores in a simple way. Mass loss by stellar wind is also taken into account. Evolutionary sequences are computed up to pre-central He-ignition stages for original 20, 60 and $100M_\odot$ stars. Finally, model results are compared with the observational data for supergiant stars.

064.048 Collisional and radiative excitation of SiO masers.   V. Bujarrabal, Nguyen-Q-Rieu.
Astron. Astrophys., Vol. 102, 65 - 72 (1981).

The authors analyse collisional and radiative pumping of SiO masers by performing multilevel calculations. They propose a radiative pumping model in which SiO is located in the inner circumstellar envelope. An analysis of IR and SiO data suggests that the authors' model is efficient enough to explain the observed SiO maser emission.

064.049 Colours and effective temperatures of extreme helium stars.   U. Heber, D. Schönberner.
Astron. Astrophys., Vol. 102, 73 - 80 (1981).

Johnson and Strömgren colours including reddening lines have been computed for extreme helium stars using a grid of (unblanketed) model atmospheres. Effective temperatures are determined from the observed colours, for all of the known extreme helium stars. From the statistics of the effective temperatures, it is concluded that the extreme helium stars evolve to higher effective temperatures and finally cool down to become white dwarfs. Their masses are then between 0.7 and $1.0M_\odot$.

064.050 An analytical model for stellar coronae.   P. C. H. Martens.
Astron. Astrophys., Vol. 102, 156 - 164 (1981).

A simple semi-analytical model for stationary stellar coronae is developed. The model is based on the assumption of an isothermal corona and a transition region of constant pressure. It is shown that the temperature structure of the transition region is described by one parameter differential equation. This leads to one parameter set of solutions for the temperature structure of the transition region. Only one of these solutions is in agreement with the boundary condition at infinity.

064.051 Line formation in turbulent media with finite correlation lengths. II.   J. Stahlberg.
Astron. Nachr., Band 302, 139 - 144 (1981).

The influence of stochastic velocity fields on line formation is investigated. The range of validity of the method in part I (see Abstr. 26.064.049) is extended to long correlation lengths. Methods for long and short correlation lengths are combined to give a better approach to the real character of the turbulence. Results for discontinuous velocity fields are derived and compared with continuous ones.

064.052 Radiation transfer in extended atmospheres.   E. Simonneau.
Bull. Soc. R. Sci. Liège, Vol. 49, 281 - 203 (1980). In French. Abstr. in Phys. Abstr., Vol. 84, Abstr. 88891 (1981).

064.053 Formation of P Cygni profiles with limb darkening of the stellar core.   J. Surdej.
Astrophys. Space Sci., Vol. 79, 213 - 227 (1981).

In the framework of the Sobolev approximation, the author investigates the effects of limb darkening of the stellar core onto the formation of line profiles in rapidly expanding envelopes.

064.054 **Stellar atmospheres.**   E. Simonneau.
Ann. Physique, Vol. 6, (see 012.026), p. 107 - 113
(1981). In French. – Abstr. in Phys. Abstr., Vol. 84, Abstr.
Abstr. 94314 (1981).

064.055 **Spicule-dominated coronae and late-type giant**
         evolution.   S. G. Wallenhorst.
Astrophys. J., Vol. 249, 176 - 184 (1981).
    The mass-loss rates and coronal temperatures of stars
that have evolved away from the main sequence are examined,
using an energy-balance model in which the coronal energy
losses due to hot, downflowing spicular material are included.
It is postulated that the coronal energy input flux remains
constant as a star evolves up the red giant branch. Stable solutions for a corona of this type are found to exist only below a
critical stellar radius. The locus of these radii on the H-R
diagram is identified with the Temperature Dividing Line of
Linsky and Haisch.

064.056 **Infrared atomic hydrogen line formation in**
         luminous stars.   J. H. Krolik, H. A. Smith.
Astrophys. J., Vol. 249, 628 - 636 (1981).
    Infrared atomic hydrogen lines observed in luminous
stars, generally attributed to compact circumstellar H II
regions, can also be formed in the winds likely to emanate
from these stars. Implications are discussed for the class of
obscured infrared point sources showing these lines, and an
illustrative model is derived for the BN object in Orion. Such
stellar winds should also produce weak, but detectable, radio
emission.

064.057 **Accretion discs in astrophysics.**   J. E. Pringle.
Annu. Rev. Astron. Astrophys., Vol. 19,
(see 003.012), 137 - 162 (1981).
    Contents: Introduction. Initial developments. Steady
discs. Viscosity. Radiation mechanisms. Time dependence.
Instability. Variability. Inner boundary. Summary.

064.058 **A stellar wind model for bipolar nebulae.**
         J. F. Barral, J. Cantó.
Rev. Mexicana Astron. Astrofis., Vol. 5, 101 - 108 (1981).
    The steady flow pattern around a wind-producing star
immersed in an interstellar disk-shaped cloud is analyzed. The
wind is taken as isotropic. In particular, the authors apply the
model to an infinite, isothermal and self-gravitating disk. The
radial velocity of the emitting material is also presented. The
model is found to account satisfactorily for the observed
shapes of optical bipolar nebulae.

064.059 **Chemical evolution of galaxies. I. Constraints im-**
         posed by the ΔY/ΔZ ratio.
A. Serrano, M. Peimbert.
Rev. Mexicana Astron. Astrofis., Vol. 5, 109 - 124 (1981).
    The authors have computed chemical evolution models
considering: a) mass loss and He production in intermediate
mass stars based on models of asymptotic giant branch evolution; b) several rates of mass loss for massive stars; c) various
initial mass functions, including that by Serrano (1978), and
d) evolution with and without instantaneous recycling. From
these models they have derived the $\Delta Y/\Delta Z$ abundance ratio
and concluded that the observed $\Delta Y/\Delta Z$ ratio provides strong
restrictions for galactic and stellar evolution models.

064.060 **Rosseland mean opacities for late-type stars.**
         T. Kipper, J. Sitska.
Tartu Astrofüüs. Obs., Teated, Nr. 64, p. 15 - 23 (1981).

064.061 **Line blanketing in the Lyman-alpha wings.**
         I. Hubený.
Upper main sequence chemically peculiar stars, (see 012.033),
p. 373 - 378 (1981).
    A schematic study of the formation of lines in the Lyman-

alpha wings in A and late B stars is presented. It is demonstrated that lines situated in the Lyman-alpha wings have very
complex profiles, and in several cases they can appear strongly
in emission. This emission tends to increase with increasing
abundance and with decreasing distance from the Lyman-alpha
centre. It is tentatively suggested that this effect can explain
the anomalous brightness variations in the Lyman-alpha wings
of α^2 CVn and similar stars.

064.062 **Spectral variability of Ap star model induced by**
         electromagnetic forces.   J. Madej.
Upper main sequence chemically peculiar stars, (see 012.033),
p. 379 - 382 (1981).
    Set of chemically homogeneous model stellar atmospheres
in LTE, radiative and hydrostatic equilibrium is constructed in
order to demonstrate influence of electromagnetic forces on
their spectra. Numerical results show that spectral differences
between the models ($T_{eff} = 12000$ K) spread over the surface
of a single Ap star can be significant agents of complex spectral variability of the star.

064.063 **The influence of an abundance stratification on the**
         manganese curve of growth.   G. Alecian.
Upper main sequence chemically peculiar stars, (see 012.033),
p. 383 - 388 (1981).
    Theoretical curves of growth of Mn with an inhomoge-
neous abundance distribution are compared to observational
curves of growth for the Hg - Mn star v Her. Results are consistent with the existence of a manganese cloud above
$\tau_{5000} \approx 10^{-4}$.

064.064 **Radiative forces and the abundance of beryllium in**
         Ap Hg-Mn stars.
J. Borsenberger, G. Michaud, I. Radiman, F. Praderie.
Upper main sequence chemically peculiar stars, (see 012.033),
p. 389 - 394 (1981).

064.065 **Silicon diffusion and the magnetic structure of Ap**
         stars.   S. Vauclair, G. Alecian.
Upper main sequence chemically peculiar stars, (see 012.033),
p. 395 - 396 (1981).

064.066 **On the infrared excess in magnetic peculiar stars.**
         O. Havnes.
Upper main sequence chemically peculiar stars, (see 012.033),
p. 403 - 408 (1981).
    The author considers three different mechanisms for pro-
ducing the infrared excesses observed in magnetic peculiar
stars. Of these three, synchrotron radiation from precipitating
electrons and thermal radiation from dust can probably be
discarded while free-free radiation from (outstreaming) gas
above the magnetic polar regions appears as the most likely
explanation.

064.067 **Motion and evaporation of grains in the vicinity of**
         magnetic peculiar stars.
J. Leiknes, O. Havnes.
Upper main sequence chemically peculiar stars, (see 012.033),
p. 409 - 413 (1981).
    The authors have calculated the motion and evaporation
of spherical grains of iron and silicates around stars of temperatures from 5800 to 20000°K. A considerable amount of grains
will approach stars with $T_* < 16000$°K and evaporate within
distances corresponding to the expected magnetosphere radius
of such stars. This results in capture of the evaporated grain
material. Above ~ 16000°K the stellar radiation pressure
becomes large and practically no grains of radius < 10 μm approach the stars sufficiently close to evaporate.

064.068 **Progress on accretion disks.**   J. C. Wheeler.
         Nature, Vol. 294, 213 (1981).

064.069 Innermost parts of accretion disks are thermally and secularly stable.   M. A. Abramowicz.
Nature, Vol. 294, 235 - 236 (1981).

It is widely held that all accretion disks must be thermally and secularly unstable in their innermost parts, which seems to be in direct conflict with observations. Much effort has been made to find a stabilizing mechanism operating in the innermost parts of the disks. Here the author shows that such a mechanism does exist. It is of general relativistic origin, is purely mechanical — operating independently of viscosity and other microphysical processes — and is similar to the mass loss caused by Roche lobe overflow in close binaries.

064.070 A supercritical accretion disk model for SS 433.   M. Calvani, L. Nobili.
Astrophys. Space Sci., Vol. 79, 387 - 395 (1981).

The authors describe a model for SS 433 based on an accretion disk around a black hole. Due to the very high mass transfer rate in the system, the disk must be geometrically thick. Two narrow funnels are formed around the rotation axis and radiation pressure in the funnels accelerates matter to relativistic velocities in the form of two opposite jets. The X and optical luminosities are evaluated and they agree well with the experimental data.

064.071 The collimation of particle beams from thick accretion discs.   M. Sikora, D. B. Wilson.
Mon. Not. R. Astron. Soc., Vol. 197, 529 - 541 (1981).

The acceleration and collimation of particle beams in the funnel of thick accretion discs is studied in the approximation that the flow is optically thin. Such flows can be collimated to within ~ 0.1 radians by sufficiently thick discs. The flow cannot convert more than a small fraction of the disc's (super-Eddington) luminosity into the energy flow of a narrow beam without being optically thick.

064.072 The influence of mass loss on the observed X-ray spectra of early-type stars.
G. C. Stewart, A. C. Fabian.
Mon. Not. R. Astron. Soc., Vol. 197, 713 - 720 (1981).

The observed X-ray spectra of early-type stars have been a major problem in the determination of the location of the emission source. The soft X-rays must propagate through a considerable column density of wind material. The authors have investigated the photo-ionization state of such a wind and find that it is very sensitive to the mass loss rate. Acceptable spectra may be obtained from X-rays emitted at the base of the wind, where radiative efficiency is highest, for mass loss rates slightly lower than those inferred from UV, infrared and radio measurements.

064.073 Free-free emission from extended envelopes. II. The mass loss and the envelope ionization.
M. Felli, N. Panagia.
Astron. Astrophys., Vol. 102, 424 - 430 (1981).

The problem of the gas ionization in the flow from stars undergoing mass loss has been considered in the general case of an accelerated wind. A critical value of recombinations per unit time in the entire envelope is found such that only if the stellar supply of ionizing photons is greater than this limit can the envelope be fully ionized. Collisional ionization of the wind is also considered but found to be of minor importance. Application of the theory to OB stars shows that, while for main sequence stars the stellar Lyman-continuum photon flux is always greater than the "mean" critical value, this may not be the case for late supergiants. The case of P Cygni is considered in detail. For T Tauri stars, the analysis points out that the "ionization luminosity" is about the same order as that of the "standard stellar luminosity" and therefore must be considered for a correct evaluation of the total energy output of the star. Finally, the importance of the constraints posed by the existence of a critical value of recombinations is stressed for the case of BN-like sources and ultra-compact H II regions.

064.074 Pulsations, grain condensation, and mass loss in long-period variable stars.
T. W. Jones, E. P. Ney, W. A. Stein.
Astrophys. J., Vol. 250, 324 - 326 (1981).

The role of pulsations on the rate of dust condensation and associated mass loss in red giants is discussed. Pulsations result in an increase of the atmospheric scale height which can enhance mass loss by several orders of magnitude. The authors stress that such effects should be included in attempts to model this mass loss mechanism.

064.075 Radiation from optically thin accretion discs.   R. Tylenda.
Acta Astron., Vol. 31, 127 - 152 (1981).

Accretion discs in cataclysmic variables with low rates of mass transfer, $\dot{M} \lesssim 10^{16} \, \mathrm{g \, s^{-1}}$, have outer regions optically thin in continuum. A simple approach that allows to calculate the radiation spectra from such discs is presented. A great number of disc models has been obtained in order to study influence of various parameters of discs on the outgoing continuous spectra, emission lines and the UBV colours.

064.076 Comments on the spectrum and polarization of relativistic accretion disks.   S. Pineault.
J. R. Astron. Soc. Canada, Vol. 75, 251 (1981). — Abstract.

064.077 Pure helium model atmospheres of red degenerate stars.   S. Kapranidis.
Publ. Astron. Soc. Pacific, Vol. 93, 548 (1981). — Abstract.

064.078 Rosseland mean opacities of stellar atmospheres.   J.-h. Jin.
Acta Astron. Sinica, Vol. 22, 273 - 278 (1981). In Chinese.

064.079 Mass loss from very luminous OB stars and the Cygnus superbubble.
D. C. Abbott, J. H. Bieging, E. Churchwell.
Astrophys. J., Vol. 250, 645 - 659 (1981).

Using the VLA, the authors detected radiation at λ 6 cm from five members of the Cyg OB2 association. They also reobserved three other stars, confirming detections of P Cyg and 9 Sgr, but not that of λ Cep. The flux densities for P Cyg and possibly 9 Sgr differ sufficiently to suggest variability on a time scale of months. Radio positions for six stars are of astrometric accuracy. Mass loss rates are derived by combining these radio fluxes with estimates for the terminal velocities of the winds and the distance to Cyg OB2. These new rates support previous conclusions that, for very luminous OB stars: (1) mass loss scales with stellar luminosity as $M \propto L^\gamma$, where $1.3 < \gamma < 2.0$, and (2) the degree of emission of the Hα line is not a reliable indicator of the rate of mass loss.

064.080 Mass loss rates from O stars in OB associations.   C. D. Garmany, G. L. Olson, P. S. Conti, M. E. Van Steenberg.
Astrophys. J., Vol. 250, 660 - 676 (1981).

The authors present mass loss rates for 31 O type stars in clusters or associations. The rates are derived from IUE observations of both ground-level resonance lines and lines from excited levels which show evidence for mass loss: C IV, N IV, N V, O IV, and O V. Not enough ions are observed in the spectral region covered by IUE to warrant the detailed calculation of the ionization balance of many elements; therefore, an alternative method is developed for determining mass loss rates. Two-thirds of the stars analyzed here are unevolved main-sequence stars, a group whose mass loss rates cannot be determined reliably by other techniques. There are indications that the problem of understanding the ionization balance in these stars may be due to the underestimation of line blanketing in the far-ultraviolet flux computed by standard model atmospheres.

064.081 Stellar chromospheres and coronae.
R. G. Athay.
Astrophys. J., Vol. 250, 709 - 718 (1981).

The basic structural features of the solar chromosphere and corona are identified either with physical properties of the plasma or with properties of the energy input. It is argued that those properties identified with plasma physics should be relatively invariant whereas those identified with the energy input should vary from star to star. Energy input (ergs $cm^{-3} s^{-1}$) to the solar atmosphere has at least two maxima, one in the low chromosphere and a second in the corona. It is assumed that the same is true of other stars.

064.082 Thick supercritical accretion disk as a model for SS 433. M. Calvani, L. Nobili.
Vistas Astron., Vol. 25, (see 012.040), 173 - 176 (1981).

064.083 Remarks on accretion-disk models of SS 433.
W. R. Stoeger.
Vistas Astron., Vol. 25, (see 012.040), 177 - 184 (1981).

The author discusses the different types of accretion-disk models which may be relevant to SS 433, especially in terms of their relation to the regions emitting the strongly red and blue shifted $H\alpha$ and $H\beta$ lines so peculiarly characteristic of this object. He describes in a qualitative way the various thick supercritical disks which may be connected with the formation of relativistic jets in SS 433.

064.084 Outer atmospheres of cool stars. VIII. IUE observations and chromospheric models for the supergiant stars β Draconis, ϵ Geminorum, and α Orionis.
G. S. Basri, J. L. Linsky, K. Eriksson.
Astrophys. J., 251, 162 - 180 (1981).

The authors extend their program of semiempirical modeling of stellar chromospheres to a previously unstudied portion of the H-R diagram – the late-type supergiants. These models were computed to match high-resolution absolute flux profiles of the Ca II K and Mg II h and k lines. In IUE ultraviolet spectra of ϵ Gem and α Ori the authors find no evidence for emission lines formed at temperatures hotter than $\sim 10^4$ K, and on this basis they compute chromospheric models which extend to $m = 10^{-6}$ g cm^{-2} at temperatures rising to 6500 K and 7000 K, respectively. Upper limits on the surface flux of the C IV $\lambda 1549$ emission feature in ϵ Gem are 0.1 that of the quiet Sun, and in α Ori the upper limits are 0.002 that of the quiet Sun, providing upper limits on the amount of 10^5 K plasma in these stars. By contrast, β Dra shows strong emission lines of C II–IV, Si IV, He II, and N V. The authors tentatively extend the β Dra chromospheric model up to 16,000 K at $P_0 = 2n_e kT = 0.012$ dynes cm^{-2}. However, density-sensitive line ratios suggest $P_0 = 0.3$ dynes cm^{-2} at 60,000 K, and the authors discuss possible explanations for the discrepancy.

064.085 Model for the circumstellar gas around α Orionis.
M. Jura, M. Morris.
Astrophys. J., Vol. 251, 181 - 189 (1981).

The authors present new models for the physical conditions in the outer circumstellar shell of the well-studied red supergiant α Ori. By modeling both optical (K I) and radio (CO) observations, the authors derive much more precise results than are possible by considering either of the alone. Good agreement is found between the models and the observations if the CO and potassium loss rates are $7 \times 10^{39} s^{-1}$ and $8 \times 10^{37} s^{-1}$. Assuming that most of the ejected carbon is in the form of CO, the authors find that the potassium to carbon ratio in α Ori is greater by about a factor of 25 than the solar value. This result is discussed in detail.

064.086 Photospheric flow and stellar winds.
E. N. Parker.
Astrophys. J., Vol. 251, 266 - 270 (1981).

The dwarf main-sequence stars, such as the sun, have strongly bound coronas whose extended temperatures cause them to expand and to form tenuous stellar winds. There is a slight mean upward motion of the gas in the photosphere of the star to replenish the mass lost through the coronal expansion. Formal solution of the time-dependent hydrodynamic equations provides an illustration of the effect of the photospheric velocity on the corona and on the wind at large distances. Using the sun as an example, the effect of blocking the upward flow in the photosphere is calculated. The solar wind velocity would be reduced by a few cm s^{-1}, and the solar wind density would decline about 1% over the next three centuries. It appears that the photospheric flow has no directly sensible effects on the wind.

064.087 Mass loss from O subdwarfs. W.-R. Hamann, J. Gruschinske, R. P. Kudritzki, K. P. Simon.
Astron. Astrophys., Vol. 104, 249 - 255 (1981).

Four subluminous O-stars were observed in the UV with IUE. P Cygni profiles of the N V resonance doublet indicate mass loss from the two more luminous sample stars, while the N V lines of the other two stars and the C IV doublets of all four stars are photospheric and fit with non-LTE calculations. Lower limits of mass loss for HD 49798 and HD 128220 B are derived, and upper limits for all four objects are estimated. The mass loss rates agree reasonably with the predictions of the fluctuation theory.

064.088 Accretion in two phases. S. A. Glasner.
Proceedings of the 5th Göttingen-Jerusalem-Symposium on Astrophysics, (see 012.041), p. 181 - 188 (1981).

064.089 Observable properties of accretion disks.
S. Hayakawa.
Plasma astrophysics, (see 012.042), p. 325 - 327 (1981).

The accretion of matter onto stars and galaxies is regarded to play important roles in a number of phenomena which require a vast amount of energy release. The matter can be accreted onto a star from interstellar matter through which the star moves slowly, known as the Bondi accretion, or from a companion star of a binary system through its stellar wind or Roche lobe overflow. These accretion processes have been investigated extensively in connection with X-ray and γ-ray sources, as the gravitational energy released by accretion may power the emission of X-rays and γ-rays.

064.090 Bondi accretion. D. Summers.
Plasma astrophysics, (see 012.042), p. 329 - 330 (1981).

The Bondi accretion model (1952) is analysed, and it is pointed out that the model is still not well understood. While the Bondi (critical) accretion rate is widely quoted its conditions of validity are in doubt. Various research avenues are suggested, including a reexamination of the stability of the Bondi (critical) solution, the shocked Bondi solutions, and the wholly subsonic solutions.

064.091 Formation of an extended envelope surrounding a pulsating star.
A. V. Tutukov, Yu. A. Fadeev.
Nauchn. Inf., Vyp. (No.) 49, (see 003.017), p. 48 - 63 (1981).
In Russian.

Pulsations of FG Sagittae, which is the core of a planetary nebula, are investigated by means of radiative hydrodynamics numerical methods. Initial model parameters: mass $M = 1 M_\odot$, luminosity $L = 3200 L_\odot$, effective temperature $T_{eff} = 5500$ K. Calculations show the formation of an extended optically thin envelope during amplitude growth. The interaction between the pulsating star and its extended envelope causes the periodical rise of a powerful shock wave which causes mass loss from the envelope.

064.092 On the acoustic and magnetoacoustic heating of the outer atmosphere of stars. P. Ulmschneider.
Space Sci. Rev., Vol. 29, (see 012.043), 355 - 356 (1981).
Observational and theoretical evidence indicates that chromospheres are very likely heated by acoustic and by slow mode magnetohydrodynamic waves. The acoustic heating of coronae which has recently been disclaimed still appears to be an important possibility.

064.093 Influence of radiative scattering on the continuous spectra of late-type stars. J.-I. Straume.
Investigations of the sun and red stars. 12, (see 003.022), p. 31 - 44 (1981). In Russian.
By Sobolev's (1975) method the continuous stellar spectrum is calculated for a two-level model stellar atmosphere. With the obtained results the energy distribution in the continuous spectra of late-type stars was calculated for effective temperature T_e = 2500, 3000, 3500 and 4000 K.

064.094 On molecular dissociative equilibria in late-type stellar atmospheres. J.-I. Straume.
Investigations of the sun and red stars. 13, (see 003.023), p. 56 - 76 (1981). In Russian.
On the basis of equations of dissociative equilibrium, partial pressures of the most abundant 14 molecules in late-type stellar atmospheres have been calculated. Calculations were made for effective temperatures T_e = 2500, 3000, 4500 K, log g = 0, 2, 4, and T_e = 3500, 4000 K, log g = 1, 3, 5 for solar chemical composition.

064.095 On laws of limb darkening of stars. M. I. Lavrov.
Tr. Kazan. Gorod. Astron. Obs., Vyp. 45, p. 40 - 57 (1980). In Russian.

064.096 Emission line profiles in nonstationary objects (Influence of the combined effect of rotation and expansion of the envelope on resolved line profiles).
L. S. Nazarova.
Tr. Kazan. Gorod. Astron. Obs., Vyp. 46, p. 54 - 65 (1980). In Russian.

064.097 On some regularities of spectral line formation in stellar atmospheres. A. A. Sapar.
Use of model atmospheres for the interpretation of stellar spectra, (see 012.054), p. 74 - 76 (1980). In Russian.

064.098 Formulae for temperature and electron density in stellar atmospheres. A. A. Sapar, L. Yu. Sapar.
Use of model atmospheres for the interpretation of stellar spectra, (see 012.054), p. 76 - 80 (1980). In Russian.

064.099 X-ray irradiated accretion disk. S. Hayakawa.
Publ. Astron. Soc. Japan, Vol. 33, 365 - 372 (1981).
The present paper describes the structure of the accretion disk irradiated by X-rays from the central star and discusses its relevance to the optical emission from X-ray sources. General properties of the accretion disk are presented. It is shown how the disk blows up to form a wall. Optical emission from the X-ray irradiated disk is discussed.

064.100 On the effective temperature of a shock wave moving in a stellar envelope.
I. A. Klimishin, O. A. Kostyuk, I. D. Lysko.
Astron. Tsirk., No. 1149, p. 2 - 4 (1981). In Russian.

Radiation transfer and stellar atmospheres.
See Abstr. 003.148.

Hot and noisy stars. See Abstr. 011.012.

Conference on stellar atmospheres, June 1981, Brno.
See Abstr. 011.036.

Digital method for solving equations of the state of matter of a stellar atmosphere. See Abstr. 021.055.

Partition functions for uranium at elevated temperature. See Abstr. 022.054.

The Spacelab Lyman alpha and white light coronagraphs program. See Abstr. 032.559.

Neutrino and photon emission from a dense, high temperature atmosphere. See Abstr. 061.051.

Effects of mass transfer, free-convection currents and heat sources on the Stokes' problem for an infinite vertical plate. See Abstr. 062.046.

Wave generation. See Abstr. 062.049.

MHD instabilities of atmospheres with magnetic fields. See Abstr. 062.076.

Diffusion models for magnetic Ap−Bp stars.
See Abstr. 062.087.

Higher order fluid equations for multicomponent nonequilibrium stellar (plasma) atmospheres and star clusters. II. Effects of nonzero relative flow velocities and skewing of velocity distribution functions. See Abstr. 062.091.

On the global structure of stellar magnetospheres with stellar winds. See Abstr. 062.125.

On the importance of convective transport of excited atoms in stellar atmospheres. See Abstr. 063.013.

Solutions to radiative transfer problems using approximate lambda operators. See Abstr. 063.025.

The radiation field in an inhomogeneous atmosphere.
See Abstr. 063.026.

Radiation transfer in circumstellar envelopes with spherical symmetry − Henyey's method of solution.
See Abstr. 063.044.

Evolution of massive stars with mass loss and formation of WR stars. See Abstr. 065.008.

Violent core helium flashes and mass loss.
See Abstr. 065.016.

Red giants and their descendants − new light on old stars. See Abstr. 065.021.

Stellar variability and individuality: observations and implications. See Abstr. 065.022.

Mass loss and evolution of massive stars.
See Abstr. 065.023.

Evolution of a 30 $M_\odot$ star: the interplay of nuclear burning and mass loss. See Abstr. 065.024.

On the significance of mass loss for the evolution of massive stars. See Abstr. 065.025.

The influence of mass loss by stellar wind on the evolution of massive helium burning stars.
See Abstr. 065.026.

Stellar evolution with SMC chemical abundances.
See Abstr. 065.027.

Mass loss from metal-poor stars.
See Abstr. 065.031.

Evolutionary effects of mass loss in low-mass stars.   See Abstr. 065.032.

Miras, mass loss, and the origin of planetary nebulae.   See Abstr. 065.033.

On the consequences of mass loss from intermediate-mass stars.   See Abstr. 065.034.

Effects of stellar evolution.
See Abstr. 065.046.

Galactic ring nebulae associated with Wolf-Rayet stars. I. Introduction and classification.
See Abstr. 065.049.

Review of the theories on SS 433.
See Abstr. 065.078.

Meridional circulation in radiative zones of hot stars.
See Abstr. 065.097.

Effect of accretion on the evolution of very-low-mass young objects.   See Abstr. 065.101.

Supercritical, steady-state, spherically symmetric accretion into a black hole.
See Abstr. 066.001.

An accreting black hole model for SS433.
See Abstr. 066.135.

SS 433: background for a relativistic model.
See Abstr. 066.153.

The optical continuum of solar and stellar flares.
See Abstr. 073.019.

Chromospheric heating.   See Abstr. 073.040.

Mechanical heating in the transition region.
See Abstr. 073.041.

Coronal heating.   See Abstr. 074.039.

Coronal loops in the sun and in the stars.
See Abstr. 074.083.

The sun as a star.   See Abstr. 080.023.

New insight into the physics of atmospheres of early-type stars.   See Abstr. 112.010.

The dependence of mass loss on the basic stellar parameters.   See Abstr. 112.021.

Mass loss from cool stars.
See Abstr. 112.034.

Line formation in the wind of Alpha Cygni.
See Abstr. 112.038.

Stellar winds and mass-loss rates from Be stars.
See Abstr. 112.045.

High sensitivity molecular line observations of IRC+10216.   See Abstr. 112.052.

Outer atmospheres of cool stars. IX. A survey of ultraviolet emission from F−K dwarfs and giants with *IUE*.
See Abstr. 114.002.

The chromosphere and corona of Procyon (α CMi, F5 IV - V).   See Abstr. 114.004.

A model for V 1016 Cyg based on the ultraviolet spectrum.   See Abstr. 114.006.

IUE observations of eight OB stars in NGC 2244: ultraviolet continua and extinction.
See Abstr. 114.015.

Carbon, nitrogen, and oxygen abundances in G and K giants.   See Abstr. 114.016.

UV observations of the intermediate helium star $CPD-46°3093$.   See Abstr. 114.020.

Circumstellar shells of luminous supergiants. I. Carbon monoxide in Rho Cassiopeiae and HR 8752.
See Abstr. 114.021.

The chemical composition, gravity, and temperature of Sirius.   See Abstr. 114.025.

The Mg II h and k lines in a sample of dMe and dM stars.   See Abstr. 114.035.

IUE spectra of F and late A stars.
See Abstr. 114.036.

The chromospheric and transition layer emission of stars with different metal abundances.
See Abstr. 114.037.

UV chromospheric and circumstellar diagnostic features among F supergiant stars.   See Abstr. 114.038.

High resolution absolute flux profiles of the Mg II h & k lines in evolved F8 to M5 stars.
See Abstr. 114.039.

IUE − ultraviolet and optical chromospheric studies of late-type giants in the Hyades cluster.
See Abstr. 114.040.

The unusual outer atmosphere of 56 Pegasi (K0 IIp).   See Abstr. 114.054.

Narrow components in UV line profiles as evidence for a two component stellar wind for O and B stars.
See Abstr. 114.085.

The effective temperature scale.
See Abstr. 114.106.

Transition region structure in F dwarfs.
See Abstr. 114.162.

Chemical composition of Procyon. Model analysis.
See Abstr. 114.193.

A contribution to the determination of the mass of Arcturus.   See Abstr. 115.003.

The radii of the Wolf-Rayet stars and the extent of their chromosphere-corona formation.   See Abstr. 115.010.

Stellar rotation in lower main-sequence stars measured from time variations in H and K emission-line fluxes. I. Initial results.   See Abstr. 116.029.

Interacting stellar winds in a binary system. See Abstr. 117.048.

SS 433, X-ray binaries and stellar evolution. See Abstr. 117.089.

SS 433 may not be as peculiar as it looks. See Abstr. 117.091.

A theoretical review of SS 433. See Abstr. 117.093.

Equilibrium figures for Beta Lyrae type disks. See Abstr. 117.094.

Accretion disks in close binary systems. See Abstr. 117.106.

The visual surface brightness relation and the absolute magnitudes of RR Lyrae stars. I. Theory. See Abstr. 122.099.

A dynamical model of type I supernova atmosphere with the velocity gradient.   See Abstr. 125.018.

The interaction of the radiation from a type II supernova with a circumstellar shell.   See Abstr. 125.053.

Electron scattering in the atmospheres of hot DA white dwarfs.   See Abstr. 126.001.

Determination of atmospheric parameters for DB white dwarfs.   See Abstr. 126.023.

'High entropy' accretion onto the surface of a white dwarf.   See Abstr. 126.029.

The evolution of protostars. III. The accretion envelope.   See Abstr. 131.026.

The dynamical effects of hypersonic stellar winds on interstellar gas.   See Abstr. 131.176.

Formation of a planetary nebula by continuous mass loss.   See Abstr. 135.033.

Effects of mass loss on the formation of planetary nebulae.   See Abstr. 135.034.

On stellar wind accretion in widely separated X-ray binaries, and the nature of 4U0115+63. See Abstr. 142.049.

Spectra of accreting X-ray pulsars. See Abstr. 142.157.

A model of the rapid burster. See Abstr. 142.158.

Optical bursts from X-ray irradiated accretion disk. See Abstr. 142.159.

Possible links between supersonic stellar winds and the origin of cosmic rays.   See Abstr. 143.032.

Cosmic ray acceleration by stellar winds and self-confinement in giant H II regions.   See Abstr. 143.080.

Stellar mass loss and galactic chemical evolution. See Abstr. 155.022.

The evolution of flows of stellar mass loss in active galaxies.   See Abstr. 158.200.

065 Stellar Structure and Evolution

065.001 An iterative theory of modal selection.
N. R. Simon.
Astrophys. J., Vol. 247, 594 - 606 (1981).

The iterative pulsation theory is employed to extend the pulsational work integral to fourth order in the amplitudes for the case of two simultaneously energized models. Explicit expressions are given for the extended integral. A geometric analysis is then performed. A brief discussion of transition lines and their relationship to the problem of double-mode pulsation closes the investigation.

065.002 The violent phase of the core helium flash.
P. W. Cole, R. G. Deupree.
Astrophys. J., Vol. 247, 607 - 613 (1981).

The calculations described previously by Cole and Deupree were continued until the core helium flash reached a violent phase. The authors find a deflagration wave, not a detonation wave as previous investigators of violent core flashes have found. They attribute this difference primarily to the nonspherical nature of the burning in our two-dimensional code. Velocities of core material reach 2000 km s^{-1}, well below either the sound speed or the escape speed. It is highly likely that some mass loss will occur.

065.003 The evolution of intermediate-mass stars from the zero-age main sequence to the base of the asymptotic giant branch as a function of mass and composition.
S. A. Becker.
Astrophys. J., Suppl. Ser., Vol. 45, 475 - 505 (1981).

Detailed numerical tabulations are provided for 20 models of intermediate mass whose evolution extends form the zero-age main sequence (ZAMS) to the base of the asymptotic giant branch. The models span the ranges of $0.02 \leq Y \leq 0.36$, $0.001 \leq Z \leq 0.03$, and $3 M_\odot \leq M_* \leq 11 M_\odot$. A total of seven different compositions are investigated with two to four different model masses considered for each composition. Quantities tabulated are the model age, effective surface temperature, luminosity, radius, surface gravity, central composition, size of the convective core, the location of the base of the convective envelope, and the center of energy production of the hydrogen- and helium-burning shells. The results show that for a given model mass the evolutionary behavior depends strongly on its initial composition.

065.004 Convective overshooting in the evolution of very massive stars. R. Stothers, C.-w. Chin.
Astrophys. J., Vol. 247, 1063 - 1069 (1981).

Two important instances of possible convective overshooting in stars of 30 - 120 $M_\odot$ are considered here: (1) the possible merger between the convective core and the fully convective intermediate zone in the envelope at the end of the main phase of core hydrogen burning and (2) the possible penetration by the outer convection zone into the hydrogen-shell region when the star is a red supergiant.

065.005 Instability and evolution of stars.
L. V. Mirzoyan.
Izv. Akad. Nauk Armyansk. SSR, Fiz., Tom 16, 126 - 133 (1981). In Russian. – Paper presented on the scientific session of the Department of Phys.-Math. Sciences of the Armenian Academy of Sciences, 1981, Jan. 26 - 29.

065.006 The thick helium-burning shell phase and the second blue loop. S. A. Becker.
Astrophys. J., Vol. 248, 298 - 310 (1981).

The evolutionary behavior of models undergoing the thick helium-burning shell phase is explored in detail as a function of mass and composition with a full quasi-static evolutionary code. Eighteen models are studied which span the ranges $3 M_\odot \leq M_* \leq 11 M_\odot$, $0.20 \leq Y \leq 0.36$, and $0.001 \leq Z \leq 0.03$. Reasons for the occurrence of the second blue loop are discussed. Finally, tentative confirmation of the existence of the second blue loop is provided by the behavior of four Cepheids showing large period decreases.

065.007 On the entropy of mixing, with particular reference to its effect on dredge-up during helium shell flashes.
P. R. Wood.
Astrophys. J., Vol. 248, 311 - 314 (1981).

It is shown that there is no energy absorption associated with the advance of a convective zone into a region containing material of a different mean molecular weight. Stellar evolution codes used in computations of dredge-up during helium shell flashes should take this fact into account. A scheme for the treatment of mixing and gravitational-internal energy production in a stellar evolution code is suggested.

065.008 Evolution of massive stars with mass loss and formation of WR stars. A.Noels, M. Gabriel.
Astron. Astrophys., Vol. 101, 215 - 222 (1981).

Evolution with mass loss of stars with initial masses 60, 80 and 100 $M_\odot$ is computed up to the end of core He-burning. The authors have investigated the plausibility of a scheme of forming massive WR stars. In this picture, WR stars have O stars of masses between 60 $M_\odot$ and 100 $M_\odot$ as progenitors. However, it seems difficult to obtain at the same time a WR/O ratio and an average mass for WR stars compatible with the more recent observations.

065.009 On stellar collapse: continual or oscillatory? A short comment. P. T. Leung.
Ann. Inst. Henri Poincaré, Sect. A, Vol. 33, 205 - 208 (1980). Abstr. in Phys. Abstr., Vol. 84, Abstr. 66923 (1981).

065.010 Shear instability of differential rotation in stars.
M. Watson.
Geophys. Astrophys. Fluid Dyn., Vol. 16, 285 - 298 (1981). Abstr. in Phys. Abstr., Vol. 84, Abstr. 66931 (1981).

065.011 The origin of supernovae.
J. C. Wheeler.
Rep. Prog. Phys., Vol. 44, No. 2, p. 85 - 138 (1981). – Abstr. in Phys. Abstr., Vol. 84, Abstr. 71223 (1981).

065.012 On stabilizing effects of relativity in cold spheric stars with a phase transition in the interior.
B. Kampfer.
Phys. Lett. B, Vol. 101B, 366 - 368 (1981). – Abstr. in Phys. Abstr., Vol. 84, Abstr. 79786 (1981).

065.013 Numerical computation of certain three-dimensional stellar structures using a semidiscrete pseudospectral method. M. J. Miketinac, S. V. Parter.
Z. Angew. Math. Phys., Vol. 32, 204 - 228 (1981). – Abstr. in Phys. Abstr., Vol. 84, Abstr. 79815 (1981).

065.014 Thermonuclear plasma conditions in stellar interiors.
R. J. Tayler.
Philos. Trans. R. Soc. London, Ser. A, Vol. 300, 641 - 648 (1981). – Abstr. in Phys. Abstr., Vol. 84, Abstr. 83613 (1981).

065.015 Highly-evolved stars. S. R. Heap.
The Universe at ultraviolet wavelengths, (see 012.009), p. 415 - 434 (1981).

065.016 Violent core helium flashes and mass loss.
R. G. Deupree.
Bull. American Astron. Soc., Vol. 13, 527 (1981). – Abstract.

065.017 Rotational modes in a slowly and uniformly rotating star.
P. Smeyers, D. Craeynest, L. Martens.
Astrophys. Space Sci., Vol. 78, 483 - 501 (1981).
A perturbation method is derived for r-modes in a slowly and uniformly rotating star. In contrast to previous studies, the perturbation of the gravitational potential is included in the perturbation method.

065.018 Analytic solutions of the radial pulsation equation for rotating and magnetic star models.
R. Stothers.
Mon. Not. R. Astron. Soc., Vol. 197, 351 - 361 (1981).
The wave equation describing small radial perturbations of spherically symmetric, gaseous stars has been generalized to include the effects of axial rotation and of tangled magnetic fields. Solutions in closed form have been obtained for the adiabatic pulsation periods of five analytic stellar models in two special cases, which have considerable astrophysical interest. Non-adiabatic stability criteria have been determined by means of the one-zone stellar model. Applications are made to the case of classical cepheids and other variable giant stars.

065.019 A study of simple polytropes. I. Fundamentals and classification of solutions. H. Kimura.
Publ. Astron. Soc. Japan, Vol. 33, 273 - 298 (1981).
The structure of a "simple" polytrope is considered, which the author defines as a (1) non-relativistic, (2) non-rotating, (3) self-gravitating m-dimensional sphere ($m = 1, 2,$ or 3), (4) being in a hydrostatic equilibrium, and (5) characterized by the single polytropic law. For a given set of system parameters n (polytropic index) and m, one finds three or more types of solutions to occur depending on the different conditions at the inner boundary. A classification of solutions by a topological method is presented and general features of the "simple polytrope" sequence are clarified.

065.020 A study of simple polytropes. II. Stability against homologous contraction. H. Kimura.
Publ. Astron. Soc. Japan, Vol. 33, 299 - 312 (1981).
The stability of simple polytropes against homologous contraction is discussed in terms of the state-plane geometry and the criterion for instability is obtained in a simpler form than those previously available; due to this simplification it becomes possible to examine, thoroughly and easily, the stability of various types of configurations. Some applications to the interstellar problem are discussed briefly.

065.021 Red giants and their descendants – new light on old stars. B. Zuckerman.
Nature, Vol. 293, 513 - 514 (1981).

065.022 Stellar variability and individuality: observations and implications. R. Costero, V. Doazan, R. Stalio, R. N. Thomas.
Effects of mass loss on stellar evolution, (see 012.015), p. 131 - 153 (1981).
The phenomena of large-amplitude variability in times short compared with evolutionary ones, and of individuality (two stars of the same taxonomic class having different atmospheric distributions of T_e and density) invalidate static evolutionary calculations. The authors summarize the evidence for such large-amplitude variability in Be and Ia supergiant B stars. They also summarize the evidence for individuality as exhibited by observations: of OVI in OB stars; of X-ray luminosity across the HR diagram; of far-UV spectra of O stars; and of visual and far-UV spectra of Be stars. These observational results require nonthermal fluxes of mass and nonradiative

energy to be imposed from below by the subatmosphere; which implies a nonthermal structure of subatmosphere and at least some part of the interior. Such nonthermal structure must then be included in evolutionary calculations.

065.023 Mass loss and evolution of massive stars.
C. Chiosi.
Effects of mass loss on stellar evolution, (see 012.015), p. 229 - 253 (1981).
In the past few years both growing observational evidence and theoretical understanding have shown that mass loss by stellar wind is a common occurrence in the evolutionary history of many types of stars. The author concentrates on those observational and theoretical aspects of the problem that demand further investigation. He is concerned with stars in the approximate range of initial mass 10 $M_\odot$ to 100 $M_\odot$, during their evolution from central H-burning to later phases.

065.024 Evolution of a 30 $M_\odot$ star: the interplay of nuclear burning and mass loss.
H. J. Falk, R. Mitalas.
Effects of mass loss on stellar evolution, (see 012.015), p. 261 - 263 (1981).
Evolutionary tracks for a 30 $M_\odot$ star with various mass loss rates (MLR) were evolved to core He exhaustion. The "overluminosity" of mass losing stars is explained in terms of the well known mass-luminosity law. A critical ZAMS MLR above which mass loss leads to evolution to fainter luminosities is derived. Two tracks showed reversals in their direction of evolution across the HR diagram. These have been shown to be a consequence of mass loss dominating over the effects of the shell source. An analytic criterion for this condition has been derived.

065.025 On the significance of mass loss for the evolution of massive stars. L. R. Yungelson (*Yungel'son*), A. G. Massevitch (*Masevich*), A. V. Tutukov.
Effects of mass loss on stellar evolution, (see 012.015), p. 265 - 270 (1981).
It is shown that mass loss by stellar wind with rates observed in O, B-stars cannot change qualitatively their evolution in the core hydrogen-burning stage. The effects, that are usually attributed to the mass loss, can be explained by other causes: e. g., duplicity or enlarged chemically homogeneous stellar cores.

065.026 The influence of mass loss by stellar wind on the evolution of massive helium burning stars.
D. Vanbeveren.
Effects of mass loss on stellar evolution, (see 012.015), p. 275 - 278 (1981).
Helium burning stars with masses between 10 $M_\odot$ and 40 $M_\odot$ are evolved up to core helium exhaustion including mass loss by stellar wind at rates between 10^{-5} $M_\odot$/yr and 10^{-4} $M_\odot$/yr appropriate for WR stars. Different $\dot{M}$ formalisms were used.

065.027 Stellar evolution with SMC chemical abundances.
P. Hellings, D. Vanbeveren.
Effects of mass loss on stellar evolution, (see 012.015), p. 279 - 282 (1981).
Evolutionary computations are presented for massive stars between 20 $M_\odot$ and 100 $M_\odot$ with chemical abundances holding for the Small Magellanic Cloud, i.e. X = 0.76 and Z = 0.003. Mass loss by stellar wind is taken into account during core hydrogen burning. After core hydrogen burning some models are considered as members of close binary systems and are followed during their Roche lobe overflow stage according to an early case B of mass transfer. During the core helium burning stage of the RLOF remnants mass loss rates comparable to WR stars are included in order to study the

formation and the evolution of WR stars. Comparison with similar galactic computations is made.

065.028 **Massive stars burning helium: the numbers of WR stars and red supergiants in galaxies.**
A. Maeder.
Effects of mass loss on stellar evolution, (see 012.015), p. 283 - 287 (1981).

065.029 **Thermal instability of hydrogen burning shells in very massive stars.**
Y. Tanaka, N. Arimoto, M. Takeuti.
Effects of mass loss on stellar evolution, (see 012.015), p. 289 - 292 (1981).

065.030 **Effects of a stochastic initial mass function on the upper main sequence band.**
C. Chiosi, L. Greggio.
Effects of mass loss on stellar evolution, (see 012.015), p. 293 - 296 (1981).

065.031 **Mass loss from metal-poor stars.**
C. Chiosi, G. Bertelli, E. Nasi, L. Greggio.
Effects of mass loss on stellar evolution, (see 012.015), p. 297 - 300 (1981).
It is essential to consider the effect of mass loss to understand the distribution of supergiant stars in the HR diagram. This research concerns the evolution of massive stars with $X = 0.700$ and $Z = 0.001$ during the phases up to central He-exhaustion with the inclusion of mass loss. Such low value of Z has been chosen in order to allow a comparison with the supergiant stars of SMC.

065.032 **Evolutionary effects of mass loss in low-mass stars.**
A. Renzini.
Effects of mass loss on stellar evolution, (see 012.015), p. 319 - 338 (1981).
The effects of mass loss on the evolution of low-mass stars (actual mass smaller than 1.4 $M_\odot$) are reviewed. The case of globular cluster stars is discussed in some detail, and it is shown that evolutionary theory sets quite precise limits to the mass-loss rate in population II red giants. The effect of mass loss on the final evolutionary stages of stars producing white dwarfs is also discussed. In particular, the interaction of the wind from the hot central star with the surrounding planetary nebula is considered. Finally, the problem of the origin of hydrogen-deficient stars is briefly discussed.

065.033 **Miras, mass loss, and the origin of planetary nebulae.** L. A. Willson.
Effects of mass loss on stellar evolution, (see 012.015), p. 353 - 356 (1981).

065.034 **On the consequences of mass loss from intermediate-mass stars.** I. Iben, Jr.
Effects of mass loss on stellar evolution, (see 012.015), p. 373 - 387 (1981).

065.035 **Self-similar motions of a self-gravitating gas in stars.**
O. I. Bogoyavlenskij.
Tr. semin. im. I. G. Petrovskogo. MGU, 1981, No. 6, p. 3 - 37. In Russian. – Abstr. in Ref. zh., 51. Astron., 8.51.159 (1981).

065.036 **La structure, l'énergie et l'évolution des étoiles.**
A. Maeder.
Histoire de l'univers, (see 003.007), p. 53 - 98 (1980).
Contents: La formation et l'évolution des protoétoiles. La structure et l'évolution nucléaire des étoiles. L'origine des éléments et la nucléosynthése stellaire. Supernovae et astres condensés: naines blanches, étoiles à neutrons, trous noirs.

065.037 **Luminosity variation due to neutrino angular momentum in stellar collapse.** N. Chaubey.
Bull. Astron. Soc. India, Vol. 9, 66 (1981). – Abstract.

065.038 **Massive configurations with constant proper and rest observed densities.**
M. C. Durgapal, A. K. Pande, K. Pandey.
Bull. Astron. Soc. India, Vol. 9, 79 (1981). – Abstract.

065.039 **Structural parameters of hot stars.**
M. K. Das, J. Kar, J. N. Tandon.
Bull. Astron. Soc. India, Vol. 9, 85 - 86 (1981). – Abstract.

065.040 **Aspetti evolutivi delle stelle massicce.**
C. Firmani, G. F. Bisiacchi.
Mem. Soc. Astron. Italiana, Vol. 52, (see 012.019), 49 - 56 (1981).

065.041 **The evolution of intermediate-mass stars.**
A. Renzini.
Mem. Soc. Astron. Italiana, Vol. 52, (see 012.019), 57 (1981). Summary.

065.042 **Evolution and nucleosynthesis in massive stars with mass loss: the yields in helium and heavy elements and constraints on the past star formation rate.** A. Maeder.
Astron. Astrophys., Vol. 101, 385 - 396 (1981).
Three grids of models of stellar evolution in the range $9-170\,M_\odot$ have been computed from the main-sequence to the end of the carbon-burning phase. One grid concerns evolution at constant mass, the other two including mass loss according to observational rates. The synthesis of helium and heavy elements is examined.

065.043 **The gravo-thermal catastrophe and stellar evolution.**
G. Barbaro, G. Bertelli, J. Perdang, L. Pigatto.
Astron. Astrophys., Vol. 102, 109 - 115 (1981).
The stability discussion of the classical isothermal gas sphere due to Lynden-Bell and Wood (1968) is extended to isothermal structures obeying a realistic stellar equation of state. The results of this analysis offer a theoretical framework for the interpretation of a series of numerical experiments on stellar evolution described in this paper, which were performed by suppressing the nuclear energy generation. A gravo-thermal catastrophe is found to show up in the later evolutionary stages, provided that the mass of the model exceeds the Chandrasekhar limit.

065.044 **The effect of differential rotation on the internal structure of low mass stars.** W.-Y. Law.
Astron. Astrophys., Vol. 102, 178 - 190 (1981).
Axisymmetric, differentially rotating main sequence models for stars of $0.9\,M_\odot$ and homogeneous composition ($X = 0.7999$, $Z = 0.0001$) have been constructed. The effects of rotation on the internal structure of the stellar models have been studied with different amounts and different distributions of angular momentum. The angular velocity distribution is assumed to be constant on surfaces of constant density.

065.045 **Nacimiento, vida y muerte de las estrellas.**
A. D. Neto, F. J. Jablonski.
R Muscae, Vol. 6, No. 1/2, p. 22 - 27 (1981).

065.046 **Effects of stellar evolution.**
C. de Loore.
Ann. Physique, Vol. 6, (see 012.026), p. 149 - 166 (1981). In French, English. – Abstr. in Phys. Abstr., Vol. 84, Abstr. 94316 (1981).

065.047 **Nitrogen in halo stars.**
B. Barbuy.
Ann. Physique, Vol. 6, (see 012.026), p. 121 - 126 (1981). In

French. – Abstr. in Phys. Abstr., Vol. 84, Abstr. 94331 (1981).

065.048   Evolving dust shells.
H. P. Siegel.
Phys. Rev. D, Vol. 23, 2835 - 2849 (1981). – Abstr. in Phys. Abstr., Vol. 84, Abstr. 94375 (1981).

065.049   Galactic ring nebulae associated with Wolf-Rayet stars. I. Introduction and classification.   Y.-H. Chu.
Astrophys. J., Vol. 249, 195 - 200 (1981).
   The author presents a new list of 15 galactic ring nebulae associated with Wolf-Rayet stars. The nebulae are selected so as to ensure that no stars other than WR stars are responsible for exciting them. For these nebulae he has obtained photographic and detailed kinematic data, which will be presented in subsequent papers of this series. The author uses these data to determine the formation mechanisms of the nebulae and classifies them accordingly.

065.050   Stellar core collapse. I. Infall epoch.
K. A. Van Riper, J. M. Lattimer.
Astrophys. J., Vol. 249, 270 - 289 (1981).
   The authors report on simulations of the collapse of the central iron core of a 15 $M_\odot$ spherically symmetric star. In this paper they consider the infall epoch, between the onset of collapse and core bounce. The models use the recent equation of state of Lamb, Lattimer, Pethick and Ravenhall and general-relativistic hydrodynamics.

065.051   Mass loss during the core helium flash.
R. G. Deupree, P. W. Cole.
Astrophys. J., Lett., Vol. 249, L35 - L38 (1981).
   The 2-D hydrodynamic calculations of the core helium flash by Cole and Deupree have been improved by the inclusion of a small nuclear reaction network, individual α capture reactions, and an expanding coordinate system. These nuclear modifications allow inclusion of neon photodisintegration, which limits the peak flash temperature to roughly 2 billion degrees, while the expanding coordinate system permits following the flash for about 50 s after the violent runaway begins.

065.052   Collapse of degenerate iron stellar cores and model of a supernova.
L. N. Ivanova, V. M. Chechetkin.
Astron. Zh., Tom 58, 1028 - 1042 (1981). In Russian. English translation in Soviet Astron., Vol. 25, No. 5.
   The collapse of iron stellar cores with envelopes of thermonuclear fuel due to neutronization of matter is considered. Depending on the temperature of matter of the envelope before the beginning of the hydrodynamic contraction, a compact remnant similar to a neutron star or to a white dwarf is obtained. It is shown that in the thermonuclear supernova model the maximum energy of the thrown-off shell resulting from a thermonuclear explosion is $(5 - 7) \times 10^{49}$ erg.

065.053   Circular polarization of electromagnetic radiation at stellar collapse.
S. A. Rozenshtejn, M. Yu. Khlopov.
Pis'ma Astron. Zh., Tom 7, 623 - 626 (1981). In Russian. English translation in Soviet Astron. Lett., Vol. 7.
   The interaction of neutrino radiation with matter in the outer layers of collapsing stars produces hard (X and γ) radiation. Owing to the parity nonconservation in neutrino interactions the electromagnetic radiation is circularly polarized.

065.054   Normal modes of oscillation for rotating stars. I. The effect of rigid rotation on four low-order pulsations.   M. J. Clement.
Astrophys. J., Vol. 249, 746 - 760 (1981).
   Some normal modes of rotating stars can be computed numerically without having to assume a slow angular velocity. This can be done by solving directly on a two-dimensional grid the linearized dynamical equations governing adiabatic oscillations. In this paper, the effects of rigid rotation on four axisymmetric modes are found for several equilibrium systems including polytropes and a 15 $M_\odot$ stellar model. It is shown that even a small rotation changes significantly the space dependence of the eigenfunctions.

065.055   The origin and structure of the fields of the upper main sequence magnetic stars.   D. Moss.
Upper main sequence chemically peculiar stars, (see 012.033), p. 309 - 319 (1981).
   After a brief survey of the observational evidence relating to the upper main sequence magnetic stars, an account is given of current theoretical problems. Particular attention is paid to the rival fossil and dynamo theories for the origin of the field. Some comments are made about evolutionary effects.

065.056   Does γ Equ have a magnetic cycle like the sun?
F. Krause, G. Scholz.
Upper main sequence chemically peculiar stars, (see 012.033), p. 323 - 331 (1981).

065.057   Hydromagnetic rotational deceleration of magnetic Ap stars.   R. C. Fleck, Jr.
Upper main sequence chemically peculiar stars, (see 012.033), p. 341 - 342 (1981).
   It is suggested that the magnetic Ap stars can be rotationally decelerated to long periods by the braking action of the associated magnetic field on time scales of order 10^7 to 10^{10} years depending on whether the star's dipole field is aligned perpendicular or parallel to the rotation axis. It is not necessary to postulate mass loss or mass accretion for this purely hydromagnetic braking effect.

065.058   Comments on theories of the chemically-peculiar stars of the upper main sequence.
W. K. Bonsack.
Upper main sequence chemically peculiar stars, (see 012.033), p. 345 - 354 (1981).

065.059   A parameter free model for HgMn stars.
G. Michaud.
Upper main sequence chemically peculiar stars, (see 012.033), p. 355 - 363 (1981).
   Detailed radiative accelerations have been calculated, over the last few years in the context of a parameter free model for HgMn stars. The abundance anomalies of He, B, Si, Ca, Mn and Sr are explained except that in 2 stars, boron is present whereas it is not expected to be. The model fails for Be. It is shown that for equatorial velocities below 90 km s^{-1}, the meridional circulation is slow enough to allow the settling of He and the disappearance of the He II convection zone. This acts as a triggering mechanism. Once the He II convection zone disappears, the overabundances appear with no more correlation expected with the rotational velocity. This is in agreement with observations. However the parameter free model can only explain the envelope of the anomalies. It cannot explain the variations in the anomalies from star to star.

065.060   Transfer equation of arbitrarily polarized radiation in the presence of a magnetic field.
G. Mathys.
Upper main sequence chemically peculiar stars, (see 012.033), p. 365 - 371 (1981).
   The authors present a new method of derivation of the line transfer equation of an arbitrarily polarized beam of radiation in presence of a magnetic field. It is based upon quantum electrodynamics and allows the study of non-local thermodynamic equilibrium.

065.061 **Diffusion in the presence of a magnetic field.**
C. Megessier, G. Michaud.
Upper main sequence chemically peculiar stars, (see 012.033), p. 397 - 401 (1981).

The authors show how diffusion in the presence of a magnetic field can lead to surface inhomogeneities similar to those deduced from the spectroscopic observations of Ap stars. Here they will present the principle of that effect and its application to the Ap star 53 Cam.

065.062 **Advanced evolutionary phase of a first-generation star.** D. Eryurt-Ezer.
Astrophys. Space Sci., Vol. 79, 265 - 287 (1981).

The theoretical evolution of a first-generation star of $3 M_\odot$ after the core helium-exhaustion phase has been investigated. The star displays the character of a double shell burning model. Prior to carbon-burning phase, the mass of the complete hydrogen-exhausted region is $1.14 M_\odot$ and that of complete helium-exhausted region is $0.83 M_\odot$. A carbon-oxygen core of about $0.87 M_\odot$ has developed within the star in which the ratio of carbon to oxygen is about 0.85, but decreases down to a value of 0.50 near the boundary of the core.

065.063 **Rotational evolution of solar-type stars.**
V. S. Geroyannis, G. A. Antonakopoulos.
Astrophys. Space Sci., Vol. 79, 367 - 378 (1981).

The aim of the present investigation has been to consider rotational evolution of solar-type stars simulated by a polytropic model that possesses differential rotation of Clement's type. The present treatment is based upon the general Eulerian equation, governing nonuniform (i.e., non-rigid- body) rotation, which has been set up in a previous investigation. Nonconservative terms, arising when stellar wind torque is under consideration, are taken into account. Data available for the viscosity of the Sun are used to construct a plausible viscosity model.

065.064 **On the core mass luminosity relation.**
R. Kippenhahn.
Astron. Astrophys., Vol. 102, 293 - 295 (1981).

The luminosity of a hydrogen burning shell surrounding a core of high density depends on the core mass only. Although they appear rather different at first glance, the two relations obtained by Refsdal and Weigert (1970) and Paczyński (1970) are two limiting cases of a unique relation.

065.065 **Helium diffusion in horizontal-branch-star evolutionary models.**
P. Giannone, L. Rossi.
Astron. Astrophys., Vol. 102, 386 - 390 (1981).

The authors studied the inward diffusion of helium in the atmosphere and subatmosphere of horizontal-branch stars in globular clusters. The process of gravitational diffusion was taken into account together with the stellar evolution due to the helium burning in the core and hydrogen burning in an intermediate shell. An improved procedure for the treatment of the induced semiconvection was also included in the models. The surface depletion of helium appears to be rather efficient and might explain the observed deficiency of helium in blue HB stars.

065.066 **Grids of evolutionary models for the upper part of the HR diagram. Mass loss and the turning of some red supergiants into WR stars.** A. Maeder.
Astron. Astrophys., Vol. 102, 401 - 410 (1981).

Results concerning the evolutionary tracks in the HR diagram and the associated lifetimes are given for three sets of models in the 9 - 120 $M_\odot$ range from the MS to the end of the carbon-burning phase: one set for constant mass evolution and the other two with mass loss based on observed rates. These grids provide an outlook of the effects of mass loss in the HR diagram up to presupernova models. The conditions of mass loss rates and luminosity for the turning of RSG (*red supergiants*) into WR stars are examined. For mass loss rates in the observed range, this turning may occur for initial $M \gtrsim 30 M_\odot$. The numerical models show that amazingly large changes of the ratio of the lifetimes t_R / t_B (red to blue) and above all of t_R / t_{WR} (red to WR stars) result from differences in mass loss. As an example, an increase of mass loss rates by a factor of 2 (in the MS and RSG stages) increases the average lifetime in the WR stage by a factor 17. This extreme sensitivity of the lifetimes to mass loss is likely responsible for the strong anticorrelation of RSG and WR stars in the Galaxy, the LMC and SMC.

065.067 **Late stages of stellar evolution: central stars of planetary nebulae.** D. Schönberner.
Astron. Astrophys., Vol. 103, 119 - 130 (1981).

The evolution of the central stars of planetary nebulae is studied by means of an observed correlation between their absolute visual magnitudes and the corresponding nebular radii as an indicator of their ages. The observed evolutionary behaviour can be well explained by post-AGB stars with typical masses near $0.6 M_\odot$, which burn hydrogen in a shell under quiet conditions. This implies that the formation of a nebula occurs during the quiescent hydrogen burning phase between successive thermal pulses and not when the star enters its peak luminosity during a thermal pulse.

065.068 **Computer simulations of stellar collapse and shock wave propagation.** W. Hillebrandt, E. Müller.
Astron. Astrophys., Vol. 103, 147 - 153 (1981).

Utilizing recent progress in deriving realistic equations of state of matter at high densities and temperatures, the authors compute the gravitational collapse of cores of massive stars through core-bounce at neutron star densities. In particular they analyze the sensitivity of the results (i.e. the question of whether or not the core-bounce gives rise to a supernova explosion of the stellar envelope) with respect to details of the equation of state, neutrino emissivities in the shock region and properties of the hydrocode. The authors find that in none of the cases considered is the core-bounce followed by an explosion of the stellar mantle. Although a shock always forms, it is never strong enough to accelerate matter beyond escape velocity. This result is independent of both the details of the equation of state and the assumptions about neutrino losses from the shocked matter.

065.069 **The collapse of rotating stellar cores.**
E. Müller, W. Hillebrandt.
Astron. Astrophys., Vol. 103, 358 - 366 (1981).

The collapse of the rotating cores of massive stars ($M \gtrsim 15 M_\odot$) is considered. The dynamical evolution of the core is followed by an explicit two-dimensional hydrodynamic code. A realistic, finite-temperature equation of state is used in the calculations. Rigidly and differentially rotating initial models are both studied. The authors find that in all cases considered the collapse proceeds up to nearly nuclear matter density before it is stopped. In none of the examined models is the core-bounce followed by an explosion. The dynamical behaviour of the slowly rotating models is very similar to that of non-rotating models. The rapidly rotating cores already show large non-spherical effects before core-bounce.

065.070 **Vibrational stability of first generation stars.**
A. Ibrahim, A. Boury, A. Noels.
Astron. Astrophys., Vol. 103, 390 - 392 (1981).

The vibrational stability towards radial oscillations of homogeneous main sequence stars initially composed of 80% of hydrogen and 20% of helium in mass and no metals is studied. During the pre-main sequence phase, enough carbon is accumulated to trigger the CN-cycle. The critical mass for vibrational stability is found to be $123 M_\odot$.

065.071   **Topics in nuclear astrophysics.**   D. Arnett.
Nuclear astrophysics, (see 012.036), p. 159 - 160
(1981).
The fundamental factors determining the evolution of
abundances of nuclei in stars were reviewed. The collapse of
the iron core was examined analytically, and the neutrino
trapping, entropy change and size of the homologous inner
core were derived. The behaviour of luminosity, effective
temperature and photospheric velocity were presented
analytically for exploding stars.

065.072   **Equation of state of hot, dense stellar matter: finite
temperature nuclear Thomas-Fermi approach.**
M. Barranco, J.-R. Buchler.
Phys. Rev. C, Vol. 24, 1191 - 1202 (1981). – Abstr. in Phys.
Abstr., Vol. 84, Abstr. 108197 (1981).

065.073   **Achieving population inversions in the lasing recom-
bination ring model of SS 433.**   W. R. Stroeger.
Nuovo Cimento, Lett., Ser. 2, Vol. 31, 501 - 504 (1981).
Abstr, in Phys. Abstr., Vol. 84, Abstr. 108235 (1981).

065.074   **Kelvin–Helmholtz instabilities and their application
to B-type variables.**   H. Ando.
Mon. Not. R. Astron. Soc., Vol. 197, 1139 - 1152 (1981).
A Kelvin–Helmholtz instability, formed from the
differential rotation in the narrow region between the core
and envelope, is proposed as a promising mechanism
responsible for the excitation of pulsations in B-type variables
(53 Per variables and β Cep stars), in which the unstable inertia
wave resulting from this instability resonates with an eigen-
mode of the non-radial oscillation of the whole star.

065.075   **Multimode stellar pulsations: a new approach.**
O. Regev, J. R. Buchler.
Astrophys. J., Vol. 250, 769 - 775 (1981).
A multi-time formalism is adapted to the problem of
multimode stellar pulsation with slow thermal evolution. The
original system of differential equations in time t of order $3N$,
where N is the number of zones, is reduced to a system of
order $N+p$ in the slow (thermal) time variable $\tau \equiv \epsilon t$, where p is
the number of simultaneously excited normal modes. In parti-
cular, the formalism is expected to be fruitful for the problems of modal selection, as well as for the search for stable
double-mode pulsators.

065.076   **The search for double-mode pulsator models.**
J. R. Buchler, O. Regev.
Astrophys. J., Vol. 250, 776 - 781 (1981).
A novel mathematical formalism is described, which
permits a search for and an analysis of multimode pulsators.
An application is made to a simple analytical model. On the
basis of these results arguments are given why a conventional
search for double-mode pulsational models is unlikely to be
successful.

065.077   **The volume and surface area of a uniformly rotating
polytrope.**   A. P. Linnell.
Astrophys. Space Sci., Vol. 80, 501 - 511 (1981).
This paper develops algebraic expressions for the volume
and surface area of a uniformly rotating polytrope. The expressions depend on an analytic theory for boundary shape
developed in previous papers. A comparison with the calculations of James indicates the present theory improves on the
original Chandrasekhar theory by a factor 10 or more.

065.078   **Review of the theories on SS 433.**
M. Milgrom.
Vistas Astron., Vol. 25, (see 012.040), 141 - 151 (1981).
The author discusses the various models proposed for ex-
plaining SS 433 with particular emphasis on the moving-lines
phenomenon and its 164^d periodicity. Observational constraints

on the individual models are outlined. It is concluded that observational evidence strongly suggests a double-outflowing-
beam picture. The problem of the central source and mechanisms of particle acceleration and confinement are reviewed.

065.079   **Evolution of massive population II stars with semi-
convection.**   D. S. P. Dearborn, V. Trimble.
Nukleonika, Vol. 25, 1441 - 1446 (1980). – Abstr. in Phys.
Abstr., Vol. 85, Abstr. 3254 (1982).

065.080   **Theoretical aspects of early type magnetic stars.**
M. Schüssler.
Nukleonika, Vol. 25, 1425 - 1440 (1980). – Abstr. in Phys.
Abstr., Vol. 85, Abstr. 3261 (1982).

065.081   **Statistical theory of non-local convection in
chemically inhomogeneous stars.**   D. Xiong.
Sci. Sinica, Vol. 24, 1406 - 1417 (1981).
The present paper investigates the non-local convection in
spherically symmetric stars. The rotation and magnetic field
are not considered.

065.082   **Low-mass evolution: zero-age main sequence to
asymptotic giant branch.**   K. H. Despain.
Astrophys. J., Vol. 251, 639 - 653 (1981).
The evolution of a $0.6\,M_\odot$ stellar model ($X=0.9, Z=10^{-3}$)
from the zero-age main sequence (ZAMS) to the asymptotic
giant branch (AGB) is described. The red giant evolution has
been done without artificial shifting of the hydrogen-burning
shell. The core flash and a sequence of 12 subsequent shell
flashes have been followed until the star reaches the horizontal
branch (HB). The HB and post-HB evolution is followed
through the first major shell on the AGB. An approximate
stability criterion, suitable for both core and shell flashes, is
also presented.

065.083   **Helium shell burning in intermediate mass stars.**
Z. Barkat.
Proceedings of the 5th Göttingen-Jerusalem-Symposium on
Astrophysics, (see 012.041), p. 167 - 172 (1981).

065.084   **Comments on the equation of state of supernova
matter.**   M. F. El Eid.
Proceedings of the 5th Göttingen-Jerusalem-Symposium on
Astrophysics, (see 012.041), p. 235 - 244 (1981).

065.085   **Hydrodynamic collapse calculations for nonrotating
and rotating supernova models.**   W. Hillebrandt.
Proceedings of the 5th Göttingen-Jerusalem-Symposium on
Astrophysics, (see 012.041), p. 245 - 255 (1981).

065.086   **On the collapse instability of rapidly rotating pre-
supernova cores having finite entropy.**
W. Glatzel.
Proceedings of the 5th Göttingen-Jerusalem-Symposium on
Astrophysics, (see 012.041), p. 261 - 269 (1981).
The author establishes the marginally unstable configurations according to the present knowledge on the thermodynamic structure of highly evolved stellar cores assuming Maclaurin
spheroids for the description of the rotational state. The
equilibrium and stability of the Maclaurin spheroids are briefly
described. The thermodynamic structure of core matter is
specified. The resulting initial conditions for core collapse are
summarized and discussed.

065.087   **The evolution and nucleosynthesis of massive
pop III stars.**   W. W. Ober.
Proceedings of the 5th Göttingen-Jerusalem-Symposium on
Astrophysics, (see 012.041), p. 271 - 278 (1981).

065.088   **Stellar rotation and magnetic stars.**
D. Moss, R. C. Smith.

Rep. Prog. Phys., Vol. 44, 831 - 891 (1981). − Abstr. in Phys. Abstr., Vol. 85, Abstr. 6590 (1982).

065.089 Investigation of a carbon flash in degenerate cores of mean mass stars with convection taken into account. A. D. Kudryashov.
Nauchn. Inf., Vyp. (No.) 47, (see 003.016), p. 15 - 30 (1981). In Russian.

The influence of convection (in the mixing length approximation) on the development of the carbon flash in a degenerate core of mass $1.4 M_{\odot}$ has been investigated. At low effectiveness of convection the numerical results agree with the hypothesis on hydrostatic evolution of the core after the onset of carbon burning in the center. In case of more effective convection burning times of successive mass zones become shorter than the core's hydrodynamic time. Then the carbon burning might start with rapidly spreading thermal instability. Connection between the convective and neutrino mechanisms of burning propagation is discussed.

065.090 Influence of additional mixing on the evolution of massive stars. A. G. Masevich, E. I. Popova, A. V. Tutukov, L. R. Yungel'son.
Nauchn. Inf., Vyp. (No.) 47, (see 003.016), p. 31 - 37 (1981). In Russian.

The influence of additional mixing, e. g. caused by the convective overshooting from the core, on the evolution of stars with $M \geqslant 10 M_{\odot}$ before helium ignition in the core is studied. It is shown that the growth of the convective zone leads to significant widening of the main-sequence strip in the H−R diagram.

065.091 On structure and evolution of stars. II. Method for computation of stellar models. T. Angelov.
Vasiona, Année 29, 44 - 49 (1981). In Croatian.

065.092 On the nature of WR-type stars with ring nebulae. T. A. Lozinskaya, A. V. Tutukov.
Nauchn. Inf., Vyp. (No.) 49, (see 003.017), p. 21 - 30 (1981). In Russian.

The main characteristics of WR stars with ring nebulae are analysed in the frame of different theoretical scenarios of their formation. The supposition that such objects are formed in the course of massive close binaries evolution gives the best explanation of these characteristics.

065.093 On the evolution of carbon-oxygen cores of mean mass stars. Yu. L. Frantsman.
Nauchn. Inf., Vyp. (No.) 49, (see 003.017), p. 74 - 79 (1981). In Russian.

The evolution of matter on the external boundary of the carbon-oxygen cores of $3 - 7 M_{\odot}$ stars is considered. After the core mass reaches about $1 M_{\odot}$ its further evolution is the same for all stars in the considered mass range. The degeneration of matter in the centre is reached considerably earlier than on the outer boundary of the core. A jump of the temperature on the outer boundary of the core and simultaneously short rise of luminosity and decrease of radius occurs. After the mass of the core reaches $1 M_{\odot}$, the rate of the core increase becomes equal for the stars of mass range considered, and it depends only on the mass of the core.

065.094 Influence of accretion on the gravitational contraction of very low-mass degenerate dwarfs.
A. V. Fedorova, E. P. Kruglova.
Nauchn. Inf., Vyp. (No.) 49, (see 003.017), p. 87 - 94 (1981). In Russian.

The evolution of a degenerate hydrogen dwarf with initial mass $0.01 M_{\odot}$ on the stage of gravitational contraction is calculated for the case of accretion of matter of a dense cloud in which a young cluster is embedded. The accretion rate is $10^{-8} M_{\odot}/$year. The resulting mass of the star is $0.015 M_{\odot}$. The evolution of a star with constant mass of $0.01 M_{\odot}$ is calculated too for the time interval 5×10^8 years. The conclusion is made that the accretion can strongly increase the rate of cooling of the central parts of the star.

065.095 On structure and evolution of stars. III. Stellar models. T. Angelov.
Vasiona, Année 29, 70 - 74 (1981). In Croatian.

065.096 On the possibility of applying the Monaghan-Roxburgh method in the second approximation.
V. P. Merezhin.
Tr. Kazan. Gorod. Astron. Obs., Vyp. 45, p. 34 - 39 (1980). In Russian.

065.097 Meridional circulation in radiative zones of hot stars. V. P. Merezhin.
Tr. Kazan. Gorod. Astron. Obs., Vyp. 46, p. 3 - 54 (1980). In Russian.

065.098 Nonradial oscillations for stars on the left-hand side of the Cepheid instability strip.
H. Shibahashi, Y. Osaki.
Publ. Astron. Soc. Japan, Vol. 33, 427 - 448 (1981).

Vibrational stability against nonradial oscillations of high harmonic index l has been examined systematically for stars on the left-hand side of the Cepheid instability strip in the HR diagram. It is confirmed that regions for instability against nonradial oscillations extend far to the bluer side than those for instability against radial oscillations of the Cepheid strip. Unstable modes are confined to nonradial f-modes in this region of the HR diagram, and higher overtone p-modes are found all stable.

065.099 Distribution of energy in stellar non-radial oscillations. R. Scuflaire.
Bull. Soc. R. Sci. Liège, 49e année, 164 - 177 (1980) = Inst. Astrophys. Univ. Liège, Cointe-Ougrée (Belgique), Coll. 8°, No. 680.

The author considers the kinetic energy and the potential energy of a non-radial adiabatic oscillation of a star. The kinetic energy can be divided into a fraction associated to the radial motion and a fraction associated to the horizontal motion and the potential energy can be decomposed into several terms, each one being related to a force involved in the problem. The author uses these different terms to describe the non-radial modes of a star.

065.100 Thermal instability of hydrogen-burning shells in very massive stars.
Y. Tanaka, N. Arimoto, M. Takeuti.
Sci. Rep. Tôhoku Univ., Ser. 8, Vol. 2, 16 - 28 (1981).

The one-zone model has been modified to include the nuclear energy sources and Sackmann's geometrical measure m_s. The deviation of physical quantities from the equilibrium values for entropy perturbations are analytically obtained mainly in terms of m_s and the exponent of the temperature in the nuclear energy production rate n. A mechanism for Beta Cephei pulsations is suggested.

065.101 Effect of accretion on the evolution of very-low-mass young objects. A. V. Federova.
Planetary interiors, (see 012.061), p. 241 - 243 (1981).

The evolution of a degenerate dwarf of an initial mass $0.01 M_{\odot}$ accreting matter at a rate $10^{-6} M_{\odot}/$year is calculated. The final mass of the star is $0.015 M_{\odot}$. The evolution of the star of constant mass $0.01 M_{\odot}$ is calculated too. The conclusion is made, that the accretion increases two orders of magnitude the rate of the cooling of the degenerate dwarf.

Physics and evolution of stars.
See Abstr. 003.100.

Instationarity and evolution of stars.
See Abstr. 003.104.

Kernexplosionen an Sternoberflächen.
See Abstr. 011.022.

Polytropic, differentially rotating cylinders.
See Abstr. 014.006.

Annihilations of stationary particles on a lattice.
See Abstr. 022.047.

On the evolution of the homogeneous ellipsoidal figures. II. Gravitational collapse and gravitational radiation. See Abstr. 042.071.

Current rate of nucleosynthesis and its implications. See Abstr. 061.026.

Bulk viscosity of a reactive (ν_e npe⁻) system. See Abstr. 061.034.

Some aspects of the nucleosynthesis of the light elements.   See Abstr. 061.040.

Synthesis of the heavy elements.
See Abstr. 061.041.

s-process branch at ^{176}Lu.
See Abstr. 061.046.

Neutron cross sections of importance to astrophysics.   See Abstr. 061.049.

s-process nucleosynthesis, stellar abundances, and galactic evolution.   See Abstr. 061.053.

On neutrino thermal conduction and viscosity in stellar collapse.   See Abstr. 061.055.

Electron capture as supernova killer.
See Abstr. 061.056.

Linear series and "catastrophies" in astronomy. See Abstr. 061.057.

On some trends in the development of astrophysics. See Abstr. 061.058.

Internal motions in magnetic stars.
See Abstr. 062.005.

The potentials for the g-, p-, and the toroidal-modes of self-gravitating fluids.   See Abstr. 062.015.

On the construction of models of rotating stars and stellar systems.   See Abstr. 062.026.

Explosive mass loss in binary stars: the two timescale method.   See Abstr. 064.009.

Stellar convection theory. III. Dynamical coupling of the two convection zones in A-type stars by penetrative motions.   See Abstr. 064.015.

Theoretical evidence of mass loss from globular cluster stars.   See Abstr. 064.039.

Effect of mass gain on stellar evolution.
See Abstr. 064.040.

Mass loss and overshooting in massive stars.
See Abstr. 064.047.

Colours and effective temperatures of extreme helium stars.   See Abstr. 064.049.

Scale covariant gravitation. VI. Stellar structure and evolution.   See Abstr. 066.032.

The Einsteinian star model in Treder's tetrad theory of gravitation.   See Abstr. 066.047.

Exterior spacetimes for rotating stars.
See Abstr. 066.090.

Cooling of young neutron stars and the Einstein X-ray observations.   See Abstr. 066.512.

Some aspects of hydrogen burning in a neutron star envelope.   See Abstr. 066.521.

Theory of the solar cycle.   See Abstr. 072.033.

Is a stellar wind inherent in WR-stars throughout the whole of their evolution?   See Abstr. 112.024.

Mass loss from hot stars below the main sequence. See Abstr. 112.031.

The Wolf-Rayet stars.   See Abstr. 114.203.

Masses of Magellanic Wolf-Rayet stars: mass loss and evidence for a WR subclass vs. mass relation. See Abstr. 115.007.

The initial/final mass relation for stellar evolution with mass loss.   See Abstr. 115.009.

Broad-band linear polarization and magnetic intensification in rotating magnetic stars.   See Abstr. 116.017.

Double stars.   See Abstr. 117.015.

SS 433, X-ray binaries and stellar evolution.
See Abstr. 117.089.

Evolutionary scenario for close binary systems of low and moderate masses.   See Abstr. 117.108.

$l = 0, 1, 2$, and 3 pulsation constants for evolutionary models of δ Scuti stars.   See Abstr. 122.080.

Gravitational-collapse supernovae.
See Abstr. 125.067.

Formation of a planetary nebula by continuous mass loss.   See Abstr. 135.033.

Planetary nebulae and stellar evolution.
See Abstr. 135.042.

X-ray binaries and stellar evolution.
See Abstr. 142.132.

Nucleosynthesis in OB associations and its effect upon cosmic rays, peculiar stars and isotopic anomalies in meteorites.   See Abstr. 152.006.

The cyanogen distribution of M4 and the possible connection between horizontal branch morphology and chemical inhomogeneity.   See Abstr. 154.012.

The fraction of O-type supergiants in our galaxy, in the LMC and in the SMC: an evidence of the correlation between mass loss rate and chemical abundance.
See Abstr. 155.020.

Peculiarities in the distribution of galactic Wolf-Rayet stars: constraints on evolutionary scenarios?
See Abstr. 155.021.

Chemical evolution in the solar neighborhood. IV. Some revised general equations and a specific model.
See Abstr. 155.050.

Nuclear astrophysics and particles.
See Abstr. 162.118.

066 Relativistic Astrophysics, Gravitation Theory, Background Radiation, Black Holes, Neutron Stars

066.001 Supercritical, steady-state, spherically symmetric accretion into a black hole. D. Freihoffer.
Astron. Astrophys., Vol. 100, 178 - 182 (1981).

The author studies supercritical, spherically symmetric and time independent accretion into nonrotating black holes, which are embedded in a fully ionized H II region. It is shown that if one uses general relativistic equations to describe fluid and radiative transfer, there arises, besides the usual critical point due to the sound speed, a second one, which determines the luminosity produced by compression-work in the optically thick part of the flow. The author discusses models for black hole masses M between 1 and 1000 $M_\odot$ and various accretion rates $\dot{M}$. In all cases the resulting luminosities are very small compared to the Eddington luminosity.

066.002 Le teorie gravitazionali: Parte prima: fondamenti. R. Bedogni.
Coelum, Vol. 50, 101 - 111 (1981).

066.003 On the mutual coherence degree of images formed by a gravitational lens. A. V. Mandzhos.
Pis'ma Astron. Zh., Tom 7, 387 - 389 (1981). In Russian. English translation in Soviet Astron. Lett., Vol. 7.

A general formula for the mutual coherence degree of images formed by a gravitational lens with arbitrary mass distribution is obtained. Numerical estimates show that multiple quasar images formed by gravitator-galaxies are mutually incoherent. Only for quasar radiation splitted by stars of intervening galaxies the coherence degree can reach observable magnitude.

066.004 Magnetic field generation by rotating black holes. D. A. Leahy, A. Vilenkin.
Astrophys. J., Vol. 248, 13 - 17 (1981).

A new mechanism of cosmic magnetic field generation is discussed. Neutrinos asymmetrically emitted by rotating black holes scatter on protons and produce a proton current which generates the magnetic field. It is shown that this mechanism can in principle produce a seed field sufficiently strong to account for present galactic fields.

066.005 Near-millimeter spectrum of the microwave background. D. P. Woody, P. L. Richards.
Astrophys. J., Vol. 248, 18 - 37 (1981).

A complete description is given of a measurement of the spectrum of the cosmic microwave background (CMB). A fully calibrated, liquid helium cooled, balloon borne spectrophotometer was used to measure the emission spectrum of the night sky over the frequency range from 1.7 to 40 cm^{-1}. The apparatus was calibrated using a blackbody source with a variety of temperatures before and after the flight. The atmospheric emission was subtracted by fitting the data to the spectrum computed from an atmospheric model. The resulting spectral measurements show that the spectrum of the CMB peaks at 6 cm^{-1} and is approximately that of a blackbody out to several times that frequency. Some deviations from blackbody shape are observed. The measured flux is equivalent to that from a blackbody in the temperature range from 2.88 to 3.09 K. These measurements are combined with previous observations of the CMB to produce a set of data covering three decades in frequency. This data set is compared with various models of the CMB.

066.006 Distortions of the microwave background spectrum by dust.
J. Negroponte, M. Rowan-Robinson, J. Silk.

Astrophys. J., Vol. 248, 38 - 46 (1981).

Exact calculations of the distortion of the microwave background spectrum by dust in the early universe are presented. A range of dust materials and grain formation epochs, z_f, are investigated and the dust temperature history calculated subject to the condition that the melting temperature of the grains not be exceeded. It is found that distortions similar to those observed by Woody and Richards can be produced by several grain materials if $z_f \gtrsim 200$. The distortions found in the exact treatment are more pronounced than those predicted in an earlier approximate calculation.

066.007 Resource letter BH-1: black holes. S. Detweiler.
American J. Phys., Vol. 49, 394 - 400 (1981). − Abstr. in Phys. Abstr., Vol. 84, Abstr. 63114 (1981).

066.008 Acceleration of a static observer near the event horizon of a static isolated black hole. N. A. Doughty.
American J. Phys., Vol. 49, 412 - 416 (1981). − Abstr. in Phys. Abstr., Vol. 84, Abstr. 63116 (1981).

066.009 Conformal geometry and spatially homogeneous cosmology. R. T. Jantzen.
Ann. Inst. Henri Poincaré, Sect. A, Vol. 33, 121 - 146 (1980). Abstr. in Phys. Abstr., Vol. 84, Abstr. 63206 (1981).

066.010 Contributions to the theory of a Weber-type gravitational antenna. C. Mantea, A. Corciovei, V. Barsan.
Rev. Roumaine Phys., Vol. 25, 1069 - 1084 (1980). − Abstr. in Phys. Abstr., Vol. 84, Abstr. 63220 (1981).

066.011 Statistical mechanics and the gravothermal catastrophe. P. Cally, J. J. Monaghan.
J. Math. Phys., Vol. 22, 348 - 351 (1981). − Abstr. in Phys. Abstr., Vol. 84, Abstr. 63222 (1981).

066.012 Stretching a black hole. W. J. Wild, R. M. Kerns, W. F. Drish, Jr.
Phys. Rev. D, Vol. 23, 829 - 831 (1981). − Abstr. in Phys. Abstr., Vol. 84, Abstr. 63224 (1981).

066.013 Gravitational clumping and the annihilation of monopoles. T. Goldman, E. W. Kolb, D. Toussaint.
Phys. Rev. D, Vol. 23, 867 - 875 (1981). − Abstr. in Phys. Abstr., Vol. 84, Abstr. 63227 (1981).

066.014 Massive spin-1 field in expanding universes and the quantum equivalence principle. A. Foussats, R. Laura, O. Zandron.
J. Math. Phys., Vol. 22, 357 - 370 (1981). − Abstr. in Phys. Abstr., Vol. 84, Abstr. 63228 (1981).

066.015 Macroscopical consequences of a propagating torsion potential. V. de Sabbata, M. Gasperini.
Nuovo Cimento, Lett., Ser. 2, Vol. 30, 503 - 506 (1981). Abstr. in Phys. Abstr., Vol. 84, Abstr. 63229 (1981).

066.016 Supergravity. P. van Nieuwenhuizen.
Phys. Rep., Vol. 68, 189 - 398 (1981). − Abstr. in Phys. Abstr., Vol. 84, Abstr. 63232 (1981).

066.017 **Current measurements of the gravitational 'constant' as a function of the mass separation.**
D. R. Long.
Nuovo Cimento B, Ser. 11, Vol. 62B, 130 - 138 (1981).
Abstr. in Phys. Abstr., Vol. 84, Abstr. 63235 (1981).

066.018 **Experimental tests of Newtonian gravity at relativistic velocities.** A. C. Melissinos.
Nuovo Cimento B, Ser. 11, Vol. 62B, 190 - 202 (1981).
Abstr. in Phys. Abstr., Vol. 84, Abstr. 63236 (1981).

066.019 **Spacecraft-Doppler gravity-wave detection. I. Theory.**
R. W. Hellings.
Phys. Rev. D, Vol. 23, 832 - 843 (1981). – Abstr. in Phys. Abstr., Vol. 84, Abstr. 63238 (1981).

066.020 **Spacecraft-Doppler gravity-wave detection. II. Results.** R. W. Hellings, P. S. Callahan,
J. D. Anderson, A. T. Moffet.
Phys. Rev. D, Vol. 23, 844 - 851 (1981). – Abstr. in Phys. Abstr., Vol. 84, Abstr. 63239 (1981).

066.021 **The gravitational constant G and its measurement.**
P.-h. Zhang.
Wuli, Vol. 9, 361 - 365 (1980). In Chinese. – From Phys. Abstr., Vol. 84, Abstr. 63240 (1981).

066.022 **The total four momentum – its relation to transformations of coordinates and frames of reference.**
U. Kasper.
Acta Phys. Polonica B, Vol. B12, 189 - 202 (1981). – Abstr. in Phys. Abstr., Vol. 84, Abstr. 67279 (1981).

066.023 **Quadratic Lagrangians and the Reissner-Nordström-de Sitter metric.** J. A. Leiro.
Int. J. Theor. Phys., Vol. 19, 873 - 876 (1980). – Abstr. in Phys. Abstr., Vol. 84, Abstr. 67288 (1981).

066.024 **A dilemma in the physics of gravitational fields.**
R. A. Vera.
Int. J. Theor. Phys., Vol. 20, 19 - 50 (1981). – Abstr. in Phys. Abstr., Vol. 84, Abstr. 67289 (1981).

066.025 **Exact spatially inhomogeneous cosmologies.**
J. Wainwright.
J. Phys. A, Vol. 14, 1131 - 1147 (1981). – Abstr. in Phys. Abstr., Vol. 84, Abstr. 67291 (1981).

066.026 **Black holes and elementary particles.**
P. F. Gonzalez-Diaz.
Nuovo Cimento, Lett., Ser. 2, Vol. 31, 39 - 42 (1981). – Abstr. in Phys. Abstr., Vol. 84, Abstr. 67623 (1981).

066.027 **The astrophysical sources of gravitational waves.**
Y. Li.
Acta Sci. Nat. Univ. Sunyatseni, No. 4, p. 65 - 79 (1980). In Chinese. – Abstr. in Phys. Abstr., Vol. 84, Abstr. 71071 (1981).

066.028 **Deflection of massive neutrinos by gravitational fields.** D. Fargion.
Nuovo Cimento, Lett., Ser. 2, Vol. 31, 49 - 52 (1981). – Abstr. in Phys. Abstr., Vol. 84, Abstr. 71072 (1981).

066.029 **Constraint on the planetary scale value of the Newtonian gravitational constant from the gravity profile within a mine.** F. D. Stacey, G. J. Tuck,
S. C. Holding, A. R. Maher, D. Morris.
Phys. Rev. D, Vol. 23, 1683 - 1692 (1981). – Abstr. in Phys. Abstr., Vol. 84, Abstr. 71074 (1981).

066.030 **On the transformation of gravitational radiation into electromagnetic radiation.**
D. Papadopoulos, F. P. Esposito.
Astrophys. J., Vol. 248, 783 - 789 (1981).
The equations which determine the response of a charged particle moving in a uniform magnetic field to an incident gravitational wave are derived in the linearized approximation to general relativity. The response is best understood when it is expressed in terms of the intensity of the magnetic bremsstrahlung emitted, where effects due to the interaction with the gravitational wave are easily identified. The authors briefly discuss several applications of the derived formulae.

066.031 **Scale covariant gravitation. V. Kinetic theory.**
S.-H. Hsieh, V. M. Canuto.
Astrophys. J., Vol. 248, 790 - 800 (1981).
The authors construct a scale covariant kinetic theory for particles and photons. The mathematical framework of the theory is given by the tangent bundle of a Weyl manifold. The Liouville equation is then derived. Solutions corresponding to equilibrium distributions are presented and shown to yield thermodynamic results identical to the ones obtained previously.

066.032 **Scale covariant gravitation. VI. Stellar structure and evolution.** V. M. Canuto, S.-H. Hsieh.
Astrophys. J., Vol. 248, 801 - 812 (1981).
The authors study stellar structure and evolution in the framework of the scale covariant theory of gravitation. A radiative transfer equation is derived from the kinetic theory presented in Paper V. This equation, combined with the hydrostatic equilibrium equation, yields a relation between the stellar luminosity and β, the scaling function directly related to a variable G. Application of the above relations allows to study the evolution of the solar luminosity, its effects on the history of the surface temperature of the earth, and finally the age of globular clusters.

066.033 **Is the gravitational constant changing?**
T. C. Van Flandern.
Astrophys. J., Vol. 248, 813 - 816 (1981).
Although the question posed by the title is not yet conclusively resolved, the most recent determinations of the lunar tidal acceleration are in reasonable accord with each other, suggesting that $\dot{n}_{tidal} = (-28.8 \pm 1.5)''$ century^{-2}. Recent lunar occultation solutions are now in accord with laser ranging results in indicating that $\dot{n}_{total} = (-23.2 \pm 1.2)''$ century^{-2}. The difference between these results gives an excess of possible cosmological origin of $\dot{n}/n = (+3.2 \pm 1.1) \times 10^{-11}$ yr^{-1} which is also consistent with very preliminary results from planetary radar ranging. Interpreted with the scalar covariant cosmology of Canuto and Hsieh, these results indicate that $\dot{G}/G = (-6.4 \pm 2.2) \times 10^{-11}$ yr^{-1}. Implications for relativity and cosmology are briefly discussed.

066.034 **On embeddings of the Kerr geometry.** N. A. Sharp.
Canadian J. Phys., Vol. 59, 688 - 692 (1981).
Abstr. in Phys. Abstr., Vol. 84, Abstr. 71514 (1981).

066.035 **Theoretical advances in general relativity.**
S. W. Hawking.
Some strangeness in the proportion. Centennial symposium to celebrate the achievements of Albert Einstein, (see 003.001), p. 145 - 152 (1980). – Abstr. in Phys. Abstr., Vol. 84, Abstr. 71518 (1981).

066.036 **Quantum gravity and supergravity.** Y. Ne'eman.
Some strangeness in the proportion. Centennial symposium to celebrate the achievements of Albert Einstein, (see 003.001), p. 429 - 443 (1980). – Abstr. in Phys. Abstr., Vol. 84, Abstr. 71526 (1981).

066.037 **Gravitational wave dispersion in condensed matter systems.**
A. Widom, G. Megaloudis, T. D. Clark, R. J. Prance.
J. Phys. A, Vol. 14, L213 - L215 (1981). — Abstr. in Phys. Abstr., Vol. 84, Abstr. 71531 (1981).

066.038 **Time evolution of pure gravitational waves.**
S. M. Miyama.
Prog. Theor. Phys., Vol. 65, 894 - 909 (1981). — Abstr. in Phys. Abstr., Vol. 84, Abstr. 75010 (1981).

066.039 **On the existence of solutions to Einstein's equation with non-zero Bondi news.**   A. Ashtekar, T. Dray.
Commun. Math. Phys., Vol. 79, 581 - 599 (1981). — Abstr. in Phys. Abstr., Vol. 84, Abstr. 75381 (1981).

066.040 **Gravitational radiation by the thermal phonons of a solid.**   G. Schäfer, H. Dehnen.
Phys. Rev. D, Vol. 23, 2129 - 2137 (1981). — Abstr. in Phys. Abstr., Vol. 84, Abstr. 75383 (1981).

066.041 **Gauge theories, black hole evaporation and cosmic censorship.**   P. C. W. Davies.
Phys. Lett. B, Vol. 101B, 399 - 400 (1981). — Abstr. in Phys. Abstr., Vol. 84, Abstr. 75387 (1981).

066.042 **Supergravity.**
P. Van Nieuwenhuizen.
Some strangeness in the proportion. Centennial symposium to celebrate the achievements of Albert Einstein, (see 003.001), p. 444 - 452 (1980). — Abstr. in Phys. Abstr., Vol. 84, Abstr. 75397 (1981).

066.043 **Experimental challenges posed by the general theory of relativity.**   I. I. Shapiro.
Some strangeness in the proportion. Centennial symposium to celebrate the achievements of Albert Einstein, (see 003.001), p. 115 - 136 (1980). — Abstr. in Phys. Abstr., Vol. 84, Abstr. 75404 (1981).

066.044 **The variability of the gravitational constant.**
T. L. Chow.
Nuovo Cimento, Lett., Ser. 2, Vol. 31, 119 - 120 (1981). Abstr. in Phys. Abstr., Vol. 84, Abstr. 79714 (1981).

066.045 **Simple presentation of effect of gravity on clocks.**
W. Hirsch.
Phys. Teach., Vol. 19, 247(1981). — Abstr. in Phys. Abstr., Vol. 84, Abstr. 79909 (1981).

066.046 **Homothetic solutions of Einstein's equations and shock waves.**   G. Moschetti.
J. Math. Phys., Vol. 22, 830 - 834 (1981). — Abstr. in Phys. Abstr., Vol. 84, Abstr. 80008 (1981).

066.047 **The Einsteinian star model in Treder's tetrad theory of gravitation.**   S. Gottlöber.
Ann. Physik, Band 38, 67 - 72 (1981). — Abstr. in Phys. Abstr., Vol. 84, Abstr. 80021 (1981).

066.048 **Non-monotonicity of the mass distribution and existence of the gravitational phase transition.**
J. Messer.
Phys. Lett. A, Vol. 83A, 304 - 306 (1981). — Abstr. in Phys. Abstr., Vol. 84, Abstr. 83477 (1981).

066.049 **Quantum wormholes. I. Choice of the classical solution.**   P. Hajicek.
Nucl. Phys. B, Vol. B185, 254 - 268 (1981). — Abstr. in Phys. Abstr., Vol. 84, Abstr. 83672 (1981).

066.050 **Gravitational lensing and the relation between QSO and galaxy magnitude-number counts.**
J. A. Tyson.
Astrophys. J., Lett., Vol. 248, L89 - L93 (1981).
The recent availability of a complete set of data on galaxy counts over a wide range of apparent magnitudes reveals a peculiar numerical coincidence and suggests a new test for the gravitational lens origin of quasars: The observed bright quasar number-magnitude relation $d\log N_{QSO} / dm = 0.9$ results from simple consideration of statistical gravitational lensing of unresolved Seyfert nuclei, by galaxies and clusters of galaxies, combined with the observed galaxy number-magnitude relation and estimates of the K-correction for Seyferts.

066.051 **On gravitational lenses and the cosmological evolution of quasars.**   Y. Avni.
Astrophys. J., Lett., Vol. 248, L95 - L99 (1981).
A heuristic model for the effect of gravitational lenses on the apparent cosmological evolution of quasars is considered. The model satisfies the requirement of average flux conservation and has no net mean amplification. This requirement is shown to be numerically important in studying the effect. On the basis of the values of the evolution indicators calculated from the model, it is concluded that it is premature to assert that lensing plays an important role in affecting the apparent evolution. A qualitative, model independent observational test for the effect is suggested.

066.052 **Another possible case of a gravitational lens.**
B. Paczyński, K. Gorski.
Astrophys. J., Lett., Vol. 248, L101 - L104 (1981).
A cluster of three quasars at redshift 2.05 discovered by Burbidge et al. may be a triple image of a single quasar produced by two spherical clusters of galaxies acting as a gravitational lens. The core radii and velocity dispersions required for the clusters by the authors' model are not unreasonable, but a positive cosmological constant would bring the velocity dispersion within clusters down, closer to that commonly observed.

066.053 **A diffraction limit on the gravitational lens effect.**
R. J. Bontz, M. P. Haugan.
Astrophys. Space Sci., Vol. 78, 199 - 210 (1981).
The focussing of gravitational radiation by the interior and exterior gravitational field of a Newtonian gravitational lens is considered. A graphical method for determining the caustic structure of a Newtonian gravitational lens is presented and the caustic structure of a solar type gravitational lens is discussed. Estimates of the amplitude magnification in the caustic region indicate that waves with frequencies less than a critical cutoff frequency ω_c are not amplified significantly. For a lens of mass M this cutoff frequency is $\omega_c \approx (10^{-1} \pi M)^{-1}$; for the Sun $\omega_c \approx 10^4$ s^{-1}.

066.054 **Superluminal velocities: a gravitational lens effect.**
J. M. Barnothy.
Bull. American Astron. Soc., Vol. 13, 529 - 530 (1981). Abstract.

066.055 **Simulating the transparent gravitational lens.**
C. C. Dyer, R. C. Roeder.
Bull. American Astron. Soc., Vol. 13, 533 (1981). — Abstract.

066.056 **Optically thick relativistic accretion onto a black hole.**   P. A. Vitello.
Bull. American Astron. Soc., Vol. 13, 534 (1981). — Abstract.

066.057 **Scale-invariant gravity: a reformulation and an astrophysical test.**   P. S. Wesson.
Mon. Not. R. Astron. Soc., Vol. 197, 157 - 165 (1981).
The field equations of the scale-invariant theory of gravity (Dirac, Hoyle/Narlikar, Canuto et al.) are reformulated

in a simpler form than hitherto. Attention is focused on the physically most interesting case, namely that having spherical symmetry, a time-dependent gauge function and matter with the properties of a perfect fluid (with values for the velocity, pressure and cosmological constant that can all be non-zero). For this case, a set of equations for the density, pressure and mass is given. This form of the field equations yields new physical insight into the theory and provides an astrophysical test of it based on mass loss from stars.

066.058 Generation of gravitational radiation in the laboratory. F. Romero B., H. Dehnen.
Z. Naturforsch., Band 36a, 948 - 955 (1981).

The generation of gravitational radiation by coherent excitation of a long row of oscillators is investigated. In particular the authors have considered diatomic linear chains as an idealization of thin piezoelectric crystals. They find a highly focussed superradiant beam of gravitational radiation in direction of the row and a total radiation power larger than the incoherent superposition of the oscillator radiation.

066.059 On observable consequences of the three body problem in Einstein's gravitation theory.
L. M. Chechin.
Kazakhsk. univ. Alma-Ata, 1981. 11 pp. In Russian. – Abstr. in Ref. zh., 51. Astron., 7.51.68 (1981).

066.060 Relativistic effects in a planetary reference system.
L. M. Chechin.
Kazakhsk. univ. Alma-Ata. 1981. 17 pp. In Russian. – Abstr. in Ref. zh., 51. Astron., 7.51.69 (1981).

066.061 Can extended bodies move along a Riemannian space-time geodetic?
V. I. Denisov, A. A. Logunov, M. A. Mestvirishvili.
Inst. yader. issled. AN SSSR. Prepr., 1981, No. 0193, 38 pp. In Russian. – Abstr. in Ref. zh., 51. Astron., 7.51.133 (1981).

066.062 Condensed structures with arbitrarily large redshifts and total mass. M. C. Durgapal, P. S. Rawat.
Bull. Astron. Soc. India, Vol. 9, 79 - 80 (1981). – Abstract.

066.063 Relativistic clusters with distribution function of classical polytropes.
M. C. Durgapal, P. S. Rawat, R. Banerjee.
Bull. Astron. Soc. India, Vol. 9, 86 (1981). – Abstract.

066.064 Application of the dynamical interpretation of general relativity. E. Deumens.
Acta Cosmologica, Zesz. 10, 25 - 32 (1981).

The paper argues that the gravitational field seen in the fully dynamical way, described here, is a useful tool for understanding some fundamental results in a coherent general relativistic way.

066.065 The manifold model for space-time. M. Heller.
Acta Cosmologica, Zesz. 10, 33 - 51 (1981).

Physical processes happen on a space-time arena. It turns out that all contemporary macroscopic physical theories presuppose a common mathematical model for this arena, the so-called manifold model of space-time. The present study is an heuristic introduction to the concept of a smooth manifold, starting with the intuitively more clear concepts of a curve and a surface in the Euclidean space.

066.066 Relativistic model for space-time. M. Heller.
Acta Cosmologica, Zesz. 10, 53 - 69 (1981).

The Lorentz structure of the space-time model is reviewed and critically discussed.

066.067 The importance of general relativity for astrophysics. S. A. Teukolsky.

News Lett. Astron. Soc. N. Y., Vol. 1, No. 10, p. 28 - 37 (1981).

066.068 Scale-invariant gravity: a simple formulation.
P. S. Wesson.
Astron. Astrophys., Vol. 102, 45 - 52 (1981).

Using the Cosmological Principle as justification, it is suggested that the scale-invariant theory of gravity be based on a Conspiracy Hypothesis. This hypothesis yields a formulation of the scale-invariant theory that is simpler than other versions of it in which the Newtonian gravitational parameter G is treated as a field variable. This simple formulation of scale-invariant gravity agrees with a recent reformulation of the (Perfect) Cosmological Principle. It also agrees with observations that have been made to date.

066.069 A new geometrical gravitational theory.
T. Obata, H. Oshima, J. Chiba.
Gen. Relativ. Gravitation, Vol. 13, 313 - 334 (1981).

066.070 Equation of motion including the reaction of gravitational radiation.
A. Papapetrou, B. Linet.
Gen. Relativ. Gravitation, Vol. 13, 335 - 359 (1981).

066.071 Rotating black holes as dissipative spin-thermodynamical systems. A. Curir.
Gen. Relativ. Gravitation, Vol. 13, 417 - 423 (1981).

A black hole is interpreted as an "open system" in the Prigogine sense. From the point of view of spin transformations, the existence of a phase transition corresponding to an extreme Kerr hole is recognized. The role played by the spin entropy into superradiance of a rotating black hole is investigated.

066.072 Two kinds of rotation: an argument for torsion.
P. B. Yasskin.
Gen. Relativ. Gravitation, Vol. 13, 463 - 471 (1981).

There are now many theories of gravity with a torsion field as well as the usual metric field. One of the arguments for allowing torsion is based upon a gauge theory analogy. The purpose of this paper is to clarify exactly which symmetries are being gauged in this process. The principal observation is that special relativity is invariant under two different kinds of Lorentz transformations. The first type rotate the fields and move them from one point to another in space-time. The second type merely rotate the fields at each point without changing their location. To gauge both types of rotations requires a torsion field as well as a metric field.

066.073 Relativistic celestial mechanics of binary stars.
N. Spyrou.
Gen. Relativ. Gravitation, Vol. 13, 473 - 485 (1981).

The author presents the results of a systematic study of the dynamics of realistic binary systems in the post-Newtonian approximation (PNA) of general relativity. He proposes definitions valid in the PNA for the self-angular-momenta of the binary's members, as well as for the angular momentum of their relative orbital motion, and he examines under which conditions they can be considered as constant in the PNA.

066.074 Relativistic effects in many-body systems of finite size, internal structure, and internal motions. II. The determination of the inertial and rest masses of binary stars.
N. Spyrou.
Gen. Relativ. Gravitation, Vol. 13, 487 - 493 (1981).

The author evaluates the absolute orbits of the members of a close binary system, the corresponding orbital elements and apsidal motions, and generalizes the concept of the classical mass function. Moreover he proposes star models expressing the inertial mass of a star in terms of its rest mass. So the rest mass, which for stars of equal inertial masses depends on the star's nature, can in principle be determined from the observational data concerning the inertial masses. An application to the binary pulsar PSR 1913 + 16 shows that the star models proposed are quite satisfactory for distinguishing the rest from the inertial masses with the present accuracy of the observational data.

066.075   Eddington's cosmological number, Einstein's number criterion and Rydberg's rational system of dimensions.   H.-J. Treder.
Astron. Nachr., Band 302, 115 - 125 (1981). In German.
    The discussions about the meaning of the "hierarchy of interactions" and in connection with this about the role of Eddington's "cosmological number" imply the question of the "big numbers" in physics. According to Einstein's and Bridgman's criteria such "big numbers" are hints at unsolved problems in the foundations of physics. Eddington gives a theory of the big numbers like cosmological quantities. – A new point of view on this question may be to remember Rydberg's suggestion on independent physical dimensions of lengths L, surfaces S, and volumina V, and to remember Dällenbach's suggestion to introduce a new universal constant α which describes the operational connections between the quadrate of lengths L^2 and the surface S in microphysics.

066.076   Inertia-free mechanics and the bi-metric procedure.   D.-E. Liebscher.
Astron. Nachr., Band 302, 127 - 132 (1981).
    The author considers the bi-metric procedure to implement field-theoretical concepts into the mechanical models developed so far in inertia-free gravodynamics. The main result is the fact, that he is able to hide the absolute time underlying the mechanical models for non-gravitational fields by the equivalence principle, but not for the gravitation theory, which reveals the instantaneous gravitational action in one parameter of the post-Newtonian approximation.

066.077   Properties of extended bodies in spacetimes admitting isometries.   R. Schattner, M. Streubel.
Ann. Inst. Henri Poincaré, Sect. A, Vol. 34, 117 - 143 (1981). Abstr. in Phys. Abstr., Vol. 84, Abstr. 83906 (1981).

066.078   The connection between local and asymptotic structures for isolated gravitating systems with isometries.   M. Streubel, R. Schattner.
Ann. Inst. Henri Poincaré, Sect. A, Vol. 34, 145 - 152 (1981). Abstr. in Phys. Abstr., Vol. 84, Abstr. 83907 (1981).

066.079   Scattering of a plane gravitational wave by a magnetic dipole field in the Schwarzschild metric.
T. Elster.
Ann. Inst. Henri Poincaré, Sect. A, Vol. 34, 163 - 172 (1981). Abstr. in Phys. Abstr., Vol. 84, Abstr. 83929 (1981).

066.080   Evolution of the interior of a charged black hole.   W. A. Hiscock.
Phys. Lett. A, Vol. 83A, 110 - 112 (1981). – Abstr. in Phys. Abstr., Vol. 84, Abstr. 83930 (1981).

066.081   A new exact solution of classical supergravity.   L. F. Urrutia.
Phys. Lett. B, Vol. 102B, 393 - 396 (1981). – Abstr. in Phys. Abstr., Vol. 84, Abstr. 83937 (1981).

066.082   Quantum field theory, horizons and thermodynamics.   D. W. Sciama, P. Candelas, D. Deutsch.
Adv. Phys., Vol. 30, 327 - 366 (1981). – Abstr. in Phys. Abstr., Vol. 84, Abstr. 83938 (1981).

066.083   On the thermodynamics of a collapsed object.   N. Chaubey, V. De Sabbata.
Nuovo Cimento, Lett., Ser. 2, Vol. 31, 281 - 284 (1981). Abstr. in Phys. Abstr., Vol. 84, Abstr. 88971 (1981).

066.084   Gravitational deflection of light at 1 1/2 PPN order.   G. W. Richter, R. A. Matzner.
Astrophys. Space Sci., Vol. 79, 119 - 127 (1981).
    The authors compute in parametrized form the effect of source-velocity terms in the gravitational deflection of light. Internal rotation corresponds to sources of order ϵ^3 where $\epsilon^2 \sim GM_\odot/R_\odot c^2 \sim (v/c)^2$. They find for rigidly rotating models of the Sun and of Jupiter that, in each case, this effect is of order 10^{-6} arc sec.

066.085   Gravitational lens.   J. Higbie.
American J. Phys., Vol. 49, 652 - 655 (1981).
Abstr. in Phys. Abstr., Vol. 84, Abstr. 89184 (1981).

066.086   Surface properties of Kerr-Newman black holes.   N. A. Doughty.
American J. Phys., Vol. 49, 720 - 724 (1981). – Abstr. in Phys. Abstr., Vol. 84, Abstr. 89199 (1981).

066.087   On singularities of general relativity and gravitational equations of fourth order.
H.-H. von Borzeszkowski.
Ann. Physik, Band 38, 239 - 248 (1981). – Abstr. in Phys. Abstr., Vol. 84, Abstr. 89361 (1981).

066.088   Hypersurfaces of constant mean extrinsic curvature.   S. M. Stumbles.
Ann. Physics, Vol. 133, 28 - 56 (1981). – Abstr. in Phys. Abstr., Vol. 84, Abstr. 89362 (1981).

066.089   The general theory of relativity with spin and torsion.   H. Knutsen.
Fra. Fys. Verden, Vol. 43, No. 1, p. 6 - 8 (1981). In Norwegian. – Abstr. in Phys. Abstr., Vol. 84, Abstr. 89365 (1981).

066.090   Exterior spacetimes for rotating stars.   B. C. Xanthopoulos.
J. Math. Phys., Vol. 22, 1254 - 1259 (1981). – Abstr. in Phys. Abstr., Vol. 84, Abstr. 89369 (1981).

066.091   High-frequency waves in gravitational theories with fourth-order derivative equations.
H. H. von Borzeszkowski.
Ann. Physik, Band 38, 231 - 238 (1981). – Abstr. in Phys. Abstr., Vol. 84, Abstr. 89380 (1981).

066.092   Field of a charged particle in Brans-Dicke theory of gravitation.
D. R. K. Reddy, V. U. M. Rao.
J. Phys. A, Vol. 14, 1973 - 1976 (1981). – Abstr. in Phys. Abstr., Vol. 84, Abstr. 89386 (1981).

066.093   Path-integral evaluation of Feynman propagator in curved spacetime.   J. D. Bekenstein, L. Parker.
Phys. Rev. D, Vol. 23, 2850 - 2869 (1981). – Abstr. in Phys. Abstr., Vol. 84, Abstr. 89392 (1981).

066.094   Calibration of gravitational radiation antenna by dynamic Newton field.
T. Suzuki, K. Tsubono, K. Kuroda, H. Hirakawa.
Japanese J. Appl. Phys., Vol. 20, L498 - L500 (1981). – Abstr. in Phys. Abstr., Vol. 84, Abstr. 89397 (1981).

066.095   Comparison of numerical methods for the integration of black hole geodesic equations.
N. A. Sharp.
J. Comput. Phys., Vol. 41, 295 - 308 (1981). – Abstr. in Phys. Abstr., Vol. 84, Abstr. 94390 (1981).

066.096   Complete description of photon trajectories in the Kerr-Newman space-time.

M. Calvani, R. Turolla.
J. Phys. A, Vol. 14, 1931 - 1942 (1981). – Abstr. in Phys. Abstr., Vol. 84, Abstr. 94391 (1981).

066.097 Some properties of static general relativistic stellar models. II. L. Lindblom.
J. Math. Phys., Vol. 22, 1324 - 1326 (1981). – Abstr. in Phys. Abstr., Vol. 84, Abstr. 94392 (1981).

066.098 Classical and quantum restrictions on geodesic motion near a Schwarzschild black hole.
F. R. Tangherlini.
Nuovo Cimento B, Ser. 11, Vol. 63B, 588 - 600 (1981). Abstr. in Phys. Abstr., Vol. 84, Abstr. 94395 (1981).

066.099 Models of evaporating black holes. I.
W. A. Hiscock.
Phys. Rev. D, Vol. 23, 2813 - 2822 (1981). – Abstr. in Phys. Abstr., Vol. 84, Abstr. 94396 (1981).

066.100 Models of evaporating black holes. II. Effects of the outgoing created radiation. W. A. Hiscock.
Phys. Rev. D, Vol. 23, 2823 - 2827 (1981). – Abstr. in Phys. Abstr., Vol. 84, Abstr. 94397 (1981).

066.101 Self-forces and atoms in gravitational fields.
L. Parker.
Phys. Rev. D, Vol. 24, 535 - 537 (1981). – Abstr. in Phys. Abstr., Vol. 84, Abstr. 94398 (1981).

066.102 On the metric tensor of empty space in general relativity. V. Benza, P. Caldirola.
Nuovo Cimento B, Ser. 11, Vol. 62B, 327 - 336 (1981). Abstr. in Phys. Abstr., Vol. 84, Abstr. 94580 (1981).

066.103 Toward a realistic nebular gravitational lens.
C. C. Dyer, R. C. Roeder.
Astrophys. J., Vol. 249, 290 - 296 (1981).
The authors apply the results of optical scalar techniques to a static, spherical distribution of matter in order to derive properties of the images of a distant radiating source. The multiple image theorem is briefly mentioned, and a method is outlined which allows for possible variation in the mass density traversed by each beam of radiation. The fact that the source has finite size is shown to produce a damping effect on any possible large amplification introduced by the star nearest the particular beam being considered.

066.104 Still another "black hole"? M. Eliāss.
Zvaigžnotā debess, 1980. gada rudens, p. 24 - 25.
In Latvian.

066.105 Vacuum counterexamples to the cosmic censorship hypothesis. B. D. Miller.
J. Math. Phys., Vol. 22, 1452 - 1456 (1981). – Abstr. in Phys. Abstr., Vol. 84, Abstr. 94710 (1981).

066.106 Solutions to wave equations on black hole geometries. W. E. Couch.
J. Math. Phys., Vol. 22, 1457 - 1462 (1981). – Abstr. in Phys. Abstr., Vol. 84, Abstr. 94711 (1981).

066.107 All non-twisting Ns with cosmological constant.
A. Garcia Diaz, J. F. Plebanski.
Group theoretical methods in physics, (see 012.030), p. 413 - 417 (1980). – Abstr. in Phys. Abstr., Vol. 84, Abstr. 94719 (1981).

066.108 Quantum conformal fluctuations near the classical space-time singularity. J. V. Narlikar.
Found. Phys., Vol. 11, 473 - 492 (1981). – Abstr. in Phys. Abstr., Vol. 84, Abstr. 94727 (1981).

066.109 Wheeler-Einstein-Mach spacetimes.
J. A. Isenberg.
Phys. Rev. D, Vol. 24, 251 - 256 (1981). – Abstr. in Phys. Abstr., Vol. 84, Abstr. 94729 (1981).

066.110 Can quantum effects prevent spacetime collapse?
M. J. Gotay, J. A. Isenberg.
Group theoretical methods in physics, (see 012.030), p. 418 - 423 (1980). – Abstr. in Phys. Abstr., Vol. 84, Abstr. 94731 (1981).

066.111 The equation of state of a cosmological self-gravitating fluid influenced by thermal or turbulent motion and the primeval background radiation.
E. Nowotny, M. G. Corona.
Astrophys. Space Sci., Vol. 79, 443 - 461 (1981).
The usual dispersion relation of a self-gravitating cosmological fluid $\omega^2 = k^2 v_s^2 - 4\pi G\rho$ is analyzed within thermodynamical theory. Restricting to a static universe an equation of state for the substratum is found including gravitational pressure. Non-thermal influences (turbulence) on the equation of state are also discussed.

066.112 Spectral shifts near compact objects.
K. Lake, E. Myra.
Astrophys. Space Sci., Vol. 79, 515 - 519 (1981).
The authors show that radiation emitted from material freely falling toward a black hole or neutron star cannot be blue-shifted as recently claimed by Cohen and Struble. The relativistic corrections to the classical apparent limb angle are given explicitly for spherical sources in collapse.

066.113 Fluctuations in the microwave background at intermediate angular scales.
F. Melchiorri, B. O. Melchiorri, C. Ceccarelli, L. Pietranera.
Astrophys. J., Lett., Vol. 250, L1 - L4 (1981).
An important distortion of the dipole pattern in the microwave background has been detected by two groups (Fabbri, Melchiorri, Guidi, and Natale; Boughn, Cheng, and Wilkinson). Peebles has suggested that this irregularity may arise from large-scale gradients in gravitational potential due to the irregular distribution of mass in clusters of galaxies. His theory anticipates anisotropies in the microwave background at smaller angular scales. The authors present the experimental evidence for microwave background fluctuations at an angular scale of 6°, with an amplitude of $(1.1 \pm 0.2) \times 10^{-4}$ K.

066.114 The gravitational quadrupole formulae and gravitationally bound matter systems. G. Schäfer.
Astrophys. J., Lett., Vol. 250, L5 - L8 (1981).
For gravitationally bound matter systems, a new approach to the gravitational quadrupole radiation problem is presented. It simultaneously gives the energy loss rate at infinity and the radiation-reaction force on the matter system, without using singular asymptotic expansions and without encountering long-range divergences. The quadrupole formulae are confirmed.

066.115 On an internal solution of general relativistic equations. A. N. Krishtal'.
Inst. fiz. AN USSR. Prepr., 1981, No. 6, 27 pp. In Russian. Abstr. in Ref. zh., 51. Astron., 10.51.216 (1981).

066.116 On a representation of the gravitational field of simplest bodies by means of point mass attraction.
V. N. Fomin.
Astron. i geod., Tomsk, 1980, No. 8, p. 102 - 110. In Russian. Abstr. in Ref. zh., 51. Astron., 10.51.217 (1981).

066.117 Unified relativistic theory of motion of the inner planets of the solar system. Relativistic effects in determination of planetary orbits using radar location data.
M. D. Kislik, Yu. F. Kolyuka, V. A. Kotel'nikov, G. M. Petrov,

V. F. Tikhonov.
Usp. fiz. nauk, Tom 134, 165 - 167 (1981). In Russian.
Abstr. in Ref. zh., 51. Astron., 10.51.218 (1981).

066.118 General relativistic connection between orbital and rotational planetary motions.
V. I. Antonov.
Izv. vuzov. Fiz., Tom 24, No. 3, p. 11 - 15 (1981). In Russian.
Abstr. in Ref. zh., 51. Astron., 10.51.219 (1981).

066.119 On an experiment for measuring the velocity of the earth's motion relative to absolute space.
E. V. Borovko.
Redkol. zh. Izv. vuzov. Fiz. Tomsk, 1981. 9 pp. In Russian.
Abstr. in Ref. zh., 51. Astron., 10.51.220 (1981).

066.120 Black holes, thermodynamics and information.
D. A. Kirzhnits, V. P. Frolov.
Priroda, 1981, No. 11, p. 2 - 14. In Russian.

066.121 On some possible additional relativistic effect in SS433.   R. Ruffini, Doo Jong Song, L. Stella.
Astron. Astrophys., Vol. 103, L7 - L9 (1981).

The 163.8 days period observed in the shifted lines of SS433, if due to Lense-Thirring precession, should change with time as a consequence of matter accreting into the black hole and changing its angular momentum. The propagation of light signals within a few Schwarzschild radii of the black hole surface, in the ring models, may lead to the observation of further relativistic effects.

066.122 On the observability of gravitational scintillation.
J. M. Hameury, M. Perault, S. Bonazzola, J. L. Puget.
Astron. Astrophys., Vol. 103, 63 - 68 (1981).

The authors study analytically the possibility of gravitational scintillation which would lead to an estimate of the missing mass. The approximation used is the thin screen which seems to be rather good as the analytical expression derived for the r.m.s. angular deviation fits rather well with numerical simulations of Press and Gunn (1973). They investigate other phenomenons than the formation of double images: scintillation in time and frequency; the authors conclude that there is no detectable radio-source small enough to give observable effects, but from the observed angular size of quasars, they are able to give constraints on the mass distribution.

066.123 Gravitational waves from cosmos.
V. N. Rudenko.
Zemlya Vselennaya, 1981, No. 6, p. 28 - 32. In Russian.

066.124 On the Nijenhuis tensor in relativistic electrodynamics.   L. N. Katkar.
Nuovo Cimento B, Ser. 11, Vol. 64B, 9 - 19 (1981). − Abstr. in Phys. Abstr., Vol. 84, Abstr. 98993 (1981).

066.125 On quantum theory of general-relativistic many-particle systems. I. Formalism.   I. Ichinose.
Prog. Theor. Phys., Vol. 65, 2023 - 2037 (1981). − Abstr. in Phys. Abstr., Vol. 84, Abstr. 99005 (1981).

066.126 General relativistic collapse of axially symmetric stars leading to the formation of rotating black holes.   T. Nakamura.
Prog. Theor. Phys., Vol. 65, 1876 - 1890 (1981). − Abstr. in Phys. Abstr., Vol. 84, Abstr. 102035 (1981).

066.127 Structure of superposed two Kerr metrics.
K. Oohara, H. Sato.
Prog. Theor. Phys., Vol. 65, 1891 - 1900 (1981). − Abstr. in Phys. Abstr., Vol. 84, Abstr. 102149 (1981).

066.128 Hiding of the conserved (anti)baryonic charge into black holes.   A. D. Dolgov.
Phys. Rev. D, Vol. 24, 1042 - 1044 (1981). − Abstr. in Phys. Abstr., Vol. 84, Abstr. 102150 (1981).

066.129 Comment on 'Collision of plane gravitational waves without singularities'.   Y. Nutku.
Phys. Rev. D, Vol. 24, 1040 - 1041 (1981). − Abstr. in Phys. Abstr., Vol. 84, Abstr. 102283 (1981).

066.130 Black holes and trapped points.   A. Krolak.
Acta Phys. Polonica B, Vol. B12, 643 - 649 (1981).
Abstr. in Phys. Abstr., Vol. 84, Abstr. 98810 (1981).

066.131 Interacting quantum fields around a black hole.
S. W. Hawking.
Commun. Math. Phys., Vol. 80, 421 - 442 (1981). − Abstr. in Phys. Abstr., Vol. 84, Abstr. 98811 (1981).

066.132 The atmospheric Cherenkov technique in searches for exploding primordial black holes.
S. Danaher, D. J. Fegan, N. A. Porter, T. C. Weekes.
Philos. Trans. R. Soc. London, Ser. A, Vol. 301, 665 - 667 (1981). − Abstr. in Phys. Abstr., Vol. 84, Abstr. 98812 (1981).

066.133 Metric and electromagnetic perturbations of the Reissner-Nordström black hole.
B. C. Xanthopoulos.
Proc. R. Soc. London,Ser. A, Vol. 378, 73 - 88 (1981).
Abstr. in Phys. Abstr., Vol. 84, Abstr. 98813 (1981).

066.134 Quantization of electromagnetic and gravitational perturbations of·a Kerr black hole.
P. Candelas, P. Chrzanowski, K. W. Howard.
Phys. Rev. D, Vol. 24, 297 - 304 (1981). − Abstr. in Phys. Abstr., Vol. 84, Abstr. 98814 (1981).

066.135 An accreting black hole model for SS433.
L. Z. Fang.
Irish Astron. J., Vol. 14, (see 012.035), 153 - 155 (1980).

066.136 Generalised Kerr-Schild space-times.
A. H. Taub.
Ann. Physics, Vol. 134, 326 - 372 (1981). − Abstr. in Phys. Abstr., Vol. 84, Abstr. 102429 (1981).

066.137 The equivalence principle and spatial curvature.
D. E. Dugdale.
Eur. J. Phys., Vol. 2, No. 1, p. 48 - 51 (1981). − Abstr. in Phys. Abstr., Vol. 84, Abstr. 102433 (1981).

066.138 On a canonical treatment of the Einstein equations in an expanding Universe. Isotropic coordinate conditions.   M. Kihara, H. Nariai.
Prog. Theor. Phys., Vol. 65, 1613 - 1620 (1981). − Abstr. in Phys. Abstr., Vol. 84, Abstr. 102441 (1981).

066.139 Gravitational field of a pencil of light.
N. V. Mitskievich *(Mitskievič).*
Exp. Tech. Phys., Vol. 29, 213 - 215 (1981). − Abstr. in Phys. Abstr., Vol. 84, Abstr. 102445 (1981).

066.140 Free-electron generation of gravitational radiation.
J.-Y. Vinet.
Nuovo Cimento A, Ser. 11, Vol. 64A, 163 - 178 (1981).
Abstr. in Phys. Abstr., Vol. 84, Abstr. 102446 (1981).

066.141 Gravitational radiation from the propulsion of bodies in a line.   F. I. Cooperstock, D. W. Hobill.
Phys. Lett. A, Vol. 84A, 297 - 300 (1981). − Abstr. in Phys. Abstr., Vol. 84, Abstr. 102447 (1981).

066.142 **Comment on colliding plane gravitational waves.**
M. Halilsoy.
Phys. Lett. A, Vol. 84A, 359 - 360 (1981). – Abstr. in Phys.
Abstr., Vol. 84, Abstr. 102448 (1981).

066.143 **Restrictions on relativistically rotating fluids.**
J. Schendel, J. Winicour.
J. Math. Phys., Vol. 22, 1696 - 1702 (1981). – Abstr. in Phys.
Abstr., Vol. 84, Abstr. 102451 (1981).

066.144 **Remarks on a possible relation between gravitational
instantons and the spin thermodynamics of a Kerr
black hole.** A. Curir.
Nuovo Cimento, Lett., Ser. 2, Vol. 31, 517 - 520 (1981).
Abstr. in Phys. Abstr., Vol. 84, Abstr. 102453 (1981).

066.145 **Electromagnetic and gravitational perturbations of
the Kerr metric.**
K. R. Pechenick, M. W. Kearney, J. M. Cohen.
Nuovo Cimento B, Ser. 11, Vol. 64B, 453 - 470 (1981).
Abstr. in Phys. Abstr., Vol. 84, Abstr. 102454 (1981).

066.146 **Hamiltonian formulation for the gauge theory of the
gravitational coupling.**
R. Giachetti, R. Ricci, E. Sorace.
J. Math. Phys., Vol. 22, 1703 - 1710 (1981). – Abstr. in Phys.
Abstr., Vol. 84, Abstr. 102458 (1981).

066.147 **The quantum era.** A. Casher, F. Englert.
Phys. Lett. B, Vol. 104B, 117 - 120 (1981). – Abstr.
in Phys. Abstr., Vol. 84, Abstr. 102482 (1981).

066.148 **On black holes in magnetic universes.**
W. A. Hiscock.
J. Math. Phys., Vol. 22, 1828 - 1833 (1981). – Abstr. in Phys.
Abstr., Vol. 84, Abstr. 108248 (1981).

066.149 **Optical observations from within a black-hole: a
large-scale viewpoint.** G. Debney, D. Farnsworth.
Nuovo Cimento B, Ser. 11, Vol. 64B, 367 - 382 (1981).
Abstr. in Phys. Abstr., Vol. 84, Abstr. 108252 (1981).

066.150 **On the variation of G and scale invariant gravitation.**
A. K. Kembhavi, M. D. Pollock.
Mon. Not. R. Astron. Soc., Vol. 197, 1087 - 1091 (1981).
It is shown that with the scale-breaking part of the
matter Lagrangian neglected, the theory of Canuto et al. is not
consistent with a varying 'constant' of gravitation G. The
effect of the scale-breaking terms cannot be obtained by
imposing external conditions.

066.151 **On the variation of G and scale invariant gravitation–
a reply.** V. M. Canuto.
Mon. Not. R. Astron. Soc., Vol. 197, 1093 - 1095 (1981).
The scale invariant gravitational Lagrangian presented by
the author several years ago, must be supplemented by a
matter Lagrangian L_m. It is shown that the form of L_m
proposed by Kembhavi & Pollock (KP) far from being
'inescapable' or 'the only possible form' as the authors claim,
is instead a very particular case that contains already im-
printed all the results KP believe they have discovered. The
conclusions of KP are therefore erroneous.

066.152 **Cosmic thermalization and the microwave background radiation.** N. C. Rana.
Mon. Not. R. Astron. Soc., Vol. 197, 1125 - 1137 (1981).
The author has suggested here a different origin of the
microwave background radiation (MBR) in view of some of
the difficulties associated with the standard interpretation.
Extensive stellar-type nucleosynthesis could provide radiation
with the requisite energy density of the MBR and its spectral

features are guaranteed by adequate thermalization of the
above radiation by an ambient intergalactic dust medium.

066.153 **SS 433: background for a relativistic model.**
L. Z. Fang, R. Ruffini, L. Stella.
Vistas Astron., Vol. 25, (see 012.040), 185 - 211 (1981).
A detailed relativistic model for SS 433 is developed by
assuming an accretion disk around a slowly rotating Kerr black
hole with mass $M \approx 10 M_{\odot}$. The structure of the accretion disk
is characterized by a nearly toroidal magnetic field. In this
model the "moving emission lines" originate in two emitting
regions of a near circular ring of material moving in Keplerian
orbits in the inner parts of the disk. A possible observational
test of general relativity with SS 433 is briefly outlined.

066.154 **Self-focussed electromagnetic waves and SS 433.**
G. Benford, A. Ferrari, S. Massaglia.
Vistas Astron., Vol. 25, (see 012.040), 213 - 216 (1981).
Double jets of matter emitted by neutron stars (or black
holes) have been invoked to explain the periodicity in wavelength shifts in SS 433. The authors attempt to show that the
jets need not be beams of particles, but could instead be self-
focussed, large-amplitude electromagnetic waves.

066.155 **On certain aspects of the relativistic frequency shift
formula.** S. L. Bazanski.
Nukleonika, Vol. 25, 1393 - 1404 (1980). – Abstr. in Phys.
Abstr., Vol. 85, Abstr. 39 (1982).

066.156 **Asymptotic symmetries of de Sitter space-time.**
P. T. Chrusciel.
Acta Phys. Polonica B, Vol. B12, 719 - 737 (1981). – Abstr.
in Phys. Abstr., Vol. 85, Abstr. 81 (1982).

066.157 **World vectors, Jacobi vectors and Jacobi one-forms
on a manifold with a linear symmetric connection.**
R. Schattner, M. Trümper.
J. Phys. A, Vol. 14, 2345 - 2352 (1981). – Abstr. in Phys.
Abstr., Vol. 85, Abstr. 84 (1982).

066.158 **On the regularization of the cosmological singularity
in general relativity.** D.-E. Liebscher.
Nukleonika, Vol. 25, 1387 - 1391 (1980). – Abstr. in Phys.
Abstr., Vol. 85, Abstr. 86 (1982).

066.159 **Nonlinear effects of gravitational and electromagnetic
radiation on the propagation of light.** B. Linet.
J. Phys. A, Vol. 14, 2673 - 2678 (1981). – Abstr. in Phys.
Abstr., Vol. 85, Abstr. 87 (1982).

066.160 **The far-field of a gravitating source in relation to its
interior.** N. Spyrou.
Gen. Relativ. Gravitation, Vol. 13, 551 - 559 (1981).
The author proves that in the context of the first post-
Newtonian perfect fluid theory it is possible to express the
far-field metric tensor of an arbitrary perfect-fluid source in
terms of parameters, which are known as three-dimensional
volume integrals over the source's interior. Such a description
for an arbitrary source is not known in the context of the
exact theory of gravity. It seems that the approximate description presented here is the only and the more general one valid
consistently to the post-Newtonian approximation.

066.161 **Electromagnetic-gravitational energy systems.**
K. H. Schatten.
Gen. Relativ. Gravitation, Vol. 13, 561 - 567 (1981).
Two methods are considered to "tap" the earth's rotational energy. This ancient "collapsed gravitational energy" exceeds the earth-lunar binding energy. One involves an orbiting
"electromagnetic-gravitational" coupling system whereby the
earth's rotation, with its nonuniform mass distribution, first
uses gravity to add orbital energy to a satellite, similar to a

planetary "flyby". The second stage involves enhanced satellite "drag" as current-carrying coils withdraw the added orbital energy as they pass through the earth's nonuniform magnetic field. A second more direct method couples the earth's rotational motion using conducting wires moving through the noncorotating part (ionospheric current systems) of the geomagnetic field. These methods, although not immediately feasible, are considerably more efficient than using pure gravitational coupling to earth-moon tides.

066.162   **Coordinate-dependent 3 + 1 formulation of the general relativity equations for a perfect fluid.**
R. A. Nelson.
Gen. Relativ. Gravitation, Vol. 13, 569 - 580 (1981).
      This paper defines mass, momentum, and energy densities for a perfect fluid, and derives a coordinate-dependent 3 +1 decomposition of the equation of motion in terms of a scalar potential $\psi \equiv c^2 [(-g_{44})^{1/2} -1]$ and a vector potential $A_i \equiv cg_{4i}/(-g_{44})^{1/2}$. The momentum equation has the form of the Euler equation except there is an additional force proportional to the vector potential and the rate of change of kinetic energy per unit volume. The momentum and energy equations are integrated to obtain the equations previously derived for a particle. The momentum equation is solved for the total acceleration of a fluid element. The equations are exact and do not depend on the choice of coordinate system.

066.163   **Double-dual solutions of generalized theories of gravitation.**
I. M. Benn, T. Dereli, R. W. Tucker.
Gen. Relativ. Gravitation, Vol. 13, 581 - 589 (1981).
      Solutions to the Stephenson-Yang theory of gravity and its generalizations are discussed. By considering the inclusion of a cosmological term in the action spherically symmetric static solutions are presented that do not fall into the vacuum Einstein class. A simple double-duality ansatz is responsible for all the solutions that are discussed.

066.164   **Galaxies as gravitational lenses: realistic models.**
C. C. Dyer, R. C. Roeder.
Gen. Relativ. Gravitation, Vol. 13, 591 - 597 (1981).
      The gravitational lens effects associated with a transparent mass distribution are quite different from those of the well-known opaque sphere. The authors have shown that any spherical galaxy whose mass distribution, when projected onto the plane of the sky, decreases outward from the center of the galaxy and diverges less rapidly than $1/h$ as $h \rightarrow 0$, must always produce an odd number of images, usually one or three, of a source located behind the galaxy. Using optical scalar techniques, the amplification of each image can readily be obtained. For a given source and galaxy, the authors can define a dimensionless focal length, a function of impact parameter, and a dimensionless distance factor, depending on lens parameters and source distance. The central value of the ratio of these quantities determines the multiplicity of the images. The mass distribution of the galaxy is a crucial function, and the authors show in some detail how this affects the various focal lengths.

066.165   **A charged particle in bimetric general relativity.**
D. Falik, N. Rosen.
Gen. Relativ. Gravitation, Vol. 13, 599 - 604 (1981).
      The gravitational field of a charged particle is investigated on the basis of the bimetric general relativity theory. It is found that the field differs from the Reissner-Nordström field only very close to the sphere $R = m + (m^2 - Q^2)^{1/2}$. This sphere is impenetrable, and its interior is unphysical.

066.166   **Primordial black holes.**   E. Bettwieser.
      Proceedings of the 5th Göttingen-Jerusalem-Symposium on Astrophysics, (see 012.041), p. 129 - 139 (1981).

066.167   **Accretion onto massive black holes: effects of $e^- e^+$ pairs.**   R. Z. Yahel.
Proceedings of the 5th Göttingen-Jerusalem-Symposium on Astrophysics, (see 012.041), p. 141 - 147 (1981).

066.168   **Collective gravitational effects.**   G. E. Tauber.
      Plasma astrophysics, (see 012.042), p. 365 - 368 (1981).

066.169   **The deviation of timelike geodesics in space-time.**
R. F. Crade, G. S. Hall.
Phys. Lett. A, Vol. 85A, 313 - 315 (1981). — Abstr. in Phys. Abstr., Vol. 85, Abstr. 3436 (1982).

066.170   **Reaction force of gravitational radiation on a continuous medium. I. Perturbation of a quasi-Newtonian motion.**   B. Linet.
Ann. Inst. Henri Poincaré Sect. A, Vol. 34, 419 - 425 (1981). In French. — Abstr. in Phys. Abstr., Vol. 85, Abstr. 3442 (1982).

066.171   **Reaction force of gravitational radiation on a continuous medium. II. Perturbation of a quasi-Minkowskian motion.**   B. Linet.
Ann. Inst. Henri Poincaré Sect. A, Vol. 34, 427 - 435 (1981). In French. — Abstr. in Phys. Abstr., Vol. 85, Abstr. 3443 (1982).

066.172   **Quantum detection on the vacuum by non-uniformly accelerated observers.**   N. Sanchez.
Phys. Lett. B, Vol. 105B, 375 - 380 (1981). — Abstr. in Phys. Abstr., Vol. 85, Abstr. 3444 (1982).

066.173   **Perturbations around a slowly rotating and slightly charged black hole.**   C. H. Lee.
Prog. Theor. Phys., Vol. 66, 180 - 194 (1981). — Abstr. in Phys. Abstr., Vol. 85, Abstr. 3446 (1982).

066.174   **Stability of massive objects in a new scalar-tensor theory.**
G. Schmidt, W. Greiner, U. Heinz, B. Müller.
Phys. Rev. D, Vol. 24, 1484 - 1490 (1981). — Abstr. in Phys. Abstr., Vol. 85, Abstr. 3455 (1982).

066.175   **Metric from matter.**
D. Amati, G. Veneziano.
Phys. Lett. B, Vol. 105B, 358 - 362 (1981). — Abstr. in Phys. Abstr., Vol. 85, Abstr. 3462 (1982).

066.176   **Symmetry restoration in conformally flat metrics.**
G. Denardo, E. Spallucci.
Nuovo Cimento A, Ser. 11, Vol. 64A, 15 - 26 (1981). — Abstr. in Phys. Abstr., Vol. 85, Abstr. 3737 (1982).

066.177   **n-dimensional complex Riemann-Einstein spaces with O(n−1, C) as the symmetry group.**
A. Krasinski, J. Plebanski.
Rep. Math. Phys., Vol. 17, 217 - 248 (1980). — Abstr. in Phys. Abstr., Vol. 85, Abstr. 6532 (1982).

066.178   **On the frequency shift in radiation from a source orbiting a black hole.**   R. C. Kapoor.
Bull. Astron. Soc. India, Vol. 9, 232 - 242 (1981).
      An analytical study has been carried out of the non-tangential emission of photons in the orbital plane from a monochromatic source of radiation orbiting a black hole, using the method of geometrical optics. The frequency shift in radiation, which is a combination of Doppler effect and gravitational redshift, is studied for various angles of emission.

066.179   **Gravitational lens in the universe.**
V. Mukhanov.

Nauk. i zhizn', 1981, No. 5, p. 28 - 32. In Russian. − Abstr. in Ref. zh., 51. Astron., 11.51.1008 (1981).

066.180 Generalities on geometric theories of gravitation.
A. Trautman.
Cosmology and gravitation, (see 012.051), p. 1 - 4 (1980).

066.181 The motion of test-particles in non-Riemannian space-times.
D.- E. Liebscher.
Cosmology and gravitation, (see 012.051), p. 125 - 137 (1980).

066.182 Experimental gravitation with measurements made from within a planetary system.
R. D. Reasenberg.
Cosmology and gravitation, (see 012.051), p. 317 - 357 (1980).

The first two parts deal with solar-system tests of relativity that have been or are currently being conducted at MIT by the radio physics group: 1) solar-system data and the corresponding mathematical models; 2) analysis, estimation, and results. The third part deals with three possible future missions: The Solar Probe is a proposed NASA mission to fly within a few solar radii of the Sun; VOIR is a Venus radar mapping mission that may provide high-quality Earth-Venus delay measurements: POINTS, Precision Optical INTerferometry in Space, is an instrument concept in an early stage of development at MIT and the Charles Stark Draper Laboratory. POINTS is intended to measure the second-order bending of light by solar gravity.

066.183 Bimetric general relativity theory.
N. Rosen.
Cosmology and gravitation, (see 012.051), p. 383 - 405 (1980).

066.184 Graviton photoproduction in static electromagnetic fields and some astrophysical applications.
S. R. Valluri.
Cosmology and gravitation, (see 012.051), p. 467 - 476 (1980).

A review of the photoproduction processes that give rise to gravitons is presented. Some astrophysical applications are also considered.

066.185 Relativistic spherical polytropes: an analytical approach. J. P. Sharma.
Gen. Relativ. Gravitation, Vol. 13, 663 - 667 (1981).

Using the technique of Padé (2, 2) approximant the author presents an approximate analytical solution to the field equations of general relativity for time-independent, spherically symmetric systems in which the pressure P and density ρ are related by a polytropic equation of state: $P = K\rho^{1+1/n}$.

066.186 Quantum fluctuations in the conformally flat and the Schwarzschild space-times.
T. Padmanabhan, J. V. Narlikar.
Gen. Relativ. Gravitation, Vol. 13, 669 - 679 (1981).

A general technique is described for dealing with the quantum fluctuations between conformally flat space-times. The second part of the paper deals with the Schwarzschild space-time. It is shown there that this space-time is stable against fluctuations of mass, but transitions between two space-times of different masses can be obtained via conformal fluctuations.

066.187 Spherically symmetric fields in Rosen's bimetric theories of gravitation. M. Israelit.
Gen. Relativ. Gravitation, Vol. 13, 681 - 688 (1981).

By means of bimetric Killing vectors two spherically symmetric fields are investigated: (1) the time-dependent one in Rosen's flat-background bimetric theory; and (2) the energy-preserving in Rosen's cosmological-background bimetric theory with $k = 1$. In the first case a wave behavior of the field is present. In the second case a time evolution is obtained for fields, created by insular systems of constant energy. These phenomena are typical for bimetric theories of gravitation.

066.188 Post-Newtonian spin and angular momentum of bounded systems. A. Caporali, N. Spyrou.
Gen. Relativ. Gravitation, Vol. 13, 689 - 701 (1981).

The authors generalize the Newtonian expressions for the orbital angular momentum of a two-body system, and for the spin of each body, by introducing corresponding definitions in the post-Newtonian approximation of fully conservative theories of gravity. Using this definition of the spin and assuming that the bodies rotate rigidly and that the equations of motion are Hamiltonian, they show that in fully conservative theories of gravity the spin of each body undergoes a relativistic precession about the direction of the orbital angular momentum, as a consequence of the local equations of motion for a perfect fluid.

066.189 A new argument in favour of the existence of gravitational waves? A. Balklavs.
Zvaigžņota debess, 1980/81. gada ziema, p. 21 - 24. In Latvian.

066.190 On a possibility of detecting gravitational waves of extraterrestrial origin from their action on the earth's ionosphere. V. I. Bashkov, V. B. Konstantinov, Z. G. Murzakhanov, E. V. Parshin.
Tr. Kazan. Gorod. Astron. Obs., Vyp. 45, p. 66 - 74 (1980). In Russian.

066.191 Los agujeros negros y sus implicancias astrofísicas. Parte I. D. L. Block.
Rev. Astron., Tomo 53, No. 218, p. 2 - 5 (1981).

066.192 Null geodesics in the Kerr-Newman metric.
Z. Stuchlík.
Bull. Astron. Inst. Czechoslovakia, Vol. 32, 366 - 373 (1981).

A complete discussion of the radial motion of photons in the Kerr-Newman metric is given. The qualitative differences between the Kerr-Newman metric and the Kerr metric occur only below the inner horizon and around naked singularities.

066.193 Tests of general relativity at the quantum level.
E. Fischbach.
Cosmology and gravitation, (see 012.051), p. 359 - 373 (1980).

The author considers several tests of general relativity at the quantum level. One of these is based on the observation that an effect of a gravitational field on a hydrogen atom is to admix states of opposite parity such as $2S_{1/2}$ and $2P_{1/2}$. This leads to a circular polarization P_γ of emitted radiation which could be sufficiently large to be detected in white dwarfs or in certain binary systems. Since the magnitude of P_γ varies from theory to theory, it is possible that a study of this effect could provide a feasible means of testing general relativity at the quantum level.

066.194 Conformal anomaly and massless particle production by a conformally-flat metric. A. D. Dolgov.
Zh. ehksp. i teor. fiz., Tom 81, 417 - 428 (1981). In Russian. Abstr. in Ref. zh., 51. Astron., 12.51.1015 (1981).

066.195 On Soldner's value of Newtonian deflection of light.
H.-J. Treder, G. Jackisch.
Astron. Nachr., Band 302, 275 - 277 (1981).

066.196 The gravitational lens. E. Leibowitz.
Mada, Vol. 25, 85 - 87 (1981). In Hebrew.

066.197 Spherical accretion of massive black holes: a model for galactic nuclei.
L. Maraschi, G. C. Perola, A. Treves.
High-energy astrophysics, (see 012.067), p. 67 - 70 (1981).

The possibility of explaining the continuous emission of active galactic nuclei in the frame of a model of spherical accretion onto a massive black hole is discussed. Cool inhomogeneities within the accretion flow could be responsible for the broad line emission if half of the accreting matter is in the dense phase. A crucial test of this hypothesis is the expected correlation between the ratio of the luminosity in lines to the total luminosity and the hardness of the continuous spectrum.

066.198 **The special theory of relativity and systems of two or a great amount of bodies.**
Published by J. C. Proost-Thoden van Velzen, Thorbeckelaan 46, 1412 BR Naarden, The Netherlands. 3 + 40 pp. (1980/81).

066.199 **On the role of ion viscosity in accretion disks around black holes.**
O. P. Medvedeva, N. I. Shakura.
Astron. Tsirk., No. 1148, p. 3 - 6 (1981). In Russian.

066.200 **Rocket measurement of the cosmic background submillimeter spectrum.**   H. P. Gush.
Phys. Rev. Lett., Vol. 47, 745 - 748 (1981).
    Measurements of the spectrum of the night sky have been made in the wave-number region 5 to 35 cm^{-1} with use of a liquid-helium–cooled rocket-borne interferometer. The observed spectrum deviates from a 2.7-K Planck spectrum, there being a deficit of energy between 5 and 11 cm^{-1} and an excess at higher wave numbers.

066.201 **Quantum limit on the Michelson interferometer used for gravitational-wave detection.**   R. Loudon.
Phys. Rev. Lett., Vol. 47, 815 - 818 (1981).
    The quantum limitations on measurements of the relative positions of the two mirrors in a Michelson gravity-wave detector are derived. The effects of intensity fluctuations in the two light beams and radiation pressure on the mirrors are modeled in a unified calculation. The radiation-pressure uncertainty in mirror position differs in general from earlier work, but gives agreement for coherent input light.

066.202 **Indirect evidence for quantum gravity.**
    D. N. Page, C. D. Geilker.
Phys. Rev. Lett., Vol. 47, 979 - 982 (1981).
    An experiment gave results inconsistent with the simplest alternative to quantum gravity, the semiclassical Einstein equations. This evidence supports (but does not prove) the hypothesis that a consistent theory of gravity coupled to quantized matter should also have the gravitational field quantized.

066.203 **Approximate effective action for quantum gravity.**
    B. S. DeWitt.
Phys. Rev. Lett., Vol. 47, 1647 - 1650 (1981).
    A new gauge-invariant effective action is proposed for quantum gravity, based on older results that go beyond finite-order perturbation theory. Expressed in coordinate space rather than momentum space it should find important applications in theory of the early universe.

Elementary general relativity.   See Abstr. 003.044.

The edge of infinity: naked singularities and the destruction of spacetime.   See Abstr. 003.051.

Invitation to physics.   See Abstr. 003.125.

The science of space-time.   See Abstr. 003.130.

Spaziotempo.   See Abstr. 003.142.

Relativistic theory of gravity. Sources and formation. 1900 - 1915.   See Abstr. 003.155.

Space, time, and gravity. The theory of the big-bang and black holes.   See Abstr. 003.157.

Relativistische Astrophysik. Ein Bericht über das X. Texas Symposium in Baltimore.   See Abstr. 011.016.

Relativistic astrophysics: the view from Texas in Baltimore.   See Abstr. 011.024.

Modern theoretical and experimental problems of relativity theory.   See Abstr. 012.049.

Superspace and supergravity. Proceedings of a workshop, Cambridge, England, July 1980.   See Abstr. 012.070.

Pressure broadening of oxygen and its implications for cosmic background measurements.
See Abstr. 022.024.

Beyond the black hole.
See Abstr. 022.046.

Simulating the transparent gravitational lens.
See Abstr. 022.116.

Suppression of the firehouse instability by hard energy spectra.   See Abstr. 022.124.

A study of the external-noise input in Weber-type gravitational-wave antennas.   See Abstr. 034.013.

The cryogenic detector of gravitational waves in Frascati.   See Abstr. 034.034.

Sensitivity analysis of a resonant mass gravitational wave antenna with resonant transducer.
See Abstr. 034.047.

Enhanced sensitivity of a gravitational wave detector.
See Abstr. 034.048.

The Roman experiment for the detection of gravitational waves.   See Abstr. 034.062.

Relativistic perturbations of planetary orbits in the generalized three-parametric Schwarzschild metric. The case of Mercury.   See Abstr. 042.001.

Comment on 'General relativity and satellite orbits: the motion of a test particle in the Schwarzschild Metric' by D. P. Rubincam.   See Abstr. 042.039.

On the evolution of the homogeneous ellipsoidal figures. II. Gravitational collapse and gravitational radiation.
See Abstr. 042.071.

On the new method of determining the gravitational constant.   See Abstr. 043.003.

A new mechanism for energy release in astrophysical objects.   See Abstr. 061.024.

Time delay between gravitational waves and neutrino bursts from a supernova explosion: a test for the neutrino mass.
See Abstr. 061.048.

Self-consistent solutions for electromagnetic plane waves in a relativistic plasma.   See Abstr. 062.011.

Physical conditions in an optically thin relativistic gas irradiated by γ-rays.   See Abstr. 062.012.

Generation of a d.c. field by nonlinear electromagnetic waves in relativistic plasmas. See Abstr. 062.068.

The effects of *in situ* particle acceleration on the appearance of relativistically moving plasmons. See Abstr. 062.073.

Compton scattering in a converging fluid flow — III. Spherical supercritical accretion. See Abstr. 063.001.

On the transport and propagation of relativistic electrons in galaxies. See Abstr. 063.046.

A new look at the dynamics of twisted accretion disks. See Abstr. 064.001.

Structure and stability of heated accretion flow onto compact objects. See Abstr. 064.027.

A supercritical accretion disk model for SS 433. See Abstr. 064.070.

Thick supercritical accretion disk as a model for SS 433. See Abstr. 064.082.

Remarks on accretion-disk models of SS 433. See Abstr. 064.083.

La structure, l'énergie et l'évolution des étoiles. See Abstr. 065.036.

Computer simulations of stellar collapse and shock wave propagation. See Abstr. 065.068.

The collapse of rotating stellar cores. See Abstr. 065.069.

Review of the theories on SS 433. See Abstr. 065.078.

Internal gravity waves in the solar atmosphere. I. Adiabatic waves in the chromosphere. See Abstr. 080.034.

Independence of the gravitational constant from gross Earth data. See Abstr. 081.061.

The equivalence principle and anomal motions of bodies in the solar system. See Abstr. 091.003.

Interpretation of observations supporting a magnetospheric model of SS 433. See Abstr. 117.092.

Axisymmetric gravitational collapse of rotating interstellar gas clouds. See Abstr. 131.117.

Collapse of gas with embedded stars. See Abstr. 131.187.

Broad emission line profiles from discs and quasars. See Abstr. 141.069.

Triple quasar — effect of a gravitational lens? See Abstr. 141.112.

Quasar reddening, intergalactic dust, and the microwave background. See Abstr. 141.117.

Observations of a decrement in the microwave background radiation toward the distant cluster of galaxies 0016 + 16. See Abstr. 141.155.

Gravitational waves from an orbiting pulsar. See Abstr. 141.527.

Pulsar theory. See Abstr. 141.562.

Anisotropy: high energy cosmic rays and the thermal background radiation. See Abstr. 143.062.

Odd-parity perturbations of spherically symmetric star clusters in general relativity. See Abstr. 151.001.

Stability of contracting pancakes. See Abstr. 151.017.

N-body simulations of instantaneous mass loss during dissipationless collapse. See Abstr. 151.036.

Galaxy mergers and active galactic nuclei. See Abstr. 151.066.

Extragalactic variable sources and cosmic-ray acceleration near massive black holes. See Abstr. 158.032.

Supermassive binaries in active galactic nuclei. See Abstr. 158.143.

Evolution of the hot gas cores of clusters of galaxies, $n(\theta)$ counts and anisotropy of the cosmic background radiation. See Abstr. 160.043.

Intergalactic shells at large redshift. See Abstr. 161.006.

New pathways in gravitational research. See Abstr. 162.010.

Why is the cosmological constant so small? See Abstr. 162.021.

Physical uniformity of the Universe. See Abstr. 162.041.

Quelques grands problèmes cosmologiques actuels. See Abstr. 162.053.

Mach–Poincaré gravodynamics and condensation phenomena in an expanding universe. See Abstr. 162.072.

Matter-antimatter separation in the early universe by rotating black holes. See Abstr. 162.088.

A note on a second-order differential equation with infinitely many solutions. See Abstr. 162.091.

Cosmological black hole production in grand unified theories. See Abstr. 162.111.

A cold big bang with small black holes. See Abstr. 162.120.

Cosmology in bimetric Machian gravitation. See Abstr. 162.125.

Elementary growth rates of local inhomogeneities in a flat Robertson-Walker background with a relativistic equation of state. See Abstr. 162.130.

The formation of galaxies from massive neutrinos. See Abstr. 162.133.

Stokesian fluids and cosmology.
See Abstr. 162.139.

Local inhomogeneities in a Robertson-Walker background. III. Elementary growth rates in a flat background with a relativistic equation of state.   See Abstr. 162.143.

Matter annihilation in the late Universe.
See Abstr. 162.149.

Torsion and strong gravity in the realm of elementary particles and cosmological physics.
See Abstr. 162.154.

Neutron Stars

066.501   **Configuration of a neutron star with axially polarized spin.**   A. P. Yefremov (*Efremov*).
Acta Phys. Polonica B, Vol. B12, 185 - 188 (1981). — Abstr. in Phys. Abstr., Vol. 84, Abstr. 71226 (1981).

066.502   **Stellar thermal expansion effects on the cooling of neutron stars.**   G. Baym.
Astrophys. J., Vol. 248, 767 - 770 (1981).
In the cooling of degenerate neutron stars thermal contraction produces a correction to the net luminosity of order $(\pi T/T_f)^2$, where T is the temperature and T_f the Fermi temperature of the supporting matter. Thus, except in the atmosphere, or at very early times, thermal effects on the neutron star structure can be neglected in computing the cooling of neutron stars.

066.503   **Energy and magnetic susceptibility of neutron matter. II. Variational calculations by Pandharipande's LOCV method.**   O. Forseth, E. Ostgaard.
Ark. Fys. Semin. Trondheim, No. 1, p. 1 - 43 (1981). — Abstr. in Phys. Abstr., Vol. 84, Abstr. 75159 (1981).

066.504   **Hot and cold, nuclear and neutron matter.**   B. Friedman, V. R. Pandharipande.
Nucl. Phys. A, Vol. A361, 502 - 520 (1981). — Abstr. in Phys. Abstr., Vol. 84, Abstr. 80500 (1981).

066.505   **Thermonuclear flashes on accreting neutron stars: a systematic study.**   S. Ayasli, P. C. Joss.
Bull. American Astron. Soc., Vol. 13, 527 - 528 (1981). Abstract.

066.506   **Synchrotron neutrino-pair radiation in neutron stars.**   D. G. Yakovlev, R. Tschaepe.
Astron. Nachr., Band 302, 167 - 176 (1981).
The neutrino-pair radiation by electrons in a non-quantizing magnetic field B is investigated. Under typical neutron star conditions at $B \sim 10^{13}$ G neutrino synchrotron radiation appears to be one of the most effective mechanisms of neutrino energy loss in the envelopes of neutron stars; this mechanism may also compete with other known neutrino production mechanisms in the neutron star cores if pion condensate or quark matter is absent.

066.507   **Kernmaterie in Neutronensternen.**   B. Kämpfer.
Sterne, 57. Band, 212 - 217 (1981).

066.508   **Proton $^1S_0$-dominant superfluidity with two-dimensional character under well-developed π^0 condensation (*neutron star matter*).**   T. Takatsuka, R. Tamagaki.
Prog. Theor. Phys., Vol. 65, 1333 - 1349 (1981). — Abstr. in Phys. Abstr., Vol. 84, Abstr. 94385 (1981).

066.509   **The coldest neutron star.**   G. Feinberg.

Phys. Rev. D, Vol. 23, 3075 (1981). — Abstr. in Phys. Abstr., Vol. 84, Abstr. 94386 (1981).

066.510   **Neutrino oscillations in neutron star matter.**   H. J. Haubold.
Astron. Nachr., Band 302, 223 - 226 (1981). In German.
From the theory of neutrino oscillations of B. Ponte-Korvo (1967) the author derives the oscillation length L of neutrinos in vacuum and the characteristic oscillation length L^* of neutrinos taking into consideration the refraction index n_e of neutron star matter. The comparison of both oscillation lengths shows that for electron densities, characteristic of neutron star matter, the oscillation length L is considerably larger than the oscillation length L^*. Therefore neutrino oscillations cannot influence the scenario for neutrino emission of the neutron star.

066.511   **Uncertainty in the saturation density of nuclear matter and neutron star models.**   P. Haensel, M. Kutschera, M. Prószyński.
Astron. Astrophys., Vol. 102, 299 - 302 (1981).
Standard minimal requirement for a theory of dense matter is that it should reproduce empirical saturation parameters of nuclear matter. The authors derive equations of state using the relativistic mean field theory of dense matter, the phenomenological parameters of which are determined by imposing this minimal condition. Then they study how the uncertainty in the saturation density (n_0) of nuclear matter may influence masses, radii and moments of inertia of neutron star models. The authors obtain maximum allowable mass of neutron stars ranging from 2.6 $M_\odot$ to 3.15 $M_\odot$ and maximum allowable moment of inertia from 4.05×10^{45} g cm^2 to 7.14×10^{45} g cm^2.

066.512   **Cooling of young neutron stars and the Einstein X-ray observations.**   K. Nomoto, S. Tsuruta.
Astrophys. J., Lett., Vol. 250, L19 - L23 (1981).
Cooling of neutron stars is calculated using an exact stellar evolution code. The fully general relativistic version of the stellar structure equations are solved with the best physical input currently available. For neutron stars with a stiff equation of state, the authors find that the deviation from isothermality in the interior is significant and that it takes at least a few thousand years to reach the isothermal state. By comparing the most recent theoretical and observational results, they conclude that for Cas A, SN 1006, and probably Tycho, "standard" cooling is inconsistent with the results from the Einstein Observatory, if neutron stars are assumed to be present in these objects. On the other hand, the "detection" points for RCW 103 and the Crab are consistent with these theoretical results.

066.513   **Equations of state of neutron stellar matter and parameters of neutron stars.**   V. I. Rejzlin, V. A. Filimonov.
Nejtron. fiz. Mater. 5-j Vses. konf. po nejtron. fiz. Kiev, 1980, Ch. 1. Moskva, 1980, p. 196 - 199. In Russian. — Abstr. in Ref. zh., 51. Astron., 10.51.417 (1981).

066.514   Neutron stars.   J. M. Irvine.
Nuclear astrophysics, (see 012.036), p. 191 - 213 (1981).
Contents: Pulsars, bursters and neutron stars. Neutron star structure. Equations of state.

066.515   Nuclear explosions on neutron star surfaces.
M. Ruderman.
Nuclear astrophysics, (see 012.036), p. 215 - 234 (1981).
After their formation, neutron stars may slowly increase their masses for various reasons. Many will be in close binaries. Mass transfer from their companions can be accomplished by a stellar wind. Especially large neutron star accretion is expected when the companion expands to fill and then overflow its Roche lobe, pushing matter into the neutron star's gravitational potential well. Isolated neutron stars, very many of which were once pulsars, may accrete from the ambient interstellar medium. Conditions under which accretion is expected to be explosive, the expected observable consequences of the explosions, and comparisons with observed X-ray and γ-ray bursts are discussed.

066.516   Structure and stability of warm cores in neutron stars.   J. M. Ibáñez Cabanell.
Mon. Not. R. Astron. Soc., Vol. 197, 845 - 864 (1981).
Relativistic equations of structure are solved using Lamb's equations of state for warm neutron degenerate matter. The stability of isothermal cores in neutron stars is discussed and also the possible compatibility of the results obtained with experimental evidence is shown.

066.517   Transport properties of dense matter. III. Analytic formulae for thermal conductivity.
E. Flowers, N. Itoh.
Astrophys. J., Vol. 250, 750 - 752 (1981).
Analytic formulae are presented for the thermal conductivity of the dense matter. The results are particularly useful for the calculations of the cooling of neutron stars.

066.518   The deleptonization and heating of proton-neutron stars.   A. Burrows, T. J. Mazurek, J. M. Lattimer.
Astrophys. J., Vol. 251, 325 - 336 (1981).
The authors derive the half-life for excess lepton loss from the proton-neutron stars is ~0.5 s whether or not convection obtains. The results indicate that a positive lepton gradient forms in the outer core. In addition, the authors show that heating always accompanies deleptonization. In pure diffusion models most of the heat is deposited in the outer core, and in convective models most of the heat is deposited in the inner core. In all cases, the increase in the pressure due to heating is insufficient to counter the decrease in the pressure due to the loss of lepton degeneracy pressure, and the core contracts. However, the heating of the shocked mantle by the core is likely to cause it to expand on diffusion time scales. Since a lot of energy ($>10^{52}$ ergs) is deposited in the mantle by the core, the subsequent expansion of the mantle may push an accretion shock outward and aid in the production of supernovae.

066.519   The tearing mode instability in an accretion disc of a neutron star and a rapid X-ray burst.
D. Wang, Q. Qu.
Sci. Sinica, Vol. 24, 1400 - 1405 (1981).
In the case that the rotation axis and the axis of the dipolar magnetism of a neutron star are aligned, in the magnetic field of the accretion disc of neutron star there appears a neutron sheet configuration. This article studies the tearing mode instability in the neutral sheet with plasma flow velocity and outside source, and the result is used to explain the burst mechanism of the rapid X-ray burst source MXB 1730-335.

066.520   Superfluidity in neutron stars.
J. Shaham.
Proceedings of the 5th Göttingen-Jerusalem-Symposium on Astrophysics, (see 012.041), p. 211 - 231 (1981).

066.521   Some aspects of hydrogen burning in a neutron star envelope.   A. D. Kudryashov.
Nauchn. Inf., Vyp. (No.) 49, (see 003.017), p. 64 - 73 (1981). In Russian.
Burning of matter ($X = 0.68$, $X_C^{12} = X_O^{16} = 0.01$) at high temperatures $T_9 \geqslant 0.1$ and under isothermal and adiabatic conditions is investigated. It is shown that at typical densities of the neutron star envelope of $\sim 10^6$ g/cc the use of a nuclear network restricted by magnesium is adequate only for $T_9 < 0.5$.

066.522   Cooling of neutron stars and X-ray observations.   K. Nomoto, S. Tsuruta.
Space Sci. Rev., Vol. 30, (see 012.044), 265 - 271 (1981).
Cooling of neutron stars is calculated using an exact stellar evolution code. The full general relativistic version of the stellar structure equations are solved, with the best physical input currently available. For neutron stars with a stiff equation of state, the authors find that the deviation from the isothermality in the interior is significant and that it takes at least a few thousand years to reach the isothermal state. By comparing the most recent theoretical and observational results, they conclude that for Cas A, SN1006, and probably Tycho, "standard" cooling is inconsistent with the results from the Einstein Observatory.

066.523   Models for accretion flow on to neutron stars and X-ray photon-electron interactions in ultra-high magnetic fields.   S. Cecchini, A. J. Evans, R. Lieu, J. J. Quenby.
Space Sci. Rev., Vol. 30, (see 012.044), 335 (1981).
Abstract.

066.524   Observability of magnetically strongly shifted iron line emission from X-ray pulsars.   G. Wunner, H. Ruder, H. Herold, J. Trümper.
Space Sci. Rev., Vol. 30, (see 012.044), 337 - 340 (1981).
As an application of extensive calculations of energies and intensities of atomic lines in very intense magnetic fields of the order of 10^{11} - 10^{13} G the authors discuss the possibility of observing magnetically strongly shifted iron lines in spectra of pulsating X-ray sources. Careful estimates of the relevant parameters lead to the conclusion that it would be profitable to look for magnetically shifted iron line emission in magnetic neutron stars of low luminosity using spectrometers working in the energy range 10 - 100 keV with sensitivities of 10^{-4} cm^{-2} s^{-1} and energy resolutions E/ΔE $\sim$ 10 - 100.

066.525   Magnetohydrodynamic instabilities in a magnetic accretion disk.
R. Horiuchi, T. Kadonaga, A. Tomimatsu.
Prog. Theor. Phys., Vol. 66, 172 - 179 (1981). – Abstr. in Phys. Abstr., Vol. 85, Abstr. 6628 (1982).

066.526   Nonstationary processes in neutron stars.
Yu. A. Berezin, O. E. Dmitrieva, N. N. Yanenko.
Inst. teor. i prikl. mekh. SO AN SSSR. Prepr., 1981, No. 15, 36 pp. In Russian. – Abstr. in Ref. zh., 51. Astron., 12.51.593 (1981).

066.527   Cyclotron radiation of magnetized degenerate dwarfs and neutron stars: I – theories.
I. G. Mitrofanov, G. G. Pavlov.
High-energy astrophysics, (see 012.067), p. 103 - 116 (1981).
Many degenerate dwarfs and neutron stars have the strongest known magnetic fields B = $10^6 - 10^9$ G and $\geqslant 10^{11}$ G, respectively. The effects of strong magnetic fields on the emission and absorption of radiation are investigated and various

approximate solutions are discussed. By solving the equation of radiative transfer the parameters of the emitting plasma can be related to the characteristics of the emitted radiation.

066.528   The γ-ray bursts as a result of the proper activity of the neutron star.
G. S. Bisnovatyi-Kogan (*Bisnovatyj-Kogan*), V. M. Chechetkin.
High-energy astrophysics, (see 012.067), p. 153 - 159 (1981).

A model of the γ-ray burst is developed which is based on the nuclear explosion in the neutron star envelope because of the chain fission reaction of the superheavy nuclei. It is shown that the main part of the γ-ray burst radiation must have thermal origin, but a nonthermal component and γ-ray lines may be present. The parameters of the strong burst of 5 March 1979 are evaluated.

Quark matter in the Hartree-Fock approximation.
See Abstr. 022.026.

Magnetic susceptibility of relativistic Fermi gas.
See Abstr. 022.089.

The quark matter.   See Abstr. 061.039.

Neutrino and photon emission from a dense, high temperature atmosphere.   See Abstr. 061.051.

On charge neutrality and dissipative effects in pulsar systems.   See Abstr. 062.002.

X-ray and gamma-burst radiation stress limits and diamagnetic accretion.   See Abstr. 062.044.

Annihilation radiation from a hot e^+-e^- plasma.
See Abstr. 062.083.

Directionality effects in the transfer of X-rays from an accreting magnetized neutron star: beam and pulse shapes.
See Abstr. 063.023.

Directionality effects in the transfer of X-rays from an accreting magnetized neutron star: beam and pulse shapes.
See Abstr. 063.048.

Spectral shifts near compact objects.
See Abstr. 066.112.

Distinguishing between a white dwarf and a neutron star in an X-ray binary.   See Abstr. 117.006.

SS 433 may not be as peculiar as it looks.
See Abstr. 117.091.

White dwarfs and supernovae.
See Abstr. 126.027.

Cyclotron radiation of magnetized degenerate dwarfs and neutron stars: II — observations.
See Abstr. 126.031.

Pair creation above pulsar polar caps: steady flow in the surface acceleration zone and polar cap X-ray emission.
See Abstr. 141.512.

A revised calculation of the bremsstrahlung cross-section in the high magnetic field of pulsars.
See Abstr. 141.536.

Pulsar and spin-spin interaction.
See Abstr. 141.541.

A model of the normal and null states of pulsars.
See Abstr. 141.543.

Pulsar disk systems.   See Abstr. 141.548.

Pulsars: polar pumps, interpolar currents, and induced Landau radiation.   See Abstr. 141.549.

The high energy X-ray spectrum of 4U 0900—40 observed from OSO 8.   See Abstr. 142.061.

Some remarks on the rotational state of pulsating X-ray sources.   See Abstr. 142.081.

Diagnosis of neutron star and its environments with X-ray astronomy.   See Abstr. 142.083.

A 4.2 second period in the gamma-ray burst of 1977 October 29.   See Abstr. 142.501.

Gamma ray bursts and neutron star accretion of a solid body.   See Abstr. 142.509.

Gravitational scattering of asteroids onto neutron stars as a cause of γ-ray bursts.   See Abstr. 142.523.

Radiation from an asteroid–neutron star collision.
See Abstr. 142.524.

Gamma-ray burst spectra.
See Abstr. 142.542.

Sun

071 Photosphere, Spectrum

071.001 The third central moment of photospheric lines as a measure of velocity gradients and line shifts.
C. Marmolino, G. Severino.
Astron. Astrophys., Vol. 100, 191 - 193 (1981).
The meaning of the third central moment (M_3) of photospheric line profiles is analysed. Whereas for infinite resolution data M_3 is a nearly linear measure of velocity gradients, at finite resolution it is essentially determined by the different weights of the shifted granular and intergranular line components. Some indications on the possibilities of disentangling velocity gradients and horizontal integration of inhomogeneities are given.

071.002 Absorption feature observed on the H Lyman-alpha solar line: an interpretation. G. Artzner, S. Cazes, C. Emerich, J. C. Vial, P. Lemaire.
Astron. Astrophys., Vol. 100, 205 - 208 (1981).
A narrow absorption feature on the red part of the solar H Lyman α profile has been observed by two instruments in 1975 and attributed to atomic hydrogen of the nearby interplanetary gas. This hypothesis implies that the absorption feature shifts with the position of the earth on its orbit. New measurements above active regions at other seasons rule out this hypothesis. On the other hand, appropriate laboratory calibrations indicate that the absorption feature is likely due to a ghost image of the geocoronal absorption.

071.003 On the establishment of internally consistent solar scales of oscillator strengths and abundances of chemical elements. II. On the errors of the oscillator strengths of Fe I lines in the Kurucz-Peytremann gf-scale.
E. (*Eh.*) A. Gurtovenko, R. I. Kostik (*Kostyk*).
Astron. Astrophys., Vol. 101, 132 - 133 (1981).
It is shown that the errors on the oscillator strengths of the Kurucz-Peytremann gf-scale depend in a regular way upon the excitation potential and gf-value. Oscillator strengths with small values of gf and large excitation potentials are underestimated, and vice versa. Especially strong underestimations (by $\approx$ 2 orders of magnitude) of the oscillator strengths take place for the lines with excitation potentials EPL > 4 eV and oscillator strengths log $gf <$ −4.0.

071.004 On the history of the study of the telluric spectrum of the sun. II. 1900 - 1970.
O. A. Mel'nikov, E. D. Khilov.
Tr. Astron. Obs., Leningrad, Tom 36 = Uch. Zap. Leningr. Univ., No. 402 = Ser. mat. nauk, Vyp. 58, 153 - 220 (1981). In Russian.
A review of the history of study of the telluric spectrum of the sun 1900 - 1970 is given. Some 250 publications on the problem are briefly discussed.

071.005 On the problem of functions of contribution and calculation of mean optical depths of formation of Fraunhofer lines. B. T. Babij, R. E. Rykalyuk.
Astron. Zh., Tom 58, 825 - 837 (1981). In Russian. English translation in Soviet Astron., Vol. 25, No. 4.
The functions of contribution to emission and depression as well as the so-called response functions are considered.

Basing on the physical conditions in the solar atmosphere, an attempt to formulate the demands on the contribution functions used in the calculation of mean optical depths of formation of Fraunhofer lines is made. The mean optical depths of different parts of some Fraunhofer lines of Fe I and O l are calculated as an example.

071.006 The transition layer from the photosphere to the chromosphere under solar flares. Eh. E. Dubov.
Astron. Zh., Tom 58, 904 - 906 (1981). In Russian. English translation in Soviet Astron., Vol. 25, No. 4.
The transition layer from the photosphere to the chromosphere (minimum temperature layer) under a flare is hotter and deeper than in the quiet sun. This fact should not necessarily be explained by substantial energy supply from the flare, but as a natural result of radiative cooling mechanism change (for instance because of additional ionization caused by the flare).

071.007 Periodic shifts of Fraunhofer lines in the spectrum of the center of the solar disk.
N. A. Drake, V. G. Nikiforov, Yu. A. Solonskij.
Soln. Dannye 1980 Byull., No. 12, p. 106 - 110 (1981). In Russian.
Spectra of the solar disk center obtained during 130 minutes are used to derive the parameters (periods, amplitudes and initial phases) of the vertical oscillations in the solar photosphere. Two fundamental periods are revealed − about five and sixty minutes. Variation of the parameters of the 5-minutes oscillation with depth in the photosphere was not found. The parameters of the 60-minutes oscillations vary with photospherical depth by a small amount.

071.008 Photometric profile of a granule and horizontal derivatives of the brightness temperature in granulation. V. N. Karpinskij.
Soln. Dannye 1981 Byull., No. 1, p. 88 - 91 (1981). In Russian.

071.009 Improved identification of some lines in the solar spectrum from 4145 Å to 4190 Å.
G. A. Porfir'eva.
Soln. Dannye 1981 Byull., No. 1, p. 105 - 112 (1981). In Russian.
In the region from 4145 Å to 4190 Å the identification of more than 20 solar lines is revised. Some weak cyanogen lines are identified.

071.010 On the asymmetry of Fraunhofer line profiles.
E. K. Kokhan, V. A. Krat.
Pis'ma Astron. Zh. Tom 7, 497 - 499 (1981). In Russian. English translation in Soviet Astron. Lett., Vol. 7.
Profiles of unblended Fraunhofer lines are analysed. Two faint satellites have been detected in the red and blue wing of the lines. It is suggested that the satellites are formed by regular up and down plasma motions over granules and an intergranular dark net.

071.011 The aluminum I autoionization doublet in the quiet solar spectrum.

J. N. Heasley, D. Roussel-Dupré, H. C. McAllister, C. Beerman.
Astrophys. J., Vol. 248, 352 - 357 (1981).

Observations are presented of the Al I autoionization
doublet λ1932 and λ1936 in the quiet solar spectrum. The
observed profiles are compared with theoretical spectra computed for the Harvard·Smithsonian Reference Atmosphere and
the Vernazza, Avrett, and Loeser solar models. The authors
find that nonlocal thermodynamic equilibrium effects are
important in the line-formation problem and the synthetic
spectra are in good agreement with the data.

071.012 On the temporal variation of the solar continuous
         brightness fluctuations. Time dependence of the
spatial power spectra.   B. Hadjebi.
Sol. Phys., Vol. 73, 25 - 35 (1981) = Mitt. Kiepenheuer-Inst.
Nr. 188.

From a sequence of white-light photographs of solar
granulation at the centre of the disk, obtained by Spectro-
Stratoscope on May 17, 1975, two-dimensional spatial power
spectra of photospheric intensity fluctuations were deduced.
These show periodicities of ~1000 s, 250–450 s (5-min
oscillation), and shorter ones in the range 30–120 s.

071.013 rms-value and power spectrum of the photospheric
         intensity fluctuations.
W. Schmidt, M. Knölker, E. H. Schröter.
Sol. Phys., Vol. 73, 217 - 231 (1981) = Mitt. Kiepenheuer-
Inst. Nr. 190.

With this investigation, based on granulation photographs
of excellent quality, the authors demonstrate that the rms
intensity fluctuations of the solar granulation at λ = 550 nm
is higher than 7.2% and does very likely not exceed 12% and
is certainly less than 13%.As the most probable rms-contrast
for this wavelength the authors obtained 10.5%. They also
present the most probable 'true' power spectrum of the
granular intensity fluctuations.

071.014 Identification of new solar OH lines in the 10 -
         12 micron region.
A. Goldman, F. J. Murcray, J. R. Gillis, D. G. Murcray.
Astrophys. J., Lett., Vol. 248, L133 - L135 (1981).

High-resolution ($0.02\,cm^{-1}$) infrared solar spectra obtained with a balloon-borne interferometer reveal new solar absorption features, which appear as regularly spaced quartets, in the
825 - 960 cm^{-1} region. The lines are interpreted as high N'' (25 -
33) pure rotation lines of solar OH. An effective amount of
~ 8×10^{15} molecules cm^{-2} of OH is estimated from the spectra.

071.015 On the variability of Lyman-alpha with solar activity.
L. Bossy, M. Nicolet.
Planet. Space Sci., Vol. 29, 907 - 914 (1981).

After taking into account the changes in the calibration,
or in the sensitivity, of the instruments used to measure solar
irradiance of Lyman-alpha, a critical analysis is performed for
the relation between the Lyman-alpha irradiance and the solar
flux at 10.7 cm.

071.016 On the spatial scale of brightness fluctuations in the
         solar photosphere.   Eh. E. Dubov.
Pis'ma Astron. Zh., Tom 7, 559 - 560 (1981). In Russian.
English translation in Soviet Astron. Lett., Vol. 7.

Photospheric brightness fluctuation scales evaluated using
one-dimensional photometric scans are shown to reflect the
supergranular structure of the photosphere.

071.017 Detection of a temperature deficit in magnetic
         faculae at the solar photosphere.
P. Foukal, T. Duvall, Jr., B. Gillespie.
Bull. American Astron. Soc., Vol. 13, 551 (1981). – Abstract.

071.018 Photospheric limb darkening as a ground-based
         diagnostic for variations in the solar effective tem-

perature.   W. A. Rosen, P. V. Foukal, R. Kurucz,
A. K. Pierce.
Bull. American Astron. Soc., Vol. 13, 551 (1981). – Abstract.

071.019 Dynamics of the solar photosphere.
         J. M. Beckers.
The sun as a star, (see 003.004), p. 11 - 64 (1981).

A model of the solar atmosphere and convection zone is
described and, briefly, the methods used in solar velocity field
observations. The inferences drawn from integrated sun observations the velocity fields of the quiet sun, sunspots, and other
magnetic structures are described.

071.020 Time-dependent interaction of granules with mag-
         netic flux tubes.
P. Venkatakrishnan, S. S. Hasan.
J. Astrophys. Astron., Vol. 2, 133 - 139 (1981).

The time-dependent interaction of the granulation
velocity field with a magnetic flux tube is investigated. It is
seen that when a magnetic field line is displaced normal to
itself so as to simulate the buffeting action of granules, a flow
of gas is initiated along the field. By choosing a lateral velocity
field which is consistent with observations of granules, it is
found that the resulting gas motion is a downward flow with a
velocity compatible with the observed downflow in isolated
photospheric flux tubes.

071.021 Combination of absorption and scattering as
         mechanism of Fraunhofer line formation in the
solar spectrum.   B. T. Babij, M. M. Koval'chuk.
Problems of cosmic physics. Vyp. 16, (see 003.005), p. 34 - 37
(1981). In Russian.

The profiles of absorption lines of moderate strength and
faint intensity have been investigated. It is marked that distinction of the physical processes on absorption and scattering
leads to discrepancies of theory with observation. It is shown
that the mechanism formation of absorption lines is the
combination of real absorption and coherent scattering.

071.022 On the barium abundance in the solar atmosphere.
         B. T. Babij, M. B. Girnyak.
Problems of cosmic physics. Vyp. 16, (see 003.005), p. 38 - 40
(1981). In Russian.

The abundance of barium is determined using equivalent
line widths of ionized barium in the spectrum of the solar
photosphere.

071.023 Minimizing the calculation scope of spectrum
         interpretation with fine structure of the solar plasma
taken into account.   E. A. Rudenchik.
Fiz. soln. aktivnosti. Moskva, 1980, p. 163 - 176. In Russian.
Abstr. in Ref. zh., 51. Astron., 8.51.433 (1981).

071.024 On wave channels on the solar surface.
         M. B. Kerimbekov.
Izv. AN AzSSR. Ser. fiz.–tekh. i mat. nauk. 1980, No. 2,
p. 75 - 80. In Russian. – Abstr. in Ref. zh., 51. Astron.,
8.51.436 (1981).

071.025 Determination of the limb darkening of the solar
         disc in the visible and near ultraviolet.
J. C. Bhattacharyya, R. Chandramohan, A. Sundareswaran,
A. Charles.
Observations of the total solar eclipse of 16 February 1980,
(see 003.009), p. 4 (1981).

071.026 Isotopes of nickel in the sun.
         J. W. Brault, H. Holweger.
Astrophys. J., Lett., Vol. 249, L43 - L46 (1981).

High-resolution Fourier transform spectrometer spectra
of near-infrared Ni I lines reveal isotopic structure both in a
laboratory source and in the spectrum of the quiet photo-

sphere. The solar $^{58}Ni/^{60}Ni$ ratio agrees with that found in terrestrial and meteoritic matter.

071.027 Non-resonance lines of neutral calcium in the spectra of the Sun and Procyon.   G. Smith.
Astron. Astrophys., Vol. 103, 351 - 357 (1981).
Neutral calcium lines of moderately high excitation (2.5–3.0 eV) in the solar spectrum and in the spectrum of Procyon have been analysed using precise oscillator strengths and estimates of hydrogen damping parameters based on measured parameters for broadening by helium. A conventional analysis based on an LTE model atmosphere and using all relatively unblended lines for which precise atomic data are now available leads to extremely consistent values for calcium abundance and microturbulence.

071.028 The use of spectral emission lines in the diagnostics of hot solar plasmas.   U. Feldman.
Phys. Scr., Vol. 24, 681 - 711 (1981).

071.029 Photospheric situation before flares.   Z. Krušina.
Říše hvězd, Vol. 62, 201 - 203 (1981). In Czech.

071.030 The analysis of solar limb observations. II. Geometrical smearing.
C. J. Durrant, F. Kneer, G. Maluck.
Astron. Astrophys., Vol. 104, 211 - 214 (1981).
Centre-of-disk and limb observations of photospheric brightness fluctuations demonstrate a systematic suppression of the values at the limb. This is a geometrical effect. The authors look at this effect in the light of recent granular temperature models using a simple model incorporating the dominant granular scale. They find the geometry to have little influence on the deep-seated granular brightness field but to strongly suppress the brightness fluctuations due to the upper photospheric temperature field. The significance of these results for the restoration of limb observations is discussed.

071.031 An analysis of the profiles of "dashes" in the Ca II K and Hβ lines.   A. N. Demidova.
Soln. Dannye 1981 Byull., No. 6, p. 83 - 88 (1981). In Russian.
The profiles of "dashes" in the Ca II K and Hβ lines observed over the limb on the 27th of August and the 1st of September 1980 are studied. In most profiles two components were detected using the Gauss-analysis. These components can be interpreted as corresponding to the downward flows of matter from the top of the flare loop prominence seen edge-on.

071.032 A local approach to the problem of two-dimensional radiative energy transfer in the photosphere near inhomogeneities. P- and H-approximations.
Yu. A. Nagovitsin.
Soln. Dannye 1981 Byull., No. 7, p. 104 - 109 (1981). In Russian.
The Eddington and diffusion approaches to the problem of the radiation transfer near inhomogeneities of the solar photosphere are critisized. A local approach to the problem is proposed consisting in a consideration of the field of an underphotospheric point source. Solutions have been obtained in the P (point) and H (horizontal) approximations which asymptotically limit the exact solution.

071.033 Improved identification of some lines in the solar spectrum from 4128 Å to 4145 Å.
G. A. Porfir'eva.
Soln. Dannye 1981 Byull., No. 7, p. 109 - 116 (1981). In Russian.
In the region from 4128 Å to 4145 Å the identification

of more than 40 solar lines is revised. Some weak cyanogen lines are identified.

071.034 Magnetic fine-structures and granular velocities.
B. Caccin, R. Falciani, M. T. Gomez, C. Marmolino, G. Roberti, G. Severino, L. A. Smaldone.
Space Sci. Rev., Vol. 29, (see 012.043), 373 - 374 (1981).
In the last years the authors have gained some experience in the diagnostics of small-scale structures, both on the interpretative and on the observational point of view. They report here the conclusions and the suggestions for future developments attained in two main fields of interest.

071.035 Additional identification of some cyanogen lines in the solar spectrum from 4128 Å to 4145 Å.
G. A. Porfir'eva.
Astron. Tsirk., No. 1121, p. 6 - 8 (1980). In Russian.

071.036 On silicon abundance in the solar photosphere.
B. T. Babij, M. B. Girnyak, R. E. Rykalyuk.
Tsirk. Astron. Obs., L'vov, No. 54, p. 27 - 33 (1979). In Russian.

071.037 Quelques résultats récents en spectroscopie solaire.   M. Migeotte.
C. R. Acad. Sci. Paris, Vie Acad., Tome 293, (see 012.056), IX - XI, XIV (1981).

071.038 Hydrodynamical and electrodynamical interactions between magnetic features in the active-region photosphere.   W. Unno, K. Tanaka, M. Semel.
Publ. Astron. Soc. Japan, Vol. 33, 495 - 504 (1981).
Theoretical interpretation is attempted for hydrodynamical and electrodynamical phenomena in an active region observed by various means in the Meudon-Mitaka Solar Collaboration Observations in 1979. Some pieces of observational evidence are quantitatively examined, showing that hydrodynamical interactions are likely to be effective between magnetic features in an active region. There is also evidence which shows the existence of the dynamo action in the active-region photosphere.

071.039 The appearance of bright regions in the photosphere and their connection with flares.
L. F. Lazareva, E. N. Sidorkina.
Astron. Tsirk., No. 1152, p. 6 - 7 (1981). In Russian.

Solar spectrum synthesis. I. A sample atlas from 224 to 300 nm.   See Abstr. 002.071.

Oscillator strengths for Zr I and Zr II and a new determination of the solar abundance of zirconium.
See Abstr. 022.037.

Measurement of the oscillator strengths and autoionization widths of the neutral-aluminum multiplet $3s^2 3p\,^2P^o-3s3p^2\,^2P$.   See Abstr. 022.050.

On the establishment of internally consistent solar scales of oscillator strengths and abundances of chemical elements. I. Oscillator strengths for 865 Fe I lines iron abundance.   See Abstr. 022.105.

Atomic calculation for Fe XXIII, UV, and X-ray lines.   See Abstr. 022.110.

On the mechanism of formation of weak lines of rare-earth elements.   See Abstr. 022.168.

Source functions for weak lines of rare-earth elements.   See Abstr. 022.169.

Speckle interferometric techniques applied to the observation of the solar photosphere.   See Abstr. 031.527.

A comparison of optical and digital Fourier transformation of solar granulation.   See Abstr. 031.563.

The analysis of solar limb observations. I. Restoration of data in a tilted reference frame.
See Abstr. 031.598.

Ein Vergleich zwischen optischer und digitaler Fouriertransformation von Sonnengranulationsaufnahmen.
See Abstr. 031.617.

A two-dimensional solar spectrometer.
See Abstr. 034.037.

Penetrative convection.   See Abstr. 062.048.

Detection of a temperature deficit in magnetic faculae at the solar photosphere.   See Abstr. 072.032.

Physical properties of the solar chromosphere deduced from optically thick lines: I. Observations, data

reduction, and modeling of an average plage.
See Abstr. 073.077.

Diffuse separation of charges and electrical fields in the solar plasma.   See Abstr. 073.095.

Transient plasmas in the solar transition zone.
See Abstr. 074.075.

Solar luminosity variation. III. Calcium K variation from solar minimum to maximum in cycle 21.
See Abstr. 080.036.

Effects of diffusion and mass flows on C IV and Si IV lines formed in the solar atmosphere.
See Abstr. 080.053.

Investigation of the empirical damping constant and iron abundance in the solar atmosphere from spectral line profiles.   See Abstr. 080.066.

072 Sunspots, Faculae, Activity Cycles, Solar Patrol

072.001 A unified working model for the atmospheric structure of large sunspot umbrae.
J. Staude.
Astron. Astrophys., Vol. 100, 284 - 290 (1981).

Horizontally averaged working models have been derived for both the umbra of a large stable sunspot and the mean undisturbed environment for reference. The present sunspot model provides the lower boundary condition for an attempt to model the transition layer and lower corona above large umbrae. Possibilities for such an extension of the sunspot model are also shortly outlined here.

072.002 Sunspot dynamics: gravitational draining – a cooling mechanism. K. H. Schatten.
Astrophys. J., Lett., Vol. 247, L139 - L142 (1981).

Inward and downward flow of cooled material beneath sunspots, as suggested by Meyer et al. and Parker, is viewed as a key to understanding their stability, temperature, and heat flow. In this view, the magnetic field of pores, magnetic knots, and sunspots provides a conduit wherein cooled gases can return more easily to their heat source at the base of the convective zone by gravitational draining. Calculations of the physics involved suggest downflow material to be roughly 1,100 K cooler than the photosphere and downflow velocities to be several km s^{-1}. These velocities are not observed within the visible layer of spots, however, they are seen in knots and inferred in pores.

072.003 A comparative study of brightness distribution in an active region on hydrogen and calcium filtergrams.
I. Photometry of active regions. G. M. Kornienko.
Soln. Dannye 1981 Byull., No. 1, p. 112 - 118 (1981). In Russian.

The relationship of brightness in an active region in the K Ca II line center and that in the center and symmetric wings ($\Delta\lambda = \pm 0.5$ Å) of the Hα line is considered. It is found that the local decrease of brightness in the K Ca II line accompanies a local brightness increase at the boundaries of the active region in H$\alpha \pm$.

072.004 On bright points in a sunspot umbra.
L. D. Parfinenko.
Soln. Dannye 1981 Byull., No. 2, p. 96 - 101 (1981). In Russian.

Fifty bright points in a sunspot's umbra were studied. The mean contrast of bright points as referred to the photosphere is equal to 0.69. The maximum observed contrast is 1.0%. The contrast of bright points on the boundary with the penumbra is higher and reaches 1.14. The brightness of bright points may vary, their lifetime is 180 min.

072.005 Characteristics of fluctuations of Wolf numbers for the ascending branch of the 21st solar cycle.
Yu. I. Vitinskij.
Soln. Dannye 1981 Byull., No. 2, p. 101 - 103 (1981). In Russian.

Values of the fluctuation index and indices of "perturbations" for the Wolf number during 1976 - 1979 are given. Also their positive and negative strong fluctuations for 1965 - 1979 are listed. Some peculiarities of these characteristics are discussed.

072.006 Some characteristics of the space distribution of solar proton active regions. B.-s. Tang.
Acta Astron. Sinica, Vol. 22, 188 - 190 (1981). In Chinese.

072.007 Nitrate ion in Antarctic firn as a marker for solar activity. E. J. Zeller, B. C. Parker.
Geophys. Res. Lett., Vol. 8, 895 - 898 (1981).

The paper presents the first comparison of nitrate ion (NO$_3^-$) concentrations in firn cores collected from South Pole

(1978 - 79) and Vostok (1979 - 80) stations, covering a continuous sequence of approximately 1200 years.

072.008 The case of the missing sunspots.
M. M. De Lancy.
Astronomy, Vol. 9, No. 2, p. 66 - 71 (1981). – Abstr. in Phys. Abstr., Vol. 84, Abstr. 71154 (1981).

072.009 Solar activity studies in the past using cosmogenic isotopes. P. Povinec.
Acta Fac. Rerum Nat. Univ. Comenianae Phys., Vol. 20, 171 - 183 (1980). – Abstr. in Phys. Abstr., Vol. 84, Abstr. 75098 (1981).

072.010 Preliminary observation of the missing energy flux of sunspots. T. Hirayama, T. Okamoto.
Sol. Phys., Vol. 73, 37 - 43 (1981).

The authors have searched for the missing energy flux of sunspots by measuring white-light photographs over a region around the sunspots exceeding ten times of the diameter of spots. It was found that the excess brightness amounts to 0.3–0.8% of the local continuum intensity immediately beyond the penumbra and smoothly decreases to zero at the edge of the inspected radii of 130000–160000 km.

072.011 The parameters of solar cycle No. 21.
M. Waldmeier.
Sol. Phys., Vol. 73, 207 (1981).

072.012 Notes on the current sunspot cycle. F. E. Cook.
J. Electr. Electron. Eng. Aust., Vol. 1, No. 1, p. 29 - 32 (1981). – Abstr. in Phys. Abstr., Vol. 84, Abstr. 83585 (1981).

072.013 Additional measurements of the high-latitude sunspot rotation rate. D. A. Landman, J. T. Takushi.
Sol. Phys., Vol. 73, 379 - 382 (1981).

The authors report measurements of the sunspot rotation rate at high sunspot latitudes for the years 1966–1968. On the average the authors find a sidereal rotation rate of 13.70 ± 0.07 deg day^{-1} at 31.05 ± 0.01 deg.

072.014 Gravitational draining: a cooling mechanism for sunspots and pores. K. H. Schatten.
Bull. American Astron. Soc., Vol. 13, 542 (1981). – Abstract.

072.015 New information on the spatial distribution of active regions.
S. F. Martin, L. M. Hermans, W. H. Marquette.
Bull. American Astron. Soc., Vol. 13, 551 - 552 (1981). Abstract.

072.016 Dynamic phenomena in the visible layers of sunspots.
R. L. Moore.
Space Sci. Rev., Vol. 28 (see 012.014), 387 - 421 (1981).

The empirical properties of the various dynamic phenomena are reviewed and interrelated with emphasis on recent observational results. It is proposed from the observations that umbral dots and penumbral grains are essentially the same phenomenon, and that the observational goal of highest priority with respect to both the origin of the periodic phenomena and the problem of the missing heat flux is to better determine the nature of these elementary bright features.

072.017 Magnetohydrodynamics of sunspots.
H. C. Spruit.
Space Sci. Rev., Vol. 28, (see 012.014), 435 - 448 (1981).
Current theories are reviewed concerning the pressure

equilibrium and thermal balance of sunspots, their hydro-magnetic stability, energy transport mechanisms, and propagation of waves.

072.018   On the structure of a sunspot's magnetic field.
A. V. Baranov.
Soln. Dannye 1981 Byull., No. 4, p. 103 - 107 (1981). In Russian.

A potential model of the magnetic field of the upper atmospheric layers is considered. The model explains quantitatively and qualitatively the following observational facts: decrease of the gradient of the field with distance from the sunspot center, decrease of the gradient of the field with height, the distribution of H and γ along the sunspot radius within the errors of measurement. A method for an experimental check of the field potential is presented.

072.019   Longitudinal distribution of sunspots at the beginning of the 21st solar cycle.
V. S. Berdichevskaya.
Soln. Dannye 1981 Byull., No. 5, p. 108 - 111 (1981). In Russian.

The longitudinal distribution of new polarity sunspots at the beginning of the 21st solar cycle is considered. The first 30 sunspot groups appeared in July 1976 - July 1977, 58% of their total area falling on the Carrington longitudes 120 - 150°. During the following year the pattern of the sunspot distribution changed essentially and became in fact uniform.

072.020   On large fluctuations of monthly Wolf numbers.
Yu. I. Vitinskij.
Soln. Dannye 1981 Byull., No. 5, p. 111 - 116 (1981). In Russian.

The main parameters of large positive and negative fluctuations of Wolf numbers are given for the years 1755 - 1878. A basic difference is found between the distributions of positive and negative fluctuations with the phase of the 11 year solar cycle. It can be explained by the suggestion that in the origin of the former there are burst processes while in the origin of the latter there are dissipative processes.

072.021   The boundary problem for magnetic fields of active regions.
E. A. Kornitskaya, M. M. Molodenskij.
Inst. zemn. magn., ionos. i rasprostr. radiovoln AN SSSR. Prepr., 1981, No. 3, 20 pp. In Russian. — Abstr. in Ref. zh., 51. Astron., 8.51.457 (1981).

072.022   Evolution and flare activity of the McMath 15403 — 1978 active region.
V. N. Ishkov, Z. B. Korobova, Eh. I. Mogilevskij, L. I. Starkova, V. G. Utrobin.
Fiz. soln aktivnosti. Moskva, 1980, p. 48 - 79. In Russian. Abstr. in Ref. zh., 51. Astron., 8.51.463 (1981).

072.023   Longitude distribution of solar activity in the 18th to the 20th cycles.
G. A . Bazilevskaya, E. S. Vernova, M. I. Tyasto, T. N. Charakhch'yan.
Izv. AN SSSR. Ser. fiz., Tom 45, 617 - 620 (1981). In Russian. Abstr. in Ref. zh., 51. Astron., 8.51.506 (1981).

072.024   Model dependence of the rotational temperature of SiO in sunspots.
L. M. Punetha, G. C. Joshi, M. C. Pande.
Bull. Astron. Soc. India, Vol. 9, 74 (1981). — Abstract.

072.025   The two-component nature of the size-distribution of sunspot-groups in a solar cycle.
M. H. Gokhale, K. R. Sivaraman.
Bull. Astron. Soc. India, Vol. 9, 86 (1981). — Abstract.

072.026   Types of sunspots, magnetic classification and flare yield in cycle No. 20.
Š. Knoška, L. Křivský.
Bull. Astron. Inst. Czechoslovakia, Vol. 32, 292 - 303 (1981).

The yield of flare activity and some other characteristics were investigated in the years 1965 - 1976 with regard to the types of sunspot groups of the Zurich and magnetic classification. From the point of view of flare yield of spotgroups it has been found that in the course of cycle No. 20 the change of flare yields demonstrates another course in the relation to the sunspot number R. The largest flare yield was displayed by the F and E types and by the magnetic δ-and γ-types.

072.027   The asymmetry of Wilson's effect.
S. O. Obashev, R. Kh. Gajnullina, T. M. Minasyants.
Vestn. AN KazSSR, 1981, No. 3, p. 36 - 42. In Russian. Abstr. in Ref. zh., 51. Astron., 9.51.361 (1981).

072.028   General characteristics of the active region McMath 15974 according to spectral-polarimetric observations with high resolution.
Sh. B. Akhmedov, V. M. Bogod, N. G. Peterova.
13-ya Vses. konf. po radioastron. issled. soln. sistemy, Kiev, 1981. Tez. dokl. Kiev, 1981, p. 14 - 15. In Russian. — From Ref. zh., 51. Astron., 9.51.378 (1981).

072.029   On the magnetic field gradient over sunspots according to optical and radioastronomical measurements.
V. M. Bogod, G. F. Vyal'shin, G. B. Gel'frejkh, N. G. Peterova.
13-ya Vses. konf. po radioastron. issled. soln. sistemy, Kiev, 1981. Tez. dokl. Kiev, 1981, p. 10 - 11. In Russian. — From Ref. zh., 51. Astron., 9.51.386 (1981).

072.030   Die Sonnenfleckenhäufigkeit im elfjährigen Zyklus.
W. Schulze.
Sterne, 57. Band, 203 - 211 (1981).

072.031   Structure of the sunspot penumbra.
R. L. Moore.
Astrophys. J., Vol. 249, 390 - 393, plates 1 - 5 (1981).

The purpose of this paper is to present an exceptionally highly resolved sunspot photograph which reveals the fine-scale structure of the photospheric penumbra down to 0."2 (150 km). This photograph shows that some of the dark striations in the penumbra are elevated fibrils.

072.032   Detection of a temperature deficit in magnetic faculae at the solar photosphere.
P. Foukal, T. Duvall, Jr., B. Gillespie.
Astrophys. J., Vol. 249, 394 - 398, plates 6, 7 (1981).

The authors show that faculae can be clearly observed in continuum across the entire solar disc. They use exceptionally clean continuum and reduce granular noise by subtracting simultaneous spectroheliograms taken at widely separated points of the photospheric radiation curve. The dark magnetic faculae revealed near sun center through this photometric technique appear to contain material significantly cooler than the mean photosphere. Their observations also provide relatively model-independent evidence that the presence of both a "hot cloud" and a hot interior wall are important to the physical description of small-diameter magnetic flux tubes.

072.033   Theory of the solar cycle.   M. Stix.
Sol. Phys., Vol. 74, (see 012.029), 79 - 101 (1981) = Mitt. Kiepenheuer-Inst. No. 194.

The properties of kinematic $\alpha\omega$-dynamos are briefly reviewed. The mean field concept, including turbulent diffusivity, is defended against recent criticism. It is pointed out that although the Maunder minimum cannot be explained by kinematic dynamo theory alone, this does not invalidate dynamo theory in general. A special discussion is devoted to attempts to evaluate the coefficients of the mean field induction equation in the case of very large conductivity. The field then behaves intermittent, in the form of locally concentrated flux tubes, and the α-effect and the turbulent diffusivity may be determined by asymptotic techniques or with the help of an exact solution of the non-dissipative induction equation in Lagrangian co-ordinates. Magnetic cycles of main sequence stars other than the Sun are briefly discussed.

072.034 **Sunspot populations and their relation with the solar cycle.**   J. I. García de la Rosa.
Sol. Phys., Vol. 74, (see 012.029), 117 - 123 (1981).

The joint consideration of theoretical and observational arguments is used to conclude that two different spot populations co-exist in the Sun.

072.035 **Possible use of (a) solar faculae and (b) the interplanetary magnetic field as heralds of a solar cycle peak.**   G. M. Brown.
Sol. Phys., Vol. 74, (see 012.029), 125 - 129 (1981).

Two independent methods of predicting the magnitude of the peak of a forthcoming sunspot cycle are summarized. One is based on considerations of the development of spots relative to the area of the faculae within which they form during the early stages of the cycle in question, and gives a lead-time of about 2 years. The other uses measurements of the quiet-day variations of the Earth's magnetic field at the time of the preceding sunspot minimum and allows predictions to be made a half-cycle ahead. A possible extension of this technique to the use of data on the component of the interplanetary magnetic field normal to the ecliptic plane is suggested. References to fuller details of both methods are given.

072.036 **On the need for space observations of the umbra/photosphere intensity ratio.**
F. Albregtsen, P. Maltby.
Sol. Phys., Vol. 74, (see 012.029), 147 - 151 (1981).

The authors draw attention to the possibility of distinguishing between different sunspot theories by observing: (1) the umbra/photosphere intensity ratio as a function of spot size and (2) the morphology and time evolution of sunspot inhomogeneities such as umbral dots. In arguing the need for space observations of sunspot intensities the authors discuss the corrections for stray light for ground based and space observations. The opportunity to use the November 13, 1986 Mercury transit as an in situ calibration event is pointed out.

072.037 **Quasiperiodic variations of solar activity, physical characteristics of the solar wind, and cosmic ray intensity and anisotropy.**
L. Kh. Shatashvili, T. V. Dzhapiashvili, B. D. Naskidashvili.
Cosmic rays, (see 012.025), 1980, p. 234 - 240. In Russian.
Abstr. in Ref. zh., 51. Astron., 10.51.400 (1981).

072.038 **On the structure of the magnetic field and its evolution around sunspots.**
S. I. Gopasyuk, L. G. Kartashova.
Izv. Krymskoj Astrofiz. Obs., Tom 63, 25 - 45 (1981). In Russian. English translation in Bull. Crimean Astrophys. Obs., Vol. 63.

The structure of the magnetic field and its evolution around a single big sunspot has been studied. It is shown that there are characteristic directions corresponding to the transition of the spot field without sign change into an extended region of the same polarity and coinciding with extended (100 000 - 300 000 km) systems of filamentary feature chains of the fine chromospheric structure in an active region.

072.039 **Structure of the magnetic field and flare activity in the active region McMath 9740 (October 1968).**
M. B. Ogir', T. T. Tsap.
Izv. Krymskoj Astrofiz. Obs., Tom 63, 46 - 65 (1981). In Russian. English translation in Bull. Crimean Astrophys. Obs., Vol. 63.

The active region McMath N 9740 during the period from October 26 to November 2, 1968 is studied. Longitudinal and transversal components of the solar magnetic field were recorded in the Fe I λ5250 Å line. 28 magnetic maps were obtained for this period. The magnetic field observations were accompanied by simultaneous Hα cinematography of this region. The results are presented in detail.

072.040 **Line profiles and magnetic field in penumbral fine structures.**   G. Stellmacher, E. Wiehr.
Astron. Astrophys., Vol. 103, 211 - 215 (1981).

Spectra at high spatial resolution are taken in a sunspot penumbra at disc center. The magnetic low sensitive lines Fe^+ 5264.8 (g = 0.1) and Ti 5222.7 (g = 0) as well as the large split line Fe 6302.3 are analyzed by comparison with model calculations. No systematic fluctuations of the magnetic field strength and inclination are found in the bright and dark spectral streaks, b and d. It is shown that this result is still compatible with actual field fluctuations of ≤ 1000 Gauß between the individual penumbral bright and dark regions, BR and DR.

072.041 **Weiteres Material gegen die Eddysche Interpretation des Maunder-Minimums.**   O. Hein, H. Kastl.
Sterne Weltraum, Jahrg. 20, 459 - 461 (1981).

072.042 **Alfvén waves and magneto-sensitive spectral lines.**   S.-h. Ye, J.-h. Jin.
Acta Astron. Sinica, Vol. 22, 265 - 272 (1981).

Under the assumption of the propagation of Alfvén waves the authors have numerically solved the system of equations of transfer of Stokes parameters and calculated the effects of Alfvén waves on the magneto-sensitive line Fe I λ6302.499 in the spectra of sunspots.

072.043 **Visual observation of the sun in Czechoslovakia in the year 1980.**   L. Schmied.
Říše hvězd, Vol. 62, 162 - 164 (1981). In Czech.

072.044 **The distribution of the spiral spot on the solar disk.**   Y.-j. Ding, H.-z. Wang, Q.-f. Hong.
Acta Astrophys. Sinica, Vol. 1, 264 - 272 (1981). In Chinese.

072.045 **Propagation of waves in an atmosphere in the presence of a magnetic field. IV. Alfvén waves in sunspot umbrae.**   N. Bel, B. Leroy.
Astron. Astrophys., Vol. 104, 203 - 206 (1981).

The propagation of Alfvén waves in a sunspot is studied. The authors find that the outgoing energy flux density reaching the corona is less than 10^{-5} of the input energy flux density. This clearly shows that the cooling of sunspots by ascending Alfvén waves is not possible.

072.046 **Coupling equations for flow-wave field applied to turbulent heating of solar faculae and corona.**
X.-q. Li, M.-t. Song, Y.-g. Zhang, Z.-d. Zhang.
Plasma astrophysics, (see 012.042), p. 415 (1981).

072.047 **Slow magnetohydrodynamic shocks as transformer of the magnetic energy in solar active regions.**
L. B. Tsirul'nik.
Tsirk. Shemakhinsk. Astrofiz. Obs., No. 68, p. 28 - 32 (1981). In Russian.

Slow hydromagnetic shocks are the only mechanism of magnetic field dissipation in a plasma of infinite conductivity. On the basis of numerical calculations of the motion of such a wave in the solar atmosphere the conclusion has been drawn that the wave propagation may play a considerable part in the origination of flares and outbursts.

072.048 **The active sun.**   C. Jordan.
Space Sci. Rev., Vol. 29, (see 012.043), 363 - 365

(1981).

A brief summary is given of observations which will be required to investigate further the structure and energy balance of active regions.

072.049 **Simultaneous observations of AR 2490 performed in the X-ray, UV, optical, and radio wavelength domain.**   F. Chiuderi Drago.
Space Sci. Rev., Vol. 29, (see 012.043), 439 - 440 (1981).

072.050 **Global velocity fields of the Sun and the activity cycle.**   R. Howard.
American Sci., Vol. 69, 28 - 36 (1981). – Abstr. in Phys. Abstr., Vol. 85, Abstr. 6576 (1982).

072.051 **Ionized molecules in facula models.**
B. M. Tripathi, V. P. Gaur, M. C. Pande.
Bull. Astron. Soc. India, Vol. 9, 227 - 231 (1981).

Equivalent widths for the lines of ionized molecules have been calculated in five facular and two photospheric models.

072.052 **Cosmogenic radio carbon and the nature of the Maunder minimum.**
V. A. Vasil'ev, V. A. Dergachev.
Izv. An SSSR. Ser. fiz., Tom 45, 1189 - 1194 (1981). In Russian. – Abstr. in Ref. zh., 51. Astron., 11.51.399 (1981).

072.053 **On multi-dimensional classification of the 11-year solar activity cycles.**   A. M. Ehjgenson.
Vestn. L'vov. univ. Ser. astron., 1981, No. 56, p. 22 - 24. In Russian. – Abstr. in Ref. zh., 51. Astron., 11.51.496 (1981).

072.054 **Using a method of dynamical image recognition for the problem of solar activity forecast.**
B. T. Kabulov.
Vopr. kibernet., Tashkent, 1981, No. 113, p. 25 - 28. In Russian. – Abstr. in Ref. zh., 51. Astron., 11.51.504 (1981).

072.055 **Solar observations at the Astronomical Observatory of the Lvov University in 1979.**
M. B. Girnyak, M. M. Koval'chuk, I. S. Laba, P. A. Olijnyk.
Vestn. L'vov. univ. Ser. astron., 1981, No. 56, p. 9 - 13. In Russian. – Abstr. in Ref. zh., 51. Astron., 11.51.511 (1981).

072.056 **Provisional sunspot-numbers for June - November 1981.**
Yamamoto Circ., Nos. 1960, 1961, 1963 - 1965, 1967 (1981).

072.057 **Solar activity prognosis for the end of the 20th and beginning of the 21st centuries.**   A. Balklavs.
Zvaigžņota debess, 1980/81. gada ziema, p. 24 - 25. In Latvian.

072.058 **Sonnenfleckenrelativzahlen des S.I.D.C. (Sunspot Index Data Center).**
Sterne Weltraum, Jahrg. 20, 304, 349, 391, 436, 485 (1981).

072.059 **Solare Beobachtungsergebnisse. Solar data. Solar radio emission.** 1981 January - August.
HHI Sol. Data, Vol. 32, A - G, 1 - 131 (1981).

072.060 **Daily maps of the sun and magnetic fields of sun-spots.**
Soln. Dannye 1980 Byull., No. 12, p. 1 - 80; Soln. Dannye 1981 Byull., No. 1, p. 1 - 81; No. 2, p. 1 - 80; No. 3, p. 1 -ʻ92; No. 4, p. 1 - 102; No. 5, p. 1 - 82; No. 6, p. 1 - 77; No. 7, p. 1 - 85; No. 8, p. 1 - 101 (1980/1981). In Russian.

072.061 **The solar one-year oscillation found by Chistyakov and possible errors in the values of I and Ω.**
J. Tuominen.

Proceedings of the Third Finnish-Soviet Astronomical Symposium, (see 012.057), p. 71 - 72 (1981). – Abstract.

072.062 **Solar maps and activity,** 1981 March - 1981 August.
F. J. Heyden, V. L. Badillo.
Manila Obs., Sol. Div.

072.063 **Solar observations.** Nos. 58 - 60, 1980 April - 1980 December. Solar group.
Publ. Astron. Obs. Trieste, Nos. 724, 749, 764 (1980).

072.064 **Active regions and flares.**
A. T. Altyntsev, V. G. Banin, S. I. Gopasyuk, V. P. Maksimov, B. V. Somov, V. E. Stepanov, V. M. Tomozov.
Issled. po geomagn., aehron. i fiz. Solntsa, Moskva, 1981, No. 56, p. 39 - 56. In Russian. – Abstr. in Ref. zh., 51. Astron., 12.51.494 (1981).

072.065 **On the periodicity of solar activity.**
M. N. Gnevyshev, A. I. Ol'.
Issled. po geomagn., aehron. i fiz. Solntsa, Moskva, 1981, No. 56, p. 3 - 11. In Russian. – Abstr. in Ref. zh., 51. Astron., 12.51.532 (1981).

072.066 **On the nonstationarity of the 22-yearly variation of Wolf numbers.**   Yu. V. Rivin.
Issled. po geomagn., aehron. i fiz. Solntsa, Moskva, 1981, No. 56, p. 88 - 94. In Russian. – Abstr. in Ref. zh., 51. Astron., 12.51.533 (1981).

072.067 **A comparison of some mathematic methods of forecast of active spot groups.**
V. M. Efimenko, S. V. Pasechnik, V. V. Tel'nyuk-Adamchuk.
Issled. po geomagn., aehron. i fiz. Solntsa, Moskva, 1981, No. 56, p. 123 - 129. In Russian. – Abstr. in Ref. zh., 51. Astron., 12.51.541 (1981).

072.068 **On the possible use of panoramic magnetograms for solar activity prediction.**
V. M. Grigor'ev, N. V. Klochek, V. S. Peshcherov.
Issled. po geomagn., aehron. i fiz. Solntsa, Moskva, 1981, No. 56, p. 95 - 98. In Russian. – Abstr. in Ref. zh., 51. Astron., 12.51.542 (1981).

072.069 **Actividad solar en 1979.** I. Números relativos de Wolf; II. Estadística de manchas y superfioie de las mismas.   M. Lopez Arroyo.
Bol. Astron. Obs. Madrid, Vol. 10, No. 4, p. 1 - 77 (1979).

072.070 **Observations solaires. Rotations 1690 - 1702, 28 décembre 1979 - 16 décembre 1980.**
E. Țifrea, V. Dinulescu, A. Dimitriu, S. Dinulescu, G. Mariș.
Cent. Astron. Sci. Spat., Acad. Repub. Socialiste România, Bucarest. 79 pp. Price Lei 4 (1981).

072.071 **Sunspots: sunspot relative-numbers and sunspot-areas.**
Q. Bull. Sol. Act., Vol. 21, Part I, p. 1 - 4 (1980).

072.072 **Quasistationary outflow of gases from solar active regions.**   E. R. Mustel (*Eh. R. Mustel'*).
Cosmic rays in the heliosphere, (see 012.059), p. 15 - 23 (1981).

It is suggested that the origin of outflow of plasma from "quiet" active regions and from active regions with enhanced flare activity is the same and is due to some continuous non-stationary processes in the active regions. The velocity of gases in all these streams grows with increasing continuous flare activity in the active regions. It is concluded that quasistationary corpuscular streams from active regions with enhanced flare activity are important sources of cosmic rays from the sun.

072.073   Swinging sun, 79-year cycle, and climatic change.
T. Landscheidt.
J. Interdisciplinary Cycle Res., Vol. 12, No. 1, p. 3 - 19 (1981).

072.074   Wilson effect and axial inclination of a sunspot tube
to the solar surface.
R. Kh. Gajnullina, T. M. Minasyants.
Astron. Tsirk., No. 1150, p. 5 - 7 (1981). In Russian.

072.075   On peculiarities of rapid variations of magnetic
fields of sunspots.   G. F. Vyal'shin.
Astron. Tsirk., No. 1159, p. 3 - 5 (1981). In Russian.

072.076   Bispiral structure of transversal magnetic field
maps of sunspots.   Eh. V. Kandrashov.
Astron. Tsirk., No. 1161, p. 4 - 7 (1981). In Russian.

072.077   Investigation of the vertical gradient of the
magnetic field in sunspots with different area.
A. V. Baranov, V. S. Loskutnikov, N. A. Mosunova.
Astron. Tsirk., No. 1162, p. 3 - 5 (1981). In Russian.

072.078   An analysis of the r_v parameters of Fe I lines in the
spectrum of a sunspot.
A. V. Baranov, N. A. Mosunova, S. G. Mozharskij.
Astron. Tsirk., No. 1163, p. 6 - 7 (1981). In Russian.

072.079   Sunspot numbers.
Sky Telesc., Vol. 62, 179, 291, 390, 510, 627
(1981).

072.080   L'activité solaire.   M.- J. Martres, G. Zlicaric.
Astronomie, Vol. 95, 358 - 359, 414 - 415, 464 -
466, 518 - 519, 570 - 571 (1981).

Graphic catalogue of spot groups, Carrington
longitude – time – latitude, for cycle No. 20. I. Interaction of
groups and flare occurrence with type II radio bursts.
See Abstr. 002.007.

Physics of solar activity.   See Abstr. 003.107.

Solar physics at Oxford.   See Abstr. 011.001.

The physics of sunspots.   See Abstr. 011.004.

The MHD of sunspots.   See Abstr. 011.015.

Enträtselung der Sonnenaktivität und ihrer Auswirkungen – ein internationales Forschungsprogramm.
See Abstr. 013.036.

Real-time monitoring of solar activity: the NOAA
Space Environment Laboratory.   See Abstr. 013.043.

Broad-band polarization in molecular spectra.
See Abstr. 022.022.

A method for unambiguous determination of
starspot temperatures and areas: application to II Pegasi,
BY Draconis, and HD 209813.   See Abstr. 031.579.

Solar activity and Earth's rotation.
See Abstr. 044.001.

Magnetic flux tubes.   See Abstr. 062.050.

The Schwarzschild criterion for convection in the
presence of a magnetic field.   See Abstr. 062.070.

Free and forced oscillations of a flux tube.
See Abstr. 062.096.

On the variability of Lyman-alpha with solar activity.
See Abstr. 071.015.

Detection of a temperature deficit in magnetic
faculae at the solar photosphere.   See Abstr. 071.017.

The transition region and corona associated with
sunspots.   See Abstr. 073.047.

Links of Hα-emission features with the underlying
elements of sunspot fine structures in some flares.
See Abstr. 073.056.

Chromospheric and photospheric evolution of an
extremely active solar region in solar cycle 19.
See Abstr. 073.057.

The chromosphere above sunspot umbrae.
III. Spatial and temporal variations of chromospheric lines.
See Abstr. 073.059.

Magnetic transients in flares.
See Abstr. 073.080.

Coronal condensations and proton flares in the
20th solar cycle.   See Abstr. 074.027.

Solar magnetism: a new look.
See Abstr. 075.005.

Solar magnetic structure and the solar activity
cycle. Review of observational data.   See Abstr. 075.008.

Surface magnetic fields during the solar activity
cycle.   See Abstr. 075.009.

Bound oscillations on thin magnetic flux tubes:
convective instability and umbral oscillations.
See Abstr. 075.011.

Relations between the toroidal magnetic field in the
solar convection zone and the bipolar magnetic field of sunspots.   See Abstr. 075.016.

Comparative magnetospherology. Part 11. A model
of heliomagnetospheric excursion in every sunspot declining-
minimum phase.   See Abstr. 075.020.

Comparative magnetospherology. Part 12. Evidences
to support the two-hemisphere model on rotational reversing
of the heliodipole in sunspot maximum phase.
See Abstr. 075.021.

Thermal radio emission of solar active regions derived from quantitative analysis of Skylab X-ray pictures and
compared with observation.   See Abstr. 077.003.

Excess emission of local sources of bipolar sunspot
groups.   See Abstr. 077.058.

Temporal variations of the cosmic ray intensity and
the magnetic configurations of the heliosphere.
See Abstr. 078.023.

Solar activity and solar neutrino flux.
See Abstr. 080.005.

Le Soleil.   See Abstr. 080.029.

The age dependence of photospheric tracer rotation.
See Abstr. 080.047.

Positions of sunspot groups and solar rotation.
See Abstr. 080.048.

Variation of the coefficients of velocities of photoionization of the neutral components of the planetary upper
atmospheres with solar activity.   See Abstr. 091.070.

Large-scale variations of cosmic ray intensity and
solar activity.   See Abstr. 143.031.

Conversion coefficients for determination of the
three-dimensional cosmic ray anisotropy in different periods
of solar activity.   See Abstr. 143.090.

Galactic cosmic ray gradients in the ecliptic plane
and at high latitudes during two solar cycles (meteorite data).
See Abstr. 143.112.

Preliminary search for cosmic radiation and solar-
terrestrial parameters correlated with the reversal of the solar
magnetic field.   See Abstr. 143.113.

073 Chromosphere, Flares, Prominences

073.001   New ionization fractions for the lithium- and helium-like ionization stages of calcium and iron.
J. G. Doyle, J. C. Raymond.
Mon. Not. R. Astron. Soc., Vol. 196, 907 - 910 (1981).

The authors have re-interpreted the high resolution X-ray spectra of Ca XIX and Fe XXV observed during a solar flare on 1979 March 25 using new ionization fractions for Ca XVIII, Ca XIX, Fe XXIV and Fe XXV. These new calculations substantially change the interpretation of the spectra, implying the flare to be ionizing during the rise phase and recombining during the decay phase.

073.002   Structure of the solar chromosphere. III. Models of the EUV brightness components of the quiet sun.
J. E. Vernazza, E. H. Avrett, R. Loeser.
Astrophys. J., Suppl. Ser., Vol. 45, 635 - 725, plate 28 (1981).

Skylab observations of the quiet sun in the EUV wavelength range 40 - 140 nm show the inhomogeneous structure of the chromosphere. The authors use these observations to determine separate chromospheric models for six observed brightness components, ranging from a dark cell center to a very bright network element. In each case they solve the non-LTE radiative transfer, statistical equilibrium, and hydrostatic equilibrium equations to determine the temperature-density stratification that produces a calculated spectrum in best agreement with the observed one. The authors' hydrogen calculations are carried out with an atomic model consisting of eight explicit levels and supplementary levels 9 - 12.

073.003   Determination of the complete vector magnetic field in solar prominences, using the Hanle effect.
V. Bommier, J. L. Leroy, S. Sahal-Bréchot.
Astron. Astrophys., Vol. 100, 231 - 240 (1981).

In this paper the authors investigate various methods which are liable to provide the three components of the prominence magnetic field. The methods consist in adding a supplementary information to those which result from the Hanle effect analysis of linear polarization measurements of an emission line. The analysis of two prominences helium lines λ 5876 Å and λ 10,830 Å is given as an example.

073.004   Flares in the heliocentric galactic coordinate system.
A. A. Shpital'naya.
Soln. Dannye 1980 Byull., No. 12, p. 95 - 106 (1981). In Russian.

The distribution of large flares along galactic parallels and meridians is characterized by the following peculiarities: 1) Maxima of longitudinal distributions approximately coincide with the direction center-anticenter of the Galaxy. 2) The number of flares in the longitudinal interval $0° \leqslant l^{II} \leqslant 179°$ is statistically significantly larger than in the interval $180° \leqslant l^{II} \leqslant 359°$. 3) The flare activity is symmetric relative to the galactic equator. 4) The flare activity is greater at high galactic latitudes.

073.005   Thermal flares in the corona: possible mechanism of origin.
V. P. Vasil'ev, V. I. Kucherov, V. I. Lapshin.
Soln. Dannye 1980 Byull., No. 12, p. 110 - 114 (1981). In Russian.

A thermalisation mechanism of the fast fragments of the chromospheric ejections in the magnetic field of the corona is investigated.

073.006   A determination of chromospheric spicule diameters from observations of their occultation by the moon during a partial solar eclipse.
G. K. Ajmanova, A. K. Ajmanov, R. A. Gulyaev.

Soln. Dannye 1981 Byull., No. 1, p. 85 - 87 (1981). In Russian.

Cinematographic observations of the spicule occultation by the moon were carried out during the partial solar eclipse of October 2, 1978. The relative brightness distribution across two spicules was obtained in the Hα + 0.8 Å light. The diameters of the spicules were found to be equal to 1 200 and 280 ± 40 km.

073.007   Spectral observations of the Hα spicules at various heights of the solar chromosphere.
V. I. Kulidzhanishvili, Eh. V. Khutsishvili.
Soln. Dannye 1981 Byull., No. 2, p. 81 - 86 (1981). In Russian.

Almost simultaneous height sequences of 61 spicules in the Hα line are studied. Radial velocities, total intensities or equivalent widths, full widths at half-maximum of intensity at all heights are determined (about 450 profiles of the Hα line). It is stated that: 1. Absolute values of radial velocities increase linearly with the height. 2. Variation of the sign of the radial velocity along single spicules was never observed.

073.008   Temporal variations of the asymmetry of the line profile in solar flares.   T. P. Bushueva.
Soln. Dannye 1981 Byull., No. 3, p. 113 - 118 (1981). In Russian.

Time variations of the asymmetry of Hα emission on the basis of visual measurements of red and blue Hα line wings of 88 solar flares are studied. It is shown that the number of flares with asymmetric profiles is decreasing limbwards. A classification of the flares of time variations of their asymmetry is given.

073.009   A class of analytic solutions for the thermally balanced magnetostatic prominence sheet.
B. C. Low, S. T. Wu.
Astrophys. J., Vol. 248, 335 - 343 (1981).

The paper is a theoretical study of the nonlinear interplay between magnetostatic equilibrium and energy balance in a Kippenhahn-Schlüter type prominence sheet. The basic effects are illustrated explicitly with an analytic model in which a radiative loss proportional to $\rho^2 T$ balances against wave heating proportional to ρ, with thermal conduction confined along magnetic field lines, where ρ and T denote the plasma density and temperature, respectively. The particular choices of heat sink and source enable the authors to integrate the governing equations exactly while they are of the basic mathematical forms to simulate radiative loss in an optically thin plasma which is heated by wave dissipation. The steady solutions exhibit three different basic behaviors. The implications of the steady solutions for the formation of prominences are discussed.

073.010   Spatial and temporal structures of impulsive bursts from solar flares observed in UV and hard X-rays.
C.-C. Cheng, E. Tandberg-Hanssen, E. C. Bruner, L. Orwig, K. J. Frost, P. J. Kenny, B. E. Woodgate, R. A. Shine.
Astrophys. J., Lett., Vol. 248, L39 - L43, plates L3 - L4 (1981).

The authors present new observations of impulsive UV and hard X-ray bursts in two solar flares obtained with instruments on Solar Maximum Mission. The UV bursts were observed in the Si IV and O IV emission lines, whose intensity ratio is density-sensitive. By comparing the spatially resolved Si IV/O IV observations with the corresponding hard X-ray observations, the authors are able to study their spatial and temporal relationships.

073.011   Continuum emission in the 1980 July 1 solar flare.
H. Zirin, D. F. Neidig.

Astrophys. J., Lett., Vol. 248, L45 - L48, plates L5 - L6 (1981).

Comparison of continuum measurements of the 1980 July 1 flare at Big Bear Solar Observatory and Sacramento Peak Observatory show strong blue emission kernels with the ratio of Balmer continuum (Bac): $\lambda 3862$ continuum: continuum above 4275 Å to be about 10:5:1. The blue continuum at 3862 Å is too strong to be explained by unresolved lines. The Bac intensity was 2.5 times the photosphere and the strongest $\lambda 3862$ continuum was 2 times the photosphere. The brightest continuum kernel occurred late in the flare, after the hard X-ray peak and related in time to an isolated peak in the 2.2 MeV line, suggesting that that continuum was excited by protons above 20 MeV.

073.012 Downward shift of the acceleration/injection region during solar flares.   S. R. Kane, A. Raoult.
Astrophys. J., Lett., Vol. 248, L77 - L82 (1981).

Simultaneous observations of impulsive hard X-rays and type III radio bursts associated with two recent solar flares have been analyzed in order to study the characteristic increase in the starting frequency of fast-drift type III bursts during the impulsive phase. The time variation of the starting frequency has been found to be similar to that of the ~35 keV X-rays, especially during the increasing phase. A similar variation has been observed for the low frequency cutoff of these bursts. The observations are consistent with a systematic downward shift of the electron acceleration/injection region during the impulsive phase of a solar flare.

073.013 A time dependent model for spicule flow.
S. S. Hasan, P. Venkatakrishnan.
Sol. Phys., Vol. 73, 45 - 57 (1981).

A time dependent model for the flow of gas in a spicule is studied. In this model, the flow occurs in a magnetic flux sheath. Starting from hydrostatic equilibrium, the flux sheath is allowed to collapse normal to itself. The collapse induces a flow of gas along the magnetic field and this flow is identified as a spicule. It is proposed that the initial rapid collapse occurs during an 'impulsive spicule' phase and it is the subsequent gradual relaxation of the flow which is observed as a spicule.

073.014 The He I 10830 Å chromosphere and filament associated structures.
M. K. McCabe, D. L. Mickey.
Sol. Phys., Vol. 73, 59 - 66 (1981).

Full disk, He I 10830 Å solar spectrograms were developed for the detection of coronal holes, and have been compared with nearly simultaneous Hα and Ca K filtergrams. Areas of reduced helium absorption have been noted in the neighborhood of filaments and neutral zones in the longitudinal solar magnetic field. The existence of these 'helium lanes' is discussed in terms of their relationship to Hα filament channels or to the coronal cavities which surround prominences.

073.015 Role of plasma flow in determining structure of the chromosphere-corona transition zone of the sun.
W. M. Glencross.
Sol. Phys., Vol. 73, 67 - 79 (1981).

Transfer of material between the chromosphere and corona of the sun must occur whenever the geometry of any interconnecting magnetic structure changes, and there will also be a flow of plasma along field lines caused by any pressure difference between the two feet of each arch. This study shows that the term associated with flow is comparable to the radiation-loss term in the energy budget of the transition zone if the plasma speed at the base of the corona reaches about 3 km s^{-1}. This value is probably exceeded within most flux tubes during some period of their development, and speeds an order of magnitude higher can occur in favourable regions. This paper also examines limits to the temperature

gradient of the transition zone set by the requirement of continuity of plasma flow.

073.016 Hydrodynamic response of the solar chromosphere to an elementary flare burst. I: Heating by accelerated electrons.
B. V. Somov, S. I. Syrovatskii (*Syrovatskij*), A. R. Spektor.
Sol. Phys., Vol. 73, 145 - 155 (1981).

The authors investigate non-steady gasdynamic plasma motion in the case of chromospheric heating by energetic electrons.

073.017 Solar flare plasmas.   A. H. Gabriel.
Philos. Trans. R. Soc. London, Ser. A, Vol. 300, 497 - 504 (1981). − Abstr. in Phys. Abstr., Vol. 84, Abstr. 83587 (1981).

073.018 Mass motions in the transition region.   G. Poletto.
Sol. Phys., Vol. 73, 233 - 256 (1981).

Following previous order of magnitude estimates (Poletto, 1980), the possibility that hot downflowing motions in the solar transition region could be ascribed to spicular matter returning to the chromosphere after being heated by compression, is more thoroughly investigated. The equations describing the one-dimensional non stationary motion of the spicular plasma during the heating process are analytically solved, and the temporal profiles of temperature, density and velocity are given for a set of representative situations. The results are finally compared with available data.

073.019 The optical continuum of solar and stellar flares.
M. A. Livshits, O. G. Badalyan, A. G. Kosovichev, M. M. Katsova.
Sol. Phys., Vol. 73, 269 - 288 (1981).

A further development of the Kostyuk-Pikelner's model is presented. The response of the chromosphere heated by nonthermal electrons of the power-law energy spectrum has been studied on the basis of the numerical solution of the one-dimensional time-dependent equations of gravitational gas dynamics. The ionization and energy loss for the emissions in the Lyman and Balmer lines have been determined separately for the optically thin and thick Lα-line layers. In their essential features, the gas dynamic processes during the flares in red dwarf atmospheres are the same as those in the solar atmosphere. However, the high atmospheric densities, smaller height scale in red dwarf atmospheres, and greater energy of this processes in stellar flares, give rise, in practice, to the regular generation of optical continuum.

073.020 Thermal nonequilibrium: a trigger for solar flares?
A. W. Hood, E. R. Priest.
Sol. Phys., Vol. 73, 289 - 311 (1981).

The authors suggest that a solar flare may be friggered by a lack of thermal equilibrium rather than by a magnetic instability. The possibility of such a thermal nonequilibrium (or catastrophe) is demonstrated by solving approximately the energy equation for a loop under a balance between thermal conduction, optically thin radiation and a heating source.

073.021 The recurrent surges in McMath 9760 and associated surge brightenings.   R. Začh, V. Bar.
Sol. Phys., Vol. 73, 331 - 339 (1981).

The unique surge activity that occurred in McMath 9760 region during November 7 to November 16, 1968 is described and discussed. The surge activity was unusually intense. Some surges were accompanied by surge brightenings. The region was studied at Hα −0.5 Å. In spite of the enhanced activity, the region appeared to be highly stable.

073.022 Analysis of a series of solar flare X-ray spectra.
D. L. McKenzie, P. B. Landecker.
Astrophys. J., Vol. 248, 1117 - 1125 (1981).

The authors analyze 13 X-ray line spectra (7.8–23.0 Å) acquired during a solar flare that peaked at 2326 UT on 1979 March 31. The data include fluxes of lines excited throughout the temperature range $2-18 \times 10^6$ K, the entire range present in the coronal flare. The coronal flare emission measure began to decrease 500 s or less after the broad-band X-ray flux peaked. The authors construct the differential emission measure $\epsilon(T)$ for $2-18 \times 10^6$ K from the data by using collision strengths now available for a small number of the observed lines. The total emission measure for T greater than 2×10^6 K, the radiative energy loss rate, the density at 2×10^6 K, and the conductive cooling time are derived from the data, but lack of information on electron densities prevents a detailed study of the flare energetics.

073.023   Models of a hot flare region based on X-ray photometric data of the Interkosmos 11 satellite.
M. Vandas.
Bull. Astron. Inst. Czechoslovakia, Vol. 32, 215 - 222 (1981).

Soft X-ray flare data are used to determine some physical conditions in the flare region. A model is suggested with a Gaussian dependence of the temperature and electron density on coordinates and compared with the constant temperature model. In order to justify the Gaussian approximation a simple model for studying flare cooling is created and the obtained results are shown.

073.024   On the possibility to observe current sheets on the sun.   V. A. Ostapenko.
Pis'ma Astron. Zh., Tom 7, 561 - 565 (1981). In Russian.
English translation in Soviet Astron. Lett., Vol. 7.

The compact arc model with a current sheet is shown to provide a description of moustaches spectral line profiles. Plasma flows from the current sheet define extended wings of moustache lines. The current sheet is manifested apparently by a stripe of continuous emission.

073.025   X-ray observations of two different systems of "post-flare" loops.
Z. Svestka, H. W. Dodson-Prince, O. C. Mohler, S. F. Martin, R. L. Moore, J. T. Nolte, R. D. Petrasso.
Bull. American Astron. Soc., Vol. 13, 542 (1981). – Abstract.

073.026   The general non-thermal models of impulsive phase of solar flares.   J. Leach, V. Petrosian.
Bull. American Astron. Soc., Vol. 13, 543 (1981). – Abstract.

073.027   Flare densities from iron Kα lines.
J. R. Lemen, K. J. H. Phillips.
Bull. American Astron. Soc., Vol. 13, 543 (1981). – Abstract.

073.028   Helium excitation in quiescent prominences.
D. A. Landman, A. P. Bernat, J. M. Pasachoff.
Bull. American Astron. Soc., Vol. 13, 552 (1981). – Abstract.

073.029   Mass and energy balance in the 1973 August 9 flare.
J. W. Cook, K. P. Dere.
Bull. American Astron. Soc., Vol. 13, 554 (1981). – Abstract.

073.030   Evidence for delayed second phase acceleration in solar flares.   J. B. Willett, J. C. Ling,
G. R. Riegler, W. A. Mahoney, A. S. Jacobson.
Bull. American Astron. Soc., Vol. 13, 555 (1981). – Abstract.

073.031   SMM observations of soft X-ray line spectra.
J. W. Leibacher, K. J. H. Phillips, C. J. Wolfson.
Bull. American Astron. Soc., Vol. 13, 555 (1981). – Abstract.

073.032   Solar flare magnetohydrodynamics – introduction.
E. R. Priest.
Solar flare magnetohydrodynamics, (see 003.002), p. 1 - 46 (1981).

Contents: Basic description of a flare. Recent advances in understanding. An overview of the flare phenomenon. Earlier theories for the solar flare. The magnetohydrodynamic equations. Summary of the book.

073.033   Flare observations.   Z. Švestka.
Solar flare magnetohydrodynamics, (see 003.002), p. 47 - 137 (1981).

Contents: Introduction. Flare observations in the Hα line. Chromospheric flare spectra. Thermal flares in soft X-rays. Impulsive hard X-ray and microwave bursts. Energy transfer through the transition layer. Metric radio bursts. Flare ejecta. Production and effects of high-energy particles in flares.

073.034   Two-ribbon flares: magnetostatic equilibria.
J. Birn, K. Schindler.
Solar flare magnetohydrodynamics, (see 003.002), p. 337 - 378 (1981).

Contents: Introduction. Basic equations and assumptions of the quasi-static theory. Existence of equilibrium solutions in two dimensions. Special solutions. Stability. Slow passive evolution. Discussion.

073.035   Two-ribbon flares: (post)-flare loops.
G. W. Pneuman.
Solar flare magnetohydrodynamics, (see 003.002), p. 379 - 428 (1981).

Contents: Introduction. Observations. Loop prominences as a reconnection process. Magnetic field dynamics. Mass supply. Heating and cooling of the loops. Summary and conclusions.

073.036   Particle acceleration in solar flares.   J. Heyvaerts.
Solar flare magnetohydrodynamics, (see 003.002), p. 429 - 555 (1981).

Contents: Observational evidence for accelerated particles. Energetic particles at 1 au. Hard X-ray emission in flares. Implications of X-ray observations: how many electrons? Microwave impulsive bursts. Gamma-ray emission. Elemental and isotopic abundances of solar cosmic rays. Fast particle impact on the chromosphere and photosphere from EUV and Hα bursts. Directed flow of fast particles to the chromosphere or hot fireball observations. Directed flow of fast particles to the chromosphere or hot fireball: theory. First-phase acceleration process: an overview. Runaways through turbulence. Current-driven micro-instabilities: weak field (B ≈ 0). Current-driven micro-instabilities: magnetized medium. Production of a Langmuir wave spectrum suitable for acceleration. Strong turbulence effects. Proposed microturbulent acceleration models. Double-layers. Approaching microturbulent conditions: a difficulty. Tearing modes and fast particles. Second-phase acceleration: an overview. Second-phase Fermi acceleration is compatible with observational data. Fermi mechanism and resonant MHD processes. First-order Fermi mechanism near shocks. Nuclei preheating. Acceleration in and near shock fronts.

073.037   Activity and outer atmosphere of the sun.
G. L. Withbroe.
Activity and outer atmospheres of the sun and stars, (see 012.012), p. 1 - 88 (1981).

The atmospheric layers that the author is concerned with are the chromospheres, chromospheric-coronal transition region and corona. A schematic illustrating the radial temperature and density distribution in a representative region of the solar atmosphere is given. The author discusses some aspects of observations of the outer solar atmosphere, its structure and important phenomena that occur there such as flares.

073.038   The chromosphere and transition region.
R. G. Athay.

The sun as a star, (see 003.004), p. 85 - 133 (1981).

A simple empirical definition of a chromosphere is adopted and various general characteristics of the chromosphere are described.

073.039 **Solar flare observations and their interpretations.**
J. C. Brown, D. F. Smith, D. S. Spicer.
The sun as a star, (see 003.004), p. 181 - 227 (1981).

This review is split into three main parts: the main observational aspects of the flare problem; the present interpretations of these observations; and the present understanding of hard X-rays produced during the first phase particle acceleration.

073.040 **Chromospheric heating.**   S. D. Jordan.
The sun as a star, (see 003.004), p. 301 - 319 (1981).

Certain general features of the solar chromospheric heating problem also apply to many stellar chromospheres. These general features are reviewed.

073.041 **Mechanical heating in the transition region.**
G. Withbroe.
The sun as a star, (see 003.004), p. 321 - 329 (1981).

In this paper, attention is focused on the energy balance in the transition region and the role that mechanical heating plays in determining the temperature density structure of this region in a stellar atmosphere. The main conclusion is that, because of its role as the interface layer through which mass and energy flow between the chromosphere and corona, direct deposition of mechanical energy is a relatively unimportant factor in the overall energy balance in the transition region, except in the uppermost layers where the temperature approaches coronal values.

073.042 **Solar flare theory.**   D. S. Spicer, J. C. Brown.
The sun as a star, (see 003.004), p. 413 - 470 (1981).

The authors discuss the physical mechanisms important to flares, which, simultaneously, introduce various physical concepts useful for understanding the physics of the flare mechanism. There is a critical examination of the question of flare model requirements and flare triggers, which also gives a brief summary of the flare models currently in vogue, which demonstrate how each mechanism can be used in the context of a model. This review is limited to the first acceleration phase of the flare.

073.043 **Analysis of the solar loop flare on March 27, 1979.**
Q. Su, T. Cao, D. Wang.
Sci. Sinica, Vol. 24, 656 - 661 (1981).

073.044 **A calculation of the curvature vector of magnetic field lines near filaments.**
Den O Gym, E. A. Kornitskaya.
Inst. zemn. magn., ionos. i rasprostr. radiovoln AN SSSR. Prepr., 1981, No. 2, 15 pp. In Russian. − Abstr. in Ref. zh., 51. Astron., 7.51.302 (1981).

073.045 **Flare ejection on the sun according to cinematographic observations in Hα.**
O. E. Den, M. M. Molodenskij, B. P. Filippov.
Dokl. AN SSSR, Tom 257, 305 - 309 (1981). In Russian. Abstr. in Ref. zh., 51. Astron., 7.51.312 (1981).

073.046 **Long term variations in solar flare activity.**
J. N. Goswami, R. Jha, D. Lal.
J. Astrophys. Astron., Vol. 2, 201 - 212 (1981).

The authors present an intercomparison between the contemporary flare data (1956−1973) based on rocket and spacecraft experiments and the long term ($>10^4$ yr) averaged solar flare data, based on stable- and radio-nuclide studies of lunar samples.

073.047 **The transition region and corona associated with sunspots.**   F. Q. Orrall.
Space Sci. Rev., Vol. 28, (see 012.014), 423 - 434 (1981).

To the present time, no structure has been identified immediately above the chromosphere in sunspots that is invariably present and that thus might be called the transition region and corona over the spot. But the magnetic flux tubes emerging from spots give rise to many of the plasma filled loops that characterize the active region corona. These emit strongly from ions characteristic of the transition region, or the corona, but seldom both simultaneously. This paper presents an overview of the morphology, evolution and theory of these structures.

073.048 **Emission energy of optical solar flares. I.**
L. N. Kuròchka, L. A. Stasyuk.
Soln. Dannye 1981 Byull., No. 5, p. 83 - 91 (1981). In Russian.

The emission energy is calculated for all lines and continua of hydrogen series under two extreme Balmer decrements for each class of flare. It is shown that the main energy loss of large flares takes place in the Lα line and the Lyman continuum. The coefficients are obtained for flares of different importance. They permit to find the total emission energy in hydrogen spectra with data on the intensity of only the Hα line.

073.049 **Characteristics of the D_3 He I line in spicules at various heights.**   Eh. V. Khutsishvili.
Soln. Dannye 1981 Byull., No. 5, p. 116 - 119 (1981). In Russian.

A series of spectrograms of spicules in the D_3 He I line at eight heights of the solar chromosphere is obtained. The series begins at 4980 km and goes up to 9450 km. Forty-eight spicules were chosen to be treated. The radial velocities, half-widths, and equivalent widths are determined for each height. It is found that the radial velocities increase linearly along the spicules, the half-widths slightly decrease with height and as for the spicules brightness it drops exponentially with height.

073.050 **Energetics and phenomenology of large solar flares.**
Eh. I. Mogilevskij.
Fiz. soln aktivnosti. Moskva, 1980, p. 3 - 47. In Russian.
Abstr. in Ref. zh., 51. Astron., 8.51.465 (1981).

073.051 **Dynamical phenomena of the limb flare of September 12, 1966.**
V. N. Ishkov, B. Rompolt.
Fiz. soln. aktivnosti. Moskva, 1980, p. 80 - 92. In Russian.
Abstr. in Ref. zh., 51. Astron., 8.51.475 (1981).

073.052 **On ways of future investigation of solar flares being rich in helium-3.**
G. E. Kocharov, L. G. Kocharov, Yu. E. Charikov.
Izv. AN SSSR. Ser. fiz., Tom 45, 579 - 587 (1981). In Russian.
Abstr. in Ref. zh., 51. Astron., 8.51.497 (1981).

073.053 **Evidence for prominent enhancement in solar flare activity on a million year time scale.**
S. K. Bhattacharya, R. Jha, D. Lal.
Bull. Astron. Soc. India, Vol. 9, 73 (1981). − Abstract.

073.054 **Solar flare and solar wind activity at present and in the past.**   M. N. Rao.
Bull. Astron. Soc. India, Vol. 9, 73 (1981). − Abstract.

073.055 **A time dependent model for spicule flow.**
S. S. Hasan, P. Venkatakrishnan.
Bull. Astron. Soc. India, Vol. 9, 74 - 75 (1981). − Abstract.

073.056 **Links of Hα-emission features with the underlying elements of sunspot fine structures in some flares.**
V. Bumba, J. Suda, V. N. Ishkov.
Bull. Astron. Inst. Czechoslovakia, Vol. 32, 286 - 291 (1981).
The positions of individual flare-emission features

penetrating into the umbrae of sunspots are compared with the underlying elements of these umbrae. Five flares in three large and complex active regions are studied. The obtained results demonstrate that the flare emission observed in the wings of the hydrogen Hα-line correlates well with the photospheric-like sunspot light-bridges and gulfs of umbrae.

073.057 **Chromospheric and photospheric evolution of an extremely active solar region in solar cycle 19.**
S. M. P. McKenna-Lawlor.
NASA Contract. Rep., NASA CR 3377. 12 + 218 pp. (1981).
   A comprehensive investigation was made of phenomena attending the disk passage, July 07-21, 1959, of active solar center HAO-59Q. At the photospheric level this comprised an aggregate of groups of sunspots of which one group, Mt. Wilson 14284, showed all the attributes deemed typical of solar regions associated with the production of major flares. A special characteristic of 59Q was its capability to eject dark material. Two apparently different kinds of flare were identified in 59Q; namely, Prominence Flares (which comprised brightenings within part of the suspended dark prominence) and Plage Flares (which comprised brightenings within part of the chromospheric network). At least 87.5 percent, and probably all, impulsive brightenings in 59Q began directly above minor spots, many of which were satellites to major umbrae.

073.058 **Dynamics in the filaments. I. Oscillations in a quiescent filament.**
J. M. Malherbe, B. Schmieder, P. Mein.
Astron. Astrophys., Vol. 102, 124 - 128 (1981).
   A time sequence of a quiescent filament has been obtained with the MSDP spectrograph operating on the Meudon Solar Tower, the duration was 720 s, the time step 30 s. The Fourier analysis of the radial velocities measured in the Hα line shows that the steady velocities are principally upward in the filament, the chromospheric oscillations are almost undetectable inside the filament and reduced around it. Steady dynamical structures are most consistent with the theoretical model of filaments proposed by Unno and Ribes.

073.059 **The chromosphere above sunspot umbrae. III. Spatial and temporal variations of chromo-**spheric lines.   F. Kneer, W. Mattig, M. v. Uexküll.
Astron. Astrophys., Vol. 102, 147 - 155 (1981) = Mitt. Kiepenheuer-Inst. No. 203.
   The authors analyse photographic spectrograms of the lines Ca II H, K, 8542 Å, Na I D_1 and D_2, Ni I 5893 Å, and Hα, which were obtained from a sunspot with the Vacuum Tower Telescope at the Sacramento Peak Observatory. The results for the umbra are presented.

073.060 **On the differential rotation of solar filaments.**
V. Dermendjiev.
Astrofiz. issled., NRB, Vol. 3, 103 - 107 (1981). In Russian. Abstr. in Ref. zh., 51. Astron., 9.51.331 (1981).

073.061 **Physical nature of solar flares.**
S. I. Syrovatskij.
Probl. soln.-zemn. svyazej. Dokl. simpoz. KAPG, Ashkhabad, 1979. Ashkhabad, 1981, p. 21 - 41. In Russian. – Abstr. in Ref. zh., 51. Astron., 9.51.405 (1981).

073.062 **Helium-3 rich solar flare studies: state of the art and prospects.**   L. G. Kocharov.
Cosmic rays, (see 012.005), 1980, p. 190 - 198. In Russian. Abstr. in Ref. zh., 51. Astron., 9.51.407 (1981).

073.063 **On active longitudes of quiescent prominences.**
V. Dermendjiev.
Astrofiz. issled., NRB, Vol. 3, 108 - 112 (1981). In Russian. Abstr. in Ref. zh., 51. Astron., 9.51.417 (1981).

073.064 **A study of active regions generating flares with high-energy protons and their forecast.**
B. O. Berlyand.
Tr. Inst. prikl. geofiz. Gos. kom. SSSR po gidrometeorol. i kontrolyu prirod. sredy, 1981, No. 50, p. 27 - 34. In Russian. Abstr. in Ref. zh., 51. Astron., 9.51.422 (1981).

073.065 **Chromospheric flash spectrum.**   M. C. Pande, K. R. Bondal, V. P. Gaur, K. Sinha, G. C. Joshi.
Observations of the total solar eclipse of 16 February 1980, (see 003.009), p. 42 - 43 (1981).

073.066 **Determination of the abundance of helium in the solar chromosphere and structure of the transition region between the corona and the chromosphere.**
T. Hirayama, M. Iris, K. Kumagai.
Observations of the total solar eclipse of 16 February 1980, (see 003.009), p. 44 - 47 (1981).

073.067 **Chromosphere-corona transition region models with magnetic field and fluid flow.**   R. G. Athay.
Astrophys. J., Vol. 249, 340 - 348 (1981).
   Energy balance models of the chromosphere-corona transition region are computed for a segment of average Sun network using assumed magnetic field geometry. The energy fluxes considered include radiation, conduction, enthalpy, and gravitational potential energy, but do not include mechanical heating. Two classes of models are considered: conduction driven and flow driven.

073.068 **High resolution solar flare X-ray spectra: the temporal behavior of electron densitiy, temperature,** and emission measure for two class M flares.
G. A. Doschek, U. Feldman, P. B. Landecker, D. L. McKenzie.
Astrophys. J., Vol. 249, 372 - 382 (1981).
   High resolution soft X-ray flare spectra recorded by Naval Research Laboratory (NRL) and Aerospace Corporation Bragg crystal spectrometers flown on an orbiting spacecraft (P78-1) are combined and analyzed. The authors analyze the spectra of two flares which occurred on 1980 April 8 and May 9. Temporal coverage is fairly complete for both flares, including the rise and decay phases. Measurements of electron density N_e with rather high time resolution (about 1 minute) have been obtained throughout most of the lifetimes of the two flares. These measurements were obtained from the O VII lines and pertain to flare plasma at temperatures near $2 \times 10^6 K$.

073.069 **The Lyman-α/Hα ratio in solar flares and quasars.**
R. C. Canfield, R. C. Puetter, P. J. Ricchiazzi.
Astrophys. J., Vol. 249, 383 - 389 (1981).
   The authors summarize the most recent observational data on the Lyα/Hα ratio in solar flares and quasars. They calculate solar flare models comparable to their previous quasar emission line cloud models. They discuss what the Lyα/Hα ratio implies and their conclusions are summarized.

073.070 **On the importance of reverse current ohmic losses in electron-heated solar flare atmospheres.**
A. G. Emslie.
Astrophys. J., Vol. 249, 817 - 820 (1981).
   The author considers the passage of a beam of nonthermal electrons through the flaring solar atmosphere, paying particular attention to the requirement that the beam be stable to the generation of plasma turbulence. He then computes the ratio of energy losses due to reverse current ohmic heating and heating by Coulomb collisions, respectively, for the greatest flux which can pass stably through the atmosphere.

073.071 **Simultaneous measurements of EUV and soft X-ray solar flare emission.**
D. M. Horan, R. W. Kreplin.

Sol. Phys., Vol. 74, (see 012.029), 265 - 272 (1981).

Broadband sensors aboard the Naval Research Laboratory's SOLRAD 11 satellites measured solar emission in the 0.5 to 3 Å, 1 to 8 Å, 8 to 20 Å, 100 to 500 Å, 500 to 800 Å, and 700 to 1030 Å bands between March 1976 and October 1979. The authors present measurements of the X-ray and EUV emission from several flares with special emphasis on the relative EUV response associated with flares in different categories determined by 1 to 8 Å soft X-ray flux. An example of a flare exhibiting an impulsive (nonthermal) phase is included.

073.072 **Intensity of astrophysical lines in the transition region.**   S. Chandra.
Physica B, C, Vol. 106 B + C, 305 - 307 (1981). − Abstr. in Phys. Abstr., Vol. 84, Abstr. 98775 (1981).

073.073 **The outer layers of the Sun.**
C. Jordan.
Sci. Prog., Vol. 67, 1 - 26 (1981). − Abstr. in Phys. Abstr., Vol. 84, Abstr. 98776 (1981).

073.074 **Observations of polarized radiation of prominences in the radiowave range.**
Sh. B. Akhmedov, I. A. Ipatova.
13-ya Vses. konf. po radioastron. issled. soln. sistemy, Kiev, 1981. Tez dokl., Kiev, 1981, p. 16. In Russian. − From Ref. zh., 51. Astron., 10.51.353 (1981).

073.075 **Energetics of solar flares.**   B. V. Somov.
Fiz. inst. AN SSSR. Prepr., 1981, No. 15. In Russian. Abstr. in Ref. zh., 51. Astron., 10.51.378 (1981).

073.076 **On thermal radiation of a solar flare during impulsive heating.**   V. A. Kovalev.
Pis'ma Astron. Zh., Tom 7, 696 - 700 (1981). In Russian. English translation in Soviet Astron. Lett., Vol. 7.

On the basis of the hydrodynamic model of a solar flare the thermal gyromagnetic radiation of an individual magnetic tube with inhomogeneous field is calculated. The flare's temperature transition region determines the dynamic characteristics of the microwave continuum.

073.077 **Physical properties of the solar chromosphere deduced from optically thick lines: I. Observations, data reduction, and modeling of an average plage.**
P. Lemaire, P. Gouttebroze, J. C. Vial, G. E. Artzner.
Astron. Astrophys., Vol. 103, 160 - 176 (1981).

Simultaneous Ca II H and K, Mg II h and k, and H I Lα and Lβ profiles have been recorded by the OSO-8/LPSP spectrometer on a quiet Sun area and a part of plage, McMath 13738. After data reduction and filtering, a full set of calibrated profiles is obtained. Using a code developed at LPSP (Gouttebroze et al., 1978), profiles computed from VAL III model C (Vernazza et al., 1981) are compared to observations. A specific plage model has been computed to fit the observed profiles. This model reproduces successfully Ca II H and K, and H I Lα and Lβ, but partly fails with Mg II h and k. Atmospheric inhomogeneities and dynamical effects may contribute to this discrepancy, but also, the authors propose to increase the level number entering in the Mg II computation to take account of a likely fluorescence induced by the H I Lβ radiation field.

073.078 **Runaway acceleration in a radio flare.**
J. Kuijpers, P. van der Post, C. Slottje.
Astron. Astrophys., Vol. 103, 331 - 338 (1981).

Radio observations of a solar flare are explained by induced electric fields in several small regions within a flaring flux tube. In each acceleration region runaway electrons are produced which lead to a pulsed production of high-frequency plasma waves. The model is used for an accurate determination of the physical conditions in the flare. During the runaway process an essential fraction of the runaway energy is put into plasma waves. The required electric field strength is of order $E/E_c \cong 0.1$ (E_c is twice the Dreicer field) and the ratio of electron cyclotron to plasma frequency is of order unity or larger.

073.079 **Observations of transitions of hydrogen-like Fe XXVI in solar flare spectra.**
A. N. Parmar, J. L. Culhane, C. G. Rapley, E. Antonucci, A. H. Gabriel, M. Loulergue.
Mon. Not. R. Astron. Soc., Vol. 197, 29P - 34P (1981).

Observations of solar flare spectra over the wavelength range 1.769−1.796 Å with the Bent Crystal Spectrometer on the NASA Solar Maximum Mission satellite are presented. The $^2P_{3/2}$, $^2P_{1/2}$ and associated satellite transitions of Fe XXVI are identified, their intensities and widths estimated and the results compared with theoretical predictions.

073.080 **Magnetic transients in flares.**
H. Zirin, K. Tanaka.
Astrophys. J., Vol. 250, 791 - 795, plates 21 - 26 (1981).

The authors present new data on magnetic transients (mgtr's) observed in flares on 1980 July 1 and 5 with the Big Bear videomagnetograph (VMG). The 1980 July 1 event was a white light flare in which a strong bipolar mgtr was observed, and a definite change in the sunspots occurred at the time of the flare. In the 1980 July 5 flare, a mgtr was observed in only one polarity, and, although no sunspot changes occurred simultaneously with the flare, major spot changes occurred in a period of hours. Late in the 1980 July 1 flare, the radio burst position shifted with the optical emission to a new kernel associated with a secondary peak in the 2.2 MeV line.

073.081 **Eruptive prominences.**   V. Rušin.
Kozmos, Vol. 12, 161 - 162 (1981). In Slovak.

073.082 **Neon composition in solar flares.**
T. R. Venkatesan, C. M. Nautiyal, M. N. Rao.
Geophys. Res. Lett., Vol. 8, 1143 - 1146 (1981).

To examine whether the long-term averaged solar flare Ne is "planetary" type, as observed in contemporary solar flares, the authors have studied etched feldspar separates of several lunar soils and three samples from different depths of two well-documented lunar rocks. The results of noble gas elemental and isotopic ratios suggest that the long-term average solar flare Ne composition is more similar to "solar" type than the "planetary" type.

073.083 **Observation and preliminary analysis of the chromospheric flash spectrum obtained during the total solar eclipse of Feb. 16, 1980.**
L.-x. Shen, Q.-d. Li, J.-q. Yu, R.-q. Sun.
Acta Astrophys. Sinica, Vol. 1, 169 - 187 (1981). In Chinese.

A sucessful observation of the flash spectrum was made by a slitless spectrograph during the total solar eclipse of Feb. 16, 1980. The authors have identified 1042 chromospheric lines ranging from 4799 Å to 5845 Å. The absolute photometric calibration was made by the spectra of the limb obtained by the slitless spectrograph before the second contact and the spectra of the disc center before the day of the eclipse. The authors have measured and calculated the intensity of each spectrum line near the base of the chromosphere.

073.084 **Cross-correlation analysis between solar flares and the southward components of the interplanetary magnetic field.**   G.-h. Chang.
Acta Astrophys. Sinica, Vol. 1, 285 - 290 (1981). In Chinese.

073.085 **Quiescent prominence spectrophotometry: sodium $D_{1,2}$, helium D_3, and calcium$^+$ λ8498.**
D. A. Landman.
Astrophys. J., Vol. 251, 768 - 780 (1981).

The results of a series of high-dispersion observations of the lines Na $D_{1,2}$, He D_3, and $Ca^+ \lambda 8498$ are presented. The measurements were made using the Haleakala 25 cm coronograph/coudé spectrograph with a Si vidicon detector system. The Na D line profiles are satisfactorily analyzed in terms of an isothermal model. The D_3 and $\lambda 8498$ profiles, on the other hand, require a two-component model. The measured cool-component $\lambda 8498$ Doppler widths range from values comparable to, to considerably less than, the widths predicted from the corresponding D_3 and $D_{1,2}$ profiles. Thus, the assumption of a common cool-emission region for these lines appears to have limited validity.

073.086 Impulsive phase of solar flares. I. Characteristics of high energy electrons.
J. Leach, V. Petrosian.
Astrophys. J., Vol. 251, 781 - 791 (1981).

The authors investigate the variation along a magnetic field line of the energy and pitch angle distribution of high energy electrons injected into a cold hydrogen plasma containing either an open or closed magnetic field structure. The problem is formulated as a time independent Fokker-Planck equation for the electron number distribution as a function of the electron energy, electron pitch angle, and the structure of the magnetic field. They present a simple analytic solution valid in the small pitch angle regime and for a slowly varying magnetic field. For the more general situations they use a numerical code for solving the Fokker-Planck equation and they show that for most practical applications the analytic expression instead of lengthy numerical computations is sufficient.

073.087 On ionization equilibrium in solar flares as determined from X-ray emission lines of Ca XVIII and Ca XIX.
G. A. Doschek, U. Feldman.
Astrophys. J., Vol. 251, 792 - 796 (1981).

X-ray emission lines of Ca XIX and Ca XVIII are used for diagnostics of high-temperature flare plasmas ($T_e > 10^7$ K). One of the Ca XVIII lines (q), originating from the transition $1s(2s2p^3 P)^2 P_{3/2} \to 1s^2 2s^2 S_{1/2}$ is more intense than expected from theoretical calculations. The authors show that a blend with a line of Ar XVII and possibly another blend with two lines that belong to Ni XXVII may remove part of the discrepancy. After consideration of the accuracy of various ionization equilibrium calculations, it appears that flare plasmas are in or near ionization equilibrium, at least up to helium-like ionization stages.

073.088 Properties of solar flare electrons, deduced from hard X-ray and spatially resolved microwave observations.
K. A. Marsh, G. J. Hurford, H. Zirin, G. A. Dulk, B. R. Dennis, K. J. Frost, L. E. Orwig.
Astrophys. J., Vol. 251, 797 - 804 (1981).

The multiply impulsive class 1B flare which occurred in AR 2339 on 1980 March 23 during 16 : 57 - 17 : 15 UT was observed by the VLA at 15.1 GHz with up to 0″.2 spatial resolution, the Owens Valley three-element solar interferometer at 10.6 GHz with up to 6″.5 resolution, and the Hard X-ray Burst Spectrometer aboard the Solar Maximum Mission. The microwave bursts occurred along a highly sheared neutral line (at different locations in some cases). Analysis of the two most prominent bursts showed that the sources were ~ 6″ in extent at both 10.6 and 15.1 GHz, and oppositely polarized on either side of the neutral line. The latter result is consistent with emission near the top of the magnetic loop. The spatial and spectral data argue against the production of microwaves and hard X-rays by a common population of electrons with either a Maxwellian or power-law distribution of electron energies.

073.089 A new component of hard X-rays in solar flares.
R. P. Lin, R. A. Schwartz, R. M. Pelling, K. C. Hurley.
Astrophys. J., Lett., Vol. 251, L109 - L114 (1981).

The authors present high resolution (~ 1 keV FWHM) spectral measurements from 13 to 300 keV of a solar flare hard X-ray burst observed on 1980 June 27 by a balloon-borne array of cooled germanium planar detectors. At energies below ~ 35 keV they identify a new component of solar flare hard X-rays. This component is characterized by an extremely steep spectrum which fits closely to that from a Maxwellian electron distribution with a maximum temperature of ~ 34 × 10⁶ K and an emission measure of 2.9 × 10⁴⁸ cm⁻³.

073.090 Steady flows in the solar transition region observed with SMM.
K. B. Gebbie, F. Hill, J. Toomre, L. J. November, G. W. Simon, J. B. Gurman, R. A. Shine, B. E. Woodgate, R. G. Athay, E. C. Bruner, Jr., R. A. Rehse, E. A. Tandberg-Hanssen.
Astrophys. J., Lett., Vol. 251, L115 - L118, plates L4 - L5 (1981).

Steady flows in the quiet solar transition region have been observed with the Ultraviolet Spectrometer and Polarimeter (UVSP) experiment on the Solar Maximum Mission (SMM) satellite. The persistent vertical motions seen at disk center have spatial rms amplitudes of 1.4 km s⁻¹ in the C II line, 3.9 km s⁻¹ in Si IV, and 4.2 km s⁻¹ in C IV. The amplitudes of the more horizontal flows seen toward the limb tend to be somewhat higher. Plots of steady vertical velocity versus intensity seen at disk center in Si IV and C IV show two distinct branches.

073.091 Preacceleration in collapsing magnetic neutral sheets and anomalous abundances of solar flare particles.
D. J. Mullan, R. H. Levine.
Astrophys. J., Suppl. Ser., Vol. 47, 87 - 102 (1981),

The authors summarize the relevant aspects of collapsing magnetic neutral sheets in the solar atmosphere. They describe the procedure they have used to calculate the ϕ_x values in a magnetic loop structure. Numerical results are presented. Comparison with empirical Q values and discussion are given.

073.092 Quasistatic evolution of preflare equilibria.
W. Zwingmann.
Plasma astrophysics, (see 012.042), p. 397 - 400 (1981).

The newly-emerging-flux-model describes the quasistatic evolution occurring before the onset of a solar eruptive process (e.g. flare) as the formation of a new magnetic structure within an otherwise stationary equilibrium. This model is analysed quantitatively in the framework of two-dimensional magnetohydrostatic theory.

073.093 Energy of optical solar flares. II. III.
L. N. Kurochka, V. M. Rossada.
Soln. Dannye 1981 Byull., No. 6, p. 78 - 83; No. 7, p. 95 - 100 (1981). In Russian.

Approximate expressions which describe the meaning of some parameters for flares of different importance are proposed. A method of calculation of the total energy radiated during flares in all the lines and the continuum of hydrogen series is given.

073.094 Optical thickness of a flare in the Hα line.
N. S. Shilova.
Soln. Dannye 1981 Byull., No. 6, p. 99 - 107 (1981). In Russian.

The results of the reduction of observations of flares on the 4, 16 and 20th of June 1979 in Hα are given. A table gives the main results of the determination of the optical thickness of the observed flares at various distances from the center of Hα.

073.095 Diffuse separation of charges and electrical fields in the solar plasma.
V. N. Krivodubskij.
Soln. Dannye 1981 Byull., No. 7, p. 86 - 95 (1981). In

Russian.

Expressions are obtained for the electric intensities caused by diffuse separation of charges under the effect of non-uniformities of pressure, temperature and electron concentration in the partially ionized plasma. The electric intensities in the chromosphere, photosphere and convective zone of the sun caused by stationary radial gradients of the physical parameters of the solar plasma are calculated. Sign reversion of the total electrostatic field in the chromosphere is discovered.

073.096 **Physical conditions in the limb flare of August 11, 1972.** V. A. Ostapenko.
Soln. Dannye 1981 Byull., No. 7, p. 100 - 104 (1981).

The total system of the equations of statistical equilibrium has been solved for a homogeneous solar structure consisting of hydrogen atoms. The physical conditions have been determined in the faint limb flare of 11 August, 1972. The observed flare radiation has been provided completely with a thermal excitation mechanism.

073.097 **On a method to determine the physical parameters of solar flares.** A. A. Rustamov.
Tsirk. Shemakhinsk. Astrofiz. Obs., No. 68, p. 3 - 7 (1981). In Russian.

Physical parameters (n_e, T_e, τ_0) for 9 moments of three flares have been determined by comparing observed and theoretical intensities.

073.098 **Hope yet for a chromospheric model!**
C. J. Durrant.
Space Sci. Rev., Vol. 29, (see 012.043), 357 - 361 (1981).

Current chromospheric models are briefly reviewed with particular emphasis on the inadequacy of the dynamical description. Whilst an improvement in models of the gross structure seems possible with present knowledge, a fully self-consistent model, of the upper chromosphere in particular, requires the better empirical understanding that awaits a space observing facility.

073.099 **Neutrino generation in solar flares.**
G. A. Koval'tsov.
Izv. AN SSSR. Ser. fiz., Tom 45, 1151 - 1154 (1981). In Russian. – Abstr. in Ref. zh., 51. Astron., 11.51.453 (1981).

073.100 **An interacting loop model for solar flare bursts.**
A. G. Emslie.
Astrophys. Lett., Vol. 22, 171 - 177 (1981).

A schematic model is presented which attempts to explain the quasi-periodic behavior (on a timescale $\lesssim 10$ s) frequently observed in solar hard X-ray bursts. As a result of the strong heating produced during a solar flare burst, the local gas pressure can transiently attain very large values in regions corresponding to the upper preflare chromosphere. The effectiveness of the surrounding magnetic field at confining this high pressure plasma is therefore reduced and the flaring loop becomes free to expand laterally. In so doing it may drive magnetic field lines into neighboring, non-flaring, loops in the same active region, causing magnetic reconnection to take place and triggering another flare burst. The features of this interacting loop model are found to be in good agreement with the energetics and time structure of flare-associated solar hard X-ray bursts.

073.101 **Solar flares.** N. Cimahovića, A. Spektor.
Zvaigžņota debess, 1981. gada vasara, p. 2 - 9.
In Latvian.

073.102 **Study of Hα emission from solar limb prominences using Fabry-Perot interferometry.**
A. H. Jarrett, J. Stapelberg, A. H. Lategan.

South African J. Sci., Vol. 77, 81 - 83 (1981) = Boyden Obs. Repr. No. 45.

073.103 **Interferometry of solar limb prominences at 656.3 nm. Part 1.**
A. H. Lategan, A. H. Jarrett, J. Stapelberg.
South African J. Phys., Vol. 4, 49 - 57 (1981) = Boyden Obs. Repr. No. 47.

The technique of obtaining Hα Fabry-Perot interferograms from solar limb prominences is discussed. Measurements of the fringe half-widths show that there is no general dependence on apparent height above the solar limb.

073.104 **Interferometry of solar limb prominences at 656.3 nm. Part 2. Interpretation of line widths in terms of temperature and turbulence.**
A. H. Lategan, A. H. Jarrett.
South African J. Phys., Vol. 4, 58 - 62 (1981) = Boyden Obs. Repr. No. 48.

The deconvoluted full widths at half maximum determined for Hα line profiles appearing on Fabry-Perot interferograms obtained from four quiescent limb prominences are interpreted in terms of thermal and microturbulent velocities.

073.105 **Numerical methods of calculations of spectral line profiles and empirical models of the chromosphere.**
R. B. Teplitskaya, Eh. A. Baranovskij.
Issled. po geomagn., aehron. i fiz. Solntsa, Moskva, 1981, No. 56, p. 23 - 28. In Russian. – Abstr. in Ref. zh., 51. Astron., 12.51.422 (1981).

073.106 **Motion in flare knots and magnetic fields in the 1979, October 6 flare.**
V. E. Stepanov, L. V. Ermakova, V. E. Merkulenko, L. Eh. Palamarchuk, V. I. Polyakov, N. V. Klochek.
Issled. po geomagn., aehron. i fiz. Solntsa, Moskva, 1981, No. 56, p. 98 - 123. In Russian. – Abstr. in Ref. zh., 51. Astron., 12.51.496 (1981).

073.107 **Eruptions chromosphériques brillantes.**
P. Simon.
Q. Bull. Sol. Act., Vol. 21, Part III, p. 1 - 80 (1980).

073.108 **Direct observation of charge state abundances of energetic He, C, O, and Fe emitted in solar flares.**
D. Hovestadt, G. Gloeckler, H. Höfner, B. Klecker, C. Y. Fan, L. A. Fisk, F. M. Ipavich, J. J. O'Gallagher, M. Scholer.
Cosmic rays in the heliosphere, (see 012.059), p. 61 - 64 (1981).

The ionic charge states of helium, carbon, oxygen, and iron have been determined for three solar particle enhancements by an electrostatic deflection analyzer. The observations were obtained during the periods September 23 to 29, 1978, June 6 to 8, 1979, and September 15 to 26, 1979, with an instrument on board the ISEE–3 spacecraft. The most surprising feature is the observation of a small ($\sim 10\%$) contribution of singly ionized helium.

073.109 **On compositional variations of heavy ions during solar particle events.** B. Klecker, M. Scholer, D. Hovestadt, C. Y. Fan, L. A. Fisk, G. Gloeckler, F. M. Ipavich, J. J. O'Gallagher.
Cosmic rays in the heliosphere, (see 012.059), p. 65 - 68 (1981).

Intensity-time profiles of protons, alpha particles, and heavy ions (C, O, Fe) in the MeV/nucleon energy range have been analyzed for one solar particle event following the solar flare on September 23, 1978. The data have been obtained with the wide angle double dE/dx–E sensor of the Max-Planck-Institut/University of Maryland experiment on board ISEE–3. The authors found time variations in the iron to helium ratio and of the O/He ratio during this event, whereas the C/O-ratio appears to be time independent. The ratio changes are probably caused by multiple injection at the sun.

073.110 **Energy spectra of high energy electrons and hard X rays as observed on board the space probe Venera 11 during the solar flare event of April 13, 1979.**
E. I. Daibog (*Dajbog*), E. A. Devicheva, S. V. Golenetskii (*Golenetskij*), Yu. A. Guryan (*Gur'yan*), V. G. Kurt, Yu. I. Logachev, A. V. Nogteva, V. G. Stopovskii (*Stolpovskij*), A. Varga.
Cosmic rays in the heliosphere, (see 012.059), p. 73 - 76 (1981).

The event was observed on board the space probe Venera 11 at a heliolongitude close to 57°. Electron spectra in the energy range from 60 to 2100 keV are determined and compared with X-ray spectra. As a result it was found that conditions of the "thin target" model were realized in the April 13, 1979 flare. Estimates of the total number of accelerated electrons and the energy of the flare are presented.

073.111 **Solar flare particle injection spectra and spectra of flux maxima at the point of observation.**
M. Scholer.
Cosmic rays in the heliosphere, (see 012.059), p. 121 - 124 (1981).

Numerical models of impulsive solar flare particle events usually assume the radial diffusion coefficient to be independent of energy per nucleon, T, although the observations indicate a $T^{0.5}$ dependence (constant mean free path). The assumption of a constant diffusion coefficient results in a preservation of a power law injection spectrum at all radial distances throughout the event. The author investigates the effect of an energy dependent diffusion coefficient on the spectrum of flux maxima at a fixed point in interplanetary space. This spectrum is harder than that of initial differential number densities close to the sun. Furthermore, the spectrum hardens with increasing radial distance which seems to be at variance with observations.

073.112 **Analysis of the physical conditions in a strong X-ray flare.** B. Sylwester, J. Jakimiec, J. Sylwester, B. Valníček.
High-energy astrophysics, (see 012.067), p. 239 - 242 (1981).

The temperature distribution of the hot plasma emission measure in a large but slowly developing flare has been investigated using the following data obtained from the INTERCOSMOS 4 satellite: (1) the X-ray spectra in the range 1.7 - 1.9 Å, (2) the hard X-ray fluxes in the range 10 - 40 keV. It has been found that all the data can be explained by a consistent thermal model of the emitting region.

073.113 **X-ray and gamma-ray observations of a white-light flare.** H. S. Hudson.
High-energy astrophysics, (see 012.067), p. 247 - 250 (1981).

HEAO-1 observed hard radiations (X- and gamma-rays) from a major solar flare on 11 July 1978. The observations showed gamma-ray line and continuum emission extending to the highest energy observed. The lines are identified with the 2.2 MeV line of deuterium formation and the 4.4 MeV line of inelastic scattering on ^{12}C. The 11 July flare was identified as a white-light flare by observations at Debrecen. It thus provides the first opportunity for a detailed examination of white-light flare theories that depend upon proton heating of the photosphere.

073.114 **Structural development of the X-ray limb flare of 30 April 1980.** C. de Jager, R. Fryer, P. Hoyng, H. Lafleur, A. Schadee, G. M. Simnett, Z. Svestka, H. F. van Beek, W. van Tend.
High-energy astrophysics, (see 012.067), p. 251 - 254 (1981).

The authors describe the development of the limb flare of 30 April 1980, 20:20 UT, as observed by the Hard X-ray

Imaging Spectrometer aboard the Solar Maximum Mission. It consisted of a short-lived bright nucleus (FWHM < 10,000 km), just inside the Sun's limb; a longer lasting tongue, extending to a height of ~30,000 km, and a more complicated feature, approximately situated at the Sun's limb.

073.115 **Cycles of solar flares.** T. Landscheidt.
9th International Congress of Biometeorology, D. Overdieck, J. Mueller, H. Lieth (Editors), Osnabrück 1981. ICR-12. – Abstract.

073.116 **Solar flares of 9 and 10 November, 1979.**
S. O. Obashev, R. S. Minasyants, Eh. Ya. Vil'kovskij.
Astron. Tsirk., No. 1162, p. 1 - 3 (1981). In Russian.

On the first extra-eclipse observations of solar prominences. See Abstr. 004.078.

Measurement of the absolute photoionization cross section of the $3p^3 P^0$ term of neutral magnesium. See Abstr. 022.080.

Unresolved dielectronic satellites of the resonance line of heliumlike iron (Fe XXV). See Abstr. 022.187.

A new method for studying celestial emission spectrum. See Abstr. 031.539.

A method to use ground-based observations to determine the Lα flux in prominences. See Abstr. 031.553.

The contribution of OSO-8 to our knowledge of the chromosphere and transition region. See Abstr. 032.519.

A discussion of theoretical ionization equilibrium calculations based on solar flare X-ray spectra. See Abstr. 062.006.

Comment on 'First phase acceleration mechanisms and implications for hard X-ray burst models in solar flares'. See Abstr. 062.033.

Experiment and m.h.d. theory of stability and relaxation in toroidal discharges. See Abstr. 062.035.

A model of flare-produced magneto-radiative shock with increasing energy. See Abstr. 062.037.

Current sheets. See Abstr. 062.047.

On refraction of acousto-gravity waves in the solar atmosphere. See Abstr. 062.119.

Stellar chromospheres and coronae. See Abstr. 064.081.

The transition layer from the photosphere to the chromosphere under solar flares. See Abstr. 071.006.

The appearance of bright regions in the photosphere and their connection with flares. See Abstr. 071.039.

Types of sunspots, magnetic classification and flare yield in cycle No. 20. See Abstr. 072.026.

Active regions and flares. See Abstr. 072.064.

Quasistationary outflow of gases from solar active regions. See Abstr. 072.072.

Solar flare acceleration of solar wind: influence of active region magnetic field.   See Abstr. 074.013.

Kinematical analysis of flare spray ejecta observed in the corona.   See Abstr. 074.024.

Coronal condensations and proton flares in the 20th solar cycle.   See Abstr. 074.027.

Simple-loop flares: magnetic instabilities. See Abstr. 074.036.

Simple-loop flares: thermal evolution. See Abstr. 074.037.

Calculation of the parameters of the transition layer and of the corona of the quiet sun. See Abstr. 074.072.

Solar flare proton release from coronal magnetic traps and strong Alfvén turbulence in the corona. See Abstr. 074.117.

On the computation of constant α force-free magnetic field.   See Abstr. 075.002.

June–July 1974 proton-flare region. I. Individual stages of its background and local magnetic field development. See Abstr. 075.022.

Nucleosynthesis in solar flares as a source of solar gamma radiation.   See Abstr. 076.002.

Observations of soft X-ray line emissions from solar flares by Tansei 4 satellite.   See Abstr. 076.019.

High-resolution ultraviolet solar observations from sounding rockets and Spacelab.   See Abstr. 076.021.

Gamma ray lines from solar flares and cosmic transients.   See Abstr. 076.022.

Solar Maximum Mission experiment: early results of the hard X-ray imaging experiment.   See Abstr. 076.028.

Solar Maximum Mission experiment: early results from the soft X-ray polychromator experiment. See Abstr. 076.029.

Spatial correlation of radio granulation and the calcium chromospherical network.   See Abstr. 077.037.

Identification and characteristics of source flares emitting ^{3}He or Fe emriched cosmic rays in the period September to December 1977.   See Abstr. 078.001.

Proton flares in 1978: quantitative diagnostics from radio bursts and results of direct measurements with the Meteor satellite.   See Abstr. 078.017.

Long lasting energetic particle injection from a weak flare.   See Abstr. 078.022.

Identification and characteristics of source flares emitting 3-He or Fe enriched cosmic rays in the period September to December 1977.   See Abstr. 078.024.

Internal structure of reconnecting current sheets and the emerging flux model for solar flares. See Abstr. 080.008.

Temperature minimum heating in solar flares by resistive dissipation of Alfvén waves.   See Abstr. 080.012.

Lectures on theoretical aspects of magnetic energy storage and conversion: their relationship to solar activity. See Abstr. 080.020.

On a method for investigation of solar fluctuations on the basis of flares.   ·See Abstr. 080.024.

Le Soleil.   See Abstr. 080.029.

Internal gravity waves in the solar atmosphere. I. Adiabatic waves in the chromosphere. See Abstr. 080.034.

Measurements of a solar flare-generated shock wave at 13.1 R_0.   See Abstr. 106.006.

On strong suppression of interplanetary shock waves when they interact in complex streams from flare series. See Abstr. 106.011.

Interplanetary dust collected in the earth's stratosphere: the question of solar flare tracks. See Abstr. 106.016.

Dependence of ESP (*energetic storm particle*) intensity on collisionless shock wave characteristics. See Abstr. 106.047.

Interplanetary acceleration of low-energy protons as observed during the 25 September 1978 shock event. See Abstr. 106.048.

A cosmic-ray-mediated shock in the solar system. See Abstr. 143.006.

074 Corona, Solar Wind

074.001 The energy balance in coronal holes and average quiet-Sun regions. J. C. Raymond, J. G. Doyle.
Astrophys. J., Vol. 247, 686 - 691 (1981).

The authors construct emission measure curves for average coronal hole and quiet-Sun spectra taken during the Skylab mission. These curves are then used to discuss the implications for the energy balance in each region. Models of constant cross section coronal loops with radiative losses balanced by the divergence of the thermal conduction flux predict an emission measure rise between 10^5 K and 10^6 K.

074.002 Closed coronal structures. III. Comparison of static models with X-ray, EUV, and radio observations.
R. Pallavicini, G. Peres, S. Serio, G. S. Vaiana, L. Golub, R. Rosner.
Astrophys. J., Vol. 247, 692 - 706, plates 17 - 19 (1981).

The authors have used numerical models of static coronal loops in energy balance for comparison with soft X-ray, extreme ultraviolet, and centimetric observations of a solar active region. They find that simple static models with thermal conductive flux vanishing at the loop base are able to reproduce satisfactorily the observed properties in the upper portion ($T \gtrsim 10^5$ K) of loop structures ranging from compact, high-pressure loops in the core of the region, to more extended, fainter loops, and to large-scale loops interconnecting different active regions. They find some discrepancy between model predictions and observations for the lower sections of loop structures (at $T \lesssim 10^5$ K).

074.003 Solar wind helium and hydrogen structure near the heliospheric current sheet: a signal of coronal streamers at 1 AU.
G. Borrini, J. T. Gosling, S. J. Bame, W. C. Feldman, J. M. Wilcox.
J. Geophys. Res., Vol. 86, 4565 - 4573 (1981).

Examination of a large body of solar wind plasma data at 1 AU has revealed a characteristic pattern of variation associated with polarity reversals in the interplanetary magnetic field. A superposed epoch analysis of the data using well-defined sector boundaries as key dates adequately demonstrates this pattern and establishes the average 1-AU characteristics of the plasma surrounding the current sheet which encircles the sun. Among the more interesting average features of the current sheet is a well-developed minimum in helium abundance A(He), which must be a consequence of physical processes occurring in the corona. It appears that A(He) variations can be used to study such processes and to help distinguish the coronal origins of various types of solar wind flows.

074.004 Wave-electron interactions in the high speed solar wind.
S. J. Schwartz, W. C. Feldman, S. P. Gary.
J. Geophys. Res., Vol. 86, 4574 - 4578 (1981).

The authors investigate the role of plasma wave-particle interactions in modifying the electron velocity distributions of the high speed solar wind. Both electrostatic and electromagnetic fluctuations are considered. The authors find that interactions between thermal electrons and a small fraction of the observed wave spectrum provide a momentum exchange rate strong enough to compete with Coulomb collisions.

074.005 Solar wind pickup of ionized Venus exosphere atoms.
S. A. Curtis.
J. Geophys. Res., Vol. 86, 4715 - 4720 (1981).

The author shows that the inclusion of thermal effects, namely, the finite velocity spreads of both the solar wind/ionosheath plasma and the planetary ion plasma, has strong effects on the estimated pickup times of exospheric atoms which are ionized in the plasma flow past Venus. Electromagnetic in-

stability growth rates at low and intermediate real frequencies are strongly affected by the inclusion of thermal effects, and both the dispersion relation for real frequencies and the growth rates are strongly modified for electrostatic instabilities at intermediate real frequencies. Using these growth rates and dispersion relations corrected for thermal effects, the author has estimated the pickup times by the solar wind.

074.006 The solar origins of solar wind interstream flows: near-equatorial coronal streamers.
W. C. Feldman, J. R. Asbridge, S. J. Bame, E. E. Fenimore, J. T. Gosling.
J. Geophys. Res., Vol. 86, 5408 - 5416 (1981).

Vela heavy ion and IMP solar wind data are used to identify the coronal origins of the interstream, low-speed solar wind as well as to understand the causes of the long-term trends in solar wind densities and electron temperatures observed at 1 AU. Several lines of evidence suggest a strong association between interstream flows and the extensions of the near equatorial band of coronal streamers into interplanetary space.

074.007 Organization of solar wind plasma properties in a tilted, heliomagnetic coordinate system.
X.- P. Zhao, A. J. Hundhausen.
J. Geophys. Res., Vol. 86, 5423 - 5430 (1981).

The authors have used a superposed epoch analysis to examine the variation in solar wind properties observed in 1974 in a "heliomagnetic" coordinate system tilted with respect to the solar equator. A tilt of $30° \pm 10°$ was found to produce the best "organization" of these properties in such a coordinate system. The solar wind speed increased with heliomagnetic latitude, while the proton density and the proton flux density decreased. These variations are qualitatively consistent with those inferred from coronal hole and other interplanetary observations.

074.008 Coronal streamers in the solar wind at 1 AU.
J. T. Gosling, G. Borrini, J. R. Asbridge, S. J. Bame, W. C. Feldman, R. T. Hansen.
J. Geophys. Res., Vol. 86, 5438 - 5448 (1981).

074.009 Kalte Schauer von der Sonne. R. Schwenn.
Umschau, 81. Jahrg., 438 - 439 (1981).

074.010 Solar wind electron densities from _Viking_ dual-frequency radio measurements.
D. O. Muhleman, J. D. Anderson.
Astrophys. J., Vol. 247, 1093 - 1101 (1981).

Simultaneous phase coherent, two-frequency measurements of the time delay between the Earth station and the _Viking_ spacecraft have been analyzed in terms of the electron density profiles from 4 solar radii ($R_\odot$) to 200 $R_\odot$. The measurements were made during a period of solar activity minimum (1976 - 1977) and show a strong solar latitude effect. The data were analyzed with both a model independent, direct numerical inversion technique and with model fitting, yielding essentially the same results.

074.011 Measurement of systematic outflow (?) from the solar transition region underlying a coronal hole.
G. J. Rottman, F. Q. Orrall, J. A. Klimchuk.
Astrophys. J., Lett., Vol. 247, L135 - L138 (1981).

This Letter presents measurements of small Doppler shifts in the line center position of $\lambda629$ O V obtained with a new high-resolution EUV spectrometer flown aboard a sounding rocket. A major result is the detection of an apparent systematic outflow (relative to the quiet Sun) of ~ 3 km s^{-1} average and 5 km s^{-1} maximum in the solar transition region underlying a well-defined low-latitude coronal hole. This is reminiscent of a similar apparent outflow observed by Cushman and Rense in the coronal line λ303 Si XI. The hypothesis that this is evidence for acceleration of the high speed solar wind deep in the transition region and inner corona is explored briefly.

074.012   Soviet-French cooperative study of the solar corona.
     3. The structure and some dynamical peculiarities of the corona on June 30, 1973.
S. K. Vsekhsvyatskij, N. I. Dzyubenko, V. I. Ivanchuk,
O. S. Popov, G. A. Rubo, S. Koutchmy, O. Koutchmy,
G. Stellmacher.
Astron. Zh., Tom 58, 810 - 824 (1981). In Russian. English translation in Soviet Astron., Vol. 25, No. 4.
     General, detailed and fine coronal structure up to $r \cong 3.0$ - $4.0\,R_\odot$ is studied. A comparison has been made between the coronal structure and formations on the solar surface, and some dynamic changes are traced for 80 - 90 min. The structure of the outer corona rays up to $12\,R_\odot$ is studied.

074.013   Solar flare acceleration of solar wind: influence of active region magnetic field.
H. Lundstedt, J. M. Wilcox, P. H. Scherrer.
Science, Vol. 212, 1501 - 1502 (1981).
     The direction of the photospheric magnetic field at the site of a solar flare is a good predictor of whether the flare will accelerate solar wind plasma. If the field has a southward component, high-speed solar wind plasma is usually observed near the earth about 4 days later. If the field has a northward component, such high-speed solar wind is almost never observed.

074.014   Thermal iron ions in high speed solar wind streams. 2. Temperatures and bulk velocities.
D. G. Mitchell, E. C. Roelof, W. C. Feldman, S. J. Bame,
D. J. Williams.
Geophys. Res. Lett., Vol. 8, 827 - 830 (1981).
     Thermal velocities and magnetic field-aligned bulk flow velocity increments are estimated using the recently discovered sensitivity to iron ions in high speed solar wind streams. Comparison with LASL hydrogen plasma measurements and GSFC magnetometer data in several recurrent streams in 1974 is performed.

074.015   Propagation of low energy solar electrons.
K. A. Anderson, J. P. McFadden, R. P. Lin.
Geophys. Res. Lett., Vol. 8, 831 - 834 (1981).
     Electrons of solar origin in the energy range 2 to 20 keV frequently appear near Earth in impulsive bursts of one to several hours duration. The angular distributions of these particles vary greatly from one event to the next. The authors discuss four such events.

074.016   Energetic particles in the heliosphere — results from the ISEE-3 spacecraft.
A. Balogh, R. J. Hynds, J. J. van Rooijen, G. A. Stevens,
T. R. Sanderson, K.-P. Wenzel.
ESA Bull., No. 27, p. 4 - 12 (1981).

074.017   Generation of coronal electric currents due to convective motions on the photosphere.
T. Sakurai, R. H. Levine.
Astrophys. J., Vol. 248, 817 - 829 (1981).
     Generation of electric currents in a magnetized plasma overlying a dense convective layer is studied, assuming that the magnetic field perturbation is small and satisfied the force-free equation. Currents are produced by rotational motions on the boundary in the case of a uniform equilibrium field. In a simple two-dimensional bipolar configuration, however, both irrotational and incompressible motions give rise to currents, and the current density has a peak at the magnetic neutral line. Scaling laws for the current density as well as for the

stored magnetic energy are derived, and the possibility of heating the solar corona through the dissipation of coronal currents generated in this way is discussed.

074.018   Solar corona magnetohydrostatics.   J. Heyvaerts.
     Ann. Physique, Vol. 5, (see 012.006), 315 - 336 (1980). In French. — Abstr. in Phys. Abstr., Vol. 84, Abstr. 75085 (1981).

074.019   MHD stability of coronal structures.   J. Heyvaerts.
     Ann. Physique, Vol. 5, (see 012.006), 337 - 377 (1980). In French. — Abstr. in Phys. Abstr., Vol. 84, Abstr. 75086 (1981).

074.020   Recent development in coronal structure thermodynamics and stability.   C. Chiuderi.
Ann. Physique, Vol. 5, (see 012.006), 425 - 445 (1980). In French. — Abstr. in Phys. Abstr., Vol. 84, Abstr. 75087 (1981).

074.021   Thermal equilibrium and non equilibrium in coronal loops.   A. Hood.
Ann. Physique, Vol. 5, (see 012.006), 447 - 451 (1980). In French. — Abstr. in Phys. Abstr., Vol. 84, Abstr. 75088 (1981).

074.022   Current confinement in solar coronal loops.
     C. Chiuderi, G. Einaudi.
Sol. Phys., Vol. 73, 89 - 103 (1981).
     Solar coronal loops are regarded as regions of localized current flows. The main purpose is to investigate the consequences of current confinement rather than to produce a model. The physical and observational basis for this assumption are presented as well as the connection with previous studies on loop structure.

074.023   Forerunners: early coronal manifestations of solar mass ejection events.   B. V. Jackson.
Sol. Phys., Vol. 73, 133 - 144 (1981).
     Coronal ejection transients viewed with the white light coronagraph on Skylab are studied from the times of their very earliest manifestations for clues to their origin. The observations place severe constraints on different solar mass ejection mechanisms because they spread the process responsible for the ejection over a larger region of the corona and over a longer period of time than normally considered. The observations suggest the corona is an active participant in the ejection that begins with the acceleration of the outer portion of a preexisting structure and ends with the obvious surface manifestation.

074.024   Kinematical analysis of flare spray ejecta observed in the corona.   D. F. Webb, B. V. Jackson.
Sol. Phys., Vol. 73, 341 - 361 (1981).
     The mass ejection event on 17 January 1974 was a classic spray associated with a flare from an over the limb region. The structure of the accompanying coronal transient was typical of well-observed mass ejections, with coronal loops and a forerunner racing ahead of the rising prominence. Observations in Hα, soft X-ray, white light and radio wavelengths allowed to track both cool ($T_e \sim 10^4$ K) and hot ($T_e > 10^6$ K) material from limb de-occultation to $6R_\odot$. The authors determined the kinematics and thermodynamics of the internal material, and the overall mass and energy budget of the event. The observations are most consistent with magnetic propulsion models of coronal transients.

074.025   Energy balance of the corona and the origin of quasi-stationary high-speed solar wind streams.
V. A. Kovalenko.
Sol. Phys., Vol. 73, 383 - 403 (1981).
     The energy balance of open-field regions of the corona and solar wind and the influence of the flow geometry in the corona upon the density and temperature, are analyzed. It is

found that the energy flux arriving at the corona is constant for the corona's open regions with different flow geometries. For the waves heating the corona and solar wind, the dependence of the absorption coefficient on the corona's plasma density is found to be within the range of distances $r = 1.05 - 1.5 R_\odot$. Theoretical models of coronal holes and the question of why the high-speed solar wind streams are precisely flowing out of coronal holes, are discussed.

074.026 A new interpretation of James's solar radar echoes involving lower-hybrid waves. D. G. Wentzel.
Astrophys. J., Vol. 248, 1132 - 1143 (1981).

James observed solar radar echoes at 38 MHz that were surprising in their large bandwidth and in their intensity. The radar beam is scattered by density fluctuations high in the corona where $n_e \approx 10^6$ cm^{-3}. At that height, the plasma wave mode most suitable for interaction with the radar is the lower-hybrid mode. Such waves are driven at sites of steep density gradients and associated electrical currents in the corona. The required coronal activity and energy supply is reasonable only on the hypothesis that the lower-hybrid waves are coherent. The radar may then be considered as a probe for coherent plasma waves in the corona.

074.027 Coronal condensations and proton flares in the 20th solar cycle. V. Rušin.
Bull. Astron. Inst. Czechoslovakia, Vol. 32, 193 - 214 (1981).

The question of correlation between limb coronal condensations (observations of the coronal emission line 569.4 nm) and proton flares, connected with the arrival of particles at the Earth, in the course of solar cycle 20 are discussed. Direct time correlation, i. e. that a proton flare occurs after a limb condensation has been observed, is relatively low (7 : 36). The distribution of occurrence of the investigated phenomena is nonuniform not only as regards the two hemispheres, but also as regards the heliographic longitudes of the individual hemispheres.

074.028 Differences between solar wind plasmoids and ideal magnetohydrodynamic filaments.
J. Lemaire, M. Roth.
Planet. Space Sci., Vol. 29, 843 - 849 (1981).

Plasma irregularities present in the solar wind are plasmoids, i.e. plasma-magnetic field entities. These actual plasmoids differ from ideal magnetohydrodynamic (MHD) filaments. Indeed, (1) their "skin" is not infinitely thin but has a physical thickness which is determined by the gyromotion of the thermal ions and electrons, (2) they are of finite extent and their magnetic flux is interconnected with the interplanetary magnetic flux, (3) when they penetrate into the magnetosphere their magnetic field lines become rooted in the ionosphere, (4) the external Lorentz force acting on their boundary surface depends on the orientation of their magnetic moment with respect to the external magnetic field, (5) when their mechanical equilibrium is disturbed, hydromagnetic oscillations can be generated.

074.029 Solar wind structure beyond Jupiter, as seen by Pioneer - 10 and - 11. S. E. Kayser, A. Barnes.
Bull. American Astron. Soc., Vol. 13, 511 (1981). – Abstract.

074.030 The structure of a force-free coronal loop.
K. A. Wear, S. K. Antiochos, A. G. Emslie, P. A. Sturrock.
Bull. American Astron. Soc., Vol. 13, 542 (1981). – Abstract.

074.031 The ionization state of the solar wind.
S. P. Owocki.
Bull. American Astron. Soc., Vol. 13, 544 (1981). – Abstract.

074.032 Rotation of the solar electron corona during solar cycle 20. G. D. Parker.
Bull. American Astron. Soc., Vol. 13, 552 - 553 (1981). Abstract.

074.033 Spectral broadening and solar wind speed measurements under conditions of strong coronal scattering.
H. M. Bradford, D. Routledge.
Bull. American Astron. Soc., Vol. 13, 554 (1981). – Abstract.

074.034 On the thermal stability of coronal loop plasma.
S. K. Antiochos, A. G. Emslie.
Bull. American Astron. Soc., Vol. 13, 555 (1981). – Abstract.

074.035 Density gradients in the solar plasma observed by interplanetary scintillation.
G. R. Gapper, A. Hewish.
Mon. Not. R. Astron. Soc., Vol. 197, 209 - 216 (1981).

A new technique is described which overcomes the limitation set by Fresnel filtering in previous IPS studies of the small-scale density irregularities in the solar plasma. Phase gradients introduced by irregularities larger than the Fresnel limit cause transverse displacements of the small-scale scintillation pattern. In the presence of the solar wind, such refraction effects may be revealed by simultaneous measurements of intensity scintillation at two radio frequencies. Observations show that the structure corresponding to temporal frequencies ~0.02 Hz is in agreement with an extrapolation of the Kolmogorov spectrum derived from spacecraft data at lower frequencies.

074.036 Simple-loop flares: magnetic instabilities.
G. van Hoven.
Solar flare magnetohydrodynamics, (see 003.002), p. 217 - 275 (1981).

Contents: Introduction. Physical conditions. Equilibrium magnetic fields. Infinite-conductivity stability. Resistive magnetic instability. Attempts at a loop-flare model.

074.037 Simple-loop flares: thermal evolution. I. J. D. Craig.
Solar flare magnetohydrodynamics, (see 003.002), p. 277 - 336 (1981).

Contents: Introduction. Observational background. The hydrodynamic equations. Quasi-static models of the solar atmosphere. Analytic conduction models. Hydrodynamic models of a flaring loop. Summary: theory and observation revisited.

074.038 The solar corona and the solar wind.
J. B. Zirker.
The sun as a star, (see 003.004), p. 135 - 162 (1981).

This paper begins with an observational overview of the corona and the wind. It then discusses in more detail two contrasting classes of coronal regions: those with closed magnetic fields (loops) and those whose open magnetic fields extend out into interstellar space (the coronal holes). For each type of region those observable properties that relate to the problems of heating, acceleration, variability, and the role of the magnetic field are discussed.

074.039 Coronal heating. D. G. Wentzel.
The sun as a star, (see 003.004), p. 331 - 353 (1981).

This paper summarizes the theoretical arguments for coronal heating that take coronal structure into account. The ingredients of these theories should also apply to other stellar coronae.

074.040 The energy balance of the solar wind.
J. V. Hollweg.
The sun as a star, (see 003.004), p. 355 - 372 (1981).

The purpose of this paper is to examine the effects of modifying some of the "classical" assumptions underlying many of the solar wind models constructed over the past 20 years, in an effort to obtain both a better fit with the observations and a deeper understanding of the relevant physical processes.

074.041   Heating and acceleration of the solar wind.
R. A. Kopp.
The sun as a star, (see 003.004), p. 373 - 384 (1981).
This paper treats mainly the dynamics of the expanding corona, emphasizing, in particular, the physical processes which accelerate the plasma as it flows away from the sun. Lately it has become increasingly apparent, primarily because of the identification of coronal holes as the localized source regions of recurrent high speed streams, that the solar wind plays a dominant role in the energy balance of the part of the corona in which it originates. Thus, in studying the wind acceleration processes one may be dealing, in large part, with the mechanisms that transport energy to the corona in the first place.

074.042   On a new simple method for inferring solar wind
parameters from ground-based geomagnetic data.
V. V. Shelomentsev, L. P. Sergeeva.
Issled. po geomagn., aehron. i fiz. Solntsa. Moskva, 1981, No. 53, p. 191 - 199. In Russian. – From Ref. zh., 51. Astron., 7.51.326 (1981).

074.043   Measurement of polarization of the Fe XIV λ5303
line in coronal loops.
K. I. Nikol'skaya, Eh. I. Tetruashvili.
Fiz. soln. aktivnosti. Moskva, 1980, p. 145 - 151. In Russian. Abstr. in Ref. zh., 51. Astron., 8.51.448 (1981).

074.044   Fast magnetic reconnection and transient
phenomena with particle acceleration in the solar
corona.   B. V. Somov.
Izv. AN SSSR. Ser. fiz., Tom 45, 576 - 578 (1981). In Russian. – Abstr. in Ref. zh., 51. Astron., 8.51.456 (1981).

074.045   Photometry of the solar corona of 1980 February 16.
K. A. Raju.
Bull. Astron. Soc. India, Vol. 9, 68 (1981). – Abstract.

074.046   Coronal temperature measured during the total solar
eclipse of 1980 February 16.
J. N. Desai, T. Chandrasekhar, H. Bhatt, N. M. Ashok, D. B. Vaidya, P. D. Angreji.
Bull. Astron. Soc. India, Vol. 9, 68 - 69 (1981). – Abstract.

074.047   Coronal electron density distribution from white
light totality photographs taken through a polaroid.
K. C. Sahu, J. N. Desai, V. B. Kamble, V. Metha.
Bull. Astron. Soc. India, Vol. 9, 69 (1981). – Abstract.

074.048   Coronal temperatures as measured from
[Fe X] 6374 Å.   J. Singh, A. K. Saxena.
Bull. Astron. Soc. India, Vol. 9, 69 (1981). – Abstract.

074.049   Interpretation of Hα emission from solar corona.
P. K. Raju.
Bull. Astron. Soc. India, Vol. 9, 69 (1981). – Abstract.

074.050   Polarisation of the solar corona of 1980 February 16.
K. R. Sivaraman, M. Jayachandran, G. S. D. Babu.
Bull. Astron. Soc. India, Vol. 9, 70 (1981). – Abstract.

074.051   Temperature and flux distribution along a solar
coronal loop in presence of a source of heating.
U. Narain, H. P. Mital.
Bull. Astron. Soc. India, Vol. 9, 75 (1981). – Abstract.

074.052   Coronal holes according to observations on BPR.
N. G. Peterova.
13-ya Vses. konf. po radioastron. issled. soln. sistemy, Kiev, 1981. Tez. dokl. Kiev, 1981, p. 19 - 20. In Russian. – From Ref. zh., 51. Astron., 9.51.360 (1981).

074.053   Temporal changes in coronal fine structures.
K. Adachi.
Observations of the total solar eclipse of 16 February 1980, (see 003.009), p. 1 - 2 (1981).

074.054   Polarization of the corona.   G. S. D. Babu,
M. Jayachandran, A. P. Jayarajan, L. Peter.
Observations of the total solar eclipse of 16 February 1980, (see 003.009). p. 3 (1981).

074.055   Determination of coronal temperature distribution
by measuring the Doppler widths of the green
coronal line 5303 Å (Fe XIV) at different points in the corona.
J. N. Desai, T. Chandrasekhar, N. M. Ashok, D. B. Vaidya.
Observations of the total solar eclipse of 16 February 1980, (see 003.009), p. 5 - 6 (1981).

074.056   Determination of the electron density distribution
in the corona using white light pictures at 4 polaroid
positions.   J. N. Desai, K. C. Sahu, Kamble, P. Panchal.
Observations of the total solar eclipse of 16 February 1980, (see 003.009), p. 7 (1981).

074.057   Monochromatic picture of the corona in λ5303 Å.
J. N. Desai, H. Bhatt, P. D. Angreji, P. Patel,
P. Pandya.
Observations of the total solar eclipse of 16 Februar 1980, (see 003.009), p. 8 (1981).

074.058   Multicolour photometry and polarimetry of the solar
corona.   J. Dürst, A. Zelenka.
Observations of the total solar eclipse of 16 February 1980, (see 003.009), p. 9 - 10 (1981).

074.059   Experiments of the German group at Palem/India.
H.- J. Proll, D. Staps, S. Wenzel.
Observations of the total solar eclipse of 16 Februar 1980, (see 003.009), p. 11 - 15 (1981).

074.060   Kitt Peak coronal velocity experiment.
W. Livingston, J. Harvey, L. A. Doe, B. Gillespie,
G. Ladd.
Observations of the total solar eclipse of 16 Februar 1980, (see 003.009), p. 19 - 22 (1981).

074.061   White light coronal photography.   M. C. Pande,
K. R. Bondal, V. P. Gaur, K. Sinha, G. C. Joshi.
Observations of the total solar eclipse of 16 February 1980, (see 003.009), p. 23 (1981).

074.062   Spectrophotometry of the inner corona.
R. Rajamohan, K. C. A. Raheem, A. P. Jayarajan.
Observations of the total solar eclipse of 16 February 1980, (see 003.009), p. 24 (1981).

074.063   Coronal photography at H alpha.
K. E. Rangarajan, S. Giridhar.
Observations of the total solar eclipse of 16 February 1980, (see 003.009), p. 25 (1981).

074.064   Search for a ring around the sun.   U. R. Rao,
T. K. Alex, V. S. Iyengar, M. S. Phadke, P. D. Prasad, K. Kasturirangan, T. M. K. Marar, R. S. Mathur, D. P. Sharma.
Observations of the total solar eclipse of 16 February 1980, (see 003.009), p. 26 - 27 (1981).

074.065   Study of the solar corona.   V. Rusin, J. Sykora.
Observations of the total solar eclipse of 16 February 1980, (see 003.009), p. 28 - 30 (1981).

074.066   Heterogeneous structures of the lower corona.
S. Saito, H. Kurokawa, Y. Ogimachi.
Observations of the total solar eclipse of 16 February 1980,
(see 003.009), p. 31 - 33 (1981).

074.067   Multislit spectrophotometry of the corona.
A. K. Saxena, J. Singh, F. Gabriel.
Observations of the total solar eclipse of 16 February 1980,
(see 003.009), p. 34 (1981).

074.068   Broad-band photometry of the corona.
K. K. Scaria, S. P. Bagare, A. P. Jayarajan.
Observations of the total solar eclipse of 16 February 1980,
(see 003.009), p. 35 (1981).

074.069   Eclipse observations of coronal Fe XIV emission.
R. N. Smartt, J. B. Zirker, H. A. Mauter.
Observations of the total solar eclipse of 16 February 1980,
(see 003.009), p. 36 - 38 (1981).

074.070   Shape and structure of maximum corona.
M. Waldmeier.
Observations of the total solar eclipse of 16 February 1980,
(see 003.009), p. 39 - 41 (1981).

**074.071   Intensities of the coronal emission line 637.4 nm
observed at Lomnický Štít in the years 1965 - 1976.**
V. Rušin, M. Rybanský, L. Scheirich.
Contrib. Astron. Obs. Skalnaté Pleso, Vol. 10, 7 - 37 (1981).

**074.072   Calculation of the parameters of the transition layer
and of the corona of the quiet sun.**
V. N. Soshnikov.
Astron. Vestn., Tom 15, 127 (1981). In Russian.

**074.073   On the modeling of the three-fluid structure of the
quiet solar wind.**
S. Cuperman, N. Metzler, M. Dryer.
Astrophys. Space Sci., Vol. 79, 67 - 75 (1981).
The reciprocal influence of the electrons and protons, on
one side, and the α-particles, on the other side in the quiet
solar wind is investigated within the framework of a conductive
three-fluid model (with frictional forces included). For this
purpose two mathematical methods are used, namely: I. Simultaneous solution of the fluid equations for all three species;
and II. Solution of two-fluid equations (for electrons and protons) followed by that of a "modified" one-fluid equation for
the α-particles (in which the two-fluid solutions are used for
electrons and protons).

**074.074   Radio wave scattering observations of the solar
corona: first-order measurements of expansion
velocity and turbulence spectrum using *Viking* and *Mariner 10*
spacecraft.**   G. L. Tyler, J. F. Vesecky, M. A. Plume,
H. T. Howard, A. Barnes.
Astrophys. J., Vol. 249, 318 - 332 (1981).
The authors' emphasis in this paper is on (1) data reduction and analysis methods, (2) interpretation of signal amplitude fluctuation spectra in terms of coronal expansion velocity
V and turbulence spectrum $\Phi_N(\kappa)$, and (3) the implications of
these measurements for the solar wind acceleration mechanism.

074.075   Transient plasmas in the solar transition zone.
K. P. Dere, J.-D. F. Bartoe, G. E. Brueckner,
M. D. Dykton, M. E. VanHoosier.
Astrophys. J., Vol. 249, 333 - 339 (1981).
Observations have been obtained which show the highly
transient nature of solar transition zone emission at high
spatial resolution. This indicates that static and steady-state
atmospheric models are probably inadequate for describing
these regions. A simplified model is used to explain some of the
known features of the solar atmosphere with special reference

to the observed C IV emissions. It is suggested that this radiation is produced in radiatively cooling plasmas that are a result
of thermal radiative instabilities in the solar corona. Such
plasmas are shown to be capable of producing rapid variations
in typical transition zone lines.

**074.076   Restoring polarized emission of local sources and
the structure of coronal magnetic fields.**
N. G. Peterova, B. I. Ryabov.
Astron. Zh., Tom 58, 1070 - 1077 (1981). In Russian. English
translation in Soviet Astron., Vol. 25, No. 5.
The present paper continues the study of the structure of
solar coronal magnetic fields, using the effect of sign inversion
of the polarized radio emission of active regions. The case of
an asymmetric distribution of the magnetic flux in sunspot
groups is investigated. A simple numerical model is shown to be
in accordance with the radio observations if one takes into account the magnetic fields of plages.

074.077   On the theory of coronal heating mechanisms.
M. Kuperus, J. A. Ionson, D. S. Spicer.
Annu. Rev. Astron. Astrophys., Vol. 19, (see 003.012), 7 - 40
(1981).
The authors stress the importance of a unifying and
consistent theory of the formation and heating of coronal
structures that eventually bridges the gap between the photospheric "driver" (e.g. the velocity and magnetic fields as they
should be observed in great detail in the photosphere) and the
observed chromospheric and coronal radiation structures. A
first attempt to consider coronal structures as electrodynamic
circuits seems to lead to a promising and original way of describing the outer solar atmosphere.

**074.078   Solutions of the two-fluid solar wind equations:
adiabatic and conduction dominated solutions.**
D. P. Rowse, I. W. Roxburgh.
Sol. Phys., Vol. 74, (see 012.029), 169 - 177 (1981).
The equations governing the two-fluid spherically symmetric models of the solar wind have been solved numerically
for a wide range of base conditions. As predicted from an
asymptotic analysis the authors find a whole domain of solutions which are asymptotically adiabatic with the proton and
electron temperatures tending to equality and varying like
$r^{-4/3}$. In these 4/3 solutions the electron and proton heat conduction is asymptotically negligible and if it is neglected the
resulting equations can be integrated analytically and shown to
have the 4/3, 4/3 behaviour.

074.079   Microinstabilities and models of the solar wind.
D. P. Rowse, I. W. Roxburgh, S. J. Schwartz.
Sol. Phys., Vol. 74, (see 012.029), 179 - 185 (1981).
The collision-dominated two-fluid plasma models of the
solar wind are shown to become collisionless and subject to
microinstabilities at a few solar radii. Assuming that once the
plasma is unstable it stays close to marginal stability models of
the solar wind are constructed including waves and proton
heating. The resulting models have higher velocities and proton
temperatures than the collision dominated two fluid models.

**074.080   Preferred Bartels days of high-speed plasma streams in
the solar wind.**   B. A. Lindblad.
Sol. Phys., Vol. 74, (see 012.029), 187 - 196 (1981).
An analysis of 346 high-speed solar wind streams observed
at 1 AU during 1964 - 75 is presented. The analysis shows that
a two-sector structure was the dominant feature of the interplanetary magnetic field associated with the high-speed solar
wind plasma. The high-speed streams occurred at preferred
Bartels days: positive polarity streams were most frequent near
Bartels day 4, negative polarity streams were most frequent
near Bartels day 17. The observations are explained in terms of
a tilted dipole model of the solar-interplanetary field.

074.081   Energetics of coronal hole expansion.   X.-P. Zhao.
J. Geophys. Res., Vol. 86, 7480 - 7486 (1981).

Based on the assumption of the undamped propagation of Alfvén waves between 2 and 5 R_S, the possible range of Alfvén wave energy flux density at the base of a polar coronal hole has been deduced. It is concluded that extended heating above 2 R_S is needed to maintain the temperature maximum above 2 R_S and to supply energy for the formation of high-speed wind streams. The energy contribution of extended heating to high-speed wind streams might be greater than that of Alfvén waves by direct momentum addition.

074.082   The expected pulsations of the heliosphere relevant to cosmic ray variations.
L. I. Dorman, V. S. Ptuskin.
Astrophys. Space Sci., Vol. 79, 397 - 404 (1981).

The possible natural large-scale pulsations of the solar wind cavity are examined. The period of the pulsations $T \sim 2\pi R/a$ (R is the size of the cavity, a is the sound velocity in the interstellar space near the solar system) may vary from a year to tens of years. The relevant new type of cosmic ray variations is predicted.

074.083   Coronal loops in the sun and in the stars.
M. Landini, B. C. Monsignori Fossi.
Astron. Astrophys., Vol. 102, 391 - 400 (1981).

The balance of energy and momentum has been discussed in detail for stationary solutions of coronal loops. The general case of loops in gravity fields including mass motions and different shapes of energy deposition has been developed and a set of dimensionless variables are introduced to discuss the main properties of the model from the point of view of the temperature dependence of the conductive flux, of the temperature and pressure profiles, of the generation of differential emission measure functions. The application of the model to coronae of other stars generates a scaling law of the total emission measure for the stars which is proportional to $g_*^4 R_*^5$ and appears to be consistent with the observed X-ray luminosity of dwarfs and early giants.

074.084   Are coronal loops stable?
I. J. D. Craig, A. N. McClymont.
Nature, Vol. 294, 333 - 334 (1981).

The simple fact that coronal loops exist stably over long periods demonstrates the inadequacy of the current theory. The authors point out here that a crucial ingredient is missing from the theory and hence reconcile the existence of coronal loops with the disruptive effect of the radiative instability.

074.085   Resonance of Alfvén waves in coronal arches.
Yu. D. Zhugzhda, V. Lotsans.
13-ya Vses. konf. po radioastron. issled. soln. sistemy, Kiev, 1981. Tez. dokl., Kiev, 1981, p. 24. In Russian. − From Ref. zh., 51. Astron., 10.51.355 (1981).

074.086   Solar wind motion within 30 $R_\odot$: spacecraft radio scintillation observations.
J. W. Armstrong, R. Woo.
Astron. Astrophys., Vol. 103, 415 - 421 (1981).

The authors report here new remote-sensing observations of the solar wind motion within about 30 $R_\odot$, using multiple-station intensity scintillations of spacecraft signals. The spacecraft used in this study were the Viking orbiters and Helios A/B; the observations were taken in the ecliptic plane.

074.087   Velocity of the drift of inhomogeneities in the region of solar wind acceleration from data of radio occultation experiments.   A. I. Efimov.
Geomagn. Aehron., Tom 21, 769 - 774 (1981). In Russian.

074.088   Effects of viscous interaction of the solar wind with the plasma layer of the magnetospheric tail.
V. G. Pivovarov.
Geomagn. Aehron., Tom 21, 876 - 879 (1981). In Russian.

074.089   Parametric interaction and spatial collapse of beam-driven Langmuir waves in the solar wind.
D. A. Gurnett, J. E. Maggs, D. L. Gallagher, W. S. Kurth, F. L. Scarf.
J. Geophys. Res., Vol. 86, 8833 - 8841 (1981).

The paper presents observations of the parametric decay and spatial collapse of Langmuir waves driven by an electron beam streaming into the solar wind from the Jovian bow shock. High-resolution frequency-time spectrograms from Voyager 1 and 2 show that long wavelength Langmuir waves upstream of the bow shock are very effectively converted into short wavelength Langmuir waves which are no longer in resonance with the beam. This conversion is shown to be the result of a nonlinear interaction involving the beam-driven pump, a sideband emission and a low level of ion-acoustic turbulence which always appears to be present in the solar wind.

074.090   Coronal density and the solar wind speed at all latitudes.   D. G. Sime, B. J. Rickett.
J. Geophys. Res., Vol. 86, 8869 - 8876 (1981).

Data from essentially all heliographic latitudes (50°S to 60°N) are used to demonstrate directly the anticorrelation between the solar wind speed and the coronal density to which it maps. The data, coronameter observations of density and interplanetary scintillation measurements of wind speed, are well suited for comparison, since they have similar selection effects, both responding most accurately to large-scale structures. Their relationship is demonstrated on time scales from 2 days to 1 year for the years 1972 - 1975, during which the structures were corotating or slowly evolving. The study makes a quantitative point-by-point comparison of coronal density and solar wind velocity.

074.091   Multi-spacecraft observations of heliographic latitude-longitude structure in the solar wind.
E. J. Rhodes, Jr., E. J. Smith.
J. Geophys. Res., Vol. 86, 8877 - 8892 (1981).

The heliographic latitude-longitude structure of high speed solar wind streams observed prior to the maximum of solar cycle 20 has been studied with a multi-spacecraft comparison. Plasma and magnetic field data obtained near the earth by Explorers 34 and 35 during mid-1967 were mapped radially inward to the location of Mariner 5, which was making simultaneous measurements at lower heliographic latitudes. For latitude separations less than 3.5°, cross-correlations of the Explorer-Mariner velocities showed only the previously reported systematic increase in velocity with latitude. However, for latitude separations ranging from 3.5° to 6.2°, significant differences in the high speed stream structure of up to 250 km/s were seen. For these stream structures, the multi-spacecraft observations made possible the separation of spatial from temporal variations with little ambiguity.

074.092   Minor ions in the low corona.   J. V. Hollweg.
J. Geophys. Res., Vol. 86, 8899 - 8905 (1981).

The ability of Coulomb friction to drag minor ions out of the subsonic region of the low corona is examined analytically. With some assumptions, the author obtains new analytical expressions for the "minimum proton flux" that is required to drag minor ions out of the corona and for the velocity, relative to the protons, at which the minor ions are dragged out. The author uses these new results to suggest that the positive n_α/n_p versus v_p correlation and the negative n_α/n_p versus $n_p v_p$ correlation observed for solar wind helium at 1 AU are due to variations in the temperature of the low corona. He also suggests that the charge dependence of the Coulomb friction may lead to errors in determining the coronal temperature from measurements of the ionization state at 1 AU.

074.093 Pronounced proton core temperature anisotropy, ion differential speed, and simultaneous Alfvén wave activity in slow solar wind at 0.3 AU.
E. Marsch, K.-H. Mühlhäuser, H. Rosenbauer, R. Schwenn, K. U. Denskat.
J. Geophys. Res., Vol. 86, 9199 - 9203 (1981).

In the Helios 2 perihelion (0.3 AU) in May 1978, pronounced ion differential speeds $\Delta v_{a_p} > 0$ and large proton temperature anisotropies $T_{\|p}/T_{\perp p} < 1$ have been observed in slow solar wind, in distinct contrast to earlier observations during the phase of solar activity minimum. These features of ion distributions occurred simultaneously with high Alfvénic waves activity. Coulomb friction between the two ion species can be shown to play a minor role, thus obviously favouring the establishment of a large differential ion speed by preferentially accelerating wave forces. The observations may also indicate that strong waves sometimes only heat the protons without creating a fast solar wind.

074.094 Critical conditions for magnetic instabilities in force-free coronal loops.   A. W. Hood, E. R. Priest.
Geophys. Astrophys. Fluid Dyn., Vol. 17, 297 - 318 (1981).
Abstr. in Phys. Abstr., Vol. 84, Abstr. 102085 (1981).

074.095 Structure of the solar corona.
V. B. Bhatia, R. Dubey.
Indian J. Phys.,Part B, Vol. 55B, 1 - 7 (1981). − Abstr. in Phys. Abstr., Vol. 84, Abstr. 102087 (1981).

074.096 Measurements of the coronal spectral broadening of spacecraft radio signals and the solar wind speed.
H. M. Bradford.
J. R. Astron. Soc. Canada, Vol. 75, 246 (1981). − Abstract.

074.097 Observations of solar wind stream with high abundance of heavy ions and relation with coronal conditions.   G. N. Zastenker, Yu. I. Yermolaev (*Ermolaev*).
Planet. Space Sci., Vol. 29, 1235 - 1240 (1981).

Long intervals, during which heavy ions were detected in the high energy tail of the energy spectra of solar wind ions, were recorded by the plasma spectrometer SCS onboard the Prognoz 7 satellite. In particular, such a region with unusual features−low velocity, high density, low temperature of protons and, especially, low temperature of α-particles−was observed during 10 - 13 December 1978. The time dependence of these parameters makes it possible to recognize this event as "noncompressive density enhancement". In this region heavy ions such as O^{+6}, O^{+7}, Si^{+7}, Si^{+8}, Si^{+9} and a group of iron from Fe^{+6} to Fe^{+13} were identified by the electrostatic analyzer. The abundance of these ions relative to protons was about ten times higher than had previously been observed. The coronal temperature, estimated from the ratios of the ion fluxes with different ionizing states, is higher than that estimated earlier for the oxygen ions.

074.098 Heating processes of the solar corona.
M. Kuperus.
Plasma astrophysics, (see 012.042), p. 113 - 128 (1981).

Coronal heating processes are discussed in the context of the magnetic structure of the solar corona. Existing theories are divided into acoustic theories, where the magnetic field plays a passive role and electrodynamic theories where the magnetic field plays an active role in the sense that transversal waves are possible. The dissipation mechanisms are shock wave dissipation in the acoustic theory and viscous and Joule dissipation of Alfvén waves in the electrodynamic theory. The dissipation of Alfvén waves appears to be much more effective in a structured medium than in a homogeneous medium. Anomalous Joule heating is likely to be unimportant as a steady heating mechanism.

074.099 Inferences about the solar wind dynamics from observed distributions of electrons and ions.
S. Olbert.
Plasma astrophysics, (see 012.042), p. 135 - 143 (1981).

A review of recent measurements of the solar wind − mostly from Helios and Voyager − is presented. The observational material is compiled into a series of radial profiles for various macroscopic plasma parameters. Moreover, several examples of experimentally determined 3-D distribution functions of protons are shown. A brief survey of the physical properties of the solar-wind electrons is given. The driving mechanism of the solar wind is discussed. The problem of the energy balance in "pure" high-velocity streams emanating from solar coronal holes is outlined. A suggestion is made that the suprathermal electrons may play an important role in driving the wind. Some numerical estimates referring to the lower corona are offered in support of this hypothesis.

074.100 Wave heating and acceleration of solar wind ions by cyclotron resonance.   E. Marsch.
Plasma astrophysics, (see 012.042), p. 405 - 406 (1981).

A model is presented, which describes the radial evolution of solar wind ion temperatures and speeds within a fluid-type approach. Resonant wave-particle interactions are taken into account and the ion-wave dynamic equilibrium is calculated self-consistently. Numerical results are briefly discussed.

074.101 Comments on polytropic solar wind models.
D. Summers.
Plasma astrophysics, (see 012.042), p. 407 (1981).

A formal analysis of the two-fluid, polytropic solar wind model is presented, and the applicability of polytropic solutions to the solar wind is discussed.

074.102 The yellow coronal line in the active region on October 6, 1979.
G. M. Nikol'skij, N. F. Tyagun, M. Rybanský.
Soln. Dannye 1981 Byull., No. 6, p. 88 - 91 (1981).

The results of observations of the emission of the yellow coronal line for about 2 hours on the 6th of October 1979 are given. The characteristics of the line profile, nonthermal random velocities in the region of its emission, the total number of electrons along the line of sight passing through the magnetic region knot and the emission measure are determined.

074.103 Scattering of magnetohydrodynamic waves on solar wind turbulence.   I. V. Chashej, V. I. Shishov.
Geomagn. Aehron., Tom 21, 961 - 967 (1981). In Russian.

074.104 Modelling of nonstationary processes in the solar wind.   I. S. Veselovskij.
Geomagn. Aehron., Tom 21, 968 - 972 (1981). In Russian.

074.105 Thermal instability of solar loop structures.
R. Rosner.
Nature, Vol. 294, 611 - 612 (1981).

074.106 Loops, fields and all that.   C. Chiuderi.
Space Sci. Rev., Vol. 29, (see 012.043), 349 - 353 (1981).

074.107 A search of short-period coronal waves.
S. Koutchmy.
Space Sci. Rev., Vol. 29, (see 012.043), 375 - 376 (1981).

In 1973, during the total solar eclipse, an experiment aboard the Concorde supersonic airliner was made in order to investigate the possible presence of white-light coronal waves. The experiment failed to detect any significant effects, so it became clear that such waves should be searched for by use of finer, i.e., spectroscopic methods.

074.108   A mechanism for heating the solar corona.
D. L. Book.
Comments Plasma Phys. Controlled Fusion, Vol. 6, 193 - 198
(1981). – Abstr. in Phys. Abstr., Vol. 85, Abstr. 6578 (1982).

074.109   Solar wind streams originating under the action of gravitation of planets.   V. D. Reshetov.
Kosm. Issled., Tom 19, 896 - 901 (1981). In Russian.

074.110   Numerical modelling of conditions of quasi-transverse propagation of electromagnetic waves in the solar corona.   B. Ryabov.
Investigations of the sun and red stars. 12, (see 003.022),
p. 44 - 56 (1981). In Russian.
    Many important details of the structure of local sources
of solar radio emission can be understood only on the basis of
quantitative analysis of both generation and propagation conditions of electromagnetic waves in the solar atmosphere. To
solve such a problem a computer program was made for
modelling generation by a thermal magneto-bremsstrahlung
mechanism and propagation of radio waves through a QT
region.

074.111   Interference of Alfvén waves in coronal arches.
J. Žugžda, V. Locāns.
Investigations of the sun and red stars. 13, (see 003.023),
p. 98 - 125 (1981). In Russian.
    Coronal arches are interference filters for Alfvén waves
with particular resonant frequencies. Interference of waves in
filters-arches influences the income and absorption of Alfvén
waves in the solar corona. The time of establishment of
resonance in coronal arches after "switching on" the flow of
waves from the photosphere can be observable because of the
low resonant frequencies of the filters-arches. The absorption
of Alfvén waves with the resonant frequencies can make a
considerable contribution to the energy balance of the coronal
arch.

074.112   On the temperature of solar coronal holes.
A. Balklavs.
Zvaigžņota debess, 1981. gada vasara, p. 19 - 20. In Latvian.

074.113   Temperatures and differential line-of-sight velocities observed in the solar corona during a total solar eclipse.
T. Chandrasekhar, J. N. Desai, P. D. Angreji.
Appl. Opt., Vol. 20, 2172 - 2173 (1981).

074.114   On rotational forces in the solar wind.
J. V. Hollweg, P. A. Isenberg.
J. Geophys. Res., Vol. 86, 11463 (1981).
    The authors present a new and simpler derivation of the
rotational forces on minor ions in the solar wind. They show
that the rotational forces can be interpreted as potential forces
affecting all particles equally. As such, they do not involve
interactions between different particle species and do not
represent an equilibrating process.

074.115   Study of the mechanism of hot region formation in the active solar corona.
V. P. Vasil'ev, V. I. Kucherov, V. I. Lapshin.
Probl. yader. fiz. i kosm. luchej, Khar'kov, 1981, No. 14,
p. 86 - 92. In Russian. – Abstr. in Ref. zh., 51. Astron.,
12.51.493 (1981).

074.116   Intensité de la couronne solaire. en lumière monochromatique, selon des angles de position variant de 5° en 5°.
Q. Bull. Sol. Act., Vol. 21, Part IV, p. 1 - 30 (1980).

074.117   Solar flare proton release from coronal magnetic traps and strong Alfvén turbulence in the corona.

B. I. Meerson, P. V. Sasorov.
Cosmic rays in the heliosphere, (see 012.059), p. 77 - 81
(1981).
    Drift instabilities arising when accelerated protons are
trapped by coronal magnetic fields of active regions are investigated theoretically. The problem of release of flare produced
high-energy protons ($E_p \gtrsim 10$ MeV) into interplanetary space
is communicated.

074.118   Corotating events in the energy range 4–13 MeV as observed on board Helios 1 and 2 in 1975 and 1976.
F. Zöllich, G. Wibberenz, H. Kunow, G. Green.
Cosmic rays in the heliosphere, (see 012.059), p. 89 - 92
(1981).
    Intensity increases of 4 - 13 MeV protons correlated with
recurrent high speed solar wind streams were observed on
board Helios 1 and 2 at heliocentric distances between 0.3 and
1.0 AU. Radial gradients of corotating events from February
till April 1976 were investigated. The observations showed
that probably there were no stationary corotating structures
in interplanetary space. The authors suggest that this effect is
related to local disturbances of solar origin.

074.119   Shock acceleration of energetic particles in corotating interaction regions.   L. A. Fisk, M. A. Lee.
Cosmic rays in the heliosphere, (see 012.059), p. 93 - 96
(1981).
    A simple shock model for the acceleration of energetic
particles in corotating interaction regions (CIR) in the solar
wind is presented. Particles are accelerated at the forward and
reverse shocks which bound the CIR by being compressed between the shock fronts and magnetic irregularities upstream
from the shocks, or by being compressed between upstream
irregularities and those downstream from the shocks. Particles
also suffer adiabatic deceleration in the expanding solar wind.

074.120   Propagation of upstream protons in the near-earth solar wind.
V. Domingo, T. R. Sanderson, K.-P. Wenzel.
Cosmic rays in the heliosphere, (see 012.059), p. 125 - 128
(1981).
    The propagation of energetic protons (35–1600 keV)
from the earth's magnetosphere to the ISEE-3 spacecraft
located about 240 earth radii upstream in the solar wind is
used as a tool to study the interaction between these protons
and the solar wind.

074.121   Solar wind and its interaction with the magnetosphere: measured parameters.   R. Schwenn.
The mesosphere and thermosphere, (see 012.066), p. 3 - 17
(1981).
    Taking into account the interdependence of the solar
wind parameters in context with the underlying solar phenomena, the author may be able to pinpoint the mechanism which
controls the action of the solar wind on the magnetosphere.

    **A catalogue of high-speed plasma streams in the
solar wind.   See Abstr. 002.051.**

    On the application of numerical models to the inverse mapping of solar wind flow structures.
See Abstr. 031.513.

    **Measuring electron density in coronal active regions.
I: A K-coronameter with a reflex monitor at λ5303 Å.**
See Abstr. 032.006.

    **The contribution of OSO-8 to our knowledge of the
chromosphere and transition region.   See Abstr. 032.519.**

Solar physics assessment of future high-resolution observations in the grazing-incidence domain. See Abstr. 032.553.

The Spacelab Lyman alpha and white light coronagraphs program.   See Abstr. 032.559.

A solar extreme ultraviolet telescope and spectrograph for Shuttle/Spacelab.   See Abstr. 032.560.

The Coronal Helium Abundance Experiment on Spacelab 2.   See Abstr. 032.561.

Sky photometer for ground-based coronal photography.   See Abstr. 034.012.

Resistive stability.   See Abstr. 062.028.

Kinetic effects on the propagation of surface waves and their relevance to the heating of the solar corona. See Abstr. 062.032.

A model of flare-produced magneto-radiative shock with increasing energy.   See Abstr. 062.037.

Temperature anisotropy instabilities driven by Whang's ion model.   See Abstr. 062.079.

On refraction of acousto-gravity waves in the solar atmosphere.   See Abstr. 062.119.

A transient MHD model applicable for the source of solar cosmic ray acceleration.   See Abstr. 062.126.

Stellar chromospheres and coronae. See Abstr. 064.081.

Photospheric flow and stellar winds. See Abstr. 064.086.

A unified working model for the atmospheric structure of large sunspot umbrae.   See Abstr. 072.001.

Quasiperiodic variations of solar activity, physical characteristics of the solar wind, and cosmic ray intensity and anisotropy.   See Abstr. 072.037.

Quasistationary outflow of gases from solar active regions.   See Abstr. 072.072.

Thermal flares in the corona: possible mechanism of origin.   See Abstr. 073.005.

The He I 10830 Å chromosphere and filament associated structures.   See Abstr. 073.014.

Role of plasma flow in determining structure of the chromosphere-corona transition zone of the sun. See Abstr. 073.015.

Mass motions in the transition region. See Abstr. 073.018.

Two-ribbon flares: (post)-flare loops. See Abstr. 073.035.

Activity and outer atmosphere of the sun. See Abstr. 073.037.

The chromosphere and transition region. See Abstr. 073.038.

Mechanical heating in the transition region. See Abstr. 073.041.

The transition region and corona associated with sunspots.   See Abstr. 073.047.

Solar flare and solar wind activity at present and in the past.   See Abstr. 073.054.

Chromospheric flash spectrum. See Abstr. 073.065.

Determination of the abundance of helium in the solar chromosphere and structure of the transition region between the corona and the chromosphere. See Abstr. 073.066.

Chromosphere-corona transition region models with magnetic field and fluid flow.   See Abstr. 073.067.

Intensity of astrophysical lines in the transition region.   See Abstr. 073.072.

The outer layers of the Sun. See Abstr. 073.073.

Steady flows in the solar transition region observed with *SMM*.   See Abstr. 073.090.

On the computation of constant α force-free magnetic field.   See Abstr. 075.002.

Modelling coronal magnetic fields. See Abstr. 075.010.

Eruptive solar magnetic fields. See Abstr. 075.015.

Magnetostatic atmospheres in a spherical geometry and their application to the solar corona. See Abstr. 075.024.

Solar abundances from X-ray flare observations. See Abstr. 076.011.

High-resolution ultraviolet solar observations from sounding rockets and Spacelab.   See Abstr. 076.021.

Solar radio bursts of spectral type II, coronal shocks, and optical coronal transients.   See Abstr. 077.017.

Radio evidence for shock acceleration of electrons in the solar corona.   See Abstr. 077.068.

Dynamics of solar cosmic ray events: processes at large heliocentric distances ($\geqslant 1$ AU).   See Abstr. 078.021.

The sun.   See Abstr. 080.082.

Excitation of magnetospheric hydromagnetic waves by solar-flare–induced change in ionospheric conductivity. See Abstr. 083.069.

Disappearing solar filaments: a useful predictor of geomagnetic activity.   See Abstr. 084.001.

Solar wind control of auroral zone geomagnetic activity.   See Abstr. 084.015.

ULF geomagnetic power near $L = 4.6$. Relationship to upstream solar wind quantities.   See Abstr. 084.052.

Solar wind flow about the terrestrial planets.
1. Modeling bow shock position and shape.
See Abstr. 091.074.

Magnetospheres and the interplanetary medium.
See Abstr. 091.096.

On the possibility of detection of ions from Venus
at 1 AU.   See Abstr. 093.002.

The role of charge exchange in the solar wind
absorption by Venus.   See Abstr. 093.070.

Solar-wind interactions with the Moon: nature and
composition of nitrogen compounds.   See Abstr. 094.030.

Mars satellite Deimos interaction with the solar wind
and its influence on flow around Mars.
See Abstr. 097.008.

Compression of Jupiter's magnetosphere by the
solar wind: reexamination via MHD simulation of evolving
corotating interaction regions.   See Abstr. 099.029.

Interplanetary scintillation in Jupiter's decametric
radiation.   See Abstr. 099.206.

Interaction of comets with the interplanetary
medium.   See Abstr. 102.019.

Plasma dynamics in the tail of
Comet Kohoutek 1973 XII.   See Abstr. 103.242.

Three-dimensional interaction of interplanetary
shock waves with the bow shock and magnetopause: a comparison of theory with ISEE observations.
See Abstr. 106.004.

Temporary spectra of scintillations with account
for solar wind velocity distribution. Theory.
See Abstr. 106.010.

The coronal and interplanetary current sheet in
early 1976.   See Abstr. 106.031.

Cross-correlation skewness of interplanetary
scintillations.   See Abstr. 106.039.

Spin-down effect on an interplanetary dust grain.
See Abstr. 106.040.

Acceleration theory for 5–40 keV ions at interplanetary shocks.   See Abstr. 106.046.

Dependence of ESP (*energetic storm particle*)
intensity on collisionless shock wave characteristics.
See Abstr. 106.047.

Determination of solar proton fluxes and energies
at high solar latitudes by UV radiation measurements.
See Abstr. 106.051.

Solar cycle modulation of galactic cosmic rays:
speculation on the role of coronal transients.
See Abstr. 143.003.

Solar wind conditioned galactic cosmic ray modulation and energy dependence of the diffusion coefficient.
See Abstr. 143.093.

Solar wind and cosmic ray density variations.
See Abstr. 143.094.

Galactic cosmic ray modulation by the solar wind
from selected heliographic latitudes.
See Abstr. 143.095.

Prediction of energetic particle behavior at high
solar latitudes.   See Abstr. 143.110.

Erratum

074.901   Erratum: 'Dynamics of coronal structures: magnetic
              field related heating and loop energy balance'
[Astrophys. J., Vol. 240, 301 - 308 (1980)].
A. A. Galeev, R. Rosner, S. Serio, G. S. Vaiana.
Astrophys. J., Vol. 248, 882 (1981). – See Abstr. 29.074.020.

075 Magnetic Fields

075.001 **Solar surface magnetic fields: a model.**
E. Knobloch.
Astrophys. J., Lett., Vol. 247, L93 - L96 (1981).

A simple model of nonlinear, three-dimensional turbulent convection in an imposed magnetic field is presented. The model describes the turbulent formation of small-scale photospheric flux tubes and gives results that are compatible with recent solar observations.

075.002 **On the computation of constant α force-free magnetic field.** C. E. Alissandrakis.
Astron. Astrophys., Vol. 100, 197 - 200 (1981).

The author presents solutions of the boundary value problem of constant α, force-free magnetic field in plane geometry, in terms of Fourier transforms. These solutions are easily adaptable to computations far from the center of the solar disk. The author also discusses the numerical computation of the field with fast Fourier transforms, which make the computation about 100 times faster compared to computations using Green's functions; in particular he discusses the effects of aliasing.

075.003 **Solar cycle Lorentz force waves and the torsional oscillations of the sun.** H. Yoshimura.
Astrophys. J., Vol. 247, 1102 - 1112 (1981).

A hypothesis is proposed and analyzed that the longitudinal component of the Lorentz force of the dynamo waves which well simulate the observed solar cycle can drive the recently observed torsional oscillations by Howard and LaBonte. The force component, which is reduced to a correlation between the toroidal and poloidal magnetic fields, consists of a nonwave part and a wave part. Only the nonwave part remains in the deep regions of the convection zone where the dynamo waves can propagate freely. The wave part, called here the Lorentz force waves, emerges only near the surface where the propagating dynamo waves are piled up and their wave profiles are deformed.

075.004 **Large-scale distribution of magnetic fields on active latitudes of the sun.** P. A. Golub.
Soln. Dannye 1981 Byull., No. 3, p. 102 - 106 (1981). In Russian.

Synoptic maps of the background solar magnetic fields Mount Wilson data were reduced obtaining Fourier spectra. The existence of their gigantic structure was confirmed. The mean period of the evolution of the magnetic field distribution is shown to last 3 - 4 solar rotations.

075.005 **Solar magnetism: a new look.** L. Golub.
Astronomy, Vol. 9, No. 3, p. 66 - 71 (1981).
Abstr. in Phys. Abstr., Vol. 84, Abstr. 83583 (1981).

075.006 **Flux tubes and the solar magnetic field spectrum.**
E. Knobloch.
Astrophys. J., Vol. 248, 1126 - 1131 (1981).

Improved spatial resolution of solar surface magnetograms makes the study of the solar surface magnetic field spectrum an increasingly promising possibility. The author explains how the spectrum can be calculated from flux-tube models and discusses in detail the nature of the information about the spatial and size distribution of the flux tubes that may be extracted from the observed spectrum. On the basis of a number of plausible assumptions, specific predictions of the spectrum can be obtained from existing models.

075.007 **Measurements of magnetic fields in the solar atmosphere.** C. W. Querfeld.
Bull. American Astron. Soc., Vol. 13, 560 (1981). − Abstract.

075.008 **Solar magnetic structure and the solar activity cycle. Review of observational data.**
C. Zwaan.
The sun as a star, (see 003.004), p. 163 - 179 (1981).

This review focuses on observational data on solar magnetism that may offer clues for understanding stellar magnetism in general. It concentrates on magnetic phenomena in the photosphere and low chromosphere, where the magnetic structure can be studied in detail.

075.009 **Surface magnetic fields during the solar activity cycle.** R. Howard, B. J. LaBonte.
Sol. Phys., Vol. 74, (see 012.029), 131 - 145 (1981).

We examine magnetic field measurements from Mount Wilson that cover the solar surface over a $13\frac{1}{2}$ year interval, from 1967 to mid-1980. Seen in long-term averages, the sunspot latitudes are characterized by fields of preceding polarity, while the polar fields are built up by a few discrete flows of following polarity fields. Weak field plots show essentially the same pattern as the stronger fields, and both data indicate that the large-scale field patterns result only from fields emerging at active region latitudes. The total magnetic flux over the solar surface varies only by a factor of about 3 from minimum to a very strong maximum (1979). Magnetic flux is highly concentrated toward the solar equator and appears at the solar surface at a rate which is sufficient to create all the flux that is seen at the solar surface within a period of only 10 days. Flux can spread relatively rapidly over the solar surface from outbreaks of activity. This is presumably caused by diffusion. In general, magnetic field lines at the photospheric level are nearly radial.

075.010 **Modelling coronal magnetic fields.**
D. P. Rowse, I. W. Roxburgh.
Sol. Phys., Vol. 74, (see 012.029), 165 - 168 (1981).

The "hairy ball" model of coronal magnetic fields has a spherical source surface separating potential and radial magnetic fields. In the present model the source surface is chosen such that the wind speed equals the Alfvénic speed at selected points on the source surface. Results have been obtained for a dipole base field and an isothermal corona.

075.011 **Bound oscillations on thin magnetic flux tubes: convective instability and umbral oscillations.**
J. V. Hollweg, B. Roberts.
Astrophys. J., Vol. 250, 398 - 407 (1981).

The possibility that "tube waves" can be trapped on slender solar magnetic flux tubes is investigated. For rigid isothermal flux tubes, the authors find that the flux tube geometry can by itself lead to waves which are trapped on the part of the tube that expands with height. Some geometries lead to trapped modes with eigenperiods near 180 s, if parameters appropriate to sunspot umbrae are chosen. It is possible that the umbral oscillations are a manifestation of such trapped waves, if sunspot umbrae consist of an assembly of slender flux tubes, as in the spaghetti model of Parker. For flux tubes which have a constant ratio of Alfvén speed to sound speed, they find that it is primarily the variation of temperature with height which determines whether trapped waves can exist. Certain temperature profiles lead to disturbances for which $\omega^2 < 0$, corresponding to convective instability of Rayleigh-Taylor instability.

075.012 **Annual variation of the solar mean magnetic field.**
V. A. Kotov, L. S. Levitskij, N. N. Stepanyan.
Izv. Krymskoj Astrofiz. Obs., Tom 63, 3 - 14 (1981). In Russian. English translation in Bull. Crimean Astrophys. Obs., Vol. 63.

Measurements of the mean magnetic field of the sun seen as a star have been made at the three observatories Crimean, Mount Wilson and Stanford, during the interval 1968 - 1978. The present analysis of these data clearly shows an annual variation in the solar mean magnetic field with an amplitude of about 0.1 G. An analogous annual wave was found also in the background magnetic field inferred from synoptic Hα-charts. Arguments are given that this one-year periodicity can be explained neither by instrumental nor by seasonal causes.

075.013 **Enhanced cosmic ray anisotropies and the extended solar magnetic field.**
D. B. Swinson, T. Saito, S. Mori.
J. Geophys. Res., Vol. 86, 8845 - 8851 (1981).

Saito's two-hemisphere model for the three-dimensional magnetic structure of the inner heliomagnetosphere is used to determine the orientation of the two solar magnetic hemispheres. This orientation, as viewed from the earth, varies throughout the year. The orientations during 1974 are presented and are confirmed by satellite data for the interplanetary magnetic field. These data suggest a role for the field component perpendicular to the ecliptic plane B_z in giving rise to cosmic ray anisotropies detected at the earth. It is shown that an enhanced solar diurnal variation in cosmic ray intensity at the earth can arise from the constructive interference of three cosmic ray anisotropies, two of which depend on the direction of the interplanetary magnetic field. This is demonstrated by using cosmic ray data.

075.014 **Magnetic loops in the sun's atmosphere.**
P. Foukal.
Sky Telesc., Vol. 62, 547 - 550 (1981).

075.015 **Eruptive solar magnetic fields.**   B. C. Low.
Astrophys. J., Vol. 251, 352 - 363 (1981).

The paper considers the quasi-steady evolution of solar magnetic fields in response to gradual photospheric changes. Special interest is taken in the threshold of a sudden eruption in the solar atmosphere. The formal model of an evolving, force-free field dependent on two Cartesian coordinates has been treated previously, and the author extends it to a field which is not force free but in static equilibrium with plasma pressure and gravity. The basic physics is illustrated by the evolution of a loop-shaped electric current sheet enclosing a potential bipolar field with footpoints rooted in the photosphere. A free-boundary problem is posed and solved for the equilibrium configuration of the current sheet in a hydrostatically supported isothermal atmosphere. The results are discussed with a particular interest in the triggering of the white-light coronal transient.

075.016 **Relations between the toroidal magnetic field in the solar convection zone and the bipolar magnetic field of sunspots.**   D.-q. Zhou.
Acta Astrophys. Sinica, Vol. 1, 188 - 196 (1981). In Chinese.

075.017 **The Hanle effect applied to magnetic field diagnostics.**   S. Sahal Bréchot.
Space Sci. Rev., Vol. 29, (see 012.043), 391 - 401 (1981).

The Hanle effect is the modification by the local magnetic field of the linear polarization parameters of a spectral line. A brief summary of the theory is presented and it is shown how the complete magnetic vector field can be recovered from the interpretation of the Hanle effect in spectral lines polarized by resonance scattering.

075.018 **Wave propagation in solar magnetic tubes.**
P. Venkatakrishnan.
Bull. Astron. Soc. India, Vol. 9, 214 - 226 (1981).

The work done on propagation of waves in magnetic tubes is reviewed. It is seen that majority of the results obtained so far are based on the thin tube approximation. It is shown that

estimates of the mechanical energy flux based on such an approximation could be grossly in error. The need for studying the propagation of waves in finite tubes is emphasised.

075.019 **On sign inversion of the solar polar magnetic field in the 21st solar cycle.**
V. I. Makarov, M. P. Fat'yanov.
Astron. Tsirk., No. 1130, p. 1 - 3 (1980). In Russian.

075.020 **Comparative magnetospherology. Part 11. A model of heliomagnetospheric excursion in every sunspot declining-minimum phase.**   T. Saito.
Proceedings of the 14th ISAS Lunar and Planetary Symposium, (see 012.055), p. 82 - 90 (1981).

075.021 **Comparative magnetospherology. Part 12. Evidences to support the two-hemisphere model on rotational reversing of the heliodipole in sunspot maximum phase.**
T. Saito, R. Howard.
Proceedings of the 14th ISAS Lunar and Planetary Symposium, (see 012.055), p. 91 - 102 (1981).

075.022 **June–July 1974 proton-flare region. I. Individual stages of its background and local magnetic field development.**   V. Bumba, L. Hejna.
Bull. Astron. Inst. Czechoslovakia, Vol. 32, 349 - 359 (1981).

The main evolutionary characteristics of the large-scale magnetic field pattern producing the June–July 1974 proton-flare region are described. Its space and time relation to the intersection of two different magnetically active longitudes is estimated, its individual development stages demonstrated. The process of successive complication of local magnetic field producing the proton-flare region and its growth of complexity in frame of the background magnetic field continuous redistribution is studied.

075.023 **Effets du champ magnétique solaire sur l'intensité semi annuelle de la raie verte coronale.**
J. Xanthakis, B. Petropóulos, H. Mavromichalaki.
Res. Cent. Astron. Appl. Math., Acad. Athens, Contrib. Ser. I (Astron.), No. 82.

The authors have studied the effects of the intensity of the solar magnetic field of the corona on the semi-annual intensity of the green coronal line at 5303 Å. They give a relation which can be used to compute the semi-annual coronal intensity in function of the index of solar activity, the number of proton events and the components of the intensity of the magnetic field of the corona. This relation is valid for the 19 and 20 solar cycle.

075.024 **Magnetostatic atmospheres in a spherical geometry and their application to the solar corona.**
J. R. Hundhausen, A. J. Hundhausen, E. G. Zweibel.
J. Geophys. Res., Vol. 86, 11117 - 11126 (1981).

The formalism for deriving 'two-dimensional' magnetostatic equilibria is extended to spherical coordinates and applied to magnetic fields that are functions of radius and polar angle. A family of analytic solutions is readily found. The basic properties of these solutions are displayed for a dipole magnetic field at the base of the atmosphere and for physical parameters appropriate to the solar corona.

075.025 **Synoptic charts of solar magnetic fields. Mount Wilson Observatory.**   R. Howard.
Q. Bull. Sol. Act., Vol. 21, Part II, p. 1 - 8 (1980).

Evolution of magnetic field and atmospheric responses. II. Formulation of proper boundary equations.
See Abstr. 062.004.

Recent development in the theory of perfect MHD waves in magnetically structured media. See Abstr. 062.029.

Dynamically consistent nonlinear dynamos driven by convection in a rotating spherical shell. See Abstr. 062.036.

Solutions for force-free magnetic fields. See Abstr. 062.045.

Magnetic flux tubes.   See Abstr. 062.050.

The present status of dynamo theory. See Abstr. 062.077.

Ideal MHD fluids.   See Abstr. 062.105.

New magnetodynamic principles in cosmic physics. See Abstr. 062.127.

Time-dependent interaction of granules with magnetic flux tubes.   See Abstr. 071.020.

Sunspot dynamics: gravitational draining – a cooling mechanism.   See Abstr. 072.002.

Types of sunspots, magnetic classification and flare yield in cycle No. 20.   See Abstr. 072.026.

Detection of a temperature deficit in magnetic faculae at the solar photosphere.   See Abstr. 072.032.

Theory of the solar cycle.   See Abstr. 072.033.

Line profiles and magnetic field in penumbral fine structures.   See Abstr. 072.040.

Determination of the complete vector magnetic field in solar prominences, using the Hanle effect. See Abstr. 073.003.

On the possibility to observe current sheets on the sun.   See Abstr. 073.024.

Chromosphere-corona transition region models with magnetic field and fluid flow.   See Abstr. 073.067.

Magnetic transients in flares. See Abstr. 073.080.

Preacceleration in collapsing magnetic neutral sheets and anomalous abundances of solar flare particles. See Abstr. 073.091.

The solar corona and the solar wind. See Abstr. 074.038.

A comparison of type III metric radio bursts and global solar potential field models.   See Abstr. 077.015.

On the presence of electric currents in the solar atmosphere. I: A theoretical framework. See Abstr. 080.009.

Recent global scale solar oscillations and magnetic field observations.   See Abstr. 080.015.

Lectures on theoretical aspects of magnetic energy storage and conversion: their relationship to solar activity. See Abstr. 080.020.

Global circulation and the solar dynamo. See Abstr.. 080.022.

On the non-symmetric solar dynamo. See Abstr. 080.045.

Fast motions in the solar atmosphere and magnetic field variations.   See Abstr. 080.068.

Champs de vitesses et champs magnétiques à petite échelle dans l'atmosphère solaire.   See Abstr. 080.072.

Comparative magnetospherology. Part 13. Effects of the rotating solar magnetosphere on disturbances of the earth's magnetosphere.   See Abstr. 085.056.

The coronal and interplanetary current sheet in early 1976.   See Abstr. 106.031.

Prediction of energetic particle behavior at high solar latitudes.   See Abstr. 143.110.

Preliminary search for cosmic radiation and solar-terrestrial parameters correlated with the reversal of the solar magnetic field.   See Abstr. 143.113.

076 UV, X, Gamma Radiation

076.001   Relationship between a soft X-ray Long Duration Event and an intense metric noise storm.
P. Lantos, A. Kerdraon, G. G. Rapley, R. D. Bentley.
Astron. Astrophys., Vol. 101, 33 - 38 (1981).

On March 30, 1980 an event was observed in active region 2363 both by the soft X-ray Polychromator experiment on the Solar Maximum Mission and by the Nançay Metric Radioheliograph (France). The soft X-ray observations reveal a typical Long Duration Event: an extensive system of loops forms over the site of a disappearing Hα filament, with Hα ribbons locating the loop footpoints. A centimetric gradual rise and fall burst and a coronal white light loop transient accompany the event. At metric wavelengths the onset of a strong type 1 noise storm is observed in the absence of a type IV burst.

076.002   Nucleosynthesis in solar flares as a source of solar gamma radiation.   B. M. Kuzhevskij.
Astron. Zh., Tom 58, 859 - 861 (1981). In Russian. English translation in Soviet Astron., Vol. 25, No. 4.

Nucleosynthesis in a flare region is one of the main processes resulting in gamma-ray emission in corresponding lines. In principle this allows to determine the emission measure, the value of the emitting volume, the effectiveness of the synthesis of nuclei in the flare as well as to find out if the preferential acceleration of heavy nuclei took place during the flare.

076.003   Rocket measurements of the solar spectral irradiance during solar minimum, 1972–1977.
G. J. Rottman.
J. Geophys. Res., Vol. 86, 6697 - 6705 (1981).

Five sounding rocket experiments conducted between December 1972 and March 1977, a period spanning solar minimum between cycles 20 and 21, provide full disc solar irradiance data in the spectral range 120-190 nm. The five measurements have been combined to give a solar minimum reference table. This spectrum is compared with other measurements obtained during the same time period. A table of intensities for the strong emission lines at wavelengths between 120 and 190 nm is presented.

076.004   A mechanism for solar ultraviolet flux variability.
K. H. Schatten, D. F. Heath.
Sol. Phys., Vol. 73, 13 - 23 (1981).

Solar UV emission observed by a filter photometer on Nimbus IV from 1969 to 1973 is examined in an attempt to understand the short term (27 day) and secular variability. Two models are discussed to explain the variations – a calcium plage model and a chromospheric network (faculae and spicule) structure model. Both relate to the remnant magnetic fields of active regions. An association between UV brightenings and the large scale magnetic field has been found consistent with the network model.

076.005   Analysis of the high-resolution X-ray spectra obtained aboard the INTERCOSMOS 16 satellite.
I: Identification of the lines in the 9.14–9.33 Å spectral region.
V. V. Krutov, V. V. Korneev, U. I. Karev, V. M. Lomkova, S. N. Oparin, A. M. Urnov, I. A. Zhitnik, G. Bromboszcz, M. Siarkowski, J. Sylwester, S. Vasha.
Sol. Phys., Vol. 73, 105 - 119 (1981).

A Bragg-type, flat ADP crystal spectrometer has been used for measurements of the X-ray spectra emitted from solar active region plasmas. About 60 spectra have been registered. In the present paper using a spectrum averaged over 20 scans, the authors measured the wavelengths corresponding to the statistically significant spectral features seen in this spectrum in the wavelength range 9.14–9.33 Å. By comparison with the calculated line wavelengths and intensities

predicted in the framework of the thermal model of the 'average' active region, the authors performed the identification of these features.

076.006   Height structure of thermal hard X-ray sources on the sun.   J. C. Brown, J. Hayward.
Sol. Phys., Vol. 73, 121 - 132 (1981).

The height distribution of hard X-ray bremsstrahlung is predicted for dissipative thermal models, involving rapid heating of many small (tearing mode) islands near the top of a magnetic arch.

076.007   Extreme ultraviolet emission of O IV and N IV from the sun.
K. P. Dere, J.-D. F. Bartoe, G. E. Brueckner.
Bull. American Astron. Soc., Vol. 13, 542 (1981). – Abstract.

076.008   Fine temporal structure in solar hard X-ray bursts and its implications.   A. G. Emslie.
Bull. American Astron. Soc., Vol. 13, 543 (1981). – Abstract.

076.009   ATM observations of H I Lyman α.
D. Roussel-Dupré.
Bull. American Astron. Soc., Vol. 13, 554 (1981). – Abstract.

076.010   Giant X-ray arches associated with post-flare radio noise storms.   Z. Svestka, M. Pick,
C. G. Rapley, R. T. Stewart, B. Woodgate.
Bull. American Astron. Soc., Vol. 13, 555 (1981). – Abstract.

076.011   Solar abundances from X-ray flare observations.
N. J. Veck, J. H. Parkinson.
Mon. Not. R. Astron. Soc., Vol. 197, 41 - 55 (1981).

The absolute coronal abundances of Si, Ar and Ca are derived from line and continuum X-ray flare observations. The lines used in the analysis are from either H- or He-like ions and allowance has been made for the presence of satellite lines and other unresolved line blends. In addition the calculations of line and continuum intensities are presented in the form of equivalent width versus temperature plots for use in analysing cosmic X-ray spectra.

076.012   The solar irradiance from 200 to 330 nm.
J. E. Mentall, J. E. Frederick, J. R. Herman.
J. Geophys. Res., Vol. 86, 9881 - 9884 (1981).

The full disk solar spectrum over the wavelength range 200–330 nm has been measured with a spectral resolution of 0.12 nm. The measurement was performed by using a double monochromator pointed at the sun from a high altitude parachute. Only small corrections were required to account for atmospheric absorption. Good agreement was obtained with previously reported measurements.

076.013   Solar irradiance below 120 nm and its variations.
G. Schmidtke.
Sol. Phys., Vol. 74, (see 012.029), 251 - 263 (1981).

Due to technological improvements of space and laboratory instrumentations, an almost complete set of data for the solar irradiance below 120 nm has been obtained from 1972 to date. These observations exhibit strong differences of the flux variations from solar cycle 20 to 21. For the theoretical and for semi-empirical treatments of many aeronomic processes controlled by the solar EUV radiation, its adequate representation e. g. as indices is required. The problems involved and possible solutions are discussed. Results from some relevant aeronomically oriented computations based on variable solar EUV fluxes are presented.

076.014   **Solar irradiance between 120 and 400 nm and its variations.**   P. C. Simon.
Sol. Phys., Vol. 74, (see 012.029), 273 - 291 (1981).

The solar ultraviolet irradiance measurements in the 120 - 400 nm wavelength range are reviewed and compared showing still important discrepancies between the irradiance values deduced from the most recent observations. The possible variations of the solar ultraviolet irradiances with the 27-day rotation period of the Sun and with the 11-year activity cycle are presented and discussed on the basis of the available irradiation fluxes obtained during the rising phase of solar cycle 21. The spectral features of both kinds of variation are clearly related to the solar atmospheric layer from which the corresponding radiation is emitted.

076.015   **Measurements of the complete flow of ultraviolet radiation of the sun aboard AES.**
S. I. Avdyushin, L. L. Bukusova, S. M. Gutkevich, D. A. Gonyukh, V. E. Davydov, F. L. Dlikman, G. S. Ivanov-Kholodnyj, T. V. Kazachevskaya, Yu. M. Kulagin, B. P. Seredin, I. S. Yudkevich.
Geomagn. Aehron., Tom 21, 925 - 927 (1981). In Russian.

076.016   **The solar spectral irradiance 1200 - 3184 Å near solar maximum: July 15, 1980.**
G. H. Mount, G. J. Rottman.
J. Geophys. Res., Vol. 86, 9193 - 9198 (1981).

Full-disk solar spectral irradiances near solar maximum were obtained in the spectral range 1200 - 3184 Å at a spectral resolution of approximately 1 Å from rocket observations. Comparison with measurements made in 1979 and during solar minimum confirms a large increase at solar maximum in the solar irradiance near 1200 Å. Irradiances in the range 1900 - 2100 Å are in excellent agreement with previous measurements, and those in the 2100 to 2500 Å range are lower than the Broadfoot results. The authors find agreement with previous values for 2600 - 2900 Å and then a decrease below those values for 2900 - 3184 Å.

076.017   **Observational, reference and model data on solar EUV, from measurements on AE-E.**
H. E. Hinteregger, K. Fukui, B. R. Gilson.
Geophys. Res. Lett., Vol. 8, 1147 - 1150 (1981).

076.018   **Observation of the solar H Lyman-α radiation by a ion-chamber with an O_2 gas filter.**
N. Watanabe, I. Higashino, T. Oshio.
Bull. Inst. Space Aeronaut. Sci. Univ. Tokyo B, Vol. 16, 1431 - 1443 (1980). In Japanese. − Abstr. in Phys. Abstr., Vol. 85, Abstr. 3225 (1982).

076.019   **Observations of soft X-ray line emissions from solar flares by Tansei 4 satellite.**   K. Tanaka, T. Watanabe, K. Nishi, F. Moriyama, H. Miyazaki.
Bull. Inst. Space Aeronaut. Sci. Univ. Tokyo B, Vol. 16, 1445 - 1468 (1980). In Japanese. − Abstr. in Phys. Abstr., Vol. 85, Abstr. 3226 (1982).

076.020   **Large loop thermal models of solar hard X-ray bursts.**   D. F. Smith, D. W. Harmony.
Plasma astrophysics, (see 012.042), p. 401 - 404 (1981).

The results of Smith and Harmony (1981) for small loop thermal models of hard X-ray bursts are extended to large loops.

076.021   **High-resolution ultraviolet solar observations from sounding rockets and Spacelab.**   G. E. Brueckner.
Space Sci. Rev., Vol. 29, (see 012.043), 407 - 418 (1981).

High spatial (∼1″) and temporal (20s) resolution UV spectroscopy of the Sun has been carried out with a new instrument flown on sounding rockets. These observations reveal a multitude of new high energetic phenomena in the outer solar atmosphere which may play a decisive role in the mechanical energy balance of the chromosphere, transition zone and corona.

076.022   **Gamma ray lines from solar flares and cosmic transients.**
R. Ramaty, R. E. Lingenfelter, B. Kozlovsky.
Gamma-ray transients, (see 012.046), 19 pp. (1981).

The authors review the theory of solar gamma ray line production and present results of new numerical calculations based on more detailed and accurate nuclear cross sections. They illustrate the application of the theory by considering the June 7, 1980, flare. The authors discuss the physical processes responsible for gamma ray line production in cosmic transients, in particular, positron production and annihilation and nuclear line emission. They review the possible origin of the gamma ray lines observed in a longer duration transient which is a particularly strong gamma ray line emitter.

076.023   **Gamma-line emission from the quiet sun.**
G. A. Koval'tsov.
Izv. AN SSSR. Ser. fiz., Tom 45, 1155 - 1158 (1981). In Russian. − Abstr. in Ref. zh., 51. Astron., 11.51.413 (1981).

076.024   **On the measurement of solar hard X-ray emission.**
A. V. Baskakov.
Izv. AN SSSR. Ser. fiz., Tom 45, 1159 - 1173 (1981). In Russian. − Abstr. in Ref. zh., 51. Astron., 11.51.444 (1981).

076.025   **Le spectre ultraviolet du soleil et ses variations en fonction de l'activité solaire.**   M. Nicolet.
C. R. Acad. Sci. Paris, Vie Acad., Tome 293, (see 012.056), XXI - XXII, XXIV (1981).

076.026   **The variability of the sun's ultraviolet radiation.**
G. E. Brueckner.
Planetary aeronomy and astronomy, (see 012.063), p. 101 - 115 (1981).

The intensity of continua and emission lines which form the solar UV spectrum below 2100 Å is variable. Continua and emission lines originating from different layers in the solar atmosphere show a different degree of variability. Typical time-scales of solar UV variability are minutes (flare induced), days (birth of active regions), 27 days (solar rotation), 11 years (solar cycle) and perhaps centuries, caused by long-term changes of the solar activity. Plages are the main contributor to the solar UV variability.

076.027   **Representations of solar EUV fluxes for aeronomical applications.**   H. E. Hinteregger.
The mesosphere and thermosphere, (see 012.066), p. 39 - 52 (1981).

076.028   **Solar Maximum Mission experiment: early results of the hard X-ray imaging experiment.**
G. M. Simnett, A. Boelee, C. P. Charlton, C. de Jager, A. Duijveman, R. Fryer, M. Galama, R. Hoekstra, P. Hoyng, J. P. Imhof, H. Lafleur, M. E. Machado, H. V. A. M. Maseland, W. A. Mels, A. Schadee, J. Schrijver, Z. Svestka, H. F. van Beek, P. van Rens, J. J. M. van der Laan, W. van Tend, F. Werkhoven, G. Wiersma, A. P. Willmore, J. W. G. Wilson, W. Zandee.
High-energy astrophysics, (see 012.067), p. 255 - 266 (1981).

The authors have selected four widely different flares from the early period of operations of the Hard X-ray Imaging Spectrometers (HXIS) on SMM to illustrate the characteristic imaging properties of this experiment. Data are reported for the small flare of April 4, 1980, an extensive weak flare of April 6, a 1B flare of April 7, and a 1N flare of April 10 with particularly strong hard X-ray bursts.

076.029   **Solar Maximum Mission experiment: early results**

from the soft X-ray polychromator experiment.
A. H. Gabriel, J. L. Culhane, L. W. Acton, E. Antonucci,
R. D. Bentley, C. Jordan, J. W. Leibacher, A. N. Parmar,
K. J. H. Phillips, C. G. Rapley, C. J. Wolfson, K. T. Strong,
High-energy astrophysics, (see 012.067), p. 267 - 273 (1981).

The X-ray polychromator experiment has been in operation on the SMM satellite for over three months. It is observing flares and active regions in the wavelength range 1 Å to
23 Å using a number of different modes. These include poly-
chromatic imaging, high resolution line profiles, high dispersion spectra, and light curves with high time-resolution. Data
are described together with some of the preliminary analysis
and interpretation.

X-ray astronomy.   See Abstr. 003.049.

Solar observations and atomic data for the
$3s^2 \, ^1S_0 - 3s3p \, ^3P_1$ transition in S V.   See Abstr. 022.120.

A high precision solar ultraviolet spectral irradiance
monitor for the wavelength region 120 - 400 nm.
See Abstr. 032.542.

The ESA project for a Grazing Incidence Solar
Telescope (GRIST).   See Abstr. 032.544.

High resolution grazing incidence telescopes for the
EUV-regime.   See Abstr. 032.562.

Photoelectric array detectors for use at XUV
wavelengths.   See Abstr. 032.565.

Solar Maximum Mission experiment: ultraviolet
spectroscopy and polarimetry on the Solar Maximum Mission.
See Abstr. 032.631.

Comment on 'First phase acceleration mechanisms
and implications for hard X-ray burst models in solar flares'.
See Abstr. 062.033.

Simultaneous observations of AR 2490 performed
in the X-ray, UV, optical, and radio wavelength domain.
See Abstr. 072.049.

New ionization fractions for the lithium- and helium-
like ionization stages of calcium and iron.
See Abstr. 073.001.

Downward shift of the acceleration/injection region
during solar flares.   See Abstr. 073.012.

Analysis of a series of solar flare X-ray spectra.
See Abstr. 073.022.

Models of a hot flare region based on X-ray photometric data of the Interkosmos 11 satellite.
See Abstr. 073.023.

X-ray observations of two different systems of
"post-flare" loops.   See Abstr. 073.025.

Evidence for delayed second phase acceleration in
solar flares.   See Abstr. 073.030.

SMM observations of soft X-ray line spectra.
See Abstr. 073.031.

Flare observations.   See Abstr. 073.033.

Solar flare observations and their interpretations.
See Abstr. 073.039.

High resolution solar flare X-ray spectra: the
temporal behavior of electron density, temperature, and
emission measure for two class M flares.
See Abstr. 073.068.

Simultaneous measurements of EUV and soft X-ray
solar flare emission.   See Abstr. 073.071.

Intensity of astrophysical lines in the transition
region.   See Abstr. 073.072.

On ionization equilibrium in solar flares as determined from X-ray emission lines of Ca XVIII and Ca XIX.
See Abstr. 073.087.

Properties of solar flare electrons, deduced from
hard X-ray and spatially resolved microwave observations.
See Abstr. 073.088.

A new component of hard X-rays in solar flares.
See Abstr. 073.089.

Steady flows in the solar transition region observed
with *SMM*.   See Abstr. 073.090.

An interacting loop model for solar flare bursts.
See Abstr. 073.100.

Energy spectra of high energy electrons and hard
X rays as observed on board the space probe Venera 11 during
the solar flare event of April 13, 1979.
See Abstr. 073.110.

Analysis of the physical conditions in a strong X-ray
flare.   See Abstr. 073.112.

X-ray and gamma-ray observations of a white light
flare.   See Abstr. 073.113.

Structural development of the X-ray limb flare of
30 April 1980.   See Abstr. 073.114.

The energy balance in coronal holes and average
quiet-Sun regions.   See Abstr. 074.001.

Closed coronal structures. III. Comparison of static
models with X-ray, EUV, and radio observations.
See Abstr. 074.002.

Thermal radio emission of solar active regions derived from quantitative analysis of Skylab X-ray pictures and
compared with observation.   See Abstr. 077.003.

Energetic electrons, type III radio bursts, and
impulsive solar flare X-rays.   See Abstr.077.004.

Interpretation of microwave and X-ray bursts observed with the VLA and P78–1.
See Abstr. 077.022.

VLA observations of positions of 6 centimeter burst
peaks associated with hard X-ray burst spikes.
See Abstr. 077.054.

On a complex model of solar microwave radiation
consistent with soft and hard X-ray burst emission.
See Abstr. 077.067.

Identification and characteristics of source flares
emitting ^{3}He or Fe enriched cosmic rays in the period
September to December 1977.   See Abstr. 078.001.

Some consequences of the application of solar gamma-astronomy to the study of solar cosmic rays. See Abstr. 078.013.

The sun. See Abstr. 080.082.

Stratospheric chemical and thermal response to long-term variability in solar UV irradiance. See Abstr. 082.010.

The E and F region ionospheric response to solar flares: 1. Effects of approximations of solar flare EUV fluxes. See Abstr. 083.001.

Modulation of Venus ion densities associated with solar variations. See Abstr. 093.011.

An attempt to determine the solar Lyα flux independently of instrument calibration. See Abstr. 106.033.

Interpretations and implications of γ-ray lines from solar flares, the galactic centre and γ-ray transients. See Abstr. 142.522.

077 Radio, Infrared Radiation

077.001 **Observation of new emission lines in the infrared solar spectrum near 12.33, 12.22, and 7.38 microns.**
F. J. Murcray, A. Goldman, F. H. Murcray, C. M. Bradford, D. G. Murcray, M. T. Coffey, W. G. Mankin.
Astrophys. J., Lett., Vol. 247, L97 - L99 (1981).
High resolution infrared solar spectra reveal new emission features at 811.575, 818.058, and 1356.182 cm⁻¹. The features width is $\lesssim 0.02$ cm⁻¹ and their intensity is ~10% above the continuum. Coincidences with energy level differences of ionized light atoms are noted, but the identification and excitation mechanism are not yet firmly established.

077.002 **Solar type I radio bursts: shock model.**
D. G. Wentzel.
Astron. Astrophys., Vol. 100, 20 - 23 (1981).
The brief intense emission of type I bursts is taken to be evidence for localized strong electrical currents in the corona. The radiation mechanism is considered to be the combination of upper-hybrid and lower-hybrid waves. The upper-hybrid waves, driven by the loss-cone of trapped fast electrons, are highly directional and can yield the observed high brightness temperature. The lower-hybrid waves provide a high radiative opacity and are most probably generated by shocks.

077.003 **Thermal radio emission of solar active regions derived from quantitative analysis of Skylab X-ray pictures and compared with observation.**
G. Elwert, W. Villing, J. Vorpahl, R. M. Broussard.
Astron. Astrophys., Vol. 101, 150 - 157 (1981).
From X-ray pictures of the Sun taken by the Aerospace/Marshall Space Flight Center experiment during the Skylab mission in August 1973, distributions of temperature T and differential emission measure E of hot coronal active regions are calculated. From T and E the thermal radio emission of these active regions at 2.8 cm wavelength is determined and compared with radio maps of the Sun which were scanned at the same days with the Effelsberg radio dish. There are indications that the radio emission is not totally thermal bremsstrahlung and that there is an additional contribution of magnetic bremsstrahlung.

077.004 **Energetic electrons, type III radio bursts, and impulsive solar flare X-rays.**   S. R. Kane.
Astrophys. J., Vol. 247, 1113 - 1121 (1981).
Observations of impulsive hard X-ray and type III radio bursts made during the maximum of the last solar activity cycle have been analyzed for a statistical study of the relationship between these two solar flare phenomena. Spectral measurements of 10 - 68 keV X-rays, which covered 7068 hr of observation time and the range 10^{-8} to 10^{-5} ergs cm⁻² s⁻¹ of flux of X-rays $\gtrsim 20$ keV, were made with the University of California (Berkeley) experiment aboard the *OGO 5* satellite. The radio data consisted of copies of the original spectral records as well as tabulated data.

077.005 **Spectral characteristics of current sheets in the radio range.**   V. D. Kuznetsov.
Soln. Dannye 1980 Byull., No. 12, p. 81 - 84 (1981). In Russian.
Spectral characteristics in the radio range are presented for a steady and pre-flare current sheet. The latter may be observed as a hot feature at shorter wavelengths and as a cold one at longer wavelengths.

077.006 **On a model of the gyroresonance emission of local sources of solar radio emission.**
Sh. B. Akhmedov, B. I. Ryabov.
Soln. Dannye 1980 Byull., No. 12, p. 85 - 94 (1981). In Russian.
The variation of the form of spectra of local sources during their evolution is explained in terms of a multicomponent gyroresonance model of the source. The distribution of the temperature and density of electrons is assumed irrespective of height within the region of gyroresonance emission, and the distribution of magnetic fields is found. A comparison of the calculations with the parameters of the observed local sources shows good agreement of the calculated and observed spectra.

077.007 **The fine structure of the 1978 July 10 and 11 compound radio bursts.**
A. A. Gnezdilov, A. K. Markeev, G. P. Chernov.
Soln. Dannye 1981 Byull., No. 1, p. 93 - 104 (1981). In Russian.
The fine structure in the meter range of the compound radio bursts July 10 and 11, 1978, connected with proton geoeffective flares is presented. The flux density and frequency spectrum of the two radio bursts were typical for proton events: stripes in emission and absorption (fiber bursts and zebra patterns) and second and minute pulsations are observed.

077.008 **High-resolution observations of the radio burst on December 28, 1977.**
S. I. Boldyrev, V. N. Ikhsanova.
Soln. Dannye 1981 Byull., No. 2, p. 103 - 109 (1981). In Russian.
The results are given of the observations of radio emission of the McMath 15081 active region where a radio burst emerged at 9^h15^m on December 28, 1977 which accompanied the chromospheric flare McMath 967. The variation of the emitting region after the burst is discussed.

077.009 **Some peculiarities of the variation of the flux of solar microwave radio emission preceding a burst.**
V. M. Plotnikov, V. L. Ermoshenko.
Soln. Dannye 1981 Byull., No. 3, p. 93 - 98 (1981). In Russian.
It is considered how the solar radio emission flux varies before a microwave burst. A statistical analysis at the frequency 6100 MHz allowed to detect four main types of the variation. Considering the dynamical spectrum of the variations their complex dependence on the wavelengths is found.

077.010 **Expected brightness distribution of the radio lines of excited hydrogen in the solar atmosphere.**
A. F. Dravskikh, Z. V. Dravskikh.
Soln. Dannye 1981 Byull., No. 3, p. 99 - 102 (1981). In Russian.
It is shown that the recombination lines of excited hydrogen (at waves shorter than 1 mm) may be discovered on the sun in the radiation. In these lines a bright ring may be observed on the solar limb at a height of about 1800 km above the chromospheric level. The computations were done for a homogenous chromosphere and taking spicules into consideration.

077.011 **Some effects in the bremsstrahlung radio emission of nonstationary solar sources.**   V. A. Kovalev.
Pis'ma Astron. Zh., Tom 7, 493 - 496 (1981). In Russian.
English translation in Soviet Astron. Lett., Vol. 7.
It is shown that an excess in the rate of the circularly polarized radiation change above the rate of the intensity change in the microwave range is a criterion (sufficient condition) of the existence of a longitudinal magnetic field time variation in a nonstationary solar source. Heating of the optically thin source in a magnetic loop can lead to depression of the radiation.

077.012 **On the night-time reception of solar radio bursts.**
J. J. Riihimaa.
Observatory, Vol. 101, 117 - 119 (1981).
There are indications that the propagation of decametric solar radio bursts into the night-time hemisphere of the Earth

may take place rather often. In fact, under certain conditions, decametric solar bursts can apparently be observed only at night. If the usual concepts of propagation are applied, the conditions turn out to be extremely marginal.

077.013   The polarization of second harmonic radio emission in type III bursts.   E. Ya. Zlotnik.
Astron. Astrophys., Vol. 101, 250 - 258 (1981).

The conversion of plasma waves into magnetoionic modes in the solar corona is discussed as a possible origin of observed polarization of the second harmonic radio emission in type III bursts. Observational data are in good agreement with the theory provided that the plasma waves excited by coronal electron streams and responsible for type III bursts have nearly one-dimensional angular distribution.

077.014   Solar limb brightening at 350 microns.
C. Lindsey, R. H. Hildebrand, J. Keene, S. E. Whitcomb.
Astrophys. J., Vol. 248, 830 - 835 (1981).

The authors have observed the intensity profile of the quiet solar limb in the 300–400 μm continuum. They find a significant resolved brightening of several percent over the outer 60″ of the solar limb in this band. More-recent ground-based observations indicate that the magnitude of solar limb brightening at 800 μm and at 1.3 mm is stronger than that at 350 μm. This may be regarded as an indication that the hot material which produces the brightening at the extreme limb, thought to consist in part of chromospheric spicules, is optically thin in the 350 μm continuum.

077.015   A comparison of type III metric radio bursts and global solar potential field models.
B. V. Jackson, R. H. Levine.
Sol. Phys., Vol. 73, 183 - 190 (1981).

The authors compare evidence of coronal magnetic fields from polarized metric type III radio bursts with (a) global potential field models, (b) direct averages of the observed photospheric magnetic field, and (c) Hα synoptic charts. The comparison clearly indicates both that the principal aspects of type III burst radiation are understood and that global potential field models are a significantly more accurate representation of coronal magnetic field structure than either the large-scale photospheric field or Hα synoptic charts.

077.016   Langmuir-wave conversion as the explanation of moving type IV solar meter-wave radio outbursts.
R. A. Duncan.
Sol. Phys., Vol. 73, 191 - 204 (1981).

The properties of moving type IV solar radio outbursts observed with the Culgoora radio-heliograph and radiospectrograph suggest that these outbursts − even those which rise to great coronal heights − originate through Langmuir-wave conversion. This implies that coronal densities are sometimes greatly enhanced.

077.017   Solar radio bursts of spectral type II, coronal shocks, and optical coronal transients.
A. Maxwell, M. Dryer.
Sol. Phys., Vol. 73, 313 - 329 (1981).

The association of solar radio bursts of spectral type II and coronal shocks with solar flare ejecta observed in Hα, the green coronal line, and white-light coronagraphs is examined. Rather than identifying fast-moving optical coronal transients with outward-travelling shock waves that generate type II radio bursts, as has been suggested in some earlier papers, the authors suggest that, for the most part, such transients should probably be identified with piston-type phenomena well behind the shock. The authors discuss a general model, consisting of three main velocity regimes, in which the authors relate type II radio bursts and coronal shocks to optically-observed ejecta.

077.018   Observations and interpretation of the slowly varying component of solar radio emission at decameter wavelengths.   C. V. Sastry, K. S. Dwarkanath, R. K. Shevgaonkar, V. Krishan.
Sol. Phys., Vol. 73, 363 - 377 (1981).

The authors have observed the slowly varying component of solar radio emission at a frequency of 34.5 MHz. It is found that the observed brightness temperatures vary within the limits of 0.3×10^6 K to 1.5×10^6 K, and the average half power widths of the brightness distribution on the sun is about $3 R_\odot$. Thermal emission from coronal regions of various electron densities and temperatures with and without the magnetic field has been computed and compared with the observed results.

077.019   Solar type III radio emission process.
D. G. Wentzel.
Bull. American Astron. Soc., Vol. 13, 543 - 544 (1981).
Abstract.

077.020   Correlations of solar microwave bursts with prompt $E > 20$ MeV proton events.   S. W. Kahler.
Bull. American Astron. Soc., Vol. 13, 544 (1981). − Abstract.

077.021   High spectral resolution observations of solar microwave emission.
G. J. Hurford, K. A. Marsh, H. Zirin.
Bull. American Astron. Soc., Vol. 13, 553 (1981). − Abstract.

077.022   Interpretation of microwave and X-ray bursts observed with the VLA and P 78 − 1.
E. J. Schmahl, M. R. Kundu, P. B. Landecker, D. L. McKenzie.
Bull. American Astron. Soc., Vol. 13, 553 (1981). − Abstract.

077.023   Microwave observations of the solar atmosphere.
H. Zirin.
Bull. American Astron. Soc., Vol. 13, 559 (1981). − Abstract.

077.024   Infrared observations of the solar atmosphere.
T. A. Clark.
Bull. American Astron. Soc., Vol. 13, 560 (1981). − Abstract.

077.025   On the interpretation of radio emission of local sources on the sun.
S. I. Agureev, A. Kurbanov, L. V. Yasnov.
Izv. TadzhSSR. Otd. fiz.-mat., khim. i geol. nauk, 1980, No. 4, p. 12 - 16. In Russian. − Abstr. in Ref. zh., 51. Astron., 7.51. 306 (1981).

077.026   On generation of subthermal radio emission of the sun by tail electrons.
V. A. Kovalev, V. I. Petviashvili.
Inst. zemn. magn., ionos. i rasprostr. radiovoln AN SSSR. Prepr., 1981, No. 4, 15 pp. In Russian. − Abstr. in Ref. zh., 51. Astron., 7.51.318 (1981).

077.027   On the development of S-component local sources of solar radio emission at 6.6 cm wavelength.
N. P. Stasyuk.
Problems of cosmic physics. Vyp. 16, (see 003.005), p. 41 - 53 (1981). In Russian.

The results of processing of stripscans of the sun at wavelength 6.6 cm during the period of minimum solar activity in 1964 are discussed. The identification of local sources of the S-component with an active region that was separated from the processing is carried out. The development of some concrete local radio sources is described. It is investigated which relation of parameters determines the properties of radio sources with characteristics of an active region.

077.028   Two-dimensional image of local solar radio sources using consecutive aperture synthesis with

RATAN 600.
O. A. Golubchina, V. N. Ikhsanova, V. M. Bogod,
G. S. Golubchin.
Soln. Dannye 1981 Byull., No. 4, p. 108 - 118 (1981). In Russian.

A new method of observations with RATAN 600 is proposed. It permits to obtain a two-dimensional image of radio sources and to guide them during 10 - 15 hours. Two series of observations of solar radio emission were made. Two-dimensional images of local sources were obtained. Their heliographic coordinates were determined. "Radio isophotes" were constructed. Dynamical spectra of a rapidly developing local source are presented.

077.029   Depression of the integral solar radio flux preceding a burst.   V. M. Plotnikov, N. G. Peterova.
Soln. Dannye 1981 Byull., No. 5, p. 92 - 96 (1981). In Russian.

From observations with low and high resolution in the range ~6.5 GHz an outburst is studied. It is shown that the effect is observed in 80% of bursts near the limb and depression reaches 50% in the emission flux of the local source in the active region where the burst occurred. The above effect may be associated with the current sheet formed before the appearance of the burst.

077.030   Study of the power spectrum of solar radio emission with the maximum entropy method.
Sh. Sh. Gusejnov, V. G. Utrobin.
Fiz. soln. aktivnosti. Moskva, 1980, p. 118 - 130. In Russian.
Abstr. in Ref. zh., 51. Astron., 8.51.450 (1981).

077.031   Microwave flux measurements during 1980 February 16 total solar eclipse.
S. K. Alurkar, S. S. Degaonkar, R. V. Bhonsle, O. P. N. Calla,
G. Raju, S. S. Rana, B. Lokanadham, B. V. Gopal.
Bull. Astron. Soc. India, Vol. 9, 70 (1981). – Abstract.

077.032   Radio observations of the solar eclipse of 1980 February 16 at Ootacamund.
A. Pramesh Rao, D. S. Bagri.
Bull. Astron. Soc. India, Vol. 9, 70 (1981). – Abstract.

077.033   Observation of 10.4 cm radionoise during the solar eclipse of 1980 February 16.
S. Raghavan, T. R. Sivaramakrishnan, V. M. Varadarajan.
Bull. Astron. Soc. India, Vol. 9, 71 (1981). – Abstract.

077.034   Intense microwave bursts observed at 2.8, 10 and 19.3 GHz on 1980 June 4.   S. S. Degaonkar,
O. P. N. Calla, S. K. Alurkar, R. V. Bhonsle, B. Lokanadham.
Bull. Astron. Soc. India, Vol. 9, 75 (1981). – Abstract.

077.035   Observations and interpretation of the slowly varying component of solar radio emission at decameter wavelengths.
C. V. Sastry, K. S. Dwarkanath, R. K. Sheygaonkar, V. Krishan.
Bull. Astron. Soc. India, Vol. 9, 75 - 76 (1981). – Abstract.

077.036   Radio spectrum of the quiet sun in the 1 - 20 mm range.   S. A. Pelyushenko.
13-ya Vses. konf. po radioastron. issled. soln. sistemy, Kiev, 1981. Tez. dokl., Kiev, 1981, p. 3 - 4. In Russian. – From Ref. zh., 51. Astron., 9.51.346 (1981).

077.037   Spatial correlation of radio granulation and the calcium chromospherical network.
G. B. Gel'frejkh, Eh. V. Kononovich, S. L. Ovchinnikov,
O. B. Smirnova, S. V. Startsev.
13-ya Vses. konf. po radioastron. issled. soln. sistemy, Kiev, 1981. Tez. dokl., Kiev, 1981, p. 7. In Russian. – From Ref. zh., 51. Astron., 9.51.357 (1981).

077.038   Observations of rising local radio sources on the RATAN-600 radio telescope and some possibilities of forecasting several characteristics of sunspot groups using radio data.   S. A. Andrianov.
13-ya Vses. konf. po radioastron. issled. soln. sistemy, Kiev, 1981. Tez. dokl. Kiev, 1981, p. 12 - 13. In Russian. – From Ref. zh., 51. Astron., 9.51.387 (1981).

077.039   An attempt of measuring local polarization of emission of local sources according to observations on RATAN-600.   A. N. Korzhavin.
13-ya Vses. konf. po radioastron. issled. soln. sistemy, Kiev, 1981. Tez. dokl. Kiev, 1981, p. 13. In Russian. – From Ref. zh., 51. Astron., 9.51.388 (1981).

077.040   Some properties of spectral direction of several radio components of solar local sources according to observations on the RATAN-600 radio telescope.
B. I. Lubyshev.
13-ya Vses. konf. po radioastron. issled. soln. sistemy, Kiev, 1981. Tez. dokl. Kiev, 1981, p. 14. In Russian. – From Ref. zh., 51. Astron., 9.51.389 (1981).

077.041   Some properties of radio emission of a local source over a large spot according to observations on the RATAN-600 radio telescope.
V. M. Bogod, A. N. Korzhavin, B. I. Lubyshev.
13-ya Vses. konf. po radioastron. issled. soln. sistemy, Kiev, 1981. Tez. dokl. Kiev, 1981, p. 15. In Russian. – From Ref. zh., 51. Astron., 9.51.390 (1981).

077.042   Characteristics of solar radio sources according to automatically reduced radio image data.
M. I. Ryabov, S. K. Panishko.
13-ya Vses. konf. po radioastron. issled. soln. sistemy, Kiev, 1981. Tez. dokl. Kiev, 1981, p. 28. In Russian. – From Ref. zh., 51. Astron., 9.51.391 (1981).

077.043   Structure of the radio emission spectrum of local sources on the sun in the 5 - 12 GHz range.
N. S. Kaverin.
13-ya Vses. konf. po radioastron. issled. soln. sistemy, Kiev, 1981. Tez. dokl. Kiev, 1981, p. 20 - 22. In Russian. – From Ref. zh., 51. Astron., 9.51.392 (1981).

077.044   On some peculiarities in radio spectra of local sources in the 8 - 12 GHz range.
N. S. Kaverin, A. I. Korshunov, V. V. Shushunov.
13-ya Vses. konf. po radioastron. issled. soln. sistemy, Kiev, 1981. Tez. dokl. Kiev, 1981, p. 21. In Russian. – From Ref. zh., 51. Astron., 9.51.393 (1981).

077.045   On polarization measurements of solar radio radiation in the 5 - 7 GHz range with high-frequency resolution.   Yu. V. Tikhomirov, E. I. Lebedev.
13-ya Vses. konf. po radioastron. issled. soln. sistemy, Kiev, 1981. Tez. dokl. Kiev, 1981, p. 21 - 22. In Russian. – From Ref. zh., 51. Astron., 9.51.394 (1981).

077.046   On fluctuations of radio emission of active regions on the sun.
I. I. Berulis, A. I. Kozlovskij, B. Ya. Losovskij,
N. G. Franchuk, L. V. Yasnov.
13-ya Vses. konf. po radioastron. issled. soln. sistemy, Kiev, 1981. Tez. dokl. Kiev, 1981, p. 26 - 27. In Russian. – From Ref. zh., 51. Astron., 9.51.395 (1981).

077.047   Quasi-periodic fluctuations of microwave radiation of active regions during minimum solar activity years.   V. G. Zandanov.
13-ya Vses. konf. po radioastron. issled. soln. sistemy, Kiev,

1981. Tez. dokl. Kiev, 1981, p. 27 - 28. In Russian. – From Ref. zh., 51. Astron., 9.51.396 (1981).

077.048 Quasi-periodic pulsations of radio emission of a proton region on the sun in July 1974 at 3.5, 2.5 and 1.9 cm.
V. I. Abramenko, N. N. Eryushev, L. I. Tsvetkov.
13-ya Vses. konf. po radioastron. issled. soln. sistemy, Kiev, 1918. Tez. dokl. Kiev, 1981, p. 28. In Russian. – From Ref. zh., 51. Astron., 9.51.397 (1981).

077.049 Oscillating characteristics of local solar radio sources in the cm-range. N. P. Stasyuk.
13-ya Vses. konf. po radioastron. issled. soln. sistemy, Kiev, 1981. Tez. dokl. Kiev, 1981, p. 29 - 30. In Russian. – From Ref. zh., 51. Astron., 9.51.398 (1981).

077.050 On some results of fluctuation studies of solar radio emission at f = 755 and 612 MHz for June - September 1979.
E. A. Aver'yanikhina, G. Ozolins, M. Paupere, M. Eliass.
13-ya Vses. konf. po radioastron. issled. soln. sistemy, Kiev, 1981. Tez. dokl. Kiev, 1981, p. 23. In Russian. – From Ref. zh., 51. Astron., 9.51.399 (1981).

077.051 On the possible role of "superthermal" electrons in the origin of fluctuations of cm-radio radiation of the sun. S. I. Arbuzov, O. A. Shejner.
13-ya Vses. konf. po radioastron. issled. soln. sistemy, Kiev, 1981. Tez. dokl. Kiev, 1981, p. 31. In Russian. – From Ref. zh., 51. Astron., 9.51.400 (1981).

077.052 On a possibility of using spectral peculiarities of S-component sources in the mm-range for forecast of the state of an active centrum. M. N. Ryabov.
13-ya Vses. konf. po radioastron. issled. soln. sistemy, Kiev, 1981. Tez. dokl. Kiev, 1981, p. 22 - 23. In Russian. – From Ref. zh., 51. Astron., 9.51.424 (1981).

077.053 Solar radio flux measurements during the total solar eclipse of 16 February 1980.
R. V. Bhonsle, S. K. Alurkar, S. S. Degaonkar, O. P. N. Calla, G. Raju, S. S. Rana, B. Lokanadham, B. Vijay Gopal.
Observations of the total solar eclipse of 16 February 1980, (see 003.009), p. 80 - 83 (1981).

077.054 VLA observations of positions of 6 centimeter burst peaks associated with hard X-ray burst spikes.
M. R. Kundu, M. Bobrowsky, T. Velusamy.
Astrophys. J., Vol. 251, 342 - 351 (1981).
Using the Very Large Array with a spatial resolution of 3″, the authors have produced two-dimensional, 10 second synthesis maps of four 6 cm λ bursts associated with multiply-spiked hard X-ray bursts (20–300 keV) and studied their positional, structural, and polarization characteristics.

077.055 Energetic electrons and plasma waves associated with a solar type III radio burst.
R. P. Lin, D. W. Potter, D. A. Gurnett, F. L. Scarf.
Astrophys. J., Vol. 251, 364 - 373 (1981).
The authors present detailed in situ observations from the ISEE 3 spacecraft of energetic electrons, plasma waves, and radio emission for the type III solar radio burst of 1979 February 17. The reduced one-dimensional distribution function $f(v)$ of the electrons is constructed as a function of time and discussed in detail.

077.056 Correlation between bandwidth and frequency drift velocity of intermediate drift bursts.
Ø. Elgarøy, O. Soldal.
Astron. Astrophys., Vol. 104, 99 - 100 (1981).
Bandwidth and frequency drift velocity has been mea-

sured on intermediate drift bursts observed at Oslo Solar Observatory. The two quantities are correlated. Possible implications of this finding are discussed.

077.057 Radio emission of the quiet sun and structure of local sources at 10 cm wavelength observed during the December 24, 1973 solar eclipse.
O. Alvarez, Yu. F. Yurovskij.
Soln. Dannye 1981 Byull., No. 6, p. 92 - 98 (1981). In Russian.

077.058 Excess emission of local sources of bipolar sunspot groups. Sh. B. Akhmedov, G. B. Gel'frejkh.
Soln. Dannye 1981 Byull., No. 6, p. 107 - 112 (1981).
An analysis of a large series of spectra of radio local sources in the centimeter wavelength region was made. It was found that the sources over a bipolar structure are capable of excess emission whose spectrum is of a growing pattern in this region ($\lambda \geqslant 4.5$ cm). Estimations show that this emission may be caused by synchrotron radiation of nonthermal electrons whose density is $n_e \leqslant 1 - 2$ cm^{-3}.

077.059 The accuracy of measurement while investigating the fine structure of the S-component spectrum of solar radio radiation in the 1 - 4 GHz region.
A. M. Gusejnov.
Tsirk. Shemakhinsk. Astrofiz. Obs., No. 68, p. 8 - 12 (1981). In Russian.
An analysis of the accuracy of spectrographic measurements of the fine structure of the S-component spectrum of the solar radio emission in the region 1 - 4 GHz is given. The main factors which influence the accuracy of measurements and methods of their elimination are shown.

077.060 Methods and results of investigations of the fine structure of the S-component spectrum of solar radio emission in the 8.9 - 9.8 GHz region.
A. R. Abbasov, A. M. Gusejnov, N. M. Svikhnushin.
Tsirk. Shemakhinsk. Astrofiz. Obs., No. 68, p. 17 - 23 (1981). In Russian.
The present paper is devoted to methods of investigations of the fine structure of the S-component spectrum of the solar radio emission in the region 8.9 - 9.8 GHz. The observational results obtained with a spectrograph are given.

077.061 Fast fine structures of a microwave outburst on May 16, 1981. R. Zhao, S. Jin.
Astron. Circ., No. 9, 3 pp. (1981). In Chinese and English.

077.062 The solar radio outbursts on April 27, 1981 observed at Purple Mountain Observatory.
Z.-y. Gu, J.-z. Sun, A.-h. Zhou.
Astron. Circ., No. 10, 4 pp. (1981). In Chinese and English.

077.063 Type-III solar bursts and radio emission of the circumterrestrial space. S. I. Musatenko.
Astron. Tsirk., No. 1140, p. 2 - 4 (1980). In Russian.

077.064 On solar centimeter burst observations with radio telescopes for measuring the shift of the effective radio emission centre.
M. Alibegov, Yu. Barabanshchikov, N. Cimahovic̆a.
Investigations of the sun and red stars. 11, (see 003.021), p. 43 - 61 (1980). In Russian.

077.065 A model of the emission mechanism of type I solar radio bursts. I. Suzuki.
Publ. Astron. Soc. Japan, Vol. 33, 505 - 519 (1981).
As an emission mechanism of type I solar radio bursts, the author proposes the decay process of Langmuir waves propagating along a static magnetic field into two circularly polarized electromagnetic waves. Wave kinetic equations

describing the decay process are obtained in random-phase approximation. With use of these equations, it is shown that such a mechanism gives very high brightness temperature of the order of 10^{11} K with acceptable parameters in type I bursts.

077.066 **Prediction of the radio emission indices of the sun in the frequency range $1000 \leqslant F \leqslant 3750$ MHz.**
J. N. Xanthakis, C. Poulakos.
Sol.-Terr. Predictions Proc., Vol. 3: Sol. Act. Predictions, p. A77 - A95 (1980) = Res. Cent. Astron. Appl. Math., Acad. Athens, Contrib. Ser. I (Astron.), No. 70.

Predicted mean annual and mean monthly values of the solar radiation flux at different wavelengths for the 21st solar cycle are presented. The forecast is derived from analytical relations between the relative sunspot numbers, R, and the radio-emission of the sun in the frequency range 1000 MHz $\leqslant F \leqslant 3750$ MHz. Two samples of predicted values of the Zürich relative sunspot numbers for the 21st solar cycle are used in this paper. The first is derived with the help of analytical relations between the Zürich relative sunspot numbers and the time of rise, T_R, while the second one is derived from a regression analysis of the cycles N = 8 to N = 20. The relationship between the Zürich numbers, R, and the radio flux densities at 2800 MHz and 3750 MHz - on a daily basis investigation - is briefly discussed.

077.067 **On a complex model of solar microwave radiation consistent with soft and hard X-ray burst emission.**
A Böhme, F. Fürstenberg, J. Hildebrandt, P. Hoyng, A. Krüger, O. Saal, G. A. Stevens.
HHI-STP-Rep. No. 6, 48 pp. (1976).

A two component (core-halo) emission model has been applied to reconcile the observed hard and soft X-ray burst emissions with the impulsive micorwave burst radiation. Electron energy distributions in the form of a Maxwellian distribution the tail of which was partly replaced by a power law in the core region and emission measures are derived from X-ray measurements and used as input quantities for model calculations of the resulting radio spectra. Information about the magnetic field in the source region required for the interpretation of both, X-ray and radio measurements is derived.

077.068 **Radio evidence for shock acceleration of electrons in the solar corona.** H. V. Cane, R. G. Stone,
J. Fainberg, R. T. Stewart, J. L. Steinberg, S. Hoang.
Geophys. Res. Lett., Vol. 8, 1285 - 1288 (1981).

A new class of kilometer wavelength solar radio bursts has been observed with the ISEE-3 Radio Astronomy Experiment. These events resemble groups of ordinary type III bursts but have some unique properties. It is proposed that this new class of bursts are the long wavelength continuation of herringbone structure and it seems probable that the electrons producing the radio emission are accelerated by shocks. The new type of events will be referred to as shock accelerated (SA) events. The characteristics of SA events are discussed.

077.069 **Diagnostics of current sheets on the sun by radio observations.**
V. D. Kuznetsov, S. I. Syrovatskij.
Izv. vuzov. Radiofiz., Tom 24, 395 - 400 (1981). In Russian.
Abstr. in Ref. zh., 51. Astron., 12.51.504 (1981).

077.070 **Some peculiarities of statistical distributions of solar radio bursts.** I. E. Pogodin.
Astron. Tsirk., No. 1152, p. 5 (1981). In Russian.

077.071 **Observation of cm-radio emission of the sun with the Josephson radiometer.**
V. S. Ehtkin, V. A. Il'in, K. Z. Fatykhov, V. M. Charugin.
Astron. Tsirk., No. 1153, p. 1 - 2 (1981). In Russian.

077.072 **Solar radio observations at 32 GHz by a large single dish.** H. Kumagai, S. Isozaki, C. Ouchi, K. Ohbu,
T. Isobe, E. Ouchi, T. Nishino, T. Hori, F. Takahashi.
Publ. Astron. Soc. Japan, Vol. 33, 485 - 494 (1981).

Solar radio observations at millimeter wavelength 9.5 mm (32 GHz) using a large steerable antenna have started at Hiraiso Branch, Radio Research Laboratories. The instruments for the observations are composed of a 10-m diameter Cassegrain antenna, low-noise parametric amplifiers, and a real-time data processing system. Two kinds of observational modes are available; one is the two-dimensional scanning mode of the solar disk and the other is the continuous tracking mode of an arbitrarily selected spot on the disk. The main objective of this paper is to present an overall feature of the hardware and the software systems developed. Further, some examples of the observational results are illustrated and brief discussions about some features on the solar radio map and the burst events at millimeter wavelength are given.

Graphic catalogue of spot groups, Carrington longitude – time – latitude, for cycle No. 20. I. Interaction of groups and flare occurrence with type II radio bursts.
See Abstr. 002.007.

Verzeichnis der Publikationen zur solaren Radio-astronomie 1954 - 1979. See Abstr. 002.085.

The Air Force RSTN *(Radio Solar Telescope Network)* **system.** See Abstr. 013.005.

On some possibilities to use bispectral analysis in the treatment of observations of solar radio emission fluctuations. See Abstr. 021.054.

The microwave solar radar experiment. I. Observations. See Abstr. 031.592.

Automatic method for determination of the variability index of the solar radio flux. See Abstr. 031.630.

On an improved method for measuring the variability of solar radio emission. See Abstr. 031.634.

Collisionless perpendicular shocks: applications to solar type II radio bursts and the Antares (α Sco) B radio emission. See Abstr. 062.013.

Identification of new solar OH lines in the 10 - 12 micron region. See Abstr. 071.014.

Simultaneous observations of AR 2490 performed in the X-ray, UV, optical, and radio wavelength domain. See Abstr. 072.049.

Solare Beobachtungsergebnisse. Solar data. Solar radio emission. 1981 January - August. See Abstr. 072.059.

Downward shift of the acceleration/injection region during solar flares. See Abstr. 073.012.

Flare observations. See Abstr. 073.033.

Runaway acceleration in a radio flare. See Abstr. 073.078.

Properties of solar flare electrons, deduced from hard X-ray and spatially resolved microwave observations. See Abstr. 073.088.

Closed coronal structures. III. Comparison of static

models with X-ray, EUV, and radio observations.
See Abstr. 074.002.

A new interpretation of James's solar radar echoes involving lower-hybrid waves.　See Abstr. 074.026.

Radio wave scattering observations of the solar corona: first-order measurements of expansion velocity and turbulence spectrum using *Viking* and *Mariner 10* spacecraft. See Abstr. 074.074.

Solar wind motion within 30 $R_\odot$: spacecraft radio scintillation observations.　See Abstr. 074.086.

Relationship between a soft X-ray Long Duration Event and an intense metric noise storm. See Abstr. 076.001.

Giant X-ray arches associated with post-flare radio noise storms.　See Abstr. 076.010.

Erratum

077.901　Erratum: 'Possible slowdown of the plasma wave sources associated with Type III radio bursts', [Astron. Astrophys., Vol. 99, L18 - L21 (1981)] G. V. de Genouillac, D. F. Escande. Astron. Astrophys., Vol. 101, 276 (1981). – See Abstr. 29.077.065 (1981).

078 Cosmic Radiation

078.001　Identification and characteristics of source flares emitting ^{3}He or Fe enriched cosmic rays in the period September to December 1977.　F. Fárník, S. Fischer, L. Křivský, B. Valníček, O. B. Likin, V. N. Lutsenko. Bull. Astron. Inst. Czechoslovakia, Vol. 32, 228 - 238 (1981).
During the observation of solar cosmic rays on the Prognoz 6 and Helios 1 and 2 spacecrafts, several events with anomalous composition of accelerated particles (higher abundance of ^{3}He or Fe nuclei) occurred. The source flares were identified and their characteristics described. The character of the X-ray emission accompanying the emission of the accelerated particles with an anomalous composition shows no pronounced difference from other flares.

078.002　On ion heating at cyclotron frequency as a cause of anomalies in solar cosmic ray composition. L. G. Kocharov, M. Slivka. Izv. AN SSSR. Ser. fiz., Tom 44, 2489 - 2495 (1980). In Russian. – Abstr. in Ref. zh., 51. Astron., 7.51.332 (1981).

078.003　On solar cosmic ray acceleration in impulsed electric fields arising during explosive reconnection of magnetic force lines.　S. V. Bulanov. Izv. AN SSSR. Ser. fiz., Tom 45, 571 - 575 (1981). In Russian. Abstr. in Ref. zh., 51. Astron., 8.51.476 (1981).

078.004　Formation of the spectrum and energetics of solar cosmic rays.　L. I. Miroshnichenko. Izv. AN SSSR. Ser. fiz., Tom 45, 588 - 591 (1981). In Russian. Abstr. in Ref. zh., 51. Astron., 8.51.495; 62. Issled. kosm. prostranstva, 8.62.374 (1981).

078.005　Event in solar particles from January 1 - 5, 1978. V. G. Kurt, Yu. I. Logachev, V. G. Stolpovskij, N. F. Pisarenko, M. Gros, L. Treguer. Izv. AN SSSR. Ser. fiz., Tom 45, 609 - 612 (1981). In Russian. Abstr. in Ref. zh., 51. Astron., 8.51.496 (1981).

078.006　Investigation of iron-rich events in the AYAKS experiment (22.IX.1977 - 26.I.1978). A. A. Kolchin, V. V. Lebedev, V. F. Levchenko, A. I. Repin, G. P. Skrebtsov, V. L. Shubin. Izv. AN SSSR. Ser. fiz., Tom 45, 592 - 608 (1981). In Russian. Abstr. in Ref. zh., 51. Astron., 8.51.498 (1981).

078.007　Solar and cosmic radiation 4.6×10^9 years ago. G. S. Anufriev. Izv. AN SSSR. Ser. fiz., Tom 45, 539 - 546 (1981). In Russian. Abstr. in Ref. zh., 62. Issled. kosm. prostranstva, 8.62.362 (1981).

078.008　Solar (flare) cosmic ray proton fluxes in the recent past.　T. R. Venkatesan, C. M. Nautiyal, J. T. Padia, M. N. Rao. Proc. Eleventh Lunar Planet. Sci. Conf., (see 012.020), p. 1271 - 1284 (1980).

078.009　Catalogue of solar events enriched with helium-3. L. G. Kocharov, G. E. Kocharov. Fiz.-tekh. inst. AN SSSR. Prepr., 1981, No. 706, 15 pp. In Russian. – Abstr. in Ref. zh., 51. Astron., 9.51.423 (1981).

078.010　Penetration of solar protons and alpha particles with energies above 1 MeV/nucleon into the polar caps. N. A. Vlasova, T. A. Ivanova, M. I. Panasyuk, Eh. N. Sosnovets. Kosm. Issled., Tom 19, 551 - 558 (1981). In Russian.

078.011　Spectrum of flare protons in the low-energy region. E. I. Dajbog, V. G. Kurt, V. G. Stolpovskij. Kosm. Issled., Tom 19, 704 - 711 (1981). In Russian.

078.012　On the quantitative correspondence of the solar proton flux measured outside and inside the earth's magnetosphere.　V. V. Klimenko. Kosm. Issled., Tom 19, 775 - 777 (1981). In Russian.

078.013　Some consequences of the application of solar gamma-astronomy to the study of solar cosmic rays. B. M. Kuzhevskij. Cosmic rays, (see 012.025), 1980, p. 263 - 267. – Abstr. in Ref. zh., 51. Astron., 10.51.377 (1981).

078.014　Solar cosmic rays in the system of solar-terrestrial relations.　L. I. Miroshnichenko. Probl. soln.-zemn. svyazej. Dokl. simpoz. KAPG, Ashkhabad, 1979. Ashkhabad, 1981, p. 42 - 62. In Russian. – Abstr. in Ref. zh., 51. Astron., 10.51.389 (1981).

078.015　On the possibility of detecting ^{2}H and ^{3}H isotopes in the composition of solar energetic particles.

V. N. Lutsenko, N. S. Nikolaeva.
Cosmic rays, (see 012.025), 1980, p. 251 - 262. In Russian.
Abstr. in Ref. zh., 51. Astron., 10.51.390 (1981).

078.016   Solar and cosmic radiation 4.6×10^9 years ago.
G. S. Anufriev.
Izv. AN SSSR. Ser. fiz., Tom 45, 539 - 546 (1981). In Russian.
From Ref. zh., 51. Astron., 10.51.394 (1981).

078.017   Proton flares in 1978: quantitative diagnostics from
          radio bursts and results of direct measurements with
the Meteor satellite.
S. I. Avdyushin, V. D. Kozlovskij, M. N. Nazarova,
N. K. Pereyaslova, I. E. Petrenko, S. T. Akin'yan,
V. V. Fomichev, I. M. Chertok.
Fiz. soln. aktivnosti, Moskva, 1980, p. 93 - 117. In Russian.
From Ref. zh., 51. Astron., 10.51.402 (1981).

078.018   Survey on experimental data for heavy nuclei-rich
          solar events in the 1971 - 1978 period.
A. V. Orishchenko.
Izv. AN SSSR. Ser. fiz., Tom 45, 1174 - 1188 (1981). In
Russian. – Abstr. in Ref. zh., 51. Astron., 11.51.481 (1981).

078.019   Cosmogenic nuclides as an indirect method for
          cosmic ray investigation.   P. Povinec.
Izv. AN SSSR. Ser. fiz., Tom 45, 1141 - 1150 (1981). In
Russian. – Abstr. in Ref. zh., 51. Astron., 11.51.483 (1981).

078.020   On the influence of the interplanetary magnetic
          field on solar cosmic ray propagation.
A. T. Filippov.
Cosmic ray variations and solar wind, (see 003.013), p. 66 - 76
(1980). In Russian. – From Ref. zh., 51. Astron., 11.51.487
(1981).

078.021   Dynamics of solar cosmic ray events: processes at
          large heliocentric distances ($\geqslant 1$ AU).
D. C. Hamilton.
Cosmic rays in the heliosphere, (see 012.059), p. 25 - 40
(1981).
    Four spacecraft have now travelled beyond the orbit of
Jupiter – Pioneer 10/11 and Voyager 1/2 – and are producing
a growing body of distant observations of solar cosmic ray
events. Initial studies using Pioneer 10/11 data out to ~6 AU
interpreted flare particle observations in terms of a diffusion
model. The fact that large events at very large distances can
last up to two solar rotations implies that solar wind stream
structure will also play a role in the event dynamics.

078.022   Long lasting energetic particle injection from a
          weak flare.   V. G. Kurt, Yu. I. Logachev,
V. G. Stolpovskii (Stolpovskij), G. A. Trebukhovskaya,
T. I. Gombosi, K. Kecskeméty, A. J. Somogyi.
Cosmic rays in the heliosphere, (see 012.059), p. 69 - 72
(1981).
    A relatively weak solar cosmic ray event registered at
the Earth orbit following the flare of December 17, 1976 is
discussed. The main feature of the event is the existence of a
prolonged unusually high proton and electron anisotropy. The
durations of proton and electron anisotropies were different.
Time-intensity and anisotropy profiles of electrons and protons are fitted by a diffusive model including prolonged particle injection at the Sun.

078.023   Temporal variations of the cosmic ray intensity and
          the magnetic configurations of the heliosphere.
H. S. Ahluwalia.
Cosmic rays in the heliosphere, (see 012.059), p. 151 - 154
(1981).
    Data from worldwide network of neutron and muon

detectors are used to study 11-year variations of cosmic rays,
over four solar activity cycles. The author finds that the recovery of the cosmic ray intensity follows one of the two
distinct modes. During odd cycles recovery is completed in
6 to 8 years, but during even cycles complete recovery occurs
in 2 to 3 years. Two model magnetic configurations of the
heliosphere are proposed to understand these recovery modes.
Implications of these models are also discussed.

078.024   Identification and characteristics of source flares
          emitting 3-He or Fe enriched cosmic rays in the
period September to December 1977.   F. Fárník, S. Fischer,
L. Křivský, B. Valníček, O. B. Likin, V. N. Lutsenko.
High-energy astrophysics, (see 012.067), 243 - 246 (1981).
    During the observation of solar cosmic rays on the
Prognoz 6 and Helios 1 and 2 spacecrafts, several events with
anomalous composition of accelerated particles (higher abundance of 3-He or Fe nuclei) occurred. The authors found seven
such events from the period September to December 1977 for
which data from the Prognoz 6 solar X-ray photometer are
available. This material together with published optical and
radio data from terrestrial observatories enabled them to
identify more reliably the source flares and describe their
characteristics.

          A transient MHD model applicable for the source
of solar cosmic ray acceleration.   See Abstr. 062.126.

          Quasistationary outflow of gases from solar active
regions.   See Abstr. 072.072.

          A study of active regions generating flares with
high-energy protons and their forecast.
See Abstr. 073.064.

          On compositional variations of heavy ions during
solar particle events.   See Abstr. 073.109.

          Corotating events in the energy range 4–13 MeV as
observed on board Helios 1 and 2 in 1975 and 1976.
See Abstr. 074.118.

          Correlations of solar microwave bursts with prompt
$E > 20$ MeV proton events.   See Abstr. 077.020.

          A 1600 year long record of solar change derived
from atmospheric ^{14}C levels.   See Abstr. 085.040.

          Detection of stopped solar flare helium in lunar
rock 68815.   See Abstr. 094.503.

          Record of the solar corpuscular radiation in minerals
from lunar soils: a comparative study of noble gases and
tracks.   See Abstr. 094.560.

          Interplanetary propagation of < 1 MeV protons in
nonimpulsive energetic particle events.
See Abstr. 106.003.

          Numerical study of solar flare particle propagation
in the heliosphere.   See Abstr. 106.050.

          Cosmic ray intensity variations and two types of
high speed solar streams.   See Abstr. 143.029.

          Time-dependent Green's functions of the cosmic ray
equation of transport.   See Abstr. 143.073.

          Propagation modes of energetic charged particles in
the heliosphere.   See Abstr. 143.109.

079 Solar Eclipses

079.001 **Observing the sun during eclipses.**
M. R. Kundu.
Mercury, Vol. 10, 108 - 111, 126 (1981).

079.002 **In de schaduw van de maan.** J. Houtgast.
Zenit, 8. Jaarg., 444 - 452 (1981).

Solar eclipse 1972 July 10

079.101 **Observation of the extreme solar limb at 3.9 μm during the partial solar eclipse of 10 July, 1972.**
T. A. Clark, R. W. Clay.
Astron. Astrophys., Vol. 100, 254 - 257 (1981).
  A simple spatially-chopping near-infrared photometer was used to monitor the 4th contact of the partial solar eclipse of July 10, 1972 at 3.9 μm. The eclipse curve agrees with that derived from the data of Léna (1970) and Johnson (1971) to $\mu = \cos\theta = 0.15$, but departs rapidly from their curves, showing extreme limb darkening to $\mu = 0.10$. The results also show tentative evidence for an enhanced peak of near infrared energy within 4″ of the limb, the reality of which is discussed with respect to other such observations.

Solar eclipse 1983 June 11

079.201 **Predictions for the Indonesian total solar eclipse of 1983.** F. Espenak.
Strolling Astron., Vol. 29, 30 - 38 (1981).

Solar eclipse 1980 February 16

079.301 **Experiments during the total solar eclipse of 1980 February 16.**
A. Paranjpye, K. T. Ramesh, M. Rao, B. N. Shridhar.
Bull. Astron. Soc. India, Vol. 9, 72 (1981). — Abstract.

079.302 **Observation of the total solar eclipse of 16 February 1980.** A. Kubicela.
Observations of the total solar eclipse of 16 February 1980, (see 003.009), p. 16 - 18 (1981).

079.303 **Observations of timings of the four contacts.**
A. Bandyopadhyay, A. K. Bhatnagar,
S. Sridharan, B. Dasgupta.
Observations of the total solar eclipse of 16 February 1980, (see 003.009), p. 48 (1981).

079.304 **Contact times.** T. Mori, T. Kanazawa.
  Observations of the total solar eclipse of 16 February 1980, (see 003.009), p. 49 (1981).

079.305 **Precise observations of contact times.** K. Sato.
  Observations of the total solar eclipse of 16 February 1980, (see 003.009), p. 50 (1981).

079.306 **Report on eclipse observation.**
R. K. Bhattacharyya.
Observations of the total solar eclipse of 16 February 1980, (see 003.009), p. 105 - 106 (1981).

079.307 **Photographing the coronal and prominent features of solar eclipse of 16 February 1980 in colour.**
H. Gunawardena.
Observations of the total solar eclipse of 16 February 1980, (see 003.009), p. 107 (1981).

079.308 **Observation of the total solar eclipse on 16 February 1980.** R. Lalrosanga.
Observations of the total solar eclipse of 16 February 1980, (see 003.009), p. 108 - 109 (1981).

079.309 **Variation in the brightness of the sun.**
J. E. S. Singh, J. Edwards, R. V. Singh, R. Singh.
Observations of the total solar eclipse of 16 February 1980, (see 003.009), p. 110 - 112 (1981).

Symposium on the total solar eclipse of 1980 February 16. See Abstr. 011.020.

Tagebuch einer Sonnenfinsternisexpedition nach Kenia. See Abstr. 011.051.

Observation and preliminary analysis of the chromospheric flash spectrum obtained during the total solar eclipse of Feb. 16, 1980. See Abstr. 073.083.

Solar radio flux measurements during the total solar eclipse of 16 February 1980. See Abstr. 077.053.

Study of variations in mesospheric structure during the total solar eclipse of 16 February 1980. See Abstr. 082.039.

Observations of atmospheric scattering in the ultra-violet during the solar eclipse. See Abstr. 082.040.

Study of atmospheric ozone variations in the stratosphere and mesosphere during the solar eclipse. See Abstr. 082.041.

Observation of the ionospheric total electron content during the solar eclipse of February 16, 1980. See Abstr. 083.003.

The behaviour of F_2 layer above Ruili during the solar eclipse of February 16, 1980. See Abstr. 083.004.

Ionospheric E region instabilities produced by a solar eclipse. See Abstr. 083.005.

Ionospheric effects of the total solar eclipse of 16 February 1980. See Abstr. 083.028.

Ionospheric electron content observations during the total solar eclipse of 16 February 1980. See Abstr. 083.029.

Detection of ionospheric polarization scintillations. See Abstr. 083.030.

Study of variations in the electron density profiles and the electrojet irregularities caused by the solar eclipse. See Abstr. 083.031.

Triggering of ionospheric irregularities by the solar eclipse. See Abstr. 083.032.

Effects of the solar eclipse of 16 February 1980 on the change in phase height and absorption at 2.2 MHz. See Abstr. 083.033.

Ionospheric effects of the total solar eclipse of 16 Feb. 1980 observed over Ahmedabad.
See Abstr. 083.035.

Analysis of the geomagnetic effect during the February 16, 1980 solar eclipse in Yunnan Province.
See Abstr. 084.011.

Solar eclipse effect on short period geomagnetic field variations.   See Abstr. 084.043.

A note on the geomagnetic observations made at Hyderabad and Etaiyapuram during the solar eclipse of 16 February 1980.   See Abstr. 084.044.

Solar eclipse effect on short-period geomagnetic field variations.   See Abstr. 084.084.

Solar eclipse 1980 February 4

079.401   Die tasmanische Sonnenfinsternis.
F. Dorst.
Sterne Weltraum, Jahrg. 20, 385 - 387 (1981).

Annual solar eclipse observed for solar radius determination.   See Abstr. 080.016.

Solar eclipse 1981 July 31

079.501   Darkness at midday.   G. Lovi, A. Seltzer.
Sky Telesc., Vol. 62, 319 - 321 (1981).

079.502   Reflections: solar eclipse 1981.   B. J. Bok.
Sky Telesc., Vol. 62, 322 - 323 (1981).

079.503   Sonnenfinsternis über Sibirien. Kein Anzeichen für eine schrumpfende Sonne.   M. Waldmeier.
Umschau, 81. Jahrg., 631 (1981).

079.504   Solar eclipse.   E. Levitan.
Nauk. i zhizn', 1981, No. 4, p. 99 - 103. In Russian.
Abstr. in Ref. zh., 51. Astron., 9.51.100 (1981).

079.505   Totale zonsverduistering in Rusland.
D. W. Jannink.
Zenit, 8. Jaarg., 526 (1981).

Solar eclipse 1968 September 22

079.601   Observations of the 22 September, 1968 solar eclipse in Xinjiang.
Acta Astron. Sinica, Vol. 22, 254 - 264 (1981). In Chinese.

080 Atmosphere, Figure, Internal Constitution, Neutrinos, Rotation, etc.

080.001 **A new measure of the solar rotation.**
D. O. Gough.
Mon. Not. R. Astron. Soc., Vol. 196, 731 - 745 (1981).

It is suggested that the fluctuations in the Princeton solar oblateness data, recently reanalysed by Dicke, may result from rotational splitting of non-axisymmetrical oscillations. If this hypothesis is correct, the splitting frequency provides an integral measure of the internal rotation of the Sun. Upper and lower bounds to the solar gravitational quadrupole moment are computed under the assumption that a single mode of oscillation predominates. The results depend on the nature of that mode: if it is a p mode the upper bound may be greater than the direct estimates obtained from the value of the oblateness of the solar image; but if it is a g mode, the upper bound is considerably lower than the value Dicke and Goldenberg have claimed.

080.002 **Chaotic solar oscillations?**
S. Blacher, J. Perdang.
Mon. Not. R. Astron. Soc., Vol. 196, 109P - 113P (1981).

A numerical experiment on Hamiltonian oscillations demonstrates the existence of chaotic motions which satisfy the property of phase coherence. It is observed that the low-frequency end of the power spectrum of such motions is remarkably similar in structure to the low-frequency SCLERA spectra. Since the smallness of the observed solar amplitudes is not a sufficient mathematical ground for inefficiency of non-linear effects the possibility of chaos among solar oscillations cannot be discarded a priori.

080.003 **Solar oscillations and limb darkening fluctuations.**
R. Yerle.
Astron. Astrophys., Vol. 100, L23 - L25 (1981), with a correction, Vol. 103, 428 (1981).

High speed scanning of the apparent solar limb darkening profile shows fluctuations of the maximum brightness gradient with frequencies found by other investigators. These results cannot be interpreted as terrestrial atmospheric transparency but rather as a brightness redistribution of the apparent limb darkening.

080.004 **Solar spin variation.**   D. W. Hughes.
Nature, Vol. 292, 668 (1981).

080.005 **Solar activity and solar neutrino flux.**
L. J. Lanzerotti, R. S. Raghavan.
Nature, Vol. 293, 122 - 124 (1981).

Among discussions of the implications of the Davis experiment to detect solar neutrinos, there have been several suggestions that manifestations of solar activity are related to the flux of the neutrinos. These suggestions have been speculative because the available solar neutrino data covered only a fraction of a solar cycle. The authors now re-examine the proposals using data over a significantly longer time base and find no evidence for either of these possibilities.

080.006 **Transformation of magnetogravitational waves in the solar atmosphere.**
Yu. D. Zhugzhda, N. S. Dzhalilov.
Astron. Zh., Tom 58, 838 - 847 (1981). In Russian. English translation in Soviet Astron., Vol. 25, No. 4.

The theory of mutual transformation of all types of magnetogravitational waves in an isothermal atmosphere is developed. The transformation coefficients allowed to compute the spectra of a slow wave in the case when the spectrum of sound noise in the convective zone is given. The transformation leads to an increase of the directivity of the slow waves and to the appearance of reflected slow modes of the calculat-

ed value of the horizontal wavelength, for which the transformation coefficient is maximum.

080.007 **Solar rotation measurements at Mount Wilson.**
II. Systematic instrumental effects and the absolute rotation rate.   B. J. LaBonte, R. Howard.
Sol. Phys., Vol. 73, 3 - 12 (1981).

Possible sources of systematic error in solar Doppler rotational velocities are examined. Scattered light is shown to affect the Mount Wilson solar rotation results, but this effect is not enough to bring the spectroscopic results in coincidence with the sunspot rotation. Interference fringes at the spectrograph focus at Mount Wilson have in two intervals affected the rotation results. It has been possible to correlate this error with temperature and thus correct for it. A misalignment between the entrance and exist slits is a possible source of error, but for the Mount Wilson slit configuration the amplitude of this effect is negligibly small. Rapid scanning of the solar image also produces no measurable effect.

080.008 **Internal structure of reconnecting current sheets and the emerging flux model for solar flares.**
A. M. Milne, E. R. Priest.
Sol. Phys., Vol. 73, 157 - 181 (1981).

The authors present a steady-state model for reconnecting current sheets, which relates the central values of temperature, density and pressure within the sheet to the prescribed external values of these parameters as well as the magnetic field strength and inflow velocity (or reconnection rate). The simplifying feature of the authors' model is the assumption of quasi-one-dimensionality so that only variations across the sheet at the centre of symmetry are considered in detail. The results are applied to the emerging flux model for solar flares.

080.009 **On the presence of electric currents in the solar atmosphere. I. A theoretical framework.**
M. Hagyard, B. C. Low, E. Tandberg-Hanssen.
Sol. Phys., Vol. 73, 257 - 268 (1981).

The authors present a simple framework of analysis aimed at deducing information on the electric currents in the solar atmosphere based on data of the photospheric magnetic vector field. The basic ideas are described with theoretical illustrations.

080.010 **Solar radius variations over the past 265 years.**
R. L. Gilliland.
Astrophys. J., Vol. 248, 1144 - 1155 (1981).

Analysis of five different data sets including meridian circle observations, timings of transits of Mercury, and durations of total solar eclipses consistently suggests the presence of a 76 year modulation of the solar radius. The last solar radius maximum occurred in about A.D. 1911, and the half-amplitude of variation is $\sim 0''.2$. The 76 year radius variation is negatively correlated with the Gleissberg cycle of the mean sunspot numbers. A more significant correlation between the solar radius and sunspot cycle is found at 11 years. Again the correlation is negative. The half-amplitude of the 11 year variation is $\sim 0''.1$. A secular decrease of $\sim 0''.1$ per century over the last 265 years is also likely from an objective analysis of available data.

080.011 **Rapid rotation of the solar interior.**
A. Claverie, G. R. Isaak, C. P. McLeod,
H. B. van der Raay, T. Roca Cortes.
Nature, Vol. 293, 443 - 445 (1981).

A measurement of the splitting of the discrete lines in the 5-min oscillations of the solar surface produced experimental evidence for the rapid internal rotation of the sun. These data demonstrate that the sun does not rotate uniformly but that the 'core' of the sun rotates 2 - 9 times as rapidly as the observed surface rotation. The number of components into which the lines are split also allows unambiguous identification of 11 examples of each of the $l = 0$, $l = 1$ and $l = 2$ modes in the frequency range 2.40 - 3.85 mHz.

080.012   **Temperature minimum heating in solar flares by resistive dissipation of Alfvén waves.**
A. G. Emslie, P. A. Sturrock.
Bull. American Astron. Soc., Vol. 13, 543 (1981). — Abstract.

080.013   **Mixing and low-Z solar models.**   W. B. Fechner.
Bull. American Astron. Soc., Vol. 13, 544 (1981). Abstract.

080.014   **Solar rotation.**   B. J. LaBonte.
Bull. American Astron. Soc., Vol. 13, 548 (1981). Abstract.

080.015   **Recent global scale solar oscillations and magnetic field observations.**   P. H. Scherrer.
Bull. American Astron. Soc., Vol. 13, 548 - 549 (1981). Abstract.

080.016   **Annual solar eclipse observed for solar radius determination.**   A. D. Fiala, D. Herald, D. W. Dunham.
Bull. American Astron. Soc., Vol. 13, 552 (1981). — Abstract.

080.017   **Solar radius variations over the past 265 years.**
R. L. Gilliland.
Bull. American Astron. Soc., Vol. 13, 553 - 554 (1981). Abstract.

080.018   **Effects of diffusion and mass flows on C IV, Si IV, He I and He II resonance lines formed in the solar atmosphere.**   R. Roussel-Dupré.
Bull. American Astron. Soc., Vol. 13, 554 (1981). — Abstract.

080.019   **Solar diameter variations.**   T. P. Caudell.
Bull. American Astron. Soc., Vol. 13, 559 - 560 (1981). — Abstract.

080.020   **Lectures on theoretical aspects of magnetic energy storage and conversion: their relationship to solar activity.**   D. S. Spicer.
Activity and outer atmospheres of the sun and stars, (see 012.012), p. 89 - 155 (1981).
Contents: Introduction. Global electrodynamic coupling and magnetic energy storage. Magnetic free energy dissipation mechanisms. The preflare state and flare triggers.

080.021   **Pulsations and oscillations.**   F.-L. Deubner.
The sun as a star, (see 003.004), p. 65 - 84 (1981).
This paper deals with the g-modes (gravitational restoring force), p-modes (pressure restoring force), and relatively high frequency acoustic waves that have been observed, or reportedly observed, in the solar atmosphere.

080.022   **Global circulation and the solar dynamo.**
P. A. Gilman.
The sun as a star, (see 003.004), p. 231 - 252 (1981).
This paper considers the latest theoretical efforts to model the solar, large-scale differential rotation and convection and also the latest models for the solar dynamo, which is thought to be responsible for generating the observed magnetic field.

080.023   **The sun as a star.**   R. J. Rutten, L. E. Cram.
The sun as a star, (see 003.004), p. 473 - 512 (1981).
There is a new solar-stellar connection, emphasizing astrophysical processes. It differs from the solar-stellar connection of old, which concentrated on the equilibrium structure of stellar photospheres. The authors first review the old connection before outlining the new connection and its observational and interpretational consequences.

080.024   **On a method for investigation of solar fluctuations on the basis of flares.**   A. I. Laptukhov.
Issled. kosm. plazmy. 2-ya Nauchn. konf. IZMIRAN, 1979. Moskva, 1980, p. 12 - 17. In Russian. — Abstr. in Ref. zh., 51. Astron., 7.51.291 (1981).

080.025   **On fluctuations of the solar atmosphere.**
V. I. Afanas'eva, A. I. Laptukhov, P. P. Pavlov.
Issled. kosm. plazmy. 2-ya Nauchn. konf. IZMIRAN, 1979. Moskva, 1980, p. 5 - 11. In Russian. — Abstr. in Ref. zh., 51. Astron., 7.51.292 (1981).

080.026   **On short-period global oscillations of the sun.**
B. A. Ioshpa.
Fiz. soln. aktivnosti. Moskva, 1980, p. 152 - 162. In Russian. Abstr. in Ref. zh., 51. Astron., 8.51.427 (1981).

080.027   **Oscillator mechanism of thermal pulsations of the sun.**   E. A. Gavryuseva, Yu. S. Kopysov.
Kratk. soobshch. po fiz., 1981, No. 2, p. 14 - 19. In Russian. Abstr. in Ref. zh., 51. Astron., 8.51.428 (1981).

080.028   **The rotational effect of light pressure and disintegration of circumsolar matter.**
V. V. Radzievskij, A. V. Artem'ev.
Gor'kov. gos. ped. inst. Gor'kij, 1981, 11 pp. In Russian. Abstr. in Ref. zh., 51. Astron., 8.51.101 (1981).

080.029   **Le Soleil.**   M. Pick.
Histoire de l'univers, (see 003.007), p. 151 - 173 (1980).
Contents: L'atmosphère solaire. L'activité solaire: les régions actives. L'activité solaire: les éruptions.

080.030   **Solar oscillations as evidence for neutrino mass.**
L. Paternò.
Mem. Soc. Astron. Italiana, Vol. 52, (see 012.023), 471 - 473 (1981).

080.031   **Problems with solar oscillations.**   D. Gough.
Nature, Vol. 293, 703 - 704 (1981).

080.032   **The missing solar neutrinos.**
H. P. Trautvetter.
Comments Nucl. Part. Phys., Vol. 10, 123 - 129 (1981). Abstr. in Phys. Abstr., Vol. 84, Abstr. 88871 (1981).

080.033   **Calibration of the color-temperature relations. The (B—V) color of the Sun.**   M.-N. Perrin.
Ann. Physique, Vol. 6, (see 012.026), 115 - 120 (1981). In French. — Abstr. in Phys. Abstr., Vol. 84, Abstr. 94296 (1981).

080.034   **Internal gravity waves in the solar atmosphere. I. Adiabatic waves in the chromosphere.**
B. W. Mihalas, J. Toomre.
Astrophys. J., Vol. 249, 349 - 371 (1981).
This paper begins with an investigation of the propagation of linear and adiabatic internal gravity waves in a realistic solar atmosphere. Then the authors consider nonlinear processes in these waves, and develop, in terms of their linear results, criteria that will be used to estimate wave-breaking heights in the chromosphere.

080.035   **Konvektion in der Sonnenatmosphäre.**
C. J. Durrant, A. Nesis.
Sterne Weltraum, Jahrg. 20, 405 - 408 (1981).

080.036 Solar luminosity variation. III. Calcium K variation from solar minimum to maximum in cycle 21.
O. R. White, W. C. Livingston.
Astrophys. J., Vol. 249, 798 - 816 (1981).

Measurement of the full disk Ca II H and K profiles has now been completed from minimum to maximum in the current solar cycle 21. The central intensity of the K line increases by 30% on the average, but a peak change of +40% was recorded near the maximum of solar activity in late 1979. The 1 Å K index shows a corresponding but smaller increase of 18%. These changes in the Sun viewed as a star appear to be directly attributable to the occurrence of solar plages on the visible solar hemisphere. Similar measurements of a 1' $\times$ 3' quiet region at disk center show no significant systematic variability. This suggests that the quiet network has no long-term variability in the rising phase of the solar cycle. Ca II K line widths and asymmetries also show systematic changes associated with the degree of solar activity. In general, spectral features formed above the temperature minimum are all closely correlated in their variability, but they are not well related to changes in the strength of narrow photospheric lines and the K_1 wings. Ca II variability correlates very closely with the plage index, the Zurich sunspot number, and the Ottawa 10 cm flux measurements.

080.037 Reflections on solar variability.   C. de Jager.
Sol. Phys., Vol. 74, (see 012.029), 11 - 17 (1981).

The sun is a variable star in many respects: there are secular variations related to the general solar evolution, and — as discovered in recent years — there are many short period variations of which the 5 min and 160 min pulsations are the most important ones. Magnetic fields in the outer convective mantle play a dominant role in the 22 years cycle. The origin of solar magnetic variability is a key problem in astrophysics. The relation between solar variability and changes in the Earth's climate is a crucial element in climatology.

080.038 Solar interior structure and luminosity variations.
D. O. Gough.
Sol. Phys., Vol. 74, (see 012.029), 21 - 34 (1981).

The assumptions of standard solar evolution theory are mentioned briefly, and the principle conclusions drawn from them are described. The result is a rationalization of the present luminosity and radius of the Sun. Because there is some uncertainty about the interior composition of the Sun, a range of models is apparently acceptable.

080.039 Quasi-biennial periodicity in the solar neutrino flux and its relation to the solar structure.
K. Sakurai.
Sol. Phys., Vol. 74, (see 012.029), 35 - 41 (1981).

By analysing the observed results on the neutrino flux from the Sun for the years 1970 - 1978, it is shown that the production rate of the neutrinos at the central core of the Sun had been varying with a period almost equal to 26 months for these years. This so-called "quasi-biennial" periodicity in this rate suggests that the physical state of the central core of the Sun must have been modulated with this period through the variation of physical parameters as temperature and the chemical composition at the central core of the Sun. An idea to interpret this observed periodicity is thus proposed by taking the variations of these parameters into consideration. Some supporting evidence on this periodicity can be found on the variations of the solar activity as the relative sunspot numbers and the equatorial rotation speed of the Sun.

080.040 Solar oscillations: past, present, and future.
G. R. Isaak.
Sol. Phys., Vol. 74, (see 012.029), 43 - 49 (1981).

Observation of global oscillations of the Sun constitutes a primitive seismology of the solar interior. After a brief historical review emphasizing global velocity spectroscopy an account is given of the present status of the observations of global oscillations in the range of periods of 3 to 160 min. Finally the future capabilities of the observational techniques and their resultant potential is discussed.

080.041 Structure of the 5-minute solar oscillations 1976 - 1980.
A. Claverie, G. R. Isaak, C. P. McLeod, H. B. van der Raay, T. Roca Cortès.
Sol. Phys., Vol. 74, (see 012.029), 51 - 57 (1981).

The discrete structure in the 5 min velocity oscillations of the solar surface has been confirmed by a re-analysis of data obtained between 1976 and 1979, and in addition a preliminary analysis of 1980 data show excellent consistency of the determined frequencies over the five year period. It is further shown that atmospheric transparency, as measured by the power in the solar intensity fluctuations, shows no correlation with the measured amplitude of the velocity fluctuations, over 2 orders of magnitude.

080.042 Solar pulsations observed from the geographic South Pole: initial results.
E. Fossat, G. Grec, M. Pomerantz.
Sol. Phys., Vol. 74, (see 012.029), 59 - 63 (1981).

Initial results of round-the-clock observations of solar oscillations at the South Pole are briefly summarized.

080.043 Present state of the study of the 160-minutes solar oscillations.
A. B. Severny (*Severnyj*), V. A. Kotov, T. T. Tsap.
Sol. Phys., Vol. 74, (see 012.029), 65 - 71 (1981).

Global oscillation of the Sun with a period of 160 min were first discovered in 1974 and since observed in Crimea during the last 6 years. The average amplitude of the oscillation is about 0.5 m s^{-1}. The phase shows remarkable stability at the period 160.010 min. The oscillation is probably accompanied by synchronous fluctuations in the IR brightness and radio-emission of the Sun, and exhibits a dependence of the amplitude on the phase of solar rotation (with a peak of power at 27.2 days). The authors also find evidence in favour of a discrete spectrum within the 5 min global oscillations of the Sun, with an average splitting of about 69.5 μHz in frequency. Strict gas-dynamical equations being solved in the adiabatic approximation for a polytropic sphere $n = 3$ display the pattern of radial oscillations with wave packets separated by 120 min time-intervals filled with high frequency (and split by 117 μHz) oscillations implying a similarity with the observed pattern.

080.044 Short-period intensity fluctuations of integral sunlight.   A. Claverie, G. R. Isaak, C. P. McLeod.
Sol. Phys., Vol. 74, (see 012.029), 73 - 78 (1981).

An attempt has been made to detect short-period solar luminosity fluctuations in the vicinity of 5 min, analogous to the observed velocity oscillation. Using silicon photodiodes to monitor integral sunlight, an upper limit for the amplitude of the intensity fluctuations of 3×10^{-5} rms was found.

080.045 On the non-symmetric solar dynamo.
I. K. Csada.
Sol. Phys., Vol. 74, (see 012.029), 103 - 105 (1981).

The external field of the solar magnetohydrodynamic dynamo is expressed in terms of spherical harmonics and in powers of $1/r$. The non-symmetric dynamo is stabilized by a φ-dependent rotational oscillation which interacts with the magnetic field, thus compensating for Ohmic loss. As a consequence, the axis of a dipole wave is found to move on a great circle, with revolution time equal to the magnetic cycle.

080.046 The influence of the angular velocity distribution on the energy transport in the Sun's convection zone.
G. Belvedere, G. La Rosa, L. Paternò.
Sol. Phys., Vol. 74, (see 012.029), 107 - 110 (1981).

This contribution is a purely exploratory search to investigate the way the distribution of angular velocity inside the Sun's convection zone affects the energy transport. This is related with problems concerning the magnetic activity, whose appearance at the Sun's surface depends on the shape of the isorotation surfaces, and the latitudinal variations in flux. The proposed model is non-linear and axisymmetric.

080.047   **The age dependence of photospheric tracer rotation.**
M. Ternullo, R. A. Zappalà, F. Zuccarello.
Sol. Phys., Vol. 74, (see 012.029), 111 - 115 (1981).
From the analysis of the motions of sunspot groups recorded at Catania Astrophysical Observatory over a 7-years period from 1972 to 1978 the mean angular velocity as a function of latitude and age is calculated. The results suggest that the age of photospheric tracer (sunspot groups) affects the rotation curve slope. The implications of this result are discussed.

080.048   **Positions of sunspot groups and solar rotation.**
J. Tuominen, J. Kyröläinen.
Sol. Phys., Vol. 74, (see 012.029), 153 - 163 (1981).
The question is studied whether the one-year solar oscillation found by V. F. Chistyakov for the years 1965 - 1973 can be traced in the observations of sunspots of 1874 - 1971 published by Greenwich Observatory. The result is negative. But the study leads to the following two conclusions: (1) The average observable centres of gravity of spot groups are variably displaced towards the central meridian or towards the limb, the time scale of this variability being of the order of 70 years. Thus the angular velocity should be determined from recurrent groups in transit of the central meridian only. (2) The angular velocity will be smaller when determined from older spots.

080.049   **Solar radiation and its variation in time.**
C. Fröhlich, R. W. Brusa.
Sol. Phys., Vol. 74, (see 012.029), 209 - 215 (1981).
In order to assess the variability of the solar radiation, the record of determinations of the total and spectral solar irradiance of the last 15 years is analysed. Although the datapoints for the period before 1969 suggest a slight decrease, the uncertainties of these determinations are too large to render this statement significant. Together with the results of the following period (1969 - 1980) which show that within the uncertainty no change is detectable, it can be concluded, that the solar constant has not changed during the last 15 years. The same result is found from the spectral distribution record.

080.050   **Solar total irradiance observations by Active Cavity Radiometers.**   R. C. Willson.
Sol. Phys., Vol. 74, (see 012.029), 217 - 229 (1981).
Pyrheliometry, definition of the radiation scale in the International System of Units and monitoring the variability of solar total irradiance have been a focus of research at the Jet Propulsion Laboratory since the mid 1960's. A series of automated, electrically self-calibrating, cavity pyrheliometers known as Active Cavity Radiometers (ACR's) was developed as part of this program. ACR flight experiments have been conducted to determine the 1 AU total solar irradiance and monitor its variability in time.

080.051   **Improved data of solar spectral irradiance from 0.33 to 1.25 μ.**   H. Neckel, D. Labs.
Sol. Phys., Vol. 74, (see 012.029), 231 - 249 (1981).
The conversion of the authors' centre of disk intensities published in 1968/70 into mean disk intensities has been repeated, using more accurate data for the centre-to-limb variation of both continuous radiation and strong absorption lines. The random observational mean error of the new irradiance data very likely is not larger than 1.5% in the UV and not larger than 1% in the visible and infrared. Comparison with the fluxes of Sun-like stars observed by Hardorp (1980) confirms these errors and seems to exclude the possibility of a systematic, wavelength-dependent scale error which would correspond to a temperature difference larger than 50 K. The resulting integral value of the irradiance between 0.33 and 1.25 μ is 1.060, the corresponding value of the solar constant lies between 1.368 and 1.377 kW m^{-2}.

080.052   **Realistic calculations of solar-neutrino oscillations.**
V. Barger, K. Whisnant, R. J. N. Philips.
Phys. Rev. D, Vol. 24, 538 - 541 (1981). − Abstr. in Phys. Abstr., Vol. 84, Abstr. 98653 (1981).

080.053   **Effects of diffusion and mass flows on C IV and Si IV lines formed in the solar atmosphere.**
R. Roussel-Dupré, C. Beerman.
Astrophys. J., Vol. 250, 408 - 422 (1981).
A model for the transition region is derived from an initial interpretation of EUV observations, assuming ionization equilibrium and constant elemental abundance with height. The effects of diffusion and mass flows are then included in the initial model and the emergent profiles of several C IV and Si IV lines are computed. The authors find that diffusion and mass flows have a strong effect on both the emergent intensity and spectral shape of these lines. Diffusion acts to deplete the transition region of heavy ions to an extent which depends on the detailed temperature and density structure.

080.054   **On quasi-periodic oscillations in the solar atmosphere.**
A. G. Kosovichev, Yu. P. Popov.
Izv. Krymskoj Astrofiz. Obs., Tom 63, 15 - 24 (1981). In Russian. English translation in Bull. Crimean Astrophys. Obs., Vol. 63.
The propagation of oscillations in the convective zone and the external atmosphere, chromosphere and corona of the sun is studied by numerical solution of the nonlinear equations of one-dimensional gas dynamics. It is shown that the inhomogeneity of the solar atmosphere in the gravity field leads to generation of atmospheric oscillations with 300 s period. The influence of the nonlinearity on the oscillations is discussed and the results obtained are compared with observational data.

080.055   **Properties of a nonlinear solar dynamo model.**
N. I. Kleeorin, A. A. Ruzmaikin (*Ruzmajkin*).
Geophys. Astrophys. Fluid Dyn., Vol. 17, 281 - 296 (1981). Abstr. in Phys. Abstr., Vol. 84, Abstr. 102084 (1981).

080.056   **Some wave reflection problems in solar physics.**
J. A. Adam.
Irish Astron. J., Vol. 14, (see 012.034), 133 - 137 (1980).

080.057   **Solar neutrinos.**   J. N. Bahcall.
Nuclear astrophysics, (see 012.036), p. 111 - 123 (1981).
The author presents an overview of the subject of solar neutrinos, a brief summary of the theory of stellar evolution, a description of the main sources of solar neutrinos, a brief summary of the results of the Brookhaven ^{37}Cl experiment, an analysis of the principal new solar neutrino experiments that have been proposed, a discussion of how solar neutrino experiments can be used to detect the collapse of stars in the Galaxy, and finally, a description of how the proposed ^{71}Ga experiment can be used to test for charge non-conversation. The most important fact about the subject the author is reviewing is that there is a serious discrepancy between the standard theory and observation.

080.058   **The solar neutrino capture cross section for ^{81}Br.**
W. C. Haxton.
Nucl. Phys. A, Vol. A367, 517 - 525 (1981). − Abstr. in Phys. Abstr., Vol. 84, Abstr. 103173 (1981).

080.059   Seismology and geodesy of the Sun: solar geodesy.
R. H. Dicke.
Proc. Natl. Acad. Sci. USA, Vol. 78, 1309 - 1312 (1981).
Abstr. in Phys. Abstr., Vol. 84, Abstr. 108160 (1981).

080.060   Seismology and geodesy of the Sun: low-frequency oscillations.   R. H. Dicke.
Proc. Natl. Acad. Sci. USA, Vol. 78, 1989 - 1993 (1981).
Abstr. in Phys. Abstr., Vol. 84, Abstr. 108161 (1981).

080.061   Measurement of solar radius changes.
B. J. LaBonte, R. Howard.
Science, Vol. 214, 907 - 909 (1981).
Photoelectric solar radius measurements since 1974 at Mount Wilson show no change in the solar radius, with a limit of about 0.1 arc second (1 standard deviation), over the interval. The limit is set by residual systematic effects.

080.062   An improved search for large-scale convection cells in the solar atmosphere.
B. J. LaBonte, R. Howard, P. A. Gilman.
Astrophys. J., Vol. 250, 796 - 798 (1981).
A reanalysis of Mount Wilson solar velocity observations was made to search for giant cellular patterns. The reanalysis avoids several errors made in a previous search. No cells are detected with sensitivity of 3 to 12 m s^{-1} depending upon wavenumber. The observed amplitudes do not conflict with recent model predictions.

080.063   Currents on the solar surface.   M. Šolc.
Vesmír, Vol. 60, 298 - 301 (1981). In Czech.

080.064   Comparison of observed solar whole-disk oscillation frequencies with the predictions of a sequence of solar models.   J. Christensen-Dalsgaard, D. O. Gough.
Astron. Astrophys., Vol. 104, 173 - 176 (1981).
The frequencies of oscillation of whole-disk spectrum line shifts reported recently by Grec et al. (1980) and Claverie et al. (1980) and the frequencies of whole-disk intensity fluctuations reported by Deubner (1981) are compared with the eigenfrequencies ν_{Ai}, ν_{Bi}, ν_{Ci} of the sequence of three solar models discussed by Christensen-Dalsgaard et al. (1979). The comparison suggests that the sun is not deficient in heavy elements.

080.065   Quiet sun.   P. Delache.
Space Sci. Rev., Vol. 29, (see 012.043), 367 - 372 (1981).
The author underlines the diagnostic strength of recent observations of the oscillating quiet sun. While high quality (k, ω) power spectra permit a better knowledge of the convection zone, long and continuous survey of oscillations of the integrated sun provides an efficient sounding of the inner solar body.

080.066   Investigation of the empirical damping constant and iron abundance in the solar atmosphere from spectral line profiles.   B. T. Babij, L. I. Ben'ko.
Vestn. L'vov. univ. Ser. astron., 1981, No. 56, p. 3 - 8. In Russian. − Abstr. in Ref. zh., 51. Astron., 11.51.402 (1981).

080.067   Abundances of rare earth elements in the solar atmosphere.   M. M. Koval'chuk.
Vestn. L'vov. univ. Ser. astron., 1981, No. 56, p. 14 - 21. In Russian. − Abstr. in Ref. zh., 51. Astron., 11.51.403 (1981).

080.068   Fast motions in the solar atmosphere and magnetic field variations.   B. P. Filippov.
Pis'ma Astron. Zh., Tom 7, 748 - 751 (1981). In Russian. English translation in Soviet Astron. Lett., Vol. 7.
Models of solar flare matter ejections in the chromosphere and corona arising due to electromagnetic forces action are compared in respect of the parameters characterizing the magnetic field evolution.

080.069   Energy transfer by Alfvén waves in the solar atmosphere.   J. Žugžda, V. Locāns.
Investigations of the sun and red stars. 11, (see 003.021), p. 62 - 89 (1980). In Russian.
The propagation of Alfvén waves in an inhomogeneous atmosphere with inhomogeneous magnetic field is considered. General expressions for phase velocities, velocity of the energy transfer and the energy flux density for linear polarized waves have been found. The general expression for the energy transmission coefficient of Alfvén waves through a multi-layered atmosphere is obtained. Calculations of the energy transmission coefficient of Alfvén waves from the convection zone into the corona above a sunspot and the quiet atmosphere with magnetic field are presented. The multi-layered approximation is used for a model of the sunspot atmosphere and the quiet atmosphere.

080.070   Quasi-biennial periodicity in solar phenomena and its possible connection with internal processes in the sun.   K. Sakurai.
Proceedings of the 14th ISAS Lunar and Planetary Symposium, (see 012.055), p. 112 - 117 (1981).
The author analyses available observed solar neutrino fluxes and claims evidence for a quasi-biennial variation in the flux. He states that the fundamental period for the physical processes in the interior of the sun seems to be about 26 months.

080.071   Oscillations et sismologie solaire.   P. Ledoux.
C. R. Acad. Sci. Paris, Vie Acad., Tome 293, (see 012.056), I - V, VIII (1981).

080.072   Champs de vitesses et champs magnétiques à petite échelle dans l'atmosphère solaire.   J.-C. Pecker.
C. R. Acad. Sci. Paris, Vie Acad., Tome 293, (see 012.056), XV - XVII, XX (1981).

080.073   Il sole: davvero sempre più piccolo?   G. Picchio.
Orione, Vol. 2, 263 - 267 (1981).

080.074   Global oscillations of the sun.
A. B. Severnyj, V. A. Kotov, T. T. Tsap.
Issled. po geomagn., aehron. i fiz. Solntsa, Moskva, 1981, No. 56, p. 11 - 23. In Russian. − Abstr. in Ref. zh., 51. Astron., 12.51.432 (1981).

080.075   Free oscillations of the sun and giant planets.
S. V. Vorontsov, V. N. Zharkov.
Usp. fiz. nauk, Tom 134, 675 - 710 (1981). In Russian.
Abstr. in Ref. zh., 51. Astron., 12.51.566 (1981).

080.076   Observation of solar oscillations for periods close to 160 mn.   P. Delache.
C. R. Acad. Sci. Paris, Tome 293, Sér. II, 949 - 951 (1981). In French.
The observed spectrum of five-minute solar oscillations enables an almost complete prediction of the power spectrum that has been attributed, up to now, to longer period oscillations.

080.077   Two-year pulsations of the solar neutrino flux and seismic activity of the sun.   Yu. S. Kopysov.
Pis'ma v ZhEhTF, Tom 34, 289 - 292 (1981). In Russian.
Abstr. in Ref. zh., 51. Astron., 1.51.426 (1982).

080.078   Solar Maximum Mission experiment: initial observations by the Active Cavity Radiometer.
R. C. Willson, H. S. Hudson.
High-energy astrophysics, (see 012.067), p. 285 - 288 (1981).

The Active Cavity Radiometer on board the SMM is providing high-quality measurements of the solar irradiance. The observed standard deviation is in the range 10 - 15 parts per million in a 96-minute integration. Measurable solar variations occur on time scales of a few minutes to a few days. The total amplitude of the variations in the daily averages from February 16 to March 31, 1980, was 0.10% based upon 96-minute averages.

080.079 Solar oscillation, flare pattern, rise of leucopenia, and cycle of locust plagues.   T. Landscheidt.
9th International Congress of Biometeorology, D. Overdieck, J. Mueller, H. Lieth (Editors), Osnabrück 1981. ICR-13. Abstract.

080.080 On the problem of the structure of the stationary convective zone of the sun. I.
N. I. Kozhevnikov.
Astron. Tsirk., No. 1151, p. 6 - 8 (1981). In Russian.

080.081 On the problem of the structure of the stationary convective zone of the sun. II.
N. I. Kozhevnikov.
Astron. Tsirk., No. 1159, p. 5 - 7 (1981). In Russian.

080.082 The sun.   J. A. Eddy.
The new solar system, (see 003.026), p. 11 - 22 (1981).

Daytime star: the story of our sun.
See Abstr. 003.105.

Notre univers.   See Abstr. 003.114.

In the light of the Sun.   See Abstr. 003.160.

Landolt-Börnstein.   See Abstr. 003.172.

Future directions in ground-based optical observations.   See Abstr. 011.014.

Optical polarimeters for solar research.
See Abstr. 031.521.

Searching for $l = 1$ modes of solar oscillations.
See Abstr. 034.038.

Observations du soleil en 1979 à l'astrolabe du CERGA.   See Abstr. 041.011.

DISCO. Re-assessment study.
See Abstr. 051.052.

Direct capture cross sections at low energy.
See Abstr. 061.018.

Gravitational energy release induced by the nuclear energy generation processes: the resolution of the solar neutrino dilemma.   See Abstr. 061.027.

Dynamically consistent nonlinear dynamos driven by convection in a rotating spherical shell.
See Abstr. 062.036.

Penetrative convection.   See Abstr. 062.048.

Wave generation.   See Abstr. 062.049.

On parametric generation of acoustic-gravitational waves in the solar atmosphere.   See Abstr. 062.063.

Modulation instabilities in astrophysics.
See Abstr. 062.065.

Compressible convection in a rotating spherical shell. IV. Effects of viscosity, conductivity, boundary conditions, and zone depth.   See Abstr. 062.098.

Mechanical wave-energy flux in magnetoatmospheres: discrete and continuous spectra.   See Abstr. 064.028.

Oscillations and pulsations.   See Abstr. 064.031.

Periodic shifts of Fraunhofer lines in the spectrum of the center of the solar disk.   See Abstr. 071.007.

Photospheric limb darkening as a ground-based diagnostic for variations in the solar effective temperature. See Abstr. 071.018.

Dynamics of the solar photosphere.
See Abstr. 071.019.

Sunspot dynamics: gravitational draining – a cooling mechanism.   See Abstr. 072.002.

Propagation of waves in an atmosphere in the presence of a magnetic field. IV. Alfvén waves in sunspot umbrae.   See Abstr. 072.045.

Global velocity fields of the Sun and the activity cycle.   See Abstr. 072.050.

Swinging sun, 79-year cycle, and climatic change.
See Abstr. 072.073.

Diffuse separation of charges and electrical fields in the solar plasma.   See Abstr. 073.095.

Solar cycle Lorentz force waves and the torsional oscillations of the sun.   See Abstr. 075.003.

Bound oscillations on thin magnetic flux tubes: convective instability and umbral oscillations.
See Abstr. 075.011.

Finite bandwidth and scattered light effects on the radiometric determination of atmospheric turbidity and the solar constant.   See Abstr. 082.089.

Mass loss from cool stars.
See Abstr. 112.034.

On a connection of the north-south anisotropy of galactic cosmic rays with the sign of the general field of the sun.   See Abstr. 143.084.

On an analogy of some nonstationary processes in nuclei of extragalactic sources and on the sun. 2.
See Abstr. 158.256.

Erratum

080.901 Erratum: 'The solar gravitational figure $-J_2$ and J_4' [Astrophys. J., Vol. 246, 985 - 988 (1981)].
R. K. Ulrich, G. W. Hawkins.
Astrophys. J., Vol. 249, 831 (1981). – See Abstr. 29.080.060.

Earth

081 Structure, Figure, Gravity, Orbit, etc.

081.001 Geophysical evidence for non-newtonian gravity.
F. D. Stacey, G. J. Tuck.
Nature, Vol. 292, 230 - 232 (1981).
Large-scale measurements of G are important because the validity of the inverse square law of gravity at short range is being questioned. The authors have made such a series of measurements and have found other data sets in the literature that suffice for the estimation of G. They also report here a statistical analysis of 1,100 km² of overlapping sea floor and sea surface gravity data from the Gulf of Mexico. All these estimates of G give values that are higher than the conventional, laboratory-determined one.

081.002 Structure of the earth's inner core.
D. R. Fearn, D. E. Loper, P. H. Roberts.
Nature, Vol. 292, 232 - 233 (1981).
The authors propose that the interface separating the inner and outer core is dendritic and argue that the region in which freezing takes place may extend throughout the entire inner core.

081.003 Grand unification magnetic monopoles inside the earth.
F. J. Lowes, with a reply by R. A. Carrigan, Jr.
Nature, Vol. 292, 273 (1981).

081.004 Eruption mechanics on the Earth, Moon and Mars.
G. Wadge.
Nature, Vol. 292, 493 - 494 (1981).

081.005 The motion of the Earth-Moon system between 1700 and 2100 in Newcomb's theory and in JPL-ephemerides. P. Stumpff.
Astron. Astrophys., Vol. 101, 52 - 71 (1981).
The motion of the Earth-Moon barycenter (EMB) in Newcomb's (1898) theory is compared with the JPL ephemerides DE102 (relativistic) and DE28 (newtonian) with-in the 400-year interval 1700 - 2100. In the analysis, both the conventional and the new (IAU, 1976) precession theories are applied.

081.006 Eine astro-gravimetrische Berechnung des Quasigeoids für die Bundesrepublik Deutschland.
D. Lelgemann, D. Ehlert, H. Hauck.
Deutsche Geod. Komm. Bayerisch. Akad. Wiss., Reihe A, Höhere Geod., Heft Nr. 92, 67 pp. (1981).
Characteristics, computation formulas, data and results of a combined astro-gravimetric determination of the quasigeoid are described. The result of a least-squares adjustment are 786 coefficients of a series of so-called analytic kernel functions, which approximate the quasigoid in the area of the Federal Republic of Germany.

081.007 Intercomparisons of earth models by means of lumped coefficients.
J. Klokočník, L. Pospíšilová.
Planet. Space Sci., Vol. 29, 653 - 671 (1981).
A new procedure for comparing and testing the earth gravity field models by the order of their harmonic geo-

potential coefficients is explained. The differences in the lumped coefficients computed from the different earth models are presented as a set of 20 figures (for order of $6 \leqslant \beta \leqslant 15$ and the orbital inclination of $30° \leqslant I \leqslant 140°$). A statistical description of those dispersions is added, too.

081.008 Structure of the inner core inferred from observations of its spheroidal shear modes.
G. Masters, F. Gilbert.
Geophys. Res. Lett., Vol. 8, 569 - 571 (1981).

081.009 Modern geodetic Earth reference models.
B. H. Chovitz.
EOS Trans. American Geophys. Union, Vol. 62, No. 7, p. 65 - 67 (1981). – Abstr. in Phys. Abstr., Vol. 84, Abstr. 83103 (1981).

081.010 On the geoid determination by combination of gravity anomalies and a spherical harmonic model using integral formulae. H.-G. Wenzel.
Z. Vermessungswes., Vol. 106, 102 - 111 (1981). – Abstr. in Phys. Abstr., Vol. 84, Abstr. 83107 (1981).

081.011 Thermal regime of the Earth's outer core.
W. Ullmann, U. Walzer.
Pure Appl. Geophys., Vol. 119, No. 1, p. 59 - 79 (1980/81). Abstr. in Phys. Abstr., Vol. 84, Abstr. 83166 (1981).

081.012 Local and global gravity field representation.
E. Groten.
Rev. Geophys. Space Phys., Vol. 19, 407 - 414 (1981).

081.013 Stability investigations of various representations of the gravity field. M. Gerstl, R. Rummel.
Rev. Geophys. Space Phys., Vol. 19, 415 - 420 (1981).

081.014 Sampling function and finite element method representation of the gravity field.
H. F. Schmidt.
Rev. Geophys. Space Phys., Vol. 19, 421 - 436 (1981).
After introducing the fundamental method of interpolation and approximation of functions using sampling functions, some examples of one-dimensional models are demonstrated. Then a method of sampling functions for the representation and determination of the earth's gravity field – based on spherical harmonics and an arrangement of sampling points on a sphere – is described and discussed by investigating its mathematical implication. Additionally, this method is compared with the finite element method, with a view to the interpolation and modeling of the gravity field of the earth.

081.015 Free oscillations of the earth. R. Buland.
Annu. Rev. Earth Planet. Sci., Vol. 9, (see 003.003), 385 - 413 (1981).

081.016 Long wavelength gravity and topography anomalies.
A. B. Watts, S. F. Daly.
Annu. Rev. Earth Planet. Sci., Vol. 9, (see 003.003), 415 - 448 (1981).

The authors show that gravity and topography anomalies on the earth's surface may provide useful new information on deep processes occurring in the earth, such as those associated with mantle convection.

081.017 **Galactic cyclicity of geological processes.**
Yu. A. Bagdasarov.
Priroda, 1981, No. 8, p. 57 - 59. In Russian.

081.018 **Hypothesis of neutron irradiation of terrestrial proto-matter and isotopic ratios of some light elements.**   R. L. Kharus.
Yaderno-geofiz. metod. pri poiskakh i razvedke tverdykh polezn. iskopaemykh Urala. Sverdlovsk, 1980, p. 73 - 75. In Russian. – Abstr. in Ref. zh., 51. Astron., 8.51.220 (1981).

081.019 **A possible cosmic cause of large-scale disasters and catastrophes in the history of the earth.**
G. G. Polyakov.
Astrakhan. gos. ped. inst. Astrakhan', 1980. 20 pp. In Russian. Abstr. in Ref. zh., 51. Astron., 8.51.221 (1981).

081.020 **La planète Terre.**   V. Courtillot.
Histoire de l'univers, (see 003.007), p. 175 - 181 (1980).
Contents: Quelques caractéristiques. L'activité de la Terre. L'évolution de la Terre.

081.021 **Optimal condition for the study of the earth's motion by laser observations of Starlette.**
E. Proverbio, I. Orru'.
Mem. Soc. Astron. Italiana, Vol. 52, 249 - 273 (1981).

081.022 **Composition of the earth's upper mantle–II: Volatile trace elements in ultramafic xenoliths.**
J. W. Morgan, G. A. Wandless, R. K. Petrie, A. J. Irving.
Proc. Eleventh Lunar Planet. Sci. Conf., (see 012.020), p. 213 - 233 (1980).

081.023 **On the compensation of geoid anomalies due to subconducting slabs.**   D. C. McAdoo.
NASA Tech. Memo., NASA TM 82157, 22 pp. (1981).
Candidate models of the forces which oppose the sinking of slabs are all constrained to produce results consistent with the following observation: relative geoid highs, which one assumes are due to slabs, characteristically occur over subduction zones. This study has extended a published model of viscous corner flow in subduction zones in order to demonstrate that it can – in certain cases – produce the requisite geoid highs.

081.024 **Elastic and anelastic structure of the deep interior of the Earth.**   A. M. Dziewonski.
Mem. Soc. Astron. Italiana, Vol. 52, (see 012.023), 401 - 405 (1981).

081.025 **Difficulties in determining thermodynamic properties in the Earth's interior.**
F. Mulargia, E. Boschi.
Mem. Soc. Astron. Italiana, Vol. 52, (see 012.023), 455 (1981). Abstract.

081.026 **Chemistry of the earth and the role of primary and secondary objects for the formation of the inner planets and meteorite parent bodies.**   H. Wänke.
Mem. Soc. Astron. Italiana, Vol. 52, (see 012.023), 541 - 545 (1981).

081.027 **Seismic surface waves and free oscillations in a regionalized earth model.**
J. H. Woodhouse, A. M. Dziewonski.

Mem. Soc. Astron. Italiana, Vol. 52, (see 012.023), 559 - 562 (1981).

081.028 **Use of Cauchy integral analogs in the geopotential field theory.**   M. S. Zhdanov.
Ann. Géophys., Vol. 36, 447 - 458 (1980). – Abstr. in Phys. Abstr., Vol. 84, Abstr. 88256 (1981).

081.029 **Is the gravitational force constant?**
Yu. D. Bulanzhe.
Zemlya Vselennaya, 1981, No. 4, p. 10 - 14. In Russian.

081.030 **The chemical and biological evolution of the earth.**
E. Siliņš.
Zvaigžņotā debess, 1980. gada rudens, p. 10 - 17. In Latvian.

081.031 **On a self-consistent representation of earth models, with an application to the computing of internal flattening.**   C. Denis, A. Ibrahim.
Bull. Géod., Vol. 55, 179 - 195 (1981).
The authors establish algorithmic formulae, which may be used to compute all the mechanical properties of an earth model in an entirely consistent way, once the density as well as P– and S– wave velocities are known. They then use this formulation to integrate Clairaut's equation in a very efficient way, and thus obtain the hydrostatic flattening to the first order in smallness at any point inside the model. For most geodynamic purposes, one may suffice with this approximation.

081.032 **Picone's theorem and the convergence of the expansion in spherical harmonics of the gravitational potential of the Earth in the external space.**   K. Arnold.
Boll. Geofis. Teor. Appl., Vol. 22, 95 - 103 (1980). – Abstr. in Phys. Abstr., Vol. 84, Abstr. 98364 (1981).

081.033 **Investigation of the earth as a planet with the methods of astronomy, geodesy and geophysics.**
A. O. Korsun'.
Visnik AN URSR, 1981, No. 3, p. 83 - 85. In Ukrainian. Abstr. in Ref. zh., 51. Astron., 10.51.31 (1981).

081.034 **Present state and prospects of planetary geodynamical investigations.**   L. P. Pellinen.
Sovrem. dvizheniya zmn. kory. Teor., metod., prognoz. Moskva, 1980, p. 35 - 45. In Russian. – Abstr. in Ref. zh., 52. Geod. Aehrosemka, 10.52.76 (1981).

081.035 **Solving the shape of the earth by using digital density models.**   M. Heikkinen.
Rep. Finnish Geod. Inst., 81:2, 69 pp. (1981). ISBN 951-711-069-3, ISSN 0355-1962.

081.036 **A new method of computing geopotential fields.**
W. Moon.
Geophys. J. R. Astron. Soc., Vol. 67, 735 - 746 (1981).
A new method is proposed for the geopotential field computation and gravitational attraction modelling. The usual method is to use a uniform density discrete numerical integration to represent either the gravitational potential or the gravitational attraction from a given density configuration. In this paper, an interpolation scheme is explained, using a piecewise continuous basis function to represent the arbitrarily varying density configuration in one, two and three dimensions. This new approach greatly simplifies the potential integrations and, in certain cases where symmetry exists, analytical evaluation of the integrals is also possible. Numerical tests and examples are given for a hypothetical salt dome, a vertical dyke with varying density structure and the hydrostatic ellipticity of earth model 1066B.

081.037 **The major element chemistry of Libyan Desert Glass and the mineralogy of its precursor.**
R. F. Fudali.
Meteoritics, Vol. 16, 247 - 259 (1981).

081.038 **Determination of the Earth potential and possible improvements.** C. Reigber.
Ann. Géophys., Vol. 37, (see 012.037), 15 - 23 (1981).
Abstr. in Phys. Abstr., Vol. 84, Abstr. 107565 (1981).

081.039 **Least squares combination of satellite and terrestrial data in physical geodesy.** L. Sjoberg.
Ann. Géophys., Vol. 37, (see 012.037), 25 - 30 (1981).
Abstr. in Phys. Abstr., Vol. 84, Abstr. 107566 (1981).

081.040 **Models of the Earth's core.** D. J. Stevenson.
Science, Vol. 214, 611 - 619 (1981).

081.041 **Planetary tides as a part of deformations of the Earth's body.** M. Burša.
Říše hvězd, Vol. 62, 247 - 250 (1981). In Czech.

081.042 **Tilts due to the free nutation of the Earth's axis in the results of geodetic observations.**
A. Zeman.
Stud. Geophys. Geod., Vol. 25, 284 - 288 (1981).

081.043 **Statistics of narrow structures of the gravity field of the earth. 1. General theory.** B. Bertotti.
J. Geophys. Res., Vol. 86, 10835 - 10842 (1981).

081.044 **A simple physical model for the terrestrial dynamo.**
P. Olson.
J. Geophys. Res., Vol. 86, 10875 - 10882 (1981).
The strength of the earth's magnetic field results from an equilibration between rates of buoyant energy production and Ohmic dissipation. Changes in magnetic field, in particular the long term changes in dipole moment, provide an indication of changes in core energy sources, and so become critical data for understanding the evolution of both the core and deep mantle. A simple physical model is proposed to establish a connection between dipole moment behavior and production of buoyancy within the core.

081.045 **A dynamic objection to the inversion of the Earth on its spin axis.** V. J. Slabinski.
J. Phys. A, Vol. 14, 2503 - 2507 (1981). – Abstr. in Phys. Abstr., Vol. 85, Abstr. 2984 (1982).

081.046 **Supernovae and nitrate in the Greenland Ice Sheet.**
T. Risbo, H. B. Clausen, K. L. Rasmussen.
Nature, Vol. 294, 637 - 639 (1981).
Nitrate concentration in the absolutely dated Greenland ice core from Crête has been measured for six time intervals, five surrounding the time of appearance of the well established historical supernova during the past 1,000 yr and one during the Maunder minimum of solar activity, to look for a possible correlation between supernovae and nitrate concentration. The findings of Rood et al. (1979) of nitrate spikes corresponding to the appearance of the historical supernovae and a pronounced minimum in nitrate contents during the Maunder minimum are not confirmed. A very regular annual variation of nitrate concentration is observed superimposed on a constant background. The authors show here that both these signals seem unaffected by the known variations in the solar activity for the periods analysed.

081.047 **Gravity slippage and the indifferent equilibrium shape of the Earth (IES-geoid).** G. P. Gregori.
Ann. Geophys., Vol. 37, (see 012.037), 79 - 90 (1981). In French. – Abstr. in Phys. Abstr., Vol. 85, Abstr. 6298 (1982).

081.048 **Marées terrestres.**
P. Melchior (Editor).
Bull. Inf., (Obs. R. Belgique, Bruxelles), No. 86, p. 5498 - 5565 (1981).

081.049 **Le dimensioni della terra.** R. Sala.
Orione, Vol. 2, 276 - 281 (1981).

081.050 **Parameterized thermal convection in a layered region and the thermal history of the earth.**
D. McKenzie, F. M. Richter.
J. Geophys. Res., Vol. 86, 11667 - 11680 (1981).

081.051 **Rotation of the inner core.** D. Gubbins.
J. Geophys. Res., Vol. 86, 11695 - 11699 (1981).

081.052 **On the volumetric gravitational differentiation of the earth.** A. S. Monin, O. G. Sorokhtin.
Dokl. AN SSSR, Tom 259, 1076 - 1079 (1981). In Russian.
Abstr. in Ref. zh., 51. Astron., 12.51.301 (1981).

081.053 **On an interconnection between the gravitational field and topography of a planet (earth, Mars, moon)**
A. L. Tserklevich, Eh. M. Evseeva.
Geod.,kartogr. i aehrofotosemka, L'vov, 1981, No. 33, p. 85 - 94. In Russian. – Abstr. in Ref. zh., 52. Geod. Aehrosemka, 12.52.306 (1981).

081.054 **An effect of the earth ellipticity and inertial forces is visible from M_2 and O_1 tidal gravity measurements in the Trans World Profiles.** P. Melchior.
Obs. R. Belgique Commun., Sér. A, No. 63, p. 1 - 9 (1981).
Paper presented at the 9th International Symposium on Earth Tides, New York, August 1981.

081.055 **On the influence of polar shift upon tectonic plate boundaries.** G.-x. Song.
Ann. Shanghai Obs. Acad. Sinica, No. 1, p. 15 - 19 (1979).

081.056 **On the use of point mass models of the geopotential for orbit predictions.** G. A. Meshcheryakov,
A. N. Marchenko, S. K. Tatevian (Tatevyan), N. A. Sorokin.
Satellite perturbations and orbital determination, (see 012.060), p. 1 - 6 (1981).
Along with the classical expansion of the geopotential by means of spherical functions, various point-mass models have been developed. In this paper several preliminary results of investigation of the possibility to utilize point-mass models for predictions of satellite orbits during different time intervals are summarized. The model considered is an intermediate one between general point-mass models and the representation of geopotential by means of expansion on spherical harmonics.

081.057 **Earth and ocean tides from long-term analysis of satellite orbits.** E. M. Gaposchkin.
Satellite perturbations and orbital determination, (see 012.060), p. 13 - 28 (1981).
Precision laser tracking data on three close-earth satellites (Geos 1, Geos 2, and Lageos) are analyzed to obtain measurements of ocean tides, core-mantle resonance, and other mass displacements. Mean elements are computed for 6.0 years for Geos 1, 4.9 years for Geos 2, and 2.1 years for Lageos. Ocean tides T2, S2, K2, P1, K1, and ϕ1 are measured, along with the core-mantle resonance at the K1 and P1 frequencies, as predicted by Molodensky. Also measured are the Sa and Ssa tides and radiation-pressure area-to-mass ratios.

081.058 **Recent results of the study of physical background of the geoidal figure.** G. Barta, A. Hajósy.
Planetary interiors, (see 012.061), p. 195 - 202 (1981).
The paper is summing up hypotheses concerning the internal structure of the Earth.

081.059 Information theory lateral density distribution for earth inferred from global gravity field.
D. P. Rubincam.
NASA Tech. Memo., NASA TM 83825, 62 pp. (1981).

081.060 Hydromagnetic stability of the Kennedy—Higgins model of the Earth's core.   I. A. Eltayeb.
Phys. Earth Planet. Inter., Vol. 27, P1 - P5 (1981).

081.061 Independence of the gravitational constant from gross Earth data.   H. W. S. McQueen.
Phys. Earth Planet. Inter., Vol. 26, P6 - P9 with a correction Vol. 27, 233 (1981).

Determination of the Newtonian gravitational constant G from free oscillation and travel-time data for a spherically symmetric non-rotating Earth is shown to be inherently impossible as long as G is spatially constant. Since Earth densities are normalised by the laboratory value of G, one consequence is that the accuracy of the density and elastic parameters derived from inversion of such data is limited to the accuracy of G. Another is the derivation of closer geophysical limits on any intermediate range variation of G than previously reported. Although these limits are not reliable if G changes in the 10 - 1000 km range, a significant variation of G in this range should be detectable from free oscillation and travel-time inversions.

Bibliographie générale des marées terrestres. Supplément V, 1980 - 1981.   See Abstr. 002.068.

Le viscere della terra — dinamica e struttura dell'-interno terrestre.   See Abstr. 003.058.

Geophysik in Heidelberg.   See Abstr. 003.083.

The earth: its birth and growth. See Abstr. 003.120.

Systematics of global cycles of natural processes. Geological aspects.   See Abstr. 003.123.

On the origin of the results of determining the size of the earth mentioned by Aristotle and Archimedes. See Abstr. 004.075.

The development of science at the Astronomical Observatory of the Leningrad University. Celestial mechanics, geodesy and gravimetry.   See Abstr. 009.015.

Cosmochemistry and the origin of life. See Abstr. 011.002.

A versatile interactive computer program for computation and automatic optimization of gravity models. See Abstr. 021.042.

New method for measuring Earth albedo. See Abstr. 031.534.

The use of micro-accelerometers for space geodetic experiments.   See Abstr. 032.577.

On some applications of the problem of many fixed centres to geophysics.   See Abstr. 042.064.

Relation between the secular variation of longitude and the plate motion.   See Abstr. 044.037.

On the cause of a wobble of polar motion oscillating with a half Chandler period.   See Abstr. 045.008.

Space geodesy, paleogeodesy and paleogeophysics. See Abstr. 046.004.

A new trajectory concept for exploring the Earth's geomagnetic tail.   See Abstr. 051.016.

A SIRIO-2 laser ranging experiment. See Abstr. 052.015.

An analytical singularity-free orbit predictor for near-earth satellites.   See Abstr. 052.019.

Precision orbit analyses in support of the Seasat altimeter experiment.   See Abstr. 052.034.

Analysis of Lageos' altitude decrease. See Abstr. 052.062.

On an experiment for measuring the velocity of the earth's motion relative to absolute space. See Abstr. 066.119.

Electromagnetic-gravitational energy systems. See Abstr. 066.161.

Research results from analysis of satellite orbits. See Abstr. 082.035.

Earth radiation budgets.   See Abstr. 082.052.

Dynamo theory of the Earth's varying magnetic field.   See Abstr. 084.050.

Revised age estimates of Brunhes palaeomagnetic events: support for a link between geomagnetism and eccentricity.   See Abstr. 084.067.

Strange attractor character of large-scale non-linear dynamos.   See Abstr. 084.095.

Comparison of Cretaceous and present Earth albedos: implications for the causes of paleoclimates. See Abstr. 085.020.

Chemical composition of Earth, Venus, and Mercury.   See Abstr. 092.002.

Carbon dioxide within Venus and the Earth. See Abstr. 093.077.

On the influence of the secular change in the eccentricity of the earth's orbit on the motion of the moon. See Abstr. 094.011.

Meteorites and the origin of water and reduced carbon on the Earth.   See Abstr. 105.011.

082 Atmosphere (Refraction, Scintillation, Extinction, Airglow, Site Testing)

082.001 **Observations of neutral iron emission in twilight spectra.**
C. A. Tepley, J. W. Meriwether, Jr., J. C. G. Walker, J. D. Mathews.
J. Geophys. Res., Vol. 86, 4831 - 4835 (1981).

The authors present a method for the analysis of twilight airglow spectra that may be contaminated by atmospheric continuum emission of unknown brightness. The necessity of correcting for this continuum emission when measuring weak airglow features in twilight is illustrated by application of the method to the neutral iron line at 3860 Å.

082.002 **Comment on "Airglow observations of the OI 7774 Å multiplet at Arecibo during a magnetic storm" by Roger G. Burnside, John W. Meriwether, and James C. G. Walker.**
Y. Sahai, J. A. Bittencourt, N. R. Teixeira, with a reply by R. G. Burnside, J. W. Meriwether, Jr., J. C. G. Walker.
J. Geophys. Res., Vol. 86, 4839 - 4841 (1981). – See Abstr. 27.082.004.

082.003 **Comment on 'Simultaneous spectroscopic measurements of stratospheric species: O_3, CH_4, CO, CO_2, N_2O, H_2O, HCl, and HF at northern and southern midlatitudes' by C. B. Farmer et al.**
H. K. Roscoe, J. R. Eyre, with a reply by C. B. Farmer, O. F. Raper.
J. Geophys. Res., Vol. 86, 5383 - 5384 (1981). – See Abstr. 27.082.015.

082.004 **Simultaneous measurements of OH (9, 4), (8, 3), (7, 2), (6, 2) and (5, 1) bands in the airglow.**
H. Takahashi, P. P. Batista.
J. Geophys. Res., Vol. 86, 5632 - 5642 (1981).

The authors present simultaneous measurements of the OH (9, 4), (8, 3), (7, 2), (6, 2) and (5, 1) bands, the NaD and OI 5577 Å emissions. Nocturnal and day to day variations of the OH band intensities and the rotational temperatures are presented, and an analysis is made of the correlations between these parameters. The OH vibrational populations and the initial excitation rates are calculated. The OH emission profiles are also calculated for the individual bands, using the obtained initial excitation rates.

082.005 **Rocket measurements of O_2 atmospheric and OH Meinel bands in the airglow.**
T. Watanabe, M. Nakamura, T. Ogawa.
J. Geophys. Res., Vol. 86, 5768 - 5774 (1981).

Altitude distributions of the (0, 0) 7619-Å band of O_2 atmospheric system ($b^1\Sigma_g^+ - X^3\Sigma_g^-$) and of the (7, 2) 6863-Å band of OH ($X^2\Pi$) in the night airglow emissions were measured at 31°N latitude. In spite of the different seasons and local times of the observations, all three measurements present the peak emission rate occurring at an altitude of 94 ± 1 km for the O_2 band and at 90 ± 2 km for the OH band. The former emitting altitude is in basic agreement with the atomic oxygen density profile as measured by using an OI 1300-Å resonance lamp, but the latter is considerably higher than expected from relevant theoretical models. After examining various possible causes for this discrepancy, it seems that the continuum airglow may contaminate the authors' observations of the OH airglow.

082.006 **Stratospheric aerosols properties from earth limb photography.**
M. Ackerman, C. Lippens, C. Muller.
Nature, Vol. 292, 587 - 591 (1981).

Balloon-borne observation at three wavelengths of visible sunlight scattered by the earth limb allows the determination of aerosols abundances and size distributions at various altitudes in the stratosphere. The stratospheric aerosols are apparently still under the influence of the Mount St. Helens volcanic eruption five months after its occurrence on 18 May 1980.

082.007 **A self-consistent evaluation of the rate constants for the production of the OI 6300 Å airglow.**
R. Link, J. C. McConnell, G. G. Shepherd.
Planet. Space Sci., Vol. 29, 589 - 594 (1981).

The quenching rate k_{N_2} of $O(^1D)$ by N_2 and the specific recombination rate α_{1D} of O_2^+ leading to $O(^1D)$ are re-examined in light of available laboratory and satellite data.

082.008 **Stratospheric observations of NO_3 and its experimental and theoretical distribution between 20 and 40 km.**
J. P. Naudet, D. Huguenin, P. Rigaud, D. Cariolle.
Planet. Space Sci., Vol. 29, 707 - 712 (1981).

A simultaneous night-time observation of NO_3 and O_3 has been made by means of a balloon-borne spectrophotometer pointing at the rising planet Venus. The spectrum recorded between 642 and 672 nm makes it possible to determine the NO_3 and O_3 absorptions in the 662 nm band and the Chappuis bands, respectively. The NO_3 vertical distribution is deduced, and is found to reach a peak of $(3.4 \pm 0.4) \times 10^7$ molecules cm^{-3} at 35 km. Such an observational result can be interpreted in terms of a theoretical profile deduced from a one-dimension time-dependent photochemical model which takes account of the night-time stratospheric NO_2, NO_3 and N_2O_5 constituents and the latest kinetic and photochemical data for the rate constants.

082.009 **Average nocturnal and seasonal variations of sodium nightglow at 23°S, 46°W.**
V. W. J. H. Kirchhoff, B. R. Clemesha, D. M. Simonich.
Planet. Space Sci., Vol. 29, 765 - 766 (1981).

The seasonal variation of the nocturnal average intensity, and the typical variation of the sodium airglow intensity during the night have been deduced, using a data set that has been accumulated over a period of 5 years. The maximum intensity, which occurs at the equinoxes, is about 2–3 times as large as the intensity during winter, when the intensity of the D_2 line is about 30 R. The nocturnal variation is symmetrical about midnight, where the intensity is some 30% less than just after (before) dusk (dawn) values.

082.010 **Stratospheric chemical and thermal response to long-term variability in solar UV irradiance.**
G. Brasseur, P. C. Simon.
J. Geophys. Res., Vol. 86, 7343 - 7362 (1981).

A theoretical analysis of the chemical response of the stratosphere to possible long-term variability of solar ultraviolet irradiance has been performed, taking into account the thermal feedback effect on the reaction rates. Numerical values of ultraviolet and visible irradiation fluxes used in this work are given for aeronomic modeling purposes and a possible variability related to the 11-year solar cycle is suggested on the basis of recent and reliable observations of solar ultraviolet irradiance. This variability has been introduced in a stratospheric two-dimensional model which simulates the zonally averaged distribution of the chemical species related to the oxygen, hydrogen, nitrogen and chlorine families.

082.011 **Measurement of atomic oxygen and related airglows in the lower thermosphere.**
R. J. Thomas, R. A. Young.
J. Geophys. Res., Vol. 86, 7389 - 7393 (1981).

Instruments on-board a sounding rocket were used to make simultaneous observations of atomic oxygen density and airglow emissions between 80 and 120 km. Atomic oxygen was measured with a resonance lamp and was found to have a peak density of 6×10^{11} at 94 km. Similar structure is seen in the oxygen-density profile on both uplegs and downlegs. The following airglow emissions were measured by using vertical-viewing photometers: Herzberg I bands near 300 nm; O(^1S) green line at 557.7 nm; background at 566 nm; O$_2$($^1\Delta_g$) bands at 1.27 μm; and OH ($X^2\pi$) Meinel bands near 1.7 μm.

082.012 Seasonal variations in aerosols and atmospheric transmission at Mauna Loa Observatory.
B. A. Bodhaine, B. G. Mendonca, J. M. Harris, J. M. Miller.
J. Geophys. Res., Vol. 86, 7395 - 7398 (1981).
Aerosol light scattering and atmospheric transmission data taken at Mauna Loa Observatory show seasonal variations that are strongly correlated and in phase. Total ozone above Mauna Loa is also in phase but accounts for only about 20% of the seasonal variation in atmospheric transmission while aerosols account for the other 80%. Total precipitable water vapor above Mauna Loa is out of phase with atmospheric transmission and cannot account for its seasonal variation. Long-range atmospheric trajectory analyses indicate that tropospheric transport of aerosols from the direction of the Asian continent is most likely responsible for the seasonal variations in atmospheric transmission and aerosol optical properties.

082.013 The atmospheric neutral sodium layer. 1. Recent modeling compared to measurements.
V. W. J. H. Kirchhoff, B. R. Clemesha, D. M. Simonich.
J. Geophys. Res., Vol. 86, 6892 - 6898 (1981).

082.014 Site testing on Hawaii, Madeira and the Canary Islands.   B. McInnes.
Q. J. R. Astron. Soc., Vol. 22, 266 - 271 (1981).

082.015 Microthermal fluctuations and their relation to seeing conditions at Roque de los Muchachos Observatory, La Palma.   M. Hartley, B. McInnes, F. G. Smith.
Q. J. R. Astron. Soc., Vol. 22, 272 - 278 (1981).
Rapid fluctuations in air temperature were recorded at the Roque de los Muchachos Observatory, La Palma, over a period of six months during 1975 for comparison with night-time tests of seeing conditions. The results show that bad seeing is usually caused by large-scale atmospheric conditions. It does not seem advantageous to build telescopes at large heights above ground for night-time observations on this site.

082.016 The atmospheric extinction at SAAO Sutherland, and the red leak of the Schott BG18 filter in the St. Andrews Scanner.   J. H. Spencer Jones.
Mon. Notes Astron. Soc. South Africa, Vol. 39, 89 - 93 (1980).
The St. Andrews Scanner has been used to determine the atmospheric extinction at SAAO Sutherland. The observations have been utilised further to give a measure of the red leak in the second order of the Schott BG18 order-separating filter.

082.017 Site testing in Binchuan (I). Observations.
Acta Astron. Sinica, Vol. 22, 195 - 206 (1981).
In Chinese.

082.018 Solar Mesosphere Explorer to study ozone.
C. A. Barth.
Nature, Vol. 293, 259 - 260 (1981).

082.019 The detection of very thin aerosol layers by the measurement of spectral sky radiance.
K. H. J. Adams.
Atmos. Environ., Vol. 15, 371 - 380 (1981). – Abstr. in Phys. Abstr., Vol. 84, Abstr. 74914 (1981).

082.020 A three-dimensional time-dependent global model of the thermosphere.
T. J. Fuller-Rowell, D. Rees.
J. Atmos. Sci., Vol. 37, 2545 - 2567 (1980). – Abstr. in Phys. Abstr., Vol. 84, Abstr. 74970 (1981).

082.021 Investigation of a theoretical description of distribution of the main neutral component of the upper atmosphere.   M. N. Vlasov, V. E. Davydov.
Geomagn. Aehron., Tom 21, 683 - 688 (1981). In Russian.

082.022 Atmospheric tides – a review.   S. Kato.
J. Atmos. Terr. Phys., Vol. 43, (see 012.008), 491 - 493 (1981).

082.023 Radar studies of gravity waves and tides in the middle atmosphere: a review.   P. K. Rastogi.
J. Atmos. Terr. Phys., Vol. 43, (see 012.008), 511 - 524 (1981).

082.024 Chemistry of middle atmospheric ionization – a review.   A. P. Mitra.
J. Atmos. Terr. Phys., Vol. 43, (see 012.008), 737 - 752 (1981).
The author describes the observations and chemistry of the ionization of the earth's middle atmosphere, here defined as the region between 100 km and the tropopause, consisting of the D-region (50 - 100 km) and the stratosphere. Observational data on electron and ion distributions and on ion composition, both for the D-region and the stratosphere, are discussed.

082.025 Particles above the tropopause: measurements and models of stratospheric aerosols, meteoric debris, nacreous clouds, and noctilucent clouds.
O. B. Toon, N. H. Farlow.
Annu. Rev. Earth Planet. Sci., Vol. 9, (see 003.003), p. 19 - 58 (1981).

082.026 Nucleation and growth of stratospheric aerosols.
A. W. Castleman, Jr., R. G. Keesee.
Annu. Rev. Earth Planet. Sci., Vol. 9, (see 003.003), 227 - 249 (1981).

082.027 Determination of the refraction angles on extended tracks.   A. V. Alekseev, M. V. Kabanov.
Geod. i fotogrammetr. Rostov n/D., 1980, p. 25 - 33. In Russian. – Abstr. in Ref. zh., 52. Geod. Aehrosemka, 7.52.53 (1981).

082.028 The influence of the atmosphere's ellipticity on the angles of total and astronomical refraction.
I. F. Kushtin.
Geod. i fotogrammetr. Rostov n/D., 1980, p. 3 - 24. In Russian. → Abstr. in Ref. zh., 52. Geod. Aehrosemka, 7.52.72 (1981).

082.029 Experimental determinations of the angles of astronomical refraction.   N. N. Redichkin.
Geod. i fotogrammetr. Rostov n/D., 1980, p. 52 - 65. In Russian. – Abstr. in Ref. zh., 52. Geod. Aehrosemka, 7.52.73 (1981).

082.030 Review of energetic (> 20 keV) bremsstrahlung X-ray measurements from satellites.
W. L. Imhof.
Space Sci. Rev., Vol. 29, 201 - 217 (1981).
The present status of satellite bremsstrahlung X-ray measurements is reviewed, some of the key results summarized and suggestions made for future improvements in instrumentation. The review is limited to X-rays > 20 keV and therefore to the associated precipitation of only the more energetic electrons.

082.031 On a possibility of mathematical modelling of radar
         observations in active cosmic experiments.
R. I. Mojsya, A. R. Kolomiets, G. I. Kolomiets,
I. I. Slyusarenko, Yu. V. Chumak.
Problems of cosmic physics. Vyp. 16, (see 003.005), p. 60-65
(1981). In Russian.

082.032 Radio reflections from artificial plasma formations
         in the Zarnitsa 2 experiment.
R. I. Mojsya, A. R. Kolomiets, G. I. Kolomiets, I. A. Zhulin,
I. I. Slyusarenko, V. S. Dokukin, Yu. Ya. Ruzhin.
Problems of cosmic physics. Vyp. 16, (see 003.005), p. 66-74
(1981). In Russian.

082.033 Cosmogenic radio carbon in the earth's atmosphere.
         V. A. Dergachev.
Izv. AN SSSR. Ser. fiz., Tom 45, 507 - 521 (1981).
In Russian. – Abstr. in Ref. zh., 51. Astron., 8.51.514 (1981).

082.034 Soviet-Indian investigations of cosmic radiation on
         high-altitude balloons.
A. M. Gal'per, V. G. Kirillov-Ugryumov, Yu. D. Kotov,
L. V. Kurnosova, A. V. Kurochkin, N. G. Lejkov,
V. I. Logachev, B. I. Luchkov, L. A. Razorenov, Yu. I. Smirnov,
M. I. Fradkin, Yu. T. Yurkin, V. N. Yurov, G. S. Kokkhal,
S. V. Damle, P. K. Kunte, B. V. Sreekantan.
Izv. AN SSSR. Ser. fiz., Tom 45, 644 - 648 (1981). In Russian.
Abstr. in Ref. zh., 62. Issled. kosm. prostranstva, 8.62.402
(1981).

082.035 Research results from analysis of satellite orbits.
         D. G. King-Hele.
Spacecraft flight dynamics, (see 012.021), p. 3 - 11 (1981).
       The paper describes some of the discoveries about the
Earth's upper atmosphere and the gravitational field resulting
from the analysis of changes in satellite orbits.

082.036 Comets and the photochemistry of the paleoatmo-
         sphere.   J. S. Levine, T. R. Augustsson,
R. E. Boughner, M. Natarajan, L. J. Sacks.
Comets and the origin of life, (see 012.022), p. 161 - 190
(1981).
       The photochemistry/chemistry of atmospheric O_3 in the
paleoatmosphere is investigated using a one-dimensional photochemical model that includes the chemistry of the oxygen,
nitrogen, hydrogen, carbon, and chlorine gases. The role of
cometary influx of H_2O on the photochemistry of the paleoatmosphere is also examined.

082.037 Site-Untersuchungen für ein neues Sonnenobserva-
         torium auf den Kanarischen Inseln.
P. N. Brandt, H. Wöhl.
Sterne Weltraum, Jahrg. 20, 359 - 366 (1981).

082.038 Comments on "A meteor-ablation model of the
         sodium and potassium layers" by D. M. Hunten.
B. R. Clemesha, V. W. J. H. Kirchhoff, D. M. Simonich.
Geophys. Res. Lett., Vol. 8, 1023 - 1025 (1981). – See Abstr.
29.082.085.

082.039 Study of variations in mesospheric structure during
         the total solar eclipse of 16 February 1980.
B. H. Subbaraya, S. Lal, Y. B. S. Acharya, K. S. Modh, R. N. Misra.
Observations of the total solar eclipse of 16 February 1980,
(see 003.009), p. 101 (1981).

082.040 Observations of atmospheric scattering in the ultra-
         violet during the solar eclipse.
B. H. Subbaraya, S. Lal, R. N. Misra, Y. B. Acharya.
Observations of the total solar eclipse of 16 February 1980,
(see 003.009), p. 102 (1981).

082.041 Study of atmospheric ozone variations in the
         stratosphere and mesosphere during the solar eclipse.
B. H. Subbaraya, S. Lal, Y. B. Acharya, S. K. Banerjee,
R. N. Misra.
Observations of the total solar eclipse of 16 February 1980,
(see 003.009), p. 103 (1981).

082.042 Quadrupole transitions of the 1←0 band of N_2
         observed in a high resolution atmospheric spectrum.
C. Camy-Peyret, J.-M. Flaud, L. Delbouille, G. Roland,
J. W. Brault, L. Testerman.
J. Phys. Lett., Vol. 42, L279 - L283 (1981). – Abstr. in Phys.
Abstr., Vol. 84, Abstr. 88560 (1981).

082.043 H_2O^+ ions in the upper atmosphere.
         G. Herzberg.
Ann. Géophys., Vol. 36, 605 (1980). – Abstr. in Phys. Abstr.,
Vol. 84, Abstr. 88619 (1981).

082.044 Rocket observations of near infrared airglow.
         S. Hayakawa, K. Ito, T. Matsumoto, H. Murakami,
K. Noguchi, T. Ono, K. Uyama.
J. Geomagn. Geoelectr., Vol. 32, 691 - 707 (1980). – Abstr. in
Phys. Abstr., Vol. 84, Abstr. 88621 (1981).

082.045 The green light of the night sky.   D. R. Bates.
         Planet. Space Sci., Vol. 29, 1061 - 1067 (1981).
       The development of the theory of the 5577 Å emission
from the upper atmosphere is reviewed. Evidence from both
aeronomy and chemical kinetics shows that the Barth
mechanism is a much more important source of $O(^1S)$ than in
the Chapman process. The molecular oxygen state involved is
probably $c^1\Sigma_u^-$ (the upper state of the Herzberg II band
system).

082.046 The silver-blue cloudlets again: nucleation and
         growth of ice in the mesosphere.   M. Gadsden.
Planet. Space Sci., Vol. 29, 1079 - 1087 (1981).
       The existence of noctilucent clouds probably indicates
that there are submicron ice crystals in the upper atmosphere
at altitudes of 80–85 km. In this paper, the results of numerical calculations of crystal sizes are discussed to provide a guide
to the interpretation of observational data.

082.047 Non-thermal O (^{1}D) produced by dissociative
         recombination of O_2^+: a theoretical model and
observational results.
G. A. Schmitt, V. J. Abreu, P. B. Hays.
Planet. Space Sci., Vol. 29, 1095 - 1099 (1981).
       Thermal and non-thermal $O(^1D)$ number density profiles
are calculated. The two populations are assumed to be coupled
by a thermalization cross-section which determines the loss and
production in the non-thermal and thermal populations,
respectively. The sources, sinks and transport of the two
populations are used to model volume emission rate profiles at
6300 Å. The 6300 Å brightness measured by the Visible
Airglow Experiment is then used to establish the presence of
the non-thermal population and to determine the thermalization cross-section.

082.048 Influence of turbulence and infrared radiation on
         the thermal regime of the earth's thermosphere.
B. F. Gordiets, Yu. N. Kulikov.
Kosm. Issled., Tom 19, 539 - 550 (1981). In Russian.

082.049 Analysis of modern semi-empirical models of the
         thermosphere on the basis of theoretical calcula-
tions and different rocket measurements of atomic oxygen.
M. N. Vlasov, V. E. Davydov, A. I. Livshits.
Kosm. Issled., Tom 19, 638 - 641 (1981). In Russian.

082.050 Rocket measurements of density and distribution of aerosol particles in the atmosphere according to sizes.   T. I. Orishich, A. V. Cherkasov.
Kosm. Issled., Tom 19, 641 - 645 (1981). In Russian.

082.051 Tunguska meteor fall of 1908: effects on stratospheric ozone.   R. P. Turco, O. B. Toon, C. Park, R. C. Whitten, J. B. Pollack, P. Noerdlinger.
Science, Vol. 214, 19 - 23 (1981).

In 1908, when the giant Tunguska meteor disintegrated in the earth's atmosphere over Siberia, it may have generated as much as 30 million metric tons of nitric oxide (NO) in the stratosphere and mesosphere. The photochemical aftereffects of the event have been simulated using a comprehensive model of atmospheric trace composition. Calculations indicate that up to 45 % of the ozone in the Northern Hemisphere may have been depleted by Tunguska's nitric oxide cloud early in 1909 and large ozone reductions may have persisted until 1912.

082.052 Earth radiation budgets.
G. L. Stephens, G. G. Campbell, A. T. Vonder Haar.
J. Geophys. Res., Vol. 86, 9739 - 9760 (1981).

The paper presents the annual and seasonal averaged earth atmosphere radiation budgets derived from the most complete set of satellite observations available in late 1979. The budgets are derived from a composite of 48 monthly mean radiation budget maps. The annual, global average emitted infrared flux is 234 W m^{-2}, planetary albedo is 0.30, and the net flux is zero within measurement uncertainty. The annual cycle of net flux is also studied in detail, and the observed globally averaged net flux displays an annual cycle that is of similar magnitude and phase to the annual cycle imposed by the influence of sun-earth distance variations on solar radiation input into the atmosphere. A study of the geographical distribution of the annual variability of the net flux is presented.

082.053 Stratospheric observations of the attenuated solar irradiance in the Schumann-Runge band absorption region of molecular oxygen.
J. E. Frederick, R. D. Hudson, J. E. Mentall.
J. Geophys. Res., Vol. 86, 9885 - 9890 (1981).

A spectrometer flown on the first Solar Absorption Balloon Experiment (SABE-1) observed the attenuated solar irradiance between 184 and 202 nm from an altitude near 40 km. These measurements provide a check on the absorption cross sections of molecular oxygen in the spectral region of the Schumann-Runge bands. Comparison of the measurements with calculations based on cross sections derived from laboratory data shows a general agreement although the irradiance measurements have large error bars near the centers of the absorption bands. The results imply that the 184–200 nm solar irradiance that penetrates to the stratosphere can be computed to an accuracy of ±30% or better by using presently available cross sections.

082.054 High-sensitivity detection of negative ions in the stratosphere.   J. L. McCrumb, F. Arnold.
Nature, Vol. 294, 136 - 139 (1981).

The authors report in situ negative ion composition measurements which led to the detection of previously unobserved ion species.

082.055 Simultaneous measurements of meteor winds and green line intensity variations: gravity waves and planetary waves.   H. Teitelbaum, M. Massebeuf, J. L. Fellous, M. Petitdidier, J. Christophe, F. Blanco.
J. Geophys. Res., Vol. 86, 7767 - 7770 (1981).

The authors provide evidence of the influence of planetary waves on the intensity variations of the 557.7-nm airglow emission.

082.056 Observations of the Ca$^+$ twilight airglow from intermediate layers of ionization.   C. A. Tepley, J. D. Mathews, J. W. Meriwether, Jr., J. C. G. Walker.
J. Geophys. Res., Vol. 86, 7781 - 7786 (1981).

082.057 Absorption of infrared radiation by atmospheric water vapor in the region of 4.3 - 5.5 micron – preliminary measurements.
A. Ben-Shalom, A. D. Devir, S. G. Lipson, U. P. Oppenheim.
Proc. Soc. Photo-Opt. Instrum. Eng., Vol. 253, 261 - 265 (1980). – Abstr. in Phys. Abstr., Vol. 84, Abstr. 98564 (1981).

082.058 Investigations of the twilight layer of the earth's atmosphere from space.
G. M. Grechko, N. B. Divari, G. M. Nikol'skij, Yu. V. Romanenko, S. A. Savchenko.
Kosm. Issled., Tom 19, 786 - 790 (1981). In Russian.

082.059 Cosmogenic radio carbon in the earth's atmosphere.
V. A. Dergachev.
Cosmic rays, (see 012.025), 1980, p. 122 - 125. – Abstr. in Ref. zh., 51. Astron., 10.51.413 (1981).

082.060 Satellite measurements of H_2O fluorescence in the mesosphere.   J. R. Drummond, C. T. Mutlow.
Nature, Vol. 294, 431 - 432 (1981).

The authors report the observation of resonant fluorescent radiation from water vapour in the 50-95 km region by the Stratospheric and Mesospheric Sounder on Nimbus 7 satellite. Their preliminary data suggest that the mixing ratio of water vapour decreases with increasing height over this region with a value of $\sim 1 \times 10^{-6}$ v/v at 75-85 km.

082.061 On the mechanism of N and NO formation in the lower thermosphere and mesosphere.
M. N. Vlasov, V. V. Medvedev.
Geomagn. Aehron., Tom 21, 857 - 862 (1981). In Russian.

082.062 On irregular variations of the parameters of the semi-daily tide.
O. V. Kajdalov, Yu. I. Portnyagin.
Geomagn. Aehron., Tom 21, 871 - 875 (1981). In Russian.

082.063 Observation of new emission bands λ 6150 and λ 6190 Å at high latitudes.
G. F. Totunova, N. I. Fedorova.
Geomagn. Aehron., Tom 21, 949 - 950 (1981). In Russian.

082.064 General formulation of optical paths for large zenith angles in the Earth's curved atmosphere.
P.-H. Wang, A. Deepak, S.-S. Hong.
J. Atmos. Sci., Vol. 38, 650 - 658 (1981). – Abstr. in Phys. Abstr., Vol. 84, Abstr. 107871 (1981).

082.065 Horizontal thermal structure of the mesosphere from observations of OH(8–3) band emissions.
C. A. Tepley, R. G. Burnside, J. W. Meriwether, Jr.
Planet. Space Sci., Vol. 29, 1241 - 1249 (1981).

Two computerized tilting-filter photometers and a programmable dual axis mirror system have been used to produce maps of OH rotational temperature and intensity. Each map consists of a square array of 121 sky positions. Significant horizontal structure is not generally observed in mesospheric OH(8–3) rotational temperature at Arecibo. However, there is evidence for the occasional occurrence of a thermal wave just after evening twilight.

082.066 Studies of the motion of equatorial 630.0 nm airglow depletions.
D. P. Sipler, M. A. Biondi, R. D. Hake, Jr.
Planet. Space Sci., Vol. 29, 1267 - 1272 (1981).

An all-sky-mapping filter photometer has been used to

study equatorial 630.0 nm nightglow intensity depletions. A generally eastward drift of the N-S aligned depletions was observed, with velocities ranging from ~240 m/sec to ~0 m/sec. At times two airglow depletions moving at different velocities were observed simultaneously; in one case they merged. Simultaneous interferometric determinations of doppler shifts in the 630.0 nm line indicated that the velocity of the neutral atmosphere does not appear to be correlated with the drift of the airglow depletions. The results are compared with other observations.

082.067 Nitric oxide delta band emission in the earth's atmosphere: comparison of a measurement and a theory.
D. W. Rusch, W. E. Sharp.
J. Geophys. Res., Vol. 86, 10111 - 10114 (1981).
    The altitude dependent emission rate in the δ-bands of nitric oxide has been measured in the earth's atmosphere at night by a scanning ultraviolet spectrometer. The reaction responsible is the two-body association of nitrogen and oxygen atoms. The measurements reveal a vertical intensity beneath the layer for the δ-band system of 19 R. The horizontal emission rate increases from 70 R at 117 km to 140 R at 150 km. The data have been analyzed with a one-dimensional, time-dependent, vertical-transport model of odd nitrogen photochemistry. The diurnal variation of δ-band emission, $N(^4S)$, and NO densities resulting from model calculations are presented and discussed.

082.068 Site testing in Binchuan (II) – discussion and estimation.
Acta Astron. Sinica, Vol. 22, 305 - 309 (1981). In Chinese.

082.069 A new method of deducing the atomic oxygen density profiles in the lower thermosphere using ground-based night airglow observations.
M. N. M. Rao, G. S. N. Murty.
J. Atmos. Terr. Phys., Vol. 43, 1253 - 1264 (1981).
    A semi-empirical method has been developed to deduce the atomic oxygen density profile in the altitude range of 90 to 120 km at any instant at night, making use of the ground-based night airglow observations of (OI) 5577 Å, (OI) 6300 Å and OH band emission zenith intensities.

082.070 NO γ-band airglow measurement with a self-absorption gas cell: an error analysis.
N. Iwagami, T. Ogawa.
Bull. Inst. Space Aeronaut. Sci. Univ. Tokyo B, Vol. 16, 1287 - 1299 (1980). In Japanese. – Abstr. in Phys. Abstr., Vol. 85, Abstr. 3114 (1982).

082.071 Middle atmosphere dynamics. J. Gregory.
Nature, Vol. 294, 519 - 520 (1981).

082.072 Investigation of refractive characteristics of the atmosphere by absolute observations with the Engelhardt Observatory's meridian circle.
A. Yu. Yatsenko.
Astrometric investigations, (see 012.048), p. 100 - 102 (1981). In Russian. – Abstr. in Ref. zh., 51. Astron., 11.51.177 (1981).

082.073 Refraction in observations of a geodetic mark and stars. S. V. Korzhinskaya,
G. S. Tyuterev, A. F. Kantorov, N. A. Sharkovskij.
Astrometric investigations, (see 012.048), p. 103 - 107 (1981). In Russian. – Abstr. in Ref. zh., 51. Astron., 11.51.178 (1981).

082.074 Studies of near-earth temperature layers in daytime determinations of declinations.
V. P. Sibilev.
Astrometric investigations, (see 012.048), p. 107 - 110 (1981). In Russian. – Abstr. in Ref. zh., 51. Astron., 11.51.179 (1981).

082.075 Chromatic refraction and its influence on determination of declinations from observations with the Pulkovo FVK.
B. K. Bagil'dinskij, E. G. Zhilinskij, S. P. Pulyaev.
Astrometric investigations, (see 012.048), p. 111 - 117 (1981). In Russian. – Abstr. in Ref. zh., 51. Astron., 11.51.180 (1981).

082.076 On determination of the astronomical refraction from measured meteorological elements and zenith distances. D. I. Maslich, L. S. Khizhak.
Geod., kartogr. i aehrofotosemka, L'vov, 1981, No. 33, p. 39 - 43. In Russian. – Abstr. in Ref. zh., 52. Geod. Aehrosemka, 11.52.100 (1981).

082.077 On the frequency spectrum of temperature fluctuations in the night-time atmospheric surface layer at Mt. Maydanak. A. Eh. Gur'yanov.
Astron. Tsirk., No. 1115, p. 3 - 4 (1980). In Russian.

082.078 On the influence of the night-time atmospheric surface layer on the astronomical "seeing" at Mt. Maydanak. A. Eh. Gur'yanov.
Astron. Tsirk., No. 1115, p. 5 - 7 (1980). In Russian.

082.079 On the morning minimum of astronomical "seeing" effects caused by the atmospheric surface layer at Mt. Dushak-Erekdagh. A. Eh. Gur'yanov.
Astron. Tsirk., No. 1123, p. 4 - 6 (1980). In Russian.

082.080 Observations of ozone depletion associated with solar proton events.
R. D. McPeters, C. H. Jackman, E. G. Stassinopoulos.
J. Geophys. Res., Vol. 86, 12071 - 12081 (1981).
    Data from the backscattered ultraviolet instrument (BUV) on Nimbus 4 show depletion of ozone following solar proton events in January and September 1971 and in August 1972. The direct effect of high energy protons on the BUV instrument was determined by comparing nighttime BUV counting rates with particle data from IMP 6. The instrumental effects were then subtracted to allow accurate calculation of ozone profiles during the three events.

082.081 Detection of nighttime atmospheric scattering of lunar UV radiances by the Nimbus 4 BUV instrument.
E. G. Stassinopoulos, R. A. Goldberg, R. D. McPeters, J. M. Barth.
J. Geophys. Res., Vol. 86, 12082 - 12086 (1981).
    Morphological studies of nighttime dark current data from the Nimbus 4 backscattered UV experiment clearly indicate enhancements during the period of the full moon.

082.082 Total ozone seasonal and interannual variations derived from the 7 year Nimbus-4 BUV data set.
E. Hilsenrath, B. M. Schlesinger.
J. Geophys. Res., Vol. 86, 12087 - 12096 (1981).
    Backscattered ultraviolet (BUV) data from the Nimbus-4 spacecraft for the period 1970-1977 have been recently processed to final form. This paper is based upon an analysis of all the total ozone data available in the 7-year data set. The seasonal and interannual variations of total ozone over the globe are examined, using daily zonal means of 10° latitude bands and a time latitude cross section. A harmonic analysis was performed on the daily zonal means, and the amplitude, phase, and percentage of variance were computed for the annual, semiannual, and higher harmonics for several years and for individual years.

082.083 Rocket measurements of the distribution of water vapor in the stratosphere at high latitudes.
R. S. O'Brien, W. F. J. Evans.
J. Geophys. Res., Vol. 86, 12101 - 12107 (1981).
    Two measurements of the altitude distribution of strato-

spheric water vapor have been made with a rocket-borne infrared photometer, which measures long path atmospheric absorption of the solar beam in the 2.7-μm spectral region.

082.084   Structure of new twilight emissions near 6200 and 7000 Å.
V. I. Krasovskij, A. I. Semenov.
Astron. Tsirk., No. 1124, p. 1 - 3 (1980). In Russian.

082.085   PSM measured r_0 distribution for two mountains in Middle Asia and a free atmosphere.
P. V. Shcheglov.
Astron. Tsirk., No. 1124, p. 3 - 4 (1980). In Russian.

082.086   On 1978 - 1979 PSM seeing data histograms at Mt. Dushak-Erekdagh.   Yu. V. Khan.
Astron. Tsirk., No. 1124, p. 6 - 8 (1980). In Russian.

082.087   On the behaviour of the H_2O^+ emission in the upper atmosphere.
V. I. Krasovskij, A. I. Semenov, V. G. Sobolev, N. N. Shefov.
Astron. Tsirk., No. 1143, p. 6 - 7 (1980). In Russian.

082.088   Wavelength variation of visible and near-infrared resolution through the atmosphere: dependence on aerosol and meteorological conditions.
N. S. Kopeika, S. Solomon, Y. Gencay.
J. Opt. Soc. America, Vol. 71, 892 - 901 (1981).

082.089   Finite bandwidth and scattered light effects on the radiometric determination of atmospheric turbidity and the solar constant.   M. A. Box.
Appl. Opt., Vol. 20, 2215 - 2219 (1981).
Multispectral solar radiometric measurements are routinely performed at a large number of sites, using equipment of varying degrees of sophistication. From the standard Langley plot technique, one may extract the total optical thickness of the atmosphere (and hence the aerosol component) plus the extraterrestrial solar flux. In this paper, the author analytically examines the effects of finite filter bandwidth (in the absence of spectral lines) and finds them to be less than one part in a thousand. This is compared with earlier results on the effects of scattered light, which turns out to be typically an order of magnitude larger.

082.090   Nitrogen-broadening coefficient of vibration-rotation lines of carbon monoxide.
J. N.-P. Sun, P. R. Griffiths.
Appl. Opt., Vol. 20, 2332 - 2333 (1981).

082.091   Validated band model for the NO fundamental.
J. H. Pierluissi, K. Tomiyama, F. X. Kneizys.
Appl. Opt., Vol. 20, 2517 - 2521 (1981).
A previously reported transmission model for the 5.3 μm band of NO, whose defining parameters had been developed with line-by-line calculated spectra, is now presented in validated and upgraded form through the use of measured transmittance data. The model consists of a double-exponential function, which approximates homogeneous-path transmittance at 5 cm^{-1} intervals with a spectral resolution of 20 cm^{-1}. The use of the proposed model parameters in transmittance calculations yielded an average standard deviation of 0.28% and an overall maximum deviation of 1.48% from the measured transmittance data.

082.092   Nitric oxide γ band airglow radiometer with a self-absorbing gas cell.
N. Iwagami, T. Ogawa.
Appl. Opt., Vol. 20, 2522 - 2527 (1981).
A radiometer which makes use of a self-absorbing gas cell was developed for rocket measurement of the NO γ band airglow. The error due to the incompleteness fo the self-absorp-

tion was analyzed theoretically and found to be unimportant compared with other experimental errors.

082.093   Quartz-glass chopping method of measuring solar irradiance.   P. Schlyter, G. Witt.
Appl. Opt., Vol. 20, 2528 - 2533 (1981).
A solar irradiance photometer was launched on a pair of ESRO Centaure rockets. The instrument, a solar irradiance photometer used six wavelength channels. Three of these channels were used to measure mesospheric ozone by the occultation method. Two other channels were equipped with 214 nm Fabry-Perot filters to monitor the absorption of sunlight required for the excitation of NO γ band fluorescence, which was measured by another instrument in the same payload.

082.094   Finite sun effect on the interpretation of solar aureole.   M. A. Box, A. Deepak.
Appl. Opt., Vol. 20, 2806 - 2810 (1981).
Although it is usually assumed that solar radiation falls on the earth's atmosphere in the form of plane waves, the finite angular size of the solar disk contradicts this assumption. For most purposes, this finite sun effect on computed or measured radiation quantities is negligible. However, in the region of the solar aureole, which is dominated by aerosol diffraction scattering, measurable effects may be obtained. In this paper, the authors show that the finite sun effect is related to derivatives of the scattering phase function and that a 1% effect may be obtained close to the sun if enough large particles are present in the atmosphere.

082.095   5 cm^{-1} band model option to LOWTRAN5.
D. C. Robertson, L. S. Bernstein, R. Haimes, J. Wunderlich, L. Vega.
Appl. Opt., Vol. 20, 3218 - 3226 (1981).
Modifications to the atmospheric transmission and radiation code LOWTRAN5 are presented which include (1) an increase in the spectral resolution from 20 to 5 cm^{-1}, (2) the addition of temperature-dependent molecular absorption coefficients, (3) the use of a multiparameter Doppler-Lorentz band model for calculation of molecular transmittance, and (4) the use of the Curtis-Godson approximation for multilayered paths.

082.096   Scintillation at two optical frequencies.
W. B. Hubbard, H. J. Reitsema.
Appl. Opt., Vol. 20, 3227 - 3232 (1981).
Stellar scintillation data were obtained on a single night at a variety of zenith distances and azimuths, using a photon-counting photometer recording at 100 Hz simultaneously at wavelengths of 0.475 μm and 0.870 μm. Orientable apertures of 42 cm diam separated by 1 m were used to establish the average upper atmosphere wind direction and velocity. Although there is clear evidence of a complicated velocity field, scintillation power was predominantly produced by levels at pressures of 130 ± 30 mbar.

082.097   Nonthermal line profiles from O(^{1}S) in the thermospheric nightglow.   T. L. Killeen, P. B. Hays.
Appl. Opt., Vol. 20, 3314 - 3317 (1981).
Generalized composite nonthermal line profiles are calculated for the thermospheric nighttime emission from O(^{1}S) metastable atoms based on an extension to the simple relaxation model of Hays and Walker. It is shown that the Dynamics Explorer Fabry-Perot interferometer will enable accurate values to be obtained for the branching ratio of the O(^{1}S) production reaction and for the excitation exchange collision frequency.

082.098   Atmospheric hydrogen cyanide absorption near 14 μm.   M. T. Coffey, A. Goldman.
Appl. Opt., Vol. 20, 3480 - 3481 (1981).

082.099 Modeling of growth and evaporation effects on the extinction of 1.0-μm solar radiation traversing stratospheric sulfuric acid aerosols. G. K. Yue, A. Deepak.
Appl. Opt., Vol. 20, 3669 - 3675 (1981).

The effects of growth and evaporation of sulfuric acid aerosols in the stratosphere on the extinction of solar radiation traversing such an aerosol medium are reported in this paper. Extinction of 1.0-μm solar radiation was studied since this wavelength was used to monitor the aerosol extinction properties by two recent satellite experiments: Stratospheric Aerosol Measurement II (SAM II) and Stratospheric Aerosol and Gas Experiment (SAGE).

082.100 Intensity of the hydrogen peroxide $\nu_6(b)$ band around 1266 cm^{-1}.
F. P. J. Valero, D. Goorvitch, F. S. Bonomo, R. W. Boese.
Appl. Opt., Vol. 20, 4097 - 4101 (1981).

082.101 Stratospheric measurements of continuous absorption near 2400 cm^{-1}. C. P. Rinsland,
M. A. H. Smith, J. M. Russell III, J. H. Park, C. B. Farmer.
Appl. Opt., Vol. 20, 4167 - 4171 (1981).

Solar occultation spectra obtained with a balloon-borne interferometer have been used to study continuous absorption by N_2 and CO_2 near 2400 cm^{-1} in the lower stratosphere. Synthetic continuum transmittances, calculated from published coefficients for far-wing absorption by CO_2 lines and for pressure-induced absorption by the fundamental band of N_2, are in fair agreement with the observed stratospheric values. The continuum close to the ν_3 R-branch band head of CO_2 is sensitive to the CO_2 far-wing line shape.

082.102 Zenith polarization and color ratio during twilight.
F. E. Volz.
Appl. Opt., Vol. 20, 4172 - 4174 (1981).

The excellent data of zenith polarization and color ratio (CR) during twilight obtained by Coulson at the Mauna Loa Observatory, Hawaii are subjected to a reinterpretation, especially with regard to the frequent deviations from the clear average. It is shown that a moderate lift of the earth's shadow by distant clouds (or by low level haze) will reduce the red/green CR, while greater lift shifts the CR peak to a smaller solar depression.

082.103 New tendencies of the research into the astronomical refraction. G. Teleki.
Bull. Obs. Astron. Belgrade, No. 131, p. 3 - 8 (1981).

A review is given of the more recent researches into the refractional influences, with special emphasis on those new tendencies in studying or eliminating the influence of the astronomical refraction (prevention, three-dimensional refraction, multi-wave methods, the use of the artificial satellites) of which it is expected to considerably promote the knowledge of this phenomenon.

082.104 On the identification of planetary-scale waves during disturbed conditions. G. W. Prölss.
Veröff. Astron. Inst. Bonn, Nr. 94, 12 pp. (1981).

This study reinvestigates thermospheric density variations previously interpreted as propagating gravity waves. It is shown that these variations are primarily spatial in nature and should be attributed to the quasi-stationary daytime geomagnetic activity effect.

082.105 Determination of stratospheric H_2O and O_3 column densities from balloon altitude far infrared absorption spectra by a curve of growth method.
D. A. Naylor, T. A. Clark, R. T. Boreiko.
Infrared Phys., Vol. 21, 271 - 281 (1981).

Pure rotational lines of H_2O, Q branches of O_3 and single lines of O_3 in the far infrared absorption spectrum of the stratosphere, taken at an altitude of 29.1 km above Texas on 19 June, 1978 from a balloon-borne solar telescope equipped with a rapid-scanning Michelson interferometer, have been analyzed by a curve-of-growth method to yield good values of vertical column densities for these important minor constituents. In this mid-latitude atmosphere, vertical column densities of H_2O and O_3 were found to be $1.32(\pm 0.39) \times 10^{18}$ and $1.72(\pm 0.55) \times 10^{18}$ molecules cm^{-2}, respectively.

082.106 Spectroscopy of the extreme ultraviolet dayglow during active solar conditions.
E. P. Gentieu, P. D. Feldman, R. W. Eastes, A. B. Christensen.
Geophys. Res. Lett., Vol. 8, 1242 - 1245 (1981).

The report focuses on a comparison of the column emission rates of principal airglow features under conditions of vastly increased solar activity as well as a qualitative consideration of newly identified atomic and molecular transitions between 830 and 1250 Å.

082.107 The airglow spectrum from 5500–7000 Å.
A. R. Walker.
South African Astron. Obs. Circ., No. 6, p. 101 - 109 (1981).

Identifications are given for 55 airglow lines in the wavelength range $\lambda\lambda 5500$ to 7000. Accurate wavelengths are tabulated.

082.108 Optical performance data for the UK Infrared Telescope; atmospheric seeing effects in the infrared.
C. M. Humphries, T. E. Perkins.
UKIRT Rep., No. 16, 11 pp. (1980).

082.109 HEAO 3 measurements of the atmospheric positron annihilation line.
W. A. Mahoney, J. C. Ling, A. S. Jacobson.
J. Geophys. Res., Vol. 86, 11098 - 11104 (1981).

All spectra measured with the High Energy Resolution Gamma-Ray Spectroscopy Experiment (HEAO C-1) on the third High Energy Astronomy Observatory (HEAO 3) contain a strong line at 511 keV resulting from positron annihilation. This line originates in the instrument itself, the earth's atmosphere, and cosmic sources, possibly including the diffuse cosmic background. In order to understand the emission from cosmic sources, the atmospheric positron annihilation line emission has been determined as a function of geomagnetic latitude and zenith angle. The atmospheric line has an energy of 511.07 ± 0.10 keV and a net width of 2.29 ± 0.30 keV FWHM. Characteristics of the instrument background have allowed an upper limit of 9.4×10^{-3} photons/cm^2-sec-sr to be placed on any narrow ($\lesssim 3$ keV) diffuse cosmic emission at 511 keV.

082.110 Study of amplitude-frequency scintillation parameters of stars.
I. F. Najbauer, V. S. Fedorishina.
Uzhgorod. univ. Uzhgorod, 1981, 4 pp. In Russian. – Abstr. in Ref. zh., 51. Astron., 12.51.1028 (1981).

082.111 Results of visual and photographic observations of noctilucent clouds with the scientific orbital station Salyut 6.
O. A. Avaste, Ch. Willmann, G. M. Grechko, Yu. V. Romanenko.
Atmos.-opt. yavleniya po nablyud. s orbital'n. stantsii "Salyut", Tartu, 1981, p. 139 - 146. In Russian. – Abstr. in Ref. zh., 62. Issled. kosm. prostranstva, 12.62.482 (1981).

082.112 On the astronomical refraction (II). C. Sugawa.
Proc. Int. Latitude Obs. Mizusawa, No. 20, p. 1 - 20 (1981). In Japanese.

082.113 Effects of meteorological environments on the astrometries of the Mizusawa Observatory (I).
N. Kikuchi, T. Goto, E. Onodera.

Proc. Int. Latitude Obs. Mizusawa, No. 20, p. 111 - 120(1981). In Japanese.

082.114 **Observations of the underground temperature and the meteorology at Nobeyama Radio Observatory.**
S. Hata, A. Tojō, K. Akabane.
Tokyo Astron. Obs. Rep. (No. 74), Vol. 19, 323 - 335 (1981). In Japanese.

082.115 **Numerical computations of refraction.**
R. Fukaya, S. Suzuki.
Tokyo Astron. Obs. Rep. (No. 74), Vol. 19, 466 - 474 (1981). In Japanese.

082.116 **The temperature characteristic for the zenith distance correction of the Danjon astrolabe − discussion of a method for studying the characteristic curve.**
H.-j. Yan, Y.-f. Chen, X.-m. Tong.
Ann. Shanghai Obs. Acad. Sinica, No. 1, p. 44 - 49 (1979).

082.117 **The correction for tropospheric refraction effect in satellite Doppler measurement.** G.-l. Zhu.
Ann. Shanghai Obs. Acad. Sinica, No. 1, p. 101 - 105 (1979).

082.118 **Earth as viewed in the ultraviolet.**
C. B. Opal.
Planetary aeronomy and astronomy, (see 012.063), p. 117 - 125 (1981).

A great variety of atmospheric phenomena on the Earth can be observed in the 300 to 4000 Å region. These include simple absorption; Rayleigh scattering; resonance fluorescence from atoms, molecules, and ions; electron impact excitation; recombination radiation; chemiluminescence; and emission from precipitating ions and neutrals. Observations at these wavelengths from space have been carried out primarily with grating spectrometers and scanning filter photometers. Discussed are variations in reflected sunlight due to ozone absorption, resonant-scattering by geocoronal atoms, resonant-scattering by ions in the plasmasphere, excitation of ultraviolet emissions by electrons in the dayglow and aurora, and the tropical ultraviolet arcs caused by O^+ recombination.

082.119 **Measurement of minor species (H_2, Cl, O_3, NO) in the earth's atmosphere by the occultation technique.**
S. K. Atreya.
Planetary aeronomy and astronomy, (see 012.063), p. 127 - 141 (1981).

This paper provides an overview of the observations of terrestrial minor gases by the stellar occultation technique. The shortcomings of the technique are discussed, and the potential for future observations of significant trace species by such spacecraft as IUE and ST is explored.

082.120 **On an estimate of density variations in the stratosphere from observations of scintillations of stars made from space.** V. Kan.
Izv. AN SSSR. Fiz. atmos. i okeana, Tom 17, 867 - 869 (1981). In Russian. − Abstr. in Ref. zh., 62. Issled. kosm. prostranstva, 1.62.629 (1982).

082.121 **Hard X-ray latitude effect measured during a trans-atlantic balloon flight.**
F. Frontera, F. Fuligni, E. Morelli, G. Ventura.
Scientific ballooning−II, (see 012.065), p. 107 - 110 (1981).

The hard X-ray latitude effect, as measured by two X-ray scintillator detectors, is reported in the range of McIlwain shell parameter L from 1.4 to 2.6 (earth radius units). The data analysis method used to measure the flux from weak X-ray sources is also given.

082.122 **Empirical models of the thermosphere and requirements for improvements.** L. G. Jacchia.

The mesosphere and thermosphere, (see 012.066), p. 81 - 86 (1981).

The different types of variation in the thermosphere are briefly examined and the solar-activity effect is singled out for special attention.

082.123 **Some recent advances in thermospheric models.**
G. Kockarts.
The mesosphere and thermosphere, (see 012.066), p. 197 - 211 (1981).

082.124 **The CO_2 greenhouse effect and the thermal history of the atmosphere.** G. Marx, F. Miskolci.
Life sciences and space research XIX, (see 012.068), p. 5 - 18 (1981).

The influence of the expected rise of CO_2 content in our atmosphere upon terrestrial temperature is uncertain. In order to clarify this problem, and to contribute to the understanding of the CO_2 greenhouse effect on Venus the authors have computed the absorption spectrum of CO_2 for a wide range of atmospheric concentrations. Their results agree with the experimental data of F. W. Taylor. The estimated increase in surface temperature does not reach the boiling point of water.

082.125 **Guide lines for minimizing urban sky glow near astronomical observatories.**
R. Cayrel, F. G. Smith, A. J. Fisher, J. B. de Boer.
Published by International Astronomical Union, Commission Internationale de l'Éclairage, Publ.IAU/CIE No. 1, 4 + 24 pp. (1980).

082.126 **Secular non-random variations of cosmogenic carbon-14 in the terrestrial atmosphere.**
A. Neftel, H. Oeschger, H. E. Suess.
Earth Planet. Sci. Lett., Vol. 56, 127 - 147 (1981).

Exploration of the polar upper atmosphere. See Abstr. 003.054.

Dynamics of the upper atmosphere. See Abstr. 003.082.

Atmospherically-optical phenomena from observations with the Salyut scientific orbital stations. See Abstr. 003.133.

Atmospheric refraction and extinction near the horizon. See Abstr. 004.031.

Infrared spectroscopy of interplanetary dust in the laboratory. See Abstr. 022.152.

Remote investigation of tropospheric turbulence by two-dimensional analysis of stellar scintillation. See Abstr. 031.535.

On the influence of scanning speed in turbulence degraded one-dimensional stellar images. See Abstr. 031.538.

A study of sky noise, 1.5 μm - 5 μm. See Abstr. 031.540.

On the possibility of using a sodar in site testing. See Abstr. 031.627.

Maximum entropy estimation of spread function in astronomical imagery. See Abstr. 031.633.

Alternate approach to the analysis of solar photometer data. See Abstr. 031.635.

Effects of atmospheric refraction on locating and tracking of a telescope.   See Abstr. 032.062.

The neutral mass spectrometer on Dynamics Explorer B.   See Abstr. 032.582.

Synopsis of troposphere and lower stratosphere. See Abstr. 032.621.

Atmospheric sampling.   See Abstr. 032.622.

Vertical sounding balloons for stratospheric photochemistry.   See Abstr. 032.623.

Airglow photography with hypersensitized infrared film.   See Abstr. 036.011.

Variations of UT1−TAI and atmospheric excitations.   See Abstr. 044.032.

Synoptic studies of the middle atmosphere using balloons.   See Abstr. 051.054.

Modelling of orbital perturbations due to radiation pressure for high earth satellites.   See Abstr. 052.017.

The variations of the orbit of 1976 - 87A and determination of air density.   See Abstr. 052.051.

Influence of atmospheric density fluctuations on the accuracy of determination and prediction of orbits of artificial earth satellites.   See Abstr. 052.052.

On the history of the study of the telluric spectrum of the sun. II. 1900 - 1970.   See Abstr. 071.004.

OI 6300 and 7774 Å airglow measurements of equatorial plasma depletions.   See Abstr. 083.024.

Equatorial night-time F-region events: a survey of 6300Å airglow intensity maps at Arecibo. See Abstr. 083.025.

Possible conditions and causes of polarized radiation appearance in the spectrum of polar aurorae and night sky luminescence.   See Abstr. 084.082.

Coordinated ionospheric and magnetospheric observations from the ISIS 2 satellite by the ISIS 2 experimenters. Volume 4. A. Large storms. B. Airglow and related measurements. C. VLF observations.   See Abstr. 084.096.

Increase in the response of the Earth's atmosphere to the sunspot cycle with height above sea level. See Abstr. 085.032.

The Sun's rotation and perturbations of geopotential height and temperature fields in the stratosphere. See Abstr. 085.034.

A dynamical mechanism through which variations in solar ultraviolet radiation can influence tropospheric climate. See Abstr. 085.035.

Climate and paleoclimate: what we can learn about solar luminosity variations.   See Abstr. 085.038.

Solar variability and stochastic effects on climate. See Abstr. 085.039.

The dynamics of a rapidly escaping atmosphere: applications to the evolution of Earth and Venus. See Abstr. 091.094.

Comparative aspects of Venus and terrestrial meteorology.   See Abstr. 093.014.

An evaluation of the sky Lyman alpha emission in IUE spectra.   See Abstr. 106.012.

Connection of the components of the interplanetary magnetic field vector with vertical currents at high latitudes of the northern hemisphere.   See Abstr. 106.036.

Dissipation of the primordial terrestrial atmosphere due to irradiation of the solar far-UV during T Tauri stage. See Abstr. 107.024.

Errata

082.901 Erratum: 'Changes in the planetary heat balance with chemical changes in air' [Planet Space Sci., Vol. 28, 1011 - 1018 (1980)].   J. W. Chamberlain. Planet. Space Sci., Vol. 29, 713 (1981). − See Abstr. 28.082.063.

082.902 Correction to 'Analyses of atomic oxygen, the green line, and Herzberg bands in the lower thermosphere' [J. Geophys. Res., Vol. 86, 206 - 210 (1981)].   R. J. Thomas. J. Geophys. Res., Vol. 86, 11469 (1981). − See Abstr. 29.082.003.

083 Ionosphere

083.001   The *E* and *F* region ionospheric response to solar
flares: 1. Effects of approximations of solar flare
EUV fluxes.   J. T. Mariska, E. S. Oran.
J. Geophys. Res., Vol. 86, 5868 - 5872 (1981).
    SOLRAD and many other satellite systems have provided
a large data base showing the time-dependent behavior of
broadband solar fluxes in the X-ray and EUV spectral regions.
These bands are broad in the sense that one band may contain
many ionospherically important spectral lines. The authors present results of tests performed to determine how this information can best be used to predict the effects of a solar flare on
the ionosphere.

083.002   Global ionospheric electric field measurements in
April 1978.   R. Holzworth, J. Wygant, F. Mozer,
C. Gonzales, R. Greenwald, M. Blanc, J. Vickrey, A. Kishi.
J. Geophys. Res., Vol. 86, 6859 - 6868 (1981).
    The paper presents an introduction to a global campaign
of simultaneous quasistatic electric field measurements from
radars, balloons, and satellites at various places within the
earth's environment for April 8—14, 1978.

083.003   Observation of the ionospheric total electron
content during the solar eclipse of February 16,
1980.   J. Li, L.-b. Li, Z.-h. Wu, B.-k. Wang, Q.-s. Yang.
Acta Geophys. Sinica, Vol. 24, 252 - 256 (1981). In Chinese.

083.004   The behaviour of F_2 layer above Ruili during the
solar eclipse of February 16, 1980.
Z.-x. Tan, X.-y. Huang.
Acta Geophys. Sinica, Vol. 24, 257 - 262 (1981). In Chinese.
    The morphology of F_2 layer during the solar eclipse is
examined, and the mechanism of its formation is discussed
theoretically.

083.005   Ionospheric E region instabilities produced by a
solar eclipse.   X.-j. Zhang, N.-l. Xiong.
Acta Geophys. Sinica, Vol. 24, 263 - 268 (1981). In Chinese.

083.006   Lorentz polarisation term and ionospheric wave
propagation.   R. Misra, B. Chakravarty.
Indian J. Theor. Phys., Vol. 28, 177 - 190 (1980). – Abstr. in
Phys. Abstr., Vol. 84, Abstr. 66788 (1981).

083.007   Ionospheric storm of 4 - 6 December 1958 in the
Indian equatorial region.   J. H. Sastri.
Indian J. Radio Space Phys., Vol. 9, 209 - 213 (1980). – Abstr.
in Phys. Abstr., Vol. 84, Abstr. 66789 (1981).

083.008   HF Doppler observations during the total solar
eclipse of 16 February 1980.
C. S. V. Ramana Murthy, A. S. Mohana Rao, B. R. Rao.
Indian J. Radio Space Phys., Vol. 9, 230 - 233 (1980). – Abstr.
in Phys. Abstr., Vol. 84, Abstr. 66792 (1981).

083.009   Joule heating of the ionosphere.
P. N. Khosa, R. N. Singh.
Indian J. Phys., Part B, Vol. 53B, 199 - 212 (1979). – Abstr.
in Phys. Abstr., Vol. 84, Abstr. 74976 (1981).

083.010   Investigation of the role of various mechanisms in
forming the F2 region of the ionosphere with a
two-dimensional model.
A. G. Kolesnik, I. A. Golikov.
Geomagn. Aehron., Tom 21, 612 - 616 (1981). In Russian.

083.011   The phenomenon of delay of formation of the
sporadic layer and meteor activity.

O. Alimov, G. G. Novikov, L. N. Rubtsov.
Geomagn. Aehron., Tom 21, 732 - 733 (1981). In Russian.

083.012   Ultrawarm ions in the outer ionosphere of the earth.
A. P. Yaichnikov.
Geomagn. Aehron., Tom 21, 733 - 735 (1981). In Russian.

083.013   Global structure of ionospheric disturbances of
convective origin.
N. P. Ben'kova, Yu. M. Berezin, G. V. Bukin, N. K. Osipov.
Geomagn. Aehron., Tom 21, 735 - 737 (1981). In Russian.

083.014   Sizes of sources and mechanisms of generation of
sporadic radio radiation in the auroral ionosphere in
the VHF range.
N. K. Osipov, V. V. Klimenko, A. A. Chernov.
Geomagn. Aehron., Tom 21, 738 - 740 (1981). In Russian.

083.015   Complex geophysical experiment in the near-
equatorial lower ionosphere.
G. V. Givishvili, S. V. Pakhomov.
Geomagn. Aehron., Tom 21, 745 - 747 (1981). In Russian.

083.016   The equatorial ionospheric electric fields. A review.
B. G. Fejer.
J. Atmos. Terr. Phys., Vol. 43, (see 012.008), 377 - 386
(1981).

083.017   The *F*-region dynamo.   H. Rishbeth.
J. Atmos. Terr. Phys., Vol. 43, (see 012.008) 387 -
392 (1981).
    The paper reviews the theory of the *F*-region dynamo
which drives about 10 - 15% of the total mid-latitude ionospheric current by day, and the major part at night. Polarization fields associated with the dynamo cause marked effects in
the night-time *F*-region, notably the mean estward wind. The
paper also discusses the equipotentiality of geomagnetic field
lines and the question of location of *Sq* and *L* current systems.

083.018   Internal structure of the equatorial ionospheric
dynamo.   K.-I. Maeda.
J. Atmos. Terr. Phys., Vol. 43, (see 012.008), 393 - 401 (1981).

083.019   Electromagnetic interactions between high and low
latitudes shown by computer simulation movie.
S. Matsushita, Y. Kamide.
J. Atmos. Terr. Phys., Vol. 43, (see 012.008), 403 - 410 (1981).
    In order to examine the transient behavior of the electro-
magnetic interactions in the ionosphere between high and low
latitudes, the ionospheric electric fields and currents during a
substorm life cycle are computed in relation to downward and
upward pairs of field-aligned currents with a use of realistic
electric conductivity model. The simulated results are compiled in the form of a 16 mm color movie which compresses real
time by a factor of 1/180.

083.020   Penetration of high-latitude electric fields.
Y. Kamide, S. Matsushita.
J. Atmos. Terr. Phys., Vol. 43, (see 012.008), 411 - 425 (1981).
    Simulation studies of ionospheric electric fields with
special emphasis placed on the electrical coupling between
high and low latitudes are presented by means of the algorithm
developed by Kamide and Matsushita (1979) to derive the
horizontal electric fields in the global ionosphere generated by
field-aligned currents in auroral latitudes.

083.021 *D*-region research at Arecibo.   J. D. Mathews.
J. Atmos. Terr. Phys., Vol. 43, (see 012.008), 549 -
556 (1981).

083.022 Modeling the ambient, low latitude *F*-region ionosphere – a review.   D. N. Anderson.
J. Atmos. Terr. Phys., Vol. 43, (see 012.008), 753 - 762 (1981).
The paper describes some recent theoretical modeling efforts which primarily study the effects of **E** × **B** drift, both
vertical and horizontal, on the equatorial ionosphere.

083.023 Longitudinal structure of the *F*-region equatorial
anomaly – a review.   G. O. Walker.
J. Atmos. Terr. Phys., Vol. 43, (see 012.008), 763 - 774 (1981).

083.024 OI 6300 and 7774 Å airglow measurements of
equatorial plasma depletions.
J. G. Moore, E. J. Weber.
J. Atmos. Terr. Phys., Vol. 43, (see 012.008), 851 - 858 (1981).
Simultaneous measurements of the equatorial airglow in
the region of the Appleton anomaly in OI 6300 and 7774 Å
radiation are reported. Peak electron densities in the *F*-region,
both within and outside field-aligned plasma depletions are
estimated from measurements of the OI 7774 Å airglow. The
derived density changes are consistent with satellite and
ionosonde measurements and show the value of optical measurements in determining physical conditions in *F*-region
plasma depletions.

083.025 Equatorial night-time *F*-region events: a survey of
6300 Å airglow intensity maps at Arecibo.
F. A. Herrero, J. W. Meriwether, Jr.
J. Atmos. Terr. Phys., Vol. 43, (see 012.008), 859 - 866 (1981).

083.026 Lunar and planetary influences upon the peak electron density of the ionosphere.
E. Harnischmacher, K. Rawer.
J. Atmos. Terr. Phys., Vol. 43, 643 - 648 (1981).
Observable effects of semidiurnal tidal character caused
by the sun (radiation and gravity) and moon (gravity only)
seem to be influenced by the planet Jupiter and, during
approach, by Venus. These effects can be identified during
summer for suitably selected planetary configurations. The
importance of the lunar effect is found to depend on the solar
hour, season and planetary constellation.

083.027 Circumterrestrial space and ionospheric inhomogeneities.   Yu. S. Vardanyan.
Izv. Akad. Nauk Armyansk.SSR, Fizika, Tom 16, 196 - 199
(1981). In Russian.
Taking into account the ambipolar diffusion in the upper
layers of the ionosphere as well as the terrestrial conductivity,
ionospheric inhomogeneities caused by the motion of neutral
gas in the dynamo-domain are studied.

083.028 Ionospheric effects of the total solar eclipse of
16 February 1980.   H. Chandra, G. Sethia,
G. D. Vyas, M. R. Deshpande, H. O. Vats.
Observations of the total solar eclipse of 16 February 1980,
(see 003.009), p. 57 - 60 (1981).

083.029 Ionospheric electron content observations during the
total solar eclipse of 16 February 1980.
A. Das Gupta, A. Maitra, S. K. Das, S. K. Sen, M. K. Das Gupta.
Observations of the total solar eclipse of 16 February 1980,
(see 003.009), p. 61 - 64 (1981).

083.030 Detection of ionospheric polarization scintillations.
M. R. Deshpande, H. O. Vats, G. Sethia, H. Chandra,
K. N. Iyer, A. V. Janve.
Observations of the total solar eclipse of 16 February 1980,
(see 003.009), p. 65 - 66 (1981).

083.031 Study of variations in the electron density profiles
and the electrojet irregularities caused by the solar
eclipse.   S. P. Gupta, V. K. Parmar, R. C. Shah,
D. K. Chakravarty, B. H. Subbaraya.
Observations of the total solar eclipse of 16 February 1980,
(see 003.009), p. 71 (1981).

083.032 Triggering of ionospheric irregularities by the solar
eclipse.   S. P. Gupta, R. C. Shah,
V. K. Parmar, R. N. Misra, B. H. Subbaraya.
Observations of the total solar eclipse of 16 February 1980,
(see 003.009), p. 72 (1981).

083.033 Effects of the solar eclipse of 16 February 1980 on
the change in phase height and absorption at
2.2 MHz.   N. N. Purkait, M. K. Das Gupta.
Observations of the total solar eclipse of 16 February 1980,
(see 003.009), p. 74 - 76 (1981).

083.034 The influence of the ionosphere-plasmasphere coupling upon the latitude variations of ionospheric
parameters.   I. A. Krinberg, A. V. Tashchilin.
Ann. Géophys., Vol. 36, 537 - 548 (1980). – Abstr. in Phys.
Abstr., Vol. 84, Abstr. 88624 (1981).

083.035 Ionospheric effects of the total solar eclipse of
16 Feb. 1980 observed over Ahmedabad.
H. Chandra, G. Sethia, G. D. Vyas, M. R. Deshpande,
H. O. Vats.
Indian J. Radio Space Phys., Vol. 10, 45 - 48 (1981). – Abstr.
in Phys. Abstr., Vol. 84, Abstr. 88629 (1981).

083.036 Planetary scale waves in the low latitude ionosphere.
U. V. G. Kumar, T. A. Mani, R. V. Narayana,
K. V. V. Ramana.
Indian J. Radio Space Phys., Vol. 10, 58 - 60 (1981). – Abstr.
in Phys. Abstr., Vol. 84, Abstr. 88631 (1981).

083.037 Influence of sporadic-E irregularities on the equatorial lower ionospheric absorption.
K. Parameswaran, B. V. Krishnamurthy.
Indian J. Radio Space Phys., Vol. 10, 61 - 64 (1981). – Abstr.
in Phys. Abstr., Vol. 84, Abstr. 88632 (1981).

083.038 A study of F-region characteristics at a low latitude
station.   B. A. Rao, C. Jogulu.
Indian J. Radio Space Phys., Vol. 10, 65 - 69 (1981). – Abstr.
in Phys. Abstr., Vol. 84, Abstr. 88633 (1981).

083.039 The earth's ionosphere.
Eh. S. Kazimirovskij, Eh. K. Solomatina.
Zemlya Vselennaya, 1981, No. 5, p. 42 - 46. In Russian.

083.040 Variations of the charged particle concentration in
the high-latitude ionosphere during magnetically-
ionospheric disturbances in September 1977 according to
Cosmos 900 data.
G. L. Gdalevich, I. S. Vsekhsvyatskaya, V. D. Ozerov,
T. N. Soboleva.
Kosm. Issled., Tom 19, 718 - 725 (1981). In Russian.

083.041 Forecast of f_o F2 and h_m F2 on the basis of a calcu-
lation of the F2 region and comparative estimate of
its precision for a period of solar activity increase.
A. V. Mikhajlov, G. N. Novikova, G. I. Ostrovskij,
B. E. Serebryakov.
Geomagn. Aehron., Tom 21, 828 - 832 (1981). In Russian.

083.042 Electromagnetic disturbances in the mid-latitude
ionosphere.
Yu. M. Berezin, N. K. Osipov, S. I. Yakhimovich.
Geomagn. Aehron., Tom 21, 833 - 840 (1981). In Russian.

083.043 **Seasonal variations of the ion composition of the D region.**   A. G. Simonov, A. D. Danilov.
Geomagn. Aehron., Tom 21, 932 - 934 (1981). In Russian.

083.044 **Simultaneous rocket measurements of the electron concentration of the ionospheric D region of polar, mean and equatorial latitudes.**   S. V. Pakhomov.
Geomagn. Aehron., Tom 21, 934 - 936 (1981). In Russian.

083.045 **Moving disturbances (stratifications) in the ionospheric E layer.**   K. N. Vasil'ev.
Geomagn. Aehron., Tom 21, 936 - 939 (1981). In Russian.

083.046 **On the depth of the "valley" between the E and F layers of the ionosphere.**   G. N. Tkachev.
Geomagn. Aehron., Tom 21, 939 - 941 (1981). In Russian.

083.047 **Comparison of model and experimental relative ion concentrations in the outer ionosphere.**
A. P. Yaichnikov.
Geomagn. Aehron., Tom 21, 941 - 942 (1981). In Russian.

083.048 **Winter polar ionosphere in the periods of moderate magnetic activity (a mathematical model).**
N. K. Osipov.
Geomagn. Aehron., Tom 21, 942 - 944 (1981). In Russian.

083.049 **On forecast of the development of an ionospheric disturbance.**
V. P. Kuleshova, E. V. Lavrova, L. N. Lyakhova.
Geomagn. Aehron., Tom 21, 948 - 949 (1981). In Russian.

083.050 **Electron temperature models for the F-region and topside ionosphere.**   V. K. Pandey, K. K. Mahajan.
Indian J. Radio Space Phys., Vol. 10, No. 3, p. 85 - 94 (1981).
Abstr. in Phys. Abstr., Vol. 84, Abstr. 107970 (1981).

083.051 **The influence of aurora on ionospheric electron content.**   P. A. Forsyth, J. A. Fulford, J. Hofstee, G. F. Lyon.
Canadian J. Phys., Vol. 59, 1089 - 1096 (1981). – Abstr. in Phys. Abstr., Vol. 85, Abstr. 3116 (1982).

083.052 **Ionospheric electron content measurements during 'Waterhole'.**
J. W. MacDougall, J. A. Fulford, P. A. Forsyth.
Canadian J. Phys., Vol. 59, 1170 - 1174 (1981). – Abstr. in Phys. Abstr., Vol. 85, Abstr. 3126 (1982).

083.053 **Thermal self-focusing instability of plasma waves in a resonance region.**
V. V. Vas'kov, A. V. Gurevich, A. N. Karashtin.
Geomagn. Aehron., Tom 21, 973 - 980 (1981). In Russian.

083.054 **Influence of ionospheric plasma on the general circulation in the upper atmosphere. I.**
A. G. Khantadze, A. I. Gvelesiani.
Geomagn. Aehron., Tom 21, 988 - 992 (1981). In Russian.

083.055 **Influence of nonstationary convection on the distribution of cold plasma in the ionosphere and protonosphere of the earth.**
V. V. Klimenko, A. A. Namgaladze.
Geomagn. Aehron., Tom 21, 993 - 998 (1981). In Russian.

083.056 **Estimate of the anisotropy of the electron temperature in the mid-latitude ionosphere and plasmasphere of the earth.**   Yu. V. Konikov, G. V. Khazanov.
Geomagn. Aehron., Tom 21, 999 - 1003 (1981). In Russian.

083.057 **Using satellite sounding data for investigation of wave motions and the inhomogeneous structure of**
the outer ionosphere.
G. K. Solodovnikov, A. S. Bakaj, V. M. Russkin, M. N. Fatkullin.
Geomagn. Aehron., Tom 21, 1009 - 1117 (1981). In Russian.

083.058 **Precipitation waves in the auroral ionosphere.**
P. A. Bespalov.
Geomagn. Aehron., Tom 21, 1018 - 1022 (1981). In Russian.

083.059 **On quasi-periodic disturbances in the F-region of the ionosphere.**   L. P. Vorob'eva, V. A. Luganin.
Geomagn. Aehron., Tom 21, 1023 - 1028 (1981). In Russian.

083.060 **On a connection of sporadic E_{sr} formations with the B_z-component of the interplanetary magnetic field and electrojet activity.**
R. S. Kukushkina, A. M. Lyatskaya.
Geomagn. Aehron., Tom 21, 1029 - 1033 (1981). In Russian.

083.061 **The ion composition of the lower ionosphere.**
M. N. Vlasov, V. V. Medvedev.
Geomagn. Aehron., Tom 21, 1034 - 1038 (1981). In Russian.

083.062 **On peculiarities of the daily variation of the He^+ and N^+ ion concentrations in the outer ionosphere.**
A. P. Yaichnikov.
Geomagn. Aehron., Tom 21, 1120 - 1121 (1981). In Russian.

083.063 **On colliding instabilities of the longitudinal current in the polar ionosphere.**
A. V. Kustov, V. A. Liperovskij.
Geomagn. Aehron., Tom 21, 1121 - 1124 (1981). In Russian.

083.064 **An association between the equatorial counter-electrojet and the azimuthal component of the IMF.**   N. S. Sastri, B. N. Bhargava, B. R. Arora.
Indian J. Radio Space Phys., Vol. 10, 109 - 112 (1981).
Abstr. in Phys. Abstr., Vol. 85, Abstr. 6496 (1982).

083.065 **Motion of neutral gas in the E-layer of the ionosphere and the electrodynamic state of the circumterrestrial space.**   Yu. S. Vardanyan.
Izv. Akad. Nauk Arm. SSR, Fizika, Tom 16, 268 - 272 (1981). In Russian.
Taking into account the ambipolar diffusion in the upper layers of the ionosphere as well as the conductivity and the horizontal component of the terrestrial magnetic field, the electrodynamic state of the circumterrestrial space due to the motion of neutral gas in the dynamo-domain of the ionosphere is studied.

083.066 **Extraatmospheric visual observations of the nightglow of the earth's ionosphere.**
G. M. Grechko, Yu. V. Romanenko, 9. A. Savchenko.
Atmos.-opt. yavleniya po nablyud. s orbital'n. stantsii "Salyut", Tartu, 1981; p. 147 - 169. In Russian. – Abstr. in Ref. zh., 62. Issled. kosm. prostranstva, 12.62.483 (1981).

083.067 **Energy sources for the ionosphere – a survey.**   K. Rawer.
The mesosphere and thermosphere, (see 012.066), p. 87 - 100 (1981).
An outline is given of early aeronomical ideas about the formation of the ionosphere by solar wave radiation, and its development under the impetus of increasing basic knowledge.

083.068 **Rapid conversion of electromagnetic waves to electrostatic waves in the ionosphere.**
A. Y. Wong, G. J. Morales, D. Eggleston, J. Santoru, R. Behnke.
Phys. Rev. Lett., Vol. 47, 1340 - 1343 (1981).

083.069   Excitation of magnetospheric hydromagnetic waves
        by solar-flare-induced change in ionospheric
conductivity.
T. J. Rosenberg, P. B. Morris, L. J. Lanzerotti.
Phys. Rev. Lett., Vol. 47, 1343 - 1346 (1981).

Observations are reported of magnetospheric hydromagnetic (Alfvén) waves excited by rapid changes in ionospheric conductivity induced by solar flare X-ray and extreme-
ultraviolet fluxes. The observed wave frequencies (~0.01–0.02
(~0.01–0.02 Hz) are those expected for fundamental eigen-
oscillations of the geomagnetic field at middle latitudes. The
measurements confirm a natural hydromagnetic wave generation mechanism not associated with the solar wind or geomagnetic storms. These results have implications for Alfvén
wave generation in other cosmic plasmas.

On a method of estimating the spectral characteristics of the electromagnetic field scattered on ionospheric inhomogeneities.   See Abstr. 022.180.

Spectral characteristics of the electromagnetic field
scattered on ionospheric inhomogeneities.
See Abstr. 022.181.

The Dynamics Explorer Langmuir probe instrument.   See Abstr. 032.587.

The Ion Drift Meter for Dynamics Explorer-B.
See Abstr. 032.588.

Dynamics Explorer program: an overview.
See Abstr. 051.034.

Dynamics Explorer spacecraft and ground operations systems.   See Abstr. 051.035.

Inductive interaction of electroconductive bodies
with ionospheric and interplanetary plasma.
See Abstr. 062.030.

Physical mechanism of the lower-hybrid-drift
instability in a collisional plasma.   See Abstr. 062.040.

Inverse scattering – exact solution of the Gel'fand-
Levitan equation (*ionosphere structure measurement*).
See Abstr. 063.035.

On a possibility of detecting gravitational waves of
extraterrestrial origin from their action on the earth's ionosphere.   See Abstr. 066.190.

Power transmission from the solar wind–magnetosphere dynamo to the magnetosphere and to the ionosphere:
analysis of the IMS Alaska meridian chain data.
See Abstr. 084.006.

A formula for calculating theoretical photoelectron
fluxes resulting from the He^+ 304 Å solar spectral line.
See Abstr. 085.015.

Modulations in Jovian decametric spectra: propagation effects in terrestrial ionosphere and Jovian environment.   See Abstr. 099.193.

Influence of electrostatic fields of the ionosphere on
the dynamics of an ionized meteor trace.
See Abstr. 104.020.

Spectra of scintillations on inhomogeneities of the
ionosphere and interplanetary plasma and the possibility of
their resolution in the decameter radiowave range.
See Abstr. 106.045.

Scintillations of cosmic radio sources in the
decametre waveband. I. Spectra of scintillations due to ionospheric and interplanetary plasma fluctuations and the possibility of their separation.   See Abstr. 141.092.

Interaction of cosmic rays with the magnetosphere
and ionosphere.   See Abstr. 143.063.

084 Aurorae, Geomagnetic Field, Magnetosphere

084.001   Disappearing solar filaments: a useful predictor of
geomagnetic activity.
J. A. Joselyn, P. S. McIntosh.
J. Geophys. Res., Vol. 86, 4555 - 4564 (1981).

A search for possible solar sources of geomagnetic storms
from June 1976 through June 1979 has revealed that a significant number of the storms, including the two largest, can only
be associated with filament disappearances. This result is supported by the many recent papers studying SKYLAB and
other observations of coronal transients which always find a
strong correlation between those transients and eruptive prominences. By analyzing the physical characteristics of those disappearances which precede magnetic storms and those which
do not, some tentative guidelines for forecasting geomagnetic
disturbances have been developed based on evidence of a significant restructuring of the implied coronal magnetic field
which could release solar wind plasma favorably positioned to
impact the earth.

084.002   Alfvén wave resonances in a realistic magnetospheric
magnetic field geometry.
H. J. Singer, D. J. Southwood, R. J. Walker, M. G. Kivelson.
J. Geophys. Res., Vol. 86, 4589 - 4596 (1981).

084.003   Long-period geomagnetic secular variations since
12,000 yr BP.   K. M. Creer.
Nature, Vol. 292, 208 - 212 (1981).

A comparison of secular variation records of the geomagnetic field inclination obtained from palaeomagnetic
studies of European and North American post-Glacial and
late-Glacial sediments suggests that geomagnetic westward
drift has persisted through the past 12,000 yr. Between 12,000
and 2,500 yr BP the overall average drift rate was between
0.14 and 0.16 deg yr^{-1}. At ~2,500 yr BP it is tentatively
suggested that the drift rate increased and rose to a maximum
of nearly 0.5 deg yr^{-1} before decreasing to historically observed values of ~0.25 deg yr^{-1}.

084.004   Towards the numerical magnetosphere.
M. Brown.
Nature, Vol. 292, 290 - 291 (1981).

084.005   The mechanism of magnetospheric substorm and
the MHD waves of the solar wind.   W.-R. Hu.
Planet. Space Sci., Vol. 29, 695 - 702 (1981).

084.006   Power transmission from the solar wind–magnetosphere dynamo to the magnetosphere and to the
ionosphere: analysis of the IMS Alaska meridian chain data.
S.-I. Akasofu, Y. Kamide, J. R. Kan, L. C. Lee, B.-H. Ahn.
Planet. Space Sci., Vol. 29, 721 - 730 (1981).

084.007   Configuration of the auroral oval and the interplanetary magnetic field.   T. Yeh, Y. Kamide.
Planet. Space Sci., Vol. 29, 731 - 734 (1981).

The poleward boundary of the auroral oval, whose
footline forms the periphery of the polar cap, is calculated,
based on a model in which the geomagnetic field is inter-
permeated with the interplanetary field. It is shown that the
calculated auroral oval size varies with the strength and
direction of the interplanetary magnetic field, in agreement
with recent observations of the location of large-scale nightside auroras.

084.008   Some properties of geomagnetic field pulsations in
the 0.1–1.0-Hz frequency range: a quantitative
description and comparison with the satellite and ground-
based observations.

A. E. Antonova, D. S. Fligel (*Fligel'*), B. V. Dovbnja (*Dovbnya*).
Planet. Space Sci., Vol. 29, 793 - 801 (1981).

084.009   Auroral X-ray images.
B. H. Mauk. J. Chin, G. Parks.
J. Geophys. Res., Vol. 86, 6827 - 6835 (1981).

A new X-ray 'pin-hole' camera has been developed to
study the spatial and temporal characteristics of the energetic
component of auroral electron precipitation. The camera
obtains spectral information on bremsstrahlung X rays with
energies between 22 and 110 keV, and it details structure with
scale lengths between 10 and 80 km ($9°-62°$ cone angles). The
camera has been deployed by a balloon into several auroral
break-up associated events. In this report preliminary images
are shown from these events, and several types of observed
phenomena are discussed.

084.010   Optical emissions and ionization profiles during an
intense pulsating aurora.
R. D. Sears, R. R. Vondrak.
J. Geophys. Res., Vol. 86, 6853 - 6858 (1981).

084.011   Analysis of the geomagnetic effect during the
February 16, 1980 solar eclipse in Yunnan Province.
C.-f. Liu, H.-l. Huang, C.-c An, Y.-h. Wang, X.-l. Wang, J. Li.
Acta Geophys. Sinica, Vol. 24, 269 - 278 (1981). In Chinese.

084.012   Alfvén fluctuation and the earth's magnetospheric
substorm.   W.-r. Hu.
Acta Geophys. Sinica, Vol. 24, 279 - 286 (1981). In Chinese.

A mechanism of the earth's magnetospheric substorm is
discussed. Alfvén waves with energy greater than 10^{18} ergs/sec
can pass through from the magnetosheath into the magnetotail, and the wave energy is dissipated in the plasma sheet. The
particles are heated and accelerated in the plasma sheet, and
then injected into the near earth space, and produce the ring
current and auroral substorm. The propagation process of the
Alfvén wave in a shear velocity field is calculated, and the
dissipation of Alfvén waves in the magnetosphere is given.

084.013   Saturation and energy-conversion efficiency of
auroral kilometric radiation.
C. S. Wu, S. T. Tsai, M. J. Xu, J. W. Shen.
Astrophys. J., Vol. 248, 384 - 391 (1981).

A quasi-linear theory is used to study the saturation level
of the auroral kilometric radiation. The investigation is based
on the assumption that the emission is due to a cyclotron
maser instability as suggested by Wu and Lee and Lee et al. The
thermodynamic bound on the radiation energy is also
estimated separately. The energy-conversion efficiency of the
radiation process is discussed. The results are consistent with
observations.

084.014   ISEE-1, -2 and -3 observation of the interaction between an interplanetary shock and the earth's
magnetosphere: a rapid traversal of the magnetopause.
D. Winterhalter, M. G. Kivelson, C. T. Russell, E. J. Smith.
Geophys. Res. Lett., Vol. 8, 911 - 914 (1981).

084.015   Solar wind control of auroral zone geomagnetic
activity.   C. R. Clauer, R. L. McPherron, C. Searls,
M. G. Kivelson.
Geophys. Res. Lett., Vol. 8, 915 - 918 (1981).

An empirical analysis of solar wind-magnetosphere energy
coupling functions is reported. Using the technique of linear
prediction filtering with 2.5 minute data, the authors examine
the relationship of auroral zone geomagnetic activity to solar
wind power input functions which depend on the solar wind

quantities VB^2, VB_s, or $V^2 B_s$. In this analysis a least squares prediction filter or impulse response function which relates a solar wind power function to an auroral zone geomagnetic index is designed directly from the data.

084.016 **The auroral plasma cavity.** W. Calvert.
Geophys. Res. Lett., Vol. 8, 919 - 921 (1981).

084.017 **An overview of the external magnetic field with regard to magnetic surveys.**
R. D. Regan, P. Rodriguez.
Geophys. Surv., Vol. 4, 255 - 296 (1981). − Abstr. in Phys. Abstr., Vol. 84, Abstr. 83113 (1981).

084.018 **Nature of geomagnetic field variation on AQDs** *(abnormal quiet days)* **at equatorial latitudes.**
J. H. Sastri.
Indian J. Radio Space Phys., Vol. 10, 35 - 38 (1981). − Abstr. in Phys. Abstr., Vol. 84, Abstr. 83116 (1981).

084.019 **Equatorial counter electrojet & interplanetary magnetic field.** R. G. Rastogi.
Indian J. Radio Space Phys., Vol. 10, No. 1, p. 1 - 15 (1981). Abstr. in Phys. Abstr., Vol. 84, Abstr. 83402 (1981).

084.020 **Magnetospheric plasmas.** J. W. Dungey.
Philos. Trans. R. Soc. London, Ser. A, Vol. 300, 489 - 496 (1981). − Abstr. in Phys. Abstr., Vol. 84, Abstr. 83412 (1981).

084.021 **Equilibrium motion of quiet auroral arcs.**
V. B. Lyatskij, S. V. Leont'ev.
Geomagn. Aehron., Tom 21, 668 - 673 (1981). In Russian.

084.022 **Electric fields and electron acceleration above discrete aurorae.** A. P. Kropotkin.
Geomagn. Aehron., Tom 21, 689 - 692 (1981). In Russian.

084.023 **On a possible mechanism of generation of the earth's magnetic field. I.** Yu. A. Brodskij.
Geomagn. Aehron., Tom 21, 720 - 726 (1981). In Russian.

084.024 **Estimate of the ground-level effect of currents on the magnetopause and on the magnetospheric tail.**
I. T. Matveenkov.
Geomagn. Aehron., Tom 21, 747 - 749 (1981). In Russian.

084.025 **Linear dependence of the intensity of geomagnetic variations in the near-polar region on the value of the southern and northern components of the interplanetary magnetic field.**
V. O. Papitashvili, O. A. Troshichev, A. N. Zajtsev.
Geomagn. Aehron., Tom 21, 752 - 754 (1981). In Russian.

084.026 **Exploiting geomagnetic data.**
S. R. C. Malin, D. R. Barraclough.
Nature, Vol. 293, 337 (1981).

084.027 **Dependence of the geometry of the region of open field lines on the interplanetary magnetic field.**
S.-I. Akasofu, D. N. Covey, C.-I. Meng.
Planet. Space Sci., Vol. 29, 803 - 807 (1981).
The geometry of the open flux area in the polar region is computed by superposing a uniform interplanetary magnetic field with various orientation angles to a model of the magnetosphere.

084.028 **Asymmetry effects associated with the x-component of the IMF in a magnetically open magnetosphere.** S. W. H. Cowley.
Planet. Space Sci., Vol. 29, 809 - 818 (1981).
The origin of magnetospheric asymmetry effects associated with the equatorial plane component of the interplanetary

magnetic field (IMF) is discussed in terms of the forces exerted on open flux tubes mapping into the solar wind.

084.029 **Annual and semiannual variations of the geomagnetic field at equatorial locations.** W. H. Campbell.
J. Atmos. Terr. Phys., Vol. 43, (see 012.008), 607 - 616 (1981).

084.030 **Lunar and solar daily variations of the geomagnetic field at Italian stations.** A. Palumbo.
J. Atmos. Terr. Phys., Vol. 43, 633 - 642 (1981).

084.031 **Source effects and the interpretation of geomagnetic sounding data at sub-auroral latitudes.**
M. Mareschal.
Geophys. J. R. Astron. Soc., Vol. 67, 125 - 136 (1981).

084.032 **A differencing method for the determination of the lunar daily geomagnetic variation from short data series.** R. Sellek.
Geophys. J. R. Astron. Soc., Vol. 67, 229 - 233 (1981).

084.033 **The influence of injection conditions on the low-edge altitude of artificial auroral rays.**
G. P. Milinevskij.
Problems of cosmic physics. Vyp. 16, (see 003.005), p. 74 - 79 (1981). In Russian.

084.034 **TV observations of auroral active forms and their spectra.** N. D. Galinskij, A. M. Evtushevskij, G. P. Milinevskij, I. V. Myasishcheva, V. A. Kravchenko, O. S. Popov.
Problems of cosmic physics. Vyp. 16, (see 003.005), p. 79 - 86 (1981). In Russian.
The results of TV observations of auroral active forms and their spectra are discussed.

084.035 **Relations between $\lambda\lambda$ 5577 and 6300 Å [O I] emissions from patrol spectral observations of aurorae in the southern hemisphere.**
N. N. Bliznyuk, N. I. Dzyubenko.
Problems of cosmic physics. Vyp. 16, (see 003.005), p. 87 - 89 (1981). In Russian.
From the materials of auroral observations on patrol spectrographs during the IGY in the coordinates invariant altitude − invariant time a picture of spatial-temporal variations for the ratio I_{5577}/I_{6300} of oxygen lines intensities in the southern hemisphere is built.

084.036 **Magnetic signals from the core of the earth and secular variation.** L. R. Alldredge.
J. Geophys. Res., Vol. 86, 7957 - 7965 (1981).
An oscillating, radial magnetic dipole source was assumed to exist in the core of the earth, 100 km beneath the core-mantle boundary. As an approximation, electromagnetic propagation was assumed in the core in lieu of hydromagnetic propagation, which could not be used because of unknown internal fields. Using Debye potentials, the radial and horizontal components of the surface fields were calculated using various assumed conductivity parameters in the core and in the mantle. It is concluded that most spherical harmonic models of the earth's magnetic field do not include enough terms to describe properly the field of core sources with periods of tens of years. These short-period variations are especially important in describing the secular variation.

084.037 **Capture of electrons with E > 100 MeV generated in a low-density residual atmosphere.**
N. L. Grigorov, L. V. Kurnosova, L. A. Razorenov, M. I. Fradkin.
Izv. AN SSSR. Ser. fiz., Tom 45, 642 - 643 (1981). In Russian. Abstr. in Ref. zh., 62. Issled. kosm. prostranstva, 8.62.401 (1981).

084.038 Excitation of the auroral kilometric radiation by whistler mode instability.   S. Sharma.
Bull. Astron. Soc. India, Vol. 9, 64 (1981). – Abstract.

084.039 On the planetary nature of the secular variation jump in 1969–1970.
H. D. Chau, J. Ducruix, J.-L. Le Mouël.
C. R. Acad. Sci. Paris, Tome 293, Sér. II, 157 - 160 (1981). In French.
It is shown that the secular acceleration impulse of 1969–1970 occurred all over the earth. From 1950 to 1979 the secular variation of the East component of the geomagnetic field can be represented in a first approximation as the product of a space function by a function of time with a discontinuous derivative.

084.040 The aurora 1979.   D. Gavine.
J. British Astron. Assoc., Vol. 91, 583 - 587 (1981).

084.041 The spectral hardening associated with the westward travelling surge.   Å. Steen, G. Gustafsson.
Planet. Space Sci., Vol. 29, 1011 - 1017 (1981).
Photometer recordings of auroral emissions from N_2^+ and OI have demonstrated that the ratios between the emissions show a characteristic plateau just prior to the passage of a westward travelling surge. The results have been interpreted in terms of an instability that is switched on during the time of the plateau and then turned off when the electron density is increased above a certain limit.

084.042 Self-exciting dynamos and geomagnetic polarity changes.   R. Hide.
Nature, Vol. 293, 728 - 729 (1981).
Palaeomagnetic and archaeomagnetic data might be expected to show evidence that departures from axial symmetry are systematically less during the decay phase of a geomagnetic polarity "reversal" or "excursion" than during the growth or recovery phase. The recent proof that neither fluid compressibility nor thermoelectric effects of the Nernst-Ettinghausen type can prevent the ohmic decay of axisymmetric magnetic fields is also discussed.

084.043 Solar eclipse effect on short period geomagnetic field variations.
G. K. Rangarajan, A. V. S. Murty.
Observations of the total solar eclipse of 16 February 1980, (see 003.009), p. 51 - 52 (1981).

084.044 A note on the geomagnetic observations made at Hyderabad and Etaiyapuram during the solar eclipse of 16 February 1980.
B. J. Srivasatava, D. Pandurangam, T. S. Sastry, H. Abbas.
Observations of the total solar eclipse of 16 February 1980, (see 003.009), p. 53 - 56 (1981).

084.045 Method of natural orthogonal components applied to equatorial geomagnetic variations.
M. Rajaram.
Ann. Géophys., Vol. 36, 599 - 603 (1980). – Abstr. in Phys. Abstr., Vol. 84, Abstr. 88273 (1981).

084.046 The Kelvin-Helmholtz instability at the magnetopause and the solar wind-magnetosphere interaction.
J. Ziemkiewicz-Dabrowska.
Acta Geophys. Polonica, Vol. 28, 267 - 265 (1980). – Abstr. in Phys. Abstr., Vol. 84, Abstr. 88658 (1981).

084.047 Severe geomagnetic disturbance of 19 Dec. 1980.
G. K. Rangarajan, R. S. Ramteke, G. V. Krishnamoorty.
Indian J. Radio Space Phys., Vol. 10, 76 - 77 (1981). – Abstr. in Phys. Abstr., Vol. 84, Abstr. 88662 (1981).

084.048 Magnetospheric substorms: a newly emerging model.
S.-I. Akasofu.
Planet. Space Sci., Vol. 29, 1069 - 1078 (1981).
A surge of progress in magnetospheric substorm studies is expected by the following three recent developments, (1) the finding of the solar wind-magnetosphere energy coupling function ϵ, (2) the determination of the Pedersen current distribution over the entire polar region, and (3) a new understanding of the auroral potential structure. In this paper, the significance of the three developments and the newly emerging model of magnetospheric substorms is described.

084.049 On the use of geomagnetic observatory results in geoelectric field work.   J. C. Miletits, J. Vero.
Magy. Geofiz., Vol. 22, No. 2, p. 62 - 71 (1981). In Hungarian. Abstr. in Phys. Abstr., Vol. 84, Abstr. 93763 (1981).

084.050 Dynamo theory of the Earth's varying magnetic field.   D. Rittenhouse Inglis.
Rev. Mod. Phys., Vol. 53, 481 - 496 (1981). – Abstr. in Phys. Abstr., Vol. 84, Abstr. 93765 (1981).

084.051 The 1980 geomagnetic reference field.
D. R. Barraclough.
Nature, Vol. 294, 14 - 15 (1981).

084.052 ULF geomagnetic power near $L = 4, 6$. Relationship to upstream solar wind quantities.
A. Wolfe, A. Meloni.
J. Geophys. Res., Vol. 86, 7507 - 7512 (1981).
Geomagnetic energy densities, computed at hourly intervals in three frequency bands from magnetometer data at Pittsburg, New Hampshire, ($L \sim 3.5$) for 28 days during July and August 1975, are correlated with upstream solar wind parameters obtained from instruments on the Imp J spacecraft.

084.053 IMF sector effects on the polar geomagnetic field in winter.   S. Matsushita, W.-Y. Xu.
J. Geophys. Res., Vol. 86, 7733 - 7743 (1981).

084.054 Generation of electric fields in the magnetospheric tail in the presence of a conducting ionosphere.
V. T. Pivovarov.
Kosm. Issled., Tom 19, 726 - 732 (1981). In Russian.

084.055 Emission of λ 6300 Å [O I] in auroral rays.
L. V. Moiseeva.
Geomagn. Aehron., Tom 21, 863 - 866 (1981). In Russian.

084.056 On a possible mechanism of generation of the earth's magnetic field. II.   Yu. A. Brodskij.
Geomagn. Aehron., Tom 21, 890 - 897 (1981). In Russian.

084.057 X-ray scanning of overhead aurorae from rockets.
J. R. Barcus, R. A. Goldberg, L. H. Gesell.
J. Atmos. Terr. Phys., Vol. 43, 1003 - 1013 (1981).

084.058 Global properties of the magnetosphere during a substorm growth phase: a case study.
D. N. Baker, E. W. Hones, Jr., P. R. Higbie, R. D. Belian, P. Stauning.
J. Geophys. Res., Vol. 86, 8941 - 8956 (1981).
It is the purpose of the paper to present detailed evidence for the point of view that the magnetosphere often stores energy for a substantial period of time and begins to release this energy suddenly at the substorm expansion phase onset.

084.059 Mechanism for the auroral red lower border.
W. Benesch.
J. Geophys. Res., Vol. 86, 9065 - 9072 (1981).
A model is proposed for the generation of the red lower

border of type B auroras that results from the effect of increased collision frequency at the lower altitudes on the distribution of excitation within the nitrogen molecule. There are two characteristic spectral distributions for the red sequence ($\Delta\nu = 3$) of the molecular nitrogen first-positive system, one for normal auroras and one for laboratory (higher pressure) discharges. The laboratory spectrum is shifted in wavelength by the intersystem collisional transfer of excitation with respect to the auroral spectrum, and such a shift manifests itself to the human eye as a two-fold increase in luminous flux in the red. Recent investigations have suggested that it is the $\lambda 5577$ Å cut-off that is responsible for the red border and have placed the lower limit of the green line emission at about 100 km. The present enhancement mechanism, however, is consistent with the onset of added bright red emission at 85 km.

084.060 A campaign to study pulsating auroras.
D. J. McEwen, C. N. Duncan.
Canadian J. Phys., Vol. 59, 1029 - 1033 (1981). – Abstr. in Phys. Abstr., Vol. 84, Abstr. 107938 (1981).

084.061 Energies of precipitating electrons during pulsating aurora events derived from ionosonde observations.
J. W. MacDougall, J. Hofstee, J. A. Koehler.
Canadian J. Phys., Vol. 59, 1070 - 1076 (1981). – Abstr. in Phys. Abstr., Vol. 84, Abstr. 107939 (1981).

084.062 TV morphology of some episodes of pulsating auroras.   A. Vallance Jones, R. L. Gattinger.
Canadian J. Phys., Vol. 59, 1077 - 1082 (1981). – Abstr. in Phys. Abstr., Vol. 84, Abstr. 107940 (1981).

084.063 Electron energy measurements in pulsating auroras.
D. J. McEwen, E. Yee, B. A. Whalen, A. W. Yau.
Canadian J. Phys., Vol. 59, 1106 - 1115 (1981). – Abstr.in Phys. Abstr., Vol. 84, Abstr. 107941 (1981).

084.064 Pulsating auroras in relation to proton and electron auroras.   F. Creutzberg, R. Gattinger, F. Harris, A. Vallance Jones.
Canadian J. Phys., Vol. 59, 1124 - 1130 (1981). – Abstr. in Phys. Abstr., Vol. 84, Abstr. 107942 (1981).

084.065 Satellite observations of spatial and temporal fluctuations in post-midnight auroras.
C. D. Anger, J. S. Murphree, L. L. Cogger.
Canadian J. Phys., Vol. 59, 1137 - 1142 (1981). – Abstr. in Phys. Abstr., Vol. 84, Abstr. 107943 (1981).

084.066 Statistics of pulsating auroras on the basis of all-sky TV data from five stations. I. Occurrence frequency.
T. Oguti, S. Kokubun, K. Hayashi, K. Tsuruda, S. Machida, T. Kitamura, O. Saka, T. Watanabe.
Canadian J. Phys., Vol. 59, 1150 - 1157 (1981). – Abstr. in Phys. Abstr., Vol. 84, Abstr. 107944 (1981).

084.067 Revised age estimates of Brunhes palaeomagnetic events: support for a link between geomagnetism and eccentricity.   M. R. Rampino.
Geophys. Res. Lett., Vol. 8, 1047 - 1050 (1981).
The timing of the geomagnetic events seems to coincide with times of peak eccentricity of the earth's orbit, suggesting a causal connection.

084.068 Auroral kilometric radiation: a theoretical review.
C. L. Grabbe.
Rev. Geophys. Space Phys., Vol. 19, 627 - 633 (1981).
A number of theories have been proposed in recent years to explain auroral kilometric radiation. These include an anisotropic velocity distribution instability, mode conversion of electron cyclotron waves to ordinary mode radiation, soliton radiation, beam-driven instability of electromagnetic

waves via low-frequency turbulence, a loss cone instability, beating of coherent electrostatic waves, and beam amplification of electromagnetic waves via coherent density fluctuations. These will all be reviewed, and comparisons of prediction made with observations. Emphasis will be placed on the three recent proposals.

084.069 The influence of the interplanetary magnetic field and thermal pressure on the position and shape of the magnetopause.   H. C. Zhuang, C. T. Russell, R. J. Walker.
J. Geophys. Res., Vol. 86, 10009 - 10021 (1981).
An ellipsoidal model is used to represent the location of the dayside magnetopause and to study the influences of the interplanetary magnetic field and thermal pressure on its location.

084.070 Polar aurora of 12 and 13 April 1981 and preceding solar activity.   L. Křivský.
Říše hvězd, Vol. 62, 185 - 186 (1981). In Czech.

084.071 Latitudinal and temporal characteristics of pulsating auroras.   C. N. Duncan, F. Creutzberg, R. L. Gattinger, F. R. Harris, A. Vallance Jones.
Canadian J. Phys., Vol. 59, 1063 - 1069 (1981). – Abstr. in Phys. Abstr., Vol. 85, Abstr. 3115 (1982).

084.072 Auroral electron energies: comparisons of in situ measurements with spectroscopically inferred energies.   D. J. McEwen, C. N. Duncan, R. Montalbetti.
Canadian J. Phys., Vol. 59, 1116 - 1123 (1981). – Abstr. in Phys. Abstr., Vol. 85, Abstr. 3117 (1982).

084.073 Latitudinally propagating on-off switching aurorae and associated geomagnetic pulsations: a case study of an event of February 20, 1980.   T. Oguti, S. Kokubun, K. Hayashi, K. Tsuruda, S. Machida, T. Kitamura, O. Saka, T. Watanabe.
Canadian J. Phys., Vol. 59, 1131 - 1136 (1981). – Abstr. in Phys. Abstr., Vol. 85, Abstr. 3118 (1982).

084.074 Rotational temperature variations in pulsating auroras.
R. A. Koehler, M. M. Shepherd, G. G. Shepherd, K. V. Paulson.
Canadian J. Phys., Vol. 59, 1143 - 1149 (1981). – Abstr. in Phys. Abstr., Vol. 85, Abstr. 3119 (1982).

084.075 Preliminary results from Project Waterhole – an auroral modification experiment.   B. A. Whalen, A. W. Yau, F. Creutzberg, R. L. Gattinger, F. R. Harris, A. Vallance Jones, G. A. McNamara, M. B. Pongratz, G. M. Smith, P. A. Forsyth, J. A. Koehler.
Canadian J. Phys., Vol. 59, 1175 - 1182 (1981). – Abstr. in Phys. Abstr., Vol. 85, Abstr. 3120 (1982).

084.076 Plasma measurements in pulsating auroras.
A. G. McNamara.
Canadian J. Phys., Vol. 59, 1083 - 1088 (1981). – Abstr. in Phys. Abstr., Vol. 85, Abstr. 3125 (1982).

084.077 Formation of ion conics and lower hybrid acceleration in the supraauroral region.
T. Chang, B. Coppi.
Plasma astrophysics, (see 012.042), p. 379 - 381 (1981).

084.078 Identification of the discontinuities at the magnetopause.   G.-c. Zhou.
Plasma astrophysics, (see 012.042), p. 413 - 414 (1981).

084.079 On a calculation of the characteristics of magnetospheric plasma in a collisionless region.
S. A. Mart'yanov.
Geomagn. Aehron., Tom 21, 1039 - 1043 (1981). In Russian.

084.080 **Influence of the structure of the magnetospheric boundary layer on the Kelvin-Helmholtz instability.**
A. G. Morozov, V. V. Mishin.
Geomagn. Aehron., Tom 21, 1044 - 1047 (1981). In Russian.

084.081 **Variations of the ring current field in time of being of the earth in high-speed solar wind.**
M. S. Bobrov.
Geomagn. Aehron., Tom 21, 1048 - 1052 (1981). In Russian.

084.082 **Possible conditions and causes of polarized radiation appearance in the spectrum of polar aurorae and night sky luminescence.**   E. G. Berezhko.
Geomagn. Aehron., Tom 21, 1131 - 1132 (1981). In Russian.

084.083 **First remote sensing of the plasmapause by terrestrial myriametric radiation.**   D. Jones.
Nature, Vol. 294, 728 - 730 (1981).

084.084 **Solar eclipse effect on short-period geomagnetic field variations.**
G. K. Rangarajan, A. V. S. Murty.
Curr. Sci., Vol. 50, 185 - 186 (1981). — Abstr. in Phys. Abstr., Vol. 85, Abstr. 6307 (1982).

084.085 **Ionosonde observations of pulsating auroras.**
J. W. MacDougall, J. A. Koehler, J. Hofstee, D. J. McEwen.
Canadian J. Phys., Vol. 59, 1049 - 1055 (1981). — Abstr. in Phys. Abstr., Vol. 85, Abstr. 6487 (1982).

084.086 **An auroral torch structure as an activity center of pulsating auroras.**   T. Oguti, S. Kokubun, K. Hayashi, K. Tsuruda, S. Machida, T. Kitamura, O. Saka, T. Watanabe.
Canadian J. Phys., Vol. 59, 1056 - 1062 (1981). — Abstr. in Phys. Abstr., Vol. 85, Abstr. 6488 (1982).

084.087 **Low-frequency bursts in the earth's plasmasphere.**
V. P. Grigor'eva.
Astron. Tsirk., No. 1115, p. 1 - 2 (1980). In Russian.

084.088 **La magnétosphère terrestre.**   M. Roth.
Ciel, Vol. 43, 269 - 273 (1981).

084.089 **Three-dimensional radial trajectories of low-frequency electromagnetic waves in the earth's magnetosphere.**   V. I. Aksenov, A. V. Moshkov.
Kosm. Issled., Tom 19, 876 - 883 (1981). In Russian.

084.090 **Radio emission at mean latitudes during the sudden commencement of a magnetic storm.**
S. I. Musatenko.
Astron. Tsirk., No. 1140, p. 1 - 2 (1980). In Russian.

084.091 **Parallax and atmospheric scattering effects on the inversion of satellite auroral observations.**
V. J. Abreu, P. B. Hays.
Appl. Opt., Vol. 20, 2203 - 2209 (1981).
A technique for the inversion of satellite auroral brightness observations is developed, which takes into account the backscattering of light from the snow-covered ground and atmospheric scattering. The theory includes parallax effects. Parallax arises when a point in the aurora is observed from different angles against a background with a variable brightness. It is shown that observations from a spinning satellite at any given angle from nadir are sufficient to recover the auroral form.

084.092 **Vibrational enhancement and the excitation of N_2^+ and the first negative system in the high-altitude red aurora and the dayside cusp.**   V. Degen.
J. Geophys. Res., Vol. 86, 11372 - 11378 (1981).
Broadfoot's (1967) theory for fluorescent scattering of sunlight by N_2^+ in high-altitude auroral rays is updated with new radiative and electron impact cross sections and extended to include the analogous process of fluorescence induced by low energy electrons. The two mechanisms account successfully for the observed vibrational enhancement of the first negative system in both the sunlit and non-sunlit phases of the great type-A red aurora of February 10, 1958, without recourse to proton precipitation as has commonly been suggested, and permit a number of conclusions regarding vibrational temperatures, ion lifetimes, and the role of electron fluxes. Recalculated g-values for solar scattering and corresponding arrays of band intensities of the first negative and Meinel systems are given.

084.093 **Paleomagnetism and Mesozoic-Cenozoic paleocontinental maps.**
J. C. Briden, A. M. Hurley, A. G. Smith.
J. Geophys. Res., Vol. 86, 11631 - 11656 (1981).
A method of making paleocontinental reconstructions based on ocean floor magnetic anomaly data and land-based paleomagnetic data is briefly described, and its limitations are assessed. Synthetic polar wander paths for the period 20–200 Ma are presented for Australia, Antarctica, India, Africa, Eurasia, North America, and South America. Comparison is made with observed paths, except for Antarctica, for which the data are too sparse. There are no significant differences between observed and synthetic paths.

084.094 **Zonal harmonic models of reversal transition fields.**
I. Williams, M. Fuller.
J. Geophys. Res., Vol. 86, 11657 - 11665 (1981).

084.095 **Strange attractor character of large-scale non-linear dynamos.**   F. Krause, P. H. Roberts.
Planetary interiors, (see 012.061), p. 231 - 240 (1981).
Solutions showing the characteristics of strange attractors are proved to exist for the non-linear equations of strongly simplified dynamo models. Particular attention is given to reversals associated with toroidal oscillations within the Earth's core.

084.096 **Coordinated ionospheric and magnetospheric observations from the ISIS 2 satellite by the ISIS 2 experimenters. Volume 4. A. Large storms. B. Airglow and related measurements. C. VLF observations.**
J. R. Burrows, L. L. Cogger, H. G. James (Coordinators).
National Space Science Data Center/World Data Center A for Rockets and Satellites, National Aeronautics and Space Administration, Goddard Space Flight Center, Greenbelt, Maryland, NSSDC/WDC-A-R & S 81-01, 6 + 177 pp. (1981).
Contents: List of ISIS 2 experimenters. Satellite description. Instrument descriptions and data processing. Data format descriptions. Geophysical data set: large storms; airglow and related measurements; VLF observations. List of contributors to the Alouette-ISIS program.

084.097 **Perspectives of auroral X-ray measurements in the 1980s.**   G. Kremser.
Scientific ballooning—II, (see 012.065), p. 39 - 47 (1981).

084.098 **A decade of balloon observations of auroral X-rays.**
D. Venkatesan, K. K. Vij.
Scientific ballooning—II, (see 012.065), p. 49 - 62 (1981).

084.099 **Observations of the aurora in the far ultraviolet from "Cosmos-900".**   K. I. Gringauz, L. Martini, N. M. Shutte (Shyutte), A. I. Puolokainen.
The mesosphere and thermosphere, (see 012.066), p. 77 - 80 (1981).
The results from observations of auroral emissions

within the wavelength band 115 - 135 nm are presented. The experiment was carried out on board the satellite "Cosmos-900", launched on March 30, 1977, to an almost circular polar orbit. The authors assume that the precipitating fluxes of protons and electrons were the sources of excitation, according to the theory.

084.100 SI and Gaussian CGS units, conversions and equations for use in geomagnetism.   M. A. Payne.
Phys. Earth Planet Inter., Vol. 26, P10 - P16 with a correction Vol. 27, 233 (1981).
A resumé is given of the fundamental equations and units of electromagnetism, and their conversion from CGS to SI equivalents.

084.101 Geomagnetic and solar data.
J. V. Lincoln, H. E. Coffey (Editors).
J. Geophys. Res., Vol. 86, 4842, 5902, 6960, 7799, 9235, 10130, 11470 (1981).

Die Magnetosphäre der Erde und ihre Dynamik.
See Abstr. 003.139.

The auroral 2145 Å feature.
See Abstr. 022.021.

Laboratory experiments and space phenomena.
See Abstr. 022.145.

The supra-thermal plasma analysers on the ESA GEOS satellites.   See Abstr. 032.526.

Global auroral imaging instrumentation for the Dynamics Explorer mission.   See Abstr. 032.579.

The energetic ion composition spectrometer (EICS) for the Dynamics Explorer-A.   See Abstr. 032.583.

High-Altitude Plasma Instrument for Dynamics Explorer-A.   See Abstr. 032.584.

The Low Altitude Plasma Instrument (LAPI).
See Abstr. 032.585.

The Retarding Ion Mass Spectrometer on Dynamics Explorer-A.   See Abstr. 032.586.

Instrumentation for vector electric field measurements from DE-B.   See Abstr. 032.589.

An all-sky imaging photometer.
See Abstr. 034.014.

A new trajectory concept for exploring the Earth's geomagnetic tail.   See Abstr. 051.016.

Dynamics Explorer program: an overview.
See Abstr. 051.034.

Dynamics Explorer spacecraft and ground operations systems.   See Abstr. 051.035.

The effects of cold plasma on the Kelvin-Helmholtz instability.   See Abstr. 062.009.

Modulation instabilities in astrophysics.
See Abstr. 062.065.

Stabilization of the electrostatic Kelvin-Helmholtz instability in high β plasmas.   See Abstr. 062.088.

Plasma processes at collisionless shock waves.
See Abstr. 062.107.

Electron acceleration in magnetosonic shock fronts.
See Abstr. 062.109.

Comments on high Mach number magnetosonic shocks.   See Abstr. 062.113.

Differences between solar wind plasmoids and ideal magnetohydrodynamic filaments.   See Abstr. 074.028.

On a new simple method for inferring solar wind parameters from ground-based geomagnetic data.
See Abstr. 074.042.

Effects of viscous interaction of the solar wind with the plasma layer of the magnetospheric tail.
See Abstr. 074.088.

Propagation of upstream protons in the near-earth solar wind.   See Abstr. 074.120.

Solar wind and its interaction with the magnetosphere: measured parameters.   See Abstr. 074.121.

Penetration of solar protons and alpha particles with energies above 1 MeV/nucleon into the polar caps.
See Abstr. 078.010.

On the quantitative correspondence of the solar proton flux measured outside and inside the earth's magnetosphere.   See Abstr. 078.012.

On the identification of planetary-scale waves during disturbed conditions.   See Abstr. 082.104.

Earth as viewed in the ultraviolet.
See Abstr. 082.118.

The influence of aurora on ionospheric electron content.   See Abstr. 083.051.

On a connection of sporadic E_{sr} formations with the B_z-component of the interplanetary magnetic field and electrojet activity.   See Abstr. 083.060.

Mean-field theories of planetary magnetism.
See Abstr. 091.083.

Analytical two-dimensional model for a pole-on magnetosphere.   See Abstr. 101.012.

Three-dimensional interaction of interplanetary shock waves with the bow shock and magnetopause: a comparison of theory with ISEE observations.
See Abstr. 106.004.

Orbital dynamics of magnetospherically trapped lunar ejecta.   See Abstr. 106.054.

Submicron lunar ejecta in the magnetosphere associated with meteor showers.   See Abstr. 106.055.

Interaction of cosmic rays with the magnetosphere and ionosphere.   See Abstr. 143.063.

085 Solar-terrestrial Relations

085.001   Asymmetric access and temporal variations of energetic solar protons at the polar caps.
S. O. Ifedili.
J. Geophys. Res., Vol. 86, 5397 - 5407 (1981).
    The paper presents measurements of solar protons over the north and south polar caps for the events of November 2, December 18, and December 19, 1969. Data are from the charged particle counters on Ogo 6, which was launched on June 5, 1969, into a nearly polar orbit with a perigee of 400 km and an apogee of 1100 km. The measurements are used to furnish further evidence of asymmetric access of solar protons to the earth's polar caps as well as to study the magnetospheric and interplanetary field interactions. The author also provides a new and clear evidence of periodic temporal variations of the proton flux over the polar caps.

085.002   Effect of sunspot activity on the occurrence of M and N echoes in the equatorial region.
J. H. Sastri.
Curr. Sci., Vol. 49, 898 - 899 (1980). – Abstr. in Phys. Abstr., Vol. 84, Abstr. 79675 (1981).

085.003   Statistical dependences of geomagnetic pulsations on Wolf numbers.
A. I. Bilinskij, F. I. Sedova.
Geomagn. Aehron., Tom 21, 760 - 762 (1981). In Russian.

085.004   Connection between the PA index of aurorae and solar phenomena and the parameters of the interplanetary medium.   V. V. Viskov, V. N. Obridko, E. V. Ivanov, L. Yu. Konstantinova, T. A. Khviyuzova.
Polyarn. siyaniya i svechenie nochn. neba, Moskva, 1981, No. 28, p. 53 - 59. In Russian. – Abstr. in Ref. zh., 51. Astron., 7.51.340 (1981).

085.005   Recurrence of auroral phenomena during a solar activity cycle.   T. A. Khviyuzova.
Polyarn. siyaniya i svechenie nochn. neba, Moskva, 1981, No. 28, p. 59 - 62. In Russian. – Abstr. in Ref. zh., 51. Astron., 7.51.341 (1981).

085.006   Some problems in studying solar-terrestrial relations.
    G. V. Vasil'ev, V. I. Dmitriev, L. A. Emel'yanova, Yu. V. Kushnerevskij, M. V. Kulagin, V. V. Migulin, M. D. Fligel', I. P. Khar'kov, Yu. N. Shaulin, L. D. Shoya, A. F. Yakovlev.
Dokl. AN SSSR, Tom 257, 316 - 318 (1981). In Russian.
Abstr. in Ref. zh., 51. Astron., 7.51.342 (1981).

085.007   Solar activity indices and the ionospheric F region.
    M. Berkeliev.
Izv. AN TurkmSSR. Ser. fiz.-tekh., khim. i geol. nauk, 1981, No. 1, p. 126 - 128. In Russian. – From Ref. zh., 51. Astron., 7.51.343 (1981).

085.008   Effects of the solar eclipse of 22nd September 1968 on the conjugate ionosphere.
G. V. Bukin, L. E. Kolokolov.
Fiz. i struktura ehkvatorial'n. ionos. Moskva, 1981, p. 152 - 155. In Russian. – Abstr. in Ref. zh., 51. Astron., 7.51.344 (1981).

085.009   Numerical investigation of the interaction between the solar wind and the earth's magnetosphere in the framework of a homogeneous-layer model.
N. V. Erkaev, V. G. Pivovarov.
Vychisl. tsentr. SO AN SSSR. Krasnoyarsk, 1981. 32 pp. In Russian. – Abstr. in Ref. zh., 62. Issled. kosm. prostranstva, 7.62.342 (1981).

085.010   A relation of disastrous weather events with solar activity.   N. A. Lebedeva.
Soln. Dannye 1981 Byull., No. 5, p. 97 - 100 (1981). In Russian.
    The results of the combination and analysis of the data of 1964 - 1976 concerning the relation of disastrous weather events with solar activity are given for the territory of the European part of the USSR.

085.011   Solar radiation and air temperature measurement during the solar eclipse of 1980 February 16.
T. R. Sivaramakrishnan.
Bull. Astron. Soc. India, Vol. 9, 71 (1981). – Abstract.

085.012   Ionospheric effects at Ahmedabad of the total solar eclipse of 1980 February 16.   H. Chandra, G. Sethia, G. D. Vyas, M. R. Deshpande, H. O. Vats.
Bull. Astron. Soc. India, Vol. 9, 71 (1981). – Abstract.

085.013   Effects of the total solar eclipse of 1980 February 16 on total electron content at low latitudes.
M. R. Deshpande, H. Chandra, G. Sethia, H. O. Vats, K. N. Iyer, A. V. Janve.
Bull. Astron. Soc. India, Vol. 9, 72 (1981). – Abstract.

085.014   Solar activity and tropospheric circulation.
M. M. Abdel-Wahab, Z. Metwally.
J. Astron. Soc. Egypt, Vol. 2, 1 - 5 (1980).

085.015   A formula for calculating theoretical photoelectron fluxes resulting from the He^+ 304 Å solar spectral line.   P. G. Richards, D. G. Torr.
Geophys. Res. Lett., Vol. 8, 995 - 998 (1981).
    The authors present a method for evaluating the ionospheric photoelectron fluxes resulting from the 304 Å solar radiation. This method can be used to verify the full theoretical calculation of these specific photoelectron fluxes. The formula may also be useful in monitoring variations in the 304 Å solar flux whenever direct solar flux measurements are unavailable. The authors have also performed photoelectron flux calculations using two different sets of inelastic electron collision cross sections and compared with observations.

085.016   The solar spectral irradiance and its action in the atmospheric photodissociation processes.
M. Nicolet.
Planet. Space Sci., Vol. 29, 951 - 974 (1981).
    A critical analysis has been made of solar irradiance in the spectral region covering wavelengths from 100 nm upwards; the absorption characteristics of molecules of oxygen and ozone have been taken into account with a view to the direct application of the results to atmospheric photochemistry.

085.017   Solar radio pulsations before proton flares and evidence of the existence of analogous pulsations in ionizing radiation appearing in geomagnetic field fluctuations.   M. M. Kobrin, V. I. Malygin, S. D. Snegirev.
13-ya Vses. konf. po radioastron. issled. soln. sistemy, Kiev, 1981. Tez. dokl. Kiev, 1981, p. 32 - 33. In Russian. – From Ref. zh., 51. Astron., 9.51.425 (1981).

085.018   Study of earth's atmospheric emissions.
    V. V. Agashe, S. M. Rathi, S. K. Pechad.
Observations of the total solar eclipse of 16 February 1980, (see 003.009), p. 87 - 88 (1981).

085.019 A survey of possible effects of long-lasting absence of solar activity on climate and some speculations on possible mechanisms.   G. Molnar.
Clim. Change, Vol. 3, 189 - 203 (1981). — Abstr. in Phys. Abstr., Vol. 84, Abstr. 88549 (1981).

085.020 Comparison of Cretaceous and present Earth albedos: implications for the causes of paleoclimates.
S. L. Thompson, E. J. Barron.
J. Geol., Vol. 89, 143 - 167 (1981). — Abstr. in Phys. Abstr., Vol. 84, Abstr. 88551 (1981).

085.021 Variability of solar EUV fluxes and exospheric temperatures.   G. Schmidtke, N. Borsken, G. Sunder.
J. Geophys., Vol. 49, 146 - 148 (1981). — Abstr. in Phys. Abstr., Vol. 84, Abstr. 88617 (1981).

085.022 Vertical drift velocity in the topside F layer during Aug.7, 1972 flare.   M. P. Singh, S. N. Ghosh.
Indian J. Phys., Part B, Vol. 53B, 266 - 272 (1979). — Abstr. in Phys. Abstr., Vol. 84, Abstr. 88636 (1981).

085.023 On the geoefficiency of the solar wind parameters.
M. I. Pudovkin, M. A. Shukhtina, D. I. Ponyavin, S. A. Zaitseva (*Zajtseva*), O. U. Ivanov.
Ann. Géophys., Vol. 36, 549 - 553 (1980). — Abstr. in Phys. Abstr., Vol. 84, Abstr. 88655 (1981).

085.024 Solar wind streams, sector boundary passage at low latitude geomagnetic field.
G. K. Rangarajan, B. R. Arora.
Indian J. Radio Space Phys., Vol. 10, 54 - 57 (1981). — Abstr. in Phys. Abstr., Vol. 84, Abstr. 88661 (1981).

085.025 Heliophysical foundations of solar-biological relations.   M. N. Gnevyshev.
Zemlya Vselennaya, 1981, No. 4, p. 22 - 25. In Russian.

085.026 How the solar activity influences the biosphere.
B. M. Vladimirskij.
Zemlya Vselennaya, 1981, No. 4, p. 26 - 28. In Russian.

085.027 On the stability of the effect of growth of tropospheric instability after the earth's entering into the solar corpuscular stream.
V. E. Chertoprud, N. B. Mulyukova, Eh. R. Mustel'.
Astron. Zh., Tom 58, 1063 - 1069 (1981). In Russian. English translation in Soviet Astron., Vol. 25, No. 5.

085.028 Long-term variations in the stratosphere of the northern hemisphere during the last two sunspot cycles.   B. Naujokat.
J. Geophys. Res., Vol. 86, 9811 - 9816 (1981).
Stratospheric geopotential height and temperature data of three levels are examined for the period 1957–1979. By using a low-pass filter for each time series at a 10°-by-10° latitude-longitude grid, long-term variations over the northern hemisphere are obtained, whose amplitudes and phases depend on latitude and longitude as well as on the height level. Analyses of differences between sunspot maximum and minimum are made, looking for a possible solar relationship.

085.029 Effects of solar variations on the upper atmosphere.   G. Kockarts.
Sol. Phys., Vol. 74, (see 012.029), 295 - 320 (1981).
The primary objective of this paper is to stress the importance of reliable solar ultraviolet data for a quantitative understanding of the physical phenomena occurring in the upper atmosphere. As a consequence the behavior of the upper atmosphere as a function of solar activity is discussed in the frame-work of the limited amount of solar ultraviolet data presently available.

085.030 The response of ozone to solar activity variations: a review.   G. M. Keating.
Sol. Phys., Vol. 74, (see 012.029), 321 - 347 (1981).
Observational evidence and theoretical predictions of the response of ozone to solar variations are reviewed. Short-term solar proton effects, possible effects of galactic cosmic rays modulated by the Sun, and the effects of 27-day solar rotation and 11-year solar cycle variations are discussed.

085.031 Chemical response of the middle atmosphere to solar variations.
P. de Baets, G. Brasseur, P. C. Simon.
Sol. Phys., Vol. 74, (see 012.029), 349 - 353 (1981).
The possible variation of the trace species concentration in the middle atmosphere related to long term solar irradiance variability is estimated by means of a one-dimensional numerical model.

085.032 Increase in the response of the Earth's atmosphere to the sunspot cycle with height above sea level.
H. Schwentek, W. Elling.
Sol. Phys., Vol. 74, (see 012.029), 355 - 372 (1981).
The average, longterm behaviour of many parameters P of the Earth's atmosphere and ionosphere depends on the solar activity cycle. In many cases it can be roughly described by a linear approximation of the form $P=P_0(1+kR)$, where R is an average sunspot number. It is shown that coefficient k considerably depends on season, and also slightly on the hemisphere considered and on geographic latitude, and that it increases with height above the sea level.

085.033 Similar periodicities in the range 12 to 150 days in solar, ionospheric and atmospheric time series.
W. Elling, H. Schwentek.
Sol. Phys., Vol. 74, (see 012.029), 373 - 384 (1981).
Data series for the same time interval of characteristic solar parameters (sunspot number R; flux at 2.8 GHz), ionospheric parameters (critical frequency of the E-region) and atmospheric parameters (stratospheric and tropospheric temperatures T) have been analysed by the maximum-entropy method in order to study the occurrence of periodicities in those parameters in the range from 12 to 150 days.

085.034 The Sun's rotation and perturbations of geopotential height and temperature fields in the stratosphere.
A. Ebel, B. Schwister.
Sol. Phys., Vol. 74, (see 012.029), 385 - 398 (1981).
The morphology of a solar activity effect apparently connected with the Sun's rotation and showing up in 25-day and 13.6-day oscillations of stratospheric geopotential and temperature fields is analysed in this study. The direct physical mechanisms of the Sun-climate relationship are not yet clear, but it can be concluded that atmospheric dynamics is an important factor for its morphology .

085.035 A dynamical mechanism through which variations in solar ultraviolet radiation can influence tropospheric climate.   J. R. Bates.
Sol. Phys., Vol. 74, (see 012.029), 399 - 415 (1981).
Variations in solar UV radiation can lead to changes in the mean temperature and wind distributions in the stratosphere and, through modification of the ozone photochemistry, to changes in the damping rate of temperature perturbations about the mean. Such changes can influence the stratospheric propagation characteristics of planetary waves generated in the troposphere. In particular, the poleward heat transfer in the troposphere can be strongly modified, thus providing a mechanism that can influence climate. The dynamics of the mechanism is presented in a simple form.

085.036 **Tropospheric effects of variable solar activity.**
C. J. E. Schuurmans.
Sol. Phys., Vol. 74, (see 012.029), 417 - 419 (1981).
Within 12 hr after strong solar flares at middle to high latitudes a cooling at tropopause level is observed along with changes in temperature and height of isobaric levels in the troposphere. Delayed tropospheric effects are reported to occur at 2 - 4 days after a flare. It is suggested that the early effect through changing the baroclinic instability conditions of the troposphere is responsible for the occurrence of the late effects.

085.037 **On the nature of the apparent response of the vorticity area index to the solar magnetic field.**
J. M. Wilcox, P. H. Scherrer.
Sol. Phys., Vol. 74, (see 012.029), 421 - 432 (1981).
The apparent response of the vorticity area index to the solar magnetic field is confined to tropospheric regions of intense circulation. Discussions and calculations that include large volumes of the troposphere would not be expected to show a significant Sun-weather effect. Analysis of the effect in time intervals outside the original 1963 - 73 is also discussed. An assessment of this Sun-weather effect at the present time is given.

085.038 **Climate and paleoclimate: what we can learn about solar luminosity variations.**   T. M. L. Wigley.
Sol. Phys., Vol. 74, (see 012.029), 435 - 471 (1981).
The Earth's climate is not constant, and has experienced major changes in the past on all timescales. The causes of these changes vary according to the timescale considered. Some of the most important causal mechanisms include continental drift, changes in the Earth's orbital parameters, volcanic activity and solar variations. Solar variations have been invoked to explain climatic change on almost all timescales from 1 to 10^9 yr. This paper reviews past changes in climate and proposed causal mechanisms on timescales of from 1 to 10^9 yr. The evidence for solar activity-climate links is discussed.

085.039 **Solar variability and stochastic effects on climate.**
C. Nicolis.
Sol. Phys., Vol. 74, (see 012.029), 473 - 478 (1981).
The effect of fluctuations on a simple energy balance model is examined. A new, long characteristic time scale referring to the passage between different stable climatic states is identified. It is shown that a weak external forcing whose period is comparable to this scale enables the system to switch between different states with a high probability. The connection with glaciation cycles is pointed out.

085.040 **A 1600 year long record of solar change derived from atmospheric ^{14}C levels.**
M. Stuiver, P. D. Quay.
Sol. Phys., Vol. 74, (see 012.029), 479 - 481 (1981).
Tree-ring ^{14}C measurements indicate the long-term solar variations as modulations of the cosmic ray flux.

085.041 **Energetics of some solar-terrestrial relations.**
I. V. Kovalevskij.
Probl. soln.-zemn. svyazej. Dokl. simpoz. KAPG, Ashkhabad, 1979. Ashkhabad, 1981, p. 63 - 92. In Russian. – Abstr. in Ref. zh., 51. Astron., 10.51.410 (1981).

085.042 **Solar activity and concentration of cosmogenic radio nuclides in the earth's atmosphere.**
V. Yu. Luyanas, V. I. Styro.
Cosmic rays, (see 012.025), 1980, p. 380 - 386. In Russian. Abstr. in Ref. zh., 51. Astron., 10.51.414 (1981).

085.043 **Long-term relationships between sunspots, solar faculae and the ionosphere.**
P. A. Smith, J. W. King.

J. Atmos. Terr. Phys., Vol. 43, 1057 - 1063 (1981).
Evidence is presented which shows that the relationship between the ionospheric critical frequencies and the sunspot numbers has changed significantly on a secular time-scale.

085.044 **Daily variation of geomagnetic field at low latitudes associated with stable solar wind flow.**
G. K. Rangarajan.
Proc. Indian Acad. Sci. Earth Planet. Sci., Vol. 90, 55 - 62 (1981). – Abstr. in Phys. Abstr., Vol. 84, Abstr. 101808 (1981).

085.045 **Correlations between solar wind parameters and auroral kilometric radiation intensity.**
D. L. Gallagher, N. D'Angelo.
Geophys. Res. Lett., Vol. 8, 1087 - 1089 (1981).

085.046 **The stimulation of auroral kilometric radiation by type III solar radio bursts.**   W. Calvert.
Geophys. Res. Lett., Vol. 8, 1091 - 1094 (1981).

085.047 **Relationships between the solar wind electric field and the magnetospheric convection electric field.**
L. Wu, R. Gendrin, B. Higel, J. Berchem.
Geophys. Res. Lett., Vol. 8, 1099 - 1102 (1981).

085.048 **An estimation of the geomagnetic activity maximum in the twenty-first solar cycle,**   M.-c. Gao.
Acta Geophys. Sinica, Vol. 24, 455 - 460 (1981). In Chinese.

085.049 **On forecasting the fields of meteorological elements with account for solar activity.**
A. A. Dmitriev.
Soln. Dannye 1981 Byull., No. 8, p. 102 - 107 (1981). In Russian.
An algorithm of a forecast for two-dimensional fields with account for the solar activity is proposed.

085.050 **Long-term and seasonal sign inversion of solar-terrestrial relationships.**
B. G. Sherstyukov.
Soln. Dannye 1981 Byull., No. 8, p. 107 - 111 (1981). In Russian.
Reference data of geoactive solar longitudes are detected. Dependence of the effect of geoactive longitudes in the atmospheric circulation on the season and polarity of the general magnetic field of the sun is shown.

085.051 **The Sun, the Moon and the weather.**
J. Gribbin.
New Scientist, Vol. 90, 754 - 766 (1981). – Abstr. in Phys. Abstr., Vol. 85, Abstr. 6425 (1982).

085.052 **Analysis of solar-terrestrial connections by the radar method of meteors.**
B. V. Kal'chenko, B. L. Kashcheev.
Meteorn. issled., Moskva, 1981, No. 7, p. 5 - 12. In Russian. Abstr. in Ref. zh., 51. Astron., 11.51.351 (1981).

085.053 **On the connection between the ionospheric, strato-tropospheric parameters and solar activity.**
M. Shirmammedov.
Izv. AN TurkmSSR. Ser. Fiz.-tekh., khim. i geol. nauk, 1981, No. 2, p. 46 - 51. In Russian. – Abstr. in Ref. zh., 51. Astron., 11.51.515 (1981).

085.054 **Amplification and regulation mechanism in solar-atmospherical relations.**   N. G. Skryabin.
Dinamich. protsessy i struktura polyarn. ionos. Apatity, 1980, p. 112 - 122. In Russian. – Abstr. in Ref. zh., 51. Astron., 11.51.520 (1981).

085.055 **Position of the earth relative to flattened flare-generated streams and some peculiarities of solar-terrestrial relations.**   K. G. Ivanov.
Astron. Zh., Tom 58, 1262 - 1267 (1981). In Russian.
English translation in Soviet Astron., Vol. 25, No. 6.

A suggestion on strongly flattened flare-generated interplanetary streams is developed. A method of determination of the earth's position relative to the streams is specified. The method is used for an interpretation of several paradoxical peculiarities which cannot be explained from the traditional point of view.

085.056 **Comparative magnetospherology. Part 13. Effects of the rotating solar magnetosphere on disturbances of the earth's magnetosphere.**   T. Saito, K. Yumoto.
Proceedings of the 14th ISAS Lunar and Planetary Symposium, (see 012.055), p. 103 - 110 (1981).

085.057 **Solar cycles and tree rings: eocene to present.**   R. Ammons, R. B. Ammons, A. Ammons.
J. American Assoc. Variable Star Obs., Vol. 10, 35 (1981). Abstract.

085.058 **Influence of solar proton event on upper stratospheric temperatures.**   J. Xanthakis, C. Zerefos, S. Sehra, C. Repapis, C. Poulakos.
Praktika Akad. Athens, Tom. 55, 362 - 371 (1980) = Res. Cent. Astron. Appl. Math. Acad. Athens, Contrib. Ser. I (Astron.) No. 79.

085.059 **Possible periodicities of the annually released planetary seismic energy (M $\geqslant$ 7.8) during the period 1898–1977.**   J. Xanthakis.
Praktika Akad. Athens, Tom. 56, 242 - 256 (1981) = Res. Cent. Astron. Appl. Math. Acad. Athens, Contrib. Ser. I (Astron.), No. 87.

085.060 **On causes of geomagnetic storms during increased solar activity.**
I. Ya. Plotnikov, L. P. Shadrina.
Variatsii kosm. luchej i soln. veter. Yakutsk, 1980, p. 86 - 94. In Russian. – From Ref. zh., 51. Astron., 12.51.546 (1981).

Investigations on solar-terrestrial physics in Turkmenistan.   See Abstr. 013.015.

Swinging sun, 79-year cycle, and climatic change. See Abstr. 072.073.

Solar cosmic rays in the system of solar-terrestrial relations.   See Abstr. 078.014.

Solar oscillation, flare pattern, rise of leucopenia, and cycle of locust plagues.   See Abstr. 080.079.

Stratospheric observations of the attenuated solar irradiance in the Schumann-Runge band absorption region of molecular oxygen.   See Abstr. 082.053.

Secular non-random variations of cosmogenic carbon-14 in the terrestrial atmosphere. See Abstr. 082.126.

The Kelvin-Helmholtz instability at the magnetopause and the solar wind-magnetosphere interaction. See Abstr. 084.046.

On the use of geomagnetic observatory results in geoelectric field work.   See Abstr. 084.049.

Planetary System

091 Physics of the Planetary System (Dynamics, Figure, Rotation, Interiors, Atmospheres, Magnetic Fields, etc.)

091.001 Schwerefelder von Planeten und Mond – zur Entwicklung des modernen Kenntnisstandes. III.
L. Ballani.
Sterne, 57. Band, 172 - 181 (1981) = Mitt. Zentralinst. Phys. Erde, Nr. 845 (2).

091.002 Solution of the problem of rotation of planets in the statistical theory of accumulation.
A. V. Vityazev, G. V. Pechernikova.
Astron. Zh., Tom 58, 869 - 878 (1981). In Russian. English translation in Soviet Astron., Vol. 25, No. 4.
An analytical theory is derived for the rate of rotation acquired by a planet as it grows. The values of regular and random components of angular momentum are determined by a time-dependent mass and velocity distribution function of the bodies. It is shown that large obliquities of planets and retrograde rotation of some of them are due to the large random component of angular momentum. For planets which were not slowed down by tidal friction the theoretical dependence of the specific angular momentum on the mass of the planet is in agreement with the observed one.

091.003 The equivalence principle and anomal motions of bodies in the solar system.
N. I. Kolosnitsyn, A. V. Osipova.
Astron. Zh., Tom 58, 888 - 897 (1981). In Russian. English translation in Soviet Astron., Vol. 25, No. 4.
Anomal secular and periodic motions of the solar system's planets and of the moon which are caused by a violation of the equivalence principle for massive bodies are calculated. The greatest effects arise from Jupiter and from the neighbouring planets.

091.004 The positions of secular resonance surfaces.
J. G. Williams, J. Faulkner.
Icarus, Vol. 46, 390 - 399 (1981).
The surfaces for the three strongest secular resonances have been located as function of proper semimajor axis, eccentricity, and inclination for semimajor axes between 1.25 and 3.5 AU. The results are presented graphically. The v_5 resonance only occurs at high inclinations ($\gtrsim 23°$). The v_6 resonance passes through both the main belt and Mars-crossing space. The v_{16} resonance starts near the inner edge of the belt and, at low inclinations at least, folds around a portion of the Mars-crossing space until it runs nearly parallel with the Earth-crossing boundary.

091.005 Tectonic evolution of the terrestrial planets.
J. W. Head, S. C. Solomon.
Science, Vol. 213, 62 - 76 (1981).
The article discusses the tectonic style of each terrestrial planet as inferred from its surface features, compares these styles with models of planetary thermal history, and reviews the major factors responsible for, and the processes governing, planetary tectonic evolution.

091.006 On the trail of the "Jupiter effect".
L. G. Thompson.
Sky Telesc., Vol. 62, 220 (1981).

091.007 Solitary Rossby waves in the presence of vertical shear.   P. D. Weidman, L. G. Redekopp.
J. Atmos. Sci., Vol. 37, 2243 - 2247 (1980). – Abstr. in Phys. Abstr., Vol. 84, Abstr. 66867 (1981).

091.008 New insight into the physical state of solar system objects.   P. D. Feldman.
The Universe at ultraviolet wavelengths, (see 012.009), p. 21 - 25 (1981).
The application of IUE to observations of solar system objects is summarized and a brief survey of new discoveries made during the first two years of IUE operation is given.

091.009 Planetary crusts: a comparative review.
J. V. Smith.
The lunar highlands crust, (see 012.013), p. 441 - 456 (1980).
This paper brings up to date the references and ideas in the review "Mineralogy of the planets: a voyage in space and time", and selects properties of planetary crusts which have special significance for the origin of the lunar highlands crust. Particular emphasis is placed on the effects caused by absence of atmophile volatiles in lunar rocks, and on the use of lunar impact chronology to place constraints on development of crusts on other planets.

091.010 Interrelation between the value of gravitational acceleration on planets, the distance of planets from the sun, and the size of planets.   Yu. P. Kozlov.
Kazan. khim. tekhnol. inst. Kazan', 1981. 14 pp. In Russian. Abstr. in Ref. zh., 51. Astron., 7.51.136 (1981).

091.011 On the specific character of seismic investigations of the terrestrial planets.
N. I. Galkin, A. V. Nikolaev, M. K. Rozhdestvenskij, V. I. Serbin, O. B. Khavroshkin, V. V. Tsyplakov.
Issled. Zemli nevzryvn. sejsmich. istochnikami. Moskva, 1981, p. 64 - 75, 333. In Russian. – Abstr. in Ref. zh., 51. Astron., 7.51.143 (1981).

091.012 On the vertical statistical structure of planetary atmospheres.   G. M. Bogolyubov.
Vestn. LGU, 1981, No. 4, p. 46 - 52. In Russian. – Abstr. in Ref. zh., 51. Astron., 7.51.144 (1981).

091.013 How high can a mountain be?   P. A. G. Scheuer.
J. Astrophys. Astron., Vol. 2, 165 - 169 (1981).
The possible height of a mountain on a solid self-gravitating object such as a planet or a neutron star is limited by the strength of the rock. Estimates of the limiting height and conditions for their validity are discussed.

091.014 Noble gases in the terrestrial planets.
M. B. McElroy, M. J. Prather.

Nature, Vol. 293, 535 - 539 (1981).

Abundances of primordial noble gases are lower for Mars than for Earth, but are higher for Venus. The data for Venus are attributed to implantation of solar wind in small preplanetary particles. Results for Mars are explained by escape of gas from planetesimals with radius between 5 and 100 km which form within the first 10^7 yr of the solar system. Volatile loss is associated with melting caused by short-lived radioisotopes such as ^{26}Al.

091.015 Groupements de planètes. B. Junod.
Orion, 39. Jahrg., 146 - 148 (1981).

091.016 The influence of "external" sources and dissipative processes on the abundance of rare gases in the atmospheres of terrestrial planets. A. K. Pavlov.
Izv. AN SSSR. Ser. fiz., Tom 45, 621 - 625 (1981). In Russian. Abstr. in Ref. zh., 51. Astron., 8.51.212 (1981).

091.017 Le système solaire. G. Israël.
Histoire de l'univers, (see 003.007), p. 183 - 241 (1980).
Contents: Les planètes et leurs satellites. Les comètes. Les météores. Géophysique et planétologie. La second vague de l'exploration planétaire. L'origine et l'histoire du système solaire.

091.018 La riscoperta dei pianeti. M. Fulchignoni.
Mem. Soc. Astron. Italiana, Vol. 52, (see 012.019), 137 - 151 (1981).

091.019 Predictions of mineral assemblages in planetary interiors. E. Stolper.
Proc. Eleventh Lunar Planet. Sci. Conf., (see 012.020), p. 235 - 250 (1980).

091.020 On the early global melting of the terrestrial planets. C. J. Hostetler, M. J. Drake.
Proc. Eleventh Lunar Planet. Sci. Conf., (see 012.020), p. 1915 - 1929 (1980).

091.021 Near surface magma movement. O. H. Muller, M. R. Muller.
Proc. Eleventh Lunar Planet. Sci. Conf., (see 012.020), p. 1979 - 1985 (1980).

091.022 Equations of state in planet interiors. O. L. Anderson, J. R. Baumgardner.
Proc. Eleventh Lunar Planet. Sci. Conf., (see 012.020), p. 1999 - 2014 (1980).

091.023 Formation of bowl-shaped craters. A. J. Piekutowski.
Proc. Eleventh Lunar Planet. Sci. Conf., (see 012.020), p. 2129 - 2144 (1980).

091.024 Control of crater morphology by gravity and target type: Mars, Earth, Moon. R. J. Pike.
Proc. Eleventh Lunar Planet. Sci. Conf., (see 012.020), p. 2159 - 2189 (1980).

091.025 A comparison of secondary craters on the Moon, Mercury, and Mars. P. H. Schultz, J. Singer.
Proc. Eleventh Lunar Planet. Sci. Conf., (see 012.020), p. 2243 - 2259 (1980).

091.026 Some thermodynamic relationships governing the behavior of permafrost and frozen ground. D. M. Anderson.
Mem. Soc. Astron. Italiana, Vol. 52, (see 012.023), 327 - 332 (1981).

091.027 Geologic overview of the terrestrial planets. R. E. Arvidson.
Mem. Soc. Astron. Italiana, Vol. 52, (see 012.023), 333 - 348 (1981).
The terrestrial moons and planets of our solar system are composed largely of rocky and metallic materials, and preserve geologic records. The moons of the giant planets, especially the larger moons, can be thought of as terrestrial bodies. In this paper, the Moon, our most primitive object is discussed first, followed by Mars, then Venus, and the paper ends with a discussion of the evolution of the Galilean satellites.

091.028 The relationship of cooling, subsidence, and thermal stress to the topography and tectonics of multi-ringed basins. S. R. Bratt, S. C. Solomon, J. W. Head.
Mem. Soc. Astron. Italiana, Vol. 52, (see 012.023), 353 - 357 (1981).
Models for the thermal evolution of large impact structures suggest that thermal displacement and stress can be significant contributors to the topography and tectonics of basins. Thermal modelling of impact basins can provide useful constraints on the actual quantity and distribution of heat associated with the basin forming process.

091.029 Formation of terrestrial planets: primordial phases. A. Coradini, C. Federico, G. Magni.
Mem. Soc. Astron. Italiana, Vol. 52, (see 012.023), 383 - 387 (1981).

091.030 The shape of small solar system bodies: gravitation versus solid state interactions. P. Farinella, P. Paolicchi, F. Ferrini, A. Milani, A. M. Nobili, V. Zappalà.
Mem. Soc. Astron. Italiana, Vol. 52, (see 012.023), 407 - 408 (1981).

091.031 Aeolian activity as a planetary process. R. Greeley.
Mem. Soc. Astron. Italiana, Vol. 52, (see 012.023), 409 - 418 (1981).
Many physical and chemical processes modify planetary surfaces. Any planet or satellite having a dynamic atmosphere and a solid surface is subject to aeolian, or wind processes. A survey of the Solar System shows that Earth, Mars, Venus, and possibly Titan meet these criteria. In this paper aeolian activity is reviewed and wind-related features on the planets are described.

091.032 Volcanism on the terrestrial planets. J. E. Guest, R. Lopes.
Mem. Soc. Astron. Italiana, Vol. 52, (see 012.023), 419 - 421 (1981).

091.033 An approach to the radial distribution of planetary orbits. A. Kyrala.
Mem. Soc. Astron. Italiana, Vol. 52, (see 012.023), 435 - 438 (1981).

091.034 Magnetic fields in the interplanetary medium: observational and theoretical aspects of a comparative magnetoplanetology. F. Mariani.
Mem. Soc. Astron. Italiana, Vol. 52, (see 012.023), 445 - 446 (1981). - Abstract.

091.035 Basin tectonics on the terrestrial planets: Moon, Mars, and Mercury. T. A. Maxwell.
Mem. Soc. Astron. Italiana, Vol. 52, (see 012.023), 449 - 453 (1981).

091.036 Solar system cratering chronology and dating of the surface structures of the terrestrial-type planets. G. Neukum.
Mem. Soc. Astron. Italiana, Vol. 52, (see 012.023), 461 - 466 (1981).

091.037 **Impact cratering mechanics.**   J. Pohl.
Mem. Soc. Astron. Italiana, Vol. 52, (see 012.023), 487 - 494 (1981).

091.038 **Rayed craters on planetary surfaces: compositional implications for the upper crust of the outer icy satellites.**   M. Poscolieri, P. H. Schultz.
Mem. Soc. Astron. Italiana, Vol. 52, (see 012.023), 495 - 497 (1981).

091.039 **The stability of the solar system.**   A. E. Roy.
Mem. Soc. Astron. Italiana, Vol. 52, (see 012.023), 505 - 512 (1981).
The empirical stability criteria approach to the stability of hierarchical dynamical n-body systems is described. Application of this approach is made to the problem of the long-term stability of the solar system's hierarchical systems.

091.040 **Multiple scattering and photochemistry in planetary atmospheres.**   G. Visconti.
Mem. Soc. Astron. Italiana, Vol. 52, (see 012.023), 535 - 539 (1981).

091.041 **A comparison of some explosive volcanic eruption processes on the Earth, Moon, Mars, Venus and Io.**
L. Wilson, J. W. Head III.
Mem. Soc. Astron. Italiana, Vol. 52, (see 012.023), 553 - 557 (1981).

091.042 **Photogeologic mapping of planetary surfaces.**
Planetary Geology Group, Arizona State University.
Mem. Soc. Astron. Italiana, Vol. 52, (see 012.023), 567 - 585 (1981).

091.043 **Interiors of the giant planets.**   W. B. Hubbard.
Science, Vol. 214, 145 - 149 (1981).
Unlike the terrestrial planets, the giant planets — Jupiter, Saturn, Uranus, and Neptune — have retained large amounts of the carbon, nitrogen, and oxygen compounds that were present in their zone of formation. A smaller fraction of the available hydrogen and helium was retained. The distribution and relative amounts of these components in the interiors of the Jovian planets can be inferred from theoretical and experimental data on equations of state and from the planets' hydrostatic equilibrium response to rotation.

091.044 **Excitation of oxygen emissions in the night airglow of the terrestrial planets.**
V. A. Krasnopolsky (*Krasnopol'skij*).
Planet. Space Sci., Vol. 29, 925 - 929 (1981).
The excitation, energy transfer and quenching of O_2 ($A^3 \Sigma_u^+$, $C^3 \Delta_u$, $c^1 \Sigma_u^-$) and $O(^1S)$ are discussed, taking into account laboratory measurements and observations on the airglow of the Earth, Venus and Mars.

091.045 **Aspects of the physics of planetary interiors.**
G. H. A. Cole.
Contemp. Phys., Vol. 22, 397 - 442 (1981). – Abstr. in Phys. Abstr., Vol. 84, Abstr. 88784 (1981).

091.046 **Collisionless Saturnian rings.**   F. C. Michel.
Planet. Space Sci., Vol. 29, 1137 - 1142 (1981).
The point of the letter is to note that ring particles can assume spatial configurations in which their motions, imposed by the finite thickness of the disk, are coordinated so that they do not collide. Whether or not Saturn's rings might be so configured is a testable hypothesis.

091.047 **Kinetic inhibition of CO and N_2 reduction in circumplanetary nebulae: implications for satellite composition.**   R. G. Prinn, B. Fegley, Jr.
Astrophys. J., Vol. 249, 308 - 317 (1981).

In contrast to the solar nebula, the conversion of CO to CH_4 and of N_2 to NH_3 in the circumplanetary nebulae of the Jovian planets is fast enough relative to radial mixing and nebula cooling rates that CO and N_2 are minor constituents in the circumplanetary nebulae. Thus, although the Jovian planets may have accreted carbon and nitrogen from the solar nebula mainly in the form of CO and N_2, these species were then reprocessed within the circumplanetary nebulae to form mainly CH_4 and NH_3. Satellites of the Jovian planets which accreted in sufficiently cool parts of their circumplanetary nebula are therefore predicted to retain large amounts of NH_3 and CH_4 in the form of clathrate hydrates and also very small but chemically important amounts of HCN.

091.048 **Radiotechnical methods for investigation of the circulation of the atmospheres of planets.**
V. A. Arkhangel'skij, V. V. Kerzhanovich, B. P. Trusov.
Kosm. Issled., Tom 19, 574 - 590 (1981). In Russian.

091.049 **Free oscillations of giant planets. Influence of rotation and ellipticity.**
S. V. Vorontsov, V. N. Zharkov.
Astron. Zh., Tom 58, 1101 - 1114 (1981). In Russian. English translation in Soviet Astron., Vol. 25, No. 5.
A second-order perturbation theory is developed for the calculation of the influence of all effects of rotation-Coriolis forces, centrifugal forces and ellipticity on the free oscillations of a gaseous-liquid planet or a star. The perturbation theory is constructed taking into account the coupling between different modes of oscillations caused by rotation. Theoretical spectra of the free oscillations of Jupiter and Saturn are computed.

091.050 **Magnetfelder im Sonnensystem. Teil 1.**
H. Volland.
Sterne Weltraum, Jahrg. 20, 409 - 414 (1981)

091.051 **Ejecta patterns diagnostic of planetary rotations.**
A. Dobrovolskis.
Icarus, Vol. 47, 203 - 219 (1981).
This paper describes a new scheme for reckoning a body's rotational history from the geologic record. Trajectories of ejecta on a rotating object are affected by the Coriolis effect, in a manner characteristic of the direction and speed of the object's spin. This information is registered on the surface in the form of debris deposits and ray patterns, and can be interpreted by comparing photographs of the planet's surface with the paradigms given in this article.

091.052 **Third order theory of the four large planets.**
J. L. Simon, G. Francou.
Astron. Astrophys., Vol. 103, 223 - 243 (1981). In French.
The authors present the construction of a theory of the four large planets. They have obtained the perturbations of the first, second and third order with respect to the masses. They use the variables a, λ, e, $\tilde{\omega}$, γ and Ω. The metrical elements (a, e, γ) contain secular terms at the first order and Poisson terms at the second and the third order. The solutions are analytical functions of the mean longitudes of the four planets and are numerical with respect to the other constants of integration. Their conventional set of integration constants is presented.

091.053 **Schwerefelder von Planeten und Mond – Zur Entwicklung des modernen Kenntnisstandes. IV.**
L. Ballani.
Sterne, 57. Band, 301 - 311 (1981) = Mitt. Zentralinst. Phys. Erde, Nr. 845.

091.054 **On a formalism accounting for distances of planets.**
R. Louise.
Moon Planets, Vol. 25, 389 - 396 (1981). In French.
The well-known Titius-Bode law giving distances of

planets from the Sun was improved by Basano and Hughes (1979) who found: $a_n = 0.285 \times 1.523^n$; a_n being the semi-major axis expressed in astronomical units, of the n-th planet. The integer n is equal to 1 for Mercury, 2 for Venus etc. In this paper, the author tries to find a formalism accounting for the Basano-Hughes law. It is based on the turbulence, assumed to be responsible of accretion of matter within the primeval nebula.

091.055 **Rings in the solar system.**
J. B. Pollack, J. N. Cuzzi.
Sci. American, Vol. 245, No. 5, p. 78 - 84, 86, 89 - 93 (1981).
Three of the giant planets are now known to have them, and the rings around Saturn are now known to consist of myriad ringlets. The form of the rings is maintained by a complex interplay of sculpturing forces.

091.056 **A method of defining topographic datums of planetary bodies.**   S. S. C. Wu.
Ann. Géophys., Vol. 37, (see 012.037), 147 - 160 (1981).
Abstr. in Phys. Abstr., Vol. 84, Abstr. 108084 (1981).

091.057 **Gravity field and rotation of planets. A review from the point of view of planetary geodesy.**
G. Balmino.
Ann. Géophys., Vol. 37, (see 012.037), 161 - 172 (1981).
Abstr. in Phys. Abstr., Vol. 84, Abstr. 108085 (1981).

091.058 **Magnetfelder im Sonnensystem. Teil 2.**
H. Volland.
Sterne Weltraum, Jahrg. 20, 446 - 451 (1981).

091.059 **Correlation between optical and thermophysical surface properties of solar system objects.**
K. D. Pang, K. Lumme, E. Bowell, M. Pontanen.
Bull. American Astron. Soc., Vol. 13, 710 (1981). – Abstract.

091.060 **Effects of subsurface volume scattering on planetary microwave brightness temperature measurements.**
S. J. Keihm.
Bull. American Astron. Soc., Vol. 13, 710 - 711 (1981). Abstract.

091.061 **How to identify a planet's past rotation.**
A. R. Dobrovolskis.
Bull. American Astron. Soc., Vol. 13, 722 (1981). – Abstract.

091.062 **Equilibration of para and ortho hydrogen in the major planets.**   S. T. Massie, D. M. Hunten.
Bull. American Astron. Soc., Vol. 13, 736 (1981). – Abstract.

091.063 **Accretion of volatile satellites.**
G. A. Ransford, S. K. Croft.
Bull. American Astron. Soc., Vol. 13, 741 (1981). – Abstract.

091.064 **About the terrestrial planets.**   K. Beneš.
Vesmir, Vol. 60, 197 - 199 (1981). In Czech.

091.065 **Satellites of Jupiter, Saturn and Neptune.**
J. Bouška.
Říše hvězd, Vol. 62, 182 - 185 (1981). In Czech.

091.066 **The role of electrostatic charging of small and intermediate sized bodies in the solar system.**
D. A. Mendis.
Investigating the universe, (see 003.014), p. 353 - 384 (1981).
The role of electrostatic charging of small and intermediate sized bodies in the solar system is reviewed. These bodies include planetary, interplanetary and cometary dust as well as cometary nuclei (at large heliocentric distances), asteroids and the larger bodies in the Saturnian ring system. The main physical consequences for the small grains are electrostatic erosion

("chipping") and disruption, whereas for the larger bodies they include electrostatic levitation and blow-off of fine loose dust from their surfaces. A large variety of solar system phenomena, recently observed by the Pioneer and Voyager deep space probes as well as the HEOS-2 earth satellite, are explained in terms of these processes. Certain peculiar features observed in the dust tails of comets as well as the spatial orientation of the zodiacal dust cloud may also be explained along these lines. The possible electrostatic erosion of the dust mantles of new comets as well as the electrostatic "polishing" of the smaller asteroids are also discussed.

091.067 **Effects of the tides on the evolution of natural satellites.**   F. Mignard.
Ann. Géophys., Vol. 37, (see 012.037), 173 - 178 (1981). In French. – Abstr. in Phys. Abstr., Vol. 85, Abstr. 6550 (1982).

091.068 **Schwerefelder von Planeten und Mond – Zur Entwicklung des modernen Kenntnisstandes. V.**
L. Ballani.
Sterne, 57. Band, 342 - 346 (1981) = Mitt. Zentralinst. Phys. Erde Nr. 845 (4).

091.069 **On the interpretation of radio emission of giant planets.**   K. P. Gajkovich, A. P. Naumov.
13-ya Vses. konf. po radioastron. issled. soln. sistemy, Kiev, 1981. Kiev, 1981, p. 73. In Russian. – From Ref. zh., 51. Astron., 11.51.296 (1981).

091.070 **Variation of the coefficients of velocities of photo-ionization of the neutral components of the planetary upper atmospheres with solar activity.**
A. V. Pavlov.
Kosm. Issled., Tom 19, 907 - 912 (1981). In Russian.

091.071 **Free oscillations of giant planets. Influence of differential rotation.**   S. V. Vorontsov.
Astron. Zh., Tom 58, 1275 - 1285 (1981). In Russian. English translation in Soviet Astron., Vol. 25, No. 6.
A perturbation theory is developed for the calculation of the influence of differential rotation on the free oscillations of a gaseous-liquid planet or a star. Effects of Coriolis forces, centrifugal forces and ellipticity are investigated simultaneously using a second-order perturbation theory. Theoretical spectra of free oscillations of Jupiter and Saturn are computed for simple models of differential rotation distributions.

091.072 **CH_4 photochemistry in the outer planets.**
O. Ashihara.
Proceedings of the 14th ISAS Lunar and Planetary Symposium, (see 012.055), p. 42 - 45 (1981). – Abstract.

091.073 **Planetary rotation and accretion.**
T. Matsui, S. Okubo.
Proceedings of the 14th ISAS Lunar and Planetary Symposium, (see 012.055), p. 302 - 305 (1981).
Assuming tidal drag force as a promoting agent, dynamical evolution of rotation of each terrestrial planet is estimated. Original rotation states of the terrestrial planets are shown not to be preserved except for Mars, as far as we assume a plausible dissipation factor such as the tidal $Q = 10 \sim 100$. Mars probably preserves its original rotation state. Accretion process of Mars is, therefore, suggested to be very different from those of other terrestrial planets.

091.074 **Solar wind flow about the terrestrial planets. 1. Modeling bow shock position and shape.**
J. A. Slavin, R. E. Holzer.
J. Geophys. Res., Vol. 86, 11401 - 11418 (1981).
General technique for modeling the position and shape of planetary bow waves are reviewed. A three-parameter method was selected to model the near portion of the Venus,

earth, and Mars bow shocks and the results compared with existing models using 1 to 6 free variables. By limiting consideration to the forward part of the bow wave, only the region of the shock surface that is most sensitive to obstacle shape and size was examined.

091.075 **A theoretical interpretation of the phase curves of atmosphereless solar system bodies.**
K. Lumme, E. Bowell.
Proceedings of the Third Finnish-Soviet Astronomical Symposium, (see 012.057), p. 81 - 88 (1981).

091.076 **Hetegonic orbit structure of planets and moons.**
D. T. F. Möhlmann.
Planetary interiors, (see 012.061), p. 61 - 65 (1981).
A scheme is proposed, ordering the orbital parameters of regular satellites orbiting around central masses. First attempts are discussed to understand the supposed physics behind these schemes.

091.077 **The importance of latest discoveries in the solar system.**
S. K. Vsekhsvyatsky (*Vsekhsvyatskij*).
Planetary interiors, (see 012.061), p. 75 - 82 (1981).
The discoveries of recent years in the solar system made with the help of cosmic probes are discussed which completely confirmed the author's predictions and conclusions on the existence of rings around giant planets and powerful volcanic processes on them and on their satellites. These great discoveries contributed to comprehension of the true cosmogony of the solar system and eruptive evolution of planetary bodies. In this connection the solar system astronomy faces urgent problems of revealing all the peculiarities of eruptive activity of planets like the Earth and Venus.

091.078 **State of matter and internal structure of planets.**
H. Stiller, S. Franck.
Planetary interiors, (see 012.061), p. 85 - 102 (1981).
The paper reviews investigations on structure and composition of planets which are related to properties of planetary material at high pressures and temperatures. The authors list important observations that contain information on the interior of a planet and sketch some theoretical methods of high-pressure physics. Planetary matter may be devided into three groups: gases, ices, and rocky material. The authors discuss the state of these materials at thermodynamical conditions relevant for the interior of planets. Such states include phenomena as solid-solid phase transitions, melting, metallization, critical points, miscibility, and convection. Remarks are given on the importance of high-pressure investigations for problems in planetary cosmogony.

091.079 **Interiors of the giant planets: recent advances.**
R. Smoluchowski.
Planetary interiors, (see 012.061), p. 103 - 115 (1981).
The broad features of the interiors of the giant planets have been elucidated over the last few years with considerable success. The main new results described in this paper are based on improved equations of state and better observational data.

091.080 **Interior structure and comparative analysis of the terrestrial planets.**   V. N. Zharkov,
S. V. Kozlovskaya, I. Ya. Zasurskii (*Zasurskij*).
Planetary interiors, (see 012.061), p. 117 - 129 (1981).
The three-layer models of Mercury, Venus and Mars are constructed. Particularly, the model of Venus is investigated in detail. It is possible to construct the model of Venus with the same total iron content as in the Earth. High and low temperature models of Mercury and Mars are considered. The surface density of iron and silicates in the formation zones of the terrestrial planets is calculated and the results are compared

with the surface density of the condensate in the formation zones of the giant planets.

091.081 **The internal structure of the planets Mercury, Venus, Mars and Jupiter according to the Savić-Kašanin theory.**   P. Savić.
Planetary interiors, (see 012.061), p. 131 - 146 (1981).

091.082 **Using the thermal pressure to compute the physical properties of terrestrial planets.**
J. R. Baumgardner, O. L. Anderson.
Planetary interiors, (see 012.061), p. 159 - 176 (1981) =
Publ. 2056, Inst. Geophys. Planet. Phys., Univ. Calif.,
Los Angeles (USA).
The density and temperature profiles of a planet's interior are calculated using a set of differential equations defining the major thermal and mechanical variables of a planet. One of these equations, the equation of state (EOS), is perhaps unique, because the density and volume effects are separated into two independent functions. The temperature-dependent part of the EOS is found from high-temperature properties measured on minerals at one atmosphere. The volume-dependent part of the EOS requires parameters determined by high-pressure experiments at room temperature.

091.083 **Mean-field theories of planetary magnetism.**
K.-H. Rädler.
Planetary interiors, (see 012.061), p. 219 - 229 (1981).
A survey is given on the basic features of the mean-field approach to dynamo models for the Earth and the planets. Several types of dynamo mechanisms are possible.

091.084 **Magnetic fields of planets and the features of their rotation.**   Sh. Sh. Dolginov.
Planetary interiors, (see 012.061), p. 245 - 256 (1981).
Some evidence has been presented for the importance of the acceleration due to planetary precession (the Poincare acceleration) in the generation mechanism of the planetary magnetic fields.

091.085 **Planetary magnetism.**
C. T. Russell.
Planetary interiors, (see 012.061), p. 257 - 263 (1981).
Several reviews of planetary magnetism have recently appeared. Instead of duplicating the material included therein, the author concentrates on a discussion of the latest results in this area especially those results that have appeared since the earlier reviews were written.

091.086 **The magnetic fields of Mercury, Venus and Mars.**
C. T. Russell.
Progress in planetary exploration, (see 012.062), p. 3 - 20 (1981).
Just as clearly as Mariner 10 established that Mercury has an intrinsic magnetic field, the Pioneer Venus orbiter has established that Venus has no significant intrinsic field. This is perhaps the opposite of what might be expected. Mercury, a small planet might be expected to cool rapidly and its internal dynamo to cease, while Venus, which is almost as large as the Earth, should not have lost much heat. On the contrary the source of energy of the Mercury dynamo appears to be extant whereas that of Venus appears to be extinct. The existence of a Martian magnetic field is controversial. No unambiguous signature of a Martian magnetic field has been reported. If the field on the nightside of Mars is of planetary rather than solar origin the Russian Mars spacecraft observations indicate the Martian dipole lies near the planetary equator rather than its pole.

091.087 **Comparative analysis of chemical composition and optical properties of gaseous and disperse phases of the earth-group planetary atmospheres.**

K. Ya. Kondratyev (*Kondrat'ev*), N. I. Moskalenko.
Progress in planetary exploration, (see 012.062), p. 53 - 58 (1981).

The models of chemical composition and structure of the Earth-type planetary atmospheres are offered. The optical properties of gaseous and disperse phases of the atmospheres are investigated.

091.088 **The correlation of the planetary atmosphere structural characteristics with their optical properties and radiative heat exchange peculiarities.**
K. Ya. Kondratyev (*Kondrat'ev*), N. I. Moskalenko.
Progress in planetary exploration, (see 012.062), p. 59 - 64 (1981).

The structure characteristics of the Earth-type planetary atmospheres are calculated solving the radiative heat exchange problem. Their correlations with the surface reflection characteristics and optical properties of their atmospheres are considered.

091.089 **Comparative atmospheric dynamics for terrestrial planets.** G. S. Golitsyn.
Progress in planetary exploration, (see 012.062), p. 141 - 149 (1981).

A review is presented of theoretical ideas on the general circulations in the atmospheres of Earth, Mars and Venus and also of results of their theoretical modelling. The role of various factors is discussed in the formation of the circulations. These results are compared with observational data obtained by different means. Data of direct local measurements of meteorological parameters in the atmospheres of Venus and Mars are discussed including those obtained at their surfaces.

091.090 **Dynamic systems as factors determining the structure, transformations and evolution of planets exemplified by the terrestrial planets.** O. Wolczek.
Progress in planetary exploration, (see 012.062), p. 217 - 221 (1981).

A more precise definition of a planet is proposed, based on the existence of dynamic planetary systems. Four basic planetary systems: the intraplanetary, the atmospheric, the magnetospheric and the biospheric one are discussed taking into account interactions of internal and external (cosmic) factors.

091.091 **Raman scattering as a probe of planetary atmospheres.** W. D. Cochran.
Planetary aeronomy and astronomy, (see 012.063), p. 143 - 153 (1981).

The observable effects of Raman scattering on the spectra of the giant planets may provide new information on the composition and structure of these atmospheres. Satellite observations have already shown the influence of Raman scattering on the UV continuum albedo. A cross correlation technique is presented for detecting rotational and vibrational transitions of the Raman active gases in the atmosphere. This technique has been applied to ground-based visible spectra of Venus, Jupiter, Saturn and Uranus. Extension of this method into the UV would improve the detectability of the Raman lines.

091.092 **Developments in the use of spectral signature for mapping the surface of solar system objects other than the earth.** T. B. McCord.
Sessions on remote sensing 1980. Proceedings of the Topical Meeting of the COSPAR Interdisciplinary Scientific Commission A (Sessions A1 and A2) of the COSPAR Twenty-third Plenary Meeting held in Budapest, Hungary, 2 - 14 June 1980.
A. B. Kahle, G. Weill, W. D. Carter (Editors).
Adv. Space Res., Vol. 1, No. 10, p. 3 - 17 (1981).

Reflection spectroscopy and multispectral mapping, especially when used in conjunction with other remote sensing data, have proved to be very important tools for studying the structure and composition of surface units of solar system objects. Several example applications are discussed to illustrate both the methods and the rewards.

091.093 **The submillimeter spectra of the planets: narrowband photometry.** C. T. Cunningham, P. A. R. Ade, E. I. Robson, I. G. Nolt, J. V. Radostitz.
Icarus, Vol. 48, 127 - 139 (1981).

Multicolor submillimeter observations of Jupiter, Saturn, Mars, and Uranus are reported. Narrow-band filters are used to define three passbands between 300 and 1600 μm to give accurate spectrophotometric data. A method for determining the atmospheric opacity from a single set of multicolor observations of a known source is shown to be consistent with secant plot data taken during very stable observing conditions. Such extinction data show that in the submillimeter region significant fluctuations in opacity occur over periods as short as 1 hr. The disc emission spectra for Saturn, Jupiter, and Uranus are compared with various atmospheric models and other observations.

091.094 **The dynamics of a rapidly escaping atmosphere: applications to the evolution of Earth and Venus.**
A. J. Watson, T. M. Donahue, J. C. G. Walker.
Icarus, Vol. 48, 150 - 166 (1981).

A simple, idealized model for the rapid escape of a hydrogen thermosphere provides some quantitative estimates for the energy-limited flux of escaping particles. The model assumes that the atmosphere is "tightly bound" by the gravitational field at lower altitudes, that diffusion through the lower atmosphere does not limit the flux, and that the main source of heating is solar euv. Applied to the Earth and Venus, the results suggest that the escape of hydrogen from these planets would have been energy-limited if their primordial atmospheres contained total hydrogen mixing ratios exceeding a few percent.

091.095 **Vertical-structure effects on planetary microwave brightness temperature measurements: application to the lunar regolith.** S. J. Keihm, J. A. Cutts.
Icarus, Vol. 48, 201 - 229 (1981).

The effects of vertical variations in density and dielectric constant on nadir-viewing microwave brightness temperatures are examined. Stratification models as well as models of a continuous increase in density with depth are analyzed. Specific applications address the vertical structure of the lunar frontside regolith, utilizing combined constraints from Apollo data, bistatic radar signatures, and Earth-based measurements of the lunar microwave brightness temperature.

091.096 **Magnetospheres and the interplanetary medium.**
J. A. Van Allen.
The new solar system, (see 003.026), p. 23 - 32 (1981).

091.097 **The collision of solid bodies.**
E. M. Shoemaker.
The new solar system, (see 003.026), p. 33 - 44 (1981).

091.098 **Surfaces of the terrestrial planets.**
J. W. Head, III.
The new solar system, (see 003.026), p. 45 - 56 (1981).

091.099 **Atmospheres of the terrestrial planets.**
J. B. Pollack.
The new solar system, (see 003.026), p. 57 - 70 (1981).

091.100 **Planetary rings.** J. A. Burns.
The new solar system, (see 003.026), p. 129 - 142 (1981).

091.101   **Small bodies and their origins.**   W. K. Hartmann.
The new solar system, (see 003.026), p. 197 - 204
(1981).

Mapping the planets and moons.
See Abstr. 002.014.

Bibliography of articles concerning the solar system
from journals received in 1980.   See Abstr. 002.024.

Planetary mission data available at NSSDC *(National
Space Science Data Center).*   See Abstr. 002.063.

Atmospheres.   See Abstr. 003.030.

Climate of the planets.   See Abstr. 003.034.

The planets, a decade of discovery.
See Abstr. 003.063.

Pictorial guide to the planets.
See Abstr. 003.079.

Nuclear-physical investigations of the moon and
planets.   See Abstr. 003.086.

Planets of the solar system.   See Abstr. 003.096.

Notre univers.   See Abstr. 003.114.

Earthlike planets.   See Abstr. 003.115.

Determination of the positions of major planets
with the photographic method.   See Abstr. 003.119.

Systematics of global cycles of natural processes.
Geological aspects.   See Abstr. 003.123.

Orbiting the sun: planets and satellites of the solar
system.   See Abstr. 003.161.

Landolt-Börnstein.   See Abstr. 003.172.

Significant achievements in the Planetary Geology
Program, 1981.   See Abstr. 011.046.

XIIIth All-Union conference on radio astronomical
investigations of the solar system. Kiev, 20 - 23 April 1981.
See Abstr. 012.024.

Tracking the planets.   See Abstr. 021.007.

Aspects of isochronism in Pb isotope systematics –
application to planetary evolution.   See Abstr. 022.014.

High-resolution infrared spectrum and analysis of the
ν_9 band of ethane at 12.17 μm.   See Abstr. 022.034.

High temperature crystal field spectra of transition
metal-bearing minerals: relevance to remote-sensed spectra of
planetary surfaces.   See Abstr. 022.064.

The effects of scattering geometry on the spectro-
photometric properties of powdered material.
See Abstr. 022.066.

A possible mechanism for the magnification of
magnetic fields during crater formation.
See Abstr. 022.073.

Comparison of band model and integrated line-by-
line synthetic spectra for methane in the 2.3 μm region.
See Abstr. 022.087.

Experimental impacts: recent progress on regolith
velocity distributions.   See Abstr. 022.132.

Laser induced photodissociation of cometary and
planetary molecules: the production and quenching of
$NH\,(A^3\,\pi_i)$.   See Abstr. 022.133.

Strength and temperature dependence of infrared
spectrum of methane.   See Abstr. 022.134.

Laboratory spectra of cold methane from 1.0 to
2.7 μ with abundances up to 1 km-amagat.
See Abstr. 022.135.

Laboratory experiments and space phenomena.
See Abstr. 022.145.

keV neutron capture cross sections for the s-process
isotopes of Se, Br and Kr and the abundance of krypton in the
solar system.   See Abstr. 022.148.

Adiabats and phase boundaries in planets.
See Abstr. 022.177.

Millimeter-wavelength continuum calibration
sources.   See Abstr. 031.599.

On the possibility of using cosmic radio emission
sources for investigation of the moon and planets by a bistatic
radar method.   See Abstr. 031.609.

TV pictures of another world.
See Abstr. 032.532.

Astrometric observations of planets, minor planets,
and satellites: 1976–1980.   See Abstr. 041.004.

Construction of reference networks on planets
and their satellites.   See Abstr. 041.012.

On periodic flybys of the Moon.
See Abstr. 042.007.

Construction of a theory of the outer planets
through an iterative method.   See Abstr. 042.033.

Estimate of the error in numerical integration of
the N-body problem.   See Abstr. 042.078.

Note on the numerical expressions for precession
quantities.   See Abstr. 043.004.

The golden age of solar system exploration.
See Abstr. 051.061.

A new look at the discrete ordinate method for
radiative transfer calculations in anisotropically scattering
atmospheres.   See Abstr. 063.028.

The role of scattering in planetary surface reflectance
spectra.   See Abstr. 063.040.

Radiative transfer in the surfaces of atmosphereless
bodies. I. Theory.   See Abstr. 063.051.

Computing diffuse reflection from particulate
planetary surface with a new function.
See Abstr. 063.060.

Unified relativistic theory of motion of the inner planets of the solar system. Relativistic effects in determination of planetary orbits using radar location data. See Abstr. 066.117.

General relativistic connection between orbital and rotational planetary motions.   See Abstr. 066.118.

Solar wind streams originating under the action of gravitation of planets.   See Abstr. 074.109.

Free oscillations of the sun and giant planets. See Abstr. 080.075.

Chemistry of the earth and the role of primary and secondary objects for the formation of the inner planets and meteorite parent bodies.   See Abstr. 081.026.

Lunar and planetary influences upon the peak electron density of the ionosphere.   See Abstr. 083.026.

On the nature of ring photo anomalies fixed by remote sensing.   See Abstr. 094.009.

Morphology of shock craters on the moon and planets.   See Abstr. 094.022.

Planetological considerations on the geologic intermediate stage of Mars between Moon and Earth and observations on the Jovian satellites.   See Abstr. 097.011.

Radiative transfer in the surfaces of atmosphereless bodies. II. Interpretation of phase curves. See Abstr. 098.073.

Ultraviolet emissions from the upper atmospheres of the planets.   See Abstr. 099.221.

On the stability of Saturn's rings. See Abstr. 100.048.

On some peculiarities of the structure of impact craters on planets and satellites of the solar system. See Abstr. 105.066.

Origin of satellite systems.   See Abstr. 107.006.

Interaction of ocean-atmosphere with planetary interior.   See Abstr. 107.027.

Interstellar gas near and within the solar system. See Abstr. 131.015.

Erratum

091.901 Correction to 'The atmospheres of the outer planets and satellites' [Rev. Geophys. Space Phys., Vol. 19, 43 - 89 (1981)]. L. Trafton.
Rev. Geophys. Space Phys., Vol. 19, 690 (1981). – See Abstr. 29.091.007.

092 Mercury

092.001 Mercury: magnetospheric processes and the atmospheric supply and loss rates.
B. E. Goldstein, S. T. Suess, R. J. Walker.
J. Geophys. Res., Vol. 86, 5485 - 5499 (1981).
    Magnetospheric processes at Mercury are investigated to determine how they affect the source and loss rates of the neutral He and H atmosphere. The atmospheric source rate caused by direct impact of the solar wind on the planetary surface is estimated by using a model that included the effect of the large planetary core upon the compressibility of the magnetosphere. As the nonconducting planetary surface could inhibit Birkeland currents that cause erosion of the dayside magnetosphere, the impact of the solar wind on the surface is modeled with and without the effects of erosion.

092.002 Chemical composition of Earth, Venus, and Mercury.   J. W. Morgan, E. Anders.
Proc. Natl. Acad. Sci. USA, Vol. 77, 6973 - 6977 (1980).
Abstr. in Phys. Abstr., Vol. 84, Abstr. 79731 (1981).

092.003 A color-ratio map of Mercury.
    B. Hapke, C. Christman, B. Rava, J. Mosher.
Proc. Eleventh Lunar Planet. Sci. Conf., (see 012.020), p. 817 - 821 (1980).
    A color-ratio map of the portion of Mercury imaged by Mariner 10 has been constructed from orange and UV frames. In order to prevent blemish-induced spurious color, at least two independent color-ratio images were used for each region.

Color differences appear to be somewhat smaller than on the moon. Many fresh-appearing craters and their ray systems tend to be bluer than their surroundings. Little correlation with the geology or topography is evident. Several regions of interesting color contrast are pointed out.

092.004 Central peaks in mercurian craters: comparisons to the moon.   W. Hale, J. W. Head.
Proc. Eleventh Lunar Planet. Sci. Conf., (see 012.020), p. 2191 - 2205 (1980).

092.005 Ridge systems of Caloris: comparison with lunar basins.   T. A. Maxwell, A. W. Gifford.
Proc. Eleventh Lunar Planet. Sci. Conf., (see 012.020), p. 2447 - 2462 (1980).

092.006 Stratigraphy of the Caloris basin, Mercury.
    J. F. McCauley, J. E. Guest, G. G. Schaber, N. J. Trask, R. Greeley.
Icarus, Vol. 47, 184 - 202 (1981).
    The 1300-km-diameter Caloris impact basin is surrounded by well-defined ejecta units that can be recognized from more than 1000 km, radially outward from the basin edge. A formal rock stratigraphic nomenclature is proposed for the Caloris ejecta units, which are collectively called the Caloris Group. Each of the individual formations within the Group are described and compared to similar rock units associated with the

lunar Imbrium and Orientale basins. A crater degradation chronology is also proposed.

092.007 On the termal regime and radio emission of Mercury.   V. D. Krotikov, O. B. Shchuko:
13-ya Vses. konf. po radioastron. issled. soln. sistemy, Kiev, 1981. Kiev, 1981, p. 74. In Russian. – From Ref. zh., 51. Astron., 11.51.264 (1981).

092.008 Seismic exploration of Mercury: an assessment.   J. Koyama, T. Matsui.
Proceedings of the 14th ISAS Lunar and Planetary Symposium, (see 012.055), p. 364 - 371 (1981).

Seismic activity in Mercury is studied through comparison with those in the earth and in the moon. Seismic activity in Mercury may resemble the lunar one, because of similarities between Mercury and the moon of the object's size, the surface geology, and the tidal situation. The level of activity, however, may be far higher in Mercury than in the moon, because of the large amount of tidal energy dissipated in Mercury. A seismometer package is proposed for the exploration, and expected quality and the deployment of the packages are also discussed.

092.009 Measurement accuracies required for the determination of a Mercurian liquid core.   S. J. Peale.
Icarus, Vol. 48, 143 - 145 (1981).

The discovery of an intrinsic magnetic field on Mercury (Ness et al., 1974) implies that either Mercury's core has a sufficiently extensive molten region to support an active dynamo, or that remnant magnetism in the crust (Stephenson, 1976) is the source of the observed dipole moment. The authors consider here the precision required in the measurement of the several parameters which allow a liquid core size to be inferred and discuss some possible schemes by which measurements of the required accuracy can be obtained.

092.010 Some aspects of the solar radiation incident at the top of the atmospheres of Mercury and Venus.
E. Van Hemelrijck, J. Vercheval.
Icarus, Vol. 48, 167 - 179 (1981).

A formalism has been developed for the calculation of the insolation on the planets Mercury and Venus neglecting any atmospheric absorption. For Mercury, the instantaneous insolation curves are repeated in a 2-tropical-year cycle, the distribution of the solar radiation being perfectly symmetric between both hemispheres. In addition to latitudinal variations, one observes a longitudinal effect. The small obliquity of Venus results in a nearly symmetric solar radiation distribution with respect to the equator except at the poles, where an important seasonal effect has been found. It has to be noted that no longitudinal dependence exists.

Radio investigations of the moon and terrestrial planets.   See Abstr. 003.015.

Relativistic perturbations of planetary orbits in the generalized three-parametric Schwarzschild metric. The case of Mercury.   See Abstr. 042.001.

A comparison of secondary craters on the Moon, Mercury, and Mars.   See Abstr. 091.025.

093 Venus

093.001 **Ion temperature anisotropy and heat flow in the Venus lower ionosphere.**
R. W. Schunk, J.-P. St.-Maurice.
J. Geophys. Res., Vol. 86, 4823 - 4827 (1981).

Motivated by the recent observations of supersonic ion flow in the Venus ionosphere near the terminator, the authors have studied the extent to which such a flow can induce an ion temperature anisotropy and a diffusion-thermal heat flow.

093.002 **On the possibility of detection of ions from Venus at 1 AU.** C. T. Russell, M. Neugebauer.
J. Geophys. Res., Vol. 86, 5895 - 5997 (1981).

The possibility that singly ionized helium enhancements observed near 1 AU are due to ion pickup from Venus is examined and found to be implausible. It is still possible that Venus ions are present in the solar wind with sufficiently high fluxes to be detectable by earth-orbiting spacecraft near the time of inferior conjunction.

093.003 **Chemical composition of the atmosphere of Venus.**
V. A. Krasnopolsky (*Krasnopol'skij*), V. A. Parshev.
Nature, Vol. 292, 610 - 613 (1981).

Measurements onboard the Venera 11, 12 and Pioneer Venus spacecraft stimulated the authors to study the chemical composition of the subcloud atmosphere of Venus in terms of thermochemical equilibrium calculations, comparison of typical mixing and chemical times and a rule of height-independent element mixing ratio in the absence of condensation. The photochemistry of the atmosphere down to 50 km was calculated using transport effects and number densities of CO_2, H_2O, HCl, SO_2 and CO at the lower boundary and rate coefficients of 102 reactions.

093.004 **The atmosphere of Venus.**
G. Schubert, C. Covey.
Sci. American, Vol. 245, No. 1, p. 44 - 52 (1981).

A decade of exploration by spacecraft now shows that the atmosphere of Venus consists almost entirely of carbon dioxide. Its clouds of sulfuric acid are driven by winds that attain a speed of 360 kilometers per hour.

093.005 **Venusian saltation.** B. R. White.
Icarus, Vol. 46, 226 - 232 (1981).

Estimates of the trajectories of saltating particles on Venus show the level of saltation to be low when compared to either Earth or Mars. Particles in saltation on Venus obtain maximum heights of only 1 cm over a wide range in particle size and surface wind speeds. Their path lengths are only a few centimeters at the wind speed of 1 and 2 m/sec. The entire saltation process and particle trajectories are insensitive to changes in surface pressure over the range from 70 to 100 bars and to changes in surface temperature over the range from 600 to 900°K. The net rate of surface material transport due to saltation on Venus is small when compared to Earth or Mars.

093.006 **An electrodynamic model of electric currents and magnetic fields in the dayside ionosphere of Venus.**
P. A. Cloutier, T. F. Tascione, R. E. Daniell, Jr.
Planet. Space Sci., Vol. 29, 635 - 652 (1981).

The electric current configuration induced in the ionosphere of Venus by the interaction of the solar wind has been calculated in previous papers for average steady-state solar wind conditions and interplanetary magnetic field. This model is generalized to include the effects of (a) plasma depletion and magnetic field enhancement near the ionopause, (b) velocity-shear-induced MHD instabilities of the Kelvin-Helmholtz type within the ionosphere, and (c) variations in solar wind parameters and interplanetary magnetic field.

093.007 **Hot oxygen atoms in the upper atmosphere of Venus.**
A. F. Nagy, T. E. Cravens, J.-H. Yee, A. I. F. Stewart.
Geophys. Res. Lett., Vol. 8, 629 - 632 (1981).

The energy distribution of "hot" oxygen atoms in the upper atmosphere of Venus was calculated using two different numerical methods. The two different approaches were based on 1) 2-stream transport and 2) diffusion equation considerations. Dissociative recombination of molecular oxygen ions and charge exchange of atomic oxygen ions with the neutral hydrogen and oxygen gas were the source terms considered; the calculations show that the dissociative recombination term is the dominant one. The altitude distribution of the hot oxygen atoms above the exobase was also calculated and compared with measurements by the Pioneer Venus Orbiter Ultraviolet Spectrometer.

093.008 **The altitude distribution of the Venus ultraviolet nightglow and implications on vertical transport.**
J. C. Gérard, A. I. F. Stewart, S. W. Bougher.
Geophys. Res. Lett., Vol. 8, 633 - 636 (1981).

The altitude distribution of the nitric oxide nightglow has been measured with the ultraviolet spectrometer on board Pioneer Venus. Limb profiles obtained near periapsis with a ~5 km vertical resolution have been fitted using a method independent of the spacecraft attitude data. It is found that the altitude of the maximum of the layer is 115 ± 2 km and the topside scale height $\simeq 3$ km. It is shown that downward transport by diffusion alone is insufficient. If vertical motion is parametrized by eddy diffusion, an eddy diffusion coefficient $K \simeq 8^{+8}_{-4} \times 10^{12}/(M)^{1/2}$ cm^2 s^{-1} is deduced from the altitude of the layer.

093.009 **Continental rifting and the origin of Beta Regio, Venus.**
G. E. McGill, S. J. Steenstrup, C. Barton, P. G. Ford.
Geophys. Res. Lett., Vol. 8, 737 - 740 (1981).

Topographic maps based on Pioneer Venus altimetry suggest that Beta Regio, an elevated feature centered at 27°N., 282°E., is analogous to domes associated with continental rift systems on Earth. This interpretation is consistent with the commonly quoted analogy between the East African rift system and the topography of the region from Beta Regio southward to Phoebe Regio. If Beta Regio is a dome, major structural uplift of the crust of Venus is implied, suggesting a more dynamic upper mantle than would be the case if Beta Regio were simply a large volcanic construct.

093.010 **Effects of lateral resolution on the identification of volcanotectonic provinces on Earth and Venus.**
R. E. Arvidson, G. F. Davies.
Geophys. Res. Lett., Vol. 8, 741 - 744, with a correction p. 1119 - 1120 (1981).

Degradation of topographic data sets for the continental U. S. and the north Pacific to simulate the lateral and vertical resolution of the Pioneer-Venus altimeter provides an indication of what types of volcanotectonic features would be discernible on Venus.

093.011 **Modulation of Venus ion densities associated with solar variations.** S. J. Bauer, H. A. Taylor.
Geophys. Res. Lett., Vol. 8, 840 - 842 (1981).

The dayside-ion concentrations in the Venus ionosphere obtained by the Pioneer Venus orbiter ion mass spectrometer exhibit a modulation corresponding to the 27-day solar variation. Comparisons were made of the amplitudes of modulation of CO_2^+, C^+, and O_2^+, with the amplitudes of the 27-day variation in the 10.7 cm solar radio flux and the simultaneously measured EUV fluxes at He II (304 Å) and Lyman β (1026 Å), together with a theoretical analysis of the effects of solar variability on the ionosphere and neutral atmosphere of Venus. This analysis leads the authors to the conclusion that the observed modulation of dayside ion densities is primarily due to the variability in the ionizing EUV radiations and, to a much lesser extent, the result of the variability with solar activity of the neutral atmosphere via the variability in exospheric temperature. The authors also show, theoretically, why the percentage variation of exospheric temperature on Venus for a given variation in $F_{10.7}$, is only half of the exospheric temperature variation for Earth.

093.012   **The distant bow shock and magnetotail of Venus: magnetic field and plasma wave observations.**
C. T. Russell, J. G. Luhmann, R. C. Elphic, F. L. Scarf.
Geophys. Res. Lett., Vol. 8, 843 - 846 (1981).
   An examination of the magnetic field and plasma wave data obtained by the Pioneer Venus orbiter in the wake region behind Venus discloses a well developed bow shock whose location is similar to that observed on previous missions in contrast to the dayside bow shock. Venus also has a well developed magnetotail in which the field strength is enhanced over magnetosheath values and in which the magnetic field is aligned approximately with the solar wind direction. The boundary between magnetosheath and magnetotail is also marked by a change in the plasma wave spectrum.

093.013   **The atmospheres of Venus and Jupiter.**
P. L. Sowerby.
J. British Astron. Assoc., Vol. 91, 501 - 503 (1981).

093.014   **Comparative aspects of Venus and terrestrial meteorology.**   F. W. Taylor, L. S. Elson, D. J. McCleese, D. J. Diner.
Weather, Vol. 36, No. 2, p. 34 - 41 (1981). – Abstr. in Phys. Abstr., Vol. 84, Abstr. 71105 (1981).

093.015   **A world revealed: Venus by radar.**
S. Nozette, P. Ford.
Astronomy, Vol. 9, No. 3, p. 6 - 15 (1981). – Abstr. in Phys. Abstr., Vol. 84, Abstr. 83513 (1981).

093.016   **Venus gravity: a high-resolution map.**
R. D. Reasenberg, Z. M. Goldberg, P. E. MacNeil, I. I. Shapiro.
J. Geophys. Res., Vol. 86, 7173 - 7179 (1981).
   The Doppler data from the radio tracking of the Pioneer Venus Orbiter (PVO) have been used in a two-stage analysis to develop a high-resolution map of the gravitational potential of Venus, represented by a central mass and a surface mass density. The two-stage procedure invokes a Kalman filter-smoother to determine the orbit of the spacecraft, and a stabilized linear inverter to estimate the surface mass density. The resultant gravity map is highly correlated with the topographic map derived from the PVO radar altimeter data. However, the magnitudes of the gravity variations are smaller than would be expected if the topography were uncompensated, indicating that at least partial compensation has taken place.

093.017   **High-resolution observation of the Venus dayglow spectrum 1250–1430 Å.**
S. T. Durrance, R. R. Conway, C. A. Barth, A. L. Lane.
The Universe at ultraviolet wavelengths, (see 012.009), p. 27 - 31 (1981).
   The spectrum of the dayglow of Venus between 1250 and 1430 Å has been measured in high-resolution with the International Ultraviolet Explorer. The atomic oxygen lines at 1302.2, 1304.9, 1306.0, and 1355.6 Å are present. In addition, the (14,3) and (14,4) bands of the carbon monoxide fourth positive systems at 1317 and 1354 Å respectively are identified. These bands are compared with synthetic spectra, showing the

excitation mechanism to be fluorescent scattering of solar Lyman alpha radiation.

093.018   **On Rossby-Blinova waves on Venus.**   A. S. Monin.
Dokl. AN SSSR, Tom 257, 309 - 312 (1981). In Russian. – Abstr. in Ref. zh., 51. Astron., 7.51.147 (1981).

093.019   **The atmosphere of Venus.**   V. I. Moroz.
Space Sci. Rev., Vol. 29, 3 - 127 (1981).
   Contents: The planet Venus – basic astronomical data. Chemical composition. Temperature, pressure, density (from 0 to 100 km). Clouds. Thermal regime and greenhouse effect. Dynamics. Chemical processes. Upper atmosphere. Origin and evolution. Problems for future studies.

093.020   **4-day waves in the Venus atmosphere.**
C. Covey, G. Schubert.
Icarus, Vol. 47, 130 - 138 (1981).
   Ultraviolet albedo contrasts in the Venus atmosphere are probably large-scale atmospheric waves propagating slowly with respect to the rapid cloud-top zonal winds. Using a simple theoretical model and profiles of mean wind and thermal structure based on Pioneer Venus data, the authors find planetary-scale gravity waves with phase velocities matching the speeds of the UV markings. They propose an upward-propagating wave and waves trapped at cloud levels as candidates to explain the observed UV features.

093.021   **Calculation on thermal radiative transfer in the subcloudy atmospheres of Venus.**
T. G. Adiks, A. P. Gal'tsev, V. M. Osipov, V. P. Shari.
Inst. prikl. mat. AN SSSR. Prepr., 1980 (1981), No. 15, 26 pp. In Russian. – Abstr. in Ref. zh., 51. Astron., 8.51.217 (1981).

093.022   **Venus ontsluierd.**   J. van Diggelen.
Zenit, 8. Jaarg., 396 - 402 (1981).

093.023   **The mineral composition of Venus surface rocks: a preliminary prediction.**   V. L. Barsukov, V. P. Volkov, I. L. Khodakovsky (*Khodakovskij*).
Proc. Eleventh Lunar Planet. Sci. Conf., (see 012.020), p. 765 - 773 (1980).

093.024   **On the search for an intrinsic magnetic field at Venus.**
C. T. Russell, R. C. Elphic, J. G. Luhmann, J. A. Slavin.
Proc. Eleventh Lunar Planet. Sci. Conf., (see 012.020), p. 1897 - 1906 (1980) = Inst. Geophys. Planet. Phys., Univ. Calif., Publ. No. 2036.
   Magnetic field observations obtained by the Pioneer Venus orbiter at low altitude are now available for two sets of orbits in the Venus wake. Data from these 130 orbits are examined for possible surface correlated features or any intrinsic magnetic moment. No surface correlated magnetic fields are observed. The upper limit of the present day Venus moment is less than 4×10^{-5} of the terrestrial moment.

093.025   **The physical and chemical properties of the surface of Venus.**   S. Nozette.
Mem. Soc. Astron. Italiana, Vol. 52, (see 012.023), 467 - 469 (1981).

093.026   **Planet Venus – discoveries and riddles.**
R. Z. Sagdeev, V. I. Moroz.
Nauka v SSSR, 1981, No. 1, p. 13 - 16. In Russian. – From Ref. zh., 51. Astron., 9.51.188 (1981).

093.027   **The albedo of Venus in the 0.2 - 4.0 μm range.**
V. I. Moroz.
Kosm. Issled., Tom 19, 591 - 598 (1981). In Russian.

093.028 **Spectrophotometric experiment aboard the launched instruments of Venera 11 and Venera 12. 3. Results of photometric measurements.**   V. I. Moroz, Yu. M. Golovin, B. E. Moshkin, A. P. Ehkonomov.
Kosm. Issled., Tom 19, 599 - 612 (1981). In Russian.

093.029 **Three-dimensional form and position of the shock wave near Venus.**   V. N. Smirnov, O. L. Vajsberg, S. A. Romanov, J. A. Slavin, C. T. Russell, D. S. Intriligator.
Kosm. Issled., Tom 19, 613 - 623 (1981). In Russian.

093.030 **On the field configuration in the magnetic tail of Venus.**   Sh. Sh. Dolginov, Eh. M. Dubinin, E. G. Eroshenko, P. L. Izrajlevich, I. M. Podgornyj, S. I. Shkol'nikova.
Kosm. Issled., Tom 19, 624 - 633 (1981). In Russian.

093.031 **Veneras and Pioneers tell about Venus. 1. 2.**   E. Mūkins.
Zvaigznota debess, 1980. gada vasara, p. 19 - 23; gada rudens, p. 33 - 41. In Latvian.

093.032 **Charge exchange in the Venus ionosphere as the source of the hot exospheric hydrogen.**   R. R. Hodges, Jr., B. A. Tinsley.
J. Geophys. Res., Vol. 86, 7649 - 7656 (1981).
   A global Monte Carlo model of the exosphere of Venus, simulating the normal exospheric processes, as well as the production of a 'hot' hydrogen component by charge exchange of H^+ with H and O, has been computed. The resulting altitude profiles of atomic hydrogen concentration over both the day and night hemispheres are in reasonable agreement with Mariner 5 and Mariner 10 observations of Lyman α, showing that the ionospheric charge exchange reactions are a significant source of 'hot' hydrogen, possibly the dominant source.

093.033 **Model of the ion composition of the Venus day ionosphere.**   M. N. Izakov, O. P. Krasitskij, A. V. Pavlov.
Kosm. Issled., Tom 19, 733 - 748 (1981). In Russian.

093.034 **On the nature of small aerosol particles in the upper part of the Venus cloud layer.**   D. V. Titov.
Kosm. Issled., Tom 19, 794 - 797 (1981).In Russian.

093.035 **Scattering of ultraviolet radiation in the He I 584 Å and O II 834 Å resonance lines in the upper atmosphere of Venus.**   M. S. Burgin, V. S. Sholokhov.
Kosm. Issled., Tom 19, 797 - 799 (1981). In Russian.

093.036 **The inverse problem of interpretation of optical measurements in the Venus atmosphere aboard the automatic interplanetary station Venera 10.**   N. V. Konovalov, N. L. Lukashevich.
Inst. prikl. mat. AN SSSR. Prepr., 1981, No. 45, 26 pp. In Russian. – Abstr. in Ref. zh., 51. Astron., 10.51.237 (1981).

093.037 **Investigation of aerosols of the Venus cloud layer with the Venera 12 automatic interplanetary station (preliminary results).**   Yu. A. Surkov, F. F. Kirnozov, V. I. Gur'yanov, V. N. Glazov, A. G. Dunchenko, S. S. Kurochkin, V. A. Rasputnyj, Eh. G. Kharitonova, L. P. Tatsij, V. L. Gimadov.
Geokhimiya, 1981, No. 1, p. 3 - 9. In Russian. – Abstr. in Ref. zh., 62. Issled. kosm. prostranstva, 10.62.249 (1981).

093.038 **The carbon monoxide fourth positive bands in the Venus dayglow. 1. Synthetic spectra.**   S. T. Durrance.
J. Geophys. Res., Vol. 86, 9115 - 9124 (1981).

   Spectral observations of the Venus dayglow from 1200 - 1800 Å were made with the Pioneer Venus ultraviolet spectrometer, a sounding rocket telescope and spectrometer and the International Ultraviolet Explorer. Emissions due to H I, C I, O I, and the CO fourth positive system are identified. Emission rate factors for the fourth positive system are calculated using a high-resolution solar flux. The importance of solar emission line excitation to this fluorescence spectrum is shown. In particular, several bands of the $v' = 14$ progression are identified and their excitation mechanism is shown to be fluorescent scattering of solar Lyman α radiation. The CO (14, 5) band at 1392 Å is a prominent unblended spectral feature in the Venus dayglow. The (14, 3) and (14, 4) bands at 1317 and 1354 Å are blended with the O I 1304- and 1356 Å features in lower-resolution spectra. This identification resolves the problem of the interpretation of the ratio of the O I 1304/1356 Å emission from Venus. It also provides an additional remote sensing technique to determine the density distribution of CO in the upper atmosphere of Venus.

093.039 **On the maintenance of the Venus nightside ionosphere: electron precipitation and plasma transport.**   K. Spenner, W. C. Knudsen. R. C. Whitten, P. F. Michelson, K. L. Miller, V. Novak.
J. Geophys. Res., Vol. 86, 9170 - 9178 (1981).
   Suprathermal integral electron spectra between 5 and 45 eV measured by the Pioneer Venus orbiter RPA are presented for the Venus nightside ionosphere. The observed integral electron flux is relatively constant with time and altitude. The simultaneously measured plasma density is much more variable and not correlated with the electron flux. For a typical electron spectrum the ionization rates and the ion density height profiles for O^+ and O_2^+ are calculated for 10 and 90° magnetic dip angle. The O^+ and O_2^+ ion density height profiles are also calculated for a downward flux of O^+ ions at 10, 30, and 90° magnetic dip angle. Comparison of the numerical modeling results with median profiles of O^+ and O_2^+ ions measured by the RPA is performed.

093.040 **Elastic thickness of the Venus lithosphere estimated from topography and gravity.**   A. Cazenave, K. Dominh.
Geophys. Res. Lett., Vol. 8, 1039 - 1042 (1981).
   The authors investigate the state of isostatic compensation of topographic features of lesser lateral extent ($\leq$ 1000 km), located in the rolling plains province of the equatorial region of Venus, applying lithospheric flexure theory.

093.041 **A source of plasma turbulence at the ionopause of Venus.**   R. E. Daniell, Jr.
J. Geophys. Res., Vol. 86, 10094 - 10104 (1981).
   The simultaneous measurements of suprathermal ions and low frequency electric field signals at the ionopause of Venus by Pioneer Venus instruments suggests that they may be causally related. Linear Vlasov theory is used to examine both parallel and perpendicular propagating waves for instability in the presence of planetary ions added to the flowing ionosheath plasma.

093.042 **Observation of an anomalous CO_2 absorption profile on the night side of Venus.**   D. Deming, F. Espenak, D. Jennings, T. Kostiuk, M. Mumma.
Bull. American Astron. Soc., Vol. 13, 711 - 712 (1981). Abstract.

093.043 **Venus: distribution of radar reflectivity.**   P. G. Ford, G. H. Pettengill.
Bull. American Astron. Soc., Vol. 13, 713 (1981). – Abstract.

093.044 **Gravity in the vicinity of Beta Regio from PVO Doppler tracking data.**

R. D. Reasenberg, Z. M. Goldberg, I. I. Shapiro.
Bull. American Astron. Soc., Vol. 13, 713 (1981). – Abstract.

093.045   Constraints on the subsurface structure near Beta Regio.
R. D. Reasenberg, Z. M. Goldberg, I. I. Shapiro.
Bull. American Astron. Soc., Vol. 13, 713 (1981). – Abstract.

093.046   Deducing the age of the dense Venus atmosphere.
R. Kahn.
Bull. American Astron. Soc., Vol. 13, 713 (1981). – Abstract.

093.047   A dynamic scale model of aeolian transport on Venus.   S. Nozette.
Bull. American Astron. Soc., Vol. 13, 714 (1981). – Abstract.

093.048   Model sensitivity studies of the solar related circulation in the stratosphere of Venus.   L. Elson.
Bull. American Astron. Soc., Vol. 13, 714 (1981). – Abstract.

093.049   The relationship between the temperature structure in the Venus atmosphere and its general circulation pattern.   A. J. Kliore, I. R. Patel.
Bull. American Astron. Soc., Vol. 13, 714 (1981). – Abstract.

093.050   Waves in the middle atmosphere of Venus.
J. Apt, J. Leung.
Bull. American Astron. Soc., Vol. 13, 714 (1981). – Abstract.

093.051   Venus SO_2 abundance from interferometric measurement of the limb darkening at 3.4 mm.
J. C. Good, F. P. Schloerb, W. J. Welch.
Bull. American Astron. Soc., Vol. 13, 714 - 715 (1981). Abstract.

093.052   Temporal and spatial variations of sulfur dioxide on Venus.   L. W. Esposito.
Bull. American Astron. Soc., Vol. 13, 715 (1981). – Abstract.

093.053   An analysis of Barker's UV spectrum of SO_2 on Venus: implication for Io and for laboratory measurements.   M. J. S. Belton, C.-Y. R. Wu.
Bull. American Astron. Soc., Vol. 13, 715 (1981). – Abstract.

093.054   Photochemical models for the stratosphere of Venus.
Y. L. Yung, W. B. DeMore.
Bull. American Astron. Soc., Vol. 13, 715 (1981). – Abstract.

093.055   Clouds of Venus: sulfuric acid by the Lead Chamber Process.   G. T. Sill.
Bull. American Astron. Soc., Vol. 13, 715 (1981). – Abstract.

093.056   Band-model interpretation of Venera 11 spectra.
A. T. Young, L. G. Young.
Bull. American Astron. Soc., Vol. 13, 715 - 716 (1981). Abstract.

093.057   Microphysical composition of the lower cloud deck on Venus.   J. Cimino.
Bull. American Astron. Soc., Vol. 13, 716 (1981). – Abstract.

093.058   Temporal monitoring of the Venus CO J = 1 → 0 rotational transition.   F. P. Schloerb, J. C. Good.
Bull. American Astron. Soc., Vol. 13, 724 (1981). – Abstract.

093.059   Cloud structure of Venus north polar region. Coincident infrared and polarimetry data from the Pioneer orbiter.   H. Nebel, L. Elson, D. Diner, L. Travis.
Bull. American Astron. Soc., Vol. 13, 725 (1981). – Abstract.

093.060   Temporal variations in the cloud and temperature structure of Venus' north polar region.

D. J. Diner.
Bull. American Astron. Soc., Vol. 13, 725 - 726 (1981). Abstract.

093.061   Temporal variability of UV cloud albedo features on Venus.   A. del Genio, W. Rossow.
Bull. American Astron. Soc., Vol. 13, 743 (1981). – Abstract.

093.062   Turbulent accumulation of interplanetary magnetic fluxes at the Venus ionopause.
H. Pérez-de-Tejada.
Geophys. Res. Lett., Vol. 8, 1151 - 1154 (1981).
The accumulation of interplanetary magnetic fluxes around the Venus ionopause is examined in the light of the PVO magnetometer and plasma wave measurements. It is argued that convective electric currents within the ionosphere are not sufficient to account for the observed geometry of the magnetic profiles. It is suggested that turbulent flow conditions forced in the ionosheath by the interaction process with the ionospheric plasma should produce an anomalous resistivity in the local flow and an effective lag in the transport of the interplanetary magnetic fluxes around the ionospheric obstacle. The enhanced magnetic fluxes seen above the ionopause should result from effective turbulent currents generated throughout the region of interaction.

093.063   Radar maps of Venus.   E. Mūkins.
Zvaigžņota debess, 1981. gada vasara, p. 10 - 17.
In Latvian.

093.064   Cloud models of Venus from the point of view of circular polarization.   Y. Kawata.
Proceedings of the 14th ISAS Lunar and Planetary Symposium, (see 012.055), p. 11 - 16 (1981).
The author demonstrates in this paper that the circular polarization is a potentially useful tool for studying the atmosphere of Venus, particularly if it is combined with measurements of linear polarization and intensity. Contour diagrams (as a function of particle size and phase angle) of the circular polarization for spherical particles with refractive index $n_r = 1.44$ are computed for Venus.

093.065   Theoretical studies on the structure of the dayside ionopause of Venus.   H. Oya, S. Watanabe.
Proceedings of the 14th ISAS Lunar and Planetary Symposium, (see 012.055), p. 46 - 72 (1981).
The Venus ionopause has been numerically calculated on the basis of the microscopic approach using the Maxwell-Vlasov equations. These results of numerical calculations have been compared with the observations by the Pioneer-Venus missions. The good agreement of the theoretical results and the observation results suggests that the Venus ionopause is mainly maintained by the current generated by the ions motion emanating from the Venus ionosphere.

093.066   L'atmosphère de Vénus.   J. Blamont.
C. R. Acad. Sci. Paris, Vie Acad., Tome 293, (see 012.056), XXV - XXVI, XXVIII - XXIX (1981).

093.067   I. The physical parameters of the Venus atmosphere computed for different chemical compositions containing SO_2.   C. J. Macris, B. C. Petropoulos.
Praktika Akad. Athens, Tom. 54, 125 - 145 (1979) = Res. Cent. Astron. Appl. Math., Acad. Athens, Contrib. Ser. I (Astron.), No. 75.
The authors propose a standard model for the Venus atmosphere from 0 to 200 km, based on Venera 9 and 10 measurements. The physical parameters have been computed for different quantities of SO_2.

093.068   II. Physical parameters of the upper atmosphere of Venus, computed for different chemical composi-

tions of $CO_2 \cdot CO_2{}^+$.   C. J. Macris, B. C. Petropoulos.
Praktika Akad. Athens, Tom. 55, 312 - 335 (1980) = Res.
Cent. Astron. Appl. Math., Acad. Athens, Contrib. Ser. I
(Astron.), No. 81.

093.069 Quantitative tests for plate tectonics on Venus.
W. M. Kaula, R. J. Phillips.
Geophys. Res. Lett., Vol. 8, 1187 - 1190 (1981).

093.070 The role of charge exchange in the solar wind absorption by Venus.   T. I. Gombosi, M. Horanyi, T. E. Cravens, A. F. Nagy, C. T. Russell.
Geophys. Res. Lett., Vol. 8, 1265 - 1268 (1981).
The amount of solar wind absorbed due to charge exchange processes in the dayside ionosheath of Venus is calculated. The calculations indicate the existence of a lower
limiting ionopause altitude, below which all solar wind
particles are removed from the flow by charge exchange. It is
suggested that the cold and slow new ions, resulting from
this interaction, play an important role in building the magnetic barrier observed just outside the ionopause, and also in
creating the dayside "mantle" and downstream "penumbra"
regions. The total absorption caused by charge exchange is
typically 2-5%, but it can reach values as high as 16%, when
the solar wind dynamic pressure is very high.

093.071 Venus nightside ionospheric holes: the signatures of parallel electric field acceleration regions?
J. M. Grebowsky, S. A. Curtis.
Geophys. Res. Lett., Vol. 8, 1273 - 1276 (1981).
Recent observations by the instruments onboard the
Pioneer Venus spacecraft have revealed the existence of
"holes" or regions of density depletion in the nightside Venus
ionosphere associated with regions of radial magnetic fields.
The properties of the electrons within the core of these holes
are suggestive of an acceleration process along the magnetic
field lines as are the Venera 9 and 10 observations of energetic
ions in the Venus tail. Given the observational information,
the authors attribute these Venusian plasma depletions to the
presence of parallel electric fields similar to those observed in
the terrestrial auroral ionosphere.

093.072 Evidence for the acceleration of ionospheric O^+ in the magnetosheath of Venus.
J. D. Mihalov, A. Barnes.
Geophys. Res. Lett., Vol. 8, 1277 - 1280 (1981).
Energetic ions ($E/q \gtrsim 4$ kV) in the magnetosheath near
Venus, in the vicinity of the planet's terminator plane, have
been detected by the plasma analyzer of the Pioneer-Venus
Orbiter. These ions have energy per unit charge consistent
with O^+, the most abundant ionospheric ion, moving at a speed
somewhat slower than or comparable to the local magnetosheath speed. These ions are slowest and most intense at low
altitudes. The authors interpret these observations as indicating that O^+ is picked up by the magnetosheath plasma and
accelerated up to the ambient proton flow speed within one
Venus radius of the surface of the planet, and subsequently
convected away through the wake of the planet.

093.073 The ionospheric peak on the Venus dayside.
T. E. Cravens, A. J. Kliore, J. U. Kozyra, A. F. Nagy.
J. Geophys. Res., Vol. 86, 11323 - 11329 (1981).
Many electron density profiles of the dayside ionosphere
of Venus have been measured by the Pioneer Venus orbiter
between December 1978 and October 1980, using the dual-
frequency radio occultation technique. The peak electron
density as a function of solar zenith angle can be described by
a simple Chapman layer theory with proper normalization;
however, it does not predict adequately the height variation
of the electron density peak. In order to interpret these radio
occultation results, the authors have constructed a theoretical
model describing the ion composition and electron density in

the vicinity of the ionospheric peak. Good agreement between the model and the measurements was obtained only
with a specific choice of the neutral densities, the electron
temperature, and the level of solar activity.

093.074 The Venus ionopause current sheet: thickness length scale and controlling factors.   R. C. Elphic, C. T. Russell, J. G. Luhmann, F. L. Scarf, L. H. Brace.
J. Geophys. Res., Vol. 86, 11430 - 11438 (1981).
Data from the Pioneer Venus orbiter magnetometer,
plasma wave detector and Langmuir probe are used to investigate the scale thickness and controlling factors associated
with the current sheet at the Venus ionopause, the boundary
between the ionosphere and the solar wind.

093.075 Refraction effects in the Venus atmosphere from bistatic radar data.
A. G. Pavel'ev, A. I. Kucheryavenkov, O. E. Milekhin.
Izv. vuzov. Radiofiz., Tom 24, 401 - 407 (1981). In Russian.
Abstr. in Ref. zh., 51. Astron., 12.51.291; 62. Issled. kosm.
prostranstva, 12.62.263 (1981).

093.076 On the possible existence of condensates of chlorides of a series of metals and elementary sulphur in the Venus troposphere.
V. L. Barsukov, I. L. Khodakovskij, V. P. Volkov,
Yu. I. Sidorov, V. A. Dorofeeva, N. E. Andreeva.
Dokl. AN SSSR, Tom 259, 1215 - 1218 (1981). In Russian.
Abstr. in Ref. zh., 62. Issled. kosm. prostranstva, 12.62.269
(1981).

093.077 Carbon dioxide within Venus and the Earth.
T. J. Ahrens.
Planetary interiors, (see 012.061), p. 177 - 187 (1981).
The maximum inventories of CO_2 and H_2O in the terrestrial and Venus mantles are calculated on the basis of shock
wave data for magnesite and dunite to be $\sim 10^3$ times the observed atmospheric inventories.

093.078 The upper atmosphere of Venus: implications for aeronomy.   D. M. Hunten.
Progress in planetary exploration, (see 012.062), p. 139 - 140
(1981).

093.079 Equatorial cloud properties on Venus from Pioneer Orbiter infrared observations.   F. W. Taylor.
Progress in planetary exploration, (see 012.062), p. 151 - 154
(1981).
Infrared observations of Venus from the Pioneer Orbiter
have been used to study the limb darkening properties of the
cloud tops at wavelengths and spatial resolutions not previously attained. The preliminary results show evidence for an extensive haze feature over the equatorial morning terminator
and for small amounts of a far-infrared absorber concentrated
near local noon, also near the equator. The evidence for these
features is reviewed and their possible origins briefly discussed.

093.080 Absorbers seen near the Venus cloud tops from Pioneer Venus.   L. W. Esposito.
Progress in planetary exploration, (see 012.062), p. 163 - 166
(1981).
Spin-scan images from the Pioneer Venus Orbiter UV
Spectrometer and the Cloud Photopolarimeter provide a set
of planetary contrast measurements in the wavelength range
1990A to 3650A and phase angles from 33°- 130°.

093.081 The optical parameters of the atmosphere of Venus.
J. M. Dlugach (*Zh. M. Dlugach*), E. G. Yanovitskij
(*Eh. G. Yanovitskij*).
Progress in planetary exploration, (see 012.062), p. 167 - 170
(1981).
This work is devoted to the derivation of the optical

properties of the Venus atmosphere from "Venera-10" optical measurements. Within the framework of a two-layer model of Venus atmosphere it is found that in the spectral interval 0.52 - 0.85 μm the optical thickness of the upper cloud layer is ≈ 50 and the optical parameters of the lower layer are similar to the Rayleigh ones. Comparison is made between the measurements of radiation field within the atmosphere and the results of strict calculations. A preliminary conclusion is suggested that there are considerable numbers of aerosol particles with a radius $\lesssim 0.03$ μm in the lower layer. The altitude of the upper boundary of the cloud layer is estimated to be ≈ 70 km.

093.082 **The upper atmosphere of Venus: implications for aeronomy.**   D. M. Hunten.
Planetary aeronomy and astronomy, (see 012.063), p. 3 - 4 (1981).

093.083 **The physical properties of the clouds of Venus.**   J. Blamont.
Planetary aeronomy and astronomy, (see 012.063), p. 5 - 11 (1981).

Brief description of the particles constituting the Venus clouds: size distribution, index of refraction, shape, bulk optical properties, possible chemical composition.

093.084 **Vertical distribution of SO_2 in upper cloud layer of Venus and origin of UV absorption.**
L. V. Zasova, V. A. Krasnopolsky (*Krasnopol'skij*), V. I. Moroz.
Planetary aeronomy and astronomy, (see 012.063), p. 13 - 16 (1981).

It is shown that decline of spherical albedo of Venus toward the ultraviolet can be explained by the presence of two absorbing agents: (1) SO_2, for which abundance is 10^{11} cm^{-3} at height 68 km and scale height is about 1 km; (2) some unknown aerosol absorbent, possibly a 1% $FeCl_3$ admixture in a sulfuric acid concentrated solution. A mechanism of aerosol formation is proposed.

093.085 **Comparative analysis of Venusian ionosphere dual-frequency radio soundings with the satellites Venera-9, 10 and Pioneer-Venus.**   N. A. Savich.
Planetary aeronomy and astronomy, (see 012.063), p. 17 - 25 (1981).

A comparison of the Venusian ionosphere electron density profiles obtained by the dual-frequency radio occultation method for the satellites Venera-9, 10 (1975) and the Pioneer-Venus orbiter (1978 - 1979) has been carried out. It is shown that the general nature of the profiles, the main maximum heights and electron densities, the ionopause height positions determined with the satellites Venera-9, 10 on the one hand and Pioneer-Venus orbiter on the other hand are in good agreement.

093.086 **On ionization sources in the night ionosphere of Venus.**
T. K. Breus, A. S. Volokitin, J. V. (*E. V.*) Mishin.
Planetary aeronomy and astronomy, (see 012.063), p. 27 - 32 (1981).

Radio occultation observations of Venus revealed an unusual variability of the planet's night ionosphere. In some cases two distinct narrow ionization peaks were observed. The purpose of the paper is to show that the double-peaked profiles observed in the night ionosphere of Venus can be explained if a 0.5 - 1 keV electron beam is assumed to be one of the sources of ionization and if collective beam interactions are taken into account together with collisional interaction.

093.087 **Model calculations of the dayside ionosphere of Venus.**   T. E. Cravens, T. I. Gombosi, A. F. Nagy.
Planetary aeronomy and astronomy, (see 012.063), p. 33 - 36 (1981).

Model calculations of the dayside ionosphere of Venus are presented. The coupled continuity and momentum equations were solved for O_2^+, O^+, CO_2^+, C^+, N^+, He^+, and H^+ density distributions, which are compared with measurements from the Pioneer Venus ion mass spectrometer. The agreement between the model results and the measurements is good for some species, such as O^+, and rather poor for others, such as N^+.

093.088 **Temporal and spatial variations observed in the ionospheric composition of Venus: implication for empirical modelling.**   H. A. Taylor Jr., S. J. Bauer, R. E. Daniell, H. C. Brinton, H. G. Mayr, R. E. Hartle.
Planetary aeronomy and astronomy, (see 012.063), p. 37 - 51 (1981).

In situ measurements of the thermal ion composition of the ionosphere of Venus have been obtained for a period of two Venus years from the Bennett ion mass spectrometer on the Pioneer Venus Orbiter. The distributions of several ion species in this data base have been sorted to identify temporal and spatial variations, and to determine the feasibility of an analytical representation of the experimental results. The influence of solar wind perturbations upon the ion distributions is also indicated.

093.089 **Magnetic flux ropes in the Venus ionosphere: in situ observations of force-free structures?**
R. C. Elphic, C. T. Russell, J. G. Luhmann, L. H. Brace.
Planetary aeronomy and astronomy, (see 012.063), p. 53 - 58 (1981).

Force-free magnetic structures with cylindrical geometry appear under a variety of conditions in nature. Filamentary helical magnetic structures are observed to be associated with prominences and flares in the solar atmosphere, and can arise in superconductors and laboratory plasmas. Another example of cylindrical quasi-force-free configurations appears to exist in the Venus ionosphere. Magnetic flux ropes with diameters of ~20 - 30 km have been observed by the Pioneer Venus Orbiter to be a nearly ubiquitous feature of the dayside Venus ionosphere.

093.090 **Iron in the clouds of Venus.**
I. V. Petryanov, B. M. Andrejchikov, B. N. Korchuganov, E. I. Ovsyankin, B. I. Ogorodnikov, V. I. Skitovich, V. K. Khristianov.
Dokl. AN SSSR, Tom 260, 834 - 836 (1981). In Russian. Abstr. in Ref. zh., 62. Issled. kosm. prostranstva, 1.62.386 (1982).

093.091 **Characteristics of the surface and features of the propagation of radio waves in the atmosphere of Venus from data of bistatic radiolocation experiments using Venera 9 and 10 satellites.**   M. A. Kolosov, O. I. Yakovlev, A. G. Pavelyev (*Pavel'ev*), A. I. Kucheryavenkov (*Kucheryavenkov*), O. E. Milekhin.
Icarus, Vol. 48, 188 - 200 (1981).

The results of the investigation of two regions of Venus by bistatic radiolocation are presented. The experiments were carried out at wavelength $\lambda_0 = 32$ cm. Maps of the distribution of reflectivity were obtained and characteristics of the relief, dielectric permittivity, soil density, and refraction attenuation in the atmosphere were measured. There are some features on the reflectivity maps. Some of these features may correspond to mountain slopes. Corresponding changes of relief heights are contained in the interval 0.8 to 2.6 km. The results of measurements of the refraction angle and the refraction attenuation of radio waves are in good agreement with the parameters of the atmosphere of Venus received from the Soviet landers.

Radio investigations of the moon and terrestrial planets.   See Abstr. 003.015.

A new interpretation of Venera 11 spectra. I. Comparison with laboratory data.   See Abstr. 022.129.

Laboratory measurements of the microwave opacity of cloud related gases under simulated conditions for the Venus atmosphere.   See Abstr. 022.130.

Laboratory measurements of the microwave opacity of sulfur dioxide and other cloud-related gases under simulated conditions for the middle atmosphere of Venus. See Abstr. 022.183.

Restoration of the spectra of solar emission scattered in the Venus atmosphere (according to Venera 11 and Venera 12 data).   See Abstr. 031.652.

VOIR mission: the global mapping of Venus. See Abstr. 051.005.

Balloons on other planets.   See Abstr. 051.055.

Role of scattering in emergent planetary radiation. See Abstr. 063.053.

Solar wind pickup of ionized Venus exosphere atoms. See Abstr. 074.005.

The CO_2 greenhouse effect and the thermal history of the atmosphere.   See Abstr. 082.124.

The dynamics of a rapidly escaping atmosphere: applications to the evolution of Earth and Venus. See Abstr. 091.094.

Chemical composition of Earth, Venus, and Mercury.   See Abstr. 092.002.

Some aspects of the solar radiation incident at the top of the atmospheres of Mercury and Venus. See Abstr. 092.010.

Radioactivity of the moon and planets. See Abstr. 094.620.

Occultation de la planète Vénus par la Lune du 20 Janvier 1980 observée à Belgrade. See Abstr. 096.019.

Solar wind sputtering effects in the atmospheres of Mars and Venus.   See Abstr. 097.024.

Erratum

093.901   Erratum: "Venus. I. Carbon monoxide distribution and molecular-line searches" [Icarus, Vol. 45, 624 - 637 (1981)].   W. J. Wilson, M. J. Klein, R. K. Kakar, S. Gulkis, E. T. Olsen, P. T. P. Ho. Icarus, Vol. 48, 344 (1981). – See Abstr. 29.093.038.

094 Moon (Dynamics, General Aspects, Local Properties)

Moon (Dynamics, General Aspects)

094.001 Lunar seismology: the internal structure of the moon. N. R. Goins, A. M. Dainty, M. N. Toksöz.
J. Geophys. Res., Vol. 86, 5061 - 5074 (1981).

The arrival times of direct P and S waves, measured on seismograms recorded from natural lunar seismic events, have been analyzed using linearized inversion and parameter search methods to simultaneously determine event locations, origin times, and structural parameters (seismic velocities). Polarization-filtered record section plots are correlated with theoretical travel time curves to identify later phases and obtain additional structural and velocity information. Shear wave amplitudes plotted as a function of distance provide data on the existence and magnitude of seismic velocity gradients in the interior. These studies are used to delineate the structure of the moon below the crust to a depth of about 1100 km.

094.002 The relationship between the albedo and the polarization properties of the moon. Fresnel component of reflected light. Yu. G. Shkuratov.
Astron. Zh., Tom 58, 862 - 866 (1981). In Russian. English translation in Soviet Astron., Vol. 25, No. 4.

The possibility of mapping the new lunar optical parameter is confirmed. This parameter describes the regional distribution of deviations from the well-known correlation between the albedo and the maximum polarization coefficient. A two-component model of light reflection is used for establishing the relationship of the new parameter with the refractive index and with the structure characteristics of the lunar soil.

094.003 Extreme perigees and apogees of the moon. J. Meeus.
Sky Telesc., Vol. 62, 110 - 111 (1981).

094.004 Tidal dissipation in the moon. R. J. Cappallo, C. C. Counselman III, R. W. King, I. I. Shapiro.
J. Geophys. Res., Vol. 86, 7180 - 7184 (1981).

Euler's equations of motion, modified to include elasticity and solid friction, were used to study the rotation of the moon. Two heuristic models for the anelasticity were considered: Q independent of frequency and Q inversely proportional to the frequency of the strain oscillation. Parameters in each model were estimated by weighted least squares from 9 years of lunar laser range observations. The root mean square of the postfit range residuals was 19 cm in each case. For a strain period of 1 month, the estimates of Q obtained with the two models were similar and surprisingly low: 23 ± 6.

094.005 On the center of Watts' datum for the lunar marginal zone. J. D. Mulholland.
Bull. American Astron. Soc., Vol. 13, 570 (1981). — Abstract.

094.006 Early lunar petrogenesis, oceanic and extraoceanic. P. H. Warren, J. T. Wasson.
The lunar highlands crust, (see 012.013), p. 81 - 99 (1980).

An attempt has been made to ascertain which (if any) pristine nonmare rocks, other than KREEPy ones, are not cumulates from the magma ocean.

094.007 The cordierite- to spinel-cataclasite transition: structure of the lunar crust. C. T. Herzberg, M. B. Baker.
The lunar highlands crust, (see 012.013), p. 113 - 132 (1980) = Lunar Planet. Inst. Contrib. No. 402.

094.008 Dropping stones in magma oceans: effects of early lunar cratering. W. K. Hartmann.
The lunar highlands crust, (see 012.013), p. 155 - 171 (1980).

A new methodology is used to calculate the accumulation rate of megaregolith materials for two models of early lunar cratering, both with and without episodes of late cataclysmic cratering. Results show that the pulverization of early rock layers was an important process competing with the formation of a coherent rock lithosphere at the surface of the hypothetical lunar magma ocean. Furthermore, the results show that intense brecciation and pulverization of rock materials must have occurred to a depth of at least tens of kilometers in the first few hundred years of lunar history regardless of whether a "terminal lunar cataclysm" occurred around 4.0 G.y. ago.

094.009 On the nature of ring photo anomalies fixed by remote sensing.
V. Yu. Zajchenko, O. L. Kuznetsov, G. P. Popsuj-Shapko.
Sov. geol., 1981, No. 1, p. 98 - 106. In Russian. — Abstr. in Ref. zh., 51. Astron., 7.51.142 (1981).

094.010 Photometric mapping of the moon according to data of the automatic interplanetary stations Zond 6 and Zond 8. N. P. Lavrova, V. A. Psarev.
Khar'kov. univ. Khar'kov, 1981, 98 pp. In Russian. — Abstr. in Ref. zh., 62. Issled. kosm. prostranstva, 7.62.296 (1981).

094.011 On the influence of the secular change in the eccentricity of the earth's orbit on the motion of the moon. J. Vondrák.
Bull. Astron. Inst. Czechoslovakia, Vol. 32, 315 - 317 (1981).

The influence of the quadratic term in the eccentricity of earth's orbit round the sun on the motion of the moon is derived. It is shown that this effect, especially in the case of the moon's longitude, cannot be neglected.

094.012 Shallow moonquakes: how they compare with earthquakes. Y. Nakamura.
Proc. Eleventh Lunar Planet. Sci. Conf., (see 012.020), p. 1847 - 1853 (1980).

094.013 Focal mechanism of deep moonquakes. J. Koyama, Y. Nakamura.
Proc. Eleventh Lunar Planet. Sci. Conf., (see 012.020), p. 1855 - 1865 (1980).

094.014 Lunar polar wandering. S. K. Runcorn.
Proc. Eleventh Lunar Planet. Sci. Conf., (see 012.020), p. 1867 - 1877 (1980).

The modelling of lunar magnetic anomalies, observed by Apollo 15 and 16 subsatellite magnetometers, gives the direction of magnetization of areas of the lunar crust. On the hypothesis of an ancient lunar magnetic field generated by a core dynamo, pole positions are calculated. Polar wandering of 90° in the early moon is found and explained by changes in the moment of inertia due to early impact.

094.015 The first few hundred years of evolution of a moon of fission origin. A. B. Binder.
Proc. Eleventh Lunar Planet. Sci. Conf., (see 012.020), p. 1931 - 1939 (1980).

094.016 Time-dependent lunar density models. F. Herbert.
Proc. Eleventh Lunar Planet. Sci. Conf., (see 012.020), p. 2015 - 2030 (1980).

094.017 On constraining lunar mantle temperatures from gravity data. S. Pullan, K. Lambeck.
Proc. Eleventh Lunar Planet. Sci. Conf., (see 012.020), p. 2031 - 2041 (1980).

094.018 The bulk composition of the moon based on geophysical constraints. W. R. Buck, M. N. Toksöz.
Proc. Eleventh Lunar Planet. Sci. Conf., (see 012.020), p. 2043 - 2058 (1980).

094.019 Lunar gravity field determination. W. Flury.
Spacecraft flight dynamics, (see 012.021), p. 37 - 45 (1981).
The current status of the lunar gravity field determination is reviewed and then followed by a discussion of the POLO (= Polar Orbiting Lunar Observatory) gravity experiment.

094.020 The jump of the second derivative of the Moon's elongation. A. T. Fomenko.
Celestial Mech., Vol. 25, 33 - 40 (1981).
The author presents some results of new calculations of $D''(t)$ − the second derivative of the Moon's elongation as a function of time. The paper contains an explanation of the well-known R. Newton's effect − the rapid decline in $D''(t)$ from about 700 yr to about 1300 yr. The new graph of D'' is based on the revised dates of the ancient eclipses and has a qualitatively different character; in particular, the decline in $D''(t)$ vanishes completely and $D''(t)$ oscillates at a roughly constant value, which coincides with the modern one.

094.021 Lunar composition based on geophysical data and evolutionary models. W. R. Buck, M. N. Toksöz.
Mem. Soc. Astron. Italiana, Vol. 52, (see 012.023), 359 - 363 (1981).

094.022 Morphology of shock craters on the moon and planets. A. T. Bazilevskij, N. N. Grebennik.
Kosmogen. struktury Zemli. Mater. semin. Moskva, 1980, p. 13 - 17. In Russian. − Abstr. in Ref. zh., 62. Issled. kosm. prostranstva, 9.62.400 (1981).

094.023 The enigma of lunar magnetism.
L. L. Hood.
EOS Trans. American Geophys. Union, Vol. 62, 161 - 163 (1981). − Abstr. in Phys. Abstr., Vol. 84, Abstr. 88780 (1981).

094.024 Lunar floor-fractured craters: evidence for viscous relaxation of crater topography.
J. L. Hall, S. C. Solomon, J. W. Head.
J. Geophys. Res., Vol. 86, 9537 - 9552 (1981).
The authors evaluate quantitatively the hypothesis that topographic modification of floor-fractured craters on the moon was accomplished predominantly by viscous relaxation. Adopting the simple assumption that the moon may be modeled as having a uniform Newtonian viscosity, the authors compare the observed topographic profiles for a number of floor-fractured craters with the profiles predicted from the viscous relaxation of topography of fresh craters of similar diameter. Despite the simplicity of the rheological model, the comparison is quite good.

094.025 Pressure dependence of the velocity of lunar soil: the velocity/depth variation in the shallow lunar crust.
A. F. Gangi.
J. Geophys. Res., Vol. 86, 9562 - 9566 (1981).
Some recent measurements (Johnson et al., 1981) of the velocity variation with pressure (up to 2.0 bars) for lunar soil are compared with results from the Hertzian-contact theory. Contrary to the original contention of Johnson et al. (1981), the data are shown to be consistent with the results of the

Hertz theory when the effects of nonrecoverable compaction are taken into consideration.

094.026 On the properties of the second derivative of the lunar elongation and connected statistical regularities. A. T. Fomenko.
Vopr. vychisl. i prikl. mat. Tashkent, 1981, No. 63, p. 136 - 150. In Russian. − Abstr. in Ref. zh., 51. Astron., 10.51.135 (1981).

094.027 Mapping of the far side of the moon on the basis of space images.
V. D. Bol'shakov, N. P. Lavrova, V. N. Deputatova, B. V. Krasnopevtseva, N. I. Konstantinova, V. V. Usova.
Nauchn. chteniya po aviats. i kosmonavt. 10-e Gagarinsk. chteniya, 1980. Moskva, 1981, p. 291 - 292. In Russian. Abstr. in Ref. zh., 51. Astron., 10.51.282 (1981).

094.028 The Earth-figure perturbations in the lunar theory. J. Henrard.
Celestial Mech., Vol. 25, 417 - 425 (1981).
The author computes the perturbations on the motion of the Moon due to the shape of the Earth. The zonal terms in J_2, J_3, and J_4 are considered. The accuracy is estimated at $3'' \times 10^{-5}$ and the results compared with previous theories.

094.029 Comparison of ELP-2000 to a JPL numerical integration. J. Chapront, M. Chapront-Touzé.
Astron. Astrophys., Vol. 103, 295 - 304 (1981). In French.
A solution for the orbital motion of the Moon has been built by the authors which is named ELP-2000, the epoch of reference being J 2000. It is a semi-analytical solution, its structure being quite similar to Brown-Eckert's one, as it appears in the Improved Lunar Ephemeris. The main purpose of this work is to present the results of a comparison of a provisional but complete solution, to an external numerical integration, LE-51, built at JPL, and fitted to lunar laser ranging data. The JPL numerical integration is regarded as an "observational model". It is a first attempt to compare as a whole, a new lunar ephemeris, derived from a semi-analytical theory, to observations, via a numerical integration.

094.030 Solar-wind interactions with the Moon: nature and composition of nitrogen compounds.
N. R. Mukherjee.
Moon Planets, Vol. 25, 451 - 463 (1981).
The solar-wind corpuscles interact directly with the lunar surface material. The dominant nature of the interaction is essentially complete absorption of the solar-wind particles by the lunar surface grains resulting in no upstream bowshock, but a cavity downstream. On a long-term basis the interaction products are the main source of species in the lunar atmosphere. In this paper the author discusses the long-term concentrations of the solar-wind derived nitrogen species in the atmosphere and lunar surface grains.

094.031 The development of central peaks in lunar craters. J. W. Bond.
Moon Planets, Vol. 25, 465 - 476 (1981).
From a consideration of equations describing the supersonic impact of a solid body on to a solid target, the difference between final crater depth and distance vertically below the original impact at which the rarefaction wave front, resulting from the reflection of the backward propagating shock wave in the meteorite, first intersects the forward travelling shock wave front in the target has been determined. It is proposed that the intersection of these two wave fronts locally inhibits the ejection of material from behind the shock front during the excavation phase of crater formation, leading to the appearance of a centrally located peak of uplifted material.

094.032 **What is happening on the moon? Lunar transient phenomena.**   C. Rutkowski.
J. R. Astron. Soc. Canada, Vol. 75, 237 - 241 (1981).

094.033 **High resolution radar maps of the moon at 70 cm wavelength.**   T. W. Thompson.
Bull. American Astron. Soc., Vol. 13, 710 (1981). — Abstract.

094.034 **Remote sensing studies of lunar dark-halo craters.**   B. R. Hawke, J. F. Bell.
Bull. American Astron. Soc., Vol. 13, 712 (1981). — Abstract.

094.035 **On the relation between the internal structure of the moon and the moonquakes.**   G.-x. Song.
Acta Astrophys. Sinica, Vol. 1, 197 - 202 (1981). In Chinese.
This paper deals with the relation between the internal structure of the moon and the radial distribution of the focuses of moonquakes. It has been shown from lunar probes that most of the focuses of moonquakes located at the moon mantle near the possible liquid core of the moon.

094.036 **Optical detection and ranging of the moon.**   V. K. Abalakin, Yu. L. Kokurin.
Usp. fiz. nauk, Tom 134, 526 - 535 (1981). In Russian. Abstr. in Ref. zh., 51. Astron., 11.51.52 (1981).

094.037 **Numerical model of the lunar indicatrix.**   V. A. Psarev.
Astron. Tsirk., No. 1130, p. 3 - 6 (1980). In Russian.

094.038 **Multi-spectral photography of the moon on Mt. Majdanak.**   V. V. Novikov, S. I. Ivanov.
Astron. Tsirk., No. 1133, p. 6 - 7 (1980). In Russian.

094.039 **On a probability approach to the analysis of accuracy of describing the lunar gravitational field.**   R. A. Kashcheev.
Tr. Kazan. Gorod. Astron. Obs., Vyp. 45, p. 75 - 80 (1980). In Russian.

094.040 **Derivation of the parameters of physical libration from photographic observations of the moon with stars.**   F. A. Garaev.
Tr. Kazan. Gorod. Astron. Obs., Vyp. 46, p. 133 - 141 (1980). In Russian.

094.041 **Parameters of the physical libration of the moon derived from Kazan heliometric observation series of the moon.**   G. M. Stolyarov.
Tr. Kazan. Gorod. Astron. Obs., Vyp. 46, p. 141 - 165 (1980). In Russian.

094.042 **Photometric and polarimetric observations of the Moon's surface using the multichannel polarimeter installed at the Dodaira Station of the Tokyo Astronomical Observatory.**   N. Sekiguchi.
Proceedings of the 14th ISAS Lunar and Planetary Symposium, (see 012.055), p. 131 - 137 (1981).
The brief description of the multichannel polarimeter installed at the Dodaira Station of the Tokyo Astronomical Observatory, and an example of the observational results (preliminary values) are mentioned.

094.043 **The wavelength-comparative reflectivity may change from detail to detail on the lunar surface.**
N. Kumagai, T. Shimasaka.
Proceedings of the 14th ISAS Lunar and Planetary Symposium, (see 012.055), p. 138 - 140 (1981).
Along the same diameters of two developed photographic films of the image of the Moon in red and green colors taken near the Full Moon by an Earth-based telescope, the density of the films was measured along the diameters by scanning with a densitometer and from the results the ratio of the densities in the two colors was found. The results suggest that the wavelength-comparative reflectivity, a material constant of the lunar surface, may change from detail to detail on the lunar surface.

094.044 **Attempt to predict the gravitational anomalies of the moon.**   G. A. Meshcheryakov, V. E. Zinger, P. M. Zazulyak, V. V. Kirichuk.
Geod., kartogr. i aehrofotosemka, L'vov, 1981, No. 33, p. 43 - 50. In Russian. — Abstr. in Ref. zh., 52. Geod. Aehrosemka, 12.52.307 (1981).

094.045 **Heating of the moon by heterogeneous accretion.**   G. A. Ransford, W. M. Kaula.
J. Geophys. Res., Vol. 85, 6615 - 6627 (1980).
Monte Carlo models of lunar accretion are constructed by randomly sampling mass and velocity distributions of objects in the solar nebula in circumsolar and circumterrestrial orbits. The thermal effect of an impact is expressed as the volume of melt produced. The melt volumes produced by each Monte Carlo case are sorted into regions to produce percentage melt estimates as a function of radius. One hundred cases are calculated to obtain a range of plausible thermal histories for lunar accretion.

094.046 **Materialien zur Dynamik des Erde-Mond-Systems. Librationen des Mondes – Grundbegriffe.**   K. Goller, H. Gleixner, M. Schneider.
Mitt. Inst. Astron. Phys. Geod. Tech. Univ. München, Nr. 155, 42 pp. (1981).

094.047 **Seismic codas on the Earth and the Moon: a comparison.**   A. M. Dainty, M. N. Toksöz.
Phys. Earth Planet. Inter., Vol. 26, 250 - 260 (1981).

094.048 **Progress in modeling the distribution of elastic inhomogeneities in the lunar crust.**   P. E. Malin, Y. Nakamura.
Phys. Earth Planet. Inter., Vol. 26, 261 - 263 (1981).
This short note outlines the current progress in modeling the elastic inhomogeneities that exist in the lunar crust. The data for the study are the seismic codas generated by impacts near the Apollo Passive Seismic Experiment sites.

094.049 **Investigations of local Bouguer gravity anomalies of the Apennines and Taurus Mountains of the Moon.**   P. Janle.
Phys. Earth Planet Inter., Vol. 27, 47 - 59 (1981) = Contrib. Inst. Geophys. Univ. Kiel, No. 212.

094.050 **The moon.**   B. M. French.
The new solar system, (see 003.026), p. 71 - 82 (1981).

The moon – our sister planet.   See Abstr. 003.042.

Nuclear-physical investigations of the moon and planets.   See Abstr. 003.086.

Some results and prospects of an investigation of physical properties of the moon.   See Abstr. 003.173.

Bemerkenswerte Daten aus der Geschichte himmelskundlicher Forschung: Der Mond.   See Abstr. 004.048.

The moon from different points of view.
See Abstr. 011.030.

On the construction of a compiled selenodetic catalogue.   See Abstr. 021.063.

Using computers for literal developments: the lunar disturbing function arising from the sun.
See Abstr. 021.069.

On the possibility of using cosmic radio emission sources for investigation of the moon and planets by a bistatic radar method.   See Abstr. 031.609.

Beitrag zur geodätischen und astronomischen Nutzung von Sternbedeckungen durch den Mond.
See Abstr. 031.649.

Is the gravitational constant changing?
See Abstr. 066.033.

Eruption mechanics on the Earth, Moon and Mars.
See Abstr. 081.004.

On an interconnection between the gravitational field and topography of a planet (earth, Mars, moon).
See Abstr. 081.053.

Basin tectonics on the terrestrial planets: Moon, Mars, and Mercury.   See Abstr. 091.035.

Effects of subsurface volume scattering on planetary microwave brightness temperature measurements.
See Abstr. 091.060.

Vertical-structure effects on planetary microwave brightness temperature measurements: application to the lunar regolith.   See Abstr. 091.095.

Phobos, Deimos, and the moon: comparison of ejecta patterns.   See Abstr. 097.059.

Orbital dynamics of magnetospherically trapped lunar ejecta.   See Abstr. 106.054.

Submicron lunar ejecta in the magnetosphere associated with meteor showers.   See Abstr. 106.055.

Capture origin of the Moon.
See Abstr. 107.022.

Moon (Local Properties)

094.501   **Solar-flare produced ^{3}He in lunar samples.**
A. Yaniv, with a reply by M. N. Rao,
T. R. Venkatesan.
Nature, Vol. 292, 866 (1981).

094.502   **Report on the A. L. P. O.-LTP** *(Lunar Transient Phenomena)* **observing program.**
W. S. Cameron.
Strolling Astron., Vol. 29, 15 - 24 (1981).

094.503   **Detection of stopped solar flare helium in lunar rock 68815.**   A. Yaniv, K. Marti.
Astrophys. J., Lett., Vol. 247, L143 - L146 (1981).
The authors detected steep increases in the concentration profiles of ^{3}He and possibly of ^{21}Ne in the near-surface layers of lunar rock 68815, which was exposed on the lunar surface for 2 Myr. The concentration profile of ^{3}He is inconsistent with either spallation or solar wind components, but it agrees with expected ranges of stopped solar flare particles. Although the ^{3}He retention in this rock is determined to be very low (~ 3% at the surface), the authors can estimate a long-term average solar flare ratio ^{4}He/^{3}He of 12 to within a factor of about 3. This ratio is two orders of magnitude smaller than the same ratio observed in the solar wind, but it is consistent with solar flare data from recent satellite observations.

094.504   **Lunar craters with radar bright ejecta.**
T. W. Thompson, S. H. Zisk, R. W. Shorthill,
P. H. Schultz, J. A. Cutts.
Icarus, Vol. 46, 201 - 225 (1981).
A small fraction of the lunar impact craters with diameters of 1 km and greater have extensive enhanced 3.8 cm radar echoes associated with their ejecta deposits. The physical properties of these ejecta deposits and the ages of the central craters have been characterized via various infrared, radar, and optical signatures.

094.505   **The lunar crater Birt.**   M. Savill, R. McKay.
J. British Astron. Assoc., Vol. 91, 463 - 472 (1981).

094.506   **Mission objectives for geological exploration of the Apollo 16 landing site.**
W. R. Muehlberger, F. Hörz, J. R. Sevier, G. E. Ulrich.
The lunar highlands crust, (see 012.013), p. 1 - 49 (1980) = Lunar Planet. Inst. Contrib. No. 403.
This document is a committee report which outlines scientific objectives and specific rationales developed for the exploration of the Apollo 16 site, our most important mission to the lunar highlands. Although premission hypotheses erroneously favored a volcanic origin for the Cayley plains as well as Descartes Mountains, the major mission objectives were successfully accomplished: to delineate the nature and origin of two major physiographic units of the lunar central highlands. A summary of currently viable hypotheses concerning the origin of Cayley and Descartes materials is given.

094.507   **Recommended classification and nomenclature of lunar highland rocks – a committee report.**
D. Stöffler, H.-D. Knöll, U. B. Marvin, C. H. Simonds,
P. H. Warren.
The lunar highlands crust, (see 012.013), p. 51 - 70 (1980).

094.508   **On the origins of lunar pristine crustal rocks.**
A. B. Binder.
The lunar highlands crust, (see 012.013), p. 71 - 79 (1980).

094.509   **Candidate samples for the earliest lunar crust.**
S. Jovanovic, G. W. Reed, Jr.
The lunar highlands crust, (see 012.013), p. 101 - 111 (1980).
A group of non-mare samples has a Cl/P_2O_5 ratio that is much lower than in other lunar samples. Two clusters of samples with order(s) of magnitude differences in minor and trace element contents make up the group. It is proposed that these samples could be relics from an original lunar crustal layer which evolved more or less independently but concurrently with the differentiation of an underlying deep magma ocean from which all other samples were eventually derived.

094.510 A comparison of fractionation trends in the lunar crust and the Stillwater Complex.
L. D. Raedeke, I. S. McCallum.
The lunar highlands crust, (see 012.013), p. 133 - 153 (1980) = Lunar Planet. Inst. Contrib. No. 404.

094.511 Cratering in the lunar highlands: some problems with the process, record and effects.
R. A. F. Grieve.
The lunar highlands crust, (see 012.013), p. 173 - 196 (1980) = Contrib. Earth Phys. Branch No. 839.

094.512 Orientations of central peaks in lunar craters: implications for regional structural trends.
W. Hale.
The lunar highlands crust, (see 012.013), p. 197 - 209 (1980).
    The purposes of this paper are: 1) to examine a large number of craters with linear central peaks of all ages over a broad diameter range, to determine if any preferential orientation(s) exist; 2) to compare directions of elongation to downrange direction of the impacting body, where this can be determined by asymmetrically emplaced ejecta deposits, in order to define the role of impact direction in the formation of linear central peaks; and 3) to assess the degree of change in preferential peak orientations over geologic time.

094.513 An experimental study of the thermal history of fragment-laden "basalt" 77115.
C. R. Thornber, J. S. Huebner.
The lunar highlands crust, (see 012.013). p. 233 - 252 (1980).

094.514 Model thermal history of 77115 and implications for the origin of fragment-laden basalts.
R. F. Sanford, J. S. Huebner.
The lunar highlands crust, (see 012.013), p. 253 - 269 (1980).

094.515 Lunar highland melt rocks: chemistry, petrology and silicate mineralogy.   D. T. Vaniman, J. J. Papike.
The lunar highlands crust, (see 012.013), p. 271 - 337 (1980).

094.516 Silicate liquid immiscibility, evolved lunar rocks and the formation of KREEP.
G. J. Taylor, R. D. Warner, K. Keil, M.-S. Ma, R. A. Schmitt.
The lunar highlands crust, (see 012.013), p. 339 - 352 (1980).

094.517 Volcanic rocks in the lunar highlands.
G. Ryder, P. Spudis.
The lunar highlands crust, (see 012.013), p. 353 - 375 (1980).

094.518 Petrology of 60035: evolution of a polymict ANT breccia.
R. D. Warner, G. J. Taylor, K. Keil.
The lunar highlands crust, (see 012.013), p. 377 - 394 (1980).

094.519 Genesis of highland basalt breccias: a view from 66095.
J. R. Garrison, Jr., L. A. Taylor.
The lunar highlands crust, (see 012.013), p. 395 - 417 (1980).

094.520 Fission xenon in troctolite 76535.
    C. M. Hohenberg, B. Hudson, B. M. Kennedy,
F. A. Podosek.
The lunar highlands crust, (see 012.013), p. 419 - 439 (1980).

094.521 Geochemical anomalies on the eastern limb and farside of the moon.   B. R. Hawke, P. D. Spudis.
The lunar highlands crust, (see 012.013), p. 467 - 481 (1980).
    The purpose of this paper is to investigate the geology and geochemistry of selected anomalous highlands regions on the eastern limb and farside and attempt to understand the processes responsible for their formation.

094.522 Infrared and radar signatures of lunar craters: implications about crater evolution.
T. W. Thompson, J. A. Cutts, R. W. Shorthill, S. H. Zisk.
The lunar highlands crust, (see 012.013), p. 483 - 499 (1980) = Planet. Sci. Inst. Contrib. No. 124.

094.523 On the mare basalt magma source region.
    A. B. Binder.
Proc. Eleventh Lunar Planet. Sci. Conf., (see 012.020), p. 1 - 22 (1980).

094.524 The geology and petrology of the Apollo 11 landing site.   D. W. Beaty, A. L. Albee.
Proc. Eleventh Lunar Planet. Sci. Conf., (see 012.020), p. 23 - 35 (1980) = Div. Geol. Planet. Sci., Calif. Inst. Technol., Contrib. No. 3434.

094.525 The petrology and chemistry of basaltic fragments from the Apollo 11 soil: drive tubes 10004 and 10005.   M.-S. Ma, R. A. Schmitt, D. W. Beaty, A. L. Albee.
Proc. Eleventh Lunar Planet. Sci. Conf., (see 012.020), p. 37 - 47 (1980).

094.526 Chemistry of Apollo 11 low-K mare basalts.
    J. M. Rhodes, D. P. Blanchard.
Proc. Eleventh Lunar Planet. Sci. Conf., (see 012.020), p. 49 - 66 (1980).

094.527 Petrology of mare-type basalt clasts from consortium breccia 73255.
O. B. James, J. J. McGee.
Proc. Eleventh Lunar Planet. Sci. Conf., (see 012.020), p. 67 - 86 (1980).

094.528 Aluminous mare basalts: new data from Apollo 14 coarse fines.   R. D. Warner, G. J. Taylor, K. Keil, M.-S. Ma, R. A. Schmitt.
Proc. Eleventh Lunar Planet. Sci. Conf., (see 012.020), p. 87 - 104 (1980).

094.529 The formation conditions of the high-magnesium olivines from the monomineralic fraction of Luna 24 regolith.   A. V. Sobolev, L. V. Dmitriev, V. L. Barsukov, V. N. Nevsorov, A. B. Slutsky (*Slutskij*).
Proc. Eleventh Lunar Planet. Sci. Conf., (see 012.020), p. 105 - 116 (1980).

094.530 Armalcolite: an oxygen fugacity indicator.
    F. T. Stanin, L. A. Taylor.
Proc. Eleventh Lunar Planet. Sci. Conf., (see 012.020), p. 117 - 124 (1980).

094.531 Cl, P_2O_5, U and Br associated with mineral separates from a low and a high Ti mare basalt.
S. Jovanovic, G. W. Reed, Jr.
Proc. Eleventh Lunar Planet. Sci. Conf., (see 012.020), p. 125 - 134 (1980).

094.532 Comparison of basaltic clasts in lunar and eucritic polymict breccias.
H. Takeda, M. Miyamoto, T. Ishii.
Proc. Eleventh Lunar Planet. Sci. Conf., (see 012.020), p. 135 - 147 (1980).

094.533 Classification, experimental petrology and possible volcanic histories of the Apollo 11 high-K basalts.
T. L. Grove, D. W. Beaty.
Proc. Eleventh Lunar Planet. Sci. Conf., (see 012.020), p. 149 - 177 (1980).

094.534 Experimental phase relations of olivine vitrophyres from breccia 14321: the temperature- and pressure-

dependence of Fe-Mg partitioning for olivine and liquid in a
highlands melt-rock.
J. E. Grover, D. H. Lindsley, A. E. Bence.
Proc. Eleventh Lunar Planet. Sci. Conf., (see 012.020),
p. 179 - 196 (1980).

094.535   On the barrier to crystal nucleation in lunar
          glasses.   H. Yinnon, A. Roshko, D. R. Uhlmann.
Proc. Eleventh Lunar Planet. Sci. Conf., (see 012.020),
p. 197 - 211 (1980).

094.536   Chemistry and liquidus phase relations of Apollo 15
          red glass: implications for the deep lunar interior.
J. W. Delano.
Proc. Eleventh Lunar Planet. Sci. Conf., (see 012.020),
p. 251 - 288 (1980).

094.537   A model of early lunar differentiation.   J. Longhi.
          Proc. Eleventh Lunar Planet. Sci. Conf., (see
012.020), p. 289 - 315 (1980).

094.538   Geochemical constraints on the igneous evolution
          of the lunar crust.   M. D. Norman, G. Ryder.
Proc. Eleventh Lunar Planet. Sci. Conf., (see 012.020),
p. 317 - 331 (1980).

094.539   Redistribution of volatiles during lunar metamor-
          phism.   E. H. Cirlin, R. M. Housley.
Proc. Eleventh Lunar Planet. Sci. Conf., (see 012.020),
p. 349 - 364 (1980).

094.540   Rocks of the early lunar crust.   O. B. James.
          Proc. Eleventh Lunar Planet. Sci. Conf., (see
012.020), p. 365 - 393 (1980).

094.541   A synthesis of lunar highlands compositional data.
          R. L. Korotev, L. A. Haskin, M. M. Lindstrom.
Proc. Eleventh Lunar Planet. Sci. Conf., (see 012.020),
p. 395 - 429 (1980).

094.542   Further foraging for pristine nonmare rocks:
          correlations between geochemistry and longitude.
P. H. Warren, J. T. Wasson.
Proc. Eleventh Lunar Planet. Sci. Conf., (see 012.020),
p. 431 - 470 (1980).

094.543   The distinction of pristine from meteorite-contami-
          nated highlands rocks using metal compositions.
G. Ryder, M. D. Norman, R. A. Score.
Proc. Eleventh Lunar Planet. Sci. Conf., (see 012.020),
p. 471 - 479 (1980).

094.544   The meteoritic contamination of terrestrial and
          lunar impact melts and the problem of indigenous
siderophiles in the lunar highland.   H. Palme.
Proc. Eleventh Lunar Planet. Sci. Conf., (see 012.020),
p. 481 - 506 (1980).

094.545   A pristine eucrite-like gabbro from Descartes and
          its exotic kindred.   U. B. Marvin, P. H. Warren.
Proc. Eleventh Lunar Planet. Sci. Conf., (see 012.020),
p. 507 - 521 (1980).

094.546   Petrology and mineral chemistry of 67667, a unique
          feldspathic lherzolite.
E. C. Hansen, J. V. Smith, I. M. Steele.
Proc. Eleventh Lunar Planet. Sci. Conf., (see 012.020),
p. 523 - 533 (1980).

094.547   Spinel cataclasites in 15445 and 72435: petrology
          and criteria for equilibrium.
M. B. Baker, C. T. Herzberg.

Proc. Eleventh Lunar Planet. Sci. Conf., (see 012.020),
p. 535 - 553 (1980).

094.548   Lunar highland rocks: element partitioning among
          minerals. II: Electron microprobe analyses of
Al, P, Ca, Ti, Cr, Mn and Fe in olivine.
J. V. Smith, E. C. Hansen, I. M. Steele.
Proc. Eleventh Lunar Planet. Sci. Conf., (see 012.020),
p. 555 - 569 (1980).

094.549   Ion microprobe analysis and petrogenetic interpreta-
          tions of Li, Mg, Ti, K, Sr, Ba in lunar plagioclase.
I. M. Steele, I. D. Hutcheon, J. V. Smith.
Proc. Eleventh Lunar Planet. Sci. Conf., (see 012.020),
p. 571 - 590 (1980).

094.550   Significance of major and minor element variations
          in plagioclase in sodic ferrogabbro and breccia
matrix in lunar highlands sample 67915.
P. W. Weiblen, W. C. Day, J. D. Miller, Jr.
Proc. Eleventh Lunar Planet. Sci. Conf., (see 012.020),
p. 591 - 610 (1980).

094.551   Comparative thermal histories of matrix from
          Apollo 17 Boulder 7 fragment-laden melt rocks:
an analytical transmission electron microscopy study.
J. J. McGee, G. L. Nord, Jr., M.-V. Wandless.
Proc. Eleventh Lunar Planet. Sci. Conf., (see 012.020),
p. 611 - 627 (1980).

094.552   Noble gas component organization in 14301.
          T. J. Bernatowicz, C. M. Hohenberg, B. Hudson,
B. M. Kennedy, J. C. Laul, F. A. Podosek.
Proc. Eleventh Lunar Planet. Sci. Conf., (see 012.020),
p. 629 - 668 (1980).

094.553   Enhanced orbital geochemical images by the
          Laplacian subtraction method.   E. Schonfeld.
Proc. Eleventh Lunar Planet. Sci. Conf., (see 012.020),
p. 677 - 688 (1980).

094.554   Lunar highland crustal models based on iron con-
          centrations: isostasy and center-of-mass displace-
ment.   E. L. Haines, A. E. Metzger.
Proc. Eleventh Lunar Planet. Sci. Conf., (see 012.020),
p. 689 - 718 (1980).

094.555   Effects of overlapping optical absorption bands of
          pyroxene and glass on the reflectance spectra of
lunar soils.
T. G. Farr, B. A. Bates, R. L. Ralph, J. B. Adams.
Proc. Eleventh Lunar Planet. Sci. Conf., (see 012.020),
p. 719 - 729 (1980).

094.556   The lunar regolith: comparative petrology of the
          Apollo sites.   T. C. Labotka, M. J. Kempa,
C. White, J. J. Papike, J. C. Laul.
Proc. Eleventh Lunar Planet. Sci. Conf., (see 012.020),
p. 1285 - 1305 (1980).

094.557   The lunar regolith: comparative chemistry of the
          Apollo sites.   J. C. Laul, J. J. Papike.
Proc. Eleventh Lunar Planet. Sci. Conf., (see 012.020),
p. 1307 - 1340 (1980).

094.558   The Apollo 16 regolith: a petrographically-con-
          strained chemical mixing model.
M. J. Kempa, J. J. Papike, C. White.
Proc. Eleventh Lunar Planet. Sci. Conf., (see 012.020),
p. 1341 - 1355 (1980).

094.559 Cosmic ray production curves below reworking
zones.   G. E. Blanford.
Proc. Eleventh Lunar Planet. Sci. Conf., (see 012.020),
p. 1357 - 1368 (1980).

094.560 Record of the solar corpuscular radiation in minerals
from lunar soils: a comparative study of noble
gases and tracks.
R. Wieler, P. Etique, P. Signer, G. Poupeau.
Proc. Eleventh Lunar Planet. Sci. Conf., (see 012.020),
p. 1369 - 1393 (1980).

094.561 The Apollo 17 drill core: chemistry of size fractions
and the nature of the fused soil component.
J. C. Laul, J. J. Papike.
Proc. Eleventh Lunar Planet. Sci. Conf., (see 012.020),
p. 1395 - 1413 (1980).

094.562 The depositional history of the Apollo 17 deep drill
core: a reappraisal.
Y. Langevin, J. S. Nagle.
Proc. Eleventh Lunar Planet. Sci. Conf., (see 012.020),
p. 1415 - 1434 (1980).

094.563 Solar and cosmogenic nitrogen in the Apollo 17
deep drill core.
M. H. Thiemens, R. N. Clayton.
Proc. Eleventh Lunar Planet. Sci. Conf., (see 012.020),
p. 1435 - 1451 (1980).

094.564 Irradiation history of the lunar regolith at the
Apollo 14, 15 and 17 sites: additional insights.
G. Crozaz.
Proc. Eleventh Lunar Planet. Sci. Conf., (see 012.020),
p. 1453 - 1462 (1980).

094.565 Depth scales for Apollo 15, 16, and 17 drill cores.
J. H. Allton, S. R. Waltz.
Proc. Eleventh Lunar Planet. Sci. Conf., (see 012.020),
p. 1463 - 1477 (1980).

094.566 Possible rim crest deposits in cores 12027 and
15008: some interpretations and problems for
future research.   J. S. Nagle.
Proc. Eleventh Lunar Planet. Sci. Conf., (see 012.020),
p. 1479 - 1496 (1980).

094.567 Recent depositional history of Apollo 16 and 17
cores.   J. C. Evans, J. S. Fruchter, J. H. Reeves,
L. A. Rancitelli, R. W. Perkins.
Proc. Eleventh Lunar Planet. Sci. Conf., (see 012.020),
p. 1497 - 1509 (1980).

094.568 Depositional and irradiational history of the
Hadley Rille core 15010/11.
D. D. Bogard, R. V. Morris, W. C. Hirsch, H. V. Lauer.
Proc. Eleventh Lunar Planet. Sci. Conf., (see 012.020),
p. 1511 - 1529 (1980).

094.569 Lunar core 15010/11: grain size, petrology, and
implications for regolith dynamics.
D. S. McKay, A. Basu, G. Nace.
Proc. Eleventh Lunar Planet. Sci. Conf., (see 012.020),
p. 1531 - 1550 (1980).

094.570 Distribution of volatile and non volatile elements in
grain-size fractions of Apollo 17 drive tube 74001/2.
U. Krähenbühl.
Proc. Eleventh Lunar Planet. Sci. Conf., (see 012.020),
p. 1551 - 1564 (1980).

094.571 Double drive tube 74001/2: composition of noble
gases trapped 3.7 AE ago.
O. Eugster, N. Grögler, P. Eberhardt, J. Geiss.
Proc. Eleventh Lunar Planet. Sci. Conf., (see 012.020),
p. 1565 - 1592 (1980).

094.572 Multispectral imaging of the lunar regolith core
samples: preliminary results for 74002.
C. M. Pieters, B. R. Hawke, P. Butler, S. Waltz, J. S. Nagle.
Proc. Eleventh Lunar Planet. Sci. Conf., (see 012.020),
p. 1593 - 1608 (1980).

094.573 Apollo 17 drive tube 76001: modal petrology.
J. J. Papike, J. Wyszynski. .
Proc. Eleventh Lunar Planet. Sci. Conf., (see 012.020),
p. 1609 - 1621 (1980).

094.574 Luna 24 core spherules: microparticle impact
crater and accreta populations as indicators of the
past surface environment.   R. J. Allison, J. A. M. McDonnell.
Proc. Eleventh Lunar Planet. Sci. Conf., (see 012.020),
p. 1623 - 1634 (1980).

094.575 The Apollo 16 regolith: Comparative petrology of
the $> 20 \mu m$ and $20-10 \mu m$ soil fractions, lateral
transport and differential volatilization.
M. J. Kempa, J. J. Papike.
Proc. Eleventh Lunar Planet. Sci. Conf., (see 012.020),
p. 1635 - 1661 (1980).

094.576 K/Ar dating of lunar soils. IV: Orange glass from
74220 and agglutinates from 14259 and 14163.
E. C. Alexander, Jr., M. R. Coscio, Jr., J. C. Dragon, K. Saito.
Proc. Eleventh Lunar Planet. Sci. Conf., (see 012.020),
p. 1663 - 1677 (1980).

094.577 Experimental shock metamorphism of lunar soil.
R. B. Schaal, F. Hörz.
Proc. Eleventh Lunar Planet. Sci. Conf.,
p. 1679 - 1695 (1980).

094.578 Origins and size distribution of metallic iron
particles in the lunar regolith.   R. V. Morris.
Proc. Eleventh Lunar Planet. Sci. Conf., (see 012.020),
p. 1697 - 1712 (1980).

094.579 Metallic particles in the glassy constituents of three
lunar highland samples 65315, 67435 and 78235.
S. Mehta, J. I. Goldstein.
Proc. Eleventh Lunar Planet. Sci. Conf., (see 012.020),
p. 1713 - 1725 (1980).

094.580 Petrography of lunar soil 15601.   A. Basu,
D. S. McKay, G. Nace, S. A. Griffiths.
Proc. Eleventh Lunar Planet. Sci. Conf., (see 012.020),
p. 1727 - 1741 (1980).

094.581 Light elements in lunar soils revisited: carbon,
nitrogen, hydrogen and helium.   R. H. Becker.
Proc. Eleventh Lunar Planet. Sci. Conf., (see 012.020),
p. 1743 - 1761 (1980) = Div. Geol. Planet. Sci., Calif. Inst.
Technol., Contrib. No. 3439.

094.582 Lunar regolith investigated by heating techniques:
surface deposits and volatilities of trace elements.
F. Wegmüller, J. Sörensen, U. Krähenbühl, H. R. von Gunten.
Proc. Eleventh Lunar Planet. Sci. Conf., (see 012.020),
p. 1763 - 1776 (1980).

094.583 Thermal diffusivity of two Apollo 11 samples,
10020,44 and 10065,23: effect of petrofabrics on
the thermal conductivity of porous lunar rocks under vacuum.

K. Horai, J. L. Winkler, Jr.
Proc. Eleventh Lunar Planet. Sci. Conf., (see 012.020),
p. 1777 - 1788 (1980).

094.584 Comparisons of magnetic paleointensity methods
         using a lunar sample.
N. Sugiura, D. W. Strangway.
Proc. Eleventh Lunar Planet. Sci. Conf., (see 012.020),
p. 1801 - 1813 (1980).

094.585 Compressive strength, seismic Q, and elastic
         modulus.
B. R. Tittmann, V. A. Clark, T. W. Spencer.
Proc. Eleventh Lunar Planet. Sci. Conf., (see 012.020),
p. 1815 - 1823 (1980).

094.586 Bulk magnetization properties of the Fra Mauro
         and Reiner Gamma formations.   L. L. Hood.
Proc. Eleventh Lunar Planet. Sci. Conf., (see 012.020),
p. 1879 - 1896 (1980).

094.587 The lunar magma ocean: a transient lunar
         phenomenon?   J. W. Minear.
Proc. Eleventh Lunar Planet. Sci. Conf., (see 012.020),
p. 1941 - 1955 (1980).

094.588 Dynamic properties of mare basalts: relation of
         equations of state to petrology.
T. J. Ahrens, J. P. Watt.
Proc. Eleventh Lunar Planet. Sci. Conf., (see 012.020),
p. 2059 - 2074 (1980).

094.589 Variations in interior morphology of 15-20 km lunar
         craters: implications for a major subsurface dis-
continuity.   R. A. De Hon.
Proc. Eleventh Lunar Planet. Sci. Conf., (see 012.020),
p. 2207 - 2219 (1980).

094.590 Cratering flow fields: implications for the excava-
         tion and transient expansion stages of crater forma-
tion.   S. K. Croft.
Proc. Eleventh Lunar Planet. Sci. Conf., (see 012.020),
p. 2347 - 2378 (1980).

094.591 The geology and morphology of Ina.
         P. L. Strain, F. El-Baz.
Proc. Eleventh Lunar Planet. Sci. Conf., (see 012.020),
p. 2437 - 2446 (1980).

094.592 Lunar cold traps and their influence on argon-40.
         R. R. Hodges, Jr.
Proc. Eleventh Lunar Planet. Sci. Conf., (see 012.020),
p. 2463 - 2477 (1980).

094.593 Mare ridge arrangements of lunar Oceanus Procellarum.
         J. Raitala.
Mem. Soc. Astron. Italiana, Vol. 52, (see 012.023), 499 - 503
(1981).

094.594 The color of lunar regions.
         Yu. G. Shkuratov.
Astron. Vestn., Tom 15, 69 - 79 (1981). In Russian.
    Color-difference photographs and digital colorimetric
photomaps of some lunar regions are given and compared with
the color-map of Evyukov (1973) and spectrometric measurements.

094.595 Lithium in Luna-24 samples.
         S. V. S. Murty, P. N. Shukla, P. S. Goel.
Advances in lunar research. Luna-24 samples, (see 003.010),
p. 1 - 4 (1979). — Abstr. in Phys. Abstr., Vol. 84, Abstr.
94209 (1981).

094.596 Potassium-rich globules in the Luna-20 soil.
         N. Bhandari, V. G. Shah.
Advances in lunar research. Luna-24 samples, (see 003.010),
p. 5 - 6 (1979). — Abstr. in Phys. Abstr., Vol. 84, Abstr.
94210 (1981).

094.597 Surface exposure duration of individual lunar soil
         grains: results from Luna-24 drill core samples.
J. N. Goswami, V. G. Shah.
Advances in lunar research. Luna-24 samples, (see 003.010),
p. 7 - 16 (1979). — Abstr. in Phys. Abstr., Vol. 84, Abstr.
94211 (1981).

094.598 Noble gas investigations of Luna-24 drill core soil
         samples.
N. B. Bhai, J. T. Padia, M. N. Rao, T. R. Venkatesan.
Advances in lunar research. Luna-24 samples, (see 003.010),
p. 17 - 28 (1979). — Abstr. in Phys. Abstr., Vol. 84, Abstr.
94212 (1981).

094.599 Total nitrogen in Luna-24 samples.
         S. V. S. Murty, P. N. Shukla, P. S. Goel.
Advances in lunar research. Luna-24 samples, (see 003.010),
p. 29 - 31 (1979). — Abstr. in Phys. Abstr., Vol. 84, Abstr.
94213 (1981).

094.600 Natural radioactivity of Luna-24 and Apollo-16
         soils.   M. B. Potdar, N. Bhandari.
Advances in lunar research. Luna-24 samples, (see 003.010),
p. 32 - 38 (1979). — Abstr. in Phys. Abstr., Vol. 84, Abstr.
94214 (1981).

094.601 Thermoluminescence of Luna-24 core samples.
         B. D. Bhasin, C. M. Sunta.
Advances in lunar research. Luna-24 samples, (see 003.010),
p. 39 - 48 (1979). — Abstr. in Phys. Abstr., Vol. 84, Abstr.
94215 (1981).

094.602 Chemical studies of two Luna-24 regolith samples.
         A. V. Murali, P. B. Pawaskar, G. R. Reddy.
Advances in lunar research. Luna-24 samples, (see 003.010),
p. 49 - 58 (1979). — Abstr. in Phys. Abstr., Vol. 84, Abstr.
94216 (1981).

094.603 Thermogravimetric and X-ray diffraction analyses
         of Luna-24 regolith samples.
V. V. Deshpande, S. R. Dharwadkar, V. S. Jakkal.
Advances in lunar research. Luna-24 samples, (see 003.010),
p. 59 - 61 (1979). — Abstr. in Phys. Abstr., Vol. 84, Abstr.
94217 (1981).

094.604 Particle track records in the Luna-24 drill core soil
         samples.   J. N. Goswami.
Advances in lunar research. Luna-24 samples, (see 003.010),
p. 62 - 69 (1979). — Abstr. in Phys. Abstr., Vol. 84, Abstr.
94218 (1981).

094.605 Investigation of lunar glass with the method of
         X-ray spectroscopy.   I. A. Brytov, Yu. P. Dikov,
Yu. N. Romashchenko, O. A. Bogatikov, K. I. Konashenok,
L. B. Vasil'eva.
Apparatura i metod. rentgen. analiza, Leningrad, 1981, No. 25,
p. 175 - 178. In Russian. — Abstr. in Ref. zh., 62. Issled.
kosm. prostranstva, 10.62.240 (1981).

094.606 Lunar volcanic glasses and their constraints on mare
         petrogenesis.   J. W. Delano, K. Livi.
Geochim. Cosmochim. Acta, Vol. 45, 2137 - 2149 (1981).
    Volcanic glasses from the Apollo 11, 14, 15, and 16
landing sites have been analyzed for major elements and Ni by
electron microprobe. The 19 varieties of volcanic glass define
two distinct chemical arrays that provide new insights into (a)

the petrogenesis of mare basalts and (b) the structure of the deep lunar interior. A simple model is proposed whereby mare basaltic liquids may have been derived from two isolated, cumulate systems occurring at depths of ~300 km and ≳ 400 km.

094.607   Telescopic remote sensing of a suggested lunar comet impact site.   J. F. Bell, B. R. Hawke.
Bull. American Astron. Soc., Vol. 13, 699 (1981). – Abstract.

094.608   Primary ejecta in crater rays: spectral evidence from Copernicus.   C. M. Pieters, J. W. Head.
Bull. American Astron. Soc., Vol. 13, 709 (1981). – Abstract.

094.609   Multispectral unit mapping of the Aristarchus region of the moon.
P. Lucey, B. R. Hawke, C. M. Pieters, T. B. McCord.
Bull. American Astron. Soc., Vol. 13, 711 (1981). – Abstract.

094.610   Moon: near-infrared spectral reflectance, a first good look.   T. B. McCord, R. N. Clark, B. R. Hawke, L. A. McFadden, P. D. Owensby, C. M. Pieters, J. B. Adams.
J. Geophys. Res., Vol. 86, 10883 - 10892 (1981).
A positive identification of the minerals olivine, plagioclase, and several types of pyroxenes were made at several locations on the lunar surface by using remote measurements. These determinations were possible because the reflectance spectra for 10–20 km diameter lunar areas have been measured for the first time in the IR spectral region (0.65–2.5 μm) with sufficient spectral resolution and photometric precision to define mineral electronic absorption bands. The reflectance for all lunar regions observed (over 100 to present) continues to increase toward longer wavelengths to at least 2.5 μm, and several mineral absorption bands appear. The telescopic spectra are of similar quality and contain features similar to laboratory spectra of lunar samples. The absorption features in several spectra have been quantitatively analyzed using newly developed computer processing techniques.

094.611   On the structure of a Luna-24 core.
G. A. Lejkin, E. V. Zabalueva, L. P. Yaroslavskij.
Nauchn. Inf., Vyp. (No.) 43, p. 88 - 97 (1980). In Russian.
A statistic analysis of size-depth distribution of fragments in a drill core is presented. It is shown that the core consists of some tens of layers. The local fragment clustering observed just beneath the surface appears to have lost fine-size fraction.

094.612   X-ray spectroscopy of lunar glass.
I. A. Brytov, Yu. P. Dikov, Yu. N. Romashchenko, O. A. Bogatikov, K. I. Konashenok, L. B. Vasil'eva.
Apparatura i metody rentgen. anal., Leningrad, 1981, No. 25, p. 175 - 178. In Russian. – Abstr. in Ref. zh., 51. Astron., 11.51.344 (1981).

094.613   Investigation of distributions of elements in the regolith returned by Luna 16.
R. A. Kuznetsov, B. G. Lur'e, V. Ya. Minevich, V. I. Styuf, V. B. Pankratov.
Geokhimiya, 1981, No. 3, p. 323 - 328. In Russian. – Abstr. in Ref. zh., 62. Issled. kosm. prostranstva, 11.62.215 (1981).

094.614   Optical characteristics and relative density of lunar soil.   M. V. Goryachev, V. V. Novikov.
Astron. Tsirk., No. 1137, p. 3 - 5 (1980). In Russian.

094.615   Microtextures due to shock events observed in a lunar achondrite-like gabbro and resetting of the radiogenic clock.   H. Takeda, H. Mori, M. Miyamoto.
Proceedings of the 14th ISAS Lunar and Planetary Symposium, (see 012.055), p. 227 - 234 (1981).
In order to see the shock effects in submicroscopic scale, exsolution, inversion, melting and shock features were studied

by analytical transmission electron microscope (ATEM) in addition to single crystal X-ray diffraction and electron microprobe techniques. The exsolution textures of pyroxene crystals from a lunar eucrite-like gabbro (61223,47) were obscured by a later shock-heating event, which might have produced minute patches of pigeonite and Mg-poor and Al-rich glass detected by the ATEM. A solvage glass produced at interface between plagioclase and pyroxene by shock process are injected into the crystal.

094.616   Inelastic neutron scatter iron concentrations of the moon from orbital gamma ray data.
P. A. Davis, Jr., M. J. Bielefeld.
J. Geophys. Res., Vol. 86, 11919 - 11926 (1981).
A set of regional Fe values were obtained from the inelastic scatter peak of the Apollo 15 and 16 orbital gamma ray spectra. These Fe $(n, n'\gamma)$ values were determined from a reaction that does not depend upon thermal neutrons and should therefore be a test of thermal neutron flux depression. No consistent effects of thermal neutron flux depression are apparent. The majority of the higher energy Fe (n, γ) values are substantiated by the Fe $(n, n'\gamma)$ results; differences of more than one sigma between iron $(n, n'\gamma)$ values and Fe (n, γ) values are attributed to the difficulty in stripping the variation in natural radioactivity from the inelastic scatter Fe band. The Fe (n, γ) data set is, therefore, shown to be reliable and to reflect the actual Fe variation on the lunar surface.

094.617   Lunar near-surface shear wave velocities at the Apollo landing sites as inferred from spectral amplitude ratios.
P. Horvath, G. V. Latham, Y. Nakamura, H. J. Dorman.
J. Geophys. Res., Vol. 85, 6572 - 6578 (1980).
The authors reexamined the horizontal-to-vertical amplitude ratios of the long-period seismograms to determine the shear wave velocity distributions at the Apollo 12, 14, 15 and 16 lunar landing sites. Average spectral ratios, computed from a number of impact signals, were compared with spectral ratios calculated for the fundamental mode Rayleigh waves in media consisting of homogeneous, isotropic, horizontal layers.

094.618   Stratigraphy of Oceanus Procellarum basalts: sources and styles of emplacement.
J. L. Whitford-Stark, J. W. Head III.
J. Geophys. Res., Vol. 85, 6579 - 6609 (1980).
The purpose of the paper is to outline the surface geologic units visible in Procellarum by using a variety of remote sensing data, to define their stratigraphic relationships, to discuss their sources and styles of emplacement, and to correlate the Procellarum basalts with other lunar mare units.

094.619   Crystallographic and special physical investigations of the Soviet lunar material.
R. Wäsch, P. Bankwitz, E. Bankwitz, A. Zedler.
Planetary interiors, (see 012.061), p. 67 - 73 (1981).
In this paper some results of X-ray and special physical investigations of the Soviet lunar material are published. For two Luna 16 olivines the results of the crystal structure determination by X-ray analysis are presented and discussed. The Fe-cation occupations of the M 1– and M 2 sites of this olivines differ from the Apollo 11 and Apollo 12 olivines. This seems to be due to different "luno" – thermical conditions of crystallization.

094.620   Radioactivity of the moon and planets.
Yu. A. Surkov.
Progress in planetary exploration, (see 012.062), p. 21 - 38 (1981).
In this report the main results of the study of radioactivity of the solar system bodies are considered. The radioactivity of the moon and planets was measured by means of

vehicles in situ. The radioactivity of the lunar samples, brought to the Earth was studied with laboratory equipment.

094.621 Study of the moon and comparative planetology: problems and perspectives.
V. L. Barsukov, C. P. Florensky (*K. P. Florenskij*).
Progress in planetary exploration, (see 012.062), p. 41 - 47 (1981).

On the basis of the comparative planetologic study of the moon and terrestrial planets two fundamental features of their history and structure have been established. Firstly, shell-like structure of the terrestrial planets could be understood only in the terms of the heterogeneous accretion theory. Secondly, there are two types of the crust on the planetary surface. Their formation is considered to be independent and differing in the geological time.

094.622 Graphite in the Luna-24 regolith.
N. A. Ashikhmina, O. A. Bogatikov, T. L. Evstigneeva, N. D. Samotoin, V. A. Stepanchikov, D. I. Frikh-Khar.
Dokl. AN SSSR, Tom 260, 989 - 992 (1981). In Russian. Abstr. in Ref. zh., 62. Issled. kosm. prostranstva, 1.62.359 (1982).

094.623 On the zonal structure of soils on the moon.
M. P. Lysenko, G. N. Katterfel'd, A. I. Melua.
Izv. Vses. geogr. o-va, Tom 113, 438 - 441 (1981). In Russian. Abstr. in Ref. zh., 62. Issled. kosm. prostranstva, 1.62.362 (1982).

094.624 Isotopic and REE studies of lunar basalt 12038: implications for petrogenesis of aluminous mare basalts. L. E. Nyquist, J. L. Wooden, C.-Y. Shih, H. Wiesmann, B. M. Bansal.
Earth Planet. Sci. Lett., Vol. 55, 335 - 355 (1981).

The authors report Sr, Nd, and Sm isotopic studies of lunar basalt 12038, one of the so-called aluminous mare basalts. A precise internal Rb-Sr isochron yields a crystallization age of 3.35 ± 0.09 AE. They have modeled the evolution of the Sr and Nd isotopic compositions and the REE abundances within the framework of their earlier model for Apollo 12 olivine-pigeonite and ilmenite basalts. The isotopic and trace element features of 12038 can be modeled as produced by partial melting of a cumulate mantle source which crystallized from a lunar magma ocean with a chondrite-normalized REE pattern of constant negative slope.

094.625 Sm-Nd age of lherzolite 67667: implications for the processes involved in lunar crustal formation.
R. W. Carlson, G. W. Lugmair.

Earth Planet. Sci. Lett., Vol. 56, 1 - 8 (1981).

One apparently unique sample of a group of lunar crustal rocks, the feldspathic lherzolite 67667, was studied utilizing the Sm-Nd radiometric system in an attempt to define its age and the implications of that age for the evolution of the lunar highlands. Data for 67667 precisely define an isochron corresponding to an age of 4.18 ± 0.07 AE.

Bibliography of articles concerning the solar system from journals received in 1980. See Abstr. 002.024.

14 MeV neutron activation analysis of geological and lunar samples. See Abstr. 022.044.

Laboratory verification of the lunar orbital X-ray fluorescence experiment: initial results. See Abstr. 022.058.

Solar activity studies in the past using cosmogenic isotopes. See Abstr. 072.009.

Neon composition in solar flares. See Abstr. 073.082.

Solar (flare) cosmic ray proton fluxes in the recent past. See Abstr. 078.008.

Composition of the earth's upper mantle–II: Volatile trace elements in ultramafic xenoliths. See Abstr. 081.022.

Planetary crusts: a comparative review. See Abstr. 091.009.

A comparison of secondary craters on the Moon, Mercury, and Mars. See Abstr. 091.025.

Central peaks in mercurian craters: comparisons to the moon. See Abstr. 092.004.

Ridge systems of Caloris: comparison with lunar basins. See Abstr. 092.005.

Stratigraphy of the Caloris basin, Mercury. See Abstr. 092.006.

Deep seated target materials in the continuous deposits of the Ries Crater, Germany. See Abstr. 105.020.

Refractory and moderately volatile element abundances in the earth, moon and meteorites. See Abstr. 105.031.

095 Lunar Eclipses

095.001 De maan verduisterd. G. Schilling.
Zenit, 8. Jaarg., 492 - 496 (1981).

095.002 Zur Statistik der Mondfinsternisse. J. Meeus.
Sternenbote, 24. Jahrg., 178 - 182 (1981).

096 Lunar and Planetary Occultations

096.001   L'occultation de Sigma Sagittaire par Vénus le 17
novembre 1981.   J. Meeus.
Astronomie, Vol. 95, 329 - 332 (1981).

096.002   De rakende bedekking van Aldebaran.
D. Schmidt.
Zenit, 8. Jaarg., 379 - 381 (1981).

096.003   L'observation des occultations d'étoiles par les
petites planètes.   R. Boninsegna.
Ciel Terre, Vol. 97, 297 - 302 (1981).
Occultations of stars by minor planets are new vistas
open to amateur astronomers as well as to professional ones.
Visual careful timings of these events lead to data of great
value.

096.004   Lunar occultations of the Hyades. II. August 1980.
D. M. Peterson, R. L. Baron, E. Dunham, D. Mink.
Astron. J., Vol. 86, 1090 - 1097 (1981).
The authors present the results of observations of lunar
occultations of Hyades cluster members during the 5 August
1980 passage. Timings for seven events are given. Four
binaries have been resolved, three of which (θ^1 Tau, θ^2 Tau,
and ϕ 342) have been previously resolved, while the other
(71 Tau) appears to have been resolved for the first time.
The authors have also obtained an angular diameter for
θ^1 Tau of unusually high precision.

096.005   Occultation of Nunki by Venus, our last chance to
see the central flash.   D. W. Dunham.
Occultation Newsl., Vol. 2, 161 - 164, 176 (1981).

096.006   More asteroidal occultations during 1981.
A. Lowe, D. W. Dunham.
Occultation Newsl., Vol. 2, 164 - 166, 169 - 176 (1981).

096.007   The graze of Iota Aquarii.   D. W. Dunham.
Occultation Newsl., Vol. 2, 166 - 167 (1981).

096.008   Suggested radio observations of the forthcoming
lunar occultations of the Crab nebula.
F. P. Maloney, S. T. Gottesman.
Publ. Astron. Soc. Pacific, Vol. 93, 518 - 520 (1981).
The moon will occult the Crab nebula in a series of
events between 1981 and 1983. Lunar occultation techniques
can reveal the low-radio-frequency structure within the nebula.
The authors draw attention to, and describe the visibility of
the events, possible observing schemes and procedures, and
the astrophysical implications of the results. The authors urge
the formation of an international campaign to observe these
events at meter wavelengths and they are prepared to establish
a cooperative network.

096.009   Photoelectric observations of lunar occultations.
XII.   D. S. Evans, D. A. Edwards.
Astron. J., Vol. 86, 1277 - 1287 (1981).
Occultation observations of 414 events observed between
6 July 1979 and 4 September 1980 with the 0.76 m telescope
of McDonald Observatory are reported. Timings are given in
Table I, data on double stars in Table II.

096.010   Lunar occultation of Jupiter at 327 MHz.
D. S. Bagri, A. K. Singal.
Bull. Astron. Soc. India, Vol. 9, 74 (1981). − Abstract.

096.011   Occultations rasantes en France en 1982.
J. Meeus.
Astronomie, Vol. 95, 445 - 448 (1981).

096.012   Ocultaciones por planetas menores. 1981 - 1982.
D. Wallentinsen.
R Muscae, Vol. 6, No. 1/2, p. 36 - 45, Sep. R Muscae, Vol. 6,
No. 1/2, p. 1 - 32 (1981).

096.013   Discovery of a probable third satellite of Neptune.
H. J. Reitsema, W. B. Hubbard, L. A. Lebofsky,
D. J. Tholen.
Bull. American Astron. Soc., Vol. 13, 721 (1981). − Abstract.

096.014   Negative results of occultation searches for
Neptunian rings.
W. B. Hubbard, G. H. Rieke, M. J. Lebofsky, B. H. Zellner,
H. J. Reitsema, L. A. Lebofsky, H. Campins, F. Vilas,
D. J. Tholen, M. Alvarez.
Bull. American Astron. Soc., Vol. 13, 728 (1981). − Abstract.

096.015   No rings detected around Neptune: limits from oc-
cultation observations.
J. L. Elliot, D. J. Mink, J. H. Elias.
Bull. American Astron. Soc., Vol. 13, 729 (1981). − Abstract.

096.016   Future occultations by Uranus, Neptune, and their
satellites.   D. J. Mink.
Bull. American Astron. Soc., Vol. 13, 732 - 733 (1981).
Abstract.

096.017   Planetary occultation predictions for 1982.
D. W. Dunham.
Occultation Newsl., Vol. 2, 178 - 182 (1981).

096.018   Occultations of stars by the moon observed at the
Belgrade Astronomical Observatory in the years
1978 and 1979.   V. M. Protitch-Benishek.
Bull. Obs. Astron. Belgrade, No. 131, p. 38 - 39 (1981).

096.019   Occultation de la planète Vénus par la Lune du
20 Janvier 1980 observée à Belgrade.
V. Protitch-Benishek.
Bull. Obs. Astron. Belgrade, No. 131, p. 40 - 42 (1981).

096.020   Occultations of stars by the moon observed at the
Cracow Astronomical Observatory in the years
1976 - 1979.   T. Z. Dworak.
Astron. Rep., Vol. 5, 15 - 20 (1981).

096.021   Beitrag zur geodätischen und astronomischen
Nutzung von Sternbedeckungen durch den Mond.
D. Böhme.
Geod. Geophys. Veröff., Reihe III, Heft 42, 84 pp. (1980) =
Mitt. Lohrmann-Obs. Tech. Univ. Dresden, Nr. 38.
From the material of photoelectric and visual observations
of star occultations having been collected so far at the Lohr-
mann Observatory, those of 1964 to 1968 were selected for an
analysis to derive the ephemeris time and to determine the
geodetic position of the observational place.

096.022   Occultations of stars by the moon observed at the
Poltava Observatory in 1975.9 - 1980.0.
B. F. Sincheskul, V. N. Sincheskul.
Vrashchenie i priliv. deformatsii Zemli, Kiev. 1981, No. 13,
p. 58 - 94. In Russian. − Abstr. in Ref. zh., 51. Astron.,
1.51.116 (1982).

Predictions of occultations by the Earth as seen from
a geostationary satellite.   See Abstr. 021.005.

Analysis of lunar occultations – II. Personal equation.
See Abstr. 031.502.

Analysis of lunar occultations – III. Systematic corrections to Watts' limb-profiles for the Moon.
See Abstr. 032.503.

Une méthode pour estimer l'équation personnelle lors de chaque observation d'occultation d'étoile par la lune.
See Abstr. 031.515.

Beitrag zur geodätischen und astronomischen Nutzung von Sternbedeckungen durch den Mond.
See Abstr. 031.649.

Occultations of stars by the four largest minor planets, 1981 - 1989.   See Abstr. 098.004.

Mögliche Bedeckung des Sterns SAO 78931 durch den Kleinplaneten (110) Lydia.   See Abstr. 098.007.

Occultations of stars by solar system objects. II. Occultations of catalog stars by asteroids in 1982 and 1983.
See Abstr. 098.076.

The Uranus occultation of August 15, 1980.
See Abstr. 101.032.

The angular diameter of Aldebaran at 8540 Å.
See Abstr. 115.002.

Lunar occultation stellar angular diameter measurements. II.   See Abstr. 115.011.

The angular diameter of Regulus from the 28 March 1980 CTIO occultation.   See Abstr. 115.026.

Lunar occultation of the CO emission from the biconical nebula LkHα 208.   See Abstr. 134.003.

Erratum

096.901   Erratum: "Illinois occultation summary. I. 1977 - 1978" [Astron. J., Vol. 85, 1053 - 1061 (1980)].
R. Radick, D. Lien.
Astron. J., Vol. 86, 1774 (1981). – See Abstr. 28.096.002.

097 Mars, Mars Satellites

097.001   **Spectroscopy of the Cameron bands in the Mars airglow.**   R. R. Conway.
J. Geophys. Res., Vol. 86, 4767 - 4775 (1981).

Mars airglow spectra obtained by the ultraviolet spectrometer on board the Mariner 9 spacecraft were analyzed by using a high-resolution synthesis of the observed emissions. These bright limb observations were made of altitudes between 88 and 180 km. The brightest features are the carbon monoxide Cameron bands in the wavelength region 1800 - 2600 Å. The shape of these bands cannot be characterized by a single rotational temperature but is best described by temperatures of $1600°K$ for lower J values and $10,000°K$ for higher J values. The observed zenith intensity of 16.7 kR is in good agreement with an excitation theory in which electron-impact dissociation is the most important mechanism.

097.002   **Origin of Martian outflow channels: the eolian hypothesis.**   J. A. Cutts, K. R. Blasius.
J. Geophys. Res., Vol. 86, 5075 - 5102 (1981).

The authors examine rates of eolian transport on Mars in the context of eolian channel formation and develop a model for the formation of an erosional channel by sandblasting with materials transported by wind out of chaotic terrains. The authors apply this model to the interpretation of topographic measurements made in the chaotic terrain source area and channel of Aromatum Vallis, a small well-preserved channel system southwest of the Chryse basin, and develop constraints on the eolian mechanism.

097.003   **The Martian twilight.**
R. Kahn, R. Goody, J. Pollack.
J. Geophys. Res., Vol. 86, 5833 - 5838 (1981).

The changing sky brightness during the Martian twilight as measured by the Viking lander cameras is shown to be consistent with data obtained from sky brightness measurements. An exponential distribution of dust with a scale height of 10 km, equal to the atmospheric scale height, is consistent with the shape of the light curve. Multiple scattering resulting from the forward scattering peak of large particles makes a major contribution to the intensity of the twilight. The spectral distribution of light in the twilight sky may require slightly different optical properties for the scattering particles at high levels from those of the aerosol at lower levels.

097.004   **Comment on 'Development and analysis of a twelfth degree and order gravity model for Mars' by**
E. J. Christensen and G. Balmino.   K. Lambeck, with a reply by E. J. Christensen.
J. Geophys. Res., Vol. 86, 6382 - 6383 (1981). – See Abstract 27.097.016.

097.005   **What is new on Mars – Martian 1979 - 1980 apparition report II.**
C. F. Capen, D. C. Parker.
Strolling Astron., Vol. 29, 38 - 41 (1981).

097.006   **Viking bistatic radar experiment: summary of first-order results emphasizing north polar data.**
R. A. Simpson, G. L. Tyler.
Icarus, Vol. 46, 361 - 389 (1981).

Bistatic radar observations of Mars' north polar region during 1977 - 1978 showed surface rms slope σ_β ranging from 1 to 6°; these values apply to horizontal scales of 1 - 100 m. The permanent north polar cap is relatively uniform with $2.5 \lesssim \sigma_\beta \lesssim 3.0°$. Considerable structure has been found in echo spectra, indicating a heterogeneous and perhaps anisotropic scattering surface. Estimates of surface roughness and dielectric constant in the equatorial region are consistent with results from Earth-based measurements.

097.007   **Thermal tides in the dusty Martian atmosphere: a verification of theory.**
R. W. Zurek, C. B. Leovy.
Science, Vol. 213, 437 - 439 (1981).

Major features of the daily surface pressure oscillations observed by the Viking landers during the two great dust storms on Mars in 1977 can be explained in terms of the classical atmospheric tidal theory developed for the earth's atmosphere. The most dramatic exception is the virtual disappearance of only the diurnal tide at Viking Lander 1 just before the second storm. This disappearance is attributed to destructive interference between the usually westward-traveling tide and an eastward-traveling diurnal Kelvin mode generated by orographically induced differential heating.

097.008   **Mars satellite Deimos interaction with the solar wind and its influence on flow around Mars.**
A. V. Bogdanov.
J. Geophys. Res., Vol. 86, 6926 - 6932 (1981).

Data are given indicating the possibility of strong solar wind Deimos interaction. Plasma measurements on board the Soviet interplanetary probe Mars 5 showed a drop in ion density and distortion of ion energy spectra in the region behind Deimos at a distance of about 20,000 km. The dimension of Deimos is smaller than the ion gyroradius, and the effect observed may exist in the case of Deimos outgassing. Ion spectra characteristics indicate that outgassing activity is probable. The existence of strong solar wind-Deimos interaction may result in influence on the Martian magnetosphere and bow shock. The dimension of the magnetosphere is maximal when Mars is shaded by Deimos.

097.009   **Local Mars dust storm generation mechanism.**
J. A. Ryan, R. D. Sharman, R. D. Lucich.
Geophys. Res. Lett., Vol. 8, 899 - 901 (1981).

On $L_s = 340°$, first Mars year of Viking on the surface, a local dust strom was observed at the Viking Lander # 1 site by Viking Orbiter A. The storm lasted less than one martian day (sol) with the dust raised affecting the site for about three sols. It is concluded that this storm was caused by baroclinic waves and that the threshold wind speed for saltation was $25 - 30$ m sec^{-1}.

097.010   **A local dust storm in the Chryse region of Mars: Viking orbiter observations.**
P. B. James, N. Evans.
Geophys. Res. Lett., Vol. 8, 903 - 906 (1981).

A local dust storm was observed near the Viking Lander 1 site by Viking orbiter 1 in September, 1977, when the areocentric longitude of the sun, L_s, was 340° (shortly before vernal equinox). The orbiter observations, which consisted of a time sequence of pictures, show that the storm moved at about 50 m/sec to the ENE from the Lunae-Planum region into the Chryse basin. Both baroclinic waves and topography may have been associated with the generation of the storm.

097.011   **Planetological considerations on the geologic intermediate stage of Mars between Moon and Earth and observations on the Jovian satellites.**   P. Leonardi.
Atti Accad. Naz. Lincei, Rend. Ser. Ottava, Vol. 66, 558 - 562 (1979).

097.012   **Geological maps of Mars.**   Yu. M. Klejner.
Priroda, 1981, No. 7, p. 105 - 106. In Russian.

097.013 **Seasonal observations of Mars.** R. R. Conway, S. T. Durrance, C. A. Barth, A. L. Lane.
The Universe at ultraviolet wavelengths, (see 012.009), p. 33 - 37 (1981).

IUE has detected the Hartley bands of ozone in the spectrum of Mars. Seasonal observations show a variation in the north consistent with the measurements of Mariner 9. New observations during Martian late fall in the south were made.

097.014 **Delta VLBI observations of Mars Viking Lander I.** F. F. Donivan, X X Newhall.
Bull. American Astron. Soc., Vol. 13, 555 (1981). − Abstract.

097.015 **The secular acceleration in the mean orbital motion of Phobos.** E. G. Reuning.
Bull. American Astron. Soc., Vol. 13, 572 (1981). − Abstract.

097.016 **Spectral properties (0.40 to 0.75 microns) of soils exposed at the Viking 1 landing site.**
E. A. Guinness.
J. Geophys. Res,, Vol. 86, 7983 - 7992 (1981).

The bidirectional reflectance and photometric function (Hapke, 1981) was determined for seven patches of soil located near the Viking Lander 1 spacecraft. The soil photometric function is strongly backscattering and has a prominent opposition effect such that the ratio of reflectances at $1°$ and $10°$ phase angles averages 1.25, 1.24, and 1.19 in blue, green, and red wavelengths, respectively. The reflectance of the soil also exhibits a wavelength dependence as a function of phase angle. There is little need to call upon mineralogical variations to explain the subtle color variations exposed at the landing sites. Rather, brightness and color variations within the soil can be correlated with particle size, with finer-grained soil being brighter and redder than coarser-grained soil. When compared to earth−based reflectance data, the soil at the Viking 1 site is most like Martian bright areas. Such a result is consistent with the Lander soil being part of a globally homogenized soil unit.

097.017 **Liquid water on Mars: an energy balance climate model for CO_2/H_2O atmospheres.**
M. I. Hoffert, A. J. Callegari, C. T. Hsieh, W. Ziegler.
Icarus, Vol. 47, 112 - 129 (1981).

The authors develop a simple climate model for a CO_2/H_2O Mars atmosphere including water vapor-longwave opacity feedback in the atmosphere and temperature-albedo feedback at surface icecaps, under the assumption that once the Martian surface pressure was $p_s \geqslant 1$ atm CO_2. Longwave flux to space is computed as a function of the surface temperature T_s and p_s using bandabsorption models for the effect of the 15 μm fundamental, and the 10 and 15μm hot bands, of the CO_2 molecule: as well as the pure rotation bands and e continuum of H_2O. The derived global radiative balance predicts a global mean surface temperature of 283°K at 1 atm CO_2.

097.018 **Investigation of the relief, tectonics and volcanism of Mars according to data of automatic interplanetary stations.** Ya. G. Kats, V. V. Kozlov, N. V. Makarova, E. D. Sulidi-Kondrat'ev.
Nauchn. chteniya po aviats. i kosmonavt. 10-e Gagarinsk. chteniya, 1980. Moskva, 1981, p. 293. In Russian. − Abstr. in Ref. zh., 62. Issled. kosm. prostranstva, 8.62.332 (1981).

097.019 **Geochemical interpretation of the results of measuring gamma-radiation of Mars.** Yu. A. Surkov, L. P. Moskalyova (*Moskaleva*), O. S. Manvelyan, A. T. Basilevsky (*Bazilevskij*), V. P. Kharyukova.
Proc. Eleventh Lunar Planet. Sci. Conf., (see 012.020), p. 669 - 676 (1980).

Analysis of the data on the natural radioactivity of Martian surface material and a comparison with analogous data on Earth, Venus and Moon indicates that there are two fundamentally different types of crustal material on these bodies. Martian volcanic formations with a basalt-like morphology are similar to the basalts on Earth, Moon and probably Venus. The ancient formations of Martian terrae differ markedly in thorium and uranium content from the granitic rocks of the Earth's continents and from the Venus ancient terrae.

097.020 **Characterization of martian surface materials from Earth-based radar: The Memnonia Fossae region.**
P. J. Mouginis-Mark, S. H. Zisk, G. S. Downs.
Proc. Eleventh Lunar Planet. Sci. Conf., (see 012.020), p. 823 - 838 (1980).

The Memnonia Fossae region of Mars is used as a test area for the characterization of surface materials using the 1971 and 1973 Goldstone radar data. In general, a good correlation exists between radar topography and surface features, although a tendency for the radar measurements to break down over very rugged terrain has been observed.

097.021 **Martian double ring basins: new observations.**
C. A. Wood.
Proc. Eleventh Lunar Planet. Sci. Conf., (see 012.020), p. 2221 - 2241 (1980).

097.022 **Mars Tharsis region: volcanotectonic events in the stratigraphic record.** D. H. Scott, K. L. Tanaka.
Proc. Eleventh Lunar Planet. Sci. Conf., (see 012.020), p. 2403 - 2421 (1980).

097.023 **Estimation of the thickness of the Tharsis lava flows and implications for the nature of the topography of the Tharsis plateau.**
J. B. Plescia, R. S. Saunders.
Proc. Eleventh Lunar Planet. Sci. Conf., (see 012.020), p. 2423 - 2436 (1980).

097.024 **Solar wind sputtering effects in the atmospheres of Mars and Venus.**
C. C. Watson, P. K. Haff, T. A. Tombrello.
Proc. Eleventh Lunar Planet. Sci. Conf., (see 012.020), p. 2479 - 2502 (1980).

097.025 **Viking site selection and certification.**
H. Masursky, N. L. Crabill.
Scientific and Technical Information Branch, National Aeronautics and Space Administration, Washington, D.C., NASA SP-429. 5 + 34 pp. (1981).

This paper presents a brief account of how the sites were selected and certified and also includes suggestions based on this experience that might be applied to future missions.

097.026 **Daily and seasonal Viking observations of martian bore wave systems.**
G. E. Hunt, A. O. Pickersgill, P. B. James, N. Evans.
Nature, Vol. 293, 630 - 633 (1981).

A martian atmospheric phenomenon called a 'bore wave' has been observed by the Viking imaging system during late spring and early summer of two martian years, in the Tharsis Ridge region of the planet. The authors present the observational data and offer a tentative explanation for the occurrence of this feature, formed by airflow and which behaves like a thermally induced diurnal katabatic breeze.

097.027 **The climatic record in polar layered deposits on Mars.** J. A. Cutts, K. R. Blasius.
Mem. Soc. Astron. Italiana, Vol. 52, (see 012.023), 389 - 396 (1981).

097.028 **Aeolian activity in Tharsis−Viking Orbiter observations over a martian year.** S. W. Lee.

Mem. Soc. Astron. Italiana, Vol. 52, (see 012.023), 439 - 443 (1981).

097.029   Morphology and network patterns of Martian valleys.
D. C. Pieri.
Mem. Soc. Astron. Italiana, Vol. 52, (see 012.023), 483 - 486 (1981).

097.030   A Martian general circulation experiment with large topography.
J. B. Pollack, C. B. Leovy, P. W. Greiman, Y. Mintz.
J. Atmos. Sci., Vol. 38, 3 - 29 (1981). — Abstr. in Phys. Abstr., Vol. 84, Abstr. 94224 (1981).

097.031   Observations of Martian tides over two annual cycles.   C. B. Leovy.
J. Atmos. Sci., Vol. 38, 30 - 39 (1981). — Abstr. in Phys. Abstr., Vol. 84, Abstr. 94225 (1981).

097.032   An examination of the formation of linear lee waves generated by giant Martian volcanoes.
A. O. Pickersgill, G. E. Hunt.
J. Atmos. Sci., Vol. 38, 40 - 51 (1981). — Abstr. in Phys. Abstr., Vol. 84, Abstr. 94226 (1981).

097.033   Seasonal motions of air masses in the Martian atmosphere.   V. I. Aleshin.
Astron. Zh., Tom 58, 1078 - 1084 (1981). In Russian. English translation in Soviet Astron., Vol. 25, No. 5.
The thickness of the snow layer in the Martian polar caps is calculated. The variations of the mass of the snow are accompanied by seasonal variations of the atmospheric pressure. On the basis of the solution of primitive meteorological equations, the picture of motion of air masses caused by condensation and evaporation of the polar caps' snow is obtained.

097.034   Laser in the Martian atmosphere.
Priroda, 1981, No. 10, p. 103. In Russian.

097.035   Midlatitude disturbances in the Martian atmosphere: a second Mars year.   J. R. Barnes.
J. Atmos. Sci., Vol. 38, 225 - 234 (1981). — Abstr. in Phys. Abstr., Vol. 84, Abstr. 98705 (1981).

097.036   Astrolabe observations of Mars.
E. M. Standish, S. Débarbat, M. Sanchez.
Astron. Astrophys., Vol. 102, 371 - 374 (1981).
A previously reported bias in the right ascension residuals of astrolabe observations of Mars is removed by correcting for equinox motion and by relating the observations to JPL's planetary ephemeris, DE 111. The remaining residuals show a "phase effect" whose magnitude is solved for and compared with that previously found for meridian transit observations.

097.037   Mars.   P. Campbell.
Nature, Vol. 294, 303 (1981).

097.038   The geophysics of Mars: whence the Tharsis plateau?   S. C. Solomon.
Nature, Vol. 294, 304 - 305 (1981).

097.039   Volcanism on Mars.   J. W. Head III.
Nature, Vol. 294, 305 - 307 (1981).

097.040   Martian geology.   M. H. Carr.
Nature, Vol. 294, 307 - 308 (1981).

097.041   Mars climate change: where are the petroglyphs?
F. P. Fanale.
Nature, Vol. 294, 308 - 310 (1981).

097.042   The martian lower atmosphere.   C. B. Leovy.
Nature, Vol. 294, 310 - 311 (1981).

097.043   Exploration of the upper atmosphere and ionosphere of Mars.   C. T. Russell, A. F. Nagy.
Nature, Vol. 294, 311 - 312 (1981).

097.044   Main characteristics of the tectonic structure of Mars.   E. E. Milanovskij, A. M. Nikishin.
Vestn. MGU. Geol., 1981, No. 3, p. 15 - 28. In Russian.
Abstr. in Ref. zh., 51. Astron., 10.51.241 (1981).

097.045   Vertical distribution of the condensate and water vapours in the Martian atmosphere.
Yu. N. Kulikov, M. V. Rykhletskij.
Inst. prikl. mat. AN SSSR. Prepr., 1981, No. 44, 23 pp. In Russian. — Abstr. in Ref. zh., 51. Astron., 10.51.242 (1981).

097.046   Ionization coefficients and profiles of the electron concentration in the Martian atmosphere.
B. P. Komitov, S. H. Spassov, M. M. Gogoshev.
Dokl. Bolg. AN, Vol. 34, 35 - 38 (1981). — Abstr. in Ref. zh., 51. Astron., 10.51.243 (1981).

097.047   Investigation of the chemical composition of the ground of planets with automatic means.
V. I. Chesnokov.
Fiz.-tekh. inst. AN SSSR. Prepr., 1980, No. 692, 14 pp. In Russian. — Abstr. in Ref. zh., 62. Issled. kosm. prostranstva, 10.62.49 (1981).

097.048   Martian CO abundance from the $J = 1 \to 0$ rotational transition: evidence for temporal variations.
J. C. Good, F. P. Schloerb.
Icarus, Vol. 47, 166 - 172 (1981) = Contrib. No. 461 Five Coll. Astron. Dep.
Mars was observed in the CO ($J = 1 \to 0$) 2.6-mm wavelength line between 29 March and 1 April, 1980. The data were analyzed using a model atmosphere based on Viking measurements. A least-squares fit of the model to the observed line profile yielded an average CO mixing ratio of $(3.2 \pm 1.1) \times 10^{-3}$. Models of the Martian atmospheric chemistry indicate that temporal variation could easily exist and that it would be due primarily to variations in the abundance of H_2O.

097.049   Analysis of condensates formed at the Viking 2 lander site: the first winter.   S. D. Wall.
Icarus, Vol. 47, 173 - 183 (1981).
A thin light-colored ground covering appeared on the surface of Mars near the Viking 2 lander from $L_s = 230°$ to $L_s = 16°$, a total of 249 Mars days, during the lander's first winter on the surface. Appearance and disappearance rates, spectral reflectance, and photometric data all tend to confirm an earlier proposal that the covering was a combination of H_2O and CO_2, which fell already condensed onto dust particles brought northward by the season's first major dust storm.

097.050   The unusual dynamical environment of Phobos and Deimos.
D. R. Davis, K. R. Housen, R. Greenberg.
Icarus, Vol. 47, 220 - 233 (1981).
The nonintuitive dynamical environment of Phobos and Deimos is explored using a three-dimensional numerical model. Surface gravity, escape speeds, and ejecta impact contours are calculated, both for the satellites at their present orbit distances and for orbit distances they may have had in the past. Impact loci for Stickney ejecta are computed and compared with the observed groove locations in order to evaluate a possible secondary impact origin for the grooves on Phobos. Possible effects of the dynamical environment on shaping the satellites' surfaces are discussed.

097.051 **Regolith carbon dioxide: influence on Mars' atmosphere and climate.**
F. P. Fanale, W. B. Banerdt, R. S. Saunders.
Bull. American Astron. Soc., Vol. 13, 707 (1981). – Abstract.

097.052 **Seasonal buffering of atmospheric H_2O and CO_2 by the Martian regolith.**   B. M. Jakosky.
Bull. American Astron. Soc., Vol. 13, 707 - 708 (1981). Abstract.

097.053 **Mars: atmospheric opacity at the Viking lander sites.**
T. Z. Martin, R. W. Zurek.
Bull. American Astron. Soc., Vol. 13, 708 (1981). – Abstract.

097.054 **A closer look at a developing Martian dust storm: analysis of Viking images.**   L. J. Martin.
Bull. American Astron. Soc., Vol. 13, 708 (1981). – Abstract.

097.055 **Global patterns in cloud forms on Mars: results from the complete Viking and Mariner 9 image sets.**
P. Gierasch, R. Kahn.
Bull. American Astron. Soc., Vol. 13, 708 (1981). – Abstract.

097.056 **The stability of permafrost in the equatorial region of Mars.**   S. M. Clifford.
Bull. American Astron. Soc., Vol. 13, 708 - 709 (1981). Abstract.

097.057 **Splosh craters within craters in the near equatorial region of Mars: evidence for ground ice replenishment?**   S. M. Clifford.
Bull. American Astron. Soc., Vol. 13, 709 (1981). – Abstract.

097.058 **Global patterns of crater ejecta morphology on Mars.**   K. R. Blasius, J. A. Cutts.
Bull. American Astron. Soc., Vol. 13, 709 (1981). – Abstract.

097.059 **Phobos, Deimos, and the moon: comparison of ejecta patterns.**
S. Lee, P. Thomas, J. Veverka.
Bull. American Astron. Soc., Vol. 13, 710 (1981). – Abstract.

097.060 **Effect of resonance passage on the tidal evolution of Phobos' orbit.**   C. F. Yoder.
Bull. American Astron. Soc., Vol. 13, 710 (1981). – Abstract.

097.061 **Theoretical interpretation of the Martian limb brightness profile.**
K. Lumme, L. J. Martin, L. H. Wasserman, W. A. Baum.
Bull. American Astron. Soc., Vol. 13, 712 (1981). – Abstract.

097.062 **Measurements and quantitative analysis of limb brightening of the natural CO_2 laser emission on Mars.**   M. Mumma, D. Buhl, G. Chin, D. Deming,
F. Espenak, T. Kostiuk, D. Zipoy.
Bull. American Astron. Soc., Vol. 13, 713 (1981). – Abstract.

097.063 **Models of climate control of polar layer deposits on Mars.**   J. A. Cutts, K. R. Blasius.
Bull. American Astron. Soc., Vol. 13, 742 (1981). – Abstract.

097.064 **Pole and prime meridian expressions for Phobos and Deimos.**   T. C. Duxbury, J. D. Callahan.
Astron. J., Vol. 86, 1722 - 1727 (1981).
Simple trigonometric expressions are derived for the right ascensions and declinations of the spin axes of Phobos and Deimos as well as for their prime meridians. Simple expressions are possible since both satellites are in synchronous rotation about Mars and since the orbits of both satellites are accurately modeled as precessing ellipses.

097.065 **A plasma wake for Mars' satellite Deimos.**
A. Johnstone.
Nature, Vol. 294, 610 - 611 (1981).

097.066 **On radiometric heterogeneity of Mars in the millimeter wave range.**
S. O. Kuz'min, B. Ya. Losovskij.
13-ya Vses. konf. po radioastron. issled. soln. sistemy, Kiev, 1981. Kiev, 1981, p. 70. In Russian. – From Ref. zh., 51. Astron., 11.51.278 (1981).

097.067 **The rotation of water on Mars.**
V. I. Aleshin.
13-ya Vses. konf. po radioastron. issled. soln. sistemy, Kiev, 1981. Kiev, 1981, p. 74 - 75. In Russian. – From Ref. zh., 51. Astron., 11.51.288 (1981).

097.068 **Effect of dissipation processes and exogenic sources on the $^{15}N/^{14}N$ isotopic ratio in the Martian atmosphere.**   A. K. Pavlov.
Izv. AN SSSR. Ser. fiz., Tom 45, 1330 - 1334 (1981). In Russian. – Abstr. in Ref. zh., 51. Astron., 11.51.290 (1981).

097.069 **L'histoire des canaux de Mars.**   M. Gabriel.
Ciel, Vol. 43, 259 - 264 (1981).

097.070 **Spectroscopic estimate of the abundance of carbon monoxide in the upper atmosphere of Mars.**
V. A. Krasnopol'skij.
Kosm. Issled., Tom 19, 902 - 906 (1981). In Russian.

097.071 **Seasonal change of the Martian north polar cap, 1979 - 80 (II).**
K. Iwasaki. Y. Saito, T. Akabane.
Proceedings of the 14th ISAS Lunar and Planetary Symposium, (see 012.055), p. 17 - 22 (1981).
The regression curve for the north polar cap of Mars is extracted from the photographic observations of Mars at the Kwasan Observatory and at the Hida Observatory during the 1979 - 80 apparition. Dimensions of the north polar cap were measured on the selected red, green, and yellow filter negatives taken at the best seeing conditions.

097.072 **Mars: a white cloud over Nix Olympica.**
T. Akabane, K. Iwasaki, Y. Saito.
Proceedings of the 14th ISAS Lunar and Planetary Symposium, (see 012.055), p. 23 - 30 (1981).
During the 1980 apparition of Mars, the authors observed a white cloud over Nix Olympica. The dimension of the cloud is $5 \times 10^5 \, km^2$, and it showed diurnal variation. The total water vapor is 22 μm precipitable water.

097.073 **The Martian global dust storm and thermal tides.**
S. Moriyama.
Proceedings of the 14th ISAS Lunar and Planetary Symposium, (see 012.055), p. 31 - 41 (1981).
Thermal tides, which are largely excited in the dusty Martian atmosphere, are discussed using a classical tidal theory. Thermal tidal winds, pressure and temperature are calculated using various atmospheric models, i.e., vertical and latitudinal distribution of dust, vertical temperature profiles and dissipation.

097.074 **Gravity field over the Martian region of Tharsis and its mass anomaly.**   M. Burša, Z. Šíma.
Bull. Astron. Inst. Czechoslovakia, Vol. 32, 373 - 380 (1981).
The largest anomalous region of Tharsis in the Martian gravitational field is interpreted using higher derivatives of the areopotential. The vertical-horizontal derivatives of gravity are represented over the whole Martian surface.

097.075   **Mars atmospheric opacity effects observed in the northern hemisphere by Viking orbiter imaging.**
T. E. Thorpe.
J. Geophys. Res., Vol. 86, 11419 - 11429 (1981).
Observations of changing Mars contrasts by Viking orbiter television cameras have provided a description of changing atmospheric opacity in the southern hemisphere (Thorpe, 1979). This report extends those measurements into the northern hemisphere over a greater time period and provides a detailed description of photometric changes at the lander sites as seen from orbit. Reflectivity changes compared with optical depth increases produced by four dust storms indicate mean particle scattering changes with storm evolution as well as substantial opacity in the northern hemisphere for the duration of nearly a Martian year.

097.076   **Radar measurement of heterogeneous small-scale surface texture on Mars: Chryse.**
R. A. Simpson, G. L. Tyler.
J. Geophys. Res., Vol. 85, 6610 - 6614 (1980).
Radar measurements from Chryse Planitia in 1978 confirm earlier estimates of 4°-5° rms slope (on meter scales) over 45°-50°W longitude. The new measurements also confirm a skewing of the measured spectra. This has now been interpreted as a combined result of large-scale surface tilt and a side effect of the decrease in small-scale roughness as the subradar point moves from the basin floor toward the west.

097.077   **Photographic position observations of Mars and Deimos at the Main Astronomical Observatory of the Ukrainian Academy of Sciences in 1975 - 1976.**
E. M. Sereda.
Glav. astron. obs. AN USSR. Kiev, 1981. 10 pp. In Russian.
Abstr. in Ref. zh., 51. Astron., 12.51.206 (1981).

097.078   **What is new on Mars - Martian 1979 - 1980 apparition report II.**   C. F. Capen, D. C. Parker.
Strolling Astron., Vol. 29, 51 - 60 (1981).

097.079   **On the origin of the grooves on Phobos.**
E. Illés, A. Horváth.
Progress in planetary exploration, (see 012.062), p. 49 - 52 (1981).
L-grooves are the consequence of layered structure of Phobos, which are made up of parallel layers of different composition or hardness.

097.080   **Spectra of fresh photoelectrons in the Martian atmosphere.**   M. M. Gogoshev, B. P. Komitov.
Progress in planetary exploration, (see 012.062), p. 155 - 160 (1981).
In this paper the latest neutral and ionospheric models of the upper Martian atmosphere, created on the Viking data are used. For the calculations of the photoelectron spectra the Hinteregger's measurements of the solar flux together with the theoretical photoelectron distribution are also used.

097.081   **Martian cratering revisited: implications for early geologic evolution.**   M. Gurnis.
Icarus, Vol. 48, 62 - 75 (1981).
Improved crater statistics from varied Martian terrains are compared to lunar crater populations. The distribution functions for the average Martian cratered terrain and the average lunar highlands over the diameter range 8–2000 km are quite similar. The Martian population is less dense by approximately 0.70 from 8 to 256 km diameter and diverges to proportionally lower densities at greater diameters. Monte Carlo simulations were performed to constrain plausible mechanisms of crater obliteration.

097.082   **North–south asymmetry of eolian features in Martian polar regions: analysis based on crater-related wind markers.**   P. Thomas.
Icarus, Vol. 48, 76 - 90 (1981).
A comparison of crater-related wind markers in the north and south polar (40–90° latitude) regions of Mars has been made. Wind streaks show that present wind activity is most effective in both north and south in the southern spring and summer. This asymmetry is consistent with the present asymmetry of climate. The more massive intracrater dune fields are also oriented with the presently strongest winds.

097.083   **Martian channels and valleys: their characteristics, distribution, and age.**   M. H. Carr, G. D. Clow.
Icarus, Vol. 48, 91 - 117 (1981).
All Martian channels and valleys visible at a resolution of 125 to 300 meters between 65°N and 65°S were mapped at a scale of 1 : 5,000,000 and the maps then digitized. Correlations of valley presence with other surface features show that almost all valleys are in the old cratered terrain. The simplest explanation of the correlations and the restriction of valley networks to old terrain is that the channels themselves are old, and that the climatic conditions necessary for their formation did not prevail for long after the decline in the cratering rate around 3.9 billion years ago. Two types of outflow channel are distinguished.

097.084   **Mars: far-infrared spectra and thermal-emission models.**   J. P. Simpson, J. N. Cuzzi,
E. F. Erickson, D. W. Strecker, A. T. Tokunaga.
Icarus, Vol. 48, 230 - 245 (1981).
Spectra of Mars from 100 to 360 cm^{-1} were obtained during three different observation periods from NASA's Kuiper Airborne Observatory. Also, a new thermal model was constructed for the surface of Mars, and synthetic spectra were computed from the models to compare with the observations. The models include the effects of a dusty atmosphere which absorbs, scatters, and reradiates energy.

097.085   **Planetary-scale wave structure in the Martian atmosphere.**   B. J. Conrath.
Icarus, Vol. 48, 246 - 255 (1981).
Wave-like perturbations are found in the Mariner 9 IRIS atmospheric temperature data during late Northern Hemisphere winter in a latitude band between 45°N and 65°N. The nature of the data base prevents a unique separation of spatial and temporal behavior, but Fourier analysis of the data constrains the waves to discrete combinations of planetary wavenumber and period.

097.086   **Mars.**   H. Masursky.
The new solar system, (see 003.026), p. 83 - 92 (1981).

097.087   **Life on Mars?**   G. A. Soffen.
The new solar system, (see 003.026), p. 93 - 96 (1981).

Viking Lander imaging investigation during extended and continuation automatic missions. Volume I – Lander 1 Picture Catalog of Experiment Data Record, Volume II – Lander 2 Picture Catalog of Experiment Data Record.
See Abstr. 002.019.

Radio investigations of the moon and terrestrial planets.   See Abstr. 003.015.

Nomenclature of features of the Martian relief.
See Abstr. 003.040.

The satellites of Mars.   See Abstr. 003.141.

From the history of discovery of the Martian satellites.   See Abstr. 004.076.

Biological studies of Martian soil analogues. See Abstr. 015.048.

Near-infrared spectral reflectance of mineral mixtures: systematic combinations of pyroxenes, olivine, and iron oxides.   See Abstr. 022.063.

Amorphous gels as possible analogs to martian weathering products.   See Abstr. 022.065.

Impact cratering in viscous targets: laboratory experiments.   See Abstr. 022.068.

Rotacion de la tierra año 1980. See Abstr. 044.035.

Balloons on other planets.   See Abstr. 051.055.

Eruption mechanics on the Earth, Moon and Mars. See Abstr. 081.004.

On an interconnection between the gravitational field and topography of a planet (earth, Mars, moon). See Abstr. 081.053.

A comparison of secondary craters on the Moon, Mercury, and Mars.   See Abstr. 091.025.

Radioactivity of the moon and planets. See Abstr. 094.620.

Evolution of the structure of the interplanetary medium within the Martian orbit.   See Abstr. 106.038.

098 Minor Planets

098.001   **Physical studies of asteroids IV: photoelectric observations of the asteroids 47, 95, 431.**
M. Carlsson, C.- I. Lagerkvist.
Astron. Astrophys., Suppl. Ser., Vol. 45, 1 - 4 (1981).
Photoelectric lightcurves are presented. The synodic period of rotation of asteroid 95 was found to be $0\overset{d}{.}3620 \pm 0\overset{d}{.}0005$. The asteroids 47 and 431 showed no significant brightness varia - tion throughout the observing runs of 8 hours.

098.002   **Minor planets' positions obtained in May - June 1980 at the GPO telescope of ESO La Silla — two discoveries.**
H. Debehogne, R. R. De Freitas Mourao, O. C. Tavares, M. Nunes.
Astron. Astrophys., Suppl. Ser., Vol. 45, 79 - 83 (1981).
This paper contains results of photographic observations of asteroids obtained by means of the GPO of the ESO La Silla ($f = 4$ m, $D = 40$ cm), using the 4 m focal length telescope ($D = 40$ cm) as guide. By clear sky, magnitude 18 could be attained.

098.003   **Positions of selected minor planets (1979 - 80).**
V. Zappalà, G. De Sanctis, W. Ferreri.
Astron. Astrophys., Suppl. Ser., Vol. 45, 93 - 96 (1981).
343 precise positions of 22 minor planets, observed during the period September 1979 - July 1980 at the Observatory of Torino, are given.

098.004   **Occultations of stars by the four largest minor planets, 1981 - 1989.**   G. E. Taylor.
Astron. J., Vol. 86, 903 - 905 (1981).
Possible occultations of stars by the four largest minor planets are predicted for the years 1981 - 1989 as part of a continuing program to obtain accurate sizes of these bodies.

098.005   **Photoelectric photometry of the asteroids 404 Arsinoe and 628 Christine.**
H. J. Schober.
Astron. Astrophys., Vol. 100, 311 - 313 (1981).
The asteroids 404 Arsinoe and 628 Christine were ob-served during their opposition approaches in 1979. For 404 Arsinoe a period of rotation could be derived with $P = 8\overset{h}{.}93 \pm 0\overset{h}{.}02$ ($0\overset{d}{.}372 \pm 0\overset{d}{.}001$) which rules out the previous value $P = 6\overset{h}{.}0$? given by Tedesco (1979). The double wave lightcurve has a total amplitude of $\Delta m = 0.36$. 628 Christine was observed for more than seven hours, showing only one minimum. The period could be expected to be $14^h < P < 21^h$; the minimum amplitude of the lightcurve is $\Delta m \geqslant 0.40$. Absolute magnitudes were computed and UBV colors were measured frequently.

098.006   **14 Irene: a puzzling asteroid.**
F. Scaltriti, V. Zappalà, H. J. Schober, A. Hanslmeier, A. Sudy, J. Piironen, C. Blanco, S. Catalano.
Astron. Astrophys., Vol. 100, 326 - 329 (1981).
Observations in different oppositions are analyzed in order to deduce the rotational properties of the asteroid 14 Irene. Two hypotheses, equally well supported by several physical considerations, were made giving the following values for the synodic period: $P_1 = 9\overset{h}{.}35$ and $P_2 = 18\overset{h}{.}71$. In both cases the total amplitude should reach a maximum of 0.10 mag at 90° of aspect, while for smaller angles it becomes 0.03 mag only. The magnitude-phase relation was also obtained and gave $\beta_v = (0.046 \pm 0.001)$ mag/degree and $V_0 (1, 0, 90°) = (6.55 \pm 0.02)$ mag.

098.007   **Mögliche Bedeckung des Sterns SAO 78931 durch den Kleinplaneten (110) Lydia.**
K. Meisenheimer, A. M. Quetsch.
Sterne Weltraum, Jahrg. 20, 301 - 302 (1981).

098.008   **The perturbations of the orbital elements of Trojan asteroids.**   B. Érdi.
Celestial Mech., Vol. 24, 377 - 390 (1981).
An asymptotic solution for the cylindrical coordinates of Trojan asteroids is derived by using a three-variable expansion method in the elliptic restricted three-body problem. The perturbations of the orbital elements are obtained from this solution by applying the formulas of the two-body problem. The main perturbations of the mean motion are studied in detail.

098.009   **Photoelectric lightcurves and rotation periods of the asteroids 46 Hestia and 115 Thyra.**
F. Scaltriti, V. Zappalà, A. W. Harris.
Icarus, Vol. 46, 275 - 280 (1981).
Results of photoelectric observations of the asteroids 46 Hestia and 115 Thyra are presented. The rotation periods and the maximum amplitudes are $P_{syn} = 21\overset{h}{.}04 \pm 0\overset{h}{.}01$, amplitude = 0.12 mag and $P_{syn} = 7\overset{h}{.}241 \pm 0\overset{h}{.}001$, amplitude = 0.20 mag, for Hestia and Thyra, respectively. The multiple scattering factors inferred from the phase relation data are 0.054 ± 0.003 and 0.058 ± 0.002 for Hestia and Thyra, respectively.

098.010   **The diameter and albedo of 1943 Anteros.**
G. J. Veeder, E. F. Tedesco, D. J. Tholen, A. Tokunaga, C. Kowal, K. Matthews, G. Neugebauer, B. T. Soifer.
Icarus, Vol. 46, 281 - 284 (1981).
The authors report the results of broadband visual and infrared photometry of the Apollo-Amor asteroid 1943 Anteros during its 1980 apparition. By means of a radiometric model, they calculate a diameter of 2.3 ± 0.2 km and a visual geometric albedo of 0.13 ± 0.03. The albedo and reflectance spectrum of Anteros imply that it is a type S asteroid. Thus, Anteros may have a silicate surface similar to other Apollo - Amor asteroids as well as some stony-iron meteorites.

098.011   **A possible satellite of 9 Metis.**
S. Wang, Y. Wu, M. Bao, L. Deng, S. Wu.
Icarus, Vol. 46, 285 - 287 (1981).
The authors attempted by means of photographic obser-vations to search for satellites of asteroids in 1979 and 1980. An elongation of the image of 9 Metis has been detected on plates taken during 6 nights. The data suggest that there is a possible satellite with an orbital period of approximately 4.61 days and a mean distance of approximately 1100 km. The authors tabulate orbital parameters as well as predicted and ob-served position angles. A comparison with the secondary oc-cultation results taken on 11.35 December 1979 is discussed.

098.012   **Asteroidal agglutinate formation and implications for asteroidal surfaces.**   F. Hörz, R. B. Schaal.
Icarus, Vol. 46, 337 - 353 (1981).
A large number of shock recovery experiments that address the ease of impact melt formation as a function of peak shock pressure lead to the conclusion that impacts at 5 km/sec into fragmental, porous surfaces will produce agglutinate-type glasses; no shock melts are produced at these velocities in dense silicate target rocks. The authors propose that spall processes at a target's free surface play a major role in asteroidal surface evolution. As a result asteroidal surface deposits should be more coarse grained and less shocked than lunar ones — consistent with meteorite evidence and remote-sensing observations.

098.013   Orientation of the rotational axis and UBV photom-
           etry of the asteroid 354 Eleonora.
D. F. Lupishko, F. P. Velichko, F. A. Tupieva, G. P. Chernova.
Pis'ma Astron. Zh., Tom 7, 437 - 441 (1981). In Russian.
English translation in Soviet Astron. Lett., Vol. 7.
    Light curves, magnitude and colour-phase relations of the
asteroid 354 Eleonora have been obtained from observations
made during 31 nights in June-September 1979.

098.014   Kleine Planeten – 2. Halbjahr 1981: Eine sehr gün-
           stige Opposition von (433) Eros.   L. D. Schmadel.
Sterne Weltraum, Jahrg. 20, 335 - 338 (1981).

098.015   The return of a celestial maverick – the minor
           planet Ra - Shalom.   C. L. Townsend.
J. British Astron. Assoc., Vol. 91, 488 - 490 (1981).
    Brief details are given of minor planet 2100, Ra-Shalom
an unusual Apollo-type object.

098.016   Lightcurves of various asteroids (IV).
           Y. C. Chang, X.-h. Zhou, X.-y. Yang, Y.-y. Zhang,
X.-q. Li, Z.-x. Wu.
Acta Astron. Sinica, Vol. 22, 169 - 173 (1981). In Chinese.
    Lightcurves of eleven asteroids were obtained from
photoelectric observations in 1965 and in 1978. They are
(1) Ceres, (2) Pallas, (14) Irene, (15) Eunomia, (16) Psyche,
(18) Melpomene, (26) Proserpina, (32) Pomona, (39) Laetitia,
(115) Thyra, and (675) Ludmilla. For seven of them light-
variable periods have been determined.

098.017   Physical studies of asteroids. V: Photoelectric obser-
           vations of the asteroids 70, 101, 369 and 432.
C.-I. Lagerkvist, H. Rickman.
Astron. Astrophys., Suppl. Ser., Vol. 45, 177 - 179 (1981).
    Observations of asteroids 70, 101, 369 and 432 are
reported. Photoelectric lightcurves, rotation periods, maximum
amplitudes, absolute magnitudes and UBV colours are presented. The following synodic periods of rotation and maximum
amplitudes were derived: 0^d965 and 0^m13 for 101 Helena, and
0^d3453 and 0^m15 for 432 Pythia.

098.018   The asteroids: physical properties and evolutionary
           processes.   C. R. Chapman.
Bull. American Astron. Soc., Vol. 13, 525 - 526 (1981).
Abstract.

098.019   Multiple Apollo asteroids from comets?
           G. L. Strobel, T. J. Wdowiak.
Bull. American Astron. Soc., Vol. 13, 568 (1981). – Abstract.

098.020   Computer simulation of formation of the structure of
           the asteroid belt.   B. B. D'yakov, B. I. Reznikov.
Fiz.-tekh. inst. AN SSSR. Prepr., 1981, No. 698. 27 pp. In
Russian. – From Ref. zh., 51. Astron., 7.51.141 (1981).

098.021   Observations of asteroid 4 Vesta.
           M. Geffert, M. Hoffmann.
Minor Planet Bull., Vol. 8, 17 - 19 (1981).

098.022   Photoelectric photometry of the asteroid
           980 Anacostia.   T. G. McFaul.
Minor Planet Bull., Vol. 8, 19 - 21 (1981).

098.023   Minor planet photometry: late 1981 program.
           D. Wallentinsen, A. C. Porter.
Minor Planet Bull., Vol. 8, 21 - 22 (1981).

098.024   General report of position observations by the
           A. L. P. O. Minor Planets Section for the year 1980.
F. Pilcher.
Minor Planet Bull., Vol. 8, 22 - 27 (1981).

098.025   Elements of unnumbered minor planets.
           T. Kurosaki.
Minor Planet Bull., Vol. 8, 27 - 31 (1981).

098.026   Photoelectric photometry of 68 Leto.
           S. S. Mims, D. Wallentinsen, R. W. James.
Minor Planet Bull., Vol. 8, 27, 32 (1981).

098.027   Recovery of long-lost 843 Nicolaia.
           L. D. Schmadel.
Minor Planet Bull., Vol. 8, 32 - 33 (1981).

098.028   1915 Quetzálcoatl observation.   B. Hudgens.
Minor Planet Bull., Vol. 8, 33 (1981).

098.029   A note concerning an effect which arose during the
           origin of the asteroids.   V. Banfi.
Mem. Soc. Astron. Italiana, Vol. 52, 217 - 231 (1981).
    The existence of a mechanism is proved due to action of
the two primaries, Jupiter and the Sun, together with the
resisting medium in the neighbourhood of the Jupiter orbit,
which would explain the obstacling and scattering action with
respect to the process of accretion of a small unique planet in
that region.

098.030   Investigation of the motion of some peculiar
           asteroids.
T. V. Bordovitsyna, V. A. Shefer, B. T. Kharin.
Determination of the coordinates of celestial bodies, Riga,
1981, (see 003.008), p. 77 - 86. In Russian. – Abstr. in Ref.
zh., 51. Astron., 9.51.74 (1981).

098.031   On the motion of minor planets in the vicinity of
           Kirkwood gaps.   M. A. Dirikis.
Determination of the coordinates of celestial bodies, Riga,
1981, (see 003.008), p. 87 - 95. In Russian. – Abstr. in Ref.
zh., 51. Astron., 9.51.75 (1981).

098.032   Studies of small asteroids. I: positions of asteroids
           obtained during August 1979 with the ESO Schmidt
telescope.   C.-I. Lagerkvist.
Astron. Astrophys.,Suppl. Ser., Vol. 46, 21 - 24 (1981).
    792 positions of 140 asteroids, mostly unnumbered ones,
obtained during August 1979 with the ESO Schmidt telescope,
are presented.

098.033   Beobachtung und Bahnverbesserung des kleinen
           Planeten (2035) Stearns.
F. Seiler, W. Landgraf.
Sterne, 57. Band, 238 - 242 (1981).

098.034   Observations of minor planets at Dresden Lohrmann
           Observatory.   E. Asenjo, D. Böhme.
Astron. Nachr., Band 302, 233 - 234 (1981) = Mitt. Lohrmann-
Obs. Tech. Univ. Dresden, Nr. 47.
    Observed positions of minor planets from 1979 and 1980
at the Zeiss-Sonnefeld astrograph 300/1500 of the Lohrmann
Observatory at Dresden-Gönnsdorf are communicated.

098.035   Evolution of orbits of asteroids not belonging to the
           main belt.   M. A. Vashkov'yak.
Kosm. Issled., Tom 19, 528 - 538 (1981). In Russian.

098.036   List of minor planets again supplemented.
           M. Dīriķis.
Zvaigžnotā debess, 1980. gada pavasaris, p. 21 - 22. In Latvian.

098.037   New names for minor planets.   M. Dīriķis.
Zvaigžnotā debess, 1980. gada vasara, p. 16. In
Latvian.

**098.038 Minor planets and related objects. XXIX.
Asteroid 29 Amphitrite.**
E. F. Tedesco, R. E. Sather.
Astron. J., Vol. 86, 1553 - 1558 (1981).
UBV photometry and lightcurves of asteroid
29 Amphitrite from all observed apparitions (March 1956 -
May 1977) are combined to obtain a phase function, absolute
magnitude, color-phase variation, and preliminary pole loca-
tion and shape for Amphitrite.

098.039 Observations of minor planets.
    Minor Planet Circ., (M.P.C.), Nos. 6070 - 6098,
6147 - 6188, 6227 - 6284, 6350 - 6413, 6436 - 6463, 6490 -
6513 (1981).
    Observations made at the following stations are published:
Abastumani, Bergedorf, Byurakan, Catalina Stat. Lunar Planet.
Lab., Catania Obs. Serra La Nave Stat., Caussols, Cerro Calan,
Cerro el Roble, Cerro Tololo, Crimea (44th - 45th report),
El Leoncito, European Southern Obs., Geisei, Goethe Link
Obs., Göttingen, Hemingford Abbots, Kambah (Canberra),
Kleť, Kvistaberg, La Seyne sur Mer, Leiden Southern Stat.,
Lick Obs., Lincoln Lab., Lowell Obs., Lowell Obs. Anderson
Mesa Stat., Meschede, Mt. Hopkins, Mt. John Obs.,
Mt. Palomar, Mt. Wilson, Nantucket, Nice, Oak Ridge Obs.,
Ojima, Perth, Reintal, S. Vittore (Bologna), Siding Spring,
Stakenbridge, Steward Obs., Sydney, Tautenburg, Tokai,
Tokyo Obs. Kiso Stat., Turku, Uccle, Union Obs., Uppsala,
Uppsala Southern Stat., Wise Obs., Zimmerwald.

**098.040 Identifications and identification changes of minor
planets.**
Minor Planet Circ., (M.P.C.), Nos. 6066 - 6067, 6145, 6221 -
6222, 6348 - 6349, 6425 - 6429, 6482 - 6486 (1981).

098.041 Ephemerides of minor planets and comets.
    Minor Planet Circ., (M.P.C.), Nos. 6112 - 6144,
6210 - 6220, 6307 - 6344, 6422 - 6424, 6476 - 6480, 6532 -
6572 (1981).

098.042 New names of minor planets.
    Minor Planet Circ., (M.P.C.), Nos. 6207 - 6210,
6421 - 6422, 6530 - 6532 (1981).

098.043 Orbital elements of one-opposition minor planets.
    Minor Planet Circ., (M.P.C.), Nos. 6098 - 6099,
6188 - 6190, 6285 - 6287, 6413 - 6414, 6463 - 6464, 6514
(1981).

098.044 Orbital elements of numbered minor planets.
    Minor Planet Circ., (M.P.C.), Nos. 6065 - 6572
(1981).
    The minor planets are listed according to their definitive
number. Newly numbered objects are indicated by an asterisk.
The names of the computers are given behind the respective
M.P.C. numbers. (2) 6464 G. Sitarski; (168) 6190, (211) 6190,
(250) 6190, (259) 6190, (409) 6190 P. Herget; (452) 6301
L. K. Kristensen; (457) 6464 K. Ziolkowski; (488) 6190,
(536) 6191, (617) 6191 P. Herget; (649) 6464 K. Ziolkowski;
(843) 6191 L. D. Schmadel, 6414 L. K. Kristensen; (1019)
6100 W. Landgraf; (1038) 6465, (1161) 6465, (1162) 6465,
(1297) 6465 K. Ziolkowski; (1457) 6290 B. G. Marsden;
(1537) 6100, (1538) 6301 L. K. Kristensen; (1565) 6290
B. G. Marsden; (1685) 6100 W. Landgraf; (2142) 6301 - 6302,
(2145) 6302, (2160) 6415, (2162) 6302, (2163) 6302 - 6303,
(2173) 6303, (2175) 6303, (2178) 6415, (2190) 6303 - 6304,
6415, (2199) 6304, (2203) 6416, (2214) 6416, (2234) 6304,
(2289) 6304 - 6305 L. D. Schmadel; (2396)* 6100 - 6101
S. Nakano, T. Urata; (2397)* - (2404)* 6101 - 6104
B. G. Marsden; (2405)* - (2413)* 6106 - 6110 C. M. Bardwell;
(2414)* 6191 L. K. Kristensen; (2415)* - (2416)* 6192
S. Nakano, T. Urata; (2417)* - (2421)* 6194 - 6196
B. G. Marsden; (2422)* - (2436)* 6198 - 6203 C. M. Bardwell;

(2437)* - (2442)* 6288 - 6290 P. Herget; (2443)* - (2452)*
6290 - 6294 B. G. Marsden; (2453)* - (2471)* 6296 - 6299,
6416 - 6419 C. M. Bardwell; (2472)* - (2474)* 6419 - 6420
B. G. Marsden; (2475)* - (2478)* 6465 - 6467 S. Nakano,
T. Urata; (2479)* - (2482)* 6468 - 6469 B. G. Marsden;
(2483)* - (2495)* 6470 - 6474 C. M. Bardwell; (2496)* -
(2499)* 6514 - 6516 S. Nakano, T. Urata; (2500)* - (2511)*
6519 - 6523 B. G. Marsden; (2512)* - (2525)* 6525 - 6530
C. M. Bardwell.

098.045 Orbital elements of unnumbered minor planets.
    Minor Planet Circ., (M.P.C.), Nos. 6065 - 6572
(1981).
    The unnumbered minor planets are listed according to
their preliminary designation. Objects from the Palomar-Leiden
Survey are sorted by number. The names of the computers are
given behind the respective M.P.C. numbers. [1937 TD] 6110
C. M. Bardwell; [1938 DW_1] 6196 B. G. Marsden; [1938 GC]
6110, [1939 TM] 6110, [1940 GG] 6111 C. M. Bardwell;
[1941 FN] 6104 B. G. Marsden; [1942 TJ] 6203, [1949 PK]
6299 - 6300 C. M. Bardwell; [1952 UZ_1] 6305, [1953 TC_1]
6101 S. Nakano, T. Urata; [1964 TR_2] 6475 C. M. Bardwell;
[1965 UB_1] 6196 B. G. Marsden; [1966 BW] 6467 S. Nakano,
T. Urata; [1967 UO] 6294 B. G. Marsden; [1969 TH_6] 6111
C. M. Bardwell; [1970 AF_1] 6192, [1971 TZ] 6467 S. Nakano,
T. Urata; [1971 UX] 6294 B. G. Marsden; [1972 RV_3] 6204
C. M. Bardwell; [1973 QY_1] 6104 B. G. Marsden; [1973 SJ_4]
6204 C. M. Bardwell; [1974 MH] 6294 - 6295 B. G. Marsden;
[1974 QL] 6300, [1974 SG_1] 6111 C. M. Bardwell; [1975 BX]
6196 B. G. Marsden; [1975 EE_3] 6204, [1975 VD_2] 6204,
[1975 VF_2] 6205 C. M. Bardwell; [1975 VB_0] 6104
B. G. Marsden; [1975 WK_1] 6300 C. M. Bardwell; [1975 WO_1]
6193 S. Nakano, T. Urata; [1975 XB] 6295 B. G. Marsden;
[1975 XQ] 6300 C. M. Bardwell; [1976 GZ_2] 6196 - 6197,
6469 B. G. Marsden; [1976 GN_3] 6101 S. Nakano, T. Urata;
[1976 GB_8] 6419, [1976 JQ_2] 6419 C. M. Bardwell; [1976 QD]
6104 - 6105 B. G. Marsden; [1976 QF_1] 6205, [1976 YU_3]
6205 C. M. Bardwell; [1977 EB_2] 6197, [1977 NR] 6105
B. G. Marsden; [1977 PA_2] 6111 C. M. Bardwell; [1977 QB_1]
6523, [1977 QY_2] 6469 - 6470, [1977 TB_1] 6197, [1978 GB]
6470, [1978 PS_3] 6295, [1978 PB_4] 6105 B. G. Marsden;
[1978 QW_2] 6205, [1978 RF] 6205 - 6206, [1978 RJ_2] 6206
C. M. Bardwell; [1978 RR_5] 6516 S. Nakano, T. Urata;
[1978 RU_5] 6111 - 6112 C. M. Bardwell; [1978 RC_6] 6467
S. Nakano, T. Urata; [1978 TA] 6524 B. G. Marsden;
[1978 TM_7] 6516 S. Nakano, T. Urata; [1978 UJ_2] 6524
B. G. Marsden; [1978 UQ_2] 6516, [1978 WM_{14}] 6305
S. Nakano, T. Urata; [1979 DE] 6197 B. G. Marsden;
[1979 FG_2] 6516 - 6517, [1979 FJ_2] 6517, [1979 FT_2] 6517,
[1979 KX] 6517, [1979 ML_3] 6305 S. Nakano, T. Urata;
[1979 MF_4] 6206, [1979 MO_6] 6112 C. M. Bardwell;
[1979 OM_{15}] 6517, [1979 QU_2] 6518 S. Nakano, T. Urata;
[1979 TH] 6197, [1980 EC] 6198, [1980 EG] 6524
B. G. Marsden; [1980 FN_3] 6475, [1980 GC] 6300 - 6301
C. M. Bardwell; [1980 HB] 6420 - 6421, [1980 KN] 6105
B. G. Marsden; [1980 LO] 6306 S. Nakano, T. Urata;
[1980 LP] 6295, [1980 MA] 9294 - 6295, [1980 PG] 6198
B. G. Marsden; [1980 PJ] 6112, [1980 PV] 6206 C. M. Bardwell;
[1980 RB] 6198, [1980 RR] 6524, [1980 RX] 6525
B. G. Marsden; [1980 TN] 6306, [1980 TD_4] 6518 S. Nakano,
T. Urata; [1980 VJ] 6106 B. G. Marsden; [1981 CN] 6467,
[1981 EK] 6306, [1981 EL] 6307, [1981 EY] 6307 S. Nakano,
T. Urata; [1981 FN] 6112 C. M. Bardwell; [1981 JA] 6193
S. Nakano, T. Urata; [1981 JP] 6206 - 6207, [1981 JX] 6207,
[1981 LD] 6475 C. M. Bardwell; [1981 LK] 6468 S. Nakano,
T. Urata; [1981 QA] 6296, 6470, 6525, [1981 QB] 6296,
6470, 6525 B. G. Marsden; [1981 QH] 6475, [1981 QC_2]
6476, [1981 QG_2] 6530, [1981 RU] 6530 C. M. Bardwell;
[1981 VA] 6476 C. M. Bardwell, 6525 B. G. Marsden;
[2221 P-L] 6421 B. G. Marsden; [2563 P-L] 6207
C. M. Bardwell; [4008 P-L] 6106 B. G. Marsden; [4579 P-L]
6207 C. M. Bardwell; [9086 P-L] 6106 B. G. Marsden.

098.046 The discovery of an unusual Apollo asteroid
(1979 VA). E. F. Helin.
Mercury, Vol. 10, 134, 149 (1981).

098.047 Surface material variegation on asteroids.
M. J. Gaffey.
Bull. American Astron. Soc., Vol. 13, 711 (1981). – Abstract.

098.048 Spectrophotometry from 3000 to 10500 Å of a
selected group of asteroids.
R. P. Binzel, A. L. Cochran, W. D. Cochran, E. S. Barker.
Bull. American Astron. Soc., Vol. 13, 712 (1981). – Abstract.

098.049 Physical observations of near-earth asteroids.
E. F. Tedesco, D. J. Tholen, B. Zellner,
G. J. Veeder, Jr., J. G. Williams.
Bull. American Astron. Soc., Vol. 13, 712 (1981). – Abstract.

098.050 Radar detection of Iris, Psyche, Klotho, Apollo,
and Quetzalcoatl.
S. J. Ostro, D. B. Campbell, I. I. Shapiro.
Bull. American Astron. Soc., Vol. 13, 716 (1981). – Abstract.

098.051 Asteroid radiometry: a recalibration and some new
results.
R. H. Brown, D. Morrison, C. M. Telesco, W. Brunk.
Bull. American Astron. Soc., Vol. 13, 716 - 717 (1981).
Abstract.

098.052 JHK infrared photometry of asteroids.
G. J. Veeder, D. L. Matson, C. Kowal.
Bull. American Astron. Soc., Vol. 13, 717 (1981). – Abstract.

098.053 JHK photometry of selected Trojan and Hilda
asteroids. R. Shorthill, P. Johnson, D. Smith.
Bull. American Astron. Soc., Vol. 13, 717 (1981). – Abstract.

098.054 Spectra of selected asteroids, 2.4 - 4.2 μm.
D. P. Cruikshank, R. R. Howell.
Bull. American Astron. Soc., Vol. 13, 717 (1981). – Abstract.

098.055 Highlights from the eight-color asteroid survey.
B. Zellner, E. F. Tedesco, D. J. Tholen.
Bull. American Astron. Soc., Vol. 13, 717 (1981). – Abstract.

098.056 Diagnostic spectral properties of ordinary
chondritic assemblages: implications for the
S-asteroids. M. J. Gaffey.
Bull. American Astron. Soc., Vol. 13, 717 (1981). – Abstract.

098.057 Near-earth asteroids: 1981 perspectives based on
reflectance spectroscopy. L. A. McFadden.
Bull. American Astron. Soc., Vol. 13, 718 (1981). – Abstract.

098.058 Asteroid 1 Ceres: evidence for structural water in
clay minerals.
L. A. Lebofsky, M. A. Feierberg, H. P. Larson, A. T. Tokunaga.
Bull. American Astron. Soc., Vol. 13, 718 (1981). – Abstract.

098.059 Trends in the distribution of taxonomic types across
the asteroid belt. E. F. Tedesco, J. Gradie.
Bull. American Astron. Soc., Vol. 13, 718 (1981). – Abstract.

098.060 Relationships between observable asteroids and
meteorite yields. C. R. Chapman, R. Greenberg.
Bull. American Astron. Soc., Vol. 13, 718 (1981). – Abstract.

098.061 Collisional generation of asteroid rotations: does the
baby depart with the bathwater?
J. A. Burns, A. R. Dobrovolskis.
Bull. American Astron. Soc., Vol. 13, 719 (1981). – Abstract.

098.062 Photometric determination of asteroid shapes and
spin axial directions.
K. Lumme, M. Poutanen, E. Bowell.
Bull. American Astron. Soc., Vol. 13, 719 (1981). – Abstract.

098.063 Speckle interferometric observations of Ceres, Vesta,
Hebe, and Victoria. E. Bowell, B. L. Morgan,
K. Lumme, J. C. Dainty, H. A. Vine, M. Poutanen.
Bull. American Astron. Soc., Vol. 13, 719 - 720 (1981).
Abstract.

098.064 An observing program to study configurations of
large, rapidly spinning asteroids.
D. R. Davis, S. J. Weidenschilling, C. R. Chapman,
R. Greenberg.
Bull. American Astron. Soc., Vol. 13, 725 (1981). – Abstract.

098.065 A physically plausible ellipsoidal model of Hektor?
M. Poutanen, E. Bowell, K. Lumme.
Bull. American Astron. Soc., Vol. 13, 725 (1981). – Abstract.

098.066 Distribution of asteroids: orbital elements, types,
rotational frequencies, diameters, albedos and
colors. S. F. Dermott, C. D. Murray.
Bull. American Astron. Soc., Vol. 13, 744 (1981). – Abstract.

098.067 Asteroid lightcurve photometry at Table Mountain
Observatory. A. W. Harris, J. W. Young.
Bull. American Astron. Soc., Vol. 13, 744 (1981). – Abstract.

098.068 Physical studies of asteroids. VI. Asteroid
201 Penelope, a fast rotator.
C.-I. Lagerkvist, H. Rickman, F. Scaltriti, V. Zappalà.
Astron. Astrophys., Vol. 104, 148 - 149 (1981).
Photoelectric UBV lightcurves of the CMEU-type asteroid
201 Penelope, observed during September and October 1980
at the European Southern Observatory and at the Torino
Observatory, are presented. A large amplitude (~ 0.56 mag),
not observed to change with phase angle, and a rapid spin rate
($P = 3^h.7474 \pm 0^h.0001$) are found. The parameters characterizing the magnitude-phase relation of the asteroid according to
the Lumme and Bowell (1981) radiative transfer theory are
found to be $Q = 0.075 \pm 0.015$ and $V(0°) = 8.21 \pm 0.01$ mag.

098.069 Analysis of the spin rate distribution of asteroids.
P. Farinella, P. Paolicchi, V. Zappalà.
Astron. Astrophys., Vol. 104, 159 - 165 (1981).
A new analysis of the spin rate distribution of asteroids is
performed; it attempts to take into account the principal selection effects and studies separately several different subsamples.
The results show strong statistical evidence for a non-Maxwellian character of the overall distribution. The average spin rate
increases with size, while different taxonomic types present
distinct features for their distributions. The results can be
interpreted in terms of a complex and size-dependent collisional history of asteroids.

098.070 Variable asteroids. R. P. Binzel.
I. A. P. P. P. Commun., No. 5, p. 19 - 23 (1981).

098.071 Positions d'astéroïdes obtenues au GPO de 40 cm de
l'ESO, La Silla, décembre 1979.
H. Debehogne, L. Machado, E. Netto, J. Caldeira, G. Vieira.
Astron. Astrophys., Suppl. Ser., Vol. 46, 371 - 374 (1981).
116 precise positions of minor planets observed at the
GPO ($f = 4$ m, $d = 40$ cm) of the European Southern Observa-
tory, La Silla, Chile, during December 15 through December
30, 1979 are given. All plates were measured on the Ascorecord
Zeiss, Jena, measuring machine of the Valongo Observatory,
UFRJ, Rio de Janeiro. The positions were computed by means
of dependences and of least squares, with the computer
Burroughs B6700 of the Núcleo de Computação Eletrônica –

NCE, of the Universidade Federal do Rio de Janeiro. The reference stars, corrected for proper motions, are taken from SAO Star Catalogue, at the equinox (1950.0).

098.072   Evidence for color variations on the surface of 3 Juno: new photoelectric *UBV*-observations.
A. Schroll, H. J. Schober, C. I. Lagerkvist.
Astron. Astrophys., Vol. 104, 296 - 299 (1981).

3 Juno was observed in 1980 on Feb. 7 - 9 and on March 12, 13 at the European Southern Observatory, La Silla, Chile. The results show that $B-V$ varies in that sense that Juno becomes redder just before the primary minimum of the lightcurve and bluer before the primary maximum. The amplitude of this color variation during the rotation is about $0.^m025$. The rotation period is found to be $7.^h209 \pm 0.^h003$, in agreement with the published values. The phase dependences for the mean colors $B-V$ and $U-B$ for 3 Juno were also derived.

098.073   Radiative transfer in the surfaces of atmosphereless bodies. II. Interpretation of phase curves.
K. Lumme, E. Bowell.
Astron. J., Vol. 86, 1705 - 1721 (1981) = Contrib. Five Coll. Obs., No. 438.

Phase curves for 74 asteroids, the Galilean satellites, four Saturn satellites, Deimos, the Moon, and Mercury have been interpreted using a generalized theory of radiative transfer in rough and porous surfaces. The phase curves can be generated from a single phase function by varying a parameter termed the multiple-scattering factor, and contain accessible information on the surface roughness, volume density, and single-particle phase function (characterized by the asymmetry factor). The volume density appears to be similar for all bodies considered. For low- and moderate-albedo objects the surface roughness and asymmetry factor differ from object to object by modest amounts. Phase integrals and Bond albedos are calculated, and rigorous and approximate equations are given for calculating the brightness of an atmosphereless body of any albedo observed at any phase angle.

098.074   Surface materials on unusual planetary object Chiron.
W. K. Hartmann, D. P. Cruikshank, J. Degewij, R. W. Capps.
Icarus, Vol. 47, (see 012.045), 333 - 341 (1981).

The authors present the first JHK near-infrared colorimetry giving some indication of the surface materials on the peculiar object 2060 Chiron. The colors appear to exclude relatively clean bright ice surfaces. The colors are consistent with those of numerous outer solar system asteroids. These and additional VJHK colorimetric data suggest that the spectrally dominant surface is C-type material. If Chiron has the low albedo common to such material on known interplanetary bodies, its diameter would be about 310 to 400 km.

098.075   On the implausibility of a cometary origin for most Apollo-Amor asteroids.
B. J. (*B. Yu.*) Levin, A. N. Simonenko.
Icarus, Vol. 47, (see 012.045), 487 - 491 (1981).

As the rate of replenishment of short-lived Apollo-Amor asteroids from the main belt seems to be insufficient to compensate for their losses, the idea was put forward that most of them are inactive cometary nuclei. There is good evidence that Apollo-Amor asteroids represent the last parent bodies of most, or even all, classes of meteorites. But meteorites cannot be formed within cometary nuclei having a constitution like Whipple's classical model, and alternative models seem to be unsatisfactory. Therefore, it is concluded that the cometary origin of most Apollo-Amor asteroids is implausible.

098.076   Occultations of stars by solar system objects. II. Occultations of catalog stars by asteroids in 1982 and 1983.   L. H. Wasserman, E. Bowell, R. L. Millis.

Astron. J., Vol. 86, 1974 - 1979 (1981).

Results of a computer search for occultations of AGK3 and SAO stars by 91 of the larger asteroids are presented. Thirty-eight events occurring in 1982 and 39 in 1983 are identified. Nominal ground tracks are plotted for the more promising occultations.

098.077   (843) Nicolaia.
IAU Circ., No. 3617 (1981).

098.078   1979 BA.
IAU Circ., No. 3628 (1981).

098.079   1981 QB.
IAU Circ., Nos. 3631, 3633 - 3635, 3637 - 3639 (1981).

098.080   1981 QA.
IAU Circ., Nos. 3629 - 3631, 3634, 3639 (1981).

098.081   1981 UA.
IAU Circ., Nos. 3642, 3643 (1981).

098.082   Occultation of SAO 187124 by (88) Thisbe.
IAU Circ., Nos. 3642, 3652 (1981).

098.083   1981 VA.
IAU Circ., Nos. 3644 - 3646, 3650 (1981).

098.084   Occultation of SAO 145972 by (18) Melpomene.
IAU Circ., No. 3650 (1981).

098.085   1978 DA.
IAU Circ., No. 3653 (1981).

098.086   1981QA.
Yamamoto Circ., Nos. 1962, 1963.

098.087   1981 QB.
Yamamoto Circ., Nos. 1962 - 1964 (1981).

098.088   1981 VA.
Yamamoto Circ., Nos. 1965, 1966 (1981).

098.089   Occultation of SAO 145972 by (18) Melpomene.
Yamamoto Circ., No. 1967 (1981).

098.090   On the discovery of faint asteroids and some properties of their statistical distribution.
S. G. Zhuravlev, V. N. Kiryushenkov.
Astron. Zh., Tom 58, 1291 - 1295 (1981). In Russian. English translation in Soviet Astron., Vol. 25, No. 6.

A process of the discovery of faint asteroids of the Palomar-Leiden survey is shortly described. A simple statistical analysis of the distribution of faint asteroids with respect to orbital elements is presented. A comparison of the distributions with those of numbered asteroids is given.

098.091   On the period of axial rotation of Eros.
N. I. Koshkin.
Astron. Tsirk., No. 1125, p. 7 - 8 (1980). In Russian.

098.092   On stability of asteroidal orbits near the 2/3 commensurability with Jupiter.   A. M. Kazantsev.
Astron. Tsirk., No. 1140, p. 4 - 6 (1980). In Russian.

098.093   New minor planets.   M. Dīriķis.
Zvaigžņotā debess, 1981. gada pavasaris, p. 8 - 10. In Latvian.

098.094 **Some new asteroids.**   M. Dīriķis.
Zvaigžņota debess, 1981. gada vasara, p. 23 - 24.
In Latvian.

098.095 **Internal structure of small bodies in the solar system.**   K. Yomogida, T. Matsui.
Proceedings of the 14th ISAS Lunar and Planetary Symposium, (see 012.055), p. 235 - 242 (1981).

Thermal histories of small bodies in the solar system are calculated taking into account simultaneously the effect of changing consolidated state on thermal conductivity. Very low thermal conductivity of particulate matter causes even the small bodies with radius less than 100 km to be heated up to sintering temperature only by the chondritic abundance of long-period radioactive elements. The maximum temperature is determined primarily by the sintering temperature.

098.096 **Reconstruction of asteroid families.**
A. Fujiwara.
Proceedings of the 14th ISAS Lunar and Planetary Symposium, (see 012.055), p. 278 - 283 (1981).

Themis, Eos and Koronis families are believed to be formed by collisions. The pre-existing bodies of these families are reconstructed. Total masses of the families, total ejection energies of the fragments are estimated and the destruction type is considered using the data of the laboratory impact experiments. It is suggested that three parent bodies were completely shattered.

098.097 **The motion of two asteroids, 1:1 commensurable case.**   G.-i. Hori.
Proceedings of the 14th ISAS Lunar and Planetary Symposium, (see 012.055), p. 330 - 333 (1981).

1:1 commensurable case of the motion of two asteroids is treated by a canonical perturbation theory.

098.098 **The size of the minor planet 65 Cybele.**
G. E. Taylor.
J. British Astron. Assoc., Vol. 92, 13 - 15 (1981).

The mean diameter of the minor planet 65 Cybele has been determined as 230 ± 16 km, from an analysis of observations of its occultation of the star AGK3 +19°0599 on 1979 October 17.

098.099 **Positions of minor planets obtained in 1979.**
Z. Knežević.
Bull. Obs. Astron. Belgrade, No. 131, p. 33 - 37 (1981).

098.100 **Dispersion of the minor planet inclinations as a function of time.**   A. M. Michel.
Stud. Soc. Sci. Torunensis, Toruń, Poland, Sect. F (Astron.), Vol. 6, No. 3, p. 39 - 50 (1981) = Biul. Obs. Astron. Uniw. M. Kopernika Toruniu, Nr. 63.

The time dependence of dispersion of minor planet inclinations has been investigated by means of computer technics. Calculations have been carried out for thirty asteroids with random initial parameters.

098.101 **Observaciones fotográficas de pequeños planetas años 1974, 1975 y 1976.**
Inst. Obs. Marina, San Fernando (Cádiz), Ser. B, No. 5, 2 + 39 pp. (1981).

098.102 **Computation of special perturbations for 12 minor planets of the Flora group.**   Z.-r. Liu, D.-s. Pan.
Ann. Shanghai Obs. Acad. Sinica, No. 2, p. 80 - 87 (1980).

098.103 **Amateurs among the asteroids.**   J. U. Gunter.
Strolling Astron., Vol. 29, 61 - 64 (1981).

098.104 **The minor planets: as interesting as ever.**
A. Porter.
Strolling Astron., Vol. 29, 64 - 69 (1981).

098.105 **The flux of earth-crossing and moon-cratering interplanetary bodies.**
L'. Kresák.
Progress in planetary exploration, (see 012.062), p. 85 - 90 (1981).

Methods of determining the present flux and total number of kilometer-sized earth-crossing objects are discussed, including (1) probability considerations based on the frequency of chance rediscoveries of the lost objects, (2) evaluation of large-scale photographic surveys for the detection of fast moving objects, and (3) evaluation of close encounters of interplanetary bodies with the Earth. The results are interfaced with the lunar and terrestrial cratering history.

098.106 **On the shape of rapidly rotating asteroids.**
P. Farinella, P. Paolicchi, V. Zappalà.
Progress in planetary exploration, (see 012.062), p. 187 - 189 (1981).

By means of a statistical analysis of the rotational properties of asteroids, the authors define a class of large amplitude and short period objects (LASPA). A possible interpretation of their collisional evolution and present physical status, in terms of ellipsoidal figures of equilibrium, is proposed.

098.107 **Radar observations of Apollo.**
R. M. Goldstein, R. F. Jurgens, D. K. Yeomans.
Icarus, Vol. 48, 59 - 61 (1981).

Radar observations of the asteroid Apollo, at 3.5 cm wavelength, indicate a radius of 600 m and a rotation period of 3.0 hr. The data are consistent with a rough surface of either hard ice or of regolith softened rock.

098.108 **The albedo and diameter of 1862 Apollo.**
L. A. Lebofsky, G. J. Veeder, G. H. Rieke,
M. J. Lebofsky, D. L. Matson, C. Kowal, C. G. Wynn-Williams,
E. E. Becklin.
Icarus, Vol. 48, 335 - 338 (1981).

The authors report infrared thermal emission measurements of 1862 Apollo, which is the type example of an Earth-crossing asteroid. They derive a geometric albedo of 0.21 ± 0.02 which is within the albedo range of the S class of asteroids. The effective diameter was observed to vary with rotation from 1.2 ± 0.1 to 1.5 ± 0.1 km.

098.109 **Asteroids.**   C. R. Chapman.
The new solar system, (see 003.026), p. 97 - 104 (1981).

Monte Carlo simulations of the charged-particle irradiation of asteroidal regoliths.   See Abstr. 022.131.

A new asteroid observation and search technique.
See Abstr. 031.589.

Dynamics of the asteroids.   See Abstr. 042.042.

An algorithm for numerical study of the motion of peculiar minor planets based on double regularization of the equations of motion.   See Abstr. 042.045.

Radiative transfer in the surfaces of atmosphereless bodies. I. Theory.   See Abstr. 063.051.

L'observation des occultations d'étoiles par les petites planètes.   See Abstr. 096.003.

More asteroidal occultations during 1981.
See Abstr. 096.006.

Ocultaciones por planetas menores. 1981 - 1982.
See Abstr. 096.012.

Planetary occultation predictions for 1982.
See Abstr. 096.017.

The Voyager 1/Saturn encounter and the cosmo-gonic shadow effect.   See Abstr. 100.068.

On the discovery of the fine structure of Saturn's ring and the asteroid belt.   See Abstr. 100.142.

Do comets have satellites?
See Abstr. 102.059.

Meteorites.   See Abstr. 105.123.

Character of eccentricity excitation by moving Jovian commensurability resonances. Connection with the asteroids and the Kirkwood gaps.   See Abstr. 107.014.

Erratum

098.901   Erratum: "Physical studies of asteroids III: the rotation period of 85 Io" [Astron. Astrophys., Suppl. Ser., Vol. 44, 401 - 404 (1981)].
C.-I. Lagerkvist, H. J. Schober.
Astron. Astrophys., Suppl. Ser., Vol. 45, 175 (1981). — See Abstr. 29.098.074.

099 Jupiter, Jupiter Satellites

099.001 Raman scattering in the Jovian atmosphere.
W. D. Cochran, L. Trafton, W. Macy, Jr.,
J. H. Woodman.
Astrophys. J., Vol. 247, 734 - 740 (1981).

Raman scattering by H_2 in the atmosphere of Jupiter is detected by a correlation technique using high-resolution spectra of Jupiter with high signal-to-noise ratio in the 3800 - 5000 Å region. The pure rotational H_2 $S(0)$ and $S(1)$ lines are detected. The ratio of the relative number of Raman scattered photons in the $S(0)$ and $S(1)$ features indicate that the H_2 in the Jovian atmosphere is in the equilibrium, rather than the normal state. Therefore some sort of nonradiative process is responsible for transitions between the ortho and para states of H_2.

099.002 Composition and thermal profiles of the Jovian upper atmosphere determined by the Voyager ultraviolet stellar occultation experiment.
M. C. Festou, S. K. Atreya, T. M. Donahue, B. R. Sandel,
D. E. Shemansky, A. L. Broadfoot.
J. Geophys. Res., Vol. 86, 5715 - 5725 (1981).

Occultation of the star Regulus − α Leo − by the Jovian atmosphere was monitored by the Voyager 2 spacecraft on July 9, 1979. The absorption recorded in the 910 - 1200 Å range was caused primarily by the H_2-Lyman and Werner bands. These data provide the first complete measurements of atmospheric density and temperature profiles between 330 and 830 km above the ammonia cloud tops.

099.003 Viscosity of high-pressure ice VI and evolution and dynamics of Ganymede.
J. P. Poirier, C. Sotin, J. Peyronneau.
Nature, Vol. 292, 225 - 227 (1981).

The authors have measured the viscosity of the high-pressure ice VI at room temperature and pressures of 1.1−1.2 GPa in a sapphire anvil cell. The low value they found for the viscosity ($\eta \simeq 10^{14}$ P) suggests that solid state convection might have taken place during the early evolution of Ganymede, thus preventing melting and differentiation.

099.004 Jupiter tail phenomena upstream from Saturn.
F. L. Scarf, W. S. Kurth, D. A. Gurnett, H. S. Bridge,
J. D. Sullivan.
Nature, Vol. 292, 585 - 586 (1981).

Voyager 2 plasma wave and plasma probe measurements from February 1981 suggest that phenomena associated with a well defined tail of Jupiter have been detected at a distance of about 6,200 R_J. This indicates that Saturn's magnetosphere will be affected by the jovian tail and that by comparing Voyager 1 and 2 observations information on the physics of Saturn's magnetosphere can be obtained.

099.005 Latitudinal beaming and local time effects in the decametre-wave radiation from Jupiter observed at the Earth and from Voyager. C. H. Barrow.
Astron. Astrophys., Vol. 101, 142 - 149 (1981).

Occurrence probabilities and frequency characteristics of the decametre-wave jovian sources have been computed from over 20 yr of Earth-based observations. These have been used as a guide for comparing short periods of simultaneous Earth-based observations with the decametric observations made by the Planetary Radio Astronomy experiment on board Voyager 1 during the 30-d periods immediately before and after encounter. The results have been assessed with respect to possible jovian local time effects and latitudinal beaming. The effects of terrestrial ionospheric absorption have been considered in relation to characteristic decametric source intensities and frequency ranges.

099.006 Jupiter in 1979 - 80: rotation periods.
P. W. Budine.
Strolling Astron., Vol. 29, 1 - 10 (1981).

099.007 Some notes on the terminology of Jupiter.
W. H. Haas.
Strolling Astron., Vol. 29, 10 - 12 (1981).

099.008 Voyagers naar Jupiter. O. Namba.
Zenit, 8. Jaarg., 284 - 290 (1981).

099.009 De dampkring van een gasreus.
Zenit, 8. Jaarg., 291 - 294 (1981).

099.010 Gejaagd door de wind.
Zenit, 8. Jaarg., 295 - 301 (1981).

099.011 De manen in kaart gebracht.
Zenit, 8. Jaarg., 302 - 303 (1981).

099.012 De kleinste manen.
Zenit, 8. Jaarg., 304 - 307 (1981).

099.013 De hel van Io.
Zenit, 8. Jaarg., 308 - 314 (1981).

099.014 De ijsmanen van Galileï.
Zenit, 8. Jaarg., 315 - 321 (1981).

099.015 Jupiter: présentation 1980. F. Jetzer.
Orion, 39. Jahrg., 116 - 118 (1981).

099.016 A survey of Io's potassium cloud.
L. Trafton.
Astrophys. J., Vol. 247, 1125 - 1140 (1981).

Io's potassium cloud exhibits spatial and temporal variations to those observed for Io's sodium cloud. Spectra from five apparitions show that the potassium cloud is elongated so that it extends forward from Io's leading, inner hemisphere and makes an angle with Io's orbit of 10°−30°, slightly less than the angle for the sodium cloud. The potassium cloud is a long-lived phenomenon which undergoes periodic fluctuations in response to solar radiation pressure and the ionizing influence of Jupiter's plasma torus. The similarities with the sodium cloud suggest that both sodium and potassium are ejected from nearly the same regions of Io by the same physical mechanism.

099.017 Voyager photometry of surface features on Ganymede and Callisto.
S. W. Squyres, J. Veverka.
Icarus, Vol. 46, 137 - 155 (1981).

The photometric properties of selected surface features on Ganymede and Callisto have been studied using Voyager images over phase angles from 10 to 124° taken with the clear filter (effective wavelength ~0.5 μm). Normal reflectances on Ganymede average 0.35 for the cratered terrain and 0.44 for the grooved terrain. The value for the ubiquitous cratered terrain on Callisto is 0.18. The photometric properties of these regions are described closely by a simple scattering function.

099.018 The topography of Ganymede's grooved terrain.
S. W. Squyres.
Icarus, Vol. 46, 156 - 168 (1981).

Using the technique of photoclinometry, topographic profiles across areas of grooved terrain and several other features on Ganymede have been constructed. The grooved terrain examined consists of subparallel grooves spaced 3 - 10 km apart. Topographic amplitudes are typically 300 - 400 m, with a maximum of about 700 m. Slopes are very gentle and tend to be primarily concave upward. Very few major positive relief features exist on Ganymede. The most important of these is a broad, gently sloping dome-shaped feature 260 km in diameter and over 2 km high.

099.019 Study of the ammonia ice cloud layer in the equatorial region of Jupiter from the infrared interferometric experiment on Voyager.
A. Marten, D. Rouan, J. P. Baluteau, D. Gautier,
B. J. Conrath, R. A. Hanel, V. Kunde, R. Samuelson,
A. Chedin, N. Scott.
Icarus, Vol. 46, 233 - 248 (1981).

Spectra from the Voyager 1 infrared interferometer spectrometer obtained near the time of closest approach to Jupiter were analyzed for the purpose of inferring ammonia cloud properties associated with the equatorial region. The observed spectra were compared with synthetic spectra computed from a radiative transfer formulation that includes multiple scattering.

099.020 Latitudinal variations in Jovian stratospheric temperature.   R. D. Cess, B. E. Carlson, J. Caldwell,
I. G. Nolt, F. C. Gillett, A. T. Tokunaga.
Icarus, Vol. 46, 249 - 255 (1981).

Ground-based observations of Jupiter show that the planet's stratospheric and tropospheric thermal emission are anticorrelated. The observations can possibly be explained by latitudinal variations in cloud altitude. These variations cause differential stratospheric heating by sunlight which is reflected off the clouds and then absorbed within the stratosphere by visible and near-infrared bands of methane.

099.021 Tidal evolution of the Galilean satellites: a linearized theory.   R. Greenberg.
Icarus, Vol. 46, 415 - 423 (1981).

The Laplace resonance among the Galilean satellites Io, Europa, and Ganymede is traditionally reduced to a pendulum-like dynamical problem by neglecting short-period variations of several orbital elements. When tidal effects are included, the resulting evolution is substantially the same as was indicated by the pendulum approach, except that evolution out of deep resonance is found to be somewhat slower than suggested by extrapolation of the pendulum results.

099.022 Erosion of Galilean satellite surfaces by Jovian magnetosphere particles.
R. E. Johnson, L. J. Lanzerotti, W. L. Brown, T. P. Armstrong.
Science, Vol. 213, 1027 - 1030 (1981).

The Galilean satellites of Jupiter - Io (J1), Europa (J2), Ganymede (J3), and Callisto (J4) - are embedded in the intense ion and electron fluxes of the Jovian magnetosphere. The effect of these particles on the icy surfaces of the outer three satellites depends on the flux and the efficiency of the sputtering of water ice by such particles. Recent laboratory measurements provided data on the erosion of water ice by energetic particles and showed that it occurs much faster than would be expected from normal sputtering theory. The Voyager spacecraft encounters with Jupiter provided the first measurements of ion fluxes (energies $\gtrsim$ 30 kiloelectron volts) in the vicinity of the Galilean satellites. Using the laboratory sputtering data together with particle measurements from the Voyager 1 low-energy charged particle experiment, the effects of erosion on the surfaces of J2 to J4 are estimated.

099.023 1979J3: discovery of a previously unknown satellite of Jupiter.   S. P. Synnott.
Science, Vol. 212, 1392 (1981).

099.024 An upper bound to the lightning flash rate in Jupiter's atmosphere.
F. L. Scarf, D. A. Gurnett, W. S. Kurth, R. R. Anderson,
R. R. Shaw.
Science, Vol. 213, 684 - 685 (1981).

099.025 Relations between turbulent regions of interplanetary magnetic field and Jovian decametric radio wave emissions from the main source.   H. Oya, A. Morioka.
Planet. Space Sci., Vol. 29, 783 - 791 (1981).

Jovian decametric radio wave emissions that were observed at Goddard Space Flight Center, U.S.A. for a period from 1 October to 31 December, 1974 and data obtained at Mt Zao Observatory, Tohoku University, Japan, for a period from 14 July to 6 December, 1975 have been used to investigate the relationship of the occurrence of the Jovian decametric radio waves, from the main source, to the geomagnetic disturbance index, ΣK_p.

099.026 Io: could SO_2 condensation/sublimation cause the sometimes reported post-eclipse brightening?
F. P. Fanale, W. B. Banerdt, D. P. Cruikshank.
Geophys. Res. Lett., Vol. 8, 625 - 628 (1981).

Io' sometimes-reported post eclipse brightening may be caused by SO_2 frost formation during eclipse and sublimation on emergence. The known properties of SO_2 can be used to explain quantitatively the amplitude of brightening and its duration, and (qualitatively) its variability.

099.027 The origin of microscopic planetary ring particles.
J. F. Bates.
Geophys. Res. Lett., Vol. 8, 835 - 836 (1981).

The microscopic planetary ring particles around Jupiter are possibly the result of internal void formation and fracture of initially larger particles. The void formation is thought to be the result of high electron and proton fluences attained by the particles during their lifetimes in orbit. The constant irradiation produces defect species which coalesce to form voids, which provide fracture paths for the particles.

099.028 Sunlight absorption by aerosols in Jupiter's upper atmosphere.   R. A. West.
Geophys. Res. Lett., Vol. 8, 847 - 849 (1981).

The amount of sunlight deposited in the Jovian upper atmosphere is estimated from reflectivity measurements at 2400 Å by the Voyager 2 Photopolarimeter experiment and at visible and near-IR wavelengths observed by Pioneer 10 and ground-based instruments. Zero to about one percent of the incident energy is absorbed at altitudes above the 100 mbar level in models with mean values for haze optical depth and single scattering albedo. Several percent of the incident energy could be absorbed, if limiting values are used, and if an additional absorbing layer is incorporated below the high altitude haze in Pioneer models. Maximum absorption occurs at the edge of the polar regions near ±65° latitude.

099.029 Compression of Jupiter's magnetosphere by the solar wind: reexamination via MHD simulation of evolving corotating interaction regions.   Z. K. Smith,
M. Dryer, R. W. Fillius, E. J. Smith, J. H. Wolfe.
J. Geophys. Res., Vol. 86, 6773 - 6780 (1981).

The authors examine the major changes in the solar wind before, during, and after the Pioneer 10 and 11 encounters with the Jovian magnetosphere during 1973 and 1974, respectively.

099.030 Sputter ejection of matter from Io.
P. K. Haff, C. C. Watson, Y. L. Yung.
J. Geophys. Res., Vol. 86, 6933 - 6938 (1981).

The direct collisional interaction of magnetospheric particles with Io will lead to sputtering of atoms and molecules from the satellite into circum-Jovian space. The authors consider the possible efficiencies of sputtering processes for several models of the interaction of the particle flux with Io.

099.031 The Great Red Spot as synoptic vortex in the Jovian atmosphere.
R. Z. Sagdeev, V. D. Shapiro, V. I. Shevchenko.
Pis'ma Astron. Zh., Tom 7, 505 - 509 (1981). In Russian.
English translation in Soviet Astron. Lett., Vol. 7.

A solution of hydrodynamic equations for an ideal incompressible fluid is obtained in the form of a Rossby solitary wave developing in a zonal plane flow with shear of velocity. A qualitative agreement between the obtained solution and characteristics of the Great Red Spot on Jupiter is shown.

099.032 Limits on large-crater production and obliteration on Callisto. A. Woronow, R. G. Strom.
Geophys. Res. Lett., Vol. 8, 891 - 894 (1981).

By comparing results of Monte Carlo simulations of the crater population on Callisto with the observed surface, the authors demonstrate that the relative dearth of large craters on Ganymede and Callisto, compared with the terrestrial planets, can not be totally ascribed either to craters relaxing or to craters piercing a thin icy crust. Consequently, the population of objects responsible for the heavy bombardment of the jovian system differed markedly from that responsible for the late heavy bombardment of the terrestrial planets.

099.033 The moons of Jupiter.
L. Rasmussen.
Fys. Tidsskr., Vol. 78, 119 - 140 (1980). In Swedish. − Abstr. in Phys. Abstr., Vol. 84, Abstr. 66865 (1981).

099.034 The satellites of Jupiter and Saturn.
P. Masson, A. Brahic.
Recherche, No. 120, p. 290 - 303 (1981). In French. − Abstr. in Phys. Abstr., Vol. 84, Abstr. 71119 (1981).

099.035 The Jovian magnetosphere: a post-Voyager view.
T. W. Hill.
EOS Trans. American Geophys. Union, Vol. 62, No. 3, p. 25 - 27 (1981). − Abstr. in Phys. Abstr., Vol. 84, Abstr. 75038 (1981).

099.036 Structure of the source of jovian decametric emission and interplanetary scintillation.
F. Genova, A. Boischot.
Nature, Vol. 293, 382 -383 (1981).

Previous theories have assumed that the source emission takes place along magnetic field lines, close to the local gyrofrequency. The authors propose a method to test this hypothesis by studying interplanetary scintillations which modulate the emission when received on Earth, and quantify the predicted effects. Preliminary results indicate that the emission is probably spatially distributed and could occur along field lines, and that Io controlled A and B sources are on opposite sides of Jupiter.

099.037 Jupiter and Galilean satellites' positions obtained in December 1978 at Uccle with the double astrograph of 40 cm (f = 2 m). H. Debehogne, L. E. Machado.
Astron. Astrophys., Suppl. Ser., Vol. 45, 183 - 185 (1981).

In December 1978, the authors observed Jupiter and the Galilean satellites at the double astrograph (f = 2 m, D = 40 cm) of Uccle. Measures and reductions were performed at the Royal Observatory of Belgium with the Ascorecord measuring machine (0.1 μ) and by means of five reference stars on the computer Burrough at the Universidade Federal do Rio de Janeiro. The SAO Catalogue, the Least Squares and Dependences Methods were used.

099.038 Spectrometry of Jupiter at selected locations on the disk during the 1979 apparition.
A. L. Cochran, L. M. Trafton, W. D. Cochran, E. S. Barker.
Astron. J., Vol. 86, 1101 - 1107 (1981).

The authors measured Jupiter's reflectivity as a function of wavelength between 3000 and 10500 Å on 26 February and 1 March 1979 in order to permit a calibration of the Voyager images and to establish differences from previous apparitions. The observations were taken in the prominent belts and zones and both polar caps along the central meridian using a slit of dimensions 2.33 × 2.5 arcsec². The authors measured the equivalent widths of the 6190- and 7270-Å CH_4 bands and found significantly less CH_4 absorption in the north tropical zone than occurred during the 1976 apparition.

099.039 Sequential development of grooved terrain and polygons on Ganymede.
M. P. Golombek, M. L. Allison.
Geophys. Res. Lett., Vol. 8, 1139 - 1142 (1981).

Grooves on Ganymede are analyzed in a manner similar to the way a structural geologist analyzes brittle fracture features on the earth. By observing simple terminating (or "T") intersections and crosscutting relationships, the relative ages of the two intersecting grooves or groove sets can be determined. These straightforward determinations of relative age are non-genetic, assuming only that grooves are brittle fracture features. The analysis reveals that grooved terrain formed by a step-by-step process; furthermore, each step can be observed because the process stopped at different stages in various places on the surface on Ganymede.

099.040 Spatial imaging of UV emission from Jupiter and Saturn. J. T. Clarke, H. W. Moos.
The Universe at ultraviolet wavelengths, (see 012.009), p. 39 - 44 (1981).

099.041 Observations of polar aurora on Jupiter.
J. T. Clarke, H. W. Moos, S. K. Atreya, A. L. Lane.
The Universe at ultraviolet wavelengths, (see 012.009), p. 45 - 48 (1981).

North-south spatial maps of Jupiter were obtained with the SWP camera in IUE observations on 10 December 1978, 19 May 1979, and 7 June 1979. Bright auroral emissions were detected from the north and south polar regions at H Ly α (1216 Å) and in the H_2 Lyman bands (1250−1608 Å) on 19 May 1979; yet no enhanced polar emission was detected on the other days. The relationship between the IUE observing geometry and the geometry of the Jovian magnetosphere are discussed.

099.042 Observations of the Io plasma torus.
H. W. Moos, J. T. Clarke, S. K. Atreya, A. L. Lane.
The Universe at ultraviolet wavelengths, (see 012.009), p. 49 - 53 (1981).

The short wavelength spectrograph on the IUE satellite has been used to obtain spectra of the plasma torus near the orbit of Io about Jupiter. Three exposures of about 8 hours each taken in March and May 1979 show emission features due to S II, S III, and O III. The absence of features at other wavelengths permits upper limits to be set on other species in the torus.

099.043 Observation of outer planets at Lyman alpha.
J. Darius, K. H. Fricke.
The Universe at ultraviolet wavelengths, (see 012.009), p. 85 - 88 (1981).

099.044 On the origin of the Great Red Spot of Jupiter.
B. I. Luchkov.
Pis'ma Astron. Zh., Tom 7, 566 - 569 (1981). In Russian.
English translation in Soviet Astron. Lett., Vol. 7.

The origin of the Great Red Spot (GRS) is connected with the structure of the magnetic field and the radiation belt of Jupiter. A comparison of the GRS with the earth field negative anomaly (Brasil anomaly) is given. It is qualitatively shown that the GRS arised in the region of a jovian magnetic

field anomaly and is constantly supplied by high energy particle precipitation from the radiation belt.

**099.045 Identification of radio emission from the Io flux
tube.**   A. C. Riddle.
Bull. American Astron. Soc., Vol. 13, 510 (1981). – Abstract.

099.046 Studies of Io's sodium cloud.
    B. A. Goldberg, R. W. Carlson, T. V. Johnson,
S. K. LaVoie, D. L. Matson, J. W. Young.
Bull. American Astron. Soc., Vol. 13, 511 (1981). – Abstract.

099.047 A new type of planetary nebula: the Io torus.
    C. B. Pilcher.
Bull. American Astron. Soc., Vol. 13, 525 (1981). – Abstract.

099.048 Jupiter's and Saturn's atmospheres.
    A. P. Ingersoll.
Bull. American Astron. Soc., Vol. 13, 525 (1981). – Abstract.

**099.049 [S III] and [S II] observations of the Jupiter plasma
torus.**   R. J. Oliversen, F. L. Roesler,
F. Scherb, J. M. Lattis, K. P. Jaehnig.
Bull. American Astron. Soc., Vol. 13, 555 (1981). – Abstract

**099.050 Preliminary analysis of mutual satellite events in
1979/80.**   K. Aksnes, F. A. Franklin.
Bull. American Astron. Soc., Vol. 13, 572 (1981). – Abstract.

**099.051 Vertical distribution of NH_3 in the upper jovian
atmosphere from IUE observations.**
M. Combes, T. Encrenaz, K. H. Fricke.
Second European IUE Conference, (see 012.011), p. 7 - 9
(1980).
    Ammonia has been identified in the upper atmosphere of
Jupiter from the observation of individual NH_3 bands between
2000 and 2300 Å, on an IUE high resolution spectrum. The
shape of these NH_3 bands implies that the ammonia distribution has to be strongly reduced by photolysis in the upper
jovian atmosphere.

099.052 Monografías sobre planetas: Júpiter.
    A. J. Camponovo.
Rev. Astron., Vol. 53, No. 217, p. 7 - 16 (1981).

099.053 The tides of Io.   C. F. Yoder, S. J. Peale.
Icarus, Vol. 47, 1 - 35 (1981).
    The Galilean satellites Io, Europa, and Ganymede interact
through several stable orbital resonances. A theory of origin
and subsequent evolution of these resonances is described in
detail. From an initially quasi-random distribution of the orbits
the resonances are assembled through differential tidal expansion of the orbits. Io is driven out most rapidly and the
first two resonance variables above are captured into libration
about 0 and 180° respectively. The third resonance variable
and simultaneously the Laplace angle are captured into libration with probability ~ 0.9. The tidal dissipation in Io is vital
for the rapid damping of the libration amplitudes and for the
establishment of a quasi-stationary orbital configuration. A
generalized Hamiltonian theory of the resonances in which
third-order terms in eccentricity are retained is developed to
evaluate the hypothesis that the resonances were of primordial
origin.

099.054 Io: thermal models and chemical evolution.
    G. J. Consolmagno.
Icarus, Vol. 47, 36 - 45 (1981).
    A combined thermal and chemical evolution model of Io
is presented, outlining limits on the possible starting materials,
heating history, chemical history, and present state of Io. The
best scenario starts with Io being accreted from material in a
proto-Jovian nebula which condensed between 400 - 600° K.

Radionuclides and tidal heating would lead to large-scale convection within Io and chemical reactions leading to the outgassing of water and methane. Reactions between Fe^0 - FeS
and water, at least near the surface, go to completion, resulting in all Fe being oxidized with elemental sulfur producing a
low-conductivity crust. In the deep interior, these reactions
may not completely exhaust Fe metal, and an FeS-rich core
may be formed.

099.055 Internal structures of the Galilean satellites.
    G. Schubert, D. J. Stevenson, K. Ellsworth.
Icarus, Vol. 47, 46 - 59 (1981).
    New models for the interiors of Io, Ganymede, and
Callisto are proposed. The model of Io consists of a thin, high-
rigidity outer layer separated from a solid interior by a thin,
molten or partially molten shell. The model of Ganymede consists of an ice outer layer, a shell of undifferentiated, primordial ice-silicate mixture, and a rock core. Callisto's heavily
cratered surface strongly suggests that relatively little, if any,
ice-rock differentiation has occurred in its interior.

099.056 Voyager search for posteclipse brightening on Io.
    J. Veverka, D. Simonelli, P. Thomas, D. Morrison,
T. V. Johnson.
Icarus, Vol. 47, 60 - 74 (1981).
    Observations of three eclipse reappearances of Io were
made during the two Voyager encounters. No posteclipse
brightening of the type reported by some Earth-based observers – a brightening by some 10% just after eclipse which
gradually disappears on a time scale of 10 to 15 min – was
detected.

099.057 Effects of Io ejecta on Europa.
    A. Eviatar, G. L. Siscoe, T. V. Johnson, D. L. Matson.
Icarus, Vol. 47, 75 - 83 (1981).
    The authors examine the effects of Io ejecta on the sur-
face and environment of Europa. They find that the observed
sulfur on the trailing side of Europa, when interpreted as a
deposit in equilibrium between implanation of, and sputtering
by, corotating Io ejecta, implies a slow loss of material from
Europa by sputtering. From this the authors infer that the
spectrum of particles sputtered from water ice is soft. The
quantity of observed sulfur and its confinement to the trailing
hemisphere appear to exclude significant implantation and
sputtering by energetic heavy ions. The authors also conclude
that the contribution from Europa to the magnetospheric
plasma (even at Europa itself) is negligible compared to the
matter ejected from Io.

**099.058 Viscous relaxation of impact craters on icy planetary
surfaces: determination of viscosity variation with
depth.**   E. M. Parmentier, J. W. Head.
Icarus, Vol. 47, 100 - 111 (1981).
    Spacecraft images show that the icy Galilean satellites
have surfaces with very low topographic relief. Impact craters
on Ganymede and Callisto are anomalously shallow and are
characterized by sharp well-defined rims and domed floors.
These morphological characteristics can be explained by viscous relaxation of topography on an icy crust in which the
viscosity is uniform or decreases with depth. Under these conditions, large craters relax more rapidly than small craters,
therefore explaining a possible underabundance of large
craters. Viscous relaxation on an icy crust that is thin compared to the crater diameter or on a thick icy crust in which
viscosity increases with depth could not produce this crater
morphology and would result in the more rapid relaxation of
small craters rather than large craters.

**099.059 Thermal evolution of Ganymede and Callisto:
effects of solid-state convection and constraints
from Voyager imagery.**
    C. H. Thurber, A. T. Hsui, M. N. Toksöz.

Proc. Eleventh Lunar Planet. Sci. Conf., (see 012.020), p. 1957 - 1977 (1980).

099.060 Fractures on Europa: possible response of an ice crust to tidal deformation.
P. Helfenstein, E. M. Parmentier.
Proc. Eleventh Lunar Planet. Sci. Conf., (see 012.020), p. 1987 - 1998 (1980).

The surface of Europa contains a planetwide system of low albedo lineaments which have been interpreted as fractures in an icy crust. The pattern of fractures on the surface consists of radial and concentric fractures having the general appearance of tension cracks within a region near the antipode of the sub-Jupiter point. The orientation of this pattern on the surface suggests that a principal axis of the deformation that produced the fractures was approximately radial to Jupiter. Fracturing may thus be consistent with an origin due to cyclical tidal deformation resulting from orbital eccentricity.

099.061 Atomic clouds as distributed sources for the Io plasma torus.   R. A. Brown, W.-H. Ip.
Science, Vol. 213, 1493 - 1495 (1981).

Several recent developments have implications for the neutral particle environment of Jupiter. Very hot sulfur ions have been detected in the Io torus with gyrospeeds comparable to the corotation speed, a phenomenon that would result from a neutral sulfur cloud. Current evidence supports the hypothesis that extensive neutral clouds of oxygen and sulfur exist in the Jupiter magnetosphere and that they are important sources of ions and energy for the Io torus.

099.062 The present picture of Io's electrodynamic coupling with the magnetosphere of Jupiter.
M. Dobrowolny.
Mem. Soc. Astron. Italiana, Vol. 52, (see 012.023), 397 - 398 (1981).

099.063 La planète Jupiter en 1977 - 1978.
R. Néel.
Astronomie, Vol. 95, 423 - 434 (1981).

099.064 A catalogue of Jupiter's decametric emission observed by Voyager-1 and by Voyager-2 in the range 15-40 MHz.   C. H. Barrow.
Astron. Astrophys.,Suppl. Ser., Vol. 46, 111 - 114 (1981).

The catalogue lists Jupiter's decametric emissions recorded by the Voyager Planetary Radio Astronomy Experiment during periods in 1979 adjacent to each encounter.

099.065 Positions of Jupiter, Galilean satellites and Pluto obtained in May 1980 with GPO of the ESO, La Silla.   H. Debehogne, L. E. Machado, J. F. Caldeira, G. G. Vieira, E. R. Netto.
Astron. Astrophys.,Suppl. Ser., Vol. 46, 131 - 134 (1981).

099.066 Zonal spectrophotometric characteristics of Jupiter's cloud cover.
V. G. Tejfel', G. A. Kharitonova, G. I. Khudyaeva.
Astron. Vestn., Tom 15, 95 - 104 (1981).In Russian.

The latitudinal and longitudinal colour effects variations of Jupiter's cloud complexes were studied using central meridian spectrograms of Jupiter made in 1979.

099.067 The surfaces of Jupiter satellites.
Zh. F. Rodionova.
Zemlya Vselennaya, 1981, No. 5, p. 47 - 50. In Russian.

099.068 Encounters with Jupiter: the low energy charged particle results of Voyager.
J. F. Carbary, S. M. Krimigis.
APL Tech. Dig., Vol. 1, 60 - 63 (1980).

099.069 The surprising world of Jupiter's moon Io.
U. Dzērvītis.
Zvaigžnotā debess, 1980. gada vasara, p. 12 - 14. In Latvian.

099.070 Nitrogen on Jupiter: a deep atmospheric source.
R. G. Prinn, E. P. Olaguer.
J. Geophys. Res., Vol. 86, 9895 - 9899 (1981).

A study of irreversible reactions involving molecular nitrogen on Jupiter indicates that vertical motions are sufficiently rapid in the deep atmosphere to transport large amounts of N_2 from the $900° - 1700°K$ levels where it is stable up to the cold visible regions. Both homogeneous gas-phase and heterogeneous iron-catalyzed reactions between N_2 and H_2 were considered. The authors predict N_2 mixing ratios of 0.6 - 3 ppmv if catalysis is effective and up to 10 ppmv if it is not. Thus N_2 may be the most abundant nonequilibrium species in Jupiter's troposphere and potentially detectable by the neutral mass spectrometer which will be on board the 1986 Galileo Entry Probe.

099.071 Further observational support for the limited-latitude magnetodisc model of the outer Jovian magnetosphere.   M. F. Thomsen, C. K. Goertz.
J. Geophys. Res., Vol. 86, 7519 - 7526 (1981).

The distinction between the magnetic anomaly model of the outer Jovian magnetosphere (Dessler and Vasyliunas, 1979; Vasyliunas and Dessler, 1981) and the magnetodisc model (as, for example, discussed most recently by Goertz (1981)) is reviewed, and a further observational comparison of the two models is pointed out. The comparison involves the latitudinal variation of the intensity of energetic charged particles confined near the magnetic equatorial plane.

099.072 The magnetic field of Jupiter: a generalized inverse approach.   J. E. P. Connerney.
J. Geophys. Res., Vol. 86, 7679 - 7693 (1981).

The estimation of planetary magnetic fields from observations of the magnetic field gathered along a spacecraft flyby trajectory is examined with the aid of generalized inverse techniques, with application to the internal magnetic field of Jupiter. Model non-uniqueness resulting from the limited spatial extent of the observations and noise on the data is explored and quantitative estimates of the model parameter resolution are found. The presence of a substantial magnetic field of external origin due to the currents flowing in the Jovian magnetodisc is found to be an important source of error in estimates of the internal Jovian field, and new models explicitly incorporating these currents are proposed. New internal field models are derived using the vector helium magnetometer observations and the high field fluxgate observations of Pioneer 11, and knowledge of the external current system gained from the Pioneer 10 and Voyagers 1 and 2 encounters.

099.073 Studies of Jupiter's lower ionospheric layers.
R. H. Chen.
J. Geophys. Res., Vol. 86, 7792 - 7794 (1981).

Theoretical calculations are made of the fluxes of heavy ions necessary to form the layers of enhanced ionization observed in the lower ionosphere of Jupiter. The results show that if the layers are formed from sodium or sulfur ions of Galilean satellite origin injected into the Jovian atmosphere then, in order to match the Pioneer 10 observations of the L_6 layer, the Na^+ flux must be about 3×10^4 cm^{-2} s^{-1} and the S^+ flux must be about 4×10^3 cm^{-2}s^{-1}. From the shape of the ionization layers estimates of ion drift velocities and neutral wind speeds may be made.

099.074 Comment on 'Azimuthal magnetic field at Jupiter' by J. L. Parish, C. K. Goertz, and M. F. Thomsen.
J. E. P. Connerney, with a reply by M. F. Thomsen, C. K. Goertz.

J. Geophys. Res., Vol. 86, 7796 - 7798 (1981). − See Abstr. 28.099.016.

099.075 Images of Jupiter from the Pioneer 10 and Pioneer 11 infrared radiometers: a comparison with visible and 5-μm images.

G. S. Orton, A. P. Ingersoll, R. J. Terrile, S. R. Walton.
Icarus, Vol. 47, 145 - 158 (1981).

All of the data acquired at Jupiter by the infrared radiometers on board Pioneers 10 and 11 are presented in the form of images with geometric control. The images are compared with 5-μm and visible images taken in the same time frame. The association of dark (blue or brown) and light (white or red) areas with warm and cool areas (at 5, 20, and 45 μm) respectively, extends to nearly all features observed on the planet.

099.076 Éclipse annulaire d'Io par Europe.

J.-E. Arlot.
Astronomie, Vol. 95, 488 (1981).

099.077 Energetic particle events (⩾30 keV) of Jovian origin observed by Voyager 1 and 2 in interplanetary space.

R. D. Zwickl, S. M. Krimigis, J. F. Carbary, E. P. Keath, T. P. Armstrong, D. C. Hamilton, G. Gloeckler.
J. Geophys. Res., Vol. 86, 8125 - 8140 (1981).

Short-lived and long-lived ion flux increases ($E \geqslant 30$ keV) of Jovian origin have been observed by the low energy charged particle instrument on the Voyager 1 and 2 spacecraft. The short-lived events are observed more than $860\,R_J$ upstream and more than $1500\,R_J$ downstream of Jupiter. Observations of long-lived events appear to be confined to $\lesssim 200\,R_J$ upstream of Jupiter. The short-lived events last from a few minutes to a couple of hours, while the long-lived events last from 8 to 21 hours. Both types of events have sharp onsets and decays, are usually confined to energies below 1 MeV total energy, and show a large general enrichment of $Z \geqslant 6$ particles relative to proton and helium particles when compared with energetic particle events of solar or interplanetary origin. Many of the events have a noticeable peak in the energy spectrum above 100 keV after the main portion of the event. The authors conclude that a significant fraction of the particles observed during Jovian ion events originate from within the magnetosphere of Jupiter and simply leak out into the magnetosheath. If the interplanetary magnetic field favorably connects to the bow shock, particles can leak out into the interplanetary medium.

099.078 Observations of the magnetic field and plasma flow in Jupiter's magnetosheath.

R. P. Lepping, L. F. Burlaga, L. W. Klein, J. M. Jessen, C. C. Goodrich.
J. Geophys. Res., Vol. 86, 8141 - 8155 (1981).

The paper presents a comprehensive description of the Jovian magnetosheath magnetic fields and speculates briefly on possible causes of these phenomena. The authors emphasize their Voyager 1 and 2 magnetic field observations and their relations to the plasma observations, but they also show that the same phenomena are present in the Pioneer 10 magnetic field data.

099.079 A survey of the plasma electron environment of Jupiter: a view from Voyager.

J. D. Scudder, E. C. Sittler, Jr., H. S. Bridge.
J. Geophys. Res., Vol. 86, 8157 - 8179 (1981).

A survey of the plasma environment within Jupiter's bow shock is presented in terms of the in situ, calibrated electron plasma measurements made between 10 eV and 5.95 keV by the Voyager Plasma Science Experiment. These measurements have been analyzed and corrected for spacecraft potential variations; the data have been reduced to nearly model independent macroscopic parameters of the local electron density and temperature.

099.080 Measurements of plasma wave spectra in Jupiter's magnetosphere.

F. L. Scarf, D. A. Gurnett, W. S. Kurth.
J. Geophys. Res., Vol. 86, 8181 - 8198 (1981).

The authors present compressed plots of E field averages for all of the 16-channel spectrum analyzer data from the Voyager 1 and 2 magnetosphere traversals to provide an overall framework for the discussion. The authors illustrate the importance of considering peaks as well as averages by using 16-channel measurements from the first inbound and last outbound bow shock for Voyager 2. They also present selected wideband measurements from the waveform receivers to demonstrate how many important wave bursts are variable in times less than or comparable to the 4-s scan period of the 16-channel analyzer. In addition, the authors show how the continuous frequency coverage of the waveform data link provides extremely valuable information on the complex spectra of Jovian plasma waves.

099.081 Determination of Jupiter's electron density profile from plasma wave observations.

D. A. Gurnett, F. L. Scarf, W. S. Kurth, R. R. Shaw, R. L. Poynter.
J. Geophys. Res., Vol. 86, 8199 - 8212 (1981).

The paper summarizes the electron density measurements obtained in the Jovian magnetosphere from the plasma wave instruments on the Voyager 1 and 2 spacecraft. Three basic techniques are discussed for determining the electron density: (1) local measurements from the low-frequency cutoff of continuum radiation, (2) local measurements from the frequency of upper hybrid resonance emissions, and (3) integral measurements from the dispersion of whistlers. The limitations and advantages of each technique are critically reviewed.

099.082 Radio Jupiter after Voyager: an overview of the planetary radio astronomy observations.

A. Boischot, A. Lecacheux, M. L. Kaiser, M. D. Desch, J. K. Alexander, J. W. Warwick.
J. Geophys. Res., Vol. 86, 8213 - 8226 (1981).

The authors present an overview of Jupiter's low-frequency radio emission morphology as observed by the planetary radio astronomy (PRA) instrument onboard the Voyager spacecraft. The PRA measurement capabilities and limitations are summarized. As a direct consequence of the PRA spacecraft observations at least three previously-unrecognized emission components have been discovered: broadband and narrow-band kilometric emission and the lesser-arc decametric emission. Their properties are reviewed. In addition, the fundamental structure of the decameter wavelength and hectometer wavelength emission, which is now believed to be almost exclusively in the form of complex but repeating arc structures in the frequency-time domain, is described. Dramatic changes in the emission morphology of some components as a function of the sun-Jupiter-spacecraft angle (local time) are described. Finally, the PRA in situ measurements of the Io plasma torus hot-to-cold electron density and temperature ratios are summarized.

099.083 Characteristics of hot plasma in the Jovian magnetosphere: results from the Voyager spacecraft.

S. M. Krimigis, J. F. Carbary, E. P. Keath, C. O. Bostrom, W. I. Axford, G. Gloeckler, L. J. Lanzerotti, T. P. Armstrong.
J. Geophys. Res., Vol. 86, 8227 - 8257 (1981).

The authors report and interpret measurements of the intensities, energy spectra, angular variations, and composition characteristics of the low-energy ion populations (~30 keV to ~4 MeV) obtained by both Voyager spacecraft in the outer ($\gtrsim 10\,R_J$) Jovian magnetosphere. Some of the energetic electron measurements will also be shown. By using the spectral and angular ion measurements, the authors construct density and pressure profiles in the magnetosphere and compare these with results reported by the plasma wave and plasma science investigations (density), and the magnetic field investigation (pressure).

099.084 Overview of the Voyager ultraviolet spectrometry results through Jupiter encounter.
A. L. Broadfoot, B. R. Sandel, D. E. Shemansky, J. C. McConnell, G. R. Smith, J. B. Holberg, S. K. Atreya, T. M. Donahue, D. F. Strobel, J. L. Bertaux.
J. Geophys. Res., Vol. 86, 8259 - 8284 (1981).

The Voyager ultraviolet spectrometers (UVS) have been making almost continuous observations, in the 500-Å to 1700-Å wavelength range, of sources in the solar system and galaxy since launch in 1977. On Jupiter's disc, the emissions detected to date are H Lyman α, H Lyman β, He (584 Å), and the H_2 Lyman and Werner bands. Jupiter's auroral region is clearly delineated by intense emissions of H and H_2 bands on both dayside and nightside of the planet. Emission from He is also present in the auroral regions. At Jupiter, the atmosphere was also probed by means of solar and stellar occultation experiments. Strong EUV emission from a plasma torus at the orbit of Io has been observed in transitions of sulfur and oxygen ions with a possible small contribution from potassium. The appearance of the entire sky in the outer solar system has been mapped in the emission lines of He (584 Å) and H Lyα (1216 Å) arising from resonant scattering of the solar lines by neutral interstellar hydrogen and helium entering the solar system. Diffuse galactic EUV emission has been measured in a number of selected directions. Stellar photospheric emissions shortward of the Lyman limit of atomic hydrogen at 912 Å have been measured. Finally, spectral images, in several emission lines, have been obtained of the Cygnus Loop supernova remnant.

099.085 Ion anisotropies in the outer Jovian magnetosphere.
J. F. Carbary, S. M. Krimigis, E. P. Keath, G. Gloeckler, W. I. Axford, T. P. Armstrong.
J. Geophys. Res., Vol. 86, 8285 - 8299 (1981).

The authors present results from the Voyager 1 and 2 low-energy charged particle measurements of ion anisotropies in the outer Jovian magnetosphere ($R \gtrsim 20\,R_J$). These anisotropies represent the first observed from an instrument rotating in the spin plane of Jupiter.

099.086 Composition of nonthermal ions in the Jovian magnetosphere.
D. C. Hamilton, G. Gloeckler, S. M. Krimigis, L. J. Lanzerotti.
J. Geophys. Res., Vol. 86, 8301 - 8318 (1981).

The paper discusses in some detail several aspects of the nonthermal ion composition in the Jovian magnetosphere. The authors include energy spectra, abundance ratios and their variation with radial distance, the relationship of the abundance ratio changes to spectral changes, as well as the characteristics of flux increases of particles which have apparently escaped from the nightside magnetosphere.

099.087 Positive ion observations in the middle magnetosphere of Jupiter.
R. L. McNutt, Jr., J. W. Belcher, H. S. Bridge.
J. Geophys. Res., Vol. 86, 8319 - 8342 (1981).

The authors consider the positive ion data gathered by the Voyager Plasma Science experiment in the middle magnetosphere of Jupiter. The experiment measures positive ions with energies per charge between 10 and 5950 V. The observations are analyzed to obtain the mass and charge densities, velocity components, and temperatures of the low-energy plasma population. The reduced data set is discussed in the context of the outstanding questions concerning this plasma population and its dynamics.

099.088 Low-energy charged particle observations in the 5-20 R_J region of the Jovian magnetosphere.
T. P. Armstrong, M. T. Paonessa, S. T. Brandon, S. M. Krimigis,

L. J. Lanzerotti,
J. Geophys. Res., Vol. 86, 8343 - 8355 (1981).

Ion (>0.5 MeV) and electron (> 30 keV) measurements made by the low-energy charged particle instrument during the Voyager 1 and 2 traversals of the 5-20 R_J region of the Jovian magnetosphere are presented. The spatial morphology of particle intensities, energy spectra, and composition is emphasized. Diffusive radial transport is also discussed.

099.089 Broadband electrostatic noise and field-aligned currents in Jupiter's middle magnetosphere.
D. D. Barbosa, F. L. Scarf, W. S. Kurth, D. A. Gurnett.
J. Geophys. Res., Vol. 86, 8357 - 8369 (1981).

Voyager 1 plasma wave observations have revealed the presence of an impulsive electrostatic emission localized to the Jovian middle magnetosphere $10 < R < 30\,R_J$ that appears on the edges of the plasma sheet. This plasma mode has the same spectral and morphological characteristics of an emission that has been extensively studied in the earth's magnetosphere and has been associated with the presence of field-aligned currents. The authors present the results of a detailed study of the properties of this Jovian emission by using comparisons with terrestrial observations as a basis for mode identification. The occurrence regions of the waves are compared with the measured magnetic field configuration to establish a correspondence with the plasma sheet.

099.090 Modeling the Jovian current sheet and inner magnetosphere.
J. E. P. Connerney, M. H. Acuña, N. F. Ness.
J. Geophys. Res., Vol. 86, 8370 - 8384 (1981).

Voyager 1 and 2 magnetic field observations confirm and extend the earlier Pioneer 10 detection of the Jovian magnetodisc, a region of enhanced charged particles and plasma and reduced magnetic field intensity located near the magnetic equatorial plane. Modeling of the azimuthal current sheet by a finite thickness annulus of inner radius 5 R_J, 5 R_J thickness, and extending to ~50 R_J provides detailed fits of the vector magnetic field perturbations observed in relation to the planetary field for distances less than 30 R_J. Field line geometry is also investigated, and better insight into the phenomena of charged particle absorption by the Galilean satellites is obtained which provides improved explanations of observed effects due to Ganymede.

099.091 The Jovian magnetotail and its current sheet.
K. W. Behannon, L. F. Burlaga, N. F. Ness.
J. Geophys. Res., Vol. 86, 8385 - 8401 (1981).

The authors present a summary of the results of analysis of the magnetic field measurements in Jupiter's magnetic tail and in the region of the tail current sheet by both Voyager 1 and 2.

099.092 Voyager observations of Jupiter's distant magnetotail.
W. S. Kurth, D. A. Gurnett, F. L. Scarf, R. L. Poynter, J. D. Sullivan.
J. Geophys. Res., Vol. 86, 8402 - 8412 (1981).

Observations of nonthermal continuum radiation by Voyager 1 and 2 at large distances from Jupiter have led to the identification of brief encounters with the Jovian magnetosphere at distances greater than 700 R_J and in directions substantially far from the Jupiter-sun line. In addition, a number of examples of continuum radiation apparently trapped in local density depressions in the solar wind are observed. Simultaneous measurements by the Voyager plasma instruments have verified the distant magnetotail crossings and are used to correlate the occurrence of trapped continuum radiation events within solar wind density rarefactions. The Voyager observations of the distant Jovian magnetotail are compared with observations in the distant terrestrial magnetosphere and also with observations of the plasma tails of comets.

099.093 **Energetic particles in the predawn magnetotail of Jupiter.**
A. W. Schardt, F. B. McDonald, J. H. Trainor.
J. Geophys. Res., Vol. 86, 8413 - 8428 (1981).

A detailed account is given of the energetic electron and proton populations as observed with Voyager 1 and 2 during their passes through the dawn magnetotail of Jupiter. In general, the results of the Pioneer 10 pass at 90° to the Jupiter-sun line have been confirmed and extended.

099.094 **The orientation and motion of the predawn current sheet and Jupiter's magnetotail.**   C. K. Goertz.
J. Geophys. Res., Vol. 86, 8429 - 8434 (1981).

From the observed times when Pioneer 10 and Voyager 1 observe particle flux maxima and minima and when the magnetic field strength is a minimum the author derives a geometric description of the current sheet in Jupiter's magnetosphere.

099.095 **The magnetic-anomaly model of the Jovian magnetosphere: a post-Voyager assessment.**
V. M. Vasyliunas, A. J. Dessler.
J. Geophys. Res., Vol. 86, 8435 - 8446 (1981).

The authors reexamine the three predictions that they previously put forth as tests for the magnetic-anomaly model (in which the anomalously weak magnetic field region in the northern hemisphere of Jupiter influences the outer Jovian magnetosphere by one or more plasma interaction processes), taking into account the Voyager and other recent observations.

099.096 **Direct plasma measurements in the Io torus and inner magnetosphere of Jupiter.**
F. Bagenal, J. D. Sullivan.
J. Geophys. Res., Vol. 86, 8447 - 8466 (1981).

The dominant feature of the inner magnetosphere of Jupiter is the dense torus of plasma associated with the satellite Io. The paper discusses the details of positive ion measurements made in the inner magnetosphere of Jupiter; the analysis of those measurements to obtain plasma composition, flow speeds, and temperatures; and the assumptions made in the analysis. These results for the positive ions are then combined with the direct measurements of plasma electrons between 5.7 and 9 R_J and with a theoretical distribution of plasma along dipolar magnetic field lines to construct a two-dimensional model of the plasma torus. The implications of the model about the nature of the plasma source at Io and about subsequent transport are discussed. Finally, in a preliminary investigation of the interaction of Io with the plasma torus, the speed of propagation for Alfven waves near Io has been calculated.

099.097 **Dispersive Doppler measurement of the electron content of the torus of Io.**   G. S. Levy,
D. W. Green, H. N. Royden, G. E. Wood, G. L. Tyler.
J. Geophys. Res., Vol. 86, 8467 - 8470 (1981).

As Voyager 1 made its swing-by of Io, it passed through and behind the satellite's plasma torus. The phase paths of the coherent 13 cm and 3.6-cm wavelength signals transmitted from the spacecraft were shortened differentially by the plasma, resulting in the observation of a dispersive Doppler signature. Ray path integrations through three different models of the electron distribution of the torus of Io have been performed earlier. The results of the integrations are compared with the dispersive Doppler data.

099.098 **Centrifugally driven diffusion of Iogenic plasma.**
G. L. Siscoe, D. Summers.
J. Geophys. Res., Vol. 86, 8471 - 8479 (1981).

The plasma distribution around Io as measured by Voyager 1 displays an asymmetric discontinuity at Io's orbit that has been suggested to be the signature of centrifugally driven interchange diffusion fed by plasma derived from Io. The authors explore this hypothesis further and find it to be valid. The particular form for the diffusion coefficient appropriate to centrifugally driven turbulence is derived. The nonlinear character of this kind of diffusion is thereby made explicit. Solutions to the nonlinear, time-independent and linearized, time-dependent diffusion equations are given.

099.099 **Ring current impoundment of the Io plasma torus.**
G. L. Siscoe, A. Eviatar, R. M. Thorne,
J. D. Richardson, F. Bagenal, J. D. Sullivan.
J. Geophys. Res., Vol. 86, 8480 - 8484 (1981).

A newly discovered feature in the Io plasma formation that may be described as a ramp separating a high-density plasma ledge on its Jupiterward side from the lower-density radially distended Io plasma disc on its anti-Jupiterward side is observed to coincide with a marked inward decrease in the ring current population. The spatial congruency of the counter-directed maximal gradients in both plasma bodies reveals a profound coupling between them.

099.100 **Factors governing the ratio of inward to outward diffusing flux of satellite ions.**
J. D. Richardson, G. L. Siscoe.
J. Geophys. Res., Vol. 86, 8485 - 8490 (1981).

The primary loss process for the ionized component of satellite debris in Jupiter's magnetosphere appears to be radial transport, which carries the satellite-derived ions away from the source satellite. Cross L (interchange) diffusion is a likely candidate for the radial transport mechanism. Many consequences are predicted to result from satellite ions that reach regions close to and remote from Jupiter. The ratio of the inward to outward flux of diffusing ions from a given satellite is therefore important as an input to the estimates of the magnitudes of expected consequences. In the formal analysis of cross L diffusion, the flux ratio depends on poorly known boundary conditions and on a poorly determined diffusion coefficient. The authors present an analysis of the sensitivity of the flux ratio to the uncertainties in the mentioned parameters.

099.101 **Ion and electron angular distributions in the Io torus region of the Jovian magnetosphere.**
L. J. Lanzerotti, C. G. Maclennan, T. P. Armstrong,
S. M. Krimigis, R. P. Lepping, N. F. Ness.
J. Geophys. Res., Vol. 86, 8491 - 8496 (1981).

099.102 **Observations of electron gyroharmonic waves and the structure of the Io torus.**   T. J. Birmingham,
J. K. Alexander, M. D. Desch, R. F. Hubbard, B. M. Pedersen.
J. Geophys. Res., Vol. 86, 8497 - 8507 (1981).

Narrow-banded emissions were observed by the planetary radio astronomy experiment on the Voyager 1 spacecraft as it traversed the Io plasma torus. The observed frequencies always include the component near f_{uhr}, the upper hybrid resonant frequency, but the distribution of the other observed emissions varies in a systematic way with position in the torus. A detailed discussion of the observations is presented. A refined model of the electron density variation, based on identification of the f_{uhr} line, is also included.

099.103 **Plasma observations of the Alfvén wave generated by Io.**
J. W. Belcher, C. K. Goertz, J. D. Sullivan, M. H. Acuña.
J. Geophys. Res., Vol. 86, 8508 - 8512 (1981).

Measurements of positive ions near the Io flux tube are consistent with the detection of velocity perturbations in the magnetospheric flow due to the southward propagating Alfvén wave generated by Io.

099.104 **Standing Alfvén wave current system at Io: Voyager 1 observations.**
M. H. Acuña, F. M. Neubauer, N. F. Ness.
J. Geophys. Res., Vol. 86, 8513 - 8521 (1981).

The Voyager 1 spacecraft encounter with the planet Jupiter occurred on March 5, 1979, with a closest approach distance of 4.9 R_J. A preliminary report and analysis of quick look data obtained at the time of the encounter has been given by Ness et al. (1979). These preliminary analyses did not consider the ramifications of the observations reported by complementary investigations aboard the spacecraft. It is the purpose of this paper to present a more detailed and expanded report of the magnetic field observations and the results of analyses carried out taking into account a more accurate physical description of the interaction between Io, its environment and the Jovian magnetosphere and using updated and much more accurate spacecraft trajectory and especially attitude information.

099.105 Ray tracing in the Io plasma torus: application to the PRA (*planetary radio astronomy*) observations during Voyager 1's closest approach. A. Lecacheux.
J. Geophys. Res., Vol. 86, 8523 - 8528 (1981).

An attempt is made to simulate propagation of the low-frequency radio waves in the vicinity of the Io plasma torus near Jupiter. In absence of a definitive and detailed electron density model for this torus, a number of minimal hypotheses were used. Nevertheless, the anomalous radio dynamic spectrum observed during the day of Voyager 1's closest approach could agree with an emission source situated around the planet in the southern hemisphere and related to the local gyrofrequency. The major conclusion is that the entire low frequency (up to a few megahertz) dynamic radio spectrum of Jupiter could be modified by refraction inside and around the Io torus. The magnitude of the effects depends on the location of the source and on the plasma parameters distribution inside the torus.

099.106 Synoptic observations of Jupiter's radio emissions: average statistical properties observed by Voyager.
J. K. Alexander, T. D. Carr, J. R. Thieman, J. J. Schauble, A. C. Riddle.
J. Geophys. Res., Vol. 86, 8529 - 8545 (1981).

Observations of Jupiter's low-frequency radio emissions collected over one-month intervals before and after each Voyager encounter have been analyzed to provide a synoptic view of the average statistical properties of the emissions. Compilations of occurrence probability, average power flux density, and average sense of circular polarization are presented as a function of central meridian longitude, phase of Io, and frequency. The results are compared with ground-based observations.

099.107 On the arc structure of the DAM Jupiter emission. Y. Leblanc.
J. Geophys. Res., Vol. 86, 8546 - 8560 (1981).

An analysis of the dynamic spectra of the Jovian DAM emission (1.3–40 MHz) has been made from Voyager data; it appears that the different Jovian 'sources' can be defined by spectral characteristics, rather than by occurrence probability. The non-Io emission consists of two families: vertex early arcs (VEA) and vertex late arcs (VLA). These two families are superimposed at all longitudes, but one is always more intense than the other. The characteristics of the two families are specified. The Io-controlled emission consists of the four sources already known from the ground-based observations in addition to a new source, (Io-A')sp, identified by its dynamic spectrum alone. All of the sources are partially superimposed on non-Io emission.

099.108 The Jovian decametric arcs as an interference pattern. A. Boischot, M. Aubier.
J. Geophys. Res., Vol. 86, 8561 - 8563 (1981).

It is proposed that the nested arc structures, observed in the decametric Jovian radio emission by the planetary radio astronomy experiment, are due to the interference of two rays from a single source at each frequency, the relative phase changing with rotation of the planet. This idea can explain at the same time the shape of the arcs and their repetitivity. But the nature of the two interfering rays is not understood yet.

099.109 The Jovian S burst sources. Y. Leblanc, F. Genova.
J. Geophys. Res., Vol. 86, 8564 - 8568 (1981).

By using the high resolution observations of Nançay observatory, the authors have been able to identify the S burst emission on the planetary radio astronomy (PRA) records of Voyager. It is shown that the S bursts occur in two regions of the Φ_{Io}–CML plane (S-IoB and S-IoA'C regions). In these regions the S burst emission is arranged into a pattern of repetitive features, drifting negatively. These features could be incomplete vertex late arcs. The authors show that the S burst pattern is distinct from the pattern of the Io-controlled emission. These results are discussed in the frame of Goldstein and Thieman's arc model.

099.110 The formation of arcs in the dynamic spectra of Jovian decameter bursts.
M. L. Goldstein, J. R. Thieman.
J. Geophys. Res., Vol. 86, 8569 - 8578 (1981).

A model is presented that can account for several features of the dynamic spectral arcs observed at decameter wavelengths by the planetary radio astronomy experiment on Voyagers 1 and 2.

099.111 A heuristic model for Jovian decametric arcs. J. B. Pearce.
J. Geophys. Res., Vol. 86, 8579 - 8580 (1981).

A field line emission cone model is proposed that fits the frequency-time profiles of decametric arcs emitted by the Jupiter system. Four parameters (latitude of the observer, longitude of the excited flux tube, emission cone angle, and field line L shell value) can be varied to fit most of the observed radiation.

099.112 Character of the Jovian decametric arcs. D. H. Staelin.
J. Geophys. Res., Vol. 86, 8581 - 8584 (1981).

The planetary radio astronomy (PRA) experiment on the Voyager 1 and Voyager 2 spacecraft reveals strong radiation in the form of arcs when the data are displayed in time-frequency coordinates. The vertex frequencies of these arcs, i.e., the central frequencies at which the arcs are first or last observed, are correlated with the magnetic field strength at the foot of the $L = 6$ shell magnetic flux tubes that emitted the arcs, provided that the emission is conical with a cone angle that varies slightly in a prescribed way. This interpretation further supports the association of the left circularly polarized arcs with the southern hemisphere, where the relation between vertex frequency and magnetic field strength is preserved. One way to produce a frequency dependent cone angle is described.

099.113 Models for Jupiter's decametric arcs. J. W. Warwick.
J. Geophys. Res., Vol. 86, 8585 - 8592 (1981).

Several papers attempt to explain the remarkable arc-shaped structures that dominate Jupiter's decametric emissions as recorded by Voyagers 1 and 2. The present paper understands these arcs in terms of a magnetic fine structure undetected by Pioneer 11 on account of its rather large closest approach distance (50 or 60 times the smallest scales in the structures involved).

099.114 Volcanic eruptions on Io.
R. G. Strom, N. M. Schneider, R. J. Terrile, A. F. Cook, C. Hansen.
J. Geophys. Res., Vol. 86, 8593 - 8620 (1981).

The paper supplements and updates the plume observations reported earlier by Smith et al. (1979), Cook et al. (1979) and Strom et al. (1979) and discusses the implications for Io volcanism from these new data.

099.115   Spatial color variations in the volcanic plume at Loki, on Io.   S. A. Collins.
J. Geophys. Res., Vol. 86, 8621 - 8626 (1981).

Multicolor Voyager 1 photographs of the Loki volcanic plume, on Io, have been analyzed to determine the nature and quantity of the scattering material within the plume. This work indicates that there are two particle populations. The first population consists of particles with radius of 0.001–0.01 μm, while the second population, which is concentrated near the source, comprises particles with radius greater than 1 μm. The population of smaller particles includes most of the particulate mass in the plume. This work increases the previously estimated particulate mass (Johnson et al., 1979) and suggests that the SO_2 gas spectroscopically identified by Pearl et al. (1979) may represent transient flow from the volcano instead of an atmosphere in stable equilibrium with the local surface.

099.116   High resolution albedo measurements on Io from Voyager 1.   R. T. Clancy, G. E. Danielson.
J. Geophys. Res., Vol. 86, 8627 - 8634 (1981).

The photometric properties of the surface of Io were investigated at high spatial resolution by a choice of 220 sample regions from the four-color, 8 km/line pair resolution photomosaic of Io taken by Voyager 1. The mosaic longitudinal coverage extends from ~200°W to 350°W (phase angle ~10.5°). The regions were categorized on the basis of their visual color in the color print. Categories include: white, yellow, orange, red, brown (polar), and black regions. The photometrically corrected data were plotted as a function of intensity versus photometric angles for each of the color regions in all four filters (orange, blue, violet, UV) using a Minnaert function. The plots of these color regions show large scatter about the least squares fitted lines.

099.117   Coordinates of features on the Galilean satellites.   M. E. Davies, F. Y. Katayama.
J. Geophys. Res., Vol. 86, 8635 - 8657 (1981).

Control nets of the four Galilean satellites have been established photogrammetrically from pictures taken by the two Voyager spacecraft during their flybys of Jupiter in 1979. Coordinates of 504 points on Io, 112 points on Europa, 1547 points on Ganymede, and 439 points on Callisto are listed. Selected points are identified on U.S. Geological Survey maps of the satellites. Measurements of these points were made on 234 pictures of Io, 115 pictures of Europa, 282 pictures of Ganymede, and 200 pictures of Callisto. The systems of longitude were defined by craters on Europa, Ganymede, and Callisto. Preliminary solutions have been found for the directions of the axes of rotation of the Galilean satellites. New mean radii have been determined as 1815±5 km for Io, 1569±10 km for Europa, 2631±10 km for Ganymede, and 2400±10 km for Callisto.

099.118   Crater populations on Ganymede and Callisto.   R. G. Strom, A. Woronow, M. Gurnis.
J. Geophys. Res., Vol. 86, 8659 - 8674 (1981).

The discovery of heavily cratered surfaces on Ganymede and Callisto by Voyager 1 shows that like the inner Solar System, a period of heavy bombardment also occurred in the outer Solar System. Comparisons among the crater size/density curves of Ganymede, Callisto and the terrestrial planets show several striking features. The overall crater density of the most heavily cratered terrain on Ganymede is down by a factor of about 3 compared to Callisto, and when allowance is made for the difference in crater production rate due to the influence of Jupiter's gravity field it is down by a factor of nearly 6. This result is discussed in detail.

099.119   Amalthea: Voyager imaging results.
J. Veverka, P. Thomas, M. Davies, D. Morrison.
J. Geophys. Res., Vol. 86, 8675 - 8682 (1981).

Voyager images of Amalthea reveal a very irregular satellite with dimensions of some 270 × 165 × 150 km, in synchronous rotation relative to Jupiter. Its surface is scarred by large craters, sharp ridges, and other prominent topography suggestive of a long history of cosmic battering. The normal reflectance of the surface is very low (5–6%); the color is very red. The authors find that Amalthea's mean opposition magnitude is $V = +14$ and that the magnitude difference between eastern and western elongation does not exceed +0.1. The phase coefficient between phase angles of 0.8° and 42° of 0.042 ± 0.004 mag/deg indicates that the phase integral does not exceed 0.3 and that the Bond Albedo is less than 0.02. Laboratory simulations show that the combination of low albedo and red color probably requires contamination of the surface by sulfur, the most likely source of which is Io. A noteworthy mystery concerning Amalthea is the composition of the material in several prominent bright spots. These isolated spots (typically 10–50 km across) occur preferentially on local slopes and ridges, have albedos several times higher than the background, and have a greenish color.

099.120   The Jovian ring.
D. C. Jewitt, G. E. Danielson.
J. Geophys. Res., Vol. 86, 8691 - 8697 (1981).

The results of further measurements of the Jovian ring system are presented. The system has three major components: the bright ring, the faint sheet, and the out-of-plane halo. The bright ring has an outer radius of 1.81 ± 0.01 R_J, an inner radius 1.72 ± 0.01 R_J, an eccentricity not greater than 0.003 and a normal optical depth 3×10^{-5}. The faint sheet extends from the inner edge of the bright ring to the surface of Jupiter. Its optical depth is approximately 7×10^{-6}. Three arguments are presented to show that a halo of material envelops the above two rings and extends 10^4 km above the ring plane. A simple model is invoked to account for the halo by means of interactions between the Jovian magnetic field and charged ring particles less than 0.5 μm in diameter. The source of small particles is probably within the bright ring itself and may be due to micrometeorite impact into larger ring bodies. Small particles evolve in towards Jupiter under Poynting Robertson and other drag forces. The outer edge of the ring system is defined by the satellite 1979J1.

099.121   Radio occultation of Jupiter's ring: bounds on optical depth and particle size and a comparison with infrared and optical results.
G. L. Tyler, E. A. Marouf, G. E. Wood.
J. Geophys. Res., Vol. 86, 8699 - 8703 (1981).

The Jovian ring is not detectable in Voyager 1 radio occultation data, setting the bounds on its optical thickness of 2×10^{-4} and 5×10^{-4} at 13- and 3.6-cm wavelengths, respectively. Comparison of results at radio, infrared, and optical wavelengths suggests a population density that either falls more rapidly than the inverse square of the linear size or is sharply bounded in maximum particle size. A fragmentation power law, $\sim a^p$ for $a_{min} < a < a_{max}$, where a is the particle radius, and power index p between −3 and −4 lead to a minimum size estimate $a_{min} \simeq 1$–2 μm. A specific model which is consistent with all observations consists of a power law distribution of lossless or very slightly absorbent Mie scatterers with refractive index near 1.62, power law index p approximately −3.5, and a minimum particle size of about 1.5 μm. In this model the upper size limit a_{max} is not critical, provided that it is greater than about 4 μm.

099.122   Albedo, internal heat, and energy balance of Jupiter: preliminary results of the Voyager infrared investigation.   R. A. Hanel, B. J. Conrath, L. W. Herath, V. G. Kunde, J. A. Pirraglia.

J. Geophys. Res., Vol. 86, 8705 - 8712 (1981).

Full disk measurements recorded 31 days before the Voyager 1 encounter with Jupiter by the radiometer $(0.4-1.7 \mu m)$ of the infrared instrument Iris indicate a geometric albedo of 0.274 ± 0.013. Combining this measurement with the Pioneer-derived phase integral of 1.25 of Tomasko et al. (1978) yields a Jovian Bond albedo of 0.343 ± 0.032. Infrared spectra recorded at the same time by the Michelson interferometer $(4-55 \mu m)$, along with a model extrapolation to low wave numbers not covered by the instrument, yield a thermal emission of $1.359 \pm 0.014 \times 10^{-3} W cm^{-2}$. This corresponds to an equivalent blackbody temperature of $124.4 \pm 0.3 K$. From these measurements the internal heat flux of Jupiter is estimated to be $5.444 \pm 0.425 \times 10^{-4} W cm^{-2}$, and the energy balance defined as the ratio of emitted thermal to absorbed solar energy is 1.668 ± 0.085.

099.123 The helium abundance of Jupiter from Voyager.
D. Gautier, B. Conrath, M. Flasar, R. Hanel, V. Kunde, A. Chedin, N. Scott.
J. Geophys. Res., Vol. 86, 8713 - 8720 (1981).

The helium abundance in the Jovian atmosphere is derived from Voyager 1 data by two methods. The first method uses only infrared spectra from selected locations on the planet while the second method uses a thermal profile independently derived from radio occultation measurements and infrared spectra recorded near the occultation point. A hydrogen mole fraction of 0.897 ± 0.030 is obtained from the first method, while the second method gives 0.880 ± 0.036, corresponding to helium mass fractions of 0.19 ± 0.05 and 0.21 ± 0.06, respectively. The results are consistent with a uniform mix of hydrogen and helium within Jupiter's interior, but a modest amount of helium depletion ($\Delta Y \leqslant 0.05$) cannot be excluded.

099.124 The atmosphere of Jupiter: an analysis of the Voyager radio occultation measurements.
G. F. Lindal, G. E. Wood, G. S. Levy, J. D. Anderson, D. N. Sweetnam, H. B. Hotz, B. J. Buckles, D. P. Holmes, P. E. Doms, V. R. Eshleman, G. L. Tyler, T. A. Croft.
J. Geophys. Res., Vol. 86, 8721 - 8727 (1981).

Coherently related S (2.3 GHz) and X band (8.4 GHz) signals transmitted from Voyager 1 and 2 have been used to probe the Jovian atmosphere during occultations of the spacecraft by Jupiter. The observations have yielded profiles in height of the gas refractivity, molecular number density, pressure, temperature, and microwave absorption in the troposphere and stratosphere of Jupiter at latitudes ranging from $0°$ to about $70°S$. The data cover a pressure range from 1000 to 1 mbar over a height interval of 160 km.

099.125 A search for the radio occultation flash at Jupiter.
J. M. Martin, G. L. Tyler, V. R. Eshleman, G. E. Wood, G. F. Lindal.
J. Geophys. Res., Vol. 86, 8729 - 8732 (1981).

The 'evolute flash', a focusing effect caused by the curvature of a planet's limb, was sought in the radio data taken during the occultation of Voyager 1 by Jupiter, using a modified matched-filter technique. The expected frequency structure of the flash signal is double branched, while the intensity structure is highly localized in time. The search for the signal was carried out over a 6.4 s period. The signal parameters were varied to span the uncertainties introduced by imperfect knowledge of the orbit of the spacecraft and the shape of Jupiter. Several peaks at the 8 standard deviation level were present in the filter output. However, these peaks were separated in time by up to 3.3 s, and none could be identified as the flash. The consequences of this negative result are discussed.

099.126 Interaction of eddies and mean zonal flow on Jupiter as inferred from Voyager 1 and 2 images.

A. P. Ingersoll, R. F. Beebe, J. L. Mitchell, G. W. Garneau, G. M. Yagi, J.-P. Müller.
J. Geophys. Res., Vol. 86, 8733 - 8743 (1981).

Voyager 1 and 2 narrow-angle frames were used to obtain displacements of features at resolutions of 130 km over time intervals of 1 Jovian rotation.

099.127 Oscillations and velocity structure of a long-lived cyclonic spot.
A. Hatzes, D. D. Wenkert, A. P. Ingersoll, G. E. Danielson.
J. Geophys. Res., Vol. 86, 8745 - 8749 (1981).

Dark brown cyclonic spots ('barges') at $14°N$ were studied by using Voyager 1 and 2 images of Jupiter. Movie sequences were made to study the spots' behavior over intervals of 50 days and longer. These movies revealed that the length and width vary by $\pm 9\%$ with a period of about 15 days, while the area remains approximately constant. The horizontal velocity field was investigated for an interval of about 1 day. Flow around the largest barge occurs as a ring current. The vorticity is about $2^1/_2$ times that of the ambient cyclonic zonal circulation, and about one-half the value of the local planetary vorticity. Length and width variations appear to be associated with a nonzero horizontal divergence field.

099.128 Flow fields within Jupiter's Great Red Spot and White Oval BC.
J. L. Mitchell, R. F. Beebe, A. P. Ingersoll, G. W. Garneau.
J. Geophys. Res., Vol. 86, 8751 - 8757 (1981).

Using sequences of Voyager 1 high-resolution images of Jupiter's Great Red Spot (GRS) and White Oval BC the authors map the flow fields within the GRS and Oval BC. They compute relative vorticity within these features as a function of semi-major axis length and position angle in a coordinate system consisting of concentric ellipses of equal eccentricity.

099.129 Thermal structure and dynamics of the Jovian atmosphere. 1. The Great Red Spot.
F. M. Flasar, B. J. Conrath, J. A. Pirraglia, P. C. Clark, R. G. French, P. J. Gierasch.
J. Geophys. Res., Vol. 86, 8759 - 8767 (1981).

One of the objectives of the Voyager infrared spectroscopy investigation (IRIS) is the study of the thermal structure and dynamics of the Jovian atmosphere, on both global and local scales. One of the more interesting features studied has been the Great Red Spot (GRS), which has been continually observed for the past 100 years and probably has existed 300 years or longer. In this paper the authors present the temperature fields associated with the GRS, derived from the IRIS data for atmospheric pressures $500 > p$ (mbar) > 3. The authors discuss the tangential winds implied by these temperatures from the thermal wind relation. They summarize the structure observed at $5 \mu m$. The authors then examine the dynamics implied by the observed thermal structure. Finally, they discuss their results in the context of several theoretical models of the GRS. Latent heat release as an energy source for the GRS is considered, and the anlogies with and differences from tropical cyclones on earth are discussed.

099.130 Thermal structure and dynamics of the Jovian atmosphere. 2. Visible cloud features.
B. J. Conrath, F. M. Flasar, J. A. Pirraglia, P. J. Gierasch, G. E. Hunt.
J. Geophys. Res., Vol. 86, 8769 - 8775 (1981).

Because of the relatively large spatial dimensions of the Great Red Spot (GRS), it was possible to resolve its internal thermal structure, and a detailed investigation of this feature has been carried out in the preceding paper. It is the purpose of the present paper to extend the investigation of thermal structure to include other cloud features in order that the comparison of the dynamic properties of the features may be further pursued. The features examined include the major white ovals, 'hot spot' regions of strong 5-μm emission, the dark brown elongated features dubbed 'barges', and a belt-zone pair. The observations are presented, and the limitations on the data are discussed. The implications of the observed thermal structure in the upper troposphere and lower stratosphere are then considered in terms of the vorticity of the features, the implied vertical motion fields, and vertical mixing. Next the thermal structure of the upper stratosphere is discussed, and various possible models are examined, including radiative heating, vertical wave propagation, and thermally indirect meridional cells. Finally, conclusions are drawn concerning the dynamical similarities and differences of the various features.

099.131 **Visible and infrared observations of Jovian plumes during the Voyager encounter.**
G. E. Hunt, B. J. Conrath, J. A. Pirraglia.
J. Geophys. Res., Vol. 86, 8777 - 8781 (1981).

Observations by the Voyager imaging instrument have shown an organized train of features moving in a westerly current at 9°N with a zonal speed of 100–120 m s^{-1}. The measurement reports show that the number of plumes varied between 13 at the time of the Voyager 1 encounter, and 11 at the time of the second spacecraft passage. The infrared interferometer spectrometer (Iris) measurements suggest that the effect of these plume features propagates throughout the Jovian troposphere. A mechanism is suggested for the origin of the plumes in terms of wave interactions with the Jovian flow.

099.132 **Near-ultraviolet scattering properties of Jupiter.**
R. A. West, C. W. Hord, K. E. Simmons,
D. L. Coffeen, M. Sato, A. L. Lane.
J. Geophys. Res., Vol. 86, 8783 - 8792 (1981).

The Voyager 2 photopolarimeter experiment obtained photometric observations of Jupiter at four phase angles between 20° and 80° and at various illumination angles from limb to terminator. These data were analyzed to determine the altitude distribution and scattering properties of the material responsible for Jupiter's low geometric albedo at 2400 Å.

099.133 **Visible aurora in Jupiter's atmosphere?**
A. F. Cook II, A. Vallance Jones, D. E. Shemansky.
J. Geophys. Res., Vol. 86, 8793 - 8796 (1981).

The darkside limb pictures obtained by the imaging experiment on Voyager 1 have been reexamined. It is concluded that the observed luminosity is very likely due at least in part to Io torus aurora. If the effective wavelength of the emission lies in the 4000- to 5000-Å region, the slant intensity is estimated to be about 20 kR. The observed double structure may be due to a number of causes such as horizontal structure in auroral emission, aurora plus twilight or photochemical airglow plus aurora.

099.134 **Colors on Jupiter.**   T. Owen, R. J. Terrile.
J. Geophys. Res., Vol. 86, 8797 - 8814 (1981).

The colors present in the clouds of Jupiter at the time of the Voyager encounters are described as they appeared in high resolution images. Although many different tints are present in these pictures, some cloud units can be characterized by very discrete colors. The authors can show that latitude, altitude, and dwell-time are all critical factors in determining which colors appear where, but the identities of the responsible chromophores remain unestablished. Simultaneous ground-based 5 μm observations permit the determination of the relative altitudes of these cloud systems. Despite the turbulence of the Jovian atmosphere, correlations between cloud color and certain latitudes have been maintained for decades, suggesting the importance of the internal energy source and the deep circulation in generating some of the observed chromophores.

099.135 **A fireball in Jupiter's atmosphere.**
A. F. Cook, T. C. Duxbury.

J. Geophys. Res., Vol. 86, 8815 - 8817 (1981).

One fireball was photographed during two encounters with Jupiter. Its total luminosity was 1.2 X 10^5 0 mag s (at standard range 100 km). If the luminous efficiency proposed by Cook et al. (1981) is employed for slip flow of a meteoroid in its own vapors the authors obtain an estimated mass of 11 kg. A rough absolute magnitude is −12.5. The authors estimate a number density near Jupiter of 7 X 10^{-28} cm^{-3} for masses of meteoroids of 3 kg and greater. This value is about a factor of 6 smaller than a rough upper limit reached from an extrapolation from terrestrial observations of meteors and comets.

099.136 **Energetic oxygen and sulfur in the Jovian magnetosphere.**
N. Gehrels, E. C. Stone, J. H. Trainor.
J. Geophys. Res., Vol. 86, 8906 - 8918 (1981).

The paper reports measurements made by the cosmic ray subsystem onboard Voyager 1 and 2 in the Jovian magnetosphere. Energy spectra of oxygen ions in the energy range 1 - 20 MeV/nuc between 5 and 20 R_J are presented and used to calculate phase space densities. There is a steep positive radial gradient in the phase space density of the energetic oxygen ions in this region, indicating an inward diffusive flow. Solutions of the diffusion equation are fit to the radial phase space density profile of oxygen ions with magnetic moments of 680 MeV/nuc-G. On the basis of the diffusion coefficient upper limit obtained from these fits, the upper limit on the rate at which oxygen ions with > 400 MeV/nuc-G diffuse inward across 10 R_J is $5 \times 10^{21 \pm 1}$ ions s^{-1}. The observations suggest that oxygen and sulfur ions in the Io plasma torus diffuse radially outward, are nonadiabatically accelerated in some region outside 17 R_J and then diffuse inward and outward from the acceleration region.

099.137 **On the injection and scattering of protons in Jupiter's magnetosphere.**   D. D. Barbosa.
J. Geophys. Res., Vol. 86, 8981 - 8990 (1981).

The author investigates a model by which 10 keV protons accelerated on Jovian auroral field lines are scattered and localized to the magnetic equator. The criterion that the beam density be sufficiently large to generate magnetosonic waves leads to the evaluation of proton beam fluxes consistent with recent Voyager observations of a hot high β plasma sheet in the magnetosphere. A model of the Alfven speed in the plasma sheet for both inbound and outbound legs of the Voyager 1 encounter is presented. This model confirms Piddington's limit of departure from rigid corotation beyond the point where the rigid body corotation speed equals the Alfven speed. Finally, the author explores the possibility of nonadiabatic scattering of super-Alfvenic ions by magnetohydrodynamic waves throughout the plasma sheet as a basis for high-energy tail formation of ambient particle distributions.

099.138 **Corotating magnetospheric convection.**
T. W. Hill, A. J. Dessler, L. J. Maher.
J. Geophys. Res., Vol. 86, 9020 - 9028 (1981).

The longitudinal asymmetry of the Io plasma torus, as predicted by the magnetic-anomaly model and observed by Earth-based optical astronomy, provides a driving mechanism for a corotating convection system in Jupiter's magnetosphere. The authors deduce some qualitative properties of this convection system from the general equations that govern a steady state corotating convection system. The corotating convection system appears capable of providing both the dominant radial transport mechanism and the dominant mechanism for extracting energy from Jupiter's rotation (at a rate ~10^{15} W) for driving a wide variety of magnetospheric phenomena. A similar corotating convection system may occur in other rotation-dominated magnetospheres, for example, those of pulsars and Saturn.

099.139 **The Voyager 1 EUV spectrum of the Io plasma torus.**
D. E. Shemansky, G. R. Smith.
J. Geophys. Res., Vol. 86, 9179 - 9192 (1981).

The Voyager 1 EUV spectrum of the hot Io plasma torus
has been analyzed by using a collisional model. Comparison of
the model calculations with other spectroscopic observations
that can be clearly identified as originating in the hot plasma
torus shows satisfactory agreement. However, the EUV observations show strong morphological differences with recent
ground-based observations of S II emission, raising questions
concerning excitation mechanisms and radial dependence of
the observed characteristics. The ion densities indicated by the
EUV data analysis show distinct differences with the in situ
Voyager 1 plasma science estimates.

099.140 **Interaction of S- and L-bursts in Jupiter's decametric
radio spectra.**   J. J. Riihimaa, T. D. Carr.
Moon Planets, Vol. 25, 373 - 387 (1981) = Contrib. No. 36
Dep. Astron., Univ. Florida.

Dynamic spectra of Jupiter's decametric emission are
observed with a high-resolution radio spectrograph. Certain
events observed in region Io-B display interaction effects
between the S- and L-emissions. The effect appears as a gap in
the L-emission after a passage of an S-burst.

099.141 **Charged dust in the outer planetary magnetospheres.
III. Satellite impact geometries.**
J. R. Hill, D. A. Mendis.
Moon Planets, Vol. 25, 427 - 436 (1981).

The authors consider the problem of impact geometries
of charged dust fragments on the Galilean satellites and they
estimate their impact rates on the leading and trailing hemispheres of the satellites. They show that this charged dust
impaction, taken independently or together with plasma
impaction, could explain the observed brightness asymmetries.

099.142 **The weather on Jupiter.**
G. Hunt.
New Scientist, Vol. 90, 485 - 487 (1981). – Abstr. in Phys.
Abstr., Vol. 84, Abstr. 108122 (1981).

099.143 **Dissipation and turbulent heating of plasma in
Jupiter's magnetosphere.**   D. D. Barbosa.
Geophys. Res. Lett., Vol. 8, 1111 - 1114 (1981).

Voyager 1 observations of plasma waves in the dayside
Jovian magnetosphere which show a correlation with measurements of localized concentrations of cool thermal plasma are
presented. This moderately intense broadband electrostatic
noise is shown to be of sufficient intensity to accelerate superthermal ions to energies ~ 1 keV and higher. This process can
account for the extensive heating of plasma in the magnetosphere and can energize a fraction of heavy ions to injection
threshold for a high-energy second stage acceleration mechanism. A brief discussion of the relation of this noise to Jovian
magnetospheric dynamics is included.

099.144 **Joule heating of the Jovian ionosphere by corota-
tion enforcement currents.**
A. Nishida, Y. Watanabe.
J. Geophys. Res., Vol. 86, 9945 - 9952 (1981).

Heat is deposited to the Jovian ionosphere in the course
of enforcement of corotation to the outer magnetosphere. The
Joule heating rate is estimated for several possible causes of
departure from the corotation by making use of a simple
model of the magnetodisc.

099.145 **Plasma near Io: estimates of some physical parameters.**   M. G. Kivelson, D. J. Southwood.
J. Geophys. Res., Vol. 86, 10122 - 10126 (1981).
Models of Io's interaction with the Jovian plasma are examined
to assess whether conditions hazardous to a spacecraft flying
by Io at about 1000 km above its surface should be anticipated.

Only two models are regarded as consistent with most of the
presently available data. The 'ionospheric' model allows field-
aligned currents to close through Io and its ionosphere. The
'magnetospheric' model allows the currents to close on an
Ionian magnetopause and across a tail neutral sheet. Although
only the latter model provides an explanation for the pitch
angle distribution of energetic electrons measured by the
LECP detector on Voyager 1, the implications of both models
for near-Io plasma are explored.

099.146 **Ground-based observations of HCN on Jupiter and
Titan.**   A. T. Tokunaga, S. C. Beck,
T. R. Geballe, J. H. Lacy, G. Serabyn.
Bull. American Astron. Soc., Vol. 13, 701 (1981). – Abstract.

099.147 **Accretion of Ganymede and Callisto in a gaseous
nebula.**   J. I. Lunine, D. J. Stevenson.
Bull. American Astron. Soc., Vol. 13, 721 (1981). – Abstract.

099.148 **The rings of Jupiter, Saturn, and Uranus.**
A. F. Cook, F. A. Franklin, B. A. McIntosh,
R. J. Terrile.
Bull. American Astron. Soc., Vol. 13, 728 (1981). – Abstract.

099.149 **Limit on possible narrow rings around Jupiter.**
E. W. Dunham, J. L. Elliot, D. J. Mink.
Bull. American Astron. Soc., Vol. 13, 729 (1981). – Abstract.

099.150 **Ultraviolet observations of the Io torus.**
H. W. Moos, S. T. Durrance, P. D. Feldman,
J. T. Clarke.
Bull. American Astron. Soc., Vol. 13, 729 - 730 (1981).
Abstract.

099.151 **CCD images of the Jovian [S II] - [S III] nebula:
photometry and line ratios.**   J. T. Trauger.
Bull. American Astron. Soc., Vol. 13, 730 (1981). – Abstract.

099.152 **Density variations in the Io plasma torus.**
J. S. Morgan.
Bull. American Astron. Soc., Vol. 13, 730 (1981). – Abstract.

099.153 **Ionic composition in the Io plasma torus: direct
measurements by Voyager 1.**   F. Bagenal.
Bull. American Astron. Soc., Vol. 13, 730 (1981). – Abstract.

099.154 **Discovery of an Io-correlated energy source for the
plasma torus.**   B. R. Sandel.
Bull. American Astron. Soc., Vol. 13, 730 (1981). – Abstract.

099.155 **Radio emission from Io.**
I. de Pater, W. J. Jaffe, R. A. Brown, G. L. Berge.
Bull. American Astron. Soc., Vol. 13, 730 (1981). – Abstract.

099.156 **Sodium remote from Io.**
N. M. Schneider, R. A. Brown.
Bull. American Astron. Soc., Vol. 13, 730 - 731 (1981).
Abstract.

099.157 **Io's sodium cloud: recent modeling advances.**
W. H. Smyth.
Bull. American Astron. Soc., Vol. 13, 731 (1981). – Abstract.

099.158 **A movie of the Io plasma torus.**
C. B. Pilcher, J. S. Morgan, J. H. Fertel, C. C. Avis.
Bull. American Astron. Soc., Vol. 13, 731 (1981). – Abstract.

099.159 **Corotating convection – a unifying concept for
Jupiter's magnetosphere.**   A. J. Dessler.
Bull. American Astron. Soc., Vol. 13, 731 (1981). – Abstract.

099.160 **Effects due to charged particle impact on the icy satellites of Jupiter and Saturn.**
R. E. Johnson, E. Sieveka, L. Barton, L. J. Lanzerotti, W. L. Brown, V. Pirronello.
Bull. American Astron. Soc., Vol. 13, 731 (1981). – Abstract.

099.161 **On the nature of the Jovian "blue" regions.**
W. R. Thompson, C. Sagan.
Bull. American Astron. Soc., Vol. 13, 734 - 735 (1981). Abstract.

099.162 **Spectroscopy of the Jovian [S II] nebula: 1981 apparition.** J. T. Trauger.
Bull. American Astron. Soc., Vol. 13, 735 (1981). – Abstract.

099.163 **Thermal infrared morphology of the Jovian atmosphere in 1981.**
G. Orton, J. Caldwell, A. Tokunaga.
Bull. American Astron. Soc., Vol. 13, 735 (1981). – Abstract.

099.164 **The lower atmosphere composition of Jupiter's North Equatorial Belt from Voyager 1 IRIS.**
V. Kunde, R. Hanel, W. Maguire, D. Gautier, J.-P. Baluteau, A. Marten, A. Chedin, N. Husson, N. Scott.
Bull. American Astron. Soc., Vol. 13, 735 (1981). – Abstract.

099.165 **High resolution observations of Jupiter in the 1100 to 1200 cm^{-1} spectral range.** R. F. Knacke, S. Y. Kim, S. T. Ridgway, A. T. Tokunaga.
Bull. American Astron. Soc., Vol. 13, 735 (1981). – Abstract.

099.166 **H_2O and PH_3 abundances for Jovian belt regions from 5 μm measurements.**
G. Bjoraker, U. Fink, H. P. Larson, V. Kunde.
Bull. American Astron. Soc., Vol. 13, 735 - 736 (1981). Abstract.

099.167 **The spectrum of the Jovian aurora, 1150 - 1700 Å.**
S. T. Durrance, P. D. Feldman, H. W. Moos.
Bull. American Astron. Soc., Vol. 13, 736 (1981). – Abstract.

099.168 **The UV transmission of the Jovian atmosphere to auroras.** G. R. Gladstone, Y. L. Yung.
Bull. American Astron. Soc., Vol. 13, 736 (1981). – Abstract.

099.169 **Properties of NH_3 gas and ice in the Jovian atmosphere from investigation of the far infrared spectrum.** J. Appleby, G. Orton, J. Martonchik, H. Aumann.
Bull. American Astron. Soc., Vol. 13, 736 (1981). – Abstract.

099.170 **Jupiter and Saturn: a reduced NH_3 abundance at high latitudes?** I. de Pater, J. R. Dickel.
Bull. American Astron. Soc., Vol. 13, 736 - 737 (1981). Abstract.

099.171 **Preliminary global compositional unit maps of the Galilean satellites.** T. B. McCord, R. N. Clark, A. Meloy, T. V. Johnson, D. Matson, J. A. Mosher, L. Soderblom.
Bull. American Astron. Soc., Vol. 13, 737 (1981). – Abstract.

099.172 **Voyager photometry of Europa.**
B. Buratti, J. Veverka.
Bull. American Astron. Soc., Vol. 13, 737 (1981). – Abstract.

099.173 **Is Amalthea too hot?**
D. Simonelli, P. Thomas, J. Veverka.
Bull. American Astron. Soc., Vol. 13, 737 (1981). – Abstract.

099.174 **Evidence for mobility of water ice on Callisto.**
J. R. Spencer, P. R. Maloney.

Bull. American Astron. Soc., Vol. 13, 737 - 738 (1981). Abstract.

099.175 **A volcanic flow model for Io and pertinent observations.** W. Sinton.
Bull. American Astron. Soc., Vol. 13, 738 (1981). – Abstract.

099.176 **The C/H ratio in Jupiter from the Voyager experiment.** D. Gautier, B. Bezard, A. Marten, J.-P. Baluteau, A. Chedin, N. Husson, N. Scott, V. Kunde, R. Hanel.
Bull. American Astron. Soc., Vol. 13, 738 (1981). – Abstract.

099.177 **Heterodyne observations of 12 μm emission from the Jovian south polar region.** T. Kostiuk, M. Mumma, D. Deming, F. Espenak, D. Jennings, D. Zipoy.
Bull. American Astron. Soc., Vol. 13, 738 (1981). – Abstract.

099.178 **Further observations of 8 μm polar brightenings on Jupiter.** J. Caldwell, A. T. Tokunaga, G. S. Orton.
Bull. American Astron. Soc., Vol. 13, 738 (1981). – Abstract.

099.179 **Longitudinal variability of 8900 Å methane absorption on the disk of Jupiter.**
C. R. Stoker, R. A. West, C. W. Hord.
Bull. American Astron. Soc., Vol. 13, 738 - 739 (1981). Abstract.

099.180 **Investigations of the ultraviolet albedo of Jupiter with the IUE.**
P. Winkelstein, J. Caldwell, T. Owen, S. J. Kim.
Bull. American Astron. Soc., Vol. 13, 739 (1981). – Abstract.

099.181 **Investigation of the Jovian cloud systems from Voyager imaging and infrared spectrometer high spatial resolution data.**
B. Bezard, J. P. Baluteau, A. Marten.
Bull. American Astron. Soc., Vol. 13, 739 (1981). – Abstract.

099.182 **A comparison of the dynamics of convective features in the atmospheres of Jupiter and Saturn using Voyager images.**
G. E. Hunt, D. Godfrey, R. F. T. Barrey, J.-P. Muller.
Bull. American Astron. Soc., Vol. 13, 739 (1981). – Abstract.

099.183 **Power spectral analysis of turbulence and waves in the Jovian atmosphere from Voyager images.**
J.-P. Muller, R. N. Gorley, G. E. Hunt, R. F. T. Barrey.
Bull. American Astron. Soc., Vol. 13, 739 (1981). – Abstract.

099.184 **New models of Jupiter and Saturn: contrasting interiors?** D. J. Stevenson.
Bull. American Astron. Soc., Vol. 13, 740 (1981). – Abstract.

099.185 **Spatial distribution of SO_2 frost on Io.**
R. R. Howell, D. P. Cruikshank.
Bull. American Astron. Soc., Vol. 13, 740 (1981). – Abstract.

099.186 **Io's atmosphere: complications and more complications.** D. L. Matson, D. B. Nash, T. V. Johnson.
Bull. American Astron. Soc., Vol. 13, 740 (1981). – Abstract.

099.187 **Geomorphology of Ra Patera, Io: a quantitative approach to sulfur volcanism.**
D. C. Pieri, S. M. Baloga, R. M. Nelson, C. Sagan.
Bull. American Astron. Soc., Vol. 13, 741 (1981). – Abstract.

099.188 **Jovian circulation from Voyager 2 cylindrical mosaics.** S. S. Limaye, V. E. Suomi, L. A. Sromovsky, R. J. Krauss, D. S. Santek, H. E. Revercomb.
Bull. American Astron. Soc., Vol. 13, 742 (1981). – Abstract.

099.189   Effects of sampling strategy on Jovian eddy transport measurements.   L. A. Sromovsky, H. E. Revercomb, V. E. Suomi, S. S. Limaye, R. J. Krauss.
Bull. American Astron. Soc., Vol. 13, 743 (1981). – Abstract.

099.190   A post-Voyager view of Jupiter's magnetosphere.
S. M. Krimigis.
Endeavour New Ser., Vol. 5, No. 2, p. 50 - 60 (1981). – Abstr. in Phys. Abstr., Vol. 85, Abstr., 3211 (1982).

099.191   Das Wetter auf dem Jupiter.   K. Keil.
Naturwissenschaften, 68. Jahrg., 567 - 568 (1981).

099.192   Jupiter and Saturn.   A. P. Ingersoll.
Sci. American, Vol. 245, No. 6, p. 66 - 69, 72, 75 - 80 (1981).
Competing models seek to describe the sun's two giant companions. In one model the winds are confined to a thin layer at the surface; in another the winds extend through the fluid depths of each planet.

099.193   Modulations in Jovian decametric spectra: propagation effects in terrestrial ionosphere and Jovian environment.   F. Genova, M. G. Aubier, A. Lecacheux.
Astron. Astrophys., Vol. 104, 229 - 239 (1981).
A study of wide band Jovian decametric spectra observed in Nançay (France) allows to distinguish clearly three kinds of lane-like structures, each with definite spectral characteristics. One is of terrestrial ionospheric origin. The two others, the well known "modulation lanes" and a new class of modulations ("high frequency lanes") are of Jovian origin. The authors show the wide band spectral properties and occurrence in the Φ_{Io}-CML diagram as well as in function of hour and season. They give a tentative interpretation by propagation effects either in terrestrial ionosphere F-zone or in Jovian plasma torus.

099.194   Plasma dynamics in the rapidly rotating magnetosphere of Jupiter.   J. W. Belcher.
Plasma astrophysics, (see 012.042), p. 129 - 134 (1981).
The author reviews the major Voyager findings concerning the low energy plasma in the Jovian magnetosphere.

099.195   Positions of Jupiter and Galilean satellites in 1978.   H. Debehogne, L. E. Machado, E. R. Netto, J. F. Caldeira, G. G. Vieira.
Astron. Astrophys., Suppl. Ser., Vol. 46, 171 - 172 (1981).
77 positions of Jupiter and Galilean satellites are given as obtained from photographic observations made at the Observatoire Royal de Belgique during December 1978. The Double Astrograph Zeiss (D = 40 cm, f = 2 m) was used. Measurements were performed at the Observatório do Valongo with the Ascorecord measuring machine and reductions made by means of five reference stars on the Univac computer of Uccle.

099.196   Calculation of the parameters of the Jovian decimeter radio emission.   A. G. Solov'ev.
13-ya Vses. konf. po radioastron. issled. soln. sistemy, Kiev, 1981. Kiev, 1981, p. 70 - 71. In Russian. – From Ref. zh., 51. Astron., 11.51.300 (1981).

099.197   Observations of the radio emission of Jupiter with RATAN-600.   M. G. Mingaliev.
13-ya Vses. konf. po radioastron. issled. soln. sistemy, Kiev, 1981. Kiev, 1981, p. 72. In Russian. – From Ref. zh., 51. Astron., 11.51.301 (1981).

099.198   On the origin of the fine structure of Jupiter's decametric radio emission.
V. V. Zajtsev, E. Ya. Zlotnik, V. E. Shaposhnikov.
13-ya Vses. konf. po radioastron. issled. soln. sistemy, Kiev, 1981. Kiev, 1981, p. 72. In Russian. – From Ref. zh., 51. Astron., 11.51.302 (1981).

099.199   Fine structure of L-burst dynamical spectra and polarization effects in decametric emission diffusion in the Jovian magnetosphere.
V. V. Kocharovskij, Vl. V. Kocharovskij, V. E. Shaposhnikov.
13-ya Vses. konf. po radioastron. issled. soln. sistemy, Kiev, 1981. Kiev, 1981, p. 72 - 73. In Russian. – From Ref. zh., 51. Astron., 11.51.303 (1981).

099.200   Une nouvelle mission vers Jupiter.   P. Ringoet.
Ciel, Vol. 43, 182 - 185 (1981).

099.201   Photochemistry of phosphine and Jupiter's Great Red Spot.   N. Noy, M. Podolak, A. Bar-Nun.
J. Geophys. Res., Vol. 86, 11985 - 11988 (1981).
The authors present the results of an experimental investigation of the ultraviolet photolysis of mixtures of phosphine in hydrogen. Under no circumstances could red phosphorus be produced in this system. Rather, the product was always yellow. The measured indices of refraction of the phosphorus produced were used to fit the variation of reflectivity of Jupiter's Great Red Spot with wavelength. It was found that the phosphorus particles have to be small (0.05 μm) and their layer should have an optical depth of about one at λ = 0.4 μm.

099.202   "Disappearance" of the Red Spot.
Lj. Jovanović.
Vasiona, Année 29, 52 - 54 (1981). In Croatian.

099.203   Voyager and Pioneer encountered Jupiter and Saturn.   E. Mūkins.
Zvaigžņota debess, 1980/81. gada ziema, p. 33 - 43. In Latvian.

099.204   Joule heating of the Jovian ionosphere by corotation enforcement currents.
A. Nishida, Y. Watanabe.
Proceedings of the 14th ISAS Lunar and Planetary Symposium, (see 012.055), p. 73 (1981). – Abstract.

099.205   Field-aligned particle acceleration due to Io-induced drift Alfvén waves.   T. Tamao.
Proceedings of the 14th ISAS Lunar and Planetary Symposium, (see 012.055), p. 74 - 81 (1981).
The possibility of coherent acceleration of energetic particles due to the field-aligned electric potential difference associated with the Io-induced standing shear Alfvén waves is discussed. The author has estimated the magnitude of the field-aligned potential difference between Jupiter's ionosphere and Io (Io's magnetopause). He obtains the potential difference to be 0.18 Φ_0 and the accelerated electron energy flux at the ionospheric level, 40 ergs/cm^2 sec, where Φ_0 is the primary induced potential across the Io diameter due to its relative orbital motion in Jupiter's magnetosphere.

099.206   Interplanetary scintillation in Jupiter's decametric radiation.   K. Maeda.
Publ. Astron. Soc. Japan, Vol. 33, 465 - 484 (1981).
The intensity fluctuations of the 21.860-MHz radio emission from Jupiter are analyzed to investigate the characteristics of the plasma irregularities in the solar wind. The effect of ionospheric scintillation on the observed intensity fluctuations is investigated by using the spaced-receiver observations with separations of 0.1–4.5 km.

099.207   Multiply reflected standing Alfvén waves in the Io torus: Pioneer 10 observations.
R. Walker, M. Kivelson.
Geophys. Res., Lett., Vol. 8, 1281 - 1284 (1981).

Observations from the Voyager 1 pass by the Io flux tube strongly indicate that large amplitude standing Alfvén waves are generated as Jovian plasma flows past Io. Recently Gurnett and Goertz (1981) have proposed that this Io generated standing Alfvén wave system extends an appreciable way around the Io L-shell. In this paper, the authors present observations of magnetic perturbations found in the Pioneer 10 record as it crossed the Io L-shell. The field perturbations are qualitatively consistent with the passage of the spacecraft through a standing Alfvén wave pattern. The observations imply an Alfvén Mach number of 0.03, which is 1/5 of the value inferred from Voyager 1 observations. This implies a lower plasma density at the time of the Pioneer 10 flyby.

099.208 **On the charge exchange loss of energetic charged particles in the Jovian magnetosphere.** W.-H. Ip.
J. Geophys. Res., Vol. 86, 11246 - 11250 (1981).

Following the suggestion that the magnetospheric charged particles may suffer charge exchange loss in the vicinity of the Io torus owing to the presence of neutral atomic and molecular clouds emitted from Io, some quantitative estimates are given to the resulting variation in the pitch angle distribution as well as the lossy radial diffusion between 5 and 8 R_J. Comparison of the theoretical calculations with the Voyager observations indicates that the low-energy particle measurements are compatible with the existence of neutral clouds of molecules and atoms distributed between 5.6 R_J and 6 R_J (and beyond) with a total number density of $\lesssim 180$ cm^{-3}. This result reinforces the deduction from other observations that the neutral atoms and molecules ejected from Io into the Jovian magnetosphere could be quite extensive and effective in regulating the overall magnetospheric processes.

099.209 **Polarimetric observations of Io and Europa.**
O. R. Bolkvadze.
Proceedings of the Third Finnish-Soviet Astronomical Symposium, (see 012.057), p. 73 - 80 (1981).

099.210 **Numerical and analytical theories of motion of the outer satellites of Jupiter.** T. V. Bordovitsyna, T. S. Boronenko, L. E. Bykova, A. M. Chernitsov.
Proceedings of scientific lectures on cosmonautics, (see 012.058), 1980, p. 191 - 197. In Russian. − Abstr. in Ref. zh., 51. Astron., 12.51.132 (1981).

099.211 **Ephemeris of the system III (1965) longitude of the central meridian of Jupiter.** V. Meiller.
United States Naval Obs. Circ., No. 161, 18 pp. (1980).

In this Circular, an ephemeris of the central meridian of the geometric disk in system III, with the correction for phase to reduce to the illuminated disk, is given for the years 1965 to 1980.

099.212 **Observations of planets with the Danjon astrolabe of the Shanghai Observatory during 1978.12−1979.4.** Z.-f. Li, H.-j. Yan, Z.-g. Yang, Y.-f. Chen, J.-w. Zhang.
Ann. Shanghai Obs. Acad. Sinica, No. 2, p. 57 - 59 (1980).

099.213 **Free oscillations of the giant planets.**
S. V. Vorontsov, V. N. Zharkov.
Planetary interiors, (see 012.061), p. 189 - 193 (1981).

The authors present the results of theoretical calculations of adiabatic oscillations of Jupiter and Saturn. The influence of differential rotation is analyzed using a second-order perturbation theory.

099.214 **Dust observations in the Jovian system.**
A. F. Cook, G. E. Danielson, D. C. Jewitt, T. C. Owen.

Progress in planetary exploration, (see 012.062), p. 99 - 101 (1981).

Two observations of dust in the Jovian system have been made from the two Voyager spacecrafts. Jewitt and Danielson report observations of dust enveloping the newly discovered ring of Jupiter.

099.215 **The magnetic fields of Jupiter and Saturn.**
N. F. Ness.
Progress in planetary exploration, (see 012.062), p. 171 - 176 (1981).

It is the purpose of this talk to summarize what is presently known about the magnetic fields of these planets and the characteristics of their magnetospheres, which are formed by interaction with the solar wind.

099.216 **On the shape of Amalthea.**
F. Ferrini, A. Milani, A. M. Nobili.
Progress in planetary exploration, (see 012.062), p. 191 - 197 (1981).

Voyager's photographs show a highly prolated and asymmetrical Amalthea. The authors tried to fit these data into the equations for an approximately equipotential surface and conclude that its density must be 1.8 g cm^{-3} if it is formed on a synchronous orbit around Jupiter.

099.217 **The meteorology of Jupiter's atmosphere.**
G. E. Hunt.
Planetary aeronomy and astronomy, (see 012.063), p. 61 - 68 (1981).

Observations of the atmosphere of Jupiter by the imaging and infrared instruments on the Voyager spacecraft have been analysed to provide new insight into the meteorology of Jupiter. Like the Earth, the atmosphere of Jupiter appears to behave in a quasi-geostrophic manner. For a period to the Voyager 1 encounter, the analysis on imaging data indicated that the eddy momentum transfer into the mean zonal flow was a major driving mechanism for the motions. The initial analysis shows that the meteorologies of the Earth and Jupiter have more in common than was previously thought.

099.218 **Upper atmosphere of Jupiter − a post Voyager perspective.** S. K. Atreya.
Planetary aeronomy and astronomy, (see 012.063), p. 69 - 73 (1981).

099.219 **Heavy ions in Jupiter's environment.**
R. A. Brown.
Planetary aeronomy and astronomy, (see 012.063), p. 75 - 82 (1981).

The extended atmosphere of the Jupiter system consists of heavy-element atoms and ions. This material originates on the satellite Io. Energy is lost from the thermal plasma in collisionally-excited optical and ultra-violet emission. The juxtaposition of Earth and spacecraft measurements provide insight concerning the underlying processes of particle transport and energy supply.

099.220 **Interpretation of 6 and 21 cm observations of Jupiter.** I. de Pater.
Planetary aeronomy and astronomy, (see 012.063), p. 83 - 85 (1981).

Observations of Jupiter's radio emission were made at a frequency of 1412 MHz (21 cm) in 1977 and at 4885 MHz (6 cm) in 1978 using the Westerbork Synthesis Radio telescope in the Netherlands. The thermal disk temperatures at these wavelengths yield values of 320 ± 20 K at 21 cm and 220 ± 8K at 6 cm. The first value requires an overabundance of ammonia by a factor of three with respect to the solar value, the second one is in agreement with the value expected for a convective model atmosphere with solar abundances for all chemical elements. The data indicate there is a layered rather than a uniformly mixed atmosphere.

099.221 Ultraviolet emissions from the upper atmospheres of the planets.   H. W. Moos.
Planetary aeronomy and astronomy, (see 012.063), p. 155 - 164 (1981).

A large number of atoms and molecules have strong emission lines in the vacuum ultraviolet. As a result, this spectral region is particularly suited to studying the upper atmospheres of the planets. The observed emissions not only identify the constituents, but also provide information on the solar and magnetospheric excitation processes. Long term monitoring of these emissions can elucidate the effects of variations in the solar input as well as changes in magnetospheric conditions. A modest beginning in planetary upper atmospheric studies from earth orbit has been made using orbiting observatories. As examples of the power of this technique, some recent results are reviewed with an emphasis on Jupiter and the Io torus.

099.222 A differential measurement of the Lyman alpha emission from the giant planets using IUE.
K. H. Fricke, U. von Zahn, M. Combes, T. Encrenaz.
Planetary aeronomy and astronomy, (see 012.063), p. 165 - 168 (1981).

The intensity of the resonantly scattered Ly-α line of the giant planets depends on the scattering column length of atomic hydrogen above the methane layer and on the incident solar flux. The authors have obtained measurements of the Ly-α brightness of Jupiter and Saturn on December 19, 1979, with a time difference of 111 minutes, which is only slightly longer than the additional travel time for solar photons scattered at Saturn compared to those from Jupiter. This observational technique eliminates two major uncertainties − the use of different instruments and solar variability − affecting previous determinations of the relative brightness of the planets.

099.223 Vertical distribution of NH_3 in the upper Jovian atmosphere from IUE observations.   M. Combes,
R. Courtin, J. Caldwell, T. Encrenaz, K. H. Fricke, V. Moore, T. Owen, P. S. Butterworth.
Planetary aeronomy and astronomy, (see 012.063), p. 169 - 175 (1981).

The first unambiguous identification of ammonia in the upper atmosphere of Jupiter has been obtained from the observation of individual NH_3 bands in an IUE high resolution spectrum in the 2100−2400 Å spectral range. The variation with wavelength of the strengths of these NH_3 bands implies that the NH_3 abundance has to be strongly reduced by photolysis. Preliminary analysis by means of scattering models shows that the ammonia mixing ratio cannot be constant with altitude.

099.224 Observations of the Galilean satellites of Jupiter from earth orbit.   P. S. Butterworth.
Planetary aeronomy and astronomy, (see 012.063), p. 177 - 184 (1981).

In April 1972 OAO-2 obtained broadband filter measurements of the Galilean satellites from 2100 to 4300 Å. All four bodies were shown to have low albedos declining towards shorter wavelengths. The vast data return from Voyager spacecraft has not removed the need for continuing long time-base observations of the former. Since January 1978, IUE has repeatedly obtained Galilean spectra within the range 1150 to 3200 Å. Observations of Io have placed an upper limit on the global abundance of SO_2 in its atmosphere. Spectral variations with phase have allowed spatial mapping of surface reflectance in the case of Io, and may enable volcanic activity to be monitored.

099.225 Ultraviolet observations of Jupiter from earth-orbiting satellites.   V. Moore.
Planetary aeronomy and astronomy, (see 012.063), p. 185 - 191 (1981).

The International Ultraviolet Explorer (IUE) has provided both improved spectral resolution and some spatial resolution for UV observations of Jupiter. Previous satellite observations have produced albedo curves for Jupiter showing the influence of Rayleigh scattering, and of some absorber(s) shortward of 2500 Å on the UV spectrum. Constraints on the abundance of several minor constituents of the Jovian atmosphere were derived from the OAO-2 data. The IUE low dispersion data make it possible to detect individual molecular features. A series of C_2H_2 absorptions have been identified, and indications of NH_3 absorptions are present.

099.226 Rossby vortex as a model for Jupiter's Great Red Spot.   A. S. Volokitin, V. V. Krasnosel'skikh.
Dokl. AN SSSR, Tom 260, 588 - 591 (1981). In Russian.
Abstr. in Ref. zh., 51. Astron., 1.51.241 (1982).

099.227 Derivation of the collision probability between orbiting objects: the lifetimes of Jupiter's outer moons.   D. J. Kessler.
Icarus, Vol. 48, 39 - 48 (1981).

Equations are derived which relate orbital parameters to the probability of collision between orbiting objects. The equations are used in this paper to calculate the collision lifetime of Jupiter's eight outer satellites.

099.228 Irradiation of NH_3 −CH_4 mixtures as a model of photochemical processes in the Jovian planets and Titan.   J. P. Ferris, J. Y. Morimoto.
Icarus, Vol. 48, 118 - 126 (1981).

Photolysis of NH_3 in the presence of CH_4 with a 185-nm light source results in the generation of hot hydrogen atoms that abstract hydrogen from the CH_4 to produce CH_3. The extent of reaction of CH_4 was measured by the ratio of the moles of CH_4 reacted per mole of NH_3 decomposed ($\Delta CH_4/\Delta NH_3$). This ratio increases with diminishing NH_3 pressure at constant CH_4 pressure but it remains constant if $CH_4/NH_3 \gtrsim 3$. The $\Delta CH_4/\Delta NH_3$ ratio is independent of temperature in the range 298−156°K, suggesting that hot hydrogen atoms were responsible for the reaction of CH_4.

099.229 Stability of zonal flows on Jupiter.
B. J. Conrath, P. J. Gierasch, N. Nath.
Icarus, Vol. 48, 256 - 282 (1981).

Jupiter's troposphere is bounded below by a deep layer of fluid which is probably extremely close to adiabatic. This paper explores the effect of this new boundary condition on the stability of atmospheric jets.

099.230 The detection of HCN on Jupiter.
A. T. Tokunaga, S. C. Beck, T. R. Geballe,
J. H. Lacy, E. Serabyn.
Icarus, Vol. 48, 283 - 289 (1981).

The authors report the detection of HCN on Jupiter. They suggest that the Jovian HCN most likely originates from the photolysis of CH_4 and NH_3 in the lower stratosphere and upper troposphere. In addition, an upper limit was established for the column density of HCN on Saturn.

099.231 Fast-drift shadow events in Jupiter's decametric radio spectra.
J. J. Riihimaa, T. D. Carr, R. S. Flagg, W. B. Greenman,
P. P. Gombola, G. R. Lebo, J. A. Levy.
Icarus, Vol. 48, 298 - 307 (1981) = Contrib. No. 37 Dep. Astron. Univ. Florida.

High-resolution dynamic spectra of Jovian S-bursts frequently reveal sloping gaps crossing bands of L-burst emission with drift rates comparable to those of S-bursts. These "fast-drift shadow" events are often sharply bounded on one edge by an S-burst, and sometimes on both edges by a pair of S-bursts emanating from a common vertex. It is suggested that the investigation of such S- and L-burst interactions may provide new insights of considerable importance in the search for the Jovian decametric emission mechanism.

099.232 **Jupiter and Saturn.**   A. Ingersoll.
        The new solar system, (see 003.026), p. 117 - 128
(1981).

099.233 **The Galilean satellites.**   T. V. Johnson.
        The new solar system, (see 003.026), p. 143 - 160
(1981).

A catalogue of Jovian decametric radio observations from January 1978 to December 1979.
See Abstr. 002.047.

Anomalous $^{12}CH_4 : ^{13}CH_4$ strengths in $3\nu_3$.
See Abstr. 022.002.

Rossby soliton.   See Abstr. 022.059.

Spectral properties of ice mineral mixtures: implications on the composition of the Galilean satellites and other icy bodies.   See Abstr. 022.125.

Sulfur-containing glass: a possible colorant on Io and Amalthea.   See Abstr. 022.137.

Reflection spectra of molten sulfur: could sulfur liquid be on Io's surface?   See Abstr. 022.138.

Critical temperatures for dielectric-metal transition in hydrogen and implications for Jupiter and Saturn.
See Abstr. 022.176.

Rectification and enhancement of three severely distorted images of Jupiter's north polar region.
See Abstr. 031.636.

Radiometric performance of the Voyager cameras.
See Abstr. 032.550.

Rotacion de la tierra año 1979.
See Abstr. 044.034.

Rotacion de la tierra año 1980.
See Abstr. 044.035.

About the calculation of phenomena predictions of Jupiter's satellites.   See Abstr. 047.031.

Satellites galiléens de Jupiter. Phénomènes et configurations pour 1982.   See Abstr. 047.037.

The Voyager mission through the Jupiter encounters.
See Abstr. 051.022.

The Voyager encounters.   See Abstr. 051.062.

Role of scattering in emergent planetary radiation.
See Abstr. 063.053.

Parametric interaction and spatial collapse of beam-driven Langmuir waves in the solar wind.
See Abstr. 074.089.

Free oscillations of giant planets. Influence of rotation and ellipticity.   See Abstr. 091.049.

Rings in the solar system.
See Abstr. 091.055.

Free oscillations of giant planets. Influence of differential rotation.   See Abstr. 091.071.

The atmospheres of Venus and Jupiter.
See Abstr. 093.013.

An analysis of Barker's UV spectrum of SO_2 on Venus: implication for Io and for laboratory measurements.
See Abstr. 093.053.

Planetological considerations on the geologic intermediate stage of Mars between Moon and Earth and observations on the Jovian satellites.   See Abstr. 097.011.

Expected immersion of Saturn's magnetosphere in the jovian magnetic tail.   See Abstr. 100.002.

Thermal structure and dynamics of Saturn and Jupiter.   See Abstr. 100.004.

A comparison of characteristic times for satellite absorption of energetic protons trapped in the Jovian and Saturnian magnetic fields.   See Abstr. 100.065.

Spatially resolved studies of Saturn and Jupiter with the IUE.   See Abstr. 100.119.

On the orbital evolution of short period comets having low velocity encounters with Jupiter.
See Abstr. 102.025.

Variations of the orbit of comet P/Gehrels 3: temporary satellite captures by Jupiter.   See Abstr. 103.121.

Erratum

099.901 Erratum: "The surface of Io: geologic units,
         morphology, and tectonics" [Icarus, Vol. 43, 302 -
333 (1980)].   G. G. Schaber.
Icarus, Vol. 46, 292 (1981). – See Abstr. 28.099.152.

100 Saturn, Saturn Satellites

100.001 Collisional amplification of density fluctuations in Saturn's rings. J. Lukkari.
Nature, Vol. 292, 433 - 435 (1981).
The encounter of Voyager 1 with Saturn in November 1980 revealed the ringlet structure of its rings. As the theoretical examination of the collisional evolution of keplerian systems had predicted such a structure in dense matter as a consequence of amplified fluctuations in density, some new computer simulations have now been carried out to check this effect. These simulations actually led to a strong, irreversible growth in the density maximum.

100.002 Expected immersion of Saturn's magnetosphere in the jovian magnetic tail.
S. Grzędzielski, W. Macek, P. Oberc.
Nature, Vol. 292, 615 - 616 (1981).

100.003 Saturn's rings.
Nature, Vol. 292, 668 (1981).

100.004 Thermal structure and dynamics of Saturn and Jupiter.
J. A. Pirraglia, B. J. Conrath, M. D. Allison, P. J. Gierasch.
Nature, Vol. 292, 677 - 679 (1981).
High resolution Voyager IRIS measurements for Saturn and Jupiter are assembled in meridional cross-sections of the retrieved upper tropospheric temperatures. The calculated thermal wind shear in the upper troposphere is highly correlated on both planets with the cloud top winds derived from imaging data. In contrast, temperatures below ~ 300 mbar are not simply related to the zonal jet structure. The upper tropospheric temperatures seem to have been more consistently correlated with cloud top winds than with major albedo features at the time of the Voyager encounters.

100.005 Morphology of Saturn's aurora.
B. R. Sandel, A. L. Broadfoot.
Nature, Vol. 292, 679 - 682 (1981).
Aurorally-excited emissions of atomic and molecular hydrogen come from a narrow circumpolar band near $80°$ north and south latitude on Saturn. The aurorae, which lie near the edge of the polar cap region, are continuously excited in both the north and south. If the strong variations observed in the auroral intensity are temporal, rather than longitudinal, they may be related to the periodic structure in the Saturn kilometric radiation.

100.006 Saturn ionosphere: theoretical interpretation.
S. K. Atreya, J. H. Waite, Jr.
Nature, Vol. 292, 682 - 683 (1981).
Voyager 1 high latitude and Pioneer 11 equatorial ionospheric structure indicate a solar EUV-controlled ionosphere with a possible molecular ion in the topside. Vibrationally excited H_2 in the high latitudes may be an important loss mechanism. Dynamical effects are expected to be important for determining the peak density and its location.

100.007 C_3H_8 and C_3H_4 in Titan's atmosphere.
W. C. Maguire, R. A. Hanel, D. E. Jennings, V. G. Kunde, R. E. Samuelson.
Nature, Vol. 292, 683 - 686 (1981).
Four bands of propane C_3H_8 and two of methyl acetylene C_3H_4 have been identified in the Voyager IR spectrum of Titan. Stratospheric abundances of 2×10^{-5} for C_3H_8 and 3×10^{-8} for C_3H_4 have been determined for the mid-latitude region. A feature at $1,154$ cm^{-1}, previously assigned solely to CH_3D, is now identified at least in part due to C_3H_8.

100.008 C_4H_2, HC_3N and C_2N_2 in Titan's atmosphere.
V. G. Kunde, A. C. Aikin, R. A. Hanel, D. E. Jennings, W. C. Maguire, R. E. Samuelson.
Nature, Vol. 292, 686 - 688 (1981).
The compounds C_4H_2, HC_3N, and C_2N_2 have been detected in trace amounts in the stratosphere of Titan. The identification of two compounds containing nitrogen, in addition to HCN, provides further evidence for the abundance of free N_2 on Titan.

100.009 Mean molecular weight and hydrogen abundance of Titan's atmosphere.
R. E. Samuelson, R. A. Hanel, V. G. Kunde, W. C. Maguire.
Nature, Vol. 292, 688 - 693 (1981).
The $200 - 600$ cm^{-1} continuum opacity in the troposphere and lower stratosphere of Titan is inferred from thermal emission spectra from the Voyager 1 IR spectrometer (IRIS). The surface temperature and mean molecular weight are 94 K $< T_G < 97$ K and $28.3 < M < 29.2$ AMU, respectively. The mole fraction of molecular hydrogen is 0.002 ± 0.001, which is equivalent to an abundance of $\sim 0.2 \pm 0.1$ km amagat.

100.010 Titan's atmosphere: temperature and dynamics.
F. M. Flasar, R. E. Samuelson, B. J. Conrath.
Nature, Vol. 292, 693 - 698 (1981).
The authors present data obtained in three spectral intervals which sample the lower troposphere and surface, tropopause, and the upper stratosphere. They discuss the meridional and longitudinal thermal structure implied by the data. Finally, they estimate radiative time constants at various levels of Titan's atmosphere and examine a class of dynamical flows which are consistent with the data.

100.011 Implications of Titan's north-south brightness asymmetry.
L. A. Sromovsky, V. E. Suomi, J. B. Pollack, R. J. Krauss, S. S. Limaye, T. Owen, H. E. Revercomb, C. Sagan.
Nature, Vol. 292, 698 - 702 (1981).
Voyager 1 images of Titan, when normalized to remove limb darkening, reveal an axially symmetric brightness pattern with significant north-south asymmetry. This interhemispheric contrast seems to be a response to seasonal solar heating variations resulting from Titan's inclined spin axis. The contrast has a significant effect on Titan's disk-integrated brightness as seen from Earth, and probably accounts for most of the observed long term variation, with solar UV variations accounting for the remainder.

100.012 Density waves in Saturn's rings.
J. N. Cuzzi, J. J. Lissauer, F. H. Shu.
Nature, Vol. 292, 703 - 707 (1981).
Certain radial brightness variations in the outer Cassini division of Saturn's rings may be spiral density waves driven by Saturn's large moon Iapetus, in which case a value of ~ 16 g cm^{-2} for the surface density is calculated in the region where the waves are seen. The kinematic viscosity in the same region is ~ 170 cm^2 s^{-1} and the vertical scale height of the ring is estimated to be a maximum of ~ 40 m.

100.013 Moonlets in Saturn's rings?
J. J. Lissauer, F. H. Shu, J. N. Cuzzi.
Nature, Vol. 292, 707 - 711 (1981).
The brightness structure within Cassini's division in Saturn's rings is explained in terms of perturbations produced by moonlets embedded within an optically thin disk of smaller ring particles. The moonlets exert gravitational torques on neighbouring ring particles and create gaps; diffusion acts to fill the gaps. A new explanation is offered for the inner edge of

the Cassini division being located at the 2 : 1 resonance with Mimas.

100.014   Distribution of neutral gas and dust near Saturn.
E. C. Sittler, Jr., J. D. Scudder, H. S. Bridge.
Nature, Vol. 292, 711 - 714 (1981).

The distribution of neutral gas and dust within the magnetosphere of Saturn has been inferred from the electron velocity distribution functions measured by the Voyager 1 plasma science experiment. Substantial enhancements of neutral material near Titan and in the vicinity of Enceladus are found. The E ring is also shown to be larger than previously thought.

100.015   Low-frequency plasma waves near Saturn.
B. M. Pedersen, M. G. Aubier, J. K. Alexander.
Nature, Vol. 292, 714 - 716 (1981).

Voyager planetary radio astronomy observations of low frequency emissions detected around the time of closest approach to Saturn and near the outbound ring plane crossing are presented. Near the ring plane an electron density of between 5 and 20 electrons cm^{-3} at distances of $\sim 6R_S$ is estimated.

100.016   Impulsive radio discharges near Saturn.
D. R. Evans, J. W. Warwick, J. B. Pearce, T. D. Carr, J. J. Schauble.
Nature, Vol. 292, 716 - 718 (1981).

An unexpected type of emission was observed during the Voyager 1 Saturn encounter. This consisted of periodic episodes of many impulsive discharges throughout the frequency range (20.4 kHz to 40.2 MHz) of the planetary radio astronomy experiment.

100.017   X-ray and energetic neutral particle emission from Saturn's magnetosphere.
E. Kirsch, S. M. Krimigis, W. H. Ip, G. Gloeckler.
Nature, Vol. 292, 718 - 721 (1981).

Although Voyager 1 was not equipped for the detection of X-rays and neutral particles, its low energy charged particle detector (LECP) records suggest a significant flux of these radiations. X-rays could be due to substantial precipitating electron fluxes in the auroral region or the rings whereas energetic neutrals could be due to charge-exchange between trapped ions and Saturn's neutral hydrogen disk.

100.018   Topology of Saturn's main magnetic field.
M. H. Acuña, J. E. P. Connerney, N. F. Ness.
Nature, Vol. 292, 721 - 724 (1981).

The Voyager 1 magnetic field observations at Saturn confirm the principally dipolar topology of the planetary magnetic field and suggest the need for more general models which incorporate non-potential field sources external to the planet and within the planetary magnetosphere.

100.019   Saturn's ring current and inner magnetosphere.
J. E. P. Connerney, M. H. Acuña, N. F. Ness.
Nature, Vol. 292, 724 - 726 (1981).

The Voyager 1 magnetic field observations at Saturn reveal an equatorial system of (eastward) azimuthal currents, very similar in certain respects to that responsible for the jovian magnetodisk.

100.020   Arc structure in Saturn's radio dynamic spectra.
A. Boischot, Y. Leblanc, A. Lecacheux, B. M. Pedersen, M. L. Kaiser.
Nature, Vol. 292, 727 - 728 (1981).

The dynamic spectra of Saturn's kilometric radiation show arc structures whose main characteristics are described and compared with those observed in the decametre emission of Jupiter. The origin of the arcs is probably similar for the two planets.

100.021   Arcs in Saturn's radio spectra.
J. R. Thieman, M. L. Goldstein.
Nature, Vol. 292, 728 - 731 (1981).

Arcs appearing in the dynamic spectra of Saturn radio emission have been modelled using the same techniques as applied to jovian arcs. Their properties suggest a source region several tenths of a Saturn radius above the cloud tops in the late morning local time sector.

100.022   Saturnian kilometric radiation: statistical properties and beam geometry.
M. L. Kaiser, M. D. Desch, A. Lecacheux.
Nature, Vol. 292, 731 - 733 (1981).

Analysis of the average properties of Saturn's kilometre wavelength radio emission suggests that the source region is near the moon meridian in the northern auroral zone and/or the polar cusp.

100.023   Narrowband electromagnetic emissions from Saturn's magnetosphere.
D. A. Gurnett, W. S. Kurth, F. L. Scarf.
Nature, Vol. 292, 733 - 737 (1981).

A series of narrowband electromagnetic emissions were detected by the plasma wave instrument on board Voyager 1 coming from the inner region of Saturn's magnetosphere in the frequency range 3 - 30 kHz. These emissions have many similarities to continuum radiation detected in the Earth's magnetosphere and narrowband kilometric radiation in the jovian magnetosphere. The observed frequency spacing suggests that the emissions are being generated near Tethys, Dione and Rhea, probably in regions of large plasma density gradients associated with boundaries of the plasma sheet.

100.024   Saturn's radio emissions: rotational modulation.
C. K. Goertz, M. F. Thomsen, W.-H. Ip.
Nature, Vol. 292, 737 - 739 (1981).

The unexpected rotational modulation of the Saturn kilometric radiation and Saturn electrostatic discharges, as revealed by Voyager 1 observations, are discussed in terms of a ring-current system and the geometry of the dipole field.

100.025   Saturn's kilometric radiation: satellite modulation.
M. D. Desch, M. L. Kaiser.
Nature, Vol. 292, 739 - 741 (1981).

There is an episodic 66-h modulation of the Saturn kilometric radiation which is both frequency and Dione-phase dependent. The behaviour is significantly different from the way in which Io modulates the jovian emission.

100.026   Control of Saturn's kilometric radiation by Dione.
W. S. Kurth, D. A. Gurnett, F. L. Scarf.
Nature, Vol. 292, 742 - 745 (1981).

Voyager 1 observations of Saturn's kilometric radio emissions reveal a strong but apparently transitory control by the orbital phase of Dione. This may be a geometrical effect and a time-variable plasma torus associated with Dione could explain most of the observed details of the Dione modulation by creating a shadow zone near the equatorial plane.

100.027   Pre-encounter distributions of Saturn's low frequency radio emission.
T. D. Carr, J. J. Schauble, C. C. Schauble.
Nature, Vol. 292, 745 - 747 (1981).

An analysis of Voyager 1 pre-encounter data is presented in which one-month averages of flux density from Saturn are determined as functions of both central meridian lonitude and frequency. Comparisons of corresponding distributions for two one-month intervals seven months apart yield information on their stability, and a redetermination of the magnetospheric rotation period.

100.028 Plasma wave turbulence at planetary bow shocks.
F. L. Scarf, D. A. Gurnett, W. S. Kurth.
Nature, Vol. 292, 747 - 750 (1981).

Voyager 1 observations of plasma wave turbulence at Saturn's bow shock are discussed and compared with corresponding data from Jupiter, Earth, and Venus. The results suggest that the plasma instabilities that develop at the lower Mach number bow shocks of the terrestrial planets differ from those found at the high Mach number bow shocks of the outer planets.

100.029 Surface waves on Saturn's magnetopause.
R. P. Lepping, L. F. Burlaga, L. W. Klein.
Nature, Vol. 292, 750 - 753 (1981).

Voyager 1 magnetometer data have shown that small-amplitude surface waves occurred on Saturn's dayside magnetopause, causing multiple inbound crossings of this boundary. These waves were travelling approximately parallel to Saturn's equatorial plane along the magnetopause ("tailward"), suggesting that they were driven by the rotation of Saturn's magnetosphere. Hydromagnetic waves (possibly slow mode) were observed in the adjacent magnetosheath.

100.030 Saturn's magnetic tail: structure and dynamics.
K. W. Behannon, J. E. P. Connerney, N. F. Ness.
Nature, Vol. 292, 753 - 755 (1981).

Voyager 1 magnetic field observations have provided evidence of a saturnian magnetic tail. Tail current system distributions are inferred through comparison of the observations with a realistic magnetotail current system model. Temporal variations observed in the tail were probably produced by solar wind variations.

100.031 New statistical measurements of Saturn's rings.
E. Sassone-Corsi, P. Sassone-Corsi.
Strolling Astron., Vol. 29, 24 - 27 (1981).

100.032 A simple model of Saturn's rings. M. Hénon.
Nature, Vol. 293, 33 - 35 (1981).

The distribution of particle sizes in Saturn's rings is probably continuous over many orders of magnitude; there is no "typical particle size". A model based on this view accounts for a number of observed properties of the rings: apparent thickness, radar and radio observations, number and size distribution of the gaps discovered by Voyager 1, and optical thickness.

100.033 Formation of organic molecules on Titan.
L. A. Capone, S. S. Prasad, W. T. Huntress,
R. C. Whitten, J. Dubach, K. Santhanam.
Nature, Vol. 293, 45 - 46 (1981).

Voyager 1 has verified that molecular nitrogen is the major constituent on Titan and that HCN is also present. Based on these data, the authors propose that even more complex organic-nitrogen molecules such as ethyl cyanide (CH_3CH_2CN), vinyl cyanide (CH_2CHCN), and cyanoacetylene (HCCCN) may be formed efficiently in the lower atmosphere of Titan.

100.034 Der Planet Saturn. F. Renner.
Phys. unserer Zeit, 12. Jahrg., 99 - 106 (1981).

100.035 Voyager 2 ontmoet Saturnus. O. Namba.
Zenit, 8. Jaarg., 353 - 357 (1981).

100.036 Infrared reflectance spectra of Hyperion, Titania, and Triton.
L. A. Lebofsky, G. H. Rieke, M. J. Lebofsky.
Icarus, Vol. 46, 169 - 174 (1981).

Medium-resolution infrared (1 - 2.5 μm; $\Delta\lambda/\lambda \sim 0.05$) photometry of Triton, Titania, and Hyperion and medium-resolution (1.5 - 2.4 μm; $\Delta\lambda/\lambda \lesssim 0.01$) spectroscopy of Triton are presented. Hyperion and Titania have spectra roughly similar to the laboratory spectrum of water frost, while the spectrum of Triton is inconsistent with the spectra of frosts likely to be major surface constituents.

100.037 The 1966 observations of the coorbiting satellites of Saturn, S10 and S11.
S. M. Larson, B. A. Smith, J. W. Fountain, H. J. Reitsema.
Icarus, Vol. 46, 175 - 180 (1981).

Most of the positions of faint satellite images obtained during the 1966 Saturn ring plane crossing fit the period of the coorbital satellites 1980 S1 and 1980 S3. In 1966 the satellites were separated by 137° in orbital longitude. Until the mutual interaction of the satellites is understood and applied to derive the precise orbital motion, the 1966 and 1980 observations cannot be linked.

100.038 Observations of Saturn's outer ring and new satellites during the 1980 edge-on-presentation.
P. L. Lamy, N. Mauron.
Icarus, Vol. 46, 181 - 186 (1981).

Observations of Saturn's satellites and external rings during the 1980 edge-on-presentation were obtained with a focal coronograph. A faint satellite traveling in the orbit of Dione and leading it by 72° has been detected, together with the two inner satellites already suspected (cf. J. W. Fountain and S. M. Larson, 1978). The external ring has been observed on both east and west sides; it may extend up to $\cong$ 8.3 Saturn radii, and appears structured.

100.039 Saturn's rings: azimuthal variations, phase curves, and radial profiles in four colors.
W. T. Thompson, K. Lumme, W. M. Irvine, W. A. Baum,
L. W. Esposito.
Icarus, Vol. 46, 187 - 200 (1981) = Contrib. Five Coll. Astron. Dep., No. 450.

Four-color photographic photometry of Saturn for the 1977 - 1979 apparitions has been analyzed to determine the dependence of ring brightness on wavelength, solar phase angle, ring particle orbital phase angle (azimuthal effect), declination of the Earth relative to the ring plane (tilt angle), and radial distance from Saturn.

100.040 High-spectral-resolution imagery of Saturn.
W. H. Smith, T. B. McCord, W. Macy.
Icarus, Vol. 46, 256 - 262 (1981).

High-spectral-resolution images of Saturn over the wavelengths of the 6196.8 Å feature of CH_4 were obtained and used to extract line profiles for the observed feature over the disk of Saturn and to determine the spatial variation of the equivalent width at the time of the observations. The result of the observations is consistent only with a single class of models for the vertical structure of the atmosphere of Saturn, namely, the homogeneous scattering models.

100.041 The infrared spectrum of Rhea.
R. N. Clark, P. D. Owensby.
Icarus, Vol. 46, 354 - 360 (1981).

A new infrared spectrum of the leading side of Rhea is presented in the 0.65- to 2.5-μm region. The spectrum is very similar to the spectrum of the leading side of Ganymede in the 0.6- to 2.5-μm region. The Rhea spectrum is also very similar to laboratory spectra of water frost on ice blocks rather than that of an optically thick frost. The strong water ice absorption features, high albedo, and little downturn in reflectance toward shorter wavelengths from 0.6 to 0.4 μm all indicate a surface of nearly pure water ice.

100.042 Measurement of stratospheric aerosol on Saturn using an eclipse of Titan.
D. W. Smith, R. W. Shorthill, P. E. Johnson, E. Budding,
A. S. Asaad.
Icarus, Vol. 46, 424 - 428 (1981).

An eclipse of Titan by Saturn was observed on December 20, 1979, to measure the aerosol content in the atmosphere of Saturn. The measurements were made in the bandpass 6300 - 7300 Å and extend to ~5 magnitudes of eclipse darkening. The faint portion of the lightcurve unambiguously requires the presence of aerosol in the lower stratosphere of Saturn. The aerosol extends to at least 20 km above the tropopause.

100.043 **The F-ring of Saturn.**   R. Smoluchowski.
Geophys. Res. Lett., Vol. 8, 623 - 624, with a correction p. 946 (1981).

One of the startling discoveries made by the Voyager Saturn flyby is the existence of a split and twisted F-ring located outside of the well known A-ring of this planet. It is proposed that the twist is the result of the action of Saturn's magnetic field on charged particles in the F-ring. A quantitative argument shows that small particles will be deflected away from a circular orbit maintained by the larger particles. The two "shepherding" satellites prevent the small particles from being lost entirely from the ring.

100.044 **On the radial structure of Saturn's rings.**
W. R. Ward.
Geophys. Res. Lett., Vol. 8, 641 - 643 (1981).

Radial diffusion in a particulate Keplerian disc under the influence of a viscosity law $\nu(\sigma)$ that decreases with surface density σ more steeply than σ^{-1} will promote the breakdown of the system into an ensemble of thin rings. Possible application to Saturn's rings is considered.

100.045 **Après le survol de Saturne par Voyager 1 ...**
quelques photographies des anneaux et des satellites.
Ciel Terre, Vol. 97, 275 - 283 (1981).

100.046 **Voyager: science at Saturn.**
R. Berry.
Astronomy, Vol. 9, No. 2, p. 6 - 22 (1981). − Abstr. in Phys. Abstr., Vol. 84, Abstr. 71113 (1981).

100.047 **A study of ethane on Saturn in the 3 micron region.**
G. L. Bjoraker, H. P. Larson, U. Fink, H. A. Smith.
Astrophys. J., Vol. 248, 856 - 862 (1981).

The authors have detected C_2H_6 in absorption in high altitude spectroscopic observations of Saturn in the 3 μm region. Based upon comparisons with laboratory spectra of C_2H_6, the abundance estimate is $\eta a = 7.5 \pm 3.5$ cm-amagat, equivalent to a column abundance of 3.0 ± 1.4 cm-amagat. This value agrees well with predictions of models of CH_4 photolysis in the stratospheres of the outer planets. Compared to previous observational studies of gaseous hydrocarbons at thermal infrared wavelengths ($\lambda \sim 10~\mu$m), the abundance is very much less dependent upon modeling the detailed atmospheric temperature profile and the vertical distribution of C_2H_6 in Saturn's atmosphere. The observed column abundance is used as a constraint in the Strobel and the Yung and Strobel photochemical models to determine the vertical distribution of ethane on Saturn for two profiles of the eddy diffusion coefficient $K(z)$.

100.048 **On the stability of Saturn's rings.**
D. N. C. Lin, P. Bodenheimer.
Astrophys. J., Lett., Vol. 248, L83 - L86 (1981) = Lick Obs. Bull., No. 890.

The authors show that collision-dominated particle disks around planets may be unstable against a "pinch" instability induced by the nature of viscous diffusion. The authors propose that the existence of ringlets around Saturn is a manifestation of this instability.

100.049 **On the possible detection of CH_3D on Titan and Uranus.**
B. L. Lutz, C. de Bergh, J.-P. Maillard, T. Owen, J. Brault.
Astrophys. J., Lett., Vol. 248, L141 - L145 (1981).

Based on the analysis of a new band of CH_3D near 6425 cm^{-1}, possible identification of it in Fink and Larson's (1979) spectra of Titan and Uranus is proposed.

100.050 **Voyager 1 in the world of Saturn.**   G. A. Burba.
Priroda, 1981, No. 7, p. 92 - 98. In Russian.

100.051 **Saturn satellite observations and orbits from the 1980 ring plane crossing.**
P. K. Seidelmann, R. S. Harrington, D. Pascu, W. A. Baum, D. Currie, J. A. Westphal, G. E. Danielson.
Bull. American Astron. Soc., Vol. 13, 572 (1981). − Abstract.

100.052 **A method for determining the masses of the co-orbiting Saturnian satellites S10 and S11.**
C. F. Peters.
Bull. American Astron. Soc., Vol. 13, 573 (1981). − Abstract.

100.053 **The dynamics of the Saturnian satellites 1980S1 and 1980S3.**
R. S. Harrington, P. K. Seidelmann.
Bull. American Astron. Soc., Vol. 13, 573 (1981). − Abstract.

100.054 **On the structure of Saturn's rings.**
B. B. Kadomtsev.
Pis'ma v ZhEhTF, Tom 33, 361 - 363 (1981). In Russian.
Abstr. in Ref. zh., 51. Astron., 7.51.176; 62. Issled. kosm. prostranstva, 7.62.317 (1981).

100.055 **Saturn redux: the Voyager 2 mission.**
M. M. Waldrop.
Science, Vol. 213, 1236 - 1237, 1240 (1981).

100.056 **Voyager 2 at Saturn.**
Spaceflight, Vol. 23, 286 - 290 (1981).

100.057 **Low energy charged particles at Saturn.**
J. F. Carbary, S. M. Krimigis.
APL Tech. Dig., Vol. 2, 87 - 89 (1981).

Voyager 1 observations of low energy charged particles (electrons and ions) in the magnetosphere of Saturn are described. The ions consist primarily of protons. Molecular hydrogen and a low concentration of helium are also present.

100.058 **Saturn's E ring. I. CCD observations of March 1980.**
W. A. Baum, T. Kreidl, J. A. Westphal, G. E. Danielson, P. K. Seidelmann, D. Pascu, D. G. Currie.
Icarus, Vol. 47, 84 - 96 (1981).

The tenuous E ring of Saturn is found to commence abruptly at 3 Saturn radii, to peak sharply in the vicinity of the orbit of the satellite Enceladus (about 4 radii), and to spread out thinly to more than 8 radii. This distribution strongly suggests it to be associated with Enceladus and perhaps to be material ejected from Enceladus. The spread of E-ring material above and below the ring plane is greater in its tenuous outskirts than in its denser inner region, suggesting that the E ring may be at an early stage in its evolution.

100.059 **The dynamics of the Saturnian satellites 1980S1 and 1980S3.**
R. S. Harrington, P. K. Seidelmann.
Icarus, Vol. 47, 97 - 99 (1981).

The orbits of the Saturnian satellites 1980S1 and 1980S3 have approximately the same semimajor axes, but the difference in longitude librates between values of approximately +6° and −6° in 3000 days. Thus the satellites never approach extreme proximity and the orbits appear to be stable for extended periods of time.

100.060 **Saturn: "An even better look".**   J. K. Beatty.
Sky Telesc., Vol. 62, 329 - 332 (1981).

100.061 **Location of faint objects in the orbits of Tethys and Dione.** C. Veillet.
Astron. Astrophys., Vol. 102, L5 - L7 (1981).

An observing run on the Danish-ESO 1.5-m reflector (La Silla) provided in April 1981 a series of positions of three faint satellites of Saturn on the L_4 Lagrangian point of Saturn: Dione and the L_4 and L_5 points of Saturn: Tethys. This series permitted to determine an accurate position of the Tethys L_4 (1981 S1) and L_5 (1981 S2) objects and to discover a periodic variation of the Dione L_4 object (1980 S6, often called Dione B).

100.062 **Voyager at Saturn, act II.**   J. K. Beatty.
Sky Telesc., Vol. 62, 430 - 444 (1981).

100.063 **The puzzle that is Saturn.**   M. M. Waldrop.
Science, Vol. 213, 1347 - 1351 (1981).

100.064 **On a magnetic anomaly origin for the braids of Saturn's F-ring.**   A. Kyrala.
Mem. Soc. Astron. Italiana, Vol. 52, (see 012.023), 431 - 433 (1981).

100.065 **A comparison of characteristic times for satellite absorption of energetic protons trapped in the Jovian and Saturnian magnetic fields.**   L. L. Hood.
Geophys. Res. Lett., Vol. 8, 976 - 979 (1981).

The relative symmetry of the Saturnian magnetic field with respect to the rotational equatorial plane results in characteristic times for satellite absorption of trapped energetic protons that are typically one to three orders of magnitude smaller than the corresponding Jovian satellite absorption times. The maximum difference occurs for nearly equatorially mirroring particles. Assuming that the rates of radial diffusion are comparable within the two magnetospheres, the inner Saturnian satellites are more efficient absorbers of inwardly diffusing ions than their Jovian counterparts.

100.066 **Organic synthesis in the atmosphere of Titan.**
S. Gupta, E. Ochiai, C. Ponnamperuma.
Nature, Vol. 293, 725 - 727 (1981).

To gain some insight into the organic chemistry on Titan, in the light of the recent data provided by Voyager, the authors carried out several simulation experiments. The main objective was to assess the possible contributions to the organic chemistry from different possible energy sources which may be available on the satellite.

100.067 **Voyager 2 an Saturn vorbei – ein Bildbericht.**
H. W. Köhler.
Sterne Weltraum, Jahrg. 20, 402 - 404 (1981).

100.068 **The Voyager 1/Saturn encounter and the cosmogonic shadow effect.**   H. Alfvén.
Astrophys. Space Sci., Vol. 79, 491 - 505 (1981).

The Voyager 1/Saturn results demonstrate that the macro-structure of the Saturnian ring system can be explained as a result of a cosmogonic shadow effect working at the formation of the system. A similar analysis of the asteroidal belt shows that its macro-structure can also be explained by the cosmogonic shadow effect. The observational results demonstrate that during their formation both the Saturnian ring and the asteroidal belt passed a plasma state dominated by electromagnetic effects.

100.069 **Perturbations in the motion of the Saturn satellites. Part 3.**   I. G. Chugunov.
Mordovsk. gos. univ. Saransk, 1981. 15 pp. In Russian. – Abstr. in Ref. zh., 51. Astron., 10.51.137 (1981).

100.070 **Orbital elements of the Saturn satellites.**
I. G. Chugunov.
Mordovsk. gos. univ. Saransk, 1981. 15 pp. In Russian. – Abstr. in Ref. zh., 51. Astron., 10.51.138 (1981).

100.071 **Saturn satellite observations and orbits from the 1980 ring plane crossing.**
P. K. Seidelmann, R. S. Harrington, D. Pascu, W. A. Baum, D. G. Currie, J. A. Westphal, G. E. Danielson.
Icarus, Vol. 47, 282 - 287 (1981).

The ground-based observations of the recently discovered Saturnian satellites have been collected from the IAU Circulars and identified with and fit to four orbital groups: (1) the inner pair of coorbital librating satellites, (2) the satellite known as "Dione B" near the L_4 point of Dione-Saturn, (3) the satellites associated with the L_4 and L_5 points of Tethys-Saturn or, alternatively, one satellite uncomfortably near the orbit of Tethys, and (4) the F-ring satellites observed by Voyager I.

100.072 **Observations of the Saturn E ring and a new satellite.**
S. M. Larson, J. W. Fountain, B. A. Smith, H. J. Reitsema.
Icarus, Vol. 47, 288 - 290 (1981).

The faint E ring of Saturn appears as a narrow ring 246,000 ± 4,000 km from the center of Saturn on photographs taken when the ring-plane inclination was $5°.4$. The apparent brightness of the ring was uniform at all observed orbital longitudes and permits an estimate of the normal optical thickness. A faint satellite (1981S1) was observed near the L_4 triangular libration point of Tethys and is probably the same object as 1980S13.

100.073 **Behavior of Titan's atmosphere during a total eclipse.**
M. Combes, T. Encrenaz, N. Epchtein, T. Owen, J. Lecacheux.
Icarus, Vol. 47, 291 - 298 (1981).

The edge-on presentation of Saturn's rings and satellites system has provided a rare opportunity to observe total eclipses of Titan. During its emersion from the Saturnian shadow (1980, June 28), Titan has been observed simultaneously in the visible and the infrared ranges (6000 - 9000 Å, 11.8 μm and 20 μm). No change has been recorded in these three spectral ranges. The observations tend to support the thick-atmosphere model, which has been shown to be valid by Voyager a few months later.

100.074 **Saturnus-Voyager 2.**   O. Namba, G. Schilling.
Zenit, 8. Jaarg., 509 - 512 (1981).

100.075 **Saturn's youngest ring.**
W. A. Baum.
Publ. Astron. Soc. Pacific, Vol. 93, 544 - 545 (1981). Abstract.

100.076 **Monochromatic spatial scans of the disk of Saturn at ultraviolet and visual wavelengths.**
W. D. Cochran, L. Trafton.
Bull. American Astron. Soc., Vol. 13, 699 - 700 (1981). Abstract.

100.077 **Photoelectric photometry of Titan, 1972 - 1981.**
D. T. Thompson, G. W. Lockwood.
Bull. American Astron. Soc., Vol. 13, 700 (1981). – Abstract.

100.078 **The atmosphere of Titan: an analysis of the Voyager 1 radio occultation measurements.**
G. F. Lindal, H. B. Hotz, D. N. Sweetnam, V. R. Eshleman, G. L. Tyler.
Bull. American Astron. Soc., Vol. 13, 700 - 701 (1981). Abstract.

100.079 **Identification of organic compounds in Titan's atmosphere.**   W. C. Maguire, A. C. Aikin,

R. A. Hanel, D. E. Jennings, V. G. Kunde, R. E. Samuelson.
Bull. American Astron. Soc., Vol. 13, 701 (1981). − Abstract.

100.080   **The organic clouds of Titan.**
C. Sagan, B. N. Khare.
Bull. American Astron. Soc., Vol. 13, 701 (1981). − Abstract.

100.081   **Infrared opacity of Titan's stratosphere from Voyager 1 limb measurements.**
R. E. Samuelson, R. A. Hanel, V. G. Kunde, W. C. Maguire.
Bull. American Astron. Soc., Vol. 13, 701 (1981). − Abstract.

100.082   **The possible detection of Raman scattering in Titan's atmosphere: implications for atmospheric bulk.**   L. Trafton.
Bull. American Astron. Soc., Vol. 13, 701 - 702 (1981). Abstract.

100.083   **The photochemistry of methane and nitrogen on Titan.**
Y. L. Yung, M. Allen, J. P. Pinto, E. P. Gardner.
Bull. American Astron. Soc., Vol. 13, 702 (1981). − Abstract.

100.084   **Ion chemistry in the atmosphere of Titan.**
W. Huntress, V. Anicich.
Bull. American Astron. Soc., Vol. 13, 702 (1981). − Abstract.

100.085   **Ion-molecule and nitrogen chemistry in the upper atmosphere of Titan.**
M. Allen, Y. L. Yung, W. Huntress.
Bull. American Astron. Soc., Vol. 13, 702 (1981). − Abstract.

100.086   **The wavelength dependence of the albedos of Titan, Uranus, and Neptune.**
J. S. Neff, J. T. Bergstralh, E. S. Barker, A. Cochran, W. D. Cochran.
Bull. American Astron. Soc., Vol. 13, 702 (1981). − Abstract.

100.087   **The abundance of CH_3D in the atmosphere of Titan, derived from 8 to 14 μm thermal emission.**
S. J. Kim, J. Caldwell, A. T. Tokunaga.
Bull. American Astron. Soc., Vol. 13, 702 - 703 (1981). Abstract.

100.088   **Detection of CH_3D and measurements of the CH_3D/CH_4 ratio on Titan and Uranus.**
C. de Bergh, B. L. Lutz, G. W. Lockwood, T. Owen, J. Brault, J. C. Buriez.
Bull. American Astron. Soc., Vol. 13, 703 (1981). − Abstract.

100.089   **High phase angle Voyager images of Titan's main aerosol layer.**   K. A. Rages, J. B. Pollack.
Bull. American Astron. Soc., Vol. 13, 703 (1981). − Abstract.

100.090   **The microwave spectrum of Titan: compatibility with post-Voyager atmospheric models.**
W. R. Thompson, C. Sagan.
Bull. American Astron. Soc., Vol. 13, 703 (1981). − Abstract.

100.091   **Titan's turbulent atmosphere observed by Voyager radio occultation.**   D. P. Hinson, G. L. Tyler.
Bull. American Astron. Soc., Vol. 13, 703 (1981). − Abstract.

100.092   **The dark side of Iapetus.**
D. P. Cruikshank, R. Howell, C. Beerman, J. F. Bell, M. J. Gaffey, R. H. Brown, M. Rognstad.
Bull. American Astron. Soc., Vol. 13, 703 - 704 (1981). Abstract.

100.093   **Both sides of the Iapetus story.**
D. W. Smith, E. S. Barker, A. L. Cochran,

W. D. Cochran, P. E. Johnson, R. W. Shorthill.
Bull. American Astron. Soc., Vol. 13, 704 (1981). − Abstract.

100.094   **Origin of Saturn's satellites: explanation of compositional differences.**   S. J. Weidenschilling.
Bull. American Astron. Soc., Vol. 13, 704 (1981). − Abstract.

100.095   **Thermal history models of Saturn's icy satellites.**
K. Ellsworth, G. Schubert.
Bull. American Astron. Soc., Vol. 13, 704 (1981). − Abstract.

100.096   **Voyager photometry of Saturn's satellites.**
J. Veverka, P. Thomas, J. Gradie, T. V. Johnson, D. Morrison.
Bull. American Astron. Soc., Vol. 13, 720 (1981). − Abstract.

100.097   **UBV rotational lightcurves of the Saturn satellites Mimas, Enceladus, Tethys, Dione, and Rhea.**
O. G. Franz, R. L. Millis.
Bull. American Astron. Soc., Vol. 13, 720 (1981). − Abstract.

100.098   **Photometry and topography of Saturn's small satellites from Voyager data.**
P. Thomas, J. Veverka, D. Morrison, M. E. Davies.
Bull. American Astron. Soc., Vol. 13, 720 - 721 (1981). Abstract.

100.099   **Progress on the control nets of the satellites of Saturn.**   M. E. Davies.
Bull. American Astron. Soc., Vol. 13, 721 (1981). − Abstract.

100.100   **The acetylene and ethane abundances and the phosphine distribution in Saturn's atmosphere from the Voyager 1 − IRIS experiment.**
R. Courtin, D. Gautier, A. Marten, W. Maguire.
Bull. American Astron. Soc., Vol. 13, 722 (1981). − Abstract.

100.101   **The helium abundance on Saturn obtained from Voyager 1 infrared observations.**
B. J. Conrath, D. Gautier, R. A. Hanel, J. Hornstein.
Bull. American Astron. Soc., Vol. 13, 722 - 723 (1981). Abstract.

100.102   **New medium resolution 5 μm Saturn spectra.**
U. Fink, H. P. Larson, J. R. Johnson, G. Bjoraker.
Bull. American Astron. Soc., Vol. 13, 723 (1981). − Abstract.

100.103   **Temporal variations in the brightness of the atmosphere of Saturn.**
J. W. Fountain, H. J. Reitsema, B. A. Smith.
Bull. American Astron. Soc., Vol. 13, 723 (1981). − Abstract.

100.104   **Methane band images of Saturn.**   R. A. West, M. G. Tomasko, B. A. Smith, M. P. Wijesinghe, L. R. Doose, H. Reitsema, S. Larson.
Bull. American Astron. Soc., Vol. 13, 723 (1981). − Abstract.

100.105   **Polarimetry of Saturn at large phase angles.**
M. G. Tomasko, L. R. Doose.
Bull. American Astron. Soc., Vol. 13, 723 (1981). − Abstract.

100.106   **Optical polarimetry of Saturn globe: aerosols in upper Saturn atmosphere.**
R. Santer, A. Dollfus.
Bull. American Astron. Soc., Vol. 13, 723 - 724 (1981). Abstract.

100.107   **Voyagers' measurements of Saturn's main magnetic field.**   J. E. P. Connerney, M. H. Acuna, N. F. Ness.
Bull. American Astron. Soc., Vol. 13, 724 (1981). − Abstract.

100.108   **Gravitational measurements at Saturn with Voyager Doppler tracking data.**
J. D. Anderson,   J. K. Campbell.
Bull. American Astron. Soc., Vol. 13, 724 (1981). — Abstract.

100.109   **Radio occultation of Saturn's rings with Voyager 1: preliminary results.**   G. L. Tyler, E. A. Marouf.
Bull. American Astron. Soc., Vol. 13, 726 (1981). — Abstract.

100.110   **Diffraction by features in Saturn's rings: implications on edge sharpness and edge thickness.**
E. A. Marouf, G. L. Tyler.
Bull. American Astron. Soc., Vol. 13, 726 (1981). — Abstract.

100.111   **Voyager UVS observations of Saturn's rings.**
J. B. Holberg.
Bull. American Astron. Soc., Vol. 13, 726 (1981). — Abstract.

100.112   **Broadband spectrophotometry of Saturn's E ring.**
S. M. Larson, H. J. Reitsema, B. A. Smith.
Bull. American Astron. Soc., Vol. 13, 726 - 727 (1981).
Abstract.

100.113   **Photometry of the outer ring E of Saturn.**
A. Dollfus, S. Brunier.
Bull. American Astron. Soc., Vol. 13, 727 (1981). — Abstract.

100.114   **Phase curve and particle properties of Saturn's F ring.**   J. B. Pollack.
Bull. American Astron. Soc., Vol. 13, 727 (1981). — Abstract.

100.115   **Thickness of Saturn's A and B rings.**
W. A. Baum, T. J. Kreidl.
Bull. American Astron. Soc., Vol. 13, 727 (1981). — Abstract.

100.116   **The structure of Cassini's division.**   J. J. Lissauer.
Bull. American Astron. Soc., Vol. 13, 727 (1981).
Abstract.

100.117   **A morphological model for spoke formation in Saturn's rings.**
R. J. Terrile, G. Yagi, A. F. Cook, C. C. Porco.
Bull. American Astron. Soc., Vol. 13, 728 (1981). — Abstract.

100.118   **A numerical investigation of Saturn's F ring.**
M. R. Showalter.
Bull. American Astron. Soc., Vol. 13, 728 (1981). — Abstract.

100.119   **Spatially resolved studies of Saturn and Jupiter with the IUE.**
V. Moore, G. E. Hunt, J. Caldwell, T. Owen.
Bull. American Astron. Soc., Vol. 13, 739 (1981). — Abstract.

100.120   **Surface composition and radius of Hyperion.**
D. P. Cruikshank, R. H. Brown.
Bull. American Astron. Soc., Vol. 13, 741 (1981). — Abstract.

100.121   **Ejection! Vapor entrainment during cratering, and erosion of the Saturnian satellites.**
W. B. McKinnon.
Bull. American Astron. Soc., Vol. 13, 741 (1981). — Abstract.

100.122   **Atmospheric velocities of Saturn.**   R. Beebe,
G. Hunt, A. Ingersoll, A. Bunker, G. Garneau.
Bull. American Astron. Soc., Vol. 13, 742 (1981). — Abstract.

100.123   **Voyager 2 and Saturn.**   J. Bouška.
Říše hvězd, Vol. 62, 223 - 225 (1981). In Czech.

100.124   **Radius and limb darkening of Titan from speckle imaging.**
P. Nisenson, J. Apt, R. Goody, P. Horowitz.

Astron. J., Vol. 86, 1690 - 1693 (1981).
The authors have used a two-dimensional speckle-imaging technique to determine the radius (R) of Titan from ground-based visible-light observations. The value of R depends somewhat upon the limb-darkening exponent (α) used to model the observed disk; the probable range extends from $R = 2680 \pm 100$ km for $\alpha = 0.6$ to $R = 2900 \pm 100$ km for $\alpha = 1.3$, and is in good agreement with both Voyager 1 and Pioneer Saturn observations.

100.125   **Liquid water on the surface of Titan.**
Eh. M. Drobyshevskij.
Astron. Tsirk., No. 1118, p. 6 - 7 (1980). In Russian.

100.126   **Satellites of Saturn.**
IAU Circ., Nos. 3619, 3627, 3651 (1981).

100.127   **Saturn IX.**
IAU Circ., No. 3629 (1981).

100.128   **Saturn VII (Hyperion).**
IAU Circ., No. 3654 (1981).

100.129   **News about Saturn.**   Z. Borjan.
Vasiona, Année 29, 26 - 30 (1981). In Croatian.

100.130   **Voyager 2 passed near Saturn.**
N. Čabrić.
Vasiona, Année 29, 61 - 62 (1981). In Croatian.

100.131   **Voyager encounters with Saturn.**   B. A. Smith.
J. Opt. Soc. America, Vol. 71, 1591 (1981).
Abstract.

100.132   **Voyager 1 at Saturn.**   E. Mūkins.
Zvaigžņota debess, 1981. gada vasara, p. 26 - 35.
In Latvian.

100.133   **Saturn 1979 - 1980.**   A. W. Heath.
J. British Astron. Assoc., Vol. 92, 22 - 32 (1981).

100.134   **Saturn from Voyager 2.**
J. British Astron. Assoc., Vol. 92, 33 - 34 (1981).

100.135   **The visual harvest of Voyager 2.**
T. V. Johnson, R. J. Terrile.
Opt. Spectra, Vol. 15, No. 12, p. 50 - 53 (1981).

100.136   **Pioneer fly-by of Saturn and its rings.**
T. Gehrels, L. Esposito.
Progress in planetary exploration, (see 012.062), p. 67 - 71 (1981).
The authors report results from analysis of data from Pioneer Saturn's Imaging Photopolarimeter. These include the discovery of a new ring and satellite, the structure of the atmosphere of Saturn and Titan, the inhomogeneous nature of Saturn's rings, and a model for the rings' formation and bimodal particle size distribution.

100.137   **Scientific results from the Pioneer Saturn Infrared Radiometer.**   G. S. Orton, A. P. Ingersoll,
L. Froidevaux, G. Neugebauer, G. Münch, S. C. Chase.
Progress in planetary exploration, (see 012.062), p. 179 - 182 (1981).
The Pioneer 11 Infrared Radiometer instrument made observations of Saturn and its rings in broadband channels centered at 20 and 45 μm and obtained whole-disk information on Titan. A planetary average effective temperature of 96.5 ± 2.5 K implies a total emission 2.8 times the absorbed sunlight. Correlation with radio science results implies that the molar fraction of H_2 is $90 \pm 3\%$ (assuming the rest is He).

100.138   **Physical characteristics of the upper layers of Saturn's atmosphere.**
O. I. Bugaenko, A. V. Morozhenko.
Progress in planetary exploration, (see 012.062), p. 183 - 186 (1981).

Analysis of polarimetric observations of Saturn was carried out. In the long wavelength spectral range ($\lambda > 0.5$ μm) polarimetric observations do not contradict the model of spherical or irregular randomly oriented particles. In the short wavelength spectral interval ($\lambda < 0.5$ μm) it is necessary to take into account the scattering by oriented particles.

100.139   **Titan on the eve of Voyager encounter.**
J. Caldwell.
Planetary aeronomy and astronomy, (see 012.063), p. 87 - 91 (1981).

A decade of intense scientific study of Titan is reviewed. The atmosphere is not well understood at the time of this writing, but it is confidently expected that great progress will be made by the Voyager spacecraft now en route to the Saturn System.

100.140   **The marvels and rings of Saturn.**   K. Pines.
Mada, Vol. 25, 188 - 193 (1981). In Hebrew.

100.141   **On excess thermal fluxes of Titan and Saturn.**
Eh. M. Drobyshevskij.
Fiz.-tekh. inst. AN SSSR. Prepr., 1981, No. 715, 28 pp. In Russian. – Abstr. in Ref. zh., 51. Astron., 1.51.265 (1982).

100.142   **On the discovery of the fine structure of Saturn's ring and the asteroid belt.**
Yu. K. Gulak, V. I. Eremin, I. A. Mysyk, A. P. Rudenko, A. N. Timoshenko, P. M. Fedij.
Astron. Tsirk., No. 1165, p. 5 - 8 (1981). In Russian.

100.143   **The dynamics of tadpole and horseshoe orbits. II. The coorbital satellites of Saturn.**
S. F. Dermott, C. D. Murray.
Icarus, Vol. 48, 12 - 22 (1981).

The coorbital satellites of Saturn, 1980S1 and 1980S3, are shown to be librating in horseshoe orbits. The authors derive an accurate relation between the sum of the satellite masses and (a) their minimum angular separation, (b) the variation of their angular separation with time, and (c) the libration period. Observations of (b) and (c) are the most practical methods of determining the satellite masses. The orbits of the coorbital satellites of Dione and Tethys are discussed. The authors demonstrate the possibility of calculating a new value for the mass of Dione. The origin of coorbital satellites and the stability of their orbits are discussed.

100.144   **The libration of the Saturnian satellite Dione B.**
H. J. Reitsema.
Icarus, Vol. 48, 23 - 28 (1981).

Previously published positions of Dione B are combined with new observations reported in this paper to determine the motion. The data are fit to the analytic solution of Erdi (1978) to derive the parameters of the libratory motion. The satellite oscillates about the leading equilateral libration point (L_4) of Dione with a period of 785.0 ± 0.5 days. The maximum separation in orbital longitude from Dione of 76.7° was reached on UT 1980 April 20.3 ± 1 days; the minimum separation is 46.7°.

100.145   **Observations of Saturn's inner satellites and the orbit of Janus in 1980.**   A. Dollfus, S. Brunier.
Icarus, Vol. 48, 29 - 38 (1981).

Systematic observations of faint satellites were conducted at Pic-du-Midi with a focal coronograph from 1980 March 20 to 24; in addition, the 1966 discovery plates of satellite S 10 Janus were reexamined together with other 1966 observa-

tions. Among the 1980 observations, differences in magnitudes indicate that it is satellite 1980 S1 which corresponds to Janus. An object of magnitude 15–16 was seen not detached from the ring; it could be a condensation in the external part of the rings or an additional faint inner satellite.

100.146   **Orbits of the Tethys Lagrangian bodies.**
H. J. Reitsema.
Icarus, Vol. 48, 140 - 142 (1981).

The 1980 observations of the Saturn system have revealed objects at both the preceding (L_4) and following (L_5) triangular libration points of Tethys (S4). The observations indicate a small (~2°) libration amplitude for the L_4 body while the data on the L_5 object are insufficient to define its libration amplitude.

100.147   **Clearing the Cassini division.**   M. P. Schwarz.
Icarus, Vol. 48, 339 - 342 (1981).

A numerical simulation shows that the field of Mimas will clear a narrow gap in Saturn's rings just outside the 2 : 1 resonance with its orbital motion. This most likely corresponds to the 170-km-wide gap discovered by Voyager 1, rather than the much wider region referred to as the Cassini division.

100.148   **Titan.**   J. B. Pollack.
The new solar system, (see 003.026), p. 161 - 166 (1981).

A Monte Carlo approach to surface and interior scattering of tri-axial ellipsoids.   See Abstr. 021.045.

Anomalous $^{12}CH_4 : ^{13}CH_4$ strengths in $3\nu_3$.
See Abstr. 022.002.

Reflection spectra of simulated Titan organic clouds.
See Abstr. 022.126.

Critical temperatures for dielectric-metal transition in hydrogen and implications for Jupiter and Saturn.
See Abstr. 022.176.

Geometric rectification of spin-scan images from Pioneer 11.   See Abstr. 031.536.

Diffusion of ring particle orbits by a monolayer of gravitational scatterers.   See Abstr. 042.027.

Theory of satellite orbit – orbit resonance.
See Abstr. 042.036.

Numerical integration of the satellites of the outer planets.   See Abstr. 042.074.

Éphémérides des huit premiers satellites de Saturne pour 1982.   See Abstr. 047.038.

Entry and landing probe for Titan.
See Abstr. 051.020.

The Voyager encounters.   See Abstr. 051.062.

How Voyager 2 has been reprogrammed.
See Abstr. 053.001.

Voyager Saturn: preview of the August encounter.
See Abstr. 053.002.

Collisionless Saturnian rings.   See Abstr. 091.046.

Free oscillations of giant planets. Influence of rotation and ellipticity.   See Abstr. 091.049.

Rings in the solar system.
See Abstr. 091.055.

Free oscillations of giant planets. Influence of differential rotation.   See Abstr. 091.071.

Jupiter tail phenomena upstream from Saturn.
See Abstr. 099.004.

The satellites of Jupiter and Saturn.
See Abstr. 099.034.

Spatial imaging of UV emission from Jupiter and Saturn.   See Abstr. 099.040.

Observation of outer planets at Lyman alpha.
See Abstr. 099.043.

Jupiter's and Saturn's atmospheres.
See Abstr. 099.048.

Preliminary analysis of mutual satellite events in 1979/80.   See Abstr. 099.050.

Ground-based observations of HCN on Jupiter and Titan.   See Abstr. 099.146.

The rings of Jupiter, Saturn, and Uranus.
See Abstr. 099.148.

Effects due to charged particle impact on the icy satellites of Jupiter and Saturn.   See Abstr. 099.160.

Jupiter and Saturn: a reduced NH_3 abundance at high latitudes?   See Abstr. 099.170.

A comparison of the dynamics of convective features in the atmospheres of Jupiter and Saturn using Voyager images.
See Abstr. 099.182.

New models of Jupiter and Saturn: contrasting interiors?   See Abstr. 099.184.

Jupiter and Saturn.   See Abstr. 099.192.

Voyager and Pioneer encountered Jupiter and Saturn.   See Abstr. 099.203.

Observations of planets with the Danjon astrolabe of the Shanghai Observatory during 1978.12–1979.4.
See Abstr. 099.212.

Free oscillations of the giant planets.
See Abstr. 099.213.

The magnetic fields of Jupiter and Saturn.
See Abstr. 099.215.

A differential measurement of the Lyman alpha emission from the giant planets using IUE.
See Abstr. 099.222.

Irradiation of NH_3-CH_4 mixtures as a model of photochemical processes in the Jovian planets and Titan.
See Abstr. 099.228.

The detection of HCN on Jupiter.
See Abstr. 099.230.

Jupiter and Saturn.   See Abstr. 099.232.

The outer solar system   See Abstr. 101.040.

Voyager and the formation of the Saturnian satellite system.   See Abstr. 107.016.

The Voyager 1/Saturn encounter and the cosmogonic shadow effect.   See Abstr. 107.017.

101 Uranus, Neptune, Pluto, Transplutonian Planets

101.001 The ice layer in Uranus and Neptune – diamonds in the sky?   M. Ross.
Nature, Vol. 292, 435 - 436 (1981).
Many of the current models of Uranus and Neptune postulate a three-layer structure, consisting of an inner rocky core, a middle 'ice' layer of fluid, H_2O, CH_4, NH_3 and an outer hydrogen-helium layer of solar composition. The author points out that shockwave experiments on these liquids, as well as theoretical studies, imply that the H_2O and NH_3 in the ice layer are almost totally ionized and the CH_4 has been pyrolysed to carbon, possibly in the metallic or diamond form.

101.002 The periods of Neptune: evidence for atmospheric motions.
M. J. S. Belton, L. Wallace, S. Howard.
Icarus, Vol. 46, 263 - 274 (1981).
An extended photometric time series in the J and K bands of Neptune has a complex appearance which appears to require the simultaneous presence of three periodicities plus related harmonics in the $(J-K)$ color. The most apparent of the fundamental periods is N1 = 17.73 hr. The two others are at N2 = 18.56 and N3 = 18.29 hr. The presence of multiple period-

icity is indicating that distinct systems of zonal winds exist on the planet.

101.003 Spectrophotometry and upper limit of gaseous CH_4 for Triton.
J. R. Johnson, U. Fink, B. A. Smith, H. J. Reitsema.
Icarus, Vol. 46, 288 - 291 (1981).
Spectra of Triton with a CCD spectrometer yielded a relative spectral reflectivity curve from 0.56 to 1.05 μm at a resolution of 25 Å. Using low-temperature band model parameters from Fink et al. (1980), an upper limit for the one-way path gaseous CH_4 abundance of 1 m-am was derived.

101.004 Some thoughts on planet "X".   P. Moore.
J. British Astron. Assoc., Vol. 91, 483 - 487 (1981).
The discovery of Pluto is discussed and the difficulties mentioned of supposing it to be the disturbing force acting on Neptune. The suggestion that there is a large unknown planet still to be detected is examined and an estimated position is given.

101.005 Observations of Uranus with the International Ultraviolet Explorer.   J. Caldwell, T. Owen,

A. R. Rivolo, V. Moore, P. S. Butterworth, G. E. Hunt.
The Universe at ultraviolet wavelengths, (see 012.009), p. 83
(1981). — Abstract.

101.006   **Perturbations of a trans-Neptunian planet.**
          T. C. Van Flandern, K. F. Pulkkinen. E. J. Santoro,
P. K. Seidelmann, R. S. Harrington.
Bull. American Astron. Soc., Vol. 13, 568 (1981). — Abstract.

101.007   **A redetermination of the orbit of Triton.**
          A. W. Harris.
Bull. American Astron. Soc., Vol. 13, 573 (1981). — Abstract.

101.008   **The period of Charon, derived from the observed**
          barycentric motion of Pluto.
T. C. Van Flandern, K. F. Pulkkinen, J. Ries, R. L. Duncombe.
Bull. American Astron. Soc., Vol. 13, 573 (1981). — Abstract.

101.009   **Neptune's rings fading.**   R. A. Kerr.
          Science, Vol. 213, 1240 (1981).

101.010   **Upper limit of the gaseous CH_4 abundance on**
          Triton.
M. Combes, T. Encrenaz, J. Lecacheux, C. Perrier.
Icarus, Vol. 47, 139 - 141 (1981).
          A spectrum of Triton between 6000 and 9000 Å was
recorded in June 1980 at the ESO 1.52 m telescope. From
these data, an upper limit of 3.5 m-am is derived for the CH_4
gaseous abundance on Triton.

101.011   **The core and the magnetic field of Uranus.**
          M. Torbett, R. Smoluchowski.
Proc. Eleventh Lunar Planet. Sci. Conf., (see 012.020),
p. 1907 - 1913 (1980).
          The presence of a magnetic field on Uranus can be ex-
plained in a natural way using the recent three-layer model of
this planet because, in contrast to earlier models, the core is
metallic, mostly liquid and gravitational differentiation
produces enough heat in it to drive a hydromagnetic dynamo.

101.012   **Analytical two-dimensional model for a pole-on**
          magnetosphere.
H. Biernat, N. Kömle, H. Rucker.
Planet. Space Sci., Vol. 29, 1101 - 1107 (1981).
          Magnetospheres, which result from a solar wind flow
parallel to the magnetic dipole, are expected for Uranus at its
solstitial points and perhaps for the Palaeo-Earth during the
times of field reversal. The authors present a two-dimensional
analytical model for a pole-on magnetopause and the magnetospheric magnetic field line configuration, employing a
conformal mapping method.

101.013   **Unidentified features in the spectrum of Triton.**
          G. H. Rieke, L. A. Lebofsky, M. J. Lebofsky,
E. F. Montgomery.
Nature, Vol. 294, 59 - 60 (1981).
          The authors have used the Multiple Mirror Telescope to
obtain an improved IR spectrum of Triton. Their observations
show features that are not in detailed agreement with the identification of methane, although the general spectral behaviour
that led to this identification is confirmed. A satisfactory identification of the surface and/or atmospheric composition on
Triton does not yet seem possible.

101.014   **At the borders of the solar system: Pluto and**
          Charon.   F. Mignard, D. Bonneau.
Recherche, No. 123, p. 738 - 740 (1981). In French. — Abstr.
in Phys. Abstr., Vol. 84, Abstr. 98758 (1981).

101.015   **The rotation period of Neptune's upper atmosphere.**
          R. H. Brown, D. P. Cruikshank, A. T. Tokunaga.
Icarus, Vol. 47, 159 - 165 (1981).

Significant variations in the near-infrared brightness of
Neptune during July and August 1980 were observed. These
observations show a well-defined, large-amplitude variation in
Neptune's $J-K$ color, with a period of 17.73 ± 0.1 hr and are
interpreted as diurnal variations resulting from the 17.73-hr
rotation period of the upper atmosphere of Neptune in the
presence of inhomogeneous weather. A new 5-μm measurement of Uranus is also reported.

101.016   **On the origin of the Pluto–Charon system.**
          D. N. C. Lin.
Mon. Not. R. Astron. Soc., Vol. 197, 1081 - 1085 (1981).
          It is shown that if the Pluto–Charon system was once a
double satellite of Neptune, Pluto and Charon's combined
tidal interaction would have driven them to merge with each
other before their orbits around Neptune had evolved
significantly. It is proposed that the Pluto–Charon system was
formed by the binary fission of a rapidly rotating object.
Dynamical constraints indicate that a mass ratio between
Charon and Pluto should be less than 0.25 and greater than
0.05.

101.017   **Near-infrared spectrophotometry of the Uranian**
          satellites.   R. H. Brown, D. P. Cruikshank.
Bull. American Astron. Soc., Vol. 13, 721 (1981). — Abstract.

101.018   **VLA observations of Uranus.**
          W. Jaffe, G. Berge, J. Caldwell, T. Owen.
Bull. American Astron. Soc., Vol. 13, 732 (1981). — Abstract.

101.019   **Asymmetrical profiles of the H_2 (4,0) quadrupole**
          lines in the spectrum of Uranus.
J. T. Trauger, J. T. Bergstralh.
Bull. American Astron. Soc., Vol. 13, 732 (1981). — Abstract.

101.020   **Changes in the geometrical albedo of Uranus since**
          1961.
G. W. Lockwood, B. L. Lutz, D. T. Thompson, A. Warnock III.
Bull. American Astron. Soc., Vol. 13, 732 (1981). — Abstract.

101.021   **Evidence for thermal inversions in the lower strato-**
          spheres of Uranus and Neptune.
G. S. Orton, A. T. Tokunaga, J. Caldwell.
Bull. American Astron. Soc., Vol. 13, 732 (1981). — Abstract.

101.022   **The temperature of the upper atmosphere of Uranus.**
          B. Sicardy, M. Combes.
Bull. American Astron. Soc., Vol. 13, 732 (1981). — Abstract.

101.023   **Neptune: photoelectric slit scans of the disk.**
          M. J. Price, O. G. Franz.
Bull. American Astron. Soc., Vol. 13, 733 (1981). — Abstract.

101.024   **Continued studies of the rotation of Neptune.**
          R. R. Howell, R. H. Brown, D. P. Cruikshank,
J. S. Morgan, E. Shaya.
Bull. American Astron. Soc., Vol. 13, 733 (1981). — Abstract.

101.025   **Further evidence for rotation periods of Uranus**
          and Neptune.   D. Slavsky, H. J. Smith.
Bull. American Astron. Soc., Vol. 13, 733 (1981). — Abstract.

101.026   **Seasonal variations in the global dynamical regime of**
          Pluto's atmosphere.   S. A. Stern, L. Trafton.
Bull. American Astron. Soc., Vol. 13, 734 (1981). — Abstract.

101.027   **The periods of Neptune: evidence for atmospheric**
          motions.   M. J. S. Belton, L. Wallace, S. Howard.
Bull. American Astron. Soc., Vol. 13, 735 (1981). — Abstract.

101.028   200 years since the discovery of Uranus.
I. Molnár.
Kozmos, Vol. 12, 98 - 101 (1981). In Slovak.

101.029   Why don't the rings of Uranus decay?
V. Pohánka.
Kozmos, Vol. 12, 99 - 102 (1981). In Slovak.

101.030   The mean elements of Nereid.   F. Mignard.
Astron. J., Vol. 86, 1728 - 1729 (1981).
The mean elements of Nereid are determined from the 44 known observations. The gravitational theory used included the solar disturbances. As a by-product the mass of Neptune is found to be $m^{-1} = 19\,402 \pm 110$.

101.031   No evidence of rings around Neptune.
J. L. Elliot, D. J. Mink, J. H. Elias, R. L. Baron,
E. Dunham, J. E. Pingree, R. G. French, W. Liller,
P. D. Nicholson, T. J. Jones, O. G. Franz.
Nature, Vol. 294, 526 - 529 (1981).
Observations of two stellar occultations by Neptune were used to search for possible equatorial and polar rings. No ring occultation events were identified, and an upper limit of 0.07 can be placed on the optical depth of any equatorial rings greater than 5 km wide with radii greater than 31,400 km. Any ring system of Neptune must be much less extensive than the ring systems of Uranus and Saturn, but a jovian-type ring of low optical depth would have escaped detection by this search.

101.032   The Uranus occultation of August 15, 1980.
P. Bouchet, C. Perrier, A. Brahic, J. Lecacheux,
B. Sicardy.
Messenger, No. 26, p. 18 - 21 (1981).

101.033   The discovery of the third satellite of Neptune.
M. Jeličić.
Vasiona, Année 29, 62 - 63 (1981). In Serbo-Croatian.

101.034   Portrait of Pluto 1980.   E. Mūkins.
Zvaigžņotā debess, 1981. gada pavasaris, p. 13 - 14.
In Latvian.

101.035   Unidentified features in the spectrum of Triton.
G. H. Rieke, L. A. Lebofsky, M. J. Lebofsky,
E. F. Montgomery.
Prepr. Steward Obs., No. 333, 5 pp. (1981).
The infrared spectrum of Triton, the largest satellite of Neptune, has some resemblance to the spectrum of methane or methane frost, but this identification is not consistent with all available data. The authors have used the Multiple Mirror Telescope (MMT) to obtain an improved infrared spectrum of Triton. These observations show features that are not in detailed agreement with the identification of methane, although the general spectral behavior that led to this identification is confirmed. A satisfactory identification of the surface and/or atmospheric composition on Triton does not seem possible at present.

101.036   Speckle interferometric observations of Pluto and
Charon.   E. K. Hege, E. N. Hubbard,
J. D. Drummond, P. A. Strittmatter, S. P. Worden, T. Lauer.
Prepr. Steward Obs., No. 355, 26 pp. (1981).
The authors report speckle interferometric observations of Pluto and its moon Charon obtained on 5 June 1980 with a single 1.8 meter mirror of the Multiple Mirror Telescope. The observations yield a separation of $0.''31$ ($\pm 0.''05$) between Pluto and Charon at position angle 285° ($\pm 7°$) for JD 2444395.75. This result and other direct observations indicate an adjustment of -4.0 hours to the orbital epoch of Harrington and Christy (1981). The authors' observation which represents the first resolution of the system near mini-

mum separation, also suggests that the inclination of the orbit to the plane of the sky should be increased by 3°.

101.037   Geocentric ephemeris of Pluto 1975 - 1985.
G. H. Kaplan.
United States Naval Obs. Circ., No. 162, 26 pp. (1980).

101.038   Pluto's atmospheric bulk near perihelion.
L. Trafton.
Planetary aeronomy and astronomy, (see 012.063), p. 93 - 97 (1981).
The detection of CH_4 frost on Pluto's surface implies a significant atmosphere for Pluto. Although Pluto's mass is small the rapid escape of gaseous CH_4 can be prevented by the presence of a heavy gas mixed with the CH_4. The resulting slow escape of CH_4 can be accomodated by sublimation of the surface CH_4 frost so that an atmosphere exists in the steady state. Pluto is currently near perihelion where the CH_4 component of the atmosphere may be 500 times denser than at aphelion.

101.039   Can magnetic fields be generated in the icy mantles
of Uranus and Neptune?
R. Smoluchowski, M. Torbett.
Icarus, Vol. 48, 146 - 148 (1981).
It has been shown by the authors previously that a hydromagnetic dynamo can operate in the core of Uranus but probably not on Neptune. A similar analysis is made for the "icy" liquid mantles of both planets. It is concluded that pressure ionization and the associated increased conductivity of water is probably not enough to satisfy the necessary conditions for a dynamo on Uranus and that it is marginal for Neptune. On the other hand the expected presence of metallic water in a thick layer around the core of Neptune makes the operation of a dynamo on this planet plausible. A similar layer on Uranus might be too thin to play the same role.

101.040   The outer solar system.
D. Morrison, D. P. Cruikshank.
The new solar system, (see 003.026), p. 167 - 176 (1981).

A note of Galileo's observation of Neptune.
See Abstr. 004.012.

Bicentennial of the discovery of Uranus.
See Abstr. 004.049.

La recherche d'une planète trans-neptunienne ou la
découverte de Pluton.   See Abstr. 004.055.

L'observation de la planète Pluton à l'Observatoire
de Meudon en 1930.   See Abstr. 004.056.

Planetary science with Space Telescope.
See Abstr. 032.620.

Rings in the solar system.
See Abstr. 091.055.

Discovery of a probable third satellite of Neptune.
See Abstr. 096.013.

Negative results of occultation searches for
Neptunian rings.   See Abstr. 096.014.

No rings detected around Neptune: limits from oc-
cultation observations.   See Abstr. 096.015.

Future occultations by Uranus, Neptune, and their
satellites.   See Abstr. 096.016.

Observation of outer planets at Lyman alpha.
See Abstr. 099.043.

Positions of Jupiter, Galilean satellites and Pluto obtained in May 1980 with GPO of the ESO, La Silla. See Abstr. 099.065.

The rings of Jupiter, Saturn, and Uranus. See Abstr. 099.148.

Irradiation of $NH_3 - CH_4$ mixtures as a model of photochemical processes in the Jovian planets and Titan. See Abstr. 099.228.

Infrared reflectance spectra of Hyperion, Titania, and Triton.   See Abstr. 100.036.

On the possible detection of $CH_3 D$ on Titan and Uranus.   See Abstr. 100.049.

The wavelength dependence of the albedos of Titan, Uranus, and Neptune.   See Abstr. 100.086.

Detection of $CH_3 D$ and measurements of the $CH_3 D/CH_4$ ratio on Titan and Uranus. See Abstr. 100.088.

102 Comets (Origin, Structure, Atmospheres, Dynamics)

102.001 **Statistical test of the distribution of perihelion points of long-period comets – II.**
S. Yabushita, I. Hasegawa.
Mon. Not. R. Astron. Soc., Vol. 196, 353 - 356 (1981).

Long-period comets are classified by their absolute magnitude, and a statistical test is made as to whether or not their perihelion points can be regarded as uniformly distributed in the northern and southern hemispheres. It is shown that the perihelia of faint comets (absolute magnitude, $H_0 \geqslant 7.0$) are more or less uniformly distributed in each hemisphere, but that the perihelia of bright comets ($H_0 \leqslant 6.9$) cannot be regarded as uniformly distributed. It may be concluded that the apparent deviation from uniform distribution is due to bright comets, and that it cannot be ascribed to observational selection.

102.002 **A Monte Carlo investigation of Jovian perturbations on short-period comet orbits.**
C. Froeschlé, H. Rickman.
Icarus, Vol. 46, 400 - 414 (1981).

The authors present statistical distributions of Jovian perturbations on short-period comet orbits resulting from accurate numerical integrations. The sample of 60,000 cometary orbits with low inclinations and random orientations is characterized by perihelia between 0 and 7 AU and aphelia between 4 and 13 AU. The perturbations considered are those experienced because of Jupiter's gravitation per orbital revolution by the comets.

102.003 **How a cometary nucleus turns on.**
I. R. Ferrin, E. Guzman.
Sky Telesc., Vol. 62, 103 - 104 (1981).

102.004 **Comets.**
M. Festou, P. Lamy.
Recherche, No. 118, p. 46 - 56 (1981). In French. – Abstr. in Phys. Abstr., Vol. 84, Abstr. 71129 (1981).

102.005 **On the microwave emission from comets.**
D. M. Gibson, R. W. Hobbs.
Astrophys. J., Vol. 248, 863 - 866 (1981).

The authors have derived a formula which can be used to predict the microwave flux density from the icy grain halo of a new comet. This model can account for all of the radio continuum observations of comets made to date, provided the gas production rate, Z, is allowed to vary about its nominally accepted mean value by about a factor of 5. An implied consequence of this model is that the thermal radio emission arises from grains which are approximately a few cm in size, not ~ 1 mm as given in earlier models.

102.006 **Where is the ice in comets?**
M. F. A'Hearn, E. Dwek, A. T. Tokunaga.
Astrophys. J., Lett., Vol. 248, L147 - L151 (1981).

$JHKL$ photometry of comets P/Tuttle, Meier 1980q, P/Stephan-Oterma, and Bowell 1980b has yielded $J-H$ and $H-K$ colors, uncontaminated by thermal emission, which are nearly identical for all four comets. The colors are inconsistent with the reflection spectrum from a cloud of icy particles, if the particles are composed primarily of any of the ices (H_2O, CO_2 CH_4, or NH_3) commonly assumed to be present in the nuclei of comets. Circular variable filter spectra of comet Stephan-Oterma confirm the absence of any absorption features due to these ices. The reflection spectra of these comets appear most like those of C- and S-type asteroids and the ring of Jupiter. Mie calculations for core-mantle particles suggest that ices could be present as thin mantles on inherently reddish, refractory cores, such as magnetite.

102.007 **Interplanetary gas. XXVII. A catalogue of disconnection events in cometary plasma tails.**
M. B. Niedner, Jr.
Astrophys. J., Suppl. Ser., Vol. 46, 141 - 157, plates 1 - 4 (1981).

A catalog of 72 disconnection events (DEs) in cometary plasma tails is presented. The DEs span the period 1892–1976, which covers nearly the entire era of cometary photography, as well as eight solar cycles. Reduction of the observations consists primarily of measurements of the distance of rejected tails from the cometary head, from which several kinematical properties have been derived. The geometrical circumstances of each DE, as well as a brief description of each event, are also given.

102.008 **On the observed excess of retrograde orbits among long-period comets.** J. A. Fernández.
Mon. Not. R. Astron. Soc., Vol. 197, 265 - 273 (1981).

The distribution of orbital inclinations of the observed long-period comets is analysed. An excess of retrograde orbits is found which increases with the perihelion distance, except for the range $1.1 < q < 2$ AU. It is suggested that the excess of retrograde orbits among long-period comets is related to an already existent excess among the incoming new comets. Using theoretical considerations and a numerical model it is proposed that an important fraction of the so-called new comets are actually repeating passages through the planetary region. An important consequence of the presence of these comets among the new ones is the production of an excess of retrograde orbits in the whole sample.

102.009 **Ultraviolet spectroscopy of comets.**
J. Rahe.
Second European IUE Conference, (see 012.011), p. XV - XXVI (1980).

In January 1970, the first ultraviolet observations of a comet, Comet Tago-Sato-Kosaka (1969 IX), were obtained by the Orbiting Astronomical Observatory (OAO−2). On October 15, 1978, the IUE satellite was successfully used for the first time for cometary observations, and the first UV spectra of a comet, Comet Seargent 1978m, were obtained two weeks after the comet's discovery. The second comet observed with IUE was Comet Bradfield 1979l. The comets were observed at various heliocentric distances in the low and high dispersion mode. The UV observations were complemented by optical and radio measurements. The main results of these observations are presented.

102.010 **Rotation and precession of cometary nuclei.**
Z. Sekanina.
Annu. Rev. Earth Planet. Sci., Vol. 9, (see 003.003), 113 - 145 (1981).

This review is concerned with a recent breakthrough in the investigation of polar-axis orientations and spin rates, and with the implications for the surface structure of cometary nuclei. The rotation data for twelve comets are summarized.

102.011 **Study of comets from space: new aims.**
O. V. Dobrovol'skij, S. Ibadov.
Dokl. AN TadzhSSR, Tom 23, 634 - 637 (1980). In Russian. Abstr. in Ref. zh., 51. Astron., 7.51.189 (1981).

102.012 **Outbursts of cometary brightness.**
D. A. Andrienko, V. N. Vashchenko.
Problems of cosmic physics. Vyp. 16, (see 003.005), p. 3 - 20 (1981). In Russian.

The article deals with results of the treatment of a catalogue compiled by the authors. The shapes of mean photometric curves of cometary brightness outbursts are obtained. The distributions of frequency, amplitude, duration of forward and backward fronts as well as mean heliocentric distances of cometary outbursts at the phase of the 11-year solar cycle are given. The peculiarities of cometary outburst activity at odd and even cycles are noted.

102.013 **A generalized method for solution of the inverse problem of the mechanical theory of cometary forms.**   Kh. I. Ibadinov.
Dokl. AN TadzhSSR, Tom 24, 22 - 27 (1981). In Russian.
Abstr. in Ref. zh., 51. Astron., 8.51.261 (1981).

102.014 **The nature of comets.**
F. L. Whipple.
Comets and the origin of life, (see 012.022), p. 1 - 20 (1981).

The paper quickly reviews the general evidence concerning the orbits of comets, relative "ages", the relationship with the Oort cloud, non-gravitational forces, sizes, and rotation of the nuclei. Following a brief discussion of the composition of the nuclei, some of the evidence regarding homogeneity and possible development into earth-crossing asteroids is presented. The discussion then turns to problems of the place and nature of origin, and concerns next the chemical and physical structure, particularly the evidence for amorphous ices and cosmic-ray damage in new comets. Review of more evidence regarding bursts, splitting, jets, halos, dust and the basic structure of comets will center around the question of radioactive heating, possible exothermic reactions, and internal structure. This is relevant to suggestions concerning the development of life forms in comets.

102.015 **Comet nucleus: some characteristics and a hypothesis on origin and structure.**   B. Donn.
Comets and the origin of life, (see 012.022), p. 21 - 29 (1981).

The author gives a brief discussion of some spectroscopic properties of comets. These points will be followed by presentation of the problem of cosmic ray irradiation of comets in the long time they survive in the Oort Cloud. The concluding section is a hypothesis on the origin and structure of cometary nuclei.

102.016 **Ultraviolet spectroscopy of comets.**
P. D. Feldman.
Comets and the origin of life, (see 012.022), p. 31 - 41 (1981).

Atomic carbon and oxygen were discovered in rocket observations of comet Kohoutek in 1974, but comprehensive spectra spanning the wavelength range from 1150–3100 Å have been obtained only for comets West (1976 VI), Seargent (1978 XV) and Bradfield (1979 X). The observations of comet Bradfield, made with the orbiting International Ultraviolet Explorer, were the first to span a wide range of heliocentric distance (0.7 to 1.5 a.u.). The results of these observations and their interpretation are reviewed in terms of their contribution to our understanding of both the chemistry and physics of the coma and the composition of the cometary ice.

102.017 **Statistical equilibrium calculations of cometary molecules.**   K. S. Krishna Swamy.
Comets and the origin of life, (see 012.022), p. 43 - 51 (1981).

Some of the problems presented by the observations of the C_2 molecule in comets is discussed in terms of the proposed model. The model is shown to explain satisfactorily all the available observations at the present time. The results of resonance fluorescence calculations for many other molecules observed in the UV region from rockets and satellites are also discussed.

102.018 **Chemical abundances in comets.**
M. F. A'Hearn.
Comets and the origin of life, (see 012.022), p. 53 - 61 (1981).

The author discusses some of the results of a program to obtain homogeneous, quantitative data on the chemical abundances in comets.

102.019 **Interaction of comets with the interplanetary medium.**   D. A. Mendis.
Comets and the origin of life, (see 012.022), p. 71 - 89 (1981).

The interaction of the solar wind and solar radiation with a comet, as the comet moves around the sun, is discussed.

102.020 **Chemical kinetics in the coma.**
W. F. Huebner.
Comets and the origin of life, (see 012.022), p. 91 - 103 (1981).

Physical and chemical conditions in the coma of a bright "new" comet are related to the composition of the nucleus. Chemical and photolytic processes are described and correlated to distance in the coma above the nucleus and to heliocentric distance of the comet. Two classes for the composition of comets are considered based on their place of origin: in the environs of the giant planets or in a companion fragment of the presolar nebula. Comparison of model results with coma observations are discussed.

102.021 **On the mechanism of ray closure in comet tails.**
A. I. Ershkovich.
Comets and the origin of life, (see 012.022), p. 105 - 109 (1981).

The folding phenomenon of the comet tail rays is explained by means of an electric drift due to convectional electric fields. This mechanism results in an angular rate of closure which reduces to that obtained by Ness and Donn (1966) if the velocity profile across the tail is linear and the plasma conductivity is ideal. Observations of both the ray closure and the disconnection events point to the phenomenon of anomalous resistivity. Magnetic field of about $30-40\ \gamma$ in the coma and of $10\ \gamma$ in the distant tail (at 1 AU) is estimated from the MHD momentum equation.

102.022 **Comets and the origin of life – the stable isotope approach.**   P. I. Abell, A. E. Fallick,
N. J. McNaughton, C. T. Pillinger.
Comets and the origin of life, (see 012.022), p. 129 - 139 (1981).

To establish links between comets and the origin of life, a greatly improved knowledge of the carbon isotopic composition of cometary bodies will be necessary. Some suggestions are presented as to how the required data might be obtained from presumed terrestrial cometary debris or from in situ measurements by a cometary probe.

102.023 **Are comets connected to the origin of life?**
A. H. Delsemme.
Comets and the origin of life, (see 012.022), p. 141 - 159 (1981).

The orbital evolution of comets and their connections with the origin of the solar system are discussed first. Comet chemistry and its relations to interstellar molecules and to the biosphere of the earth are discussed later.

102.024 **Cometary material and the origins of life on earth.**
A. Lazcano-Araujo R., J. Oró.
Comets and the origin of life, (see 012.022), p. 191 - 225 (1981).

First the organic synthetic pathways that occur in dense interstellar clouds of our galaxy and in comets are examined, together with the possibility that more complex organic molecules than those detected up to now may exist in these environments. The authors then review the different estimates of the amount of terrestrial volatiles that may be of cometary origin, and compare them to other extraterrestrial sources of organogenic elements. They also discuss the prebiological events that might have followed the collision of cometary nuclei accreted by the primitive earth. Finally, some of the current ideas on prebiological organic synthesis of biochemical monomers and their polymers on the primitive earth are discussed, together with the emergence of precellular systems and their further evolution, which lead to the appearance of terrestrial life.

102.025 **On the orbital evolution of short period comets having low velocity encounters with Jupiter.**
A. Carusi, G. B. Valsecchi,
Mem. Soc. Astron. Italiana, Vol. 52, (see 012.023), 365 - 367 (1981).

102.026 **On the accuracy of orbit determination of short-period comets.**
V. V. Emel'yanenko, N. Yu. Emel'yanenko.
Determination of the coordinates of celestial bodies, Riga, 1981, (see 003.008), p. 19 - 27. In Russian. – Abstr. in Ref. zh., 51. Astron., 9.51.80 (1981).

102.027 **Determination of the mean-square value of a transversal impulse in the problem of long-period comets.**   A. L. Salitis.
Determination of the coordinates of celestial bodies, Riga, 1981, (see 003.008), p. 28 - 34. In Russian. – Abstr. in Ref. zh., 51. Astron., 9.51.81 (1981).

102.028 **On calculation of the mean velocity change in the problem of long-period comets.**   A. L. Salitis.
Determination of the coordinates of celestial bodies, Riga, 1981, (see 003.008), p. 35 - 39. In Russian. – Abstr. in Ref. zh., 51. Astron., 9.51.82 (1981).

102.029 **Model of stellar motion for determining perturbations from stars in cometary motion.**
K. A. Šteins, A. L. Salitis.
Determination of the coordinates of celestial bodies, Riga, 1981, (see 003.008), p. 40 - 48. In Russian. – Abstr. in Ref. zh., 51. Astron., 9.51.83 (1981).

102.030 **Families of nearly parabolic comets connected with Mercury and Venus.**   V. P. Tomanov.
Determination of the coordinates of celestial bodies, Riga, 1981, (see 003.008), p. 49 - 55. In Russian. – Abstr. in Ref. zh., 51. Astron., 9.51.84 (1981).

102.031 **Remark on the hypothesis of eruptive origin of comets.**   V. P. Tomanov.
Determination of the coordinates of celestial bodies, Riga, 1981, (see 003.008), p. 56 - 60. In Russian. – Abstr. in Ref. zh., 51. Astron., 9.51.85 (1981).

102.032 **Application of millimeter-wave remote sensing to the investigation of comets.**
R. W. Hobbs, J. C. Brandt, S. P. Maran, J. S. Chitwood, R. K. Larsen, C. Katz, S. Becker, R. M. Rudish, R. K. Hendricks.
Proc. Soc. Photo-Opt. Instrum. Eng., Vol. 259, p. 158 - 165 (1980). – Abstr. in Phys. Abstr., Vol. 84, Abstr. 94272 (1981).

102.033 **Secret of the birth of shaggy celestial bodies.**
V. V. Radzievskij.
Zemlya Vselennaya, 1981, No. 4, p. 56 - 60. In Russian.

102.034 **On the electrostatic charging of the cometary nucleus.**   D. A. Mendis, J. R. Hill,
H. L. F. Houpis, E. C. Whipple, Jr.
Astrophys. J., Vol. 787 - 797 (1981).
The authors consider, in detail, the charging of the distant cometary nucleus as it moves within the region of the heliosphere dominated by the supersonic flow of the solar wind. The authors also discuss some important physical consequences of this charging, such as the levitation, transport, and blow-off of loose, fine dust that may be present on the nuclear surface.

102.035 **Cometary mass determination.**
D. K. Yeomans, M. Ananda, W. L. Sjogren, L. J. Wood.
J. Astronaut. Sci., Vol. 29, 19 - 33 (1981). – Abstr. in Phys. Abstr., Vol. 84, Abstr. 98763 (1981).

102.036 **On a possible mechanism of the origin of periodic active comets.**   V. D. Davydov.
Kosm. Issled., Tom 19, 749 - 762 (1981). In Russian.

102.037 **The forbidden oxygen lines in comets.**
M. C. Festou, P. D. Feldman.
Astron. Astrophys., Vol. 103, 154 - 159 (1981).
High spatial resolution observations of comet Bradfield (1979X) with the IUE spectrograph had allowed the authors to tentatively identify a feature at $\cong$ 2972 Å as the $^1S-^3P$ forbidden transition of oxygen atoms. It is shown here that the most likely source of this emission is the photodissociation of water molecules, mainly by solar Lyman-alpha photons, but that a significant contribution of CO_2 cannot be excluded.

102.038 **When we may and need to use barycentric orbit of a comet?**   B. Todorovic-Juchniewicz.
Acta Astron., Vol. 31, 191 - 196 (1981).
It is studied at which distances of a comet to the Sun the perturbation caused by the planets may practically be neglected. Remarks concerning practical calculation when changing the heliocentric system into a barycentric one are given.

102.039 **The role of collisions with interplanetary particles in the physical evolution of comets.**
J. A. Fernández.
Moon Planets, Vol. 25, 507 - 519 (1981).
Effects of collisions of comets with interplanetary particles are investigated. To this purpose, collision probabilities for comets with different orbital elements are computed. It is found that collisions may have a non-negligible effect on the physical evolution of comets. Catastrophic collisions may contribute to prevent long-period comets in retrograde orbits from reaching short-period orbits by orbital diffusion. Collisions may also produce irregularities of the nucleus brightness. Collisional effects for comet Halley, for which a continuous surveillance is planned, are evaluated.

102.040 **A calculation of cometary coma abundances with H_2O-dominated and CO_2-dominated volatile mixtures.**   G. F. Mitchell, M. B. Swift.
J. R. Astron. Soc. Canada, Vol. 75, 250 (1981). – Abstract.

102.041 **Magnitude-frequency relations of comets.**
T. Gehrels.
Bull. American Astron. Soc., Vol. 13, 705 (1981). – Abstract.

102.042 **Do comets have satellites?**   T. Van Flandern.
Bull. American Astron. Soc., Vol. 13, 705 (1981). Abstract.

102.043 **Terrestrial impact rates for long and short-period comets.**   P. R. Weissman.
Bull. American Astron. Soc., Vol. 13, 705 (1981). – Abstract.

102.044 **Relative abundance of OH in recent comets.**
M. F. A'Hearn, R. L. Millis.
Bull. American Astron. Soc., Vol. 13, 707 (1981). – Abstract.

102.045 **On the dust zoning of rapidly rotating cometary nuclei.** H. L. F. Houpis, D. A. Mendis.
Astrophys. J., Vol. 251, 409 - 414 (1981).

As a dirty-ice cometary nucleus approaches the sun, some of the nuclear dust is entrained by the sublimating gas, while some (the larger particles) continuously collect on the surface. The authors consider the effects of nuclear rotation on this surface dust (specifically of a 1 km radius, H_2O-dominated, comet at 1 AU).

102.046 **Motion of a cometary particle with variable mass under the central field of forces.**
P. Žáček, V. Vanýsek.
Acta Univ. Carolinae Math. Phys., Vol. 22, No. 1, p. 51 - 61 (1981).

The equations for trajectory, energy, radius and gas flow rate of a small cometary particle with variable mass are derived in this paper. These equations are valid for the central field of forces caused by the Sun and for zero initial particle velocity relative to the nucleus of comet. The influence of the nucleus on a particle is neglected.

102.047 **Exploration of comets by means of space probes.**
Ľ. Kresák.
Kozmos, Vol. 12, 134 - 135 (1981). In Slovak.

102.048 **Im Schwung um die Sonne. Neue Impulse für die Kometenforschung.** K. Jockers.
Umschau, 81. Jahrg., 738 - 741 (1981).

102.049 **Comet showers and the steady-state infall of comets from the Oort cloud.** J. G. Hills.
Astron. J., Vol. 86, 1730 - 1740 (1981).

The author develops analytical procedures for investigating the effects of stars passing nearby on orbits of comets in the Oort cloud and the subsequent interaction with Jupiter and Saturn if the comets are deflected into the inner solar system. He infers that the apparent inner boundary of Oort's cloud at a semimajor axis $a_c \sim 2 \times 10^4$ AU is probably an observational selection effect. Comets with $a < a_c$ will be observable only during a relatively brief time after a star has come close enough to the sun to deflect some of them into orbits penetrating the planetary system. In the much longer time intervall between such stellar encounters only comets with $a > a_c$ can enter the planetary system. This suggests that most comets may actually have semimajor axes $a < a_c$ with the Oort cloud being only the outer halo of a more massive comet cloud. The author estimates that the present-day mass of this inner comet-cloud could be two orders of magnitude greater than that of the Oort cloud. He discusses a possible formation model for this comet-cloud and eventual observational consequences.

102.050 **Electrostatic disruption of a charged conducting spheroid (comets).** J. R. Hill, D. A. Mendis.
Canadian J. Phys.. Vol. 59, 897 - 901 (1981). – Abstr. in Phys. Abstr., Vol. 85, Abstr. 6568 (1982).

102.051 **Thermal modeling of cometary nuclei.**
P. R. Weissman, H. H. Kieffer.
Icarus, Vol. 47, (see 012.045), 302 - 311 (1981).

A new model of the sublimation of volatile ices from a cometary nucleus has been developed which includes the effects of diurnal heating and cooling, rotation period and pole orientation, and thermal properties of the ice and subsurface layers. The model also includes the contribution from coma opacity, scattering, and thermal emission, where the properties of the coma are derived from the integrated rate of volatile production by the nucleus. The model is applied to the specific case of the 1986 apparition of Halley's comet.

102.052 **Heat content and evolution of cometary nuclei.**
R. Smoluchowski.

Icarus, Vol. 47, (see 012.045), 312 - 319 (1981).

The heat flux into cometary nuclei before and after perihelion has been investigated for the isothermal case and for the fixed subsolar point. It turns out that this heat flux may be a large fraction of the incident solar heat input, so that the surface temperature and the associated rate of evaporation are lower than usually calculated. The effect is strongly dependent on the porosity of the nucleus. An explanation of the splitting of comets as far as 9 AU from the Sun is suggested in terms of heating of a CO_2-rich inclusion in a nucleus.

102.053 **Some consequences of a phase transition of water ice on the heat balance of comet nuclei.**
J. Klinger.
Icarus, Vol. 47, (see 012.045), 320 - 324 (1981).

The author considers spheres of water ice of about 1 km in radius moving on three different orbits with a common perihelion distance of 8 AU. The surface temperature has been numerically calculated for two extreme situations: (1) the spheres are composed of amorphous ice with a heat conduction to the interior presumed to be negligible. (2) The spheres are composed of compact hexagonal ice with a heat conduction coefficient known from laboratory experiments. The case of active ice bodies is also discussed.

102.054 **On the dectectability of icy grains in the comae of comets.**   M. S. Hanner.
Icarus, Vol. 47, (see 012.045), 342 - 350 (1981).

Evaporation of icy grains over the distance scale of the visible cometary coma sets very specific limits on their temperature. Unless the grains are very pure water ice, the maximum size of an icy grain halo will be limited to a few hundred kilometers at heliocentric distances $\lesssim 2.5$ AU. It is unlikely that the 1.5- or 2-μm ice band could be detected in the scattering by icy grains. Detection of the 3-μm ice band might be possible in comets which display a coma at large heliocentric distances.

102.055 **Far-UV, visible, and near-IR reflectance spectra of frosts of H_2O, CO_2, NH_3 and SO_2.**
B. Hapke, E. Wells, J. Wagner, W. Partlow.
Icarus, Vol. 47, (see 012.045), 361 - 367 (1981).

The reflectance spectra of frosts of several volatiles of interest for comet nuclei and outer-solar system objects have been measured over the range 0.1 - 2.5 μm. Most of the features in the frost spectra correspond to bands of absorption lines in the gas phase.

102.056 **Models of the cometary coma in which abundances are calculated for various heliocentric distances.**
M. B. Swift, G. F. Mitchell.
Icarus, Vol. 47, (see 012.045), 412 - 430 (1981).

One-dimensional radial models of the chemistry in cometary comae have been constructed for heliocentric distances ranging from 2 to 0.125 AU. The coma's opacity to solar radiation is included and photolytic reaction rates are calculated. A parent volatile mixture similar to that found in interstellar molecular clouds is assumed. Profiles through the coma of number density and column density are presented for H_2O, OH, O, CN, C_2, C_3, CH, and NH_2. Whole-coma abundances are presented for NH_2, CH, C_2, C_3, CN, OH, CO^+, H_2O^+, CH^+, N_2^+, and CO_2^+.

102.057 **Millimeter-wave experiments for cometary space missions.**
R. W. Hobbs, J. C. Brandt, S. P. Maran, J. M. Hollis.
Icarus, Vol. 47, (see 012.045), 434 - 440 (1981).

Predicted brightness temperatures for a variety of cometary nucleus models, consisting of homogeneous layers comprised of mixtures of water ice and refractory grains, are presented as functions of wavelength. These illustrative spectra are computed using simple radiative transfer techniques. It appears that millimeter-wave sensing from an interplanetary spacecraft is an effective means for distinguishing between alternate models of the nucleus.

102.058 Ultraviolet absorption studies of H_2O and other species in comet comae with satellite telescope-spectrometers.
P. L. Smith, J. H. Black, M. Oppenheimer.
Icarus, Vol. 47, (see 012.045), 441 - 448 (1981).

Ultraviolet absorption by H_2O and other species in the comae of comets could be detected by studying, with satellite telescope-spectrometers, the occultation of hot stars by comets. Such observations could produce the first direct detection of H_2O and give measures of molecular level populations. The first instrument suitable for such observations will be the High Resolution Spectrograph on Space Telescope and the authors consider its capabilities.

102.059 Do comets have satellites?
T. C. van Flandern.
Icarus, Vol. 47, (see 012.045), 480 - 486 (1981).

This paper explores several puzzling features of comets which do not fit easily into conventional cometary models, but which can be satisfactorily explained if it is assumed that comets have a full range of gravitationally bound masses, from dust size to the size of the nucleus, in orbit around the principal nucleus. This discussion also implies a higher probability of destruction of a spacecraft near a comet than is usually assumed.

102.060 Observing chemical abundances in comets.
A. H. Delsemme.
Modern observational techniques for comets, (see 012.047), p. 5 - 13 (1981).

The atomic resonance lines of the major elements have been observed in the atmospheres of a few comets, by using vacuum ultraviolet spectrographs on board rockets or orbiting observatories. Dust-to-gas ratios have also been deduced for two comets through a Finson-Probstein's analysis of their dust-tail isophotes.

102.061 Observational data needs useful for modeling the coma.   W. F. Huebner, P. T. Giguere.
Modern observational techniques for comets, (see 012.047), p. 14 - 18 (1981).

The present status of the authors' computer model of comet comae is described; results from assumed composition of frozen gases are summarized and compared to coma observations. Restrictions on relative abundance of some frozen constituents are illustrated. Modeling,when tightly coupled to observational data, can be important for comprehensive analysis of observations, for predicting undetected molecular species and for improved understanding of coma and nucleus.

102.062 The nucleus structure of a comet from systematic observations of dust features in the coma.
Z. Sekanina.
Modern observational techniques for comets, (see 012.047), p. 19 - 20 (1981).

102.063 Observational data needs for plasma phenomena.
M. B. Niedner, Jr.
Modern observational techniques for comets, (see 012.047), p. 21 - 32 (1981).

Bright comets display a rich variety of interesting plasma phenomena which occur over an enormous range of spatial scales, and which require different observational techniques to be studied effectively. Wide-angle photography of high time resolution is probably the best method of studying the phenomenon of largest known scale: the plasma tail disconnection event (DE), which has been attributed to magnetic reconnection at interplanetary sector boundary crossings.

102.064 Cometary ephemerides – needs and concerns.
D. K. Yeomans.
Modern observational techniques for comets, (see 012.047), p. 46 - 49 (1981).

102.065 Spectrophotometry of faint comets: the asteroid approach.   J. Degewij.
Modern observational techniques for comets, (see 012.047), p. 63 - 69 (1981).

This is a short description of observing programs at optical (0.35–0.8 micron) and near-infrared (1.1–2.4 micron) wavelengths, directed at the acquisition of reflection spectra of faint and distant comets. The ultimate goal is to obtain spectrophotometric measurements of comets for which a significant part of the light is expected to be reflected by the solid surface of the nucleus.

102.066 Millimeter wave radiometry as a means of determining cometary surface and subsurface temperatures.
R. W. Hobbs, J. C. Brandt, S. P. Maran.
Modern observational techniques for comets, (see 012.047), p. 96 - 103 (1981).

Thermal emission spectra for a variety of cometary nucleus models are evaluated by a radiative transfer technique adapted from modeling of terrestrial ice and snow fields. It appears that millimeter wave sensing from an interplanetary spacecraft is the most effective available means for distinguishing between alternate models of the nucleus and for evaluating the thermal state of the layer – which is below the instantaneous surface – where modern theories of the nucleus indicate that sublimation of the cometary volatiles actually occurs.

102.067 The spectroscopy of comets: introductory remarks.
A. H. Delsemme.
Modern observational techniques for comets, (see 012.047), p. 107 - 109 (1981).

102.068 A systematic program of cometary spectroscopy.
S. M. Larson, B. Donn.
Modern observational techniques for comets, (see 012.047), p. 110 - 114 (1981).

102.069 Imaging of coma and tail. Introductory remarks.
F. D. Miller.
Modern observational techniques for comets, (see 012.047), p. 169 - 170 (1981).

102.070 The JOCR program.
J. C. Brandt.
Modern observational techniques for comets, (see 012.047), p. 171 - 184 (1981).

The principal goal of the Joint Observatory for Cometary Research (JOCR) is to obtain observational data on large-scale plasma structures in comets. This data is of value in (1) analyzing the interaction between the solar wind and comets, (2) using comets as solar wind probes, and (3) using comets as an astrophysical plasma laboratory with the CO^+ plasma serving as tracers of the magnetic field.

102.071 On observing comets for nuclear rotation.
F. L. Whipple.
Modern observational techniques for comets, (see 012.047), p. 191 - 201 (1981).

The purpose of the present paper is to encourage measurements of cometary coma with the hope that more measurements will be made and that they will be better standardized, utilizing the full potential of modern energy sensing devices and analytical techniques.

102.072 Existing cometary data and future needs.
J. Rahe.

Modern observational techniques for comets, (see 012.047), p. 213 - 215 (1981).

102.073 **Probability of discovery of comets as a criterion of their origin.**   V. V. Radzievskij.
Astron. Zh., Tom 58, 1286 - 1290 (1981). In Russian.
English translation in Soviet Astron., Vol. 25, No. 6.

On the basis of the observed distribution of parabolic comets with respect to the perihelion distance q and the well-known formula of the probability of the discovery of comets their true distribution is found. It is shown that this distribution is incompatible with the hypothesis of Oort of capture and eruption of the comets by giant planets.

102.074 **A new outlook on the origin of cometary nuclei.**
B. A. Vorontsov-Vel'yaminov.
Astron. Tsirk., No. 1136, p. 7 - 8 (1980). In Russian.

102.075 **Statistics of perihelia of orbits of short-period comets of the Jupiter family.**   G. S. Guliev.
Astron. Tsirk., No. 1140, p. 6 - 8 (1980). In Russian.

102.076 **Infrared line emission from constituent molecules of cometary nuclei.**   T. Yamamoto.
Proceedings of the 14th ISAS Lunar and Planetary Symposium, (see 012.055), p. 111 (1981). — Abstract.

102.077 **The lifetimes and disappearance of periodic comets.**
L. Kresák.
Bull. Astron. Inst. Czechoslovakia, Vol. 32, 321 - 339 (1981).

Among the 117 short-period comets discovered so far, there are 30 objects which have not been observed at their last returns to the Sun. All of these cases are analyzed for possible reasons of the loss. Even without presuming a progressive decrease of their absolute brightness, it is found that most of these comets failed to be rediscovered simply because their geocentric configurations at the missed returns were much less favourable than at the observed perihelion passages. The only cases where the loss was almost definitely due to an ultimate extinction are P/Biela and P/Brorsen. For two additional objects — P/Westphal and P/Neujmin 2 — a complete extinction appears probable, and for a few others possible. These results are used to estimate the mean active lifetimes of comets. For the comets of Jupiter's family these amount to 2500–3000 years, or to 400 revolutions in a short-period. For the comets of Halley type the mean lifetime seems to be of the order of 200 revolutions, or 10 000–15 000 years. The corresponding slow rate of aging is consistent with the requirements of the formation of meteor streams but inconsistent with a rapid secular brightness decrease of comets.

102.078 **On the nature of comets.**   C. Arpigny.
Proceedings of The Robert A. Welch Foundation Conferences on Chemical Research XXI. Cosmochemistry, November 7 - 9, 1977, Houston, Texas = Inst. Astrophys. Univ. Liège, Cointe-Ougrée (Belgique), Coll. 8°, No. 678.

102.079 **The capture of interstellar comets.**
M. J. Valtonen, K. A. Innanen.
Rep. Ser., Dep. Phys. Sci., Univ. Turku, Turku-FTL-R20, 34 pp. (1981). ISBN 951-642-028-1 = Turku Univ. Obs. Informo, No. 49. — Submitted to Astrophys. J.

102.080 **A method for the determination of the inclination angle between the cometary tail and the radius**
vector.   I. Tóth.
Astron. Nachr., Band 302, 301 - 309 (1981).

If the angle between the radius vector and the axis of the comet tail is known one can determine the velocity component of the solar wind pointing into the direction of the radius vector. This paper describes a geometrical method to determine the orientation of cometary tails.

102.081 **On the peculiar dust tails of some comets and the history of one investigation.**   Z. Sekanina.
Int. Comet Q., Vol. 3, 95 - 98 (1981).

102.082 **The contemporary model of cometary nucleus and the prospects of its improvement by space research.**
L. M. Shulman (*Shul'man*).
Progress in planetary exploration, (see 012.062), p. 91 - 98 (1981).

No cometary nucleus has ever been observed directly. A model is deduced from ground-based and space data on cometary atmospheres. The main features of the chemical composition of cometary nuclei and the estimation of their sizes are described.

102.083 **On a possible mechanism of X-ray generation in the heads of comets.**   S. Ibadov.
Dokl. AN TadzhSSR, Tom 24, 164 - 168 (1981). In Russian.
Abstr. in Ref. zh., 51. Astron., 1.51.293 (1982).

102.084 **On the first close encounters of comets with Jupiter.**   A. Guliev, A. Demenko.
Komet. Tsirk., Kiev, No. 279 (1981). In Russian.

102.085 **Eruption of the heated layer of cometary nuclei.**
E. A. Kajmakov, I. S. Lizunkova.
Komet. Tsirk., Kiev, No. 280 (1981). In Russian.

102.086 **Comets and problems of the eruptive evolution of protoplanets.**   S. K. Vsekhsvyatskij.
Komet. Tsirk., Kiev, No. 280 (1981). In Russian.

102.087 **Variation of the arguments of the perihelia of orbits of periodic comets of the Neptune family.**
P. T. Veleshchuk.
Komet. Tsirk., Kiev, No. 282 (1981). In Russian.

102.088 **The possibly nearest comet to the sun.**
Komet. Tsirk., Kiev, No. 283 (1981). In Russian.

102.089 **Comets.**   J. C. Brandt.
The new solar system, (see 003.026), p. 177 - 186 (1981).

Comets and the origin of life: bibliography.
See Abstr. 002.018.

Comets. Readings from Scientific American.
See Abstr. 003.036.

Introduction to comets.   See Abstr. 003.039.

Observing facilities at the European Southern Observatory (ESO) in Chile for cometary observations.
See Abstr. 009.042.

Demonstrating the motion of comets.
See Abstr. 014.007.

Comets — a vehicle for panspermia.
See Abstr. 015.020.

Extreme environments: are there any limits to life?
See Abstr. 015.021.

Possibility of the existence of comet-like bodies with small average density in the solar system.
See Abstr. 022.061.

Modelling the interaction between solar radiation and cometary matter.   See Abstr. 022.062.

Laser induced photodissociation of cometary and planetary molecules: the production and quenching of NH ($A^3 \pi_i$).   See Abstr. 022.133.

Statistical equilibrium in cometary C_2. III. Triplet-singlet, Phillips, Ballik-Ramsay, and Mulliken bands. See Abstr. 022.140.

Laboratory measurements of cometary photo-chemical phenomena.   See Abstr. 022.153.

Introductory remarks. Photometry section. See Abstr. 031.602.

Correlated ground-based and IUE observations. See Abstr. 031.603.

Use of an image dissector scanner for spectro-photometry of faint comets.   See Abstr. 031.604.

Observations of faint comets at McDonald Observa-tory: 1978 - 1980.   See Abstr. 031.605.

Spectral imagery: recent results with the SPIFI and their implications for cometary atmospheric studies. See Abstr. 031.606.

A possible technique for cometary studies with high angular and spectral resolution.   See Abstr. 031.607.

Laboratory research.   See Abstr. 031.608.

Cometary dust observations by optical *in-situ* methods.   See Abstr. 031.651.

An opportunity for the observations of comets with wide-field cameras aboard the Saliout space station. See Abstr. 032.590.

Anticipated results from dust experiments on cometary missions.   See Abstr. 051.036.

New problems of cometary observations from space. See Abstr. 051.037.

Results to be expected from light scattering dust analyzer during a rendezvous mission.   See Abstr. 051.038.

A European probe to comet Halley. See Abstr. 051.051.

Comets and the photochemistry of the paleoatmo-sphere.   See Abstr. 082.036.

New insight into the physical state of solar system objects.   See Abstr. 091.008.

Multiple Apollo asteroids from comets? See Abstr. 098.019.

On the implausibility of a cometary origin for most Apollo-Amor asteroids.   See Abstr. 098.075.

The flux of earth-crossing and moon-cratering interplanetary bodies.   See Abstr. 098.105.

On the visibility of nuclei of dusty comets. See Abstr. 103.806.

Is the interplanetary dust cloud maintained by comets?   See Abstr. 106.022.

Interplanetary dust – its physical nature and entry into the atmosphere of terrestrial planets. See Abstr. 106.053.

Dynamical evolution of a cometary swarm in the outer planetary region.   See Abstr. 107.018.

On the ejection of planetesimals from the protosolar system.   See Abstr. 107.023.

Chemical evolution of interstellar dust – a source of prebiotic material?   See Abstr. 131.108.

Comets, interstellar molecules, and the origin of life.   See Abstr. 131.209.

103 Comets (Individual Objects)

103.001 Les comètes de l'année 1979.
   C. Bertaud.
Astronomie, Vol. 95, 403 - 413 (1981).

103.002 De kometen van 1980.   R. J. Bouma.
   Zenit, 8. Jaarg., 472 - 473 (1981).

103.003 Observations des membres de la commission des
   comètes en 1980.   C. Bertaud.
Astronomie, Vol. 95, 503 - 508 (1981).

103.004 Drie periodieke kometen. Waarnemingsresultaten
   van Encke, Tuttle en Stephan-Oterma.
H. Feijth.
Zenit, 8. Jaarg., 497 - 502 (1981).

103.005 Roman numeral designations of comets.
   Minor Planet Circ., (M.P.C.), No. 6487 (1981).

103.006 Observations of comets.
   Minor Planet Circ., (M.P.C.), Nos. 6067 - 6070, 6145 - 6147, 6222 - 6227, 6350, 6429 - 6436, 6487 - 6490 (1981).
   Observations made at the following stations are published: Calar Alto, Catalina Stat. Lunar Planet. Lab., Centro Astron. Yebes, Cerro el Roble, Chamberlin Obs., El Leoncito, Fabra Obs., Geisei, Hemingford Abbots, Kleť, La Seyne sur Mer, Lick Obs., Lowell Obs. Anderson Mesa Stat., McDonald Obs., Mt. John Obs., Mt. Palomar, Nantucket, Nice, Northern Ariz. Univ., Oak Ridge Obs., Ondřejov, Perth, Pino Torinese, Pulkovo, Stakenbridge, Steward Obs., Tokai, Tokyo Obs. Kiso Stat., Uccle, Wise Obs., Zimmerwald. Concerning observations of the following comets: 1937 II Wilk, 1960 IV Väisälä 1, 1961 X van Houten, 1965 VIII Ikeya-Seki, 1966 II Barbon, 1966 V Kilston, 1967 II Rudnicki, 1967 IV Seki, 1967 VI Arend, 1967 X Tempel 2, 1968 I Ikeya-Seki, 1968 V Whitaker-Thomas, 1969 IV Churyumov-Gerasimenko, 1969 VI Faye,

1969 VII Fujikawa, 1969 IX Tago-Sato-Kosaka, 1970 III Kohoutek, 1970 X Suzuki-Sato-Seki, 1970 XV Abe, 1971 V Toba, 1972 VII Swift-Gehrels, 1972 VIII Heck-Sause, 1972 IX Sandage, 1972 XI Kearns-Kwee, 1973 II Kojima, 1973 X Sandage, 1973 XII Kohoutek, 1974 II Schwassmann-Wachmann 1, 1974 VIII Cesco, 1974 XIII Schwassmann-Wachmann 2, 1975 VII Smirnova-Chernykh, 1975 VIII Lovas, 1975 XII Mori-Sato-Fujikawa, 1976 III Gunn, 1976 VI West, 1976 X Klemola, 1977 X Tsuchinshan, 1977 XII Sanguin, 1977 XIV Kohler, 1978 I Schuster, 1978 II Tempel 1, 1978 XIII Machholz, 1978 XVII Comas Solá,1978 XXI Meier, 1978 XXIII Clark, 1978 XXIV van Biesbroeck, 1979 II Kowal 2, 1979 VII Bradfield, 1979 VIII Schwassmann-Wachmann 3, 1979 X Bradfield, 1979 XI Howard-Koomen-Michels, 1980 I Honda-Mrkos-Pajdušáková, 1980 II Torres, 1980 III Russell 2, 1980 IV Černis-Petrauskas, 1980 VI Forbes, 1980 VII Wild 3, 1980 VIII Reinmuth 1, 1980 IX Brooks 2, 1980 X Stephan-Oterma, 1980 XI Encke, 1980 XII Meier, 1980 XIII Tuttle, 1980 XIV Harrington, 1980 XV Bradfield, 1980b Bowell, 1980i Borrelly, 1980j Kohoutek, 1980*l* Russell, 1980n Reinmuth 2, 1980r West-Kohoutek-Ikemura, 1980u Panther, 1981a Longmore, 1981b Bus, 1981c Elias, 1981d Bus, 1981f Gehrels 2, 1981g González, 1981i Slaughter-Burnham, 1981k Howell.

103.007   Ephemerides of minor planets and comets.
Minor Planet Circ., (M.P.C.), Nos. 6112 - 6144, 6210 - 6220, 6307 - 6344, 6422 - 6424, 6476 - 6480, 6532 - 6572 (1981).
Concerning ephemerides of the following comets: 1960 IV Väisälä 1, 1967 VI Arend, 1969 IV Churyumov-Gerasimenko, 1972 IV Neujmin 3, 1974 II Schwassmann-Wachmann 1, 1974 XIII Schwassmann-Wachmann 2, 1976 III Gunn, 1976 XI d'Arrest, 1977 VI Grigg-Skjellerup, 1977 VII Gehrels 3, 1980 XII Meier, 1980b Bowell, 1980u Panther, 1981a Longmore, 1981c Elias, 1981d Bus, 1981g González, 1981k Howell.

103.008   Orbital elements of comets.
Minor Planet Circ., (M.P.C.), Nos. 6065 - 6572 (1981).
The comets are listed according to their Roman numeral designation or preliminary designation. The names of the authors are given behind the respective M.P.C. numbers. 1937 II Wilk 6099, 1939 IX Friend 6099 W. Landgraf; 1977 VI Grigg-Skjellerup 6193 G. Sitarski; 1978 XIII Machholz 6518, 1980 VII Wild 3 6518, 1980 XII Meier 6518, 1980 XV Bradfield 6519, 1980b Bowell 6193, 1980u Panther 6194, 6519, 1981b Bus 6194, 1981c Elias 6519, 1981d Bus 6194, 6519, 1981g González 6290, 1981k Howell 6290, 6519 B. G. Marsden.

103.009   Low activity comets in 1981 - 1982.
J. Degewij, E. M. Shoemaker, R. F. Wolfe.
Bull. American Astron. Soc., Vol. 13, 705 (1981). – Abstract.

103.010   Combined visual and near infra-red observations of comets.
D. C. Jewitt, G. E. Danielson, B. T. Soifer, G. Neugebauer, K. Matthews.
Bull. American Astron. Soc., Vol. 13, 706 (1981). – Abstract.

103.011   Analysis of CCD comet spectra from 5600 to 10300 Å.
J. R. Johnson, U. Fink, S. Larson, M. Buie.
Bull. American Astron. Soc., Vol. 13, 706 (1981). – Abstract.

103.012   Reflectances of the dust comae of comets Bowell, P/Brooks, P/Schwassmann-Wachmann 2 and P/Stephan-Oterma.
E. S. Barker, A. L. Cochran, W. D. Cochran.
Bull. American Astron. Soc., Vol. 13, 706 (1981). – Abstract.

103.013   Broad-band spectrophotometry of cometary comae.
D. J. Tholen, E. F. Tedesco, S. M. Larson.
Bull. American Astron. Soc., Vol. 13, 707 (1981). – Abstract.

103.014   Comets in the year 1980.   J. Bouška.
Vesmír, Vol. 60, 285 (1981). In Czech.

103.015   IUE observations of faint comets.
H. A. Weaver, P. D. Feldman, M. C. Festou, M. F. A'Hearn, H. U. Keller.
Icarus, Vol. 47, (see 012.045), 449 - 463 (1981).
Comets P/Encke (1980), P/Tuttle (1980h), P/Stephan-Oterma (1980g), and Meier (1980q) were observed during November - December 1980 with the International Ultraviolet Explorer (IUE) satellite observatory, while comets P/Borrelly (1980i) and Panther (1980u) were observed with IUE on 6 March 1981. The spectra of these comets are compared with those of comet Bradfield (1979 X), studied extensively earlier in 1980 with IUE, as well as with each other. All of the cometary spectra are remarkably similar, which suggests that these comets may have a common composition and origin.

103.016   Observations of the OH radical in comets at 18 cm wavelength.
D. Bockelée-Morvan, J. Crovisier, E. Gerard, I. Kazès.
Icarus, Vol. 47, (see 012.045), 464 - 469 (1981).
The authors present observations of the OH radical at λ 18 cm with the Nançay radio telescope in comets Meier (1978 XXI), Bradfield (1979 X), Meier (1980q), P/Encke (1980), and Bradfield (1980t). The analysis of the OH radio line shape is a powerful tool to study the kinematics of the coma. An excitation model by UV pumping and fluorescence of the OH radical and an application of Haser's model lead to the production rate of the parent molecule of OH. There is a close correlation between this gas production rate and the visual brightness of the comet.

103.017   Infrared observations of comets.
R. W. Hobbs.
Modern observational techniques for comets, (see 012.047), p. 79 - 82 (1981).

103.018   Infrared observations of faint comets.
H. Campins, J. Gradie, M. Lebofsky, G. Rieke.
Modern observational techniques for comets, (see 012.047), p. 83 - 89 (1981).
Infrared observations of the periodic comets Encke, Stephan-Oterma and Chernykh indicate that the dusty component in this class of comets is not radically different from the dusty component found in nonperiodic comets. The differences in the infrared behavior among these three comets suggests that a range of behaviors rather than a single behavior typifies the cometary activity. The range in albedo (0.02 to 0.10) of the dust calculated for the periodic comets is similar to the range in albedos seen among the asteroids.

103.019   Radio observations of comets.
L. E. Snyder.
Modern observational techniques for comets, (see 012.047), p. 93 - 95 (1981).

103.020   Ground-based cometary spectroscopy.
S. Wyckoff.
Modern observational techniques for comets, (see 012.047), p. 129 - 137 (1981).
The return of comet Halley presents a rallying point for astronomers to discuss data on recent comets in the light of new developments in the field of cometary astronomy. The observational problems presented by bright comets near perihelion are discussed. High and low resolution spectra (3100–8000 Å) of the bright comets Kohoutek, Kobayashi-Berger-

Milon, West and d'Arrest are presented. Digital reduction of calibrated photographic spectra to relative intensity versus wavelength can provide useful information. The reduction of comet spectra to absolute intensities involves, however, large uncertainties and should be interpreted cautiously.

103.021 Ultraviolet spectroscopy of comets using sounding rockets, IUE and Spacelab. P. D. Feldman.
Modern observational techniques for comets, (see 012.047), p. 141 - 147 (1981).

103.022 Possible comet Shcherbanovskij.
IAU Circ., No. 3655 (1981).

103.023 Differential corrections of comet orbits.
L. Buffoni, A. Manara, M. Scardia.
Astron. Nachr., Band 302, 299 - 300 (1981).
From Smithsonian Observatory circulars observations the orbits of seven comets have been calculated by a program which provides differential correction of the orbits.

103.024 Tabulation of comet observations.
Int. Comet Q., Vol. 3, 77 - 89, 104 - 114 (1981).
Concerning comets: 1968 VI Honda, 1974 II Schwassmann-Wachmann 1, 1977 VI Grigg-Skjellerup, 1977 XIV Kohler, 1978 IV Chernykh, 1978 VII Bradfield, 1978 XI Wild 2, 1978 XIV Ashbrook-Jackson, 1978 XV Seargent, 1978 XX Haneda-Campos, 1978 XXI Meier, 1979 IX Meier, 1979 X Bradfield, 1979k Schwassmann-Wachmann 2, 1980b Bowell, 1980g Stephan-Oterma, 1980h Tuttle, 1980i Borrelly, 1980q Meier, 1980t Bradfield, 1980u Panther, 1981j Swift-Gehrels, P/Encke.

103.025 Research on comets from space.
J. Blamont.
Planetary aeronomy and astronomy, (see 012.063), p. 193 - 197 (1981).
A review of spectrometric and photometric measurements of recent comets from rockets and satellites is presented.

103.026 Comet digest. J. E. Bortle.
Sky Telesc., Vol. 62, 125 - 126, 222, 307, 411, 540 (1981).

Astronomical data bases and retrieval systems.
See Abstr. 002.066.

Premières photographies d'une comète et de son spectre. See Abstr. 004.009.

Faint comet searching. See Abstr. 031.601.

Some procedures for comet discovery.
See Abstr. 031.647.

A review of visual comet observing techniques – III.
See Abstr. 031.648.

Earth-orbit-approaching comets and their theoretical meteor radiants. See Abstr. 104.022.

Comet 1979 X Bradfield

103.101 Comète Bradfield (1979l). C. Bertaud.
Astronomie, Vol. 95, 343 - 346 (1981).

103.102 Analysis of IUE observations of CS in Comet Bradfield (1979l).
W. M. Jackson, J. Halpern, P. D. Feldman, J. Rahe.

The Universe at ultraviolet wavelengths, (see 012.009), p. 55 - 64 (1981).
The CS high-resolution and low-resolution IUE data obtained on Comet Bradfield (1979l) have been analyzed. The high resolution rotational band profiles can be fitted with theoretical band profiles which are derived using a Boltzmann temperature of 70 K. A very rapid variation with heliocentric distance, for the CS brightness has been found. The implications of these results for models of the coma along with the origin of the CS species are discussed.

103.103 Water production models for Comet Bradfield (1979l).
H. A. Weaver, P. D. Feldman, M. C. Festou.
The Universe at ultraviolet wavelengths, (see 012.009), p. 65 - 72 (1981).
IUE observations of Comet Bradfield (1979l) made from 10 January 1980 to 3 March 1980 permit a detailed study of water production for this comet. Brightness measurements are presented for all three water dissociation products, H, O, and OH, and comparisons are made with model predictions. The heliocentric variation of the water production rate is derived.

103.104 IUE observations of comet Bradfield (1979l).
C. Arpigny, P. Benvenuti, B. Donn, P. D. Feldman, M. Festou, M. F. A'Hearn, G. Hunt, W. M. Jackson, H. U. Keller, A. J. Meadows, J. Rahe, A. M. Smith, S. P. Tarafdar, H. A. Weaver.
Second European IUE Conference, (see 012.011), p. 25 (1980). – Abstract.

103.105 Comet Bradfield 1979 X: the gassiest comet?
M. F. A'Hearn, R. L. Millis, P. V. Birch.
Astron. J., Vol. 86, 1559 - 1566 (1981).
The authors present the results of narrowband filter photometry of Comet Bradfield 1979 X for postperihelion, heliocentric distances from 0.57 to 1.65 AU. The emission-to-continuum ratio was found to be larger than that for any other comet measured by them. For the first time, the authors present OH production rates obtained contemporaneously with the production rates of the species more usually observed from the ground. All molecular species show a very steep variation with heliocentric distance, averaging about $r_H^{-3.2}$. This appears inconsistent with simple models of vaporization equilibrium. There appear to be changes in the relative production rates with heliocentric distance.

103.106 An Einstein search for X-ray emission from Comet Bradfield (1979l).
H. S. Hudson, W.-H. Ip, D. A. Mendis.
Planet. Space Sci., Vol. 29, 1373 - 1376 (1981).
X-radiation may result from active plasma phenomena in the interactions of comets with the solar wind. The authors have carried out a limited but sensitive search for soft X-radiation from Comet Bradfield (1979l), on 1980 Feb. 5. No X-radiation was detected at a level of 1.7×10^{-13} erg $(cm^2 \ sec \ keV)^{-1}$ in the 0.2–4.0 keV range. This corresponds to a limit on the power dissipated in the comet by non-thermal electrons of approximately 10^{19} ergs sec^{-1}, averaged over the 2568-sec exposure to the comet. This energy deposition is near the magnitude suggested by simple theoretical ideas.

103.107 Water production models for comet Bradfield (1979 X).
H. A. Weaver, P. D. Feldman, M. C. Festou, M. F. A'Hearn.
Astrophys. J., Vol. 251, 809 - 819 (1981).
IUE observations of comet Bradfield (1979 X) made in early 1980 allow a comprehensive study of the production of water by this comet. All three water dissociation products, H, O, and OH, were observed simultaneously with a spatial resolution of ~1000 km and over a range of heliocentric distances from 0.71 to 1.55 AU. By comparing the observations to the predictions of two water models of the coma (Haser and

vectorial), it is determined that these measurements support the idea of a comet composed principally of water ice. The vaporization of the water has a rather peculiar heliocentric variation, decreasing as $r^{-3.7}$ over the entire range of observations.

103.108　**The gassiest comet?**　D. W. Hughes.
Nature, Vol. 294, 696 (1981).

103.109　**A search for millimeter-wave emission from HCN and other molecules in Comet Bradfield (1979*l*).**
L. Ekelund, W. M. Irvine, C. Andersson, F. P. Schloerb, S. E. Robinson.
Icarus, Vol. 47, (see 012.045), 431 - 433 (1981) = Contrib. Five Coll. Astron. Depart., No. 471.
　　Millimeter-wave emission from HCN, CS, CH_3OH, and two unidentified lines (previously observed in Comet Kohoutek (1973 XII)) was sought and not detected from Comet Bradfield (1979*l*) after perihelion passage. Limits on column densities and production rates are derived. In the case of HCN, the production rate is less than that reported for Comet Kohoutek, even after scaling relative to the observed OH emission from each comet.

103.110　**Ground-based photometry of comets in the spectral interval 3000 to 3500 Å.**
R. L. Millis, M. F. A'Hearn.
Modern observational techniques for comets, (see 012.047), p. 57 - 62 (1981).
　　This paper summarizes the authors' efforts to measure the strength of the 0–0 band of OH in the spectrum of Comet Bradfield (1979*l*) using the 0.6–meter Planetary Patrol telescope at Mauna Kea Observatory.

103.111　**The apparition of comet Bradfield 1979 X.**
S. J. O'Meara, D. W. E. Green,
Strolling Astron., Vol. 29, 45 - 51 (1981).
　　Visual photometric observations of comet Bradfield 1979 X (= 1979*l*) are discussed. The brightness behavior of this comet is fairly well represented by the following magnitude formula, found through a least-squares regression analysis of 117 selected observations: $m_1 = 8.10 + 5 \log \Delta + 7.69 \log r$, where m_1 is the aperture-corrected, total visual magnitude of the comet, and Δ and r are the comet's geocentric and heliocentric distance respectively.

Ultraviolet spectroscopy of comets.
See Abstr. 102.009.

The forbidden oxygen lines in comets.
See Abstr. 102.037.

Fluorescence equilibrium in the ultraviolet spectra of Comets Seargent (1978m) and Bradfield (1979*l*).
See Abstr. 103.501.

Periodic comet Gehrels 3

103.121　**Variations of the orbit of comet P/Gehrels 3: temporary satellite captures by Jupiter.**
H. Rickman, A. M. Malmort.
Astron. Astrophys., Vol. 102, 165 - 170 (1981).
　　The authors attempt to strengthen the evidence for a temporary satellite capture of comet P/Gehrels 3 by integrating backward a large number of modified starting orbits in the threebody problem (Sun-Jupiter-comet). Moreover, they extend the variations of the starting orbit far outside the range permitted by the observations of the comet in order to investigate some conditions for temporary satellite capture in the Sun-Jupiter model. This general approach is followed

with particular emphasis on the simultaneous variation of (a) the eccentricity and semimajor axis of the starting orbit, and (b) the starting mean anomaly and the longitude of perihelion.

Great March comet 1843 I

The great comet of 1843.　See Abstr. 004.007.

Periodic comet Gehrels 2

103.161　**Periodic comet Gehrels 2 (1981f).**
IAU Circ., No. 3615 (1981).

103.162　**Periodic comet Gehrels 2 (1981f).**
Yamamoto Circ., No. 1960 (1981).

103.163　**Periodic comet Gehrels 2 (1981f).**
Komet.Tsirk., Kiev, No. 280 (1981). In Russian.

Periodic comet Väisälä 1

103.181　**Periodic comet Väisälä 1 (1981*l*).**
IAU Circ., No. 3654 (1981).

Comet 1975 X Suzuki-Saigusa-Mori

103.201　**The apparition of Comet Suzuki-Saigusa-Mori 1975 X.**
D. W. E. Green.
Strolling Astron., Vol. 29, 12 - 18 (1981).

Periodic comet Shajn-Schaldach

103.211　**The past orbital evolution of periodic comets Shajn-Schaldach and Whipple.**　E. M. Pittich.
Bull. Astron. Inst. Czechoslovakia, Vol. 32, 340 - 345 (1981).
　　Using improved orbital elements of the comets P/Shajn-Schaldach and P/Whipple their orbital evolution is integrated 250 years backward. The present similarity of their orbits is shown to be merely a result of recent close encounters of these comets with Jupiter. Prior to 1875 the comets were moving in quite different orbits. The comets are not of a common origin as it was suggested on the basis of less accurate data available before.

Comet 1980b Bowell

103.221　**Dust emission from comets at large heliocentric distances. 1. The case of comet Bowell (1980b).**
H. L. F. Houpis, D. A. Mendis.
Moon Planets, Vol. 25, 397 - 412 (1981).
　　Possible mechanisms for the production of the extended dust coma observed in comet Bowell (1980b) at the large heliocentric distance of 7.17 AU are considered. It is concluded that a plausible mechanism is electrostatic blow-off of fine, loose dust from an electrically charged H_2O dominated nucleus, as recently proposed by Mendis et al. (1981). Of all the other processes considered, dust entrainment by a chemical species more volatile than H_2O is considered the most plausible.

103.222 A search for frost in comet Bowell (1980b).
H. Campins, G. H. Rieke, M. Lebofsky, L. Lebofsky.
Bull. American Astron. Soc., Vol. 13, 705 (1981). – Abstract.

103.223 **Infrared thermal emission from comet Bowell.**
M. S. Hanner, G. J. Veeder, D. L. Matson.
Bull. American Astron. Soc., Vol. 13, 705 - 706 (1981).
Abstract.

103.224 **Infrared photometry of comets Bowell and P/Stephan-Oterma.**
G. J. Veeder, M. S. Hanner.
Icarus, Vol. 47, (see 012.045), 381 - 387 (1981).
  The authors report broadband infrared photometry of comets P/Stephan-Oterma and Bowell between 1 and 20 μm. Their JHK colors are similar to P/Meier and P/Tuttle and are compatible with scattering of sunlight by micron-sized grains. The thermal emission from P/Stephan-Oterma showed an effective temperature significantly higher than that expected from a blackbody in equilibrium. The thermal emission can be fit by models of the dust coma consisting of micron-sized grains.

103.225 **Comet Bowell (1980b).**
IAU Circ., No. 3632 (1981).

103.226 **Comet Bowell (1980b).**
Yamamoto Circ., No. 1960 (1981).

103.227 **Comet Bowell (1980b).**
Komet. Tsirk., Kiev, Nos. 280, 282, 283 (1981). In Russian.

Orbital elements of comets.
See Abstr. 103.008.

Comet 1980/ Russell

103.231 **Comet Russell (1980/).**
Komet.Tsirk., Kiev, No. 279 (1981). In Russian.

Comet 1973 XII Kohoutek

103.241 **Spectrophotometric scan of Comet 1973 XII Kohoutek.**
H. S. Ishii, T. Yamamoto, S. Tamura.
Moon Planets, Vol. 25, 437 - 450 (1981).
  Photoelectric spectrophotometric scans of Comet 1973 XII Kohoutek were made on November 25.85 and 28.85 UT, 1973. The vibrational temperature of C_2, the total numbers of CN and of C_2, and the production rates for CN and C_2 are derived from the observed fluxes.

103.242 **Plasma dynamics in the tail of Comet Kohoutek 1973 XII.**
K. Jockers.
Icarus, Vol. 47, (see 012.045), 397 - 411 (1981).
  Photographs of Comet Kohoutek 1973 XII from the period 1974, Jan. 19, 0 UT to Jan. 21, 3 UT, are studied. A large-scale tail disturbance was observed which coincides with the passage of a high-speed solar wind stream and an interplanetary sector boundary. The direction of the solar wind adjacent to the tail is deduced. During the large-scale disturbance the overall plasma density seems to be enhanced.

Comet 1975 IX Kobayashi-Berger-Milon

103.261 **Comet Kobayashi-Berger-Milon (1975 IX).**
IAU Circ., No. 3618 (1981).

Periodic comet Slaughter-Burnham

103.281 **Periodic comet Slaughter-Burnham (1981i).**
IAU Circ., No. 3621 (1981).

103.282 **Periodic comet Slaughter-Burnham (1981i).**
Yamamoto Circ., No. 1961 (1981).

103.283 **Short-period comet Slaughter-Burnham (1981i).**
Komet. Tsirk., Kiev, Nos. 281, 282 (1981). In Russian.

Periodic comet Encke

103.301 **Radar observations of the nucleus of the comet P/Encke.**   P. G. Kamoun, D. B. Campbell,
S. J. Ostro, G. H. Pettengill, I. I. Shapiro.
Bull. American Astron. Soc., Vol. 13, 705 (1981). – Abstract.

103.302 **The ultraviolet spectrum of periodic comet Encke.**
P. D. Feldman, H. A. Weaver, M. C. Festou,
H. U. Keller.
Bull. American Astron. Soc., Vol. 13, 707 (1981). – Abstract.

How a cometary nucleus turns on.
See Abstr. 102.003.

Comet 1937 II Wilk

Orbital elements of comets.
See Abstr. 103.008.

Comet 1939 IX Friend

Orbital elements of comets.
See Abstr. 103.008.

Periodic comet Grigg-Skjellerup

Orbital elements of comets.
See Abstr. 103.008.

Periodic comet Swift-Gehrels

103.381 **Periodic comet Swift-Gehrels (1981j).**
IAU Circ., Nos. 3622, 3642, 3650 (1981).

103.382 **Periodic comet Swift-Gehrels (1981j).**
Yamamoto Circ., Nos. 1961, 1962, 1965, 1967 (1981).

Periodic comet Ashbrook-Jackson

103.401　Application of the Trepied-Metcalf method to observations of the Periodic Comet Ashbrook-Jackson (1977g) in the European Southern Observatory at La Silla, Chile, April 1978.　H. Debehogne, L. E. Machado. An. Acad. Brasil. Cienc., Vol. 52, 459 - 465 (1980). In Portuguese. − Abstr. in Phys. Abstr., Vol. 84, Abstr. 71122 (1981).

Comet 1978 XIII Machholz

Orbital elements of comets.
See Abstr. 103.008.

Periodic comet Wild 3

Orbital elements of comets.
See Abstr. 103.008.

Comet 1980 XII Meier

103.461　Comet Meier (1980q).
Komet. Tsirk., Kiev, Nos. 279 - 282 (1981). In Russian.

Orbital elements of comets.
See Abstr. 103.008.

Comet 1980 XV Bradfield

103.481　Comet Bradfield (1980t).
IAU Circ., No. 3621 (1981).

103.482　Comet Bradfield (1980t).
Komet.Tsirk., Kiev, No. 282 (1981). In Russian.

Orbital elements of comets.
See Abstr. 103.008.

Comet 1978 XV Seargent

103.501　Fluorescence equilibrium in the ultraviolet spectra of Comets Seargent (1978m) and Bradfield (1979l). M. F. A'Hearn, D. G. Schleicher, B. Donn, W. M. Jackson. The Universe at ultraviolet wavelengths, (see 012.009), p. 73 - 81 (1981).

　　The authors have carried out detailed fluorescence calculations for OH including the Swings effect. These calculations have been used to reproduce the high resolution spectra of Comets Seargent and Bradfield taken with IUE. The calculations also provide the OH fluorescence efficiencies (g factors), as a function of heliocentric radial velocity, which are needed to derive OH abundances from measured fluxes. A close examination of the spectra shows no sign of the corresponding emission bands of OD. Preliminary attempts to reproduce the CO^+ band structure by fluorescence are also discussed.

Ultraviolet spectroscopy of comets.
See Abstr. 102.009.

Comet 1980 u Panther

103.521　Note sur le spectre de la comète 1980 u. C.-C. Huang. Astron. Astrophys., Suppl. Ser., Vol. 46, 369 (1981).
　　The author presents a spectrum of Comet Panther (1980 u) obtained on December 29, 1980, and showing various molecular bands.

103.522　Comet Panther (1980u).
Komet.Tsirk., Kiev, Nos. 279 - 281 (1981). In Russian.

Orbital elements of comets.
See Abstr. 103.008.

Comet 1981b Bus

103.541　Comet Bus (1981b).
Komet. Tsirk., Kiev, Nos. 279, 282 (1981). In Russian.

Orbital elements of comets.
See Abstr. 103.008.

Possible comet Stättmayer

103.561　Possible comet Stättmayer.
IAU Circ., Nos. 3638, 3642 (1981).

103.562　Possible comet Stättmayer.
Komet.Tsirk., Kiev, No. 283 (1981). In Russian.

Comet 1979 XI Howard-Koomen-Michels

103.581　Comet Howard-Koomen-Michels (1979 XI).
IAU Circ., Nos. 3640, 3647 (1981).

103.582　Comet Howard-Koomen-Michels (1979 XI).
Yamamoto Circ., No. 1966 (1981).

Comet 1976 VI West

103.601　The estimation of the dust content in the atmosphere of comet West 1975n. M. Šolc, V. Vanýsek, J. Svatoš, Pham Tien Duc. Acta Univ. Carolinae Math. Phys., Vol. 22, No. 2, p. 65 - 76 (1981).

　　Two components were distinguished in the spectra of comet West 1975n on the basis of multicolour photometry − the solar light scattered by cometary dust and the proper thermal emission of the dust. The temperature and mass of the dust in the cometary atmosphere are estimated depending on the position of the comet on its orbit, under the assumption of isothermal dust and spherical shape of dust particles. It is shown that more solid material is supplied to the cometary coma from the nucleus in the period following the transit through the perihelion than in the identical heliocentric distances in the period preceding the perihelion.

103.602　Substances of cometary grains estimated from evaporation and radiation pressure mechanisms. K. Saito, S. Isobe, K. Nishioka, T. Ishii. Icarus, Vol. 47, (see 012.045), 351 - 360 (1981).

The shape and intensity distribution of tails for comets West (1976 VI), Ikeya-Seki (1965 VIII) and Seki-Lines (1962 III) are estimated on the basis of grain properties in the solar radiation field.

103.603 Optical polarimetry of Comet West 1976 VI.
J. J. Michalsky.
Icarus, Vol. 47, (see 012.045), 388 - 396 (1981).

The polarization of the continuum of Comet West 1976 VI was measured in four narrowband filters spanning the wavelengths 440 - 850 nm. The postperihelion observations indicated wavelength independent linear polarization. The magnitude of the polarization as a function of scattering angle, the wavelength independence, and the infrared and optical photometric properties suggest that dirty silicates ($n_i \sim 0.05$) with radii smaller than 5 μm but approaching this size may be responsible. No circular polarization was detected.

103.604 Narrow passband imagery of comets.
T. R. Gull.
Modern observational techniques for comets, (see 012.047), p. 185 - 189 (1981).

During an emission-line survey of the Milky Way, Comet West was accidently imaged through four different narrow passbands with a wide-field, image-intensified camera. Three passbands recorded very similar head plus tail structure. The fourth passband shows an additional large, diffuse component around the head. It was serendipitous that such was recorded as the filters, being selected for studies of emission nebulae, are not particularly suited for studies of comets. However the imagery, plus subsequent studies, encourages to suggest that much can be learned about the structure of comets using narrow passband imagery simultaneously with long slit spectroscopy.

103.605 Comet West 1976 VI: observations of the great comet of 1976.   D. Wallentinsen.
Strolling Astron., Vol. 29, 69 - 79 (1981).

Ultraviolet spectroscopy of comets.
See Abstr. 102.009.

Comet 1981c Elias

103.621 Comet Elias (1981c).
Komet. Tsirk., Kiev, Nos. 279, 281 (1981). In Russian.

Orbital elements of comets.
See Abstr. 103.008.

Comet 1981d Bus

103.641 Comet Bus (1981d).
Komet.Tsirk., Kiev, Nos. 279, 281, 282 (1981). In Russian.

Orbital elements of comets.
See Abstr. 103.008.

Periodic comet Finlay

103.651 Comet Finlay (1981e).
Komet.Tsirk., Kiev, No. 279 (1981). In Russian.

Comet 1981g González

103.661 Comet González (1981g).
IAU Circ., Nos. 3617 - 3619, 3623, 3627 (1981).

103.662 Comet González (1981g).
Yamamoto Circ., Nos. 1961, 1962 (1981).

103.663 Possible new comet González (1981g).
Komet.Tsirk., Kiev, Nos. 280 - 282 (1981). In Russian.

Orbital elements of comets.
See Abstr. 103.008.

Comet 1981k Howell

103.681 Comet Howell (1981k).
IAU Circ., Nos. 3631, 3633, 3635, 3636 (1981).

103.682 Comet Howell (1981k).
Yamamoto Circ., Nos. 1962 - 1964 (1981).

103.683 New comet Howell (1981k).
Komet.Tsirk., Kiev, Nos. 281, 283 (1981). In Russian.

Orbital elements of comets.
See Abstr. 103.008.

Periodic comet Whipple

The past orbital evolution of periodic comets Shajn-Schaldach and Whipple.   See Abstr. 103.211.

Periodic comet Schwassmann-Wachmann 1

103.701 Comète P/Schwassmann-Wachmann 1.
C. Bertaud.
Astronomie, Vol. 95, 413 (1981).

103.702 The color and CO⁺ emission of comet Schwassmann-Wachmann 1.
A. L. Cochran, W. D. Cochran, E. S. Barker.
Bull. American Astron. Soc., Vol. 13, 724 - 725 (1981).
Abstract.

103.703 Periodic comet Schwassmann-Wachmann 1 in the year 1981.   J. Bouška.
Říše hvězd, Vol. 62, 204 - 206 (1981). In Czech.

103.704 P/Schwassmann-Wachmann 1.
Yamamoto Circ., No. 1965 (1981).

103.705 On the outbursts of P/Comet Schwassmann-Wachmann 1 during 1980 - 1981.   C.-Y. Shao.
Int. Comet. Q., Vol. 3, 76 (1981).

103.706 Recent outbursts of P/Schwassmann-Wachmann 1.
M. Huruhata.
Int. Comet Q., Vol. 3, 77 (1981).

103.707 An ephemeris for periodic Comet Schwassmann-Wachmann 1.   D. W. E. Green.
Int. Comet Q., Vol. 3, 92 (1981).

103.708   Comet Schwassmann-Wachmann 1.
Komet. Tsirk., Kiev, Nos. 279, 280, 282 (1981). In
Russian.

103.709   Explanation of the activity of comet Schwassmann-
Wachmann 1 on the basis of a new model of the
comet.   E. A. Kajmakov, I. S. Lizunkova.
Komet. Tsirk., Kiev, No. 283 (1981). In Russian.

Periodic comet Schwassmann-Wachmann 3

103.721   Comet Schwassmann-Wachmann 3 (1930 VI =
1979g).   N. A. Belyaev, S. D. Shaporev.
Determination of the coordinates of celestial bodies, Riga,
1981, (see 003.008), p. 61 - 69. In Russian. — Abstr. in Ref.
zh., 51. Astron., 9.51.86 (1981).

Periodic comet Schwassmann-Wachmann 2

103.731   The apparition of P/Schwassmann-Wachmann 2
(1979k).   C. S. Morris.
Int. Comet Q., Vol. 3, 103 (1981).

Periodic comet Stephan-Oterma

103.741   Narrowband photometry of comet P/Stephan-
Oterma during the 1980/81 apparition.
R. L. Millis, D. T. Thompson, M. F. A'Hearn.
Bull. American Astron. Soc., Vol. 13, 706 (1981). — Abstract.

Infrared photometry of comets Bowell and
P/Stephan-Oterma.   See Abstr. 103.224.

Periodic comet Tuttle

103.751   Comet Tuttle (1980h).
Komet. Tsirk., Kiev, No. 282 (1981). In Russian.

Periodic comet Swift-Tuttle

103.761   Distribution and activity of discrete emission areas
on the nucleus of periodic comet Swift-Tuttle.
Z. Sekanina.
Astron. J., Vol. 86, 1741 - 1773 (1981).
Periodic comet Swift-Tuttle (1862 III), the parent comet
of the Perseid meteor stream with a revolution period of 120a,
was extensively observed in 1862. From these data the author
presents an interpretation of the jet, envelope, and tail evolution in terms of dust ejections from discrete active regions on
the rotating nucleus. An emission model is developed to determine the rotational properties of the comet; the distribution,
lifetime and size of the activity sources; their gas production
variations with time; the mode of emission; the time lag in
sublimation versus the Sun's altitude above the local horizon;
and the relation between the particle ejection velocity and the
particle acceleration by solar radiation pressure.

103.762   The next return of the comet of the Perseid meteors.
B. G. Marsden.
Int. Comet Q., Vol. 3, 69 - 71 (1981).
The prospects for recovery of comet 1862 III (P/Swift-

Tuttle) are discussed, based upon its 1862 apparition and
upon possible identifications with previous comets.

Periodic comet Tempel 2

Some consequences of a phase transition of water
ice on the heat balance of comet nuclei.
See Abstr. 102.053.

Periodic comet Halley

103.801   The long-term motion of comet Halley.
D. K. Yeomans, T. Kiang.
Mon. Not. R. Astron. Soc., Vol. 197, 633 - 646 (1981).
The orbital motion of comet Halley has been numerically integrated back to 1404 BC. The non-gravitational forces
were assumed due to the rocket effect of an outgassing water
ice nucleus. The dynamic model used to compute the long-
term motion of comet Halley assumed the comet's non-
gravitational forces remained constant from one apparition to
the next. Hence it seems likely that comet Halley's spin axis
direction and ability to outgas has also remained relatively
constant with time.

103.802   On the magnetic field in the tail of Comet Halley.
A. I. Ershkovich.
Moon Planets, Vol. 25, 521 - 522 (1981).
The method of evaluation of the cometary magnetic field
proposed by Podgorny et al. (1980) is shown not to be self-
consistent. An alternative method is discussed.

103.803   Comet Halley's motion, 1404 B. C. - 1986 A. D.
D. K. Yeomans, T. Kiang.
Bull. American Astron. Soc., Vol. 13, 704 (1981). — Abstract.

103.804   Infrared spectroscopy of comet Halley.
T. Encrenaz, M. Combes, J. F. Crifo, E. Gérard,
J. P. Bibring, J. Crovisier.
Bull. American Astron. Soc., Vol. 13, 707 (1981). — Abstract.

103.805   Comet Halley 1986 as a target of space probes.
V. Vanýsek.
Vesmír, Vol. 60, 366 - 367 (1981). In Czech.

103.806   On the visibility of nuclei of dusty comets.
R. Hellmich, H. U. Keller.
Icarus, Vol. 47, (see 012.045), 325 - 332 (1981).
The ratio of radiation scattered in the dust column above
the surface and that reflected from the cometary nucleus
determines the visibility of features on the nuclear surface. A
contrast parameter characterizing the ratio of radiation coming from the nuclear surface and that of the nuclear vicinity
describes the visibility of the full nucleus against the dust fore-
and background. These quantities and the intensity distribution of scattered solar radiation across the nucleus and its
vicinity are calculated for the case of comet Halley at a heliocentric distance of 0.9 AU after perihelion.

103.807   Photographic observations of comets at Lowell
Observatory.   H. L. Giclas.
Modern observational techniques for comets, (see 012.047),
p. 202 - 212 (1981).
The author briefly reviews the 1910 observations of
Halley's comet and illustrates just one morning's observations
as an example of utilizing every possible observational resource
available at the Observatory at that time. It also exemplifies
the comprehension that the opportunity to observe objects of
this nature from a ground base is limited and, therefore, had to

be maximized at the opportune time. He then describes a few observational improvements he has developed since and some suggestions for the coming return of Halley's comet.

103.808 Outburst and nuclear breakup of comet Halley – 1910. H. J. Wood, R. Albrecht.
Modern observational techniques for comets, (see 012.047), p. 216 - 219 (1981).
Computer processing of five plates of comet Halley taken during the 1910 apparition shows that on May 24 strong asymmetric (with respect to the tail axis) fountain-like parabolic plumes had developed on the sunward side of the nucleus. Visual observations showed that after an initial fading while passing in front of the sun, the brightness increased to about magnitude 1. On the plates taken May 31 the nucleus is clearly divided into at least three parts of nearly equal brightness. However, the last plate on June 3 shows a symmetrical coma with a small stellar-like nucleus.

103.809 The interactive astronomical data analysis facility – image enhancement techniques applied to comet Halley. D. A. Klinglesmith III.
Modern observational techniques for comets, (see 012.047), p. 223 - 231 (1981).

103.810 A review and recalculation of Bobrovnikoff's photometric power-law solutions for P/Comet Halley 1910 II. C. S. Morris, D. W. E. Green.
Int. Comet. Q., Vol. 3, 100 - 103 (1981).

The International Halley Watch.
See Abstr. 013.025.

The international Halley watch: a program of coordination, cooperation and advocacy.
See Abstr. 013.037.

The infrared experiment of the Venera-Halley mission. See Abstr. 032.552.

Halley and contemporary space science.
See Abstr. 051.008.

Missions to comets: the perspective in 1980.
See Abstr. 051.012.

Near-perihelion observations of comet Halley from Shuttle Orbiter. See Abstr. 051.039.

Plans for comet Halley. See Abstr. 051.040.

The ESA mission to comet Halley.
See Abstr. 051.041.

Rendezvous with comet Halley.
See Abstr. 051.045.

The role of collisions with interplanetary particles in the physical evolution of comets. See Abstr. 102.039.

Thermal modeling of cometary nuclei.
See Abstr. 102.051.

Some consequences of a phase transition of water ice on the heat balance of comet nuclei.
See Abstr. 102.053.

Comet 1965 VIII Ikeya-Seki

Substances of cometary grains estimated from evaporation and radiation pressure mechanisms.
See Abstr. 103.602.

Comet 1962 III Seki-Lines

Substances of cometary grains estimated from evaporation and radiation pressure mechanisms.
See Abstr. 103.602.

Periodic comet Tempel-Tuttle

Comet Tempel-Tuttle and the Leonid meteors.
See Abstr. 104.021.

Periodic comet Gunn

103.871 Periodic comet Gunn.
Komet. Tsirk., Kiev, No. 282 (1981). In Russian.

Comet 1980 IV Černis-Petrauskas

103.881 Comet Černis-Petrauskas (1980k).
V. Straižys.
Bull. Vilnius Astron. Obs., Nr. 56, p. 46 - 50 (1981). In Russian.

103.882 New comet Černis-Petrauskas (1980k).
A. Salītis.
Zvaigžņotā debess, 1981. gada pavasaris, p. 12. In Latvian.

Periodic comet Honda-Mrkos-Pajdušaková

103.901 An attempt to observe an anti-tail for P/Honda-Mrkos-Pajdušaková in 1980.
I. Halliday, B. A. McIntosh, A. F. Cook.
Modern observational techniques for comets, (see 012.047), p. 70 - 71 (1981).

Periodic comet Borrelly

103.921 Periodic comet Borrelly (1980i).
IAU Circ., No. 3616 (1981).

103.922 Periodic comet Borrelly (1980i).
Yamamoto Circ., No. 1963 (1981).

103.923 The tenth apparition of P/Comet Borrelly.
C. S. Morris, D. W. E. Green.
Int. Comet Q., Vol. 3, 115 (1981).

103.924 Periodic comet Borrelly (1980i).
Komet.Tsirk., Kiev, No. 280 (1981). In Russian.

Periodic comet Kearns-Kwee

103.941   **Periodic comet Kearns-Kwee (1981h).**
IAU Circ., Nos. 3618, 3625, 3653 (1981).

103.942   **Periodic comet Kearns-Kwee (1981h).**
Yamamoto Circ., No. 1961 (1981).

103.943   **Periodic comet Kearns-Kwee (1981h).**
Komet.Tsirk., Kiev, No. 282 (1981). In Russian.

104 Meteors, Meteor Streams

104.001 Case study using arrays of infrasonic microphones to detect and locate meteors and meteorites.
A. J. Bedard, Jr., G. E. Greene.
J. Acoust. Soc. America, Vol. 69, 1277 - 1279 (1981).
Abstr. in Phys. Abstr., Vol. 84, Abstr. 79746 (1981).

104.002 Cooling of electrons in meteor traces.
S. M. Levitskij, N. Abdrakhmanov.
Geomagn. Aehron., Tom 21, 674 - 678 (1981). In Russian.

104.003 Results of investigating meteoric matter on the Interkosmos 14 satellite.
I. Zacharov, Yu. A. Surkov, A. K. Rybakov, Z. V. Vasyukova
(Vas'yukova), Yu. D. Vasil'ev, I. Apati, I. Semirej.
Bull. Astron. Inst. Czechoslovakia, Vol. 32, 239 - 242 (1981).

The results of recording micro-meteoric bodies on the Interkosmos 14 satellite are studied. It is shown that flux density of particles with mass $m \geqslant 6 \times 10^{-14}$ in the neighbourhood of the Earth at an altitude of 1000 km is not more than 5 times higher than the density in interplanetary space. A correlation was observed between the fluxes of particles of the sporadic background and of the particles combined into groups. The results of the experiments are compared with the data of other investigators.

104.004 Excitation temperature of meteor atoms according to the instantaneous spectrum.
P. B. Babadzhanov, V. S. Getman.
Dokl. AN TadzhSSR, Tom 23, 703 - 705 (1980). In Russian.
Abstr. in Ref. zh., 51. Astron., 7.51.217 (1981).

104.005 Density change in meteor streams at the 1 : 3 commensurability of their orbits with Jupiter.
A. M. Kazantsev, L. M. Sherbaum.
Problems of cosmic physics. Vyp. 16, (see 003.005), p. 30 - 33 (1981). In Russian.

On the basis of model calculations of the motion of particles in the vicinity of the commensurability 1/3 with Jupiter the change of the linear density along the orbit is obtained.

104.006 Simultaneous radar meteor observations at Ondřejov and Budrio.   A. Hajduk, G. Cevolani.
Bull. Astron. Inst. Czechoslovakia, Vol. 32, 304 - 310 (1981).

Radar meteor observations carried out simultaneously by the Ondřejov (Czechoslovakia) and Budrio (Italy) radar equipment during the Orionid shower period in October 1978 were used to obtain the variation of shower activity. The resulting peak rate is placed at $\lambda_\odot = 208.1 \pm 0.3$. A rapid increase of meteor activity, exceeding the shower peak rate, was found at $\lambda_\odot = 214.6$ and is ascribed to a filament of the stream.

104.007 A possibility of using optimization methods for determination of meteor parameters. I.
K. V. Kostylev, K. K. Kostylev.
Meteor. rasprostranenie radiovoln, Kazan', 1980, No. 15, p. 3 - 12. In Russian. − Abstr. in Ref. zh., 51. Astron., 9.51.265 (1981).

104.008 A method of statistical analysis of density variation of radiants of sporadic meteors over the celestial sphere.   Yu. A. Pupyshev.
Meteor. rasprostranenie radiovoln, Kazan', 1980, No. 15, p. 13 - 20. In Russian. − Abstr. in Ref. zh., 51. Astron., 9.51.266 (1981).

104.009 Maps of distribution of the apparent density of radiants of sporadic meteors over the whole
celestial sphere.
Yu. A. Pupyshev, T. K. Filimonova, T. V. Kazakova.
Meteor. rasprostranenie radiovoln, Kazan', 1980, No. 15, p. 21 - 41. In Russian. − Abstr. in Ref. zh., 51. Astron., 9.51.267 (1981).

104.010 On the range-finder method for determination of the position of a meteor train.
G. I. Lyutershtejn.
Meteor. rasprostranenie radiovoln, Kazan', 1980, No. 15, p. 42 - 44. In Russian. − Abstr. in Ref. zh., 51. Astron., 9.51.268 (1981).

104.011 Determination of the initial radius of a meteor train in the characteristic height of the radio line.
A. A.-K. Gajdaev, A. M. Zulliev, O. I. Bel'kovich.
Izv. Sev.-Kavkaz. nauchn. tsentr. vyssh. shkoly. Estestv. nauk., 1981, No. 1, p. 43 - 44. In Russian. − Abstr. in Ref. zh., 51. Astron., 9.51.269 (1981).

104.012 Evolution of variable size meteoroids under the influence of non-gravitational forces.
P. B. Babadzhanov, Yu. V. Obrubov.
Dokl. AN TadzhSSR, Tom 24, 17 - 21 (1981). In Russian.
Abstr. in Ref. zh., 51. Astron., 9.51.278 (1981).

104.013 Orionid meteor shower: activity and magnitude distribution.   J. Stohl, V. Porubčan.
Contrib. Astron. Obs. Skalnaté Pleso, Vol. 10, 39 - 51 (1981).

A series of visual observations of the Orionid meteor shower, comprising over 2500 records of meteors observed during 6 different returns of the shower in 1944 - 1950 at the Skalnaté Pleso Observatory, is analysed and discussed.

104.014 The Eta Aquarid meteor shower.
A. Hajduk.
Contrib. Astron. Obs. Skalnaté Pleso, Vol. 10, 125 - 133 (1981).

On the basis of 240 000 radar echoes observed during the Eta Aquarid meteor shower period in 1958 - 1967 at the Springhill Meteor Observatory the activity of the shower is examined and hourly rates are tabulated.

104.015 Velocities of meteoroids from radar observations. I. Velocity distributions.
Yu. I. Voloshchuk, B. L. Kashcheev, A. A. Tkachuk.
Astron. Vestn., Tom 15, 118 - 126 (1981). In Russian.

The results of velocity measurements of individual meteors up to $+12^m$ from Kharkov observations carried out in 1973 - 1974 are analyzed with a computerized radar of high sensitivity. A new method for obtaining the true velocity distribution of meteoroids in the neighbourhood of the earth's orbit is proposed. The variation of the parameters of distribution of the meteoroids versus the velocity outside the earth's atmosphere and the heliocentric velocity for various sections of the celestial sphere are investigated by means of this method.

104.016 The bolid which created meteorite Tsarev.
R. L. Khotinok.
Zemlya Vselennaya, 1981, No. 4, p. 74. In Russian.

104.017 Results of an investigation of meteor matter aboard the Intercosmos 14 artificial earth satellite and
accompanying ground-based observations of meteors.
I. Apathy, Z. V. Vasyukova, B. L. Kashcheev, A. K. Rybakov, Yu. A. Surkov, A. A. Tkachuk, I. Zacharov, A. Hajduk, M. Šimek.
Kosm. Issled., Tom 19, 790 - 794 (1981). In Russian.

104.018   **Numerical modelling of the Draconid meteor shower.**
E. A. Reznikov.
Chelyabinsk. politekh. inst. Chelyabinsk, 1981. 15 pp. In
Russian. – Abstr. in Ref. zh., 51. Astron., 10.51.303 (1981).

104.019   **Three fireballs observed on 28, 29 and 30 January**
1981 in Czechoslovakia.   Z. Ceplecha.
Říše hvězd, Vol. 62, 160 - 162 (1981). In Czech.

104.020   **Influence of electrostatic fields of the ionosphere**
on the dynamics of an ionized meteor trace.
Z. M. Ioffe, L. N. Rubtsov.
Geomagn. Aehron., Tom 21, 1129 - 1131 (1981). In Russian.

104.021   **Comet Tempel-Tuttle and the Leonid meteors.**
D. K. Yeomans.
Icarus, Vol. 47, (see 012.045), 492 - 499 (1981).
The distribution of dust surrounding periodic comet
Tempel-Tuttle has been mapped by analyzing the associated
Leonid meteor shower data over the 902 - 1969 interval. It
appears that radiation pressure and planetary perturbations,
rather than ejection processes, control the dynamic evolution
of the Leonid particles. Significant Leonid meteor showers are
possible roughly 2500 days before or after the parent comet
reaches perihelion but only if the comet passes closer than
0.025 AU inside or 0.010 AU outside the Earth's orbit. As a
by-product of this study, the orbit of comet Tempel-Tuttle
has been redetermined for the 1366 - 1966 observed interval.

104.022   **Earth-orbit-approaching comets and their theoreti-**
cal meteor radiants.   J. D. Drummond.
Icarus, Vol. 47, (see 012.045), 500 - 517 (1981).
Sixteen comets produce recognizable meteor showers
that are found in A. F. Cook's (1973) working list of meteor
streams. This paper presents a comprehensive listing of theoretical meteor radiants from Earth-orbit-approaching comets
as an aid for those observers who seek to define further relationships between comets and meteors. The seasonal distribution of these theoretical radiants is discussed, the non-meteor-
producing comets are addressed and some speculations are
concluded.

104.023   **Heigh variations of meteor train drifts.**
V. V. Zhukov, B. V. Kal'chenko, B. L. Kashcheev.
Meteorn. issled., Moskva, 1981, No. 7, p. 13 - 22. In Russian.
Abstr. in Ref. zh., 51. Astron., 11.51.352 (1981).

104.024   **The effect of selectivity of radio observations on**
meteor velocity distribution.
A. A. Tkachuk.
Meteorn. issled., Moskva, 1981, No. 7, p. 28 - 36. In Russian.
Abstr. in Ref. zh., 51. Astron., 11.51.353 (1981).

104.025   **Investigation of time series of faint radio meteor**
rates.
Yu. I. Voloshchuk, M. I. Malynyak, N. B. Nazarenko.
Meteorn. issled., Moskva, 1981, No. 7, p. 37 - 43. In Russian.
Abstr. in Ref. zh., 51. Astron., 11.51.354 (1981).

104.026   **Forecasting of faint radio meteor activity.**
M. I. Malynyak.
Meteorn. Issled., Moskva, 1981, No. 7, p. 44 - 55. In Russian.
Abstr. in Ref. zh., 51. Astron., 11.51.355 (1981).

104.027   **Dependence of the average height on meteor veloc-**
ity.   V. N. Olejnikov.
Meteorn. issled., Moskva, 1981, No. 7, p. 56 - 59. In Russian.
Abstr. in Ref. zh., 51. Astron., 11.51.356 (1981).

104.028   **Selection of duration of recording individual**
meteor radio echoes.
V. V. Lizogub, V. N. Olejnikov.

Meteorn. issled., Moskva, 1981, No. 7, p. 60 - 62. In Russian.
Abstr. in Ref. zh., 51. Astron., 11.51.357 (1981).

104.029   **Regions of visibility of meteor radiants.**
A. A. Tkachuk, V. V. Matsenko.
Meteorn. issled., Moskva, 1981, No. 7, p. 63 - 73. In Russian.
Abstr. in Ref. zh., 51. Astron., 11.51.358 (1981).

104.030   **On the phase radar method for determination of the**
vectors of meteoroid velocities.
N. S. Andrianov.
Meteorn. rasprostr. radiovoln, Kazan', 1980, No. 16, p. 63 - 68.
In Russian. – Abstr. in Ref. zh., 51. Astron., 11.51.361
(1981).

104.031   **On the scale of meteor masses.**
E. N. Kramer, I. S. Shestaka.
Astron. Tsirk., No. 1130, p. 7 - 8 (1980). In Russian.

104.032   **On the structure of the Geminid meteor stream 1976.**
G. V. Andreev, G. O. Ryabova, A. E. Epishova,
L. N. Rubtsov.
Astron. Tsirk., No. 1131, p. 5 - 7 (1980). In Russian.

104.033   **On identification of spectral lines in a TV spectro-**
gram of meteors in the infrared region.
S. Mukhamednazarov, K. Kurbanmuradov.
Astron. Tsirk., No. 1125, p. 6 - 7 (1980). In Russian.

104.034   **Taking into account the astronomical factor in**
statistics of meteor orbits.   V. A. Bronshtehn.
Astron. Tsirk., No. 1141, p. 3 - 7 (1980). In Russian.

104.035   **Single wavelength measurements of the initial radii**
of radio meteor ionization columns.
W. J. Baggaley.
Bull. Astron. Inst. Czechoslovakia, Vol. 32, 345 - 349 (1981).

104.036   **A statistical survey of photographed telescopic**
meteors. I. Louisiana State University Observatory.
S. S. Mims.
Contrib. No. 5 James-Mims Obs., 2 pp. (1980).

104.037   **Observations of meteor streams.**
H. Korpikiewicz.
Astron. Rep., Vol. 4, 30 - 35 (1979).

104.038   **A new radiant of meteors in Libra.**
H. Korpikiewicz.
Astron. Rep., Vol. 4, 36 (1979).

104.039   **On the spatial structure of the Geminid meteor**
shower.
Yu. A. Pupyshev, A. M. Stepanov, R. M. Zakiev,
T. K. Filimonova.
Meteor. rasprost. radiovoln, Kazan', 1980, No. 16, p. 3 - 9. In
Russian. – Abstr. in Ref. zh., 51. Astron., 1.51.298 (1982).

104.040   **On the scale of meteor masses. II.**
E. N. Kramer, I. S. Shestaka.
Astron. Tsirk., No. 1157, p. 5 - 7 (1981). In Russian.

104.041   **A possible meteor stream.**
Komet. Tsirk., Kiev, No. 279 (1981). In Russian.

104.042   **Which fireballs are meteorites? A study of the**
Prairie Network photographic meteor data.
G. W. Wetherill, D. O. ReVelle.
Icarus, Vol. 48, 308 - 328 (1981).
Plausible meteorite mass distributions imply that in the
Prairie Network data there must be many fainter fireballs
produced by meteorites with physical properties that, except

for mass, are very similar to the recovered ordinary chondrite (*H* 5) Lost City. Four criteria are proposed for identifying these other meteorites among the fireballs. By use of the criteria, 27 Prairie Network fireballs are identified as being meteorites comparable to or greater in strength and density to Lost City, most of these should be ordinary chondrites. The orbital element distributions of these objects span a wide range.

Physics of meteor phenomena.
See Abstr. 003.037.

Distribution of meteor bodies near the earth's orbit.
See Abstr. 003.156.

Die Beobachtung sporadischer Meteore und Meteor-schauer mittels radiotechnischer Meßeinrichtungen.
See Abstr. 031.510.

Analysis of multi-ray methods for radio observations of meteor trains.   See Abstr. 031.619.

A multi-ray method for meteor radar observations with electronic commutation of receiving antennas.
See Abstr. 033.014.

Meteor position and velocity in the original orbit.
See Abstr. 042.085.

Particles above the tropopause: measurements and models of stratospheric aerosols, meteoric debris, nacreous clouds, and noctilucent clouds.   See Abstr. 082.025.

The phenomenon of delay of formation of the sporadic layer and meteor activity.
See Abstr. 083.011.

Analysis of solar-terrestrial connections by the radar method of meteors.   See Abstr. 085.052.

The next return of the comet of the Perseid meteors.
See Abstr. 103.762.

Submicron lunar ejecta in the magnetosphere associated with meteor showers.   See Abstr. 106.055.

105 Meteorites, Meteorite Craters

105.001 **Diminished tektite ablation in the wake of a swarm.**
P. Sepri, K. K. Chen, J. A. O'Keefe.
J. Geophys. Res., Vol. 86, 5103 - 5111 (1981).
   Observations of ablation markings on tektite surfaces reveal that a large variation in aerodynamic heating must have occurred among the members of a swarm during atmospheric entry. Calculations indicate that the observed ablation variations are actually possible for swarm entry at greater than escape velocity. This aerodynamic conclusion provides support for the arguments favoring extraterrestrial origin of tektites.

105.002 **Microanalysis by Raman spectroscopy of carbon in the Tieschitz chondrite.**
M. Christophe Michel-Levy, A. Lautie.
Nature, Vol. 292, 321 - 322 (1981).
   The authors report that the carbon constituent is a highly disordered graphite with a progressive ordering as shown by samples heated to temperatures between 300 and 600°C. Tieschitz could not have been heated to more than 300-350°C.

105.003 **Estimates of total quantity of meteorites in the East Antarctic ice cap.**   E. J. Olsen.
Nature, Vol. 292, 516 - 518 (1981).
   Since 1969 ~5,000 meteorite fragments have been recovered from two regions, the Yamato Mountains and Victorialand, on opposite sides of the East Antarctic ice cap. Based on a steady-state model for the ice cap, and current estimates of meteorite influx, a model is developed here which predicts that the steady-state number of meteorites being carried in and on the ice is at least 760,000.

105.004 **How "unique" can meteorites be?**
R. Hutchison.
Nature, Vol. 293, 11 (1981).

105.005 **Solar flare irradiation records in Antarctic meteorites.**
J. N. Goswami.
Nature, Vol. 293, 124 - 125 (1981).
   The author reports the first observation of solar flare heavy nuclei tracks in Antarctic meteorite samples. Two interior specimens of sample 77216, an L-3 chondrite, contain track-rich grains indicating their exposure to solar flare irradiation before compaction of this meteorite. Preliminary noble gas data also indicate the presence of solar-type gases. Results of nuclear track studies of other Antarctic meteorite samples are presented.

105.006 **Solidification zoning and metallographic cooling rates of chondrites.**   J. Willis, J. I. Goldstein.
Nature, Vol. 293, 126 - 127 (1981).
   The authors have investigated the effect of zoning produced during solidification on the formulation of the Wood model, and report that solidification zoning is erased during kamacite growth and has no influence on the resultant taenite Ni gradients for all cases in which the cooling rate is sufficiently slow to allow Ni to diffuse to the taenite grain centre.

105.007 **The isotopic composition of titanium in the Allende and Leoville meteorites.**
F. R. Niederer, D. A. Papanastassiou, G. J. Wasserburg.
Geochim. Cosmochim. Acta, Vol. 45, 1017 - 1031 (1981).
   The authors report determinations of the isotopic composition of titanium in samples of terrestrial, lunar, and meteoritic materials and present a description of the chemical and mass spectrometric procedures developed for this purpose. The intent of this study was to establish a reliable technique for the high precision measurement of the Ti isotope abundances

and to search for variations in these abundances which might cast light on nuclear-astrophysical processes.

105.008 **The compositional classification of chondrites–I. The carbonaceous chondrite groups.**
G. W. Kallemeyn, J. T. Wasson.
Geochim. Cosmochim. Acta, Vol. 45, 1217 - 1230 (1981).
   Twenty carbonaceous chondrites were analyzed by instrumental and radiochemical neutron activation analysis. Analysis of 2 or more samples of all but 2 chondrites has helped yield a high precision that allowed the resolution of numerous previously unrecognized trends.

105.009 **The Antarctic achondrite ALHA 76005: a polymict eucrite.**   L. Grossman, E. Olsen, A. M. Davis,
T. Tanaka, G. J. MacPherson.
Geochim. Cosmochim. Acta, Vol. 45, 1267 - 1279 (1981).

105.010 **Volatilization of sodium from silicate melt spheres and its application to the formation of chondrules.**
A. Tsuchiyama, H. Nagahara, I. Kushiro.
Geochim. Cosmochim. Acta, Vol. 45, 1357 - 1367 (1981).
   The rates of volatilization of Na from liquid spheres of chondrule compositions have been determined as functions of time, temperature, partial pressure of oxygen, and sizes of the spheres.

105.011 **Meteorites and the origin of water and reduced carbon on the Earth.**   H. Roy-Poulsen, L. Larsen,
N. O. Roy-Poulsen, L. Vistisen, J. M. Knudsen.
Phys. Scr., Vol. 23, 1113 - 1117 (1981).
   The oxidation state of iron in some carbonaceous chondrites has been determined by means of Mössbauer spectroscopy, and it is demonstrated that there is a correlation between the oxidation state of iron and the content of water and reduced carbon in the meteorites. Possible consequences of this fact for the origin of water and reduced carbon on the Earth are discussed.

105.012 **Host phases and origin of noble gases in meteorites.**
B. Srinivasan.
Naturwissenschaften, 68. Jahrg., 341 - 353 (1981).

105.013 **Il bolide del 18 aprile 1977.**   S. Ghedini.
Astronomia, N. 2, p. 8 - 27 (1981).

105.014 **Isotopically anomalous xenon in meteorites: a new clue to its origin.**   R. S. Lewis, E. Anders.
Astrophys. J., Vol. 247, 1122 - 1124 (1981).
   The "CCF" xenon component in primitive meteorites, which has been attributed either to fission of a superheavy element or to nucleosynthesis in a supernova, does not show the large enrichment in Xe^{129} (from decay of 16 Myr I^{129}) expected for supernova ejecta. Although this problem can be circumvented by ad hoc assumptions, a fission origin of CCFXe seems more likely.

105.015 **Study of microinclusions in minerals and microtextures of meteorites.**   Z. Li, F. Sun, M. Li.
Sci. Sinica, Vol. 24, 975 - 981 (1981).

105.016 **The significance of unique or rare meteorites.**
R. Hutchison.
Nature, Vol. 293, 260 (1981).

105.017 **Uranium determination by ^{133}Xe in meteorites; comparative study with fission track and/or ^{239}Np methods and thermal release character of ^{133}Xe.**

K. Sakamoto, K. Itoh, T. Takashima, S. Yokoyama, K. Yamazaki.
J. Radioanal. Chem., Vol. 60, 323 - 342 (1980). — Abstr. in Phys. Abstr., Vol. 84, Abstr. 79748 (1981).

105.018 **Terrestrial ages of meteorites,** D. W. G. Sears.
Nature, Vol. 293, 433 (1981).

105.019 **The distribution of trace elements in carbonaceous chondrites.** H.-J. Knab.
Geochim. Cosmochim. Acta, Vol. 45, 1563 - 1572 (1981).

12 carbonaceous chondrites, amongst them representa-.ives of nearly all known petrologic types were analyzed for twenty trace elements by spark source mass spectrography combined with the isotope dilution method. Data on different element groups show that the distribution of the trace elements in the carbonaceous chondrites can be well explained by Anders' two-component model.

105.020 **Deep seated target materials in the continuous deposits of the Ries Crater, Germany.**
F. Hörz, G. S. Banholzer, Jr.
The lunar highlands crust, (see 012.013). p. 211 - 231 (1980).

105.021 **Pulverized and impactite meteoritic matter in the Kaali crater field.** A. Aaloe, R. Tiirmaa.
Izv. AN EhstSSR. Geol., Tom 30, 20 - 27 (1981). In Russian. Abstr. in Ref. zh., 51. Astron., 7.51.281 (1981).

105.022 **Relict grains in chondrules.** E. R. Rambaldi.
Nature, Vol. 293, 558 - 561 (1981).

The author presents important information with regard to the chondrule precursor materials and the process of chondrule formation which was acquired by a compositional and textural study of three of the most unequilibrated type 3 ordinary chondrites.

105.023 **Chinese meteorites.** B. Depei.
Meteoritics, Vol. 16, 115 - 128 (1981).

105.024 **Grier(b), a "monomict", brecciated chondrite.**
K. Fredriksson, C. G. R. Reid, B. J. Fredriksson.
Meteoritics, Vol. 16, 129 - 137 (1981).

105.025 **Thermal history of the Abee enstatite chondrite II; thermal measurements and heat flow calculations.**
M. L. Rudee, J. M. Herndon.
Meteoritics, Vol. 16, 139 - 140 (1981).

105.026 **Unusual weathering products of oldhamite parentage in the Norton County enstatite achon-**
drite. A. Okada, K. Keil, G. J. Taylor.
Meteoritics, Vol. 16, 141 - 152, with a correction p. 283 (1981).

105.027 **The Innisfree meteorite fall: a photographic analysis of fragmentation, dynamics and luminosity.**
I. Halliday, A. A. Griffin, A. T. Blackwell.
Meteoritics, Vol. 16, 153 - 170 (1981).

105.028 **Irghizites and zhamanshinites: Zhamanshin crater, USSR.**
V. Bouška, P. Povondra, P. V. Florenskij, Z. Řanda.
Meteoritics, Vol. 16, 171 - 184 (1981).

105.029 **The Lahrauli ureilite.**
N. Bhandari, V. G. Shah, A. Graham.
Meteoritics, Vol. 16, 185 - 191 (1981).

105.030 **The Meteoritical Bulletin, No. 59.**
Sponsored by The Meteoritical Society.
A. L. Graham (Editor).

Meteoritics, Vol. 16, 193 - 199 (1981).

Place of fall/find, class and type, number of individual specimens, total weight and circumstances of fall/find of some meteorites are given.

105.031 **Refractory and moderately volatile element abundances in the earth, moon and meteorites.**
S. R. Taylor.
Proc. Eleventh Lunar Planet. Sci. Conf., (see 012.020), p. 333 - 348 (1980).

105.032 **Vacuum ultraviolet reflectance spectra of groups L, LL, and E chondrites and of achondrites.**
J. K. Wagner, A. J. Cohen, B. W. Hapke, W. D. Partlow.
Proc. Eleventh Lunar Planet. Sci. Conf., (see 012.020), p. 775 - 797 (1980).

105.033 **Structures and textures of the Murchison and Mighei carbonaceous chondrite matrices.**
I. D. R. Mackinnon.
Proc. Eleventh Lunar Planet. Sci. Conf., (see 012.020), p. 839 - 852 (1980).

105.034 **Origin of volatile-rich H chondrites with light/dark structures.** H. Y. McSween, Jr., M. E. Lipschutz.
Proc. Eleventh Lunar Planet. Sci. Conf., (see 012.020), p. 853 - 864 (1980).

105.035 **Flow behavior of droplet chondrules in the Manych (L-3) chondrite.**
L. C. Klein, B. Fasano, R. H. Hewins.
Proc. Eleventh Lunar Planet. Sci. Conf., (see 012.020), p. 865 - 878 (1980).

105.036 **The neon-alphabet game.**
D. D. Sabu, O. K. Manuel.
Proc. Eleventh Lunar Planet. Sci. Conf., (see 012.020), p. 879 - 899 (1980).

105.037 **Bulk compositions of chondrules in the Allende meteorite.** S. B. Simon, S. E. Haggerty.
Proc. Eleventh Lunar Planet. Sci. Conf., (see 012.020), p. 901 - 927 (1980).

105.038 **Refractory metal particles in refractory inclusions in the Allende meteorite.**
L. H. Fuchs, M. Blander.
Proc. Eleventh Lunar Planet. Sci. Conf., (see 012.020), p. 929 - 944 (1980).

105.039 **XPS and STEM studies of Allende acid insoluble residues.** R. M. Housley, D. R. Clarke.
Proc. Eleventh Lunar Planet. Sci. Conf., (see 012.020), p. 945 - 958 (1980).

105.040 **$^{39}Ar/^{40}Ar$ systematics of Allende inclusions.**
G. F. Herzog, A. E. Bence, J. Bender, G. Eichhorn, H. Maluski, O. A. Schaeffer.
Proc. Eleventh Lunar Planet. Sci. Conf., (see 012.020), p. 959 - 976 (1980).

105.041 **Proportional retention of I and radiogenic ^{129}Xe in preheated Allende.**
W. Rison, A. Zaikowski.
Proc. Eleventh Lunar Planet. Sci. Conf., (see 012.020), p. 977 - 994 (1980).

105.042 **Experimental studies bearing on the magnetite-alloy-sulfide association in the Allende meteorite:**
constraints on the conditions of chondrule formation.
B. M. McMahon, S. E. Haggerty.

Proc. Eleventh Lunar Planet. Sci. Conf., (see 012.020), p. 1003 - 1025 (1980).

105.043 **Allan Hills A77219, the first Antarctic mesosiderite.**
W. N. Agosto, R. H. Hewins, R. S. Clarke, Jr.
Proc. Eleventh Lunar Planet. Sci. Conf., (see 012.020), p. 1027 - 1045 (1980).

105.044 **Stress histories retained in olivines from pallasite meteorites.**
T. Matsui, S.-i. Karato, T. Yokokura.
Proc. Eleventh Lunar Planet. Sci. Conf., (see 012.020), p. 1047 - 1054 (1980).

105.045 **Modal studies of mesosiderites and related achondrites, including the new mesosiderate ALHA 77219.**
M. Prinz, C. E. Nehru, J. S. Delaney, G. E. Harlow, R. L. Bedell.
Proc. Eleventh Lunar Planet. Sci. Conf., (see 012.020), p. 1055 - 1071 (1980).

105.046 **Olivine clasts from mesosiderites and howardites: clues to the nature of achondritic parent bodies.**
J. S. Delaney, C. E. Nehru, M. Prinz.
Proc. Eleventh Lunar Planet. Sci. Conf., (see 012.020), p. 1073 - 1087 (1980).

105.047 **Comparative petrology and origin of Governador Valadares and other nakhlites.**
J. L. Berkley, K. Keil, M. Prinz.
Proc. Eleventh Lunar Planet. Sci. Conf., (see 012.020), p. 1089 - 1102 (1980).

105.048 **Howardites: samples of the regolith of the eucrite parent-body: petrology of Frankfort, Pavlovka, Yurtuk, Malvern, and ALHA 77302.**
T. C. Labotka, J. J. Papike.
Proc. Eleventh Lunar Planet. Sci. Conf., (see 012.020), p. 1103 - 1130 (1980).

105.049 **Clouding of pyroxene and plagioclase in eucrites: implications for post-crystallization processing.**
G. E. Harlow, R. Klimentidis.
Proc. Eleventh Lunar Planet. Sci. Conf., (see 012.020), p. 1131 - 1143 (1980).

105.050 **Oxygen isotopic compositions of aubrites and some unique meteorites.**
T. K. Mayeda, R. N. Clayton.
Proc. Eleventh Lunar Planet. Sci. Conf., (see 012.020), p. 1145 - 1151 (1980).

105.051 **Controlled cooling and crystallization of a eucrite: microprobe studies.**
M. A. Powell, D. Walker, J. F. Hays.
Proc. Eleventh Lunar Planet. Sci. Conf., (see 012.020), p. 1153 - 1168 (1980).

105.052 **Silver isotopic anomalies in iron meteorites: cosmic-ray production and other possible sources.**
R. C. Reedy.
Proc. Eleventh Lunar Planet. Sci. Conf., (see 012.020), p. 1169 - 1178 (1980).

105.053 **Carbon-14 and argon-39 in ALHA meteorites.**
E. L. Fireman.
Proc. Eleventh Lunar Planet. Sci. Conf., (see 012.020), p. 1215 - 1221 (1980).

105.054 **The Antarctic environment and its effect upon the total carbon and sulfur abundances in recovered meteorites.**   E. K. Gibson, Jr., F. F. Andrawes.

Proc. Eleventh Lunar Planet. Sci. Conf., (see 012.020), p. 1223 - 1234 (1980).

105.055 **Magnetic classification of Antarctic meteorites.**
T. Nagata.
Proc. Eleventh Lunar Planet. Sci. Conf., (see 012.020), p. 1789 - 1799 (1980).

105.056 **Meteor crater: energy of formation – implications of centrifuge scaling.**   R. M. Schmidt.
Proc. Eleventh Lunar Planet. Sci. Conf., (see 012.020), p. 2099 - 2128 (1980).

105.057 **Computer code simulations of the formation of Meteor Crater, Arizona: calculations MC-1 and MC-2.**   D. J. Roddy, S. H. Schuster, K. N. Kreyenhagen, D. L. Orphal.
Proc. Eleventh Lunar Planet. Sci. Conf., (see 012.020), p. 2275 - 2308 (1980).

105.058 **Impact melt generation and transport.**
D. L. Orphal, W. F. Borden, S. A. Larson,.
P. H. Schultz.
Proc. Eleventh Lunar Planet. Sci. Conf., (see 012.020), p. 2309 - 2323 (1980).

105.059 **Calculational investigation of impact cratering dynamics: material motions during the crater growth period.**   M. G. Austin, J. M. Thomsen, S. F. Ruhl, D. L. Orphal, P. H. Schultz.
Proc. Eleventh Lunar Planet. Sci. Conf., (see 012.020), p. 2325 - 2345 (1980).

105.060 **The search for Antarctic meteorites.**   U. B. Marvin
Sky Telesc., Vol. 62, 423 - 427 (1981).

105.061 **Comments on "Metallic minerals, thermal histories and parent bodies of some xenolithic, ordinary chondrite meteorites" by E. R. D. Scott and R. S. Rajan (1980).**   A. W. R. Bevan, H. J. Axon, with a reply by E. R. D. Scott and R. S. Rajan.
Geochim. Cosmochim. Acta, Vol. 45, 1957 - 1959 (1981).

105.062 **Do the age differences given by relative or absolute chronologies of the most ancient meteorites correspond to real age differences?**   P. Pellas.
Mem. Soc. Astron. Italiana, Vol. 52, (see 012.023), 475 - 482 (1981).

105.063 **Refractory spherules in the Murchison meteorite: are they chondrules?**   J. D. Macdougall.
Geophys. Res. Lett., Vol. 8, 966 - 969 (1981).
  Refractory spherules in the Murchison carbonaceous chondrite show a range of compositions similar to those of coexisting irregular inclusions. However, the shape and internal texture of the spherules are suggestive of formation from a liquid. The coexistance of these two types of refractory inclusions suggests that the spherules were formed by melting of previously condensed irregular inclusions. If so, they can be properly termed chondrules.

105.064 **Atmospheric ablation in meteorites: a study based on cosmic ray tracks and neon isotopes.**
N. Bhandari, D. Lal, R. S. Rajan, J. R. Arnold, K. Marti, C. B. Moore.
Nucl. Tracks Methods Instrum. Appl., p. 213 - 262 (1980).
Abstr. in Phys. Abstr., Vol. 84, Abstr. 88855 (1981).

105.065 **Space erosion of meteorites and the secular variation of cosmic rays (over 10^9 years).**
O. A. Schaeffer, K. Nagel, H. Fechtig, G. Neukum.
Planet. Space Sci., Vol. 29, 1109 - 1118 (1981).

The space erosion of stony meteorites has been determined to be 650 μm $10^6 y^{-1}$, while that of iron meteorites has been determined to be 22 μm $10^6 y^{-1}$. The erosion rates are based on flux and size distributions of small particles in the solar system, meteoroid orbitals and the relation, determined by laboratory experiments, between excavated volume due to a collision and the size and velocity of the impacting small particle. Neither multiple collision or space erosion can explain the difference in cosmic ray exposure ages based on ^{40}K and those based on ^{36}Cl, ^{39}Ar and ^{10}Be. It is concluded that there is a long term cosmic ray variation.

105.066   On some peculiarities of the structure of impact craters on planets and satellites of the solar system.
A. T. Bazilevskij.
Dokl. AN SSSR, Tom 258, 323 - 325 (1981). In Russian. Abstr. in Ref. zh., 51. Astron., 10.51.234; 62. Issled. kosm. prostranstva, 10.62.234 (1981).

105.067   Search for tracks of superheavy cosmic nuclei in olivines from meteorites.
V. P. Perelygin, S. G. Stetsenko, D. Lkhagvasurehn, O. Otgonsurehn, D. S. Yadav, P. Pellas, K. Perro, T. Taneva, P. Fater, B. Jakupi.
Cosmic rays, (see 012.025), 1980, p. 207 - 216. In Russian. Abstr. in Ref. zh., 51. Astron., 10.51.331 (1981).

105.068   Thermal history of mesosiderites revisited.
G. Crozaz, D. R. Tasker.
Geochim. Cosmochim. Acta, Vol. 45, 2037 - 2046 (1981).

The results of a fission track study of uranium-rich merrillite crystals in a number of mesosiderites are clearly incompatible with cooling rates as low as $\sim 0.1°C/Myr$ which have previously been inferred for these meteorites. Classical metallographic techniques thus appear to give erroneous results when applied to this type of meteorites. There appears to be no reason to advocate exotic scenarios to explain the thermal history of mesosiderites.

105.069   Cosmic-ray exposure ages of the ordinary chondrites and their significance for parent body stratigraphy.
J. Crabb, L. Schultz.
Geochim. Cosmochim. Acta, Vol. 45, 2151 - 2160 (1981).

105.070   Frombork, ein neuer Meteoritenkrater in der VR Polen?   J. Classen.
Sterne, 57. Band, 296 - 297 (1981).

105.071   Windows to early solar system processes: refractory inclusions in the CV and CM chondrites.
J. D. MacDougall, J. N. Goswami.
Proc. Indian Acad. Sci. Earth Planet. Sci., Vol. 90, 1 - 26 (1981). – Abstr. in Phys. Abstr., Vol. 84, Abstr. 102076 (1981).

105.072   Impact and impact-like structures in Algeria. Part II. Multi-ringed structures.   P. Lambert,
J. F. McHone, Jr., R. S. Dietz, M. Briedj, M. Djender.
Meteoritics, Vol. 16, 203 - 227 (1981).

105.073   Tierra Blanca: an unusual achondrite from West Texas.   E. A. King, E. Jarosewich, F. W. Daugherty.
Meteoritics, Vol. 16, 229 - 237 (1981).

105.074   Composition and origin of the unusual Oktibbeha County iron meteorite.
A. Kracher, J. Willis.
Meteoritics, Vol. 16, 239 - 246 (1981).

105.075   Complex shock-induced Fe-Ni-S-Cr-C melts in the Haig (IIIA) iron meteorite.

A. W. R. Bevan, J. Kinder, H. J. Axon.
Meteoritics, Vol. 16, 261 - 267 (1981).

105.076   Meteorite falls and finds: some statistics.
D. W. Hughes.
Meteoritics, Vol. 16, 269 - 281 (1981).

The statistics of meteorite falls and finds are presented. Histograms give the distribution of falls as a function of year, month and time of day. The distributions of the retrieved masses of fallen and found meteorites are given, as is also their distribution over the Earth's surface.

105.077   A photometric study of the Innisfree meteorite fall.
I. Halliday, A. A. Griffin, A. T. Blackwell.
J. R. Astron. Soc. Canada, Vol. 75, 247 - 248 (1981). Abstract.

105.078   Interstellar chemistry: polycyanoacetylene formation.
W. D. Langer, F. P. Schloerb, R. L. Snell, J. S. Young, with a reply by E. Anders, R. Hayatsu.
Science, Vol. 214, 688 - 689 (1981). – See Abstr. 28.105.037.

105.079   Systematics of r-process enrichment factors for barium, neodymium, and samarium isotopic anomalies in the Allende meteorite.
G. J. Mathews, W. A. Fowler.
Astrophys. J., Lett., Vol. 251, L45 - L48 (1981).

The decomposition of Ba, Nd, and Sm isotopic anomalies in the Allende meteorite into s-process and r-process enrichment is computed with experimental cross section data for all of the stable isotopes involved. The uncertainties in this decomposition are analyzed. It is concluded that these data may reveal a previously unobserved systematic enrichment relative to solar system r-process material which favors the population of the lighter isotopes of each element. Some possible explanations for these systematics are discussed.

105.080   Internal chemical energy: heating of parent bodies and oxygen isotopic anomalies.   D. D. Clayton.
Nukleonika, Vol. 25, 1477 - 1490 (1980). – Abstr. in Phys. Abstr., Vol. 85, Abstr. 3223 (1982).

105.081   On the nature of the isotopic anomalies in meteorites.   A. K. Lavrukhina.
Nukleonika, Vol. 25, 1495 - 1515 (1980). – Abstr. in Phys. Abstr., Vol. 85, Abstr. 3224 (1982).

105.082   Deuterium/hydrogen ratios in unequilibrated ordinary chondrites.
N. J. McNaughton, J. Borthwick, A. E. Fallick, C. T. Pillinger.
Nature, Vol. 294, 639 - 641 (1981).

The authors report high and variable deuterium enrichments in two of the most unequilibrated (unmetamorphosed) type LL3 ordinary chondrites Bishunpur and Semarkona. Water released from bulk unseparated material by pyrolysis has δD values up to $\sim 3,000°/_{00}$ (SMOW) or in absolute terms a D/H ratio of $\sim 6 \times 10^{-4}$.

105.083   Constancy of ^{244}Pu distribution in chondritic whitlockite.   P. Mold, R. K. Bull, S. A. Durrani.
Nucl. Tracks Methods Instrum. Appl., Vol. 5, 27 - 31 (1981). Abstr. in Phys. Abstr., Vol. 85, Abstr. 6574 (1982).

105.084   Some parameters involved in the interpretation of meteorite ages: a review of the present status.
S. A. Durrani, R. K. Bull, P. F. Green.
Nucl. Tracks Methods Instrum. Appl., Vol. 5, 223 - 228 (1981). – Abstr. in Phys. Abstr., Vol. 85, Abstr. 6575 (1982).

105.085   Mineralogy, petrology, and trace element geochemistry of the Johnstown meteorite: a brecciated

orthopyroxenite with siderophile and REE-rich components.
R. J. Floran, M. Prinz, P. F. Hlava, K. Keil, B. Spettel,
H. Wänke.
Geochim. Cosmochim. Acta, Vol. 45, 2385 - 2391 (1981).

The report presents, compares, and evaluates new mineralogical, bulk compositional, and trace element data for Johnstown in order to understand possible genetic relationships between unbrecciated clasts and brecciated matrix.

105.086 **Identification of the projectile at the Brent crater,
and further considerations of projectile types at
terrestrial craters.** H. Palme, R. A. F. Grieve, R. Wolf.
Geochim. Cosmochim. Acta, Vol. 45, 2417 - 2424 (1981) =
Earth Phys. Branch, Contrib. No. 939.

105.087 **Noble gases in E-chondrites.**
J. Crabb, E. Anders.
Geochim. Cosmochim. Acta, Vol. 45, 2443 - 2464 (1981).

105.088 **Further ^{40}Ar/^{39}Ar evidence for the multi-collisional
heating of the Kirin chondrite.**
T. M. Harrison, S. Wang.
Geochim. Cosmochim. Acta, Vol. 45, 2513 - 2517 (1981).

105.089 **Non-miscibility in melts as a factor of early dif-
ferentiation of meteorites and planets.**
N. S. Gorbachev, E. G. Osadchij.
Dokl. AN SSSR, Tom 255, 693 - 697 (1980). In Russian.
Abstr. in Ref. zh., 51. Astron., 11.51.370 (1981).

105.090 **Indications of meteoritic impact in the Sobolevskij
crater.** L. P. Khryanina.
Izv. AN SSSR. Ser. geol., 1980, No. 11, p. 32 - 40. In Russian.
Abstr. in Ref. zh., 51. Astron., 11.51.389 (1981).

105.091 **Effects of low reflectance materials on the spectral
reflectance of meteorites.**
A. Mito, M. Miyamoto, Y. Takano.
Proceedings of the 14th ISAS Lunar and Planetary Symposium,
(see 012.055), p. 141 - 148 (1981).

The authors performed spectral measurements of meteorite samples in which various amounts of low reflectance material were mixed to examine the effects of low reflectance material.

105.092 **Analyses of large-sized spherules and core-nucleus
spheres from marine sediments.**
K. Yamakoshi.
Proceedings of the 14th ISAS Lunar and Planetary Symposium,
(see 012.055), p. 181 - 185 (1981).

From marine sediments the authors gathered large-sized, black magnetic spherules and also silver-white magnetic spheres. Size dependences of (Co/Fe), (Ir/Fe) and (Au/Fe) components could be seen. With increasing size of the spherules, these contents decrease rapidly. From the silver-white spheres they could detect not only Fe, Ni, Co, Ir, Au but also W, Cu, Ga, Sb and As.

105.093 **Preliminary report on the isotopic ratio of nickel in
iron meteorites.** J. Okano, H. Nishimura.
Proceedings of the 14th ISAS Lunar and Planetary Symposium,
(see 012.055), p. 186 - 192 (1981).

The isotopic ratio of nickel (^{58}Ni/^{60}Ni) has been studied for the two iron meteorites Odessa and Canyon Diablo. The results show that the higher isotope ^{58}Ni has been enriched in taenite veins when compared with the kamacite phase.

105.094 **Characteristics of the unique achondrite Chassignite.**
T. Fukuoka.
Proceedings of the 14th ISAS Lunar and Planetary Symposium,
(see 012.055), p. 193 - 194 (1981). – Abstract.

105.095 **On parent bodies of the unique achondrites,
Nakhlites and Chassignites: Martian origin hypothesis.** N. Nakamura.
Proceedings of the 14th ISAS Lunar and Planetary Symposium,
(see 012.055), p. 195 - 196 (1981). – Abstract.

105.096 **Thermal history of a howardite parent body.**
M. Miyamoto, H. Mori, H. Takeda, N. Fujii.
Proceedings of the 14th ISAS Lunar and Planetary Symposium,
(see 012.055), p. 203 - 210 (1981).

A layered crust model of the parent body for a howardite achondrite is discussed. Its thermal history is outlined by estimating cooling rates and assuming either external heating by the surrounding primordial solar nebula or internal heating by the radionuclide ^{26}Al.

105.097 **Thermal stress in ordinary chondrite parent body:
a possible cause of its lithification.**
N. Fujii, Y. Hamano, M. Miyamoto.
Proceedings of the 14th ISAS Lunar and Planetary Symposium,
(see 012.055), p. 211 - 218 (1981).

The distribution of thermal stresses in parent bodies of chondrites must have had large influence on the solidification of chondrite material. An analysis may provide important clues to the question whether the parent bodies were heated internally or externally during their thermal evolution. Preliminary results are presented.

105.098 **Differences of cutting rates among chondrites and
simulated planetesimals.**
N. Fujii, M. Miyamoto, Y. Kobayashi, K. Ito.
Proceedings of the 14th ISAS Lunar and Planetary Symposium,
(see 012.055), p. 219 - 226 (1981).

The purpose of this study is to define quantitatively the mechanical strength of chondritic material and loosely consolidated aggregates. This parameter is important in understanding accretion and impact phenomena occurring with planetesimals in the primordial solar system.

105.099 **Chondrules as secondary products and their cooling
history.** H. Nagahara.
Proceedings of the 14th ISAS Lunar and Planetary Symposium,
(see 012.055), p. 243 - 248 (1981).

105.100 **The destruction of meteorites.**
H. Hasegawa.
Proceedings of the 14th ISAS Lunar and Planetary Symposium,
(see 012.055), p. 284 (1981). – Abstract.

105.101 **The conditions for meteorite fall.**
M. Ogasawara, H. Hasegawa.
Proceedings of the 14th ISAS Lunar and Planetary Symposium,
(see 012.055), p. 285 - 292 (1981).

The relation between the pre-atmospheric mass of meteorites and the terminal mass of meteorites is investigated. The numerical experiments of the meteorite fall are performed for various conditions. The authors found the residual rate of a meteorite to be much smaller than previously believed. Generally the residual rate of an ordinary chondrite becomes 0.2 – 0.0 and of carboneceous chondrite becomes 0.08 – 0.00 for an initial mass range of $10^2 \sim 10^5$ kg. They also found the residual rate to depend strongly on the entry velocity of the meteorites.

105.102 **A presence of a 2 AE period in the meteorite flux
on the earth and moon.**
M. Kumazawa, H. Mizutani.
Proceedings of the 14th ISAS Lunar and Planetary Symposium,
(see 012.055), p. 313 - 321 (1981).

The presence of a series of meteorite impact swarms with a period of ~2 AE is suggested on the basis of existing data on the frequency of terrestrial and lunar craters. The change of mass distribution around the solar system in the Galaxy caused by the tidal force of the Magellanic Clouds is supposed to disturb the quasi-stable orbital motions of bodies in the solar system, and to induce the postulated swarms of meteorite impact.

105.103 The distribution of meteoric dust around the meteorite crater at Frombork.
H. Korpikiewicz.
Astron. Rep., Vol. 5, 7 - 11 (1981).

105.104 Meteoric dust in different regions of Poland.
H. Korpikiewicz.
Astron. Rep., Vol. 5, 13 - 14 (1981).

105.105 Problem of excess lead in meteorites.
G. V. Ovchinnikova, L. A. Nejmark.
Geokhimiya, 1981, No. 8, p. 1091 - 1102. In Russian.
Abstr. in Ref. zh., 51. Astron., 12.51.399 (1981).

105.106 Crystals of chromite from the Mar'yalakhti meteorite.
V. D. Kolomenskij, G. I. Dolivo-Dobrovol'skaya, I. A. Yudin.
Zap. Vses. mineral. o-va, Tom 110, 465 - 467 (1981). In Russian. – Abstr. in Ref. zh., 51. Astron.,12.51.419 (1981).

105.107 Genesis of tektites as cause of unity of their composition and structure.
P. V. Florenskij, Yu. P. Dikov.
Geokhimiya, 1981, No. 6, p. 809 - 819. In Russian. – Abstr. in Ref. zh., 51. Astron., 12.51.425 (1981).

105.108 Estimate of the pre-catastrophic composition of the Tunguska meteorite body.
S. P. Golenetskij, V. V. Stepanok, D. A. Murashov.
Astron. vestn., Tom 15, 167 - 173 (1981). In Russian.
Abstr. in Ref. zh., 51. Astron., 1.51.335 (1982).

105.109 Composition and classification of clasts in the St. Mesmin LL chondrite breccia.
C.-L. Chou, D. W. Sears, J. T. Wasson.
Earth Planet. Sci. Lett., Vol. 54, 367 - 378 (1981).

105.110 Al-correlated ^{26}Mg excess in a large Ca-Al-rich inclusion of the Leoville meteorite.
W. Stegmann, F. Begemann.
Earth Planet. Sci. Lett., Vol. 55, 266 - 272 (1981).

Minerals from a large ($\sim$ 1.5 cm diameter) Type B inclusion of the C3V-carbonaceous chondrite Leoville have been analyzed mass spectrometrically for their Mg isotopic composition. ^{25}Mg/^{24}Mg ratios were found to be indistinguishable in all cases from those of terrestrial Mg.

105.111 The cosmic abundance of molybdenum.
H. Palme, W. Rammensee.
Earth Planet. Sci. Lett., Vol. 55, 356 - 362 (1981).

Fifteen carbonaceous chondrites were analysed for Mo and Ir by neutron activation analysis combined with a metal extraction method. The results of two Orgueil analyses gave a mean concentration of 915 ppb Mo. A constant Mo/Ir ratio is found for C1, C2, and C3V chondrites; C3Os have variable Mo/Ir ratios. These variations are due to variable Ir concentrations. Micron-sized grains enriched in Ir but not in Mo are presumably responsible for these variations. The Mo content of Karoonda is nearly a factor of four lower than that of C3V chondrites.

105.112 Nitrogen isotopes in the Allende meteorite.
M. H. Thiemens, R. N. Clayton.
Earth Planet. Sci. Lett., Vol. 55, 363 - 369 (1981).

105.113 Spalogenic rare gases in iron meteorites with anomalous silver.
I. M. Villa, J. C. Huneke, G. J. Wasserburg.
Earth Planet. Sci. Lett., Vol. 56, 9 - 18 (1981).

He, Ne, and Ar have been measured in seven iron meteorites for which anomalous Ag isotopic compositions were reported, in order to determine if ^{107}Ag excesses could be related to galactic cosmic-ray bombardment of these meteorites. The results show that no correlation exists between ^{107}Ag excess and either the fluence or the energy spectrum of the particles producing spallogenic rare gases.

105.114 New kind of type 3 chondrite with a graphite-magnetite matrix.
E. R. D. Scott, A. E. Rubin, G. J. Taylor, K. Keil.
Earth Planet. Sci. Lett., Vol. 56, 19 - 31 (1981).

The authors have discovered four clasts in three ordinary-chondrite regolith breccias which are a new kind of type 3 chondrite. Like ordinary and carbonaceous type 3 chondrites, they have distinct chondrules, some of which contain glass, highly heterogeneous olivines and pyroxenes, and predominantly monoclinic low-Ca pyroxenes. But instead of the usual fine-grained, Fe-rich silicate matrix, the clasts have a matrix composed largely of aggregates of micron- and submicron-sized graphite and magnetite. This new kind of chondrite is probably the source of the abundant graphite-magnetite inclusions in ordinary-chondrite regolith breccias, and may be more common than indicated by the absence of whole meteorites made of chondrules and graphite-magnetite.

105.115 The interstellar dust as a precursor of Ca, Al-rich inclusions in carbonaceous chondrites. J. A. Wood.
Earth Planet. Sci. Lett., Vol. 56, 32 - 44 (1981).

The observed anomalies in meteoritic oxygen isotope compositions are not due to an incomplete mixing of several dust or gas-plus-dust components in the solar nebula. The Ca, Al-rich inclusions (CAI's) in carbonaceous chondrites are unlikely to be condensates, but instead are distillation residues. If so, the observed depletion of super-refractory elements in the Group II CAI's cannot have been accomplished by fractional condensation in the solar nebula. Then this depletion, and a number of other properties of the components of primitive meteoritic material, must be relics of pre-solar system fractionations among different populations of interstellar dust grains.

105.116 On the distribution of noble gases in Allende: a differential oxidation study.
U. Frick, R. O. Pepin.
Earth Planet. Sci. Lett., Vol. 56. 45 - 63 (1981).

105.117 Microanalysis of nitrogen isotope abundances: association of nitrogen with noble gas carriers in Allende. U. Frick, R. O. Pepin.
Earth Planet. Sci. Lett., Vol. 56, 64 - 81 (1981).

105.118 Cristobalite-pyroxene in an L6 chondrite: implications for metamorphism.
E. J. Olsen, T. K. Mayeda, R. N. Clayton.
Earth Planet. Sci. Lett., Vol. 56, 82 - 88 (1981).

105.119 ^{87}Rb–^{87}Sr dating of LL chondrites.
J. F. Minster, C. J. Allègre.
Earth Planet. Sci. Lett., Vol. 56, 89 - 106 (1981).

^{87}Rb–^{87}Sr analyses of LL chondrites have been made in 10 whole rock meteorites, chondrules from Chainpur (LL3) and Soko Banja (LL4), density separates and chondrules from Guidder (LL5) and density separates from Jelica (LL6) and Ensisheim (LL6). Whole rocks define an isochron of age 4.486 ± 0.020 Ga. Analyses for chondrules from Soko Banja yield a very good isochron of age 4.452 ± 0.020 Ga. A more poorly defined isochron is obtained for Jelica; the age is

4.423 ± 0.041 Ga. These results are discussed in relation with thermal metamorphism in the LL chondrite parent bodies.

105.120 **Primitive ultrafine matrix in ordinary chondrites.**
E. R. Rambaldi, B. J. Fredriksson, K. Fredriksson.
Earth Planet. Sci. Lett., Vol. 56, 107 - 126 (1981).

105.121 **Viscous magnetization of stony meteorites.**
T. Nagata.
Phys. Earth Planet. Inter., Vol. 26, 125 - 133 (1981).

105.122 **New magnetic results from Allende C3(V).**
P. Wasilewski.
Phys. Earth Planet. Inter., Vol. 26, 134 - 148 (1981).

105.123 **Meteorites.**   J. A. Wood.
The new solar system, (see 003.026), p. 187 - 196 (1981).

Brazilian stone meteorites.   See Abstr. 003.068.

Stones from the stars: the unresolved mysteries of meteorites.   See Abstr. 003.092.

Impactites.   See Abstr. 003.095.

Geology of astroblemes.   See Abstr. 003.099.

Meteoritics – more facts, more complexity.
See Abstr. 011.025.

Uranium determinations by ^{133}Xe in terrestrial zircon, apatite and chromite; comparative study of thermal releases of fissiogenic xenon from terrestrial and meteoritic minerals.
See Abstr. 022.045.

Laboratory studies of actinide metal-silicate fractionation.   See Abstr. 022.067.

Hypervelocity impact craters in leucitite targets.
See Abstr. 022.072.

Hypervelocity impact cratering experiments.
See Abstr. 022.074.

Etch a meteorite.   See Abstr. 022.118.

Microtektite spherules vs. volcanic shards: a question of foamability.   See Abstr. 022.128.

Performance test of the high velocity shock gun with a novel sabot stopper.   See Abstr. 022.166.

Nucleosynthesis of neutron-rich heavy nuclei during explosive helium burning in massive stars.
See Abstr. 061.005.

Some aspects of the nucleosynthesis of the light elements.   See Abstr. 061.040.

Tunguska meteor fall of 1908: effects on stratospheric ozone.   See Abstr. 082.051.

The collision of solid bodies.
See Abstr. 091.097.

Diagnostic spectral properties of ordinary chondritic assemblages: implications for the S-asteroids.
See Abstr. 098.056.

Relationships between observable asteroids and meteorite yields.   See Abstr. 098.060.

On the implausibility of a cometary origin for most Apollo-Amor asteroids.   See Abstr. 098.075.

The bolid which created meteorite Tsarev.
See Abstr. 104.016.

Which fireballs are meteorites? A study of the Prairie Network photographic meteor data.
See Abstr. 104.042.

Solar-ion penetration in the early solar nebula.
See Abstr. 107.004.

Superheavy elements: an early solar system upper limit for elements 107 to 110.   See Abstr. 107.008.

Some aspects of space and global petrology.
See Abstr. 107.010.

Growth stress in a chondritic parent body as an accreting planetesimal.   See Abstr. 107.019.

A first roadmap for kryptology.
See Abstr. 125.033.

Origin of Ca-Al-rich inclusions. II. Sputtering and collisions in the three-phase interstellar medium.
See Abstr. 131.169.

Grain formation behind shocks and the origin of isotopically anomalous meteoritic inclusions.
See Abstr. 131.180.

Galactic cosmic ray gradients in the ecliptic plane and at high latitudes during two solar cycles (meteorite data).
See Abstr. 143.112.

Nucleosynthesis in OB associations and its effect upon cosmic rays, peculiar stars and isotopic anomalies in meteorites.   See Abstr. 152.006.

Errata

105.901 **Erratum: "Li^6/Li^7 variations in meteorites"**
[Meteoritics, Vol. 15, 267 (1980)].
J-L Birck, B. Zanda, C. J. Allègre.
Meteoritics, Vol. 16, 201 (1981). – See Abstr. 29.105.027.

105.902 **Erratum: 'Impact and impact-like structures in Algeria. Part I. Four bowl-shaped depressions'**
[Meteoritics, Vol. 15, 157 - 179 (1980)].
P. Lambert, J. F. McHone, Jr., R. S. Dietz, M. Houfani.
Meteoritics, Vol. 16, 283 (1981). – See Abstr. 28.105.029.

106 Interplanetary Matter, Interplanetary Magnetic Field, Zodiacal Light

106.001 On the enhancement of the IMF magnitude during 1978 - 1979.   J. H. King.
J. Geophys. Res., Vol. 86, 4828 - 4830 (1981).
The magnitude of the interplanetary magnetic field (IMF) exhibits an enhancement during 1978 - 1979 relative to all years back to 1963. It is shown that IMF magnitude variations over the 1966 - 1979 period represent the combined effect of variations in both the radial flux density of the IMF and the degree of spiraling of the IMF, consistent with the theoretical model of Parker. The 1978 - 1979 IMF magnitude enhancement is due to an enhancement of radial flux which was in turn related to an increase of magnetic flux leaving solar active regions. It is also shown that during the corotating stream dominated years 1973 - 1976, the IMF was less wound up than it was during other years, and that 1973 - 1974 were years of enhanced radial flux.

106.002 IMF enhancements associated with streams: an analysis of magnetic field behavior.   U. Villante.
J. Geophys. Res., Vol. 86, 5417 - 5422 (1981).
The author examined the interplanetary magnetic field behavior in the field enhancement regions associated with the high velocity stream fronts observed during the primary mission of Helios 2 (January - April 1976). The principal conclusions of the investigation are the following: (1) The relative variation of the field magnitude ranges between 1.3 and 1.8 and is independent of heliocentric distance at least between 0.5 and 1 AU; (2) close to the sector boundaries the field enhancements are consistent with a field perturbation occurring in a plane perpendicular to both the solar equator and the radial direction; (3) the field line has occasionally the form of a cylindrical helix rather than a simple planar geometry; (4) there is no compelling evidence to relate, in the inner solar system and in the period of interest, the field enhancements with the shape and location of the sector boundary surface.

106.003 Interplanetary propagation of < 1 MeV protons in nonimpulsive energetic particle events.
R. D. Zwickl, E. C. Roelof.
J. Geophys. Res., Vol. 86, 5449 - 5471 (1981).

106.004 Three-dimensional interaction of interplanetary shock waves with the bow shock and magnetopause: a comparison of theory with ISEE observations.
H. C. Zhuang, C. T. Russell, E. J. Smith, J. T. Gosling.
J. Geophys. Res., Vol. 86, 5590 - 5600 (1981).
The propagation of shock waves in the solar wind and their three-dimensional interaction with the terrestrial bow shock, magnetosheath, and magnetopause is examined theoretically and compared with observations made by the ISEE spacecraft.

106.005 What causes the warp in the heliospheric current sheet?   J. M. Wilcox, P. H. Scherrer, with a reply by B. T. Thomas, E. J. Smith.
J. Geophys. Res., Vol. 86, 5899 - 5901 (1981). – See Abstr. 28.106.088.

106.006 Measurements of a solar flare-generated shock wave at 13.1 R_0.   R. Woo, J. W. Armstrong.
Nature, Vol. 292, 608 - 610 (1981).
Radio-scattering observations consisting of spectral broadening, mean phase and amplitude scintillations were made on 18 August 1979, 13.1R_0 east of the sun and near the ecliptic plane using the Voyager 1 2.3- and 8.4-GHz radio signals. The results show that the shock wave had a shock speed of ~3,500 km s^{-1}.

106.007 Magnetic loop behind an interplanetary shock: Voyager, Helios, and IMP 8 observations.
L. Burlaga, E. Sittler, F. Mariani, R. Schwenn.
J. Geophys. Res., Vol. 86, 6673 - 6684 (1981).
The paper investigates the configuration of the magnetic field in a flow behind a shock observed at 2 AU on January 6, 1978, and at ~1 AU on January 3. It is shown that the magnetic field in the 'driver' gas closely resembles that of an extended loop with ordered fields as proposed by Cocconi et al. (1958) and Piddington (1958). However, there is a region (a sheath) between the shock and the stream in which the field is strong and turbulent. The authors discuss data from five spacecraft, Voyager 1 and 2, Helios 1 and 2, and IMP 8.

106.008 On a method of finding the anisotropy of the interplanetary medium.   N. S. Petrova.
Soln. Dannye 1981 Byull., No. 3, p. 107 - 113 (1981). In Russian.
The method proposed in "Solnechnye Dannye" NN 1 - 2, 1975 is criticized.

106.009 On the mechanism of origin of the spectrum of the interplanetary plasma turbulence.
I. V. Chashej, V. I. Shishov.
Pis'ma Astron. Zh., Tom 7, 500 - 504 (1981). In Russian.
English translation in Soviet Astron. Lett., Vol. 7.
It is shown that the stationary spectrum of solar wind turbulence is maintained due to balance near the outer scale between relative amplification of the turbulence power due to inhomogeneities of the medium and energy outpumping to high frequencies by nonlinear interactions. The consequences from the model are in good agreement with experimental results.

106.010 Temporary spectra of scintillations with account for solar wind velocity distribution. Theory.
N. A. Lotova.
Geomagn. Aehron., Tom 21, 593 - 597 (1981). In Russian.

106.011 On strong suppression of interplanetary shock waves when they interact in complex streams from flare series.   K. G. Ivanov.
Geomagn. Aehron., Tom 21, 750 - 752 (1981). In Russian.

106.012 An evaluation of the sky Lyman alpha emission in IUE spectra.   K. H. Fricke, O. Ojanguren.
Second European IUE Conference, (see 012.011), p. 3 - 6 (1980).
An emission line at 1216 Å is noticeable in IUE spectra for exposure times longer than about 5 minutes in low dispersion. For a set of seven sky-only spectra the expected Lyman alpha column emission rate was calculated from standard models of geocoronal and interplanetary hydrogen. The geocorona contributes between 5% and 50% of the signal, depending on the observing geometry. The calculated emission rates are higher than the observed ones by approximately 19% with a scatter of 11%. This discrepancy is smaller than the combined errors of observation and calculation.

106.013 Some parameters of the sector structure of the interplanetary magnetic field (1964 - 1973).
V. V. Tel'nyuk-Adamchuk.
Problems of cosmic physics. Vyp. 16, (see 003.005), p. 53 - 59 (1981). In Russian.
On the basis of an analysis of the direct measurement of interplanetary magnetic fields the characteristic time of the existence of sector structure is found to be 1.5 - 2.0 years. The distribution of sector duration is studied.

106.014   The interplanetary magnetic field in the plane of the ecliptic during the First International Polar Year.   P. V. Sumaruk, V. M. Litinskij.
Soln. Dannye 1981 Byull., No. 5, p. 100 - 108 (1981). In Russian.
    Mean hour values of the azimuth component of the interplanetary magnetic field are calculated for 6 - 24 hours U. T. in August - September 1932 and June - July 1933.

106.015   Maps of interplanetary plasma according to observations in August - October 1979.
V. I. Vlasov.
Fiz. inst. AN SSSR. Prepr., 1980, No. 114, 20 pp. In Russian.
Abstr. in Ref. zh., 51. Astron., 8.51.485 (1981).

106.016   Interplanetary dust collected in the earth's stratosphere: the question of solar flare tracks.
P. Fraundorf, G. J. Flynn, J. Shirck, R. M. Walker.
Proc. Eleventh Lunar Planet. Sci. Conf., (see 012.020), p. 1235 - 1249 (1980).

106.017   Hypervelocity impacts on Skylab IV/Apollo windows.
U. S. Clanton, H. A. Zook, R. A. Schultz.
Proc. Eleventh Lunar Planet. Sci. Conf., (see 012.020), p. 2261 - 2273 (1980).

106.018   Some properties of unstable atmospheric-geomagnetic relationships and geoefficiency of the interplanetary plasma.   V. N. Plakhotnyuk.
Tr. Inst. prikl. geofiz. Gos. kom. SSSR po gidrometeorol. i kontrolyu prirod. sredy, 1981, No. 50, p. 41 - 58. In Russian.
Abstr. in Ref. zh., 51. Astron., 9.51.418 (1981).

106.019   Interplanetary Alfvénic fluctuations: a stochastic model.   A. Barnes.
J. Geophys. Res., Vol. 86, 7498 - 7506 (1981).
    The strong alignment of the average directions of minimum magnetic variance and mean magnetic field in interplanetary Alfvénic fluctuations is inconsistent with the usual wave-propagation models. The author investigates the concept of minimum variance for nonplanar Alfvénic fluctuations in which the field direction varies stochastically. It is found that the tendency of the minimum variance and mean field directions to be aligned may be purely a consequence of the randomness of the field direction. The probability distribution of directions of a randomly fluctuating field of constant magnitude is calculated. A new approach for observational studies of interplanetary fluctuations is suggested.

106.020   Plasma – dust interactions in the solar vicinity and their observational consequences.
H. J. Fahr, H. W. Ripken, G. Lay.
Astron. Astrophys., Vol. 102, 359 - 370 (1981).
    A solution for the local velocity distribution function of zodiacal dust particles on the basis of a given function of their density distribution is derived under the assumption of circular dust particle orbits without net losses along these orbits. The processes which are of importance for the interception of solar wind plasma by dust grain surfaces are discussed. A dominance of dust-generated neutral densities over the corresponding interstellar values is found in regions inside of about 0.5 AU(H) and 0.05 AU(He). Solar photons resonantly scattered at these dust-generated constituents thus give an additional contribution to interplanetary resonance radiations at 121.6 nm and 58.4 nm. Shown in this paper are both integrated and spectral intensities of the radiation components which are due to dust-related and interstellar neutrals.

106.021   Multiple scattering of solar resonance radiation in the nearby interstellar medium. II.
H. U. Keller, K. Richter, G. E. Thomas.
Astron. Astrophys., Vol. 102, 415 - 423 (1981).

    The authors have obtained numerical Monte Carlo solutions for the problem of solar Lyman-α (1216 Å) photons scattered from interplanetary hydrogen. They show that the previous assumption of an optically-thin medium underestimates the radial sky background intensity by 5 - 35%, depending upon the angle θ which the line of sight makes with the interstellar wind vector. Furthermore, the line width of the scattered radiation is enhanced, yielding an effective temperature approximately 20% smaller than that deduced from the assumed optically-thin Gaussian profiles. The results indicate a dependence on θ of the multiple scattering correction factor q but a very weak radial dependence. This yields a simple expression for the calculation of the total radial intensity everywhere inside 40 AU.

106.022   Is the interplanetary dust cloud maintained by comets?   E. Grün.
Cosmic rays, (see 012.025), 1980, p. 172 - 186. – Abstr. in Ref. zh., 51. Astron., 10.51.304 (1981).

106.023   Interplanetary plasma maps from observations in August - October 1980.   V. I. Vlasov.
Fiz. inst. AN SSSR. Prepr., 1981, No. 63, 34 pp. In Russian.
Abstr. in Ref. zh., 51. Astron., 10.51.409 (1981).

106.024   The photometric center of the Gegenschein.
N. Y. Misconi.
Icarus, Vol. 47, 265 - 269 (1981).
    Model calculations are used to evaluate two factors which determine the position of the photometric center of the Gegenschein: the increased scattering efficiency of the interplanetary dust near backscattering (scattering angles $\theta \sim 165 - 180°$), and the spatial density distribution of the dust. It is found that the effect of the enhanced scattering of light by dust in the backscattering region overrides the effect of the spatial-density distribution of the dust. As a result, the photometric center should be observed at the antisolar point nearly at all times.

106.025   Collisional processes among interplanetary dust grains: an unlikely origin for the β meteoroids.
L. B. Le Sergeant D'Hendecourt, P. L. Lamy.
Icarus, Vol. 47, 270 - 281 (1981).
    The question of the collisional production of the β meteoroids is reexamined incorporating recent experimental results. The collisional model yields a flux of fragments supported by the conservation of mass flux which does not account by far for the observed flux of submicron grains. Particles larger than about 100 μm will be destroyed by collisions inside 1 AU, well before they can get near the Sun. The existence of two independent populations of interplanetary dust grains appears reinforced.

106.026   The zodiacal light from 1.0 to 0.3 A.U. as observed by the Helios space probes.
C. Leinert, I. Richter, E. Pitz, B. Planck.
Astron. Astrophys., Vol. 103, 177 - 188 (1981).
    The authors summarize the average properties of the zodiacal light as obtained from several years of observations from the Helios space probes. The zodiacal light experiments on Helios 1 and Helios 2 give consistent results on intensity, colour and polarization of the zodiacal light between 1.0 and 0.3 A.U. The intensity increase towards the sun as a function of heliocentric distance is $I(R) \sim {}^{-2.3 \pm 0.05}$ where the upper and lower limits give the deviations observed at small and large elongations, respectively, from the sun. The colour is independent of heliocentric distance, slightly reddened with respect to the sun, particularly for small elongations. The polarization is somewhat higher than found from ground-based observations. It decreases towards 0.3 A.U. to about 2/3 of its value near the earth. No convincing explanation for this effect is found.

106.027 **Interplanetary shock waves from observations of scintillations of radio sources.** V. I. Vlasov.
Geomagn. Aehron., Tom 21, 927 - 929 (1981). In Russian.

106.028 **Inversion of the brightness integral along zodiacal and coronal lines of sight.**
R. Dumont, B. Pelletanne.
C. R. Acad. Sci. Paris, Tome 293, Sér. II, 377 - 380 (1981). In French.

106.029 **Shock-associated low-energy ion enhancements observed by Voyagers 1 and 2.**
R. B. Decker, M. E. Pesses, S. M. Krimigis.
J. Geophys. Res., Vol. 86, 8819 - 8831 (1981).

Observations of shock-associated $\geqslant 30$ keV ion enhancements are presented using data from the Low Energy Charged Particle (LECP) experiment on Voyagers 1 and 2, launched on days 248 and 232, 1977, respectively. The observations include examples of energetic storm particle (ESP) events associated with flare-produced shocks and examples of corotating particle events (CPE) associated with forward and reverse shocks that bound corotating interaction regions in the outer heliosphere.

106.030 **On the origin of the MeV energy nucleon flux associated with CIRs.** S. P. Christon.
J. Geophys. Res., Vol. 86, 8852 - 8868 (1981).

Measurements of recurrent enhancements of interplanetary nucleon flux in the MeV energy range, detected at widely separated points in the heliosphere by charged particle instruments on board the Pioneer 10 and 11, IMP 7 and 8, and Mariner 10 spacecraft, are presented and interpreted. These features recur at the solar rotation period in association with stream-stream plasma interaction regions corotating with the sun (CIRs) and are not directly produced by solar flaring activity.

106.031 **The coronal and interplanetary current sheet in early 1976.**
L. F. Burlaga, A. J. Hundhausen, X.-p. Zhao.
J. Geophys. Res., Vol. 86, 8893 - 8898 (1981).

A comparison of Helios 1 and 2 observations of the interplanetary sector pattern in early 1976 with the maximum brightness curves in the K coronameter data at $1.5\,R_s$ shows that the latter may be identified with the footprints of the sector boundary surface to an accuracy of $\approx 10°$. The neutral line computed by Wilcox et al (1980) from a potential field model is similar in shape to the K coronameter maximum brightness curves but extends to higher latitudes. The Helios observations give better agreement with the K coronameter curves for the one solar rotation on which a test of the latitude extent of the neutral line was possible. The K coronameter results and the Helios data show that the sector boundary surface probably extended to $\approx 15°$ in the northern hemisphere and to $\approx 30°$ in the southern hemisphere, with little change between $1.5\,R_s$ and 1 AU. The surface was warped appreciably from a single tilted plane (a dipole configuration) suggesting a significant magnetic quadrupole contribution.

106.032 **A search for interplanetary He II, 304-A emission.**
F. Paresce, H. Fahr, G. Lay.
J. Geophys. Res., Vol. 86, 10038 - 10048 (1981).

A region of the sky around the antisolar direction in the earth's shadow was systematically surveyed for the first time at high sensitivity and angular resolution by the extreme ultraviolet telescope on the Apollo-Soyuz mission. The signal from the 170-620 A channel of the instrument, which is sensitive primarily to the He II, 304-A and He I, 584-A resonance lines, shows a sharp decrease in intensity as the line of sight sweeps into the shadow region. Once the line of sight is completely contained within the shadow, a small residual signal above instrumental background and independent of view direction

and spacecraft position is observed. Its intensity is consistent with the expected signal from the interplanetary He I, 584-A emission measured by the 500-780 A channel of the extreme ultraviolet telescope. An upper limit at the 2σ confidence level to a possible He II, 304-A component that can escape detection by the instrument corresponds to 0.02 R or to a column density of 10^9 ions cm^{-2} scattering at line center. This flux conceivably could be generated by a number of emission sources whose characteristics the authors investigate in detail.

106.033 **An attempt to determine the solar Lyα flux independently of instrument calibration.**
S. Cazes, C. Emerich, A. Vidal-Madjar, R. R. Meier.
Astron. Astrophys., Vol. 104, 10 - 14 (1981).

From comparison of geocoronal absorption of the Lyα interplanetary emission profiles observed from 500 km in two directions perpendicular to the ecliptic plane, the ratio between Lyα radiation pressure and solar gravitation is deduced independently from absolute calibration, and found to be equal to 0.62 ± 0.18, for the April 1971 period. This determination provides an absolute value for the solar Lyα flux at the center of the line allowing an increase in accuracy by a factor of two of the corresponding OSO 5 measurements.

106.034 **On the definition of albedo and application to irregular particles.**
M. S. Hanner, R. H. Giese, K. Weiss, R. Zerull.
Astron. Astrophys., Vol. 104, 42 - 46 (1981).

The various definitions of albedo used in planetary astronomy are reviewed. In particular, the Bond albedo, which refers only to the reflected and refracted components, is not applicable to small particles or highly irregular particles, where diffraction is not restricted to a well-defined lobe at small scattering angles. Measured scattering functions for irregular particles are presented in a normalized form and are applied to the case of zodiacal light.

106.035 **Measurement of the interplanetary magnetic field by a Japanese artificial planet.**
T. Saito, M. Seto, K. Yumoto, I. Aoyama.
Bull. Inst. Space Aeronaut Sci. Univ. Tokyo B, Vol. 16, 1391 - 1403 (1980). In Japanese. – Abstr. in Phys. Abstr., Vol. 85, Abstr. 3171 (1982).

106.036 **Connection of the components of the interplanetary magnetic field vector with vertical currents at high latitudes of the northern hemisphere.**
R. G. Afonina, B. A. Belov, A. E. Levitin, M. Yu. Markova, Ya. I. Fel'dshtejn.
Geomagn. Aehron., Tom 21, 1053 - 1063 (1981). In Russian.

106.037 **Small minded – the characterization of interplanetary dust by electron microscopy.** C. Pillinger.
Nature, Vol. 294, 517 - 518 (1981).

106.038 **Evolution of the structure of the interplanetary medium within the Martian orbit.**
G. Ya. Vasil'eva, P. M. Fedorov.
Izv. AN SSSR. Ser. fiz., Tom 45, 1335 - 1345 (1981). In Russian. – Abstr. in Ref. zh., 51. Astron., 11.51.479 (1981).

106.039 **Cross-correlation skewness of interplanetary scintillations.** N. A. Lotova, I. V. Chashej.
Astron. Zh., Tom 58, 1268 - 1274 (1981). In Russian. English translation in Soviet Astron., Vol. 25, No. 6.

The form of the temporal cross-correlation function of scintillations is considered for a power law irregularities spectrum taking into account the velocity dispersion in the scattering medium. The velocity dispersion is shown to give rise to the skewness of the cross-correlation function relative to its maximum as in the case of a gaussian spectrum. Estimates of the instantaneous relative velocity dispersion in the solar wind are obtained using the measured values of the skewness parameter.

106.040 Spin-down effect on an interplanetary dust grain.
T. Mukai.
Proceedings of the 14th ISAS Lunar and Planetary Symposium, (see 012.055), p. 168 - 174 (1981).

Two spin-down effects on an interplanetary dust grain are examined, i.e. (1) spin angular momentum loss due to isotropic thermal radiation from a grain surface and (2) that due to collisions of solar wind particles. It is found that the former is negligibly small compared with the orbital angular momentum loss by the Poynting-Robertson effect.

106.041 A study of the inferred interplanetary magnetic field polarity periodicities.   J. Xanthakis, V. P. Tritakis, C. Zerefos.
J. Interdiscipl. Cycle Res., Vol. 12, 205 - 215 (1981) = Res. Cent. Astron. Appl. Math., Acad. Athens, Contrib. Ser. I. (Astron.), No. 84.

106.042 Visual observations of the libration clouds in the earth–moon-system in the year 1974.
J. Kordylewska.
Cracow Obs., Repr. No. 129, 4 pp. (1981).

106.043 The structure and dynamics of the heliospheric current sheet.   B. T. Thomas, E. J. Smith.
J. Geophys. Res., Vol. 86, 11105 - 11110 (1981).

Several authors have suggested that the sector structure observed in the interplanetary magnetic field may be interpreted in terms of a warped equatorial current sheet in the heliosphere. In this paper the authors seek to investigate this suggestion and to provide a clear picture of the topology of the current sheet. They present an analysis of the magnetic field data obtained by the Pioneer 10 and 11 spacecraft between 1972 and 1976 that covers a range of heliocentric distances between 1 and 8.5 AU and reaches a heliographic latitude of 16°.

106.044 Acceleration of electrons by interplanetary shocks.
D. W. Potter.
J. Geophys. Res., Vol. 86, 11111 - 11116 (1981).

106.045 Spectra of scintillations on inhomogeneities of the ionosphere and interplanetary plasma and the possibility of their resolution in the decameter radiowave range.
V. P. Bovkun, I. N. Zhuk.
Dokl. AN USSR, 1981, A, No. 6, p. 68 - 71. In Russian.
Abstr. in Ref. zh., 62. Issled. kosm. prostranstva, 12.62.305 (1981).

106.046 Acceleration theory for 5–40 keV ions at interplanetary shocks.   M. A. Forman.
Cosmic rays in the heliosphere, (see 012.059), p. 97 - 100 (1981).

Power-law spectra $f(E) \propto E^{-2.7}$ of < 40 keV suprathermal ions within $\sim 10^7$ km of propagating interplanetary shocks are explained by diffusive scattering near a plane shock. The theory fits the 25 November 1977 event with a mean free path perpendicular to the shock is 0.01 AU in front of the shock and less than 0.0003 AU behind it, for 1 keV ions. The theory predicts a steepening spectrum at higher energies.

106.047 Dependence of ESP (*energetic storm particle*) intensity on collisionless shock wave characteristics.
K. Kecskeméty, S. Pintér.
Cosmic rays in the heliosphere, (see 012.059), p. 101 - 104 (1981).

Flux variations of 1 - 5 MeV protons are studied in energetic storm particle events with respect to the preshock solar wind plasma parameters and to the thickness of the collisionless interplanetary shock wave. It is found that the peak intensity in ESP events depends on pre-shock plasma density and on the thickness of the transition region. These relations predict, in agreement with recent observations, the increase of ESP events at larger heliocentric distances.

106.048 Interplanetary acceleration of low-energy protons as observed during the 25 September 1978 shock event.   K.-P. Wenzel, R. Reinhard, T. R. Sanderson.
Cosmic rays in the heliosphere, (see 012.059), p. 105 - 108 (1981).

ISEE-3 observations of a long-lasting low-energy proton intensity increase during the 25 September 1978 shock event are presented as an example for interplanetary particle acceleration in association with shock waves. The observations are discussed in the light of current models for particle acceleration. The particular shape of the time intensity behaviour of the particle intensity increase, the existence of a shock spike and the observed particle distributions indicate that the particles are accelerated at the shock by the induced electric field $\vec{E} = -1/c \ \vec{V} \times \vec{B}$.

106.049 Interplanetary acceleration of low energy protons in association with shock waves: the 11 February 1979 event.   A. Balogh, G. A. Stevens.
Cosmic rays in the heliosphere, (see 012.059), p. 109 - 112 (1981).

The problem of interplanetary acceleration of low energy protons in association with shock waves is examined in the context of the specific event observed on 11 February 1979 on board the ISEE-3 spacecraft. This event has been selected for special study as it apparently was not associated with a solar flare event. The low energy proton telescope system on ISEE-3 measures the proton distribution function with good spectral, directional and temporal resolution from $E_p = 35$ keV. The evolution of the anisotropies and of the energy spectrum during the event are consistent with particle acceleration taking place in the vicinity of the shock wave.

106.050 Numerical study of solar flare particle propagation in the heliosphere.   T. I. Gombosi, A. J. Owens.
Cosmic rays in the heliosphere, (see 012.059), p. 115 - 119 (1981).

Numerical solutions are presented for the propagation of solar cosmic rays in the interplanetary space, including the effects of pitch-angle scattering and adiabatic focusing. The intensity-time profiles can be well fitted by a simple radial spatial diffusion equation with scattering mean-free path λ_{fit}. For low-rigidity particles the radial mean-free path so obtained is significantly larger than the mean-free path calculated from the scattering coefficient due to the inapplicability of the diffusive approximation early in the event. The well-known discrepancy between λ_{fit} and the theoretical predictions may be resolved by these calculations.

106.051 Determination of solar proton fluxes and energies at high solar latitudes by UV radiation measurements.
N. Witt, J. M. Ajello, P. W. Blum.
Cosmic rays in the heliosphere, (see 012.059), p. 129 - 132 (1981).

The latitudinal variation of the solar proton flux and energy causes a density increase at high solar latitudes of the neutral gas penetrating the heliosphere. Measurements of the neutral density by UV resonance radiation observations from interplanetary spacecraft thus permit deductions on the dependence of the solar proton flux on heliographic latitude. The values of the solar proton fluxes and energies at polar heliographic latitudes are determined for several cases of interest. The Mariner 10 analysis, together with IPS results, indicate a significant decrease of the solar proton flux at polar latitudes.

106.052 How does the global structure of the interplanetary magnetic field affect cosmic ray modulation?
J. Kóta.
Cosmic rays in the heliosphere, (see 012.059), p. 135 - 138 (1981).

Force-field theory is studied in 3-dimensions using the full diffusion tensor incorporating drift effects. An analytical approximate solution is deduced under some assumptions which include a flat neutral sheet and a non-uniform density distribution at the outer boundary. By contrast with the usual force-field theory, the author's solution gives a large and charge dependent latitudinal gradient and near perfect isotropy, even corotation disappears. The results are in general agreement with the numerical calculations of Jokipii and Kopriva.

106.053 Interplanetary dust − its physical nature and entry into the atmosphere of terrestrial planets.
D. E. Brownlee.
Comets and the origin of life, (see 012.022), p. 63 - 70 (1981).

Studies of extraterrestrial particles collected in the stratosphere and from the sea floor indicate that the majority of interplanetary dust is a black, fine grained material similar to carbonaceous chondrites but different from known meteorites in mineralogy and structure. If the analyzed samples are typical interplanetary particles then they are probably materials similar to the vast numbers of cometary particles which have continuously entered the Earth's atmosphere for the lifetime of the solar system. If comets do not contain large strong rocks capable of producing conventional meteorites then dust is the only form in which organic materials from comets can be accreted by the Earth.

106.054 Orbital dynamics of magnetospherically trapped lunar ejecta. J. D. Corbin, W. M. Alexander.
Progress in planetary exploration, (see 012.062), p. 103 - 106 (1981).

In situ measurements by dust experiments on HEOS II showed significant enhancement of fluxes for submicron particles. Recent studies have shown that lunar ejecta in this size range can, in a highly simplified model, be trapped in the earth's magnetosphere. The present work is a more detailed study of the dynamics of lunar ejecta in the magnetosphere. The particle size ranges for which the guiding center approximation is valid, for which corotation is negligible, and for which electromagnetic forces dominate gravitational forces have been calculated. Temporal details of charge acquisition by ejecta in the plasmasphere are considered.

106.055 Submicron lunar ejecta in the magnetosphere associated with meteor showers.
W. M. Alexander, J. D. Corbin.
Progress in planetary exploration, (see 012.062), p. 107 - 110 (1981).

Recent studies of the lunar ejecta from lunar impacts of interplanetary dust particles indicate that during favorable lunar phases, over 80% of the submicron ejecta enters the earth's magnetosphere. This "pulse" of lunar ejecta produced by the sporadic meteor background will follow the random variations of the sporadic flux. An additional enhancement of this flux can be related to major meteor showers.

106.056 Zodiacal light and space observation of faint objects.
R. Dumont, A. C. Levasseur-Regourd.
Progress in planetary exploration, (see 012.062), p. 127 - 130 (1981).

Zodiacal light is examined as a "foreground noise" limiting the space photometry of faint objects. Emphasis is given to the ways of increasing the signal to noise ratio by an appropriate choice of observational epoch. In the case of the Space Telescope, predictions of average values of this ratio for the extreme faintness case V = 28 are derived from the expected performances announced by NASA and from the recent table of zodiacal brightnesses, as obtained from observations at Tenerife.

106.057 The zodiacal light. The first chapter of V. G. Fesenkov's dissertation published in 1914. Historically-critical survey with a supplement by N. B. Divari, and a survey on the zodiacal light from Arabic sources by B. A. Rozenfel'd.
Researches in the history of astronomy. 15th issue, (see 003.025), p. 233 - 292 (1980). In Russian.

Infrared spectroscopy of interplanetary dust in the laboratory. See Abstr. 022.152.

Formation of Ca-Al mineral grains through gas-solid reactions. See Abstr. 022.164.

Low-velocity impact experiment on rocks − comparison with high-velocity impact experiment. See Abstr. 022.165.

Calibration and in-flight performance of the zodiacal light experiment on Helios. See Abstr. 032.525.

An infrared sensor designed to measure the diffuse zodiacal light. See Abstr. 032.536.

Two-dimensional analyzer with dE-dE silicon detectors for cosmic ray experiments in space. See Abstr. 032.619.

Influence of the sector structure of the interplanetary magnetic field on the regime of the daily rotation of the earth. See Abstr. 044.012.

Results to be expected from light scattering dust analyzer during a rendezvous mission. See Abstr. 051.038.

Recent developments in space-borne zodiacal light photometry. See Abstr. 051.050.

Inductive interaction of electroconductive bodies with ionospheric and interplanetary plasma. See Abstr. 062.030.

Kinetic consideration of charged particle acceleration process in shearing flows of collisionless plasma. See Abstr. 062.075.

Study of interstellar scattering by interplanetary scintillations. See Abstr. 063.014.

Absorption feature observed on the H Lyman-alpha solar line: an interpretation. See Abstr. 071.002.

Possible use of (a) solar faculae and (b) the interplanetary magnetic field as heralds of a solar cycle peak. See Abstr. 072.035.

Cross-correlation analysis between solar flares and the southward components of the interplanetary magnetic field. See Abstr. 073.084.

Solar flare particle injection spectra and spectra of flux maxima at the point of observation. See Abstr. 073.111.

Solar wind helium and hydrogen structure near the heliospheric current sheet: a signal of coronal streamers at 1 AU. See Abstr. 074.003.

Energetic particles in the heliosphere – results from the ISEE-3 spacecraft.   See Abstr. 074.016.

Density gradients in the solar plasma observed by interplanetary scintillation.   See Abstr. 074.035.

Preferred Bartels days of high-speed plasma streams in the solar wind.   See Abstr. 074.080.

Corotating events in the energy range 4–13 MeV as observed on board Helios 1 and 2 in 1975 and 1976. See Abstr. 074.118.

Enhanced cosmic ray anisotropies and the extended solar magnetic field.   See Abstr. 075.013.

Comparative magnetospherology. Part 11. A model of heliomagnetospheric excursion in every sunspot declining-minimum phase.   See Abstr. 075.020.

Comparative magnetospherology. Part 12. Evidences to support the two-hemisphere model on rotational reversing of the heliodipole in sunspot maximum phase. See Abstr. 075.021.

On the influence of the interplanetary magnetic field on solar cosmic ray propagation. See Abstr. 078.020.

Temporal variations of the cosmic ray intensity and the magnetic configurations of the heliosphere. See Abstr. 078.023.

On a connection of sporadic E_{sr} formations with the B_z-component of the interplanetary magnetic field and electro-jet activity.   See Abstr. 083.060.

An association between the equatorial counter-electrojet and the azimuthal component of the IMF. See Abstr. 083.064.

The mechanism of magnetospheric substorm and the MHD waves of the solar wind.   See Abstr. 084.005.

Configuration of the auroral oval and the inter-planetary magnetic field.   See Abstr. 084.007.

ISEE-1, -2 and -3 observation of the interaction be-tween an interplanetary shock and the earth's magnetosphere: a rapid traversal of the magnetopause.   See Abstr. 084.014.

Equatorial counter electrojet & interplanetary mag-netic field.   See Abstr. 084.019.

Linear dependence of the intensity of geomagnetic variations in the near-polar region on the value of the southern and northern components of the interplanetary magnetic field. See Abstr. 084.025.

Dependence of the geometry of the region of open field lines on the interplanetary magnetic field. See Abstr. 084.027.

Asymmetry effects associated with the x-com-ponent of the IMF in a magnetically open magnetosphere. See Abstr. 084.028.

IMF sector effects on the polar geomagnetic field in winter.   See Abstr. 084.053.

The influence of the interplanetary magnetic field and thermal pressure on the position and shape of the magnetopause.   See Abstr. 084.069.

Connection between the PA index of aurorae and solar phenomena and the parameters of the interplanetary medium.   See Abstr. 085.004.

The role of electrostatic charging of small and intermediate sized bodies in the solar system. See Abstr. 091.066.

Magnetospheres and the interplanetary medium. See Abstr. 091.096.

Turbulent accumulation of interplanetary magnetic fluxes at the Venus ionopause.   See Abstr. 093.062.

Relations between turbulent regions of interplane-tary magnetic field and Jovian decametric radio wave emis-sions from the main source.   See Abstr. 099.025.

Structure of the source of jovian decametric emission and interplanetary scintillation.   See Abstr. 099.036.

Energetic particle events ($\geqslant 30$ keV) of Jovian origin observed by Voyager 1 and 2 in interplanetary space. See Abstr. 099.077.

Overview of the Voyager ultraviolet spectrometry results through Jupiter encounter.   See Abstr. 099.084.

Interplanetary scintillation in Jupiter's decametric radiation.   See Abstr. 099.206.

On the electrostatic charging of the cometary nucleus.   See Abstr. 102.034.

The role of collisions with interplanetary particles in the physical evolution of comets.   See Abstr. 102.039.

Existing cometary data and future needs. See Abstr. 102.072.

Results of investigating meteoric matter on the Interkosmos 14 satellite.   See Abstr. 104.003.

Interstellar gas near and within the solar system. See Abstr. 131.015.

On the ionization ratio of the local interstellar medium.   See Abstr. 131.220.

Scintillations of cosmic radio sources in the decametre waveband. I. Spectra of scintillations due to iono-spheric and interplanetary plasma fluctuations and the possibi-lity of their separation.   See Abstr. 141.092.

A cosmic-ray-mediated shock in the solar system. See Abstr. 143.006.

Correlation of the cosmic-ray intensity with solar-terrestrial parameters.   See Abstr. 143.011.

The role of particle drifts in solar modulation. See Abstr. 143.015.

Comment on "The role of particle drift in solar modulation" by Lee and Fisk.   See Abstr. 143.016.

Acceleration and modulation of cosmic rays in the interplanetary space.   See Abstr. 143.044.

The solar modulation of galactic cosmic rays in the outer heliosphere.   See Abstr. 143.058.

Relativistic cosmic rays and corotating interaction regions.   See Abstr. 143.059.

Remote sensing of muon variation spectra and interplanetary inhomogeneities.   See Abstr. 143.060.

Effects of interaction of the heliomagnetosphere with the galactic field in cosmic rays.   See Abstr. 143.068.

Time-dependent Green's functions of the cosmic ray equation of transport.   See Abstr. 143.073.

Spectral and compositional variations of low energy ions during an energetic storm particle event. See Abstr. 143.075.

On the motion of cosmic rays in the interplanetary magnetic field.   See Abstr. 143.089.

50–200 GeV cosmic rays at various heliolatitudes. See Abstr. 143.111.

Changes in cosmic ray propagation induced by corotating interaction regions.   See Abstr. 143.114.

107 Cosmogony

107.001   The origin of the solar system and stochastic electro-dynamics.   M. Surdin.
Nuovo Cimento C, Ser. 1, Vol. 3C, 626 - 634 (1980). – Abstr. in Phys. Abstr., Vol. 84, Abstr. 66853 (1981).

107.002   Comets and the problems of protoplanetary evolu-tion.   S. K. Vsekhsvyatskij.
Problems of cosmic physics. Vyp. 16, (see 003.005), p. 20 - 25 (1981). In Russian.
Basic conclusions and new arguments of the theory of eruptive evolution of planets are given. Some new possibilities that appeared after the discoveries of rings around the giant planets revealing the history of planets are pointed out.

107.003   Cosmogenic isotopes in the early solar system.   A. K. Lavrukhina.
Izv. AN SSSR. Ser. fiz., Tom 45, 522 - 538 (1981). In Russian. Abstr. in Ref. zh., 51. Astron., 8.51.203 (1981).

107.004   Solar-ion penetration in the early solar nebula.   K. R. Housen, L. L. Wilkening.
Proc. Eleventh Lunar Planet. Sci. Conf., (see 012.020), p. 1251 - 1269 (1980).
In order to quantify ideas concerning an early irradiation of solar nebular material, the authors compute the probability that ions, emitted by the sun, survive passage through the absorbing nebula to a distance of 3 AU and so produce radia-tion effects observed in meteorites. Gas densities and pressures, assumed in models of the early nebula, are typically several orders of magnitude greater than those needed to cause complete absorption of ions. The irradiation of meteoritic material could occur only after nearly all gas and small dust particles had been removed from the inner solar system.

107.005   Early thermal history: melting and convection during the accumulation.
A. Coradini, C. Federico, P. Lanciano.
Mem. Soc. Astron. Italiana, Vol. 52, (see 012.023), 379 - 382 (1981).

107.006   Origin of satellite systems.   S. J. Weidenschilling.
Mem. Soc. Astron. Italiana, Vol. 52, (see 012.023), 547 - 551 (1981).

107.007   Model of the accumulation process of formation of planetary systems. I. Numerical experiments.
T. M. Ehneev, N. N. Kozlov.
Astron. Vestn., Tom 15, 80 - 94 (1981). In Russian.
The evolution of a planar protoplanetary cloud consisting of many gravitationally interacting and under collisions com-bining bodies (protoplanets) which are moving in the field of a central massive body (sun or a planet) is considered.

107.008   Superheavy elements: an early solar system upper limit for elements 107 to 110.
S. Nozette, W. V. Boynton.
Science, Vol. 214, 331 - 333 (1981).
The abundance of samarium-152 in the Santa Clara iron meteorite is found to be 108×10^7 atoms per gram. This quantity, if attributed to fission of a superheavy element with atomic number 107 to 109, limits the amount of super-heavy elements in the early solar system to 1.7×10^{-5} times the abundance of uranium-238. For element 110, the limit is 3.4×10^{-5}.

107.009   Computer simulation of the evolution of flat rings of gravitating particles moving around the sun.
S. I. Ipatov.
Astron. Zh., Tom 58, 1085 - 1094 (1981). In Russian. English translation in Soviet Astron., Vol. 25, No. 5.
The evolution of flat rings of gravitating particles (bodies or material points) moving around the sun is investigated. An investigation of the used algorithm of modeling the mutual gravitational influence of the particles is carried out. The parameters of rings of gravitating bodies coagulating under every collision correspond to the feeding zones of terrestrial planets. The density of the bodies is similar to the present plan-etary density.

107.010   Some aspects of space and global petrology.   A. A. Markushev, N. I. Bezmen.
Vestn. MGU. Geol., 1980, No. 5, p. 3 - 18. In Russian. Abstr. in Ref. zh., 51. Astron., 10.51.327; 62. Issled. kosm. prostranstva, 10.62.245 (1981).

107.011   Solar nebula dispersal and the stability of the planetary system. I. Scanning secular resonance theory.   W. R. Ward.
Icarus, Vol. 47, 234 - 264 (1981).
Secular resonances in the early solar system are studied in an effort to establish constraints on the time scale and/or method of solar nebula dispersal. Simplified nebula models and dispersal routines are employed to approximate changes in an assumed axisymmetric nebula potential. These changes, in turn, drive an evolutionary sequence of Laplace-Lagrange

solutions for the secular variations of the solar system. A general feature of these sequences is a sweep of one or more giant planet resonances through the inner solar system.

107.012 The solar system evolution.
G. P. Gladyshev, V. P. Budtov.
Moon Planets, Vol. 25, 413 - 425 (1981).
   The concepts on the spatially-periodic condensation in the solar system have been considered in the light of the general theory of the evolution of the solar system. It has been shown that as proto-disk arise and compress, the role of hydromagnetic effects weakens. After the stage of spatially-periodic condensation and accretion, the concentration of gas in protodisks decreases and the role of hydromagnetic effects increases again. Specific features of the formation of planets near the Sun and satellites near the planets can be explained if these peculiarities of the evolution are taken into account.

107.013 On the formation of the Earth and Moon by gravita-tional accretion in a dust disc.   N. Schofield.
Mon. Not. R. Astron. Soc., Vol. 197, 1031 - 1047 (1981).
   The formation of the Earth by accretion from a uniform Keplerian dust disc is numerically investigated, with particular reference to the consequences of particle interactions within the planetary sphere of influence. For a given set of parameter values a planet of Earth mass is seen to form in a short time with a prograde angular momentum of magnitude close to the present-day Earth-Moon system. Furthermore the planet passed through a rotationally unstable phase which would result in fission into two parts, the smaller being of near lunar mass.

107.014 Character of eccentricity excitation by moving Jovian commensurability resonances. Connection with the asteroids and the Kirkwood gaps.
M. Torbett, R. Smoluchowski.
Bull. American Astron. Soc., Vol. 13, 719 (1981). – Abstract.

107.015 Hydrodynamical models of presolar nebula formation.
A. P. Boss.
Bull. American Astron. Soc., Vol. 13, 722 (1981). – Abstract.

107.016 Voyager and the formation of the Saturnian satel-lite system.   A. J. R. Prentice.
Bull. American Astron. Soc., Vol. 13, 743 (1981). – Abstract.

107.017 The Voyager 1/Saturn encounter and the cosmo-gonic shadow effect.   H. Alfvén.
Plasma astrophysics, (see 012.042), p. 11 - 21 (1981).
   If an electrically conducting medium (e.g. a dusty plasma) rotates around a gravitating central body, which possesses an axisymmetric dipole field, the medium is supported to two-third by the centrifugal force and to one-third by electro-magnetic forces under the condition that the magnetic field is strong enough to control the motion. If the electromagnetic forces disappear –e.g. by a deionisation of the dusty plasma – the medium will fall down to two-thirds of its original central distance. The result of this process will be a "cosmogonic shadow effect" which is described in some detail. The Voyager 1/Saturn results demonstrate that the macro-structure of the Saturnian ring system can be explained as a result of this effect working at the formation of the system. A similar analysis of the asteroidal belt shows that its macro-structure can also be explained by the cosmogonic shadow effect.

107.018 Dynamical evolution of a cometary swarm in the outer planetary region.
J. A. Fernández, W.- H. Ip.
Icarus, Vol. 47, (see 012.045), 470 - 479 (1981).
   The main aim of this study is to analyze the interrela-tions between the accretion of Uranus and Neptune with other processes of cosmogonical importance as, for example, the

formation of a cometary reservoir from bodies placed into near-parabolic orbits by planetary perturbations and the scattering of bodies to the region of the terrestrial planets. Neptune is found to be the most important contributor of comets to the cometary reservoir.

107.019 Growth stress in a chondritic parent body as an accreting planetesimal.
M. Kumazawa, A. Fujimura, M. Kato.
Proceedings of the 14th ISAS Lunar and Planetary Symposium, (see 012.055), p. 197 - 202 (1981).
   Certain growth stress develops in a quietly accreting and self-gravitating planetesimal. The type of growth stress in an elastic sphere with and without a core is calculated for two different accretional modes; instantaneous and continuous accretions. If the oriented texture observed in chondrites is assumed to be a result of deformation due to the growth stress, the necessary condition for the chondritic parent body is shown to be the presence of an incompressible core.

107.020 Growth of planetesimals in the primordial solar nebula.   Y. Nakagawa, C. Hayashi.
Proceedings of the 14th ISAS Lunar and Planetary Symposium, (see 012.055), p. 301 (1981). – Abstract.

107.021 Scattering and collisional processes of planetesimals by protoplanets.   S. Nishida.
Proceedings of the 14th ISAS Lunar and Planetary Symposium, (see 012.055), p. 306 - 310 (1981).
   The growth process of protoplanets by encounter with planetesimals is studied. Since both protoplanet and planetesi-mal are in Keplerian orbits around the protosun, treating the collision process as a two-body encounter in free space is inadequate. The author tries to solve the scattering and colli-sion problem for two bodies in Keplerian orbits numerically. Preliminary results are reported.

107.022 Capture origin of the Moon.
K. Nakazawa, T. Komuro, C. Hayashi.
Proceedings of the 14th ISAS Lunar and Planetary Symposium, (see 012.055), p. 311 - 312 (1981). – Abstract.

107.023 On the ejection of planetesimals from the protosolar system.   T. Nakamura.
Proceedings of the 14th ISAS Lunar and Planetary Symposium, (see 012.055), p. 322 - 329 (1981).
   Formation of Oort's cometary cloud is examined from the following viewpoints. First effects of the gas drag of the protosolar nebula on ejection of the planetesimals are calculat-ed and it is shown that the efficiency of capture into the Oort cloud is $5 \sim 8$ times better than that in the case without gas drag. Second a maximum perihelion distance which planetesi-mals acquire after ejection is calculated based on the dynam-ical conservation laws. This value is evidently insufficient to maintain the Oort cloud against the planetary perturbations over 4.5×10^9 yr.

107.024 Dissipation of the primordial terrestrial atmosphere due to irradiation of the solar far-UV during T Tauri stage.   M. Sekiya, C. Hayashi, K. Nakazawa.
Proceedings of the 14th ISAS Lunar and Planetary Symposium, (see 012.055), p. 341 (1981). – Abstract.

107.025 On the thermomechanical theory of the evolution of planets.   K. M. Magomedov.
Dokl. AN SSSR, Tom 259, 817 - 822 (1981). In Russian.
Abstr. in Ref. zh., 51. Astron., 12.51.285 (1981).

107.026 Origin of solar system.   H. Alfvén.
Planetary interiors, (see 012.061), p. 5 - 20 (1981).
   As the theory of the origin and evolution of the solar system should constitute the general background for the

science of planetary interiors, the author gives a summary of some of the main problems in this field of research.

107.027 **Interaction of ocean-atmosphere with planetary interior.**   G. Arrhenius.
Planetary interiors, (see 012.061), p. 37 - 48 (1981).

107.028 **On the role of the magnetic field and turbulence in the evolution of the presolar nebula.**
T. V. Ruzmaikina *(Ruzmajkina).*
Planetary interiors, (see 012.061), p. 49 - 53 (1981).

107.029 **Thermal dissipation of gas from the protoplanetary cloud.**   G. V. Pechernikova, A. V. Vitjazev
(Vityazev).
Planetary interiors, (see 012.061), p. 55 - 60 (1981).

Most authors assumed that gas was removed from the solar system by solar wind when the young sun passes through T-Tauri stage. Another possible mechanism — the thermal escape from the hot exosphere of disk-like cloud is considered. It is shown that the solar UV radiation could heat the gas in the exosphere up to temperature sufficient for removal of gas. The theory of thermal escape for rapidly rotating atmosphere is developed. The calculations show that the time of dissipation of gas from the nebula with mass $\sim 10^{-2}_\odot$ is about 10^8 years. The authors' model predicts the possibility of violent regime of evaporation.

107.030 **The problems of simulation of planetary systems accumulation processes.**
T. M. Eneev, N. N. Kozlov.
Progress in planetary exploration, (see 012.062), p. 201 - 215 (1981).

Evolution of a system consisting of a great number of bodies that are gravitationally interacting and aggregating in contacts is considered. Body motions take place in the gravitational field of a central massive body (Sun or planet) in the same plane and at the initial time of system evolution orbits of all bodies are circular. It is shown that during evolution of the protoplanetary cloud, ring zones of matter rarefaction and condensation develop. Development of the condensation zones leads to the formation of planets, the most part of which acquire a direct rotation about their axes. In the case under consideration, approximate agreement between the law of planetary distances and that of geometric progression takes place as it is observed in planetary and satellite systems. The formation of the terrestrial planets and Jovian planets has been simulated. The principal numerical results have been obtained through digital simulation of planetary accumulation.

107.031 **On the origin of the solar system and the remarkable position of the sun in the Galaxy.**
L. S. Marochnik.
Inst. kosm. issled. AN SSSR. Prepr., 1981, No. 650, 32 pp. In Russian. — Abstr. in Ref. zh., 51. Astron., 1.51.201 (1982).

107.032 **Putting it all together.**   J. S. Lewis.
The new solar system, (see 003.026), p. 205 - 211 (1981).

Probleme der modernen Kosmogonie.
See Abstr. 003.028.

Evolution kosmischer, biologischer und geistiger Strukturen.   See Abstr. 003.152.

On the foundation of the method of virtual contacts.   See Abstr. 022.075.

Sr/Ca–Ba/Ca systematics on eucrite parent body, the moon and the earth.   See Abstr. 022.167.

Quasi-equilibrium in collisional systems.
See Abstr. 042.070.

Another numerical treatment of the evolution of circumplanetary disks.   See Abstr. 042.072.

Experimental gravitation with measurements made from within a planetary system.   See Abstr. 066.182.

Composition of the earth's upper mantle–II: Volatile trace elements in ultramafic xenoliths.
See Abstr. 081.022.

Solution of the problem of rotation of planets in the statistical theory of accumulation.
See Abstr. 091.002.

Formation of terrestrial planets: primordial phases.
See Abstr. 091.029.

Kinetic inhibition of CO and N_2 reduction in circumplanetary nebulae: implications for satellite composition.   See Abstr. 091.047.

On a formalism accounting for distances of planets.
See Abstr. 091.054.

Planetary rotation and accretion.
See Abstr. 091.073.

The importance of latest discoveries in the solar system.   See Abstr. 091.077.

State of matter and internal structure of planets.
See Abstr. 091.078.

Internal structure of small bodies in the solar system.   See Abstr. 098.095.

Reconstruction of asteroid families.
See Abstr. 098.096.

Accretion of Ganymede and Callisto in a gaseous nebula.   See Abstr. 099.147.

The Voyager 1/Saturn encounter and the cosmogonic shadow effect.   See Abstr. 100.068.

Origin of Saturn's satellites: explanation of compositional differences.   See Abstr. 100.094.

Comets and problems of the eruptive evolution of protoplanets.   See Abstr. 102.086.

The neon alphabet game.   See Abstr. 105.036.

Windows to early solar system processes: refractory inclusions in the CV and CM chondrites.
See Abstr. 105.071.

Internal chemical energy: heating of parent bodies and oxygen isotopic anomalies.   See Abstr. 105.080.

The interstellar dust as a precursor of Ca, Al-rich inclusions in carbonaceous chondrites.   See Abstr. 105.115.

Cosmogonic aspects of the evolution of planetary systems.   See Abstr. 118.037.

Fluxes of energetic particles and the ionization rate in protostars and the primitive solar nebula.
See Abstr. 131.206.

Stars

111 Parallaxes, Proper Motions, Radial Velocities, Space Motions, Distances

111.001 Une étoile à CH très intense et à très grande vitesse radiale située dans la direction du Grand Nuage de Magellan.   C. Fehrenbach, M. Duflot.
Astron. Astrophys., Vol. 101, 226 - 227 (1981).

A high radial velocity star (+440 km s^{-1}) with very strong CH features has been found in the direction of LMC. The authors give a photometric profile and identify absorption bands.

111.002 Progress on the Lick proper motion program.
A. R. Klemola.
Bull. American Astron. Soc., Vol. 13, 516 (1981). – Abstract.

111.003 The precision of acceleration parallaxes.
H. Eichhorn.
Bull. American Astron. Soc., Vol. 13, 569 (1981). – Abstract.

111.004 The distances of cataclysmic variables.
J. Bailey.
Mon. Not. R. Astron. Soc., Vol. 197, 31 - 39 (1981).

A method is described for determining the distance of a cataclysmic variable by making use of the $K(2.2\ \mu m)$ magnitude of the secondary star. This is achieved by calibrating surface brightness in this band as a function of effective temperature. The method is relatively insensitive to the temperature and evolutionary state of the secondary. The author applies the method to determine the distances of a number of dwarf novae and nova-like variables as well as the recurrent nova T CrB and the AM Her type binary VV Pup.

111.005 An RV digital measuring method applied to Lac OB1: preliminary results.
A. Bijaoui, M. Lacoarret, P. Granes.
Astron. Astrophys., Suppl. Ser., Vol. 45, 483 - 486 (1981).

A digital method to determine the radial velocity of stars is described. It has been used on plates photographed with a Schmidt telescope equipped with a normal field objective prism, and measured with the CDCA microdensitometer PDS 1010A connected with PDP 11/40 computer. A preliminary study was made on the southern group Lac OB1 association whose radial velocity histogram is presented.

111.006 Proper motions of bright red giants in globular clusters. IV. The globular cluster M5.
N. Spasova, N. Mikhnevski.
Astrofiz. issled., NRB, Vol. 3, 70 - 75 (1981). In Russian.
Abstr. in Ref. zh., 51. Astron., 8.51.819 (1981).

111.007 Studies of late-type dwarfs. IV. Kinematics of a sample of faint red dwarfs at the north galactic pole.
P. C. Dawson.
Astron. J., Vol. 86, 1200 - 1203 (1981).

Radial velocities have been obtained for 20 of the stars in Sanduleak's catalog. These are combined with photometric parallaxes and published proper motions to obtain space velocities and velocity dispersions. The space density is also estimated. The velocity dispersions and the space density appear to be "normal", i. e., there is no evidence for an excess density of low-velocity M dwarfs.

111.008 The nature of the Giclas + 4 stars.
D. Weistrop.
Astron. J., Vol. 86, 1220 - 1227 (1981).

Broadband $BVRI$ photometry has been obtained for 130 stars in the Lowell Observatory proper motion survey with color class + 4. The sample is complete for the 75 + 4 class stars between R. A. = $2^h 31^m$ and R. A. = $12^h 20^m$ in the northern hemisphere. The absolute magnitudes and distances of the stars are estimated using the M_R -$(R-I)$ relation. Most of the stars are normal late-type dwarf stars. The $B-V$ color distribution for the + 4 class stars is similar to that for + 3 class stars, suggesting there is no significant difference between the two color classes. Several individual stars which may be subluminous or otherwise of interest have been identified, as have several stars within 25 pc which are not included in published catalogs of nearby stars.

111.009 Early-type high-velocity stars in the solar neighborhood. I. List of candidates.   P. B. Stetson.
Astron. J., Vol. 86, 1337 - 1359 (1981).

Studies by several authors indicate that a large fraction of the early-type stars found far from the Galactic Plane resemble Population I stars in every observable respect except in their location. The identification of significant numbers of early-type high-velocity stars in the solar neighborhood is both an extremely difficult task, and one which is important for understanding these apparent exceptions to the correlation between stellar ages, metal abundances, and kinematic properties. This paper identifies suspected nearby high-velocity stars of spectral types B and A by their proper motions, which are generally too small for the stars to have been included in previous catalogs of high-velocity stars. The same selection process is extended to stars of spectral type F to provide a comparison sample and to investigate whether the hypothetical high-velocity, metal-rich population continues to later spectral types. A list of 371 stars selected by this process and with confirmed proper motions is presented; 168 of them are of spectral types A and B. Photometric and spectroscopic observations of these stars are required to eliminate stars with misclassified spectra, and to convert the proper motions to space velocities.

111.010 Grand Nuage de Magellan. Troisième liste d'étoiles membres du Grand Nuage de Magellan et liste d'étoiles galactiques.   C. Fehrenbach, M. Duflot.
Astron. Astrophys., Suppl. Ser., Vol. 46, 13 - 19 (1981).

The authors give two new lists of stars in the direction of the Large Magellanic Cloud. 167 in the first list are Large Magellanic Cloud members, 489 in the second one are galactic stars.

111.011 Systematic effects in trigonometric parallaxes – II. Inter-observatory comparisons.
T. E. Lutz, R. B. Hanson, A. H. Marcus, W. L. Nicholson.
Mon. Not. R. Astron. Soc., Vol. 197, 393 - 412 (1981).

The relative parallaxes determined by five major General Catalogue observatories (Allegheny, McCormick, Cape, Yale and Greenwich) have been inter-compared to investigate their systematic differences. Most systematic differences between observatories are not statistically significant, or are likely to be due to selection effects. The classical "observatory corrections" are neither statistically nor physically justified. The inter-comparisons yield external error estimates which agree with most recent determinations.

111.012 Proper motion survey with the 48-inch Schmidt telescope. LVI. More bed-time stories from Lick.
W. J. Luyten.
Sep. print Univ. Minnesota, Minneapolis, Minnesota, 11 pp. (1981).

111.013 Ein Stern auf Kollisionskurs zur Sonne?
W. Gliese.
Sterne Weltraum, Jahrg. 20, 445 (1981).

111.014 Dispersion of stellar velocities.
V. K. Drofa, A. A. Molotaj.
Astrometric investigations, (see 012.048), p. 53 - 58 (1981). In Russian. – Abstr. in Ref. zh., 51. Astron., 11.51.147 (1981).

111.015 Determination of proper motions of circumpolar region stars from observations with the zonal astrograph at the Pulkovo and Nikolaev Observatories.
V. I. Voronenko.
Astrometric investigations, (see 012.048), p. 90 - 93 (1981). In Russian. – Abstr. in Ref. zh., 51. Astron., 11.51.155 (1981).

111.016 Early-type high-velocity stars in the solar neighborhood. II. Photometry for 78 candidates.
P. B. Stetson.
Astron. J., Vol. 86, 1882 - 1895 (1981).
Four-color and Hβ photometry are presented for 74 of the early-type high-velocity candidates listed in a previous paper in this series (20% of that sample), for four additional early-type high-velocity stars identified by Eggen, and for three optical companions of these stars. At least one-third of the main-sequence A stars and nearly all of the F stars appear to be true high-velocity objects and several of the high velocity F stars appear to have solar metal abundances. Thus these data support recent findings that some apparently young, metal-rich, high-velocity stars exist. This population may crudely be estimated to contribute about one A star to every thousand in the solar neighborhood.

111.017 The degree of completeness of nearby stars and the stellar luminosity function.
A. R. Upgren, T. E. Armandroff.
Astron. J., Vol. 86, 1898 - 1908 (1981).
The space distribution of the nearby stars has been analyzed in order to determine whether they are complete or whether a statistically significant number of yet unrecognized nearby stars may exist. By using UBV photometry of almost all stars earlier than spectral class M0 known to within 22 pc of the sun a luminosity function for these stars is produced. This luminosity function differs significantly in the range of absolute visual magnitude (M_v) +6 to +9 from the commonly accepted smoothed "standard" luminosity function. The authors infer that the deviation is not caused by a number of unrecognized nearby stars, but that the deficiency for the above magnitude range is real. This suggests either that stars of masses equivalent to this range of absolute magnitude are rarer than stars of somewhat greater mass, or that the mass-luminosity relation is not as linear as is presently believed.

111.018 Determining the proper motions of the Trapezium in Orion. B.-c. Qian, Z.-h. Xu.
Ann. Shanghai Obs. Acad. Sinica, No. 1, p. 66 - 73 (1979).
The proper motions of the Trapezium in Orion relative to their centre of mass have been determined. The first epoch data are the data of ten groups of double stars and one group of photographic observations obtained from 1836 to 1925. The second epoch observations were made in 1975 and 1977.

111.019 Proper motions of stars in the Orion association. B.-c. Qian, Z.-h. Xu.
Ann. Shanghai Obs. Acad. Sinica, No. 2, p. 107 - 116 (1980).
The proper motion of the early O, B type stars and the late-type emission stars of the Sword Group in the Orion Association have been determined. The results show that the two kinds of stars are characterized by the proper motion. The expansion and rotation of the early O, B type stars have been found, but similar motions are not discernible for the late-type stars.

111.020 Corrections of the components of solar motion, of the galactic rotation and variations of corrections for the precessional constant in improved systems of proper motions of the GC and N30 star catalogues.
V. V. Vityazev, E. V. Vityazeva.
Vestn. LGU. Ser. mat., mekh., astron. Leningrad, 1981. 9 pp. In Russian. – Abstr. in Ref. zh., 51. Astron., 1.51.111 (1982).

Refined data for parallax stars.
See Abstr. 002.025.

Absolute proper motions of 117 O-type stars.
See Abstr. 002.050.

Right ascension and proper motion in R. A. of 904 bright stars. See Abstr. 002.080.

A catalogue of proper motions of RR Lyrae variables. See Abstr. 002.081.

On the computation of acceleration parallaxes.
See Abstr. 031.506.

L'observation et les propriétés des étoiles.
See Abstr. 031.560.

Parallax and proper motion analyses of 21 plate series taken with the Sproul 61-cm refractor.
See Abstr. 031.562.

The N30 catalogue $\Delta\mu_\alpha$ proper motion corrections for 2305 stars. See Abstr. 041.009.

On the N30 catalogue $\Delta\mu_\alpha$ proper motion corrections. Comparison of the improved system of proper motions of the N30 catalogue with the FK4 and GC systems.
See Abstr. 041.010.

Improvement of declinations and proper motions of the latitude programme stars compiled at the Engelhardt Astronomical and the Poltava Gravimetrical Observatories.
See Abstr. 041.022.

The colours, magnitudes and parallaxes of the nearby stars. See Abstr. 113.017.

Spectral types and radial velocities of southern OB+ stars. See Abstr. 114.079.

HD 86590: a short-period RS Canum Venaticorum binary system. See Abstr. 117.054.

Radial velocities of binary and proper-motion stars.
See Abstr. 118.018.

Large proper motions of the Herbig-Haro objects HH 1 and HH 2. See Abstr. 121.019.

Hubble ratio and solar motion from 200 spiral galaxies having distances derived from the luminosity index. See Abstr. 158.047.

111.902 Erratum: "Systematic motions of fundamental stars" [Astron. Astrophys., Vol. 99, 311 - 319 (1981)].   P. Brosche, H. Schwan.
Astron. Astrophys., Vol. 103, 427 (1981). – See Abstr. 29.111.023.

Errata

111.901 Erratum: "Summary of U. S. Naval Observatory parallaxes" [Astron. J., Vol. 85, 454 - 465 (1980)].
R. S. Harrington, C. C. Dahn.
Astron. J., Vol. 86, 1414 (1981). – See Abstr. 27.111.017.

112 Circumstellar Matter (Shells, Dust, Masers, Stellar Winds, etc.)

112.001 **Detection of the SiO ($v=2, J=2 \rightarrow 1$) maser.**
H. Olofsson, O. E. H. Rydbeck, A. P. Lane,
C. R. Predmore.
Astrophys. J., Lett., Vol. 247, L81 - L84 (1981).

Weak maser emission from the $v = 2, J = 2 \rightarrow 1$ transition of SiO has been detected for the first time toward the cool S-type Mira variable χ Cyg. The intensity did not show a positive correlation with the visual phase of the star during the observation period. No SiO ($v=2, J = 2 \rightarrow 1$) emission was seen at two different epochs toward Ori A and a number of Miras and semiregular variables.

112.002 **The size of a Wolf-Rayet star's dust shell measured by speckle interferometry.**
D. A. Allen, J. R. Barton, P. T. Wallace.
Mon. Not. R. Astron. Soc., Vol. 196, 797 - 800 (1981).

The authors have resolved the WC 9 star Ve 2 - 45 at a wavelength of 2.2 μm. The shell of circumstellar dust is optically thin, and has a characteristic radius of 0.04 arcsec. This is in agreement with simple models of the dust shell, and will help to constrain more elaborate models, when they become available.

112.003 **Ultraviolet observations of 27 Canis Majoris, π Aquarii and 48 Librae.**
A. E. Ringuelet, J. M. Fontenla, M. Rovira.
Astron. Astrophys., Vol. 100, 79 - 90 (1981).

27 CMa, π Aqr, and 48 Lib, Be stars and V/R variables of luminosity class not higher than main sequence and with high rotational velocity, have been analyzed in order to determine the physical conditions in the gaseous structure that surrounds the stars. According to the two different types of absorption lines that can be recognized, namely, the narrow and deep profiles and the broad and shallower profiles, the authors divide the gaseous structure into two regions: (1) A geometrically thin region that rotates with the star. (2) A very extended cold region which does not rotate with the star.

112.004 **Observations of stellar OH masers with the VLA.**
M. J. Reid, J. M. Moran, K. J. Johnston.
Astron. J., Vol. 86, 897 - 902 (1981).

The authors observed the OH maser emission from four M-type stars with the VLA. Using narrowband filters, they isolated the two emission features that are characteristic of stellar OH masers. The observed dependence of fringe amplitude with baseline length can be explained if the emission comes from a random distribution of point components. The OH positions are in good agreement with H_2O maser, optical, and infrared positions. The authors' results strongly support models in which the OH maser emission originates in the expanding circumstellar envelopes of late-type stars.

112.005 **Influence of the envelope in the variations of transitory type Be stars.**   A. Peton.
Astron. Astrophys., Vol. 101, 96 - 100 (1981). In French.

The author considers the spectral variations of the Be stars of transitory type, those which are alternately B or Be. It is shown that the Rojas-Herman classification can be used for a quantitative study of the envelope variations. The author proposes a relation between the $(U-B)_0$ index and the equivalent width of Balmer emission lines, and the author concludes that the $(U-B)_0$ index must be divided into two parts: the first one $(U-B)_*$ is connected with the underlying star and is sensitive to the long time variations, the second one $(U-B)_{env}$ is connected with the envelope and is sensitive to the short time variations. Both indices are additive ones, which imposes that the observed luminous flux is a convolution of the flux emitted by the star and the shell.

112.006 **Recent developments in the spectra of shell stars, I. The stable shell stars.**   A. F. Gulliver.
Astrophys. J., Vol. 248, 222 - 227 (1981).

Two of the three previously known stable shell stars, HD 193182 and HD 195325, are described. A third such star, HD 179343, is identified and a marked intensity change is observed in an otherwise stable shell, HD 192954. No changes in the shell velocities or spectra of the stable shell stars have been observed for up to half a century. The underlying stellar spectra are restricted to spectral types B6 to B9. The individual shell spectra range in intensity but are qualitatively similar. A very broad Ca II K shell line component is found in several stars.

112.007 **Circumstellar winds in globular cluster giants.**
R. C. Peterson.
Astrophys. J., Lett., Vol. 248, L31 - L34 (1981).

Spectra of luminous giants in several different globular clusters, obtained at a resolution of better than 15 km s^{-1}, show Na D and Hα line profiles with cores shifted by 11 km s^{-1} and 6 km s^{-1}, respectively, and with a steep blue wing and a gradual red wing. These profiles have a striking resemblance to both theoretical circumstellar profiles and to those observed for M giants and supergiants of the field.

112.008 **Detection of the $J = 1 \rightarrow 0$ and $J = 2 \rightarrow 1$ rotational lines of SiS in the molecular envelope of IRC+10216.**
M. Grasshoff, E. Tiemann, C. Henkel.
Astron. Astrophys., Vol. 101, 238 - 240 (1981).

Searches for J = 1 → 0 transition in the v = 0,1 states as well for J = 2 → 1 transition in the ground vibrational state are presented. The J = 1 → 0, J = 2 → 1, v = 0 lines were detected toward IRC+10216. The line temperature of the J = 2 → 1 transition agrees with the value predicted by Morris (1975), but the spikes in the line wings of the J = 1 → 0 profile are larger than the model value. Two possible explanations are: (1) The region emitting the J = 1 → 0 line is more extended than previously thought or (2) maser emission is present.

112.009 **IUE spectra of RS Puppis and HD 20722: stars in symmetric dusty nebulae.**   H. M. Johnson.
Publ. Astron. Soc. Pacific, Vol. 93, 285 - 287 (1981).

One classical cepheid and one K giant are candidates for special circumstances of mass ejection by virtue of central position in symmetric nebulae. Low-dispersion near-ultraviolet spectra of these stars are reproduced and discussed. No anomalies are found in comparison with apparently similar stars that all lack nebulae. A high-dispersion study of the Mg II doublet is needed for further information about possibly current mass loss.

112.010 **New insight into the physics of atmospheres of early-type stars.**   H. J. G. L. M. Lamers.
The Universe at ultraviolet wavelengths, (see 012.009), p. 93 - 103 (1981).

The insight into the physics of atmospheres of early-type stars obtained from IUE observations is discussed. The paper is concentrated on the phenomenon of mass loss and stellar winds from hot stars. The mass loss rate of early-type stars increases by about a factor of 10^2 to 10^3 during their evolution. This seems incompatible with the radiation-driven wind models and may require another explanation for the mass loss from early-type stars. The winds of early-type stars are strongly variable and the stars may go through active phases. Eclipses in binary systems by the stellar winds can be used to probe the winds.

112.011 **Mass-loss rates from early-type stars.**
P. S. Conti, C. D. Garmany.

The Universe at ultraviolet wavelengths, (see 012.009), p. 115 - 120 (1981).

The authors have derived mass-loss rates for a number of unevolved O-type stars and a few WN stars from high dispersion IUE spectra of their P Cygni profiles. When combined with other published mass-loss rates, the authors find that the relationship between log $\dot{m}$ and M_{BOL} is a broad band rather than a linear relation, suggesting that the line radiation driven wind theory may not be sufficient to explain mass loss. The mass-loss rates for the WN stars, while more uncertain, confirm that these stars lose mass about 100 times faster than O-stars.

112.012   Detection and mapping of the N = 2 - 1 line of CN in IRC + 10216.
A. Wootten, S. Lichten, R. Sahai, P. Wannier.
Bull. American Astron. Soc., Vol. 13, 508 (1981). – Abstract.

112.013   VLA observations of circumstellar OH and H_2O masers associated with four late-type variable stars.   J. H. Spencer, P. F. Bowers, K. J. Johnston.
Bull. American Astron. Soc., Vol. 13, 513 (1981). – Abstract.

112.014   Evidence for mass loss at polar latitudes in the Be stars ω Ori and 66 Oph.   G. J. Peters.
Bull. American Astron. Soc., Vol. 13, 526 - 527 (1981). Abstract.

112.015   A comparison of circumstellar gas and dust in M giants and supergiants.
W. Hagen, D. F. Dickinson, R. M. Humphreys, R. E. Stencel.
Bull. American Astron. Soc., Vol. 13, 548 (1981). – Abstract.

112.016   Stellar winds in Be stars.
J. Dachs.
Second European IUE Conference, (see 012.011), p. 139 - 142 (1980).

High-dispersion IUE spectrograms were obtained for eight bright early-type Be and shell stars. Indications for stellar winds are found in all program stars from asymmetric or violet-shifted ultraviolet resonance line profiles.

112.017 T_e-structure of the wind in γ Cas.
V. Doazan, P. Selvelli, R. Stalio, R. N. Thomas.
Second European IUE Conference, (see 012.011), p. 145 - 146 (1980).

UV observations of the wind in γ Cas, via measure of line-displacements, delineate 2 distinct and non-overlapping velocity regions where absorbing ions for each of Si IV, C IV, N V are concentrated: at about -100 km s^{-1} and -1400 km s^{-1}. Absence of appreciable number of these ions absorbing at intermediate velocities suggests the presence of a corona too hot to support such ions in those atmospheric regions where such velocities occur; observed soft X-ray emission at coronal intensities from γ Cas supports this picture.

112.018   Variable wind-velocity in 59 Cyg.
V. Doazan, L. V. Kuhi, J. M. Marlborough,
T. P. Snow, R. N. Thomas.
Second European IUE Conference, (see 012.011), p. 151 - 154 (1980).

The authors summarize combined UV and visual observations of 59 Cyg in 1972, 1975, late 1978, and at several epochs in 1979. Visual observations suggest a similar pattern of long-term variability for 59 Cyg and γ Cas, within which the authors place the UV data on 59 Cyg. During the 1978 - 79 epoch of small but increasing Hα emission, the C IV and N V lines reach maximum expansion velocities of some 750 km s^{-1}, whereas the Si IV lines show neither appreciable displacements nor asymmetry. Line displacements of N V vary by a factor 2, and appear to oscillate, but the authors have not identified, unambiguously, line components that correspond to violet displacements $<$ 100 km s^{-1}. This 1978 - 79 situation is unlike the situation of strong Be phase of 59 Cyg in 1972 and 1975, and of γ Cas in 1978 - 79, where expansion at $<$ 100 km s^{-1} was observed.

112.019   Observational evidence for an accretion disk around the B star companion of δ Sge (M 2 II).
D. Reimers, R. P. Kudritzki.
Second European IUE Conference, (see 012.011), p. 229 - 232 (1980).

The authors observed δ Sge with the aim to determine the M star's mass-loss rate from possibly present circumstellar absorption lines in the B star's UV continuum. To the authors' surprise, they detected instead of CS lines numerous emission lines.

112.020   Observations of stellar winds in early type stars.
P. S. Conti.
Effects of mass loss on stellar evolution, (see 012.015), p. 1 - 18 (1981).

112.021   The dependence of mass loss on the basic stellar parameters.   H. J. G. L. M. Lamers.
Effects of mass loss on stellar evolution, (see 012.015), p. 19 - 25 (1981).

The author determined the dependence of mass loss on the stellar parameters for O and B stars of various luminosities. He used four homogeneous sets of mass loss rates derived by different authors from the radioflux, the infrared excess, the UV lines and Hα emission.

112.022   The velocity characteristics of WR stellar winds.
A. J. Willis.
Effects of mass loss on stellar evolution, (see 012.015), p. 27 - 33 (1981).

With the advent of IUE many WR stars have become accessible to high resolution UV spectroscopy for the first time, and this paper briefly reports a first analysis of the UV (and visible) P-Cygni profiles observed in 10 WR stars.

112.023   The iron curtain of the WC 9 star HD 164270.
K. A. van der Hucht, P. S. Conti.
Effects of mass loss on stellar evolution, (see 012.015), p. 35 - 37 (1981).

The emission lines in the optical and UV spectrum of the WC 9 star HD 164270 indicate a terminal velocity of its wind of $v_\infty \approx$ 1400 km s^{-1}. In the UV spectrum Fe III absorption lines appear from transitions with metastable lower levels. Because they are displayed over only 830 km s^{-1}, it is suggested that they are formed in the decelerating part of the wind. A radius of 1.8×10^4 R$_\odot$ is found for the circumstellar Fe III shell. This value is within the range of radii calculated by Cohen et al. (1975) for the dust shells of WC 9 stars.

112.024   Is a stellar wind inherent in WR-stars throughout the whole of their evolution?   R. E. Gershberg.
Effects of mass loss on stellar evolution, (see 012.015), p. 39 - 40 (1981).

112.025   Wind characteristics of the O7 n star HD 217086 in the Cep OB 3 association.
M. Perinotto, N. Panagia.
Effects of mass loss on stellar evolution, (see 012.015), p. 41 - 44 (1981).

The O7 n star HD 217086 which provides the ionization of the H II region S 155 A and is the brightest member of the Cep OB 3 association, has been observed in the ultraviolet with IUE. From an analysis of the UV spectra the authors determine a terminal velocity of 3560 ± 100 km s^{-1} and a mass loss rate of $(4.2 + 2.8/- 1.7) \times 10^{-7}$ M$_\odot$ yr^{-1}. A comparison is made with the stars of similar spectral type.

112.026 **Mass loss from central stars of planetary nebulae.**
M. Perinotto, P. Benvenuti, C. Cacciari.
Effects of mass loss on stellar evolution, (see 012.015),
p. 45 - 50 (1981).

From a high resolution spectrum taken with IUE, the
central star of the planetary nebula IC 2149 is found to
exhibit a wind with edge velocity of 1440 ± 100 km s^{-1}. A
preliminary evaluation of the associated mass loss rate gives
$10^{-8} M_\odot$ yr^{-1}. Other planetary nebulae nuclei are studied with
low resolution IUE spectra and indications are found of mass
loss rates consistent with the above value.

112.027 **Mass loss rates of OB stars derived from infrared
observations.**
E. G. Tanzi, M. Tarenghi, N. Panagia.
Effects of mass loss on stellar evolution, (see 012.015),
p. 51 - 56 (1981).

The authors report briefly on a study of the mass loss of
early type stars in the infrared. Up to now near infrared
$(1.25 - 4.8\ \mu)$ broad band photometry of 70 southern OB
stars of various luminosity classes has been secured. 37 stars are
found to exhibit emission in excess over a blackbody photospheric continuum, which is interpreted in terms of gas ejected
in the form of an accelerated wind. By means of model calculations the corresponding mass loss rates are derived. The data
show that mass loss rates increase with luminosity and are a
decreasing function of surface gravity.

112.028 **Radio observations and the mass flow rate of
α Cyg (A2Ia).**
B. Wolf, O. Stahl, W. J. Altenhoff.
Effects of mass loss on stellar evolution, (see 012.015),
p. 61 - 64 (1981).

112.029 **Mass loss rates for twenty-one Wolf-Rayet stars.**
M. J. Barlow, L. J. Smith, A. J. Willis.
Effects of mass loss on stellar evolution, (see 012.015),
p. 65 (1981). – Abstract.

112.030 **Mass outflow in AG Carinae and a comparison
with P Cygni.**   S. Bensammar, S. Gaudenzi,
C. Rossi, H. M. Johnson, P. S. Thé, E. J. Zuiderwijk, R. Viotti.
Effects of mass loss on stellar evolution, (see 012.015),
p. 67 - 69 (1981).

AG Car is a variable supergiant surrounded by a small
ring nebula. Its ultraviolet and optical spectrum, and UV to IR
energy distribution are similar to those of P Cygni with
$v_\infty = -290$ km s^{-1}. Also the strength of the 2200 A band is
the same and suggests E(B–V) = 0.55 for both stars. The
possible variability of the infrared flux is discussed.

112.031 **Mass loss from hot stars below the main sequence.**
R. Viotti, L. Rossi, F. D'Antona.
Effects of mass loss on stellar evolution, (see 012.015),
p. 71 - 74 (1981).

The evolutionary implications of mass loss from post-
asymptotic giant branch stars are discussed, with reference to
the UV observations of sdO's.

112.032 **Observational evidences of stellar winds.**
S. P. Tarafdar.
Effects of mass loss on stellar evolution, (see 012.015),
p. 79 - 82 (1981).

Observations mainly from the IUE spectra of
HD 152236 (B1 Ia) have been used to provide support to the
various aspects of the theory of stellar winds and its interaction with the interstellar medium.

112.033 **The radial velocity variations in IC 418.**
R. H. Méndez, A. D. Verga.
Effects of mass loss on stellar evolution, (see 012.015),
p. 83 - 86 (1981).

112.034 **Mass loss from cool stars.**
A. K. Dupree.
Effects of mass loss on stellar evolution, (see 012.015),
p. 87 - 110 (1981).

Recently obtained spectroscopic observations indicating
mass loss in cool stars are reviewed with analogies to the solar
atmosphere. Spectral diagnostics of mass loss are discussed
with new theoretical calculations of chromospheric line profiles. A general picture of mass loss from cool stars is developed and related to chromospheric and coronal emissions
measured by IUE and the HEAO-2 Observatory. These winds
range in characteristics from the hot $(10^6$ K) and fast wind
with low mass loss found in the dwarf stars to the warm
$(\sim 10^5$ K), moderate speed winds present in hybrid luminous
supergiants, and the coolest massive winds emerging from the
latest type supergiants exhibiting lowest thermal velocities
and circumstellar shells. Evidence for stellar surface inhomogeneity and variability of outflow is briefly discussed.

112.035 **Outflow of matter in the chromosphere of
α Orionis.**   L. Goldberg.
Effects of mass loss on stellar evolution, (see 012.015),
p. 111 - 112 (1981).

112.036 **Mass loss from α Ori.**
T. Tanabe, F. Kamijo.
Effects of mass loss on stellar evolution, (see 012.015),
p. 113 - 115 (1981).

High dispersion spectrograms of α Ori are obtained.
Several strong absorption lines which have circumstellar
components are measured. Assuming the plane-parallel
envelope and thermal equilibrium, the mass loss rate is estimated as $1 \times 10^{-6} M_\odot$/yr.

112.037 **On possible mass loss from the supergiant
Rho Cassiopeia.**   M. Boyarchuk.
Effects of mass loss on stellar evolution, (see 012.015),
p. 117 - 118 (1981).

112.038 **Line formation in the wind of Alpha Cygni.**
P. B. Kunasz, F. Praderie.
Effects of mass loss on stellar evolution, (see 012.015),
p. 159 - 160 (1981). – Abstract.

112.039 **Sternwinde in der Cygnus OB 2 Assoziation.**
C. Leitherer.
Diplomarb., Landessternw. Heidelberg-Königstuhl, F. R.
Germany. 5 + 59 pp. (1981).

Spectroscopic and photometric (HKLMN) observations
of Cyg OB 2 stars were carried out. For most program stars
the author derived infrared excesses due to circumstellar
free-free radiation of expanding envelopes. By combining the
infrared data with recent radio observations of five Cyg OB 2
stars the velocity law of the stellar wind from O stars was
determined. Using this empirical velocity law, the author
derived mass-loss rates for Cyg OB 2 stars with measured
infrared excess.

112.040 **Evolution of the OH maser emission from
U Orionis.**
P. R. Jewell, J. C. Webber, L. E. Snyder.
Astrophys. J., Vol. 249, 118 - 123 (1981).

The evolution of the U Orionis OH masers has been
followed for several years. The 1612 MHz maser emission has
continued to decline in the damped oscillator mode previously
reported. The main lines masers, in contrast, have become
stronger in recent years and are varying with the stellar cycle.
Their emission curves show an unusual, sharply peaked form,
however. Observations obtained by other investigators are
plotted with the authors' ones to form a comprehensive
display of the behavior of these masers. An interpretation of

the evolution is given, and the implications of this source for OH masers and circumstellar shells in general are discussed.

112.041 Infrared excess and mass-loss rate of the extreme Of star HD 108.
M. Ferrari-Toniolo, P. Persi, G. L. Grasdalen.
Publ. Astron. Soc. Pacific, Vol. 93, 633 - 635 (1981).

Infrared observations of the extreme Of star HD 108 show an IR excess already at 2.3 microns. This infrared excess has been interpreted as due to a thermal ff+bf emission from an expanding ionized gas surrounding HD 108. Between 5 and 10 microns the IR excess shows a distribution law of the type $S_\nu \propto \nu^{1.0 \pm 0.4}$, indicating that the $10\,\mu$ radiation originates from a region where the stellar wind is still undergoing acceleration. From the observed $10\,\mu$ ff+bf flux a mass-loss rate for HD 108 of $\dot{M} \simeq 2 \times 10^{-5}\,\nu_\infty (M_\odot\,y^{-1})$ with ν_∞ in $10^3 km\,s^{-1}$ is derived. This value indicates that radiation pressure only is insufficient to drive the wind in HD 108.

112.042 Galactic ring nebulae associated with Wolf-Rayet stars. III. H II region-type nebulae.
Y.-H. Chu, R. R. Treffers.
Astrophys. J., Vol. 250, 615 - 620 (1981).

The authors have obtained Fabry-Perot data for five galactic ring nebulae associated with Wolf-Rayet stars. The amount of turbulence in these nebulae is comparable to that in an ordinary H II region. It seems that these nebulae are only being illuminated by the central star. There is no evidence that these nebulae are formed from either material ejected from the stars or by stellar winds interacting with the interstellar medium.

112.043 4–8 micron spectrophotometry of OH 0739–14.
B. T. Soifer, S. P. Willner, R. W. Capps, R. J. Rudy.
Astrophys. J., Vol. 250, 631 - 635 (1981).

Spectrophotometry of the dust-embedded late-type star OH 0739–14 shows an absorption feature at 6.0 μm characteristic of H_2O ice at temperatures significantly lower than 150 K, confirming the identification of H_2O ice in the circumstellar shell in this source. The differences in the infrared spectra of OH 0739–14 and embedded molecular cloud sources are attributed to the different cloud lifetimes and temperature regimes in which the molecules are formed. A lower limit to the mass loss rate of $10^{-4} M_\odot/yr^{-1}$ is derived, based on the column density of ice and the size and the expansion velocity of the circumstellar cloud.

112.044 Extended OH maser emission from circumstellar shells and the envelope around OH 26.5 + 0.6.
B. Baud.
Astrophys. J., Lett., Vol. 250, L79 - L83 (1981).

The extended diffuse component of the 1612 MHz OH maser emission from circumstellar envelopes around OH/IR stars has been studied with the VLA spectral line system. The first results show extended emission, in some cases distributed over several arc seconds, with maser line brightness temperatures of $10^4 - 10^6$ K. The emission from OH 26.5 + 0.6 is considered in detail.

112.045 Stellar winds and mass-loss rates from Be stars.
T. P. Snow, Jr.
Astrophys. J., Vol. 251, 139 - 151 (1981).

The Copernicus scans of Si III and Si IV resonance lines in the spectra of 22 early B stars, 19 of which are known to have Be characteristics, have been analyzed for effects of stellar winds. All but one of the stars show line asymmetries to which theoretical profiles were fitted, yielding wind parameters that were then used to calculate mass-loss rates. The mass-loss rates do not appear to correlate with any other stellar parameters, possibly a result, in part, of time variability in the winds. Evidence of equatorial gravity darkening is seen in the relative velocity widths of ultraviolet and visible photo-

spheric lines. The rates of mass loss from the Be stars are probably insufficient to affect either their own evolution, since they typically lose only $\sim 10^{-3} M_\odot$ during their lifetimes, or that of the galaxy, since their winds do not contribute significantly to enrichment of the interstellar medium.

112.046 H₂O masers in the direction of southern nebular objects.
E. Scalise, Jr., G. F. Gahm, G. Sandell.
Astron. Astrophys., Vol. 104, 166 - 168 (1981).

The authors have searched 13 Orion population stars and 4 H-H objects for H_2O maser emission on two different observing runs in 1979 and 1980. They detected H_2O masers toward two stars: RU Lupi and vBH 65b. Both H_2O masers are variable, vBH 65b was below the detection limit during the second observing period. Whether these H_2O masers originate in the dust envelopes surrounding these two T Tauri stars is not clear, because the accurate maser positions have not yet been determined.

112.047 Velocity fields in the shell of the M giant star Beta Pegasi.   A. M. Boesgaard.
Astrophys. J., Vol. 251, 564 - 570 (1981).

The M2.5 II - III star, β Pegasi, has been found to have exceptionally strong Fe II emission lines in the 3100 Å region and exceptionally strong circumstellar absorption at Ca II H and K. Nine spectrograms at 6.7 Å mm^{-1} (0.13 Å spectral resolution) were obtained over a 4 yr interval with the 2.24 m telescope at Mauna Kea in the 3100 - 4000 Å region to determine radial velocities and line profile shapes to analyze the velocity field and extent of the shell. The Fe II emission is blueshifted and asymmetric in the sense to indicate outflowing material. A specific model of an extended moving envelope produces emission-line profiles which match the observed Fe II $\lambda 3228$ feature when the material in a $2R_*$ shell expands from 0 to 25 km s^{-1} and has a characteristic turbulent velocity of 10 km s^{-1}.

112.048 An expanding motion in the ionized envelope of HBV 475.   S. Tamura.
Astrophys. Lett., Vol. 22, 165 - 170 (1981).

Expansion velocities in the ionized envelope of HBV 475 are obtained from the line profiles of Hα, He I 6678, 7065 and [Fe VII] 6087. In particular two components of [Fe VII] 6087 show evidence of outward gas motion with -350 and -200 km/s from the central star. HBV 475 is thought to be a precursor of a planetary nebula. Line profiles are interpreted according to a model where a high-velocity stellar wind from a hot central star interacts with a remnant shell produced during the red giant stage of the star.

112.049 OH emission in the direction of TV Gem and BI Cyg.
L. W. Brown, R. W. Hobbs, A. G. Michalitsianos, M. Kafatos.
Astron. J., Vol. 86, 1926 - 1929 (1981).

OH maser main-line emission at 1665 MHz has been detected in the direction of two late-type supergiants, TV Gem (M1Iab) and BI Cyg (M3Iab), with the 43-m telescope and 18-cm receiver at NRAO-Green Bank. Detection of two new OH maser sources here is particularly interesting in view of the relatively small number of late-type supergiants that are known as molecular radio sources or associated in close proximity with OH emission clouds. The infrared, ultraviolet, and radio properties of these stars are discussed briefly.

112.050 Supergiant OH/IR stars.   P. F. Bowers.
Astron. J., Vol. 86, 1930 - 1934 (1981).

Approximately 70 M supergiants have been searched for Type I OH emission. OH was detected toward six stars but it is probably not associated with the stars in most cases. The results of this survey, combined with earlier searches for Type II OH supergiants, indicate that the OH luminosities are less for Type I supergiants than for Type II supergiants. There is also a trend for the OH luminosity to increase with the infrared color index. Similar effects have been noted for OH Miras, though these stars generally have smaller OH luminosities and color indexes. The results are consistent with OH supergiants having thicker dust envelopes than most OH Miras.

112.051   **The correlation of SiO maser flux with stellar flux.**
     J. H. Cahn.
Astron. J., Vol. 86, 1935 - 1938 (1981).
     A survey of available SiO maser observations of the 43.122-GHz, $v = 1, J = 1 \rightarrow 0$ transition has been made in order to further investigate the question of a correlation of maser flux with bolometric flux. This correlation has been shown to exist both for data which have been phase shifted to mean maximum and for those which have not.

112.052   **High sensitivity molecular line observations of IRC+10216.**   H. Olofsson, L. E. B. Johansson, Â. Hjalmarson, Nguyen-Quang-Rieu.
Res. Lab. Electron. Onsala Space Obs., Res. Rep. No. 142, 2 + 54 pp. (1981).
     The authors present high-sensitivity observations of a number of molecular millimeter wave transitions. The goal is to study the elemental abundances and the chemistry in the stellar atmosphere, the mass loss rate and its possible evolution with time, the kinetics, the thermal structure, and the chemistry of the gas/dust envelope, and finally to understand the detailed radiative transfer and excitation processes in the molecular shell.

112.053   **A spectrographic study of the shell star EW Lac (217050).**   G. Scholz.
Astron. Nachr., Band 302, 279 - 284 (1981).
     Investigation of 38 Zeeman spectrograms of the Be star EW Lac obtained in the three successive years, 1978, 1979, and 1980, yields no hints at the occurrence of a global magnetic field larger than 150 Gauss. The temporal variations of the radial velocities of the Balmer and metallic absorption lines indicate a disturbance travelling through the envelope of EW Lac from the inner to the outer parts. The observed variations of the line widths and equivalent widths of the metallic lines suggest that the physical parameters change remarkably. These results together with the measured variations of the radial velocity and intensity of the emission features of H_β can be explained by a thick elliptical ring surrounding EW Lac.

A mechanism for producing ground vibrational state SiO masers.   See Abstr. 022.097.

The infrared spectrum of SiO near 1240 cm⁻¹ and its relation to the circumstellar SiO maser.
See Abstr. 022.099.

The 11.5 micrometer emission from carbon stars: comparison with IR spectra of submicrometer-sized silicon carbide grains.   See Abstr. 022.102.

On the problem of detection of the circumstellar clouds from 2800 Mg II observations.   See Abstr. 031.581.

One-dimensional infrared speckle interferometry.
See Abstr. 031.594.

Mass-loss rates in early-type stars determined by fitting Balmer alpha profiles.   See Abstr. 064.014.

On the stellar gravity and effective temperature dependence of the ratio of terminal to escape velocities in stellar winds.   See Abstr. 064.032.

Properties of optically thick winds driven by radiation pressure.   See Abstr. 064.042.

Collisional and radiative excitation of SiO masers.
See Abstr. 064.048.

Mass loss from very luminous OB stars and the Cygnus superbubble.   See Abstr. 064.079.

Mass loss rates from O stars in OB associations.
See Abstr. 064.080.

Stellar chromospheres and coronae.
See Abstr. 064.081.

Infrared photometry of the X-ray stars in Cygnus OB2 (VI Cygni): preliminary results.
See Abstr. 113.052.

Far-UV wind line profile changes in the O-type star HD 175754.   See Abstr. 114.005.

Spectral energy distributions of hot stars with circumstellar dust.   See Abstr. 114.008.

A far-infrared emission feature in carbon-rich stars and planetary nebulae.   See Abstr. 114.014.

Circumstellar shells of luminous supergiants. I. Carbon monoxide in Rho Cassiopeiae and HR 8752.
See Abstr. 114.021.

Search for optical coronal line emission from the X-ray sources Epsilon Orionis (B0 Ia) and Kappa Orionis (B0.5 Ia).   See Abstr. 114.023.

IUE observations of variability in winds from hot stars.   See Abstr. 114.028.

IUE and ground-based observations of mass loss in the Magellanic Clouds.   See Abstr. 114.031.

IUE observations of circumstellar emission from the late type variable R Aqr (M7 + pec).   See Abstr. 114.041.

The chromospheric and coronal structure of αCMi (F5 IV–V).   See Abstr. 114.065.

Long term changes in ultraviolet resonance lines in γ Cas.   See Abstr. 114.070.

Coordinated ultraviolet, optical and infrared observations of X Per in December 1979: preliminary results.
See Abstr. 114.073.

Photospheric molecular line profiles in cool stars.
See Abstr. 114.084.

The ultraviolet to infrared spectrum of the large mass loss LMC supergiant S22 = HD 34664.
See Abstr. 114.086.

Spectroscopic and photometric observations of the Be star 69 Orionis.   See Abstr. 114.129.

Paschen lines in Be stars. II. Study of Paschen emission lines.   See Abstr. 114.135.

Ultraviolet observations of two extreme Population II stars: detection of chromospheric emission and mass loss.   See Abstr. 114.137.

The UV resonance lines of HD 175362, a helium variable with a strong magnetic field.
See Abstr. 116.010.

Radio observations of O-type stars.
See Abstr. 116.012.

First detection of nonflare microwave emission from the coronae of single late-type dwarf stars.
See Abstr. 116.030.

Search for colliding stellar winds in Plaskett's star (HD 47129).　See Abstr. 117.025.

The active chromosphere and corona of HD 4502 = ζ And.　See Abstr. 117.036.

IUE spectroscopy of hot binary stars HD 108, HD 149404, and HD 163181.　See Abstr. 117.077.

Is this diagram an argument for binary orbital evolution due to mass loss?　See Abstr. 118.014.

The 1979 - 1980 eclipse of Zeta Aurigae. II. The emission spectrum.　See Abstr. 119.066.

Photometry of the reflection nebulosity at V1057 Cygni.　See Abstr. 121.022.

Über einige Eigenschaften zirkumstellarer Hüllen von T-Tauri-Sternen.　See Abstr. 121.038.

Observation of mass loss in R CrB during the visual light minimum.　See Abstr. 122.061.

IUE and ground-based spectroscopic observations of the S Dor-type LMC variable R 71 during minimum state.
See Abstr. 122.103.

Detection of 6 centimeter OH emission from the Mira variable AU Geminorum.　See Abstr. 122.122.

Z Camelopardalis at standstill.
See Abstr. 122.123.

The UV spectra of two old novae (HR Del and V 603 Aql).　See Abstr. 124.103.

Graphite grains, carbon depletion and the 2200 Å feature.　See Abstr. 131.069.

Time variation of SiO maser emissions. I. Velocity structure of the Orion SiO maser envelope.
See Abstr. 131.078.

A southern survey of OH masers at 1612 MHz.
See Abstr. 131.107.

Masers.　See Abstr. 131.132.

High velocity molecular emission in Orion: a case for stellar winds.　See Abstr. 131.166.

High velocity molecular gas near Herbig-Haro objects HH 7−11.　See Abstr. 131.167.

Luminosity effects in interstellar spectral features.
See Abstr. 131.215.

Stellar mass loss and H II region morphology in Magellanic irregular galaxies.　See Abstr. 132.030.

Wolf-Rayet stars associated to giant regions of star formation.　See Abstr. 132.060.

The radial velocity of IRC + 10420.
See Abstr. 133.003.

On the radio properties of V645 Cygni.
See Abstr. 133.004.

Ionization effects in stellar winds of massive X-ray binaries.　See Abstr. 142.043.

On stellar wind accretion in widely separated X-ray binaries, and the nature of 4U0115+63.
See Abstr. 142.049.

High-dispersion spectroscopy of the most luminous F- and G-type supergiants in the Large Magellanic Cloud and the Milky Way.　See Abstr. 159.014.

Erratum

112.901　Erratum: 'The structure of OH masers around late-type stars' [Astrophys. J., Vol. 242, 1088 - 1101 (1980)].　P. F. Bowers, M. J. Reid, K. J. Johnston, J. H. Spencer, J. M. Moran.
Astrophys. J., Vol. 248, 880 (1981). − See Abstr. 28.112.054.

113 Photometric Properties

113.001 Infrared observations of southern bright stars.
D. Engels, W. A. Sherwood, W. Wamsteker,
G. V. Schultz.
Astron. Astrophys., Suppl. Ser., Vol. 45, 5 - 9 (1981).

Infrared magnitudes in the standard bands J, H, K, L, and M for 61 southern bright stars and 26 solar type stars are presented. K-magnitudes range between -3^m and $+7^m$. The magnitude scale is defined by Johnson's JKL-system. The bright stars provide a standard star system with an accuracy of $0^m_{.}03$ magnitudes in all bands.

113.002 Infrared photometry of southern Wolf-Rayet stars.
P. M. Williams, E. Antonopoulou.
Mon. Not. R. Astron. Soc., Vol. 196, 915 - 919 (1981).

Near infrared (JHK) photometry is presented for twelve WC5-7 and WN7-8 stars. From discussion of their reddening and infrared colours, it is suggested that the intrinsic $(b-v)_0$ of WC5-7 stars is about 0.15 mag bluer than previously believed. An intermediate sequence WN6-C7 star has colours similar to those of the WN stars.

113.003 Ultraviolet intrinsic colours of early type stars.
F. Llorente de Andrés, C. Morales,
J. A. Ruiz del Arbol, J. Pérez Mollá.
Astron. Astrophys., Vol. 100, 138 - 142 (1981).

Ultraviolet intrinsic colours of early type stars observed by the S2/68 Sky Survey Telescope in the TD-1A satellite are presented. These ultraviolet intrinsic colours are computed in the spectral region from 1380 to 2540 Å in wavelength intervals of 20 Å and in the photometric band of about 330 Å width centered at 2740 Å. A comparison shows that giants and dwarfs exhibit the same flux distribution, while supergiants are fainter with regard to those. Further the authors compare this set of intrinsic colours to that computed from the grid of stellar model atmospheres.

113.004 Spectrophotometry of V1016 Cygni.
A. P. Ipatov, B. F. Yudin.
Pis'ma Astron. Zh., Tom 7, 432 - 436 (1981). In Russian.
English translation in Soviet Astron. Lett., Vol. 7.

Spectrophotometry of V1016 Cyg in the range 3300 - 7500 Å was made simultaneously with $JHKN$ photometry on September 10 and 11, 1980. There are some changes in the spectrum of V1016 Cyg as compared with the middle seventies. From analysis of the observational data V1016 Cyg is to be classified as a symbiotic star.

113.005 Some notes on transformations between photometric systems in the visual, red and near-infrared regions.
A. W. J. Cousins.
Mon. Notes Astron. Soc. South. Africa, Vol. 39, 80 - 81 (1980).

Linear, and sometimes two-line, transformations are often acceptable for VRI photometry but they are not rigorous.

113.006 Colour equations and mean wavelengths.
A. W. J. Cousins.
Mon. Notes Astron. Soc. South. Africa, Vol. 39, 93 - 96 (1980).

Approximate relations between colour equations and mean reciprocal wavelengths are given for bands in the red and near infrared.

113.007 On the variability of Vega. J. D. Fernie.
Publ. Astron. Soc. Pacific, Vol. 93, 333 - 337 (1981).

Differential photometric observations of Vega with respect to ϵ^1 Lyr have been made on 14 nights spanning almost four months in 1980. All but three of these show constancy to within 0.006 mag. On two nights Vega was brighter than average by about 0.015 mag, and on one night by

0.041 ± 0.005 mag. No plausible explanation other than a real brightening of Vega can be found for these observations, although they do not seem to be pulsational in character. Simultaneous observations between ϵ^1 Lyr and ϵ^2 Lyr taken on all these nights, including the three unusual ones, show constancy to within 0.003 mag.

113.008 Infrared variability and spectrum of SS 433.
T. A. Clark, E. F. Milone.
Publ. Astron. Soc. Pacific, Vol. 93, 338 - 343 (1981) = Univ. Calgary Rothney Astrophys. Obs. Publ. No. 13.

Infrared monitoring of SS 433 from 1979 April 13 UT to April 17 UT inclusive at J, H, K, and L wavelengths demonstrated the variability of this source over time scales of days with tentative evidence of much shorter period fluctuations. These limited data show a substantially different pattern of variation to that expected from the inferred light curve of Giles et al. (1980), casting doubt upon the validity of this light curve as a representation of infrared variations, at least over the above time period. Color-index changes appear to be related to J intensity, the source becoming redder as the intensity decreases. The spectrum when corrected for interstellar reddening fits reasonably well to a blackbody curve, with a significant excess at L (3.5 μm) possibly indicating the existence of thermal emission from circumstellar material at a temperature below 1000 K.

113.009 Photometric UBV period study of eight Ap stars.
E. W. Burke, Jr., T. H. Barr.
Publ. Astron. Soc. Pacific, Vol. 93, 344 - 350 (1981).

Eight Ap stars for which periods had been determined by J. E. Winzer were the subject of this independent period determination. All were stars which he found to be short-period variables and which showed the typical small amplitude variation associated with short periods. For six of these, his period was confirmed. For one, HR 9017, a different and better period was determined and for the remaining star, HR 4430, the authors did not observe periodic variation.

113.010 A search for light variations in F- and G-type supergiants II. J. R. Percy, D. L. Welch.
Publ. Astron. Soc. Pacific, Vol. 93, 367 - 375 (1981).

Fifteen F- and G-type supergiants were surveyed for photometric variability. Eleven were found to be constant, including five bright stars (α Per, μ Per, β Cam, ϵ Gem, ϵ Leo) previously suspected of variability. The other four (HD 161796, 89 Her, HR 8752, and ρ Cas) were already known to be variable. The present observations confirm the variability, which is characterized by irregularity, small amplitude ($0^m_{.}1$ to $0^m_{.}3$), and long period (40 to 400 days).

113.011 Photometric scans of field horizontal-branch stars.
A. G. D. Philip, D. S. Hayes.
Bull. American Astron. Soc., Vol. 13, 515 (1981). – Abstract.

113.012 M and C stars in the central region of Omega Cen.
L. G. Palmer, R. F. Wing.
Bull. American Astron. Soc., Vol. 13, 546 (1981). – Abstract.

113.013 Short-period photometric variability of the Be star 28 Cygni.
G. G. Spear, J. Mills, Jr., S. A. Snedden.
Publ. Astron. Soc. Pacific, Vol. 93, 460 - 463 (1981).

Observations in the UBV system are reported for the Be star 28 Cyg. The short-period variability initially reported in 1976 has apparently increased in amplitude and in July 1978 a $0^d_{.}7$ periodicity was detected during four days of observations. It is speculated that this quasi-periodic behavior may be related to the axial rotation of the star.

113.014   UBV photographic standards in the region of the nebula IC 5070.   M. Tsvetkov, K. Tsvetkova.
Astrofiz. issled., NRB, Vol. 3, 81 - 88 (1981). In Russian.
Abstr. in Ref. zh., 51. Astron., 8.51.563 (1981).

113.015   Four-color and H beta photometry of field horizontal-branch stars.
A. G. D. Philip, P. Dubois.
News Lett. Astron. Soc. N. Y., Vol. 1, No. 10, p. 41 - 42 (1981). — Abstract.

113.016   Up-to-date *UBVRI* values for the E-region standard stars.   N. Vogt, H. S. Geisse, S. Rojas.
Astron. Astrophys.,Suppl. Ser., Vol. 46, 7 - 11 (1981).
A table with revised *UBVRI* standard values of the E-region stars of Cousins and Stoy (1962) is given.

113.017   The colours, magnitudes and parallaxes of the nearby stars.   M. Grenon, F. Rufener.
Astron. Astrophys.,Suppl. Ser., Vol. 46, 25 - 40 (1981).
This catalogue presents the apparent visual magnitudes and colours measured in the Geneva system of 1 077 nearby stars belonging to the 1969 edition of Gliese's catalogue. The program limiting magnitude is m_v 11.0 or m_{pg} 12.5. All parallaxes published since the Jenkins catalogue (1963) have been compiled and the stars are given with their trigonometric absolute magnitude and an up-to-date spectral classification.

113.018   Simultaneous spectroscopic and photometric observations of 2A0311-227.
J. van Paradijs, F. Verbunt, E. P. J. van den Heuvel, T. J. van der Linden, J. Brand, F. van Leeuwen.
Astron. Astrophys.,Suppl. Ser., Vol. 46, 89 - 92 (1981).
Results of photoelectric five-colour photometry (Walraven system) and simultaneous spectroscopic observations (ESO 3.6 m/IDS) of the AM Her-type system 2A0311-227 are presented.

113.019   Note on photometric properties of red dwarfs in the Cousins *VRI* system.
P. S. Thé, C. Karman, G. Alcaino.
Astron. Astrophys.,Suppl. Ser., Vol. 46, 105 - 110 (1981).
Colour-colour and colour-magnitude relations of red dwarfs are studied in Cousins's photometric system. The effect of TiO blanketing on the shape of the *V-R, R-I* two-colour diagram has been studied qualitatively. For statistical studies based on space densities of M dwarfs the colour-magnitude relations M_V, *V-I*, M_R, *V-I* or M_I, *V-I* are the most suitable. The dispersion in absolute magnitude which should be employed in these statistical studies is approximately ±0^m 4.

113.020   A photoelectric investigation of light variability in Ap stars.   H. Hensberge, H. M. Maitzen, G. Deridder, M. Gerbaldi, F. Delmas, P. Renson, C. Doom, W. W. Weiss, N. Morguleff.
Astron. Astrophys.,Suppl. Ser., Vol. 46, 151 - 170 (1981).
The light variability of a number of peculiar A stars has been studied mainly in Strömgren colours u and v, and in peculiarity index Δa. Periods of light variability are proposed for HD 5601 (1^d11), HD 19712 (2^d19), HD 30849 (15^d86), HD 38823 (8^d64), HD 53116 (11^d98 or 18^d10), HD 56022 (0^d92) and HD 81009 (33^d97).

113.021   Further observations of a possible star-nebula system.   W. I. Hartkopf, K. M. Yoss.
Astrophys. Lett., Vol. 21, 111 - 114 (1981).
Additional photometric and spectrographic observations are presented for the high galactic latitude K giant HD 20722, which is suspected to be either a newly-formed planetary

nebula or a recently dismembered binary star. The present observations find the K giant to be a normal star, consistent with a binary model but also with the possibility that the observed nebulosity is due to a plate flaw or a background galaxy. The authors also place some additional constraints on the surface brightness of the nebula.

113.022   Ultraviolet photometry of stellar populations in galaxies.   J. M. Deharveng.
Ann. Physique, Vol. 6, (see 012.026), p. 9 - 16 (1981). In French. — Abstr. in Phys. Abstr., Vol. 84, Abstr. 94423 (1981).

113.023   Energetics of stellar populations.
A. Renzini.
Ann. Physique, Vol. 6, (see 012.026), p. 87 - 102 (1981). In French. — Abstr. in Phys. Abstr., Vol. 84, Abstr. 94424 (1981).

113.024   Discovery of microvariability in normal stars. Analysis of six main sequence stars.
A. G. Totochava, B. E. Zhiljaev (*Zhilyaev*).
Astron. Nachr., Band 302, 219 - 222 (1981).
To study the microvariability of the normal main se-quence stars, six stars of spectral type A2V — K1V have been observed by means of the method of photon counting using the 48-cm telescope of the Abastumani Astrophysical Observatory. Microvariability with an amplitude of about 0.001 - 0.003 mag. seems to be specific for the majority of normal stars.

113.025   Results of a photometric investigation of SS 433 in 1979 - 1980.   S. A. Gladyshev.
Pis'ma Astron. Zh., Tom 7, 594 - 599 (1981). In Russian.
English translation in Soviet Astron. Lett., Vol. 7.
Results of photometric observations of SS 433 (= V1343 Aql) carried out in 1979 - 1980 are considered. Basic regularities in the light variability of SS 433 are described. Interpretation of these effects is based on the model of SS 433 as an eclipsing binary system with a precessing accretion disc around the neutron component.

113.026   A search for variability of the object SS 433 by the method of normalized factorial moments.
V. S. Lebedev, A. A. Pimonov.
Pis'ma Astron. Zh., Tom 7, 600 - 604 (1981). In Russian.
English translation in Soviet Astron. Lett., Vol. 7.
From independent observations with the 6 m telescope the relative power of the variable component in the optical radiation of the object SS 433 has been limited in the period range from 10^{-4} to 1 sec. The method of analysis is presented.

113.027   Thirteen-color photometry of subdwarf stars. IV. HD 25329 and HD 122563.   W. J. Schuster.
Rev. Mexicana Astron. Astrofis., Vol. 5, 69 - 78 (1981).
The unusual ultraviolet excesses of HD 25329 and HD 122563 are explained by their nitrogen abundances and by Rayleigh scattering. For 45 - 63 $\gtrsim$ +1.0 eight-color photome-try can be used to identify (metal poor) dwarf stars with nitrogen overabundances. The difference curve of the extremely metal-poor giant HD 122563 shows very distinctly the increased importance of Rayleigh scattering with respect to the H$^-$ processes as a source of continuum opacity.

113.028   Ultraviolet and optical photometry of the chemically peculiar stars of the upper main sequence.
S. J. Adelman.
Upper main sequence chemically peculiar stars, (see 012.033), p. 13 - 38 (1981).

113.029   Photometry of peculiar stars around λ4200.
M. Gerbaldi, N. Morguleff.

Upper main sequence chemically peculiar stars, (see 012.033), p. 39 - 50 (1981).

Photometric indices around the region λ4200 are defined to study the intense line blocking present in the spectra of Bp and Ap stars. Combining the authors' observations with those of Adelman, it is shown that photometric indices alone cannot measure any broad continuum features in the spectra of Ap stars. The indices ΔC(J) permit to show that stars belonging to the group Si, Sr - Cr, and Sr - Cr - Eu exhibit an increase of their line blocking with the increase of the rotational period.

113.030 CP-stars in open clusters.   H. M. Maitzen.
Upper main sequence chemically peculiar stars, (see 012.033), p. 51 - 53 (1981).

A report is given on a project to investigate the occurrence of CP-stars in open clusters using a photoelectric technique (Maitzen, 1976) instead of spectroscopic classification.

113.031 Ap stars detected in open clusters by the Geneva photometry.   P. North, N. Cramer.
Upper main sequence chemically peculiar stars, (see 012.033), p. 55 - 59 (1981).

The capability of the Geneva photometric system to separate the magnetic Ap stars was used for their detection in thirty-six open clusters. The frequency of the Ap stars is found to be in agreement with previous works, while the strength of their photometric peculiarity shows no trend as a function of age over the range considered.

113.032 Photometric classification of B and Ap stars with an application to 3600 stars.
N. Cramer, A. Maeder.
Upper main sequence chemically peculiar stars, (see 012.033), p. 61 - 65 (1981).

The authors present here the properties and first results of a method of photometric analysis suitable for the hotter Ap stars, defined in the Geneva system. This reddening-free analysis of the various classes of peculiarity is carried out in the general framework, given by the X, Y and Z parameters, in which calibrations in T_{eff}, gravity, absolute magnitude, age and mass have been established. The parameter Z, in particular, is shown to be related to the intensity of the mean surface magnetic field of Ap stars within the range of 0 to 5 kG. A calibration is established and applied to Rufener's (1981) photometric data.

113.033 The peculiar Rosetta stone: Beta Coronae Borealis.
S. J. Adelman, B. A. Boytim, D. M. Pyper, S. N. Shore.
Upper main sequence chemically peculiar stars, (see 012.033), p. 109 - 114 (1981).

The authors discuss the photometric and spectroscopic variability of β CrB with respect to the oblique rotator model. They present new spectrophotometric results and a possible method for improving the comparison of model atmospheres and spectrophotometry. This method is also immediately applicable to statistical abundance determinations.

113.034 Photométrie UV des étoiles Am et Delta Delphini.
C. Van't Veer, R. Faraggiana, C. Burkhart.
Upper main sequence chemically peculiar stars, (see 012.033), p. 465 - 473 (1981).

The photometric study of A and Am stars in the UV range (Van't Veer et al 1980, 1981) is extended to δ Del stars. The observed colours are compared with those computed from Kurucz's models. The flux distribution of the evolved Am star HD 25425, which shows strong UV peculiarities, is derived from IUE spectra.

113.035 Search for variability among the Am stars measured in the Geneva system.   B. Hauck, D. Lovy.
Upper main sequence chemically peculiar stars, (see 012.033), p. 485 - 489 (1981).

Nearly 400 Am stars have been measured in the Geneva photometric system. Using Deeming and Renson's methods the authors have investigated whether the stars with the most measurements present light variations. They have also studied the distribution of the standard deviation for the V filter for various samples (AV, Am, Ap, δ Scuti stars). In both cases, the behaviour of Am stars is similar to that of the normal AV stars.

113.036 Ultraviolet colours of early-type stars.   R. Barbier.
Astron. Astrophys., Vol. 102, 307 - 312 (1981).

The properties of an ultraviolet system of reddening-free colour indices based on the TD1A S2/68 fluxes are investigated for stars of spectral types O, B and early A. Theoretical indices are computed from Kurucz's (1979) blanketed models. A set of stars of known spectral types is used to determine the empirical relation between spectral classification and mean ultraviolet colour indices. A quantitative comparison between observed and computed main-sequence indices is also performed.

113.037 Infrared studies of the two stellar populations in 30 Doradus.   P. J. McGregor, A. R. Hyland.
Astrophys. J., Vol. 250, 116 - 134 (1981).

Broad band J, H, and K magnitudes are presented for 33 blue and red supergiant stars within the central 12' of the 30 Doradus complex. Band strength indices for the 2.3 μm CO bands of 10 of the red stars are also given. These more detailed data confirm an earlier conclusion that at least two bursts of star formation have occurred in the 30 Doradus region in the last ~5 × 10^7 yr, with the most recent event occurring ≲ 10^6 yr ago. Mass loss rates derived for the 30 Dor Wolf-Rayet stars are comparable with the highest rates found by other workers for galactic Wolf-Rayet stars. These high rates of mass loss place severe limits on the age of the blue population. The broad band colors of the late-type supergiants differ from the colors of galactic stars of the same luminosity. The CN molecular opacity is shown to be important in determining the infrared colors of these luminous supergiants and to be sensitive to metal abundance. The observed differences can therefore be plausibly interpreted as evidence for metal deficiency in the 30 Dor supergiants.

113.038 Relations between some photometric temperature parameters.   G. Meylan, B. Hauck.
Astron. Astrophys., Suppl. Ser., Vol. 46, 281 - 303 (1981).

The relations between temperature parameters of various photometric systems are given. The photometries concerned are the following: $UBVB_1B_2V_1G$ (Geneva), UBV (Johnson), $uvby\beta$ (Strömgren), $UPXYZVS$ (Vilnius), $UBVRIJKLMNH$ (Johnson), $UVBGRI$ (Stebbins and Whitford). The effects due to the different luminosity classes and to blanketing (Am, Ap and population II dwarf stars) are taken into account.

113.039 Comparison between the observed intrinsic colour $(b-y)_0$ and the calculated theoretical index $(b-y)$ for A-type stars.   C. Burkhart, C. Van't Veer, R. Faraggiana.
Astron. Astrophys., Vol. 103, 145 - 146 (1981).

The observed intrinsic colour $(b-y)_0$ (Crawford, 1979) corresponds to the computed theoretical index $(b-y)$ (Kurucz, 1979) in the region of A-type stars calibrated by Crawford; this correlation is not valid in the temperature range 7000–8500 K when the metal abundances are greater than or equal to 3 times the solar abundances. Therefore for Am stars, the m_1 index cannot be quantitatively expressed in metal overabundances.

113.040 Photometry of metal weak stars.   J. F. Dean.
Mon. Notes Astron. Soc. South. Africa, Vol. 40, 14 - 20 (1981).

$UBVRI$ photometry was obtained for 89 metal weak

stars. The DDO photometry obtained for 49 of these stars indicates that most of the later type stars are giants. The colour differences, $(V-R)-(R-I)$, when plotted versus $V-I$, fall below Cousins' sequences for normal dwarfs and giants possibly due to a difference in metal blanketing.

113.041 Pulsating Be stars?   J. R. Percy.
J. R. Astron. Soc. Canada, Vol. 75, 251 (1981).
Abstract.

113.042 Les plus brillantes étoiles de M81.
R. Racine, M. M. Shara.
J. R. Astron. Soc. Canada, Vol. 75, 251 - 252 (1981). --Abstract.

113.043 Absolute spectrophotometry of Wolf-Rayet stars.
P. Massey.
Publ. Astron. Soc. Pacific, Vol. 93, 549 (1981). – Abstract.

113.044 High-speed photometry of, and speculations on, the central star of M 1 - 2.   A. D. Grauer, H. E. Bond.
Publ. Astron. Soc. Pacific, Vol. 93, 630 - 632 (1981) =
Louisiana State Univ. Obs. Contrib., No. 166.

Minkowski 1 - 2 is a planetary nebula whose central star spectroscopically resembles a G-type supergiant. Although the central star has recently been suggested to be an eclipsing binary with a period of four hours, no eclipses or other variability during nearly ten hours of high-speed photometric monitoring on two consecutive nights have been found. If M 1 - 2 contains a binary system at all, this photometry indicates that its orbital period is very unlikely to be less than several days. Spectroscopic evidence suggests a period in excess of half a year. The authors tentatively suggest that the central star is a rather wide binary containing the G star and a hot companion with the G-type "supergiant" being a less luminous giant with heavy-element enhancements similar to those seen in marginal barium stars.

113.045 Narrow-band photometry of bright Be stars.
J. J. Clariá, L. A. Escosteguy.
Publ. Astron. Soc. Pacific, Vol. 93, 636 - 640 (1981).

Narrow-band filters centered at the Hα and Hβ Balmer lines are used to separate emission and nonemission stars from a sample of early-type stars. By comparing the present observations with earlier data, significant changes in the emission-line strengths of five Be stars are found.

113.046 Is HR 2947 a variable?
N. K. Rao, N. M. Ashok, P. V. Kulkarni.
Bull. Astron. Soc. India, Vol. 9, 144 - 150 (1981).

113.047 Spectrophotometry of bright southern stars.
H. J. Trodahl, D. J. Sullivan, R. G. Gibb.
Mon. Not. R. Astron. Soc., Vol. 197, 941 - 948 (1981).

The authors have completed careful intermediate resolution (10–18 nm) multicolour photometric measurements on 28 bright stars in the southern winter sky at wavelengths from 339 to 700 nm. The spectra are tied to the calibration of α Lyr through secondary standards. Relative spectra are accurate to better than 0.01 mag over the entire range and the authors estimate that the absolute spectra are accurate to 0.02 mag.

113.048 Optical continuum from SS 433 – star or accretion disk?   P. Murdin.
Vistas Astron., Vol. 25, (see 012.040), 165 - 168 (1981).

The continuum (0.2 - 2 μ) from SS 433 appears to be black-body in form. A luminous reddened star fits the data. Constraints on accretion disk spectra are discussed.

113.049 Field population II blue stragglers.
B. W. Carney, R. C. Peterson.
Astrophys. J., Vol. 251, 190 - 200 (1981).

The authors have undertaken a program of high-dispersion spectroscopy of blue stragglers as a means of addressing the question of rotation, as well as radial velocity variability and abundance anomalies that might hint at extensive mass transfer. Photometry and high-resolution spectra are presented and discussed of metal-poor stars hotter than the turn-off of globular clusters.

113.050 Photoelectric observation of unusual suspected variable stars in M4.
B.-a. Yao, J.-s. Yin, Z.-h. Guo.
Acta Astrophys. Sinica, Vol. 1, 311 - 316 (1981). In Chinese.

113.051 Observations of late-type dwarfs of the young and old disk population using DDO photometry.
A. R. Upgren, D. W. Dawson, P. K. Lü.
Astrophys. J., Vol. 251, 557 - 563 (1981).

Intermediate-band DDO photometry is presented for 52 dwarf stars which cover the spectral range from K2 to M2. The stars are taken from a sample of 145 dwarfs for which parallaxes and proper motions have recently been determined at the Van Vleck Observatory. These stars all have radial velocity determinations and have been divided into young and old disk populations on the basis of their space motions. All of the stars observed here also have measures of the emission intensities of the H and K lines of calcium in their spectra. Thirty of the stars appear to be young stars on the bases of both their motions and their calcium emission intensities; the rest are members of the old disk population. The DDO photometry of these stars defines the locus of K and early M dwarfs more clearly than had been done previously. It also shows a small separation between the young and old stars which is interpreted as a small difference in surface gravity.

113.052 Infrared photometry of the X-ray stars in Cygnus OB2 (VI Cygni): preliminary results.
P. Persi, M. Ferrari-Toniolo, G. L. Grasdalen.
Space Sci. Rev., Vol. 30, (see 012.044), 207 - 212 (1981).

The Cygnus OB2 association (also known as VI Cygni) is very young and includes several giants and supergiants earlier than O7, one of which is an O3If. The authors report the preliminary results of infrared photometry from 2.3 up to 10 and 20 microns of the most luminous stars in VI Cygni. The observations were carried out with the Ge-multifilter system during an observational campaign on September 1980 at the 2.3 m Wyoming Infrared Telescope.

113.053 The surface brightness parameter in relation to the VRI$_c$ system.   A. W. J. Cousins.
Mon. Notes Astron. Soc. South. Africa, Vol. 40, 37 - 40 (1981).

The surface brightness parameter, F_v, can be obtained if VRI$_c$ colours are transformed to V–R by means of the equation V–R = 0.587 (V–R$_c$) + 0.413 (V–I$_c$) + 0.03. A table gives the relation between F_v and V–I$_c$.

113.054 The new calibration of the Vilnius photometric system in spectral types and absolute magnitudes. I.
V. Straižys, E. Jodinskienė.
Bull. Vilnius Astron. Obs., Nr. 56, p. 3 - 24 (1981). In Russian.

The reddening-free diagrams Q, Q of the seven-color Vilnius photometric system are calibrated in spectral types and absolute magnitudes. Five diagrams Q, Q make possible two-dimensional classification of stars of all spectral types and luminosities of solar chemical composition. The intrinsic color indices, the ratios E_{B-V}/E_{Y-V} and the ratios $R = A_V/E_{B-V} = R_0 + qE_{B-V}$ for different spectral types and absolute magnitudes are also tabulated.

113.055 Estimation of optimality of the Vilnius photometric system bandpasses X, Y, and Z for stellar classification.   V. Straižys, G. Kurilienė.

Bull. Vilnius Astron. Obs., Nr. 56, p. 25 - 37 (1981). In Russian.

The positions and half-widths of the response curves X, Y, and Z of the Vilnius photometric system are investigated with respect to their optimality to determine spectral classes and absolute magnitudes. The investigation is based on two-color and reddening-free Q, Q diagrams calculated theoretically.

113.056 **Wide-band photometric systems in the red and near infrared.**
V. Straižys, G. Kurilienė, Z. Sviderskienė.
Bull. Vilnius Astron. Obs., Nr. 57, p. 9 - 34 (1981). In Russian.

The study deals with the Kron, Johnson, Cousins-Bessell and Washington systems. The authors compare observed and calculated indices and estimate the reliability of the response curves. They present the color excess ratios E_{R-I}/E_{B-V} of different systems as a function of spectral type for different luminosities. Some considerations on the selection of an optimum international red-infrared system are presented. The Cousins-Bessell, Washington and Palomar systems are among the candidates to become international.

113.057 **Photoelectric observations of Pleione (BU Tau) from September 1977 to March 1980.**
A. S. Sharov, V. M. Lyutyj.
Astron. Tsirk., No. 1119, p. 1 - 2 (1980). In Russian.

113.058 **Observations of the nova-like variable FY Per.**
S. Yu. Shugarov.
Astron. Tsirk., No. 1119, p. 3 - 5 (1980). In Russian.

113.059 **Photoelectric measurements of three stars in M67.**
A. V. Mironov, E. N. Pastukhova.
Astron. Tsirk., No. 1119, p. 5 - 7 (1980). In Russian.

113.060 **On the problem of Vega's variability.**
V. S. Kozyreva, V. G. Moshkalev, Kh. F. Khaliullin.
Astron. Zh., Tom 58, 1241 - 1248 (1981). In Russian.
English translation in Soviet Astron., Vol. 25, No. 6.

Photoelectric $WBVR$ observations of Vega made during three months in 1980 and aimed to search for its variability suspected by Guthnick did not show light variations in significant excess of the measurement errors ($\sigma = 0^m006$). However, the analysis of these data shows the possibility of (quasi) periodic microvariability of Vega with amplitude $\cong 0^m02$ and period about one hour.

113.061 **Photometric observations of symbiotic stars in the $UBVRJHKLMN$ system. 2. Z Andromedae.**
O. G. Taranova, B. F. Yudin.
Astron. Zh., Tom 58, 1249 - 1256 (1981). In Russian.
English translation in Soviet Astron., Vol. 25, No. 6.

Photometry of Z And in the $UBVRJHKLMN$ system in 1978 - 1980 when the star was in a quiet state revealed a periodic variation in the ultraviolet. The period and the phase are the same as in the active state. The infrared flux showed small fluctuations uncorrelated with the ultraviolet flux. A variable infrared excess near 10 μm is found which is also uncorrelated with the U flux. The authors estimated the physical parameters of three components of Z And: a hot star, a cool star and ionized gas.

113.062 **The discovery of 6.8 minute oscillations in α Cir.**
D. W. Kurtz, M. S. Cropper.
Inf. Bull. Variable Stars, No. 1987, 3 pp. (1981).

113.063 **UBV observations of 88 Her (V 744 Her).**
L. Baldinelli, A. Ferri, S. Ghedini.
Inf. Bull. Variable Stars, No. 1993, 2 pp. (1981).

113.064 **Nouvelle recherche de périodes d'étoiles Ap observées à l'ESO-VI.**
J. Manfroid, P. Renson.
Inf. Bull. Variable Stars, No. 2004, 3 pp. (1981).

113.065 **Variations photometriques et periode de HR 2971.**
J. Manfroid, P. Renson.
Inf. Bull. Variable Stars, No. 2032, 2 pp. (1981) – see Abstr. 113.064.

113.066 **BD-8°4232, a new Delta Scuti star?**
D. Kilkenny.
Inf. Bull. Variable Stars, No. 2034, 2 pp. (1981).

113.067 **Photometric behaviour of T Tauri stars in the T-associations T1 and T3 Tau. I. Variations at various wavelengths.**   U. A. Nurmanova.
Astron. Tsirk., No. 1127, p. 1 - 2 (1980). In Russian.

113.068 **Photometric behaviour of T Tauri stars in the T-associations T1 and T3 Tau. II. U–B, B–V diagram for excess radiation.**   U. A. Nurmanova.
Astron. Tsirk., No. 1127, p. 2 - 5 (1980). In Russian.

113.069 **Photoelectric BVR observations of SS 433 from June - October 1980.**   S. A. Gladyshev.
Astron. Tsirk., No. 1138, p. 1 - 6 (1980). In Russian.

113.070 **Further observations of the carbon star RW LMi (CIT 6) in the years 1977/78.**
A. Alksnis, I. Eglītis.
Investigations of the sun and red stars. 11, (see 003.021), p. 18 - 26 (1980). In Russian.

U-, B-, V-, R(0.63)- and I(0.81)-magnitudes obtained photographically in 1977/78 are presented for the infrared carbon variable star RW LMi = CIT 6.

113.071 **Slopes of the reddening lines and ratios of total to selective absorption for carbon stars in the Vilnius and UBV systems.**   U. Dzērvitis, O. Paupers.
Investigations of the sun and red stars. 13, (see 003.023), p. 12 - 21 (1981). In Russian.

Published spectrophotometric curves of the spectral energy distribution of carbon stars have been used to calculate slopes of the reddening lines and ratios of total to selective absorption in the Vilnius and UBV systems. Calculated characteristics are compared with similar ones for red giants with normal chemical composition.

113.072 **Results of electrophotometric investigations of carbon and zirconium stars in the Vilnius photometric system.**   U. Dzērvitis, O. Paupers.
Investigations of the sun and red stars. 13, (see 003.023), p. 22 - 47 (1981). In Russian.

Results of electrophotometric observations of carbon and zirconium stars in the Vilnius photometric system are given. It has been found that the normal colour indices of carbon stars are much more scattered than those of K-M stars – they actually fill a large region on the two-colour diagrams.

113.073 **Normal colour indices, slopes of the reddening lines and ratios of total to selective absorption for a black body in the Vilnius photometric system.**   U. Dzērvitis.
Investigations of the sun and red stars. 13, (see 003.023), p. 48 - 55 (1981). In Russian.

Normal colour indices, slopes of the reddening lines and ratios of total to selective absorption for the black body in the Vilnius photometric system have been calculated. It has been stated that the reddening lines form an angle with normal-colour lines only in those three-colour diagrams where one of the colours is placed to the blue side from the break of the interstellar absorption curve.

113.074 Carbon stars – possible members of open clusters. IV. Four carbon stars in the surroundings of the cluster NGC 1528 = C011 + 511.
Z. Alksne, A. Alksnis.
Investigations of the sun and red stars, 14, (see 003.024), p. p. 28 - 43 (1981). In Russian.

Four carbon stars were photographically monitored for light variations in the R(0.63), V and B passbands at the time interval JD 2443400-2444300. The photometric results are given in a table and individual observations are discussed.

113.075 Photoelectric photometry of cyanogen and barium stars in the Vilnius system.
U. Dzĕrvītis, L. Duncāns, O. Paupers.
Investigations of the sun and red stars, 14, (see 003.024), p. 59 - 66 (1981). In Russian.

Results of photoelectric observations of 34 cyanogen and 7 barium stars in the seven-colour Vilnius photometric system are presented. Methods of observation and reduction of colour-indices in a standard system are given.

113.076 Study of the errors in the light curve of artificial celestial bodies due to stars.   E. B. Vovchik.
Tsirk. Astron. Obs., L'vov, No. 54, p. 44 - 48 (1979). In Russian.

113.077 Near-infrared photometry of carbon stars.
K. Noguchi, K. Kawara, Y. Kobayashi, H. Okuda, S. Sato, M. Oishi.
Publ. Astron. Soc. Japan, Vol. 33, 373 - 397 (1981).

An extensive survey of near-infrared photometric spectra and the depth of the 3-μm absorption was carried out for 138 carbon stars. Infrared color indices of carbon stars correlate well with one another. The depth of the 3-μm absorption also shows good correlations with color indices, suggesting a temperature dependence of the 3-μm absorption. Dilution of the 3-μm absorption by thermal emission due to circumstellar dust is found for extremely red carbon stars.

113.078 Photoelectric measures of AAVSO comparison star sequences – II.   R. H. Stanton.
J. American Assoc. Variable Star Obs., Vol. 10, 1 - 8 (1981).

Three related comparisons between photoelectric and visual magnitudes are made: (1) photoelectric V vs m_v for 260 stars listed in the Revised Harvard Photometry, (2) photo-electric magnitude (measured by author) vs chart, magnitude for 183 stars on preliminary AAVSO charts, and (3) photoelectric magnitude (from published literature) vs comparison star magnitude for 124 stars on standard AAVSO charts.

113.079 Response functions in the red and near infrared.
A. W. J. Cousins.
South African Astron. Obs. Circ., No. 6, p. 4 - 9 (1981).

This paper uses magnitudes obtained by numerical integration, with response functions for the VRI$_c$ system and flux distributions for MK types, to obtain colour indices and compare them with those obtained observationally.

113.080 DDO observations of southern stars.   J. F. Dean.
South African Astron. Obs. Circ., No. 6, p. 10 - 27 (1981).

113.081 VI photometry of selected SAO stars.
E. R. Craine, W. W. G. Scharlach.
Prepr. Steward Obs., No. 349, 11 pp. (1981).

Johnson V and I band photoelectric photometry has been obtained for 158 SAO stars in a continuing program in support of the Near Infrared Photographic Sky Survey. These data are utilized in the calibration of the authors survey photographs and are presented here to assist other programs reliant upon access to photometry in the photographic infra-red.

113.082 UBV-Photometrie der jungen Be/Ae-Sterne HDE 250550, BD + 65° 1637 und BD + 46° 3471.
S. Rößiger.
Mitt. Veränderl. Sterne (MVS), Band 9, 1 - 3 (1981).

Photoelectric UBV observations of three Herbig Be/Ae stars associated with nebulosity over a period of some years are presented. Each star has only a very small variation in brightness. Probably the objects are situated at the end of their pre-main-sequence state.

113.083 Star colours: hit, not myth.   P. Murdin.
Q. J. R. Astron. Soc., Vol. 22, 353 - 360 (1981).
Naked-eye observations of bright star colours see truthfully on average, and can measure colour index ±0.24 mag and spectral type ±one letter class.

113.084 Mean intrinsic U−B and B−V indices for population II stars as functions of temperature.
V. D. Malyuto, P. A. Traat.
Astron. Tsirk., No. 1158, p. 2 - 3 (1981). In Russian.

113.085 Possible stars of the red horizontal branch of the galactic field.
V. Straižys, A. Bartkevičius, J. Sperauskas
Astron. Tsirk., No. 1161, p. 1 - 3 (1981). In Russian.

Thirteen stars are suspected as belonging to the red horizontal branch on the basis of their seven-color photometry in the Vilnius photometric system. Their spectral types are within F5 - G5 and luminosities are between III and I classes. Most of them are metal-deficient.

Third catalogue of stars measured in the Geneva Observatory Photometric System.   See Abstr. 002.008.

Documentation for the machine-readable version of the Thirteen-Color Photometry of 1380 Bright Stars. See Abstr. 002.021.

Photometric data for the nearby stars. See Abstr. 002.026.

A photometric catalogue of stars in the direction of the bright cloud B in Sagittarius.   See Abstr. 002.045.

The early-type chemically peculiar stars in the catalogue of stellar groups.   See Abstr. 002.052.

Atlas and catalogue of stellar magnitudes and photoelectric standards.   See Abstr. 002.076.

A catalogue of homogeneous photometry of bright stars on the DDO system.   See Abstr. 002.077.

The perils of high-speed stellar photometry. See Abstr. 031.514.

Magnitude-image diameter calibration of the UKSTU J-survey films.   See Abstr. 031.541.

Photometric observations of faint objects and their processing on the M-220M computer. See Abstr. 031.622.

A quick method of determining monochromatic absorption corrected stellar magnitudes. See Abstr. 031.625.

Two-star high-speed photometry. See Abstr. 034.015.

Early-type high-velocity stars in the solar neighborhood. I. List of candidates.   See Abstr. 111.009.

Early-type high-velocity stars in the solar neighborhood. II. Photometry for 78 candidates.
See Abstr. 111.016.

A spectroscopic study of the light variable peculiar star HD 188136.   See Abstr. 114.007.

Scanner measurements of NH, CN, CH, and C_2 bands.   See Abstr. 114.051.

Photometric carbon and nitrogen abundances in G and K stars.   See Abstr. 114.052.

Simultaneous spectroscopic (UV Mg II and Al II lines) and photometric variations of ζ^1 Sco (HD 152236).
See Abstr. 114.062.

On the estimation of photometric spectral types.
See Abstr. 114.081.

Metallicity indicators for the stellar and integrated light spectra.   See Abstr. 114.099.

A spectroscopic and photometric study of the Bp star HR 3413.   See Abstr. 114.110.

Spectroscopic and photometric observations of the Be star 69 Orionis.   See Abstr. 114.129.

On some extreme metal-deficient giants.
See Abstr. 114.161.

Carbon stars with very large infrared colour indices.
See Abstr. 114.188.

On the opacity of Be star envelopes in the Balmer continuum.   See Abstr. 114.198.

The luminosity function and the colours of the solar neighbourhood.   See Abstr. 115.013.

Bolometric luminosities and infrared properties of carbon stars in the Magellanic Clouds and the Galaxy.
See Abstr. 115.014.

Infrared luminosities of M supergiants and their use as distance indicators.   See Abstr. 115.015.

Synchronous changes of polarization, brightness and radial velocities of o Andromedae.
See Abstr. 116.044.

Optical photometry of SS 433: 200 nights of data 1979 - 1980.   See Abstr. 117.086.

High resolution monitoring of rest Hα in SS 433.
See Abstr. 117.087.

On the photometric behavior of SS 433 in 1979 and 1980.   See Abstr. 117.115.

The 1976-1978 eclipse of VV Cephei.
See Abstr. 119.033.

Photometric observations of the eclipsing variable QX Carinae.   See Abstr. 119.109.

Near-infrared photometry of some Herbig emission-line stars.   See Abstr. 121.004.

Photometric classification of pulsating variables with periods between one and three days.
See Abstr. 122.058.

UBVRI photometry of FK Comae.
See Abstr. 122.130.

A discussion on new *VBLUW* observations of the X-ray binary Sk. 160 = SMC X-1.   See Abstr. 142.007.

UBV photometry of the star V1357 Cyg (Cyg X-1). The mean light curves.   See Abstr. 142.057.

The X-ray source A0538-66 in optical quiescence.
See Abstr. 142.124.

Blue stragglers in M67.   See Abstr. 153.002.

The region of NGC 2287 and Cr 121.
See Abstr. 153.003.

A photoelectric investigation of Ap-stars in open clusters.   See Abstr. 153.004.

Geneva photometric boxes. III. Distances and reddenings for 43 open clusters.   See Abstr. 153.028.

Photoelectric photometry of the open clusters IC 4665, NGC 6633 and NGC 7092 (M39) in the Vilnius photometric system.   See Abstr. 153.029.

On the giant, asymptotic and horizontal branches of globular clusters – I. Photographic photometry of M5.
See Abstr. 154.002.

Photometry of faint stars in globular clusters. IV. B-system luminosity function for the globular cluster NGC 6366.   See Abstr. 154.042.

A (B, V) photoelectric sequence of stars in Omega Centauri.   See Abstr. 154.044.

RGU photometry of a field in the Large Sagittarius Cloud (Sgr III).   See Abstr. 155.032.

Carbon stars in the Carina dwarf spheroidal galaxy.
See Abstr. 158.283.

114 Spectra, Temperatures, Chemical Composition, etc.

114.001 Radial velocities for different spectral lines of B and A supergiants in our Galaxy and in the Large Magellanic Cloud.   E. Kontizas, M. Kontizas.
Astron. Astrophys., Suppl. Ser., Vol. 45, 121 - 128 (1981).

Coudé spectra of 23 early-type supergiants (16 of them belong to our Galaxy and 7 to the Large Magellanic Cloud (LMC)) were measured to study the radial velocities of all metallic and non-metallic lines in relation to their EP + IP (high excitation + ionization potential).

114.002 Outer atmospheres of cool stars. IX. A survey of ultraviolet emission from F−K dwarfs and giants with *IUE*.   T. R. Ayres, N. C. Marstad, J. L. Linsky.
Astrophys. J., Vol. 247, 545 - 559 (1981).

The authors report preliminary results of an ultraviolet survey of cool-star emission properties with *IUE*. They present 1150–2000 Å spectra of representative F−K dwarfs and giants and construct correlation diagrams that compare chromospheric ($T \lesssim 10^4$ K) and transition-region ($T \approx 10^5$ K) emission line strengths, and broad-band coronal ($T \gtrsim 10^6$ K) soft X-ray fluxes.

114.003 Abundances in 11 field stars with large metal deficiencies.   E. M. Leep, G. Wallerstein.
Mon. Not. R. Astron. Soc., Vol. 196, 543 - 556 (1981).

Equivalent widths from high dispersion spectra in the red region are used to derive abundances of elements from oxygen to barium in 11 weak-lined stars. Values of Fe/H from 1/10 to 1/200 of the solar value are found. The ratio of oxygen-to-iron is found to vary from one to about five times the solar value. Calcium and titanium are found to be deficient in most of these metal-poor stars confirming an odd-even effect among Si, Ca, Sc and Ti that has been recognized in other samples of metal-deficient stars. The authors find that barium shows an over-deficiency in stars with [Fe/H] <-1.5 as seen in other metal-poor field stars but not in the globular clusters M15 and M92 with similarly large iron deficiencies.

114.004 The chromosphere and corona of Procyon (α CMi, F5 IV - V).
A. Brown, C. Jordan.
Mon. Not. R. Astron. Soc., Vol. 196, 757 - 779 (1981).

Observations of Procyon (α CMi), an F5 IV - V star, have been obtained with the IUE satellite at low and high spectral resolution. Emission line fluxes are used to find the emission measure distribution. The range of likely electron pressure is discussed, based in part on consideration of line opacities. Models are made for several boundary values of the electron pressure.

114.005 Far-UV wind line profile changes in the O-type star HD 175754.   L. Carrasco, R. Costero, R. Stalio.
Astron. Astrophys., Vol. 100, 183 - 185 (1981).

The authors report resonance line profile changes in far-UV spectra of HD 175754, obtained with the IUE satellite. They interpret these changes in terms of variations in the dynamics and density-ionization structure of the star's wind.

114.006 A model for V 1016 Cyg based on the ultraviolet spectrum.   H. Nussbaumer, H. Schild.
Astron. Astrophys., Vol. 101, 118 - 131 (1981).

The authors interpret the ultraviolet spectrum of the symbiotic star V 1016 Cyg with a single star model. They arrive at a high density planetary nebula type object, with a central star of $T^* = 160,000$ K, a radius of $R^* = 0.06$ $R_\odot$, and $\log (L^*/L_\odot) = 3.3$. The mass of the ionisation bounded H^+ shell is $M = 2.8 \times 10^{-4} M_\odot$. The electron density in the shell is approximately 3×10^6 cm^{-3} and the electron temperature of

the H^+ region varies from 8000 K to 27,000 K. From the reddening the authors derive a distance of 2.2 kpc. The spectra also show the presence of a large neutral hydrogen region with a lower expansion velocity than that of the H^+ region.

114.007 A spectroscopic study of the light variable peculiar star HD 188136.   G. Wegner.
Astrophys. J., Vol. 247, 969 - 974 (1981).

A curve-of-growth analysis of the element abundances in the atmosphere of Kurtz's light-variable, peculiar, metallic-line star, HD 188136, is reported. The low-dispersion appearance of this star, according to Houk and Cowley, is similar to Fm δ Del stars, except for the outstanding strength of Sr II lines. The present study reveals that HD 188136 shows a rare earth enhancement intermediate between that of the cool Ap stars and the other light, variable, peculiar star, HD 101065 and, hence, larger than usual in the Am and δ Del stars. The simultaneous presence of light variability and the strong over-abundances of heavy metals in both of these objects suggests that they are related, despite differences in the lengths of their periods and the rapid decline of the heavy rare earths with atomic weight in HD 188136 compared to HD 101065. Possible evolutionary states of HD 188136 and its relation to metal enhancement mechanisms for peculiar stars are discussed, and it is concluded that HD 188136 could be an evolved Ap star.

114.008 Spectral energy distributions of hot stars with circumstellar dust.   M. L. Sitko.
Astrophys. J., Vol. 247, 1024 - 1038 (1981).

Spectral energy distribution curves of hot stars with circumstellar dust shells, combining data obtained with the *IUE* satellite and a variety of ground-based optical and infrared telescope systems, are presented. The data illustrated cover the wavelength range ~1300 Å to 12.5 µm. For a few objects, longer wavelength data are also shown. The combined ultraviolet, visual, and infrared data indicate that the dust surrounding these stars may be different from that seen in the diffuse interstellar medium.

114.009 A study of CNO elements in barium stars.   C. Sneden, D. L. Lambert, C. A. Pilachowski.
Astrophys. J., Vol. 247, 1052 - 1062 (1981).

Carbon, nitrogen, and oxygen abundances are presented for seven mild barium stars and two classical barium stars. The mild barium stars do not show the carbon enhancement typical of the classical Ba II stars. The CNO abundances of the mild barium and normal G and K giants are identical. An s-process enhancement is confirmed for some of the mild barium stars.

114.010 The FK Comae stars.   B. W. Bopp, R. E. Stencel.
Astrophys. J., Lett., Vol. 247, L131 - L134 (1981).

The authors present *IUE* observations of three very rapidly rotating G−K giants ($v \sin i = 100$ km s^{-1}). The UV spectra show strong chromospheric and transition region emission lines similar to (and in excess of) the RS CVn binaries. These stars show no evidence for radial velocity variations in excess of ± 3 to ± 20 km s^{-1}, arguing against duplicity. As a class, they lend support to the rotation-activity hypothesis. Coalesced W UMa binaries, rather than single stars, are the possible progenitors for these FK Com variables.

114.011 Correlation between spectrum characteristics and photometric behaviour of Be stars.
G. A. Ponomareva.
Pis'ma Astron. Zh., Tom 7, 414 - 416 (1981). In Russian. English translation in Soviet Astron. Lett., Vol. 7.

A comparison between spectrum characteristics and

photometric behaviour of 140 bright Be stars has been done. The percentage of variability is 44% for all stars in the sample. Spectral features which most closely correlate with light variations are narrow metallic absorption lines. On the basis of this correlation variability for the stars HD 50138 and HD 193182 can be expected.

114.012 **Spectrum of FG Sagittae in 1980.** T. A. Kipper. Pis'ma Astron. Zh., Tom 7, 428 - 431 (1981). In Russian. English translation in Soviet Astron. Lett., Vol. 7.

6-m telescope spectrograms of FG Sge obtained in 1980 are analysed. Spectral type (K0Ib − K2Ib) and abundances of elements are estimated. The evolution of the spectrum of FG Sge towards later types has ceased in recent years.

114.013 **The reddest red dwarf?** A. Walker. Mon. Notes Astron. Soc. South. Africa, Vol. 39, 96 - 100 (1980).

A spectrum of the star van Biesbroeck 8 shows that it is an M dwarf of very late type, consistent with its extremely large colour index. Comparisons are made with other cool M dwarfs.

114.014 **A far-infrared emission feature in carbon-rich stars and planetary nebulae.** W. J. Forrest, J. R. Houck, J. F. McCarthy. Astrophys. J., Vol. 248, 195 - 200 (1981).

The 16 − 30 μm spectra of several carbon stars and the planetary nebulae IC 418 and NGC 6572 have been obtained using the NASA C-141 Kuiper Airborne Observatory. A newly observed emission feature appears in the spectrum of IRC +10216 and several other carbon stars at wavelengths greater than 24 μm. The feature is interpreted as resulting from a solid-state resonance in the dust grains which have condensed around these stars. A similar feature appears in the spectra of IC 418 and NGC 6572, implying that the same type of dust is present. Since the dust probably condensed from a carbon-rich gas, this indicates an evolutionary link between carbon stars and these planetary nebulae.

114.015 **IUE observations of eight OB stars in NGC 2244: ultraviolet continua and extinction.** D. Massa, P. S. Conti. Astrophys. J., Vol. 248, 201 - 213 (1981).

The authors present low-resolution IUE ultraviolet continuum observations of O and B stars in the young open cluster NGC 2244. They find that the ultraviolet energy distributions of the O stars have significantly cooler ultraviolet color temperatures than the cluster early B stars. The effect does not appear to be a property of the B stars because their observed continua can be well represented by model atmospheres and ultraviolet extinction characteristic of the local interstellar medium. Thus, the effect seems to originate with the O star continua and can be due to either interstellar effects or stellar effects. Current data do not allow a definitive distinction to be made between these two alternatives, and so the possible origins and consequences of each are considered.

114.016 **Carbon, nitrogen, and oxygen abundances in G and K giants.** D. L. Lambert, L. M. Ries. Astrophys. J., Vol. 248, 228 - 248 (1981).

C, N, and O abundances are presented for 32 G and K giants and subgiants. The primary abundance indicators are C_2 lines at 5086 and 5135 Å, the [O I] lines at 6300 and 6363 Å, and CN red system lines. The equivalent widths are obtained from high-resolution low-noise spectra and analyzed with line-blanketed model atmospheres.

114.017 **Observations of Wolf-Rayet stars in the emission-line galaxy Tololo 3.** D. Kunth, W. L. W. Sargent. Astron. Astrophys., Vol. 101, L5 - L8 (1981).

The authors have found a strong broad He II 4686 emis-sion-line in the spectrum of the dwarf galaxy Tol 3. This emission originates from Wolf-Rayet stars mostly of WN type.

114.018 **Monitoring line profile changes in κ Orionis, B0.5 Ia.** R. Stalio, G. Sedmak, L. Rusconi. Astron. Astrophys., Vol. 101, 168 - 173 (1981).

Line profile variability has been detected in the UV and visual spectra of the B0.5 Ia supergiant κ Orionis. Several UV observations of the O VI and N V doublets, of the Si III − λ1206 resonance line and of the C III−λ1175 metastable line, obtained with the highest resolution mode of the Copernicus satellite, and quasi-simultaneous Hα coudé observations have been compared.

114.019 **Effective temperature of Ap stars.** M. Floquet. Astron. Astrophys., Vol. 101, 176 - 183 (1981).

To determine the effective temperature of a sample of 69 mostly cool Ap stars, the intensity of the CaII K line was taken as a temperature indicator. Effective temperatures deduced from the K line intensity, T_{eff} CaII are lower than those generally admitted for Ap stars. T_{eff} CaII are in agreement with T_{eff} deduced from ultraviolet photometric index Δm 2100 and with those deduced from (U−B) and (B−V) corrected of line blocking effects.

114.020 **UV observations of the intermediate helium star CPD−46° 3093.** U. Heber, K. Hunger. Astron. Astrophys., Vol. 101, 269 - 272 (1981).

UV spectrograms of CPD −46° 3093 have been obtained with IUE in low resolution. No evidence for mass loss can be found at this resolution. A synthetic spectrum, calculated in LTE and including 800 atomic transitions matches the observations. Also the UV Si III/Si IV ionization equilibrium is reproduced as well as the Strömgren colours. The atmospheric parameters are T_{eff} = 26000 K, log g = 3.95 and n_{He}/n_H = 0.37.

114.021 **Circumstellar shells of luminous supergiants I. Carbon monoxide in Rho Cassiopeiae and HR 8752.** D. L. Lambert, K. H. Hinkle, D. N. B. Hall. Astrophys.J., Vol. 248, 638 - 650 (1981).

High resolution (λ/Δλ ∼ 32,000) spectra have been obtained of the 2.3 μm CO Δυ = 2 region in two F Ia supergiants, HR 8752 and ρ Cas. Both stars are known to have extensive circumstellar shells without infrared excesses, indicating circumstellar gas with little dust. The present spectra reveal strong CO Δυ = 2 lines of purely circumstellar origin at kinetic temperatures near 2000 K. The 2 μm observations are complemented by new spectra at shorter wavelengths (0.6 μm < λ < 0.9 μm). The atmospheric structures of the two supergiants are discussed in detail.

114.022 **Lithium abundances, K line emission and ages of nearby solar type stars.** D. K. Duncan. Astrophys. J., Vol. 248, 651 - 669 (1981) = Lick Obs. Bull. No. 877.

Li abundances and chromospheric emission fluxes measured in the core of the Ca II K line have been determined in over 100 field F5–G5 dwarfs and subgiants. Although both quantities are known statistically to decrease in older stars, the correlation between them is not good. In particular, there are a number of anomalous solar type stars which show high Li abundances and very little chromospheric flux; the converse is rare. This might be understood if the intensity of chromospheric emission undergoes a sudden decrease when stars reach an age of 1 to 2 × 10^9 years, before much Li depletion occurs. Consistent time scales for Li depletion are fitted to Pleiades, Ursa Major cluster, Hyades, and solar data, and are applied to the field stars. The resulting age distribution indicates a slightly decreasing rate of star formation during the last 4–5 × 10^9 years. Many of the earliest type stars observed (F5–F6) show significant Li depletion, which cannot be explained by the same process which depletes Li in the sun. Different mechanisms are discussed which may explain Li depletion in the sun and in earlier type stars.

114.023   Search for optical coronal line emission from the X-ray sources Epsilon Orionis (B0 Ia) and Kappa Orionis (B0.5 Ia).   K. H. Nordsieck, J. P. Cassinelli, C. M. Anderson.
Astrophys. J., Vol. 248, 678 - 683 (1981).

High signal-to-noise observations have been made at the wavelengths of the lines of [Fe X] 6574 Å and [Fe XIV] 5303 Å to search for evidence of a coronal region at the base of the winds of ϵ Ori (B0 Ia) and κ Ori (B0.5 Ia). These two stars have been detected as soft X-ray sources and both stars also show anomalously strong O VI lines in the UV spectra, which has been explained as a product of Auger ionization in cool stellar wind. The nondetection of the iron coronal lines places new upper limits on the emission measure. The measured upper limits on the equivalent width of the Fe X and Fe XIV lines are 1 mÅ for narrow lines and 2.5 mÅ for lines assumed to be rotationally broadened. This constrains the coronal temperatures to $T_c > 1 \times 10^6$ for κ Ori and $T_c > 2 \times 10^6$ for ϵ Ori. For ϵ Ori the constraints are then no longer compatible with the energy distribution measured in the X-ray region. It is suggested that at least some of the X-rays arise not from the base corona, but from source regions farther out in the wind.

114.024   Far-ultraviolet observations of the variable radio star LSI +61°303.
L. Maraschi, E. G. Tanzi, A. Treves.
Astrophys. J., Vol. 248, 1010 - 1014 (1981).

The radio star LSI +61°303 has been observed with the IUE in 1980 September. One short wavelength ($\lambda\lambda 1200-1950$) and two long wavelength ($\lambda\lambda 1900-3200$) spectra have been obtained. Assuming $E_{B-V} = 0.9$ as given by the rectification of the $\lambda 2200$ absorption dip, the ultraviolet continuum is well fitted by a model atmosphere corresponding to a supergiant of subclass b and spectral type around B1.5, at a distance of ~4 kpc. With respect to short wavelength observations made in 1978, significant variations of the emission features are found.

114.025   The chemical composition, gravity, and temperature of Sirius.   R. A. Bell, L. A. Dreiling.
Astrophys. J., Vol. 248, 1031 - 1042 (1981).

A number of flux-constant, line blanketed model stellar atmospheres have been computed for Sirius. The stellar effective temperature has been found to be 10,150±300 K. Balmer line profiles give $\log g = 4.3 \pm 0.1$. The Ti II curve of growth gives $\log N(\text{Ti}) = 5.5$ on the scale $\log N(\text{H}) = 12.0$. The Fe II and Fe I lines give $\log N(\text{Fe}) = 8.1$ and 8.0, respectively. The fluxes of the model $T_{\text{eff}} = 10,100$ K, $\log g = 4.3$, [A/H] = 0.5 are presented for the wavelength interval $0.12-30 \, \mu m$, along with the line blocking for the $0.12-1.2 \, \mu m$ interval.

114.026   Far-ultraviolet fluorescence of carbon monoxide in the red giant Arcturus.
T. R. Ayres, H. W. Moos, J. L. Linsky.
Astrophys. J., Lett., Vol. 248, L137 - L140 (1981).

The authors present evidence that many of the weak features observed with the International Ultraviolet Explorer (IUE) in the far-ultraviolet (1150 - 2000 Å) spectrum of the archetype red giant Arcturus (K2 III) are A - X fourth positive bands of carbon monoxide excited by chromospheric emissions of O I, C I, and H I. The appearance of fluorescent CO bands near the wavelengths of commonly used indicators of high-temperature ($T > 2 \times 10^4$ K) plasma, such as C II $\lambda 1335$ and C IV $\lambda 1548$, introduces a serious ambiguity in diagnosing the presence of hot material in the outer atmospheres of the cool giants by means of low-dispersion IUE spectra.

114.027   High resolution profiles of chromospheric lines in M dwarf stars.
S. P. Worden, T. J. Schneeberger, M. S. Giampapa.
Astrophys. J., Suppl. Ser., Vol. 46, 159 - 175, plate 5 (1981).

The authors show 46 hydrogen and sodium line profiles for 17 selected dwarf M and dwarf M emission line stars. These profiles, derived from echelle spectrograms, typically have 0.25 Å spectral resolution. Most of the dMe stars show hydrogen emission lines with central reversals. Central emission features in the cores of the Na D lines are often observed, and the presence of these emission features is strongly correlated with hydrogen line emission.

114.028   IUE observations of variability in winds from hot stars.   C. A. Grady, T. P. Snow, Jr.
The Universe at ultraviolet wavelengths, (see 012.009), p. 121 - 125 (1981).

Observations of variability in stellar winds or envelopes provide an important probe of their dynamics. For this purpose a number of O, B, Be, and Wolf-Rayet stars have been repeatedly observed with the IUE satellite in high-resolution mode. In the course of analysis, instrumental and data handling effects were found to introduce spurious variability in many of the spectra. Software has therefore been developed to partially compensate for these effects, but limitations remain on the type of variability that can be identified from IUE spectra. With these constraints, preliminary results of multiple observations of two OB stars, one Wolf-Rayet star, and a Be star are discussed.

114.029   Anomalous ionization seen in the spectra of B supergiants.   J. P. Cassinelli, D. C. Abbott.
The Universe at ultraviolet wavelengths, (see 012.009), p. 127 - 134 (1981).

An IUE survey of B supergiants has been conducted to study the persistence with spectral type of the ultraviolet resonance lines of N V, C IV, and Si IV. N V is seen as late as B2.5 Ia, C IV until B6 Ia and Si IV throughout the range from B1.5 to B9. This is in fairly good agreement with the Auger ionization model of Cassinelli and Olson (1979). The terminal velocities are derived for the 20 stars in the sample.

114.030   Absolute energy curves from late B-type supergiants. A. B. Underhill.
The Universe at ultraviolet wavelengths, (see 012.009), p. 135 - 140 (1981).

Absolute energy curves for six late B and early A-type supergiants have been determined from IUE data and from other ultraviolet and ground-based photometry. Effective temperatures and angular diameters are presented as well as estimates of the outflow velocity of the wind. All six stars show a strong Balmer continuum in emission; the Ia supergiants also show an infrared excess which reaches into the visible range. Evidence is found for the presence of a warm mantle as well as for wind from the Ia stars.

114.031   IUE and ground-based observations of mass loss in the Magellanic Clouds.
F. Macchetto, P. Benvenuti, S. D'Odorico, N. Panagia.
The Universe at ultraviolet wavelengths, (see 012.009), p. 141 - 148 (1981).

Ground-based and IUE observations of hot stars in the Large and Small Magellanic Clouds have been carried out to investigate the mass loss process in these objects and to search for differences with galactic hot stars. Preliminary results show that in a large proportion of the stars observed the mass-loss process is taking place. A mechanism for acceleration of the wind in OB stars is proposed.

114.032   The UV resonance lines of three hot Ap stars HD 133029, HD 175362 and HD 219749.
K. D. Rakos.

The Universe at ultraviolet wavelengths, (see 012.009), p. 167 - 176 (1981).

High resolution spectra of the Ap stars HD 133029, HD 175362 and HD 219749 have been obtained during two 16 hour shifts with the IUE satellite. Stellar wind, extended atmosphere in rigid corotation with the stellar surface and the influence of the strong magnetic field on the upper part of the atmosphere would explain the shape and the strength of the resonance lines. The resonance lines of HD 175362 in particular show very peculiar behavior. The necessary driving forces for the expanding envelope are not compatible with the diffusion theory of the Ap atmospheres.

114.033   IUE observations of blue halo high luminosity stars.
M. Hack, M. L. Franco, R. Stalio.
The Universe at ultraviolet wavelengths, (see 012.009), p. 193 - 199 (1981).

Two high luminosity population II blue stars of high galactic latitude, BD+33°2642 and HD 137569 have been observed at high resolution. The stellar spectra show the effect of mass loss in BD+33°2642 and abnormally weak metallic lines in HD 137569. The interstellar lines in the direction of BD+33°2642, which lies at a height z⩾6.2 kpc from the galactic plane, are split into two components. No high ionization stages are found at the low velocity component; nor can they be detected in the higher velocity clouds.

114.034   The emission/absorption Fe II spectrum of HD 45677.   R. Stalio, P. L. Selvelli.
The Universe at ultraviolet wavelengths, (see 012.009), p. 201 - 207 (1981).

The complex behavior of the emission/absorption spectrum of Fe II is analyzed. The far UV spectrum is characterized almost solely by absorption lines, while, in the near UV, strong emissions are predominant. Radiative excitation from the ground to the highest levels ($\chi \simeq 10$ eV) with re-emission in the near UV, visible and I.R. seems to be the main mechanism capable of explaining the observed spectral features.

114.035   The Mg II h and k lines in a sample of dMe and dM stars.   M. S. Giampapa, P. L. Bornmann,
T. R. Ayres, J. L. Linsky, S. P. Worden.
The Universe at ultraviolet wavelengths, (see 012.009), p. 279-286 (1981).

The authors present observed Mg II h and k line fluxes for a sample of 4 dMe and 3 dM stars obtained with the IUE satellite in the long wavelength, low dispersion mode. The observed fluxes are converted to stellar surface flux units and the importance of chromospheric non-radiative heating in this sample of M dwarf stars is intercompared. In addition, the authors compare the net chromospheric radiative losses due to the Ca II H and K lines in those stars in the sample for which calibrated Ca II H and K line data exist. Moreover, the authors estimate active region filling factors which likely give rise to the observed optical and ultraviolet chromospheric emission. Finally, the authors briefly discuss the implications of the results for homogeneous, single-component stellar model chromospheres analyses.

114.036   IUE spectra of F and late A stars.
J. L. Linsky, N. C. Marstad.
The Universe at ultraviolet wavelengths, (see 012.009), p. 287 - 295 (1981).

The authors report on IUE spectra of α CMi (F5 IV–V), β Cas (F2 IV), α Car (F0 Ib), and γ Boo (A7 III) in the context of the question as to whether chromospheres disappear in the early F-late A portions of the HR diagram. Both α CMi (Procyon) and β Cas show bright emission line spectra indicative of chromospheres and transition regions, but neither α Car (Canopus) nor γ Boo show any evidence of emission in their SWP spectra or at the Mg II lines, despite very deep

exposures. These results are consistent with those recently published by Böhm-Vitense and Dettmann.

114.037   The chromospheric and transition layer emission of stars with different metal abundances.
E. Böhm-Vitense.
The Universe at ultraviolet wavelengths, (see 012.009), p. 303 - 309 (1981).

The author reports preliminary results on observations of chromospheric and transition layer emission of stars with different metal abundances. Metal deficient stars generally show reduced emission in the Mg II resonance lines and also in the other chromospheric and transition layer emission lines. This is interpreted as showing that energy fluxes other than acoustic fluxes must at least be co-responsible for the coronal and transition layer heating.

114.038   UV chromospheric and circumstellar diagnostic features among F supergiant stars.
R. E. Stencel, S. P. Worden, M. S. Giampapa.
The Universe at ultraviolet wavelengths, (see 012.009), p. 311 - 316 (1981).

The authors undertook a survey of F supergiant stars to evaluate the extension of chromospheric and circumstellar characteristics commonly observed in the slightly cooler G, K and M supergiant spectra. In the optical regions, the usual diagnostic spectral features are swamped by the brighter photospheric light in F stars. Therefore, an ultraviolet survey was elected since UV features of Mg II and Fe II might persist in revealing outer atmosphere phenomena even among F supergiants. The survey encompassed spectral types F0 to G0, and luminosity classes Ib, Ia and Ia-0.

114.039   High resolution absolute flux profiles of the Mg II h & k lines in evolved F8 to M5 stars.
R. E. Stencel, D. J. Mullan, G. S. Basri, J. L. Linsky.
The Universe at ultraviolet wavelengths, (see 012.009), p. 317 - 324 (1981).

The authors present the central results of a survey of the Mg II resonance line emission in a sample of over 50 evolved late-type stars, including spectral-luminosity types F8-M5 and Ia-IV. Observed and surface fluxes have been derived and correlations noted.

114.040   IUE – ultraviolet and optical chromospheric studies of late-type giants in the Hyades cluster.
S. L. Baliunas, L. Hartmann, A. K. Dupree.
The Universe at ultraviolet wavelengths, (see 012.009), p. 325 - 331 (1981).

The authors present ultraviolet and optical observations of four bright, late-type giants in the Hyades cluster detected with IUE in order to study chromospheric and coronal activity in stars of the same age.

114.041   IUE observations of circumstellar emission from the late type variable R Aqr (M7 + pec).
R. W. Hobbs, A. G. Michalitsianos, M. Kafatos.
The Universe at ultraviolet wavelengths, (see 012.009), p. 355 - 365 (1981).

As part of a program to observe circumstellar emission from late type stars, IUE observations of R Aqr (M7 + pec) have been obtained in low dispersion. Strong permitted, semi-forbidden and forbidden emission lines are seen, superimposed on a bright ultraviolet continuum. The authors deduce that the strong emission line spectrum arises from a dense compact nebula the size of which is comparable to the orbital radius of the binary system of which R Aqr is the primary star. The secondary is probably a white dwarf, comparable to or somewhat brighter than the sun. The authors attribute the UV continuum to Balmer recombination from the dense nebula and not to blackbody emission from the hot companion.

114.042 **IUE observations of two late type stars BX Mon (M4 + pec) and TV Gem (M1 Iab).**
A. G. Michalitsianos, R. W. Hobbs, M. Kafatos.
The Universe at ultraviolet wavelengths, (see 012.009), p. 367 - 375 (1981).

IUE observations of two late type stars BX Mon and TV Gem have been obtained that reveal the emission properties in the ultraviolet of subluminous companions.

114.043 **Horizontal-branch stars, and galactic and Magellanic Cloud globular clusters.**   K. S. de Boer.
The Universe at ultraviolet wavelengths, (see 012.009), p. 527 - 531 (1981).

Seven blue horizontal branch stars in the field have been observed, and a few HB stars have been isolated in globular clusters. Energy distributions are compared to assess possible differences, also in comparison with model atmospheres. Observed energy distributions of HB stars in NGC 6397 are used to estimate the total number of HB stars which produced the integrated fluxes as observed by ANS. Preliminary   results are given for colors of globular clusters observed in the Magellanic Clouds and for their extent, based on the Washburn IUE extraction.

114.044 **Improvements to the accuracy of the IUE wavelength scales in high dispersion.**
B. E. Turnrose, C. A. Harvel, R. C. Bohlin.
The Universe at ultraviolet wavelengths, (see 012.009), p. 795 - 800 (1981).

The data base of Pt-Ne emission lines used to calibrate the IUE high dispersion wavelength scales has been scrutinized to improve the internal consistency of the adopted laboratory wavelength values and provide a homogeneous, documented line list, which IUE Guest Observers may use to evaluate quantitatively those Pt-Ne spectra taken to calibrate their data.

114.045 **Far-ultraviolet fluorescence of carbon monoxide in the red giant Arcturus.**
T. R. Ayres, H. W. Moos, J. L. Linsky.
Bull. American Astron. Soc., Vol. 13, 515 (1981). – Abstract.

114.046 **IUE observations of stellar lines in OB stars.**
L. W. Kamp, C. J. Needham, L. A. York.
Bull. American Astron. Soc., Vol. 13, 515 - 516 (1981). Abstract.

114.047 **Spectral classifications for northern Kapteyn Selected Areas.**   W. Buscombe.
Bull. American Astron. Soc., Vol. 13, 517 - 518 (1981). Abstract.

114.048 **Indications of IUE spectral artifacts in observations of flat-continuum sources.**
R. L. Hackney, K. R. Hackney.
Bull. American Astron. Soc., Vol. 13, 532 (1981). – Abstract.

114.049 **Brackett-alpha emission from LkHα 101.**
M. Daehler, H. A. Smith, D. P. McNutt, S. H. Knowles.
Bull. American Astron. Soc., Vol. 13, 542 (1981). – Abstract.

114.050 **Ultraviolet spectra of BL Ori.**
H. R. Johnson, G. T. O'Brien.
Bull. American Astron. Soc., Vol. 13, 546 (1981). – Abstract.

114.051 **Scanner measurements of NH, CN, CH, and C_2 bands.**   R. F. Wing, G. L. Lange.
Bull. American Astron. Soc., Vol. 13, 546 (1981). – Abstract.

114.052 **Photometric carbon and nitrogen abundances in G and K stars.**   G. L. Lange.
Bull. American Astron. Soc., Vol. 13, 546 (1981). – Abstract.

114.053 **High-resolution spectra of five late-type dwarfs and giants obtained with the IUE satellite.**
T. R. Ayres, G. S. Basri, R. C. Henry, W. Landsman, J. L. Linsky, H. W. Moos, R. E. Stencel.
Bull. American Astron. Soc., Vol. 13, 546 - 547 (1981). Abstract.

114.054 **The unusual outer atmosphere of 56 Pegasi (K0 IIp).**
M. Schindler, R. Stencel, J. Linsky, D. Helfand, G. Basri.
Bull. American Astron. Soc., Vol. 13, 547 (1981). – Abstract.

114.055 **Scanner observations of the Be stars ν Gem, κ Ori, β Mon and ω CMa.**   P. S. Goraya.
Astrophys. Space Sci., Vol. 78, 419 - 425 (1981).

The photoelectric spectrophotometric scans of the Be stars ν Gem, κ Ori, β Mon and ω CMa have been analyzed to find out a few stellar parameters. The absolute energy distributions of these stars in the wavelength range λλ350–750 nm have been given. Their effective temperatures and gravities have been estimated from comparisons with non-LTE model atmospheres. The evolutionary aspects of these stars are discussed and their masses have been estimated.

114.056 **The near ultraviolet stellar spectra of α Lyrae and β Orionis.**   E. Michelson.
Mon. Not. R. Astron. Soc., Vol. 197, 57 - 74 (1981).

A comparison is made of high resolution, near ultraviolet observations of α Lyrae and β Orionis with synthetic spectra. The observations, taken on a balloon-borne spectrograph, are in the wavelength region 2730–2880 Å and have a resolution of about 0.1 Å. The model atmospheres used are in LTE, both unblanketed and blanketed.

114.057 **High dispersion observations of LMC/SMC stars.**   B. D. Savage, K. S. de Boer.
Second European IUE Conference, (see 012.011), p. XLI (1980).

114.058 **Ultraviolet observations and the nature of the Wolf-Rayet stars.**   A. J. Willis.
Second European IUE Conference, (see 012.011), p. IL - LVII (1980).

Ultraviolet observations of WR stars obtained prior to and with IUE are reviewed. The new IUE data allow accurate interstellar reddening corrections and colour temperatures to be determined for a large sample of stars. WR stellar wind terminal velocities show a strong correlation with WN subclass in both galactic and LMC stars. A revised mass loss rate scale for WR stars is discussed.

114.059 **Ultraviolet observations of a Bp-star at high galactic latitude.**   J. Berger, A. M. Fringant, M. Gerbaldi, N. Morguleff.
Second European IUE Conference, (see 012.011), p. 29 - 30 (1980).

The high-velocity metal-poor star HD 214539 (A0Ib) has been analysed for MgII lines around 2800 Å. It is possible to separate the interstellar components.

114.060 **Ultraviolet variation of the Ap star 21 Com.**
M. Gerbaldi, N. Morguleff, C. Megessier.
Second European IUE Conference, (see 012.011), p. 31 - 34 (1980).

High resolution spectra permit to identify lines of elements responsible for the large absorption features observed at low resolution. With a series of low resolution spectra covering the period one is able to identify probable continuum windows which seem not to be variable. At short wavelength, the authors confirm the large amplitude of periodical variations due to numerous metallic lines.

114.061 Fe II in the IUE spectra of emission line stars.
R. Viotti, A. Giangrande, A. Altamore,
A. Cassatella, M. Friedjung, G. Muratorio, O. Ricciardi.
Second European IUE Conference, (see 012.011), p. 39 - 41 (1980).

The authors study the behaviour of the Fe II lines in the ultraviolet spectra of the emission line stars Boss 1985, η Car, and S 22 in the Large Magellanic Cloud. The authors discuss in particular the dependence of the Fe II lines emission as a function of the excitation potential, line strength, and wavelength.

114.062 Simultaneous spectroscopic (UV Mg II and Al II lines) and photometric variations of ζ^1 Sco
(HD 152236).   A. Heck, G. Burki, L. Bianchi, A. Cassatella, J. Clavel.
Second European IUE Conference, (see 012.011), p. 43 - 46 (1980).

Repeated IUE observations of ζ^1 Sco show variations in the line profiles of the two Mg II doublets at about 2800 Å and of the Al II λ1670 line. A series of puffs in the expanding envelope have been identified, implying a non-stationary mass loss process. A phase of particularly discontinuous loss of matter has been correlated with a sharp decrease of the V magnitude.

114.063 The chromosphere of the A7V star α Aql.
C. Blanco, S. Catalano, E. Marilli.
Second European IUE Conference, (see 012.011), p. 63 - 66 (1980).

High resolution spectrograms of α Aql (A7V) obtained with the IUE satellite in the short and long wavelength regions are presented. Evidence of chromospheric emission in the h and k lines of Mg II has been found. The width $\Delta\lambda \simeq 1.4$ Å fits very closely the linear relationship between the absolute visual magnitude (M_V) and Mg II emission width. The hydrogen Lα emission has been detected and a flux of $(5.58 \pm 1.0) \times 10^5$ erg/cm^2 sec at the stellar surface has been estimated.

114.064 Chromospheres and transition regions in F dwarfs.
M. Saxner.
Second European IUE Conference, (see 012.011), p. 67 - 70 (1980).

Low resolution spectra in the wavelength region 1150 - 2000 Å have been obtained for five F stars using the SWP camera on the IUE satellite. In four of these spectra emission lines from O I, C II, Si IV, C IV, and N V are visible. These lines are also prominent in the UV spectra of the quiet sun and cooler stars with chromospheric activity. The lines, with the exception of the O I line, originate in a transition region between the chromosphere and the corona. It is found that the surface flux in the emission lines of the stars observed is comparable to that in late type stars with chromospheric activity.

114.065 The chromospheric and coronal structure of αCMi (F5 IV−V).   A. Brown, C. Jordan.
Second European IUE Conference, (see 012.011), p. 71 - 75 (1980).

Observations of αCMi (Procyon) obtained with the IUE satellite have been analysed to determine the structure of the outer atmosphere. This analysis is used to examine the terms of the energy balance equation. Radiation losses exceed those from comparable regions of the solar atmosphere and dominate the energy balance of the atmosphere. The coronal temperature and pressure are $\sim 3 \times 10^5$ K and $\sim 2 \times 10^{14}$ cm^{-3} K. Line profiles from high resolution spectra are used to determine the

non-thermal energy flux. Comparisons made between the required energy input and that expected from heating by acoustic waves show that, from the authors' observations and models, energy deposition by acoustic waves should be sufficient to account for the radiative losses.

114.066 IUE observations of the star HD93521.
C. Morossi, M. Ramella.
Second European IUE Conference, (see 012.011), p. 95 (1980). Abstract.

114.067 Observations of high velocity components.
D. Giaretta, B. Bates, R. E. L. Bankhead,
W. Brown-Kerr, J. A. McQuoid.
Second European IUE Conference, (see 012.011), p. 115 - 118 (1980).

Several stars which, from earlier investigations, are known to show well separated high velocity components, have been observed with IUE at high dispersions. The physical conditions in these components are discussed.

114.068 Coordinated ground-based and IUE observations of hot stars in the Magellanic Clouds.
F. Macchetto, P. Benvenuti, S. D'Odorico, N. Panagia.
Second European IUE Conference, (see 012.011), p. 119 - 122 (1980).

Ground-based and IUE observations of hot stars in the Large and Small Magellanic Clouds have been carried out to investigate the mass-loss process in these objects and to search for differences with galactic hot stars. Preliminary results show that in a large proportion of the stars observed the mass-loss process is taking place. A mechanism for acceleration of the wind in OB stars is proposed.

114.069 Ultraviolet observations of 27 Canis Majoris, π Aquarii and 48 Librae.
A. E. Ringuelet, J. M. Fontenla, M. Rovira.
Second European IUE Conference, (see 012.011), p. 143 (1980). − Abstract.

114.070 Long term changes in ultraviolet resonance lines in γ Cas.
H. F. Henrichs, G. Hammerschlag-Hensberge,
H. J. G. L. M. Lamers.
Second European IUE Conference, (see 012.011), p. 147 - 149 (1980).

High-resolution ultraviolet spectra of the B 0.5 IVe star γ Cas, obtained with IUE between April 1978 and October 1979 show spectacular changes in the profiles of the resonance doublets of C IV, N V and Si IV. In addition to the broad asymmetric low-velocity absorption lines which seem to be always present, narrow, blue-shifted absorption components, with velocities up to 1500 km s^{-1}, are observed at 4 occasions. The time-dependent high-velocity features are interpreted in terms of (presumed) spherical high-density "shells", which travel outwards with a terminal velocity not very much exceeding 1500 km s^{-1}.

114.071 Ultraviolet emission line spectrum of peculiar stars: HD 45677.   P. L. Selvelli, R. Stalio.
Second European IUE Conference, (see 012.011), p. 155 - 157 (1980).

SWP and LWR IUE high resolution spectra of HD 45677 have been studied in order to analyse the complex atmospheric structure revealed by the visible and infrared observations. The star has a particularly rich spectrum of both emission and absorption lines.

114.072 Simultaneous ultraviolet and optical observations of HDE 245770/A 0535+26.   F. Giovannelli,
M. Ferrari Toniolo, A. Giangrande, P. Persi, C. Bartolini,
A. Guarnieri, A. Piccioni, S. M. Rucinski.

Second European IUE Conference, (see 012.011), p. 159 - 164 (1980).

The authors present high and low dispersion UV spectra of HD 245770, which is the optical counterpart of the recurrent transient X-ray pulsar A 0535+26. UBV photometric measurements were simultaneously carried out and some medium dispersion optical spectra were obtained close to the UV ones. In this paper is shown a preliminary analysis of the UV spectra.

114.073 **Coordinated ultraviolet, optical and infrared observations of X Per in December 1979: preliminary results.**   R. Viotti, M. Ferrari-Toniolo, A. Giangrande, P. Persi, L. Bianchi, G. Grasdalen, P. Kalv, R. Stalio.
Second European IUE Conference, (see 012.011), p. 165 - 167 (1980).

Coordinated ultraviolet, optical and infrared observations of the X-ray binary system X Per were made in December 1979. During the observations, X Per was in a phase of rapidly increasing luminosity. The ultraviolet high resolution spectrum shows violet shifted broad absorption lines with the violet wings extending to −250/650 km s^{-1}. The double peaked Hα line profile seems to be variable on a time scale of days. An infrared excess probably due to an expanding envelope surrounding the OBe star, was observed during the campaign.

114.074 **Observations of late-type stars with 'hot' companions.**   D. P. Gilra, P. R. Wesselius, N. Kameswara Rao.
Second European IUE Conference, (see 012.011), p. 227 - 228 (1980).

IUE and ANS observations of the 'hot' companions of o Ceti, ε Aur, and π Pup are briefly discussed. Observations of HD 62001, the central star of the nebula VV 1−7 are also reported. The nebula VV 1−7 which is present on the Palomar blue print is not present on the ESO blue prints.

114.075 **Ultraviolet energy distribution of Wolf-Rayet stars.**   W. Schmutz, L. J. Smith.
Second European IUE Conference, (see 012.011), p. 249 - 251 (1980).

Low resolution spectra of 15 Wolf-Rayet stars (10 WN and 5 WC stars) have been analysed to investigate their UV continuum distribution. The reddening observed for Wolf-Rayet stars is found to follow the general galactic interstellar extinction law. Colour excesses for each star are deduced from the observed strength of the 2200Å feature. The combined de-reddened UV-visible continua are compared with black body energy distributions to deduce colour temperatures. Four stars are analysed on the bases of Cassinelli and Hartmann's theory for extended emitting regions.

114.076 **UV-observations of two extremely helium rich stars.**   U. Heber, D. Schönberner.
Second European IUE Conference, (see 012.011), p. 327 - 332 (1980).

UV-spectrograms of BD +10°2179 and BD −9°4395 have been obtained with the IUE satellite in the low resolution mode. No evidence for mass loss can be found (at least at low resolution). The observations are analysed by comparison with detailed LTE-model fluxes including 800 atomic transitions. The results confirm earlier analyses of visual spectra.

114.077 **The ultraviolet spectrum of the anomalous EUV-source halo star HD 192273.**
G. E. Bromage, A. H. Gabriel, D. W. Sciama.
Second European IUE Conference, (see 012.011), p. 23, 353 - 356 (1980).

The proposed EUV source HD 192273 has been observed with IUE at both low and high resolution. Some results of an analysis of the stellar and interstellar spectrum are presented. Strengths of interstellar absorption lines give a value for the column density of neutral hydrogen atoms, $N_H = 2.0 \pm 0.5 \times 10^{20}$ cm^{-2}, making the proposed EUV identification untenable. The star appears to be a normal B2 V star with $E_{B-V} = 0.04$ but an approximate distance of 0.5 kpc is derived from the IUE data and this leaves the star 4 magnitudes too faint for B2 V.

114.078 **IUE observations of HD 269546.**
M. Grewing, E. Schulz-Lüpertz.
Second European IUE Conference, (see 012.011), p. 129, 357 - 359 (1980).

114.079 **Spectral types and radial velocities of southern OB+ stars.**   J. S. Drilling, C. L. Perry.
Astron. Astrophys., Suppl. Ser., Vol. 45, 439 - 442 (1981) = Contrib. Louisiana State Univ. Obs. No. 162.

MK spectral types have been determined for 69 stars designated as OB+ in the catalog of Stephenson and Sanduleak. All but 11 of these stars turned out to be O-type stars or B-type supergiants. Radial velocities are presented for 31 of the stars and for the interstellar K-line in the spectra of 25 of the stars. The radial velocities indicate that both the stars and the interstellar Ca II gas are associated primarily with the Sagittarius arm, and in some cases with the Norma-Scutum arm, of the Galaxy.

114.080 **Spectral classification in the MK system of 167 northern HD stars.**   K. S. Jensen.
Astron. Astrophys., Suppl. Ser., Vol. 45, 455 - 458 (1981).

Spectral classifications in the MK system of 167 northern HD stars are presented. The spectra (102 Å/mm at H$_\gamma$, width 0.60 mm) are from objective prism plates obtained with the Schmidt telescope of the CUO, Brorfelde. Most of the stars have no previous MK classification.

114.081 **On the estimation of photometric spectral types.**   E. Oblak, M. Chareton.
Astron. Astrophys., Suppl. Ser., Vol. 45, 459 - 471 (1981).

The authors have estimated a photometric spectral type based on indices of the $uvby\beta$ photometry for the normal stars of the Hauck and Mermilliod (1975) compilation. In this sample: 1563 stars have no MK spectral types; for 440 stars it is difficult or impossible to estimate a spectral type from the photometry; for 436 stars having an estimated photometric spectral type the authors have found an MK spectral type in the literature which allowed a comparative study. The absolute magnitudes for the MK and photometric spectral types are presented.

114.082 **Photographic spectrophotometry of selected bright B and Be stars.**   E. M. Hendry, J. D. R. Bahng.
J. Astrophys. Astron., Vol. 2, 141 - 159 (1981).

Coudé spectra taken in the red region of 49 bright B and Be stars were examined for the behaviour of Hα and the He I λλ 5876 and 6678 line profiles. Evidence is presented for the existence of drastic changes in the line profile and the radial velocity in some of the stars on a time scale of fractions of a day.

114.083 **Profiles de la raie Hα par télévision analogique.**   Y. Andrillat, C. Fehrenbach.
Effects of mass loss on stellar evolution, (see 012.015), p. 57 - 60 (1981).

High resolution profiles of the Hα line are derived for about thirty Be stars and several bright stars by means of the analogical television system. The profiles obtained with this method fairly agree with the theoretical ones (α Lyrae).

114.084 **Photospheric molecular line profiles in cool stars.**   S. T. Ridgway, E. D. Friel.
Effects of mass loss on stellar evolution, (see 012.015), p. 119 - 124 (1981).

Spectral lines of the $\Delta V = 2$ rotation vibration bands of CO are well suited for study of photospheric motions and the mass ejection process in cool stars. The authors have obtained high spectral resolution (1.8 km/sec) and high signal-to-noise ($>10^2$) line profiles for a selection of K and M giants. These profiles are being studied for evidence of gas motions in the photosphere and near circumstellar regions.

114.085 Narrow components in UV line profiles as evidence for a two component stellar wind for O and B stars.
H. J. G. L. M. Lamers.
Effects of mass loss on stellar evolution, (see 012.015), p. 181 - 186 (1981).

The UV resonance lines of early type stars show narrow absorption components, with a width of the order of 300 km s^{-1}, superimposed on the wider P Cygni profiles. The author studied the characteristics of these narrow components in the Copernicus spectra.

114.086 The ultraviolet to infrared spectrum of the large mass loss LMC supergiant S22 = HD 34664.
G. Muratorio, S. Bensammar, A. Cassatella, M. Friedjung, R. Viotti.
Effects of mass loss on stellar evolution, (see 012.015), p. 271 - 274 (1981).

114.087 The hydrogen/helium ratio on the surface of Wolf-Rayet stars.   P. S. Conti, P. Massey.
Effects of mass loss on stellar evolution, (see 012.015), p. 311 - 314 (1981).

114.088 Continua of Wolf-Rayet stars.   V. I. Stebnev.
Kazanskij inzh.-stroitel'nyj inst., Kazan', 1981.
17 pp. In Russian. – Abstr. in Ref. zh., 51. Astron., 8.51.625 (1981).

114.089 Short-period variations in the moving line spectrum of SS 433.   G. H. Newsom, G. W. Collins II.
Astron. J., Vol. 86, 1250 - 1258 (1981).

The authors present evidence for the existence of short-period variations in the wavelengths of the moving lines of SS 433. Specific attention is paid to the possibility that the effects result from the temporal sampling of the data, and they find that this cannot be the case. The geometrical representation is generalized to encompass the light time effects for arbitrary orientation and location of the velocity vectors of the emitting regions consistent with the assumption that they are stationary and antiparallel in a rotating frame. From the best fit to the residuals in the data compared to this assumption, they conclude that the velocity vectors vary in a 6.06-day period, but the "blue" and "red" beams remain nearly antiparallel throughout this variation.

114.090 Scanner observations of the Be stars ν Gem, κ Ori, β Mon and ω CMa.   P. S. Goraya.
Bull. Astron. Soc. India, Vol. 9, 82 (1981). – Abstract.

114.091 Medium resolution spectra of G, K and M stars in the near infrared.   C. Barbieri, C. Bonoli, F. Bortoletto, S. di Serego, R. Falomo.
Mem. Soc. Astron. Italiana, Vol. 52, 195 - 216 (1981).

This paper describes the medium-resolution spectra of a small sample of G, K and M stars in the near infrared region. The resolution of 6 Å permits a discussion of atomic and molecular features useful for spectral classification while preserving at the same time the information content of the continuum.

114.092 Applications of an observational grid of low-metallicity stellar spectra.   W. Tobin, K. H. Nordsieck.
Astron. J., Vol. 86, 1360 - 1376 (1981).
The authors present low-resolution spectrophotometric

observations over a large wavelength interval of a selection of stars that spans a large range of spectral type, luminosity, and (low) metal content. The recent commissioning of a fast Reticon detector on the Pine Bluff Observatory 0.9 m telescope (Nordsieck et al. 1981) made feasible the rapid acquisition of this large number of high signal-to-noise spectra. Three dozen absorption features are identified in the group of metal-poor stars. The authors use a regression technique to devise a mathematical model which describes the luminosity, color, and metallicity behavior of each feature. The models are then used to attack two sample problems of astrophysical interest. The first problem is that of the rapid determination of stellar metallicity. The second problem is an attempt to detect mass segregation in a globular cluster using luminosity-sensitive features and integrated light spectrophotometry.

114.093 A spectral description and non-LTE analysis of 6 central stars of planetary nebulae.
R. H. Méndez, R. P. Kudritzki, J. Gruschinske, K. P. Simon.
Astron. Astrophys., Vol. 101, 323 - 331 (1981).

The authors present an improved spectral description of 6 central stars of planetary nebulae; several absorption and emission features are identified. A non-LTE analysis of hydrogen and helium line profiles is carried out, in order to derive T_{eff}, log g, and helium abundance in the photospheres of these stars. The masses of the 6 objects are in a narrow interval around 0.55 $M_\odot$, which is in close agreement with the mass distribution of white dwarfs.

114.094 Carbon-to-iron ratio in extreme population II stars.
B. Barbuy.
Astron. Astrophys., Vol. 101, 365 - 368 (1981).

The synthetic spectrum technique applied to the molecular feature of the CH ($A^2\Delta - X^2\pi$) band near $\lambda4300$ Å was employed to obtain carbon abundances in some very old, extremely metal deficient halo stars. The unevolved stars in the current sample show the null-hypothesis result $[C/M] \cong 0.0$, whereas for the evolved ones carbon depletions appear as a signature of mixing effects.

114.095 Analysis of the far ultraviolet emission lines in late type stars.   E. de Castro, M. J. Fernández-Figueroa, M. Rego, D. Ponz.
Astron. Astrophys., Vol. 102, 207 - 211 (1981).

Far ultraviolet emission lines in a sample of seven stars of spectral type later than F5 and different luminosity class have been analyzed to calculate the electron density and pressure in the formation region of the lines. Three different methods were used. The authors obtain electron pressures larger than solar for all dwarf stars of the sample. The low resolution observations were made with the IUE satellite.

114.096 Spectrophotometric investigation of the Hα line in the spectrum of AR Lac.
M. B. Babaev, L. Kh. Gasanalizade.
Izv. AN AzSSR. Ser. fiz.-tekh. i mat. nauk, Tom 180, No. 2, p. 81 - 85. In Russian. – Abstr. in Ref. zh., 51. Astron., 9.51.681 (1981).

114.097 Chemical composition of the atmosphere of the bright component of TX Leo.
M. L. Evtikhieva, V. V. Leushin.
Rostov. n/D. univ. Rostov n/D, 1981. 25 pp. In Russian. Abstr. in Ref. zh., 51. Astron., 9.51.682 (1981).

114.098 A compilation of physical parameters of Ap and Am stars as derived from energy distribution studies.
G. S. D. Babu, B. S. Shylaja.
Astrophys. Space Sci., Vol. 79, 243 - 255 (1981).

The physical parameters, such as the effective temperatures, radii, bolometric magnitudes and bolometric corrections of 87 Ap and Am stars out of the 125 compiled here are essentially based on the work done by the authors at the Kavalur Observatory. The results of the rest of the stars are from the earlier work done by one of the authors, which have been included for the sake of completion. All the results are derived from the observed energy distribution curves of the individual stars. The standardised magnitudes at various wavelengths also have been listed.

114.099 Metallicity indicators for the stellar and integrated light spectra.   M. Grenon.
Ann. Physique, Vol. 6, (see 012.026), 127 - 147 (1981) = Publ. Obs. Genève, Sér. A, Fasc. 84. In French.

The definitions, the physical meaning and the intrinsic limitations of selected photometric metallicity indicators are summarized in the stellar case and in that of integrated light of clusters and galaxies. Various calibrations are given and their accuracy is discussed.

114.100 SS 433: a new extraordinary object in astrophysics.   F. Ciatti.
Europhys. News, Vol. 12, No. 4, p. 8 - 11 (1981). − Abstr. in Phys. Abstr., Vol. 84, Abstr. 94372 (1981).

114.101 Spectrophotometry of two luminous variable stars in the Andromeda galaxy.
J. S. Gallagher, S. J. Kenyon, E. K. Hege.
Astrophys. J., Vol. 249, 83 - 92 (1981).

The authors present spectrophotometry of AF Andromedae and A-1 Andromedae, two S Doradus variables in M31. Emission lines of H Balmer, He I, Fe II, and [Fe II] dominate the spectra, and only Na D appears in absorption.

114.102 Constancy of the low resolution spectrum of SS 433.
C. B. Stephenson, N. Sanduleak.
Astrophys. J., Lett., Vol. 249, L19 - L20 (1981).

At the spectroscopic resolution from which the authors originally reported $H\alpha$ emission in SS 433 on plates taken in 1959, the appearance of the spectrum was the same in 1975 and 1981. The only well-observed line emission is the strongest $H\alpha$ component which does not partake of the large Doppler variations. A 10th−11th magnitude star 1ʹ5 from SS 433 is a K dwarf listed in the Bruce Proper Motion Survey General Catalogue.

114.103 SS 433 − a unique relativistic object.   U. Dzērvītis.
Zvaigžnotā debess, 1980. gada vasara, p. 16 - 18. In Latvian.

114.104 Lithium abundance in two halo stars.   M. Spite, F. Spite.
C. R. Acad. Sci. Paris, Tome 293, Sér. II, 299 - 301 (1981). In French.

114.105 Spectroscopic evidence for a wide range in abundances among faint subgiant stars in the globular cluster Omega Centauri.   R. A. Bell, G. L. H. Harris, J. E. Hesser, R. D. Cannon.
Astrophys. J., Vol. 249, 637 - 646 (1981).

Intermediate dispersion spectra have been obtained for 11 stars with $16.7 < B < 18.4 (+2 < M_V < 3.5)$ on the lower subgiant branch of ω Cen. Five of the stars, generally the redder ones, have radial velocities incompatible with cluster membership, while the remaining six stars are probably all cluster members. Five of the probable members have abundances $-1.0 \leqslant [M/H] \leqslant -1.5$, with some evidence for carbon depletion. The sixth has $[M/H] \approx -0.5$, as high an abundance as is known for any of the more highly evolved stars in ω Cen. This object also has an enhanced nitrogen abundance, the enhancement being possibly as much as a factor of 3. Possible reasons for the variation in the nitrogen abundance are discussed.

114.106 The effective temperature scale.   E. Böhm-Vitense.
Annu. Rev. Astron. Astrophys., Vol. 19, (see 003.012), 295 - 318 (1981).

The author finds that the agreement between radiative equilibrium model atmosphere $T_{eff}(B-V)$ relation and observed direct determinations is generally rather good showing, that we may understand the structure of most atmospheres reasonably well. There are still some problems in the main-sequence O and F stars and in the M star regions. The red giants and supergiants also pose some problems. The author does not know whether Rayleigh scattering alone can explain the red colors for giants around 3600 K. Model atmosphere colors for such stars have not been published. The metal-poor giants seem to be well represented by the model calculations; for the metal-poor F stars there are still some uncertainties.

114.107 Analysis of the spectrum of Arcturus.   T. Kipper, M. Kipper, J. Sitska.
Tartu Astrofüüs. Obs., Teated, Nr. 64, p. 3 - 14 (1981).

Some basic parameters of Arcturus including the effective temperature, the surface gravity, and the abundance of some elements are derived.

114.108 The chemically peculiar stars: an approach via the method of multiple working hypotheses.
C. R. Cowley.
Upper main sequence chemically peculiar stars, (see 012.033), p. 5 - 9 (1981).

114.109 The spectra of the chemically peculiar stars.   M. Hack.
Upper main sequence chemically peculiar stars, (see 012.033), p. 79 - 108 (1981).

The main results obtained from the study of the continuous and line spectra of the CP stars are reviewed: a) classification systems based on visual and ultraviolet spectrophotometry; b) flux distribution and comparison with normal stars; c) magnetic fields and spotted distribution of the elements; d) surface chemical composition; e) membership of CP stars in binary systems and clusters; f) X-ray emission from CP stars.

114.110 A spectroscopic and photometric study of the Bp star HR 3413.
E. N. Walker, C. Lloyd, C. D. Pike, D. J. Stickland, M. M. Dworetsky.
Upper main sequence chemically peculiar stars, (see 012.033), p. 115 - 123 (1981).

The Si variable star HR 3413 has been investigated through simultaneous spectroscopic and photometric observations. The Si II lines exhibit marked equivalent width and radial velocity variations which have been used together with the photometric variations to model the stellar surface. Rough average surface chemical abundances have been obtained.

114.111 A spectroscopical study of the star Iota Cassiopeiae.   A. Woszczyk, M. Jasiński, M. Muciek.
Upper main sequence chemically peculiar stars, (see 012.033), p. 125 - 130 (1981).

The variations of radial velocity and equivalent widths with phase are found. The curve of growth analysis leads to the determination of the variation of Fe II abundance from 8.1 to 8.6 ($\log N_H = 12$) and indicates that the microturbulence varies on the surface of the star. A map of abundance distribution of iron is computed − iron lies in a band along a great circle.

114.112 The ultraviolet spectrum of α^2 CVn.   J. Van Santvoort, H. Hensberge.
Upper main sequence chemically peculiar stars, (see 012.033), p. 131 - 134 (1981).

The authors use the Wavelength Coincidence Statistics

method to identify some ions in a middle ultraviolet spectrum of α^2 CVn obtained with the balloon BUSS.

114.113 Preliminary results from UV spectrum of 78 Vir.
F. Castelli, R. Faraggiana, F. A. Catalano,
H. M. Maitzen.
Upper main sequence chemically peculiar stars, (see 012.033),
p. 135 - 139 (1981).

The flux distribution of 78 Vir derived from the UV and visual spectrum is compared with Kurucz's models: the best agreement is found for the model $T_e = 9500$ K, log g = 3.5 and 10 times the solar abundances. A preliminary synthetic spectrum is computed and compared with some selected ranges of the long-wave high-resolution IUE spectrum.

114.114 Calibration of IUE high dispersion spectra of magnetic stars and the variation of spectral lines in HD 219749.   M. Barylak, K. D. Rakosch.
Upper main sequence chemically peculiar stars, (see 012.033),
p. 141 - 147 (1981).

With the help of the IUE flux calibrated low dispersion spectra the high dispersion spectra were calibrated by a method described in this paper. In these flux calibrated spectra some spectral lines were investigated for variability.

114.115 Recent results from an IUE survey of mercurymanganese stars.   J. M. Jacobs, M. M. Dworetsky.
Upper main sequence chemically peculiar stars, (see 012.033),
p. 153 - 158 (1981).

The authors present some new results of analyses of high resolution IUE spectra of 11 HgMn stars and 3 normal stars of similar effective temperatures. An effective-temperature-dependent overabundance of copper has been found in several HgMn stars. Prominent ultraviolet lines of Hg III have been searched for without success, implying severe upper limits on Hg III transition probabilities. The gallium abundance anomaly in many HgMn stars is confirmed by the identification of several Ga II lines, and a preliminary abundance analysis of the ultraviolet Ga II lines is presented.

114.116 Abundance peaks in Hg-Mn stars.
A. Magazzù, V. Pirronello, G. Strazzulla.
Upper main sequence chemically peculiar stars, (see 012.033),
p. 159 - 162 (1981).

114.117 Gallium in Hg-Mn stars.   M. Takada, J. Jugaku.
Upper main sequence chemically peculiar stars,
(see 012.033), p. 163 - 168 (1981).

Quantitative analyses of the Ga abundances in seven Hg-Mn stars are made with the resonance lines of Ga II and Ga III in the ultraviolet region observed with the IUE. The main conclusions are as follows: (1) The Ga overabundance is confirmed and can be regarded as a genuine anomaly in Hg-Mn stars. (2) The Ga overabundance shows a dependence on the stellar effective temperature. (3) The ionization equilibrium between Ga II and Ga III is well realized in the cooler stars, while not in the hotter stars.

114.118 On the iron abundances in CP stars.
C. R. Cowley.
Upper main sequence chemically peculiar stars, (see 012.033),
p. 169 - 175 (1981).

Iron abundances in late B and A stars range from about one order of magnitude below to slightly more than one order of magnitude above the solar value. A review of iron abundances in a sample of CP stars is given. It is suggested that some processes may limit the maximum galactic iron abundance.

114.119 Quantitative analysis of the spectrum of the Ap(SiTi) star HD 66318.   J. Dachs.
Upper main sequence chemically peculiar stars, (see 012.033),
p. 177 - 181 (1981).

Extremely strong lines of Si, Ti, Cr, Fe, Sr, La and Eu are found on coudé spectrograms of the V = $9^m.65$ Ap star HD 66318 obtained with a dispersion of 20 Å/mm. Results of a coarse analysis of the atmosphere of the star are reported. A large overabundance of lanthanum (4.2 dex) relative to normal solar composition poses serious problems for diffusion theory to explain the anomalies. The star is a member of the open cluster NGC 2516; its evolutionary age is about 100 million years.

114.120 Binary and single Hg-Mn stars compared.
B. N. G. Guthrie.
Upper main sequence chemically peculiar stars, (see 012.033),
p. 189 - 193 (1981).

Hg-Mn stars are classified according to their binary characteristics. Short-period spectroscopic binaries and other stars are compared with regard to rotation, the abundances of Sc, Ti, Y, and Zr, and the isotopic compositions of Hg. The short-period binaries rotate more slowly and have lower abundances of Sc. The heavy isotopic compositions of Hg in double-lined binaries are noted.

114.121 New transition probabilities for Y II and Zr II and the Sr-Y-Zr abundance patterns in CP stars.
N. Grevesse, E. Biemont, P. Hannaford, R. M. Lowe.
Upper main sequence chemically peculiar stars, (see 012.033),
p. 211 - 222 (1981).

Accurate transition probabilities have been derived from measurements of lifetimes and branching ratios for Y II and Zr II lines present in the spectra of chemically peculiar stars of the upper main sequence. These new data are used to normalize the Sr-Y-Zr abundances in a large number of CP stars.

114.122 Mapping of chemical elements over surfaces of Ap stars.   A. V. Goncharski (*Goncharskij*),
A. G. Jagola (*Yagola*), V. V. Stepanov, V. L. Khokhlova.
Upper main sequence chemically peculiar stars, (see 012.033),
p. 263 (1981).

114.123 Hot peculiar stars.   C. Jaschek, M. Jaschek.
Upper main sequence chemically peculiar stars,
(see 012.033), p. 417 - 427 (1981).

114.124 Analysis via synthetic spectrum of the atmosphere of the star Feige 86 in the UV region.
C. Morossi, M. Ramella, M. Hack, F. Castelli.
Upper main sequence chemically peculiar stars, (see 012.033),
p. 445 - 449 (1981).

The chemical composition of the atmosphere of Feige 86 was studied in the UV spectral region. A synthetic spectrum was constructed from a blanketed model of $T_{eff} = 17500°$K and log g = 4.2 and from the Kurucz and Peytremann (1975) line list. The IUE Hi. Res. data were elaborated with a new procedure. Preliminary estimates of the chemical abundances are reported.

114.125 Am stars.   V. L. Khokhlova.
Upper main sequence chemically peculiar stars,
(see 012.033), p. 457 - 463 (1981).

114.126 The abundance of the lighter elements in metallicline stars.   J. B. Lester, M. C. Lane.
Upper main sequence chemically peculiar stars, (see 012.033),
p. 475 - 479 (1981).

As part of a much larger project to reexamine the physical properties of the Am stars, the authors have determined abundances for several of the lighter elements which occur before calcium in the periodic table. The abundances of several of these elements were very poorly known previously because of the lack of suitable features in the visual spectral region.

114.127 A high resolution IUE spectrum of the G0-G5Ia supergiant HR8752.
D. J. Stickland, D. L. Lambert.
Astron. Astrophys., Vol. 102, 296 - 298 (1981).

A high-resolution long-wavelength IUE spectrum of the G0-G5Ia supergiant HR8752 and its B main sequence companion is discussed. The supergiant, which dominates for $\lambda \gtrsim 2700$ Å, appears to have a normal spectrum.

114.128 Carbon, nitrogen, and oxygen abundances in main-sequence stars. II. 20 F and G stars.
R. E. S. Clegg, D. L. Lambert, J. Tomkin.
Astrophys. J., Vol. 250, 262 - 275 (1981).

High-resolution Reticon spectra of red and near-infrared C I, N I, and O I lines have been analyzed to determine C, N, and O abundances in a sample of 20 F and G main-sequence stars. Their iron abundances, which have been determined from analysis of additional Reticon spectra of red Fe I lines, cover the range $-0.9 \leqslant$ [Fe/H] $\leqslant$ +0.4. Sulfur abundances have also been obtained. The authors find that the variations of the carbon and sulfur abundances closely follow those of iron. The oxygen abundance varies much more slowly than iron ([O/Fe] = -0.48 (±0.07) [Fe/H]). This result confirms an earlier conclusion that oxygen is overabundant in more metal-deficient stars. The behavior of the nitrogen abundance appears to be similar to that of iron ([N/Fe] = 0.0 ± 0.2). However, the unavailability of nitrogen abundances for the most metal-deficient stars in the sample makes this result less certain than the results for the other elements. These results are discussed in the light of current theories of stellar nucleosynthesis of the elements.

114.129 Spectroscopic and photometric observations of the Be star 69 Orionis.
M. Bossi, G. Guerrero, L. Mantegazza, L. Rusconi, M. Scardia, G. Sedmak.
Astron. Astrophys., Suppl. Ser., Vol. 46, 173 - 177 (1981).

Spectroscopic and photometric observations of the Be star 69 Ori are reported. The structure and evolution of the stellar envelope are discussed qualitatively via the study of the H_α line. There are indications of a motion of matter towards the observer during the period of intensive activity and of a successive decrease of the rotation of the envelope as the emission began to disappear.

114.130 Observations spectroscopiques de CI Cygni.
C. Fehrenbach, C.-C. Huang.
Astron. Astrophys., Suppl. Ser., Vol. 46, 257 - 261 (1981).

Spectroscopic observations of the symbiotic star CI Cyg were performed between July 1980 and October 1980. A fair amount of [Fe V] and very strong emission lines of high excitation, including [O III], [Ne III], He II, O III, . . ., are the main spectral characteristics of CI Cyg. The emission lines H_α, H_β and H_γ are double. There are two groups of radial velocity, one is positive for the forbidden lines, the other is negative for the permitted lines. There is a regression of radial velocity for the lines of Balmer.

114.131 Spectrophotometric study of three supergiants of class F8.
A. A. Boyarchuk, M. E. Boyarchuk.
Izv. Krymskoj Astrofiz. Obs., Tom 63, 66 - 85 (1981). In Russian. English translation in Bull. Crimean Astrophys. Obs., Vol. 63.

The spectra of the supergiants α UMi, γ Cyg and ρ Cas have been studied. The equivalent widths of 800 absorption lines were measured. The analysis of the atmospheres has been carried out with the curve-of-growth method. The physical parameters of the stellar atmospheres, the excitation temperatures, the electron densities and the turbulent velocities were determined. It was found that the chemical composition of the investigated stars does not differ significantly from that of the sun, except for C and Na.

114.132 Spectrophotometric investigation of the T Tauri-like star AS 353.
V. I. Krasnobabtsev.
Izv. Krymskoj Astrofiz. Obs., Tom 63, 93 - 103 (1981). In Russian. English translation in Bull. Crimean Astrophys. Obs., Vol. 63.

Identification of emission lines and a comparative analysis of the equivalent widths and intensities in spectra of AS 353 were carried out. From the analysis of Fe II emission lines the amount of interstellar absorption is determined. The estimated spectral type is F0 - G5. The physical conditions in the circumstellar envelope are considered. Abnormally intensification of some Fe II and Ti II emission lines that occurs simultaneously with intensification of the hydrogen Hα and Hβ emission lines has been found.

114.133 On the behaviour of the lines 2905 Si II and 2932 Mg II in the spectra of B−A stars.
G. A. Gurzadyan, S. S. Rustambekova.
Astrophys. Space Sci., Vol. 80, 231 - 235 (1981).

On the basis of the Orion-2 observational data it is shown that the ratio of the intensities of absorption lines 2905(2904 + 2906) Si II and 2932(2928 + 2936)Mg II varies strongly from B0 stars to A0. This circumstance may be used as a sensitive indicator for the determination of the effective temperatures of hot stars.

114.134 Infrared lines of O I and Ca II in Be stars with Paschen emission lines.
D. Briot.
Astron. Astrophys., Vol. 103, 1 - 4 (1981).

Emission lines of elements other than hydrogen in the near infrared, that is OI λ 8446 and the infrared triplet of Ca II are studied for Be stars with Paschen emission lines. In agreement with the prediction by Bowen, the author finds that the correlation between the equivalent widths of the OI λ 8446 and Hα emission lines is better than the correlation between the equivalent widths of the OI λ 8446 and Hβ emission lines. There is a slightly larger infrared excess on the average for stars with the infrared triplet of CaII in emission than for stars without this.

114.135 Paschen lines in Be stars. II. Study of Paschen emission lines.
D. Briot.
Astron. Astrophys., Vol. 103, 5 - 18 (1981).

Intensities of the Paschen and Balmer emission lines are determined for a sample of 15 Be stars, with spectral types ranging from B0e to B5e. The observational results are compared with predictions based on Sobolev's theory. No agreement is obtained between observations and theoretical calculations when electron collisions in the circumstellar envelope are neglected. Agreement can be obtained between the observations of the hottest stars (B0e, B1e) and the theoretical calculations when electron collisions are taken into account. However none of these theoretical calculations agrees with the intensities of the Paschen and Balmer emission lines observed for Be stars of later spectral sub-types.

114.136 Detailed analysis of a G supergiant in the Small Magellanic Cloud.
R. Foy.
Astron. Astrophys., Vol. 103, 135 - 139 (1981).

The author reports the detailed analysis of the G type supergiant AZ 369 in the Small Magellanic Cloud. It is based on high dispersion spectrograms obtained with the Lallemand-Duchesne electronic camera installed at the échelle spectrograph of the ESO 1.52 m telescope. Spectra are interpreted using model atmospheres. The author derives an abundance -0.4 ± 0.3 dex for Fe, Cr, and Ti with respect to the Sun. He discusses the possible difference between this result and the abundance derived for oxygen from gaseous nebulae.

114.137 Ultraviolet observations of two extreme Population II stars: detection of chromospheric emission and mass loss.
M. Spite, V. Caloi, F. Spite.

Astron. Astrophys., Vol. 103, L11 - L13 (1981).

The authors describe some spectroscopic features in the ultraviolet and visible ranges for two extreme Population II stars. These features can be interpreted as indicating mass loss.

114.138 La misura dei raggi stellari.
G. Longo, M. Rigutti.
Coelum, Vol. 50, 202 - 213, 230 - 249 (1981).

114.139 Balloon and IUE spectra of early-type stars.
B. Bates, D. L. Giaretta, D. J. McCartney,
J. A. McQuoid.
Irish Astron. J., Vol. 14, (see 012.034), 114 - 121 (1980).

114.140 R 136a: une étoile 2000 fois plus massive que le soleil?   A. Noels.
Ciel Terre, Vol. 97, 371 - 374 (1981).

Recent observations with the International Ultraviolet Explorer of the central object (R 136a) of the Tarantula Nebula in the Large Magellanic Cloud, the most luminous H II region in the Local Group of galaxies, seem to support the hypothesis of a supermassive star of about 2000 solar masses, with a luminosity of about 10^8 times the luminosity of the Sun and a radius of about 100 solar radii.

114.141 Abundance fluctuations in superficially normal A stars.
C. R. Cowley, R. L. Sears, G. C. L. Aikman, K. Sadakane.
J. R. Astron. Soc. Canada, Vol. 75, 246 - 247 (1981).
Abstract.

114.142 The chromospheric Mg II k_2 emission in stars with different rotational velocities and different metal abundances.   E. Böhm-Vitense.
Publ. Astron. Soc. Pacific, Vol. 93, 545 - 546 (1981).
Abstract.

114.143 Problems in the spectral classification of the red giants.   P. C. Keenan.
Publ. Astron. Soc. Pacific, Vol. 93, 548 (1981). – Abstract.

114.144 A study of radiative losses in calcium H and K and infrared lines in stars with active chromospheres.
R. A. Mielbrecht, A. Young.
Publ. Astron. Soc. Pacific, Vol. 93, 549 (1981). – Abstract.

114.145 Abundance analyses of 21 Pegasi (B9.5 V) and HR 7338 (A0 III).   K. Sadakane.
Publ. Astron. Soc. Pacific, Vol. 93, 587 - 593 (1981).

Abundances of 15 elements are determined in two superficially normal stars 21 Peg (B9.5 V) and HR 7338 (A0 III) with model atmosphere techniques. Abundances of metals are nearly solar or slightly underabundant in 21 Peg except for Mn, Y, and Ba, which may be overabundant. Nearly all the metals are underabundant in HR 7338.

114.146 Line identification in the spectrum of HD 213918.
S. Malaroda.
Publ. Astron. Soc. Pacific, Vol. 93, 614 - 620 (1981).

The spectrum of the star HD 213918 has been studied. Atomic species present on three Coudé spectrograms of the star are identified. Fe in the first three stages of ionization, P II, P III, Cl II, Sr II, Hg II are present. The star has a temperature around 14800 K. Its spectrum is compared with similar silicon and He-weak stars.

114.147 SS 433 – le mystère du siècle?
M. Verdenet, E. Schweitzer.
Bull. AFOEV, No. 18, p. 1 - 5 (1981).

114.148 Far UV observations of late K and M type stars.
D. J. Stickland, F. Sanner.
Mon. Not. R. Astron. Soc., Vol. 197, 791 - 798 (1981).

The far UV spectra of 15 late K and M type giant and supergiant stars have been recorded with *IUE*. They exhibit rather flat continua which change strength with spectral type and appear to originate in the low chromosphere. A number of chromospheric emission lines have also been measured and identifications are suggested for many of them.

114.149 Spectroscopic observations of two radio-quiet BL Lac candidates.
D. Wills, B. J. Wills.
Mon. Not. R. Astron. Soc., Vol. 197, 27P - 28P (1981).

The optical objects near the positions of the radio sources 1210 + 121 and 1620 + 103 are galactic stars, rather than radio-quiet BL Lacertae objects. Whether radio-quiet BL Lac objects exist remains an open question at present.

114.150 The resonance line of B II in *IUE* spectra of chemically peculiar stars.   D. S. Leckrone.
Astrophys. J., Vol. 250, 687 - 700 (1981).

High-dispersion *IUE* spectroscopic observations of the B II resonance line at 1362.46 Å are presented for six HgMn stars (κ Cnc, μ Lep, 46 Dra, ι CrB, HR 4072, and χ Lup), two magnetic peculiar stars (α^2 CVn and HD 32633), and one normal B9.5 V star (ν Cap). The B II line is present in great strength in κ Cnc. It is absent from the five other HgMn stars. It is probably present, though unresolved and weak, in the magnetic stars. Its strength in ν Cap is consistent with a normal boron abundance.

114.151 The spectra of two new intermediate helium stars.
J. S. Drilling.
Astrophys. J., Vol. 250, 701 - 703, plates 19, 20 (1981) =
Contrib. Louisiana State Univ. Obs. No. 165.

The stars designated as LSS 2394 and LS II +36°37 in the Case-Hamburg Luminous Stars surveys are found to be intermediate helium stars. The environment and spectrum of LSS 2394 (CPD −62°2124) appear to be quite similar to those of σ Ori E in a number of respects, including close proximity to two other OB stars, membership in a rich O association, spectrum variability, rapid rotation, and broad Hα emission wings. A low-resolution *IUE* spectrum for this star is presented and discussed.

114.152 The visible spectra of three B stars: π Cet, 36 Lyn, 134 Tau.   L. Crivellari, U. Flora, M. Mercanti,
C. Morossi, L. Rusconi, G. Sedmak.
Astrophys. Space Sci., Vol. 80, 425 - 436 (1981).

The authors present an identification list for the visible spectra of three B stars: π Cet (B7V), 36 Lyn B8 IIIp (?), and 134 Tau (B9 IV). Equivalent widths have been measured on medium dispersion plates (7.0 and 12.4 Å mm^{-1}) taken with the 152 cm coudé reflector at the Observatoire de Haute Provence (France). These results are also presented. The photographic plates were digitized by using a PDS 1010A microdensitometer. The spectroscopic data were reduced by means of a dedicated software package and an attempt was made to compute equivalent widths in a homogeneous way. T_{eff} and log g parameters were estimated by using the computer to compare automatically the spectroscopic data with the value of theoretical models.

114.153 On the nebulosities associated with the extreme Of star HD 148937.   F. C. Bruhweiler, T. R. Gull,
K. G. Henize, R. D. Cannon.
Astrophys. J., Vol. 251, 126 - 132, plates 2 - 6 (1981).

The extreme Of star HD 148937 and its associated nebulae, including NGC 6164-5, a 2° diameter H II region, and an outer dust shell, plus a newly identified interstellar bubble, are studied to understand their interrelations and thereby to understand the origin of HD 148937. Information from far-ultraviolet spectroscopy of the central star, narrow passband emission-line photography, and broad-band survey plates is assembled into a self-consistent model describing the interaction of HD 148937 with the surrounding interstellar bubble-H II region complex. The star HD 148937 is found to be quite young,less than 350,000 years in age. Constraints on its luminosity suggest it is still in the pre-main-sequence contraction phase.

114.154　A Fourier analysis of the spectral lines of Procyon.
D. F. Gray.
Astrophys. J., Vol. 251, 152 - 154 (1981).

The mean residual Fourier transform is obtained by combining the results of 34 individual lines. The analysis of this mean indicates a radial-tangential macroturbulence dispersion of 7.0 ± 0.1 km s^{-1} and a projected rotation rate of 2.8 ± 0.3 km s^{-1}. The individual residual transforms show differences from the mean that are correlated with excitation potential. This is interpreted as an effect of microturbulence, possibly a depth dependence.

114.155　An upper limit for the deuterium abundance in Canopus.
M. Peimbert, G. Wallerstein, C. A. Pilachowski.
Astron. Astrophys., Vol. 104, 72 - 74 (1981).

Canopus is a supergiant star, F0 Ib, that probably did not destroy its original deuterium during contraction to the main sequence. From observations of Hα an upper limit for the deuterium abundance has been obtained. This upper limit is comparable to the ratio observed in the interstellar medium and less than the upper end of the spread of interstellar deuterium values. Possible explanations for this result are discussed in terms of mass loss and mixing in stars with masses near $10\,M_\odot$.

114.156　The Sr-Y-Zr abundance peak in HR 6127.
V. Pirronello, G. Strazzulla.
Astron. Astrophys., Vol. 104, 80 (1981).

The relative element abundance in the Sr-Y-Zr peak on HR 6127 is the only feature that contradicts the classification of this star as a cool Ap. The authors show that using more appropriate gf-values the abundances of Y and Zr are considerably reduced and become of the same order as those observed in Am stars rather than in cool Ap stars.

114.157　Ultraviolet observations of the Be star and X-ray binary 4U 1145–61 (= HD 102567 = Hen 715) obtained with the IUE.
C. de Loore, M. Burger, H. Hensberge, E. L. Van Dessel.
Astron. Astrophys., Vol. 104, 150 - 154 (1981).

Two high resolution spectrograms in the wavelength region λλ 1150 - 2050 have been obtained using the IUE satellite. From these spectrograms line lists of the stellar and interstellar absorption have been compiled. The most important ions in the spectrum are He$^+$, one, two, and three times ionized C and Si, two and three times ionized O, three times ionized Mn and two, three and four times ionized N. From the resonance lines of C IV, Si IV, and N V, a mass loss rate of $3 \times 10^{-10}\,M_\odot$ yr^{-1} is derived. Eight interstellar molecular bands of $^{12}C^{16}O$ of the fourth positive system ($A^1\pi - X^2\Sigma^+$) are present.

114.158　Spectra of the red (2,0) CN band in 31 G and K giant stars.
P. Kjaergaard, G. A. H. Walker, S. Yang.
Astron. Astrophys., Suppl. Ser., Vol. 46, 375 - 388 (1981).

As part of a project aimed at determining C, N and O abundances in a number of G and K giant stars, spectra of the red (2,0) CN band have been observed. The observations were carried out with the 1.2 meter telescope at the Dominion Astrophysical Observatory, Victoria. A 120 Å region around 7980 Å was observed with a 1024 Reticon diode array refrigerated by liquid nitrogen at a spectral resolution of 0.25 Å.

The spectra were calibrated for wavelength, corrected for the influence of telluric water vapour lines, and rectified to a continuum. Tables of equivalent widths for selected ^{12}CN and atomic lines are given. The error in an equivalent width is typically ± 8 mÅ when compared with independent measurements by others. As a corollary to the water vapour correction procedure, high precision radial velocities (± 0.5 km/s) have been derived.

114.159　Asymmetries in the spectral lines of Procyon.
D. F. Gray.
Astrophys. J., Vol. 251, 583 - 584 (1981).

Asymmetries are measured in the spectral line profiles of the F5 IV-V star Procyon. A blueshifted core asymmetry is seen in all of the lines. Amplitudes of the asymmetry are proportional to line strength being 12% of the central depth. The characteristic widths of the asymmetry change only modestly with line strength, ranging from ~2.5 km s^{-1} for the weakest lines to ~3.5 km s^{-1} for the strongest lines. These traits differ markedly from those characterizing the line asymmetries of Arcturus.

114.160　"P Cygni" profiles in P Cygni.　L. Goldberg.
Astron. Astrophys., Vol. 104, L7 - L9 (1981).

Underhill has proposed that two strong emission features near λ1306 and λ1309 in the spectrum of P Cygni are the result of dielectronic recombination through the autoionizing levels 3s3p3d $^2F^0$ of Si II, giving rise to multiplet UV 13.04. The lines of this multiplet are closely blended with resonance lines of O I, Si II, and to some extent N I. It is shown that the observed spectrum may be accounted for by P Cygni profiles of the resonance lines, and that in the absence of supporting qualitative calculations the dielectronic recombination hypothesis is unjustified.

114.161　On some extreme metal-deficient giants.
A. Bartkevičius, V. Straižys.
Astron. Astrophys., Vol. 104, 215 - 217 (1981).

The temperatures, metallicities, and color excesses of five giants with extreme metal-deficiency are discussed. Their observations in the Vilnius photometric system after correction for interstellar reddening yield the following [Fe/H] values: -3.0 for BD $-18°5550$ and HD 4306, -2.8 for CD $-25°13871$ and HD 115444, and -2.6 for HD 126587. Their effective temperatures are within 4800–4950 K.

114.162　Transition region structure in F dwarfs.
M. Saxner.
Astron. Astrophys., Vol. 104, 240 - 248 (1981).

Strong transition region line emission has been detected in the IUE spectra of two F dwarfs. The corresponding surface fluxes are comparable to those of active binaries, such as RS CVn-systems. Line flux ratios formed from the multiplets Si IV λ 1394 + λ 1403, C IV λ 1549, and N V λ 1240 have been used to derive the structure of the transition region in the F dwarfs and other active stars. It is proposed that the large surface fluxes and pressures are a consequence of the rapid rotation of these stars.

114.163　A classification system for O – B2 stars based on the Si IV and C IV resonance lines.
K. G. Henize, J. D. Wray, S. B. Parsons.
Astron. J., Vol. 86, 1658 - 1670 (1981).

Low-dispersion ultraviolet spectra from Skylab experiment S-019 are used to explore the variations of Si IV and C IV line strengths with temperature and luminosity. These considerations lead to a classification system in which the Si/C ratio is used to discriminate luminosity among the O stars and temperature among the O9 – B2 stars of lower luminosity. Stars falling in these two regimes may be distinguished either by the presence of C IV emission or on the basis of C IV absorption strength. The log (Si IV/C IV) vs C IV diagram is proposed as a primary tool in such a classification system. The rapid variation in the Si IV/C IV ratio from less than 1/10 at O9 to greater than 10 at B1.5 for luminosity class III−V stars appears to be an especially useful criterion for the temperature classification of stars in this spectral range.

114.164 Hydrogen lines in the spectrum of the Ap star HD 184905. D. Kolev.
Nauchn. Inf., Vyp. (No.) 43, p. 3 - 16 (1980). In Russian.

Variations of the Balmer-line intensities are investigated using 16 spectrograms of the Ap star HD 184905. The mean electron density in the atmosphere is estimated. From the Hγ profiles the mean LTE model atmosphere is obtained. The spectral class derived from hydrogen lines and UBV photometry is B7pV. The spectrophotometric errors are estimated.

114.165 Spectrophotometric investigation of the Ap star HD 184905. D. Kolev.
Nauchn. Inf., Vyp. (No.) 43, p. 17 - 47 (1980). In Russian.

An investigation of spectral variability of the Ap star HD 184905 is carried out. The mean surface abundances of Mg, Si, Ca, Ti, Cr, Fe, Sr, and Eu are estimated in the phases of maximum and minimum line intensities. A separation of Mg, Si, Sr, and Eu lines into components is carried out. Some geometric and kinematic parameters of the star are obtained.

114.166 Determination of carbon and magnesium abundances in the atmospheres of Ap stars.
D. A. Ptitsyn, T. A. Ryabchikova.
Nauchn. Inf., Vyp. (No.) 43, p. 48 - 64 (1980). In Russian.

Theoretical equivalent widths of the Ca II lines $\lambda\lambda$ 3920.68, 4267.27 and the Mg II lines $\lambda\lambda$ 4390.56, 4481.33 in stellar spectra are calculated for the set of Mihalas' model atmospheres with $0.25 < \theta_e < 0.60$, lg g = 4.0 and various abundances of elements and microturbulent velocities. The carbon and magnesium abundances in the atmospheres of 62 Ap stars are determined using these calculations. The results are compared with those obtained by other authors.

114.167 Interrelation between the parameters of analytical approximation of a line profile in a stellar spectrum.
V. M. Pavlova, V. L. Khokhlova.
Nauchn. Inf., Vyp. (No.) 43, p. 65 - 69 (1980). In Russian.

The convenience of Minnaert's formula for the analytical approximation of absorption line profiles computed by numerical integration of the transfer equation is shown. The parameters of the analytical approximation are determined and their interconnection is considered.

114.168 Distribution of Si on the surface of the silicon Ap star CU Virginis and interpretation of the light curves in the v, b, y bands.
A. A. Krivosheina, T. A. Ryabchikova, V. L. Khokhlova.
Nauchn. Inf., Vyp. (No.) 43, p. 70 - 80 (1980). In Russian.

Observations of equivalent widths and radial velocity variations of Si II lines in the spectrum of the Si star CU Vir are presented. Theoretical light curves in the v, b, y bands are calculated.

114.169 Rapid neutron capture process and abundances of heavy elements in Ap stars.
D. A. Ptitsyn, B. Kuchowicz.
Nauchn. Inf., Vyp. (No.) 43, p. 81 - 87 (1980). In Russian.

A calculation of the process of neutron capture on a rapid time scale at constant temperature and neutron density assuming various durations of neutron exposure is carried out. The abundances of heavy elements observed in Ap stars are compared with those resulting from the r-process calculation.

114.170 Age-metallicity correlation determined spectroscopically for disk population stars.
S. V. Vereshchagin.
Nauchn. Inf., Vyp. (No.) 47, (see 003.016), p. 51 - 58 (1981). In Russian.

The average rate of enrichment of the galactic disk by heavy elements has been determined by three original methods. The group of A−K dwarfs from the catalogue by Morel et al. (1976) with values [Fe/H] determined spectrally is used. The value ∂[Fe/H]/∂t = 3.0 ±0.8 per 10^{11} years^{-1} is obtained.

114.171 HR 465 returning to rare-earth-maximum phase.
C. R. Cowley, J. B. Rice.
Nature, Vol. 294, 636 - 637 (1981).

The authors report three 2.4 Å mm^{-1} spectrograms of HR 465 taken at the Dominion Astrophysical Observatory in August 1981 which show that the spectrum now very closely resembles that of the early 1960s − the rare-earth-maximum phase has returned.

114.172 The astrophysical interest of EUV observations of extra-solar objects. A. K. Dupree.
Space Sci. Rev., Vol. 29, (see 012.043), 479 - 487 (1981).

The feasibility of observation of EUV sources is assessed. Many stars have been detected in the EUV range ($\lambda\lambda$100−1000); line fluxes from others can be predicted. Selected astrophysical problems are reviewed that can benefit from EUV spectroscopy. Included among them are the physics and dynamics of stellar coronae, confirmation of nuclear surface burning on cataclysmic variables, evolutionary properties of white dwarfs, the helium abundance in the interstellar medium, and spectroscopic signatures of neutrino oscillations.

114.173 Intrinsic properties of carbon stars. II. Spectra, colours, and the HR diagram of cool carbon stars.
T. Tsuji.
J. Astrophys. Astron., Vol. 2, 253 - 276 (1981).

On the basis of the effective temperature scale proposed previously for cool carbon stars, other intrinsic properties of them are examined in detail. It is shown that the major spectroscopic properties of cool carbon stars can most consistently be understood on the basis of the author's new effective temperature scale. Various photometric indices of cool carbon stars also appear to be well correlated with the new effective temperatures. As effective temperatures of some 30 carbon stars are now obtained, the calibration of any photometric index is straightforward, and some examples of such a calibration are given. An observational HR diagram of red giant stars, including carbon stars as well as K−M giant stars, is obtained on the basis of the author's colour index-effective temperature calibrations and the best estimations of luminosities. Such an observational HR diagram of red giant stars shows rather a poor agreement with the current stellar evolution models.

114.174 The "continuous" central stars of planetary nebulae − are their spectra really continuous?
R. P. Kudritzki, K. P. Simon, R. H. Méndez.
Messenger, No. 26, p. 7 - 9 (1981).

114.175 Systematization of late-type giant stars.
V. Straižys.
Bull. Vilnius Astron. Obs., Nr. 56, p. 38 - 45 (1981).

Different types of late-type giants are represented in three-dimensional space with the coordinates [Fe/H], [s/Fe] and O/C. The low temperature group (T < 3800 K) includes oxygen-rich giants of types M and S and carbon-rich giants of type N. The higher temperature group (T = 5000 −3800 K) includes oxygen-rich G−K giants and subgiants with different metallicity and carbon-rich barium stars, CH- and R-type stars as well as different transition cases.

114.176 Energy distribution in stellar spectra of different spectral types and luminosities. IV. The metal-deficient stars. A. Bartkevičius, Z. Sviderskienė.

Bull. Vilnius Astron. Obs., Nr. 57, p. 35 - 61 (1981). In Russian.

Energy distribution curves in the spectra of 10 extreme subdwarfs and 5 metal-deficient giants are obtained using the spectrophotometric and photometric data of different authors in the interval λ 3000 - 11 000. The relation between observed and calculated color indices in the Vilnius photometric system is shown.

114.177 **MK classifications for F- and G-type stars. IV.**
E. A. Harlan.
Astron. J., Vol. 86, 1896 - 1897 (1981) = Lick Obs. Bull., No. 901.

MK spectral classifications are given for 112 stars brighter than $m_v = 7.5$, having HD types F2–G5. The classifications were made on slit spectrograms of dispersion 75 Å mm⁻¹ at Hγ.

114.178 **Abundance determinations of field horizontal-branch stars.** S. C. Danford, S. M. Lea.
Astron. J., Vol. 86, 1909 - 1915 (1981).

Atmospheric metal abundances are rederived for five A-type field horizontal-branch stars. New equivalent widths for HD 2857 (−6°86) and HD 60778 give [Fe/H] = −1.8, −0.5 (±0.3), respectively. Abundances for HD 86986, 109995, and 161817, rederived from previously published equivalent widths, are [Fe/H] = −1.0, −0.5, −1.0. No evidence is seen for an overdeficiency of s-process elements for the most metal-poor stars, nor for a dependence of [Al/Fe] on the iron-to-hydrogen ratio.

114.179 *σ* **Andromedae.**
IAU Circ., No. 3650 (1981).

114.180 **HDE 245770.**
IAU Circ., No. 3655 (1981).

114.181 **The spectrum of the suspected variable HD 33331.**
J. Andersen, B. Nordström.
Inf. Bull. Variable Stars, No. 2024, 2 pp. (1981).

114.182 **On variation of the Hα and H͞α profiles and intensities in the spectrum of SS 433.**
V. P. Arkhipova, V. F. Esipov, N. Yarovenko.
Astron. Tsirk., No. 1131, p. 3 - 5 (1980). In Russian.

114.183 **Energy distribution in the spectra of five stars of the cluster NGC 6913 from photometric estimates through UBVRI filters.** R. M. Raznik.
Astron. Tsirk., No. 1128, p. 5 - 7 (1980). In Russian.

114.184 **Standard chemical composition of the atmospheres of A1 - G0 spectral class stars.**
Z. N. Fenina, Yu. S. Romanov.
Astron. Tsirk., No. 1144, p. 1 - 2 (1980). In Russian.

114.185 **The characteristics of cool stars.**
N. S. Komarov.
Astron. Tsirk., No. 1144, p. 3 - 4 (1980). In Russian.

114.186 **Unique object SS433.**
P. I. Kolykhalov, Yu. Eh. Lyubarskij.
Priroda, 1981, No. 12, p. 19 - 23. In Russian.

114.187 **The secret of SS 433.** M. S. Dimitrijević.
Vasiona, Année 29, 31 - 34 (1981). In Croatian.

114.188 **Carbon stars with very large infrared colour indices.**
A. Alksnis.
Investigations of the sun and red stars. 11, (see 003.021), p. 5 - 17 (1980). In Russian.

Observational data – colour indices K–O and K–L, L-magnitude range, equatorial coordinates and, partly,

V-magnitudes for 28 IRC and (or) AFGL/CRL objects identified as carbon stars with colour index K−O ⩾ 4ᵐ9 − are compiled from different sources. References to surrounding charts are given. V-magnitude variations between 15ᵐ6 and 17ᵐ4 for the star identified as the optical counterpart of AFGL 2699 have been found.

114.189 **New carbon stars BC 186 - BC 200.** I. Platais.
Investigations of the sun and red stars. 12, (see 003.022), p. 19 - 24 (1981). In Russian.

In several fields of the sky 15 new carbon stars have been found. Equatorial and galactic coordinates along with finding charts are given.

114.190 **New carbon stars BC 201 - BC 217.**
Z. Alksne, A. Alksnis.
Investigations of the sun and red stars. 12, (see 003.022), p. 24 - 30 (1981). In Russian.

17 new carbon stars were found on photographs. Their equatorial and galactic coordinates as well as spectral types and finding charts are given.

114.191 **New carbon stars at galactic longitudes 128°- 140°.**
Z. Alksne, A. Alksnis, I. Eglitis.
Investigations of the sun and red stars. 13, (see 003.023), p. 5 - 11 (1981). In Russian.

17 new carbon stars have been found in three fields at galactic latitude +7° and longitude 128°- 140°. Their equatorial and galactic co-ordinates are given. The location of the carbon stars is shown on identification charts.

114.192 **WR stars: continuous spectra.** V. I. Stebnev.
Use of model atmospheres for the interpretation of stellar spectra, (see 012.054), p. 83 - 87 (1980). In Russian.

114.193 **Chemical composition of Procyon. Model analysis.**
V. V. Leushin, L. I. Snezhko, V. V. Sokolov.
Use of model atmospheres for the interpretation of stellar spectra, (see 012.054), p. 88 - 89 (1980). In Russian.

114.194 **Radiation of the gaseous envelope of β Lyrae in 1976 - 1977.**
Ya. T. Blagodyr, M. Yu. Skul'skij.
Tsirk. Astron. Obs., L'vov, No. 54, p. 21 - 26 (1979). In Russian.

114.195 **Abundance analysis of the helium-weak star 20 Tauri.** M. Mon, R. Hirata, K. Sadakane.
Publ. Astron. Soc. Japan, Vol. 33, 413 - 426 (1981).

An abundance analysis of the helium-weak star 20 Tauri is performed with a fully line-blanketed model atmosphere. A value of $\log N(\text{He})/N(\text{H}) = -1.7$ is found from the average of six He I lines. Mg, Si, Ca, and Ni are underabundant, while P and Mn are overabundant. The abundances of C, Ti, Cr, and Fe coincide with the solar values. The observed abundance pattern in 20 Tau is quite different from those in other helium-weak stars, while it shows a mild characteristic of Mn-Hg stars.

114.196 **Time series infrared spectroscopy of o Ceti.**
K. H. Hinkle.
J. American Assoc. Variable Star Obs., Vol. 10, 35 (1981). Abstract.

114.197 **A spectroscopic study of the supergiant ρ Cas.**
M. E. Boyarchuk.
Proceedings of the Third Finnish-Soviet Astronomical Symposium, (see 012.057), p. 7 - 15 (1981).

114.198 **On the opacity of Be star envelopes in the Balmer continuum.** I. N. (*Ya. N.*) Chkhikvadze.
Proceedings of the Third Finnish-Soviet Astronomical

Symposium, (see 012.057), p. 31 - 42 (1981).

Some spectral and photometric characteristics of the 14 Be stars known to have anomalously large Balmer jump are considered. It appears that almost all the objects with large jumps can be classified as shell stars. It is concluded that the increase of the jumps in all the stars considered is of the same nature, namely it is connected with a large optical depth of the envelope in the Balmer continuum.

114.199 **Oxygen abundance in the red giants HD 148897 and α Boo.** I. Tuominen, J. Kyröläinen.
Proceedings of the Third Finnish-Soviet Astronomical Symposium, (see 012.057), p. 61 - 66 (1981).

114.200 **An investigation of the infrared radiation of stars with anomalous chemical composition.**
Yu. K. Melik-Alaverdian (*Melik-Alaverdyan*).
Proceedings of the Third Finnish-Soviet Astronomical Symposium, (see 012.057), p. 67 - 69 (1981).

114.201 **Spectral classification of barium stars.** Y. Yamashita, Y. Norimoto.
Ann. Tokyo Astron. Obs., Second Ser., Vol. 18, 125 - 141 (1981).

Low dispersion spectra of 48 certain barium stars and 28 marginal barium stars selected from the stars found by the Michigan Spectral Survey of the southern sky were obtained at Okayama with the use of the Cassegrain spectrograph (73 Å mm^{-1} at Hγ) of the 91-cm telescope. These spectra were classified by the spectral type (temperature class), the luminosity class and the Ba-abundance parameter. A third of the barium stars observed here shows weakening of metallic lines.

114.202 **On the oxygen abundance of the metal deficient dwarfs σ Bootis and μ Cassiopeiae.**
H. Yokoo, M. Simoda.
Bull. Tokyo Gakugei Univ., Sect. IV, Vol. 33, 203 - 206 (1981).

The oxygen triplet $\lambda\lambda$ 7772 - 7775 Å of σ Boo and μ Cas was observed and analyzed relative to α CMi and the sun, respectively. It is found that σ Boo has roughly the same deficiency in oxygen as in metals and the oxygen to hydrogen ratio for μ Cas is smaller than that for the sun.

114.203 **The Wolf-Rayet stars.** Lectures delivered at the Collège de France, Paris, on February 19, 21, 26 and 27 and March 4, 11, 18 and 25, 1980. J. Sahade.
Published by Collège de France, Chaire d'Astrophysique Théorique, Paris, France. 162 pp. (1980).
Contents: Observational features. Wolf-Rayet binaries. The O stars. Objects with Wolf-Rayet spectra. Evolutionary considerations.

114.204 **Optical continuum of V1016 Cygni.**
B. F. Yudin.
Astron. Tsirk., No. 1160, p. 4 - 7 (1981). In Russian.

114.205 **Spectral energy distributions of Oe- and Oef-type stars.** U. Hänni.
Academy of Sciences of the Estonian SSR, Prepr. A-4, p. 1- 59 (1981).

Scanner observations of the spectral energy distributions of O, Oe and OBpe stars combined with TD-1 and OAO-2 ultraviolet spectrophotometry have been compared with predictions of extended model atmospheres. Pronounced flattening of the energy distribution curves of the Oe dwarfs HD 39680, HD 60848 and of the OBpe star X Per proved to be in evidence.

A catalogue of [Fe/H] determinations. (Magnetic tape version, updated to 31.12.1980).
See Abstr. 002.001.

The Sixth Catalogue of galactic Wolf-Rayet stars, their past and present. See Abstr. 002.012.

A bibliography on galactic Wolf-Rayet literature 1867–1980. Appendix to the Sixth Catalogue of galactic Wolf-Rayet stars, their past and present. See Abstr. 002.013.

Analysis of IUE spectra using the Interactive Data Language. See Abstr. 021.008.

Problems and programming for analysis of IUE high resolution data for variability. See Abstr. 021.009.

Density sensitive C II lines in cool stars of low gravity. See Abstr. 022.003.

New spectroscopic data for the iron group elements. See Abstr. 022.101.

Washburn extraction and width of IUE point spread function. See Abstr. 031.543.

Results of basic improvements to the extraction of spectra from IUE images. See Abstr. 031.544.

Effects of temperature fluctuations on IUE data quality. See Abstr. 031.546.

The inapplicability of the Fourier convolution method to the analysis of late-type stellar spectra. See Abstr. 031.549.

L'observation et les propriétés des étoiles. See Abstr. 031.560.

On recording of black body radiation. See Abstr. 031.573.

Digital image processing of Ap-star coudé Zeeman plates. See Abstr. 031.575.

Automatic quantitative spectral classification of stars. I. Observational data reduction. See Abstr. 031.653.

Outlook for ultraviolet astronomy. See Abstr. 064.016.

An IUE's eye view of cool-star outer atmospheres. See Abstr. 064.019.

The structure of chromospheres around late-type giants and supergiants. See Abstr. 064.020.

Empirical wind models from detailed UV-line fits: Tau Scorpii. See Abstr. 064.034.

Mass loss rates from O stars in OB associations. See Abstr. 064.080.

Outer atmospheres of cool stars. VIII. IUE observations and chromospheric models for the supergiant stars β Draconis, ϵ Geminorum, and α Orionis. See Abstr. 064.084.

Mass loss from O subdwarfs. See Abstr. 064.087.

Influence of radiative scattering on the continuous spectra of late-type stars. See Abstr. 064.093.

Highly-evolved stars. See Abstr. 065.015.

Non-resonance lines of neutral calcium in the spectra of the Sun and Procyon.   See Abstr. 071.027.

Overview of the Voyager ultraviolet spectrometry results through Jupiter encounter.   See Abstr. 099.084.

Une étoile à CH très intense et à très grande vitesse radiale située dans la direction du Grand Nuage de Magellan. See Abstr. 111.001.

Early-type high-velocity stars in the solar neighborhood. I. List of candidates.   See Abstr. 111.009.

Ultraviolet observations of 27 Canis Majoris, π Aquarii and 48 Librae.   See Abstr. 112.003.

Influence of the envelope in the variations of transitory type Be stars.   See Abstr. 112.005.

IUE spectra of RS Puppis and HD 20722: stars in symmetric dusty nebulae.   See Abstr. 112.009.

New insight into the physics of atmospheres of early-type stars.   See Abstr. 112.010.

Mass-loss rates from early-type stars. See Abstr. 112.011.

Stellar winds in Be stars.   See Abstr. 112.016.

Variable wind-velocity in 59 Cyg. See Abstr. 112.018.

Observational evidence for an accretion disk around the B star companion of δ Sge (M 2 II).   See Abstr. 112.019.

Observational evidences of stellar winds. See Abstr. 112.032.

Mass loss from cool stars. See Abstr. 112.034.

Mass loss from α Ori. See Abstr. 112.036.

Galactic ring nebulae associated with Wolf-Rayet stars. III. H II region-type nebulae.   See Abstr. 112.042.

4–8 micron spectrophotometry of OH 0739–14. See Abstr. 112.043.

Spectrophotometry of V1016 Cygni. See Abstr. 113.004.

Infrared variability and spectrum of SS 433. See Abstr. 113.008.

The peculiar Rosetta stone: Beta Coronae Borealis. See Abstr. 113.033.

Photométrie UV des étoiles Am et Delta Delphini. See Abstr. 113.034.

Absolute spectrophotometry of Wolf-Rayet stars. See Abstr. 113.043.

Spectrophotometry of bright southern stars. See Abstr. 113.047.

Field population II blue stragglers. See Abstr. 113.049.

Slopes of the reddening lines and ratios of total to selective absorption for carbon stars in the Vilnius and UBV systems.   See Abstr. 113.071.

Fundamental stellar parameters derived from the evolutionary tracks.   See Abstr. 115.022.

Magnetic structure in cool stars. I. The Ca II H and K emission from giants.   See Abstr. 116.003.

The UV resonance lines of HD 175362, a helium variable with a strong magnetic field.   See Abstr. 116.010.

Liens entre la métallicité et la vitesse de rotation des étoiles Am.   See Abstr. 116.026.

The rotational profile of II Pegasi (HD 224085). See Abstr. 116.032.

Zeta Puppis: an O-type oblique rotator? See Abstr. 116.040.

Changes in the high state of AM Herculis: a simultaneous X-ray, optical, polarimetric, and spectroscopic study. See Abstr. 117.008.

Absolute spectrophotometry of υ Sgr. See Abstr. 117.072.

The ultraviolet spectrum of UW Canis Majoris. See Abstr. 119.024.

Fast variability of emission lines of β Lyrae. See Abstr. 119.071.

BD + 34°4216: a composite spectrum binary system.   See Abstr. 120.006.

High-dispersion spectra of RU Lup. See Abstr. 121.023.

Spectroscopic observations of pre-main-sequence stars in the Pleiades.   See Abstr. 121.028.

IUE observations of eight dwarf novae: a study of the outburst cycle from 0.12 to 3.5 microns. See Abstr. 122.001.

Spectroscopic study of the infrared Ca II triplet in S-type Mira variable stars.   See Abstr. 122.024.

Spectrum variation and chemical compositions of the Cr type Ap star ε Ursae Majoris. See Abstr. 122.085.

The distribution of iron and chromium over the surface of ε UMa.   See Abstr. 122.086.

IUE spectroscopy of cataclysmic variables. See Abstr. 122.096.

V923 Aquilae.   See Abstr. 122.116.

Ultraviolet spectra of dwarf solar neighbourhood stars – I.   See Abstr. 122.119.

Far UV spectrophotometry of SNR. See Abstr. 125.028.

The white dwarf companion of the Ba II star ζ Cap.   See Abstr. 126.008.

The ultraviolet spectrum of the O-type subdwarf HD 49798. See Abstr. 126.021.

The young Of star HD 148937 and its associated interstellar bubble − H II region. See Abstr. 131.045.

Anomalous far-UV extinction in the WN6 star HD 147419. See Abstr. 131.071.

Observations of the exciting stars of NGC 2023 and M43 − interstellar extinction and abundances. See Abstr. 131.074.

High resolution observations of interstellar lines in the IUE spectra of LMC stars. See Abstr. 131.076.

Interstellar diffuse bands in the spectra of six stars previously reported anomalous. See Abstr. 131.144.

A fine analysis of stellar and interstellar lines towards four halo B stars. See Abstr. 131.157.

Central object of the 30 Doradus Nebula, a supermassive star. See Abstr. 132.008.

H II regions and Wolf-Rayet stars in M33. See Abstr. 132.022.

On the far ultraviolet flux distribution of the Orion Nebula. See Abstr. 132.036.

Untersuchungen an extragalaktischen H II Gebieten unter besonderer Berücksichtigung des Riesen-H II-Gebietes NGC 604 in M33. See Abstr. 132.051.

High resolution spectrophotometry of the O I line (8446 Å) towards Cyg OB2 No. 12. See Abstr. 133.013.

Far infrared spectrophotometry of evolved objects. See Abstr. 135.023.

The ultraviolet spectrum of the planetary nebula NGC 2371 and its exciting star. See Abstr. 135.041.

Spectroscopic investigation of the Cygnus X-1 optical component. See Abstr. 142.011.

Far ultraviolet observations of Cyg X-2. See Abstr. 142.041.

Ultraviolet spectral variations in LMC X-4 and SMC X-1. See Abstr. 142.042.

X-ray emission from Of stars and OB supergiants. See Abstr. 142.065.

Lanning 10 and 33: the X-ray, UV, and optical fluxes. See Abstr. 142.077.

X-ray and optical observations of HD 155638: a remarkably active cool star. See Abstr. 142.080.

Coronal activity in F-, G-, and K-type stars. See Abstr. 142.098.

First UV observations of the optical counterpart of the X-ray source 2S0114+650. See Abstr. 142.101.

Spectroscopic studies of stars in Ori OB1 (belt). See Abstr. 152.002.

Abundances in globular cluster red giants. IV. M22 and Omega Centauri. See Abstr. 154.007.

Carbon and nitrogen abundances in the giant stars of the globular clusters M3 and M13. See Abstr. 154.037.

Étude de la structure galactique dans une région de la Poupe. See Abstr. 155.013.

Wolf-Rayet stars and giant H II regions in M33: casual associations or meaningful relationships? See Abstr. 158.160.

The metal abundance range in the Ursa Minor dwarf galaxy. See Abstr. 158.212.

Ultraviolet interstellar extinction in the Large Magellanic Cloud using observations with the International Ultraviolet Explorer. See Abstr. 159.002.

Distribution of hot stars and hydrogen in the Large Magellanic Cloud. See Abstr. 159.003.

Errata

114.901 Erratum: "A joint optical-radio study of SS 433", [Astron. Astrophys., Vol. 95, 177 - 183 (1981)]. F. Ciatti, A. Mammano, C. Bartolini, A. Guarnieri, A. Piccioni, A. J. B. Downes, D. T. Emerson, C. J. Salter. Astron. Astrophys., Vol. 100, 330 (1981). − See Abstr. 29.114.013.

114.902 Erratum: 'Absorption lines in the visual spectrum of the "continuous" central star of the planetary nebula NGC 3242', [Astron. Astrophys., Vol. 99, L15 - L17 (1981)]. R. P. Kudritzki, R. H. Méndez, K. P. Simon. Astron. Astrophys., Vol. 101, 276 (1981). − See Abstr. 29.114.171.

114.903 Corrigendum: 'Intrinsic properties of carbon stars. I. Effective temperature scale of N-type carbon stars' [J. Astrophys. Astron., Vol. 2, 95 - 113 (1981)]. T. Tsuji. J. Astrophys. Astron., Vol. 2, 213 (1981). − See Abstr. 29.114.099.

114.904 Erratum: 'A new weak-banded, probable carbon star, having bright Hα' [Publ. Astron. Soc. Pacific, Vol. 92, 653 (1980)]. C. B. Stephenson. Publ. Astron. Soc. Pacific, Vol. 93, 535 (1981). − See Abstr. 28.114.151.

114.905 Erratum: "R 81: P Cygni of the LMC." [Astron. Astrophys., Vol. 99, 351 - 361 (1981)]. B. Wolf, O. Stahl, M. J. H. de Groot, C. Sterken. Astron. Astrophys., Vol. 103, 427 (1981). − See Abstr. 29.114.174.

114.906 Erratum: "Forbidden lines of four and five times ionized iron and nickel in the spectra of RR Telescopii and Eta Carinae" [Astrophys. J., Vol. 243, 217 - 222 (1981)]. A. J. J. Raassen, J. E. Hansen. Astrophys. J., Vol. 251, 415 (1981). − See Abstr. 29.114.037.

115 Luminosities, Masses, Diameters, HR and other Diagrams

115.001   The absolute magnitudes of the Ap stars.
S. Grenier, M. Jaschek, A. E. Gomez, C. Jaschek, A. Heck.
Astron. Astrophys., Vol. 100, 24 - 27 (1981).
The mean absolute magnitude of the Ap stars has been determined by statistical parallaxes. The authors' results given as a function of UBV photometric indices, have been checked against the absolute magnitudes derived from Ap stars in clusters and in visual binaries. The agreement is good when the errors are taken into account. The authors conclude that the Ap stars do have the same absolute magnitude as the main sequence stars of the same color.

115.002   The angular diameter of Aldebaran at 8540 Å.
R. R. Radick, J. L. Africano.
Astron. J., Vol. 86, 906 - 908 (1981).
Two occultation observations of Aldebaran at 8540 Å yield best-fit angular diameters of 22.2 ± 3.4 and 23.1 ± 2.3 milliarcsec for the uniform disk.

115.003   A contribution to the determination of the mass of Arcturus.   M. Spite, P. Martin.
Astron. Astrophys., Vol. 101, 265 - 268 (1981).
In this paper the authors study the influence of the choice of the type of model for the atmosphere of Arcturus on the determination of its surface gravity and thus of its mass. They show that taking fixed effective temperature and helium abundance, the value of the mass deduced from a model to another can change by a factor of four and that all the models used give practically the same abundance ratio [Ti/Fe].

115.004   Absolute magnitude of a helium-rich star.
B. Bohannan.
Bull. American Astron. Soc., Vol. 13, 526 (1981). − Abstract.

115.005   The stellar luminosity function at the galactic poles.
S. L. Kipp.
Bull. American Astron. Soc., Vol. 13, 539 (1981). − Abstract.

115.006   The theoretical HR diagram for a cluster of stars with mass accretion.
S. K. Bhattacharjee, I. P. Williams.
Mon. Not. R. Astron. Soc., Vol. 197, 75 - 79 (1981).
It is assumed that stars form in a cluster in a sequence due to the passage of some triggering mechanism such as a shock wave. Each protostar initially has about the same mass but gains mass by accretion from the ambient cloud. A theoretical HR diagram is obtained for a cluster consisting of stars evolving in this manner and this is found to be in excellent agreement with the observed HR diagram for the young cluster NGC 2264.

115.007   Masses of Magellanic Wolf-Rayet stars: mass loss and evidence for a WR subclass vs. mass relation.
A. F. J. Moffat.
Effects of mass loss on stellar evolution, (see 012.015), p. 301 - 305 (1981).
New spectroscopic observations in the LMC/SMC combined with published data on Galactic WR stars lead to a correlation between mass ratio, M_{WR}/M_O, and WR subclasses. As a consequence of this and their high mass loss rates WR stars probably evolve from cool to hot ionization class in a way which depends on the metallicity.

115.008   How massive are the Wolf-Rayet stars?
V. S. Niemelä.
Effects of mass loss on stellar evolution, (see 012.015), p. 307 - 310 (1981).

115.009   The initial/final mass relation for stellar evolution with mass loss.   V. Weidemann.
Effects of mass loss on stellar evolution, (see 012.015), p. 339 - 344 (1981).
The relation between initial and final masses is discussed under consideration of changing theoretical concepts and new empirical data on masses of white dwarfs and nuclei of planetary nebulae. It is concluded that presently adopted schemes of evolution need revision, and that no universal relation exists.

115.010   The radii of the Wolf-Rayet stars and the extent of their chromosphere-corona formation.
J. Sahade, J. Zorec.
Mem. Soc. Astron. Italiana, Vol. 52, (see 012.019), 23 - 35 (1981).
The radii of a number of Wolf-Rayet stars have been computed for different effective temperatures.

115.011   Lunar occultation stellar angular diameter measurements. II.
W. I. Beavers, R. R. Cadmus, Jr., J. J. Eitter.
Astron. J., Vol. 86, 1404 - 1409 (1981).
Results are reported for the analyses of 14 diameter candidate lunar occultation observations. Included are two measurements (μ Gem and 45 Ari) with large angular diameters, seven cases with small or barely resolvable diameters, two point sources, and three cases where the records are too noisy to permit an angular diameter measurement.

115.012 UBV and H_β observations of stars towards M8.
R. Chini, T. Neckel.
Astron. Astrophys., Vol. 102, 171 - 174 (1981).
New and more accurate UBV magnitudes have been obtained for 106 stars with known proper motions in the field of the emission nebula M8. The colour-magnitude diagram of the embedded cluster NGC 6530 consists of a main-sequence from B 1 to about B 9 with a distance modulus of 11.4 mag. Spectral types earlier than B 1 and later than B 9 fall above the main sequence. Evidence is given that the so-called "pre-main sequence" of NGC 6530 is not related to this cluster but consists of field stars. H_β observations strongly support previous results that many cluster members are intrinsic Be stars.

115.013   The luminosity function and the colours of the solar neighbourhood.   M. L. Malagnini.
Ann. Physique, Vol. 6, (see 012.026), p. 63 - 72 (1981). In French, English. − Abstr. in Phys. Abstr., Vol. 84, Abstr. 94312 (1981).

115.014   Bolometric luminosities and infrared properties of carbon stars in the Magellanic Clouds and the Galaxy.   J. G. Cohen, J. A. Frogel, S. E. Persson, J. H. Elias.
Astrophys. J., Vol. 249, 481 - 503 (1981).
Broad band J, H, K photometry and narrow band CO and H_2O indices have been obtained for 89 luminous red stars in the Large Magellanic Cloud and 21 in the Small Magellanic Cloud. Most are known to be carbon stars, and their infrared properties are compared with new observations of 33 galactic carbon stars. The bolometric luminosity distributions of an unbiased sample of Magellanic Cloud carbon stars are compared with those predicted from evolutionary calculations by Renzini and Voli for double shell burning stars under-

going He shell flashes. The observed and theoretical distributions disagree markedly: nearly all the observed stars have lower luminosities than even the faintest theoretical carbon star.

115.015 Infrared luminosities of M supergiants and their use as distance indicators.
J. H. Elias, J. A. Frogel, R. M. Humphreys, S. E. Persson.
Astrophys. J., Lett., Vol. 249, L55 - L59 (1981).

Infrared photometry of the brightest red supergiants in five Local Group galaxies shows that their K (2.2 μm) luminosities range from about -12 mag in the Milky Way and the LMC to approximately -11.4 mag in the SMC, NGC 6822, and IC 1613. This dispersion contrasts with the situation for the visual luminosities of the brightest M supergiants in the same galaxies which are nearly constant at $M_V = -8$ mag. This spread in the brightest K luminosity is due to variations in the spectral type distribution of the M supergiant population which may depend on the metallicity of the parent galaxy. Thus without further information on the spectral type of the star, the large range in M_K limits the usefulness of the M supergiants as distance indicators in the infrared.

115.016 The structure of the upper main sequence and the subgroups of the Ap stars. R. L. Stratford.
Upper main sequence chemically peculiar stars, (see 012.033), p. 67 - 75 (1981).

The structure of the upper main sequence between spectral types B0 and F1 is discussed using composite (β, [u - b]) and ([m_1], [c_1]) diagrams of open star clusters. It is found that the distribution of stars along the main sequence is not uniform but shows a complex "gap-and-clump" structure. The gaps in the upper main sequence are tabulated. The regions of the two photometric diagrams occupied by the different types of chemically peculiar stars (from He strong stars to Am stars) are found to be bounded by these gaps.

115.017 The absolute magnitude of δ Scuti and δ Del type stars.
S. Grenier, A. E. Gomez, C. Jaschek, M. Jaschek, A. Heck.
Upper main sequence chemically peculiar stars, (see 012.033), p. 491 - 492 (1981).

115.018 The cool half of the H-R diagram in soft X-rays.
T. R. Ayres, J. L. Linsky, G. S. Vaiana, L. Golub, R. Rosner.
Astrophys. J., Vol. 250, 293 - 299 (1981).

The authors report results of an Einstein Guest Observing program to map the occurrence of soft X-ray emission, which is a signature of hot stellar coronae ($T > 10^6$ K), in the cool half of the Hertzsprung Russell (H-R) diagram. They detect X-rays from F-M dwarfs and late F through early K giants, but not from the cooler giants, other than the spectroscopic binary ϵ Car (K0 II+B), or from any supergiants, other than Canopus (F0 Ib–II). The empirical separation of the cool half of the H-R diagram into a region where stellar soft X-ray emission is a common phenomenon, and a region where hot coronae are rare, if present at all among single stars, is similar to that found previously by Linsky and Haisch for C IV λλ1548,1551 emission ($T \approx 10^5$ K) and by Stencel and Mullan for the onset of rapid mass loss in strong, cool ($T \lesssim 10^4$ K) stellar winds. The authors discuss the energy balance in the outer atmospheres of the coronal stars, the likely absorption of X-ray emission by cool winds in the "hybrid-spectrum" supergiants, a rotation-activity connection among the G dwarfs, and possible evolutionary origins of the structure seen in the cool half of the X-ray H-R diagram.

115.019 Empirical bolometric corrections for the main sequence. G. M. H. J. Habets, J. R. W. Heintze.
Astron. Astrophys., Suppl. Ser., Vol. 46, 193 - 237 (1981).
By using the most reliable data for visual/astrometric

binaries and for eclipsing binaries, which are at the same time double-lined spectroscopic binaries, empirical bolometric corrections were derived for the main-sequence as a function of mass, spectral type and effective temperature. An empirical relation between mass and bolometric correction was found for the whole spectral range of main-sequence stars. This relation agrees for large masses ($> 7 M/M_\odot$) and for $1.35 M/M_\odot$ nicely with the theoretical results of Buser and Kurucz (1978) and with the empirical results of Code et al. (1976), whenever their relations are transformed as a function of mass.

115.020 Stellar diameter measurements by two-telescope interferometry in optical wavelengths.
D. Bonneau, L. Koechlin, J. L. Oneto, F. Vakili.
Astron. Astrophys., Vol. 103, 28 - 34 (1981).

Angular diameters of α Cas, β And, γ And, α Per and α Cyg are measured using the two-telescope stellar interferometer at C.E.R.G.A. Stellar diameters and effective temperatures derived from T_{eff} – bolometric correction empirical relations are compared with determinations based on infrared photometry and stellar model atmospheres. For α Cas and β And the authors obtain good agreement with the relation T_{eff}-(V-K) color established from lunar occultation diameters and infrared photometry. They also give estimates of the stellar linear radii.

115.021 Multiplicity and absolute magnitudes of Wolf-Rayet stars in the Large Magellanic Cloud.
M. L. Prévot-Burnichon, L. Prévot, E. Rebeirot, J. Rousseau, N. Martin.
Astron. Astrophys., Vol. 103, 83 - 93 (1981).

Photographic surveys of the Large Magellanic Cloud made at the ESO with the 40 cm astrograph, the 1 m Schmidt telescope and partly with the 3.6 m telescope allowed the inspection of images of the LMC Wolf-Rayet stars in the bands U, B, V, and R. The V astrographic images revealed a "WC effect" from which it could be assessed that 50% of the WC stars are probably single, or connected with an underluminous companion. The inspection of other images in order to find the visual companions of all WR stars, has led to a study of their apparent multiplicity: 33% of WC stars and 16% of WN stars have companions closer than 3".

115.022 Fundamental stellar parameters derived from the evolutionary tracks.
V. Straižys, G. Kuriliene.
Astrophys. Space Sci., Vol. 80, 353 - 368 (1981).

The surface gravities and radii of stars are calculated for different MK spectral types using the masses of stars determined from their evolutionary tracks in the HR diagram and the most reliable values of effective temperatures and absolute bolometric magnitudes. MK spectral types are calibrated in absolute visual magnitudes using the studies of M_V published since 1965. The calibration of MK types in temperatures is based on the newest investigations including the results both from the ultraviolet and the infrared. The obtained masses, gravities, and the mass-luminosity relationship show reasonable agreement with independent observational data.

115.023 On a gap in the stellar mass distribution.
M. A. Giannuzzi.
Astron. Astrophys., Vol. 104, 81 - 82 (1981).

Some evidence has been collected about a gap in stellar mass distribution near 3.5 $M_\odot$. Probably some differences between theory and observation for stars in advanced evolutionary stages could be connected with this fact. The question arises whether the star formation is bimodal.

115.024 Absolute magnitudes and intrinsic colours of OB stars. E. I. Vega, J. C. Muzzio.
Observatory, Vol. 101, 211 - 213 (1981).
The authors use spectral types obtained previously to

derive mean absolute magnitudes and intrinsic colours of OB stars classified on thin-objective-prism plates. These results are compared with those obtained for stars classified on objective-prism plates at larger dispersion.

115.025 **Space density of stars and interstellar extinction near h and χ Persei (Perseus I).**
W. Becker, W. H. Wooden II.
Astron. Astrophys., Suppl. Ser., Vol. 46, 355 - 364 (1981).

Perseus I, a star field near h and χ Persei, measuring 0.155 square degrees and containing 1281 stars down to the limiting magnitude $G = 18\overset{m}{.}5$ has been studied photometrically in the RGU-system. The two-colour-diagram of all stars with $G < 18^m$ is given. The evaluation of the two-colour-diagrams for consecutive intervals in G leads to an interstellar reddening produced by two screens in distances of 0.6 kpc and 1.5 kpc from the sun, with a total reddening of $E(G-R) = 0\overset{m}{.}88$. The density-functions have been determined for different intervals of absolute magnitude. There is a rather large number of late-type gaints at the left side of the main-sequence. In addition to them, with $G > 18^m$, further late-type giants are found at the right side of the main-sequence which are possibly metal-poor. The luminosity-functions have been determined for two adjoining distance-intervals.

115.026 **The angular diameter of Regulus from the 28 March 1980 CTIO occultation.**   R. R. Radick.
Astron. J., Vol. 86, 1685 - 1689 (1981).

On 28 March 1980 an occultation of Regulus (=α Leo) was observed in two colors from Cerro Tololo Inter-American Observatory. The best-fit uniform disk angular diameters (in milliarcseconds) derived from these data are 1.32 ± 0.12 (yellow channel, λ = 5768 Å) and 1.37 ± 0.11 (blue channel, λ = 4356 Å), in excellent agreement with the measurement obtained through intensity interferometry and in reasonable agreement with a daytime occultation observed at Kitt Peak National Observatory.

115.027 **Spectroscopic determination of stellar masses: mene, mene, tekel, Arcturus.**
V. Trimble, R. A. Bell.
Q. J. R. Astron. Soc., Vol. 22, 361 - 379 (1981).

The determination of stellar masses (i.e. surface gravities) from spectroscopic data requires, in principle, a complete understanding of stellar atmospheric structure, including the effects of turbulence and convection, departures from LTE and plane parallel structure, and variations in abundances of individual elements. Where very accurate data are available, as for Arcturus, it becomes clear that the uncertainties in derived effective temperatures, surface gravities, and abundances are both rather larger and somewhat more interdependent than is usually supposed. As a result, it takes surprisingly large error bars, e.g. $M = 0.7 \pm 0.5\,M_\odot$, to bracket all modern determinations of the mass of Arcturus. The authors discuss why this is so and what, if anything, can be done about it.

115.028 **The mass-luminosity relation for red dwarfs.**
M. Joeveer.
Astron. Tsirk., No. 1163, p. 1 - 3 (1981). In Russian.

115.029 **On luminosity calibration of O and B stars.**
K. A. Barkhatova, A. V. Loktin.
Astron. Tsirk., No. 1164, p. 2 - 5 (1981). In Russian.

Catalogue of apparent diameters and absolute radii of stars (CADARS).   See Abstr. 002.004.

L'observation et les propriétés des étoiles.
See Abstr. 031.560.

On recording of black body radiation.
See Abstr. 031.573.

Effects of a stochastic initial mass function on the upper main sequence band.   See Abstr. 065.030.

Mass loss from metal-poor stars.
See Abstr. 065.031.

The degree of completeness of nearby stars and the stellar luminosity function.   See Abstr. 111.017.

The surface brightness parameter in relation ot the VRI_c system.   See Abstr. 113.053.

Lithium abundances, K line emission and ages of nearby solar type stars.   See Abstr. 114.022.

Absolute energy curves from late B-type supergiants.
See Abstr. 114.030.

Scanner observations of the Be stars ν Gem, κ Ori, β Mon and ω CMa.   See Abstr. 114.055.

A compilation of physical parameters of Ap and Am stars as derived from energy distribution studies.
See Abstr. 114.098.

La misura dei raggi stellari.   See Abstr. 114.138.

A classification system for O − B2 stars based on the Si IV and C IV resonance lines.   See Abstr. 114.163.

Intrinsic properties of carbon stars. II. Spectra, colours, and the HR diagram of cool carbon stars.
See Abstr. 114.173.

The characteristics of cool stars.
See Abstr. 114.185.

Absolute dimensions and masses of eclipsing binaries. II. YZ Cassiopeiae.   See Abstr. 119.065.

An estimate of the mass of zero metal stars.
See Abstr. 131.159.

Luminosity effects in interstellar spectral features.
See Abstr. 131.215.

Relations among stellar X-ray emission observed from Einstein, stellar rotation and bolometric luminosity.
See Abstr. 142.014.

Low luminosity galactic X-ray sources.
See Abstr. 142.096.

The color-magnitude diagram for stars in the central part of the globular cluster M 15.   See Abstr. 154.004.

Superluminous giants in Magellanic Cloud clusters.
See Abstr. 159.011.

116 Magnetic Fields, Polarization, Figure, Rotation, Radio Radiation

116.001 **Multislit photoelectric magnetometer observations of Cepheids and supergiants: probable detections of weak magnetic fields.** E. F. Borra, J. M. Fletcher, R. Poeckert.
Astrophys. J., Vol. 247, 569 - 576 (1981).
The authors have assembled an instrument that measures the circular polarization in 230 spectral lines simultaneously. The instrument is capable of measuring longitudinal magnetic fields in bright stars with standard deviations of the order of 1 gauss. The authors have obtained several detections at the 2, 3, and up to 4,6 standard deviations level in the Cepheid variables α UMi and δ Cep and in the F8 Ib supergiant γ Cyg, indicating longitudinal fields from 10 to 30 gauss. The observations support the hypothesis advanced by Stothers to reconcile pulsational and evolutionary masses.

116.002 **Starlight polarization in the direction of two H I complexes.** R. Morras.
Astron. J., Vol. 86, 875 - 880 (1981).
Observational data from starlight polarization in the direction of two H I complexes are given. It is shown that the Sun is found in a region of low density of material, and that the dust, which produces the starlight polarization, is mixed with the H I gas. This is useful in order to estimate the distance for two H I complexes.

116.003 **Magnetic structure in cool stars. I. The Ca II H and K emission from giants.**
F. Middelkoop, C. Zwaan.
Astron. Astrophys., Vol. 101, 26 - 32 (1981).
Wilson's (1976) eye estimates of the Ca II H and K emission from evolved cool stars are reanalysed. G-type giants show a large spread in H and K emission; giants with broadened spectral lines tend to produce enhanced emission. The H and K emission from stars of metallicities deviating from the solar metallicity is relatively small. For spectroscopic binaries the authors find a critical period (between 120 and 200 days) below which the orbits tend to be circular, indicating that rotation and orbital motion are probably synchronized. For periods less than the critical period the emission strength is enhanced; the emission is stronger according as the orbital period is shorter. The data are consistent with the hypothesis that Ca II H and K emission depends on the dynamo action in the convective envelope, the dynamo efficiency decreasing with decreasing rotation rate.

116.004 **Le stelle magnetiche.** F. A. Catalano.
G. Astron., Vol. 7, 97 - 115 (1981).

116.005 **Polarization of the radiation of stars, nebulae and galaxies.** O. S. Shulov.
Tr. Astron. Obs., Leningrad, Tom 37 = Uch. Zap. Leningr. Univ., No. 406 = Ser. mat. nauk, Vyp. 59, 181 - 208 (1981). In Russian.

116.006 **Magnetic structure in cool stars. II. Observational evidence for transverse magnetic fields.**
J. Tinbergen, C. Zwaan.
Astron. Astrophys., Vol. 101, 223 - 225 (1981).
The apparently intrinsic linear polarization of the light from nearby F, G, K and M stars is attributed to transverse magnetic fields near the stellar limb in activity belts on either side of the stellar equator. The polarization results from the π components of strong spectral lines being more saturated than the σ components (Leroy, 1962). The proposed mechanism is compatible with existing observational data. Stringent tests and prospects for further observations are indicated.

116.007 **Rotation of dwarf star chromospheres in the ultraviolet.** K. L. Hallam, C. L. Wolff.
Astrophys. J., Lett., Vol. 248, L73 - L76 (1981).
Periodic variations in the ultraviolet chromospheric fluxes of H I, Si II, and Mg II have been found in six main-sequence F, G, and K stars with periods as long as 47 days. The authors interpret the modulation as being caused by rotation and, if true, this new technique offers as much as an order of magnitude increase in the length of rotation periods detectable in dwarf stars. The periodicity persists over many rotational cycles, but the period and time dependence of the modulated spectral flux is not identical from one ionic species to another in the same star, presumably because of differential stellar rotation and differing global distributions and excitation conditions of the chromosphere.

116.008 **The rotation of horizontal branch stars.**
T. D. Tarbell, R. C. Peterson.
Bull. American Astron. Soc., Vol. 13, 516 - 517 (1981). Abstract.

116.009 **Intrinsic polarization variations of the symbiotic star R Aqr and the single-star hypothesis.**
J. Svatoš, M. Šolc.
Astrophys. Space Sci., Vol. 78, 503 - 504 (1981).
Due to similarity of the intrinsic polarization changes near phase 0.8 and due to other similar physical features observed both in o Ceti and R Aqr it is suggested that the activity in R Aqr system is caused by a flaring region on a single star.

116.010 **The UV resonance lines of HD 175362, a helium variable with a strong magnetic field.**
K. D. Rakos.
Second European IUE Conference, (see 012.011), p. 35 - 37 (1980).
High resolution spectra of the Ap stars HD 133029, HD 219749, HD 168733 and HD 175362 have been obtained during two 16 hours shifts with the IUE satellite at Goddard Space Flight Center. The resonance lines of HD 175362 in particular show very peculiar behaviour. Stellar wind, extended atmosphere in rigid corotation with the stellar surface and the influence of the strong magnetic field on the upper part of the atmosphere would explain the shape and the strength of the resonance lines. The necessary driving forces for the expanding envelope are not compatible with the diffusion theory of the Ap atmospheres.

116.011 **On polarization variations of V CVn.**
T. A. Polyakova.
Vestn. LGU, 1981, No. 1, p. 105 - 110. In Russian. – Abstr. in Ref. zh., 51. Astron., 7.51.443 (1981).

116.012 **Radio observations of O-type stars.**
M. Felli, N. Panagia.
Effects of mass loss on stellar evolution, (see 012.015), p. 179 - 180 (1981).
Eight O-type giants and supergiants, selected for being relatively isolated in the sky, have been observed at 5 GHz. Most of the stars (6 out of 8) were not detected. Possible detection has been achieved for the two supergiants HD 225160 and HD 30614 both with a flux density of 0.15 mJy. The mass loss rates have been computed by using the formulae of Panagia and Felli (1975).

116.013 **Infrared polarization of stars in the Taurus cloud.**
A. Moneti, J. L. Pipher.
News Lett. Astron. Soc. N. Y., Vol. 1, No. 10, p. 38 (1981). Abstract.

**116.014 11.1- and 3.7-cm variations in the flux density of
SS 433.   K. J. Johnston, N. J. Santini,**
J. H. Spencer, W. J. Klepczynski, G. H. Kaplan, F. J. Josties,
P. E. Angerhofer, D. R. Florkowski, D. N. Matsakis.
Astron. J., Vol. 86, 1377 - 1383 (1981).

The peculiar galactic object SS 433 has been observed at
radio wavelengths of 11.1 and 3.7 cm from JD 2444067 until
JD 2444487. These observations (with a baseline of 2.4 km)
show that the radio radiation displays (1) a quiescent phase in
which there is little change in the radio flux density and (2) an
active flaring phase in which the radio flux density varies by
30% - 100% in an interval of one to five days. No dominant
periodicities are present in the data, although over short periods
(~ 100 days) there may be some weak correlations.

**116.015 Magnetic structure in cool stars. III. Ca II H and K
emission and rotation of main-sequence stars.**
F. Middelkoop.
Astron. Astrophys., Vol. 101, 295 - 298 (1981).

Ca II H and K fluxes were measured in 31 rapidly rotating main-sequence stars of spectral type F and in 7 single-line
main-sequence spectroscopic binaries with periods less than
10 days. Most of the rapidly rotating stars observed appear to
have relatively high fluxes compared to the slowly rotating
stars of the same spectral type. All main-sequence binaries
with periods less than 9 days show eccentricities smaller than
0.1, which indicates that in these systems the rotational and
orbital motions are probably synchronized.

**116.016 Study of stellar polarization with the CERGA
interferometer.   F. Vakili.**
Astron. Astrophys., Vol. 101, 352 - 355 (1981).

An attempt has been made to detect scattering effects
in stellar atmospheres with the interferometer at CERGA.
Fringes have been observed on Alpha Lyrae both in natural
and linearly polarized light. Several polarization effects are
investigated and it is shown that the next generation interferometer with single apertures of 1.5 m should provide new
kinds of information on the atmospheres and magnetic fields
of stars.

**116.017 Broad-band linear polarization and magnetic intensi-
fication in rotating magnetic stars.**
M. Landi Degl'Innocenti, G. Calamai, E. Landi Degl'Innocenti,
P. Patriarchi.
Astrophys. J., Vol. 249, 228 - 237 (1981).

Magnetic intensification is proposed as a mechanism to
explain the general features of the variable broad-band linear
polarization emerging from rotating magnetic stars. This
mechanism is studied in detail, and some efforts are made to
investigate the wide variety of polarization diagrams that can
result from it. Theoretical results are compared with direct
observations of the variable magnetic star 53 Cam to determine its geometric and magnetic configuration.

**116.018 Decaying stellar magnetic fields, magnetic braking:
evidence from magnetic observations in Orion OB1.**
E. F. Borra.
Astrophys. J., Lett., Vol. 249, L39 - L42 (1981).

The author has observed the magnetic fields of 13 Ap
and helium-weak stars in the Orion OB1 association. The
magnetic fields observed are compared to those of older
magnetic stars. It is concluded that the magnetic stars in Orion
have magnetic fields stronger (by a factor of 3) than the older
stars, in agreement with a fossil origin of the fields. The
e-folding time of decay is estimated to be of the order of
10^8 years. This short time scale is seen as evidence that
internal mass motions are present in upper-main-sequence
stars, with potentially important effects on their evolution.
Although the periods of variation are uncertain, they suggest
that young magnetic stars rotate faster than their older
counterparts, implying magnetic braking on the main sequence.

**116.019 A magnetic survey of Ap stars in young clusters –
preliminary results.**
D. N. Brown, J. D. Landstreet, I. Thompson.
Upper main sequence chemically peculiar stars, (see 012.033),
p. 195 - 198 (1981).

The authors have undertaken a magnetic survey of Ap stars
in young clusters. The results will contribute to a description
of the magnetic, chemical and rotational evolution of the
chemically peculiar stars. The data obtained to date include
several new magnetic identifications and display the character
of the survey but are not yet sufficient to support any firm
evolutionary conclusions.

**116.020 Symmetric rotator models with inhomogeneous
distribution of elements.**
L. Oetken, F. Krause.
Upper main sequence chemically peculiar stars, (see 012.033),
p. 271 - 272 (1981).

**116.021 Non uniform surface distribution of abundances
and magnetic field intensity variations.**
F. A. Catalano, A. Del Bufalo.
Upper main sequence chemically peculiar stars, (see 012.033),
p. 273 - 275 (1981).

**116.022 Axisymmetric and mixed poloidal-toroidal magnetic
fields for HD 215441 and 53 Cam.**
M. Goossens, W. Van Assche.
Upper main sequence chemically peculiar stars, (see 012.033),
p. 277 - 281 (1981).

The oblique rotator model is adopted, and axisymmetric,
mixed poloidal-toroidal magnetic fields are considered for the
description of the periodic magnetic variations observed in
many Ap-stars. A schema is presented for the determination of
the surface structure of the magnetic field from H_e and H_s observations. This schema has been succesfully applied to the
magnetic observations of HD 215441 and 53 Cam. It was
found that several magnetic surface structures fit the observational data.

**116.023 The intrinsic linear polarization of the magnetic
variable HD 71866.**
V. Piirola, I. Tuominen.
Upper main sequence chemically peculiar stars, (see 012.033),
p. 283 - 288 (1981).

Period changes in the degree and position angle of the
linear polarization of the magnetic variable HD 71866 are reported. The period of the variations is compatible with the
period of photometric and magnetic field variation, and the
mean degree of polarization is smaller than reported earlier by
other observers. The polarization parameters $p_x = p \cos 2\theta$,
$p_y = p \sin 2\theta$ form double waves with a total amplitude of
0.08 ± 0.01 % over one period. The changes of polarization
could be interpreted as a combination of an interstellar component and an intrinsic polarization vector rotating with the
star.

116.024 Evidence for intrinsic variation in Ap-magnetic fields.
W. K. Bonsack.
Upper main sequence chemically peculiar stars, (see 012.033),
p. 289 - 294 (1981).

**116.025 The non-axisymmetric surface field structure of
magnetic stars.   M. J. Stift.**
Upper main sequence chemically peculiar stars, (see 012.033),
p. 333 - 336 (1981).

**116.026 Liens entre la métallicité et la vitesse de rotation des
étoiles Am.**
C. Burkhart, C. Van't Veer, R. Faraggiana.
Upper main sequence chemically peculiar stars, (see 012.033),
p. 481 - 484 (1981).

116.027 Stellar dynamo and the galactic X-ray sources.
G. Belvedere, D. Molteni.
Astron. Astrophys., Vol. 102, 283 - 286 (1981).

After the discovery of X-ray emission from many isolated stars and the explanation of their emission in terms of coronal activity, the authors examine the possibility that also a fraction of those X-ray sources, usually depicted as accreting binary systems, may be interpreted as active stars supplied by the $\alpha - \omega$ dynamo mechanism. It is shown that the dynamo mechanism can produce high X-ray luminosity in sufficiently fast rotating late main sequence stars. The hypothesis of an $\alpha - \omega$ dynamo mechanism in faint late type stars seems interesting also in view of a more careful appealing to the accretion mechanism particularly for those systems which exhibit unresolved features.

116.028 Periodic changes in the compact radio structure of SS 433. A. E. Niell, T. G. Lockhart,
R. A. Preston.
Astrophys. J., Vol. 250, 248 - 253 (1981).

VLBI observations of SS 433 at 2.3 GHz made on 12 days between 1979 May and 1980 August yield the following results: (1) The position angle of the radio "jet" of angular size ~0″.1 varies approximately sinusoidally about a mean value of $100°.2 \pm 1°.7$ with an amplitude of $19°.3 \pm 3°.4$ for a period fixed at 163.6 days. This resolves the ambiguity in the two angles of the optical model of Abell and Margon and assigns the inclination of the axis of the precession cone to the 79° value. (2) The position angle of the radio structure on this scale lags that of the optical model by 17.4 ± 1.6 days. (3) At least some of the radio emitting material appears to propagate away from the core in blobs. Using the rate of change of angular separation of the blobs from the core obtained from measurements on four different days, and assuming that the radio emission is traveling at the speed derived for the optical jets from the kinematic model, the authors obtain a distance to SS 433 of 5.1 ± 0.5 kpc.

116.029 Stellar rotation in lower main-sequence stars measured from time variations in H and K emission-line fluxes. I. Initial results. A. H. Vaughan,
S. L. Baliunas, F. Middelkoop, L. W. Hartmann, D. Mihalas,
R. W. Noyes, G. W. Preston.
Astrophys. J., Vol. 250, 276 - 283 (1981).

Fluxes at 1 Å bands at the centers of the H and K lines in 46 lower main-sequence field stars, and in eight selected subgiants and giant stars, have been measured at nightly intervals in the course of a nearly continuous 14-week observing run. In 19 stars the authors have found clear evidence of rotational modulation, from which values of the rotational periods can be assigned by inspection. In nine others, periods have been found by an autocorrelation analysis of the flux records. The periods obtained imply rotation velocities that are in good accord with spectroscopically determined values of $V \sin i$ in the literature for 13 of the stars they have observed.

116.030 First detection of nonflare microwave emission from the coronae of single late-type dwarf stars.
D. E. Gary, J. L. Linsky.
Astrophys. J., Vol. 250, 284 - 292 (1981).

The authors report on an observing program with the VLA in its C configuration to detect microwave radiation from the coronae of nearby late-type dwarf stars which are not members of close binary systems and do not have large winds. Six stars, chosen on the basis of strong apparent X-ray flux, were observed during a 24 hour period, and two stars were detected. χ^1 Orionis (G0 V) was detected as a 0.6 mJy source (S/N $\approx$ 7) at 6 cm, and the authors obtained an upper limit at 2 cm. The flare star UV Cet (dM 5.5e) was detected as a steady 1.55 mJy source (S/N $\approx$ 17) at 6 cm during a 2.5 hour observation.

116.031 The magnetic field of Zeta Puppis.
P. K. Barker, J. D. Landstreet, J. M. Marlborough,
I. Thompson, J. Maza.
Astrophys. J., Vol. 250, 300 - 305 (1981).

The authors report a null magnetic field measurement for the O4ef star ζ Pup, with $\sigma \approx 100$ gauss. The observations do not eliminate a corotating magnetic wind model for the star because a surface field of over 2 kilogauss could escape detection. However, theoretical arguments do not support the suggestion that Type III P Cygni profiles may result from corotation of the stellar wind. If an expanding envelope is forced into solid body rotation by a global magnetic field, the resultant line profiles are expected to be of Type VI (Be-like) and not Type III.

116.032 The rotational profile of II Pegasi (HD 224085).
S. M. Ruciński, M. S. Staniucha.
Acta Astron., Vol. 31, 163 - 173 (1981).

A 277 Å long segment of the red spectrum obtained with the Reticon detector is analysed by the Fourier transform method: the transforms for II Peg and non-rotating stars are divided and then transformed back to the wavelength domain. The rotational profile can be interpreted by the solid rotation with $V \sin i = 26.4 \pm 1.8$ (m. e.) km/s and the fully darkened stellar disc. Hα and λ6707 lithium lines are discussed in more detail.

116.033 Decaying stellar magnetic fields, magnetic braking: evidence from magnetic observations in Orion OB1.
E. F. Borra.
J. R. Astron. Soc. Canada, Vol. 75, 245 (1981). – Abstract.

116.034 The periodic radio star LSI + 61°303.
A. R. Taylor, P. C. Gregory.
J. R. Astron. Soc. Canada, Vol. 75, 252 (1981). – Abstract.

116.035 A preliminary interpretation of stellar chromospheric Ca II emission variations within the framework of stellar dynamo theory.
B. R. Durney, D. Mihalas, R. D. Robinson.
Publ. Astron. Soc. Pacific, Vol. 93, 537 - 543 (1981).

Recent stellar rotation measurements, combined with Wilson's stellar-cycle observations, are interpreted in terms of simple dynamo concepts. The authors derive an empirical expression for the chromospheric H and K emission flux $\langle F \rangle$ as a function of the dynamo number N_D, and of $(B-V)$ color index. They suggest that the Vaughan-Preston gap (in a plot of log $\langle S \rangle$ vs. $(B-V)$) is a consequence of a discontinuity in the magnetic-field morphology caused by a transition from large to small dynamo numbers.

116.036 Magnetic fields in late-type stars.
E. Knobloch, R. Rosner, N. O. Weiss.
Mon. Not. R. Astron. Soc., Vol. 197, 45P - 49P (1981).

Observations show that magnetic activity in late-type stars is correlated with rotation rates and that there is a discontinuous change in behaviour at a critical rotation period. This can be explained as a consequence of a transition from convection in rolls parallel to the rotation axis to normal convection cells as the angular velocity is decreased.

116.037 Intrinsic polarization of SS 433: an electron scattering model. I. S. McLean, S. Tapia.
Vistas Astron., Vol. 25, (see 012.040), 45 - 50 (1981).

The intrinsic polarization of SS 433 is clearly modulated with the 164 day cycle. Short-term variations, possibly related to the underlying 13 day period, are also observed. An excellent sinusoidal fit to the U Stokes parameter is obtained which suggests that the source of polarization may be consistent with a simple model involving optically thin electron scattering in jets or a highly inclined disk.

116.038   Radio properties of SS 433.   E. R. Seaquist.
Vistas Astron., Vol. 25, (see 012.040), 79 - 85
(1981).
SS 433 is a variable non-thermal radio source with a
spectral index near $\alpha = -0.6$. There is definite evidence for a
low frequency turnover both in the quiescent and the flaring
component. VLA synthesis maps show an unresolved core
$(\Theta < 0\rlap{.}''2)$ plus a "jet" aligned with the lobes on the SNR W50.
Definite changes occur in the jet on a time scale of months
which may be related to the flaring activity.

116.039   408 MHz monitoring of SS 433 radioemission.
S. R. Bonsignori Facondi, A. Braccesi.
Vistas Astron., Vol. 25, (see 012.040), 91 - 94 (1981).
Total power radio observations at 408 MHz of SS 433 in
the period May - September 1980 are presented. The preliminary results seem to indicate flux variations of the source in a
time-scale of a day and possibly a few minutes.

116.040   Zeta Puppis: an O-type oblique rotator?
A. F. J. Moffat, G. Michaud.
Astrophys. J., Vol. 251, 133 - 138 (1981).
Extensive spectrophotometric observations of the
P Cygni type $H\alpha$ profile from the O4ef star, ζ Puppis, show
periodic variability in the depth of the nearly central absorption reversal. The 5.075 day period derived implies that the
$H\alpha$-forming part of the stellar wind corotates with the photosphere, for which $2\pi R/(v \sin i) \approx 4.8$ days. The periodic variations can most simply be explained by a rotating magnetic
multipole, as observed in Ap stars.

116.041   Rotation and turbulence in G giant stars.
D. F. Gray.
Astrophys. J., Vol. 251, 155 - 161 (1981).
Rotation and turbulence measurements have been made
for five G5 III stars. Comparison with evolutionary models
indicate that a rotational brake is needed. The braking appears
to occur abruptly during the G giant phase bringing the rotation down from ~ 25 km s^{-1} to ~ 5 km s^{-1}. It is suggested
that the brake is a dynamo generated magnetic field coupled
with mass loss and that ϵ UMi is currently undergoing rapid
rotational slowdown. The radial-tangential macroturbulence
dispersions are ~ 7 km s^{-1}. When compared with values
~ 4 km s^{-1} for K giants, the temperature dependence is in
accord with that expected for nonradial oscillations.

116.042   Regular variations in the $H\alpha$ profile of FK Comae.
L. W. Ramsey, H. L. Nations, S. C. Barden.
Astrophys. J., Lett., Vol. 251, L101 - L104 (1981).
The authors present a series of observations of the rapidly
rotating G type giant FK Comae. The unusually broad $H\alpha$ profile is seen to vary regularly in the sense that the violet-to-red
emission ratio is correlated with photometric phase and presumably with rotation. These observations are described by an
excretion disk model with substantial density inhomogeneities
driven at the angular velocity of the star by an imbedded magnetic field. That sufficient magnetic flux is available is suggested by both the high observed $v \sin i$ of ~ 120 km s^{-1} and photometric evidence of starspots.

116.043   On the rotation of stars.   V. P. Merezhin.
Tr. Kazan. Gorod. Astron. Obs., Vyp. 45, p. 3 - 33
(1980). In Russian.

**116.044   Synchronous changes of polarization, brightness
and radial velocities of o Andromedae.**
J. Arsenijević.
Bull. Obs. Astron. Belgrade, No. 131, p. 13 - 14 (1981).

**116.045   Variable intrinsic polarization in the peculiar star
CH Cygni.**   V. Piirola.

Proceedings of the Third Finnish-Soviet Astronomical Symposium, (see 012.057), p. 127 - 132 (1981).

**116.046   Polarizacija zračenja nekih hladnih superdžinova.
(Polarization of radiation of some cool supergiants).**
J. Arsenijević.
Publ. Obs. Astron. Beograd, No. 28, 56 pp. (1980).

116.047   Outer atmospheres and rotation of late-type stars.
M. M. Katsova.
Astron. Tsirk., No. 1154, p. 1 - 4 (1981). In Russian.

**116.048   An explanation of the radio flux mystery of
HD 192163.**   T. Nugis.
Academy of Sciences of the Estonian SSR, Prepr. A-5, p. 1 - 30
(1981).
It is suggested that the radio flux paradox of the star
HD 192163 (WN6) is the effect of a special ionization structure. Considering the fact that helium is left only in the
neutral stage beginning at some distance from the core, the
radio flux is due to the f-f radiation of H$^+$ and N$^+$. Good agreement with the observed UV, IR and radio flux values and with
line intensities was achieved.

Broad-band polarization in molecular spectra.
See Abstr. 022.022.

**Fourier analysis of differential rotation in late-type
stars.**   See Abstr. 031.519.

Measurements of stellar magnetic field.
See Abstr. 031.577.

**A method for unambiguous determination of star-
spot temperatures and areas: application to II Pegasi,
BY Draconis, and HD 209813.**   See Abstr. 031.579.

Internal motions in magnetic stars.
See Abstr. 062.005.

**Collisionless perpendicular shocks: applications to
solar type II radio bursts and the Antares (α Sco) B radio
emission.**   See Abstr. 062.013.

The present status of dynamo theory.
See Abstr. 062.077.

Diffusion models for magnetic Ap–Bp stars.
See Abstr. 062.087.

**On the global structure of stellar magnetospheres
with stellar winds.**   See Abstr. 062.125.

**Variable mass loss and magnetic topology in cool
giant stars.**   See Abstr. 064.024.

**Rotational modes in a slowly and uniformly rotating
star.**   See Abstr. 065.017.

Does γ Equ have a magnetic cycle like the sun?
See Abstr. 065.056.

Diffusion in the presence of a magnetic field.
See Abstr. 065.061.

Theoretical aspects of early type magnetic stars.
See Abstr. 065.080.

Stellar rotation and magnetic stars.
See Abstr. 065.088.

Solar magnetic structure and the solar activity cycle. Review of observational data.   See Abstr. 075.008.

Radio observations and the mass flow rate of α Cyg (A2Ia).   See Abstr. 112.028.

A spectrographic study of the shell star EW Lac (217050).   See Abstr. 112.053.

Photometric *UBV* period study of eight Ap stars. See Abstr. 113.009.

The peculiar Rosetta stone: Beta Coronae Borealis. See Abstr. 113.033.

Far-ultraviolet observations of the variable radio star LSI +61°303.   See Abstr. 114.024.

The spectra of the chemically peculiar stars. See Abstr. 114.109.

Hot peculiar stars.   See Abstr. 114.123.

A Fourier analysis of the spectral lines of Procyon. See Abstr. 114.154.

The cool half of the H-R diagram in soft X-rays. See Abstr. 115.018.

Changes in the high state of AM Herculis: a simultaneous X-ray, optical, polarimetric, and spectroscopic study. See Abstr. 117.008.

Spectroscopy of the unique degenerate binary star LSI +61°303.   See Abstr. 117.051.

Interpretation of observations supporting a magnetospheric model of SS 433.   See Abstr. 117.092.

A possible nonrotational periodic magnetic variation in the Ap star HR 710.   See Abstr. 120.025.

A linear polarization survey of T Tauri stars. See Abstr. 121.013.

Microwave emission from flare star AT Mic. See Abstr. 122.008.

Search for linear polarization in solar neighborhood flare stars and spotted stars.   See Abstr. 122.011.

The variability of Bp and Ap stars. See Abstr. 122.084.

The principal frequency of the rapidly oscillating Ap star α Cir.   See Abstr. 122.177.

On rapid variations of the magnetic star 53 Cam spectrum.   See Abstr. 122.203.

The magnetic fields of white dwarfs. See Abstr. 126.003.

Radio continuum and carbon monoxide observations of V645 Cygni (GL 2789).   See Abstr. 131.083.

Einstein X-ray identification of the variable radio star LSI +61°303.   See Abstr. 142.002.

Relations among stellar X-ray emission observed from Einstein, stellar rotation and bolometric luminosity. See Abstr. 142.014.

Low luminosity galactic X-ray sources. See Abstr. 142.096.

Coronal activity in F-, G-, and K-type stars. See Abstr. 142.098.

Variable linear polarization in the X-ray binaries HD 77581 and HD 153919.   See Abstr. 142.153.

Polarization observations of X-ray binaries. See Abstr. 142.154.

117 Close Binaries (Observations, Theory)

117.001 **M5 V101: a close binary system in a globular cluster.**
B. Margon, R. A. Downes, J. E. Gunn.
Astrophys. J., Lett., Vol. 247, L89 - L92 (1981).
Spectrophotometric and photographic observations of
the large amplitude variable V101 in the globular cluster M5
show it to have the characteristics of a dwarf nova. In particular, a spectrum obtained with the Hale 5 m reflector of the
object in quiescence ($V \sim 20$) shows strong, broad Balmer and
He I emission. It is suggested that M5 V101 is the strongest
case for the existence of close binary systems in globular clusters.

117.002 **Parametric excitation of stellar modes in close
binary systems.**   J. Papaloizou, J. E. Pringle.
Mon. Not. R. Astron. Soc., Vol. 196, 371 - 379 (1981).
The authors discuss the parametric instabilities that occur
in stars in close binary systems, when they are subjected to a
periodic disturbance. They relax the constraint inherent in
previous work that the star is nearly spherical. The authors
show that, even when the star nearly fills its Roche lobe,
parametric destabilization of modes is possible.

117.003 **Spectroscopy of the AM Herculis type binary
2A 0311 − 227.**   J. Bailey, M. Ward.
Mon. Not. R. Astron. Soc., Vol. 196, 425 - 434 (1981).
The authors present time resolved spectroscopy of
2A 0311 − 227 with 2 Å resolution. The observations show
large variations in the emission line profiles with orbital phase.
They observe a double peaked structure at certain phases, and
this appears to move with a larger radial velocity amplitude
than the base of the line. The authors discuss several possible
models to explain the double structure. Comparison with
other spectroscopic studies of 2A 0311 − 227 indicate changes
in the behaviour of the system on a time-scale of months.

117.004 **RZ Gru − a new cataclysmic binary?**
B. D. Kelly, D. Kilkenny, J. A. Cooke.
Mon. Not. R. Astron. Soc., Vol. 196, 91P - 94P (1981).
Recent spectroscopic and photometric observations of
RZ Gru (= S5150 Gruis) indicate that it is an interacting bina-
ry star, possibly a cataclysmic system. Earlier observations by
Siedel and Hoffmeister suggest that it may also be an eclipsing
binary.

117.005 **Magnetic braking in low-mass X-ray binaries.**
F. Verbunt, C. Zwaan.
Astron. Astrophys., Vol. 100, L7 - L9 (1981).
It is pointed out that the cool main-sequence components
of cataclysmic binaries and low-mass X-ray binaries will undergo rotational braking by a magnetically coupled stellar wind, in
a similar way as single main-sequence stars. This leads to a loss
of orbital angular momentum from the system, and to an enhanced mass-transfer rate, sufficient to power the brightest low-
mass X-ray binaries for the secondary mass $M_2 = 0.8\,M_\odot$.

117.006 **Distinguishing between a white dwarf and a neutron
star in an X-ray binary.**   J. Patterson.
Nature, Vol. 292, 810 - 811 (1981).
It is proposed that the single most useful clue to the
nature of the accreting compact star in an X-ray binary is its
ratio of X-ray to visual luminosity $L_x(2 - 10\,\text{keV})/L_v$. Present
data indicate that typical L_x/L_v ratios are ~ 4 for white dwarfs,
and ~ 5000 for neutron stars.

117.007 **A quest for the red companion in six cataclysmic
binaries.**   P. Young, D. P. Schneider.
Astrophys. J., Vol. 247, 960 - 968 (1981).
Near-infrared CCD spectra of six cataclysmic binaries

have been obtained in an attempt to detect their red dwarf
components. Similar observations of M dwarfs yield a classification scheme, based on the TiO bands, which allows one to
describe the red dwarfs in cataclysmic binaries despite their
dilution by light from an accretion disk of unknown character.
Observations of AM Her while in its low state found a red
dwarf of type M4$^+$ V and a distance of 80 pc, confirming the
authors' previous work.

117.008 **Changes in the high state of AM Herculis: a simultaneous X-ray, optical, polarimetric, and spectroscopic
study.**   L. Crosa, P. Szkody, G. Stokes, J. Swank,
G. Wallerstein.
Astrophys. J., Vol. 247, 984 - 993 (1981).
Simultaneous optical photometry, polarimetry, and
spectroscopy during an Einstein X-ray observation of AM Her
at the high state are reported. The circular polarization at this
time was always negative and had the smallest peak to peak
amplitude yet reported. No correlation between the optical
photometry and 1−10 keV X-ray intensity were found on
flickering time scales. The combined results of these observations are discussed within the framework of models with accretion onto one or both magnetic poles of the white dwarf.

117.009 **SS 433: enigma of the century.**   R. A. Schorn.
Sky Telesc., Vol. 62, 100 - 102 (1981).

117.010 **HD 192163 − a possible close binary Wolf-Rayet
star with a relativistic component.**
A. A. Aslanov, A. M. Cherepashchuk.
Pis'ma Astron. Zh., Tom 7, 482 - 487 (1981). In Russian.
English translation in Soviet Astron. Lett., Vol. 7.
Spectroscopic observations of the WN6 star HD 192163
are presented to confirm the periodic ($P = 4^{\text{d}}5$) spectroscopic
variability of this star suspected by Koenigsberger et al. (1980).
In the frame of a close binary system model the mass function
of the star is $\sim 10^{-2} \cdot 10^{-3}\,M_\odot$. The mass of the second com-
ponent − probably a neutron star − seems to be of $\sim 1\,M_\odot$.
HD 192163 can be considered in its evolution as a close post
X-ray binary system.

117.011 **A photometric study of the close binary Delta
Orionis A.**   R. H. Koch, B. J. Hrivnak.
Astrophys. J., Vol. 248, 249 - 255 (1981).
Green and blue photoelectric light curves show the
historical intrinsic variability of the δ Ori A close binary superposed on the interaction and eclipse effects. There is a considerable measure of agreement between spectrographic and
photometric determinations of the rate of apsidal advance.
The determinacy of orbital eccentricity, however, is confused
because few minima of indifferent precision exist to check the
spectrographic value. After numerous trials, a less-than-perfect
theoretical representation of the light curve was achieved and
shows the system to be detached. The absolute stellar parameters make clear that both components have evolved substantially. A mean stellar structure k_2 is derived but cannot be
compared usefully to existing theoretical values. The
importance of the recently discovered visual companion, hz 42,
is emphasized.

117.012 **Gravitational radiation and the evolution of cataclysmic binaries.**
B. Paczyński, R. Sienkiewicz.
Astrophys. J., Lett., Vol. 248, L27 - L30 (1981).
Model evolutionary computations were done for a low-
mass secondary component filling its Roche lobe. Mass
transfer from the secondary onto a primary component was
driven by angular momentum loss caused by gravitational

radiation. The secondary was initially a lower-main-sequence star, and it gradually became degenerate as a result of mass loss. The orbital period initially decreased to a minimum value close to 80 minutes and then increased. The observed hydrogen-rich cataclysmic binaries have a short-period cutoff at 81 minutes. This indicates that gravitational radiation is the main driving force for the evolution of such short-period systems.

117.013 Time-resolved spectroscopy of cataclysmic variables: U Geminorum.   R. J. Stover.
Astrophys. J., Vol. 248, 684 - 695 (1981).

Spectroscopic observations of the dwarf nova U Gem are presented. The author also presents new measurements of the radial-velocity variations of the accretion disk emission lines. With these new measurements, the author derives the masses of the white dwarf primary and the red dwarf secondary. Analysis of the emission-line profiles show that the disk in U Gem cannot be approximated by a model in which the disk particles are assumed to follow simple periodic orbits described by the three-body approximation. Furthermore, the line profiles indicate the presence of turbulent velocities in the disk.

117.014 Radial velocity and line profile variations in the dwarf nova EM Cygni.
R. J. Stover, E. L. Robinson, R. E. Nather.
Astrophys. J., Vol. 248, 696 - 704 (1981).

Spectroscopic observations of the dwarf nova EM Cygni are presented. From these spectra the authors measure the radial velocity variations of the accretion disk emission lines and the secondary star absorption lines. The emission line profiles show that the accretion disk was very unstable during the nights of observation, leading to uncertainties in the emission line radial velocities. In spite of the uncertainties the authors can estimate the masses of the accreting white dwarf and the mass-losing secondary star. They find that the secondary star is at least 10% larger than a zero-age main sequence star of the same mass.

117.015 Double stars.   C. de Loore.
Recherche, No. 121, p. 430 - 440 (1981). In French.
Abstr. in Phys. Abstr., Vol. 84, Abstr. 83700 (1981).

117.016 A spectroscopic study of the binary Be star φ Persei.
R. Poeckert.
Publ. Astron. Soc. Pacific, Vol. 93, 297 - 317, with a correction p. 535 - 536 (1981).

High-dispersion high signal-to-noise ratio spectrograms of φ Per have been obtained. The radial velocities of various spectral features have been measured and an orbit solution obtained. The φ Persei system is found to consist of a 21 $M_\odot$ star, whose $V \sin i$ is 450 km s^{-1}, and a 3.4 $M_\odot$ secondary. The secondary is peculiar in that He II λ4686 emission arises from within its vicinity. It is suggested that the secondary is the helium core of a once more massive star, its mass having been transferred to the primary. Both stars in the system have circumstellar disks and the dimensions of these disks are discussed.

117.017 The short-term variation of period and the mass transfer in the semidetached binary system TW Draconis.   J. Tremko, J. M. Kreiner.
Bull. Astron. Inst. Czechoslovakia, Vol. 32, 242 - 253 (1981).

The period changes of the semidetached binary system TW Dra were studied. The short term changes of period are explained by the variable rate of the mass transfer from the less massive subgiant to the more massive primary component, accompanied by the exchange between the orbital momentum and the rotational momenta of the star and that of the disk, respectively.

117.018 Determination of parameters of W UMa systems. I: AE Phe, AQ Tuc, 44 i Boo.
C. Maceroni, L. Milano, G. Russo, C. Sollazzo.

Astron. Astrophys., Suppl. Ser., Vol. 45, 187 - 192 (1981).

Multicolour lightcurves of three W UMa systems have been solved using the Wilson-Devinney computer code, modified in order to run on a medium-size minicomputer. The solution confirms the contact hypothesis for these systems, one of which (AQ Tuc) has turned out to be of Binnendijk's A-type (deeper minimum produced by the eclipse of the more massive star), and the other two (AE Phe and 44 i Boo) of W-type. A comparison with previous solutions is made.

117.019 Observations of the gas stream in the mass transfer binary HR 2142.   G. J. Peters.
The Universe at ultraviolet wavelengths, (see 012.009), p. 157 - 165 (1981).

The mass transfer binary system HR 2142 has been observed at selected phases with the high resolution spectrograph on IUE. The observations were scheduled throughout the interval $0.91 < \phi < 0.00$ in order to view the light of the primary star through the gas stream as it presents different orientations to the line of sight. Numerous UV lines formed in the gas stream have been identified. The strengths and velocity variations displayed by these lines are compared with those observed in the ground-based spectral region.

117.020 IUE observations and interpretation of the symbiotic star RW Hya.
M. Kafatos, A. G. Michalitsianos, R. W. Hobbs.
The Universe at ultraviolet wavelengths, (see 012.009), p. 349 - 354 (1981).

IUE observations of the high excitation symbiotic star RW Hya (gM2 + pec) have been obtained. Analysis of the intense UV continuum observed between 1100 Å to 2000 Å suggests that this star is a binary system in which the secondary is identified as a hot subdwarf with $T_{eff} \sim 10^5$ K. The authors deduce a distance to the system of ~ 1000 pc. The UV spectrum consists of mainly semi-forbidden and allowed transition lines of which the CIV (1548 Å, 1550 Å) emission lines are particularly strong, and UV continuum at both shorter and longer wavelengths. Strong forbidden lines seem to be absent suggesting the presence of a nebula of high densities, in the approximate range $10^8 - 10^9$ cm^{-3}. Tidal interaction between the red giant primary and the hot subdwarf is suggested as a likely means to form the observed nebula. A general discussion of elemental and ionic abundances in the nebula is also presented.

117.021 IUE spectra of a flare in HR 5110: a flaring RS CVn or Algol system?
T. Simon, J. L. Linsky, F. H. Schiffer III.
The Universe at ultraviolet wavelengths, (see 012.009), p. 435 - 441 (1981).

Ultraviolet spectra of the RS CVn-type binary system HR 5110 have been obtained with IUE on May 31, 1979 during a period of intense radio flaring of this star. High temperature transition region lines are present, but are not enhanced above observed quiescent strengths. The similarities of HR 5110 to the Algol system, AS Eri, suggest that the 1979 May-June flare may involve mass exchange rather than annihilation of coronal magnetic fields.

117.022 Analysis of the symbiotic star AG Pegasi.
C. D. Keyes, M. J. Plavec.
The Universe at ultraviolet wavelengths, (see 012.009), p. 443 - 450 (1981).

High and low dispersion IUE data are analyzed in conjunction with coincident ground-based spectrophotometric scans and supplementary infrared photometry of the symbiotic object AG Pegasi. The IUE observations yield an improved value of E(B–V) = 0.12. The two stellar components are easily recognized in the spectra. The cool component may be an M1.7 III star and the hot component appears to have T_{eff} of approximately 30000 K. The emission lines observed in the ultraviolet indicate two or three distinct emitting regions. Nebular component ultraviolet intercombination lines suggest an electron density of several times 10^{10} cm^{-3}.

117.023   IUE observations of symbiotic stars.
J. Sahade, E. Brandi.

The Universe at ultraviolet wavelengths, (see 012.009), p. 451 - 454 (1981).

The IUE observations suggest that the symbiotic stars can be placed in two broad groups. One of the groups is characterized by strong, narrow emissions arising from a wide range of excitation energies, while the other one typically shows a strong continuum with absorption lines and very few or no emissions at all. Both broad groups appear to suggest that we are dealing with binary systems and that they probably differ in the characteristics and extent of the chromosphere-corona formation that is present in the system.

117.024   Ingress observations of the 1980 eclipse of the symbiotic star CI Cygni.   R. E. Stencel,
A. G. Michalitsianos, M. Kafatos, A. A. Boyarchuk.

The Universe at ultraviolet wavelengths, (see 012.009), p. 459 - 460 (1981).

Two symbiotics, AR Pav and CI Cyg are thought to be eclipsing binaries, and the authors have begun IUE observations during the predicted 1980 eclipse of CI Cygni.

117.025   Search for colliding stellar winds in Plaskett's star (HD 47129).   S. R. Heap.
The Universe at ultraviolet wavelengths, (see 012.009), p. 485 - 493 (1981).

High-dispersion spectra of Plaskett's star (HD 47129) were obtained with the short-wavelength spectrograph on IUE at five phases of the binary cycle. The unsaturated wind profiles, particularly those of Si IV λ 1400, show complex, phase-dependent structure. Two interpretations for the structure are suggested, neither of which is entirely satisfactory: (1) the structure is a consequence of directed streams, and (2) the structure is a consequence of colliding winds from the primary and secondary.

117.026   Cyclotron emission from AM Herculis.
G. Chanmugam.

The Universe at ultraviolet wavelengths, (see 012.009), p. 515 - 519 (1981).

The cyclotron absorption coefficients, in the ordinary and extraordinary modes, are calculated for the shock heated region of AM Her. The equations of radiative transfer are solved and the intensity of the emitted UV radiation determined as a function of angle. The average spectrum is shown to have deviations from the previously predicted Rayleigh-Jeans spectrum and the magnetic field of AM Her is deduced to be roughly 5×10^7 gauss.

117.027   Spectrophotometry of HM Sge.
A. P. Ipatov, B. F. Yudin.

Pis'ma Astron. Zh., Tom 7, 554 - 558 (1981). In Russian. English translation in Soviet Astron. Lett., Vol. 7.

Spectrophotometric observations of HM Sge in the spectral range 3300 - 7500 Å and photometric observations in the $JHKN$ system were carried out in September 1980. During the last four years the flux in the optical continuum of HM Sge has decreased by a factor of about 1.4. At the same time the equivalent width of the H β line ($W_\lambda \sim 760$ Å) practically has not changed. Significant increase of the He II λ4686 line flux during the last two years characterises the growth of excitation in the gaseous envelope of HM Sge. It is shown that the gaseous envelope incompletely absorbs Lc radiation and likely is of disc shape. This circumstance is in favor of the symbiotic nature of HM Sge.

117.028   M5 V101: a close binary system in a globular cluster.
B. Margon, R. A. Downes, J. E. Gunn.

Bull. American Astron. Soc., Vol. 13, 513 (1981). – Abstract.

117.029   Evolution of contact binary stars with magnetic braking.   S. W. Mochnacki.
Bull. American Astron. Soc., Vol. 13, 513 - 514 (1981). Abstract.

117.030   Recent light curves of V471 Tauri.
J. D. Himer, R. M. Robb, E. F. Milone.

Bull. American Astron. Soc., Vol. 13, 514 (1981). – Abstract.

117.031   Light curves and solutions for ST Cen.
B. J. Hrivnak, E. F. Milone.

Bull. American Astron. Soc., Vol. 13, 524 (1981). – Abstract.

117.032   Nebular emission line diagnostics in symbiotic stars.
N. Oliversen, C. M. Anderson, K. H. Nordsieck.

Bull. American Astron. Soc., Vol. 13, 524 (1981). – Abstract.

117.033   Problems of low-mass binary evolution.   O. Vilhu.
Astrophys. Space Sci., Vol. 78, 401 - 418 (1981).

Some problems connected with low-mass binary evolution (from contact binaries to cataclysmic variables and origin of bursters) are considered. Most attention is given to contact W UMa-stars and to (still unclear) scenarios where the angular momentum loss by magnetic braking may, at least partly, control the contact binary evolution.

117.034   Mass loss and mass transfer in the binary star HD 41335.   F. Paterson-Beeckmans.
Second European IUE Conference, (see 012.011), p. 51 - 54 (1980).

The Be star HD 41335 is a prototype of interacting binary. IUE high resolution spectra obtained at 5 different phases of the binary cycle show prominent variations due to the transfer of mass between the secondary and the primary star (B2V). Redshifted (V = +50 km/s) narrow lines are characteristics of the phase 0 (primary seen through the mass flow). Around phase 0.5, blueshifted shell lines (V = −65 km/s) indicate a mass flow through the external Lagrangian point. The Al III (1), Si IV (1), C IV (1), Fe III (34, 48) multiplets present at all the phases strong and asymmetric absorption features, with outward velocities up to 550 km/s (C IV), larger than the escape velocity for this rapidly rotating star (V sin i = 415 km/s).

117.035   Simultaneous UV and optical observations of σ Gem, an RS CVn-like star: preliminary results.
F. Giovannelli, S. M. Rucinski, C. Bartolini, A. Guarnieri, S. Guadenzi, C. Rossi, A. Giangrande, A. Piccioni.

Second European IUE Conference, (see 012.011), p. 59 - 60 (1980).

In the framework of the IUE programme FG 176 (Rucinski et al. 1978) "The UV Excesses of RS CVn-Type Binaries" the authors present some UV spectra of σ Gem and discuss them shortly. The UBV photometric measurements and optical spectra carried out simultaneously are also commented on.

117.036   The active chromosphere and corona of HD 4502 = ζ And.   D. Reimers.
Second European IUE Conference, (see 012.011), p. 77 - 79 (1980).

Low resolution IUE spectra covering the range 1200 to 3200 Å show that the binary ζ And (K2 II) has an emission line spectrum of chromospheric and transition layer lines similar to that observed in stars like λ And and HR 1099. The underlying continuum flux between 1600 Å and 3000 Å is due to the K giant. Also in the UV the companion is not visible. The UV lines are typically an order of magnitude stronger than in the sun. The NV 1240 line is much stronger than in the otherwise similar λ And spectrum which indicates that there is more coronal material with T ≃ 2 × 10⁵ K. Si IV 1402/C III 1909 and Si III 1892/C III 1909 ratios yield a transition layer pressure log $N_e \times T_e \simeq 15.5$.

117.037 Mass loss from the interacting contact binary SV Cen. H. Drechsel, J. Rahe, B. Wolf.
Second European IUE Conference, (see 012.011), p. 213 - 218 (1980).

High resolution IUE (SWP and LWR) spectrograms of the close eclipsing binary SV Cen (B1 III + B4) were taken near orbital phase 0.4 on 1979, November 3. The dominant spectral features are resonance lines of ionized metals (Al II, Mg II, Si IV, C IV, etc.), and Fe III lines originating from metastable lower levels.

117.038 Simultaneous observations of V861 Sco.
C. de Loore, M. Burger, I. D. Howarth, R. Wilson, B. S. Carter, J. W. Menzies, G. Roberts, P. A. Whitelock, E. L. van Dessel, M. C. W. Sandford.
Second European IUE Conference, (see 012.011), p. 237 - 242 (1980).

High dispersion spectra (Δλ ~ 0.2Å) of V861 Sco (HD 152667) taken with IUE at several orbital phases are presented, as well as simultaneous optical spectroscopy, and uvby and infrared photometry. The UV spectrum is normal for a B0Ia supergiant; a number of lines show P Cygni profiles with in several cases sharp multiple absorption components.

117.039 A model of the Algol type close binary TT Hydrae.
A. G. Kulkarni, K. D. Abhyankar.
J. Astrophys. Astron., Vol. 2, 119 - 132 (1981).

UBV photometric observations and elements of TT Hydrae obtained by Kulkarni and Abhyankar (1980) are combined with the radial velocity curve of Popper (1979) to derive the absolute dimensions and a model of this important Algol system. The system shows an ultraviolet excess of 0.5 to 0.6 magnitudes during primary eclipse, which is attributed to an asymmetric circumstellar distribution of matter around the primary. The evolutionary status of the secondary, which does not appear to fill its Roche lobe completely, is discussed.

117.040 The influence of mass loss on the evolution of binaries. C. de Loore.
Effects of mass loss on stellar evolution, (see 012.015), p. 405 - 430 (1981).

Contents: Introduction. Mass loss in not evolved binaries. Stellar wind mass loss rates for binaries. The mass transfer phase. Mass and/or angular momentum loss from binary systems. Binary evolutionary computations with mass and/or angular momentum loss. Mass loss and the production of WR stars. Non-conservative evolution. Change of the chemical abundances of the outer layers in binaries. The effect of mass loss on the mass ratio distribution. The lifetime of the secondary after accretion. Different evolutionary scenarios for massive systems with mass ratios ~1 and extreme mass ratios. Conclusions.

117.041 Mass loss from interacting close binary systems.
M. J. Plavec.
Effects of mass loss on stellar evolution, (see 012.015), p. 431 - 456 (1981).

117.042 Non conservative massive binary evolution. How much mass leaves the binary during the evolution from OB + OB to WR + OB. D. Vanbeveren.
Effects of mass loss on stellar evolution, (see 012.015), p. 461 - 463 (1981).

117.043 On the evolutionary time scale of the accreting component in massive close binaries: consequences

for the supernova event. C. Doom, J. P. de Grève.
Effects of mass loss on stellar evolution, (see 012.015), p. 465 - 468 (1981).

The remaining core hydrogen burning lifetime after a case B of mass exchange is computed for the mass gaining component in massive close binaries. Effects of stellar wind mass loss during Roche lobe overflow are included. Consequences for the evolutionary scenario are discussed.

117.044 Non-conservative evolutionary scenario for 100 Algols. F. Mardirossian, G. Giuricin.
Effects of mass loss on stellar evolution, (see 012.015), p. 473 - 475 (1981).

The authors have examined the observational data of 100 Algols in order to check the validity of several simple models of non-conservative mass transfer. Strong evidence of mass and angular momentum loss have been found at least in about 20% of the authors' Algols. Case B mass exchange is favoured for low-mass Algols, while case A predominates, though not so widely as expected, in Algols of higher total mass.

117.045 Infrared observations and mass loss of the binary system V861 Sco.
E. G. Tanzi, L. Maraschi, M. Tarenghi, A. Treves.
Effects of mass loss on stellar evolution, (see 012.015), p. 477 - 480 (1981).

The authors report on a series of photometric observations of V861 Sco in the infrared (from 1.25 to 4.8 μ) taken at various orbital phases which improve and extend previous measurements by Tanzi et al. (1979). The results give evidence of a phase modulated infrared excess which can be interpreted in terms of a non isotropic mass flow in the system or, alternatively, of a contribution from a colder secondary component.

117.046 The nature of V861 Sco (=HD 152667).
I. D. Howarth, R. Wilson.
Effects of mass loss on stellar evolution, (see 012.015), p. 481 - 486 (1981).

The physical parameters of the components of V861 Sco are derived from light-curve analysis and published spectroscopy. Good agreement with evolutionary models is obtained. The stellar wind is investigated using IUE data; the results include no large phase dependence of the mass loss rate and insensitivity of the velocity of the wind to changes in the photospheric velocity, even near the base of the wind. However, small random changes in velocity near the base of the wind are amplified to larger changes further out.

117.047 Ultraviolet and optical observations of the mass-losing contact binary SV Centauri.
H. Drechsel, H. D. Radecke, J. Rahe, G. Rupprecht, W. Wargau, B. Wolf.
Effects of mass loss on stellar evolution, (see 012.015), p. 487 - 489 (1981).

117.048 Interacting stellar winds in a binary system.
S. Kwok.
Effects of mass loss on stellar evolution, (see 012.015), p. 499 - 502 (1981).

The author describes a binary system in which two stellar winds are interacting with dramatic effects.

117.049 Evolutionary computations for intermediate mass close binary systems. T. J. van der Linden.
Effects of mass loss on stellar evolution, (see 012.015), p. 503 - 506 (1981).

Numerical simulations of the evolution of close binaries with primary mass between 3 and 15 M⊙ and mass ratios of ~1.5 were carried out. These results show that after mass transfer the system consists of a bright main-sequence star together with a faint He-star in a rather wide orbit (p> 60ᵈ).

Evolution of this kind predicts isotope anomalies for the CNO-elements with respect to their solar values.

117.050 **The ultraviolet spectrum of the twin-degenerate interacting binary G61-29.**
D. L. Lambert, M. H. Slovak.
Publ. Astron. Soc. Pacific, Vol. 93, 477 - 480 (1981).

A low-resolution IUE spectrum (1200 Å−1900 Å) of the twin-degenerate white-dwarf binary star G61-29 is presented. A comparison is made to similar IUE spectra of the old novae DQ Her and V603 Aql. G61-29 shows N V 1240 Å and He II 1640 Å in emission. Unlike the old novae, the C IV 1550 Å doublet is not present. It is proposed that the mass-losing He white dwarf ($M \cong 0.02\,M_\odot$) is the H-exhausted, He-rich core of an old star in which the original CNO nuclei have been converted to ^{14}N by the CNO-cycles.

117.051 **Spectroscopy of the unique degenerate binary star LSI +61°303.**   J. B. Hutchings, D. Crampton.
Publ. Astron. Soc. Pacific, Vol. 93, 486 - 489 (1981).

IUE ultraviolet, and ground-based spectroscopic data are presented and discussed. The primary is a rapidly rotating B0 main-sequence star, with a stable shell. The radial velocities indicate binary motion in a 26.5-day orbit (in agreement with the radio flaring period), with a secondary of mass $1.1-1.5\,M_\odot$. The broad Balmer emission appears to arise principally near the primary, and the line of sight is close to equatorial. These considerations bear on models involving X and γ radiation from the system.

117.052 **Convective auto-oscillations in the components of binary systems.**   G. I. Burdeh.
Izv. AN SSSR. Mezh. zhidkosti i gaza, 1981, No. 2, p. 149 - 155. In Russian. − Abstr. in Ref. zh., 51. Astron., 8.51.542 (1981).

117.053 **Evidence for apsidal motion in the Cygnus X-1 system and a third-body interpretation.**
R. E. Wilson, R. K. Fox.
Astron. J., Vol. 86, 1259 - 1266 (1981).

The authors have compared optical light curves of the HDE 226868 − Cyg X-1 system, made in six consecutive observing seasons, with model light curves and have found evidence for rotation of the elliptical orbit with a period of 4.5 yr. They discuss reasons to account for earlier findings that any apsidal motion must be much slower. The rotation, if real, is far too fast to be due to the finite mass distribution of the optical star, because that star is highly evolved and must have a high degree of central condensation. The authors favor an explanation in terms of a perturbing third body, and use the relations given by Lidov and Ziglin to find the expected mass-distance relation.

117.054 **HD 86590: a short-period RS Canum Venaticorum binary system.**
C. T. Bolton, Z. Aslan, K. W. Kamper, R. W. Lyons.
Astron. J., Vol. 86, 1267 - 1276 (1981).

The authors have derived spectroscopic orbital elements for the single-line binary HD 86590 from radial velocities measured on moderate- and high-dispersion spectrograms taken at the Kottamia Observatory in 1966 and the David Dunlap Observatory in 1975 - 1979. The orbital period is 1.070 354 days. The primary is a K0V star, and the secondary is almost certainly a white dwarf. The primary shows evidence of enhanced solar activity similar to that seen in many RS CVn binaries.

117.055 **On the frequency distribution of mass ratios in close binary stellar systems and its explanation based on mass transfer.**   A. G. Kulkarni.
Bull. Astron. Soc. India, Vol. 9, 84 (1981). − Abstract.

117.056 **AS 296, a symbiotic star of very high radial velocity.**
E. W. Brugel, G. Wallerstein.
Observatory, Vol. 101, 164 - 166 (1981).

As part of a larger spectroscopic study of suspected symbiotic stars the authors have obtained several 20 Å/mm plates of AS 296 with the three-stage image tube on the 1.8 m Cassegrain spectrograph of the Dominion Astrophysical Observatory. The spectrum of AS 296 is clearly composite, showing *TiO* bands and emission lines of various ions. A systemic velocity of +117 km/s with respect to the LSR is obtained from the [*O* III] lines. It is argued that AS 296 is more likely to be a disk star of high velocity than an unusually luminous symbiotic star.

117.057 **Symbiotic stars as an old disk population.**
G. Wallerstein.
Observatory, Vol. 101, 172 - 174 (1981).

The author has assembled a table of all symbiotic stars in the *General Catalogue of Variable Stars* and its supplements and discusses their radial velocities. A velocity dispersion of 63 ± 14 km/s is found for all the stars and a value of 58 ± 14 km/s is established if the probable halo star, AG Dra, is omitted. The space distribution is similar to that of an old disk population. Some implications of low masses for the symbiotic stars are discussed, and some suggestions are made regarding possibly useful observations.

117.058 **Ultraviolet photometry from ANS: chromospheric emission of W Ursae Majoris and 44 i Bootis.**
J. A. Eaton, C.- C. Wu.
Astron. J., Vol. 86, 1387 - 1393 (1981).

The authors present photometry of two W-type systems obtained with the ANS satellite in 1974 and 1975 in wavelength regions which sample chromospheric and transition region emission of solar-type stars. One star, 44 i Boo, which is contaminated seriously by light of a third component, has a marginal ultraviolet excess associated with chromospheric emission. The other star, W UMa, has strong ultraviolet excesses at both 1550 and 1800 Å.

117.059 **On the spin-up of the mass accreting component in a close binary system.**   W. Packet.
Astron. Astrophys., Vol. 102, 17 - 19 (1981).

Using a simplified model it is shown that during mass-exchange in a close binary system enough angular momentum is transferred towards the mass accreting star to spin it up to its critical rotation velocity after gaining only a few percent of its original mass. The consequences of this result for the evolution of close binary systems are briefly discussed.

117.060 **Evolution of W UMa systems and angular momentum loss.**   T. Rahunen.
Astron. Astrophys., Vol. 102, 81 - 90 (1981).

The evolution of contact binaries is investigated in an attempt to understand the structure and evolutionary state of W UMa stars. The observed periods, mass ratios and colours of W UMa systems are compared with the properties of the models. This comparison shows that the division of W UMa systems into W-type and A-type systems is probably related to the convective or radiative nature of the common envelope at the inner contact surface.

117.061 **Über den W UMa-Stern SW Lacertae.**
K. Wälke.
BAV Rundbrief, 30. Jahrg., 79 - 82 (1981).

117.062 **Reanalysis of the light curve of ES Lib.**
G. Giuricin, F. Mardirossian, S. Ferluga.
Astron. Nachr., Band 302, 187 - 188 (1981).

Using Wood's (1972) model the authors have reanalyzed Bartolini et al.'s (1973) photoelectric light curve − as yet only crudely studied − of the single-lined eclipsing binary ES Lib.

Their photometric elements considerably differ from the previous ones. For a plausible value of the mass ratio ($q = 0.4$) the hotter (A2−3) component fills its Roche lobe, whereas the (K-type) secondary is detached. Nevertheless, in view of the large sum of the fractional radii of the two components, ES Lib can be related to the contact systems, for which broken-contact phases may occur.

117.063 The massive contact system LY Aur.
R. Margoni, R. Stagni, A. Mammano.
Astrophys. Space Sci., Vol. 79, 145 - 152 (1981).

From 44 spectra obtained at Asiago in 1969 - 1971, the authors infer the orbital elements of the double line system LY Aur. The mass-functions indicate rather massive components, with mass ratio 0.58 ± 0.03. This is in good agreement with a contact photometric solution found by Eaton (1978). Combination of spectroscopic and photometric data yields the following absolute elements: $m_1 = 13$, $m_2 = 22$, separation 34.5, $R_1 = 11$, $R_2 = 15$, in solar units. A short comment on the large number of massive contact systems is given.

117.064 S 10830 Trianguli − an interesting cataclysmic variable.
G. A. Richter, P. Notni, F. Börngen, V. Afanasjev (*Afanas'ev*), I. D. Karachentsev, A. Kopylov.
Astron. Nachr., Band 302, 211 - 218 (1981). In German.

With the aid of slit spectrograms gained at the 6 meter Selenchukskaja telescope and at the 2 meter Tautenburg telescope a brightness outburst (the only one so far observed) could be followed up spectroscopically. The spectral behaviour resembles SS Cygni, but the presence of CIII and NIII emission is not typical of dwarf novae. Possibly the object is related to the anomalous recurrent nova WZ Sge. The definite determination of its belonging to a known subclass of cataclysmic variables will be possible only after obtainment of extensive photometric material.

117.065 Photometric observations of symbiotic stars in the UBVRJHKLMN system. CI Cygni.
O. G. Taranova, B. F. Yudin.
Astron. Zh., Tom 58, 1051 - 1056 (1981). In Russian. English translation in Soviet Astron., Vol. 25, No. 5.

Photometric observations of CI Cyg in the UBVRJHKLMN system during 1978 - 1980 are reported.

117.066 SS 433: gewoon model voor zeer ongewone ster.
F. Verbunt.
Zenit, 8. Jaarg., 453 - 457 (1981).

117.067 A radial-velocity study of the dwarf nova RU Pegasi.
R. J. Stover.
Astrophys. J., Vol. 249, 673 - 679 (1981).

The author has obtained a series of short-exposure spectra of RU Peg and, from these spectra, he has measured the radial-velocity variations of both the absorption and emission lines. The author has also obtained an improved value for the orbital period, $P = 0\overset{d}{.}3746 \pm 0\overset{d}{.}0002$. He finds an apparent phase shift between the emission and absorption velocity curves. The author suggests that this phase shift and similar shifts found in several other systems may be due to an asymmetry in the emission distribution in the accretion disk.

117.068 Spectroscopy of 2A 0526−328: a triple periodic cataclysmic variable.
J. B. Hutchings, D. Crampton, A. P. Cowley, J. R. Thorstensen, P. A. Charles.
Astrophys. J., Vol. 249, 680 - 686 (1981).

The cataclysmic variable X-ray binary 2A 0526−328 has a spectroscopic (orbital) period of 0.228600 days, a photometric (rotation) period of 0.21631 days, and a beat (photometric) period of 4.024 days. Radial velocity behavior is complex, suggesting orbital motions, a disk hot spot, and a spiraling gas stream. Emission line variations are large and not generally periodic. The system is discussed in terms of a non-

synchronously rotating white dwarf of the DQ Her type. The system presents several unique phenomena for this type of binary.

117.069 The X-ray and optical characteristics of the cataclysmic variable V794 Aquilae.
P. Szkody, L. Crosa, G. D. Bothun, R. A. Downes, R. A. Schommer.
Astrophys. J., Lett., Vol. 249, L61 - L64 (1981).

The authors report the Einstein detection of V794 Aql as an X-ray source with a 0.1 - 4.5 keV band flux of 4.5×10^{-12} ergs cm^{-2} s^{-1} during its high state. Ground-based optical observations show blue colors, strong H emission with a flat Balmer decrement, and variability of 0.5 mag on hourly time scales and of 3 mag on nightly time scales. These properties argue for the inclusion of V794 Aql in the general class of cataclysmic variables.

117.070 Mass, angular momentum, and energy transfer in close binary stars.
F. H. Shu, S. H. Lubow.
Annu. Rev. Astron. Astrophys., Vol. 19, (see 003.012), 277 - 293 (1981).

The article discusses some aspects of the dynamics and energetics of close binary stars which have heretofore not received a detailed review. The authors concentrate their attention to those processes of mass, angular momentum, or energy transfer which they believe have potentially important consequences for the evolutionary history of close binary systems. The presentation proceeds principally along theoretical lines. The authors consider in turn detached binaries, semidetached binaries, and contact binaries.

117.071 Analytical method for calculating orbital elements of spectrally double stars.
V. I. Oumlenski.
Dokl. Bolg. AN, Vol. 34, 157 - 158 (1981). − Abstr. in Ref. zh., 51. Astron., 10.51.585 (1981).

117.072 Absolute spectrophotometry of υ Sgr.
V. I. Burnashev.
Izv. Krymskoj Astrofiz. Obs., Tom 63, 104 - 117 (1981). In Russian. English translation in Bull. Crimean Astrophys. Obs., Vol. 63.

Observations of the absolute energy distribution and narrow-band photometry of the close binary system υ Sgr were carried out in 1973 - 1978. Interstellar absorption $A_V = 0\overset{m}{.}75 - 1\overset{m}{.}2$ has been estimated. The spectral type of the invisible companion has been estimated to be O9 V. Irregular variabilities of light and color of several percent during few days were observed. Emission of the binary system in the Hα-line was nearly constant during all phases of the orbital period.

117.073 Mass loss in close binary systems.
M. A. Giannuzzi.
Astron. Astrophys., Vol. 103, 111 - 118 (1981).

It is now believed that in close binary systems the process of mass transfer between the component stars is, generally, not conservative. Therefore various cases of mass and momentum loss have been investigated and the trends of period P against mass ratio μ have been calculated. A comparison of theoretical values with the observed periods of the semidetached systems of Algol type has been carried out.

117.074 Is II Pegasi (HD 224085) a new spot-cycle activity star?
E. Bohusz, A. Udalski.
Acta Astron., Vol. 31, 185 - 190 (1981).

New observations of II Pegasi (HD 224085) are presented. They are used together with the earlier ones to study the photometric period variations. Basing on the spot model the photometric period changes can be explained by solar-like spot activity with the period 8 - 10 years.

117.075 **Etoiles Wolf-Rayet de hautes vitesses et compagnon compact.**
R. Lamontagne, A. F. J. Moffat, W. Seggewiss.
J. R. Astron. Soc. Canada, Vol. 75, 250 (1981). – Abstract.

117.076 **IUE spectra of the hot close binary V Puppis.**
R. H. Koch, D. H. Bradstreet, P. M. Perry, R. J. Pfeiffer.
Publ. Astron. Soc. Pacific, Vol. 93, 621 - 625 (1981).
The early B-type close binary, V Pup, has been observed with the IUE spacecraft and both high- and low-dispersion spectra have been obtained. A temperature of 25,000 K determined from a solar-abundance model atmosphere fits most of the ground-based and spacecraft data. The line spectrum of C, Si, and Al ions is discussed and it is shown that the equivalent widths of the silicon lines do not agree with the predictions of a modern theory for them. It is suggested that the complexities of both the continuum and the absorption spectrum are due to contamination by circumstellar gas. Several newly-observed interstellar lines are used to supplement the line-of-sight absorptions already presented by previous workers.

117.077 **IUE spectroscopy of hot binary stars HD 108, HD 149404, and HD 163181.**
J. B. Hutchings, J. van Heteren.
Publ. Astron. Soc. Pacific, Vol. 93, 626 - 629 (1981).
High-dispersion UV spectroscopy is presented of the stars at several orbital phases. Changes in the spectrum of HD 108 are seen, but its binary nature is in doubt. The other two binaries do not show stellar wind changes with phase, but their binary motions are seen in weak lines. An expanding Fe III shell is indicated in all spectra. Stellar temperatures and extinction are derived.

117.078 **Evolution and period changes in binary stars.**
K. D. Abhyankar.
Bull. Astron. Soc. India, Vol. 9, 99 - 111 (1981).

117.079 **On the evolutionary time scale of the accreting component in massive close binaries.**
C. Doom, J. P. de Grève.
Astrophys. Space Sci., Vol. 80, 369 - 395 (1981).
The remaining core hydrogen burning lifetime after mass exchange is computed for the mass gaining component in massive close binary systems. Results are obtained for various modes of mass accretion, for a wide range of initial masses and mass ratios, and for different values of mass loss from the system. The effect of mass loss by stellar wind is included. By use of general properties of massive close binary evolution, the position in time of four events is determined: the end of core hydrogen burning of the primary and the secondary, and the occurrence of the supernova of the primary and the secondary. It is found that the two supernova events occur rather close in time, compared to the main sequence time scales of the components.

117.080 **Hydrogen burning on a white dwarf accreting hydrogen in a binary system.**
S. A. Glasner, G. Rakavy, Y. Tuchman.
Astrophys. Space Sci., Vol. 80, 397 - 403 (1981).
The authors follow the accretion of hydrogen-rich matter onto the surface of a white dwarf in a close binary system. Two phases of accretion are assumed. First – slow accretion from the interstellar clouds, second – fast accretion from the companion. Hydrogen is ignited at the interface between the fast and slowly accreted layers. After a short runaway burning continues in the form of quasistationary deflagration front propagating inwards. The features of this front are discussed. A possibility of mass loss is indicated.

117.081 **54 Cam: a new variable star.**
J. A. Eaton, D. S. Hall, G. W. Henry, C. A. Vaucher,

H. Louth, D. R. Skillman.
Astrophys. Space Sci., Vol. 80, 405 - 410 (1981).
Differential photometry of the RS CVn-type binary 54 Cam in 1978, 1979, and 1980 shows its light to be variable with a period of $10^d163 \pm 0^d009$ and an amplitude in V (max. to min.) which increased from 0^m03 to 0^m06 between 1979.19 and 1980.82. An epoch of light minimum was JD 2444529.7. The 9% difference between P(phtm.) = 10^d163 and P(orb.) = 11^d0764, a much larger difference than is characteristic of other RS CVn binaries, is suggested as an explanation for the radio emission.

117.082 **SS 433: how do the moving lines move?**
S. A. Grandi.
Vistas Astron., Vol. 25, (see 012.040), 7 - 11 (1981).
A progress report is given on the analysis of an ever-increasing set of moving line radial velocity measurements of SS 433. Model fits to the "twin-beam" model are discussed, and an assessment of the quality of the fit is made. A preliminary investigation is conducted into the source of the "random" scatter of the data points around the best-fit model. Finally, a discussion of moving line bullets and light-travel-time effects is made with reference to a spectral montage of SS 433.

117.083 **Optical observations of the "stationary" emission and absorption line systems.**
D. Crampton, J. B. Hutchings.
Vistas Astron., Vol. 25, (see 012.040), 13 - 21 (1981).
Continued spectroscopic observations of SS 433 confirm and refine the discovery by Crampton, Cowley and Hutchings (1980) that SS 433 is a spectroscopic binary with P = 13.1 days Both the absorption and emission line systems exhibit similar small amplitude (K ~ 70 km s^{-1}) velocity variations that are precisely in phase with each other. A 164 day variation is found in the intensities of both emission and absorption lines, and in the profile shape of Hβ. A small 164-day variation is also present in the radial velocities of lines in the "stationary" system.

117.084 **A confirmation of the 13^d binary period in SS 433.**
R. Bedogni, A. Braccesi, B. Marano, A. Messina, D. C. Crampton.
Vistas Astron., Vol. 25, (see 012.040), 23 - 25 (1981).
Spectroscopic observations of SS 433 made in the same period at different longitudes confirm the 13^d periodicity in the velocity of the stationary lines. The alias values, of the order of 1^d, can be definitively ruled out. A systemic velocity of +90 km/s is derived from the present data.

117.085 **The moving features in SS 433: new observations and analysis.**
F. Ciatti, A. Mammano, A. Vittone.
Vistas Astron., Vol. 25, (see 012.040), 27 - 30 (1981).
Spectroscopic observations of the extremely peculiar object SS 433 have been carried out at the Asiago Observatory since the time of its identification as the optical counterpart of a radio/X-ray source. The results obtained during the year 1980 are here presented. The authors also briefly describe both the morphological properties of the blue- and red-shifted features, and a kinematical analysis of the system.

117.086 **Optical photometry of SS 433: 200 nights of data 1979 - 1980.**
J. C. Kemp, M. S. Barbour, G. N. Kemp, D. M. Hagood.
Vistas Astron., Vol. 25, (see 012.040), 31 - 43 (1981).
The authors report the results of over 200 nights of photoelectric photometry of SS 433, over the period July 1979 - October 1980. The data include 199 nightly points in unfiltered light (S-20); 166 points in V band; 26 points in a far-red band; and 46 points in a narrow (12 Å wide) interference-filter band centered on stationary Hα, at 6567 Å. They find: (1) improved evidence for an underlying 164^d light variation,

with peak-to-peak amplitudes 0^m50 in V and 0^m7 in λ6567; the object is brightest during maximum separation of the "moving" lines. (2) A binary-like variation with period $13^d074 \pm 0^d020$, with peak-to-peak amplitudes (as averaged coherently over the entire data sets) of 0^m50 in V and 0^m60 in λ6567. (3) Additional irregular behaviour, with occasional changes of more than 1 mag in 2 - 3 days.

117.087   High resolution monitoring of rest Hα in SS 433.
M. A. Dopita, A. M. Cherepashchuk.
Vistas Astron., Vol. 25, (see 012.040), 51 - 59 (1981).

The authors have simultaneously monitored SS 433 in V and the Hα line at $\Delta\lambda = 0.6$ Å. There was an optical outburst on JD 2444435.5, correlated with a sudden broadening of the Hα line. They also conclude that SS 433 has a massive accretion disk which contributes nearly all the optical luminosity of the object. It has $L = 10^{40}$ erg s^{-1}, $T_e = 30000°K$ and $R = 6 \times 10^{12}$ cm.

117.088   SS 433 – an October 1980 view. Concluding remarks to the Rome conference on SS 433.
J. Shaham.
Vistas Astron., Vol. 25, (see 012.040), 217 - 233 (1981).

The author reviews the body of observational data for SS 433 accumulated until late 1980 with particular emphasis on the "moving line phenomenon". The various theoretical models proposed for the emission mechanism and the nature of the central source are critically evaluated. An extensive bibliography is given.

117.089   SS 433, X-ray binaries and stellar evolution.
E. P. J. van den Heuvel.
Vistas Astron., Vol. 25, (see 012.040), 95 - 108 (1981).

It is argued from the stationary optical and I. R. spectrum, absolute luminosity and the optical light- and radial-velocity curves, that the SS 433 system consists of a luminous early-type star which is overflowing its Roche lobe ($R > 20 R_\odot$), together with a large luminous accretion disk. Both components have roughly similar luminosities. The star transfers mass to the disk at a rate of some $10^{-4} M_\odot$/yr. The entire system is embedded in a dense spherically symmetric outflowing atmosphere, resembling that of a Wolf-Rayet star; the outflow velocities are $\sim 1 - 2 \times 10^3$ km/sec and the wind mass loss rate is $\sim 10^{-5} - 10^{-4} M_\odot$/yr. The presence of the beams suggests that the object at the center of the disk is a neutron star or a black hole. It appears likely that the system is a later evolutionary state of a massive X-ray binary.

117.090   Does SS 433 have cousins?
L. Maraschi, A. Treves.
Vistas Astron., Vol. 25, (see 012.040), 137 - 139 (1981).

The authors speculate whether SS 433 belongs with Cir X-1, Cyg X-3, GT 0236+610 to a specific class of objects representing an evolutionary state immediately following supernova explosion and neutron star formation in a close binary system.

117.091   SS 433 may not be as peculiar as it looks.
W. Kundt.
Vistas Astron., Vol. 25, (see 012.040), 153 - 164 (1981).

It is argued that SS 433 is a massive binary consisting of a B-star and a young neutron star plus accretion disk, at a distance of (3 ± 0.5) kpc. W50 is the $4 \times 10^{4\pm0.3}$ yr old supernova remnant that was ejected at the neutron star's birth. The radio ansae of W50 have been blown out by a pair of extremely relativistic jets driven by the fast-spinning neutron star, and confined by the wind of the B-star. The system is rare because the accretion disk was already assembled some 10^4 yr after the neutron star's birth, probably because of almost equal component masses of the progenitor system.

117.092   Interpretation of observations supporting a magnetospheric model of SS 433.
G. W. Collins II, G. H. Newsom.
Vistas Astron., Vol. 25, (see 012.040), 169 - 172 (1981).

The authors present a Fourier analysis of the average "velocity" of the moving line spectra of SS 433. An interpretation of the Fourier co-efficients in terms of a very general geometrical model and the light travel time effects required by the finite size of the system is presented. It is found that these effects are present and are statistically significant at the 1.6σ level. Implications of this effect are discussed.

117.093   A theoretical review of SS 433.
J. A. Petterson.
High-energy astrophysics, (see 012.067), p. 49 - 61 (1981).

The "twin jet model" for SS 433 is briefly described, and its likely superiority to other models demonstrated. It is then used as a framework within which the implications of the main observational results are discussed. Many theoretical questions about SS 433 are raised, but only few are answered.

117.094   Equilibrium figures for Beta Lyrae type disks.
R. E. Wilson.
Astrophys. J., Vol. 251, 246 - 258 (1981).

The author computes equilibrium figures for self-gravitating disks around stars in binary systems as a start toward eventual computation of complete disk models. A key role is played by centrifugally limited rotation of the central star, which would naturally arise late in the rapid phase of mass transfer. Beta Lyrae is thus postulated to be a double-contact binary, which makes possible nonarbitrary separation of star and disk into separate structures. The computed equilibrium figures are three-dimensional, as the gravitation of the second star is included. The author computes the total potential and locates equipotential surfaces. The centrifugal potential is written in a particularly convenient form. Experiments leading to a case much like β Lyrae are described. Departures from axial symmetry resulting from the presence of the second star are illustrated. The conditions which must be satisfied by a satisfactory potential field are given. Values for the main physical variables within the disk are roughly estimated. Some of the main problems for the next stage of model computations, such as energy sources, are briefly discussed.

117.095   Binary nature of the barium stars.
Z. Mikulášek.
Říše hvězd, Vol. 62, 245 - 247 (1981). In Czech.

117.096   Evolution of close binaries.   O. Obůrka.
Říše hvězd, Vol. 62, 186 - 189 (1981). In Czech.

117.097   Image tube spectroscopic studies of rapid variables. IV. Spectroscopic and photometric observations of AE Aquarii.   G. Chincarini, M. F. Walker.
Astron. Astrophys., Vol. 104, 24 - 32 (1981).

Photoelectric and time-resolved spectroscopic observations of the short-period eruptive binary AE Aqr are presented. The following spectroscopic elements of the system are derived: $p = 0^d4116537$. $K_1 = 135$ km s^{-1} and $K_2 = 159$ km s^{-1} for the emission and absorption-line components, respectively, from which $q = 1.18$ and $a \sin i = 1.66 \times 10^{11}$ cm. The spectral type of the absorption-line component is found to be K5V, from which $M_1 = 0.82 M_\odot$, $M_2 = 0.69 M_\odot$ and $i = 64°$. The type of mass transfer process proposed by Plavec and Sehnal (1964) is possible.

117.098   The Algol-type systems.
A. H. Batten.
Investigating the universe, (see 003.014), p. 207 - 225 (1981).

The development of our ideas about Algol-type systems over the past quarter of a century is discussed with special reference to the hypothesis of mass-transfer between the components, and to the role of circumstellar matter in the evolution of these systems.

117.099   **An analytical approach to contact binary stability.**
          J. Hazlehurst.
Investigating the universe, (see 003.014), p. 227 - 240 (1981).

It is argued that the Laplace transformation affords a convenient means of investigating the secular stability of contact binaries.

117.100   **The Roche dimensions of rotating gaseous disks in close binary systems.**   M. Kitamura.
Investigating the universe, (see 003.014), p. 241 - 270 (1981).

The problem of rotating emission disks (or rings) in close binary systems is discussed from a simple hydrodynamic point of view for the inviscid gas flow with the aid of the Roche coordinates. Assuming, as a sufficient approximation, that the rotating gas flows around one component in a close binary system move along zero-velocity curves in the orbital plane, a certain simple relation can be deduced between the flow velocity and the outward density gradient. The author can easily evaluate, with the relation, the dimension of the rotating disk from the velocity. Application is made to emission disks in Algol-type close binaries and also to accretion disks in cataclysmic variables.

117.101   **Close binary systems of short period.**
          E. Budding.
Investigating the universe, (see 003.014), p. 271 - 319 (1981).

Information on 406 close binary systems with periods less than 1.5 days for which the primary is of known spectral type, is assembled from recent sources. The organization of the presentation depends on a light curve descriptor which takes the values EA1, EA2, EA3, EB, EW and E 11. Within these groups the distribution in the spectral type:period plane is examined. An additional distribution of this kind is presented for 311 EA2 candidate stars (evolved Algols), with periods of up to 15 days. Recent empirical information on main-sequence stars is presented for comparison. Concerning the EW (W UMa) group, detailed investigation of a particular case reveals the extent of parameter indeterminacy. A recommendation is made for a simple but more quantitative means of defining light curve type.

117.102   **Discovery of a large-amplitude photometric wave in the RS CVn binary BD +61°1211 (=DM UMa) and constraints on starspot models for the observed variation.**
R. A. Kimble, S. M. Kahn, S. Bowyer.
Astrophys. J., Vol. 251, 585 - 590 (1981).

Photometric observations were made in 1979 May and June of BD +61°1211 (= DM UMa), which has been identified as the optical counterpart of the X-ray source 2A 1052 + 606. These observations show the existence of a large-amplitude photometric wave (0.32 mag in V), confirming the classification as an RS CVn system. Model fitting to the light curve using distributions of cool starspots on the surface of the active star confirms the applicability of the canonical starspot model of RS CVn activity to this system. Constraints are derived on the surface area covered by spots, the asymmetry of the spot distribution, and the temperature difference between spots and photosphere. An unusually large spot covering fraction is inferred.

117.103   **The SS 433 binary system.**
          D. Crampton, J. B. Hutchings.
Astrophys. J., Vol. 251, 604 - 610 (1981).

Continued observations of SS 433 reveal that the Hβ emission profile and the intensity of the He I emission and the He I and Fe II absorption lines vary during the 164 day period in a manner consistent with origin in a precessing accretion disk. The radial velocities of Hβ, He I, and Fe II exhibit a 13 day velocity variation during epochs when radio observa-

tions indicate that the source is quiescent, but the variations either disappear or are masked by irregular variations when the object is active. The photometric variability may be similarly correlated with the radio variability. A velocity variation with $K \sim 195$ km s^{-1} has been detected in the He II 4686 emission feature which, if interpreted as orbital motion, indicates that SS 433 may be an eclipsing binary and be related to the massive X-ray binaries.

117.104   **On the elusive cause of cataclysmic variable outbursts.**   F. Meyer, E. Meyer-Hofmeister.
Astron. Astrophys., Vol. 104, L10 - L12 (1981).

The authors show that a transition from convective to radiative structure in the accretion disk around the white dwarf provides a natural explanation for the cataclysmic variable outbursts.

117.105   **On the period of the interacting binary UW Canis Majoris.**   T. Herczeg, H. Drechsel, J. Rahe.
Astron. Astrophys., Vol. 104, 256 - 259 (1981).

The period of the massive interacting binary UW CMa is rediscussed; the value proposed here is 4.^{d}39336. A marked discrepancy between the photometrically and spectroscopically determined epochs is pointed out and commented upon.

117.106   **Accretion disks in close binary systems.**
          G. Hensler.
Proceedings of the 5th Göttingen-Jerusalem-Symposium on Astrophysics, (see 012.041), p. 189 - 200 (1981).

A numerical pseudo-particle method is developed which should allow three-dimensional MHD investigations of accretion disks in close binary systems in the presence of magnetic fields. First test calculations are presented for thin (2-D) accretion disks. Density and temperature distributions for the stationary disk nearly filling the whole Roche-area are presented. The primary accretes 95% of the infalling matter.

117.107   **Short-time-scale Doppler and intensity variations in the spectrum of SS 433.**
R. M. Wagner, G. H. Newsom, C. B. Foltz, P. L. Byard.
Astron. J., Vol. 86, 1671 - 1678 (1981).

Spectroscopic observations of SS 433 were obtained on ten consecutive nights in 1981 May. Measurements of the equivalent widths of the spectral lines and Doppler shifts of the moving lines are presented. Not only does the 6-day period in the wavelengths of the moving lines appear quite clearly, but the data strongly suggest that this period is decreasing at a rate of $(3.0 \pm 1.1) \times 10^{-5}$ days/day. In addition, large equivalent width variations of the stationary lines are observed.

117.108   **Evolutionary scenario for close binary systems of low and moderate masses.**
A. V. Tutukov, L. R. Yungel'son.
Nauchn. Inf., Vyp. (No.) 49, (see 003.017), p. 3 - 20 (1981).
In Russian.

The evolution of low and moderate mass close binary systems ($M_1 \lesssim (8 - 10) M_\odot$) is discussed from initial main-sequence stars up to formation of compact objects. The place of Algol-type stars, novae, dwarf novae, symbiotic stars, binary nuclei of planetary nebulae, low-mass X-ray sources in the evolutionary sequence is shown.

117.109   **Photoelectric observations of the eclipsing variable EY Orionis.**   M. M. Zakirov.
Perem. Zvezdy, Tom 21, 227 - 235 (1979). In Russian.

An observational study of the close binary system EY Ori has been carried out. Photoelectric UBVR observations of EY Ori were obtained in 1976 - 1978. The individual observations in each color are listed in a table. The new period is equal to 16.^{d}787832.

117.110 **AS Camelopardalis: an outstanding detached eclipsing binary.** R. A. Botsula.
Perem. Zvezdy, Tom 21, 237 - 245 (1979). In Russian.

Light curve variations of the detached eclipsing system AS Cam are explained by assuming the existence of obscuring clouds above underlying bright spots on the surface of the more massive star. The spots' temperature is about 14400°, that is 2000° more than the star's temperature. Possibly magnetic fields appear on the star's surface.

117.111 **Variability of soft X-ray emission of EX Hydrae observed with Einstein Observatory.**
A. Kruszewski, R. Mewe, J. Heise, T. Chlebowski, W. van Dijk, R. Bakker.
Space Sci. Rev., Vol. 30, (see 012.044), 221 - 229 (1981).

The cataclysmic variable EX Hydrae has been observed in X-rays with Einstein Observatory. A 67 min periodic X-ray modulation has been found in the energy range 0.1 - 2 keV, whereas the source is approximately constant above 2 keV. The modulation is approx. in phase with the stable 47 min modulation in the optical brightness.

117.112 **Optical observations of the X-ray source 2S0921−630.** G. Branduardi-Raymont, R. Corbet, A. N. Parmar, P. G. Murdin, K. O. Mason.
Space Sci. Rev., Vol. 30, (see 012.044), 279 - 286 (1981).

The authors report photometric observations of the optical counterpart of the X-ray source 2S0921−630. The data, obtained at the South African Astronomical Observatory during 3 weeks in 1980 and 1981, are consistent with a 17.9 day periodic modulation of the flux in the B band. Correlated variability of the (B−V) and (U−B) colour indices with the B mag. is demonstrated and quantified. The observed B mag. and colours at maximum and minimum light are used to compute some of the system parameters. A model of 2S0921−630 is proposed in terms of a binary system in which the variable inclination of a luminous accretion disc produces the long-term modulation of the optical flux.

117.113 **Ultraviolet study of V1341 Cyg = Cyg X−2. Observation of an accretion disk.**
L. Chiappetti, L. Maraschi, E. G. Tanzi, A. Treves.
Space Sci. Rev., Vol. 30, (see 012.044), 287 - 292 (1981).

The authors discuss observations with I.U.E. in the range 1200 - 3000 Å, with particular reference to the continuum.

117.114 **The enigma of SS433.** G. W. Collins,II.
Astron. Q., Vol. 4, No. 13, p. 5 - 23 (1981).

117.115 **On the photometric behavior of SS433 in 1979 and 1980.** T. Mazeh, E. M. Leibowitz, O. Lahav.
Astrophys. Lett., Vol. 22, 185 - 191 (1981).

Results and analysis of photometric measurements of SS433 are presented. The light of the star in the V and the B photometric bands varies with a period of either 6.55 or 6.43 days. The periodicity of ~164 days is also apparent in the photometric data, with its first and third harmonics. Fluctuations of the order of half a magnitude within a few hours have also been recorded. The B-V color shows no dependence on the phase of the periodic variations, while it does change by up to 0.25 mag in an apparent association with the short time scale fluctuations.

117.116 **AM Herculis.**
IAU Circ., Nos. 3615, 3638, 3643 (1981).

117.117 **PG 1550+191.**
IAU Circ., No. 3616 (1981).

117.118 **HZ Herculis.**
IAU Circ., No. 3629 (1981).

117.119 **AM Herculis.**
Yamamoto Circ., No. 1960 (1981).

117.120 **AR Aurigae − triple system?**
J. Zverko, D. Chochol, K. Juza, J. Žižňovský.
Inf. Bull. Variable Stars, No. 1997, 3 pp. (1981).

117.121 **A dramatic change in the light curve of V 711 Tau (HR 1099).**
C. Blanco, S. Catalano, E. Marilli, M. Rodono, F. Scaltriti.
Inf. Bull. Variable Stars, No. 2000, 4 pp. (1981).

117.122 **Note on the period of the W UMa star AH Vir.**
U. Hopp, S. Witzigmann.
Inf. Bull. Variable Stars, No. 2005, 2 pp. (1981).

117.123 **UBVR photometry of UV Psc.**
M. Zeilik, R. Elston, G. Henson, P. Smith.
Inf. Bull. Variable Stars, No. 2006, 4 pp. (1981).

117.124 **5 Ceti = HR 14: a new and puzzling variable star.**
R. D. Lines, D. S. Hall.
Inf. Bull. Variable Stars, No. 2013, 2 pp. (1981).

117.125 **The period of the symbiotic star AG Pegasi.**
L. Meinunger.
Inf. Bull. Variable Stars, No. 2016, 2 pp. (1981).

117.126 **R photometry of RW Com.**
T. J. Davidge, J. D. Himer, E. F. Milone.
Inf. Bull. Variable Stars, No. 2037, 3 pp. (1981) = Rothney Astrophys. Obs. Publ., Ser. B, No. 6.

117.127 **Spectroscopic observations of symbiotic stars in July and September 1981.** R. Gravina.
Inf. Bull. Variable Stars, No. 2041, 4 pp. (1981).

117.128 **Photoelectric light curves of CK Bootis (HD 128141).**
Z. Aslan, M. Gören, E. Derman.
Inf. Bull. Variable Stars, No. 2043, 3 pp. (1981).

117.129 **The actual photometric behaviour of CH Cygni.**
U. Hopp, S. Witzigmann.
Inf. Bull. Variable Stars, No. 2048, 4 pp. (1981).

117.130 **Photoelectric observations of CI Cyg.**
G. H. Gevorkian (*Gevorkyan*).
Inf. Bull. Variable Stars, No. 2056, 3 pp. (1981).

117.131 **Recent photometry of several RS CVn stars.**
G. W. Henry.
J. American Assoc. Variable Star Obs., Vol. 10, 35 - 36 (1981). Abstract.

117.132 **An interactive binary model for eruptive symbiotic systems.** L. A. Willson.
J. American Assoc. Variable Star Obs., Vol. 10, 36 (1981). Abstract.

117.133 **On the origin and decay of contact binaries.**
F. van't Veer.
Proceedings of the Third Finnish-Soviet Astronomical Symposium, (see 012.057), p. 89 - 95 (1981).

117.134 **Angular momenta and evolution of W UMa systems.**
T. Rahunen.
Proceedings of the Third Finnish-Soviet Astronomical Symposium, (see 012.057), p. 97 - 111 (1981).

The author has studied the angular momenta of 13 A-type and 16 W-type W UMa systems using their known periods and mass ratios. It is shown that the W-type systems can be understood as zero-age systems, whereas most of the A-type systems are probably evolved. Some of the A-type systems can be explained by zero-age thermal non-equilibrium models which are losing angular momentum on a timescale of $\sim 5 \times 10^8$ yr. The author concludes that the problem of angular momentum loss may be crucial to the evolution of W UMa systems.

117.135   CH Cygni – M-giant and a white dwarf binary with transient accretion disk.   L. Luud.
Proceedings of the Third Finnish-Soviet Astronomical Symposium, (see 012.057), p. 117 - 124 (1981).

117.136   Problems of low-mass binary evolution.
O. Vilhu.
Proceedings of the Third Finnish-Soviet Astronomical Symposium, (see 012.057), p. 125 (1981). – Abstract.

117.137   The 1979 minimum state of AN Ursae Majoris.
J. Liebert, S. Tapia, H. E. Bond, A. D. Grauer.
Prepr. Steward Obs., No. 328, 27 pp. (1981)
The authors report spectrophotometry, photometry and circular polarimetry of the magnetic binary AN UMa during 1979 February/March.

117.138   PG 1550+191: a new AM-Herculis type binary system.   J. Liebert, H. S. Stockman,
R. E. Williams, S. Tapia, R. F. Green, D. Rautenkranz, D. H. Ferguson, P. Szkody.
Prepr. Steward Obs., No. 350, 38 pp. (1981).
The authors report the discovery of a new AM-Her type magnetic variable. The orbital period, as defined by linear and circular polarimetry, is 113.56 min. The maximum of linear polarization was generally weak ($\sim$2–5%), while the circular polarization varies between 0% and -12%. This is the first object to show substantial circular polarization at the time of the linear pulse, yet the circular polarization does reach zero at phase 0.5. Simultaneous UBVR photometry shows mild flickering and smooth variations which are roughly synchronous with the polarization. Infrared (JHK) photometry indicates an energy distribution similar to AM Her.

117.139   Infrared variability of V1329 Cygni.
B. F. Yudin.
Astron. Tsirk., No. 1160, p. 8 (1981). In Russian.

117.140   Evidence on high rate mass loss in the eclipsing variable VW Cep from ultraviolet spectra.
I. Pustyl'nik, A. Sapar.
Academy of Sciences of the Estonian SSR, Prepr. A-3, p. 1 - 20 (1981).
The results of an analysis of two ultraviolet spectra obtained for the contact binary VW Cep with the aid of the IUE satellite are presented. Mass loss from the system with a rate of up to $10^{-6} M_\odot$ per year has been estimated from the Mg II 2795, 2802 Å emission feature.

Symbiotic stars.   See Abstr. 011.003.

High resolution image reconstruction of binary stars and other objects.   See Abstr. 031.548.

On the construction of Roche harmonics.
See Abstr. 042.067.

The sizes and edge temperatures of optically thick accretion discs.   See Abstr. 064.002.

Explosive mass loss in binary stars: the two time-scale method.   See Abstr. 064.009.

Large macroturbulence in the chromosphere of an RS CVn star.   See Abstr. 064.021.

Response of low-mass main sequence stars to accretion.   See Abstr. 064.041.

Innermost parts of accretion disks are thermally and secularly stable.   See Abstr. 064.069.

A supercritical accretion disk model for SS 433.
See Abstr. 064.070.

Numerical computation of certain three-dimensional stellar structures using a semidiscrete pseudospectral method.   See Abstr. 065.013.

Stellar evolution with SMC chemical abundances.
See Abstr. 065.027.

Review of the theories on SS 433.
See Abstr. 065.078.

On the nature of WR-type stars with ring nebulae.
See Abstr. 065.092.

Optical continuum from SS 433 – star or accretion disk?   See Abstr. 113.048.

IUE observations of two late type stars BX Mon (M4 + pec) and TV Gem (M1 Iab).   See Abstr. 114.042.

The rotational profile of II Pegasi (HD 224085).
See Abstr. 116.032.

Intrinsic polarization of SS 433: an electron scattering model.   See Abstr. 116.037.

EX Hydrae: modulated mass transfer or intermediate polar?   See Abstr. 119.002.

Photometric orbit of the massive system RY Scuti.
See Abstr. 119.003.

BB Peg: A W UMa-W system with a high degree of overcontact.   See Abstr. 119.009.

The 1979–1980 eclipse of Zeta Aurigae. I. The circumstellar envelope.   See Abstr. 119.011.

Revised photometric elements of BF Vir.
See Abstr. 119.014.

The impact of IUE on binary star studies.
See Abstr. 119.017.

O'Connell effect correlations with eclipsing binary star parameters.   See Abstr. 119.018.

Mass transfer and stellar wind effects in the eclipsing binary RT Andromedae.   See Abstr. 119.027.

Variations in the light curve of V 505 Sagittarii and their interpretation.   See Abstr. 119.034.

WY Hya: a main sequence detached binary system with nearly equal members.   See Abstr. 119.046.

The eclipsing variable HD 134518.
See Abstr. 119.047.

Rapid oscillations in cataclysmic variables. VI. Periodicities in erupting dwarf novae.   See Abstr. 122.007.

Observations of cataclysmic variables with IUE.
See Abstr. 122.038.

Ultraviolet observations of Mira Ceti.
See Abstr. 122.053.

Stepanian's star: the energy distribution reveals a nontypical cataclysmic variable.   See Abstr. 122.063.

V 1329 Cyg: a nova-like object with recurrent outbursts?   See Abstr. 122.076.

Rapid and extreme spectral variability of the peculiar emission-line object BE Ursae Majoris.
See Abstr. 122.113.

BE Ursae Majoris (PG 1155 + 492): a unique cataclysmic-variable-like object.   See Abstr. 122.124.

MV Lyrae: spectrophotometric properties of minimum light; or on MV Lyrae off.   See Abstr. 122.129.

UV observations of MV Lyrae.
See Abstr. 122.146.

Novae and related stars at quiescence.
See Abstr. 124.003.

Generation of an external ring during the 1978 outburst of WZ Sagittae.   See Abstr. 124.006.

Coordinated IUE, Einstein and optical observations of accreting degenerate dwarfs.   See Abstr. 126.009.

White dwarfs and supernovae.
See Abstr. 126.027.

The collapse of a rotating non-axisymmetric isothermal cloud.   See Abstr. 131.141.

Fragmentation in a rotating protostar: a re-examination of comparison calculations.   See Abstr. 131.142.

A photometric study of 2A 0526-328.
See Abstr. 142.006.

Ultraviolet observations of LMC X-4 and SMC X-1.
See Abstr. 142.016.

IUE observations of X-ray binaries.
See Abstr. 142.040.

Spin-reversed accretion as the cause of intermittent spindown in slow X-ray pulsars.   See Abstr. 142.051.

Identification of 4U 1849−31 with V1223 Sagittarii and discovery of optical pulsations.
See Abstr. 142.055.

X-ray observations of symbiotic stars.
See Abstr. 142.060.

The X-ray spectrum of AM Herculis from 0.1 to 150 keV.   See Abstr. 142.066.

HD 36705: a new bright X-ray emitting RS CVn star.
See Abstr. 142.075.

Hard X-ray balloon observations of compact galactic and extragalactic X-ray sources.
See Abstr. 142.103.

High energy X-ray spectrum of Her X-1.
See Abstr. 142.104.

Hard X-rays observation of Her X-1.
See Abstr. 142.105.

Optical properties of low-mass X-ray binaries.
See Abstr. 142.110.

New optical observations of 2S 0921−630 and its relation to other X-ray binaries.   See Abstr. 142.111.

X-ray binaries and stellar evolution.
See Abstr. 142.132.

A model for the cosmic γ-ray burst event on March 5, 1979.   See Abstr. 142.510.

Supermassive binaries in active galactic nuclei.
See Abstr. 158.143.

118 Visual Binaries, Multiple Stars, Astrometric Binaries

118.001   Discovery of X-ray emission lines from the RS CVn binary σ CrB.
P. C. Agrawal, G. R. Riegler, N. E. White.
Mon. Not. R. Astron. Soc., Vol. 196, 73P - 79P (1981).

X-ray observations of the RS CVn binary σ CrB carried out with the solid state spectrometer on board the Einstein Observatory are reported. X-ray emission lines due to Mg XI, Si XIII, S XV and a blend of L-transition lines due to Fe XVII and Fe XVIII are clearly detected in the spectrum of σ CrB. The observed X-ray spectrum is best described by a two-temperature coronal X-ray emission model. Intensity variations over a period of a few hours were also recorded during the observations. Implications of these results are discussed in terms of the constant-pressure coronal loop model.

118.002   Visual multiples. VII. MK classifications.
H. A. Abt.
Astrophys. J., Suppl. Ser., Vol. 45, 437 - 456 (1981).

Classifications are given for 865 components of visual multiples; they show no systematic differences from the MK system, and the random errors are one subclass in type and two-thirds of a luminosity class. It is found that at least 1% of the F-type IV and V stars are weak-lined, 32% of the A4-F1 IV and V stars are Am, and 5% of the A0-A3 IV and V stars are early-type Am. Attention is called to the large fraction (55%) of the A3-A9 III-V stars that are of luminosity classes III or IV, unlike the percentage (16%) at neighboring types.

118.003   Orbits of 15 visual binaries.   W. D. Heintz.
Astrophys. J., Suppl. Ser., Vol. 45, 559 - 564 (1981).
Elements and masses for the pairs ADS 896, 2336, 6315, 7054, 7629, 8092, 8555, 8739, 13987, 16185, Rst 1658, 3906, 3972, 4529, and Jsp 691 have been determined with recent reobservations.

118.004   The nearby quadruple system G 107 − 69/70.
R. S. Harrington, J. W. Christy, K. Aa. Strand.
Astron. J., Vol. 86, 909 - 911 (1981).
The binary system G 107 − 69/70 is nearby, having a relative parallax of 0.0896 ± 0.0014. The primary (G 107 − 69) is a newly discovered unresolved astrometric binary, with a best-period estimate of approximately 0.94 yr, suggesting that the faint component might be a low-mass red dwarf. The secondary (G 107 − 70) is a partially resolved binary, having a period of 20.5 yr, and consisting of two low-mass white dwarfs.

118.005   HD 202908: a young, solar-type triple system.
F. C. Fekel, Jr.
Astrophys. J., Vol. 248, 670 - 677 (1981).
HD 202908 is a solar-type, spectroscopic-visual triple system. High-dispersion spectroscopic observations show that the primary visual component is a double-lined spectroscopic binary with a period of 3.966 days and a mass ratio of 1.0. Velocity changes due to motion in the long-period, 76 year orbit are now perceptible. Contrary to expectation, the short-period and long-period orbits are not coplanar. The large lithium abundances, strength of the Ca II emission lines, and rotational velocities all suggest that this multiple system has an age similar to that of the Hyades cluster.

118.006   The "orbit" of β^1 Cygni.   E. M. Hendry.
Publ. Astron. Soc. Pacific, Vol. 93, 323 - 327 (1981).
The resolution of β^1·Cyg by the technique of speckle interferometry (and later by visual means) has led to speculations of a recent increase in separation between the two stars of the composite spectrum. Radial velocities from recently obtained spectra as well as those already in the literature have provided relatively complete coverage from 1898 to 1980 and show no essential change in the radial velocity of the K star or its B-star companion during this time, leading to the conclusion that either the period of β^1 Cyg is considerably longer than suspected or that the two stars are physically unrelated. Inconsistencies between the speckle observations are commented upon.

118.007   The binary system 20 Persei.   W. D. Heintz.
Publ. Astron. Soc. Pacific, Vol. 93, 328 - 329 (1981).
The orbital motion, with P = 31.37 yr, and other properties of the subgiant system are rediscussed.

118.008   Astrometric orbits, mass ratios, and masses of ζ Herculis and 85 Pegasi from plates taken with the Sproul 61-cm refractor.   S. L. Lippincott.
Publ. Astron. Soc. Pacific, Vol. 93, 376 - 380 (1981).
Sproul photographic data covering two periods of revolution for each of these well-known visual binaries provide independent orbital elements.

118.009   Astrometric analysis of the unresolved binary Mu Cassiopeiae from photographs taken with the Sproul 61 centimeter refractor.   S. L. Lippincott.
Astrophys. J., Vol. 248, 1053 - 1058 (1981).
Mu Cassiopeiae, a high-velocity Population II subdwarf, is an astrometric binary which has been on the Sproul Observatory astrometric program since 1937. The data yield P = 21.43 yr, with a photocentric semiaxis major, α = 0.186±0.001 (p.e.) and a relative parallax, π = +0.130±0.001. The best tentative values of Δm and separation so far found suggest $M_A \approx 0.7\,M_\odot$, and $M_B \approx 0.2\,M_\odot$

with $\Delta m \sim 4.5$, which indicate higher He content for μ Cas A than for the sun.

118.010   Computation of orbits of visual binaries ADS: 1223, 2459, 1321.   A. Valbousquet.
Astron. Astrophys., Suppl. Ser., Vol. 45, 181 - 182 (1981).
The orbit of the visual binaries ADS 1223 & 2459 are recomputed sice their previous orbits do not agree with the measurements made afterwards. A first orbit for ADS 1321 is given.

118.011   Preliminary orbital elements for ADS 14893.
F. R. West, H. A. McAlister.
Bull. American Astron. Soc., Vol. 13, 569 - 570 (1981).
Abstract.

118.012   The ultraviolet spectra of the two components of the visual binary system Δ199.
M. Hack, P. S. Thê, H. R. E. Tjin A Djie.
Second European IUE Conference, (see 012.011), p. 219 - 222 (1980).
High resolution spectra of the two components of the visual binary system Δ199 have been obtained with IUE. The main interest of this system is that HR 5999 is a nebular variable whose variability is mainly due to obscuration by circumstellar dust clouds and whose age is estimated at about 7×10^5 y; while the companion, HR 6000, is a Bp star with strong phosphorus overabundance. This association establishes a lower limit for the age of chemically peculiar stars.

118.013   Eléments orbitaux des étoiles doubles visuelles ADS 1737, ADS 2446, ADS 2612 et ADS 2799.
M. Scardia.
Astron. Astrophys., Suppl. Ser., Vol. 45, 431 - 437 (1981).
The orbital elements of the visual binary stars ADS 1737, ADS 2446, ADS 2612 and ADS 2799 are given. The provisional elements obtained by the Kowalsky method have been corrected with the differential method of Hellerich. Finally, the dynamical parallaxes and total masses of the systems have been calculated.

118.014   Is this diagram an argument for binary orbital evolution due to mass loss?   J. Dommanget.
Effects of mass loss on stellar evolution, (see 012.015), p. 507 - 513 (1981).
In 1963, a diagram has been established by the author, for visual and spectroscopic binaries, showing the existence of a real correlation between total mass and eccentricity. The consideration of an up-dated material leads today to an identical diagram confirming the reality and the stability of the mentioned correlation. This diagram is discussed with the assumption of a substantial secular mass loss of the components of the considered binaries.

118.015   A spectroscopic and photometric study of the triple system HD 7215.   D. P. Hube.
Publ. Astron. Soc. Pacific, Vol. 93, 490 - 494 (1981).
HD 7215 is a visual binary (ADS 988) in which the visual primary is itself a double-lined spectroscopic binary. The visual components have been observed spectroscopically and photometrically, and a solution for the spectroscopic orbit of the primary found. The magnitudes, color indices, and spectral types are most consistent with a model in which the visual secondary is also a binary with similar components. All three (four?) stars are A-type dwarfs.

118.016   Nauwe dubbelsterren en planetenstelsels.
F. van't Veer.
Zenit, 8. Jaarg., 403 - 407 (1981).

118.017   Visual double stars measurements.   P. J. Morel.
Astron. Astrophys., Suppl. Ser., Vol. 46, 3 - 6 (1981).

The author gives 563 micrometer measurements of 174 visual binaries made at Nice Observatory in 1969-71 and 1980.

118.018   Radial velocities of binary and proper-motion stars.
W. D. Heintz.
Astrophys. J., Suppl. Ser., Vol. 46, 247 - 253 (1981).
Coudé plates have been obtained at Kitt Peak for radial velocities of 127 program stars.

118.019   Statistical method for calculating parallaxes and masses of binaries with unknown orbits.
P. Couteau.
Astron. Astrophys., Vol. 102, 313 - 315 (1981). In French.

A method is given for the computation of the orbital constant of a binary, a''^3/P^2, which can be applied as soon as the couple is in measurable motion. About forty close and rapid binaries are computed with this method. The stars are well distributed over the main sequence.

118.020   Discovery of duplicity of three bright B stars.
G. A. Starikova, A. A. Tokovinin.
Pis'ma Astron. Zh., Tom 7, 692 - 695 (1981). In Russian.
English translation in Soviet Astron. Lett., Vol. 7.

Observations with a photoelectric interferometer revealed duplicity of the bright B stars ζ Dra, o Her and ι Lyr. These new pairs are likely to have short periods.

118.021   Alpha Centauri.   E. Hantzsche.
Sterne, 57. Band, 267 - 282 (1981).

118.022   Spectroscopic observations of ADS 11579.
A. H. Batten, J. M. Fletcher, R. D. McClure, C. D. Scarfe, G. Wallerstein.
J. R. Astron. Soc. Canada, Vol. 75, 245 (1981). – Abstract.

118.023   Near-infrared observations of trapezium-type multiple systems. Catalogue of observations and a new determination of the reddening law.   M. Tapia.
Mon. Not. R. Astron. Soc., Vol. 197, 949 - 965 (1981).

Results are presented from a large scale near-infrared survey of small sky areas of the Milky Way containing trapezium-type multiple star systems located within young galactic clusters, H II regions or dark nebulosities. Photometric maps at wavelength 2.2μm complemented by broadband JHKL photometry of the majority of the sources detected and of the star members of the systems were obtained. Analyses of the infrared two-colour diagrams provide determinations of the values of colour excess ratios.

118.024   Trapezium-type multiple systems with variable stars as their components.   G. N. Salukvadze.
Soobshch. AN GruzSSR, Tom 101, 573 - 576 (1981). In Russian. – Abstr. in Ref. zh., 51. Astron., 11.51.906 (1981).

118.025   Existe-t-il d'autres systèmes planétaires?
G. Mathys.
Ciel, Vol. 43, 209 - 222 (1981).

118.026   Orbites nouvelles.
Circ. Inf., No. 85 (1981).

118.027   Etoiles doubles nouvelles.
G. M. Popovic, P. Couteau, P. Muller.
Circ. Inf., No. 85 (1981).

118.028   HD 12180: a new quadruple star system.
C. R. Chambliss.
Inf. Bull. Variable Stars, No. 2058, 2 pp. (1981).

118.029   Results of speckle interferometry with the 6 m telescope.   V. N. Dudinov, V. N. Erokhin,

V. V. Konichek, S. G. Kuz'menkov, V. S. Rylov, V. S. Tsvetkova.
Astron. Tsirk., No. 1134, p. 4 - 6 (1980). In Russian.

118.030   Orbits of five visual binaries.   D. J. Zulević.
Bull. Obs. Astron. Belgrade, No. 131, p. 15 - 19 (1981).

118.031   Les orbites de deux étoiles doubles visuelles (ADS 2301 = A 2414 et ADS 2531 = A 829).
V. Erceg.
Bull. Obs. Astron. Belgrade, No. 131, p. 20 - 22 (1981).

118.032   New double stars discovered in Belgrade with the Zeiss refractor 65/1055 cm, Supplement VI.
G. M. Popović.
Bull. Obs. Astron. Belgrade, No. 131, p. 23 - 26 (1981).

118.033   Micrometer measures of double stars (Series 32).
D. J. Zulević.
Bull. Obs. Astron. Belgrade, No. 131, p. 27 - 32 (1981).

118.034   On the uniqueness of solution of a certain incorrect astronomical problem.   A. Kh. Amirov.
Priblizhen. metod. resheniya i vopr. korrektnosti obratn. zadach. Novosibirsk, 1981, p. 9 - 16. In Russian. – Abstr. in Ref. zh., 51. Astron., 12.51.781 (1981).

118.035   Micrometer observations of 82 double stars.
L.-s. Yan, M.-z. Xu, Z.-y. Chu, D.-t. Tan.
Ann. Shanghai Obs. Acad. Sinica, No. 1, p. 74 - 82 (1979).

118.036   Eléments orbitaux provisoires des étoiles doubles visuelles ADS 10279 et ADS 12889.
M. Scardia.
Astron. Nachr., Band 302, 291 - 298 (1981).

The orbital elements of the visual binary stars ADS 10279 and ADS 12889 are given and the dynamical parallaxes and total masses of the systems have been calculated.

118.037   Cosmogonic aspects of the evolution of planetary systems.   H. Stiller, H.-J. Treder, D. Möhlmann.
Planetary interiors, (see 012.061), p. 21 - 36 (1981).

The fundamental approaches to the problem of formation of planetary and satellite systems are discussed. Especially the hetegony principle and its proposed generalization to a formation principle including the central body, and a restricted actualistic principle are supposed to guide further approaches. Relevant characteristic parameters to identify the formation processes and possible future planetological tasks have been derived on this basis.

118.038   γ and τ Centauri.   I. N. Latyshev.
Astron. Tsirk., No. 1151, p. 8 (1981). In Russian.

Preliminary orbital parallax catalog.
See Abstr. 002.074.

L'observation des etoiles doubles visuelles.
See Abstr. 003.047.

Observing visual double stars.
See Abstr. 003.048.

On meridian observations according to the international programme (double stars).   See Abstr. 041.028.

Photoelectric observations of lunar occultations. XII.
See Abstr. 096.009.

The problems of simulation of planetary systems accumulation processes.   See Abstr. 107.030.

Observations of late-type stars with 'hot' companions.   See Abstr. 114.074.

Empirical bolometric corrections for the main sequence.   See Abstr. 115.019.

AR Aurigae − triple system?
See Abstr. 117.120.

Erratum

118.901   Erratum: 'Low-mass unseen companions to two
nearby red dwarfs, CC 1228 and Wolf 922'
[Publ. Astron. Soc. Pacific, Vol. 91, 784 - 788 (1979/80)].
S. L. Lippincott.
Publ. Astron. Soc. Pacific, Vol. 93, 535 (1981). − See Abstr.
27.118.014.

119 Eclipsing Binaries

119.001 Revised photometric data for nine eclipsing binaries.
G. Giuricin, F. Mardirossian.
Astron. Astrophys., Suppl. Ser., Vol. 45, 85 - 91 (1981).

Using Wood's (1972) model the authors have reanalyzed the lightcurves of nine eclipsing binaries in order to obtain homogeneous photometric elements. It has been found that V346 Aql, CV Car, TZ CrA, SZ Her, and RS Lep are very probably common semidetached systems with lobe-filling subgiant (or giant) secondaries; K0 Aql appears to be an sd-d system; AW Cam and WX Eri, having close − though not exactly in contact − members, are unlikely to be ordinary main sequence detached systems. It has been pointed out that RR TrA shows an anomalous behaviour.

119.002 EX Hydrae: modulated mass transfer or intermediate polar?   B. Warner, J. T. McGraw.
Mon. Not. R. Astron. Soc., Vol. 196, 59P - 63P (1981).

The morphology of total eclipses in the dwarf nova EX Hya shows that the 67-min cycle in this system is caused by periodic variations in the luminosity of the bright spot. This can be explained either by modulated mass transfer arising from pulsations of the cool component of the binary system, or by beamed emission from the white dwarf if it has a rotation period of 40 min.

119.003 Photometric orbit of the massive system RY Scuti.
L. Milano, A. Vittone, F. Ciatti, A. Mammano, R. Margoni, G. Strazzulla.
Astron. Astrophys., Vol. 100, 59 - 65 (1981).

The UBV lightcurves of RY Scuti were analysed using the Wilson and Devinney approach. The system is found to be an evolved overcontact system, embedded in a gas and dust envelope, revealed by forbidden lines and IR excess. It is suggested that RY Scuti has early-type components of 39 and 49 $M_\odot$, making it one of the most massive contact systems known so far. A possible variation in the period is analysed.

119.004 Four-colour photometry of eclipsing binaries, XIII A. Photometric elements and absolute dimensions of TY Pyxidis.
J. Andersen, J. V. Clausen, B. Nordström, B. Reipurth.
Astron. Astrophys., Vol. 101, 7 - 15 (1981).

The previously published *uvby* light curves (Andersen et al., 1981) of the double-lined eclipsing binary TY Pyx have been analysed. TY Pyx is a unique RS CVn system in the sense that its components have quite identical mass, luminosity and spectral type. The distortions typical of RS CVn systems are present in the light curves, but are very small with a total range of only $0^m.03$. Photometric elements have been determined by means of the binary model and the light curve simulation programme WINK described by Wood (1971, 1972). Combined with the spectroscopic data by Andersen and Popper (1975), accurate absolute dimensions have been obtained.

119.005 A lightcurve analysis for the massive binary RY Sct.
G. Giuricin, F. Mardirossian.
Astron. Astrophys., Vol. 101, 138 - 141 (1981).

Using Wood's (1972) model the authors have analyzed Ciatti et al.'s (1980) *UBV* lightcurves − as yet unexplored − of the massive eclipsing binary RY Sct. They have found that the eclipses are partial (the inclination angle is about 75°) and that the primary minimum is an occultation. Further the authors discuss the evolutionary stage of RY Sct, which is likely to evolve into a Wolf-Rayet binary system.

119.006 SAO 072799 − ein neuer heller Bedeckungsveränderlicher?   M. Fernandes.
BAV Rundbrief, 30. Jahrg., 49 - 51 (1981).

119.007 Distribution of masses of components and semimajor axes of orbits of eclipsing binaries.
Z. T. Krajcheva, E. I. Popova, A. V. Tutukov, L. R. Yungel'son.
Pis'ma Astron. Zh., Tom 7, 488 - 492 (1981). In Russian.
English translation in Soviet Astron. Lett., Vol. 7.

The distribution of primary masses and semi-major axes of orbits for 482 eclipsing binaries with known spectra and luminosity classes is investigated.

119.008 The history of U Coronae Borealis − a sequel.
G. A. Bakos, J. Tremko.
J. R. Astron. Soc. Canada, Vol. 75, 124 - 131 (1981).

Recent determinations of epochs of primary minima of U CrB point to a periodic nature of the $O-C$ diagram. In the authors' interpretation it indicates a light-time orbit of the eclipsing pair about a distant companion.

119.009 BB Peg: a W UMa-W system with a high degree of overcontact.
M. Cerruti-Sola, L. Milano, F. Scaltriti.
Astron. Astrophys., Vol. 101, 273 - 275 (1981).

V and B photoelectric lightcurves of the eclipsing binary BB Peg are analysed by the Wilson-Devinney computer code. Though the system might be classified as a W UMa system of W type a degree of overcontact greater than usual is obtained from lightcurves analysis. The adopted solution is poor because of either a variable degree of asymmetry from the yellow to blue lightcurve or a phase shift of the secondary minimum. This morphology of the lightcurves is not well explainable by the current theory on W UMa's.

119.010 The nature of the eclipsing binary V822 Aquilae.
D. M. Popper.
Publ. Astron. Soc. Pacific, Vol. 93, 318 - 319 (1981).

The five-day eclipsing binary, V822 Aql, heretofore classified as B8, contains components of estimated types B3 and B9 and may be a semidetached system.

119.011 The 1979−1980 eclipse of Zeta Aurigae. I. The circumstellar envelope.   R. D. Chapman.
Astrophys. J., Vol. 248, 1043 - 1052, plate 17 (1981).

The resonance lines of Mg^+ and C^{+++} have been studied in the spectrum of ζ Aurigae during 1979 and 1980. A model of the K-star wind far from the K star and its interaction with the B star in the system has been derived. The data suggest a mass loss rate from the K star of $2 \times 10^{-8} M_\odot$ yr^{-1}. The rate of accretion by the B star of material from the K supergiant is such that the matter accreted in the course of about 10 years is of the order of the total mass of the photosphere of the B star.

119.012 Some aspects of mass loss and mass transfer in Algol variables.   G. Giuricin, F. Mardirossian.
Astrophys. J., Suppl. Ser., Vol. 46, 1 - 26 (1981).

The authors have examined the observational data of 102 Algol variables in order to check the validity of several simple models of nonconservative mass transfer used in the literature and to clarify their implications on the evolutionary scenario of Algol variables.

119.013 Photoelectric elements of AR Lacertae.
R. K. Srivastava.
Astrophys. Space Sci., Vol. 78, 123 - 140 (1981).

Geometrical elements of the system AR Lacertae have been obtained and its colour has been discussed. The absolute dimensions have been obtained on the basis of the spectroscopic elements given by Sanford (1951). The primary component lies fairly close to the Main Sequence on the log m - log R

plot and the secondary falls away from the Main Sequence. The values of the Roche constants indicate that AR Lacertae is a detached system.

119.014 Revised photometric elements of BF Vir.
G. Russo, C. Sollazzo.
Astrophys. Space Sci., Vol. 78, 141 - 144 (1981).

The B light curve of BF Vir published by Mallama and Witt (1976) has been re-analyzed using the Wilson and Devinney (1971) approach and the absolute dimensions have been computed. The system turns out to be a semi-detached system, with an A2V primary and an evolved G2 undermassive secondary which fills its Roche lobe; the primary, even if detached, is not far from contact.

119.015 IUE ultraviolet observations of W UMa stars.
A. K. Dupree, S. Preston.
The Universe at ultraviolet wavelengths, (see 012.009), p. 333 - 340 (1981).

Four W UMa eclipsing binary systems have been observed with IUE: 44 Boo, VW Cep, W UMa, and ε CrA. They generally show large surface fluxes of high temperature lines (C II, C IV, N V, Si IV) which may result from the high rotational velocities forced by synchronous rotation. High dispersion spectra of the 44 Boo system in the Mg II line enable the individual stellar components to be identified. The line widths and phase variations are consistent with the optically determined spectroscopic orbit. Circumstellar absorption of Mg II may be present at selected phases.

119.016 The 1979 eclipse of Zeta Aurigae.
R. D. Chapman.
The Universe at ultraviolet wavelengths, (see 012.009), p. 377 - 380 (1981).

Observations of the system ζ Aurigae made around primary eclipse are described, and their significance is discussed in a preliminary fashion.

119.017 The impact of IUE on binary star studies.
M. J. Plavec.
The Universe at ultraviolet wavelengths, (see 012.009), p. 397 - 413 (1981).

Every class of binary stars can be profitably studied with IUE, and for most of them, such observations are of fundamental importance, and have already yielded extremely valuable results or new surprising facts. Some of the classes are discussed in the review.

119.018 O'Connell effect correlations with eclipsing binary star parameters.   T. J. Davidge, E. F. Milone.
Bull. American Astron. Soc., Vol. 13, 514 (1981). – Abstract.

119.019 Preliminary results for the eclipsing binary A0 Cam.
D. H. Piggott, E. F. Milone, S. L. Morris.
Bull. American Astron. Soc., Vol. 13, 514 (1981). – Abstract.

119.020 SX Cassiopeiae: spectral energy distribution and a model.
M. J. Plavec, J. J. Dobias, J. L. Weiland, R. H. Koch.
Bull. American Astron. Soc., Vol. 13, 523 (1981). – Abstract.

119.021 RX Cassiopeiae and SX Cassiopeiae: the case of the dubious twins.
M. J. Plavec, J. L. Weiland, J. J. Dobias, R. H. Koch.
Bull. American Astron. Soc., Vol. 13, 523 - 524 (1981). Abstract.

119.022 Ground based and UV observations of the spot and chromospheric activity of the short period eclipsing binary XY Ursae Majoris.   E. H. Geyer, M. Hoffmann.
Second European IUE Conference, (see 012.011), p. 81 - 84 (1980).

Six LWR-low resolution IUE spectrograms were obtained of the spotted eclipsing binary XY UMa. They show a strong chromospheric continuum rising from λλ 240 nm to λλ 310 nm. On it are superimposed a large number of emission and a few absorption features. The most prominent ones are those of the MgII resonance doublet and of FeII emission line blends. Both, the continuum and the emission features, show orbital phase dependent variations. The comparison with nearly simultanious optical photometric and spectroscopic observations exhibits the interaction of the photospheric and chromospheric activity, the latter of which is much enhanced whenever the optical light curve shows depressions.

119.023 IUE observations of Zeta Aurigae.
R. Faraggiana, M. Hack.
Second European IUE Conference, (see 012.011), p. 223 - 226 (1980).

High resolution IUE spectra of Zeta Aur have been obtained near orbital phase 0.91, covering the 1175 - 3000 A region. The spectral type of the B companion is estimated as B7 V from its energy distribution, which gives $T_e = 12000$ K, and the strength of Si III lines. The absorption profiles of the strong lines are complex, indicating the presence of several components. High excitation features like the resonance lines of N V, C IV and Si IV and λ 1640 He II are present in absorption.

119.024 The ultraviolet spectrum of UW Canis Majoris.
H. Drechsel, J. Rahe, Y. Kondo, G. E. McCluskey, Jr.
Astron. Astrophys., Suppl. Ser., Vol. 45, 473 - 481 (1981).

High resolution spectra of the eclipsing binary UW CMa have been obtained in the wavelength range 950 to 3200 Å with the COPERNICUS (OAO-3) and IUE satellites. UW CMa is an interacting system, consisting of an O7f Ia primary and an O-B secondary. The spectra have been used to compile a list of stellar and interstellar lines. Identifications together with position and laboratory wavelengths are given. The spectra are dominated by envelope (P Cygni) lines. Numerous, but less conspicuous photospheric absorption lines from the O7f component are present. In addition, a wealth of interstellar lines appear.

119.025 Revised photometric elements of five eclipsing binaries.   G. Giuricin, F. Mardirossian.
Astron. Astrophys., Suppl. Ser., Vol. 45, 499 - 505 (1981).

Using Wood's (1972) model, the authors have reanalyzed the photoelectric lightcurves – as yet no adequately explored – of five eclipsing binaries. For each system they have derived photometric elements appreciably different from the earlier ones, thus allowing a more consistent interpretation of the binaries studied.

119.026 A spectroscopic study of VV Cephei during the 1976—78 eclipse. II. Structure of the Hα emission envelope around the early-type component.   K. Saijo.
Publ. Astron. Soc. Japan, Vol. 33, 351 - 364 (1981).

Profiles of the Hα emission line obtained at various phases of the 1976—78 eclipse of VV Cep are analyzed, by taking into account the atmospheric eclipse of the M-type component, to find out the structure of the emitting region. It is shown that the structure can be approximately expressed by a simple model of an axisymmetric flat disk around the early-type component.

119.027 Mass transfer and stellar wind effects in the eclipsing binary RT Andromedae.
G. A. Bakos, J. Tremko.
Effects of mass loss on stellar evolution, (see 012.015), p. 491 - 494 (1981).

119.028 Further period changes in U Cephei.
E. C. Olson, R. C. Crawford, D. S. Hall, H. Louth,

N. L. Markworth, V. Piirola.
Publ. Astron. Soc. Pacific, Vol. 93, 464 - 469 (1981).

The authors report 19 new photoelectric determinations of times of minima for U Cep. Most of these eclipses were photometrically undisturbed and gave true times of stellar conjunction. They combine these times with earlier ones to discuss period changes during the past nine years. Two period decreases and one increase occurred in this time, separated by intervals of constant period. There is no evidence of a gradually changing period such as would be produced by a steady tidal interaction.

119.029 SW Lacertae − a quadruple system.
T. Panchatsaram, K. D. Abhyankar.
Bull. Astron. Soc. India, Vol. 9, 31 - 39 (1981).

Orbital period study using photoelectric times of minima of SW Lac reveals that the period change is caused by the orbital motion of a triple system around a fourth body. The orbital periods of the third and the fourth body are 19.67 yr and 70.25 yr respectively, and their masses are close to 1 $M_\odot$ each.

119.030 UBV photometry of XX Cassiopeiae.
R. K. Srivastava.
Bull. Astron. Soc. India, Vol. 9, 83 - 84 (1981). − Abstract.

119.031 Three colour photometry of BZ Eridani.
R. K. Srivastava, B. K. Sinha.
Bull. Astron. Soc. India, Vol. 9, 84 (1981). − Abstract.

119.032 UBV photometry and absolute dimensions of DI Pegasi.   U. S. Chaubey.
Bull. Astron. Soc. India, Vol. 9, 85 (1981). − Abstract.

119.033 The 1976-1978 eclipse of VV Cephei.
L. Baldinelli, A. Ferri, S. Ghedini, S. Marmi.
Mem. Soc. Astron. Italiana, Vol. 52, 275 - 303 (1981).

The eclipsing binary system VV Cephei has been observed photoelectrically and photographically during the 1976−1978 eclipse. The observations show the usual semiregular variations and a suspected short period flickering in the U light. From both the B and U light curves the dimensions of the two components have been estimated. As a consequence the primary M-type supergiant component seems to fill its Roche lobe and to produce gas streaming between the two stars.

119.034 Variations in the light curve of V 505 Sagittarii and their interpretation.   K. Walter.
Astron. Astrophys., Vol. 101, 369 - 376 (1981).

UBV observations of V 505 Sgr are discussed. For the epoch 1976−78 conspicuous deformations and an increase of the orbital period were found. It seems probable that also the increased brightness of the system which was observed in B and U in 1978 was connected with the active phase of increased mass transfer. The deviations of the single observations from the mean light curves showed clear variations with a period of somewhat more than 300^d. This period is interpreted to be the precessional period of the rotational axis of the primary component.

119.035 Der „Franksche" Bedeckungsveränderliche SAO 072799.   M. Fernandes.
BAV Rundbrief, 30. Jahrg., 82 (1981).

119.036 Photoelectric observations of the Am eclipsing binary AN Andromedae.
G. A. Bakos, J. Tremko.
Contrib. Astron. Obs. Skalnaté Pleso, Vol. 10, 53 - 88 (1981).

Photoelectric observations in the V band of the eclipsing binary AN And have been obtained during the period 1967 - 1977 at the Skalnaté Pleso Observatory and the University of

Waterloo Observatory. During this period 3690 observations have been obtained. The mean error of these observations is ±0^{m}0067.

119.037 Photoelectric photometry of the close binary SZ Camelopardalis in the intermediate passband filters.   D. Chochol.
Contrib. Astron. Obs. Skalnaté Pleso, Vol. 10, 89 - 124 (1981).

Photoelectric observations of the eclipsing binary SZ Cam, obtained in the years 1972 - 1976 at the Skalnaté Pleso Observatory, in two intermediate passband filters, with the mean wavelength 472 and 527 nm and half width 19.5 nm, are given.

119.038 Estimate of absolute elements of ER Cephei − an eclipsing variable in NGC 188.
L. F. Istomin.
Stellar aggregates, Sverdlovsk, 1980, (see 003.006), p. 80 - 83. In Russian. − Abstr. in Ref. zh., 51. Astron., 9.51.673 (1981).

119.039 The eccentric orbit of the short-period eclipsing binary V 1010 Oph.
R. Margoni, R. Stagni, A. Mammano, E. Illés-Almár.
Astrophys. Space Sci., Vol. 79, 159 - 163 (1981).

Radial velocity observations of V 1010 Oph obtained at Asiago from 23 spectra combined with those published by Guinan and Koch (1977) reveal a significant eccentricity ($e \cong 0.20 \pm 0.03$) with a periastron angle close to 90° at 1σ level. This underlines the need for a new analysis of the light curve for photometric elements.

119.040 Flare activity of the RS CVn star SV Cam.
L. Patkós.
Astrophys. Lett., Vol. 22, 131 - 133 (1981).

The flare activity of the short-period eclipsing binary system SV Cam was observed. Analysis of the light variations shows that the flare-active region is connected with the suspected spotted region on the surface of the secondary component.

119.041 Is the riddle of ε Aurigae stellar systems solved?
Z. Alksne.
Zvaigžnota debess, 1980. gada pavasaris, p. 18 - 20. In Latvian.

119.042 Light curves and solutions for ST Centauri.
E. F. Milone, B. J. Hrivnak.
Astron. J., Vol. 86, 1546 - 1552 (1981) = Publ. Rothney Astrophys. Obs., No. 14.

Results of the first UBV photometry, classification spectroscopy, and synthetic light curve analysis of the 1^{d}2-period eclipsing binary ST Cen are presented. The existence of an asymmetry in the light curve is confirmed. There is no strong evidence of Ca H and K emission or of light curve variability, and the nature of the asymmetry is not understood at present.

119.043 Photometry and elements of GW Geminorum.
P. Broglia, P. Conconi.
Astron. Astrophys., Suppl. Ser., Vol. 46, 185 - 191 (1981).

Photoelectric observations of the eclipsing system GW Gem made during the years 1978 and 1979 enable the authors to derive nine epochs of minimum light. The period of the binary star appears to be constant. The light curves are analyzed referring to the Roche model. The brighter component appears to be an A8 star and the secondary G3-G4 subgiant fills its Roche lobe.

119.044 Photometric observations of the Algol variable V 505 Sagittarii.   K. Walter.
Astron. Astrophys., Suppl. Ser., Vol. 46, 263 - 275 (1981).

UBV observations of the typical Algol system V 505 Sgr obtained between 1973 and 1979 at the European Southern Observatory are reported. Times of primary minima and deviations of the observations from the mean light curves are derived.

119.045 The variable lightcurves of RT Andromedae.
L. Milano, G. Russo, S. Mancuso.
Astron. Astrophys., Vol. 103, 57 - 62 (1981).

All presently available photoelectric lightcurves of the short-period eclipsing binary RT Andromedae, belonging to a group of binaries with some properties similar to the RS CVn group, have been analysed in a homogeneous way using two different methods, viz. the Wood and the Wilson-Devinney methods. A unique solution has been found for all the epochs, thus eliminating the inconsistencies generated by the use of the rectification procedure (Mancuso et al., 1979 c). The presence of a "migration wave" on the lightcurves, with a period of about 22-years, is ascertained by the analysis of the residuals with respect to the synthetic lightcurves and compared with the results obtained by means of the usual procedure. Although a complete analysis needs continuous observation of the system, yet some trends indicate that this group of the binaries has much more in common to the RS CVn one, than presently assumed.

119.046 WY Hya: a main sequence detached binary system with nearly equal members.
G. Giuricin, F. Mardirossian, M. Mezzetti.
Astron. Astrophys., Vol. 103, 349 - 350 (1981).

Using Wood's (1972) model the authors have analyzed Carr's (1971) three-colour observations — as yet unexplored — of the short-period eclipsing binary WY Hya. Photometric elements have been obtained. This binary turns out to consist of two detached components of practically equal temperatures, sizes, masses, and luminosities.

119.047 The eclipsing variable HD 134518.
A. Przybylski.
Acta Astron., Vol. 31, 221 - 229 (1981).

The ninth magnitude star HD 134518 is a β Lyrae type eclipsing binary with a period of 1.154553 days. The principal minimum is 0.24 magnitudes deep, the secondary minimum does not exceed 0.06 magnitudes. HD 134518 is probably a contact binary. Because of its fairly high proper motion (0".090 p. a.) it may be a Population II star.

119.048 The spectroscopic orbit of the eclipsing binary AR Cas revisited.   M. Gaida, W. Seggewiss.
Acta Astron., Vol. 31, 231 - 239 (1981).

The authors have determined a new spectroscopic orbit for the eclipsing binary AR Cas from spectrograms taken at Hoher List Observatory in 1969/70 and 1978. The new orbital elements agree well with other determinations between 1934 and 1969.

119.049 Ephemerides of eclipsing binaries for the year 1982.
P. Flin.
Rocznik Astronomiczny Obserwatorium Krakowskiego 1982, (see 047.025), p. 1 - 101 (1981).

119.050 Period changes of AB Cassiopeiae.   P. Flin.
Rocznik Astronomiczny Obserwatorium Krakowskiego 1982, (see 047.025), p. 102 - 106 (1981).

119.051 Ephemerides of eclipsing binaries among cataclysmic variables for the year 1982.   J. M. Kreiner.
Rocznik Astronomiczny Obserwatorium Krakowskiego 1982, (see 047.025), p. 107 - 111 (1981).

119.052 HR 7551: a new supergiant eclipsing binary.
P. Mayer, D. Chochol.
Publ. Astron. Soc. Pacific, Vol. 93, 608 - 613 (1981).

The binary nature of HR 7551, type B0.5 Ib has been known since 1931. New radial velocities have enabled the authors to determine a period of 13^{d}374 and the elements of the spectroscopic orbit. The UBV light curve displays eclipses at phases in agreement with the spectroscopy. The system seems to be unaffected by mass transfer.

119.053 A study of the period of U Ophiuchi.
T. Panchatsaram.
Bull. Astron. Soc. India, Vol. 9, 139 - 143 (1981).

With the available photoelectric times of minima, the variations in the orbital period of U Oph can be interpreted in terms of the presence of a third body.

119.054 The southern RS CVn binary, HD 5303.
A. C. Collier, J. B. Hearnshaw, R. R. D. Austin.
Mon. Not. R. Astron. Soc., Vol. 197, 769 - 778 (1981).

Spectroscopic and photometric ($BVRI$) observations are presented for the southern RS Canum Venaticorum-type spectroscopic and eclipsing binary, HD 5303. These are used to obtain both the radial velocity and light curves for this system, which are analysed to obtain an improved ephemeris and orbital parameters. The out-of-eclipse light curve can be interpreted with a model that has a single large spot on the corotating cooler star, as well as tidal distortion of both stars in the binary system.

119.055 Brightness variations of the cool subgiants of totally eclipsing Algol binaries.   E. C. Olson.
Astrophys. J., Vol. 250, 704 - 708 (1981).

In a monitoring program lasting several years, the author has searched for eclipse-to-eclipse variations in the light levels of totality in 10 Algol-like binaries. Significant variations were found in U Cep, RW Tau, X Tri, U Sge, and RV Oph. Five-color observations of the first three are adequate to show that the cool contact subgiants were responsible for these variations. He shows that brightness increases of the subgiants in U Cep and RW Tau are correlated with eruptive transfer events, and therefore are directly related to instabilities in the lobe-filling subgiants.

119.056 On the accuracy of the interpretation of the light curves of eclipsing variables.
A. C. De Landtsheer.
Astrophys. Space Sci., Vol. 80, 349 - 352 (1981).

Error estimates resulting from light curve interpretations are often too optimistic, as is shown with the example of YZ Cas.

119.057 Preliminary results from the spectra of V471 Tauri.
E. Hamzaoğlu.
Astrophys. Space Sci., Vol. 80, 517 - 520 (1981).

The findings about this unusual eclipsing variable which possesses a white-dwarf component are reported. It was found that the H and K absorption lines of Ca II are strongly pressure-broadened, and exhibit two double-reversal emission components. The radial velocities derived from the central strong emission components of Ca II (H and K) are in agreement with the mean radial velocity curve obtained from the metallic absorption lines of the existing elements, whereas Hγ ($\lambda\lambda$4340.47 Å) does not conform with the mean radial velocity curve. In certain orbital phases the Hα (6562.82Å) line occurs in emission.

119.058 Simultaneous photoelectric and single-trail spectroscopic observations of V 471 Tauri (BD +16°516).
E. Hamzaoğlu.
Astron. Astrophys., Vol. 104, 65 - 68 (1981).

Spectroscopic observations of the white dwarf eclipsing binary V 471 Tauri (BD +16°516) are reported. The behaviour of Hγ (λ4340.47 Å, in absorption) and H, K lines of Ca II (λ 3968.47, λ 3933.66 Å in emission) are investigated. The radial velocities obtained from Hγ (4340.47 Å) do not conform with the radial velocity curve obtained from the existing absorption lines and the strong central H, K emission lines of the K 0 V component. The observation of Hα in emission may suggest that some material exists between the components.

119.059   Observing project: U Cephei.
      D. S. Hall, E. C. Olson.
I. A. P. P. P. Commun., No. 5, p. 12 - 14 (1981).

119.060   5 Ceti: a puzzling, newly discovered variable star.
      R. D. Lines, D. S. Hall.
I. A. P. P. P. Commun., No. 5, p. 17 - 18 (1981).

119.061   UX Comae: a wild eclipsing binary.
      D. M. Popper.
I. A. P. P. P. Commun., No. 6, p. 22 - 23 (1981).

119.062   Photoelectric observations of ζ Aurigae during the 1979 - 1980 eclipse.   Z.-h. Guo, X.-f. Liu.
Acta Astrophys. Sinica, Vol. 1, 203 - 212 (1981). In Chinese.
      Photoelectric observations of ζ Aur during the 1979 - 1980 eclipse were obtained. The observations were made in six narrow-band and the standard UBV systems. The central wavelengths of the narrow-band filters are in λλ 3520, 3940, 4240, 5020, 6595 w and 6560 n Å, respectively. The photometric feature of the narrowband λ 3940 is consistent with that of spectroscopic observations of Ca II K line. The photoelectric observations in λ 6595 w and 6560 n during totality indicate that it seems to be a pulsation.

119.063   The radial velocity of ζ Aur in its eclipse 1979 to 1980.
      H.-s. Tan, S.-q. Peng, Y.-l. Guo, X.-f. Liu.
Acta Astrophys. Sinica, Vol. 1, 317 - 321 (1981). In Chinese.

119.064   Photoelectric photometry of the eclipsing binary V 338 Cephei.   F. Gieseking.
Astron. Astrophys., Suppl. Ser., Vol. 46, 365 - 368 (1981).
      More than 350 photoelectric V- and B-magnitudes of the eclipsing binary V 338 Cep are presented yielding the first photoelectric light curve of this relatively bright system. Preliminary results are given.

119.065   Absolute dimensions and masses of eclipsing binaries. II. YZ Cassiopeiae.   C. H. Lacy.
Astrophys. J., Vol. 251, 591 - 596 (1981).
      Absorption lines of the F2 V secondary star of YZ Cas have been detected for the first time in high signal-to-noise ratio digital spectral scans obtained with the Reticon spectrometer of the 2.7 m reflecting telescope at McDonald Observatory. Radial velocities of high precision have been obtained for both the primary and secondary stars in this well known large light-ratio eclipsing binary system. Spectroscopic orbits derived from these data have been combined with a photometric orbit derived from Kron's photoelectric light curve to give very accurate masses and radii: $(2.31 \pm 0.01\,M_\odot, 2.53 \pm 0.03\,R_\odot)$ for the primary and $(1.35 \pm 0.01 M_\odot, 1.35 \pm 0.02\,R_\odot)$ for the secondary. Theoretical evolutionary tracks indicate an age of about 4×10^8 years based on the observed masses and radii. The primary star is near core hydrogen exhaustion and is rotating about 20% faster than the synchronous rotational velocity. The secondary is rotating synchronously.

119.066   The 1979 - 1980 eclipse of Zeta Aurigae. II. The emission spectrum.
R. E. Stencel, R. D. Chapman.
Astrophys. J., Vol. 251, 597 - 603, plate 14 (1981).
      The spectra in the UV of Zeta Aur, obtained with the *IUE* satellite near times of second contact and mid-eclipse in 1979, are discussed. Emission line flux and identifications are provided. Changes in the fluxes and line profiles are used to infer the geometric configuration of the interaction region between the B8 V and K-type supergiant in the system. It is argued that the high speed flows seen in the UV spectra originate from the B star side of the supergiant and are driven by the action of local surface heating. Several shock fronts exist, including an accretion shock around the B star and a terminator shock around the supergiant.

119.067   Is TY Coronae Australis an eclipsing system?
      V. I. Kardopolov, G. K. Filip'ev.
Perem. Zvezdy, Tom 21, 195 - 198 (1979). In Russian.
      The results of photoelectric UBV observations of TY CrA in 1976 - 77 are given. A decrease of the stellar brightness of about 0ᵐ4 during two nights was found. In the rest of nights the data show brightness constancy. The photoelectric data are in very good accordance with light curve elements as JD 2422527ᵈ75 + 2ᵈ8888·E by Innes (1925). So TY CrA may be classified as an eclipsing variable star.

119.068   The eclipsing variable θ^1 Orionis A in Trapezium.
      M. M. Zakirov.
Perem. Zvezdy, Tom 21, 223 - 226 (1979). In Russian.
      Photoelectric UBVR observations of θ^1 Ori A were obtained in the early 1978. The individual observations in each color are listed in a table. The amplitude of the variations decreases from U (1ᵐ02) to R (0ᵐ82). The light curves in V were solved by the Russell-Merrill method.

119.069   On the orbital periods of the eclipsing binaries CM Lacertae, AB Andromedae and YY Eridani.
T. Panchatsaram, K. D. Abhyankar.
Bull. Astron. Soc. India, Vol. 9, 243 - 248 (1981).
      A study of all the available photoelectric times of minima of CM Lac, AB And and YY Eri does not reveal any changes in their orbital periods.

119.070   Emission of circumstellar matter of β Lyrae in 1969 and 1978.   V. S. Kaserkevich, M. Yu. Skul'skij.
Vestn. L'vov. univ. Ser. astron., 1981, No. 56, p. 28 - 37. In Russian. – Abstr. in Ref. zh., 51. Astron., 11.51.737 (1981).

119.071   Fast variability of emission lines of β Lyrae.
      V. S. Kaserkevich, M. Yu. Skul'skij.
Vestn. L'vov. univ. Ser. astron., 1981, No. 56, p. 38 - 44. In Russian. – Abstr. in Ref. zh., 51. Astron., 11.51.738 (1981).

119.072   Improved period of AM Herculis and its possible variations.
I. L. Andronov, S. V. Vasil'eva, V. P. Tsesevich.
Astron. Tsirk., No. 1122, p. 1 - 3 (1980). In Russian.

119.073   89th - 90th list of minima of eclipsing binaries.
      Compiled by M. Andrakakou, G. Boistel,
R. Boninsegna, R. Diethelm, D. P. Elias, R. Germann,
R. Leyman, K. Locher, P. Matagne, D. Mourikis, E. Nezry,
I. Nikolaou, A. Parris, H. Peter, E. Poretti, P. Ralincourt,
J. Vialle, L. Capol, N. Contopoulos, C. Maranta,
G. Mavrofridis, C. Pampaloni, N. Stoikidis, W. Zwing.
BBSAG Bull., No. 56, p. 1 - 7, No. 57, p. 1 - 5 (1981).

119.074   V 1933 Sagittarii: the minimum brightness and duration.   K. Locher.
BBSAG Bull., No. 56, p. 7 (1981).

119.075   DP Cephei: discordant results.
      K. Locher.
BBSAG Bull., No. 56, p. 7 (1981).

119.076   V 456 Ophiuchi: revised elements.
      R. Diethelm.
BBSAG Bull., No. 57, p. 6 (1981).

119.077   EX Andromedae: evidence of an overrounding against the GCVS ephemeris.   K. Locher.
BBSAG Bull., No. 57, p. 6 (1981).

119.078 Revised Svetchnikov classification of 6 binaries
based on recent BBSAG results.   K. Locher.
BBSAG Bull., No. 57, p. 7 (1981).

119.079 RU Canis Maioris: the minimum duration and
magnitude.   K. Locher.
BBSAG Bull., No. 57, p. 7 (1981).

119.080 A photoelectric time of minimum of TV Cas in
four colours.   A. C. de Landtsheer.
Inf. Bull. Variable Stars, No. 1985, 2 pp. (1981).

119.081 Photoelectric observations of BV Draconis.
P. Rovithis, H. Rovithis-Livaniou.
Inf. Bull. Variable Stars, No. 1994, 2 pp. (1981).

119.082 Infrared photometry of Beta Lyrae.
M. Zeilik, P. Heckert, G. Henson, P. Smith.
Inf. Bull. Variable Stars, No. 1995, 4 pp. (1981).

119.083 HD 224113 – a new eclipsing, double-lined binary.
R. Haefner.
Inf. Bull. Variable Stars, No. 1996, 4 pp. (1981).

119.084 Search for the period of the binary shell star V 505
Monocerotis.
D. Chochol, A. Kučera.
Inf. Bull. Variable Stars, No. 1998, 4 pp. (1981).

119.085 Light elements of W Gru.
M. A. Cerruti, M. A. de Laurenti.
Inf. Bull. Variable Stars, No. 2012, 4 pp. (1981).

119.086 Period variability and new ephemeris of
RU Leporis.   Z. Kviz, F. Rufener.
Inf. Bull. Variable Stars, No. 2014, 4 pp. (1981).

119.087 An ultraviolet photoelectric lightcurve of the
eclipsing binary BB Peg.
M. Cerruti Sola, F. Scaltriti.
Inf. Bull. Variable Stars, No. 2023, 3 pp. (1981).

119.088 A light curve of the eclipsing binary LU Lacertae.
M. Hoffmann.
Inf. Bull. Variable Stars, No. 2025, 2 pp. (1981).

119.089 Photoelectric minima of U Pegasi.
H. Rovithis-Livaniou, P. Rovithis.
Inf. Bull. Variable Stars, No. 2026, 2 pp. (1981).

119.090 Photoelectric minima and light curves of the
eclipsing binary VZ Librae.
J. J. Clariá, E. Lapasset.
Inf. Bull. Variable Stars, No. 2035, 4 pp. (1981).

119.091 1 Per: a new eclipsing binary with a long period and
an elliptical orbit.
P. North, F. Rufener, A. Figer, L. Maurin.
Inf. Bull. Variable Stars, No. 2036, 7 pp. (1981).

119.092 Two colour photoelectric light curves of WW Dra.
Z. Tunca, C. Ibanoğlu, M. Kurutaç, S. Evren,
O. Tümer, A. Y. Ertan.
Inf. Bull. Variable Stars, No. 2040, 3 pp. (1981).

119.093 The eclipsing binary HD 124195.
P. R. Wesselius.
Inf. Bull. Variable Stars, No. 2044, 3 pp. (1981).

119.094 Photoelectric minima observations of the eclipsing
binary ST Carinae.
J. J. Claria, E. Lapasset.
Inf. Bull. Variable Stars, No. 2049, 4 pp. (1981).

119.095 The light curve and elements for EM Cephei.
C. Cristescu, G. Oprescu, M. D. Suran.
Inf. Bull. Variable Stars, No. 2050, 3 pp. (1981).

119.096 Spectroscopy of V1425 Cygni and AH Cephei.
C. R. Chambliss.
Inf. Bull. Variable Stars, No. 2051, 2 pp. (1981).

119.097 Light elements of GG Vel.
M. A. Cerruti.
Inf. Bull. Variable Stars, No. 2052, 4 pp. (1981).

119.098 SAO 072799, a new bright eclipsing binary.
M. Fernandes, P. Frank.
Inf. Bull. Variable Stars, No. 2053, 3 pp. (1981).

119.099 Optical observations of the primary minimum of
the solar-type binary AI Phe with the IUE satellite.
E. F. Milone, B. J. Hrivnak, T. A. Clark, O. Kjeldseth Moe,
J. C. Blades, I. Shelton.
Inf. Bull. Variable Stars, No. 2060, 3 pp. (1981) = Rothney
Astrophys. Obs. Publ., Ser. B, No. 8.

119.100 New Algol-type variable SVS 2342.
N. E. Kurochkin.
Astron. Tsirk., No. 1131, p. 7 - 8 (1980). In Russian.

119.101 On the light curve of AM Herculis and its variations.
I. L. Andronov, S. Vasil'eva, V. P. Tsesevich.
Astron. Tsirk., No. 1142, p. 5 - 8 (1980). In Russian.

119.102 Photometry and orbital period of SS 433
(V1343 Aql).
S. A. Gladyshev, V. P. Goranskij, N. E. Kurochkin,
A. M. Cherepashchuk.
Astron. Tsirk., No. 1145, 8 pp. (1980). In Russian.

119.103 Principal regularities in light variations of SS 433 =
V1343 Aql.   S. A. Gladyshev,
V. P. Goranskij, N. E. Kurochkin, A. M. Cherepashchuk.
Astron. Tsirk., No. 1146, 8 pp. (1980). In Russian.

119.104 The precessing accretion disk in the eclipsing binary
system SS 433.   N. G. Bochkarev,
E. A. Karitskaya, N. E. Kurochkin, A. M. Cherepashchuk.
Astron. Tsirk., No. 1147, 10 pp. (1980). In Russian.

119.105 A photoelectric light curve and elements of the
eclipsing binary EO Aurigae.   P. Hartigan.
J. American Assoc. Variable Star Obs., Vol. 10, 13 - 20 (1981).
       EO Aurigae is an early-type eclipsing binary of large
amplitude. A photoelectric light curve consisting of 316 V
observations is presented. The period does not seem to be
changing significantly. A solution for the elements of the
system is derived and a theoretical light curve constructed
from these elements is found to fit the data well. These new
elements differ considerably from those published previously

119.106 The 1978 eclipse of R Aquarii.
L. A. Willson, P. Garnavich, J. A. Mattei.
J. American Assoc. Variable Star Obs., Vol. 10, 34 (1981).
Abstract.

119.107 Visual minima of eclipsing binaries.
H. Korpikiewicz, G. Sitarski.
Astron. Rep., Vol. 4, 37 - 38 (1979).

119.108   **Visual observations of the eclipsing binary RZ Cas.**
J. Speil.
Astron. Rep., Vol. 4, 39 - 41 (1979).

119.109   **Photometric observations of the eclipsing variable**
QX Carinae.   A. W. J. Cousins.
South African Astron. Obs. Circ., No. 6, p. 1 - 3 (1981).

119.110   **Long light curves of VW Cep.**
J.-E. Solheim, K. E. Egge.
Proceedings of the Third Finnish-Soviet Astronomical
Symposium, (see 012.057), p. 113 - 115 (1981).

119.111   **Wolf-Rayet stars in the Magellanic Clouds. II. The**
peculiar eclipsing binary HD 5980 in the SMC.
J. Breysacher, A. F. J. Moffat, V. S. Niemelae.
ESO Sci. Prepr., No. 177, 27 pp. (1981). − Submitted to
Astrophys. J.

119.112   **Analisis de la curva de luz del sistema binario**
eclipsante S Velorum.   A. Gimenez.
Rev. Real Acad. Cienc., Madrid, Tomo 73, 577 - 592 (1979) =
Univ. Complutense - Fac. Cienc., Madrid, Semin. Astron.
Geod., Publ. núm. 107.

119.113   **He I line emission and the helium abundance in**
cataclysmic variables.
R. E. Williams, D. H. Ferguson.
Prepr. Steward Obs., No. 336, 2 + 46 pp. (1981).
   Time-resolved spectroscopy of a sample of eclipsing
cataclysmic variables indicates that the He I emission lines
are formed with the Balmer lines in the outer regions of the
accretion disk. Additional observations of lower inclination
non-eclipsing systems demonstrate that the He I emission is
frequently characterized by relatively high singlet to triplet
intensity ratios and an inverted line decrement, an unusual
situation which can be explained in terms of emission from
an optically thick (in the lines) region in LTE. Calculations
of steady-state accretion disk structure show that the
observed characteristics of the He I and H I lines can be reproduced in normal disk models, but only if the helium
abundance is assumed to be very high. Enhanced helium
abundances are thus implied for some CV secondaries, suggesting that they are either highly evolved cores or that they
have an outer layer of processed material which was accreted
from the white dwarf during earlier evolution of the system.

119.114   **Visual minima of eclipsing binaries.**   P. Filin.
Mitt. Veränderl. Sterne (MVS), Band 9, 17 - 19
(1981).

119.115   **Verbesserte Elemente von V 393 Cygni.**
T. Berthold.
Mitt. Veränderl. Sterne (MVS), Band 9, 19 - 20 (1981).

119.116   **Photoelektrische Messungen des langperiodischen**
Bedeckungssternes EE Cephei.
L. Meinunger, W. Pfau.
Mitt. Veränderl. Sterne (MVS), Band 9,   38 - 40 (1981).

119.117   **Light curve synthesis for the eclipsing binary**
BB Peg.   G. Giuricin, F. Mardirossian, M. Mezzetti.
Astron. Nachr., Band 302, 285 - 286 (1981).
   Using Wood's (1972) model the authors reanalyzed
Cerruti-Sola and Scaltriti's (1980) photoelectric observations
of the eclipsing binary BB Peg. The photometric elements,
which turn out to be considerably different from the earlier
preliminary results, describe BB Peg as a contact system, in
which the component eclipsed at the deeper minimum is
smaller than its companion. BB Peg appears to conform to
the general properties of the W-type subclass of W UMa
systems.

119.118   **On systems of the W UMa-type.**
L. F. Istomin.
Astron. Tsirk., No. 1151, p. 1 - 4 (1981). In Russian.

119.119   **On the period of QU Persei.**
V. P. Tsesevich.
Astron. Tsirk., No. 1152, p. 7 - 8 (1981). In Russian.

119.120   **On the eclipses of SY Cep.**
V. I. Kardopolov, G. K. Filip'ev.
Astron. Tsirk., No. 1158, p. 4 - 5 (1981). In Russian.

119.121   **Photoelectric observations of BS Dra.**
R. A. Botsula.
Astron. Tsirk., No. 1158, p. 5 - 6 (1981). In Russian.

119.122   **The moments of minima of eclipsing binaries in**
Delphinus.   V. G. Karetnikov.
Astron. Tsirk., No. 1162, p. 5 - 6 (1981). In Russian.

119.123   **Photometric elements of RT UMi.**
M. I. Lavrov, N. V. Lavrova.
Astron. Tsirk., No. 1165, p. 1 - 2 (1981). In Russian.

119.124   **The apsidal motion in the eclipsing binary system**
RR Lyn.   M. I. Lavrov, N. V. Lavrova.
Astron. Tsirk., No. 1165, p. 3 - 4 (1981). In Russian.

   A study of the convergent solutions of Kopal's
iterative method for solving the light curves of eclipsing
binaries.   See Abstr. 031.597.

   Photoelectric BVR observations of SS 433 from
June - October 1980.   See Abstr. 113.069.

   Spectrophotometric investigation of the Hα line
in the spectrum of AR Lac.   See Abstr. 114.096.

   Observations spectroscopiques de CI Cygni.
See Abstr. 114.130.

   Radiation of the gaseous envelope of β Lyrae in
1976 - 1977.   See Abstr. 114.194.

   Empirical bolometric corrections for the main
sequence.   See Abstr. 115.019.

   Spectroscopy of the AM Herculis type binary
2A 0311 − 227.   See Abstr. 117.003.

   A photometric study of the close binary Delta
Orionis A.   See Abstr. 117.011.

   Time-resolved spectroscopy of cataclysmic variables:
U Geminorum.   See Abstr. 117.013.

   Radial velocity and line profile variations in the
dwarf nova EM Cygni.   See Abstr. 117.014.

   Ingress observations of the 1980 eclipse of the
symbiotic star CI Cygni.   See Abstr. 117.024.

   Mass loss from the interacting contact binary
SV Cen.   See Abstr. 117.037.

   A model of the Algol type close binary TT Hydrae.
See Abstr. 117.039.

   Non-conservative evolutionary scenario for
100 Algols.   See Abstr. 117.044.

Ultraviolet photometry from ANS: chromospheric emission of W Ursae Majoris and 44 i Bootis. See Abstr. 117.058.

The massive contact system LY Aur. See Abstr. 117.063.

IUE spectra of the hot close binary V Puppis. See Abstr. 117.076.

The Algol-type systems. See Abstr. 117.098.

Evidence on high rate mass loss in the eclipsing variable VW Cep from ultraviolet spectra. See Abstr. 117.140.

Statistical models for spectroscopic and for eclipsing binary stars.   See Abstr. 120.015.

Revised photometric elements of the eclipsing systems BS Draconis and CD Tauri.   See Abstr. 120.019.

Photoelectric observation of δ Capricorni. See Abstr. 120.030.

Period changes in Z Cha.   See Abstr. 122.003.

On the nature of the two supergiant components in the system of V 810 Cen = HR 4511 = HD 101947. See Abstr. 122.005.

Rapid oscillations in cataclysmic variables. VI. Periodicities in erupting dwarf novae.   See Abstr. 122.007.

Notes on the early-type components of W Cep, o Cet, CH Cyg, AR Mon, and BL Tel. See Abstr. 122.035.

Ultraviolet spectroscopy of old novae and symbiotic stars.   See Abstr. 124.002.

The nature of V - V 1 - 7 and its central star HD 62001.   See Abstr. 135.038.

The X-ray modulation of Cygnus X-3. See Abstr. 142.047.

The high energy X-ray spectrum of 4U 0900−40 observed from OSO 8.   See Abstr. 142.061.

Optical eclipses and the precession effects in the X-ray binary system HD 153919 = 4U 1700−37. See Abstr. 142.149.

Identification of the X-ray source H 0850 + 13 with the eclipsing binary AC Cancri.   See Abstr. 142.165.

New probable planetary nebulae and VV Cephei stars in the Small Magellanic Cloud.   See Abstr. 159.008.

Erratum

119.901   Erratum: 'IUE ultraviolet spectra of the interacting binary U Cephei' [Astrophys. J., Vol. 247, 202 - 209 (1981)].   Y. Kondo, G. E. McCluskey, Jr., C. A. Harvel. Astrophys. J., Vol. 249, 402 (1981). − See Abstr. 29.119.115.

120 Spectroscopic Binaries

120.001   Ultraviolet and optical studies of binaries with luminous cool primaries and hot companions. I. Intrinsic colors and newly recognized systems. S. B. Parsons. Astrophys. J., Vol. 247, 560 - 568 (1981).

Late-type, noneclipsing giant and supergiant stars with essentially unresolved ($\lesssim 1''$) hot companion stars are being analyzed by means of far-ultraviolet spectra and the total energy distributions from far-UV to near-IR. Spectral types and visual magnitudes of the companions are estimated from the UV fluxes. Statistics of F and G supergiants show about 20% with hot companions.

120.002   Infrared and X-ray observations of the binary system V 861 Sco.   E. G. Tanzi, L. Maraschi, A. Treves, M. Tarenghi. Astron. Astrophys., Vol. 100, 68 - 71 (1981).

Infrared (1.25 - 4.8μ) observations at various orbital phases of the single line spectroscopic binary V 861 Sco are presented. Emission in excess over the photospheric continuum of the primary is found and interpreted as due to a strong stellar wind ($\dot{M} \simeq 2 \times 10^{-6} M_\odot yr^{-1}$). Modulation of the colour excess with orbital phase is discussed. X-ray data obtained within a month from the infrared observations yield a 3σ upper limit F(0.4 - 4.5 keV) $\lesssim 3 \times 10^{-12} erg\, cm^{-2} s^{-1}$.

120.003   A spectroscopic, photometric, and magnetic study of the starspot on II Pegasi.   S. S. Vogt. Astrophys. J., Vol. 247, 975 - 983 (1981) = Lick Obs. Bull. No. 882.

Photometric observations of II Peg (HD 224085) in 1977 revealed V-band variations of 0.43 mag range accompanied in phase by well-defined $(V−R)$ variations of 0.06 mag range, attributable to a large, cool starspot or spot group on the star at this epoch. A series of high-dispersion Reticon spectra showed a strong and asymmetrical Hα emission feature whose profile was remarkably constant with rotational phase but whose strength correlated with spot visibility, demonstrating that the Hα emission is preferentially spatially associated with the cool spot. Asymmetry of the Hα feature is attributed to differential mass motions in the chromosphere. A high quality Reticon spectrum of the $\lambda6707$ region was also obtained to confirm published reports of lithium in this star, an indicator of relative stellar youth.

120.004   Spectroscopic binary orbits from photoelectric radial velocities. Paper 39: HD 27144. R. F. Griffin. Observatory, Vol. 101, 115 - 116 (1981).

120.005   The evolutionary status of Upsilon Sagittarii (= HD 181615) as derived from ultraviolet and visual observations. P. Hellings, C. de Loore, M. Burger, H. J. G. L. M. Lamers. Astron. Astrophys., Vol. 101, 161 - 167 (1981).

The line content of the ultraviolet spectrum of Upsilon

Sagittarii is presented in the range λλ2025–2190 and 2830–3220 (BUSS spectrum) and in the range λλ1350–2550 (S2/68 spectrum). The possible spectral type of the secondary according to the S2/68 data is discussed. Stellar radii and masses are derived for both components of the system. It is shown that 90% of the matter expelled during the Roche lobe overflow phase of the primary has left the system. Observations and evolutionary computations indicate case C evolution.

120.006 BD + 34°4216: a composite spectrum binary system.
S. A. Naftilan, E. F. Milone.

Publ. Astron. Soc. Pacific, Vol. 93, 364 - 366 (1981).

The star BD + 34°4216 has a composite spectrum. Spectroscopy, photometry, and spectrophotometry of this object show that it is a physical binary consisting of an early-type dwarf and a late-type giant. The variable radial velocity of each component confirms this conclusion.

120.007 Visible-band and IUE observations of μ Sagittarii.
J. D. Dorren, E. F. Guinan, E. M. Sion.

The Universe at ultraviolet wavelengths, (see 012.009), p. 381 - 388 (1981).

Hα and u band photometry and IUE spectra of the binary system μ Sagittarii are discussed. An estimate of mass-loss is made from the observed P Cygni profiles. There are indications of pulsation in the supergiant B8 component.

120.008 Chromospheric plages on II Pegasi (HD 224085).
M. Rodonò, G. Romeo, G. Strazzulla.

Second European IUE Conference, (see 012.011), p. 55 - 57 (1980).

Low resolution IUE spectra in the long wavelength region of II Peg, which were obtained in July 1979 simultaneously with groundbased UBV observations at Catania Observatory, are presented.

120.009 Hα photometry of UX Arietis during 1979–80.
E. F. Guinan, G. P. McCook, J. L. Fragola, W. C. O'Donnell, A. G. Weisenberger.

Publ. Astron. Soc. Pacific, Vol. 93, 495 - 499 (1981).

The 6^d438 RS CVn-type binary UX Ari was observed photoelectrically on 17 nights from October 1979 through February 1980. The observations were made with a pair of narrow- and intermediate-band interference filters centered near the wavelength of the Balmer Hα line. An α index was formed from the data yielding a measure of the net Hα line strength. The light curve formed from the λ6585 intermediate-band data is well defined and has an amplitude of about 0^m04 with minimum light occurring near 0.87 phase computed with the ephemeris given by Hall, Montle, and Atkins. The value of the α index indicates the presence of Hα emission at all orbital phases with a possible weak correlation of Hα emission strength with phase. A study of all available photometry of the system reveals significant changes in the light amplitude and mean light level as well as changes in the phase at which light minimum occurs. The observed variations can be attributed to changes in both the area covered by star spots and in their location on the surface of the active member of the system.

120.010 HD 15558: an extremely luminous O-type binary star.
C. D. Garmany, P. Massey.

Publ. Astron. Soc. Pacific, Vol. 93, 500 - 503 (1981).

The authors present an orbital solution for the single-line binary HD 15558, an O5(f) star in the association IC 1805. The period is 440^d, the longest of any O-type spectroscopic binary, and e = 0.54. This star is one of the most luminous stars in the Galaxy and in the theoretical H-R diagram it lies near the track corresponding to stars with initial masses of 100 $M_\odot$.

120.011 41 Sextantis – a metallic line spectroscopic binary.
C. Raghavender Rao, S. Sreedhar Rao, R. Swaminathan.

Bull. Astron. Soc. India, Vol. 9, 85 (1981). – Abstract.

120.012 Spectroscopic binary orbits from photoelectric radial velocities. Paper 40: HR 551.
R. F. Griffin.

Observatory, Vol. 101, 175 - 177 (1981).

120.013 The ellipsoidal variable 33 Tauri. I. Spectroscopic orbit.
D. P. Hube.

Astron. J., Vol. 86, 1394 - 1396 (1981).

Orbital elements are presented for the spectroscopic binary, and recently recognized ellipsoidal variable, 33 Tau.

120.014 Speckle interferometry of Tau Persei.
H. A. McAlister.

Astron. J., Vol. 86, 1397 - 1400 (1981) = Astron. Contrib. Georgia State Univ., No. 53.

Six speckle interferometric observations of the 4.15-yr spectroscopic binary star τ Per are combined with the spectroscopic orbit of Colacevich (1941) to determine the elements of the apparent orbit. The new elements are $\Omega = 101°$, $i = 95°.0 \pm 2°.4$, and $a'' = 0''.050 \pm 0''.002$. Adoption of reasonable masses for the G5III primary and the A5V secondary lead to astrophysically consistent distance and luminosity values. It is possible that the system may exhibit a partial eclipse at one phase of its orbit.

120.015 Statistical models for spectroscopic and for eclipsing binary stars.
J. L. Halbwachs.

Astron. Astrophys., Vol. 102, 191 - 196 (1981).

The relations between the statistical features of spectroscopic and eclipsing binaries and the statistical properties of binaries as a whole were investigated. Possible distributions of the semimajor axes and of the mass ratios were selected; then certain values accessible from the observations – the proportion of spectroscopic and eclipsing binaries and the distributions of the K velocities and of the depths of eclipse – were computed for each case.

120.016 The true period of rotational light variations of the Ap star HD 219749 = ET And.
G. Hildebrandt, A. Hempelmann.

Astron. Nachr., Band 302, 155 - 160 (1981). In German.

The photoelectric measurements of the Ap star HD 219749 were analysed. The authors estimate the true period of light variation to $P = 1.61883$ days. The other periods which are given in the literature could be excluded.

120.017 Evidence for a starspot cycle on BD +26°730.
L. Hartmann, B. W. Bopp, M. Dussault, P. V. Noah, A. Klimke.

Astrophys. J., Vol. 249, 662 - 665 (1981).

The authors demonstrate that the dK5e star BD +26°730 exhibits a smooth variation in mean light suggestive of a 60 year spot cycle. Photographic data from the Harvard archival plate collection indicate a total amplitude of at least 0.5 mag, the largest yet observed in a BY Dra star. Spectroscopic observations show that the star is a single-lined spectroscopic binary with a period of ~2 days and velocity amplitude of ~20 km s⁻¹. These data indicate that the primary star is viewed nearly pole-on from earth. The authors suggest that this geometry reduces the rotational modulation of photospheric light, so that the long-term starspot cycle is more clearly apparent in the photometric data.

120.018 Chemically peculiar stars among spectroscopic binaries.
W. Seggewiss.

Upper main sequence chemically peculiar stars, (see 012.033), p. 183 - 188 (1981).

This paper presents a statistical investigation of peculiar stars (Am, Hg-Mn, Ap) among spectroscopic binaries. The fraction of Am stars among all binaries in the comparable spectral range is 33%, that of Hg-Mn stars 19% and that of Ap stars 7%. – Selection effects are briefly discussed. – The distributions of orbital elements (P, e, f(m)) for the sets of peculiar and non-peculiar binaries are compared.

120.019 **Revised photometric elements of the eclipsing systems BS Draconis and CD Tauri.**
G. Russo, L. Milano, A. D'Orsi, S. Marcozzi.
Astrophys. Space Sci., Vol. 79, 359 - 366 (1981).
   The recently published two-colour photoelectric light curves of the double-lined spectroscopic binaries CD Tauri and BS Draconis have been analysed. The absolute elements, computed using Popper's spectroscopic data, show that both binaries have normal Main Sequence components, and the comparison of their relative radii with the Roche lobes allows to classify them as geometrically detached.

120.020 **Bright new spectroscopic binaries discovered from composite spectra. III. HR 233, HR 676, HR 1129, ξGem, HR 7573, and HD 213470-1.**   E. M. Hendry.
Astron. J., Vol. 86, 1540 - 1545 (1981).
   Results of a continuing search for spectroscopic binaries among known composite spectra are presented. Radial velocities and preliminary orbital elements for four of these stars are given. ξ Gem and HD 213470-1 are found to be constant in velocity over the time period considered.

120.021 **Spectroscopic observations of HD 206267 as a possible counterpart of the X-ray source Cep X-4.**
T. S. Galkina.
Izv. Krymskoj Astrofiz. Obs., Tom 63, 86 - 92 (1981). In Russian. English translation in Bull. Crimean Astrophys. Obs., Vol. 63.
   To confirm the probability of the X-ray source Cep X-4 associated with the binary system HD 206267, the spectrum of the spectroscopic binary system HD 206267 is analysed. These observations show no evidence for the connection of HD 206267 with the X-ray source Cep X-4.

120.022 **A model for the expanding C III envelope of the Wolf-Rayet spectroscopic binary HD 152270.**
W. Neutsch, H. Schmidt, W. Seggewiss.
Acta Astron., Vol. 31, 197 - 205 (1981).
   A model is presented to explain the behaviour of the broad depression feature on top of the C III λ5696 emission line of the Wolf-Rayet (WR) spectroscopic binary HD 152270. In the authors' model the expanding C III envelope of the WR star is perturbed in the section adjacent to the hot O-type companion, causing an emission-free region and in turn the depression on the line top. The model is able to explain the variability of the depression feature and gives the basic physical parameters of the binary components.

120.023 **The spectroscopic orbit of HD 225292.**
R. F. Griffin.
J. R. Astron. Soc. Canada, Vol. 75, 222 - 226 (1981).
   The sixth-magnitude star HD 225292, whose radial velocity was found to be variable by Heard at the David Dunlap Observatory, is shown to be a spectroscopic binary with a rather eccentric orbit and a period of 954 days.

120.024 **The line-profile variations of Spica.**
   G. A. H. Walker, K. Moyles, S. Yang, G. G. Fahlman.
J. R. Astron. Soc. Canada, Vol. 75, 254 (1981). – Abstract.

120.025 **A possible nonrotational periodic magnetic variation in the Ap star HR 710.**   W. K. Bonsack.
Publ. Astron. Soc. Pacific, Vol. 93, 546 (1981). – Abstract.

120.026 **Spectroscopic binary orbits from photoelectric radial velocities. Paper 41: HR 7135.**
R. F. Griffin.
Observatory, Vol. 101, 208 - 211 (1981).

120.027 **A spectroscopic orbit for HD 115968.**
   R. F. Griffin.
J. Astrophys. Astron., Vol. 2, 309 - 313 (1981).
   Photoelectric radial-velocity measurements show that the eighth-magnitude star HD 115968 is a spectroscopic binary with a period of 16.195 days. The star has a large proper motion, and is unlikely to have the luminosity corresponding to the spectral type of G8 III favoured by Zaitseva. It is most probably a late-G dwarf.

120.028 **The spectroscopic triple system η Orionis A.**
   E. R. Zizka, W. R. Beardsley.
Astron. J., Vol. 86, 1944 - 1952 (1981).
   A reinvestigation of the spectroscopic triple system η Ori A is presented combining radial velocity measures from plates made at five observatories. The final elements derived from the combined set of observations show the periods of the short- and long-period orbits to be 7.989 268 days and 9.219 yr, respectively. The spectrum is double lined, and a weakening of the secondary hydrogen spectrum is observed when redshifted. Attempts have been made to minimize the distortion of the radial velocity curves resulting from pair blending. A large external error for the secondary orbit may be evidence for the existence of rapid pulsation. Minimum masses for the primary and secondary stars are $14.9 M_\odot$ and $12.3 M_\odot$, respectively. Assuming coplanarity, the minimum mass for the third star is $13.6 M_\odot$.

120.029 **V1343 Aquilae = SS 433.**
   IAU Circ., Nos. 3626, 3649 (1981).

120.030 **Photoelectric observation of δ Capricorni.**
   S. Ohmori.
Inf. Bull. Variable Stars, No. 2011, 5 pp. (1981).

120.031 **BVR photoelectric observations of ER Vu1 June, July, 1981.**   T. H. Kadouri.
Inf. Bull. Variable Stars, No. 2057, 5 pp. (1981).

120.032 **On the statistical nature of Be and Ae binaries.**
   T. Kogure.
Publ. Astron. Soc. Japan, Vol. 33, 399 - 411 (1981).
   The period distribution and the period-rotational velocity relation are examined for spectroscopic Be and Ae binaries whose MK spectral types lie in the range of B0-B9, V-III and A0-A9, V-III, respectively. It is found that Be and Ae binaries are divided into two groups of short period (P < 30 d) and long period (P > 30 d) in their orbital motion. The former is closely related to Algol binaries and may be mass-exchanging binaries, whereas the latter group is characterized by rapid stellar rotation favoring a strong similarity to classical single Be stars.

Preliminary orbital parallax catalog.
See Abstr. 002.074.

Lunar occultations of the Hyades. II. August 1980.
See Abstr. 096.004.

The radial velocity variations in IC 418.
See Abstr. 112.033.

Observations of late-type stars with 'hot' companions.   See Abstr. 114.074.

Binary and single Hg-Mn stars compared.
See Abstr. 114.120.

The Wolf-Rayet stars.   See Abstr. 114.203.

Magnetic structure in cool stars. I. The Ca II H and K emission from giants.   See Abstr. 116.003.

Magnetic structure in cool stars. III. Ca II H and K emission and rotation of main-sequence stars. See Abstr. 116.015.

Image tube spectroscopic studies of rapid variables. IV. Spectroscopic and photometric observations of AE Aquarii. See Abstr. 117.097.

HD 202908: a young, solar-type triple system. See Abstr. 118.005.

Is this diagram an argument for binary orbital evolution due to mass loss?   See Abstr. 118.014.

A spectroscopic and photometric study of the triple system HD 7215.   See Abstr. 118.015.

HR 7551: a new supergiant eclipsing binary. See Abstr. 119.052.

The southern RS CVn binary, HD 5303. See Abstr. 119.054.

Simultaneous photoelectric and single-trail spectroscopic observations of V 471 Tauri (BD + 16°516). See Abstr. 119.058.

5 Ceti: a puzzling, newly discovered variable star. See Abstr. 119.060.

Spectroscopy of V 1425 Cygni and AH Cephei. See Abstr. 119.096.

Photometry of HD 45088: a new bright BY Draconis variable.   See Abstr. 122.065.

A search for light-time effects in binary cepheids: AW Persei.   See Abstr. 122.110.

The ultraviolet spectrum of the O-type subdwarf HD 49798.   See Abstr. 126.021.

The binary central star of NGC 2346 and the extinction puzzle.   See Abstr. 135.050.

X-ray spectroscopic investigation of the coronal structure of Capella.   See Abstr. 142.097.

Erratum

120.901 Corrigendum: 'A spectroscopic orbit for 26 Comae' [J. Astrophys. Astron., Vol. 2, 115 - 118 (1981)]. R. F. Griffin. J. Astrophys. Astron., Vol. 2, 213 (1981). – See Abstr. 29.120.015.

121 Early-stage Stars (T Tauri Stars, Herbig-Haro Objects, etc.)

121.001 Two young stars in L 43. W. Herbst, J. W. Warner. Astron. J., Vol. 86, 885 - 891 (1981).
Optical and infrared photometry has been obtained for the two stars (RNO 90 and RNO 91) illuminating reflection nebulae in the small, filamentary dust cloud L 43. Spectrograms of RNO 90 in the blue and red have also been obtained, which show that it is a T Tauri star. Both stars have strong infrared excesses, placing them firmly in the "dust" position of the $H - K, K - L$ diagram. L 43 is located near the star-forming end of the Sco OB2 association, but is relatively isolated from the major center of activity, the ρ Oph cloud.

121.002 A study of the peculiar T Tauri star V 1331 Cygni. C. Chavarría-K.
Astron. Astrophys., Vol. 101, 105 - 117 (1981).
New photometric and spectroscopic observations of V 1331 Cygni and of selected objects in its vicinity are discussed. A distance and an interstellar reddening similar to that of the T associations in NGC 7000/IC 5070 are derived. The observations of this star are explained best with an early type star (F0 - A8) of fairly high luminosity (= 86 $L_\odot$). Its location in the Hertzsprung Russell diagram is between the domains of the T Tauri stars and of the more luminous Herbig's Ae and Be stars. Using these new results the mass loss rate estimates from earlier authors are revised. The variability of the star is discussed and constraints for theoretical interpretations of this peculiar object are given. In addition the young Orion population stars GG 2–2 and LkHα 324 in the vicinity of V 1331 Cygni are briefly discussed.

121.003 Extinction properties for T Tauri stars. U. A. Nurmanova.
Astron. Zh., Tom 58, 762 - 764 (1981). In Russian. English translation in Soviet Astron., Vol. 25, No. 4.
The amount of extinction toward T Tauri stars in the T-associations T1 and T3 Tauri are evaluated. Dependencies observed are the following: stars of earlier spectral types are located in more obscured regions; the value of the ratio of total to selective extinction is higher for earlier-type stars; more active variables are embedded in regions with higher extinction.

121.004 Near-infrared photometry of some Herbig emission-line stars.   M. A. Pogodin.
Astron. Zh., Tom 58, 796 - 800 (1981). In Russian. English translation in Soviet Astron., Vol. 25, No. 4.
The results of a narrow-band photometry in the near infrared of 9 Herbig emission-line stars are reported. The variability of Z CMa and HD 259431 is noted. In the spectra of 5 objects absorption features coinciding in wavelengths with water vapour absorption bands are observed.

121.005 Discovery of three X-ray luminous pre-main-sequence stars.
E. D. Feigelson, G. A. Kriss.
Astrophys. J., Lett., Vol. 248, L35 - L38, plate L2 (1981).
The authors present an optical study of three X-ray sources found serendipitously in the Taurus—Auriga cloud. All three sources are identified as pre-main-sequence (PMS) stars of very low activity, exhibiting only weak emission lines and no optical variability. The authors suggest that the en-

hanced X-ray emission from PMS stars is not dependent on the conditions (i.e., stellar wind or infall) responsible for their unusual spectroscopic properties.

121.006 Nearly simultaneous optical and infrared photometry of T Tauri stars.
A. E. Rydgren, F. J. Vrba.
Astron. J., Vol. 86, 1069 - 1075 (1981).

Nearly simultaneous *UBVri* and *JHKL* photometry is presented for 17 T Tauri and related young stars in the Taurus and NGC 2264 regions. The seven stars in the authors' sample with extremely weak line emission have *B−V* colors which average about 0.1 bluer than expected from their *V−i* colors, assuming luminosity class V intrinsic colors and normal reddening. The intrinsic locus of the authors' program stars in the (*J−H,H−K*) infrared two-color diagram is quite well defined and supports the dust-shell model for the infrared excess. Maximum dust temperatures in the approximate range 1000−1500 K are inferred from the intrinsic locus of the authors' stars in the (*H−K, K−L*) diagram. Infrared excesses similar to those found in typical T Tauri stars are clearly present in several late-type young stars which have extremely weak emission-line spectra. Some *BVri* photometry for known T Tauri stars in the southern part of the Taurus complex is also presented.

121.007 IUE observations of young variables.   G. F. Gahm.
The Universe at ultraviolet wavelengths, (see 012.009), p. 105 - 114 (1981).

New insight to the physics and behavior of young variables have been provided by observations with the IUE satellite. These results are briefly reviewed.

121.008 The ultraviolet variability of the T Tauri star RW Aurigae.   C. L. Imhoff, M. S. Giampapa.
The Universe at ultraviolet wavelengths, (see 012.009), p. 185 - 191 (1981).

Between 1978 and 1979 the visible brightness of RW Aurigae increased by 0^m9. During this time (1) CIV and SiIV increased by factors of 2 to 4 while the lower ionization lines remained unchanged, (2) the fluorescent OI line increased by a factor of 8, (3) the shell spectrum changed from emission to absorption, and (4) the ultraviolet continuum brightened by 2^m3. On a time scale of a week the continuum varied by as much as 0^m8 but the MgII emission lines showed no variability over 10%. The authors hypothesize an active chromosphere, transition region, and envelope cooled by mass loss in order to explain the ultraviolet observations of RW Aur.

121.009 Mass spectra of young stars.   R. B. Larson.
Bull. American Astron. Soc., Vol. 13, 539 - 540 (1981). − Abstract.

121.010 High velocity molecular gas near Herbig-Haro objects HH 7 - 11.   R. L. Snell, S. Edwards.
Bull. American Astron. Soc., Vol. 13, 540 (1981). − Abstract.

121.011 High velocity molecular gas around T Tauri stars.
S. Edwards, R. L. Snell.
Bull. American Astron. Soc., Vol. 13, 540 (1981). − Abstract.

121.012 The smothered coronae of T Tauri stars.
F. M. Walter, L. V. Kuhi.
Bull. American Astron. Soc., Vol. 13, 540 - 541 (1981). Abstract.

121.013 A linear polarization survey of T Tauri stars.
P. Bastien.
Bull. American Astron. Soc., Vol. 13, 541 (1981). − Abstract.

121.014 Far-infrared observations of Herbig-Haro objects.
P. M. Harvey, R. Levreault, M. Cohen.
Bull. American Astron. Soc., Vol. 13, 541 (1981). − Abstract.

121.015 B35 and FU Ori: dust temperatures and energetics from 60 - 250 μm photometry.
H. A. Smith, C. Lada, H. A. Thronson, D. A. Harper,
J. Keene, R. F. Loewenstein, J. Smith.
Bull. American Astron. Soc., Vol. 13, 541 (1981). − Abstract.

121.016 UV spectrograms of T Tauri stars.
I. Appenzeller, C. Chavarria K., J. Krautter,
R. Mundt, B. Wolf.
Second European IUE Conference, (see 012.011), p. 209 - 211 (1980).

The IUE satellite has been used to observe the ultraviolet spectra (1200 $\leqslant \lambda \leqslant$ 3200 Å, resolution − 6 Å) of the T Tauri stars DR Tauri, CoD −35°10525, and AS 205. All observed UV spectra show emission lines of a great variety of different atoms and ions. The ion of highest ionisation stage observed is N V. Two of the observed T Tauri stars show a strong UV continuum and UV envelope absorption lines.

121.017 Do T Tauri stars have extensive coronae?
J. Krautter, G. Gahm.
Messenger, No. 25, p. 19 - 21 (1981).

121.018 Spectropolarimetry of Herbig-Haro objects and the exciting star of HH 30.
M. Cohen, G. D. Schmidt.
Astron. J., Vol. 86, 1228 - 1231 (1981).

Spectrophotometric and spectropolarimetric observations are presented of three HH objects: HH 11, 30, and 43. Although null polarization results are obtained for HH 11 and 43, the unusually strong continuum of HH 30 is linearly polarized by 3%. The emission lines in this nebula are unpolarized. The electric vector position angle appropriate to the continuum polarization, and the presence of Fe II emission lines apparently from the photosphere of a T Tauri-like star, indicate HL Tau as the exciting/illuminating star for HH 30. Evidence is adduced for a highly anisotropic distribution of circumstellar obscuration around this star.

121.019 Large proper motions of the Herbig-Haro objects HH 1 and HH 2.   G. H. Herbig, B. F. Jones.
Astron. J., Vol. 86, 1232 - 1244 (1981) = Lick Obs. Bull. No. 887.

The authors describe the large cross motions that they have discovered in the original objects HH 1 and HH 2, review current ideas on the nature of such isolated HH objects, and offer their own interpretation of the new observations.

121.020 Are we beginning to understand T Tauri stars?
M. Cohen.
Sky Telesc., Vol. 62, 300 - 303 (1981).

121.021 Light curve peculiarities and the circumstellar envelope of the YY Ori star S CrA.
V. I. Kardopolov, G. K. Filip'ev.
Astron. Zh., Tom 58, 801 - 809 (1981). In Russian. English translation in Soviet Astron., Vol. 25, No. 4.

The results of BV photoelectric measurements of the YY Ori star S CrA in 1976 - 79 are presented. The observed light curve peculiarities of S CrA are assumed that they are caused by several different mechanisms. The age of the YY Ori phenomenon of S CrA is estimated to be some hundred years.

121.022 Photometry of the reflection nebulosity at V1057 Cygni.
D. K. Duncan, E. A. Harlan, G. H. Herbig.
Astron. J., Vol. 86, 1520 - 1525 (1981) = Lick Obs. Bull.

No. 895.

The decline in brightness of V1057 Cyg since its 1969 - 1970 flareup has been accompanied by a fading of the reflection nebulosity which was first seen shortly after the outburst. Calibrated photometry of Crossley direct plates taken in the 6100 - 6700 Å region now shows that the surface brightness of the nebula has in fact decreased in proportion to the brightness of the star at the same wavelength. Except for nebulosity very near the star, whose photometry was not possible, at the present rate of fading of V1057 Cyg the nebula should fall below the detection threshold of the present material about 1985.

121.023 **High-dispersion spectra of RU Lup.**
R. D. Schwartz, R. W. Heuermann.
Astron. J., Vol. 86, 1526 - 1539 (1981).

The authors report the results of echelle spectra of the southern T Tauri star RU Lup. At minimum light, a photospheric absorption spectrum was detected for the first time in this star which usually suffers strong continuum veiling. The general characteristics of the spectrum, with particular emphasis upon the radial velocities obtained from the photospheric absorption lines and the permitted and forbidden emission lines are given.

121.024 **The smothered coronae of T Tauri stars.**
F. M. Walter, L. V. Kuhi.
Astrophys. J., Vol. 250, 254 - 261 (1981).

The authors present Einstein IPC observations of 10 fields containing pre-main-sequence stars. They report detection of significant soft X-ray flux from eight cataloged T Tauri stars plus the optical identification of two new T Tauri-like stars associated with X-ray sources. The authors demonstrate that an inverse correlation between the Hα equivalent width and the detection of X-ray flux exists. They interpret this in terms of a model wherein the X-ray emission arises in a corona similar to that of active late-type stars, which is small compared to the extent of the circumstellar envelope. The authors postulate that the lack of correlation of X-ray flux (or Hα) with A_v is due to a lack of grains in the envelopes of the more active T Tauri stars.

121.025 **The young pre-main-sequence Ae-type star HR 5999.**
P. S. Thé.
Irish Astron. J., Vol. 14, (see 012.034), 122 - 132 (1980).

121.026 **The origin of blue continua in Herbig-Haro objects: two-photon continua in low-velocity shocks.**
M. A. Dopita, R. D. Schwartz.
Publ. Astron. Soc. Pacific, Vol. 93, 546 (1981). – Abstract.

121.027 **Infrared photometry of the probable exciting source for Herbig-Haro objects 7 - 11.** R. D. Schwartz.
Publ. Astron. Soc. Pacific, Vol. 93, 549 - 550 (1981). Abstract.

121.028 **Spectroscopic observations of pre-main-sequence stars in the Pleiades.** R. Mundt, U. Bastian.
Publ. Astron. Soc. Pacific, Vol. 93, 601 - 607 (1981).

Medium-resolution blue and red spectrograms of three Pleiades pre-main-sequence stars are presented. All three stars show the lower Balmer lines and the Ca II H and K lines in emission. In the spectrum of the star H-134 the emission lines are broad enough to be resolved. The emission-line width observed for this star (FWHM $\approx$ 110 km s^{-1}) is considerably higher than for the older field dMe stars (FWHM $\approx$ 20 km s^{-1}). The widths of emission lines in Me stars may therefore be an important age indicator for this type of stars. The broad emission lines observed for H-134 may be a relic of its T Tauri phase. A conservative upper limit for the Li abundance log N(Li) = 0.6 has been determined for the three observed stars.

121.029 **Intrinsic polarization changes and the Hα and Ca II emission features in T-Tauri stars.**
J. Svatoš, M. Šolc.
Astrophys. Space Sci., Vol. 80, 473 - 476 (1981).

On grounds of the correlation between polarization and emission features observed in some T-Tauri stars, it is concluded that flaring effects associated with UV and/or X-ray irradiation and with increased magnetic field are the cause of the intrinsic polarization changes in T-Tauri stars.

121.030 *IUE* **observations of pre-main-sequence stars.**
I. Mg II and Ca II resonance line fluxes for T Tauri stars. M. S. Giampapa, N. Calvet, C. L. Imhoff, L. V. Kuhi.
Astrophys. J., Vol. 251, 113 - 125 (1981).

The authors present absolute flux measurements of the Mg II and Ca II resonance lines in a sample of T Tauri stars, obtained with the International Ultraviolet Explorer satellite and the image tube scanner of the 3 m Lick reflector. They discuss the Mg II h and k line emission and the Ca II H and K line emission within the context of stellar chromospheres, and they present the corroborative evidence for the chromospheric origin of these particular resonance lines in the T Tauri stars.

121.031 **Herbig-Haro objects: recent observational and theoretical developments.** J. Cantó.
Investigating the universe, (see 003.014), p. 95 - 124 (1981).

A brief summary of our present observational and theoretical knowledge of Herbig-Haro objects is given. Special emphasis is given to the recent observational results in the radio frequency range and on the most recent models advanced to explain these objects. It is concluded that HH-objects are most probably intrinsic emission nebulae produced by the cooling of shock-heated material. The source of energy is likely to be a strong wind from a new born star which is displaced ~ 0.1 pc typically from the optical object.

121.032 **HH 103 – an unpolarized Herbig-Haro object.**
D. J. King, S. M. Scarrott.
Observatory, Vol. 101, 197 - 200 (1981).

121.033 **Emission line spectra of Herbig-Haro objects.**
E. W. Brugel, K. H. Böhm, E. Mannery.
Astrophys. J., Suppl. Ser., Vol. 47, 117 - 138 (1981).

Spectrophotometric data have been obtained for 12 Herbig-Haro nebulae with the multichannel spectrometer on the Mt. Palomar 5.08 m telescope and with the image intensified dissector scanner on the Kitt Peak 2.13 m telescope. Empirical two-component density models have been constructed to interpret the emission line spectra of the five brightest condensations. Slightly less satisfactory homogeneous models are presented for the remaining five objects.

121.034 **Flare activity of T Tau-type stars: DR and DQ Tau.**
N. E. Kurochkin.
Astron. Tsirk., No. 1134, p. 1 - 3 (1980). In Russian.

121.035 **On the absence of coronal line emission from Orion population stars.**
G. F. Gahm, J. Krautter.
ESO Sci. Prepr., No. 171, 2 + 12 pp. (1981). – Submitted to Astron. Astrophys.

121.036 **Observations of rapid line profile variability in the spectra of T Tauri stars.**
R. Mundt, M. S. Giampapa.
Prepr. Steward Obs., No. 339, 38 pp. (1981).

The authors present high spectral and temporal resolution observations of 6 T Tauri stars, obtained with the Multiple Mirror Telescope echelle spectrograph and Reticon detector. They discuss the rapid line profile variability mani-

fested by RW Aur within the context of two physical processes, namely (1) variable mass infall into an accretion shock,
and (2) flare activity. Besides presenting evidence for the
existence of rapid line profile variabiltiy in the spectrum of
at least one T Tauri star, the authors also discuss properties
of the line spectrum of RW Aur in the vicinity of the Hα,
Hγ and Na D lines.

121.037 Eruptive events in T Tauri stars.   R. Mundt.
           Prepr. Steward Obs., No. 354, 10 pp. (1981).
The NaD line profiles of about 20 T Tauri stars have
been observed at high spectral resolution (Δλ = 0.2 Å) with
high signal-to-noise (S/N = 30–100). In seven T Tauri stars
the broad NaD emission lines showed relatively narrow
(FWHM = 10–40 km/s) blueshifted absorption components
with velocity shifts of about −50 to −150 km/s (that are all
of circumstellar origin). The observed narrow and blueshifted
absorption features are discussed in terms of shells ejected
during outbursts of these objects. These oubursts are probably the low energetic analogs of the outburst observed in the
FU Orionis objects.

**121.038 Über einige Eigenschaften zirkumstellarer Hüllen
           von T-Tauri-Sternen.   W. Götz.**
Veröff. Sternw. Sonneberg, Band 9, 237 - 305 (1980).
   By assembling and analyzing results of statistical investigations of T-associations and young open clusters general
statements on behaviour and properties of circumstellar gas
and dust shells at T Tauri stars are derived. They permit adequate reduction of observational data at single objects and
provide theory with a complex portrayal of the shell phenomenon and its influence on the total brightness.

**121.039 Photoelektrische Beobachtungen des T-Tauri-Sterns
           V 410 Tauri = BD+28°637.   S. Rößiger.**
Mitt. Veränderl. Sterne (MVS), Band 9, 35 - 38 (1981).
   The results of photoelectric UBV observations of the
T Tauri star V 410 Tau over a period of some years are
presented. The star shows irregular light variation with an
amplitude in V of about 0.35 mag.

**121.040 Photoelectric and polarimetric observations of
           T Tauri.**
N. P. Red'kina, A. V. Zubarev, G. P. Chernova.
Astron. Tsirk., No. 1149, p. 6 - 7 (1981). In Russian.

121.041 On the possible duplicity of RY Tau.
           U. A. Nurmanova.
Astron. Tsirk., No. 1150, p. 1 - 3 (1981). In Russian.

**121.042 On the extinction of light toward T Tauri-type
           stars.   U. A. Nurmanova.**
Astron. Tsirk., No. 1155, p. 1 - 3 (1981). In Russian.

**H_2O masers in the direction of southern nebular
objects.   See Abstr. 112.046.**

**Photometric behaviour of T Tauri stars in the
T-associations T1 and T3 Tau. I. Variations at various wavelengths.   See Abstr. 113.067.**

**Photometric behaviour of T Tauri stars in the
T-associations T1 and T3 Tau. II. U−B, B−V diagram for
excess radiation.   See Abstr. 113.068.**

**Spectral energy distributions of hot stars with
circumstellar dust.   See Abstr. 114.008.**

Balloon and IUE spectra of early-type stars.
See Abstr. 114.139.

Infrared polarization of stars in the Taurus cloud.
See Abstr. 116.013.

**High velocity molecular gas near Herbig-Haro
objects HH 7−11.   See Abstr. 131.167.**

Optical observations of interstellar shockwaves.
See Abstr. 131.175.

**Star formation in the λ Orionis region: 1. The
distribution of Hα emission objects.   See Abstr. 131.213.**

**An X-ray active region in Orion: X-rays from a
Herbig-Haro object?   See Abstr. 142.021.**

X-ray and ultraviolet observations of NGC 2264.
See Abstr. 153.009.

Erratum

**121.901 Erratum: "New upper limit to the coronal line emis-
           sion from the T Tauri star RU Lupi" [Mon. Not. R.**
Astron. Soc., Vol. 195, 59P - 62P (1981)].
G. F. Gahm, M. T. V. T. Lago, M. V. Penston.
Mon. Not. R. Astron. Soc., Vol. 196, 1087 (1981). − See
Abstr. 29.121.019.

122 Intrinsic Variables (Pulsating Variables, Spectrum Variables, etc.)

122.001 *IUE* **observations of eight dwarf novae: a study of the outburst cycle from 0.12 to 3.5 microns.**
P. Szkody.
Astrophys. J., Vol. 247, 577 - 589 (1981).

IUE observations of eight dwarf novae at various stages of their outburst cycles are joined to optical and IR data to provide a continuum flux distribution from 0.12 to 3.5 μm. These distributions have slopes ranging from λ^{-1} to λ^{-4} and, for the most part, do not appear consistent with simple steady state accretion disk models.

122.002 **Beat Cepheid studies – I. Temperatures of beat Cepheids.** S. L. Barrell.
Mon. Not. R. Astron. Soc., Vol. 196, 357 - 369 (1981).

Effective temperatures, based on the Hα absorption line profile, have been obtained for 10 southern beat Cepheid variables, yielding results which are similar to those for single-mode Cepheids of comparable period. The temperatures do not fall exclusively at either edge of the Instability Strip, but rather define a narrow domain related to a predicted region of complex, possibly mixed-mode behaviour. Evidence suggests that the beat Cepheids are not mode switching but rather have a pulsational status, not yet theoretically defined, where continuing mixed-mode behaviour is possible.

122.003 **Period changes in Z Cha.**
M. C. Cook, B. Warner.
Mon. Not. R. Astron. Soc., Vol. 196, 55P - 57P (1981).

The authors have analysed eclipse data on Z Cha over the period 1972 - 81, and have found that the times of white dwarf eclipse can no longer be represented by a linear ephemeris. A quadratic fit to the data implies an evolution time-scale $P/\dot{P} = 9.6 \times 10^6$ yr.

122.004 **The infrared spectrum of the dwarf nova EX Hydrae.**
J. Frank, A. R. King, M. R. Sherrington, A. B. Giles, R. F. Jameson.
Mon. Not. R. Astron. Soc., Vol. 196, 921 - 925 (1981).

Infrared photometry and spectrophotometry of the dwarf nova EX Hydrae shows that its flux increases for wavelengths > 2 μm. This is interpreted in terms of a circumstellar dust cloud resulting from mass loss during the accretion process. The authors comment on the variability of the near-infrared spectrum, and report the detection of the Bγ line of hydrogen.

122.005 **On the nature of the two supergiant components in the system of V 810 Cen = HR 4511 = HD 101947.**
A. M. van Genderen.
Astron. Astrophys., Vol. 100, 175 - 177 (1981).

V 810 Cen = HR 4511 = HD 101947 is a yellow variable supergiant on the blue side of the Cepheid Strip. Photometric characteristics and recent IUE satellite spectra indicated the presence of a blue supergiant companion. On account of spectroscopic, photometric and evolutionary considerations, the companion has likely a T_{eff} of 25,000–30,000 K. With this temperature and the $V-B/B-L$ diagram of the *VBLUW* system it is tried to determine the range of M_v, T_{eff}, log g and unreddened photometric parameters of both stars. The spectral type of the cool supergiant could be F 7.5.

122.006 **The Delta Scuti variable BD +28°1494.**
P. Broglia, P. Conconi.
Astron. Astrophys., Vol. 100, 201 - 204 (1981).

B and V light curves which show the δ Scuti variability of BD +28°1494 are given. Period analysis of the observations obtained on seven nights in 1978 and 1979 are presented. A beat phenomenon with $P_1/P_0 = 0.78$ can be seen during one

night, one pulsation only with the same value for $P_1 = 0\overset{d}{.}0553$, but with slightly different amplitudes appears on four nights, whilst on the two remaining nights no significant power appears above the noise level.

122.007 **Rapid oscillations in cataclysmic variables. VI. Periodicities in erupting dwarf novae.**
J. Patterson.
Astrophys. J., Suppl. Ser., Vol. 45, 517 - 539 (1981).

The author reports an extensive study of the coherent oscillations observed in high-speed photometry of dwarf novae during eruption. The oscillations are in all cases singly periodic and sinusoidal to the limits of measurement. The detection of oscillations in 14 separate eruptions of AH Her and SY Cnc enables a general study of period variations. The stars trace out characteristic loops in the period-intensity plane. New detections are also reported for SS Cyg, EM Cyg, and HT Cas.

122.008 **Microwave emission from flare star AT Mic.**
O. B. Slee, I. R. Tuohy, G. J. Nelson, C. J. Rennie.
Nature, Vol. 292, 220 - 221 (1981).

The authors describe the most convincing detection (7σ) yet made of a microwave flare during simultaneous optical and 5-GHz observations of the dMe 4.5 flare star AT Mic on 25 October 1980.

122.009 **Beobachtungen an den Delta Scuti-Sternen δ Del und ϵ Cep im Jahr 1980.** M. Fernandes.
BAV Rundbrief, 30. Jahrg., 42 - 48 (1981).

122.010 **RR Lyrae stars: a theoretical study of Bailey types** *a* **and** *b***.** R. Stothers.
Astrophys. J., Vol. 247, 941 - 948 (1981).

The full-amplitude behaviour of theoretical models of RR Lyr stars pulsating in the fundamental mode has been explored with the use of Carson's radiative opacities. A detailed comparison of the models with metal-poor RR Lyr stars of Bailey types *a* and *b* suggests that these stars have properties that lie somewhere in the following ranges: $M/M_\odot = 0.55 - 0.65$, $Y = 0.2 - 0.3$, and log $(L/L_\odot) = 1.6 - 1.7$. In obtaining these results, the only data that are used are velocity curves and light curves. Independent information, however, seems to corroborate the values found here. The use of Los Alamos opacities gives somewhat similar results, although the derived masses are smaller by about 20%.

122.011 **Search for linear polarization in solar neighborhood flare stars and spotted stars.**
B. R. Pettersen, J.-C. Hsu.
Astrophys. J., Vol. 247, 1013 - 1023 (1981).

Multifilter polarimetric measurements of 19 solar neighborhood flare stars and spotted stars outside of flares are presented. The stars cover a large range of luminosity among the red dwarfs. No linear polarization is detected in any of the broad-band filters used. A detailed study of simultaneous polarimetric and photometric measurements of two spotted flare stars (BY Dra and EV Lac) reveals no variability in polarization data when starspot groups move across the stars as they rotate.

122.012 **Study of IR-light and polarization variability of** χ **Cygni.** L. V. Larionova.
Tr. Astron. Obs., Leningrad, Tom 36 = Uch. Zap. Leningr. Univ., No. 402 = Ser. mat. nauk, Vyp. 58, 40 - 48 (1981). In Russian.

The results of photometric (R, I, H, K) and polarimetric (R, I, K) observations of χ Cyg made in 1969 - 1978 are presented. Variations of the polarization angle and short-scale variations of brightness are found.

122.013 Photoelectric observations and analysis of variability of the β Cephei-type star KP Persei.
T. Jarzębowski, M. Jerzykiewicz, M. Ríos Herrera, M. Ríos Berumen.
Rev. Mexicana Astron. Astrofis., Vol. 5, 61 - 68 (1981).

Photoelectric observations of the β Cephei-type variable, KP Persei, made in 1977 and 1978 are presented. The photometry was performed mainly in the B system and some nights also in the UB or UBV. A frequency analysis of the blue photometry was carried out. Three sine-wave components with periods $P_1 = 0^d201779$, $P_2 = 0^d198085$, and $P_3 = 0^d227099$ were found.

122.014 Magnitude damping in the Blazhko effect of XZ Cygni.   P. O. Taylor.
J. American Assoc. Variable Star Obs., Vol. 9, 57 - 58 (1980).

The amplitude of secondary cycle magnitude maxima appears to have decreased abruptly, coincident with the increase in primary period that occurred in 1979.

122.015 On the period of RR Canum Venaticorum.
E. Belserene, W. Larson.
J. American Assoc. Variable Star Obs., Vol. 9, 61 - 63 (1980).

Observations of the RR Lyrae star, RR Canum Venaticorum, from 1932 to 1980 are satisfied by a constant basic pulsation period on which are superimposed fluctuations due to changes in the shape and amplitude of the light curve.

122.016 The period of AA Comae Berenices.
T. P. Greene.
J. American Assoc. Variable Star Obs., Vol. 9, 73 (1980).

122.017 The periodicities of Z Ursae Majoris.
M. K. Suchko.
J. American Assoc. Variable Star Obs., Vol. 9, 74 - 80 (1980).

A Fourier analysis of the AAVSO visual light curve of Z Ursae Majoris shows that the cycle of variation is currently 195.5 days with an unstable superposed period of about 205 days. There is no sign of the reported 1560 day period.

122.018 Period revisions for AX Andromedae and WX Canis Minoris.   H. B. Hammel.
J. American Assoc. Variable Star Obs., Vol. 9, 81 - 84 (1980).

122.019 A revised period for the long-period variable, Y Capricorni.   E. O. Waagen.
J. American Assoc. Variable Star Obs., Vol. 9, 88 - 91 (1980).

122.020 Nonradial pulsations in the zero-age main-sequence star Upsilon Orionis (O9.5 V).   M. A. Smith.
Astrophys. J., Vol. 248, 214 - 221 (1981).

High-resolution Reticon observations of the zero-age main-sequence star (ZAMS) υ Orionis over 1976–1979 show periodic line profile variations interpreted in terms of small-amplitude nonradial pulsations. Except for one interval in which a period of 23.52 hr was twice observed, υ Ori usually exhibits a period of ~12.3 hr. Viewed in the context of other line profile variable B stars (53 Per–type), υ Ori shows an average period length and a comparatively high period stability. The most significant aspect of the study is that a ZAMS B star may pulsate nonradially. On theoretical grounds, it is argued that this fact makes the stellar core an unlikely site for the excitation of these oscillations.

122.021 Image-tube spectroscopic studies of rapid variables. III. Hydrogen bombs in SS Cygni.   M. F. Walker.
Astrophys. J., Vol. 248, 256 - 267, plates 8 - 10 (1981) = Lick Obs. Bull., No. 871.

Radial-velocity observations of SS Cyg in 1970 and 1971 are discussed. Time-resolved spectra in 1970 provide additional examples of the "ultraviolet flares" observed previously by Walker and Chincarini, and detailed analysis indicates that these flares are caused by the brief appearance of extremely broad H and sometimes Ca II and He I emission features with total widths of ~94 Å, often accompanied by a much smaller increase in continuum emission. Inspection of similar time-resolved spectra of AE Aqr, T CrB, MV Lyr, RW Tri, and UX UMa indicates that this type of broad-line hydrogen emission flare occurs in at least two other nova-like systems, RW Tri and AE Aqr.

122.022 The color of the fast oscillations of AH Herculis.
R. H. Hildebrand, E. J. Spillar, R. F. Stiening.
Astrophys. J., Vol. 248, 268 - 273 (1981).

The authors have measured the color of the fast coherent oscillations of AH Herculis by comparing simultaneous records of the photon arrival rate in two wavelength bands, 355±30 nm and 795±80 nm. They find that the color temperature of the coherent oscillation detected in these two wavelength bands is $28,000 \text{ K} < T < 73,000 \text{ K}$. The color temperature of the nonoscillating luminosity is much lower. The authors have also correlated the arrival times of photons in these two bands and find that the time interval between the long-wavelength coherent signal and the ultraviolet coherent signal is 0.26±0.5 seconds. These results suggest that it is the surface of the white dwarf or the hot inner portion of the accretion disk which is emitting the coherent signal.

122.023 The structural properties of Cepheid light curves.
N. R. Simon, A. S. Lee.
Astrophys. J., Vol. 248, 291 - 297 (1981).

Fourier decompositions are performed for the V magnitude variations of a sample of 57 Cepheids ranging in period from 2 to 17 days. The coefficients have been tabulated to allow rapid reconstruction of the light curves. It is shown that combinations of the low-order coefficients quantitatively describe the progression of curve shape with period. The pulsation amplitude plays only a minor role. Sharp breaks in the progression, occurring at around 10 days, are attributed to the resonance $P_2/P_0 = 0.5$. It is suggested that the Fourier decompositions provide a straightforward method for resolving a number of controversies concerning theoretical interpretations of the Hertzsprung sequence. The paper closes with a brief discussion of three short-period stars which seem to occupy a unique position in the sample.

122.024 Spectroscopic study of the infrared Ca II triplet in S-type Mira variable stars.
M. E. Contadakis, J. Solf.
Astron. Astrophys., Vol. 101, 241 - 246 (1981).

The Ca II infrared triplet in the spectra of 22 S-type Mira variable stars has been observed at various phases of the light cycle. The strength as well as the duration of the Ca II infrared emission are correlated with the period of the Mira stars. These results are explained referring to an interpretation, according to which the Ca II line emission originates within a hot layer deep in the stellar atmosphere which is moving upwards. The authors show that their observations are consistent with the assumption of the periodic shock wave producing such a layer in its relaxation region.

122.025 On the physical nature of Delta Delphini stars.
M. Saez, M. Auvergne, J.-C. Valtier, A. Baglin, P. Morel.
Astron. Astrophys., Vol. 101, 259 - 264 (1981).

In this paper it is proposed that the envelopes of δ Del stars are not completely deprived from helium. A partial mixing would produce intermediate He content, allow for the high metals abundances and restore a normal abundance of calcium. Models computed with recent physical data show that for $\log T < 3.9$ variability is obtained for $Y = 0.18$.

122.026 Synchronous extreme spectral variability of BE Ursae Majoris.
B. Margon, R. A. Downes, J. I. Katz.
Nature, Vol. 293, 200 - 202 (1981).

The authors report observations which demonstrate that the spectrum of BE UMa is dramatically variable on a time scale of hours and days, in synchrony with the 2.29-day periodic light variations. They interpret the system as a detached binary consisting of an extremely hot ($T \sim 10^5$ K) white dwarf and a cool M star, whose upper layers are irradiated by the intense ionizing flux of the hot companion.

122.027 Note on the nature of HD 179315.
J. D. Fernie, R. F. Garrison.
Publ. Astron. Soc. Pacific, Vol. 93, 330 - 332 (1981).

The authors have carried out MK classification and *UBVRI* photometry of HD 179315 to help decide the question of whether this star is a cepheid or an ellipsoidal binary. The spectral type near maximum proves to be G1 I*b* and not K2 as given in the HD catalog, so that the star is almost certainly a cepheid.

122.028 A survey of variable yellow supergiants in the southern Milky Way. A. Arellano Ferro.
Publ. Astron. Soc. Pacific, Vol. 93, 351 - 360 (1981).

Forty-three supergiants of spectral type F0 - G8, including the RV Tauri star U Mon and the small-amplitude cepheid HR 4768 have been monitored in brightness for about a month. Three new variables are announced. HR 6109 – a comparison star and a member of a spectroscopic binary with $P = 40$ days (Eggen 1973) – is suspected to be a δ Scuti star. HR 4912 shows a range of $0^m.32$ in *B*; the period is between 44 and 68 days. The cepheid nature of HR 4912 cannot be established on the basis of the author's data. HR 4110 – the central star of the galactic cluster IC 2581 – shows a range in *B* of $0^m.07$ and its most likely period is about 59 days. Arguments are offered against the possible cepheid interpretation of HR 4110. The variability of HR 2910, Hr 3026, and HD 67458 is suspected. The star R Pup which has been claimed and disclaimed as a variable for a century, did not show significant variation during the author's observing period.

122.029 A photometric study of the dwarf nova WZ Sagittae in outburst.
J. Patterson, J. T. McGraw, L. Coleman, J. L. Africano.
Astrophys. J., Vol. 248, 1067 - 1075 (1981).

The authors report the results of high-speed photometry of WZ Sagittae during its 1978 eruption. Periodic humps are seen in the light curve at all times. A variation at the orbital period appears to be present throughout the eruption; during the first 10 days, the phase of the periodic hump was displaced from that of the orbital humps at quiescence. If this variation arises from a mass-transfer bright spot, then at maximum light the spot must have migrated ~60° downstream from its position at quiescence. During the latter part of the eruption, a more prominent modulation ("superhump") appeared with a period 0.8% greater than the orbital period. The star is identified as an SU UMa type dwarf nova. The light curves are in fair agreement with predictions of the mass-transfer instability theory, but are not easily explained if instability originates from the accretion disk or the white dwarf.

122.030 The Oosterhoff period groups and the age of globular clusters. I. Photometry of cluster variables in M15. A. Sandage, B. Katem, M. Sandage.
Astrophys. J.,Suppl. Ser., Vol. 46, 41 - 74 (1981).

Three-color photometry of 60 RR Lyrae stars in M15 permits study of the distribution of colors, amplitudes, light curve shapes, and periods within the instability strip. Period-amplitude, period-rise time, period-color, and color-amplitude relations for this Oosterhoff group II cluster are compared

with similar data for M3 which is in group I. The mean age of M3 and M15 is $(16.4 \pm 3.3) \times 10^9$ yr.

122.031 Variable stars in the bulge of the galaxy in a field around NGC 6304. F. D. A. Hartwick, D. J. Barlow, J. E. Hesser.
Astron. J., Vol. 86, 1044 - 1056 (1981).

Forty-two new variables have been discovered in the field surrounding the globular cluster NGC 6304. Periods have been determined for 40 RR Lyrae stars in the field. The period-frequency histogram is similar to that for the RR Lyrae variables in Baade's "window" around NGC 6522 and is consistent with the interpretation that the stars are only mildly metal deficient. The space distribution of the RR Lyrae stars is apparently different from that of the system of globular clusters, thereby suggesting a different dynamical history for these stars.

122.032 Variations in the UV spectrum of α Centauri.
R. P. Fahey.
The Universe at ultraviolet wavelengths, (see 012.009), p. 177 - 184 (1981).

Equivalent width and central depth measurements of the ultraviolet absorption lines from the spectrum of the He I variable star α Cen [HD 125823] are presented. The measurements are from 53 short-wavelength, and 45 long-wavelength, high-resolution spectrograms taken with the International Ultraviolet Explorer (IUE) satellite over a 25 day period during July and August of 1978. Central depth measurements were of lines of Cr III, Mn III, V III, C III, Si II, and S II seen in the region between 1245 and 1255 Å. Equivalent widths are of the three He I 2^3S $-$ n^{3}P^0 transitions found at 2945, 2829 and 2764 Å. The period for cyclic intensity variations in He I is determined using a four-parameter, least-squares fit to a sinusoid.

122.033 The detection of companion stars to the Cepheid variables Eta Aquilae and T Monocerotis.
J. T. Mariska, G. A. Doschek, U. Feldman.
The Universe at ultraviolet wavelengths, (see 012.009), p. 209 - 215 (1981).

The authors have obtained ultraviolet spectra with IUE of the classical Cepheid variables η Aql and T Mon at several phases in their periods. For η Aql significant ultraviolet emission is detected at wavelengths less than 1600 Å, where little flux is expected from classical Cepheids. Furthermore, the emission at wavelengths less than about 1600 Å does not vary with phase. Comparison with model atmosphere flux distributions shows that the nonvariable emission is consistent with the flux expected from a main-sequence companion star with an effective temperature of about 9500 K (A0 V – A1 V). For T Mon a nonvarying component to the ultraviolet emission is observed for wavelengths less than about 2600 Å. Comparison with model atmosphere flux distributions suggests that the companion has an effective temperature of around 10,000 K (A0) and is near the main sequence.

122.034 Line strength variations in β Cephei.
D. Fischel, W. M. Sparks.
The Universe at ultraviolet wavelengths, (see 012.009), p. 217 - 224 (1981).

The line strength variations of the resonance line of C IV (1550 Å, 2s ^{2}S – ^{2}P) observed by OAO-II have been confirmed by IUE observations. In addition, the N V resonance line (1204 Å, 2s ^{2}S – ^{2}P), the Si III line (1206 Å, 3p ^{1}P – ^{1}D, multiplet 11) and the Si IV resonance line (1395 Å, 3s ^{2}S – ^{2}P) all vary in line strength essentially in phase with the C IV variation. The (preliminary) period of the variation is 6.02/12.04 days.

122.035 Notes on the early-type components of W Cep, o Cet, CH Cyg, AR Mon, and BL Tel.

R. F. Wing, K. G. Carpenter.
The Universe at ultraviolet wavelengths, (see 012.009), p. 341 - 347 (1981).

Low-resolution IUE spectra in both spectral regions have been used to clarify the nature of the warmer components of several binary systems. W Cep, the primary of which is a luminous K-type supergiant, shows an ultraviolet absorption spectrum of type B0 or B1; this system is heavily reddened. The hot companion of Mira (o Cet) is surprisingly faint in the short-wavelength region, but it excites a rich emission spectrum from the surrounding gas. The ultraviolet-active M7 giant CH Cyg is shown to be a binary with a hot companion. The eclipsing binaries AR Mon and BL Tel are shown not to have hot companions.

122.036   **On the nature of the nova-like variable CD−42°14462.**   E. F. Guinan, E. M. Sion.
The Universe at ultraviolet wavelengths, (see 012.009), p. 471 - 475 (1981).

Low-dispersion long and short wavelength IUE spectra of the nova-like system CD −42°14462 were obtained on August 24 U.T. 1979. The short wave spectrum exhibits absorption features due to C III (λ1175), Lα (λ1216), N V (λ1240), He II (λ1640), Si IV (λ1394), N IV (λ1785) with C IV (λ1550) as a P Cygni feature with blue-shifted absorption suggesting the presence of material leaving the system. Possible interpretations of this object are discussed.

122.037   **Orbital phase dependent IUE spectra of the nova-like binary TT Arietis.**
E. F. Guinan, E. M. Sion.
The Universe at ultraviolet wavelengths, (see 012.009), p. 477 - 484 (1981).

The authors have obtained nine low-dispersion IUE spectra of the nova-like binary TT Ari over its $3^h 17^m$ orbital period. Four short-wave spectra and five long-wave spectra exhibit marked changes in line strength and continuum shape with orbital phase. The short wave spectra show the presence, in absorption, of C III, Lyman α, Si III, N V, Si IV, C IV, He II, Al III, and N IV. C IV shows a P Cygni profile on two of the spectra. Implications of these spectra for the nature of nova-like variables are discussed.

122.038   **Observations of cataclysmic variables with IUE.**
L. Hartmann, J. Raymond.
The Universe at ultraviolet wavlengths, (see 012.009), p. 495 - 500 (1981).

The authors report observations of the cataclysmic variables AN UMa, 2AO311-227, VV Pup, DQ Her, and GK Per made with IUE. The authors have been able to detect continuum emission in the short-wavelength (λ1180-1950) region in DQ Her. This object exhibits a quasi-blackbody ($\sim \nu^2$) spectrum at short wavelengths; such blackbody components are a common property of the variables AM Her, SS Cyg, and U Gem, suggesting an underlying similarity in the activity of these diverse systems.

122.039   **The corona and chromosphere of Proxima Centauri during flare and quiescent times.**
B. M. Haisch, P. L. Bornmann, J. L. Linsky, R. E. Stencel, O. B. Slee.
Bull. American Astron. Soc., Vol. 13, 515 (1981). − Abstract.

122.040   **The cepheid period-luminosity relation: a new application of infrared photometry.**
B. F. Madore, R. McGonegal, R. A. McLaren, C. McAlary.
Bull. American Astron. Soc., Vol. 13, 522 (1981). − Abstract.

122.041   **Properties of type II cepheid variables.**
H. C. Harris.
Bull. American Astron. Soc., Vol. 13, 522 (1981). − Abstract.

122.042   **Analysis of improved cepheid light curves.**
C. G. Davis, N. R. Simon.
Bull. American Astron. Soc., Vol. 13, 522 (1981). − Abstract.

122.043   **Theoretical velocity and light curves for classical bump cepheids.**   S. W. Hodson, A. N. Cox.
Bull. American Astron. Soc., Vol. 13, 523 (1981). − Abstract.

122.044   **The chromospheres of classical cepheids.**
E. G. Schmidt, S. B. Parsons.
Bull. American Astron. Soc., Vol. 13, 523 (1981). − Abstract.

122.045   **Ultraviolet observations of R CrB variables.**
A. V. Holm, C.-C. Wu, L. R. Doherty.
Bull. American Astron. Soc., Vol. 13, 523 (1981). − Abstract.

122.046   **Differential photometry of HDE 310376.**
B. B. Bookmyer, J. R. Kern.
Bull. American Astron. Soc., Vol. 13, 524 (1981). − Abstract.

122.047   **Two dwarf anti-novae.**   E. P. Belserene.
Bull. American Astron. Soc., Vol. 13, 524 (1981). − Abstract.

122.048   **On the amplitudes of Delta Scuti stars.**
E. Antonello, M. Fracassini, L. Pastori.
Astrophys. Space Sci., Vol. 78, 435 - 441 (1981).

There is a clear relation between the amplitude of the visual light curve, the period and the luminosity of the low amplitude δ Scuti stars.

122.049   **The pulsating helium star BD + 13°3224.**
P. W. Hill, D. Kilkenny, D. Schönberner, H. J. Walker.
Mon. Not. R. Astron. Soc., Vol. 197, 81 - 94 (1981).

The period of the small-amplitude photometric variable extreme hydrogen-deficient star BD + 13°3224 is improved to 0.1079950 ± 0.0000002 day. Measurements of radial velocity show it to vary with the same period as the light and provide for the first time clear evidence for pulsation of a hot hydrogen-deficient star. With gravity determined from helium line profiles, the derived mass is 0.9 $M_\odot$ The observed variation is compared with theoretical models of pulsation in helium stars.

122.050   **Dwarf novae.**
J. A. J. Whelan.
Second European IUE Conference, (see 012.011), p. XXXIX - XL (1980).

An outline is given of a long-term programme to study dwarf novae spectra at ultraviolet and optical wavelengths. The programme has recently started and is now producing its first results.

122.051   **Pulsation of the outer atmosphere of BW Vul.**
M. Burger, C. de Jager.
Second European IUE Conference, (see 012.011), p. 47 - 49 (1980).

Observations of the far UV spectrum of the β Cephei variable BW Vul show that the pulsation consists of three phases: (a) a phase of photospheric rest; (b) a sudden upward acceleration of the photosphere and outer atmosphere; the maximum acceleration being approximately equal to the gravitational value; (c) descend of the whole atmosphere with downward acceleration $\sim$ 1/4 of the gravitational value. At the end of phase (b) a temporal stellar wind occurs with maximum velocity of $\sim$ 500 km s^{-1}.

122.052   **The UV spectrum of WZ Sge at nova maximum.**
M. Friedjung, B. Rocca-Volmerange, G. Debève.
Second European IUE Conference, (see 012.011), p. 85 - 86 (1980).

Profiles of relatively unblended lines on a short wavelength region high dispersion spectrum, have been measured. Both line absorption and line emission are seen. The former is stellar and not interstellar. The profiles have been classified, and their regions of formation are considered. The C IV resonance blend can be understood as consisting of emission components having a similar profile to that of the He II 1640 A, and black absorption lines. The form of the He II 1640 A profile suggests moreover the existence of an outer excretion disk, in addition to the well known accretion disk.

122.053 Ultraviolet observations of Mira Ceti.
A. Cassatella, J. Clavel, D. Gilra, D. Reimers, D. J. Stickland.
Second European IUE Conference, (see 012.011), p. 233 - 236 (1980).

IUE observations of Mira Ceti show a wealth of emission lines, a continuum gradually increasing to long wavelengths and a number of absorption features due to Fe II, Mg II and perhaps Fe I and CO. These spectral components exhibit some variability and can be explained with a model of accretion into a disk around Mira B (a white dwarf) of material from Mira A (M6 III e).

122.054 Changes in IUE postmaximum spectra of the Mira variable Chi Cygni. A. Cassatella, A. Heck, F. Querci, M. Querci, D. J. Stickland.
Second European IUE Conference, (see 012.011), p. 243 - 245 (1980).

IUE long-wavelength (2500 - 3350 Å) spectra of the long-period variable χ Cygni have been obtained at phases ϕ = 0.04, 0.18 and 0.22 following the 1979 light maximum. Low spectral resolution was used except for the last phase studied where a high resolution spectrum has been secured. The observations are interpreted on the basis of a shock-wave model.

122.055 Frequency analysis of photometric observations of the β Cephei star ν Eridani.
J. Cuypers, M. Goossens.
Astron. Astrophys., Suppl. Ser., Vol. 45, 487 - 497 (1981).

A phase dispersion minimization method is used to perform a frequency analysis of Van Hoof's photometric observations of the β Cephei star, ν Eridani. The frequency analysis reveals the existence of four oscillations in the light variability of ν Eridani.

122.056 Les variables du type R CrB. M. Verdenet.
Bull. AFOEV, No. 17, p. 106 - 115 (1981).

122.057 The Hubble-Sandage variable HDE 269006: a hot supergiant with a cool envelope.
I. Appenzeller, O. Stahl, B. Wolf.
Messenger, No. 25, p. 2 - 5 (1981).

122.058 Photometric classification of pulsating variables with periods between one and three days.
R. Diethelm.
Messenger, No. 25, p. 29 - 31 (1981).

122.059 A determination of the characteristics of Cepheids from B-type companions.
D. H. McNamara, K. A. Feltz, Jr.
Effects of mass loss on stellar evolution, (see 012.015), p. 389 - 395 (1981).

Photometry ($uvby\beta$) of five Cepheids with B-type companions has been utilized to derive the spectral types and apparent magnitudes of the companions. By utilizing the period-luminosity relation it is possible to estimate the absolute magnitudes of the companions. The spectral types and absolute magnitudes of the B-type components indicate they are all evolved from the zero-age main sequence. By utilizing the ages of the B stars the authors can derive the ages of the Cepheids. A period-age relation of the form $\log t_{age} = 8.60 - 0.83 \log P$ is found. Minimum masses of the Cepheids can also be estimated from the data. No evidence of large mass loss is evident in the intermediate-band photometry of Cepheids.

122.060 What "masses" for Cepheids?
C. G. Davis.
Effects of mass loss on stellar evolution, (see 012.015), p. 397 - 400 (1981).

To understand the evolution of giant stars, it is important to pin down the masses for Cepheids. The 7- to 10-day "bump" Cepheids imply lower than evolutionary mass (60%). Recent theoretical work, though, indicates that for Cepheids with periods of 15 to 16 days, the best understanding of the light curves results from using evolutionary masses.

122.061 Observation of mass loss in R CrB during the visual light minimum. N. K. Rao.
Effects of mass loss on stellar evolution, (see 012.015), p. 469 - 471 (1981).

122.062 A search for violet-displaced cores in the optical spectrum of Mira. G. Wallerstein.
Publ. Astron. Soc. Pacific, Vol. 93, 453 - 455 (1981).

A search for circumstellar lines in the spectrum of Mira from $\lambda 3900$ to $\lambda 8700$ reveals only the sodium D lines and possibly the Ca II K line to be present. As suggested by Yamashita and Maehara (1978) the commonly seen circumstellar lines in red giants arising from various neutral species as well as Ca II and Sr II are weakened by ionization by the hot companion of Mira.

122.063 Stepanian's star: the energy distribution reveals a nontypical cataclysmic variable. P. Szkody.
Publ. Astron. Soc. Pacific, Vol. 93, 456 - 459 (1981).

Einstein, IUE, optical multichannel spectrophotometry, and IR observations of Stepanian's star are discussed in terms of other known cataclysmics. While the X-ray flux and IUE emission-line data are similar to that of dwarf novae, the total continuum flux distribution from uv-IR is cooler (peaking near a 10,000 K blackbody) and is unlike either a stellar component or a classic steady-state disk. The IR data show no evidence for a late-type component.

122.064 A new radial-velocity curve for the cepheid. η Aquilae. T. S. Jacobsen, G. Wallerstein.
Publ. Astron. Soc. Pacific, Vol. 93, 481 - 485 (1981).

A new radial-velocity curve based upon 21 10 Å mm^{-1} red spectrograms of the bright cepheid η Aql is presented. The shape of the velocity curve is unchanged since 1923 but a small increase (0.4 km s^{-1}) in amplitude may have occurred since 1898. The systematic velocity appears to have decreased by 0.74 km s^{-1} during the same interval which, if real, may be associated with orbital motion around the companion recently discovered by Mariska et al. (1980). The increase in period (since about 1825) that had previously been noted by various authors is confirmed and extended. The Hα absorption velocity curve shows the same bump seen in metallic lines except about 0.16 cycle later.

122.065 Photometry of HD 45088: a new bright BY Draconis variable. B. W. Bopp, D. S. Hall, G. W. Henry, P. Noah, A. Klimke.
Publ. Astron. Soc. Pacific, Vol. 93, 504 - 506 (1981).

The K-dwarf SB2 system HD 45088, predicted to be a BY Dra variable on the basis of its orbital period and emission-line characteristics, is found to be photometrically variable. The amplitude is significantly variable, having ranged between $\Delta V = 0^m.02$ during February–April 1980 to $\Delta V = 0^m.045$ during September–October 1980. The photometric period, shown to be 7.36 days, differs appreciably from the 6.99-day orbital period. Possible explanations for this difference are (1) nonsynchronization in a young binary, (2) spot development at high latitudes on a differentially rotating star, and (3) imperfect synchronization in a binary with an eccentric orbit.

122.066 Cepheid instability strip investigation with U−B intrinsic colours.
N. S. Nikolov, G. R. Ivanov.
Astrofiz. issled., NRB, Vol. 3, 3 - 17 (1981). − Abstr. in Ref. zh., 51. Astron., 8.51.551 (1981).

122.067 Parallel two-colour observations of flare stars in the region of Cygnus.
M. Tsvetkov, L. K. Erastova.
Astrofiz. issled., NRB, Vol. 3, 76 - 80 (1981). In Russian. Abstr. in Ref. zh., 51. Astron., 8.51.614 (1981).

122.068 Investigation of the possibility of existence of short periods for a few long-period Ap stars.
K. Panov.
Astron. issled., NRB, Vol. 3, 48 - 56 (1981). In Russian. Abstr. in Ref. zh., 51. Astron., 8.51.633 (1981).

122.069 Photometric abundances and colors of Cepheids in the Small Magellanic Cloud.   H. C. Harris.
Astron. J., Vol. 86, 1192 - 1199 (1981).

Washington-system colors of 45 SMC Cepheids are compared to colors of Galactic Cepheids to investigate the metal abundances of the Cepheids. A mean value relative to the Sun of $[A/H] = -0.54$ is found, with weak evidence that short-period Cepheids are more metal poor, but with no correlation with projected position in the SMC. The consistency of their colors, temperatures, abundances, and reddenings relative to Galactic Cepheids is discussed. The small reddenings found in the SMC imply a low dust-to-gas ratio and a high value of R.

122.070 The variable stars in the globular cluster NGC 6101.
M. H. Liller.
Astron. J., Vol. 86, 1204 - 1209 (1981).

Fifteen variable stars have been discovered in the region of the globular cluster NGC 6101. Eleven of these appear to be RR Lyrae stars with magnitudes appropriate for cluster membership. A distance of 15 ± 3 kpc is derived for NGC 6101.

122.071 Near-infrared observations of R Coronae Borealis.
N. M. Ashok, P. V. Kulkarni, N. Kameswara Rao.
Bull. Astron. Soc. India, Vol. 9, 67 - 68 (1981). − Abstract.

122.072 A new cepheid variable, HD 200925.
T. D. Padalia, S. K. Gupta.
Bull. Astron. Soc. India, Vol. 9, 82 - 83 (1981). − Abstract.

122.073 HV 1369, a cepheid at a possible depth of 32 kpc in the Small Magellanic Cloud.
A. M. van Genderen.
Astron. Astrophys., Vol. 101, 289 - 291 (1981).

HV 1369 ($P = 31^d$) in the SW part of the SMC appeared to be far too underluminous for its period in a photometric (VBLUW system) study (van Genderen, 1977). New VBLUW photometry seems to support its long period. A comparison with the brightnesses of ten other cepheids distributed all over the SMC with similar periods, shows that HV 1369 is possibly 32 ± 7 kpc more distant than these ones.

122.074 On the possibility of detecting companions to cepheids and their effect on the CORS method.
G. Russo, C. Sollazzo, M. Coppola.
Astron. Astrophys., Vol. 102, 20 - 24 (1981).

The idea of Fernie (1980) to detect the presence of companions to cepheids from the phase-shift between the light and colour curves has been applied to the extensive set of observations in the VBLUW photometric system. Approximately 25% of the cepheids considered have possible companions. The study of the effect of (unresolved) companions on the cepheid mean radius computed by means of the CORS method of Caccin et al. (1980, 1981) has shown the stability of this method.

122.075 UBV observations and period determination for the Delta Scuti star HD 177392.
W. Schöneich, E. Żelwanowa.
Astron. Nachr., Band 302, 181 - 186 (1981). In German.

UBV observations of the Delta Scuti type star HD 177392 were carried out in 7 nights. The three periods 0^d1215, 0^d11094, and $0^d109414$ were found. These periods can not be explained by radial pulsation.

122.076 V 1329 Cyg: a nova-like object with recurrent outbursts?   T. Iijima, A. Mammano.
Astrophys. Space Sci., Vol. 79, 55 - 66 (1981).

Spectrophotometric observations in optical region ($\lambda\lambda$ 3800 ∼ 8000 Å) on the symbiotic object V 1329 Cyg (HBV 475) were carried out with the 122 cm telescope of the Astrophysical Observatory of Asiago in 1979 and 1980. Significant changes of intensity of emission lines with a time-scale of a few weeks were observed. The results of these observations suggest that probably V 1329 Cyg is a nova-like object with recurrent outbursts at intervals of about 950 days. The properties of this object seem to agree with those of a white dwarf which is accreting mass rapidly.

122.077 Radii and luminosities of classical cepheids.
G. R. Ivanov.
Astrophys. Space Sci., Vol. 79, 107 - 117 (1981).

The radii of 17 classical cepheids are determined. The linear surface brightness-colour relation $S_v = b(B-V)_0 +$ const is accepted. The present method permits the determination of the absolute magnitude of cepheids. The period-luminosity relation is approximately the same as the one obtained by van den Bergh (1976) for cepheids in open clusters.

122.078 BV photometry of the RR$_c$ variable star DH Pegasi.
U. Hopp.
Astrophys. Space Sci., Vol. 79, 239 - 242 (1981).

Revised light elements of the RR Lyrae variable star DH Pegasi are presented together with a new BV light curve.

122.079 Discovery and observation of BY Draconis variables.
B. W. Bopp, P. V. Noah, A. Klimke, J. Africano.
Astrophys. J., Vol. 249, 210 - 217 (1981).

The authors describe a spectroscopic technique for finding BY Dra variables that requires only a single observation with a telescope of moderate aperture. The success of the technique is demonstrated by the authors' discovery of four new BY Dra variables with $V \sim 8$ (as bright as the prototype variable or somewhat brighter). As a further result of this spectroscopic survey, they establish the existence of an important difference in the character of stellar surface activity that is revealed by the emission lines of Ca II and hydrogen.

122.080 $l = 0, 1, 2$, and 3 pulsation constants for evolutionary models of δ Scuti stars.   W. S. Fitch.
Astrophys. J., Vol. 249, 218 - 227 (1981).

Evolutionary sequences at $M/M_\odot = 1.5$, 2.0, and 2.5 were calculated with a version of the Eggleton stellar evolution code for an initial composition $X = 0.70$, $Z = 0.02$. From these sequences 14 models spanning the lower part of the Cepheid instability strip were selected for pulsation calculations. All the linear adiabatic m-degenerate eigenvalues with $l \leqslant 3$ for pulsation constants $Q [= P(\langle\rho\rangle/\langle\rho_\odot\rangle)^{1/2}]$ in the range $0^d0096 \leqslant Q \leqslant 0^d067$ were obtained with a modified version of Dziembowski's second pulsation code, and the author's code uses a variant of the Cowling approximation for the nonradial modes.

122.081 The zero point of the period-luminosity relation and kinematics of cepheids.
D. K. Karimova, E. D. Pavlovskaya.
Astron. Zh., Tom 58, 984 - 995 (1981). In Russian. English translation in Soviet Astron., Vol. 25, No. 5.

75 classical cepheids with known radial velocities, UBV photoelectric magnitudes and proper motions were used for a kinematical investigation. The correction to the zero point of the period – luminosity relation by the statistical parallax method was found to be $-0^m.1$. The study of the cepheid velocity field based on their space velocities, shows that the motions of 11 cepheids in the region $90° < l < 180°$ and $r > 1.5$ kpc are in disagreement with the adopted kinematical models and therefore the angular velocity of the galactic rotation was found to be $\omega_0 \approx 13$ km/s kpc. After excluding these stars, ω_0 increases to 19.3 km/s kpc. The elements of the cepheid box orbits are calculated.

122.082 The unusual Cepheid HR 7308 with a variable amplitude. M. Breger.
Astrophys. J., Vol. 249, 666 - 672 (1981).

HR 7308 is a classical Cepheid with the shortest known period. Photoelectric observations taken during 1966–1969 are presented and combined with recent work. The primary period of oscillation is found to be $1^d.49077 \pm 0^d.00001$. The amplitude of HR 7308 varies by over a factor of 5 with a period of about 1210^d. Except for the variable amplitude, no secondary periods can be found in the data. The author has applied various models utilizing two close periods as well as models of variable amplitudes superposed on a constant oscillation.

122.083 The high resolution ultraviolet spectrum of the CP star HR 6000.
F. Castelli, M. Cornachin, M. Hack.
Upper main sequence chemically peculiar stars, (see 012.033), p. 149 - 152 (1981).

The results of a preliminary quantitative analysis of the CP star HR 6000 are given. The method of the synthetic spectrum has been used. The main peculiarity is the defect of silicon by a factor of about 1000.

122.084 The variability of Bp and Ap stars.
W. Schöneich.
Upper main sequence chemically peculiar stars, (see 012.033), p. 235 - 255 (1981).

122.085 Spectrum variation and chemical compositions of the Cr type Ap star ε Ursae Majoris.
K. Sadakane.
Upper main sequence chemically peculiar stars, (see 012.033), p. 257 - 262 (1981).

Periodic spectrum variations and chemical compositions in ε UMa are examined. Intensity of the Ca II K line increases sharply from minimum to maximum between phase 0.0 and 0.3 and then decreases gradually to the next minimum. Lines of Mn, Cr, and Sr become strongest at phase 0.0, while those of Ti and Fe remain nearly constant. ε UMa shows a unique abundance pattern among cool Ap stars. Cr, Mn and Sr are overabundant, Mg and Fe are normal, while Al, Si, Ca, Sc, V, Ti, Ni, and Ba are underabundant compared to the normal A0 IV star γ Gem.

122.086 The distribution of iron and chromium over the surface of ε UMa.
J. Rice, W. Wehlau, V. L. Khokhlova, N. E. (N. Eh.) Piskunov.
Upper main sequence chemically peculiar stars, (see 012.033), p. 265 - 270 (1981).

The distribution of Fe and Cr on the surface of ε UMa has

been determined using high precision profiles of Fe I, Fe II, Cr I and Cr II lines obtained with a Reticon. The mathematical methods previously described by Goncharski et al (1977, 1981) have been used to derive maps of local equivalent width from the observational data. Both elements are concentrated in two large regions situated near the rotational equator. There is an indication that the spots for the neutral states are displaced in longitude from the spots for the singly ionized states.

122.087 Short timescale spectrum variations in Ap stars – a progress report on HD 108945 (21 Com).
W. W. Weiss, R. G. Tull, H. Peterlik.
Upper main sequence chemically peculiar stars, (see 012.033), p. 295 - 296 (1981).

122.088 Rapidly oscillating Ap stars. D. W. Kurtz.
Upper main sequence chemically peculiar stars, (see 012.033), p. 297 (1981). – Summary.

122.089 The influence of a magnetic field on the oscillations of early-type stars.
D. Biront, A. Cousens, M. Goossens, L. Mestel.
Upper main sequence chemically peculiar stars, (see 012.033), p. 337 - 340 (1981).

Some early-type pulsationally unstable stars almost certainly have an intrinsic magnetic field (Weiss et al., 1981) and the authors attempt to discover what effects the coupling of the magnetic field and the pulsation have. Subsequent work in this field could explain the long period variability of the light curves of some stars (Blazhko effect) as a modulation caused by the coupling between the pulsation under the influence of the magnetic field and the oblique rotation of the star.

122.090 IUE observations of three helium rich stars.
S. N. Shore, S. J. Adelman.
Upper main sequence chemically peculiar stars, (see 012.033), p. 429 - 434 (1981).

The authors report UV observations of the helium rich spectrum variables HD 37017, σ Ori E, and HD 37776. In all three, He I variations are in antiphase with the ions of C, Si, and Al; all vary periodically on the rotational timescale. Evidence is presented for weak stellar winds which are controlled by the magnetic field. The line variations and wind structure support a unified model for the class as extensions of the magnetic peculiar stars of the upper main sequence.

122.091 The helium-rich variable star HD 37776.
D. Groote, J. P. Kaufmann.
Upper main sequence chemically peculiar stars, (see 012.033), p. 435 - 438 (1981).

122.092 The helium-rich star W66 in NGC 6530.
A. P. Odell.
Upper main sequence chemically peculiar stars, (see 012.033), p. 439 - 443 (1981).

Spectra were taken on two nights with the CTIO 4m Echelle Spectrograph of the reportedly helium-rich star Walker 66 in the young cluster NGC 6530. Comparison of spectral line equivalent widths with published helium rich models yields $N_{He} = 0.58$ on one night and 0.18 on the other, typical of a helium-rich star. The cluster diagram indicates an age for the cluster of $< 10^6$ years, which puts a limit on the timescale for the development of the helium rich phenomenon. A strong feature in the spectra at 4725 Å remains to be identified.

122.093 Study of the variability of the Delta Scuti stars. II. Photometric observations of HR 5492.
M. Bossi, G. Guerrero, L. Mantegazza, M. Scardia.
Astrophys. Space Sci., Vol. 79, 463 - 468 (1981).

Photoelectric observations of the suspected Delta Scuti star HR 5492 are given. The present data do not show any light variation greater than 0.01 mag. In spite of the very small light variations, the data analysis shows the presence of two periodicities, which are statistically significant. While the longest period (1^d296) cannot be justified with the present data, the shorter one (0^d044) is in excellent agreement with the hypothesis of an F4V star that is radially pulsating in the fundamental mode.

122.094   The luminosities of red supergiant variables in the Small Magellanic Cloud.   R. M. Catchpole, M. W. Feast.
Mon. Not. R. Astron. Soc., Vol. 197, 385 - 391 (1981).

Infrared photometry for 22 red supergiant variables in the SMC is used to derive bolometric magnitudes. Except for two anomalous objects these fit the period—luminosity relation for massive, core helium burning stars which was previously shown to represent well the LMC red supergiant variables. The $J-K$ colours of the LMC and SMC variables are similar.

122.095   Outburst spectra of UZ Serpentis.
J. Echevarria, D. H. P. Jones, R. E. Wallis, S. K. Mayo, B. J. M. Hassall, J. E. Pringle, J. A. J. Whelan.
Mon. Not. R. Astron. Soc., Vol. 197, 565 - 570 (1981).

The authors present observations of UZ Ser in outburst and approaching quiescence made in 1980 August. They discuss spectral fits to simultaneous ultraviolet and optical data in the wavelength range $\lambda\lambda$ 1250—8000 Å. An upper limit of the reddening $E_{B-V} = 0.1$ mag is found. The ultraviolet spectrum at outburst shows broad, shallow absorption lines.

122.096   IUE spectroscopy of cataclysmic variables.
J. Krautter, G. Klare, B. Wolf, H. W. Duerbeck, J. Rahe, N. Vogt, W. Wargau.
Astron. Astrophys., Vol. 102, 337 - 346 (1981).

Spectra of the classical novae V 603 Aql, HR Del, and RR Pic, the recurrent nova T CrB, the dwarf nova RU Peg, and the peculiar object TT Ari, obtained with the IUE in the range 1150 - 3200 Å, are presented. Based on the continuum fluxes, the authors determined interstellar extinction values and temperatures in the range of 25,000 - 45,000 K. The spectra of all variables (except for TT Ari) are characterized by strong emission features of highly ionized and/or highly excited ions of N, C, Si, Al, and He. The old novae exhibit pronounced P Cygni characteristics. From the P Cygni profiles of the unresolved C IV doublet mass loss rates were derived for HR Del ($\dot{M} = 2.6 \times 10^{-11} M_\odot \mathrm{yr}^{-1}$) and for TT Ari ($\dot{M} = 4.8 \times 10^{-11} M_\odot \mathrm{yr}^{-1}$). The UV observations of the line spectrum and continuum fluxes agree well with the properties of theoretical accretion disk models of cataclysmic binaries.

122.097   RR Lyrae populations in the galactic field.
V. Castellani, C. Maceroni, M. Tosi.
Astron. Astrophys., Vol. 102, 411 - 414 (1981).

The pulsational properties of field RR Lyrae have been studied analyzing the data relevant to about 2000 stars. The characteristics of three different populations membering the galactic bulge, the halo and the disk are reported and similarity between the first two and globular cluster pulsators is stressed. A particular discussion is devoted to short period pulsators in the solar neighbourhood. The authors show that the observational data of these stars can be interpreted in terms of a metal abundance of $Z \cong 10^{-2}$ and a He-content similar to that of galactic globular clusters.

122.098   On the intrinsic polarization of red dwarfs.
G. C. Clayton, P. G. Martin.
Astron. J., Vol. 86, 1518 - 1519 (1981).

High-precision polarimetric measurements of five UV Ceti flare stars, including the spotted flare star, BY Dra, are reported. No linear or circular polarization was detected in any of the stars.

122.099   The visual surface brightness relation and the absolute magnitudes of RR Lyrae stars. I. Theory.
A. Manduca, R. A. Bell.
Astrophys. J., Vol. 250, 306 - 311 (1981).

A theoretical counterpart to the Barnes-Evans relation between stellar surface brightness and $V-R$ color has been calculated from model atmospheres for parameters appropriate to RR Lyrae stars. The theoretical relation is in reasonable agreement with the empirical relation of Barnes, Evans, and Moffett but reveals small but significant sensitivities to gravity and metal abundance. This relation will be used in subsequent papers to derive radii, distances, and absolute magnitudes for RR Lyrae stars by the method of Barnes et al. The method is discussed in detail and shown to compare very favorably to the variations of the Baade-Wesselink technique currently in use.

122.100   The visual surface brightness relation and the absolute magnitudes of RR Lyrae stars. II. RR Lyrae and X Arietis.   A. Manduca, R. A. Bell, T. G. Barnes III, T. J. Moffett, D. S. Evans.
Astrophys. J., Vol. 250, 312 - 323 (1981).

The theoretical surface brightness relation for RR Lyrae stars derived in Paper I has been combined with new VR photometry to determine distances, radii, and absolute magnitudes for X Arietis and RR Lyrae. For X Arietis, the authors obtain a pulsational variation in the radius from $5.06\,R_\odot$ to $5.96\,R_\odot$, a distance of 495 ± 50 pc, and an intensity mean absolute magnitude $\langle M_v \rangle = +0.59 \pm 0.25$. The corresponding quantities for RR Lyrae are $4.81-5.67\,R_\odot$, 263 ± 39 pc, and $\langle M_v \rangle = +0.61 \pm 0.35$. The larger errors quoted for RR Lyrae are due to uncertainties arising from the Blazhko effect. The authors' results are discussed in the context of the dependence of RR Lyrae absolute magnitudes on metal abundance.

122.101   Abundance peculiarities in the atmosphere of SU Tauri.   M. Ya. Orlov, M. H. Rodriguez.
Pis'ma Astron. Zh., Tom 7, 689 - 691 (1981). In Russian. English translation in Soviet Astron. Lett., Vol. 7.

Quantitative analysis of the spectrum of SU Tau has been carried out by the differential curve-of-growth method. The region $\lambda\lambda$ 3900 - 5000 Å has been studied. The spectral class, microturbulent velocity, and electron pressure are estimated.

122.102   Inverse Compton effect and the colorimetric characteristics of flare stars.   G. A. Gurzadyan.
Astrophys. Space Sci., Vol. 80, 211 - 230 (1981).

It is shown that the observed color diagrams $(U-B)_f \sim (B-V)_f$ for pure flare emission of UV Cet type flare stars may be explained within the framework of a fast electron hypothesis. Under the real conditions of the generation of flares around the star the frequency transformation law at the photon-electron interaction reads $\nu = n\gamma^2\nu_0$, where n may take values from 0.15 up to 4. A possible role of reflection effect at the generation of stellar flares is outlined.

122.103   IUE and ground-based spectroscopic observations of the S Dor-type LMC variable R 71 during minimum state.   B. Wolf, I. Appenzeller, O. Stahl.
Astron. Astrophys., Vol. 103, 94 - 102 (1981).

Using the IUE satellite the authors obtained high resolution UV spectrograms ($1200 < \lambda < 3200$ Å) of the S Dor-type variable R 71 in the LMC. The IUE observations were supplemented by coordinated ground-based high dispersion spectroscopy and by photometric observations. From these observations the authors derive for the minimum state of R 71 the following stellar parameters: $L = 2.0 \times 10^5 L_\odot$, $R \approx 81\,R_\odot$, $T_{eff} \approx 13600$ K. For the expanding envelope they find a surprisingly low temperature of only about 6000 K and an apparently decelerated velocity field with a maximum outflow velocity of ~ 127 km s^{-1}. The minimum state mass loss rate is of the order of $3 \times 10^{-7} M_\odot$ yr^{-1}. Their results support the suggestion that the visual light variations of the S Dor-type variables are produced by strong density variations of the expanding envelopes of these objects.

122.104 G 255-2: a new ZZ Ceti variable star.
G. Vauclair, N. Dolez, M. Chevreton.
Astron. Astrophys., Vol. 103, L17 - L19 (1981).

The authors report the discovery of a new ZZ Ceti variable selected from its G-R color. G255-2 has been found to vary with two main periods: 830s and 685s of similar amplitude of about 4% of star luminosity. Two more periods at 450s and 380s of smaller amplitude may be present. Both periods and amplitudes seem to have remained stable during the six months interval between the two observing runs.

122.105 BD + 16°2356, an RRc Lyr variable. T. Oja.
Astron. Astrophys., Vol. 103, 339 - 341 (1981).

UBV observations show the star BD + 16°2356 to be an RR Lyr variable of Bayley type c with a period of 0.449 d and a mean visual magnitude of $\langle V \rangle = 10.5$. The star is one of the brightest RRc variables in the sky and definitely the brightest representative of the RRc variables with periods longer than 0.4 d. $\Delta S = 6$ indicates a metal abundance of [Fe/H] = -1.2 relative to the Sun.

122.106 On the excitation mechanism in β Cephei variables.
W. Dziembowski, M. Kubiak.
Acta Astron., Vol. 31, 153 - 161 (1981).

The driving effect of He$^+$ ionization edge suggested by Stellingwerf is investigated for both radial and nonradial modes. For the nonradial modes the driving effect is almost identical as for radial modes of the same frequency. The effect as calculated on the basis of the present opacity data is insufficient to cause instability, but the assumption that it is actually responsible for excitation offers a natural explanation of all major properties of β Cephei stars.

122.107 Photoelectric observations and analysis of AM CVn.
G. Kovács.
Acta Astron., Vol. 31, 207 - 212 (1981).

New photoelectric data of the ultrashort-period variable AM CVn obtained in 1980 were analysed in order to determine minima of the ~ 17.5 min light variation and to look for short-periodic oscillations. Using the new minima the author derived a mean period of 1051.19 ± 0.01 s. He has not found clear evidence for the presence of long-lived coherent oscillations above 0.001 mag in the range 0.002 - 0.04 Hz.

122.108 FG Sagittae: combination of the existing photometric and spectroscopic data.
J. Jurcsik, L. Szabados.
Acta Astron., Vol. 31, 213 - 219 (1981).

The radial velocity and the effective temperature curves during a pulsation period were constructed utilizing the available spectroscopic data. Both the spectroscopic and the photometric behaviour of FG Sge indicate that the mixing episode in 1970 took place within a few months.

122.109 The astrophysical significance of variable stars.
B. Szeidl.
Fiz. Sz., Vol. 31, 121 - 127 (1981). In Hungarian. – Abstr. in Phys. Abstr., Vol. 84, Abstr. 108233 (1981).

122.110 A search for light-time effects in binary cepheids: AW Persei. N. R. Evans.
J. R. Astron. Soc. Canada, Vol. 75, 247 (1981). – Abstract.

122.111 The cepheid period-luminosity relation – a new application of infrared photometry.
R. McGonegal, R. A. McLaren, B. F. Madore, C. W. McAlary.
J. R. Astron. Soc. Canada, Vol. 75, 250 (1981). – Abstract.

122.112 Ephemerides of RR Lyrae-type variables for the year 1982. W. P. Zessewitsch (*V. P. Tsesevich*), B. N. Firmaniuk (*Firmanyuk*), J. M. Kreiner.
Rocznik Astronomiczny Obserwatorium Krakowskiego 1982, (see 047.025), p. 113 - 130 (1981).

122.113 Rapid and extreme spectral variability of the peculiar emission-line object BE Ursae Majoris.
B. Margon, R. Downes.
Publ. Astron. Soc. Pacific, Vol. 93, 548 (1981). – Abstract.

122.114 Fourier decomposition of RR Lyrae pulsations.
N. R. Simon, T. J. Teays.
Publ. Astron. Soc. Pacific, Vol. 93, 550 (1981). – Abstract.

122.115 Recent high-dispersion observations of CH Cygni.
G. Wallerstein.
Publ. Astron. Soc. Pacific, Vol. 93, 577 - 580 (1981).

In 1979 - 80 CH Cyg showed an increasingly intense emission spectrum. Lines of [O I] are now present and indicate a systemic velocity of -63.4 km s^{-1}. The Na I D lines show inverted P Cygni profiles. Large numbers of permitted and forbidden lines of singly ionized metals are present. Four components of circumstellar Ca II absorption lines are present and two components of Na I can be recognized. Broad, shallow absorption lines of He I are visible; they are likely to come from a hot star or an accretion disk. Soft X-rays are not observed.

122.116 V923 Aquilae.
A. E. Ringuelet, J. Sahade.
Publ. Astron. Soc. Pacific, Vol. 93, 594 - 600 (1981).

Spectrographic observations of V923 Aql carried out at the Cerro Tololo Inter-American Observatory have disclosed absorption and emission components in the spectrum of the shell that make it possible to describe it in terms of different regions characterized by different velocities, velocity gradients, and/or line-broadening mechanisms at work.

122.117 New high speed photometry of EH Lib and a binary interpretation of its period change.
S.-y. Jiang, Z.-z. Yang.
Acta Astron. Sinica, Vol. 22, 279 - 284 (1981). In Chinese.

122.118 The DDO properties of cepheids and supergiants.
J. F. Dean.
Mon. Not. R. Astron. Soc., Vol. 197, 779 - 790 (1981).

Galactic and LMC cepheids and supergiants have been observed at the SAAO in the DDO photometric system. Their comparative properties are studied here using de-reddened two-colour diagrams with reddening ratios derived from early-type stars.

122.119 Ultraviolet spectra of dwarf solar neighbourhood stars – I.
C. J. Butler, P. B. Byrne, A. D. Andrews, J. G. Doyle.
Mon. Not. R. Astron. Soc., Vol. 197, 815 - 827 (1981).

Short-wavelength *IUE* spectra of three nearby M-type dwarfs are presented. Two of these stars, Gliese 867A and AU Mic, are active UV Ceti flare stars and their UV spectra exhibit strong emission lines. The third star, Gliese 825, is a low-activity flare star. A spectrum of the flare light is extracted and compared with the predictions of various flare models.

122.120 On the nature of WX Ceti.
R. A. Downes, B. Margon.
Mon. Not. R. Astron. Soc., Vol. 197, 35P - 38P (1981).

WX Ceti is a large amplitude (8 mag) variable which has undergone at least four outbursts during this century. Bailey has suggested that this object may be related to WZ Sge, and is part of a small but distinct subgroup of dwarf novae. The authors' data confirm the cataclysmic variable nature of WX Cet, and indicate that the proposed association with WZ Sge is plausible.

122.121 The metal abundances of RR Lyrae stars in the globular clusters NGC 6712 and NGC 6723.
H. A. Smith.
Astrophys. J., Vol. 250, 719 - 722 (1981).

The spectroscopic metal abundance parameter ΔS has been measured for RR Lyrae stars in the globular clusters NGC 6712 and NGC 6723. Mean ΔS values are 2.6 ± 0.4 for NGC 6712 and 2.8 ± 0.5 for NGC 6723, corresponding to [Fe/H] ~ -0.7 on Butler's abundance scale. This iron abundance is considerably higher than those found for 47 Tuc and M71 in recent analyses of high dispersion spectra of red giant stars.

122.122 Detection of 6 centimeter OH emission from the Mira variable AU Geminorum.
M. J. Claussen, J. D. Fix.
Astrophys. J., Lett., Vol. 250, L77 - L78 (1981).

Fourteen stars which are known to emit ground-state OH maser radiation were examined for excited-state emission from the $^2\pi_{1/2}$, $J = 1/2$ state at 6 cm. One star, AU Gem, was found to have such emission at the 4750 MHz transition, the only main line ($\Delta F = 0$) in the multiplet. The authors believe that AU Gem is the first star from which 6 cm OH emission has been observed.

122.123 Z Camelopardalis at standstill.
P. Szkody, R. A. Wade.
Astrophys. J., Vol. 251, 201 - 204 (1981).

Coudé spectra of the Hα region in Z Cam are used to study the line structure and velocity variations during the current extended standstill of Z Cam. Orbital elements and fine structure in the line profiles are found to be consistent with results of Robinson, who studied Z Cam in quiescence. The standstill therefore does not alter the velocity structure of the inner parts of the accretion disk nor that of a possible circumstellar shell. Absolute fluxes from multichannel spectrophotometry of Z Cam are used to derive mass loss rates from the system of $\sim 6 \times 10^{-10} M_\odot yr^{-1}$ in both quiescence and standstill.

122.124 BE Ursae Majoris (PG 1155 + 492): a unique cataclysmic-variable-like object.   D. H. Ferguson,
J. Liebert, R. F. Green, J. T. McGraw, H. Spinrad.
Astrophys. J., Vol. 251, 205 - 213 (1981).

BE Ursae Majoris (PG 1155 + 492) is a unique spectrum variable with similarities to the cataclysmic variables (CVs). Its emission-line spectrum shows the high-excitation CN 4650 Å blend as the strongest optical feature, with weaker helium and hydrogen lines and a strong Balmer jump; it also shows an absorption spectrum dominated by helium. The object shows no evidence for photometric variability on very short (6–200 s) or very long time scales. The luminosity is dominated by an ultraviolet (1200–3000 Å) ν^{+1} power-law component, though the 4000–10000 Å continuum is flat ($\nu^{-1/4}$). Despite the spectroscopic variability, the object's photometric behavior, energy distribution, and long period do not readily support the hypothesis that BE UMa is an active mass transfer binary.

122.125 On the existence of hysteresis effects in pulsating stars.   M. Auvergne, A. Baglin, P.-J. Morel.
Astron. Astrophys., Vol. 104, 47 - 56 (1981).

A critical review of observations of Cepheids in the Magellanic Clouds and in our Galaxy shows that identical (same M, L, T_e) stars are seen pulsating at the same period with two different amplitudes. The classical interpretations of this phenomenon are discussed. The analogy of the amplitude-period relation with the resonance curves of non linear oscillators has led the authors to build a simple model mimicking the behaviour of the outer layers of a pulsating star.

122.126 Radial-velocity observations of two low-amplitude cepheids.   L. A. Balona.
Observatory, Vol. 101, 205 - 207 (1981).

Radial-velocity observations are presented for two recently-discovered low-amplitude cepheids, HD 129708 and HD 179315. The former is probably a binary with a blue companion.

122.127 A simultaneous photometric and radial velocity study of short-period southern Cepheids. I. The radial velocities.   W. Gieren.
Astrophys. J., Suppl. Ser., Vol. 46, 287 - 293 (1981).

Simultaneous photoelectric radial velocity and $UBV(RI)_{KC}$ photometric observations of 15 southern classical Cepheids with pulsation periods between 3^d and 7^d have been carried out. The results of the radial velocity observations are presented. For all of the variables, complete radial velocity curves could be obtained within two to four consecutive pulsation cycles. Subsequent papers will present the results of the photometry and investigate radii, pulsation modes, and distances of the observed Cepheids, as well as their possible binary character.

122.128 Chromospheric lines in red dwarf flare stars. I. AD Leonis and GX Andromedae.
B. R. Pettersen, L. A. Coleman.
Astrophys. J., Vol. 251, 571 - 582 (1981).

Line profiles with a spectral resolution of 0.45 Å have been obtained for Hα(λ6563), the Na D lines (λλ5890, 5896), He I lines (λλ5876, 6678), and the Ca II infrared triplet lines (λλ8498, 8542, 8662) in the flare stars AD Leonis and GX Andromedae. AD Leo shows strong flare activity whereas GX And is more than an order of magnitude less active. Chromospheric emission lines are discussed in detail and correlations between spectral line features and flare activity are investigated.

122.129 MV Lyrae: spectrophotometric properties of minimum light; or on MV Lyrae off.
E. L. Robinson, E. S. Barker, A. L. Cochran, W. D. Cochran, R. E. Nather.
Astrophys. J., Vol. 251, 611 - 619 (1981).

The nova-like variable MV Lyr is normally at maximum light near $B \sim 12.5$, but it occasionally fades to minimum light near $B \sim 17.3$. The authors have obtained photometric and spectrophotometric observations of MV Lyr at maximum light in 1969 and at minimum light in 1980. They show that minimum light is caused by a total cessation of mass transfer from the late-type star to the white dwarf in the system. The distribution of orbital periods of the cataclysmic variables has a gap at orbital periods between 2 hr and 3 hr, and MV Lyr is at the long-period edge of the gap. The authors argue that the cataclysmic variables do evolve through the gap, but that they cease mass transfer while in the gap.

122.130 UBVRI photometry of FK Comae.
S. M. Rucinski.
Astron. Astrophys., Vol. 104, 260 - 263 (1981).

The light curves of FK Comae in five Johnson photometry colours are presented. The period of photometric variability equal 2.400 d is confirmed. The flare activity increased the noise of the light curves, especially at shorter wavelengths. The light variation of about 0.08 in V can be explained by a longitudinally asymmetric region of dark spots covering no less than 10% of the hemisphere. The $V-R$ and $R-I$ colours at maximum suggest an effective temperature of about 4900 K.

122.131 20 CVn, a monoperiodic Delta Scuti star.
J. H. Peña, S. González B.

Astron. J., Vol. 86, 1679 - 1684 (1981).

Differential photoelectric photometry of the δ Scuti star HR 5017 has been carried out. Period determination of the data shows one frequency at 8.214 c/d that gives high statistical coefficients and reproduces the observations well. This result indicates that this star has maintained a stable period for a time span of 11 yr.

122.132 Theoretical Hertzsprung progression and mass-luminosity relation of classical cepheids.
Yu. A. Fadeev.
Nauchn. Inf., Vyp. (No.) 47, (see 003.016), p. 38 - 50 (1981). In Russian.

Results of nonlinear calculations of classical cepheid pulsations are presented. The masses of models make 3.1 to 7.0 $M_\odot$, and the pulsation periods 4.4 to 27 days. The results were compared with observational diagrams for the period-bump phase and period-light curve amplitude.

122.133 Distribution of population I pulsating stars according to their evolutionary mass.
Ts. G. Tsvetkov.
Nauchn. Inf., Vyp. (No.) 47, (see 003.016), p. 59 - 64 (1981). In Russian.

The theoretical distribution of population I pulsating stars according to their evolutionary mass has been derived. It agrees qualitatively with the observed one, however, for the cepheids it is shifted to smaller values of mass. The possible character of the initial mass function (monotonous or discrete) is discussed.

122.134 Carbon stars – possible members of open clusters. I. Variable carbon star BC 89 in the cluster NGC 744.
Z. Alksne, A. Alksnis.
Nauchn. Inf., Vyp. (No.) 47, (see 003.016), p. 65 - 70 (1981). In Russian.

The carbon star BC 89 situated in the cluster NGC 744 is found to be a long-period variable star with a period of 470^d. The V-magnitude range observed during the time interval JD 2442450 - 2444175 is $11^m4 - 14^m5$. The light curves in R (0.63) and V are given.

122.135 Search for variability of four η Carinae-like objects.
N. E. Kurochkin, A. S. Sharov, R. I. Noskova.
Perem. Zvezdy, Tom 21, 149 - 151 (1979). In Russian.

Results of search for variability of four η Car-like objects, MWC 645, MWC 819, IRC + 10420 and MWC 349 (= V 1478 Cyg), are presented.

122.136 Use of the method of principal components for the problem of classification of RR Lyrae-type stars.
M. B. Bogdanov.
Perem. Zvezdy, Tom 21, 153 - 159 (1979). In Russian.

The method of principal components was used for an analysis of an RR Lyrae-type stars ensemble. This method is based on search of linear combinations of object parameters, and this linear combination takes into account summary dispersion of parameters. For every star seven parameters were considered: period, amplitude of light variation, amplitude of color index variation, asymmetry of light curve, metal abundance index, surface gravity index and index of effective temperature. Diagrams drawing on the plane of principal components shows that among RRab-type stars there are perhaps three groups of objects.

122.137 Statistical investigation of period changes of RR Lyrae variables in globular clusters.
T. I. Gryzunova.
Perem. Zvezdy, Tom 21, 161 - 168 (1979). In Russian.

The possibility of application of mathematical statistics methods to the period changes of an individual pulsating RR Lyrae variable star and all the stars of the same type in a globular cluster is examined. A mathematical model of the period change number of an individual pulsating star and all the stars of the same type in a globular cluster is constructed. It is shown that the observed changes of period are described by a Markov process with continuous time. The changes of periods of RR Lyrae variables in the globular clusters NGC 5024, NGC 5272, NGC 5466, NGC 5904, NGC 6171, NGC 7078 are considered.

122.138 Variable stars in the globular cluster M13. II.
R. M. Rusev, T. S. Ruseva.
Perem. Zvezdy, Tom 21, 169 - 174 (1979). In Russian.

The investigation results on nine variable stars in the globular cluster M13 are given. More accurate data for the elements of the light curves in the photographic B system are obtained. It is shown that V4 is an RRc type variable star with $P = 0^d298827$ and V12 is a variable star with $P = 5^d21753$.

122.139 On the optical activity of rapid irregular variables.
A. Ya. Filin.
Perem. Zvezdy, Tom 21, 175 - 182 (1979). In Russian.

The activity of light variations of an irregular variable star may be characterized by the index of optical activity k which is the mean light variation rate per unit time. This method for the calculation of the activity index was applied to the variables SU Aur, UX and BN Ori.

122.140 Photometric observations of XX Oph from 1964 through 1978.
T. S. Belyakina, G. E. Erleksova, G. V. Zajtseva.
Perem. Zvezdy, Tom 21, 183 - 190 (1979). In Russian.

The photographic light curve of XX Oph from 1964 through 1978 is shown. Photoelectric UBV measurements of this star for the period 1970 - 78 demonstrate that in the average the brightness in the V system is connected with reddening of the colour index (U−B). From the observations it follows that the behaviour of XX Oph during 1964 - 1978 was not changed as compared with that of 1890 - 1946. There were rather rapid light variations of XX Oph with the amplitude close to $0^m1 - 0^m2$.

122.141 V379 Cassiopeiae is a pulsating variable star.
V. I. Kardopolov, G. K. Filip'ev.
Perem. Zvezdy, Tom 21, 191 - 194 (1979). In Russian.

The results of photoelectric UBV observations of V379 Cas in 1976 - 77 are presented. It is shown that V379 Cas is an s-type cepheid variable star with the light curve elements Max = JD $2442958^d27 + 4^d3077 \cdot E$.

122.142 Photoelectric observations of the super-shortperiod cepheid EH Librae.
V. G. Karetnikov, Yu. A. Medvedev.
Perem. Zvezdy, Tom 21, 199 - 210 (1979). In Russian.

From 1146 photoelectric BV observations of EH Lib 22 moments of maxima have been determined and the light behaviour of the variable at maximum has been studied. It is shown that the moment of light at maximum proceeds in B and V simultaneously. However, the light itself varies both in magnitude and in time at maximum. EH Lib seems to show Blazhko effect. In the light curve of EH Lib about the phase of 0^p2 the presence of a "bump" changing its light from 0 to 0^m15 is suspected.

122.143 RW Arietis.
V. P. Goranskij, S. Yu. Shugarov.
Perem. Zvezdy, Tom 21, 211 - 217 (1979). In Russian.

Wiśniewski's conclusion that the RR Lyrae variable RW Ari is simultaneously an eclipsing binary star with period 3^d1754 is not confirmed by a large array of photographic and photoelectric observations carried out in August 1976 - February 1978. New more precise elements were determined.

122.144   **Variable H42 in the M31 Andromeda nebula.**
A. S. Sharov, P. N. Kholopov.
Perem. Zvezdy, Tom 21, 219 - 222 (1979). In Russian.
Results of the investigation of the variable H42 in M31 are given. The star probably is a cepheid with the longest period in this galaxy.

122.145   **Rapid oscillations in SS Cygni.**
F. Giovannelli.
Space Sci. Rev., Vol. 30, (see 012.044), 213 - 219 (1981).
High speed photometry of the dwarf nova SS Cygni during an outburst reveals the presence of rapid oscillations detected over an interval of four days. The period of the oscillations ranged from a maximum of 9.91 s to a minimum of 8.96 s. The results are discussed from the viewpoint of several current models.

122.146   **UV observations of MV Lyrae.**
L. Chiappetti, L. Maraschi, E. G. Tanzi, A. Treves.
Space Sci. Rev., Vol. 30, (see 012.044), 231 - 233 (1981).
Abstract

122.147   **The discovery of a new SU UMa star.**
B. Stolz.
Messenger, No. 26, p. 16 - 17 (1981).

122.148   **Study of period changes for 38 RR Lyrae variables in the globular cluster M15.**
H. A. Smith, A. Sandage.
Astron. J., Vol. 86, 1870 - 1881 (1981).
Plate material spanning the interval 1896−1978 has been used to study period changes for 38 RR Lyrae stars in the globular cluster M15. Abrupt period changes have occurred in some stars. For others, no decision can be made between abrupt changes or a linear period change with time. For stars whose period change is either certain or probable, five times as many have increased their periods as decreased. The mean rate of period change for the sample is larger than can be explained by stellar evolution. No correlation exists between the observed period change, neither in sign nor amount, and a star's position in the instability strip, hence the cause of the period changes remains unknown.

122.149   **V 697 Tauri − a member of the Hyades?**
V. P. Tsesevich.
Astron. Tsirk., No. 1115, p. 7 - 8 (1980), with a correction in No. 1152, p. 8 (1981). In Russian.

122.150   **Investigation of the form of curves of O−C residual variations for RR Lyrae-type stars of the galactic field.**   B. N. Firmanyuk.
Astron. Tsirk., No. 1118, p. 1 - 2 (1980). In Russian.

122.151   **On the instability of periods of light variation of RR Lyrae-type stars of the galactic field.**
B. N. Firmanyuk.
Astron. Tsirk., No. 1118, p. 3 - 5 (1980). In Russian.

122.152   **A new variable star in Draco.**
G. O. Gevorkyan.
Astron. Tsirk., No. 1118, p. 7 - 8 (1980). In Russian.

122.153   **Variable RR Lyr-type star SVS 596 = CSV 1902.**
V. P. Tsesevich.
Astron. Tsirk., No. 1119, p. 7 - 8 (1980). In Russian.

122.154   **UBV photometry of SS Psc.**
G. A. Garbuzov.
Astron. Tsirk., No. 1120, p. 6 - 7 (1980). In Russian.

122.155   **Observations of the variable ST Psc.**
V. P. Smykov.
Astron. Tsirk., No. 1120, p. 8 (1980). In Russian.

122.156   **On the Blazhko effect in DM Cygni.**
L. E. Lysova, B. N. Firmanyuk.
Astron. Tsirk., No. 1122, p. 3 - 5 (1980). In Russian.

122.157   **New Mira Ceti-type variable star SVS 2315.**
S. V. Ignatov.
Astron. Tsirk., No. 1122, p. 5 - 6 (1980). In Russian.

122.158   **On the period of AW Tauri.**
V. P. Tsesevich.
Astron. Tsirk., No. 1122, p. 6 - 7 (1980). In Russian.

122.159   **New light elements of NO Cas.**
V. P. Smykov, L. I. Shakun.
Astron. Tsirk., No. 1122, p. 8 (1980). In Russian.

122.160   **RR Lyrae variables in the globular cluster M3.**
Z. I. Kadla, A. N. Gerashchenko.
Astron. Tsirk., No. 1123, p. 1 - 3 (1980). In Russian.

122.161   **SU Cygni.**
IAU Circ., No. 3615 (1981).

122.162   **SU Cygni.**
Yamamoto Circ., No. 1960 (1981).

122.163   **Observations of SU UMa before and during a super-outburst.**   H. Barwig, R. Schoembs.
Inf. Bull. Variable Star, No. 1989, 4 pp. (1981).

122.164   **Photoelectric observations of VZ Cas.**
J. B. Srivastava, C. D. Kandpal.
Inf. Bull. Variable Stars, No. 1990, 2 pp. (1981).

122.165   **Photometric variability of 10 Lacertae.**
A. J. Delgado, R. Garrido.
Inf. Bull. Variable Star, No. 1992, 2 pp. (1981).

122.166   **Seven colour photoelectric photometry of the red variable Gliese 83.3.**
U. Dzervitis, O. Paupers.
Inf. Bull. Variable Stars, No. 2002, 2 pp. (1981).

122.167   **V 68 in the globular cluster M 3 (NGC 5272) is a double mode RR Lyrae type star.**
V. P. Goranskij.
Inf. Bull. Variable Stars, No. 2007, 3 pp. (1981).

122.168   **HD 200356: a possible new δ Scuti star.**
S. F. Gonzalez-Bedolla, J. H. Peña.
Inf. Bull. Variable Stars, No. 2008, 3 pp. (1981).

122.169   **V 1334 Cyg.**
C. Bartolini, A. Dapergolas, A. Piccioni.
Inf. Bull. Variable Stars, No. 2009, 2 pp. (1981).

122.170   **The Delta Scuti star Beta Leonis.**
C. Bartolini, A. Dapergolas, A. Piccioni.
Inf. Bull. Variable Stars, No. 2010, 3 pp. (1981).

122.171   **Note on published times of maxima of DY Pegasi.**
R. Quigley, J. L. Africano.
Inf. Bull. Variable Stars, No. 2017 (1981).

122.172   **Photoelectric observations of the flare star EV Lac in 1980.**
L. N. Mavridis, G. Asteriadis, K. P. Panov, M. K. Tsvetkov.
Inf. Bull. Variable Stars, No. 2022, 4 pp. (1981).

122.173 **The maximum times and the light elements of 28 Andromedae.**
Z. Tunca, S. Evren, C. Ibanoglu, O. Tümer, A. Y. Ertan.
Inf. Bull. Variable Stars, No. 2027, 2 pp. (1981).

122.174 **The maximum times and new light elements of 44 Tauri.**
O. Tümer, A. Y. Ertan, S. Evren, Z. Tunca, C. Ibanoglu.
Inf. Bull. Variable Stars, No. 2028, 2 pp. (1981).

122.175 **TU Men, the first SU UMa star beyond the gap.**
B. Stolz, R. Schoembs.
Inf. Bull. Variable Stars, No. 2029, 4 pp. (1981).

122.176 **Standstill of γ CrB.**
B. Vetó, G. Kovács.
Inf. Bull. Variable Stars, No. 2030, 4 pp. (1981).

122.177 **The principal frequency of the rapidly oscillating Ap star α Cir.**
D. W. Kurtz, S. Allen, M. S. Cropper.
Inf. Bull. Variable Stars, No. 2033, 2 pp. (1981).

122.178 **Synchronous UBV flare observations on UV Ceti.**
N. D. Melikian (*Melikyan*), I. Jankovics,
N. N. Kiljachkov (*Kilyachkov*), V. S. Shevchenko,
O. I. Stalbovsky (*Stal'bovskij*).
Inf. Bull. Variable Stars, No. 2038, 3 pp. (1981).

122.179 **Period changes of AC And.**
I. Guman.
Inf. Bull. Variable Stars, No. 2046, 4 pp. (1981).

122.180 **New observations of the Delta Scuti variables Sigma Octantis and B Octantis.**
D. W. Coates, L. Halprin, T. T. Moon, K. Thompson.
Inf. Bull. Variable Stars, No. 2047, 3 pp. (1981).

122.181 **On the suspected variable CSV 96.**
W. P. Bidelman, K. Cudworth.
Inf. Bull. Variable Stars, No. 2055 (1981).

122.182 **Broad emissions in the spectra of R CrB stars.**
A. Eh. Rozenbush.
Astron. Tsirk., No. 1131, p. 1 - 2 (1980). In Russian.

122.183 **On causes of anticorrelations of maser and infrared radiation from long period variables.**
V. S. Strel'nitskij.
Astron. Tsirk., No. 1125, p. 1 - 3 (1980). In Russian.

122.184 **Long-period variations of RY Tauri.**
G. V. Zajtseva, N. E. Kurochkin.
Astron. Tsirk., No. 1126, p. 3 - 7 (1980). In Russian.

122.185 **New semiregular variable star SVS 2381.**
S. V. Ignatov.
Astron. Tsirk., No. 1135, p. 7 - 8 (1980). In Russian.

122.186 **Determination of radii and luminosities of classical cepheids using Wesselink's method.**
G. R. Ivanov.
Astron. Tsirk., No. 1136, p. 1 - 3 (1980). In Russian.

122.187 **Moments of maxima of SS Psc.**
G. A. Garbuzov.
Astron. Tsirk., No. 1138, p. 7 - 8 (1980). In Russian.

122.188 **Velocities of cepheids included in complexes.**
D. K. Karimova, E. D. Pavlovskaya.
Astron. Tsirk., No. 1139, p. 1 - 3 (1980). In Russian.

122.189 **Brightness decrease of XX Ophiuchi in 1979 - 1980: photometric and spectral observations.**
T. S. Belyakina, G. V. Zajtseva, E. A. Kolotilov.
Pis'ma Astron. Zh., Tom 7, 736 - 740 (1981). In Russian. English translation in Soviet Astron. Lett., Vol. 7.
Results of spectral and photometric observations of the variable star XX Oph obtained in 1979 - 1980 mostly during the brightness minimum are presented.

122.190 **Photometric investigations of some type RCB(?) variable stars.** I. Platais, A. Rozenbush.
Investigations of the sun and red stars. 11, (see 003.021), p. 27 - 42 (1980). In Russian.
Photographic U, B, V, R(0.63), R'(0.66) observations of CT Vul and V1405 Cyg are presented. The shape of light curves for these stars is not similar to that of RCB variables.

122.191 **Carbon stars – possible members of open clusters. II. The carbon star CCS 65 near the cluster NGC 457**
Z. Alksne, A. Alksnis.
Investigations of the sun and red stars. 12, (see 003.022), p. 5 - 19 (1981). In Russian.
According to photographic photometry in R(0.63)-, V- and B-magnitudes the carbon star No. 65 in Stephenson's General Catalogue of cool carbon stars (CCS 65) has cyclic variations with an amplitude of $0^m.8$ in V and a period of about 425 days.

122.192 **Carbon stars – possible members of open clusters. III. Variable red giants near the cluster NGC 7789 = C2354 + 564.** Z. Alksne, I. Daube.
Investigations of the sun and red stars, 14, (see 003.024), p. 5 - 27 (1981). In Russian.
On the basis of photographic photometry in the R(0.63)-, V-, and B-passbands at the time interval JD 2440095 −2444242 the known characteristics of the light variations for the C-spectral type stars V532 Cas and V533 Cas, the S-type star WY Cas and M-type star V542 Cas have been improved.

122.193 **Carbon stars – possible members of open clusters. V. HN Aur – a carbon star in the centre of the cluster NGC 1664 = C0447 + 436.** A. Alksnis.
Investigations of the sun and red stars, 14, (see 003.024), p. 44 - 50 (1981). In Russian.
The carbon star HN Aur has been monitored photographically for light variations in the R(0.63)-, V- and B-passbands at the time interval between 1971 and 1980. The star had semi-regular light variations with a mean cycle length of 160^d-170^d superposed to very slow variations, probably also semi-regular with cycle length of about 1400^d. The individual magnitudes are given.

122.194 **Carbon stars – possible members of open clusters. VI. Irregular variable carbon star CCS 414 near the cluster NGC 2099 = C0549 + 325.**
Z. Alksne, I. Daube.
Investigations of the sun and red stars, 14, (see 003.024), p. 51 - 58 (1981). In Russian.
At the time interval JD 2441429-2444327 the carbon star CCS 414 has been photographically monitored for light variations in the R(0.63)-, V- and B-passbands. It is stated that CCS 414 belongs to irregular variable stars. The cluster membership of the star is discussed.

122.195 **La variabile Rho Cassiopeiae.**
F. Gambino, M. Vallone.
Orione, Vol. 2, 268 - 271 (1981).

122.196 **V801 Cygni.** M. Yarlott.
J. American Assoc. Variable Star Obs., Vol. 10, 9 - 10 (1981).
Linear and parabolic light elements have been derived for

the RR Lyr variable V801 Cygni using published data and new epochs for the interval 1967 - 1980 derived from photographs taken at the Maria Mitchell Observatory.

122.197   A possible change in the period of HS Aquilae.
K. D. Thompson.
J. American Assoc. Variable Star Obs., Vol. 10, 11 - 12 (1981).

122.198   A study of UU Aquilae.
J. F. Davis, J. A. Mattei.
J. American Assoc. Variable Star Obs., Vol. 10, 28 - 32 (1981).

122.199   Photographic (B, V) photometry of Magellanic Cloud cepheids. I. − Observational data.
W. L. Martin, Y. Thomas, B. S. Carter, H. E. Davies.
South African Astron. Obs. Circ., No. 6, p. 31 - 95 (1981).
   Photographic (B, V) observations of 213 LMC and 181 SMC cepheids have been secured and are presented here together with the relevant light curve parameters. Comparison of this work with stars in common with those which have photoelectric photometry indicate that there are no significant systematic differences.

122.200   Multicolour photoelectric photometry of Magellanic Cloud cepheids. IV: B, V observations of 20 short period cepheids.   W. L. Martin.
South African Astron. Obs. Circ., No. 6, p. 96 - 100 (1981).
   B and V observations of ten LMC and ten SMC cepheids in the period range 1 - 5 days are presented. In addition, from the respective light curves, the relevant B and V parameters have been derived for 18 of these cepheids.

122.201   A search for AlO emission in the spectra of Mira variables.   T. A. Kipper, M. A. Kipper.
Proceedings of the Third Finnish-Soviet Astronomical Symposium, (see 012.057), p. 25 - 29 (1981).

122.202   A search for rapid brightness variations in stars.
V. P. Zalinian (*Zalinyan*), Yu. K. Melik-Alaverdian (*Melik-Alaverdyan*), S. E. Nersissian (*Nersisyan*), H. M. Tovmassian (*G. M. Tovmasyan*).
Proceedings of the Third Finnish-Soviet Astronomical Symposium, (see 012.057), p. 45 - 47 (1981).

122.203   On rapid variations of the magnetic star 53 Cam spectrum.   N. S. Polosukhina, K. K. Chuvaev,
V. P. Malanushenko.
Proceedings of the Third Finnish-Soviet Astronomical Symposium, (see 012.057), p. 49 - 60 (1981).

122.204   Model envelopes for BQ Serpentis: a double-mode cepheid.   K. Uji-iye.
Sci. Rep. Tôhoku Univ., Ser. 8, Vol. 1, 155 - 160 (1980).
   Inhomogeneous model envelopes with helium enriched outer layer for BQ Ser, a double-mode cepheid, are constructed with T_{eff} = 6450 K estimated from the observed spectral type. Periods of those models are compared with the observed period and period ratio. The mass of model satisfying its periods and the resonance condition is less than that derived from the evolution theory.

122.205   Observational study on line weakening in the spectra of Mira Ceti during 1965−1980.
Y. Yamashita, H. Maehara, Y. Norimoto.
Ann. Tokyo Astron. Obs., Second Ser., Vol. 18, 142 - 163 (1981).
   Plate material obtained for Mira Ceti during the period from 1965 to 1980 and consisting of about 150 spectrograms with various dispersions is used for a statistical study on the general line-weakening phenomenon.

122.206   Photoelektrische UBV-Photometrie des Sternes RV Equulei.   S. Rößiger.
Mitt. Veränderl. Sterne (MVS), Band 9, 3 - 5 (1981).
   Photoelectric UBV observations from 1975 to 1979 show that RV Equulei is a semiregular variable star of the type SRb. It does not belong to the RW Aurigae class.

122.207   Beobachtungen von V 13 im Kugelhaufen M3.
I. Meininger.
Mitt. Veränderl. Sterne (MVS), Band 9, 6 - 7 (1981).

122.208   Die Periodenänderung des Mira-Sterns UZ Cam.
B. Fuhrmann.
Mitt. Veränderl. Sterne (MVS), Band 9, 8 - 10 (1981).

122.209   AG Ser - ein langperiodischer Mira-Stern.
B. Fuhrmann.
Mitt. Veränderl. Sterne (MVS), Band 9, 11 - 12 (1981).

122.210   Der Lichtwechsel des Veränderlichen TT Ari.
B. Fuhrmann.
Mitt. Veränderl. Sterne (MVS), Band 9, 14 - 15 (1981).

122.211   Veränderlichkeit einiger C-Sterne mit sehr langen Perioden.   A. Alksnis.
Mitt. Veränderl. Sterne (MVS), Band 9, 25 - 34 (1981).
   Basing on the photovisual Sky Patrol plates of the plate collection of Sonneberg Observatory the light variations for the six long period variable carbonstars S Aur, UW Aur, CT Lac, KL Cyg, V 659 Cyg, and SY Per have been studied. Elements of the light curve for the primary periodic variations have been determined or improved, and secondary very slow variations searched for.

122.212   Beobachtungen von V 1 im Kugelhaufen M 3.
I. Meininger.
Mitt. Veränderl. Sterne (MVS), Band 9, 40 - 43 (1981).

122.213   Mehrfarben-Beobachtungen von V 79 im Kugelhaufen M 3.   I. Meininger.
Mitt. Veränderl. Sterne (MVS), Band 9, 44 - 47 (1981).

122.214   On the infrared excess temperature variability of R Coronae Borealis.   T. S. Khruzina.
Astron. Tsirk., No. 1150, p. 3 - 5 (1981). In Russian.

122.215   On the photometric and spectral variability of Kuwano's object.   Ya. N. Chkhikvadze.
Astron. Tsirk., No. 1150, p. 7 - 8 (1981). In Russian.

122.216   Two giants in projection to the large Orion nebula region.
M. M. Zakirov, V. I. Kardopolov, V. S. Shevchenko.
Astron. Tsirk., No. 1156, p. 7 - 8 (1981). In Russian.

122.217   Visual observations of DY Pegasi.
I. N. Atroshchenko, A. S. Gadun, S. V. Kudlinkov, S. N. Mel'nikov, A. M. Chausovskij.
Astron. Tsirk., No. 1158, p. 6 - 7 (1981). In Russian.

122.218   Observations of four classical cepheids.
E. N. Makarenko.
Astron. Tsirk., No. 1159, p. 7 - 8 (1981). In Russian.

122.219   On the possible cyclicity in variations of SU Aurigae.
L. V. Timoshenko.
Astron. Tsirk., No. 1164, p. 5 - 7 (1981). In Russian.

A catalogue of proper motions of RR Lyrae variables.   See Abstr. 002.081.

A method for unambiguous determination of starspot temperatures and areas: application to II Pegasi, BY Draconis, and HD 209813. See Abstr. 031.579.

Pulsations, grain condensation, and mass loss in long-period variable stars. See Abstr. 064.074.

Radiation from optically thin accretion discs. See Abstr. 064.075.

Formation of an extended envelope surrounding a pulsating star. See Abstr. 064.091.

An iterative theory of modal selection. See Abstr. 065.001.

The thick helium-burning shell phase and the second blue loop. See Abstr. 065.006.

Analytic solutions of the radial pulsation equation for rotating and magnetic star models. See Abstr. 065.018.

Miras, mass loss, and the origin of planetary nebulae. See Abstr. 065.033.

Kelvin–Helmholtz instabilities and their application to B-type variables. See Abstr. 065.074.

Multimode stellar pulsations: a new approach. See Abstr. 065.075.

The search for double-mode pulsator models. See Abstr. 065.076.

Nonradial oscillations for stars on the left-hand side of the Cepheid instability strip. See Abstr. 065.098.

Thermal instability of hydrogen-burning shells in very massive stars. See Abstr. 065.100.

The distances of cataclysmic variables. See Abstr. 111.004.

Detection of the SiO ($\nu = 2, J = 2 \rightarrow 1$) maser. See Abstr. 112.001.

Ultraviolet observations of 27 Canis Majoris, π Aquarii and 48 Librae. See Abstr. 112.003.

IUE spectra of RS Puppis and HD 20722: stars in symmetric dusty nebulae. See Abstr. 112.009.

The correlation of SiO maser flux with stellar flux. See Abstr. 112.051.

Photometric UBV period study of eight Ap stars. See Abstr. 113.009.

A search for light variations in F- and G-type supergiants II. See Abstr. 113.010.

Pulsating Be stars? See Abstr. 113.041.

Is HR 2947 a variable? See Abstr. 113.046.

Photoelectric observations of Pleione (BU Tau) from September 1977 to March 1980. See Abstr. 113.057.

Photometric observations of symbiotic stars in the $UBVRJHKLMN$ system. 2. Z Andromedae. See Abstr. 113.061.

BD-8°4232, a new Delta Scuti star? See Abstr. 113.066.

Further observations of the carbon star RW LMi (CIT 6) in the years 1977/78. See Abstr. 113.070.

Carbon stars – possible members of open clusters. IV. Four carbon stars in the surroundings of the cluster NGC 1528 = C011 + 511. See Abstr. 113.074.

Spectrum of FG Sagittae in 1980. See Abstr. 114.012.

IUE observations of circumstellar emission from the late type variable R Aqr (M7 + pec). See Abstr. 114.041.

Ultraviolet variation of the Ap star 21 Com. See Abstr. 114.060.

Time series infrared spectroscopy of o Ceti. See Abstr. 114.196.

The absolute magnitude of δ Scuti and δ Del type stars. See Abstr. 115.017.

Multislit photoelectric magnetometer observations of Cepheids and supergiants: probable detections of weak magnetic fields. See Abstr. 116.001.

Intrinsic polarization variations of the symbiotic star R Aqr and the single-star hypothesis. See Abstr. 116.009.

Regular variations in the Hα profile of FK Comae. See Abstr. 116.042.

Variable intrinsic polarization in the peculiar star CH Cygni. See Abstr. 116.045.

Polarizacija zračenja nekih hladnih superdžinova. (Polarization of radiation of some cool supergiants). See Abstr. 116.046.

M5 V101: a close binary system in a globular cluster. See Abstr. 117.001.

A quest for the red companion in six cataclysmic binaries. See Abstr. 117.007.

Time-resolved spectroscopy of cataclysmic variables: U Geminorum. See Abstr. 117.013.

Radial velocity and line profile variations in the dwarf nova EM Cygni. See Abstr. 117.014.

Simultaneous observations of V861 Sco. See Abstr. 117.038.

A radial-velocity study of the dwarf nova RU Pegasi. See Abstr. 117.067.

Spectroscopy of 2A 0526–328: a triple periodic cataclysmic variable. See Abstr. 117.068.

The X-ray and optical characteristics of the cataclysmic variable V794 Aquilae. See Abstr. 117.069.

54 Cam: a new variable star. See Abstr. 117.081.

Variability of soft X-ray emission of EX Hydrae observed with Einstein Observatory. See Abstr. 117.111.

CH Cygni — M-giant and a white dwarf binary with transient accretion disk.   See Abstr. 117.135.

EX Hydrae: modulated mass transfer or intermediate polar?   See Abstr. 119.002.

Flare activity of the RS CVn star SV Cam. See Abstr. 119.040.

He I line emission and the helium abundance in cataclysmic variables.   See Abstr. 119.113.

Chromospheric plages on II Pegasi (HD 224085). See Abstr. 120.008.

Evidence for a starspot cycle on BD + 26°730. See Abstr. 120.017.

Flare activity of T Tau-type stars: DR and DQ Tau. See Abstr. 121.034.

Novae and related stars at quiescence. See Abstr. 124.003.

Two new variable white dwarfs: G185-32 and G191-16.   See abstr. 126.025.

The variable infrared emission of cosmic dust sources See Abstr. 131.207.

Miras and planetary nebula formation. See Abstr. 135.058.

The X-ray cataclysmic variable 1E0634.0−1648. See Abstr. 142.018.

The ultraviolet spectrum of the X-ray source 2A0526−33.   See Abstr. 142.050.

Lanning 10 and 33: the X-ray, UV, and optical fluxes.   See Abstr. 142.077.

X-ray observations of cataclysmic variables. See Abstr. 142.078.

Photographic observations of the newly discovered X-ray emitting dwarf nova 1E 0643.0 - 1648. See Abstr. 142.150.

Early-type stars in the field of the 16-day Cepheid X Cygni.   See Abstr. 153.006.

The Oosterhoff period groups and the age of globular clusters. II. Properties of RR Lyrae stars in six clusters: the P-L-A relation.   See Abstr. 154.011.

RR Lyrae stars in Magellanic Cloud clusters. See Abstr. 159.013.

123 Variable Stars (Surveys, Lists of Observations, Charts, etc.)

123.001   AV Pegasi.   K. Wenske.
BAV Rundbrief, 30. Jahrg., 52 - 53 (1981).

123.002   AE Ursae Majoris — ein "schneller" Stern.
H. Grzelczyk.
BAV Rundbrief, 30. Jahrg., 54 (1981).

123.003   Suspected variable CSV 3775 found and confirmed.
E. Velasquez.
J. American Assoc. Variable Star Obs., Vol. 9, 59 - 60 (1980).

123.004   Intermediate-type supergiant variables: request for photoelectric observations.   J. R. Percy.
J. American Assoc. Variable Star Obs., Vol. 9, 64 - 70 (1980).
   This paper lists and describes a group of about 20 supergiant variables, with spectral types F, G, and K, ranges of $0.^m1$ to $1.^m0$, time scales of 40 to 300 days, and semiregular light curves. In order to understand the variations in these stars, it is necessary to observe them regularly throughout the observing season, preferably for several seasons.

123.005   The period of GQ Aurigae.   H. B. Hammel.
J. American Assoc. Variable Star Obs., Vol. 9, 85 - 87 (1980).

123.006   Tableaux des observations faites par les sociétaires de l'AFOEV en avril, mai et juin 1981.
Bull. AFOEV, No. 17, p. 117 - 134 (1981).

123.007   Tableaux des observations faites par les observateurs de l'AFOEV en juillet, août et septembre 1981.
Bull. AFOEV, No. 18, p. 13 - 49 (1981).

123.008   A request for photoelectric observations of Cepheid variables.   R. Diethelm.
I. A. P. P. P. Commun., No. 6, p. 7 - 11 (1981).

123.009   Why not observe a Mira variable or two?
B. F. Marino.
I. A. P. P. P. Commun., No. 6, p. 16 - 18 (1981).

123.010   New variable stars in the field around MV Lyrae.
G. Romano.
Inf. Bull. Variable Stars, No. 1999, 2 pp. (1981).

123.011   Two new variable stars in the bright star catalogue.
B. Reipurth.
Inf. Bull. Variable Stars, No. 2015, 2 pp. (1981).

123.012   Flare stars in Orion.
N. D. Melikian (Melikyan).
Inf. Bull. Variable Stars, No. 2018 (1981).

123.013   Flare stars in the Pleiades region.
N. D. Melikian (Melikyan), I. Jankovics, J. Kelemen.
Inf. Bull. Variable Stars, No. 2019, 2 pp. (1981).

123.014   August 1981 faintening of V 348 Sgr.
A. Heck.
Inf. Bull. Variable Stars, No. 2020 (1981).

123.015   Variability in the Wolf-Rayet star HD 164270.
I. Lundström, B. Stenholm.
Inf. Bull. Variable Stars, No. 2021 (1981).

123.016 **66th name-list of variable stars.**
P. N. Kholopov, N. N. Samus, N. P. Kukarkina,
G. I. Medvedeva, N. B. Perova.
Inf. Bull. Variable Stars, No. 2042, 7 pp. (1981).

123.017 **Additional and newly-named variables in two objective-prism M-emission-star lists.**
W. P. Bidelman.
Inf. Bull. Variable Stars, No. 2054, 4 pp. (1981).

123.018 **List of variable stars discovered in the USSR and preliminary SVS designations.**
Astron. Tsirk., No. 1126, p. 8 (1980). In Russian.

123.019 **Variable stars in the globular cluster M13.**
Z. I. Kadla, A. N. Gerashchenko, N. V. Yablokova.
Astron. Tsirk., No. 1136, p. 5 - 7 (1980). In Russian.

123.020 **Photographic observations at Pulkovo of the variable X Leo.** A. N. Deutsch.
Astron. Tsirk., No. 1137, p. 8 (1980). In Russian.

123.021 **Observations of variable stars.** January - December 1980.
Nederlandse Ver. Weer Sterrenkunde. Kapteyn Astron. Lab.
Groningen, Netherlands, Rep. No. 35, 18 pp. (1981).

123.022 **V 645 Cyg – ein unregelmäßiger, langsamer Veränderlicher.** B. Fuhrmann.
Mitt. Veränderl. Sterne (MVS), Band 9, 12 - 13 (1981).

123.023 **Nova-verdächtiges Objekt V 605 Aql - nicht sichtbar auf Sonneberger Platten.** B. Fuhrmann.
Mitt. Veränderl. Sterne (MVS), Band 9, 13 (1981).

123.024 **Bearbeitung von 74 Veränderlichen am Südhimmel (Feld β Hydri).** H. Geßner.
Mitt. Veränderl. Sterne (MVS), Band 9, 16 - 17 (1981).

123.025 **Visuelle Beobachtungen von SS Cygni durch den Arbeitskreis "Veränderliche Sterne".**
E. Zische.
Mitt. Veränderl. Sterne (MVS), Band 9, 21 (1981).

123.026 **Beobachtungen von R. Coronae Borealis 1978 bis 1980.**
Mitt. Veränderl. Sterne (MVS), Band 9, 22 - 23 (1981).

123.027 **Visuelle Beobachtungen des roten Kohlenstoffsterns VY Ursae Maioris 1977 bis 1980.**
K. Reichenbächer, M. Rätz.
Mitt. Veränderl. Sterne (MVS), Band 9, 23 - 24 (1981).

123.028 **Der Lichtwechsel des Veränderlichen BI Cru auf Sonneberger Platten.** B. Fuhrmann.
Mitt. Veränderl. Sterne (MVS), Band 9, 48 - 49 (1981).

123.029 **Visuelle Lichtkurve 1970 bis 1980 von V Bootis.**
E. Zische.
Mitt. Veränderl. Sterne (MVS), Band 9, 50 - 51 (1981).

123.030 **Beobachtungen veränderlicher Sterne auf Sonneberger Überwachungsplatten.** D. Böhme.
Mitt. Veränderl. Sterne (MVS), Band 9, 51 - 55 (1981).

123.031 **Sonneberger δ-Cephei-Sterne im Bereich der Kleinen Magellanschen Wolke.** H. Geßner.
Mitt. Veränderl. Sterne (MVS), Band 9, 55 - 57 (1981).

123.032 **3 neue Veränderliche im Feld R Lyrae.**
H. Geßner.
Mitt. Veränderl. Sterne (MVS), Band 9, 57 - 58 (1981).

123.033 **Beobachtungsergebnisse der Berliner Arbeitsgemeinschaft für Veränderliche Sterne e.V. (BAV).**
W. Braune, E. Mundry.
BAV-Mitt. Nr. 32, 6 pp. (1981).

123.034 **Maxima of variable stars.** V. P. Smykov.
Astron. Tsirk., No. 1158, p. 8 (1981). In Russian.

123.035 **New red variable stars.** I. Daube.
Astron. Tsirk., No. 1161, p. 7 - 8 (1981). In Russian.

123.036 **New variables in Coma Berenices.**
G. V. Zhukov, V. Ya. Solov'ev.
Astron. Tsirk., No. 1164, p. 7 - 8 (1981). In Russian.

123.037 **Variable star notes.** J. A. Mattei.
J. R. Astron. Soc. Canada, Vol. 75, 164 - 168 (1981).

I.A.U. archives of unpublished observations of variable stars: 1979–81 data. See Abstr. 002.015.

The machine-readable version of the General Catalogue of Variable Stars, third edition.
See Abstr. 002.034.

Charts for southern variables. Series No. 13.
See Abstr. 002.069.

An atlas of southern and equatorial dwarf novae.
See Abstr. 002.075.

Le stelle variabili. See Abstr. 003.136.

L'exploitation des observations d'étoiles variables.
See Abstr. 031.557.

89th - 90th list of minima of eclipsing binaries.
See Abstr. 119.073.

124 Novae

124.001   A spectrogram of the recurrent nova U Scorpii.
    A. D. Mallama, B. Starosta.
J. American Assoc. Variable Star Obs., Vol. 9, 71 - 72 (1980).
    The spectrum of U Scorpii, observed five days after its
outburst in 1979, consists of strong, broad emission lines of
hydrogen, helium, carbon, nitrogen, oxygen, and iron superposed on a weak continuum.

124.002   Ultraviolet spectroscopy of old novae and symbiotic
        stars.   D. L. Lambert, M. H. Slovak, G. A. Shields,
G. J. Ferland.
The Universe at ultraviolet wavelengths, (see 012.009), p. 461 -
467 (1981).
    Low-dispersion, short-wavelength IUE spectra are
presented for two old novae, DQ Herculis (1934) and
V603 Aquilae (1918) = HD 174107, and for two symbiotic
variables, AG Pegasi = HD 207757 and CI Cygni. Prominent
emission line spectra are revealed as is a continuum whose
appearance is effected by the system inclination. These data
provide evidence for hot companions in the symbiotic stars,
making plausible the binary model for these peculiar stars.

124.003   Novae and related stars at quiescence.
    H. W. Duerbeck, G. Klare, J. Krautter, B. Wolf,
W. C. Seitter, W. Wargau.
Second European IUE Conference, (see 012.011), p. 91 - 94
(1980).
    UV spectra of the novae V603 Aql, HR Del, RR Pic, the
recurrent nova T CrB, the dwarf nova RU Peg and the novalike
variable TT Ari are presented. Interstellar extinction values,
temperatures and sizes are derived; implications are discussed
in terms of the cataclysmic binary/accretion disk model.

124.004   Classical novae: the thermonuclear runaway model.
    J. W. Truran.
Nuclear astrophysics, (see 012.036), p. 177 - 190 (1981).
    A brief overview of the evolution of a classical nova
system through outburst is presented. The critical role played
by the properties of the CNO cycle nuclear reaction sequences
is elaborated.

124.005   Transition and the infrared development of novae.
    M. F. Bode, A. Evans.
Mon. Not. R. Astron. Soc., Vol. 197, 1055 - 1065 (1981).
    The authors consider the infrared development of novae
from the point of view of the behaviour of the underlying
system. They describe the various transition-related phenomena and show that novae surrounded prior to outburst by small
graphite grains begin to radiate significantly at infrared wavelengths when the photospheric flux in the visual has declined
by ~4 mag from maximum, and that this gives rise to the
coincidence between infrared development and transition.

124.006   Generation of an external ring during the 1978 out-
        burst of WZ Sagittae.
E. M. Leibowitz, T. Mazeh.
Astrophys. J., Vol. 251, 214 - 220 (1981).
    It has been proposed that the emission lines observed in
the spectrum of WZ Sge during its recent outburst originated
in a circumstellar gaseous ring surrounding this entire binary
system. The authors suggest a possible mechanism for the
formation of this ring. The luminosity of the outburst, near
or at the surface of the white dwarf component of the system,
stops for a few days the mass transfer from the secondary onto
a disk around the primary through the Lagrangian point L_1.
The radiation pressure causes instead a flow of mass through
the L_2 point to a new ring around the system. A few relevant

observations during the outburst episode are discussed and
found to be consistent with this model.

124.007   Ultraviolet spectroscopy of the recurrent nova
        U Scorpii during outburst.   R. E. Williams,
W. M. Sparks, J. S. Gallagher, E. P. Ney, S. G. Starrfield,
J. W. Truran.
Astrophys. J., Vol. 251, 221 - 229 (1981).
    Observations of the recurrent nova U Sco during the 1979
outburst have been obtained with the IUE satellite. The
spectral evolution has been found to differ from that of other
recurrent novae. The spectra are dominated by emission lines,
though broad resonance absorption is present during the first
week after outburst, and the strong forbidden-line emission
characteristic of seemingly similar systems such as T CrB and
RS Oph is conspicuously absent. A method for determining the
masses of nova shells is outlined, and a mass $\sim 10^{-8} - 10^{-7} M_\odot$
is inferred for the ejecta. Analysis of the emission lines reveals
the nova ejecta to be rich in He relative to H and shows an
enrichment in N relative to C and O. Optical spectra of U Sco
obtained follwing its return to quiescence show predominantly
He II emission lines, suggesting an enrichment of the preout-
burst gas in He and thus the presence of a highly evolved companion.

124.008   Radiatively accelerated nova envelopes.
    G. J. Ferland, J. W. Younger.
Astrophys. J., Lett., Vol. 251, L17 - L20 (1981).
    Photoionization calculations which describe the thermal
structure, ionization, and radiative acceleration of the dense
envelope surrounding an active nova are described. Assuming
conditions appropriate for the period near maximum light,
the authors show that the envelope will be strongly accelerated
outward when the nova has a luminosity at or even below the
classical Eddington limit. Such a luminosity is effectively
super-Eddington in the outer layers of the envelope because of
the increased opacity of the gas due to photoelectric and line
absorption. The calculations suggest terminal velocities and
mass loss rates close to those observed, and may provide a
mechanism to produce the discrete velocity systems usually
observed near maximum light.

124.009   The maximum luminosity-rate of decline relation
        and the problem of distance determination of novae.
A. S. Sharov.
Perem. Zvezdy, Tom 21, 141 - 147 (1979). In Russian.
    Data on novae discovered in the region of M31 especially
at large distances from its center are discussed. It is concluded
that either the statistical relation between maximum luminosity
and rate of decline has a large dispersion or we may consider a
number of novae as very distant objects from the galaxies.

124.010   PU Vulpeculae (object Honda - Kuwano 1979)-
        78.1 days period.
D. Chochol, L. Hric, J. Papoušek.
Inf. Bull. Variable Stars, No. 2059, 4 pp. (1981).

124.011   On a possible nova in Cygnus.
    G. N. Kimeridze, R. Ya. Inasaridze.
Astron. Tsirk., No. 1160, p. 7 - 8 (1981). In Russian.

    A spirit of search.   See Abstr. 031.637.

    On the evolution of thermal disturbances in stars.
See Abstr. 062.052.

    Observations of the nova-like variable FY Per.
See Abstr. 113.058.

S 10830 Trianguli – an interesting cataclysmic variable.   See Abstr. 117.064.

V 1329 Cyg: a nova-like object with recurrent outbursts?   See Abstr. 122.076.

X-ray observations of cataclysmic variables.
See Abstr. 142.078.

Novae and the extra-galactic distance scale.
See Abstr. 162.057.

Nova Delphini 1967 = HR Delphini

124.101 Direct photography of the nebular remnant of nova Delphini 1967.   L. Kohoutek.
Mon. Not. R. Astron. Soc., Vol. 196, 87P - 89P (1981).

Direct photographs of N Del 1967 taken in the light of the [O III] 5007 Å line show an oval nebular remnant: angular size 3.7 × 2.5 arcsec, the position angle of the large (probably equatorial) axis is 43°. The nebular expansion method yields a distance of 850 ± 50 pc.

124.102 Variation of the structure of emission lines in the spectrum of nova HR Delphini.   M. B. Babaev.
Pis'ma Astron. Zh., Tom 7, 422 - 427 (1981). In Russian.
English translation in Soviet Astron. Lett., Vol. 7.

Variations in the structure of emission lines with evolution of the spectrum of the nova HR Del are discussed. A comparison is made between intensities of the internal and external components of Balmer hydrogen lines, lines of helium, nitrogen, oxygen, and neon. It is concluded that there are some time variations of physical conditions in the envelope of HR Del.

124.103 The UV spectra of two old novae (HR Del and V 603 Aql).   D. Dultzin-Hacyan, Y. Andrillat, J. Audouze, M. Friedjung, C. Gordon, B. Rocca-Volmerange, G. Stasinska.
Second European IUE Conference, (see 012.011), p. 87 - 89 (1980).

Two old novae have been observed by IUE (short and long wavelength regions). The reddening has been evaluated by the disappearance of the absorption feature at 2200 A. This determination enables the authors to evaluate the UV fluxes and to fit the HR Delphini flux to an approximate 40000 K black body curve and a power law for far UV wavelengths. V 603 Aql is better fitted by two black body curves of about 50000 K and 8000 K for far UV wavelengths. An evaluation of the total emitted energy is given. In addition a P Cygni profile has been found for the C IV doublet in the HR Del spectrum and allows the authors to obtain a minimum mass loss rate and provides some information on the nature of the wind.

124.104 Light variability of nova Delphini 1967 in 1980.   L. Kohoutek, R. Pauls, H.-M. Steinbach.
Inf. Bull. Variable Stars, No. 1991, 4 pp. (1981).

Nova LMC 1977b

124.121 Spectral properties of Nova LMC 1977b.   R. Canterna, L. F. Thompson.
Publ. Astron. Soc. Pacific, Vol. 93, 581 - 586 (1981).

Spectra obtained of Nova LMC 1977b have been reduced and analyzed. An expansion velocity of 625 km sec^{-1} from the absorption lines of the principal spectrum was observed 1.16 days after maximum. From the emission profiles of Hα and [O I] λ6300 an expansion velocity of 1400 km sec^{-1} was observed 13.5 days after maximum. Estimates of mass loss, photospheric radius and total energy release are given.

Nova Aquilae 1918 = V603 Aquilae

124.201 A photometric study of the old nova V603 Aquilae.   M. H. Slovak.
Astrophys. J., Vol. 248, 1059 - 1066 (1981).

High-speed photometry is presented for the old nova V603 Aquilae. No evidence for regular eclipses or other periodic features was found in the light curves. The periodic variations discovered by Rahe et al. using the IUE satellite remain a puzzle. It is suggested these variations may arise from the formation of transient features in the accretion disk. Recent IUE and X-ray data support the low inclination ($i \approx 15°$) previously inferred from spectroscopic studies.

124.202 Nova Aql 1918: a nude old nova.   P. L. Selvelli, A. Cassatella.
Effects of mass loss on stellar evolution, (see 012.015), p. 515 - 522 (1981).

IUE observations at high and low resolution of Nova Aql 1918 show neither evidence of outflow nor the presence of nebular lines. This indicates that the shell ejected at the time of the outburst and surrounding the system for many years has by now disappeared. The high excitation spectrum presents rapid variations in the far UV and eclipse effects in the near UV that seem well correlated with the orbital phase. The observations can be interpreted in terms of phenomena occurring in or near the accretion disk surrounding a white dwarf.

124.203 Periodic light variations of V 603 Aql.   R. Haefner.
Inf. Bull. Variable Stars, No. 2045, 4 pp. (1981).

The UV spectra of two old novae (HR Del and V 603 Aql).   See Abstr. 124.103.

Nova Cygni 1980

124.221 Honda's variable in Cygnus.
IAU Circ., No. 3628 (1981).

124.222 Objekt Honda 1980.   H. Geßner.
Mitt. Veränderl. Sterne (MVS), Band 9, 15 (1981).

Nova in Large Magellanic Cloud

124.241 Nova in Large Magellanic Cloud.
IAU Circ., Nos. 3641, 3648 (1981).

Nova RR Telescopii

124.301 IUE and other new observations of the slow nova RR Tel.   M. V. Penston, P. Benvenuti, A. Cassatella, A. Heck, P. Selvelli, D. Ponz, F. Macchetto, C. Jordan, N. Cramer, F. Rufener, J. Manfroid.
The Universe at ultraviolet wavelengths, (see 012.009), p. 469 - 470 (1981). – Abstract.

Nova Pictoris 1925 = RR Pictoris

124.321   **Rapid oscillations in RR Pictoris
(Nova Pictoris 1925).**
R. Schoembs, B. Stolz.
Inf. Bull. Variable Stars, No. 1986, 4 pp. (1981).

Nova Persei 1901 = GK Persei

124.341   **The orbital period of the old nova GK Per (1901).**
A. Bianchini, F. Sabbadin, E. Hamzaoglu.
Inf. Bull. Variable Star, No. 1988, 2 pp. (1981).

Nova Cygni 1978 = V1668 Cygni

124.401   **Nova Cygni 1978 – I. The nebular phase.**
D. J. Stickland, C. J. Penn, M. J. Seaton,
M. A. J. Snijders, P. J. Storey.
Mon. Not. R. Astron. Soc., Vol. 197, 107 - 138 (1981).

UV observations of Nova Cygni 1978, which were obtained using IUE on 17 dates between 1978 September 12 and
1979 July 8 inclusive, will be discussed in the present series
of papers. Paper I is mainly concerned with the interpretation,
during the nebular stage, of the UV observations together with
optical and IR data from other observers.

124.402   **The distance and visual lightcurve of Nova
Cygni 1978.**   S. S. Mims, R. W. James.
Contrib. No. 6 Inst. Astron. Res., 3 pp. (1981).

Nova Coronae Austrinae 1981

124.421   **On the location of nova Coronae Austrinae 1981.**
N. Brosch.
Inf. Bull. Variable Stars, No. 2003, 3 pp. (1981).

Nova BD Pavonis

124.441   **Detection of periodic light variations of the old
nova BD Pav.**
H. Barwig, R. Schoembs.
Inf. Bull. Variable Stars, No. 2031, 3 pp. (1981).

Nova Sagittarii 1978 = V3876 Sagittarii

124.501   **Spectral features of Nova Sagittarii 1978.**
M. F. McCarthy, B. M. Lasker, T. D. Kinman.
Publ. Astron. Soc. Pacific, Vol. 93, 470 - 473 (1981).

Finding charts and spectroscopic data are presented for
Nova Sgr 1978. The outstanding spectroscopic feature is the
$\lambda 6085$ line of the highly ionized species [Fe VII].

Nova Vulpeculae 1968 No. 1 = LV Vulpeculae

124.601   **General characteristics of nova LV Vulpeculae.**
D. Rajkova.
Astrofiz. issled., NRB, Vol. 3, 41 - 47 (1981). In Russian.
Abstr. in Ref. zh., 51. Astron., 8.51.656 (1981).

Nova Trianguli 1963

124.701   **S 10828 Trianguli: an unusual nova in M33.**
G. A. Richter, F. Börngen.
Astrophys. Lett., Vol. 21, 101 - 102 (1981).

S 10828 = Nova Trianguli 1963 belongs to the extreme
halo of M33 at a distance from the centre greater than 17 kpc.

Nova Cygni 1975 = V1500 Cygni

124.801   **Once more about nova Cygni 1975.**
Z. Alksne.
Zvaigžnotā debess, 1980. gada rudens, p. 25 - 28. In Latvian.

124.802   **Photometric observations of Nova V1500 Cygni in
the years 1977 - 1979.**
H. H. Lanning, I. Semeniuk.
Acta Astron., Vol. 31, 175 - 183 (1981).

Photometric observations of V1500 Cygni in the time
interval November 1977 - July 1979 are presented. The 3 hour
variations are still present 4 years after the outburst. The
amplitude of light variability stays at the level of 0.5 mag. The
upper limit on the period change $\dot{P}/P$ is 4.2×10^{-5} year^{-1}.

124.803   **Simultaneous TV observations of nova
Cygni 1975 (V1500 Cyg) in some parts of the
spectrum.**   A. N. Abramenko, E. P. Pavlenko.
Astron. Tsirk., No. 1129, p. 1 - 3 (1980). In Russian.

125 Supernovae, Supernova Remnants

125.001 Cooling and evolution of adiabatic blast waves in a dusty medium. E. Dwek.
Astrophys. J., Vol. 247, 614 - 627 (1981).

The author presents detailed numerical calculations of the cooling and destruction of collisionally heated interstellar dust grains that are swept up by adiabatic supernova blast waves. He calculates the infrared luminosity that is emitted from the interior of the supernova cavity, studies the effect of the energy loss on the evolution of the remnant, and briefly discusses recent infrared observations of several young supernova remnants.

125.002 Supernova remnants in M31.
W. P. Blair, R. P. Kirshner, R. A. Chevalier.
Astrophys. J., Vol. 247, 879 - 893, plate 28 (1981).

Image-tube photographs of eight fields in M31, using interference filters that isolate Hα + [N II] and [S II], have revealed nebulae that may be supernova remnants in this galaxy. Spectroscopic observations show that 11 of these nebulae resemble galactic supernova remnants (SNRs). The [O III] temperatures determined for four of these SNRs are in the range of 37,000 - 65,000 K. The [S II] ratios indicate densities in the S^+ zone in the range of 150 - 730 cm^{-3}.

125.003 Supernova remnant evolution in an inhomogeneous medium. I. Numerical models.
L. L. Cowie, C. F. McKee, J. P. Ostriker.
Astrophys. J., Vol. 247, 908 - 924 (1981).

The authors present the first numerical simulations of supernova remnant evolution in an inhomogeneous gas. Evolution in the lowest density substrate (the intercloud) is assumed to be spherically symmetric with a large intercloud filling factor and many dense regions (clouds) within the remnant; however, mass momentum and energy transfer between cloud and intercloud are included and the position and morphology of individual clouds tracked.

125.004 On the acceleration of cosmic rays in supernova explosions. V. N. Fedorenko.
Astron. Zh., Tom 58, 790 - 795 (1981). In Russian. English translation in Soviet Astron., Vol. 25, No. 4.

It is shown that the radio emission from supernova remnants with the age $t_L \gtrsim 10^2$ yrs can result from relativistic electrons accelerated at the stage $t_L \sim 10^2$ yrs. A model is developed in which the acceleration process occurs in the shock wave front and also by pulsar magnetic dipole radiation. A possible explanation is given of the lack of the radio emission of supernova remnants at the stage $t_L \lesssim 30$ yrs.

125.005 Spatial distribution of type I and type II supernovae in spiral galaxies. D. Yu. Tsvetkov.
Pis'ma Astron. Zh., Tom 7, 463 - 465 (1981). In Russian. English translation in Soviet Astron. Lett., Vol. 7.

In spiral galaxies the thickness of the layer of type I supernova formation is shown to be less than the thickness of the dust layer. No difference has been found in the values of the surface density gradient for type I and type II supernovae.

125.006 Radio-emitting supernovae are young plerions.
I. S. Shklovskij.
Pis'ma Astron. Zh., Tom 7, 479 - 481 (1981). In Russian. English translation in Soviet Astron. Lett., Vol. 7.

A hypothesis is substantiated that the radio emission of supernovae is due to young plerions forming immediately after the outburst. When the plerion expands its power decreases approximately as R^{-1}. It is predicted that radio emission can be attributed only to type II supernovae outbursts.

125.007 The infrared emission from supernova condensates. E. Dwek, M. W. Werner.
Astrophys. J., Vol. 248, 138 - 151 (1981).

The authors examine the possibility of detecting grains formed in supernovae by observations of their emission in the infrared. The basic processes determining the temperature and infrared radiation of grains in supernovae environments are analyzed, and the results are used to estimate the infrared emission from the highly metal enriched "fast moving knots" in Cas. A. The predicted fluxes lie within the reach of current ground-based facilities at 10 μm, and their emission should be detectable throughout the infrared band with cryogenic space telescopes.

125.008 Two X-ray supernova remnants: G296.1−0.7 and 1E 1149.4−6209. T. H. Markert, R. C. Lamb, R. C. Hartman, D. J. Thompson, G. F. Bignami.
Astrophys. J.,Lett., Vol. 248, L17 - L21, plate L1(1981).

Using the imaging X-ray detectors on the Einstein Observatory, the authors have discovered what appear to be two overlapping galactic supernova remnants; one of which is clearly identified with the previously cataloged radio remnant G296.1−0.7. The other feature has no radio or optical counterpart. It is nearly complete ring with a diameter of $20' - 25'$, designated 1E 1149.4−6209. Because of its morphology, because there is evidence for supernova events in the vicinity, and because there seems to be no plausible alternative, the authors classify 1E 1149.4−6209 as a supernova remnant.

125.009 High resolution X-ray and radio images of the Crab-like supernova remnant G21.5−0.9.
R. H. Becker, A. E. Szymkowiak.
Astrophys. J., Lett., Vol. 248, L23 - L26 (1981).

The Crab-like supernova remnant G21.5−0.9 has been imaged at 6 cm wavelength with the VLA and between 0.1−4 keV with the Einstein Observatory. Both the X-ray and radio images show a centrally peaked distribution of extended emission with no indication of a pointlike component. In contrast, the linearly polarized radio emission is distributed in a ring with a distinct central minimum. The electric vectors of the polarized radio emission are oriented circumferentially. The data support a Crab-like model for G21.5−0.9, but one in which a lower percentage of the radiated flux is emitted as X-rays relative to the Crab Nebula.

125.010 Density, velocity, and temperature profiles for the extended envelope model of type I supernovae.
G. Lasher.
AIP Conf. Proc., No. 63, (see 012.001), p. 1 - 6 (1980).

The early light curve of a type I supernova has been fitted by a model in which the low density envelope of a supergiant is exploded by the sudden release of energy at its center. This paper supplements previous publications on this model by giving information about the model which is necessary for the computation of the emitted spectrum, namely, the density, temperature and velocity profiles of the expanding shell.

125.011 The light curve of type I supernovae.
S. A. Colgate, A. G. Petschek, J. T. Kriese.
AIP Conf. Proc., No. 63, (see 012.001), p. 7 - 14 (1980).

Calculations of the intermediate and late time luminosity of type I supernovae based on 100% efficiency for optical emission of energy deposited by the Ni^{56} decay chain give good agreement with observations provided $M_{ej} v^{-2} = (2.2 \pm 0.5) \times 10^{17} M_\odot s^2 cm^{-2}$ where M_{ej} is the ejected mass and v is the expansion velocity. Account must be taken of the escape of both gamma rays and positrons. These two escape processes as well as the early luminosity peak, as calculated by Colgate and

McKee, are all consistent with the same value of M_{ej}/v^2.

125.012 Supernova models and light curves.
T. A. Weaver, S. E. Woosley.
AIP Conf. Proc., No. 63, (see 012.001), p. 15 - 32 (1980).

The paper reviews the current status of understanding of type II supernovae with particular emphasis on the processes responsible for the emission of electromagnetic radiation. A relatively novel evolutionary scenario that appears to lead to a type I supernova explosion is presented.

125.013 Type I supernovae: an observer's view.
R. P. Kirshner.
AIP Conf. Proc., No. 63, (see 012.001), p. 33 - 37 (1980).

125.014 Synthetic spectra of supernovae.   D. Branch.
AIP Conf. Proc., No. 63, (see 012.001), p. 39 - 48 (1980).

The predicted characteristics of individual line profiles and blends formed by resonance scattering in supernova envelopes are described. Synthetic spectra consisting of blended scattering profiles on an underlying continuum are compared with McDonald Observatory spectra of a type I and a type II supernova. Some rough constraints on the density profiles and chemical compositions of the observable layers of the envelopes are given.

125.015 The excitation of spectra in the envelopes of supernovae at late times by the deposition of positrons and γ-rays.   R. E. Meyerott.
AIP Conf. Proc., No. 63, (see 012.001), p. 49 - 74 (1980).

The excitation of spectra in the envelopes of supernovae at late times by the deposition of positrons and γ-rays is analyzed. In type I supernovae envelopes at late times, the principal source of cooling is forbidden line radiation. In type II supernovae envelopes, charge and energy transfer collisions with H^+ and He^+ may be important in determining the excitation and ionization states of the minor species.

125.016 Some statistical properties of Type I supernovae.
D. Branch.
Astrophys. J., Vol. 248, 1076 - 1080 (1981).

A list of measurements of the wavelength of the 6130 Å absorption feature in the spectra of 16 Type I supernovae (SN I) has been compiled from the literature. The corresponding matter velocities at the photosphere are used to test the validity of proposed subclassifications of SN I. The "fast" and "slow" subclasses defined by Barbon and his collaborators on the basis of light curve properties do have significantly different expansion velocities. The data suggest, however, that SN I form a continuous sequence, rather than two discrete subclasses, and confirm Pskovskii's conclusion that expansion velocity and peak absolute magnitude are correlated with the rate of decline of the postpeak light curve.

125.017 An oxygen-rich young supernova remnant in the Small Magellanic Cloud.
M. A. Dopita, I. R. Tuohy, D. S. Mathewson.
Astrophys. J., Lett., Vol. 248, L105 - L108 (1981).

The second brightest X-ray source in the Small Magellanic Cloud (SMC), 1E 0102.2 – 7219, has been identified as a bright, optically emitting supernova remnant (SNR) by means of narrow-band imaging with the Anglo-Australian Telescope. The remnant is invisible at Hα but prominent in its [O III] emission and so belongs to the class of young SNRs resulting from the explosion of a Population I type star. This is the first such remnant discovered in the SMC.

125.018 A dynamical model of type I supernova atmosphere with the velocity gradient.
V. S. Imshennik, D. K. Nadyozhin (*Nadezhin*), V. P. Utrobin.
Astrophys. Space Sci., Vol. 78, 105 - 122 (1981).

A compact structure of a low-mass type I presupernovae is assumed to be an essential feature of the hydrodynamical problem dealing with the supernova type I (SNI) envelope outbursts. The structure of the SNI atmosphere expanding with the velocity gradient can be employed for an interpretation of the observed SNI spectra. The SNI light curves are considered to occur due to an additional slow (with time-scale 10^6 - 10^7 s) release of the bulk of the SNI energy, $W \cong 10^{51}$ erg. A short ($\Delta t \cong 10^{-2}$ s) burst of soft (2 - 10 keV) X-rays with total radiated energy of about 10^{40} erg is found to appear 10 - 20 days before the SNI optical maximum.

125.019 Supernova remnants and H II regions in M31.
M. Dennefeld, D. Kunth.
Astron. J., Vol. 86, 989 - 997 (1981).

New spectroscopic data obtained on SNR candidates and H II regions in M31 are used to confirm the SNR diagnostic and to derive abundances. With all data available in the literature, confirmation is obtained for 12 SNR from a detailed study of some characteristic line ratios. In the absence of a direct temperature determination for the H II regions and the SNR, empirical methods are used to derive oxygen and nitrogen abundances. While N/H and N/O gradients are present with values comparable to the ones found in our Galaxy, no evidence exists for one in oxygen, in view of the large uncertainties in the abundance determinations. The SN rate is found to be about one in 200 yr, at variance with the higher rate (one every 40 yr) found from SN statistics. The difference between the two determinations is found to be smaller in M33.

125.020 New meter-wavelength observations of S 147.
P. E. Angerhofer, M. R. Kundu.
Astron. J., Vol. 86, 1003 - 1009 (1981).

The authors present recent high-resolution 430-MHz (69.8 cm) continuum observations of the old galactic supernova remnant S 147. These new low-frequency measurements indicate the presence of a well defined radio shell, in which are apparently embedded numerous small-diameter radio sources. Included are four 4C sources, several B2 sources, and three anonymous sources. The nature of these small-diameter radio sources and their association with the supernova remnant are discussed. The 430-MHz flux density (97 ± 20 Jy) reported here supports the suggestion of a break in the flux density spectrum.

125.021 A compressed cloud in the Vela supernova remnant.
E. B. Jenkins, G. Wallerstein, E. M. Leep, J. Silk.
The Universe at ultraviolet wavelengths, (see 012.009), p. 589 - 594 (1981).

To elucidate the nature of the interstellar medium in the vicinity of the Vela SNR, the authors have undertaken an extensive study with the IUE of interstellar absorption lines toward 35 stars in the vicinity of the Vela SNR. Observations of interstellar absorption, in particular of CI, towards one of these stars, HD 72350 (type B4 III), are reported in this paper.

125.022 IUE observations of supernova remnants.
J. C. Raymond.
The Universe at ultraviolet wavelengths, (see 012.009), p. 595 - 599 (1981).

The author discusses the UV emission spectra of several filaments in the Cygnus Loop and the Vela Supernova Remnant, including several which are anomalously bright in the optical O III lines. The effects of internal and interstellar resonance line scattering are estimated. Shock velocities, elemental abundances, and in some cases filament ages are given.

125.023 A search for young optical pulsars in the galaxies NGC 4647 and NGC 4321.

G. M. Beskin, V. S. Lebedev, S. I. Neizvestnyj,
V. L. Plakhotnichenko, V. F. Shvartsman.
Pis'ma Astron. Zh., Tom 7, 537 - 542 (1981). In Russian.
English translation in Soviet Astron. Lett., Vol. 7.

The results of a search for optical variability of supernovae (SN) exploded in 1979 in the galaxies NGC 4647 and
NGC 4321 are presented. There is no variability in the whole
investigated time range from 10^{-6} to 100 s. The optical luminosity of pulsars with periods $10^{-3} - 10^{-1}$ s does not exceed
3×10^{40} erg/s for SN in NGC 4647 and 2×10^{39} erg/s for SN in
NGC 4321 on the 60th and 300th day after the brightness
maximum, respectively. Possible interpretations of the negative
results of search for extragalactic pulsars are discussed.

125.024 **Supernova remnants and abundance gradients in M31.**
W. P. Blair, R. P. Kirshner, R. A. Chevalier.
Bull. American Astron. Soc., Vol. 13, 518 - 519 (1981).
Abstract.

125.025 **Soft X-ray observation of supernova remnant IC443.**
C. M. F. Galas, D. Venkatesan, G. Garmire.
Bull. American Astron. Soc., Vol. 13, 550 (1981). – Abstract.

125.026 **Sequential explosions of supernovae in an OB**
association and formation of a superbubble.
K. Tomisaka, A. Habe, S. Ikeuchi.
Astrophys. Space Sci., Vol. 78, 273 - 285 (1981).

From the standpoint of view that the early type stars are
formed sequentially at an OB association, it is expected that
the supernova explosions will also occur sequentially. The
authors study the expansion law of a supernova remnant,
which is formed by sequential explosions of supernovae. The
superbubbles and supershells with the radii 200 ~ 1000 pc are
naturally explained by this model. Assuming that the sequential explosion of supernovae occurs at ever OB association, the
authors deduce the star formation rate in our Galaxy.

125.027 **Observations of supernovae.**
N. Panagia.
Second European IUE Conference, (see 012.011), p. XXVII -
XXXI (1980).

The IUE observations of supernovae are reviewed. The
well studied case of the bright supernova 1979c in the galaxy
M 100 is discussed in some detail. The most important findings
for SN 1979c are presented.

125.028 **Far UV spectrophotometry of SNR.**
P. Benvenuti, S. D'Odorico, M. Dopita.
Second European IUE Conference, (see 012.011), p. 11 (1980).
Abstract.

125.029 **A relationship between SNR G109.1−1.0 and the**
molecular cloud of Sh2−152?
M. Heydari-Malayeri, C. Kahane, R. Lucas.
Nature, Vol. 293, 549 - 550 (1981).

The authors report recent observations of the complex
Sh2−147/Sh2−153 in the molecular line ^{13}CO $(J = 1 \rightarrow 0)$.
Their results are compared with those of Israel (1980) in ^{12}CO.
Several lines of evidence point to a physical association of
SNR G109.1−1.0 and the X-ray source GF2259+586 with the
molecular cloud of Sh2−152.

125.030 **Spectra and internal motions of supernova remnants**
and ring nebulae in the Magellanic Clouds.
B. M. Lasker.
Publ. Astron. Soc. Pacific, Vol. 93, 422 - 427 (1981).

New data are presented and interpreted for supernova
remnants (SNR) and for other large ring nebulae in the Magellanic Clouds. Shock velocities determined for the SNRs by
different methods are reasonably consistent; however, densities
determined by different methods are slightly discordant, probably because of the effects of preionization, magnetic fields,

or some source of excitation. Emission from the large rings,
N9 and N70, may be characterized by a single temperature
(~12,000 K), which is more like the situation expected in an
H II region than in a recombination zone behind a shock.

125.031 **A study of a supernova remnant.**
K. S. Dwarakanath, C. V. Sastry.
Bull. Astron. Soc. India, Vol. 9, 77 (1981). – Abstract.

125.032 **What stars make supernovae?**
R. K. Kochhar, T. P. Prabhu.
Bull. Astron. Soc. India, Vol. 9, 83 (1981). – Abstract.

125.033 **A first roadmap for kryptology.**
D. Heymann, M. Dziczkaniec.
Proc. Eleventh Lunar Planet. Sci. Conf., (see 012.020),
p. 1179 - 1213 (1980).

The authors have calculated isotopic compositions of
xenon and krypton which issue forth from a number of
specific locations of a supernova from a star with zero-time
main-sequence mass of 25 solar masses. They surmise that
isotopic variations of Kr and Xe in carbonaceous chondrites
may be excellent monitors for conditions in the deep interiors
of massive stars during the last few days before as well as
during the supernova explosion.

125.034 **Neutrino mass and detection of neutrino supernova**
bursts. T. Piran.
Phys. Lett. B, Vol. 102B, 299 - 302 (1981). – Abstr. in Phys.
Abstr., Vol. 84, Abstr. 88962 (1981).

125.035 **Exploding stars and their remnants.**
I. S. Shklovskij.
Zemlya Vselennaya, 1981, No. 4, p. 15 - 21; No. 5, p. 32 - 37.
In Russian.

125.036 **Formation of supernova remnants: the pre-blast-**
wave phase.
E. M. Jones, B. W. Smith, W. C. Straka.
Astrophys. J., Vol. 249, 185 - 194 (1981).

The effect of stellar structure on supernova remnant
formation is studied with a series of computer models of a
10^{51} erg explosion in a 15 $M_\odot$ star. The authors find that
immediately after the explosion shock wave travels down a
steep density gradient, the material in the gradient goes into
free expansion, forming a collapsible piston. At the outer edge
of such a piston are two shock waves: the expanding supernova shock and a reverse shock moving back into the
collapsible piston. Until the piston is completely collapsed it is
Rayleigh-Taylor stable, but after collapse the inner material
behaves as a massive piston and the interface is R-T unstable.
If there is a significant mass in an external density gradient,
the material between the supernova shock moving out through
the interstellar medium and the reverse shock will be a significant source of X-rays during the pre–blast-wave phase of
remnant formation.

125.037 **Estimate of the ionization and mass of SN I enve-**
lopes based on the radioactive hypothesis.
I. S. Shklovskij.
Astron. Zh., Tom 58, 1017 - 1021 (1981). In Russian. English
translation in Soviet Astron., Vol. 25, No. 5.

Analysis of SN spectra observations made in 1972 fully
confirms the hypothesis of radioactive ^{56}Ni decay being responsible for the "exponential" portion of the SN type I light
curves. The interaction of positrons generated in the β-decay of
^{56}Co with matter ejected in the explosion of a supernova envelope has been studied, which permits ionization of the latter
to be estimated. The conclusion follows that neutron stars
form in SN I outbursts as well as in SN II outbursts. In the
Universe nucleosynthesis of iron, however, occurs only in
SN I outbursts.

125.038   Summary of radial velocities of fast-moving condensations of Cassiopeia A.   Yu. P. Pskovskij.
Astron. Zh., Tom 58, 1022 - 1027 (1981). In Russian. English translation in Soviet Astron., Vol. 25, No. 5.

A summary list of radial velocities and coordinates of 141 fast-moving condensations in Cassiopeia A is compiled on the basis of the works by Minkowski, van den Bergh and Kirshner-Chevalier. In addition, 41 condensations of these also have proper motion estimates according to Kamper – van den Bergh.

125.039   A broad line in the meter wave range spectrum of the Cassiopeia A supernova remnant.
M. I. Agafonov, K. S. Stankevich.
Pis'ma Astron. Zh., Tom 7, 612 - 616 (1981). In Russian. English translation in Soviet Astron. Lett., Vol. 7.

A spectral line has been detected in the spectrum of Cassiopeia A radio emission with a maximum at the frequency of 114 MHz and a width of 35 MHz at half-power level. At the line maximum the intensity exceeds the known synchrotron spectrum of the source by 18%. The line is interpreted as a maser radiation in the magnetosphere of a neutron star inside the supernova remnant.

125.040   Analytical methods for the hydrodynamical evolution of supernova remnants. II. Arbitrary form of boundary conditions.   B. Gaffet.
Astrophys. J., Vol. 249, 761 - 786 (1981).

The paper is an extension of an earlier work in which the author presented an analytical method for computing the hydrodynamical evolution of supernova remnants. The author introduces a generalized formalism so as to obtain formulae which are valid for the most general type of boundary conditions at the outer edge. In addition, the higher orders of the approximation are constructed in a more rigorous way, based on an iteration method. The second-order solution then gives the pressure distribution with a relative accuracy of about 10^{-4} in self-similar cases with parameters which may be considered as typical of supernova remnants. The author further discusses the accuracy to be expected in non-self-similar cases.

125.041   The radial velocity field of the optical filaments associated with the SNR W63.
M. Rosado, J. González.
Rev. Mexicana Astron. Astrofis., Vol. 5, 93 - 99, plates 2, 3 (1981).

The authors obtained the radial velocity field of some optical filaments of the SNR W63. They estimated for its expansion velocity values of 35 ± 12 km s^{-1} and 70 ± 30 km s^{-1} depending on whether the SNR is thought to be the non-thermal radio source W63 ($< 1°$) or the whole filamentary complex shown on Palomar Plates ($\sim 4°$). The authors derived the SNR ratio E_0/n_0 for both cases and found upper limits for the pre-shock density.

125.042   Supernovae and star formation in clusters of galaxies.
C. N. Caldwell, A. Oemler, Jr.
Astron. J., Vol. 86, 1424 - 1428 (1981).

It has recently been suggested that all supernovae, including those in early-type galaxies, are the products of recent star formation. If this is so, and if elliptical and S0 galaxies in the cores of dense clusters are swept of gas by the intracluster medium, one should expect these galaxies to have a lower supernova rate and redder colors than similar galaxies in the field. The authors demonstrate the existence of both effects, with amplitudes consistent with the predictions of the star-formation hypothesis.

125.043   High-resolution Hα spectra of an outer filament in the Cygnus Loop.   R. R. Treffers.
Astrophys. J., Vol. 250, 213 - 215 (1981).

Spectra of the Hα line of a faint filament at the western edge of the Cygnus Loop are presented. This filament is interesting because it shows only hydrogen line emission. The data are best fit by two approximately equal components: one with a FWHM of ≈ 35 km s^{-1}, and a second broader component with a FWHM of > 130 km s^{-1}. These components may represent pre- and postshock kinetic temperatures of $\approx 10^4$ and $> 4 \times 10^5$ K, and they may imply a shock velocity greater than or equal to 130 km s^{-1}.

125.044   Soft X-ray observation of supernova remnant IC 443.
C. M. F. Galas, D. Venkatesan, G. Garmire.
Astrophys. J., Vol. 250, 216 - 221 (1981).

The low-energy X-ray spectrum (0.15–3.0 keV) and X-ray centroid position of the supernova remnant IC 443 have been determined using data obtained by the HEAO-1 A-2 experiment. The spectrum exhibits a low-energy excess over and above that predicted by an extrapolation of the higher-energy spectrum determined by Parkes et al. The excess can be attributed to line emission, predominantly from high-ionization states of iron from a plasma at a temperature of $7.7^{+0.6}_{-4.3} \times 10^6$ K. Other models which fit the data are discussed, in particular, a lower-temperature component required to explain the Fe X optical emission.

125.045   Kinematics of the supernova remnant G65.2 + 5.7 in Cygnus.   M. Rosado.
Astrophys. J., Vol. 250, 222 - 226, plate 17 (1981).

The radial-velocity field of some filaments of the SNR G65.2 + 5.7 has been obtained by means of Fabry-Perot interferometry. The author finds that these specific filaments are expanding with a velocity of about 90 km s^{-1}. Inhomogeneities in the interstellar medium seem to be required in order to explain the optical data. The type of inhomogeneities that could be present and their observational consequences are also discussed.

125.046   Near-infrared spectroscopy of northern supernova-remnants.   M. Dennefeld, Y. Andrillat.
Astron. Astrophys., Vol. 103, 44 - 49 (1981).

Spectroscopy of bright northern SNR in the spectral range 8500 Å-11000 Å reveals the presence of lines of [S II], [S III], [CI] and He I. Comparison with shock-ionization models suggests a carbon deficiency of about 2 with respect to cosmic values for the Cygnus Loop and at least 150 for the fast-moving knots in Cas A in good agreement with previous independent determinations. A lower limit of 0.27 in number is derived for the C/O ratio in the Crab nebula, however a firm value cannot be given unless models are produced which explain satisfactorily the very presence of the [C I] lines in this object.

125.047   Further radio observations of W 50: total intensity and linear polarization measurements at 1.7 and 2.7 GHz.   A. J. B. Downes, T. Pauls, C. J. Salter.
Astron. Astrophys., Vol. 103, 277 - 287 (1981).

Radio continuum observations at 1.7 and 2.7 GHz are presented of the supernova remnant W 50 which is associated with the unusual object SS 433. Both total intensity and linear polarization were measured at the two frequencies, and parameters for the Faraday rotation and depolarization are derived. The remnant shows polarization of up to 10% of the total intensity at 1.7 GHz and up to 40% at 2.7 GHz. The S 74 loop is completely unpolarized, as would be expected from a thermal source. The part of the remnant nearest S 74 is also unpolarized, suggesting that S 74 lies nearer to us than does W 50, and may be causing depolarization of that region of W 50. This would put a lower limit of 2.3 kpc on the distance of W 50. SS 433 was strong during both sets of measurements.

125.048   The point radio source in the supernova remnant G 78.2 + 2.1.   L. A. Higgs, R. S. Roger,

T. L. Landecker, S. R. Spangler, J. M. Cordes, J. M. Dickey.
Astron. Astrophys., Vol. 103, 370 - 373 (1981).

Observations of a point source, 2020 + 401, in the eastern rim of the supernova remnant G 78.2 + 2.1 have been made at 20 cm, 6 cm, and 2 cm using the Very Large Array. This source was previously thought to be variable, but no variability was detected over a two-month time interval. The radio properties of the source give no reason to associate it with the supernova remnant. It is a point source (< 0".05) and has a spectrum typical of many quasars, and is therefore probably a background extragalactic source. Two other small-diameter sources within the remnant are possibly related to it. However, the number of such sources found is consistent with extragalactic source counts.

125.049 **Spectral and polarization characteristics of the supernova remnant CTA1.**
W. Sieber, C. J. Salter, C. J. Mayer.
Astron. Astrophys., Vol. 393 - 404 (1981).

The spectral characteristics of the continuum emission from the supernova remnant CTA 1 have been studied over the frequency range from 151.5 to 2695 MHz. A point-by-point analysis of the spectral index shows that there exist only small deviations from the mean value $\langle \alpha \rangle = 0.47 (S \sim \nu^{-\alpha})$. Polarization measurements at 2695 and 1720 MHz indicate a magnetic field which is tangential to the shell at those places where shell structure is well defined. The physical parameters of the remnant including possible X-ray emission are discussed as are the characteristics of some point sources in the field.

125.050 **Heavy neutrinos and supernova explosions.**
C. W. Kim, K. Uehara.
Phys. Rev. D, Vol. 24, 1023 - 1026 (1981). − Abstr. in Phys. Abstr., Vol. 84, Abstr. 102140 (1981).

125.051 **On the properties and the nature of W50.**
N. Panagia, K. W. Weiler.
Vistas Astron., Vol. 25, (see 012.040), 87 - 89 (1981).

A consideration of the supernova remnant W50 suggests that it may consist of a normal shell at least partially surrounding a "plerion", i. e. a filled-center supernova remnant resembling the Crab Nebula. The suggestion is put forward that filled-center type supernova remnants, built and maintained by a central energy source, may represent locally "in miniature" the mechanism for the conversion of rotational and gravitational energy into relativistic synchrotron electrons seen in the cores of extragalactic radio sources.

125.052 **Supernova remnants and their relationship to W50 and SS 433.**   S. van den Bergh.
Vistas Astron., Vol. 25, (see 012.040), 109 - 118 (1981).

Both the high nitrogen content of W50 and the fact that SS 433 is a binary are consistent with a scenario in which these objects were formed by the explosion of a massive WN star. Attention is drawn to a deficiency of OB stars near the centres of the 38 presently known optical supernova remnants. It is suggested that this deficiency is due to the explosion of the OB stars that formed these SNR's.

125.053 **The interaction of the radiation from a type II supernova with a circumstellar shell.**
R. A. Chevalier.
Astrophys. J., Vol. 251, 259 - 265 (1981).

Detailed models for the light curves and photospheric properties of type II supernovae have shown that these events are likely to be the explosions of massive, red supergiant stars. Observations of red supergiants show that they have winds with velocities of the order of 10 km s⁻¹. The interaction of the supernova radiation with the circumstellar shell created by this wind can be important for ultraviolet, radio, and infrared observations of type II supernovae, and it is the purpose of this paper to explore the expected interaction.

125.054 **Are we overdue for a galactic supernova?**
D. H. Clark, P. J. Andrews, R. C. Smith.
Observatory, Vol. 101, 203 - 205 (1981).

125.055 **Observations of the expansion of the optical remnant of SN 1006 (Lupus).**
J. E. Hesser, S. van den Bergh.
Astrophys. J., Vol. 251, 549 - 551, plates 12 - 13 (1981).

Intercomparison of plates obtained in 1976 and 1981 shows that the optical remnant of the supernova of 1006 is expanding with $\mu = 0".39 \pm 0".06$ (m. e.) yr⁻¹. Comparison with the value $\mu = 0".33$ yr⁻¹ expected for a Sedov solution shows that the remnant is presently in the adiabatic phase of its evolution.

125.056 **Supernova hydrodynamics.**   S. A. Colgate.
Plasma astrophysics, (see 012.042), p. 291 - 296 (1981).

The explosion of a star, a supernova, occurs at the end of its evolution when the nuclear fuel in its core is almost, or completely, consumed. The star may explode due to a small residual thermonuclear detonation, type I SN or it may collapse, type I and type II SN leaving a neutron star remnant. The type I progenitor is thought to be an old accreting white dwarf, $1.4\,M_\odot$, with a close companion star. A type II SN is thought to be a massive young star $6\text{-}10\,M_\odot$. The mechanism of explosion is still a challenge to our ability to model the most extreme conditions of matter and hydrodynamics that occur presently and excessively in the universe.

125.057 **Halo around the Crab Nebula.**
P. Murdin, D. H. Clark.
Nature, Vol. 294, 543 - 544 (1981).

The Crab Nebula as usually depicted consists of filaments of total mass $2-3\,M_\odot$ and a pulsar of $1.4\,M_\odot$. If the precursor of SN 1054 was heavier than $\sim 4\,M_\odot$, its outer envelope could be detectable as an extended halo around the Crab Nebula. The authors have detected such an outer halo, whose size implies that the progenitor of SN 1054 was indeed a massive star.

125.058 **A new view of supernova remnants in the Magellanic Clouds.**   D. Clark.
Nature, Vol. 294, 690 (1981).

125.059 **X-ray images of the supernova remnant Puppis A.**
E. Pfeffermann, B. Aschenbach, H. Bräuninger, J. Trümper.
Space Sci. Rev., Vol. 30, (see 012.044), 251 (1981).
Abstract.

125.060 **The structure of the Cygnus Loop at 34.5 MHz.**
C. V. Sastry, K. S. Dwarakanath, R. K. Shevgaonkar.
J. Astrophys. Astron., Vol. 2, 339 - 347 (1981).

The authors have observed the large supernova remnant Cygnus Loop at 34.5 MHz with the low frequency radio telescope at Gauribidanur, India. A radio map of the region with a resolution of 26 arcmin × 40 arcmin ($\alpha \times \delta$) is presented.

125.061 **Neutral hydrogen in the vicinity of galactic radio sources. Supernova remnant W 44.**
A. P. Venger, I. V. Gosachinskij, V. G. Grachev, T. M. Egorova, N. F. Ryzhkov, V. K. Khersonskij.
Astron. Zh., Tom 58, 1187 - 1194 (1981). In Russian. English translation in Soviet Astron., Vol. 25, No. 6.

The results of the observations of emission and absorption in the H I line near the radio source W 44 obtained with the radio telescope RATAN-600 are presented. The absorption line observations show that the nearby H II region NRAO 584 cannot be physically connected with the supernova remnant. Six H I clouds are found in absorption and emission that probably are parts of a patchy envelope around W 44. Three of them coincide with a molecular cloud near W 44. The total H I mass of these clouds is about 220 $M_\odot$. The age of the supernova remnant and the physical parameters of the H II region NRAO 584 are estimated.

125.062 **Rate of type II supernovae in the Galaxy.**
         M. A. Smirnov, F. Kh. Sakhibov.
Astron. Tsirk., No. 1128, p. 8 (1980). In Russian.

125.063 **Shock-wave model for the optical emission from**
         oxygen-rich supernova ejecta. II. Precursor region.
H. Itoh.
Publ. Astron. Soc. Japan, Vol. 33, 521 - 530 (1981).
    A numerical model is constructed for the preshock structure and emission of a plane-parallel, steady-state, ionizing shock wave in a rarefied pure oxygen gas, the emphasis being placed upon the application to the [O III] filament of Cassiopeia A.

125.064 **Radio emission from supernova remnants in the**
         galaxy M 33.
S. D'Odorico, W. M. Goss, M. A. Dopita.
ESO Sci. Prepr., No. 167, 1 + 13 pp. (1981). — Submitted to Mon. Not. R. Astron. Soc.

125.065 **The radio morphology of supernova remnants.**
         P. A. Shaver.
ESO Sci. Prepr., No. 173, 24 pp. (1981). — Submitted to Astron. Astrophys.

125.066 **Supernova remnants.**
         R. A. Chevalier.
High-energy astrophysics, (see 012.067), p. 71 - 81 (1981).
    Optical observations of young supernova remnants can give information on the abundance structure of the supernova ejecta, from which properties of the supernova explosion can be deduced. Some young remnants are promising candidates for particle acceleration in shock waves, while in other remnants a central compact object is probably responsible for particle acceleration.

125.067 **Gravitational-collapse supernovae.**
         R. I. Epstein.
High-energy astrophysics, (see 012.067), p. 83 - 94 (1981).
    The gravitational collapse of the core of a massive star releases ample energy for a supernova explosion. The mechanisms by which this energy can be imparted to the outer parts of the stars to actually yield an explosion is outlined, and some of the related physical processes are examined. In particular the generation of entropy and the loss of neutrinos during the collapse of the core and the generation of lepton-driven buoyancy instabilities are discussed in some detail.

125.068 **On the morphology of supernova remnants.**
         V. Radhakrishnan, G. Srinivasan.
High-energy astrophysics, (see 012.067), p. 95 - 98 (1981).
    The difference in morphology between filled and shell type supernova remnants is attributed to differences in the activity of the neutron stars left by the supernovae. Pulsar activity leads to centrally concentrated remnants similar to the Crab. Non-activity as a pulsar results in all of the rotational energy loss going into dipole radiation. The pressure of this radiation creates shell-like objects with hollow interiors such as Cas A.

125.069 **Stellar clustering as induced by a supernova.**
         R. Baierlein, E. Schwing, W. Herbst.
Icarus, Vol. 48, 49 - 58 (1981).
    When the shock wave from a supernova expands, it sweeps up not only interstellar matter but also magnetic field. The field is greatly amplified by compression and will provide the dominant pressure during the cool radiative phase of an expanding supernova shell. The authors examine a hydromagnetic instability in this system and find that it will concentrate gas at intervals of the order of parsecs. The length and time scales make the instability promising as an explanation of the stellar clustering that is seen in Canis Major R1.

Supernovae.   See Abstr. 011.008.

Requirements for future SN observations: γ-ray,
X-ray, UV, visible, IR.   See Abstr. 011.037.

Atomic physics and spectroscopic data needs for
improved hydrodynamic predictions of composition, tempera-
tures, and densities of SN envelopes.
See Abstr. 011.040.

Recent advances in charged particle energy deposi-
tion and applications to supernova spectra.
See Abstr. 022.032.

Charge and energy transfer in heavy particle col-
lisions.   See Abstr. 022.033.

Nucleosynthesis of neutron-rich heavy nuclei during
explosive helium burning in massive stars.
See Abstr. 061.005.

Effects of neutrino degeneracy on the transport
properties of presupernovae.   See Abstr. 061.007.

Equations of state for hot, dense nuclear matter.
See Abstr. 061.032.

Neutrinos from electron-positron pair annihilation
in supernovae.   See Abstr. 061.044.

Time delay between gravitational waves and neutrino
bursts from a supernova explosion: a test for the neutrino mass.
See Abstr. 061.048.

The opacity of an expanding medium.
See Abstr. 062.025.

Plasma diagnostics in an astrophysical setting.
See Abstr. 062.039.

On the evolution of thermal disturbances in stars.
See Abstr. 062.052.

Effects of scattering on continuum radiation from
supernovae and determination of their distances.
See Abstr. 063.036.

The origin of supernovae.
See Abstr. 065.011.

La structure, l'énergie et l'évolution des étoiles.
See Abstr. 065.036.

Collapse of degenerate iron stellar cores and model
of a supernova.   See Abstr. 065.052.

Comments on the equation of state of supernova
matter.   See Abstr. 065.084.

Hydrodynamic collapse calculations for nonrotating
and rotating supernova models.   See Abstr. 065.085.

On the collapse instability of rapidly rotating pre-
supernova cores having finite entropy.   See Abstr. 065.086.

Cooling of young neutron stars and the Einstein X-ray observations.   See Abstr. 066.512.

The deleptonization and heating of proton-neutron stars.   See Abstr. 066.518.

Cooling of neutron stars and X-ray observations. See Abstr. 066.522.

Supernovae and nitrate in the Greenland Ice Sheet. See Abstr. 081.046.

Overview of the Voyager ultraviolet spectrometry results through Jupiter encounter.   See Abstr. 099.084.

Radio properties of SS 433. See Abstr. 116.038.

Does SS 433 have cousins? See Abstr. 117.090.

White dwarfs and supernovae. See Abstr. 126.027.

On the 26.13 MHz absorption line towards Cassiopeia A.   See Abstr. 131.017.

A compressed cloud in the Vela supernova remnant. See Abstr. 131.029.

High velocity gas associated with supernova remnants.   See Abstr. 131.075.

The 26.13 MHz absorption line in the direction of Cassiopeia A.   See Abstr. 131.133.

Optical observations of interstellar shockwaves. See Abstr. 131.175.

New insights into the physical state of gaseous nebulae.   See Abstr. 132.015.

Very high-resolution observations of compact radio sources in the directions of supernova remnants. See Abstr. 141.036.

Radio patrol of the northern Milky Way: a survey for variable sources.   See Abstr. 141.044.

The point radio source in the supernova remnant G 78.2 + 2.1.   See Abstr. 141.052.

Scintillations of cosmic radio sources in the decametre waveband. II. Compact sources in supernova remnants of the Crab Nebula and Cassiopeia A. See Abstr. 141.093.

Radio observations of W50 and the surrounding area.   See Abstr. 141.139.

CTB80: a galactic radio source with directed jets. See Abstr. 141.140.

A search for galactic compact radio sources. I. Two-frequency method and scintillating sources in the vicinity of supernova remnants.   See Abstr. 141.175.

On synchrotron nebulae and pulsars. See Abstr. 141.553.

Expanding shells of young pulsars as sources of high-energy neutrinos.   See Abstr. 141.563.

The Gemini-Monoceros X-ray enhancement: a giant X-ray ring.   See Abstr. 142.013.

An X-ray pulsar in SNR G109.1−1.0. See Abstr. 142.017.

The discovery of an X-ray pulsar in the SNR G109.1 - 1.0.   See Abstr. 142.024.

X-ray observations of SS 433 with the Einstein Observatory.   See Abstr. 142.069.

Cosmic X-ray observations performed with a gas scintillation spectrometer.   See Abstr. 142.099.

Inverse comptonization and the nature of the March 1979 γ-ray burst event.   See Abstr. 142.505.

A constraint on prompt supernova cosmic ray production from γ-ray observations.   See Abstr. 143.030.

On the origin of antiprotons observed in cosmic rays.   See Abstr. 143.057.

A search for expanding supershells of gas around OB associations.   See Abstr. 152.004.

Large-scale formations of matter in the region $l^{II} = 15°- 45°$, $b^{II} = 0°- 15°$ according to 102.5 MHz observations at the large synphase array of the Lebedev Physical Institute.   See Abstr. 156.008.

A study of M100 in X-rays. See Abstr. 158.236.

A soft X-ray study of the Large Magellanic Cloud. See Abstr. 159.004.

Supernova in NGC 4321 (1979c)

125.101 Observations of supernova 1979c in M 100.
        N. Panagia.
The Universe at ultraviolet wavelengths, (see 012.009), p. 521-525 (1981).
    The IUE observations of supernova 1979c in M 100 are presented and discussed. The main results are: (1) The bulk of the energy is in the form of continuous emission which is radiated by the main SN envelope. (2) The absorption features originate mostly in both the disks and the haloes of our Galaxy and M 100. (3) The emission lines are produced in a highly ionized shell which has a radius greater than twice the radius of the main envelope and consists of compressed circumstellar material in which the abundance ratio N/C is about 30 times higher than solar.

125.102 The ultraviolet spectrum of supernova 1979c in
        NGC 4321.   N. Panagia.
Mem. Soc. Astron. Italiana, Vol. 52, (see 012.019), 87 - 94 (1981).

Observations of supernovae. See Abstr. 125.027.

Observations of M100 with the Einstein Observatory shortly after the explosion of its fourth supernova, SN 1979c. See Abstr. 158.009.

Supernova in NGC 1316 No. 1

125.201   **Infrared light curves of Type I supernovae.**
J. H. Elias, J. A. Frogel, J. A. Hackwell, S. E. Persson.
Astrophys. J., Lett., Vol. 251, L13 - L16 (1981).

Two Type I supernovae in NGC 1316 and one in NGC 4536 have been observed at J, H, and K. These light curves are characterized by a double maximum; the effect is strongest at J. This is probably due to the development of a transient absorption feature during the intervening minimum. The three sets of infrared light curves are very similar, and their dispersion in absolute magnitude is less than the uncertainty in relative distances of the supernovae.

125.202   **Supernovae in NGC 1316.**
IAU Circ., No. 3624 (1981).

Spectra of two recent supernovae.
See Abstr. 125.601.

Interstellar Ca II in supernova Wischnjewsky in NGC 1316 (Fornax A).   See Abstr. 131.005.

Supernova in NGC 1316 No. 2

Infrared light curves of Type I supernovae.
See Abstr. 125.201.

Supernova in MCG +07 −29 −043

125.241   **A supernova in MCG + 07−29−043.**
N. Sanduleak.
Inf. Bull. Variable Stars, No. 2061 (1981).

Supernova in M 31

125.301   **Bigourdan's forgotten observations of SN 1885 (S Andromedae) in M31.**
G. de Vaucouleurs, R. Buta.
Publ. Astron. Soc. Pacific, Vol. 93, 294 - 296 (1981).

The light curve of SN 1885 in M31 from 1885 September 6 to October 26 is derived from 18 visual observations made by G. Bigourdan with a 30-cm refractor at Paris Observatory and reduced with respect to newly determined photoelectric magnitudes for five comparison stars. These observations had been overlooked in the variable stars literature.

Supernova in NGC 4536

125.401   **Supernova outburst in the galaxy NGC 4536.**
V. G. Surdin.
Priroda, 1981, No. 7, p. 105. In Russian.

125.402   **Supernova in the galaxy NGC 4536.**
D. Yu. Tsvetkov.
Zemlya Vselennaya, 1981, No. 5, p. 38. In Russian.

Infrared light curves of Type I supernovae.
See Abstr. 125.201.

Intergalactic supernova 1980 I

125.501   **The spectrum of the intergalactic supernova 1980I.**
H. A. Smith.
Astron. J., Vol. 86, 998 - 1002 (1981).

The supernova 1980I cannot be clearly associated with a single galaxy. Its apparent position is several arcminutes from the centers of three bright galaxies−NGC 4374, NGC 4387, and NGC 4406. Blue and red spectra, obtained shortly after discovery, show 1980I to be a supernova of Type I, but one possessing an unusual absorption feature near 6680 Å. The outlying location of 1980I suggests that at least some Type I supernovae have low-mass progenitors, or that star formation is occurring at large (~ 50 kpc) distances from the centers of the neighboring galaxies.

Supernova in NGC 7448

125.521   **Supernova in NGC 7448.**
R. Ya. Inasaridze.
Astron. Tsirk., No. 1143, p. 8 (1980). In Russian.

Supernova in NGC 6946

125.601   **Spectra of two recent supernovae.**   T. P. Prabhu.
Bull. Astron. Soc. India, Vol. 9, 60 - 63 (1981).

Supernova in NGC 3913

125.621   **Observations of the supernova in the galaxy NGC 3913.**   M. V. Gyulzadyan.
Astron. Tsirk., No. 1149, p. 7 - 8 (1981). In Russian.

Supernova in NGC 1073 (1962*l*)

125.701   **Peculiar type-I supernova 1962*l* in the galaxy NGC 1073.**   B. A. Sandler.
Nauchn. Inf., Vyp. (No.) 43, p. 98 - 104 (1980). In Russian.

The spectrum of SN 1962*l* has been identified on the basis of the absorption hypothesis.

Supernova in NGC 1532

125.801   **Supernova in NGC 1532.**
IAU Circ., No. 3621 (1981).

Supernova in ESO 356−G20

125.821   **Possible supernova in ESO 356-G20.**
IAU Circ., Nos. 3627, 3628 (1981).

Erratum

125.901   **Erratum: "Intensity and spectrum of the continuum gamma ray emission from supernovae"** [Astron. Astrophys., Vol. 88, 367 - 369 (1980)].
C. Cavallo, F. Pacini.
Astron. Astrophys., Vol. 101, 159 (1981). − See Abstr. 28.125.011.

126 Low-luminosity Stars, Subdwarfs, White Dwarfs, Degenerate Stars

126.001 Electron scattering in the atmospheres of hot DA white dwarfs.   F. Wesemael.
Astrophys. J., Vol. 247, 590 - 593 (1981).

The process of high-frequency emission from the photosphere of hot, hydrogen-rich white dwarfs is reexamined in the light of the recent result of Böhm and Kapranidis that electron scattering has a strong influence on the energy distribution of these stars at extreme-ultraviolet (EUV) wavelengths. Although electron scattering generally dominates the source function over a large range of depths, detailed model atmosphere calculations indicate that the EUV continuum is formed much deeper in the atmosphere.

126.002 Discovery of strong ultraviolet absorption in the spectrum of the DC white dwarf G 33−49.
G. Vauclair, V. Weidemann, D. Koester.
Astron. Astrophys., Vol. 100, 113 - 115 (1981).

Strong ultraviolet absorption features have been observed by IUE in the spectrum of the DC white dwarf G 33−49. The spectrum between 1400 Å and 1960 Å is quite similar to the one obtained for the C_2 weak white dwarf L145−141, and shows strong absorptions at 1910 Å (FWHM 70 Å) and between 1550 Å and 1680 Å. Identification with carbon absorption features is discussed.

126.003 The magnetic fields of white dwarfs.
J. R. P. Angel, E. F. Borra, J. D. Landstreet.
Astrophys. J., Suppl. Ser., Vol. 45, 457 - 474 (1981).

In a survey program carried out over the past decade, more than 100 white dwarfs have been observed for magnetic fields by continuum circular polarization measurements. Twelve white dwarfs have been measured with greater accuracy by Zeeman measurements in absorption lines. These observations are reported in full in this paper.

126.004 The least luminous star.   D. W. Hughes.
Nature, Vol. 293, 9 - 10 (1981).

126.005 High-frequency stellar oscillations. XIV. Null results from a search for white-dwarf variability between 0.13 and 6.8 seconds.
B. M. Lasker, S. B. Bracker, J. E. Hesser.
Publ. Astron. Soc. Pacific, Vol. 93, 320 - 322 (1981).

A survey of 7 DA (including ZZ Ceti), 4 DC, 1 DF, 1 DG, and 2 C_2 white dwarfs for "white light" photometric vaiability in the range 0.13 to 6.8 sec was made using an on-line Fourier technique; no indications of coherent variability in excess of 0.01 mag were discovered.

126.006 The interstellar medium and the highly ionized species observed in the spectrum of the nearby white dwarf G191 − B2B.   F. C. Bruhweiler, Y. Kondo.
Astrophys. J., Lett., Vol. 248, L123 - L127 (1981).

High-resolution spectra of the nearby (48 pc) white dwarf G191 − B2B obtained with the International Ultraviolet Explorer (IUE) reveal sharp resonance lines of N V, C IV, and Si IV. The origin of these features is most likely linked to the white dwarf, possibly being formed in an expanding halo around the star. Interstellar lines of C II, N I, Mg II, Si II, and Fe II are also seen in the spectrum. Analysis of these features indicates an average neutral hydrogen number density, $n_{HI} = 6.4 \times 10^{-3}$, for this line of sight. In combination with the recent EUV and soft X-ray results, the authors interpret this to mean that the interstellar medium in the most immediate solar vicinity is of the "normal" density ($n \approx 0.1$ cm^{-3}) of lower ionization, while just beyond it, at least in some directions, is a hot, lower density plasma. These results are apparently in conflict with the model of the interstellar medium by McKee and Ostriker in its present form.

126.007 Observations of strong ultraviolet absorptions in the spectrum of the DC white dwarf G218-8.
G. Wegner.
Astrophys. J., Lett., Vol. 248, L129 - L132 (1981).

The spectrum of the 14th magnitude DC white dwarf G218-8 (EG 245) is described. In the visible spectral region the star shows a featureless spectrum. However, the ultraviolet observations using the IUE satellite show that strong absorption lines attributable to neutral carbon are present. G218-8 has an effective temperature near 11,500 K and a helium dominated atmosphere with a carbon-to-helium abundance ratio, C:He $\lesssim 5 \times 10^{-4}$.

126.008 The white dwarf companion of the Ba II star ζ Cap.   E. Böhm-Vitense.
The Universe at ultraviolet wavelengths, (see 012.009), p. 455 - 458 (1981).

The Ba II star ζ Cap has a white dwarf companion. Its T_{eff} is determined to be 22000 K, its mass M ~ 1 $M_\odot$. The importance of this finding for the explanation of abundance peculiarities is discussed.

126.009 Coordinated IUE, Einstein and optical observations of accreting degenerate dwarfs.
G. Fabbiano, J. E. Steiner, L. Hartmann, J. Raymond.
The Universe at ultraviolet wavelengths, (see 012.009), p. 501 - 514 (1981).

Three binary systems believed to be composed of a white dwarf and a late type star, AM Her, SS Cyg and U Gem, were observed simultaneously in the UV X-ray and optical wavelengths. AM Her was in its customary high state at the time of the observations, while SS Cyg and U Gem were in a low state. In all three cases, a significant UV black body component with KT $\gtrsim 10$ eV was found. The flux in this component is in excess of the amount predicted by current scenarios of gravitational energy release. The authors compare their observations of these objects with the data available in the literature, and they suggest an alternative scenario that would explain their general behavior.

126.010 Spectroscopy of four subdwarf O stars.
A. R. Walker.
Mon. Not. R. Astron. Soc., Vol. 197, 241 - 246 (1981).

Spectra of four subdwarf O stars, three of them new discoveries, are discussed. Two of the stars lie in the galactic plane enabling reasonably accurate limits to their distance, and hence absolute magnitudes, to be calculated.

126.011 The nitrogen and silicon abundance of sdO-stars.
K. P. Simon, J. Gruschinske, K. Hunger, R. P. Kudritzki.
Second European IUE Conference, (see 012.011), p. 305 - 306 (1980).

The abundance of nitrogen and silicon are determined by means of multi-level-calculations for three helium-rich subluminous O-stars, HD 49798, HD 127493 and BD+75°325. The analysis of the line spectra of N III and Si IV yields three results: First, microturbulence is small, i.e. less than 10 km/s, second, the silicon abundance is solar, and third, nitrogen is strongly overabundant.

126.012 Non-LTE analysis of the sdO-star BD+75°325.
R. P. Kudritzki, K. Hunger, J. Gruschinske, K. P. Simon, J. P. Kaufmann.
Second European IUE Conference, (see 012.011), p. 307 - 310

(1980).

Effective temperature, gravity and helium abundance of the sdO-star BD+75°325 are determined by a NLTE analysis of high resolution IUE spectra and visual coudé spectra. The evolutionary status of BD+75°325, as well as of other sdO-stars, which have recently been analyzed by the authors, is discussed by comparing the positions in the (log g, log T_{eff})-plane with evolutionary tracks computed for stars with similar H-He mixture.

126.013   The carbon abundance of sdO-stars.
J. Gruschinske, K. Hunger, R. P. Kudritzki, K. Simon.
Second European IUE Conference, (see 012.011), p. 311 - 314 (1980).

High resolution IUE spectra of the three sdO-stars HD 127493, HD 49798 and BD +75°325 are used to derive an upper limit for the carbon abundance. The CIV resonance line doublet at λ1548/1551 is analyzed by means of non-LTE line formation in the two level approximation. All three sdO-stars are found to be carbon-deficient by at least a factor of ten (relative to the sun).

126.014   The far ultraviolet spectrum of the OB-type subdwarf HD 149 382.
B. Baschek, R.-P. Kudritzki, M. Scholz.
Second European IUE Conference, (see 012.011), p. 315 - 318 (1980).

High-resolution IUE spectra of the OB-type subdwarf HD 149 382 have been obtained from λ = 1160 to 3230 Å. The stellar spectrum, which only shows hydrogen, helium and sulfur lines in the blue, exhibits more than 1500 absorption features with central depths $\gtrsim 0.2$ in the far ultraviolet. He II 1640 is the only helium line found in this spectral region. The spectrum is dominated by the doubly and triply ionized stages of carbon, nitrogen and twelve heavier elements which are certainly or probably present. Oxygen and eleven further elements are possibly present. The Si IV resonance doublet at 1394 and 1403 Å is not found. Preliminary atmospheric parameters $T_{eff} \simeq 35\,000$ K, $g \simeq 10^{5.5}$ cm s^{-2}, log $N_{He}/N_H \simeq -1.4$, have been derived.

126.015   The far ultraviolet spectrum of the B-type subdwarf HD 205 805.
B. Baschek, R.-P. Kudritzki, M. Scholz.
Second European IUE Conference, (see 012.011), p. 319 - 322 (1980).

High-resolution IUE spectra of HD 205 805 (sdB, $m_v = 10.2$, $T_{eff} = 26500$ K, $g = 10^5$ cm s^{-2}) have been obtained in the range 1170–3230 Å. In contrast to the very few lines found in the blue, this range is characterized by numerous absorption lines, some 1150 line with central depths ≥ 0.2 have been measured. The radial velocity of -80 km s^{-1} permits the distinction between stellar and interstellar lines. The stellar spectrum exhibits many lines of doubly ionized elements, especially of C, Si, S, Ti, Cr, Mn and Fe. Preliminary estimates of selected abundances are given by comparison with the normal B star γ Peg.

126.016   Analysis of the UV spectrum of three hot subdwarfs.
L. Rossi, R. Viotti, J. Darius, F. D'Antona.
Second European IUE Conference, (see 012.011), p. 323 - 325 (1980).

The authors present IUE high resolution observations of three O-subdwarfs (BD+37°442, BD+37°1977 and BD+48°1777). The spectra are characterized by very high excitation photospheric lines. Intense interstellar lines are present, whose possible circumstellar origin is suggested in view of the absence of the 2200Å interstellar band.

126.017   A strong absorption feature near 1900 Å in the spectrum of the C_2 white dwarf LP 145 141.
D. Koester.
Second European IUE Conference, (see 012.011), p. 333 - 337 (1980).

An absorption feature of exceptional strength has been detected in the spectrum of the cool degenerate "λ4670" or "C_2" star LP 145 - 141. This feature is tentatively identified as a dissociation continuum of C_2. Theoretical calculations of dissociation and ionization equilibria confirm that molecule formation is strongly favored in the helium-rich, high pressure atmosphere of this object, with CO, C_2, or O_2 being dominant, depending on the He/C and C/O ratios.

126.018   Lanning 14, a new DO white dwarf: spectrophotometry and atmospheric parameters.
J. Liebert, R. F. Green, F. Wesemael, B. Margon.
Astron. J., Vol. 86, 1384 - 1386 (1981).

Optical spectrophotometry of Lanning 14, recently classified as a DO white dwarf, is presented and analyzed. This object has one of the strongest He II 4686 absorption line of any known star ($W_\lambda \sim 7$ Å), with strong He II Brackett lines and weak He I. Use of a new grid of hot, high-gravity, mixed-composition models permits to derive $T_{eff} = 55\,000 \pm 2500$ K, log g near 8, and log $[N(He)/N(H)] \gtrsim 2$; thus Lanning 14 is both hotter and more helium-rich than the prototype DO star HZ 21.

126.019   New spectral classifications for two peculiar white dwarfs.   G. Wegner.
Astron. Astrophys., Vol. 102, 223 - 224 (1981).

Two white dwarfs that were previously given the spectral type of DC have been re-examined spectroscopically and photometrically. The first, G171−27, has been found to have Swan bands of C_2, while the second, G188−27, has weak neutral helium lines and appears to be among the coolest DB stars yet found.

126.020   Mass limits for non-degenerate white dwarfs.
G. Schmidt, H.-T. Elze, J. Rafelski.
Nucl. Phys. A, Vol. A364, 527 - 532 (1981). – Abstr. in Phys. Abstr., Vol. 84, Abstr. 88916 (1981).

126.021   The ultraviolet spectrum of the O-type subdwarf HD 49798.
F. C. Bruhweiler, Y. Kondo, G. E. McCluskey.
Astrophys. J., Suppl. Ser., Vol. 46, 255 - 265 (1981).

The bright O-type subdwarf (O6p) single-lined spectroscopic binary HD 49798, with an orbital period 1.5477, has been observed with the IUE satellite. Four high resolution (~0.1 Å) spectra in the far-ultraviolet region (λλ 1150–2000) and two high resolution (~0.2 Å) spectra in the mid-ultraviolet region (λλ 2000–3200) were obtained. In the spectra of this sharp-lined star, a large number of absorption lines have been identified. Most numerous are the lines attributable to Fe V. This finding has serious consequences on the model atmospheres computed for stars of spectral types O6 to B0 since the existing models underestimate the blanketing effect of the Fe IV and Fe V lines. Mass loss from the subdwarf is also discussed.

126.022   Strange abundances in an OB-subdwarf based on IUE observations.
B. Baschek, R.-P. Kudritzki, M. Scholz, K. P. Simon.
Upper main sequence chemically peculiar stars, (see 012.033), p. 451 - 454 (1981).

Element abundances in the atmosphere of the OB-type subdwarf HD 149382, largely based on high-dispersion far ultraviolet spectra obtained with the IUE satellite, are presented and discussed.

126.023   Determination of atmospheric parameters for DB white dwarfs.

D. Koester, H. Schulz, G. Wegner.
Astron. Astrophys., Vol. 102, 331 - 336 (1981).

New spectroscopic observations of 25 DB white dwarfs are reported and analysed together with Strömgren colours, using a new set of model atmospheres for He-rich degenerate stars. The DB seem to follow a line of constant log g – implying constant mass – in the temperature range from 13,000 K to at least 20,000 K. Straightforward evaluation of gravities from the line spectrum and use of the mass-radius relation give a mean mass of 0.44 $M_\odot$ and a mass range (1σ) of $\pm$ 0.10 $M_\odot$.

126.024 **Gravity modes instability in DA white dwarfs.**
N. Dolez, G. Vauclair.
Astron. Astrophys., Vol. 102, 375 - 385 (1981).

A linear non adiabatic stability analysis of gravity modes for chemically stratified DA white dwarf envelopes is presented. Two slightly overlapping instability strips are found, according to the amount of hydrogen in the outer envelope. For "hydrogen poor" DAs the instability strip is located in the temperature range from 11,500 K to ~ 13,500 K. The Kappa-mechanism in the helium partial ionisation zone is responsible for driving the pulsations. For "hydrogen rich" DAs the instability strip is located between ~ 10,000 K and 11,500 K and the Kappa-mechanism in the hydrogen partial ionisation zone drives the pulsations. The properties of the modes, their periods and stability, are discussed in detail. A physical interpretation in terms of mode propagation is provided. From this analysis, the DA character of ZZ Ceti stars is quite fundamental as the amount of hydrogen in the envelope must be large enough so that the Kappa-mechanism acting in the hydrogen partial ionisation zone may drive pulsations in the observed range of effective temperature.

126.025 **Two new variable white dwarfs: G185-32 and G191-16.** J. T. McGraw, G. Fontaine, D. S. P. Dearborn, J. Gustafson, P. Lacombe, S. G. Starrfield.
Astrophys. J., Vol. 250, 349 - 354 (1981).

The authors report the discovery of two new ZZ Ceti variables, originally selected as candidates for variability on the basis of their (G–R) colors. The star G191-16 is a stable, large-amplitude variable showing several periods and significant nonlinear effects in its light curve. The star G185-32 is a much smaller amplitude, multiply periodic variable. The period structures of both stars act as counterexamples to the accepted correlation for the ZZ Ceti variables between pulsation amplitude and period-structure complexity. The implication of this result with respect to the structure of hydrogen-rich white dwarf (DA) stars is discussed.

126.026 **Polarimetric observations of white-dwarf stars.**
Yu. S. Efimov.
Izv. Krymskoj Astrofiz. Obs., Tom 63, 118 - 139 (1981). In Russian. English translation in Bull. Crimean Astrophys. Obs., Vol. 63.

A survey of linear polarization of a sample of 85 white-dwarf stars and some subdwarfs in $UBVOR$ bands is given. An intrinsic polarization twice large as the root-mean-square errors of measurements is revealed at more than 50% of the sample. There is evidence for polarization variations in some objects (for example GD 299, EG 191, Feige 110 and EG 159).

126.027 **White dwarfs and supernovae.**
R. Canal, J. Isern, J. Labay.
C. R. Acad. Sci. Paris, Tome 293, Sér. II, 441 - 443 (1981). In French.

The authors show that the long-term evolution of accreting C – O white dwarfs in close binary systems may be a source of neutron stars and/or SNI explosions. They take into account the effects of solidification of the star's core on the speed of propagation of thermonuclear burning. They also consider the possible separation of oxygen from carbon as a result of solidification.

126.028 **White dwarfs with carbon features in their spectra.**
G. Wegner.
Publ. Astron. Soc. Pacific, Vol. 93, 550 (1981). – Abstract.

126.029 **'High entropy' accretion onto the surface of a white dwarf.** A. Glasner, Y. Tuchman.
Astrophys. Space Sci., Vol. 80, 281 - 287 (1981).

The accretion of hydrogen rich matter onto the surface of a white dwarf, assuming that the accreted matter keeps a small part of its gravitational falling energy, is investigated. The influence of this energy excess on the hydrogen ignition and the thermal instability that follows is described and discussed.

126.030 **Ariel 6 observations of the hot white dwarf HZ43.**
R. E. Cole, A. M. Cruise, J. P. Osborne, D. Bedford, E. Carpenter, C. Goodall, A. M. T. Pollock.
Space Sci. Rev., Vol. 30, (see 012.044), 201 - 205 (1981).

The hot white dwarf HZ43 was observed by the soft X-ray experiment on the Ariel 6 satellite from 1980 March 1.8 to March 8.7. The pulse height spectrum has been fit to an input black body spectrum of temperature 140 000 K. An upper limit of 10% has been placed on the source variability on time scales of minutes.

126.031 **Cyclotron radiation of magnetized degenerate dwarfs and neutron stars: II – observations.**
I. G. Mitrofanov, G. G. Pavlov.
High-energy astrophysics, (see 012.067), p. 117 - 120 (1981).

Pure helium model atmospheres of red degenerate stars. See Abstr. 064.077.

Effect of accretion on the evolution of very-low-mass young objects. See Abstr. 065.101.

Cyclotron radiation of magnetized degenerate dwarfs and neutron stars: I – theories. See Abstr. 066.527.

Thirteen-color photometry of subdwarf stars. IV. HD 25329 and HD 122563. See Abstr. 113.027.

Distinguishing between a white dwarf and a neutron star in an X-ray binary. See Abstr. 117.006.

The ultraviolet spectrum of the twin-degenerate interacting binary G61-29. See Abstr. 117.050.

Hydrogen burning on a white dwarf accreting hydrogen in a binary system. See Abstr. 117.080.

On the elusive cause of cataclysmic variable outbursts. See Abstr. 117.104.

CH Cygni – M-giant and a white dwarf binary with transient accretion disk. See Abstr. 117.135.

EX Hydrae: modulated mass transfer or intermediate polar? See Abstr. 119.002.

Simultaneous photoelectric and single-trail spectroscopic observations of V 471 Tauri (BD +16°516). See Abstr. 119.058.

Period changes in Z Cha. See Abstr. 122.003.

G 255-2: a new ZZ Ceti variable star. See Abstr. 122.104.

New subdwarfs. III. On obtaining the vertical galactic metallicity gradient from the kinematics of nearby stars. See Abstr. 155.055.

Interstellar Matter, Nebulae

131 Interstellar Matter, Star Formation

131.001 **Observations of $^{14}N/^{15}N$ in the galactic disk.**
P. G. Wannier, R. A. Linke, A. A. Penzias.
Astrophys. J., Vol. 247, 522 - 529 (1981).

A new survey has been made of $HC^{15}N$ and $H^{13}CN$ which includes 16 giant molecular clouds and one circumstellar cloud of expelled material. The observed double isotope ratio $(^{12}C/^{13}C) \times (^{15}N/^{14}N)$ is shown to be a good tracer of equilibrium CNO processing. The authors determine that the $^{15}N/^{14}N$ ratio shows a small difference between the galactic center region and the rest of the galactic disk but has no appreciable radial gradient throughout the outer Galaxy.

131.002 **The gas-to-dust ratio in the Rho Ophiuchi cloud.**
D. C. B. Whittet.
Mon. Not. R. Astron. Soc., Vol. 196, 469 - 472 (1981).

The reported anomaly in the gas-to-dust ratio in the ρ Ophiuchi dark cloud is investigated. It is found that the anomaly appears to be restricted to the line of sight to the star ρ Oph itself, and is not a general feature of the dark cloud at low optical depth. The ratio $N(H)/A_V$ is deduced to be 1.9×10^{21} atom cm^{-2} mag^{-1} for $A_V \lesssim 2$ mag, which is identical to the result for the general interstellar medium.

131.003 **Spectrum of the interstellar plasma turbulence in the direction of the pulsar PSR 0329 + 54.**
A. Wolszczan, N. Bartel, W. Sieber.
Mon. Not. R. Astron. Soc., Vol. 196, 473 - 480 (1981).

A set of bandwidth-averaged time-autocorrelation functions of intensity has been computed from interstellar scintillation data for PSR 0329 + 54 and compared with the corresponding theoretical functions for a Gaussian and a Kolmogorov spectrum of density irregularities in the interstellar plasma. It is shown that the results of these comparisons provide evidence that the Kolmogorov spectrum adequately describes the plasma turbulence in the direction of this pulsar.

131.004 **Analysis of *IUE* observations of the galactic corona and the existence of molecular clouds in the halo.**
T. W. Hartquist, A. Tallant.
Mon. Not. R. Astron. Soc., Vol. 196, 527 - 532 (1981).

The authors re-analyse *IUE* observations of ions in the galactic corona and confirm Savage & de Boer's (1979) conclusion that the detected hot phase has $T \sim 8 \times 10^4$ K. The cold phase must be at $T < 2.4 \times 10^4$ K and probably contains substantial amounts of molecular material. Molecules, such as CH^+, OH and SH, may exist in detectable abundances in the transition regions between the hotter gas and the cold gas on the far sides of halo molecular clouds.

131.005 **Interstellar Ca II in supernova Wischnjewsky in NGC 1316 (Fornax A).** J. C. Blades.
Mon. Not. R. Astron. Soc., Vol. 196, 65P - 71P (1981).

Results are presented of a high-resolution spectrum of interstellar Ca II in the 1980 supernova in the cD galaxy NGC 1316 (Fornax A). Three interstellar clouds are detected, with heliocentric velocities of +18, +1364 and +1886 km s^{-1}. The lowest velocity cloud, which is also the strongest $(W_\lambda(Ca\ II\ K) = 0.21$ Å), originates in our Galaxy, probably in local (< 3 kpc) gas. The other two clouds are much weaker and are attributed to gas in NGC 1316, although the +1364 km s^{-1} component could be unbound from it. No high-velocity halo clouds in our Galaxy or in the intergalactic medium to NGC 1316 are detected to a limit of $N(Ca\ II) \lesssim 1-2 \times 10^{11}$ cm^{-2}.

131.006 **A high-velocity H I stream interacting with the galactic disc.** R. J. Cohen.
Mon. Not. R. Astron. Soc., Vol. 196, 835 - 844 (1981).

21-cm line observations have been made of the high-velocity H I stream near $l = 160°$, $b = -50°$, $V = -110$ km s^{-1} discovered by van Kuilenburg. The observations have an angular resolution of $0°.55$, a velocity resolution of 7.3 km s^{-1} and an rms noise of ~ 25 mK. The -110 km s^{-1} stream is at least 25° long, and is coincident with a filament in the local H I gas at a velocity of -10 km s^{-1}. It is suggested that the -110 km s^{-1} stream is colliding with the galactic H I disc, and that the -10 km s^{-1} filament is shocked galactic gas. The distance from the Sun is estimated to be between 100 and 300 pc.

131.007 **Ice mantles and the anomalous ultraviolet extinction of HD 29647.**
D. C. B. Whittet, M. F. Bode, A. Evans, I. Butchart.
Mon. Not. R. Astron. Soc., Vol. 196, 81P - 85P (1981).

HD 29647, a reddened early-type star in the Taurus dark cloud, was shown by Snow & Seab to have anomalously weak absorption in the λ 2200 feature. In this contribution, complementary infrared (1 - 4 μm) observations are presented. Results indicate that the anomaly cannot be explained by ice mantle formation on the grains. Chemical processing of graphite grains is proposed as an alternative explanation.

131.008 **An atlas of models of H^+ blisters.** V. Icke.
Astrophys. J., Suppl. Ser., Vol. 45, 585 - 602 (1981).

Early-type stars apparently form near the periphery of massive neutral clouds. The ionizing radiation of these stars interacts with the neutral gas and, because of the density gradient in the cloud, forms a "blister" of ionized gas. The author presents a sequence of models of these blisters, based on an earlier theoretical approximation of the transfer of the ionizing radiation. The models are presented as contour maps showing the predicted appearance of blisters around O5 V, O7 V, and O9 V stars in the 10 GHz radio and 30 μm, 50 μm, and 100 μm infrared bands. The characteristic systematic shift in the position of the emission maxima is again evident.

131.009 **The giant outburst of the 8 km s^{-1} water maser feature in Orion.**
Z. Abraham, N. L. Cohen, R. Opher, J. C. Raffaelli, S. H. Zisk.
Astron. Astrophys., Vol. 100, L10 - L13 (1981).

The authors report synoptic observations of the giant outburst of the 8 km s^{-1} water maser source near Orion-KL. The light curve is characterized by flux oscillations of the order of several months, with the degree of linear polarization apparently correlated with the flux variations. The line width has increased steadily with time. The flare connotes isolated and energetic events in the neighbourhood of the Orion-KL region which may be due to the action of an emerging protostar or a region whose density and temperature now favor enhanced masering. The high degree of linear polarization favors a saturated maser.

131.010 SiO isotopes in Orion A.
H. Olofsson, Å. Hjalmarson, O. E. H. Rydbeck.
Astron. Astrophys., Vol. 100, L30 - L33 (1981).

Time variable features have been observed in the Orion ^{29}SiO (v = 0, J = 2 - 1) emission, presumably of maser origin. The corresponding ^{30}SiO line displays a very different line shape and could also be influenced by maser effects. No signs of variability have been seen in the ^{28}SiO (v = 0, J = 2 - 1) emission, which appears to be partly saturated. A mapping of the ^{28}SiO (v = 0, J = 2 - 1) line revealed three separable emission regions within an area considerably smaller than the 43″ beam.

131.011 Detection of the CO J = 4 → 3 transition from the Kleinmann-Low Nebula. A. H. F. van Vliet,
T. de Graauw, T. J. Lee, S. Lidholm, H. v. d. Stadt.
Astron. Astrophys., Vol. 101, L1 - L3 (1981).

The CO J = 4 → 3 rotational transition has been detected in the Kleinmann-Low Nebula in the Orion Molecular Cloud from a ground based site. Using several of the previously measured transitions a minimum gas kinetic temperature is derived of the high velocity "Plateau Source" of ~600 K along with a minimum CO column density of $\sim 7 \times 10^{18}$ cm^{-2}.

131.012 A comparison of visual extinction with H_2CO and H I absorption in Heiles Cloud 2.
W. A. Sherwood, T. L. Wilson.
Astron. Astrophys., Vol. 101, 72 - 78 (1981).

The visual extinction, A_{vn}, has been determined over Heiles Cloud 2 ($l = 175°$, $b = -14°$) for a scale size which matches the resolution of H_2CO and H I data. A minimum visual extinction is required before H I self absorption and H_2CO are detected. The size of the cloud in the visual is larger than that of the H I self-absorption region, which in turn is larger than the size of the H_2CO region.

131.013 The newest stars in Orion.
C. G. Wynn-Williams.
Sci. American, Vol. 245, No. 2, p. 30 - 39 (1981).

Clouds of dust and gas in the familiar constellation emit radiation at infrared and radio wavelengths. Some of the denser clouds appear to conceal new stars that are expelling fast-moving streams of gas.

131.014 Die Geburtsstätten der Sterne. M. Gaida.
Umschau, 81. Jahrg., 474 - 475 (1981).

54 organic and inorganic molecules have been detected in interstellar dense clouds. Their formation in many of these regions leads to the conclusion of protostellar objects. From this point biological evolution seems to be a continuation of physical and chemical evolution in the interstellar medium. It is hoped that laboratory simulations of the dark and dusty clouds will help us to understand the generation process of molecules.

131.015 Interstellar gas near and within the solar system.
M. S. Burgin.
Comments Astrophys., Vol. 9, 157 - 168 (1981).

The picture of the interaction between the local interstellar medium (LISM) and the solar environment developed in recent years is described, and prospects for obtaining complete information about the LISM are discussed. The author focuses his attention on the neutral component of the LISM, and particularly on the results of observations of the UV radiation scattered from hydrogen and helium atoms penetrating into the solar system from interstellar space.

131.016 Proper motions and distances of H_2O maser sources. II. W51 MAIN. R. Genzel, D. Downes,
M. H. Schneps, M. J. Reid, J. M. Moran, L. R. Kogan,

V. I. Kostenko, L. I. Matveyenko (Matveenko), B. Rönnäng.
Astrophys. J., Vol. 247, 1039 - 1051 (1981).

The authors report VLBI measurements of relative proper motions of H_2O maser features in W5 MAIN. They have achieved relative positional accuracies of ~100 micro-arcsec from the analysis of fringe phases and from aperture synthesis maps. The observed proper motions are typically 1 milli-arcsec per year and probably represent kinematic motions of the maser cloudlets. The kinematics of the maser features in W51 MAIN are more complex than in Orion–KL. The transverse motions of most features in W51 MAIN appear to be random and can be interpreted best as turbulent motions created when a strong stellar wind interacts with density inhomogeneities in the surrounding molecular cloud.

131.017 On the 26.13 MHz absorption line towards Cassiopeia A. A. A. Konovalenko, L. G. Sodin.
Pis'ma Astron. Zh., Tom 7, 402 - 405 (1981). In Russian.
English translation in Soviet Astron. Lett., Vol. 7.

New observational data are reported confirming that the absorption line detected earlier towards the radio source Cas A at the frequency of 26.13 MHz is not a nitrogen hyperfine structure line but a carbon recombination line. Some relevant consequences are discussed.

131.018 Interpretation of the 1.2 cm methanol emission in Orion. V. S. Strel'nitskij.
Pis'ma Astron. Zh., Tom 7, 406 - 413 (1981). In Russian.
English translation in Soviet Astron. Lett., Vol. 7.

Possible pumping mechanisms for CH_3OH masers in OMC-1 ($\Delta J = 0$, $k = 2$ −1 transitions at $\lambda = 1.2$ cm) are analysed. It is shown that the radiative-radiative pumping through vibrational transitions requires the unlikely combination of low gas density and high methanol content. Collisional-radiative pumping with the energy sink on the cold dust is suggested.

131.019 Unveiling interstellar clouds.
G. Righini-Cohen, M. Simon. M. Felli.
Sky Telesc., Vol. 62, 225 - 227 (1981).

131.020 High-resolution profiles for the diffuse interstellar band at 6196 Å.
W. H. Smith, T. P. Snow, M. Jura, W. D. Cochran.
Astrophys. J., Vol. 248, 128 - 131 (1981).

Observations of the 6196.0 Å diffuse feature in a number of directions toward moderately and little reddened stars have been made at high spectral resolution and signal-to-noise ratio. The authors find that the 6196.0 Å feature is frequently a symmetric feature as narrow as 0.4 Å FWHM, while in some stars it is much broader, apparently due to blending in several clouds in the line of sight. While no fine structure is apparent in the 6196.0 Å feature, a new, equally narrow feature was found at 6194.6 Å. The feature at 6202.6 Å is seen to be highly asymmetric. The observations are interpreted as strong support for a molecular origin for the carrier of the 6196.0 Å diffuse feature.

131.021 The largest molecules in space. II.
J. M. Greenberg.
Nederlands Tijdschr. Natuurkd. A, Vol. A47, No. 1, p. 24 - 26 (1981). In Dutch. − Abstr. in Phys. Abstr., Vol. 84, Abstr. 71290 (1981).

131.022 On the origin and distribution of C IV and Si IV ions in the neighboring interstellar medium.
L. L. Cowie, W. Taylor, D. G. York.
Astrophys. J., Vol. 248, 528 - 541 (1981).

The authors present data on the C IV, Si IV, and N V absorption lines in 46 distant stars observed with the IUE spectrometer at FWHM resolution of 30 km s^{-1}. Sharp strong Si IV and C IV lines are regularly detected, while N V is not generally observed. Most of the Si IV and C IV lines probably arise in the H II regions surrounding the observed star or near neighbors. A simple analytic theory is given for these contributions, including the effects of ionization by a soft X-ray background. For some of the most distant field stars a photoionized contribution may be ruled out, but C IV absorption lines are detected. These particular lines are also unusual in that Si IV is not detected and that the lines are shallow and broad. The authors suggest that in these cases they are detecting contributions from the hot interstellar gas responsible for the O VI absorption lines.

131.023   **Ultraviolet interstellar extinction toward 1367 stars observed by ANS.**
D. M. Meyer, B. D. Savage.
Astrophys. J., Vol. 248, 545 - 562 (1981).

The character of interstellar extinction toward 1367 stars of spectral type B3 and earlier is investigated using ANS five-band UV photometry. The observations produce a galactic average $E(\lambda-V)/E(B-V)$ = 5.11, 4.78, 6.52, 4.10, and 1.95 at 1550, 1800, 2200, 2500, and 3300 Å, respectively. Regional plots of $E(22-33)$ and $E(15-33)$ versus $E(B-V)$ isolate many stars whose extinction bump at 2200 Å or far-UV extinction at 1550 Å deviates significantly from the mean. Differences in the strengths of the UV extinction features of these stars appear to be independent; objects with either strong or weak 2200 Å extinction can have strong normal, or weak far-UV extinction. This independence supports the idea that different dust grain species within the interstellar medium are responsible for extinction at 2200 Å and in the far-UV. Further investigation reveals that some of the deviant stars are embedded in localized regions whose average extinction curves diverge dramatically from the galactic mean.

131.024   **A semiempirical model for heavy element depletion in the interstellar medium.**
B. M. P. Trivedi, J. W. Larimer.
Astrophys. J., Vol. 248, 563 - 568 (1981).

One important process operating in interstellar space is grain sputtering which preferentially destroys grains with weak binding energies and returns the constituent elements to the gas. An empirical relationship has been derived which reproduces the observed depletion factors in interstellar gas by relating them to some preshock distribution factor and the binding energy of the elements in appropriate compounds.

131.025   **The collapse to equilibrium of rotating, adiabatic spheroids. I. Protostars.**   J. E. Tohline.
Astrophys. J., Vol. 248, 717 - 726 (1981).

The role that rotation plays during the dynamic collapse of a protostellar gas cloud is examined under the assumptions that the cloud conserves angular momentum and evolves along a fixed adiabat. An analytic model is presented that predicts what the final radius, density, temperature, and rate of rotation of the protostellar cloud will be for any chosen adiabatic exponent of the gas from the knowledge only of what the cloud's initial ratio of thermal to gravitational energy (α_0) and the initial ratio of rotational to gravitational energy (β_0) are. Assuming that all gas clouds are susceptible to fragmentation if they collapse to configurations having $\beta \geqslant 0.274$, the model also predicts, from initial conditions alone, when a protostellar object will fragment into a multiple stellar system.

131.026   **The evolution of protostars. III. The accretion envelope.**   S. W. Stahler, F. H. Shu, R. E. Taam.
Astrophys. J., Vol. 248, 727 - 737 (1981).

The authors consider the radiation gas dynamics of the rapidly inflowing material during the main accretion phase of protostellar evolution. The authors follow in detail the momentum and energy transfer by radiation and matter, the dissociation of the molecular gas, and the thermochemical de-

struction of graphite grains. Considerable physical insight is gained by taking advantage of the vastly disparate time and length scales associated with the various processes. Different computational techniques are used in different portions of the accretion flow: the optically thick dust envelope, the thermalization layer of stellar photons, the opacity gap, and the radiative precursor.

131.027   **A high-resolution optical survey of interstellar absorption lines toward globular clusters and extragalactic objects. II. Further data.**
A. Songaila, L. L. Cowie, D. G. York.
Astrophys. J., Vol. 248, 956 - 962 (1981).

An ongoing survey of narrow-line visual absorption components is expanded to include lines of sight to the quasar 3C 273, the nucleus of M87, and the cores of the globular clusters NGC 2808, NGC 1851, and 47 Tuc. In addition, observations of three additional LMC stars are presented. The number density of components in the halo is calculated to be between one-third and one per 100 kpc, though uncertainties are substantial. This is entirely consistent with the galactic halo interpretation of the origin of high-column–density, quasar absorption-line systems.

131.028   **Interstellar grain size. II. Infrared photometry and polarization in Orion.**
M. Breger, R. D. Gehrz, J. A. Hackwell.
Astrophys. J., Vol. 248, 963 - 976 (1981).

New observations, including 0.36–0.82 μm linear polarimetry and extensive infrared photometry, are presented for stars within a degree of the Orion Nebula. The authors use infrared colors, wavelength dependence of the polarization, and criteria for cluster membership to examine the interstellar extinction law in the Orion region. An analysis of nonpeculiar cluster members yields $(m-M)_0$ = 8.00 mag, $E(B-V)$ = 0.054 mag, $E(V-K)$ = 0.18 mag, and p = 0.25% for the majority of cluster stars which lie in front of the thick intracluster dust clouds. The stars inside the dust clouds are examined on a star-by-star basis. The different intracluster grain size indicators are studied in detail. In Orion, large interstellar grain sizes are found in, and restricted to, regions of nebulosity.

131.029   **A compressed cloud in the Vela supernova remnant.**
E. B. Jenkins, J. Silk, G. Wallerstein, E. M. Leep.
Astrophys. J., Vol. 248, 977 - 983 (1981).

Observations by the IUE of stars behind the Vela supernova remnant show that HD 72350 (type B4 III) has exceptionally strong interstellar absorption lines of C I arising from the two excited fine-structure levels. A curve of growth analysis yields column densities for many neutron atomic species and the CO molecule. An analysis of the excited-level populations of C I provides limits on the local temperature and pressure. The authors discuss this unusually high pressure cloud in terms of shock compression by the Vela supernova blast wave. The relationship of this kind of cloud compression to star formation and to the origin of the characteristic filamentary emission arcs seen in Vela and in other supernova remnants is outlined.

131.030   **An investigation of the neutral and ionized gas in M16.**   S. L. Mufson, W. F. Fountain, G. A. Gary, W. E. Howard III, C. R. O'Dell, M. T. Wolff.
Astrophys. J., Vol. 248, 992 - 1009 (1981).

The authors present the results of new mapping observations of the neutral and ionized hydrogen in M16. In addition, they report observations of ^{12}CO from selected positions within the H II regions. The aim of the study is to understand the physical and kinematic conditions in this site of active star formation.

131.031   **The increasing chemical complexity of the Taurus dark clouds: detection of CH_3CCH and C_4H.**

W. M. Irvine, B. Höglund, P. Friberg, J. Askne, J. Elldér.
Astrophys. J., Lett., Vol. 248, L113 - L117 (1981).

The authors report the first detection in cold dark clouds of methyl acetylene (CH_3CCH) and the butadiynyl radical (C_4H). The $2_0 \rightarrow 1_0$ and $2_1 \rightarrow 1_1$ transitions of CH_3CCH and the hyperfine-resolved $N = 3 \rightarrow 2$ transitions of C_4H, which occur at frequencies near 34 and 29 GHz, respectively, are all seen astronomically for the first time. The distribution of CH_3CCH in TMC-1 appears to correlate with that of the cyanopolyynes, and not with NH_3 or CN. Abundance ratios at the TMC-1 HC_5N peak include $[CH_3C_2H] / [HC_2CN] \approx 1$, $[C_4H] / [C_2H] \approx 1$, and $[C_4H] / [HC_4CN] \approx 4$.

131.032 Isocyanic acid in the Taurus Molecular Cloud 1.
R. L. Brown.
Astrophys. J., Lett., Vol. 248, L119 - L122 (1981).

The author reports the detection of the 1_{01}- 0_{00} line of isocyanic acid (HNCO) in the Taurus Molecular Cloud 1 (TMC-1). The hyperfine structure of the transition is clearly resolved in TMC-1, but the intensities of the hyperfine components are not in the LTE ratio 5:3:1; rather the author finds the ratios to be approximately 5:2:1.

131.033 Absorption bands and extinction in small interstellar particles.   W. W. Duley.
Astrophys. Space Sci., Vol. 78, 145 - 156 (1981).

The production of discrete line and broad-band extinction by small interstellar oxide and silicate particles is discussed quantitatively. Restrictions on particle size and refractive index that are required to produce "pure" absorption features are reviewed. The relationship between optical depth in interstellar extinction and absorption coefficients for bulk materials is used to reach some general conclusions concerning the diffuse interstellar features, VUV extinction and the composition of interstellar dust. It is noted that charge transfer bands of ions such as Fe^{3+} may be detectable in the VUV spectrum of dust.

131.034 Axisymmetric gravitational collapse of an isothermal rotating interstellar gas cloud. A semi-analytical approach.   N. Virgopia, F. Ferraioli.
Astrophys. Space Sci., Vol. 78, 211 - 233 (1981).

The gravitational collapse mechanism of a non-magnetized isothermal rotating interstellar gas cloud is discussed analytically. The study predicts the formation of final flattened disk-like configurations in the central region of the infalling uniform gas cloud. The existence of critical times indicates the possibility that, beyond these times, discrepancies can arise in the numerical calculations. It seems therefore very likely that the ring-like structures, as suggested by earlier numerical calculations of hydrodynamic models, would be caused by pressure effects which arise through the introduction of higher order terms in the linearized theory.

131.035 New discoveries from the interstellar laboratory.
E. N. Rodríguez Kuiper.
Nature, Vol. 293, 339 - 340 (1981).

131.036 The nearby interstellar medium.   P. C. Frisch.
Nature, Vol. 293, 377 - 379 (1981).

The high dispersion spectrometer on board the International Ultraviolet Explorer (IUE) satellite was used to observe interstellar absorption lines in Rasalhague (α Oph). The satellite results, combined with previous interstellar line data and observations of the 'local interstellar wind' and soft X-ray emission, support earlier suggestions that the sun is immersed in a supernova remnant which may be an extension of the 'Loop I' or 'North Polar Spur' supernova remnant seen in the Scorpius − Ophiuchus region of the sky.

131.037 Deviations of galactic reddening from the cosecant relation.   T. W. Noonan.
Astron. J., Vol. 86, 1042 - 1043 (1981).

The scatter in the reddening diagram of $E(B-V)$ vs $\csc |b|$ for galactic globular clusters is found to be proportional to $(\csc |b|)^{1/2}$ and is due to cloudiness of the reddening layer. An asymmetry between the north and south galactic hemispheres suggests that the Sun may be south of the mid-plane of the Galaxy.

131.038 Ammonia-line observations of the Mon R2 molecular cloud.
R. F. Willson, F. J. Folch-Pi.
Astron. J., Vol. 86, 1084 - 1089 (1981).

Line emission of the $(J,K)=(1,1)$ and $(2,2)$ transition of ammonia, NH_3, has been mapped across the Mon R2 molecular cloud with an angular resolution of 1.5′ and a velocity resolution of 0.1 km s^{-1}. The observations show that the NH_3 emission is contained within an ellipsoidal region whose major axis lies nearly along the galactic plane. The line velocities vary systematically along the major axis, suggesting that the cloud is rotating along the galactic plane with a period of about 8×10^6 yr. The authors failed to observe any systematic variations of the linewidths across the cloud, indicating that the denser interior of the cloud outlined by the NH_3 emission is not collapsing or expanding. Kinetic temperatures of about 15 K, hydrogen densities of about 3×10^3 cm^{-3}, and a cloud mass of 300 $M_\odot$ are inferred from the observations.

131.039 Mode of variation of the scattering index parameter in the interstellar medium.   P. N. Okeke.
Astron. J., Vol. 86, 1110 - 1112 (1981).

Using recent data obtained for decorrelation frequency f_ν for 20 pulsars with dispersion measure DM $\leqslant 50$ pc cm^{-3}, the author investigates whether there is any systematic variation of the scattering index parameter, $(\langle \Delta N_e^2 \rangle / L)^{1/2} = \epsilon$, in the interstellar medium (ISM). It is found that ϵ appears to vary uniformly in certain regions of the ISM, defined approximately by $110 \leqslant z < 420$ pc and with corresponding galactic coordinates given by $26° < |b| < 87°$. Elsewhere, ϵ is found to have a very random variation. The results for ϵ arising from values of f_ν obtained earlier by other authors are compared with the author's result. It is observed that more data are required before a definite conclusion on the mode of variation of ϵ can be made.

131.040 Observations of the interstellar medium with IUE.
E. B. Jenkins.
The Universe at ultraviolet wavelengths, (see 012.009), p. 541 - 556 (1981).

131.041 Interstellar abundance determination using IUE data.
C. Joseph, T. P. Snow, Jr.
The Universe at ultraviolet wavelengths, (see 012.009), p. 567 - 569 (1981).

Analyses of the silicon interstellar abundances have been made for more heavily reddened lines-of-sight than were accessible to the Copernicus satellite.

131.042 A search for interstellar molecules in the spectra of highly reddened stars.   D. Lien, D. Buhl, R. M. Crutcher, B. Donn, A. M. Smith, L. E. Snyder, L. J. Stief.
The Universe at ultraviolet wavelengths, (see 012.009), p. 581 - 588 (1981).

The dark cloud in the line of sight towards X Persei has been searched for the molecular species OH, CH_2, HCl, CO and C_2 using the IUE satellite. The authors have detected CO, with a column density of 5×10^{15} cm^{-2} and 1×10^{14} cm^{-2} for ^{12}CO and ^{13}CO, respectively, and have placed upper limits on the log of the column densities of OH, CH_2 and HCl of 14.0, 12.8, and 12.3, respectively.

131.043 Ultraviolet absorption by interstellar gas at large distances from the galactic plane.

B. D. Savage, K. S. de Boer.
The Universe at ultraviolet wavelengths, (see 012.009), p. 611 - 612 (1981).

131.044 **Superbubbles.**   T. R. Gull, F. C. Bruhweiler,
M. Kafatos, S. Sofia.
The Universe at ultraviolet wavlengths, (see 012.009), p. 679 - 685 (1981).

Individual massive stars with $M_{bol} < -6$ have huge stellar winds that create interstellar bubbles. Stars with masses greater than $8 M_\odot (4 M_\odot ?)$ are considered supernova progenitors. These massive stars are numerous in OB associations where few supernova remnants are detected. Model calculations describing the evolution of an association show: (1) that large, hot cavities are formed by pushing the ambient gas into neutral shells; (2) that the shell radii change with galactocentric radius, (3) that only thirty percent of the interstellar medium is in the form of supercavities and (4) that a consequence is that only a small fraction of supernovae form supernova remnants. These results have strong bearing on interpretation of interstellar studies being done by IUE and by HEAO-B.

131.045 **The young Of star HD 148937 and its associated interstellar bubble – H II region.**
F. C. Bruhweiler, T. R. Gull.
The Universe at ultraviolet wavelengths, (see 012.009), p. 687 - 691 (1981).

HD 148937 and nebulosities surrounding the star are found to be closely inter-related. IUE spectroscopy of HD 148937 shows the star to be a young Of star with low mass loss. Properties of the surrounding interstellar bubble and the H II region support the implied youth of HD 148937.

131.046 **H I kinetic temperatures in shell, filaments, and elsewhere.**   C. Heiles.
Bull. American Astron. Soc., Vol. 13, 507 - 508 (1981). Abstract.

131.047 **Ammonia and water observations in the southern hemisphere.**   D. F. Dickinson, S. Gulkis,
M. J. Klein, T. B. H. Kuiper, M. Batty, F. F. Gardner, D. Jauncey, J. B. Whiteoak.
Bull. American Astron. Soc., Vol. 13, 508 (1981). – Abstract.

131.048 **Coupling of the magnetic field and rotation in the dark cloud B5.**
P. F. Goldsmith, W. D. Langer, J. S. Young, R. W. Wilson.
Bull. American Astron. Soc., Vol. 13, 508 (1981). – Abstract.

131.049 **Statistical properties of the radiation fields from H_2O masers.**   J. M. Moran.
Bull. American Astron. Soc., Vol. 13, 508 (1981). – Abstract.

131.050 **Absorption of the soft X-ray background by the Coalsack Nebula.**   W. T. Sanders.
Bull. American Astron. Soc., Vol. 13, 509 (1981). – Abstract.

131.051 **A survey of interstellar K I absorption lines. I. Observations.**   F. H. Chaffee, Jr., R. E. White.
Bull. American Astron. Soc., Vol. 13, 509 (1981). – Abstract.

131.052 **VLBI observations of H_2O masers in W49N.**
R. C. Walker, D. N. Matsakis, J. A. Garcia-Barreto.
Bull. American Astron. Soc., Vol. 13, 511 (1981).

131.053 **Collisionally determined mass spectra of interstellar clouds.**   M. A. Hausman.
Bull. American Astron. Soc., Vol. 13, 511 - 512 (1981). Abstract.

131.054 **On the possibility of star formation behind interstellar shocks.**   G. Welter, J. Schmid-Burgk.
Bull. American Astron. Soc., Vol. 13, 512 (1981). – Abstract.

131.055 **The density distribution in the core of the ρ Oph and R CrA molecular clouds.**
R. B. Loren, A. Sandqvist, A. Wootten.
Bull. American Astron. Soc., Vol. 13, 512 (1981). – Abstract.

131.056 **Molecular line mapping of the Orion Molecular Cloud.**
N. Z. Scoville, F. P. Schloerb, P. F. Goldsmith.
Bull. American Astron. Soc., Vol. 13, 513 (1981). – Abstract.

131.057 **Star formation in the outer Galaxy.**
M. R. Viner, V. A. Hughes, P. D. Jackson, J. R. Sewall.
Bull. American Astron. Soc., Vol. 13, 513 (1981). – Abstract.

131.058 **High velocity H_2 in NGC 2071.**   S. E. Persson,
T. R. Geballe, T. Simon, C. J. Lonsdale, F. Baas.
Bull. American Astron. Soc., Vol. 13, 513 (1981). – Abstract.

131.059 **Cyanopolyyne molecular absorption in the direction of Cassiopeia A.**
P. A. Feldman, M. B. Bell, H. E. Matthews.
Bull. American Astron. Soc., Vol. 13, 519 (1981). – Abstract.

131.060 **Detection of deuterated cyanodiacetylene (DC_5N) in the molecular cloud TMC 1.**
J. M. MacLeod, L. W. Avery, N. W. Broten.
Bull. American Astron. Soc., Vol. 13, 519 (1981). – Abstract.

131.061 **Detection of deuteriocyanobutadiyne in the interstellar cloud TMC 1.**
F. P. Schloerb, R. L. Snell, J. S. Young, W. D. Langer.
Bull. American Astron. Soc., Vol. 13, 519 - 520 (1981). Abstract.

131.062 **Molecule depletion in the cores of dense clouds.**
W. Boland, T. de Jong.
Bull. American Astron. Soc., Vol. 13, 520 (1981). – Abstract.

131.063 **A survey of interstellar potassium absorption. II. Abundance and velocities.**
R. E. White, F. H. Chaffee, Jr., E. K. Knittle.
Bull. American Astron. Soc., Vol. 13, 537 (1981). – Abstract.

131.064 **Molecular clouds in the outer Galaxy.**
M. L. Kutner, K. N. Mead.
Bull. American Astron. Soc., Vol. 13, 538 (1981). – Abstract.

131.065 **Cloud-cluster associations: the B0 phenomenon reexamined.**   D. P. Clemens, N. Z. Scoville.
Bull. American Astron. Soc., Vol. 13, 539 (1981). – Abstract.

131.066 **A second region of energetic activity associated with star formation in Orion; NGC 2071.**   J. Bally.
Bull. American Astron. Soc., Vol. 13, 540 (1981). – Abstract.

131.067 **CO J = 3 → 2 observations of galactic star formation regions.**   G. J. White, J. P. Phillips, G. D. Watt.
Bull. American Astron. Soc., Vol. 13, 540 (1981). – Abstract.

131.068 **Collisional excitation of interstellar molecules due to H_2. II. Excited rotational-vibrational states of SiO.**
S. C. Mehrotra.
Astrophys. Space Sci., Vol. 78, 463 - 465 (1981).

Collision induced vibrational rates of SiO due to H_2 are computed using the normalized perturbative, semiclassical approach as described in Paper I (Mehrotra, 1980). The rates are very small as compared to the rates for rotational excitation and are more sensitive to temperature.

131.069 Graphite grains, carbon depletion and the 2200 Å feature. A. Blanco, E. Bussoletti.
Astrophys. Space Sci., Vol. 78, 467 - 471 (1981).

In a recent paper Millar has shown that if one assumes that the carbon depleted from the gas phase is all tied up into small graphite grains the observational data raise serious doubts against the hypothesis that these particles are responsible for the 2200 Å extinction hump. In the present paper it is shown that this problem may be overcome if the presence in the interstellar space of graphite grains with sizes greater than 0.02 μ is taken into account. The derived ratios between the masses of large grains, which do not contribute to the ultraviolet extinction hump, and those of the small ones are consistent with those evaluated in the circumstellar shells of carbon stars.

131.070 The temperature of interstellar oxide grains. T. J. Millar.
Astrophys. Space Sci., Vol. 78, 505 - 508 (1981).

Grain temperatures have been derived for small oxide grains in a typical interstellar radiation field. Detailed Mie calculations have been used to determine Q_{abs} from 1000 Å to 30 μm. Longward of 30 μm, there is no optical data available but the fact that such grains are likely highly defected implies that a λ^{-1} dependence in the far-infrared may be appropriate.

131.071 Anomalous far-UV extinction in the WN6 star HD 147419. A. J. Willis, D. J. Stickland.
Mon. Not. R. Astron. Soc., Vol. 197, 1P - 9 P (1981).

IUE observations of two WN stars in ring nebulae, HD 147419 and HD 89358, show that in the former star, the level of far UV interstellar extinction, $\lambda > 1600$ Å, is far higher than expected from the mean galactic extinction law. The results imply the existence of an abnormal abundance of small grains at some location in the line of sight to HD 147419, possibly in the ring nebula RCW 104 surrounding the star.

131.072 The rate of star formation in galaxies. B. F. Madore.
The structure and evolution of normal galaxies, (see 012.010), p. 239 - 249 (1981).

The paper is a topical review of the many types of star formation mechanisms suggested to date but set in a speculative extragalactic context.

131.073 Abundances and physical conditions in a high-velocity cloud. S. R. Pottasch, P. R. Wesselius, E. M. Arnal.
Second European IUE Conference, (see 012.011), p. 13 - 17 (1980).

The physical conditions in the 'high-velocity cloud' in the direction of HD 175754 are discussed, based on the high-dispersion IUE measurements. It is found that the abundances are very close to solar. Although the oxygen (and presumably also the hydrogen) is mostly neutral, the ionisation is higher than one would expect from a neutral cloud. Consequences for the ionisation and heating of the interstellar medium in general are discussed.

131.074 Observations of the exciting stars of NGC 2023 and M43 – interstellar extinction and abundances.
D. P. Gilra, E. M. Arnal, P. R. Wesselius, S. R. Pottasch, J. de Vries.
Second European IUE Conference, (see 012.011), p. 19 - 20 (1980).

The authors present preliminary results on extinction curves for two stars in dusty H II regions – HD 37903 and HD 37061. HD 37903 is the exciting star of NGC 2023, a nebula in the Orion belt region. HD 37061 is the exciting star of M43 – the north eastern companion of the Orion Nebula,

and is about 9' NE of the trapezium. These stars were also observed at high resolution to study the interstellar gaseous abundances and investigate if these were correlated with the shape of the extinction curves in the UV.

131.075 High velocity gas associated with supernova remnants. P. M. Gondhalekar, A. P. Phillips.
Second European IUE Conference, (see 012.011), p. 109 - 113 (1980).

High velocity gas associated with supernova remnants has been studied by obtaining high dispersion IUE spectra of early-type background stars. The authors present the analysis of absorption lines originating in gas associated with the remnant S147 and the detection of gas associated with the remnant IC443. In the star HD 36665 behind S147, as well as high velocity lines, low velocity interstellar CO has been detected probably originating in gas associated with the remnant.

131.076 High resolution observations of interstellar lines in the IUE spectra of LMC stars.
P. M. Gondhalekar, A. J. Willis, D. H. Morgan, K. Nandy.
Second European IUE Conference, (see 012.011), p. 123 - 127 (1980).

Velocity profiles of numerous interstellar lines observed in high resolution IUE spectra of several stars in the LMC are discussed. In all lines of sight, well defined velocity components are seen at heliocentric velocities of +0 km s^{-1}, +70 km s^{-1}, +150 km s^{-1} and +200 - 300 km s^{-1}, similar to those reported for the Ca II k interstellar line observed in the spectrum of HD 38268: the central star of the 30 Doradus complex. The extended blue asymmetries at LMC velocities observed in the Ca II k line in HD 38268 are observed in many of the UV interstellar lines in LMC stars well removed from 30 Doradus, favouring an origin for this feature as absorption in gas in a halo around the LMC, as proposed by de Boer and Savage.

131.077 The extinction and the H I content of the dark cloud complex Khavtasi 141. T. Saito, H. Ohtani, Y. Tomita.
Publ. Astron. Soc. Japan, Vol. 33, 327 - 340 (1981).

The dark cloud complex Khavtasi 141 is examined by both star counts and an analysis of the self-absorption in the 21-cm line of neutral hydrogen. From an extinction analysis, the mass of the complex is found to be $\gtrsim 1 \times 10^4 M_\odot$, while the mass of neutral hydrogen is estimated to be $\sim 4 \times 10^2 M_\odot$.

131.078 Time variation of SiO maser emissions. I. Velocity structure of the Orion SiO maser envelope.
N. Ukita, N. Kaifu, Y. Chikada, T. Miyaji, K. Miyazawa.
Publ. Astron. Soc. Japan, Vol. 33, 341 - 350 (1981).

The authors report monitoring observations of the SiO $v = 1$, $J = 2-1$ transition with high frequency resolution. Velocity structures in the double emission feature tend to shift inward recurrently. The authors discuss the velocity structure, the size, and the nature of the central star of the Orion SiO maser envelope.

131.079 Interstellar masers: the influence of the geometrical shape on the radiation properties.
E. Bettwieser.
J. Astrophys. Astron., Vol. 2, 187 - 199 (1981).

The relation between the minimum and the maximum of the energy density of the microwave field is important in determining the internal physical conditions in a maser source and is directly connected with the size of the emission spot. This relation is investigated for models of homogeneous maser clouds for three different geometries: a thin tube, a thin disk and a sphere. For substantial degrees of saturation, an approximate analytical calculation scheme is presented.

131.080 **More absorption toward Cygnus X-3.**
T. C. Weekes, H. P. Murphy, R. Schild, H. Gursky,
J. Geary, T. Stephenson.
Publ. Astron. Soc. Pacific, Vol. 93, 474 - 476 (1981).
   Deep CCD images in the I band show a limiting magni-
tude for Cygnus X-3 of $I > 20^m.3$; when this is combined with
measurements of H and K it indicates an absorption
$A_v = 28^m.7$. This is compatible with a greater distance to the
source with a consequent increase in the already high X-ray
and gamma-ray luminosities.

131.081 **Search for and investigation of condensations of star
formation in dark galactic nebulae.**
I. I. Zinchenko, A. G. Kislyakov.
Spektr. issled. kosm. i atmos. izluch. Gor'kij, 1979, p. 34 - 66.
In Russian. – Abstr. in Ref. zh., 51. Astron., 8.51.736 (1981).

131.082 **CO emission in the direction of some continuum
sources at millimeter radio waves.**
A. G. Kislyakov, M. A. Gordon.
Spektr. issled. kosm. i atmos. izluch. Gor'kij, 1979, p. 67 - 83.
In Russian. -- Abstr. in Ref. zh., 51. Astron., 8.51.751 (1981).

131.083 **Radio continuum and carbon monoxide observations
of V645 Cygni (GL 2789).**
L. F. Rodríguez, J. M. Torrelles, J. M. Moran.
Astron. J., Vol. 86, 1245 - 1249 (1981).
   The authors report 6-cm VLA observations and 3-mm
CO observations of the region. Their main objectives were to
search for a compact H II region, in order to determine precisely the location of the exciting star of the system, and to
determine accurately the spatial position of the CO peak
first detected by Harvey and Lada (1980).

131.084 **Panel on dynamic vs. static models of the interstellar
medium.**   Introductory remarks by L. Spitzer, Jr.
The phases of the interstellar medium, (see 012.018), p. 1 - 5
(1981).

131.085 **Numerical simulations of the interstellar medium.**
T. M. Bania.
The phases of the interstellar medium, (see 012.018), p. 7 - 13
(1981).

131.086 **The growth of molecular clouds.**
V. Icke.
The phases of the interstellar medium, (see 012.018),
p. 15 - 22 (1981).

131.087 **Static vs. dynamic models: concluding remarks.**
V. Icke.
The phases of the interstellar medium, (see 012.018),
p. 23 - 24 (1981).

131.088 **Magnetic coupling between various phases of
interstellar matter.**   B. G. Elmegreen.
The phases of the interstellar medium, (see 012.018),
p. 25 - 34 (1981).

131.089 **Studies of optical interstellar absorption lines.**
L. M. Hobbs.
The phases of the interstellar medium, (see 012.018),
p. 35 - 39 (1981).

131.090 **Radio observations of neutral hydrogen and molecules in diffuse clouds.**   J. Crovisier.
The phases of the interstellar medium, (see 012.018),
p. 41 - 44 (1981).
   The paper will report on statistical properties of H I
clouds derived from 21-cm absorption measurements; observations of small-scale fluctuations of the 21 cm line; studies of

correlations between H I and molecules (OH and CO) in H I
clouds.

131.091 **Radio spectroscopy of diffuse clouds.**
J. M. Dickey.
The phases of the interstellar medium, (see 012.018),
p. 45 - 50 (1981).

131.092 **Neutral hydrogen spin temperatures.**
E. E. Salpeter.
The phases of the interstellar medium, (see 012.018),
p. 51 - 52 (1981).

131.093 **Comments related to the state of diffuse interstellar
clouds.**   W. D. Watson.
The phases of the interstellar medium, (see 012.018),
p. 53 - 56 (1981).
   The author's comments will deal with aspects of the
state of the classical "diffuse interstellar clouds" – (1) new
data and a reconciliation of old data about the temperature,
density and pressure of the "prototypical" diffuse cloud
toward Zeta Ophiuchi, and (2) the likelihood that shock
waves are a common phenomenon in these clouds.

131.094 **Recent progress toward an understanding of interstellar chemistry.**   S. R. Federman.
The phases of the interstellar medium, (see 012.018),
p. 57 - 59 (1981).
   Observational results and theoretical predictions now
appear to be converging toward a comprehensive understanding of many physical processes occurring in interstellar clouds.
The fractionation of ^2H and ^{13}C and the chemistry of simple
molecules like CO and CH^+ are examples of such processes.
The enhancement of deuterium in HCO^+ has been discussed in
great detail before, so the author will concentrate on the latter
processes.

131.095 **Galactic systematics of dark nebulae.**
B. T. Lynds.
The phases of the interstellar medium, (see 012.018),
p. 61 - 63 (1981).

131.096 **Opacities of dark clouds in spiral galaxies.**
D. M. Elmegreen.
The phases of the interstellar medium, (see 012.018),
p. 65 - 68 (1981).

131.097 **Morphology and evolution of molecular clouds: a
brief review.**   B. E. Turner.
The phases of the interstellar medium, (see 012.018),
p. 69 - 85 (1981).
   I. Classification of molecular clouds: Traditional classification, modern picture, complexes and GMC, cloud masses and
star masses. II. Structure and dynamics of group A clouds.
III. Structure and dynamics of group B clouds. IV. The galactic
distribution of giant complexes. Do they lie in spiral arms?
Evidence from local complexes, evidence from galactic CO
surveys. V. Theories of star formation and evolution of giant
complexes: Purely kinematic models, formation of GMC from
dust lanes in spiral shocks, modified coalescence models,
formation by Parker instability triggered by galactic shocks,
some general comments.

131.098 **New observational evidence regarding the lifetime of
giant molecular cloud complexes.**   L. Blitz.
The phases of the interstellar medium, (see 012.018),
p. 87 - 91 (1981).

131.099 **The phase of the interstellar medium which produces
the very low surface brightness Hα background.**
R. J. Reynolds.
The phases of the interstellar medium, (see 012.018),

p. 109 - 120 (1981).

There is strong evidence for a phase of the interstellar medium consisting of wide-spread regions of nearly fully ionized hydrogen. These regions have an electron density of about 0.2 cm^{-3}, a temperature near 10^4 K, and occupy about 10-20% of the interstellar volume. While this warm, ionized phase seems to be concentrated or has higher gas densities in the spiral arms, it appears to be present along every line of sight, even toward the galactic pole. The ionization mechanism for this gas has not been determined, but it must be wide-spread and capable of nearly fully ionizing hydrogen, while not producing detectable [O III]λ5007 emission. The Hα surface brightness at high galactic latitudes requires an ionization rate in the vicinity of the Sun of 3 $\times 10^6$ hydrogen ionizations per second per cm^2 of galactic disk.

131.100 Constraints on the ISM phases from pulsar observations.   B. J. Rickett.
The phases of the interstellar medium, (see 012.018), p. 121 - 125 (1981).

131.101 Radio recombination lines that are not from bright H II regions.   F. J. Lockman.
The phases of the interstellar medium, (see 012.018), p. 127 - 129 (1981).

131.102 X-ray observations of the hot interstellar medium. W. T. Sanders.
The phases of the interstellar medium, (see 012.018), p. 133 - 134 (1981).

131.103 Observational constraints on the properties of the coronal interstellar gas.   E. B. Jenkins.
The phases of the interstellar medium, (see 012.018), p. 139 - 144 (1981).

131.104 Origin of the narrow mixed ion quasar absorption lines.   B. D. Savage.
The phases of the interstellar medium, (see 012.018), p. 181 - 186 (1981).

131.105 Le milieu interstellaire.   J. Guibert.
Histoire de l'univers, (see 003.007), p. 99 - 122 (1980).

Contents: La découverte du milieu interstellaire. Les principaux constituants du milieu interstellaire. L'émission, l'absorption et la propagation du rayonnement dans le milieu interstellaire. Les molécules interstellaires. La distribution et la cinématique de la matière interstellaire dans la Galaxie. Les interactions et les relations des étoiles et de la matière interstellaire. Les perspectives d'avenir de l'étude du milieu interstellaire.

131.106 On the determination of the UV albedo of interstellar dust.
J. Wolf, T. Herter, H. L. Helfer, J. L. Pipher.
News Lett. Astron. Soc. N. Y., Vol. 1, No. 10, p. 23 (1981).
Abstract.

131.107 A southern survey of OH masers at 1612 MHz.
J. L. Caswell, R. F. Haynes, W. M. Goss, U. Mebold.
Australian J. Phys., Vol. 34, 333 - 355 (1981).

A search for OH has been made along the galactic plane from longitude 340° to the galactic centre, yielding 78 emission sources. Of these sources 55 are masers of the variety showing two intensity peaks spaced in velocity – a characteristic of OH/IR stars. The velocity and spatial distributions of these new OH/IR stars (which are not as yet identified in the optical or infrared) are discussed, with special reference to their kinematic properties and population type.

131.108 Chemical evolution of interstellar dust – a source of prebiotic material?   J. M. Greenberg.
Comets and the origin of life, (see 012.022), p. 111 - 127 (1981).

As a result of laboratory simulation of the photoprocessing of the grains in interstellar space, the author finds that the order of 10% of all the available oxygen, carbon and nitrogen mixtures is converted in 10^7 years into large molecules whose infra-red absorption spectra are characteristic of carboxylic acid and amino groups. The material does not evaporate at temperatures less than 400–500 K and is soluble in water. The molecular weight of one of the samples which is called the "yellow stuff" was measured and found to be 514 a.m.u. The author considers the above properties to be strongly indicative of prebiotic organic molecules. Following the thermal evolution of a typical comet, he finds that comets entering from the Oort cloud should contain at least as much organic molecule material as originally contained in the interstellar dust out of which they accreted. The deposit of prebiotic molecules on a primitive earth is considered from several points of view.

131.109 The multiple-point fringe-rate method of mapping spectral-line VLBI sources with application to H$_2$O masers in W3-IRS5 and W3(OH).   R. C. Walker.
Astron. J., Vol. 86, 1323 - 1331 (1981).

Astrophysical masers often have spatial structures which consist of a small number of well separated, compact features at each frequency. VLBI observations of such structures produce relative fringe-rate spectra which have multiple peaks. Giuffrida has shown that these peaks can be used to produce maps while avoiding some of the worst calibration difficulties of aperture synthesis. The current multiple-point fringe-rate mapping method, along with some of its advantages and limitations, is described in this paper. The method has been used to map the H$_2$O maser emission in W3-IRS5 and W3(OH). The additional details seen in these maps, compared to previously published observations, make division of the sources into centers of activity difficult.

131.110 Cyanopolyyne absorption in the direction of Cassiopeia A.
M. B. Bell, P. A. Feldman, H. E. Matthews.
Astron. Astrophys., Vol. 101, L13 - L16 (1981).

The long-chain linear molecules HC$_5$N and HC$_7$N have been detected at 10 GHz in absorption by interstellar clouds in the Orion arm against the strong continuum source Cas A. The velocity components generally correspond with those found for atomic hydrogen and light molecules. The inferred total column densities are $\sim 2 \times 10^{11}$ cm^{-2} for HC$_5$N and $\gtrsim 1 \times 10^{11}$ cm^{-2} for HC$_7$N. These column densities are unexpectedly high relative to HCN if these molecules occur in diffuse cloud material ($n_{HI} \sim 10$ cm^{-3}).

131.111 On the association of the 1720 MHz OH masers with the H$_2$CO masers in NGC 7538 (IRS1).
S. Guilloteau, R. Lucas.
Astron. Astrophys., Vol. 101, L19 - L21 (1981).

The 1720 MHz OH masers associated with the ultracompact H II region NGC 7538 IRS1 have been observed with high spectral resolution (0.03 km s^{-1}). The spectrum displays about ten circularly polarized velocity features, with LSR velocities ranging from -59.4 to -56.0 km s^{-1}. This suggests that these masers are not intimately related with the H$_2$CO maser existing in this region. Implications are discussed for the pumping mechanisms of both masers, showing that 1720 MHz OH masers imply densities at least one order of magnitude higher than expected for H$_2$CO masers.

131.112 H I fine structure in a high-velocity cloud (HVC AI).   U. J. Schwarz, J. H. Oort.
Astron. Astrophys., Vol. 101, 305 - 314 (1981).
H I observations with the Westerbork Radio Telescope

were made of a field of 40′ diameter of a component (HVC 139 + 28 − 190) of the chain A of high-velocity clouds. A general survey of the fine-structure is given. Several features have a filamentary form. The dense features are distributed in an irregular fashion in velocity as well as position.

131.113 Statistics of neutral hydrogen absorption toward pulsars.   J. M. Dickey, J. M. Weisberg, J. M. Rankin, V. Boriakoff.
Astron. Astrophys., Vol. 101, 332 - 341 (1981).
This paper presents new spectra of the absorption of pulsar radiation by interstellar H I observed from Arecibo. The abundance of 21-cm absorption lines of various optical depths is found to be the same toward pulsars as toward extragalactic background sources. Variations of the ratio of the integral of 21-cm optical depth to dispersion measure as a function of pulsar distance reveal the statistical properties of the inhomogeneities of both the absorbing gas and the dispersive medium.

131.114 Observations of CO in H I clouds: correlations with H I and OH.   I. Kazès, J. Crovisier.
Astron. Astrophys., Vol. 101, 401 - 408 (1981).
The CO 2.6-mm line was searched for in 76 directions showing strong H I 21-cm absorption. It was detected in 21 directions, most of them corresponding to diffuse clouds. The CO and H I lines roughly coincide in velocity, but there is no apparent correlation between CO line intensities and H I optical depths, column densities or spin temperatures. The CO velocities are strongly correlated with those of the OH absorption lines.

131.115 On the origin of diffuse interstellar lines.
H. D. Breuer, H. Petry.
Naturwissenschaften, 68. Jahrg., 520 - 521 (1981).

131.116 Lifetimes of certain interstellar organic molecules.
K. K. Ghosh, S. N. Ghosh.
Indian J. Radio Space Phys., Vol. 10, 78 - 79 (1981). − Abstr. in Phys. Abstr., Vol. 84, Abstr. 89043 (1981).

131.117 Axisymmetric gravitational collapse of rotating interstellar gas clouds.
N. Virgopia, F. Ferraioli.
Astrophys. Space Sci., Vol. 79, 129 - 144 (1981).
Numerical calculations have been made of the gravitational axisymmetric collapse of isothermal gas clouds endowed with angular momentum. The evolutionary study is based on the so-called Fluid-in-Cell method coupled to an efficient algebraic algorithm which allows the Poisson equation to be integrated by means of block tri-diagonal matrices. The results, at ages slightly larger than the initial free-fall time, indicate that flattened disk-shaped structures are formed in the central region of the clouds − in good agreement with the previous analytical results predicted by the authors.

131.118 Star formation tracers.
E. Falgarone.
Ann. Physique, Vol. 6, (see 012.026), p. 103 - 105 (1981). In French. − Abstr. in Phys. Abstr., Vol. 84, Abstr. 94313 (1981).

131.119 Ultraviolet interstellar extinction toward stars in the Orion Nebula and toward HD 147889.
R. C. Bohlin, B. D. Savage.
Astrophys. J., Vol. 249, 109 - 117 (1981).
Measurements of UV interstellar extinction toward two interesting regions are presented. One region is the Orion Nebula and the other is toward the star HD 147889, a heavily reddened star in the ρ Oph reflection nebula. The UV extinction curves obtained are significantly different from the average galactic curve of Savage and Mathis (1979).

131.120 Proper motions and distances of H_2O maser sources. III. W51 NORTH.   M. H. Schneps, A. P. Lane, D. Downes, J. M. Moran, R. Genzel, M. J. Reid.
Astrophys. J., Vol. 249, 124 - 133 (1981).
The authors present the results of a proper motion experiment for the water maser source W51 NORTH which is associated with a very compact H II region in the radio continuum complex G49.5−0.4. They compare the positions of the water masers obtained in two experiments separated by about 2 years and identify components which have moved. These observations are used to estimate the distance to the H II region. The kinematics of the masers in W51 MAIN and W51 NORTH are found to differ from those observed for Orion.

131.121 Magnetic fields in molecular clouds: OH Zeeman observations.
R. M. Crutcher, T. H. Troland, C. Heiles.
Astrophys. J., Vol. 249, 134 - 137 (1981).
The authors have carried out sensitive OH Zeeman observations of the absorption lines produced in interstellar dust clouds toward 3C 133, 3C 123, and W51. Conservative (3 σ) upper limits to the magnetic field strengths at each position are 15, 25, and 30 microgauss respectively.

131.122 Hot-gas cold-dust pumping for water masers associated with H II regions.   S. Deguchi.
Astrophys. J., Vol. 249, 145 - 151 (1981) = Contrib. 3445 Div. Geol. Planet. Sci., Calif. Inst. Technol.
A collisional pump with an internal sink is proposed for the water masers associated with H II regions, where the population inversion occurs due to the absorption by cold ice-mantle grains in a highly dusty cloud of the far-infrared line radiation of hot water vapor. A new escape probability method is developed to calculate the transfer of line radiation in dusty medium. The pump mechanism explains the power of usual maser sources associated with H II regions and the enormous power of the sources associated with W49 N and external galaxies.

131.123 Structure of molecular clouds. V. Detailed models and observational characteristics.
L. G. Stenholm, T. W. Hartquist, G. E. Morfill.
Astrophys. J., Vol. 249, 152 - 160 (1981).
A detailed model of centrally condensed clouds containing regions of active star formation is derived from observations. The model has nearly equal turbulent and systematic collapse velocities and indicates that such clouds are not clumpy. Alfvénic waves can explain the turbulent velocities if the magnetic field varies as $n_{H_2}{}^{1/2}$. The model indicates the importance of chemical isotopic fractionation in CO, and suggests that metals play an important rolé in the ionization balance.

131.124 Suppositions for the comparison of infrared spectra from terrestrial silicates with interstellar spectra and the selection of suitable silicates.
R. Schmidt.
Astron. Nachr., Band 302, 235 - 245 (1981) = Mitt. Univ.-Sternw. Jena, Nr. 149.
The influence of potassium bromide on the bands near 10 μm, 18 μm, and 33 μm was calculated for any silicate. The wavelength shift, the deepening, and the broadening of the bands were calculated for different band depths of each band. The results were applied on the investigation of an extensive catalogue of silicate spectra, which were got on the basis of the KBr pressing technique too. So it was possible to pick out those silicates, which are probable candidates for the interstellar silicate component.

131.125 On the relative abundances of silicon isotopes in the interstellar medium.   A. A. Penzias.

Astrophys. J., Vol. 249, 513 - 517 (1981).

Measurements of interstellar silicon monoxide indicate that the isotopes of interstellar silicon have relative abundances equal to those in the solar system. These results serve to modify the results of a previous study which had suggested a different interstellar value. The interpretation of the earlier data involved the assumption that the observed SiO spectra were optically thin. The present work shows this assumption to have been invalid.

131.126   The isotopic abundances of interstellar oxygen.
A. A. Penzias.
Astrophys. J., Vol. 249, 518 - 523 (1981).

The relative abundance of ^{17}O and ^{18}O is found to be remarkably uniform in the giant molecular clouds of our Galaxy, varying by less than ~5% between the galactic center and the most distant (~12 kpc) sources. This uniformity is seen to reflect the close similarity in the galactic distributions of the processes by which these two isotopes are introduced into the interstellar medium, via novae explosions and mass loss from low-to-intermediate mass stars, respectively. There are two small-scale deviations from this general uniformity. First, the $^{18}O/^{17}O$ ratio in the envelope of the evolving star IRC + 10216 is found to be an order of magnitude below the galactic value. While this circumstance is comfortably consistent with the present understanding of nucleosynthesis, the second exception, the marked (~40%) $^{18}O/^{17}O$ enhancement in the solar system, remains an enigma.

131.127   Determination of the HNC to HCN abundance ratio in giant molecular clouds.   P. F. Goldsmith, W. D. Langer, J. Ellder, W. Irvine, E. Kollberg.
Astrophys. J., Vol. 249, 524 - 531 (1981).

The authors have used the rare isotopes of HCN and HNC to measure the relative abundance of these two chemical isomers in four giant molecular clouds. The lines of $H^{13}CN$, $HC^{15}N$, $HN^{13}C$, and $H^{15}NC$ are all sufficiently weak to be little affected by saturation. A comparison of intensities measured with different beamsizes indicates that the sources are not resolved even with the 42″ beamsize used in the present work. The abundance ratio [HNC]/[HCN] varies by more than an order of magnitude, ranging from 0.015 to 0.40. These low values are not consistent with HNC and HCN production solely from various sources of the precursor ion H_2CN^+. The authors suggest that neutral atom-molecule reactions may make an important contribution to the HCN production rate in such regions.

131.128   Properties of the galactic molecular cloud ensemble from observations of ^{13}CO.
H. S. Liszt, X. Delin, W. B. Burton.
Astrophys. J., Vol. 249, 532 - 549 (1981).

The authors have observed galactic ^{13}CO over the region $b = 0°$, $l = 28° - 40°$, at 3′ spacings of the 36 foot (11 m) telescope antenna pattern. The major results of interpretation of these observations are as follows: (1) The size distribution of molecular clouds, inferred from correcting measured sizes of features for the bias introduced by sampling only at $b = 0°$, is characterized by the moments $\langle D \rangle \lesssim 25$ pc, $\langle D^2 \rangle^{1/2} \lesssim 27$ pc, $\langle D^3 \rangle^{1/3} \lesssim 29$ pc. (2) The smallest mean density of hydrogen molecules derived for the region $R \approx 5$ kpc consistent with available constraints is $\langle n_{H_2} \rangle \approx 2.5$ cm^{-3}. (3) The radial abundance variations of the ^{13}CO emissivity follows the general behavior established earlier from ^{12}CO, but with prominent enhancements at $R \approx 9-10$ kpc and $R \approx 7-8$ kpc. (4) Comparison of the longitudinal variation of terminal velocities measured in ^{13}CO and in H I indicates that the galactic kinematics of the densecentered molecular clouds are essentially identical to those found for H I. (5) Comparison of ^{13}CO and H I integrated intensities indicates that one may find substantial regions over each of which the atomic and molecular intensities may be strongly positively or anticor-

related, or statistically independent. Reasons for this behavior are summarized.

131.129   High-velocity CO wings and CO self-reversals.
R. B. Loren.
Astrophys. J., Vol. 249, 550 - 562 (1981).

The redshifted and blueshifted CO line wings are found to occur in two oppositely directed lobes in a number of molecular clouds associated with recent star formation. These CO lobes may have their origin in nonisotropic mass outflow from young protostellar objects embedded in the cloud. Away from the cloud's dense core, the velocity extent of the CO wings decreases with projected radial distance (R), following a $R^{-0.5}$ law indicating deceleration of the stellar wind material as it encounters the surrounding molecular cloud. CO profiles with enhanced wings are found beyond the boundaries of the compact H II regions located within the dense cloud core. Carbon recombination lines occurring at the H II region-molecular cloud interface have line widths narrower than the ^{13}CO lines and much narrower than the broad CO profiles. This suggests that carbon recombination line emission and the broad wings of CO lines arise in substantially different regimes in the cloud.

131.130   Far-infrared observations of star-forming regions.
A. I. Sargent, R. J. van Duinen, C. V. M. Fridlund, H. L. Nordh, J. W. G. Aalders.
Astrophys. J., Vol. 249, 607 - 621 (1981).

The authors discuss observations at 85 μm and 150 μm of the molecular clouds NGC 2071, Mon R2, ON 1, and IC 5146. They investigate the distribution of dust emission in these clouds, the nature of their luminosity sources, and their large-scale energetics. Mapping of areas of order 20′ × 20′ permits a detailed scrutiny of the energetics of the clouds, not only near the dense cores but also in the more diffuse regions. Combining the results with other observations, the authors conclude that massive stars form at different depths within molecular clouds. Their apparent tendency to form near the edges of dense molecular clouds would seem to be the result of observational selection effects.

131.131   Detection of the $S(8), S(12), S(13), S(14)$, and $S(15), v = 0 \rightarrow 0$ rotation lines of molecular hydrogen in Orion.   R. F. Knacke, E. T. Young.
Astrophys. J., Lett., Vol. 249, L65 - L69 (1981).

The molecular hydrogen rotation lines, $v = 0 \rightarrow 0$, $S(8)$, $S(12), S(13), S(14)$, and $S(15)$, have been detected in the shocked region of the Orion molecular cloud. Line frequencies agree with predicted values to experimental accuracy. The H_2 energy levels are in LTE at 2000 ± 400 K up to the $J = 17$ level at 1.86 eV above the ground state. The 2.1 μm extinction at peak 1 is found to be between 2.2 and 4.0 mag. The authors find evidence for variations of H_2 emission intensity, apart from reddening, and set limits on the amount of gas hotter than 2000 K.

131.132   Masers.   M. J. Reid, J. M. Moran.
Annu. Rev. Astron. Astrophys., Vol. 19, (see 003.012), 231 - 276 (1981).

Contents: Introduction. Interstellar masers. Stellar masers. Maser theory. Interstellar scattering. Distance measurements.

131.133   The 26.13 MHz absorption line in the direction of Cassiopeia A.   A. A. Konovalenko, L. G. Sodin.
Nature, Vol. 294, 135 - 136 (1981).

It has recently been suggested that the absorption line $v = 26.13$ MHz that the authors had detected (1980) in the direction of Cassiopeia A might be a recombination line due either to carbon (C631α) present in H I clouds or to heavier elements of hot gas. Observations reported here taken in February 1981 suggest that the low-frequency attenuation in the spectrum of Cas A is due to the presence of a cold cloud of ionized carbon.

131.134 A search for high-velocity water masers in regions of star formation.   J. A. García-Barreto, L. F. Rodríguez, J. M. Moran, B. F. Burke.
Rev. Mexicana Astron. Astrofis., Vol. 5, 87 - 91 (1981).

Recently, several masers have been detected with all or most of the maser emission substantially shifted with respect to the velocity of the associated molecular cloud. To test if high-velocity water masers were actually a common phenomenon overlooked by an instrumental selection, the authors searched for high-velocity H_2O maser emission in 42 suspected sites of star formation. New masers related to S88 and W42 were detected.

131.135 On the determination of the Doppler shifts of interstellar lines.   I. Pustyl'nik.
Tartu Astrofüüs. Obs., Teated, Nr. 65, p. 11 - 15 (1981).

The aim of this communication is to show how one can improve the accuracy of measuring the Doppler shift of a sufficiently strong interstellar line with purely instrumental profile, provided that one posesses a uniform set of spectral measurements with comparatively high spectral resolution.

131.136 Determination of the structure of Orion A cloud from spectral data for carbon monoxide.
B. Bhattacharyya, S. Ray, A. K. Barua.
Indian J. Phys., Part B, Vol. 54B, 209 - 214 (1980). − Abstr. in Phys. Abstr., Vol. 84, Abstr. 98851 (1981).

131.137 Molecules in interstellar space.
R. K. Asundi.
Proc. Indian Natl. Sci. Acad., Part A, Vol. 46, 423 - 434 (1980). − Abstr. in Phys. Abstr., Vol. 84, Abstr. 98856 (1981).

131.138 The formation of cosmic grains. An experimental and theoretical study.   B. Donn, J. Hecht, R. Khanna, J. Nuth, D. Stranz, A. B. Anderson.
Surf. Sci., Vol. 106, 576 - 581 (1981). − Abstr. in Phys. Abstr., Vol. 84, Abstr. 98859 (1981).

131.139 Diffuse interstellar absorption bands between 2.9 and 4.0 μm.   D. A. Allen, D. T. Wickramasinghe.
Nature, Vol. 294, 239 - 240 (1981).

IRS7, the obscured M supergiant near the galactic centre, is an excellent source against which to study interstellar absorption bands near 3-μm wavelength. The authors report data between 2.9 and 4.0 μm, and find a broad absorption trough with numerous discrete features stretching from 2.9 to 3.6 μm. Although no definite identification of the material producing this absorption is possible, there is some indication that complex organic molecules may be involved.

131.140 Chemistry in low and medium density interstellar clouds.   J. B. Pickles, D. A. Williams.
Mon. Not. R. Astron. Soc., Vol. 197, 429 - 449 (1981).

The sensitivity of chemical modelling of low and medium density interstellar clouds with respect to variations in many astrophysical and chemical parameters is explored. Results are particularly sensitive to cloud density, temperature, size and shape and to the value adopted for the elemental depletions.

131.141 The collapse of a rotating non-axisymmetric isothermal cloud.   R. A. Gingold, J. J. Monaghan.
Mon. Not. R. Astron. Soc., Vol. 197, 461 - 475 (1981).

The collapse of an isothermal cloud which, initially, is uniformly rotating and non-axisymmetric, has been followed using the particle method SPH. The initial model is as similar as possible to that evolved by Boss & Bodenheimer. The SPH calculations agree qualitatively with the finite difference

calculations up to 1.7 initial free-fall times. Beyond 1.7 free-fall times the present calculations diverge from those of Boss & Bodenheimer who found the final state consisted of a binary system. In the present calculation the two condensations eventually coalesce.

131.142 Fragmentation in a rotating protostar: a re-examination of comparison calculations.
P. Bodenheimer, A. P. Boss.
Mon. Not. R. Astron. Soc., Vol. 197, 477 - 485 (1981).

The self-gravitating collapse of a rotating, isothermal protostellar cloud has been recalculated with two independent fluid-dynamic computer codes, in three space dimensions, with improved spatial resolution compared to previous calculations. The results again predict fragmentation and formation of a binary protostellar system with properties similar to those obtained in the lower-resolution calculations. The results are in disagreement with those obtained by Gingold & Monaghan (1981).

131.143 H_2CO mapping toward DR21 and W58 (K3 # 50).
J. R. Forster, W. M. Goss, H. R. Dickel, H. J. Habing.
Mon. Not. R. Astron. Soc., Vol. 197, 513 - 527 (1981).

The authors have mapped the distribution of 4.8-GHz H_2CO absorption against the galactic H II regions DR21 and W58 (K3 # 50) with 8 arcsec resolution. Opacity variations on scales less than 18 arcsec (0.2 pc) are seen against the extended continuum emission toward DR21. Extreme variations are observed toward the various radio components in W58. The linewidths are comparable to single-dish values for both DR21 and W58. H_2 column densities derived from the H_2CO equivalent widths are compared to the visual extinction. Large variations in the apparent H_2CO/H_2 abundance ratio are derived for W58.

131.144 Interstellar diffuse bands in the spectra of six stars previously reported anomalous.
J. C. Blades, W. B. Somerville.
Mon. Not. R. Astron. Soc., Vol. 197, 543 - 551 (1981).

Optical observations are presented of six stars for which anomalies had previously been reported in the strengths of certain of the unidentified diffuse interstellar bands. For four stars of low reddening, HD 25558, 26912, 35708 and 166182, anomalously strong λ 4430 had been reported. It is shown that this disappears when adequate spectroscopic resolution is used. For HD 13476 and 19820 anomalously strong λ 5780 and λ 5797 had been reported. These anomalies also are not confirmed.

131.145 Numerical studies of shocked magnetic gas clouds − I.
J. Nittmann.
Mon. Not. R. Astron. Soc., Vol. 197, 699 - 712 (1981).

A numerical technique is used for the analysis of nonsteady flow fields generated by shocks in interstellar gas. The governing set of Eulerian, partial differential equations describing the system is integrated using a finite difference scheme. This paper investigates the influence of large scale interstellar magnetic fields on the early evolution (time $\leqslant 3.0 \times 10^6$ yr) of a high density gas cloud (mass $\sim 700 M_\odot$) which is hit by a strong shock wave.

131.146 Observations of CO $J = 3 \rightarrow 2$ emission from molecular clouds.
G. J. White, J. P. Phillips, G. D. Watt.
Mon. Not. R. Astron. Soc., Vol. 197, 745 - 767 (1981).

Extensive observations in the $J = 3 \rightarrow 2$ transition of carbon monoxide at 345 GHz have been obtained using an indium antimonide heterodyne spectral-line receiver. Mapping observations of up to 1/2 deg^2 areas are reported towards the sources NGC 1333, 2023, 2024 and 2068. High-quality spectra have been obtained towards the self-absorbed sources NGC 1333 IRS-1, NGC 2071, Mon R2, AFGL 961 and DR 21.

Spectral observations are also reported for L1551/HH-29. The present data support the authors' previous model for NGC 2071 and show it to be a rotating cloud.

131.147 Observations of interstellar lithium toward σ Scorpii, β¹ Scorpii, and 55 Cygni.
R. L. Snell, P. A. Vanden Bout.
Astrophys. J., Vol. 250, 160 - 162 (1981).

The λ6708 doublet of interstellar Li I has been observed toward σ Sco, β¹ Sco, and 55 Cyg with measured equivalent widths of 0.63 ± 0.10, ≤ 0.20, and 1.63 ± 0.18 mÅ, respectively. Neutral and total lithium column densities are calculated and compared with those of potassium, sodium,and hydrogen nuclei. The Li/K ratio is roughly constant toward the six stars with detected interstellar lithium lines, with a value double that found in Type I carbonaceous chondrites. Comparison with Li/H and Li/Na ratios suggests that all the alkalis are depleted, but the potassium depletion is twice that of lithium. No significant limit can be placed on the ^{6}Li abundance by these spectra.

131.148 A new search for interstellar H_2O absorption in the spectrum of ζ Ophiuchi.
T. P. Snow, Jr., W. H. Smith.
Astrophys. J., Vol. 250, 163 - 165 (1981).

Intensive Copernicus scans of the $1_{11}-0_{00}$ line of the $\tilde{F}-\tilde{X}$ band of H_2O were carried out, resulting in a 2 σ upper limit of 1.04 mÅ for this line, at 1114.225 Å. The recently published f-value implies a column density limit of 5.3×10^{12} cm^{-2}, marginally above the level expected from recent chemical model calculations. A 1.8 σ feature appears at the expected position of the H_2O line, implying a column density for the ground rotational state of 2.8×10^{12} cm^{-2}.

131.149 The energetics of molecular clouds. IV. The S88 molecular cloud.
N. J. Evans II, G. N. Blair, P. Harvey, F. Israel, W. L. Peters III, M. Scholtes, T. de Graauw, P. Vanden Bout.
Astrophys. J., Vol. 250, 200 - 212 (1981).

The S88 molecular cloud has been observed in several molecular lines and at infrared wavelengths from 1 to 100 μm. The CO emission has a single, sharp peak which is near the Hα emission region S88 B and centered on a compact H II region observed in the radio continuum. The infrared observations indicate that the principal luminosity source is located near the radio continuum peak and is hidden behind substantial extinction, presumably from the molecular cloud.

131.150 The molecular cloud associated with NGC 7538.
H. R. Dickel, J. R. Dickel, W. J. Wilson.
Astrophys. J., Lett., Vol. 250, L43 - L47 (1981).

CO maps and preliminary H_2S and H_2CO data for the molecular cloud associated with NGC 7538 are used to compare the molecular distributions with those of ionized, neutral, and molecular hydrogen and dust. South of the H II regions is a ridge of high ^{13}CO column density with cold, self-absorbed H I just beyond it. A dense clump within the ridge is found adjacent to the H II region in the southeast and a lower density region of expanding gas is seen next to the H II region in the southwest. NGC 7538 is the first region which appears to show observational evidence for a molecular dissociation wave.

131.151 Observations of the interstellar cyanoacetylene J = 4 → 3 rotational transition.
A. M. Tolmachev, R. L. Sorochenko.
Pis'ma Astron. Zh., Tom 7, 682 - 688 (1981). In Russian. English translation in Soviet Astron. Lett., Vol. 7.

The J = 4 → 3 rotational transition (ν = 36.4 GHz) of interstellar cyanoacetylene (HC_3N) has been observed in the directions of 6 galactic sources. A comparison is made between the intensities of this line and of other rotational lines of this molecule. On the basis of available observational data the density, kinetic temperature, and HC_3N content are estimated for the sources.

131.152 Far UV radiation transfer and H_2CO lifetime in dense interstellar clouds.
S. Aiello, L. Morbidelli, L. Ulivi.
Astrophys. Space Sci., Vol. 80, 173 - 187 (1981).

The UV radiation transfer within spherical interstellar dust clouds is analyzed using the method of successive scatterings. The results are used to determine the lifetime of interstellar H_2CO against photo-destruction. The effectiveness of this process is compared with those of chemical mechanisms.

131.153 H_2 production in dense molecular clouds.
V. Pirronello, G. Strazzulla, G. Foti.
Astron. Astrophys., Vol. 103, L5 - L6 (1981).

Recent experimental results about the molecular character of the particles released from ice targets bombarded by energetic ion beams have been applied to the interaction between cosmic rays and frozen mantles on grains. The authors show that this interaction becomes a very effective mechanism for the H_2 production in dense molecular clouds.

131.154 A survey of the H I self-absorption in the ρ Ophiuchi region.
Y. K. Minn.
Astron. Astrophys., Vol. 103, 269 - 276 (1981).

The 21-cm H I self-absorption line has been mapped over an area of 3° × 2° in the ρ Ophiuchi dark cloud with an 8.7 resolution and a 10' spacing. The distribution of visual extinction in the same region is also obtained by the star counting method. The comparisons of the angular distribution of the cold H I with those of extinction and molecules show that they do not coincide.

131.155 The distribution of emission in TMC-1 from the strong unidentified line at 85339 MHz.
L. W. Avery, N. W. Broten, J. M. MacLeod.
J. R. Astron. Soc. Canada, Vol. 75, 245 (1981). − Abstract.

131.156 M82 − a superb laboratory for star formation bursts.
P. P. Kronberg, P. Biermann, F. R. Schwab.
J. R. Astron. Soc. Canada, Vol. 75, 249 (1981). − Abstract.

131.157 A fine analysis of stellar and interstellar lines towards four halo B stars.
F. P. Keenan, C. D. McKeith, P. L. Dufton, J. C. Blades.
Mon. Not. R. Astron. Soc., Vol. 197, 799 - 813 (1981).

High resolution observations of Ca II and Na I interstellar line profiles towards four halo B stars are analysed in terms of multicloud models. The interstellar line profiles were analysed to derive information on the radial velocities, internal velocity dispersions and column densities within individual interstellar clouds. The stellar CNO lines in the halo stars imply a composition similar to that found for unevolved B stars in the galactic plane.

131.158 Surface chemistry on interstellar oxide grains.
P. Denison, D. A. Williams.
Mon. Not. R. Astron. Soc., Vol. 197, 835 - 844 (1981).

Detailed calculations are made to test the predictions of Duley, Millar & Williams (1978) concerning the chemical reactivity of interstellar oxide grains. A method is established for calculating interaction energies between atoms and the perfect crystal with or without surface vacancy sites. The outline by Duley, Millar & Williams (1979) of interstellar oxide grain growth and destruction is justified by these calculations.

131.159 An estimate of the mass of zero metal stars.
R. G. Carlberg.
Mon. Not. R. Astron. Soc., Vol. 197, 1021 - 1029 (1981).

The temperature of a cloud of hydrogen−helium gas uncontaminated by metals is followed from post recombination conditions to hydrogen ionization, assuming the collapse takes place on a free-fall time-scale. A metal production yield of 10 per cent combined with the minimum metal abundance for ordinary stars implies that the zero metal stars comprised at most 10^{-4} of the gas mass of the Universe.

131.160 The abundance ratio $[^{16}OH]/[^{18}OH]$ in Sgr A and Sgr B2.   J. B. Whiteoak, F. F. Gardner.
Mon. Not. R. Astron. Soc., Vol. 197, 39P - 44P (1981).

Isotope abundance ratios $[^{16}OH]/[^{18}OH]$ of 277±20 and 261±20 have been obtained for the +40 km s^{-1} molecular cloud near Sgr A and the +60 km s^{-1} cloud in Sgr B2. For the +40 km s^{-1} cloud the satellite-line anomalies of the ^{16}OH and ^{18}OH transitions are similar. However, with the +60 km s^{-1} cloud the difference between the satellite lines is less marked for the ^{18}OH than for the ^{16}OH transitions.

131.161 Star formation in M33 and in our Galaxy. M. Kaufman.
Astrophys. J., Vol. 250, 534 - 550 (1981).

It is assumed that in the disks of M33 and our Galaxy, star formation results from a galactic spiral density wave plus successive generations of stars born via sequential star formation. For M33, orbit calculations and the observed distributions of giant H II regions and OB associations are used to calculate the reproductivity y for sequential star formation and the percentage of O stars formed by the spiral density wave. For each galaxy, the following results pertain to O stars in the annulus where the classical density wave operates. In both M33 and our Galaxy, the radial distribution of young stars agrees with the radial gradient predicted for a spiral density wave.

131.162 High-temperature methyl cyanide in Orion molecular cloud 1.
R. B. Loren, L. Mundy, N. R. Erickson.
Astrophys. J., Vol. 250, 573 - 578 (1981).

The $K=0$ through $K=6$ lines of the $J=13-12$ and $J=12-11$ rotational transitions of methyl cyanide (CH_3CN) at 239 GHz and 221 GHz have been observed toward OMC-1. The relative strengths of the $\Delta K=0$ transitions have been used to obtain the excitation temperatures of the CH_3CN emitting region.

131.163 New H_2O masers associated with far-infrared sources.   D. T. Jaffe, R. Güsten, D. Downes.
Astrophys. J., Vol. 250, 621 - 630 (1981).

The authors have searched for H_2O masers in the direction of far-infrared sources found in a recent survey. There are 25 H_2O masers near the 42 far-IR sources, of which 13 masers are new detections. Many of the far-IR sources are in the general vicinity of compact H II regions, and are thus likely to be young objects. The results are consistent with a 50–100% coincidence of H_2O masers with far-IR sources, suggesting that such objects are even more suitable candidates for maser searches than the compact H II regions. The high coincidence rate implies a lifetime of more than 2×10^5 years for the maser phase.

131.164 Collapse and fragmentation of rotating, adiabatic clouds.   A. P. Boss.
Astrophys. J., Vol. 250, 636 - 644 (1981).

A numerical hydrodynamics code has been used to calculate the collapse of rotating, adiabatic clouds. The three-dimensional nature of the calculation allows the clouds to fragment in the dynamic collapse phase. Clouds with adiabatic exponents of 7/5 and initial cos (2ϕ) density variations fragment into binary systems if the initial ratio of thermal to gravitational energy is small ($\sim$0.05). Clouds with higher thermal energy, however, damp the density variation, and form near-equilibrium ellipsoids, with ratios of rotational to gravitational energy less than the critical value for dynamic growth of

nonaxisymmetry in Maclaurin spheroids. Even with an adiabatic pressure law, dynamic fragmentation of a collapsing cloud is possible, implying for star formation theory that the low thermal energy fragments produced in isothermal collapse calculations may undergo a subsequent dynamic fragmentation in the nonisothermal regime.

131.165 The C^+, C, CO network in interstellar clouds. J. B. Pickles, D. A. Williams.
Astrophys. Space Sci., Vol. 80, 337 - 347 (1981).

A chemical network determining the formation and destruction of C^+, C, and CO and other species is set up and applied to spherical clouds in the normal interstellar radiation field. The spherical geometry adopted gives results which are different from those for slab models. The sensitivity of chemical species to chemical and astronomical parameters is explored. The relevance of this work to the observations is discussed.

131.166 High velocity molecular emission in Orion: a case for stellar winds.   T. B. H. Kuiper, B. Zuckerman, E. N. R. Kuiper.
Astrophys. J., Vol. 251, 88 - 102 (1981).

The authors report sensitive observations of a variety of molecules in the high velocity souce ("plateau") of the Orion molecular cloud. They also report a search for high velocity emission in other molecular clouds. On the basis of these data and data in the literature, the authors propose that the high velocity source consists of clumps of dense molecular gas which are being accelerated by the high velocity stellar wind of a pre-main-sequence star. The H_2, some of the H_2O maser, and the Herbig-Haro–type emissions are attributed to shocks associated with the breakup of the protostellar cocoon.

131.167 High velocity molecular gas near Herbig-Haro objects HH 7–11.   R. L. Snell, S. Edwards.
Astrophys. J., Vol. 251, 103 - 107 (1981).

Observations of the $J = 2-1$ and $J = 1-0$ transitions of ^{12}CO and ^{13}CO reveal the presence of high velocity molecular gas associated with a low luminosity infrared source in the vicinity of the Herbig-Haro objects HH 7–11. The blueshifted and redshifted wings show peak intensities spatially separated by 1.5 (0.2 pc), suggesting an energetic bipolar outflow of gas from a young low mass star. The mass loss rate implied by these observations is $8 \times 10^{-6} M_\odot$ yr^{-1}.

131.168 Polarization of scattered light in globular clusters. P. G. Martin, S. J. Shawl.
Astrophys. J., Vol. 251, 108 - 112 (1981).

The authors present a simple model to show that, under certain circumstances, polarization of light scattered from dust particles (Rayleigh scatterers) in patches in globular clusters should produce observable polarization. A search for polarization of radiation from dark patches in M3, M13, and M15 provides a tentative detection in M15 of 0.22% ± 0.06%.

131.169 Origin of Ca-Al-rich inclusions. II. Sputtering and collisions in the three-phase interstellar medium.
D. D. Clayton.
Astrophys. J., Vol. 251, 374 - 386 (1981).

The theory introduced by Clayton for the formation of the Ca-Al-rich inclusions within C3 meteorites is extended to an evolutionary history in a three-phase interstellar medium. Widespread supersonic turbulence in the hot interstellar medium is maintained by supernova shock waves, causing heavy sputtering of the refractory dust. Subsequent reaccumulation with varying dust/gas ratios or varying particle size leads to isotopically fractionated Ca-Al-rich accumulates. A new theoretical explanation of the correlation of mass-fractionated isotopes with unidentified nuclear anomalies follows from this picture, as do many other intriguing details.

131.170 Detection of deuterated cyanodiacetylene (DC$_5$N) in Taurus Molecular Cloud 1.
J. M. MacLeod, L. W. Avery, N. W. Broten.
Astrophys. J., Lett., Vol. 251, L33 - L36 (1981).

Deuterated cyanodiacetylene (DC$_5$N) has been observed for the first time in an interstellar cloud. The $J = 10 \to 9$ and $J = 9 \to 8$ transitions have been detected in emission in TMC 1. The abundance ratio DC$_5$N/HC$_5$N is found to lie in the range 0.006 - 0.016. From these observations, it is not possible to rule out the formation of HC$_5$N on grain surfaces or by dissociation of larger molecules.

131.171 Detection of deuteriocyanobutadiyne (DC$_5$N) in the interstellar cloud TMC-1.
F. P. Schloerb, R. L. Snell, W. D. Langer, J. S. Young.
Astrophys. J., Lett., Vol. 251, L37 - L41 (1981).

The authors report the first detection of DC$_5$N in interstellar space via its $J = 9 \to 8$ rotational transition at 22.878955 GHz. The ratio of the abundances of DC$_5$N to HC$_5$N, R(DC$_5$N), is 0.016 ± 0.003, assuming that the lines are optically thin. This large fractionation shows that ion-molecule reactions are active in the Taurus clouds, but it is not sufficient to prove that HC$_5$N is formed by ion-molecule chemistry. Comparison of R(DC$_5$N) with R(DC$_3$N) shows that R(DC$_5$N) < R(DC$_3$N), which is consistent with ion-molecule models for their formation.

131.172 Discovery of interstellar rubidium.
M. Jura, W. H. Smith.
Astrophys. J., Lett., Vol. 251, L43 - L44 (1981).

The authors have discovered interstellar rubidium by observing the resonance line of Rb I at 7800 Å toward ζ Oph. This is the heaviest element yet detected within the interstellar medium, and the detection of this element is important for understanding the history of nucleosynthesis in the solar neighborhood.

131.173 Where are the blue supergiants born?   M. Šolc.
Kozmos, Vol. 12, 175 - 177 (1981). In Czech.

131.174 Observations of HCO$^+$, H^{13}CO$^+$, ^{13}CO, and C^{18}O in Taurus cloudlets.
A. Baudry, J. Cernicharo, J. Pérault, J. de la Noë, D. Despois.
Astron. Astrophys., Vol. 104, 101 - 115 (1981).

In this work the authors have observed the $J = 1 - 0$ transition of HCO$^+$, H^{13}CO$^+$, ^{13}CO, and C^{18}O in the direction of several Taurus dark clouds extending from Barnard 18 and Taurus Molecular Cloud 2 to Barnard 212 and 210. Maps of HCO$^+$ emission show that this molecule can be used to delineate the extent of molecular cloudlets in obscured areas. There is no simple systematic cloud motion in the Taurus Molecular Cloud 2 region.

131.175 Optical observations of interstellar shockwaves.
M. A. Dopita.
Investigating the universe, (see 003.014), p. 29 - 59 (1981).

Although shocks are observed optically in novae, supernovae, accretion disks and columns, ring nebulae about WR stars and in H II regions and planetary nebulae, the author considers only three classes of object: supernova remnants, Herbig-Haro objects and active galactic nuclei.

131.176 The dynamical effects of hypersonic stellar winds on interstellar gas.   J. E. Dyson.
Investigating the universe, (see 003.014), p. 125 - 150 (1981).

Early type stars have energetic winds which impinge hypersonically on interstellar gas. The author reviews this interaction with particular emphasis on the resulting dynamical effects. The basic flow pattern and physical state of the gas are first briefly discussed and the results then applied to the competition between a wind driven shell and the ionization front produced by the stellar radiation field. The author then examines a variety of topics: the dynamical structure of galactic and extragalactic H II regions, the production and maintenance of density fluctuations within nebulae, and the formation of large scale loop structures in the plane of the galaxy. The emphasis throughout the review is on simple semi-qualitative models of various wind related phenomena.

131.177 The exchange process of energy in the molecular cloud W 40.
L.-p. Xu, J. Xing, Y.-f. Wu, S.-d. Xie, J.-z. Zhao.
Acta Astrophys. Sinica, Vol. 1, 234 - 242 (1981). In Chinese.

131.178 Abundance of atomic carbon (C I) in dense interstellar clouds.   T. G. Phillips, P. J. Huggins.
Astrophys. J., Vol. 251, 533 - 540 (1981).

The abundance of interstellar neutral atomic carbon (C I) is investigated by means of its ground state fine-structure line emission at 492 GHz (609 μ). Since this particular submillimeter line lies in a relatively opaque region of the Earth's atmospheric spectrum, the investigation was carried out from the 91.5 cm telescope of NASA's Kuiper Airborne Observatory. Atomic carbon is found to be very abundant in dense interstellar (molecular) clouds. Its column densities of about 10^{19} cm^{-2} seem comparable to those for CO in directions towards the cloud centers. By contrast, current theories of carbon chemistry in dense clouds suggest that C I should be the dominant species only near cloud edges giving column densities of $\sim 10^{17}$ cm^{-2}, and should be a negligible constituent for deep regions where the UV field cannot penetrate. It is suggested that these observations of considerably greater column densities imply that the physical conditions in the clouds are not as simple as assumed in the chemical models.

131.179 New interstellar molecular transitions in the 2 millimeter range.
J. M. Hollis, L. E. Snyder, D. H. Blake, F. J. Lovas, R. D. Suenram, B. L. Ulich.
Astrophys. J., Vol. 251, 541 - 548 (1981).

The authors have detected interstellar line emission in the 2 mm wavelength region from SO, ^{34}SO, SO$_2$, CH$_3$OH, CH$_3$CCH, CH$_3$CH$_2$CN, HC$_3$N, and nine unidentified transitions. The specific transitions involved in these detections have not been previously reported in interstellar molecular clouds. The authors derive a Sgr B2 kinetic temperature of ~47 K based on observations of the K components of the 9_K-8_K transitions of CH$_3$CCH. In addition, they attempted to confirm the existence of interstellar HNO, searched for interstellar HCP, and attempted to detect sulfuric acid, formic anhydride, and the ^{18}O isotopic form of carbon dioxide in interstellar clouds and in the Venusian atmosphere, but could not detect these species.

131.180 Grain formation behind shocks and the origin of isotopically anomalous meteoritic inclusions.
B. G. Elmegreen.
Astrophys. J., Vol. 251, 820 - 833 (1981).

The growth of grains behind dense, snowplowing shock fronts is studied. Grain growth by condensation and coagulation is studied to determine the maximum radii of the grains that form behind the shock front before these grains emerge or the shock stops. The results are applied to the shocks created by supernova ejecta that hit giant molecular clouds. In such clouds large grains (~ 1 cm) or stones should form during the 10^7 years when an OB association forms and supernovae explode nearby. An example shows that behind a typical supernova shock large grains can coagulate into cm-size chondrules similar in size, structure and isotopic anomalies to inclusions found in Allende and other carbonaceous chondrites. Thus, during the formation of the solar system, the isotopically anomalous meteoritic inclusions could have formed behind nearby supernova shock fronts.

131.181   Coupling of the magnetic field and rotation in the dark cloud B5.
J. S. Young, W. D. Langer, P. F. Goldsmith, R. W. Wilson.
Astrophys. J., Lett., Vol. 251, L81 - L84 (1981).
The authors present CO observations which indicate that the central region of the dark cloud B5 is rotating in the opposite sense to that of the bulk of this $\sim 300\,M_\odot$ region. The direction of elongation of the cloud coincides with its axis of rotation; both are perpendicular to the galactic plane and to the direction of the magnetic field in the region. This configuration, together with the unusual rotation curve of the cloud, suggests that magnetic braking has played an important role in its dynamical evolution.

131.182   High velocity H_2 line emission in the NGC 2071 region.
S. E. Persson, T. R. Geballe, T. Simon, C. J. Lonsdale, F. Baas.
Astrophys. J., Lett., Vol. 251, L85 - L89 (1981).
Emission in the $v = 1 \rightarrow 0\, S(1)$ line of H_2 at 2.12 μm has been measured at 20 km s^{-1} resolution in the vicinity of the microwave maser and infrared continuum sources north of NGC 2071. The line profile observed near the peak of the extended H_2 emission region has full width at zero intensity of ~ 100 km s^{-1}, with a prominent blue wing. The NGC 2071 region thus represents the second detection, after OMC-1, of high velocity H_2 emission in a region showing signs of ongoing star formation. The width of the line may result from a supersonic outflow of gas from within the molecular cloud.

131.183   Far-infrared and submillimeter observations of Barnard 35: heat sources for bright-rimmed molecular clouds.   C. J. Lada, H. A. Thronson, Jr., H. A. Smith, D. A. Harper, J. Keene, R. F. Loewenstein, J. Smith.
Astrophys. J., Lett., Vol. 251, L91 - L95, plate 2 (1981).
The authors have obtained far-infrared and submillimeter continuum observations of a region of enhanced gas temperature in the bright-rimmed molecular cloud B35. A dust temperature of $T_d = 10 \pm 5$ K was determined by fitting optically thin blackbody functions to the observed spectrum. They estimate the total dust luminosity of the cloud to be $\sim 69\,L_\odot$ and show that this can easily be supplied by stellar radiation from the background interstellar radiation field and from the λ Ori cluster which excites the ionized gas in the bright rim at one border of the molecular cloud. Observations of CO molecular line emission at 2.7 mm were also obtained to determine an accurate gas temperature at the position of the infrared observations. From these observations a gas temperature of $T_g = 23.4 \pm 3.5$ K was derived.

131.184   Ammonia in Orion. II. The gas in and around OMC-1.   L. M. Ziurys, R. N. Martin, T. A. Pauls, T. L. Wilson.
Astron. Astrophys., Vol. 104, 288 - 295 (1981).
Results of ammonia observations of the Orion region made with an angular resolution of 40″ are presented. Within 1′ of the most intense center of NH_3 emission in the region, OMC-1, spectra of the (1,1), (2,2), (3,3), and (4,4) lines were taken with a 20″ spacing on an (α, δ) grid.

131.185   Star formation in molecular clouds and the formation of compact H II regions.   H. W. Yorke.
Proceedings of the 5th Göttingen-Jerusalem-Symposium on Astrophysics, (see 012.041), p. 3 - 30 (1981).
Theoretical aspects of the main evolutionary phases of star formation are briefly reviewed: the collapse and fragmentation of molecular clouds, the collapse and subsequent evolution of a protostellar cloud, the structure and spectral appearance of protostellar envelopes, the formation of a compact H II region and its subsequent expansion. Emphasis is placed on 1-D (spherically symmetric) hydrodynamic and radiation transfer calculations of the formation of stars of mass $M \geqslant 10\,M_\odot$.

131.186   Amplification of microwaves in the interstellar medium.   E. Bettwieser.
Proceedings of the 5th Göttingen-Jerusalem-Symposium on Astrophysics, (see 012.041), p. 31 - 42 (1981).

131.187   Collapse of gas with embedded stars.   K. J. Fricke.
Proceedings of the 5th Göttingen-Jerusalem-Symposium on Astrophysics, (see 012.041), p. 43 - 48 (1981).
The author finds the observations of 10^5 - $10^6\,M_\odot$ of ionized gas in elliptical galaxies consistent with the collapse properties of such gas in the nuclei of these galaxies. The cloud collapse is expected to end up in the formation of one or several massive objects at the center of the Galaxy where they might become the engines for more violent activity as observed in radio galaxies and quasars.

131.188   Model calculations for the star formation in the galactic center and other galactic nuclei.
H.-H. Loose.
Proceedings of the 5th Göttingen-Jerusalem-Symposium on Astrophysics, (see 012.041), p. 61 - 72 (1981).

131.189   Interstellar matter.   S. Hayakawa.
Plasma astrophysics, (see 012.042), p. 37 - 42 (1981).
Several components of the interstellar medium coexist by thermal balance and/or pressure balance. The two-phase stability is explained by the temperature and density dependences of the heating and cooling rates. The cooling rates of a thin hot plasma are given under the ionization equilibrium for the interstellar media of solar abundances and of metal depleted abundances. Thermal instabilities are discussed with regard to the temperature dependence of the cooling rate. It is pointed out that a deviation from the ionization equilibrium and from the electron-ion equipartition may take place in a tenuous plasma. Numerical examples are given for the non-steady behavior. The generation of a hot plasma by supernova blast waves is discussed on the basis of evolution models of the supernova remnant. Observational results indicating the distribution of the hot plasma component are shown by reference to soft X-ray and non-thermal radio maps.

131.190   Stability of the equilibrium states of the interstellar gas and fields system.   C. J. Cesarsky.
Plasma astrophysics, (see 012.042), p. 51 - 59 (1981).
Possible static equilibrium configurations of the interstellar gas and fields system (including cosmic rays) are described, and tested for stability.

131.191   On the chemical equilibrium abundance in cold cosmic dust.
V. I. Gol'danskij, L. V. Gurvich, V. V. Muzylev, V. S. Strel'nitskij.
Nauchn. Inf., Vyp. (No.) 47, (see 003.016), p. 3 - 8 (1981).
In Russian.
The method of free energy minimization is used to calculate the chemical equilibrium of a medium consisting of H, C, N, O at $T = 0$. It has been stated that such organic molecules as CH_3COOH and $(NH_2)_2CO$ are produced in rather large amounts, if hydrogen is significantly deficient (as compared with standard "cosmic" abundance). This confirms the possibility of organic synthesis in cold cosmic objects (interstellar dust grains, comets).

131.192   Pumping of a cosmic H_2O maser in a two-temperature gas.   G. T. Bolgova.
Nauchn. Inf., Vyp. (No.) 47, (see 003.016), p. 9 - 14 (1981).
In Russian.
The possibility of pumping an H_2O maser is investigated in a medium with different temperatures of neutral particles and electrons. The observational criteria permitting to reveal sources in which this pumping mechanism works, and a pos-

sible method of estimating the degree of ionization in a source from the observed relations of line intensities are discussed.

131.193 Non-equilibrium processes in interstellar molecules.
V. S. Strel'nitskij.
Nauchn. Inf., Vyp. (No.) 49, (see 003.017), p. 95 - 113 (1981). In Russian.

The problem of interpretation of non-equilibrium emission and absorption in radio lines of cosmic molecules is discussed. Basic observational data on these sources are summarized. The necessity of maser amplification in compact emission sources is strictly proved (taking H_2O sources as an example). The general thermodynamic and kinetic requirements to the pumping models of cosmic masers and "anti-masers" are formulated. An informative classification of the pumping mechanisms, indicating the types of the source and the sink of energy, is proposed. Several pumping models of different type are critically discussed.

131.194 Hydrodynamical collapse of interstellar clouds. I. Numerical method of calculation.
A. E. Dudorov, Yu. V. Sazonov.
Nauchn. Inf., Vyp. (No.) 49, (see 003.017), p. 114-134 (1981). In Russian.

The problem of numerical calculations of the magneto-hydrodynamical evolution of self-gravitating interstellar clouds is considered. The explicit two-step numerical method by Lax-Vendroff was used to compute the collapse of clouds in spherical polar coordinates with respect to a Lagrangian grid. The applicability of the numerical method to the star formation problem is discussed. The isothermal collapse of a non-magnetic cloud of $10^4 \ M_\odot$ is investigated as a numerical test of the suggested method.

131.195 La polvere interstellare.   E. Bussoletti.
G. Astron., Vol. 7, 133 - 147 (1981).

131.196 Star formation in Bok globules.
B. Reipurth.
Messenger, No. 26, p. 2 - 4 (1981).

131.197 Modelling of pulse broadening due to multi-ray enhancement in the interstellar medium.
A. A. Bocharov.
Inst. kosm. issled. AN SSSR. Prepr., 1981, No. 637, 31 pp. In Russian. – Abstr. in Ref. zh., 51. Astron., 11.51.696 (1981).

131.198 Chemistry of molecules in interstellar clouds.
A. A. Rejtblat.
Inst. kosm. issled. AN SSSR. Prepr., 1981, No. 623, 74 pp. In Russian. – Abstr. in Ref. zh., 51. Astron., 11.51.764 (1981).

131.199 Interstellar abundances derived from IUE data. I. Silicon depletions toward reddened stars.
T. P. Snow, Jr., C. L. Joseph.
Astron. J., Vol. 86, 1916 - 1922 (1981).

The International Ultraviolet Explorer (IUE) has the capability of acquiring interstellar line data for a large number of stars, reaching substantially greater column densities than previous instruments, and accordingly the authors are undertaking wide-ranging surveys of abundances for species particularly well suited to IUE's wavelength coverage and spectral resolution. In the process, a complex package of interactive computer programs has been developed to facilitate the efficient analysis of data on large numbers of stars. This program package, along with results on Si II column densities towards two moderately reddened stars, is described. For both stars, substantial silicon depletions are inferred.

131.200 Distances to 14 molecular clouds (including two associated with supernova remnants) by a new technique.   T. E. Armandroff, W. Herbst.
Astron. J., Vol. 86, 1923 - 1925 (1981).

The authors illustrate a new technique for determining distances to dense molecular clouds. Star counts on the POSS blue print are used to establish the foreground star density for 14 clouds. From this quantity, the distance is derived using the empirical relation given by Herbst and Sawyer (1981). The method is reliable for clouds with $A_V \gtrsim 7$ mag over areas in excess of 20 arcmin^2. Except for spectroscopic parallax of associated stars, it is probably the most accurate method for determining distances to clouds within ~3200 pc of the Sun. The authors show that the North American Nebula has a distance of ~1 kpc, not ~200 pc as has recently been suggested. They also derive distances for two supernova remnants, W28 and W44, which are known to be interacting with molecular clouds.

131.201 High-sensitivity survey of NH_3 in the southern hemisphere.   E. Scalise, Jr., R. E. Schaal, Y. Bakor, J. W. S. Vilas Boas, P. C. Myers.
Astron. J., Vol. 86, 1939 - 1943 (1981).

Microwave emission associated with the para form of ammonia, in the (1, 1) line, was searched for in the direction of 46 southern sky positions where strong H_2CO absorption or strong H_2O maser emission was observed previously. The results of those observations are presented with 17 new ammonia sources discovered.

131.202 Ammoniak in der OMC1-OMC2-Region.
W. Batrla, P. Bastien, T. L. Wilson, K. Ruf, T. Pauls, R. N. Martin.
Mitt. Astron. Ges., Nr. 54, (see 012.050), p. 278 - 280 (1981).

131.203 Interstellar extinction in the region of the dark cloud Khavtassi 286+287 in Taurus and its distance.
V. Straižys, E. Meištas.
Astron. Tsirk., No. 1121, p. 3 - 5 (1980). In Russian.

131.204 Giant bubble of hot gas in the galactic interstellar medium.   V. S. Imshennik.
Priroda, 1981, No. 12, p. 33 - 35. In Russian.

131.205 Heterogeneous condensation of ice in the cosmic environments.   J. Seki, H. Hasegawa.
Proceedings of the 14th ISAS Lunar and Planetary Symposium, (see 012.055), p. 175 - 180 (1981).

Astrophysical problems of water ice condensation are investigated. It is derived that most of H_2O molecules condense into ice mantles heterogeneously on dust grains of high temperature minerals. The authors calculate the temperature of heterogeneous condensation of ice for a wide range of astrophysical conditions.

131.206 Fluxes of energetic particles and the ionization rate in protostars and the primitive solar nebula.
T. Umebayashi, T. Nakano.
Proceedings of the 14th ISAS Lunar and Planetary Symposium, (see 012.055), p. 293 - 300 (1981).

Solving the one-dimensional transport equation the authors obtain intensities and ionization rates of the primary and secondary cosmic-ray particles in a protostar and the primordial solar nebula. It is found that energetic particles are quickly attenuated with increasing column density and that in the inner regions of the nebula ionization by radioactive elements is more important than that by energetic particles.

131.207 The variable infrared emission of cosmic dust sources.
M. Bode.
J. British Astron. Assoc., Vol. 92, 1 - 6 (1981).

Many objects are known to possess circumstellar dust shells whose composition can be inferred by observations at wavelengths from the far ultraviolet to the far infrared. Here the author discusses work carried out recently into what happens to the long wavelength emission from a dust shell when the central source of heating radiation is intrinsically variable.

131.208   **On a simple model of chemical evolution and abundance of carbon, nitrogen and oxygen in interstellar matter.**   D. L. Dimitrov.
Bull. Astron. Inst. Czechoslovakia, Vol. 32, 359 - 365 (1981).

A simple model of chemical evolution in the solar vicinity is investigated. Using equations governing the evolution of primary and secondary isotopes, abundances of ^{12}C, ^{13}C, ^{14}N and ^{16}O are derived. The model does not predict a large enrichment of ^{13}C relative to ^{12}C over the last 4.6×10^9 years.

131.209   **Comets, interstellar molecules, and the origin of life.**   W. M. Irvine, Å. Hjalmarson.
Res. Lab. Electron. Onsala Space Obs., Res. Rep. No. 143, 31 pp. (1981).

131.210   **Das Massenspektrum bei der Sternentstehung.**   H. Zinnecker.
Max-Planck-Inst. Phys. Astrophys., Inst. Extraterr. Phys., Garching b. München. MPI-PAE/Extraterr. 167, 10 + 282 pp. (1981). ISSN 0340-8922.

131.211   **Global pressure instability in diffuse molecular clouds due to optical effect.**   Y. Kannari.
Sci. Rep. Tôhoku Univ., Ser. 8, Vol. 2, 1 - 15 (1981).

There is a possibility that an interstellar cloud will become unstable as a whole when the cloud is compressed slightly, provided that UV radiation is attenuated enough for molecule formation to proceed in the cloud, and if the new molecules work as an efficient coolant and decrease the gas pressure toward the center of the cloud. It is shown that the formation of H_2 molecules on the surface of dust grains does not induce such instabilities.

131.212   **Far infrared and submillimeter observations of Barnard 35: heat sources for bright-rimmed molecular clouds.**   C. J. Lada, H. A. Thronson, Jr., H. A. Smith, D. A. Harper, J. Keene, R. F. Lowenstein, J. Smith.
Prepr. Steward Obs., No. 331, 2 + 18 pp. (1981).

The authors have obtained far infrared and submillimeter continuum observations of a region of enhanced gas temperature in the bright-rimmed molecular cloud B35. A dust temperature of $T_d = 10 \pm 5$ K was determined by fitting appropriately weighted, optically thin blackbody functions to the observed spectrum. The authors estimate the total dust luminosity of the cloud to be ~ 69 $L_\odot$ and show that this can easily be supplied by stellar radiation from the background interstellar radiation field and from the λ Ori cluster which excites the ionized gas in the bright rim at one border of the molecular cloud. Observations of CO molecular line emission at 2.7 millimeters were also obtained to determine an accurate gas temperature at the position of the infrared observations. From these observations a gas temperature of $T_g = 23.4 \pm 3.5$ was derived. The observed difference in gas and dust temperatures confirms earlier suggestions (Lada and Wilking 1980) that the gas and dust are not in thermal equilibrium and that gas-dust collisions are not the heat source for the molecular gas in this source.

131.213   **Star formation in the λ Orionis region: 1. The distribution of Hα emission objects.**
R. Duerr, C. L. Imhoff, C. J. Lada.
Prepr. Steward Obs., No. 340, 38 pp. (1981).

The authors present the results of the first thoroughly sampled survey of the λ Orionis OB association for Hα emission objects. They also present photographic photometry of the entire list of previously known and newly discovered Hα emission objects.

131.214   **A study of interstellar extinction in the direction of carbon stars. Part I. The fields of WZ Cas, U Cyg and TT Tau.**   S. Krawczyk, J. Krempeć-Krygier, J. Gertner.
Stud. Soc. Sci. Torunensis, Toruń, Poland, Sect. F (Astron.), Vol. 6, No. 3, p. 3 - 22 (1981) = Biul. Obs. Astron. Uniw. M. Kopernika Toruniu, Nr. 63.

Based on photographic observations performed at the Toruń Observatory with the method of colour excesses applied, the interstellar extinction around two carbon stars, namely U Cyg and TT Tau have been determined. A revision of Ampel's galactic structure study in the Cassiopeia region has given an opportunity to estimate interstellar extinction for the third carbon star, i.e. WZ Cas. By means of introduced interstellar extinction curves gradient values of reddening for these stars were quantitatively evaluated and compared.

131.215   **Luminosity effects in interstellar spectral features.**   J. Krełowski, A. Strobel.
Stud. Soc. Sci. Torunensis, Toruń, Poland, Sect. F (Astron.), Vol. 6, No. 3, p. 23 - 38 (1981) = Biul. Obs. Astron. Uniw. M. Kopernika Toruniu, Nr. 63.

The paper summarizes the results of the investigations of local effects in interstellar spectral lines, mostly diffuse ones. The intensities of these spectral features seem to correlate with E_{B-V}, when the luminosity class of an observed star is taken as a parameter. The possible explanation is that probably most of the massive, hot stars are still embedded in relics of their parent clouds.

131.216   **Fragmentating of a rotating toroidal configuration and the origin of the Trapezium.**
B.-c. Qian, R.-y. Wang.
Ann. Shanghai Obs. Acad. Sinica, No. 2, p. 95 - 99 (1980).

131.217   **Interstellar dust in the neighbourhood of the sun.**   J. Dorschner.
Progress in planetary exploration, (see 012.062), p. 75 - 84 (1981).

The distribution of interstellar dust within 500 pc from the sun obtained from recent investigations is described. Statistical properties of dust clouds in the neighbourhood of the sun and individual data of two near clouds in high galactic latitudes are discussed. The present knowledge of the chemical composition of the interstellar dust grains is outlined. Possible relations between solar system solids and interstellar solids are indicated.

131.218   **Ionization degree of matter in the dust clouds Heiles No. 2 and L 1630.**
E. A. Abramenkov, V. V. Krymkin, A. A. Rejtblat, V. I. Slysh.
Inst. kosm. issled. AN SSSR, Prepr., 1981, No. 654, 20 pp. In Russian. − Abstr. in Ref. zh., 51. Astron., 1.51.683 (1982).

131.219   **First detection of line emission from the hot interstellar medium with solid state detectors.**
H. W. Schnopper, J. P. Delvaille, R. Rocchia, C. Blondel, C. Chéron, J. C. Christy, R. Ducros, L. Koch, R. Rothenflug.
High-energy astrophysics, (see 012.067), p. 121 - 123 (1981).

X-ray observations of the interstellar medium near the North-Galactic Polar region with Si(Li) detectors were performed during a rocket flight from the White Sands Missile Range on March 22, 1980. Emission lines of highly ionized carbon (CV, CVI) and oxygen (OVII, OVIII) were detected. These lines are direct evidence for the thermal origin of the emission and confirm the presence of a hot ($\sim 10^6$ K) component in the interstellar medium.

131.220   **On the ionization ratio of the local interstellar medium.**   P. W. Blum, S. Grzedzielski, N. Witt.
High-energy astrophysics, (see 012.067), p. 197 - 200 (1981).

The abundance ratio of neutral hydrogen to neutral helium, as deduced from interplanetary observations of

Lyman-alpha and He 584 A radiation by Mariner 10, is significantly lower than the cosmic abundance ratio of these elements, thus showing that the local interstellar medium (LISM) is partly ionized. It is shown that an important source of ionization of the LISM can be thermal collisions, yielding an ionization degree of about 50% for the hydrogen component.

131.221 **Fragmentation processes.** W. M. Tscharnuter.
Rôle des nuages moléculaires dans la formation des étoiles, (see 012.069), 21 pp.
The current status of theoretical investigations of the fragmentation mechanism in collapsing interstellar clouds is reviewed. Consequences of angular momentum and magnetic fields for the gravitational collapse are outlined and a "magnetic" Jeans-criterion is derived. Some results of recent numerical calculations are presented.

131.222 **Models of molecular clouds.** W. Boland.
Rôle des nuages moléculaires dans la formation des étoiles, (see 012.069), 11 pp.
In order to study the physical circumstances in dark molecular clouds the author has calculated hydrostatic cloud models which are in thermal and in chemical equilibrium.

131.223 **Processus de formation et de destruction des molécules dans les nuages interstellaires. Rôle des grains de poussière.** Y. Viala.
Rôle des nuages moléculaires dans la formation des étoiles, (see 012.069), 30 pp.
A review of chemical processes in dense interstellar clouds is given. The importance of grain surfaces as catalytic agents in the formation of molecular hydrogen and more complex molecules is discussed. The mechanisms of molecule destruction by stellar UV radiation and cosmic ray particles are described and a detailed survey of the various types of gas-phase chemical reactions in a molecular cloud is given. In particular the role of radiative association of free radicals in the formation of complex molecules is emphasized.

131.224 **La formation des étoiles chaudes dans les galaxies de types avancés.** F. Viallefond.
Rôle des nuages moléculaires dans la formation des étoiles, (see 012.069), 5 pp.

131.225 **Quelques aspects du rôle joué par le champ magnétique dans l'évolution des nuages inter-stellaires.** N. Bel.
Rôle des nuages moléculaires dans la formation des étoiles, (see 012.069), 12 pp.

131.226 **Rôle du rayonnement cosmique.** C. J. Cesarsky.
Rôle des nuages moléculaires dans la formation des étoiles, (see 012.069), 7 pp.

131.227 **Low-dispersion spectral survey of the regions Khavtasi 189, 194.**
O. M. Kurtanidze.
Astron. Tsirk., No. 1158, p. 1 - 2 (1981). In Russian.

Catalogue of star formation regions in the Galaxy: Part 1. Observational data. Supplement to the collection "Nauchnye Informatsii" iss. 47 [Vyp. 47].
See Abstr. 002.065.

Accurate ab initio calculation of the HCS^+ interstellar ion. See Abstr. 022.007.

Vibrational disequilibrium in low pressure clouds. See Abstr. 022.011.

Rotational excitation of molecular ions by electron impact under interstellar conditions. See Abstr. 022.012.

$^{14}N/^{15}N$ **isotope fractionation in the reaction $N_2H^+ + N_2$: interstellar significance.** See Abstr. 022.013.

A laboratory study of the reaction $H_3^+ + HD \rightleftharpoons H_2D^+ + H_2$: the electron densities and the temperatures in interstellar clouds. See Abstr. 022.023.

Laboratory measurements of millimeter and sub-millimeter transitions of sodium hydride and sodium deuteride. See Abstr. 022.025.

Rotational cross sections and rate coefficients for e-CO and e-HCN collisions under interstellar conditions. See Abstr. 022.043.

A measurement of the hyperfine structure of $C^{17}O$. See Abstr. 022.049.

Radiative lifetimes of excited electronic states in molecular ions. See Abstr. 022.051.

Some positive ion reactions with H_2: interstellar implications. See Abstr. 022.057.

Detection of the infrared spectrum of cyanobuta-diyne, HC_5N. See Abstr. 022.076.

Collisional excitation of OH by H_2: transitions within the groundstate Λ doublet. See Abstr. 022.081.

Rotational excitation of OH by H_2 at thermal energies. See Abstr. 022.082.

The extinction coefficients in mid- and far-infrared of silicate and iron-oxide minerals of interest for astronomical observations. See Abstr. 022.086.

Observation of the infrared spectrum of the triatomic molecular ion H_2D^+. See Abstr. 022.091.

Hydroxycarbene (HCOH) and protonated formaldehyde: two potentially observable interstellar molecules. See Abstr. 022.096.

Oscillator strengths for lines of the $\widetilde{F}(0,0,0) - \widetilde{X}(0,0,0)$ band of H_2O at 111.5 nanometers and the abundance of H_2O in diffuse interstellar clouds. See Abstr. 022.104.

Can silicon monoxide grains be responsible for the interstellar 9.7 micrometer absorption band? See Abstr. 022.108.

HOC^+**: an observable interstellar species? A comparison with the isomeric and isoelectronic HCO^+, HCN and HNC.** See Abstr. 022.111.

Reactions of Si^+ with H_2O and O_2 and SiO^+ with H_2 and D_2. See Abstr. 022.112.

Dust-sensitive forbidden line ratios. See Abstr. 022.119.

Laboratory millimeter and submillimeter spectra of CO^+. See Abstr. 022.122.

Laboratory millimeter and submillimeter spectrum of CCH. See Abstr. 022.142.

The mid-infrared spectrum of high temperature silicate minerals cooling on low temperature. See Abstr. 022.162.

Thermal metamorphism of synthetic silicates. See Abstr. 022.163.

Formation of Ca-Al mineral grains through gas-solid reactions.   See Abstr. 022.164.

Local gas without reddening: the contribution of stray radiation to 21 centimeter line measurements. See Abstr. 031.501.

Automated star counts in the dark cloud L1454. See Abstr. 031.643.

One-dimensional infrared speckle interferometry. See Abstr. 034.103.

Diffuse clouds and Alfvén waves. See Abstr. 062.053.

The effect of losses on acceleration of energetic particles by diffusive scattering through shock waves. See Abstr. 062.071.

Kinetic consideration of charged particle acceleration process in shearing flows of collisionless plasma. See Abstr. 062.075.

Convective instability in bounded uniform self-gravitating spherical clouds.   See Abstr. 062.090.

Chemical evolution of galaxies. I. Constraints imposed by the $\Delta Y/\Delta Z$ ratio.   See Abstr. 064.059.

Mass loss from very luminous OB stars and the Cygnus superbubble.   See Abstr. 064.079.

La structure, l'énergie et l'évolution des étoiles. See Abstr. 065.036.

Interstellar chemistry: polycyanoacetylene formation.   See Abstr. 105.078.

The interstellar dust as a precursor of Ca, Al-rich inclusions in carbonaceous chondrites.   See Abstr. 105.115.

Multiple scattering of solar resonance radiation in the nearby interstellar medium. II. See Abstr. 106.021.

Galactic ring nebulae associated with Wolf-Rayet stars. III. H II region-type nebulae.   See Abstr. 112.042.

H_2O masers in the direction of southern nebular objects.   See Abstr. 112.046.

Infrared studies of the two stellar populations in 30 Doradus.   See Abstr. 113.037.

Observations of high velocity components. See Abstr. 114.067.

Space density of stars and interstellar extinction near h and χ Persei (Perseus I).   See Abstr. 115.025.

Starlight polarization in the direction of two H I complexes.   See Abstr. 116.002.

Two young stars in L 43.   See Abstr. 121.001.

High velocity molecular gas near Herbig-Haro objects HH 7 - 11.   See Abstr. 121.010.

High velocity molecular gas around T Tauri stars. See Abstr. 121.011.

B35 and FU Ori: dust temperatures and energetics from 60 - 250 μm photometry.   See Abstr. 121.015.

Emission line spectra of Herbig-Haro objects. See Abstr. 121.033.

Cooling and evolution of adiabatic blast waves in a dusty medium.   See Abstr. 125.001.

The infrared emission from supernova condensates. See Abstr. 125.007.

Sequential explosions of supernovae in an OB association and formation of a superbubble. See Abstr. 125.026.

A relationship between SNR G109.1−1.0 and the molecular cloud of Sh2−152?   See Abstr. 125.029.

Supernovae and star formation in clusters of galaxies. See Abstr. 125.042.

Stellar clustering as induced by a supernova. See Abstr. 125.069.

The interstellar medium and the highly ionized species observed in the spectrum of the nearby white dwarf G191− B2B.   See Abstr. 126.006.

A high-resolution search for small-scale structure in Sharpless H II regions at 4.995 GHz. II. General properties of the entire sample.   See Abstr. 132.004.

A high-resolution search for small-scale structure in Sharpless H II regions at 4.995 GHz. III. Description of selected sources.   See Abstr. 132.005.

Polarization of starlight in W 3. See Abstr. 132.006.

Aperture synthesis observations of a giant H II region in Cygnus X.   See Abstr. 132.007.

Wolf-Rayet stars in the giant H II region NGC 604. See Abstr. 132.011.

CO ($J = 2 \rightarrow 1$) observations of southern H II regions.   See Abstr. 132.035.

N70: a mass-loss bubble within a massive collapsing H I cloud.   See Abstr. 132.043.

Star formation and extinction in extragalactic H II regions.   See Abstr. 132.048.

Untersuchungen an extragalaktischen H II Gebieten unter besonderer Berücksichtigung des Riesen-H II-Gebietes NGC 604 in M33.   See Abstr. 132.051.

Wolf-Rayet stars associated to giant regions of star formation.   See Abstr. 132.060.

Comparison between optical and ultraviolet interstellar lines formed in the Carina Nebula (NGC 3372). See Abstr. 132.106.

An extended far-infrared emission complex at IC 1318b and IC 1318c.   See Abstr. 133.001.

Infrared objects near to H_2O masers in regions of active star formation. II. Survey and $1-20$ μm observations of southern sources.   See Abstr. 133.006.

A search for the infrared counterpart of type II OH masers – I. A model for the IR background source confusion. See Abstr. 133.010.

The location of the hot molecular core in Orion. See Abstr. 133.012.

On the ice content of KL nebula in Orion. See Abstr. 134.023.

Dust in planetary nebulae.   See Abstr. 135.008.

Planetary nebulae and their influence on the interstellar medium.   See Abstr. 135.037.

Observations of the H_2O radio line in the sources W 49 and Orion A.   See Abstr. 141.125.

An Effelsberg-Green Bank galactic H I absorption line survey. I. The observations.   See Abstr. 141.147.

An X-ray active region in Orion: X-rays from a Herbig-Haro object?   See Abstr. 142.021.

γ rays from the cosmic ray irradiation of local molecular clouds.   See Abstr. 142.506.

Cosmic gamma rays from the Orion molecular cloud complex.   See Abstr. 142.520.

Contribution of cosmic ray-irradiated molecular clouds to the number of apparent γ-ray sources. See Abstr. 142.535.

On cosmic ray acceleration in the interstellar medium.   See Abstr. 143.065.

The propagation of galactic cosmic rays. See Abstr. 143.072.

Confinement of cosmic rays in molecular clouds. See Abstr. 143.081.

Lifetime of molecular clouds and spiral structure. See Abstr. 151.006.

The generality of our local environment. See Abstr. 151.026.

Star formation and evolution of galaxies. See Abstr. 151.034.

Chemical evolution with inhibited star formation rate.   See Abstr. 151.042.

The effects of induced star formation on the evolution of the galaxy. I. One-zone models. See Abstr. 151.046.

High-velocity interstellar C IV and Si IV toward two stars in the I Per OB association.   See Abstr. 152.001.

A search for HI towards SY, SZ and TT Phoenicis. See Abstr. 152.003.

The Vulpecula OB1 complex – some thoughts on the star formation process in a rich association. See Abstr. 152.005.

The star-forming history of the young cluster NGC 2264.   See Abstr. 153.008.

On the effects of radiation pressure in the cores of globular clusters.   See Abstr. 154.025.

Star formation in globular clusters at initial stages of their evolution.   See Abstr. 154.027.

Kinematical and chemical evolution of the galactic disk near the Sun.   See Abstr. 155.002.

The extinction toward the galactic center from observations of interstellar lines.   See Abstr. 155.005.

High latitude H I shells in the Galaxy. I. See Abstr. 155.006.

Extreme positive velocities of neutral hydrogen in the Southern Milky Way.   See Abstr. 155.016.

The large scale H I distribution in the outer Galaxy. See Abstr. 155.017.

The distribution of interstellar C IV in the Galaxy. See Abstr. 155.019.

^{13}CO near the galactic center. See Abstr. 155.024.

Is the galactic corona hot or cool? See Abstr. 155.026.

A corona and a wind in our Galaxy. See Abstr. 155.027.

Molecular clouds outside the solar circle in the first quadrant of our Galaxy.   See Abstr. 155.036.

Abundances in stellar populations and the interstellar medium in galaxies.   See Abstr. 155.039.

Ammonia in the neighbourhood of the galactic center.   See Abstr. 155.045.

Dynamics of the galactic fountain. See Abstr. 155.052.

Determination du taux de formation d'étoiles et de la fonction initiale de masse dans la Galaxie. See Abstr. 155.062.

Nuages moléculaires et structure spirale. See Abstr. 155.063.

Structure of the magnetic field in the cloud of interstellar matter associated with the Perseus OB2 association.   See Abstr. 156.011.

The γ-ray emissivity of the local interstellar medium from correlations with gas at intermediate latitudes. See Abstr. 157.011.

Low-latitude galactic γ-ray emission: a probe, not a proof.   See Abstr. 157.012.

Prognoz-6 data about ultraviolet sky background in dark and Milky Way regions of the sky. See Abstr. 157.019.

Blue compact dwarf galaxies. I. Neutral hydrogen observations of 115 galaxies.   See Abstr. 158.034.

The CO distribution in M 51. See Abstr. 158.092.

CO observations of the Sc galaxy NGC 3628. See Abstr. 158.093.

CO in M 31.   See Abstr. 158.094.

CO radial distributions in two Scd galaxies: IC 342 and NGC 6946.   See Abstr. 158.095.

The selection of galaxies in the Lick survey and the relationship between galaxy counts and galactic absorption. See Abstr. 158.098.

Bursts of star formation in the central regions of spiral galaxies.   See Abstr. 158.130.

Dust clouds in the stellar spheroids of Local Group galaxies.   See Abstr. 158.134.

Distribution of molecular gas in three face-on galaxies.   See Abstr. 158.167.

Preliminary stellar photographic photometry in the Sculptor Dwarf Irregular Galaxy (SDIG). See Abstr. 158.189.

The giant spiral galaxy M 101. VII. Associations of H I concentrations and H II complexes. See Abstr. 158.217.

Neutral hydrogen in M31. I. The distribution of H I gas and spiral arms.   See Abstr. 158.269.

Interstellar extinction in the Large Magellanic Cloud. See Abstr. 159.001.

Ultraviolet interstellar extinction in the Large Magellanic Cloud using observations with the International Ultraviolet Explorer.   See Abstr. 159.002.

Distribution of hot stars and hydrogen in the Large Magellanic Cloud.   See Abstr. 159.003.

The velocity structure of gas in the lines of sight to the Magellanic Clouds.   See Abstr. 159.005.

Discovery of a protostar in the Large Magellanic Cloud.   See Abstr. 159.015.

A strong 1665-MHz OH maser in the Large Magellanic Cloud.   See Abstr. 159.016.

The young phenomena within the Large Magellanic Cloud.   See Abstr. 159.018.

Studies of the Magellanic Clouds. III. Colours, gas and past star formation rate.   See Abstr. 159.019.

Neutral hydrogen streams in groups of galaxies. I. Observations.   See Abstr. 160.030.

H I observations in the Virgo cluster area. See Abstr. 160.044.

Errata

131.901   Erratum: 'Detection of HC$_5$N in four dark clouds' [Astrophys. J., Lett., Vol. 242, L87 - L91 (1980)]. P. J. Benson, P. C. Myers. Astrophys. J., Lett., Vol. 248, L87 (1981). – See Abstr. 28.131.302.

131.902   Erratum: "The first stars" [Irish Astron. J., Vol. 14, 41 - 49 (1979)]. W. H. McCrea. Irish Astron. J., Vol. 14, 198 (1980). – See Abstr. 28.131.003.

132 H I, H II Regions

132.001 Magnetic field strengths in the H II regions S117, S119, and S264.
C. Heiles, Y.- H. Chu, T. H. Troland.
Astrophys. J., Lett., Vol. 247, L77 - L80 (1981).

The authors derive magnetic field strengths in three H II regions from the Faraday rotation measures of linearly polarized extragalactic radio sources that lie behind the H II regions. The observations were obtained with the Very Large Array, using the technique of Heiles and Chu (1980) and a bandwidth of 3.125 MHz.

132.002 Recombination lines (76α) of hydrogen, helium, carbon and one other element from high-emission-measure H II regions.
R. X. McGee, L. M. Newton.
Mon. Not. R. Astron. Soc., Vol. 196, 889 - 905 (1981).

Twenty-five southern H II regions of high emission measure were selected as a first step in a recombination line survey of helium, carbon and other elements made with a 14.7-GHz receiver and the Parkes 64-m radio telescope. Spectra are presented together with observational and derived parameters. Electron temperatures and the helium abundance ratios $[N(He^+)/N(H^+)]$ and their distribution with galactic radius are discussed. Carbon recombination lines were detected in 20 sources. Lines of another element, assumed to be sulphur, were detected in four sources.

132.003 A comparison of the dynamics of the ionized and neutral gas in NGC 281.
P. G. Johnson, N. J. White, A. Pedlar.
Mon. Not. R. Astron. Soc., Vol. 196, 995 - 1003 (1981).

Splitting of the [N II] line of up to 26 km s^{-1} has been found over much of the H II region NGC 281. These observations have been correlated with distinct features in the associated neutral clouds mapped by recent radio observations. The intimate relationship of the ionized and neutral gas associated with NGC 281 is discussed and the possibility of the dynamics being explained by a stellar-wind-driven shell is proposed.

132.004 A high-resolution search for small-scale structure in Sharpless H II regions at 4.995 GHz. II. General
properties of the entire sample.   M. Felli, R. H. Harten.
Astron. Astrophys., Vol. 100, 28 - 41 (1981).

The optical and radio properties of a heterogeneous sample of 75 H II regions in the longitude range $50° < l < 235°$ are presented and discussed. Instrumental and/or selection effects that may help to define the statistical significance of the results are also examined. The nebulae are divided into five classes according to optical morphology and presence of obscuration.

132.005 A high-resolution search for small-scale structure in Sharpless H II regions at 4.995 GHz. III.
Description of selected sources.   M. Felli, R. H. Harten.
Astron. Astrophys., Vol. 100, 42 - 58 (1981).

A detailed discussion of 22 H II regions with small structure in their radio emission is presented. Of the 22 objects discussed, 12 can be classified as compact H II regions with an internal source of excitation. The remaining small diameter radio features are found in more extended H II regions and can be attributed to increases in the electron density at ionization fronts located at the edges of dense molecular clouds. Their ionization can be accounted for by the same early type star which ionizes the diffuse extended H II region in which they are located. The results imply that, in the sample, the expansion of an H II region plays little role as the trigger of the next generation star formation.

132.006 Polarization of starlight in W 3.
R. Lenzen, A. Schulz, T. Schmidt.
Astron. Astrophys., Vol. 100, 249 - 253 (1981).

Continuing the analysis of Schulz et al. (1978), linear polarization measurements of a number of heavily reddened early type stars in the W 3 region and its environs are presented. The wavelength dependence of the degree of polarization has been found to agree in all cases with that of the general interstellar field, as established by Serkowski et al. (1975) and extended by Wilking et al. (1980). The wavelengths of maximum polarization have been found to be correlated with the corresponding visual extinctions. The ratios of polarization to extinction found in W 3 imply a very efficient polarization mechanism.

132.007 Aperture synthesis observations of a giant H II region in Cygnus X.
J. W. M. Baars, H. J. Wendker.
Astron. Astrophys., Vol. 101, 39 - 48 (1981).

The area around IC 1318 b and c of Cygnus X was observed with the SRT Westerbork mainly at 1415 MHz. Each of the fields is dominated by more than one very extended component. A method is presented which obtains previously missing short baseline information thereby giving a consistent picture for single dish and interferometer continuum data. It turns out that a hierarchy of structures exists. The major parts of these features are the genuine components of a typical so-called giant H II region. A few features are interpreted as recently-ionized substructures of neutral clouds of the interstellar medium which point to a typical scale length of 1 to 2 2 pc for neutral density fluctuations in this area.

132.008 Central object of the 30 Doradus Nebula, a supermassive star.
J. P. Cassinelli, J. S. Mathis, B. D. Savage.
Science, Vol. 212, 1497 - 1501 (1981).

R136 (HD 38268) is the central object of the 30 Doradus Nebula, a giant region of ionized hydrogen in the Large Magellanic Cloud. Observations of R136 at low and high spectral resolution with the International Ultraviolet Explorer reveal a peculiar hot object with a massive stellar wind. The bulk of the observed ultraviolet radiation must come from R136a, the brightest and bluest component of R136. Its absolute visual magnitude and observed temperature imply a luminosity about 10^8 times that of the sun. Most of the ionizations produced in 30 Doradus are provided by this peculiar object. If R136a is a dense cluster of very hot stars, about 30 stars of classes O3 and WN3 exist in a region estimated to have a diameter of less than 0.1 parsec. This is inconsistent with the ultraviolet line spectrum and the evidence for optical variability.

132.009 Isolated extragalactic H II regions − invisible galaxies rendered visible.   A. P. Fairall.
Mon. Notes Astron. Soc. South. Africa, Vol. 39, 100-102 (1980).

Isolated extragalactic H II regions are dwarf galaxies undergoing a burst of star formation. In their dormant stage they are almost certainly too faint to detect − their elliptical profiles may suggest a relation to dwarf spheroidal systems. Their spatial distribution suggests that these otherwise "invisible" systems prefer open spaces between clusters or inhabit poor groupings of galaxies.

132.010 Formation of OB clusters: VLA observations.
P. T. P. Ho, A. D. Haschick.
Astrophys. J., Vol. 248, 622 - 637 (1981).

Three ultracompact H II regions associated with OH masers are studied with the VLA. Multiple structures are found in each region and are interpreted as clusters of OB stars. This interpretation is supported in the case of G10.6−0.4 by a good agreement between the distribution of spectral types and the initial mass function of field stars. If the OB cluster extends to lower mass stars, the authors find important implications about the far-infrared characteristics of these regions and, in particular, about the deduced properties of the dust in the H II regions. Size scales and spatial distribution of the OB stars suggest that collision processes between cloud fragments must have been important in the evolution toward stars.

132.011   **Wolf-Rayet stars in the giant H II region NGC 604.**
S. D'Odorico, M. Rosa.
Astrophys. J., Vol. 248, 1015 - 1020, plate 16 (1981).
The authors have detected strong Wolf-Rayet emission bands in NGC 604 at all positions where the stellar continuum is strong. The W-R stars are comparable in number with the massive hot stars ionizing the H II region. The presence of about 50 W-R stars with a mean type of WN 7 is interpreted as the result of a single burst of star formation some 4×10^6 years ago in the core of NGC 604.

132.012   **Detection of [O I] 63 micron emission from the galactic center.**   D. F. Lester, M. W. Werner, J. W. V. Storey, D. M. Watson, C. H. Townes.
Astrophys. J., Lett., Vol. 248, L109 - L112 (1981).
The detection of the 63 μm line of [O I] is reported for three positions in the H II region complex Sgr A at the galactic center. Velocity resolution of the line indicates that the emitting material has both rotational and radial motion of magnitude similar to that of the ionized gas in the core and that a substantial amount of the emitting material lies within the central few parsecs of the Galaxy. A model in which [O I] is collisionally excited by neutral hydrogen, either from the warm region ahead of an ionization front or behind a shock, is proposed and gives a total mass of hot, neutral gas within the central 3 pc of the Galaxy of between 10 and $10^3 M_\odot$. A limit on the flux of this line has been set for Sgr B2.

132.013   **A systematic radial velocity survey of the H II regions IC 1318a and HS 191.**
K. L. Dixon, P. G. Johnson, R. Songsathaporn.
Astrophys. Space Sci., Vol. 78, 189 - 198 (1981).
The results of a systematic radial velocity survey of two H II regions in the Cygnus-X complex are presented. The [N II] emission line components obtained at many positions over each object have been compared and correlated with other phenomena. It is suggested that they are probably connected objects and may be part of a giant shell or cylinder.

132.014   **The calibrated map of the H II region S237 in Hα emission.**
S. Mizuno, K. Sakka, T. Sasaki, T. Kogure.
Astrophys. Space Sci., Vol. 78, 235 - 242 (1981).
The calibrated Hα-map of S237 and its profiles in section through the nebular center are presented. The core-envelope structure and central open shell appearance are noticed. Physical parameters such as emission measure and mass of ionized gas are deduced and compared with results of radio observations.

132.015   **New insights into the physical state of gaseous nebulae.**   M. Peimbert.
The Universe at ultraviolet wavelengths, (see 012.009), p. 557 - 565 (1981).
The impact of our knowledge of H II regions, planetary nebulae and supernova remnants due to IUE is briefly examined. Some of the more relevant aspects related to the physical conditions of gaseous nebulae are reviewed. The analysis of IUE data is under process and already significant results have been obtained on the following properties of gaseous nebulae: (1) density and temperature distribution, (2) ionization structure, (3) chemical composition, (4)

internal dust, and (5) shock velocity for supernova remnants. The CNO abundances of planetary nebulae are compared with stellar evolution models.

132.016   **Comparisons between optical and ultraviolet interstellar lines formed in the Carina Nebula (NGC 3372).**   J. E. Hesser, N. R. Walborn.
The Universe at ultraviolet wavelengths, (see 012.009), p. 571 - 579 (1981).
Discovery of complex Ca II H and K interstellar line profiles towards stars embedded in the giant H II region surrounding Eta Carinae led the authors to undertake a reconnaissance of the richer ultraviolet interstellar line spectrum. Single IUE spectra were secured for those stars exhibiting the greatest variety of structure in the optical interstellar lines, namely, HD 93130, 93160, 93162, 93204, 93205, 93206 and HDE 303308.

132.017   **The carbon abundance in two H II regions of the Small Magellanic Cloud.**
R. J. Dufour, R. J. Talbot, Jr., G. A. Shields.
The Universe at ultraviolet wavelengths, (see 012.009), p. 671 - 678 (1981).
Observations of the ultraviolet spectra of two locations in the H II region NGC 346 and of the entire H II region IC 1644 in the Small Magellanic Cloud (SMC) were made using the International Ultraviolet Explorer (IUE) satellite. From measurements of the C III] λ 1909 lines, the abundance of carbon in the nebulae was derived using theoretical model analysis combined with ground-based spectrophotometry of other emission lines. The abundance of C relative to H in the SMC was found to be lower by −0.9 dex compared with the sun and lower by −0.8 dex compared with the Orion Nebula. This C deficiency is discussed in detail.

132.018   **The spiral structure of our Galaxy and kinematics of H II regions.**   E. M. Grivnev.
Pis'ma Astron. Zh., Tom 7, 543 - 546 (1981). In Russian. English translation in Soviet Astron. Lett., Vol. 7.
The kinematics of H II regions is investigated supposing the existence of density waves in our Galaxy. It is proved that there are H II regions near the inner edge of the Sagittarius spiral arm and near the outer edge of the Perseus arm. The distance of the corotation circle is 11 kpc. The group velocity of the density wave is directed from the corotation circle.

132.019   **H II regions in external galaxies.**   P. W. Hodge.
Bull. American Astron. Soc., Vol. 13, 518 (1981). Abstract.

132.020   **Models of H II regions in M101.**
L. H. Aller, K. Sedwick.
Bull. American Astron. Soc., Vol. 13, 518 (1981). − Abstract.

132.021   **IUE observations of carbon in SMC and LMC H II regions.**
R. J. Dufour, G. A. Shields, R. J. Talbot.
Bull. American Astron. Soc., Vol. 13, 518 (1981). − Abstract.

132.022   **H II regions and Wolf-Rayet stars in M33.**
P. Massey, P. S. Conti.
Bull. American Astron. Soc., Vol. 13, 532 (1981). − Abstract.

132.023   **The z distribution of H II regions in the Galaxy.**
M. Fich, L. Blitz.
Bull. American Astron. Soc., Vol. 13, 539 (1981). − Abstract.

132.024   **The dynamics and structure of the ionized and neutral gas in the 30 Doradus nebula.**   N. J. White.
Astrophys. Space Sci., Vol. 78, 443 - 461 (1981).
A variety of new optical observations have been made over the bright core of the supermassive H II region 30 Doradus

and the ionized filamentary material surrounding this object. In addition, a more detailed analysis of previously published interstellar absorption and H I emission profiles has been undertaken. The velocity and density structure of this complex region is discussed and the new results analysed here shown to be compatible to the model of 30 Doradus presented in Cantó et al. (1980) and Meaburn (1980).

132.025 Ionization front interactions and the formation of globules.   P. W. J. L. Brand.
Mon. Not. R. Astron. Soc., Vol. 197, 217 - 233 (1981).

It is assumed that an H II region has evolved inside a molecular cloud. The interactions that result from the expanding shell of compressed molecular gas reaching the edge of the cloud are calculated, and the instability of the ionization front to the formation of globules is investigated. If the shell phase of H II region evolution has proceeded significantly, then globules of up to a fraction of a solar mass may be formed in an H II region caused by a star with an ionizing luminosity of 10^{49} photon s^{-1} in a molecular cloud of density 10^3 cm^{-3}.

132.026 Ultraviolet spectra of extragalactic H II regions.
P. M. Gondhalekar, D. H. Morgan, K. Nandy, R. Wilson.
Second European IUE Conference, (see 012.011), p. 131 - 132 (1980).

H II regions in the galaxies NGC 2366 and IC 2574 have been observed with the IUE at low resolution. The spectra show the 2200 Å absorption feature and many emission lines.

132.027 The exciting stars of giant H II regions in M 33 and M 101.   M. Rosa.
Second European IUE Conference, (see 012.011), p. 135 (1980). – Abstract.

132.028 Observations of H II regions in the Magellanic Clouds.
D. P. Gilra, M. Dennefeld, S. R. Pottasch.
Second European IUE Conference, (see 012.011), p. 205 - 206 (1980).

IUE observations of 5 positions in the 30 Doradus region and one in N79A in the Large Magellanic Cloud and, N81 and N66 in the Small Magellanic Cloud were obtained. The C III] line at 1909 Å is present in almost all the objects. Dust-scattered light is detected in 30 Dor and N66. A preliminary analysis of the abundances in N81 shows that carbon is underabundant by about a factor of 10.

132.029 The interstellar extinction law in some dusty HII regions.   T. Neckel, R. Chini.
Astron. Astrophys., Suppl. Ser., Vol. 45, 451 - 454 (1981).

UBVRI data for 62 O and B stars in dusty HII regions and for 9 field stars are presented. The data are used to investigate the reddening law in the range BVRI. For the field stars and the HII regions M8, M20, S5, S66, S67 the normal interstellar extinction law has been confirmed. However, in NGC 6334, NGC 6357 and M16 unusually high colour excess ratios E_{V-R}/E_{B-V} and E_{V-I}/E_{B-V} indicating abnormal values of the ratio of total to selective extinction, $R = 3.8$ for NGC 6334/6357 and $R = 4.0$ for M16, have been found.

132.030 Stellar mass loss and H II region morphology in Magellanic irregular galaxies.
M. A. Dopita, I. R. Wilson.
Effects of mass loss on stellar evolution, (see 012.015), p. 523 - 534 (1981).

The problem of the interaction of a strong stellar wind with a massive collapsing neutral hydrogen cloud has been considered and a new class of steady flow solutions found. The detailed comparison of these models with one filamentary shell, N70 in the LMC, shows that they successfully explain all

the observed properties of the H II region and enable mean mass-loss rates to be derived for the central stars.

132.031 Opportunities offered by optical region observations of H II regions.   L. H. Aller.
The phases of the interstellar medium, (see 012.018), p. 99 - 103 (1981).

132.032 Shell structures in compact H II regions.
J. H. van Gorkom.
The phases of the interstellar medium, (see 012.018), p. 105 - 107 (1981).

Compact H II regions are by definition H II regions with sizes smaller than 0.5 pc and electron densities greater than 10^4 cm^{-3}. It has been known for some time that the filling factor in these regions increases with decreasing density, indicating that the more compact sources have fine structure, which smooths out as they expand. Some recent aperture synthesis observations of this fine structure in the radio continuum and recombination line emission are discussed.

132.033 Infrared and radio observations of the compact H II region DR - 22.   T. Herter.
News Lett. Astron. Soc. N. Y., Vol. 1, No. 10, p. 39 (1981). Abstract.

132.034 Measurements of the equivalent width of the H_β emission line and age determination of H II regions of the LMC and SMC.   H. A. Dottori, E. L. D. Bica.
Astron. Astrophys., Vol. 102, 245 - 249 (1981).

The equivalent width of the emission line H_β was measured photoelectrically in 29 H II regions of the LMC and 2 of the SMC respectively. The age of these regions was obtained through a calibration of $W_{H\beta}$. A relation was found between $W_{H\beta}$ and the ratio of the H II region radius to that of the embedded stellar association.

132.035 CO $(J = 2 \to 1)$ observations of southern H II regions.   T. de Graauw, S. Lidholm, B. Fitton, J. Beckman, F. P. Israel, H. Nieuwenhuijzen, J. Vermue.
Astron. Astrophys., Vol. 102, 257 - 264 (1981).

Observations of the ^{12}CO $(J = 2 \to 1)$ line in the direction of the Carina Nebula (RCW 53), G 333.6−0.2, G 333.3−0.4, G 333.1−0.4 (RCW 106), G 332.8−0.6 and G 332.7−0.6 are presented. An extended molecular cloud appears to be wrapped around the brightest portion of the Carina Nebula (known as Car I and Car II). The situation is consistent with the Elmegreen-Lada model for sequential star formation; a new heating center may have been found northwest of Car I, indicating the next step in the star formation sequence. At the position of the other sources observed, a giant molecular complex, at least 75 pc long and 20 pc wide is found. The complex consists of at least four large clouds. A velocity shift of 5 km s^{-1} between the northern and southern end of the complex is interpreted as caused by the presence of two cloud groups within the complex, each associated with its own H II regions. The molecular complex coincides accurately with a distinct minimum in the H I distribution. A large scale triggering factor for the recent star formation in the complex appears necessary.

132.036 On the far ultraviolet flux distribution of the Orion Nebula.   G. R. Carruthers, H. M. Heckathorn.
Astrophys. Lett., Vol. 22, 135 - 141 (1981).

The authors' previously reported sounding rocket far-UV absolute flux distribution for the $\theta^{1,2}$ Ori star cluster is shown to have a significant nebular contribution on the basis of IUE observations of individual component stars. In fact, the integrated far-UV brightness of the Orion Nebula is quite comparable to the directly observed flux from the central stars – confirmation that the scattering efficiency of the dust particles in the Orion region is high.

132.037 **The largest H II regions in M101.**
L. Blitz, F. P. Israel, G. Neugebauer, I. Gatley,
T. J. Lee, D. H. Beattie.
Astrophys. J., Vol. 249, 76 - 82 (1981).

Coordinated millimeter, infrared, and ultraviolet observations of the five brightest H II regions in M101 are presented. A CO complex has been found to be associated with NGC 5461 which is much more massive than any Milky Way counterpart. A narrow line width has been observed which suggests that the molecular complex may consist of numerous fragments with a volume filling fraction $\sim 10^{-2}$. NGC 5461 also shows a 10 μm and 20 μm excess which is similar to that of galactic H II regions. The ultraviolet observations show that the 2200 Å dust feature is greatly attenuated in all of the H II regions but least of all in NGC 5461. The near-infrared flux densities and Bγ line strengths are consistent with the hypothesis that a significant fraction of the infrared emission is free-free.

132.038 **Internal motions in H II regions. IX. The bipolar nebula S106.**   P. Pişmiş, I. Hasse.
Rev. Mexicana Astron. Astrofis., Vol. 5, 79 - 85, plate 1 (1981).

The authors have determined radial velocities by photographic Fabry-Pérot interferometry, at 236 points in and around the small (1 × 3 arcmin) bipolar H II region Sharpless 106. They believe the distance of S106 to be less than 1 kpc, more nearly 600 pc, in agreement with the suggestion of Eiroa et al. (1979) based on physical arguments on the ionization source of S106.

132.039 **Highly excited OH in W3 (OH).**   A. Baudry,
C. M. Walmsley, A. Winnberg, T. L. Wilson.
Astron. Astrophys., Vol. 102, 287 - 292 (1981).

The $^2\pi_{3/2}$, $J = 7/2$, $F = 4 \rightarrow 4$ and $F = 3 \rightarrow 3$ transitions of OH have been observed in absorption against the continuum radio spectrum of the compact H II region W3 (OH). The data give a temperature of 160 ± 30 K between the $J = 7/2$ and 9/2 rotational levels. It is estimated that the excitation of the OH gas to these high levels is dominated by infrared continuum radiation ($\lambda 65$ μm) emitted by dust grains at a temperature of $\sim 150 - 200$ K. The narrow emission component (probably maser amplified) in the $J = 7/2$, $F = 4 \rightarrow 4$ transition has not changed significantly in intensity since its detection. It is . shown that it splits into one right and one left circularly polarized component. If this splitting is caused by the Zeeman effect the frequency separation corresponds to a magnetic field of 6.0 ± 0.4 mG directed away from us.

132.040 **Physics of the peripheral zones of H II regions. I. Border enhancements of [O III]/Hβ and He I/Hβ in H II regions.**   M. Heydari-Malayeri.
Astron. Astrophys., Vol. 102, 316 - 320 (1981).

This paper presents the results of high-resolution bidimensional studies of the intensity ratios [O III] λ 5007/Hβ, He I λ 5876/Hβ and [O II] $\lambda\lambda$ 3727 + 3729/Hβ for the Galactic H II regions Sh2 − 152 and 156 situated in the Perseus arm. The author shows that the ratios are enhanced unexpectedly in the faint peripheral zones of the H II regions. He discusses the physical conditions that might give rise to these phenomena. The author finds that these nebulae are not overabundant in heavy elements.

132.041 **Frequency distribution of H II region diameters.**
S. van den Bergh.
Astron. J., Vol. 86, 1464 - 1467 (1981).

It is shown that the frequency distribution of the largest H II regions in galaxies is well represented by the relation $N(D) = N_0 e^{-D/D_0}$, where $N(D)$ is the number of H II regions with diameter larger than D.

132.042 **Very high-velocity H I clouds: an intergalactic population?**   R. Giovanelli.
Astron. J., Vol. 86, 1468 - 1479 (1981).

The very high-velocity clouds, found only in the galactic quadrant $l < 180°$, $b < 0°$, are considered vis-à-vis the galaxies of the Local Group and the Magellanic Stream, in both their space and velocity distributions. This comparison does not favor the hypothesis that the very high-velocity clouds are intergalactic clouds in the Local Group. Structural parameters derived from new observations of some clouds, and observations of the Magellanic Stream presented by Giovanelli and Haynes (1981), confirm this conclusion and favor a picture of a fragmenting Magellanic Stream, where the very high-velocity clouds are shreds of Magellanic material precipitating toward the galactic disk.

132.043 **N70: a mass-loss bubble within a massive collapsing H I cloud.**   M. A. Dopita, V. L. Ford,
P. J. McGregor, D. S. Mathewson, I. R. Wilson.
Astrophys. J., Vol. 250, 103 - 115 (1981).

From a wide variety of observational data, it is concluded that N70, a giant shell H II region in the LMC, is a mass-loss bubble confined by the ram pressure of a massive, collapsing H I cloud, along the lines of the model described by Dopita (1981). From this model, the age, the parameters of the H II region and the H I cloud, the mass-loss rate, and the other properties of the exciting stars are determined. A supernova origin of N70, often suggested for this nebulosity, is shown to be untenable.

132.044 **Abundances of argon, sulfur, and neon in six galactic H II regions from infrared forbidden lines.**
T. Herter, H. L. Helfer, J. L. Pipher, W. J. Forrest,
J. McCarthy, J. R. Houck, S. P. Willner, R. C. Puetter,
R. J. Rudy, B. T. Soifer.
Astrophys. J., Vol. 250, 186 - 199 (1981).

Airborne measurements of the [Ar II] (6.99 μm) and [S III] (18.71 μm) lines for six compact H II regions are presented, as well as ground-based 2–4 μm and 8–13 μm spectroscopy if not already published. From these data and radio data, the authors deduce lower limits to the elemental abundances of Ar, Ne, and S. G29.9–0.0 at 5 kpc from the galactic center is overabundant in all these elements. The other five regions (at distances 6–13 kpc from the center) mainly appear to be consistent with standard abundances, with the exception of G75.84 + 0.4 at 10 kpc from the galactic center, which is overabundant in S.

132.045 **High-resolution radio observations of G82.6 + 0.4.**
G. S. Rossano, R. W. Russell.
Astrophys. J., Vol. 250, 227 - 231 (1981).

The region of AFGL 2636, a multicomponent infrared source from which the unidentified 3.3 μm emission feature is observed, has been mapped at 20 cm with a resolution of 1.″5, at 9 cm at a resolution of 4.′6, and at 6 cm with a resolution of 0.″5. The authors find that the large-scale distribution of the ionized gas and the extent of the heated dust are well correlated. They place an upper limit of 3 kpc on the distance to this region. Nine cm observations of AFGL 437 and AFGL 3053, which also exhibit the 3.3 μm feature, are also discussed.

132.046 **Giant [C II] halos around H II regions.**
R. W. Russell, G. Melnick, S. D. Smyers, N. T. Kurtz,
T. R. Gosnell, M. Harwit, M. W. Werner.
Astrophys. J., Lett., Vol. 250, L35 - L38 (1981).

The authors have mapped the submillimeter (157 μm) [C II] emission from the M17 complex. The [C II] emission extends over at least 1/4° in the sky. The regions emitting the C II radio recombination lines contribute only in a minor way to the total 157 μm [C II] flux. The total [C II] luminosity of M17 exceeds 2×10^3 $L_\odot$.

132.047 Synthesis observations of the radio continuum radiation of the H II region NGC 7822 (W1).
R. H. Harten, W. M. Goss, H. E. Matthews, F. P. Israel.
Astron. Astrophys., Vol. 103, 50 - 56 (1981).

Observations of the large evolved H II region W1 (NGC 7822, S171) have been made using the Westerbork Synthesis Radio Telescope at 0.6, 1.4 and 5.0 GHz. The angular resolutions range from 7″ to 56″. The source contains two bright components. The eastern component G118.6+4.8 consists of a ridge of emission with an arc-like structure in the center. The western component G118.1+5.0 consists of a diffuse plateau and an ionization front on the western edge. Based on the structural evidence, no indication of a supernova component exists. The authors discuss the suggestion that some of the exciting stars of W1 may not yet be identified and that more than one mechanism of star formation may be operating in the region.

132.048 Star formation and extinction in extragalactic H II regions. J. Lequeux, M. Maucherat-Joubert, J. M. Deharveng, D. Kunth.
Astron. Astrophys., Vol. 103, 305 - 318 (1981).

Various aspects of star formation — distribution of stellar masses at birth, history of the rate and efficiency of star formation — are investigated in extensively studied extragalactic H II regions. Evolutionary models of far UV luminosity, production of ionizing photons, effective temperature and ejection of heavy elements for an ionizing star cluster are compared with properties of extragalactic H II regions; the sample under study includes CM 39 in the irregular galaxy NGC 4449 (new IUE observations are presented), NGC 604 in M 33, NGC 5471 and NGC 5461 in M 101, NGC 2363 in the irregular galaxy NGC 2366, 30 Dor in the Large Magellanic Cloud and the blue compact galaxies IZw 18 and IIZw 70, for which IUE observations are available.

132.049 Radio astronomy confronts elementary particle cosmology: radio measurements of helium abundances. J. Schmid-Burgk.
Nuclear astrophysics (see 012.036), p. 295 - 303 (1981).

Radio lines, emitted as a consequence of H^+ and He^+ recombinations in H II regions, provide nearly ideal tools for the determination of He/H abundance ratios all across the galaxy. After applying appropriate corrections for some unseen He^0, and taking into account the post-big bang fraction of He derived from radio and optical observations of oxygen, recent 100-m radio telescope measurements give a primordial He mass fraction Yp of between 0.21 and 0.234. This puts stringent limits on some elementary particle scenarios in standard cosmology.

132.050 Spectrophotometry of H II regions in the spiral galaxy M101. K. E. Sedwick, L. H. Aller.
Proc. Natl. Acad. Sci. USA, Vol. 78, 1994 - 1997 (1981).
Abstr. in Phys. Abstr., Vol. 84, Abstr. 108304 (1981).

132.051 Untersuchungen an extragalaktischen H II Gebieten unter besonderer Berücksichtigung des Riesen-H II-Gebietes NGC 604 in M 33. M. Rosa.
Diss. Naturwiss.-Math. Gesamtfak. Ruprecht-Karls-Univ. Heidelberg. 5 + 155 pp. (1981).

Spectra of the ionizing OB clusters of giant H II regions in nearby galaxies have been obtained in the IUE UV and visual wavelength regions. UV extinctions and the luminosity functions of the stellar populations are discussed. Special emphasis is laid on NGC 604. Numerous WR stars dominate the kinematics of this nebula. High resolution maps, photographed in the light of $H\beta$, [O III] and the stellar continuum are presented for NGC 604 and NGC 5471.

132.052 Far-infrared [O III] and [N III] line emission from galactic H II regions and planetary nebulae.

D. M. Watson, J. W. V. Storey, C. H. Townes, E. E. Haller.
Astrophys. J., Vol. 250, 605 - 614 (1981).

The authors report observations of the [O III] 51.8 μm and [N III] 57.3 μm lines in various galactic H II regions and planetary nebulae. The results are combined with their previous [O III] 88.4 μm data (Storey, Watson, and Townes 1979) to yield electron densities and N^{++}/O^{++} density ratios where possible.

132.053 VLA observations of DR 21 NH_3 (1, 1) absorption: direct evidence for clumping.
D. N. Matsakis, Å. Hjalmarson, P. Palmer, A. C. Cheung, C. H. Townes.
Astrophys. J., Lett., Vol. 250, L85 - L89 (1981).

The (1, 1) transition of NH_3 was observed in absorption against DR 21 with a synthesized beam of 2″ size at the VLA. The DR 21 continuum is resolved by the VLA into components distributed within a range of about 30″, but only the brightest (and most southerly) component is absorbed by NH_3. The indicated filling factor and optical depth of the absorption feature are consistent with a model suggested by emission results, ammonia existing in optically thick thermalized clumps. One of these would be producing the observed absorption.

132.054 The equivalent width of the $H\beta$ emission line and the evolution of the H II regions. H. A. Dottori.
Astrophys. Space Sci., Vol. 80, 267 - 280 (1981).

The synthetic equivalent width ($W_{H\beta}$) of the line $H\beta$ in emission is obtained for H II regions opaque to the Lyman photon flux, with embedded OB associations with different initial chemical compositions and initial mass functions. The variation of $W_{H\beta}$ as a function of the evolution of the ionizing stars is analysed. The observations of $W_{H\beta}$ for M33, M101, and M51 by Searle (1971) are discussed.

132.055 RCW 58: a remarkable H II region around a WN 8 star. M. C. Lortet, G. Testor.
Messenger, No. 26, p. 21 - 22 (1981).

132.056 On the dynamics of H I supershells.
V. G. Surdin.
Astron. Tsirk., No. 1126, p. 1 - 2 (1980). In Russian.

132.057 Anomalous motions of H I clouds. P. A. Shaver, V. Radhakrishnan, K. R. Anantharamaiah,
D. S. Retallack, W. Wamsteker, A. C. Danks.
ESO Sci. Prepr., No. 166, 16 pp. (1981). — Submitted to Astron. Astrophys.

132.058 Wolf-Rayet stars in extragalactic H II regions: discovery of a peculiar WR in IC 1613/#3.
S. D'Odorico, M. Rosa.
ESO Sci. Prepr., No. 175, 1 + 10 pp. (1981). — Submitted to Astron. Astrophys.

132.059 On the statistics of galactic H I clouds.
K. R. Anantharamaiah, V. Radhakrishnan, P. A. Shaver.
ESO Sci. Prepr., No. 176, 1 + 34 pp. (1981). — To be published in Proceedings of the Second Asian-Pacific Regional Meeting of the IAU, held in Bandung, Indonesia, Aug. 24 - 29, 1981.

132.060 Wolf-Rayet stars associated to giant regions of star formation. S. D'Odorico, M. Rosa.
ESO Sci. Prepr., No. 180, 4 pp. (1981). — To appear in the I.A.U. Symposium 99 on Wolf-Rayet stars.

132.061 Spektralphotometrie der galaktischen H II Region S 106 im Wellenlängenbereich 8 - 13 μm.
E. Hölzle.

Diss. Naturwiss.-Math. Gesamtfak. Ruprecht-Karls-Univ., Heidelberg. 5 + 74 pp. (1981).

Beobachtungen von H II-Gebieten mit verschiedenen Fabry-Perot-Interferometern.   See Abstr. 031.613.

Galactic ring nebulae associated with Wolf-Rayet stars. I. Introduction and classification. See Abstr. 065.049.

Galactic ring nebulae associated with Wolf-Rayet stars. III. H II region-type nebulae.   See Abstr. 112.042.

R 136a: une étoile 2000 fois plus massive que le soleil?   See Abstr. 114.140.

On the nebulosities associated with the extreme Of star HD 148937.   See Abstr. 114.153.

UBV and H_β observations of stars towards M8. See Abstr. 115.012.

Near-infrared observations of trapezium-type multiple systems. Catalogue of observations and a new determination of the reddening law.   See Abstr. 118.023.

Supernova remnants and H II regions in M31. See Abstr. 125.019.

Supernova remnants and abundance gradients in M31. See Abstr. 125.024.

Further radio observations of W 50: total intensity and linear polarization measurements at 1.7 and 2.7 GHz. See Abstr. 125.047.

Neutral hydrogen in the vicinity of galactic radio sources. Supernova remnant W 44.   See Abstr. 125.061.

An investigation of the neutral and ionized gas in M16.   See Abstr. 131.030.

The young Of star HD 148937 and its associated interstellar bubble — H II region.   See Abstr. 131.045.

Star formation in the outer Galaxy. See Abstr. 131.057.

Observations of the exciting stars of NGC 2023 and M43 — interstellar extinction and abundances. See Abstr. 131.074.

Radio continuum and carbon monoxide observations of V645 Cygni (GL 2789).   See Abstr. 131.083.

On the association of the 1720 MHz OH masers with the H_2CO masers in NGC 7538 (IRS1). See Abstr. 131.111.

Proper motions and distances of H_2O maser sources. III. W51 NORTH.   See Abstr. 131.120.

Hot-gas cold-dust pumping for water masers associated with H II regions.   See Abstr. 131.122.

H_2CO mapping toward DR21 and W58 (K3#50). See Abstr. 131.143.

The molecular cloud associated with NGC 7538. See Abstr. 131.150.

Star formation in M33 and in our Galaxy. See Abstr. 131.161.

The dynamical effects of hypersonic stellar winds on interstellar gas.   See Abstr. 131.176.

Star formation in molecular clouds and the formation of compact H II regions.   See Abstr. 131.185.

High-sensitivity survey of NH_3 in the southern hemisphere.   See Abstr. 131.201.

La formation des étoiles chaudes dans les galaxies de types avancés.   See Abstr. 131.224.

Infrared objects near to H_2O masers in regions of active star formation. II. Survey and $1-20\,\mu m$ observations of southern sources.   See Abstr. 133.006.

The nature of NGC 2024: near-infrared spectroscopy of IRS 1 and IRS 2.   See Abstr. 133.007.

Near infrared high resolution spectrophotometry of forbidden [C I] in the Orion Nebula.   See Abstr. 134.018.

The QSO B234 and the irregular galaxy NGC 4861: examples of isolated extragalactic H II regions with low helium abundances.   See Abstr. 141.042.

The distance to G316.8−0.1.   See Abstr. 141.087.

Cosmic ray acceleration by stellar winds and self-confinement in giant H II regions.   See Abstr. 143.080.

The abundance of argon at the galactic center. See Abstr. 155.008.

Does the Galaxy have four spiral arms? See Abstr. 155.049.

On the width and profile of nuclear emission lines in galaxies.   See Abstr. 158.021.

Evidence of the ionized interstellar hydrogen structure in the M33 galaxy. Observations owing to the 6 m telescope of Zelentchuk.   See Abstr. 158.031.

NGC 3344: unusual abundances for an Sbc galaxy? See Abstr. 158.054.

IUE spectra of clumpy irregular galaxies. See Abstr. 158.115.

The size distribution of H II regions as a new variant to determine the distances of galaxies.   See Abstr. 158.157.

Wolf-Rayet stars and giant H II regions in M33: casual associations or meaningful relationships? See Abstr. 158.160.

The giant spiral galaxy M 101. VII. Associations of H I concentrations and H II complexes.   See Abstr. 158.217.

Extragalactic and galactic UV observations owing to the balloon borne SCAP telescope at the wavelength 2000 Å.   See Abstr. 158.291.

Erratum

132.901 Erratum: "Extinction of extragalactic H II regions" [Astrophys. Lett., Vol. 21, 1 - 9 (1980)]. F. P. Israel, R. C. Kennicutt. Astrophys. Lett., Vol. 21, 129 (1981). — See Abstr. 28.132.032.

133 Infrared Sources

133.001 An extended far-infrared emission complex at IC 1318b and IC 1318c.
M. F. Campbell, W. F. Hoffmann, H. A. Thronson, Jr.
Astrophys. J., Vol. 247, 530 - 539 (1981).

A new map of extended 90 μm emission with 15′ resolution is presented for a complex of sources centered on IC 1318b and c. The map shows very strong correlation with radio continuum emission. The extended sources apparently have their far-infrared emitting dust inside the ionized region, unlike compact far-infrared sources. Far-infrared emission due to dust grains is associated with the nonthermal component of the radio source DR 4 which is due to a supernova remnant.

133.002 Discovery of radio brightening in AFGL 618.
S. Kwok, P. A. Feldman.
Astrophys. J., Lett., Vol. 247, L67 - L71 (1981).

Recent radio continuum observations of AFGL 618 have shown that the free-free emission flux density has increased by approximately a factor of 2 over a 2 - 3 yr interval, whereas the spectrum has remained optically thick up to at least 12 GHz. This is interpreted as the result of expansion of a compact H II region within the molecular/dust envelope of AFGL 618. The central star of the nebula is likely to be a 1 $M_\odot$ star now rapidly evolving into the planetary nebula stage.

133.003 The radial velocity of IRC + 10420.
J. D. Fix.
Astrophys. J., Vol. 248, 542 - 544 (1981).

The radial velocity (with respect to the local standard of rest) of IRC + 10420 at visible wavelengths has been found to be 97 ± 3 km s^{-1} for lines of calcium silicon, and iron and 50 ± 3 km s^{-1} for the sodium D lines. The D line absorption is attributed to interstellar sodium atoms with an average distance of about 4 kpc, thus placing a lower limit on the distance to IRC + 10420. The radial velocity of the remaining lines is about 30 km s^{-1} larger than the stellar velocity inferred from spectra of the 18 cm OH maser lines. The discrepancy between the OH velocity and the visible velocity of IRC +10420 is accounted for on the basis of a thick, expanding particle shell which produces redshifts in the wavelengths of the emergent stellar photons.

133.004 On the radio properties of V645 Cygni.
S. Kwok.
Publ. Astron. Soc. Pacific, Vol. 93, 361 - 363 (1981).

Upper limits of 0.5 mJy at 5 GHz and 4 mJy at 15 GHz have been obtained for the continuum flux densities of V645 Cyg. These low upper limits imply that any associated H II region must be extremely dense and compact, with angular size < 0″.06 and density > 3 × 10^6 cm^{-3}. Alternatively, there may be a stellar wind associated with the star with a mass loss rate < 10$^{-5} M_\odot$ yr^{-1}.

133.005 Polarimetry from 1 to 5 μm of compact infrared sources. P. A. Heckert, M. Zeilik, II.
Astron. J., Vol. 86, 1076 - 1083 (1981).

The authors present near-infrared (1−5-μm) polarimetry data for 38 compact infrared sources, mostly in regions of massive star formation. For 17 of these sources, they have multicolor polarimetry. The authors make a statistical argument to show that there are two distinctly different mechanisms causing the polarization in the sources in this collection. One possible mechanism is grain alignment by the interstellar magnetic field; the other is anisotropic scattering off a nonspherical dust cloud surrounding the infrared source.

133.006 Infrared objects near to H$_2$O masers in regions of active star formation. II. Survey and 1−20 μm observations of southern sources.
A. F. M. Moorwood, P. Salinari.
Astron. Astrophys., Vol. 102, 197 - 206 (1981).

The authors continue their discussion of the positional association of infrared objects, H$_2$O and OH masers and H II regions and present a similar statistical analysis to that in Paper I on the degree of IR/H$_2$O association within both this sample and their combined sample of 90 H$_2$O masers. The authors also discuss the nature of those infrared objects studied in most detail and comment on their failure to establish any obvious relationship between their evolutionary phase and the appearance of the associated H$_2$O emission spectra.

133.007 The nature of NGC 2024: near-infrared spectroscopy of IRS 1 and IRS 2. R. I. Thompson, H. A. Thronson, Jr., B. G. Campbell.
Astrophys. J., Vol. 249, 622 - 627 (1981).

New 1.4−2.4 μm moderate-resolution infrared spectra of NGC 2024 IRS 1 and IRS 2 have been obtained in an effort to determine the exciting source for NGC 2024. These spectra, combined with existing optical, infrared, and radio data, argue that neither IRS 1 nor IRS 2 is the exciting source. It is shown that the existing data point toward a single exciting star with a temperature of 34,000 K on the zero-age main sequence. The most likely location of this star is in the southern mid- and far-infrared peaks which are located near the centroid of the radio emission.

133.008 Preliminary results of the Air Force Infrared Sky Survey. S. G. Kleinmann, F. C. Gillett, R. R. Joyce.
Annu. Rev. Astron. Astrophys., Vol. 19, (see 003.012), 411 - 456 (1981).

Contents: Introduction. The Air Force Survey. Comparison with the Two-Micron Sky Survey. The new AFGL sources: isolated stars. New AFGL sources − objects associated with interstellar clouds. Consensus of the AFGL catalog.

133.009 Number densities of cosmic long wavelength infrared (LWIR) sources. J. E. Davis, T. D. Fay.
Proc. Soc. Photo-Opt. Instrum. Eng., Vol. 256, (see 012.031), p. 48 - 54 (1980). − Abstr. in Phys. Abstr., Vol. 84, Abstr. 98891 (1981).

133.010 A search for the infrared counterpart of type II OH masers − I. A model for the IR background source confusion.
T. J. Jones, M. Ashley, A. R. Hyland, A. Ruelas-Mayorga.
Mon. Not. R. Astron. Soc., Vol. 197, 413 - 428 (1981).

A simple exponential disc model for the distribution of stars and extinction (dust) in the galaxy is developed for the purpose of predicting the infrared stellar luminosity function along an arbitrary line of sight at any given wavelength and apparent magnitude limit. The model luminosity function is used to determine the extent to which the stellar field will contaminate searches for the infrared counterpart of type II OH masers.

133.011 High-velocity gas in the Orion BN/KL region: observations of the carbon monoxide (2−1) and sulfur dioxide ($13_{1,13}−12_{0,12}$) lines.
G. R. Knapp, T. G. Phillips, P. J. Huggins, R. O. Redman.
Astrophys. J., Vol. 250, 175 - 185 (1981).

The high-velocity gas in the region of the BN/KL cluster in Orion has been observed at high spatial resolution (25″) in the 230 GHz CO and 251 GHz SO$_2$ lines. Carbon monoxide emission is detected to about 100 km s^{-1}. The source radius is ~25", close to that of the inner edge of the H_2 line emission region. The SO_2 emission region is smaller in both velocity and angular extent than is the CO emission. The evidence suggests a roughly spherical outflow with velocity $V \propto r$ and a rapid density fall off: the authors' observations give $n \approx r^{-3--4}$. The CO(2−1) line is optically thin at high velocities. The $^{12}C/^{13}C$ isotope ratio is found to be 75 ± 20. The mass of high-velocity gas is 5 $M_\odot$ and the energy ~2.5×10^{46} ergs. These values suggest that the source of out-flowing gas is the wind of a massive O star. The outflow is characterized by velocities ~100 km s^{-1} and mass loss rates ~$10^{-3}-10^{-2} M_\odot$yr^{-1}, similar to those found in η Carinae.

133.012 The location of the hot molecular core in Orion.
B. Zuckerman, M. Morris, P. Palmer.
Astrophys. J., Lett., Vol. 250, L39 - L42 (1981).

The authors mapped the hot molecular core in the Kleinmann-Low (KL) infrared nebula in Orion in the (7, 6) ammonia inversion transition. The source is small ($\lesssim$ 20"), and the position of its centroid agrees best with the compact infrared source IRc4 and the strong H_2O maser "Source A." Profiles of the (4, 3) and (6, 5) inversion transitions with high signal-to-noise ratios were also obtained. A comparison of the temperatures of the hot molecular core and IRc4 suggests that the latter is a protostar that is very heavily reddened by obscuration local to itself.

133.013 High resolution spectrophotometry of the O I line (8446 Å) towards Cyg OB2 No. 12.
K. V. K. Iyengar, F. Strafella, C. B. Cosmovici.
Astron. Astrophys., Vol. 103, 382 - 385 (1981).

By means of a Fabry-Perot Tilting Filter Spectrophotometer, high resolution observations ($\lambda/\Delta\lambda \cong 10^4$) of the $(3s^3 S^0 - 3p^3 P)$ O I line at 8446 Å were carried out in the direction of Cyg OB2 No. 12, with the 182 cm Asiago telescope. The line is of stellar origin and appears in absorption with an equivalent width of (40 ± 12) mÅ. This value is at least 25 times lower than expected from an absorption line correlated to its stellar spectral type and luminosity. The authors interpret the observed small equivalent width of the 8446 Å line in Cyg OB2 No. 12 as due to the contribution from emission arising as a result of the resonant excitation of the O I 3 $^3D^0$ state by Ly-β photons.

133.014 Identification of objects discovered in the AFGL four-color infrared sky survey.
W. P. Bidelman.
Publ. Astron. Soc. Pacific, Vol. 93, 545 (1981). – Abstract.

133.015 Velocity resolved spectroscopy of the Brackett gamma line emission of CRL 490 and M17 IRS 1.
M. Simon, G. Righini-Cohen, J. Fischer, L. Cassar.
Astrophys. J., Vol. 251, 552 - 556 (1981).

High spectral resolution observations of the BN-like objects CRL 490 and M17 IRS 1 show that the Brγ line of CRL 490 is ~150 km s^{-1} wide at half-power and is blueshifted by ~30 km s^{-1} with respect to the CO velocity of the surrounding molecular cloud. The Brγ line of M17 IRS 1 is ~135 km s^{-1} wide at half-power and is blueshifted by ~24 km s^{-1} with respect to its molecular cloud. Line emission of both objects extends to a full width of at least 300 km s^{-1} and probably arises in gas flow associated with mass loss. The mass loss rates inferred exceed $10^{-7} M_\odot$ yr^{-1}.

133.016 Les sources infrarouges compactes. P. Léna.
Rôle des nuages moléculaires dans la formation des étoiles, (see 012.069), 20 pp.

A compilation of 50 compact infrared sources is presented and their observed properties are evaluated. Most of the objects are either suspected to be "protostars" or are thought to be extremely young early-type stars. However, some highly evolved objects are also included. The author stresses the importance of spatial interferometry in elucidating the nature of these sources.

Infrared Astronomical Data Base and Catalog of Infrared Observations.   See Abstr. 002.037.

A search for S stars in the IRC and AFGL Catalogues. See Abstr. 002.054.

One-dimensional infrared speckle interferometry. See Abstr. 031.594.

Automated star counts in the dark cloud L1454. See Abstr. 031.643.

Infrared atomic hydrogen line formation in luminous stars.   See Abstr. 064.056.

Detection of the $J = 1 \to 0$ and $J = 2 \to 1$ rotational lines of SiS in the molecular envelope of IRC+10216. See Abstr. 112.008.

Infrared excess and mass-loss rate of the extreme Of star HD 108.   See Abstr. 112.041.

4−8 micron spectrophotometry of OH 0739−14. See Abstr. 112.043.

Far-infrared observations of star-forming regions. See Abstr. 131.130.

Observations of CO $J = 3 \to 2$ emission from molecular clouds.   See Abstr. 131.146.

The energetics of molecular clouds. IV. The S88 molecular cloud.   See Abstr. 131.149.

New H_2O masers associated with far-infrared sources.   See Abstr. 131.163.

High velocity H_2 line emission in the NGC 2071 region.   See Abstr. 131.182.

The variable infrared emission of cosmic dust sources. See Abstr. 131.207.

High-resolution radio observations of G82.6 + 0.4. See Abstr. 132.045.

Far-infrared [O III] and [N III] line emission from galactic H II regions and planetary nebulae. See Abstr. 132.052.

Spektralphotometrie der galaktischen H II Region S 106 im Wellenlängenbereich 8 - 13 μm. See Abstr. 132.061.

The Red Rectangle: its polarization and structure. See Abstr. 134.002.

The distance to G316.8−0.1. See Abstr. 141.087.

Compact radio source 1413 + 135 is a far-IR extragalactic object.   See Abstr. 141.088.

Further observations of 3C 273 for the 3.3-μm dust feature.   See Abstr. 141.128.

IR observations of a new X-ray globular cluster and galactic bulge X-ray sources.   See Abstr. 142.027.

Mass models of the galactic bulge derived from the distribution of OH/IR stars. See Abstr. 155.042.

Distribution of near infrared sources in the galactic disk. See Abstr. 156.001.

Distribution of infrared sources in the galactic plane. See Abstr. 156.005.

134 Emission Nebulae, Reflection Nebulae

134.001 **The Toby Jug nebula (IC 2220): a bipolar and biconical nebula.**
H. G. Perkins, D. J. King, S. M. Scarrott.
Mon. Not. R. Astron. Soc., Vol. 196, 403 - 408 (1981).
An optical linear polarization map of IC 2220, the nebula surrounding the cool red giant HD 65750, is presented. The nebula appears to be bipolar and biconical in structure. The mass of the nebula is estimated to be 0.01 $M_\odot$ and is consistent with the nebula being formed from the current mass loss stage of the central star.

134.002 **The Red Rectangle: its polarization and structure.**
H. G. Perkins, S. M. Scarrott, P. Murdin, R. G. Bingham.
Mon. Not. R. Astron. Soc., Vol. 196, 635 - 639 (1981).
Two colour polarization maps of the Red Rectangle are presented which show that the object is a reflection nebula with the dust concentrated on a hollow biconical surface centred on HD 44179. The polarizations in the two wavebands (red and blue) imply that the broad red feature centred near Hα in the spectrum of the nebulosity originates in the nebular medium itself and is not reflected radiation from the central object.

134.003 **Lunar occultation of the CO emission from the biconical nebula LkHα 208.**
J. Good, N. Scoville, F. P. Schloerb, J. Bally.
Astron. J., Vol. 86, 892 - 896 (1981).
The biconical reflection nebula LkHα 208 was observed in the 2.6-mm CO line during the lunar occultation of 19 April 1980. The object is situated near a dark cloud which has an abrupt edge along the line of sight to the reflection nebula and tapers off gradually on the opposite side. The data show possible local heating of the cloud edge by LkHα 208 with an enhancement of 16 K above the ambient cloud temperature of 11 K. No definite evidence is seen for a compact source in LkHα 208.

134.004 **Peculiar optical spectrum of the Red Rectangle.**
R. F. Warren-Smith, S. M. Scarrott, P. Murdin.
Nature, Vol. 292, 317 - 319 (1981).
In high resolution spectra of the nebula the authors have identified a narrow-line component (Na D, Ca H and K, Hα) from a low excitation plasma and they have resolved broader features presumably from molecular bands. They give constraints on the molecules responsible, and propose as a candidate carbyne. The Red Rectangle may thus be a factory for the production of the raw material from which some interstellar molecules are made.

134.005 **On the variable cometary nebula GM 1 - 29 in Cepheus.** T. Yu. Magakyan.
Pis'ma Astron. Zh., Tom 7, 398 - 401 (1981). In Russian.
English translation in Soviet Astron. Lett., Vol. 7.
Changes of shape of the variable cometary nebula GM 1 - 29 in 1950 - 1980 are described. Three brightness in-

creases in 1950 - 1956, 1968 - 1971 and 1975 - 1980 are traced. The present-day brightness of the object is decreasing.

134.006 **Observations of the magnitude and color variation of the "bN? " object in Perseus.**
B.-a. Yao, J.-s. Yin, Y.-l. Sun.
Acta Astron. Sinica, Vol. 22, 174 - 179 (1981). In Chinese.

134.007 **Studies of bipolar nebulae. VII. The exciting star of OH0739-14 (= OH231.8 + 4.2).** M. Cohen.
Publ. Astron. Soc. Pacific, Vol. 93, 288 - 290 (1981).
The red spectrum of the bipolar reflection nebula associated with OH0739-14 implies a spectral type of M9 III for the underlying star, making it the coolest star known to occur in a bipolar system.

134.008 **Detection of new filamentary nebulosity near the giant non-thermal radio loop II.** A. M. I. Osman.
Astrophys. Space Sci., Vol. 78, 169 - 174 (1981).
IIIaJ survey plates from the 1.2 m SRC Schmidt have been high contrast copied in search for nebulosity at high galactic latitudes in the vicinity of the giant radio loop II. Some filamentary nebulosity, possibly reflection, has been certainly detected.

134.009 **Surface mapping of selected regions in the Orion Nebula.** P. M. Perry, B. E. Turnrose, C. A. Harvel, R. W. Thompson, A. D. Mallama.
The Universe at ultraviolet wavelengths, (see 012.009), p. 601 - 610 (1981).
Low-dispersion, large-aperture, ultraviolet spectra ($\lambda\lambda$ 1135–3255 Å) of selected regions in the Orion Nebula were obtained with the International Ultraviolet Explorer (IUE) scientific instrument. Spectra obtained at 35 contiguous locations defining a mosaic within the nebula were used to generate monochromatic images of high spatial resolution at the wavelengths of the ultraviolet emission lines C III] λ 1909, C II λ 2326, and [O II] λ 2470. Image-processing techniques were utilized to generate and analyze these ultraviolet surface maps.

134.010 **Preliminary report on IUE spectra of the Crab Nebula.** K. Davidson, T. R. Gull, S. P. Maran, T. P. Stecher, M. Kafatos, V. L. Trimble.
The Universe at ultraviolet wavelengths, (see 012.009), p. 693 - 700 (1981).
The Crab Nebula is marginally observable with the IUE. Observations of the optically brightest filamentary regions, made with IUE in August 1979, show the C IV λ1549, He II λ1640, and C III] λ1909 emission lines. The intensities of these lines have been compared with visual-wavelength data. It appears that carbon is not overabundant in the Crab; carbon/oxygen is approximately "normal" and oxygen is slightly scarcer than "normal" as a fraction of the total mass.

134.011 **The far-UV flux distribution of the Orion Nebula.**
H. M. Heckathorn, G. R. Carruthers.

Bull. American Astron. Soc., Vol. 13, 508 - 509 (1981).
Abstract.

**134.012 New Einstein observations of the Eta Carinae
Nebula.**   T. Chlebowski, F. D. Seward.
Bull. American Astron. Soc., Vol. 13, 512 (1981). — Abstract.

**134.013 Helium abundances and mass in the Crab Nebula
filaments.**   R. Henry, G. M. MacAlpine.
Bull. American Astron. Soc., Vol. 13, 512 (1981). — Abstract.

**134.014 A spectroscopic study of the high excitation nebula
NGC 6302.**   L. H. Aller, J. E. Ross, B. J. O'Mara,
C. D. Keyes.
Mon. Not. R. Astron. Soc., Vol. 197, 95 - 106 (1981).

The wonderfully intricate irregular gaseous nebula
NGC 6302 shows a great range of excitation from [S I] to
[Fe VII]. Data obtained with the International Ultraviolet
Explorer (IUE) covering the spectral range $\lambda 1200-3203$ and
with the image-photon counting system at the Anglo-Australian
telescope emphasizing the region $\lambda 3200-4880$ Å are used for
a new determination of the chemical composition of this object.

134.015 Ultraviolet extinction in the Orion nebula.
P. Patriarchi, M. Perinotto.
Second European IUE Conference, (see 012.011), p. 201 - 203
(1980).

The authors have obtained extinction curves in the UV
range in the direction of θ^1 Ori C in the central brightest area
of the nebula and toward θ^2 Ori A which is located 2.3 arc
minutes south-east from the Trapezium. The two extinction
curves show significant differences amounting to about 2 magnitudes, the one toward θ^1 Ori C running below and the one
toward θ^2 Ori A above the mean Orion curve of Bless and
Savage (1972).

**134.016 Globules in the Orion nebula. II. Polarization
observations.**   S. Isobe, K. Okida.
Publ. Astron. Soc. Japan, Vol. 33, 313 - 325 (1981).

Polarization observations were performed for the Orion
nebula within an angular distance 1′ from the exciting star
HD 37022 at five wavelengths in continuum light and at four
wavelengths of emission lines. Position angles in magnetic
vector of these polarizations are distributed in the radial
direction at all five continuum wavelengths, and it shows
that these continuum wavelenths are due to light scattered by
dust grains.

**134.017 Physical conditions and abundances of some ions in
diffuse nebulae.**
I. Yankulova, G. Petrov, V. Golev.
Astrofiz. issled., NRB, Vol. 3, 89 - 94 (1981). In Russian.
Abstr. in Ref. zh., 51. Astron., 8.51.731 (1981).

**134.018 Near infrared high resolution spectrophotometry
of forbidden [C I] in the Orion Nebula.**
C. B. Cosmovici, F. Strafella, T. Iijima.
Astron. Astrophys., Vol. 101, 397 - 400 (1981).

High resolution ($\lambda/\Delta\lambda \sim 10^4$) near infrared observations
of the [C I] $- {}^1S_0 - {}^1D_2$ transition (8727 Å) were carried out
at 18 positions across the Orion Nebula (M 42) by using a
tilting filter Fabry-Perot photometer. The absolute fluxes
derived here are discussed on the basis of possible line emission in the boundary zones between H I and H II regions and
are compared with other experimental results.

**134.019 Laboratory produced visible spectral emission
features correlate with those of the Red Rectangle.**
T. J. Wdowiak.
Nature, Vol. 293, 724 - 725 (1981).

The author reports that matrix isolation experiments
directed towards understanding the origin of the diffuse inter-

stellar bands also show spectral emission features having
characteristics similar to those found in the visible spectrum
of the Red Rectangle nebulosity. These experiments suggest
that the observed Red Rectangle emissions have their origins
in fluorescence phenomena of species that are matrix-isolated
in grain mantles.

**134.020 The ultraviolet properties of dust in the Orion
Nebula.**   J. S. Mathis, M. Perinotto,
P. Patriarchi, F. H. Schiffer III.
Astrophys. J., Vol. 249, 99 - 108 (1981).

The continuum in the Orion Nebula was observed with
the *IUE* at 16 positions varying from 30″ to 5′ in angular
separation from the central star θ^1 Orionis C. The atomic
continuum was estimated from the Hβ brightnesses. The
remaining scattered starlight was analyzed by multiple-
scattering models in which the main parameters are: (1) the
albedo ϖ_λ of the dust; (2) the scattering phase function
parameter g_λ, the averaged cosine of the angle of scattering;
and (3) the density of dust grains at small distances from the
star.

**134.021 Models for the structure and origin of bipolar
nebulae.**   M. Morris.
Astrophys. J., Vol. 249, 572 - 585 (1981).

The known characteristics of bipolar nebulae are summarized. It is argued that observations to date of bipolar
nebulae are consistent with a dust and gas distribution that
decreases monotonically with latitude above the equatorial
plane. The details of the proposed mass ejection mechanism
are presented, including the results of numerical three-body
calculations which simulate that mechanism both with and
without stellar pulsation. The ultimate fate of a bipolar nebula
depends on, among other things, the mass ratio and the initial
separation. With appropriate initial conditions, bipolar nebulae
can be the logical precursors of common envelope binaries,
planetary nebulae, and cataclysmic binaries.

**134.022 Galactic ring nebulae associated with Wolf-Rayet
stars. II. M1-67: a nebula braked by the interstellar
medium.**   Y.-H. Chu, R. R. Treffers.
Astrophys. J., Vol. 249, 586 - 591 (1981).

Narrow band interference filter photographs and high
resolution (15 km s^{-1}) Fabry-Perot spectra of Hα are
presented for the nebula M1-67. These spectra show two components: a narrow component at $V_\odot$ = +185 km s^{-1} which is
concentrated near the central star, and a broader component
at +150 km s^{-1} which is more extended. The kinematics of
M1-67 is not consistent with that of a stellar wind-blown
bubble. It appears that this nebula consists of material ejected
from the central star 209 BAC at two distinct epochs, about
10^4 to 10^5 yr ago. The ejected material has been subsequently
slowed down and displaced by the interstellar medium.

134.023 On the ice content of KL nebula in Orion.
R. Papoular.
Astron. Astrophys., Vol. 104, L1 - L3 (1981).

During the Osiris II campaign in 1978, OMC 1 was observed in the infrared, with an airborne telescope in an attempt to detect the 45 μm band of water-ice. The obtained
data were included in a simple emission model for the
Kleinmann-Low nebula.

134.024 Polarimetric observations of S 106.
M. G. Lacasse, D. Boyle, R. Levreault, J. L. Pipher,
S. Sharpless.
Astron. Astrophys., Vol. 104, 57 - 64 (1981).

Optical and near infrared polarimetric observations of the
bipolar nebula S 106 have revealed the presence of significant
reflection nebulosity associated with this source at these wavelengths. The radial pattern of the polarization in all of the authors' data is consistent with the hypothesis that S 106 #3 is

the source of illumination for the nebula. The authors' analysis also reveals that the optical depth at 2.2 μm to the star S 106 # 3 is probably approximately 2.5 which, in conjunction with other data, implies that the distance to S 106 is between 200 pc and 750 pc.

134.025 Ultraviolet observations of the X-ray sources in the Eta Carinae nebula. R. Viotti, A. Giangrande, A. Cassatella, F. Macchetto.
Space Sci. Rev., Vol. 30, (see 012.044), 235 - 239 (1981).

η Car is surrounded by a dense nebula ejected during the last 150 years. The Einstein satellite recently detected intense X-ray emission from η Car and its nebula indicating the existence of efficient gas heating processes. Ultraviolet observations with IUE confirmed the presence of hot gas in the condensation S of the nebula and in η Car. Possible interpretations of the results are discussed.

134.026 Abundances of some ions in diffuse nebulae. I. Galactic diffuse nebulae.
G. T. Petrov, V. K. Golev, I. M. Yankulova.
Astron. Tsirk., No. 1133, p. 1 - 2 (1980). In Russian.

134.027 Abundances of some ions in diffuse nebulae. II. Diffuse nebulae in nearby galaxies.
V. K. Golev, I. M. Yankulova, G. T. Petrov.
Astron. Tsirk., No. 1133, p. 2 - 4 (1980). In Russian.

134.028 On luminosity sources of the reflection nebula IC 2118. D. A. Rozhkovskij.
Astron. Tsirk., No. 1149, p. 4 - 6 (1981). In Russian.

134.029 Polarimetry of the faint nebula IC 2118.
Sh. N. Sabitov.
Astron. Tsirk., No. 1153, p. 6 - 8 (1981). In Russian.

Interference between ordinary and higher-order amplitudes in O II magnetic dipole transitions.
See Abstr. 022.084.

The Orion nebulae in color. See Abstr. 036.007.

A stellar wind model for bipolar nebulae.
See Abstr. 064.058.

Suggested radio observations of the forthcoming lunar occultations of the Crab nebula.
See Abstr. 096.008.

UBV and H$_\beta$ observations of stars towards M8.
See Abstr. 115.012.

Polarization of the radiation of stars, nebulae and galaxies. See Abstr. 116.005.

IUE observations and interpretation of the symbiotic star RW Hya. See Abstr. 117.020.

Halo around the Crab Nebula.
See Abstr. 125.057.

Interstellar grain size. II. Infrared photometry and polarization in Orion. See Abstr. 131.028.

Ultraviolet interstellar extinction toward stars in the Orion Nebula and toward HD 147889.
See Abstr. 131.119.

High-velocity gas in the Orion BN/KL region: observations of the carbon monoxide (2–1) and sulfur dioxide $(13_{1,13}-12_{0,12})$ lines. See Abstr. 133.011.

Recombination spectrum of the C III ion for nebulae. See Abstr. 135.062.

Hard X-ray structure of the Crab Nebula.
See Abstr. 142.100.

On the evidence for high energy γ-ray emission from the Orion Nebula stemming from COS-B observations.
See Abstr. 142.519.

135 Planetary Nebulae

135.001 **A planetary nebula with high oxygen abundance in the galactic bulge.**   C. M. Price.
Astrophys. J., Vol. 247, 540 - 544 (1981).

Extensive photoelectric spectra of H1-55, a planetary nebula in the galactic bulge, have been obtained. The temperature and density in the nebula are estimated. The logarithmic oxygen abundance is found to be 9.50, several times the normal abundances in planetary nebulae in the disk. The abuandances of other heavy elements are also enhanced. These high abundances are consistent with the nebula's exceptionally low electron temperature of less than 5600 K.

135.002 **Detection of six new extended planetary nebulae by means of interference filter photography.**
R. Weinberger, F. Sabbadin.
Astron. Astrophys., Vol. 100, 66 - 67 (1981).

Interference filter photographs in H_α + [N II] and 5007 Å ([O III]) enabled the authors to find six new very faint planetary nebulae, tentatively identified on the POSS. In two nebulae central stars were found; one might be a variable.

135.003 **Molecular hydrogen ion (H_2^+) absorption in planetary nebulae.**
W. A. Feibelman, A. Boggess, C. W. McCracken, R. W. Hobbs.
Astron. J., Vol. 86, 881 - 884 (1981).

Several IUE spectra of planetary nebulae show an absorption feature shortward of λ 1500 which is believed to be due to H_2^+. The nebulae are excited by stars of spectral type O3 - O7 or continuum, and all but one have double-shell structure.

135.004 **UV spectroscopy of planetary nebulae.**
M. Perinotto, P. Benvenuti.
Astron. Astrophys., Vol. 100, 241 - 248 (1981).

The planetary nebulae IC 2149, NGC 2440, 3242, and 6826 are investigated with spectra obtained with the IUE satellite in the spectral range λ 1150 - 3200 Å at a resolution of 7 Å. The physical conditions in the nebulae and the abundances of C, N, O, and Ne are derived. A faint continuum is observed in NGC 2440, and a very faint nebular continuum is seen in NGC 3242. The behaviour of the stellar continuum observed in IC 2149, NGC 3242, and NGC 6826 is briefly discussed.

135.005 **The planetary nebula NGC 7009.**
M. Perinotto, P. Benvenuti.
Astron. Astrophys., Vol. 101, 88 - 95 (1981).

The properties of the planetary nebula NGC 7009 have been investigated using spectroscopic observations in the UV obtained with the IUE satellite. A combination of optical and ultraviolet data yields the physical conditions prevailing in the relevant ionization zones. The UV data then provide good information on the chemical abundances of C, N, O, and Ne. A continuum emitted by the nebula is observed. It is due to atomic processes with no evidence of light scattered by dust particles. The spectrum of the central star from λ 1200 to λ 3100 Å is presented.

135.006 **A "variable" stellar object in a variable blue nebula V-V 1-7.**   N. Kameswara Rao, D. P. Gilra.
Observatory, Vol. 101, 108 - 110 (1981).

135.007 **The 5 GHz flux density of the planetary nebula K648 in M 15.**
M. Birkinshaw, A. J. B. Downes, G. G. Pooley.
Observatory, Vol. 101, 120 (1981).

135.008 **Dust in planetary nebulae.**
A. Natta, N. Panagia.
Astrophys. J., Vol. 248, 189 - 194 (1981).

Near-infrared and far-infrared data of 10 planetary nebulae are analyzed and compared with corresponding radio data. Evidence is found for the presence of dust grains mixed with the ionized gas. The dust-to-gas mass ratio, the grain size, and the grain number density are found to vary systematically with the nebular radius. These results can be explained either as due to progressive grain fragmentation and partial destruction with time or as reflecting a higher concentration of dust grains in the inner part of the planetary nebula envelope when it was ejected from the parent star. Possible evidence and tests which may support or disprove either possibility are briefly reviewed.

135.009 **Kinematics of planetary nebulae. II.**
A. Purgathofer, M. Perinotto.
Astron. Astrophys., Vol. 101, 247 - 249 (1981).

In a program of study of radial velocities of planetary nebulae, eighty-four spectra of eight planetary nebulae in the direction of the galactic anticenter have been obtained at the L. Figl Observatory near Vienna with an image tube spectrograph giving a reciprocal dispersion of 26 Å mm^{-1}. With this material the authors have studied the kinematical behaviour of the objects and shown that most of them deviate significantly from circular motion in the Galaxy.

135.010 **The optical and ultraviolet spectrum of the planetary nebula NGC 2440.**
G. A. Shields, L. H. Aller, C.D. Keyes, S. J. Czyzak.
Astrophys. J., Vol. 248, 569 - 583 (1981).

New measurements of the optical and ultraviolet emission-line intensities of the high-excitation planetary nebula NGC 2440 in the wavelength range 1240–8578 Å are analyzed with the aid of photoionization models. The observed [O III] and [N II] temperatures (13,800 K and 10,000 K, respectively) differ by more than the models predict. Inclusion of charge-transfer reactions at published rate improves agreement between calculated and observed abundances of many ions, but worsens the agreement for several highly ionized ions. Nitrogen shows a larger overabundance $N(N)/N(O) \approx 1.0$, whereas ultraviolet lines give a carbon abundance close to the solar value. The refractory elements magnesium, calcium, and iron have similar gas-phase depletions of ~1.5 dex. The abundances of oxygen, neon, sodium, sulfur, chlorine, potassium, and argon are roughly solar.

135.011 **Near-infrared spectroscopy of possible precursors to planetary nebulae: HM Sagittae.**
H. A. Thronson, Jr., P. M. Harvey.
Astrophys. J., Vol. 248, 584 - 590 (1981).

Two 1.5–2.3 μm spectra of the candidate protoplanetary nebula HM Sge are presented and discussed. A number of the Brackett series of hydrogen are seen in emission, allowing an estimate of source excitation and extinction. In addition, one spectrum shows the photospheric absorption features that are characteristic of a late-type giant star. The second spectrum shows no such features. Using previous models and observations, the infrared line-emitting region is found to be extremely small—in agreement with the idea that the source has recently begun a period of rapid mass loss or exchange. It is suggested that the recently observed brightening of HM Sge was due to partial unveiling of the source due to the motion of a dusty shell. The object is found to lie on the Harman-Seaton sequence, as appropriate for young evolving protoplanetaries.

135.012 **Near-infrared spectroscopy of possible precursors to planetary nebulae: AFGL 618.**

H. A. Thronson, Jr.
Astrophys. J., Vol. 248, 984 - 991 (1981).

A spectrum in the range $1.9-2.5\,\mu m$ of the suspected protoplanetary nebula AFGL 618 is presented. It shows at least 11 members of the H_2, $v = 1 \to 0$ and $v = 2 \to 1$ rotation-vibration branches. These lines appear to arise in a dusty region of shocked gas, and the line intensities are used to estimate shock velocity, preshock density, H_2 excitation temperature, and extinction to the emitting region. The observations are consistent with the suggested geometry of a double-lobed reflection nebula, banded about the center with a ring of dusty material from which the H_2 lines are emitted.

135.013 Planetary nebulae. I. S. Shklovskij.
Priroda, 1981, No. 7, p. 3 - 13. In Russian.

135.014 Electron densities for six planetary nebulae and HM Sge derived from the C III] λ 1907/1909 ratio.
W. A. Feibelman.
The Universe at ultraviolet wavelengths, (see 012.009), p. 613 - 621 (1981).

Electron densities for IC 418, NGC 6572, IC 1297, NGC 3242, NGC 6818, NGC 3211, and HM Sge derived from high-dispersion IUE C III] spectrograms are consistently higher than those derived from either surface brightness measurements of forbidden line intensity ratios in the visible. The nebulae were selected for a range of excitation classes from 3 to 9. Line splitting due to expansion velocities is observed for three objects. The great width of the λ1909 C III] line in HM Sge suggests large expansion velocities.

135.015 Silicon and magnesium in planetary nebulae.
J. P. Harrington, P. A. Marionni.
The Universe at ultraviolet wavelengths, (see 012.009), p. 623 - 631 (1981).

The IUE satellite spectra of some planetary nebulae show features due to silicon and magnesium: Si III] λλ 1883, 1892; Si IV λλ 1394, 1403; Mg II λλ 2796, 2804 and [Mg V] λλ 2784, 2929. In addition to previous observations of NGC 7662 and IC 418, the authors now have data for NGC 2440, Hu 1-2, IC 2003 and IC 2165. Silicon appears depleated by up to an order of magnitude relative to the sun. Large variations of magnesium abundance are found, which are likely to reflect differing degrees of depletion due to grain formation.

135.016 Elemental abundances in high-excitation planetary nebulae. P. A. Marionni, J. P. Harrington.
The Universe at ultraviolet wavelengths, (see 012.009), p. 633 - 639 (1981).

The IUE satellite has been used to obtain low dispersion spectra of the high excitation planetary nebulae IC 351, IC 2003, NGC 2022, IC 2165, NGC 2440, Hu 1-2, and IC 5217. Numerical modeling has been undertaken to determine the chemical composition of these objects with particular emphasis on obtaining elemental carbon and nitrogen abundances. Preliminary results for several nebulae suggest large variations in the C/N ratio from object to object.

135.017 The high-excitation planetary nebulae NGC 3918 and IC 2448.
S. Torres-Peimbert, M. Peña, E. Daltabuit.
The Universe at ultraviolet wavelengths, (see 012.009), p. 641 - 647 (1981).

The authors present IUE observations of NGC 3918 and IC 2448. Combining these observations with data in the optical range and computed model structures the authors derive the chemical composition for these objects.

135.018 Analysis of high excitation planetary nebulae.
L. H. Aller, C. D. Keyes.
The Universe at ultraviolet wavelengths, (see 012.009), p. 649 - 656 (1981).

Combination of extensive ground-based spectroscopic observation of high excitation planetary with IUE data permit determination not only of improved diagnostics but also better abundances for elements such as C and N that are well-represented in the ultraviolet spectra, and also C, Ar and metals Na, Ca and K whose lines appear in the λ 3200-8100 Å region. The authors summarize some of their principal results.

135.019 Discovery of the molecular hydrogen ion (H_2^+) in the planetary nebulae.
S. R. Heap, T. P. Stecher.
The Universe at ultraviolet wavelengths, (see 012.009), p. 657 - 661 (1981).

Low-dispersion spectra of fifteen planetaries and hot subdwarfs were obtained with the SWP camera on IUE and continuous flux distributions corrected for interstellar extinction were derived. Several planetaries, particularly the young planetaries of high surface-brightness, show anomalous flux distributions. The most anomalous case is NGC 6210. The authors suggest that these anomalies may be explained as absorption by H_2^+ in the nebula. For the case of NGC 6210, the authors derive a column density, $N(H_2^+) = 8 \times 10^{16}$ cm^{-2}.

135.020 An atlas of emission line fluxes of planetary nebulae in the 1150-3200 Å region.
A. Boggess, W. A. Feibelman, C. W. McCracken.
The Universe at ultraviolet wavelengths, (see 012.009), p. 663 - 669 (1981).

Emission line fluxes in units of 10^{-12} erg cm^{-2} sec^{-1} for 28 planetary nebulae are presented. The nebulae were chosen to cover a wide range of excitation classes, apparent diameters, location in the sky, and types of central stars. All objects were observed in the low-dispersion mode of the IUE spectrographs, using the large entrance aperature.

135.021 Chemical enrichment in halo planetary nebulae.
S. Torres-Peimbert, J. F. Rayo, M. Peimbert.
Bull. American Astron. Soc., Vol. 13, 512 (1981). – Abstract.

135.022 IUE observations of the planetary nebulae NGC 7662, NGC 7027, NGC 6572, NGC 5315 and BD + 30°3639. S. Torres-Peimbert, M. Peña.
Bull. American Astron. Soc., Vol. 13, 519 (1981). – Abstract.

135.023 Far infrared spectrophotometry of evolved objects.
H. Moseley, R. F. Silverberg.
Bull. American Astron. Soc., Vol. 13, 519 (1981). – Abstract.

135.024 The [S II] electron density distribution over the planetary nebula NGC 7009.
J. Meaburn, J. R. Walsh.
Astrophys. Space Sci., Vol. 78, 473 - 481 (1981).

Electron densities have been measured from [S II] 6716/6731 Å line ratios for a grid of points over the surface of the planetary nebula NGC 7009 using a photon counting detector. The radial dependence of the electron density has been modelled, and the relationship provides possible evidence that the planetary nebula shell is driven by a strong stellar wind.

135.025 On nitrogen abundances of planetary nebulae.
B. J. Wilkes, G. J. Ferland, D. Hanes, J. W. Truran.
Mon. Not. R. Astron. Soc., Vol. 197, 1 - 6 (1981).

Observations of N II recombination lines in the planetary nebula NGC 3242 are presented which allow the first temperature independent measurement of the nitrogen abundance of a nebula. Model nebula calculations are utilized to identify a reliable N/H indicator.

135.026 Ultraviolet spectra of planetary nebulae – V. The C II λ 1335 dielectronic recombination lines in IC 418. J. Clavel, D. R. Flower, M. J. Seaton.

Mon. Not. R. Astron. Soc., Vol. 197, 301 - 311 (1981).

The C II lines λ 1334.53 and λλ 1335.66, 1335.71 (blended) are observed in high-dispersion spectra of IC 418, and have profiles of P Cygni type. The calculated ratio of the flux in C II λ 1335 relative to that in C II λ 4267 is found to be about twice the observed ratio. This discrepany is probably due to some absorption of the C II λ 1335 resonance lines by dust internal to the nebula.

135.027 A spatial–kinematical model for the planetary nebula NGC 650–1.
F. Sabbadin, E. Hamzaoglu.
Mon. Not. R. Astron. Soc., Vol. 197, 363 - 368 (1981).

The velocity field of the emitting gas observed within the planetary nebula NGC 650–1 has been used to derive a three-dimensional model for this nebula. The results indicate that the brighter part has a ringlike shape (expanding at a maximum velocity of 42 (±2) km s^{-1} in [N II]), oriented almost edge-on to the line of sight and surrounded by two half-shells of less dense material expanding faster than the gas contained in the ring.

135.028 IUE observations of planetary nebulae.
H. Nussbaumer.
Second European IUE Conference, (see 012.011), p. XLIII - XLVIII (1980).

135.029 C, N and O in the planetary nebula NGC 2371.
S. R. Pottasch, D. P. Gilra, A. Natta, A. Preite-Martinez, P. R. Wesselius.
Second European IUE Conference, (see 012.011), p. 185 - 186 (1980).

Low resolution IUE spectra have been taken of NGC 2371, some centered on the central star, while other spectra include only the nebula. The resultant abundances are discussed.

135.030 IUE observations of planetary nebulae: nebular continuum and mass loss from central stars.
P. Benvenuti, M. Perinotto.
Second European IUE Conference, (see 012.011), p. 187 - 190 (1980).

IUE low resolution spectra of eight planetary nebulae from low to fairly high excitation have been taken under ESA project MP 28. The authors report here about two particular topics: the nebular continuum and the mass loss from central stars. A nebular continuum has been measured in NGC 7662 and shown to be essentially of atomic origin. A significant stellar wind from the nuclei of most of the planetaries has been detected with velocities around 1000 - 3000 km/s. A crude estimate of the associated mass loss has been made.

135.031 High dispersion EUV observations of planetary nebulae.
J. Köppen, R. Wehrse.
Second European IUE Conference, (see 012.011), p. 191 - 195 (1980).

The three high excitation planetary nebulae NGC 3242, NGC 7009 and NGC 6210 have been observed in the short wavelength range with high resolution. The most prominent emission lines are the He II 1640 and C III 1909 Å lines. The C IV 1550 lines of NGC 3242 and NGC 7009 are in emission, whereas in NGC 6210 they are in absorption. This may suggest that the central star of NGC 6210 is a subdwarf O.

135.032 A search for absorption in the fourth positive system of CO in the spectrum of the planetary nebula IC 418.
J. Clavel, D. Flower.
Second European IUE Conference, (see 012.011), p. 197 - 200 (1980).

The continuous emission of IC 418 has been observed at high dispersion with the IUE satellite and the spectrum searched for the $A^1\Pi - X^1\Sigma^+$ bands of CO. It was not possible to unequivocally detect these CO bands in IC 418. However, an upper limit to the absorption by CO was determined, giving a column density which is much less than the value implied by the $J = 1 \rightarrow 0$ rotational transition at 2.6 mm.

135.033 Formation of a planetary nebula by continuous mass loss.
A. Harpaz, A. Kovetz.
Effects of mass loss on stellar evolution, (see 012.015), p. 345 - 346 (1981).

The evolution of a 1.2 M$_\odot$ star along the asymptotic branch with continuous mass loss is presented, showing that this mass loss leads to the formation of a PN with a typical central star in its center.

135.034 Effects of mass loss on the formation of planetary nebulae.
S. Kwok.
Effects of mass loss on stellar evolution, (see 012.015), p. 347 - 351 (1981).

135.035 A giant halo around the planetary nebula NGC 3242.
H. E. Bond.
Publ. Astron. Soc. Pacific, Vol. 93, 429 - 430 (1981).

The planetary nebula NGC 3242 is surrounded by an extremely large (18 × 24 arc minutes) low–surface-brightness shell, making NGC 3242 the planetary nebula with the second largest known angular diameter. This "giant halo" appears to have been ejected from the central star approximately 10^5 years ago. The bright inner planetary nebula is itself a double-shell nebula, with diameters of 15″ and 38″ and ages of a few thousand years. The central star therefore appears to have undergone three separate phases of mass ejection.

135.036 The spectrum of planetary nebula K 1-27.
K. G. Henize, A. P. Fairall.
Publ. Astron. Soc. Pacific, Vol. 93, 435 - 436 (1981).

The spectrum of K 1-27 shows He II λ4686 stronger than either Hβ or [O III] λ4959. K 1-27 therefore appears to belong to the small group of old, very hot planetary nebulae which also includes NGC 246, NGC 4361, and Abell 36.

135.037 Planetary nebulae and their influence on the interstellar medium.
Y. Terzian.
The phases of the interstellar medium, (see 012.018), p. 131 - 132 (1981). – Abstract.

135.038 The nature of V - V 1 - 7 and its central star HD 62001.
N. Kameswara Rao, D. P. Gilra.
Bull. Astron. Soc. India, Vol. 9, 82 (1981). – Abstract.

135.039 Kinematics of planetary nebulae, and the galactic rotation curve.
S. E. Schneider, Y. Terzian.
News Lett. Astron. Soc. N. Y., Vol. 1, No. 10, p. 25 (1981). Abstract.

135.040 Spectral variations and evidence for edge and/or line locking mechanism(s) in the low-excitation planetary nebula HD 138403.
A. Surdej, J. Surdej, J. P. Swings.
ESO Sci. Prepr., No. 168, 36 pp. (1981). – Submitted to Astron. Astrophys.

135.041 The ultraviolet spectrum of the planetary nebula NGC 2371 and its exciting star.
S. R. Pottasch, R. Gathier, D. P. Gilra, P. R. Wesselius.
Astron. Astrophys., Vol. 102, 237 - 244 (1981).

Low resolution IUE spectra have been taken of the planetary nebula NGC 2371. Some were centered on the exciting star while other spectra measure only nebula emission. The authors discuss the nebula line and continuum spectrum in terms of the physical state of the nebula gas and its abundance. The stellar spectrum yields information on the atmospheric temperature and mass loss rate.

135.042 **Planetary nebulae and stellar evolution.**
J. B. Kaler.
Mercury, Vol. 10, 114 - 120 (1981).

135.043 **The structure of a typical double-ring planetary nebula NGC 2392.**   R. Louise.
Astrophys. Space Sci., Vol. 79, 229 - 237 (1981).

Monochromatic photographs in Hα, [N II] λ 6584 and [O III] λ 5007 Å show many different details in the morphological structure of the Eskimo Nebula (NGC 2392). A photometric study gives various geometrical parameters. The classical stratification structure in planetary nebulae is not clearly observed in NGC 2392. This fact is probably a consequence of the peculiar structure in the geometry of the nebula. A model consisting of an inner toroid surrounded by a spherical shell is proposed to account for both photometric and spectroscopic observations.

135.044 **Carbon in planetary nebulae.**   J. B. Kaler.
Astrophys. J., Vol. 249, 201 - 209 (1981).

Carbon abundances, and in particular C/O ratios, are calculated for 53 planetary nebulae from the strength of the optical λ4267 C II line. Where necessary, the intensity of λ4267 is corrected for systematic error for each nebula by comparing it to nearly observed λ4541 He II, λ4363 [O III], and λ4471 He I lines whose true intensities are known from measurements of λ4686 He II, λ5007 [O III], or from accurate photometry of λ4471.

135.045 **The abundances of neon, sulfur, and argon in planetary nebulae.**   S. C. Beck, J. H. Lacy, C. H. Townes, L. H. Aller, T. R. Geballe, F. Baas.
Astrophys. J., Vol. 249, 592 - 601 (1981).

Infrared observations of [Ne II], [S IV], and [Ar III] are used with optical observations to discuss the abundances of Ne, S, and Ar in 18 planetary nebulae. In addition, infrared observations of 18 other nebulae are presented. The derived abundances of S and Ar are each slightly enhanced relative to previous studies.

135.046 **Two contrasting Abell planetary nebulae.**
J. B. Kaler, W. I. Hartkopf.
Astrophys. J., Vol. 249, 602 - 606 (1981).

The authors present line fluxes, radial velocities, and analyses of two large Abell-type planetaries that show marked contrasts. Abell 43 is a high-excitation galactic-disk, or Population I, nebula that is optically thin in the hydrogen Lyman continuum. Abell 50 is a medium-excitation, optically thick, Population II (high velocity) object. Although the two nebulae are of similar size (radius > 0.2 pc), which in a simple theory might imply a similar evolutionary state, the central star of A43 appears to be much more luminous than that of A50. The set of large nebulae does not constitute a homogeneous evolutionary class in which the stars can be linked by a monotonic evolutionary track. The comparisons between these two objects illustrate and perhaps epitomize very large differences among large planetaries which probably involve differences in population type, and in stellar and nebular masses.

135.047 **An analysis of the planetary nebula NGC 2867.**
L. H. Aller, C. D. Keyes, J. E. Ross, B. J. O'Mara.
Mon. Not. R. Astron. Soc., Vol. 197, 647 - 658 (1981).

NGC 2867 is a compact planetary nebula of moderately high surface brightness. Spectroscopic observations secured with the Anglo-Australian Telescope are combined with data obtained with the International Ultraviolet Explorer to provide material for a theoretical analysis. Except for [S II], line intensities in the optical region can be represented reasonably well by a simple theoretical model. The derived chemical composition is compared with the Sun and other planetaries. Some possible nucleosynthesis scenarios in progenitor stars are briefly described.

135.048 **A search for nebulosities associated with planetary nebulae.**
R. Louise, F. Michel, J. C. Mevolhon.
Astron. Astrophys., Vol. 102, 303 - 306 (1981).

In order to detect faint nebulosities around planetary nebulae, long exposure photographs are made with a large bandwidth (Δλ=50 Å) interference filter centered on Hα. The authors found only one nebula presenting "secondary structures" not previously observed on six selected objects. It is the low excitation nebula NGC 40 associated with a very active Wolf-Rayet nucleus.

135.049 **Radio observations of compact planetary nebulae.**
S. Kwok, C. R. Purton, D. W. Keenan.
Astrophys. J., Vol. 250, 232 - 239 (1981).

An attempt has been made to identify very young planetary nebulae from radio observations. Details of the observations at the Algonquin Radio Observatory and the Very Large Array are presented. Four nebulae which were previously designated as stellar are resolved using aperture synthesis techniques with angular resolution as high as 0.″2. Two of the observed nebulae, SwSt-1 and M 3-38, exhibit characteristics expected of very young planetary nebulae. These two nebulae are compared to other planetary nebulae which are believed to be young, and the possibility that young planetary nebulae may have different spectral behavior in the radio is discussed.

135.050 **The binary central star of NGC 2346 and the extinction puzzle.**
R. H. Méndez, V. S. Niemelae.
Astrophys. J., Vol. 250, 240 - 247 (1981).

New observations of the A-type central star of the planetary nebula NGC 2346 have confirmed that it is a single-lined spectroscopic binary with a period of 16 days. The γ-velocity of the system is shown to coincide with the nebular velocity; therefore, the authors have good reason to assume that the binary and the nebula are associated. On the other hand, there is observational evidence that the nebula is more reddened than the A-type star. An explanation based on the bipolar structure of the nebula is discussed.

135.051 **Large high-excitation planetary nebulae.**
J. B. Kaler.
Astrophys. J., Lett., Vol. 250, L31 - L34, plate L2 (1981).

Absolute total and relative emission-line fluxes are given for: (1) the two planetary nebulae, Abell 30 and 78, that exhibit zones of nearly pure helium near their cores, and (2) a list of other large, high-excitation nebulae that are outwardly similar to A30 and A78, and which are prime candidates for further investigation. All the nebulae are optically thin in the hydrogen Lyman continuum and have luminous central stars. They are morphologically similar, most being spherical shells.

135.052 **Optical positions for northern stellar planetary nebulae.**   S. R. Blackwell, C. R. Purton.
Astron. Astrophys., Suppl. Ser., Vol. 46, 181 - 183 (1981).

Accurate optical positions are presented for 84 stellar (unresolved) planetary nebulae north of declination −19°. Substantial offsets have been measured from many previously published positions.

135.053 **On the pseudo-resonance absorption lines in planetary nebulae.**   G. A. Gurzadyan.
Astrophys. Space Sci., Vol. 80, 189 - 195 (1981).

The possibility of the excitation of a new class of spectral lines, so called pseudo-resonance absorption lines in the conditions of planetary nebulae, is predicted. These lines are formed by permitted atomic transitions from the metastable levels to the upper subordinate, but not metastable, levels. To observe pseudo-resonance lines of 'nebular' origin is possible only in the form of absorption lines. A preliminary list of pseudo-resonance lines is given.

135.054   Line widths in peculiar emission line objects.
J. P. Swings, Y. Andrillat.
Astron. Astrophys., Vol. 103, L3 - L4 (1981).

Measurements of the line widths of Hα, HeI and [N II] in a series of peculiar emission line objects tend to support the interacting winds theory for the formation of planetary nebulae.

135.055   Stellar nucleogenesis and planetary nebulae compositions.   L. H. Aller, S. J. Czyzak.
Publ. Astron. Soc. Pacific, Vol. 93, 544 (1981). – Abstract.

135.056   *IUE* observations of four planetary nebulae.
H. M. Johnson.
Astrophys. J., Vol. 250, 590 - 595 (1981).

Low dispersion spectra of NGC 6905 and NGC 5189 are reproduced. These are nebulae with "O VI sequence" nuclei. Line features are identified and classified according to site in the nucleus or the nebula; their flux densities are measured above continuum and presented before and after correction for interstellar extinction. The behavior of the line spectra, including line width, $\Delta\lambda/\lambda$, is summarized. The extinction parameter c is further discussed in connection with establishing the characteristics of the continua.

135.057   The optical and ultraviolet spectra of the high excitation planetary nebula, CD −23°12238=Me 2−1.
L. H. Aller, C. D. Keyes, S. J. Czyzak.
Astrophys. J., Vol. 250, 596 - 604 (1981).

The bright, high-excitation planetary CD −23°12238, Me 2−1 or (342+27°1) has been studied in both optical and ultraviolet spectral regions with an image dissector scanner at the Shane 3 m telescope and with the *IUE*. Analysis is carried out with the aid of a theoretical model which is used as an interpolation device to obtain the nebular chemical composition. Carbon and nitrogen appear to be enhanced with respect to solar values, but probably not much, if any, compared with an average galactic planetary.

135.058   Miras and planetary nebula formation.
Y. Tuchman.
Proceedings of the 5th Göttingen-Jerusalem-Symposium on Astrophysics, (see 012.041), p. 173 - 180 (1981).

135.059   The effect of nucleus duplicity of a planetary nebula on its emission line spectrum.
V. V. Golovatyj, O. S. Yatsyk.
Vestn. L'vov. univ. Ser. astron., 1981, No. 56, p. 25 - 27. In Russian. – Abstr. in Ref. zh., 51. Astron., 11.51.751 (1981).

135.060   A possible duplicity of the nucleus of the planetary nebula NGC 6826.   R. I. Noskova.
Astron. Tsirk., No. 1128, p. 1 - 3 (1980). In Russian.

135.061   Light curve of the nucleus of the planetary nebula Abell 46.   N. E. Kurochkin.
Astron. Tsirk., No. 1143, p. 3 - 6 (1980). In Russian.

135.062   Recombination spectrum of the C III ion for nebulae.   T. Kh. Feklistova.
Use of model atmospheres for the interpretation of stellar spectra, (see 012.054), p. 98 - 100 (1980). In Russian.

135.063   Study of physical conditions in gaseous nebulae. III. Determination of temperatures and radii of central stars of planetary nebulae.
V. V. Golovatyj, O. S. Yatsyk.
Tsirk. Astron. Obs., L'vov, No. 54, p. 3 - 10 (1979). In Russian.

135.064   Study of physical conditions in gaseous nebulae. IV. On the determination of relative helium abundance in planetary nebulae.   O. S. Yatsyk.
Tsirk. Astron. Obs., L'vov, No. 54, p. 11 - 14 (1979). In Russian.

135.065   8 - 13 μm spectrophotometry of compact planetary nebulae and emission line objects.
D. K. Aitken, P. F. Roche.
Anglo-Australian Obs. Prepr. No. 154, 26 pp. (1981). – Submitted to Mon. Not. R. Astron. Soc.

The ESO/Uppsala survey of the ESO (B) Atlas of the southern sky. IX.   See Abstr. 002.055.

The emission spectrum of H_2 from associative detachment and ultraviolet pumping.   See Abstr. 022.092.

Red giants and their descendants – new light on old stars.   See Abstr. 065.021.

Miras, mass loss, and the origin of planetary nebulae.   See Abstr. 065.033.

Late stages of stellar evolution: central stars of planetary nebulae.   See Abstr. 065.067.

Mass loss from central stars of planetary nebulae.   See Abstr. 112.026.

The radial velocity variations in IC 418.   See Abstr. 112.033.

An expanding motion in the ionized envelope of HBV 475.   See Abstr. 112.048.

Further observations of a possible star-nebula system.   See Abstr. 113.021.

High-speed photometry of, and speculations on, the central star of M 1 - 2.   See Abstr. 113.044.

A model for V 1016 Cyg based on the ultraviolet spectrum.   See Abstr. 114.006.

A far-infrared emission feature in carbon-rich stars and planetary nebulae.   See Abstr. 114.014.

A spectral description and non-LTE analysis of 6 central stars of planetary nebulae.   See Abstr. 114.093.

The "continuous" central stars of planetary nebulae – are their spectra really continuous?   See Abstr. 114.174.

FG Sagittae: combination of the existing photometric and spectroscopic data.   See Abstr. 122.108.

Anomalous far-UV extinction in the WN6 star HD 147419.   See Abstr. 131.071.

New insights into the physical state of gaseous nebulae.   See Abstr. 132.015.

Far-infrared [O III] and [N III] line emission from galactic H II regions and planetary nebulae.   See Abstr. 132.052.

Discovery of radio brightening in AFGL 618.   See Abstr. 133.002.

Galactic ring nebulae associated with Wolf-Rayet stars. II. M1-67: a nebula braked by the interstellar medium. See Abstr. 134.022.

New probable planetary nebulae and VV Cephei stars in the Small Magellanic Cloud.   See Abstr. 159.008.

Planetary nebula progenitors in Magellanic Cloud clusters.   See Abstr. 159.012.

Radio Sources, X-ray Sources, Cosmic Radiation

141 Radio Sources, Quasars, Pulsars

Radio Sources, Quasars

141.001 Low frequency variable sources 5 year monitoring program at 408 MHz.
C. Fanti, R. Fanti, A. Ficarra, F. Mantovani, L. Padrielli, K. W. Weiler.
Astron. Astrophys., Suppl. Ser., Vol. 45, 61 - 78 (1981).

Results of a 5 year monitoring program with the East-West arm of the Northern Cross (Bologna) at 408 MHz are given. They concern 114 extragalactic radiosources measured approximately once a month. The authors find 44 radiosources variable or possibly at low frequencies. Among these sources, 8 belong to a homogeneous sample of 32 small angular diameter radiosources, selected on the basis of scintillation studies, and 23 to a homogeneous sample of 45 flat spectrum radiosources. In these two samples, the percentage of variable sources is therefore 25% and 51% respectively.

141.002 Distance-independent evidence for relativistic motion in the quasar NRAO 140.
A. P. Marscher, J. J. Broderick.
Astrophys. J., Lett., Vol. 247, L49 - L52 (1981).

The authors have made VLBI radio and X-ray observations of the bright X-ray quasar NRAO 140. The radio structure consists of two very compact components plus a more diffuse jet. Using standard synchrotron theory, the authors find that the Compton X-ray emission from one of these compact components should exceed the observed value by three orders of magnitude. Since this result is independent of distance and since they find no evidence for coherent emission, they conclude that the component is moving toward us at a relativistic speed with a Lorentz factor exceeding 4.

141.003 Observations of six flat spectrum sources from the 5 GHz survey.
P. Biermann, H. Duerbeck, A. Eckart, K. Fricke, K. J. Johnston, H. Kühr, J. Liebert, I. I. K. Pauliny-Toth, H. Schleicher, H. Stockman, P. A. Strittmatter, A. Witzel.
Astrophys. J., Lett., Vol. 247, L53 - L56 (1981).

The authors report observations at radio, optical, and X-ray wavelengths of six sources from the Bonn-NRAO 5 GHz survey. The sources were selected on the basis of their flux densities and spectral indices at 5 GHz. All have been shown to contain compact radio cores and to emit strongly at X-ray wavelengths; five are strongly polarized at optical wavelengths. The measured flux densities suggest that the sources are of comparable luminosity (per fractional bandwidth) in the X-ray and optical regions. The interpretation of these results in terms of a synchrotron-self-Compton mechanism is briefly discussed.

141.004 Extended radio emission aligned with compact nuclear sources in normal galaxies.
D. L. Jones, R. A. Sramek, Y. Terzian.
Astrophys. J., Lett., Vol. 247, L57 - L61 (1981).

Observations of eight galaxies with the VLA showed five to have extended radio emission which in general appears asymmetric and on a scale of only a few seconds of arc. Comparison of this structure with VLBI models of the milli-arcsec scale structure in the nuclei revealed close alignments between the position angles, suggesting a long-term dynamical connection between these different regions. This is very similar to the situation found in many active radio galaxies and quasars. If a common physical mechanism is indeed responsible for the similar radio structures seen in all of these objects, then it must be capable of operating over a range of at least 10^6 in radio luminosity.

141.005 Bending of electromagnetic beams and head-tail radio sources. G. Bodo, A. Ferrari, S. Massaglia.
Mon. Not. R. Astron. Soc., Vol. 196, 481 - 489 (1981).

The authors present an interpretation of bridge bending in head-tail radio sources in the framework of an electromagnetic beam model. The physical effect responsible for the structural distortion is proposed to be the refraction of a large-amplitude wave in a medium with a density gradient perpendicular to the wave propagation vector; this gradient is consistently produced by the relative motion of the beam source in the surrounding medium with a velocity higher than the speed of sound. These effects are calculated in some detail and a quantitative fit of model parameters to the typical radio source associated with NGC 1265 is discussed.

141.006 Measurements of the linear and circular polarization of some compact radio sources at 2.7 GHz.
M. Ryle, A. C. Brodie.
Mon. Not. R. Astron. Soc., Vol. 196, 567 - 581 (1981).

Observations of linear polarization at 2.7 and 15 GHz and circular polarization at 2.7 GHz with the 5-km telescope are described for some compact radio sources; they are discussed, together with earlier measurements at 5 GHz using the same method, in relation to theories of the origin of circular polarization. There is evidence for possible short time-scale variations (~ 1 month) in circular polarization for two of the sources.

141.007 Models of radio source evolution − II. The 2700-MHz source count.
J. V. Wall, T. J. Pearson, M. S. Longair.
Mon. Not. R. Astron. Soc., Vol. 196, 597 - 610 (1981).

The authors' technique for deriving cosmological evolution from source counts and identifications is applied to data at 2700 MHz. The analysis is carried out on the assumption that two populations with different evolutionary behaviours appear in surveys at this frequency: 'steep-spectrum' sources with extended radio structures, and 'non-steep-spectrum' sources with compact structures, the majority of which are identified with QSOs. The 2700-MHz data add constraints to the evolution deduced for the 'steep-spectrum' sources from low-frequency data; in particular, of the two types of model obtained in the authors' analysis of the 408-MHz data, only one now appears tenable. The present results for 'non-steep-spectrum' sources agree with the results from luminosity − volume tests on samples of 'flat-spectrum' QSOs − the change in space density with epoch appears less dramatic than for the powerful radio sources with steep spectra and extended radio structures.

141.008 **Multifrequency models for the cosmological evolution of extragalactic radio sources.**
J. A. Peacock, S. F. Gull.
Mon. Not. R. Astron. Soc., Vol. 196, 611 - 633 (1981).

This paper presents new models for the epoch dependence of the luminosity function of extraglactic radio sources. These models have been derived on the assumption that radio sources may be divided into two distinct populations: the extended steep-spectrum sources and the compact flat-spectrum sources. By using the correlation between luminosity and spectral index for the steep-spectrum sources, the luminosity functions defined at different frequencies may be related, allowing a single model to be tested against all available source counts, luminosity distributions and identification data.

141.009 **Optical and radio properties of 4C galaxies.**
A. Wilkinson, R. G. Hine, W. L. W. Sargent.
Mon. Not. R. Astron. Soc., Vol. 196, 669 - 693, Microfiche MN 196/1 (1981).

The relations between the optical and radio properties of a sample of 4C radio galaxies of intermediate radio luminosity have been investigated. Optical spectra were taken with the Hale 5-m telescope and radio maps of a subset of the galaxies were made with the Cambridge 1-mile and 5-km telescopes.

141.010 **The spectrum of the QSO 0805 +046 (4C 05.34) at intermediate dispersion.**
J.-s. Chen, D. C. Morton, B. A. Peterson, A. E. Wright, D. L. Jauncey.
Mon. Not. R. Astron. Soc., Vol. 196, 715 - 730 (1981).

Spectra of the radio QSO PKS 0805 + 046 (z_{em} = 2.8772) have been obtained with the AAT at 2 Å resolution from 3300 to 6100 Å. Two absorption line systems at z_{abs} = 2.87717 and 2.47568 are certain and three more at 1.01422, 0.95915 and 0.70280 are possible. These systems leave about 80 per cent of the absorption lines unidentified shortward of Lα emission and 26 per cent longward. A strong peak in the cross-correlation corresponding to Lα - Lβ pairs and the explicit identification of many of these pairs supports the hypothesis that these lines originate in neutral hydrogen clouds.

141.011 **Identification of southern radio sources – IV.**
A. Savage, A. E. Wright.
Mon. Not. R. Astron. Soc., Vol. 196, 927 - 932 (1981).

The authors suggest identifications for 32 radio sources drawn from the Parkes 2700-MHz survey, 25 with QSOs and seven with galaxies. The identifications were made using the SRC United Kingdom Schmidt Telescope Unit (UKSTU) IIIa-J objective prism plates, together with deep, direct IIIa-J survey plates and some direct IIIa-F plates.

141.012 **The symmetry, misalignment and kinematic evolution of double radio sources.** J. T. Macklin:
Mon. Not. R. Astron. Soc., Vol. 196, 967 - 986 (1981).

The symmetry properties of a carefully selected sample of 76 double radio sources have been examined. Computer simulations show that, assuming a random distribution of source axes, the mean intrinsic misalignment of these sources is 0.038, and this result does not depend on the intrinsic size of a source. Sources which are most misaligned tend to have the highest values of D, the ratio of hot-spot separations from the nucleus. Simulations in which the major contribution to the D distribution is independent of orientation and is correlated with misalignment give the best explanation of this result. Such a correlation might arise if the supply from the nucleus is intermittent and changing in direction. This correlation also implies that the main contribution to the observed misalignment does not come from motion of the parent galaxy relative to the external medium.

141.013 **Variability of extragalactic sources at 2.7 GHz. IV. Evidence for weak extended emission and for rapid variability.** J. F. C. Wardle, A. H. Bridle, M. J. L. Kesteven.
Astron. J., Vol. 86, 848 - 853 (1981).

The results of two programs to monitor the λ 11.1-cm flux densities of extragalactic radio sources are compared. The programs were carried out between 1972 and 1974 on the NRAO 300-ft telescope and the NRAO three-element interferometer. The two sets of data are generally in excellent agreement.

141.014 **The 5-GHz strong source surveys. V. Survey of the area between declinations 70° and 90°.**
H. Kühr, I. I. K. Pauliny-Toth, A. Witzel, J. Schmidt.
Astron. J., Vol. 86, 854 - 863 (1981).

The north pole region between declinations 70° and 90° has been surveyed at 4.9 GHz using the MPI 100-m telescope. A total of 476 sources with flux densities above 50 mJy were found in this survey, which covers 0.401 sr of the sky and is essentially complete above 250 mJy. It is thus the most sensitive of the S-surveys. Radio positions accurate to between 0.2 and 6 arcsec are given in this catalog, as well as accurate flux densities at 2.7, 5.0, and 10.7 GHz, and suggested optical identifications for the stronger sources.

141.015 **Changes in the radio structure of SS 433.**
W. S. Gilmore, E. R. Seaquist, J. T. Stocke, P. C. Crane.
Astron. J., Vol. 86, 864 - 870 (1981).

The authors report aperture synthesis observations of SS 433 with the VLA at 4.9 and 15 GHz on two different epochs, spaced to show nearly maximum structural changes according to the ephemeris of Margon et al. (1980). Changes were clearly seen in the structure of the jets. Comparison of the radio maps with the helical pattern expected from twin precessing jets shows some evidence for a helical pattern, although this evidence from the authors' data alone is inconclusive. Linear polarization is clearly associated with the jets and is variable in structure. The degree of polarization is $\geq$ 10% in at least some parts of the radio source, indicative of a highly ordered magnetic field.

141.016 **Spectrophotometry of Paschen and Balmer lines in PKS 0312-77 and 3C 109.**
W. Kollatschny, K. J. Fricke.
Astron. Astrophys., Vol. 100, L4 - L6 (1981).

J, H, K photometry and narrow band infrared photometry of Pα in the K (2.2μ) band has been performed for the QSO PKS 0312-77 and for the N galaxy 3C 109 together with IDS spectrophotometry in the optical region. The authors determined Pα/Hβ ratios for both objects. The infrared continuum of the QSO proved to be variable from a comparison with measurements of this source a year ago.

141.017 **A 408 MHz all-sky continuum survey. I. Observations at southern declinations and for the north polar region.**
C. G. T. Haslam, U. Klein, C. J. Salter, H. Stoffel, W. E. Wilson, M. N. Cleary, D. J. Cooke, P. Thomasson.
Astron. Astrophys., Vol. 100, 209 - 219 (1981).

The observation, data reduction and calibration of two large area radio continuum surveys at 408 MHz are described in detail. The resolution of both is close to 0.°85. One covers the complete southern sky, the other the north polar regions and, together with existing observations, they form the data-base for the first all-sky continuum survey of better than one degree resolution. Sample maps of astronomical interest are presented and possible uses of the 408 MHz survey are discussed. The full set of maps will appear in a companion all-sky atlas.

141.018 **DA 240 and 3C 236: spatial variation in their radio spectra and its implications.**
R. G. Strom, J. R. Baker, A. G. Willis.
Astron. Astrophys., Vol. 100, 220 - 230 (1981).

From 49 cm observations made with the WSRT and 6 cm measurements obtained with the 100 m telescope, the authors have determined the spectral index distribution for the large radio galaxies DA 240 and 3C 236. The methods of data reduction and sources of error are considered in some detail.

141.019   IR photometry of flat spectrum radio sources.
C. D. Impey, P. W. J. L. Brand.
Nature, Vol. 292, 814 - 816 (1981).

18 Parkes flat spectrum sources have been observed in the IR without regard to optical morphology, and the red sources are found to be just the tail of the normal distribution of quasar colours. The results imply that a considerable fraction of flat spectrum sources from complete samples may have properties similar to the BL Lac objects.

141.020   Emission line regions and stellar associations in extended extragalactic radio sources.
D. S. De Young.
Nature, Vol. 293, 43 - 44 (1981).

Recent observations of two radio galaxies, 3C277.3 and Centaurus A, reveal the coexistence of emission line filaments and the extended radio-emitting regions. Coincident associations of young stars are also found in Cen A. A model is proposed that can account for the presence of both emission line regions and young stars as arising from the entrainment of interstellar matter by the outflowing gas which powers the radio source.

141.021   Compact radio sources at declinations > 67°.
E. Waltman, K. J. Johnston, J. H. Spencer,
I. Pauliny-Toth, J. Schraml, A. Witzel.
Astron. Astrophys., Vol. 101, 49 - 51 (1981).

The authors present a list of compact radio sources at 5 GHz with declinations > 67°. Further observations at other frequencies may establish them as calibrators for radio astrometric observations. Since these sources are extragalactic, they may also be used in establishing an almost absolute reference frame. These sources are circumpolar at latitudes north of 23°.

141.022   Optical morphology of 13 QSOs.
J. B. Hutchings, D. Crampton, B. Campbell,
C. Pritchet.
Astrophys. J., Vol. 247, 743 - 749 (1981).

Faint, underlying structures have been detected and mapped around all (10) QSOs with $z < 0.3$ for which good, red, sensitive plates were obtained at the prime focus of the Canada-France-Hawaii Telescope (CFHT). Surface photometry of the objects shows that the radial intensity distribution falls off exponentially in the outer regions, while in the inner regions it is impossible to distinguish between a spheroidal, $r^{-1/4}$, distribution or an exponential disk. An analysis with simple models confirms that the intensity distribution of a disk plus point source is indistinguishable from a spheroidal plus point source on the basis of such data.

141.023   Resolution of quasar images.
S. Wyckoff, P. A. Wehinger, T. Gehren.
Astrophys. J., Vol. 247, 750 - 761, plates 19 - 26 (1981).

The purpose of this paper is to present imaging data and the results of an image analysis of 15 low redshift ($z \lesssim 0.6$) quasars. Evidence is presented for extended nebulosity around all except two of the program quasars. Statistical methods are used for testing the hypothesis that quasars are the luminous nuclei of distant galaxies. The observed characteristics of the resolved nebulosity in the sample studied are consistent with the nebulosity being identified with a galaxy associated with the quasar.

141.024   The three-dimensional distribution of quasars in the CTIO surveys.   P. S. Osmer.
Astrophys. J., Vol. 247, 762 - 773 (1981).

The availability of confirmed redshifts for the 174 high-latitude quasars in the CTIO surveys, together with their large angular extent and high surface density, permits a first look at the spatial distribution of redshift 2 quasars on scales of 100 - 3000 Mpc (present epoch coordinates for $H_0 = 50$). Visual inspection of the data in graphical form shows a number of close groups and pairs with small Δz in the 4 m survey. Although they turn out to be chance fluctuations, they should be useful in other studies. Three statistical techniques from the field of galaxy clustering, binning analysis, the nearest neighbor test, and the correlation function, are applied to the survey data. After modification for the cosmological, edge, and selection effects in the samples, the techniques indicate no evidence for clustering or other deviation from randomness on the 100 - 3000 Mpc scale.

141.025   Milli-arcsec structure of 3C 345.
M. H. Cohen, S. C. Unwin, R. S. Simon,
G. A. Seielstad, T. J. Pearson, R. P. Linfield, R. C. Walker.
Astrophys. J., Vol. 247, 774 - 779 (1981).

The quasar 3C 345 has been studied at eight epochs between 1977.56 and 1979.92 using very-long-baseline interferometry, alternating between frequencies of 10.65 and 5.01 GHz. Hybrid maps for each epoch show that the source comprises a bright, barely-resolved "core" with an inverted spectrum, and an elongated region of emission extending ~ 5 milli-arcsec from its western side. The position angle of this "jet" changes systematically from P.A. $\approx -85°$ close to the core to P.A. $\approx -75°$ at a radius of 3 milli-arcsec. Components within this "jet" change in position and brightness with time, but they do not show clear indications of superluminal expansion. No emission is found on the eastern side of the core above a level of 5% of the peak brightness.

141.026   Extremely red compact radio sources: the empty field objects.   C. A. Beichman, S. H. Pravdo,
G. Neugebauer, B. T. Soifer, K. Matthews, H. A. Wootten.
Astrophys. J., Vol. 247, 780 - 786 (1981).

10 μm radiation has been detected from 1413 + 135, one of the very red objects discovered by Rieke, Lebofsky, and Kinman at near-infrared wavelengths. The spectrum of this object flattens at wavelengths longer than 2.2 μm. Upper limits are also given for the 10 μm emission from 2255 + 14, 0026 + 34, and 0406 + 121. Photometry between 1.25 and 2.2 μm confirms the variability of 1413 + 135, 2255 + 41, and 0406 + 121. Five percent resolution spectra of 1413 + 135 and 0406 + 121 between 1.5 and 2.4 μm show no emission or absorption lines. The spectral data rule out the possibility that 1413 + 135 is a quasar with normal line strengths and a redshift $1.3 < z < 4$. The lack of features of the $1.5 - 2.4$ μm spectra, the rapid variability, and the overall shape of the radio, infrared, and X-ray energy distributions are consistent with a BL Lac nature for these objects.

141.027   Investigation of BL Lac-type radio sources at the frequency of 102 MHz.
V. S. Artyukh, Yu. N. Vetukhnovskaya.
Astron. Zh., Tom 58, 689 - 700 (1981). In Russian. English translation in Soviet Astron., Vol. 25, No. 4.

Measurements of flux densities and angular dimensions of BL Lac-type radio sources, based on interplanetary scintillation observations at the wavelength 3 m, are reported. Radio emission has been detected from 40 BL Lac-type objects.

141.028   On the possibility of observing X-ray absorption lines in the spectra of quasars.
M. M. Basko, B. V. Komberg, E. I. Moskalenko.
Astron. Zh., Tom 58, 701 - 705 (1981). In Russian. English translation in Soviet Astron., Vol. 25, No. 4.

The authors discuss the possibility to detect absorption features in the X-ray spectra of quasars, formed in hot plasmas of cosmic objects that happen to lie on the line of sight. For typical parameters of the hot gas in rich clusters and giant galaxy halos the equivalent widths of these features are shown to be less than 50 eV. For the most abundant heavy elements the most promising transitions are pointed out for which it is not unreasonable in some cases to undertake a searching procedure just now with the aid of the Einstein X-ray telescope.

141.029   A survey of the sky at the frequencies 8.7 and 14.4 GHz between the declinations 0° and +4°.
V. R. Amirkhanyan, A. G. Gorshkov, A. A. Kapustkin, V. K. Konnikova, A. N. Lazutkin, M. G. Larionov, A. S. Nikanorov, V. N. Sidorenkov, L. S. Ugol'kova, O. I. Khromov.
Astron. Zh., Tom 58, 717 - 724 (1981). In Russian. English translation in Soviet Astron., Vol. 25, No. 4.

A survey at 8.7 and 14.4 GHz of 0.33 steradian of the sky at $\delta = 0°- 4°$ has revealed 82 and 24 sources respectively.

141.030   A model of the radio source Sagittarius B2.
I. V. Gosachinskij, V. K. Khersonskij.
Astron. Zh., Tom 58, 743 - 753 (1981). In Russian. English translation in Soviet Astron., Vol. 25, No. 4.

A dynamical model of the gas cloud around the radio source Sagittarius B2 is suggested. This model describes the kinematic features of the gas in this source: contraction of the core and rotation of the envelope. The parameters of the primary cloud and some features of the collapse are calculated. The conditions in the core at the moment of its fragmentation into masses of stellar order are established.

141.031   Scattering of radio emission in the envelopes of compact sources.   V. I. Altunin.
Astron. Zh., Tom 58, 754 - 761 (1981). In Russian. English translation in Soviet Astron., Vol. 25, No. 4.

It is shown that in the scattering of radio emission in an envelope surrounding a compact source, a considerable role may be played by inhomogeneities of electron density created by high-frequency plasma waves.

141.032   VLBI observations of compact radio sources at 18 cm.
L. I. Matveenko, V. I. Kostenko, A. Kh. Papatsenko, N. Bartel, M. Massi, J. D. Romney, K. W. Weiler, A. Ficarra, F. Mantovani, L. Padrielli, I. G. Moiseev, L. Baath, G. Nicolson.
Pis'ma Astron. Zh., Tom 7, 470 - 478 (1981). In Russian. English translation in Soviet Astron. Lett., Vol. 7.

On February 14 - 15, 1980 structures of 33 compact radio sources were investigated with the global network of radio interferometers at 18 cm. The source brightness distributions are given.

141.033   The milli-arcsecond structure of a complete sample of radio sources. I. VLBI maps of seven sources.
T. J. Pearson, A. C. S. Readhead.
Astrophys. J., Vol. 248, 61 - 81 (1981).

The sample of 51 sources between declinations 35° and 70° with galactic latitude $>10°$ and flux density $\geqslant 1.3$ Jy at 5 GHz contains 36 objects that are suitable for study by very long baseline interferometry (VLBI). Observations have been made of seven of these with a VLBI array of four U.S.A. antennas at a frequency of 5 GHz with full hour-angle coverage. The results show a wide variety of morphologies.

141.034   Theoretical quasar emission-line ratios. III. Flux divergence and photon escape.
R. C. Canfield, R. C. Puetter, P. J. Ricchiazzi.
Astrophys. J., Vol. 248, 82 - 86 (1981).

The authors develop a computationally useful version of the probabilistic first-order differential radiative transfer equation of Frisch and Frisch. This approximate radiative transfer equation is especially appropriate for the evaluation of radiative transfer effects in multilevel atomic systems due to its extreme computational efficiency and reasonably accurate description of the physics of radiation transfer. In particular, it recognizes the distinction between the flux divergence coefficient, ρ, and the photon escape probability, p_e. The authors show that this distinction is crucial for calculations that attempt to construct self-consistent energy balance models since substitution of p_e for ρ in such models leads to large errors (of sign as well as magnitude) in the local cooling rate.

141.035   Magnetic fields in extragalactic radio sources.
R. A. Laing.
Astrophys. J., Vol. 248, 87 - 104 (1981).

Distributions of total intensity and linear polarization have been calculated for synchrotron radiation from a variety of magnetic fields which may occur in extragalactic radio sources, assuming that the energy distributions of radiating electrons are uniform and isotropic. Analytical solutions are derived for helical and transverse self-similar fields of the type suggested for radio jets by Chan and Henriksen and compared with results for a field which has no component along the jet axis, but is otherwise random (as is expected to develop in an expanding jet containing a tangled field). A comparison with observation is performed.

141.036   Very high-resolution observations of compact radio sources in the directions of supernova remnants.
B. J. Geldzahler, D. B. Shaffer.
Astrophys. J., Vol. 248, 132 - 137 (1981).

Compact radio sources whose positions lie within the outlines of supernova remnants may be the stellar remnants of supernova explosions and, if they are related to the supernova remnants, may be used to explore the nature of any morphological connection between the Galactic and extragalactic radio sources. Three such compact sources, G127.11+0.54, CL 4, and 2051+433, have been observed at 10.65 GHz with an array of very long baseline interferometers.

141.037   Speckle interferometry observations of the triple QSO PG 1115+08.   E. K. Hege, E. N. Hubbard, P. A. Strittmatter, S. P. Worden.
Astrophys. J., Lett., Vol. 248, L1 - L3 (1981).

Speckle interferometric observations of the 16.2 mag component A of PG 1115+08 show that it is resolved into two components, separated by $0.''54$ in position angle 20°. The data suggest that the two components are approximately equally bright and are unresolved at a resolution of $\sim 0.''15$. These results are consistent with those inferred from long exposure images.

141.038   An improved optical position of 3C 273B in the FK4-system.   C. de Vegt, U. K. Gehlich.
Astron. Astrophys., Vol. 101, 191 - 193 (1981).

For the quasar 3C 273B, which has been recommended as R. A. zeropoint in radio astrometric work, a new precise optical position in the FK4-system has been obtained from 8 plates taken with the 23 cm-astrograph and 5 plates from the 60 cm-refractor of the Hamburg Observatory.

141.039   The structure of a sample of low luminosity radio galaxies.   R. D. Ekers, R. Fanti, C. Lari, P. Parma.
Astron. Astrophys., Vol. 101, 194 - 214 (1981).

New maps for eight radio galaxies from the B2 catalogue, obtained with the Westerbork synthesis radio telescope, at frequencies of 0.6, 1.4, and 5 GHz, are presented here. Together with other observations already published by other authors, they provide a complete sample of radio galaxies with linear size larger than 50 kpc. A discussion of the radio properties of this sample is presented.

141.040 Can the twin-exhaust model explain radio jets?
M. D. Smith, L. Smarr, M. L. Norman, J. R. Wilson.
Nature, Vol. 293, 277 - 279 (1981).

The authors conclude that high-velocity, high-powered jets cannot be produced in gravitational potentials produced by stellar clusters. However, the more general notion of the twin-exhaust mechanism, in which the central potential can be cusp-like near a massive black hole, is not ruled out for the high-powered radio sources.

141.041 Detection of 21 centimeter absorption at $z \approx 1.94$ in the QSO PKS 1157 + 014.
A. M. Wolfe, F. H. Briggs, D. L. Jauncey.
Astrophys. J., Vol. 248, 460 - 467 (1981).

The authors report the detection of 21 cm absorption by neutral hydrogen in front of PKS 1157+014, a QSO with emission redshift $z_{em} = 1.978 \pm 0.008$. The absorption line arises at 482.537 ± 0.002 MHz, corresponding to a redshift $z_{abs} = 1.94362 \pm 0.00001$ that is identical, within observational errors, to the previously detected optical redshift of absorption system d. The presence of several other absorption systems, along with this one, suggests that this system, along with the others, is ejected from QSO. However, the determination of the spin temperature T_s from a comparison of Lyα and 21 cm absorption spectra argues against this interpretation. The upper limit of $T_s < 470$ K places system d farther than 10 kpc from the QSO to avoid excitation effects of the 21 cm continuum. The authors suggest that system d is part of an H I disk of a galaxy gravitationally bound to a cluster containing the QSO.

141.042 The QSO B234 and the irregular galaxy NGC 4861: examples of isolated extragalactic H II regions with low helium abundances.   H. B. French, J. S. Miller.
Astrophys. J., Vol. 248, 468 - 471 (1981) = Lick Obs. Bull., No. 884.

Line fluxes are presented for the QSO B234 and for the nucleus of the peculiar galaxy NGC 4861 and are used to derive physical parameters for the excited gas and abundances of helium, oxygen, nitrogen, neon, and sulfur relative to hydrogen. The values derived are found to be similar to those of typical "isolated extragalactic H II regions" (IEHRs) and well below what is considered to be normal. The helium abundance of both objects is below the average for the IEHRs and close to the primordial value derived by French in his study of these objects. The line profiles and line ratios of B234 are typical of IEHRs and at variance with those of QSOs. The authors conclude that B234 is a luminous, distant IEHR rather than a QSO.

141.043 VLA observation of radio/optical knots in 3C 277.3 = Coma A.   A. H. Bridle, E. B. Fomalont, J. J. Palimaka, A. G. Willis.
Astrophys. J., Vol. 248, 499 - 503 (1981).

The radio galaxy 3C 277.3 has been mapped at ~1″ resolution using the VLA at 4885 MHz. Its structure resembles that of the inner ~50″ of the radio emission from M87. A bridge or jet linking the unresolved central radio core to the southern lobe contains two bright radio knots. One of these coincides with a blue polarized continuum knot in the envelope of the galaxy. Minimum energy calculations for this knot demand in situ particle deposition there, as in the M87 optical-radio jet. If the flow velocity through the knots is of the same order as the 200−300 km s^{-1} peculiar radial velocity of an optical emission-line system detected in their vicinity by Miley et al., two interpretations are possible.

141.044 Radio patrol of the northern Milky Way: a survey for variable sources.
P. C. Gregory, A. R. Taylor.
Astrophys. J., Vol. 248, 596 - 605 (1981).

Repeated observations of the entire northern galactic plane are being carried out at a wavelength of 6 cm and resolution of 2″.7 in a survey for variable radio sources with the NRAO 91 m transit telescope. Details of the patrol program are described together with results on the first five variable sources discovered, including one periodic variable with a period of 26^d.52. Accurate radio positions for four of the sources have been obtained with the Very Large Array, and possible optical counterparts are discussed. The survey data base is also being used to generate a very sensitive differential map of the galactic plane, and a new map of the supernova remnant G127+0.5 is presented. The central source in G127+0.5 was found to be constant on time scales of a few days to one year.

141.045 A model for extragalactic radio sources and quasi-stellar objects.   P. F. Browne.
Phys. Lett. A, Vol. 83A, 406 - 410 (1981). − Abstr. in Phys. Abstr., Vol. 84, Abstr. 79860 (1981).

141.046 Faint blue objects at high galactic latitude. I. Palomar Schmidt field centered on Selected Area 57.
P. D. Usher.
Astrophys. J., Suppl. Ser., Vol. 46, 117 - 139 (1981).

Objects have been selected for relative ultraviolet excess from a Palomar Schmidt field centered on Kapteyn Selected Area 57, near the north galactic pole. Object selection to $B = 20$ mag has been made by a novel semiquantitative technique, permitting a two parameter color classification. All known quasars in the field have been rediscovered. Four new quasars are reported in the paper. The integral number counts of the listed objects are given as a function of their assigned color classes. The population of objects belonging to color classes 1A, 1 and 1B should be comprised primarily of quasars and white dwarfs and should be virtually complete to $B = 18.5$ mag.

141.047 A VLA survey of strong radio sources.
J. Ulvestad, K. Johnston, R. Perley, E. Fomalont.
Astron. J., Vol. 86, 1010 - 1035 (1981).

The VLA has been used to survey the structures of 360 strong high-frequency radio sources at 1480 and 4900 MHz. Of those sources, 250 are found to contain at least 90% of their flux density in an unresolved component ($\lesssim 1$ arcsec), while the remaining 110 sources are considered to be resolved. Positions and structural information for these sources are given, and maps of 20 of the resolved objects are shown. The VLA data have been combined with other structural information available for sources listed in the Parkes and S surveys in order to study a complete sample of 444 extragalactic objects having flux densities greater than 1 Jy at 5 GHz.

141.048 Sco X-1: a galactic radio source with an extragalactic radio morphology.   B. J. Geldzahler, E. B. Fomalont, K. Hilldrup, B. E. Corey.
Astron. J., Vol. 86, 1036 - 1041 (1981).

VLA observations of the radio emission at 1465 and 4885 MHz of Sco X-1 confirm the collinear triple structure of the object. Simultaneous VLBI observations at 5010 MHz were also made. The authors present strong evidence that the three components of Sco X-1 are physically associated. They also discuss the likelihood that a physical phenomenon similar to that occurring in extragalactic radio sources occurs in Sco X-1 but at a much lower energy level. A previously unknown radio source 20 arcmin away from Sco X-1 has been found to lie nearly along the radio axis formed by the components of Sco X-1. This source is probably unrelated to Sco X-1.

141.049 Fast correlated variability of OI 090.4 (PKS 0754 + 100) in optical and radio wavelengths.
S. A. Pustil'nik, V. M. Lyutyj, S. I. Neizvestnyj.

Pis'ma Astron. Zh., Tom 7, 547 - 553 (1981). In Russian. English translation in Soviet Astron. Lett., Vol. 7.

Fast optical and radio emission variability with a characteristic time of the order of a few days is detected for the radio object with continuous optical spectrum OI 090.4 (PKS 0754 + 100). Flux variations at 8.2 cm wavelength are evidently correlated with a delay of two days with variations of the optical R-band flux.

141.050 Observations of absorption troughs in broad absorption-line QSOs.
V. T. Junkkarinen, E. M. Burbidge, H. E. Smith.
Bull. American Astron. Soc., Vol. 13, 520 (1981). − Abstract.

141.051 Optical Fe II emission in radio-quiet QSOs.
C. B. Foltz, B. M. Peterson.
Bull. American Astron. Soc., Vol. 13, 520 - 521 (1981). Abstract.

141.052 The point radio source in the supernova remnant G 78.2 + 2.1. L. A. Higgs, R. S. Roger,
T. L. Landecker, S. R. Spangler, J. M. Cordes, J. M. Dickey.
Bull. American Astron. Soc., Vol. 13, 522 (1981). − Abstract.

141.053 A simple formula for the surface brightness of turbulent radio jets. R. N. Henriksen.
Bull. American Astron. Soc., Vol. 13, 528 (1981). − Abstract.

141.054 The polarization of the jet in NGC 6251.
R. A. Perley, A. G. Willis.
Bull. American Astron. Soc., Vol. 13, 528 (1981). − Abstract.

141.055 The intensity and polarization structure of the jets in 3C449.
T. J. Cornwell, R. A. Perley, A. G. Willis.
Bull. American Astron. Soc., Vol. 13, 528 - 529 (1981). Abstract.

141.056 Direct imaging of low redshift QSOs.
J. B. Hutchings, D. Crampton, B. Campbell.
Bull. American Astron. Soc., Vol. 13, 529 (1981). − Abstract.

141.057 VLBI observations of Circinus X-1 and Centaurus A.
R. A. Preston, D. D. Morabito, A. E. Wehrle,
L. J. Skjerve, D. L. Jauncey, J. G. Ables, M. J. Batty,
J. G. Bolton, D. J. Cooke, R. F. Haynes, A. E. Wright,
G. D. Nicolson.
Bull. American Astron. Soc., Vol. 13, 529 (1981). − Abstract.

141.058 The periodicity in the distribution of quasar redshifts and the density perturbation in the early universe. L. Z. Fang, Y. Q. Chu.
Bull. American Astron. Soc., Vol. 13, 530 (1981). − Abstract.

141.059 Detailed maps of a quasar with a radio jet and with optical emission in one of its radio lobes.
F. D. Ghigo, L. Rudnick, P. A. Wehinger, S. Wyckoff.
Bull. American Astron. Soc., Vol. 13, 530 (1981). − Abstract.

141.060 Multifrequency VLA observations of 3C388 − evidence for an intermittent jet?
J. O. Burns, A. Christiansen, D. H. Hough.
Bull. American Astron. Soc., Vol. 13, 530 (1981). − Abstract.

141.061 Spectra of compact radio sources in galactic nuclei.
D. L. Jones, R. A. Sramek, Y. Terzian.
Bull. American Astron. Soc., Vol. 13, 531 (1981). − Abstract.

141.062 He II λ4686 as a diagnostic for quasar and Seyfert galaxy-emission-line analyses. G. M. MacAlpine.
Bull. American Astron. Soc., Vol. 13, 531 (1981). − Abstract.

141.063 The 3000 Å bump in quasars. S. A. Grandi.
Bull. American Astron. Soc., Vol. 13, 531 - 532 (1981). − Abstract.

141.064 Photoelectric comparison sequences in the fields of four extragalactic radio sources. H. R. Miller.
Bull. American Astron. Soc., Vol. 13, 532 (1981). − Abstract.

141.065 Structure of seven 3C sources at metre wavelengths.
M. N. Joshi.
Mon. Not. R. Astron. Soc., Vol. 197, 7 - 17 (1981).

Accurate positions and high-resolution brightness profiles derived from lunar occultations at 327 MHz are presented for seven 3C sources. The results are combined with those of IPS and interferometric observations available in the literature in order to derive models for the fine structure of these sources at metre wavelengths.

141.066 Radio synthesis observations of 3C 296, 3C 442 A and 3C 449 at 0.4, 1.4 and 2.7 GHz.
M. Birkinshaw, R. A. Laing, J. A. Peacock.
Mon. Not. R. Astron. Soc., Vol. 197, 253 - 263 (1981).

The low-luminosity radio galaxies 3C 296, 442 A and 449 have been observed at 0.4 and 1.4 GHz with the Cambridge One-Mile telescope and at 2.7 GHz with the 5-km telescope. The high-resolution maps show that both 3C 296 and 449 possess radio jets. Models for the magnetic-field structure are considered: a partially-ordered configuration which leads to predictions of the jet inclinations is advocated. All three sources are identified with galaxies that possess close companions, but it is argued that orbital interactions are not important in determining their morphologies.

141.067 The radio jet in NGC 6251. R. Saunders,
J. E. Baldwin, G. G. Pooley, P. J. Warner.
Mon. Not. R. Astron. Soc., Vol. 197, 287 - 300 (1981).

High-resolution observations are presented of the 300-kpc jet in the giant radio galaxy NGC 6251. The width of the jet is resolved over most of its length, and the axis of the jet wiggles with an amplitude increasing linearly with distance from the nucleus. Polarization data are used to derive densities of cold matter in the jet. The authors estimate the speed of the jet as $c/20$.

141.068 A statistical study of flat-spectrum radio sources at 966 MHz. P. K. Moore, I. W. A. Browne,
E. J. Daintree, R. G. Noble, D. Walsh.
Mon. Not. R. Astron. Soc., Vol. 197, 325 - 337 (1981).

A sample of 36 flat-spectrum radio sources has been observed at 966 MHz with a resolution of ~0.5 arcsec. Sixty-seven per cent of them showed evidence of extended emission several arcsec from an unresolved nucleus. The results show that most of the sources have asymmetric double structure. Interpreting this in terms of the relativistic jet model requires bulk relativistic motions up to several hundred kpc from the nucleus. The presence of extended emission, variability and spectral type are found to be correlated, and possible explanations for this are sought in terms of the relativistic jet model.

141.069 Broad emission line profiles from discs and quasars.
D. J. Raine, A. Smith.
Mon. Not. R. Astron. Soc., Vol. 197, 339 - 350 (1981).

Numerical computations of the line profiles from a 'scattering illumination' black hole accretion disc model of quasars and active galactic nuclei are reported. At low inclinations (edge on), when Doppler broadening dominates any internal broadening mechanisms, the model yields the observed range of 'slightly asymmetric' line profiles. At moderate inclinations the red−blue asymmetry of the lines is much reduced, but the range of basic profiles is otherwise unchanged. The effect of including some internal broadening by electron scattering is to improve the agreement with the observed 'logarithmic' profiles.

141.070   Ultraviolet observations of quasars.
           M. A. J. Snijders.
Second European IUE Conference, (see 012.011), p. LXXI - LXXX (1980).

Ultraviolet observations of QSOs with the IUE satellite are reviewed. In particular the author discusses observations of QSOs for $\lambda_o < 912$ Å and the covering factor of the central source, the hydrogen emission line formation problem and the possibility of extinction in and along the line of sight towards QSOs, as well as the sources of the narrow absorption lines seen in many QSO spectra.

141.071   IUE observations of the twin QSOs 0957+561 A, B.
           P. M. Gondhalekar, R. Wilson.
Second European IUE Conference, (see 012.011), p. 295 - 298 (1980).

The ultraviolet spectra of the twin quasars 0957+561 A, B have been obtained with IUE. These quasars are suspected to be the images of a single object formed by an intervening galaxy acting as a gravitational lens. The ultraviolet data give an intensity ratio for the two objects which is the same, within the errors, as that derived at radio frequencies. This shows that differential reddening between the two light paths is negligible and also gives support to the gravitational lens hypothesis.

141.072   Far ultraviolet observations of the Q.S.O.
           PG1351+64.   L. Maraschi, E. G. Tanzi, A. Treves.
Second European IUE Conference, (see 012.011), p. 299 - 301 (1980).

PG1351 is a bright low redshift QSO (V = 14.8, Z = 0.088. Its optical luminosity is 10^{45} erg s^{-1} and its X-ray luminosity in the 0.5 − 4.5 keV band is 10^{43} erg s^{-1}. The optical spectrum has not been published yet. The object was observed with IUE on Oct 15, 1979, using the blind off-set technique. The spectrum in the short and long wavelength range is reported.

141.073   UBV photometry of quasars.   G. Adam.
           Messenger, No. 25, p. 23 - 26 (1981).

141.074   Stars with zero proper motion and the number of
           faint QSOs.   R. G. Kron, L.-T. G. Chiu.
Publ. Astron. Soc. Pacific, Vol. 93, 397 - 404 (1981).

The authors survey the field of SA 57 for faint QSOs by studying the properties of stars which lack detectable proper motion. In principle, the method is capable of avoiding selection effects intrinsic to other search techniques. The zero-proper motion criterion is generally successful, but all of the QSOs that the authors studied would have been found with some other conventional technique, so that evidently no significant population of stellar-appearing extragalactic objects has remained undiscovered. After combining the results of several techniques, they arrive at a surface density of QSOs which supersedes, but is completely consistent with, the previous point on the log N-m diagram at B = 21.4.

141.075   Dual-frequency VLBI observations of extragalactic
           sources with distinctive radio spectra.
S. R. Spangler, J. M. Benson, J. M. Cordes, R. B. Hall, T. W. Jones, K. J. Johnston.
Astron. J., Vol. 86, 1155 - 1164 (1981).

Very Long Baseline Interferometer observations, utilizing a continental baseline interferometer at 6 and 18 cm, have been obtained of ten sources whose spectra indicate that they may consist of a single component. The VLBI observations were made to determine whether the shapes of the integrated spectra were due to the spectral properties of a single, dominant component, or to the superposition of two or more subcomponents becoming optically thick at different frequencies.

141.076   High-resolution radio observations of the X-ray galaxy NGC 3862 (3C 264) in Abell 1367.
A. H. Bridle, J. P. Vallée,
Astron. J., Vol. 86, 1165 - 1174 (1981).

The radio source 3C 264, identified with the X-ray emitting elliptical galaxy NGC 3862 in the rich cluster Abell 1367, has been mapped with resolution from 2 to 8 arcsec at 1465 MHz using 14 antennas of the VLA. The new data delineate three main structural components: a small-diameter core in the galactic nucleus, a jet-like feature about 900 pc long emanating from the core, and an amorphous emission plateau about 11 × 6 kpc radius around, but not centered on, the nucleus.

141.077   Relativistic beaming in the central components of extended radio sources.
V. K. Kapahi, D. J. Saikia.
Bull. Astron. Soc. India, Vol. 9, 77 - 78 (1981). − Abstract.

141.078   Extragalactic radio sources with asymmetric extended structure.   V. K. Kapahi.
Bull. Astron. Soc. India, Vol. 9, 78 (1981). − Abstract.

141.079   Structure of 3C2 and 3C279 at metre wavelengths.
           M. N. Joshi.
Bull. Astron. Soc. India, Vol. 9, 78 (1981). − Abstract.

141.080   Redshift distributions of extragalactic radio sources and optical identification statistics.
C. R. Subrahmanya.
Bull. Astron. Soc. India, Vol. 9, 78 - 79 (1981). − Abstract.

141.081   De optische straling van radiomelkwegstelsels.
           P. D. Barthel, G. K. Miley.
Zenit, 8. Jaarg., 427 - 430 (1981).

141.082   The double quasar 0957 + 561 AB as a probe of quasar structure.   G. Gilmore.
Observatory, Vol. 101, 170 - 171 (1981).

If the individual components of the double quasar 0957 + 561 AB exhibit uncorrelated time-variations, it will become possible to estimate the size and number of clouds in the region of the quasar in which the emission lines are formed.

141.083   The large- and small-scale structures of 3C 293.
           A. H. Bridle, E. B. Fomalont, T. J. Cornwell.
Astron. J., Vol. 86, 1294 - 1305 (1981).

The radio galaxy 3C 293 has been mapped with the VLA at 1.465 and 15.035 GHz with resolutions of 6″ and 0″.2, respectively, and with the Multi-Element Radio Linked Interferometer Network (MERLIN) at 1.666 GHz with a resolution of 0″.25. The VLA 1.465-GHz map shows that the source has a two-sided Z-shaped structure whose physical association with the galaxy VV 5-33-12 is now clear.

141.084   31.4 and 89.6 GHz flux density measurements of extragalactic radio sources.
B. J. Geldzahler, A. Witzel.
Astron. J., Vol. 86, 1306 - 1311 (1981).

Measurements at 31.4 GHz are presented for 333 radio sources from the 2.7 and 5 GHz surveys. In addition, the authors present the results of observations at 89.6 GHz of a few selected objects known or suspected to have inverted radio spectra between 5 and 31.4 GHz. While many of the sources show steepening radio spectra with increasing radio frequency, when these results are combined with those of the authors' previous study, they find that more than half of the sources they measured have flat radio spectra between 5 and 31.4 GHz, which is similar to the result for spectra between 2.7 and

5 GHz. These flat-spectrum sources are expected to contain components which are opaque at short-centimeter wavelengths.

141.085 Extragalactic radio sources with very steep decimetrewave spectrum.
Gopal-Krishna, H. Steppe.
Astron. Astrophys., Vol. 101, 315 - 319 (1981).

A sample of 61 sources showing very steep spectra between 327 and 2700 MHz has been selected from 577 sources originally observed in the Ooty lunar occultation survey at 327 MHz. Compared to sources having normal spectra, the very steep spectrum sources are found to show a considerably smaller optical identification percentage and smaller angular sizes. Their angular sizes look very similar to those of the 3C/4C quasars. Possible cosmological implications are briefly discussed.

141.086 Westerbork 5 GHz observations of head-tail radio sources in A 2022, A 2256, and A 2462.
E. A. Valentijn.
Astron. Astrophys., Vol. 102, 53 - 58 (1981).

5 GHz radio maps are presented for 5 head-tail radio sources detected in the three clusters A 2022, A 2256, and A 2462. For A 2022 it is noted that the observed widening of the tail is close to what one would expect from the local Hubble expansion, for a relatively low galaxy velocity of 75 km s^{-1}. The head-tail radio source C in A 2256 is still unresolved at 5 GHz. The complex radio source in A 2462 has been identified with a dominant E galaxy and a compact object, close to the galaxy on a photographic plate, has been detected as well.

141.087 The distance to G316.8–0.1.
P. A. Shaver, D. S. Retallack, W. Wamsteker, A. C. Danks.
Astron. Astrophys., Vol. 102, 225 - 229 (1981).

The authors present new radio and infrared observations which, together with other data, indicate that G316.8–0.1 is located at its near kinematic distance, 2.9 kpc, rather than the far kinematic distance (12.1 kpc) which is usually assumed. At the near distance it is a typical H II region requiring a single O6 star for its excitation, and one of the infrared sources detected near the radio peak may be the exciting star.

141.088 Compact radio source 1413 + 135 is a far-IR extragalactic object.   C. A. Beichman, G. Neugebauer, B. T. Soifer, H. A. Wootten, T. Roellig, P. M. Harvey.
Nature, Vol. 293, 711 - 713 (1981).

The radio source 1413 + 135 is shown to be one of the strongest known emitters of millimetre radiation. The energy distribution of this object measured between metre and X-ray wavelengths reveals that most of the power emitted by this object comes out at millimetre and far-IR wavelengths. If the emission at 1 mm is due to incoherent synchrotron radiation, then the source must be very compact, with a size around 25 μ arc s, and contain a magnetic field larger than 10 G.

141.089 Multifrequency observations of the red QSO 1413 + 135.
J. N. Bregman, M. J. Lebofsky, M. F. Aller, G. H. Rieke, H. D. Aller, P. E. Hodge, A. E. Glassgold, P. J. Huggins.
Nature, Vol. 293, 714 - 717 (1981).

The extremely red QSO 1413 + 135 shows characteristics typical of BL Lac objects – no emission lines – is embedded in a luminous galaxy, is subject to rapid radio and IR variability, has an inverted radio spectrum and shows substantial IR polarization. The rapid steepening of the nonthermal spectrum at $\lambda < 5\,\mu$m is interpreted as synchrotron emission from an electron distribution that ends sharply at some maximum energy. The X-ray emission is probably inverse Compton radiation.

141.090 The binary quasar QSO 0957+561 A, B – a gravitational lens?   V. F. Mukhanov.
Usp. fiz. nauk, Tom 133, 729 - 732 (1981). In Russian.
Abstr. in Ref. zh., 51. Astron., 9.51.895 (1981).

141.091 Spatial distribution, luminosity function and cosmological evolution of quasars.   G. Mathez.
Ann. Physique, Vol. 6, 295 - 315 (1981). In French. – Abstr. in Phys. Abstr., Vol. 84, Abstr. 89114 (1981).

141.092 Scintillations of cosmic radio sources in the decametre waveband. I. Spectra of scintillations due to ionospheric and interplanetary plasma fluctuations and the possibility of their separation.
V. P. Bovkoon (*Bovkun*), I. N. Zhouck (*Zhuk*).
Astrophys. Space Sci., Vol. 79, 165 - 180 (1981).

The effect of fluctuations of the interplanetary plasma and the ionosphere upon the scintillation spectra of radio sources at decametre waves is considered with due regard for the finite antenna aperture, fluctuation anisotropy, and the direction of their drift in space. It has been shown that scintillation due to interplanetary plasma, can be reliably separated from the ionospheric scintillation background at decametre wavelengths.

141.093 Scintillations of cosmic radio sources in the decametre waveband. II. Compact sources in supernova remnants of the Crab Nebula and Cassiopeia A.
V. P. Bovkoon (*Bovkun*), I. N. Zhouck (*Zhuk*).
Astrophys. Space Sci., Vol. 79, 181 - 189 (1981).

By observing the scintillation of radio emission due to interplanetary plasma fluctuations, the authors discovered a compact source in Cassiopeia A at decametre wavelengths. The parameters of the compact source in the Crab Nebula are also given. In both supernova remnants the spectrum without account of the compact source shows some flattening.

141.094 Observed colors of QSOs in relation to their redshifts.   D. Basu.
Astrophys. Lett., Vol. 21, 85 - 86 (1981).

The relation between observed colors of QSOs and their redshifts is studies with updated data. It is concluded that the relation is non-linear. The rather complex variation of colors with redshifts is due to the effect of emission lines on the observed values of color indices.

141.095 Flat spectrum multicomponent radio sources: cosmic conspiracy or geometry?
A. G. Pacholczyk.
Astrophys. Lett., Vol. 21, 87 - 92 (1981).

It is suggested that the characteristic overall flatness of multicomponent spectra of a certain class of compact radio sources (which seems to include BL Lacertae type objects) may be related to a specific symmetry in their radio structure, namely to a predominantly linear, one-dimensional evolution of radio radiating material involving structures such as beams, jets or plasmons (plasmoids) in a channel.

141.096 Optical variability and the redshift of quasars.
C. M. Gaskell.
Astrophys. Lett., Vol. 21, 103 (1981).

The apparent dependency of the degree of optical variability on redshift reported by Basu (1980) is shown to be due to the effects of time dilation.

141.097 The variable radio structure of 3C 446.
R. L. Brown, K. J. Johnston, F. H. Briggs, A. M. Wolfe, S. G. Neff, R. C. Walker.
Astrophys. Lett., Vol. 21, 105 - 110 (1981).

The radio source associated with the violently variable quasar 3C 446 consists of an extremely compact core and an asymmetric 1″ extension which lies along the apparent axis of the source magnetic field. From VLBI observations made on 4 epochs following a pronounced optical outburst we see the radio flux density increase as the morphology of the radio core changes from that of a single resolved (0.4 mas) source to that of a resolved source plus an elongated structure with the ~ 1.1 milli-arcsecond elongation again found along the magnetic axis. Since the timescale for these variations is of order one year while spatial extent over which they occur is ~ 20 light-years the authors infer that relativistic effects, perhaps occurring in separate centers of activity, play a deciding role in 3C 446.

141.098 One millimeter continuum observations of high red-shift quasars.
D. J. Ennis, B. T. Soifer, G. Neugebauer, M. Werner.
Astrophys. Lett., Vol. 22, 143 - 150 (1981).

Upper limits to the one-millimeter continuum flux densities of the high redshift quasars B2 1225+31, Ton 490, and PHL 957 are presented. The upper limit to the power observed from these quasars at 1 mm is, on average, $^{1}/_{2}$ the observed power in the continuum at $L\alpha$. These observations are used to constrain the temperature of a hypothetical dust shell which reddens the quasar line and continuum emission by an extinction optical depth sufficient to account for the anomalously low $L\alpha/H\alpha$ emission line ratio observed in each of these quasars.

141.099 Superluminal quasar 3C179 with double radio lobes.
R. W. Por as.
Nature, Vol. 294, 47 - 49 (1981).

VLBI observations of the quasar 3C179 reveal that its two milli arc second components have an apparent relative velocity of 7.6 times the velocity of light. This is the fifth radio source in which "superluminal" motion has been reported but the first which also exhibits double lobe structure on the arc second scale. Statistical arguments which apply to such sources cause difficulties for explanations based on motion in a relativistic jet.

141.100 Is a bi-stable beam responsible for the complex radio structure of 3C133?   D. W. Robson.
Nature, Vol. 294, 57 - 59 (1981).

Recent high resolution observations of the radio structure of 3C133 reveal a complex structure in which rotation of the source axis may have uncovered evidence for a bi-stable nature of the energy supply. There are, however, several difficulties in such an interpretation.

141.101 The C III] λ1909 effective wavelength-redshift relationship in quasars.   G. J. Ferland.
Astrophys. J., Vol. 249, 17 - 22 (1981).

The author reexamines one of the mechanisms suggested by Wills (1980) as the cause of the shift in effective wavelength, viz., that λ1906.7 emitted by low-density gas may make a significant contribution to the C III] profile. He compiles a mean narrow-line spectrum and then constructs a grid of photoionization models to deduce the physical conditions in the narrow-line region (NLR) gas. Results of these calculations are then used to construct a synthetic C III] λ1909 profile whose centroid is in very good agreement with the Wills results. The implications of the Wills correlation for high-redshift quasars are discussed.

141.102 Radio jet refraction in galactic atmospheres with static pressure gradients.
R. N. Henriksen, J. P. Vallée, A. H. Bridle.
Astrophys. J., Vol. 249, 40 - 47 (1981).

A theory of double radio sources which has a Z or S morphology is proposed, based on the refraction of radio jets in the extended atmosphere of an elliptical galaxy. The model describes a collimated jet of supersonic material bending self-consistently under the influence of external static pressure gradients. Gravity and magnetic fields are

neglected in the simplest case except insofar as they determine the static pressure distribution. The calculation is a straightforward extension of the method used by Begelman, Rees, and Blandford to calculate a ram-pressure model for twin radio trails (C morphology). It may also be described as a continuous-jet version of the buoyancy model proposed by Gull and Northover.

141.103 The extragalactic nature of CL4.
B. Margon, R. A. Downes, J. E. Gunn.
Astrophys. J., Lett., Vol. 249, L1 - L4 (1981).

Spectroscopy of the faint optical counterpart of the variable radio source CL4 shows a single broad emission line at 6480 Å. The object is almost surely a quasar, thus settling a decade old dispute of whether this source, near the center of the Cygnus Loop, is galactic or extragalactic.

141.104 Linear polarization of extragalactic radio sources at 3.71 and 11.1 centimeters.
M. Simard-Normandin, P. P. Kronberg, S. Button.
Astrophys. J., Suppl. Ser., Vol. 46, 239 - 245 (1981).

The three element interferometer of the National Radio Astronomy Observatory has been used to measure the linear polarization of 261 radio galaxies and quasars. Integrated polarization values are presented for 259 sources at 11.1 cm and 142 sources at 3.71 cm. Redshifts are known for 72% of the sources.

141.105 X-ray and VLBI radio observations of the quasars NRAO 140 and NRAO 530.
A. P. Marscher, J. J. Broderick.
Astrophys. J., Vol. 249, 406 - 414 (1981).

The authors have detected the quasars NRAO 140 and NRAO 530 with the Einstein X-ray Observatory, thereby confirming the previously reported detection by HEAO 1. The low-energy X-ray flux of NRAO 140 is consistent with the high-energy results from the earlier satellite. However, either the X-ray emission from NRAO 530 decreased by an order of magnitude from 1978.0 to 1979.7, or a second bright source was present within the HEAO 1 field. The authors have observed each of these two sources at 2.8 cm using a four station intercontinental VLB array, in order to determine whether the X-ray emission is caused by Compton scattering within the radio components.

141.106 The origin of a new absorption system discovered in both components of the double QSO Q0957+561.   P. Young, W. L. W. Sargent, A. Boksenberg, J. B. Oke.
Astrophys. J., Vol. 249, 415 - 421 (1981).

Ultraviolet spectroscopy at 1.2 Å resolution of the two gravitationally lensed images of Q0957+561 has led to the discovery of a new absorption-line system at z_{abs} = 1.1249. This is visible as a C IV $\lambda\lambda$1548, 1550 doublet in both images Q0957+561A,B. The absorption lines have the same strengths in both images and have a velocity difference Δv_{abs} = -8 ± 11 km s^{-1}. The velocity dispersion of the absorbing cloud is $\sigma_v < 26$ km s^{-1}. Interpreted as an ejection, the observed value of $z_{em} - z_{abs}$ for this system corresponds to a velocity of 37,500 km s^{-1}. Alternatively, if the new system were due to an intervening galaxy or intergalactic cloud, it would lie at a distance of roughly 300 Mpc from the QSO. The authors show that the small velocity difference observed in the absorption lines in the two images strengthens the ejection hypothesis for this particular system.

141.107 Two-phase models of quasar emission line regions.
J. H. Krolik, C. F. McKee, C. B. Tarter.
Astrophys. J., Vol. 249, 422 - 442 (1981).

Previous studies of quasar emission line regions have established that the emitting gas is in clouds occupying a small fraction of the available volume and have suggested that these clouds are confined by a hot intercloud medium. The authors demonstrate that the clouds must be confined, provided only that the heating mechanisms are strong enough to drive the temperature of the low density intercloud gas above a few 10^7 K. The authors have analyzed the thermal properties of the cloud-intercloud system under the assumption that it is in thermal balance (heating equals cooling) although they do not rule out the possibility that the intercloud gas is so hot (possibly relativistic) that it is far from equilibrium.

141.108 Abundances of refractory elements in quasars.
C. M. Gaskell, G. A. Shields, E. J. Wampler.
Astrophys. J., Vol. 249, 443 - 448 (1981) = Lick Obs. Bull., No. 889.

New observations of iron, silicon, aluminum, magnesium, and carbon lines in quasars are presented. From comparison of these and previous observations with theoretical models, it is found that the gas-phase abundances of these refractory elements cannot be much less than solar, and in particular that they do not show the order of magnitude depletions that are found in planetary nebulae and the interstellar medium. Because of this lack of depletion of refractory elements it is argued that the broad emission-line clouds are probably devoid of dust.

141.109 Multiwavelength VLBI observations of the galactic center. K. Y. Lo, M. H. Cohen, A. C. S. Readhead, D. C. Backer.
Astrophys. J., Vol. 249, 504 - 512 (1981).

VLBI observations of the galactic center compact radio source at 6, 3.6, 2.8, and 1.35 cm were carried out in an attempt to determine the source structure. At 6 cm and 3.6 cm, the sizes were approximately $0''.05$ and $0''.015$, respectively. The $0''.001$ core component reported by Kellermann et al. at 3.8 cm was not detected at 3.6 cm. The observations do not allow a distinction between the suggestions that the apparent size of the radio source is controlled by interstellar electron scattering or free-free self-absorption. Thus, the source structure is not yet known despite the measured sizes, but an absolute upper limit of 10^{15} cm can be set to the linear size. This poses a severe constraint on possible physical models of the underlying energy source. The plausibility of various models is discussed.

141.110 Absorption lines in the spectra of quasistellar objects. R. J. Weymann, R. F. Carswell, M. G. Smith.
Annu. Rev. Astron. Astrophys., Vol. 19, (see 003.012), 41 - 76 (1981).

Contents: Introduction. Classification scheme for QSO absorption lines. Techniques. Properties of sharp, metal-containing systems. Lyman alpha systems. Origin of sharp absorption line systems. QSOs with broad or complex absorption lines. Miscellaneous problems. Concluding remarks.

141.111 Compact radio sources.
K. I. Kellermann, I. I. K. Pauliny-Toth.
Annu. Rev. Astron. Astrophys., Vol. 19, (see 003.012), 373 - 410 (1981).

Contents: Introduction. Some basic relations. Spectra. Variability. Structure. Structural variations. Spatial distribution. Summary.

141.112 Triple quasar – effect of a gravitational lens?
V. G. Surdin.
Priroda, 1981, No. 10, p. 101 - 102. In Russian.

141.113 High-resolution observations of faint radio sources and the angular size–flux density relation.
A. J. B. Downes, M. S. Longair, M. A. C. Perryman.
Mon. Not. R. Astron. Soc., Vol. 197, 593 - 626 (1981).

High-resolution radio observations are presented of 40 5C sources which make up two samples complete to 15 mJy at 1407 MHz and 55 mJy at 408 MHz respectively. The observations are used to define the angular size distribution of radio sources at low flux densities. The values are compared with predictions of evolution models from the literature. New flux density measurements at 2.7, 5 and 10.6 GHz are used to define the spectra of some sources over a wide range of frequency.

141.114 Extragalactic double radio sources with asymmetric jets. D. J. Saikia.
Mon. Not. R. Astron. Soc., Vol. 197, 11P - 15P (1981).

Several possible solutions to the problem of an apparently asymmetric jet supplying energy to a roughly symmetric double source have been suggested. Here the author discusses a simple consequence of one of the attractive alternatives in which the apparent asymmetry is due to the relativistic bulk velocity of the radiating material, which enhances the flux density of the approaching jet and diminishes that of the receding one.

141.115 Morphologie et photometrie de 3C 120.
G. Wlérick, P. Bouchet, V. Cayatte, D. Michet.
Astron. Astrophys., Vol. 102, L17 - L20 (1981).

In 3C 120 there exists a jet like structure made of four segments of different orientations. It emits mostly in the continuum. Using H = 60 km s^{-1} Mpc^{-1} its projected length is 11 kpc and its absolute magnitude $M_V = -18.5 \pm 0.3$. These properties are intermediate between the jets of Virgo A and 3C 273. An electronographic plate, taken in the continuum (530 - 660 nm) shows a very simplified structure; in addition to the nucleus and jet, one observes a smooth nebulosity with isophotes similar to those of a giant elliptical galaxy.

141.116 The Fe II spectrum of Seyfert 1 galaxies and quasars. M. Joly.
Astron. Astrophys., Vol. 102, 321 - 330 (1981).

The author determines the physical conditions of the region emitting the permitted Fe II lines observed in the spectrum of Seyfert 1 galaxies and quasars. The exact transfer equations in all the lines are simultaneously solved with the statistical equilibrium equations for a 14 level atom. It is shown that the optical thickness of the lines is the main parameter governing the ratio of the total UV resonance lines intensity over the total subordinate visible lines intensity. The results are compared to the observed spectrum of I Zw 1, PHL 1092, II Zw 136, 3 C 273. It is found that the Fe II emission region has a high electron density $10^{10} \lesssim n_e \lesssim 10^{12}$ cm^{-3}, a low temperature 7500 K $\lesssim T_e \lesssim$ 10,000 K and a very high optical thickness in the UV lines ($\tau_{UV3} \sim 10^5$).

141.117 Quasar reddening, intergalactic dust, and the microwave background. E. L. Wright.
Astrophys. J., Vol. 250, 1 - 6 (1981).

Quasar reddening and distortion of the microwave background are calculated assuming that dust in intervening galaxies and any possible intergalactic dust follow the extinction curve for dust in the Galaxy. Quasar reddening implies that the average line of sight has a visual extinction $A_V = 0.85 \pm 0.51$ at $z = 3$; and this amount of dust is enough to distort the microwave background (given a very bright initial population of stars) but not enough to thermalize the added energy at wavelengths > 3 cm.

141.118 Thermal instabilities in radiatively driven winds– application to emission line clouds of quasars and active galactic nuclei. M. Beltrametti.
Astrophys. J., Vol. 250, 18 - 30 (1981).

It is shown that radiatively driven optically thin winds from active galactic nuclei are thermally unstable provided that the mass loss rates are not smaller than $\sim 50\, M_\odot$ yr^{-1}. Clouds form at distances of $\lesssim 1$ pc, with electron densities

$\gtrsim 10^8$ cm^{-3}, temperatures of $\gtrsim 10^4$ K, and radii between 3×10^{14} and 10^{15} cm. These values agree with the values deduced from observations. Since the clouds are formed in a high velocity wind, this model (as does the model by Eilek and Caroff) avoids problems of disruption inherent in using radiation pressure to accelerate clouds from rest to velocities up to $0.1c$ needed to explain the emission line widths. The thermal balance of the gas is discussed, the criteria for thermal instability are given, and a perturbation analysis is made.

141.119 Quasars near companion galaxies. H. Arp.
            Astrophys. J., Vol. 250, 31 - 42, plates 1 - 11 (1981).
Twenty-two new quasars close to galaxies are reported. Most of them are so close to companion galaxies that the probability of accidental occurrence is less than 0.01. One section of the sky between R.A. = $7^h 47^m$ and $10^h 15^m$ (1950), decl. = $-4°$ to $+90°$, has been most completely analyzed. All nonelliptical galaxies brighter than $m_H = 12.8$ with apparent companion galaxies have been examined. There are 34 candidate companion galaxies in this list. Near 13 of these companions are found quasars with accidental probability less than 0.01. Some companions still remain to be investigated, but already the compound probability of obtaining this result by chance is only $\sim 10^{-17}$.

141.120 Fluid dynamical twisting of the radio jets in 3C 449.
            P. E. Hardee.
Astrophys. J., Lett., Vol. 250, L9 - L13 (1981).
The jet structure of the double jets in the radio galaxy 3C 449 can be understood as resulting from helical instability of an initially cylindrically symmetric straight jet. A growing helical wave is convected out along the jets and reaches large amplitude at the position of the radio lobes disrupting well-directed flow. Adiabatic expansion of the thermalized jet material produces the radio lobes. The extended diffuse emission outside the present radio lobes is the remnant of earlier more luminous jets. Conditions in the expanding inner jets required for this mechanism to operate are almost the same as those inferred by Perley, Willis, and Scott from their VLA observations.

141.121 Optical positions of benchmark radio sources south
            of +5° declination. R. M. West, H. G. Walter.
Astron. Astrophys., Suppl. Ser., Vol. 46, 277 - 279 (1981).
Optical positions of 39 radio sources, mainly in the southern hemisphere have been measured relative to reference stars from the Perth 70 Catalogue. These objects form a subset of extragalactic radio sources, provisionally selected by IAU Commission 24, for the establishment of a reference frame. Typical internal accuracies are at the $0.''2$ level in both coordinates.

141.122 Further spectroscopic observations of 25 quasi-
            stellar objects. J. Surdej, J. P. Swings.
Astron. Astrophys., Suppl. Ser., Vol. 46, 305 - 309 (1981).
Twenty-five quasars reported either variable or with P Cygni profiles, and for the majority of which the redshift is high ($Z_e \gtrsim 2$), were observed spectrographically at the Cassegrain focus of the ESO 3.6 m telescope. For most of them the emission and absorption redshifts are confirmed: in a few cases new lines are detected, as well as new absorption redshift(s).

141.123 Television photometric observations of ten extra-
            galactic peculiar objects.
L. P. Metik, V. V. Prokof'eva.
Izv. Krymskoj Astrofiz. Obs., Tom 63, 140 - 145 (1981). In Russian. English translation in Bull. Crimean Astrophys. Obs., Vol. 63.
During March - April and September, 1976 the brightness of extragalactic objects was measured in a photometric system close to BV using a high-sensitive television system. Five qua-

sars and five BL Lac-type objects having magnitudes in the range from 14^m5 to 18^m5 were observed. Three of these objects (3C279, OJ287 and OY091) showed light variations in the scale time of several days.

141.124 Far separated "pairs" of quasistellar objects –
            possible members of clusters of galaxies or their
groups.   B. V. Komberg.
Pis'ma Astron. Zh., Tom 7, 643 - 647 (1981). In Russian. English translation in Soviet Astron. Lett., Vol. 7.
An analysis of a sample of 61 pairs of quasistellar objects shows that at least 18 of them with $M_{eff} \leqslant 10^{16} M_\odot$ are members of distant pairs of clusters of galaxies.

141.125 Observations of the H_2O radio line in the sources
            W 49 and Orion A.
A. P. Venger, I. V. Gosachinskij, T. M. Egorova, G. N. Il'in, R. A. Kandalyan, N. A. Yudaeva.
Pis'ma Astron. Zh., Tom 7, 677 - 681 (1981). In Russian. English translation in Soviet Astron. Lett., Vol. 7.
Profiles of the H_2O radio line at 1.35 cm wavelength obtained in May - June 1981 on the RATAN-600 radio telescope are presented.

141.126 Optical position and "proper motion" of the radio
            source OQ 208. P. Brosche, M. Geffert.
Astron. Astrophys., Vol. 103, 78 - 82 (1981).
From plates with epochs around 1916 and 1979 the position of OQ 208 has been derived with respect to different catalogue systems. A vanishing "proper motion" was obtained in an extragalactic reference system, whereas the "proper motion" is several $0.''1/100$ yr in the AGK 3 system.

141.127 A simple formula for the surface brightness of
            turbulent radio jets. R. N. Henriksen.
J. R. Astron. Soc. Canada, Vol. 75, 248 (1981). – Abstract.

141.128 Further observations of 3C 273 for the 3.3-μm dust
            feature.
R. J. Rudy, P. D. LeVan, J. M. Rodriguez-Espinosa.
Publ. Astron. Soc. Pacific, Vol. 93, 558 - 559 (1981).
The authors have reobserved the QSO 3C 273 for the 3.3-μm dust emission feature detected by Allen. They find no evidence for this feature in excess of 7% of the surrounding continuum, in contrast to the 40% feature reported previously by Allen. If source variability is to account for the difference in the two sets of observations, the two-year separation implies that the feature is either associated with a compact region or originates nearer the central source than originally thought.

141.129 Positions and flux densities at 5000 MHz for
            1133 weak radio sources.
L. Binette, C. Carignan, J. G. Bolton, A. E. Wright.
Australian J. Phys., Vol. 34, 407 - 444 (1981).
Flux densities and positions have been measured at 5000 MHz for all the radio sources in the Parkes 2700 MHz catalogue with $S_{2700} \geqslant 0.25$ Jy for which no 5000 MHz data have previously been published. The new positions of the sources, which are of considerably higher accuracy than those measured at 2700 MHz, have been examined on the available sky survey material: 17 previously suggested identifications are rejected, but alternative identifications are proposed for 4 of these; identifications are proposed for 118 other sources.

141.130 New optical identifications for weak sources from
            the Parkes 2700 MHz catalogue.
J. G. Bolton, J. Trett, C. Carignan, L. Binette.
Australian J. Phys., Vol. 34, 445 - 470 (1981).
Optical identifications are suggested for 122 weak radio sources from the Parkes 2700 MHz catalogue. Optical positions

have been measured for the proposed counterparts and for 81 previously suggested identifications.

141.131   Flux densities at 8.87 GHz of 347 small-diameter radio sources.   A. J. Shimmins, J. G. Bolton.
Australian J. Phys., Vol. 34, 471 - 491 (1981).

Flux densities at 8.87 GHz ($\lambda = 3.4$ cm) have been determined with the Parkes 64 m telescope for 347 radio sources, using an improved on-off integration method. The sources were selected from the Parkes 2700 MHz catalogues as those having estimated flux densities at 8.87 GHz greater than 0.5 Jy ($\equiv 0.5 \times 10^{-26}$ Wm^{-2} Hz^{-1}) and relatively small angular sizes. A total of 156 of the sources are identified with QSOs and 58 with galaxies, and 133 have not been optically identified. Of the 347 sources 101 have been previously measured at this frequency at Parkes. Some of these repeats were made to check variability and others to compare the flux density scale with that of previous measurements.

141.132   Interferometric observations of compact components in extragalactic radio sources.
A. J. Kerr, P. Birch, R. G. Conway, R. J. Davis, D. Stannard.
Mon. Not. R. Astron. Soc., Vol. 197, 921 - 930 (1981).

Interferometric observations at $\lambda 31$ and 73 cm with resolutions of 2.7 and 1.2 arcsec respectively have been used to study the linear polarization of compact components in extended radio sources.

141.133   The spectral index—luminosity relationship for steep-spectrum cores in extragalactic radio sources.
D. J. Saikia.
Mon. Not. R. Astron. Soc., Vol. 197, 1097 - 1101 (1981).

The author has investigated the spectral index—luminosity relationship for steep-spectrum cores in galaxies and quasars, and finds that his sample of galaxies supports earlier suggestions of a strong correlation, while there is weak evidence for a similar relationship for the quasars. He shows that a strong spectral index—luminosity correlation can be used to set an upper limit to the velocities of the radio-emitting material which is expelled from the nucleus.

141.134   A precessing jet model of compact radio sources.
R. Linfield.
Astrophys. J., Vol. 250, 464 - 468 (1981).

An attempt is made to explain the structure of compact radio sources with a precession-like motion of a relativistic jet. It is found that the curvature of these sources can be readily explained in this way. In addition, the knots which are often revealed by VLBI observations arise naturally in such a model. The main problem with the model is that it cannot by itself explain the arcsecond structure of asymmetric radio sources.

141.135   A spectrophotometric classification of low-redshift quasars and active galactic nuclei.
J. E. Steiner.
Astrophys. J., Vol. 250, 469 - 477 (1981).

Current classifications for quasars and active galactic nuclei are based on apparent properties. The author proposes a spectrophotometric classification which is based on intrinsic properties by separating objects into three groups: Class A — objects with Fe II emission lines; Class B — objects without Fe II and with permitted lines broader than forbidden ones; and Class C — objects without Fe II and with permitted lines and forbidden lines having the same width. He studies the properties of these groups using intrinsic parameters like line ratios, line width, and luminosities. Class A objects have the property that the ratio Hβ/[O III] is proportional to the absolute luminosity (either L_χ or L_{opt}) of the object.

141.136   The formation of emission lines in quasars and Seyfert nuclei.   J. Kwan, J. H. Krolik.
Astrophys. J., Vol. 250, 478 - 507 (1981).

The authors calculate the photoionization and heating throughout a quasar emission-line cloud which is very optically thick at the Lyman edge. Conditions of ionization, thermal, and pressure equilibrium are imposed at each point. Photoionization and collisional ionization from excited states of hydrogen are included; these processes maintain a substantial electron fraction even after the exhaustion of Lyman continuum photons halts ground-state photoionization. In determining the thermal equilibrium, the authors include, besides the hydrogen lines and continua, all lines of helium and heavy elements generally observed in quasar spectra.

141.137   The correlation of X-ray emission with strong millimeter activity in extragalactic sources.
F. N. Owen, D. J. Helfand, S. R. Spangler.
Astrophys. J., Lett., Vol. 250, L55 - L58 (1981).

A sample of 25 strong extragalactic millimeter sources have been observed at ~ 1 keV with the Imaging Proportional Counter on the Einstein X-ray Observatory and at 90 GHz with the NRAO 11 m telescope. A strong correlation between the millimeter and X-ray flux densities is found. This result is consistent with a synchrotron self-Compton model in which the X-ray emission arises from the Compton scattering of millimeter radio photons off the relativistic electrons which created them.

141.138   Structure investigations of 3C 196 and 3C 280 in the decameter waveband.   V. P. Bovkoon (*Bovkun*), I. N. Zhouck (*Zhuk*), A. V. Megn (*Men'*).
Astrophys. Space Sci., Vol. 80, 411 - 423 (1981).

Observational results are presented concerning the structure of the quasar 3C 196 and the radio galaxy 3C 280 at 20 and 25 MHz, obtained by the scintillation method with the URAN-1 interferometer. Angular dimensions of the scintillating components and extended regions of the sources have been evaluated. In the case of the quasar 3C 196, the effective angular size of the scintillating component equals $2 \pm 1\rlap.{''}5$ and that of the extended region 18 X 25$''$. The contribution of the compact component into the total radiation flux is 0.46 ± 0.20. Spectra of the structural formations in 3C 196 have been obtained in the range 20–5000 MHz.

141.139   Radio observations of W50 and the surrounding area.
A. J. B. Downes, T. Pauls, C. J. Salter.
Vistas Astron., Vol. 25, (see 012.040), 75 - 77 (1981).

SS 433 lies at the centre of the galactic non-thermal radio source W50. The authors present observations of W50 and the surrounding area at 1.7 and 2.7 GHz in both total intensity and linear polarization made with the Effelsberg 100-m telescope.

141.140   CTB 80: a galactic radio source with directed jets.
J. R. Dickel, P. E. Angerhofer, R. G. Strom, M. D. Smith.
Vistas Astron., Vol. 25, (see 012.040), 127 - 135 (1981).

The galactic radio source CTB 80 shows several remarkable jets which can be traced outward to over 30 pc from a point less than 0.1 pc within a central core. If the object is associated with a supernova of 1408 then the jets must move out at a velocity of 0.2 c but the morphology of the whole source suggests a significant proper motion of the central object which argues for an older age. A model involving a supernova explosion ejecting several massive path-cutting fragments is quantitatively explored.

141.141   The V/V_m test for quasars: a new interpretation.
M. R. S. Hawkins, N. J. Stewart.
Astrophys. J., Vol. 251, 1 - 3 (1981).

The V/V_m test is discussed, and it is suggested that if certain problems of misidentification are not taken into account, the expected value for $\langle V/V_m \rangle$ without evolution is not $^1/_2$, but depends on the quasar luminosity function and is compatible with currently observed values of about 0.68. A complete sample of quasars is examined, and it appears that beyond a redshift of about 0.5, the evidence for evolution is not strong. Differences in the value of $\langle V/V_m \rangle$ for steep and flat spectrum quasars can be simply explained as differences in luminosity function.

141.142   Observations of optical Fe II emission in the spectra of radio-quiet quasi-stellar objects.
B. M. Peterson, C. B. Foltz, P. L. Byard.
Astrophys. J., Vol. 251, 4 - 7 (1981).

The optical spectra of four radio-quiet QSOs have been examined. In each of these spectra, emission from the optical multiplets of Fe II appears strongly. In contrast, a study by Phillips indicates that Fe II emission is not a general feature of QSOs which are strong radio sources. The similarity between the optical spectra of Seyfert 1 galaxies and radio-quiet QSOs suggests that the latter may simply be more distant examples of the former. The differences between radio sources and radio-quiet objects may be more fundamental than the differences between QSOs and broad-line emission galaxies.

141.143   The quasar 2141 + 174: not a case of self-absorption in the Balmer lines.   C. M. Gaskell.
Astrophys. J., Vol. 251, 8 - 9 (1981) = Lick Obs. Bull., No. 894.

The apparent self-absorption in the Balmer line profiles of the quasar 2141 + 174 is shown not to be real. The unusual Balmer line profiles are due to a large difference in redshift between the broad lines and the narrow lines in this quasar.

141.144   The discovery and observed properties of QSOs at large redshifts – an update.   M. G. Smith.
Investigating the universe, (see 003.014), p. 151 - 206 (1981).

Contents: Techniques, surveys and selection effects. The distribution of QSOs on the sky. The continuous spectrum of QSOs, and the nature of the central source of energy. The emission-line spectrum. Absorption lines in QSO spectra. Gravitational lenses.

141.145   Magnetic field distribution in extended radio sources and halo source features.
E. Ma, F. Durret, B. Chen.
Sci. Sinica, Vol. 24, 1545 - 1552 (1981).

By comparison between X-ray and radio contour maps of several radio galaxies and clusters of galaxies, the distribution of the lower limit of magnetic field and related physical quantities have been calculated according to the theory of inverse Compton scattering. On the basis of this comparison and calculated distribution, the role that the inverse Compton scattering plays in the X-ray emission has been discussed for different kinds of sources. It is shown that clusters of galaxies can be a combination of two components: one or several radio-galaxies and an extended halo source.

141.146   Some possible clusters of quasars.
Y.-z. Liu, S.-d. Jiang, Y.-l. Bian, X.-y. Tang, S.-l. Cao, X.-h. Xiao.
Acta Astrophys. Sinica, Vol. 1, 243 - 248 (1981). In Chinese.

141.147   An Effelsberg-Green Bank galactic H I absorption line survey. I. The observations.
U. Mebold, A. Winnberg, P. M. W. Kalberla, W. M. Goss.
Astron. Astrophys., Suppl. Ser., Vol. 46, 389 - 419 (1981).

Observations of galactic H I absorption in the continuum spectra of 69 radio sources have been carried out using the NRAO 3-element interferometer and the MPIfR 100-m telescope. A special effort was made to obtain high velocity resolution and good signal-to-noise ratio. Emission profiles were obtained for all sources using either the 100-m telescope at Effelsberg, the 91-m telescope at Green Bank or the 64-m telescope at Parkes. In the present paper the observations are published.

141.148   Compact and extended structure in B2 radio sources of intermediate strength.
L. Padrielli, V. K. Kapahi, J. K. Katgert-Merkelijn.
Astron. Astrophys., Suppl. Ser., Vol. 46, 473 - 481 (1981).

A complete sample of sources for which high-resolution data obtained at Jodrell Bank at 408 MHz are available has been observed with the Westerbork Synthesis Radio Telescope (WSRT) at 5 GHz. A catalogue of positions, structures and flux densities as derived from the WSRT observations is given, together with the results of an optical identification programme. A comparison of the high-resolution (i. e. up to 0.3 arcsec) data and the WSRT data (with a resolution of $\lesssim 8$ arcsec) is made.

141.149   Fe II emission in quasars.   S. A. Grandi.
Astrophys. J., Vol. 251, 451 - 464 (1981).

In an attempt to understand Fe II emission in quasars, observations of Fe II blends at $\lambda 2100$, $\lambda 2500$, $\lambda 2950$, $\lambda 3200$, $\lambda 4570$, $\lambda\lambda 5190$, 5320 as well as Mg II $\lambda 2798$, the Balmer lines and continuum, and the power law spectral index are reported for 38 quasars and high redshift Seyfert 1 galaxies. Considering only relative strengths among the UV Fe II blends, which should be insensitive to column density effects, the observed blend strengths agree well with predicted strengths from optically thick collisional excitation models for Fe II emission. However, synthetic spectra based on two extensive Fe II calculations (due to Phillips and to Kwan and Krolik) disagree with observed quasar spectra.

141.150   On He II $\lambda 4686$ emission and the question of reddening in quasars and Seyfert galaxies.
G. M. MacAlpine.
Astrophys. J., Vol. 251, 465 - 470 (1981).

The He II $\lambda 4686/H\beta$ intensity ratio in quasars or class 1 Seyfert galaxies provides an important test for calculations which attempt to explain ratios of Lyman lines to Balmer lines by recourse to collisional excitation of the Balmer lines without any reddening. With the assumption that $\lambda 4686$ arises from recombination and radiative cascading, the models fail to produce enough emission for this line; and a comparison of calculated and observed equivalent widths suggests that the continua are significantly altered by reddening. Other possible contributing processes for $\lambda 4686$ emission have been examined to determine their importance in comparison with recombination.

141.151   On the energetics and morphology of radio tail galaxies.
W. A. Christiansen, A. G. Pacholczyk, J. S. Scott.
Astrophys. J., Vol. 251, 518 - 522 (1981).

The energetics of the multiple ejection plasmon (plasmoid) model of a radio tail source is in full agreement with the observed luminosity of the entire tail for time scales required by observations of typical tail sources such as, e. g., NGC 1265. Furthermore, the inclusion of the effects of an interstellar medium as suggested by Jones and Owen assures the agreement between the observed morphological structure and that predicted by the multiple ejection plasmon model.

141.152   Detection of radio emission from the jet in Centaurus A.
E. J. Schreier, J. O. Burns, E. D. Feigelson.
Astrophys. J., Vol. 251, 523 - 529 (1981).

Radio emission from the X-ray jet in Centaurus A has been detected at 20 cm and 6 cm using the VLA. The radio morphology is very similar to that of the X-ray jet. The spectral index between radio and X-ray is about 0.85; a flatter spectrum is consistent with the radio data, but a spectral break is required below the X-ray band. The authors suggest that the same population fo relativistic electrons is responsible for both radio and X-ray synchroton emission; in situ acceleration of the electrons in the knots would be mandatory. Alternatively, the relativistic beam may heat the surrounding gas, leading to thermal X-ray emission. The presence of the ambient hot gas in the galaxy appears sufficient to statically confine the knots of the jet. The nucleus of the galaxy has an inverted spectrum in the radio, and, at low frequencies, the jet is as bright as the nucleus.

141.153 On the superluminal motion in the quasar 3C 345.
J. Schraml, I. I. K. Pauliny-Toth, A. Witzel, K. I. Kellermann, K. J. Johnston, J. H. Spencer. Astrophys. J., Lett., Vol. 251, L57 - L59 (1981).

The authors report VLBI observations of the quasar 3C 345 at 2.8 and 6 cm during the period 1977 - 1980. The simplest interpretation is that the overall structure appears to have increased in size during the period 1974 - 1980 due to the separation of a low surface brightness region from a bright double core with an apparent angular velocity of ~ 0.3 mas yr^{-1}. During this same period the separation of the core components has remained close to 1.3 mas. However, alternative models cannot be ruled out.

141.154 A complex picture of 3C 345.
J. H. Spencer, K. J. Johnston, I. I. K. Pauliny-Toth, A. Witzel. Astrophys. J., Lett., Vol. 251, L61 - L63 (1981).

The authors present 1980.1 epoch VLBI observations at 5011 MHz of 3C 345 which show it to have complex structure on the milliarcsecond scale, indicating that it is not a simple double source. This complex source structure appears to cloud the interpretation of superluminal expansion in this source. The structure appears to be consistent with a model of a bent beam that extends to the arcsecond scale structure.

141.155 Observations of a decrement in the microwave background radiation toward the distant cluster of galaxies 0016 + 16.
M. Birkinshaw, S. F. Gull, A. T. Moffet. Astrophys. J., Lett., Vol. 251, L69 - L73 (1981).

A cooling of the microwave background radiation of -1.36 ± 0.28 mK has been detected at 10.7 GHz in the direction of the distant cluster of galaxies 0016 + 16. The central gas temperature and pressure implied by this measurement and the X-ray data of White, Silk, and Henry are ~ 14 keV and 2.5×10^{-11} N m^{-2}. Only a lower limit (10 km s^{-1} Mpc^{-1}) to the Hubble constant can be derived from the present data.

141.156 Hydrogen line spectrum in quasars. I. Approximation procedures for line transfer versus an exact treatment.
S. Collin-Souffrin, P. Delache, S. Dumont, H. Frisch. Astron. Astrophys., Vol. 104, 264 - 275 (1981).

Approximation procedures frequently used to handle self-absorption effects in hydrogen emission lines of quasars are discussed and compared to an exact numerical treatment of line transfer. The model used is a finite slab with prescribed density and temperature.

141.157 The MPI-NRAO 6 cm surveys for extragalactic radio sources.
A. Witzel. Proceedings of the 5th Göttingen-Jerusalem-Symposium on Astrophysics, (see 012.041), p. 99 - 110 (1981).

141.158 VLA observations of 1400 + 162 and other BL Lacertae candidates.
P. Hintzen, F. Owen. Astron. J., Vol. 86, 1577 - 1584 (1981).

Twenty-centimeter VLA maps are presented for six radio sources which have been identified in the literature as probable BL Lacertae objects. The source 0957 +227, whose optical

counterpart shows no emission lines, is a double with heavily resolved components. Also observed were four objects identified as BL Lacertids from polarization measurements of suggested optical counterparts. The radio positions indicate that three of these four identifications are incorrect. The correctly identified object, 2201 +171, consists of a single source with a 4″ jet extending to the west. Finally, the source 1400 + 162, identified by Baldwin et al. as a BL Lacertae object in a group of galaxies, was found to be a bent, heavily resolved triple radio source.

141.159 The structure of DA 344 at 1.67 GHz.
R. L. Mutel, R. B. Phillips, R. Skuppin. Astron. J., Vol. 86, 1600 - 1603 (1981) = Contrib. No. 86 Iowa Radio Astron. Group.

The radio galaxy DA 344 has been mapped at 1.67 GHz using a five-telescope intercontinental VLBI array. The structure consists of two well separated, nearly equal simple lobes ~ 10 milliarcsec in size, separated by 55 milliarcsec. This morphology is very similar to several previous mapped compact objects such as CTD 93 and 3C 395, and evidently results from similar emission processes.

141.160 318-MHz variability of complete samples of extragalactic radio sources. II.
B. Dennison, J. J. Broderick, J. E. Ledden, S. L. O'Dell, J. J. Condon. Astron. J., Vol. 86, 1604 - 1618 (1981).

The authors report the remainder of two- and three-epoch 318-MHz observations of extragalactic sources in samples complete to 3 Jy at 1400 MHz and 1 Jy at 5000 MHz. From analysis of this low-frequency variability survey, they find that steep-spectrum ($\alpha \geqslant 0.5$) sources do not appear to vary, but about 40% of all flat-spectrum ($\alpha < 0.5$) sources exhibit low-frequency variability exceeding 8% over ~ 5 yr. Among the flat-spectrum sources, those with inverted spectra show the largest fractional variations. They also find that the incidence of low-frequency variability is strongly correlated with the determination that a source is an optically violent variable. These statistical properties are consistent with models invoking relativistic beaming of radio and optical emission.

141.161 Extragalactic radio sources.
R. D. Blandford. Plasma astrophysics, (see 012.042), p. 207 - 214 (1981).

Recent observational and theoretical developments in the study of extragalactic radio sources are reviewed. Over seventy radio sources have been shown to possess jets. It is argued that these jets are conduits along which flow mass, momentum, energy and magnetic flux. Particular attention is paid to recent work on symmetric sources, magnetic collimation and instabilities.

141.162 Extra-galactic double radio sources.
R. V. E. Lovelace, H. A. Scott. Plasma astrophysics, (see 012.042), p. 215 - 221 (1981).

A discussion is given of different aspects of a unified model of double radio sources in which the accretion disc of a massive black hole acts as a unipolar induction dynamo producing oppositely directed beams of relativistic particles. The hydrodynamics of accretion flows with angular momentum is discussed.

141.163 Production and propagation of jets.
M. J. Rees. Plasma astrophysics, (see 012.042), p. 267 - 271 (1981).

The paper deals with the processes in active galactic nuclei which may give rise to the large-scale jet structures in radio galaxies.

141.164 A doughnut model of quasars.
P. Allan. Plasma astrophysics, (see 012.042), p. 349 - 350 (1981).

The author proposes a model for quasars which over-comes the problem with accretion disc of not knowing the correct viscosity. These models have the shape of doughnuts, and can be usefully compared with thick accretion disc models.

141.165 **Plasma beams from doughnuts.**   P. M. Allan.
Plasma astrophysics, (see 012.042), p. 351 - 352 (1981).
The author proposes a model to explain the jets seen in radio galaxies and SS433, in which gas is accelerated by radia-tion pressure from a doughnut.

141.166 **Large scale Kelvin-Helmholtz instabilities in extra-galactic radio jets.**   A. Ferrari, E. Trussoni.
Plasma astrophysics, (see 012.042), p. 353 - 357 (1981).
The relation of the morphological features detected in radio jets from active galactic nuclei with the development of large scale Kelvin-Helmholtz instabilities in supersonic flows is discussed. In particular it is pointed out that, depending on the Mach number and the beam velocity, different damping processes are expected to occur which dissipate the unstable modes and saturate their non linear evolution. Applications of this theoretical scheme to peculiar radio jets are shortly commented.

141.167 **The unsteady beam.**   M. Nepveu.
Plasma astrophysics, (see 012.042), p. 359 (1981).

141.168 **Radio jet of 3C273.**
R. G. Conway, R. J. Davis, A. R. Foley, T. P. Ray.
Nature, Vol. 294, 540 - 542 (1981).
Most radio sources are two-sided, but a minority appear one-sided, 3C273 being the first-known and brightest example. The authors report here new radio observations at 408 MHz of 3C273 which show that the brightness of the postulated counter-jet is $<1/100$ of the brightness of the visible jet. If this ratio is due to Doppler beaming, the source must be seen almost end-on, and the whole jet must be moving at a quasi-relativistic speed ($>0.7c$) into an ambient medium with num-ber density <0.6 m^{-3}. Because several arguments suggest that such a density is implausibly low, the jet of 3C273 cannot be identified with the radio lobe of a normal double source.

141.169 **Intrinsic variations of the double quasar 0957+56 AB.**
C. Lloyd.
Nature, Vol. 294, 727 - 728 (1981).
The author reports observations that show a variation of ~ 1 mag in both components and behaviour typical of in-trinsically variable quasars with similar radio structure. The near constancy of the magnitude difference between the com-ponents at several epochs despite overall variations favours all the variations being intrinsic to the quasar and also supports the gravitational lens hypothesis.

141.170 **A Doppler theory of quasars.**
J. V. Narlikar, M. G. Edmunds.
J. Astrophys. Astron., Vol. 2, 289 - 307 (1981).
The authors examine a Doppler theory of quasars in which it is assumed that a fraction of the total population of quasars are fired from centres of explosion with moderate cosmological redshifts. It is argued that the substantial part of the redshift of a typical high redshift quasar could be of Doppler origin.

141.171 **Prediction of multiplicity of QSS is confirmed.**
B. A. Vorontsov-Vel'yaminov.
Astron. Tsirk., No. 1120, p. 5 - 6 (1980). In Russian.

141.172 **On brightness temperatures of extragalactic variable radio sources.**

V. N. Kuril'chik, M. N. Yakovlev.
Astron. Tsirk., No. 1121, p. 1 - 3 (1980). In Russian.

141.173 **Radio outbursts in 3C 345, 3C 454.3 and Q0241 + 622.**
IAU Circ., No. 3637 (1981).

141.174 **Q0957 + 561A, B.**
IAU Circ., No. 3644 (1981).

141.175 **A search for galactic compact radio sources. I. Two-frequency method and scintillating sources in the vicinity of supernova remnants.**
A. V. Pynzar', V. A. Udal'tsov.
Astron. Zh., Tom 58, 1177 - 1186 (1981). In Russian.
English translation in Soviet Astron., Vol. 25, No. 6.
A new two-frequency correlation method of observation of scintillating sources is outlined. The sensitivity of this meth-od is higher than that of the known one-frequency method. A selected search for scintillating sources in the vicinity of super-nova remnants in the galactic latitudes region ±10° has been carried out with the 20 000 m² effective area array at 102.5 MHz. Fourteen new scintillating sources have been de-tected in the galactic anticentre region. The positions, scintil-lating component flux densities and total flux densities of the sources were determined.

141.176 **Large-scale density distribution in thermal radio sources.**   S. A. Gulyaev, A. B. Men'shchikov.
Astron. Zh., Tom 58, 1207 - 1212 (1981). In Russian.
English translation in Soviet Astron., Vol. 25, No. 6.
A simple method for finding the density distribution in the outer regions of thermal radio sources from the appearance of its continuum spectra at low frequencies is proposed. The method is applied to the radio sources W 49A, ON 1 and the Orion nebula. The possibility of investigation of the galactic center region by the method proposed is discussed.

141.177 **On the Kellermann and Pauliny-Toth criterion.**
Z. I. Tsvetanov, V. M. Charugin.
Astron. Tsirk., No. 1135, p. 3 - 5 (1980). In Russian.

141.178 **A survey of the sky between 2° and 4° declination.**
V. R. Amirkhanyan, A. G. Gorshkov,
A. A. Kapustkin, V. K. Konnikova, A. N. Lazutkin,
M. G. Larionov, A. S. Nikanorov, V. N. Sidorenkov,
L. S. Ugol'kova, O. I. Khromov.
Astron. Tsirk., No. 1137, p. 5 - 7 (1980). In Russian.

141.179 **Observations of radio sources at 3.6 GHz.**
V. R. Amirkhanyan. A. G. Gorshkov.
M. G. Larionov, L. S. Ugol'kova.
Astron. Tsirk., No. 1139, p. 7 - 8 (1980). In Russian.

141.180 **New quasistellar objects. I.**
B. E. Markarian, D. A. Stepanyan,
V. A. Lipovetskij.
Astron. Tsirk., No. 1141, p. 1 - 3 (1980). In Russian.

141.181 **New quasistellar objects. II.**
B. E. Markarian, D. A. Stepanyan, V. A. Lipovetskij.
Astron. Tsirk., No. 1142, p. 1 - 3 (1980). In Russian.

141.182 **Polarization survey of radio objects with continuous optical spectrum. I.**
G. M. Beskin, Yu. S. Efimov, S. I. Neizvestnyj, S. A
S. A. Pustil'nik, N. M. Shakhovskoj.
Pis'ma Astron. Zh., Tom 7, 707 - 713 (1981). In Russian.
English translation in Soviet Astron. Lett., Vol. 7.
Results of a search for optical linear polarization of 12 radio objects with continuous optical spectrum are reported.

Polarization of 11 of them is detected and its values are from 1 to 24%.

141.183 Imaging observations of optical emission from jets.
H. R. Butcher.
Optical jets in galaxies, (see 012.052), p. 23 - 24 (1981).

A ground-based survey for optical counterparts to radio jets is described. Present detection limits are consistent with all radio jets having continuous spectra into the optical with spectral indices near -0.7 (i.e., $S_\nu \sim \nu^{-0.7}$).

141.184 Spectroscopic evidence for collimated outflow.
H. R. Butcher.
Optical jets in galaxies, (see 012.052), p. 39 - 41 (1981).

Recent optical observations of several nearby radio sources are discussed. At present, optical evidence for collimated outflow is mostly indirect, involving the results of interactions of radio emitting plasma with interstellar material. One case of simultaneous ejection of thermal and relativistic plasmas has apparently been found, however.

141.185 Jets in 3C 273 and other extragalactic objects.
H. Arp.
Optical jets in galaxies, (see 012.052), p. 53 - 61 (1981).

Photographic observations of optical jets in 3C 273, M 87, NGC 5128, 3C 120, 3C 66B and NGC 4448 are presented. The morphology and possible origins of the jets are discussed.

141.186 Optical emission from the extended radio source 3C 277.3 (Coma A).   W. van Breugel.
Optical jets in galaxies, (see 012.052), p. 63 - 67 (1981).

Optical emission has been detected both from the radio jet and from the outer lobe regions of the radio galaxy 3C 277.3 (Coma A) at a redshift of 0.0857. Two optical continuum knots are seen in the jet, the brighter of which is found to be 12% polarized with the magnetic vector along the jet. Intense optical emission lines have been observed from the knot region of the jet and from regions near the peripheries of the radio lobes.

141.187 Recent observations of large angular size radio jets.
A. G. Willis.
Optical jets in galaxies, (see 012.052), p. 71 - 76 (1981).

Various properties of radio jets having projected angular sizes of more than a few arcseconds are summarized. Their intrinsic projected sizes range up to $\sim$290 kpc. Their radio spectra tend to be flatter than the spectra of the integrated emission of the sources with which they are associated. Jet wiggles, gaps, magnetic field structures, collimation and alignment are discussed. The bulk flow velocity of material in these jets is probably non-relativistic.

141.188 Radio jets in flat-spectrum sources.
R. A. Perley.
Optical jets in galaxies, (see 012.052), p. 77 - 81 (1981).

VLA-maps at 6 cm of high-luminosity quasars are presented. The appearance of jet-like structures is discussed in the context of the need of continuous energy supply for the large extended radio lobes of these objects.

141.189 The large radio galaxy DA 240: evidence for jets?
R. G. Strom, A. G. Willis.
Optical jets in galaxies, (see 012.052), p. 83 - 86 (1981).

Although there appears to be no radio jet in the large double source DA 240, there is evidence for collimated structure in its outer components. The authors note a probable morphological connection between the outer radio emission and the optical jet in the central galaxy MCG 9−13−57. This system and NGC 5128/Cen A have striking features in common including a strong indication that the gas flowing in the jet contains thermal material and is moving nonrelativistically.

141.190 Observations of core-dominated sources with MTRLI.   I. W. A. Browne, M. J. L. Orr.
Optical jets in galaxies, (see 012.052), p. 87 - 89 (1981).

Maps of core-dominated sources have been made at 408 and 1666 MHz with the Jodrell Bank MTRLI. These maps show that many such sources have clearly defined jets, in addition to compact cores and regions of diffuse emission. The authors argue that, if Doppler beaming models offer a true explanation of the core properties, then a) the jets are moving relativistically, and b) core-dominated and classical double quasars are one and the same type of object.

141.191 Jets in a sample of low luminosity radio galaxies.
R. Fanti, P. Parma.
Optical jets in galaxies, (see 012.052), p. 91 -95 (1981).

The authors discuss a sample of low luminosity radiogalaxies selected from the B2 catalogue. In a large number of these sources jets were detected (9 out of 17). They discuss in more detail the low luminosity radiosource B2 1321+31. In some regions in the jet the brightness and polarization data suggest that the flow velocity along the jet decreases.

141.192 VLBI observations of jets and active nuclei.
E. Preuss.
Optical jets in galaxies, (see 012.052), p. 97 - 105 (1981).

Some important observational facts from VLBI relevant to the discussion on "jets" are summarized. Lists of objects and samples observed by VLBI and published since 1975 are attached.

141.193 Jet-like radiostructure in the core of 3C 236.
R. T. Schilizzi, G. K. Miley, F. L. J. Janssen, P. N. Wilkinson, T. J. Cornwell, E. B. Fomalont.
Optical jets in galaxies, (see 012.052), p. 107 - 108 (1981).

Preliminary conclusions regarding the small scale radio structure of 3C 236 are reported.

141.194 X-ray observations of jets.
E. J. Schreier.
Optical jets in galaxies, (see 012.052), p. 109 - 114 (1981).

The current status of X-ray emission from jets, as observed by the Einstein X-ray Observatory, is reviewed. X-rays are detected from the jets of 3C 273, M 87, and Centaurus A. Preliminary results from VLA observations of the Centaurus A jet are also discussed. In all three cases, the data indicate synchrotron radiation, with the spectral index breaking in the range of $10^{14} - 10^{15}$ Hz; in situ acceleration of the relativistic electrons is required, since the electron lifetimes are significantly shorter than the light travel times from the nucleus.

141.195 X-ray emission from the radiogalaxy 3C 66B.
M. Tarenghi, D. Maccagni.
Optical jets in galaxies, (see 012.052), p. 115 - 116 (1981).
Abstract.

141.196 Optical jets and beam models of extended radio sources.   A. Ferrari, E. Trussoni, L. Zaninetti.
Optical jets in galaxies, (see 012.052), p. 119 - 123 (1981).

Theoretical models about the morphology and optical nonthermal emission of galactic jets are discussed, mainly referring to the case of M 87. It is shown that observed physical features can be related to the dynamical interaction of the jet moving through the intracluster gas. Dense clouds or large scale unstable perturbations can explain the structure of the blobs, while nonthermal high frequency radiation can be due to synchrotron emission by electrons accelerated in turbulent plasmas or in shock fronts.

141.197 Collimated ejection of radio emitting material in Seyfert galaxies.   A. S. Wilson.
Optical jets in galaxies, (see 012.052), p. 125 - 130 (1981).

VLA maps of six Seyfert galaxies show double (or triple)

radio structures, with two radio components straddling the optical continuum nucleus and a third, when present, coincident with it. The scales of the double sources lie in the range a few hundred parsecs to a few kiloparsecs. It is argued that these doubles represent ejection from the inner nucleus, either as plasmoids or a particle beam, followed by stopping or beam disruption by surrounding interstellar gas in the inner region of the galaxy. The broadened forbidden optical emission lines in these galaxies may originate in interstellar clouds accelerated by the hypothetical particle beam.

141.198   An infrared study of quasars.
A. R. Hyland, D. A. Allen.
Anglo-Australian Obs. Prepr. No. 152, 21 pp. (1981). – Submitted to Mon. Not. R. Astron. Soc.

141.199   Quasars in a control field far from bright galaxies.
H. Arp. J. Surdej.
ESO Sci. Prepr., No. 164, 18 pp. (1981). – Submitted to Astron. Astrophys.

141.200   On the quasar surface density.
P. Véron, M. P. Véron.
ESO Sci. Prepr., No. 165, 1 + 18 pp. (1981). – Submitted to Astron. Astrophys.

141.201   The 1919+479 radio tail, a moving galaxy within an accumulated gaseous halo.   E. A. Valentijn.
ESO Sci. Prepr., No. 172, 1 + 4 pp. (1981). – To be published in the Proc. of the IAU Symp. No. 97: 'Extragalactic radio sources', D. S. Heeschen, C. M. Wade (Editors). D. Reidel Publishing Company, Dordrecht, Holland–Boston, U.S.A.–London, England.

141.202   20 and 25 MHz observation of a compact source in Cas A.   V. P. Bovkun, S. Ya. Braude, A. V. Men'.
Dokl. AN SSSR, Tom 259, 811 - 813 (1981). In Russian. Abstr. in Ref. zh., 51. Astron., 12.51.864 (1981).

141.203   Multifrequency observations of the red QSO 1413+135.   J. N. Bregman, M. J. Lebofsky, M. F. Aller, G. H. Rieke, H. D. Aller, P. E. Hodge, A. E. Glassgold, P. J. Huggins.
Prepr. Steward Obs., No. 334, 24 pp. (1981).
The red QSO 1413+135 is particularly suited to detailed study because it is one of the brightest members of its class at radio and infrared wavelengths. By determing the redshift and monitoring the infrared and radio flux, the authors have obtained essential information on the size of the emitting region. They have also obtained nearly simultaneous observations of the continuous spectrum in the radio, infrared, and X-ray bands. The infrared observations provide a good determination of a cutoff in the spectrum. The composite spectrum permits the parameters of a simple theoretical model to be determined, i.e. size, magnetic field, and bulk velocity and the maximum energy of the synchrotron electrons. The key element in the analysis is that the unusual shape of the steeply falling infrared spectrum of 1413+135 places strong constraints on theoretical models of the source of its continuous emission.

141.204   Time dependent energy supply in radio sources and morphology of radio lobes.
W. A. Christiansen, A. G. Pacholczyk, J. S. Scott.
Prepr. Steward Obs., No. 337, 3 pp. (1981).

141.205   37 GHz-radio observations of quasar-like objects at Metsähovi, Finland.   H. Lehto, P. Teerikorpi, S. Haarala, E. Valtaoja, A. Sillanpää, M. Valtonen, H. Teräsranta, E. Salonen, S, Urpo, M. Tiuri.
Turku Univ. Obs. Informo, No. 52 (1981). – Abstract.

141.206   Statistical investigation of QSO's distribution.
J.-l. Zhao.
Ann. Shanghai Obs. Acad. Sinica, No. 1, p. 50 - 55 (1979).
The distribution of redshifts and magnitudes of over 630 QSO's is analyzed statistically by χ^2-test, and it is shown that the redshift distribution of QSO's is significantly anisotropic, while the QSO's magnitude distribution appears isotropic.

141.207   Analysis of periodicity in the distribution of QSO's redshifts.   W.-x. Lu, B.-c. Qian, Z.-x. Xu, L.-L. Mi.
Ann. Shanghai Obs. Acad. Sinica, No. 1, p. 56 - 59 (1979).

141.208   QSO's continuum energy distribution.
R.-y. Wang, B.-c. Qian.
Ann. Shanghai Obs. Acad. Sinica, No. 1, p. 60 - 65 (1979).
The continuum energy distribution of QSO's is obtained, using the data of reliable three colour photometry and the redshifts of 235 QSO's.

141.209   Scintillation observations of compact sources in the Crab nebula and in Cassiopeia A in the decametric radio range.   V. P. Bovkun, I. N. Zhuk.
Dokl. AN USSR, 1981, A, No. 7, p. 57 - 60. In Russian. Abstr. in Ref. zh., 51. Astron., 1.51.705 (1982).

141.210   A new attempt of absorption line identification in quasar spectra with molecular lines.   I. E. Val'ts.
Inst. kosm. issled. AN SSSR. Prepr., 1981, No. 640, 37 pp. In Russian. – Abstr. in Ref. zh., 51. Astron., 1.51.794 (1982).

141.211   The angular distribution of the cosmic background radiation.   R. Fabbri, B. Melchiorri, F. Melchiorri.
High-energy astrophysics, (see 012.067), p. 19 - 32 (1981).
The authors review the measurements of the cosmic background anisotropies at all angular scales. Special attention is devoted to experimental problems and questions concerning the data analysis.

141.212   Extragalactic radio sources.
R. D. Blandford.
High-energy astrophysics, (see 012.067), p. 39 - 48 (1981).
Some recent observational results on extended and compact extragalactic radio sources are described. Theoretical interpretations are critically examined.

141.213   High-energy neutrinos from powerful radio galaxies.
R. Silberberg, M. M. Shapiro.
High-energy astrophysics, (see 012.067), p. 63 - 66 (1981).
In order to explore mechanisms for the production of radio lobes from radio galaxies, the authors propose observational tests involving neutrinos at $E \geqslant 4$ TeV. Among the mechanisms that have been suggested are: an explosive burst of energy that has been stored in or near the galactic nucleus; diffusive escape of particles from the vicinity of the galactic nucleus into plasmons; a beam from the galactic core that interacts with the circumgalactic medium; and black holes or spinars ejected from the galactic nucleus by a gravitational slingshot mechanism. As an example, the authors estimate neutrino fluxes from Cen A.

141.214   Radio objects with continuum optical spectrum showing rapid variability at cm wavelengths.
S. A. Pustil'nik.
Astron. Tsirk., No. 1152, p. 1 - 2 (1981). In Russian.

141.215   A survey of the sky between declinations 4°- 6°.
V. R. Amirkhanyan, A. G. Gorshkov, A. A. Kapustkin, V. K. Konnikova, A. N. Lazutkin, M. G. Larionov, A. S. Nikanorov, V. N. Sidorenkov,

L. S. Ugol'kova, O. I. Khromov.
Astron. Tsirk., No. 1157, p. 2 - 5 (1981). In Russian.

141.216   Flare of the H₂O maser radio source U Orionis.
         E. E. Lekht, G. M. Rudnitskij, R. L. Sorochenko.
Astron. Tsirk., No. 1162, p. 7 - 8 (1981). In Russian.

141.217   On the hydrogen spectrum of 3C 273B.
         L. Luud, M. Ilmas.
Academy of Sciences of the Estonian SSR, Prepr. A-6, p. 1 - 18
(1980). In Russian.

The former results obtained by the authors on hydrogen
line formation in the spectrum of 3C 273B have been revised
using new observational data. The observed intensity ratios
and fluxes of hydrogen lines refer to a moderate interstellar
and/or intraquasar reddening. The slightly diluted ionizing
radiation field is probably due to local sources with
$T \approx 18000 - 20000$ K.

A catalogue of extragalactic radio sources having
flux densities greater than 1 Jy at 5 GHz.
See Abstr. 002.011.

Neutral hydrogen observations of a large sample of
galaxies.   See Abstr. 002.064.

Radio galaxies. Radiative transfer, dynamics,
stability and evolution of a synchrotron plasmon.
See Abstr. 003.121.

Le stelle variabili.   See Abstr. 003.136.

Quasars.   See Abstr. 010.023.

Extragalactic jets: facts and fancies.
See Abstr. 011.011.

The Tucson workshop on active galaxies.
See Abstr. 011.049.

The future of radio astronomy.
See Abstr. 013.021.

Dust-sensitive forbidden line ratios.
See Abstr. 022.119.

A new method for making maps with unstable radio
interferometers.   See Abstr. 031.504.

Analysis of cryogenic FTS as a survey instrument
to discover new quasars.   See Abstr. 031.550.

Reduction of double-beam observations of extended
radio sources.   See Abstr. 031.565.

Recursive estimation of the reduction parameters
of an astrometric plate.   See Abstr. 031.566.

Selection effects in spectral searches for quasars.
See Abstr. 031.578.

Extended radio sources: a method for baseline
improvement.   See Abstr. 031.583.

Millimeter-wavelength continuum calibration
sources.   See Abstr. 031.599.

Reseau astrometry with Palomar Schmidt plates:
position-coincidence optical identification of radio sources.
See Abstr. 041.014.

Peculiarities of solutions of hyperbolic equations
and the problem of twins in astronomy.
See Abstr. 061.028.

Stability of galactic radio jets.
See Abstr. 062.016.

Ion pressure-supported accretion tori and the origin
of radio jets.   See Abstr. 062.110.

Formation and transfer of permitted Si II emission
lines in Seyfert I galaxies and quasars.
See Abstr. 063.017.

Inverse Compton scattering in anisotropic synchrotron sources.   See Abstr. 063.030.

Theoretical quasar emission line ratios, IV. General
asymptotic escape probabilities and the effects of linear Stark
broadening.   See Abstr. 063.047.

Gravitational lensing and the relation between QSO
and galaxy magnitude-number counts.   See Abstr. 066.050

On gravitational lenses and the cosmological evolution of quasars.   See Abstr. 066.051.

Another possible case of a gravitational lens.
See Abstr. 066.052.

Superluminal velocities: a gravitational lens effect.
See Abstr. 066.054.

On the observability of gravitational scintillation.
See Abstr. 066.122.

Restrictions on relativistically rotating fluids.
See Abstr. 066.143.

The Lyman-α/Hα ratio in solar flares and quasars.
See Abstr. 073.069.

Interplanetary shock waves from observations of
scintillations of radio sources.   See Abstr. 106.027.

Radio properties of SS 433.
See Abstr. 116.038.

408 MHz monitoring of SS 433 radioemission.
See Abstr. 116.039.

SS 433: enigma of the century.
See Abstr. 117.009.

High resolution X-ray and radio images of the Crab-
like supernova remnant G21.5−0.9.   See Abstr. 125.009.

New meter-wavelength observations of S 147.
See Abstr. 125.020.

The radial velocity field of the optical filaments associated with the SNR W63.   See Abstr. 125.041.

Further radio observations of W 50: total intensity
and linear polarization measurements at 1.7 and 2.7 GHz.
See Abstr. 125.047.

The point radio source in the supernova remnant
G 78.2 + 2.1.   See Abstr. 125.048.

On the properties and the nature of W50.
See Abstr. 125.051.

Neutral hydrogen in the vicinity of galactic radio sources. Supernova remnant W 44.   See Abstr. 125.061.

A high-resolution optical survey of interstellar absorption lines toward globular clusters and extragalactic objects. II. Further data.   See Abstr. 131.027.

Radio continuum and carbon monoxide observations of V645 Cygni (GL 2789).   See Abstr. 131.083.

Origin of the narrow mixed ion quasar absorption lines.   See Abstr. 131.104.

Statistics of neutral hydrogen absorption toward pulsars.   See Abstr. 131.113.

Synthesis observations of the radio continuum radiation of the H II region NGC 7822 (W1). See Abstr. 132.047.

Radio astronomy confronts elementary particle cosmology: radio measurements of helium abundances. See Abstr. 132.049.

An extended far-infrared emission complex at IC 1318b and IC 1318c.   See Abstr. 133.001.

Discovery of radio brightening in AFGL 618. See Abstr. 133.002.

On the radio properties of V645 Cygni. See Abstr. 133.004.

Detection of new filamentary nebulosity near the giant non-thermal radio loop II.   See Abstr. 134.008.

Radio observations of compact planetary nebulae. See Abstr. 135.049.

Optical studies of X-ray selected quasars and active galaxies.   See Abstr. 142.026.

The contribution of quasars to the 2 keV–100 MeV background radiation and the X-ray source counts at 2 keV. See Abstr. 142.039.

HEAO A-2 observations of non-Abell Zwicky clusters containing extended radio sources. See Abstr. 142.054.

A second SS 433?   See Abstr. 142.071.

X-ray astronomy in the Einstein era. See Abstr. 142.085.

An X-ray survey of flat spectrum radio sources. See Abstr. 142.086.

"Einstein" observations of BL Lacertae objects. See Abstr. 142.087.

An X-ray selected list of QSOs: implications for the QSO luminosity function and the origin of the diffuse X-ray background.   See Abstr. 142.088.

Physical processes for X-ray emission in galactic nuclei.   See Abstr. 142.092.

Some recent results on extragalactic X-ray sources from the Einstein Observatory.   See Abstr. 142.155.

Orbital motion of the head-tail radio galaxy IC 708. See Abstr. 151.052.

Luminosity evolution of quasars and active galaxies: theoretical models of the evolving mass supply rate. See Abstr. 151.062.

Radio structures of Seyfert galaxies. II. See Abstr. 158.003.

The dynamics of the S0 galaxy IC 5063. See Abstr. 158.014.

The rotation axes of five radio galaxies. See Abstr. 158.018.

High-resolution maps of the hotspots of several class II radio galaxies.   See Abstr. 158.028.

NGC 7714: the prototype star-burst galactic nucleus. See Abstr. 158.042.

Radio polarization rotators: BL Lacertae and 0727–115.   See Abstr. 158.043.

New insight into the physical state of galaxies and quasars.   See Abstr. 158.064.

Correlations between infrared, optical, and X-ray fluxes from Seyfert galaxies and quasars. See Abstr. 158.081.

Ultraviolet excess of Seyfert 1 galaxies and QSOs. See Abstr. 158.084.

VLA observations of the nucleus of NGC 1365 in the 6- and 20-cm continuum.   See Abstr. 158.085.

The inactive radio galaxy in Abell 566. See Abstr. 158.086.

Six-centimeter survey of Arakelian galaxies. See Abstr. 158.087.

Radio continuum emission from galaxies. See Abstr. 158.107.

Observations of the star-like BL Lac object OI 090.4.   See Abstr. 158.116.

IUE observations of two (non-Seyfert) radio galaxies.   See Abstr. 158.120.

The structure of radio emission in the inner few hundred parsecs of spiral galaxies.   See Abstr. 158.126.

La découverte et l'exploration de l'univers extragalactique.   See Abstr. 158.128.

Les galaxies "actives" et les quasars. See Abstr. 158.129.

Spectra of compact radio sources in galactic nuclei. See Abstr. 158.131.

Optical spectroscopic and electronographic observations of the radio galaxy IC 5063.   See Abstr. 158.139.

The ionized gas in NGC 5128: evidence for a shock-heated component.   See Abstr. 158.165.

Gamma ray constraints on annihilation in active galaxies.   See Abstr. 158.166.

Galactic absorption line coronae.
See Abstr. 158.171.

The appearance of broad emission lines in the spectrum of the BL Lac object PKS 0521-36.
See Abstr. 158.185.

Observations of the head-tail radio galaxy NGC 3862 (3C 264) at 0.6, 1.4, and 5.0 GHz.
See Abstr. 158.186.

Importance of the Doppler Differential Effect in the interpretation of active nuclei spectra. I. The hydrogen spectrum.   See Abstr. 158.187.

Puzzling ejections from Centaurus A.
See Abstr. 158.191.

Photoelectric comparison sequences in the fields of four radio sources.   See Abstr. 158.198.

What is Markarian 3?   See Abstr. 158.201.

The infrared continua of active galaxies.
See Abstr. 158.202.

X-ray emission around radio galaxies in non-Abell clusters: a possible physical link between environment and nonthermal radio emission.   See Abstr. 158.203.

VLBI and X-ray observations of compact nuclei in pairs of galaxies.   See Abstr. 158.207.

The X-ray structure of Centaurus A.
See Abstr. 158.211.

Information deduced from the spectra of quasars and Seyfert galaxies.   See Abstr. 158.221.

The new BL Lac object 0716 + 71.
See Abstr. 158.223.

21-cm line studies of spiral galaxies. I. Observations of the galaxies NGC 5033, 3198, 5055, 2841, and 7331.
See Abstr. 158.239.

Jets and the Space Telescope – an introduction.
See Abstr. 158.258.

Jets optiques dans les galaxies elliptiques, les galaxies N et les objets de type BL Lacertae.
See Abstr. 158.260.

Optical jets in active galaxies.
See Abstr. 158.264.

The ratio of hydrogen line intensities in Seyfert galaxies and quasars.   See Abstr. 158.267.

A polarization burst in the BL Lac object AO 0235+164.   See Abstr. 158.281.

On the motion of gas in the envelopes of active nuclei.   See Abstr. 158.292.

The Ursa Major supercluster – I. The optical field and the 5C10 radio survey.   See Abstr. 160.004.

Observation of the cluster of galaxies A 401 at 11 cm.   See Abstr. 160.005.

Low-frequency radio observations of poor clusters of galaxies.   See Abstr. 160.008.

408 MHz observations of clusters of galaxies. I. Halo sources in the Coma-A1367 supercluster.
See Abstr. 160.011.

Low frequency extended radio sources – magnetic field strength and classification.   See Abstr. 160.034.

Optical emission-line gas associated with dominant cluster galaxies.   See Abstr. 160.042.

H I observations in the Virgo cluster area.
See Abstr. 160.044.

VLA observations of the radio continuum emission from Stephan's Quintet.   See Abstr. 160.054.

Radio observations of Abell clusters and a comparison with certain Einstein observations.   See Abstr. 160.055.

The physical state of primordial intergalactic clouds.
See Abstr. 161.007.

Pulsars

141.501 Linear polarization of optical radiation from the Crab pulsar.
D. H. P. Jones, F. G. Smith, P. T. Wallace.
Mon. Not. R. Astron. Soc., Vol. 196, 943 - 953 (1981).
    The light curve of the Crab Nebula pulsar has been observed using an aperture of 1.3 arcsec with the Anglo-Australian Telescope. The background was measured in an annulus of 2 arcsec radius, contamination from the pulsar being estimated from the pulsed component. The resultant light curve has a minimum intensity 0.6 per cent of the peak. Around minimum the light is ~70 per cent linearly polarized, and at other phases the authors' measures accord with earlier work.

141.502 Frequency dependence of the P_2 and P_3 periods in four pulsars.
A. Wolszczan, N. Bartel, W. Sieber.
Astron. Astrophys., Vol. 100, 91 - 96 (1981).
    An analysis of the frequency dependence of the subpulse drift parameters P_2 and P_3 has been carried out for four pulsars. The frequency behaviour of the P_2 period in PSR 0809 + 74 and PSR 1919 + 21 may indicate that for these two objects the radiation received at frequencies above 1 GHz originates close to the neutron star surface in a non-dipolar magnetic field. The P_3 period is shown to be generally fre-

quency independent, in accord with the predictions of the polar cap models of the pulsar emission.

141.503   Depolarization of pulsar integrated pulse profiles.
D. Morris, D. A. Graham, W. Sieber.
Astron. Astrophys., Vol. 100, 107 - 112 (1981).
The rate of depolarization of the linear polarization of the average pulse profile is a function of pulsar period. Short period pulsars depolarize less rapidly and show a variation of depolarization rate across their pulse profiles. The predominant depolarization mechanism is apparently different for the long and short period pulsars. The majority of pulsars (> 70% in this sample) show depolarization of their circular polarization at high frequency. There are indications that this is period dependent.

141.504   Pulsar slow-down epochs.
H. Heintzmann.
Nature, Vol. 292, 811 - 814 (1981).
The author presents a new model for pulsar slow-down. The present model predicts four different slow-down epochs from evolutionary changes of the magnetosphere, which are dominated by different braking mechanisms. The model assumes that the masses, magnetic moments and initial rotation periods of all neutron stars are equal. The author shows that no direct relationship can exist between the "slow-down age" and the true age of a pulsar and that the pulsar birth rate is one per 100 yr.

141.505   Pulsar birthrates.
W. Kundt, with a reply by R. Narayan,
M. Vivekanand.
Nature, Vol. 292, 865 - 866 (1981).

141.506   Inertial development of vorticity in pulsar magneto-spheres.   R. R. Burman.
Australian J. Phys., Vol. 34, 91 - 95 (1981).
It is pointed out that inertial drift of a species will become important, corresponding to development of vorticity, as its number density becomes small. Hence, regions of inertial development of vorticity will form natural boundary layers to domains in which a species is concentrated.

141.507   Sub-rotation and super-rotation in pulsar magneto-spheres.   R. R. Burman.
Australian J. Phys., Vol. 34, 97 - 103 (1981).
For magnetospheric species that are nonrelativistic at the stellar surface in the general oblique rotator model, it is shown that a nondissipative flow branch which is sub-rotating when near the star remains sub-rotating everywhere, but that one which is super-rotating when near the star can become sub-rotating inside the light cylinder. Hence, both flow branches can cross the light cyclinder and remain valid outside it.

141.508   Orientation of the accretion disk in binary X-ray pulsars.
V. M. Lipunov, E. S. Semenov, N. I. Shakura.
Astron. Zh., Tom 58, 765 - 770 (1981). In Russian. English translation in Soviet Astron., Vol. 25, No. 4.
The interaction of the accretion disk with the magnetic field of an X-ray pulsar is considered. It is shown that under real conditions both magnetic forces and viscosity tend to turn the inner parts of the disk along the equator of rotation of the neutron star. At distances of the order of the Alfvén radius the accretion disk lies practically in the equatorial plane.

141.509   Statistical analysis of the temporal fine structure of the pulsar PSR 0809 + 74.
V. A. Soglasnov, T. V. Smirnova, M. V. Popov, A. D. Kuz'min.
Astron. Zh., Tom 58, 771 - 778 (1981). In Russian. English translation in Soviet Astron., Vol. 25, No. 4.
A statistical analysis of the microstructure of

PSR 0809 + 74 was performed. Two types of microstructure have been detected, with "short" (< 500 μs) and "long" (0.6 - 9 ms) time-scales. The microstructure is often of periodic character. The distribution of periods $P_{\star}$ is also divided into two components with "short" ($50 \leqslant P \leqslant 350 \mu$s) and "long" ($1 < P < 10$ ms) periods. The obtained distribution of scales and periods of microstructure is discussed.

141.510   Radio radiation characteristics of pulsars and the magnetic dipole angle.   Z.-r. Wang, Y. Chu.
Acta Astron. Sinica, Vol. 22, 191 - 194 (1981). In Chinese.

141.511   Electron-positron cascade in pulsar outer gaps.
A. Ray, G. Benford.
Phys. Rev. D, Vol. 23, 2142 - 2150 (1981). – Abstr. in Phys. Abstr., Vol. 84, Abstr. 75156 (1981).

141.512   Pair creation above pulsar polar caps: steady flow in the surface acceleration zone and polar cap X-ray emission.   J. Arons.
Astrophys. J., Vol. 248, 1099 - 1116 (1981).
The existence of steady, electrically driven flow including pair creation above the polar caps of a strongly magnetic, rotating neutron star is reconsidered. The free emission of electrons from the stellar surface is assumed. It is shown that pair creation by the trapped positrons does not disrupt the extraction of the primary stream of electrons, while continued pair creation above the top of the acceleration zone generates a dense, approximatively force-free column of pair plasma which streams away from the star. The result is a steady flow of plasma along polar field lines. Qualitative arguments are given which suggest the flow in the acceleration zone is stable. Bombardment of the polar caps by positrons trapped within the acceleration zone implies EUV–soft X-ray emission from each polar cap. When applied to the Vela pulsar, the theory predicts a modulated luminosity $L_{cap} \sim 4 \times 10^{30}$ ergs s^{-1} and $T_{eff} \sim 10^6$ K.

141.513   Radio spectra of pulsars. I. Observations of flux densities at meter wavelengths and analysis of the spectra.   V. A. Izvekova, A. D. Kuzmin (*Kuz'min*), V. M. Malofeev, Yu. P. Shitov.
Astrophys. Space Sci., Vol. 78, 45 - 72 (1981).
The results of flux pulsar radioemission measurements at meter wavelengths, made at Pushchino Radio Astronomical Observatory are presented. Spectra of 52 pulsars were plotted. In practically all investigated pulsars the authors have detected a turn-over frequency at which the flux density of pulsar radioemission attained its maximum. Its mean value is 130 ± 80 MHz. Averaged on many pulsars, the spectral index is negative in the 39 - 61 MHz frequency range and passes through zero at frequencies of about 100 MHz, becoming positive in the 100 - 400 MHz frequency range. Using the spectra, more precise radio luminosities of pulsars have been computed.

141.514   Radio spectra of pulsars. II. The interpretation.
I. F. Malov, V. M. Malofeev.
Astrophys. Space Sci., Vol. 78, 73 - 83 (1981).
The interpretation of the average spectra of pulsar radiation is given. It is shown that the observed relationship between the part of the pulsar kinetic energy, which is transformed into radio emission, and the basic period can explain the lack of very long period pulsars. The statistical relationships between the characteristic frequencies rule out effective operating of the Melrose's mechanism in pulsar magnetospheres. Two possible reasons of the high-frequency cut-off are analyzed. The values of the energy of emitting electrons are obtained.

141.515   A search for interpulse emission in pulsars.
J. M. Weisberg, V. Boriakoff, D. C. Ferguson,
P. R. Backus, J. M. Cordes.
Astron. J., Vol. 86, 1098 - 1100 (1981).

Using the Arecibo telescope, the authors examined 28 pulsars for evidence of interpulses or bridges of emission. Two objects, PSR 0940+16 and PSR 1530+27, were discovered to have components separated from the main pulse by 90 and 52 deg, respectively. Sensitive upper limits on such emission from the 26 other pulsars are listed. Improved dispersion measures are also tabulated for 25 of the pulsars, and improved positions are given for eight sources.

141.516 IUE observations of the Crab Pulsar.
P. Benvenuti, L. Bianchi, A. Cassatella, J. Clavel, J. Darius, A. Heck, H. V. Penston, F. Macchetto, P. L. Selvelli, J. Zamorano.
The Universe at ultraviolet wavelengths, (see 012.009), p. 701 (1981). – Abstract.

141.517 Pulsar disk systems.   F. C. Michel, A. J. Dessler.
Bull. American Astron. Soc., Vol. 13, 557 (1981). Abstract.

141.518 Single pulses from the binary pulsar PSR 1913 + 16.
D. C. Ferguson, V. Boriakoff.
Bull. American Astron. Soc., Vol. 13, 557 (1981). – Abstract.

141.519 Electromagnetic cascades in pulsars.
A. K. Harding, J. K. Daugherty.
Bull. American Astron. Soc., Vol. 13, 557 (1981). – Abstract.

141.520 IUE observations of the Crab pulsar.
P. Benvenuti, L. Bianchi, A. Cassatella, J. Clavel, J. Darius, A. Heck, M. V. Penston, F. Macchetto, P. L. Selvelli, J. Zamorano.
Second European IUE Conference, (see 012.011), p. 339 - 342 (1980).

The UV spectrum of the Crab Nebula pulsar has been detected with the IUE long wavelength spectrograph. The data are in good agreement with the optical spectrum. The interstellar extinction seems to deviate from the standard law in the 2200 Å region.

141.521 Pulsar as star with a core as pulse thermonuclear reactor.
B. V. Alekseev, V. S. Vinogradov, V. M. Razorenov.
Mosk. aviats. inst. Moskva, 1981. 8 pp. In Russian. – Abstr. in Ref. zh., 51. Astron., 8.51.667 (1981).

141.522 Initial period and age of a pulsar.
S. A. Gerasimov, Yu. I. Neshpor, A. A. Stepanyan.
Izv. AN SSSR. Ser. fiz., Tom 45, 629 - 632 (1981).
In Russian. – Abstr. in Ref. zh., 51. Astron., 8.51.672 (1981).

141.523 Corotation and poloidal flow in pulsar magnetospheres.   R. R. Burman.
Australian J. Phys., Vol. 34, 303 - 316 (1981).

The purpose of the previous work was not to propose any particular model but to reach an understanding of the essential features of the dissipation-free flow dynamics for the general oblique rotator. The author concentrates on magnetospheric regions in which the particles have azimuthal velocity components that are close to the local speed of corotation.

141.524 Boundary-layer acceleration and particle mirroring in pulsar magnetospheres.   R. R. Burman.
Australian J. Phys., Vol. 34, 317 - 331 (1981).

The purpose of this paper is to initiate study of the physics of the boundary layers separating the electron and ion corotation zones, and to consider the consequences of that physics. The poloidal accelerations and azimuthal drift velocities generated in those layers will be estimated for both ions and electrons. The behaviour of the accelerated ions and electrons as they subsequently penetrate into the electron and ion

zones, respectively, will be deduced. Implications of the results for model building will be discussed.

141.525 Simultaneous five colour photometry of the double period optical pulsar H 2254–033.
C. Motch, M. W. Pakull.
Astron. Astrophys., Vol. 101, L9 - L12 (1981).

Fast photometry of the optical counterpart of the 805 second pulsating X-ray source H 2254–033 performed simultaneously in the 5 pass bands of the Walraven system ranging from 3200 Å to 5500 Å confirms the presence of the two reported optical periodicities at 805 and 859 s and of a modulation with the 3.6 hour orbital period. The amplitude of the two pulsations are comparable in the ultraviolet whereas the 859 second pulsation strongly dominates in the visual. Spectrophotometric data show that the He II 4686 flux is also modulated likely with the 859 second period.

141.526 Mode-changing and quantized subpulse drift-rates in pulsar PSR 2319+60.
G. A. E. Wright, L. A. Fowler.
Astron. Astrophys., Vol. 101, 356 - 361 (1981).

The authors report observations of the long-period pulsar PSR 2319+60 at 1415 MHz. When not in a null state, the pulsar is found to select from three different "modes" of emission, each corresponding to a different subpulse drift-rate. A comparison with other pulsars suggests that the conventional definition of mode-changing might be reconsidered. The authors attempt to indicate the directions in which the polar cap model for pulsar emission must be developed if these observations are to be understood theoretically.

141.527 Gravitational waves from an orbiting pulsar.
J. M. Weisberg, J. H. Taylor, L. A. Fowler.
Sci. American, Vol. 245, No. 4, p. 66 - 74 (1981).

Einstein's 1915 prediction that an accelerating mass should radiate energy in the form of gravitational waves is supported by evidence that a pulsar's orbit around a companion star is slowly shrinking.

141.528 Inverse Compton emission of gamma rays near the pulsar surface.   M. Morini.
Astrophys. Space Sci., Vol. 79, 203 - 211 (1981).

The physical conditions near pulsar surface that might give rise to gamma ray emission from Crab and Vela pulsars are not yet well understood. The author suggests that, in the context of the vacuum discharge mechanism proposed by Ruderman and Sutherland (1975), gamma rays are produced by inverse Compton scattering of secondary electrons with the thermal radiation of the star surface as well as for curvature and synchrotron radiation.

141.529 Microstructure in the pulsar 0950+08 interpulse at radio wavelengths.   T. H. Hankins, V. Boriakoff.
Astrophys. J., Vol. 249, 238 - 240 (1981).

The authors have made high time resolution measurements of the PSR 0950+08 interpulse intensity. They find that the interpulse microstructure is similar to the main pulse microstructure, which implies that the emission mechanisms for the two regions must be similar. The characteristic microstructure time scale of 85 μs which they obtained from autocorrelation analysis compares with 130 μs obtained for the main pulse.

141.530 Interpulse emission from pulsar 0950+08: how many poles?   T. H. Hankins, J. M. Cordes.
Astrophys. J., Vol. 249, 241 - 253 (1981).

The authors discuss the relationship of the main pulse and interpulse by considering the frequency dependence of the average pulse profile over 2 decades of frequency (100 to 5000 MHz) and by investigating the average polarization behavior and pulse-to-pulse fluctuations. The data do not unequivocally support either the single-pole or double-pole models.

141.531 Giant glitches and pinned vorticity in the Vela and other pulsars.   M. A. Alpar, P. W. Anderson, D. Pines, J. Shaham.
Astrophys. J., Lett., Vol. 249, L29 - L33 (1981).

The authors call attention to a number of hitherto unexamined regularities in the Vela pulsar timing data and show how these find a natural explanation in a theory of giant glitches in the Vela and other pulsars as the dynamic consequence of catastrophic unpinning events in the pinned crustal neutron superfluid, with postglitch behavior resulting from glitch-induced vortex creep.

141.532 Observations of the polarization of average pulsar profiles at high frequency.
D. Morris, D. A. Graham, W. Sieber, N. Bartel, P. Thomasson.
Astron. Astrophys., Suppl. Ser., Vol. 46, 421 - 472 (1981).

Average pulse profiles are given with full polarization information for 64 pulsars. Most of these pulsars have been observed at 1720 MHz (58) and 2650 MHz (52), several at 8700 MHz (15) and a few at 14.8 GHz (5). Tables list average intensity and polarization. The reduced prominence at high frequency of the central components of multicomponent pulse profiles which has been observed for some pulsars may be a consequence of a non central cut of the line of sight across nested emission cones whose dimensions are frequency dependent. With two possible exceptions (PSR 1237 + 25 and PSR 2045 − 16) most pulsars have less average linear polarization at high frequency. In general the average circular polarization is less at high frequency.

141.533 JPL pulsar timing observations. I. The Vela pulsar.
G. S. Downs.
Astrophys. J., Vol. 249, 687 - 697 (1981).

Twelve years of arrival time data for PSR 0833−45 are analyzed by fitting a simple model of the pulse period to the data in a least-squares sense. There exist regions of exceptionally stable behavior in which rms residuals (measured minus predicted arrival times) are typically ~40 μs, each exhibiting a significant $\dot{P}$ which changes little within a particular interjump era. Changes ΔP, $\Delta \dot{P}$, and $\Delta \ddot{P}$ in the period P and its derivatives introduced by each of the four large observed jumps in period are recomputed in a detailed analysis.

141.534 The gamma-ray spectra of radio pulsars.
S. Ayasli.
Astrophys. J., Vol. 249, 698 - 703 (1981).

A general expression for the γ-ray spectra of pulsars is derived for models where the radiation is produced by particles accelerated along the open field lines of a neutron star. The accelerating potentials are assumed to be limited by pair production discharges and to decrease linearly with increasing number of electron-positron pairs created during these discharge processes. The corresponding energy distribution of particles is derived both for the case of strong radiation reaction and for that of no radiation reaction. The resultant formula for the γ-ray spectrum fits well to the Crab and the Vela pulsar spectra assuming that radiation reaction is negligible in these two fast pulsars.

141.535 Time asymmetries in pulsar signals.
D. Stinebring, J. M. Cordes.
Astrophys. J., Vol. 249, 704 - 719 (1981).

The authors present a technique for analyzing time asymmetries of stochastic processes and apply it to high time resolution data from pulsars PSR 0950+08 and PSR 2016+28. Subpulses and average waveforms show similar time asymmetries. This is consistent with subpulses arising from beams of radiation rather than temporal modulations. Micropulses are, on average, time symmetric. It is possible that individual micropulses are asymmetric, but over a data set of several hundred pulses, there is no preferred sense of asymmetry.

141.536 A revised calculation of the bremsstrahlung cross-section in the high magnetic field of pulsars.
R. Lieu.
Astrophys. Space Sci., Vol. 80, 157 - 172 (1981).

It has been postulated that electron bremsstrahlung in a strong external magnetic field is the dominant radiation mechanism within the accretion plasma near the magnetic polar regions of binary X-ray sources. The present work uses the simple structure of the propagator obtained in a previous calculation for mildly relativistic electrons occupying no more than the first few Landau levels. Typical behaviour for two linear polarization modes is illustrated for forward and backward electron scattering. At higher frequencies resonances are present irrespective of the polarization of the emitted radiation.

141.537 Effect of long-term intensity variations on pulsar searches and the pulsar luminosity function.
S. Krishnamohan.
Mon. Not. R. Astron. Soc., Vol. 197, 497 - 506 (1981).

Long-term intensity data for five pulsars have been used to obtain probability density distribution of intensities for each pulsar. The author finds that χ^2-distributions describe them well and shows from these χ^2 distributions that the luminosity function derived from a single survey is not affected significantly by the intensity variations. He also describes a method of deriving the luminosity function by combining different searches of a given area on the basis of a probabilistic approach to the evaluation of selection effects.

141.538 Pulsar altitude distribution as a clue to their mean velocity and lifetime.
M. Arnaud, R. Rothenflug.
Astron. Astrophys., Vol. 103, 263 - 268 (1981).

The pulsar altitude distribution in the galaxy depends on: (1) the gravitational field, (2) the pulsar lifetime τ; the pulsar creation rate is assumed to be constant in time, (3) the pulsar initial velocity distribution and the corresponding mean value $\langle |V_z| \rangle$, (4) the altitude distribution of the pulsar progenitors, characterized by their scale-height H_0. In this paper the authors compute the pulsar altitude distributions in a model including all the above ingredients and using as parameters $\langle |V_z| \rangle$, τ, and H_0.

141.539 The stability of the pulse intensity of the X-ray pulsar in the Crab nebula.
M. Meidav, D. Sadeh.
Astron. Astrophys., Vol. 103, 367 - 369 (1981).

The large area detector on the H.E.A.O.-A satellite and the fast time resolution allowed the detection of individual Crab pulses. A statistical analysis performed on those pulses shows that although there are pulse-to-pulse variations, they fall within the random expectation.

141.540 A conjectured pulsar magnetosphere model.
R. Burman.
Speculations Sci. Technol., Vol. 4, No. 1, p. 91 - 97 (1981).
Abstr. in Phys. Abstr., Vol. 84, Abstr. 98808 (1981).

141.541 Pulsar and spin-spin interaction.
A. P. Yefremov (Efremov).
Exp. Tech. Phys., Vol. 29, 217 - 219 (1981). − Abstr. in Phys. Abstr., Vol. 84, Abstr. 108245 (1981).

141.542 On the magnetic field of a pulsar.
A. P. Yefremov (Efremov).
Exp. Tech. Phys., Vol. 29, 221 - 222 (1981). − Abstr. in Phys. Abstr., Vol. 84, Abstr. 108246 (1981).

141.543 A model of the normal and null states of pulsars.
    P. B. Jones.
Mon. Not. R. Astron. Soc., Vol. 197, 1103 - 1124 (1981).

A solvable three-dimensional polar cap model of pair creation and charged particle acceleration has been derived. There are no free parameters of significance apart from the polar surface magnetic flux density. Solutions of the model exist for both normal and null states of a pulsar, and the instability in the normal state leading to the normal to null transition has been identified. The predicted necessary condition for the transition is entirely consistent with observation.

141.544 Propagation of microwaves in pulsar magnetospheres.
    G. Bodo, A. Ferrari, S. Massaglia.
Astrophys. Space Sci., Vol. 80, 261 - 266 (1981).

The authors discuss the dispersion relation of linearly-polarized waves, propagating along a strong background magnetic field embedded in an electron-positron plasma. The results are then applied to the study of the propagation conditions of coherent curvature radio radiation inside neutron stars magnetospheres, as produced by electric discharges following current pulsar models.

141.545 The frequency dependence of micropulse separation.
    J. Gil.
Astron. Astrophys., Vol. 104, 69 - 71 (1981).

It is demonstrated that within an angular beaming model of microstructure, the longitude of micropulses should be independent of frequency, whereas the component separation can vary in the frequency band in which the micropulses are visible.

141.546 A new analysis of the pulsar distribution in the Galaxy.
    M. Morini.
Astron. Astrophys., Vol. 104, 75 - 79 (1981).

As discussed here it is difficult to derive the properties of the distribution of pulsars in the Galaxy from the current observations. In this work the author presents some Monte-Carlo simulations of the pulsar observations starting from proposed galactic distributions. A large pulsar population concentrated in the inner part of the Galaxy as proposed by Taylor and Manchester (1977) could agree with the observations only if selection effects bias the surveys more than the author expects.

141.547 On the distribution of pulsars in the galactic plane.
    A. del Romero, J. Gómez-González.
Astron. Astrophys., Vol. 104, 83 - 87 (1981).

The spatial distribution of the projections of pulsars on the galactic plane within distances less than 5 kpc has been studied by looking for a correlation of these objects with the projections of the H II regions of best known distances. The resulting spatial correlation of pulsars with the galactic spiral arms in the vicinity of the Sun indicates that pulsars are generated from young objects, and allows to estimate an upper limit for their mean lifetimes, which is an agreement with previous values.

141.548 Pulsar disk systems.
    F. C. Michel, A. J. Dessler.
Astrophys. J., Vol. 251, 654 - 664 (1981).

The authors argue that the radio pulsars and the X-ray pulsars differ mainly in the fact that the latter are surrounded by an accretion disk, while the former are surrounded by a fossil collapse disk presumably left over from the formation event. They attribute the difference between these two types of pulsars to a strong interaction (enforced accretion) of the X-ray pulsars with their disks as opposed to a relatively weak interaction (and negligible accretion) in the case of the radio pulsars. A number of observational problems (e. g., role of alignment, ion confinement, nulling, drifting subpulses, braking index, residuals and the supernova association) are readily addressed in terms of the disk model. Rough estimates suggest that pulsars with disks could function with magnetic fields at the neutron star surface as low as 10^9 gauss, far below that often assumed.

141.549 Pulsars: polar pumps, interpolar currents, and induced Landau radiation.   E. A. Jackson.
Astrophys. J., Vol. 251, 665 - 673 (1981).

Three dynamical effects are discussed which permit the formulation of a fully self-consistent theory of pulsar atmospheres in the case of aligned magnetic and rotational axes. The basic effect is the pumping action near the poles caused by the electron-lattice interactions and the slippage between the stellar crust and the plasma in the force-free atmosphere. This pumping action produces a continuous injection of particles into the atmosphere near both poles. The second effect concerns the possibility of circulations involving interpolar currents associated with each individual pole. Such pole-to-pole circulation requires outer regions containing only one sign of the charge. The third effect concerns a radiative dissipative mechanism stimulated by the nonrelativistic poloidal circulation. This dissipation is by synchrotron radiation resulting from transitions between Landau levels. These dynamical elements are combined with previous self-consistent, force-free models to suggest a complete, qualitative picture of the aligned atmospheric behaviour.

141.550 Pulsar electrodynamics.   H. Ardavan.
    Astrophys. J., Vol. 251, 674 - 686 (1981).

It is shown that with the quasi-static constraint $\partial/\partial t = -\omega \partial/\partial \varphi \neq 0$, there exist no retarded solutions to the inhomogeneous Maxwell's equations for which the electromagnetic fields are nonsingular at the light cylinder $r = c/\omega$, unless at all points of this surface the densities of both electric charge and electric current are zero (ω is a constant angular frequency, c is the speed of light, t is time, and r and φ are the radius and the azimuthal angle in cylindrical coordinates).

141.551 Pulsar theory: particle acceleration and photon emission in the polar flux tube.   J. Arons.
Plasma astrophysics, (see 012.042), p. 273 - 289 (1981).

Particle acceleration and photon emission in the polar field lines of a pulsar are considered. Observations of pulsar pulse, subpulse and micropulse morphology, spectra and polarization, as well as high frequency radiation, are reviewed and used to suggest a steady streaming model with pair creation as the seat of the emission phenomenon. The energetics of polar beam flow are described, with emphasis on acceleration in a slot gap. A possible global model of the magnetosphere is outlined. The plasma content of the flow is described, then used to discuss several transfer processes which may be associated with subpulse morphology and polarization and with the formation of micropulses. Various ideas about possible emission processes are discussed with emphasis on emission from shear flows in the boundary layers of the pair plasma flow. Finally, a new model of the origin of marching subpulses is suggested, involving a rotating drift wave over the polar cap.

141.552 Pulsar magnetosphere potential structure.
    D. F. Smith, L. A. Muth, J. Arons.
Plasma astrophysics, (see 012.042), p. 333 - 335 (1981).

The electric fields in the polar cap region of the magnetosphere of a pairless orthogonal rotator are being calculated using a modified Poisson's equation for the potential Φ. This is the first step in a program to determine whether return currents are formed in these models and the resulting global current flow patterns.

141.553 On synchrotron nebulae and pulsars.
    D. J. Helfand, R. H. Becker, R. Novick.

Space Sci. Rev., Vol. 30, (see 012.044), 263 (1981).
Abstract.

141.554 Pulsar magnetospheres: some fundamental considerations.   K. C. Westfold.
Australian J. Phys., Vol. 34, 595 - 607 (1981).

The equations governing a pulsar magnetosphere under quasi-static conditions are presented in a vector form from which the theory can be developed in a systematic manner. In particular, integrals applicable in the cylindrical and axisymmetric cases are obtained direct from these vector equations.

141.555 A new look at pulsar statistics − birthrate and evidence for injection.
M. Vivekanand, R. Narayan.
J. Astrophys. Astron., Vol. 2, 315 - 337 (1981).

The authors make a statistical analysis of the periods P and period-derivatives $\dot{P}$ of pulsars using a model-independent theory of pulsar flow in the $P-\dot{P}$ diagram. Using the available sample of P and $\dot{P}$ values, the authors estimate the current of pulsars flowing unidirectionally along the P-axis, which is related to the pulsar birthrate. The birthrate estimated here is consistent with the supernova rate. The authors further conclude that a large majority of pulsars make their first appearance at periods greater than 0.5 s. This "injection", which runs counter to present thinking, is probably connected with the physics of pulsar radio emission.

141.556 Observed and derived parameters for 330 pulsars.   R. N. Manchester, J. H. Taylor.
Astron. J., Vol. 86, 1953 - 1973 (1981).

A compilation of the principal observational parameters of 330 pulsars known at the time of writing is presented. References to the sources of information are included. Commonly used parameters derived from the observational data, including galactic coordinates, distance, characteristic age, radio luminosity, and surface magnetic field, are given in a separate table.

141.557 PSR 0833-45.
IAU Circ., No. 3644 (1981).

141.558 Parameters of pulsars of the second Molonglo survey and their luminosity function.
O. Kh. Gusejnov, F. K. Kasumov, I. M. Yusifov.
Astron. Zh., Tom 58, 1195 - 1206 (1981). In Russian.
English translation in Soviet Astron., Vol. 25, No. 6.

The main parameters for 224 pulsars of the second Molonglo survey were calculated. The pulsar luminosity function is investigated on the basis of the given statistical sample.

141.559 On the theory of orthogonal modes in pulsar radio emission.   O. G. Onishchenko.
Pis'ma Astron. Zh., Tom 7, 731 - 735 (1981). In Russian.
English translation in Soviet Astron. Lett., Vol. 7.

The effect of a magneto-active relativistic electron-positron plasma in a pulsar magnetosphere on the propagation and the character of polarization of the radio emission is considered. It is shown that characteristic double structure and orthogonal polarization of subpulses may be accounted for as a result of anomalous double refraction while radiation is propagating through a dense magnetospheric pulsar plasma.

141.560 Another double-star pulsar found?
U. Dzērvītis.
Zvaigžņota debess, 1980/81. gada ziema, p. 20 - 21. In Latvian.

141.561 Galactic asymmetry of physical characteristics of pulsars.   A. M. Ehjgenson.
Tsirk. Astron. Obs., L'vov, No. 54, p. 19 - 20 (1979). In Russian.

141.562 Pulsar theory.
V. Usov.
High-energy astrophysics, (see 012.067), p. 125 - 139 (1981).

The detection of pulsed gamma-rays from the Crab nebula and the Vela pulsar confirms the presence of relativistic plasma in pulsars. Processes in the ultrarelativistic plasmas of pulsar magnetospheres are discussed and mechanisms for the emission of nonthermal radiation are outlined.

141.563 Expanding shells of young pulsars as sources of high-energy neutrinos.   M. M. Shapiro, R. Silberberg.
High-energy astrophysics, (see 012.067), p. 141 - 144 (1981).

Ultra-high energy protons ($E \geqslant 10^{15}$ eV) accelerated by a young pulsar that was formed in a recent supernova explosion should interact with the supernova shell and generate cascades of mesons. The mesons decay by emitting neutrinos. Neutrino fluxes as function of the initial energy input are estimated and the feasibility of detecting these neutrinos with the DUMAND experiment are discussed.

141.564 The luminosity function and radial distribution of pulsars in the Galaxy.   I. M. Yusifov.
Astron. Tsirk., No. 1164, p. 1 - 2 (1981). In Russian.

IAU symposium on pulsars.
See Abstr. 011.019.

A method for searching for optical pulsars.
See Abstr. 031.582.

Variations in the angular velocity of the earth's rotation and the rate of change of the pulsar period.
See Abstr. 044.014.

On charge neutrality and dissipative effects in pulsar systems.   See Abstr. 062.002.

Laboratory study of coherent curvature radiation as a pulsar emission mechanism.   See Abstr. 062.089.

On the magnetospheric structure of pulsars.
See Abstr. 062.092.

On the possibility of existence of large amplitude waves around pulsars.   See Abstr. 062.108.

Neutron stars.   See Abstr. 066.514.

A search for young optical pulsars in the galaxies NGC 4647 and NGC 4321.   See Abstr. 125.023.

Exploding stars and their remnants.
See Abstr. 125.035.

Spectrum of the interstellar plasma turbulence in the direction of the pulsar PSR 0329 + 54.
See Abstr. 131.003.

Radio spectroscopy of diffuse clouds.
See Abstr. 131.091.

Constraints on the ISM phases from pulsar observations.   See Abstr. 131.100.

Temporal variability of the Crab pulsar in hard X-rays.   See Abstr. 142.025.

Hard X-ray structure of the Crab Nebula.
See Abstr. 142.100.

High-energy γ-rays from the direction of the Crab pulsar.   See Abstr. 142.517.

On the nature of the galactic 2CG γ-ray sources. See Abstr. 142.526.

Detection of pulsed γ-rays at energies above 300 GeV from pulsars—TIFR experiments. See Abstr. 142.529.

Distribution of the electron concentration in the Galaxy.   See Abstr. 155.037.

Galactic γ-ray emission from pulsars. See Abstr. 157.001.

Errata

141.901 Corrigendum: 'The structure of integrated pulse profiles' [J. Astrophys. Astron., Vol. 1, 119 - 128 (1980)].   M. Vivekanand, V. Radhakrishnan.
J. Astrophys. Astron., Vol. 2, 213 (1981). – See Abstr. 28.141.540.

141.902 Erratum: 'The average quasar spectrum in the wavelength range 0.1−0.6 μm' [Mon. Not. R. Astron. Soc., Vol. 195, 831 - 837 (1981)].
J. E. Cheney, M. Rowan-Robinson.
Mon. Not. R. Astron. Soc., Vol. 197, 1153 (1981). – See Abstr. 29.141.130.

142 UV Sources, X-ray Sources, X-ray Background, Gamma-ray Sources, Gamma-ray Background

UV Sources, X-ray Sources, X-ray Background

142.001 Masses of X-ray bursters and super-Eddington luminosities. R. Hoshi.
Astrophys. J., Vol. 247, 628 - 631 (1981).

The author shows that the upper limit to the mass of X-ray bursters and the minimum of the ratio of peak luminosity to Eddington luminosity, k_{min}, can be determined as functions of the apparent peak luminosity of a burst and the corresponding blackbody temperature. Observations support the conclusion that the upper limit of the mass of these sources is $\sim 1 M_\odot$, and the ratio k_{min} is distinctly greater than 5 for an assumed distance of 10 kpc.

142.002 Einstein X-ray identification of the variable radio star LSI +61°303. G. F. Bignami, P. A. Caraveo, R. C. Lamb, T. H. Markert, J. A. Paul.
Astrophys. J., Lett., Vol. 247, L85 - L88 (1981).

Einstein X-ray observations are presented for the star LSI +61°303 associated with the variable radio source GT 0236 +61. Both the IPC and HRI instruments were used, yielding an accurate positional identification and a flux evaluation, as well as some spectral information for the X-ray source coincident with the B star. No evidence for variability is found, or any evidence for diffuse emission surrounding the X-ray source. The possible connection between the star, now a recognized X-ray source, with the high-energy *COS B* gamma-ray source 2CG 135 + 01 is discussed on the basis of overall energetic arguments.

142.003 Photometric and spectroscopic observations of H2252 −035 (=3A 2254 −033). B. Warner, D. O'Donoghue, A. P. Fairall.
Mon. Not. R. Astron. Soc., Vol. 196, 705 - 713 (1981).

Photometric observations of H2252 −035 with a time resolution of 5 s are reported. Fourier analysis shows the presence of an 805.2-s periodicity which agrees with the observed X-ray period and is thought to arise from a rotating magnetic white dwarf. The predominant light variation is, however, an 858.68-s periodicity which is the "reflected" pulse from the secondary component of this close double star. Slow brightness variations, synchronized with the 3.59-hr orbital period, are found. Rapid flickering and occasional bursts of quasi-periodic flares with periods near 100s are reported.

142.004 The X-ray source A0538 - 66 in optical quiescence. P. Murdin, G. Branduardi-Raymont, A. N. Parmar.
Mon. Not. R. Astron. Soc., Vol. 196, 95P - 99P (1981).

Photometric observations of A0538 - 66 in quiescence show that the optical star is an object of equivalent spectral type B7eII. Spectroscopic data taken simultaneously confirm this spectral classification and are consistent with the reported periodic variations seen in the emission lines. It is unlikely that the optical outburst is due to X-rays reprocessed in the atmosphere of the primary star or elsewhere.

142.005 The X-ray pulsars 2S1145–619 and 1E1145.1–6141: optical identifications and a nearby supernova remnant. J. B. Hutchings, D. Crampton, A. P. Cowley.
Astron. J., Vol. 86, 871 - 874 (1981).

High Resolution Imager X-ray pictures from the Einstein Observatory identify the two pulsars optically with Hen 715 and a 12^m reddened B1 supergiant, respectively. Optical spec-

troscopic data are discussed for each object. Hen 715 is a main-sequence B0 star whose properties are consistent with the 188-day X-ray binary period proposed by Watson et al. (1981). However, no optical periodicity is evident.

142.006 A photometric study of 2 A 0526-328. C. Motch.
Astron. Astrophys., Vol. 100, 277 - 283 (1981).

UBV and fast photometry of the X-ray source 2A 0526-328 optical counterpart confirms its cataclysmic-variable nature. The light curve shows a periodic hump ~ 0.27 mag full amplitude in B and V at $P = 5.18$ h, average brightness variations of 0.33 mag on a time scale of 2 d and ~ 0.1 mag flickering on a time scale of some minutes. The hump is interpreted as due to a bright spot on an accretion disc; fast colour behaviour gives evidence for both disc and hot-spot emissivity. A strong hump phase shift depending on the system's average luminosity is explained by the hot spot motion on a disc changing size with accretion rate.

142.007 A discussion on new *VBLUW* observations of the X-ray binary Sk. 160 = SMC X-1. A. M. van Genderen, E. van Groningen.
Astron. Astrophys., Vol. 101, 101 - 104 (1981).

VBLUW photometry (Walraven system) of Sk. 160 = SMC X-1 made in 1973, 1975, and 1977 are discussed. They are compared with those of van Paradijs (1977) made in 1976. Using all available photometric and X-ray mid-eclipse data, a new period is determined by a weighted least-squares solution: $P = 3^d89235 \pm 0^d00002$ m. e.

142.008 A 5.57 hour modulation of the X-ray flux from 4U 1822-37. N. E. White, R. H. Becker, E. A. Boldt, S. S. Holt, P. J. Serlemitsos, J. H. Swank.
Astrophys. J., Vol. 247, 994 - 1002 (1981).

The optical counterpart of 4U 1822-37 is a 16th magnitude blue star (Griffiths et al. 1978) whose flux is modulated at a period of 5.57 hr with a color independent peak-to-peak amplitude of 1 mag (Mason et al. 1980). The duty cycle of the modulation suggests the occultation of an extended region of optical emission (e.g., an accretion disk). The authors have made observations with the HEAO 1 and Einstein observatories which demonstrate that the X-ray flux of this source is also modulated with the 5.57 hr period. This result is presented here along with high quality X-ray spectra and a discussion of the probable nature of 4U 1822-37 and its similarity to Cyg X-3.

142.009 Pulse-timing observations of Hercules X-1. J. E. Deeter, P. E. Boynton, S. H. Pravdo.
Astrophys. J., Vol. 247, 1003 - 1012 (1981).

The authors present results from pulse-timing analysis of Hercules X-1, based on satellite observations spanning 7 years. The binary orbital elements are redetermined and the near circularity of the orbit is confirmed with a new 2 σ upper limit of 0.0003 on the eccentricity.

142.010 X-ray detection of the symbiotic star AG Draconis. C. M. Anderson, J. P. Cassinelli, W. T. Sanders.
Astrophys. J., Lett., Vol. 247, L127 - L130 (1981).

The authors report the detection of the yellow symbiotic star AG Draconis by the *Einstein* Observatory Imaging Proportional Counter. The object is an intense source of very soft X-rays. The X-ray luminosity is estimated to be $L_x \approx 10^{32}$ ergs s^{-1} with a temperature of less than 2×10^6 K. If one assumes an interstellar column density of 3×10^{20} cm^{-2}, then the emission measure is deduced to be greater than 3×10^{55} cm^{-3} from the X-ray data and less than 3×10^{59} cm^{-3} from optical data. The observation is discussed in the context of various models of the symbiotic stars.

142.011 Spectroscopic investigation of the Cygnus X-1 optical component.   O. Eh. Aab, L. V. Bychkova, I. M. Kopylov, R. N. Kumajgorodskaya, E. L. Chentsov.
Pis'ma Astron. Zh., Tom 7, 417 - 421 (1981). In Russian. English translation In Soviet Astron. Lett., Vol. 7.

Results of an investigation of the Cygnus X-1 optical component from spectrograms obtained with the 6-m telescope are presented. Equivalent widths, central depths and differential shifts of lines vary with orbital phase. The supergiant's parameters are quantitatively determined.

142.012 Infrared observations of galactic X-ray sources.   Yu. N. Gnedin, G. V. Khozov, V. M. Larionov.
Pis'ma Astron. Zh., Tom 7, 466 - 469 (1981). In Russian. English translation in Soviet Astron. Lett., Vol. 7.

Observations of the X-ray binary systems 4U 0115 + 634 and 4U 0535 + 262 in the H and K bands show the presence of noticeable infrared excesses which are likely to be of intrinsic nature. The ratio of IR to X-ray fluxes is $\sim 10^{-2}$. The IR excesses can hardly be explained by free-free emission of a hot gas in a binary system. One cannot exclude the existence of a dust envelope.

142.013 The Gemini-Monoceros X-ray enhancement: a giant X-ray ring.   J. A. Nousek, L. L. Cowie, E. Hu, C. J. Lindblad, G. P. Garmire.
Astrophys. J., Vol. 248, 152 - 160, plates 6 - 7 (1981).

A $1^\circ 5$ spatial resolution map of the 1/4 keV diffuse X-ray background enhancement in the Gemini and Monoceros constellations shows a striking, circular, ring-shaped emission feature with a diameter of about 20°. At a distance of 300 pc, the region has a radius of 50 pc and, for a spectral temperature of 3×10^6 K, an emitting electron density of 0.01 cm^{-3}. The authors discuss this feature and, also, the possible X-ray contribution from the Mon OB1 association and neighboring supernova remnants.

142.014 Relations among stellar X-ray emission observed from Einstein, stellar rotation and bolometric luminosity.   R. Pallavicini, L. Golub, R. Rosner, G. S. Vaiana, T. Ayres, J. L. Linsky.
Astrophys. J., Vol. 248, 279 - 290 (1981).

The authors have determined the correlation between observed stellar X-ray luminosities, bolometric luminosities, and projected rotational velocities for stars of various spectral types and luminosity classes observed by the Einstein Observatory. Early type stars (O3–A5) have X-ray luminosities which are proportional to bolometric luminosity. The proportionality constant is independent of luminosity class. In contrast, late type stars (G to M) have X-ray luminosities strongly dependent on rotation rate and independent of bolometric luminosity; this relation for late type stars is again found to be independent of luminosity class. F stars as a class are intermediate. The location of RS CVn stars as a class is discussed with respect to the dependence of X-ray luminosity on rotation.

142.015 On the maximum luminosity in X-ray bursts.   J. van Paradijs.
Astron. Astrophys., Vol. 101, 174 - 175 (1981).

The average maximum luminosity of X-ray bursts is larger than the Eddington limit by a factor of ~ 3. A qualitative model is suggested to explain the burst behaviour of 1608-52 observed by Murakami et al. (1980) and the observed variability in the peak luminosity of X-ray bursts.

142.016 Ultraviolet observations of LMC X-4 and SMC X-1.   J. M. Bonnet-Bidaud, S. A. Ilovaisky, M. Mouchet, G. Hammerschlag-Hensberge, M. van der Klis, W. M. Glencross, A. J. Willis.
Astron. Astrophys., Vol. 101, 184 - 186 (1981).

Low-resolution IUE spectra of the massive X-ray binaries LMC X-4 and SMC X-1 have been obtained at both short and long wavelengths. The continua are consistent with expected early-type model atmosphere fluxes showing small amounts of reddening and exhibit the characteristic double-wave light curves seen at visible wavelengths.

142.017 An X-ray pulsar in SNR G109.1-1.0.   G. G. Fahlman, P. C. Gregory.
Nature, Vol. 293, 202 - 204 (1981).

The unusual extended X-ray source G109.1-1.0 looks like a supernova remnant (SNR) and is of particular interest because it contains a point source 1E2259+586. The authors have analysed the photon arrival times for the point source. They conclude that the point source is an X-ray pulsar.

142.018 The X-ray cataclysmic variable 1E0643.0–1648.   J. Bailey, J. H. Hough, A. Schwarzenberg-Czerny, I. Gatley, R. F. Jameson, D. A. Hanes.
Nature, Vol. 293, 204 - 206 (1981).

The authors have used a new simultaneous IR/optical high-speed photometer on the UK IR telescope to study the recently discovered X-ray cataclysmic variable 1E0643.0–1648. The light curve of this object obtained by the American Association of Variable Star Observers has shown it to be a dwarf nova with a recurrence time scale of 15 days. The authors report here that their photometry, obtained during the decline from an outburst, shows slow flickering, with the optical and IR light curves correlated with no delay. They have also obtained moderate resolution spectra of the star.

142.019 Hard X-ray spectrum of Cygnus X-1.   P. L. Nolan, D. E. Gruber, F. K. Knight, J. L. Matteson, R. E. Rothschild, F. E. Marshall, A. M. Levine, F. A. Primini.
Nature, Vol. 293, 275 - 277 (1981).

The authors have measured the low-state spectrum on 26 October to 18 November 1977, over a wide energy range (3 keV–8 MeV). This spectrum agrees well with a single temperature comptonization model at low energy, but shows a significant excess at high energy ($E > 300$ keV).

142.020 A model for the extended X-ray source surrounding M87.   M. Takahara, F. Takahara.
Prog. Theor. Phys., Vol. 65, 369 - 373 (1981). – Abstr. in Phys. Abstr., Vol. 84, Abstr. 67133 (1981).

142.021 An X-ray active region in Orion: X-rays from a Herbig-Haro object?
S. H. Pravdo, F. E. Marshall.
Astrophys. J., Vol. 248, 591 - 595, plates 11 - 12 (1981).

The authors have observed with the Imaging Proportional Counter on the Einstein X-ray Observatory a $a \sim 1^\circ$ square region in Orion centered near α (1950) = 5^{h}34^m, δ = $-6^\circ47'$. Some 25 stellar X-ray sources are found, and several are identified with stars which have X-ray luminosities at the upper end of the distribution of X-ray luminosity versus stellar type. In addition there is evidence for X-ray emission associated with HH 1 which can provide support for models which include strong stellar winds interacting with the interstellar medium.

142.022 Evidence for an X-ray period in the 'Sco-like' source 2S0614 + 091.   N. Marshall, J. M. Millit.
Nature, Vol. 293, 379 - 382 (1981).

A class of galactic X-ray source exists whose members show properties which have many similarities to those of Sco X-1. The authors report the results of a detailed analysis for one such source, 2S0614 + 091, which undergoes significant variations in its X-ray intensity consistent with a period of 5.2 ± 0.3 days. This is the first discovery of a periodic modulation in the X-ray emission from this class of source.

142.023 Optical studies of empty field X-ray sources with a charge-coupled device (CCD) detector.
G. R. Ricker, M. Bautz, D. Dewey, S. S. Meyer.
Bull. American Astron. Soc., Vol. 13, 529 (1981). – Abstract.

142.024 The discovery of an X-ray pulsar in the SNR G109.1 - 1.0.
G. G. Fahlman, P. C. Gregory.
Bull. American Astron. Soc., Vol. 13, 533 - 534 (1981). Abstract.

142.025 Temporal variability of the Crab pulsar in hard X-rays.
M. S. Strickman, W. N. Johnson, J. D. Kurfess.
Bull. American Astron. Soc., Vol. 13, 534 (1981). – Abstract.

142.026 Optical studies of X-ray selected quasars and active galaxies. G. A. Kriss, C. R. Canizares,
C. J. Berg, R. A. Remillard, P. F. Winkler.
Bull. American Astron. Soc., Vol. 13, 549 (1981). – Abstract.

142.027 IR observations of a new X-ray globular cluster and galactic bulge X-ray sources.
P. Hertz, J. E. Grindlay.
Bull. American Astron. Soc., Vol. 13, 549 (1981). – Abstract.

142.028 The nuclear X-ray source in M 81.
M. Elvis, L. Van Speybroeck.
Bull. American Astron. Soc., Vol. 13, 550 (1981). – Abstract.

142.029 Power density spectra of X-ray sources exhibiting aperiodic variability.
H. Bradt, J. Doty, J. G. Jernigan, R. Kelley, E. Morgan,
L. Petro, R. Remillard, B. Schaefer.
Bull. American Astron. Soc., Vol. 13, 558 (1981). – Abstract.

142.030 High energy X-ray studies of the 35d cycle of Her X-1.
A. Gorecki, M. Bautz, F. Lang, A. Levine, F. Primini,
W. H. G. Lewin, D. Gruber, W. Baity, R. Rothschild.
Bull. American Astron. Soc., Vol. 13, 558 (1981). – Abstract.

142.031 X-ray timing observations of 4U1626−67.
D. Leahy, W. Darbro, R. F. Elsner, M. C. Weisskopf,
P. G. Sutherland, J. E. Grindlay, S. M. Kahn.
Bull. American Astron. Soc., Vol. 13, 558 (1981). – Abstract.

142.032 Einstein X-ray observations of the 5.2 hour X-ray binary 4U2129+47.
J. E. McClintock, R. A. London, H. E. Bond, A. D. Grauer.
Bull. American Astron. Soc., Vol. 13, 558 (1981). – Abstract.

142.033 Tentative identification of the flaring X-ray source H0323+02.
R. E. Doxsey, J. E. McClintock, L. Petro, R. Remillard,
D. A. Schwartz.
Bull. American Astron. Soc., Vol. 13, 558 (1981). – Abstract.

142.034 Precise locations of HEAO-1 X-ray sources.
D. A. Schwartz, M. Garcia, J. Patterson, E. Ralph,
W. Roberts, R. Doxsey, M. D. Johnston, K. Wood, J. Meekins,
D. Yentis, R. Bleach, J. McClintock.
Bull. American Astron. Soc., Vol. 13, 558 - 559 (1981). Abstract.

142.035 IUE observations of globular cluster X-ray sources.
J. Grindlay, L. Hartmann.
Bull. American Astron. Soc., Vol. 13, 559 (1981). – Abstract.

142.036 Spectral analysis of the 1975 May transition of Cygnus X-1. L. Chiappetti, R. J. Blissett,
G. Branduardi-Raymont, S. J. Bell Burnell, J. C. Ives,
A. N. Parmar, P. W. Sanford.
Mon. Not. R. Astron. Soc., Vol. 197, 139 - 150 (1981).
X-ray spectra from Ariel V of the 1975 May active state and the subsequent transition of Cyg X-1 are examined. The presence of a black-body soft component ($kT \sim 0.3$ keV) in the spectrum is revealed. The data are found to be in general agreement with the predictions of inverse Compton models for Cyg X-1. An estimate of the electron temperature and the optical thickness for Thomson scattering of the emitting region is attempted.

142.037 The X-ray spectrum of SMC X-1 observed from the Ariel V satellite. M. J. Coe, S. J. Bell Burnell,
A. R. Engel, A. J. Evans, J. J. Quenby.
Mon. Not. R. Astron. Soc., Vol. 197, 247 - 251 (1981).
The X-ray binary pulsator, SMC X-1, has been observed simultaneously over the photon range 2−1200 keV with the Ariel V satellite. The spectrum obtained shows many similarities with comparable sources in our own Galaxy (Her X-1 and Cen X-3), exhibiting a high-energy cut-off around 20 keV. Its extreme X-ray luminosity (in excess of the Eddington limit) is confirmed and a measure of its magnetic field obtained.

142.038 Observations and models of H2252−035.
B. J. M. Hassall, J. E. Pringle, M. J. Ward,
J. A. J. Whelan, S. K. Mayo, J. Echevarria, D. H. P. Jones,
R. E. Wallis, D. A. Allen, A. R. Hyland.
Mon. Not. R. Astron. Soc., Vol. 197, 275 - 286 (1981).
The authors present observations of H2252−035 made in 1980 August and November, which span the wavelength range 1200−22000 Å. The continuum is well represented by a black-body of 12500 K over most of this range. The authors discuss and reject a model for the optical flux in terms of reprocessing of high-energy radiation from the secondary, and propose a model in terms of reprocessing from the disc and bright-spot region.

142.039 The contribution of quasars to the 2 keV−100 MeV background radiation and the X-ray source counts at 2 keV. J. E. Cheney, M. Rowan-Robinson.
Mon. Not. R. Astron. Soc., Vol. 197, 313 - 323 (1981).
The authors estimate the contribution of quasars to the X- and γ-ray background, using the rates of evolution they have previously found necessary to explain the optical source counts of quasars. With plausible assumptions about the optical−X-ray spectral index, α_{ox}, the authors are able to present a self-consistent picture in which the background intensity (and spectral shape) from 2 keV−100 MeV is produced almost entirely by quasars and a significant fraction of the counts of unidentified extragalactic X-ray sources at 2 keV are explained.

142.040 IUE observations of X-ray binaries.
G. Hammerschlag-Hensberge.
Second European IUE Conference, (see 012.011), p. LIX - LXV (1980).
The author reviews UV observations of massive X-ray binaries obtained by IUE. Most early-type X-ray binaries show continuum and line variations which are correlated with binary phase. High dispersion spectra of the X-ray binary HD 77681/Vela X-1 provide a unique opportunity to study the ionization structure of the stellar wind under influence of the X-ray source. The author compares these results with the line profile variations of Si IV and C IV in low-resolution spectra obtained for three other early-type X-ray binaries. It is shown that the intensity of the X-ray source and the density of the stellar wind strongly influence the observed line variations. He gives also a short discussion of the UV observations of

three low-mass X-ray binaries which were observed with IUE: Sco X-1, Her X-1 and AM Her.

142.041 Far ultraviolet observations of Cyg X-2.
L. Maraschi, E. G. Tanzi, A. Treves.
Second European IUE Conference, (see 012.011), p. 169 - 171 (1980).

Two far UV spectra (1150 - 1950 Å) of the optical counterpart of the binary X-ray source Cyg X-2 are presented. The results are compared with the expectations of a model where the X-rays are produced by an accreting neutron star and the optical and UV fluxes derive from the X-ray heated atmosphere of the primary.

142.042 Ultraviolet spectral variations in LMC X-4 and SMC X-1.
J. M. Bonnet-Bidaud, W. M. Glencross, A. J. Willis, G. Hammerschlag-Hensberge, M. van der Klis, S. A. Ilovaisky, M. Mouchet.
Second European IUE Conference, (see 012.011), p. 173 - 175 (1980).

Several low-resolution IUE spectra of the massive X-ray binaries LMC X-4 and SMC X-1 have been obtained at both short and long wavelengths. Highly ionized resonance lines of N V, C IV and Si IV show marked changes with orbital phase, being particularly weak when the X-ray source is in front of the primary; the effect is most striking in LMC X-4. The continuum in each case shows the characteristic double wave light variation seen at visible wavelengths and with similar amplitude.

142.043 Ionization effects in stellar winds of massive X-ray binaries.
G. Hammerschlag-Hensberge.
Effects of mass loss on stellar evolution, (see 012.015), p. 457 - 460 (1981).

High resolution ultraviolet spectroscopy of the early-type X-ray binary Vela X-1/HD 77581 provides a unique example to study the ionization structure of the stellar wind which varies under influence of the X-ray source. New results for other early-type X-ray binaries are presented and compared with the Vela X-1 model. These results show that the observed variations strongly depend on the intensity of the X-ray source and on the density structure of the stellar wind.

142.044 Structural studies in the soft diffuse X-ray background.
K. P. Singh, P. C. Agrawal, R. K. Manchanda, S. Naranan, B. V. Sreekantan.
Bull. Astron. Soc. India, Vol. 9, 65 (1981). — Abstract.

142.045 The X-ray luminosity function of normal stars.
J.-P. Caillault.
News Lett. Astron. Soc. N. Y., Vol. 1, No. 10, p. 26 - 27 (1981). — Abstract.

142.046 The origin of the X-ray background.
G. Setti.
Mem. Soc. Astron. Italiana, Vol. 52, (see 012.019), 159 - 164 (1981).

142.047 The X-ray modulation of Cygnus X-3.
J. M. Bonnet-Bidaud, M. van der Klis.
Astron. Astrophys., Vol. 101, 299 - 304 (1981).

Results are presented of three ~40 d COS-B X-ray observations of Cyg X-3. The amplitude of the 4.8 h modulation is shown to vary more than proportional to the average source level. The shape of the light curve is seen to be generally very stable when averaged over many cycles. It is shown that the correlated behaviour of amplitude, average level and mean shape of the light curve can be explained in terms of an eclipsing binary surrounded by scattering gas. A variation is found in the phase of arrival at the minimum of the light curve, suggestive of a ~20 d periodicity.

142.048 The optical counterpart of A0538−66.
M. Pakull, A. Parmar.
Astron. Astrophys., Vol. 102, L1 - L4 (1981).

The authors present optical and X-ray observations of the recurrent X-ray transient A0538−66. They show that this source is a member of the LMC and consequently among the most luminous stellar sources known. The optical counterpart is a slightly evolved late O to early B type star with episodes of Be type behaviour. The authors suggest that the transient nature of A0538−66 is a result of time dependent accretion.

142.049 On stellar wind accretion in widely separated X-ray binaries, and the nature of 4U0115+63.
Y. Avni, I. Goldman.
Astron. Astrophys., Vol. 102, 12 - 16 (1981).

The authors point out that widely separated, eccentric, pulsating X-ray binaries are appropriate systems for studying the basic properties of stellar wind accretion. They set up a framework of four observational tests to study the characteristics of wind accretion, and to distinguish between different models for the acceleration of the wind. They apply the tests to the source 4U0115+63.

142.050 The ultraviolet spectrum of the X-ray source 2A0526−33.
M. Mouchet, J. M. Bonnet-Bidaud, S. A. Ilovaisky, C. Chevalier.
Astron. Astrophys., Vol. 102, 31 - 34 (1981).

Low resolution spectra of the X-ray source 2A0526-33 have been obtained with IUE. The ultraviolet spectrum of 2A0526-33 shows strong highly ionized resonance lines in emission, with a slight indication of a P-Cygni profile for the C IV line. The observed weak ultraviolet interstellar absorption indicates a source closer than 500 pc. The continuum flux is consistent with a steady emission of an optically thick disc dominating at $\lambda < 2100$ A, and a hot spot of temperature $T \sim 9000$ K contributing mostly in the optical.

142.051 Spin-reversed accretion as the cause of intermittent spindown in slow X-ray pulsars.
Y.-M. Wang.
Astron. Astrophys., Vol. 102, 36 - 44 (1981).

It is shown that the magnitude and sign of the angular momentum captured by an X-ray pulsar from its mass-losing binary companion are very sensitive to the distribution of the relative flow velocity (as well as the matter density) across the capture cross-section. Small fluctuations in the mass transfer process can thus lead to a reversal in the direction of the accretion torque, providing an explanation for the spinup-spindown behaviour of some of the slow X-ray pulsars.

142.052 An analysis of the pulse profiles of the binary X-ray pulsars.
Y.-M. Wang, G. L. Welter.
Astron. Astrophys., Vol. 102, 97 - 108 (1981).

The pulse profiles of the majority of the known binary X-ray pulsars are analyzed in an attempt to derive information about the beaming geometries and the structure of the plasma magnetosphere. Evidence for a bias of the magnetic and rotation axes of the underlying neutron stars toward alignment is presented.

142.053 Galactic X rays observed with X-ray astronomy satellite 'Hakucho'.
S. Hayakawa.
Space Sci. Rev., Vol. 29, 221 - 290 (1981).

Highlights of the results obtained with Japanese X-ray astronomy satellite 'Hakucho' are reviewed. After a brief account of instrumentation, some new features of non-bursting, non-pulsating objects are presented. The main part of the present review is devoted for X-ray bursts which are found more complex than one might have thought. The observation of X-ray pulsars, including a change of spin rate of Vela X-1, is described. The main results obtained in the first two years are summarized.

142.054 *HEAO A-2 observations of non-Abell Zwicky clusters containing extended radio sources.*
G. D. Holman, J. D. McKee.
Astrophys. J., Vol. 249, 35 - 39 (1981).

Burns and Owen's sample of 25 4C radio sources which coincide with Zwicky clusters of galaxies has been searched for X-ray emission using the *HEAO 1 A-2* experiment. X-ray emission was detected from five sources at the 3 σ level, two of which exceeded 5 σ. Positions and 90% X-ray error boxes are given for each of the five sources detected. The clusters surveyed are non-Abell clusters which would generally not be expected to be detectable with the *A*-2 experiment. The search for X-ray emission was prompted by the knowledge of the existence of extended radio sources in the clusters. The relaxed structure of these sources implies the presence of a relatively dense intracluster medium which is expected to produce thermal bremsstrahlung X-ray emission.

142.055 **Identification of 4U 1849–31 with V1223 Sagittarii and discovery of optical pulsations.**
J. E. Steiner, D. A. Schwartz, F. J. Jablonski, I. C. Busko, M. G. Watson, J. P. Pye, I. M. McHardy.
Astrophys. J., Lett., Vol. 249, L21 - L24, plate L2 (1981).

On the basis of a precise HEAO 1 modulation collimator X-ray position uniquely selected by the Ariel 5 sky survey location, the authors identify 4U 1849–31 (3A 1851–312) with V1223 Sgr, a 13 mag star previously classified as an irregular variable. Optical spectroscopy and photometry demonstrate that V1223 Sgr is a cataclysmic variable which displays a photometric period of 13.2 minutes. This pulsation is interpreted as being due to the rotation of a degenerate dwarf. The system shows remarkable similarities to H2252–035.

142.056 **The 805 second X-ray pulsar H2252–035.**
N. E. White, F. E. Marshall.
Astrophys. J., Lett., Vol. 249, L25 - L28 (1981).

The authors demonstrate that the X-ray flux from the 3.6 hr binary system H2252–035 is modulated at a period of 805 s. The spectrum is consistent with either a 1.4 photon index power law or a >20 keV thermal model. A 560 $\pm$ 350 eV equivalent width iron line is seen at $\sim$6.7 keV. They discuss the possibility that this system contains a slowly rotating neutron star.

142.057 *UBV* photometry of the star V1357 Cyg (Cyg X-1). **The mean light curves.**
Kh. F. Khaliullin, A. I. Khaliullina.
Astron. Zh., Tom 58, 1043 - 1050 (1981). In Russian. English translation in Soviet Astron., Vol. 25, No. 5.

The results of individual *UBV* measurements of V1357 Cyg obtained in June - October 1974 are presented. The mean light curves are plotted which combine all the published photoelectric data for this star.

142.058 **On optical eclipses in the system Cyg X-1.**
N. I. Balog, A. V. Goncharskij, A. M. Cherepashchuk.
Pis'ma Astron. Zh., Tom 7, 605 - 611 (1981). In Russian. English translation in Soviet Astron. Lett., Vol. 7.

Results of solving the inverse problem in interpretation of the average high-accurate *UBV* light curves of the Cyg X-1 system are presented for assumed existence of an accretion disk around a probable black hole. The most probable values of the system parameters are derived.

142.059 **Is the riddle of the X-ray bursters solved?**
W. H. G. Lewin.
Priroda, 1981, No. 10, p. 86 - 99. In Russian. Translated from English into Russian by Yu. Eh. Lyubarskij.

142.060 **X-ray observations of symbiotic stars.**
D. A. Allen.
Mon. Not. R. Astron. Soc., Vol. 197, 739 - 743 (1981).

Observations of 19 symbiotic stars made with the image proportional counter of the Einstein Observatory are reported. Three were detected as soft X-ray sources. All three have shown slow-nova eruptions in the past 40 years. The data are interpreted as support for a model for slow novae involving thermonuclear events on white dwarfs which accrete from M giant companions. Symbiotic stars in their steady state, not being detected X-ray sources, are presumed to be powered by the accretion process alone.

142.061 **The high energy X-ray spectrum of 4U 0900–40 observed from OSO 8.**   J. F. Dolan, D. C. Ellison, C. J. Crannell, B. R. Dennis, K. J. Frost, L. E. Orwig.
Astrophys. J., Vol. 250, 355 - 361 (1981).

The X-ray source 4U0900–40 (= Vela XR-1) was observed with the high-energy X-ray spectrometer on OSO 8 for one week in 1976 and three weeks in 1978. Spectra of the source are presented above 16 keV. No systematic difference exists between the X-ray eclipse centers the authors observe during the 2 years and the eclipse centers predicted from optical ephemerides. Short period "intrinsic" variability in the system's X-ray intensity may be related to changes in the Compton scattering optical depth in the system and does not require sporadic mass transfer via Roche lobe overflow. The 282 s modulation in the source's X-ray flux above 21 keV consists of two essentially similar pulses per period, most easily interpreted as arising from the two different magnetic poles of a rotating neutron star. White dwarf and black hole models for the secondary can be ruled out by the observational properties of the system. The secondary appears to be a spherically accreting, magnetic neutron star.

142.062 **On the variation of the period of 4.8 hr of the source Cyg X-3.**
Yu. I. Neshpor, Yu. L. Zyskin.
Izv. Krymskoj Astrofiz. Obs., Tom 63, 157 - 160 (1981). In Russian. English translation in Bull. Crimean Astrophys. Obs. Vol. 63.

The data of observations of gamma-ray quanta with energy 2×10^{12} eV from Cygnus X-3 obtained during 1972 - 1978 have been analysed. It is shown that a period of $\sim$4.8 hr apparently varies with time.

142.063 **Average properties of low-mass X-ray binaries.**
J. van Paradijs.
Astron. Astrophys., Vol. 103, 140 - 144 (1981).

It is shown that the absolute magnitudes and colours of low-mass X-ray binaries are distributed within narrow ranges: $M_V = 1.0 \pm 1.0$ (s.d.), $(B-V)_0 = 0.0 \pm 0.25$ (s.d.), and $(U-B)_0 = -1.0 \pm 0.25$ (s.d.). Also the distribution of the ratio of optical to X-ray luminosity is fairly narrow (± 1.1 mag s.d.). It is proposed that this uniformity in their optical properties is the consequence of a confinement of the X-ray luminosity function of low-mass X-ray binaries to the luminosity range between $\sim 10^{36}$ and $\sim 10^{38}$ erg.

142.064 **Extragalactic X-ray astronomy.**
J. L. Culhane.
Sci. Prog., Vol. 67, 223 - 244 (1981). – Abstr. in Phys. Abstr., Vol. 84, Abstr. 108342 (1981).

142.065 **X-ray emission from Of stars and OB supergiants.**
J. P. Cassinelli, W. L. Waldron, W. T. Sanders, F. R. Harnden, Jr., R. Rosner, G. S. Vaiana.
Astrophys. J., Vol. 250, 677 - 686 (1981).

The result of a survey of X-ray emission from luminous early-type stars is reported in which observations were made using the imaging proportional counter on the Einstein Observatory. The survey suggests that all Of stars and OB supergiants earlier than B1 I are X-ray sources with luminosities $\gtrsim 10^{32}$ ergs s^{-1} and that some later B supergiants have X-ray luminosities $\gtrsim 10^{31}$ ergs s^{-1}. The X-ray fluxes are sufficient to explain the anomalous ionization seen in the ultraviolet spectra of the winds of OB supergiants.

142.066   The X-ray spectrum of AM Herculis from 0.1 to 150 keV.   R. E. Rothschild, D. E. Gruber, F. K. Knight, J. L. Matteson, P. L. Nolan, J. H. Swank, S. S. Holt, P. J. Serlemitsos, K. O. Mason, I. R. Tuohy.
Astrophys. J., Vol. 250, 723 - 732 (1981).

AM Herculis was viewed by the *HEAO 1* satellite in March and April of 1978. The large energy range spanned by the A2 and A4 instruments provided simultaneous coverage from 0.1 to 150 keV. No significant flux at 100 keV was detected in these observations, nor in several *OSO 7* observations. The spectrum above 2 keV can be fitted by a composite thermal bremsstrahlung model which includes an approximation to the albedo expected from the white dwarf.

142.067   Soft X-ray spectral variations in Scorpius X-1.
S. M. Kahn, P. A. Charles, S. Bowyer, R. J. Blissett.
Astrophys. J., Vol. 250, 733 - 738 (1981).

The authors present soft X-ray spectra of Scorpius X-1 obtained with the low-energy detectors of the A-2 experiment on *HEAO 1*. The raw count spectra are deconvolved using the Kahn and Blissett technique to reveal the presence of oxygen absorption in the range 0.5–0.7 keV. The strength of this feature is shown to vary on a time scale of order hours. These results are interpreted as evidence for variable X-ray photo-ionization of circumsource material in the system. An alternative model, involving variable Compton broadening of an oxygen edge, is also discussed.

142.068   X-ray burst sources near the galactic center and their burst peak luminosities.
H. Inoue, K. Koyama, K. Makishima, M. Matsuoka, T. Murakami, M. Oda, Y. Ogawara, T. Ohashi, N. Shibazaki, Y. Tanaka, I. Kondo, S. Hayakawa, H. Kunieda, F. Makino, K. Masai, F. Nagase, Y. Tawara, S. Miyamoto, H. Tsunemi, K. Yamashita.
Astrophys. J., Lett., Vol. 250, L71 - L75 (1981).

X-ray bursts from the GCX bursters, GX 3 + 1, Terzan 1, Terzan 5, and MXB 1728 - 34, all within 0.1 rad from the galactic center, were observed with the Hakucho burst monitor system. Distinctly high concentration of burst sources to the galactic center direction and the frequent association of burst sources with globular clusters imply that those burst sources the authors observed are located spatially near the galactic center. For the probable source distances of ~ 10 kpc, the burst peak luminosities for these sources are found to exceed the Eddington limit for a ~ $1.4\,M_\odot$ neutron star by large factors.

142.069   X-ray observations of SS 433 with the Einstein Observatory.   E. R. Seaquist.
Vistas Astron., Vol. 25, (see 012.040), 61 - 69 (1981).

The author discusses X-ray observations of SS 433 made with the Einstein Observatory. The data reveal that SS 433 contains a variable compact X-ray source with L ~ 10^{35} erg s^{-1} and a diffuse component with L ~ 10^{34} erg s^{-1} which is aligned with the bulges of the SNR W50. The alignment of the features confirms the association between SS 433 and W50. A comparison between X-ray and radio variations suggests that the X-ray emission is nonthermal.

142.070   X-ray observations of SS 433, 1974 - 1980.
M. J. Ricketts, R. Hall, C. G. Page, K. A. Pounds, M. R. Sims.
Vistas Astron., Vol. 25, (see 012.040), 71 - 74 (1981).

The X-ray source A1909 + 04 was reported from analysis of data obtained by the Leicester Sky Survey Instrument on the Ariel V satellite. These data, from the period 1974 October to 1975 July, were sufficient to suggest an association between the X-ray source and W50, and to indicate that the 2 - 15 keV flux was variable in the range 2 - 5 SSI count s^{-1}.

142.071   A second SS 433?   P. C. Gregory, G. G. Fahlman.
Vistas Astron., Vol. 25, (see 012.040), 119 - 125 (1981).

An extraordinary new X-ray source has been discovered with the Einstein Observatory. The source exhibits many parallels with SS 433 and its related SNR, W50. Preliminary results on the associated radio structure observed with the VLA at 20 cm and 6 cm are also presented. The observations are discussed in terms of a precessing beam model.

142.072   The asymmetric 4.8 hour X-ray modulation of Cygnus X-3: model light curves and inferred orbital parameters.   P. Ghosh, R. F. Elsner, M. C. Weisskopf, P. G. Sutherland.
Astrophys. J., Vol. 251, 230 - 245 (1981).

The authors show that the 4.8 hour X-ray light curve of Cygnus X-3, as observed by the Einstein X-Ray Observatory in 1978 December, can be quantitatively described by several models proposed for this source, under the assumption that the binary orbit of Cyg X-3 is elliptic. The authors infer orbital parameters for Cyg X-3 from these models. They find that the cocoon model for Cyg X-3 implies an apparent size for the X-ray star's companion which is larger than that of its critical lobe; they suggest that the emission of a dense stellar wind by the companion could lead to this situation, and they discuss some characteristics of such a hybrid wind-cocoon model.

142.073   Rapid X-ray and optical flares from Scorpius X-1.
L. D. Petro, H. V. Bradt, R. L. Kelley, K. Horne, R. Gomer.
Astrophys. J., Lett., Vol. 251, L7 - L11 (1981).

The authors report the discovery of 1 s time structure in X-ray flares from Sco X-1 and simultaneous optical flares with time structure no faster than 20 s. This decreases by a factor of 10 the shortest reported time scale in the X-ray emission of Sco X-1, but is consistent with the parameters of the standard plasma model for Sco X-1. Different mechanisms for the production of X-ray flares and optical flares are suggested by the disparity in the fastest X-ray and optical time scales. The production of optical flares by reprocessing on the surface of the binary companion of Sco X-1 is ruled out by the X-ray/optical flare time delay.

142.074   Correlation between X-ray and high energy gamma-ray emission from Cygnus X-3.
T. C. Weekes, S. Danaher, D. J. Fegan, N. A. Porter.
Astron. Astrophys., Vol. 104, L4 - L6 (1981).

In May - June 1980, the 4.8 hour modulated X-ray flux from Cygnus X-3 underwent a significant change in the shape of the light curve; this change correlates with the peak in the high-energy (E > 2×10^{12} eV) gamma ray emission at the same epoch.

142.075   HD 36705: a new bright X-ray emitting RS CVn star.   M. W. Pakull.
Astron. Astrophys., Vol. 104, 33 - 36 (1981).

The seventh magnitude K1 IV - V star HD 36705 has been detected with the Einstein observatory to be a flaring X-ray source. Subsequent optical photometry carried out between 1980 February and October reveals a migrating wave with a period of 0.514 d. The data are consistent with the assumption that HD 36705 is a new bright RS CVn system.

142.076   X-ray pulsars – a possible γ-burst source.
A. Balklavs.
Zvaigžnotā debess, 1980. gada rudens, p. 20 - 24. In Latvian.

142.077   Lanning 10 and 33: the X-ray, UV, and optical fluxes.   P. Szkody, L. Crosa.

Astrophys. J., Vol. 251, 620 - 625 (1981).

The authors have used UV spectra from *IUE*, optical spectroscopy, broad band photometry, and upper limits on the X-ray flux from Einstein to specify the characteristics of Lanning 10 and Lanning 33. They find the data on Lanning 33 to be consistent with a 22,500 K Be star at a distance of 5 to 7 kpc, while Lanning 10 appears to include an F5 − G0 star plus a hot component ($T > 35,000$ K) at a distance of 0.9 to 1.3 kpc. There are several characteristics which distinguish Lanning 10 from a typical cataclysmic variable.

142.078   **X-ray observations of cataclysmic variables.**
          R. H. Becker.
Astrophys. J., Vol. 251, 626 - 629 (1981).

Twelve cataclysmic variables (CVs) have been observed with the Imaging Proportional Counter (IPC) on the Einstein Observatory. These objects share many of the X-ray characteristics common to previously observed CVs, namely time variability over intervals from minutes to months, relatively hard X-ray spectra, and X-ray luminosities L_x between 10^{29} and 10^{32} erg s^{-1}. The results on these 12 objects have been combined with results for 20 other CVs from two other surveys to determine the mean X-ray properties of the subclasses of CVs. The X-ray properties of old nova systems can be used either to constrain their space densities or to infer an evolution in their X-ray luminosities.

142.079   **A search for apsidal motion in 4U0115 +63.**
          R. L. Kelley, S. Rappaport, M. J. Brodheim,
L. Cominsky, R. Stothers.
Astrophys. J., Vol. 251, 630 - 638 (1981).

The authors have carried out a pulse arrival-time analysis of the archival Uhuru data from the 1971 transient outburst of the binary X-ray pulsar 4U0115 +63. The 3.6 s X-ray pulsations are clearly present in the data, and they show that the average fractional rate of change in pulse period over the 7 yr interval 1971 - 1978 corresponds to $\dot{P}/P = -2 \times 10^{-6}$ yr^{-1}. This spin-up rate is consistent with an average source luminosity ~ 20 times less than that observed during its flare state. The pulse arrival times were tracked for $\sim 7^d$, and by combining these data with the 1978 *SAS 3* orbital determination, they place a limit on the advance of periastron in the 4U0115 +63 system of $\dot{\omega} \lesssim 2°1$ yr^{-1} (95% confidence). The analysis also yields an improved value for the orbital period, $P_{orb} = 24^d3162$.

142.080   **X-ray and optical observations of HD 155638: a**
          remarkably active cool star.
R. A. Stern, J. A. Nousek, J. J. Nugent, P. C. Agrawal,
G. R. Riegler, A. Rosenthal, S. H. Pravdo, G. P. Garmire.
Astrophys. J., Lett., Vol. 251, L105 - L108, plate L3 (1981).

The authors report X-ray observations with the Einstein Observatory and spectroscopic investigation of the star HD 155638. The spectra are consistent with a spectral type G 8 IV - V. They show evidence of Ca II H and K core emission reversals and a filled-in Hα line. The X-ray luminosity is estimated to be $\sim 10^{30}$ - 10^{32} erg s^{-1}. It is suggested that the X rays are of coronal origin, indicating that this star has a remarkably active chromosphere and corona. The similarity of the X-ray properties to those of the highly active RS CVn binaries is pointed out.

142.081   **Some remarks on the rotational state of pulsating**
          X-ray sources.   G. Börner.
Proceedings of the 5th Göttingen-Jerusalem-Symposium on Astrophysics, (see 012.041), p. 201 - 210 (1981).

A rotating neutron star with a strong magnetic field surrounded by an accretion disk is adopted as a model for a regularly pulsating X-ray source. In the context of this model it is the interaction between the stellar magnetic field and the plasma in the accretion disk which determines the rotational state of the neutron star. Two different configurations are discussed, and resulting period changes are compared with the ob-

servations. Finally the possibility that an X-ray source can turn off itself is investigated.

142.082   **Evolution of X-ray astronomy.**   B. Rossi.
          Plasma astrophysics, (see 012.042), p. 225 - 249
(1981).

The evolution of X-ray astronomy up to the launching of the Einstein Observatory, proceeded through the following major steps. (1) The discovery in 1962 of an extrasolar X-ray source, Sco X-1. (2) The identification of a strong X-ray source with the Crab Nebula. (3) The identification of Sco X-1 with a faint, peculiar optical object. (4) The demonstration that X-ray stars are binary systems. (5) The discovery of X-ray "bursts". (6) The discovery of exceedingly strong X-ray emission from active galaxies, quasars and clusters of galaxies. (7) The demonstration that in these the principal X-ray source is a hot gas filling the space between galaxies.

142.083   **Diagnosis of neutron star and its environments with**
          X-ray astronomy.   M. Oda.
Plasma astrophysics, (see 012.042), p. 251 - 263 (1981).

Among a number of questions with regard to the neutron star, its atmosphere and its environments raised by observational results in X-ray astronomy, some selected topics are discussed. The behavior of the X-ray pulsar Vela X-1, observed with the X-ray astronomy satellite "Hakucho", is discussed in relation to the nature of the neutron star and its interaction with an accretion disk. X-ray bursts, also observed with the satellite, are a rich source of information for physics of the atmosphere and the plasma environment of the neutron star. The hard X-ray imaging of the Crab Nebula undertaken by a collaborative balloon observation may reveal the internal structure of the Crab Nebula.

142.084   **A four-hour orbital period of the X-ray burster**
          4U/MXB1636−53.
H. Pedersen, J. van Paradijs, W. H. G. Lewin.
Nature, Vol. 294, 725 - 727 (1981).

The authors report variations in the persistent optical flux of 4U/MXB1636−53 which seem to vary periodically with a period near 4 h.

142.085   **X-ray astronomy in the Einstein era.**
          R. Giacconi.
Space Sci. Rev., Vol. 30, (see 012.044), 3 - 32 (1981).

A review of X-ray observations of extragalactic objects emphasizing their role as a tool for investigating the evolution of large-scale structure in the universe is given. In particular the progress is discussed that has been brought about by the Einstein mission in the understanding of clusters of galaxies, quasars and the X-ray background.

142.086   **An X-ray survey of flat spectrum radio sources.**
          P. Biermann.
Space Sci. Rev., Vol. 30, (see 012.044), 33 - 37 (1981).

Flat spectrum radio sources from the MPIfR north pole radio survey at 5 GHz were observed with the Einstein X-ray observatory, at optical, mm, and radio wavelengths. The sources show rather extreme properties in all wavelength regions.

142.087   **"Einstein" observations of BL Lacertae objects.**
          D. Maccagni, M. Tarenghi.
Space Sci. Rev., Vol. 30, (see 012.044), 55 - 58 (1981).

Nine fields containing BL Lacertae objects have been observed with the Imaging Proportional Counter of the "Einstein" Observatory. The authors have detected seven BL Lacertae objects and eight serendipitous sources.

142.088   **An X-ray selected list of QSOs: implications for the**
          QSO luminosity function and the origin of the
          diffuse X-ray background.

B. Margon, G. Chanan, R. Downes.
Space Sci. Rev., Vol. 30, (see 012.044), 59 (1981).
Abstract.

142.089 Optical identification of X-ray sources in the Einstein Observatory medium and deep surveys.
R. E. Griffiths, T. Maccacaro, S. S. Murray, R. Giacconi.
Space Sci. Rev., Vol. 30, (see 012.044), 67 - 73 (1981).

Methods are discussed for establishing the optical identification of X-ray sources in the 'medium' and 'deep' X-ray surveys of the Einstein Observatory. Of the 63 X-ray sources with a statistical significance of $\gtrsim 5\sigma$ in the medium survey, optical identification work is summarized for 51, of which identifications have been made with 30 active galactic nuclei. The Einstein deep survey of Pavo is used to illustrate the problems and methods used for securing optical identifications for X-ray sources in the deep survey fields.

142.090 The Einstein Observatory medium sensitivity survey.
T. Maccacaro, E. D. Feigelson, R. Giacconi,
I. M. Gioia, R. E. Griffiths, J. Liebert, S. S. Murray, J. Stocke, G. Zamorani.
Space Sci. Rev., Vol. 30, (see 012.044), 75 - 80 (1981).

Results are presented from an X-ray survey of ~50 square degrees of the high galactic latitude sky at sensitivities in the range $7 \times 10^{-14} - 5 \times 10^{-12}$ erg/cm^2 sec (0.3–3.5 keV) carried out with the Imaging Proportional Counter (IPC) aboard the Einstein Observatory. The extragalactic sample consists of 48 sources which have been used to determine the number flux relation. The content of the sample is analyzed in terms of types of sources and is found to be significantly different from the content of similar samples selected at higher fluxes.

142.091 On the properties of 21 X-ray globular clusters in M31.
P. Battistini, F. Bonoli, R. Buonanno, C. Cacciari, C. E. Corsi, F. Fusi Pecci.
Space Sci. Rev., Vol. 30, (see 012.044), 81 - 84 (1981).

B and V plates for 19 fields covering a $4° \times 4°$ area centered on M31 have been obtained with the Loiano 152 cm Ritchey-Chrétien F/8 telescope. A search for globular clusters is in progress, yielding in the central field 240 candidates. 21 candidates of the quoted sample have been identified with 17 HRI and 4 IPC X-ray sources derived from Einstein Observatory data.

142.092 Physical processes for X-ray emission in galactic nuclei.
M. J. Rees.
Space Sci. Rev., Vol. 30, (see 012.044), 87 - 99 (1981).

The 2 - 10 keV X-ray luminosities of galactic nuclei range from $L_X \leqslant 10^{36}$ erg s^{-1} for the Galactic Nucleus to $L_X \geqslant 10^{46}$ erg s^{-1} for quasars. The author reviews possible emission mechanisms in central continuum sources, the conditions in hot and "transrelativistic" plasmas and outlines the implications of rapid variability. Constraints on models of the ultimate energy source are discussed and the analogies between active galactic nuclei and compact galactic objects are pointed out.

142.093 The X-ray background 2–50 keV: a superposition of known sources, or evidence of a new component?
A. Cavaliere, L. Danese, G. De Zotti, A. Franceschini.
Space Sci. Rev., Vol. 30, (see 012.044), 101 - 105 (1981).

The contribution of the canonical quasars to the X-ray background 2–50 keV is not likely to exceed 50%. A sensible estimate for the intensity ratio L_X/L_O results in a further reduction of the percentage. In addition, recent counts of radio-loud and radio-quiet quasars definitely limit the number of faint sources. A "missing component" with definite spectral properties should comprise a sizeable fraction of the XRB.

142.094 X-rays from active galactic nuclei – hard component.
S. Tsuruta, F. Takahara, S. Ichimaru.

Space Sci. Rev., Vol. 30, (see 012.044), 123 - 127 (1981).

A model for the emission of hard X-rays in active galactic nuclei is proposed by assuming the existence of a supermassive black hole with $M \sim 10^8 M_\odot$ and sufficient mass accretion $(M \sim 10^{26}$ g s$^{-1})$ in the center. The hard X-ray component would be due to the Comptonization of self-absorbed cyclotron photons.

142.095 Is PKS 2155 an extragalactic source?
L. Maraschi, E. G. Tanzi, M. Tarenghi, A Treves.
Space Sci. Rev., Vol. 30, (see 012.044), 129 - 133 (1981).

The X-ray source H2155–304, identified with the radio-source PKS 2155 and with a 14 mag star-like object is commonly believed to be a BL Lac object. The authors present observations in the far ultraviolet (1200 - 3000 Å) obtained with I.U.E. The presence of weak variable emission features is discussed and the extragalactic nature of the object is questioned.

142.096 Low luminosity galactic X-ray sources.
G. S. Vaiana.
Space Sci. Rev., Vol. 30, (see 012.044), 151 - 179 (1981).

A comprehensive review of stellar X-ray emission based on Einstein Observatory results is given. X-ray luminosity functions for various spectral types are discussed. Correlations between X-ray luminosity and other stellar parameters (bolometric luminosity, surface rotation) are studied. Theoretical models of stellar surface activity, including coronal activity, stellar winds and accretion shocks are outlined. It is concluded that X-ray emission from stars over a large range in the H–R diagram is very similar and most probably represents solar coronal surface activity modulated by stellar parameters which govern stellar magnetic activity.

142.097 X-ray spectroscopic investigation of the coronal structure of Capella.
E. H. B. M. Gronenschild, R. Mewe, N. J. Westergaard, J. Heise, F. D. Seward, T. Chlebowski, N. P. M. Kuin, A. C. Brinkman, J. H. Dijkstra, H. W. Schnopper.
Space Sci. Rev., Vol. 30, (see 012.044), 185 - 189 (1981).

The binary system Capella (G6 III + F9 III) has been observed on 1979 March 15 and on 1980 March 15 - 17 with the Objective Grating Spectrometer onboard the Einstein Observatory. The spectrum measured covers the range 5 - 30 Å with a resolution < 1 Å. Line emissions from O VIII, Fe XVII, Fe XVIII, Fe XXIV and Ne X have been identified. The data are interpreted in terms of emission from hot static coronal loops rather similar to the magnetic arch structures found on the sun.

142.098 Coronal activity in F-, G-, and K-type stars.
R. Mewe, C. J. Schrijver, C. Zwaan.
Space Sci. Rev., Vol. 30, (see 012.044), 191 - 199 (1981).

Soft X-ray (0.3 - 3.5 keV) observations with the Imaging Proportional Counter (IPC) onboard Einstein Observatory are presented for a sample of some 20 cool stars of luminosity classes III-V; The results are compared with the Ca II H and K emission, which had served as a selection criterion. The specific X-ray flux F_X is an increasing function of the specific Ca II H and K line-core flux F_{H+K}. The results are interpreted in the framework of magnetic activity. The X-ray emission and the excess Ca II H and K flux are attributed to magnetic structure in the corona and chromosphere, the magnetic features emerging from the stellar convective envelope, where they are generated by dynamo action.

142.099 Cosmic X-ray observations performed with a gas scintillation spectrometer.
R. D. Andresen, G. Manzo, A. Peacock, B. G. Taylor, R. C. Catura, J. L. Culhane, R. Berthelsdorf.
Space Sci. Rev., Vol. 30, (see 012.044), 243 - 249 (1981).

Observations of the spectrum of Cas-A in the X-ray

energy range 3 to 10 keV have been made using two gas scintillation proportional counters flown on an Aries rocket. Analysis of the data from the 180 second observation has resolved the emission features at ~ 7 keV, which results from transitions in highly ionised iron.

142.100   **Hard X-ray structure of the Crab Nebula.**
K. Makishima, Y. Ogawara, M. Matsuoka, M. Oda, S. Miyamoto, R. M. Pelling, L. E. Peterson, W. S. Paciesas. Space Sci. Rev., Vol. 30, (see 012.044), 259 - 262 (1981).

Two-dimensional maps of the Crab Nebula have been synthesized in 22 - 64 keV range through the modulation collimator experiment. The effective angular resolution is about 15″. The result indicates that the Crab morphology is strictly controlled by the pulsar.

142.101   **First UV observations of the optical counterpart of the X-ray source 2S0114+650.**   L. Bianchi.
Space Sci. Rev., Vol. 30, (see 012.044), 273 - 277 (1981).

The X-ray source 2S0114+650 discovered by SAS–3 in 1978 has been associated with a highly reddened star showing broad H_α emission. The author observed this star with the LWR spectrograph on board the I.U.E. Satellite on Sep. 1980 and Feb. 1981. An ultraviolet flux was detected longward of the 2200 Å bump, consistent with the B0.5 spectral classification and the values of V=11 and E(B–V) = 1.48. The UV extinction in the region around 2200 Å seems to deviate from the standard interstellar extinction law.

142.102   **Soft X-ray sources and their optical counterparts in the error box of the COS-B source 2CG 135+01.**
P. A. Caraveo, G. F. Bignami, J. A. Paul, B. Marano, G. P. Vettolani.
Space Sci. Rev., Vol. 30, (see 012.044), 301 - 306 (1981).

A field containing the error boxes of the COS-B source 2CG 135+01 and covering 6 square degrees in the galactic plane has been surveyed with the IPC instrument of the Einstein observatory. The five sources detected were identified by optical spectroscopy with one quasar and four stars of spectral types B, F, G.

142.103   **Hard X-ray balloon observations of compact galactic and extragalactic X-ray sources.**   R. Staubert, E. Kendziorra, W. Pietsch, R. J. Proctor, C. Reppin, H. Steinle, J. Trümper, W. Voges.
Space Sci. Rev., Vol. 30, (see 012.044), 311 - 323 (1981).

A balloon program in hard X-ray astronomy (20 - 200 keV) is jointly pursued by the Astronomisches Institut der Universität Tübingen (AIT) and the Max-Planck-Institut für Extraterrestrische Physik in Garching (MPE). Since 1973 nine successful balloon flights have been performed from Texas and Australia. Here results on Centaurus A and on several galactic binary X-ray sources are summarized. In particular the high energy photon spectrum of Hercules X-1 and the evidence for the cyclotron line feature which was discovered by the authors in 1976 is reviewed.

142.104   **High energy X-ray spectrum of Her X-1.**
A. Scheepmaker, F. A. Jansen, A. J. M. Deerenberg, G. R. Ricker, J. E. Ballintine, J. V. Vallerga, W. H. G. Lewin. Space Sci. Rev., Vol. 30, (see 012.044), 325 (1981). Abstract.

142.105   **Hard X-rays observation of Her X-1.**
P. Ubertini, A. Bazzano, C. La Padula, V. F. Polcaro, G. Vialetto, R. K. Manchanda.
Space Sci. Rev., Vol. 30, (see 012.044), 327 - 330 (1981).

A hard X-rays (15 - 170 keV) measurement of the spectrum of Her X-1, during a mid turn on is presented. The presence of an emission line at about 53 keV during the mid-on state is confirmed by the present measurement.

142.106   **New hard X-rays sources at 38° declination.**
P. Ubertini, A. Bazzano, C. La Padula, V. F. Polcaro. Space Sci. Rev., Vol. 30, (see 012.044), 331 - 334 (1981).

The authors report the detection of three new hard X-rays sources emitting in the range 15 - 150 keV. Their observation was carried out by means of a balloon borne payload, consisting of two large area high spectral resolution Multiwire Spectroscopic Proportional Counters.

142.107   **Magnetic alignment and counter-alignment in binary X-ray pulsars.**   Y.-M. Wang.
Space Sci. Rev., Vol. 30, (see 012.044), 341 - 346 (1981).

The magnetic torque exerted on a binary X-ray pulsar acts to decrease the angle between the dipole and spin axes during braking episodes, but to increase it whenever spinup occurs. The likely effect on the X-ray pulse profiles is considered.

142.108   **Ariel VI observations of Sco X-1.**
A. M. T. Pollock, C. V. Goodall, G. F. Carpenter, D. K. Bedford, A. M. Cruise, R. E. Cole, J. P. Osborne, J. L. Culhane.
Space Sci. Rev., Vol. 30, (see 012.044), 347 - 351 (1981).

Two of the soft X-ray detectors on Ariel VI have been used to observe Sco X-1 for 7 min near JD 244 4332.98 and for 8 min near JD 244 4333.05. A 32 channel pulse height spectrum covering the range 0.1 to 1.5 keV was recorded.

142.109   **Periodic behaviour in three galactic bulge sources.**
T. J. Ponman.
Space Sci. Rev., Vol. 30, (see 012.044), 353 - 360 (1981).

Evidence for periodicities of several days in the three bright galactic bulge sources GX349+2, GX17+2 and Ser X-1 has been found in Ariel V RMC data collected over 4 years.

142.110   **Optical properties of low-mass X-ray binaries.**
J. van Paradijs, F. Verbunt.
Space Sci. Rev., Vol. 30, (see 012.044), 361 - 364 (1981).

An important class of bright galactic X-ray sources (comprising the bright galactic bulge sources, X-ray bursters, globular cluster sources and the soft X-ray transients) consist of a neutron star and a low-mass (< 1 M☉) Roche-lobe filling star.

142.111   **New optical observations of 2S 0921–630 and its relation to other X-ray binaries.**
C. Chevalier, S. A. Ilovaisky.
Space Sci. Rev., Vol. 30, (see 012.044), 365 - 366 (1981). Abstract.

142.112   **Ariel 6 observations of Cyg X-1 in the high state.**
C. G. Page, A. J. Bennetts, M. J. Ricketts.
Space Sci. Rev., Vol. 30, (see 012.044), 369 - 371 (1981).

Cygnus X-1 was observed by the Leicester University instrument on Ariel 6 during the period 1980 June 24 to July 6. The source was in transition from the high to the low state. The main aim of the observation was to obtain energy spectra with high time resolution. Most data were obtained with 8 energy channels geometrically spanning the range 1 - 50 keV and time resolution of 62.5 msec or 125 msec.

142.113   **Ariel VI soft X-ray observations of Cygnus X-1.**
D. K. Bedford, G. F. Carpenter, C. V. Goodall, A. M. T. Pollock, R. E. Cole, A. M. Cruise, J. P. Osborne. Space Sci. Rev., Vol. 30, (see 012.044), 373 - 378 (1981).

Observations of Cygnus X-1 during the high-low transition of June - July 1980 reveal an intense flux between 0.5 and 1.5 keV. Although the intensity broadly follows the 1 - 12 keV flux through the transition, there is no evidence of variations in the shape of the energy spectrum. The implications of these results and derived limits on the minute-to-minute variability are discussed.

142.114 **Power spectral density analysis of hard X-ray emission of Cyg X-1.**   F. Frontera, F. Fuligni.
Space Sci. Rev., Vol. 30, (see 012.044), 379 - 384 (1981).

The authors report on results of analysis performed on data of three balloon observations of Cyg X-1 in the range 20 - 200 keV. Evidence of periodic, even if not stable, pulsations were previously obtained at about 0.058 Hz. The continuous power spectrum, still showing a broad peak around the above frequency, can be explained on the basis of a model, similar to the shot noise one, where the single shot is replaced by a short sequence of equispaced pulses with a decaying amplitude.

142.115 **Search for fast time fluctuations in Cygnus X-1 and Herculis X-1.**   S. Bonazzola.
Space Sci. Rev., Vol. 30, (see 012.044), 385 - 388 (1981).

The author reports analysis of HEAO-1 A2 and HEAO-B SSS data in view of searching fast time fluctuations in Cyg X-1 and Her X-1.

142.116 **X-ray absorption in Circinus X-1 in the low state.**
L. Chiappetti, S. J. Bell-Burnell.
Space Sci. Rev., Vol. 30, (see 012.044), 389 - 393 (1981).

The authors report observations of Cir X-1 with the Ariel 5 spectrometer, showing that the hydrogen column density in the low state is comparable to that observed in the high state. The implications of this fact for the models of the source are discussed.

142.117 **Time variation of the pulse period of Vela X-1.**
F. Nagase.
Space Sci. Rev., Vol. 30, (see 012.044), 395 - 398 (1981).

X-ray pulsar Vela X-1 was observed with the X-ray astronomy satellite HAKUCHO on five occasions between March 1979 and March 1981. An increase of the pulsation period at an average rate of $\dot{P}/P \simeq 3.0 \times 10^{-4}\,\mathrm{yr}^{-1}$ was observed over the time span of two years. Besides, variations of the pulse period in the time scale of $\sim$10 days were resolved in superposition on the secular spin-down trend. The observed rate of change $\dot{P} \simeq 3 \times 10^{-8}$, for both spin-up and spin-down, is an order of magnitude greater than the secular spin-down rate.

142.118 **Observation of an outburst from the X-ray pulsator 0115+63.**   M. J. Ricketts, R. Hall, C. G. Page, K. A. Pounds.
Space Sci. Rev., Vol. 30, (see 012.044), 399 - 403 (1981).

The Ariel 6 X-ray instruments observed 0115+63 during 1980 December 16 - 30, when the mean flux was $\sim$ 150 millicrab. Analysis of the $\sim$ 3.6 s pulsations gave a refined orbital period of 24.3155 ± 0.0002 days and a periastron angle of $47°.15 \pm 0°.13$, setting a limit on the rate of advance of periastron since 1978 of 0.11 yr^{-1}.

142.119 **The optical light curve of LMC X-4 and the 30-day X-ray period.**   C. Chevalier, S. A. Ilovaisky, C. Motch, M. Pakull, J. Lub, J. van Paradijs.
Space Sci. Rev., Vol. 30, (see 012.044), 405 - 406 (1981).

Analysis of an extensive set of optical photometric observations of the massive X-ray binary LMC X-4 obtained from 1976 through 1981 at the European Southern Observatory, La Silla, Chile, shows a clear modulation of the 1.408-d binary light curve amplitude with the 30.48-d X-ray period. The B filter amplitude varies by a factor of 2 from a lowest value of 0.09 mag during the X-ray OFF portion of the 30.48-d cycle to a highest value of 0.18 mag during the X-ray ON part of the cycle.

142.120 **Ultraviolet and optical observations of HDE 245770/ A 0535+26 system during an X-ray flare.**
F. Giovannelli, M. Ferrari-Toniolo, A. Giangrande, P. Persi, C. Bartolini, A. Guarnieri, A. Piccioni.
Space Sci. Rev., Vol. 30, (see 012.044), 407 - 414 (1981).

The authors present high and low dispersion UV spectra of HDE 245770, which is the optical counterpart of the recurrent transient X-ray pulsar A 0535+26, during a decay of an X-ray flare. UBV photometric measurements and medium dispersion optical spectra were simultaneously obtained. The energy distribution of the star in the 0.12 - 10 micron range is compared with the models of Kurucz and Poeckert and Marlborough.

142.121 **Recent optical observations of the X-ray pulsar 4U 1626 - 67.**
S. A. Ilovaisky, C. Chevalier, C. Motch.
Space Sci. Rev., Vol. 30, (see 012.044), 415 - 417 (1981).
Abstract.

142.122 **COS-B X-ray observations of Cyg X-3; an indication for apsidal motion.**
M. van der Klis, J. M. Bonnet-Bidaud.
Space Sci. Rev., Vol. 30, (see 012.044), 419 - 421 (1981).

Results are presented of three $\sim$40 d COS-B X-ray observations of Cyg X-3. The shape of the 4.8 h light curve showed variations on all observational time scales. An $\sim$20 d periodicity was found in the arrival time of the minimum of the light curve, interpreted as apsidal motion of an orbit with an eccentricity of $\sim$0.03.

142.123 **An optical outburst from the periodic recurrent X-ray transient A0538-66.**   P. A. Charles, L. Booth, R. H. Densham, J. R. Thorstensen, A. J. Willis.
Space Sci. Rev., Vol. 30, (see 012.044), 423 - 431 (1981).

The authors present optical spectroscopy and photometry and IUE spectroscopy of the counterpart of the LMC recurrent X-ray transient A0538-66 during an outburst at the end of December 1980 which was consistent with the 16.6 day X-ray period. The optical spectra show steadily increasing Balmer and He I emission (indicative of a shell phase) superposed on a B2 IV spectrum with a substantial brightness increase of 2^m. Significant radial velocity changes have been detected but they show no correlation with the 16.6 day period. IUE spectra during a subsequent outburst show very strong and broad ($\sim$5000 km s^{-1}) emission from C IV λ1550 and He II λ1640.

142.124 **The X-ray source A0538-66 in optical quiescence.**   P. Murdin, G. Branduardi-Raymont, A. N. Parmar.
Space Sci. Rev., Vol. 30, (see 012.044), 433 - 439 (1981).

Photometric observations of A0538-66 in quiescence show that the optical star is an object of equivalent spectral type B7eIII. Spectroscopic data taken simultaneously confirm this spectral classification and are consistent with the reported periodic variations seen in the emission lines. It is unlikely that the optical outburst is due to X-rays reprocessed in the atmosphere of the primary star or elsewhere.

142.125 **The periodic X-ray transient A0538-66 and its optical counterpart.**   G. K. Skinner.
Space Sci. Rev., Vol. 30, (see 012.044), 441 - 446 (1981).

The long term light curve of the optical counterpart of A0538-66 is discussed and a revised estimate for the period obtained. Observations of the optical spectrum during outburst and preliminary results from Einstein observations of two recent X-ray flares are presented.

142.126 **Ariel 5 observations of GX 339-4.**
L. Chiappetti.
Space Sci. Rev., Vol. 30, (see 012.044), 447 - 449 (1981).

Observations of GX 339-4 with the Ariel 5 spectrometer are reported. The spectrum of the source is very steep, possibly a low temperature blackbody or Bremsstrahlung distribution. Differences and similarities with the related sources Cyg X-1 and Cir X-1 are discussed.

142.127   **Highlights of results from HAKUCHO.**
S. Hayakawa.
Space Sci. Rev., Vol. 30, (see 012.044), 453 - 456 (1981).
A small X-ray satellite named HAKUCHO after Constellation Cygnus has been active since its launch on 21 February 1979. The author reviews observational results for X-ray bursters obtained with this satellite.

142.128   **The Ariel V sky survey of fast-transient X-ray outbursts.**   I. M. McHardy, J. P. Pye.
Space Sci. Rev., Vol. 30, (see 012.044), 457 - 460 (1981).
The authors analyzed the data of the Ariel V Sky Survey Instrument (SSI) comprising ~30 000 orbits of the instrument in search of fast X-ray bursts (5σ limit). A total of 38 fast-transient sources is reported; the morphology, identification and possible nature of these sources are briefly discussed.

142.129   **Periodic X-ray outbursts from A0535+26, 2S1145−619 and GX301−2.**
R. S. Warwick, M. G. Watson, M. R. Sims.
Space Sci. Rev., Vol. 30, (see 012.044), 461 - 466 (1981).
The authors report Ariel V(SSI) observations of three X-ray pulsars A0535+26, 2S1145−619 and GX301−2 (2S1223−624). These sources exhibit X-ray outbursts which appear, on the basis of observations extending over ~5 years, to have recurrence periods of 110 days, 187.5 days and 41.4 days respectively. The optical counterparts are all early-type stars; A0535+26 and 2S1145−619 are identified with Be main sequence stars and GX301 −2 with a B2 supergiant which also shows an emission line spectrum.

142.130   **X-ray and optical observations of the November 19, 1978 gamma-ray burst source region.**   G. Pizzichini, J. Danziger, P. Grosbøl, M. Tarenghi, T. L. Cline, U. D. Desai, R. Mushotzky, B. J. Teegarden, W. D. Evans, R. W. Klebesadel, J. G. Laros, C. Barat, K. Hurley, M. Niel, G. Vedrenne, I. V. Estulin (*Ehstulin*), G. Mersov, V. Zenchenko, V. Kurt.
Space Sci. Rev., Vol. 30, (see 012.044), 467 - 470 (1981).
The November 19, 1978 gamma-ray burst has a very well determined error box, 10 square arcmin. An 8000 second IPC exposure with the Einstein Observatory detected, at a 3.4σ level, one low intensity X-ray source inside the error box. Inside the X-ray source error box there are two weak radio sources, one of them highly polarized and two m_v ~20 magnitude objects, not coincident with the radio sources, visible in the ESO/SRC J and R plates.

142.131   **Discovery of rapid optical variability in GX 339−4.**
C. Motch, S. A. Ilovaisky, C. Chevalier.
Space Sci. Rev., Vol. 30, (see 012.044), 475 - 476 (1981).
The optical counterpart of the Cygnus X-1-like X-ray source GX339−4 has recently undergone large brightness variations. Although it was fainter than 21 mag in B on March 8, 1981, it was found to be at V = 15.4 on May 28, 1981.

142.132   **X-ray binaries and stellar evolution.**
E. P. J. van den Heuvel.
Space Sci. Rev., Vol. 30, (see 012.044), 623 - 642 (1981).
Observational evidence suggests that most − if not all − binary X-ray sources are neutron stars. The evolutionary status and possible formation mechanisms of the type I (massive) and type II (low-mass) X-ray binaries are discussed. The difference between the "standard" massive X-ray binaries and the Be/X-ray binaries is ascribed to a somewhat different evolutionary history and status, and possible reasons for the existence of short- and long-period X-ray pulsars are discussed. Type II X-ray sources in globular clusters were most probably formed by capture processes; their formation rate inferred from the observations indicates that only a small fraction ($\lesssim$ 1 to 10 percent) of the originally formed neutron stars have remained in their clusters. Type II sources in the galactic bulge may also have formed from cataclysmic binaries in which a

white dwarf was driven over the Chandrasekhar limit by accretion.

142.133   **X-ray surveys with the Einstein Observatory.**
J. Danziger.
Messenger, No. 26, p. 5 - 6 (1981).

142.134   **Discovery of a very fast optical activity in the X-ray source GX 339−4.**   C. Motch, S. A. Ilovaisky, C. Chevalier.
Messenger, No. 26, p. 28 - 30 (1981).

142.135   **X-ray emission from lateral radio components of Sco X-1.**   V. M. Lipunov, E. I. Moskalenko.
Astron. Tsirk., No. 1117, p. 1 - 2 (1980). In Russian.

142.136   **V926 Scorpii.**
IAU Circ., No. 3615 (1981).

142.137   **GX 339-4.**
IAU Circ., No. 3616 (1981).

142.138   **New optical candidates for galactic X-ray sources.**
IAU Circ., No. 3620 (1981).

142.139   **GX 17+2.**
IAU Circ., Nos. 3624, 3628 (1981).

142.140   **V1333 Aquilae = 4U 1908+00.**
IAU Circ., No. 3626 (1981).

142.141   **V801 ARAE = 4U 1636-53 = MXB 1636-53.**
IAU Circ., No. 3628 (1981).

142.142   **4U 1916-05.**
IAU Circ., No. 3632 (1981).

142.143   **LMC X-4.**
IAU Circ., No. 3632 (1981).

142.144   **SS Cygni.**
IAU Circ., No. 3639 (1981).

142.145   **H0139-68.**
IAU Circ., No. 3649 (1981).

142.146   **Possible optical identification of 1E 064301-1640.8.**
IUA Circ., No. 3651 (1981).

142.147   **X-ray/optical/infrared/radio observations.**
IAU Circ., No. 3652 (1981).

142.148   **SS Cygni.**
Yamamoto Circ., No. 1964 (1981).

142.149   **Optical eclipses and the precession effects in the X-ray binary system HD 153919 = 4U 1700−37.**
A. M. Cherepashchuk, T. S. Khruzina.
Astron. Zh., Tom 58, 1226 - 1240 (1981). In Russian.
English translation in Soviet Astron., Vol. 25, No. 6.
Long-period ($P = 32\overset{d}{.}46$) variability of the amplitude and shape of the optical light curves of the X-ray binary HD 153919 has been discovered from an analysis of all published photometric data. This periodicity is presumably due to eclipses of the optical star by the "precessing" accretion structure surrounding the neutron star. Estimates of the parameters of the system are obtained. A supernova explosion in the binary system is suggested as the main cause of the precession of the rotational axis of the optical star, and related problems are discussed.

142.150 Photographic observations of the newly discovered X-ray emitting dwarf nova 1E 0643.0 - 1648.
L. Meinunger.
Inf. Bull. Variable Stars, No. 2001, 2 pp. (1981).

142.151 Long-term variability of X-ray binary systems as a result of a supernova explosion.
A. M. Cherepashchuk.
Pis'ma Astron. Zh., Tom 7, 726 - 730 (1981). In Russian.
English translation in Soviet Astron. Lett., Vol. 7.

Arguments are presented in favour of the idea that the forced precession of the rotation axis of the optical star together with related long-term optical and X-ray variability appears to be a general property of X-ray binaries and may be a result of a supernova explosion.

142.152 Einstein observations of the confused 2A 2315-428 region. P. A. Charles, M. M. Phillips.
Anglo-Australian Obs. Prepr. No. 156, 18 pp. (1981). – Submitted to Mon. Not. R. Astron. Soc.

142.153 Variable linear polarization in the X-ray binaries HD 77581 and HD 153919. T. Korhonen.
Turku Univ. Obs. Informo, No. 53 (1980). – Abstract.

142.154 Polarization observations of X-ray binaries.
T. Korhonen.
Turku Univ. Obs. Informo, No. 58 (1981). – Abstract.

142.155 Some recent results on extragalactic X-ray sources from the Einstein Observatory. P. Gorenstein.
High-energy astrophysics, (see 012.067), p. 3 - 18 (1981).

Extragalactic research studies by the Harvard/Smithsonian group with the Einstein Observatory have emphasized quasars and clusters of galaxies. More than 100 QSO's have been detected, including 20 serendipitous discoveries. The ratio of L_x/L_o for radio loud quasars is on the average 3 times that of radio quiet ones. QSO's with a large intrinsic optical luminosity have a smaller ratio of L_x/L_o. X-ray images of clusters of galaxies reveal a variety of morphological types which may correspond to different stages in their evolution. Several examples of bi-modal clusters have been discovered. An X-ray plume associated with M86 is apparently gas being stripped. From X-ray studies, a mass between $1.7 \times 10^{13} M_\odot$ and $4.0 \times 10^{13} M_\odot$ has been derived for M87.

142.156 X-ray background from evolving QSOs.
A. Cavaliere, L. Danese, G. De Zotti, A. Franceschini.
High-energy astrophysics, (see 012.067), p. 33 - 37 (1981).

An analysis of the deep HEAO-2 survey has led the authors to predict that QSO's up to $m_B = 20$ should emit $L_x = 0.2\ L_o$ if radio-quiet, as compared with $0.7\ L_o$ resulting for catalogued radio-loud; they contribute 20% of the X-ray background at 2 keV. Fainter QSO's can fill up the remaining quota but at the cost of definite and testable predictions on faint star counts and X-ray counts in deeper surveys.

142.157 Spectra of accreting X-ray pulsars.
J. Ventura.
High-energy astrophysics, (see 012.067), p. 99 - 102 (1981).

Using recent improved results on the frequency and polarization dependent opacities in a strong magnetic field the author examines the physics of the lower accretion column on a magnetic neutron star to determine the spectrum radiated at infinity. He argues that photon convection by the fast infalling material (free fall velocities $\sim c/2$) should substantially modify the spectrum radiated through the accretion column due to the frequency dependence of the opacity.

142.158 A model of the rapid burster.
S. Hayakawa.
High-energy astrophysics, (see 012.067), p. 145 - 148 (1981).

A new burst mode of MXB 1730 - 335, the rapid burster, as discovered by Hakucho in August 1979, is characterized by a train of long X-ray bursts whose behaviour is dictated by the accretion rate. In this mode the burst luminosity reaches the Eddington limit, so that the X-ray pressure controls the accretion from a reservoir in the magnetopause and accretion columns in the polar regions explain general features of the rapid burster.

142.159 Optical bursts from X-ray irradiated accretion disk.
S. Hayakawa.
High-energy astrophysics, (see 012.067), p. 149 - 152 (1981).

Optical radiation is emitted from an accretion disk which is irradiated by X-rays. In the outer part the disk is heated mainly by X-rays. The outer region in which X-ray heating is dominant gives a main contribution to optical emission. When the X-ray source emits an X-ray burst, the disk also emits an optical burst with some time delay. This explains the time profile of an optical burst with respect to that of an X-ray burst, as observed for MXB 1636 - 536.

142.160 Two years of observation with the International Ultraviolet Explorer.
C. de Jager, H. J. G. L. M. Lamers.
High-energy astrophysics, (see 012.067), p. 185 - 196 (1981).

The launching of IUE in January 1978 has provided a powerful observatory-type spacecraft allowing continuous observations of UV-spectra of celestial bodies. This review summarizes the most important results obtained for galactic objects: early-type and Wolf-Rayet stars, medium- and late-type stars, planetary nebulae, novae and dwarf-novae, X-ray binaries, supernovae, the interstellar medium and the galactic halo.

142.161 Low energy observations of Cygnus X-2 by Ariel VI.
A. M. Cruise, C. V. Goodall, D. K. Bedford, D. J. Campbell, G. F. Carpenter, R. E. Cole, J. L. Culhane, J. Osborne, A. M. T. Pollock, A. P. Willmore, J. Zarnecki.
High-energy astrophysics, (see 012.067), p. 211 - 214 (1981).

Ariel VI observations of Cygnus X-2 have revealed a rather flat spectrum between 0.1 and 1.5 keV with variable emission at low energy. Of the two conflicting interpretations of this object in terms of 1) a distant high-luminosity ($L_x \sim 10^{38}$ ergs s^{-1}) binary and 2) a nearby low-luminosity ($L_x \sim 10^{35}$ ergs s^{-1}) degenerate dwarf system, these measurements support the latter.

142.162 Rapid variability of 10 - 140 keV X-rays from Cygnus X-1. P. L. Nolan, J. P. Doty, D. E. Gruber, A. M. Levine, J. L. Matteson, L. E. Peterson, F. A. Primini, R. E. Rothschild.
High-energy astrophysics, (see 012.067), p. 215 - 218 (1981).

On five occasions in 1977 and 1978, Cygnus X-1 was observed using the Low-Energy Detectors of the UCSD/MIT Hard X-Ray and Low-Energy Gamma-Ray Experiment on the HEAO-1 satellite. Rapid $(0.08\ s \leqslant t \leqslant 1000\ s)$ variability was found in the 10 - 140 keV band. A variable accretion rate could cause the observed effects.

142.163 Evidence of a cyclotron line at 70 keV from 4U0531 + 21. V. F. Polcaro, A. Bazzano, C. D. La Padula, P. Ubertini, R. K. Manchanda.
High-energy astrophysics, (see 012.067), p. 219 - 222 (1981).

The HXR79 hard X-ray experiment was flown on August 26th 1979. The scientific payload consisted of two Multiwire Spectroscopic Proportional Counters in the hard X-ray range (16 - 180 keV). A single drift scan was performed on the Crab Nebula region. The observation gave a maximum counting rate of about 25 counts/s superimposed on a background of about 75 counts/s. The Crab power-law spectrum was observed, and these data provide evidence for a line emission enhancement around 70 keV.

142.164   Observations of A0535 + 26 transient X-ray source.
I. V. Estulin (*Ehstulin*), Sh. Yu. Rakhamimov,
B. L. Novak, N. A. Eismont.
High-energy astrophysics, (see 012.067), p. 223 - 226 (1981).
Observations of the transient X-ray A0535 + 26 source
with the SIGNE−2MP instrument on board the Soviet
"Prognoz-6" and "Prognoz-7" satellites are considered in the
paper.

142.165   Identification of the X-ray source H 0850+13 with
the eclipsing binary AC Cancri.
N. E. Kurochkin, S. Yu. Shugarov.
Astron. Tsirk., No. 1154, p. 8 (1981). In Russian.

The Ariel V (3 A) catalogue of X-ray sources − I.
Sources at low galactic latitude ($|b| < 10°$).
See Abstr. 002.061.

The Ariel V (3 A) catalogue of X-ray sources − II.
Sources at high galactic latitude ($|b| > 10°$).
See Abstr. 002.062.

X-ray astronomy.   See Abstr. 003.049.

The X-ray spectrum of a hot interstellar plasma.
See Abstr. 022.052.

Density and temperature diagnostics of X-ray
sources: line ratios for helium-like ions.   See Abstr. 022.098.

Vorschlag einer Weitwinkel-Röntgen-Optik mittlerer
Auflösung für astronomische Anwendungen.
See Abstr. 031.015.

High resolution grazing incidence telescopes for the
EUV-regime.   See Abstr. 032.562.

Röntgenastronomie mit Ballonen und Satelliten.
See Abstr. 032.593.

Experiment for the observation of short-term time
variations in the hard X-ray region of Circinus X-1.
See Abstr. 032.626.

X-ray astronomy in the future.
See Abstr. 051.031.

UV astronomy with balloons.
See Abstr. 051.056.

The European X-ray astronomy satellite EXOSAT
observer's guide.   See Abstr. 051.058.

High X-ray luminosity from dynamo stars.
See Abstr. 062.116.

On the transfer equation for the cyclotron line
in Her X-1.   See Abstr. 063.016.

Directionality effects in the transfer of X-rays from
an accreting magnetized neutron star: beam and pulse shapes.
See Abstr. 063.023.

Compton scattering of relativistic electrons in
compact X-ray sources.   See Abstr. 063.029.

Radiative transfer in a strongly magnetized plasma.
I. Effects of anisotropy.   See Abstr. 063.041.

Radiative transfer in a strongly magnetized plasma.
II. Effects of Comptonization.   See Abstr. 063.042.

The influence of mass loss on the observed X-ray
spectra of early-type stars.   See Abstr. 064.072.

Observable properties of accretion disks.
See Abstr. 064.089.

X-ray irradiated accretion disk.
See Abstr. 064.099.

Cooling of young neutron stars and the Einstein
X-ray observations.   See Abstr. 066.512.

Neutron stars.   See Abstr. 066.514.

Nuclear explosions on neutron star surfaces.
See Abstr. 066.515.

The tearing mode instability in an accretion disc of a
neutron star and a rapid X-ray burst.   See Abstr. 066.519.

Cooling of neutron stars and X-ray observations.
See Abstr. 066.522.

Observability of magnetically strongly shifted iron
line emission from X-ray pulsars.   See Abstr. 066.524.

Solar abundances from X-ray flare observations.
See Abstr. 076.011.

Hard X-ray latitude effect measured during a transatlantic balloon flight.   See Abstr. 082.121.

Simultaneous spectroscopic and photometric
observations of 2A0311-227.   See Abstr. 113.018.

Search for optical coronal line emission from the
X-ray sources Epsilon Orionis (B0 Ia) and Kappa Orionis
(B0.5 Ia).   See Abstr. 114.023.

Simultaneous ultraviolet and optical observations of
HDE 245770/A 0535+26.   See Abstr. 114.072.

Coordinated ultraviolet, optical and infrared observations of X Per in December 1979: preliminary results.
See Abstr. 114.073.

Ultraviolet observations of the Be star and X-ray
binary 4U 1145−61 (= HD 102567 = Hen 715) obtained with
the IUE.   See Abstr. 114.157.

Stellar dynamo and the galactic X-ray sources.
See Abstr. 116.027.

Periodic changes in the compact radio structure of
SS 433.   See Abstr. 116.028.

Outer atmospheres and rotation of late-type stars.
See Abstr. 116.047.

Spectroscopy of the AM Herculis type binary
2A 0311 − 227.   See Abstr. 117.003.

Magnetic braking in low-mass X-ray binaries.
See Abstr. 117.005.

SS 433: enigma of the century.
See Abstr. 117.009.

Spectroscopy of the unique degenerate binary star
LSI +61°303.   See Abstr. 117.051.

Evidence for apsidal motion in the Cygnus X-1 system and a third-body interpretation.   See Abstr. 117.053.

The X-ray and optical characteristics of the cataclysmic variable V794 Aquilae.   See Abstr. 117.069.

SS 433, X-ray binaries and stellar evolution. See Abstr. 117.089.

Does SS 433 have cousins? See Abstr. 117.090.

Variability of soft X-ray emission of EX Hydrae observed with Einstein Observatory.   See Abstr. 117.111.

Optical observations of the X-ray source 2S0921−630.   See Abstr.117.112.

Ultraviolet study of V1341 Cyg = Cyg X−2. Observation of an accretion disk.   See Abstr. 117.113.

Discovery of X-ray emission lines from the RS CVn binary σ CrB.   See Abstr. 118.001.

Spectroscopic observations of HD 206267 as a possible counterpart of the X-ray source Cep X-4. See Abstr. 120.021.

Discovery of three X-ray luminous pre-main-sequence stars.   See Abstr. 121.005.

The smothered coronae of T Tauri stars. See Abstr. 121.012.

The smothered coronae of T Tauri stars. See Abstr. 121.024.

Two dwarf anti-novae.   See Abstr. 122.047.

Two X-ray supernova remnants: G296.1−0.7 and 1E 1149.4−6209.   See Abstr. 125.008.

High resolution X-ray and radio images of the Crab-like supernova remnant G21.5−0.9.   See Abstr. 125.009.

An oxygen-rich young supernova remnant in the Small Magellanic Cloud.   See Abstr. 125.017.

Soft X-ray observation of supernova remnant IC443. See Abstr. 125.025.

Soft X-ray observation of supernova remnant IC 443.   See Abstr. 125.044.

X-ray images of the supernova remnant Puppis A. See Abstr. 125.059.

Ariel 6 observations of the hot white dwarf HZ43. See Abstr. 126.030.

Absorption of the soft X-ray background by the Coalsack Nebula.   See Abstr. 131.050.

More absorption toward Cygnus X-3. See Abstr. 131.080.

First detection of line emission from the hot interstellar medium with solid state detectors. See Abstr. 131.219.

New Einstein observations of the Eta Carinae Nebula.   See Abstr. 134.012.

Observations of six flat spectrum sources from the 5 GHz survey.   See Abstr. 141.003.

On the possibility of observing X-ray absorption lines in the spectra of quasars.   See Abstr. 141.028.

VLBI observations of Circinus X-1 and Centaurus A. See Abstr. 141.057.

The correlation of X-ray emission with strong millimeter activity in extragalactic sources. See Abstr. 141.137.

X-ray observations of jets. See Abstr. 141.194.

X-ray emission from the radiogalaxy 3C 66B. See Abstr. 141.195.

Pulsar disk systems.   See Abstr. 141.517.

Simultaneous five colour photometry of the double period optical pulsar H 2254−033.   See Abstr. 141.525.

The stability of the pulse intensity of the X-ray pulsar in the Crab nebula.   See Abstr. 141.539.

On synchrotron nebulae and pulsars. See Abstr. 141.553.

High energy X-ray observations of the COS-B gamma ray source CG135+1.   See Abstr. 142.514.

Observations of the gamma-ray flux in the 10^{12} eV region from the X-ray source Cyg X-3 at the Tian-Shan installation during 1977 - 1978.   See Abstr. 142.534.

X-rays of gamma-ray bursts observable by HEAO-B. See Abstr. 142.537.

On the nature of a recurrent source of gamma and X-ray bursts in Doradus.   See Abstr. 142.547.

High energy primary electron spectrum and diffusion X-ray source.   See Abstr. 143.070.

X-ray observations of the Pleiades cluster of stars. See Abstr. 153.007.

X-ray and ultraviolet observations of NGC 2264. See Abstr. 153.009.

Stellar coronae in the Hyades: a soft X-ray survey with the Einstein Observatory.   See Abstr. 153.023.

Picture processing of the globular cluster NGC 6441. See Abstr. 154.017.

A very rapidly expanding gas in the core of the X-ray globular cluster NGC 6624?   See Abstr. 154.020.

Photometry of the cores of globular clusters. III. NGC 6712.   See Abstr. 154.031.

Position, magnitudes and color for stars in the central part of the X-ray globular cluster M15. See Abstr. 154.038.

Observation of soft diffuse X-rays in the southern galactic hemisphere.   See Abstr. 157.003.

Spectral observation of the soft X-ray background and of the North Polar Spur with solid state spectrometers. See Abstr. 157.016.

Einstein observations of the galactic centre. See Abstr. 157.017.

X-ray emission from M87: a pressure confined cooling atmosphere surrounding a low mass galaxy? See Abstr. 158.007.

Observations of M100 with the Einstein Observatory shortly after the explosion of its fourth supernova, SN 1979c. See Abstr. 158.009.

Detection of the 3.3 μm emission feature in the nuclei of IC 4329A and NGC 5506.   See Abstr. 158.029.

NGC 7714: the prototype star-burst galactic nucleus. See Abstr. 158.042.

The detection of X-ray emission from the BL Lacertae object Markarian 180.   See Abstr. 158.053.

Discovery of a Seyfert 1 with an unusually soft X-ray spectrum.   See Abstr. 158.083.

X-ray observations of late type galaxies with the Einstein Observatory.   See Abstr. 158.097.

X rays from normal galaxies and clusters of galaxies.   See Abstr. 158.108.

UGC 10683B – a possible X-ray-emitting Seyfert galaxy.   See Abstr. 158.133.

NGC 4507: a weak Seyfert 1 and X-ray galaxy. See Abstr. 158.136.

Possible detection of far-ultraviolet line emission from a hot galactic corona.   See Abstr. 158.161.

X-ray emission around radio galaxies in non-Abell clusters: a possible physical link between environment and nonthermal radio emission.   See Abstr. 158.203.

HEAO 1 observations of high-energy X-rays from the Seyfert I galaxy Mkn 509.   See Abstr. 158.205.

VLBI and X-ray observations of compact nuclei in pairs of galaxies.   See Abstr. 158.207.

Rapid X-ray variability in the Seyfert galaxy NGC 6814.   See Abstr. 158.209.

X-rays from active galactic nuclei. See Abstr. 158.210.

The X-ray structure of Centaurus A. See Abstr. 158.211.

Discovery of a Seyfert 1 galaxy with an unusually soft X-ray spectrum.   See Abstr. 158.218.

X-ray emission from new BL Lac objects. See Abstr. 158.224.

An optical and X-ray survey of s-type Markarian galaxies.   See Abstr. 158.225.

X-ray emission from active galaxies. See Abstr. 158.229.

Ariel-6 medium energy spectral observations of active galaxies.   See Abstr. 158.230.

X-ray observations of emission line galaxies with the Einstein Observatory.   See Abstr. 158.231.

NGC 4151 and MCG 8-11-11: two X-ray Seyfert galaxies with strong soft γ-ray emission. See Abstr. 158.232.

The X-ray luminosity function and source counts for Seyferts.   See Abstr. 158.233.

X-ray emission from galactic jets. See Abstr. 158.235.

A study of M100 in X-rays. See Abstr. 158.236.

Mid-infrared observations of Seyfert 1 and narrow-line X-ray galaxies.   See Abstr. 158.278.

A soft X-ray study of the Large Magellanic Cloud. See Abstr. 159.004.

A soft X-ray study of the Large Magellanic Cloud. See Abstr. 159.020.

Optical spectrophotometry of the suspected X-ray cluster of galaxies E 1455 + 2232.   See Abstr. 160.007

Low-frequency radio observations of poor clusters of galaxies.   See Abstr. 160.008.

Low energy X-ray emission from five galaxy cluster sources.   See Abstr. 160.013.

The distribution and morphology of X-ray – emitting gas in the core of the Perseus cluster.   See Abstr. 160.016.

Soft X-ray images of the central region of the Perseus cluster.   See Abstr. 160.017.

Angular separation studies of X-ray bright rich clusters of galaxies.   See Abstr. 160.022.

Einstein observations of A1367. See Abstr. 160.026.

X-ray observations of A2256 at large radial distance. See Abstr. 160.027.

The X-ray structure of a galaxy cluster at $z = 0.54$: implications for cluster evolution and cosmology. See Abstr. 160.047.

X-ray measurements and the mass of Abell 1763 out to a radius of 1.4 Mpc.   See Abstr. 160.053.

Radio observations of Abell clusters and a comparison with certain Einstein observations.   See Abstr. 160.055.

Gamma-ray Sources, Gamma-ray Background

142.501 A 4.2 second period in the gamma-ray burst of 1977 October 29. K. S. Wood, E. T. Byram, T. A. Chubb, H. Friedman, J. F. Meekins, G. H. Share, D. J. Yentis.
Astrophys. J., Vol. 247, 632 - 638 (1981).

The A-1 experiment on HEAO 1 detected the γ-ray burst of 1977 October 29, and obtained a burst profile with high statistical precision. The overall profile can be fitted with a periodic function, giving a period of 4.2 ± 0.2 s. The limitations of fitting periods to bursts of short duration are described, and implications of period detections for neutron star models of gamma bursts are developed. Gamma bursters are shown likely to be neutron stars in binary systems.

142.502 Detection of a soft gamma-ray emission from the region of NGC 4151.
F. Perotti, A. Della Ventura, G. Villa, G. Di Cocco, L. Bassani, R. C. Butler, J. N. Carter, A. J. Dean.
Astrophys. J., Lett., Vol. 247, L63 - L66 (1981).

During a balloon flight on 1979 September 30, a region of the sky containing Seyfert galaxy NGC 4151 was studied over the photon energy range 0.02 - 19 MeV. A 4 σ excess above 260 keV was detected within an error box encompassing this galaxy. With respect to the 1977 May observation of this source with the same telescope, the γ-ray luminosity in the energy range 0.5 - 5 MeV shows a reduction at the 99% confidence level, and the ratio between the two luminosities is 4 ± 2. The emission spectrum has been evaluated, and it may be represented by a single power law over the entire range.

142.503 A deep optical search of the 1979 April 6 gamma-ray burst error box.
C. Chevalier, S. A. Ilovaisky, C. Motch, C. Barat, K. Hurley, M. Niel, G. Vedrenne, J. G. Laros, W. D. Evans, E. E. Fenimore, R. W. Klebesadel, I. V. Estulin (*Ehstulin*), V. M. Zenchenko.
Astron. Astrophys., Vol. 100, L1 - L3 (1981), with a correction, Vol. 103, 428 (1981).

The authors report preliminary results of a deep optical search of the error box for the 1979 April 6 gamma-ray burst. A 90 min direct IIIa-J plate exposed at the prime focus of the ESO 3.6 m telescope reveals several very faint objects in the box reported by Laros et al. (1981). Statistical arguments based on published faint object counts show that most of them are probably distant galaxies unrelated to the burst source.

142.504 Angular sizes of gamma-ray sources.
T. P. Li, A. W. Wolfendale.
Astron. Astrophys., Vol. 100, L26 - L29 (1981).

An analysis is made of the extent to which the data on the γ-ray sources in the 2CG catalogue of Swanenburg et al. (1981) are consistent with the authors' contention (Li and Wolfendale, 1981) that about 60% of the sources are molecular clouds irradiated by cosmic rays. Although there can be no question yet of proof the authors conclude that there is a measure of support for the idea.

142.505 Inverse comptonization and the nature of the March 1979 γ-ray burst event. E. P. T. Liang.
Nature, Vol. 292, 319 - 321 (1981).

The author points out that the observed burst spectrum seems to be most naturally and consistently interpreted as that of a synchrotron spectrum modified by inverse Compton scattering from ~MeV $e^{\pm}$-pairs. This model then allows to derive, from first principles, the intrinsic synchrotron luminos-

ity of the burst source, which agrees basically with that expected from N49 (distance ~55 kpc).

142.506 γ rays from the cosmic ray irradiation of local molecular clouds. M. R. Issa, A. W. Wolfendale.
Nature, Vol. 292, 430 - 433 (1981).

The authors show that only in a few clouds does the cosmic ray intensity need to exceed the value near the earth to explain the γ-ray fluxes. Possible reasons for the few excesses are considered.

142.507 A 928 MHz search for periodicities in 2CG195 + 04.
J. H. Seiradakis.
Astron. Astrophys., Vol. 101, 158 (1981).

An area of four square degrees around the γ-ray source 2CG195 + 04 was searched for periodicities at 928 MHz using the 76 m radio telescope at Jodrell Bank. No positive detection was made.

142.508 Gamma rays − the last frontier.
T. C. Weekes.
Mercury, Vol. 10, 78 - 84, 95 (1981).

142.509 Gamma ray bursts and neutron star accretion of a solid body. S. A. Colgate, A. G. Petschek.
Astrophys. J., Vol. 248, 771 - 782 (1981).

The sequence of events that would probably take place if a comet or asteroid were to make a direct impact with a neutron star is described. The authors have in mind an explanation of the 1979 March 5 gamma burst where a 0.1 s burst of hard X-rays was followed by a protracted 8 s pulsation.

142.510 A model for the cosmic γ-ray burst event on March 5, 1979. Q. Qu, Z. Li, D. Wang, A. Xu.
Kexue Tongbao, Vol. 26, 53 - 55 (1981). − Abstr. in Phys. Abstr., Vol. 84, Abstr. 83769 (1981).

142.511 Observations of gamma radiation between 0.4 MeV and 7 MeV at balloon altitudes using a Compton telescope. J. A. Lockwood, W. R. Webber, L. A. Friling, J. Macri, L. Hsieh.
Astrophys. J., Vol. 248, 1194 - 1201 (1981).

Results are presented from a balloon flight to measure the atmospheric and diffuse γ-ray flux in the energy range 0.4−7.0 MeV. The observations were made with a Compton telescope which included pulse-shape discrimination on the first scattering detector and a time-of-flight system between the first and second detector elements. The results are discussed and compared to other observations.

142.512 Long-term variation in the Cygnus region observed with the HEAO-3 high resolution gamma-ray spectrometer.
J. C. Ling, W. A. Mahoney, G. R. Riegler, W. A. Wheaton, A. S. Jacobson.
Bull. American Astron. Soc., Vol. 13, 549 (1981). − Abstract.

142.513 Negative visual search for CG195 + 4.5.
S. R. Williams, J. R. Dickel, M. E. Özel.
Bull. American Astron. Soc., Vol. 13, 550 (1981). − Abstract.

142.514 High energy X-ray observations of the COS-B gamma ray source CG135+1. A. Scheepmaker, F. A. Jansen, A. J. M. Deerenberg, G. R. Ricker, J. E. Ballintine, J. V. Vallerga, J. S. Kruper, W. H. G. Lewin.
Bull. American Astron. Soc., Vol. 13, 557 (1981). − Abstract.

142.515 On the possible nature of the gamma-ray source SG 195+4. B. M. Vladimirskij.
Izv. AN SSSR. Ser. fiz., Tom 45, 626 - 628 (1981). In Russian. Abstr. in Ref. zh., 51. Astron., 8.51.693 (1981).

142.516   **Observation of diffuse cosmic γ-radiation in the**
          100 - 700 MeV energy range aboard Cosmos 731.
V. L. Bokov, V. K. Bocharkin, I. F. Bugakov,
G. M. Gorodinskij, E. M. Kruglov, G. A. Pyatigorskij,
E. I. Chujkin.
Izv. AN SSSR. Ser. fiz., Tom 45, 633 - 636 (1981). In Russian.
Abstr. in Ref. zh., 62. Issled. kosm. prostranstva, 8.62.301
(1981).

142.517   **High-energy γ-rays from the direction of the Crab**
          pulsar.
T. Drikowsky, B. Grochalska, J. Gawin, J. Wdowczyk.
Cosmic rays, (see 012.005), 1980, p. 369 - 373. — Abstr. in
Ref. zh., 51. Astron., 9.51.745 (1981).

142.518   **Gamma ray astronomy and the origin of cosmic**
          rays.   R. Schlickeiser.
Fortschr. Phys., Vol. 29, 95 - 134 (1981). — Abstr. in Phys.
Abstr., Vol. 84, Abstr. 94122 (1981).

142.519   **On the evidence for high energy γ-ray emission**
          from the Orion Nebula stemming from COS-B
observations.   P. A. Caraveo.
Philos. Trans. R. Soc. London, Ser. A, Vol. 301, 569 - 571
(1981). — Abstr. in Phys. Abstr., Vol. 84, Abstr. 94483 (1981).

142.520   **Cosmic gamma rays from the Orion molecular cloud**
          complex.   M. R. Issa, A. W. Wolfendale.
J. Phys. G, Vol. 7, L187 - L191 (1981). — Abstr. in Phys.
Abstr., Vol. 84, Abstr. 94569 (1981).

142.521   **Cosmic γ-ray bursts.**   G. Vedrenne.
Philos. Trans. R. Soc. London, Ser. A, Vol. 301,
645 - 658 (1981). — Abstr. in Phys. Abstr., Vol. 84, Abstr.
94570 (1981).

142.522   **Interpretations and implications of γ-ray lines from**
          solar flares, the galactic centre and γ-ray transients.
R. Ramaty, R. E. Lingenfelter.
Philos. Trans. R. Soc. London, Ser. A, Vol. 301, 671 - 686
(1981). — Abstr. in Phys. Abstr., Vol. 84, Abstr. 94571 (1981).

142.523   **Gravitational scattering of asteroids onto neutron**
          stars as a cause of γ-ray bursts.   D. Van Buren.
Astrophys. J., Vol. 249, 297 - 301 (1981).
          The author proposes a mechanism by which asteroids
may be scattered onto neutron stars causing impulsive and
perhaps all γ-ray bursts. Asteroids which remain bound to the
neutron star after it is formed can, through close encounters
with planets, enter a region of magnetic drag near the star.
This region is larger than the star's collisional cross section; it
is $\sim 10^{10}$ cm in radius and hence produces an interesting rate if
even only a small fraction of all neutron stars are responsible
for these bursts. The author has interpreted the 1979 March 5
burst in this context and can satisfy all of the observational
constraints except for its suggested association with N49 in
the LMC. His model predicts that γ-ray bursts have a low
luminosity (10^{35} ergs s^{-1}) X-ray precursor of fairly long duration (10^4 s).

142.524   **Radiation from an asteroid–neutron star collision.**
          W. M. Howard, J. R. Wilson, R. T. Barton.
Astrophys. J., Vol. 249, 302 - 307 (1981).
          The authors calculate a zero impact parameter collision
of a 5×10^{17} g asteroid with a 1.51 $M_\odot$ neutron star using a
two-dimensional Lagrangian–Eulerian hydrodynamics code.
The radiation transfer is followed with a Planckian LTE
diffusion model, and allowed to couple to the matter through
bremsstrahlung and Compton processes. The effects self-
gravity on the asteroid, relativity, and magnetic fields are not
included. The kinetic energy of impact is converted into
radiant energy within 1 ms of impact. However, the neutron

star is rapidly (<1 ms) covered by a low-density optically
thick cloud that radiates within an order of magnitude of the
Eddington limit at an effective temperature of only a few
keV. Thus, such models, without the inclusion of confinement effects such as magnetic fields, are insufficient to
explain cosmic γ-ray bursts or the 1979 March 5 event.

142.525   **Search for gamma burst sources.**   E. Mūkins.
          Zvaigžnotā debess, 1980. gada pavasaris, p. 16 - 18.
In Latvian.

142.526   **On the nature of the galactic 2CG γ-ray sources.**
          R. Buccheri, M. Morini, B. Sacco.
Philos. Trans. R. Soc. London, Ser. A, Vol. 301, 495 - 503
(1981). — Abstr. in Phys. Abstr., Vol. 84, Abstr. 98894
(1981).

142.527   **Extended γ-ray sources and active regions in the**
          Galaxy: the Carina and Orion complexes.
T. Montmerle.
Philos. Trans. R. Soc. London, Ser. A, Vol. 301, 505 - 518
(1981). — Abstr. in Phys. Abstr., Vol. 84, Abstr. 98895
(1981).

142.528   **Extragalactic γ-rays.**
          A. J. Dean, D. Ramsden.
Philos. Trans. R. Soc. London, Ser. A, Vol. 301, 577 - 602
(1981). — Abstr. in Phys. Abstr., Vol. 84, Abstr. 98896
(1981).

142.529   **Detection of pulsed γ-rays at energies above**
          300 GeV from pulsars–TIFR experiments.
B. V. Sreekantan.
Philos. Trans. R. Soc. London, Ser. A, Vol. 301, 629 - 632
(1981). — Abstr. in Phys. Abstr., Vol. 84, Abstr. 98897
(1981).

142.530   **Do γ-ray bursts contain γ-rays of energies above**
          1 GeV?   P. N. Bhat, N. V. Gopalakrishnan,
S. K. Gupta, P. V. Ramana Murthy, B. V. Sreekantan,
S. C. Tonwar.
Philos. Trans. R. Soc. London, Ser. A, Vol. 301, 659 - 660
(1981). — Abstr. in Phys. Abstr., Vol. 84, Abstr. 98898
(1981).

142.531   **The astronomy of gamma pulses.**
          G. Vedrenne.
Recherche, No. 122, p. 536 - 545 (1981). In French. — Abstr.
in Phys. Abstr., Vol. 84, Abstr. 98899 (1981).

142.532   **Galactic origin of cosmic gamma-ray bursts.**
          E. P. Mazets, S. V. Golenetskij, R. L. Aptekar',
Yu. A. Gur'yan, V. N. Il'inskij.
Fiz.-tekh. inst. AN SSSR. Prepr., 1980, No. 686, 14 pp. In
Russian. — Abstr. in Ref. zh., 51. Astron., 10.51.555 (1981).

142.533   **On the possible nature of the gamma-source**
          SG 195 + 4.   B. M. Vladimirskij.
Izv. Krymskoj Astrofiz. Obs., Tom 63, 146 - 150 (1981). In
Russian. English translation in Bull. Crimean Astrophys. Obs.,
Vol. 63.
          It is suggested that the gamma-source SG 195 + 4 might be
a remnant of a supernova outburst in AD 437.

142.534   **Observations of the gamma-ray flux in the 10^{12} eV**
          region from the X-ray source Cyg X-3 at the Tian-
Shan installation during 1977 - 1978.   Dzh. B. Mukanov.
Izv. Krymskoj Astrofiz. Obs., Tom 63, 151 - 156 (1981). In
Russian. English translation in Bull. Crimean Astrophys. Obs.,
Vol. 63.
          The Tian-Shan high-level installation of the Lebedev
Physical Institution for detection of Cherencov flashes of ex-

tensive atmospheric showers is described. The results of observations of the X-ray source Cyg X-3 during 1977 - 1978 are presented. There is evidence of the existence of a periodic component of the γ-ray flux from this object.

142.535 **Contribution of cosmic ray-irradiated molecular clouds to the number of apparent γ-ray sources.**
Li Ti Pei, A. W. Wolfendale.
Astron. Astrophys., Vol. 103, 19 - 27 (1981).
The nature of most of the so-called γ-ray sources is very uncertain. The case is made here for a significant fraction, $(55 \pm 25)\%$ of the sources with fluxes above 100 MeV and latitude $|b| < 10°$, being due to molecular clouds which have been irradiated by the ambient cosmic ray flux.

142.536 **A model of the cosmic gamma-ray burst event on 5 March 1979.**
Q.-y. Qu, D.-y. Wang, Z.-q. Li, A.-a. Xu.
Acta Astron. Sinica, Vol. 22, 293 - 298 (1981). In Chinese.

142.537 **X-rays of gamma-ray bursts observable by HEAO-B.**
I. G. Mitrofanov.
Astrophys. Space Sci., Vol. 80, 303 - 306 (1981).
On grounds of Mazets' et al. (1980a) conclusion that all gamma-ray bursts recorded were emitted at the distance $\lesssim 1$ kpc from the Sun it is shown that bursts from the Galaxy and from other galaxies may be observable by Einstein Observatory (HEAO-B). The HEAO-B observing programs are considered and the numbers of possibly recorded events are estimated. It is concluded that the important information about sources of gamma-bursts may be obtained in this way.

142.538 **Probable optical counterpart of a γ-ray burster.**
B. E. Schaefer.
Nature, Vol. 294, 722 - 724 (1981).
The author describes 1.6×10^7 s (over half a year) of optical monitoring of three γ-ray burst positions using the collection of archival plates at the Harvard College Observatory. The search has uncovered the probable optical counterpart for the 19 November 1978 γ-ray burster on a blue emulsion plate exposed in 1928.

142.539 **Are γ-ray bursters neutron stars accreting interstellar matter?**
S. Bonazzola, J. M. Hameury, J. Heyvaerts, J. Ventura.
Space Sci. Rev., Vol. 30, (see 012.044), 471 - 474 (1981).
About 150 gamma-ray burst events have been studied and catalogued in recent years. The authors investigate a model of a neutron star accreting interstellar matter as it moves through space. It is concluded that specific low-accretion rate systems could produce gamma-rays by triggering helium detonation in the accreting shell at the neutron star surface.

142.540 **Discrete sources of cosmic gamma-rays.**
P. A. Riley, A. W. Wolfendale.
Nuovo Cimento, Riv., Ser. 3, Vol. 4, 1 - 29 (1981). – Abstr. in Phys. Abstr., Vol. 85, Abstr. 6685 (1982).

142.541 **A review of the 1979 March 5 transient.**
T. L. Cline.
Gamma-ray transients, (see 012.046), 17 pp. (1981).
The understanding of the 1979 March 5 event remains a problem of central importance for researchers in gamma ray transient astronomy. A consensus of opinion has not yet been reached regarding its possible origin in N49 at 55 kpc distance, versus in an invisible source 3 or 4 orders of magnitude closer, although interpretations favoring N49 appear to be presently gaining momentum. This presentation outlines the existing data in a review of what remains the most singular high-energy astrophysical phenomenon of the space age.

142.542 **Gamma-ray burst spectra.**
B. J. Teegarden.
Gamma-ray transients, (see 012.046), 20 pp. (1981).
A review of recent results in gamma-ray burst spectroscopy is given. Particular attention is paid to the recent discovery of emission and absorption features in the burst spectra. These lines represent the strongest evidence to date that gamma-ray bursts originate on or near neutron stars. The behavior of the continuum spectrum is also discussed. A remarkably good fit to nearly all bursts is obtained with a thermal-bremsstrahlung-like continuum.

142.543 **Search for time variations in 511 keV flux by ISEE–3 gamma-ray spectrometer.**
J. P. Norris, T. L. Cline, B. J. Teegarden.
Gamma-ray transients, (see 012.046), 6 pp. (1981).
The authors report upper limits to the variation of the galactic 511 keV flux as observed by the ISEE–3 gamma-ray spectrometer for 500 days of nearly continuous observation.

142.544 **Time variations of an absorption feature in the spectrum of the gamma-ray burst on 1980 April 19.**
B. R. Dennis, K. J. Frost, A. L. Kiplinger, L. E. Orwig, U. Desai, T. L. Cline.
Gamma-ray transients, (see 012.046), 10 pp. (1981).
The authors present high energy X-ray spectral data for a gamma-ray burst with over an order of magnitude finer time resolution than that obtained by Mazets et al. The measured photon number spectrum can be interpreted as showing a broad absorption feature at energies below ~ 100 keV similar to the features reported by Mazets et al for other events. These observations support the idea that gamma-ray bursts are produced in the vicinity of neutron stars.

142.545 **Search for high frequency pulsations in the onset of the March 5 gamma-ray burst.**
M. C. Weisskopf, R. F. Elsner, P. G. Sutherland, J. E. Grindlay.
Astrophys. Lett., Vol. 22, 179 - 184 (1981).
The onset of the March 5 gamma-ray burst was observed with microsecond time resolution by the Monitor Proportional Counter aboard the HEAO-2/Einstein Observatory. The data were analyzed for high frequency periodic pulsations which, if present, would be strong evidence for vibrations of a neutron star following the initial release of energy. The authors do not find evidence for such pulsations in the data over the period range 29.4737 to 0.0962 ms, and they set upper limits to the amplitude of 49–70%.

142.546 **High-energy gamma astronomy.**
V. G. Kirillov-Ugryumov.
Vestn. AN SSSR, 1981, No. 9, p. 12 - 19. In Russian. Abstr. in Ref. zh., 51. Astron., 1.51.607 (1982).

142.547 **On the nature of a recurrent source of gamma and X-ray bursts in Doradus.** I. G. Mitrofanov.
High-energy astrophysics, (see 012.067), p. 161 - 165 (1981).
The X-ray pulsar FXP 0520-66 is believed to be the source of the gamma-ray burst from March 5, 1977. A model implying a neutron star in a close binary system is proposed and physical model parameters are derived from observational results.

142.548 **Determination of confidence contours for gamma ray burst source locations.** G. Pizzichini.
High-energy astrophysics, (see 012.067), p. 227 - 230 (1981).
The method for determining gamma-ray burst source regions which has produced the best results until now is long-baseline wavefront triangulation. The direction of the source of the gamma rays is derived from the delay in the arrival time of the burst at widely spaced detectors on at least three spacecraft. An exact method to compute confidence contours for the source error boxes is given in the present paper.

142.549   Low energy gamma-ray observations of CG135 + 1
            and CG195 + 4.   G. Di Cocco, G. Boella,
A. Della Ventura, G. Mangia, F. Perotti, G. Villa, R. E. Baker,
R. C. Butler, A. J. Dean, R. I. Hayles, D. Ramsden.
High-energy astrophysics, (see 012.067), p. 231 - 234 (1981).
   Regions of the sky containing the high energy γ-ray
sources CG135 + 1 and CG195 + 4 were searched for X-ray
and γ-ray emission in the energy range from 20 keV to 25MeV
from balloon altitudes on October 8, 1978. A 5σ excess was
measured in the counting rate of the telescope above 120 keV
from the region of the sky containing CG135 + 1 and the spectrum of the source evaluated. No X- and γ-rays were found
from CG195 + 4.

        The second COS-B catalogue of high-energy γ-ray
sources.   See Abstr. 002.048.

        Catalog of cosmic gamma-ray bursts from the
KONUS experiment data.   See Abstr. 002.056.

        The Signe 2 Franco-Soviet interplanetary gamma
ray burst experiment network.   See Abstr. 032.523.

        Das Compton-Teleskop.   See Abstr. 032.594.

        An imaging telescope for soft gamma ray astronomy.
See Abstr. 032.627.

        Medium energy gamma ray astronomy with trans-
pacific balloon flights.   See Abstr. 032.628.

        An air shower array designed for cosmic ray varia-
tion measurements and high energy gamma ray astronomy.
See Abstr. 034.046.

        Six years of gamma-ray astronomy with Cos-B.
See Abstr. 051.033.

        X-ray and gamma-burst radiation stress limits and
diamagnetic accretion.   See Abstr. 062.044.

        Annihilation radiation from a hot e^+-e^- plasma.
See Abstr. 062.083.

        On the theory of gamma ray amplification through
stimulated annihilation radiation (grasar).
See Abstr. 063.056.

        On the theory of gamma ray amplification through
stimulated annihilation radiation (grasar).
See Abstr. 063.057.

        Nuclear explosions on neutron star surfaces.
See Abstr. 066.515.

        The γ-ray bursts as a result of the proper activity of
the neutron star.   See Abstr. 066.528.

        Gamma ray lines from solar flares and cosmic
transients.   See Abstr. 076.022.

        HEAO 3 measurements of the atmospheric positron
annihilation line.   See Abstr. 082.109.

        More absorption toward Cygnus X-3.
See Abstr. 131.080.

        Inverse Compton emission of gamma rays near the
pulsar surface.   See Abstr. 141.528.

        The gamma-ray spectra of radio pulsars.
See Abstr. 141.534.

        Einstein X-ray identification of the variable radio
star LSI +61°303.   See Abstr. 142.002.

        Correlation between X-ray and high energy
gamma-ray emission from Cygnus X-3.
See Abstr. 142.074.

        X-ray pulsar − a possible γ-burst source.
See Abstr. 142.076.

        Soft X-ray sources and their optical counterparts in
the error box of the COS-B source 2CG 135+01.
See Abstr. 142.102.

        X-ray and optical observations of the November 19,
1978 gamma-ray burst source region.   See Abstr. 142.130.

        Medium energy gamma ray sky in the Northern
Hemisphere: cross correlation analysis.
See Abstr. 157.004.

        Gamma-ray lines from the galactic center and
gamma-ray transients.   See Abstr. 157.009.

        Observations of 1−30 MeV gamma rays from the
galactic center.   See Abstr. 157.020.

        The γ-ray emission from the Seyfert galaxy
NGC 4151.   See Abstr. 158.163.

        Gamma ray constraints on annihilation in active
galaxies.   See Abstr. 158.166.

        Absorption of γ rays in active galaxies as a test of
the jet hypothesis.   See Abstr. 158.178.

        NGC 4151 and MCG 8-11-11: two X-ray Seyfert
galaxies with strong soft γ-ray emission.
See Abstr. 158.232.

Erratum

142.901   Erratum: 'An 81 minute modulation of the X-ray
            flux from 2A0311−227' [Astrophys. J., Lett.,
Vol. 244, L85 - L88 (1981)].   N. E. White.
Astrophys. J., Lett., Vol. 248, L87 (1981). − See Abstr.
28.142.037.

143 Cosmic Radiation

143.001 The power-law spectrum of shock-accelerated relativistic particles. F. C. Michel.
Astrophys. J., Vol. 247, 664 - 670 (1981).
The author develops an alternative formalism for calculating the power-law index of relativistic particles (e.g., cosmic rays) accelerated by shock waves. This formalism reproduces the results found by Bell and by Blandford and Ostriker. Moreover, higher order effects are easily calculated and examples of such generalizations are provided.

143.002 Observation of pronounced isotropic variations of cosmic ray intensity during August - September 1979.
D. Venkatesan, K. Volk, S. P. Agrawal.
J. Geophys. Res., Vol. 86, 4836 - 4838 (1981).
A study of the cosmic ray intensity data from high-latitude stations reveals an isotropic variation of ~ 30-day periodicity and a peak-to-peak amplitude of 10% during the two months August - September 1979. Such an unusual modulation of cosmic ray intensity has not been observed before. The rigidity spectra of the variations is found to be similar to that of Forbush decreases. The results are discussed in terms of solar activity. No obvious physical mechanism is readily seen, to account for this intensity variation.

143.003 Solar cycle modulation of galactic cosmic rays: speculation on the role of coronal transients.
G. Newkirk, Jr., A. J. Hundhausen, V. Pizzo.
J. Geophys. Res., Vol. 86, 5387 - 5396 (1981).
Coronal transients are believed to involve the expulsion of magnetic "bubbles" or loops into interplanetary space. The close relationship of coronal transients with other forms of solar activity suggests a variation in their frequency of occurrence by a factor of 3 - 10 over the solar cycle. As magnetic "inclusions" in the interplanetary magnetic field, transients should then lead to a solar cycle dependent scattering of cosmic rays entering the inner solar system. The power spectrum of a model of the transverse magnetic fluctuations and its solar cycle dependence is compared to the observed power spectra.

143.004 Antiprotons in the cosmic radiation. P. Kiraly, J. Szabelski, J. Wdowczyk, A. W. Wolfendale.
Nature, Vol. 293, 120 - 122 (1981).
The authors describe an improved calculation of the energy spectrum of $\bar{p}$ expected for the standard leaky box model and show that even the enhanced secondary $\bar{p}$ production in the 'closed Galaxy' model of Peters and Westergaard (1977) fails by a wide margin to reproduce the high observed flux at low energies. A more radical departure from conventional thinking is to regard the observed $\bar{p}$s as primaries; a universal baryon-symmetric model and a black hole evaporation model are considered.

143.005 Die Kosmische Strahlung im TeV-Bereich — ein unerforschtes Gebiet. D. Müller.
Phys. Bl., 37. Jahrg., 267 - 274 (1981).

143.006 A cosmic-ray-mediated shock in the solar system.
D. Eichler.
Astrophys. J., Vol. 247, 1089 - 1092 (1981).
It is pointed out that the flare-induced blast wave of 1972 Aug 4, the most violent disturbance in the solar wind on record, produced cosmic rays with an efficiency of ~50%. Such a high efficiency is predicted by the self-regulating production model of cosmic-ray origin in shocks. Most interplanetary shocks, according to simple theoretical analysis, are not strong enough to produce cosmic rays efficiently. However, if shock strength is the key parameter governing efficiency, as present interplanetary data suggest, then shocks from supernova blasts, quasar outbursts, and other violent astrophysical phenomena should be extremely efficient sources of cosmic rays.

143.007 Recent cosmic-ray antiproton measurements and astrophysical implications.
A. Buffington, S. M. Schindler.
Astrophys. J., Lett., Vol. 247, L105 - L109 (1981).
Cosmic-ray antiprotons have been detected by a new balloon-borne experiment which covers the energy range between 130 and 320 MeV. Fourteen detected events yield a measured flux of $1.7 \pm 0.5 \times 10^{-4}$ antiprotons m^{-2} sr^{-1} s^{-1} MeV^{-1}. The corresponding antiproton/proton ratio is $2.2 \pm 0.6 \times 10^{-4}$, only slightly smaller than the ratio observed by other experiments at higher energies. The measured flux is significantly larger than predicted, and the authors discuss some cosmic-ray models which could explain this result.

143.008 Cosmic-ray abundances of elements with atomic number $26 \lesssim Z \lesssim 40$ measured on *HEAO* 3.
W. R. Binns, R. K. Fickle, T. L. Garrard, M. H. Israel, J. Klarmann, E. C. Stone, C. J. Waddington.
Astrophys. J., Lett., Vol. 247, L115 - L118 (1981).
Individual elements in the cosmic radiation of even atomic number (Z) in the interval $26 \lesssim Z \lesssim 40$ have been resolved and their relative abundances measured. The results are inconsistent with a cosmic-ray source whose composition in this charge interval is dominated by r-process nucleosynthesis. The ratios of cosmic-ray source abundances to solar system abundances in this interval follow the same general correlation with first ionization potential as for the lighter elements, although there are deviations in detail.

143.009 High-resolution observations of the isotopic composition of carbon and silicon in the galactic cosmic rays. M. E. Wiedenbeck, D. E. Greiner.
Astrophys. J., Lett., Vol. 247, L119 - L122 (1981).
The authors present new high-resolution observations of the isotopic composition of galactic cosmic ray carbon and silicon and consider the relationship of these compositions to the isotopic anomalies previously reported in the elements neon and magnesium in the galactic cosmic ray source (Wiedenbeck and Greiner 1981) and to carbon and silicon isotopic anomalies found in the interstellar medium.

143.010 Cosmic ray gradients in the heliosphere and particle drifts. G. Newkirk, Jr., J. A. Lockwood.
Geophys. Res. Lett., Vol. 8, 619 - 622 (1981).
K — coronameter, solar wind, and neutron monitor data are used to estimate the latitudinal gradient of ~5 GeV protons in the heliosphere at 1 AU during 1965 and 1975. The reversal of sign of the gradient between 1965 and 1975 and the strong positive latitudinal gradient in 1975 predicted by models in which drifts are dominant are not observed.

143.011 Correlation of the cosmic-ray intensity with solar-terrestrial parameters. J. R. Jokipii.
Geophys. Res. Lett., Vol. 8, 837 - 839 (1981).
Shea and Smart (1981) have shown that the correlation between the Mt. Washington neutron monitor counting rate and the geomagnetic aa index changes significantly from the eleven-year period centered on one solar minimum to the next such period. They suggested that this may be a manifestation of cosmic-ray drift in the interplanetary magnetic field. This Letter reports the results of numerical simulations of cosmic-ray modulation, including drift, which verify that drifts can indeed produce the sense of the observed effect.

143.012 **Cosmic ray power spectral variations. 1. Simulation study.**   L. J. Lanzerotti, C. G. Maclennan, S. P. Agrawal, D. Venkatesan.
J. Geophys. Res., Vol. 86, 6951 - 6956 (1981).

The study of the daily variation of the cosmic ray intensity over a long interval, extending, for example, over the 11-year solar cycle, is complicated by other types of superposed variations. An analysis of real data can be considerably facilitated by a power spectral study using simulated input incorporating different artificial but physically based variations. The present study utilizes, in the simulation data, the known average characteristics of the daily variation as well as other variations present in real data. Power spectra are then calculated from a series of test data sets, each of which consists of hourly values for an interval of 27 days. The essential parameters relating to the peaks of the daily variation are tabulated.

143.013 **Hydromagnetic shock structure in the presence of cosmic rays.**   L. O'C. Drury, H. J. Völk.
Astrophys. J., Vol. 248, 344 - 351 (1981).

The time asymptotic structure of a shock significantly modified by the back-reaction from the diffuse acceleration of cosmic rays is investigated. Making a physically plausible assumption about the diffusion, it is shown that for given upstream conditions and shock speed only a finite odd number of shock structures are possible; an explicit method of determining these is given. The results of this nonlinear study are contrasted with those of the linear test-particle theory and shown to confirm the possibility of efficient particle acceleration in shocks.

143.014 **Cosmic-ray scintillations observed from space.**   A. J. Owens.
Geophys. Res. Lett., Vol. 8, 907 - 910 (1981).

A theoretical calculation is given for the "scintillations" or fluctuations of cosmic rays expected to be observed during quiet times by a detector in interplanetary space or on the lunar surface. Interplanetary scintillations caused by fluctuations in the interplanetary magnetic field are not sufficient to explain the power spectrum recently reported by Benson et al. (1981) from a detector on the moon. Their inferred cosmic-ray fluctuations are considerably larger than observed during undisturbed periods by ground-based and satellite experiments.

143.015 **The role of particle drifts in solar modulation.**   M. A. Lee, L. A. Fisk.
Astrophys. J., Vol. 248, 836 - 844 (1981).

The importance of gradient and curvature drifts in the description of the transport of cosmic rays in the heliospheric magnetic field is considered. The authors examine critically the work of Isenberg and Jokipii (1979) and show that their extension of the conventional drift-flux formula to static magnetic fields of arbitrary spatial variation is invalid for interplanetary cosmic-ray transport. The authors present a field configuration involving helical magnetic fields which suppresses curvature and gradient drift and which may occur in the solar wind. Finally, they present their conclusions, which may be summarized as follows: the conventional drift-flux contribution to the solar modulation of cosmic rays may, but need not, be important.

143.016 **Comment on "The role of particle drifts in solar modulation" by Lee and Fisk.**
P. A. Isenberg, J. R. Jokipii.
Astrophys. J., Vol. 248, 845 - 846 (1981).

The preceding paper by Lee and Fisk discusses many of the topics considered in an earlier paper of the authors. The authors reiterate their previous results and point out several misinterpretations and inaccurate statements.

143.017 **The abundances and energy spectra of cosmic ray iron and nickel at energies from 1 to 10 GeV per**

amu.   G. Minagawa.
Astrophys. J., Vol. 248, 847 - 855 (1981).

A scintillation-Cerenkov counter telescope has been used on high-altitude balloons to measure the absolute flux and energy spectra of the nuclear components of the primary cosmic radiation. The author presents the measurements of the abundances and the differential energy spectra of iron and nickel in the energy interval from 1 to 10 GeV per amu. The mean abundance of nickel relative to iron is found to be 0.050 ± 0.003. This ratio is consistent with abundances at the source which are similar to those of the interstellar medium as estimated both by Meyer and by Cameron. If the measured energy spectra are fitted to single power laws in total energy, the author obtains spectral indices of 2.44 ± 0.07 for iron and 2.33 ± 0.13 for nickel.

143.018 **Modulation of cosmic rays during solar minimum. I. Cosmic ray intensity survey at sea-level during 1976: experimental details.**
P. H. Stoker, A. J. van der Walt, M. S. Potgieter.
South African J. Phys., Vol. 3, No. 3-4, p. 73 - 76 (1980).
Abstr. in Phys. Abstr., Vol. 84, Abstr. 74987 (1981).

143.019 **Modulation of cosmic rays during solar minimum. II. Cosmic ray latitude distribution at sea-level during 1976.**   M. S. Potgieter, B. C. Raubenheimer, P. H. Stoker, A. J. van der Walt.
South African J. Phys., Vol. 3, No. 3-4, p. 77 - 89 (1980).
Abstr. in Phys. Abstr., Vol. 84, Abstr. 74988 (1981).

143.020 **Development of atmospheric cosmic-ray showers. II.**   G. Thornton, R. Clay.
Phys. Rev. D, Vol. 23, 2090 - 2094 (1981). – Abstr. in Phys. Abstr., Vol. 84, Abstr. 74990 (1981).

143.021 **Modulation of cosmic rays during solar minimum. III. Comparison of the latitude distributions for the periods of solar minimum during 1954, 1965 and 1976.**
M. S. Potgieter, H. Moraal, B. C. Raubenheimer, P. H. Stoker.
South African J. Phys., Vol. 3, No. 3 - 4, p. 90 - 94 (1980).
Abstr. in Phys. Abstr., Vol. 84, Abstr. 79697 (1981).

143.022 **Data photographs from the Auckland cosmic ray telescope.**   P. C. M. Yock.
AIP Conf. Proc., No. 68, Part 2, p. 820 - 823 (1980). – Abstr. in Phys. Abstr., Vol. 84, Abstr. 79698 (1981).

143.023 **Cosmic rays in the atmosphere and characteristics of high energy interactions.**
G. B. Yodh, R. W. Ellsworth, T. Stanev, T. K. Gaisser.
Nucl. Phys. B, Vol. B183, 12 - 28 (1981). – Abstr. in Phys. Abstr., Vol. 84, Abstr. 79700 (1981).

143.024 **Effects of particle drift on the transport of cosmic rays. IV. More realistic diffusion coefficients.**
J. R. Jokipii, J. M. Davila.
Astrophys. J., Vol. 248, 1156 - 1161 (1981).

New results from numerical simulations of cosmic-ray modulation by the solar wind are presented. It is argued that the scattering mean free path should be larger than the particle gyroradius in the average magnetic field. Since this constraint was violated in a previous paper, the authors discuss here simulations which incorporate the larger diffusion coefficients.

143.025 **Neutron oscillation as a source of cosmic ray anti-nucleons.**
O. Sawada, M. Fukugita, J. Arafune.
Astrophys. J., Vol. 248, 1162 - 1165 (1981).

The authors consider effects of neutron oscillation on cosmic ray antineutrons and antiprotons. Presently available experiments on the cosmic ray $\bar{p}/p$ ratio lead to the direct lower bound of the mixing time of free neutron

$(\delta m)^{-1} > 2.5 \times 10^3$ s. If $(\delta m)^{-1} \gtrsim 10^5$ s as theoretically suggested, the effect gives $(\bar{p})_{osc}/p \lesssim 2 \times 10^{-7}$ at low energies. Such an effect, if $(\delta m)^{-1} \sim 10^5$ s, would be seen in a low momentum region ($P \sim 500$ MeV/c) in future experiments.

143.026 Measurement of the cosmic-ray sidereal anisotropy near 1500 GV.
D. J. Cutler, H. E. Bergeson, J. F. Davis, D. E. Groom.
Astrophys. J., Vol. 248, 1166 - 1178 (1981).

A large, underground muon detector has been used to search for periodic variations in the cosmic-ray flux at a median rigidity of 1.5×10^{12} V. The detector has operated at an average rate of 4.47 Hz for most of a 2.7 year interval starting 1978 January 1. A peak with relative amplitude 4.1×10^{-4} is observed in the Fourier transform of data at the sidereal frequency. After correction for solar motion, the projected anisotropy is 6.4×10^{-4} at right ascension $3^h.8 \pm 0^h.7$. The authors observe a second sidereal harmonic with amplitude 2.8×10^{-4} at $\alpha = 7^h.8 \pm 1^h.0$ and a third harmonic with amplitude 2.3×10^{-4} at $\alpha = 7^h.8 \pm 1^h.4$.

143.027 A measurement of the cosmic-ray antiproton flux and a search for antihelium.
A. Buffington, S. M. Schindler, C. R. Pennypacker.
Astrophys. J., Vol. 248, 1179 - 1193 (1981).

A balloon-borne instrument has measured the cosmic-ray antiproton flux between 130 and 320 MeV and searched for antihelium between 130 and 370 MeV per nucleon. The observed 14 antiprotons yield a measured differential flux of $1.7 \pm 0.5 \times 10^{-4}$ antiprotons m^{-2} sr^{-1} s^{-1} MeV^{-1} at the top of the atmosphere. The corresponding antiproton/proton ratio is $2.2 \pm 0.6 \times 10^{-4}$, only slightly smaller than the ratio observed by other experiments at higher energies. Thus the antiprotons have a spectral shape similar to the protons, at least down to about 100 MeV. The expected flux of these particles can be calculated under the assumption that they were created by collisions of high-energy cosmic rays with the interstellar gas. Calculations using the standard leaky box model for propagation in the Galaxy predict a flux two orders of magnitude smaller than that observed. The search for cosmic-ray antihelium sets a 95% confidence level upper limit on the $\overline{He}$/He ratio of 2.2×10^{-5}.

143.028 Spectrum of cosmic ray electrons in the range of 7 - 40 MeV from measurements at the AES Prognoz 4.
R. N. Basilova, G. V. Lupenko, G. I. Pugacheva.
Geomagn. Aehron., Tom 21, 598 - 601 (1981). In Russian.

143.029 Cosmic ray intensity variations and two types of high speed solar streams.
D. Venkatesan, A. K. Shukla, S. P. Agrawal.
Bull. American Astron. Soc., Vol. 13, 544 (1981). – Abstract.

143.030 A constraint on prompt supernova cosmic ray production from γ-ray observations.
G. E. Morfill, L. O'C. Drury.
Mon. Not. R. Astron. Soc., Vol. 197, 369 - 375 (1981).

The consequences of prompt cosmic ray production intrinsic to supernovae are examined for supernova explosions occurring in dense molecular clouds. For reasonable parameters it is shown that prompt cosmic ray production cannot exceed 10^{48} erg per supernova. This suggests that cosmic ray production takes place mainly in the intercloud medium.

143.031 Large-scale variations of cosmic ray intensity and solar activity.   G. P. Lyubimov.
Izv. AN SSSR. Ser. fiz., Tom 44, 2588 - 2609 (1980). In Russian. – Abstr. in Ref. zh., 51. Astron., 7.51.335 (1981).

143.032 Possible links between supersonic stellar winds and the origin of cosmic rays.

T. Montmerle, M. Cassé, J. Paul.
Effects of mass loss on stellar evolution, (see 012.015), p. 155 - 158 (1981).

143.033 Acceleration and modulation of cosmic rays in the interplanetary medium.
B. A. Tverskoj, T. N. Charakhch'yan, M. F. Bakhareva.
Izv. AN SSSR. Ser. fiz., Tom 45, 450 - 460 (1981). In Russian. – Abstr. in Ref. zh., 51. Astron., 8.51.499 (1981).

143.034 On cosmic ray modulation in the interplanetary medium.
V. Kh. Babayan, A. V. Belov, L. I. Dorman, V. S. Ptuskin.
Izv. AN SSSR. Ser. fiz., Tom 45, 557 - 560 (1981). In Russian. Abstr. in Ref. zh., 51. Astron., 8.51.500 (1981).

143.035 A particular modulation event of galactic cosmic rays in 1973.
A. N. Charakhch'yan, Yu. I. Stozhkov, T. N. Charakhch'yan.
Izv. AN SSSR. Ser. fiz., Tom 45, 561 - 565 (1981). In Russian. Abstr. in Ref. zh., 51. Astron., 8.51. 501 (1981).

143.036 Variations of cosmic ray asymmetry during the 20th and 21st solar activity cycles.
S. I. Avdyushin, N. K. Pereyaslova, Yu. M. Kulagin, M. N. Nazarova, I. E. Petrenko.
Izv. AN SSSR. Ser. fiz., Tom 45, 613 - 616 (1981). In Russian. Abstr. in Ref. zh., 51. Astron., 8.51.508 (1981).

143.037 On the importance of ordinary stars and stellar systems in cosmic ray generation.   A. Z. Dolginov.
Izv. AN SSSR. Ser. fiz., Tom 45, 486 - 492 (1981). In Russian. Abstr. in Ref. zh., 51. Astron., 8.51.781; 62. Issled. kosm. prostranstva, 8.62.370 (1981).

143.038 Some peculiarities of propagation and acceleration of cosmic rays in the Galaxy.
L. I. Dorman, V. S. Ptuskin.
Izv. AN SSSR. Ser. fiz., Tom 45, 553 - 556 (1981). In Russian. Abstr. in Ref. zh., 51. Astron., 8.51.782; 62. Issled. kosm. prostranstva, 8.62.371 (1981).

143.039 Temporal, radial and latitudinal variations of the galactic cosmic ray gradients in the solar system.
A. K. Lavrukhina, G. K. Ustinova, M. V. Alaniya, L. I. Dorman.
Izv. AN SSSR. Ser. fiz., Tom 45, 547 - 552 (1981). In Russian. Abstr. in Ref. zh., 51. Astron., 8.51.783; 62. Issled. kosm. prostranstva, 8.62.372 (1981).

143.040 Investigation of electron and γ-quanta fluxes in the near cosmos.   A. M. Gal'per, V. M. Grachev,
V. V. Dmitrenko, V. G. Kirillov-Ugryumov, V. A. Lyakhov, V. V. Ryumin, S. E. Ulin, N. I. Shvets.
Izv. AN SSSR. Ser. fiz., Tom 45, 637 - 641 (1981). In Russian. Abstr. in Ref. zh., 62. Issled. kosm. prostranstva, 8.62.400 (1981).

143.041 An attempt to understand the observed antiprotons in cosmic radiation.   S. A. Stephens.
Bull. Astron. Soc. India, Vol. 9, 66 (1981). – Abstract.

143.042 Revised closed galaxy model for cosmic ray propagation.   S. A. Stephens.
Bull. Astron. Soc. India, Vol. 9, 66 - 67 (1981). – Abstract.

143.043 The relative abundances of the elements scandium to manganese in relativistic cosmic rays and the possible radioactive decay of manganese 54.
L. Koch, J. J. Engelmann, P. Goret, E. Juliusson, N. Petrou, Y. Rio, A. Soutoul, B. Byrnak, N. Lund, B. Peters, I. L. Rasmussen, M. Rotenberg, N. Westergaard.

Astron. Astrophys., Vol. 102, L9 - L11 (1981).

The authors report measurements of the abundances relative to iron of the elements scandium through manganese in galactic cosmic rays in the energy interval 0.7 to 18 GeV/n using data collected by a cosmic ray telescope on the HEAO-3 satellite. The variation versus energy of the abundance ratio of manganese over iron is markedly flatter than that of other iron secondaries. This difference is interpreted as being due to ^{54}Mn survival at $E > 15$ GeV/n and to its progressive beta decay at lower energies.

143.044   **Acceleration and modulation of cosmic rays in the interplanetary space.**
B. A. Tverskoj, T. N. Charakhch'yan, M. F. Bakhareva.
Cosmic rays, (see 012.005), 1980, p. 199. – Abstr. in Ref. zh., 51. Astron., 9.51.416 (1981).

143.045   **Cosmic ray origin. Some aspects of the problem.**
V. L. Ginzburg.
Cosmic rays, (see 012.005), 1980, p. 1 - 36. In Russian. Abstr. in Ref. zh., 51. Astron., 9.51.741 (1981).

143.046   **Spectrum of galactic electrons and models of their propagation.**
A. A. Gusev, G. I. Pugacheva, A. F. Titenkov.
Izv. AN SSSR. Ser. fiz., Tom 45, 566 - 567 (1981). In Russian. Abstr. in Ref. zh., 51. Astron., 9.51.742; 62. Issled. kosm. prostranstva, 9.62.420 (1981).

143.047   **The role of normal stars and stellar systems in cosmic ray generation.**   A. Z. Dolginov.
Cosmic rays, (see 012.005), 1980, p. 187 - 189. – Abstr. in Ref. zh., 51. Astron., 9.51.743 (1981).

143.048   **Cosmic ray acceleration in sources by supersonic turbulence.**   A. M. Bykov, I. N. Toptygin.
Cosmic rays, (see 012.005), 1980, p. 203 - 206. – Abstr. in Ref. zh., 51. Astron., 9.51.744 (1981).

143.049   **Investigation of primary cosmic rays of the iron group on Zond 5.**   D. G. Baranov, Yu. F. Gagarin, N. S. Ivanova, V. N. Kulikov, A. Yu. Merkur'eva.
Cosmic rays, (see 012.005), 1980, p. 226 - 233. In Russian. Abstr. in Ref. zh., 51. Astron., 9.51.749 (1981).

143.050   **Cosmic rays of ultra-high energy.**
G. D. Rochester, K. E. Turver.
Contemp. Phys., Vol. 22, 425 - 450 (1981). – Abstr. in Phys. Abstr., Vol. 84, Abstr. 88675 (1981).

143.051   **Conundrum of high energy ($> 10^{18}$ eV) cosmic ray particles.**   A. K. Dasgupta.
Astrophys. Space Sci., Vol. 79, 153 - 158 (1981).

It is shown that the observed characteristic of cosmic-ray air showers is not inconsistent with the hypothesis that the initiating primary particles cannot be charged dust grains of radii ($3 \times 10^{-6} \sim 3 \times 10^{-5}$ cm).

143.052   **The cosmic ray proton spectra at different atmospheric depths derived from the recent primary spectrum of Olejniczak et al.**   D. P. Bhattacharyya.
Fizika, Vol. 13, No. 1, p. 15 - 22 (1981). – Abstr. in Phys. Abstr., Vol. 84, Abstr. 94123 (1981).

143.053   **The local interstellar electron spectrum.**
L. C. Tan, L. K. Ng.
J. Phys. G, Vol. 7, 1123 - 1134 (1981). – Abstr. in Phys. Abstr., Vol. 84, Abstr. 94124 (1981).

143.054   **The energy spectra of cosmic-ray positrons and electrons.**   L. C. Tan, L. K. Ng.

J. Phys. G, Vol. 7, 1135 - 1148 (1981). – Abstr. in Phys. Abstr., Vol. 84, Abstr. 94125 (1981).

143.055   **Antiprotons in cosmic rays.**
Zemlya Vselennaya, 1981, No. 5, p. 23. In Russian.

143.056   **The galactic centre and origin of cosmic rays.**
V. S. Ptuskin, Ya. M. Khazan.
Astron. Zh., Tom 58, 959 - 968 (1981). In Russian. English translation in Soviet Astron., Vol. 25, No. 5.

A model of the origin of galactic cosmic rays is considered in which it is suggested that the major part of the energetic particles observed in the Galaxy is accelerated during the repeated periods of activity of the galactic nucleus at a time interval of 10^7 - 10^8 years. In each of the active periods, the total energy of the accelerated cosmic rays should be 10^{55} - 10^{56} erg. Most characteristic features of this model are considered and it is concluded that the model does not disagree with existing observational data.

143.057   **On the origin of antiprotons observed in cosmic rays.**   V. L. Ginzburg, V. S. Ptuskin.
Pis'ma Astron. Zh., Tom 7, 585 - 589 (1981). In Russian. English translation in Soviet Astron. Lett., Vol. 7.

Possibilities to explain the relatively large flux of antiprotons in primary cosmic rays near the earth are discussed. The generation of antiprotons in young supernova shells is considered in detail.

143.058   **The solar modulation of galactic cosmic rays in the outer heliosphere.**
F. B. McDonald, N. Lal, J. H. Trainor, M. A. I. Van Hollebeke, W. R. Webber.
Astrophys. J., Lett., Vol. 249, L71 - L75 (1981).

The observations of Pioneer 10 and Helios 1 and 2 of the intensity changes of galactic cosmic rays associated with the enhanced solar activity during the onset of cycle 21 over an extended range of energy and heliocentric distance provide new insight on the relative importance of the various processes involved in the long-term modulation. There is a close correspondence between changes at 1 AU and those at 23 AU for hydrogen and helium in the range of 100 - 200 MeV per nucleon. The relative decrease is of the same order at both locations with an appropriate time delay that corresponds to an outward propagation velocity of some 550 km s^{-1}. These measurements suggest that the recently discovered, moderately long-lived, radially propagating shock waves in the outer heliosphere may play a key role in the long-term modulation. At the location of Pioneer 10, the variations in the intensity of the anomalous and galactic cosmic-ray helium are well correlated.

143.059   **Relativistic cosmic rays and corotating interaction regions.**   S. P. Duggal, B. T. Tsurutani, M. A. Pomerantz, C. H. Tsao, E. J. Smith.
J. Geophys. Res., Vol. 86, 7473 - 7479 (1981).

Analyses of relativistic galactic cosmic ray intensity variations have been conducted to determine the nature of the modulations that are related to the presence of corotating interaction regions in interplanetary space.

143.060   **Remote sensing of muon variation spectra and interplanetary inhomogeneities.**
M. El-Raey, S. Gaber.
J. Geophys. Res., Vol. 86, 7744 - 7748 (1981).

The similarity between the general remote sensing integral equation and Dorman's integral equation relating secondary to primary muon variation spectra has been pointed out and exploited. Primary muon variation spectra and parameters of interplanetary field inhomogeneities are remotely sensed by a non-linear inversion technique, using two different coupling functions. Results are compared to those obtained by the spectrographic technique and interpreted in terms of interplanetary field fluctuations. Based on relationships among kernels of the integral equation, optimization of location and orientation of measuring stations become possible.

143.061   Results of measurements of relativistic particles in September 1977 with a Čerenkov detector aboard the Cosmos 900 satellite.
E. V. Gorchakov, V. A. Iozenas, M. V. Ternovskaya, V. G. Afanas'ev, K. G. Afanas'ev, N. I. Klimov, A. E. Indyukov.
Kosm. Issled., Tom 19, 778 - 780 (1981). In Russian.

143.062   Anisotropy: high energy cosmic rays and the thermal background radiation.   P. R. Marchant.
Astrophys. Space Sci., Vol. 79, 521 - 523 (1981).
The motion of a large ($\gtrsim 100$ Mpc radius) volume of local matter with respect to the frame in which the thermal cosmic background radiation appears isotropic is shown to produce a contribution to anisotropy in the arrival directions of the highest energy ($> 10^{19}$ eV) cosmic rays. The magnitude of the effect predicted is $\cong 1\%$, below that currently observed and below the sensitivity of present experiments.

143.063   Interaction of cosmic rays with the magnetosphere and ionosphere.   L. I. Dorman.
Probl. soln.-zemn. svyazej. Dokl. simpoz. KAPG, Ashkhabad, 1979. Ashkhabad, 1981, p. 108 - 131. In Russian. – Abstr. in Ref. zh., 51. Astron., 10.51.411 (1981).

143.064   On kinetic properties of the cosmic ray gas.
V. S. Ptuskin.
Pis'ma Astron. Zh., Tom 7, 671 - 673 (1981). In Russian. English translation in Soviet Astron. Lett., Vol. 7.
From an equation for the energy distribution function of cosmic rays the speed of sound and compressibility for the cosmic ray gas are found.

143.065   On cosmic ray acceleration in the interstellar medium.   V. S. Ptuskin.
Pis'ma Astron. Zh., Tom 7, 674 - 676 (1981). In Russian. English translation in Soviet Astron. Lett., Vol. 7.
The role of interstellar acceleration in the diffusion model of cosmic ray propagation is estimated. The acceleration is due to scattering on weak irregularities of the galactic magnetic field.

143.066   The first four harmonics of the daily variation of galactic cosmic rays.
A. Z. Dolginov, Yu. P. Mel'nikov.
Geomagn. Aehron., Tom 21, 775 - 780 (1981). In Russian.

143.067   A regular mechanism of acceleration of charged particles under cosmic conditions. III. Time characteristics.   V. K. Elshin, G. F. Krymskij, S. I. Petukhov, A. A. Turpanov, Yu. A. Romashchenko.
Geomagn. Aehron., Tom 21, 781 - 787 (1981). In Russian.

143.068   Effects of interaction of the heliomagnetosphere with the galactic field in cosmic rays.
G. F. Krymskij, P. A. Krivoshapkin, V. P. Mamrukova, G. V. Skripin.
Geomagn. Aehron., Tom 21, 923 - 925 (1981). In Russian.

143.069   Propagation of cosmic ray electrons in the Galaxy.
H.- Q. Zhang.
Phys. Energ. Fortis Phys. Nucl., Vol. 5, 477 - 483 (1981). In Chinese. – Abstr. in Phys. Abstr., Vol. 84, Abstr. 108021 (1981).

143.070   High energy primary electron spectrum and diffusion X-ray source.   C.-X. Xu, H.-Q. Zhang.
Phys. Energ. Fortis Phys. Nucl., Vol. 5, 484 - 488 (1981). In Chinese. – Abstr. in Phys. Abstr., Vol. 84, Abstr. 108022 (1981).

143.071   Study of tri-diurnal variation of galactic cosmic radiation.   S. P. Agrawal.
J. Geophys. Res., Vol. 86, 10115 - 10121 (1981).
Using the experimental data of high counting rate neutron and meson monitors, the solar tri-diurnal anisotropy of galactic cosmic radiation has been investigated for the period 1962–79. The enhancement of the average tri-diurnal amplitude observed by all the detectors during 1973–75 provides sufficient signal to noise ratio to obtain its variational characteristics.

143.072   The propagation of galactic cosmic rays.
A. N. Hall.
Mon. Not. R. Astron. Soc., Vol. 197, 977 - 993 (1981).
Large scale (~ 15 pc) turbulence in the interstellar medium (ISM) causes the firehose and mirror instabilities to occur. These produce small scale ($\sim 10^{-7}$ pc) magnetic irregularities, which scatter cosmic rays. The author constructs a slab model of the turbulent ISM. Then he finds the amplitudes and wavelengths of the magnetic irregularities that arise, and calculates the coefficients for the diffusion of cosmic rays along the interstellar magnetic fields. He shows that his model can account for both the lifetime of low energy cosmic rays, and the variation of their mean pathlength with energy.

143.073   Time-dependent Green's functions of the cosmic ray equation of transport.   G. M. Webb.
Astrophys. Space Sci., Vol. 80, 323 - 335 (1981).
Two spherically symmetric time-dependent Green's functions of the equation of transport for cosmic rays in the interplanetary region are derived by transform techniques. The solar wind velocity is assumed radial and of constant speed V. In the first model the radial diffusion coefficient $\kappa = \kappa_0 r$ (κ_0 constant), and in the second solution $\kappa = \kappa_0 = $ constant. The solutions are for monoenergetic, impulsive release of particles from a fixed heliocentric radius. Integration of the solutions over time t, from $t = 0$ to $t = \infty$, gives the steady-state Green's functions obtained previously.

143.074   Cosmic ray antiprotons in the closed galaxy model.
R. J. Protheroe.
Astrophys. J., Vol. 251, 387 - 392 (1981).
The author has made a calculation of the flux of secondary antiprotons expected for the leaky box model and for the closed galaxy model of Peters and Westergaard. The $\bar{p}/p$ ratio observed at several GeV is a factor of 4 higher than the prediction for the leaky box model but is consistent with that predicted for the closed galaxy model. New low energy data are not consistent with either model. The possibility of a primary antiproton component is discussed.

143.075   Spectral and compositional variations of low energy ions during an energetic storm particle event.
B. Klecker, M. Scholer, D. Hovestadt, G. Gloeckler, F. M. Ipavich.
Astrophys. J., Vol. 251, 393 - 401 (1981).
The authors study the spectral and compositional variations of protons, helium, and heavy ions (C, O, and Fe) in the energy range 0.3–20 MeV per nucleon during an energetic storm particle event on 1978 September 28 - 29. In particular, they discuss the data in terms of a first-order Fermi acceleration process. Using the time scale of the intensity increase of low energy protons ($\lesssim 0.6$ MeV) in front of the shock front, the authors obtain an estimate for the absolute value and the energy dependence of the diffusion coefficient upstream of the propagating interplanetary shock.

143.076   Indicators of nucleosynthesis and acceleration processes in the ultraheavy cosmic rays: $24 \leqslant Z \leqslant 59$.

J. B. Blake, S. H. Margolis.
Astrophys. J., Vol. 402 - 408 (1981).

The observable abundances of the ultraheavy cosmic rays in the charge range $24 \leqslant Z \leqslant 59$ have been calculated using different source abundances hypotheses. These include the standard solar-system abundances, r-process-only abundances, and these two sets modified by the first ionization potential of the elements to test for preferential acceleration effects. The propagated elemental abundance patterns that result are distinctive signatures of nucleosynthesis and acceleration. Present data seem to suggest solar-system abundances with a dependence upon the first ionization potential for the elements $30 \leqslant Z \leqslant 40$.

143.077   The isotopic composition of cosmic ray B, C, N, and O nuclei.
R. A. Mewaldt, J. D. Spalding, E. C. Stone, R. E. Vogt.
Astrophys. J., Lett., Vol. 251, L27 - L31 (1918).

The authors report new high resolution measurements of the elemental and isotopic composition of galactic cosmic ray B, C, N, and O nuclei with ~ 30 to ~ 130 MeV nucleon^{-1}. These observations place limits on the isotopic composition of the cosmic ray source and restrict possible models of cosmic ray origin and propagation. In particular, the authors find that N is significantly depleted in the cosmic ray source with respect to the solar system and local interstellar medium, a result inconsistent with models in which a majority of cosmic rays are accelerated interstellar medium material.

143.078   The study of high energy electrons using transition radiation.   A. A. Marin, D. Hasegan, C. Blaj, D. Dorcioman, M. Draghicescu, I. Tutos, N. L. Grigorov, A. A. Gusev, G. I. Pugacheva, A. F. Titenkov, V. I. Zatzepin.
Rev. Roumaine Phys., Vol. 26, 545 - 551 (1981). – Abstr. in Phys. Abstr., Vol. 85, Abstr. 3169 (1982).

143.079   High energy cosmic ray acceleration by diffusive shocks.   P. O. Lagage, C. J. Cesarsky.
Plasma astrophysics, (see 012.042), p. 317 - 318 (1981).

The mechanism of acceleration of cosmic rays by shock waves presents many attractive features, but it also encounters some difficulties. One of the problems is that it cannot accelerate particles to very high energies. This is because the mean time between shock crossings is long for high energy particles, so that the acceleration process is slow. The authors present a derivation of the mean crossing time which has the advantage, over previous methods, of being readily generalized when the diffusion coefficient is space dependent. Then, the authors evaluate an upper limit to the energy that can be attained for a supernova shock and for a stellar wind terminal shock.

143.080   Cosmic ray acceleration by stellar winds and self-confinement in giant H II regions.
T. Montmerle, C. J. Cesarsky.
Plasma astrophyiscs, (see 012.042), p. 319 - 322 (1981).

143.081   Confinement of cosmic rays in molecular clouds.
E. G. Zweibel, J. M. Shull.
Plasma astrophysics, (see 012.042), p. 371 - 373 (1981).

The authors discuss the consequences of cosmic ray production by a supernova in a molecular cloud.

143.082   Certain aspects of primary cosmic rays composition at energies $10^{14} - 10^{15}$ eV.   J. Kempa.
Plasma astrophysics, (see 012.042), p. 375 - 376 (1981).

Many aspects indicate that the chemical composition of primary cosmic rays at energies $10^{14} - 10^{15}$ eV consists of heavier particles than in the area of energy 10^{12} eV. The comparison of various components proves that the average atomic weight of the primary particles in the area mentioned above is about $10-15$.

143.083   Acceleration of cosmic rays by shock waves.
W. I. Axford.
Plasma astrophysics, (see 012.042), p. 425 - 449 (1981).

Contents: Introduction. Galactic cosmic rays. The interstellar medium. Scatter-free shock acceleration. Acceleration by shocks in scattering media. Plane shocks with energy losses and time dependence. Acceleration by non-planar shocks. Acceleration of cosmic rays by supernova blast waves. The cosmic ray diffusion coefficient in the HISM. Non-linear cosmic ray shocks. Conclusions.

143.084   On a connection of the north-south anisotropy of galactic cosmic rays with the sign of the general field of the sun.   V. G. Grigor'ev, A. T. Filippov.
Geomagn. Aehron., Tom 21, 1112 - 1113 (1981). In Russian.

143.085   On the nature of increase of the galactic cosmic ray intensity on October 27, 1977.
A. T. Filippov, V. G. Grigor'ev.
Geomagn. Aehron., Tom 21, 1113 - 1114 (1981). In Russian.

143.086   Clues for the origin of cosmic rays.   P. Meyer.
Nature, Vol. 294, 518 - 519 (1981).

143.087   A southern hemisphere search for a nonrandom component in cosmic rays.
R. W. Clay, B. R. Dawson.
Australian J. Phys., Vol. 34, 591 - 593 (1981).

A search has been made for a nonrandom component in southern hemisphere cosmic radiation with energies greater than $\sim 10^{14}$ eV. The authors find no evidence to correspond to the positive effect reported in the northern hemisphere by Bhat et al.

143.088   On the nature of the cosmic ray positron spectrum.
R. J. Protheroe.
NASA Tech. Memo., NASA TM 83834, 23 pp. (1981).

The author has made a new calculation of the flux of secondary positrons above 100 MeV expected for various propagation models. The models investigated are the leaky box or homogeneous model, a disk-halo diffusion model, a dynamical halo model and the closed galaxy model. The positron flux predicted for these models is compared with the available data. The possibility of a primary positron component is considered.

143.089   On the motion of cosmic rays in the interplanetary magnetic field.
L. I. Dorman, M. E. Kats, S. F. Nosov, M. Steglik, Yu. I. Fedorov, B. A. Shakhov.
Izv. AN SSSR. Ser. fiz., Tom 45, 1285 - 1286 (1981). In Russian. – Abstr. in Ref. zh., 51. Astron., 11.51.486 (1981).

143.090   Conversion coefficients for determination of the three-dimensional cosmic ray anisotropy in different periods of solar activity.
G. A. Gonchar, E. V. Kolomeets, N. V. Slyunyaeva.
Fiz. atom. yadra i kosm. luchej. Alma-Ata, 1980, p. 101 - 108. In Russian. – Abstr. in Ref. zh., 51. Astron., 11.51.489 (1981).

143.091   Expected effects of charged particle drift in a broad variety of cosmic ray intensity and anisotropy variations and comparison with an experiment.
M. V. Alaniya, R. A. Aslamazishvili, D. P. Bochikashvili, N. A. Nachkebiya.
Izv. AN SSSR. Ser. fiz., Tom 45, 1278 - 1284 (1981). In Russian. – Abstr. in Ref. zh., 51. Astron., 11.51.491 (1981).

143.092   On the study of galactic cosmic ray modulation.
V. P. Antonova, A. G. Zusmanovich, L. A. Mirkin, L. F. Churunova.

Izv. AN SSSR. Ser. fiz., Tom 45, 1287 - 1290 (1981). In Russian. – Abstr. in Ref. zh., 51. Astron., 11.51.492 (1981).

143.093 **Solar wind conditioned galactic cosmic ray modulation and energy dependence of the diffusion coefficient.**
B. N. Dujsenbaev, E. V. Kolomeets, V. N. Sevast'yanov, N. V. Stekol'nikov.
Izv. AN SSSR. Ser. fiz., Tom 45, 1291 - 1295 (1981). In Russian. – Abstr. in Ref. zh., 51. Astron., 11.51.493 (1981).

143.094 **Solar wind and cosmic ray density variations.**
N. P. Chirkov.
Cosmic ray variations and solar wind, (see 003.013), p. 25 - 39 (1980). In Russian. – Abstr. in Ref. zh., 51. Astron., 11.51.502 (1981).

143.095 **Galactic cosmic ray modulation by the solar wind from selected heliographic latitudes.**
Kh. Z. Aldagarova, D. S. Amankulov, E. V. Kolomeets, V. T. Pivneva.
Fiz. atom. yadra i kosm. luchej, Alma-Ata, 1980, p. 38 - 47. In Russian. – Abstr. in Ref. zh., 51. Astron., 11.51.503 (1981).

143.096 **Mass composition of primary cosmic radiation in the 10^{15} - 10^{16} eV energy range.**
I. N. Kirov, S. I. Nikol'skij, N. M. Nikol'skaya, J. N. Stamenov, S. Z. Ushev, V. D. Yanminchev.
Izv. AN SSSR. Ser. fiz., Tom 45, 1273 - 1277 (1981). In Russian. – Abstr. in Ref. zh., 51. Astron., 11.51.881 (1981).

143.097 **Chemical and isotopic composition of cosmic rays.**
N. Lund.
Cosmic rays, (see 012.025), 1980, p. 55 - 69. – Abstr. in Ref. zh., 51. Astron., 11.51.886; 62. Issled. kosm. prostranstva, 11.62.247 (1981).

143.098 **The anisotropy of very high-energy cosmic rays.**
A. W. Wolfendale.
Cosmic rays, (see 012.025), 1980, p. 70 - 76, 114 - 116. Abstr. in Ref. zh., 51. Astron., 11.51.887 (1981).

143.099 **Energetic spectrum of primary cosmic radiation in the region of superhigh energies.**
Yu. A. Fomin, G. B. Khristiansen.
Cosmic rays, (see 012.025), 1980, p. 77 - 96. In Russian. Abstr. in Ref. zh., 51. Astron., 11.51.888 (1981).

143.100 **Spectrum of high and ultrahigh energy cosmic rays.**
J. Wdowczyk.
Cosmic rays, (see 012.025), 1980, p. 97 - 113, 114 - 116. Abstr. in Ref. zh., 51. Astron., 11.51.889; 62. Issled. kosm. prostranstva, 11.62.246 (1981).

143.101 **Energetic spectra of protons and helium cores of galactic cosmic rays in the deep minimum of solar activity 1975.**
A. V. Khrimyan, N. A. Nalbandyan, A. A. Avakyan, G. V. Khrimyan, T. G. Vartanyan, M. Zh. Mkrtchyan.
Cosmic rays, (see 012.025), 1980, p. 217 - 225. In Russian. Abstr. in Ref. zh., 51. Astron., 11.51.890; 62. Issled. kosm. prostranstva, 11.62.250 (1981).

143.102 **Modulation of cosmic rays by high-velocity solar wind streams from measurements on the Prognoz 3 satellite and the Mars 7 automatic interplanetary station in January 1974.** P. V. Vakulov, N. I. Vologdin, B. M. Kuzhevskij, E. S. Spir'kova, I. P. Shestopalov.
Kosm. Issled., Tom 19, 869 - 875 (1981). In Russian.

143.103 **Observation of electrons with energies $\geqslant 10^{12}$ eV in the primary cosmic radiation.** R. A. Nymmik.
Kosm. Issled., Tom 19, 884 - 888 (1981). In Russian.

143.104 **Variability of the diurnal anisotropy of the cosmic ray intensity during the interval 1973–1978.**
S. P. Agrawal, D. Venkatesan.
J. Geophys. Res., Vol. 86, 11093 - 11097 (1981).
The diurnal anisotropy of cosmic ray intensity has been investigated over the interval 1973–1978 by using neutron monitor data from Calgary, Deep River, Goose Bay, and Inuvik. It is observed that in general, the 27-day variation in the diurnal anisotropy is greater for the component perpendicular to the nominal interplanetary magnetic field (IMF) in comparison with the parallel (or field-aligned) component, which has very little 'power'. However, the direction of the average anisotropy vector is aligned almost perpendicular (within the hour) to the IMF. The results are discussed in terms of alternate possibilities: the 'power' with 27-day period and its harmonics arise from fluctuations in the amplitude aligned in the direction (a) of the total vector itself (≈ 16 hours) and (b) perpendicular to the IMF.

143.105 **A study of the long-term variation and radial gradient of cosmic rays out to 23 AU.**
W. R. Webber, J. A. Lockwood.
J. Geophys. Res., Vol. 86, 11458 - 11462 (1981).
An examination of the long-term modulation of cosmic rays with $E > 60$ MeV from 1972-1980 using data from the Pioneer, Voyager and IMP spacecraft has revealed that (1) the long-term modulation effects are propagated outward radially from the sun with a typical speed of 350-500 km/s, (2) the average integral radial gradient during this period between 2 and 23 AU is $(2.85 \pm 0.5)\%/\text{AU}$, and (3) the boundary of the heliocentric modulation region in the ecliptic plane as derived from simple conventional modulation theory is >65 AU during this time period. The implications of these results to cosmic ray modulation theories are discussed.

143.106 **Taking into account back coupling during acceleration of charged particles by a shock wave.**
G. F. Krymskij.
Izv. AN SSSR. Ser. fiz., Tom 45, 461 - 473 (1981). In Russian. – Abstr. in Ref. zh., 51. Astron., 12.51.867 (1981).

143.107 **Cosmic ray acceleration in sources by supersonic turbulence.** A. M. Bykov, I. N. Toptygin.
Izv. AN SSSR. Ser. fiz., Tom 45, 474 - 485 (1981). In Russian. – Abstr. in Ref. zh., 51. Astron., 12.51.868 (1981).

143.108 **Cosmic ray acceleration in a medium with account for losses and diffuse sources of gamma-radiation.**
H. Völk.
Izv. AN SSSR. Ser. fiz., Tom 45, 1122 - 1136 (1981). In Russian. – Abstr. in Ref. zh., 51. Astron., 12.51.870 (1981).

143.109 **Propagation modes of energetic charged particles in the heliosphere.** B. A. Tverskoi (*Tverskoj*).
Cosmic rays in the heliosphere, (see 012.059), p. 5 - 14 (1981).
Several years ago, the anisotropic diffusion and convective transport accompanied by adiabatic deceleration were considered as the principal means for cosmic ray propagation. The theory describing the 11-year variation of galactic cosmic ray intensity and the propagation of solar cosmic rays was founded on this basis. The anomalies of the 11-year variation of galactic cosmic ray intensity in 1969 - 1971 revealed the necessity to take into account the influence of the general electromagnetic field of the heliosphere giving rise to a rapid magnetic drift of particles.

143.110 **Prediction of energetic particle behavior at high solar latitudes.** L. A. Fisk.

Cosmic rays in the heliosphere, (see 012.059), p. 41 - 55 (1981).

Emphasis is placed on predictions of the galactic cosmic-ray flux at high heliographic latitudes. Recent work on gradient and curvature drifts in the large-scale heliospheric magnetic field have modified the traditional argument that the cosmic rays should be essentially unmodulated over the solar poles. In fact, drift effects during the next solar cycle, when the International Solar Polar Mission is to fly, are predicted to cause considerable modulation in the polar regions.

143.111   **50–200 GeV cosmic rays at various heliolatitudes.**
G. Erdős, J. Kóta.
Cosmic rays in the heliosphere, (see 012.059), p. 139 - 142 (1981).

The authors consider regular motion of 50–200 GeV particles in a large-scale interplanetary magnetic field model. Numerical calculations based upon energy losses along various trajectories are carried out to obtain the predicted omni-directional density and anisotropy of cosmic rays at various solar latitudes.

143.112   **Galactic cosmic ray gradients in the ecliptic plane and at high latitudes during two solar cycles (meteorite data).**   A. K. Lavrukhina, G. K. Ustinova.
Cosmic rays in the heliosphere, (see 012.059), p. 143 - 146 (1981).

Results of investigations of cosmogenic isotope radioactivity in chondrites fallen to the earth during two solar cycles are presented. The data obtained on radial and latitudinal gradients cover the period 1955 - 1976, heliocentric distances from 1.03 AU to 3.33 AU, and heliographic latitudes from 23°S to 16°N. The dependence of radial and latitudinal gradients on the phase of solar activity is established, as well as a north-south asymmetry during a certain period after the inversion of the general solar magnetic field in 1969.

143.113   **Preliminary search for cosmic radiation and solar-terrestrial parameters correlated with the reversal of the solar magnetic field.**   M. A. Shea, D. F. Smart.
Cosmic rays in the heliosphere, (see 012.059), p. 147 - 150 (1981).

A statistical study has been made of cosmic ray intensity, as observed by a neutron monitor, and of selected solar and geophysical parameters in a search for phenomena which may be associated with the reversal of the solar magnetic field. The results reported here utilized the Zurich sunspot number and the geomagnetic aa index.

143.114   **Changes in cosmic ray propagation induced by corotating interaction regions.**
R. Gall, B. T. Thomas, G. Vidargas.
Cosmic rays in the heliosphere, (see 012.059), p. 155 - 158 (1981).

By simulating the trajectories for scatter free and diffusive propagation of relativistic cosmic rays in a model of the heliospheric magnetic fields containing a representation of corotating interaction regions the authors find that the motion is strongly affected and differs substantially from the predictions of current modulation theory.

143.115   **Analysis of galactic electron spectrum measurements.**   A. A. Gusev, G. I. Pugacheva,
A. F. Titenkov.
Cosmic rays in the heliosphere, (see 012.059), p. 159 - 162 (1981).

Results of measurements of primary electron spectra performed during the last decade are presented together with methodical peculiarities of the experiments. A possible physical interpretation of the experimental results is discussed.

143.116   **The HEAO-3 French-Danish cosmic ray spectrometer: preliminary results on the elemental abundances of cosmic ray nuclei in the iron peak.**
The Copenhagen - Saclay collaboration for HEAO.
High-energy astrophysics, (see 012.067), p. 173 - 184 (1981).

The French-Danish cosmic ray spectrometer was launched on HEAO-3 on September 20, 1979. This instrument was optimized to measure the charge composition and isotopic abundances of galactic cosmic rays with energies of about 0.5 to 7 GeV/nucleon, using the multi-Cerenkov detector technique and a flash tube hodoscope. The high charge resolution of the instrument permits complete separation of all chemical elements between Be and Ni, even at energies larger than 5 GeV/nucleon. Preliminary results are presented for relative abundances of individual elements $21 \leqslant Z \leqslant 28$.

143.117   **Results of cosmic radiation dose field measurements aboard the "Salyut-6" orbital station.**
Yu. A. Akatov, T. V. Batenchuk, A. M. Borodin, V. E. Dudkin, A. M. Marenny, V. A. Sakovich, G. G. Vasilev, A. G. Karadjov, P. P. Szabó, I. Fehér, W. Malz, G. Voigtman, D. Hasheganu.
Life sciences and space research XIX, (see 012.068), p. 67 - 71 (1981).

During the 3rd main expedition on board the "Salyut-6" orbital station in 1979 the integral characteristics of cosmic radiation were measured in various positions inside the manned modules. The dose gradient inside the manned modules of the station amounted to 70% for long intervals of time. During the experimental period the dose rate inside the station was 15 to 30 mrad per day.

143.118   **Abundance enhancements in cosmic rays produced by collisionless shocks.**
D. Eichler, K. Hainebach.
Phys. Rev. Lett., Vol. 47, 1560 - 1563 (1981).

It is shown that shocks preferentially accelerate partially ionized heavy elements over protons. Within the framework of a previously published model of injection, the spectra of different ion species are calculated from thermal to ultra-relativistic energies. For typical astrophysical parameters, the predicted enhancement is in qualitative agreement with observations for a 10^6-K preshock plasma; ions with $Z > 10$ are enhanced by about an order of magnitude or so. The continuing increase with energy of the heavy-element abundances into the air-shower regime is discussed.

High energy astrophysics. An informal introduction for students of physics and astronomy.   See Abstr. 003.093.

Problems of cosmic ray physics. Volume 1.
See Abstr. 003.165.

Physics of the atomic nucleus and cosmic rays.
See Abstr. 003.170.

Status of research on ultra-heavy nuclei in cosmic rays.   See Abstr. 013.023.

Soviet-Indian research of cosmic radiation by high altitude balloons.   See Abstr. 013.050.

Characteristics of cosmic ray cutoffs for a satellite orbiting at 400 km; the effect of the solid earth.
See Abstr. 031.650.

The UH-nuclei cosmic ray detector on the third High Energy Astronomy Observatory.   See Abstr. 032.538.

A new experiment for measuring the flux of high energy iron nuclei in cosmic rays.   See Abstr. 032.539.

Fast heavy ions in the heliosphere.
See Abstr. 032.618.

Two-dimensional analyzer with dE-dE silicon detectors for cosmic ray experiments in space.
See Abstr. 032.619.

A new thermoluminescent dosimeter system for space research.   See Abstr. 032.632.

Nature's own particle accelerator.
See Abstr. 034.005.

Measurements of characteristics of the gas proportional chamber in cosmic ray showers.
See Abstr. 034.040.

A high efficiency 4π neutron detector.
See Abstr. 034.041.

An air shower array designed for cosmic ray variation measurements and high energy gamma ray astronomy.
See Abstr. 034.046.

Elastic and inelastic scattering of the relic neutrinos by high energy cosmic rays.   See Abstr. 061.011.

Some aspects of the nucleosynthesis of the light elements.   See Abstr. 061.040.

The effect of losses on acceleration of energetic particles by diffusive scattering through shock waves.
See Abstr. 062.071.

Quasiperiodic variations of solar activity, physical characteristics of the solar wind, and cosmic ray intensity and anisotropy.   See Abstr. 072.037.

The expected pulsations of the heliosphere relevant to cosmic ray variations.   See Abstr. 074.082.

Enhanced cosmic ray anisotropies and the extended solar magnetic field.   See Abstr. 075.013.

Solar and cosmic radiation 4.6×10^9 years ago.
See Abstr. 078.007.

Solar and cosmic radiation 4.6×10^9 years ago.
See Abstr. 078.016.

Cosmogenic nuclides as an indirect method for cosmic ray investigation.   See Abstr. 078.019.

Space erosion of meteorites and the secular variation of cosmic rays (over 10^9 years).   See Abstr. 105.065.

How does the global structure of the interplanetary magnetic field affect cosmic ray modulation?
See Abstr. 106.052.

H_2 production in dense molecular clouds.
See Abstr. 131.153.

Fluxes of energetic particles and the ionization rate in protostars and the primitive solar nebula.
See Abstr. 131.206.

Rôle du rayonnement cosmique.
See Abstr. 131.226.

Gamma ray astronomy and the origin of cosmic rays.   See Abstr. 142.518.

Contribution of cosmic ray-irradiated molecular clouds to the number of apparent γ-ray sources.
See Abstr. 142.535.

Nucleosynthesis in OB associations and its effect upon cosmic rays, peculiar stars and isotopic anomalies in meteorites.   See Abstr. 152.006.

Correlation of γ-ray flux and molecular hydrogen column densities in the galactic plane.   See Abstr. 155.011.

The distribution of cosmic-ray electrons and nuclei in the Galaxy.   See Abstr. 155.012.

The diffuse galactic gamma radiation: the Compton contribution and component separation by energy interval and galactic coordinates.   See Abstr. 157.015.

Stellar Systems, Galaxy, Extragalactic Objects, Cosmology

151 Stellar Systems (Kinematics, Dynamics, Evolution)

151.001 **Odd-parity perturbations of spherically symmetric star clusters in general relativity.**
R. Semenzato, J. R. Ipser.
Astrophys. J., Vol. 247, 671 - 676 (1981).
  The theory of odd-parity nonspherical perturbations of collisionless, isotropic, spherically symmetric star clusters is developed within general relativity for $l \geqslant 2$. A variational principle is derived for the associated normal modes of oscillation. The variational expression reveals that an unstable normal mode has a pure exponentially growing time dependence and hence that a normal mode can become unstable in a smooth fashion only through zero frequency. It is shown that no instabilities can set in through zero-frequency modes along smooth sequences of models with fewer high-energy stars than low-energy stars.

151.002 **Resonant excitation of motion perpendicular to galactic planes.** J. Binney.
Mon. Not. R. Astron. Soc., Vol. 196, 455 - 467 (1981).
  Resonant coupling between oscillations perpendicular to the equatorial plane of a galaxy and periodic changes in the force towards the Galactic Centre can cause stars to move far from the plane. The variation of the force to the Galactic Centre may be caused by a non-axisymmetric potential, but instability is possible even in the axisymmetric case. The author discusses the circumstances under which these instabilities occur in highly flattened potentials and in nearly spherical systems.

151.003 **On the wind-up of galactic warps.**
A. H. Nelson.
Mon. Not. R. Astron. Soc., Vol. 196, 557 - 565 (1981).
  The implications of the existence of fast and slow corrugation waves in a galactic disc for the wind-up of warps is considered. The propagation of corrugation wavefronts is calculated using the pattern velocities derived from a local approximation dispersion relation, and it is shown that slow corrugation waves may exhibit persistent open tilts and warps in certain regions of a galaxy.

151.004 **Scontri mareali fra galassie.** M. Fulle.
Astronomia, N. 2, p. 28 - 38 (1981).

151.005 **Possible South Polar Cap group moving stars.**
P. K. Lu.
Chinese J. Phys., Vol. 18, 119 - 127 (1980). – Abstr. in Phys. Abstr., Vol. 84, Abstr. 71260 (1981).

151.006 **Lifetime of molecular clouds and spiral structure.**
J. M. Huntley, H. Gerola.
Astrophys. J., Lett., Vol. 248, L69 - L72, plates L7 - L9 (1981).
  The authors investigate the effects of finite lifetimes and varying birthsite distributions on the dynamics of ballistic "clouds" in a barred galaxy by combining two-dimensional n-body calculations with numerical gas dynamics. The authors find that, regardless of the cloud birthsite distribution, for cloud lifetimes comparable to a galactic rotation period, trailing spiral arms with spurs develop in the distribution of clouds. This result implies that sufficiently long-lived giant molecular clouds need not be born at the locations of density wave shocks in order to produce the spiral structure observed in disk galaxies.

151.007 **The central densities of spherical stellar systems with anisotropic velocity dispersions.**
F. D. A. Hartwick.
Publ. Astron. Soc. Pacific, Vol. 93, 279 - 280 (1981).
  N-body collapse calculations by various authors show evidence for an appreciable velocity anisotropy in the outer parts of the system. Using a simple stellar dynamical model of Ogorodnikov, the author considers the effects of this anisotropy on the central density obtained from the core-fitting technique of King and Minkowski. The result suggests that the central densities of clusters of galaxies may be ~ 20% less than those calculated without taking the anisotropy into account.

151.008 **Vector equation of the motion of centroids in stellar systems and some of its consequences.**
K. F. Ogorodnikov.
Pis'ma Astron. Zh., Tom 7, 570 - 571 (1981). In Russian. English translation in Soviet Astron. Lett., Vol. 7.
  By transforming three scalar Jeans equations of motion of centroids within stellar systems into a single vector equation, the dependence is derived of the second order peculiar velocity momenta tensor on the logarithm of stellar density and on the variation of the normalized peculiar velocity distribution function.

151.009 **Simulations of clusters of galaxies.**
E. M. Malumuth, D. O. Richstone.
Bull. American Astron. Soc., Vol. 13, 506 (1981). – Abstract.

151.010 **Numerical experiments in collisions of disk galaxies.**
B. F. Smith, R. H. Miller.
Bull. American Astron. Soc., Vol. 13, 517 (1981). – Abstract.

151.011 **Numerical experiments on the clustering of galaxies.**
R. H. Miller, B. F. Smith.
Bull. American Astron. Soc., Vol. 13, 533 (1981). – Abstract.

151.012 **Coherent galactic oscillations and global spiral patterns.** P. E. Seiden, L. S. Schulman.
Bull. American Astron. Soc., Vol. 13, 538 (1981). – Abstract.

151.013 **Hierarchical clustering in astrophysical systems.**
J. G. Jernigan.
Bull. American Astron. Soc., Vol. 13, 539 (1981). – Abstract.

151.014 **Riemann-like galactic bars.**
P. O. Vandervoort, D. E. Welty.
Bull. American Astron. Soc., Vol. 13, 568 (1981). – Abstract.

151.015 **Is there a unique galactic rotation curve?**
P. Pişmiş.
Bull. American Astron. Soc., Vol. 13, 568 - 569 (1981). Abstract.

151.016 **Experimental test of dynamical friction acting on a galaxy.** R. H. Miller, B. F. Smith.
Bull. American Astron. Soc., Vol. 13, 569 (1981). – Abstract.

151.017 **Stability of contracting pancakes.**
P. L. Palmer.
Mon. Not. R. Astron. Soc., Vol. 197, 167 - 178 (1981).
The stability of an isothermal, contracting disc to the growth of radial perturbations is examined. The analysis is applied to gaseous protoclusters of galaxies (pancakes) and the mass of the fragments is calculated in terms of the total mass of the protocluster and its density contrast at recombination. It is found that the fragment masses lie in a very narrow range for all initial conditions, and have comparable masses to galaxies.

151.018 *N*-body **simulations of disc–halo galaxies: isolated systems, tidal interactions and merging.**
O. E. Gerhard.
Mon. Not. R. Astron. Soc., Vol. 197, 179 - 208 (1981).
Collisions of galaxies are studied through numerical simulations of 250-particle systems. Each model galaxy consists of a flat disc, an extended live halo and a core. The evolution of such systems and possible relaxation effects are investigated in detail.

151.019 **The macroscopic dynamics of elliptical galaxies.**
J. Binney.
The structure and evolution of normal galaxies, (see 012.010), p. 55 - 66 (1981).
Contents: Equations of macroscopic dynamics. Application of the tensor virial theorem. Applications of the equations of stellar hydrodynamics.

151.020 **Galaxy mergers.** S. Tremaine.
The structure and evolution of normal galaxies, (see 012.010), p. 67 - 84 (1981).
Contents: Introduction. Mergers of galaxies of comparable mass. Mergers of satellite galaxies. Dynamical friction. The validity of the dynamical friction formula. Numerical techniques. Mergers in rich clusters. Do merged spirals make ellipticals? Conclusions.

151.021 **What amplifies the spirals?** A. Toomre.
The structure and evolution of normal galaxies, (see 012.010), p. 111 - 136 (1981).
Contents: Some remarkable responses. Nature of the conspiracy. Fate of a leading wave packet. Global modes of a Gaussian disk. What does it all mean? T/W and all that. Tidal waves after all?

151.022 **Chemical evolution of normal galaxies.**
B. E. J. Pagel.
The structure and evolution of normal galaxies, (see 012.010), p. 211 - 237 (1981).
Contents: Introduction and overview. Survey of the observational data. The standard or "cosmic" abundance distribution. Abundances in stars. Abundances in H II regions and supernova remnants. Ingredients of chemical evolution models. Initial conditions. Endproducts of stellar evolution. The initial mass function (IMF). The rate of star formation (SFR). Supplementary assumptions. Some actual models. The instantaneous recycling (IR) approximation. The simple model and its success: irregular galaxies. The simple model and its failure: the solar neighbourhood. Dynamical collapse models. Sketch of a self-consistent model for the solar neighbourhood. Abundance gradients and mass-metallicity relation in galaxies. Conclusion.

151.023 **Extended maximum likelihood analysis of apparent flattenings of S0 and spiral galaxies.**
S. Okamura, B. Takase, M. Hamabe, Y. Nakada, K. Kodaira.
Publ. Astron. Soc. Japan, Vol. 33, 223 - 245 (1981).
Apparent flattenings of S0 and spiral galaxies compiled by Sandage and van den Bergh, and those listed in the Second Reference Catalogue are analyzed by means of the extended maximum likelihood method which was recently developed in the information theory for statistical model identification. Emphasis is put on the possible difference in the distribution of intrinsic flattenings between S0's and spirals as a group, and on the apparent disagreements present in the previous results.

151.024 **Plane-parallel magnetic fields in a thin galactic gaseous disk in random motion.**
M. Fujimoto, T. Sawa.
Publ. Astron. Soc. Japan, Vol. 33, 265 - 271 (1981).
A Fokker-Planck-type equation for describing magnetic fluxes in the gaseous component in random rising and falling motion in the galactic disk is solved. If weak fields exist in the halo and if they can diffuse into the disk across its interface, the magnetic fields may be maintained stronger in the disk without recourse to a dynamo action. The authors stress that the dynamo is rather necessary in the halo of a larger extent.

151.025 **Model of the large-scale gravitational field of galaxies.**
S. A. Kutuzov, L. P. Osipkov.
Vestn. LGU, 1981, No. 1, p. 99 - 105. In Russian. – Abstr. in Ref. zh., 51. Astron., 7.51.645 (1981).

151.026 **The generality of our local environment.**
D. P. Cox.
The phases of the interstellar medium, (see 012.018), p. 195 - 200 (1981).

151.027 **L'évolution des galaxies normales.** L. Martinet.
Histoire de l'univers, (see 003.007), p. 287 - 312 (1980).
Contents: La formation des galaxies. L'apport de la dynamique stellaire. Le rôle du gaz. Modèles de formation des galaxies. La structure spirale des galaxies. L'évolution chimique des galaxies. Les interactions gravitationnelles entre galaxies.

151.028 **Asymmetry in energy transfer in a collision of galaxies.** F. Ahmed, S. M. Alladin.
Bull. Astron. Soc. India, Vol. 9, 40 - 59 (1981).
The time dependence of binding and translational energies of a pair of galaxies during head-on collisions has been studied with impulsive approximation. The two galaxies are assumed to be identical with mass distribution represented by that of a polytrope of index $n = 4$. The results indicate that more than 80 per cent of the increment in the binding energy takes place when the centres of the two galaxies are separated by a distance less than three times the median radius of a galaxy. The change in the binding energy is asymmetric with respect to the distance of closest approach, the increment in the second half of the collisions being about double that of the first half.

151.029 **Global spiral density waves in disk galaxies.**
A. Ambastha, R. K. Verma.
Bull. Astron. Soc. India, Vol. 9, 80 (1981). – Abstract.

151.030 **Tidal effects of our galaxy on a globular cluster passing through its nucleus.**
T. M. Singh, K. S. Sastry.
Bull. Astron. Soc. India, Vol. 9, 80 (1981). – Abstract.

151.031 **The formation of a ring structure during a head-on collision between a disk and a spherical galaxy.**
T. K. Chatterjee.
Bull. Astron. Soc. India, Vol. 9, 81 (1981). – Abstract.

151.032 **Tidal distortions of a close pair of galaxies from stellar orbits.**   P. V. Subrahmanyam.
Bull. Astron. Soc. India, Vol. 9, 81 - 82 (1981). — Abstract.

151.033 **Galactisch kannibalisme.**   T. Dethier.
Zenit, 8. Jaarg., 418 - 422 (1981).

151.034 **Star formation and evolution of galaxies.**   H. Gerola.
Mem. Soc. Astron. Italiana, Vol. 52, (see 012.019), 67 - 73 (1981).

151.035 **Formation of galaxies through hierarchic fragmentation.**   J. L. Sérsic.
Mem. Soc. Astron. Italiana, Vol. 52, (see 012.019), 75 - 85 (1981).
A model for hierarchic fragmentation is proposed, which describes the global properties of galaxies.

151.036 **N-body simulations of instantaneous mass loss during dissipationless collapse.**
R. G. Carlberg, F. D. A. Hartwick.
Astron. J., Vol. 86, 1410 - 1413 (1981).
N-body collapse calculations with and without mass loss have been carried out to determine whether instantaneous mass loss affects the structure of the remaining relaxed bound system. The two main results of these calculations are: (1) that amount of mass which can be lost yet still leave an appreciable bound remnant is much larger than that predicted by simple analytical models; and (2) to within the statistical uncertainties the object remaining after mass loss possesses virtually identical structural and dynamical characteristics to that of a relaxed object collapsed without mass loss.

151.037 **Some problems with the evolution of gas in clusters of galaxies.**   M. Nepveu.
Astron. Astrophys., Vol. 101, 362 - 364 (1981).
Mixing of dynamically swept galaxy and cluster gas, so as to render one-fluid calculations completely reliable, is sufficient only if the flow Mach number is much smaller than one. The gas sweeping rate from a galaxy is connected with a number of cluster and dynamical parameters, and the one dimensional gas dynamic equations governing cluster gas evolution are rederived. The outcome of numerical experiments is presented and discussed for illustration.

151.038 **The density response of a stellar disk to growing oval mass distributions. The exponential mass model in first order epicyclic approximation.**
D. Polzin, K. O. Thielheim.
Astron. Astrophys., Vol. 101, 409 - 416 (1981).
The authors calculate the shape of the density response of a not self-gravitating stellar component of a model galaxy to an imposed rotating, growing oval density perturbation. They assume exponential mass distributions for the axisymmetric disk and the oval perturbation respectively and investigate the form of the density response as dependent on the model parameters involved.

151.039 **A "symmetrical" kinematical model for elliptical galaxies. Application to the edge on bulge of M 31.**
G. Monnet, M. Rosado.
Astron. Astrophys., Vol. 102, 175 - 177 (1981).
The authors have derived a kinematical model for elliptical galaxies, under the assumptions of cylindrical symmetry of the residual velocities and the absence of third order velocity asymmetries. It is shown to be in correct agreement with known experimental data on the bulge of M 31, and to give an M/L ratio very similar to that computed for very different assumptions.

151.040 **The effects of resonances near corotation in barred galaxies.**   G. Contopoulos.
Astron. Astrophys., Vol. 102, 265 - 278 (1981).
The author studies the effects of higher order resonances in the corotation region. In this region there is an infinity of resonant families of "simple" periodic orbits closing after one revolution around the center, and n epicyclic oscillations. In the axisymmetric case these families branch off the family of circular periodic orbits x_1. When a bar is added the resonant families either branch off the (perturbed) family x_1 (n = odd), or they are produced by breaking the family x_1 (n = even).

151.041 **Formal aspects of possible hierarchies within a stellar system.**   J. B. Nugeyre, P. Bouvier.
Celestial Mech., Vol. 25, 51 - 64 (1981).
A formalism is proposed, in order to study the hierarchical structures observed in many small stellar systems. Following some descriptive terminology, it is shown that the choice of suitable hierarchical coordinates allows one to keep a simple form of the equations of motion. When the structure is of binary type, every subsystem corresponds to a single equation where the terms describing the two-body approximation for the internal motion of the system are easily separated from those pertaining to the influence of the other subsystems. This type of structure appears likely to present, for small systems, a necessary condition of stability.

151.042 **Chemical evolution with inhibited star formation rate.**   R. Caimmi.
Astrophys. Space Sci., Vol. 79, 87 - 106 (1981).
In this paper a simple model of chemical evolution is proposed, according to which active phases of star formation occurred, each followed by a quiescent phase where the formation of stars with $m \lesssim 1 \, m_\odot$ was totally inhibited. The contraction of the gaseous component of the system was not stopped during quiescent phases, according to observed distributions of chemical and dynamical parameters in objects of different populations. The author assumes that: (i) star formation is inhibited during a given active phase, at a rate which is connected to the number of supernovae born at that time; (ii) total mass is conserved in the unit comoving volume; (iii) instantaneous recycling approximation holds.

151.043 **Dynamics of galaxies and evolution of stellar populations.**   L. Martinet.
Ann. Physique, Vol. 6, (see 012.026), 73 - 86 (1981) = Publ. Obs. Genève, Sér. A, Fasc. 84. In French.
Review of recent theoretical and observational works on the dynamics of galaxies and possible influence on the evolution of stellar populations.

151.044 **Elliptical and lenticular galaxies evolution.**   L. Vigroux.
Ann. Physique, Vol. 6, (see 012.026), p. 183 - 190 (1981). In French. — Abstr. in Phys. Abstr., Vol. 84, Abstr. 94504 (1981).

151.045 **Some fluid-dynamical problems in galaxies.**   C. C. Lin, W. W. Roberts, Jr.
Ann. Rev. Fluid Mech., Vol. 13, (see 003.011), p. 33 - 55 (1981). — Abstr. in Phys. Abstr., Vol. 84, Abstr. 94536 (1981).

151.046 **The effects of induced star formation on the evolution of the galaxy. I. One-zone models.**
S. N. Shore.
Astrophys. J., Vol. 249, 93 - 98 (1981).
The effect of stellar-induced star formation is explored in the context of a one-zone model galaxy. The formalism is similar to that employed in standard models for the chemical evolution of such a system. It is shown that under a variety of assumptions, it is possible to obtain bursts of star forma-

tion for two- and three-phase models (stars, diffuse gas, and clouds). One of the controlling factors during such bursts is the rate of infall of halo and extragalactic material. The model can be generalized to include also the effects of density wave-induced collapse of clouds and formation of clouds from diffuse gas due to spiral arm shocks. A comparison is drawn with the behavior observed in stochastic models for flat rotation curve galaxies.

151.047 Investigation of the stability of general models of collisionless spherically symmetric systems.
V. L. Polyachenko, I. G. Shukhman.
Astron. Zh., Tom 58, 933 - 948 (1981). In Russian. English translation in Soviet Astron., Vol. 25, No. 5.

A matrix equation is suggested for investigating the large-scale stability of general spherical collisionless systems. The equation derived is used for a stability analysis of Idlis' one-parametric series of distribution functions. It is shown that the stability condition restricts strongly the permissible degree of anisotropy of the stellar velocity distribution.

151.048 Merging instability in groups of galaxies.
P. Carnevali, A. Cavaliere, P. Santangelo.
Astrophys. J., Vol. 249, 449 - 461 (1981).

The authors present the results and the interpretation of N-body simulations for the dynamical evolution of galaxy groups. They find that, for group sizes not more than about 20 galactic radii, slow hyperbolic encounters give rise to abrupt and extensive merging associated with overall contraction of the group as a whole. This leads ultimately to the formation of a large central object.

151.049 Spiral structure of galaxies — two alternatives.
L. S. Marochnik, A. A. Suchkov.
Astrophys. Space Sci., Vol. 79, 337 - 343 (1981).

At present the wave theory of the spiral structure of the galaxies includes two alternative viewpoints. Two types of spiral waves — short-wave and long-wave — can be excited. According to Lin and Shu (1964) the short-wave mode K_∞ is responsible for the spiral structure of the galaxies and, according to Marochnik et al. (1972), the long-wave mode (K_0). The theoretical and observational evidences in favour of the model of Marochnik et al. (1972) are given.

151.050 Dissipation and the global properties of elliptical galaxies. C. Struck-Marcell.
Mon. Not. R. Astron. Soc., Vol. 197, 487 - 496 (1981).

It is proposed that the observed relation between the luminosities and velocity dispersions of elliptical galaxies is the result of dissipative processes that are important during their formation. This hypothesis is illustrated with schematic models of galaxy formation via mergers of protogalactic subsystems. The models reproduce the observed correlations between luminosity, metallicity, and velocity dispersion in elliptical galaxies. Possible correlations with flattening are also discussed.

151.051 Velocity dispersions in the bulges of spiral and S0 galaxies. II. Further observations and a simple three-component model for spiral galaxies.
B. C. Whitmore, R. P. Kirshner.
Astrophys. J., Vol. 250, 43 - 54 (1981).

The authors have obtained velocity dispersions for 24 galaxies in the Virgo cluster to supplement their earlier results. A 2000 channel intensified Reticon scanner has again been used on the 1.3 m telescope of McGraw-Hill Observatory, and a Fourier quotient technique has been employed to yield dispersions. The authors have confirmed their earlier result that spiral bulges exhibit a relation between total luminosity and velocity dispersion with the form $L \propto \sigma^4$, but with velocity dispersions that are 17 ± 8% smaller than elliptical galaxies at the same absolute magnitude.

151.052 Orbital motion of the head-tail radio galaxy IC 708.
J. P. Vallée, A. H. Bridle, A. S. Wilson.
Astrophys. J., Vol. 250, 66 - 78 (1981).

The Very Large Array has been used to map the unusual head-tail radio galaxy IC 708 in the cluster of galaxies Abell 1314 with resolutions from 0".5 to ~3".3 at 4.89 GHz. The new data strongly suggest that the unusual structure of the source is due to a gravitational interaction between IC 708 and its neighbor IC 709 while both orbit the center of Abell 1314. The shapes of the radio trails and some details of the variation of intensity along them can be explained by projection effects arising from the orbital motion of IC 708 and by the effects of the variation of the orbital velocity of IC 708 on the motions of its ejecta. The authors have made numerical simulations of the orbiting head-tail system incorporating various models of the ejecta dynamics. The most satisfactory model for the structure is one in which a pair of continuous supersonic jets bends behind IC 708 under the ram pressure of the intracluster medium. Furthermore, the distributions of intensity and polarization over the trails of IC 708 resemble those observed in straight radio jets in several low luminosity radio galaxies. It appears that external pressures have bent similar jets in IC 708 through almost 90° without disrupting them.

151.053 Dynamics of non-spherical elliptical galaxies. I. Case of no rotation, isotropic velocity distribution and no damping. K. Aizu.
Prog. Theor. Phys., Vol. 65, 1858 - 1869 (1981). — Abstr. in Phys. Abstr., Vol. 84, Abstr. 102237 (1981).

151.054 General solution of Poisson's equation in three dimensions for dish-like galaxies.
Y. Tong, X.-t. Zheng, Q.-h. Peng.
Acta Astron. Sinica, Vol. 22, 285 - 292 (1981). In Chinese.

151.055 N-body simulations of secondary infall. I. Formation of mass profile in galactic halos.
A. Dekel, M. Kowitt, J. Shaham.
Astrophys. J., Vol. 250, 561 - 572 (1981).

The cosmological evolution of a flat bound density perturbation around a central "galaxy" is studied by means of N-body simulations with 250–500 particles. The results confirm previous theoretical studies (Gott) showing that under certain initial conditions the maximum expansion radii could be almost proportional to the mass contained within these radii. Furthermore, as suggested by Gunn, the shape is roughly preserved during the subsequent infall, in which self-similar relaxation occurs and violent relaxation is found to be ineffective. Hence, the relaxed systems can indeed show almost flat rotation curves ($M/R \propto R^{-0.25}$). Failures to obtain similar results using spherical shell models are discussed.

151.056 Diffusion of stars in a harmonic potential.
H. E. Kandrup.
Astrophys. Space Sci., Vol. 80, 443 - 455 (1981).

In a simple approximation, the evolution of a stellar system can be described in terms of the solutions to a diffusion equation for motion in a harmonic potential. This paper presents a discussion and characterization of the normal modes for this equation. These solutions are of particular interest in that they provide a simple example of the interplay between dynamical and relaxation phenomena.

151.057 The role of magnetic fields in extragalactic astronomy. J. H. Piddington.
Astrophys. Space Sci., Vol. 80, 457 - 471 (1981).

The evidence is reviewed for a universal magnetic field of strength $10^{-9} - 10^{-8}$ G; it has been extended to include the diffuse fields of galactic clusters and the extensive magnetic halos of spiral galaxies.

151.058   **Inner Lindblad resonance in galaxies. Nonlinear theory. IV. Self-consistent bars.**
G. Contopoulos.
Astron. Astrophys., Vol. 104, 116 - 126 (1981).

The self-consistency of a galaxy requires that the response density should be equal to the imposed density. A review of the various approaches to this problem is made. The author derives the non-linear self-consistency equation in the case of a bar and its linearized version. He finds that in all linear solutions the amplitude decreases considerably outwards near the inner Lindblad resonances.

151.059   **A hydrodynamic interaction between the galactic disk and halo, and its application to the origin of warping gaseous disks.**   Y. Yoshii, M. Fujimoto.
Astron. Astrophys., Vol. 104, 142 - 147 (1981).

A hydrodynamical interaction is considered between the disk and halo of a galaxy in differential rotation. A pressure jump arises in the halo gas across the disk plane, which exerts a force on the disk to go up or down. The condition is derived from the disk-halo system that the bending oscillation is unstable. The amplitude grows towards the outer parts of the disk with spiral pattern winding. The characteristics of the unstable modes may account for the large-scale asymmetric flexure of isolated galaxies.

151.060   **Interacting galaxies.**
P. Stewart.
Investigating the universe, (see 003.014), p. 385 - 422 (1981).

The paper is divided into two parts: (1) a short review of the literature on interacting galaxies, (2) a detailed account of several 20,000 body computer simulations of rotating galaxies in direct collision.

151.061   **Techniques for simulating galactic collisions.**
R. A. James.
Investigating the universe, (see 003.014), p. 423 - 441 (1981).

Numerical simulation of the evolution of two or more interacting galaxies requires a fine spatial resolution to resolve detail in individual systems. The much larger regions separating the galaxies must be treated in the same detail with current fast Fourier transform based potential solvers. The paper describes a mesh segmentation technique which permits to solve very large potential problems and to concentrate the effort on the regions where detailed potential maps are required. The author discusses the need for this method in simulating single disc galaxies, and in the study of galactic collisions.

151.062   **Luminosity evolution of quasars and active galaxies: theoretical models of the evolving mass supply rate.**
S. L. W. McMillan, A. P. Lightman, H. Cohn.
Astrophys. J., Vol. 251, 436 - 445 (1981).

Leading models for the power source of quasars (QSOs) and active galactic nuclei (AGN) involve gravitational energy release of gas. The authors propose that a comparison of theoretical models for the gas supply mechanism to the observed luminosity and luminosity evolution of QSOs and AGN may rule out or constrain some such models. In particular, they calculate the gas liberation rate $\dot{M}(t)$ resulting from (1) physical stellar collisions and (2) tidal disruptions in a dense star system surrounding a massive black hole, taking into account the dynamical evolution of the stellar system. The main conclusions are that both mechanisms (1) and (2) lead to an eventual decreasing function $\dot{M}(t)$; that for a total mass of $10^8 M_\odot$ or more, a peak value of $\dot{M}(t)$ exceeding $\sim 0.1 M_\odot \, yr^{-1}$ requires a stellar system whose dynamical evolution is dominated by physical collisions; and that $\dot{M}(t)$ in a collision-dominated system increases until the mass in gas liberated is comparable to the remaining mass in stars, after which $\dot{M}(t)$ decreases and attains an asymptotic form $\dot{M}(t) \propto t^{-2}$.

151.063   **The velocity dispersion among galaxies.**
A. R. Rivolo, A. Yahil.
Astrophys. J., Vol. 251, 477 - 484 (1981).

The galaxies in the Revised Shapley-Ames redshift catalog, which is complete to $B \approx 13$, are used to determine the velocity dispersion in the local galaxy field ($v < 4000 \, km \, s^{-1}$). By studying the velocity differences of neighboring pairs of galaxies, the authors find for the one dimensional velocity dispersion per galaxy $\sigma = 70 \pm 10 \, km \, s^{-1}$. This velocity dispersion is found in pairs of galaxies irrespective of how isolated they are, and whether or not they are each other's nearest neighbors. The interpretation of isolated galaxy pairs as binaries, whose dynamics is dominated by the two-body force, is therefore questioned.

151.064   **Globular clusters and galaxy mergers.**
W. E. Harris.
Astrophys. J., Vol. 251, 497 - 500 (1981).

Recent observations of globular cluster populations in many nearby galaxies, combined with current theory concerning galaxy mergers, suggest two principal differences between globular cluster systems located in different environments: (1) Elliptical galaxies found in small groups have noticeably fewer globular clusters per unit galaxy luminosity than do ellipticals in the Virgo cluster; and (2) Major spiral galaxies contain fewer globular clusters per unit spheroidal luminosity than do ellipticals, regardless of environment. These results are consistent with the views that galaxy mergers occur more frequently for ellipticals outside rich clusters, and that the stellar populations in the halos of disk galaxies may be fundamentally different from those in elliptical galaxies.

151.065   **Dynamics of luminous galaxies.**
E. M. Malumuth, R. P. Kirshner.
Astrophys. J., Vol. 251, 508 - 517 (1981).

The authors have used a Fourier quotient method to obtain velocity dispersions in 19 luminous E and cD galaxies to clarify the dynamical difference between these morphological types. They find that the E galaxies follow a relation between luminosity and velocity dispersion of the form $L \propto \sigma^4$. The brightest cluster galaxies, some of which are cD's, have velocity dispersions no larger than $400 \, km \, s^{-1}$ and are generally more luminous than E galaxies at the same velocity dispersion by 0.52 mag. Mass-to-light ratios determined for E galaxies averaged $M/L_v = 6.5 \pm 0.7$, while the brightest cluster members averaged $M/L_v = 9.9 \pm 0.8$. They also found that the quantity $M_\sigma - M_v$, the difference in magnitudes between a brightest cluster member and an elliptical with the same velocity dispersion, is correlated with the galaxy's core radius and central surface brightness in a manner consistent with galaxy merger models.

151.066   **Galaxy mergers and active galactic nuclei.**
N. Roos.
Astron. Astrophys., Vol. 104, 218 - 228 (1981).

The tidal disruption rate of stars near a central black hole in a galactic nucleus may be considerably enhanced during a merger of a galaxy with a smaller companion due to scattering of stars into loss-cone orbits by the perturbing gravitational field of the intruding galaxy. Using a merging rate consistent with observational data and with cosmological simulations the fraction of galaxies which are active at some level of total luminosity is estimated. The agreement between the model prediction and observed luminosity functions of active galactic nuclei is encouraging. If both merging galaxies contain a central black hole, a binary black hole system will be formed at the centre of the combined galaxy.

151.067   **Non-stationary viscous gaseous disks in stellar systems.**   W. Glatzel.
Proceedings of the 5th Göttingen-Jerusalem-Symposium on Astrophysics, (see 012.041), p. 157 - 161 (1981).

151.068 **Galactic dynamics and gravitational plasmas.**
C. C. Lin, G. Bertin, with a supplementary note by
J. Haass.
Plasma astrophysics, (see 012.042), p. 191 - 205 (1981).

The authors attempt to formulate a dynamical classification of spiral galaxies. Some realistic galaxy models are found to support discrete unstable spiral modes. Modes can be unstable or stable, and they can be either normal spiral modes or barred spiral modes depending on the properties of the equilibrium model. A tentative proposal is that these modes correspond to normal spiral galaxies (SA) and transition barred galaxies (SAB) and some barred galaxies (SB). Many issues related to such a proposal are discussed in the paper.

151.069 **Instabilities in inhomogeneous collision-free selfgravitating stellar systems.** W. Seboldt.
Plasma astrophysics, (see 012.042), p. 385 - 389 (1981).

Stability criteria for collision-free selfgravitating stellar systems are discussed with applications to elliptical galaxies. It is concluded that rotationally symmetric systems which are either sufficiently oblate or prolate are unstable against ringlike and sausagelike perturbations respectively.

151.070 **Irregular force as a random process.**
G. K. Bakhman.
Nauchn. Inf., Vyp. (No.) 47, (see 003.016), p. 71 - 100 (1981).
In Russian.

Properties of an irregular gravitational force acting on some control body with peculiar velocity $\vec{V}$ have been considered. The uniform field of gravitating bodies with given mean number density and distribution law of mass and peculiar velocities of field bodies has been accepted as a source of this force. The irregular force produced by such a field of gravitating bodies is a three-dimensional random process.

151.071 **Model of a source of irregular forces for numerical experiments on computers.**
G. K. Bakhman.
Nauchn. Inf., Vyp. (No.) 47, (see 003.016), p. 101 - 110 (1981).
In Russian.

A mathematical model of the source of irregular gravitational forces realized on computers and meant for numerically-experimental investigation of various dynamical processes caused by these forces has been worked out. As the source of irregular forces the field of gravitating bodies with any chosen number density and any chosen distribution law of mass and peculiar velocities of the field bodies has been accepted.

151.072 **Relaxation process caused by the action of an irregular force.** G. K. Bakhman.
Nauchn. Inf., Vyp. (No.) 49, (see 003.017), p. 135 - 167 (1981).
In Russian.

The change of the velocity of a body caused by the gravitational influence of the surrounding bodies (the relaxation process) is analysed by means of a new mathematical formalism. This formalism is more adequate to long-range nature of the gravitational forces than the traditional mathematical formalism based on the approximation of the problem by two-body "collision".

151.073 **Rotation of the galaxies.** A. Serrano.
Rev. Mexicana Fis., Vol. 27, 361 - 391 (1981). In Spanish. – Abstr. in Phys. Abstr., Vol. 85, Abstr. 6674 (1982).

151.074 **Escape velocities of interacting spherical galaxies.**
K. S. Sastry, S. M. Alladin.
Bull. Astron. Soc. India, Vol. 9, 249 - 254 (1981).

The dependence of the escape velocity of a pair of interacting identical spherical galaxies on their separation is studied using polytropic models for the galaxies.

151.075 **Examples of angular rotation curves of galaxies.**
A. P. Fairall, D. L. Block.
Mon. Notes Astron. Soc. South. Africa, Vol. 40, 40 - 42 (1981).

Angular rotation curves best illustrate differential rotation. Spiral arms can be seen to be regions of least differential rotation.

151.076 **Some comments on the pitch angle in spiral structure.** D. L. Block, A. P. Fairall.
Mon. Notes Astron. Soc. South. Africa, Vol. 40, 43 - 46 (1981).

The empirical behaviour of the pitch angle, taken from observational evidence, is discussed – in particular its tendency towards $\sim 73°$. Branching may be a possible method of regenerating the spiral pattern.

151.077 **I. Normal galaxies.** A. V. Zasov.
Problems of extragalactic astronomy, (see 003.018), p. 3 - 47 (1981). In Russian.

151.078 **II. Instationary phenomena in galaxies.**
Eh. A. Dibaj.
Problems of extragalactic astronomy, (see 003.018), p. 48 - 82 (1981). In Russian.

151.079 **The mass-angular momentum-diagram of astronomical objects.** P. Brosche.
Cosmology and gravitation, (see 012.051), p. 375 - 382 (1980).

151.080 **Instability of a nonlinearly pulsating model of a stellar system. I. The Einstein sphere.**
V. A. Antonov, S. N. Nuritdinov.
Astron. Zh., Tom 58, 1158 - 1166 (1981). In Russian.
English translation in Soviet Astron., Vol. 25, No. 6.

A nonequilibrium model of a stellar system as a pulsating Einstein sphere with arbitrary amplitude of oscillation is constructed. Its stability with respect to small nonsymmetric perturbations is investigated. A system of equations determining the frequency of surface oscillations is obtained.

151.081 **On the beam mechanism of generation of spiral structure.** A. G. Morozov.
Astron. Tsirk., No. 1137, p. 1 - 3 (1980). In Russian.

151.082 **Vlasov simulations of stellar systems: infinite homogeneous case.** T. Fujiwara.
Publ. Astron. Soc. Japan, Vol. 33, 531 - 540 (1981).

The splitting scheme for numerical integration of the Vlasov equation is tested on one-dimensional problems with spatially periodic boundary conditions. The problems considered are (1) damping of small and large amplitude disturbances and (2) weak and strong Jeans instabilities, in an infinite homogeneous medium. These nonlinear phenomena are reproduced with high accuracy, efficiency, and stability. This suggests that the splitting scheme is applicable to simulations of more realistic stellar systems such as stellar disks and spherical stellar systems.

151.083 **Vlasov simulations of stellar disks. I. Axisymmetric case.** Y. Watanabe, S. Inagaki, M. T. Nishida,
Y. D. Tanaka, S. Kato.
Publ. Astron. Soc. Japan, Vol. 33, 541 - 565 (1981).

Axisymmetric stellar disks are simulated by solving a set of the Vlasov and Poisson equations. The splitting scheme is used to integrate the Vlasov equation, and the Poisson equation is solved by a biorthogonal set of density and potential. The numerical code is checked and found to be sufficiently accurate. In particular, this is an excellent method to study the weak nonlinear effects.

151.084 **Vlasov simulations of stellar disks. II. Nonaxisymmetric case.** M. T. Nishida, M. Yoshizawa,
Y. Watanabe, S. Inagaki, S. Kato.

Publ. Astron. Soc. Japan, Vol. 33, 567 - 582 (1981).

The splitting scheme to solve the Vlasov equation is applied to study the time evolution of nonlinear, nonaxisymmetric (bisymmetric) perturbations on flat stellar disks. The gravitational potential for a given density distribution is obtained by solving the Poisson equation in terms of biorthogonal sets of density and potential pairs. The unperturbed equilibrium disks adopted in the simulations are Toomre's No. 1 models with Miyamoto's velocity distribution function. Perturbations imposed are fan-shaped density perturbations. The splitting scheme is found to be applicable to the present problem with sufficient accuracy. The results of calculation show that such large-scale prominent bars as noticed in the N-body simulation do not appear in Vlasov simulations.

151.085 **What is a spiral galaxy?**   K. J. Donner.
Proceedings of the Third Finnish-Soviet Astronomical Symposium, (see 012.057), p. 133 - 140 (1981).

The maintenance of a permanent global spiral pattern has to rely on some quite complex non-linear process involving both stars and gas. It is argued that the simplest scheme is one where a stellar bar gives rise to a spiral in the gas. A physical picture of the instability of a stellar disc towards bar formation is given. This allows one to estimate the angular velocity of the bar.

151.086 **The effects of resonances near corotation in barred galaxies.**   G. Contopoulos.
ESO Sci. Prepr., No. 159, 50 pp. (1981). − Submitted to Astron. Astrophys.

151.087 **Invariant surfaces and orbital behaviour in dynamical systems of 3 degrees of freedom. II.**
G. Contopoulos, L. Magnenat, L. Martinet.
ESO Sci. Prepr., No. 163, 1 + 22 pp. (1981). − Submitted to Phys. D.

151.088 **On the sizes of rings and lenses in disk galaxies.**
E. Athanassoula, A. Bosma, M. Crézé, M. P. Schwarz
ESO Sci. Prepr., No. 178, 23 pp. (1981). − Submitted to Astron. Astrophys.

151.089 **On a general form of the gravitational potential allowing to solve the problem of planar stellar orbits in elliptic integrals.**   G. A. Malasidze.
Soobshch. AN GruzSSR, Tom 102, 333 - 336 (1981). In Russian. − Abstr. in Ref. zh., 51. Astron., 12.51.876 (1981).

151.090 **On a group of conformal transformations of R^3 in stellar dynamics and in inverse kinematic problems of seismism.**   V. P. Golubyatnikov, L. N. Pestov.
Priblizhen. metod. resheniya i vopr. korrektnosti obratn. zadach. Novosibirsk, 1981, p. 35 - 43. In Russian. − Abstr. in Ref. zh., 51. Astron., 12.51.877 (1981).

151.091 **Extra nuclear clues to the origin and evolution of activity in galaxies.**   B. Balick, T. M. Heckman.
Prepr. Steward Obs., No. 352, 2 + 85 pp. (1981).

The authors consider the environment of presently active galaxies in order to understand the possible effects of cluster or group membership on the one hand, and nearby interacting companions on the other. Then, they consider the relationships between galactic activity and internal properties of galaxies such as the galaxy's stellar content, mass, dynamics, structure, and optical luminosity. The authors also examine the distributions, the motions, and the physical and chemical properties of the gas and dust in active galaxies for historical clues.

151.092 **Angular momentum orientation of spiral galaxies.**
T. Yamagata, M. Hamabe, M. Iye.
Ann. Tokyo Astron. Obs., Second Ser., Vol. 18, 164 - 174

(1981).

The distribution of spin-angular momentum orientation of 1650 spiral galaxies in the Second Reference Catalogue of Bright Galaxies is studied through their spiral winding direction. The predominance of "S" galaxies over "reversed S" galaxies is not confirmed to be statistically significant for spiral galaxies except for those of type Sbc-Sc.

151.093 **Determination of the average projection factor in calculating the mass of binary galaxies.**
Y.-q. Mao, D.-s. Ji.
Ann. Shanghai Obs. Acad. Sinica, No. 2, p. 100 - 106 (1980).

151.094 **On the relation between "volume" and "surface" adiabatic indices for gaseous subsystems of flat galaxies.**   S. M. Churilov, I. G. Shukhman.
Astron. Tsirk., No. 1157, p. 1 - 2 (1981). In Russian.

Galactic astronomy.   See Abstr. 003.103.

Galaxies: structure and evolution.
See Abstr. 003.150.

Quasi-equilibrium in collisional systems.
See Abstr. 042.070.

Orbital stability constraints on the nature of planetary systems.   See Abstr. 042.073.

Linear series and "catastrophies" in astronomy.
See Abstr. 061.057.

Disk models with convective viscosity.
See Abstr. 062.001.

On the construction of models of rotating stars and stellar systems.   See Abstr. 062.026.

Effect of a helical magnetic field on the stability of a gravitating cylinder.   See Abstr. 062.082.

Higher order fluid equations for multicomponent nonequilibrium stellar (plasma) atmospheres and star clusters. II. Effects of nonzero relative flow velocities and skewing of velocity distribution functions.   See Abstr. 062.091.

Chemical evolution of galaxies. I. Constraints imposed by the $\Delta Y/\Delta Z$ ratio.   See Abstr. 064.059.

Thermal instabilities in radiatively driven winds− application to emission line clouds of quasars and active galactic nuclei.   See Abstr. 141.118.

Fluid dynamical twisting of the radio jets in 3C 449.
See Abstr. 141.120.

Globular clusters and galaxy mergers.
See Abstr. 154.035.

Thermodynamic equilibrium hard binary models for globular star clusters.   See Abstr. 154.039.

Some numerical experiments concerning the determination of the general velocity field of the Galaxy from proper motions.   See Abstr. 155.001.

Kinematical and chemical evolution of the galactic disk near the Sun.   See Abstr. 155.002.

The post explosion shock propagation in the central region of the Galaxy.   See Abstr. 155.029.

A photometric and kinematic study of the barred spiral galaxy NGC 253. II. The velocity field. See Abstr. 158.008.

Density-wave induced star formation: a model for M81. See Abstr. 158.010.

The systematics of rotation curves and their use as luminosity discriminators. See Abstr. 158.077.

Galaxy dynamics: observations. See Abstr. 158.102.

Evolutionary synthesis of the stellar population in elliptical galaxies. III. Detailed optical spectra. See Abstr. 158.154.

Galactic absorption line coronae. See Abstr. 158.171.

Spectroscopic observations of superthin galaxies. See Abstr. 158.173.

The evolution of flows of stellar mass loss in active galaxies. See Abstr. 158.200.

The extended H I disk of the S0 galaxy NGC 4203. See Abstr. 158.206.

21-cm line studies of spiral galaxies. II. The distribution and kinematics of neutral hydrogen in spiral galaxies of various morphological types. See Abstr. 158.240.

The Magellanic stream and the interacting galaxies. See Abstr. 159.007.

Binary-galaxy-rich clusters of galaxies. See Abstr. 160.045.

Theory of galaxy formation triggered by quasar explosions. See Abstr. 162.044.

Galaxy clustering and the Rubin-Ford effect. See Abstr. 162.050.

The initial fragmentation of collapsing gaseous protoclusters. See Abstr. 162.132.

The formation of galaxies from massive neutrinos. See Abstr. 162.133.

Toward the application of a metric size function in galactic evolution and cosmology. See Abstr. 162.141.

Errata

151.901   Erratum: 'On the development of vorticity and waves in shearing media with preliminary application to the solar nebula' [Astrophys. J., Vol. 243, 1030 - 1039 (1981)]. J. H. Hunter, Jr., K. S. Schweiker. Astrophys. J., Vol. 248, 881 (1981). – See Abstr. 29.151.028.

151.902   Erratum: "A numerical model for a triaxial stellar system in dynamical equilibrium. II. Some dynamical features of the model" [Astrophys. J., Suppl. Ser., Vol. 43, 435 - 455 (1980)] D. Merritt. Astrophys. J., Suppl. Ser., Vol. 46, 115 (1981). – See Abstr. 28.151.049.

152 Stellar Associations

152.001   High-velocity interstellar C IV and Si IV toward two stars in the I Per OB association.
A. P. Phillips, P. M. Gondhalekar.
Mon. Not. R. Astron. Soc., Vol. 196, 533 - 542 (1981).
Blue-shifted interstellar C IV, Si IV and Al III lines have been detected in the *IUE* spectra of the B2Ia stars HD 14818 and HD 14143 which are members of the I Per OB association. The radial velocities and column densities of these high ion state lines are similar toward the two stars. These observations, together with previous optical and radio observations, suggest that one is observing a shell of gas expanding away from the OB association at about 25 km s^{-1}. The most likely explanation is that the expanding shell has been produced by a very large supernova event or a number of normal Type II supernovae, with stellar winds from the hot supergiant stars in the association providing a secondary energy source.

152.002   Spectroscopic studies of stars in Ori OB1 (belt).
H. H. Guetter.
Astron. J., Vol. 86, 1057 - 1068 (1981).
MK spectral classifications are presented for 120 B- and A-type stars in the Belt subgroup of the Ori OB1 association. Combined with other published spectroscopic data, and the assuming absolute magnitudes given by Schmidt-Kaler (1965)

and Lesh (1979), a mean corrected distance modulus of 8.56 ± 0.11 is computed for the eastern region, while for the remainder of the Belt a modulus of 8.20 ± 0.05 is derived. Several B9–A2 subgiants and A7–F0 main-sequence stars were observed in the western and central regions of the Belt. The ages of these stars imply a spread in formation ages for these subgroups from about 10^6 yr to at least 10^7 yr.

152.003   A search for HI towards SY, SZ and TT Phoenicis.
J. C. Cersosimo, E. M. Arnal.
Astrophys. Lett., Vol. 21, 81 - 83 (1981).
A 1° × 1° grid covering 50 square degrees centred at $\alpha = 1^h 35^m$, $\delta = -42°44'.6$ (1950) in Phoenix T1 (Kholopov, 1959) was observed in the 21 cm line. The mean value of N_H as deduced from the radio profiles is 8×10^{19} atoms cm^{-2} over the whole grid. No excess or deficiency of atomic hydrogen associated with the region around the T Tauri stars could be detected. An upper limit of 4 solar masses was obtained for the neutral hydrogen in that region, assuming a distance of 100 parsecs.

152.004   A search for expanding supershells of gas around OB associations.
L. L. Cowie, E. M. Hu, W. Taylor, D. G. York.

Astrophys. J., Lett., Vol. 250, L25 - L29 (1981).

The authors record the results of a search for giant, rapidly expanding shells of gas surrounding OB associations. Two out of thirteen nearby associations (Orion and Carina) show supershells of gas with radii of ~ 100 pc, expansion velocities of ~ 100 km s^{-1}, and ages of ~4×10^5 years. Observations of the newly discovered Carina supershell are presented. They suggest that these features are formed by supernova explosions in the associations. The birthrate of association supernovae then lies between 1.5×10^{-14} pc^{-3} yr^{-1} and 10^{-13} pc^{-3} yr^{-1}.

152.005 **The Vulpecula OB1 complex — some thoughts on the star formation process in a rich association.** D. Turner.
J. R. Astron. Soc. Canada, Vol. 75, 252 - 253 (1981). Abstract.

152.006 **Nucleosynthesis in OB associations and its effect upon cosmic rays, peculiar stars and isotopic anomalies in meteorites.**   D. N. Schramm.
Nukleonika, Vol. 25, 1543 - 1554 (1980). — Abstr. in Phys. Abstr., Vol. 85, Abstr. 3287 (1982).

152.007 **Stellar groups (review).**   V. Straižys.
Bull. Vilnius Astron. Obs., Nr. 57, p. 62 - 75 (1981). In Russian.

A review of investigations of stellar groups with common spatial motion in the Galaxy is given. The stellar groups are an important tool to test stellar evolution and kinematic theories and to calibrate photometric systems in absolute magnitudes and metallicities.

152.008 **OB-associations and stellar streams.** I. N. Latyshev.
Astron. Tsirk., No. 1141, p. 7 - 8 (1980). In Russian.

152.009 **Elements of galactic orbits of six stellar associations.** L. P. Osipkov.
Astron. Tsirk., No. 1153, p. 2 - 4 (1981). In Russian.

152.010 **New member of the T-association T3 Tau.** U. A. Nurmanova.
Astron. Tsirk., No. 1154, p. 4 - 6 (1981). In Russian.

Mass loss from very luminous OB stars and the Cygnus superbubble.   See Abstr. 064.079.

Mass loss rates from O stars in OB associations. See Abstr. 064.080.

An RV digital measuring method applied to Lac OB1: preliminary results.   See Abstr. 111.005.

Wind characteristics of the O7 n star HD 217086 in the Cep OB 3 association.   See Abstr. 112.025.

Sternwinde in der Cygnus OB 2 Assoziation. See Abstr. 112.039.

Infrared photometry of the X-ray stars in Cygnus OB2 (VI Cygni): preliminary results. See Abstr. 113.052.

Photometric behaviour of T Tauri stars in the T-associations T1 and T3 Tau. I. Variations at various wavelengths.   See Abstr. 113.067.

Photometric behaviour of T Tauri stars in the T-associations T1 and T3 Tau. II. U—B, B—V diagram for excess radiation.   See Abstr. 113.068.

Decaying stellar magnetic fields, magnetic braking: evidence from magnetic observations in Orion OB1. See Abstr. 116.018.

Decaying stellar magnetic fields, magnetic braking: evidence from magnetic observations in Orion OB1. See Abstr. 116.033.

Two young stars in L 43.   See Abstr. 121.001.

On the extinction of light toward T Tauri-type stars.   See Abstr. 121.042.

Sequential explosions of supernovae in an OB association and formation of a superbubble. See Abstr. 125.026.

Stellar clustering as induced by a supernova. See Abstr. 125.069.

Superbubbles.   See Abstr. 131.044.

Star formation in M33 and in our Galaxy. See Abstr. 131.161.

La formation des étoiles chaudes dans les galaxies de types avancés.   See Abstr. 131.224.

Untersuchungen an extragalaktischen H II Gebieten unter besonderer Berücksichtigung des Riesen-H II-Gebietes NGC 604 in M33.   See Abstr. 132.051.

High resolution spectrophotometry of the O I line (8446 Å) towards Cyg OB2 No. 12.   See Abstr. 133.013.

Étude de la structure galactique dans une région de la Poupe.   See Abstr. 155.013.

Structure of the magnetic field in the cloud of interstellar matter associated with the Perseus OB2 association.   See Abstr. 156.011.

153 Open Clusters

153.001 **Photographic photometry of the open clusters NGC 2910, NGC 2925, Ru 79 and Ru 82 in Vela II and NGC 6031 in Norma II.** L. Topaktas.
Astron. Astrophys., Suppl. Ser., Vol. 45, 111 - 117 (1981).

Four open clusters have been investigated in the Vela II and Norma II fields by means of RGU three colour photographic photometry and the cluster NGC 2910 in the Vela II field by means of UBV photometry.

153.002 **Blue stragglers in M67.**
O. J. Eggen.
Astrophys. J., Vol. 247, 503 - 506 (1981).

Intermediate band and Hβ observations of blue stragglers in M67 are used to derive the reddening $[E(b-y) = 0.037 \pm 0.022 \, (\sigma)$ mag] and modulus $[9.3 \pm 0.2 \, (\sigma)$ mag] of the cluster. A modulus as large as some recent suggestions, ~9.9 mag, seems unlikely. From a comparison of luminosity functions of the Pleiades, Hyades, and M67, there appears to be as many blue stragglers with mass near twice the solar value as there were original main-sequence stars of this mass in M67. This raises the possibility of red stragglers.

153.003 **The region of NGC 2287 and Cr 121.**
O. J. Eggen.
Astrophys. J., Vol. 247, 507 - 521 (1981).

Two hundred and seventy-five intermediate band and Hβ observations of 135 stars in the regions of the clusters NGC 2287 and Cr 121 are discussed. A luminosity calibration of photometric parameters for late G- to early K-type bright giants and supergiants is introduced. NGC 2287 is found to be 740 pc distant, very little reddened, 10^8 years old. Cr 121 is 1.17 kpc distant, also very little reddened, has an age near 1.5×10^6 years.

153.004 **A photoelectric investigation of Ap-stars in open clusters. II. NGC 6475.**
H. M. Maitzen, M. Floquet.
Astron. Astrophys., Vol. 100, 3 - 6 (1981).

Photoelectric Δa-photometry for the detection of Ap-stars (Maitzen, 1976) was obtained for the old galactic cluster NGC 6475. The three known Silicon Ap-stars could be confirmed, two other peculiar stars were found in the spectral range A5 and earlier. A reddening value E(b–y)= 0.050 was obtained from this type of photometry. The slight discordance with the colour excess of Snowden's (1976) photometry is discussed. The main reasons for this difference are an evolutionary effect in the group of B-type stars and the application of a preliminary calibration of the uvbyß-system.

153.005 **Probable members of the Orion nebula cluster.**
O. S. Shulov, E. N. Kopatskaya.
Tr. Astron. Obs., Leningrad, Tom 36 = Uch. Zap. Leningr. Univ., No. 402 = Ser. mat. nauk, Vyp. 58, 49 - 73 (1981). In Russian.

The membership probabilities based on the proper motion catalogue by Parenago (1954) are evaluated for 1267 stars in the Orion nebula cluster. 495 stars have shown probabilities larger than 0.5.

153.006 **Early-type stars in the field of the 16-day Cepheid X Cygni.** D. G. Turner.
Bull. American Astron. Soc., Vol. 13, 513 (1981). – Abstract.

153.007 **X-ray observations of the Pleiades cluster of stars.**
W. H.-M. Ku.
Bull. American Astron. Soc., Vol. 13, 526 (1981). – Abstract.

153.008 **The star-forming history of the young cluster NGC 2264.**
M. Adams, S. E. Strom, K. M. Strom.
Bull. American Astron. Soc., Vol. 13, 541 (1981). – Abstract.

153.009 **X-ray and ultraviolet observations of NGC 2264.**
T. Simon, W. Cash, T. P. Snow.
Bull. American Astron. Soc., Vol. 13, 541 - 542 (1981). Abstract.

153.010 **Photometry of Praesepe in $BVRI$ colors. II.**
E. W. Weis.
Publ. Astron. Soc. Pacific, Vol. 93, 437 - 440 (1981).

Photoelectric $BVRI$ photometry has been obtained for 65 stars thought to be possible members of the Praesepe star cluster on the basis of their proper motions. Most of the stars have no previously published photometry. From these observations, it appears that 45 of the stars are probable cluster members.

153.011 **The young open clusters Berkeley 62 and Berkeley 86.** D. Forbes.
Publ. Astron. Soc. Pacific, Vol. 93, 441 - 446 (1981).

Berkeley 62 and Berkeley 86 are two previously unstudied open clusters whose brightest probable members are OB stars. Photoelectric UBV observations show Be 62 to lie at a distance of 2.05 ± 0.24 kpc with a reddening of $E_{B-V} = 0.86 \pm 0.04$ and earliest spectral type B1. Be 86 is found to lie 1.72 ± 0.20 kpc from the sun with a reddening of $E_{B-V} = 0.96 \pm 0.07$ and earliest spectral type O9.

153.012 **uvbyβ photometry of the open cluster M52.**
S. C. Danford, J. Thomas.
Publ. Astron. Soc. Pacific, Vol. 93, 447 - 452 (1981).

Photometric $uvbyβ$ observations of the open cluster M52 (NGC 7654) are used to determine the cluster's distance modulus $(V_0 - M_v = 10.83 \pm 0.24)$ and mean cluster reddening $(E(B-V) = 0.57)$. The likelihood of cluster membership of the F7 Ib supergiant 60°2532 is verified on evolutionary grounds; its absolute magnitude is determined to be $M_v = -4.42 \pm 0.46$. The cluster M52 is shown to be similar in age to the Pleiades and the α Persei cluster.

153.013 **On the dynamics of open star clusters.**
V. M. Danilov.
Stellar aggregates, Sverdlovsk, 1980, (see 003.006), p. 103 - 113. In Russian. – Abstr. in Ref. zh., 51. Astron., 8.51.806 (1981).

153.014 **The luminosity function of six open star clusters.**
N. V. Matkin.
Stellar aggregates, Sverdlovsk, 1980, (see 003.006), p. 68 - 79. In Russian. – Abstr. in Ref. zh., 51. Astron., 8.51.808 (1981).

153.015 **The luminosity function of the open star cluster NGC 1245.** M. Popova, K. Tsvetkova.
Astrofiz. issled., NRB, 1981, Vol. 3, 57 - 62 (1981). In Russian. – Abstr. in Ref. zh., 51. Astron., 8.51.809 (1981).

153.016 **Photometry of the star cluster NGC 7160.**
P. E. Zakharova, O. I. Kazantseva.
Stellar aggregates, Sverdlovsk, 1980, (see 003.006), p. 56 - 67. In Russian. – Abstr. in Ref. zh., 51. Astron., 8.51.811 (1981).

153.017 **Photometry of the open clusters NGC 3960 and NGC 5823.** K. A. Janes.
Astron. J., Vol. 86, 1210 - 1219 (1981).

Photoelectric and photographic BV photometry and DDO photometry of NGC 3960 and NGC 5823 are used to derive color-magnitude diagrams, distances, reddenings, and metallicities of the two clusters. NGC 3960 is a Hyades-age cluster, while NGC 5823 is somewhat younger. NGC 5823 is unlikely to be associated with the nearby cluster NGC 5822.

153.018   Study of the galactic cluster NGC 6913.
U. C. Joshi, B. B. Sanwal, R. Sagar.
Bull. Astron. Soc. India, Vol. 9, 83 (1981). — Abstract.

153.019   Evolution of moving clusters in the regular field of the Galaxy.   L. P. Osipkov.
Stellar aggregates, Sverdlovsk, 1980, (see 003.006), p. 114 - 121. In Russian. — Abstr. in Ref. zh., 51. Astron., 9.51.769 (1981).

153.020   Morphological features of open clusters.
K. A. Barkhatova, O. P. Pyl'skaya.
Stellar aggregates, Sverdlovsk, 1980, (see 003.006), p. 8 - 26. In Russian. — Abstr. in Ref. zh., 51. Astron., 9.51.771 (1981).

153.021   Some problems of investigation of star clusters.
I. V. Petrovskaya.
Stellar aggregates, Sverdlovsk, 1980, (see 003.006), p. 122 - 138. In Russian. — Abstr. in Ref. zh., 51. Astron., 9.51.772 (1981).

153.022   Investigation of the open star cluster NGC 2194.
K. A. Barkhatova, P. E. Zakharova, L. P. Shashkina.
Stellar aggregates, Sverdlovsk, 1980, (see 003.006), p. 27 - 55. In Russian. — Abstr. in Ref. zh., 51. Astron., 9.51.774 (1981).

153.023   Stellar coronae in the Hyades: a soft X-ray survey with the Einstein Observatory.   R. A. Stern,
M.-C. Zolcinski, S. K. Antiochos, J. H. Underwood.
Astrophys. J., Vol. 249, 647 - 661, plate 8 (1981).
An X-ray survey of the central region of the Hyades cluster with the Einstein Observatory has demonstrated that soft X-ray emission is a common property of the stars in the cluster. Of the 85 stars surveyed, about half are detected above the sensitivity threshold. More than 80% of the F and G dwarfs are X-ray sources, as well as three Hyades giants. The brightest X-ray source is 71 Tau, a rapid rotator of spectral type F0 V. The high incidence of X-ray emission and range of observed X-ray luminosities provide convincing evidence that stellar coronae produce the observed X-ray emission. The results are discussed in the context of coronal heating theories.

153.024   Four-color and Hβ photometry for four southern open clusters.   P. B. Stetson.
Astron. J., Vol. 86, 1500 - 1517 (1981).
Four-color and Hβ observations are presented for 81 stars in the fields of the southern open clusters NGC 2437, NGC 2659, NGC 4103, and NGC 5822. Mean cluster reddenings and distance moduli are derived, and upper limits for the variation of reddening in the cluster fields are estimated. The upper main sequence of NGC 2437 consists of evolved A0-type stars. The brightest members of NGC 2659 are unevolved B and A0 stars. NGC 4103 contains slightly evolved B stars, and an extreme Be star. NGC 5822 is an intermediate-age cluster with a main-sequence turnoff in the A-star range.

153.025   The open cluster NGC 2281.   J. Glaspey.
J. R. Astron. Soc. Canada, Vol. 75, 247 (1981). Abstract.

153.026   Differential ages in open clusters derived from *uvby*, Hβ observations of B stars.   A. M. Jakobsen.
Publ. Astron. Soc. Pacific, Vol. 93, 547 - 548 (1981). Abstract.

153.027   Photometry in the ancient open cluster NGC 6791.
W. E. Harris, R. Canterna.
Astron. J., Vol. 86, 1332 - 1336 (1981).
New photoelectric UBV photometry in the field of the old open cluster NGC 6791 is reported and discussed. The foreground reddening of the cluster is determined as $E_{B-V} = 0.13$ and the distance modulus $(m - M)_v = 14.0$ $(D_\odot = 5.2$ kpc), by analysis of the color-magnitude (CM) and two-color diagrams along with published spectral types of several cluster stars.

153.028   Geneva photometric boxes. III. Distances and reddenings for 43 open clusters.   B. Nicolet.
Astron. Astrophys., Vol. 104, 185 - 197 (1981).
An independent and homogeneous method for obtaining distances and reddenings of open clusters is presented. It is based on the Geneva photometric boxes as defined by Golay et al. (1969), Nicolet (1981), and uses data from the Rufener (1981) catalogue.

153.029   Photoelectric photometry of the open clusters IC 4665, NGC 6633 and NGC 7092 (M39) in the Vilnius photometric system.
A. Gurklytė, V. Straižys.
Bull. Vilnius Astron. Obs., Nr. 57, p. 3 - 8 (1981). In Russian.
The results of photoelectric photometry of 72 stars in three open clusters are given. Spectral types in the MK system given in tables are collected from literature. When MK classification was absent the stars were classified by the Q-method in the Vilnius photometric system.

153.030   On universality of the zero-age luminosity function of open star clusters.   P. E. Zakharova.
Astron. Zh., Tom 58, 1170 - 1176 (1981). In Russian.
English translation in Soviet Astron., Vol. 25, No. 6.
Comparison of luminosity functions of 26 star clusters by statistical criteria of Pearson and Kolmogorov shows mainly random differences for very young, young and intermediate-age clusters. Such an accordance confirms the hypothesis of universality of the zero-age luminosity function of open clusters.

153.031   Revised distance modulus for the open cluster NGC 7142.
A. I. Kopylov, N. N. Samus', S. Yu. Shugarov.
Astron. Tsirk., No. 1129, p. 3 - 5 (1980). In Russian.

153.032   Galactic orbits of old open clusters.
K. A. Barkhatova, S. A. Kutuzov, L. P. Osipkov.
Astron. Tsirk., No. 1135, p. 1 - 3 (1980). In Russian.

153.033   Supercorona of the α Persei cluster.
R. B. Shatsova.
Pis'ma Astron. Zh., Tom 7, 720 - 725 (1981). In Russian.
English translation in Soviet Astron. Lett., Vol. 7.
A list of 98 B-A3 stars, most probable members of the α Per cluster supercorona, is formed on the basis of kinematic, photometric and spectroscopic data.

153.034   Probable multiple galactic star clusters.
K. A. Barkhatova, E. D. Pavlovskaya.
Astron. Tsirk., No. 1155, p. 4 - 6 (1981). In Russian.

153.035   The motion of open clusters forming the Perseus complex.
K. A. Barkhatova, S. A. Kutuzov, L. P. Osipkov.
Astron. Tsirk., No. 1155, p. 6 - 8 (1981). In Russian.

The Lund-Strasbourg catalogue of open cluster data.
See Abstr. 002.036.

Discussion on the maximum-likelihood method for the determination of the membership in open clusters. See Abstr. 031.518.

Lunar occultations of the Hyades. II. August 1980. See Abstr. 096.004.

CP-stars in open clusters.　　See Abstr. 113.030.

Ap stars detected in open clusters by the Geneva photometry.　　See Abstr. 113.031.

Photoelectric measurements of three stars in M67. See Abstr. 113.059.

IUE observations of eight OB stars in NGC 2244: ultraviolet continua and extinction. See Abstr. 114.015.

IUE – ultraviolet and optical chromospheric studies of late-type giants in the Hyades cluster. See Abstr. 114.040.

Quantitative analysis of the spectrum of the Ap(SiTi) star HD 66318.　　See Abstr. 114.119.

Energy distribution in the spectra of five stars of the cluster NGC 6913 from photometric estimates through UBVRI filters.　　See Abstr. 114.183.

The theoretical HR diagram for a cluster of stars with mass accretion.　　See Abstr. 115.006.

UBV and H_β observations of stars towards M8. See Abstr. 115.012.

Near-infrared observations of trapezium-type multiple systems. Catalogue of observations and a new determination of the reddening law.　　See Abstr. 118.023.

Spectroscopic observations of pre-main-sequence stars in the Pleiades.　　See Abstr. 121.028.

V 697 Tauri – a member of the Hyades? See Abstr. 122.149.

Cloud-cluster associations: the B0 phenomenon reexamined.　　See Abstr. 131.065.

Formation of OB clusters: VLA observations. See Abstr. 132.010.

Distribution of hot stars and hydrogen in the Large Magellanic Cloud.　　See Abstr. 159.003.

Erratum

153.901　Erratum: 'Binaries in open clusters. II. Discrimination between double and rotating stars' [Astron. Astrophys., Vol. 97, 403 - 406 (1981)].
V. L. Trimble, J. P. Ostriker.
Astron. Astrophys., Vol. 102, 142 (1981). – See Abstr. 29.153.016.

154 Globular Clusters

154.001　Infrared studies of four highly reddened globular clusters: Palomar 2, Palomar 8, Palomar 10, and NGC 6749.　　R. Canterna, L. Rosino.
Astron. Astrophys., Suppl. Ser., Vol. 45, 53 - 60 (1981).
Infrared plates of the four highly reddened globular clusters, Pal 2, Pal 8, Pal 10, and NGC 6749 have been used to obtain star count data, concentrations, tidal radii, and distance moduli of these clusters. Use of the 25 brightest stars in the infrared has been successfully applied to the authors' data and reliable distance moduli estimates have been obtained. Adopting their values of reddening and distances, the authors obtain diameters and integrated absolute visual magnitudes that are consistent with the mean values of all globular clusters of the Galaxy.

154.002　On the giant, asymptotic and horizontal branches of globular clusters – I. Photographic photometry of M5.　　R. Buonanno, C. E. Corsi, F. Fusi Pecci.
Mon. Not. R. Astron. Soc., Vol. 196, 435 - 454 (1981).
B, V photographic photometry has been obtained for stars in the globular cluster M5. All stars with $B \leqslant 18.5$ and $2 \leqslant r \leqslant 5.6$ arcmin have been measured. The reduction procedure used is based on a multicomponent bidimensional Gaussian fit to the data derived by scanning the plates with a PDS microphotometer. From the colour-magnitude diagram obtained with a photometric accuracy of about 0.04 mag down to $B = 17.5$, the following can be deduced: (a) no blue HB stars have been found fainter than $V = 16.4$; (b) the HB is steeper and slightly redder than the HB found by Arp (1962). Possible explanations for this difference and a detailed analysis

of its implications are discussed in the light of evolutionary theories.

154.003　Far-ultraviolet photometry of globular clusters with ANS – III. Globular cluster ages.
T. S. van Albada, R. J. Dickens, B. M. H. R. Wevers.
Mon. Not. R. Astron. Soc., Vol. 196, 823 - 833 (1981).
Synthesis of the far-ultraviolet colours of globular clusters in the ANS system has been carried out using isochrones and luminosity functions by Ciardullo & Demarque and theoretical ANS colours for the Kurucz-model atmospheres. For a sample of eight clusters with well-observed HR diagrams, the authors have subtracted the total horizontal branch flux from the ultraviolet colours and fitted the residual C_{18-25} colours to those of isochrones of various ages to yield ages of $11 - 14 \times 10^9$ yr. For a sample of seven low-metallicity clusters (group B) the observed scatter in the two-colour, C_{18-25}, C_{15-33} diagram leads to an upper limit for the dispersion of the distribution of ages of 3×10^8 yr. From this the authors conclude that the volume containing the low-metallicity clusters has collapsed on a dynamical time-scale.

154.004　The color-magnitude diagram for stars in the central part of the globular cluster M 15.
M. Aurière, J.-P. Cordoni.
Astron. Astrophys., Vol. 100, 307 - 310 (1981).
The color-magnitude diagram for stars in the central part of M 15 is obtained from data presented in another paper. It is compared to the color-magnitude diagram obtained by Sandage et al. for the external part of the cluster. The radial variations

of the projected density of red giants and horizontal branch stars are examined; they provide an explanation for the observed radial color variations in M 15 and a basis for the investigation of the radial distribution of stars of different masses in globular clusters.

154.005 **On the radial color distribution in the globular cluster M5 = NGC 5904.**
R. Buonanno, V. Castellani, C. E. Corsi, F. Fusi Pecci.
Astron. Astrophys., Vol. 101, 1 - 6 (1981).

Astronomical plates of the galactic globular cluster NGC 5904 (M5) have been studied by means of modern techniques of plate analysis and image reconstruction, in order to reach a more precise knowledge in the origin of the radial colour distribution observed by Chun and Freeman (1979). It is shown that the colour gradient they found is due to position and size of the measuring aperture with respect to position and colour of a few bright giants, rather than to "collective" properties of the cluster. The occurrence of spurious reddenings in the centers of other clusters is also discussed, suggesting cautions in deriving firm conclusions on colour gradients in globular clusters.

154.006 **The ages and metallicities of the globular clusters in the Fornax dwarf spheroidal galaxy.**
R. Zinn, S. E. Persson.
Astrophys. J., Vol. 247, 849 - 859 (1981).

The authors report observations that provide some information on the age-metallicity relationship in the Fornax dwarf galaxy, and they compare what is known about the age-metallicity relationships in Fornax, other dwarf spheroidal galaxies, the Magellanic Clouds, and the halo of the Galaxy.

154.007 **Abundances in globular cluster red giants. IV. M22 and Omega Centauri.** J. G. Cohen.
Astrophys. J., Vol. 247, 869 - 878 (1981).

A detailed abundance analysis has been performed for three members of M22 and five of ω Cen. The three stars in M22 are chemically identical with [Fe/H] = -1.78 dex except for a range of Na and, less certainly, Ba larger than the observational and modeling errors. The ω Cen giants show a range of abundance for all elements, with the light elements and the rare earths (plus Ba) enhanced by a larger factor than the Fe peak nuclei in the more metal-rich ω Cen stars. The two most metal-rich ω Cen stars have vastly different Na/Mg and Al/Mg ratios and Sc abundances. In the appendix the author gives information on the interstellar lines and Hα emission found in these globular cluster giants.

154.008 **The intermediate age globular cluster NGC 152 in the Small Magellanic Cloud.** P. W. Hodge.
Astrophys. J., Vol. 247, 894 - 907 (1981).

The color-magnitude diagram of the populous Small Magellanic Cloud cluster NGC 152, measured to $V \approx 22$, shows that it has an age of about 8×10^8 years and is moderately metals-deficient. Its giant branch is sparsely populated, but includes at least two carbon stars at very red colors. There is a richly populated clump of intermediate-color giants at $M_V \approx 0$. Several superluminous giants are detected, also at intermediate colors, but averaging $M_V \approx -2.0$. The C-M diagram for the nearby field has very similar properties to those of the cluster.

154.009 **Star cluster membership: separating sheep from goats.**
L. A. Marschall, L.- T. G. Chiu, W. F. van Altena.
Sky Telesc., Vol. 62, 112 - 115 (1981).

154.010 **The dynamics of globular clusters.** I. R. King.
Q. J. R. Astron. Soc., Vol. 22, 227 - 243 (1981).

154.011 **The Oosterhoff period groups and the age of globular clusters. II. Properties of RR Lyrae stars in six clusters: the P-L-A relation.** A. Sandage.
Astrophys. J., Vol. 248, 161 - 176 (1981).

Data for RR Lyrae stars in six clusters are used to test the conclusion of an earlier paper that, at every temperature, period shifts exist between variables in one cluster relative to those in another of a different Oosterhoff group. Observations for RR Lyrae stars in M3, NGC 6121, NGC 6171, NGC 6981, and ω Cen are discussed; correlations between period, temperature, amplitude, and rise times are given and a comparison is made with a model of the Oosterhoff period shifts. The intrinsic dispersion in luminosity and mass among variables in a given cluster is discussed. The data for M3 and ω Cen are used to show the existence of a period-luminosity-amplitude (P-L-A) relation. Finally, the formulation of the P-L-A relation for RR Lyrae stars is shown to be similar in principle to that required for the long period Cepheids.

154.012 **The cyanogen distribution of M4 and the possible connection between horizontal branch morphology and chemical inhomogeneity.** J. Norris.
Astrophys. J., Vol. 248, 177 - 188 (1981).

A spectroscopic survey of 45 red giants in the globular cluster M4 has been completed with a view to ascertaining whether the bimodal distribution of stars on the horizontal branch (Lee) is accompanied by a cyanogen dichotomy on the giant branch, similar to the situation found in NGC 6752. From analysis of some 118 spectra, it is concluded that the red giants in M4 (in the magnitude range $M_V \sim 0.3$ to -1.2) do show a bimodal cyanogen distribution. There appears also to be an anticorrelation between the behavior of CN and CH. A working hypothesis is proposed which will explain most of the known peculiarities of the three globular clusters 47 Tuc, M4, and NGC 6752, for which comprehensive cyanogen surveys are available. It is suggested that there is a spectrum of core rotational velocities in the main-sequence stars of globular clusters.

154.013 **The giant branch of the globular cluster NGC 3201.**
G. S. Da Costa, J. A. Frogel, J. G. Cohen.
Astrophys. J., Vol. 248, 612 - 621 (1981).

Infrared photometry has been obtained for 26 stars in the field of the globular cluster NGC 3201. For 14 of these stars, optical spectra have also been obtained. These observations show that the giant branch of this cluster possesses an intrinsic width of 0.13 mag in $(V-K)_0$. The authors conclude that variable reddening across the cluster is the most likely explanation of this result. There is a range in G band (CH) strength at constant $(V-K)_0$ that correlates with CO strength: stars with weak CH also have weak CO. Comparisons with published synthetic spectra show that star-to-star variations in carbon abundance of less than a factor of 3 are sufficient to explain the range in G-band strengths. The position of the giant branch in the infrared C-M diagram and the strengths of metal lines in the optical spectra indicate that the overall metal abundance of NGC 3201 is comparable to those of M3, M5, and NGC 6752.

154.014 **Vidicon photometry in old LMC clusters.**
W. E. Harris, J. E. Hesser, B. Atwood.
Bull. American Astron. Soc., Vol. 13, 531 (1981). − Abstract.

154.015 **Luminosity functions in globular clusters.**
E. M. Green.
Bull. American Astron. Soc., Vol. 13, 545 (1981). − Abstract.

154.016 **A comment on the metal abundance of the globular cluster M71.** R. A. Bell, B. Gustafsson.
Bull. American Astron. Soc., Vol. 13, 545 (1981). − Abstract.

154.017 **Picture processing of the globular cluster NGC 6441.**
D. H. Martins, C. A. Harvel.
Bull. American Astron. Soc., Vol. 13, 559 (1981). − Abstract.

154.018 Globular cluster systems: implications for galaxy formation. K. C. Freeman.
The structure and evolution of normal galaxies, (see 012.010), p. 251 - 260 (1981).
Contents: Globular cluster populations. Galactic distribution of globular clusters. Abundance gradients in globular cluster systems. Cluster ages and the second parameter. The abundance distribution for galactic globular clusters. Globular cluster initial mass functions. Orbital properties of globular clusters. Conclusion.

154.019 IUE observations of the galactic globular cluster M 80. V. Caloi, V. Castellani, A. Cassatella, D. Ponz.
Second European IUE Conference, (see 012.011), p. 179 - 182 (1980).
IUE observations of the globular cluster M 80 are presented and discussed; their importance in relation to problems of galactic populations and population synthesis programs is stressed.

154.020 A very rapidly expanding gas in the core of the X-ray globular cluster NGC 6624?
N. Bel, J. Clavel, R. Foy.
Second European IUE Conference, (see 012.011), p. 183 - 184 (1980).
IUE observations of the centre of the X-ray globular cluster NGC 6624 reveal an emission line spectrum. A tentative identification needs a very large velocity of -10^4 km s^{-1}. A point source emits most of the UV flux, with a continuum distribution. A resolved expanding nebula would emit the emission line spectrum.

154.021 Membership in the field of globular cluster M56.
B. E. Rishel, W. L. Sanders, R. Schröder.
Astron. Astrophys., Suppl. Ser., Vol. 45, 443 - 450 (1981).
Relative proper motions in the field of M56 are convolved with spatial information to compile a list of 39 stars out to a radius of about 17 pc, 30 of which are expected to be cluster stars.

154.022 Structure of the core of globular clusters.
M. Aurière.
Messenger, No. 25, p. 12 - 14 (1981).

154.023 Anomalous giants in Magellanic Cloud clusters.
P. W. Hodge.
J. Astrophys. Astron., Vol. 2, 161 - 164 (1981).
Many of the populous intermediate-age star clusters of the Magellanic Clouds have a few stars that occupy a 'forbidden zone' in the colour-magnitude diagrams of the clusters. The properties of these stars in 13 clusters are presented. Several hypotheses regarding their nature are discussed, and it is concluded tentatively that they represent an unpredicted advanced stage of evolution from the giant branch, though no theoretical evidence supports such a conclusion.

154.024 Revised metal abundances for galactic globular clusters. Preliminary results on the abundance gradient and on the second parameter problem. F. Caputo.
Mem. Soc. Astron. Italiana, Vol. 52, (see 012.019), 59 - 65 (1981).
New homogeneous metal abundances have been derived for 94 galactic globular clusters by using recent high-quality spectroscopic [Fe/H] values to calibrate several metallicity indicators. The results are consistent with a metal abundance gradient in the galactic halo.

154.025 On the effects of radiation pressure in the cores of globular clusters.
L. Angeletti, R. Capuzzo-Dolcetta, P. Giannone.

Astron. Astrophys., Vol. 102, 250 - 256 (1981).
The possible effects of a presence of a dust cloud in the cores of globular clusters was investigated. Two cluster models were considered together with various models of clouds. The problem of radiation transfer was solved under some simplifying assumptions. Owing to a differential absorption of the star light in the cloud, radiation pressure turned out to be inward-directed in some cloud models. This fact may lead to a confinement of some dust in the central regions of globular clusters.

154.026 New index system of heavy element abundance in globular clusters.
N. N. Samus', E. N. Pastukhova.
Stellar aggregates, Sverdlovsk, 1980, (see 003.006), p. 84 - 92. In Russian. – Abstr. in Ref. zh., 51. Astron., 9.51.785 (1981).

154.027 Star formation in globular clusters at initial stages of their evolution.
Al. A. Suchkov, An. A. Suchkov, Yu. A. Shchekinov.
Pis'ma Astron. Zh., Tom 7, 617 - 622 (1981). In Russian. English translation in Soviet Astron. Lett., Vol. 7.
It is shown that the mean density $\bar{n}$ of globular clusters can be entirely explained by thermal fragmentation of protoclusters, the clusters being formed only in the density range $0.4 < \bar{n} < 2 \times 10^3$ cm^{-3}. Thermal instability directly leads to star formation.

154.028 The chemical composition, structure, and dynamics of globular clusters. K. C. Freeman, J. Norris.
Annu. Rev. Astron. Astrophys., Vol. 19, (see 003.012), 319 - 356 (1981).
Contents: Introduction. Abundances of the heavy elements. Inhomogeneities within clusters. The second-parameter problem. The galactic globular cluster system. The structure of globular clusters.

154.029 Three *UBV* sequences in the LMC. A. J. Penny.
Mon. Not. R. Astron. Soc., Vol. 197, 693 - 698 (1981).
Photoelectric and electronographic photometry is presented for three sequences near the clusters NGC 1466, 2214 and 2257. Photoelectric measures of stars previously observed by Butler provide a tie to other photometric studies.

154.030 The main sequence of the globular cluster NGC 3201. G. Alcaino, W. Liller.
Astron. J., Vol. 86, 1480 - 1499 (1981).
The authors present photographic photometry for 1452 stars in the globular cluster NGC 3201. For the main-sequence turn-off, they have determined its position to lie at $V = 18.0 \pm 0.1$ (m. e.) and $B-V = 0.61 \pm 0.03$ (m. e.). From these values, they calculate the intrinsic values $M_v = 3.93$ and $(B-V)_0 = 0.39$. For the cluster as a whole, the authors derive a distance modulus $(m-M)_v = 14.07 \pm 0.15$ and reddening $E(B-V) = 0.22 \pm 0.03$. They deduce a range of ages from 7.9×10^9 to 12.3×10^9 yr. They confirm the existence of a deficiency of stars over ~ 0.7 mag of the subgiant branch.

154.031 Photometry of the cores of globular clusters. III. NGC 6712. D. H. Martins, C. A. Harvel.
Astrophys. J., Vol. 250, 135 - 141, plate 14 (1981).
New photometry has been done for the central region of the X-ray globular cluster NGC 6712. Of a total of 208 stars, 39 were newly observed. A color-magnitude diagram is derived, including the recently discovered sdO star (Remillard, Canizares, and McClintock), showing that there are no stars in the field as blue as the sdO object down to a limiting $V \approx 18$ mag. Systematic errors appear to exist in previous photometry of the cluster, such that the reddening estimate has been revised to $E(B-V) = 0.33$ mag and the cluster distance to 7.9 ± 0.1 kpc.

154.032   Photoelectric photometry of globular clusters in the Andromeda nebula. VII.
A. S. Sharov, V. M. Lyutyj.
Pis'ma Astron. Zh., Tom 7, 662 - 665 (1981). In Russian.
English translation in Soviet Astron. Lett., Vol. 7.

Results of photoelectric UBV observations of 110 globular clusters in the Andromeda nebula obtained in 1980 are presented.

154.033   Far ultraviolet investigation of three nuclei of globular clusters.   V. Caloi, A. Cassatella,
V. Castellani, F. Macchetto, J. Melnick.
Astron. Astrophys., Vol. 103, 386 - 389 (1981).

Spectra of the central regions of the galactic globular clusters NGC 362, NGC 5824, and NGC 6093 have been obtained in low dispersion mode with the International Ultraviolet Explorer. The peculiar cluster NGC 362 shows an UV behaviour consistent with the HB morphology and the metal content inferred from observations in the optical range. NGC 6093 appears to be a member of the "extremely blue" HB group, defined by ANS observations; results for NGC 5824 are similar, but more uncertain. UV properties of G.C. are discussed in relation to the population of HB's.

154.034   Ultraviolet spectrophotometry of the galactic globular cluster M 5.   A. Altamore,
L. Angeletti, R. Capuzzo-Dolcetta, P. Giannone.
Astron. Astrophys., Vol. 103, 424 - 426 (1981).

Ultraviolet low-resolution observations of M 5 (NGC 5904) obtained with the International Ultraviolet Explorer (IUE) Satellite are presented. General characteristics of the spectrum are discussed. A low $E(B-V)$ value was confirmed; some spectral features are identified as being of stellar origin. Broad- and narrow-band fluxes and their spatial distribution perpendicular to the direction of dispersion were computed. The importance and degree of reliability of this kind of observations are emphasized.

154.035   Globular clusters and galaxy mergers.
W. E. Harris.
J. R. Astron. Soc. Canada, Vol. 75, 248 (1981).

154.036   The galactic distribution of globular clusters.
V. Castellani, M. Melchiorri.
Astrophys. Space Sci., Vol. 80, 289 - 302 (1981).

The spatial distribution of galactic globular clusters has been studied by means of Walsh-Hadamard transforms. Consistent indications have been collected suggesting the existence, at least in a first approximation, of two distinct populations of clusters – the 'metal poor' clusters showing a spherical distribution whereas 'metal rich' clusters are flattened along the disk. Simulated cluster-halos have been computed in order to study the dependence of observed radial velocities on the eccentricity of cluster orbits, showing that only a general high value of eccentricity seems able to reproduce the observational features.

154.037   Carbon and nitrogen abundances in the giant stars of the globular clusters M3 and M13.   N. B. Suntzeff.
Astrophys. J., Suppl. Ser., Vol. 47, 1 - 32 (1981) = Lick Obs. Bull., No. 881.

Carbon and nitrogen abundances, as well as the strengths of calcium II H and K and the $\Delta v = 0$ cyanogen band, have been measured in red giant stars in the globular clusters M3 and M13. The data consist of spectrophotometric scans of low resolution (10 Å) of 29 giants in M3 and 35 giants in M13 in the wavelength region 3000–5000 Å.

154.038   Position, magnitudes and color for stars in the central part of the X-ray globular cluster M15.
M. Aurière, J.-P. Cordoni.
Astron. Astrophys., Suppl. Ser., Vol. 46, 347 - 354 (1981).

Position, magnitudes and color for 734 stars in a $1' \times 1'$ square field centered on the X-ray globular cluster M 15 are given. The observations consist of long focus (about 35 m) high spatial resolution (FWHM down to 0.55 arcsecond) unfiltered image tube and B and V classical photographs of the central part of the cluster obtained with two specially designed cameras at the 1 meter telescope of Pic du Midi Observatory. The reduction procedure is a compromise between automatic and interactive technics. Errors in the data relative to photometry as well as astrometry are discussed.

154.039   Thermodynamic equilibrium hard binary models for globular star clusters.   G. Horwitz.
Proceedings of the 5th Göttingen-Jerusalem-Symposium on Astrophysics, (see 012.041), p. 149 - 155 (1981).

A thermodynamic equilibrium (TDE) model for a highly evolved globular cluster (GC) of stars with one or more hard binaries is described. In this model there is an abrupt transition to a state of stable, thermodynamic equilibrium in which some binaries possess a significant fraction of the total energy of the GC. The TDE sequence corresponds to states of successively more tightly bound binaries, thereby releasing energy from the residual single stars. This corresponds to an evolutionary mechanism in which binary hardening competes with evaporation as the driving force of evolution.

154.040   Globular clusters in galaxies beyond the local group. I. New cluster systems in selected northern ellipticals.   W. E. Harris, S. van den Bergh.
Astron. J., Vol. 86, 1627 - 1642 (1981).

Globular cluster systems have been detected in eight northern elliptical galaxies (NGC 3226, 3377, 3379, 3607, 4278, 4472, 5813, and 5846) with redshifts in the range $740 \text{ km s}^{-1} < V_0 < 1800 \text{ km s}^{-1}$. By radial star counts around each galaxy the authors have estimated the total population and structural characteristics of their cluster systems. These data, combined with previous material for other galaxies, are used to discuss the "specific frequency" S [number of globular clusters per unit ($M_v = -15$) galaxy luminosity] in elliptical galaxies generally. They find that the majority of ellipticals have S values consistently in the range $4 \lesssim S \lesssim 10$. No strong correlation of S with galaxy luminosity is seen , through there is some hint of environmental differences.

154.041   Mass segregation in globular clusters.
K. K. Scaria, M. K. V. Bappu.
J. Astrophys. Astron., Vol. 2, 215 - 244 (1981).

Photoelectric aperture-photometry of ω Cen in U, B, V, R and I bands has established that the cluster is bluer between 2 arcmin and 4 arcmin from the centre, than it is elsewhere. Equidensitometry of ω Cen in B, V and infrared bands shows a wavelength dependence with the cluster being nearly spherical in the infrared band. It shows a maximum ellipticity around 3 arcmin from the cluster centre. The blue contribution in this zone comes from both a diffuse background of unresolved stars and an increase in the relative abundance of horizontal branch (HB) stars. The similarity between the diffuse background and the HB stars is demonstrated.

154.042   Photometry of faint stars in globular clusters. IV. B-system luminosity function for the globular cluster NGC 6366.
E. V. Naumova, N. N. Samus', S. Yu. Shugarov.
Astron. Tsirk., No. 1129, p. 5 - 7 (1980). In Russian.

154.043   Luminosity functions of bright red giant stars in globular clusters.   E. M. Green.
Univ. Texas Publ. Astron., No. 18, 13 + 305 pp. (1981).
Diss. Philos. Univ. Texas, Austin.

The current knowledge of globular cluster metallicities, helium abundances, distances, reddenings, and ages is discussed, with emphasis on the interdependence of these quantities, and the assumptions involved in their determination. Luminosity functions are shown to provide valuable and complementary information in the analysis of color-magnitude diagrams. They are free from the theoretical uncertainties associated with model radii and temperatures, and they require very few assumptions about the observed data.

154.044 A (B, V) photoelectric sequence of stars in Omega Centauri. W. L. Martin.
South African Astron. Obs. Circ., No. 6, p. 28 - 30 (1981).
 Photoelectric (B, V) observations of 55 stars in the globular cluster ω Centauri in the range $10.4 < V < 15.1$ are presented. Comparison of the photometry with that of other authors shows that there are no significant systematic differences.

154.045 A carbon star in the globular cluster Lindsay 102. A. C. Danks.
ESO Sci. Prepr., No. 170, 10 pp. (1981). – Submitted to Astron. Astrophys.

154.046 The extended giant branches of intermediate age globular clusters in the Magellanic Clouds. II.
M. Aaronson, J. Mould.
Prepr. Steward Obs., No. 332, 82 pp. (1981).
 The authors report on a study of upper asymptotic giant branch (AGB) stars in the red globular clusters of the Magellanic Clouds. A photographic near-infrared survey of the clusters is now more than half completed, and infrared (JHK) photometry of the stars so identified implies that the majority of the red clusters have extended giant branches and are therefore of intermediate age, unlike globular clusters in the Galaxy. With the assumption of a uniform luminosity function on the upper AGB, ages can be deduced for sufficiently well-populated cloud clusters. For sparser cloud clusters an upper limit to the age can be derived. Independent of the exact calibration, the very distinct distributions of AGB tips implies a very different age distribution of existing clusters in the Clouds. The intermediate-age globulars of the Large Cloud are younger than those of the Small Cloud in the mean, and their AGB tips contain the reddest and most luminous carbon stars.

154.047 On the origin of globular clusters: connection between the mass of a gaseous protocluster and the star cluster formed from it. V. G. Surdin.
Astron. Tsirk., No. 1151, p. 4 - 6 (1981). In Russian.

Simultaneous echelle spectroscopy of 15 or more giants in globular clusters using an aperture plate.
See Abstr. 031.552.

Theoretical evidence of mass loss from globular cluster stars. See Abstr. 064.039.

Evolutionary effects of mass loss in low-mass stars. See Abstr. 065.032.

Helium diffusion in horizontal-branch-star evolutionary models. See Abstr. 065.065.

Proper motions of bright red giants in globular clusters. IV. The globular cluster M5. See Abstr. 111.006.

Circumstellar winds in globular cluster giants.
See Abstr. 112.007.

M and C stars in the central region of Omega Cen.
See Abstr. 113.012.

Abundances in 11 field stars with large metal deficiencies. See Abstr. 114.003.

Horizontal-branch stars, and galactic and Magellanic Cloud globular clusters. See Abstr. 114.043.

Applications of an observational grid of low-metallicity stellar spectra. See Abstr. 114.092.

Spectroscopic evidence for a wide range in abundances among faint subgiant stars in the globular cluster Omega Centauri. See Abstr. 114.105.

M5 V101: a close binary system in a globular cluster.
See Abstr. 117.001.

M5 V101: a close binary system in a globular cluster. See Abstr. 117.028.

The Oosterhoff period groups and the age of globular clusters. I. Photometry of cluster variables in M15.
See Abstr. 122.030.

Variable stars in the bulge of the galaxy in a field around NGC 6304. See Abstr. 122.031.

The variable stars in the globular cluster NGC 6101.
See Abstr. 122.070.

RR Lyrae populations in the galactic field.
See Abstr. 122.097.

The metal abundances of RR Lyrae stars in the globular clusters NGC 6712 and NGC 6723.
See Abstr. 122.121.

Statistical investigation of period changes of RR Lyrae variables in globular clusters.
See Abstr. 122.137.

Variable stars in the globular cluster M13. II.
See Abstr. 122.138.

Study of period changes for 38 RR Lyrae variables in the globular cluster M15. See Abstr. 122.148.

RR Lyrae variables in the globular cluster M3.
See Abstr. 122.160.

V 68 in the globular cluster M 3 (NGC 5272) is a double mode RR Lyrae type star. See Abstr. 122.167.

Beobachtungen von V 1 im Kugelhaufen M 3.
See Abstr. 122.212.

Mehrfarben-Beobachtungen von V 79 im Kugelhaufen M 3. See Abstr. 122.213.

Variable stars in the globular cluster M13.
See Abstr. 123.019.

A high-resolution optical survey of interstellar absorption lines toward globular clusters and extragalactic objects. II. Further data. See Abstr. 131.027.

Deviations of galactic reddening from the cosecant relation. See Abstr. 131.037.

Polarization of scattered light in globular clusters.
See Abstr. 131.168.

IR observations of a new X-ray globular cluster and galactic bulge X-ray sources. See Abstr. 142.027.

IUE observations of globular cluster X-ray sources. See Abstr. 142.035.

On the properties of 21 X-ray globular clusters in M31.   See Abstr. 142.091.

Globular clusters and galaxy mergers. See Abstr. 151.064.

Where is Population III?   See Abstr. 155.009.

On the circular velocity of the Galaxy. See Abstr. 155.053.

The mass of M31.   See Abstr. 158.124.

The metal abundance range in the Ursa Minor dwarf galaxy.   See Abstr. 158.212.

UBV observations of globular clusters in the Magellanic Clouds.   See Abstr. 159.010.

Superluminous giants in Magellanic Cloud clusters. See Abstr. 159.011.

Planetary nebula progenitors in Magellanic Cloud clusters.   See Abstr. 159.012.

RR Lyrae stars in Magellanic Cloud clusters. See Abstr. 159.013.

155 Galaxy (Structure, Evolution)

155.001 Some numerical experiments concerning the determination of the general velocity field of the Galaxy from proper motions. J. Byl, M. W. Ovenden.
Mon. Not. R. Astron. Soc., Vol. 196, 659 - 668, Microfiche MN 196/1 (1981).

Analyses have been made of fictitious proper motions for velocity fields with, and without, spiral arm kinematics. Dispersions in velocity, distance and observational errors were included in some solutions, and a comparison was made between the mean, the median and the mode, as used for representative points. Spiral arms yield apparent values of the Oort constants that are algebraically smaller than the true values, a result compatible with recent studies of actual proper motions. The presence of dispersions increases the uncertainties in the estimation of the Oort constants, but does not produce significant systematic effects. When using representative points, the median proper motion for a field yields better solutions than either the mean or the mode.

155.002 Kinematical and chemical evolution of the galactic disk near the Sun. J. P. Vader, T. de Jong.
Astron. Astrophys., Vol. 100, 124 - 137 (1981).

The authors have constructed models of the galactic disk near the Sun by calculating the chemical and the kinematical evolution of gas and stars in a cylinder perpendicular to the galactic plane. One of the main novel features of these models is that the z-dependence of all properties is explicitly retained. Stars are formed at a rate proportional to the n-th power of the gas density. Another novel feature of the authors' evolutionary models is the explicit treatment of stellar acceleration.

155.003 The rotation curve of the neutral hydrogen subsystem in the galactic plane. I. V. Petrovskaya.
Tr. Astron. Obs., Leningrad, Tom 36 = Uch. Zap. Leningr. Univ., No. 402 = Ser. mat. nauk, Vyp. 58, 97 - 109 (1981). In Russian.

The rotation curves of the H I subsystem are obtained for the first and forth quadrants of galactic longitude.

155.004 On the ratio between the halo and disk masses in the Galaxy. A. G. Morozov.
Astron. Zh., Tom 58, 734 - 742 (1981). In Russian. English translation in Soviet Astron., Vol. 25, No. 4.

The evolution and final stationary states of model galaxies consisting of a halo with a density of $\rho(r) \sim r^{-2}$ and $\rho(r) \sim r^{-4}$ and a disk of a finite thickness with N = 200 particle stars with an exponential surface density $\sigma \sim \exp(-r/L)$ are investigated by numerical experiments.

155.005 The extinction toward the galactic center from observations of interstellar lines.
S. R. Federman, N. J. Evans II.
Astrophys. J., Vol. 248, 113 - 118 (1981).

A substantial galactic gradient in the ^{13}CO−to-dust ratio has been suggested recently by Blitz and Shu. They used the ^{13}CO emission seen toward the galactic center and a local calibration of the ^{13}CO−to-extinction ratio to predict the extinction to the galactic center. In this paper, H_2CO absorption observations against Sgr A are used to select the molecular material lying in front of the galactic center, and a similar analysis is carried out. The value for the ratio $A_v/W(H_2CO)$ is ~12 mag (km s^{-1})$^{-1}$. When this ratio is applied to published data for the galactic center an extinction of <86 mag is derived. By considering only the molecular components in front of the infrared cluster, the estimate for the extinction is lowered to 15−46 mag. Extinction estimates based on other interstellar species were also considered.

155.006 High latitude H I shells in the Galaxy. I. E. M. Hu.
Astrophys. J., Vol. 248, 119 - 127, plates 4 - 5 (1981).

A search has been made for H I shells in the 21 cm maps made by Heiles and Habing for $|b| > 10°$. A filter in the velocity dimension was used to enhance shell features over background gas. Positions, expansion velocities, column densities, and angular extents were measured for the 50 shells that were discovered. Incorporating distance estimates, masses and kinetic energies were estimated for the shells. A striking result was the failure of swept up matter in a standard model for gas in the disk to reproduce the observed shell column densities. The distribution of H I shells was compared with the distribution of other large-scale features in the Galaxy: the author finds H I shell structures to correlate with the positions of radio continuum loops I, II, and III, and to show little relation to the distribution of O and B stars.

155.007 A new determination of the halo luminosity density of the Galaxy.
D. O. Richstone, F. G. Graham.
Astrophys. J., Vol. 248, 516 - 523 (1981).

Assuming a particular form for the halo velocity distribution function near the sun, the authors have explicitly computed the distribution of tangential velocities for arbitrary halo velocity dispersions and rotation rates. For reasonable choices of these parameters, applying a lower tangential velocity limit of 250 km s^{-1} in a proper motion catalog excludes a large fraction of the halo from consideration. The authors have computed that fraction and used it to rediscuss Schmidt's estimate of the halo luminosity function.

155.008 The abundance of argon at the galactic center.
D. F. Lester, J. D. Bregman, F. C. Witteborn, D. M. Rank, H. L. Dinerstein.
Astrophys. J., Vol. 248, 524 - 527 (1981).

Measurements of [Ar II] 6.99 μm and Pfα 7.45 μm made from the Kuiper Airborne Observatory are presented for Sgr A, the H II region at the center of the Galaxy. These line strengths, when combined with ground-based measurements, suggest a factor of 2 enhancement in the Ar/H ratio in the galactic center region relative to that in the sun and in the solar neighborhood.

155.009 Where is Population III? H. E. Bond.
Astrophys. J., Vol. 248, 606 - 611 (1981).

A simple one-zone model of chemical evolution in the galactic halo predicts that the halo should contain appreciable numbers of "Population III" stars (roughly defined as stars with [Fe/H] < −3). However, unbiased samples of halo globular clusters, dwarf spheroidal systems, and high-velocity subdwarfs all show a statistically significant lack of objects with Population III metallicity. Even extensive special searches for field stars of ultralow metallicity made by the writer and others have failed to reveal more than a very small number of objects that may have [Fe/H] < −3. A chemical evolution model in which the primordial gas stars with a small but nonzero metallicity near [Fe/H] = −2.6 is shown to predict a metallicity distribution function in the galactic halo that is in good agreement with the observations.

155.010 A method for the determination of metal abundances in the Galactic nuclear bulge and nearby galaxies.
J. M. Scalo, G. E. Miller.
Astrophys. J., Lett., Vol. 248, L65 - L68 (1981).

A nonphotometric method for the determination of metal abundances in nearby galaxies is presented. The method involves a theoretical calibration of the observed strong correlation between the ratio of cool carbon stars to M giants (C/M ratio) with metal abundance, assuming that carbon stars are internally mixed during helium shell flashes. A lower limit on the mass fraction of metals in the Galactic nuclear bulge, $Z(NB) > 0.05$, is determined by this means from the observed C/M ratio.

155.011   Correlation of γ-ray flux and molecular hydrogen column densities in the galactic plane.
T.-p. Li, A. W. Wolfendale.
J. Phys. G, Vol. 7, L157 - L160 (1981). – Abstr. in Phys. Abstr., Vol. 84, Abstr. 83750 (1981).

155.012   The distribution of cosmic-ray electrons and nuclei in the Galaxy.   M. R. Issa, P. A. Riley,
A. W. Strong, A. W. Wolfendale.
J. Phys. G., Vol. 7, 973 - 994 (1981). – Abstr. in Phys. Abstr., Vol. 84, Abstr. 83751 (1981).

155.013   Étude de la structure galactique dans une région de la Poupe.   D. Peton-Jonas.
Astron. Astrophys., Suppl. Ser., Vol. 45, 193 - 205 (1981).
The present investigation was undertaken to get more information about the structure and the rotation of our Galaxy in the Puppis OB associations area, towards longitude $l'' = 244°$. The author gives spectrophotometric and kinematic results which were obtained through the measurements of plates taken at the ESO Objective Prism Astrograph. Spectral types, B magnitudes, distance moduli and distances from the galactic centre for 102 stars are presented, 74 of them having spectral types ranging from O5 to B3, the others being B5 and B6 stars.

155.014   A rotational standard of rest.   W. L. H. Shuter.
Bull. American Astron. Soc., Vol. 13, 535 (1981).
Abstract.

155.015   On the vertical structure of galactic rotation.
F. J. Lockman, T. M. Bania.
Bull. American Astron. Soc., Vol. 13, 538 (1981). – Abstract.

155.016   Extreme positive velocities of neutral hydrogen in the Southern Milky Way.
P. D. Jackson, F. J. Kerr.
Bull. American Astron. Soc., Vol. 13, 538 - 539 (1981). Abstract.

155.017   The large scale H I distribution in the outer Galaxy.
L. Blitz, S. Kulkarni, C. Heiles.
Bull. American Astron. Soc., Vol. 13, 539 (1981). – Abstract.

155.018   A test for completeness of nearby G–K dwarfs.
A. R. Upgren, T. E. Armandroff.
Bull. American Astron. Soc., Vol. 13, 570 (1981). – Abstract.

155.019   The distribution of interstellar C IV in the Galaxy.
G. E. Bromage, A. H. Gabriel, D. W. Sciama.
Second European IUE Conference, (see 012.011), p. 21, 345 - 351 (1980).
27 high-dispersion spectra of 13 halo and disc stars have been reduced and column densities of C IV are presented. The galactic distribution of C IV is discussed and, combining the data with other published measurements, a mean exponential scale height of 3^{+4}_{-1} kpc and disc space density $n_0 = 8.5 \times 10^{-9}$ cm^{-3} are derived. The results are compared with Copernicus O VI data. Strengths and profiles of Si IV, C IV and N V from IUE data are also compared for a few representative lines of sight.

155.020   The fraction of O-type supergiants in our galaxy, in the LMC and in the SMC: an evidence of the correlation between mass loss rate and chemical abundance.
G. F. Bisiacchi, C. Firmani.
Effects of mass loss on stellar evolution, (see 012.015), p. 255 - 259 (1981).
The distribution of the spectral types of the WR stars in our galaxy is different at different distances from the galactic center. This distribution is also different in all three galaxies, in our, in the LMC and in the SMC. These results have been interpreted as due to the dependence of the mass loss rate on the original chemical abundance which is known to be different in these objects.

155.021   Peculiarities in the distribution of galactic Wolf-Rayet stars: constraints on evolutionary scenarios?   A. Gomez, M.-C. Lortet, A. Pitault.
Effects of mass loss on stellar evolution, (see 012.015), p. 315 - 317 (1981).
The re-examination of the spatial distribution of 158 Wolf-Rayet stars in the Galaxy may help to define different breeds of Wolf-Rayet stars and put constraints on possible evolutionary scenarios.

155.022   Stellar mass loss and galactic chemical evolution.
A. Serrano, M. Peimbert.
Effects of mass loss on stellar evolution, (see 012.015), p. 535 - 538 (1981).
The authors summarize some of the results that they have obtained from chemical evolution models. In particular, they discuss those aspects related to stellar mass loss, helium production, a varying heavy element yield and the very light stars.

155.023   Mass loss and ΔY/ΔZ ratio.
C. Chiosi, F. Matteucci.
Effects of mass loss on stellar evolution, (see 012.015), p. 543 - 549 (1981).
The $\Delta Y/\Delta Z$ abundance ratio is studied taking into account mass loss from intermediate and massive stars, and the Y_Z vs. Z relation of Peimbert and Serrano (1980).

155.024 ^{13}CO near the galactic center.
G. M. Heiligman.
The phases of the interstellar medium, (see 012.018), p. 93 - 98 (1981).
A preliminary report on a survey of the galactic center region being done at Bell Laboratories in the 2.6 mm line of carbon-13 monoxide is given. This research differs from earlier CO observations taken with the 11-m antenna at Kitt Peak in three respects: (1) increased coverage in galactic latitude, (2) use of the less saturated ^{13}CO isotope, and (3) improved sidelobe suppression of the BTL 7-m antenna. Observations will be continued during the 1981-82 observing season.

155.025   A synoptic view of galactic processes.
F. W. Stecker.
The phases of the interstellar medium, (see 012.018), p. 151 - 161 (1981).
The author will discuss the power of using synoptic galactic surveys in many wavelength bands in order to obtain a more complete picture and a better understanding of the dynamics of the interstellar medium and to study galactic structure and evolution on a large scale. In particular, he will discuss the implications of the picture presented by mm-wave CO, far infrared and γ-ray surveys of the Galaxy.

155.026   Is the galactic corona hot or cool?
R. A. Chevalier.
The phases of the interstellar medium, (see 012.018), p. 175 - 179 (1981).

155.027   A corona and a wind in our Galaxy.
J. N. Bregman.
The phases of the interstellar medium, (see 012.018), p. 191 - 193 (1981).

155.028 **La Galaxie.**   L. Martinet.
Histoire de l'univers, (see 003.007), p. 123 - 144
(1980).
Contents: Le soleil dans la Galaxie. La structure
"verticale" de la Galaxie. La rotation différentielle. Les
mouvements des étoiles dans le disque et dans le halo. La
structure spirale du disque galactique. Les nuages gazeux à
grande vitesse et le courant magellanique. Les régions centrales
de la Galaxie. Un trou noir au centre de la Galaxie? Un
scénario possible d'évolution initiale de la Galaxie.

155.029 **The post explosion shock propagation in the**
central region of the Galaxy.
T. Bhattacharyya, B. Basu.
Bull. Astron. Soc. India, Vol. 9, 80 - 81 (1981). − Abstract.

155.030 **Galactic kinematic distances from velocity gradients.**
K. Rohlfs.
Astron. Astrophys., Vol. 102, 91 - 92 (1981).
The slope dV/dl of the radial velocity of a feature close
to the direction $l = 0°$ or $l = 180°$ is shown to be a good
measure of the distance to the galactic centre. Neither the
geometric shape of the feature nor variation of the galactic
rotation velocity with l affect this distance, only a possibly
existing gradient of a galactic expansion (or contraction)
velocity field $d\Pi/dl$ influences the result.

155.031 **A southern atlas of galactic hydrogen. III. The**
regions $320° \leqslant l \leqslant 345°, +18° \leqslant b \leqslant +26°$ **and**
$346° \leqslant l \leqslant 350°, +18° \leqslant b \leqslant +20°$.
C. A. Olano, W. G. L. Pöppel, E. R. Vieira.
Astron. Astrophys.,Suppl. Ser., Vol. 46, 41 - 47 (1981).
The authors present observational data in the 21-cm
neutral hydrogen line, which were obtained with the 30 m dish
of the IAR. The radial velocity interval extends from −100 to
+ 100 km s^{-1}.

155.032 *RGU* **photometry of a field in the Large Sagittarius**
Cloud (Sgr III).   L. Topaktas.
Astron. Astrophys.,Suppl. Ser., Vol. 46, 93 - 99 (1981).
A galactic centre field of 0.103 square degrees with 1649
stars in the Large Sagittarius Cloud (Sgr III) has been studied
photometrically in the *RGU* system down to a limiting
magnitude of 16.3 in *G*. The interstellar reddening function, the
density functions for different groups of stars and the
luminosity functions have been determined. The reddening is
caused by only one cloud with distance of about 700 pc. The
density functions of the late-type giants and of the main-
sequence stars with $0 \leqslant M(G) < 2$ show maxima between 1.2
and 1.8 kpc which agree with the distance of the next inner
spiral arm (−I).

155.033 *UBV* **surface brightness photometry of eight sections**
through the Milky Way from the Helios space
probes.   C. Leinert, I. Richter.
Astron. Astrophys.,Suppl. Ser., Vol. 46, 115 - 129 (1981).
The zodiacal light experiment on Helios measures the
brightness of the Milky Way along four strips of constant
ecliptical latitude. In extension of earlier work *UBV* brightnesses were obtained for all of the resulting eight sections
through the Milky Way. When comparing them to available
photometries the agreement is reasonably good with Pfleiderer
and Mayer in *U*, with Elsässer and Haug in *V*, and particularly
close with Classen in *B*. The Helios results support the *U-B*
and *B-V* colours predicted from Mattila's synthetic Milky Way
model. They also suggest that the intensity of the interstellar
radiation field is decreasing from *V* to *U*.

155.034 **History of star formation and enrichment in the**
Galaxy.   A. A. Suchkov.
Stellar aggregates, Sverdlovsk, 1980, (see 003.006), p. 93 -

102. In Russian. − Abstr. in Ref. zh., 51. Astron., 9.51.797
(1981).

155.035 **Estimate of the frequency of galactic rotation from**
motions of B stars.
T. P. Gerasimenko, A. V. Loktin.
Stellar aggregates, Sverdlovsk, 1980, (see 003.006), p. 143 -
145. In Russian. − Abstr. in Ref. zh., 51. Astron., 9.51.800
(1981).

155.036 **Molecular clouds outside the solar circle in the first**
quadrant of our Galaxy.
M. L. Kutner, K. N. Mead.
Astrophys. J., Lett., Vol. 249, L15 - L18 (1981).
The authors report the detection of extensive low-level
CO emission from molecular clouds well outside the solar
circle, in the longitude range 55°−95°. Three strips, at
$b = 1°3, 1°5,$ and $1°7$, were mapped at intervals of $0°1$ in l,
covering the LSR velocity range of −10 to −135 km s^{-1}.
Maps from the $1°5$ strip are presented here. Armlike concentrations are evident, including an extension of the Perseus
arm into the first quadrant and a well-defined arm 15 kpc
from the galactic center. It is estimated that the amount of
molecular material in the outer Galaxy could be comparable
to that in the "molecular ring," though spread out over a larger
area. Detailed maps of five regions show cloud complexes
60−80 pc long, containing one or more localized CO peaks.

155.037 **Distribution of the electron concentration in the**
Galaxy.
O. Kh. Gusejnov, F. K. Kasumov, I. M. Yusifov.
Astron. Zh., Tom 58, 996 - 1010 (1981). In Russian. English
translation in Soviet Astron., Vol. 25, No. 5.
In order to determine the main parameters of the pulsars
of the last Molonglo survey, the distribution of the electron
concentration in the Galaxy has been studied. The data on the
galactic distribution of various objects (H II regions, H$_2$O
masers, OB stars, etc.) influencing directly or indirectly the
magnitude of n_e, have been used for this purpose. These data,
together with general constraints on the character of the
pulsar distribution obtained earlier and the distribution of
pulsars of the most high-sensitive search in Arecibo, have permitted to define a statistically correct n_e value for the majority (∼ 90%) of pulsars of the second Molonglo survey.

155.038 **Determination of spiral structure parameters of the**
Galaxy from stellar kinematics. Nonlinear descrip-
tion.   V. G. Berman, Yu. N. Mishurov.
Pis'ma Astron. Zh., Tom 7, 590 - 593 (1981). In Russian.
English translation in Soviet Astron. Lett., Vol. 7.
Using five samples including long-period cepheids, supergiants of spectral classes B, A, F − G, M (500 objects in all) in
terms of nonlinear description of star motion disturbed by the
spiral arm gravitational field, the spiral structure parameters of
the Galaxy were obtained. The derived parameters are in agreement with data on radio observations of neutral hydrogen at
the 21-cm line.

155.039 **Abundances in stellar populations and the inter-**
stellar medium in galaxies.
B. E. J. Pagel, M. G. Edmunds.
Annu. Rev. Astron. Astrophys., Vol. 19, (see 003.012), 77 -
113 (1981).
The authors review the results of abundance determination in our own neighborhood, which is of special interest
because of the relatively large amount of detailed information
available for a sufficient number of stars to permit statistical
discussions of such important topics as the age-metallicity and
number-metallicity relations and primordial variations among
different elements. The authors then consider more large-scale
trends, i.e. abundance gradients in the disk and halo of our
Galaxy, the diffuse ISM, and molecular clouds. The remainder

of the article is devoted to other galaxies, which raise special problems in abundance determination.

155.040 A realistic model of the Galaxy.
K. Rohlfs, J. Kreitschmann.
Astrophys. Space Sci., Vol. 79, 289 - 319 (1981).

A three-component mass model of the Galaxy consisting of a modified exponential disk, a spherical inner bulge and a massive outer corona has been derived that attempts to give a valid description of observed features over a wide range of galactic distances. The new model shows the steep rise of the circular velocity close to the galactic centre with a first peak of the rotational velocity at approximately $r = 350$ pc and a deep minimum around $r = 2$ kpc. A feature of this model is a variable disk-thickness like that of an equilibrium disk with constant velocity dispersion in the z-direction.

155.041 H I observations of galaxies in the galactic plane.
J. Pfleiderer, M. D. Gruber, G. M. Gruber, L. Velden.
Astron. Astrophys., Vol. 102, L21 - L22 (1981).

Forty-one faint galaxies detected by Weinberger in the galactic plane ($|b| < 8°$) were observed for signs of H I emission in the velocity range -2000 km s^{-1} to $+6000$ km s^{-1}. Ten were detected, none of which is a possible member of the Local Group. A further ten galaxies gave rise to a suspicion of possible emission.

155.042 Mass models of the galactic bulge derived from the distribution of OH/IR stars.
R. Isaacman, M. J. A. Oort.
Astron. Astrophys., Vol. 102, 347 - 350 (1981).

A series of spheroidal power-law mass models of the galactic bulge is tested against the distribution of OH/IR stars found within 1° of the galactic center. For a power-law index of 1.8, the authors find that the mass of the bulge is $\cong 5 \times 10^5 \sigma_{OH}^2 M_\odot$, where σ_{OH} is the velocity dispersion of the OH/IR stars. σ_{OH} is poorly known because of low-velocity OH absorption, but is $\cong 130$ km s^{-1} in planetary nebulae; the bulge mass is thus $\cong 9 \times 10^9 M_\odot$. The OH/IR star distribution is less flattened than the stellar bulge and shows little sign of systematic rotation.

155.043 Chemical and dynamical evolution of the Galaxy.
A. A. Suchkov.
Priroda, 1981, No. 11, p. 56 - 63. In Russian.

155.044 New details of the local spiral structure of the Galaxy.
M. V. Dolidze.
Pis'ma Astron. Zh., Tom 7, 666 - 670 (1981). In Russian.
English translation in Soviet Astron. Lett., Vol. 7.

Groups (complexes) of regions of star formation concentrations of H I regions, SNRs, X-ray and γ-ray sources located in the local spiral arm are considered as components of more complicated interrelated objects — interarm superassociations.

155.045 Ammonia in the neighbourhood of the galactic center.
R. Güsten, C. M. Walmsley, T. Pauls.
Astron. Astrophys., Vol. 103, 197 - 206 (1981).

The molecular clouds in the immediate neighbourhood of the galactic center have been mapped in the (1,1) and (2,2) transitions of ammonia with 40″ resolution. Selected positions have also been measured in the (4,4) line in order to obtain an estimate of the kinetic temperature in the molecular clouds. The authors conclude that the temperature is in the range $50-120$ K and is fairly uniform throughout the cloud. The present data suggest that the gas temperature is higher than the temperature of the dust in the same region, and they conclude that the gas may be heated due to ionization by a large flux of low-energy cosmic rays. Some consequences of both this unusual heating source and the high gas temperature are discussed briefly. The authors compare their results with other molecular maps as well as with the distribution of late-type stars and suggest that the molecular clouds are probably about 100 pc from the center on the near side. The ammonia column density towards the center itself appears to be more than an order of magnitude smaller than in the direction of the clouds.

155.046 La forma delle galassie.   G. Galletta.
Coelum, Vol. 50, 181 - 201 (1981).

155.047 Our bigger and better Galaxy.   B. J. Bok.
Mercury, Vol. 10, 130 - 133, 158 (1981).

155.048 The Milky Way at a glance.
Sky Telesc., Vol. 62, 524 - 526 (1981).

155.049 Does the Galaxy have four spiral arms?
F. N. Bash.
Astrophys. J., Vol. 250, 551 - 560 (1981).

The Georgelin and Georgelin spiral four-arm model for the Galaxy has been examined. It is shown that a two-arm, density-wave model for the Galaxy predicts an H II region in the same direction and with the same radial velocity as the observed H II regions which the Georgelins use, if one takes the velocity dispersions into account. The results of the two-arm model indicate that the Georgelin and Georgelin kinematic distances differ from the two-arm model distances by more than a factor of 2 in 30% of the cases. This difference arises because their assumption that the H II regions move on circular orbits differs from the author's assumption that the spiral structure has a local affect on the kinematics.

155.050 Chemical evolution in the solar neighborhood. IV. Some revised general equations and a specific model.
B. M. Tinsley.
Astrophys. J., Vol. 250, 758 - 768 (1981).

Three main points are made in this paper: (1) It is shown that, contrary to common belief, extrapolation of standard data suggests that "stars" below 0.1 $M_\odot$ are most unlikely to add significantly to the local surface density. (2) The general equations of chemical evolution are revised to separate living stars explicitly from dead remnants; it is then easier to incorporate constraints based on star counts, etc., consistently into models. (3) A schematic, analytic model is proposed for the solar neighborhood: there is an initial burst of (halo) star formation, followed by a lull with no star formation, and then by evolution of the disk itself with constant rates of star formation and infall. This model crudely represents the outer regions of some dynamical models for disk formation, and it is related to two-era models by many authors, and to a recent disk model by Twarog. A new specific model is proposed, with empirical constraints based on point (1) and on Twarog's stellar ages and metallicities. Predictions of the model agree with nucleochronological ages of the elements and with the stellar age-metallicity relation.

155.051 The mass distribution within our Galaxy: a three component model.
J. A. R. Caldwell, J. P. Ostriker.
Astrophys. J., Vol. 251, 61 - 87 (1981).

A set of mass distribution models for the Galaxy is presented, wherein the relative predominance of the "dark" corona is parametrized by the escape velocity at the sun. Currently existing Galactic structure information is used to select a small range of models probably in good agreement with the true gross large-scale structure of the Galaxy. The method used is "objective" in that no assumed mass-to-light ratios are used; rather, the results are derived from a process of minimizing χ^2 given known kinematical measurements with their estimated associated errors. In the best model (minimum χ^2) the solar radius is 9.1 ± 0.6 kpc and the local disk surface density is 82 ± 12 $M_\odot$pc^{-2}. The local circular velocity is 243 ± 20 km s^{-1} and the local escape velocity is probably in the range of 550–650 km s^{-1} depending on the extent of the dark corona. Within the solar radius the relative fractions of disk, spheroid, and dark corona are approximately 44%, 33%, and 23%. The most surprising result is that at the solar radius the density of matter in the spheroidal component is $1.1 \times 10^{-3} M_\odot$ pc^{-3}, far in excess of that found by Schmidt from analysis of observed high velocity stars.

155.052 **Dynamics of the galactic fountain.**
F. D. Kahn.
Investigating the universe, (see 003.014), p. 1 - 28 (1981).

The author discusses the properties of the hot inter-cloud medium; to do so something needs to be said about the input of energy, mainly from supernova explosions, about radiative processes which cause substantial cooling, and about the dynamical consequences of having gas in the Galaxy which is so hot that its thermal velocity far exceeds the escape velocity from the galactic disk, but not from the Galaxy as a whole. There is an intricate inter-relation between these various aspects, so that the author cautiously approaches the problem by a series of successive approximations.

155.053 **On the circular velocity of the Galaxy.**
D. Lynden-Bell, C. S. Frenk.
Observatory, Vol. 101, 200 - 202 (1981).

At each point in the Galaxy we define the circular velocity V_c as that transverse velocity that would balance the radial component of gravity. The root-mean-square value of V_c at the positions of the globular star clusters may be obtained from their motions. It is $V_c = 212 \pm 16$ km/s.

155.054 **Element abundance gradients across the Galaxy as determined by radio observations.**
J. Schmid-Burgk.
Proceedings of the 5th Göttingen-Jerusalem-Symposium on Astrophysics, (see 012.041), p. 49 - 59 (1981).

155.055 **New subdwarfs. III. On obtaining the vertical galactic metallicity gradient from the kinematics of** nearby stars.   A. Sandage.
Astron. J., Vol. 86, 1643 - 1657 (1981).

Radial velocities and UBV photometry are given for 34 subdwarf candidates, 32 of them new. Space motions calculated using photometric parallaxes and known proper motions range between 39 and 403 km s^{-1}. The space motions for $\sim\frac{1}{2}$ of the stars are greater than 200 km s^{-1}. These stars plus 83 others from paper II not previously used provide $\sim$350 subdwarfs to show again that the mean reduced ultraviolet excess $\langle \delta (0.6) \rangle$ and the $|W|_0$ velocity in the plane are correlated, implying a chemical gradient within the lower halo. The distribution of metallicity expected at any height is calculated from the observed (δ, W_0) distributions in the plane using both a continuum and a two-component kinematic model.

155.056 **Evolution of the galactic disk in the solar neighbourhood. Rate of star formation and metal enrichment.**
S. V. Vereshchagin, A. Eh. Piskunov.
Nauchn. Inf., Vyp. (No.) 49, (see 003.017), p. 31 - 47 (1981). In Russian.

"Mean star metallicity – age" and "present-day mass spectrum slope – rate of star formation" relations are derived for different modern models of the galactic disk evolution in the instantaneous recycling approximation.

155.057 **Differences of the chemical composition of matter in the Galaxy.**   Yu. L. Frantsman.
Nauchn. Inf., Vyp. (No.) 49, (see 003.017), p. 80 - 86 (1981). In Russian.

According to the cepheid period distribution differences inside 5 kpc distance from the sun a conclusion about the heavy element abundance differences in the Galaxy is made. Comparing the data about the cepheid periods and star evolu-

tion calculations, the conclusion is made that in this region of the Galaxy Z changes between 0.01 and 0.04.

155.058 **On the distribution of bright stars on the sky.**
I. N. Latyshev.
Astron. Tsirk., No. 1117, p. 6 - 8 (1980). In Russian.

155.059 **Kinematical characteristics of the centroids of stellar groups at large distances from the galactic** plane.   N. V. Kharchenko.
Astron. Tsirk., No. 1135, p. 5 - 7 (1980). In Russian.

155.060 **Observation of the isotope ^{13}CO of carbon monoxide in the second galactic quadrant.**
F. Casoli, F. Combes.
C. R. Acad. Sci. Paris, Tome 293, Sér. II, 761 - 763 (1981). In French.

Millimetric observations at the Bordeaux telescope (P.O.M.) have established two ^{13}CO maps in the Orion and Perseus arms. In the second galactic quadrant, the molecular emission of spiral arms is much easier to determine than in the first quadrant, since there is neither distance ambiguity, nor line overlap. The choice of the ^{13}CO molecule yields a better estimate of densities than the existing CO observations, the CO line being generally optically thick.

155.061 **On the motion of our Galaxy.**   P. Teerikorpi.
Proceedings of the Third Finnish-Soviet Astronomical Symposium, (see 012.057), p. 141 (1981). – Abstract.

155.062 **Determination du taux de formation d'étoiles et de la fonction initiale de masse dans la Galaxie.**
G. Serra.
Rôle des nuages moléculaires dans la formation des étoiles, (see 012.069), 24 pp.

The implications of observations of diffuse galactic infrared and thermal radio continuum radiation for models of the galactic star formation rate are analyzed. The present distribution of interstellar matter and of hot, massive stars is derived and the bearing of the results upon the initial mass function of the Galaxy is described. The data are compared with the inferences from earlier star counts.

155.063 **Nuages moléculaires et structure spirale.**
F. Combes.
Rôle des nuages moléculaires dans la formation des étoiles, (see 012.069), 7 pp.

155.064 **On the application of the edge-on spiral galaxies orientation criterion to the Milky Way.**
I. I. Pasha, M. A. Smirnov, F. A. Tsitsin.
Astron. Tsirk., No. 1160, p. 3 - 4 (1981). In Russian.

155.065 **A programme of studying the main meridional section of the Galaxy from the viewpoint of** chemical evolution.
V. Malyuto, H. Eelsalu, G. Dzhimshelejshvili.
Academy of Sciences of the Estonian SSR, Prepr. A-2, p. 1 - 7 (1981).

A programme is described for conducting combined investigations in the principal meridional section of the Galaxy (in the plane perpendicular to the galactic plane and crossing the centre of the Galaxy and the sun). A joint treatment of physical and kinematic stellar data from the viewpoint of galactic evolution is planned.

A photometric catalogue of stars in the direction of the bright cloud B in Sagittarius.   See Abstr. 002.045.

La galaxie, l'univers extragalactique. Tome no. 3 de l'encyclopédie scientifique de l'univers.
See Abstr. 002.092.

The Milky Way.   See Abstr. 003.038.

Galactic astronomy.   See Abstr. 003.103.

New survey technique for halo G-dwarfs.
See Abstr. 031.551.

s-process nucleosynthesis, stellar abundances, and galactic evolution.   See Abstr. 061.053.

Early-type high-velocity stars in the solar neighborhood. I. List of candidates.   See Abstr. 111.009.

Early-type high-velocity stars in the solar neighborhood. II. Photometry for 78 candidates.
See Abstr. 111.016.

The degree of completeness of nearby stars and the stellar luminosity function.   See Abstr. 111.017.

The stellar luminosity function at the galactic poles.
See Abstr. 115.005.

The luminosity function and the colours of the solar neighbourhood.   See Abstr. 115.013.

Space density of stars and interstellar extinction near h and χ Persei (Perseus I).   See Abstr. 115.025.

Variable stars in the bulge of the galaxy in a field around NGC 6304.   See Abstr. 122.031.

Observations of $^{14}N/^{15}N$ in the galactic disk.
See Abstr. 131.001.

Analysis of *IUE* observations of the galactic corona and the existence of molecular clouds in the halo.
See Abstr. 131.004.

A high-velocity H I stream interacting with the galactic disc.   See Abstr. 131.006.

Deviations of galactic reddening from the cosecant relation.   See Abstr. 131.037.

Molecular clouds in the outer Galaxy.
See Abstr. 131.064.

Observational constraints on the properties of the coronal interstellar gas.   See Abstr. 131.103.

Origin of the narrow mixed ion quasar absorption lines.   See Abstr. 131.104.

Properties of the galactic molecular cloud ensemble from observations of ^{13}CO.   See Abstr. 131.128.

Model calculations for the star formation in the galactic center and other galactic nuclei.
See Abstr. 131.188.

Detection of [O I] 63 micron emission from the galactic center.   See Abstr. 132.012.

The spiral structure of our Galaxy and kinematics of H II regions.   See Abstr. 132.018.

The z distribution of H II regions in the Galaxy.
See Abstr. 132.023.

Kinematics of planetary nebulae, and the galactic rotation curve.   See Abstr. 135.039.

Multiwavelength VLBI observations of the galactic center.   See Abstr. 141.109.

An Effelsberg-Green Bank galactic H I absorption line survey. I. The observations.   See Abstr. 141.147.

A new analysis of the pulsar distribution in the Galaxy.   See Abstr. 141.546.

On the distribution of pulsars in the galactic plane.
See Abstr. 141.547.

Cosmic ray antiprotons in the closed galaxy model.
See Abstr. 143.074.

Resonant excitation of motion perpendicular to galactic planes.   See Abstr. 151.002.

Is there a unique galactic rotation curve?
See Abstr. 151.015.

Chemical evolution with inhibited star formation rate.   See Abstr. 151.042.

Spiral structure of galaxies – two alternatives.
See Abstr. 151.049.

Infrared studies of four highly reddened globular clusters: Palomar 2, Palomar 8, Palomar 10, and NGC 6749.
See Abstr. 154.001.

The ages and metallicities of the globular clusters in the Fornax dwarf spheroidal galaxy.   See Abstr. 154.006.

Globular cluster systems: implications for galaxy formation.   See Abstr. 154.018.

The galactic distribution of globular clusters.
See Abstr. 154.036.

Distribution of near infrared sources in the galactic disk.   See Abstr. 156.001.

Distribution of galactic synchrotron emission. II.
See Abstr. 156.010.

The stellar contribution to the galactic soft X-ray background.   See Abstr. 157.010.

Optical polarization of M82 and the local spiral arm.
See Abstr. 158.046.

Die Radialgeschwindigkeiten und Massen von Galaxien hoher Leuchtkraft und die differentielle Rotation unserer Milchstraße.   See Abstr. 158.141.

Dynamical or static halo – is there a galactic wind?
See Abstr. 158.146.

Structure of galaxies.   See Abstr. 158.214.

The velocity structure of gas in the lines of sight to the Magellanic Clouds.   See Abstr. 159.005.

The Magellanic Clouds and the galactic halo.
See Abstr. 159.023.

156 Galaxy (Magnetic Field, Radio and Infrared Radiation)

156.001 Distribution of near infrared sources in the galactic disk. S. Hayakawa, T. Matsumoto, H. Murakami, K. Uyama, J. A. Thomas, T. Yamagami.
Astron. Astrophys., Vol. 100, 116 - 123 (1981).

Near infrared surveys of the galactic plane were performed by two balloon experiments over the galactic longitude range −70° ~ 50° at a wavelength of 2.4 μm and partially at 3.4 μm. Steps and humps observed in the longitude distribution coincide with the directions tangent to major spiral arms and with star forming regions relatively near the Sun. No absorption feature expected from the concentration of dust towards the galactic plane was observed. This indicates that the near infrared sources contain a component strongly concentrated towards the plane with the scale height of about 50 pc.

156.002 Submillimeter survey of the galactic plane. M. T. Stier, R. F. Silverberg, L. H. Cheung, M. G. Hauser, T. Kelsall, D. Y. Gezari.
Bull. American Astron. Soc., Vol. 13, 537 (1981). − Abstract.

156.003 Origin of the diffuse infrared emission from the galactic plane.
L. H. Cheung, G. G. Fazio, F. W. Stecker, D. B. Sanders, P. M. Solomon.
Bull. American Astron. Soc., Vol. 13, 537 - 538 (1981). Abstract.

156.004 Galactic non-thermal radiation. R. L. Brown.
The phases of the interstellar medium, (see 012.018), p. 163 - 168 (1981).

156.005 Distribution of infrared sources in the galactic plane. A. G. Ananth, B. V. Nagaraja.
Bull. Astron. Soc. India, Vol. 9, 67 (1981). − Abstract.

156.006 Diffuse emission of the galactic disk in the radio, far-IR and near-IR ranges and stellar populations.
G. Serra.
Ann. Physique, Vol. 6, (see 012.026), p. 53 - 61 (1981). In French. − Abstr. in Phys. Abstr., Vol. 84, Abstr. 94501 (1981).

156.007 Magnetic field generation in the galactic disk. A. A. Ruzmajkin, A. M. Shukurov.
Astron. Zh., Tom 58, 969 - 978 (1981). In Russian. English translation in Soviet Astron., Vol. 25, No. 5.

The process of the large-scale galactic magnetic field generation is determined by the values of two dimensionless dynamo numbers. The authors evaluate them by employing the observed rotation curve of the Galaxy and taking into account the radial variations of the ionized gas layer thickness and of a characteristic velocity of turbulent motions.

156.008 Large-scale formations of matter in the region $l^{II} = 15°\text{-}45°$, $b^{II} = 0°\text{-}15°$ according to 102.5 MHz observations at the large synphase array of the Lebedev Physical Institute. V. I. Ariskin.
Astron. Zh., Tom 58, 979 - 983 (1981). In Russian. English translation in Soviet Astron., Vol. 25, No. 5.

Observations have been carried out and a contour map of the radio brightness distribution in the region $l^{II} = 15°\text{-}50°$, $b^{II} = 0°\text{-}15°$ was obtained with the purpose of a search for supernova remnants.

156.009 High resolution VLA observations of the galactic center. R. L. Brown, K. J. Johnston, K. Y. Lo.
Astrophys. J., Vol. 250, 155 - 159 (1981).

The authors present a 5 GHz map of the inner 1′ × 1′ of the galactic center made with the VLA. The angular resolution of the map is 2″ × 8″ (α × δ). The radio map is very similar to the 10 μm maps of this region; in particular, nearly all of the thermal 10 μm peaks have corresponding peaks in the radio brightness distribution. The authors find that the ratio of 10 μm flux density to radio flux density for these discrete regions is surprisingly high, 100−1000, a result which implies that the source of excitation is very soft−(blackbody) temperatures ~25,000 K are indicated. The 5 GHz flux density of the point nonthermal source in Sgr A West has increased by a factor ~ 2 in 2 years.

156.010 Distribution of galactic synchrotron emission. II. S. Phillipps, S. Kearsey, J. L. Osborne, C. G. T. Haslam, H. Stoffel.
Astron. Astrophys., Vol. 103, 405 - 414 (1981).

Using the all-sky radio continuum map at 408 MHz of Haslam et al. (1981), the model of the large scale distribution of synchrotron emissivity on the galactic plane derived in a previous paper (Phillipps et al., 1981) is extended to three dimensions. It is found that the observed map cannot be accounted for entirely in terms of a distribution of emissivity with distance from the plane that is the same throughout the Galaxy. A distribution whose characteristics scale length varies as exp $((R-10)/8)$ where R is the galactocentric distance gives a better fit. In addition to the radio disc a non-spherical halo of emission extending ~10 kpc from the plane with an emissivity ≲ 10% of that in the plane is required to account for the high latitude brightness temperature. The main difference between the observed and modelled distributions of radio emission is the broader observed distribution about the plane in the second quadrant of galactic longitude. This is believed to be due to a local arm which does not appear in the unfolding.

156.011 Structure of the magnetic field in the cloud of interstellar matter associated with the Perseus OB2 association. T. Markkanen.
Proceedings of the Third Finnish-Soviet Astronomical Symposium, (see 012.057), p. 17 - 24 (1981).

A provisional report about a polarization observation programme in the Perseus OB2 association is given. The amounts of polarization are found to be correlated with extinction in the association region. The magnetic field structure is found to agree with the shape of the HI cloud associated with the stellar association.

156.012 Spectrum of the galactic magnetic field. A. A. Ruzmajkin, A. M. Shukarov.
Inst. prikl. mat. AN SSSR. Prepr., 1981, No. 75, 18 pp. In Russian. − Abstr. in Ref. zh., 51. Astron., 12.51.914 (1981).

Programmes for measurement processing of linear polarized galactic radio emission by the NAIRI-K computer. See Abstr. 021.026.

A search for the infrared counterpart of type II OH masers − I. A model for the IR background source confusion. See Abstr. 133.010.

Radio patrol of the northern Milky Way: a survey for variable sources. See Abstr. 141.044.

A synoptic view of galactic processes. See Abstr. 155.025.

Element abundance gradients across the Galaxy as determined by radio observations.
See Abstr. 155.054.

Determination du taux de formation d'étoiles et de la fonction initiale de masse dans la Galaxie.
See Abstr. 155.062.

157 Galaxy (UV, X, Gamma Radiation)

157.001 Galactic γ-ray emission from pulsars.
A. K. Harding.
Astrophys. J., Vol. 247, 639 - 649 (1981).
   The contribution of pulsars to the γ-ray flux from the galactic plane is examined using data from the most recent pulsar surveys. It is assumed that pulsar γ-rays are produced by curvature radiation from relativistic particles above the polar cap and attenuated by pair production in the strong magnetic and electric fields; their luminosities can be predicted as a function of period and magnetic field strength. The distribution of pulsars in the Galaxy is determined from data on 328 pulsars detected in three surveys. A present estimate is that pulsars contribute from 15% - 20% of the total flux of γ-rays from the galactic plane.

157.002 Variable positron annihilation radiation from the galactic center region.   G. R. Riegler,
J. C. Ling, W. A. Mahoney, W. A. Wheaton, J. B. Willett, A. S. Jacobson, T. A. Prince.
Astrophys. J., Lett., Vol. 248, L13 - L16 (1981).
   Studies of 511 keV positron annihilation radiation from the vicinity of the galactic center are reported, based on data which were recorded during 1979 September/October and 1980 March/April by the HEAO 3 Cosmic Gamma-Ray Spectrometer. The 1979 fall data show unshifted, narrow 511 keV line emission of intensity $(1.85 \pm 0.21) \times 10^{-3}$ photons $cm^{-2} s^{-1}$, consistent with earlier measurements. The 1980 spring measurement showed a statistically significant reduction in 511 keV emission from this region, thus requiring that a significant fraction of the flux originate in one or more compact sources of size $\lesssim 10^{18}$ cm. The data are well satisfied by assuming that the emission originates in a single compact source at the galactic center.

157.003 Observation of soft diffuse X-rays in the southern galactic hemisphere.   K. P. Singh, R. K. Manchanda, V. S. Yengar, S. Naranan, B. V. Sreekantan.
Indian J. Radio Space Phys., Vol. 10, No. 1, p. 16 - 19 (1981).
Abstr. in Phys. Abstr., Vol. 84, Abstr. 83749 (1981).

157.004 Medium energy gamma ray sky in the Northern Hemisphere: cross correlation analysis.
A. D. Zych, J. Long, E. Zanrosso, R. S. White.
Bull. American Astron. Soc., Vol. 13, 534 (1981). – Abstract.

157.005 Solrad 11 observations of the 1220 - 1500 Å far-UV background near the galactic poles.
C. S. Weller.
Bull. American Astron. Soc., Vol. 13, 537 (1981). – Abstract.

157.006 HEAO C-1 limits on diffuse galactic plane line emission.   W. A. Mahoney, J. C. Ling,
G. R. Riegler, W. A. Wheaton, A. S. Jacobson.
Bull. American Astron. Soc., Vol. 13, 549 (1981). – Abstract.

157.007 How much coronal phase?
D. McCammon.
The phases of the interstellar medium, (see 012.018), p. 135 - 137 (1981).

157.008 Limits on the contribution of galactic halo emission to the diffuse X-ray background.
J. A. Nousek.
The phases of the interstellar medium, (see 012.018), p. 187 - 189 (1981).
   The author assesses the observations of the soft X-ray background from the standpoint of placing an upper limit on the contribution from sources distributed in a halo around our Galaxy to that background.

157.009 Gamma-ray lines from the galactic center and gamma-ray transients.
R. Ramaty, D. Leiter, R. E. Lingenfelter.
NASA Tech. Memo., NASA TM 82125, 20 pp. (1981).
   The observations and interpretations of cosmic (nonsolar) gamma-ray lines are discussed. The most prominent of these lines is the $e^+ - e^-$ annihilation line which has been observed from the galactic center and from several gamma-ray transients.

157.010 The stellar contribution to the galactic soft X-ray background.   R. Rosner, Y. Avni, J. Bookbinder,
R. Giacconi, L. Golub, F. R. Harnden, Jr., C. W. Maxson, K. Topka, G. S. Vaiana.
Astrophys. J., Lett., Vol. 249, L5 - L9 (1981).
   The authors construct log N-log S relations for stars based on median X-ray luminosities for dF, dG, and dK stars previously reported for the Einstein Observatory/Center for Astrophysics stellar survey and on a detailed X-ray luminosity function derived here for dM stars, and investigate the stellar contribution to the diffuse soft X-ray background. The principal results are that stars provide ~20% of the soft X-ray background in the 0.28–1.0 keV passband and therefore contribute significantly to the soft X-ray background in this energy range (with dM stars constituting the dominant contributing class), and that the stellar contribution to the diffuse X-ray background in the 0.15–0.28 keV passband is $\lesssim$3%.

157.011 The γ-ray emissivity of the local interstellar medium from correlations with gas at intermediate latitudes.   A. W. Strong, A. W. Wolfendale.
Philos. Trans. R. Soc. London, Ser. A, Vol. 301, 541 - 553 (1981). – Abstr. in Phys. Abstr., Vol. 84, Abstr. 98853 (1981).

157.012 Low-latitude galactic γ-ray emission: a probe, not a proof.   G. F. Bignami.
Philos. Trans. R. Soc. London, Ser. A, Vol. 301, 555 - 567 (1981). – Abstr. in Phys. Abstr., Vol. 84, Abstr. 98854 (1981).

157.013 An analysis of the longitudinal distribution of gamma rays from SAS-II data.   S. A. Stephens.
Astrophys. Space Sci., Vol. 79, 419 - 434 (1981).
   An analysis of the longitudinal distribution of gamma rays from SAS-II data has been carried out using the available information on the gas distribution in the Galaxy. The overall distribution of cosmic rays in the galactic plane can be represented by an exponential function in galactocentric distance with a scale length of ~8 kpc up to the solar circle

and ~10 kpc beyond. There is no evidence for a large gradient of the cosmic ray intensity in the outer parts of the Galaxy.

157.014 An X-ray study of the galactic center.
M. G. Watson, R. Willingale, J. E. Grindlay, P. Hertz.
Astrophys. J., Vol. 250, 142 - 154, plates 15, 16 (1981).

The authors present the results from two long observations with the Einstein Observatory Imaging Proportional Counter (IPC) of a ~ 1° × 1° field centered near the galactic nucleus. The X-ray images reveal a complex of weak sources within 20' of the galactic nucleus (Sgr A West) together with a region of apparently diffuse emission ~25' × 15' in extent. Three of the sources are tentatively identified: two with nearby galactic objects and a third positionally coincident with Sgr A West itself (within the arcminute accuracy available with the IPC). The authors discuss the nature of this source and the implications of both the high source density and presence of diffuse emission in the galactic center region.

157.015 The diffuse galactic gamma radiation: the Compton contribution and component separation by energy interval and galactic coordinates.
D. A. Kniffen, C. E. Fichtel.
Astrophys. J., Vol. 250, 389 - 397 (1981).

The diffuse high-energy galactic γ-radiation to be expected from cosmic ray interactions with matter and photons is examined. Particular emphasis is placed on the Compton emission since work in this area is hindered by the limited knowledge of the galactic photon densities. Both the photon density in and near the visible region and that in the infrared region are deduced from the estimates of the emission functions throughout the Galaxy. The blackbody radiation is also included in the estimate of the total Compton emission. The result suggests that the γ-ray Compton radiation from cosmic ray interactions with galactic visible and infrared photons is substantially larger than previously believed.

157.016 Spectral observation of the soft X-ray background and of the North Polar Spur with solid state spectrometers.
R. Rocchia, M. Arnaud, C. Blondel, C. Cheron, J. C. Christy, R. Ducros, L. Koch, R. Rothenflug, H. W. Schnopper, J. P. Delvaille.
Space Sci. Rev., Vol. 30, (see 012.044), 253 - 257 (1981).

The authors present spectral measurements of the soft X-ray (E < 1 keV) background obtained with 3 cooled Si(Li) solid state detectors during the flight of a spin-stabilized rocket. The rocket was launched from the White Sands Missile Range on 22 March 1980 at 0330 AM MST. Approximately 260 seconds of data were obtained above an altitude of 120 km.

157.017 Einstein observations of the galactic centre.
M. G. Watson, R. Willingale, J. E. Grindlay, P. Hertz.
Space Sci. Rev., Vol. 30, (see 012.044), 293 - 299 (1981).

The authors present the X-ray observations made with the Einstein Observatory Imaging Proportional Counter (IPC) of a 1 × 1 degree field centred near the galactic nucleus. This is the first observation of this region with an imaging X-ray telescope. Radio and infrared maps reveal the presence of numerous H II regions and molecular cloud complexes, and in the near infrared one sees evidence for the rapidly increasing stellar density towards the galactic nucleus. The nucleus itself is believed to be coincident with Sgr A West, a radio/infrared source which has a compact, nonthermal core of very small size.

157.018 On the origin of galactic annihilation radiation.
F. A. Agaronyan, A. M. Atoyan.

Pis'ma Astron. Zh., Tom 7, 714 - 719 (1981). In Russian. English translation in Soviet Astron. Lett., Vol. 7.

A relationship between photon emissivity of the 0.511 MeV line and continuous γ-radiation owing to suprathermal positron annihilation is obtained. The analysis of this relationship along with observational data yields a rather severe restriction imposed on the injection spectrum of positrons which produce the 0.511 MeV annihilation line observed in the direction of the galactic center.

157.019 PROGNOZ-6 data about ultraviolet sky background in dark and Milky Way regions of the sky.
A. Zvereva, A. Severny (Severnyj), C. T. Hua, P. Cruvellier, G. Courtes.
High-energy astrophysics, (see 012.067), p. 201 - 209 (1981).

Ultraviolet spectra (1100 - 1900 Å) of the sky background of 10 wide angle (6° × 6°) regions obtained between 70 000 km and 200 000 km from the Earth with the photoelectric spectrometer "GALACTIKA" on board the satellite "PROGNOZ-6" are considered. The spectral energy distribution of the sky background, after subtraction of the stellar component, is similar for regions on both sides of the Milky Way and exhibits a strong UV light contribution. In contrast, the Milky Way (1_{II} = 190° b_{II} = +6°) is less rich in far UV light; this can be related to the predominance of an expected selective absorption near the galactic plane.

157.020 Observations of 1 - 30 MeV gamma rays from the galactic center.
E. Zanrosso, J. L. Long, A. D. Zych, R. S. White.
High-energy astrophysics, (see 012.067), p. 235 - 238 (1981).

Preliminary results are reported for gamma ray observations of the galactic center region. The observations were carried out with the UCR double-scatter gamma-ray telescope at energies of 1 to 30 MeV. The observations are compatible with a galactic source of approximately equal brightness along the region 300° < l^{II} < 60°. The energy distribution joins smoothly to previous spark chamber results at energies above 30 MeV and to scintillator results below 1 MeV. It appears to be a combination of nuclear gamma ray lines superimposed on a bremsstrahlung spectrum. The $^{12}C^*$ line at 4.4 MeV appears to be present with a significance of about 16σ.

Overview of the Voyager ultraviolet spectrometry results through Jupiter encounter.   See Abstr. 099.084.

Analysis of IUE observations of the galactic corona and the existence of molecular clouds in the halo. See Abstr. 131.004.

Gamma rays – the last frontier. See Abstr. 142.508.

Interpretations and implications of γ-ray lines from solar flares, the galactic centre and γ-ray transients. See Abstr. 142.522.

A synoptic view of galactic processes. See Abstr. 155.025.

X rays from normal galaxies and clusters of galaxies.   See Abstr. 158.108.

Possible detection of far-ultraviolet line emission from a hot galactic corona.   See Abstr. 158.161.

158 Single and Multiple Galaxies, Peculiar Objects

158.001   **A 1415 MHz survey of Seyfert and related galaxies.**
II.   E. J. A. Meurs, A. S. Wilson.
Astron. Astrophys., Suppl. Ser., Vol. 45, 99 - 110 (1981).
  The authors present the second installment of a high
sensitivity survey with the Westerbork telescope for radio continuum emission at 1415 MHz from Seyfert and related galaxies. 36 Seyferts, 10 possibly Seyfert-related, and 4 other
galaxies, all with declination $\delta > +10°$, have been observed to
a flux density limit (3 σ) of about 3 mJy. 21 (58%) of the
Seyferts, 7 (70%) of the possibly Seyfert-related and 3 (75%)
of the other galaxies were detected. Several galaxies are extended and contour maps with resolution 23 × 23 cosec δ arc sec
are presented for Mark 506, 700, 871, Arak 42, NGC 6764,
MCG 8-11-11, 4C29.6 and 4C35.37. With the exception of
4C29.6 and 4C35.37, no linear polarization is found in the
detected galaxies.

158.002   **Emission-line profiles and kinematics of the narrow-**
line region in Seyfert and radio galaxies.
T. M. Heckman, G. K. Miley, W. J. M. van Breugel,
H. R. Butcher.
Astrophys. J., Vol. 247, 403 - 418 (1981).
  This paper presents results for the [O III] $\lambda5007$ line in
36 Seyfert and radio galaxies. Some parameters which describe
the profiles are extracted and tabulated. The distributions of
these parameters and their relationship to other parameters of
the galaxies in the sample are considered. Consequences of the
results for the energetics of active galaxies are discussed.

158.003   **Radio structures of Seyfert galaxies. II.**
J. S. Ulvestad, A. S. Wilson, R. A. Sramek.
Astrophys. J., Vol. 247, 419 - 442 (1981).
  High resolution radio maps of 16 Seyfert and active galaxies have been made with the Very Large Array at 1.465 and/
or 4.885 GHz. Two Seyfert galaxies have been found to contain double or triple radio sources (Mrk 6 and Mrk 78); a third
object (NGC 4151) is probably also a triple source. In the
double and triple sources, there is evidence that the major axis
of the complex of nuclear narrow line emitting clouds may be
correlated with the radio source axis.

158.004   **CO emission from Seyfert galaxies.**
J. H. Bieging, L. Blitz, C. J. Lada, A. A. Stark.
Astrophys. J., Vol. 247, 443 - 448 (1981).
  The authors report the detection of millimeter-wave CO
emission from two Seyfert galaxies, NGC 3227 and NGC 4051.
Sensitive upper limits for CO emission toward four other
Seyfert galaxies (NGC 1275, NGC 4151, NGC 4439, and
NGC 6764) are also reported. Scaling the CO detections and
upper limits by H I and 10 μm emission shows that neither of
these quantities correlates well with the CO emission.

158.005   **The C IV $\lambda1550$ profile in type 1 Seyfert galaxies.**
C.-C. Wu, A. Boggess, T. R. Gull.
Astrophys. J., Vol. 247, 449 - 457 (1981).
  The authors present C IV $\lambda1550$ line profiles for the
following type 1 Seyfert galaxies: NGC 5548, Mrk 509,
NGC 7469, and MCG−2-58-22. They consider several possible
line broadening mechanisms and the theoretical line profiles
that would result. While C IV profiles strongly favor the ballistic model suggested by Capriotti, Foltz, and Byard, both
Mrk 509 and NGC 7469 have profiles with significant asymmetry. It is also noted that for all four galaxies, the C IV lines are
at lower redshift than the Balmer lines.

158.006   **Concurrent radio, infrared, optical, and X-ray obser-**
vations of the nucleus of the Seyfert galaxy
NGC 4151.   J. H. Beall, W. K. Rose, B. R. Dennis,

C. J. Crannell, J. F. Dolan, K. J. Frost, L. E. Orwig.
Astrophys. J., Vol. 247, 458 - 463 (1981).
  The authors observed the nucleus of the Seyfert galaxy
NGC 4151 at X-ray energies from 20 to 200 keV and at radio
frequencies of 2695 and 8085 MHz. Concurrently with these
observations, data were taken in the visual, and at infrared
wavelengths of 1 to 12 μm. Observations are also available
from 2 to 6 keV during this interval. The optical data show
variability comparable to the variability in the 2 to 6 keV
X-ray flux. No variability was measured at infrared or radio
frequencies.

158.007   **X-ray emission from M87: a pressure confined cool-**
ing atmosphere surrounding a low mass galaxy?
J. Binney, L. L. Cowie.
Astrophys. J., Vol. 247, 464 - 472 (1981).
  With a conventional model for the mass of the galaxy
M87 ($5 \times 10^{11} M_\odot$ at 100 kpc), the authors show that a radiative cooling flow agrees remarkably well with spectroscopically
derived mass flow rates onto M87 ($\approx 10 M_\odot$ yr^{-1}) and surface
brightness profiles of X-ray emission around the galaxy. Values
of the density ($\sim$ few $\times 10^{-4}$ cm^{-3}) and temperature ($\sim 10^8$ K)
in the diffuse cluster are obtained. No massive halo is required.
The atmosphere of M87 contains $\sim 10^{12} M_\odot$ of gas and possesses
a temperature profile which increases outward.

158.008   **A photometric and kinematic study of the barred**
spiral galaxy NGC 253. II. The velocity field.
W. D. Pence.
Astrophys. J., Vol. 247, 473 - 483 (1981).
  The velocity field of the SAB(s)c spiral galaxy NGC 253
is derived from Hα Fabry-Perot interferometry covering most
of the main disk of H II emission. Several basic galaxy parameters are derived which are in good agreement with previous
determinations. The fit of various mass models to the rotation
curve is discussed. A two-component mass model shows that
the mass and luminosity cannot be similarly distributed in
NGC 253. A complicated pattern of noncircular motions of up
to ± 50 km s^{-1} is observed. These residual velocities cannot be
explained by the spiral density wave theory.

158.009   **Observations of M100 with the Einstein Observatory**
shortly after the explosion of its fourth supernova,
SN 1979c.   G.G.C. Palumbo, T. Maccacaro, N. Panagia,
G. Vettolani, G. Zamorani.
Astrophys. J., Vol. 247, 484 - 487, plate 14 (1981).
  The authors report the results of X-ray observations with
the Einstein Observatory of M100 and SN 1979c 63, 73, and
239 days after explosion. No X-ray flux has been detected
from SN 1979c or from three previously reported supernovae
in the same galaxy. The nucleus of M100 has been detected
with an X-ray luminosity of $\sim 1.5 \times 10^{40}$ ergs s^{-1}. A source projected on the northern spiral arm of the galaxy was also detect-
ed. The possibility of a background QSO or of a cluster of galaxies by chance covered by M100 is also discussed.

158.010   **Density-wave induced star formation: a model for**
M81.   F. N. Bash, H. C. D. Visser.
Astrophys. J., Vol. 247, 488 - 498, plates 15, 16 (1981).
  The two-armed spiral shock wave, fitted to observations
of the H I gas in M81 by Visser, is used to give initial conditions for ballistic particles launched from the spiral arms. The
model predicts the location of molecular clouds which agrees
with the one detection of CO in M81. The model agrees with
the observed distribution of H II regions near the eastern arm
and also predicts the radial velocities of, and velocity differences between, H II regions and molecular clouds near the

major and minor axes. The model also produces a galaxy which resembles a photograph of M81.

158.011 The metallicities, velocity dispersions and true shapes of elliptical galaxies.
R. Terlevich, R. L. Davies, S. M. Faber, D. Burstein.
Mon. Not. R. Astron. Soc., Vol. 196, 381 - 395 (1981).

The authors have re-analysed the relationships between velocity dispersion, line strength and absolute magnitude for normal elliptical galaxies. They find that, at fixed absolute magnitude, galaxies with high velocity-dispersions tend to have high line-strengths and those with low velocity-dispersions have low line-strengths. This implies that elliptical galaxies are at least a two-parameter family. This conclusion is based on a preliminary sample of 24 galaxies for which both types of measurements are available. Well-known apparent scale differences in velocity dispersion between various authors are shown to be due to mean line-strength differences between galaxy samples. After correction for this effect, all sources considered here agree well to within a few per cent.

158.012 Spectroscopic survey of southern compact and bright-nucleus galaxies – IV. A. P. Fairall.
Mon. Not. R. Astron. Soc., Vol. 196, 417 - 424 (1981).

Seventy galaxies with strong emission lines have been found. They include three Seyfert 1 galaxies, five Seyfert 2 galaxies and eleven near-Seyfert galaxies.

158.013 Early-type ("discless") galaxies with dust lanes.
T. G. Hawarden, R. A. W. Elson, A. J. Longmore, S. B. Tritton, H. G. Corwin, Jr.
Mon. Not. R. Astron. Soc., Vol. 196, 747 - 756 (1981).

The authors present a comprehensive homogeneous sample of 40 galaxies which contain strong dust lanes but exhibit no obvious associated luminous discs. A dearth of such galaxies in the north is attributed to the poorer resolution of the Palomar Sky Survey relative to the ESO/SRC IIIaJ Southern Sky Survey which is the authors' source material in the south. They infer that there may not be a comparably large population of galaxies with dust lanes closely confined to their central regions.

158.014 The dynamics of the S0 galaxy IC 5063.
I. J. Danziger, W. M. Goss, K. J. Wellington.
Mon. Not. R. Astron. Soc., Vol. 196, 845 - 856 (1981).

Low-dispersion spectrophotometry of the S0 galaxy IC 5063 suggests there is an intense hot source at its centre, photo-ionizing not only its immediate surroundings but also regions of the gaseous disc at large distances from the centre. Long slit spectroscopy with an IPCS reveals circular motion in the gaseous disc with an observed amplitude of 185 km s^{-1} and a real amplitude of 226 km s^{-1}. The morphology of the galaxy and spatial extension of the emission lines are consistent with the gaseous disc's being warped and optically thick to ionizing radiation from the central source. The 21 cm line of neutral hydrogen has been detected in emission from this galaxy with a systemic velocity and width almost precisely that observed for the ionized gas.

158.015 Detailed observations of NGC 4151 with *IUE* – I. Low dispersion data up to 1979 January.
M. V. Penston, A. Boksenberg, G. E. Bromage, J. Clavel, A. Elvius, P. M. Gondhalekar, C. Jordan, J. Lind, L. Lindegren, G. C. Perola, M. Pettini, M. A. J. Snijders, E. G. Tanzi, M. Tarenghi, M. H. Ulrich.
Mon. Not. R. Astron. Soc., Vol. 196, 857 - 887 (1981).

NGC 4151 is the brightest type 1 Seyfert galaxy and has attracted much observational attention at all wavebands. The purpose of this paper is to give a further discussion of data on this galaxy taken over the first year of the guest observer programme of the *IUE* by a widespread European collaboration. It covers a data description, the continuum, the absorp-

tion lines, the emission lines, and a discussion of the variability. The authors anticipate a further paper to discuss later observations and other refined analyses of the presently reported results.

158.016 Colours and morphology of spiral galaxies.
R. F. G. Wyse.
Mon. Not. R. Astron. Soc., Vol. 196, 911 - 914 (1981).

Tinsley has proposed that late-type spirals have relatively more non-luminous material than early-type spirals. A re-examination of the data indicates that this proposal is equally consistent with dark matter being more dominant in barred galaxies than in unbarred galaxies. Neither conclusion can be firm, since the dataset is far from ideal.

158.017 On the local mass to light ratio in galaxies.
M. Petrou.
Mon. Not. R. Astron. Soc., Vol. 196, 933 - 942 (1981).

Evidence is presented that there is a universal law which relates the local light with the local mass at each point in a galaxy. The implication is that when the photometric parameters of a disc are established, the distribution of mass in the dark halo is automatically established too. A method is presented of how to extend these calculations to S0s and ellipticals if more data become available.

158.018 The rotation axes of five radio galaxies.
C. R. Jenkins.
Mon. Not. R. Astron. Soc., Vol. 196, 987 - 993 (1981).

Spectroscopic results are presented on the internal dynamics of five radio galaxies. These data, taken together with those of Jenkins & Scheuer, suggest that there is no tendency for absorption-line rotation axes to align with radio structure. However, radio ellipticals seem to be dynamically peculiar compared to ordinary ellipticals, in that they often show rotation about the minor axis and so are not oblate spheroids.

158.019 The question of extinction in active galactic nuclei: infrared spectral observations of NGC 1614, NGC 7469 and NGC 1275.
D. K. Aitken, P. F. Roche, M. M. Phillips.
Mon. Not. R. Astron. Soc., Vol. 196, 101P - 107P (1981).

The authors present 8 - 13 μm spectra of the nuclei of the three active galaxies NGC 1614, 7469 and 1275. For the first time the [Ne II] fine structure line and the unidentified "11.25 μm" feature are definitively observed in a Seyfert nucleus, NGC 7469, and are also seen in the high luminosity emission-line nucleus of NGC 1614. It is shown that the observations are consistent with there being no large optical extinction of amount $A_V \gtrsim 10$ mag in any of these objects. The presence of 11.25 and 8.65 μm emission features in the type 1 Seyfert NGC 7469 and in NGC 1614 indicates a dust-emission mechanism, whereas the featureless spectrum of NGC 1275 shows no evidence for the presence of dust in this object.

158.020 UV observations of the new BL Lac object 0716 + 71.
K. J. Fricke, W. Kollatschny, H. Schleicher.
Astron. Astrophys., Vol. 100, 1 - 2 (1981).

IUE observations of the new BL Lac object 0716 + 71 are presented.

158.021 On the width and profile of nuclear emission lines in galaxies. M. P. Véron.
Astron. Astrophys., Vol. 100, 12 - 19 (1981).

The author has obtained spectra with a resolution of 100 km/s of 9 Seyfert 1, 8 Seyfert 2 galaxies and 8 nuclear H II regions. He has studied the profiles of the [O III] λ5007 line in all these galaxies and he has shown that in Seyfert 1 and 2 galaxies, the narrow emission lines have complex profiles, being usually asymmetric and having wings which can extend over 500 to 1000 km/s. The FWHM of the observed objects cover the range 200 to 530 km/s. The profiles of the emission lines of the nuclear H II regions are symmetric and can be approximated by a gaussian with a FWHM not exceeding 150 km/s.

158.022   The extended H I−envelope of NGC 5236 (M 83).
W. K. Huchtmeier, H.-D. Bohnenstengel.
Astron. Astrophys., Vol. 100, 72 - 78 (1981).

21 cm line observations of the galaxy NGC 5236 (M 83) with the 100-m radiotelescope of the Max-Planck-Institut für Radioastronomie at Effelsberg reveal one of the greatest H I−envelopes observed so far (linear extent of $\geqslant 217$ kpc), 6.5 times the optical extent. About 80% of the neutral hydrogen of $2.4 \times 10^{10} M_\odot$ is located outside the galaxy's Holmberg limits. The H I distribution is asymmetric with greater extent to the east. The outer half of the H I−envelope seems to rotate in a different plane than the galaxy itself.

158.023   Millimeter-wave and X-ray observations of a Cen-A flare.   P. Kaufmann, F. M. Strauss, M. J. Coe, G. F. Carpenter.
Astron. Astrophys., Vol. 100, 189 - 190 (1981).

Simultaneous observations of Cen-A were made in December 1979 at 22 GHz (Itapetinga) and soft X-rays (Ariel-5). A significant radio flare was detected. In phase X-ray excess emission was not significant, and is discussed.

158.024   Metric properties of the inner ring structures of galaxies.   M. Pedreros, B. F. Madore.
Astrophys. J., Suppl. Ser., Vol. 45, 541 - 557 (1981).

New data are presented for the diameters of inner ring structures in S(r) galaxies. It is found that ring diameters decrease with advancing Hubble type and increase toward higher luminosity classes. Expressions are derived which relate absolute ring dimensions to the parent galaxy parameters, and the application of the formulae is discussed in terms of distance determinations.

158.025   Accurate optical positions for Markarian galaxies 798 - 1095.   G. Kojoian, R. Elliott, H. M. Tovmassian (G. M. Tovmasyan).
Astron. J., Vol. 86, 811 - 815 (1981) = Contrib. No. 3 of Casey Obs.

Optical positions of 298 Markarian objects, Markarian 798 - 1095, were measured from the Palomar Sky Survey Prints, and are given with the accuracy of about 1.5 arcsec in each coordinate.

158.026   Accurate optical positions for Markarian galaxies 1096 - 1302.
G. Kojoian, R. Elliott, M. D. Bicay.
Astron. J., Vol. 86, 816 - 819 (1981) = Contrib. No. 4 of Casey Obs.

Optical positions of 207 Markarian objects, Markarian 1096 - 1302, were measured from the Palomar Sky Survey Prints, and are given with an accuracy of about 1.5 arcsec in each coordinate.

158.027   Accurate optical positions of Arakelian galaxies.
G. Kojoian, R. Elliott, M. D. Bicay, M. A. Arakelian (Arakelyan).
Astron. J., Vol. 86, 820 - 825 (1981) = Contrib. No. 5 of Casey Obs.

Optical positions of 591 Arakelian objects, Arak 1 −591, were measured from the Palomar Sky Survey Prints, and are given with an accuracy of about 1.5 arcsec in each coordinate.

158.028   High-resolution maps of the hotspots of several class II radio galaxies.   J. W. Dreher.
Astron. J., Vol. 86, 833 - 847 (1981).

The Very Large Array was used to map the hotspots of 3C 390.3 at 5 GHz; of 3C 33 (south only), 3C 61.1 (south only), and 3C 234 at 15 GHz; and of Cyg A at 22 GHz. The beam was generally $< 0\rlap{.}''4$ and polarization was mapped at 5 and 15 GHz. A typical hotspot has an ~ 1 kpc head with a sharper outer edge approximately perpendicular to the source axis and $\leqslant 200$ pc in width and with an opposed tail several kiloparsecs long. The hotspots are 25% to 40% polarized with the magnetic field parallel to the edges. Three of the hotspots show limb-brightening. The absence of Rayleigh-Taylor instabilities suggests that $n_e \leqslant 10^{-3}$ cm^{-3} within the heads.

158.029   Detection of the 3.3 μm emission feature in the nuclei of IC 4329A and NGC 5506.
A. F. M. Moorwood, P. Salinari.
Astron. Astrophys., Vol. 100, L16 - L19 (1981).

The authors report detections of the 3.3 μm emission feature in the nuclei of IC 4329A, an extreme Seyfert 1 galaxy and NGC 5506, a narrow line X-ray galaxy which may contain a heavily obscured Seyfert 1 nucleus. This feature probably originates in mantles on a population of small dust grains at temperatures around 300 K which may contribute substantially to the infrared continua of these galaxies. In the case of IC 4329A the authors discuss evidence provided by the large width of the observed feature that the grains responsible could be located within the broad line region. Observations of NGC 5506 made around 10 μm are compatible with the presence of significant absorption by silicate dust.

158.030   Colour-absolute magnitude relation for spiral galaxies.
N. Visvanathan.
Astron. Astrophys., Vol. 100, L20 - L22 (1981).

New photometric data at V (5500 Å) r (6738 Å) and IV (10500 Å) of the Virgo cluster and field spirals show a correlation between colours $(V - IV)_{0.5'}^C (R - IV)_{0.5'}^C$ and the IV luminosity. The colour magnitude gradient is 0.23 and 0.17 in V and r wavelengths respectively. As the IV luminosity is dominated by an old giant population, the C - M relations indicate that in r and V wavelengths we see not only the contribution from the old population but also that from the young population resulting from the recent star formation. This young population content is directly related to the mass of the old population of the galaxy.

158.031   Evidence of the ionized interstellar hydrogen structure in the M33 galaxy. Observations owing to the 6 m telescope of Zelentchuk.
G. Courtès, J. Boulesteix, J.-P. Sivan.
C. R. Acad. Sci. Paris, Tome 292, Sér. II, 1521 - 1527 (1981). In French.

The use of the world's largest optical telescope, the 6 m telescope of Zelentchuk (U.S.S.R.) leads to some new features of the real structure of the interstellar gas (H$_\alpha$ emission) in the M33 galaxy. The number of the bubble like spherical 200 pc diameter H II regions and the 400 pc extension of the filamentary structures seem to be only limited by the sensitivity and space resolution. Bubble and filaments are likely evolved states of the interstellar gas dissipation process related to a rapid expansion due to supernovae explosions and stellar winds.

158.032   Extragalactic variable sources and cosmic-ray acceleration near massive black holes.
M. Kafatos, M. M. Shapiro, R. Silberberg.
Comments Astrophys., Vol. 9, 179 - 198 (1981).

The rapid variations in the luminosity of active galactic nuclei imply that their energy is emitted from a highly compact region with dimensions of the order of light hours to light days. The energy is emitted over a broad band of frequencies, ranging from the radio to the X-ray and γ-ray parts of the spectrum. The effects of the powerful, compact source can be seen at distances that considerably exceed the dimensions of the parent galaxy: radio lobes, jets and high-velocity clouds are expelled from the center. Various models have been proposed to explain the central power source. The authors examine here the acceleration processes near a black hole. An accretion disk around a supermassive black hole in the center of an active galactic nucleus is shown to be a likely site of particle acceleration. The authors find that, depending on the confinement time for the protons, different galactic nuclei would contribute to different parts of the rigidity spectrum. Examining the magnetic field strengths in different disk models, they find that the betatron process can boost particle momenta by some four orders of magnitude. Finally, if shock waves can be sustained in an accretion disk (for example if the disk has a corona around it, or if supersonic turbulence can be sustained), such shock waves would also accelerate particles to cosmic-ray energies.

158.033 **The gaseous filaments in the northeast halo region of NGC 5128 (Centaurus A).**
J. A. Graham, R. M. Price.
Astrophys. J., Vol. 247, 813 - 822, plate 27 (1981).

Photographs, radial velocity measurements, and spectrophotometry are presented for the gaseous filaments which stretch 25' from the galaxy NGC 5128 toward its NE radio lobes. For some regions, fine structure is found on a scale of a few seconds of arc both in the velocity pattern and in emission line strengths. The authors appear to be observing regions of highly structured and turbulent gas in which velocity differences of 200 - 300 km s^{-1} are common. On the average and with respect to the main body of the galaxy, the gas is systematically moving toward the observer with a velocity of 300 - 400 km s^{-1}. The authors discuss the dominant sources for the excitation of the emission regions and conclude that the energy dissipation within a hot turbulent gas is the most likely, although they cannot completely rule out photoionization from a nonthermal source within the nucleus of NGC 5128. Possible origins of the gas are considered.

158.034 **Blue compact dwarf galaxies. I. Neutral hydrogen observations of 115 galaxies.**
T. X. Thuan, G. E. Martin.
Astrophys. J., Vol. 247, 823 - 848 (1981).

H I observations are reported for a sample of 115 blue compact dwarf ($M_B \gtrsim -18$) galaxies chosen mostly from the objective prism surveys of Markarian and Haro, with a few objects from Zwicky and other investigators. Ninety-three galaxies are detected. H I profiles, neutral hydrogen masses, total masses, and all available optical data are given for the 115 galaxies in a consistent and homogeneous system and in a useful format for statistical studies. The data are used in a companion paper to study the stochastic mode of star formation in galaxies.

158.035 **An estimate of peculiar velocities of field galaxies.**
D. Yu. Tsvetkov.
Astron. Zh., Tom 58, 725 - 733 (1981). In Russian. English translation in Soviet Astron., Vol. 25, No. 4.

For the root mean square peculiar velocity of field galaxies at a distance less than 20 Mpc (with H = 75 km/sec/Mpc) an estimate of 70 ± 40 km/sec has been obtained.

158.036 **NGC 6872: the largest known barred spiral.**
D. L. Block.
Sky Telesc., Vol. 62, 116 - 117 (1981).

158.037 **Active galaxies.** Eh. E. Khachikyan.
Izv. Akad. Nauk Armyansk. SSR, Fiz., Tom 16, 134 - 143 (1981). In Russian. – Paper presented on the scientific session of the Department of Phys.-Math. Sciences of the Armenian Academy of Sciences, 1981, Jan. 26 - 29.

The results of photometry and spectroscopy of some UV galaxies are presented. It is shown that many of them have some or other form of activity. A conclusion is drawn that the

form of activity is independent of the morphology and of the presence of a starlike nucleus.

158.038 **Rotation curves and mass-to-luminosity ratios of Seyfert galaxies.** V. L. Afanas'ev.
Pis'ma Astron. Zh., Tom 7, 390 - 393 (1981). In Russian. English translation in Soviet Astron. Lett., Vol. 7.

Masses of 12 Seyfert galaxies are determined from 6-m telescope observations of rotation curves. The difference between the mass-to-luminosity ratio for Seyfert galaxies and the mean one for normal galaxies of the same Hubble type is found to be small.

158.039 **Photoelectric surface colorimetry of the Andromeda nebula.** A. S. Sharov, V. M. Lyutyj.
Pis'ma Astron. Zh., Tom 7, 394 - 397 (1981). In Russian. English translation in Soviet Astron. Lett., Vol. 7.

Results of photoelectric surface colorimetry of the Andromeda nebula are given. It is shown that the radial color gradients in $B - V$ and $U - B$ up to a distance of 10 kpc from the centre are equal to $-0^m\!.017$ and $-0^m\!.028$ kpc^{-1} respectively. In the region of spiral structure both colors are bluer and further become again redder.

158.040 **On the difference between Seyfert 1 and 2 types of galaxies.** Eh. A. Dibaj.
Pis'ma Astron. Zh., Tom 7, 451 - 458 (1981). In Russian. English translation in Soviet Astron. Lett., Vol. 7.

Masses and bolometric luminosities of nuclei of 42 Seyfert 1 type galaxies, 14 quasi-stellar objects and 17 Seyfert 2 type galaxies are determined. The ratio of observed luminosity to the critical (Eddington) one is derived. The mean values of this ratio are equal to 0.46 ± 0.17 for Seyfert 1 galaxies and QSO's and 0.01 ± 0.003 for Seyfert 2 galaxies. The diversity of properties of active nuclei of galaxies seems to be connected with the approach to the Eddington limit (Seyfert 1 and QSO), or going away from that (Seyfert 2).

158.041 **Colour peculiarities of inner regions of Seyfert galaxies.** A. V. Zasov, V. M. Lyutyj.
Pis'ma Astron. Zh., Tom 7, 459 - 462 (1981). In Russian. English translation in Soviet Astron. Lett., Vol. 7.

Colour anomalies of nuclear regions of Seyfert galaxies found earlier are confirmed by new measurements of 11 galaxies. The inner regions of these galaxies are relatively blue in comparison with analogous regions of normal galaxies of similar morphological type. This may be caused by enhanced star formation. Resemblance of nuclear region of Seyfert galaxies and of structural "hot spot" nuclei of some normal galaxies is stressed.

158.042 **NGC 7714: the prototype star-burst galactic nucleus.**
D. W. Weedman, F. R. Feldman, V. A. Balzano, L. W. Ramsey, R. A. Sramek, C.-C. Wu.
Astrophys. J., Vol. 248, 105 - 112 (1981).

New X-ray, ultraviolet, optical, and radio observations are presented for the bright nucleus galaxy NGC 7714. All data are explainable as resulting from intense star-formation activity in this nucleus – a star burst. The X-ray and radio luminosities are explained by the supernova remnants arising from the massive stars, and ~10^4 supernova remnants are required in a volume of 280 pc radius. It is thought that NGC 7714 is typical of a significant class of emission-line galaxies. The authors point out that individual supernovae should be observable in this, or similar nuclei, because they are expected to occur at rates approaching one per year.

158.043 **Radio polarization rotators: BL Lacertae and 0727−115.** H. D. Aller, P. E. Hodge, M. F. Aller.
Astrophys. J., Lett., Vol. 248, L5 - L8 (1981).

Rotation in the polarization position angles over substantially more than 180° have been observed in BL Lac and 0727−115 at centimeter wavelengths. The broad-banded nature of the phenomenon eliminates frequency-dependent mechanisms as a possible cause, and the large range of the rotations is not consistent with relativistic-aberration type models. The most straightforward explanation for the origin of the apparent rotations is a rotating or a revolving structure in the radio emitting region.

158.044   Infrared surface brightness and absolute magnitude of spiral galaxies.   S. van den Bergh.
Astrophys. J., Lett., Vol. 248, L9 - L11 (1981).

It is shown that absolute magnitudes of spiral galaxies predicted from their infrared surface brightnesses have standard deviations of only 0.94 mag. Using M31 and M33 as calibrators for the Σ versus M_H relation in the Virgo cluster yields H(local) = 49 and 62 km s^{-1} Mpc^{-1} from M31 and M33, respectively. A similar calibration of the Aaronson et al. observations in four distant clusters yields H(global) = 92 and 115 km s^{-1} Mpc^{-1} from M31 and M33, respectively. Taken at face value, the infrared surface brightness data suggest that the Virgo cluster is at 1.86 ± 0.14 (m.e.) times the distance suggested by its redshift.

158.045   Small-scale structure of the core of M33 (NGC 598).
C. T. Hua, T. Nguyen-Trong.
Astron. Astrophys., Vol. 101, 187 - 190 (1981).

The authors present monochromatic images obtained with the Image Photon Counting System which is intended for the Canada-France-Hawaii 3.60m telescope and which is equipped with the Thomson Nocticon camera. The principal result of the present work is the detection of a double ring structure in the vicinity of the so-called nucleus at the centre of the nearest Sc galaxy M33 (NGC 598).

158.046   Optical polarization of M82 and the local spiral arm.
D. J. Axon.
Thesis Univ. Durham, England (1977). − Abstr. in Phys. Abstr., Vol. 84, Abstr. 67117 (1981).

158.047   Hubble ratio and solar motion from 200 spiral galaxies having distances derived from the luminosity index.   G. de Vaucouleurs, W. L. Peters.
Astrophys. J., Vol. 248, 395 - 407 (1981).

New solutions for the motion of the sun with respect to an all-sky sample of 200 spiral galaxies having precise distance moduli μ_0 recently derived from optical tertiary indicators lead to the following conclusions: (1) The solar motion relative to the frame of reference defined by galaxies in the distance interval $2 < \Delta < 32$ Mpc is $\langle V_s{}'' \rangle = 340 \pm 60$ km s^{-1} toward galactic coordinates $l_s{}'' = 125° \pm 16°$, $b_s{}'' = +24° \pm 8°$. (2) After correction for solar motion in the Local Group the motion of the Group is $V_G{}'' \approx 255$ km s^{-1} toward $L_G{}'' = 84°$, $B_G{}'' = -16°$, which is displaced 25° to the northeast of the Virgo cluster. (3) The mean Hubble ratio is essentially constant with $\langle H^* \rangle = 96$ km s^{-1} Mpc^{-1}. (4) The velocity dispersion $\sigma_v \approx 200$ km s^{-1} is definitely greater than in the Local Group, where $\sigma_v \approx 50$ km s^{-1}.

158.048   Hubble ratio and solar motion from 300 spirals having distances derived from H I line widths.
G. de Vaucouleurs, W. L. Peters, L. Bottinelli, L. Gouguenheim, G. Paturel.
Astrophys. J., Vol. 248, 408 - 422(1981).

New solutions for the motion of the sun with respect to an all-sky sample of 300 spiral galaxies having precise distance moduli μ_0 derived from improved versions of the Tully-Fisher relation lead to the following conclusions: (1) The solar motion relative to the frame of reference defined by galaxies in the distance interval $2 < \Delta < 29$ Mpc is $\langle V_s{}'' \rangle = 306 \pm 40$ km s^{-1} toward galactic coordinates $l_s{}'' = 117° \pm 12°$, $b_s{}'' = +30° \pm 7°$. (2) After correction for solar motion in the

Local Group the peculiar motion of the Group is $V_G{}'' = 254$ km s^{-1} toward $L_G{}'' = 96°$, $B_G{}'' = -10°$. (3) The mean Local Group apex is ~18° to the northeast of the Virgo cluster. (4) The mean Hubble ratio is essentially constant with $\langle H^* \rangle = 103 \pm 2$ km s^{-1} Mpc^{-1}. (5) The velocity dispersion $\langle \sigma_v \rangle \approx 198$ km s^{-1} is also in good agreement with the optical estimate $\langle \sigma_v \rangle \approx 192$ km s^{-1}. (6) After correction for the mean solar motion relative to the nearby galaxies, the apparent peculiar motion of the comoving nearby region−including the Local supercluster − with respect to the frame of reference defined by the 3 K background radiation is $V^* = 488$ km s^{-1} toward $L^* = 155°$, $B^* = -25°$.

158.049   The structure of giant elliptical galaxies in poor clusters of galaxies.
T. X. Thuan, W. Romanishin.
Astrophys. J., Vol. 248, 439 - 459 (1981).

Photographic surface photometry is presented for nine first brightest galaxies in poor clusters suspected to be cD galaxies by Morgan, Kayser, and White and by Albert, White, and Morgan. Exponentially truncated Hubble law and de Vaucouleurs $r^{1/4}$ law are fitted to the observed profiles to derive structural parameters. First brightest galaxies in poor clusters have profiles which are well fitted by an $r^{1/4}$ law over a range of more than 9 mag. They do not show the distinct envelope component seen in cD galaxies in rich clusters and in this sense are not "true" cD galaxies. The structural parameters of first brightest galaxies in poor clusters are compared with those of cD galaxies in rich clusters and those of normal ellipticals.

158.050   Optical spectral variability of the N galaxies 3C 382 and 3C 390.3.
H. K. C. Yee, J. B. Oke.
Astrophys. J., Vol. 248, 472 - 484 (1981).

A study of the variability of the absolute energy distribution and emission lines of the N galaxies 3C 382 and 3C 390.3 using multichannel spectrophotometry data acquired from 1969 to 1980 is presented. The continuum spectra can be decomposed into a nuclear nonthermal component of constant spectral shape and a constant elliptical galaxy component. Apparent visual magnitudes of 15.27 and 16.04 within a 10" aperture are derived for the underlying galaxies of 3C 382 and 3C 390.3 respectively. Forbidden lines and narrow Balmer lines do not change with the central continuum. The intensity variations of the broad Balmer lines, however, are correlated nonlinearly with changes in the luminosity of the central continuum source. Changes in the line profiles of the broad Balmer lines over a large range of velocity are also observed in both objects. A simple model in which a large part of the broad Balmer lines arise from an expanding shell or ring of gas clouds with a radius of a few lights years can explain adequately both the profile and emission-line luminosity changes.

158.051   On the triaxiality of elliptical galaxies.
R. Leach.
Astrophys. J., Vol. 248, 485 - 498 (1981).

Ellipticities and position angles are presented as functions of the major axis radius for 32 nearby elliptical galaxies from images obtained with a CCD camera. Comparisons with other observers and with multiple observations of some of the galaxies with a different camera configuration establish the reliability of the measurements. Ellipticity, ellipticity gradient, and position angle gradient are tabulated for each galaxy and statistically compared to projections of triaxial, coaxial, and concentric spheroids whose ellipticities vary with radius. The data are consistent with the galaxies all being triaxial with substantial ellipticity gradients and inconsistent with their all being coaxial, concentric, and axially symmetric.

158.052 A million cubic megaparsec void in Boötes?
R. P. Kirshner, A. Oemler, Jr., P. L. Schechter,
S. A. Shectman.
Astrophys. J., Lett., Vol. 248, L57 - L60 (1981).

In the course of a redshift survey of galaxies brighter than $R \approx 16.3$, 133 redshifts were measured in three fields, each separated by roughly $35°$ from the other two. If the galaxies in these fields were distributed uniformly, the combination of a galaxian luminosity function and the authors' magnitude limits predicts that the distribution of redshifts should peak near 15,000 km s^{-1}. In fact, only one galaxy of the 133 was observed with a redshift in the 6000 km s^{-1} interval centered on 15,000 km s^{-1}. One plausible interpretation is that a large volume in this region of order 10^6 Mpc^3 is nearly devoid of galaxies.

158.053 The detection of X-ray emission from the BL Lacertae object Markarian 180.
S. L. Mufson, D. J. Hutter.
Astrophys. J., Lett., Vol. 248, L61 - L64 (1981).

The authors report the detection of X-ray emission from Mrk 180, an elliptical galaxy which has a BL Lac object embedded in its core. The X-ray observations were made by the Einstein and HEAO 1 satellites. In addition, the authors present new multiaperture UBVR photometric observations which were used to separate the nonthermal optical emission from the galactic emission.

158.054 NGC 3344: unusual abundances for an Sbc galaxy?
M. L. McCall, G. A. Shields, P. M. Rybski.
Publ. Astron. Soc. Pacific, Vol. 93, 273 - 278 (1981).

Spectrophotometry of H II regions in the Sbc galaxy NGC 3344 reveals that it has a strong radial excitation gradient, with log [O III]/Hβ rising to +0.7 in the outermost arms. Such behavior has heretofore been observed only in Scd galaxies, and it contrasts with observations of a number of galaxies of type Sc or earlier.

158.055 UBV photometry of Markarian and S0 galaxies.
B. M. Peterson, K. Fricke, P. Biermann.
Publ. Astron. Soc. Pacific, Vol. 93, 281 - 284 (1981).

New UBV measurements for 36 Markarian and S0 galaxies are presented.

158.056 Photoelectric UBVRI photometry in two fields near NGC 5128.   J. A. Graham.
Publ. Astron. Soc. Pacific, Vol. 93, 291 - 293 (1981).

Photoelectric photometry in UBVRI is presented for 21 stars in two fields near the radio galaxy NGC 5128 (Centaurus A). The visual magnitudes cover a range from $7^{m}2$ to $16^{m}6$. The photometry is suitable for the calibration of two-dimensional imagery of stars and clusters associated with the galaxy itself.

158.057 2 micron spectroscopy of the nucleus of NGC 1068.
D. N. B. Hall, S. G. Kleinmann, N. Z. Scoville,
S. T. Ridgway.
Astrophys. J., Vol. 248, 898 - 905 (1981).

A 2.0–2.4 μm spectrum taken at a resolution of 5 cm^{-1} (350 km s^{-1}) with a small aperture ($3\rlap{.}''8$) centered on the nucleus of NGC 1068 exhibits both stellar and interstellar features. Stars are evident from the presence of first overtone CO absorption bands, the strengths of which suggest that late-type giant stars contribute about 25% of the 2.3 μm continuum within 2″ (200 pc) of the galactic center. The estimated velocity dispersion of the stars is $\lesssim 200$ km s^{-1}. Subtraction of this starlight from the spectrum leaves a gray body continuum component characterized by $T \sim 800$ K. The K band flux distribution of this nonstellar emission suggests an identification with the compact source dominant at $\lambda \gtrsim 3\,\mu$m and is inconsistent with extrapolation of the $\nu^{-1.85}$ power law ultraviolet source. Measured Brackett γ ($n = 7 \rightarrow 4$) and quadrupole

H$_2$ emission line fluxes are in agreement with previous, larger aperture measurements, implying that both the atomic and molecular emissions are predominantly confined to within 200 pc of the nucleus.

158.058 Spectroscopic measures of galaxies, their companions, and peculiar galaxies in the southern hemisphere.
H. Arp.
Astrophys. J., Suppl. Ser., Vol. 46, 75 - 112 (1981).

Examples of apparent association of galaxies and also of single peculiar galaxies have been drawn from the Catalogue of Southern Peculiar Galaxies and Associations (Arp and Madore, in preparation). Spectroscopic measures are reported for 75 central or peculiar galaxies and for 97 companion galaxies. Objects are identified by position and illustrated by photographic prints from the UK Schmidt (SRC) survey. Absorption and emission characteristics are tabulated for each spectrum, and heliocentric redshifts are given.

158.059 The distribution of luminosity in spiral galaxies.
T. Boroson.
Astrophys. J., Suppl. Ser., Vol. 46, 177 - 209, plates 6 - 7 (1981).

Luminosity profiles have been derived from photographic surface photometry measurements for 26 spiral galaxies. A decomposition procedure has been devised and applied to these profiles. In addition to the standard photometric parameters for the bulges and disks, accurate inclinations and true bulge flattenings were determined for some of the objects. A comparison of the actual profiles with the decomposition models is performed. The distributions of disk central surface brightness, disk scale length, and true bulge flattening are discussed in terms of previous investigations and theoretical predictions. An investigation of the Hubble sequence finds that the bulge-to-disk ratio is related to the Hubble type, but the scatter is much larger than observational uncertainties. The distribution of bulge-to-disk mass ratios and the disk properties of the spirals are compared with the results of a similar study of S0 galaxies by Burstein. The comparison of disk parameters strongly suggests that intrinsic differences between spirals and S0s existed at the time of formation.

158.060 Spiral galaxies in clusters. I. Neutral hydrogen observations in Abell 1367, Coma, and Zwicky 74-23 (1400.4+0949).   W. T. Sullivan, III, G. D. Bothun, B. Bates, R. A. Schommer.
Astron. J., Vol. 86, 919 - 942 (1981).

H I observations with the 305-m Arecibo telescope of 82 spiral galaxies, mostly types Sb–Scd, in the clusters Abell 1367, Abell 1656 (Coma), and Zwicky 74 - 23 (1400.4+0949) are reported. Reduction procedures, observational uncertainties, and problems with a comparison sample of field galaxies are fully discussed. The H I properties of these galaxies appear to correlate well with the morphological classification and X-ray properties of each cluster. Five individual galaxies of special note are discussed in detail: UGC 6697, NGC 3860, NGC 3947, NGC 4911, and NGC 4921.

158.061 Spiral galaxies in clusters. II. Neutral hydrogen observations in Cancer, Hercules (A 2151), and Pegasus I.   R. A. Schommer, W. T. Sullivan, III, G. D. Bothun.
Astron. J., Vol. 86, 943 - 952 (1981).

H I observations with the 305-m Arecibo telescope of 67 spiral galaxies in Abell 2151 (Hercules), Peg I, and the Cancer clusters are reported. Systemic velocities, linewidths, and hydrogen masses are computed for the 37 galaxies which were detected, and upper limits on the H I content are derived for nondetections.

158.062   **The distribution of the spiral galaxies in the direction of the Coma/A 1367 supercluster.**
B. A. Williams, F. J. Kerr.
Astron. J., Vol. 86, 953 - 980 (1981).

The three-dimensional distribution of spiral galaxies in a 1080-deg^2 region of the sky surrounding the Coma cluster, A 1367, and Zw 74-23 (NGC 5416 cluster) but not including the cores of Coma and A 1367 is analyzed using a sample of 233 spiral galaxies. The survey was done with the Arecibo telescope and is limited to H I observations of spiral galaxies whose angular diameter is less than 2 arcmin, with the exception of six NGC objects and one IC object. Out of a total of 87 galaxies detected, 67 of these redshifts are reported here for the first time. Most of the nondetections are believed to be above the observing range. Within the triangular region formed in the sky by these three clusters mentioned above, the authors find a population of faint isolated galaxies located as much as 30° from either Coma or A 1367. These isolated galaxies have radial velocities near 7000 km/s and could belong to the Coma/A 1367 supercluster.

158.063   **Color gradients in the spheroids of galaxies.**
A. Wirth.
Astron. J., Vol. 86, 981 - 988 (1981).

Photoelectric surface photometry designed to measure colors at low-brightness levels is presented for the stellar spheroids in galaxies of various morphological types. Radial color gradients of varying sizes are found in all galaxies examined. From a comparison of these data with integrated colors for composite stellar systems, it is concluded that metallicity is the dominant factor controlling the broadband colors of the stellar spheroids. Interpreting the color gradients as a metallicity effect, the author shows that disk-dominated spiral galaxies display the same metal-rich bulge, metal-poor halo structure seen in the Milky Way.

158.064   **New insight into the physical state of galaxies and quasars.**   R. F. Green.
The Universe at ultraviolet wavelengths, (see 012.009), p. 711 - 723 (1981).

Data from the International Ultraviolet Explorer satellite have revolutionized many concepts in extragalactic astronomy. These include the physical processes at work in the emitting gas characteristic of active objects, the nature of the continuum source itself in those objects, and the constituent hot stellar and gaseous components of normal galaxies. The review is not exhaustive, but concentrates on several problems of extragalactic research investigated with IUE.

158.065   **Observations of the nucleus of M 100.**
N. Panagia, G. Vettolani, G. G. C. Palumbo, P. Benvenuti, F. Macchetto.
The Universe at ultraviolet wavelengths, (see 012.009), p. 725 - 728 (1981).

The IUE observations of the nucleus of M 100 are presented and briefly discussed.

158.066   **IUE observations of the nuclear region of M51.**
P. Benvenuti, S. D'Odorico.
The Universe at ultraviolet wavelengths, (see 012.009), p. 729 (1981). − Abstract.

158.067   **The ultraviolet spectra of early-type galaxies.**
G. Bruzual A., H. Spinrad.
The Universe at ultraviolet wavelengths, (see 012.009), p. 731 736 (1981).

The average spectral energy distribution for a sample of bright elliptical galaxies is presented in the range λλ 2000 to 3200 Å. Spectral synthesis indicates that elliptical galaxies are most likely older than 9 Gyrs. The ultraviolet flux is consistent with a population of red horizontal branch stars, as those present in metal-rich globular clusters. Data for distant ($z \sim 1$) first-ranked cluster galaxies show indications of spectral evolution.

158.068   **IUE observations of Seyfert galaxies.**
C.-C. Wu, A. Boggess, T. R. Gull.
The Universe at ultraviolet wavelengths, (see 012.009), p. 737 - 742 (1981).

The authors discuss the following three topics: (1) Lα/Hβ ratio, (2) continuous energy distribution, (3) line profile.

158.069   **The UV variability of the Seyfert I galaxies III Zw 2 and Markarian 509.**
J. Huchra, M. Geller, D. Morton.
The Universe at ultraviolet wavelengths, (see 012.009), p. 743 - 749 (1981).

158.070   **Simultaneous observations of active galactic nuclei with IUE.**
J. N. Bregman, A. E. Glassgold, P. J. Huggins.
The Universe at ultraviolet wavelengths, (see 012.009), p. 751 - 755 (1981).

IUE observations of four active nuclei have been coordinated with radio, infrared, and X-ray measurements to obtain simultaneous determinations of their continuous spectra. The results for the BL Lac objects 0735+178 and I Zw 187 indicate sufficient UV and X-ray fluxes to ionize any gas. Comparison of the X-ray measurements with the extrapolated optical-UV continuum show a definite X-ray excess for I Zw 187 but none for the other BL Lac object.

158.071   **IUE observations of Fe II galaxies.**
M. V. Penston, M. A. J. Snijders, A. Boksenberg, J. D. J. Haskell, R. A. E. Fosbury.
The Universe at ultraviolet wavelengths, (see 012.009), p. 757 - 766 (1981).

Repeated observations of the Seyfert I galaxies I Zw 1 and II Zw 136, which have very strong Fe II emission lines in the optical region, were made at low resolution with the IUE satellite. The ultraviolet spectra are very similar: both are variable and show broad emission features of Fe II as well as the emission lines usually strong in Seyferts and quasars e.g: Lyα, Mg II, C III], C IV and N V. The data strongly support the hypothesis that the optical Fe II emission lines are primarily due to collisional excitation and that resonance fluorescence makes only minor contribution to the excitation of these lines.

158.072   **A strict method for elimination of optical pairs of galaxies.**   B. I. Fesenko.
Pis'ma Astron. Zh., Tom 7, 524 - 526 (1981). In Russian. English translation in Soviet Astron. Lett., Vol. 7.

A method for elimination of false pairs of galaxies is proposed and used. The mass-to-luminosity ratio is estimated from the motion of components by a formula which is true for an arbitrary distribution of orbital eccentricities. The existence of hidden masses is not confirmed.

158.073   **UBVR photometry of 27 galaxies.**
Eh. A. Dibaj, V. T. Doroshenko, K. A. Postnov.
Pis'ma Astron. Zh., Tom 7, 527 - 532 (1981). In Russian. English translation in Soviet Astron. Lett., Vol. 7.

UBVR photometry of 27 galaxies is carried out. Most of them are taken from Arakelyan's list of high surface brightness galaxies. Absolute magnitudes and intrinsic colours are determined. Their position on the two-colour diagram is analysed. Most of the objects investigated are shown to belong to spiral systems. The nucleus of the Seyfert 2 type galaxy NGC 6764 seems to be optically variable.

158.074 **Peculiar galaxy DDO 161.**
I. D. Karachentsev, V. E. Karachentseva,
V. A. Mineva.
Pis'ma Astron. Zh., Tom 7, 533 - 536 (1981). In Russian.
English translation in Soviet Astron. Lett., Vol. 7.
   Spectral observations of four remarkable central knots in
the southern dwarf galaxy DDO 161 have shown these knots
to be usual associations of hot stars and gas having relative
velocities of about 20 km/s. The mass-to-luminosity ratio for
DDO 161 lies in the range $(1 - 6)$ $(M/L)_\odot$.

158.075 **A new limit on the luminosity of the halo of
NGC 4565 from infrared observations.**
P. R. Saulson, S. P. Boughn, M. Seldner.
Bull. American Astron. Soc., Vol. 13, 506 (1981). – Abstract.

158.076 **Discovery of a carbon star in the Draco dwarf
spheroidal galaxy.**
M. Aaronson, J. Liebert, J. Stocke.
Bull. American Astron. Soc., Vol. 13, 506 (1981). – Abstract.

158.077 **The systematics of rotation curves and their use as
luminosity discriminators.**
N. Thonnard, V. C. Rubin, W. K. Ford, Jr.
Bull. American Astron. Soc., Vol. 13, 507 (1981). – Abstract.

158.078 **The optical warp in M31.**   K. A. Innanen,
K. Kamper, K. A. Papp, S. van den Bergh.
Bull. American Astron. Soc., Vol. 13, 507 (1981). – Abstract.

158.079 **Photometric redshift estimates for distant radio
galaxies.**   J. J. Puschell, F. N. Owen, R. A. Laing.
Bull. American Astron. Soc., Vol. 13, 507 (1981). – Abstract.

158.080 **On the redshift of PKS 2155-304.**   L. Maraschi,
E. G. Tanzi, A. Treves, M. Tarenghi.
Bull. American Astron. Soc., Vol. 13, 507 (1981). – Abstract.

158.081 **Correlations between infrared, optical, and X-ray
fluxes from Seyfert galaxies and quasars.**
N. P. Carleton, J. E. Steiner.
Bull. American Astron. Soc., Vol. 13, 520 (1981). – Abstract.

158.082 **High-resolution optical spectroscopy of Seyfert
galaxy emission lines.**
J. M. Vrtilek, N. P. Carleton, J. P. Huchra.
Bull. American Astron. Soc., Vol. 13, 521 (1981). – Abstract.

158.083 **Discovery of a Seyfert 1 with an unusually soft
X-ray spectrum.**   S. H. Pravdo, J. J. Nugent,
J. A. Nousek, K. Jensen, A. S. Wilson, R. H. Becker.
Bull. American Astron. Soc., Vol. 13, 521 (1981). – Abstract.

158.084 **Ultraviolet excess of Seyfert 1 galaxies and QSOs.**
M. Malkan.
Bull. American Astron. Soc., Vol. 13, 521 (1981). – Abstract.

158.085 **VLA observations of the nucleus of NGC 1365 in
the 6- and 20-cm continuum.**
Å. Sandqvist, S. Jörsäter, P. O. Lindblad.
Bull. American Astron. Soc., Vol. 13, 521 (1981). – Abstract.

158.086 **The inactive radio galaxy in Abell 566.**
D. E. Harris, R. G. Robertson, P. E. Dewdney,
C. H. Costain.
Bull. American Astron. Soc., Vol. 13, 522 (1981). – Abstract.

158.087 **Six-centimeter survey of Arakelian galaxies.**
G. Kojoian, D. Dickinson, M. A. Arakelian
(*Arakelyan*), M. D. Bicay.
Bull. American Astron. Soc., Vol. 13, 522 (1981). – Abstract.

158.088 **The nature of the obscuring region in Seyfert
galaxies.**   A. Lawrence, M. Elvis.
Bull. American Astron. Soc., Vol. 13, 531 (1981). – Abstract.

158.089 **CCD observations of Mkn 421 and Mkn 501.**
P. Hickson, G. G. Fahlman, J. R. Auman,
G. A. H. Walker, Z. Ninkov.
Bull. American Astron. Soc., Vol. 13, 531 (1981). – Abstract.

158.090 **IUE observations of spectral variations in the active
BL Lac Object OJ 287.**
K. R. Hackney, R. L. Hackney, R. L. Scott, R. C. Roeder,
J. T. Pollock, A. J. Pica, R. J. Leacock, A. G. Smith.
Bull. American Astron. Soc., Vol. 13, 532 (1981). – Abstract.

158.091 **1 to 10 μm observations of strong emission line
galaxies**   C. J. Lonsdale.
Bull. American Astron. Soc., Vol. 13, 532 - 533 (1981).
Abstract.

158.092 **The CO distribution in M 51.**
S. Lord, J. S. Young, N. Z. Scoville.
Bull. American Astron. Soc., Vol. 13, 535 (1981). – Abstract.

158.093 **CO observations of the Sc galaxy NGC 3628.**
L. J. Tacconi, J. S. Young, N. Z. Scoville.
Bull. American Astron. Soc., Vol. 13, 535 (1981). – Abstract.

158.094 **CO in M 31.**
A. A. Stark, R. A. Linke, M. A. Frerking.
Bull. American Astron. Soc., Vol. 13, 535 (1981). – Abstract.

158.095 **CO radial distributions in two Scd galaxies:
IC 342 and NGC 6946.**
J. S. Young, N. Z. Scoville.
Bull. American Astron. Soc., Vol. 13, 538 (1981). – Abstract.

158.096 **Spectrophotometry of carbon stars in Local Group
galaxies and in the Magellanic Stream.**
H. Richer, B. Westerlund.
Bull. American Astron. Soc., Vol. 13, 545 - 546 (1981).
Abstract.

158.097 **X-ray observations of late type galaxies with the
Einstein Observatory.**
G. Fabbiano, L. Van Speybroeck.
Bull. American Astron. Soc., Vol. 13, 550 (1981). – Abstract.

158.098 **The selection of galaxies in the Lick survey and the
relationship between galaxy counts and galactic
absorption.**   S. Phillipps, R. S. Ellis, A. W. Strong.
Mon. Not. R. Astron. Soc., Vol. 197, 151 - 156 (1981).
   The relationship between galaxy counts and galactic
absorption has been examined in the Lick astrographic survey.
It is found that results of previous workers are consistent with
the survey being limited by the isophotal size of galaxies rather
than by the apparent magnitude. The selection of galaxies by
size is also consistent with the degree of galaxy clustering
observed in the sample.

158.099 **The ultraviolet spectrum of the active elliptical
galaxy NGC 1052.**   R. A. E. Fosbury,
M. A. J. Snijders, A. Boksenberg, M. V. Penston.
Mon. Not. R. Astron. Soc., Vol. 197, 235 - 240 (1981).
   Low resolution observations of the elliptical galaxy
NGC 1052 have been obtained with both the long and short
wavelength cameras of the IUE satellite. These have been
combined with optical multi-aperture photometry to demonstrate a low reddening of $E_{B-V} \sim 0.06$. There is no evidence for
a compact source of non-thermal radiation capable of ionizing
sufficient gas to radiate the observed Balmer line flux. The
strongest UV emission lines are O II] λ 2326 and C III] λ 1909.

The energy distribution of NGC 1052, if typical, implies important revisions to the K-corrections for high redshift galaxies.

158.100 General properties of galaxies.   S. M. Fall.
The structure and evolution of normal galaxies, (see 012.010), p. 1 - 11 (1981).
Contents: Introduction. Characteristic morphologies. Intensity profiles. Luminosity functions. Stellar content. Gas content. Internal kinematics. Groups and clusters.

158.101 Photometric and dynamical properties of elliptical galaxies.   F. Bertola.
The structure and evolution of normal galaxies, (see 012.010), p. 13 - 25 (1981).
Contents: Introduction. Luminosity profiles. Ellipticity changes and the isophote twists. Dynamical properties. Rotation curves. Velocity dispersions. Masses and mass-to-light ratios. Three-dimensional shapes.

158.102 Galaxy dynamics: observations.   G. Illingworth.
The structure and evolution of normal galaxies, (see 012.010), p. 27 - 41 (1981).
Contents: Dark halos about galaxies. Elliptical galaxies. Bulges of disk systems. Implications for formation.

158.103 Shapes of unperturbed galaxies.
M. Schwarzschild.
The structure and evolution of normal galaxies, (see 012.010), p. 43 - 53 (1981).
Contents: Variety of equilibrium configurations. Levels of symmetry. Non-classical effective integrals. Numerical construction of equilibrium models. Stochastic orbits. Major families of regular orbits.

158.104 The structure of barred galaxies.   J. Kormendy.
The structure and evolution of normal galaxies, (see 012.010), p. 85 - 110 (1981).
Contents: Introduction. Morphology of barred galaxies. Distinct components in barred galaxies. Frequency of incidence of inner rings and lenses. Shapes of lenses and inner rings. Bars. Basic properties of bars: theoretical results. The principal resonances. Bars as almost-kinematic density waves. The pattern speed. The length of the bar. Stellar velocity fields. Triaxial bulge components. Lenses. Conclusion.

158.105 H I spiral structure in M31 and M33.
J. E. Baldwin.
The structure and evolution of normal galaxies, (see 012.010), p. 137 - 147 (1981).

158.106 Neutral hydrogen in galaxies.   R. Sancisi.
The structure and evolution of normal galaxies, (see 012.010), p. 149 - 168 (1981).
Some of the more recent work on the large-scale properties of the HI gas in elliptical, S0 and spiral galaxies and in the neighbourhood of these galaxies are discussed. These observations have revealed new aspects of the structure and dynamics of galaxies and constitute new evidence which may have a bearing on their formation and evolution.

158.107 Radio continuum emission from galaxies.
R. D. Ekers.
The structure and evolution of normal galaxies, (see 012.010), p. 169 - 180 (1981).
Contents: Introduction. Surveys. Emission mechanism. Disks of spiral galaxies. The Galaxy. Disk emission from spiral galaxies. The distribution of emission in z – the radio haloes. Centres of galaxies. Sagitarius A. Structure in the central region. Nuclear radio sources. Statistical properties of galaxies. Method of analysis. Correlations. Spectra. Elliptical

galaxies. Total emission. Elliptical/S0 problem – radio source ejection and dynamics.

158.108 X rays from normal galaxies and clusters of galaxies.   A. C. Fabian.
The structure and evolution of normal galaxies, (see 012.010), p. 181 - 199 (1981).
In this review the author first summarises the various classes of pointlike and extended X-ray sources in our own Galaxy in order to provide a background to the new results. Some well studied normal galaxies (the Magellanic Clouds, M31, M33, M101 and others) are then covered in some detail, together with the luminosity distribution of their point sources. The first preliminary results of surveys of other galaxies are then reported. Finally the X-ray emission from groups and clusters of galaxies is discussed with an emphasis on the role of the galaxies.

158.109 Observations of NGC 4151 with IUE.
A. Boksenberg, G. Bromage, J. Clavel, A. Elvius, A. Gabriel, P. Gondhalekar, C. Jordan, J. Lind, L. Lindegren, M. Longair, M. Penston, G. Perola, M. Perryman, M. Pettini, M. Rees, D. Sciama, M. Snijders, E. Tanzi, M. Tarenghi, M. H. Ulrich, R. Wilson.
Second European IUE Conference, (see 012.011), p. LXVII - LXX (1980).
NGC 4151 has been observed with IUE at six different epochs between May 1978 and January 1979. Complex spectral variations have been found. The variations of the UV continuum which the authors observed are slower and have smaller amplitudes than the rapid variations observed in the soft X-ray range with the Ariel V satellite. The emission lines are those normally seen in quasar spectra; the intensities and profiles of C IV and Si IV are variable. A number of absorption lines are present in the short-wavelength spectra.

158.110 Ultraviolet spectrum of dwarf blue galaxy Tololo 1924–416 (z = 0.0096).   R. F. Carswell,
P. M. Gondhalekar, D. Morgan, K. Nandy, R. Wilson.
Second European IUE Conference, (see 012.011), p. 133 - 134 (1980).
The ultraviolet spectrum of the dwarf blue galaxy Tololo 1924–416 (z = 0.0096) has been obtained with the International Ultraviolet Explorer. The spectrum has a strong 2200 Å absorption feature and the color excess obtained from the 2200 Å feature (assuming the galactic extinction law) and that obtained from the Balmer decrement are similar. The stellar composition of the galaxy has been determined and the physical state of the gas in the galaxy has been described. The $L\alpha/H\beta$ ratio in the dereddened spectrum is found to be normal.

158.111 The UV surface brightness distribution in the central region of the galaxy M 31.
J. M. Deharveng, G. Monnet, M. Maucherat-Joubert.
Second European IUE Conference, (see 012.011), p. 253 - 255 (1980).
IUE observations have been used to investigate the UV surface brightness distribution in the central region of M 31 as a possible clue for the origin of the excess UV light found there. From a comparison with previous UV photometric measurements it is found that the UV distribution is similar to the B light distribution. A possible steeper gradient in the inner bulge is also discussed. Hot stars in old stellar population are thus suggested as the major source of the excess UV light.

158.112 The UV spectrum of the elliptical galaxy NGC 3379.
F. Bertola, M. Capaccioli, J. B. Oke.
Second European IUE Conference, (see 012.011), p. 257 (1980). – Abstract.

158.113 Observations of NGC 5253.
G. C. Perola, M. Tarenghi.

Second European IUE Conference, (see 012.011), p. 259 (1980). – Abstract.

158.114 Observations of two narrow emission line galaxies, NGC 4507 and 5506.
    J. Bergeron, T. Maccacaro, G. C. Perola.
Second European IUE Conference, (see 012.011), p. 261 (1980). – Abstract.

158.115 IUE spectra of clumpy irregular galaxies.
    P. Benvenuti, C. Casini, J. Heidmann.
Second European IUE Conference, (see 012.011), p. 263 (1980). – Abstract.

158.116 Observations of the star-like BL Lac object OI 090.4.   G. E. Bromage, W. M. Burton,
B. E. Patchett, A. G. Smith.
Second European IUE Conference, (see 012.011), p. 267 - 269 (1980).
    Results of simultaneous IUE spectroscopy and UBVI photometry are presented for an apparently star-like BL Lac project, the Ohio radio source OI 090.4. The flux distribution is compared with that for two nearby BL Lac objects in elliptical galaxies, Mkn 421 and Mkn 501, as observed both in early 1978 and in March 1979. Some general comments are made concerning IUE analysis procedures for faint spectra on high-background images.

158.117 IUE observations of Seyfert galaxies and a BL Lac object.   H. Schleicher, K. J. Fricke,
W. Kollatschny.
Second European IUE Conference, (see 012.011), p. 271 - 278 (1980).
    Spectrophotometric data for four Seyfert galaxies and one BL Lac object are presented.

158.118 IUE observations of the Seyfert 1 galaxies I Zw1, II Zw 136, and Mk 231.   M. A. J. Snijders,
A. Boksenberg, J. D. J. Haskell, R. A. E. Fosbury, M. V. Penston.
Second European IUE Conference, (see 012.011), p. 279 - 287 (1980).
    The authors present ultraviolet observations for three Seyfert galaxies, with very strong optical Fe II emission lines, which have been extensively studied in the optical and infrared.

158.119 IUE observations of X-ray emitting Seyfert galaxies.
    P. Barr, A. J. Willis, R. Wilson.
Second European IUE Conference, (see 012.011), p. 289 - 292 (1980).
    IUE observations have been made of the X-ray emitting Seyfert galaxies NGC 3783, Mkn 506, Mkn 279, 3C 390.3 and NGC 5548. The continuum intensities lead to estimates of reddening consistent with local (galactic) extinction; there is no evidence for any reddening intrinsic to the source.

158.120 IUE observations of two (non-Seyfert) radio galaxies.
M. V. Penston, R. A. E. Fosbury, M. A. J. Snijders,
A. Boksenberg.
Second European IUE Conference, (see 012.011), p. 293 (1980). – Abstract.

158.121 Paschen and Balmer lines in active galactic nuclei.
    K. J. Fricke, W. Kollatschny.
Messenger, No. 25, p. 21 - 23 (1981).

158.122 Spectrophotometric investigation and the physical conditions in the nucleus of Mrk 534.
V. K. Golev, G. T. Petrov, I. M. Yankulova.
Dokl. Bolg. AN, Vol. 33, 1033 - 1036 (1980). In Russian. Abstr. in Ref. zh., 51. Astron., 7.51.685 (1981).

158.123 Observations of faint Haro galaxies.
    T. D. Kinman, P. Hintzen.
Publ. Astron. Soc. Pacific, Vol. 93, 405 - 421 (1981).
    Spectroscopic and photometric data are given for 23 faint-compact uv-excess galaxy candidates which Haro picked out by his three-color technique. Eighteen were found to be emission-line galaxies ($14.4 < V < 17.9$) with redshifts in the range 1670 to 39,450 km s^{-1}; the remaining five appear to be stars. Much of the apparent compactness of the galaxy images is caused by the finite resolution of the Survey plates. In surface brightness, and their range in color and absolute magnitude, these faint Haro galaxies are not distinguishable from the original Haro galaxies.

158.124 The mass of M31.   S. van den Bergh.
    Publ. Astron. Soc. Pacific, Vol. 93, 428 (1981).
    The projected mass method of Bahcall and Tremaine has been used to derive the mass of the core of M31. From the observed radial velocities of 43 globular clusters, the mass interior to $\langle R \rangle \sim 30'$ is found to be $M_I = (9 \pm 2) \times 10^{10} M_\odot$.

158.125 A search for neutral hydrogen in D and cD galaxies.
    J. O. Burns, R. A. White, M. P. Haynes.
Astron. J., Vol. 86, 1120 - 1125 (1981).
    Recent X-ray spectroscopic results suggest that 2 - 3 $M_\odot$ yr^{-1} of cooling gas may be falling into M87 and as much as 300 $M_\odot$ yr^{-1} into giant galaxies in rich clusters. The eventual fate of the gas in the radiatively accreting flow is an open question. In an attempt to set some bounds on the evolution of the gas, the authors performed a sensitive search for H I in a sample of eight D and cD galaxies in rich and poor clusters. Null detections in all cases set upper limits of (4 - 20) $\times 10^8 M_\odot$ of H I in the giant galaxies. Possible reasons for these nondetections and alternative evolutionary fates for the accreting gas are discussed.

158.126 The structure of radio emission in the inner few hundred parsecs of spiral galaxies.
    J. M. van der Hulst, P. C. Crane, W. C. Keel.
Astron. J., Vol. 86, 1175 - 1191 (1981).
    This paper describes observations of the central radio continuum sources in 81 galaxies, mostly spirals, obtained with the Very Large Array at λ 6 cm. The $\sim 1''$ resolution is sufficient to resolve the majority of the central sources and to reveal the global structure of the radio emission. A simple classification scheme is used to characterize the radio structure, and the dependence of radio morphology on other galaxy properties is investigated. It appears that the early Hubble-type galaxies mostly exhibit strong simple central sources, whereas the central sources in late-type galaxies are generally weak and complex or absent. There also exists a correlation with the color, or stellar population, of the center of a galaxy, in the sense that galaxies with red centers have strong, simple radio sources, and galaxies with blue centers exhibit weak complex central sources.

158.127 High-resolution UV images of galaxies.
    R. C. Bohlin, T. P. Stecher.
The phases of the interstellar medium, (see 012.018), p. 145 - 150 (1981).

158.128 La découverte et l'exploration de l'univers extra-galactique.   L. Gouguenheim, S. Collin-Souffrin.
Histoire de l'univers, (see 003.007), p. 247 - 286 (1980).
    Contents: Historique. Les propriétés des galaxies. La cinématique et la dynamique des galaxies. Les groupes et les amas de galaxies. L'espace intergalactique.

158.129 Les galaxies "actives" et les quasars.
    S. Collin-Souffrin.
Histoire de l'univers, (see 003.007), p. 313 - 341 (1980).
    Contents: Les différentes classes de galaxies actives. Les

quasars et les lacertides. Le "moteur" des quasars et des noyaux de galaxies actives.

158.130   Bursts of star formation in the central regions of spiral galaxies.   T. P. Prabhu.
Bull. Astron. Soc. India, Vol. 9, 86 (1981). – Abstract.

158.131   Spectra of compact radio sources in galactic nuclei.   D. Jones, Y. Terzian, R. Sramek.
News Lett. Astron. Soc. N. Y., Vol. 1, No. 10, p. 24 (1981). Abstract.

158.132   The activity of galactic nuclei.   F. Pacini.
Mem. Soc. Astron. Italiana, Vol. 52, (see 012.019), 153 - 157 (1981).

158.133   UGC 10683B – a possible X-ray-emitting Seyfert galaxy.   A. S. Wilson, K. Wood, M. J. Ward, R. E. Griffiths, R. F. Mushotzky.
Astron. J., Vol. 86, 1289 - 1293 (1981).
Optical spectra have been obtained of both members of an interacting pair of galaxies (UGC 10683) lying close to the error box of the HEAO-A1 unidentified high-latitude X-ray source 1H 1703 - 01 (=4U 1716 - 01?). One galaxy (UGC 10683B) is of Seyfert type I and may be related to the X-ray source.

158.134   Dust clouds in the stellar spheroids of Local Group galaxies.   J. S. Gallagher, D. A. Hunter.
Astron. J., Vol. 86, 1312 - 1322 (1981).
The central regions of the four Local Group galaxies M31, M32, NGC 185, and NGC 205 have been imaged with the KPNO video camera in the B and I_K passbands as part of a program to search for dust clouds in spheroidal stellar system. The relatively long wavelength baseline allows an approximate determination of cloud optical depths, which provide the basis for an exploratory analysis of gas content of dust clouds in NGC 185 and the M31 bulge. These two systems have different types of clouds and correspondingly different levels of star-forming activity.

158.135   The mild abundance gradient of NGC 1365.   D. Alloin, M. G. Edmunds, P. O. Lindblad, B. E. J. Pagel.
Astron. Astrophys., Vol. 101, 377 - 384 (1981).
ESO and AAT observations are combined so as to estimate O and N abundance of thirteen H II regions in the nuclear region, bar and arms of the barred spiral, hot-spot galaxy NGC 1365 which has recently been found to have a Seyfert nucleus. In agreement with an earlier result based on a smaller data sample, the radial abundance gradient is quite small. The behavior of NGC 1365 in this respect is compared with that of a few other galaxies of similar morphological type but without a bar.

158.136   NGC 4507: a weak Seyfert 1 and X-ray galaxy.   P. Véron, M. P. Véron, E. J. Zuiderwijk.
Astron. Astrophys., Vol. 102, 116 - 118 (1981).
The authors show that the faint X-ray galaxy NGC 4507 (1232−39) detected with the Einstein Observatory by Kriss et al. (1980) has a starlike nucleus and exhibits a broad Hα component. It is therefore a Seyfert 1.9 galaxy. This adds further evidence that all nuclear X-ray galaxies contain a Seyfert 1 nucleus or miniquasar. The authors also comment on the reality of a correlation between the X-ray luminosity and the full width zero intensity of the Balmer lines, which has been suggested by Kriss et al. (1980).

158.137   Light distribution, inclination, and mass distribution of M51.
G. Monnet, G. Paturel, F. Simien.
Astron. Astrophys., Vol. 102, 119 - 123 (1981).

Bulge and disk parameters of M51 have been measured by two-dimensional B photometry. A redetermination of the inclination is made, both by classical isophotometry and by using a new technique which compares mass-to-light ratios determined from emission line rotation velocities and absorption line velocity dispersions. A mean value of 35° is obtained, and the mass to luminosity ratio of the bulge and the disk are derived.

158.138   The neutral hydrogen distribution of irregular galaxies.
W. K. Huchtmeier, J. H. Seiradakis, J. Materne.
Astron. Astrophys., Vol. 102, 134 - 141 (1981).
Twenty-one relatively large irregular galaxies spread over a wide range of absolute magnitudes ($-18.9 > M > -13.7$) have been mapped with the 100 m radio telescope at Effelsberg at a wavelength of 21 cm. On the average the half power width of the H I distribution of galaxies reported here is similar to their Holmberg diameters. Eight of these twenty-one galaxies have extended H I envelopes at a low H I surface density level, while other global properties are within the normal range for this type of galaxy.

158.139   Optical spectroscopic and electronographic observations of the radio galaxy IC 5063.
I. Appenzeller, G. Gaida.
Astron. Astrophys., Vol. 102, 230 - 236 (1981).
Low resolution (171 Å/mm) IDS spectrograms, medium resolution (56 Å/mm) long slit image tube spectrograms, and electronographic direct images in the visual and blue spectral range have been obtained of the radio galaxy IC 5063. According to the results of these observations the main body of IC 5063 is a slowly rotating giant elliptical or S0 galaxy. At its center is an active nucleus which is surrounded by a rapidly rotating disk of gas and dust filaments. This system shows many similarities with NGC 5128 (Cen A). From the observations the authors conclude that the present structure of IC 5063 is most likely the result of the capture of a gas rich system by an oblate giant elliptical or S0 galaxy.

158.140   The matter of the missing mass.   W. Tucker.
Mercury, Vol. 10, 107, 125 (1981).

158.141   Die Radialgeschwindigkeiten und Massen von Galaxien hoher Leuchtkraft und die differentielle Rotation unserer Milchstraße.   D. Wiedemann.
Sterne Weltraum, Jahrg. 20, 369 - 371 (1981).

158.142   BL Lacertae-objekt.   A. Sikℓ.
Astron. Tidsskr., Årg. 14, 116 - 123 (1981).

158.143   Supermassive binaries in active galactic nuclei.   D. P. Whitmire, J. J. Matese.
Nature, Vol. 293, 722 - 724 (1981).
Recent observational evidence indicating the apparent precession of jets in compact and extended radio sources suggests the presence of a second massive body in the nuclei of the associated active galaxies. On the basis of their similarities with close binaries in the Galaxy the authors consider the possibility that the secondary object is a Roche overflowing supermassive star of mass $M_2 \sim 10^6$ - $10^8 M_\odot$ in orbit around a primary massive black hole of mass $M_1 \sim 10^7$ - $10^9 M_\odot$.

158.144   On the interpretation of the Shane-Wirtanen counts of galaxies in $10' \times 10'$ and $1° \times 1°$ squares.
G. Dautcourt.
Astrofiz. issled., NRB, Vol. 3, 95 - 102 (1981). – Abstr. in Ref. zh., 51. Astron., 9.51.834 (1981).

158.145   Some properties of Seyfert galaxies.
B. V. Komberg.

Inst. kosm. issled. AN SSSR. Prepr., 1981, No. 627, 38 pp. In Russian. – Abstr. in Ref. zh., 51. Astron., 9.51.867 (1981).

158.146 Dynamical or static radio halo – is there a galactic wind? I. Lerche, R. Schlickeiser.
Astrophys. Lett., Vol. 22, 161 - 163 (1981).

The effect of a galactic wind on a radio halo can be best observed at frequencies smaller than about 1 GHz. At higher frequencies static halo models predict the same features as dynamical halo models. External galaxies, which exhibit a break by ~0.5 in their high frequency nonthermal integral flux spectrum, are the best candidates for studying the influence of galactic winds on the formation of relativistic electron haloes around these systems. The authors present several such cases.

158.147 Diffuse Hα emission in galaxies. J. P. Sivan.
Ann. Physique, Vol. 6, (see 012.026), p. 17 - 34 (1981). In French. – Abstr. in Phys. Abstr., Vol. 84, Abstr. 94499 (1981).

158.148 Optical photometry of galaxies. G. Comte.
Ann. Physique, Vol. 6, (see 012.026), p. 35 - 52 (1981). In French. – Abstr. in Phys. Abstr., Vol. 84, Abstr. 94500 (1981).

158.149 Photometric evolution of galaxies: observables and models. B. Rocca-Volmerange.
Ann. Physique, Vol. 6, (see 012.026), p. 167 - 181 (1981). In French. – Abstr. in Phys. Abstr., Vol. 84, Abstr. 94503 (1981).

158.150 Energy distribution in elliptical galaxies. F. Bertola.
Ann. Physique, Vol. 6, (see 012.026), p. 191 - 192 (1981). In French. – Abstr. in Phys. Abstr., Vol. 84, Abstr. 94505 (1981).

158.151 A particularly anomalous Seyfert galaxy. J. H. Krolik.
Nature, Vol. 294, 13 (1981).

158.152 CCD photometry of the BL Lacertae objects 1218+304, 1219+28, and 1727+50: point sources, associated nebulosity, and broad-band spectra.
D. Weistrop, D. B. Shaffer, R. F. Mushotzky, H. J. Reitsema, B. A. Smith.
Astrophys. J., Vol. 249, 3 - 12 (1981).

The authors have obtained visual and far-red surface photometry of two X-ray emitting BL Lacertae objects, 1218+304 (2A 1219+305) and 1727+50 (I Zw 187), as well as the highly variable object 1219+28 (ON 231, W Com). The intensity distribution for 1727+50 can be modeled using a central point source plus a de Vaucouleurs intensity law for an underlying galaxy. The broad-band spectral energy distribution so derived is consistent with what is expected for an elliptical galaxy. The spectral index of the point source is $\alpha = 0.97$. New VLBI and X-ray data are also reported for 1727+50. There is nebulosity associated with the recently discovered object 1218+304.

158.153 Detection of Lyman continuum absorption in the BL Lacertae object PKS 0735+178.
J. N. Bregman, A. E. Glassgold, P. J. Huggins.
Astrophys. J., Vol. 249, 13 - 16 (1981).

Strong continuum absorption has been detected with *IUE* in the BL Lac objects PKS 0735+178 shortward of the Lyman edge associated with the optically determined absorption redshift. The size of the discontinuity leads to a lower limit for the column density of absorbing atomic hydrogen, $N(\text{H I}) \geqslant 4(17) \text{ cm}^{-2}$. An upper limit, obtained from Ly α absorption, is $\sim 2(19) \text{ cm}^{-2}$.

158.154 Evolutionary synthesis of the stellar population in elliptical galaxies. III. Detailed optical spectra.
J. E. Gunn, L. L. Stryker, B. M. Tinsley.
Astrophys. J., Vol. 249, 48 - 67 (1981).

The evolutionary synthesis technique is used to construct population models for giant elliptical galaxies, using detailed spectrophotometric data for the galaxies and for stars obtained with the Oke multichannel spectrometer on the Hale telescope. The authors find that ellipticals are well represented by an old, metal-rich population with a turnoff at $B-V \sim 0.80$ and a turnoff mass function slope $x \lesssim 1$, plus a quite significant contribution from stars above the turnoff. The nature of these objects is discussed, and it is concluded that the present data and astrophysical constraints cannot distinguish between a small young population and a blue straggler population augmented by a few O stars.

158.155 The surface brightness-axis ratio relation as a test of intrinsic shapes of elliptical galaxies.
D. W. Olson, G. de Vaucouleurs.
Astrophys. J., Vol. 249, 68 - 75 (1981).

The relation between apparent ellipticity and surface brightness for a sample of 112 normal, noninteracting, bright elliptical galaxies in the Second Reference Catalogue is used as a test of their intrinsic shapes. The observed trend is consistent with that expected from geometric projection effects for oblate spheroids, when it is assumed that intrinsic surface brightness, as viewed down the symmetry axis, is independent of the intrinsic axis ratio. Prolate models for elliptical galaxies cannot be ruled out, however, if the ad hoc hypothesis is made that a correlation strong enough to dominate the projection effects exists between intrinsic flattening and surface brightness.

158.156 Selection, statistics and photometry of compact galaxies with different instrumental equipment, part II. The cluster Abell 1775.
N. Richter, A. Schnell.
Astron. Nachr., Band 302, 247 - 250 (1981).

A field of 0.8 □° near the galactic north pole is investigated with two large photographic telescopes of very different optical constants. The field contains the cluster Abell 1775. The fraction of optical compact galaxies up to the identification magnitude $18^{m}9$ (B) is found to be more than 30% of the total number of cluster galaxies. The data for a secondary photometric sequence, which had to be determined for the photometry of the galaxies, are presented.

158.157 The size distribution of H II regions as a new variant to determine the distances of galaxies. I. A. Issa.
Astron. Nachr., Band 302, 251 - 254 (1981).

The apparent radii of some H II regions in the galaxies NGC 224, NGC 2403, NGC 300, NGC 598 and NGC 1313 were determined. The size distribution of the apparent radii of the H II regions in these galaxies were studied. Two variants are suggested to determine the distances of these galaxies. The reliability of these variants could be seen by comparison with the distances determined by other methods.

158.158 Physical characteristics of normal galaxies. V. A. Dostal', M. A. Smirnov.
Pis'ma Astron. Zh., Tom 7, 582 - 584 (1981). In Russian. English translation in Soviet Astron. Lett., Vol. 7.

For normal galaxies mean values of electron temperatures and density of ionized gas and the ratio of number of L_c-quanta to luminosity in solar units are determined.

158.159 Seyfert galaxies with weak broad Hα emission lines. D. E. Osterbrock.

Astrophys. J., Vol. 249, 462 - 470 (1981) = Lick Obs. Bull., No. 886.

Spectrophotometric measurements are presented of five galaxies with fairly strong narrow emission lines combined with weak broad Hα, and in some cases Hβ, emission components. These objects have line and continuum spectral properties intermediate between those of Seyfert 1 and Seyfert 2 galaxies, but much closer to the latter, and they are therefore called Seyfert 1.8 and Seyfert 1.9 galaxies, depending on whether or not broad Hβ emission can be seen in the scans. The physical properties of these objects are described and discussed.

158.160   **Wolf-Rayet stars and giant H II regions in M33: casual associations or meaningful relationships?**
P. S. Conti, P. Massey.
Astrophys. J., Vol. 249, 471 - 480 (1981).

The authors have discovered 14 new Wolf-Rayet (W-R) stars in the Local Group galaxy M33. Six are surrounded by relatively small H II regions; their spectra and M_v are similar to field W-R stars in our own Galaxy. Eight are found among the brightest stars in the giant H II regions NGC 588, 592, 595, and 604; these are likewise similar to those superluminous W-R stars found in 30 Dor in the Large Magellanic Cloud and HD 97950 in NGC 3603. The authors discuss the statistics of the W-R types known in M33 and conclude that selection effects still dominate the discovery process. The authors also suggest that the morphological similarities among the spectra of the superluminous W-R stars and the dynamics of the giant H II regions themselves suggest a common pattern of evolution.

158.161   **Possible detection of far-ultraviolet line emission from a hot galactic corona.**
P. D. Feldman, W. H. Brune, R. C. Henry.
Astrophys. J., Lett., Vol. 249, L51 - L54 (1981).

Rocket observations at low resolution of the spectrum of the diffuse far-ultraviolet background near the north galactic pole suggest the presence of an emission-line component to the radiation field. The lines are at the wavelengths of the emissions, from collisionally excited atoms, that have been predicted to arise from a hot galactic corona. Removal of the line emission leaves a residual uniform cosmic ultraviolet background radiation of only 150 ± 50 photons $(cm^2 s sr Å)^{-1}$, about half that previously reported.

158.162   **Milliarcsecond structure of BL Lac during outburst.**
R. L. Mutel, H. D. Aller, R. B. Phillips.
Nature, Vol. 294, 236 - 238 (1981).

The authors have mapped the radio structure of BL Lac at 5 and 10.6 GHz at three epochs during the large flux outburst of 1980. They show that comparison of the size and flux density of the core component with the flux history provides evidence for relativistic beaming effects, independent of detailed model considerations.

158.163   **The γ-ray emission from the Seyfert galaxy NGC 4151.**   L. Bassani.
Astrophys. Space Sci., Vol. 79, 469 - 481 (1981).

Light curves at X-ray, ultraviolet, optical and infrared wavelengths of NGC 4151 over the period 1977–1980 are presented together with a summary of the large amount of observational data relating to this Seyfert galaxy. Intensity variations in different wavebands are discussed and compared with recent low energy γ-ray observations of this object. The data presented are compared with the predictions of some emission mechanisms, which have been proposed to account for the gamma-radiation.

158.164   **Infrared and optical polarization of Markarian 421.**
J. Bailey, E. C. Cunningham, J. H. Hough, D. J. Axon.
Mon. Not. R. Astron. Soc., Vol. 197, 627 - 632 (1981).

The authors present observations of the polarization of the BL Lac object Markarian 421 in the V, J and K bands. They find the infrared polarization to be substantially lower than that in the optical. The drop in polarization is too large to be explained solely as due to dilution by the unpolarized infrared flux of the host galaxy, and thus indicates a true wavelength dependence in the polarization of the non-thermal source.

158.165   **The ionized gas in NGC 5128: evidence for a shock-heated component.**   M. M. Phillips.
Mon. Not. R. Astron. Soc., Vol. 197, 659 - 677 (1981).

New spectrophotometric and kinematic data for the ionized gas in the main body of the giant radio galaxy NGC 5128 are presented. An analysis of the emission-line spectra shows (1) discrete and diffuse low-ionization H II regions which are photo-ionized by the radiation of normal O and B stars, and (2) a turbulent, diffuse component which emits a spectrum more typical of shock-ionized nebulae.

158.166   **Gamma ray constraints on annihilation in active galaxies.**   A. J. Allen.
Mon. Not. R. Astron. Soc., Vol. 197, 687 - 692 (1981).

Observational constraints bearing on the possibility that active galaxies are powered by matter–antimatter annihilation are discussed. It is shown, contrary to previous work, that annihilation is not ruled out by observations.

158.167   **Distribution of molecular gas in three face-on galaxies.**   L. J. Rickard, P. Palmer.
Astron. Astrophys., Vol. 102, L13 - L16 (1981).

The $J = 1 - 0$ emission of carbon monoxide has been mapped over the optical disks of the three nearly face-on galaxies M 51, NGC 6946 and IC 342. The inferred H_2 distributions are flat disks plus central peaks. The disks are confined to the optical disks, not extended like H I. They show much structure, but no obvious spiral patterns.

158.168   **Variability of the continuum and the emission lines in the Seyfert 1 galaxy Akn 120.**
W. Kollatschny, K. J. Fricke, H. Schleicher, H. W. Yorke.
Astron. Astrophys., Vol. 102, L23 - L25 (1981).

Continuum and emission line variability of the Seyfert 1 galaxy Akn 120 is established from simultaneous optical and ultraviolet observations at two different epochs.

158.169   **CCD photometry of edge-on spirals in the Ursa Major and Cancer clusters.**   T. C. Weekes.
Astron. J. Vol. 86, 1415 - 1423 (1981).

A CCD camera was used to do R-band photometry on edge-on spirals in the Ursa Major and Cancer clusters; $R_{-0.5}$ magnitudes were derived in the same way as the $H_{-0.5}$ magnitudes used by Aaronson, Mould, and Huchra (1979) in defining a distance scale. The $R_{-0.5}$ magnitudes when used in the Tully-Fisher relationship show only slightly more scatter than $H_{-0.5}$, the slope is close to -10. The CCD R-band photometry has the advantage that the galaxy diameters are derived from the CCD images themselves. The color differences can be used to derive an inclination correction in R and B.

158.170   **Integrated magnitudes and mean colors of the DDO dwarf galaxies in the UBV system: I. Observations and catalog.**
G. de Vaucouleurs, A. de Vaucouleurs, R. Buta.
Astron. J., Vol. 86, 1429 - 1463 (1981).

The integrated magnitudes and colors of 163 DDO "dwarf" galaxies in the UBV system are derived from 542 photoelectric observations with four telescopes between 1976 and 1980. The authors derive total B_T magnitudes, effective apertures, and mean colors for 150 objects, study the mean luminosity profiles of each type T, discuss the mean effective surface brightness and the effective and isophotal diameter.

They discuss the color gradients and effective colors, and reduce the Fisher-Tully magnitudes M_{pg} to the B_T system to provide at least approximate magnitudes for the remaining 79 DDO objects as yet unobserved by the authors. The observing log and a detailed error analysis are given in two appendices.

158.171 Galactic absorption line coronae.
J. N. Bregman.
Astrophys. J., Vol. 250, 7 - 17 (1981).

The author has investigated whether gaseous coronae around galaxies give rise to the absorption systems seen in quasar spectra. In his model, gas originally located in the disk is heated to the million degree range and rises to surround the galaxy; the gas remains bound to the galaxy. Optically thin radiative cooling drives a thermal instability in the hot gas which causes cool clouds ($T \lesssim 10^4$ K) to condense out of the corona. These clouds, which follow ballistic trajectories back to the disk, are the absorption sites. A two-dimensional hydrodynamic code with radiative cooling was used to study the dynamics and thermodynamics of the corona as well as the position and rate at which clouds form.

158.172 Empirical results from a study of active galactic nuclei.   J. M. Shuder, D. E. Osterbrock.
Astrophys. J., Vol. 250, 55 - 65 (1981) = Lick Obs. Bull. No. 888.

Spectrophotometric observations are presented for emission-line galaxies. When combined with previous Lick Observatory measurements, they are used to deduce criteria that allow the Seyfert 2 phenomena to be quantitatively isolated from the narrow-emission-line galaxies. These criteria are based on the authors' finding that galaxies with [O III] λ 5007/Hβ < 3 have emission line widths that are usually considerably smaller than the widths in galaxies with [O III] λ 5007/Hβ > 3. In addition, large He II λ4686/Hβ ratios are consistent with [O III] λ 5007/Hβ > 3. The authors also compare the physical conditions in the narrow-line regions of Seyfert 1 galaxies with those in Seyfert 2 galaxies. Significant differences are found for the temperature-sensitive [O III] ratio, and for the luminosity of [Fe VII] λ 6087 relative to the luminosity of the low-ionization forbidden lines.

158.173 Spectroscopic observations of superthin galaxies.
J. W. Goad, M. S. Roberts.
Astrophys. J., Vol. 250, 79 - 86, plates 12, 13 (1981).

Superthin galaxies are distinguished by linelike images with axial ratios of up to 20 : 1 and little or no central bulge. Their masses, luminosities, and sizes are typical of the latest type spirals, in this case seen exactly edge-on. With the 4 m and 2.1 m telescopes at Kitt Peak, the authors observed four superthin galaxies: UGC 4278 (= IC 2233), 7170, 7321, and 9242. For comparison, they also observed two related edge-on systems: UGC 3697 (which has a thin disk distorted into an integral-sign shape) and UGC 4943 (which has a thin disk plus a nuclear bulge). Unlike the latter two galaxies, the four superthin systems are characterized by velocity curves of modest gradient, and by line ratios [N II]/[S II] having typically one-fourth the value seen in a normal H II region. Since [N II]/Hα is also low, these systems appear to be deficient in nitrogen.

158.174 Spectral components of NGC 4151.
G. H. Rieke, M. J. Lebofsky.
Astrophys. J., Vol. 250, 87 - 97 (1981).

Near-infrared spectrophotometry of NGC 4151 shows lines of Bγ and [Fe II]. Accurate differential photometry at 10 μm indicates no variation greater than 8% with changes in near-infrared and nonthermal flux of a factor of 2 or more. These observations are combined with previously published ones to construct a comprehensive model for the continuum of the nucleus of this galaxy. For wavelengths longer than 1 μm, the continuum is dominated by thermal reradiation by dust. Reddening by this dust strongly influences the observed properties of the nonthermal continuum and the relative emission-line strengths.

158.175 Near-infrared spectrophotometry of NGC 4151.
C. W. McAlary, R. A. McLaren.
Astrophys. J., Vol. 250, 98 - 102 (1981).

The authors have obtained 2 - 4 μm spectrophotometry of the Seyfert 1.5 galaxy NGC 4151. The Bα and Bγ lines of hydrogen have been detected with strengths of 4.2 ± 0.7 and $1.9 \pm 0.2 \times 10^{-13}$ ergs cm^{-2} s^{-1}, respectively. The Bα/Hγ ratio has been used to derive an upper limit, $E(B-V) \leqslant 0.13$ mag, for the reddening toward the line formation regions of the nucleus. The 2–4 μm continuum of NGC 4151 can be adequately fitted by two blackbodies, one representing a stellar component and the other strong thermal dust emission. No more than 10% of the total emission at 2 μm can be attributed to any nonthermal spectral distribution.

158.176 A new limit on the mass-to-light ratio of the halo of NGC 4565.
S. P. Boughn, P. R. Saulson, M. Seldner.
Astrophys. J., Lett., Vol. 250, L15 - L18, plate L1 (1981).

K-band (2.2 μm) photometry is reported for selected positions near the minor axis of the edge-on spiral galaxy NGC 4565. If the "dark matter" implied by the rotation curve is in the form of a spherical halo, its mass-to-light ratio is $M/L > 38$ (2 σ lower limit) in solar K-band units. The faintest known main-sequence star, VB 10, has a mass-to-light ratio of $M/L \approx 34$. In addition, the K-band data suggests a ($V-K$) color gradient for the central bulge of this galaxy.

158.177 New observational constraints on the M87 jet.
J. T. Stocke, G. H. Rieke, M. J. Lebofsky.
Nature, Vol. 294, 319 - 322 (1981).

New observations at 1.6–3.45 μm confirm the presence of a dramatic ($\Delta\alpha \sim 1$) break between radio–IR wavelengths and 6,000 Å, in the spectrum of the M87 jet. These data, in combination with data taken in other spectral regions, show that the individual knots in the M87 jet have nearly the same spectral indices and nearly the same large ($\Delta\alpha \sim 1$) spectral break. This large spectral break and the constancy of spectral properties between the knots pose serious constraints for models of the M87 jet.

158.178 Absorption of γ rays in active galaxies as a test of the jet hypothesis.   L. Bassani, A. J. Dean.
Nature, Vol. 294, 332 - 333 (1981).

The authors investigate the attenuation of γ rays as a result of pair-production in photon–photon collisions with X rays and show that whilst Seyfert galaxies are transparent to γ radiation up to ~1 GeV, QSOs and BL Lac objects are generally opaque above 1 MeV. However, the detection of γ rays from these latter classes of objects may be explained if beaming of the X–γ emission is assumed.

158.179 Geometrical analysis of catalogues of galaxies.
Sh. A. Guberman, A. G. Doroshkevich, Eh. V. Kotok, S. F. Shandarin.
Inst. prikl. mat. AN SSSR. Prepr., 1981, No. 42, 28 pp. In Russian. – Abstr. in Ref. zh., 51. Astron., 10.51.679 (1981).

158.180 Morphological analysis of Seyfert galaxies.
N. N. Pavlova.
Astrofiz. inst. AN KazSSR. Alma-Ata, 1981. 34 pp. In Russian. Abstr. in Ref. zh., 51. Astron., 10.51.699 (1981).

158.181 Some results of an investigation of galaxies with UV excess.
M. A. Kazaryan, A. R. Petrosyan, V. S. Tamazyan.
Pis'ma Astron. Zh., Tom 7, 648 - 650 (1981). In Russian. English translation in Soviet Astron. Lett., Vol. 7.

Preliminary results of a morphological and spectral investigation of galaxies with UV excess from lists by Kazaryan are presented. About 2/3 among 120 investigated galaxies are spirals. The distributions of absolute magnitudes as well as of emission line intensity ratios N II/Hα for these galaxies are given.

158.182 Magnetic field structure of the central source of the radio galaxy Virgo A.   V. N. Kuril'chik.
Pis'ma Astron. Zh., Tom 7, 651 - 656 (1981). In Russian. English translation in Soviet Astron. Lett., Vol. 7.

The observed peculiarities of linearly polarized emission of the central source of the radio galaxy Virgo A are in agreement with the bipolar large-scale structure of the magnetic field of this source. There is a regular growth of rotation measure in the knots of the Virgo A jet, which may result from sweeping of interstellar gas by axially expanding symmetric magnetic field loops.

158.183 Peculiarities in the optical variability of the galaxy NGC 4151.   V. M. Lyutyj, V. L. Oknyanskij.
Pis'ma Astron. Zh., Tom 7, 657 - 661 (1981). In Russian. English translation in Soviet Astron. Lett., Vol. 7.

Photographic and photoelectric observations of the optical variability of the nucleus of the Seyfert galaxy NGC 4151 during 1934 - 1980 have been analysed. The presence of a quasi-periodic component with $P = 126^d$ is confirmed. There are cyclic variations of this period with a cycle time of ~ 20 years.

158.184 Recurrent Seyfert activity in spiral galaxy nuclei.   R. H. Sanders.
Nature, Vol. 294, 427 - 429 (1981).

The author suggests a model for recurring activity in normal galactic nuclei based on the assumed presence of a massive nuclear black hole ($\sim 10^7 M_\odot$) and the observation of a very clumpy distribution of interstellar gas in the inner 200 pc of our Galaxy. The system of massive molecular clouds with low net angular momentum could provide an 'accretion event' every 10^7 yr with a duration of $\sim 10^5$ yr.

158.185 The appearance of broad emission lines in the spectrum of the BL Lac object PKS 0521-36.
M. H. Ulrich.
Astron. Astrophys., Vol. 103, L1 - L2 (1981).

The radio galaxy PKS 0521-36 has a compact continuum source which is variable and polarized in the optical range. It emits weak narrow emission lines from low excitation gas and on this basis could be classified as a BL Lac object. The observations presented here show that the line spectrum has recently undergone remarkable changes. In particular, intense broad permitted lines have appeared between 1975 and 1981. The full width at zero intensity of Hα is ~ 120 Å.

158.186 Observations of the head-tail radio galaxy NGC 3862 (3C 264) at 0.6, 1.4, and 5.0 GHz.
G. Gavazzi, G. C. Perola, W. Jaffe.
Astron. Astrophys., Vol. 103, 35 - 43 (1981).

The radio source 3C 264, identified with the elliptical galaxy NGC 3862 in the A1367 cluster, has been mapped with the WSRT (full intensity and polarization) at 0.6, 1.4 and 5.0 GHz, and with the VLA at 4.9 GHz. The WSRT maps show a head-tail morphology with a twin arm structure. Like in other well studied sources of this type, the spectrum becomes steeper away from the galaxy, the prevailing direction of the magnetic field is roughly parallel to the direction of the far tail, the density of the thermal electrons is about 10^{-3} cm^{-3}.

158.187 Importance of the Doppler Differential Effect in the interpretation of active nuclei spectra. I. The hydrogen spectrum.   C. Gordon, S. Collin-Souffrin,

D. Dultzin-Hacyan.
Astron. Astrophys., Vol. 103, 69 - 77 (1981).

The introduction of the Doppler Differential Effect (DDE) in the interpretation of Seyfert I galaxies and quasars spectra explains the observed relative intensities of the hydrogen lines using a simplified photoionized model. The DDE may occur inside one cloud optically thick of the Lyman continuum or between two different clouds of a disk or a cone.

158.188 Variations in the spectrum of the Seyfert galaxy Ak 120.   H. Schulz, P. Rafanelli.
Astron. Astrophys., Vol. 103, 216 - 219 (1981).

The intensity ratio of the emission lines [O III] λ 5007 and Hβ decreased strongly within the 13 months between November 1978 and December 1979. Variations of line profiles and equivalent widths have also been observed. The physical conditions in the emission line region and the relevant time scales are discussed. It is concluded that the observed line changes are most likely caused by variations of the photoionizing flux.

158.189 Preliminary stellar photographic photometry in the Sculptor Dwarf Irregular Galaxy (ŞDIG).
J. Lequeux, R. M. West.
Astron. Astrophys., Vol. 103, 319 - 323 (1981).

Preliminary electronographic V photometry is given for individual stars in the Sculptor Dwarf Irregular Galaxy (SDIG), in the Sculptor group of galaxies. These are the first such data available for a galaxy outside the Local Group. The brightest star has an absolute V magnitude of -6.3. The lack of brighter stars which are present in bigger galaxies is not statistically significant. The luminosity function is similar to that in the irregular galaxies IC 1613 and NGC 6822 and, like these objects, SDIG appears to be in a relatively quiescent stage of star formation.

158.190 Optical brightness variations of BL Lac objects.
H. Zekl, G. Klare, I. Appenzeller.
Astron. Astrophys., Vol. 103, 342 - 348 (1981).

Using the plate collection of the Bruce astrograph of the Landessternwarte Heidelberg-Königstuhl optical brightness variations between 1901 and 1980 have been measured for 21 BL Lac objects north of declination $\delta = -30°$. The observed magnitude index $R_B^s(\max)$, [but not the maximum brightness variation $\Delta B(\max)$] is found to be positively correlated with the spectral index at 5 GHz.

158.191 Puzzling ejections from Centaurus A.   V. G. Surdin.
Zemlya Vselennaya, 1981, No. 6, p. 24. In Russian.

158.192 Spiral galaxies.   I. I. Pasha, F. A. Tsitsin.
Zemlya Vselennaya, 1981, No. 6, p. 50 - 56. In Russian.

158.193 The chemistry of galaxies.   G. Shields.
Astronomy, Vol. 9, No. 6, p. 6 - 17 (1981). – Abstr. in Phys. Abstr., Vol. 84, Abstr. 108320 (1981).

158.194 Search for Wolf-Rayet stars in external galaxies with the CFHT.   A. F. J. Moffat, M. M. Shara.
J. R. Astron. Soc. Canada, Vol. 75, 250 - 251 (1981). Abstract.

158.195 Photometry of bright stars in the dwarf irregular galaxy Leo A.   J. Mongeon, S. Demers.
J. R. Astron. Soc. Canada, Vol. 75, 251 (1981). – Abstract.

158.196 Filamentary structure in NGC 5253.   J. A. Graham.

Publ. Astron. Soc. Pacific, Vol. 93, 552 - 553 (1981).

A photograph of the galaxy NGC 5253 in the light of [O III] λ5007 is reproduced. Ionized gas with extensive filamentary structure is associated with the galaxy. Emphasis is placed upon the resemblance between this galaxy and the peculiar galaxies Messier 82, NGC 3077, and NGC 1569 which appear in each case to be in a similar environment close to a large spiral galaxy.

158.197 A new Seyfert galaxy with asymmetric forbidden-line profiles in an interacting system.
A. J. Wasilewski.
Publ. Astron. Soc. Pacific, Vol. 93, 560 - 563 (1981).

On an exploratory IIIa-J plate taken with the Burrell Schmidt telescope equipped with a 4° objective prism, a system of three interacting galaxies (NGC 7212) has been found whose spectrum suggests the presence of a Seyfert component. The objective-prism spectrum of NGC 7212 indicates that it contains a Sy 2 nucleus, and this is confirmed by a high-resolution scan. Forbidden-line asymmetries are discussed.

158.198 Photoelectric comparison sequences in the fields of four radio sources. H. R. Miller.
Publ. Astron. Soc. Pacific, Vol. 93, 564 - 566 (1981).

Photoelectric *UBV* comparison sequences in the fields of four identified or suspected BL Lacertae objects have been calibrated to facilitate and encourage long-term photometric monitoring of these objects.

158.199 Optical positions of Seyfert galaxies.
E. D. Clements.
Mon. Not. R. Astron. Soc., Vol. 197, 829 - 834 (1981).

Positions with an accuracy of about 0.1 arcsec are presented for over 100 Seyfert galaxy nuclei observed with the 26-inch and 13-inch refractors at the RGO.

158.200 The evolution of flows of stellar mass loss in active galaxies. J. MacDonald, M. E. Bailey.
Mon. Not. R. Astron. Soc., Vol. 197, 995 - 1019 (1981).

Gas flows in elliptical galaxies fuelled by stellar mass loss have been calculated numerically for three fiducial epochs using both King and de Vaucouleurs models of NGC 3379. The authors' model explains the association of nuclear activity with giant elliptical galaxies and predicts rapid comoving density evolution as observed.

158.201 What is Markarian 3? C. R. Jenkins.
Mon. Not. R. Astron. Soc., Vol. 197, 1049 - 1053 (1981).

Although the radio galaxy Markarian 3 has been classified as a Seyfert 2 by Khachikian & Weedman, recent long-slit spectra reveal no sign of a gas disc and suggest instead that the galaxy is an elliptical. There may be some slight rotation of the gas, in which case the radio axis does not appear to be aligned with the rotation axis.

158.202 The infrared continua of active galaxies.
I. S. Glass.
Mon. Not. R. Astron. Soc., Vol. 197, 1067 - 1079 (1981).

Multi-colour, multi-aperture infrared (*JHKL*) photometry is presented for 24 Seyfert and X-ray emitting southern galaxies. The essential uniformity of the colours of the Seyfert 1 nuclei and their similarity to those of the violently variable quasars is shown for this wavelength region. The nuclear colours of the remaining X-ray emitting galaxies are found to be similar after correction for reddening. Further evidence for the universality of the underlying power sources of these galaxies and of quasars is drawn from the ratios of infrared to X-ray fluxes.

158.203 X-ray emission around radio galaxies in non-Abell clusters: a possible physical link between environ-
ment and nonthermal radio emission.
J. O. Burns, S. A. Gregory, G. D. Holman.
Astrophys. J., Vol. 250, 450 - 463 (1981).

X-ray observations of a sample of 11 4C radio galaxies in clusters with $z \lesssim 0.1$ were performed with the IPC (resolution FWHM ~1.5) on the Einstein Observatory. The sample galaxies lie in the directions of non-Abell Zwicky clusters; these clusters are generally poor, ranging in effective Abell richness class from $\lesssim 0$ to 1. In 10 out of the 11 observed fields 1 - 3 keV emission was detected. In nearly every case, the X-ray emission is centered on the radio galaxy; 5 out of 10 of the X-ray sources are resolved with HWHM of 70–300 kpc.

158.204 Temporal variations of the Balmer line profiles in the spectrum of the Seyfert 1 galaxy Arakelian 120.
C. B. Foltz, B. M. Peterson, E. R. Capriotti, P. L. Byard, R. Bertram, D. G. Lawrie.
Astrophys. J., Vol. 250, 508 - 512 (1981).

Observations of the emission-line strengths and the Hα and Hβ line profiles in the spectrum of the Seyfert 1 galaxy Arakelian 120 have been obtained with the Ohio State University image-dissector scanner. Comparison of these data with observations of comparable resolution and quality obtained with the Lick Observatory image-dissector scanner show that a dramatic change occurred in the Hβ profile between 1976 November and 1981 January. A change of the same sense is observed in Hα to a lesser extent, implying that the Balmer decrement of the component which has apparently brightened is less steep than that of the total line emission. The changes in the emission-line profiles were accompanied by decreases in both the Hα and Hβ equivalent widths and in the [O III] λ5007/Hβ flux ratio. The latter variations imply that the broad lines have increased in strength in response to an increase in the luminosity of the continuum source.

158.205 *HEAO 1* observations of high-energy X-rays from the Seyfert I galaxy Mkn 509. S. Dil, F. A. Primini, E. Basinska, M. Bautz, S. K. Howe, F. Lang, A. M. Levine, W. H. G. Lewin, D. M. Worrall, P. L. Nolan, J. L. Matteson.
Astrophys. J., Vol. 250, 513 - 516 (1981).

The results of observations of the Seyfert I galaxy Mkn 509 at high X-ray energies (14 - 140 keV) with the UCSD/MIT instrument aboard *HEAO 1* are reported. The source is detected at energies above 40 keV, and the spectrum is consistent with an extrapolation of the power-law spectrum obtained at lower X-ray energies. Observations on four separate days in 1978 October indicate that the source is varying on a time scale of days.

158.206 The extended H I disk of the S0 galaxy NGC 4203.
D. Burstein, N. Krumm.
Astrophys. J., Vol. 250, 517 - 527, plate 18 (1981).

An Arecibo 21 cm H I map of this remarkable galaxy is presented. As judged by the usual classification criteria, NGC 4203 is a fairly typical S0 galaxy, although its disk-to-bulge ratio is at the high end of that observed for S0's. Unlike most S0's, NGC 4203 is embedded in an H I disk at least 38 h^{-1} kpc in diameter, over three times the Holmberg size. The velocity field of the H I is generally indicative of a rotating disk with a flat rotation curve, but it is distorted in a manner suggesting an oval distortion or a kinematical warp. The distribution of gas in the outer parts of the galaxy is observed to be asymmetric and is inferred to be asymmetric near the center of the galaxy by combining the H I data with an inner optical rotation curve kindly provided by V. C. Rubin. Intercomparison of the H I profiles bracketing the center with the center profile indicates the gas has a central, unresolved minimum.

158.207 VLBI and X-ray observations of compact nuclei in pairs of galaxies.
P. Biermann, P. P. Kronberg, E. Preuss, R. T. Schilizzi,

D. B. Shaffer.
Astrophys. J., Lett., Vol. 250, L49 - L53 (1981).

The authors observed five pairs of isolated pairs of galaxies with radio sources from the sample of Stocke et al. with a three-element VLBI array operating at 4996 MHz (6 cm) with antennas in Germany, The Netherlands, and the United States. They have detected parsec-scale components in three of these objects (NGC 3894, Arp 102, and NGC 6500). The authors then observed these objects and UGC 10094B (= K475) with the IPC aboard the Einstein Observatory. They detected Arp 102B as a strong X-ray source with a luminosity in the 0.5 - 4.5 keV band of about 1×10^{43} erg s^{-1}. Comparing optical and X-ray flux densities, the authors find that Arp 102B is similar to Seyfert class 1 galaxies.

158.208   He I λ10830 observations of five Seyfert galaxies.

P. D. LeVan, R. C. Puetter, R. J. Rudy, H. E. Smith, S. P. Willner.
Astrophys. J., 251, 10 - 14 (1981).

Helium λ10830 line fluxes have been measured for the Seyfert 1 galaxies NGC 3516, NGC 4151, NGC 7469, and Mrk 509, and for the Seyfert 2 galaxy NGC 1068. The observed λ10830/λ5876 ratio is considerably smaller than that predicted for a high density ($n_e \gtrsim 10^4$ cm^{-3}), optically thin gas in the case of the four Seyfert 1 galaxies. This is interpreted as being due to optical depth effects in the λ10830 transition with $\tau(\lambda 10830) \gtrsim 250$. The observed line ratio for NGC 1068 is much larger. The intrinsic ratio in NGC 1068 is probably similar to the Seyfert 1 galaxies but more highly reddened by dust. Although the assumed intrinsic value could be explained by very low values of n_e and T_e for that object, it is likely that a large λ10830 optical depth is also required for NGC 1068.

158.209   Rapid X-ray variability in the Seyfert galaxy NGC 6814.   A. F. Tennant, R. F. Mushotzky, E. A. Boldt, J. H. Swank.

Astrophys. J., Vol. 251, 15 - 25 (1981).

The authors report HEAO 1 A-2 high time resolution X-ray observations of the X-ray emitting Seyfert I galaxy NGC 6814. In sharp distinction to a sample of over 30 active galactic nuclei (Tennant et al.), this object showed strong X-ray variability on time scales less than 3 hr. The mean flux on a time scale of 90 minutes varied by a factor of ~2.5 corresponding to $\Delta L_x \sim 1 \times 10^{43}$ ergs s^{-1}. An autocorrelation analysis shows a characteristic time for variability of 100 (+60/−25) seconds. There is no indication of spectral variability. The authors consider in detail the constraints of such rapid variability on a wide variety of X-ray source mechanisms.

158.210   X-rays from active galactic nuclei.

F. Takahara, S. Tsuruta, S. Ichimaru.
Astrophys. J., Vol. 251, 26 - 30 (1981).

Through detailed numerical calculations, the authors have shown that production of thermal photons at the self-absorbed cyclotron higher harmonics and Comptonization of those soft photons in a hot plasma with the electron temperature 10^9 K provide a natural mechanism to account for the dominant observed features of the X-ray emission in the active galactic nuclei, both in intensities and in spectra. As a most probable environment where the necessary conditions may be satisfied, the authors propose a hot plasma deep within the accretion flows onto supermassive black holes. The model is consistent with the rapid time variabilities observed in some of those X-ray sources.

158.211   The X-ray structure of Centaurus A.

E. D. Feigelson, E. J. Schreier, J. P. Delvaille, R. Giacconi, J. E. Grindlay, A. P. Lightman.
Astrophys. J., Vol. 251, 31 - 51, plate 1 (1981).

Detailed study of the nearby radio galaxy Cen A =

NGC 5128 with the imaging detectors on board the Einstein X-ray Observatory reveals X-ray emission associated with several components of the galaxy: the compact nucleus; an X-ray jet pointed toward the NE radio lobes; the "middle" NE radio lobe; the disk or dust lane; and diffuse emission extending several arc minutes around the nucleus. The different components are discussed in detail.

158.212   The metal abundance range in the Ursa Minor dwarf galaxy.   R. Zinn.

Astrophys. J., Vol. 251, 52 - 60 (1981).

A small sample of red giants in the Ursa Minor dwarf spheroidal galaxy has been observed with the multichannel scanner and the digital spectrograph of the Hale 5 m telescope. These observations confirm the large width of the giant branch in the color-magnitude diagram and indicate that it is caused by a range in metal abundance of ~1 dex. Ursa Minor and the Draco dwarf spheroidal appear to have very similar ranges in metal abundance. These observations and those of the other dwarf spheroidal galaxies are compared with the models of the chemical evolution of elliptical galaxies that include mass loss. The spectrophotometry of the peculiar star, K, that was discovered by Canterna and Schommer shows that it is a mild CH star. A very similar star has been discovered in the globular cluster M2. If this star is a member of M2, which seems likely, then CH stars are not found only in clusters of low central concentration, as the previous discoveries of CH stars had suggested.

158.213   Velocity dispersions of low luminosity ellipticals: $L \sim \sigma^3$?   J. L. Tonry.

Astrophys. J., Lett., Vol. 251, L1 - L5 (1981).

A complete sample of elliptical galaxies in the core of the Virgo cluster are used to derive a relation between stellar velocity dispersion and luminosity. The exponent of $L \sim \sigma^n$ is found to be $n = 3.2 \pm 0.2$ over 2 decades of luminosity, with a scatter of 0.38 mag. The discordance between this value and the slope of 4 traditionally assumed is discussed, and the possibilities of curvature in the relation or rotational support of the galaxies are examined. A derivation of a law of slope 3 is presented and compared with the derivation for a slope of 4.

158.214   Structure of galaxies.   J. Palouš.

Vesmír, Vol. 60, 234 - 239 (1981). In Czech.

158.215   Multidimensional statistical analysis of normal galaxies.

V. Bujarrabal, J. Guibert, C. Balkowski.
Astron. Astrophys., Vol. 104, 1 - 9 (1981).

A multidimensional statistical analysis of integral properties of normal galaxies has been carried out using principal component analysis applied to radio and optical data from several large (~ 100 objects) catalogues. The authors have investigated in detail the influence of distance uncertainties on statistics of galaxies.

158.216   [O III]/Hβ ratios of emission regions in the arms and disk of M 33 and luminosity functions at the fronts of the arms.

J. Boulesteix, R. Dubout-Crillon, G. Monnet.
Astron. Astrophys., Vol. 104, 15 - 23 (1981).

The authors used a Fabry-Perot interferometer to obtain a great number of coherent spectrophotometric measures of both normal H II regions and faint, diffuse emission regions in M 33. All the observational results (line ratios, early star numbers at the fronts of the arms, Hα fluxes) for the southern and northern arms of M 33 and the galactic arms near the Sun are in agreement with a very simple model.

158.217   The giant spiral galaxy M 101. VII. Associations of H I concentrations and H II complexes.

F. Viallefond, R. J. Allen, W. M. Goss.

Astron. Astrophys., Vol. 104, 127 - 141 (1981).

The authors have investigated the detailed relationship between the H I concentrations and the giant H II complexes in M 101 on a scale of about one kiloparsec. They discuss new observational data on the H I/H II associations in M101, and derive the physical parameters of the complexes. The variation of these parameters with galactocentric radius is presented and a brief discussion follows.

158.218 Discovery of a Seyfert 1 galaxy with an unusually soft X-ray spectrum.
S. H. Pravdo, J. J. Nugent, J. A. Nousek, K. Jensen, A. S. Wilson, R. H. Becker.
Astrophys. J., Vol. 251, 501 - 507 (1981).

The authors report the identification of one of the brightest soft X-ray sources in the sky, H1613+06, with a newly discovered Seyfert 1 galaxy, E1615+061. Its X-ray spectrum is steep, energy index $\alpha > 2$, and similar to the variable soft X-ray components seen in BL Lac objects and in the Seyfert 1, ESO 141- G55. A significant fraction of the presently unidentified, brightest soft X-ray sources at high galactic latitudes could be these compact extragalactic objects.

158.219 Mapping of NGC 5128 (= Centaurus A) at J, H, and K. P. Harding, T. J. Jones, A. W. Rodgers.
Astrophys. J., Vol. 251, 530 - 532, plates 10 - 11 (1981).

The central 10′ region of NGC 5128 has been mapped by drift scanning in J, H, and K. Evidence of saturation of previous estimates of A_v in the dust lane is presented, and it is shown that the near-infrared light is dominated by reddened stars of the elliptical component of the galaxy. The apparent asymmetry of the dust lane across the nucleus is confirmed for the near-infrared radiation.

158.220 The line spectrum of the Fe II Seyfert I galaxy Akn 120. W. Kollatschny, H. Schleicher, K. J. Fricke, H. W. Yorke.
Astron. Astrophys., Vol. 104, 198 - 202 (1981).

Combined UV and optical spectra for the Fe II Seyfert galaxy Akn 120 are presented. This galaxy shows unusually strong Lyα and UV Fe II emission and a jump in the continuum at 3000 Å - 4000 Å.

158.221 Information deduced from the spectra of quasars and Seyfert galaxies. H. Netzer.
Proceedings of the 5th Göttingen-Jerusalem-Symposium on Astrophysics, (see 012.041), p. 75 - 87 (1981).

158.222 Spectroscopy of Seyfert galaxies.
H. Schleicher.
Proceedings of the 5th Göttingen-Jerusalem-Symposium on Astrophysics, (see 012.041), p. 89 - 98 (1981).

158.223 The new BL Lac object 0716+71.
W. Kollatschny.
Proceedings of the 5th Göttingen-Jerusalem-Symposium on Astrophysics, (see 012.041), p. 111 - 113 (1981).

158.224 X-ray emission from new BL Lac Objects.
K. J. Fricke.
Proceedings of the 5th Göttingen-Jerusalem-Symposium on Astrophysics, (see 012.041), p. 115 - 125 (1981).

158.225 An optical and X-ray survey of s-type Markarian galaxies. D. J. Hutter, S. L. Mufson.
Astron. J., Vol. 86, 1585 - 1594 (1981).

The authors report the results of a study of 23 compact, lineless Markarian galaxies using broadband optical photometry and X-ray satellite observations. The photometry shows that the sample can be broken into four groups. In one group (Mrk 180, 421, and 501) are composite objects in which a BL Lacertae object is embedded in an elliptical galaxy. For

this group, they present the results of multiepoch X-ray observations using the HEAO-1 and -2 satellites. In addition, they use the photometry to decompose the optical emission into nonthermal and galactic components. In the second group are objects showing a small ultraviolet excess relative to normal galaxies. The X-ray survey indicates that the X-ray luminosity of objects in group 2 is much lower than those in group 1. The third and fourth groups are objects whose colors are indistinguishable from those of normal field galaxies and those of galactic stars, respectively. No X-ray emission was detected from objects in either of these groups.

158.226 Spectral variations of Markarian 335.
J. M. Shuder.
Astron. J., Vol. 86, 1595 - 1599 (1981) = Lick Obs. Bull. No. 898.

Examination of spectral scans of Mrk 335 has revealed it to be a variable Seyfert 1 galaxy. In addition to the continuum and the emission-line variability, differences were found in the 3000 Å bump between the 1974 and 1976 observations.

158.227 Radiation processes in active galactic nuclei.
A. Cavaliere.
Plasma astrophysics, (see 012.042), p. 97 - 103 (1981).

A discussion of the peculiar conditions of the plasma filling the emission regions of the powerful continuum radiation from active galactic nuclei and quasars is given. The plasma is energized by gravitation; it is radiation dominated, out of thermal equilibrium and at least weakly relativistic; it is likely to comprise a large number of positrons. Thus the exploration of a new and specific range of plasma physics is called for.

158.228 Hot plasmas in active galactic nuclei.
M. J. Rees.
Plasma astrophysics, (see 012.042), p. 297 - 301 (1981).

In the deep potential well around a compact object, plasma can retain thermal energies of 1 - 100 MeV per particle, even if it is gravitationally bound. The radiative losses are then likely to be due to synchrotron/Compton emission; pair production may be so prolific that positrons greatly outnumber the ions; the ion and electron temperatures would be unequal unless there are strong collective effects which couple them. These processes may be important in determining the radiative and dynamical properties of galactic nuclei.

158.229 X-ray emission from active galaxies.
M. J. C. Hayes, S. J. Bell Burnell, J. L. Culhane, M. J. Ward, P. Barr, J. C. Ives, P. W. Sanford.
Space Sci. Rev., Vol. 30, (see 012.044), 39 - 45 (1981).

The authors review all the Ariel V spectrometer observations of Seyfert and narrow emission line galaxies, examining the X-ray spectral properties of this class of object and using this data to investigate the nature of the source.

158.230 Ariel-6 medium energy spectral observations of active galaxies.
R. Hall, M. J. Ricketts, C. G. Page, K. A. Pounds.
Space Sci. Rev., Vol. 30, (see 012.044), 47 - 54 (1981).

X-ray spectra of the BL Lac type object Mkn 421 and several Seyfert type 1 galaxies; IIIZw2, MCG8-11-11 and NGC 4151, have been obtained using the Leicester University instrument on board the Ariel-6 satellite.

158.231 X-ray observations of emission line galaxies with the Einstein Observatory.
T. Maccacaro, G. C. Perola, M. Elvis.
Space Sci. Rev., Vol. 30, (see 012.044), p. 61 - 65 (1981).

Four emission line galaxies, NGC 1365, 2992, 5506 and 7582, previously recognized as X-ray emitters in the range $10^{42} - 10^{43}$ erg/sec, have been observed with the Einstein

Observatory Imaging Proportional Counter (IPC), in one case (NGC 2992) also with the High Resolution Imager (HRI).

158.232 NGC 4151 and MCG 8-11-11: two X-ray Seyfert galaxies with strong soft γ-ray emission.
L. Bassani, R. C. Butler, A. J. Dean, G. Di Cocco, N. A. Dipper, F. Perotti, G. Villa.
Space Sci. Rev., Vol. 30, (see 012.044), 107 - 112 (1981).

During a balloon flight of the MISO telescope on 30th September 1979, the Seyfert galaxies NGC 4151 and MGC 8-11-11 were studied in the hard X-ray range ($E_X > 20$ keV) and low-energy γ-ray range up to 19 MeV. An emission at the 4.5 σ level above 20 keV was detected in the direction of NGC 4151. γ-ray emission at the 3.9 σ level above 90 keV was also observed from the direction of MCG 8-11-11. The emission photon spectrum shows a high-energy cutoff at about 3 MeV.

158.233 The X-ray luminosity function and source counts for Seyferts.
A. C. Fabian, A. K. Kembhavi, M. J. Ward.
Space Sci. Rev., Vol. 30, (see 012.044), 113 - 118 (1981).

The authors obtain an X-ray luminosity function for Seyfert 1 nuclei using the optical luminosity function due to Veron (1979) and the ~ 2 keV X-ray data primarily due to Kriss et al. (1980). This is compared with the 2 - 10 keV X-ray luminosity function due to Piccinotti et al. (1981) which was determined using a complete X-ray survey. In the second part of the paper they use this last function to predict X-ray source counts for Seyferts and compare these with the Einstein deep X-ray survey observations of a field in Pavo.

158.234 Ultraviolet, optical and X-ray monitoring of the continuum of NGC 4151.   G. C. Perola,
A. Boksenberg, G. E. Bromage, J. Clavel, M. Elvis, A. Elvius, P. M. Gondhalekar, J. Lind, C. Lloyd, M. V. Penston, M. Pettini, M. A. J. Snijders, E. G. Tanzi, M. Tarenghi, M. H. Ulrich, R. S. Warwick.
Space Sci. Rev., Vol. 30, (see 012.044), 119 - 122 (1981).

The nucleus of the Seyfert galaxy NGC 4151 has been extensively monitored with the International Ultraviolet Explorer (IUE) in the wavelength range 1200 - 3000 Å, a region of the spectrum where the contamination by the stellar component of the nuclear light is negligible. This note is a summary of the data for the continuum, which also includes optical and X-ray observations.

158.235 X-ray emission from galactic jets.
A. Ferrari, E. Trussoni.
Space Sci. Rev., Vol. 30, (see 012.044), 135 - 140 (1981).

X-ray emission has been detected from the galactic jets in Cen A and M87. Some aspects of theoretical models interpreting this emission, e.g. acceleration of relativistic electrons, plasma confinement, Kelvin-Helmholtz instabilities are outlined.

158.236 A study of M100 in X-rays.
G. G. C. Palumbo, T. Maccacaro, N. Panagia, G. Vettolani, G. Zamorani.
Space Sci. Rev., Vol. 30, (see 012.044), 143 - 147 (1981).

During a search for X-ray emission from Supernova 1979c, the parent galaxy M100 (NGC 4321) was repeatedly observed with the IPC and HRI instruments aboard the Einstein X-ray Observatory. The X-ray data reveal two possible sources in the arms of the spiral galaxy, two components in the nuclear bulge and extended X-ray emission from the central part of the galaxy. The authors find that the extended X-ray emission cannot be explained in terms of inverse Compton effect on radio, optical or 3 K blackbody photons but rather it is likely to originate from supernova remnants.

158.237 Variability of the continuum and the emission lines in the Seyfert galaxy Arakelian 120.
K. J. Fricke, W. Kollatschny.
Messenger, No. 26, p. 9 - 11 (1981).

158.238 H I observations of strongly interacting galaxies.
J. S. Gallagher, G. R. Knapp, S. M. Faber.
Astron. J., Vol. 86, 1781 - 1790 (1981).

H I observations of small systems of galaxies in some cases reveal the presence of anomalous, sloping velocity shoulders on the integrated H I line profiles. These are interpreted as evidence for ongoing tidal interactions, most probably associated with galaxies in slow hyperbolic encounters in which extensive transfers of orbital to internal energy can produce extreme noncircular motions. The kinematics of such systems therefore provide a means to identify actively interacting galaxies and a method to probe the energetics of collisions between nonbinary galaxies.

158.239 21-cm line studies of spiral galaxies. I. Observations of the galaxies NGC 5033, 3198, 5055, 2841, and 7331.   A. Bosma.
Astron. J., Vol. 86, 1791 - 1824 (1981).

The author has mapped the galaxies NGC 5033, 3198, 5055, 2841, and 7331 in the 21-cm line with angular resolutions such that the ratio radius/beam size is of order 5−10. For four of the five galaxies the H I distribution and the velocity field cannot be described by the simple picture of a spiral having an axisymmetric disk in differential rotation. For those four galaxies the main deviations from circular motion can be described by a warping of the plane at large radii. The determination of the mass distributions for these galaxies is severely compromised by this. For all galaxies the author finds asymmetries in the H I distribution, and for most of them there are H I clouds in the outer parts which do not coincide with optical features. NGC 5055 and NGC 2841 have H I disks extending 2 and 2.5 times as far as the Holmberg radius. NGC 3198, NGC 2841, and NGC 7331 have a depression in the H I distribution in the central parts.

158.240 21-cm line studies of spiral galaxies. II. The distribution and kinematics of neutral hydrogen in spiral galaxies of various morphological types.   A. Bosma.
Astron. J., Vol. 86, 1825 - 1846 (1981).

The author has collected detailed 21-cm maps and associated velocity fields in order to study the relation between the dynamical and morphological properties of spiral galaxies. A description is given of the various deviations from axisymmetry in spirals as is evident from the velocity fields. A large number of spirals exhibit large-scale symmetric deviations, to be interpreted either as oval distortions or as kinematical warps. Most galaxies also have asymmetries, which are sometimes quite large. Despite the complications due to the irregularities in the velocity fields, crude rotation curves can be derived. Most of them are flat enough to indicate that there is a large amount of high M/L material in the outer parts. The extent of a galaxy is not well determined and there is a large variety of H I sizes.

158.241 The shapes of spiral arms along the Hubble sequence.
R. C. Kennicutt, Jr.
Astron. J., Vol. 86, 1847 - 1858 (1981).

Measurements are presented of the shapes and pitch angles of the spiral arms in a large set of nearby Sa-Sc galaxies. The question of the mathematical forms of the spirals is shown to be irrelevant for the majority of real galaxies. Measured pitch angles for 113 objects correlate weakly with arm structure and bulge-to-disk ratio, suggesting that Hubble's classification criteria are less tightly coupled than previously believed. The maximum rotation velocity in a galaxy is shown to be well correlated with arm pitch, suggesting that the shape of the spiral pattern is dictated by kinematic parameters, independent of the physical origin of the arms. The arm pitch appears to be tied to the absolute value of the rotation velocity, not to the angular velocity.

158.242 On the optical appearance of distant galaxies.
C. Pritchet, M. I. Kline.
Astron. J., Vol. 86, 1859 - 1869 (1981).

The authors have used the recent evolutionary and K-corrections of Bruzual and Kron to predict the optical appearance of galaxies spanning a wide range of magnitudes and redshifts. It is found that nearly all galaxies with $J \lesssim 25$ are resolved in 1-arcsec seeing. At fixed apparent magnitude, galaxies with large redshifts are more diffuse in appearance than those at small z. This fact causes the most distant galaxies at any magnitude level to be missed, and, depending on the measurement algorithm employed, may cause the luminosities of detected galaxies to be seriously under-estimated. Both of these effects deserve consideration when attempting to interpret number counts of faint galaxies. Observations made with the Space Telescope are expected to resolve nearly all galaxies at $J \lesssim 27.5$; however, several factors conspire to render Space Telescope observations less effective than certain ground-based CCD observations for the optical detection of distant galaxies.

158.243 Broad-line Hα sources in M31.
IAU Circ., No. 3643 (1981).

158.244 NGC 4151.
IAU Circ., No. 3648 (1981).

158.245 NGC 4151.
Yamamoto Circ., No. 1967 (1981).

158.246 Rotation and mass distribution of the spiral galaxy NGC 6643.   G. A. Kyazumov.
Astron. Zh., Tom 58, 1167 - 1169 (1981). In Russian.
English translation in Soviet Astron., Vol. 25, No. 6.

The results of spectral observations of the spiral galaxy NGC 6643 are given. The total mass of the galaxy and mass-to-luminosity ratio are $M = 1.2 \times 10^{11} M_\odot$; $M/L = 3.91$.

158.247 The identification of OI 090.4 and CSV 1180.
J. H. Baumert, K. Cudworth.
Inf. Bull. Variable Stars, No. 2039, 2 pp. (1981).

158.248 Two-component model of the elliptical galaxy M32.
J. Einasto, P. Tenjes, P. Traat.
Astron. Tsirk., No. 1132, p. 5 - 6 (1980). In Russian.

158.249 On the blue object near Markarian 123.
A. R. Petrosyan, K. A. Saakyan, Eh. E. Khachikyan.
Astron. Tsirk., No. 1132, p. 7 - 8 (1980). In Russian.

158.250 New Seyfert-type objects.
B. E. Markarian, D. A. Stepanyan, V. A. Lipovetskij.
Astron. Tsirk., No. 1125, p. 4 - 6 (1980). In Russian.

158.251 Luminosity functions of Seyfert 1 and Seyfert 2 galaxies.   A. M. Romanov.
Astron. Tsirk., No. 1127, p. 5 - 6 (1980). In Russian.

158.252 UBVR estimates of the magnitude of some Seyfert galaxies.   A. M. Romanov.
Astron. Tsirk., No. 1128, p. 3 - 5 (1980). In Russian.

158.253 Seyfert-type objects from the XIIIth and XIVth lists of galaxies with UV continuum.
B. E. Markarian, V. A. Lipovetskij, D. A. Stepanyan.
Astron. Tsirk., No. 1134, p. 6 - 8 (1980). In Russian.

158.254 Statistical research of Seyfert galaxies.
A. M. Romanov.
Astron. Tsirk., No. 1142, p. 3 - 5 (1980). In Russian.

158.255 Spectrophotometry of the nuclei of the emission-line galaxies NGC 7463, Mrk 313, Mrk 531 and III Zw 103.
G. T. Petrov, V. K. Golev, I. M. Yankulova.
Astron. Tsirk., No. 1143, p. 1 - 3 (1980). In Russian.

158.256 On an analogy of some nonstationary processes in nuclei of extragalactic sources and on the sun. 2.
Yu. A. Kovalev.
Astron. Tsirk., No. 1144, p. 6 - 8 (1980). In Russian.

158.257 Blazars.   M. S. Dimitrijević.
Vasiona, Année 29, 64 - 67 (1981). In Serbo-Croatian.

158.258 Jets and the Space Telescope – an introduction.
G. K. Miley.
Optical jets in galaxies, (see 012.052), p. 9 - 13 (1981).

Jets in active galaxies are prime targets for the ST. They occur on length scales ranging from smaller than a parsec to almost a megaparsec. There appears to be a close association between the narrow-line region and radio-jet structure on the kiloparsec scale. The angular resolution of the ST will be ideal for studying these kiloparsec-sized components. Complementary radio observations are essential if the unique capability of the ST is to be exploited fully.

158.259 The optical jet in M87 – deconvolution by maximum entropy.   M. A. C. Perryman.
Optical jets in galaxies, (see 012.052), p. 25 - 28 (1981).

Deconvolution of astronomical images by the method of maximum entropy is described and the algorithm applied to observations of the optical jet in M87. The importance of the application of such techniques to data obtained with the Space Telescope is emphasized.

158.260 Jets optiques dans les galaxies elliptiques, les galaxies N et les objets de type BL Lacertae.
G. Wlérick.
Optical jets in galaxies, (see 012.052), p. 29 - 35 (1981). In French.

Electronographic exposures in U and V of the optical jet in M87, and of the galaxies 3C 66B, 3C 234, 3C 445, PKS 0521−36 and 3C 120 have been obtained. Isophotes of the exposures are presented and a detailed U photometry of the components of the jet in M87 is given.

158.261 High resolution electrographs of the M 87 jet.
J.-L. Nieto, J.-L. Vidal, P. Laques, A. Augé, R. Despiau.
Optical jets in galaxies, (see 012.052), p. 43 - 44 (1981).

The authors present and briefly discuss high resolution electrographs of the M 87 jet obtained with the Lallemand camera at Pic du Midi observatory.

158.262 New optical and radio data on the outer parts of Centaurus A.   R. D. Cannon.
Optical jets in galaxies, (see 012.052), p. 45 - 47 (1981).

Deep photographs of NGC 5128 (Cen A) taken on the UK 1.2 m Schmidt telescope show very faint diffuse extensions aligned approximately with the radio lobes. However, the optical isophotes and radio contours do not coincide exactly. The position angle varies with distance from the nucleus for a series of X-ray, optical and radio features in the NE quadrant, perhaps indicating precession of the central source. The faint optical extensions have colours similar to those of the main body of the galaxy, and may consist either of stars or dust.

158.263 **The optical jets of NGC 1097.**
R. D. Wolstencroft.
Optical jets in galaxies, (see 012.052), p. 49 - 51 (1981).
The optical and radio properties of the four jets of this barred spiral galaxy are reviewed and their origin is discussed.

158.264 **Optical jets in active galaxies.**
M. S. Longair.
Optical jets in galaxies, (see 012.052), p. 133 - 135 (1981).
The author outlines summarily current theoretical considerations for explaining the origin and dynamical evolution of radio and optical jets in active galaxies and quasars.

158.265 **Optical properties of the jet in NGC 5128.**
I. J. Danziger.
Optical jets in galaxies, (see 012.052), p. 143 - 144 (1981).
A brief review is given of the optical observations pertaining to the presence of the jet apparently emerging from the nucleus of NGC 5128.

158.266 **Old and new observations of the M 87 jet.**
M. Tarenghi.
Optical jets in galaxies, (see 012.052), p. 145 - 151 (1981).
Photographic and spectroscopic optical observations of the jet in M 87 are reviewed. Furthermore preliminary IUE spectra of the nucleus and the jet in this galaxy are presented.

158.267 **The ratio of hydrogen line intensities in Seyfert galaxies and quasars.**   L. S. Nazarova.
Tr. Kazan. Gorod. Astron. Obs., Vyp. 45, p. 58 - 65 (1980). In Russian.

158.268 **Energetic L_C-spectrum of five Seyfert galaxies.**
V. V. Golovatyj, Yu. V. Fridel'.
Tsirk. Astron. Obs., L'vov, No. 54, p. 15 - 18 (1979). In Russian.

158.269 **Neutral hydrogen in M31. I. The distribution of H I gas and spiral arms.**   Y. Sofue, T. Kato.
Publ. Astron. Soc. Japan, Vol. 33, 449 - 464 (1981).
The authors propose a method to obtain the spatial distribution of the H I gas in a largely tilted (nearly edge-on) disk galaxy from velocity profiles of the H I line emission. The method is applied to the data from the high-sensitivity, filled aperture survey of M31 in the 21-cm H I line emission made with the 100-m radio telescope.

158.270 **List of 44 galaxies suspected in variability.**
K. Rudnicki, A. M. Tomaszewska.
Astron. Rep., Vol. 4, 7 - 22 (1979).

158.271 **The near-infrared continua of BL Lacertae objects.**
D. A. Allen, M. J. Ward, A. R. Hyland.
Anglo-Australian Obs. Prepr. No. 152, 19 pp. (1981). – Submitted to Mon. Not. R. Astron. Soc.

158.272 **The near infrared properties of Seyfert and related active galaxies.**   M. Ward, D. A. Allen,
A. S. Wilson, M. G. Smith, A. E. Wright.
Anglo-Australian Obs. Prepr. No. 152, 32 pp. (1981). – Submitted to Mon. Not. R. Astron. Soc.

158.273 **A Seyfert 2 galaxy in the Virgo cluster.**
M. M. Phillips, D. F. Malin.
Anglo-Australian Obs., Prepr. No. 153, 21 pp. (1981). – Submitted to Mon. Not. R. Astron. Soc.

158.274 **The nature of the shells of NGC 1344.**
D. Carter, D. A. Allen, D. F. Malin.
Anglo-Australian Obs. Prepr. No. 157, 6 pp. (1981). – Submitted to Nature.

158.275 **Optical absorption lines in the high redshift BL Lac object 0215+015.**
J. C. Blades, R. W. Hunstead, H. S. Murdoch, M. Pettini.
Anglo-Australian Obs. Prepr. No. 158, 42 pp. (1981). – Submitted to Mon. Not. R. Astron. Soc.

158.276 **The very large, interacting galaxy pair IC 5174/75.**
R. M. West, R. Barbier.
ESO Sci. Prepr. No. 160, 2 + 12 pp. (1981). – Submitted to Astron. Astrophys.

158.277 **21 cm line observations of cD galaxies.**
E. A. Valentijn, R. Giovanelli.
ESO Sci. Prepr., No. 169, 1 + 8 pp. (1981). – Submitted to Astron. Astrophys.

158.278 **Mid-infrared observations of Seyfert 1 and narrow-line X-ray galaxies.**
I. S. Glass, A. F. M. Moorwood, W. Eichendorf.
ESO Sci. Prepr., No. 179, 30 pp. (1981). – Submitted to Astron. Astrophys.

158.279 **Activity of galactic nuclei.**   F. Pacini.
Izv. AN SSSR. Ser. fiz., Tom 45, 1137 - 1140 (1981). In Russian. – Abstr. in Ref. zh., 51. Astron., 12.51.915 (1981).

158.280 **Physical conditions in double galaxies with emission lines. Markarian 171 a, b.**
G. T. Petrov, V. K. Golev, I. M. Yankulova.
Dokl. Bolg. AN, Vol. 34, 461 - 464 (1981). – Abstr. in Ref. zh., 51. Astron., 12.51.951 (1981).

158.281 **A polarization burst in the BL Lac object AO 0235 + 164.**   C. D. Impey, P. W. J. L. Brand,
S. Tapia.
Prepr. Steward Obs., No. 329, 16 pp. (1981). – Submitted to Mon. Not. R. Astron. Soc.
Large and variable linear polarization has been observed in AO 0235+164. Over a period of two months, the position angle of the polarization at two radio frequencies showed a remarkable linear rotation of 130°, suggesting a rotating magnetic field structure in the emitting region. The observations presented here are the first simultaneous optical/infrared polarimetry and photometry of AO 0235+164.

158.282 **Nonthermal optical-infrared emission from NGC 1052.**
G. H. Rieke, M. J. Lebofsky, J. C. Kemp.
Prepr. Steward Obs., No. 335, 12 pp. (1981).
The infrared excess of NGC 1052 is approximately 4.5% polarized. Its spectrum resembles those of the reddest BL Lac-type sources. Despite its lack of variability, the authors conclude that this source is nonthermal but of exceptionally low luminosity for its type.

158.283 **Carbon stars in the Carina dwarf spheroidal galaxy.**
J. R. Mould, R. D. Cannon, M. Aaronson,
J. A. Frogel.
Prepr. Steward Obs., No. 338, 31 pp. (1981).
A $1°1 \times 1°1$ field centered on the Carina dwarf spheroidal galaxy has been scanned on objective prism plates taken with the UK 1.2 m Schmidt Telescope. This search has yielded eight carbon star candidates. Near infrared JHK photometry has been obtained for six of the stars. The brightest of these have derived bolometric magnitudes and colors similar to carbon stars in the Fornax dwarf galaxy. Five of the Carina carbon stars lie well above the normal red giant branch tip as defined by galactic globular clusters. These stars are probably more massive (and hence younger) than classical Population II in our Galaxy, although it is possible that they have become carbon stars because of some special circumstance such as

membership of close binary systems. Comparison with the intermediate-age globular clusters of the Magellanic Clouds suggests that Carina is not purely an old stellar population. An upper limit to the most recent epoch of star formation is approximately 8 billion years.

158.284 Discovery of carbon stars in the Draco dwarf spheroidal galaxy. M. Aaronson, J. Liebert, J. Stocke.
Prepr. Steward Obs., No. 341, 29 pp. (1981).

A grism survey of the Draco dwarf spheroidal galaxy that is 97% areally complete has led to the discovery of three carbon stars. Membership of these stars in Draco is firmly established from luminosity, proper motion, and radial velocity considerations. Optical and preliminary infrared photometry suggest that the Draco carbon stars are more closely related to the CH stars in ω Cen than to the luminous carbon stars found in the Fornax and Carina dwarfs. Carbon stars have now been located in all four of the dwarf spheroidals that have been examined using grism techniques. The rarity of these stars in galactic globulars, systems to which the dwarf spheroidals are often compared, indicates a fundamental population difference the cause of which is not yet fully understood.

158.285 Infrared observations of radio galaxies. G. H. Rieke.
Prepr. Steward Obs., No. 343, 8 pp. (1981).

For technical reasons, infrared studies of active galaxies have lagged far behind optical and radio ones. This is unfortunate, since entirely new aspects of these sources are often revealed in the infrared. The complete identification of samples of radio sources will require infrared observations to supplement the optical techniques now generally employed, and the extreme properties of the sources bright in the infrared can provide new insights to conditions in extragalactic nonthermal sources. To illustrate these points, the author discusses three cases: 1.) galaxies undergoing a powerful burst of star formation, 2.) intermediate type Seyfert galaxies, and 3.) an extreme infrared identification of an extragalactic radio source.

158.286 New observational constraints on the M87 jet. J. T. Stocke, G. H. Rieke, M. J. Lebofsky.
Prepr. Steward Obs., No. 345, 20 pp. (1981).

New observations at 1.6 μm to 3.45 μm confirm the presence of a dramatic ($\Delta\alpha \sim 1$) break between radio-IR wavelengths and 6000 Å in the spectrum of the M87 jet. These data, in combination with data taken in other spectral regions, show that the individual knots in the M87 jet have nearly the same spectral indices and nearly the same large ($\Delta\alpha \sim 1$) spectral break. This large spectral break and the constancy of spectral properties between the knots pose serious constraints for models of the M87 jet.

158.287 Double galaxy investigations. II. The redshift periodicity in optically observed pairs. W. G. Tifft.
Prepr. Steward Obs., No. 348, 2 + 13 pp. (1981).

The new sample of optically observed double galaxies is tested for the 72 km s^{-1} periodicity. The periodicity is present in the 200 pair high weight sample at or above the 99% confidence level. Detection varies between subsamples in accord with variation in data uncertainty as expected. Uncertainties in ΔV inferred from the visibility of the periodicity are in close accord with directly determined internal ΔV errors. The result is not dependent upon inclusion or exclusion of the zero peak.

158.288 Near infrared spectrophotometry of four Seyfert 1 galaxies and NGC 1275. R. J. Rudy, B. Jones, P. D. LeVan, R. C. Puetter, H. E. Smith, S. P. Willner,
A. T. Tokunaga.
Prepr. Steward Obs., No. 353, 29 pp. (1981).

Low resolution spectrophotometry from 2–4 μm is reported for the four Seyfert 1 galaxies Mrk 335, 3C 120, Mrk 509, NGC 7469 and the peculiar emission line galaxy NGC 1275.

158.289 B-V vs. HM relation for ScI galaxies and the peculiar motion of the Local Group. P. Teerikorpi.
Rep. Ser., Dep. Phys. Sci., Univ. Turku, Turku-FTL-R19, 2 + 34 pp. (1981). ISBN 951-642-027-3 = Turku Univ. Obs. Informo, No. 48. – Submitted to Astron. Astrophys.

158.290 Light scattering and extinction in M82. P. Notni, H. Tiersch, W. Bronkalla.
Astron. Nachr., Band 302, 259 - 274 (1981).

Published low-resolution measurements of colour and polarisation over the face of M82 are discussed to separate the contribution of starlight and scattered light. The authors show that in all places of the middle and outer halo the scattered light comes predominantly from a central source of very high ultraviolet excess, the contribution of the disc is negligible there. The projected distributions of H$_\alpha$-light and scattered continuum are of considerable similarity. The stellar populations seen in M82 are different in the northern and in the southern halfs of the galaxy. The main body and the region of the northern "halo" consist of an old population of normal metal content (pop. I); the colours of the southern parts – which are partly considerably influenced by extinction – can be due to either metal poor F-stars (pop. II) or to young B-stars.

158.291 Extragalactic and galactic UV observations owing to the balloon borne SCAP telescope at the wavelength 2000 Å. G. Courtès, M. Golay, M. Viton, W. Bentz, J. M. Deharveng, M. Laget, J. Donas, J. P. Sivan, B. Milliard.
Scientific ballooning–II, (see 012.065), p. 81 - 88 (1981).

The use of the 2000 Å transmission window of the atmosphere permitted to observe for the first time from a balloon gondola the nearest galaxies; 30 arc/sec resolution images were obtained for M 31, spiral arms and nucleus, M 33, M 101, with their external spiral structure, M 82 and NGC 3077. Nearly three hundred more distant galaxies have been detected up to the magnitude V = 13. Several H II regions of the Galaxy have been observed, chiefly NGC 7000 and the Cygnus Loop. Numerous blue stars of the halo have been identified.

158.292 On the motion of gas in the envelopes of active nuclei. K. A. Postnov, S. N. Fabrika.
Astron. Tsirk., No. 1148, p. 1 - 3 (1981). In Russian.

158.293 On the problem of optical variability of the nuclei of galaxies. Eh. A. Dibaj, V. M. Lyutyj.
Astron. Tsirk., No. 1152, p. 2 - 4 (1981). In Russian.

158.294 Results of spectral observations of four galaxies with UV excess.
M. A. Kazaryan, Eh. L. Karapetyan, V. S. Tamazyan.
Astron. Tsirk., No. 1154, p. 6 - 7 (1981). In Russian.

158.295 On the difference between Seyfert 1- and 2-type galaxies. Eh. A. Dibaj.
Astron. Tsirk., No. 1156, p. 1 - 3 (1981). In Russian.

158.296 U, B, V, R observations of seven galaxies with UV excess.
I. G. Babkin, A. V. Didenko, G. V. Khonin.
Astron. Tsirk., No. 1157, p. 8 (1981). In Russian.

Accurate positions of Markarian galaxies.
See Abstr. 002.017.

Spectroscopic and photometric observations of galaxies from the ESO/Uppsala list. Third catalogue. See Abstr. 002.046.

UBV photometry of double galaxies. Catalogue. See Abstr. 002.049.

The ESO/Uppsala survey of the ESO (B) Atlas of the southern sky. IX.   See Abstr. 002.055.

Neutral hydrogen observations of a large sample of galaxies.   See Abstr. 002.064.

Catalogue of clusters of galaxies with measured red-shifts.   See Abstr. 002.067.

A near-infrared atlas of spiral galaxies. See Abstr. 002.073.

Double galaxy investigations I. Observations. See Abstr. 002.078.

Index of galaxy spectra.   See Abstr. 002.087.

A revised Shapley-Ames catalog of bright galaxies. See Abstr. 002.091.

Webb Society deep-sky observer's handbook. Vol. 4: Galaxies.   See Abstr. 003.081.

Galactic astronomy.   See Abstr. 003.103.

Radio galaxies. Radiative transfer, dynamics, stability and evolution of a synchrotron plasmon. See Abstr. 003.121.

The Tucson workshop on active galaxies. See Abstr. 011.049.

Surveying velocity fields in galaxies. See Abstr. 031.567.

Recherche d'objets faibles fortement polarisés. See Abstr. 031.588.

Automatic classification of galaxy images by Fourier structural analysis.   See Abstr. 031.595.

Possible existence of massive neutrino halos. See Abstr. 061.012.

Formation and transfer of permitted Si II emission lines in Seyfert I galaxies and quasars. See Abstr. 063.017.

On the transport and propagation of relativistic electrons in galaxies.   See Abstr. 063.046.

Massive stars burning helium: the numbers of WR stars and red supergiants in galaxies. See Abstr. 065.028.

Gravitational lensing and the relation between QSO and galaxy magnitude-number counts.   See Abstr. 066.050.

Galaxies as gravitational lenses : realistic models. See Abstr. 066.164.

Spherical accretion of massive black holes: a model for galactic nuclei.   See Abstr. 066.197.

Ultraviolet photometry of stellar populations in galaxies.   See Abstr. 113.022.

Les plus brillantes étoiles de M81. See Abstr. 113.042.

Observations of Wolf-Rayet stars in the emission-line galaxy Tololo 3.   See Abstr. 114.017.

Metallicity indicators for the stellar and integrated light spectra.   See Abstr. 114.099.

Spectrophotometry of two luminous variable stars in the Andromeda galaxy.   See Abstr. 114.101.

Spectroscopic observations of two radio-quiet BL Lac candidates.   See Abstr. 114.149.

Infrared luminosities of M supergiants and their use as distance indicators.   See Abstr. 115.015.

Polarization of the radiation of stars, nebulae and galaxies.   See Abstr. 116.005.

S 10828 Trianguli: an unusual nova in M33. See Abstr. 124.701.

Supernova remnants in M31. See Abstr. 125.002.

Spatial distribution of type I and type II supernovae in spiral galaxies.   See Abstr. 125.005.

Supernova remnants and H II regions in M31. See Abstr. 125.019.

A search for young optical pulsars in the galaxies NGC 4647 and NGC 4321.   See Abstr. 125.023.

Supernova remnants and abundance gradients in M31. See Abstr. 125.024.

Radio emission from supernova remnants in the galaxy M33.   See Abstr. 125.064.

Bigourdan's forgotten observations of SN 1885 (S Andromedae) in M31.   See Abstr. 125.301.

The rate of star formation in galaxies. See Abstr. 131.072.

Opacities of dark clouds in spiral galaxies. See Abstr. 131.096.

M82 – a superb laboratory for star formation bursts. See Abstr. 131.156.

Star formation in M33 and in our Galaxy. See Abstr. 131.161.

Optical observations of interstellar shockwaves. See Abstr. 131.175.

Model calculations for the star formation in the galactic center and other galactic nuclei. See Abstr. 131.188.

La formation des étoiles chaudes dans les galaxies de types avancés.   See Abstr. 131.224.

Isolated extragalactic H II regions – invisible galaxies rendered visible.   See Abstr. 132.009.

H II regions in external galaxies.
See Abstr. 132.019.

Models of H II regions in M101.
See Abstr. 132.020.

H II regions and Wolf-Rayet stars in M33.
See Abstr. 132.022.

Ultraviolet spectra of extragalactic H II regions.
See Abstr. 132.026.

The exciting stars of giant H II regions in M 33 and
M 101.   See Abstr. 132.027.

The largest H II regions in M101.
See Abstr. 132.037.

Frequency distribution of H II region diameters.
See Abstr. 132.041.

Star formation and extinction in extragalactic
H II regions.   See Abstr. 132.048.

Spectrophotometry of H II regions in the spiral
galaxy M101.   See Abstr. 132.050.

Untersuchungen an extragalaktischen H II Gebieten
unter besonderer Berücksichtigung des Riesen-H II-Gebietes
NGC 604 in M 33.   See Abstr. 132.051.

Extended radio emission aligned with compact
nuclear sources in normal galaxies.   See Abstr. 141.004.

Optical and radio properties of 4C galaxies.
See Abstr. 141.009.

Identification of southern radio sources – IV.
See Abstr. 141.011.

DA 240 and 3C 236: spatial variation in their radio
spectra and its implications.   See Abstr. 141.018.

Optical morphology of 13 QSOs.
See Abstr. 141.022.

Resolution of quasar images.
See Abstr. 141.023.

Extremely red compact radio sources: the empty
field objects.   See Abstr. 141.026.

Investigation of BL Lac-type radio sources at the
frequency of 102 MHz.   See Abstr. 141.027.

The structure of a sample of low luminosity radio
galaxies.   See Abstr. 141.039.

The QSO B234 and the irregular galaxy NGC 4861:
examples of isolated extragalactic H II regions with low
helium abundances.   See Abstr. 141.042.

VLA observation of radio/optical knots in
3C 277.3 = Coma A.   See Abstr. 141.043.

Optical Fe II emission in radio-quiet QSOs.
See Abstr. 141.051.

The polarization of the jet in NGC 6251.
See Abstr. 141.054.

Multifrequency VLA observations of 3C388 – evidence for an intermittent jet?   See Abstr. 141.060.

Spectra of compact radio sources in galactic nuclei.
See Abstr. 141.061.

He II λ4686 as a diagnostic for quasar and Seyfert
galaxy-emission-line analyses.   See Abstr. 141.062.

Radio synthesis observations of 3C 296, 3C 442 A
and 3C 449 at 0.4, 1.4 and 2.7 GHz.   See Abstr. 141.066.

IUE observations of the twin QSOs 0957+561 A, B.
See Abstr. 141.071.

High-resolution radio observations of the X-ray galaxy NGC 3862 (3C 264) in Abell 1367.
See Abstr. 141.076.

De optische straling van radiomelkwegstelsels.
See Abstr. 141.081.

The large- and small-scale structures of 3C 293.
See Abstr. 141.083.

Radio jet refraction in galactic atmospheres with
static pressure gradients.   See Abstr. 141.102.

Morphologie et photométrie de 3C 120.
See Abstr. 141.115.

The Fe II spectrum of Seyfert 1 galaxies and
quasars.   See Abstr. 141.116.

Thermal instabilities in radiatively driven winds—
application to emission line clouds of quasars and active
galactic nuclei.   See Abstr. 141.118.

Quasars near companion galaxies.
See Abstr. 141.119.

Fluid dynamical twisting of the radio jets in 3C 449.
See Abstr. 141.120.

Television photometric observations of ten extragalactic peculiar objects.   See Abstr. 141.123.

The spectral index—luminosity relationship for
steep-spectrum cores in extragalactic radio sources.
See Abstr. 141.133.

A spectrophotometric classification of low-redshift
quasars and active galactic nuclei.   See Abstr. 141.135.

The formation of emission lines in quasars and
Seyfert nuclei.   See Abstr. 141.136.

Structure investigations of 3C 196 and 3C 280 in the
decameter waveband.   See Abstr. 141.138.

On He II λ4686 emission and the question of reddening in quasars and Seyfert galaxies.
See Abstr. 141.150.

On the energetics and morphology of radio tail
galaxies.   See Abstr. 141.151.

Detection of radio emission from the jet in
Centaurus A.   See Abstr. 141.152.

VLA observations of 1400 + 162 and other
BL Lacertae candidates.   See Abstr. 141.158.

Imaging observations of optical emission from jets.
See Abstr. 141.183.

Spectroscopic evidence for collimated outflow.
See Abstr. 141.184.

Jets in 3C 273 and other extragalactic objects.
See Abstr. 141.185.

The large radio galaxy DA 240:evidence for jets?
See Abstr. 141.189.

Jets in a sample of low luminosity radio galaxies.
See Abstr. 141.191.

VLBI observations of jets and active nuclei.
See Abstr. 141.192.

X-ray observations of jets.
See Abstr. 141.194.

Collimated ejection of radio emitting material in
Seyfert galaxies.   See Abstr. 141.197.

Optical studies of X-ray selected quasars and active
galaxies.   See Abstr. 142.026.

The nuclear X-ray source in M 81.
See Abstr. 142.028.

An X-ray selected list of QSOs:implications for the
QSO luminosity function and the origin of the diffuse X-ray
background.   See Abstr. 142.088.

On the properties of 21 X-ray globular clusters in
M31.   See Abstr. 142.091.

Physical processes for X-ray emission in galactic
nuclei.   See Abstr. 142.092.

X-rays from active galactic nuclei – hard component.
See Abstr. 142.094.

Hard X-ray balloon observations of compact
galactic and extragalactic X-ray sources.
See Abstr. 142.103.

Some recent results on extragalactic X-ray sources
from the Einstein Observatory.   See Abstr. 142.155.

Detection of a soft gamma-ray emission from the
region of NGC 4151.   See Abstr. 142.502.

Lifetime of molecular clouds and spiral structure.
See Abstr. 151.006.

The macroscopic dynamics of elliptical galaxies.
See Abstr. 151.019.

What amplifies the spirals?
See Abstr. 151.021.

Chemical evolution of normal galaxies.
See Abstr. 151.022.

L'évolution des galaxies normales.
See Abstr. 151.027.

A "symmetrical" kinematical model for elliptical
galaxies. Application to the edge on bulge of M 31.
See Abstr. 151.039.

The effects of resonances near corotation in barred
galaxies.   See Abstr. 151.040.

Dynamics of galaxies and evolution of stellar populations.   See Abstr. 151.043.

Elliptical and lenticular galaxies evolution.
See Abstr. 151.044.

Some fluid-dynamical problems in galaxies.
See Abstr. 151.045.

Velocity dispersions in the bulges of spiral and
S0 galaxies. II. Further observations and a simple three-component model for spiral galaxies.   See Abstr. 151.051.

Orbital motion of the head-tail radio galaxy IC 708.
See Abstr. 151.052.

Interacting galaxies.   See Abstr. 151.060.

Luminosity evolution of quasars and active galaxies:
theoretical models of the evolving mass supply rate.
See Abstr. 151.062.

Rotation of the galaxies.
See Abstr. 151.073.

The ages and metallicities of the globular clusters in
the Fornax dwarf spheroidal galaxy.   See Abstr. 154.006.

Vidicon photometry in old LMC clusters.
See Abstr. 154.014.

Globular cluster systems: implications for galaxy
formation.   See Abstr. 154.018.

Photoelectric photometry of globular clusters in the
Andromeda nebula. VII.   See Abstr. 154.032.

Globular clusters in galaxies beyond the local group. I.
New cluster systems in selected northern ellipticals.
See Abstr. 154.040.

A method for the determination of metal abundances
in the Galactic nuclear bulge and nearby galaxies.
See Abstr. 155.010.

Abundances in stellar populations and the interstellar medium in galaxies.   See Abstr. 155.039.

H I observations of galaxies in the galactic plane.
See Abstr. 155.041.

The H I content of galaxies in the Hercules supercluster: evidence for sweeping.   See Abstr. 160.001.

Objective prism radial velocities of cluster galaxies
from UK Schmidt telescope plates.   See Abstr. 160.002.

Isophotometry of galaxies in the Leo group G 11.
See Abstr. 160.009.

The combined effect of radioemissivity and galaxy
type on redshift.   See Abstr. 160.010.

Hydrogen deficient galaxies in A262 and neighboring clusters.   See Abstr. 160.023.

Luminosity function and colors of the 3C 295
cluster of galaxies.   See Abstr. 160.046.

The luminosity function of Virgo cluster galaxies.
See Abstr. 160.049.

Galaxies and intergalactic matter.
See Abstr. 161.002.

Interaction of intergalactic-gas flow with a rigid-body, spheroidal galaxy.
See Abstr. 161.008.

On the extragalactic distance scale and the Hubble constant.
See Abstr. 162.022.

Errata

158.901   Erratum: 'On the physical environment in the nucleus of Centaurus A (NGC 5128)' [Astrophys. J., Vol. 238, 539 - 545 (1980)].   J. H. Beall, W. K. Rose.
Astrophys. J., Vol. 248, 883 (1981). − See Abstr. 27.158.323.

158.902   Erratum: 'Identification of the nucleus in the spiral galaxy NGC 4631' [Publ. Astron. Soc. Pacific, Vol. 90, 28 - 31 (1978)].   M. Aaronson.
Publ. Astron. Soc. Pacific, Vol. 93, 535 (1981). − See Abstr. 21.158.150.

159 Magellanic Clouds

159.001   Interstellar extinction in the Large Magellanic Cloud.
K. Nandy, D. H. Morgan, A. J. Willis, R. Wilson, P. M. Gondhalekar.
Mon. Not. R. Astron. Soc., Vol. 196, 955 - 966 (1981).

A systematic investigation of interstellar extinction in the ultraviolet as a function of position in the Large Magellanic Cloud has been made from an enlarged sample of reddened and comparison stars distributed throughout the cloud. The intrinsic colours $(m_\lambda - V)_0$ of the luminous cloud members as a function of spectral type have been derived, using the mean extinction law for the LMC.

159.002   Ultraviolet interstellar extinction in the Large Magellanic Cloud using observations with the International Ultraviolet Explorer.   J. Koornneef, A. D. Code.
Astrophys. J., Vol. 247, 860 - 868 (1981).

Ten early-type supergiants in the Large Magellanic Cloud (LMC) have been observed with the IUE. The spectra (1150 - 3200 Å) are shown, and their photometric properties are discussed. It is confirmed that the LMC interstellar extinction law for these stars deviates significantly from the average galactic law in the sense that the 2200 Å feature is deficient in strength and that, in the far-ultraviolet ($\lambda < 2000$ Å), the observed LMC extinction law is significantly above the galactic curve.

159.003   Distribution of hot stars and hydrogen in the Large Magellanic Cloud.   T. Page, G. R. Carruthers.
Astrophys. J., Vol. 248, 906 - 924, plates 13 - 14 (1981).

Imagery of the Large Magellanic Cloud in the wavelength ranges 1050–1600 and 1250–1600 Å, was obtained by the S201 far ultraviolet camera during the Apollo 16 mission in 1972 April. These images have been reduced to absolute far-UV intensity distributions over the area of the LMC, with $3' - 5'$ angular resolution. Comparison of the authors' far-UV measurements in the LMC with Hα and 21 cm surveys reveals that interstellar hydrogen in the LMC is often concentrated in 100 pc clouds within the 500 pc clouds detected by McGee and Milton. Furthermore, at least 25 associations of O and B stars in the LMC are outside the interstellar hydrogen clouds; four of them appear to be on the far side. Far-UV and mid-UV spectra were obtained of stars in 12 of these associations, using the International Ultraviolet Explorer (IUE). Equivalent widths of Lα and six other lines, and relative intensities of the continuum at seven wavelengths from 1300 to 2900 Å, have been measured and are discussed.

159.004   A soft X-ray study of the Large Magellanic Cloud.
K. S. Long, D. J. Helfand, D. A. Grabelsky.
Astrophys. J., Vol. 248, 925 - 944, plate 15 (1981).

A survey of the Large Magellanic Cloud (LMC) has been carried out with the imaging X-ray telescope aboard the Einstein Observatory. Ninety-seven sources were detected. The authors estimate that less than 25 of these are likely to be either foreground stars or background quasars. The remaining 75 sources which are in the LMC have luminosities ranging from 1×10^{35} to 2×10^{38} ergs s^{-1} and are heavily concentrated around the H I complex surrounding the 30 Doradus nebula. At least 25 of the sources are supernova remnants (SNRs), including six nebulae that had not been previously recognized as such. Of the remaining 50 sources, about half appear to be extended and are also likely to be SNRs. High resolution observations of selected remnants indicate that SNR diameters in the LMC have been consistently underestimated by radio and optical observers, an effect that reduces previous estimates of the SNR rate.

159.005   The velocity structure of gas in the lines of sight to the Magellanic Clouds.   A. Songaila.
Astrophys. J., Vol. 248, 945 - 955 (1981).

The results of a survey of visual absorption lines (Ca K and Na D) in the spectra of stars in the Large and Small Magellanic Clouds and in foreground objects is brought together with existing ultraviolet and visual absorption-line data in an attempt to form a coherent picture of the velocity structure of gas along this line of sight. The author confirms the presence of two absorption components first seen in the ultraviolet covering the LMC, at LSR velocities of 60 km s^{-1} and 120 km s^{-1}. The favored model for both features is a condensation, formed by radiative cooling of a coronal halo gas. It can be shown on physical grounds that this model accounts well for the shape, dimensions, mass, and velocity of the observed features. The author points out, however, that the number of features seen in this line of sight is anomalous, compared with more general extragalactic lines of sight, and that the LMC may be expected to show anomalous velocity structure. Two particular lines of sight, toward the 30 Doradus region and toward the supernova remnant N206, are singled out for special discussion.

159.006   The dynamics of giant filamentary shells in the Large Magellanic Cloud − III. N59A (DEM 241).
J. Meaburn, D. L. Terrett, J. C. Blades.
Mon. Not. R. Astron. Soc., Vol. 197, 19 - 30 (1981).

The dynamics and structure of the 40 pc diameter filamentary shell, N59A (DEM 241) in the LMC have been investigated with the Anglo-Australian telescope and SRC Schmidt camera. The presence of many velocity components, over a velocity range of 100 km s^{-1}, and its complex filamentary structure both suggest that it is far from being a single spherical shell whose radial expansion is driven by stellar winds. Alternative configurations are proposed.

159.007 The Magellanic stream and the interacting galaxies.
K. I. Tanaka.
Publ. Astron. Soc. Japan, Vol. 33, 247 - 264 (1981).

The Magellanic stream is reproduced numerically by taking account of the gravitational interaction of three galaxies, the Galaxy, and the Large and Small Magellanic Clouds. The best reproduction of the stream is obtained as a narrow band of the particles pulled out from the LMC, whose orbital plane is perpendicular to our galactic plane and whose motion is clockwise as seen from the sun with the perigalacticon at 20 kpc. The galactic sheet of test particles is not warped enough by the tidal force of the two Clouds to explain the observed bending of the galactic disk.

159.008 New probable planetary nebulae and VV Cephei stars in the Small Magellanic Cloud.
N. Sanduleak, P. Pesch.
Publ. Astron. Soc. Pacific, Vol. 93, 431 - 434 (1981).

Deep objective-prism plates have been used to detect six new faint probable planetary nebulae in the Small Magellanic Cloud. Also discussed are five suspected VV Cephei stars in the SMC.

159.009 Discovery of the first S star in the Small Magellanic Cloud.   V. M. Blanco, J. A. Frogel,
M. F. McCarthy.
Publ. Astron. Soc. Pacific, Vol. 93, 532 - 534 (1981).

A pure S-type star found in the SMC is described. The extreme rarity of such stars relative to C and M giants is shown by Magellanic Clouds surveys for red giants. For the S star, M_{bol} is -5.1 ± 0.1, a value in fair agreement with that found for the few known galactic counterparts with reliable distance moduli; and one near the faint limit expected theoretically.

159.010 UBV observations of globular clusters in the Magellanic Clouds.   S. van den Bergh.
Astron. Astrophys.,Suppl. Ser., Vol. 46, 79 - 87 (1981).

Integrated UBV photometry, including some new data, is given for 61 clusters in the SMC and for 147 clusters in the LMC. These observations are used to set up a rough age classification system for the clusters in the Magellanic Clouds. Age classifications are given for 50 SMC and 141 LMC clusters.

159.011 Superluminous giants in Magellanic Cloud clusters.
P. J. Flower.
Astrophys. J., Lett., Vol. 249, L11 - L14, plate L1 (1981).

The C-M diagram of the LMC blue, globular-like cluster NGC 1866 is known to exhibit a large number of yellow and red core helium-burning giants. Short exposure photographic plates taken with the CTIO 4 m Ritchey-Chrétien camera provide an opportunity to study the stellar content of the central regions of this cluster. In and near the core of NGC 1866, 11 superluminous giants (SLGs), apparently similar to those found in other LMC blue, globular-like clusters, have been identified. These stars lie more than 0.5 mag above the core helium-burning giants. It is proposed that SLGs are protoplanetary nebulae; i.e., stars that, having ejected their envelopes as luminous red giants, are crossing the C-M diagram toward becoming central star planetary nebulae.

159.012 Planetary nebula progenitors in Magellanic Cloud clusters.   P. J. Flower.
Publ. Astron. Soc. Pacific, Vol. 93, 546 (1981). – Abstract.

159.013 RR Lyrae stars in Magellanic Cloud clusters.
J. A. Graham.
Publ. Astron. Soc. Pacific, Vol. 93, 546 - 547 (1981).
Abstract.

159.014 High-dispersion spectroscopy of the most luminous F- and G-type supergiants in the Large Magellanic Cloud and the Milky Way.   W. Hagen, R. M. Humphreys, R. E. Stencel.
Publ. Astron. Soc. Pacific, Vol. 93, 567 - 576 (1981).

The authors have obtained high-dispersion echellograms, at 2.5 Å and 5.1 Å mm⁻¹ dispersion, of four F- and G-type LMC supergiants ($M_V \sim -9$) for the purpose of evaluating the outer atmospheres of such objects, as compared to Milky Way counterparts. In combination with IR photometry, line doubling at Na I "D", and circumstellar cores in Ca II H and K indicate extensive circumstellar envelopes and estimates of mass-loss rates in excess of $10^{-4} M_\odot$ yr⁻¹. The physical connection between layers in the outer atmospheres, in terms of chromospheric heating and the stellar-wind energy source, is discussed, and these stars are compared with galactic supergiants.

159.015 Discovery of a protostar in the Large Magellanic Cloud.
I. Gatley, E. E. Becklin, A. R. Hyland, T. J. Jones.
Mon. Not. R. Astron. Soc., Vol. 197, 17P - 21P (1981).

A near infrared search fo the H II region/molecular cloud complex N159 in the Large Magellanic Cloud has revealed a very red ($H-K = 2.1$, $K-L' = 2.7$) compact object. The location, brightness, colour and $2.1-2.4$ μm spectrum of this source suggest that it is very young, and similar to the galactic infrared 'protostars'. This is the first identification of an infrared protostar in an external galaxy. Its discovery provides direct evidence of current star formation in the Large Magellanic Cloud.

159.016 A strong 1665-MHz OH maser in the Large Magellanic Cloud.   R. F. Haynes, J. L. Caswell.
Mon. Not. R. Astron. Soc., Vol. 197, 23P - 25P (1981).

A new OH Type I maser has been discovered in the Large Magellanic Cloud in the direction of the H II region N105. The 1665-MHz emission has a peak intensity of nearly 1 Jy and is strongly polarized in the right-hand circular sense. Weaker OH emission at 1667 MHz was also found but none at 1612 or 1720 MHz. No H_2CO was detected in the direction of N105.

159.017 The Magellanic Stream and other hydrogen remnants of strong tidal disruption of the Magellanic Clouds.
I. F. Mirabel.
Astrophys. J., Vol. 250, 528 - 533 (1981).

A review of results from recent surveys of neutral-hydrogen, high-velocity clouds reveals the connection of the Magellanic Stream with a widespread population of clouds with large infalling motions toward the Galaxy. The tidal disruption of the Magellanic Clouds during their last close encounter with the Milky Way produced, along with the Magellanic Stream, a population of high-velocity clouds scattered throughout the first and second quadrants of the southern galactic hemisphere. High angular resolution observations in the northern tip of the Magellanic Stream are presented. They show that the leading head of the Stream is disintegrating as it falls toward the Milky Way. The scattering material may be observed as a conglomerate of discrete clouds surrounding the tip of the Magellanic Stream. The need for new computer simulations of the remnants of strong tidal disruption of the Magellanic Clouds is discussed.

159.018 The young phenomena within the Large Magellanic Cloud.   J. Meaburn.
Investigating the universe, (see 003.014), p. 61 - 94 (1981).

The most pertinent observations of the very youngest phenomena occurring within the Large Magellanic Cloud are presented. A picture emerges where giant (20 - 220 pc diam.) shells are formed within $\sim < 10^6$ yr by successive supernova explosions, stellar winds and radiation pressure. These chisel their supergiant counterparts out of the > 3 kpc sheets of H I within periods of 10^7 yr, by a process of stochastic self-propagating star formation. The massive 30 Doradus nebula and R136 become the nucleus of the young spiral structure formed in this way. Interstellar velocities over a range of 400 km s^{-1} within this central complex could be driven predominantly by the stellar wind and radiation pressure of R136 alone, supplemented by all other sources.

159.019   Studies of the Magellanic Clouds. III. Colours, gas and past star formation rate.
B. Rocca-Volmerange, J. Lequeux, M. Maucherat-Joubert.
Astron. Astrophys., Vol. 104, 177 - 184 (1981).

The authors show how quantitative information on the history of star formation in a galaxy can be obtained in principle by fitting photometric data, gas fraction and metallicity with an evolutionary model. They apply these results to the Magellanic Clouds, using their integrated colours from the far UV up to the red and other data, and find that their initial mass function cannot differ much from that in the solar neighbourhood, and that their evolution has been smooth, with a steady or slightly decreasing rate of star formation in the past.

159.020   A soft X-ray study of the Large Magellanic Cloud.
D. J. Helfand, K. S. Long.
Space Sci. Rev., Vol. 30, (see 012.044), 141 (1981).
Abstract.

159.021   Observations of the giant bubbles in the Large Magellanic Cloud.
Y. Georgelin, Y. Georgelin, A. Laval, G. Monnet, M. Rosado.
Messenger, No. 26, p. 11 - 14 (1981).

159.022   The gas to dust ratio and the near-infrared extinction law in the Large Magellanic Cloud.
J. Koornneef.
ESO Sci. Prepr., No. 162, 18 pp. (1981). – Submitted to Astron. Astrophys.

159.023   The Magellanic Clouds and the galactic halo.
M. J. Valtonen, K. A. Innanen, L. Tähtinen.
Rep. Ser., Dep. Phys. Sci., Univ. Turku, Turku-FTL-R21, 26 pp. (1981). ISBN 951-642-058-3 = Turku Univ. Obs. Informo, No. 50. – Submitted to Astrophys. J.

Stellar evolution with SMC chemical abundances. See Abstr. 065.027.

Mass loss from metal-poor stars. See Abstr. 065.031.

Grand Nuage de Magellan. Troisième liste d'étoiles membres du Grand Nuage de Magellan et liste d'étoiles galactiques.   See Abstr. 111.010.

Infrared studies of the two stellar populations in 30 Doradus.   See Abstr. 113.037.

Radial velocities for different spectral lines of B and A supergiants in our Galaxy and in the Large Magellanic Cloud. See Abstr. 114.001.

IUE and ground-based observations of mass loss in the Magellanic Clouds.   See Abstr. 114.031.

Horizontal-branch stars, and galactic and Magellanic Cloud globular clusters.   See Abstr. 114.043.

High dispersion observations of LMC/SMC stars. See Abstr. 114.057.

Coordinated ground-based and IUE observations of hot stars in the Magellanic Clouds.   See Abstr. 114.068.

The ultraviolet to infrared spectrum of the large mass loss LMC supergiant S22 = HD 34664. See Abstr. 114.086.

Detailed analysis of a G supergiant in the Small Magellanic Cloud.   See Abstr. 114.136.

Masses of Magellanic Wolf-Rayet stars: mass loss and evidence for a WR subclass vs. mass relation. See Abstr. 115.007.

Bolometric luminosities and infrared properties of carbon stars in the Magellanic Clouds and the Galaxy. See Abstr. 115.014.

Multiplicity and absolute magnitudes of Wolf-Rayet stars in the Large Magellanic Cloud.   See Abstr. 115.021.

Wolf-Rayet stars in the Magellanic Clouds. II. The peculiar eclipsing binary HD 5980 in the SMC. See Abstr. 119.111.

The Hubble-Sandage variable HDE 269006: a hot supergiant with a cool envelope.   See Abstr. 122.057.

Photometric abundances and colors of Cepheids in the Small Magellanic Cloud.   See Abstr. 122.069.

HV 1369, a cepheid at a possible depth of 32 kpc in the Small Magellanic Cloud.   See Abstr. 122.073.

The luminosities of red supergiant variables in the Small Magellanic Cloud.   See Abstr. 122.094.

IUE and ground-based spectroscopic observations of the S Dor-type LMC variable R 71 during minimum state. See Abstr. 122.103.

The DDO properties of cepheids and supergiants. See Abstr. 122.118.

Photographic (B, V) photometry of Magellanic Cloud cepheids. I. – Observational data. See Abstr. 122.199.

Multicolour photoelectric photometry of Magellanic Cloud cepheids. IV: B, V observations of 20 short period cepheids.   See Abstr. 122.200.

Sonneberger δ-Cephei-Sterne im Bereich der Kleinen Magellanschen Wolke.   See Abstr. 123.031.

Spectral properties of Nova LMC 1977b. See Abstr. 124.121.

Nova in Large Magellanic Cloud. See Abstr. 124.241.

An oxygen-rich young supernova remnant in the Small Magellanic Cloud.   See Abstr. 125.017.

Spectra and internal motions of supernova remnants and ring nebulae in the Magellanic Clouds. See Abstr. 125.030.

A new view of supernova remnants in the Magellanic Clouds.   See Abstr. 125.058.

High resolution observations of interstellar lines in the IUE spectra of LMC stars.   See Abstr. 131.076.

The carbon abundance in two H II regions of the Small Magellanic Cloud.   See Abstr. 132.017.

IUE observations of carbon in SMC and LMC H II regions.   See Abstr. 132.021.

Observations of H II regions in the Magellanic Clouds.   See Abstr. 132.028.

Stellar mass loss and H II region morphology in Magellanic irregular galaxies.   See Abstr. 132.030.

Measurements of the equivalent width of the H_β emission line and age determination of H II regions of the LMC and SMC.   See Abstr. 132.034.

Very high-velocity H I clouds: an intergalactic population?   See Abstr. 132.042.

N70: a mass-loss bubble within a massive collapsing H I cloud.   See Abstr. 132.043.

Ultraviolet observations of LMC X-4 and SMC X-1. See Abstr. 142.016.

The optical counterpart of A0538−66. See Abstr. 142.048.

The optical light curve of LMC X-4 and the 30-day X-ray period.   See Abstr. 142.119.

An optical outburst from the periodic recurrent X-ray transient A0538-66.   See Abstr. 142.123.

LMC X-4.   See Abstr. 142.143.

The ages and metallicities of the globular clusters in the Fornax dwarf spheroidal galaxy.   See Abstr. 154.006.

The intermediate age globular cluster NGC 152 in the Small Magellanic Cloud.   See Abstr. 154.008.

Anomalous giants in Magellanic Cloud clusters. See Abstr. 154.023.

Three *UBV* sequences in the LMC. See Abstr. 154.029.

The extended giant branches of intermediate age globular clusters in the Magellanic Clouds. II. See Abstr. 154.046.

The fraction of O-type supergiants in our galaxy, in the LMC and in the SMC: an evidence of the correlation between mass loss rate and chemical abundance. See Abstr. 155.020.

Spectrophotometry of carbon stars in Local Group galaxies and in the Magellanic Stream.   See Abstr. 158.096.

Interaction of intergalactic-gas flow with a rigid-body, spheroidal galaxy.   See Abstr. 161.008.

160 Groups of Galaxies, Clusters of Galaxies, Superclusters

160.001 **The H I content of galaxies in the Hercules supercluster: evidence for sweeping.**
R. Giovanelli, G. L. Chincarini, M. P. Haynes.
Astrophys. J., Vol. 247, 383 - 402 (1981).

Arecibo observations of the neutral hydrogen content of galaxies in the Hercules supercluster reveal clear signs of environmental dependence. The supercluster galaxies are compared with a sample of isolated galaxies characterized by the same absolute luminosity range and observed with the same telescope. A sweeping mechanism which depends on the intracluster medium density is at work. Preliminary optical data show the swept galaxies to be redder than the ones with normal M_H/L ratios.

160.002 **Objective prism radial velocities of cluster galaxies from UK Schmidt telescope plates.**
J. A. Cooke, D. Emerson, B. D. Kelly, H. T. MacGillivray, R. J. Dodd.
Mon. Not. R. Astron. Soc., Vol. 196, 397 - 402 (1981).

Radial velocities for galaxies in the cluster Abell 2670 have been measured on an objective prism plate and compared with previously published velocities from slit spectra. These indicate that the prism velocities are accurate to ~ 1800 km s^{-1}. Similar velocities have been measured for the cluster of galaxies Abell 140, the results indicating a wider dispersion in velocities than would be expected for a single cluster. The authors conclude that the technique is useful for determining the radial velocities of clusters, for testing the cluster membership of bright elliptical galaxies, and for resolving superimposed clusters.

160.003 **Observations of rich clusters of galaxies at metre wavelengths.** H. V. Cane, W. C. Erickson, R. J. Hanisch, P. J. Turner.
Mon. Not. R. Astron. Soc., Vol. 196, 409 - 415 (1981).

Observations have been made at 10 frequencies between 50 and 120 MHz of 17 rich, X-ray emitting clusters of galaxies with the 78 × 156 m dipole array at Llanherne. The observed flux densities were compared to the flux densities expected on the basis of the known discrete sources in the fields. In no case was a significant flux excess found that might have indicated the presence of a diffuse halo component of radio emission in the cluster. For those clusters in which spectral indices could be determined, the spectra all tend to be much steeper than is normal for extragalactic radio sources, although a strict correlation between the X-ray luminosity and the low-frequency radio luminosity or spectral index is not found. The occurrence of large halo sources such as that which is present in the Coma cluster seems to be quite unusual.

160.004 **The Ursa Major supercluster – I. The optical field and the 5C10 radio survey.** N. J. Schuch.
Mon. Not. R. Astron. Soc., Vol. 196, 695 - 704, Microfiche MN 196/1 (1981).

The Ursa Major supercluster is an association of 10 Abell clusters which lie within 7° of each other, and it is the richest nearby Abell supercluster in the northern hemisphere. This paper defines the optical field of the supercluster, presents the results of the 5C10 radio survey, the optical identifications of the 5C10 radio sources, and their relation to the supercluster. Two hundred and sixty-five sources are listed with $S(408$ MHz$) \geqslant 9.8$ mJy and 48 sources with $S(1407$ MHz$) \geqslant 1.7$ mJy. Twenty-six sources appear extended and 12 of these are identified with galaxies. Four of the sources have radio-trail structures. At 408 MHz a lower limit of 11 per cent for the identification of unresolved sources has been found, while the total identification rate is 18 per cent. At 1407 MHz, 41

per cent of all the sources have possible identifications. Two other papers are to follow.

160.005 **Observation of the cluster of galaxies A 401 at 11 cm.** J. Roland, H. Sol, I. Pauliny-Toth, A. Witzel.
Astron. Astrophys., Vol. 100, 7 - 11 (1981).

Observations of the Abell cluster A 401 have been made at a frequency of 2.7 GHz. When combined with other measurements, the present observations suggest that this cluster contains a radio halo, similar to Coma C, which contributes to its radio emission at low frequencies. Available data on radio halos in clusters of galaxies are examined. It is shown that they are consistent with the simple relation between the radio luminosity, the X-ray luminosity, and the velocity dispersion of the galaxies suggested by Roland (1981).

160.006 **List of clusters of galaxies with published redshifts.** T. W. Noonan.
Astrophys. J., Suppl. Ser., Vol. 45, 613 - 620 (1981).

This paper lists 439 clusters with published redshifts. The Humason-Mayall-Sandage designation is extended to all clusters in the list, and the Abell catalog and Zwicky catalog numbers are given where possible.

160.007 **Optical spectrophotometry of the suspected X-ray cluster of galaxies E 1455 + 2232.**
K. O. Mason, H. Spinrad, S. Bowyer, G. Reichert, J. Stauffer.
Astron. J., Vol. 86, 803 - 805 (1981).

The redshift of the cluster of galaxies at the position of the X-ray source E 1455 + 2232 is measured based on spectrophotometry of four member galaxies. It is found to be $z = 0.259$, significantly lower than the value of $z \sim 0.7$ proposed by Schild et al. The line-of-sight velocity dispersion is ~ 600 km s^{-1} and the X-ray luminosity in the $0.5 - 4.5$-keV energy range at the source is calculated to be 9×10^{44} erg s $^{-1}$. The redshift of the cluster A 2001, which is centered ~ 0.5 deg from the X-ray position, is also measured. It is shown that A 2001 is physically distinct from the new cluster probably associated with the X-ray source.

160.008 **Low-frequency radio observations of poor clusters of galaxies.** R. J. Hanisch, R. A. White.
Astron. J., Vol. 86, 806 - 810 (1981).

Observations have been made at the Clark Lake Radio Observatory of 16 poor clusters of galaxies at 34.3 MHz. Four of the poor clusters were detected at flux densities greater than 20 Jy. The spectra of the four detected clusters are all rather steep. Two of the detected clusters, AWM 4 and AWM 5, are also known to be X-ray sources. The authors have investigated the possibility that the X-ray-emitting gas is heated by Coulomb interactions with the relativistic electrons responsible for the radio emission, and find that the observed X-ray luminosities can be accounted for if the electron energy spectrum extends to very low energies ($\gamma \sim 1 - 10$). Collective plasma effects may increase the heating efficiency and eliminate the need to extrapolate the electron energy spectrum to such low values.

160.009 **Isophotometry of galaxies in the Leo group G 11.**
R. J. Dickson, P. W. Hodge.
Astron. J., Vol. 86, 826 - 832 (1981).

Photographic surface photometry has been carried out for 11 members of the group of galaxies in Leo, G 11. Data are presented in the form of graphs showing the variation of ellipticity and orientation with radius, and also tables of individual values. The implied three-dimensional structure of each galaxy is discussed.

160.010 The combined effect of radioemissivity and galaxy type on redshift.
M. Moles, L. Nottale.
Astron. Astrophys., Vol. 100, 258 - 270 (1981).

The question of a possible dependence of redshift on the morphological type of galaxies in clusters has been of interest. For 5 different samples, the authors find in a given cluster (mean velocities) V (ellipticals, radio) $< V$ (ellipticals, non radio) $< V$ (spirals, non radio) $< V$ (spirals, radio). The type effect in the Virgo cluster is reanalysed. A significant effect with type at 4.2σ is found for 17 clusters and groups.

160.011 408 MHz observations of clusters of galaxies. I. Halo sources in the Coma-A1367 supercluster.
B. Ballarati, L. Feretti, A. Ficarra, G. Gavazzi, G. Giovannini, M. Nanni, M. C. Olori.
Astron. Astrophys., Vol. 100, 323 - 325 (1981).

As a result of a 408 MHz survey of clusters of galaxies, made with the Bologna Northern Cross Radio Telescope, the authors present observations of three extended sources of "halo" type detected in the Coma−A1367 supercluster. Two of them are found in the Coma cluster: the well known Coma-C and a newly detected structure located next to the strong radio galaxy 3C 277.3 (Coma-A). The third is detected close to the centre of A1367, in agreement with previous observations.

160.012 The relation between velocity dispersion and central galaxy density in clusters of galaxies.
N. A. Bahcall.
Astrophys. J., Vol. 247, 787 - 791 (1981).

A correlation between cluster velocity dispersion and average central galaxy density, $N_{0.5}$, is reported. The correlation covers the range from rich clusters to small groups of galaxies, or, in terms of velocity dispersion, from $v_r \sim 1500$ to ~ 100 km s^{-1}. The correlation satisfies $N_{0.5} \approx 21(v_r/1000)^{1.2}$. This result is useful for estimating unknown velocity dispersions in clusters with the aid of the relatively easily determined parameter $N_{0.5}$. When combined with the virial theorem, the above relation also suggests that the mass-to-light ratio of galaxy systems increases with the system's velocity dispersion.

160.013 Low energy X-ray emission from five galaxy cluster sources.
G. Reichert, K. O. Mason, S. M. Lea, P. A. Charles, S. Bowyer, S. Pravdo.
Astrophys. J., Vol. 247, 803 - 812 (1981).

The authors report the detection of soft (0.2 - 2.5 keV) X-ray emission from several known cluster X-ray sources using the low energy detectors of the HEAO 1 A-2 experiment. Soft X-ray emission was observed from five clusters − the Centaurus cluster, Abell 2147, SC 1329 − 314, Abell 2319, and Abell 133. The temperature of Abell 133 is constrained to be less than 2×10^7 K, making it the coolest X-ray cluster yet detected.

160.014 Absence of the hidden mass in clusters.
B. I. Fesenko.
Tr. Astron. Obs., Leningrad, Tom 36 = Uch. Zap. Leningr. Univ., No. 402 = Ser. mat. nauk, Vyp. 58, 74 - 96 (1981). In Russian.

The problem of estimation of the radial velocity dispersion in clusters of galaxies is examined. The statistical properties of galaxies in the Virgo cluster are considered in detail.

160.015 The dynamical age of the Local Group of galaxies.
D. Lynden-Bell.
Observatory, Vol. 101, 111 - 114 (1981).

The distance to those Local Group members whose expansion has just been stopped by the gravity of the Local Group yields Mt^2 where M is the mass of the Group and t the time since expansion began. The distance and radial velocity of M 31 yield a relationship between Mt^2 and t. Thus M and t may be deduced. Sandage-Tammann distances to Local Group members yield $t = 1.6 \times 10^{10}$ years and $M = 3.6 \times 10^{12} M_\odot$. Accurate distances to outlying members of the Local Group could refine this method and make a lasting contribution to cosmogony and cosmology.

160.016 The distribution and morphology of X-ray −emitting gas in the core of the Perseus cluster.
A. C. Fabian, E. M. Hu, L. L. Cowie, J. Grindlay.
Astrophys. J., Vol. 248, 47 - 54, plates 1 - 2 (1981).

A high-resolution ($\sim$4") X-ray image of the core of the Perseus cluster obtained with the Einstein Observatory is presented. An unresolved source is found coincident with the nucleus of NGC 1275. In the surrounding extended emission, isointensity contours twist at progressively larger radii from a position angle with SE orientation at 19" to 100" to a W one at 150". Absorption in the optical features at high velocity with respect to NGC 1275, which are thought to be associated with a foreground galaxy, does not produce any detectable X-ray absorption. The emission tends to become asymmetric in the presence of the lower-velocity filaments, but there is no obvious detailed correlation between X-ray enhancements and individual filaments. The surface brightness is deprojected to obtain density and temperature profiles of the intracluster gas.

160.017 Soft X-ray images of the central region of the Perseus cluster.
G. Branduardi-Raymont, D. Fabricant, E. Feigelson, P. Gorenstein, J. Grindlay, A. Soltan, G. Zamorani.
Astrophys. J., Vol. 248, 55 - 60, plate 3 (1981).

The authors report the results of 0.5−3.0 keV X-ray observations of the central region of the Perseus cluster. In addition to the very extended thermal cluster emission and a sharply peaked component at NGC 1275 previously known, the high resolution image reveals a point source coincident with the optical nucleus of NGC 1275. The 0.5−3.0 keV luminosity of the compact source is $\sim 10^{44}$ ergs s^{-1}. The data show the cluster emission to be a factor of 1.2 more elongated east-west than north-south. The centroid of the cluster emission is found to be offset 1.7 east of NGC 1275, in the direction away from the prominent line of galaxies that stretch between NGC 1275 and IC 310. The cluster center determined from galaxy counts lies 6.4 southwest of the X-ray centroid.

160.018 Clustering of the relic neutrinos and the velocity dispersion in a cluster of galaxies.
H. Sato, F. Takahara.
Prog. Theor. Phys., Vol. 65, 374 - 377 (1981). − Abstr. in Phys. Abstr., Vol. 84, Abstr. 67126 (1981).

160.019 Groups of spiral galaxies around the Coma cluster and upper limits on its mass.
F. D. A. Hartwick.
Astrophys. J., Vol. 248, 423 - 428 (1981).

A sample of 137 spiral galaxies found by Abell centered on the Coma cluster has been analyzed. The existence of small groups of spirals is suggested on the basis of the angular two point correlation function. A virial-theorem analysis of six of the groups for which sufficient velocity data was available suggests that at least five of the 20 groups are real gravitationally bound systems. Luminosities and tidal radii were then derived and were used to compute $(M/L)_{Coma}/(M/L)_{group}$ for each group. Upper limits on the mass of the Coma cluster were then derived. The results are consistent with the hypothesis that the groups were tidally limited at an earlier epoch and that the groups are now expanding away from the cluster. The implied mass of Coma agrees with virial-theorem determinations to within a factor of 2.

160.020 On the equilibrium distribution of the elements in the gas in the Coma cluster.
F. Abramopoulos, G. A. Chanan, W. H.-M. Ku.
Astrophys. J., Vol. 248, 429 - 438 (1981).

The equations governing the equilibrium distribution of the elements for an isothermal plasma in a gravitational potential well are presented, togehter with a rapidly converging iterative procedure for determining their solution. These results are applied to the case of the intracluster gas in clusters of galaxies and are shown to have important implications (1) for the recent controversy as to whether or not the iron nuclei "settle out" in the cluster core, and (2) for the interpretation of the size and shape of the distribution of the continuum X-ray emission from clusters. The authors consider a model for the Coma cluster in which the intracluster gas is in equilibrium and is "partially processed"; that is, the elements with $Z > 2$ have abundances proportional to their cosmic values, but reduced by a factor of ~ 20.

160.021 Redshifts of multiple clusters of galaxies.
I. D. Karachentsev, A. I. Kopylov.
Pis'ma Astron. Zh., Tom 7, 515 - 523 (1981). In Russian.
English translation in Soviet Astron. Lett., Vol. 7.

Radial velocities were measured for 128 galaxies from isolated multiple clusters. Physical systems of the clusters were distinguished by statistical analysis of redshifted differences of fourteen pairs and seven triplets.

160.022 Angular separation studies of X-ray bright rich clusters of galaxies.
M. P. Ulmer, M. P. Kowalski, R. G. Cruddace.
Bull. American Astron. Soc., Vol. 13, 507 (1981). – Abstract.

160.023 Hydrogen deficient galaxies in A262 and neighboring clusters.
R. Giovanelli, M. P. Haynes, G. L. Chincarini.
Bull. American Astron. Soc., Vol. 13, 530 (1981). – Abstract.

160.024 Symmetry and distribution studies of 23 rich clusters of galaxies. W. S. Burgett.
Bull. American Astron. Soc., Vol. 13, 533 (1981). – Abstract.

160.025 Dynamical models for clusters of galaxies.
F. Occhionero. L. Veccia-Scavalli, N. Vittorio.
Bull. American Astron. Soc., Vol. 13, 533 (1981). – Abstract.

160.026 Einstein observations of A1367.
J. Bechtold, W. Forman, C. Jones, J. Schwarz, W. Tucker, L. Van Speybroeck, R. Giacconi.
Bull. American Astron. Soc., Vol. 13, 550 (1981). – Abstract.

160.027 X-ray observations of A2256 at large radial distance.
D. G. Fabricant, P. Gorenstein.
Bull. American Astron. Soc., Vol. 13, 550 (1981). – Abstract.

160.028 The galaxies of the Local Group.
S. van den Bergh.
The structure and evolution of normal galaxies, (see 012.010), p. 201 - 210 (1981).

Contents: Group membership. Distribution of Group members. The dwarf spheroidal galaxies. Local Group luminosity function. Globular cluster frequency. Dynamical friction and giant molecular clouds. Galactic distribution of the oldest open clusters. The "cloudiness" of galaxies.

160.029 Clusters of galaxies.
W. K. Huchtmeier, O.-G. Richter, J. Materne.
Messenger, No. 25, p. 8 - 11 (1981).

160.030 Neutral hydrogen streams in groups of galaxies.
I. Observations. M. P. Haynes.
Astron. J., Vol. 86, 1126 - 1154 (1981).

Nearby loose groups of galaxies have been the object of a search for neutral hydrogen appendages, referred to as H I streams, extending outward from the disks of member galaxies. Observational data obtained with the Arecibo 305-m telescope are summarized for 15 nearby groups containing at least one pair of galaxies separated by less than 250 kpc. The aggregates considered in this sample are often subsystems within more extended groupings. The restriction of group definition has been dictated by the preconception that these H I streams are likely tidal debris left over from the close encounters of neighboring galaxies. Of the 15 groups, nine show no evidence for extended neutral hydrogen appendages. In the remaining six groups, however, H I streams associated with at least one member of the aggregate are identified. In several cases, the H I contours coincide with faint optical emission. While previously recognized H I streams have been almost exclusively associated with late-type spiral galaxies, several of the newly identified ones extend in the regions of early-type galaxies. Neutral hydrogen profiles and their derived parameters are presented for 47 galaxies observed in the course of this study.

160.031 The Local Supercluster of galaxies.
G. de Vaucouleurs.
Bull. Astron. Soc. India, Vol. 9, 1 - 23 (1981).

The history of the discovery of the Local Supercluster or "Supergalaxy" is traced. Recent studies of its influence on the distribution and velocities of nearby galaxies are reviewed.

160.032 Dynamics of the Virgo supercluster.
G. L. Hoffman, E. E. Salpeter.
News Lett. Astron. Soc. N. Y., Vol. 1, No. 10, p. 40 (1981). Abstract.

160.033 A2670 – the nearest richness class 3 cluster of galaxies. L. A. Thompson.
Mercury, Vol. 10, 122 - 123 (1981).

160.034 Low frequency extended radio sources – magnetic field strength and classification.
E. Ma, F. Durret, B. Chen.
Sci. Sinica, Vol. 24, 1259 - 1270 (1981).

Using very recently published X-ray and low frequency radio data the authors have calculated the lower limit of the magnetic field and other quantities for 25 clusters of galaxies and 12 radio galaxies.

160.035 Dynamical analysis of clusters of galaxies and considerations upon the hidden mass. G. Mathez.
Ann. Physique, Vol. 6, 223 - 229 (1981). In French. – Abstr. in Phys. Abstr., Vol. 84, Abstr. 89093 (1981).

160.036 Typical effect on redshift of galaxies of the Hercules supercluster. E. Giraud.
C. R. Acad. Sci. Paris, Tome 293, Sér. II, 295 - 298 (1981). In French.

The author studies the dependance of redshift on the morphological type of galaxies in the Hercules supercluster. He shows that S-galaxies have a significant 4.72 σ excess in 105 redshift samples.

160.037 Supercluster bridge between groups of galaxy clusters.
G. Chincarini, H. J. Rood, L. A. Thompson.
Astrophys. J., Lett., Vol. 249, L47 - L50 (1981).

Redshifts were obtained for a random sample of 44 galaxies selected from a complete population of 350 galaxies brighter than $m_z \leqslant 15.7$ in a 332 sq. deg. region between the Hercules group of clusters (A2151, A2152, and A2147) and the A2199/A2197 group. A plot of redshift versus declination reveals a bridge of galaxies connecting the two groups. The bridge contains 33 of the 44 galaxies sampled. The redshifts vary smoothly along the bridge by 2000 km s^{-1} from the Hercules region to the A2199/A2197 region. The groups are separated by 44 h^{-1} Mpc (Hubble constant $H = 100h$ km s^{-1} Mpc^{-1}), and the bridge is tilted 27° off the plane of the sky. Apart from two sparsely populated associations of galaxies with dimensions $\gtrsim 15$ Mpc, the foreground volume is devoid of galaxies.

160.038 Measurements of the gas contents of clusters of galaxies by observations of the background radiation at 10.6 GHz – II.
M. Birkinshaw, S. F. Gull, K. J. E. Northover.
Mon. Not. R. Astron. Soc., Vol. 197, 571 - 592 (1981).

The final results of a 4500-hr search for anisotropies in the microwave background radiation towards 15 clusters of galaxies are presented. Significant depressions in the temperature of this radiation at 10.6 GHz were found in the directions of Abell 576 and 2218; the mean gas density and temperature at the cluster centres inferred from the data on a subset of the authors' cluster sample are about 3×10^{-24} kg m^{-3} and 3×10^8 K.

160.039 Émission de rayons-X du gaz thermique, hors des galaxies, à l'intérieur des amas de galaxies.
J. P. Vallée, A. H. Bridle.
J. R. Astron. Soc. Canada, Vol. 75, 253 (1981). – Abstract.

160.040 Distances to the galaxies Stephan's Quintet.
S. M. Kent.
Publ. Astron. Soc. Pacific, Vol. 93, 554 - 557 (1981).

Measurements of the nuclear velocity dispersions in four members of Stephan's Quintet are combined with a 21-cm line width for the fifth member to determine the absolute magnitude and distances to these galaxies. All objects are found to be at distances consistent with their redshifts. In particular, the conjecture by Arp that the high-velocity members ($z = 6700$ km s^{-1}) are nearby and at the same distance as the low-velocity member ($z = 1000$ km s^{-1}) is strongly rejected.

160.041 Dynamic analysis of Coma Berenices galaxies.
G. Des Forêts, R. Dominguez-Tenreiro, D. Gerbal, G. Mathez, A. Mazure, E. Salvador-Solé.
C. R. Acad. Sci. Paris, Tome 293, Sér. II, 579 - 582 (1981). In French.

In a preceding Note (1981) the equations of the simplest dynamical model accounting for most of available data on the Coma Cluster have been established. In the present Note the typical results of the dynamic analysis of Coma are given.

160.042 Optical emission-line gas associated with dominant cluster galaxies. T. M. Heckman.
Astrophys. J., Lett., Vol. 250, L59 - L63 (1981).

The author has surveyed a representative sample of 13 dominant cluster galaxies for the presence of the type of emission-line region seen around M87 and NGC 1275. Recent X-ray data and theoretical work suggest that emission-line filaments will form in a cooling accretion flow of the X-ray gas onto the dominant galaxy, and so should be a common feature of such galaxies. He finds that about half of his sample shows detectable emission lines with $L_{H\alpha + [N\ III]} > 10^{40.5}$ ergs^{-1} ($H_0 = 75$) and that emission at the level seen in Virgo/M87 cannot be ruled out in any of the sample galaxies. However, the NGC 1275 emission-line region remains unsurpassed in both linear extent and luminosity. The emission-line luminosity appears to be related to the X-ray luminosity and/or the cluster richness.

160.043 Evolution of the hot gas cores of clusters of galaxies, n(θ) counts and anisotropy of the cosmic background radiation. R. Fabbri, V. Natale.
Nuovo Cimento B, Ser. 11, Vol. 64B, 173 - 190 (1981).
Abstr. in Phys. Abstr., Vol. 85, Abstr. 3310 (1982).

160.044 H I observations in the Virgo cluster area.
G. Helou, C. Giovanardi, E. E. Salpeter, N. Krumm.
Astrophys. J., Suppl. Ser., Vol. 46, 267 - 286 (1981).

The authors report 21 cm H I line data obtained at Arecibo for 35 galaxies in or near the Virgo cluster. For all gal-

axies but two, they show contour plots of the spectral profile as a function of position along the major axis, with an average of six profiles per galaxy. The authors tabulate detailed information on each spectral profile and discuss improved or new procedures for gain correction, sidelobe removal, and total flux estimations. For each galaxy, they tabulate systemic velocity, total line width, and H I extent and content. The authors report determinations of the kinematic major axis for five face-on galaxies, and for most galaxies, a complete determination of the spin vector is shown for the first time. Upper limits on the H I content of 29 undetected galaxies are also given.

160.045 Binary-galaxy-rich clusters of galaxies.
M. F. Struble, H. J. Rood.
Astrophys. J., Vol. 251, 471 - 476, plates 7 - 9 (1981).

The authors report the discovery of a new, rare type of cluster of galaxies: those that contain a large number of binary galaxies among their brightest members. Counts are given within annuli of two Abell diameters for 10 prominent examples; photographs are given of three. It is suggested that the average angular separation of binaries in a cluster might be used as a distance indicator. The potential importance of these binary-galaxy-rich clusters for the mass discrepancy problem and implications for N-body simulations are briefly discussed.

160.046 Luminosity function and colors of the 3C 295 cluster of galaxies.
R. D. Mathieu, H. Spinrad.
Astrophys. J., Vol. 251, 485 - 496 (1981).

A photometric study of the cluster of galaxies around the cD galaxy 3C 295 ($z = 0.46$) has been done using two plates, one each in the F and J bandpasses, obtained at the prime focus of the KPNO 4 m Mayall reflector. The photometric data was obtained with the Berkeley PDS and the associated galaxy photographic photometry package. The results show the cluster to be relatively poor, with an F luminosity function consistent with the standard Schechter function. The color distribution has a very large number of blue galaxies, as found earlier by Butcher and Oemler. However, estimates from the surrounding field show a very high fraction of galaxies bluer than $(J-F) = 1.5$ are likely to be field galaxies. The cluster members appear to have a color distribution typical of local Bautz-Morgan type I clusters. The authors find no evidence for evolutionary effects occurring between the $z = 0.46$ epoch and the present.

160.047 The X-ray structure of a galaxy cluster at $z = 0.54$: implications for cluster evolution and cosmology.
S. D. M. White, J. Silk, J. P. Henry.
Astrophys. J., Lett., Vol. 251, L65 - L68 (1981).

The authors present high resolution X-ray observations of the rich cluster 0016 + 16 at a redshift of 0.541. The emitting gas in this cluster is hot and extremely luminous, and its structure resembles that seen in the brightest nearby cluster sources. In most of its properties, 0016 + 16 resembles a richer version of the Coma cluster, and it offers little support to the hypothesis that clusters at $z > 0.5$ differ fundamentally from nearer objects.

160.048 Multicolor photometry of the red cluster 0016 + 16 at $z = 0.54$. D. C. Koo.
Astrophys. J., Lett., Vol. 251, L75 - L79, plate 1 (1981).

Some highlights of photographic $UBVI$ photometry from 4 m plates are presented for the distant cluster known as 0016 + 16, at $z = 0.54$. This BM type II - III cluster is about twice as rich as Coma, possesses a structure elongated NE-SW with an axial ratio of ~ 0.6, and has a core radius of 0.38 Mpc. Within a 1.5 Mpc diameter circle, virtually all galaxies ($\gtrsim 90\%$) above background are intrinsically very red. The author concludes that the excess of blue galaxies found by Butcher and Oemler in other rich distant clusters is not a universal phenomenon and that star formation must have ceased in the S0 galaxies of 0016 + 16 more than 9 billion years ago.

160.049 The luminosity function of Virgo cluster galaxies.
R. C. Kraan-Korteweg.
Astron. Astrophys., Vol. 104, 280 - 287 (1981).

The luminosity function of a complete sample of 176 Virgo cluster galaxies is derived. The 160 cluster members with known redshift yield a mean radial velocity 967 ± 53 km s^{-1} with a dispersion $\sigma = 668$ km s^{-1}. Separate luminosity functions for different morphological types indicate significant differences; they are particularly significant between E and S0 galaxies. The data are consistent with an increasing number of fainter E (and S0?) galaxies.

160.050 Galactic halos/secondary peaks in clusters of galaxies.
J. Shaham.
Proceedings of the 5th Göttingen-Jerusalem-Symposium on Astrophysics, (see 012.041), p. 163 (1981).

160.051 Optical studies of southern X-ray clusters of galaxies. II. Velocity dispersions for A 119, Ser 40/6, SC 0316−444, and SC 2008−569.
J. Melnick, H. Quintana.
Astron. J., Vol. 86, 1567 - 1576 (1981).

Redshifts and new velocity dispersions are given for three rich southern galaxy clusters associated with X-ray sources. An improved value of the dispersion is given for the cluster Ser 40/6. The main optical and X-ray properties of these clusters are briefly discussed.

160.052 III. Clusters of galaxies.   M. A. Arakelyan.
Problems of extragalactic astronomy, (see 003.018), p. 83 - 123 (1981). In Russian.

160.053 X-ray measurements and the mass of Abell 1763 out to a radius of 1.4 Mpc.   J. P. Vallée.
Astrophys. Lett., Vol. 22, 193 - 198 (1981).

The radial dependence of the extended X-ray emission in the cluster of galaxies Abell 1763 is presented, as detected with the 2-arc min resolution Imaging Proportional Counter onboard the Einstein Observatory satellite. From this radial dependence of the X-ray intensity, the radial dependences of the emission measure, the thermal electron density, and the total mass density are obtained out to 1.4 Mpc from the X-ray centroid. The total mass so obtained suggests that this cluster possesses a dark component.

160.054 VLA observations of the radio continuum emission from Stephan's Quintet.
J. M. van der Hulst, A. H. Rots.
Astron. J., Vol. 86, 1775 - 1780 (1981).

Observations of the λ 20-cm radio continuum emission associated with Stephan's Quintet have been made with the VLA. Maps with an angular resolution of 2."25 and 6" are presented and show that the ridge of radio emission between the galaxies NGC 7319 and NGC 7318b resolves into a point-like source (possibly an unrelated background object) and diffuse emission that partly coincides with spiral-arm-like regions in NGC 7318b. The central source in NGC 7319 is resolved in a direction perpendicular to the bar and shows a jet-like feature. No radio emission was detected from NGC 7320. The idea that NGC 7318b is a recent interloper in the group, and that its interaction with an intergalactic medium causes the enhanced activity at its east side is briefly discussed and adopted as the still most reasonable interpretation.

160.055 Radio observations of Abell clusters and a comparison with certain Einstein observations.   H. M. Johnson.
Astrophys. J., Suppl. Ser., Vol. 47, 235 - 241 (1981).

Areas around 70 clusters have been observed at the Arecibo Observatory primarily for coordination with about 60 Einstein X-ray sources. Arecibo source coordinates and flux density are tabulated, and some identifications are made with cataloged radio sources. Statistical estimates are made of the percentage of cluster sources among all sources in certain areas and to limits on S_{430}. Arecibo and other radio data are compared with the available Einstein data of seven clusters. NRAO Green Bank interferometer observations of A401 and A2142 are reported.

160.056 Optical emission-line gas associated with dominant cluster galaxies.   T. M. Heckman.
Prepr. Steward Obs. No. 330, 18 pp. (1981).

The author has surveyed a representative sample of 13 dominant cluster galaxies for the presence of the type of emission-line region seen around M87 and NGC 1275. Recent X-ray data and theoretical work suggests that emission-line filaments will form in a cooling accretion flow of the X-ray gas onto the dominant galaxy, and so should be a common feature of such galaxies. The author finds a relationship between optical-line and radio emission, leading him to hypothesize that the radio source may help in some way to form or excite the emission-line gas. The excitation state of the detected gas is always low.

160.057 The velocity field in the Local Supercluster.
M. Aaronson, J. Huchra, R. B. Tully.
Prepr. Steward Obs., No. 351, 45 pp. (1981).

The authors consider a model for the velocity field of the Local Supercluster which includes both the deceleration of galaxies by a spherically symmetric density enhancement and a random motion of the Local Group with respect to nearby galaxies.

A catalogue of ultraviolet, optical, and H I data for 201 Virgo cluster galaxies.   See Abstr. 002.033.

Lifetime constraints on massive neutrinos from ultraviolet observations of clusters of galaxies.
See Abstr. 022.186.

Possible existence of massive neutrino halos.
See Abstr. 061.012.

Metallicity indicators for the stellar and integrated light spectra.   See Abstr. 114.099.

Supernovae and star formation in clusters of galaxies.
See Abstr. 125.042.

High-resolution radio observations of the X-ray galaxy NGC 3862 (3C 264) in Abell 1367.
See Abstr. 141.076.

Far separated "pairs" of quasistellar objects — possible members of clusters of galaxies or their groups.
See Abstr. 141.124.

Magnetic field distribution in extended radio sources and halo source features.   See Abstr. 141.145.

Observations of a decrement in the microwave background radiation toward the distant cluster of galaxies 0016 + 16.
See Abstr. 141.155.

HEAO A-2 observations of non-Abell Zwicky clusters containing extended radio sources.
See Abstr. 142.054.

X-ray astronomy in the Einstein era.
See Abstr. 142.085.

Some recent results on extragalactic X-ray sources from the Einstein Observatory.   See Abstr. 142.155.

The central densities of spherical stellar systems with anisotropic velocity dispersions.   See Abstr. 151.007.

Simulations of clusters of galaxies. See Abstr. 151.009.

Some problems with the evolution of gas in clusters of galaxies.   See Abstr. 151.037.

Merging instability in groups of galaxies. See Abstr. 151.048.

The velocity dispersion among galaxies. See Abstr. 151.063.

Dynamics of luminous galaxies. See Abstr. 151.065.

Angular momentum orientation of spiral galaxies. See Abstr. 151.092.

Colour-absolute magnitude relation for spiral galaxies. See Abstr. 158.030.

Infrared surface brightness and absolute magnitude of spiral galaxies.   See Abstr. 158.044.

Hubble ratio and solar motion from 200 spiral galaxies having distances derived from the luminosity index. See Abstr. 158.047.

Hubble ratio and solar motion from 300 spirals having distances derived from H I line widths. See Abstr. 158.048.

The structure of giant elliptical galaxies in poor clusters of galaxies.   See Abstr. 158.049.

Spiral galaxies in clusters. I. Neutral hydrogen observations in Abell 1367, Coma, and Zwicky 74-23 (1400.4+0949).   See Abstr. 158.060.

Spiral galaxies in clusters. II. Neutral hydrogen observations in Cancer, Hercules (A 2151), and Pegasus I. See Abstr. 158.061.

The distribution of the spiral galaxies in the direction of the Coma/A 1367 supercluster.   See Abstr. 158.062.

The inactive radio galaxy in Abell 566. See Abstr. 158.086.

The selection of galaxies in the Lick survey and the relationship between galaxy counts and galactic absorption. See Abstr. 158.098.

General properties of galaxies. See Abstr. 158.100.

X rays from normal galaxies and clusters of galaxies.   See Abstr. 158.108.

A search for neutral hydrogen in D and cD galaxies. See Abstr. 158.125.

La découverte et l'exploration de l'univers extra-galactique.   See Abstr. 158.128.

Dust clouds in the stellar spheroids of Local Group galaxies.   See Abstr. 158.134.

Selection, statistics and photometry of compact galaxies with different instrumental equipment, part II. The cluster Abell 1775.   See Abstr. 158.156.

CCD photometry of edge-on spirals in the Ursa Major and Cancer clusters.   See Abstr. 158.169.

X-ray emission around radio galaxies in non-Abell clusters: a possible physical link between environment and nonthermal radio emission.   See Abstr. 158.203.

Multidimensional statistical analysis of normal galaxies.   See Abstr. 158.215.

A Seyfert 2 galaxy in the Virgo cluster. See Abstr. 158.273.

B-V vs. HM relation for ScI galaxies and the peculiar motion of the Local Group.   See Abstr. 158.289.

New models for the intracluster gas. See Abstr. 161.001.

Galaxies and intergalactic matter. See Abstr. 161.002.

Dynamics of intracluster gas in aspherical clusters of galaxies and their X-ray appearances. See Abstr. 161.003.

Intergalactic matter in the groups of galaxies. See Abstr. 161.009.

On the extragalactic distance scale and the Hubble constant.   See Abstr. 162.022.

A statistical measure of the cosmological density parameter using clusters of galaxies.   See Abstr. 162.142.

Erratum

160.901   Erratum: "Diffuse radio emission in the Coma cluster and Abell 1367: observations at 430 and 1400 MHz" [Astron. J.,Vol. 85, 1565 - 1576 (1980)]. R. J. Hanisch. Astron. J., Vol. 86, 1113 (1981). – See Abstr. 28.160.069.

161 Intergalactic Matter

161.001 New models for the intracluster gas.
  A. Cavaliere, R. Fusco-Femiano.
Astron. Astrophys., Vol. 100, 194 - 196 (1981).
  To describe the disposition of the intracluster hot medium, the authors propose a new static model, that incorporates — rather than the conventional polytropic assumption — the effects of localized heating. The model passes the test against high-resolution, X-ray brightness distributions; it entails direct and predictive links between dynamics and X-ray emission.

161.002 Galaxies and intergalactic matter.   G. B. Field.
  Some strangeness in the proportion. Centennial symposium to celebrate the achievements of Albert Einstein, (see 003.001), p. 308 - 330 (1980). — Abstr. in Phys. Abstr., Vol. 84, Abstr. 75206 (1981).

161.003 Dynamics of intracluster gas in aspherical clusters of galaxies and their X-ray appearances.
  Y. Hirayama, S. Ikeuchi.
Prog. Theor. Phys., Vol. 65, 463 - 476 (1981). — Abstr. in Phys. Abstr., Vol. 84, Abstr. 79854 (1981).

161.004 Interface between a hot and cool gas and the evolution of cool clouds in the intergalactic medium.
  A. G. Doroshkevich, Ya. B. Zel'dovich.
Zh. ehksp. i teor. fiz., Tom 80, 801 - 815 (1981). In Russian. Abstr. in Ref. zh., 51. Astron., 7.51.741 (1981).

161.005 Intergalactic extinction and the deceleration parameter.   R. Meinel.
Astron. Nachr., Band 302, 177 - 179 (1981) = Mitt. Univ. Sternw. Jena Nr. 148.
  The deceleration parameter q_0 is calculated from the relation between apparent magnitudes m of the brightest galaxies in clusters and their redshifts z considering an intergalactic extinction. The calculation is valid for a Friedmann universe, homogeneously filled with dust grains, assuming the extinction to be 0.5 mag at $z = 1$ and a λ^{-1}-law of extinction (according to Oleak and Schmidt 1976). Using the m, z-values of Kristian, Sandage, and Westphal (1978) a formal value of $q_0 \approx 2.1$ is obtained instead of $q_0 \approx 1.6$ without consideration of intergalactic extinction.

161.006 Intergalactic shells at large redshift.
  J. M. Shull, J. Silk.
Astrophys. J., Vol. 249, 26 - 34 (1981).
  Galactic explosions at large redshift may produce intergalactic shells, the interiors of which cool by inverse Compton scattering off the cosmic background radiation. They radiate strongly in H I (Lyα), He II λ304, and infrared fine-structure lines of C II and Si II. The redshifted emission from individual shells is at the current level of detectability. The predicted extragalactic background emission from many shells, strongly peaked toward the UV, sets an upper limit to the number of exploding sources at $z \sim 10$. Shell absorption lines of H I, C II, Si II, and Fe II may be seen at more recent epochs in quasar spectra. Each of these observations may probe otherwise invisible explosions in the early universe.

161.007 The physical state of primordial intergalactic clouds.
  J. H. Black.
Mon. Not. R. Astron. Soc., Vol. 197, 553 - 563 (1981).
  The physical state of low-density, primordial gas in intergalactic space at redshift $z \cong 2.4$ is examined. Expressions for the rates of ionization, recombination, heating and cooling in hydrogen and helium are presented. When the rates of these processes are in steady state, the structure of the gas approaches that of a polytrope. Lα absorption lines in distant QSOs can arise in the densest parts of a smooth distribution of intergalactic matter. Some properties of intergalactic clouds are discussed with reference to recent observations and equilibrium models.

161.008 Interaction of intergalactic-gas flow with a rigid-body, spheroidal galaxy.
  S. Ikeuchi, K. Tomisaka.
Astrophys. Space Sci., Vol. 80, 483 - 500 (1981).
  The interaction of intergalactic-gas flow, which is assumed to be an incompressible fluid, with a rigid-body, spheroidal galaxy is examined analytically. The gas-flow patterns, the distributions of pressure at the surface of galaxy and some other quantities are calculated for the cases with and without viscosity. By use of results, the authors discuss the formation of H I ridges, trailing clouds and gas streams accompanying with galaxies, and the bending (or warping) of interstellar gas observed in spiral galaxies are analyzed. Some discussion on the Magellanic Stream is given.

161.009 Intergalactic matter in the groups of galaxies.
  J. Švestka.
Pokroky, Vol. 26, 41 - 45 (1981). In Czech.

  Quasar reddening, intergalactic dust, and the microwave background.   See Abstr. 141.117.

  La découverte et l'exploration de l'univers extragalactique.   See Abstr. 158.128.

  X-ray emission around radio galaxies in non-Abell clusters: a possible physical link between environment and nonthermal radio emission.   See Abstr. 158.203.

  The distribution and morphology of X-ray — emitting gas in the core of the Perseus cluster.   See Abstr. 160.016.

  On the equilibrium distribution of the elements in the gas in the Coma cluster.   See Abstr. 160.020.

  X-ray measurements and the mass of Abell 1763 out to a radius of 1.4 Mpc.   See Abstr. 160.053.

Erratum

161.901 Erratum: 'On the origin of the intergalactic magnetic field and the radio halo associated with the Coma cluster of galaxies' [Astron. Astrophys., Vol. 93, 407 - 410 (1981)].   J. Roland.
Astron. Astrophys., Vol. 102, 142 (1981). — See Abstr. 29.161.001.

162 Universe (Structure, Evolution)

162.001 **Origin of the galaxies.**   J. Silk.
Nature, Vol. 292, 409 - 411 (1981).

162.002 **Grand unification, the neutron electric dipole moment and galaxy formation.**
J. Ellis, M. K. Gaillard, D. V. Nanopoulos, S. Rudaz.
Nature, Vol. 293, 41 - 43 (1981).
Grand Unified Theories (GUTs) provide a possible solution to the long-standing cosmological problem of the apparent asymmetry between matter and antimatter. The authors emphasize that the experimental upper limit on the neutron electric dipole moment and the present cosmological baryon-to-photon ratio can be used to limit the amount of extra entropy generated after baryosynthesis. As an application of this restriction of entropy generation, the authors establish an upper limit on shear in the early Universe.

162.003 **Imprints of the damping of adiabatic perturbations.**   A. Dekel.
Astron. Astrophys., Vol. 101, 79 - 87 (1981).
The damping that adiabatic density perturbations undergo prior to and during the recombination epoch can produce a secondary peak in their two point, spatial, correlation function. It may then show up in the angle-averaged density profiles around real high density peaks. It is expected to appear at the same time in the non-linear regime of rich clusters of galaxies (length scale of a few Mpc) and in the linear regime of extended superclusters (length scale of a few tens of Mpc). Such peaks are tentatively identified in the observed spatial correlation function of galaxies and in the Virgocentric density profile. The adiabatic component is found to be relatively important on fitting a theoretical model in which the initial spectrum is a combination of adiabatic and isothermal power-laws. The power indices are found to be limited by $n_a \gtrsim -1$ and $n_b \lesssim 0$, respectively. By matching the peaks from the two length scales the author obtains $\Omega_0 \sim 0.1$ and $M_D \sim 2 \times 10^{14}\ M_\odot h_{75}^{-1}$ for the damping scale. It is proposed that the structure in the Universe evolves according to a script which is a combination of the hierarchical clustering theory and the "pancake" theory.

162.004 **Das Milchstraßensystem – Hypothesen über Ursprung und Entwicklung.**   H.-E. Fröhlich.
Astron. Schule, 18. Jahrg., 63 - 67 (1981).

162.005 **Spazio e tempo in cosmologia.**   S. Mora.
Coelum, Vol. 50, 127 - 132 (1981).

162.006 **Beiträge der Radioastronomie zum Urknallmodell der Weltentstehung. Teil I.**   P. G. Mezger.
Sterne Weltraum, Jahrg. 20, 265 - 272 (1981).

162.007 **The stage of superheavy particle dominance in the universe and primordial black holes.**
A. G. Polnarev, M. Yu. Khlopov.
Astron. Zh., Tom 58, 706 - 716 (1981). In Russian. English translation in Soviet Astron., Vol. 25, No. 4.
The stage of non-relativistic superheavy particle dominance in the early universe is predicted within the frame of the grand unified theories of elementary particles. The minimal probability of primordial black hole (PBH) formation at this stage is estimated and compared with the astrophysical upper limits on the PBH density. Astrophysical restrictions on grand unified theories are obtained.

162.008 **Inflation and the mysteries of the cosmos.**
M. M. Waldrop.
Science, Vol. 213, 121 - 122 (1981).
Modern particle theory and the inflationary scenario may explain how the universe was filled with matter and energy.

162.009 **Beiträge der Radioastronomie zum Urknallmodell der Weltentstehung. Teil II.**   P. G. Mezger.
Sterne Weltraum, Jahrg. 20, 319 - 324 (1981).

162.010 **New pathways in gravitational research.**
P. S. Wesson, R. E. Goodson.
Observatory, Vol. 101, 105 - 108 (1981).

162.011 **On the linear theory of density perturbations in a neutrino + baryon universe.**   I. Wasserman.
Astrophys. J., Vol. 248, 1 - 12 (1981).
Various aspects of the linear theory of density perturbations in a universe containing a significant population of massive neutrinos are calculated. The gravitational effects of nonneutrino species are included in calculating the maximum neutrino Jeans mass, which is found to be $[M_J(t)]_{max} \sim 10^{17} M_\odot / [m_\nu(\text{eV})]^2$, about an order of magnitude smaller than is obtained when nonneutrino species are ignored. An explicit expression for the nonviscous damping of neutrino density perturbations less massive than the maximum neutrino Jeans mass is derived. The linear evolution of density perturbations after photon decoupling is discussed.

162.012 **Cosmological consequences of a first-order phase transition in the SU$_5$ grand unified model.**
A. H. Guth, E. J. Weinberg.
Phys. Rev. D, Vol. 23, 876 - 885 (1981). – Abstr. in Phys. Abstr., Vol. 84, Abstr. 63483 (1981).

162.013 **Our Universe – and others.**
M. Rees.
New Scientist, Vol. 89, p. 270 - 273 (1981). – Abstr. in Phys. Abstr., Vol. 84, Abstr. 67139 (1981).

162.014 **Generic singularity studies revisited.**
J. D. Barrow, F. J. Tipler.
Phys. Lett. A, Vol. 82A, 441 - 445 (1981). – Abstr. in Phys. Abstr., Vol. 84, Abstr. 67140 (1981).

162.015 **The cosmological term in the Einstein theory.**
R. V. Tevikian.
Phys. Lett. A, Vol. 83A, 49 - 50 (1981). – Abstr. in Phys. Abstr., Vol. 84, Abstr. 67141 (1981).

162.016 **Einstein-Cartan-Sciama-Kibble cosmological models with spinning matter and magnetic field.**
D. Tsoubelis.
Phys. Rev. D, Vol. 23, 823 - 828 (1981). – Abstr. in Phys. Abstr., Vol. 84, Abstr. 67142 (1981).

162.017 **Bianchi type IX cosmological models with homogeneous spinor fields.**   O. Obregon, M. P. Ryan, Jr..
J. Math. Phys., Vol. 22, 623 - 630 (1981). – Abstr. in Phys. Abstr., Vol. 84, Abstr. 71339 (1981).

162.018 **Monopole production in the very early Universe in a first-order phase transition.**
M. B. Einhorn, K. Sato.
Nucl. Phys. B, Vol. B180 (FS2), 385 - 404 (1981). – Abstr. in Phys. Abstr., Vol. 84, Abstr. 71344 (1981).

162.019 **On the evolution of the primordial cosmic turbulence.**   K. Tanabe.
Prog. Theor. Phys., Vol. 65, 190 - 203 (1981). – Abstr. in Phys. Abstr., Vol. 84, Abstr. 71345 (1981).

162.020 **On the visualization of Bolyai-Lobatchevsky's geometry.**   H. V. Fagundes.
Rev. Brasil. Fis., Vol. 10, 721 - 729 (1980). − Abstr. in Phys. Abstr., Vol. 84, Abstr. 71348 (1981).

162.021 **Why is the cosmological constant so small?**   P. C. W. Davies, S. D. Unwin.
Proc. R. Soc. London, Ser. A, Vol. 377, 147 - 149 (1981). Abstr. in Phys. Abstr., Vol. 84, Abstr. 71517 (1981).

162.022 **On the extragalactic distance scale and the Hubble constant.**   G. de Vaucouleurs.
Some strangeness in the proportion. Centennial symposium to celebrate the achievements of Albert Einstein, (see 003.001), p. 416 - 422 (1980). − Abstr. in Phys. Abstr., Vol. 84, Abstr. 75207 (1981).

162.023 **The absolute and relative nature of motion; self motion; the so-called expansion of the Universe.**   I. S. Zheludev.
At. Energy Rev., Vol. 18, 1079 - 1098 (1980). − Abstr. in Phys. Abstr., Vol. 84, Abstr. 75228 (1981).

162.024 **Tilted electromagnetic Bianchi type I and type II cosmologies.**   D. Lorenz.
Phys. Lett. A, Vol. 83A, 155 - 157 (1981). − Abstr. in Phys. Abstr., Vol. 84, Abstr. 75230 (1981).

162.025 **Heavy Majorana leptons and cosmological baryon excess.**   T. Yanagida, M. Yoshimura.
Phys. Rev. D, Vol. 23, 2048 - 2054 (1981). − Abstr. in Phys. Abstr., Vol. 84, Abstr. 75231 (1981).

162.026 **Homogeneous cosmological model in general scalar-tensor theory.**   A. Banerjee, N. O. Santos.
Phys. Rev. D, Vol. 23, 2111 - 2116 (1981). − Abstr. in Phys. Abstr., Vol. 84, Abstr. 75232 (1981).

162.027 **Improved standard cosmology.**   P. S. Wesson.
Phys. Rev. D, Vol. 23, 2137 - 2141 (1981). − Abstr. in Phys. Abstr., Vol. 84, Abstr. 75233 (1981).

162.028 **The size and shape of the Universe.**   M. J. Rees.
Some strangeness in the proportion. Centennial symposium to celebrate the achievements of Albert Einstein, (see 003.001), p. 291 - 301 (1980). − Abstr. in Phys. Abstr., Vol. 84, Abstr. 75234 (1981).

162.029 **Issues in cosmology.**   D. W. Sciama.
Some strangeness in the proportion. Centennial symposium to celebrate the achievements of Albert Einstein, (see 003.001), p. 387 - 404 (1980). − Abstr. in Phys. Abstr., Vol. 84, Abstr. 75235 (1981).

162.030 **On the formation of collapsed superdense nuclei in the Universe.**   T. Ohnishi.
Nucl. Phys. A, Vol. A362, 480 - 502 (1981). − Abstr. in Phys. Abstr., Vol. 84, Abstr. 79870 (1981).

162.031 **In the beginning (matter-antimatter asymmetry).**   D. Lindley.
New Scientist, Vol. 89, 685 - 687 (1981). − Abstr. in Phys. Abstr., Vol. 84, Abstr. 79871 (1981).

162.032 **Age of a neutrino dominated universe.**   J. Bernstein, G. Feinberg.
Phys. Lett. B, Vol. 101B, 39 - 42 (1981). − Abstr. in Phys. Abstr., Vol. 84, Abstr. 79872 (1981).

162.033 **Cosmological origin of mass scales.**   H. Terazawa.
Phys. Lett. B, Vol. 101B, 43 - 47 (1981). − Abstr. in Phys. Abstr., Vol. 84, Abstr. 79873 (1981).

162.034 **The baryon asymmetry and CPT invariance in the early Universe.**   S. Barshay.
Phys. Lett. B, Vol. 101B, 155 - 158 (1981). − Abstr. in Phys. Abstr., Vol. 79874 (1981).

162.035 **Quantum effects in the early universe. V. Finite particle production without trace anomalies.**   J. B. Hartle.
Phys. Rev. D, Vol. 23, 2121 - 2128 (1981). − Abstr. in Phys. Abstr., Vol. 84, Abstr. 79876 (1981).

162.036 **Symmetry paradoxes and other cosmological comments.**   C. W. Misner.
Some strangeness in the proportion. Centennial symposium to celebrate the achievements of Albert Einstein, (see 003.001), p. 405 - 415 (1980). − Abstr. in Phys. Abstr., Vol. 84, Abstr. 79878 (1981).

162.037 **On the particle-defining modes for a free neutral scalar field in spatially homogeneous and isotropic universes.**   H. Kodama.
Prog. Theor. Phys., Vol. 65, 507 - 524 (1981). − Abstr. in Phys. Abstr., Vol. 84, Abstr. 80012 (1981).

162.038 **Interaction of the de-Sitter-Schwarzschild Universe with massless spin fields.**   U. Khanal, N. Panchapakesan.
V high energy physics symposium, (see 012.007), p. II.A2/11 (1980). − Abstr. in Phys. Abstr., Vol. 84, Abstr. 80017 (1981).

162.039 **The present mass of the universe and Dirac's large number hypothesis.**   A. J. Rutgers.
Proc. K. Nederlandse Akad. Wet., Ser. B, Vol. 84, 1 - 11 (1981). − Abstr. in Phys. Abstr., Vol. 84, Abstr. 83775 (1981).

162.040 **Primeval adiabatic perturbations: constraints from the mass distribution.**   P. J. E. Peebles.
Astrophys. J., Vol. 248, 885 - 897 (1981).
 The autocorrelation function of the mass distribution after decoupling of matter and radiation is computed under the assumption of linear primeval adiabatic perturbations using a new numerical method, and the results are compared to what is inferred from the present galaxy distribution. The computations are based on a Friedmann-Lemaître model with $\Lambda = 0$ containing radiation, zero-mass neutrinos, hydrogen, and helium.

162.041 **Physical uniformity of the Universe.**   Z. Horák.
Astrophys. Space Sci., Vol. 78, 287 - 292 (1981).
 The question of the spatial homogeneity of the Universe is re-examined from the viewpoint of the hypothesis on the physical unity of the Universe. It is shown that the demand for the universal validity of the theory of relativity implies that the average value of the Newtonian world potential is constant everywhere in the universe which is spatially homogeneous on a large scale. It turns out that Mach's principle is compatible with the special theory of relativity if the average value of the normalized world potential is exactly equal to $-c^2$.

162.042 **A direct hydrodynamical approach to the evolution of the density correlations in an expanding flat Friedmann universe.**   E. Nowotny.
Astrophys. Space Sci., Vol. 78, 427 - 433 (1981).
 A direct approach of the dynamical equation for the evolution of the two-point density correlation function ω is given in an expanding flat Friedmann universe in the Newtonian approximation. If the third and higher moments are neglected, a wave-like equation of third-order for the two-point density correlation function is found. The exact solution of this equa-

tion shows, in the large time limit, the usual Jeans instability $\omega \sim t^{4/3}$. It is suggested that the higher n-point correlation function of the density grows like $t^{2n/3}$ in the same approximation.

162.043   Estado actual de la cosmología.   L. C. Green.
Rev. Astron., Vol. 53, No. 217, p. 2 - 6 (1981).

162.044   **Theory of galaxy formation triggered by quasar explosions.**   S. Ikeuchi.
Publ. Astron. Soc. Japan, Vol. 33, 211 - 222 (1981).
A two-step process of galaxy formation is presented. As objects generated first, the author considers quasars (QSOs) formed at the cosmological redshift $z = 4-10$. The cooled shells at the shock fronts generated by QSO explosions are split into galaxies of the mass $10^{10-11} M_{\odot}$. Clusters of galaxies are formed at the collision of the cold shells, suggesting the large-scale shell-like or chain-like distributions of galaxies.

162.045   **The theory of vacuum seems to solve the puzzle of cosmology.**   Ya. B. Zel'dovich.
Usp. fiz. nauk, Tom 133, 479 - 503 (1981). In Russian.
Abstr. in Ref. zh., 51. Astron., 7.51.762 (1981).

162.046   **Neutrino and the universe.**
A. G. Doroshkevich, Ya. B. Zel'dovich, R. A. Syunyaev, M. Yu. Khlopov.
Vestn. AN SSSR, 1981, No. 4, p. 115 - 126. In Russian.
Abstr. in Ref. zh., 51. Astron., 7.51.763 (1981).

162.047   **On the theory of a fluctuating universe.**
M. F. Shirokov, Yu. R. Musin.
Izv. vuzov. Fiz., Tom 23, No. 12, p. 92 - 94 (1980). In Russian.
Abstr. in Ref. zh., 51. Astron., 7.51.765 (1981).

162.048   **The removal of a causal particle horizon in anisotropic cosmological singularities.**
V. A. Ruban, A. Yu. Ushakov, A. D. Chernin.
Zh. ehksp. i teor. fiz., Tom 80, 816 - 829 (1981). In Russian.
Abstr. in Ref. zh., 51. Astron., 7.51.767 (1981).

162.049   **Size and age of the universe.**   S. van den Bergh.
Science, Vol. 213, 825 - 830 (1981).
A number of different studies have been made of the Hubble flow in nearby regions of space. These investigations yield conflicting results, which indicate that the velocity with which the Local Group is falling into the Virgo cluster may lie anywhere in the range 0 to 500 km sec^{-1}. Until the local flow pattern is better understood, determinations of the Hubble parameter H will have to be derived from exceedingly difficult studies of distant galaxies with redshifts $\geqslant 1000$ km sec^{-1}. Presently available data indicated that $50 < H < 100$ km sec^{-1} Mpc^{-1}. The corresponding expansion time scale for the universe lies in the range 10 to 20 Gyr.

162.050   **Galaxy clustering and the Rubin-Ford effect.**
M. Clutton-Brock, P. J. E. Peebles.
Astron. J., Vol. 86, 1115 - 1119 (1981).
The solar motion relative to the Rubin-Ford sample of galaxies and the solar motion relative to the microwave background suggest that the Rubin-Ford sample is moving relative to the distant universe with a speed of perhaps 800 km s^{-1}. The authors enquire whether this effect could be produced by gravitational accelerations due to density fluctuations corresponding to the observed galaxy correlation function $\xi(r)$.

162.051   **La constante de Hubble et le paramètre d'accélération de l'univers.**   P. Véron.
Histoire de l'univers, (see 003.007), p. 349 - 364 (1980).
Contents: La découverte de l'expansion de l'univers. Les révisions de l'échelle des distances extragalactiques. La cosmologie newtonienne. Quelle est la vitesse de l'expansion?

L'univers est-il ouvert ou fermé? La valeur cosmologique des radiosources extragalactiques. Conclusion.

162.052   **L'évolution de l'univers.**   A. Maeder.
Histoire de l'univers, (see 003.007), p. 365 - 382 (1980).
Contents: Les conséquences de l'expansion et du principe cosmologique. La découverte d'un rayonnement de fond cosmologique. Les observations et les propriétés du rayonnement de fond cosmologique. Discussion sur les modèles cosmologiques. L'univers initial dominé par le rayonnement. L'interprétation du rayonnement de fond cosmologique: des photons avec un décalage spectral d'un facteur 1 000. Les phases initiales de l'univers. Questions liées au big-bang et aux singularités. La synthèse cosmologique de l'helium: un succès de la théorie du big-bang. L'abondance du deutérium: un test cosmologique. L'origine et la formation des galaxies.

162.053   **Quelques grands problèmes cosmologiques actuels.**   P. Bouvier.
Histoire de l'univers, (see 003.007), p. 383 - 403 (1980).
Contents: Le paradoxe du ciel nocturne. L'état stationnaire. La constante cosmologique et ses vicissitudes. Espace-temps et matière: le principe de Mach. La constante de la gravitation varie-t-elle? L'antimatière dans l'univers. Les ondes de gravitation.

162.054   **Conclusion. L'univers, quel avenir?**   A. Hayli.
Histoire de l'univers, (see 003.007), p. 421 - 424 (1980).

162.055   **Viscous Friedmann cosmology.**   Z. Klimek.
Acta Cosmologica, Zesz. 10, 7 - 19 (1981).
The evolution of Friedmann models with bulk viscosity in the plane "Hubble's constant"-energy density is presented. The general conclusions are: 1) Viscosity leads to intense energy production – energy density increases in spite of expansion. 2) If the above result be regarded as non-physical, the bulk viscosity can produce cosmological models without the initial singularity only for flat universes. 3) The results do not essentially depend on the equation of state.

162.056   **Note on the use of type I supernovae as cosmic clocks.**   P. Teerikorpi.
Acta Cosmologica, Zesz. 10, 21 - 24 (1981).
There is a bias influencing the cosmological expansion test based on type I supernovae. One should also be cautious of a bias when the radio variations of quasars are used in the test.

162.057   **Novae and the extra-galactic distance scale.**   S. van den Bergh.
J. R. Astron. Soc. Canada, Vol. 75, 169 - 174 (1981).
It is shown that CCD arrays can detect significant numbers of novae in galaxies at the distance of the Virgo Cluster. For the most luminous novae, such observations can, through the M_{pg} (max) versus t_2 relation, be used to determine distances to galaxies with distance moduli $(m-M)_{pg} \lesssim 31$. Such observations could significantly strengthen distance determinations to the Virgo Cluster itself and might place interesting constraints on models in which the Local Group is falling into the Virgo Super Cluster.

162.058   **Spatially homogeneous and anisotropic cosmological models in Brans-Dicke theory.**
V. B. Johri, G. K. Goswami.
Australian J. Phys., Vol. 34, 261 - 265 (1981).
Spatially homogeneous and anisotropic cosmological models corresponding to Bianchi type I solutions of Brans-Dicke theory are investigated. The physical and geometrical

properties of the models are discussed and compared with the corresponding relativistic models.

162.059 Towards a non-Friedmannian universe.
R. Fabbri, F. Melchiorri.
Gen. Relativ. Gravitation, Vol. 13, 201 - 204 (1981).

The authors present the result of an experiment of the Florence group which has detected a quadrupole anisotropy in the cosmic background radiation. They show that this result implies that the universe either has large metric perturbations outside the particle horizon, or will become largely irregular in the future.

162.060 Stability of geodesic incompleteness for Robertson-Walker space-times. J. K. Beem, P. E. Ehrlich.
Gen. Relativ. Gravitation, Vol. 13, 239 - 255 (1981).

162.061 Instabilities of certain empty Robertson-Walker space-times. R. P. A. C. Newman.
Gen. Relativ. Gravitation, Vol. 13, 257 - 282 (1981).

162.062 Neutrinos of non-zero mass in Friedmann universes.
P. S. Joshi, S. M. Chitre.
Nature, Vol. 293, 679 (1981).

162.063 Los limites del universo, según Einstein.
R. R. de Freitas Mourão.
R Muscae, Vol. 6, No. 1/2, p. 20 - 21 (1981).

162.064 Quantum cosmology and geometric quantization.
J. A. Isenberg, M. J. Gotay.
Gen.Relativ. Gravitation, Vol. 13, 301 - 306 (1981).

162.065 Nonscalar singularities in spatially homogeneous cosmologies. S. T. C. Siklos.
Gen. Relativ. Gravitation, Vol. 13, 433 - 441 (1981).

162.066 Left-flat spaces and null geodesics.
J. R. Porter.
Gen. Relativ. Gravitation, Vol. 13, 443 - 449 (1981).

Equations are derived giving the null geodesics in any left-flat space as the intersection of hypersurfaces. The connection with null geodesics given in terms of the good-cut functions for H spaces is established.

162.067 Quantum fluctuations and nonavoidance of the singularity in Bianchi type I cosmology.
T. Padmanabhan.
Gen. Relativ. Gravitation, Vol. 13, 451 - 455 (1981).

An effective metric is defined and used for analyzing the quantum fluctuations in a classical geometry. Earlier work showing that quantum (conformal) fluctuations avoid the classical singularity in the case of spherically symmetric collapse is briefly reviewed. It is shown that this result does not extend to anisotropic Bianchi type I cosmology. Here the dispersion in the fluctuations increases too slowly to quench the classical singularity. The singularity persists in the space-time described by the effective metric.

162.068 Public and private space curvature in Robertson-Walker universes. W. Rindler.
Gen. Relativ. Gravitation, Vol. 13, 457 - 461 (1981).

The question is asked: what space curvature would a fundamental observer in an ideal Robertson-Walker universe obtain by direct local spatial measurements, i. e., without reference to the motion pattern of the other galaxies? The answer is that he obtains the curvature $\tilde{K}$ of his "private" space generated by all the geodesics orthogonal to his world line at the moment in question, and that $\tilde{K}$ is related to the usual curvature $K = k/R^2$ of the "public" space of galaxies by $\tilde{K} = K + H^2/c^2$, where H is Hubble's parameter.

162.069 Creation of gravitons in a homogeneous and isotropic universe. A. D. Popova.
Inst. kosm. issled. AN SSSR. Prepr., 1980, No. 604, 32 pp. In Russian. – Abstr. in Ref. zh., 51. Astron., 9.51.935 (1981).

162.070 The problem of the "thermal death" of the universe in the history of cosmology and presence.
A. P. Trafimenka.
Izv. AN BSSR. Ser. obshchestv. nauk, 1981, No. 2, p. 26 - 33. In Belorussian. – Abstr. in Ref. zh., 51. Astron., 9.51.936 (1981).

162.071 On some properties of Gödel's nonstationary generalized cosmological model.
V. G. Agakov, N. I. Andreeva.
Redkol. zh. Izv. vuzov. Fiz., Tomsk, 1981. 7 pp. In Russian. Abstr. in Ref. zh., 51. Astron., 9.51.940 (1981).

162.072 Mach–Poincaré gravodynamics and condensation phenomena in an expanding universe.
D.-E. Liebscher, J. P. Mücket.
Astron. Nachr., Band 302, 133 - 138 (1981).

The authors discuss the condensation phenomena in a very simple model, which allows for transparent calculations. They compare the condensation times of subsystems decoupled from the cosmological expansion in different approaches to Mach–Poincaré gravodynamics with Newtonian mechanics. In all cases contracting subsystems without inner rotation collapse, only the time-scales differ.

162.073 Is the Universe expanding?
N. Rosen.
Nuovo Cimento, Lett., Ser. 2, Vol. 31, 193 - 198 (1981). Abstr. in Phys. Abstr., Vol. 84, Abstr. 83917 (1981).

162.074 CP violation and the development of cosmological baryon asymmetry. G. Senjanovic.
AIP Conf. Proc., No. 72, p. 192 - 223 (1980). – Abstr. in Phys. Abstr., Vol. 84, Abstr. 84200 (1981).

162.075 Non-vacuum cosmologies with toroidal topology of space sections. N. V. Mitskievic, Y. E. Senin.
Acta Phys. Polonica B, Vol. B12, 541 - 547 (1981). – Abstr. in Phys. Abstr., Vol. 84, Abstr. 89125 (1981).

162.076 How homogeneous is the Universe?
P. S. Wesson.
Astrophys. Lett., Vol. 21, 97 - 99 (1981).

Recent observations of the quadrupole anisotropy of the 3K microwave background indicate that the Universe may be inhomogeneous on scales larger than the present Hubble radius. Observations of galaxies and QSOs are analyzed to see what non-microwave data indicate about the homogeneity of the Universe. They give the limit $\Delta\rho/\rho \lesssim 0.4$ for the global variation of matter density. This does not satisfy the condition $\Delta\rho/\rho \ll 1$ for homogeneity. The non-microwave data are therefore in agreement with the microwave data in indicating that the Universe may be globally inhomogeneous.

162.077 Left-ring symmetry, composite models and baryon number of the universe. R. N. Mohapatra.
AIP Conf. Proc., No. 72, p. 647 - 664 (1980). – Abstr. in Phys. Abstr., Vol. 84, Abstr. 89788 (1981).

162.078 Exact solution for vaccum Bianchi type III model with a cosmological constant.
A. Moussiaux, P. Tombal, J. Demaret.
J. Phys. A, Vol. 14, L277 - L280 (1981). – Abstr. in Phys. Abstr., Vol. 84, Abstr. 94577 (1981).

162.079 Vacuum stress tensor for a slightly squashed Einstein Universe. R. Critchley.

J. Phys. A, Vol. 14, 1943 - 1955 (1981). — Abstr. in Phys. Abstr., Vol. 84, Abstr. 94578 (1981).

162.080 **Thermal Green functions in an expanding universe.**
I. T. Drummond.
Nucl. Phys. B, Vol. B190, 93 - 112 (1981). — Abstr. in Phys. Abstr., Vol. 84, Abstr. 94579 (1981).

162.081 **Creation of wormholes by first order phase transition of a vacuum in the early Universe.**
K. Sato, M. Sasaki, H. Kodama, K. Maeda.
Prog. Theor. Phys., Vol. 65, 1443 - 1446 (1981). — Abstr. in Phys. Abstr., Vol. 84, Abstr. 94581 (1981).

162.082 **Effects of spacetime curvature on a cosmological first-order phase transition.** Y. Fujii.
Phys. Lett. B, Vol. 103B, 29 - 32 (1981). — Abstr. in Phys. Abstr., Vol. 84, Abstr. 94583 (1981).

162.083 **The redshift for inhomogeneous cosmological models.** N. A. Tomimura.
Astrophys. J., Vol. 249, 23 - 25 (1981).
Calculations are presented for the redshift for certain inhomogeneous cosmological models. The method the author uses is based on the paper by Kristian and Sachs. The null geodesic equations are satisfied by an a priori solution using an inductive method. It is found that the results for the inhomogeneous models are in identical agreement with the result for the Friedmann model to first and second order over an expansion in r_A, where r_A is the corrected luminosity distance. The anisotropic differences for the redshift are seen to arise in the author's model only to third order in r_A.

162.084 **Formation of the structure in the neutrino universe.**
A. G. Doroshkevich, M. Yu. Khlopov.
Astron. Zh., Tom 58, 913 - 924 (1981). In Russian. English translation in Soviet Astron., Vol. 25, No. 5.
The evolution of perturbations and the formation of inhomogeneities in cosmological models with non-zero neutrino mass are analyzed. Detailed study of the transition from the radiation domination stage to the modern stage of massive neutrino domination provides a more accurate definition of parameters of such models. The amplitude of initial metric perturbations is estimated on the basis of the nonlinear theory of inhomogeneities formation. The nonlinear stage of the evolution of inhomogeneities is considered.

162.085 **Examples of relativity of temporal finiteness and infiniteness of cosmological models.**
L. I. Kharbediya.
Astron. Zh., Tom 58, 1131 - 1132 (1981). In Russian. English translation in Soviet Astron., Vol. 25, No. 5.
The present paper is a continuation of Zelmanov's and the author's works devoted to the problems of the relativity of spatial and temporal finiteness and infiniteness (the extension) in general relativity. Some examples on the relativity of temporal extension are given for an empty universe as well as for a non-empty one.

162.086 **Birth of a closed universe and the antropogenic principle.** Ya. B. Zel'dovich.
Pis'ma Astron. Zh., Tom 7, 579 - 581 (1981). In Russian. English translation in Soviet Astron. Lett., Vol. 7.
A scenario is proposed beginning with the quantum birth of a closed universe in the minimum of de Sitter's self-consistent cosmological solution with vacuum polarization. The point that the universe is closed and its density is always larger than the critical one is a direct consequence of the possibility of quantum birth. Probably it is the antropogenic principle which selects the universe with actually observed characteristics.

162.087 **The early universe.** E. R. Harrison.
South. Stars, Vol. 29, 2 - 20 (1981).

162.088 **Matter-antimatter separation in the early universe by rotating black holes.** D. A. Leahy.
Astrophys. J., Vol. 249, 403 - 405 (1981).
The effect of rotating black holes evaporating very early in the universe is considered. They produce oppositely directed neutrino and antineutrino currents which push matter and antimatter apart. This separation mechanism is too feeble to explain the present baryon to photon ratio of 10^{-9}, nor will it have other significant observational consequences.

162.089 **The extragalactic distance scale.**
P. W. Hodge.
Annu. Rev. Astron. Astrophys., Vol. 19, (see 003.012), 357 - 372 (1981).
Contents: Introduction. The Hubble distance scale. The Sandage-Tammann program. Kennicutt's refinements. Van den Bergh's assessment. De Vaucouleurs' reassessment. The H I line-width method. The present situation.

162.090 **Ageing and final life time of a photon in cosmological space.** A. Kipper.
Tartu Astrofüüs. Obs., Teated, Nr. 63, 59 pp. (1981). In Russian.

162.091 **A note on a second-order differential equation with infinitely many solutions.**
D. Lorenz, R. E. Zimmermann.
Nuovo Cimento,Lett., Ser. 2, Vol. 31, 603 - 606 (1981). Abstr. in Phys. Abstr., Vol. 84, Abstr. 94712 (1981).

162.092 **A static world model. I.**
S. Sundman.
Speculations Sci. Technol., Vol. 4, No. 1, p. 71 - 82 (1981). Abstr. in Phys. Abstr., Vol. 84, Abstr. 95087 (1981).

162.093 **A static world model. II.**
S. Sundman.
Speculations Sci. Technol., Vol. 4, No. 1, p. 83 - 90 (1981). Abstr. in Phys. Abstr., Vol. 84, Abstr. 95088 (1981).

162.094 **Effect of finite-temperature quantum fields on the early Universe.** B. L. Hu.
Phys. Lett. B, Vol. 103B, 331 - 337 (1981). — Abstr. in Phys. Abstr. Vol. 84, Abstr. 98901 (1981).

162.095 **Baryon nonconservation at intermediate mass scales and matter-antimatter asymmetry.**
A. Masiero, R. N. Mohapatra.
Phys. Lett. B, Vol. 103B, 343 - 348 (1981). — Abstr. in Phys. Abstr., Vol. 84, Abstr. 98902 (1981).

162.096 **Self-adjointness-based quantum field theory in de Sitter and anti-de Sitter space-time.**
H. Rumpf.
Phys. Rev. D, Vol. 24, 275 - 289 (1981). — Abstr. in Phys. Abstr., Vol. 84, Abstr. 98903 (1981).

162.097 **Cosmological scalar-particle creation rate.**
M. Castagnino, D. Harari, L. Chimento.
Phys. Rev. D, Vol. 24, 290 - 296 (1981). — Abstr. in Phys. Abstr., Vol. 84, Abstr. 98904 (1981).

162.098 **Electromagnetic waves in a Bianchi type-I Universe.**
A. Sagnotti, B. Zwiebach.
Phys. Rev. D, Vol. 24, 305 - 319 (1981). — Abstr. in Phys. Abstr., Vol. 84, Abstr. 98905 (1981).

162.099 **SO(3,2)-invariant scattering and Dirac singletons.**
L. Castell, W. Heidenreich.
Phys. Rev. D, Vol. 24, 371 - 377 (1981). − Abstr. in Phys.
Abstr., Vol. 84, Abstr. 98906 (1981).

162.100 **Expansion of the Universe as an indirect cause of the cosmological redshift: some consequences.**
P. Voráček.
Astrophys. Space Sci., Vol. 79, 507 - 508 (1981).
Consequences are discussed of the fact that it is
not possible to interpret directly the cosmological redshifts as
a Doppler effect due to a recession velocity of the source of
the radiation.

162.101 **The expansion of the metagalaxy.** B. Bonnevier.
Astrophys. Space Sci., Vol. 79, 509 - 513 (1981).
In the big-bang hypothesis the metagalaxy (or Universe)
is considered to have expanded from a state of extremely high
density. The observations of distant galaxies are usually interpreted with this picture in mind. As pointed out by Alfvén it is
important to see whether these observations do really require
the hypothesis of an early extremely high density. It is found
that with present available information, it is not possible to
conclude that the density has been higher than implied by the
Schwarzschild limit.

162.102 **On observational constraints against symmetric cosmologies.** A. J. Allen.
Mon. Not. R. Astron. Soc., Vol. 197, 679 - 686 (1981).
The problem of observational constraints on antimatter
in the Universe is re-investigated. It is shown that some
previous investigations of annihilation are in error. It is
emphasized that symmetric cosmologies in which the spectrum
of emulsion scales is predominantly given by cluster scales are
not precluded by currently derived observational limits,
whilst specific macroscopic inclusions on smaller scales which
retain their identity cannot be precluded.

162.103 **Effects of shape upon the evolution of perturbations in the post-recombination Universe.**
P. L. Palmer.
Mon. Not. R. Astron. Soc., Vol. 197, 721 - 729 (1981).
The importance of shape upon the evolution of cosmological perturbations in the post-recombination Universe is
demonstrated. The simple results for the evolution of spherical
perturbations in an $\Omega = 1$ Universe are generalized to triaxial
perturbations. The consequences of these results are mainly
for observational cosmology, and the interpretation of such
data.

162.104 **Neutrinos in the early universe.** A. D. Dolgov.
Yader. fiz., Tom 33, 1309 - 1319 (1981). In Russian.
Abstr. in Ref. zh., 51. Astron., 10.51.752 (1981).

162.105 **Do antimatter domains exist in the universe?**
V. A. Kuz'min, I. I. Tkachev, M. E. Shaposhnikov.
Pis'ma v ZhEhTF, Tom 33, 557 - 560 (1981). In Russian.
Abstr. in Ref. zh., 51. Astron., 10.51.758 (1981).

162.106 **Effect of quantum gravitational processes on the evolution of the isotropic universe.**
G. M. Vereshkov, Yu. S. Grishkan, N. M. Ivanov,
A. N. Poltavtsev.
Zh. ehksp. i teor. fiz., Tom 80, 1665 - 1676 (1981). In Russian.
Abstr. in Ref. zh., 51. Astron., 10.51.759 (1981).

162.107 **Quantum fluctuations and the "non-singular" universe.** V. F. Mukhanov, G. V. Chibisov.
Pis'ma v ZhEhTF, Tom 33, 549 - 553 (1981). In Russian.
Abstr. in Ref. zh., 51. Astron., 10.51.760 (1981).

162.108 **Quantum statistics in cosmological models.**
E. N. Rumyantseva.
Izv. vuzov. Fiz., Tom 24, No. 5, p. 61 - 65 (1981). In Russian.
Abstr. in Ref. zh., 51. Astron., 10.51.761 (1981).

162.109 **Conformal scalar field, gravitation, vacuum and cosmology.** V. N. Mel'nikov.
Gravitatsiya i teor. otnositel'n. Kazan', 1980, No. 17, p. 71 - 80.
In Russian. − Abstr. in Ref. zh., 51. Astron., 10.51.774 (1981).

162.110 **Unity of processes of evolutionary non-stability of matter in the universe.**
M. L. Stebakov, V. M. Stebakov.
Gos. NII gornokhim. syr'ya. Lyubertsy, 1981, 27 pp. In
Russian. − Abstr. in Ref. zh., 51. Astron., 10.51.775 (1981).

162.111 **Cosmological black hole production in grand unified theories.** A. F. Grillo, Y. Srivastava.
Nucl. Phys. B, Vol. B187, 271 - 284 (1981). − Abstr. in Phys.
Abstr., Vol. 84, Abstr. 102278 (1981).

162.112 **Perturbation of the de Sitter-Schwarzschild Universe with massless fields.**
U. Khanal, N. Panchapakesan.
Phys. Rev. D, Vol. 24, 829 - 834 (1981). − Abstr. in Phys.
Abstr., Vol. 84, Abstr. 102279 (1981).

162.113 **Production of massless particles in the de Sitter-Schwarzschild Universe.**
U. Khanal, N. Panchapakesan.
Phys. Rev. D, Vol. 24, 835 - 838 (1981). − Abstr. in Phys.
Abstr., Vol. 84, Abstr. 102280 (1981).

162.114 **Monopole dissociation in the early Universe.**
P. J. Steinhardt.
Phys. Rev. D, Vol. 24, 842 - 857 (1981). − Abstr. in Phys.
Abstr., Vol. 84, Abstr. 102281 (1981).

162.115 **Cosmic strings in unified gauge theories.**
A. E. Everett.
Phys. Rev. D, Vol. 24, 858 - 868 (1981). − Abstr. in Phys.
Abstr., Vol. 84, Abstr. 102282 (1981).

162.116 **On the solution to the cosmological horizon problem proposed by Zee.** M. D. Pollock.
Phys. Rev. D, Vol. 24, 1045 - 1048 (1981). − Abstr. in Phys.
Abstr., Vol. 84, Abstr. 102284 (1981).

162.117 **Baryon-to-entropy ratio and initial conditions of the Universe.** L. Parker.
Phys. Rev. D, Vol. 24, 1049 - 1052 (1981). − Abstr. in Phys.
Abstr., Vol. 84, Abstr. 102285 (1981).

162.118 **Nuclear astrophysics and particles.** R. J. Tayler.
Nuclear astrophysics, (see 012.036), p. 5 - 22 (1981).
The author discusses two topics: the physics of the early
Universe and late stages of stellar evolution. In both cases
current developments in elementary particle physics are very
important and they are also both topics which relate closely to
nuclear astrophysics. If current cosmological and astrophysical
ideas are correct, the light chemical elements which are observed today were essentially produced by nuclear reactions
during the first few minutes of the expanding Universe, while
the heavy elements were produced by nuclear fusion reactions
in stars, with the final stages of stellar evolution being
particularly important. In each case a very important role is
played by neutrinos and by the precise form of the theory of
weak interactions.

162.119 **Cosmological implications of Grand Unified Theories.**
D. V. Nanopoulos.
Nuclear astrophysics, (see 012.036), p. 23 - 67 (1981).

Some basic points concerning the structure of the standard strong and electroweak interactions prior to grand unification are presented. A detailed exposé of GUTs is attempted including their basic principles and their consequences for particle physics. The most important cosmological implications of GUTs, including baryon number generation in the early Universe; dissipative processes in the very early Universe; grand unified monopoles, etc. are given.

162.120 **A cold big bang with small black holes.**
    D. Lindley.
Nuclear astrophysics, (see 012.036), p. 279 - 283 (1981).
    The author shows how one might construct a cosmological model in which the universe starts out entirely in the form of small black holes, whose subsequent evaporation generates the presently observed cosmic entropy.

162.121 **The first order phase transition of a vacuum and baryon-number domain structure of the universe.**
K. Sato.
Nuclear astrophysics, (see 012.036), p. 311 - 317 (1981).
    If CP-nonconservation arises from spontaneous symmetry breaking in the very early universe, the universe will have a domain structure of baryon number. The author proposes a model of the early universe in which domains are stretched exponentially and the radius of domains is much greater than that of the standard big bang model, provided that the GUT phase transition is of first order. If the size of the stretched domains is sufficiently big to avoid pair annihilations of baryon and antibaryon domains, the difficulties of the baryon symmetric universe may be removed.

162.122 **Grand unification and cosmology.**
    J. Ellis, M. K. Gaillard, D. V. Nanopoulos.
Unification of the fundamental particle interactions, (see 012.038), p. 461 - 493 (1980). − Abstr. in Phys. Abstr., Vol. 84, Abstr. 102879 (1981).

162.123 **Cosmology confronts grand unification.**
    G. Steigman.
Unification of the fundamental particle interactions, (see 012.038), p. 495 - 508 (1980). − Abstr. in Phys. Abstr., Vol. 84, Abstr. 102880 (1981).

162.124 **Spherically symmetric cosmological solutions of the Lyttleton-Bondi Universe.**   A. Nduka.
Acta Phys. Polonica B, Vol. B12, 833 - 845 (1981). − Abstr. in Phys. Abstr., Vol. 84, Abstr. 108344 (1981).

162.125 **Cosmology in bimetric Machian gravitation.**
    R. Goldoni.
Nuovo Cimento, Lett., Ser. 2, Vol. 31, 481 - 486 (1981). Abstr. in Phys. Abstr., Vol. 84, Abstr. 108348 (1981).

162.126 **Vector particle creation and annihilation in a Friedmann expansion.**
S. Matarrese, S. A. Bonometto.
Nuovo Cimento A, Ser. 11, Vol. 64A, 125 - 140 (1981). Abstr. in Phys. Abstr., Vol. 84, Abstr. 108350 (1981).

162.127 **Particles in Friedmann and Einstein-Strauss universes.**
    R. T. Jantzen, R. Ruffini.
Phys. Lett. B, Vol. 104B, 373 - 376 (1981). − Abstr. in Phys. Abstr., Vol. 84, Abstr. 108351 (1981).

162.128 **Survey of cosmological models with gravitational, scalar and electromagnetic waves.**
M. Carmeli, C. Charach, S. Malin.
Phys. Rep., Vol. 76, 79 - 156 (1981). − Abstr. in Phys. Abstr., Vol. 84, Abstr. 108356 (1981).

162.129 **The production of magnetic monopoles in the very early Universe.**   M. B. Einhorn.
Unification of the fundamental particle interactions, (see 012.038), p. 569 - 582 (1980). − Abstr. in Phys. Abstr., Vol. 84, Abstr. 108359 (1981).

162.130 **Elementary growth rates of local inhomogeneities in a flat Robertson-Walker background with a relativistic equation of state.**   C. Hellaby, K. Lake.
J. R. Astron. Soc. Canada, Vol. 75, 248 (1981). − Abstract.

162.131 **On the evolution of cosmological adiabatic perturbations in the weakly non-linear regime.**
R. Juszkiewicz.
Mon. Not. R. Astron. Soc., Vol. 197, 931 - 940 (1981).
    The purpose of this paper is to investigate the weakly non-linear stage in the evolution of adiabatic density fluctuations. It is shown that tidal processes may lead to the disruption of large scale inhomogeneities into smaller units. This phenomenon may change significantly the Doroshkevich, Sunyaev & Zeldovich scenario for development of structure in the expanding Universe. Accurate N-body experiments are necessary to decide whether the approach used in this paper is reasonable.

162.132 **The initial fragmentation of collapsing gaseous protoclusters.**
B. J. T. Jones, P. L. Palmer, R. F. G. Wyse.
Mon. Not. R. Astron. Soc., Vol. 197, 967 - 975 (1981).
    The authors consider the initial fragmentation of a 'pancaking' protocluster of galaxies. Other aspects of the authors' picture of galaxy formation are also discussed.

162.133 **The formation of galaxies from massive neutrinos.**
    M. Davis, M. Lecar, C. Pryor, E. Witten.
Astrophys. J., Vol. 250, 423 - 431 (1981).
    Neutrinos with nonzero rest mass strongly influence galaxy formation in the early universe. If stable neutrinos have rest masses on the order of 100 eV, they close the universe, but they erase initial perturbations on mass scales less than $4 \times 10^{15} M_\odot$. However, if in addition there exist unstable neutrinos with rest masses on the order of 100 keV, they preserve and amplify initial perturbations on galactic mass scales ($10^{12} M_\odot$). These perturbations are picked up and further amplified by the lighter, stable neutrinos, as long as the heavy neutrinos decay somewhat after the lighter neutrinos go non-relativistic. If the heavy neutrinos decay into light neutrinos, the decay products contribute about one-half of the present mass density in a hot unclustered background. The only alternative method of retaining initial perturbations until the light neutrinos become nonrelativistic is to introduce large amplitude initial fluctuations such as primordial black holes. If the light neutrinos close the universe, black hole seeds of size $10^9 M_\odot$ would be required for galaxies of $10^{12} M_\odot$ to form.

162.134 **The growth of anisotropic structures in a Friedmann universe.**   J. D. Barrow, J. Silk.
Astrophys. J., Vol. 250, 432 - 449 (1981).
    Motivated by observational evidence for the existence of flattened or striated structures of very large scale (~100 Mpc) in the universe, the authors study the evolution of nonspherical structures. They set up equations to describe the evolution of a rotating, dust ellipsoid in an expanding universe. The authors derive solutions for the shape and density evolution as the ellipsoid expands along with a Friedmann background universe and show that deviations from spherical symmetry are preserved by adiabatic expansion. Analytic expressions are derived to describe the nonlinear evolution of pressure-free ellipsoids. The authors calculate details of "pancake" formation which are relevant for realistic theories of galaxy and cluster formation.

162.135 **Nucleogenesis: The origin of elements in the universe.**
J. Štohl.
Kozmos, Vol. 12, 163 - 166 (1981). In Slovak.

162.136 **Evolution of the universe and history of the earth.**
J. Kalvoda.
Pokroky, Vol. 26, 181 - 192 (1981). In Czech.

162.137 **Baryon asymmetry of the Universe versus left-right symmetry.** V. A. Kuzmin *(Kuz'min)*,
M. E. Shaposhnikov.
Electroweak interactions and unified theories. Proceedings of the 15th Rencontre de Moriond. Vol. 2, Les Arcs, Savoie, France, 9 - 21 March 1980. Editions Frontieres, Dreux, France (1980). p. 525 - 533. – Abstr. in Phys. Abstr., Vol. 85, Abstr. 286 (1982).

162.138 **Homogeneous radiation-filled Universe in general scalar tensor theory.**
A. Banerjee, N. O. Santos.
J. Phys. A, Vol. 14, 2829 - 2835 (1981). – Abstr. in Phys. Abstr., Vol. 85, Abstr. 3315 (1982).

162.139 **Stokesian fluids and cosmology.**
M. Novello.
Nukleonika, Vol. 25, 1405 - 1413 (1980). – Abstr. in Phys. Abstr., Vol. 85, Abstr. 3316 (1982).

162.140 **The rest mass of neutrinos and clustering in the early universe.** L.-z. Fang, Y.-z. Liu.
Acta Astrophys. Sinica, Vol. 1, 291 - 300 (1981). In Chinese.

162.141 **Toward the application of a metric size function in galactic evolution and cosmology.**
S. Djorgovski, H. Spinrad.
Astrophys. J., Vol. 251, 417 - 423 (1981).
The authors have measured the metric size function $\eta(r)$ defined by Petrosian on a limited sample of luminous distant galaxies, up to a redshift of 1.175. Although the investigation has a preliminary character, they find some evidence for the evolution of elliptical galaxies. Using the Ostriker-Hausman theoretical models, they place some constraints on the rates of dynamical evolution. The eta function appears to be an extremely sensitive probe for the investigation of galactic evolution and potentially useful in cosmology. Some problems and directions for future work are outlined.

162.142 **A statistical measure of the cosmological density parameter using clusters of galaxies.**
R. M. Wagner, S. C. Perrenod.
Astrophys. J., Vol. 251, 424 - 428 (1981).
Using complete redshift information from a recent supercluster catalog, the authors have applied a statistical test outlined by Sargent and Turner to estimate the cosmological density parameter Ω. However, rather than using individual galaxies as test particles, they use individual clusters. The test utilizes cluster redshifts and positions to detect and measure distortions from uniform Hubble expansion. These observed distortions are then compared statistically with theory for $0 < \Omega < 0.79$. The authors find the results are consistent with no distortions from uniform Hubble flow, and they indicate $\Omega \lesssim 0.1$. The limit does not apply to the contribution by any nonclustered component to the mass-energy of the universe.

162.143 **Local inhomogeneities in a Robertson-Walker background. III. Elementary growth rates in a flat background with a relativistic equation of state.**
C. Hellaby, K. Lake.
Astrophys. J., Vol. 251, 429 - 435 (1981).
The growth of a class of inhomogeneities embedded in a spatially flat Robertson-Walker background is examined. The inhomogeneities are restricted by the simplifying assumption that the isotropic pressure, as measured along the associated boundary surface, equals the local energy flux. The background is described both by means of a noninteracting mixture of blackbody photons and dust, and by means of a baryon-conserving equilibrium mixture of blackbody photons and a neutral, two-component, relativistic Maxwell-Boltzmann gas. The authors compare the resultant mass of the inhomogeneities with the Jeans and particle horizon mass scales and demonstrate the critical effect that varying the initial photon-to-baryon ratio in the background has on all masses.

162.144 **Origin of the large-scale structure of the Universe.**
I. D. Novikov.
Plasma astrophysics, (see 012.042), p. 179 - 189 (1981).
The paper deals with the plasma problems of the theory of origin of the large-scale structure of the Universe. The gravitational instability of the expanding relativistic plasma and collisionless particles (neutrinos), the numerical simulation of the non-linear stage of the evolution of inhomogeneites, and the origin of the primordial cosmological sound waves are considered.

162.145 **A characteristic initial-value problem approach to the formation of inhomogeneities in the early Universe.** D. S. Chellone.
J. Phys. A, Vol. 14, 2339 - 2343 (1981). – Abstr. in Phys. Abstr., Vol. 85, Abstr. 6690 (1982).

162.146 **Cosmology in bimetric Machian gravitation. II.**
R. Goldoni.
Nuovo Cimento, Lett., Ser. 2, Vol. 32, 50 - 54 (1981). Abstr. in Phys. Abstr., Vol. 85, Abstr. 6691 (1982).

162.147 **Symmetry breaking and restoration in the Einstein universe.** G. Denardo, E. Spallucci.
Nuovo Cimento A, Ser. 11, Vol. 64, 27 - 38 (1981). – Abstr. in Phys. Abstr., Vol. 85, Abstr. 6692 (1982).

162.148 **Upper bounds on neutrino masses from the large-scale structure of space-time.** P. S. Joshi,
S. M. Chitre.
Phys. Lett. A, Vol. 85A, 131 - 134 (1981). – Abstr. in Phys. Abstr., Vol. 85, Abstr. 6693 (1982).

162.149 **Matter annihilation in the late Universe.**
D. N. Page, M. R. McKee.
Phys. Rev. D, Vol. 24, 1458 - 1469 (1981). – Abstr. in Phys. Abstr., Vol. 85, Abstr. 6694 (1982).

162.150 **Origins of relativistic cosmology.**
M. Heller, O. Godart.
Astron. Q., Vol. 4, No. 13, p. 27 - 33 (1981).

162.151 **Large-scale structure of the universe.**
G. Chincarini.
Messenger. No. 26, p. 14 - 15 (1981).

162.152 **Baryon non-conservation: cosmological and experimental consequences.** V. A. Kuz'min.
Usp. fiz. nauk, Tom 134, 153 - 155 (1981). In Russian.
Abstr. in Ref. zh., 51. Astron., 11.51.1020 (1981).

162.153 **Elements of the large-scale structure of the universe.**
V. I. Arnol'd, Ya. B. Zel'dovich, S. F. Shandariñ.
Usp. mat. nauk, Tom 36, 244 - 245 (1981). In Russian.
Abstr. in Ref. zh., 51. Astron., 11.51.1029 (1981).

162.154 **Torsion and strong gravity in the realm of elementary particles and cosmological physics.**
V. de Sabbata, M. Gasperini.
Cosmology and gravitation, (see 012.051), p. 139 - 171 (1980).

162.155 **Self-similar cosmological models.**   W. Z. Chao.
Gen. Relativ. Gravitation, Vol. 13, 625 - 647 (1981).
The kinematics and dynamics of self-similar cosmological
models are discussed. The degrees of freedom of the solutions
of Einstein's equations for different types of models are listed.
The relation between kinematic quantities and the classifications of the self-similarity group is examined. All dust local
rotational symmetry models have been found.

162.156 **The de Sitter-Castelnuovo universe and elementary**
particle physics.   G. Arcidiacono.
Gen. Relativ. Gravitation, Vol. 13, 703 - 710 (1981).
If one studies the de Sitter universe in the geodetic
representation, one obtains the "projective relativity", valid
on the cosmic scale, for hyperdense matter and high energies.
The new theory can be applied also to quantum and elementary particle physics. In this work the author examines the
angular momentum operator, the D'Alembertian generalized
equation, and studies the electron, conceived as an oscillating
microuniverse.

162.157 **What is Hubble's constant?**
S. A. Wrona, R. Snopkowski.
Astron. Rep., Vol. 4, 23 - 29 (1979).

162.158 **Scaling hypothesis.**   M. Zabierowski.
Astron. Rep., Vol. 5, 21 - 28 (1981).

162.159 **Bouncing quantum cosmologies.**   J. Demaret.
Acad. R. Belgique, Bull. Cl. Sci., 5e Sér., Tome
66, 473 - 486 (1980) = Inst. Astrophys. Univ. Liège, Cointe-
Ougrée (Belgique), Coll. 8°, No. 681.

162.160 **Canonical quantization of diagonal Bianchi V**
models filled with a perfect fluid.   J. Demaret.
Phys. Lett., Vol. 95B, 413 - 418 (1980) = Inst. Astrophys.
Univ. Liège, Cointe-Ougrée (Belgique), Coll. 4°, No. 332.

162.161 **Steps toward the Hubble constant VIII. The global**
value.   A. Sandage, G. A. Tammann.
ESO Sci. Prepr., No. 174, 2 + 21 pp. (1981). — Submitted to
Astrophys. J.

162.162 **Structure of the universe and fundamental constants.**
I. L. Rozental'.
Inst. kosm. issled. AN SSSR. Prepr., 1981, No. 636, 12 pp. In
Russian. — Abstr. in Ref. zh., 51. Astron., 12.51.1018 (1981).

162.163 **Conformal cosmological models.**
A. V. Nosovets.
Kuban. univ. Krasnodar, 1981. 17 pp. In Russian. — Abstr. in
Ref. zh., 51. Astron., 12.51.1021 (1981).

162.164 **Spirals in the universe.**   M. Mikhajlov.
Priroda, NRB, Vol. 29, No. 5, p. 3 - 9 (1980). In
Bulgarian. — From Ref. zh., 51. Astron., 1.51.819 (1982).

162.165 **Closed model of the metagalaxy in a modified**
theory of gravitation.
O. Sh. Sharshekeev, A. D. Mukambaev.
Izv. vuzov. Fiz., Tom 24, No. 7, p. 112 - 113 (1981). In
Russian. — Abstr. in Ref. zh., 51. Astron., 1.51.832 (1982).

162.166 **The problem of model selection in modern**
cosmology.   Ch. Sh. Tsybikov.
Dialektika svyazi filos. i konkretno-nauchn. znaniya. Irkutsk,
1980, p. 98 - 106. In Russian. — Abstr. in Ref. zh., 51.
Astron., 1.51.835 (1982).

162.167 **Why is the cosmological constant so small?**
P. C. W. Davies, S. D. Unwin.
Proc. R. Soc. London, Ser. A, Vol. 377, 147 - 149 (1981).

162.168 **Cosmological baryon-number generation in grand**
unified models.   J. A. Harvey, E. W. Kolb,
D. B. Reiss, S. Wolfram.
Phys. Rev. Lett., Vol. 47, 391 - 394 (1981).
Methods for complete calculation of cosmological
baryon-number generation in the hot big-bang early universe
are outlined and are applied to several SU(5) models. Effects
of several baryon-number-nonconserving bosons and the
presence of nonthermalizing modes are treated.

The creation.   See Abstr. 003.029.

The edge of infinity: naked singularities and the
destruction of spacetime.   See Abstr. 003.051.

The cosmology of infinity.   See Abstr. 003.061.

Genesis: the origins of man and the universe.
See Abstr. 003.067.

Cosmology. The science of the universe.
See Abstr. 003.070.

Emerging cosmology.   See Abstr. 003.094.

The theory of cosmic aberration — a new inter-
pretation of the Hubble redshift.   See Abstr. 003.124.

Invitation to physics.   See Abstr. 003.125.

The isotropic universe. An introduction to cosmology.   See Abstr. 003.129.

Cosmology.   See Abstr. 003.138.

Space, time, and gravity. The theory of the big-bang
and black holes.   See Abstr. 003.157.

Por que el cielo es negro por la noche?
See Abstr. 004.027.

The early Universe.   See Abstr. 011.043.

The large-scale structure of the universe.
See Abstr. 012.071.

On relativistic kinetic theory: neutrino-antineutron
systems. II. The transport coefficients. Separation of matter
and anti-matter.   See Abstr. 022.048.

Phase picture of the SU(5) grand unified model.
See Abstr. 022.078.

Grand unified theories and proton decay.
See Abstr. 022.079.

Topological symmetry restoration.
See Abstr. 022.090.

Magnetic monopoles and grand unified theories.
See Abstr. 022.114.

The lore of large numbers: some historical back-
ground to the anthropic principle.   See Abstr. 022.178.

Lifetime constraints on massive neutrinos from
ultraviolet observations of clusters of galaxies.
See Abstr. 022.186.

Astrophysical production of fractional charge in
broken quantum chromodynamics.   See Abstr. 022.188.

Existing and potential limits to lifetimes of massive neutrinos.   See Abstr. 061.002.

Neutron-capture cross sections for osmium isotopes and the age of the Universe.   See Abstr. 061.006.

Magnetic moment of massive neutrinos and the cosmic helium abundances.   See Abstr. 061.010.

Nucleosynthesis with nonzero lepton numbers: is there a limit on the neutrino flavors?   See Abstr. 061.016.

Nucleocosmochronology.   See Abstr. 061.019.

Cosmological upper limit to neutrino magnetic moments.   See Abstr. 061.029.

Neutrinos and cosmology.
See Abstr. 061.030.

Nuclear astrophysics. Introductory note.
See Abstr. 061.037.

Nuclear and particle physics in the early universe.
See Abstr. 061.038.

Limits on neutrino degeneracy from nucleosynthesis.
See Abstr. 061.045.

Neutrinos, Sun and universe.
See Abstr. 061.052.

Magnetic field generation by rotating black holes.
See Abstr. 066.004.

Near-millimeter spectrum of the microwave background.   See Abstr. 066.005.

Distortions of the microwave background spectrum by dust.   See Abstr. 066.006.

Conformal geometry and spatially homogeneous cosmology.   See Abstr. 066.009.

Gravitational clumping and the annihilation of monopoles.   See Abstr. 066.013.

Massive spin-1 field in expanding universes and the quantum equivalence principle.   See Abstr. 066.014.

Exact spatially inhomogeneous cosmologies.
See Abstr. 066.025.

Scale-invariant gravity: a simple formulation.
See Abstr. 066.068.

Eddington's cosmological number, Einstein's number criterion and Rydberg's rational system of dimensions.
See Abstr. 066.075.

Hypersurfaces of constant mean extrinsic curvature.
See Abstr. 066.088.

The general theory of relativity with spin and torsion.   See Abstr. 066.089.

Path-integral evaluation of Feynman propagator in curved spacetime.   See Abstr. 066.093.

On the metric tensor of empty space in general relativity.   See Abstr. 066.102.

All non-twisting Ns with cosmological constant.
See Abstr. 066.107.

Can quantum effects prevent spacetime collapse?
See Abstr. 066.110.

The equation of state of a cosmological self-gravitating fluid influenced by thermal or turbulent motion and the primeval background radiation.
See Abstr. 066.111.

Generalised Kerr-Schild space-times.
See Abstr. 066.136.

On a canonical treatment of the Einstein equations in an expanding Universe. Isotropic coordinate conditions.
See Abstr. 066.138.

The quantum era.   See Abstr. 066.147.

On black holes in magnetic universes.
See Abstr. 066.148.

On the variation of G and scale invariant gravitation.
See Abstr. 066.150.

On the variation of G and scale invariant gravitation– a reply.   See Abstr. 066.151.

On the regularization of the cosmological singularity in general relativity.   See Abstr. 066.158.

Bimetric general relativity theory.
See Abstr. 066.183.

Rocket measurement of the cosmic background submillimeter spectrum.   See Abstr. 066.200.

An estimate of the mass of zero metal stars.
See Abstr. 131.159.

Radio astronomy confronts elementary particle cosmology: radio measurements of helium abundances.
See Abstr. 132.049.

Models of radio source evolution – II. The 2700-MHz source count.   See Abstr. 141.007.

Multifrequency models for the cosmological evolution of extragalactic radio sources.   See Abstr. 141.008.

The periodicity in the distribution of quasar redshifts and the density perturbation in the early universe.
See Abstr. 141.058.

The V/V_m test for quasars: a new interpretation.
See Abstr. 141.141.

The angular distribution of the cosmic background radiation.   See Abstr. 141.211.

Stability of contracting pancakes.
See Abstr. 151.017.

Galaxy mergers and active galactic nuclei.
See Abstr. 151.066.

Infrared surface brightness and absolute magnitude of spiral galaxies.   See Abstr. 158.044.

Hubble ratio and solar motion from 200 spiral galaxies having distances derived from the luminosity index. See Abstr. 158.047.

Hubble ratio and solar motion from 300 spirals having distances derived from H I line widths. See Abstr. 158.048.

The dynamical age of the Local Group of galaxies. See Abstr. 160.015.

Typical effect on redshift of galaxies of the Hercules supercluster.   See Abstr. 160.036.

The X-ray structure of a galaxy cluster at $z = 0.54$: implications for cluster evolution and cosmology. See Abstr. 160.047

Intergalactic extinction and the deceleration parameter.   See Abstr. 161.005.

Intergalactic shells at large redshift. See Abstr. 161.006.

Author Index

The authors are listed in alphabetical order according to the initial letter following the first names.

Subject Index

Starting with Volume 18 of *Astronomy and Astrophysics Abstracts,* some alterations concerning formation, arrangement, and versatility of the key words have been made. In order to provide an adequate description of a paper, specific key words are used as frequently as possible. References to a whole subject category are suppressed now. The user, therefore, has to refer to the contents at the beginning of each volume.

Whenever possible, the key words are formed in such a way that there are two different supplementary terms, e.g. the pair

interstellar matter
molecules.

An effort is made to choose preferably terms which can be inverted in order to increase the usefulness of this index. In the example given there are the two entries

interstellar matter
molecules

and

molecules
interstellar matter.

Exceptions to the rule of inversion of terms are given in all cases where the second key word is either a very specific one (e.g. Urca processes) or a general one (e.g. history). The use of substantives is preferred. In order to obtain the possibility to extend a one-term key word in a two-term one, combinations as

Mars   or   sun
 atmosphere   active regions

are changed into

Mars atmosphere   and   solar active regions,

respectively.

Starting with Volume 30 of *Astronomy and Astrophysics Abstracts,* this Index is given in a four-column arrangement. The overall size further has been optimized by suppression of a repetition of the first key word in a two-term combination. Thus, the primary term is printed only once for all the following secondary key words which belong to this header.

The user is — in any case — requested to look for synonymous entries, because further references to this topic might exist elsewhere in the Index under another current astronomical term.

ASTRONOMY AND ASTROPHYSICS ABSTRACTS

A Publication of the Astronomisches Rechen-Institut Heidelberg

Member of the Abstracting Board
of the International Council of Scientific Unions

Editors: S. Böhme, W. Fricke, I. Heinrich, W. Hofmann,
D. Krahn, V. R. Matas, D. Rosa, L. D. Schmadel, G. Zech

Published for Astronomisches Rechen-Institut by
Springer-Verlag Berlin Heidelberg New York

W. Högner, N. Richter

Isophotometric Atlas of Comets

This beautiful atlas contains a carefully selected collection of material needed for the study of the physics of comets. The authors scrutinized more than 300 photographs taken in the years 1902–1967. They applied photographic equidensities as quasiisophotes according to the method of Lau and Krug by using the Sabattier effect. Reproductions of
- the original photographs,
- their isophote diagrams,
- enlarged isophote diagrams of the cometary heads and nuclei

are presented. The most important documentary and astronomical data of the objects represented in the atlas are compiled in a special set of tables. The IAU considers the compilation of this atlas to be of "extreme value" to the astronomical community.

Part 1
1980. 90 plates, comments and tables
ISBN 3-540-09171-8

Part 2
1980. 55 plates, comments and tables
ISBN 3-540-09172-6

K.R. Lang

Astrophysical Formulae
A Compendium for the Physicist and Astrophysicist

2nd corrected and enlarged edition. 1980. 46 figures, 69 tables. XXIX, 783 pages
ISBN 3-540-09933-6

"… For astronomers, teachers and students it represents an important reference source for fundamental formulae used in astrophysics. For a student it may serve as a compact review of a familiar field or a handy aid to gain a rapid insight into the techniques of new fields of astrophysics. For a teacher the **Astrophysical Formulae** will be a useful guide in the very broad field of modern astrophysics… It is a work that everybody from us needs and that will spare much of our time and effort."
Journal of the British Interplanetary Society

Springer-Verlag
Berlin
Heidelberg
NewYork